# last YEARS 10 SOLVED PAPERS

(All Sets Included : Outside Delhi & Delhi)

## 2022 EXAMINATION

## SCIENCE STREAM (PCM)

English • Physical Education
Physics • Chemistry
Mathematics

# CBSE XII

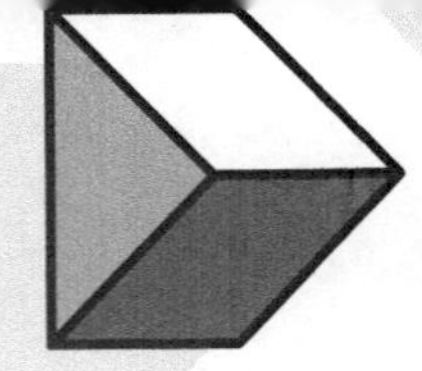

## DISCLAIMER

With the ambition of providing standard academic resources, we have exercised extreme care in publishing the content. In case of any discrepancies in the matter, we request readers to excuse the unintentional lapse and not hold us liable for the same. Suggestions are always welcome.

**EDITION :** YEAR 2021

**ISBN :** 978-93-91184-52-0

**PUBLISHED BY**

 OSWAL PUBLISHERS

 1/12, Sahitya Kunj, M.G. Road, Agra - 282002

 (0562) 2527771-4, +91 7534077222

 info@oswalpublishers.in

 www.oswalpublishers.com

The cover of this book has been designed using resources from Freepik.com

Printed At Upkar Printing Unit, Agra

In accordance with the latest syllabus prescribed by the Central Board of Secondary Education, New Delhi.

# PREFACE

With immense pleasure, we would like to present the thoroughly updated edition of CBSE 10 Last Years Solved Papers for the students of class 12th. This volume is prepared to meet the requirements of students preparing for their coming board examinations. Structured to facilitate easy last minute revision, this book gives a fair idea of the pattern of the question paper and the kind of questions asked by the board in the previous years.

This title also aims at gauging the preparation of the students for their examination. The content of this book is prepared, keeping in mind the dynamic nature of the board exam pattern. A team of experienced academicians have put in meticulous efforts provide befitting solutions, which have been framed in accordance to the marking scheme issued by the board.

Immense care has been taken to present the content with lucidity to enhance the readability of the book. We hope that this book will prove to be the quintessential tool for the students to ace their board examination.

Constructive suggestions to improve the quality of this book are always welcome.

**—Publisher**

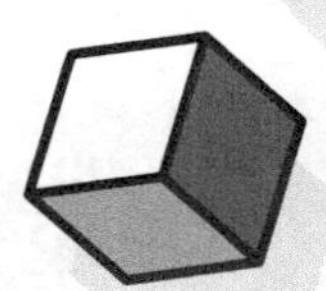

# HOW TO PRACTICE THE 3R'S AT HOME?

Shred the waste paper and use it as a packaging material.

Take your own cloth bag to shopping. Avoid taking products with extra packaging.

Change the refills not the pen.

Use the old towels and beddings as braided rugs.

Use old coffee or tea mugs and bottles for growing plants at home.

Don't sell the old furniture but use the wood to make shelves and drawers.

Use old glass jars, containers and cans as desk tools like pen holder or paper weight.

Use newspapers & old comic pages for gift wrapping.

Keep plastic lids under oil jars to keep your shelves stain-free.

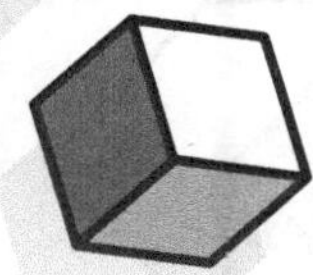

# STUDY LOG

## TOPICS I'VE MASTERED

## TOPICS I'VE TO WORK ON

## PRACTICE STRATEGY

## TIPS

1. Divide lengthy topics into smaller chunks and work on understanding one sub-topic at a time.
2. At the end of each study session, make short notes for quick revision.
3. Revisit previously learnt topics regularly from self-made notes.
4. Try to understand the topic than just mugging it up. It will stay in your memory for longer.

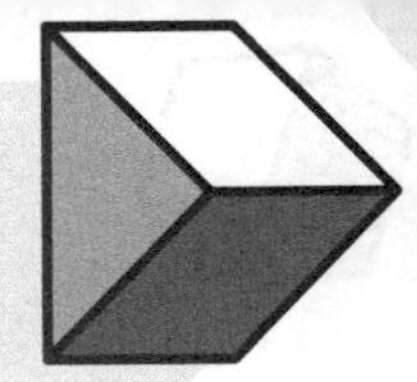

# HOW TO SAVE WATER AT HOME

Turn off the tap when you have wet the toothbrush or while applying soap to the dishes.

Take shorter showers. Prefer bucket-bath over shower one.

Soak your dishes first in warm water before cleaning them in running water. This will make the dirt come out much faster.

Leaky faucets can waste upto 20 gallons of water. Keep a close eye on them. Turn them off tightly.

Use broom instead of pipes to clean sidewalks or driveways.

Try watering plants during the early part of the day. Avoid watering them when it is too sunny or windy.

For harvesting rainwater use barrels or drums at the rooftop.

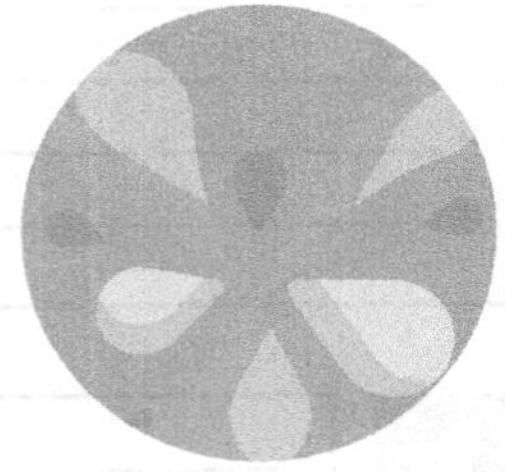

Avoid unnecessary throwing or splashing of water. Drink it upto the very last drop.

Wash your car manually instead of using a pipe.

Be conscious of the leaky washbasins/sinks.

Make use the waste water for cleaning floors.

Collected rainwater can be used for plants as it is rich in minerals.

# ENGLISH

## Outside Delhi [Set-I] 2020

**Time allowed : 3 hours** **Maximum marks : 80**

**General Instructions :**

**Read the following instructions very carefully and strictly follow them :**

(i) *The question paper comprises* **three** *Section* **A, B** *and* **C.**
*Section A – 20 marks*
*Section B – 30 marks*
*Section C – 30 marks*

(ii) *There are 10 questions in the question paper. All questions are compulsory.*

(iii) *There is no overall choice. However, an internal choice has been provided in one question in Section A, four questions in Section B, and three questions in Section C. Make your choice correctly.*

(iv) *However, separate instructions are given with each section and question wherever necessary.*

(v) *Do not exceed the prescribed word limit while answering the questions.*

### SECTION–A [20]

**(Reading)**

**1. Read the passage given below: [12]**

**Donated Organs and their Transportation**

(1) Once an organ donor's family gives its consent and the organs are matched to a recipient, medical professionals are faced with the big challenge of transporting organs while ensuring that the harvested organ reaches its destination in the shortest possible time. This is done in order to preserve the harvested organs and involves the police, especially the traffic police department.

(2) The traditional method of transporting organs by road is referred to as a "green corridor". This process entails police escorting an ambulance, so as to move around traffic – usually a specific traffic lane is chosen and all signals on the route stay green to ensure it to reach its destination in the shortest possible time. A 'green corridor' is a route cleared and cordoned off by the traffic police to ensure the smooth and steady transportation of harvested organs, on most occasions, to those awaiting a life-saving transplant. Organs tend to have a very short preservation time, such as the heart which has to be harvested and transplanted within four hours or the lungs which can be preserved for only six hours once they are harvested.

(3) The first green corridor in India was created by Chennai Traffic Police in September 2008 when they accomplished their task of enabling an ambulance to reach its destination within 11 minutes during peak hour traffic. That organ saved a nine-year-old girl whose life depended on the transplant.

(4) Similarly, such green corridors have been created by traffic police of various cities such as Pune, Mumbai, Delhi NCR etc. Personnel are stationed at selected points to divert, control, and clear the traffic giving way to the ambulance. Apart from this, a motorcade of police vehicles accompanies the ambulance ensuring that it does not face any problems. Delhi Traffic Police provided a green corridor from IGI Airport to Institute of Liver and Biliary Sciences in Vasant Kunj for transportation of a liver. The distance of 14 kms was covered in 11 minutes.

(5) Experts point out the lack of a robust system to transport organs to super-speciality hospitals in least possible time. National Organ & Tissue Transplant Organisation (NOTTO), the country's apex organ donation agency, is now framing a proposal to airlift cadaver organs and will send a report to the Union Health Ministry. "Cadaver organs have a short life and so transplant should be done within a few golden hours", Director (NOTTO) expressed, "Therefore, we are preparing a proposal for airlifting organs at any given moment".

(6) Most states do not have enough well-trained experts to retrieve or perform, transplant procedures. Also, there is an acute shortage of advanced healthcare facilities to carry out a transplant. So, it is referred to other big centres in metropolitan cities. Organs retrieved from Aurangabad, Indore, Surat, Pune are sent to Mumbai as these cities do not have super-speciality healthcare centres, informed officials.

(7) "In India about fifty thousand to one lakh patients are suffering from acute heart failure and need heart transplant at any point of time. In a private set-up, a heart transplant costs ₹ 15-20 lakhs, which is followed up by postoperative medication of about ₹ 30,000 per month lifelong."

**1.1. On the basis of your understanding of the above passage, answer any five of the following questions by choosing the most appropriate options :** **[1 × 5 = 5]**

**(a) The first green corridor in India was created in :**
(i) New Delhi (ii) Chennai
(iii) Mumbai (iv) Pune

**(b) The organization which is framing a proposal to airlift cadaver organs is :**
(i) Union Health Ministry
(ii) Regional Organ and Tissue Transplant Organisation
(iii) National Organ and Tissue Transplant Organisation
(iv) State Organ and Tissue Transplant Organisation.

**(c) The onerous task the author is talking about in Para 1 is :**
(i) finding organ donors.
(ii) finding doctors capable of performing transplants.
(iii) to carry the harvested organ in the shortest possible time.
(iv) to arrange the requisite facilities for the transplant.

**(d) Most of the people do not go for the heart transplant as :**
(i) it is very risky.
(ii) it is very painful.
(iii) it may cause death of the recipient.
(iv) the cost is prohibitive.

**(e) Most states refer organ transplant cases to big hospitals because :**
(i) they don't have well trained experts.
(ii) the patients don't trust local doctors.
(iii) the state hospitals are very crowded.
(iv) they don't have a pool of harvested organs.

**(f) Heart retrieved from a body is alive only for ................. hours.**
(i) two (ii) three
(iii) four (iv) five

Answer:
(a) (ii) Chennai
(b) (iii) National Organ and Tissue Transplant Organisation.
(c) (iii) To carry the harvested organs in the shortest possible time.
(d) (iv) The cost is prohibitive.
(e) (i) They don't have well-trained experts
(f) (iii) Four

**1.2. Answer the following questions briefly:** **[1 × 5 = 5]**

**(a) What is a 'green corridor'?**

**(b) Why is smooth transportation of the retrieved organ necessary?**

**(c) What opinion do you form of Chennai Police with regard to the transportation of the harvested heart?**

**(d) What does the author mean by 'a few golden hours'?**

**(e) How much does a heart transplant cost a patient in a private hospital?**

Answer :
(a) A green corridor is a route cleared and cordoned off by the traffic police to ensure the smooth and steady transportation of harvested organs, on most occasions, to those awaiting a life-saving transplant.
(b) A retrieved organ requires smooth transportation because it has very short preservation time post its harvest.
(c) Chennai Police did a phenomenal job by successfully transporting an organ in 11 minutes during the peak traffic hour and saving the life of a nine–year old girl.
(d) By 'a few golden hours', the author means 'short preservation time' that the harvested organs have with them.
(e) In a private hospital, a heart transplant costs ₹ 15-20 lakhs, which is followed by postoperative medication of about ₹ 30,000 per month lifelong.

**1.3. Pick out the words from the passage which mean the same as the following :** **[1 × 2 = 2]**

**(a) save (para 1)**

**(b) achieved/carried out (para 3)**

Answer :
(a) Preserve
(b) Accomplished

**2. Read the following passage:** **[8]**

(1) How does television affect our lives? It can be very helpful to people who carefully choose the shows that they watch. Television can increase our knowledge of the outside world; there are high quality programmes that help us understand many fields of study, science, medicine, the different arts and so on. Moreover, television benefits very old people, who can't leave the house, as well as patients in hospitals. It also offers non-native speakers the advantages of daily informal language practice. They can increase their vocabulary and practice listening.

(2) On the other hand, there are several serious disadvantages of television, of course, it provides us with a pleasant way to relax and spend our free time, but in some countries people watch television for an average of six hours or more a day. Many children stare at the TV screen for more hours a day than they spend on anything else, including studying

and sleeping. It's clear that TV has a powerful influence on their lives and that its influence is often negative.

**(3)** Recent studies show that after only thirty seconds of television viewing, a person's brain 'relaxes' the same way that it does just before the person falls asleep. Another effect of television on the human brain is that it seems to cause poor concentration. Children who view a lot of television can often concentrate on a subject for only fifteen to twenty minutes. They can pay attention only for the amount of time between commercials.

**(4)** Another disadvantage is that television often causes people to become dissatisfied with their own lives. Real life does not seem so exciting to these people. To many people, television becomes more real than reality and their own lives seem boring. Also many people get upset or depressed when they can't solve problems in real life as quickly as television actors seem to.

**(5)** Before a child is fourteen years old, he or she views eleven thousand murders on the TV. He or she begins to believe that there is nothing strange about fights, killings, and other kinds of violence. Many studies show that people become more violent after viewing certain programmes. They may even do the things that they see in a violent show.

**(a)** On the basis of your reading of the above passage, make notes on it using Headings and Sub-headings. Use recognisable abbreviations (minimum four) and a format you consider suitable. Supply a suitable title to it.

**(b)** Make a summary of the above passage in about **80** words.

**Answer:**

**Title : Effects of Television**

1. **Advantages of TV**
   (i) Increases knowledge
   (ii) Understanding of sci. art and medicine
   (iii) Language practice
   (iv) Relaxes mind
2. **Disadvantages of TV**
   (i) Wastage of time
   (ii) Eye strain
   (iii) Negative influence
3. **Effects on children**
   (i) Poor Conc.
   (ii) Lost in fantasy
   (iii) Normalcy towards violence
   (iv) Threat to the personality of child

**Abbreviations**
Adv.—advantages
TV—Television
Sci—Science
Conc.—concentration

## Summary

Television is the biggest evil in every household. Though a mode of entertainment for those who cannot step out, it has equal negative impacts; especially on children. It helps to expand our horizons and gain new knowledge along with an exposure to different dialects across the globe. Violence, injustice, and fantasy portrayed on TV tend to appear normal to the growing children. If used judiciously, TV can be a useful device for entertainment and knowledge but can prove to be detrimental otherwise.

## SECTION–B [30]

**(Writing Skills)**

**3. At Rohini in Delhi you have a three-bedroom flat with all modern amenities. It is fully air-conditioned and has power backup. For the sale of this flat draft a suitable advertisement in not more than 50 words to be published in a local daily. Give all the necessary details. Your contact number is 9911223344. [4]**

**OR**

**You are Secretary, Social Service League of your school. Design a poster to be displayed in your colony and in a local hospital premises inspiring people to make a pledge to donate eyes and other organs of their bodies.**

**Answer :**

**FLAT FOR SALE**

Newly constructed flat in Rohini for sale. Three bedrooms with attached washrooms and all the other modern amenities. Fully air-conditioned. Located in society with power backup and security. Interested buyers may contact Mr. Ramesh at 9911223344.

**OR**

**Donate Organs, Donate Life**

Kendriya Vidyalaya, Mahapuri
appeals one and all to sign up for the

**Organ Donation Camp**

to be organised at the basement of
Sri Jeevandan Hospital, Vikaspuri on
25th March 20XX, 11:00 AM onwards

Get a free medical check-up on successful sign up for the camp !

Secretary
Social Service League
Kendriya Vidyalaya

**4. You are Tapas/Tapasya of A-150, Mount Road, Chennai. You have seen an advertisement in the newspaper, 'The Chennai Times' for the post of Manager (Accounts) in Sundaram Westside, Chennai. Apply for the post with your complete biodata. (120-150 words) [6]**

**OR**

**Write a letter to the Editor, 'The Indian times', Jaipur highlighting the need to tap the sports talent at a young age by sports teachers, coaches, etc., so that it does not go unrecognized. Thus, we shall have a large pool of young talented sports persons who can be groomed. You are Poorva/ Paras of 78, Inderpuri Jaipur. (120-150 words)**

**Answer :**

A-105
Mount Road
Chennai

August 11, 20XX

The Manager
Sundaram Westside
Chennai

**Sub. : Application for the post of Accountant**

Sir,

In response to your advertisement in 'The Chennai Times' dated 1 August, 20XX, I wish to be considered for the position mentioned above. I feel my qualifications and experience are good enough to enable me to discharge my duties. I herewith attach attested copies of my certificates and my bio-data. If given a chance, I assure you, that I shall spare no pains in the discharge of my duties honestly. I will certainly leave no stone unturned to perform well.

Thanking you

Yours faithfully

Tapas

Encl. : Bio-data

## BIO–DATA

| | |
|---|---|
| Name | Tapas Verma |
| Father's Name | Sh. Mohan Lal Verma |
| Address | A-150, Mount Road, Chennai |
| Date of Birth | 15 September, 19XX |
| Educational Qualifications | (i) CBSE (10 + 2) in 19XX from St. Marks School with 83%<br>(ii) B.Com. (Hons.) in 19XX from Deviram College with 73%<br>(iii) CA in 19XX<br>(iv) ICWAI in 19XX |
| Experience | Working with TISCO, Bhopal since Nov. 20XX |
| Salary Expected | ₹ 9,000 p.m. + perks |
| Marital Status | Unmarried |
| Languages known | Hindi, English, Tamil, Telugu |
| References | (i) Dr. V. Anand, H.O.D. Commerce, Holker University, Indore<br>(ii) Sh. R.K. Sareen, Financial Advisor GRASIM (India) Bengaluru. |

**OR**

78, Inderpuri
Jaipur

April 5, 20XX

The Editor
The Indian Times
Jaipur

**Sub. : Need to tap sports talent at younger age**

Sir,

Through the columns of your esteemed newspaper, I wish to express my views on the need to tap sports talent at young age to create a successful sportsperson.

India has left its marks on the world by producing phenomenal sportspersons in both indoor and outdoor sports. Continuing the legacy, we should take effective measures to train the sports enthusiasts.

From a young age, teachers and coaches can create the zeal for sports and motivate the pupils for active participation so that it does not go unrecognised and also, tapping young talent will help to train them better and provide a boost to their performance.

Therefore, I request you to take the necessary measures to help the schools and communities create a sports pool and work towards acquiring many more feathers in the cap of our nation.

Yours truly
Poorva/Paras

5. **'No detention policy for classes sixth to eighth is academically very unsound.' Write a debate in 150-200 words either for or against the motion.** **[10]**

**OR**

**As per last census, the literacy rate in India was around 74%. In our day-to-day life, we find people who cannot even read or write. Looking at the gravity of the situation you decide to deliver a speech in your school's morning assembly on the topic, 'Each One, Teach One'. Write your speech in 150-200 words. You are Vinitha/Bejo.**

**Answer :**

**NO DETENTION POLICY FOR CLASSES SIXTH TO EIGHTH IS ACADEMICALLY VERY UNSOUND**

**FOR :**

Good morning everyone ! Today I, XYZ, stand before you to speak for the motion 'No detention policy for classes sixth to eight is academically very unsound'.

'No detention policy' states that no child shall be held back in any class or expelled from the school till the completion of elementary education. The step was taken to reduce stress and pressure among students, but the policy has led to poor outcomes, as there has hardly been any improvement in the academic standard.

It has encouraged negligence and reluctance of children to study. A student develops a carefree and an easy–going attitude, with an automatic promotion system up to class 8 and in class 9, he is unable to change his attitude suddenly, hence, they fail to cope with the academic pressure and end up failing their board exams.

I, therefore, firmly agree that the policy of no detention till class 8 is not in the interest of students, as it gives security of passage to the next class, regardless of a child's competence to learn heavier concepts.

Thank You.

**AGAINST :**

Good morning to one and all. Today I, xyz, stand here to speak against the motion 'No detention policy for classes, sixth to eight is academically very unsound.' No detention policy. No detention policy states that no child shall be held back in any class or expelled from the school till the completion of his elementary education. This policy will not only reduce the stress to pass in a particular grade but wall also reduce the drop-out rates especially from government schools.

Indian education system gives more importance to marks rather than achieving academic excellence through gaining knowledge. In such a scenario, passing is more crucial than learning and the 'no detention policy' thus, comes as a rescue.

A major benefit of this policy is that it has a direct impact on the dropout rates and boosts the confidence in under-performing or average students. They are able to grow and learn at par with their peers. A sense of responsibility thus, develops within them by the time they are in grade 9.

Owing to the above reasons, I strongly agree with the policy, especially due to the current educational system in the country.

Thank You.

**OR**

**EACH ONE, TEACH ONE**

Good Morning to one and all. Today, I Vinitha/ Bejo, would like to express my views on the slogan - 'Each one, Teach one'. It is important for the educated class to understand their moral and social responsibility to teach at least one illiterate person and make a difference in society.

Government policies and laws can promise many rights to the underprivileged children and make attempts to eliminate or reduce disparities, but it is a fact that many children are deprived of the right to education.

If we want children of all classes, even from poor families, to be literate and aware of their basic health and hygiene and to grow up as educated citizens, we must move forward on their behalf. It is our social responsibility to share our wealth of knowledge and learning with the least privileged section of the society.

It is true that the educated mass of the society that includes students, educated homemakers, professionals and students can play a vital role in the education of the disadvantaged children. The contribution of each individual can bring a significant positive change in the current state of illiteracy.

So, let's join hands and make India a totally literate country. Thank you and have a great day !

6. **A programme on 'Swachh Bharat Mission' was organised in your school on Mahatma Gandhi's birthday. Posters were prepared and pasted in the colony near your school. A procession was taken out. School premises and its surroundings**

**were cleaned by the students. Public was advised to make the mission successful. Write a report on the programme in 150-200 words. You are Srinivasan/Latha. [10]**

**OR**

**The word 'father' is synonymous with strict discipline, etc., but it is not completely true. The father fulfils his responsibilities affectionately for the family. Write an article in 150-200 words on the topic, 'Role of father in the family'. You are Dhruv/Deepa.**

**Answer :**

**CLEANLINESS DRIVE UNDER SWACHH BHARAT MISSION**

A Report by Srinivasan/Latha

Our school organised a cleanliness drive under Swachh Bharat Mission Programme on the occasion of Gandhi Jayanti. The main purpose of this programme was to create awareness among the students regarding cleanliness and its benefits. Under this programme, all the students from 9th to 12th had to participate. Teachers were the essential part of this drive. In this cleanliness drive, we had to clean the whole school. The sweepers of the school had to be the observers.

Our Principal started the programme by cleaning the corners of the playground. Then the teachers and students followed the suit. All the teachers and students did the dusting of the rooms. Posters were prepared and pasted around the school premises in order to create awareness. After collecting the garbage from the rooms, it was through into the dustbins.

Thereafter, our Principal delivered the speech telling us the importance of sanitation. He also honored the sweepers of the school for their noble job. In the end, we all took an oath of keeping our homes, locality, and city clean. It was a unique experience for all of us.

**OR**

**ROLE OF FATHER IN THE FAMILY**

By Deepa/Dhruv

Father is the strongest member of the family, both by choice and by compulsion. Our society is still very much patriarchal in essence, which means there is a great emphasis on the man of the family to be the strongest one, to hide their weaknesses and carry on unperturbedly in the toughest of times. That is perhaps, the greatest lesson we can learn watching our fathers as we grow.

Each time when life throws you a bad hand, your father is there to pick you up with his unquestioning love and support. He teaches you the importance of relationships and that family comes before everything, because he puts his family first all the time.

The father is, in every sense of the term, the man of the family. He holds our hand the first time when we are born and never lets it go. If there is one man we can bank on forever, he is the one. He will not be with us forever in person, but in spirit, he is eternal. So, we shouldn't hesitate to shower him with love and show him what he means in every chance we get. He might not say much, but honestly, we mean the world to him.

## SECTION–C [30]

**(Literature : Text Books)**

**7. Read the extracts given below and answer the questions that follow each of them. [4 + 4 = 8]**

**(a) He couldn't have addressed a more dazed and silent audience–no one knew what he was talking about and his accent defeated any attempt to understand what he was saying.****

**(i) Identify the chapter. Who is 'he'? [1]**

**(ii) What was 'he' in real life? [1]**

**(iii) How did the audience react to his speech? [1]**

**(iv) Why was his speech not a success? [1]**

**(b) Driving from my parent's**
**home to Cochin last Friday**
**morning, I saw my mother,**
**beside me,**
**doze, open mouthed, her face**
**ashen like that**
**Of a corpse and realised with**
**pain**
**that she was as old as she**
**looked................ .**

**(i) Where was the poet driving to? Who was sitting beside her? [1]**

**(ii) What did the poet notice about her mother? [1]**

**(iii) Which thought made the poet feel painful? [1]**

**(iv) Name the figure of speech used in the expression : 'her face ashen like that of a corpse'. [1]**

**Answer :**

**(b) (i)** The poet was driving to Cochin from her parent's home. Her mother was sitting beside her.

**(ii)** The poet noticed that her dozing mother had grown old with wrinkles on her face.

**(iii)** The poet felt painful looking at her mother's aging looks and wrinkled face.

**(iv)** Simile.

**Answers are not given due to change in present syllabus

**8. Answer any five of the following questions in 30-40 words each : [2 × 5 = 10]**

**(a) What explanation did the children offer the writer for not wearing footwear? Did she agree to it? (Lost Spring)**

**(b) How did Douglas' experience at the beach in California affect him? (Deep Water)**

**(c) Why has Raj Kumar Shukla been described as being resolute? (Indigo)**

**(d) How will 'Keeping Quiet' protect our environment?**

**(e) How did Charley 'reach' the third level of the Grand Central Station?**

**(f) What is common between Derry and Mr. Lamb?**

**(g) Why was Dr. Sadao not arrested on the charge of harbouring an enemy?**

**Answer :**

(a) The children were poor ragpickers. They always went barefoot and had become used to it. Even if they had shoes , they looked for excuses to not to wear them. Some even said that going barefoot was a tradition among them. The author disagrees with the usual explanation that is offered about it being a tradition in India to walk barefoot. She feels that this is only an excuse for lack of money. They live in a 'perpetual state of poverty'.

(b) Once Douglas' father took him to a beach in California where he was knocked down by strong waves and got breathless. It left a haunting fear lurking in his heart. He was extremely scared of water and feared going near it then onwards.

(c) Rajkumar Shukla has been described as being 'resolute' because he wanted Gandhiji to accompany him to his district named Champaran anyhow, Gandhiji was busy at that time and had several engagements. But Rajkumar Shukla never left Gandhiji's side. He followed him whenever he went. At last, Gandhiji had to find time to go with him. It shows how resolute Rajkumar Shukla was.

(d) Keeping quiet and introspecting will initiate peace and brotherhood among men, halting all destructive activities like waging wars which harm people besides damaging the environment. Thus, the environment will be protected by keeping quiet.

(e) Grand Central Station, New York, has only two levels. One evening Charley reached this station and then walked down the second level to catch an early train to his home. While he was on the second level, he strangely happened to notice a doorway down. He followed the steps and reached the third level which was never heard of or seen by anyone. In the third level, Charley experienced a hundred-year-old world and its people.

(f) Derry was a boy whose face was burnt because of acid and Mr. Lamb was an old man who had a tin leg because his leg was blown off by a bomb in the war. The common thing in both of them was that they both were physically impaired but their attitude towards life was very much opposite.

(g) Dr. Sadao, on humanitarian grounds as well as considering it his professional duty, tended to a wounded war prisoner which was officially a serious crime. However, he did not get punished for this offence as it was never revealed to anyone, except his wife, loyal but timid servants, and a General who was too self-obsessed with his own treatment that he would never let the doctor leave him.

**9. Answer the following question in 120-150 words: [6]**

**Why did the Crofter repose confidence in the peddler? How did the peddler feel after betraying the Crofter?**

**OR**

**Educating children is the responsibility of society. Justify the statement in view of 'The Last Lesson'.**

**Answer :**

The Crofter reposed confidence in the peddler because he was a lonesome man who used to live alone. He wanted someone with whom he could share his feelings, so he trusted him to the extent of showing him where he had kept his money. Despite the Crofter treating the peddler with hospitality, the peddler robbed him and was quite pleased with his smartness. However, the fear of getting caught haunted him. So, he avoided the public highway and turned into the woods. It was a big and confusing forest, and due to darkness, he lost his way. He got exhausted moving around the same place, and was filled with despair. He began to feel that the forest was like a big rattrap and the thirty kronors he had stolen were like a bait set to tempt him. He felt helpless like a rat who had no way out of the trap in which he had fallen.

**OR**

Adults in a society play a very important role in motivating the young learners to take education seriously. They have gone through this stage and can guide us in the right track through their experiences. This can also be seen in the lesson "The Last Lesson" by Alphanso Daudet. This story is of a boy, Franz who took his studies as a burden and a teacher, M. Hamel who took the noble profession like teaching quite lightly.

However, during the latter part of the story, it was only M. Hamel who gave rise to a feeling of remorse in the mind of Franz for not taking the french language seriously and went on praising the language to be the most logical and clear. With this, one can see that it was M. Hamel who motivated young Franz to think nice about the language and pay his complete undivided attention on the Last Lesson.

**10. Answer the following question in 120-150 words: [6]**

**What was the prediction of the astrologers regarding the ultimate fate of the Tiger King? How did it come to be true? Describe with reference to the story.****

**OR**

**What are phytoplankton? How are they important to our ecosystem? (Journey to the End of the Earth)****

●●

# Outside Delhi [Set-II]

**Note : Except for the following questions, all the remaining questions have been asked in previous set.**

**SECTION–B [30]**

**(Writing Skills)**

**6. The number of private vehicles is increasing in the cities. It leads to traffic jams and a rise in air pollution levels. The only solution seems to be that people start using public transport. Write an article in 150-200 words on the topic, 'Importance of Public Transport'. You are Sanjith/Ranjitha. [10]**

**OR**

**Teacher's Day was celebrated in your school. The function was presided over by the State Education Minister. A variety programme was organised. Five retired teachers were honoured. Write a report in 150-200 words on the function. You are Pritam/Nirmala.**

**Answer :**

IMPORTANCE OF PUBLIC TRANSPORT

By Ranjitha/Sanjith

Out of several issues doing rounds in the country, heavy traffic congestion accompanied with pollution is one of the biggest problems that is faced by everyone in the country. A gradual increment of traffic has been observed in major cities. However, resorting to public transport will definitely solve a lion's share of this problem.

In order to curb the rising pollution, public transport such as buses, should be actively used. With good connectivity throughout the city; this mode of transport could easily be utilised at an optimum level. Travelling by bus not only is cheaper than on your own vehicle but also is a step towards pollution control.

Moreover, citizens can use car-pooling or alternative modes of transport for commuting. Using transport such as cycling, will keep both their health and pollution in check. Being the vigilant citizens of the country, it becomes our moral obligation to avail the services offered by the government and work towards the betterment of the country.

As it says, "A developed country is not a place where the poor have cars. It's where the rich use public transport."

OR

TEACHER'S DAY CELEBRATION AT ST. MARTIN'S HIGH SCHOOL

A Report by Pritam/Nirmala

St. Martins High School, Bengaluru, celebrated Teacher's Day with zeal and zest on September 5, 20XX. There was an unusual excitement in the school which was quite contagious. The students of classes XI and XII took over the duties of the teachers for the day. A special morning assembly was held in which the students paid homage to their teachers through poems and speeches. The Principal, Mrs. R. Joseph, addressing the entire school, lauded the entire staff for their exemplary contribution to the school's phenomenal success. State Education Minister, Hon. Mr. Dilip Jadhav, graced the occasion with his presence as the Chief Guest and addressed the audience through his words of wisdom. He spoke about the importance of teachers in a student's life and that how they steer us in the right direction. There was a special celebration organised for the retiring teachers. The retiring teachers were felicitated by the Chief Guest for their long-term contribution. Other entertaining programmes which included games

**Answers are not given due to change in present syllabus

with many prizes, were also carried out followed by a grand lunch.

## SECTION–C [30]

### (Literature : Text Books)

**8. Answer any five of the following questions in 30-40 words each : [2 × 5 = 10]**

**(a) How did the Crofter entertain the peddler?**

**(b) What was Sophie's first choice of job after completing her school? How did she hope to succeed in her aim?****

**(c) Mention any two difficulties faced by the bangle sellers of Firozabad.**

**(d) What are the probable causes of the passing cars to stop near the roadside stand?****

**(e) Describe the scene at the threshing floor observed by Bama on her way back from school.****

**(f) Dr. Sadao and servants were only partly loyal to their country. Comment.**

**(g) How did Jo want the story to end?**

**Answer :**

(a) Crofter entertained the peddler with his talks and friendly gestures. The crofter was so talkative and friendly with the peddler because he was leading a lonely life in the cottage. He had no one with whom he could talk to and due to this, his life was so boring and monotonous. He was an old man who was living there without a wife or child. So, he was happy to get someone to talk to in his loneliness.

(c) The bangle sellers of Firozabad are prone to many diseases as they work in suffocation and unhygienic conditions. They also lose their eyesight early in their life as they work at high temperatures.

(f) Dr. Sadao, a Japanese surgeon and his servants gave shelter to a wounded American soldier during times of war between America and Japan. They not only treated the soldier back to health but also looked after him with care and concern. Dr. Sadao also helped the man escape from Japan under disguise. Treating the wounded enemy as a doctor but also making him leave the country, demonstrates partial loyalty by Dr. Sadao and his servants.

(g) Jo was not convinced with the ending of the story and coaxed her father to re-tell the story the next day, giving the story a pre-determined path that she had set. According to her, neither Roger Skunk nor the wizard was wrong in the story. Jo refused to accept the end where Roger Skunk's mother hits the wizard and that too without being hit back. She wanted the story to end with the wizard hitting back the mother skunk with his magic wand and chopping off her arms 'force fully'.

**9. Answer the following question in 120-150 words: [6]**

**The order from Berlin created a great enthusiasm in the school. Comment.**

**OR**

**How did Gandhiji use Satyagraha and non-violence at Champaran to achieve his goal?**

**Answer :** In the short story 'The Last Lesson' by Alphonse Daudet, the atmosphere at the school on the day of the last lesson was quite different from usual. There was a lot of sadness and solemnity in the school due to the order from Berlin which stated that only German language would be taught in the schools of Alsace and Lorraine.

The French teacher, M. Hamel, was very sad and emotional that he would have to leave the school after forty years of teaching his mother tongue over there. There was also a profound change in his behaviour and he was more polite than usual.

He displayed his zeal and enthusiasm towards the French language by calling it the most beautiful, clearest, and the logical language in the world. He asked his students to guard their mother tongue among them and never forget or lose it under any circumstances. He asked the students to take a sense of pride in their native language. He supported his views with the reason that when the people of a country are enslaved and forced to use a foreign language, their mother tongue holds the key to their prison.

He had also brought new copies for the students in which "France, Alsace, France, Alsace" was written in beautiful round handwriting.

The back benches of the classroom were occupied by the village elders who were attending the class as a mark of respect for their mother-tongue.

**OR**

Mahatma Gandhi, who was an apostle of peace and non-violence, agitated in Champaran district of Bihar in 1917 and then in Kheda district of Gujarat in 1918. Although the term 'Satyagraha' was coined in 1919, when the Anti - Rowlatt Act protests took place, Champaran witnessed this kind of non-violent protests for the first time.

The protests in Champaran took place as the British introduced farming of cash crops like indigo, which they bought at a low price for export. Opium was also extracted from the seeds

**Answers are not given due to change in present syllabus

of poppy, which was grown to export to China to fill the coffers of the imperialist rulers. The protests were against two decisions of the British, *viz.*, forcing farmers to cultivate cash crops in place of food crops, which they earlier cultivated for sustenance, and payment of abysmally low wages to the growers. Rajendra Prasad, Anugrah Narayan Sinha, Brajkishore Prasad, Acharya Kripalani and many others took part in the agitation with Gandhiji.

Satyagraha was actually a mass civil disobedience movement. Gandhiji set up an ashram and volunteers in scores joined his agitation. He then began cleaning up the villages, built hospitals and schools, induced villagers to remove untouchability and do away with the purdah system. After this, people became conscious of their rights and non-violent protests were all they would do. This culminated in the arrest of Gandhiji. He was later released by an order of the court, after the agitators agitated gathered en masse in front of the police station and the court. Thereafter, Gandhiji was called Bapu across the country. The Champaran movement brought to heel the British rulers and their cohorts, the wealthy landlords; the poor farmers were subsequently paid remunerative wages and their farming right of the desired crops was restored.

●●

# Outside Delhi [Set-III]

**Note : Except for the following questions, all the remaining questions have been asked in previous sets.**

**SECTION–B [30]**

**(Writing Skills)**

**6. Due to heavy rainfall the river near your town started overflowing its banks. As a result your town was flooded. It led to a great loss of life and property. Many N.G.O.s came forward to help the people with food, shelter, and medicines. Write a report in 150-200 words for a local newspaper on the tragedy and rescue efforts. You are Raghav/Ritika. [10]**

OR

**Teachers are regarded as nation builders. They are responsible for developing the character of students and transforming them into responsible citizens. Write an article in 150-200 words on the topic, 'Role of the Teacher in Society'. You are Gaurav/Gauri.**

**Answer :** FLOOD HITS KERALA

A Report by Raghav/Ritika

Recently one of the worst floods of the decade has hit Kerala. Owing to copious rainfall that started on August 8, 20XX, all the dams of the state were filled to their fullest capacity. The gates had to be opened to keep the dams safe. The incessant rains further deteriorated the flood situation. Approximately, 370 people have died in the floods so far. Thousands of homes have been affected and damaged. More than a million people had to take shelter in relief camps. Normal life has come to a standstill.

The entire nation has come forward to lend a helping hand to the Kerala flood victims. Central Government, State Governments, Union Territories, Multi-national Corporations, Big Business Houses, Celebrities, Sportsmen and women, schools, colleges' and common people have contributed to Kerala's Chief Minister's Relief Fund generously. Even the help from foreign countries is pouring in. It will take many months of hard-work to bring life to normalcy in the town.

OR

ROLE OF THE TEACHER IN SOCIETY

By Gauri/Gaurav

The role of a teacher is not confined within the classroom walls; it is stretched beyond and forever. Teachers steer the students in the right direction and provide the light of knowledge banishing the darkness of ignorance. The teacher inspires and encourages the student to strive for greatness, give shape to future and nurtures their talent and potential.

An effective teacher can have a positive impact on the career of a student. A good mentor can motivate the student at every point of difficulties in and out of the classroom. A person with proper vision, experience, and the passion of teaching, can enter the teaching profession.

A good teacher, in fact, becomes a role model for the students. Students tend to follow their teacher in almost every way. The teacher's caring attitude will have a major influence on the students. Student's self-esteem could be uplifted by a good teacher. A teacher gives wings to the ambitions of the student for future academic success. Hence, a teacher has one of the major roles to play in society.

**Answers are not given due to change in present syllabus

## SECTION–C [30]

### (Literature : Text Books)

**8. Answer any five of the following questions in 30-40 words each : [2 × 5 = 10]**

**(a) Why did Rudyard Kipling dislike being interviewed? Give two reasons.****

**(b) Why is Sophie attracted to Danny Casey?****

**(c) For Saheb, how was work at the tea stall different from ragpicking?**

**(d) What is the colour of 'sour cream'? Why has the poet used these words to describe the classroom walls? (An Elementary Classroom in a Slum)**

**(e) Why was Dr. Sadao not sent to the battlefield?**

**(f) Why did the landlord's man ask Bama's brother on which street he lived?****

**(g) Why was Evans not ready to remove his hat ?**

**Answer :**

(c) Saheb was a ragpicker. By chance, he got a job working at a tea stall down the road. His job at the tea stall included odd tasks like getting milk from the milk booth. Even though he was paid 800 rupees, his face lacked the carefree look. He was unhappy and felt enslaved. The steel milk canister were heavier than the plastic bags for him.

(d) The colour of 'sour cream' is off-white or yellowish. Perhaps, the poet has used this expression to describe the classroom walls in order to underline the prevalent neglect. The dull colour and the suggested bitterness, echoes the situation of the underprivileged children.

(e) Dr. Sadao was not sent to the battlefield because of two reasons firstly, he had perfected a discovery which would render wounds entirely clean. Secondly, the General might have needed some operation urgently and Dr. Sadao was the best doctor available for his need.

(g) When Evans was asked by the prison officer, Mr. Jackson, to take off his hat and he denied doing that. He was asked by the officer to smarten himself in half an hour. After half an hour, he was seen with his hat and was ordered to take it off. In reply, he said that the hat was a symbol of good luck for him and he won't take it off.

**10. Answer the following question in 120-150 words:**

**The play 'On The Face of It' depicts the unusual behaviour of the people towards the physically disabled which makes them lonely. Comment. [6]**

**OR**

**If we want to study and examine the Earth's past, present, and future, Antarctica is the place to go to. Comment.****

**Answer :**

The play 'On the Face of it' aptly depicts the loneliness and sense of alienation experienced by Derry and Mr. Lamb on account of disability. The actual pain and inconvenience caused by the disabilities is often much less than the sense of alienation felt by the disabled person. Derry suffered from severe negative complexes because of his burnt face. He became a pessimistic loner who indulged in self-pity and was always suspicious of the intention of others. His anger and frustration made him an introvert. Mr. Lamb, on the other hand, was inwardly a loner who craved company and acceptance. Though outwardly he was always jovial, outgoing, and optimistic but from inside, he was an extremely sensitive person. Derry and Mr. Lamb's physical disabilities caused pain and suffering not only to their bodies but also to their minds and souls.

**Answers are not given due to change in present syllabus

●●

# ENGLISH

## Delhi [Set-I]

## 2020

Time allowed : 3 hours

Maximum marks : 80

**Note : Except for the following questions, all the remaining questions have been asked in previous sets.**

### SECTION–B [30]

**(Writing Skills)**

3. **You are Dhruv/Nidhi, Student Editor of your school magazine, 'The Buds'. Write a notice in not more than 50 words to be placed on your school notice board, inviting short stories, articles, poems, etc., from students of all classes for the school magazine. Give all the necessary details.** [4]

**OR**

**You are Vinod/Vineeta, General Manager, Global Solutions, Meerut. You want to rent a bungalow to be used as a guest house. It should have all the modern amenities. Draft a suitable advertisement in not more than 50 words to be published in a local newspaper.**

**Answer :**

Achievers International High School,
New Delhi

**NOTICE**

25th November, 20XX

**CALL FOR COMPOSITIONS FOR "THE BUDS"**

Our annual school magazine "THE BUDS" will be published in the second week of March. Students who wish to contribute short stories, articles, poems, riddles, caricatures or even quizzes are invited to send in their entries by 14th January, 20XX. This is open to students from all classes. Kindly, submit your entries typed or neatly written on an A4 size paper to the undersigned.

Dhruv/Nidhi

Student Editor

THE BUDS

**OR**

**Guest House Wanted**

Global Solutions, Meerut, is looking for a well-designed bungalow in a renowned location in and around Meerut on company lease as a guest house. The bungalow must be equipped with all the modern amenities including uninterrupted water and electricity supply. Connectivity to the local markets, hospitals, and malls will be an add-on.

Interested owners are requested to contact :

Vineeta/Vinod

General Manager, Global Solutions, Meerut

4. **A book fair was organised in your city, Bhopal. Thousands of people including a large number of students visited the fair. It aroused a great interest in reading and buying books. You want that such book fairs are held in other cities of the state also to promote the habit of reading. Write a letter in 120-150 words to the editor of a local newspaper giving your views. You are Navtej / Navita, F-112, Malviya Nagar, Bhopal.** [6]

**OR**

**You carried a hit and run accident victim to a local hospital. You were shocked to see that in its casualty ward the conditions were chaotic. The patient was attended to after a lot of precious time was lost. You are Deepak / Deepa, A-114 Roop Nagar, Delhi. Write a letter of complaint in 120-150 words to the Chief Medical Officer of the hospital.**

**Answer :**

F-112,

Malviya Nagar,

Bhopal

Madhya Pradesh – 462011

15th February, 20XX

The Editor

Patrika

Chinar Fortune City

Bhopal

Madhya Pradesh : 462047

**Sub. : Promotion of reading habits through book fairs in Bhopal**

Sir,

Through the columns of your esteemed newspaper, I would like to talk about the need for book fairs in and around Bhopal city to promote the habit of reading in students.

Recently, XYZ school in Bhopal organised a book fair in collaboration with local publishers. This book fair housed several fiction and non-fiction books as well as books essential for various academic subjects. The two-day fair had seen a plethora of the public thronging to view and purchase the books on sale. I also had the opportunity to be a part of this book fair.

I was surprised to see young and old alike choosing books of their interest as well as providing reading suggestions. Parents and students at the event waited in long queues just to buy their favourite books. This indicated that book fairs open to the public bring about the habit of reading. I am certain that such book fairs will only increase the fervor of reading among children and help them to be less reliant on social media for their entertainment. I, hereby, request you to give my article a place in your esteemed newspaper so that the relevant authorities may express their interest and take steps to conduct such book fairs across other cities of the state to promote the habit of reading.

Thanking you

Yours sincerely
Navtej/Navita

**OR**

A-114,

Roop Nagar,

Delhi–110007

13th March, 20XX

The,

Chief Medical Officer

Roop Nagar Hospital

Delhi–110007

**Sub. : Expressing concern over the chaotic conditions in the casualty ward**

Sir,

I would like to express my anguish and pain at the chaotic conditions prevalent in the casualty ward of your esteemed hospital. Carrying a victim of a hit and run accident, I expected to see immediate action and treatment for the victim. However, I was rewarded with delay, chaos in the casualty ward and scant attentive attitude of the staff on call. This is to draw the attention of the lethargic authorities to initiate some steps to improve the situation.

While walking home, I witnessed a hit and run accident that took place in front of me. The man was bleeding profusely and despite giving him first aid, we couldn't stop the bleeding. At that stage, I took the responsibility of safely admitting the victim to the hospital. We were soon in for a rude shock. As we reached there, we did not find anyone at the emergency so much so there was no doctor or nurse at the counter. Prodding further, we found a group of nurses out on their tea break in this grave situation. The hygiene and the cleanliness were at their worst and the number of patients asking to be helped was high. As a result, treatment was extremely delayed due to which the victim unfortunately lost a limb.

This callousness of the hospital staff and chaotic conditions can cost the life of a person. What is the use of having a local hospital in a posh location when treatment cannot be given at the right time? I hope that your good-self takes heed of this situation and brings about changes in the hospital that will help the people. I am anticipating a positive reply and a change for the best.

Thanking you

Yours truly,

Deepa/Deepak

5. **Every year there are floods or droughts in the country. Victims suffer a lot. The government does a lot to help the people, but it is not enough. Society also must do its bit. Write a speech in 150-200 words to be delivered in the school morning assembly describing how the students can also help by collecting clothes, money, and medicines from their neighbourhoods. You are Arun/Aruna.** **[10]**

**OR**

**'Rain harvesting is the only solution to water crisis in the present times'. Write a debate either for or against the motion in 150-200 words. You are Sujatha/Ajith.**

**Answer :**

**GIVE, AND YOU WILL RECEIVE**

Today, our country is facing frequent problems with respect to environmental changes. Victims of flood and drought have no hope but to rely on the meagre help of the Government to fulfil their needs. Honourable Principal, Respected teachers and my dear friends, today I, Arun/Aruna, the secretary of the school's Social Involvement Program, stand before you to speak about how we can do our bit to create a difference in the lives of these people.

Drought and floods have ravaged our country from centuries, however, today we stand at such times when increasing prices makes it difficult for victims of these environmental hazards to rely on the aid provided by the government. In most cases, aid does not even reach on time leaving these poor families stranded like a headless chicken. It is our duty to help our suffering brethren and as citizens of tomorrow, we shall come together to do our bit. Collecting medicines, clothes and even small amounts of money from our neighbourhood can go a long way to help our people. Let us all consciously strive to make efforts to help our suffering brothers and sisters.

Thank You !

OR

## RAIN HARVESTING IS THE ONLY SOLUTION TO WATER CRISIS IN THE PRESENT TIMES

**FOR :**

Honourable members of the Jury, my teachers, and my contenders, greetings to each one of you gathered here. Today I, Sujtha/Ajith, am glad to have been given this opportunity to voice my views on the topic "Rain harvesting is the only solution to water crisis in the present times." I am concentrating on the facts and advantages of the elements and components of this topic.

The most common slogan "You never know the worth of water until the well runs dry" is a widely spoken topic. Although planet earth consists of three-fourths water, not all water is drinkable, resulting in scarcity of water. With the increasing population, and with major parts of the world struggling for water, the need for improvising the management of water resources is a major concern.

The solution to this water crisis lies in rainwater harvesting. Rainwater harvesting is collecting and storing rainwater for utilising rather than let it go waste. Rainwater harvesting is free from chemicals, and favourable for multiple uses like domestic use, livestock, gardening, irrigation especially, at the time of droughts.

The bright and yet simple idea of saving groundwater and using rainwater is by harvesting rainwater *i.e.,* collect, filter, store, and utilise the rainwater for any and every purpose. Rainwater harvesting setup is affordable to install, easy to maintain, useful for irrigation, and has multi-purpose usage. I hope that public makes use of this environmental-friendly resource that will benefit all.

Thank you.

**AGAINST :**

Honourable judges, teachers, and my dear friends, today I, Sujatha/Ajith, stand before you to express my views against the topic "Rain harvesting is the only solution to water crisis in the present times". Yes, this topic is a crucial one but when we look at the various drawbacks, it will surely help decide for itself.

Water harvesting is not a bad idea but, who can precisely predict the rainfall? It could rain enough only to a limited extent or low or even not at all. It is beneficial to the areas where there is a downpour of rain.

It is initially a costly affair depending on the quantity of rainfall and the complexity of the system. The harvesting system requires routine maintenance. Therefore, if no care is taken, it could be a breeding ground for algae, bacteria, and pests. Another aspect is the quality of the construction of the rainwater harvesting system and the storage limit it holds which may not be sustainable for all.

Rain water harvesting will be of no benefit to drought-affected regions, it would only benefit the places that receive heavy rainfall letting water diversions to arid lands. Therefore, I believe that rainwater harvesting is not an optimum method of water conservation.

Thank you.

**6. Rising prices of essential commodities make life difficult for the common man. You are Nalini/ Ranbir. Write an article in 150-200 words describing the causes, possible solutions and how students can help their parents to cope with the problem. [10]**

OR

**On your way to school, right in front of the school gate, you saw a verbal dispute leading to a physical fight between an auto rickshaw driver and his passenger. A teacher of your school intervened, found out the cause of the quarrel and amicably resolved the issue (problem). Write a report on the incident in 150-200 words. You are Balbir/Bimla.**

**Answer :**

## RISING PRICES OF ESSENTIAL COMMODITIES MAKE LIFE DIFFICULT FOR THE COMMON MAN

By Nalini/Ranbir

As alarming as it sounds, the rise in prices is considered to be a positive indicator of a developing economy. However, today the prices of the essential commodities have risen so much that the real worth of people's income has declined. Most of the people live from hand to mouth these days. Further, the basic necessities of the common man are sold at a great price so much so that they have doubled at an alarming rate. It is uncommon to not see businessmen or sometimes middlemen hoard goods and wait for the prices to go up, so that they can then sell the goods at exorbitant prices. The Government therefore, should ensure equal distribution of basic commodities. Further, families can take up steps to improve this situation by budgeting wisely and curbing unnecessary expenses. Students can contribute by reusing and recycling every item that they own to ensure maximum use and save on unrequired expenditure.

OR

**VERBAL SPAT TURNS TO A PHYSICAL FIGHT**

A Report by : Balbir/Bimla

A verbal spat between an auto-rickshaw driver and his passenger turned into a physical confrontation with both throwing accusations at one another. This incident occurred near Modern High School, Mall Road, Kanpur. According to the passenger, Jatin Singh, the auto driver, Sunil was talking on the mobile right from the beginning of the journey and up to the end of the auto rickshaw ride. Jatin made multiple attempts and asked Sunil to cut the call and concentrate on driving the auto, however, Sunil bluntly refused to do so. Jatin claims that Sunil was driving quite rashly and the focus was more on the call than on the road. By the time Mr. Hariram Das, a biology teacher of Modern High School intervened, a huge crowd had already gathered at the spot and traffic increased, people started honking madly. Mr. Das questioned both parties about the situation and immediately called the police who took both the parties for further questioning and helped ease the traffic on the road.

## SECTION–C [30]

**(Literature : Text Books)**

**7. Read the extracts given below and briefly answer the questions that follow each. [4 + 4 = 8]**

**(a) But just as he laid his head on the ground, he heard a sound–a hard regular thumping. There was no doubt as to what that was. He raised himself, "Those are the hammer strokes from an iron mill", he thought.**

**(i) Who is he? [1]**

**(ii) Where was 'he' at that moment? [1]**

**(iii) Why did he lay his head on the ground? [1]**

**(iv) Did he feel comfortable on hearing the thumping sound? Why? [1]**

**(b) And such too is the grandeur of the dooms**
**We have imagined for the mighty dead;**
**All lovely tales that we have heard or read;**
**An endless fountain of immortal drink,**
**Pouring unto us from the heaven's brink.**

**(i) Which two things of beauty are mentioned in these lines? [1]**

**(ii) Why are the 'lovely tales' called an endless fountain? [1]**

**(iii) Where is this fountain situated? [1]**

**(iv) Explain : 'grandeur of the dooms'. [1]**

**Answer :**

**(a)** (i) 'he' in the extract is the rattrap peddler.

(ii) At that moment, the rattrap peddler was trapped in a forest.

(iii) The rattrap peddler was stuck in the forest and could not find his way out. As darkness was settling around him, he gave up and laid his head on the ground tired.

(iv) Yes, he felt comfortable on hearing the thumping sound as it gave him hope that there was a way out from the forest and he would not remain trapped.

**(b)** (i) The things of beauty mentioned in these lines are the tales we have heard or read about our ancestors describing their sacrifices.

(ii) The 'lovely tales' are called an endless fountain as they describe the beautiful bounty of the earth.

(iii) The fountain is situated on heaven's brink.

(iv) It is the magnificence that we can imagine for our ancestors on the doomsday.

**8. Answer any five of the following questions in 30-40 words each : [2 × 5 = 10]**

**(a) How was Saheb's life at the tea-stall?**

**(b) What are some of the positive views on interviews?****

**(c) Why is Jansie so critical of Sophie?****

**(d) How do we know that children at the elementary school are coming from a slum?**

**(e) Why does Mr. Lamb leave the gate of his house always open?**

**(f) How did Zitkala – Sa feel when her hair was shingled?****

**(g) When did the 'Tiger King' decide to get married?****

**Answer :**

**(a)** The tea-stall ensured that Saheb had a stable life, with a regular wage of rupees eight hundred plus his all-time meals. He was not worried about his food but was sad with his new life as he was no longer his own master and he had to work long hours at the tea stall.

**(d)** The appearance of the children in the classroom tells us that the children at the elementary school are coming from a slum. The children's

**Answers are not given due to change in present syllabus

faces are dull and their appearances show that they are poor.

(e) Mr. Lamb always left the gate of his house open. He lived alone and did not mind if strangers entered his house or garden. As he lived alone, he hoped that someone would come and talk to him.

**9. Answer the following question in 120-150 words:**

**How did the swimming instructor build a swimmer out of Douglas?** **[6]**

**OR**

**How did the court scene at Motihari change the course of India's struggle for freedom?**

**Answer :**

Douglas was extremely scared of water so to overcome this fear, he decided to get an instructor to teach him to swim. Douglas and the instructor spent an hour each day for five days a week at the swimming pool. The instructor would put a belt around Douglas which was attached to a rope on an overhead cable; the instructor would hold one end of the rope while Douglas went back and forth in the pool. This went on hour after hour, day after day, week after week. It took nearly three months for Douglas to ease the tension of swimming across the pool. Once Douglas gained the confidence, the instructor taught Douglas to put his face underwater and exhale, and then raise his nose and inhale. For several weeks, the instructor made Douglas kick with his legs until he was able to control his legs underwater. Slowly, but steadily, the instructor made a swimmer out of Douglas.

**OR**

The episode at Champaran was a turning point for India's struggle for freedom. This was when Gandhiji decided to oppose the British rule in India. Gandhiji went to Motihari to gather complete information about the indigo plantation contracts from the peasants; he was accompanied by several lawyers. He was on his way to meet a peasant who was mistreated by some of the indigo planters; he was stopped by a messenger of the police superintendent who served him a notice which asked him to leave. However, Gandhiji ignored the notice and continued his investigation. A case was filed against him and he was arrested. On the day of the trial, unanimously, a large crowd of peasants gathered near the court demanding the release of Gandhiji. When it became impossible to control them, Gandhiji spoke to them and they all obeyed him. Gandhiji managed to convince the court that he was not there to break laws but he wanted to help the peasants. He was immediately granted bail and later, the case against him was dropped.

**10. Answer the following question in 120-150 words:** **[6]**

**Describe briefly the scene at the third level of Grand Central as seen (or seemed to be seen) by Charley.**

**OR**

**How did Dr. Sadao resolve the conflict in his mind between his loyalty to his country and his duty as a doctor?**

**Answer :**

One night, after working late at the office, Charley was in a hurry to get back to his apartment, so he decided to take the subway from Grand Central instead of taking the bus. He walked down the flight of stairs and entered an arched doorway, where he got lost in the corridors. At the end of the doorway, he went down another small flight of stairs and reached the third level. On reaching there, he was much surprised to find that everything looked old and different. There were few ticket windows that were made of wood. The architecture made him feel as if he was in the 1890s. The men had beards and fancy moustaches. The women wore long dresses and skirts with high buttoned boots. Charley also saw a Currier and Ives Locomotive and an issue of an 1894 'The World' newspaper.

**OR**

In the lesson, 'The Enemy', Dr. Sadao was a successful surgeon. He managed to strike a balance between his loyalty to his country and his duty as a doctor in a very commendable way. At that time, Japan was at war with America. On one of his walks on the beach with his wife, he found a wounded American soldier who was on the verge of dying. As a doctor, he was unable to leave the man to die on the beach when he knew that he could save him. He operated on the American soldier and saved his life as it was his moral duty to value human life. At the same time, he was a loyal citizen of his country. So, Dr. Sadao informed the General that he was treating the American soldier and was also taking care of him until he was fit. In this way, Dr. Sadao fulfilled his duty as a doctor as well as a patriot citizen.

●●

# Delhi [Set-II]

Note : Except for the following questions, all the remaining questions have been asked in previous sets.

## SECTION–B
### (Writing Skills)

3. The Environment Club of your school is organising a tree plantation drive on the World Environment Day around your school and in the neighbourhood. Write a notice in not more than 50 words inviting students to volunteer for the event. Give all the relevant details. You are Secretary of the Club.

OR

You have a 3-bedroom, ground floor flat in a posh locality. You want to sell it off as you are going abroad. Write an advertisement in not more than 50 words to be published in the classified columns of a local daily, giving all the necessary details. Your contact number is : XXX XXX...

**Answer :**

Scholars International High School, Kolkata
NOTICE

23$^{th}$ June, 20XX

TREE PLANTATION DRIVE

Sage, the environment club of our school, is organising its annual tree plantation drive on the occasion of Van Mahautsav. In a bid to encourage plantation, 1500 new trees will be planted in and around the school campus. Gardening equipments, seeds, and saplings will be distributed to the students. All students from Grade VII to XII are invited to take part in this noble event which will be inaugurated by Mr. Desai, school principal, on 10th July, 20XX. Participants are requested to assemble in the school auditorium by 8 : 15 am.

XYZ
Secretary-Sage

OR

FLAT FOR SALE

A three-bedroom flat located on the ground floor in a posh locality is available for purchase at the earliest. The flat is well equipped with all the modern amenities including a modular kitchen, uninterrupted water and electricity supply. The building also has a parking facility. Located in Delhi's Safdarjung locality, the house offers connectivity to the local markets, hospitals, and malls.

Interested parties are requested to contact :
XYZ
XXXXXXXXXX

4. You have read an advertisement in The National Times for the post of Chartered Accountant in Happy Times Industry, Kanpur. You believe you possess the requisite qualifications and experience and would prove to be an asset to the company. Apply for the post giving your complete biodata stressing your suitability. You are Amrendra Singh of 2, Kailash Puri, Delhi. (120-150 words)

OR

The main park in your locality is in a state of utter neglect. The gates of the park are broken and the greenery has almost vanished. The employees responsible for the upkeep of the park are seldom seen doing their assigned duties. Write a letter to the Director, Horticulture Department, Meerut complaining about the problem and requesting him to take appropriate action. You are Brij Kishore / Gopa, a resident of Kishore Nagar, Meerut. (120-150 words)

**Answer :**

2, Kailash Puri

Delhi

November 21, 20XX

The Manager

Happy Times Industry

Kanpur

**Sub. : Application for the post of Chartered Accountant**

Sir,

This is in response to your advertisement in The National Times dated 13 November, 20XX wherein I wish to offer my candidature for the position mentioned above.

I believe that my qualifications and work experience are good enough to enable me to discharge my duties at your reputed office. I am certain that I am the best candidate for this role. I am attaching with this letter the attested copies of my certificates and bio-data. If given an opportunity, I assure you that I shall spare no efforts in discharging my duties to the best of my ability and honouring the trust placed in me.

Thanking you

Yours faithfully

Amrendra Singh

Encl. : Bio-data

**BIO–DATA**

| Name | Amrendra Singh |
|---|---|
| Father's Name | Sh. Vijendra Singh |
| Address | 2, Kailash Puri, Delhi |
| Date of Birth | 14th May 19XX |
| Educational Qualifications | (i) CBSE (10 + 2) in the your 2002 with 91% |
| | (ii) B.Com. (Hons.) in the year 2006 with 82% |
| | (iii) CA. in 2008 |
| | (iv) CFA in 2012 |
| Experience | Happy and Sons LLP |
| Salary Expected | ₹ 30,000 p.m. + perks |
| Marital Status | Unmarried |
| Languages known | Hindi, English, Punjabi |
| References | (i) Smt. Kalika Saxena, Financial Advisor RGD (India) |
| | (ii) Dr. Khan, H.O.D. Commerce, St. Stephens College, Delhi |

**OR**

B-206
Kishore Nagar
Meerut

23rd July, 20XX

The Director
Horticulture Department
Meerut

**Sub. : Complaint regarding the derelict neighbourhood municipal park.**

Sir,

My locality boasts of a huge park that is equipped with benches and playground. The park was inaugurated in 2012 and was promoted as a place where one and all come to unwind and relax. However, the hugely promoted park is now in a state of abject disrepair. I am writing this letter to draw your attention to the horrendous conditions of the park and to initiate some steps to improve the situation.

The gates of this magnanimous park are broken and no attempts have been made to fix them. Further, for a park that is supposed to be promoting a green environment, it is bare of any greenery and what is left is slowly deteriorating away. The shocking bit is that the employees responsible for the upkeep of the park are seldom seen around and this has led to squatters occupying the park and making it their home. This situation reflects poorly on the otherwise good work done by your department.

I hope that your good-self takes heed of this situation and ensures that appropriate action is taken to bring life back into the park as many people look forward to unwinding at the park. I am anticipating a positive reply and a change for the best.

Thanking you

Yours truly

Brij Kishore/Gopa

## SECTION–C

**(Literature : Text Books)**

8. **Answer any five of the following questions in 30-40 words each :**

   (a) **Why did the elder of the village attend the last lesson?**

   (b) **What does Umberto Eco mean by the concept of 'Interstices' in our lives?**

   (c) **How was the make-up room in the Gemini Studios a fine example of national integration?****

   (d) **Describe the tigers created by Aunt Jennifer.**

   (e) **What would Sam have done in Galesbury for his living? Why? (The Third Level)**

   (f) **What efforts did Dr. Sadao and Hana make to save the American soldier?**

   (g) **Why did the prison officers call Evans, 'Evans the Break'?**

**Answer :**

(a) The elders of the village attended the last lesson to show their affection for the French language and to pay respect to the teacher, M. Hamel for his forty years of faithful services. The elders occupied the last benches of the classroom.

(b) Umberto Eco believes that just as the universe has empty spaces, our lives also have empty

**Answers are not given due to change in present syllabus

spaces or 'interstices'. They refer to the time between two tasks. For example, Umberto Eco keeps himself busy in the empty time while waiting for someone to come up through the elevator.

(d) Aunt Jennifer embroiders tigers on a tapestry. The poet, Adrienne Rich describes the tigers created by Aunt Jennifer as 'bright topaz denizens' of the forest. They are portrayed as fearless in contrast to Aunt Jennifer. They are sure of their purpose and move ahead confidently.

(e) Sam would have set up a hay, feed, and grain business in Galesbury. Charley infers this as he enquired about Sam and comes to know that he had bought old style currency worth eight hundred dollars at the stamp and coin store.

(f) Although, the American soldier was an enemy, Hana washed the soldier with warm water. She continued to assist Dr. Sadao during the operation. Dr. Sadao was in dilemma between patriotism and his duty as a doctor, but he managed to save the American soldier without betraying his country.

(g) The prison officers called Evans, 'Evans the Break' because he was a habitual jail breaker. Before being shifted to the Oxford Prison, he had tried breaking out of the previous prison thrice.

**10. Answer the following question in 120-150 words:**

**Why did the Tiger King decide to kill a hundred tigers? Describe the efforts he made to attain his target.****

**OR**

**Why is Antarctica the place to go to for the scientists today? (Journey to the End of the Earth)****

## Delhi [Set-III]

**Note : Except for the following questions, all the remaining questions have been asked in previous sets.**

### SECTION–B
### (Writing Skills)

**6. History Club of your school organized a picnic on a river bank. During the picnic you observed a small boat (with two persons in it) sinking. You saw a good swimmer jump into the river and rescue the two persons in the boat. Write a report in 150-200 words on the picnic and the rescue. You are Rohit of 2, Mall Road, Kanpur.**

**OR**

**Wearing the school uniform gives the student an identity, pride, and a feeling of discipline. Write an article in 150-200 words on, 'Importance of wearing a uniform in school'. You are Jagan/ Jagriti.**

**Answer :**

ONE ACT OF KINDNESS SAVES TWO

A Report by Rohit

**14th December, 20XX** : A picnic organised by the school history club on the river bank soon turned into an awful event. While the students brought out their picnic baskets and gathered around to celebrate their achievements at the recent inter-school history quiz, a couple on a paddling boat was soon going to face the shock of their lives. As they continued paddling around the river, they lost hold of the boat, which subsequently toppled with both the persons on it. Shouts of cry for help prompted every passerby to look up and gather around the river to try and help the sinking couple. It was one brave man who made the ultimate sacrifice by jumping into the river and rescuing the two persons on-board. The students and the crowd were fearful as it was not an easy task to save the flailing souls as it soon started to rain. A winter downpour and drowning people were not the best match but this brave man was able to save the couple and himself. The crowd gathered around the rescuer and applauded him for his brave efforts.

OR

IMPORTANCE OF WEARING A UNIFORM IN SCHOOL

By : Jagriti/Jagan.

The first thing that comes to your mind when you talk about school is the school uniform. Every individual for sure remembers the colour of the uniforms that they wore to school. It is sad to see that some of the schools are now planning to abandon the school uniforms. The importance of the school uniform is in the name itself. A school uniform is worn to ensure that there is uniformity across the school. Each school has its own particular style of uniform. A uniform is like the unique identify of the students through which one can be identified. School uniforms provide a sense of pride, whether the students belong to

**Answers are not given due to change in present syllabus

the higher class or lower class. Wearing a school uniform also make the students responsible for their behaviour inside and outside school. This sense of responsibility begins to inculcate within the student from childhood itself. Another advantage of wearing school uniform is that there is no peer pressure amongst the students when it comes to choosing the attire for the day.

## SECTION–C

### (Literature : Text Books)

**8. Answer any five of the following questions in 30-40 words each : [2 × 5 = 10]**

**(a) How is the mother tongue important to a person? What does M. Hamel, the teacher, say about it?**

**(b) Why didn't Sophie want Jansie to know about her meeting with Danny Casey?****

**(c) How bad were the living conditions in which Mukesh and his family survived?**

**(d) Why has the poet mentioned 'Merry children spilling out of their homes' in the poem? (My Mother at Sixty-six)**

**(e) How is Antarctica different from the place you live in?****

**(f) On seeing Rogar Skunk again with a very bad smell, how did the little animals react first and then later on when he had lost it?**

**(g) "I felt like sinking to the floor", said Zitkala-Sa. When did she feel so and why?****

**Answer :**

(a) M. Hamel, the french teacher, told the class that a mother tongue was the most beautiful language in the world and one must always safeguard it. When in times of captivity and enslavement, the mother language serves as a ray of hope.

(c) Mukesh, a child labourer, worked in a bangle-making factory. His family was very poor. They lived and worked in hazardous conditions. They used to work in the glass furnaces with high temperatures in cells with no light or air.

(d) The children spilling out of their homes, represents youth which is opposite to the poet's mother. It is possible that the poet has used this image to show the difference between youth and old age.

(f) On seeing Roger Skunk with a very bad smell, the little animals ran away from him at first. Roger Skunk then went to the wizard who changed his smell to the smell of roses. The little animals became Roger Skunk's friends and played with him when he smelled like roses afterwards.

**9. Answer the following question in 120-150 words:**

**How did Douglas develop a fear of water?**

**OR**

**How did Civil Disobedience triumph at Motihari? (Indigo)**

**Answer :**

When Douglas was ten or eleven years old, he was standing alone at the side of the YMCA swimming pool for people to come, when a boy who was probably eighteen years old, came to him and asked if he would liked to be ducked. Without waiting for any answer, he picked up Douglas and threw him into the deep end of the pool. Douglas landed in a sitting position inside the pool and had swallowed water. Douglas decided that soon as his feet touched the bottom of the pool, he would try and make a jump to the surface and paddle to the edge of the pool, but nothing worked and he started drowning. This experience scared Douglas and he never returned to the YMCA pool. He started fearing water. He would avoid water whenever he would get a chance.

**OR**

Gandhiji went to Motihari to gather complete information about the indigo plantation contracts from the peasants. There he was accompanied by several lawyers. He was on his way to meet a peasant who was mistreated by some of the indigo planters; but was stopped by a messenger of the police superintendent who served him a notice which asked him to leave. However, Gandhiji ignored the notice and continued his investigation. A case was filed against him and he was arrested. On the day of the trial, unanimously, a large crowd of peasants gathered near the court demanding the release of Gandhiji. When it became impossible to control them, Gandhiji spoke to them and they all obeyed him. Gandhiji managed to convince the court that he was not there to break the laws but just to help the peasants. He was immediately granted bail and later, the case against him was dropped. An official enquiry was conducted and the landlords agreed to refund the peasants. Although they did not give 50%, the landlords agreed to refund 25%. Thus, Gandhiji won the case for the peasants through Satyagraha and non-violence.

●●

**Answers are not given due to change in present syllabus

# ENGLISH
## Outside Delhi [Set-I] 2019

**Time allowed : 3 hours** **Maximum marks : 100**

### SECTION–A
### (READING) 30

1. **Read the passage given below and answer the questions that follow : [20]**

(1) The sage of science, Einstein, was sitting in a depressive and pensive mood one evening. His eyes were brimming with tears. The pain was evident on his face. He peeped out of the window of his room. The sun had set a few minutes back. The sky was filled with a reddish glow. At this sunset, he felt that it was humanity that had sunk into devilish darkness and the reddish glow in the sky was the blood of humanity spilling all over the sky from earth. With tired steps, he walked back to his chair and settled down. It was 9th of August, 1945. Three days back, he had felt the same agony as if someone had torn him apart. He was deeply hurt and depressed when he heard on the radio that America had dropped an atom bomb on the Japanese city, Hiroshima. Today, within three days another bomb was dropped on another city, Nagasaki and lakhs of people had been killed.

(2) He had heard that the blast released so much energy that it had paled all past destructions in comparison and death had played out a pitiable dance of destruction. The flames that broke out of the bomb were burning, melting and exploding buildings. Scared of the heat of the bomb, people had jumped into lakes and rivers, but the water was boiling and the people too were burnt and killed. The animals in the water were already boiled to death. Animals, trees, herbs, fragrant flowering plants were all turned into ashes. The atomic energy destruction had just not stopped there. It had entered the atmosphere there and had spread radiation that would affect people for generations to come and would also bring about destructive irreversible biological change in animals and plants.

(3) As the news of the atomic attack reached Einstein, and he became aware of the glaring horror of the abuse of atomic energy, his distress and restlessness knew no bounds. He could not control himself and picked up his violin to turn his mind on to other things. While playing the violin, he tried to dissolve his distress in its sad notes, but couldn't. He was burning on the embers of destruction; his heart was filled with an ocean of agony and tears just continued streaming uncontrollably out of his eyes. Night had fallen. His daughter came up and asked him to eat something as he had not taken anything for the last four days. His voice was restrained and he said, "I don't feel like eating."

(4) He could not sleep that night. Lying down, he was thinking how he had drawn the attention of the American President Roosevelt towards the destructive powers of an atomic bomb. He had thought that this would be used to scare Hitler and put an end to the barbarism that Hitler was up to. However, Roosevelt kept him in the dark and made false promises. Eventually, he had abused Einstein's equation of $E=mc^2$ that resulted in the destructive experiments. His actions had made science and scientists as murderers. Einstein kept on thinking for a long time. Eventually, he slipped into sleep. When he woke up at dawn, there was a new dawn in him too. The atomic threat had transformed his heart.

(5) The next day, he decided to disassociate himself from the scientific policy of the government and all governmental institutions. He decided to open educational institutions for children, adolescents and youth—institutions where along with science, spirituality will be compulsorily taught.

(6) To inaugurate this institution, he had invited two great philosophers, Bertrand Russell and Albert Schweitzer. Ten other great scientists who had own Nobel Prizes in different fields were also invited. They all saw a different Einstein, not a great scientist but a sage in him. The institution was opened by garlanding a photo of Mahatma Gandhi. While garlanding the Mahatma, he became emotional and said with a lump in his throat, "I bow down to the great man who fought for the independence of his country through non-violence. He could do so because he was a truthful man and true spiritualist."

(7) Those who teach science should be taught spirituality too. Without harmony between science and spirituality, the destruction would continue unabated. A few years after this institution was built, a Japanese delegation came to meet him. Einstein broke down in the meeting and said, "You can give me any punishment and I will accept it. Anyway, I have decided to lead my life in penitence." The Japanese were moved by this sincerity and forgot their grief.

**1.1 On the basis of your understanding of the above passage, answer each of the questions given below by choosing the most appropriate option : [1 × 5 = 5]**

**(a) Besides two great philosophers how many other scientists were invited by Einstein to inaugurate the institution where spirituality would be compulsorily taught?**

(i) Five (ii) Ten
(iii) Eight (iv) Fifteen

**(b) Which musical instrument did Einstein play when he was in grief?**

(i) Harmonium (ii) Guitar
(iii) Violin (iv) Flute

**(c) Einstein came to know that America had dropped an atom bomb on the Japanese city, Hiroshima through**

(i) television
(ii) newspaper
(iii) radio
(iv) a telephonic message

**(d) Which American President was told about the destructive power of an atomic bomb?**

(i) Kennedy (ii) Bill Clinton
(iii) Lincoln (iv) Roosevelt

**(e) Einstein said to the Japanese delegation,**

(i) "You can give me any punishment and I will accept it."
(ii) "I am not at fault."
(iii) "What could I do?"
(iv) "The President didn't agree to my advice."

**Answer :**

**(a)** (ii) Ten
**(b)** (iii) Violin
**(c)** (iii) Radio
**(d)** (iv) Roosevelt.
**(e)** (i) "You can give me any punishment and I will accept it."

**1.2 Answer the following questions briefly: [1 × 6 = 6]**

**(a) What did Einstein do to overcome his distress after getting the news of the atomic attack?**

**(b) Which event in 1945, according to Einstein, turned science and scientists into murderers?**

**(c) What did Einstein do to show his displeasure over the atomic attack?**

**(d) Whose photo was garlanded at the inauguration of Einstein's institute for children, adolescents and youth?**

**(e) Name the philosophers that Einstein invited to inaugurate the new institution.**

**(f) Why did Einstein want harmony between science and spirituality while teaching in educational institutes?**

**Answer :**

**(a)** Einstein was extremely distressed after getting the news of the atomic attack. To distract his mind from this serious news, he picked up his violin and began playing sad notes on it. He did not even eat for four days.

**(b)** In 1945, America had dropped atom bombs on the Japanese cities of Hiroshima and Nagasaki. The bombs made to scare Hitler and put an end to barbarism had been misused to kill many innocent lives. This event according to Einstein, turned science and scientists into murders.

**(c)** To show his displeasure over the atomic attack, Einstein decided to disassociate himself from the scientific policy of the government and all its institutions. He decided to open educational institutions where science and spirituality would be compulsorily taught to children, adolescents and youth.

**(d)** The photo of Mahatma Gandhi was garlanded at the inauguration of Einstein's institute for children, adolescents and youth.

**(e)** Two great philosophers, Bertrand Russell and Albert Schweitzer were invited by Einstein to inaugurate the new institution.

**(f)** Einstein's fear of destruction due to the atomic bomb attacks transformed his thinking. He felt the need for bonding between science and spirituality to bring about harmony. This could bring a decrease in destruction of life and peace would prevail in the world.

**1.3 Answer any three of the following questions in 25–30 word each : [2 × 3 = 6]**

**(a)** What did Einstein feel while looking at the sunset from his room's window?

**(b)** Give a brief description of the disaster when the atomic bomb was dropped on the Japanese city, Nagasaki.

**(c)** What did Einstein think of Mahatma Gandhi?

**(d)** What was Einstein's reaction when the Japanese delegation met him?

**Answer :**

**(a)** Einstein with tears brimming from his eyes and pain evident in his face, peeped out of the window of his room. He was deeply hurt and depressed. Looking at the sunset, he felt the sinking of humanity into a devilish darkness and the reddish glow in the sky felt like human blood spilling all over the sky from the earth. He was torn apart.

**(b)** The explosion at Nagasaki caused death, burning, melting and exploding of buildings. The water in the lakes and rivers was boiled hot leading humans as well as aquatic animals to death. All living and non-living things were turned into ashes.

**(c)** Einstein thought of Mahatma Gandhi as a great man who used non-violence as his tool while fighting for his country's independence. Einstein referred to Mahatma as a truthful and spiritualist man.

**(d)** Einstein, on meeting the Japanese delegation, broke down and announced that he was ready for any punishment given to him. He also spoke of his decision to lead the rest of his life in penitence.

**1.4 Find words/phrases from the passage which are similar in meaning to each of the following : [1 × 3 = 3]**

**(a) mental pain (para 1)**

**(b) agreement (para 7)**

**(c) regret/remorse (para 7)**

**Answer :**

**(a)** agony

**(b)** harmony

**(c)** penitence

**2. Read the passage given below carefully and answer the questions that follow : [10]**

(1) To live in harmony with oneself and the environment is the wish of every human. However, in modern times greater physical and emotional demands are constantly placed upon many areas of life. More and more people suffer from physical and mental tension such as stress, anxiety, insomnia, and there is an imbalance in physical activity and proper exercise. This is why methods and techniques for the attainment and improvement of health, as well as physical, mental and spiritual harmony, are of great importance, and Yoga meets this requirement.

(2) The word "Yoga" originates from Sanskrit and means "to join, to unite". Yoga exercises have a holistic effect and bring body, mind, consciousness and soul into a balance. In this way Yoga assists us in coping with everyday demands, problems and worries. Yoga helps to develop a greater understanding of our self, the purpose of life and our relationship with God.

(3) On the spiritual path, Yoga leads us to supreme knowledge and eternal bliss in the union of the individual Self with the universal Self. Yoga is the supreme, cosmic principle. It is the light of life, the universal creative consciousness that is always awake and never sleeps, that always was, always is, always will be.

(4) Many thousands of years ago in India, Rishis (wise men and saints) explored nature and the eosmos in their meditations. They discovered the laws of the material and spiritual realms and gained an insight into the connection within the universe. They investigated the cosmic laws, the laws of nature and the elements, life on earth and the powers and energies at work in the universe—both in the external world and on a spiritual level. The unity of matter and energy, the origin of the universe and the effects of the elementary powers have been described and explained in the Vedas. Much of this knowledge has been rediscovered and confirmed by modern science.

(5) From these experiences and insight a far-reaching and comprehensive system known as Yoga originated and gave us valuable, practical instructions for the body, breathing, concentration, relaxation and meditation. The system "Yoga in Daily Life" is taught worldwide in Yoga Centres, Adult Education Centres, Health Institutions, Fitness and Sports Clubs, Rehabilitation Centres and Health Resorts. It is suitable for all age groups—the name itself indicates that Yoga can be and should be used "in Daily Life".

(6) The exercise levels have been worked out in consultation with doctors and physiotherapists and can therefore —with observation of the stated rules and precautions—be practiced independently at home by anyone. "Yoga in Daily Life" is a holistic system, which means it takes into consideration not only the physical, but also the mental and spiritual aspects.

Positive thinking, perseverance, discipline, orientation towards the Supreme, prayer as well as kindness and understanding form the way to Self-Knowledge and Self-Realisation.

**2.1** On the basis of your understanding of the above passage, make notes on it using headings and subheadings. Use recognizable abbreviations (wherever necessary — minimum four) and a format you consider suitable. Also supply an appropriate title to it. **[5]**

**2.2** Write a summary of the above passage in about 100 words. **[5]**

**Answer :**

**TITLE : YOGA IN DAILY LIFE**

1. Need for Yoga
   (i) Decreases mental and phy. stress in daily life.
   (ii) Attnt. and improv. of phy., mental health and spiritual harmony.
2. Benefits of Yoga
   (i) Bal. of body, mind, consciousness and soul.
   (ii) Cope with daily probs.
   (iii) Supreme knwldg. of union of indvl. self and univ. self.
3. Practice of Yoga
   (i) Practical instructions for – body, breathing, concentration, relaxation, meditation.
   (ii) Taught worldwide in Yoga Centres, Health Institutions, Rehabilitation Centres etc.
4. Yoga in daily life
   (i) Use of yoga daily for positivity in life
   (ii) Holistic system, considers physical, mental and spiritual aspects

| Key to abbreviations | | |
|---|---|---|
| Phy. | – | Physical |
| Attnt. | – | Attainment |
| Improv. | – | Improvement |
| Knwldg. | – | Knowledge |
| Indvl. | – | Individual |
| Bal. | – | Balance |
| Probs. | – | Problems |
| Univ. | – | Universal |

**SUMMARY**

Today, people face greater physical and mental problems in their daily lives. These problems are related to lack of physical, mental and spiritual harmony. Yoga helps to create bridge for this gap. "Yoga" originated from a Sanskrit word that means "to join" or "to unite". Yoga exercises balance of body, mind, consciousness and soul. This in a way, helps us to tackle various problems that afflict us in our day-to-day life. The practice of Yoga has been prevalent since ancient times as prescribed in the Vedas. "Yoga in daily life" provides practical instructions for the body, breathing, concentration, relaxation and meditation. Yoga paves the way from self-knowledge to self-realization.

## SECTION–B

## (ADVANCED WRITING SKILLS) 30

**3. You are Dhruv/Deepa. Your father, Shri Dheeraj Garg of Gurugram wants you to draft an invitation to be sent to friends and relatives on the occasion of your elder sister's marriage. Prepare the invitation giving necessary details in not more than 50 words. [4]**

**OR**

**Water is precious and each one of us must stop its wastage to avoid its scarcity in the near future. Prepare a poster requesting people of your colony to adopt means to save water, prevent its wastage, etc. You are Rama/Rohan of Ashima Vihar, Saket, Ahmedabad. (50 words)**

**Answer :**

**Mrs. and Mr. Dheeraj Garg**
solicit your gracious presence on the auspicious occasion of the marriage of their daughter
**Alisha**
**With**
**Rohan**
(Son of Mrs. Kalpana and Mr. Nayak Sharma)
on
**24th April 20XX**
at 5 p.m. at
**Diamond Party Lawn, S.V. Road, Gurugram**

R.S.V.P :
Dheeraj Garg
B-1077, Green Enclave,
Gurugram — With best compliments from
Ph.:9876543210 — Deepa Sharma & All Relatives

**OR**

**THERE'S NO TIME TO WASTE**
**SAVE WATER POST-HASTE**

Save Water, Save Lives

**Every Drop Counts**

Do your part to save Water today !

Conserve Water By :

- Turning off the tap when not in use.
- Use a bucket to bath, to ensure minimum consumption of water.
- Avoid water leakage in tanks.
- Check taps repeatedly for water leakage.
- Promote rain water harvesting technique.

Issued in Public Interest by Rama/Rohan
N. Ahimsa Vihar, Saket, Ahmedabad

**4. Kerala was badly hit by floods due to heavy rains. Even today the victims are suffering financially, physically and mentally. Write a letter in 120 – 150 words to the Editor. 'The Times of Kerala' giving details of the loss of life and property. Also, make an appeal to the people and the Government to provide help and relief to the afflicted. You are Varun/Vandana, 31 Shastri Nagar, Trichi. [6]**

**OR**

**You are Shobha/Sushil from 56, A Block, Saket, Agra. Write an application along with a personal resume in 120–150 words in response to the following advertisement.**

**Wanted an experienced hockey coach having B.P.Ed Degree for our school. Apply to the Principal, Agarsen Public School, Agra by 15th March, 2019.**

**Answer :**

31 Shastri Nagar,
Trichi

3rd March 20XX

The Editor
The Times of Kerala
Thiruvandapuram

**Subject : Relief for survivors of floods in Kerala**

Sir/Madam,

Through the columns of your esteemed and widely read newspaper, I would like to highlight the present conditions of the people affected by the floods that devastated Kerala last year.

The floods that caused a lot of death and destruction in 'God's own land, has left behind a trail of victims a year after it wreaked havoc in the state. These survivors of the flood are still struggling to deal with the consequences on their own. They are dealing with the physical damage as well as the mental and emotional trauma that comes with having survived the event without those that are dear to them. These people have lost their livelihood and some have even lost mobility and structure.

I request you to highlight their plight in the newspaper. They are in need of the basic necessities and more. Any help offered by the government and the people reading this, would make a world of change in their lives.

Yours respectfully,
Varun/Vandana

**OR**

56, A Block, Saket
Agra

10th March 20XX

The Principal
Agarsen Public School
Agra

**Subject: Application for the post of hockey coach.**

Sir/Madam,

In response to your advertisement in the Indian Express dated 5th March, 20XX for the position of hockey coach, I hereby offer my candidature for the same. I believe that my experience and background fit your requirements for the position.

I have had the pleasure to be the hockey coach at St. Stephen's College, Delhi for the last five years and the pleasure of assisting the esteemed coach Pramod Srivastav at Tripti High School for four years as well. I have two Bachelor's degree, one in Physical education and another in Education. Coaching students has been my passion and I am sure that I can work well with the team. Both the teams I have worked previously have been able to play in multiple competitions statewide. I have received positive feedback from the students, the parents and the school.

You may call me for an interview on any date as per your convenience. I shall be able to join my duties at one month's notice if appointed. As requested, I have enclosed my resume with this letter.

Yours sincerely,
Shobha /Sushil

Encl. : Resume

**RESUME**

| | | |
|---|---|---|
| Name | : | Shobha / Sushil |
| Father's Name | : | Mr. Krishna |
| Age | : | 32 years |
| Address | : | 56, A Block, Saket Agra |
| Hobbies | : | Music, Reading, Researching, Photography |
| Languages Known | : | Hindi, English |
| Nationality | : | Indian |

| | | |
|---|---|---|
| Educational Qualifications | : | Graduation with First Division in B.P.Ed from Delhi University<br>B.Ed. with First Division from Delhi University |
| Experience | : | Worked as a Hockey Coach for five years at St. Stephen's and as Assistant Coach for four years at Tripti High School, Delhi |
| Expected Salary | : | Negotiable |
| References | : | 1. Mr. Pramod Srivastav (Head Coach), Tripti High School, New Delhi<br>2. Mr. Matthew (Principal), St. Stephen's, Delhi |

**5. You are concerned about the cleanliness of parks, roads and even your school premises. In spite of various measures taken by local and school administration, lack of cleanliness is still a problem. You are Namita/Keshav. Prepare a speech in 150–200 words to be delivered in the morning assembly of your school on the topic, 'Cleanliness is the Responsibility of All.' [10]**

**OR**

**Write a debate in 150 – 200 words either for or against the motion :**

**'Stray dogs should not be put to sleep'.**

**Answer :**

**CLEANLINESS IS THE RESPONSIBILITY OF ALL**

Honorable Principal, teachers and all my dear friends, Good Morning to all!

I am Keshav/Namita of class XII C. It's a great honour for me to stand before everyone and deliver a speech on the importance of cleanliness.

Today, I would like to address the idea of cleanliness in our locality. Specifically, I would like to talk about cleanliness in public spaces such as parks and roads as well as other spaces that host large groups of people simultaneously such as our school.

Local authorities have taken various measures to combat this issue of keeping an area clean by fining people for littering in public spaces and banning defecation in open spaces. The school also has rules regarding cleanliness and certain punishments for littering. However, I feel that these don't seem as useful because the job of cleaning up has been left to the custodians, both in school and in these public spaces. It is a state of sorry to see people twice our age cleaning up after the future citizens of a country. It is important for us, as students to start realizing that cleanliness of a space is not a job that we should be pushing onto other people, but something we should be doing as a personal responsibility as good citizens.

It is important that each of us as citizens do our part in maintaining the cleanliness of our surroundings. It can be as simple as picking up any piece of paper we drop or making sure that the things we are throwing away are inside the wastebaskets. These small steps can have a large impact, so to all of us, I say it is time for us to stop thinking of the idea of cleanliness and start doing what we can to make a clean and healthy environment a reality. Thank you.

**OR**

**STRAY DOGS SHOULD NOT BE PUT TO SLEEP**

**FOR :**

Honourable judges, teachers and my worthy opponents. A very good morning to all of you gathered here. Today, I am glad to have the opportunity to speak for the topic "Stray Dogs Should Not Be Put To Sleep". This topic is one that can be emotional to many. I shall be focusing on both the emotional and factual aspects of this topic.

The saying "a dog is man's best friend" has been around for centuries. For years, we have taken care of them and they have been nothing but loyal to us. They don't demand much from us except for love and attention. The stray dogs that roam in the streets should be seen from another point of view they are as the dogs that have not had the love and care that animals need to survive. Their aggressiveness comes from a place of fear for those who have hurt them, thrown stones at them and yelled at them multiple times. Had people seen these vulnerable creatures and shown them affection and care, then they would not need to be treated in such a harsh manner.

Studies have shown that most stray dogs have lesser health problems and are friendlier when given love and attention. Maintaining these breeds who are well-adjusted to the Indian climate is less costly than getting a costly breed. I believe that the issue of having strays on the roads could have been tackled with a little love and care. They wouldn't have been strays in the first place if we were more accommodate this breed. So, extermination is not the right solution. Instead, we should spread awareness about these beautiful God's creation and encourage people to adopt them.

In conclusion, I would like to point out that putting down animals that have never had the opportunity to be loved as they should have because of decisions made by humans, is a selfish act.

**AGAINST :**

Honourable judges, teachers and my worthy opponents. A very good morning to have all of you. Today, I am going to speak against the motion.

We often see stray dogs roaming here and there in the locality. Some of them can be dangerous but mostly are harmless. They remain in untidy atmosphere, eat whatever is available and breathe dirty smog. Stray dogs are certainly a threat to routine life. So, it is better to put them off to sleep. When stray dogs sleep children and elders can heave a sigh of relief. These stray dogs otherwise chase them, scare them and make their life miserable . They do not have control over their excitement and anger like the domestic dogs. They are not vaccinated and are more prone to many of the diseases, they can spread the infection if any, or diseases to humans too. Stray dogs also have a temptation to eat everything and anything as they do not get proper food and care. Their babies too catch infections and spread the same to children or people who touch them or come close to them. They are left to fend for their own which culminates into chaos. Hence, it is better to keep away from stray dogs and it is the best to put them to sleep more often.

6. **Abundance of advertisements in newspapers, magazines and even on television is a wastage of both time and money of readers and viewers. You are Shobha/Sameer. Write an article in 150 – 200 words on the topic, 'Excessive Use of Advertisements in Media'. [10]**

**OR**

**An NGO, 'Health for All' organised a health check-up camp in a slum at Amritsar. You are Ruchika/Raunak, and you visited the camp. Write a report in 150–200 words covering arrangements such as registration, check-up, tests carried out, etc.**

**Answer :**

**EXCESSIVE USE OF ADVERTISEMENTS IN MEDIA**

By Shobha/Sameer, XII A

Mark Twain once said, "Many a small thing has been made large by the right kind of advertising". Today's media platforms contain more ads than the content itself and this applies to all form of media that we choose to entertain ourselves with. To the media, advertising is more profitable than the content they provide.

Televisions today show more advertisements than the actual content and newspapers contain more ads per page than it does the actual information. Magazines are even more confusing, often giving articles that serve no purpose other than to convince the reader to buy the product advertised on the next page. These advertisements waste the money of those who have subscribed to the product or service for the sake of information. Furthermore, it is statistically shown that the advertisements make people believe that they are missing out on trends leading to excessive consumerism and materialism.

Taking everything into account, it is important to note that reducing the amount of advertising present in the media might have a positive impact. Perhaps, the space allotted to advertisements can be kept as a percentage of information and content presented in order to ensure that the content takes precedence over the advertisements.

**OR**

**HEALTH CHECK UP CAMP IN A SLUM**

By Ruchika/Raunak

**Amritsar, March 20XX :** A health camp was organised in a slum in Amritsar by the NGO 'Health for All.' The camp was set up in such a way that each person who opted to get the health checkup first had to register at the counter by giving his name and age. Once the check-up was completed, the people were requested to submit the forms at the registration desk.

Upon registration, the people were required to visit a general physician, a dentist, and an optometrist. Once the available doctors had examined them, the participants were required to get a blood test done. For most people tested, it was a basic work-up of their non-fasting blood sugar and other related tests. However, for those who suffered from other diseases, the doctors made a note of it in the form given to them, requesting specific tests be done to understand the severity of the condition. Financing for additional tests were also covered by the medical laboratories associated with the NGO.

The results of these tests shall be available to the NGO in a few weeks' time and then the doctors and volunteers will return to the slum to treat or advice as necessary to the participants that require it. It was refreshing to see a huge turn-up for this eye-opening event.

**SECTION–C**

**(LITERATURE : TEXTBOOKS AND LONG READING TEXT) 40**

7. **Read the extract given below and answer the questions that follow : [1×4=4]**

Perhaps the Earth can teach us
as when everything seems dead
and later proves to be alive.
Now I'll count up to twelve
and you keep quiet and I will go.

**(a) Name the poem and its poet.** [1]

**(b) What does the Earth teach us?** [1]

**(c) What does the poet want to achieve by counting up to twelve?** [1]

**(d) Give an example of what now seems to be dead but later on becomes alive.** [1]

**OR**

A thing of beauty is a joy forever
Its loveliness increases, it will never
Pass into nothingness; but will keep
A bower quiet for us, and a sleep
Full of sweet dreams, and health, and quiet breathing.

**(a) Name the poem and its poet.** [1]

**(b) How does a thing of beauty remain a joy forever?** [1]

**(c) Explain : 'never pass into nothingness'.** [1]

**(d) How does the bower help us in our life? Give two examples.** [1]

**Answer :**

**(a)** The poem is 'Keeping Quiet' and the poet is Pablo Neruda.

**(b)** The earth teaches us to be patient, productive and to live in tranquility.

**(c)** The poet wants to enlighten us on the importance of patience and meditation by counting up to twelve. By remaining still and through reasoning, we can achieve to be positive and overthrow the negativity within us.

**(d)** A seed when falls into the earth and dies, grows into a plant which bears fruit, gives shade when it grows into a tree, and helps the environment.

**OR**

**(a)** The poem is 'A Thing of Beauty' and the poet is John Keats.

**(b)** A thing of beauty is eternal as it revives our memories and leaves a positive influence on our minds. As time passes by, its beauty only increases with time. This only helps in providing everlasting joy to the eye of the beholder.

**(c)** Joy that lasts in beauty is everlasting, its importance never diminishes but grows with time.

**(d)** Bower is a peaceful dwelling place that helps us is our life. Under the shade of the tree, it provides us protection from the sun and rain. It is also a safe and quiet place for rest and meditation.

8. **Answer any four of the following question in 30 – 40 words each :** **[3×4=12]**

**(a) 'Garbage to them is gold.' Why does the author say so about the ragpickers of Seemapuri?**

**(b) Which two incidents in Douglas' early life made him scared of water ?**

**(c) Why was the crofter so talkative and friendly with the peddler?**

**(d) "It ate my face up. It ate me up." Who said these words? Why?**

**(e) Why did Evans not take off his hat when Jackson ordered him to do so?**

**(f) Why did Zitkala-Sa resist the shingling of her hair?****

**Answer :**

**(a)** Rag picking is a means of survival to the ragpickers of Seemapuri. They sell all that they pick from the heap of garbage in exchange for money to meet their needs. At times, they even find currency notes and coins in it. To these ragpickers, garbage is equivalent to treasures of gold.

**(b)** At the age of three or four, Douglas was knocked down by strong waves of the Yakima river in California. As a result, Douglas was almost buried under water, and got breathless. Later, at the age of ten or eleven, he was tossed in the YMCA pool by a senior where he ended up swallowing water. These two incidents played a powerful role in developing Douglas' aversion to water.

**(c)** The crofter was a lonely old man living all alone with no family. He craved for a human company. Therefore, when the peddler requested the old man for shelter, the old man, in his generosity and need for someone to talk, welcomed the peddler for supper old men offered porridge and tobacco to eat. He also played 'Majolis' with him.

**(d)** Derry showing his inferiority complex, says these words since he is often singled out by people as an ugly poor boy due to his acid-burnt face. As a result of his injury and the behavior of the people in his community, he doesn't trust people and he looks down upon him self as an outcast.

**(e)** Evans had long wavy hair and covered it with a hat as after the exam he was going to

**Answers are not given due to change in present syllabus

impersonate Mcleery and escape from the prison. Mcleery had short hair so Evans had cropped up his hair and hid it under the hat. He confided in Jackson that the hat was the only lucky charm that had brought him luck in his life and so refused to take off his hat.

**9. Answer any one of the following questions in 120 – 150 words : [6]**

**(a) Everybody during the last lesson was filled with regret. Comment.**

**(b) Why is the Champaran episode considered to be the beginning of the Indian struggle for independence?**

**(c) It is not unusual for a lower middle class girl to dream big. How unrealistic were Sophie's dreams?****

**Answer :**

**(a)** The story, "The Last Lesson" is set during the Franco-Prussian War when the French districts of Alsace and Lorraine passed into the hands of Prussia. The Prussian government then made the declaration that only German was to be taught in the schools in Alsace and Lorraine. This order affected the local public badly and each one of them was repenting for wasting time and not learning French. Each person in the class was filled with regret. Even the school appeared solemn and strange. The French language teacher, M. Hamel felt guilty for sending his students to complete his personal tasks. The parents of the students never bothered to send their children to class but instead preferred that the children work on the farm or at mills. Even the older people in the room like old Hauser, the former mayor, the former postman and other elders of the village showed regret on neither learning French themselves nor encouraging their children to learn the language.

**(b)** India was under the suzerainty of the British power for more than a hundred years before the incident at Champaran. The people of India were tired of the unjust rules and policies of the British and started the first mass movement in India. The landlords at Champaran were forced to grow indigo in 15% of the available land and then surrender the entire harvest as rent. When the peasants learned about the development of synthetic indigo, they demanded for their money. Gandhiji managed to get the amount back and provided courage to the peasants. The incident at Champaran was an attempt to alleviate the distress of large numbers of poor people. At Champaran, Gandhiji received the support of thousands of people. The peasants became aware of their rights, grew more courageous and began to contribute to the movement. Gandhiji successfully implemented the method of non-violence and non-cooperation. Therefore, it can be said that the Champaran episode is the beginning of the Indian struggle for independence. This is the successful Civil Disobedience movement.

**10. Answer any one of the following questions in 120–150 words : [6]**

**(a) Describe the efforts made by the Tiger King to achieve his target of killing a hundred tigers.****

**(b) Explain the reaction of Dr. Sadao's servants when he decided to give shelter to an enemy in the house.**

**(c) Describe Jack's art of storytelling.**

**Answer :**

**(b)** Dr. Sadao Hoki resided along the coast in Japan with his wife Hana. During the war, he was not sent along with the other troupes because the General was under his treatment and needed Dr. Sadao's services. On one occasion, Dr. Sadao saw an injured young American soldier in need of his help. He brought the soldier home and began treating him. The servants were frightened and puzzled. As a matter of fact, the old gardener declared that Dr. Sadao ought not to heal the wound of that white man because he should have been dead also as he was their enemy. He also believed that the soldier was first shot and then the sea caught him and wounded him with her rocks. If Dr. Sadao healed that shot and the sea had done, then Dr. Sadao and the servants would have to suffer. They were also skeptical because the man was their enemy and if the news of the man at the house spread, then their lives would be in danger. The servants finally chose to leave the house altogether on the seventh day.

**(c)** Storytelling is a ritual for Jack and his four-year-old daughter Jo. He aims to express his wisdom, art and creativity infused with cunningness through his stories. Jack begins his story telling sessions in the evenings and before Saturday naps. The originality of these stories are that they all revolve around a tiny creature called Roger. His stories more or less follow one pattern with minor variations depending on the theme for the day. Each story begins and describes a Roger who may be a squirrel, a fish or a chipmunk. The stories also have a wise owl and a magic wand wielding wizard. The story always starts with

**Answers are not given due to change in present syllabus

a problem. Then, they goes to wise owl for advice, owl send him to wizard. At last wizard casts a spell to solve the problem. Each story is a lesson in morals. Jack successfully manages to capture the attention of his daughter because he understands her span of attention while using his creativity to create the desired effect. He can imitate different voices and uses this positive attribute to communicate him ideas in an enthusiastic manner. Each of these stories are his own invention and he uses this talent along with his other talent of adopting the right facial expressions to capture his daughter's interest and provide life lessons along the way.

**11. Answer any one of the following questions in 120–150 words : ** [6]**

**(a) Describe the burglary that took place at Mr. Bunting's house. (The Invisible Man)**

**(b) How did Griffin move about in the village of Iping?**

**(c) Why does Silas return to Lantern Yard after having lived and prospered at Raveloe? How does his visit prove useful to him?**

**(d) How was Silas transformed completely with the arrival of a child?**

**12. Answer any one of the following questions in 120 – 150 words : ** [6]**

**(a) Describe Nancy Lammeter's personality and approach to life. (Silas Marner)**

**(b) Writer a character sketch of Godfrey Cass. (Silas Marner)**

**(c) Marval is the only character in 'The Invisible Man' who interacts with Griffin and gains something. Elaborate.**

**(d) How did Dr. Kemp use the information given to him by Griffin himself to get him arrested? Was he successful in doing so?**

●●

# Outside Delhi [Set-II]

**Note : Except for the following questions, all the remaining questions have been asked in previous sets.**

## SECTION–B
## (ADVANCED WRITING SKILLS)

**4. The elderly living alone in cities are not safe. Reports of violence against them are really very disturbing. The recent murder of two elderly persons has rocked the city. It seems the police are not taking adequate steps to make the elderly feel safe and secure. Write a letter in 120–150 words to the Editor, 'Modern Times' to draw the attention of the law and order authorities to the problem and take effective steps to deal it. You are Sohan/Sangeeta, 12, Roshan Colony, Alampur. [6]**

**OR**

**You have read an advertisement in the newspaper, 'Pune Times' for the post of software engineer in Alex Software, Pune. You believe that you possess the requisite qualifications and experience and your innovative ideas will prove an asset to the company. Write an application in 120–150 words emphasising your strong points and your suitability for the post. Also include your biodata. You are Saurabh/Shikha, 12, Mall Road, Chandigarh.**

**Answer :**

12 Roshan Colony,
Alampur

3rd March 20XX

The Editor
Modern Times
Alampur

**Subject : Safety precautions for the elderly in the city.**

Sir/Madam,

Through the columns of your esteemed newspaper, I, on behalf of the residents of Roshan Colony, wish to bring to your kind attention, the murder of the two elderly people that have taken place recently. As a resident of the area in which the attack took place, I write to point out the lack of protection that the seniors are offered.

The recent murders, while brutal, are not a surprising phenomenon considering the lack of proper precautions taken to ensure the safety of the elderly. While I mourn the loss of the two souls that were attacked, I also fear for those who are still with us. It is even more surprising to see that despite these attacks, very little is done to ensure their safety. It is important, in today's times that the authorities take proper precaution for those in our communities who are the most vulnerable *i.e.*, the elderly. I suggest to increase police vigilance and also increase the awareness for joint families and effective RWA.

I believe that highlighting it through your newspaper would bring it to the attention of the proper authorities so that they can take

**Answers are not given due to change in present syllabus

appropriate and legal steps to ensure the safety of the elderly at the earliest.

Thanking you,
Yours respectfully,
Sangeeta/Sohan

**OR**

12, Mall Road
Chandigarh

10th March 20XX

The Manager
Alex Software
Pune

**Subject : Application for the post of a Software Engineer**

Sir/Madam,

In response to your advertisement in the Pune Times dated January, 20XX for the post of a software engineer, I hereby offer my candidature for the same. I believe that my skills and qualifications match your requirements.

I have worked as a software engineer at GNC for the last eight years. It has been my dream to work in this sector which is playing such an important role in the development of society. My area of specialty is user interface and coding. I have been appreciated for my efficiency by my previous employers. If given the opportunity, I am sure I will be able to prove my capabilities to you as well.

I would be glad to get this opportunity and speak more about my work in a personal interview. I am available to come to Pune at your convenience. Thank you so much for reading my letter and for considering me for this position. I am enclosing my detailed resume for your perusal.

Sincerely,

Shikha / Saurabh

Encl. : Detailed Resume and Work Experience Certificate

**RESUME**

| | | |
|---|---|---|
| Name | : | Shikha / Saurabh |
| Father's Name | : | Mr. Amit |
| Age | : | 36 years |
| Address | : | 12, Mall Road, Chandigarh |
| Hobbies | : | Coding, Reading, Music, Photography and Sports |
| Languages Known | : | Hindi, English, Marathi and Punjabi |
| Nationality | : | Indian |
| Educational Qualifications | : | Graduation with First Division in Computer Engineering from IIT Delhi<br>Diploma in Graphic Design from Ace Design Academy |
| Professional Qualification | : | Diploma in Computers from Aptech with first division |
| Experience | : | Worked as a Manager – Technology for eight years at GNC, Mumbai and Software Engineer for two years at Tonarry Tech, New Delhi |
| Expected Salary | : | Negotiable |
| References | : | 1. Mr. Agarwal (Director), GNC, Mumbai<br>2. Mrs. Shah, Principal,IIT Delhi, New Delhi |

**5. Career conscious students focus primarily on academics and neglect participation in games and sports, not realising how important physical activities are. You are Vineet/Vinita. Prepare a speech in 150–200 words to be delivered in the morning assembly of your school on the topic, 'Balancing Academics and Sports.' [10]**

**OR**

**Write a debate in 150–200 words either for or against the motion :**

**'21st century is an age of greed'.**

**Answer :**

**BALANCING ACADEMICS AND SPORTS**

Dear audience, Good morning to all. I, Vineet/ Vinita, feel extremely happy as it's a matter of great privilege for me to deliver a speech on the topic, "Balancing Academics and Sports".

Today is a beautiful day for playing outside, and yet, many of our students and my fellow classmates, will skip the chance to play in favour of another round of revision of our study materials. This is done in hopes of improving their scores and having a better career, a better future. However, in doing so, they are ignoring one very important factor that is essential for a successful future i.e., health.

While pulling away from the books to play or participate in sports for a while might sound like a huge sacrifice, it is a necessity. Finding a balance between sports and academics ensures success in the future. It has been proven that our brain cannot process all the information given to it without breaks. Sports and other physical activities help our brain process things faster. Focusing solely on academics leads to anxiety, distress and depression in students. Another reason to neglect physical activities is parental Pressure, peer pressure and increasing desire for ambition, which leads to poor physical health.

I would like to conclude this speech by saying that health is something to be treasured, we can never regain it if lost. Hence, Sports should be an integral part of the curriculum. Focusing all our energy on academics at such a crucial age can have a huge effect that can change our lives. Thanks to all for being so generous in listening to my views so attentively and patiently.

**OR**

**21$^{ST}$ CENTURY IS AN AGE OF GREED**

**AGAINST :**

Respected chairperson, members of the jury and my worthy opponents, I, xyz of class XII am glad to have the opportunity today to argue against the motion that "21st Century is an age of greed".

The notion that the 21st century is full of greed is one that is easy to arrive at. Today we have more material comfort than we probably know what to do with and a great divide in the incomes that people earn. Yet, we are in a time when more people are giving to charity than ever before. In the past few years, those who can, have started donating generously to multiple causes. There exist NGO's and charities that help people with everything from getting back onto their feet after a natural disaster to curing terminal illnesses. What is important to note is not just that these charities exist, but that there is an increasing number of people giving generously towards various causes. Moreover, according to statistics, people want peace by involving in many activities like yoga, meditation, etc. As a matter of fact, social media has played an important role in encouraging people to donate to the right medium.

To conclude, I would like to point out that while there is an increasing number of products, technological advances and services people are paying for, there is a balanced increase in the amount that people are giving to those in need. Thank You!

**OR**

**FOR :**

Respected chair person, members of the jury and my worthy opponents, I, XYZ of class XII am glad to have the opportunity today to argue in favour of the motion-'21st century is an age of greed'.

21st century is an age of technological advancements, digitization of services, and dependence on social media. It is rightly said by Dan Robinson that, 'One of the weaknesses of our age is our apparent inability to distinguish our needs from our greed.' Greed is defined as an insatiable longing for food, money, power, or status.

The 21st century has not only satisfied our needs, but it has pushed us to the point of becoming greedy. The culture of consumerism is fostered by e-commerce websites which lure customers into buying things by glorifying the idea of spending money and by offering strategic discounts. This fulfills the greed of both the traders as they are chasing profits as well as customers who are eyeing after material goods. Moreover, quality of these goods is mostly compromised since fraud is the daughter of greed.

Next comes the greed for power. Political parties are greedy to win the majority and hence there are uprisings, protests, debates, and other forms of political manoevuvres to hold on to that power and to demonstrate it.

Everyone has two wolves inside them one representing greed and the other gratitude. It is up to each person to choose the wolf they want to feed. And the people in the 21st century are only feeding greed.

**SECTION–C**

**(LITERATURE : TEXTBOOKS AND LONG READING TEXT)**

**8. Answer any four of the following questions in 30 – 40 words each : [3×4=12]**

**(a) What are the two most admirable qualities in Mukesh ?**

**(b) What is wrong with Jansie's outlook on life?****

**(c) What does Neruda wish fishermen and warmongers to do ?**

**(d) How is the Derry we meet at the beginning different from the Derry at the end ?**

**(e) Why was the General apologetic when Sadao told him that the prisoner had escaped?**

**(f) Why did Zitkala-Sa oppose tooth and nail the shingling of her long hair ?****

**Answer :**

(a) Two admirable qualities in Mukesh are that he dares to dream and is focused on achieving them. Mukesh refuses to accept his destiny as a bangle-maker like his father and ancestors. He is ready to walk the long distance from his house to the garage where he could learn to be a motor mechanic which he aims to be.

**Answers are not given due to change in present syllabus

(c) Neruda is a lover of peace and non-violence. He wishes that the fishermen do not hunt the whales in the cold regions. He also wishes that the warmongers who normally indulge in wars and violence, walk with their brothers in some of their best clothes in the shade of trees, doing nothing.

(d) Derry at the beginning of the story is a very negative person and because of his ugly appearance, hides away from the crowds. He believes that the acid has eaten his skin and him alive. He feels that he doesn't need any second chance to live. However, Derry at the end of the story is more optimistic and practical. He decides to live happily by being friendlier with people.

(e) The General believed that Sadao was indispensable to his life as none were as skilled as Dr Sadao. When Sadao told him about the prisoner who had escaped, he was apologetic as he had forgotten about his promise of sending assassins to kill the enemy, that he made to Sadao, in his greed and selfish needs.

●●

# Outside Delhi [Set-III]

**Note : Except for the following questions, all the remaining questions have been asked in previous sets.**

## SECTION–B
## (ADVANCED WRITING SKILLS)

**5. Respect for law, concern for others and maintaining a clean environment are some of the qualities of a civilized person. It is sad that we do not pay much attention to them. You are Shoma/Sonu. Write a speech in 150–200 words to be delivered in the morning assembly on the topic, 'Importance of Civic Sense'. [10]**

**OR**

**Write a debate in 150–200 words either for or against the motion :**

**'Money is the only thing that matters in today's life'.**

**Answer :**

**IMPORTANCE OF CIVIC SENSE**

Dear Friends and Faculty, I, Shoma/Sonu of class XII-C, wish all of you a very good Morning. I am privileged to present my views on the importance of civic sense with you today.

How often have we seen people spitting on the roadside? How often have we seen people standing, watching someone else in pain? How many of us leave the lights on when we don't really need it? Many of us show disrespect for law by breaking traffic rules, vandalising public property, abusing public facilities etc.

We see these things on a daily basis. We curse the people who do these things but then we do it ourselves too. We say to ourselves, "Why should I stop when no one else is?" We forget that every single person's actions matter.

We live in a time where we stand at the edge of irreversible changes. In such a time, it becomes more important than ever that we hold on to our civic sense. If we continue to live as individuals rather than a society on this planet, just functioning based on rules that ensure our mere survival, then there will come a time when the society will never be the same. Knowing this, it becomes even more important to emphasize the importance that civic sense can have in our very lives. It is high time that we all start to live with civic sense and social mannerism for a cleaner and healthier environment. This will help us change the society at the earliest.

Thank you.

**OR**

**MONEY IS THE ONLY THING THAT MATTERS IN TODAY'S LIFE**

**FOR :**

Honourable judges, teachers and my worthy opponents, today, I am glad to receive this opportunity to present to you my side for the argument "Money is the only thing that matters in today's life".

In today's society, having money can be the difference between life and death. It has become common knowledge to know that one can get anywhere if the right people are on your side. The way our society is built, the value placed on money and having more of it has lead to the rise of Billionaires with so much wealth that they no longer are in touch with how the common man survives and in the same place, we have people who cannot access basic necessities because of the lack of funds.

This disparity does not reduce as time goes on, it grows. This is because access to good education, high-paying jobs and better accommodations are available only to those who have the money to afford it. While it is easy to say that the younger generations place less importance on money, it is only those with money that can say this. Without money, one cannot get anywhere in today's society. At the end of the day, one can't help but agree that money moves the world around.

OR

**AGAINST :**

Honourable judges, teachers and my worthy opponents, today, I am glad to receive the opportunity to present to you my side against the motion.

No, money is not the only valid thing which matters in today's era as there are other things too. Relationships and emotions are more important than anything else. People are trying to build their lost connections with their near and dears. Money is also important for several things but at the same time relationships are losing value but everyone is reaching out to their long lost bonds with their roots. Health factor is also high and people are finding ways to improve the same to march towards longevity. Yoga, gym, and exercise everything is finding relevance among the young and the old as well. People are becoming more conscious on the health point of view and hence are trying their best to remain fit and fine. Many youngsters even trying to volunteer and educate people at all levels and places to stay alert and aware about their health and make efforts to remain healthy. Health and wealth put together makes a perfect ingredient for everyone's life. There is also a lot of awareness about food and beverages for everyone's better health. Money does matter for all but it is not the only thing that matters in the current scenario.

6. **The tendency to use violence at the slightest provocation and thus to solve personal and general problems is growing among people in our society. This is not a civilized way of dealing with the problems. Violence solves no problem; it must be shunned. Write a article in 150–200 words on the topic, You are Shan/Shobha. [10]**

OR

**The Fine Arts Club of your school celebrated its tenth anniversary in a grand manner. Distinguished artists Pt. Jasraj, Hariprasad Chaurasia and the veteran Kathak dancer Shovana Narayan graced the occasion and enthralled the audience by their mesmerizing compositions. Shovana Narayan's graceful movements were a delight to watch. The function ended with two lilting songs by Shambu, a class 10$^{th}$ student of your school. Write a report in 150–200 words on the event, to be published in your school magazine. You are Sohan/Sakhi, Secretary of the Club.**

**Answer :**

**'DON'T TAKE LAW INTO YOUR OWN HANDS.'**

By Shobha / Shan

"An eye for an eye makes the whole world blind." This was stated by Mahatma Gandhi when our nation was fighting for its independence from the British Raj. Yet, today our nation faces the growing problem of violence at the slightest provocation. There is no respect for law and no fear of consequences.

All it takes for a violent attack on another person is their presence. Women have been attacked for working at officer or studying in colleges because others believe they do not have the right to be there, people of different race or colour are abused and attacked verbally or physically simply for existing in a certain space. The tendency to hate is strong. There is a growing crowd of both victims and perpetrators and sometimes one can't help but wonder if the need to rely on the laws of the land is indeed worthwhile. All these incidents lead to stressful life style. Intolerance level is growing day-by-day.

Sadly, taking matters into ones hands are not the methods through which problems are truly solved. All violence does is to breed more violence. It is important to take steps to help people find the right way to solve problems, through the use of words over actions. Schools, parents, and media can play a responsible role in spreading awareness. This can only be done if there are proper precautions in place for people who act out aggressively with the barest hint of provocation. It is time that we, as a society, look at the problem of violence more seriously and try to find a solution together.

OR

**10$^{th}$ ANNIVERSARY OF THE FINE ARTS CLUB**

By Sakhi / Sohan, Secretary of Fine Arts Club

Last Friday marked the tenth anniversary of our school's Fine Arts Club. To commemorate this achievement, a grand celebration was held to honour the milestones achieved by the club. The celebration, lasting nearly an entire day, included a variety of renowned artists and veterans of the business as well as performances by the club members. Held at the School Auditorium, the Club is proud to say that the entire hall was full of guests as well as audiences. Aside from the notable faces, the crowd also held parents of the club members as well as outsiders who had bought tickets to attend the event.

The audience was treated by performances of the distinguished artists such as Pt. Jasraj and Hariprasad Chaurasia. Their performances captivated the audience and their compositions moved the audience changing the air of the entire celebration. We were also graced with the veteran Kathak dancer Shovana Narayan who gave us the honour of performing on the stage.

In grand finale of the show, there was two songs performed by Shambu, a class 10 student and a club member. The talented and heart-wrenching performance earned a standing ovation from the audience and there were requests for an encore, which we were unable to perform due to time constraints. The President of the club presented the vote of thanks after which light refreshments were served.

## SECTION–C
## (LITERATURE : TEXTBOOKS AND LONG READING TEXT)

**8. Answer any four of the following questions in 30 – 40 words each : [3×4=12]**

**(a) What steps did Gandhiji take to remove the cultural and social backwardness in the rural areas of Champaran ?**

**(b) Why was the crofter so kind and generous to the peddler ?**

**(c) Give examples from 'Keeping Quiet' to show that Neruda was peace loving.**

**(d) How did Annan's advice to his sister Bama impact her ?****

**(e) What dilemma did Sadao and his wife face when they saw the wounded American soldier ?**

**(f) How did Jo want the story narrated by her father to end ?**

**Answer :**

(a) The social and cultural backwardness in the rural areas of Champaran caused Gandhi a great pain. His appeal for teachers was responded by several people. In about six villages, primary schools were opened. Gandhi's wife and youngest son also came to Champaran to help him. Kasturba taught personal cleanliness and community sanitation to the people. Gandhi also managed to get the help of a doctor who with three medicines, tried to fight the miserable health conditions of Champaran.

**(b)** The crofter was a lonely old man living all alone with no family. He craved for a human company. Therefore, when the peddler requested the old man for a shelter the old man, in his generosity and need for someone to talk, welcomed the peddler for supper.

**(c)** We know that the poet Neruda is a peace-loving person through several instances in the poem. He cares for both humans and nature. In the poem, Neruda wishes that fishermen do not hunt whales, he wants the man who is collecting salt to stop for a while and see the self-harm it is doing and lastly, he wishes that people who run about indulging in wars, to stop what they are doing and instead, embrace peace with their brothers.

**(e)** Dr. Sadao and his wife found the wounded American soldier on the beach near their house. The soldier needed an immediate medical attention which Dr.Sadao could give. If Dr. Sadao sheltered the American soldier, they could be arrested. However, if they handed the soldier to the police he would surely die. Thus, Dr. Sadao was in a dilemma.

**(f)** The story narrated by Jo's father ended with Roger the skunk's mother hitting the wizard for making her son smell like roses. Jo did not like this ending. She wanted the wizard to hit Roger's mother back for not letting this to be happen.

●●

**Answers are not given due to change in present syllabus

# ENGLISH

## Delhi [Set-I] 2019

**Time allowed : 3 hours** **Maximum marks : 100**

### SECTION–A
### (READING)

**1. Read the passage given below : [20]**

**(1)** All of the Earth's ocean share one thing in common : plastic pollution. Discarded plastic bags, cups, and bottles make their way into the sea. Today, it seems that no part of the ocean is safe from plastic trash. In recent years, oceanographers have searched in vain for a pristine marine environment. They have found plastic everywhere they have looked. "It is a common global problem, we can't point to a single habitat or location with no plastic."

**(2)** Plastic harms wildlife and introduces dangerous chemicals into marine ecosystems—communities of organisms interacting with their surroundings. Once plastic enters the environment, it lasts a long time. Scientists are working to prevent plastic pollution from entering the sea.

**(3)** When people litter, or when trash is not properly disposed of, things like plastic bags, bottles, straws, foam beverage cups get carried to the sea by winds and waterways. About 80 percent of ocean plastic originates on land. The rest comes from marine industries such as shipping and fishing.

**(4)** In 2015, engineer Jenna Jambeck at the University of Georgia and other researchers calculated that at least 8 million tons of plastic trash is swept into the ocean from coasts every year. That's the equivalent of a full garbage truck of plastic being dumped into the sea every minute. If current trends in plastic production and disposal continue, that figure will double by 2025. A report published by the World Economic Forum last year predicts that by 2050, ocean plastic will outweigh all the fish in the sea.

**(5)** In today's world, plastic is everywhere. It's found in shoes, clothing, household items, electronics and more. There are different types of plastics, but one thing they all have in common is that is that they're made of polymers – large molecules made up of repeating units. Their chemical structure gives them a lot of advantages : they're cheap and easy to manufacture, lightweight, water-resistant, durable, and can be moulded into nearly any shape.

**(6)** Unfortunately, some of the properties that make plastics great for consumer goods also make them a problem pollutant. Plastic's durability comes in part from the fact that unlike paper or wood, it is not biodegradable, or beak down naturally. Instead it just fragments, or breaks into tiny pieces over time. These tiny pieces, known as microplastic, can potentially stick around for hundreds or perhaps even thousands of years.

**(7)** Another problem with plastics is the other chemicals they contain, like dyes and flame retardants. When plastic isn't disposed of properly, these additives end up in the environment. Plastic also tends to absorb harmful chemicals from its surroundings. "It's like a sponge for persistent organic pollutants." These long-lasting, toxic substances include pesticides and industrial chemicals. If plastic absorbs the chemicals, and marine organisms eat the plastic, they may be exposed to higher concentrations of these contaminants.

**(8)** One of the biggest impacts of plastic pollution is its effects on sea life. Seals, sea turtles, and even whales can become entangled in plastic netting. They can starve to death if the plastic restricts their ability to move or eat. Or the plastic can cut into the animals' skin, causing wounds that develop severe infections.

**(9)** Sea turtles eat plastic bags and soda-can rings, which resemble jellyfish, their favourite food. Seabirds eat bottle caps or chunks of foam cups. Plastic pieces may make an animal feel full, so it doesn't eat enough real food to get the nutrients it needs. Plastic can also block an animal's digestive system, making it unable to eat.

**(10)** Plastic and its associated pollutant can even make it into our food supply. Scientists recently examined fish and shell-fish bought at markets in California and Indonesia. They found plastic in the guts of more than a quarter of samples purchased at both locations. In organisms that people eat whole, such as sardines and oysters, that means we're eating plastic too. In larger fish, chemicals from plastic may seep into their muscles and other tissues that people consume.

**(11)** One way to keep the ocean cleaner and healthier is through cleanup efforts. A lot of plastic waste caught in ocean currents eventually washes up on beaches. Removing it can prevent it from blowing out to sea again. Beach clean-up is ocean clean-up.

**(12)** Cleanup efforts can't reach every corner of the ocean or track down every bit of microplastic. That means it's critical to cut down on the amount of plastic that reaches the sea in the first place. Scientists are working towards new materials that are safer for the environment. For example, Jambeck and her colleagues are currently testing a new polymer that breaks down more easily in seawater.

**(13)** "Individual actions make a big difference," says Jambeck. Disposing of plastic properly for recycling or trash collection is a key step. "And simple things like reusable water bottles, mugs, and bags really cut down on waste," she says. Skipping straws or using paper ones helps too. Ocean pollution can seem overwhelming, but it's something everyone can help address. This is a problem we can really do something about.

**1.1 On the basis of your understanding of the above passage, answer each of the questions given below by choosing the most appropriate option : [1×5=5]**

**(i) Percentage of ocean plastic that originates from land is :**

(a) 20% (b) 50%
(c) 80% (d) 25%

**(ii) In which year did Jenna Jambeck and other researchers calculate that at least 8 million tons of plastic trash is swept into ocean every year?**

(a) 2018 (b) 2015
(b) 2005 (d) 2010

**(iii) Plastic is not biodegradable because it is made up of :**

(a) low atomic particles
(b) tiny particles
(c) strong big particles
(d) large molecule polymers

**(iv) Sea turtles eat :**

(a) plastic bottles
(b) plastic bags and soda-can rings
(c) bottle caps
(d) chunks of foam cups

**(v) Scientists bought fish and shell-fish for examination at markets in :**

(a) China and Russia
(b) Pakistan and Afghanistan
(c) California and Indonesia
(d) Australia and Brazil

**Answer :**

(i) (c) 80%
(ii) (b) 2015
(iii) (d) large molecule polymers
(iv) (b) Plastic bags and soda-can rings
(v) (c) California and Indonesia

**1.2 Answer the following question briefly: [1×6=6]**

**(i) Which articles made of plastic generally cause pollution in the sea ?**

**(ii) How does plastic in oceans harm marine ecosystems?**

**(iii) How is microplastic formed?**

**(iv) Why is plastic compared to a sponge?**

**(v) What is the biggest impact of plastic pollution on sea life?**

**(vi) How are scientists trying to reduce the plastic pollutants?**

**Answer :**

(i) Discarded plastic bags, cups and bottles cause pollution in the sea.

(ii) Plastic absorbs dangerous chemicals from its surroundings that include pesticides and other industrial chemicals. When marine life eats the plastic, they may be exposed to higher concentration of these contaminants.

(iii) Plastic is not biodegradable or breakdown naturally. It just breaks into tiny pieces which are known as microplastic.

(iv) Plastic tends to absorb harmful chemicals from its surroundings which is why it is compared to a sponge. It's like a sponge for persistent organic pollutants.

(v) The biggest impact of plastic pollution on the life is that seals, sea turtles and even whales can become entangled in plastic netting. They can starve to death if plastic restricts their ability to move or eat. It can cut into the animals' skin and may cause wounds.

(vi) To reduce the plastic pollutants, scientists are working towards new materials that are safer for the environment. They are testing a new polymer that breaks down more easily in seawater.

**1.3 Answer any three of the following questions in 25-30 words each : [2×3=6]**

**(i) How does plastic waste enter the ocean?**

**(ii) How is it true to say that plastic is everywhere in today's world?**

**(iii) Which property of plastic makes it a problem pollutant?**

**(iv) What has scientist Jambeck suggested for having cleaner and healthier oceans?**

**Answer :**

(i) When people litter or when trash is not properly disposed of, things like plastic bags, bottles, straws, foam, beverage cups get carried to the sea by winds and waterways. This is how plastic waste enters the oceans. About 80% of ocean plastic originates on land. The rest comes from marine industries such as shipping and fishing.

(ii) Its true to say that in today's world, plastic is everywhere as it's found in shoes, clothing, household items, electronics and more.

(iii) Some of the properties of plastic make it a problem pollutant. Unlike paper or wood, it is not biodegradable or breakdown naturally. Instead, it just fragments or breaks into tiny pieces known as microplastic which potentially sticks around for hundreds or perhaps thousands of years.

(iv) Jambeck has suggested that individual actions make a big difference. Disposing of plastics properly for recycling or trash collection is a key step. Skipping straws or using paper ones helps too.

**1.4 Pick out the words/phrases from the passage which are similar in meaning to the following: [1×3=3]**

**(i) unspoiled (para 1)**

**(ii) long lasting (para 5)**

**(iii) people working together (para 12)**

**Answer :**

(i) pristine (ii) durable (iii) colleagues

**2. Read the passage given below : [10]**

(1) Getting enough sleep is an important as taking time out to relax. A good night's sleep is essential for preserving the health of your brain and gives you the best chance to meet the coming day with a razor sharp mind. An average person needs about six to eight hour sleep a night—although it is also true that you need slightly less than this, as you grow older-another advantage of aging stress and sleep deprivation often feed on each other, since stress tends to make it harder for you to fall asleep at night and sleep deprivation in itself causes stress.

(2) Eventually, too little sleep can dramatically interfere with the performance of your memory-something you obviously want to prevent. If you are not getting enough sleep, try going to bed 30 to 60 minutes earlier than your normal bed time for a few days. Lie down on the bed and try to relax by dissociating yourself from your daily routine work. This is normally enough to catch up on any sleep deprivation.

(3) If, however, you suffer from insomnia you should seek the advice of your doctor. The chances are it is already affecting your ability to remember and recall information – and if you are struggling to improve your memory scores, this could be at the root of your problem. Prolonged periods of insufficient sleep can deplete your immune system, make you more accident prone and even cause depression—this can also reinforce a more negative outlook on life, which can contribute to your stress burden. The good news is that your memory and mood should automatically improve once you improve your sleep patterns. Tackle your sleep issues and everything else should fall into place.

(4) Because stress management is so essential to maximize your brain power, if you are not in the habit of setting aside time to relax, make it a priority to do so. Even a minute or two of deep breathing can start to work wonders. Often the best ideas and memories can come to you when you are in a state of relaxation as it is during these moments that your brain stores, processes and plays with the information it has received.

(5) Meditation has long been part of religious and spiritual life, specially in Asia. Today, more and more people are adopting it in Western countries also, for its value of developing peace of mind and lowering stress. There is some evidence that regular meditation can have real sleep gain and health benefits particularly in terms of protecting your brain against aging.

**2.1 On the basis of your understanding of the above passage, make notes on it using headings and subheadings. Use recognizable abbreviations (wherever necessary-minimum four) and a format you consider suitable. Also supply an appropriate title to it. [5]**

**2.2 Write a summary of the above passage in about 100 words. [5]**

**Answer :**

**TITLE : BENEFITS OF A GOOD SLEEP**

1. **Imp. of enough sleep**
   (i) Essential for the health of the brain
   (ii) Gives chance to meet the cmng. day with a razor sharp mind
   (iii) Avg. person reqrs. six to eight hour sleep a night
2. **Effects of little sleep or insomnia**
   (i) Interferes with memory
   (ii) Depletes immune sys.
   (iii) Makes you accident prone
   (iv) Causes depression
3. **Stress mngmnt. and meditation**
   (i) Max. brain power
   (ii) Brings best ideas and memories
   (iii) Develops peace of mind
   (iv) Protects brain against aging

| Abbreviations Used | | |
|---|---|---|
| Imp. | – | Importance |
| Cmng. | – | Coming |
| Avg. | – | Average |
| Reqrs. | – | Requires |
| Mngmnt. | – | Management |
| Max. | – | Maximise |
| Sys. | – | System |

**SUMMARY :**

Good sleep is necessary for a healthy brain. Six to eight hours sleep a night is required for an average person. Too little sleep or insomnia interferes with the performance of memory and can lead to depression. It further adds on to stress burden. Good sleep patterns help to improve mood and memory. Besides, stress management also maximizes brain power. Relaxed mind is a home for best ideas and memories. Meditation, which has been a part of religious and spiritual life particularly in Asia, develops peace of mind and lowers stress, regains sleep and health. Western countries are also adopting meditation for the same reason.

## SECTION–B
## (ADVANCED WRITING SKILLS)

3. **You are Principal of National Public School, Jaipur. You require a TGT (Maths) for your school. Drat a suitable advertisement in not more than 50 words for the 'Situations Vacant' column of 'The National Times' stating essential and desirable qualifications, experience etc. of the candidates.** [4]

**OR**

**Arts Club of your school is going to organize a drawing and painting competition. Write a notice in not more than 50 words, to be displayed on the school notice board, inviting students to participate in it. Give all the necessary details. You are Rishabh/Ridhima, Secretary, Arts Club, Sunrise Public School, Gurugram, Haryana.**

**Answer :**

**SITUATION VACANT**

National Public School requires a trained, efficient maths teacher below 35 years of age for teaching IX and X classes. The candidate must be B.Ed and possess an experience of teaching for five years in a reputed school. Walk-in-interview on 20th April from 9.00 a.m. to 11. a.m. in the Principal's office. Contact :

Principal,

National Public School,

F-16, Durgapura, Jaipur

**OR**

**SUNRISE PUBLIC SCHOOL, GURUGRAM**

**NOTICE**

5th September, 20XX

**DRAWING AND PAINTING COMPETITION**

This is to inform all the students of the school that 'Arts Club' is going to organize a drawing and painting competition for classes V to X on Saturday, 19th September in the school Crafts Room between 10.a.m to 12.00 p.m. Those who are willing to participate can give their names to their respective class teachers.

Rishabh/Ridhima

Secretary

Arts Club

4. **You have realized the necessity of education and financial independence of women for their family, society and in turn for the nation. Write a letter to the Editor, 'The National Times' highlighting your ideas on the importance of education of women leading to a better status for them. You are Tarun/Taruna, B-7/9, Mall Road, Delhi. (100-125 words)** [6]

**OR**

**You bought a refrigerator two months ago from Mohan Sales, Ashok Vihar, Bangalore. It has developed certain problems regarding its functioning. Cooling has stopped and it is making a lot of noise. Write a letter of complaint to the Manager asking him for immediate repair/**

**replacement of the same. You are Sachin/Shashi, 61 Pratap Enclave, Bangalore. (100-125 words)**

**Answer :**

B-7/9, Mall Road
Delhi

14th November, 20XX

The Editor
The National Times
Delhi

**Subject : Importance of Education for Women**

Sir,

Through the columns of your esteemed newspaper, I would like to highlight the issue of women education in our country. Education plays a vital role in improving the socio-economic conditions of our nation and educating women holds an important position in this regard. Lack of Education in women effects family and society, restricts women growth and financial independence. Earlier, women's education was forbidden in our society because of patriarchal and conservative mindset and illiteracy. But today women are highly educated and qualified professionals and scholars. There are women judges, doctors, executives, entrepreneurs, engineers, pilots and sportspersons in the country today. Women are present in every walk of life, competing with men quite successfully and this is possible only because they are given proper education facilities and opportunities. They can prove to be productive for the welfare of the society as well as the country if their education is given due importance like their male counterparts. Today, women even provide financial support to their family also.

It is said that a child's first school is her/his home and for that purpose, the mother needs to be educated enough. This not only helps the child but also helps in improving women's status and position in her family. For a developing country like India where women form half of the population, it's important to educate women for the betterment of the their family, society and in turn for the nation. Government should take steps in spreading awareness with the help of media to change patriarchal mind sets. Government and NGO's must take some incentives favouring girl child through reservations, free food, uniform, books, etc.

I hope my thoughts will persuade the people to act towards the betterment of women.

Thanking You
Yours faithfully,
Tarun/Taruna

**OR**

61, Pratap Enclave
Bangalore

10th August, 20XX

The Manager
Mohan Sales,
Ashok Vihar
Bangalore

**Subject. : Complaint regarding defective refrigerator**

Sir,

Two months ago, I purchased a double door five star Samsung refrigerator in the colour cherry red, model no. A 233408 vide cash memo no. 453, dated 10th June, 20XX from your showroom. I was extremely happy to have purchased such a branded product but to my utter disappointment, it has not been working properly for the past few days.

Though for the first six weeks, it gave a trouble-free service but after that it started having problems. Its cooling has stopped and it makes a lot of noise too. Sometimes it even gives shock and gets heated quickly.

Therefore, I request you to send your service engineer to check its functioning. In case if there is any serious technical fault, I would like you to replace it as it is under the warranty period. I hope you would do the needful.

Thanking You
Yours faithfully,
Sachin/Shashi

5. **Write a debate in 150-200 words either for or against the motion : 'Capital Punishment should be abolished.'** **[10]**

**OR**

**Regular practice of yoga is useful in maintaining good health. It is also important for good concentration and peace of mind. You are Shivam/Shabnam.**

**Write a speech in 150-200 words to be delivered in the morning assembly of your school, highlighting the impact of yoga in our life.**

**Answer :**

**CAPITAL PUNISHMENT SHOULD BE ABOLISHED**

**FOR :**

Honorable Judges, Respected Principal, Teachers and my dear friends! Good Morning. Today I feel privileged to present my views in favour of the motion 'Capital Punishment should be abolished'.

Before starting of, I would like to explain what 'Capital punishment' is? It is actually the death penalty given by the government of a country to the person who has committed any serious crime. In my opinion, 'Capital Punishment' should be abolished. By giving someone a death penalty, it doesn't mean that the crime is over or the victim is given justice. Instead, they should be punished in such a way that would make them realize and feel the pain and embarrassment from deep inside the heart. As we know that the crime graph is rising higher day by day and it's not possible to hang all the criminals to death for their heinous crimes. So the accused must be given punishment that would make others afraid and they dare not think of committing crimes in future. It is unethical to take someone's life. Life is a god's gift and may be taken by only him.

Moreover, death penalty is not for the democratic countries and civilized societies. It is not considered just and appropriate as in some cases influential and rich people hire good and reputed lawyers and could be saved if it falls on them. But on the other hand, those who cannot afford a good lawyer, succumb to death. Therefore, it is not justified. Rather imprisonment is far better to make a criminal realize his fault and to give him an opportunity to reform himself.

**AGAINST :**

Honorable Judges, Respected Principal, Teachers and my dear friends ! Good Morning. Today I stand before you to present my views against the motion 'Capital Punishment should be abolished'.

For any crime there is a punishment and certain crimes are so ghastly for which capital punishment must be given to the criminals, otherwise the justice would be denied to the victim. Besides, the criminals will get the opportunity to commit more crimes. If they wouldn't get the fear of death, they would continue their wrongdoings. If capital punishment will be abolished, the other criminals will feel free to do such things in future which is unacceptable in the society as well as in the country. When a person commits a crime, he tries to escape the punishment by offering bribe to the police officials or jail authorities or hiring an influential lawyer. Even if he gets life imprisonment, it is nothing in front of the pain he caused to the victim.

Countries like United States of America give capital punishment to the criminals to lower the crime rates. If a person commits serious crimes like murder or rape, they should be given death penalty so as to give perceptual relief to the victim's relatives. In a lot of cases, when the criminals are excused, they repeat the gruesome crimes. To avoid such situation, death penalty is must as it would refrain others from doing crimes and make the society safe. It is an appropriate punishment for the accused and also for those who continue to commit crimes even after being released from the jail.

**OR**

**IMPACT OF YOGA IN OUR LIFE**

By Shivam/Shabnam

Good Morning everybody, I am Shivam/ Shabnam, a student of class XII of your school. It is a great opportunity for me to share my views on the topic 'Impact of Yoga in our life' in today's assembly.

Yoga is not merely a physical exercise which includes stretching, breathing or twisting the body in different ways, but it results into a healthy development of mind, body and soul. Today's sedentary lifestyle has invited many diseases where man has limited time to spare for his health. It is also important for good concentration and peace of mind. With regular practice of yoga, we tend to become more sensitive to the kind of food our body asks. With the genuine effort made by our honorable Prime Minister Mr. Narendra Modi, it is celebrated every year on 21st June. Not only Indians but the whole world has accepted the power of yoga. A few minutes of yoga helps to get relief from stress. It helps detox the body and de-stress the mind. Yoga is the best way to calm a disturbed mind and increase the level of well–being in a person. On the other hand, yoga postures massage organs and strengthen muscles; breathing techniques and meditation release stress and improve immune system. Yoga and meditation work on keeping the mind happy and peaceful. When we feel drained out of energy, a few minutes of yoga can provide the freshness and energy back to our body. It also helps in improving body postures. Today, students face many problems due to the burden of studies and feel depressed but yoga acts as the stress buster for them. In many schools yoga is practiced on the daily basis as a part of curriculum. Hence, we too should add yoga to our daily routines for a healthy life.

Thank you.

**6. Hard work and punctuality are essential for a happy and successful life. They help in meeting the desired targets of our life. You are Kavya/Kanha. Write a article in 150-200 words highlighting the importance of hard work and punctuality in a student's life. [10]**

**OR**

**In your locality a blood donation camp was organized by an NGO- 'For Your Health'. Many people visited the camp and donated blood.**

**Write a report in 150-200 words for a local newspaper covering the arrangements, doctors, team, refreshment served etc.**

**Answer :**

**IMPORTANCE OF HARD WORK AND PUNCTUALITY IN STUDENT'S LIFE**

By Kanha/Kavya

Punctuality is a trait which makes a man successful in life and if it is accompanied with hard work, nothing can stop him from reaching the zenith of success. In a student's life, punctuality is must so that he/she can achieve good scores. A hard working and punctual student is respected and liked by all whether it is his teachers, society, family or friends. Students should follow the examples of great leaders like Abdul Kalam Azad in order to achieve success and fame. Being punctual refers to the quality to be on time and to complete every task in a systematic way. It encourages the student to do the work on time and keep an update of everything. This quality can be developed at any stage of life but if it is developed during childhood, it would prove to be better. Hard work and punctuality gives fruitful results in the long run. Lack of punctuality and hard work causes harmful effects. It is only the result of hard work and efforts of the scientists that the world has developed so far. Only hard work and punctuality make a student stand aloft the others and lack such trait make them lag behind. It is truly said that time is priceless. Once you have lost it, you can never get it back.

OR

**BLOOD DONATION CAMP**

By XYZ, staff reporter

**24th June, 20XX:** 'A blood donation camp' was organized by an NGO 'For Your Health' yesterday in our society. A team of eminent doctors and their efficient staff attended the camp. The main objective of organizing this camp was to emphasize the need of creating awareness among the people about the need of blood donation. People in large numbers came to donate blood whole-heartedly for this noble cause. It was so overwhelming to see their enthusiasm. The event made it clear that there are still lots of people who give their selfless contribution to save the life of others. Blood donation is ranked as the highest form of service towards mankind. Addressing the gathering, the chief of medical staff told that it was a matter of pleasure to see the enthusiastic response of the public for such a cause. He further added that donating blood regularly doesn't affect the health at all. Impressed by his speech, many people took an oath to donate blood whenever required.

At the end of the event, juice, fruits and other healthy refreshments was given to all those who donated the blood. Overall, the camp was a big success.

## SECTION-C

## (LITERATURE : TEXT BOOKS AND LONG READING TEXT)

**7. Read the extract given below and answer the questions that follow : [1×4=4]**

I looked again at her, wan, pale
as a late winter's moon and felt that old
familiar ache, my childhood's fear,
but all I said was, see you soon, Amma,
all I did was smile and smile and smile......

**(i) Name the poet and the poem.**

**(ii) What was the poet's childhood fear?**

**(iii) What is the poetic device used in line 1-2?**

**(iv) Explain : 'late winter's moon.**

OR

............The stunted, unlucky heir

Of twisted bones, reciting a father's gnarled disease.

His lesson, from his desk. At back of the dim class

One unnoted, sweet and young. His eyes live in a dream.

Of squirrel's game, in tree room, other than this.

**(i) Who is the unlucky heir?**

**(ii) What has he inherited?**

**(iii) Who is sitting at the back of the dim class?**

**(iv) How is he different from rest of the class?**

**Answer :**

**(i)** Kamala Das is the poet and the name of the poem is 'My Mother at Sixty-Six'.

**(ii)** The poet's childhood fear was that her mother would leave her alone and go away.

**(iii)** The poetic device used in lines 1-2 is simile.

**(iv)** Late winter's moon appears dull and hazy and it has no brightness. Similarly, poet's mother also looks tired and lifeless.

OR

(i) The unlucky heir is the boy with the stunted growth.

(ii) He has inherited his father's gnarled disease of twisted bones.

(iii) A young, sweet, unnoted boy is sitting at the back of the dim class.

(iv) He is different from the rest of the class as he is lost in the world of his dreams of a bright future and he isn't sad or morbid like others.

**8. Answer any four of the following questions in 30-40 word each : [3 × 4 = 12]**

**(i) Who did M. Hamel blame for the neglect of learning on the part of boys like Franz?**

**(ii) How did the instructor turn Dougla into a swimmer?**

**(iii) Why do you think Gandhiji considered the Champaran episode to be a turning point in his life?**

**(iv) What made the chief astrologer place his finger on his nose?****

**(v) How did the writer indicate that Dr. Sadao's father was a very traditional and conventional man ?**

**(vi) How does Jo want the story to end? Why?**

**Answer :**

(i) M.Hamel blamed the parents for the neglect of learning on the part of boys like Franz because their parents wanted them to work on farm to earn some money. Franz himself wanted to enjoy his time and thus, avoided going to school. M. Hamel blamed himself too for the neglect as he made the boys work in the garden which made them miss their lessons and whenever he wanted to go fishing, he gave them a holiday.

(ii) Douglas was afraid of water and he wanted to overcome his fear. So he took the help of an instructor who taught him the strokes in swimming, how to inhale and exhale and how to use his limbs and his hands in water. He put a belt around Douglas during practice and after perfecting each part, he built a complete swimmer out of Douglas.

(iii) Gandhiji considered the Champaran episode as the turning point of his life because after this episode, the peasants realized their rights and came out of the fear of British landlords. They also got 25% of their land tax as compensation money. This was also the beginning of the cultural, economic and social transformation of the peasants in India. The British were forced to leave the lands of the peasants and hence, the peasants became the owner of their own lands.

(v) Dr. Sadao's father was a very traditional and conventional man. He was a gentle and quiet person who never joked or played with his son. Sadao's education was his chief concern. He bestowed infinite pains upon Sadao, who was his only son. He was strict but a caring and loving father. He considered arranged marriage the best and believed in the purity of race.

(vi) Joe wants another ending of the story. She wants that the wizard should slap Roger Skunk's mother because of her silly action and take a revenge. By changing the end, she wants to force her views on her father.

**9. Answer any one of the following in 120–150 words : [6]**

**(i) "Seemapuri, a place on the periphery of Delhi yet miles away from it, metaphorically. " Explain.**

**(ii) 'The Rattrap' focuses on human loneliness and the need to bond with others. Comment.**

**(iii) Attempt a character sketch of Sophie as a girl who lives in her dreams.****

**Answer :**

**(i)** The story 'Lost Spring' revolves around the pathetic lives of the poor children who live in the slums and work in unhygienic conditions.

Seemapuri is a slum area on the periphery of Delhi with a settlement of more than 10,000 rag pickers. Those who live here are squatters who came from Bangladesh back in 1971. They live in the structures of mud with roofs of tin or tarpaulin, devoid of sewage systems, drainage or running water. One may think that the life of people living on the periphery of the national capital must be very easy-going, rich and good but it is all opposite. They have no identity or permits except ration cards as food is more important to them than any identity. Children grow up only to become the supporters of their parents in survival. To them, garbage means 'gold'. It is their daily bread, a roof over their heads. It is a gift for them because whenever they find some silver coin in a heap of garbage, they hope to survive on it for some days.

**(ii)** 'The Rattrap' is a bit philosophical story which tells that the world is a trap and the materialistic things are bait. In the story, the author tells that humans need someone to share their joys and sorrows. They can't live alone and that's why they are called social animals. In the story too, the peddler is lonely and so is the crofter. The peddler is a lonely man who sells rattraps to avoid his loneliness

**Answers are not given due to change in present syllabus

whereas the crofter is an old man who has neither wife nor child. Due to his loneliness, he wants someone to talk to. So, he becomes talkative and friendly with the peddler and allows him to have a conversation with him. He not only gives him a shelter but also gives him the food to eat and tobacco to smoke. He tells the peddler everything about himself and his earnings. On the other hand, the iron master and his daughter are also lonely. They have lots of money but no companion and so they request the peddler to spend Christmas with them.

**10. Answer any one of the following in 120–150 words : [6]**

**(i) Derry and Mr. Lamb both are victims of physical impairment, but their attitudes towards life are completely different. Elaborate.**

**(ii) Write a character sketch of the Governor of Oxford Prison based on the story, 'Evans Tries an O-Level'.**

**(iii) Why did Bama stroll in the market place instead of hurrying back home? Describe the sights she enjoyed seeing there.****

Answer :

(i) Derry and Mr. Lamb both are victims of physical impairment, but their attitudes towards life are completely different. Derry and Mr. Lamb have the same sense of loneliness and they both live a secluded life due to their physical disability. Mr. Lamb is disabled but calm and maintains peace with people around him. He has a tin leg as his real one was blown away in a war. He is not irritated by the children who call him 'Lamey-Lamb'. He always lives in peace. On the other hand, Derry has a burnt face as he got acid all down on his face in an unfortunate accident. He is always conscious of it and so gets alienated from the world. Before coming in contact with Mr. Lamb, Derry feels lonely due to the behavior of the people but Lamb inspires him and fills him with self-confidence and determination. His meeting with Mr. Lamb is a turning point in his life. He learns to live a respectful life. Mr. Lamb teaches him to live life to the fullest regardless of his disability. He tells him to move ahead in life as his brain and senses are competent enough. He imbibes positive attitude in him. He shows him the way to live in this world in spite of his disability because he could do much better than others.

(ii) Governor of Oxford Prison was a generous and an understanding man who allowed Evans to appear for an O-level exam. Evans was a criminal and had escaped from jail earlier twice, so governor took all the possible precautions for the smooth conduct of the examination. The examination had been arranged in the cell itself. The door was locked one day before and the security was made very tight. No visitors were allowed to see him and a microphone was fixed in the cell so that the Governor could hear every bit of a sound. Everything was checked thoroughly to ensure the ways that could help him escape. This shows that he was very wise; on the other hand he was so simple that he couldn't understand the cunningness of Evans. He didn't realize the extent of Evans, cunningness. Governor was also trusting and therefore, he was fooled by Evans again in the end.

**11. Answer any one of the following questions in 120-150 words :** [6]**

**(i) Why was Mrs. Hall happy to have a guest at 'Coach and Horses' Inn? How did the stranger behave at the inn?**

**(ii) Describe Mr. Marvel's meeting with the mariner at Port Stowe. What kind of person was the mariner?**

**(iii) How did Dunstan Cass meet his end?**

**(iv) Describe the circumstances under which Silas Marner had to leave Lantern Yard.**

**12. Answer any one of the following question in 120–150 words :** [6]**

**(i) How does the novel. 'The Invisible Man' highlight the theme of corruption of morals in the absence of social restriction?**

**(ii) Compare and contrast Griffin (the invisible man) and Dr. Kemp as scientist and also as members of society.**

**(iii) George Eliot has portrayed Godfrey as a morally weak characters. Comment.**

**(iv) In 'Siles Marner', describe the role of Dolly Winthrop.**

**Answers are not given due to change in present syllabus

# Delhi [Set-II]

**Note : Except for the following questions, all the remaining questions have been asked in previous sets.**

**SECTION-A**

**(READING)**

**1.4 Pick out the words/phrases from the passage which are similar in meaning to the following:** **[1 × 3 = 3]**

**(i) dropped/thrown (para 4)**

**(ii) unluckily (para 6)**

**(iii)look like (para 9)**

**Answer :**

(i) dumped

(ii) unfortunately

(iii) resemble

**SECTION–B**

**(ADVANCED WRITING SKILLS)**

**5. Write a debate in 150–200 words either for or against the motion : Indulgence of the youth in the use of Internet is a great bane for them. [10]**

**OR**

**Outdoor exercise or morning and evening walk during winter months has become difficult in the polluted atmosphere of the metropolitan cities. In such a situation indoor (team) games have become very important. Write a speech in 150–200 words persuading your students to make more use of facilities for basketball, badminton etc available in your school. You are Principal of your school.**

**Answer :**

**INDULGENCE OF THE YOUTH IN THE USE OF THE 'INTERNET' IS A GREAT BANE**

**FOR :**

Honorable Judges, Respected Principal, Teachers and my dear friends! Good Morning. Today, I stand before you to speak in favour of the motion 'Indulgence of the youth in the use of internet is a great bane for them'.

Today we are living in a high-technology and advance world. Modern gadgets and the internet has become an integral part of the youth's life. They live in a fascinating world where they want everything to move fast. Even while looking for the meaning of a word, the youngsters don't use dictionary, instead they switch over the internet to look for it. While scrolling through Facebook or Instagram, they feel that everyone is living a better life than them. This leads to frustration. This social media behavior is linked with major depression disorder. E-books have replaced printed books. An e-book reader uses rapidly changing technology to read books in a digital format. Then comes the mobile phones. There is no doubt that mobile phones have brought revolution in the field of communication and are considered as the biggest blessing of modern science. But it proves to be the most hazardous gadget as it distracts the youth from studies and their academic goals. Computers and internet have made every possible information accessible at one click but with lots of information and also with incorrect information. Over-dependency upon internet has made life sedentary which in turn, leads to many disorders like obesity, diabetes, heart problems and blood pressure.

Researches have proved over use of internet hampers intellectual growth, language development, innovative thinking and vulnerable to cyber crime and plagiarism. Therefore, I must say that too much indulgence in the use of internet is a great bane for the youth.

**AGAINST :**

Honorable Judges, Respected Principal, Teachers and my dear friends! Good Morning. Today, I stand before you to speak against the motion 'Indulgence of the youth in the use of internet is a great bane for them.'

In my view, indulgence of the youth in the use of internet is not a bane; rather it is a boon for them. The internet revolution has completely changed the lives of the youth today. The number of internet user has increased a lot due to easy accessibility and affordability. Whether it is homework, chatting, e-mailing, project work, information gathering or any other purpose, internet has occupied its solid space in every home. The internet provides information on any topic and youth considers it the best source of gaining knowledge as it saves their time and energy comparatively if they look into books or anywhere else. Therefore, books have taken a back seat in this regard and internet has become more reliable. It is supposed to be multitasking and provides knowledge on the basis of students' ability and understanding. It makes the routine lesson interesting, motivating and exciting for students. It is accompanied with audio-visual aids which give pleasure in learning and the students' quest to know more increases. The internet is not only the source of getting information but it is equally an efficient means of communication. The world has shrunk due to it because e-mailing and chatting have paved a path to discuss and spread education on a large scale.

Internet also opens many earning opportunities and saves time and resources. Therefore, Internet is a boon for the youth.

OR

**THE IMPORTANCE OF INDOOR GAMES**

Good Morning everybody! Today as we observe that life in metro cities has become very difficult due to the increasing rate of pollution, it's my duty to count on the hazardous consequences of the outdoor activities for all the students. Whenever we go out, we inhale smoke and poisonous gases along with oxygen. The burden of overgrowing population, unrestricted cutting of trees and unlimited urbanization has resulted in excessive pollution and so it has become impossible to do any outdoor exercise or morning and evening walk especially during harsh winters. Therefore, I would say that students must fully utilize the facilities of sports like badminton, basketball etc. available in the school as for good mental growth, we cannot deny the importance of these games. It helps to interact with each other as well. Children exchange ideas and thoughts while playing. As it is not possible to avail these facilities outside, our school has started the games within school premises. Trained coaches have been appointed for the students keeping in view the interest of the children in their respective field of interest. Students can play during zero periods and before the assembly. Provisions have been also made for evening training classes. These games are must for the children's balanced personality. Moreover, physical activities provide an outlet for the surplus energy.

Therefore, I would like to suggest all the students to make use of the sports facilities available in the school for which no extra fee would be charged.

Thank you !

## SECTION-C

## (LITERATURE : TEXT BOOKS AND LONG READING TEXT)

**8. Answer any four of the following question in 30-40 word each : [3 × 4 = 12]**

**(i) Why did the village elders occupy the back benches in the classroom to attend the last lesson delivered by M. Hamel?**

**(ii) Why did William Douglas choose Y.M.C.A. pool when he decided to learn to swim?**

**(iii) What did Gandhiji do about the social and cultural upliftment of the Champaran villages?**

**(iv) What great miracle did the ten-day-old Jilani Jung Jung Bahadur perform? How did it affect the people who listened to him?****

**(v) How did Dr. Sadao perform his duty as a patriotic Japanese ?**

**(vi) How can we say that Jack is a dominant male?**

**Answer :**

(i) The village elders occupied the back benches in the classroom to attend the last lesson delivered by M. Hamel because they felt sorry for not attending the school ever. They wanted to show their sympathy and respect for the teachers. They thanked M. Hamel for his long and devoted service of forty years to teach the children of the school.

(ii) William Douglas chose Y.M.C.A. pool when he decided to learn to swim because it was a safe and an ideal place for swimming. It was only two or three feet deep at the shallow end; and while it was nine feet at the other, the drop was gradual whereas Yakima river was dangerous and many people had drowned in it earlier.

(iii) For social and cultural upliftment of the Champaran villages, Gandhiji did many things. He fought for the justice to the sharecroppers of Champaran. The sharecroppers got freedom from the prolonged fear of the British landlords only due to him. He decided to eradicate illiteracy by opening primary schools over there. His wife Kasturba taught the lessons of cleanliness and sanitation to the people. A doctor volunteered for six months in the village to keep a check on health issues Personal and community hygiene was emphasized by Gandhiji.

(v) Dr. Sadao was a Japanese surgeon. An American prisoner of war was washed ashore in a dying state. He was thrown on the doorsteps of Dr. Sadao. Doctor was full of patriotism but as a doctor, he had to save the life of a dying man first. Therefore, he operated on the prisoner of war and as a true Japanese, he informed the General and agreed to the General's plan of assassination. This is how he performed his duty as a patriotic Japanese.

(vi) Jack was the father of a little girl Jo' to whom he used to tell the bedtime stories. When his daughter grew up she began to ask him a number of questions. Jack was unable to invent new stories and thus, all his stories were alike except the main creature. He was indeed a dominant male as he enjoyed when his daughter listened to him carefully without interrupting him and his authority. He is a male chauvinist person who doesn't like women to take anything for granted.

**Answers are not given due to change in present syllabus

**9. Answer any one of the following questions in 120–150 words : [6]**

**(i) What are the hazards of working in a glass bangle industry ?**

**(ii) What made the peddler finally change his ways ?**

**(iii) What was Sophie's dream ? How was it unrealistic ? How was she different from Jansie ?****

**Answer :**

(i) Hazards of working in a glass bangle factory are manifold. It is harmful for the eyes of the maker as the polishing of the glass bangles produces dust which may result into blindness. Working in the glass furnaces with high temperature and in dingy cells without proper ventilation and light take away the brightness of eyes. Similarly, during the welding work in the glass factory, sitting in dark rooms in front of flickering lamps is equally harmful for the eyes and many children lose their eyesight even before they grow up. Their eyes become so accustomed to that light that they cannot see the natural daylight outside. There are chances of skin burn too. Working in glass bangle industries leaves an adverse effect on the health of the workers. Besides the workers are physically, emotionally and psychologically hurt too. They are exploited by middlemen, policemen, sahukars, bureaucrats and politicians and it is impossible for them to come out of their trap.

(ii) The peddler was a rattrap seller who led a dull and monotonous life of a vagabond. He was a petty thief who stole thirty kronors from the house of crofter. Once, he was invited by an ironmaster, who thought him to be an old acquaintance of his. The peddler declined his invitation but when his daughter Edla persuaded him, he agreed. Edla showed pity on him and asked him to spend Christmas with them. Her behavior changed his conscience and he finally changed his way of living when got immense love and affection from Edla. His life was miserable but when he was shown hospitality and respect, he changed his way of living. He was full of gratitude to Edla for her kindness and generosity and started living the life of dignity and respect. While giving Edla the gift of rattrap, he asked her to return the thirty kronors of the crofter which he had stolen earlier. Thus, his life was completely transformed.

●●

# Delhi [Set-III]

**Note : Except for the following questions, all the remaining questions have been asked in previous sets.**

## SECTION–A
## (READING)

**1.4 Pick out the words/phrases from the passage which are similar in meaning to the following: [1 × 3 = 3]**

**(i) unsafe (para 2)**

**(ii) rubbish/junk (as a noun) (para 3)**

**(iii) shaped/formed (para 5)**

**Answer :**

(i) dangerous (ii) trash

(iii) moulded

## SECTION–B
## (ADVANCED WRITING SKILLS)

**6. People like to travel to their places of work in their own cars. It is comfortable and they also reach their destinations more or less in time. However, use of private vehicles leads to pollution and traffic jams and it is expensive also. Air-conditioned public buses, which are smaller in size, seating 15–20 people and tickets, reasonably priced can be a boon for the people. Write an article in 150–200 words on the topic. 'Need of better Public Transport.' You are Resham/Rehana. [10]**

**OR**

**Your school held a two-day Diwali Mela in the school grounds. Various food stalls were put up by the students. On the second day, there was a song and dance programme. Prizes were awarded to the participants. However, there were no fire works. You are Amit/Amrita. Write a report on the event in 150–200 words for your school magazine.**

**Answer :**

NEED OF BETTER PUBLIC TRANSPORT

By Resham/Rehana

We all know that travelling by a private vehicle is very convenient and takes far less time than public transport but simultaneously, it leads to pollution and traffic jams. Moreover, they prove to be extremely expensive as the petrol prices are increasing day by day. In that case, there is

**Answers are not given due to change in present syllabus

a serious need of better public transport which is comfortable and can accommodate 15 to 20 people at a time. The ticket prices should also be reasonable so that everyone can afford it. It is a known fact that in metro cities, from every house, a minimum of two to three vehicles are used by the members for going to their working places. Sometimes they remain stranded in their vehicles for long hours due to heavy traffic jams which result into excessive emission of polluted gases in the atmosphere and delay to reach their destinations.

So, it would be far better if the government provides good public transport facilities which could be accessible to all. Government should create awareness of public transport. It would create lot of convenience for the general public and daily commuters. It has many social, economic and environmental benefits. Social benefits of using public transport are multifold. It helps in developing a sense of being a community among people. They feel a connection with those who are traveling with them. It encourages people to have a healthy lifestyle as they walk to their stations, stands or stops. Moreover, it is less stressful as it saves a lot of time. Economic benefits include less expenditure of money on buying cars or spending money on its maintenance. Travelling by public transport is much cheaper than private transport. It also reduces pollution and road congestion thus, helps in environmental conservation.

**OR**

**DIWALI MELA**

By Amit/Amrita

The festival of Goddess Lakshmi and Festival of lights came alive in our school on 27th October. The school organized a two-day Diwali Mela in the school ground. The whole school wore a festive look. The building were illuminated with earthen lamps and lanterns. The reception hall and the corridors were decorated with marigold flowers which are considered auspicious. Children were very excited and had lot of fun. The main attraction was different food stalls which were set up by the students which included Indian, Continental, Italian and Chinese food. There was also a stall which displayed decorative and perfumed candles and many other decorative items made by the school children which were also meant for sale. People in large numbers purchased those candles and decorative items to motivate the children. On the second day, a cultural program was organized which stole the hearts of the viewers. Everybody appreciated the enthusiasm and endeavor of the students. The participants were awarded prizes by the Principal for their overwhelming performances.

It was an eco- friendly Diwali as there were no fire-works. Students were also asked to spread the message of celebrating eco-friendly Diwali. Our Principal briefed the students about the harmful effects of fire crackers on the environment.

## SECTION-C

## (LITERATURE : TEXT BOOKS AND LONG READING TEXT)

**8. Answer any four of the following questions in 30-40 word each : [3 × 4 = 12]**

**(i) How did M. Hamel make his last lesson a special one ? What did he emphasize in it?**

**(ii) Which two incidents in the life of William Douglas before he was ten years old created an aversion in his mind to water.**

**(iii) Why did Gandhiji not accept C.F. Andrews, help during Champaran movement ?**

**(iv) How did the Tiger King ' manage to kill the hundredth tiger?****

**(v) What was Hana's role in Dr. Sadao's life when he brought home an injured American soldier ?**

**(vi) We can't approve of Jack's attitude towards his wife. Comment.**

**Answer :**

(i) M.Hamel made his last lesson special by wearing a special outfit and talking about the French language at length. He told that it was the most beautiful language in the world – the clearest and the most logical. He asked the students that they must guard it because when people are enslaved, the language is the only key to their prison.

(ii) As a boy, Douglas was scared of water a lot. His first experience was on the sea beach when he was with his father and a powerful wave swept over him. Although, the wave receded but it left Douglas terrified. At the second time, when William Douglas was learning swimming in Y.M.C.A. pool, a strong boy threw him into the deep water of the pool. He was saved but the terror stayed with him and he couldn't go near water for years.

**Answers are not given due to change in present syllabus

**(iii)** Gandhiji did not accept C.F. Andrew's help during Champaran movement because he thought that in having an Englishman on their side was not helpful or a fair fight. That would show the weakness of their heart. He wanted Indians to be self-reliant in their struggle against injustice. He wanted Indians to rely upon themselves to win the battle.

**(v)** When Dr. Sadao brought an injured American soldier to his home, Hana played a crucial role. On the humanitarian ground, she nursed the soldier herself when she got no cooperation from her maid. She did all the household work herself gracefully. She washed the soldier's wound and fed him, which indicates her respect and obedience towards her husband. She did whatever Dr. Sadao asked her to do during the operation.

**(vi)** Jack recalled certain moments of humiliation towards himself while telling stories to his daughter. He was influenced by his own childhood experiences and failures. He wanted to help his wife but didn't make any efforts. He was a temperamental and dominant husband. He didn't want to speak or work with her and even touch her. Thus, his attitude towards his wife is not acceptable.

**9. Answer any one of the following question in 120–150 words : [6]**

**(i) What kind of life did children living in Seemapuri lead ?**

**(ii) How did the crofter tempt the peddler to steal his money ? How did it change the peddler's life ?**

**(iii) What did Sophie tell her father and her brother about her 'meeting(s)' with Danny Casey ? How did each of them react ?****

**Answer :**

**(i)** Seemapuri is a settlement of more than 10,000 rag pickers situated on the periphery of Delhi. The people living there have migrated from Dhaka, Bangladesh. They lived in the structures of mud with roofs of tin and tarpaulin. They didn't have any identity or permits. Their children also helped them in earning their livelihood. Saheb was one of the children who lived here. He was from a poor family of Bangladesh. In order to earn their livelihood, his family shifted to Delhi where they settled in Seemapuri. But when they found no work to do, they became rag pickers. They had to search the garbage like others for something to make their livelihood. Sahib, whose age was to enjoy the life, go to school and live his childhood, also looked into the heaps of garbage for some coins. It was a tedious task as they used to leave their houses barefoot early in the morning with a bag on their shoulders. They did not have chappals, so they were not in a habit to wear them. Because their fields and homes were swept away in the flood in Bangladesh, so they came to the big city with a hope but even here they led a miserable life.

**(ii)** The peddler was a lonely man who looked like a beggar. He was a homeless wanderer who used to sell rattraps. Once he knocked the door of a crofter who was equally lonely like him for a night stay. He was received by the crofter warmly and cheerfully. Crofter was proud of his cow that gave him enough milk. So he told the peddler about the thirty kronors he got by selling the milk of the cow and he used to keep his money in a leather pouch that hung from a nail in the window frame. He felt that the peddler did not believe him so he showed the money to convince him.

Having stolen the money, the peddler didn't find it safe to walk along the public highway. So he went through the woods and got lost. Here, he thought that the whole world was a rattrap and the money he had stolen was a bait and he himself was trapped in it. Later, he regretted for stealing the crofter's money so much so that when Edla invited him for Christmas, he returned the stolen money to her.

**Answers are not given due to change in present syllabus

# ENGLISH

# 2018

**Time allowed : 3 hours** **Maximum marks : 80**

## SECTION - A
## (READING) 30

**1. Read the passage given below: [12]**

(1) When you grow up in a place where it rains five months a year, wise elders help you to get acquainted with the rain early. They teach you that it is ignorant to think that it is the same rain falling every day. Oh no, the rain is always doing different things at different times. There is rain that is gentle, and there is also rain that falls too hard and damages the crops. Hence, the prayers are for the sweet rain that helps the crops to grow.

(2) The monsoon in the Naga hills goes by the native name, khuthotei (which means the rice-growing season). It lasts from May to early or mid-October. The local residents firmly believe that Durga Puja in October announces the end of rain. After that, one might expect a couple of short winter showers, and the spring showers in March and April. Finally, comes the "big rain" in May; proper rainstorms accompanied by heart-stopping lightning and ear-splitting thunder. I have stood out in storms, looking at lightning arc across dark skies, a light-and-sound show that can go on for hours.

(3) This is the season when people use the word **sezuo** or **sizu** to refer to the week-long rains, when clothes don't dry and smell of mould, when fungus forms on the floor and when you can't see the moon or the stars because of the rainclouds. But you learn not to complain. Rain, after all, is the farmer's friend and brings food to the table. Rituals and festivals centre around the agricultural rhythm of life, which is the occupation of about 70 per cent of the population.

(4) The wise learns to understand its ways. I grew up hearing my grandfather say, "It's very windy this year. We'll get good rain." If the windy season was short and weak, he worried there might not be enough rain for the crops. I learned the interconnectedness of the seasons from childhood, and marvelled at how the wind could bring rain. Another evening, many rainy seasons ago, my paternal aunt observed the new moon and worried, "Its legs are in the air, we're in for some heavy rain." She was right. That week, a storm cut off power lines and brought down trees and bamboos.

(5) Eskimos boast of having a hundred of names for snow. Norwegians in the north can describe all kinds of snow by an equal amount of names: pudder, powder snow, wet snow, slaps, extra wet snow, tight snowfall, dry snow, and at least 95 more categories of snow. Likewise, in India we have names and names for rain. Some are common, some are passing into history.

(6) The rains are also called after flowering plants and people believe that the blossoming of those plants draws out rain. Once the monsoon's set in, field work is carried out in earnest and the work of uprooting and transplanting paddy in flooded terrace fields is done. The months of hard labour are June, July and August. In August, rain, also called phrogü, is a sign that the time for cultivation is over. If any new grain seeds are sown, they may not sprout; even if they do sprout, they are not likely to bear grain. The rain acts as a kind of farmer's almanac.

(7) The urban population of school-goers and office-goers naturally dislikes the monsoon and its accompanying problems of landslides, muddy streets and periodic infections. For non-farmers, the month of September can be depressing, when the rainfall is incessant and the awareness persists that the monsoons will last out till October. One needs to have the heart of a farmer to remain grateful for the watery days, and be able to observe—from what seems to the inexperienced as a continuous downpour—the many kinds of rain. Some of the commonly known rain-weeks are named after the plants that alternately bloom in August and September. The native belief is that the flowers draw out the rain.

(8) Each rain period has a job to fulfil : October rain helps garlic bulbs to form, while kümünyo rain helps the rice bear grain. Without it, the ears of rice cannot form properly. End October is the most beautiful month in the Naga hills, as the fields turn

gold and wild sunflowers bloom over the slopes, all heralding the harvest. Prayers go up for protecting the fields from storms, and the rains to retreat because the grain needs to stand in the sun and ripen. The cycle nears completion a few weeks before the harvest, and the rain does retreat so thoroughly from the reaped furrows that the earth quickly turns hard. The months of rain become a distant memory until it starts all over again.

**1.1 On the basis of your understanding of the above passage, complete the statements given below with the help of options that follow : [1× 4 = 4]**

**(a) The rains are called after flowering plants because**

(i) heavy rains kill plants.

(ii) flowers grow in the rainy season.

(iii) it is believed that the plants bring the rain.

(iv) flowers grow all the year round.

**(b) The rain is like a calendar for farmers because**

(i) it tells them when to sow and when to harvest.

(ii) it tells them the birthdays of their children.

(iii) each month has a time for plantation.

(iv) different kinds of rain tell different things.

**(c) People who live in cities don't like rain because**

(i) it brings mud and sickness with it.

(ii) they are not bothered about the farmers.

(iii) they don't like the plants that grow during the rain.

(iv) going shopping becomes difficult.

**(d) People pray asking the rain to retreat because**

(i) the fungus and mould to dry.

(ii) children don't get a chance to play.

(iii) the crops need the sun and heat to ripen.

(iv) they like to pray.

**1.2 Answer the following questions briefly: [1×6 = 6]**

**(e) Why do the elders want you to understand the rains in the Naga hills?**

**(f) What does Durga Puja mean to the farmers of the Naga hills?**

**(g) What kind of rain is called sezuo?**

**(h) What is the occupation of more than half the population of the Naga hills?**

**(i) How is the heart of the farmer different from that of the city person?**

**(j) When does rain become a memory in the minds of the people of the Naga hills?**

**(k) Find words from the passage which mean the same as the following : [1×2 =2]**

**(i) flowering (para 6) (ii) non-stop (para 7)**

**Answer:**

(a) (iii) it is believed that the plants bring the rain.

(b) (i) it tells them when to sow and when to harvest.

(c) (i) it brings mud and sickness with it.

(d) (iii) the crops need the sun and heat to ripen.

(e) The elders want us to understand the rains in the Naga hills because rain always does different things at different times. It tells us when to sow and harvest. It can be gentle or too hard leaving the crops damaged.

(f) Durga Puja means the end of the rainy season to the farmers of the Naga hills.

(g) Sezuo refers to the week-long rains, when clothes do not dry and there is a smell of mould, when fungus forms on the floor and when we cannot see the moon or the stars because of the rain clouds.

(h) The occupation of more than half the population on the Naga hills is farming.

(i) The city person dislikes the rain because of landslides, muddy streets and periodic infections. He finds incessant rain depressing. However, a farmer remains grateful for the rainy days.

(j) Rain becomes a memory in the minds of the people of the Naga hills, when the cycle comes near its completion a few weeks before the harvest.

(k) (i) blossoming **(ii)** incessant

**2. Read the passage given below : [10]**

(1) Every morning Ravi gives his brain an extra boost. We're not talking about drinking strong cups of coffee or playing one of those mind-training video games advertised all over Facebook. "I jump into my stationary bike and cycle for 45 minutes to work," says Ravi. "When I get to my desk, my brain is at peak activity for a few hours." After his mental focus comes to a halt later in the day, he starts it with another short spell of cycling to be able to run errands.

(2) Ride, work, ride, repeat. It's a scientifically proven system that describes some unexpected benefits of cycling. In a recent study in the Journal of Clinical and Diagnostic Research, scientists found that people scored higher on tests of memory, reasoning, and

planning after 30 minutes of spinning on a stationary bike than they did before they rode the bike. They also complete the tests faster after pedalling.

**(3)** Exercise is like fertilizer for your brain. All those hours spent on exercising your muscles, create rich capillary beds not only in leg and hip muscles, but also in your brain. More blood vessels in your brain and muscles mean more oxygen and nutrients to help them work. When you pedal, you also force more nerve cells to fire. The result: you double or triple the production of these cells—literally building your brain. You also release neurotransmitters (the messengers between your brain cells) so all those cells, new and old, can communicate with each other for better, faster functioning. That's a pretty profound benefit to cyclists.

**(4)** This kind of growth is especially important with each passing birthday, because as we age, our brains shrink and those connections weaken. Exercise restores and protects the brain cells. Neuroscientists say, " Adults who exercise display sharper memory skills, higher concentration levels, more fluid thinking, and greater problem-solving ability than those who are sedentary."

**(5)** Cycling also elevates your mood, relieves anxiety, increases stress resistance, and even banishes the blues. "Exercise works in the same way as psychotherapy and antidepressants in the treatment of depression, maybe better," says Dr. Manjari. A recent study analyzing 26 years of research finds that even some exercise—as little as 20 to 30 minutes a day—can prevent depression over the long term.

**(6)** Remember : Although it's healthy, exercise itself is a stress, especially when you're just getting started or getting back into riding. When you first begin to exert yourself, your body releases a particular hormone to raise your heart rate, blood pressure, and blood glucose levels, says Meher Ahluwalia, PhD, a professor of integrative physiology. As you get fitter, it takes a longer, harder ride to trigger that same response.

**2.1 On the basis of your understanding of the passage, complete the statements given below with the help of the options that follow :**

[1×2 = 2]

**(a) Ravi gets his brain to work at peak level by**

(i) drinking three cups of coffee.

(ii) playing games that need brain activity.

(iii) cycling on a stationary bike.

(iv) taking tablets to pump up his brain.

**(b) When nerve cells work during exercise then**

(i) the body experiences stress.

(ii) the brain is strengthened by multiplying them.

(iii) you start to lose your temper.

(iv) your stationary cycle starts to beep.

**2.2 Answer the following questions briefly :**

[1×6 = 6]

**(c) How does exercise help the brain?**

**(d) Why does Ravi do a circuit of 'ride, work, ride'?**

**(e) What is the work of neurotransmitters?**

**(f) What benefits other than greater brain activity does one get from cycling?**

**(g) Why is exercise so important for adults?**

**(h) How is exercise itself a stress?**

**(i) Find words from the passage which mean the same as the following :** **[1 × 2 = 2]**

**(i) manure (para 3) (ii) inactive (para 4)**

**Answer :**

(a) (iii) cycling on a stationary bike.

(b) (ii) the brain is strengthened by multiplying them.

(c) Exercise is like the fertilizer to the brain. It creates rich capillary beds not only in legs and hips muscles, but also in the brain. More blood vessels in the brain and muscles mean more oxygen and nutrients to help them work. It doubles or triples the nerve cells and builds the brain. It also releases neurotransmitters.

(d) Ravi does a circuit of 'ride, work, ride' because it given his brain an extra boost. People who do such rigorous activities score higher on tests of memory, reasoning and planning. It also helps in completing the tests faster.

(e) Neurotransmitters are the messengers between our brain cells. They enable faster, and better communication between the brain cells.

(f) Cycling elevates our mood, relieves anxiety, increases stress resistance and even banishes the blues. It prevents depression.

(g) As we age, our brain shrinks and the connections between the cells weaken. Exercise restores and protects the brain cells. It leads to sharper memory skills, higher concentration levels, more fluid thinking and better problem solving ability.

**(h)** Exercise itself is a stress especially when one just starts an activity or gets back to it. At that point, the body releases a particular hormone which raises the heart rate, blood pressure and blood glucose levels.

**(i)** (i) fertiliser (ii) sedentary

**3. Read the passage given below : [8]**

Keeping cities clean is essential for keeping their residents healthy. Our health depends not just on personal hygiene and nutrition, but critically also on how clean we keep our cities and their surroundings. The spread of dengue and chikungunya are intimately linked to the deteriorating state of public health conditions in our cities.

The good news is that waste management to keep cities clean is now getting attention through the Swachh Bharat Mission. However, much of the attention begins and stops with the brooms and the dustbins, extending at most to the collection and transportation of the mixed waste to some distant or not so distant place, preferably out of sight.

The challenge of processing and treating the different streams of solid waste, and safe disposal of the residuals in scientific landfills, has received much less attention in municipal solid waste management than is expected from a health point of view.

One of the problems is that instead of focusing on waste management for health, we have got sidetracked into "waste for energy". If only we were to begin by not mixing the biodegradable component of solid waste (close to 60 per cent of the total) in our cities with the dry waste, and instead use this stream of waste for composting and producing a gas called methane.

City compost from biodegradable waste provides an alternative to farmyard manure (like cow-dung). It provides an opportunity to simultaneously clean up our cities and help improve agricultural productivity and quality of the soil. Organic manure or compost plays a very important role as supplement to chemical fertilisers in enriching the nutrient-deficient soils. City compost can be the new player in the field.

Benefits of compost on the farm are well-known. The water holding capacity of the soil which uses compost helps with drought-proofing, and the requirement of less water per crop is a welcome feature for a water-stressed future. By making the soil porous, use of compost also makes roots stronger and resistant to pests and decay. Farmers using compost, therefore, need less quantity of pesticides. There is also evidence to suggest that horticulture crops grown with compost have better flavour, size, colour and shelf-life.

City compost has the additional advantage of being weed-free, unlike farmyard manure which brings with it the seeds of undigested grasses and requires a substantial labour cost for weeding as the crops grow. City compost is also rich in organic carbon, and our soils are short in this. Farmers clearly recognize the value of city compost. If city waste was composted before making it available to the farmers for applying to the soil, cities would be cleaned up and the fields around them would be much more productive.

Quite apart from cleaning up the cities of biodegradable waste, this would be a major and sustainable contribution to improving the health of our soil, without further damage by excessive chemical inputs. What a marvellous change from waste to health!

The good news is that some states are regularly laying plastic roads. Plastic roads will not only withstand future monsoon damage but will also solve a city's problem of disposing of non-recyclable plastic. It is clear that if the mountains of waste from our cities were to be recycled into road construction material, it would tackle the problem of managing waste while freeing up scarce land.

**(a)** On the basis of your understanding of the above passage, make notes on it using headings and sub-headings. Use recognisable abbreviations wherever necessary (minimum four) and a format you consider suitable. Also supply and appropriate title to it. **[5]**

**(b)** Write a summary of the passage in about 80 words. **[3]**

**Answer :**

**(a)** **Waste Management**

**I.** Reasons for keeping cities clean

**1.1** Keeps surroundings clean

**1.2** Prevents the spread of dengue and chikungunya.

**1.3** Promotes overall health.

**II.** Waste Mgmt.–Problems

**2.1** No focus on waste Mgmt. from health point of view.

**2.2** Processing and treating diff. streams of solid waste.

**2.3** Safe disposal of residuals in scientific landfills.

**2.4** Sidetracked into 'waste for energy'.

**III.** Advantages—City Compost

**3.1** Provides alternative to farmyard manure.

**3.2** Opportunity to clean cities.

3.3 Improves agricultural productivity.
3.4 Better qty. of soil.
3.5 Supplements chem. fertilisers.
3.6 Enriches nutrient deficient soils.
3.7 Weed free.
3.8 Rich in organic carbon.

**IV.** Advantages—Farm Compost
4.1 $H_2O$ holding capacity—helps drought proofing.
4.2 Requires less water per crop.
4.3 Makes soil porous—roots stronger.
4.4 Use of less pesticides.
4.5 Horticulture crops—
(i) better flavour (ii) colour
(iii) shelf-life

**V.** Advantages—Plastic roads
5.1 Withstands future monsoon damage
5.2 Solves problem—disposing non-recyclable plastic.
5.3 Better waste mgmt.
5.4 Frees up scarce land.

**Key**

| | Abbre. | Word |
|---|---|---|
| 1. | attn. | attention |
| 2. | $H_2O$ | water |
| 3. | qty. | quality |
| 4. | diff. | different |
| 5. | chem. | chemical |
| 6. | & | and |
| 7. | agri. | agriculture |
| 8. | mgmt. | management |

### (b) SUMMARY

Cleanliness in cities keeps residents healthy and prevents the spread of diseases such as dengue and chikungunya. The Swachh Bharat Mission instituted by the government is beset by a lot of challenges. City compost from biodegradable waste has a lot of advantages. It improves agricultural productivity, is weed free and enriches nutrient-deficient soils as it is rich in organic carbon. Farm compost is equally beneficial. Use of such compost in soil enhances the water holding capacity and thus, there is less requirement of water per crop. It makes the soil porous and roots stronger and resistant to pests and decay. With plastic roads, there is less monsoon damage and it simultaneously solves the city's problem of disposing non-recyclable plastic. Moreover, it also tackles problem of managing waste while freeing up scarce land.

## SECTION — B
## (WRITING SKILLS) 30

**4. You are Faiz/Falak Mazumdar living at 39, Udampur Colony, Shimla. You decide to hold a dinner party to congratulate your grandparents on their golden wedding anniversary. Draft a formal invitation in not more than 50 words to all family members to attend a grand dinner at home. [4]**

**OR**

**You are Harish/Harshita of 12, Seva Nagar, Pune. You want to sell your flat as you are shifting to another city for work. Draft a suitable advertisement in not more than 50 words to be published in *The Pune Times* under the classified columns.**

**Answer :**

**Mr. & Mrs. Mazumdar**
request the pleasure of your company on the auspicious occasion of the
50th Wedding Anniversary
celebration of our
Loving Grandparents
on March 9th, 20XX
at the Gymkhana Club
at 9 p.m.

**RSVP**
**Faiz/Falak Mazumdar**
39, Udampur Colony Programme
Shimla overleaf
Ph : 012345678

**OR**

**FOR SALE**
3 bed-room flat for sale, 1200 sq. ft., attached bathroom, big balcony facing the park, modular kitchen with all modern amenities. Price negotiable, Contact : Harsh/Harshita, 12, Seva Nagar, Pune. Ph 012345678

**5. You are Neeraj/Neerja Shekhar, Principal, Vasant Public School, Pune. Your school has just started a music department. Write a letter to the Manager of Melody House, Pune, wholesale suppliers of musical instruments, placing an order for musical instruments for the school. Ask for a discount on the catalogue prices. (120-150 words) [6]**

**OR**

**Bal Vidya Public School, Bhilai, urgently requires a post-graduate teacher to teach**

**political science for which they have placed an advertisement in The Bhilai Express. You are Sanjay/Sanjana Sharma from 21, Vasant Marg, Bhilai. Draft a letter including a CV, applying for the advertised post. (120-150 words)**

**Answer :**

Vasant Public School,
Pune–040008

March 5th, 20XX

The Manager,
Melody House,
Pune–040001.

**Sub : Order for Musical Instruments.**

Sir,

To promote the latent talents of the children, the school has started a music department for the upcoming academic session. Working towards this end, we need to order more musical instruments. As the most prominent music instrument seller in the town, we would be highly obliged if you could send the following instruments by March 15th, 20XX.

| Sr. No. | Item | Quantity |
|---|---|---|
| 1. | Casio | 2 |
| 2. | Piano | 1 |
| 3. | Guitar | 4 |
| 4. | Tabla | 2 |

I hope that the discount available to all the educational institutions is also made available to us.

Yours Faithfully,
Neeraj/Neerja Shekhar
(Principal)

**OR**

21, Vasant Marg
Bhilai

March 5th, 20XX

The Principal,
Bal Vidya Public School,
Bhilai

**Subject : Applying for the post of PGT (Pol. Sc.)**

Sir,

This is with the reference to your advertisement in the Bhilai Express dated March 1st, 20XX for the vacancy of a post-graduate teacher in Political Science and I wish to apply for the same.

I have just finished my B. Ed. from the Central Institute of Education (CIE) after completing my post-graduation in Political Science from Hindu College, University of Delhi. I have 4 years of teaching experience.

I am a person who is very motivated and dedicated. Given the job, I promise to keep up the reputation of the school and work diligently and sincerely. I will be available to walk in for an interview on the day and date specified by you.

Yours Faithfully,
Sanjay/Sanjana Sharma
Encl. Bio-data

**Bio-data**

Name : Sanjay/Sanjana Sharma
Date of birth : 10/12/1990
Permanent address : 21, Vasant Marg, Bhilai
Ph. No. : 9350106576
E-mail add. : Sanjay@gmail.com/ Sanjana@gmail.com

Educational Qualifications :

| S. No. | Course | Institution | Subjects | Grade |
|---|---|---|---|---|
| 1. | 10+2 (Hum.) | St. Mark's School, Bhilai. | Eng., Com., Pol. Sc., Eco., Hist., Geog. | 90% |
| 2. | B. A. Pol. Sc. (Hons.) | Miranda College, New Delhi. | Pol. Sc. | 70% |
| 3. | M.A. Pol. Sc. (Hons.) | Hindu College. New Delhi | Pol. Sc. | 65% |
| 4. | B. Ed. | Central Institute of Education (CIE) | Pol. Sc., Eng. | 70% |

Experience :

| | Post | Institution | Tenure |
|---|---|---|---|
| 1. | T. G. T (Social Studies) | St. Mark's School, Bhilai | 3 years |
| 2. | P. G. T. (Pol Sc.) | Birla Vidya School, Bhilai | 1 year |

Other interests : Dramatics, Reading, Photography.
References :

(i) Dr. Marwah
Reader
C-E, New Delhi.
Ph. 6578267XXX
New Delhi.

(ii) Dr. Sharma
Prof. of Pol. Science
Hindu College,
Ph. 7621356XXX

6. **Recent floods in many metropolitan cities of the country during the monsoon season laid bare the hollowness of the claims of the civic**

**authorities of their preparedness. The poor had to bear the brunt of the problem while no one was ever held accountable. Write an article in 150-200 words on the common man's woes during the monsoons and the need for accountability of the officials concerned. You are Sumit/Smita Verma. [10]**

**OR**

**You are Ali/Alia, Head girl/Head boy of your school. You are deeply disturbed by the rising cases of aggressive behaviour of students in your school. You decided to speak during the morning assembly about it. Write a speech on 'Indiscipline in Schools'. (150-200 words)**

**Answer :**

**Monsoon Fury**

by Sumit/Smita Verma

Rain, rain go away- Come again another day. This little childhood rhyme portraged the prevalent mood of most of the people of the major metropolitan cities of India – Mumbai, Kolkata and Chennai. The monsoon fury was so extensive that it led to heavy losses of material and more importantly the loss of human lives. It was all unprecedented and unexpected.

There was an extensive loss of life and property. People were stranded in their own homes as they could not move out for a couple of days. The civic system came to a halt. Phone lines were disconnected and electricity lines were snapped. Life came to a standstill in busy commercial capitals like Mumbai, Chennai and Kolkata.

People from all strata of life suffered. However, the most affected were the poor people who felt the brunt of it. Most of them are daily wage earners but they could not go to work because of the weather conditions. The children and families went hungry for many days. Rations were provided immediately to the people. The floods brought the inadequate civic amenities. into light. It led to an outbreak of diseases.

It is now the high time that the authorities wake up and take remedial measures. Prior steps should be taken to repair the roadways, waterways and all sewage systems. The officers should be held responsible and the poor should be evacuated immediately. Rations and aid must reach the victims quickly in order to control the situation faster.

**OR**

**Indiscipline in Schools**

Good morning, dear principal, teachers and my fellow students assembled here. Today, I would like to talk about one disturbing aspect—the aggressive behaviour of most of the students. Aggressive behaviour among students has almost become a norm these days. Students are no longer courteous or well-behaved. They believe that impoliteness and rudeness is going to reward them and they are seriously mistaken!

Cases of students pulling pranks are or the rise. This is a matter of concern as bullies assault students and children to get into fist fights. The growing atmosphere of hate and abuse should not be tolerated any longer.

There are a number of reasons for this kind of behaviour amongst students. The influence of TV, social media and the internet brings in a lot of ill effects. Working parents are not able to look after their children and the children are left on their own to find out what is good for them.

It is the high time that some drastic change should happen to improve such situations. Counselling sessions and workshops on child behaviour must be held periodically. Value education classes must be made compulsory for students. More attention should be paid to students individually, and more PTA meetings should be conducted. Discipline should become students way of life for the growth of the nation.

Ali/Alia

Headboy/Headgirl

**7. "Academic excellence is the only requirement for a successful career". Write a debate either for or against the motion. (120-150 words) [10]**

**OR**

**MMD School, Nashik, recently organised a science symposium on the topic : 'Effect of pollution on quality of life'. You are Amit/ Amita Raazdan, editor of the school magazine. Write a report on the event for your school magazine. (120-150 words)**

**Answer :**

**Academic excellence is the only requirement for successful career**

**AGAINST :**

Good morning ladies and gentlemen. I am Avneet Kaur of Mira Modern School. I will be speaking against the topic 'Academic excellence is the only requirement for a successful career'.

If a student is good only in academic, he will be like 'a frog in the well'. Versatility is the keyword these days. One needs to excel as a whole and be an all-rounder.

The whole world has become a global village, and life's challenges have become more demanding. Higher 'EQ' is required these days than a higher 'IQ'. In places of work you need to handle several people of different temperaments and being just into academics, does not help.

One must not only be theoretical but should also be very practical in one's approach. Theoretical knowledge is not sufficient in real life situations. On the whole, I believe only academic excellence does not help. One must learn finer skills like being able to handle people and situations and only then will be successful. Thank you.

**OR**

**FOR :**

Good Morning ladies and gentlemen. I am Avneet Kaur of Mira Modern School. I will be speaking in favour of the topic 'Academic excellence is the only requirement for a successful career'.

Whenever two new people meet each other, the first thing they ask is what you do. In reality, they are waiting to judge you by your academic qualifications. Also, one's social or professional profile is incomplete without inclusion of a person's education qualifications. Therefore, it is a person's successful acquisition of academic qualifications that grants them a smooth sail in every walk of life.

Academic qualifications ensure you have the right foundation in life. If you basic grounding in Maths, Science, and Language is strong, you can sail through almost all situations in life because these subjects provide you the necessary skills to calculate, to innovate and to communicate. Also, academic qualifications are the validation of all these skills in form of degrees and certificates. Strong academic backgrounds gives people more perspectives and experiences and help make a nation both morally strong and civilised.

People might cite examples of successful people such as Bill Gates who are school dropouts. But the count of the well-placed doctors, engineers, IT professionals, and lawyers is far more and all of these people rely on their formal education for their success, So, being successful in life is the sum total of a person's academic qualifications, their hard work and their dedication.

**OR**

**The Science Symposium**

By Amit/Amita Raazdan

A Science Symposium was held in the school, on the topic "effect of pollution on the quality of life." It was held on March 4th, between 9 a.m. and 5 p.m. More than 50 schools participated in it. Leading scientists, science teachers and students from different schools, enthusiastically took part in it. The chief guest was Dr. Khurana, a scientist from CSIR.

The programme began with the lighting of the ceremonial lamp by the chief guest, accompanied by the principal and other leading dignitaries. It was followed by a powerpoint presentation on 'the effect of pollution on the quality of life. Panel discussions were held at different levels and after that the ending address was given by the chief guest. There was also a question-answer session at the end of the programme. Students from all schools participated into it willingly.

Amir Khan, one of the participants remarked, "It was a very educative session. We learnt a lot." As could be seen, such workshops broaden the perspective of the students.

## SECTION —C
## (LITERATURE : TEXTBOOKS AND LONG READING TEXT)

**8. Read the extract given below and answer the questions that follow : [1 × 4 = 4]**

.....and clear rills
That for themselves a cooling covert make
'Gainst the hot season; the mid forest brake,
Rich with the sprinkling of fair musk-rose blooms;
.......

**(a) Identify the poem and the poet.**

**(b) What is the role of the clear rills?**

**(c) How has the mid forest brake become rich?**

**(d) Name the figure of speech in 'cooling covert'.**

**OR**

.....On their slag heap, these children
Wear skins peeped through by bones and spectacles of steel
With mended glass, like bottle bits on stones.

**(a) Name the poem and the poet.**

**(b) Explain : 'slag heap'.**

**(c) What future awaits these children?**

**(d) Name the figure of speech used in the third line.**

**Answer :**

(a) 'A Thing of Beauty' by John Keats.

(b) Clear rills make the land fertile. Travellers here, find a shady place to take rest, when it is too hot.

(c) The mid forest brake has become rich because of the flowers blooming in it. There are musk roses, adding colour and beauty to the scene.

(d) Alliteration.

**OR**

(a) 'An Elementary School Classroom in a Slum' by Stephen Spender.

(b) 'Slag heap' means garbage dump. The poet is pointing at the children of the slum who are extremely starved and malnourished. They appear Skinny like a skeleton. Their world is full of dirt and garbage and they spend their life raking these slag heaps.

(c) The future of these children will be bleak and dark unless they are educated.

(d) Simile

**9. Answer any four of the following questions in 30-40 words each : [3 × 4 = 12]**

**(a) What does the poet's smile in the poem, 'My Mother at Sixty-six' show?**

**(b) "Listening to them, I see two distinct worlds....". In the context of Mukesh, the bangle maker's son, which two worlds is Anees Jung referring to?**

**(c) Though the sharecroppers of Champaran received only one-fourth of the compensation, how can the Champaran struggle still be termed a huge success and victory?**

**(d) Which article in McLeery's suitcase played perhaps the most significant role in Evans' escape and how?**

**(e) Why does Derry's mother not want him to go back to visit Mr. Lamb?**

**(f) What considerations influenced the Tiger King to get married?****

**Answer :**

(a) The poet's smile in the poem 'My Mother at Sixty-six' shows that she is trying to pretend that everything is alright. She is trying to comfort herself and her mother by putting on a brave front.

(b) There are two worlds that Anees Jung is referring to in the context of Mukesh, the bangle maker's son. One world is of the family, caught in poverty and burdened by the stigma of caste in which they are born. The other world is the vicious circle of the sahukars, middlemen, policemen, keepers of law, bureaucrats and politicians.

(c) The Champaran struggle can be termed as a huge success and victory because of the following reasons :

(1) Gandhiji wanted to break the deadlock.

(2) The amount of the refund was not important but the landlords obliged to surrender a part of the money and with it their prestige.

(3) Lastly, the peasants had developed courage and learnt to amplify themselves. They had overcome their fear.

(d) The semi-inflated rubber ring played the most significant role in Evan's escape. It was filled with pig's blood from the slaughterhouse in Kidlington. It was mixed with actual blood with 1/10th of its own volume of 3.8 per cent tri-sodium citrate. Evan covered his head and face with this blood and thus was able to hide his identity from the eyes of the prison officer.

(e) Derry's mother did not want him to go back to visit Mr. Lamb because she did not want her child Derry to wander around and get influenced by others who laugh at him.

She wanted him to think it out on his own. She is selfish and realizes that Mr. Lamb has played a more important role in Derry's life.

**10. Answer the following question in 120-150 words: [6]**

**In one's approach to life one should be practical and not live in a world of dreams. How is Jansie's attitude different from that of Sophie?****

**OR**

**Fear is something that we must learn to overcome if we want to succeed in life. How did Douglas get over his fear of water?**

**Answer :**

OR

"All we have to fear is fear itself", said Roosevelt. When the element of fear creeps in, even the best of ventures will prove to be unsuccessful.

Douglas had an aversion to water since his childhood. He was, when a bully pushed him into the water. The fear conquered his mind and he developed an aversion to water from then onwards. However, after a point, he felt upset as he could not indulge in water activities like fishing, canoeing, boating and swimming.

So, he decided to overcome his fear of water. He appointed an instructor who trained him for 5 days. The instructor put a belt around him with a rope attached to the belt which went through a pulley that ran on an overhead cable. He held on to the end of the rope and went back and forth, week after week for nearly three months. The instructor made him put his face underwater and exhale, and then raise his nose and inhale. He held him at the side of the pool and made him kick the water with his legs. Piece by piece, Douglas became a swimmer and swam the length of the pool.

However, to fully reassure himself, he swam the length of the swimming pool alone in the absence of the instructor. He swam in Lake Wentworth in New Hampshire and the warm lake until he felt truly **confident.**

**Answers are not given due to change in present syllabus

11. **Answer the following question in 120-150 words : [6]**

**At the end of the storytelling session, why does Jack consider himself 'caught in an ugly middle position'?**

**OR**

**It may take a long time for oppression to be resisted, but the seeds of rebellion are sowed early in life. How did Zitkala-Sa face oppression as a child and how did she overcome it?****

**Answer :**

Jack used to tell a story from his own imagination every Saturday night in order to make Jo asleep. He used to create a basic plot with a hero named Roger, who had a problem which had to be solved by the wizard who would send him to the wise owl, who would then guide him how to get the required pennies. Jo was a precocious child who raised a lot of questions. On that particular day, his chief character was a skunk, who wanted to have the smell of roses so that he could relate to his peer group. There was a big controversy about the ending of the story as Jo supported Roger skunk and Jack supported his mother. At the end of the story, Jo wanted the story to be altered so she refused to sleep. Jack went down to help his wife Clare. She was furious as she felt that he had spent too much time with Jo. Jack found himself to be caught in an ugly position as he felt inadequate and helpless, both as a father and a husband.

As a father he had not been able to make Jo sleep who was unhappy with the ending of the story and as a husband, his wife was angry at him and he felt like he had nothing to do with her.

12. **Answer the following questions in 120-150 words :** [6]**

**How does the perception and attitude of the villagers of Raveloe change towards Silas Marner from the beginning to the end of the novel?**

**OR**

**How do you perceive Dr. Kemp based on his interaction with Griffin?**

13. **Answer the following question in 120-150 words :** [6]**

**How is Godfrey Cass different from his younger brother Dunstan?**

**OR**

**"Misdirected and self-serving ambition essentially remains an exercise in futility." Describe Griffin's character in the light of the above statement.**

**Note :** All the sets of Outside Delhi and Delhi are same.

**Answers are not given due to change in present syllabus

# ENGLISH

## Outside Delhi [Set-I]

# 2017

**Time allowed : 3 hours** **Maximum marks : 100**

### SECTION – A

### (READING) 30

**1. Read the passage given below and answer the questions that follow : [12]**

**(1)** We sit in the last row, bumped about but free of stares. The bus rolls out of the dull crossroads of the city, and we are soon in the open countryside, with fields of sunflowers as far as the eye can see, their heads all facing us. Where there is no water, the land reverts to the desert. While still on levelground, we see in the distance the tall range of the Mount Bogda, abrupt like a shining prism laid horizontally on the desert surface. It is over 5,000 metres high, and the peaks are under permanent snow, in powerful contrast to the flat desert all around. Heaven lake lies part of the way up this range, about 2,000 meters above sea level, at the foot of one of the highest snow-peaks.

**(2)** As the bus climbs, the sky, brilliant before, grows overcast. I have brought nothing warm to wear. It is all down at the hotel in Urumqi. Rain begins to fall. The man behind me is eating overpoweringly smelly goat's cheese. The bus window leaks inhospitably but reveals a beautiful view. We have passed quickly from the desert through arable land to pasture and the ground is now green with grass, the slopes dark with pine. A few cattle drink at a clear stream flowing past moss-covered stones; it is a Constable landscape. The stream changes into a white torrent, and as we climb higher I wish more and more that I had brought with me something warmer than the pair of shorts that have served me so well in the desert. The stream (which, we are told, rises in Heaven Lake) disappears, and we continue our slow ascent. About noon, we arrive at Heaven Lake, and look for a place to stay at the foot, which is the resort area. We get a room in a small cottage, and I am happy to note that there are thick quilts on the bed.

**(3)** Standing outside the cottage, we survey our surroundings. Heaven Lake is long, sardine-shaped and fed by snowmelt from a stream at its head. The lake is an intense blue, surrounded on all sides by green mountain walls, dotted with distant sheep. At the head of the lake, beyond the delta of the inflowing stream, is a massive snow-capped peak which dominates the vista; it is part of a series of peaks that culminate, a little out of view, in Mount Bogda itself.

**(4)** For those who live in the resort, there is a small mess-hall by the shore. We eat here sometimes, and sometimes buy food from the vendors outside, who sell kabab and naan until the last buses leave. The kababs, cooked on skewers over charcoal braziers, are particularly good; highly spiced and well-done. Horse's milk is available too from the local Kazakh herdsmen, but I decline this. I am so affected by the cold that Mr. Cao, the relaxed young man who runs the mess, lends me a spare pair of trousers, several sizes too large but more than comfortable. Once I am warm again, I feel a pre-dinner spurt of energy—dinner will be long in coming—and I ask him whether the lake is good for swimming in.

**(5)** "Swimming?" Mr. Cao says. "You aren't thinking of swimming, are you?"

**(6)** "I thought I might," I confess, "What's the water like?"

**(7)** He doesn't answer me immediately, turning instead to examine some receipts with exaggerated interest. Mr. Cao, with great off-handedness, addresses the air. "People are often drowned here," he says. After a pause, he continues. "When was the last one?" This questions is directed at the cook, who is preparing a tray of mantou (squat, white steamed bread rolls), and who now appears, wiping his doughy hand across his forehead. "Was it the Beijing athlete?" asks Mr. Cao.

**1.1 On the basis of your understanding of the above passage, answer the statements given below with the help of options that follow : [1 × 4 = 4]**

**(a) One benefit of sitting in the last row of the bus was that :**

(i) the narrator enjoyed the bumps.

(ii) no one stared at him.

(iii) he could see the sunflowers.

(iv) he avoided the dullness of the city.

**(b) The narrator was travelling to :**

(i) Mount Bogda

(ii) Heaven Lake

(iii) a 200 metre high snow peak

(iv) Urumqi

**(c) On reaching the destination, the narrator felt relieved because :**

(i) he had gotten away from the desert.

(ii) a difficult journey had come to an end.

(iii) he could watch the snow peak.

(iv) there were thick quilts on the bed.

(d) **Mount Bogda is compared to :**
- (i) a horizontal desert surface.
- (ii) a shining prism.
- (iii) a Constable landscape.
- (iv) the overcast sky.

**1.2 Answer the following questions briefly :** [1 × 6 = 6]

(e) **Which two things, in the bus, made the narrator feel uncomfortable?**

(f) **What made the scene look like a Constable landscape?**

(g) **What did he regret as the bus climbed higher?**

(h) **Why did the narrator like to buy food from outside?**

(i) **What is ironic about the pair of trousers lent by Mr. Cao?**

(j) **Why did Mr. Cao not like the narrator's wish to swim in the lake?**

(k) **Find words from the passage which mean the same as the following :** [1 × 2 = 2]
**(i) sellers (para 4) (ii) increased (para 7)**

**Answer :**

(a) (ii) No one stared at him.

(b) (i) Mount Bogda

(c) (iv) There were thick quilts on the bed.

(d) (ii) A shining prism.

(e) As the bus climbs higher, the sky grows overcast and the narrator has not brought anything warm to wear. Also, the man sitting behind the narrator starts eating overpoweringly smelly goat's cheese which made the narrator uncomfortable.

(f) A few cattle were drinking at a clear stream flowing past moss-covered stones. It made the scene look like a Constable landscape.

(g) The narrator regretted that he didn't bring anything warm to wear other than of a pair of shorts.

(h) The narrator liked to buy food from outside because the kebabs cooked on skewers over Charcoal braziers were particularly good, highly spiced and well done.

(i) Mr. Cao lends the narrator a spare pair of trousers and the irony about it was that those trousers were several sizes big.

(j) Mr. Cao did not like the narrator's wish to swim in the lake because he had seen people drown in the lake.

(k) (i) vendors (ii) exaggerated

**2. Read the passage given below and answer the following questions that follow :** [10]

(1) Thackeray reached Kittur along with a small British army force and a few of his officers. He thought that the very presence of the British on the outskirts of Kittur would terrorise the rulers and people of Kittur and that they would lay down their arms. He was quite confident that he would be able to crush the revolt in no time. He ordered that tents to be erected on the eastern side for the fighting forces, and a little away from the western slopes, tents be put up for the family members of the officers who had accompanied them. During the afternoon and evening of 20th October, the British soldiers were busy making arrangements for these camps.

(2) On the 21st morning, Thackeray sent his political assistants to Kittur fort to obtain a written assurance from all the important officers of Kittur rendering them answerable for the security of the treasury of Kittur. They be accordingly, met Sardar Gurusiddappa and other officers of Kittur and asked them to comply with the orders of Thackeray. They did not know that the people were in a defiant mood. The commanders of Kittur dismissed the agent's orders as no documents could be signed without sanction from Rani Chennamma.

(3) Thackeray was enraged and sent for the commander of the Horse Artillery, which was about 100 strong, ordered him to rush his artillery into the fort and capture the commanders of the Desai's army. When the Horse Artillery stormed into the fort, Sardar Gurusiddappa, who had kept his men on full alert, promptly commanded his men to repel and chase them away. The Kittur forces made a bold front and overpowered the British soldiers.

(4) Meanwhile, the Desai's guards had shut the gates of the fort and the British Horse Artillery men, being completely overrun and routed, had to get out through the escape window. Rani's soldiers chased them out of the fort, killing a few of them until they retreated to their camps on the outskirts.

(5) A few of the Britishers had found refuge in some private residences, while some were hiding in their tents. The Kittur soldiers captured about forty persons and brought them to the palace. These included twelve children and a few women from the British officers' camp. When they were brought in the presence of the Rani, she ordered the soldiers to be imprisoned. For the women and children she had only gentleness, and admonished her soldiers for taking them into custody. At her orders, these women and children were taken inside the palace and given food and shelter. Rani came down from her throne, patted the children lovingly and assured them that no harm would come to them.

(6) She, then, sent word through a messenger to Thackeray that the British women and children

were safe and could be taken back any time. Seeing this noble gesture of the Rani, he was moved. He wanted to meet this gracious lady and talk to her. He even thought of trying to persuade her to enter into an agreement with the British to stop all hostilities in lieu of an *inam* (prize) of eleven villages. His offer was dismissed with a gesture of contempt. She had no wish to meet Thackeray. That night she called Sardar Gurusiddappa and other leading Sardars and after discussing all the issues, they came to the conclusion that there was no point of meeting Thackeray who had come with an army to threaten Kittur into submission to British sovereignty.

**2.1 On the basis of your understanding of the above passage complete the statements given below with the help of options that follow: [1 × 2 = 2]**

**(a) Thackeray was a/an :**
- (i) British tourist
- (ii) army officer
- (iii) adviser to Rani of Kittur
- (iv) treasury officer

**(b) British women and children came to Kittur to :**
- (i) visit Kittur
- (ii) enjoy life in tents
- (iii) stay in the palace
- (iv) give company to the army officers

**2.2 Answer the following questions briefly : [1 × 6 = 6]**

**(c) Why did Thackeray come to Kittur?**

**(d) Why did the Kittur officials refuse to give the desired assurance to Thackeray?**

**(e) What happened to the Horse Artillery?**

**(f) How, do we know that the Rani was a noble queen?**

**(g) How in your opinion, would the British women have felt after meeting the Rani?**

**(h) Why did the Rani refuse to meet Thackeray?**

**(i) Find words from the passage which mean the same as the following : [1 × 2 = 2]**
- **(i) entered forcibly (para 3)**
- **(ii) aggressive/refusing to obey (para 2)**

**Answer :**

**(a) (ii)** army officer

**(b) (iv)** give company to the army officers

**(c)** Thackeray came to Kittur as he believed that the presence of the Britishers on the outskirts of Kittur would terrorise the rulers and people of Kittur due to which they would lay down their arms. He was quite confident that he would be able to crush the revolt in no time.

**(d)** The Kittur officials refused to give the desired assurance because they were in a defiant mood. Moreover, no documents could be signed without sanction from Rani Chennamma.

**(e)** When the Horse Artillery stormed into the fort, Sardar Gurusidappa, who kept his men on full alert, promptly commanded his men to repel and chase them away. The Kittur forces made a bold front and overpowered the British soldiers.

**(f)** The Kittur soldiers captured about forty people and brought them to the palace. These included twelve children and a few women from the British officers' camp. The Rani ordered the soldiers to imprison all the man, whereas for the women and children, she was gentle and admonished her soldiers for taking them into custody. At her order, the women and children were taken inside the palace and given food and shelter. The Rani came down from her throne, patted the children lovingly and assured them that no harm would come to them.

So, this is how we come to know that the Rani was a noble queen.

**(g)** The British women must have felt guilty as their husbands attacked the queen who was so humble and generous to them.

**(h)** Rani refused to meet Thackeray because there wasn't any point in meeting a person who had come with an army to threaten Kittur and its people for submission to the British sovereignty.

**(i) (i)** stormed **(ii)** defiant.

**3. Read the passage given below and answer the questions that follow : [8]**

The most alarming of man's assaults upon the environment is the contamination of air, earth, rivers, and sea with lethal materials. This pollution is for the most part irrevocable; the chain of evil it initiates is for the most part irreversible. In this contamination of the environment, chemicals are the sinister partners of radiation in changing the very nature of the world; radiation released through nuclear explosions into the air, comes to the earth in rain, lodges into the soil, enters the grass or corn, or wheat grown there and reaches the bones of a human being, there to remain until his death. Similarly, chemicals sprayed on crops lie long in soil, entering living organisms, passing from one to another in a chain of poisoning and death. Or they pass by underground streams until they emerge and combine into new forms that kill vegetation, sicken cattle and harm those who drink from once pure wells.

It took hundreds of millions of years to produce the life that now inhabits the earth and reached a stage of adjustment and balance with its surroundings. The environment contained elements that were hostile as well as supporting. Even within the light of the sun, there are short wave radiations with power to injure. Given time, life has adjusted and a balance reached. For time is the essential ingredient, but in the modern world there is no time.

The rapidity of change and the speed with which new situations are created follow the heedless pace of man rather than the deliberate pace of nature. Radiation is no longer the bombardment of cosmic rays; it is now the unnatural creation of man's tampering with the atom. The chemicals to which life is asked to make adjustments are no longer merely calcium and silica and copper and all the rest of the minerals washed out of the rocks and carried in the rivers to the sea; they are the synthetic creations of man's inventive mind, brewed in his laboratories, and having no counterparts in nature.

(a) **One the basis of your understanding of the above passage make notes on it using headings and sub-headings. Use recognizable abbreviations (wherever necessary-minimum four) and a format you consider suitable. Also supply a title to it.** [5]

(b) **Write a summary of the passage in about 80 words.** [3]

**Answer :**

(a) THE DEGRADATION OF ENVIRONMENT

**I. Contamination of the environment.**

1. Pollution is for most part irrevocable
   i. Initiates an irreversible chain of evil.
2. Chemicals are the sinister partners of radiation
   i. Radiations released through nuclear exp.
   ii. Comes to Earth in F. O. rain.
   iii. Enters the body of humans and remain till death.
3. Chemical & lying in soil
   i. Passes from one to another in F. C.
   ii. kills veg.' sickens cattle etc.

**II. Establishment of life on earth.**

1. Took hundreds of millions of yrs.
2. The env. contained elements that were hostile and supporting.

**III. Changes in life on earth**

1. Life has adj. and reached balance.
2. i. Time is essential but modern world has no time.
3. The heedless pace of man
   i. Radiation no longer the bombardment of cosmic rays.
   ii. The chemicals are no longer natural.
   iii. They have turned out to be synthetic.

**Key :**

| | | | |
|---|---|---|---|
| (i) | exp. | – | explosions |
| (ii) | F. O. | – | form of |
| (iii) | F. C. | – | food chain |
| (iv) | veg. | – | vegetation |
| (v) | yrs. | – | years |
| (vi) | env. | – | environment |
| (vii) | adj. | – | adjusted |

(b) **SUMMARY**

The above passage tells us about the degradation of environment since the human life has evolved on this planet. The chain of evil initiated by pollution is for the most part, irreversible and irrevocable. The various chemicals that come to the Earth in the form of rain and are sprayed in the soil get transferred through the food chain and remain there till the death of the host organisms. The modern world does not have much time to spare as it is well-adjusted and balanced. But the rapidity in change follows the heedless pace of man rather than the deliberate pace of nature. The radiations and chemicals are no longer natural but, rather synthetic or man-made.

## SECTION – B

## (WRITING SKILLS) 30

4. **Your friend, PV. Sathish, has invited you to attend the wedding of his sister, Jaya. You find that you have an important paper of pre-board examination on the day of the wedding. Thus you cannot attend the event. Write in about 50 words a formal reply to the invitation expressing your regret. You are Puneet/Puneeta Vij, M-114, Fort Road, Chennai.** [4]

**OR**

**You are Vikram/Sonia, an electronics engineer, who has recently returned from the U.S., and are looking for a suitable job in the IT industry. Draft an advertisement in about 50 words for the Situations Wanted column of a national newspaper. Your contact number is 9193010203.**

**Answer :**

M-114
Fort Road
Chennai

14 May 20XX

**Subject : Reply to the invitation of the wedding**

Dear Sathish,

Thank you for your kind invitation for the wedding of your sister Jaya. I am sorry to say that I shall not be able to attend the wedding as I have to appear in an important Pre-Board exam on the same day. I wish her all the best for the future.

Congratulations

With warm regards

Puneet/Puneeta.

OR

**SITUATION WANTED**

B.E. (Electronical Engineer), topper of university with a rich experience of working in the U.S., seeking a job in or around Mumbai in the IT industry. Familiar with Java and $C^{++}$. Expected salary 1lac p.m. Contact: Vikram/Sonia, 342, Dadar Mumbai. Contact No. 9193010203

5. **Mountview Public School, Kalka is run by an NGO to give quality education to the children of the deprived sections of society. The Principal of the school feels that blackboards in the classrooms need to be replaced. She decides to ask the chairperson of the NGO named, 'Education for All' for funds. Write her letter in 120-150 words. Her name is Shweta Pandit. [6]**

**OR**

**National Book Trust organised a week-long book fair at Anna Grounds, Chennai. You visited the fair and bought a few books, You were pleased with the arrangements, enthusiasm of the visitors and the fact that books have not yet lost their relevance in the world of the Internet. Write a letter in 120-150 words to the editor of a local newspaper to express your feelings. You are Lalit/ Latha, 112, Mount Road, Chennai.**

**Answer :**

Mountview Public School
Kalka

6 May 20XX

The Chairperson
Education for All
New Delhi.

**Subject : Replacement of Blackboards in classrooms**

Sir,

Through this letter, I want to bring to your notice that the condition of the blackboards in the classrooms are very pathetic. Students are not able to read what is written on them properly. Hence, lessons in classrooms are not understood by the students effectively and quality teaching is hampered. Keeping in mind the inconvenience caused to the students, I request you to get them replaced. There are 25 classrooms in the school so there is a need for 25 blackboards.

It will be very kind of you to do this before the commencement of the session.

Thanking you
Yours sincerely,
Shweta Pandit

**OR**

112, Mount Road
Chennai

6 May 20XX

The Editor
The Hindustan Times
Chennai

**Subject : Visit to the book fair**

Sir,

Through the column of your esteemed newspaper, I would like to share my pleasant feelings of visiting the book fair at Anna Grounds.

I, along with my friends went to the book fair and was thrilled and overjoyed to find among so many books that I love. There were many stalls with books on different subjects. There was a rich collection of English Literature which I was looking forward to. The arrangement was superb. Though there was a big crowd of people, we could move around and search books of our interest smoothly. There was a lot of enthusiasm as well as discipline among the visitors. What was amazing is the fact that books have not yet lost their relevance in the world of internet. I bought a few books. It was a thrilling experience. I am looking forward to visit the fair again.

Yours truly,
Lalit.

6. **Every teenager has a dream to achieve something in life. What they are going to become tomorrow depends on what our youth dream today. Write an article in 150-200 words on 'What I want to be in life'. You are Simranjit/Smita. [10]**

**OR**

**History Society of Kendriya Vidyalaya, Krishna Nagar sent a group of students to visit a place of historical interest. You, Anant/Anita, were its leader. Write a report in 150-200 words for the school newsletter on the tour, describing the place, its history, how you reached there and all that you have learnt.**

**Answer :**

**WHAT I WANT TO BE IN LIFE**

By Simranjit/Smita

Ten years down the line, we will be reading books and admiring the people associated with them. I developed a habit of reading books and as a result, I happened to read about Kalpana Chawla, a girl from a small town, who touched the skies. In her message to the college students of Chandigarh, she said that it is always necessary to realise one's dream. One could get success provided he had the vision and the courage.

This germinated the seed of dream in my mind and I also aspired to soar high and achieve something big. Her struggles and achievements gave me courage to nurture my dream. I want to go into the fashion industry. I doubt whether my parents would approve my idea or not. But I have decided not to give up on my dream for any wrong reasons. I enjoy watching fashion shows and designing clothes. I am also very good at fine arts. It is my passion. I visualise myself as the most successful fashion designer in the future, designing dresses for the biggest celebrities. So, I will work hard to achieve it.

**OR**

**A VISIT TO THE TAJ MAHAL**

By Anant/Anit

I, along with a group of ten students, visited the Taj Mahal last week. It was a trip of five days organised by the History Society of our school. It was done to enhance students' interest in History and historical places.

The place was a unique marvel. Shah Jahan gave India the gift that no one has given ever. His profuse love for his wife gave birth to his eternal beauty. Its captivating charm left us stunned. More than being the wonder of the world, it is a brand. It is built on emotion and drenched in it. It is the undying marble refusing to lose its shine. Rarest talent and unrestricted investment went into the making of it.

Its stunning symmetry and enticing inlay makes every Indian proud.

When the President of USA, Bill Clinton visited the Taj Mahal in the year 2000, he was spell bounded by its beauty. After visiting the Taj Mahal, he said that there are two types of people in the world– those who saw the Taj Mahal and those who didn't. It's immensely beautiful and that's why it is reckoned as one of the seven wonders of the world.

7. **Holi is a festival of colours. It expresses pure and simple joy. Sometimes, we start throwing coloured water and that too on strangers. As the Head boy/girl of your school write a speech in 150-200 words that you will deliver in the morning assembly of your school, describing why Holi is played and how it should be played.** **[10]**

**OR**

**"It is cruel to put stray dogs to sleep." Write a debate in 150-200 words either for or against the motion.**

**Answer :**

**HOLI**

Good Morning respected teachers and my dear friends.

Today, I am here to share my views on why Holi is played and how it should be played. Holi is a festival of colours. It is celebrated every year on a full-moon day in the month of March. It is also known as the festival of love, joy, happiness, unity, renewal and triumph. We celebrate it with vibrant colours indicating joy and happiness. There is a great reason, belief and significance behind celebrating this festival.

Long ago, there was a king Hiranyakashyap; he had a sister, Holika and a son, Prahlad. Prahlad was devoted to God. He was a holy spirit. His father, however, was keen on having everyone worship him. But, his son Prahlad did not worship him. He kept worshipping God. This enraged the king and he planned to kill him by burning him. He asked his sister to sit in the fire with Prahlad in her lap. Holika had a boon that fire could not harm her. So, she was ready to do so. But she died in the fire and Prahlad was saved by the Lord. This originated the celebration of Holi.

So, we should celebrate Holi to spread joy and happiness, to celebrate the triumph of good over evil. We can enjoy the day by playing with colours, meeting near and dear ones and having lots of fun throughout the day safely.

Thank you.

**OR**

**IT IS CRUEL TO PUT STRAY DOGS TO SLEEP**

**FOR :**

Honourable Chairperson, members of the Jury and my dear friends. I am Pulkit and I am here to speak for the motion "It is cruel to put stray dogs to sleep."

It is rather perplexing to see humans torturing the poor and silent souls and putting them to permanent sleep. Is it because they cannot speak; express their pain and suffering?

I want to ask who gives this right to them? In fact, every living being has the right to survive on this planet. God has granted us the wisdom, but we steep down so low that we kill another living creature, that too for sheer pleasure. Is putting someone to sleep a pleasure? It is a shameful thing. Such an act on our part is heinous and all living beings have the right to breathe in the open air.

Instead of killing them, we should find a different solution and relocate them to a place where stray dogs could be put together without causing anyone any problems. Dogs are very faithful and loyal by nature. If we take care of them, they respond very lovingly in their own manner. Putting them to sleep, either out of fear or out of joy, is not a sensible thing. It is high time that we take it seriously and punish the offenders adequately.

OR

**AGAINST :**

Honourable chairperson, jury member and friends, I am Pulkit and I am standing front of you to speak against the motion.

We often see stray roaming here and there in the locality. Some of them can be dangerous but mostly are harmless. They remain in untidy atmosphere, eat whatever is available and breathe dirty smog. Stray dogs are certainly a threat to routine life. So, it is better to put them off to sleep. When stray dogs sleep children and elders can heave a sigh of relief. These stray dogs otherwise chase them, scare them and make their life miserable. They do not have control over their excitement and anger like the domestic dogs. They are not vaccinated and are more prone to many of the diseases, they can spread the infection if any, or diseases to humans too. Stray dogs also have a temptation to eat everything and anything as they do not get proper food and care. Their babies too catch infections and spread the same to children or people who touch them or come close to them. They are left to fend for their own that culminate into chaos. Hence, it is better to keep away from stray dogs and it is the best to put them to sleep more often.

## SECTION – C

## (LITERATURE : TEXTBOOKS AND LONG READING TEXT) 40

**8. Read the extract given below and answer the questions that follow : [1 × 4 =4]**

At back of the dim class
One unnoted, sweet and young. His eyes live in a dream,
Of squirrel's game, in tree room, other than this.

**(a) Why is the class dim?**
**(b) How is the young child different from others?**
**(c) What is he doing?**
**(d) What is a tree room?**

OR

Aunt Jennifer's fingers fluttering through her wool
Find even the ivory, needle hard to pull.
The massive weight of Uncle's wedding band
Sits heavily upon Aunt Jennifer's hand.

**(a) What is Aunt Jennifer doing with her wool?**
**(b) Why does she find it difficult to pull her ivory needle?**
**(c) What does 'wedding band' stand for?**
**(d) Describe the irony in the third line.**

**Answer :**

(a) The class is not maintained properly. It is dim-lighted and ill-equipped.

(b) The young child is not sad or morbid like the other children. He was lost in his dreams.

(c) He is lost in his dreams of having a bright future.

(d) A tree room is the hiding place of the squirrel in the hollow of a tree trunk.

OR

(a) Aunt Jennifer is weaving the image of tigers with her wool.

(b) She finds it difficult to pull the ivory needle through the tapestry more because of mental suppression by her husband than because of her physical weakness.

(c) Wedding band stands for male authority and power. She is trapped in gender oppression and feels burdened by the authority of her husband which is represented by her wedding band.

(d) The irony is that the wedding band might be small and light but it has burdened her spirit. She seems to have lost her identity under its weight.

**9. Answer any four of the following questions in 30-40 words each : [3 × 4 =12]**

**(a) Why were some elderly persons occupying the back benches that day?**
**(b) Why did Jansie discourage Sophie from having dreams?****
**(c) Having looked at her mother, why does Kamala Das look at the young children?**
**(d) How would keeping quiet affect life in and around the sea?**
**(e) Why did the Maharaja decide to get married?****
**(f) What is mother Skunk's role in the story?**

**Answer :**

(a) After the announcement that German was to be taught in the schools of Alsace and Lorraine, the elderly people came to the school that day to attend the last French lesson.They, perhaps, wanted to thank M. Hamel for his valuable service of teaching, and it was also to pay homage to their country that was not theirs anymore.

(c) Kamala Das looks at the young children as they are in contrast to her mother's pale and old face. They are full of life, energy and vitality.

(d) The fishermen cause harm to the whales and the men at sea hurt their hands in their effort to gather salt. Stopping these activities will allow whales to live a wholesome life without

**Answers are not given due to change in present syllabus

any fear of being trapped and it will bring peace, healing and tranquility to the lives of men at sea.

(f) Mother Skunk's role in the story is to keep a balance. Her act indicates that what is natural is not ungraceful. Taking any deviation is the violation of nature. Moreover, parents are quite well known about what is best for their children.

**10. Answer the following question in 120-150 words:**
**"For the children it is wrapped in wonder, for the elders it is a means of survival." What kind of life do the rag-pickers of Seemapuri lead? [6]**

**OR**

**The peddler believed that the whole world is a rattrap. How did he himself get caught in the same?**

**Answer :**

The childern of slums like Seemapuri are the offsprings of homeless ragpickers. As they live in poverty and hunger, they start picking rags at an early age. They grow up in shifting shacks, without drainage or running water.They live in mud houses with roofs of tin and tarpaulin. For elders, garbage is a means of survival but for children, it has surprises. Finding a coin or a currency note by chance, thrills them. This profession gives them a roof over their heads, even if it leaks.

OR

The peddler believed that the whole world was nothing but a big rattrap. It baited people in the form of riches and joys. The peddler could not make ends meet, so he had to indulge into begging and petty theft. As he got lured by the bait of thirty kronors, the world closed in on him and trapped him. He got lost in the forest. At that moment, he realised that his end was near.

**11. Answer the following question in 120-150 words: [6]**

**In India, the so-called lower castes have been treated cruelly for a long time. Who advised Bama to fight, when and how, against this prejudice?****

**OR**

**To choose between professional loyalty and patriotism was a dilemma for Dr. Sadao. How did he succeed in betraying neither?**

**Answer :**

OR

Dr. Sadao succeeded as a doctor as well as a true citizen. As a doctor, he took the enemy home and treated him. He even did an operation and gave a post-operative treatment to the prisoner. He did all this amidst the opposition of his servants and a risk of arrest. But as a patriot, he informed the General about the whole incident and did not allow himself to get emotionally involved with the prisoner.

**12. Answer the following question in 120-150 words:** [6]**

**Attempt a character sketch of Dr. Kemp as a law-abiding citizen.**

**OR**

**Lammeter sisters have money but not class or education. What do you think about them?**

**13. Attempt the following question in 120-150 words:** [6]**

**Why and how did Griffin burglarise the vicarage?**

**OR**

**How did Silas' treatment of Sally Oates affect his life at Raveloe?**

●●

# Outside Delhi [Set-II]

> Note : Except for the following questions, all the remaining questions have been asked in previous sets.

## SECTION – B
## (WRITING SKILLS)

**4. The Principal, Sunshine Public School, Dindigul has invited the Inspector of Police (Traffic) to deliver a lecture on 'Road Safety' in her school. Draft a notice in about 50 words informing the students to assemble in the school auditorium. [4]**

**OR**

**Publicly we proclaim that dowry is an evil. Privately we want our sons to fetch good dowries. Right from our school days we should be taught that demanding and even giving dowry is not only illegal but immoral too. Draw a poster in about 50 words highlighting dowry as a curse. You are Vikram/Sonia.**

**Answer :**

SUNSHINE PUBLIC SCHOOL, DINDIGUL
NOTICE

$10^{th}$ MAY, 20XX

ROAD SAFETY

A lecture on 'Road Safety' has been organised by the school. It will by delivered by the Inspector of Police (traffic) on $20^{th}$ May, 20XX at 9:00 AM. All the students have to assemble in the school auditorium.

**Xyz**
Principal

**Answers are not given due to change in present syllabus

OR

| DOWRY-AN EVIL |
|---|
| Demanding dowry is : |
| • Illegal |
| • Immoral |
| • Unethical |
| In India, one woman is murdered for dowry every 20 minutes and hour |
| SAY NO TO DOWRY |
| BE A MAN, NOT A BEGGER! |

**5. There is a senior secondary school running right in the middle of Manu Vihar, a middle class colony full of housing society flats. In the morning as well as in the afternoon, the road passing in front of the school is congested with school buses, mini buses, private cars, etc. Write a letter in 120-150 words to the Dy. Commissioner (Traffic) to provide at least two traffic police constables to regulate the traffic. You are Gautam/ Gita, A-21. Akashdeep Apartments, Manu Vihar, Delhi.** [6]

OR

**Sunshine Public School, Pune requires two sports coaches (one male and one female). Each should be a degree holder in physical education as well as an SAI certified coach in athletics. You have seen their advertisement and you know that you have these qualifications. Write an application in 120-150 words along with your resume. You are Praveen/Praveena, M-114, Najafgarh. Delhi.**

**Answer :**

A-21
Akashdeep Apartments
Manu Vihar
Delhi

6 May 20XX

The Deputy Commissioner
Traffic Police
Gole Market
New Delhi

**Subject : Monitoring the traffic**

Sir,

Our colony is full of housing society flats. I want to bring to your notice the inconvenience caused to the common man due to traffic jam in the area.

There is a senior secondary school running right in the middle of Manu Vihar. In the morning as well as in the afternoon, the road passing in front of the school is congested with school buses, mini buses and private cars. At that time, it gets difficult to pass through the road. Nobody cares for the traffic signal.

Much of this nuisance will be dealt with if a traffic police officer is there at that time. It will be so kind of you if you could provide at least two traffic police constables to regulate the traffic during the morning and the afternoon hours.

Thank you

Yours sincerely,

Gautam/Gita

OR

M-114
Najafgarh
Delhi

10 May 20XX

The Principal
Sunshine Public School
Pune

**Subject : Application for the job of a sports coach.**

Sir,

This is with reference to your advertisement in The Times of India, dated 2nd May, 20XX, for the job of a sports coach.

I am, at present, working in a private school in Delhi. Since my family is stationed in Pune, I want to shift to the city. Hence, I am looking for a job in Pune.

I have 3 years of experience and can assure you of my best services, if given a chance.

I am enclosing my resume for your perusal.

Yours faithfully,

Praveen/Praveena

Encl. Resume

**RESUME**

| | | |
|---|---|---|
| Name | : | Praveen/Praveena Sharma |
| Address | : | 8, Gandhi Road, Pune |
| Date of Birth | : | 30 Nov 19XX |
| Marital Status | : | Unmarried |
| Educational Qualifications | : i. | Degree in Physical Education from Mumbai, |
| | ii. | SAI certification in athletics. |
| Job experience | : | 3 years in a private school in Delhi |
| Expected salary | : | ₹ 15000 p.m. |
| References | : i. | Prof. B.M. Sharma Adarsh college, Ambala |
| | ii. | Dr. Lal Civil hospital, Delhi |

**7. After the rainy season is over, mosquitoes start breeding. They cause malaria, dengue, chikungunya, etc. These diseases can sometimes prove to be fatal. As the Principal of your school, you have decided to deliver a speech on**

**protection from mosquitoes. Write that speech in 150-200 words. [10]**

**OR**

**"A career counsellor (not you, yourself) is the best person to guide you in the choice of a career." Write a debate in 150-200 words either for or against the motion.**

**Answer :**

**PROTECTION FROM MOSQUITOES**

Good morning dear students, today I want to make you aware of the outspread of malaria, chikungunya and dengue. These may take a great toll on health if precautions are not taken. So, it is pertinent on our part, that we save ourselves by keeping the mosquitoes away.

For this, we should not allow water to stagnate at a place that becomes a breeding place for mosquitoes. We should be very alert and cautious to make sure that water is removed from our coolers, pots and from other such areas where there is a likelihood of water to stagnate. We should spray insecticides, use mosquito nets or repellents at night. We should also cover our bodies properly to avoid mosquito bites by wearing trousers, long sleeved shirts and dressing in light colours. We should use mosquito nets while sleeping. Thus, we can avert the dengue, malaria and chikungunya menace by putting a very little effort on our part. In case, we feel any symptoms of the fever, we must immediately consult the doctor.

**OR**

**A CAREER COUNSELLOR IS THE BEST PERSON TO GUIDE YOU IN THE CHOICE OF A CAREER**

**FOR :**

Honourable Chairperson, Members of Jury and my dear friends, I am Pulkit and I am here to speak for the motion "A career counsellor is the best person to guide you in the choice of a career."

The world is changing at a fast pace. New avenues are coming up; but how the young children can be made aware of the opportunities For students, it is their moulding period. They focus mainly on studies. But not all students are bright academically and it does not mean that they can't do anything. They might be gifted with another type of talent or skill like : music, atheletics, visual intelligence. The good thing in it is that there are good opportunities out in the world for such talent. A Career Counseller works as a guide to help them to understand and utilise those opportunities. Finding their worth, recognising their interests and choosing the right career path for themselves becomes too much of a burden for the students. So, the role of a career counsellor is to help them to find and acknowledge the area of their interest and guide them to the right career options. They provide career mapping support and motivation to the students and help to boost their morale. So, let us not be a parasite on the society. Let us make an intelligent and professional move and step ahead in the right direction with the help of those who are more aware and informed.

**OR**

**AGAINST :**

Honourable chairperson, Members of jury and my dear friends, I am Pulkit and I am here to speak against.

A career counsellor may be one of the people who could give suggestions about what could be the best option for you, but he or she is not the final authority to decide our career. Our qualification, interest and future prospects mainly determine what we are supposed to do in life. Our parents, friends, siblings and people who matter are more keen on our betterment so they definitely know better. We should analyse the pros and cons ourselves too to make the right choice in the career. Career choice is a very major responsibility for one's own self and also the dependents if any so it requires proper rational thinking and the consequent decisions. Career should be the one that suits our capabilities, abilities and innate talents, it should be a place where you earn, learn and grow upwards towards success and development. Therefore, it is always better to form decisions keeping in mind the avenues that any career choice will lead to. We should keep abreast of the latest in order to make the best use of the available choices. One should have a balanced head on responsible shoulders to make the right choices in life. A career counsellor might not understand such aspects of an individual.

## SECTION – C

### (LITERATURE : TEXT BOOKS AND LONG READING TEXT)

**9. Answer any four of the following questions in 30-40 words each : [3 × 4 = 12]**

**(a) "What a thunderclap these words were to me!" (Franz). What were those words and what was their effect on Franz?**

**(b) Why did Douglas fail to come to the surface of the pool as he hoped to?**

**(c) What was Kamala Das's childhood fear?**

**(d) How is the Earth a source of life when all seems dead on it?****

**(e) How does Mr. Lamb react when Derry enters his garden?**

**(f) Which problem did the Maharaja face when he had killed seventy tigers? How did he solve it?****

**Answers are not given due to change in present syllabus

**Answer :**

**(a)** Franz was thunderstruck when Mr. Hamel announced that French will not be taught any longer. He was there to teach his last lesson of French that day. They had received orders from Berlin that teaching French was being barred. Franz felt very sorry for not learning his French lessons. He felt very sad.

**(b)** Douglas failed to come to the surface of the pool because when he reached the bottom, he felt his lungs were ready to burst. He tried to spring up, but came up slowly. He could see and feel nothing but water. He felt too suffocated. Fear gripped him and his legs seemed to have been paralyzed. Before he knew it, he was sinking back to the bottom.

**(c)** Since her childhood, Kamala Das had a fear that she would lose her mother in some way or the other or she would be separated from her. Now that her mother was sixty-six years old and had a deteriorating health, the poet's fear returned with an even greater intensity.

**(e)** Derry enters Mr. Lamb's garden by climbing over the wall. When Mr. Lamb saw him entering his garden, he was startled. Derry tried to explain that he didn't intend to steal anything, but Mr. Lamb welcomed derry instead of getting angry and treated him as his friend. He also asked him to pick up the apples lying in the long grass.

●●

# Outside Delhi [Set-III]

**Note : Except for the following questions, all the remaining questions have been asked in previous sets.**

## SECTION – B
## (WRITING SKILLS)

6. **To have a fair complexion is an obsession in our society. Demand for fair brides in matrimonial columns and sale of fairness creams are evidence enough. Write an article in 150-200 words giving your views on 'Beautiful mind, better than a fair complexion.' You are Natwar/Nimmi. Use the following clues :** [10]
   - **The west, people like to be tanned**
   - **Fair complexion, only skin deep**
   - **Beautiful mind**
   - **Attiude to life**
   - **Behaviour in society, etc.**

   OR

   **Your school is situated near a road intersection. Last week, in the morning, a bus coming at a great speed overturned when it braked suddenly. Senior students of your school rushed out and did everything to save the passengers. You were part of the rescue efforts. Write a report in 150-200 words on the accident and your friends' role, for your school newsletter. You are George/Mary.**

**Answer :**

**BEAUTIFUL MIND, BETTER THAN A FAIR COMPLEXION**

By Natwar/Nimmi

Beauty is not in the face but is a light in the heart. There is much difference between looking beautiful and feeling beautiful. Fair complexion is about visual and superficial beauty Complexion preference varies from person to person. for instance people in the west like to get tanned voluntarily. The term 'beauty' refers to the inner beauty. Inner beauty is based on personality traits such as attitude and behaviour.

But, demand for brides in matrimonial columns and sale of fairness creams make it clear that to have a fair complexion is an obsession in our society. There is a clear discrimination being carried out across our country and society. It is shocking that such trend is not only ruling our mind but is being shamelessly marketed across our TV screens as well.

Our real progress will occur when we get rid of such racist mindset within the general masses of our society.

OR

**A GHASTLY BUS ACCIDENT**

By George/Mary

Death and disasters come uninvited. They have no time-table. Last week, dated 6th May, 20XX. I was at my school gate in the morning when I heard a loud screeching of brakes. A blue line bus was approaching at a dangerously fast speed while the students of our school were crossing the road intersection. The bus was moving so rapidly that the driver could not stop it and it overturned when he applied the brakes suddenly. The scene became a havoc. Passengers were badly injured. Some senior students and I rushed out and tried to save the passengers. The bus driver fled from the scene. There were people with multiple fractures and injuries. Many people came to help. Injured people were rushed to Safdar Jung Hospital where two people died on the spot. It was a tragic sight. Rash driving of a driver took away the lives of the innocent people. I wonder how long will this recklessness continue!

## SECTION – C
## (LITERATURE : TEXT BOOKS AND LONG READING TEXT)

**9. Answer any four of the following questions in 30-40 words each : [3 × 4 = 12]**

**(a) How were the parents and M. Hamel responsible for the children's neglect of the French language?**

**(b) How did Douglas' introduction to YMCA pool revive his childhood fear of water?**

**(c) In the last line of the peom, 'My Mother at Sixty-six', why does the poet use the word 'smile' repeatedly?**

**(d) How is total inactivity on the Earth in the winter months full of life?****

**(e) When did the Maharaja decide to double the land tax for a village?****

**(f) What kind of garden does Mr. Lamb have? Why does he like it?**

**Answer :**

(a) M. Hamel feels that the parents and he himself are responsible for the children's neglect of the French language. Parents preferred to have their children work on their farms or in the mills. Even Hamel used to send the children on petty errands for his convenience or would dismiss the classes when he wanted to go fishing, thus, making them miss their lessons of french language.

(b) The YMCA pool revived unpleasant memories of Douglas' childhood. He recollected the experience on the beach in California when he was three or four years old. He remembered how waves overpowered him despite his father's presence. He had nearly drowned and felt his limbs got paralysed. But, he was determined to learn swimming. So, he decided to learn it by imitating.

(c) The poet uses the word smile repeatedly to cover up her pain and fear of losing her mother. She tries to put on a brave front by hiding her fear behind a smile.

(f) Mr. Lamb's garden is a place where there is an occasional round of birdsongs and rustling of tree leaves. There are crab apple trees and Mr. Lamb shares the apples with the children who come to his garden. His philosophy is to love all the creatures of God so he likes his garden as it gives him a chance to be in contact with the nature and people.

**13. Attempt the following question in 120-150 words:** [6]**

**Describe the circumstances that led to the unveiling of Griffin's invisibility.**

**OR**

**How did Dunstan Cass meet his end?**

**Answers are not given due to change in present syllabus

# ENGLISH

## Delhi [Set-I]

**2017**

**Time allowed : 3 hours** **Maximum marks : 100**

> **Note : Except for the following questions, all the remaining questions have been asked in previous sets.**

**SECTION – B**

**(WRITING SKILLS)**

4. **You are Vikram/Sonia, an Hon's graduate in history with specialization in Medieval India.**

**You are well acquainted with places of historical interest in Delhi, Agra and Jaipur. You are looking for the job of a tourist guide. Write an advertisement in about 50 words for the situations wanted column of a local newspaper. Your contact no. is 999751234.** **[4]**

**OR**

**While walking in a park in your neighbourhood you found a small plastic bag containing some documents and some cash. Write a notice in about 50 words to be put on the park notice board asking the owner to identify and collect it from you. You are Amar/Amrita 9399123456.**

**Answer :**

SITUATION WANTED

An Hon's graduate in history, specialised in Medieval India, well acquainted tourist guide available to give service. Expert in historical places like Delhi, Agra and Jaipur. Good in communication skills, knows Hindi, English and French. Nominal charges. Please contact Vikram/ Sonia. Contact No. 999751234 Janakpuri, Delhi.

**OR**

SHIV VATIKA PARK, JANAKPURI
NOTICE

5 Feb, 20XX

LOST ! LOST ! LOST

This is to inform you all that a small black plastic bag containing some documents and cash was found in the park premises on 2nd Feb., 20XX around 10 a.m. The genuine owner who can claim it is requested to collect it from the undersigned by giving proper identification.

Amar
The Secretary

5. **In our society we do not give to our women the respect and status that they deserve. Women are stared at, stalked and even molested. We need to change the male mindset about women. Write a letter in 120-150 words to the editor of a national newspaper giving your views on the problem. You are Omar/Amna, A 114 Mall Road, Delhi.** **[6]**

**OR**

**You want to spend a week-long holiday at Shimla in the month of October decided to stay at Hotel Snow view. Write a letter in 120-150 words to the manager to book a room. Mention the dates, facilities in the room, food, sight-seeing facilities that you will need. You are Amar/Amrita M114, Lake Road, Karnal.**

**Answer :**

A - 114
Mall Road
Delhi,
12th Feb, 20XX

The Editor
The Times of India
Bahadurshah Zafar Marg, Delhi

**Subject : Respecting the Dignity of women.**

Sir,

I shall feel highly obliged if you publish the following views in the valuable columns of your esteemed paper.

Women have always been the most vulnerable in any age, time, place and society. It has been observed over quite a long time that women are humiliated, tortured, harassed, stalked and molested. The other problems include sexual harassment, domestic violence and marital discord which cause a lot of embarrassment. Though strict measures are taken to improve the status of women in the society, otherwise the problem remains where it was.

As we all know, in modern India, women are playing a major role and have significant contribution in the welfare of democratic values. So, it is necessary to change the male mindset towards women. It is only possible by spreading awareness among women and the society at large. Also, the bringing up of boys and girls should be done equally at home. This growing rate of violence against women can be stopped. Strict measures should be adopted by the society and the government in order to do so. Moreover, we should take a pledge to start campaigns against women violence and other such issues in order to increase the status and social standing of women, economically and politically.

Yours Sincerely,

Omar/Amna

OR

M-114
Lake Road,
Karnal

12th Feb., 20XX

The Manager
Snow View Hotel
Shimla

**Subject : Booking a Room in Hotel**

Sir,

It gives me an immense pleasure to inform you that I shall be visiting Shimla during the autumn break. I intend to reach there on 15th Oct., 20XX for a week-long vacation. During the stay at Shimla, you are requested to reserve a room for me in your hotel from 15th Oct. to 21st Oct., 20XX. Please see to it that the room should be well-equipped with all the modern facilities. Complimentary breakfast and amazing meals would be appreicated. I am enclosing a bank draft of ₹ 5000/- as an advance for reservation charges. It would be highly appreciated if you can provide me with a guide to show me around Shimla. Please confirm the booking so that the remaining amount shall be sent to you. I shall also need a chauffeur driven car to take me in and around Shimla.

Yours Sincerely,

Amar/Amrita

**6. Our performance in the Rio Olympics has told us that we do not pay enough attention to athletics and outdoor games. It is the time we revise our attitude. Sports should be an important part of every school's daily routine. Write an article in 150-200 words on 'Importance of Outdoor Games'. You are Sreeja/Thomas. [10]**

**OR**

**Cultural Society, Sunshine Public School, Nellore organised an adult literacy camp in its neighbourhood. Write a report in 150-200 words on the camp for your school newsletter. You are P. V. Sunitha, Secretary. Use the following clues :**

**no. of volunteers—hours spent in teaching—location of the class—chairs, blackboards—no. of people attending the camp—benefit.**

**Answer :**

IMPORTANCE OF OUTDOOR GAMES

By Sreeja/Thomas

As in the words of Muhammad Ali :

"Champions are made from something they have deep inside them, — a desire, a dream, a vision".

But our performance in Rio olympics tells a different story. A nation of more than 125 crores has not been able to produce a sportsperson, or an athlete of International standards even after 76 years of independence! What really is the reason behind the adverse condition of Indian sports? Growing craze of cellphones and video games among children are the official apathy and the lack of facilities are like an additional handicap. Also, at present, there is a lot of emphasis on academics which leads to the lack of time for the students to get indulge in sports activities. So, to brighten up the image of sports and games, indoor and outdoor games are equally important-one for relaxation and the other for physical exercise. Games such as Volleyball, Hockey, Cricket, Basketball, Football and Lawn Tennis provide recreation, develop sportsmanship and broaden the base of sports. There is no doubt that these outdoor games develop good habits, skills, knowledge and other social qualities required in a sportsperson. Sports personalities and players should be given proper training for them to be properly motivated. If we really want to revive our sports, the Government and sports authorities of India must put serious efforts into the matter. Proper facilities should be provided to the athletes and new talent should be encouraged. In order to enhance the performance, we should chalk out weekly plans, meso-cycle plans, micro-cycle and yearly plans for the sports person. Sports play an integral part in education. So we should motivate all the students of an institution to take parts in games and sports. 'A game for each and each for a game' may be considered the motto of sports.

**OR**

AN ADULT LITERACY CAMP

By P. V. Sunitha, Secretary

In order to enlighten the fruits of education, an adult literacy campaign was conducted by Cultural Society, Sunshine Public School, Nellore from 2nd Feb. to 5th Feb. 20XX in our neighbourhood. The campaign was held with the motive of ensuring education for adults and helping them to become literate. During the campaign, 50 volunteers participated and spent 4 hours in teaching them. For this noble task, the community hall was selected at Janakpuri. Both males and females were engaged and given a taste of education. They had arranged chairs, desks and blackboards for teaching them. More than 200 people attended the camp. They were provided with free copies, pencils and other necessary items. Every illiterate person was given special attention. It was hoped not only to achieve our target, but also to exceed the same. It was positively determined to make the campaign grand success. The chief-coordinator, Mr. Joshi, thanked all the volunteers, teachers and all the members who motivated the adults for gaining education, and for making the campaign a grand success.

**7. Your PGT English Ms. Geetha is a short story writer also. 'Sky is not Far' is a collection of her latest short stories. This book has won a national award. Write a speech in 150-200 words you will deliver in her honour in the morning assembly. [10]**

**OR**

**'Private cars should be banned in the congested commercial areas of the cities.' Write a debate in 150-200 words either for or against the motion.**

**Answer :**

**IN HONOUR OF MS GEETHA'S ACHIEVEMENT**

It is rightly said that :

'Surprises render more pleasure, especially when they are totally unexpected.'

A very good morning to one and all. Respected Principal, teachers and my dear friends !

I, Vinod, a student of class, XII feel extremely proud and honoured to speak about my loving teacher, Ms. Geeta P.G.T. (Eng.), who wrote the short story, 'Sky is Not far', which has won a national award. It is really a matter of pride for our school. Each one of us wants to touch the sky but only a few of us dare to chase our dreams and our English teacher fulfilled her dream with a strong determination and a keen hobby of writing stories. Her easy style of writing along with the issues linking to the everyday problems was well received by readers, particularly the youth. This short story deals with women's struggles and essentially their attainment of the desired goals and the more women relate to it, it goes on and comes up with some other 'newer issues'. This is really a great achievement for her and our school. Her story is ultimate, unmatched and an excelled one. I feel extremely honoured for her service to the society and her excellent work. She will continue to be a source of inspiration to one and all.

Thank you.

OR

**PRIVATE CARS TO BE BANNED IN CONGESTED AREAS**

**FOR :**

A very good morning to all of you here.

Respected panel of jury members, teachers and my friends !

I, Raja, stand before you today to express my views regarding; Should private cars be banned in the congested commercial areas of cities'?

I am in favour of the motion. It is causing a lot of inconvenience to the people who are forced to be stuck in congested areas. We must think unconventionally and alternatively to solve this problem. The roads leading to the commercial areas should be made one way roads during the peak traffic hours, i.e., between 8 am to 10 am and 5 pm to 8pm respectively. Heavy fine should be levied on people who wait inside their vehicles in the 'No waiting' area. More underground parking facilities need to be provided as there is a lack of parking lots. The officials should keep strict vigilance on this issue. It is noted, that a lot of cars generally hinders the efficiency of the commercial areas, because it creates traffic jams and they are the main cause of pollution. The Government. should ban cars to force the people to use the public transportation provided and to reinforce the efficiency of commerce.

OR

**AGAINST :**

A very good morning to all the jury members and audience.

I, Raja, standing in front of you to express my views against the motion.

Private cars are a convenient way to start from the comfort of your home and reach any destination with ease. In places where there is congestion, proper lanes can be assigned for vehicles and pedestrians separately. Banning the private vehicles in these areas will not be the right solution as old people and children might find it tough to reach their cars. Walking in these congested areas is also very difficult and it is better to travel by cars. We should have a long term plan for getting away with the problem. Congested areas should be planned well to have a gate and security guards so that vehicles enter and come out in a proper manner. Parking areas should also be monitored in these areas and vehicles parked in the middle or at inappropriate places should be fined so that they do not create traffic commotions. A proper check, planning and monitoring will help us to create better atmosphere at the congested places. Banning the private vehicles will not be a solution to ease traffic woes in these areas.

## SECTION – C
## (LITERATURE : TEXT BOOKS AND LONG READING TEXT)

**8. Read the extract given below and answer the questions that follow : [1 × 4 = 4]**

Far far from gusty waves these children's faces.
Like rootless weeds, the hair torn round their pallor;
The tall girl with her weighed-down head.

**(a) Who are these children?**

(b) **Which figure of speech has been used in the first two lines?**

(c) **Why is the tall girl's head weighed down?**

(d) **What does the word, 'pallor' mean?**

**OR**

Aunt Jennifer's tigers prance across a screen,
Bright topaz denizens of a world of green.
They do not fear the men beneath the tree;
They pace in sleek chivalric certainty.

(a) **Why are the tigers called Aunt Jennifer's tigers?**

(b) **How are they described here?**

(c) **How are they different from Aunt Jennifer?**

(d) **What does the word, 'chivalric' mean?**

**Answer :**

(a) These children are the victims of social injustice and acute poverty.

(b) Simile and Alliteration have been used in the first two lines of the poem,

(c) The tall girl is malnourished and depressed. That's why she has her head weighed down by poverty.

(d) Pallor means 'an unhealthy pale appearance.'

**OR**

(a) The tigers are called Aunt Jennifer's tigers because they represent her hidden aspirations and desires.

(b) They are described as the inhabitant of a dense green forest, prancing unafraid with bright eyed.

(c) They are different from Aunt Jennifer, as the animals created by her displayed chivalry and fearlessness, though, she herself is oppressed and fearful

(d) Chivalric means brave.

9. **Answer any four of the following questions in 30-40 words each : [3 × 4 = 12]**

(a) **Why did Franz not want to go school that day?**

(b) **What was Sophie's ambition in life? How did she hope to achieve that?****

(c) **What kind of pain does Kamala Das feel in 'My Mother at Sixty-six'?**

(d) **How can 'mighty dead' be things of beauty?**

(e) **Why was the Maharaja once in danger of losing his kingdom?****

(f) **What was the basic plot of each story told by Jack?**

**Answer :**

(a) Franz did not want to go to school that day because the weather was very tempting. The bright day, chirping of birds, woods and fields, free environment outside along with the drilling by Prussion soldiers all restricted him from going to school. Moreover, he did not prepare his lesson and feared Mr. M. Hamel's scolding on that day.

(c) The failing health of the old mother upset and haunts the poetess. She feels that her mother is swiftly nearing the death. She is leaving her mother for sometime and going abroad; she fears that she would not find her alive upon her return.

(d) John Keats was the youngest of all Romantic Poets. He was famous for sensuousness, describing the mighty dead as the magnificent warrior who selflessly sacrificed his life for his nation and won grandeur or greatness. So, achieving this is equal to achieving a thing of beauty.

(f) Jack always used to tell a story to his little daughter, Jo, about Roger. Usually, every story had a small creature named Roger fish, Roger squirrel or Roger chipmunk, who always had a problem, so he went to the wise owl for advice and then the wise owl directed him to go to the wizard who solved the problem for a fee. Then Roger went to the well to get some pennies to pay the fee. The wizard used his magical spell to solve the problem after which, he went home and had a family dinner.

10. **Answer the following question in 120-150 words: [6]**

**Garbage to them is gold. How do ragpickers of Seemapuri survive?**

**OR**

**The peddler thinks that the whole world is a rattrap. This view of life is true only of himself and of no one else in the story. Comment.**

**Answer :**

Garbage provides the rag-pickers a means to survive. It is there daily bread. Sometimes, they find currency notes or even silver coins in the heaps of garbage. For them, it is wrapped in wonder and this garbage is nothing less than gold for them.

Poor Ragpicker like Saheb spend their lives looking for gold in garbage dumps. The ragpickers of Seemapuri lived in structures of mud, tin and tarpaulin with no sewage or drainage system for running water. Their parents have no fixed income. They wage war against poverty and hunger. They have no dream except finding the means of survival. These people have lived there for more than thirty years without an identity or permit. They have got ration cards that enable them to buy grains and get names on the voter list. For them, food is more important for survival than an identity, rag picking is the means of their survival.

**Answers are not given due to change in present syllabus

OR

This view of life is universal. It relates to everyone who was ever a victim of circumstances. The Peddler, a poor vagabond, makes and sells rattraps to make both ends meet. He even found a parable between the world and the rattrap. He had been living a despicable life of misery, poverty and frustration, visualising the whole world to be a rattrap ready to engorge anyone who succumbed to the bots of its riches and joy. The world had been rather unkind to only him as others like crofter, Edla and Ironmaster were not tempted with anything as they already had everything, so he felt happy to think ill about it. Despite the Crofter's kindness and hospitality, he betrayed his trust and stole his money. But Edla's compassion and understanding finally brought about a transformation in his nature. So, in his own opinion, only he suffers.

11. **Answer the following question in 120-150 words : [6]**

**Untouchability is not only a crime, it is inhuman too. Why and how did Bama decide to fight against it?****

OR

**Good human values are far above any other value system. How did Dr. Sadao succeed as a doctor as well as a patriot?**

**Answer :**

OR

The story, 'The Enemy' set during the World War, deals with a dilemma. When man faces difficult choices to be a patriotic or to be a professional, he acts impatiently in both the situations. Dr. Sadao is thoroughly professional and when it comes to his patients, he becomes oblivious to other facts. So, despite the fact that the injured man is an American Soldier, and an enemy, Dr. Sadao gives top priority to save the life of the dying soldier. As a doctor, he performed his first and foremost duty which is towards the sick and the injured who needs, his immediate attention, irrespective of his nationality. But later on, being a true patriotic nationalist and a loyal citizen of Japan, he handed him over to the police by informing the General about him. But the General is totally governed by self-absorption. This ultimately set Dr. Sadao to comply with the plan for him to escape. Thus, Dr. Sadao managed to serve as a doctor as well as a patriot.

12. **Answer the following question in 120-150 words :** [6]**

**Describe Marvel as an opportunist.**

OR

**George Eliot has portrayed Godfrey as a morally weak character. Comment.**

13. **Answer the following question in 120-150 words :** [6]**

**How does Dr. Cuss's encounter with Griffin end in a disaster?**

OR

**What kind of life did Silas lead at Lantern Yard?**

●●

# Delhi [Set-II]

**Note : Except for the following questions, all the remaining questions have been asked in previous sets.**

SECTION – B**

(WRITING SKILLS)

4. **After the rains, cases of dengue, chikungunya etc., are on the rise in your city. As the Principal of Sunshine Public School, Manu Vihar, you have decided to allow your students to wear full sleeve shirts and trousers in the school for a period of one month. Write the notice in about 50 words. [4]**

OR

**You are Health Secretary, Students Council Citizens Public School, Ram Bagh, Varanasi. The Council has decided to start from the second of October a week-long cleanliness drive around the school. Draft a notice in about 50 words asking the Class XI students to enrol for the drive.**

**Answer :**

SUNSHINE PUBLIC SCHOOL, MANU VIHAR

NOTICE

16th March, 20XX

RELAXATION IN UNIFORM

This is to inform all the parents that after incessant rains, cases of dengue and chikungunya have been noticed, concluding this, the school mangement has swung into action and decided to allow the students to wear full sleeve shirts and trousers in school for a period of one month, i.e., from July to August, 20XX. Parents are requested to co-operate in this matter.

Principal

A.K. Sharma

**Answers are not given due to change in present syllabus

OR

CITIZEN PUBLIC SCHOOL, RAMBAGH
NOTICE
16th Oct, 20XX
CLEANLINESS DRIVE
The student's council of our school is launching a week long 'Cleanliness Drive' from 2nd October to 8th October, 20XX.
The Students of class XI are requested to join this mission by enrolling for the drive and make it a great success at large.
XYZ.
Health Secretary

**5. Last month you purchased a TV set from Ram Electronics, Pushp Vihar, Ambala. It is not working properly (imagine a few defects). Write a letter of complaint in 120-150 words to the Manager asking for repair or replacement. You are Amar/Amrita, M114 Model Town, Ambala. [6]**

OR

**You are living in a village along a national highway. The state roadways bus does not stop near your village. Write a letter in 120-150 words to the General Manager. State Railways requesting a bus stop for your village. Give all the reasons why you need a bus stop there. You are Omar/Amna, Gopalpur, Dist. Ramnagar.**

**Answer :**

M 114, Model Town,
Ambala
24th January, 20XX
The Manager
Ram Electronics
Pusp Vihar
Ambala

**Subject : Complaint against defective T. V. Set.**

Sir,

I bought a L.G. 21"colour T.V. set from your show-room against cash memo no. 3212 dated 15th Jan, 20XX. You assured me that your T.V. sets were far superior in function than that of other companies.
Last week, it started showing major defects like blurred vision and sound. I thought that the sound problem was due to weak signals and that it will rectify shortly but it stayed with a more stronger frequency. Now, the situation is that the sound is not clear at all. It is jarring and, the colours are fading away suddenly. Since the T.V. set is within the guarantee period of two years, you are requested to get the defects rectified by your service engineers. In case, if the set is beyond repair, kindly get it replaced.

Thank you,
Yours sincerely,
Amar/Amrita

OR

Gopalpur,
Dist. Ramnagar
16th March, 20XX
The General Manager
The State Roadways Authority
Ramnager

**Subject : Request for a bus stop.**

Sir,

I beg to draw your kind attention to the problem of transport in our village, Gopalpur. As you are aware, Gopalpur is situated along the national highway but the villagers are facing the brunt of isolation as there is no bus stop within a radius of 2 km. Since the state buses are unavailable, the autos and taxis also don't have a stand nearby. It becomes extremely difficult to commute from the place except in one's own vehicle.
I request you to start the bus service in our village at the earliest so that the inconvinence being caused to the villagers is permanently eradicated.
Looking forward to an early action by the concerned authorities.

Yours truly,
Omar/Anna

**7. Sakshi Malik won a bronze medal in the Rio Olympics; every Indian felt proud of her. You, Head girl of your school, want to deliver a speech in order to express your feelings. Write the speech in 150-200 words. Apart from your own ideas, use the following clues :**
**Wrestling mostly a male sport-Sakshi born in conservative surroundings-most unfavourable gender ratio for females-parental encouragement-one can't forget the look of determination on her face in the last three minutes of the bout. [10]**

OR

**'Our large population is not a cause of poverty but an asset, a resource'. Write a debate in 150-200 words, either for or against the motion.**

**Answer :**

SAKSHI MALIK'S ACHIEVEMENT

Good morning everyone !

Respected Principal, teachers and my dear friends. I, xyz the head girl of the school, am going to deliver a speech on Sakshi Malik. In India, where all the sisters were celebrating the day with their brothers on Raksha Bandhan, far away in Rio de Janerio, a 23-year-old wrestler, Sakshi Malik, made the nation proud by winning a bronze medal. Malik became the first female Indian wrestler who has bagged a medal in the Olympics.

Sakshi Malik, the daughter of a bus conductor, comes from a small district in Haryana. She started her training at the age of 12. Malik and her coach, Ishwar Dahiya, faced many problems during the initial stages of the former's career as the mentality of the conservative society thought that wrestling was not meant for girls; it was thought to be a male sport. There are still so many people who are living with such male-dominant thinking.

Not only has she proved to be one of the best wrestlers in the world but also fought off long odds, bureaucratic obstacles, lack of facilities, and the conservative Indian society. Sakshi taught the nation that women in the country need support more than protection. We are proud to be the citizens of the nation where women like Sakshi Malik are born.

**OR**

**LARGE POPULATION NOT A CAUSE OF POVERTY BUT AN ASSET**

**FOR :**

A very good morning to the jury members, all my dear friends and my fellow participants. Today, I am going to speak for the topic 'Our large population is not a cause of poverty but an asset, a resource'.

India has the second highest population in the world with 125 crore people and also is the biggest provider of skilled manpower in the world. Currently, India provides the maximum number of engineers to the world. It is also one of the largest markets for automobiles, high-value foods, etc.

Large population means more intelligence that can be put to good effect and also skilled population draws more projects from foreign companies as well.

I, therefore, agree with the view that our large population is not a cause of poverty but an asset as every hand can become an earning hand for the economy and not a cause of poverty.

**OR**

**AGAINST :**

Good morning the jury members and my fellow friends. Today, I am going to speak against the topic.

Large population is certainly the main factor for people falling below the poverty line. The presence of more people demands more requirement of resources which in turn needs more jobs. Too many qualified people apply for similar jobs and then they have to compromises or settle for less wages. The situation is bad because there are too many mouths to feed. People should find ways to control the number otherwise we would be facing a situation wherein we have our major part of the population below the poverty line. There are issues of living, food, earning and education because of the large population. A family having 4 to 5 children cannot educate all their children because of their financial position with the result they do not get a job and hence remain poor. Everything comes back as a whole circle and centres around one single point that we have to check the increase in number of population. Large population becomes an asset when everyone has proper education, job, shelter and food otherwise it is a burden for the family and the country at large. Therefore, we can conclude that large population is not an asset but an issue of concern which demands everyone's efforts.

## SECTION – C
## (LITERATURE : TEXT BOOKS AND LONG READING TEXT)

**9. Answer any four of the following questions in 30-40 words each : [3 × 4 = 12]**

**(a) Why was Franz not scolded for reaching the school late that day?**

**(b) Sophie was dreaming of so many things in her life. What were they?**

**(c) Why are the youngsters described as springing? (My Mother at Sixty-Six).**

**(d) In the hot season, how do man and beast get comfort? (A thing of Beauty)**

**(e) How did the Maharaja deal with a high ranking British officer who wanted to shoot a tiger?****

**(f) Having got rid of his stink, what problem did Roger Skunk face?**

**Answer :**

**(a)** Mr. H. Hamel did not scold Franz. He only said that it is not good to put off learning. He was overwhelmed with emotions as it was his last French lesson. So, he asked him to sit in his place quickly.

**(b)** Sophie dreamt of becoming as boutique owner, and actress, a manager or a fashion designer. She wanted to be a part of the sophisticated world so that she could visit exotic places that she believed her brother went to.

**(c)** While driving to the airport, in an effort to distract herself from the thoughts of her ageing mother, the poet looks at the young trees, 'sprinting'. The trees seem to be running past the moving car. The sprinting of the trees symbolises the rapidly passing years of a human's life from childhood to old age. This image, showing activity and strength, is contrasted with that of her old and weak mother who seems pale and dormant, sleeping in the car.

**Answers are not given due to change in present syllabus

The 'young' trees represent life in contrast to her mother's approaching death.

**(d)** In the hot season, both men and beast relax in a bower and walk by a cooling covert made by the rills

**(f)** He faced the problem from his mother who felt extremely annoyed. She could not stand that smell. She decided to take him right back to that 'awful' wizard and ask him to give the original smell back to her boy.

## Delhi [Set-III]

**Note : Except for the following questions, all the remaining questions have been asked in previous sets.**

**SECTION – B**

**(WRITING SKILLS)**

**4. RJ Public School is located in a Central Government employees' residential colony. Cultural Society of the School has decided to organise a fancy dress show on 25th of January in which each participant will wear the dress particular to his/her region. The aim is to show the cultural diversity of India. As Secretary, write a notice in about 50 words the names of those who want to participate. [4]**

**OR**

**An NGO has approached your school to offer book grants to the needy students. As Head Girl of Sunshine Public School, Aram Bagh, write a notice in about 50 words asking students, who are in need, to put their requests into the box kept outside the Principal's office.**

**Answer :**

R.J. PUBLIC SCHOOL

NOTICE

16th Dec. 20XX

"FANCY DRESS SHOW"

School Cultural Society is holding a fancy dress show to portray the cultural diversity of India on 25th Jan., 20XX in the school auditorium at 2 p.m. To make this gala a roaring success, the participants are advised to wear the dress particular to their region.

The interested students should give their names to the undersigned for the show positively by 15th Jan., 20XX.

xyz

The Secretary

Cultural Society

**OR**

SUNSHINE PUBLIC SCHOOL, ARAM BAGH

NOTICE

16th March, 20XX

BOOK CONTRIBUTION BY NGO

The students should feel glad to know that an NGO has offered book grants to support the students in need. Those who are in need of the help regarding the purchase of books are adviced to put their requests in the box kept outside the Principal's office as soon as possible.

xyz

Head Girl

**5. You are Principal of Kendriya Vidyalaya, Ramnagar. The number of students in the computer class is increasing. Next year, starting in April 2017, you will need 25 more computers in your lab. Write a letter in 120-150 words to the Commissioner, K. V. Sangathan for funds. [6]**

**OR**

**You are Aslisha/Nimmi Dhar B–94 fort Road, Jammu. You have read the advertisement given below. You are qualified for the job. Write an application in 120-150 words along with a resume.**

**India Chemical Industries, Delhi**

**requires**

**Accounts Officers**

**Qualification : B. Com.**

**Experience : Minimum 4 years**

**Job requirement : Maintaining books of accounts, preparation of Balance Sheet, etc.**

**Salary : Best in the Industry**

**Apply to : Managing Director, ICI,**

**B-12 Barakhamba Road, New Delhi**

**Answer :**

Kendriya Vidyalaya
Ramnagar

16th March, 20XX

The Commissioner,

K.V. Sangathan

**Subject : Request for funds regarding purchase of new computers**

Sir,

I am writing this letter in regard with the upgradation of the computer lab. The number of students in the computer class is increasing rapidly but the number of computers is not enough. Most of the computers are outdated which causes inconvenience to the students. The lab has to be renovated. More desktop computers and an adequate seating arrangement is needed. Therefore, I request for a grant of sufficient funds in order to purchase 25 more computer for the new session of the school. Kindly take a quick decision so that no inconvenience is caused to the students.

Thank you,

Yours sincerely,

R. K. Sharma

Principal

**OR**

B-94

Fort Road

Jammu.0.1

The Managing Director, ICI,

B-12 Barakhamba Road.

New Delhi.

**Subject : Application for the post of Accounts officer.**

Sir,

With reference to your advertisement published in the Times of India, dated 15th Feb., 20XX, I would like to apply for the post of accounts officer in your company.

Since, I have a keen interest in the subject, I found your offer quite alluring. I am enclosing my Bio Data to enable you to assess my suitability for the post.

So, Please consider my application and favour me with an interview.

Thanking you,

Yours Truely,

Aslisha/Nimmi Dhar

End : Bio Data.

**BIO DATA**

1. NAME : Aslisha/Nimmi Dhar
2. FATHER'S Name : Om Prakash Sharma.
3. Date of Birth : 6th Sept., 1976
4. Address : A Block, Vikaspur. New Delhi.
5. **EDUCATIONAL QUALIFICATION :**
   (i) B. Com. from D.U.
   (ii) C.A. from Institute of Chartered Accountants.
   N. D.
   (iii) ICWAI
6. **EXPERIENCE** : 4 year experience as an accountant in Sims Pvt. Ltd.
7. **SALARY EXPECTED** : Negotiable
8. **REFERENCES** :(i) Adv. Lal Singh 31-B, Civil Line Delhi
   (ii) Mr. Arun Goyal Accontant 40-C, Rambagh, Delhi.

## SECTION – C
## (LITERATURE : TEXT BOOKS AND LONG READING TEXT)

**9. Answer any four of the following questions in 30-40 words each : [3 × 4 = 12]**

**(a) What did Franz wonder about when he entered the class that day?**

**(b) What shocking experience did Douglas have at YMCA pool?**

**(c) Why does Kamala Das compare her mother to 'a pale winter's moon'?**

**(d) What rich bounty has the heaven given us? (A Thing of Beauty)**

**(e) Why did the Maharaja have to pay a bill of three lakh rupees to the British jewellers?**

**(f) What problem did Roger Skunk face he went to play with his friends. How did he solve it?**

**Answer :**

**(a)** Franz had a bundle of surprises when he entered the class room. The village people were sitting quietly on the back benches. Mr. Hamel did not get angry with Franz, rather he requested him very kindly to go to his place and sit quietly. Mr. Hamel was dressed in his best formal clothes. Everybody in the class looked very sad.

**(b)** William Douglas had a very shocking experience at the YMCA pool. A bully threw him into the deeper end of the pool. He hit the water surface in a sitting postion and swallowed a lot of water, all at once. He nearly died in this misadventure.

**(c)** The poet's mother is sixty-six years old. She is fast approaching her death. Her shrunken pale face resembles that of a corpse. She has lost her charm and the strength of youth. Similarly, the pale winter moon looks hazy and dull.

**(d)** The heaven has given us innumerable things things of beauty. They all give us a joy forever. They are like endless fountains of immortal drink, pouring in from the heaven's brink.

**(f)** Roger Skunk smelled very bad. Whenever he would go to play, all the other little animals would run away from him. Then he went to the wise old owl, who told him to go to see the wizard. The wizard solved Roger's problem with some magic spell and he got the smell of roses.

**12. Answer the following question in 120-150 words :**** **[6]**

**Mrs. Hall is greedy but efficient in her business.**

**Attempt a character sketch of Mrs. Hall.**

**OR**

**Nancy Lammeter is a typical country girl. Comment.**

●●

** Answer is not given due to the change in present syllabus

# ENGLISH

## Outside Delhi [Set-I]

## 2016

**Time allowed : 3 hours** **Maximum marks : 100**

### SECTION - A

### (READING) 30

1. **Read the passage given below :** **[12]**

(1) Maharana Pratap ruled over Mewar only for 25 years. However, he accomplished so much grandeur during his reign that his glory surpassed the boundaries of countries and time turning him into an immortal personality. He along with his kingdom became a synonym for valour, sacrifice and patriotism. Mewar had been a leading Rajput kingdom even before Maharana Pratap occupied the throne. Kings of Mewar, with the cooperation of their nobles and subjects, had established such traditions in the kingdom, as augmented their magnificence despite the hurdles of having a smaller area under their command and less population. There did come a few thorny occasions when the flag of the kingdom seemed sliding down. Their flag once again heaved high in the sky, thanks to the gallantry and brilliance of the people of Mewar.

(2) The destiny of Mewar was good in the sense that barring a few kings, most of the rulers were competent and patriotic. This glorious tradition of the kingdom almost continued for 1500 years since its establishment, right from the reign of Bappa Rawal. In fact, only 60 years before Maharana Pratap, Rana Sanga drove the kingdom to the pinnacle of fame. His reputation went beyond Rajasthan and reached Delhi. Two generations before him, Rana Kumbha had given a new stature to the kingdom through victories and developmental work. During his reign, literature and art also progressed extraordinarily. Rana himself was inclined towards writing and his works are read with reverence even today. The ambience of his kingdom was conducive to the creation of high quality work of art and literature. These accomplishments were the outcome of a long standing tradition sustained by several generations.

(3) The lives of the people of Mewar must have been peaceful and prosperous during the long span of time; otherwise such extraordinary accomplishment in these fields would not have been possible. This is reflected in their art and literature as well as their loving nature. They compensate for lack of admirable physique by their firm but pleasant nature. The ambience of Mewar remains lovely thanks to the cheerful and liberal character of its people.

(4) One may observe astonishing pieces of workmanship not only in the forts and palaces of Mewar but also in public utility buildings. Ruins of many structures which are still standing tall in their grandeur are testimony to the fact that Mewar was not only the land of the brave but also a seat of art and culture. Amidst aggression and bloodshed, literature and art flourished and creative pursuits of literature and artists did not suffer. Imagine, how glorious the period must have been when the Vijaya Stambha which is the sample of our great ancient architecture even today, was constructed. In the same fort, Kirti Stambha is standing high, reflecting how liberal the then administration was which allowed people from other communities and kingdoms to come and carry out construction work. It is useless to indulge in the debate whether the Vijay Stambha was constructed first or the Kirti Stambha. The fact is that both the capitals are standing side by side and reveal the proximity between the king and the subjects of Mewar.

(5) The cycle of time does not remain the same. Where the reign of Rana Sanga was crucial in raising the kingdom to the acme of glory, it also proved to be his nemesis. History took a turn the fortune of Mewar—the land of the brave, started waning. Rana tried to save the day with his acumen which was running against the stream and the glorious traditions for some time.

**On the basis of your understanding of the above passage, answer each of the questions given below with the help of the options that follow :** **[1 × 4 = 4]**

**(a) Maharana Pratap became immortal because :**

(i) he ruled Mewar for 25 years

(ii) he added a lot of grandeur to Mewar

(iii) of his valour, sacrifice and patriotism

(iv) both (ii) and (iii)

(b) **Difficulties in the way of Mewar were :**
   (i) lack of cooperation the nobility
   (ii) ancient traditions of the kingdom
   (iii) its small area and small population
   (iv) the poverty of the subjects

(c) **During thorny occasions :**
   (i) the flag of Mewar seemed to be lowered
   (ii) the flag of Mewar was hoisted high
   (iii) the people of Mewar showed gallantry
   (iv) most of the rulers heaved a sigh of relief

(d) **Mewar was lucky because :**
   (i) all of its rulers were competent
   (ii) most of its people were competent
   (iii) most of its rulers were competent
   (iv) only a few of its people were incompetent

**Answer the following questions briefly : [1×6=6]**

(e) **Who is the earliest King of Mewar mentioned in the passage?**

(f) **What was Rana Kumbha's contribution to the glory of Mewar?**

(g) **What does the writer find worth admiration in the people of Mewar?**

(h) **How could art and literature flourish in Mewar?**

(i) **How did the rulers show that they cared for their subjects?**

(j) **What does the erection of Vijaya Stambha and Kirti Stambha in the same fort signify?**

(k) **Find words from the passage which mean the same as each of the following : [1×2=2]**
   **(i) surprising (para 4), (ii) evidence (para 4)**

**Answer :**

(a) (ii) he added a lot of grandeur to Mewar

(b) (iii) its small area and small population

(c) (i) the flag of Mewar seemed to be lowered

(d) (iii) most of its rulers were competent

(e) Bappa Rawal is the earliest king of Mewar.

(f) Rana Kumbha had given a new stature to the kingdom through victories and development in work. During his reign, literature and art also progressed extraordinarily. Rana himself was inclined towards writing and his works are read with reverence even today.

(g) The people of Mewar are loving, cheerful, liberal, peaceful, prosperous and have a pleasant nature. According to the writer, these are the qualities that make them worth admiration.

(h) During Rana Kumbha's reign, literature and art progressed extraordinarily. The peaceful and prosperous atmosphere of his kingdom was suitable for the creation of high quality work of literature and art. The king himself contributed to literature. Even amidst wars and bloodshed, literature and art flourished.

(i) By building Vijaya Stambha and Kirti Stambha side by side, rulers revealed the proximity between the king and subjects of Mewar.

(j) Vijaya Stambha is the sample of our great ancient architecture even today. Kirti Stambha reflects how liberal the administration was, which allowed people from other communities and kingdoms to come and carry out construction work in Mewar. Along with minning wars, the ruler patronised art.

(k) (i) astonishing (ii) testimony

2. **Read the passage given below : [10]**

(1) To ensure its perpetuity, the ground is well held by the panther both in *space* and in *time.* It enjoys a much wider distribution over the globe than its bigger cousins, and procreates sufficiently profusely to ensure its continuity for all time to come.

(2) There seems to be no particular breeding season of the panther, although its sawing and caterwauling is more frequently heard during winter and summer. The gestation period is about ninety to hundred days (Whipsnade, ninety-two days). The litter normally consists of four cubs, rarely five. Of these, generally two survive and not more than one reaches maturity. I have never come across more than two cubs at the heels of the mother. Likewise, graziers in the forest have generally found only two cubs hidden away among rocks, hollows of trees, and other impossible places.

(3) Panther cubs are generally in evidence in March. They are born blind. This is a provision of Nature against their drifting away from the place of safety in which they are lodged by their mother, and exposing themselves to the danger of their being devoured by hyenas, jackals, and other predators. They generally open their eyes in about three to four weeks.

(4) The mother alone rears its cubs in seclusion. It keeps them out of the reach of the impulsive and impatient male. As a matter of fact the mother separates from the male soon after mating and forgets all about their tumultuous union. The story that the male often looks in to find out how the mother is progressing with her cubs has no foundation except in what we wish it should do at least.

(5) The mother carries its cubs about by holding them by the scruff of their necks, in its mouth. It trains them to stalk, and teaches them how to deliver the bite of death to the prey. The cubs

learn to treat all and sundry with suspicion at their mother's heels. Instinctively the cubs seek seclusion, keep to cover and protect their flanks by walking along the edge of the forest.

(6) I have never had an opportunity to watch a mother panther train its cubs. But in Pilibhit forests, I once saw a tigress giving some lessons to its little ones. I was its kill at Mala. As the sun set, the tigress materialised in the twilight behind my *machan*. For about an hour, it scanned and surveyed the entire area looking and listening with the gravest concern. It even went to the road where my elephant was awaiting my signal. The *mahout* spotted it from a distance and drove the elephant away.

(7) When darkness descended upon the scene and all was well and safe the tigress called its cubs by emitting a low *haa-oon.* The cubs, two in number and bigger than a full-grown cat, soon responded. They came trotting up to their mother and hurried straight to the kill in indecent haste. The mother spitted at them so furiously that they doubled back its heels immediately. Thereafter, the mother and its cubs sat under cover about 50 feet (15 m) away from the kill to watch, wait, look, and listen. After about half an hour's patience and fidgetless vigil the mother seemed to say 'paid for'. At this signal, the cubs cautiously advanced, covering their flanks, towards the kill. No longer did they make a beeline for it, as they had done before.

(8) The mother sat watching its cubs eat, and mounted guard on them. She did not partake of the meal.

**On the basis of your understanding of the passage complete the statements given below with the help of options that follow : [1× 2 = 2]**

**(a) To protect its cubs, the mother panther hides them :**

(i) among rocks
(ii) in the branches of the trees
(iii) behind the tree trunks
(iv) at its heels

**(b) The male panther :**

(i) is protective of its cubs
(ii) trains its cubs
(iii) watches the progress of the mother
(iv) is impulsive and impatient

**Answer the following questions briefly : [1 × 6 = 6]**

**(c) How many cubs does the mother panther rarely deliver?**

**(d) What may happen if the panther cubs are not born blind?**

**(e) Why did the mahaut drive his elephant away?**

**(f) Why did the tigress spit at its cubs?**

**(g) From the narrator's observation, what do we learn about the nature of the tigress?**

**(h) Why does the panther not face the risk of extinction?**

**(i) Find words from the passage which mean the same as each of the following: [1 × 2 = 2]**

**(i) moving aimlessly (para 3)**
**(ii) came down/fell (para 7)**

**Answer :**

**(a) (i)** among rocks

**(b) (iv)** is impulsive and impatient

**(c)** The mother panther delivers litter of four cubs, rarely five.

**(d)** If the panther cubs are not blind they drift away from the place of safety in which they are lodged by their mother.

**(e)** The mahaut drove away his elephant to protect it from the tigress by avoiding to disturb it.

**(f)** The mother tigress spitted at its cubs to call them back to its heels and watch the prey carefully, instead of running towards it in haste.

**(g)** The tigress is very protective about her cubs. She rears them alone without the help of the male. She trains them to hunt and guards them while they eat.

**(h)** The panther does not face the risk of extinction because it procreates sufficiently and profusely to ensure its continuity for all time to come.

(i) drifting, **(ii)** descended

**3. Read the passage given below : [8]**

People tend to amass possessions, sometimes without being aware of doing so. They can have a delightful surprise when they find something useful which they did not know they owned. Those who never have to change houses become indiscriminate collectors of what can only be described as clutter. They leave unwanted objects in drawers, cupboards and attics for years in the belief that they may one day need them. Old people also accumulate belongings for two other reasons, lack of physical and mental energy, and sentiment. Things owned for a long time are full of associations with the past, perhaps with the relatives who are dead, and so they gradually acquire a sentimental value.

Some things are collected deliberately in an attempt to avoid wastage. Among these are string and brown paper, kept by thrifty people when a parcel has been opened. Collecting small items can be a mania. A lady cuts out from newspapers sketches of model clothes that she would like to buy if she had money. As she is not rich, the chances are that she will never be able to afford such purchases. It is a harmless habit, but it litters up her desk.

Collecting as a serious hobby is quite different and has many advantages. It provides relaxation for leisure hours, as just looking at one's treasure is always a joy. One doesn't have to go out for amusement as the collection is housed at home. Whatever it consists of- stamps, records, first editions of books, china–there is always something to do in connection with it; from finding the right place for the latest addition to verifying facts in reference books. This hobby educates one not only in the chosen subject, but also in general matters which have some bearing on it.

There are other benefits also. One gets to meet like-minded collectors to get advice, compare notes, exchange articles, to show off one's latest finds, etc.; so, one's circle of friends grows. Soon the hobby leads to travelling, perhaps a meeting in another town, possibly a trip abroad in search of a rare specimen, for collectors are not confined to one country. Over the years, one may well become an authority on one's hobby and will probably be asked to give informal talks to little gatherings and then, if successful, to larger audiences.

**(a) On the basis of your understanding of the above passage make notes on it, using headings and subheadings. Use recognisable abbreviations (wherever necessary - minimum four) and a format you consider suitable. Also supply an appropriate title to it. [5]**

**(b) Write a summary of the passage in about 80 words. [3]**

**Answer :**

(a) **Collecting-A Hobby**

1. **Reasons**
   - (i) Ppl. can have delightful surprises
   - (ii) Become indiscriminate collectors
   - (iii) Senti values
   - (iv) Mania
2. **Advantages**
   - (i) Avoid wastage
   - (ii) Saver money
   - (iii) Educational values
   - (iv) Relaxation
   - (v) Doesn't have to go out for amusement
3. **Other benefits**
   - (i) One may meet like-minded collectors
   - (ii) Frnd. circle grows
   - (iii) Hbby. leads to travelling

| Abbreviations Used | |
|---|---|
| Ppl. | people |
| Senti. | sentimental |
| Frnd. | friend |
| hbby. | hobby |

(b) **Summary :**

People sometimes collect things due to various reasons which may be related to their need, lack of energy or sentimental values. Sometimes collection may be done to avoid wastage which later becomes a mania. But it has many advantages too; it relaxes, amuses and educates. It has additional benefits where one may meet like-minded collectors to get advice, compare notes, exchange articles, to show off one's latest findings, etc. As a result, their friend circle grows bigger. In fact, hobby further leads to travelling.

## SECTION — B
## (WRITING SKILLS) 30

**4. You are Karan Kumar/Karuna Bajaj, a leading lawyer practising in Surat. You want to buy an independent house at City Light Road to be used as office-cum-residence. Draft an advertisement in about 50 words for the classified columns of a local newspaper. You can be contacted at 45645678 [4]**

**OR**

**On 30th November your school is going to hold its annual sports day. You want Mr. Dhanraj Pillai, a noted hockey player to give away the prizes to the budding sportspersons of the school. Write a formal invitation in about 50 words requesting him to grace the occasion. You are Karuna/Karan, Sports Secretary, Sunrise Global School, Agra.**

**Answer.**

FOR PURCHASE

An independent house at City Light Road to be used as office cum residence for purchase, good location, separate entrance and exit, spacious, 24-hour water and electricity supply, parking facility, security and power backup.

Contact : Karuna Bajaj Lawyer, Surat
45635678

**OR**

Sunrise Global School, Agra

15th November, 20XX

Mr. Dhanraj Pillai

Abhilasha Complex

Chennai

**Subject : Invitation to preside over Annual Sports Day**

Dear Sir,

Our school is going to hold its annual sports day on 30th November. A number of students are likely to participate in various sports events.

We shall be highly honored, if you kindly give your consent to preside over the programme and to give away the prizes to the budding sportspersons of the school at 10 a.m. at the school playground.

With warm regards,
Karuna/Karan
(Sports Secretary)

**5. Along with air and water pollution, our cities are also under an attack of noise pollution. Marriage processions, DJs during wedding receptions, loud music from neighbourhood flats etc. are all sources of noise which is not good for the old, the ailing and students. Write a letter in 120-150 words to the editor of a local newspaper describing the problem and making a request to the concerned authorities to solve it. You are Karuna/Karan, M-114, Mall Road, Delhi. [6]**

**OR**

**In all big cities road rage has become a serious problem. A minor scratch, a little push, or a small brushing past can lead to a scuffle sometimes resulting even in murder. Write a letter in 120-150 words to the Police Commissioner giving your views on the problem and its solutions. You are Karuna/Karan, M-114, Mall Road, Delhi.**

**Answer :**

M-114, Mall Road
Delhi

20th March, 20XX

The Editor
The Times of India
New Delhi

**Subject : Time to become aware of the hazards of noise pollution**

Dear Sir,

Through the esteemed columns of your prestigious newspaper, I, a resident of New Delhi, want to bring forth the issue of increasing noise pollution to the notice of the general public and concerned authorities. In the metro cities, there has been a tremendous increase in the levels of noise pollution. It's not only the shrieking horns and roars of vehicles but also marriage processions, loud DJ's, aircrafts and machines in industrial areas which cause a lot of noise pollution.

The consequences are extremely hazardous. They lead to increasing cases of deafness, insomnia, palpitation of heart, mental fatigue, etc. Among the sufferers are the small babies and old people who can't sleep properly, there are patients who can't take rest and students who can't study. Then, there are 'Jagrans' that force the public to keep awake the whole night. The problem further adds to the fact that noise control laws are neither followed properly nor implemented completely. The culprits bribe the police and other officials and get away easily.

Hope the concerned authorities will take initiative in this regard and relieve us from the menace of noise pollution.

Yours Truly,
Karuna/Karan

**OR**

M-114, Mall Road,

Delhi

20th March, 20XX

The Commissioner
Delhi Police

**Sub : Increasing cases of Road Rage**

Dear Sir,

I am writing this letter to bring to your kind notice, the problem of increasing cases of road rage. Day by day, it is becoming a serious problem which should be sorted out as quickly as possible. We often come across many incidents like minor scratch, a little push or a small brushing past, leading to scuffles which escalate quickly and sometimes even result in death. Everyday, there are many fatal accidents, happening and the number of hit and run cases are also rising. In Delhi alone, 532 people died in road accidents in the year 2014. It is observed that people are losing patience and tolerance due to their modern lifestyle. They begin to fight over minor issues. The feeling of ignorance, irresponsibility, lawlessness and indifference is prevailing. Everybody is taking undue advantage of the freedom provided to them by the democracy. Now, it's high time to realize the need of the hour. I hope the concerned authorities would realize the gravity of the situation and put a check to these people and their activities either by imposing penalties or fines.

Yours Truly,
Karuna/Karan

**6. 'Brain drain is not a bane for a developing country like India'. Write a debate in 150-200 words either for or against the motion. [10]**

OR

**Write a speech in 150-200 words on the topic, 'Discipline shapes the future of a student'. It is to be delivered in the morning assembly. You are Karuna/Karan.**

**Answer :**

**'BRAIN DRAIN IS NOT A BANE FOR A DEVELOPING COUNTRY LIKE INDIA'**

**AGAINST :**

Good morning, respected judges and the esteemed audience, I would like to speak against the motion- 'Brain Drain is not a bane for a developing country like India'.

Brain Drain is a major problem of India that has been harmful for the country. Brain drain means the intelligent minds that do not get desirable jobs or recognition in their own country. Thus, they migrate to other countries for better prospective where they are provided with favorable working conditions. They prove their intelligence there and are recognized throughout the world. They also become the pride of the country. Every year a large number of people go abroad either for acquiring higher studies or getting good jobs. It is because their worth and intelligence is not acknowledged aptly in the country. Sometimes, the most brilliant people also leave India and prefer to stay in a well-developed country like America, England, Canada, etc. This is the most common thing in India as compared to other countries as the companies over here fail to provide necessary working conditions and acceptable salaries to their own countrymen.

But the question arises what if they show their worth in their own country? How can they be stopped from migrating or persuaded to return to India?

It is a matter of serious concern for the government of India because it is a complete loss for India. The professional training that the students receive in India and for which the government spends lakhs of rupees, goes waste; plus it is a gain for receiving countries. They can be attracted to India if the conditions here are altered to their liking; if the merit and talent is given due consideration. Therefore, sincere efforts are required to bring them back by offering jobs according to their caliber, so as to make the country more efficient in all fields.

OR

**FOR :**

Good Morning, honorable Principal, respected teachers and my dear friends. I am Karuna, student of class XII of your school. It is a great opportunity for me to share my views in the favour of the topic, 'Brain Drain is not bane for a developing country like India'.

Brain drain refers to the continual migration of highly qualified and trained people from their country to a more advanced one, in search of lucrative job opportunities and promising lifestyles.

A major advantage of this system is that in this way, the other skilled people of the native country are able to find suitable opportunities in their own country. A couple of years ago, there were a lot of people who were highly educated but still couldn't find appropriate employment. With brain drain happening actively, if some people gone outside India for work, the locals are able to get that work. This way, it helps reduce the percentage of unemployed people in India.

Secondly, the income from the emigrated people comes back to the native country as overseas remittances. India is recipient of the highest amount of foreign currency coming in as remittances from NRIs. People working outside the country send money to their parents and family members, who in turn can invest in property and other assets. Thus, it leads the development of the native country.

Finally, it helps in improving the education and infrastructure in the native country, which in turn develops the country further, because the native country realises the value of well educated people.

OR

**'DISCIPLINE SHAPES THE FUTURE OF A STUDENT'**

Good Morning, honorable Principal, respected teachers and my dear friends. I am Karuna, student of class XII of your school. It is a great opportunity for me to share my views in today's assembly, on the topic, 'Discipline shapes the future of a student'.

The students of today are the leaders of tomorrow. They are the backbone of the society. They are going to be the future leaders, administrators and nation shapers. Student life is the time when their character is formed and takes shape. But nowadays, schools and colleges have become places of complete indiscipline. Whether it is the morning assembly, annual function, sports meet or examinations, one can see complete indiscipline in it. Man is a social being and he lives and dies in a society. So, he has to follow the rules of the society for which he has to discipline himself and this begins from the school itself.

Aristotle said," Discipline is obedience to rules formed by the society for the good of all". Without discipline, one cannot acquire knowledge and as the schools and colleges are abodes of learning, it is required much over there. Students cannot gain much without discipline and the result can never be good. Thus, it is clear that the need of discipline is much required in the student life. Indiscipline leads to failure but discipline paves the road of success.

7. **Education has always been a noble profession. Our ancestors received their learning at gurukuls and ashrams. Even in the near past pathshalas (schools) were associated with places of worship. Today, education is fast becoming commercialised. Parents have to shell out a lot of money on coaching classes, tuition fees etc. Write an article in 150-200 words on 'The State of Education, Today'. You are Karan/Karuna.**

[10]

OR

**According to 2011 census, literacy rate of hundred percent or around has been achieved by only a couple of states in India. Illiteracy is found mostly among the old and the deprived sections of society. What can the youth do to spread literacy in society? Write an article in 150-200 words on 'Role of students in eradicating illiteracy'. You are Karuna/Karan.**

**Answer :**

THE STATE OF EDUCATION TODAY

By Karuna/Karan

Our ancestors have received their learning at gurukuls and ashrams. Pathshalas were regarded as the place of worship. But the scenario has completely changed now. Today, education has been totally commercialized. Whether its school or college, students have to rush to coaching classes and parents have to shell out a lot of money on them. They don't rely on classroom teaching. On the other hand, parents' expectations from their children have increased. They want them to secure highest marks so as to get the highest paid jobs. They want them to become doctors, administrative officers, engineers, etc. For this, they send their children to the most reputed schools; still they are not satisfied with their performance. They want their child to get some extra knowledge to excel in others fields. Thus, craze for the coaching classes has increased.

Another reason for joining the coaching classes is the inferiority complex which children face when they do not go for extra coaching. Although, these so called coaching classes are just like school classes because they too follow group teaching method is used but still parents as well as children think them to be far superior to classroom teaching. They should realize that these coaching classes are no better than a money-making business. The students must pay attention in class and must concentrate more on self-study in order to be successful.

OR

ROLE OF STUDENTS IN ERADICATING ILLITERACY

By Karuna/Karan

Youth plays a very important role in building the nation. With the increase in population, the responsibilities of students have also increased. They are the pivot on which the development of any nation depends. We know that India is a land of villages and most of its population is illiterate, In most of the villages, there are no schools or teaching facilities. Thus, villagers don't get any opportunity to learn. Poverty and ignorance have deprived them and the weaker section from becoming literate. Therefore, it becomes the duty of the youth to render their services to them by teaching them.They can share their valuable advice and guidance to educate them. We know that education helps in economic, social, spiritual and mental development. It broadens ones outlook towards society. Being illiterate or uneducated, is like a curse. Students can help these people by organizing evening classes and by adopting the motto' each one, teach one'. Out of their twenty-four hours, the students can spare at least one hour for the unprivileged people. They can spread awareness among the illiterates regarding the hazards of overpopulation and recommend the need of having a small family. Many NGOs and governmental agencies are there to assist these youths in this regard. They can be encouraged to do so by giving them the incentive of gaining additional marks or extra grades.

Youth, therefore, can play a major role in making the country progressive and prosperous by bringing complete change in the attitude and thinking of the illiterate by educating them. So, let's join hands to eradicate illiteracy from our country.

## SECTION — C

## (LITERATURE : TEXT BOOKS AND LONG READING TEXT) 40

8. **Read the extract given below and answer the questions that follow :** **[4]**

Now we will count to twelve
and we will all keep still.
For once on the face of the Earth
let's not speak in any language,
let's stop for one second,
and not move our arms so much.

(a) **What is the significance of the number 'twelve'?**

(b) **Which two activities does the poet want us to stop?**

(c) **What does the poet mean by let's not speak in any language'?**

(d) **Describe the pun on the word, 'arms'.**

**OR**

..........I saw my mother,
beside me,
doze, open mouthed, her face
ashen like that
of a corpse and realised with
pain..........

(a) **Who is 'I'?**

(b) **What did 'I' realise with pain?**

(c) **Why was the realisation painful?**

(d) **Identify and name the figure of speech used in these lines.**

**Answer :**

(a) The significance of the number twelve is that at twelve O' Clock, the hands in the clock unite suggesting amity and unity.

(b) The two activities which the poet wants us to stop are speaking and moving our arms.

(c) By 'let's not speak .......... language' the poet means that we should keep quiet for some time so that we can perform a self-analysis. It is necessary for creating a feeling of mutual understanding among ourselves.

(d) The pun on the word 'arms' is that if we won't move our arms then we will never lift 'arms' i.e. weapons. Symbolically by this he means no war or no violence.

**OR**

(a) 'I' is used to refer to the poet Kamala Das.

(b) The poet realised in anguish that her mother had grown old and that her end was near. This expression was borne out of the fear of her separation.

(c) The realization was painful as she knew that her mother might not live long and she might lose her some day. It brought forth her fear of separation.

(d) The figure of speech that has been used here is simile in the lines 'her face ashen like that......'.

**9. Answer any four of the following questions in 30-40 words each : [3 × 4 = 12]**

(a) **What does the reference to chappals in 'Lost Spring' tell us about the economic condition of the rag pickers?**

(b) **What do we learn about the crofter's nature from the story, 'The Rattrap'?**

(c) **How is 'Shakespeare wicked and the map a bad example' for the children of the school in a slum?**

(d) **What picture of male chauvinism (tyranny) do we find in the poem, 'Aunt Jennifer's Tigers'?**

(e) **How did the Governor, Oxford Prison describe Evans to the Secretary, Examination Board?**

(f) **At the dining table why did Zitkala-Sa begin to cry when others started eating?****

**Answer :**

(a) The reference to chappals in 'Lost Spring' tells about the miserable life of the rag pickers settled in Seemapuri. Most of the impoverished children busy in rag-picking, were not wearing footwear. The boys were barefeet as if it was a tradition to remain barefeet. It had become their habit not to wear chappals. But according to the writer-it was their excuse to conceal their poverty.

(b) The crofter was a lonely old man who neither had a wife and a child. He wanted someone to talk to, therefore, he allowed the peddler to spend the night at his house. He not only gave him shelter, but also gave him food to eat and tobacco to smoke. He started a conversation with him and told him everything about himself and his earnings.

(c) The portrait of Shakespeare and the world map in the classroom of the slum children is far from the actual condition of the classroom. Shakespeare is the synonym of knowledge and excellence but the children of the slum school have no meaning for that. The map exhibits the world with all its bounties, whereas, for these children there is no world beyond poverty and hunger. Thus, Shakespeare is wicked and the map a bad example.

(d) In the poem 'Aunt Jennifer's Tigers', we find the picture of a woman oppressed by the male chauvinism. She needs freedom from her burdensome husband, whose ring on her finger is like a handcuff which has snatched all her freedom and happiness. She has a lot of constraints in her life because of her dominating husband and from whom she wants to get free.

(e) The Governor of the Oxford Prison made a call to the Secretary of Examination Board telling him that a prisoner named Evans wanted to take the O-level examination in German. He told him that Evans in not a violent person but he is a congenital Kleptomaniac. The Governor then requested the Secretary to see whether the arrangements could be made in the prison itself.

**Answers are not given due to change in present syllabus

**10. Answer the following question in 120-150 words : [6]**

**Every teenager has a hero/heroine to admire. So many times they become role models for them. What is wrong if Sophie fantasizes about Danny Casey and is ambitious in life?****

**OR**

**Our native language is a part of our culture and we are proud of it. How does the presence of village elders in the classroom and M. Hamel's last lesson show their love for French?**

**Answers :**

**OR**

That Sunday morning, everything was quite normal. Franz's classmates and teacher were already in the class but there was no sound of opening and closing of desks. Even unison repetition of lessons and rapping of teacher's great ruler was absent. It was all due to the order from Berlin that from then onwards, only German would be taught in the schools of Alsace and Lorraine instead of French. Then only the people realized how precious their language was to them. That day, all the villagers occupied the back benches in the classroom to attend the last French lesson and to mark the respect for their French teacher, Mr. M. Hamel, who devoted his selfless service to the society. They felt very sad. They also regretted not learning French properly. Franz was shocked to hear this news and Hamel's words were like thunderclaps to him. He was so sorry French before and for having wasted his time. His books suddenly became his friends. Even Mr. Hamel talked a lot about the French language. According to him, it was the most beautiful language in the world. It was the most logical one too. He asked his students never to forget this language.

**11. Answer the following question in 120-150 words : [6]**

**Both Derry and Lamb are victims of physical impairment, but much more painful for them is the feeling of loneliness. Comment.**

**OR**

**How did Jack end the Roger Skunk story? How and why did Jo want to change it?**

**Answers :**

A disabled or a handicapped person is considered to be an outcaste. People sympathize with them but do not accompany them. They are excluded from the society. The feeling of loneliness and alienation is much more troublesome than the pain which disabled people experience due to their physical impairment. This is very well revealed in the story, 'On the face of it', where the character, Derry has an acid burnt face and people often sympathized with him but it was not a heart-felt one. This resulted in his pessimistic approach towards life and people. Derry had no association with the outer world till he met Mr. Lamb, who had Mr. Lamb's attitude and outlook towards people and life uplifted him. He learnt to live a respectful life. Mr. Lamb inspired him to face the world in spite of his own disability. Mr. Lamb successfully infused in him the courage and self-determination. He advised him to ignore the comments made by people on his physical impairment and enjoy the beauty of life and nature. Although, the loneliness of Derry dominates the play but there are evident traces of Mr. Lamb's loneliness also. The old man had no one to be with. He spent his day in the sun, reading books. By the end of the first scene, it becomes clearer that he is lonely when he mutters that no one comes back to him after meeting him for the first time.

**OR**

Jack used to tell his little daughter, Jo, bedtime stories. When she grew older, she began to ask questions to her father. Once her father told her a story of a little animal called Roger Skunk who smelled awfully bad. He, therefore, met with a wizard who solved his problem with a magical wand and made him smell like roses. But, Skunk's mother did not like it and took him back to the wizard to make him smell originally bad once again. This ending was not acceptable to Jo as she wanted the story to have another end in which the wizard should hit the mommy by his magic wand and leave Roger skunk emitting the pleasant smell of roses. Jo thought that Roger was the hero of the story so he must not look ugly or stinky. Jo's perspective on life was different from her father's. She wanted to change the end of the story to assert her views on her father. She had her own ideals and views which she wanted to be pursued by her father.

**12. Attempt the following question in 120-150 words :** [6]**

**How did Mrs. Hall show her excitement when Griffin came to stay at Coach and Horses?**

**OR**

**How do William Dane's deceit and Sarah's desertion affect Silas?**

**13. Attempt the following question in 120-150 words :** [6]**

**Marvel is the only character in 'The Invisible Man' who interacts with Griffin and gains something. Comment.**

**OR**

**Attempt a character sketch of Eppie.**

**Answers are not given due to change in present syllabus

# Outside Delhi [Set-II]

Note : Except for the following questions, all the remaining questions have been asked in previous sets.

## SECTION — B
## (WRITING SKILLS)

4. You are Karan/Karuna of M 114, Mall Road, Delhi. You are a civil engineer and have recently returned from UAE. You are looking for a suitable job in India. Draft an advertisement for the same in about 50 words. Give details of your qualifications, experience, nature of job and expected remuneration. [4]

OR

Yesterday, during lunch break, you misplaced your notes on the chemistry lectures. You want to get them back. Write a notice in about 50 words for the school notice board. You are Karuna/ Karan, a student of class XII A.

**LOOKING FOR JOB**

Recently returned from UAE, looking for a job in Civil Engineering.

Qualification : DAE Civil from IIT, Kanpur

Experience : Eight years in Flyover and building construction, Highway Construction in Dubai

Extra : Possess high level of verbal and written communication skills

Expected Remuneration : Basic Salary- ₹1,00,000/to ₹1,50,000/SR

Correspondence Address :
Karuna/Karan
M 114, Mall Road
Delhi

**Answer.**

OR

**SUNRISE GLOBAL SCHOOL, AGRA**
**NOTICE**

12th February, 20XX

**Lost ! Lost ! Lost**

Yesterday during the lunch break, I lost my notes on chemistry lectures somewhere in the school around 11:30 a.m. They are very important as the exams are approaching and I don't have enough time to write them again. Anyone who finds it, kindly restore to the undersigned or to my class teacher Mr. Rajat Sharma. I shall be highly obliged.

Karuna/Karan
Student, XII A

5. It gives you a good feeling when you read in the newspapers how patients from abroad come to hospitals in India and get themselves treated at a fraction of expenses they will have incurred elsewhere. Write a letter in 120-150 words to the editor of a national daily describing the importance of medical tourism for India. You are Karan/Karuna M 114, Mall Road, Kanpur. [6]

OR

You bought a flat from PQR Builders, Sector 55, NOIDA. Within a period of two months you have started facing a lot of problems like seepage in the walls and ceilings, wall paint peeling off, leaking sanitary fittings, lift getting stalled, etc. Write a letter of complaint in 120-150 words to the Works Manager. You are Karuna/Karan, A9D Apoorva Appartments, NOIDA.

**Answer :**

M-114, Mall Road, Kanpur

20th March, 20XX

The Editor,
The Times of India
New Delhi

**Sub- Importance of medical tourism in India**

Dear Sir,

Through the columns of your esteemed daily, I, Karuna/Karan would like to emphasis on the importance of medical tourism in India. As we know that India is the most popular tourist place, people from all over the world come to India for their treatment by the most experienced and world class physicians. The medical facilities available in the high- end hospitals of India are excellent. They are equipped with the latest technologies and the doctors are super specialized in their respective fields. With the advent of globalization, the world has shrunk. The concept of medical tourism was devised to help global citizens to enjoy their tourist sojourns while being cost effective world class medical quality. Over the last decade, India has emerged as the preferred destination for medical tourism. The city of Chennai has been termed as 'India's health capital'. India's medical tourism sector is expected to experience an annual growth rate of 30% making it a $2 Billion industry. Therefore, medical tourism should also be promoted by the government along with other tourist sectors.

Yours Sincerely,
Karuna/Karan

OR

A9D, Apoorva Appartments

Noida

12 February, 20XX

Work Manager
PQR Builders,
Sector 55 Noida

**Sub : Complaint against the poor condition of flat**

Dear Sir,

On 20th December, 20XX, I purchased a flat from PQR Builders in 'Ashiana Apartments'. The flat is on the first floor of D-Wing in sector 55, Noida. Initially, when the flat was shown to us, it was in excellent condition with walls, doors, ceiling perfectly painted and well-furnished. The kitchen and bathroom were very well-equipped with attractive fittings, seeing it we were tempted to buy it immediately.

But now, I'm very hurt to inform you that the flat which I purchased only two months before is in a very poor condition and so I, hereby, lodge a complaint against the negligence and dereliction on your part. Within a period of two months I'm facing a lot of problems like seepage in the walls and ceilings, wall paint peeling off, leaking of sanitary fittings, lift getting stalled, etc. Even the service men do not turn up on time to rectify the problems.

Please ask your mechanic and service men to check the fittings and do the required repair work.

Yours faithfully,

Karuna/Karan

**7. Increase in the number of private vehicles has caused problems like rising air pollution, traffic jams, lack of parking space, road rage etc. The solution lies in the use of public transport. Write an article in 150-200 words on 'The Importance of Public Transport'. You are Karan/Karuna. [10]**

**OR**

**In certain states of India, there is a great imbalance in the male female ratio. This is the result of special treatment given to boys in the family. Why is it so? How can we change this mindset? Write an article in 150-200 words on 'Gender discrimination in society'. You are Karuna/Karan.**

**Answer :**

**'IMPORTANCE OF PUBLIC TRANSPORT'**

By Karuna/Karan

Recently, I heard that traffic police has started a new drive against the pollution causing vehicles. Traffic police is stationed at the important intersections. Each and every vehicle is checked for 'No Pollution' certificates and they are fined ₹ 500 for the lapse. But, what is the use of all this and how long could it go because people find too many excuses for their negligence. Next, from every house, a minimum of two to three vehicles are used by the people for going to their working places. Sometimes, they are stuck with their vehicles for long hours due to heavy traffic jams.

So, it would be far better if the government provides public transport facilities. It would create a lot of convenience for the public and daily commuters. It has many social, economic and environmental benefits. Social benefits of using public transport are multi-fold. It helps in developing a sense of unity among people. They feel a social connection with those who are travelling with them. It encourages people to have a healthy lifestyle if they walk to their station, stand or stop. Moreover, it is less stressful as it saves a lot of time and fuel. Then economic benefits include less expenditure of money on buying cars or spending money on their maintenance. Travelling by public transport is much cheaper than private transport. It also reduces pollution and road congestion thus, helping in environmental conservation.

**OR**

**'GENDER DISCRIMINATION IN SOCIETY'**

By Karuna/Karan

In India, attitude towards men and women has been discriminating since generations. Although, the Constitution of India has provided equal rights to men and women, gender disparity prevails. Nature has also divided human-beings into two equal halves and both are given equal responsibilities according to their capacity, thus, stating the rights to both of them. But we humans have shown discrimination which has now become a practice in society. The powerful become aggressors and the others face the aggression. Usually women are considered to be weaker as compared to men and are considered to be incapable of doing everything just because of their fragile appearance. Gender discrimination prevails in many homes, it is generally seen in families where there are sisters and brothers and where brothers are favored over sisters. Female infanticide and female foeticide are still common in many regions of the country. Sexual harassments, eve teasing, bride burning, woman trafficking, dowry deaths and many more evils are being perpetrated. Economically, females are less privileged as they do not have control on financial matters.

In the rural areas and villages the situation, is still bad. There has been no change in the social, economic and family status of the women. They are subjected to exploitation, domination and harassment by their male counterparts. But now it's time that women should be empowered and encouraged so that the gap between men and women may be reduced.

### SECTION — C
### (LITERATURE : TEXT BOOKS AND LONG READING TEXT)

**9. Answer any four of the following questions in 30-40 words each : [3 × 4 =12]**

**(a) Why did the Crofter show the thirty kroner to the peddler?**

**(b) What deep meaning did his experience at the YMCA swimming pool have for Douglas?**

**(f) How Did Zitkala-Sa's first day in the land of apples begin?****

**Answer :**

**(a)** The crofter was very proud of his cow as it used to give enough milk to support him in his old age. When the peddler came to him, he felt so excited that he told him about the thirty kroner from the creamery as the payment of the previous month's supply. He wanted to share his joy and moreover, he thought that peddler would not believe him, so he showed him the money to convince him.

**(b)** Experience at the YMCA swimming pool had a deep meaning for William Douglas. He had nearly escaped from drowning in the pool when he was sitting alone on the edge of the pool and a strong young man picked him up and tossed him into the pool. The experience had a deeper meaning for Douglas; he had experienced the fear of dying but at the same time, he learned to be confident and determined to overcome his fear.

●●

## Outside Delhi [Set-III]

**Note : Except for the following questions, all the remaining questions have been asked in previous sets.**

### SECTION — B
### (WRITING SKILLS)

**6. 'The Internet cannot replace a classroom teacher'. Write a debate in 150-200 words either for or against the motion. [10]**

**OR**

**Write a speech in 150-200 words to be delivered in the morning assembly on the topic. Water, a valuable asset, conservation at home and at school'. You are Karuna/Karan.**

**Answer :**

**THE INTERNET CANNOT REPLACE A CLASSROOM TEACHER**

**FOR :**

Good morning, respected judges and the esteemed audience! Today, I have the privilege of speaking in support of the motion: 'The Internet cannot replace a classroom teacher'.

It is indeed a truth that the internet can solve a bulk of problems and provide all the necessary answers with illustrations to students but it cannot replace a classroom teacher as it is devoid of any kind of emotion which is required while teaching children. While teaching, teachers have to face all types of students with different mental abilities and they have to adjust their teaching pattern according to them. They motivate their students accordingly and help the weaker students to excel. There are some queries which cannot be solved by the internet but only a teacher can resolve it. The teacher understands the emotions and sentiments of the children; they realize the positive as well as the negative feelings of the students. Teachers help us to remain alert and active in the classroom during teaching whereas a machine can never be. It's not necessary that all the information provided by the internet is genuine. The question is that where has all the information or knowledge on the internet come from? It is us humans who have fed all of this information into the computers.

In the end, I would like to say that a machine can never replace a living and breathing knowledgeable person. i.e, a teacher

**OR**

**AGAINST :**

Good morning, respected judges and the esteemed jury members and audience. Today, I have the privilege of speaking in against the motion.

The present era is a time of internet and online presence. Everyone depends on the internet to get answers to all their queries, doubts and requirements. The students mainly are very keen on going online to solve their problems in learning. A student can get answers to any number of questions anytime and he or she doesn't have to wait for time. There is no need for waiting for turns or for the time and subject period to ask

**Answers are not given due to change in present syllabus

the questions. Google is so vast that you can get answers to all your questions in each and every subject. You name it and you have the answer to your query or the topic described. While looking for solutions or answers online we get away with our hesitation and can research without any hurdles. In a classroom we might not be able to do so in front of everyone. Internet is a great tool in our hand for constructive use in all the phases of life. Internet can be a replacement for any kind of classroom teaching as it does not have any constraints and can be of use to anyone or everyone.

**OR**

**'WATER, A VALUABLE ASSET, CONSERVATION AT HOME AND AT SCHOOL'**

Honorable Principal, respected teachers and dear friends! Good Morning. I am Karuna/Karan, a student of class XII. It is a great opportunity for me to share my views on the topic 'Water, a valuable asset, conservation at home and at school' in today's assembly.

Water scarcity today has become a big issue which if not dealt with in time, will turn out to be a hazard. With the increasing population, people are facing the problem of acute water shortage. People waste water in many ways which include washing cars, bathing animals, construction work, leaving taps running, etc. Apart from this, we can see leaking water pipes at many places which should be repaired at once. Children and adults should be taught not to leave the taps running while brushing the teeth or washing utensils and clothes. At home, water used for washing fruits and vegetables can be reused for gardening purpose.

To make people aware conservation of water, importance of rain water harvesting and the method to recycle the water should be explained. There is no life without water. Public should be encouraged to improve water management practices that reduce the wastage of water and enhance the beneficial use of it. Some ancient methods of water conservation should also be practiced. In case the water is not used wisely by anyone, then a heavy fine should be imposed on them.

7. **For admission to colleges, there is a lot of competition today. For one available seat there are hundreds of candidates. Such a situation leads to stress on the minds of students in the schools. However, the practice of Yoga can help them to feel calm and stress free. Write an article in 150-200 words on the topic, 'Importance of Yoga to students.' You are Karan/Karuna. [10]**

**OR**

**While watching TV what we generally dislike most are commercial breaks, yet, a number of advertisements do add to our information. Write an article in 150-200 words on. 'Advertisements as a source of information'. You are Karuna/Karan.**

**Answer:**

**IMPORTANCE OF YOGA TO STUDENTS**

By Karuna/Karan

Today's sedentary lifestyle has given invitation to many diseases as man has limited time to spare for his health. But practicing yoga helps to keep a person away from several health problems. It is the science that cures many serious diseases and helps the mind as well as body to remain sound. It helps in maintaining balance between mental and social status of a human being. With regular practice of yoga, we tend to become more sensitive to the kind of food our body demands. A few minutes of yoga helps in getting relief from stress. It helps detox the body and de-stress the mind. We all want to live in a peaceful atmosphere surrounded by nature, but we hardly realize that peace can be found within us. Yoga is the best way to calm a disturbed mind. On the other hand, yoga postures massage organs and strengthen muscles; breathing techniques and meditation release stress and improves the immune system. Yoga and meditation work on keeping the mind happy and peaceful. When we feel drained out of energy, a few minutes spend on yoga provides the freshness and energy. It also helps in improving body posture. Today, students face many problems due to the burden of studies and they feel depressed but yoga acts as a stress buster for them. In many schools, yoga has been practiced on a daily basis as a part of the curriculum. It is gaining popularity not only in India but also in European countries.

In the end, I would like to tell that yoga is a continuous process, so keep practicing!

**OR**

**ADVERTISEMENTS AS A SOURCE OF INFORMATION**

By Karuna/Karan

Advertisements have become an inseparable part of our life. We can't get away from them as they have surrounded us from all sides. When we go out of the house, we see various hoardings displaying one or the other things, when we switch on the television, they start attacking us in between the serials and movies. Even news channels are interrupted by them. In all, we can say that we can't escape from the advertisements.

But, there are certain things that can be said in favor of advertisements. The cost of production is lowered when the advertisers give money to newspapers, magazines, radio and television for mass marketing. Advertising is a way of delivering proper message to the customers. This way, we get the information as well as knowledge of various products. Advertisements help in drawing customers to the business. It benefits the consumers also as there is a competition between the manufacturers and they try to sell their products at a lower cost than the others. Buyers, on the other hand, get wider choice to select the products according to their pocket and taste. They inform us about all the new products which are launched in the market. Without them, we would not be able to know about the new things and its benefits. Therefore, advertisements seek to generate increase in consumption of a product or a service by introducing it to the target consumers.

## SECTION — C
## (LITERATURE : TEXT BOOKS AND LONG READING TEXT)

**9. Answer any four of the following questions in 30-40 words each : [3 × 4 = 12]**

**(a) All we have to fear is fear itself. When did Douglas learn this lesson?**

**(b) Why did the iron master speak kindly to the peddler and invite him home?**

**(e) What was his German teacher's opinion of Evan's proficiency in German?**

**(f) According to Zitkala-Sa what does 'eating by formula' mean?****

**Answers are not given due to change in present syllabus

**Answer:**

(a) Douglas had a 'near to death' experience of drowning and the fear of water took control of his physical and mental strength. Finally, he decided to overcome this fear because of his interest in water activities like canoeing, swimming, fishing, boating etc. He decided to get an instructor and learn to swim. He practiced an hour daily for five days a week. He learnt to exhale in water and inhale after coming out of it. He swam two miles across the Lake Wentworth.Thus, he had experienced both the sensation of dying and the terror it can produce. He was finally able to learn the lesson that 'all we have to fear is the fear itself' and overcome his fear of water.

(b) The iron master took the peddler for an old regimental comrade, Nils Olof. He considered the peddler as an old acquaintance who had fallen on evil days so he talked to him kindly and invited him to come home to help him.

(e) Evans' German tutor started giving him tuitions for six months in night classes in the prison itself. Evans wanted to take the (O')level examination in German. When he was permitted by the Governor and the Board to take the examination, his tutor came to meet him the day before. He wished him good luck but said that he hardly had any 'chance of getting through'. But Evans gave an ironical twist by saying that he would surprise everybody.

●●

# ENGLISH

## Delhi [Set-I]

## 2016

**Time allowed : 3 hours** **Maximum marks : 100**

**Note : Except for the following questions, all the remaining questions have been asked in previous sets.**

**SECTION — B**
**(WRITING SKILLS)**

4. **Principal, Sunrise Global School, Agra requires a receptionist for her school. Draft a suitable advertisement in about 50 words to be published in the classified columns of a national newspaper giving all the necessary details of qualifications and experience required in the receptionist. [4]**

**OR**

**Water supply will be suspended for eight hours (10 am to 6 pm) on 6th of March for cleaning of the water tank. Write a notice in about 50 words advising the residents to store water for a day. You are Karan Kumar/Karuna Bajaj, Secretary, Janata Group Housing Society, Palam Vihar, Kurnool.**

**Answer :**

RECEPTIONIST REQUIRED

Smart, graduate, proficient in Hindi and English, having good communication skills, not more than 25 years of age, has two years of experience, can contact the principal with complete resume within seven days.

Attractive Salary
Contact : Principal
Sunrise Global School,
Agra.

**OR**

JANTA GROUP HOUSING SOCIETY

NOTICE

25$^{th}$Feb, 20XX

**WATER SUPPLY SUSPENDED**

All the residents of Janta Group Housing Society are here by informed that on 6$^{th}$ Mar. 20XX of water suspension from 10 am to 6 pm due to cleaning of the water tank. So, you are requested to store the water for the day.

Karuna Bajaj/Karan Bajaj.
(Secretary)
Janta Group Housing Society,
PalamVihar, Kurnool

5. **Yesterday you went to Sunrise Hospital, Market Road, New Delhi taking with you the victim of a hit and run accident. There were chaotic conditions in the casualty department. The injured was attended to after a lot of precious time had been lost.**

**Write a letter of complaint in 120-150 words to the Medical Superintendent. You are Karan/Karuna, M-114, Mall Road, Delhi. [6]**

**OR**

**Lack of job opportunities in the rural areas is forcing people to migrate to cities. Every big city thus has a number of slums in it. Life in these slums is miserable.**

**Write a letter in 120-150 words to the editor of a national newspaper on how we can improve the living conditions in these slums. You are Karan/ Karuna, M-114, Mall Road, Delhi.**

**Answer :**

M-114, Mall Road
Delhi

20$^{th}$ March, 20XX

The Medical Superintendent,
Sunrise Hospital,
Market Road,
New Delhi.

**Subject : Chaotic conditions in the casualty department**

Dear Sir,

I, Karan/Karuna, a resident of New Delhi want to bring to your kind notice the conditions of the Sunrise hospital. Yesterday, while I was going to my office, I witnessed a person lying on the road in a pool of blood. There was a crowd around him. On enquiry I came to know that he was hit by an unidentified vehicle and nobody took the pain to take him to Sunrise hospital. So, I brought him to hospital where I was further shocked to see the pathetic condition of the casualty ward. The corridors were full of people where the smell of medicines prevailed and cotton dabs were thrown away on the floor. I requested the doctor on duty to attend the patient immediately to avoid any mishap but was asked to file a police complaint first. The patient was left unattended for about an hour during which he lost ample amount of blood which could have been proved to be fatal. Later, when all the formalities were completed, the doctor came to see him and by God's grace his life was saved. Then, again drugs were administered in an inappropriate way by the staff who came only after being reminded. I think this negligence could cost anyone their life someday. So, I request you to be more vigilant towards the working of the hospital.

Yours truly,

Karuna/Karan

OR

M-114, Mall Road,

Delhi.

20th March, 20XX

The Editor,

Times of India,

New Delhi.

**Sub : Improvisation of the living conditions in slums**

Dear Sir,

Through the esteemed columns of your prestigious newspaper, I would like to highlight the quality of life of the people living in slums. Each family consists of six to seven members who share a single room. Their children can be seen playing on the heaps of garbage and breeding grounds of flies, mosquitoes and cockroaches. Sincere efforts are required to improve the physical, social and economic conditions of these slum dwellers. Practical action should be taken by providing access to clean drinking water, proper health and sanitation facilities, waste management services and better housing facilities. Slum upgrading should be a major part of the government's agenda to alleviate the poor living standards of the people.

But unless government takes a quick action, there can be no sustainable solution to the housing challenges. The government is not doing much regarding this reasons which may include low incentives, less financial resources, lack of provision of any such reform due to their short-term electoral period. But NGOs,training institutes, local residents, community groups can join hands together with government to eradicate the miseries of slum dwellers.

Hope my letter would find a place in your newspaper.

Yours Truly,

Karuna/Karan

**6. 'The policy of reservation of seats for admission to the professional courses is good for the deprived sections of society.**

**Write a debate in 150-200 words either for or against the motion. [10]**

**OR**

**Write a speech in 150-200 words on 'Benefits of early rising' to be delivered by you in the morning assembly of your school. You are Karuna/Karan, Head Girl/Head Boy.**

**Answer :**

**'POLICY OF RESERVATION OF SEATS' IN PROFESSIONAL COURSES**

**AGAINST :**

Good morning, respected judges and the esteemed audience! The topic of today's debate is 'Policy of reservation of seats for professional courses is good for the deprived section of the society, I, stand before you to speak against the motion.

According to the Indian constitution, reservation in all the fields whether education, jobs or any other place, has been made to bring under-privileged classes at par with privileged or general ones. There are numerous reservations like reservation for women, reservation for physically challenged, reservation for economically backward class, reservation for scheduled castes, etc. But, I believe that this leads to further discrimination in the society. It had started with a very good objective of uplifting the socially backward society to give them equal opportunities. But with time its meaning has been changed and the aim of starting it is lost. People have started taking undue advantage of it. This has resulted in widespread protests and counter protests as it has affected thousands of jobs and admissions opportunities. Students securing 80 or 90 percent sometimes lag behind the underprivileged who do secure only 40 or 50 percent for taking admission in the vocational or professional courses. It denies the equality of opportunity to those who do not belong to backward communities. They are denied admissions even though they have secured a good percentage. This would also perpetuate caste feeling among people and hamper national unity. Also the reservation systems in India is creating a manpower which is not able to compete at the global level. India needs people who can work for the growth and development but reservation is raising undeserving candidates at higher positions. Hence, I think reservation system should be demolished.

**FOR :**

Good morning, respected judges and the esteemed audience ! I am standing before you to speak in favour of the topic, 'Policy of reservation of seats in professional courses'.

Reservation system or quota in the context of the Indian education system basically means that a fixed percentage of seats are reserved in all higher education institutions such as universities and colleges for students who belong to socially backward castes and categories.

To understand the reason for this reservation, we need to trace its roots back to the Indian independence era when a few people made the caste system and them practised it aggressively. They also created some of very unsocial norms such as untouchability and discriminated against the lower castes on every ground. Thus, not only the social life but also the economic conditions of these people worsened. This increased the gap between the rich and the poor or in other words between the upper and lower castes.

The basis of the reservation policy is not economic but social and educational backwardness. It clears the structure of hierarchy in the society which is rigid and can't be easily disseminate even if there is betterment in the economic status of those people.

So, we must understand that the purpose of reservation is only to equalise the society and not to benefit one caste over the other.

OR

'BENEFITS OF EARLY RISING'

Honourable Principal, respected teachers and my dear friends, Good Morning! I am Karan/Karuna, Head boy/Head Girl of the school; it is a great opportunity for me to share my views on the topic 'Benefits of rising early' in today's assembly.

We derive pleasure from many things, people, and places in our life. One such pleasure is when we get up early in the morning. It gives a good start to our mood. The air we breathe is so fresh that it refreshes our whole day. One should not miss the wonderful art of nature, rising of the sun from the dark color of midnight and gradually turning brighter, spreading its vibrant colors in the sky. This glory can't be explained in words. We come into close contact with nature when we hear the chirping of birds, see green trees, flowers and leaves dancing. There is no chaos in the early morning and we can easily concentrate on our studies. It is well said that whatever we study in the early hours of morning is learnt quickly and our mind retains it forever. Early morning meditation helps in keeping our body and soul fit. Breakfast, the most important meal of the day can be taken up timely. Besides, many other things, one may get enough time to exercise which keeps the body energetic and fit throughout the day. Then our productivity and capacity to work also increases. There is no distraction in the morning hours in case we want to indulge in some creative activities.

To sum up, I can say that I have full faith in the saying, "Early to bed, early to rise, makes a man healthy, wealthy and wise."

**7. India is a land of diversity. One way in which it makes us feel proud of it is the number of festivals we enjoy. Write an article in 150-200 words on 'Festivals of India'. You are Karuna/ Karan. [10]**

OR

**Rising pollution, fast and competitive lifestyle, lack of nutritious food etc. have caused health woes for a large section of our population. Providing health care used to be a charitable and ethical activity. Today it has become commercialized, a money spinning business. Write an article in 150-200 words on 'How to provide proper health care to the common man'. You are Karan/Karuna.**

**Answer :**

'FESTIVALS OF INDIA

By Karuna/Karan

India is a land of colors. It is a country populated by the people of many castes, religions, faiths, communities, languages and creeds. Therefore, many festivals, celebrations, social and cultural gatherings, fairs take place every year. India is said to be a country which is full of pleasures, joys and festivities. Throughout the year, we celebrate many festivals. They represent the grandeur of the country. These are the occasions when families gather at one place and celebrate happy moments. Festivals are celebrated throughout the year and people await these festivals with great enthusiasm. These have an important place in the hearts of Indians as they believe that fasting on these occasions purify their soul and help them to lead a problem free life. Some festivals are celebrated all over India and people celebrate them irrespective of their status and social position. These include Diwali, Holi and Dusshera. Each festival has its religious value and contains some mythology behind it. For example, Dusshera is celebrated on account of victory of good over evil and when Lord Ram killed Ravana. Similarly, Janmashtami has also a social and religious significance as Lord Krishna was born to kill all the demons on the earth. Thus, each and every festival has its own value which teaches people to come closer and forget all their differences.

OR

**'HOW TO PROVIDE PROPER HEALTHCARE TO THE COMMON MAN'**

By Karuna/Karan

A large section of our population is suffering from physical as well as mental problems like, fever, injury, epilepsy, learning disability, etc. Women have problems related to pregnancy and childbirth whereas children require immunisation against common diseases. Most of these health woes have been the result of our changing life style, rising pollution, lack of nutritive value in food. Hence, they require access to health-care facilities and treatment. Unfortunately, health services are often planned without consulting the community members. To meet the necessities of the people for accessible and affordable services, community members should be consulted and involved in their planning. Health centers should be more vigilant. Most people prefer home remedial measures and go to the hospital or refer to a doctor only when the situation is beyond their control. But it's not necessary that they will go to a well-qualified doctor, they may approach local pharmacist or traditional healers. Therefore, it is necessary that the health sector must intervene and create awareness among the people.

Health care facilities should be within the reach of the common man. Government's investment in public health care is very low. As a result, many hospitals run by the government do not provide sufficient medical facilities. Therefore, improvements are required in this sector and it is a positive thing that the government is now planning to allocate more funds for the renovation of hospitals.

## SECTION — C
## (LITERATURE : TEXT BOOKS AND LONG READING TEXT)

8. **Read the extract given below and answer the questions that follow : [4]**

Its loveliness increases, it will never
Pass into nothingness; but will keep
A bower quiet for us, and a sleep
Full of sweet dreams, and health, and quiet breathing.

(a) **Whose loveliness will keep on increasing?**

(b) **Identify the phrase which says that 'it' is immortal.**

(c) **What is a 'bower'?**

(d) **Why do we need sweet dreams, health and quiet breathing in our lives?**

OR

Old
familiar ache, my childhood's fear,
but all I said was, see you soon,
Amma,
all I did was smile and smile and
smile................

(a) **What does the phrase, 'familiar ache' mean?**

(b) **What was the poet's childhood fear?**

(c) **What do the first two lines tell us about the poet's feelings for her mother?**

(d) **What does the repeated use of the word, 'smile' mean?**

**Answer :**

(a) Loveliness of anything that is beautiful keeps on increasing. It gives us immense joy, unending and everlasting pleasure.

(b) "It will never pass into nothingness" is the phrase which says that it is immortal.

(c) 'Bower' literary means leafy shelter or a pleasant shady place under a tree.

(d) We need sweet dreams, health and quiet breathing in our lives as they refresh us and give eternal joy and remove our sorrows and sufferings. Thus, we want to remain bound to the earth.

OR

(a) By the phrase 'familiar ache' the poet means the pain which she had already felt in her childhood. The pain was due to the fear of getting separated from her mother and the realization that she had not cared enough for her. She suppressed it by smiling again and again at her and by promising to meet her soon.

(b) The poet's childhood fear was that her mother was old and had a pale and colourless face, she might not live very long and the pact might lose her someday.

(c) The poet had the childhood fear that she would lose her mother soon. She possessed the same fear while going away from her. But she did not say anything to her except the optimistic note 'see you soon Amma'. She gave her a cheerful farewell.

(d) The word 'smile' has been repeated to show that in order to hide her fearful thought of separating from her mother, she puts on a false smile. Secondly, she gives moral support to her mother by promising to meet her soon. She tries to hide her feelings by smiling again and again.

9. **Answer any four of the following in 30-40 words each: [3 × 4 = 12]**

(a) **Describe the irony in Saheb's name.**

**(b) Why was Gandhiji opposed to C.F. Andrews helping him in Champaran?**

**(e) When he was only ten days old, a prediction was made about the future of the Tiger King. What was ironic about it?****

**(f) What was his father's chief concern about Dr. Sadao?**

**Answer :**

(a) Saheb's full name was Saheb-E-Alam which means the 'Lord of the Universe'. The irony in his name was that he was a poor rag picker who did not even have chappals to wear. He walked on the streets barefeet to earn his living and looked into the garbage for gold coins or something to survive.

(b) Gandhiji was opposed to C.F. Andrews helping him in Champaran because he wanted the Indians to be self-dependent and self-reliant in their struggle against injustice. He told him that Indians were strong enough to fight their own battle and had the capability to win it. Therefore, Gandhiji refused his help.

(f) Sadao's education had been his father's chief concern. At the age of twenty-two, he sent his son to America to learn about surgery and medicine. He lined up to his father's expectations. By the time his father died, Sadao had established himself as a famous surgeon as well as a scientist.

**10. Answer the following question in 120-150 words: [6]**

**Our language is a part of our culture and we are proud of it. Describe how regretful M. Hamel and the village elders are for having neglected their native language, French.**

**OR**

**Teachers always advise their students to dream big. Yet, the same teachers in your classrooms find fault with Sophie when she dreams. What is wrong with Sophie's dreams?****

Answer : There is no doubt that our language is a part of our culture and we feel proud of it. But the true value of it can be realized only when the freedom of speaking it, is taken away from us.

The story 'The Last Lesson' depicts the pathos of the whole situation, how people feel in not learning their language and losing an asset. One day Franz, one of the students came unwillingly to the school and noticed that everything was still and as quiet as a Sunday morning and there were elder people sitting quietly at the back of the classroom. There was an order from Berlin that French would not be taught any further in the schools and instead German was to be taught. This news shocked everyone. All the villagers gathered in the classroom to attend their last French lesson and to pay their respect for the French teacher, M. Hamel, who was going to deliver his last speech in French. The adults felt sorry that they had not attended the school and thanked Hamel for his faithful service.

To honour his French lesson, M. Hamel also put on his fine Sunday clothes. His behaviour changed to kindness. He even blamed himself for neglecting french as he used to give holidays when he wanted to go fishing or for watering his flowers. He told his students that it was the most beautiful and logical language and they should not forget it.

**OR**

**11. Answer the following question in 120-150 words : [6]**

**Derry sneaked into Mr. Lamb's garden and it became a turning point in his life. Comment.**

**OR**

**How did Jo want the Roger Skunk story to end? Why?**

**Answer :** Derry sneaked into Mr. Lamb's garden and it became a turning point in his life. Before that, Derry used to live an isolated life. He did not like meeting people or socializing. The reason was his acid-burnt face, seeing which people often got terrified. He began leading a secluded life.

Meeting Mr. Lamb brought a change in his life. His attitude towards life was completely changed. Mr. Lamb gave him confidence and supported him morally as he was too physically impaired like him. He had a tin leg because the original one was blown off in the war. He told Derry that he was not affected by people's sarcastic remarks. He motivated Derry to face the challenges of life and not to lament over his burnt face. According to him, he was lucky to have two arms, two legs, ears, eyes and tongue. He could do far better than others.

All these preaching left an impression on Derry and then he started looking at things and the world differently. He began to help Mr. Lamb with crab apples and developed a positive attitude towards life.

**OR**

Jack used to tell his little daughter Jo, bedtime stories. As soon as she grew a bit older, she began to ask questions from her father. Once her father told her a story of a little animal called Roger Skunk who smelled bad. He therefore, met with a wizard who solved his problem a magic wand and made him smell like roses. But the Skunk's mother did not like it and took him back to the wizard to smell bad once again.

This ending was not acceptable to Jo as she wanted the story to have another ending where the wizard

**Answers are not given due to change in present syllabus

would hit the mommy with his magic wand and chop off her arms 'forcibly.' Jo thought that Roger was the hero of the story so he must not look ugly or stinky. Jo's perspective on life was different from her father's. She wanted the wizard to hit mommy on head as a punishment as she imposed her will on her son unnecessarily. According to Jo, Roger should be allowed to choose his own life. Moreover, she didn't want the story to end sadly. She wanted to change the end of the story to assert her views on her father. She had her own ideals and views which she wanted to pursue.

**12. Answer the following question in 120-150 words :**** **[6]**

**What do we learn about Mrs. Hall and Griffin from their first interaction at Coach and Horses inn?**

**OR**

**What kind of life did Silas lead before coming to Raveloe?**

**13. Answer the following question in 120-150 words :****

**Everyone who comes into contact with Griffin suffers. Attempt a character sketch of Griffin in the light of this remark.** **[6]**

**OR**

**Attempt a character sketch of Nancy Lammeter.**

●●

## Delhi [Set-II]

**Note : Except for the following questions, all the remaining questions have been asked in previous sets.**

### SECTION—B
### (WRITING SKILLS)

**4. You need to buy a flat. Draft a suitable advertisement in about 50 words to be published in the classified columns of a local newspaper giving all the necessary details of your requirement. You are Karan/Karuna, M 114, Mall Road, Delhi.** **[4]**

**OR**

**Sunrise Global School, Agra is going to organize a one-act play competition in the school auditorium. You have decided to invite noted stage artist, Nalini to grace the occasion. Draft a formal invitation for her in about 50 words. You are Karuna/Karan, Cultural Secretary.**

**Answer :**

FLAT WANTED

A 2 or 3 BHK flat covering in the area of Mall Road, Delhi. Covering area 800-1500 sq. ft. required should be newly-built, ventilated and centrally located: Car parking is preferred. Should be accessible to market and metro station. Cost negotiable.
Please contact.
Karan,
M114, Mall Road,
Delhi Ph : xxx xxx xxx

OR

Sunrise Global School
Agra

15th May, 20XX

Nalini
Abhilasha Vihar,
Delhi.

**Subject : Invitation to preside over one-act play competition.**

Dear Ma'am,

Our school is going to organize a one-act play competition in the school auditorium. A large number of students from classes XI and XII are likely to participate in this inter school competition. We shall be highly honoured if you will grace the occasion with your presence on the 30th of June, 20XX at 2 pm. Kindly confirm your appearance in time.

Regards
Karan/Karuna
(Cultural Secretary)
Sunrise Global School
Agra

**5. Park Lane, Chanakyapuram, Mysore is proud of having four reputed public schools in close proximity to each other. In the morning and afternoon there is a heavy traffic along the road and hundreds of student are put to a lot of inconvenience.** **[6]**

**Write a letter in 120-150 words to the Deputy Commissioner of Police (Traffic) describing the problem. Suggest a few measures to regulate the traffic. You are Karuna Shetty/Karan Kumar, Principal, Sunrise Global School.**

**OR**

**Write an application (including a resume) in 120-150 words for the post of receptionist advertised in a national newspaper by JKL Publishers,**

**Answers are not given due to change in present syllabus

**Peshwa Road, Pune. You are Karuna, M-114 Mall Road, Pune, a graduate from SNDT University and have done a Secretarial Practice Course from YWCA, Mumbai.**

**Answer :**

Sunrise Global School,
Park Lane,
Chanakyapuram,
Mysore

1st March, 20XX

The Deputy Commissioner of Police (Traffic)
Chanakyapuram,
Mysore

Dear Sir/Madam,

**Subject : Traffic congestion on the road surrounding the schools.**

This letter intends to bring to your kind notice about the serious problem of traffic mess near the school's surrounding which has caused great inconvenience. There are four reputed public schools in the ambit of the area. In the morning, the movement of the school transport, staff vehicles and the vehicles of the parents, who come to drop their children, along with the general traffic, creates a huge chaos on the road. The same chaos could be witnessed in the afternoon when the school gets off. The situation gets worse when the traffic lights don't work. The children, who come from the nearby areas, find it difficult to cross the roads and face a huge risk. The safety of the children is of utmost importance for the schools. Therefore, I would request you to appoint traffic police on different points of the roads around the schools, so as to avoid traffic jams.

Also, a regular check of the traffic lights would help in reducing inconvenience. The schools are also planning to arrange traffic rules awareness workshop for their respective staff, parents and children. I request you to look into the matter and solve the issue at the earliest.

Yours Sincerely,
Karan Kumar/Karuna Shetty
Principal
Sunrise Global School

**OR**

M-114, Mall Road
Pune

1st March, 20XX

JKL Publishers
Peshwa Road,
Pune.

**Subject : Application for the post of Receptionist.**

Dear Sir,

With reference to your advertisement of the job in the newspaper. I want to apply for post of a receptionist. I am graduate from SNDT University and have completed a Secretarial Practice Course from YWCA, Mumbai. I have been working as a receptionist with Gupta & Sons Textiles Ltd. since March, 20XX. I have a proven record of meeting my goals by accepting all challenges throughout my professional career.

My resume has been attached herewith. Please let me know, if any further information regarding my background and qualifications, is needed.

Looking forward to hear from you soon.

Yours faithfully,
Karuna

Enclosure : Resume and two passport-size photographs

**RESUME**

| | |
|---|---|
| **Name** | : Karuna |
| **Father's Name** | : Om Prakash Chauhan |
| **Date of Birth** | : 11-07-1989 |
| **Residential Address** | : M-14, Mall Road, Pune |
| **Contact Number** | : +91-xxxxxxxxxx |
| **Educational Qualifications** | : (i) Graduation (SNDT University)<br>(ii) Secretarial Practice Course (YWCA, Mumbai) |
| **Computer Proficiency** | : MS Office and Adobe Photoshop |
| **Experience** | : Working as Receptionist with Gupta & Sons Textiles Ltd. since March 20XX |
| **Salary Expected** | : ₹ 15,000 |
| **Marital Status** | : Unmarried |
| **Hobbies** | : Travelling, reading and socializing |
| **Languages Known** | : Hindi and English |
| **References** | : (i) Mr. Anil Thackrey Marketing Head Mumbai<br>(ii) Mrs. Shabha Rana HRM, Surat |

6. **'Rising prices can be controlled only by the government.' Write a debate in 150-200 words either for or against the motion. [10]**

OR

**Write a speech in 150-200 words on the topic, 'Library plays an important role at school'. It is to be delivered in the morning assembly. You are Karuna/Karan, Head Girl/Head Boy.**

**Answer :**

**'RISING PRICES CAN BE CONTROLLED ONLY BY THE GOVERNMENT'**

**AGAINST :**

A very good morning to jury members, friends and my fellow opponents! I am Karan/Karuna Head boy/Head girl. Today, I am going to speak against the motion' Rising prices can be controlled only by the government.' The role of government in an economy is to announce measures of regulation and control. A government can impose restrictions on hoarding, the export of some items and track prices of limited commodities. However, the execution vests in the custody of the people residing in the nation.

It would not be correct to put the blame wholly on the government. The consumer is no less to be blamed for inflation. Some consumers allow themselves to be robbed due to the fear of insufficient good supply and adverse price trends. Also, some consumers indulge themselves into hoarding and black marketing of the products. Such practices not only deprive the country of much-needed resources but also serve as bad examples for those who cannot afford to pay such high prices. By reducing the demand of a good, its price can be controlled by the consumers. Thus, I believe, the rising prices or inflation can be controlled by the combined effort of the government and consumers.

Thank you.

OR

**FOR :**

Good morning to jury members, friends and my fellow opponents ! I am Karan/Karuna Head boy/Head girl. Today, I am going to speak in favour of the motion Rising prices can be controlled only by the government !

The rising prices of commodities is also called inflation is not just an economic concept but they are also a political tool, often used by the opposition parties to launch attack on the ruling government. Common man is the most affected part of the society and especially the people who belongs to the lower strata of society. Moreover, if the prices of the essential items increase, the damage is more severe.

Statistically, if we observe the trends in prices in the last five years; the prices of eight essential commodities have gone up by nearly 72 percent and on the contrary the per capita income have gone up by 38 percent of average Indian in metros, according to a study undertaken by apex chamber ASSOCHAM.

This inflation can be controlled to a large extent if the government gives farmers the independence and authority to sell their agricultural produce anywhere in the country without any restriction and also ensure free movement for them without imposing any taxes on them. In the similar line of thought, it is only the government which can create policy and enforce laws which can control rising prices and help the common man get out of the vicious cycle of poverty and helplessness.

OR

**LIBRARY PLAYS AN IMPORTANT ROLE AT SCHOOL**

Good morning everyone!

Respected Principal, teachers and my dear friends, I am Karan/Karuna, the head boy/ Head Girl of your school, and I am going to give a short speech on how 'Library plays an important role at school.' A school library is a place that plays an important role in encouraging innovation and thirst for knowledge. It provides support for learning and teaching throughout the school. The school library exhibits positive impact on the academic achievement of the student, as it helps in developing the mind of the student. It is integral to the teaching and learning process.

A library at school not only provides academic references but a wide range of fiction, non-fiction, digital and print resources. It develops the habit of reading in students and helps to to enhance students' intellectual, cultural and emotional growth. The school library also provides opportunities for teachers to plan, implement and evaluate new learning programs. It also plays a critical role in assisting the students in compiling, critically analysing and organising information, and then presenting their findings.

A school library provides a safe environment that encourages curiosity and imagination. The students become users of information and start to develop a better understanding of the world.

Thank you

●●

# Delhi [Set-III]

**Note : Except for the following questions, all the remaining questions have been asked in previous sets**

## SECTION — B
## (WRITING SKILLS)

6. **'The policy of no detention till class VIII is not in the interest of students.' Write a debate in 150-200 words either in favour of or against the motion.'** **[10]**

**OR**

**Write a speech in 150-200 words on the topic, 'A student must know how to manage his time.' It is to be delivered in the morning assembly. You are Karan/Karuna, Head Boy/Head Girl of the school.**

**Answer :**

### THE POLICY OF NO DETENTION TILL CLASS VIII IS NOT IN THE INTEREST OF STUDENTS

**FOR :**

A very good morning to jury members, friends, and my fellow opponents. Today, I am going to speak for the motion 'The policy of no detention till class VIII is not in the interest of the students.' Firstly, let me tell you what the no-detention policy states. The policy states that no child shall be held back in any class or expelled from the school till the completion of elementary education. The step was taken to reduce stress and pressure among students, but the policy has led to poor outcomes, as there has hardly been any improvement in the academic standard. It has encouraged negligence and reluctance in children to study. The student develops a carefree and an easy going attitude, due to an automatic promotion system up to class VIII, and in class IX, he is unable to change his attitude. Suddenly, they fail to cope up with the academic pressure and end up performing badly in board exams.

I, therefore, firmly agree that the policy of no detention till class VIII is not in the interest of students, as it gives then the security of making it to the next class, regardless of a child's competence to learn higher concepts.

**OR**

**AGAINST :**

A very good morning to jury member, friends and my fellow opponents. Today, I am going to speak against the motion.

Small children have very tender hearts and also their learning capacities are different. They need to be molded and guided in such a way that they learn and grow in a positive way. The fear of failure in exam is a very scary aspect for these tender minds. They might try to escape the results or run away from homes even due to this fear. There is peer pressure too and they will feel low in comparison to their peers. There is pressure from parents too to get good marks and hence the child is forced from all angles to score well and clear the exams. The no detention policy till VIII standard is a very good initiative by the government to make the child get away with all negativity and come out with flying colors in the exams. The grading system does not create any differences and the children focus more on their subjects. By the time a child reaches IX standard he or she is matured enough to handle the scores and the level of learning and hence does not have any coping issues. Therefore, the no detention policy till VIII standard is in the best interest of the students.

**OR**

### A STUDENT MUST KNOW HOW TO MANAGE HIS TIME

Good morning everyone!

Respected principal, teachers and all my dear friends, I, Karan/Karuna, the head boy/head girl of your school, am going to give a short speech on the topic, 'A student must know how to manage his time'.

When life takes a highway and everything seems to be going out of control, the only thing that can bring back the required pace in life is time management because a student's time is so valuable, it is important to use it wisely. This can be done by taking advantage of an idle time that could be spent studying. This could happen during lunch, on the bus ride to and fro from school or before a class starts.

A student must prioritize what comes first in his life.

Spending time with friends, family and having fun is important to provide a healthy balance. To successfully satisfy their personal life and meet education obligations, students must learn to juggle the two effectively by managing their time well.

When you get into the habit of procrastinating as a student, you tend to do the same as an adult. Habits that are learned at an early are hard to change when you become old. Therefore, develop a sense of responsibility towards your task at an early.

Thank you.

**7. India is an emerging economy, we are rich in natural resources. Yet a significant proportion of our people is living below poverty line. One cause of this poverty is our rising population. Write an article is 150-200 words on 'How to keep a check on population.' You are Karuna/Karan. [10]**

**OR**

**Women feel unsafe on the road, at their place of work and even at home. There is an urgent need to change the male mindset in its attitude towards women.**

**Write an article in 150-200 words on 'How to make women feel safe'. You are Karan/Karuna.**

**Answer :**

HOW TO KEEP A CHECK ON POPULATION

By Karan/Karuna

An economy requires the support of its population to function smoothly. The fact remains intact that India is extremely rich in its natural resources, yet incapable of sustaining its population. We often ponder over the question. Who is responsible for the imbalance? Is it the current government? Is it the lack of facility? Or is it just ignorance?

The disproportionate distribution of wealth in the country is mostly an outcome of the humongous population, hyperbolically termed as the 'Population Explosion'. The government, at regular intervals, has introduced birth control schemes and programmes, such as 'Hum Do Hamare Do'.

Such programs have failed tremendously because of the illiteracy and religiously guided notions of the people. The only way possible to curb the population is to educate the Indian population. The beliefs and desirability of having a male child also increases the population, which can be checked by making people understand that it is better to facilitate the children in a manner to empower them without being gender biased.

The government can work effectively only when the population understands the fragility and the need of the moment.

OR

'HOW TO MAKE WOMEN FEEL SAFE'

By : Karan/Karuna

The safety of women in the modern society is increasingly becoming a grave concern by each passing day. It is no longer an issue of morals but of criminalist mentality of a majority of the male sector of our society. Ensuring security of women can never be possible by restraining them to their houses. It can never be the solution. It is said, that to solve a problem effectively and efficiently, its root needs to be targeted' and in the same way, we need to put efforts into educating our boys and quite evidently, men as well, to respect the women. Rather than teaching a girl how to live with restrictions, we should teach a boy how to respect and accept the girl's wishes. This will help to reduce the cause a lot, but in order to eradicate the problem, severe legal measures are needed to be taken. The criminals should be punished severely and the legal procedures should be shortened. Domestic violence against women should not be tolerated and programs should be set-up to support women suffering in these scenarios. The aggressors in such cases need to be adequately penalized. The women should be trained the basic self-defense at an early age. Besides, taking these measures, special forces need to be posted at all places where crimes against women are more frequent, like tourist sites and secluded places. Special help-line numbers need to be set-up and effective and quick aid should be provided to women whenever and wherever required.

## SECTION — C

## (LITERATURE : TEXT BOOKS AND LONG READING TEXT)

**9. Answer any four of the following in 30-40 words each: [12]**

**(a) How did his experience at the YMCA swimming pool affect Douglas?**

**(b) What hospitality did the peddler receive from the crofter?**

**(e) What kind of life was enjoyed by crown prince Jung Bahadur till he reached the age of twenty?****

**(f) Where, when and how did Dr. Sadao meet Hana?**

**Answer :**

(a) The experience created a terrible fear in the heart of Douglas for a long time, after his YMCA swimming pool experience when Douglas hit the pool, he started drowning when he was saved, he became weak and trembled badly. He couldn't even eat for many days as he felt sick in the stomach. This made Douglas more afraid of water. The very sight of water made him ill. Douglas decided not to play any water sports.

(b) Unlike the indifferent attitude of the others towards him, the peddler was whole heartedly welcomed by the crofter in his cottage. The crofter, at once, put the porridge pot on fire and gave him supper. He gave him a roll of tobacco for his pipe. He also played a game of cards with him.

(f) Dr. Sadao met Hana in America at the house of a professor Harley where some foreign students were living. He liked her but waited to fall in love with her until he was sure she was a pure Japanese.

**Answers are not given due to change in present syllabus

# ENGLISH

## Outside Delhi [Set-I]

# 2015

**Time allowed : 3 hours** **Maximum marks : 100**

**SECTION – A**
**(READING)** **30**

**1. Read the passage given below carefully:** **[12]**

**(1)** For four days, I walked through the narrow lanes of the old city, enjoying the romance of being in a city where history still lives – in its cobblestone streets and in its people riding asses, carrying vine leaves and palm as they once did during the time of Christ.

**(2)** This is Jerusalem, home to the sacred sites of Christianity, Islam and Judaism. This is the place, where there is a church of the Holy Sepulchre, the place where Jesus was finally laid to rest. This is also the site of Christ's crucifixion, burial and resurrection.

**(3)** Built by the Roman Emperor Constantine at the site of an earlier temple to Aphrodite, it is the most venerated Christian shrine in the world. And justifiably so. Here, within the church, are the last five stations of the cross, the $10^{th}$ station where Jesus was stripped off his clothes, the $11^{th}$ where he was nailed to the cross, the $12^{th}$ where he died on the cross, the $13^{th}$ where the body was removed from the cross, and the $14^{th}$, his tomb.

**(4)** For all this weighty tradition, the approach and entrance to the church is non descriptive. You have to ask for directions. Even to the devout Christian pilgrims walking along the Via Dolorosa – The Way of Sorrow – first nine stations look clueless. Then a courtyard appears, hemmed in by other buildings and a doorway to one side. This leads to a vast area of huge stone architecture.

**(5)** Immediately inside the entrance is your first stop. It's the stone of anointing: this is the place, according to Greek tradition, where Christ was removed from the cross. The Roman Catholics, however, believe it to be the spot where Jesus' body was prepared for burial by Joseph.

**(6)** What happened next? Jesus was buried. He was taken to a place outside the city of Jerusalem where other graves existed and there, he was buried in a cave. However, all that is long gone, destroyed by continued attacks and rebuilding; what remains is the massive – and impressive – Rotunda (a round building with a dome) that Emperor Constantine built. Under this, and right in the centre of the Rotunda, is the structure that contains the Holy Sepulchre.

**(7)** "How do you know that this is Jesus' tomb?" I asked one of the pilgrims standing next to me. He was clueless, more interested, like the rest of them, in the novelty of it all and in photographing it, than in its history or tradition.

**(8)** At the start of the first century, the place was a disused quarry outside the city walls. According to the gospels, Jesus' crucifixion occurred 'at a place outside the city walls with graves nearby……'. Archaeologists have discovered tombs from that era, so the site is compatible with the biblical period.

**(9)** The structure at the site is a marble tomb built over the original burial chamber. It has two rooms, and you enter four at a time into the first of these, the Chapel of the Angel. Here the angle is supposed to have sat on a stone to recount Christ's resurrection. A low door made of white marble, partly worn away by pilgrims' hands, leads to a smaller chamber inside. This is the 'room of the tomb', the place where Jesus was buried.

**(10)** We entered in single file. On my right was a large marble slab that covered the original rock bench on which the body of Jesus was laid. A woman knelt and prayed. Her eyes were wet with tears. She pressed her face against the slab to hide them, but it only made it worse.

**1.1. On the basis of your understanding of this passage answer the following questions with the help of given options :** **[1×4=4]**

**(a) How does Jerusalem still retain the charm of an ancient era?**

**(i)** There are narrow lanes.
**(ii)** Roads are paved with cobblestones.
**(iii)** People can be seen riding assess.
**(iv)** All of the above.

**(b) Holy Sepulchre is sacred to ______.**

**(i)** Christianity.
**(ii)** Islam.
**(iii)** Judaism.
**(iv)** Both (i) and (iii).

(c) **Why does one have to constantly ask for directions to the church?**

(i) Its lanes are narrow.

(ii) Entrance to the church is non de script.

(iii) People are not tourist-friendly.

(iv) Everyone is lost in enjoying the romance of the place.

(d) **Where was Jesus buried?**

(i) In a cave.

(ii) At a place outside the city.

(iii) In the Holy Sepulchre

(iv) Both **(i)** and **(ii)**

**Answer:**

**(a) (iv)** All of the above.

**(b) (i)** Christianity.

**(c) (ii)** Entrance to the church is non descript.

**(d) (iv)** Both **(i)** and **(ii)**.

**1.2. Answer the following questions briefly: [1×6=6]**

(e) **What is the Greek belief about the 'stone of anointing'?**

(f) **Why did Emperor Constantine build the Rotunda?**

(g) **What is the general attitude of the pilgrims?**

(h) **How is the site compatible with the biblical period?**

(i) **Why did the pilgrims enter the 'room of the tomb' in a single file?**

(j) **Why did 'a woman' try to hide her tears?**

(k) **Find words from the passage which mean the same as : [1×2=2]**

(i) **A large grave (para 3)**

(ii) **Having no interesting features/dull (para 4)**

**Answer :**

**(e)** According to the Greek belief, the stone of anointing is a place where Christ was removed from the cross.

**(f)** Emperor Constantine built Rotunda to safeguard the Holy Sepulchre and the remaining structure around it.

**(g)** Pilgrims have a very casual attitude and they are completely unaware about the history and tradition. They are more into capturing pictures and appreciating the novelty of the church.

**(h)** According to the gospels, Jesus' crucifixion occurred at a place outside the city walls with graves nearby…' and as the archaeologists have discovered tombs from the Biblical era, the site is compatible.

**(i)** 'Room of the tomb' is a very small place, hence the pilgrims entered in a single file.

**(j)** Like a true Christian she felt overwhelmed as Jesus was buried there. Her eyes were wet with tears but she tried to hide as she didn't want to he seen crying by others.

**(k) (i)** Tomb. **(ii)** Non-descript.

**2. Read the passage given below: [10]**

**(1)** We often make all things around us the way we want them. Even during our pilgrimages we have begun to look for whatever makes our heart happy, gives comfort to our body and peace to the mind. It is as if external solutions will fulfil our needs, and we do not want to make any special efforts even in our spiritual search. Our mind is resourceful – it works to find shortcuts in simple and easy ways.

**(2)** Even pilgrimages have been converted into tourism opportunities. Instead, we must awaken our conscience and souls and understand the truth. Let us not tamper with our own nature of that of the Supreme.

**(3)** All our cleverness is rendered ineffective when nature does a dance of destruction. Its fury can and will wash away all imperfections. Indian culture, based on Vedic treatises, assists in human evolution, but we are using our entire energy in distorting these traditions according to our convenience instead of making efforts to make ourselves worthy of them.

**(4)** The irony is that humans are not even aware of the complacent attitude they have allowed themselves to sink in. Nature is everyone's Amma and her fierce blows will sooner or later corner us and force us to understand this truth. Earlier, pilgrimages to places of spiritual significance were rituals that were undertaken when people became free from their worldly duties. Even now some seekers take up this pious religious journey as a path to peace and knowledge. Anyone travelling with this attitude feels and travels with only a few essential items that his body can carry. Pilgrims traditionally travelled light, on foot, eating light, dried chickpeas and fruits, or whatever was available. Pilgrims of older days did not feel the need to stay in special AC bedrooms, or travel by luxury cars or indulge themselves with delicious food and savouries.

**(5)** Pilgrims traditionally moved ahead, creating a feeling of belonging towards all, conveying a message of brotherhood among all they came across whether in small caves, ashrams or local settlements. They received the blessings and congregations of yogis and mahatmas in return while conducting the dharma of their pilgrimage. A pilgrimage is like penance of sadhana to stay near nature and to experience a feeling of oneness with it, to keep the body healthy and fulfilled with the amount of food, while seeking freedom from attachments and yet remaining happy while staying away from relatives and associates.

(6) This is how a pilgrimage should be rather than making it like a picnic by taking a large group along and living in comfort, packing in entertainment, and tampering with environment. What is worse is giving a boost to the ego of having had a special darshan. Now, alms are distributed, charity done while they brag about their spiritual experiences !

(7) We must embark on our spiritual journey by first understanding the grace and significance of a pilgrimage and following it up with the prescribed rules and rituals – this is what translates into the ultimate and beautiful medium of spiritual evolution. There is no justification for tampering with nature.

(8) A pilgrimage is symbolic of contemplation and meditation and acceptance, and is a metaphor for the constant growth or movement and love for nature that we should hold in our hearts.

(9) This is the truth !

**2.1 On the basis of your understanding of the above passage answer the questions that follow with the help of given options :** **[1 × 2 = 2]**

**(a) How can a pilgrim keep his body healthy?**

(i) By travelling light
(ii) By eating small amount of food
(iii) By keeping free from attachments
(iv) Both (i) and (ii)

**(b) How do we satisfy our ego?**

(i) By having a special darshan
(ii) By distributing alms
(iii) By Making it like a picnic
(iv) Both (i) and (ii)

**2.2. Answer the following as briefly as possible:** **[1 × 6 = 6]**

**(c) What change has taken place in our attitude towards pilgrimages?**

**(d) What happens when pilgrimages are turned into picnics?**

**(e) Why are we complacent in our spiritual efforts?**

**(f) How does nature respond when we try to be clever with it?**

**(g) In olden days with what attitude did people go on a pilgrimage?**

**(h) What message does the passage convey to the pilgrims?**

**(i) Find words from the passage which mean the same as the following :** **[1 × 2 = 2]**

**(i) made/turned (para 3)**

**(ii) very satisfied (para 4).**

**Answer : I.**

(a) (iv) Both (i) and (ii)
(b) (iv) Both (i) and (ii)

**Answer : II**

(c) Earlier pilgrimages were a penance to stay near nature and did not require luxury rooms and big cars to travel, now they have become more of a tourism opportunity where people go for a picnic.

(d) When pilgrimages turn into picnics, the entire significance of pilgrimage as sadhana is lost. It becomes more of a social gathering, a life of comfort and a boost to ego, where pilgrims don't achieve spiritual upliftment.

(e) Pilgrimages are no more of travelling on foot and living in the ashrams with basic necessities, it has become a picnic with all the comfort, AC rooms, travelling by car with a large group to have all the fun and entertainment. We have distorted traditions according to our conveniences. Hence, we are complacent in our spiritual efforts.

(f) When we try to be clever with nature it does a dance of destruction. Its anger washes away all imperfections that are ruining our bodies because of the extra comfort that we are adapting our bodies to. We are not making any efforts to make our body work and adjust in adverse conditions rather making it all the more tender. Nature is just like everyone's Amma and would act destructively by affecting our health, etc. Sooner or later nature will force us to understand this truth.

(g) In older days, pilgrimages were more of a sadhana to love people, stay close to the nature, understand it, stay healthy and eat healthy. It was also for seeking freedom from attachment. They took it as a path to peace and knowledge.

(h) It conveys the message that pilgrimages are symbolic of contemplation meditation and acceptance and undertaken to understand the realities of life, to stay close to nature and should not be taken as a tourism opportunity where you gather in a large group, enjoy amongst all the luxuries and make a life rest upon only comfort.

(i) (i) Rendered (ii) Complacent.

3. **Read the passage given below:** **[8]**

It is surprising that sometimes we don't listen to what people say to us. We hear them, but we don't listen to them. I was curious to know how hearing is different from listening. I had thought both were synonyms, but gradually, I realised there is a big difference between the two words.

Hearing is a physical phenomenon. Whenever somebody speaks, the sound waves generated to reach your ears and you definitely hear whatever is said to you. However, even if you hear something, it doesn't always mean that you actually understand whatever is being said. Paying attention to whatever you hear means you are really listening. Consciously using your mind to understand whatever is being said is listening.

Diving deeper, I found that listening is not only hearing with attention, but is much more than that. Listening is hearing with full attention, and applying our mind. Most of the time, we listen to someone, but our minds are full of needless chatter and there doesn't seem to be enough space to accommodate what is being spoken.

We come with a lot of prejudices and preconceived notions about the speaker or the subject on which he is talking. We pretend to listen to the speaker, but deep inside, we sit in judgement and are dying to pronounce right or wrong, true or false, yes or no. Sometimes, we even come prepared with a negative mindset of proving the speaker wrong. Even if the speaker says nothing harmful, we are ready to pounce on him with our own version of things.

What we should ideally do is to listen first with full awareness. Once we have done that, we can decide whether we want to make a judgement or not. Once, we do that, communication will be perfect and our interpersonal relationship will become so much better. Listening well doesn't mean one has to say the right thing at the right moment. In fact, sometimes if words are left unspoken, there is a feeling of tension and negativity. Therefore, it is better to speak out your mind, but do so with awareness after listening to the speaker with full concentration.

Let's look at this in another way. When you really listen to, you imbibe not only what is being spoken, but you also understand what is not spoken as well. Most of the time we don't really listen even to people who really matter to us. That's how misunderstandings grow among families, husbands and wives, brothers and sisters.

**(a) On the basis of your reading of the above passage make notes on it, using headings and subheadings. Use recognizable abbreviations (wherever necessary – minimum four) and a format you consider suitable. Also supply an appropriate title to it.** **[5]**

**Answer:**

**Title : Listening Vs. Hearing**

**1. Difference between the two synonyms**
- (i) Hearing but not listening
- (ii) Paying attention
- (iii) Hearing as a physical phenomenon
- (iv) Consciously using your mind
- (v) Minds are full of needless chatter
- (vi) Accommodating what is said

**2. Barriers to Listening**
- (i) Prejudices and preconceived notions
- (ii) Negative Mindset
- (iii) Pretend to listen
- (iv) Proving our ver. of things right

**3. Ideal way of listening**
- (i) Listen first
- (ii) Listen with awareness
- (iii) Think before you judge
- (iv) Words unspoken leave a feeling of tension and negativity
- (vi) Speak your mind out

**4. Listening to understand people**
- (i) Imbibe what is not spoken
- (ii) Times when we don't listen to people who matter
- (iii) Triggers misunderstanding

**5. Importance of listening**
- (i) Reduce misunderstanding
- (ii) Perfect communication
- (iii) Improve interpersonal relation ships

**List of abbreviations used**

Vs. – Versus

Ver. – Version

Don't – Do not

**(b) Write a summary of the passage in about 80 words.** **[3]**

**Answer:**

Sometimes, we hear people but do not listen to them attentively. The difference between the two synonyms is that one is a physical phenomenon while the other is a conscious use of mind.

We, as listeners, often have preconceived notions about the speaker. We come prepared with a negative mindset, tend to prove them wrong and put forward our own version of things. Ideally, we should listen to them, understand their perception and stop judging them in order to avoid misunderstandings.

## SECTION – B
## (WRITING SKILLS) 30

4. **Your school, Akash Public School, Agra needs a canteen manager. On behalf of the Principal, write an advertisement in about 50 words to be**

**published in the classified columns of a local daily. Mention the educational and professional qualifications, other qualities required in the manager, who to apply to and the last date for the receipt of applications.** [4]

OR

**Your club is going to organise an interclass singing competition. Write a notice in about 50 words inviting names of the students who want to participate in it. Give all the necessary details. You are Navtej/Navita, Secretary, Music Club, Akash Public School, Agra.**

**Answer:**

SITUATION VACANT

Akash Public School, Agra, requires a Canteen Manager for the School Canteen. The person should be above 40 years of age, graduate with at least 5 years of managerial experience in the required field and willing to work extra hours on special occasions. Interested candidates may apply to the principal of the school latest by 10th, March 20XX.

OR

AKASH PUBLIC SCHOOL, AGRA

NOTICE

5th July, 20XX

ATTENTION! MUSIC LOVERS!

The school is organizing an Interclass Singing Competition on 10th October, 20XX at the School Auditorium. All those who are interested, may appear for the audition on 22nd September, 20XX at 9 a.m. The venue, is Music hall. For further details contact the undersigned.

Navtej/Navita
Secretary
Music Club

5. **You are Navtej/Navita, Secretary, Environment Club, Akash Public School, Agra. You, along with a group of students, went on a 3-day tour to Corbett National Park. You found how the tourists abuse the available facilities and thus endanger the environment. Write a letter in 120 – 150 words to the editor of a national daily highlighting the situation.**

**Suggest ways through which the environment of the Park can be saved.** [6]

OR

**On Teacher's Day, you read in a newspaper that privately owned and managed schools in small towns or even in the suburbs of metropolitan cities exploit their teachers by paying them just a fraction of their authorised salaries. This affects their performance in the classroom and thus the lives of their students. Write a letter in 120 – 150 words to the editor of a national daily raising your voice against such exploitation. Suggest ways to solve this problem. You are Navtej/Navita, 112 Taj Road, Agra.**

**Answer:**

Akash Public School,

Agra.

5 March, 20XX

The Editor,

The Hindu,

New Delhi

**Subject : The harmful effects of tourism on the Corbett National Park.**

Dear Sir,

Through the esteemed columns of your daily newspaper, I would like to draw the attention of the concerned authority to the rampant abuse of available facilities in the Corbett National Park by tourists and the subsequent endangerment to the environment.

The tourists not only litter the place with non-biodegradable plastic packets and wrappers, but also use wood from the forests for cooking purposes. Excessive trampling of the soil has been reported to have a harmful effect on the natural ecosystem. The tranquility of the habitat, which is so crucial to the resident fauna, is habitually disturbed by the tourists playing loud music in their safari jeeps and during their camps.

The Corbett National Park is the oldest in the country, established in 1936, to protect the endangered Bengal Tiger. The public should be made aware of the fact that eco-tourism should not be facilitated at the cost of harming the environment. There should be increase in staff and a stricter implementation. Penalty should be implemented for abuse of facilities. Use of only non-polluting fuel should be allowed. It is my humble request that this issue should be immediately be taken up by your esteemed publication to generate further awareness.

Yours sincerely,

Navita/Navtej

Secretary

Environment Club

OR

112, Taj Road,
Agra

2 March 20XX
The Editor,
The Hindu
New Delhi

**Subject : Exploitation of Teachers.**

Dear Sir,

It is said that the teachers in privately owned and managed schools in small towns and metropolitan suburbs are being exploited. On the occasion of Teacher's Day, this issue was brought to light in another national daily, mentioning that the teachers employed in such schools are paid a fraction of their authorised salaries.

Such a malpractice is seen to affect the performance of the teachers in the classroom. The school trust pays these teachers much less than the amount they were promised, leading to resignations and constant changes in the faculty. This has affected students' performance and led to an atmosphere of uncertainty in the schools, which is detrimental to the cause of education.

As a nation that prides itself on revering its teachers, it must be noted that we cannot build a future for our students if we do not give due credit to our teachers. The government should ensure that such schools pay salaries as per the government's fixed pay scales. Payments should be made through banks and the government agencies should monitor the salary perks etc. Therefore, I request your esteemed publication to take up this matter in order to investigate the issue further and mobilise the public towards calling for a complete overhaul in the education system of the country.

Yours sincerely,

Navtej/Navita

6. **The government has banned the use of animals in the laboratories for the purpose of dissection. Write a debate in 150 – 200 words either for or against this decision.** **[10]**

**OR**

**Some people feel that electronic media (TV news) will bring about the end of print media (newspapers). What are your views on the issue? Write a debate in 150-200 words either for or against this view.**

* **use of visuals on TV**
* **authentic and fast**
* **not enough news for 24-hour telecast**
* **may fabricate news**
* **become repetitive and dull**
* **even scandals become news**
* **print media – time tested**
* **analysed, verified news**
* **editorial comments**
* **cater to all interests**

**Answer:**

**BAN ON THE USE OF ANIMALS**

**AGAINST :**

Good morning everyone, Today I am going to express my views against the motion "Banning the use of animals for dissection".

Dissection of animals in the school laboratories, provides a practical experience for the theoretical knowledge that students gain from books. They procure an opportunity to actually put their learning into practice. It should not be looked at as unjust practice because such an activity is meant to develop the student's knowledge of the elements and functions of the living being.

The dissection of animals has a scientific purpose; it is a mode of hands-on education for students such that they learn outside of the pages of the textbook and actually look at what they are being taught. We should give more credit to the sensitivity and intelligence of our students, so as not to believe that a dissection class will make them more prone to animal violence. Students should be taught the value of animal life by making them understand how an animal has the same kind of organs as human beings that perform similar functions. The governing philosophy behind teaching students to dissect animals should be that life, whether human or animal, is the same; each living being is tied to each other by their inner similarity.

Thus, a ban on dissection of animals in school laboratories would not be the best course of action. I hope my views on this topic were substantial enough to throw light on this topic.

Thank you.

**FOR :**

Good Morning judges, teachers and friends Today I am going to express my views in favour of the motion, 'Ban on the use of animals.'

Animals are used for various purposes such as transportation, recreation, and scientific research. It is wrong on all grounds to use animals for human benefits and this should be banned.

As PETA founder ingrid Newkirk has said, 'When it comes to pain, love, joy, loneliness, and fear, a rat is a pig is a dog is a boy. Each one values his or her life and fights the knife'.

Over 115 million animals-mice, rats, dogs, cats rabbits, monkeys, birds, among others-are killed in laboratory experiments worldwide for chemical, drug, food, and cosmetics testing every year.

Next, humans use products made from animals. It is estimated that more than 50% the fur in the US comes from China, where millions of dogs and cats are often bled to death and skinned alive for their fur.

Many animals are ill-treated, starved, die a painful death in circuses and zoos where people go for their personal entertainment. Such places do not follow the set standards of care to be given to the animals.

Also, the exotic pet trade is a multi-billion-dollar industry. Some wild pets are bred in captivity, many are taken from their native habitats and this causes a number to die prematurely.

Animal killing is against humanity. Therefore, the use of animals for these purposes should be banned.

OR

**ELECTRONIC MEDIA THE END OF PRINT MEDIA**

**AGAINST :**

Good morning everyone. Today I take the opportunity to express my argument against the motion on the belief that electronic media (TV news) will bring about the end of print media (newspaper).

TV news that strives towards authentic and fast reportage is always at risk of fabricating its content to continue being in circulation. The governing principle for the 24-hours news format is that whether or not news constitutes the element of worthiness, it should be covered on the channel. If there is nothing worth reporting then the content can be made sensational just to gain better TRP and currency. It tends to become repetitive and dull, scandals are allowed to become news and popularity becomes the main objective. The ambition of TV news is to be the soul guardian of objective truth in the field of journalism can be considered to be a tall claim as more often it becomes a site for promotion of commercial products and sensational content in the grab of news. Print media is a time tested medium of information dissemination. The time taken to publish engenders that the content of the report has been analysed and verified to offer a composite picture to the reader. Editorial commentary ensures an informed view on the issues at hand. Newspapers are far from extinction, they are still a tried and time tested mode of receiving the world at our doorstep. Thus, the electronic media can never overrule the print media.

Thank you.

**FOR :**

The means for obtaining news of the world has been a paradigm shift from newspapers to television, especially due to the arrival of the 24-hours news format, reception of news is constant and on the go. The entire news landscape has changed as we are subjected to the concept of breaking news, that is, receiving news as it breaks and without waiting for the newspaper to arrive the next day with an account of events that have already been telecasted by them. It is helpful to illiterate people. E-media is faster and environment friendly. The power of the television medium is that the image is a direct message which does not need to be read or deciphered. TV news strives towards authenticity, covering all vantage points for maximum coverage and it is delivered fresh and fast. In such a case, it contributes towards the redundancy of the print media.

7. **Ragging has raised its ugly head again. A recent incident at a prestigious school has shown that this evil has not yet come to an end. Write an article in 150 – 200 words on 'Ragging, an Evil'. You are Navtej/Navita. [10]**

* **a practice from the British era**
* **original aim, respect for hierarchy**
* **enforcing traditions, discipline**
* **Prefect – a teacher substitute**
* **misuse of authority**
* **vulgar aspect**
* **fatalities**
* **solution**

**OR**

**India is a tourist's dream destination. Give your views on the tourism potential of India in an article in 150 – 200 words. You are Navtej/Navita.**

* **places of worship – religious tourism**
* **foreigners – places of historical interest**
* **the rich – hill stations during summers**
  - **the sun-kissed beaches in winters**
  - **leisure tourism**
* **medical tourism – world class hospitals**

**Answer :**

**THE EVIL OF RAGGING IN EDUCATIONAL INSTITUTIONS**

By Navita/Navtej

Ragging is an evil heirloom, left over from the British era when it was practised in English colleges and Universities to teach the social hierarchy and instills important values since it was believed that the students were lacking in such knowledge due to their immaturity. Unfortunately, these practices gradually became prevalent in educational institutions. The history of the best educational institutions in India is blackened with traumatic cases of brutal ragging, especially in Engineering and Medical Colleges. Ragging was practiced to enforce traditions and discipline freshers into respecting their seniors and the academic hierarchy at such institutions. In a school, 'Prefect' is considered to be a teacher's substitute. However, that does not mean that he/she can misuse his/her authority and indulge in the abuse of power, which happened in a recent case at a prestigious school. Schools are centres of learning and wholesome development, but when power equations within the student, displaying its vulgar aspect, it can lead to fatalities such as continued mental trauma or even death. The solution is to sensitise students and bring awareness to stop the harmful practice of ragging. In order to have able-bodied and healthy-minded future citizens of India, we need to eradicate such practices.

**OR**

**TOURISM POTENTIAL OF INDIA**

By Navtej/Navita

India has been blessed with a geographical, cultural and linguistic diversity, which is so rich and varied that one will need a lifetime to explore each and every part of India. The tourism potential of India is immense and still to be explored fully. Each century has brought a new culture to Indian soil leading to a diversified commingling that has shaped modern India.

India is well known all over the world, for its ancient civilization and the artifacts recovered and being excavated till today, prove the genuineness of the claims.

The ancient history of our culture has left behind exquisite monuments and temples together with archaeological finds for the tourist who are interested in them. Besides this, our country has been blessed with natural scenic beauty, right from the coast lines and beaches to the majestic mountains and hill stations.

Tourism is a high-potential industry over the world and there are several economies which are totally based on the tourism industry for their survival. The amount of money involved and high value returns could be mind-boggling and India likely, would successfully develop an infrastructure conducive to the promotion of the tourism industry.

There are so many places of historical interest, heirlooms left over from the various centuries. From the mosques to the temples and churches, India is a living testament to its vast and varied religious heritage. Being a democratic country, every religion in India has full liberty to celebrate its special festivals in which the entire country participates as a whole. For those, looking to just get away from the madding crowd, hill stations during summers, the sun-kissed beaches in winter and leisure tourism are the go-to options. Over the years, medical tourism has also emerged as an extremely viable option, considering world-class facilities are made available at half the prices here, enabling more people to come to India for their healthcare.

The options enumerated so far are just some instances of India's diversified tourist potential. There is so much more waiting to be explored in what can truly be the dream destination for tourists.

## SECTION – C
## (LITERATURE : TEXT BOOKS AND LONG READING TEXT) 40

**8. Read the extract given below and answer the questions that follow : [4]**

and
looked out at young
trees sprinting, the merry children spilling
out of their homes, but after the airport's
security check, standing a few yards
away, I looked again at her, wan,
pale as a late winter's moon and felt that
old familiar ache, ... ... ...

**(a) How do the trees sprint? [1]**

**(b) Why did the poet look at her mother again? [1]**

**(c) What did she observe? [1]**

**(d) Identify the figure of speech used in these lines. [1]**

**OR**

On their slag heap, these children
Wear skins peeped through by bones and spectacles of steel
With mended glass, like bottle bits on stones.

**(a) Who are these children? [1]**

**(b) What is their slag heap? [1]**

**(c) Why are their bones peeping through their skins? [1]**

**(d) What does 'with mended glass' mean? [1]**

**Answer :**

(a) The poetess is driving to the Cochin Airport. When she looks outside, the young trees seem to be moving backwards as the car moves ahead. With the speed of the car they seem to be running fast or sprinting. The poetess represents a contrast between her 'dozing' mother and the 'sprinting' young trees.

(b) The poetess looked at her mother again because there were obvious signs that the mother would pass away soon. She was deeply attached to her mother and had the childhood fear of losing her mother. She was pained to look at her ageing, pale face after she saw the world full of life and activity.

(c) She observed the pale and colourless face of her mother that resembles the late winter moon which lacks brightness as well as strength, like her mother did at that age.

**(d)** The figures of speech used in these lines are simile: The "wan, pale" face to the mother is compared to "a late winter's moon", and personification for "trees sprinting..."

**OR**

**(a)** These are impoverished children of the slums.

**(b)** Their slag heap is the heap of waste on which they live, roam and play.

**(c)** Their bones are peeping through their skins as they are malnourished.

**(d)** 'Mended glass' means broken spectacles. This shows their poverty and inability to buy new glasses.

**9. Answer any four of the following in 30 to 40 words: [3 × 4 = 12]**

**(a) What did garbage mean to the children of Seemapuri and to their parents?**

**(b) How did Rajkumar Shukla establish that he was resolute?**

**(c) 'Life is what it is all about.....' How is keeping quiet related to life?**

**(d) Mention any four things of beauty that add joy to our life.**

**(e) The manner of his (the Tiger King's) death is a matter of extraordinary interest. Comment.****

**(f) In what condition did Dr. Sadao find the American soldier at the seashore?**

**Answer :**

**(a)** Small slum children scrounged heaps of garbage. They expected to get some coins, notes or valuable things in it. Sometimes, they would find a rupee or even a ten rupee note. This gave the hope of finding more. So, they searched it excitedly. For children, garbage was wrapped in wonder. For the elders, it was a means of survival. Thus, garbage has two different meanings.

**(b)** Rajkumar Shukla established that he was resolute as he had come all the way from Champaran District in the foothills of Himalayas to Lucknow to speak to Gandhiji. Shukla accompanied Gandhiji everywhere. He also followed him to the Ashram near Ahmedabad. For weeks, he never left Gandhi's side till Gandhiji asked him to meet at Calcutta.

**(c)** The poet says, 'Life is what it is about ...' as he advocates and suggests the necessity of quiet introspection but does not advocate total inactivity and death. Total inactivity brings death, while stillness entails rest for a moment so that one can have calm self-realisation as otherwise the people of the world are overactive and always on the move. The poet wants 'no truck with death.' Keeping quiet is related to doing activities in a purposeful and thoughtful manner.

**(d)** Everything of nature is a thing of beauty and a source of pleasure. Some of them are : the sun, the moon, old and young trees, flowers, small streams with clear water, masses of ferns, blooming musk rose and lovely tales, etc. all of these are the things of beauty.

**(f)** Dr. Sadao found the American soldier in a wounded condition at the sea shore. He was very weak and pale. He had a bullet stuck in his wound. His face looked tortured and his back was stained with blood. From his battered cap, he learnt that he was a sailor from an American Warship. He was almost at the verge of death, if Sadao had not saved him.

**10. Answer the following in 120-150 words : [6]**

**Even today so many among us believe in superstitions. An astrologer predicted about 'the Tiger King' that he would be killed by a tiger. He 'killed' one hundred tigers yet was himself 'killed' by a tiger. How did the superstitious belief 'prevail'?****

**OR**

**Dr. Sadao faced a dilemma. Should he use his surgical skills to save the life of a wounded person or hand an escaped American P.O.W. over to the Japanese police? How did he resolve this clash of values?**

**Answer :**

**OR**

A conflict of interests arises in a situation when someone in a position of trust, such as a doctor, has competing interests that make it difficult to fulfil his duties. Dr. Sadao was a Japanese surgeon who lived in Japan during the Second World War. He had spent several years in the United States where he had experienced cultural prejudice and bias. He struggled with issues of duty, wartime medicine and racism. Yet, he risked his life to save an enemy, an American and a prisoner of war. He, like a real hero stood up for what he believes in and does not calculate the repercussions. His sentimentality for the suffering and wound help him rise above narrow national prejudices and extend help and services to an enemy. He tends to the soldier, reveals the truth about him to the General and later on helped the soldier escape to safety. He can be viewed as a true hero for his bravery and professional competence.

**Answers are not given due to change in present syllabus

11. **Answer the following question in 120–150 words: [6]**

**Everybody during the last lesson is filled with regret. Comment. (The Last Lesson)**

**OR**

**Sophie lives in a world full of dreams which she does not know she cannot realise. Comment.****

**Answer :**

According to the order from Berlin, only German was to be taught in the schools of Alsace and Lorraine. This order left Mr. Hamel, the village people and Mr. Hamel's students in deep regret. The villagers regretted not having made utmost use of the French lessons that had been given by Mr. Hamel. They had engaged themselves in their profession to earn a living and had not taken their French lessons seriously. The children and the villagers sitting in the class had always thought that they had enough time to learn the language. But now, during the last French lesson, everybody realised their mistake. They realised how little they knew of their native language. They all attended the last lesson to mark their respect towards their teacher and their country, which was no longer theirs.

**OR**

12. **Answer the following in 120-150 words :** [6]**

**Describe how Silas Marner is betrayed by his friend, William Dane.**

**OR**

**Why and how did Griffin rob the Vicar's house?**

13. **Answer the following question in 120 – 150 words:** [6]**

**'Evil begets evil'. In the light of this remark, describe the character of Dunstan Cass.**

**OR**

**Attempt a character sketch of Mrs. Hall.**

## Outside Delhi [Set-II]

**Note: Except for the following questions, all the remaining questions have been asked in the previous sets.**

### SECTION – B

### (WRITING SKILLS)

4. **Mid day meal scheme in the school introduced by the government serves several purposes. Millions of children from the deprived sections of the society get nutritious food leading to good health, are attracted to the schools, kept from going astray and, thus developed into good citizens. The scheme may have some drawbacks too. Write an article in 150-200 words on all aspects of the scheme. You are Navtej/Navita. [10]**

**OR**

**Morning assembly in the school gives a cool and calm start to the otherwise hectic schedule of the day. It can be a means of passing important information to the students. What are your views on this significant event of the day? Write an article in 150-200 words on 'Importance of Morning Assembly'. You are Navtej/Navita.**

**Answer :**

MID DAY MEAL

By Navtej/Navita

The mid day meal scheme in schools was initiated with the dual objective of improving the nutritional status of children in classes 1-5 in government schools and government-aided schools also encouraging children from socially and economically disadvantaged backgrounds to attend school regularly. It is a measure taken by the government to ensure that by meeting the nutritional requirements of the students, they would also be helping them in concentrating in school activities. The aim is also to keep children in schools, enable them to devote time to their education rather than leave the school to labour for food or get involved in criminal activities. The scheme estimates that a cooked mid day meal provides a child with a minimum of 300 calories and 8-12 grams of protein. This scheme was revised to cover children in the upper primary section, from the year 2007. The programme is carried out with the help of the local authorities. However, certain issues have cropped up. There have been a lot of cases of food poisoning resulting from the partaking of these mid day meals. In some schools, funds meant for such purposes have been funnelled off into the pockets of those who are in charge of distributing the grains. The government must inquire into all aspects of the schemes, check the supply chain of food grains, impose strict measures for observance of cleanliness in the

**Answers are not given due to change in present syllabus

school kitchens and ensure that the intention with which the scheme was started, i.e., to help the students, continues to be the cornerstone of the scheme.

OR

**IMPORTANCE OF MORNING ASSEMBLY**

By : Navita/Navtej

The morning assembly in schools starts the day on a cool and calm note, a short breath, before the hectic school day begins with classes, courses and extra-curricular activities. It offers a moment of reflection before the working day begins, the school community meets together and the assembly becomes a medium of communicating matters of significance to the students. The morning assembly is a time when the whole school assembles in the field or school auditorium, prays together, shares information and begins the day jointly on a note of quiet reflection, which is so crucial in preparing for the hectic day ahead. The most important thing is that the assembly imparts a clear sense of vision that there are moments of peace, even if the day seems like there is too much on our plates.

## SECTION – C

## (LITERATURE: TEXT BOOKS AND LONG READING TEXT)

**9. Answer any four of the following in 30-40 words each: [3 × 4=12]**

**(a) What does Saheb look for in the garbage dumps?**

**(b) How was Gandhi treated at Rajendra Prasad's house?**

**(c) Why does one feel 'a sudden strangeness' on counting to twelve and keeping quiet?**

**(d) Mention any two things which cause pain and suffering (A Thing of Beauty).**

**(e) When was the Tiger King in danger of losing his throne?****

**(f) What role did the American professor play in bringing Hana and Sadao together?**

**Answer :**

**(a)** For Saheb, the garbage dump was like a mysterious package, full of wonders and magic. He looked for different kinds of treasures in the garbage, like a silver coin or other tit bits.

**(b)** When Shukla took Gandhi to Rajendra Prasad's house, he was out of town. The servants knew Shukla as a poor farmer who was pestering their master to help the indigo share croppers. Seeing Gandhi with him they presumed him to be another farmer and allowed him to stay as Shukla's companion.

**(c)** Counting to twelve and keeping quiet brings in us a sudden strangeness because we are not used to such quiet and peaceful moments. We fail to understand ourselves, and have created a threat of self-destruction, which will be reduced by counting to twelve.

**(d)** The things that cause suffering and pain include the scarcity of people with a noble temperament and the unhealthy and miserable ways in which humanity is searching for meaning in life.

**(f)** Dr. Sadao had met Hana at his American professor's house. The professor and his wife were kind people who wanted to help the foreign students and so invited them to their house. After completing their studies in America, Sadao and Hana returned to Japan and got married.

**10. Answer the following question in 120-150 words: [6]**

**How different from usual was the atmosphere at school on the day of the last lesson? (The Last Lesson)**

OR

**How different is Jansie from Sophie?****

**Answer:** On usual days, when the school started there would be a huge bustle that could be heard out in the street. Sounds of desks being opened and closed and lessons repeated in unison could be heard. While repeating the lessons, students would put their hands on their ears for better understanding of their lessons and the sound of the teacher's great ruler rapping on the table could also be heard. But, on the day of the last lesson, the school was as quiet as on a Sunday morning. The narrator saw his classmates already in their respective places and Mr. Hamel walking up and down holding the scary iron ruler under his arm. When the narrator felt frightened to enter the class Mr. Hamel kindly told him to quickly go to his place as they were about to begin the class without him. Moreover, he saw some villagers were sitting on the last benches.

●●

**Answers are not given due to change in present syllabus

# Outside Delhi [Set-III]

**Note: Except for the following questions, all the remaining questions have been asked in the previous sets.**

## SECTION – B
## (WRITING SKILLS)

**4. Consumerism is increasing day-by-day. Luxuries of yesterday have become necessities of today. The result is that the more we want the more miserable we become. Write a debate in 150-200 words on 'The only way to minimise human suffering and pain is to control our needs.' You are Navtej/Navita. [10]**

**OR**

**Migration from villages to cities has led to the spread of urban slums. People living in these slums lead a miserable life. Economic disparity leads to the problems of law and order. Write a debate in 150-200 words on the 'Solution to the problem of misery in the urban slums lies in creating jobs in the villages.' You are Navtej/Navita.**

**Answer :**

**THE ONLY WAY TO MINIMISE HUMAN SUFFERING AND PAIN IS TO CONTROL OUR NEEDS**

**FOR :**

'There is sufficiency in this world for man's needs but not for man's greed' said Mahatma Gandhi. Respected panel of jury, teachers, my dear friends and not to forget my worthy but misguided opponents. Today, I, Navita/Navtej stand before you to speak for the motion on the topic "The only way to minimise human suffering and pain is to control our needs."

We know that there is sufficiency, in this world for man's needs, but what if a man mistakes his greed to be his need? Unfortunately, this is the scenario of today. The things that were regarded luxuries by our forefathers are now termed as necessities. We cannot imagine life without certain items like a blender, a washing machine or even a mobile phone. But weren't our forefathers able to lead their lives even without these things? It is time that we understand that we have become slaves of a time when diseases like diabetes weren't even heard off, why? It was because people worked hard without taking the help of such things and so kept their body's metabolism running. Whatever be the problem, there is only one simple solution– Controlling our needs and differentiating them from our greed. The more we enslave ourselves, the more we suffer.

Thank You.

**OR**

**AGAINST :**

Good morning to everyone present here. I am standing in front of to speak against the motion.

Human beings are born with defined needs and also we have some physical and mental pains. Our needs develop and grow more with time and age. A small child may have very meager needs but a grown up individual has different set of needs. Controlling our needs is not a solution for minimizing our suffering rather we should be able to match our needs with our outreach. Mental pain is something which does not require any financial aid but it needs soothing words and comfort Physical pain comes as a result of a disease, an attack or a lifestyle disorder. We should try to calm our mind and find happiness in simple things to have a control over the pain and suffering. Life is beautiful, if we deal it with positive thoughts and a broad-minded approach. Human needs are inevitable and we need to work towards their fulfillment. Controlling the needs will make any person mentally sick and disgusted. So, instead of controlling our needs we should find ways to satisfy the needs and march towards a happy and fulfilled life.

**OR**

**SOLUTION TO THE PROBLEM OF MISERY IN THE URBAN SLUMS LIES IN CREATING JOBS IN THE VILLAGE**

**FOR :**

Respected panel of jury, teachers, my dear friends and not to forget my worthy but misguided opponents. Today, I Navita/Navtej, stand before you to speak for the motion of the topic "Solution to the problem of misery in the urban slums lies in creating jobs in the villages."

Migration is a major issue faced by cities. In search of a better and brighter future and to fulfil their dreams, people from small towns and villages leave their life back home and move to cities. Unfortunately, their dreams shatter as they get to live in the already overpopulated urban slums, where they gradually get entangled in the deplorable conditions. Besides, the economic disparity that they face in the cities lead them to the problems of law and order. But, have we ever wondered what is the root cause of migration and the problems attached to it? Yes, it is the dearth of jobs in their native place. There is a huge scope for expansion in villages and small towns but people refuse to see it. The more jobs we create in villages and small towns, the less people will migrate, thereby leading to an economically as well as medically stronger India.

Thank you

OR

**AGAINST :**

Good Morning to the jury members and my dear friends. I am standing in front of you to speak against the motion.

The slums in India have a lot of people who are jobless and aimless; the only way to approach this problem is to give proper food and education to them. Once they are healthy physically and mentally they can be trained to fit any kind of jobs. Only by creating jobs these people cannot directly fit into that but they need to be well-equipped first for the opportunities that are available. The first and foremost need is food which can make them strong and healthy and then they need basic education for them to be able to solve their miseries of living life in a slum.These slums are filled with problems like filth, water scarcity, food unavailability and non-availability of shelter. All these problems need to be addressed first and then they need to be provided jobs later. Jobs can solve their financial issues but the other miseries should be given the first priority and then proper jobs with a good salary will make them come out of their problems forever.

## SECTION – C

## (LITERATURE: TEXT BOOKS AND LONG READING TEXT)

**6. Answer any four of the following in 30-40 words each: [3×4=12]**

**(a) "It is his karam, his destiny." What is Mukesh's family's attitude towards their situation?**

**(b) What were the terms of the indigo contract between the British landlords and the Indian peasants?**

**(c) How will 'keeping quiet' protect our environment?**

**(d) Which objects of nature does Keats mention as sources of joy in his poem, 'A Thing of Beauty'?**

**(e) Why did Tiger King decide to get married?****

**(f) What was Sadao's father's dream for him? How did Sadao realise it?**

**Answer:**

(a) Mukesh's family had resigned themselves to their fate. His father was a poor bangle maker and could not send his sons to school. He could only teach them how to make bangles. His family accepted it as their fate and felt that God-given lineage could not be broken.

(b) Most of the arable land in the Champaran district was divided into large estates that were owned by Englishmen and Indian tenants worked for them. They got the agreements from the sharecroppers to pay them compensation for being released from the 15% arrangement.

(c) By 'Keeping Quiet' the poet means that we should stop all our destructive activities that harm nature. If we count up to twelve, and do not talk, the environment will be saved from us.

(d) The poet says that, a beautiful thing is a source of ever lasting happiness. The few things that add joy to our life are The Sun, The Moon, a bower of trees and a clear stream of water.

(f) Sadao's father dreamt of a bright future for his son. He wanted him to reach the sky as the limit and take Japan towards a better future too. Sadao understood that education was the most important thing to his father and at the age of twenty-two, he went to America to study Surgery and Medicine. At the age of thirty, he retuned having realised his father's dream.

**11. Answer the following question in 120-150 words? [6]**

**How did the order from Berlin change the situation in the school? (The Last Lesson)**

OR

**It is not unusual for a lower middle class girl to dream big. How unrealistic were Sophie's dreams?****

**Answer:**

The order from Berlin was received with an uncanny silence in the school. There were new copies for the students with 'France, Alsace' written on them. They looked like little flags floating everywhere in the classroom. This was in contrast to the usual scene when the school would be buzzing with activity. The sound of opening and closing of desks, lessons repeated in chorus and the teachers' great ruler banging on the table could be heard on the street. But that day, everyone was working quietly and Mr. Hamel, who was dressed formally in the attire that he never wore except on special occasions, was walking up and down with his terrible iron ruler under his arm. The only sound that was heard was, the scratching of pen over paper. The villagers had gathered in the classroom and were sitting quietly on the lock benches.

●●

**Answers are not given due to change in present syllabus

# ENGLISH

## Delhi [Set-I] 2015

**Time allowed : 3 hours** **Maximum marks : 100**

**Note: Except for the following questions, all the remaining questions have been asked in the previous sets.**

**SECTION – B**

**(ADVANCED WRITING SKILLS)**

**4. Every year in the central park of the city a flower show is held in the month of February. Your school has received a circular from the District Collector inviting your students to visit it. Write a notice in about 50 words informing the students about the show and advising them to go and enjoy it. You are Navtej/Navita, Head Boy/ Head Girl, Sunrise Public School, Surat. [4]**

**OR**

**Sarvoday Education Society, a charitable organisation is coming to your school to distribute books among the needy students. As Head Boy/Head Girl, Sunrise Public School, Surat, write a notice in about 50 words asking such students to drop the lists of books they need in the box kept outside the Principal's office. You are Navtej/Navita.**

**Answer :**

**SUNRISE PUBLIC SCHOOL, SURAT**
**NOTICE**
10 February, 20XX
**FLOWER SHOW**
All the students of the school are invited by the District Collector to the Flower Show that is being organised in the Central Park on 20th February 20XX. The event will showcase different varieties of flowers and would be very informative and interesting. The students may take their passes from the undersigned before 18 February, 20XX.
Navita/Navtej
Head Girl/Head Boy

**OR**

**SUNRISE PUBLIC SCHOOL, SURAT**
**NOTICE**
10 February, 20XX
**BOOK DISTRIBUTION**
This is to inform all those students who are in need of books that Sarvodaya Education Society, a charitable organization, is coming to our school for distribution of books. Interested students should drop the lists of books they need, in the drop box kept outside the principal's office by 12 February, 20XX.
Navita/Navtej
Head Girl/Head Boy

**5. Recently you went to your native village to visit your grandparents. You saw that some of the children in the age group 5-14 (the age at which they should have been at school) remained at home, were working in the fields or simply loitering in the streets.**

**Write a letter in 120-150 words to the editor of a national daily analyzing the problem and offering solutions to it. You are Navtej/Navita, M-114 Mount Kailash, Kanpur. [6]**

**OR**

**When cricket teams go abroad the members are allowed to take their wives, even friends along with them. Does this fact distract them or help them to focus on their game in a better way? If it is good, why don't we allow our athletes to enjoy the same privilege?**

**Write a letter to editor of a national daily in 120-150 words giving your views on the issue. You are Navtej/Navita, M-114 Mount Kailash, Kanpur.**

**Answer:**

M-114, Mount Kailash,
Kanpur

3 March, 20XX

The Editor,
The Times of India,
Kanpur

**Subject : Pathetic condition of children**

Dear Sir,

Through the columns of your esteemed newspaper, I would like to draw the attention of the government and NGOs towards the children of Hoshiarpur village who are not receiving formal education at school. They either remain at home, work in fields or loiter in the streets.

Education, which is a necessity, is still a luxury here and rather they are made to work in the fields. Poor people hesitate to send their children to school. On a recent visit to this village, I could not help but notice the sheer amount of children who should have been in the school, but were not. Children in the age group of 5-14 are supposed to attend school to make a bright future. But in the villages, they are either seen loitering around or helping their parents in the fields, which is a very painful and a depressing situation.

It is high time that the government and the NGOs should take up the issue seriously and implement measures to solve it. Besides, literate villagers can also help by starting makeshift schools to educate the children till reformation are made by the government.

Yours sincerely,
Navita/Navtej

OR

M-114, Mount Kailash,
Kanpur

3 March, 20XX

The Editor,
The Times of India,
Kanpur

**Subject : Difference in the status of Cricketers and Athletes.**

Dear Sir,

Through the columns of your esteemed newspaper, I would like to express my views on the difference in the status of Cricketers and Athletes.

Cricket is a very popular game in India and Cricketers are idolised. The public as well as the officials are willing to give special privileges to them. The extent of their love is such that rules are easily bent for them. Cricketers are allowed to take their families with them when they go on tours, irrespective of the fact that this may distract them while playing. But when it comes to other games, Indian Government becomes rather stingy and the players do not get the same treatment. Why do we have two policies? I believe, this is because Cricket and Cricketers are worshipped, while the other games and their players are ignored. Even the finances that are allotted to these games are either too less or are utilised by the officers themselves.

The perspective of India needs a revolution and all the games need equal treatment, after all they all bring glory to the country. Government needs to implement measures to keep all the games at par.

Yours sincerely,

Navtej/Navita

6. **Mobile phone of today is no longer a mere means of communication. Music lovers are so glued to it that they don't pay attention even to the traffic while crossing the roads. This leads to accidents sometimes even fatal ones.**

**Write a speech in 150-200 words to be delivered in the morning assembly advising the students to be careful in the use of this otherwise very useful gadget. Imagine you are this Principal of your school.** [10]

OR

**Power shortage has become a norm even in the metropolitan cities. One way to face this situation is by preventing the wastage of power. Write a speech in 150-200 words on the importance of power in our daily life and how to save power at school and at home. Imagine that you are the Principal of your school.**

**Answer:**

**MOBILE AND ITS ILL-EFFECTS**

"I fear the day when technology surpasses human interaction, we will have a generation of idiots".

My dear students, a very good morning to one and all.

Mobile phone is a wonderful invention. It is multipurpose instrument from which you can note only make calls and text people but it is also installed with music facility, broadcasting facilities such as radio and television, a standard camera, internet browsing facility through which information from the whole world the lies at just one click. But to much attention has this gadget gained and it becomes a menace when people forget about their surroundings being engrossed in their phones. Just the other day, I read in a newspaper that a girl in New York fell off the harbour because she was too involved in her phone to notice where she was going. Even while crossing roads, children don't seem to notice the oncoming cars and music lovers always have earphones plugged in and they don't seem to hear the blaring horns of the traffic. They should be used sparingly and judiciously without risking self or others. If they are lucky enough to survive, they have to live with a disability for the rest of their lives due to the accident.

Children, you should understand that you have a bright life ahead if you adhere to safety. Gadgets are for our help and convenience.

Thank You.

OR

**SAVING POWER**

Good morning to one and all!

A lot of power is being wasted majorly due the carelessness and indifference at home, school business premises etc. Power is the lifeblood of regulating operations in our environment. Power is responsible for our life conveniences, comfort, running equipment, appliances and gadgets which are entirely dependent on power supply.

Though electricity can be generated, it is not easy to do so and this fact needs to be recognised and realised at the earliest. We dread power failure but do nothing to save power. so that we may not face such a situation. It is high time we should make judicious use of power. Saving electricity will save fossil fuel too. We should switch off the lights, fans, geysers, ACs, TV when not in use. Students should be taught about saving electricity at school and public recognition should be given to students who take initiative in this regard. Awareness should be spread in the neighbourhoods and everyone should cooperate.

Remember, power needs to be saved and used efficiently for future use. So, children please pay heed to my advice and save electricity. Set an example for others in doing so and lead the country towards a brighter future.

Thank you and wish you a 'powerful' day.

**7. In the year to come (if you have not already done this year) you are going to celebrate your $18^{th}$ birthday. Write an article in 150-200 words on the joys and responsibilities of being eighteen. You are Navtej/Navita. [10]**

OR

**Write an article in 150-200 words on how we can make India a carefree and enjoyable place for women when they can go wherever they like to without any fear of being stared at, molested or discriminated against. You are Navtej/Navita.**

**Answer :**

RESPONSIBILITIES AND JOYS OF BEING EIGHTEEN

By Navita/Navtej

Eighteen is a magic birthday, a milestone into adulthood of unlimited freedom accompanied by great privileges as well as serious legal implications. At 18, a teen can vote, buy a house or attain a driving licenses. He can also go to jail, get sued, gamble away his/her tuition via online poker, and make terrible stock market investments.

Some teens think it's acceptable to get out of control just because they're of the legal age. But they are wrong. Becoming 18 years old one should become more responsible one and mature instead of careless and stupid.

You can start working, get your own apartment or go off to college and live alone. The age of 18 is a transition from your young adult years to adulthood. You have to be conscientious; you'll still make mistakes as you try to find your way, but don't forget everything you were taught.

The law is harsher on legal teens than minors, so you should keep that in mind. Being legal and grown up are two different things. Have fun while you're young and live life to the fullest, but make wise decisions and choices because it will affect your future.

At the age of 18, you can do whatever you want. However, if you make the wrong move, you should also be adult enough to face the consequences. One thing to remember is that if you still live with your parents, you have to abide by their rules. So, if they say no, you might want to listen to them, so you don't end up on the streets.

As long as you're safe and smart about what you decide to do when you reach the legal age, then go ahead be adventurous! Discover yourself. You only live once.

OR

HOW SAFE ARE WOMEN?

By Navtej/Navita

One woman killed for dowry, a minor raped by an acquaintance, eve-teasing in broad day light, road rage leads to death of a woman driver – these are the headlines of everyday newspaper. On one hand, we talk of women liberation and on the other, we are so cruel to them. Is this humanity? The land where women are considered to be the manifestation of Goddess Durga and Shakti, there they are worshipped in so many forms, should they witness such ghastly crime? —folk is shocking.

What is surprising is that these incidents occur in big cities, where most of the people are educated and broad-minded as compared to rural India where honouring a women is not acceptable. It's high time that awareness is created amongst people; organisation of the Police Force is done on such humanitarian issues. We should stop treating women as an object but rather look at them as home makers, mothers, sisters, colleagues and friends. With increasing pressures on family life, with increasing education and awareness, it is evident that women need to work. We all require money and comfort, then isn't it our duty to make their safety our prime concern? We must consider safety options for them, increase police patrolling in lonely places and sensitive areas like colleges, hostels and universities. Women police and personnel should be mobilised and women should be given training in Self-Defence and Martial Arts. Self-defence techniques should be taught to girls and women in schools and colleges mandatorily. Police should adopt sensitive approach in handling cases of atrocities against women, so that women can come out to lodge complaints more openly. Punishments should be given for heinous crimes like rape and dowry deaths. All these measures, if heeded can go a long way in making life of women safer and better.

## SECTION – C

## (LITERATURE: TEXT BOOKS AND LONG READING TEXTS)

**8. Read the extract given below and answer the questions that follow: [4]**

I saw my mother,
beside me,
doze, open mouthed, her face
ashen like that

of a corpse and realised with
pain
that she was as old as she
looked but soon
put that thought away, .....

**(a) What worried the poet when she looked at her mother?** [1]

**(b) Why was there pain in her realization?** [1]

**(c) Why did she put that thought away?** [1]

**(d) Identify the figure of speech used in these lines.** [1]

**OR**

Far far from gusty waves these children's faces.
Like rootless weeds, the hair torn round their pallor:
The tall girl with her weighed-down head.

**(a) Who are these children?** [1]

**(b) What does the poet mean by 'gusty waves'?** [1]

**(c) What has possibly weighed-down the tall girl's head?** [1]

**(d) Identify the figure of speech used in these lines.** [1]

**Answer :**

**(a)** The poet was worried about her mother's advancing age.

**(b)** There was pain in the poet's realisation because her mother now looked as old as she was, her bodily infirmities that come with old age were visible on her face and she was fast approaching her death.

**(c)** The poet put that thought away because she would not be able to go through with her plan of travelling away from home if she continued to dwell on her mother's old age. (as old as she looked)

**(d)** The figure of speech used is a 'simile.'

**OR**

**(a)** The children referred to in the poem are the slum children who attend an elementary school in that slum.

**(b)** "Gusty waves" refer to the sea and its waves. It is uses to denote that the children were far away from the presence of nature and sinking further and further into a hopeless mire in spite of receiving education.

**(c)** The tall girl's head is possibly weighed-down by the troubles and tribulations of living out her life in abject poverty and thinking of a future within the hopeless confines of a slum.

**(d)** The figure of speech is a 'simile.' (Like rootless weeds)

**9. Answer any four of the following in 30-40 words each :** **[3×4=12]**

**(a) Who occupied the back benches in the class room on the day of the last lesson? Why?**

**(b) Why did Douglas' mother recommend that he should learn swimming at the YMCA swimming pool?**

**(c) What will counting upto twelve and keeping still help us achieve?**

**(d) What does a thing of beauty do for us?**

**(e) Which do you think is a better ending of Roger Skunk's story, Jo's or her father's? Why?**

**(f) What could the Governor have done to securely bring Evans back to the prison from the 'Golden Lion'?**

**Answer:**

**(a)** Old Hauser the former mayor, the former postmaster and several other villagers occupied the back benches in the classroom because they had come to know about the new order that has come from Berlin. They sat down on the back benches as it was their way of thanking their teacher who had devoted his life for the welfare of the society. They also felt sorry as they had ignored the school and the lessons.

**(b)** When Douglas was ten or eleven years old, he decided to learn swimming. He chose YMCA pool because it was safe and his mother had also warned him to learn swimming in. Yakima River was full of danger and she warned Douglas against it by reminding him of every drowning incident in this river.

**(c)** Counting up to twelve and keeping still helps us to introspect. We will be able to find our inner-self, think and redetermine our future for the betterment of ourselves as well as environment in which we live. It is necessary for creating a feeling of mutual understanding among men. This way people will be together in sudden strangeness and it would be an exotic moment.

**(d)** Beautiful things are good for mind as well as soul. They give us aesthetic pleasure and permanent joy. Its beauty increases forever. They divert our mind from all the ugly things of the world. They allow us to find a ray of hope even in despair, sorrow and sufferings.

**(e)** Jo wanted the story to have another end in which the wizard should hit the mommy by his magic wand because for her the smell of roses was awful. Jo thought that Roger was the hero of the story so he must not look ugly or stinky. Jo's perspective of life was different from her father's.

Whereas, her father thought in a different way. He wanted to convey that for a mother, her son can never smell bad. The adults' world is full of hardships and they can had to face the reality of life. His ending may not be very pleasing but is realistic and he wanted to highlight the fact that for a mother, her child is always an object of love. His smell makes no difference.

Thus, the original ending is acceptable.

(f) The Governor was extra careful at every place except in the end. He took all the steps and precautions for not letting Evan escape anyway. He got success also to bring Evan back to the prison from Golden Lion but in the end acted foolishly. He could not judge Evan's cleverness. He should understand that Evans could not go out of the cell in Mcleery's disguise. McLeery was found injured and covered with blood in the cell. No one took the pain to check the identity of the injured. It was Evans himself. Further, when Evans was arrested and handcuffed, he was made to sit in the police van. But the van and driver's identification was not done and Evans was successful escaping.

**10. Answer the following in 120-150 words : [6]**
**Giving a bribe is evil practice. How did Tiger King bribe the British officer to save his kingdom? How do you view this act of his?****

**OR**

**Dr. Sadao was a patriotic Japanese as well as a dedicated surgeon. How could he honour both the values?**

**Answer:**

**OR**

Dr. Sadao Hoki was not only a trained surgeon but also a fervent patriot who dedicated himself to the cause of serving his country in wartime through Scientific Research. However, the dilemma that Sadao faced in lieu of the arrival of the wounded enemy soldier on his doorstep was a clash between his duties as a doctor and that of a citizen of a particular nation. Sadao remarked that if the man had been whole and uninjured, then he would not have faced any difficulty in turning him over to the police. However, the fact that he was wounded, complicated this issue because as a doctor, Sadao had taken the oath to put his professional duties first and serve mankind as a whole, without any discrimination on the basis of nationality. But, he was able to protect his patriotism by informing the General about the matter. In this way, he balanced both his values by tending the soldier and helping him escape at the end, while having informed the General about his presence. He did not let national prejudice override his duties as a doctor.

**11. Answer the following in 120-150 words: [6]**
**Describe the difficulties the bangle makers of Firozabad have to face in their lives.**

**OR**

**The peddler declined the invitation of the ironmaster but accepted the one from Edla. Why?**

**Answer :**

Every family in Firozabad was engaged in the task of making bangles. It was the centre of India's glass-blowing industry, where families had spent generations working around glass furnaces, with high temperature's spend long hours in dingy rooms without air and light, welding glass and making bangles for apparently all the women in the country. The circumstances that kept them in this trade were the vicious circle of middlemen who would interfere if the young men tried to form a cooperative. Moreover, these youngsters would be hauled up by the police, beaten and dragged to jail for doing something illegal, while it was the middlemen who performed such practices. There was no leader among them who could help them see things differently. Their fathers were as tired as they and continued to sink in the mire of an endless spiral of poverty, apathy, greed and injustice. The bangle seller's families were caught in a web of poverty, burdened by the stigma of the caste into which they were born and simultaneously suffered under the oppressive regime of a vicious circle of sahukars, middlemen, policemen, bureaucrats and politicians. Together they imposed a baggage on the children born into such families, from which they could not free themselves.

**OR**

Edla proved to be much more persuasive than her father in dealing with the peddler. Her gentle and kind treatment managed to effect a monumental change in the latter. Although, Edla had misgivings, she convinced her father to let the peddler stay and revelled in the opportunity to actually help a poor, hungry, homeless man, who was always chased away by everyone. She wanted him to enjoy a day of peace and partake in the Christmas festivities. She was kind, sympathetic and friendly with the stranger, taking hold of his hand and leading him to the dinner table, thus, making him a part of her family, at least for a day. It was this act of kindness that helped the peddler to change himself. Besides, before leaving, he left a Christmas present for her and signed it as Captain Von Stahle.

**12. Answer the following in 120-150 words :** [6]**
**Describe the ironical situation in which Silas Marner had to leave Lantern Yard.**

**OR**

**Within a few days of his arrival in Iping, people became suspicious of Griffin. Why?**

**13. Answer the following in 120-150 words :** [6]**
**Describe Dolly Winthrop as the most lovable character in George Eliot's 'Silas Marner'.**

**OR**

**Attempt a character sketch of Marvel.**

●●

** Answer is not given due to the change in syllabus

# ENGLISH

## Outside Delhi [Set-I]

## 2014

**Time allowed : 3 hours** **Maximum marks : 100**

**SECTION – A**

**(READING) 20**

**1. Read the following passage and answer the questions that follow : [12]**

**(1)** Too many parents these days can't say no. As a result, they find themselves raising children who respond greedily to the advertisements aimed right at them. Even getting what they want doesn't satisfy some kids; they only want more. Now, a growing number of psychologists, educators and parents think it's time to stop the madness and start teaching kids about what's really important: values like hard work, contentment, honesty and compassion. The struggle to set limits has never been tougher – and the stakes have never been higher. One recent study of adults who were overindulged as children, paints a discouraging picture of their future : when given too much too soon, they grow up to be adults who have difficulty coping with life's disappointments. They also have a distorted sense of entitlement that gets in the way of success in the work place and in relationships.

**(2)** Psychologists say that parents who overindulge their kids, set them up to be more vulnerable to future anxiety and depression. Today's parents themselves raised on values of thrift and self-sacrifice, grew up in a culture where 'no' was a household word. Today's kids want much more, partly because there is so much more to want. The oldest members of this Generation Excess were born in the late 1980s, just as PCs and video games were making their assault on the family room. They think of MP3 players and flat-screen TVs as essential utilities, and they have developed strategies to get them. One survey of teenagers found that when they crave something new, most expect to ask nine times before their parents give in. By every measure, parents are shelling out record amounts. In the heat of this buying blitz, even parents who desperately need to say no find themselves reaching for their credit cards.

**(3)** Today's parents aren't equipped to deal with the problem. Many of them, raised in the 1960s and '70s, swore they'd act differently from their parents and have closer relationships with their own children. Many even wear the same designer clothes as their kids and listen to the same music. They even work more hours; so at the end of a long week, it's tempting to buy peace with a 'yes' and not mar precious family time with conflict. Anxiety about the future is another factor. How do well intentioned parents say no to all the sports gear and arts and language lessons they believe will help their kids thrive in an increasingly competitive world? Experts agree: too much love won't spoil a child. Too few limits will.

**(4)** What parents need to find, is a balance between the advantages of an affluent society and the critical life lessons that come from waiting, saving and working hard to achieve goals. That search for balance has to start early. Children need limits because they feel better and more secure when they live within a secured structure. Older children learn self-control by watching how others, especially their parents act. Learning how to overcome challenges, is essential to becoming a successful adult. Few parents ask kids to do chores. They think their kids are already overburdened by social and academic pressure. Every individual can be of service to others, and life has meaning beyond one's own immediate happiness. That means parents eager to teach values have to take a long, hard look at their own.

**(a) Answer the following:**

**(i) What values do parents and teachers want children to learn? [2]**

**(ii) What are the results of giving the children too much too soon? [2]**

**(iii) Why do today's children want more? [2]**

**(iv) What is the balance which the parents need to have in today's world? [2]**

**(v) What is the necessity to set limits for children? [2]**

**(b) Pick out words from the passage that mean the same as the following : [3]**

**(i) A feeling of satisfaction (para 1)**

**(ii) Valuable (para 3)**

**(iii) Important (para 4)**

**Answer: (a)**

**(i)** Teachers and parents want children to learn values like hard work, contentment, honesty and compassion.

**(ii)** Giving children too much too soon brings its own ill-effects as they grow up to be adults who experience difficulty in coping up with life's disappointments. They also have a distorted sense of entitlement that hinders their success at work place and in relationships.

**(iii)** Children of present time want more because of the increasing desire. and because there are so many things to want.

**(iv)** In today's world, there is an urgent need for parents to strike a balance between the advantages of an affluent society and the critical life lessons that come from waiting, saving and working hard to achieve goals.

**(v)** There is a necessity to set limits for children because they feel better and more assured when they live within a secured structure.

**(b)** **(i)** Contentment

**(ii)** Precious

**(iii)** Essential.

**2. Read the passage carefully.** **[8]**

**(1)** I remember my childhood as being generally happy and can recall experiencing some of the most carefree times of my life. But I can also remember, even more vividly, moments of being deeply frightened. As a child, I was truly terrified of the dark and getting lost. These fears were very real and caused me some extremely uncomfortable moments.

**(2)** May be it was the strange way in which things looked and sounded familiar in my room at night that scared me so much. There was never total darkness, but a street light or passing car lights made clothes hung over a chair take on the shape of an unknown beast. Out of the corner of my eye, I saw curtains move when there was no breeze. A tiny creak in the floor would sound a hundred times louder than in the daylight and my imagination would take over, creating burglars and monsters. Darkness always made me feel helpless. My heart would pound and I would lie very still so that 'the enemy' wouldn't discover me.

**(3)** Another childhood fear of mine was that I would get lost, especially on the way home from school. Every morning, I got on the school bus right near my home – that was no problem. After school, though, when all the buses were lined up along the curve, I was terrified that I would get on the wrong one and be taken to some unfamiliar neighbourhood. I would scan the bus for the faces of my friends, make sure that the bus driver was the same one that had been there in the morning, and even then ask the others over and over again to be sure I was in the right bus. On school or family trips to an amusement park or a museum, I wouldn't let the leaders out of my sight. And of course, I was never very adventurous when it came to taking walks or hikes because I would go only where I was sure I would never get lost.

**(4)** Perhaps, one of the worst fears I had as a child was that of not being liked or accepted by others. First of all, I was quite shy. Secondly, I worried constantly about my looks, thinking people wouldn't like me because I was too fat or wore braces. I tried to wear 'the right clothes' and had intense arguments with my mother over the importance of wearing flats instead of saddled shoes to school. Being popular was very important to me then, and the fear of not being liked was a powerful one.

**(5)** One of the processes of evolving from a child to an adult is being able to recognise and overcome our fears. I have learnt that darkness does not have to take on a life of its own, that others can help me when I am lost and that friendliness and sincerity will encourage people to like me. Understanding the things that scared us, as children helps to cope with our lives as adults.

**(a) On the basis of your reading of the above passage, make notes using headings and sub-headings. Use recognisable abbreviations wherever necessary.** **[5]**

**(b) Make a summary of the passage in not more than 80 words using the notes made and also suggest a suitable title.** **[3]**

**Answer: (a)**

**Childhood Nostalgia**

**1. Recollection of childhood memories**

(i) Happy and carefree

(ii) Terrified of darkness and getting lost

**2. Childhood fears or hallucinating vibes**

(i) Feeling helpless in the dark

(ii) Strange shadows-an unknown beast

(iii) Moving curtains

(iii) Creaking sounds

(iv) Imagining burglars and monsters

(v) Lying still, with pounding heart

(vi) Fear of getting lost (on the way back home)

(vii) Scanning school bus searching familiar faces, same driver
(viii) Re-confirming the bus
(ix) Not letting leaders out of sight
(x) Avoiding adventurous act
(xi) Going only when sure of not being lost
(xii) Fear of not being liked
(xiii) Quite shy
(xiv) Worried about fat-looks, having braces, right clothes
(xv) Wearing the right clothes
(xvi) Flats vs. saddled shoes for school
(xvii) Imp. of popularity

**3. Coping with childhood fears as an adult**
(i) Undg. evolution process from a child to an adult
(ii) Recog. and overcoming fears
(iii) Accepting help from others
(iv) Role of friendliness and sincerity
(v) Undg. things that were scary

**abbreviations used**

**1.** &-and **2.** Act.-activities
**3.** Vs.-versus **4.** Imp.-importance
**5.** Undg. – understanding **6.** Recog. - recognizing

**(b) SUMMARY**

The narrator had happy and carefree moments in his childhood. However, darkness scared him with its shadows, unexpected movement of curtains and blatant creaking sounds, made him imagine about monsters and burglars. He felt helpless and was scared of getting lost. Before boarding his school bus, he looked in for familiar faces. He was shy and afraid of not being liked by others which had been his biggest phobia. As he evolved from a child to an adult, he realised that understanding things that scare us as a child helps in coping with the challenges of life.

## SECTION – B

## (WRITING SKILLS) 35

**3. You were very upset about the reports on communal riots in various parts of the country. As a concerned social worker, design a poster in not more than 50 words, highlighting the importance of communal harmony. You are Vinay/Vineeta.** [5]

OR

**The Literary Club of your school is putting up the play 'Waiting for Godot'. As secretary of the club, draft an invitation inviting the famous writer Sudesh Gupta to be the guest of honour at the function. Write the invitation in not more than 50 words. You are Govind/Gauri.**

**Answer :**

COMMUNAL HARMONY
"The Need of The Hour"
When Diwali ends with 'Ali' and 'Ramadan' starts with 'Ram' then why fight in the name of religion.

Live and
Let
Live
Let's stop
Communal Riots!

Issued in Public Interest by
Vinay/Vineeta
Social Worker

OR

The Literary Club,
XYZ School

12th Feb, 20XX

The Writer
Sudesh Gupta,

**Subject : Invitation to be the guest of honour**

Dear Sir,

The literary Club of XYZ School requests the pleasure of your esteemed presence on the occasion of Annual Theatre Festival on Friday, 19th Feb, 20XX at 9.00 a.m. at the School Auditorium.

The students of the School are putting up the play 'Waiting for Godot'. Your presence will encourage the students thereby making the occasion more meaningful and attractive. Hope our invitation will be accepted by you.

Thanks,
Gauri/Govind
Secretary
The literary Club
XYZ School

**4. Last week the newly built auditorium of your school was inaugurated. As Deepak/Deepti Saha, the head boy/head girl of the school, write a factual description of the auditorium in 125-150 words.** [10]

OR

**You had attended a workshop on personality development for students. Many eminent personalities had been present. Write a report in 125-150 words on how the workshop proved to be beneficial. You are Rajesh/Rajshree.**

**Answer:**

**NEWLY BUILT AUDITORIUM**

By Deepak Saha / Head boy/Deepti Saha/ Head girl

The newly built auditorium of our school was inaugurated last week by Mr. Anand Seth, the MP. The auditorium is quite spacious and well-ventilated. It has five gates so as to avoid any chaos or mishap while entering or leaving the auditorium. Just after entering, one can see the exhibited portraits of educationists and artists on the walls. The seating space has been designed to accommodate about 5000 students. The roof is well-designed so as to give it a lively look. The seats are comfortable and arranged properly for a better visibility of the performances on the stage. Special care has been taken to build the stage of the auditorium. The dazzling lighting system, which constitutes dramatic lights, spot lights, etc., and the digital sound system have been installed by a leading company. The highly equipped auditorium has become an instant hit among the students and was even highly appreciated by Mr. Anand Seth, the guest of honour.

**OR**

**WORKSHOP ON PERSONALITY DEVELOPMENT**

By Rajesh / Rajshree

A workshop on personality development was organised by our school on last Thursday. Students of class XI and XII participated in the workshop. The purpose of the workshop was to prepare the students for their future in order to face bigger challenges in life. Students were addressed on various topics like self-analysis, body language, life-skills and etiquettes. Several activities and group discussions were conducted in order to help the students to enhance their communication and public speaking skills. The workshop was also graced by the presence of eminent personalities like Sushmita Sen and Kiran Rao Kher. They interacted with the students and spoke about the importance of interpersonal skills. Students left the workshop beaming with confidence and optimism.

5. **You are Anand/Arti of 14, Model Town, Delhi. You have seen an advertisement in The Hindu for the post of chief chef in a 5-Star hotel. Apply for the job with complete biodata. Write in 125-150 words.** [10]

**OR**

**You are Prem/Parul of 16, TT Nagar, Bhopal. You would like to apply for the post of Marketing Manager in a reputed firm in Mumbai. Write a letter to the Public Relations Officer, Chantac Enterprises, Mumbai, applying for the job. Write the letter in 125-150 words giving your biodata.**

**Answer:**

14, Model Town

Delhi

June 15th, 20XX,

The Chairman,

The Hindu,

Gurgaon.

**Subject : Job application for the position of Chief Chef**

Dear Sir/Madam,

With reference to your advertisement in the newspaper, 'The Hindu' dated June 8th, 201XX, for the post of Chief Chef, I wish to offer my services for the same.

I am Arti/Anand, a post-graduate in Hotel Management, from B.M College, Delhi.

Please find my attached bio-data for the purpose. I would like to connote that I am a hard working girl/boy with a keen sense of interest in different cuisines flavours, herbs and essential seasoning. I am adept at the culinary skills and have mastery in cooking. I have also undergone an industrial training of ten months at 'The Grand,' Gurgaon. If selected, I assure you that I shall work with utmost diligence and sincerity.

I thank you in anticipation of an early response.

Yours Faithfully,

Arti Anand Singh

**BIO-DATA**

| | |
|---|---|
| Name: | Arti/Anand Singh |
| Father's Name: | Mr. Dilip Singh |
| Address: | 14, Model Town, Delhi |
| Phone: | 91xxxxxxxxxx |
| Date of Birth: | 5th September, 19XX |
| Marital Status: | Unmarried |
| Educational | M.Phil. in Chemistry, HM College, Delhi (20XX-XX) |
| Qualification: | Diploma in Hotel Administration,HM College, Delhi (20XX-XX) |
| Experience: | Trainee at The Grand, Gurgaon (July 20XX- April, 20XX) |
| Skills: | Excellent communication skills, knowledge of different cuisine, diligent, ability to manipulate things in need. |
| Languages known: | English, Hindi and Gujarati |

References: (1) Mr. Kapil Sharma
Head of F Food Operations
The Grand, Gurgaon
(2) Malvika Ray
Head of Department
HM College, Delhi

**OR**

16, TT Nagar,
Bhopal
February 20th, 20XX

The Public Relations Officer,
Chantac Enterprises,
Mumbai.

**Subject : Job application for the post of Marketing Manager.**

Dear Sir/Madam,

With reference to your advertisement in 'The Hindu' dated January 10th , 20XX, I, Parul/Prem take the opportunity to offer my services for the post of Marketing Manager in your prestigious firm.

I am a diligent and honest person who is passionate about marketing. I completed my MBA in 20XX, and since then, I have been working with RP Communications, as a Manager.

I am here with, enclosing my bio-data along with the photocopies of certificates and testimonials for your further perusal and query. If selected, I assure you that I shall work with utmost devotion and sincerity to your full satisfaction.

Hoping for a favourable response.

Thank You

Yours Sincerely,

Parul/Prem

Encl : Biodata

**BIO-DATA**

| | |
|---|---|
| Name: | Parul/Prem Kumar |
| Father's Name: | Mr. Rakesh Kumar |
| Address: | 16, TT Nagar, Bhopal |
| Phone: | 91xxxxxxxxxx |
| Date of Birth: | 27th September, 19XX |
| Marital Status: | Unmarried |
| Educational Qualification: | MBA in Finance, Management DT Institute, Gurgaon (20XX-XX)<br>B.Com. in Accountancy, BU (20XX-XX) |
| Experience : | Manager at RP Communications (Apr. 20XX-February 20XX) |
| Skills: | Excellent communication skills, punctual, sincere, lively, ability to convince and influence people . |
| Languages known: | English, Hindi and Spanish |
| References: | (1) Mr. D.K. Paul<br>Chairman,<br>RP Communications,<br>Bhopal<br>(2) Mr. Subrata Das<br>Head of Department,<br>Commerce, BU |

**6. Last week, as you were coming back from school you happened to see a huge plastic bag full of leftovers of food being flung into the middle of the road from a speeding car. You wondered how people can be so devoid of civic sense. Write an article in 125-150 words on why we lack civic sense and how civic sense can be inculcated in children at a very young age. You are Shiva/Shamini. [10]**

**OR**

**You saw a stray dog beaten to death by a group of boys. Their act infuriated you and you scolded them for their cruel act. You decided to write an article on cruelty to animals. Write the article in 125-150 words. You are Nikhil/Naina.**

**Answer :**

LACK OF CIVIC SENSE

By Shiva/Shamini

Negligent attitude of the people seem to be the order of the day in today's harsh world. Modern society has given birth to indifferent and offensive individuals. People no longer think of the after effects of their negligent activities. They are driven more by their selfish needs.

People seldom follow traffic rules in India and are seen littering the streets with piles of trash. They show no respect for public property. Vandalism, road rage and intolerance are rampant on the streets. Perseverance and patience are rare virtues to be seen in people behind the wheels.

Inculcating a strong civic sense is the need of the hour. Parents must instill civic sense in children at an early age. They must be taught the values of cleanliness, discipline, patience and tolerance. Parents must encourage their children to keep their surroundings clean as well as should practice what they preach to their children. Children should be taught to respect and abide by the rules as they are the future of the country. Our country can progress only if we teach right values to our children. As a civilised citizen, it becomes the duty of a parent to fathom this dire need and start off this task at the earliest, as family is the first institution for a child to learn the right thing.

OR

CRUELTY TO ANIMALS

By Naina/Nikhil

Animals should be treated with love and compassion. However, we often fail to perform this duty.

Instances of animal cruelty are on rise in our society. Poachers mercilessly kill animals for their skin, fur and teeth. This has led to the extinction of several species of animals which has affected the ecological balance of our planet. Several companies use animals for testing medicines and cosmetics. Similarly, animals are often ill-treated and also sometimes forced into performing circus tricks and other activities for our entertainment. Even 'the tiger', one of the endangered species, is on the verge of being extinct.

We must understand that co-existence and harmony between humans and animals is essential for our planet's survival. We must be more responsible and protective towards animals. Ill-treatment with animals is the act of barbarism which is beyond the limit of one's conscientiousness. We must learn to love and treat them as living creatures. Animals lack in power of expression the but they possess the feelings of pain and agony. It is time that we reform our ways towards the animals. Immediate steps should be taken by the government to ensure proper care, protection and shelter for these animals.

## SECTION-C

## (LITERATURE : TEXT BOOKS) 45

**7. Read the following extract and answer the questions that follow : [4]**

Spite of despondence, of the inhuman dearth
Of noble natures, of the gloomy days,
Of all the unhealthy and o'er-darkened ways
Made for our searching : yes in spite of all,
Some shape of beauty moves away the pall
From our dark spirits.

**(a) Name the poem and the poet. [1]**

**(b) Why are we 'despondent'? [2]**

**(c) What removes 'the pall from our dark spirits'? [1]**

OR

............ And yet, for these
Children, these windows, not this map, their world.
Where all their future's painted with a fog,
A narrow street sealed in with a lead sky
For far from rivers, capes, and stars of words.

**(a) Who are the 'children' referred to here?**

**(b) Which is their world?**

**(c) How is their life different from that of other children?**

**Answer:**

**(a)** The lines have been taken from the poem 'A Thing of Beauty' by John Keats.

**(b)** The world we live in is full of depression illness and despair as there is demisal of noble souls. These evil and dark emotions leave us despondent and despicable.

**(c)** The poet says that even though the world is not a place worth living in because of the umpteen reasons which gloom and depress, the gusto to live a happy and content life can be derived from the sight of the beautiful bounties of nature around us. It is the thing of beauty which charms us and makes this world feel like heaven. These beautiful things remove the pall from our dark spirits.

OR

**(a)** The children who study in an elementary school and live in the slum are referred to here.

**(b)** Their world comprises of what they see from the window of their classroom. The view is full of despair and their future seems blurred and vague. They are confined to the narrow streets of the slum, far away from the open sky and rivers.

**(c)** The children in the slums spend their whole life confined in 'cramped holes' like rodents. They lack the basic amenities of life like proper food, clothing, shelter and health benefits. In fact, they are also deprived of the freedom to leave the shackles of this life and enjoy the 'green fields' as other children do. As a result, their life is left on the bereft part of unpleasant chores, which is absolutely unlike the other children.

**8. Answer any two of the following in about 30-40 words each : [2 × 2 = 4]**

**(a) What was the poet's childhood fear?**

**(b) What is the sadness the poet refers to in the poem 'Keeping Quiet'?**

**(c) How are Aunt Jennifer's tigers different from her?**

**Answer :**

**(a)** As a child, the poet always feared losing her mother.

**(b)** The poet refers to the 'sadness' of failing to understand oneself in the monotonous every day existence. He also finds it sad that the humanity is moving towards its own ruins, owing to its unprecedented actions. He regrets the rush of maneuvering others that has made one forget the values of humanity.

**(c)** Aunt Jennifer's tigers are proud, free, dauntless and sure of themselves. Unlike them, Aunt Jennifer is terrified and oppressed by her chauvinist husband. She lived her life under constant pressure of duties and responsibilities of a married lady. Living a life on her own terms is a far-fetched dream for her.

**9. Answer any six of the following in about 30-40 words each : [6 × 2 = 12]**

**(a) Why did Sophie long for her brother's affection?****

**(b) Why did Edla plead with her father not to send the vagabond away?**

**(c) What job did Saheb take up? Was he happy?**

**(d) Why were the elders of the village sitting in the classroom?**

**(e) Why did the Maharaja order the Diwan to double the land tax?****

**(f) Why did Roger Skunk go to see the old owl?**

**(g) Why was Zitkala-Sa in tears on the first day in the land of apples?****

**Answer :**

(b) Edla was a kind-hearted woman. She was disheartened by the plight of the poor peddler. She requested her father to allow him to spend one day with them in peace as a respite from the struggle that he had to endure round the year.

(c) Saheb took up a job at a tea-stall. He was not happy because he lost his freedom and his 'carefree look' which he had, when he scrounged the garbage for eight-hundred rupees.

(d) The elders of the village were present in the classroom to attend the last French lesson by Mr. Hamel. They not only paid their respect and tribute to Mr. Hamel but also showed their interest towards their own language that was perhaps being taught for the last time in the village.

(f) Roger Skunk, alike his name was a nasty Skunk who had a foul body odour because of which, other small animals avoided playing with him. He too wanted to have friends to play with. Hence, to get rid of the foul smell, he went to the wise owl and requested him to make him smell like roses.

**10. Answer any one of the following in 100-125 words: [5]**

**Why is the Champaran episode considered to be the beginning of the Indian struggle for Independence?**

OR

**What was the General's plan to get rid of the American prisoner? Was it executed? What traits of the General's character are highlighted in the lesson 'The Enemy'?**

**Answer:**

The episode of Champaran was considered to be the beginning of the Indian struggle for Independence because, for the first time, the farmers and peasants rose against the Britishers. The peasants were oppressed by their British landlords, but they lacked the courage to protest. Under the leadership of Gandhiji, they became aware of their rights. A small farmer, Rajkumar Shukla, from a small district, Champaran, helped to bring about a very prominent change. He championed the cause of Champaran movement. Likewise, many other peasants from the villages fought courageously and contributed in their own way to the movement. Their combined effort eventually helped them to win the battle of Champaran and to finally free themselves from 'Share Cropping' arrangement. The success also proved, for the first time, the effectiveness of Gandhiji's method of non-violence and non-cooperation. Gandhiji exulted over them and made the villagers feel self-dependant and undaunted.

OR

The General planned to kill the American prisoner by sending his personal assassins and then removing his body from Sadao's house. The plan was not executed because the General forgot his promise to send the assassins.

The old General had a long expressionless face, which seemed to reflect the cold and calculated ruthlessness of the man as a General. He had an unusual sense of humour and seemed to generate fear in the hearts of others quite effortlessly. In his interaction with Sadao, he came across as a man who puts his self-interest above everything else. As Sadao was the only person whom he trusted with his operation, he lets Sadao off the hook. His cold and calculating streak was manifested in the meticulousness with which he latched out the plan of executing the American prisoner of war whom Sadao had given refuge.

**11. Both Derry and Lamb are physically impaired and lonely. It is the responsibility of society to understand and support people with deformities so that they do not suffer from a sense of alienation. As a responsible citizen, write in about 100 words what you would do to bring about a change in the lives of such people. [5]**

**Answer :**

In the story, both Derry and Lamb are physically impaired and lonely. Such people can only live their lives with respect and honour, if they are not ridiculed and punished with heartless pity. These people expect empathy, rather than sympathy, from others. If everyone looks down at them with a pessimistic approach, they may never be able to come out of their sorrow; consequently, they may recline to their own secluded worlds. As responsible citizens, we should understand the tremendous mental and emotional pressure these people go through. Instead of reminding them of their disabilities, we should give them the chance to live a normal life.

** Answer is not given due to the change in syllabus

●●

# Outside Delhi [Set-II]

**Note: Except for the following questions, all the remaining questions have been asked in previous sets.**

**SECTION – B**

**(WRITING SKILLS)**

**3. You are a member of the S.P.C.A. Draft a poster in not more than 50 words, to create awareness on the need to prevent cruelty to animals. You are Suhas/Suhasini. [5]**

**OR**

**As the principal of a reputed college, you have been invited to inaugurate a Book Exhibition in your neighborhood. Draft a reply to the invitation in not more than 50 words, expressing your inability to attend the function. You are Tarun/Tanvi.**

**Answer:**

**OR**

PQ Colony,

New Delhi.

1st March, 20XX

**Sub : Reply to the invitation**

Dear Mr. XYZ,

Thank you for your thoughtful invitation for the inauguration of the Book Exhibition. However, I regret to inform you that I will not be able to attend it due to a prior engagement. I extend my warm wishes to you and all those involved with the event.

Best regards,

Yours sincerely,

Tarun/Tanvi

Principal, PQR College

**6. A newspaper report on a recent terrorist attack in your locality shattered you. You decided to write an article in a local daily on terrorist attacks and the need to maintain peace and harmony in society. Write the article in 125-150 words. You are Madhav/Meena.**

**OR**

**India is a tourists' paradise. But we have not really exploited our tourism possibilities. Write an article in 125-150 words on 'Tourism in India'. You are Ravi/Rani. [10]**

**Answer:**

**CURBING TERRORISM**

By Meena/Madhav

Terrorism is one of the major threats that is faced by our country. We live under constant fear of violence and death. Common man no longer feels safe. At every nook and corner, lurks the threat of an imminent terrorist attack. It is a giant threat which the world is confronted to. Activities of terrorists have created a horrendous grief in the hearts of the common masses.

Some of the factors that have given rise to terrorism are lack of education, unemployment, poor economy and religious differences. But, it is high time that we join hands to fight and combat terrorism. India has always boasted of its unity in diversity. Keeping this in view, we need to spread the message of peace and harmony across the nation.

Strengthening the intelligence network, proper policy formation and effective implementation of Prevention of Terrorism Act (POTA) are some of the steps that can be taken up to reduce terrorist attacks. It is also necessary to educate the masses and make them aware of the safety measures. It is the need of the hour to keep a constant vigil on the safety of our very own nation.

**OR**

**TOURISM IN INDIA**

By Ravi/Rani

India is known for its rich flora and fauna, beautiful landscapes, glorious past and varied cultural trends. All these make India a major attraction for tourists from all over the world. Tourists from far-away places visit our motherland and are influenced by its scintillating beauty. However, tourism in India has been constantly suffering setbacks due to the various factors.

Pot-holed and filthy roads, rickety buses, over-charged taxis and delayed trains make travelling a nightmare in India. Moreover, the absence of decent and hygienic accommodations at a reasonable cost adds to the distress of tourists visiting India. Besides, the growing menace of terrorism is a huge deterrent and hindrance for tourists.

The Government of India needs to put concerted efforts to revamp and rejuvenate tourism in the country. Utmost emphasis should be placed on infrastructure development and increasing the security. Such effects will certainly make India the leader in the world of tourism.

●●

# Outside Delhi [Set-III]

**Note: Except for the following questions, all the remaining questions have been asked in previous sets.**

## SECTION – B
## (WRITING SKILLS)

3. **You are a fitness trainer in a health club. Design a poster in not more than 50 words, to emphasise the importance of exercise in maintaining mental and physical fitness. You are Prem/Priya.** [5]

OR

**You have received an invitation to be the judge for a literary competition in St. Ann's School. Send a reply in not more than 50 words, confirming your acceptance. You are Mohan/Mohini.**

**Answer :**

OR

**Answer:**

22/4, PQ Puri,
New Delhi

23rd Feburary, 20XX

St. Ann's School
New Delhi

**Subject : Acceptance of the invitation**

Dear Sir,

I acknowledge, with gratitude, the invitation extended to me to judge the literary competition which is to be held in St. Ann's School. I accept the invitation whole-heartedly and confirm my presence on that day and time. I will surely make it without fail.

Regards,

Mohini/Mohan

4. **Your brother is going to Chennai to write an entrance exam. You have requested your friend to meet him at the railway station and arrange accommodation for him. As your brother has not seen your friend, give a detailed factual description of your friend in 125-150 words, so that your brother can recognise him easily. You are Parvesh/Parvati.** [10]

OR

**A few teachers and students from the Edwin University had visited your school on an exchange programme. You welcomed them grandly. Write a report in 125-150 words on the celebration. You are Brijesh/Benu.**

**Answer:**

Dear Karan,

As you are leaving for Chennai to appear for your entrance examination, I have asked my friend Sunny to meet you at the railway station and he will be arranging accomodation for you. Since you have not seen Sunny, his factual description is as follows.

Sunny is about 22 years old and he is 6 feet tall. He has a wheatish complexion and has long hair, which he usually wears in a pony tail. He told me that he will be wearing a yellow T-shirt and a pair of navy blue jeans while picking you up at the station. He has big eyes and there is a scar above his left cheek. He keeps a french beard and there is a small wound on the left side of his forehead. He wears stone studded rings on three of his fingers, one of which has a bright red coloured ruby. Hope you will find him easily.

Yours lovingly,

Parvesh/Parvati

**OR**

**EXCHANGE PROGRAMME**

By Brijesh/Benu

A group of teachers of Edwin University along with their students visited our school on a two-week long exchange programme. They were given a grand welcome by the students of our school. Many of the alumini were also invited to attend the programme. The grand event began with the inaugural speech by our Principal followed by the coordinator of the exchange programme. Their elaborate speeches focused on explaining the significance and characteristics of this exchange programme. This was followed by a cultural programme having one of the main attractions as the Rajasthani folk dance which was presented by the students of class sixth. Interestingly, the cultural programme started on a hilarious note and ended with a patriotic song. The music compelled the students of Edwin University to dance along with the cheerful crowd. The celebration ended with the Indian national anthem, sung by all the participants and the audience. It was a wonderful experience to be cherished.

**5. You are Arman/Arpita of 14, MG Road, Pune. You had bought the text book. 'Vistas' for class XII from a neighbourhood bookstore. After skimming through the book, you realised that few of the pages were missing and the print overlapped on a few pages. Write a complaint letter in 125-150 words to the Manager, Dawn Books, Lawrence Road, Chennai requesting him for a replacement or refund. [10]**

**OR**

**You had bought an expensive wrist watch last week. Now you realise that it is not functioning properly. Write a letter in 125-150 words to the Manager, Edison Watch Works, New Delhi, complaining about it and requesting him to repair or replace it immediately. You are Gobind/Gopa of 68, Park Street, Kolkata.**

**Answer :**

14, MG Road,
Pune.

2nd February, 20XX

The Manager ,

Dawn Books,
Lawrence Road ,
Chennai.

**Subject : Replacement of the book 'Vistas'/ Refund for the purchase.**

Dear Sir,

I had bought the text book 'Vistas' from your bookstore against (bill No. 002223, dated 12/4/20XX). After going through the book, I found that few pages were missing. Moreover, on some pages, the print is overlapped which is making it difficult to read.

I have bought many books from your book store and have never encountered such problem before. My exams are approaching, therefore, I request you to replace the book as soon as possible or refund the money at the earliest. I have attached a copy of the bill for your reference.

Yours truly,

Arman/Arpita

OR

68, Park Street,

Kolkata.

16th June, 20XX,

The Manager,

Edison Watch Works,

New Delhi.

**Sub : Complaint regarding defective wrist watch.**

Dear Sir,

With reference to the bill no. 008453, dated 6-6-20XX, I apprise you that I bought a watch from your retail showroom. However, I regret to inform you that the watch has not been functioning properly from the day of purchase. The watch does not keep correct time and stops every few hours. I have tried resetting the time, but that was of no use. I work as a security guard and having a watch is important for my work, as I have to record the time when people are visiting or leaving my work place. The watch has turned out to be a big trouble rather than being useful to me. I request you to either repair or replace the watch at the earliest. I have enclosed a copy of the bill for your reference.

Regards,

Gobind/Gopa

**SECTION – C**

**(LITERATURE)**

**10. Answer any one of the following in 100-125 words: [5]**

**Gandhiji was not a loyalty to abstractions; it was a loyalty to living, human beings. Why did**

**Gandhiji continue his stay in Champaran even after Indigo share cropping disappeared?**

**OR**

**Explain the reactions of the servants in Dr. Sadao's house when he decided to give shelter to an enemy in the house.**

**Answer:** Gandhiji was not satisfied with mere political and economic solution so he wanted to bring about a change in the social and cultural conditions of Champaran. He noticed the unhealthy living conditions and poor sanitation in the village. He also realised the need for literacy. He decided to stay in Champaran even after the disappearance of Indigo Share Cropping. A small farmer, Rajkumar Shukla, from a small district of Champaran, helped in bring about a very prominent change. Gandhiji opened up schools in six different villages; and several of his disciples and family members volunteered as teachers. Being a staunch believer of passive resistance, Gandhiji championed the cause of the Champaran movement. His wife, Kasturba, worked on the personal cleanliness and community sanitation of the place. Gandhiji also hired a doctor for the improvement of the health conditions.

Thus, we can say that Gandhiji was not a loyalty to abstractions. It was loyalty to living, human beings. This also made him keen to eradicate social and cultural backwardness of Champaran. Gandhiji exulted over them and made the villagers feel self-dependent and undaunted.

**OR**

When Dr. Sadao decided to give shelter to the enemy, the servants at his house overtly expressed their disapproval. However, they did not disclose the secret to the world. For them, the white man was an enemy who did not deserve their help. Yumi, the nanny of the children, refused to wash the American who was bleeding profusely. The servants expressed this displeasure and they voiced their opinion loudly for Sadao and Hana to hear. The servants constantly pressurised the Hoki family to let out the secret and get the American caught by the police. Finally, their displeasure at their master's decision made them decide to leave his household.

●●

# ENGLISH

## Delhi [Set-I] 2014

Time allowed : 3 hours

Maximum marks : 100

**Note: Except for the following questions, all the remaining questions have been asked in previous sets.**

### SECTION – B

### (WRITING SKILLS)

3. **An Inter-school Kabadi Competition is organised by your school. Write a notice, in not more than 50 words, requesting the students to be present at the venue to encourage the players. Invent all the necessary details. You are Arjun, the sports captain of your school.** [5]

**OR**

**You possess an acre of land in the heart of the city. You want to dispose of the property since you have decided to buy a flat. Write an advertisement to be published in a national daily, giving all the necessary details. You are Krishan of Moti Nagar, Delhi.**

**Answer :**

ABC SCHOOL, DELHI

**NOTICE**

13th February, 20XX

**INTER-SCHOOL, KABADDI COMPETITION**

Inter-school Kabaddi Competition will be held at our playground on Friday, February 20, at 9 a.m. Students from different schools will be participating in the competition. All the students are requested to be present at the venue to motivate the participants. Refreshment will be provided by the school. Contact the undersigned for any query.

Arjun

(Sports Captain)

ABC School

**OR**

**FOR SALE**

Sale of 1-acre plot located in the heart of the city besides 100 ft. road with lush green surroundings suitable for building residential flats. Location adjacent to posh Friends colony and free from encumbrance. Interested parties can meet in person or can contact.

Krishan

Moti Nagar, Delhi 9342XXXXXX

4. **Incessant rain has caused irrecoverable damage in your area. As an active participant in the flood relief programme, write a report in 125-150 words on the different flood relief measures carried out. You are Krishan/Krishna.** [10]

**OR**

**You have visited a book exhibition in your neighbourhood. Write a report in about 125-150 words on the exhibition. You are Rohan/Rohini.**

**Answer:**

**FLOOD RELIEF PROGRAMME**

By Krishna/Krishan Staff reporter

Xyz, 13th June, 20XX, The torrential rain in the past few days flooded our area causing irrecoverable damage to the properties. A relief camp was set up to aid the victims of the flood-affected areas. Everyone made generous contributions in the form of money, clothes, food and other basic requirements. Relief-treaties also emerged as helpful groups. The Chief Minister too announced a definite amount of relief for the victims affected. The devastating effect of the over-powering rain, played havoc with the city. The volunteers also helped the victims to be shifted to the relief camps, set up on the suburbs of the city. A huge amount of donations has come from the people of the city along with various NGOs

**OR**

**BOOK EXHIBITION**

By Rohan/Rohini

13th June, 20XX, The Surya Nagar residents organised a book exhibition in the community centre of the locality. The exhibition started on June 10th and lasted for five days. The event attracted book lovers from all over Delhi. Renowned publishing houses participated in the fair. Books by well-known writers were displayed. The exhibition also offered heavy discounts on the books written by Indian authors. Apart from Hindi novels there were multifarious writers and bards dictions available like Shakespeare, Milton and John Keats. People were excited to get access to some of the rare books that were on display. The event was marked by the presence of the renowned writer and novelist Chetan Bhagat.

5. **You are the librarian of Amla Public School. You had placed an order for text books with Dhanpati & Sons. Since the books did not arrive on time, you have decided to cancel the order.**

**Write a letter to the Manager, Dhanpati & Sons, Chennai, cancelling the order. (125-150 words) [10]**

**OR**

**You are interested in doing a short-term course in computer graphics during your holidays. Write a letter to the Director, Easy Computers, enquiring about their, short-term courses and asking for all the necessary details. You are Naresh/Nandini.**

**Answer:**

Amla Public School,
New Delhi-110018

13 June, 20XX

The Manager,
Dhanpati & Sons,
Chennai

**Subject : Regarding cancellation of order number 43/20XX**

Sir,

With reference to the order number 43/20XX placed with your publishing house on May 31st, 20XX, I would like to inform you that we have not received the delivery of 12 sets of Encyclopedia from you as promised by your company on June 10, 20XX. A cheque of five thousand was duly sent to you enclosed with the order. Since your company has failed to keep its commitment, therefore, we are cancelling the order. Kindly reimburse the money to the school account at the earliest.

Thanking you

Yours sincerely,

Librarian

Amla Public School

**OR**

23, Defence Colony,

New Delhi-110022

3rd March, 20XX

The Director,
Easy Computers,
New Delhi-110032

**Subject: Request for details of the short-term courses provided**

Dear Sir,

I came across an advertisement about your institute in today's newspaper. I would like to know the details of the short-term courses offered by your institute.

I am a student and I wish to pursue a course in computer programming/computer graphics during my holidays next month. Kindly mail me the details of all short-term courses, along with their duration, fee structure and the relevance of the course material. I would be thankful if you can dispatch the prospectus of the course too.

Looking forward to your early response.

Your faithfully,

Nandini/Naresh

6. **You are Rajendra Kumar, a social worker, You read an article in *The Hindu* on 'Health care for Indian Workers'. Write a speech in 125-150 words on the importance of health care to be delivered at a public function to create awareness among the workers. [10]**

**OR**

**Media has a strong hold on society. Write a speech in 125-150 words on how media influences public opinon to be delivered in the school assembly.**

**Answer:**

**HEALTH CARE FOR INDIAN WORKERS**

Good morning to all. Today I, Rajendra Kumar, am going to deliver a speech on the topic 'Health care for Indian workers' Recently, I had read an article in the local newspaper *The Hindu* and wondered why we don't care for the health of the workers around us. We remain mute spectators to their misery trials and turbulations of life.

Often, the workers are found working in inhuman conditions, neglecting their health. They are exploited, as they are illiterate and lacking awareness of a hygienic way of life. The authorities must ensure clean and healthy working conditions for these people. It is the responsbility of the government to provide affordable healthcare services to workers and their families. Indian workers should be taught the importance of cleanliness and hygiene. Many non-governmental organisations can play a role in spreading awareness about health and hygiene.

Workers play a major role in our life. We must understand and appreciate their contribution towards society. Access to basic health facilities is their innate right. As responsible citizens, we should get together to make their lives healthy. However, we should not forget that it is the sweat and blood of these workers which has provided us with enormous comfort while we sit in our well-furnished homes.

Thank you

### OR

### IMPACT OF MEDIA

Good morning everybody. Today, I Xyz is here to voice my concerns on how Media influences public opinion?

Media acts as a powerful weapon in democracy. It keeps people abreast with the recent changes and development in the society and the world. They also help people to understand the various developments and build their opinion about the same. With the development of better modes of communications, the influence of media has increased. Consequently, its responsibility of delivering correct information has also increased. If people blindly believe the biased news which the print and electronic media tell us, our democracy will suffer badly. Media has the power to showcase both the facets of a piece of information. It should work towards helping the common man and understand the pros and cons of any issue or event.

Thus, one can clearly say that it is the responsibility of media and the media personnel to ensure that they direct their efforts towards building a constructive and balanced society as well as a nation.

Thank you

## SECTION-C

## (LITERATURE :TEXT BOOKS)

**7. Read the following extract and answer the questions that follow : [4]**

All lovely tales that we have heard or read;
An endless fountain of immortal drink.
Pouring onto us from the heaven's brink.

**(a) Name the poem and the poet. [1]**

**(b) What is the thing of beauty mentioned in these lines? [1]**

**(c) What image does the poet use in these lines? [2]**

OR

With ships and sun and love tempting them to steal.
For lives that slyly turn in their cramped holes
From fog to endless night?

**(i) Who are 'them' referred to in the first line? [1]**

**(ii) What tempts them? [1]**

**(iii) What does the poet say about 'their' lives? [2]**

**Answer:**

**(a)** The above given lines are from the poem 'A thing of Beauty' by John Keats.

**(b)** The tales of mighty men is the thing of beauty mentioned in the given lines.

**(c)** The poet used the image of 'an endless fountain of immortal drink' to describe the beautiful bounty of the Earth. The Earth, like a fountain, gives us numerous beautiful sights like the sun, the moon, flowers, rivers and greenery.

OR

**(a)** The children studying in the school are referred to as "them".

**(b)** The beauty, vastness and radiance of things tempt them.

**(c)** According to the poet, those children spent their whole lives like rodents confined in 'their cramped holes'. The undernourished bodies of those children looked almost like skeletons. Their steel-framed spectacles with repaired glasses made them look like, the broken pieces of a bottle scattered on stones. Since, their entire lives revolved around slums, their future also seemed blotted.

**8. Answer any two of the following questions in about 30-40 words : [2 × 2 = 4]**

**(a) How does Kamala Das try to put away the thoughts of her ageing mother?**

**(b) Which is the exotic moment that the poet refers to in 'Keeping Quiet'?**

**(c) What are the difficulties that Aunt Jennifer faced in her life?**

**Answer :**

**(a)** Kamala Das is troubled by the thoughts of her ageing mother and tries to console herself by looking outside at the car. The sight is full of youthful whim and vigour. She watches the trees 'sprinting' past her speeding car and the children, full of life and activity, running out of their houses to play. Her ashen like mother irks her so much that for a moment she ponders over the past memories to feel solace.

**(b)** The exotic moment the author talks about is when everyone keeps quiet and still. The moment would be extraordinarily tranquil, with no hustle-bustle of the frenzied world. Although, it may seem a little strange in the beginning, it would eventually bring us all together. Ultimately, this is the introspection which makes one united with one's soul.

**(c)** Aunt Jennifer spent her life in accordance to the rules laid down by her husband. Her life was overburdened by the demands and duties and it lacked self-expression. She was feeble and it had to face oppression by her husband which depicted the age-old turbulations of women's lives.

**11. Answer any six of the following questions in 30-40 words : [2 × 6 = 12]**

**(a) Why did Gandhiji feel that taking the Champaran case to the court was useless?**

**(b) Why did the peddler derive pleasure from his idea of the world as a rat trap?**

**(c) How is Mukesh different from the other bangle makers of Firozabad?**

**(d) What tempted Franz to stay away from school?**

**(e) Why did the Maharaja ban tiger hunting in the state?****

**(f) How was the skunk's story different from the other stories narrated by Jack?**

**(g) Which words of her brother made a deep impression on Bama?****

**Answer :**

**(a)** Gandhiji believed that the peasants were under the grip of fear. He felt that taking the Champaran's case to the court would be useless because actual relief for the peasants would come when they would be free from fear. So, his ultimate motive was to kill the fear in them.

**(b)** The peddler was never treated kindly by the world, which is why he was amused by the idea of the world to be a rat trap. He nurtured hard feelings towards the world by comparing it to a giant rat trap.

**(c)** Mukesh who hailed from the city of Dhaka was different from other bangle makers of Firozabad because he liked to dream, unlike others who followed the family lineage of making bangles. He wanted to be a motor mechanic and drive cars.

**(d)** Franz was not prepared for test on participles. The chirping birds and sight of the Prussian soldiers were far more interesting to him than his attending the school. That is why he stayed out of the doors of the school to bask in the nature's glory.

**(f)** Usually, the stories told by Jack were taken well by But she was not satisfied and convinced with the ending of the Skunk's story. She believed that the wizard should have hit back the Skunk's mommy for demanding Skunk to get his bad smell back.

**10. Answer the following in about 100-125 words. [5]**

**Has Sophie met Danny Casey? What details of her meeting with Danny Casey did she narrate to her brother?****

**OR**

**Describe the precautions taken by the prison officers to prevent Evans from escaping.**

**Answer :**

**OR**

Special precautions were taken by the prison staff to prevent Evan from escaping during his O-level German test. The test was to be taken where he was kept. A person from St. Mary Mags was called to invigilate and to keep the prisoner incommunicado during the exam. Evans was kept in the heavily guarded Recreational Block. There were two locked doors between his cell and the yard, which boasted of a high wall. All the prison officers were also on high alert. The Governor got a microphone installed in Evans' cell, while Stephens kept, an eagle eye on Evans. Two prison officers, Mr. Jackson and Mr. Stephens, thoroughly checked his cell for any sign of a possible escape. Even his razor, nail filer and male scissors were taken away.

**11. Read the following and answer the question that follows: [5]**

**The story "Deep Waters" has made you realise that with determination and preseverance one can accomplish the impossible. Write a paragraph in about 100 words on how a positive attitude and courage will aid you to achieve success in life.**

**Answer:** In the story "Deep waters", the author overcomes his childhood fear of water through determination, perseverance and hard work. Thus, the story sends the message that with positive attitude and strength of will, one can accomplish everything in life. All of us are in the grip of fear. But fear, when conquered with courage, becomes victory which has its own charm. It might appear to be a long and arduous journey, but it will definitely culminate in success. Douglas improved himself and was able to develop into a full-fledged swimmer. It gave him a feeling of victory and achievement in the end.

** Answer is not given due to the change in syllabus

# Delhi [Set-II]

**Note: Except for the following questions, all the remaining questions have been asked in previous sets.**

## SECTION – B
## (WRITING SKILLS)

3. **You are Smrithi Saran of Victoria Public School, Hyderabad.Your school has organised a Science Exhibition in connection with the death anniversary of Ramanujam. Write a notice in not more than 50 words inviting students to participate in it. Provide all the necessary details.** [5]

OR

**You want to sell off your motor bike which you have been using for five years, as you have decided to buy a car. Write an advertisement, in not more than 50 words, to be published under the classified columns of a national daily. Furnish all the necessary details.**

**Answer:**

VICTORIA PUBLIC SCHOOL, HYDERABAD
NOTICE

15th June, 20XX

SCIENCE EXHIBITION

A Science Exhibition is being organised by our school on June 18, 20XX in the school hall from 10 a.m. to 1 p.m. in order to commemorate the death anniversary of Ramanujam. Students from classes IX to XII are requested to prepare models on the theme 'India's foray into space research.' Interested students should give their names to the undersigned by 16th June, 20XX.

Smrithi Saran
Head girl

OR

FOR SALE

Availabe for sale a black coloured, 2010 model XYZ motor bike. in a good condition, scratch-less, giving a mileage of 50 km/L. It has hardly run 9000 km. Price negotiable. The Party is selling the motorcycle as it is buying a car.

Contact: 9123456XXX

4. **You are Sampreet, an educationist. You have noticed that the youth of today are often unable to cope with stress and become frustrated and bitter. Write a speech in 125-150 words to be delivered at a college function on 'Youth, their problems and solutions'.** [10]

OR

**After Independence, India has seen a steep upward trend in the graph of corruption. The public has often become disgusted and disappointed. As Venkat, a well known journalist, write a speech in 150-200 words on 'Corruption, its causes and solutions'.**

**Answer:**

YOUTH, THEIR PROBLEMS AND SOLUTIONS

Good Morning! I am Sampreet an educationist and today I am going to deliver a speech on the topic "Youth their problems and solutions"

Aspirations are now high and with this the level of competition and peer pressure among the youth has been increasing with each passing day. The youth is not yet either matured or trained enough to handle so much stress and, therefore, they ultimately get frustrated, easily commit suicide get involve in road-rage, hatred crimes, drug abuse, looting, rapes etc. At the slightest provocation, they hold out protests and demonstrations. The minimum desired level of tolerance is missing in them. They don't see any harm in violating the norms and fundamental rights.

The solution is a doddle one. The youth of today needs our untinted support, it needs to know that, irrespective of the situations they are in, they are always loved by their parents, teachers and elders. They need to develop confidence in their capabilities. We need to remove the unnecessary pressure that our education system exerts on the young minds. It needs to be eased and, CCE is a positive step in this direction. Counsellors can help the young generation to understand and solve their problems. More job opportunities need to be created, better recruitment processes are need to be developed; and professional courses that equip them with necessary skills have to be designed.

The most important thing is to impart value-based education so that they achieve high self-esteem, tolerance and patience and more importantly the zeal to create a clean and civilised society and to be a good citizen.

OR

CORRUPTION, ITS CAUSES AND SOLUTIONS

Good morning, friends! I am Venkat. As a journalist and a responsible citizen of India, it is my duty to shed light and apprise you, that how corruption exists in almost every nook and corner of our country.

Corruption can be understood as misappropriation of government, public property or money. It is an unlawful and immoral act. Corruption has had a gradual growth in our country. How and why did corruption become so common in our country? India's colonial past has contributed in flourishing corrupt practices. It is said that corruption cannot survive long in this world. Although, with the increase in scientific technology and advancement of the nation, corruption still remains unbeatable. The corruption also has a relation with scarcity and exists especially because of the vested interests and greed of the people in power.

It is the spasmodic snare of corruption which springs from all prodigality and dissipation. Whatever the reason may be, corruption is an ill practice that is seen as the major hindrance in the process of economic growth and modernisation of the nation. The need for an anti-corruption bill like Jan Lokpal has long been discerned now. But, in a democracy like India, the power of the common man should not be underestimated. Constant vigil and intervention by the civil society is also required to bring positive changes in the country. Punishment must be increased and illegally confiscated wealth must be recovered. As aware citizens and of course as in united whole, we must overcome our flaws and work towards bringing about a corruption free society.

●●

# Delhi [Set-III]

**Note : Except for the following questions, all the remaining questions have been asked in previous sets.**

## SECTION–B
## (WRITING SKILLS)

1. **You are Anoop/Arya, the cultural secretary of your school. As part of national heritage programme the school has decided to put up a show on ancient art forms. Write a notice to be put up on the school notice board inviting students to watch the show and encourage the artists. Write the notice in not more that 50 words.** **[5]**

**OR**

**You would like to let out a portion of your independent house for office use. Write an advertisement for the classified coloumns of the local newspaper giving all the necessary details. Write the advertisement in not more than 50 words.**

**Answer :**

ABC School

NOTICE

14th November, 20XX

**SHOW ON ANCIENT ART FORMS**

As a part of national heritage programme, the school has decided to put up a show on ancient art forms on 20th November, 20XX at 12 O'Clock in the school auditorium.

All the students are invited to watch the show and encourage the artists. For any queries, you may contact the undersigned.

Arya/Anoop

The School Cultural Secretary

OR

AVAILABLE OFFICE SPACE

For rent, good location, separate entrance and exit. Spacious, 24 hour water and electricity supply, parking facility, security and power backup.

Interested may Contact Mobile : 8XXXXXXXXX

4. **On Teacher's day, the badge holders of your school organised a grand celebration to honour the teachers of your school. As the head boy of the school, write a report on the celebration in 125-150 words.** **[10]**

**OR**

**As you were travelling by train. you happened to witness an accident at an unmanned level crossing. As the reporter of a newspaper, write a report on the accident, including details of the number of people injured and the extent of damage caused.**

**Answer:**

**TEACHER'S DAY CELEBRATIONS**

By Xyz Head boy

On the occasion of the birth anniversary of our former president, Dr. S. Radhakrishnan, the badge holders of our school had organised a grand celebration on 5th September to honour the teachers of our school. the celebration took place in the assembly hall. It began at around 9:30 a.m. In the morning, senior students greeted the teachers warmly in a traditional way by putting tilak on their forehead and giving them flowers. The senior students were dressed up as and played their role teachers. They counted on the teacher's efforts by praising them. After honouring all the teachers, the students performed activities like dance, drama, singing and speech. They also organised few games like musical chair, dodge ball, balloon game, etc. to make the day memorable. Teachers were made to dance and sing. In the end, teachers were provided a lavish lunch by the seniors. It was indeed a great celebration.

**OR**

**ACCIDENT AT AN UNMANNED LEVEL CROSSING**

By Xyz

A horrible sight of a train accident was witnessed at an unmanned level crossing at Udaipur, near Chittor. A bus was carrying nearly 45 people, happened to cross the railway track when suddenly the train arrived. It happened to collide with one of the bogies of the train and overturned badly. 23 people were seriously injured and they were immediately taken to a nearby hospital by the locals. The bogie was also derailed. The extent of damage caused is not yet estimated. Due to this accident, the train got delayed and the schedule of the other trains of that route was also disturbed. It resulted into a lot of inconvenience to the passangers. Government has set up an enquiry to probe into the matter. A sum of ₹50,000 was announced for all those injured.

5. **You are the Physical Education Instructor of 'All Saints School'. You had placed an order for sports goods with Bat and Ball Enterprises. When the items arrived you found that some of them were defective. Write a letter to the Manager, Bat and Ball Enterprises, asking him to replace the defective items.** **[10]**

**OR**

**You live in a crowded area in Mumbai. Unauthorised parking of vehicles in your area is causing a lot of inconvenience to the locals. Write a letter to the Police Commissioner of the district complaining against the practice. You are Rohan/ Rohini of Sakinaka, Mumbai.**

**Answer:**

All Saints School,
New Delhi
18th December, 20XX
The Manager,
Bat and Ball Enterprises,

**Subject : Replacement of the defective sports items**

Sir,

This is to bring to your kind notice that one month ago, our school placed an order for the following sports items:

5 Bats, 10 Balls, 3 Footballs, 3 Basketballs, 5 Badminton Racquets.

On the arrival of the goods, it was found that some of them were defective. The handle of the bats were cracked, the net of one of the badminton racquet was loose and the stitching of the football was not properly done. Therefore, I request you to replace the defective items as soon as possible as the sports week of the school is approaching and the children are required to do the practice. I would be obliged if you kindly understand my problem and arrange for the early delivery.

Thanking You,

Yours Sincerely,

Xyz Sharma

Physical Education Instructor

**OR**

123, Sakinaka,

Mumbai.

24th August, 20XX

District Police Commissioner,

Mumbai.

**Subject : Unauthorized Parking of Vehicles**

Dear Sir,

This is to draw your attention towards the problems faced due to unauthorized parking of the vehicles in our area. There is not a single day in our area when the place is without disorder or free from traffic jams. As it is a crowded place, there is inadequate parking space. It creates problems not only to the residents but also to the people who come to meet their relatives or for some other purpose. As it is a busy area, vehicles are parked all over in a haphazard manner. It often results in quarrel among the people who display the lack of civic sense in them also creating noise pollution. It is therefore necessary to take some quick actions and provide the proper parking area to the residents as well as do some arrangements for the visitors too.

We all hope for an immediate action to get relief from this problem.

Thanking You,

Yours Sincerely,

Rohini/Rohan

**SECTION C**

**(LITERATURE : TEXT BOOK)**

10. **What were Sophie's plan for her future? Why would you call her dreams unrealistic? (100-125 words)**** **[5]**

** Answer is not given due to the change in syllabus

**OR**

**What benefits did Derry reap from his association with Mr. Lamb?**

**Answer :**

**OR**

Derry had no association with the outer world till he met Mr. Lamb. His meeting with Mr. Lamb was a turning point in his life. He learnt to live a respectful life and also appreciated Mr. Lamb's attitude and outlook towards people and life. He realised his true worth. Mr. Lamb inspired him to face the world in spite of his disability. Derry's thinking towards people and life completely changed and he became self-confident. Mr. Lamb successfully infused in him courage and self-determination. He advised him to ignore the comments made by people on his physical impairment and enjoy beauty of life and nature. Derry went back to Mr. Lamb's garden in spite of his mother's refusal., Derry understood the importance and true meaning of life and understanding himself better than before. Only due to the association with Mr. Lamb.

●●

# ENGLISH

## Outside Delhi [Set-I] 2013

**Time allowed : 3 hours** **Maximum marks : 100**

### SECTION-A

### (READING) 20

**1. Read the passage given below and answer the questions that follow : [12]**

(1) Air pollution is an issue which concerns us all alike. One can willingly choose or reject a food, a drink or a life comfort, but unfortunately there is little choice for the air we breathe. Everything in the air is inhaled by all living in those surroundings.

(2) Air pollutant is defined as a substance which is present in an amount exceeding the normal concentrations. It could either be gaseous or a particulate matter. The important and harmful polluting gases are carbon monoxide, carbon dioxide, ozone and oxides of sulphur and nitrogen. The common particulate pollutants are the dusts of various inorganic or organic origins. Although, we often talk of the outdoor air pollution caused by industrial and vehicular exhausts, the indoor pollution may prove to be as or a more important cause of health problems.

(3) Recognition of air pollution is relatively recent. It is not uncommon to experience a feeling of 'suffocation' in a closed environment. It is often ascribed to the lack of oxygen. Fortunately, however, the composition of air is remarkably constant all over the world. There is about 79 percent nitrogen and 21 percent oxygen in the air the other gases forming a very small fraction. It is true that carbon dioxide exhaled out of lungs may accumulate in a closed and over-crowded place. But such increment is usually small and temporary unless the room is really air-tight. Exposure to poisonous gases such as carbon monoxide may occur in a closed room, heated by burning coal inside. This may also prove to be fatal.

(4) What is more common in a poorly ventilated home is a vague constellation of symptoms described as the sick-building syndrome. It is characterised by a general feeling of malaise, headache, dizziness and irritation of mucous membranes. It may also be accompanied by nausea, itching, aches, pains and depression. Sick building syndrome is getting common in big cities with the small houses, which are generally over-furnished. Some of the important pollutants whose indoor concentrations exceed to those of the outdoors include gases such as carbon monoxide, carbon dioxide, oxides of nitrogen and organic substances like spores, formaldehydes, hydrocarbon aerosols and allergens. The sources are attributed to a variety of construction materials, insulations, furnishings, adhesives, cosmetics, house dusts, fungi and the other indoor products.

(5) By-products of fuel combustion are important in houses with indoor kitchens. It is not only the burning of dried dung and fuelwood which is responsible, but also kerosene and liquid petroleum gas. Oxides of both nitrogen and sulphur are released from their combustion.

(6) Smoking of tobacco in a closed environment is an important source of indoor pollution. It may not be high quantitatively, but significantly hazardous for health. It is because of the fact that there are over 3000 chemical constituents in tobacco smoke, which have been identified. These are harmful for human health.

(7) Micro-organisms and allergens are of special significance in the causation and spread of diseases. Most of the infective illnesses may involve more persons of a family living in common indoor environment. These include viral and bacterial diseases like tuberculosis.

(8) Besides infections, allergic and hypersensitivity disorders are spreading fast. Although asthma is the most common form of respiratory allergic disorders, pneumonias are common, but more persistent and serious. These are attributed to exposures to allergens from various fungi, molds, hay and other organic materials. Indoor air ventilation systems, coolers, air-conditioners, dampness, decay, pet animals, production or handling of the causative items are responsible for these hypersensitive diseases.

(9) Obviously, the spectrum of pollution is very wide and our options are limited. Indoor pollution may be handled relatively easily by an individual. Moreover, the good work must start from one's own house.

**(a) (i) What is an air pollutant?** **[1]**

**(ii) In what forms are the air pollutants present?** **[2]**

**(iii) Why do we feel suffocated in a closed environment?** **[1]**

**(iv) What is sick building syndrome? How is it increasing?** **[2]**

**(v) How is indoor smoking hazardous?** **[1]**

**(vi) How can one overcome the dangers of indoor air pollution?** **[2]**

**Answer :**

**(a) (i)** Air pollutant is a gaseous or a particulate matter present in the air in an amount exceeding the normal concentration.

**(ii)** Air pollutants are present in gaseous form such as carbon monoxide, carbon dioxide, ozone and oxides of sulphur and nitrogen. The pollutants are also in the form of dust of various inorganic or organic origins.

**(iii)** We feel suffocated in a closed environment due to lack of oxygen. Accumulation of carbon dioxide exhaled out of lungs and more exposure to carbon monoxide are also the reasons of suffocation.

**(iv)** Sick building syndrome is characterised by a general feeling of malaise, headache, dizziness and irritation of mucous membrane. It may be accompanied by nausea, itching, aches, pains and depression.

It is increasing due to poorly ventilated and over furnished houses in big cities in which there is high concentration of carbon dioxide, carbon monoxide and oxides of nitrogen and other harmful substances.

**(v)** Indoor smoking is hazardous in the way that there are 3,000 chemical constituents in tobacco smoke which are harmful to human health.

**(vi)** The houses should be properly ventilated with more and more greenery around them. Smoking of tobacco and burning of coal inside the houses should also be avoided to overcome the dangers of indoor air pollution.

**(b) Find the words from the above passage which mean the same as the following :** **[3]**

**(i) Giddiness (para 4)**

**(ii) Constant (para 8)**

**(iii) Humidity (para 8)**

**Answer :**

**(i)** Dizziness

**(ii)** Persistent

**(iii)** Dampness.

**2. Read the passage below and answer the questions that follow :** **[8]**

The term dietary fibres refers collectively to indigestible carbohydrates present in plant foods. The importance of these dietary fibres came into the picture when it was observed that the people having diet rich in these fibres, had low incidence of coronary heart disease, irritable bowel syndrome, dental caries and gall stones.

The foodstuffs rich in these dietary fibres are cereals and grains, legumes, fruits with seeds, citrus fruits, carrots, cabbage, green vegetables with leaves, apples, melons, peaches, pears, etc.

These dietary fibres are not digested by the enzymes of the stomach and the small intestine whereas, most of the other carbohydrates like starch and sugar are digested and absorbed. The dietary fibres have the property of holding water and because of it, they get swollen and behave like a sponge as these pass through the gastrointestinal tract. The fibres add bulk to the diet and increase transit time in the gut. Some of these fibres may undergo fermentation in the colon.

In recent years, it has been considered essential to have some amount of fibres in the diet. Their beneficial effects help in preventing coronary heart disease, and decreasing cholesterol levels. The fibres like gums and pectin are reported to decrease postprandial (after meals) glucose level in blood. These types of dietary fibres are recommended for the management of certain types of diabetes. Recent studies have shown that the fenugreek (Methi) seeds, which contain 40 percent gum, are effective in decreasing blood glucose and cholesterol levels as compared to other gum containing vegetables.

Some dietary fibres increase transit time and decrease the time of release of ingested food in colon. The diet having less fibres is associated with colon cancer and the dietary fibres may play a role in decreasing the risk of it.

The dietary fibres hold water so that stools are soft, bulky and readily eliminated. Therefore, high fibre intake prevents or relieves constipation.

The fibres increase motility of the small intestine and the colon and by decreasing the transit time there is less time for exposure of the mucosa to harmful toxic substances. Therefore, there is less desire to eat and the energy intake can be maintained within the range of requirement. This phenomenon helps in keeping a check on obesity.

Another reason in helping to decrease obesity is that the high fibre diets have somewhat lower coefficients of digestibility.

The dietary fibres may have some adverse effects on nutrition by binding some trace metals like calcium, magnesium, phosphorus, zinc and others and therefore preventing their proper absorption. This may pose a possibility of nutritional deficiency especially when diets contain marginal levels of mineral elements. This may become an important constraint on increasing dietary fibres. It is suggested that an intake of 40 grams dietary fibres per day is desirable.

**(a) On the basis of your reading of the above passage, make notes in points only, using abbreviations wherever necessary. Provide a suitable title.**

**Answer : (a)**

**Title : Dietary Fibres**

**1. Importance : lowers**

(i) hrt disease (ii) irrtble bowel syndrome

(iii) dental caries (iv) gall stones

**2. Rich dietary fibers**

(i) cereals, grains, legumes

(ii) fruits

(iii) vegetables

**3. Adverse effects**

(i) bind metals like ca, mg, ph, zn

(ii) prevents their absoptn

(iii) nutnl defcncy

**Abbreviation Used**

| | | |
|---|---|---|
| Hrt. | : | heart |
| Irrtble. | : | irritable |
| Ca. | : | calcium |
| Mg. | : | magnesium |
| Ph. | : | phosphorous |
| Zn. | : | zinc |
| Absoptn. | : | absorption |
| Nutnl. | : | nutritional |
| Defcncy. | : | deficiency |

**(b) Write a summary of the above in about 80 words.** **[3]**

**SUMMARY**

Dietary fibers are indigestible carbohydrates present in plant foods. People having diet rich in fibers have low heart diseases, irritable bowel syndrome, dental caries and gall stones. Sources of dietary fibers are legumes, cereals, fruits, vegetables and grains. The fibers like gums present in fenugreek and pectin decrease glucose level. The diet which is less in fibers results in colon cancer. The dietary fibers hold water and functions like sponge, hence they prevent constipation.

But there are adverse effects also. They bind some trace metals resulting in nutritional deficiency.

## SECTION – B
## (WRITING SKILLS) 35

**3. You are Vineeta/Vikram, School pupil leader of Rani Laxmi Bai Senior Secondary School, Gwalior. Draft a notice for your school notice board in not more than 50 words inviting the names of the students who want to participate in the cultural programme organised in aid of the victims of the recent Assam floods.** **[5]**

**OR**

**You have a three bedroom flat in Dwarka, which you want to let out on rent. Draft an advertisement in not more than 50 words to be published in "The Times of India" under classified columns, Contact 2758902.**

**Answer :**

**RANI LAXMI BAI SENIOR SECONDARY SCHOOL**

**NOTICE**

30th July, 20XX

**Participation in the Cultural Programme**

To help the flood victims of Assam, a cultural programme is being organised by the school on 30th August, 20XX, in the school auditorium. The students who are interested in participating in dance, singing or dramatics can give their names to their respective class teachers by 7th August, 20XX.

Vineeta/Vikram

School Pupil Leader

**OR**

**FLAT AVAILABLE**

For rent, good accommodation, prime location, three bhk flat in Dwarka, near metro station, fully furnished, company lease preferred, AC installed in all the rooms, expected rent 30,000/- ·Interested may contact: 2758902

**4. Maxim Gorky School, Kodaikanal celebrated "Nutrition Week" from 1st September to 7th September by arranging inter-school cookery contests, oratorical, painting and poster competition, besides talks by eminent nutritionists and medical professionals. There was an overwhelming response from students and parents. Prepare a report in 100-125 words to be published in your school magazine. You are Arjun/Anita, School pupil leader.** **[10]**

OR

**You are Roshan/Rohini, School Pupil Leader of Meerut Public School, Meerut. Your school joined a campaign organised by various agencies of your city to create awareness among people to conserve water. Write a report in 100-125 words highlighting the activities such as rain water harvesting etc.**

**Answer :**

NUTRITION-WEEK

By Anita/Arjun (School Pupil Leader)

Maxim Gorky School, Kodaikanal celebrated Nutrition Week from 1st September to 7th September in the school auditorium. The celebration started with lighting of candles by our respected Principal followed by his inspirational speech on the present day's health issues and the impact of diet on it. Different schools participated in various competitions organised by the school. The competitions included inter-school cookery contests, oratorical, painting and poster competition, besides talks by eminent nutritionists and medical professionals. The students as well as their parents gave an overwhelming response because nowadays everybody is conscious about their health and well-being. The speech on the nutritional value of food by a student was praised by all. The programme was a combined effort of the teachers and the students and received a huge appreciation by present there.

In the end, the results of the competitions organised were declared and the winners were awarded prizes by Principal.

OR

WATER CONSERVATION

By Rohini/Roshan (School Pupil Leader)

Water scarcity today has become a big issue which, if not dealt with in time, will turn out to be a hazard. To make people aware to conserve water, Meerut Public School joined a campaign organised by various agencies of the city. People in large numbers attended it and were asked to conserve every drop of water and not to waste it unnecessarily. Small skits presenting the ways of conserving water were presented by the children. It was an easy approach as the venue of the campaign was the public park and the passersby also stopped to become a part of the campaign. Besides, importance of rainwater harvesting and its methods, were also explained. People were encouraged to improve water management practices that reduce the wastage of water and enhance the beneficial uses of water. Some ancient methods of water conservation should also be practiced. In case, someon is found misusing water, he/she will be penalised. Public was motivated by this, and they joined hands and took oath to use water properly and motivate others also.

5. **Write a letter to your cousin, Raj Prakash who is currently staying in Dubai explaining the process of CCE, being used by CBSE for the secondary school examination in its affiliated schools. Also mention how you and your classmate have reacted to this scheme. You are Narain/Nisha, 20 Fort Road, Mumbai.** **[10]**

OR

**Recently you travelled from Bangalore city to Vasco in Vasco Express. To your dismay, you found that the coach was infested with cockroaches. Write a letter to the editor "The Hindu" drawing the attention of the General Manager of Southern Railways, towards the prevailing unhygienic conditions and asking for remedial action. You are Saroj/Saran, 5/31 Bangalore Cantt., Bangalore.**

**Answer :**

20, Fort Road,

Mumbai

20th April, 20XX

Dear Raj Prakash,

Hope you are good and have adjusted well in the atmosphere of Dubai. How do you find your school? What pattern of examination is followed there? Here, in our schools, CBSE has started a new pattern-CCE (Continuous and Comprehensive Evaluation). This system evaluates the performance of the students in scholastic as well as co-scholastic areas. It has reduced the burden of the children and we feel that it will go a long way. It includes the formative and summative assessment. With this new system it has become possible for us to learn and do it in a fun-filled way. Learning has become a pleasure as it has taken away the fear of examination. It not only excel us in academics but helps in our holistic development also. We all are enjoying it and it is a positive approach towards learning.

Hope you too are doing fine in your studies. Convey my regards to uncle and aunty.

Yours affectionately,

Nisha/Narain

OR

5/31, Bangalore Cantt.

Bangalore

10th January, 20XX

The Editor,

The Hindu

**Subject : Unhygienic conditions in train coaches**

Dear Sir,

Through the columns of your esteemed newspaper, I would like to draw attention of the General Manager, Southern Railways towards the unhygienic conditions of train coaches. Recently, when I was travelling from Bangalore city to Vasco in Vasco Express, I noticed that the coach was infested with cockroaches. The sheets and blankets given to the passengers were not properly washed; the food was also stale and uncovered, toilets were unclean, washbasins were dirty due to which, the cockroaches are breeding and the whole place is infested with them. Therefore, immediate remedial actions are required. There should be regular check on the cleanliness of the coaches and the staff members who are given the duty, should also be taken into account. If they are not performing their duties, they should be fined so that the negligence is not repeated. I hope, this will bring a definite change.

I shall be grateful if this letter finds a place in your newspaper.

Yours Sincerely,

Saroj/Saran

**6. Write an article in 150-200 words for your school magazine on the topic. "Obesity among School Children." You are Mohini/Mohit. [10]**

**OR**

**You are Ramesh/Ruchika. Write an article in 150-200 words for your school magazine on the topic, "Life without Modern Gadgets".**

**Answer:**

**OBESITY AMONG SCHOOL CHILDREN**

By Mohini/Mohit

The most common problem found among the children in today's scenario is the growing problem of obesity. This is the result of present day lifestyle where parents are much responsible for it. Even the tastes of the children have changed a lot and they are moving towards junk food like burgers, pizza, pasta and many more. These are not only harmful for their proper growth but also affect their metabolism. As the children are confined to their rooms playing on laptops, computer games and mobiles, the outdoor activities have come to an end. As a result, more and more children are becoming obese and overweight which is alarming because it causes health problems which results, in social issues. They are likely to be teased by their peers and they tend to become inactive. Once children become obese it is very difficult to return to the original shape. Apart from this, the risk factors are very high. They are more prone to high blood pressure, weak heart, diabetes, orthopedic disorders, etc. So, the only remedy for this is the awareness about obesity less intake of junk foods, more intake of home-made food, avoiding aerated drinks and getting involved in physical activities only then this problem can be overcomed.

**OR**

**LIFE WITHOUT MODERN GADGETS**

By Ruchika/Ramesh

Science has made life easy, convenient, comfortable, better and more beautiful. Its achievements, discoveries and inventions are wonderful. Now we cannot think of our lives without gadgets as we would just be helpless without them. Every device or appliance has a unique purpose. They have revolutionised our life so much that without them we feel handicaped. Beginning from kitchen, gadgets like mixer, microwave oven, refrigerator and water purifier have proved to be a boon. Cell phones are the center of our life have huge potential, we can listen to our favourite music while travelling and need not carry heavy radios.

Our life starts with the sound of an alarm clock, so we can say that our day begins with gadgets. They work as our personal assistant. In offices, meetings, trainings and all, we are dependent on them. Even at the shops, libraries, railways, banks, airports, a computer is required for booking tickets and keeping accounts. If for a day our internet does not work, we feel paralysed. It feels as if something important is missing from our lives. Thus, these gadgets have left a remarkable impact on our lives and we cannot work or go without them.

**SECTION-C**

**(LITERATURE : TEXTBOOK) 45**

**7. Read the extract given below and answer the questions that follow : [3]**

The stunted, unlucky heir

Of twisted bones, reciting a father's gnarled disease,

His lesson, from his desk. At the back of the dim class

One unnoted, sweet and young.

**(i) Who is this 'unlucky heir' and what has he inherited? [1]**

**(ii) What is the stunted boy reciting? [1]**

**(iii) Who is sitting at the back of the dim class? [1]**

**OR**

Driving from my parents
Home to Cochin last Friday
Morning, I saw my mother, beside me,
Doze, open mouthed, her face ashen like that
Of a corpse and realised with pain
That she was as old as she looked ...

**(i) Where was the poet driving to? [1]**

**(ii) Why was her mother's face looking like that of a corpse? [1]**

**(iii) What did the poet notice about her mother? [1]**

**Answer :**

(i) The boy who has twisted bones and is sitting in the slum classroom is the unlucky heir. He has inherited the gnarled disease of twisted bones from his father.

(ii) The stunted boy is reciting his father's gnarled disease and his lessons.

(iii) At the back of the dim class, there is an unnoted, sweet and young dreamer sitting, who is dreaming of squirrel's game.

**OR**

**Answer :**

(i) The poet was driving to Cochin from her parents' home.

(ii) Her mother's face was looking like a corpse because she was old, weak, pale and lifeless.

(iii) The poet noticed that her mother's mouth was open and her face looked like that of a corpse.

**8. Answer any three of the following in 30-40 words each : [2 × 3 = 6]**

**(a) What does Keats consider an endless fountain of immortal drink and why does he call its drink immortal?**

**(b) What are the different kinds of wars mentioned in the poem? What is Neruda's attitude towards these wars?**

**(c) What lies heavily on Aunt Jennifer's hand? How is it associated with her husband?**

**Answer :**

(a) Beauty, according to Keats is immortal and is continuous source of motivation and inspiration and things of beauty are like endless fountains pouring from heaven. It is an endless fountain of beautiful things so that man can enjoy happiness and peace.

(b) The poet talks of the chemical wars, war with fire, wars with ammunition, wars with environment and green wars. According to him, these wars are the gate to destruction. Therefore, wars should be avoided, as there will be no survivors to enjoy the victory.

(c) The wedding band and the ring lie heavily on Aunt Jennifer's hand. The ring symbolises the weight of her marriage which she has to bear, dead or alive. It reminds her of the pressures of an unhappy and unpleasant marriage.

**9. Answer any three of the following in 30-40 words each : [2 × 3 = 6]**

**(a) What was the order from Berlin and what changes did it cause in the school?**

**(b) How did Douglas make sure that he conquered the old terror?**

**(c) Why was Sophie jealous of Geoff's silence?****

**Answer :**

(a) There was an order from Berlin that French would not be taught in schools anymore. Instead, German would be taught by a new teacher. That was probably their last French lesson. It was not a Sunday morning yet, there was no noise or activity in the school. The French teacher Mr. Hamel seemed to be calm and he was becoming emotional. He was wearing a special dress and all the villagers were sitting on the back benches of the class. The whole atmosphere of the school was silent and peaceful.

(b) The old terror always returned whenever Douglas was alone in the pool. He covered the length of the pool but when he was not satisfied, he went to Lake Wentworth and swam two miles to overcome his terror. The terror was now gone and he laughed at it.

**10. Answer the following in 125-150 words : [10]**

**Why did Gandhiji agree to a settlement of 25 percent refund to the farmers? How did it**

** Answer is not given due to the change in syllabus

**influence the peasant landlord relationship in Champaran?**

**Answer :**

Gandhiji agreed to the settlement of 25 percent because the amount was not important for him but to release the peasants from the indigo raising agreement was much more important. The British planters wanted to prolong the agreement but Gandhiji proved to be wiser. He fought for the poor peasants a long battle for one year and managed to get justice for them. This was a moral victory for them. Even the Britishers had to compromise with their pride and prestige. Peasants now got the courage and became aware of their rights. Their money and time was saved and within few years British planters were forced to give up their estates and these were returned to farmers. They became the owners of their own land. It also ignited the feelings of self-dependence and patriotism among the farmers.

**11. Answer the following in 30-40 words : [7]**

**Give a character sketch of the Governor of Oxford Prison based on your understanding of the story 'Evans Tries an O-level'.**

**Answer :**

The Governor of Oxford Prison was a generous man who allowed Evans to appear for the exam. Evans was a criminal and had escaped from jail twice, so the governor took all the possible precautions for the smooth conduct of his examination. The examination had been arranged in the cell itself. The door was locked one day before and the security was made tighter. Everything was checked thoroughly to ensure that there was no way that could help him to escape. This shows that he was very wise and on the other hand, he was so simple that he couldn't understand the cunningness of Evans.

**12. Answer the following in about 100 words: [5]**

**There is a saying 'kindness pays, rudeness never'. In the story 'The Rattrap' Edla's attitude towards men and matters is different from her father's attitude. How are the values of concern and compassion brought out in the story, 'The Rattrap'?**

**Answer :**

Edla, the ironman's daughter was totally different from her father. In the beginning, the iron master seemed to be very sympathetic when he mistook the stranger for one of his old acquaintance. He even sent his daughter to call for him when he did not agree to come to their house. But soon his attitude changed when he came to know the truth that the stranger was someone else. He called him dishonest and threatened to hand him over to the police whereas his daughter was entirely different from her father. She had different ways of looking at things. She was affectionate and considerate. She acted more wisely than her father. She stopped her father from behaving badly towards the stranger. She proved to be a good host and it was her, who brought a change in his behaviour. It was her generosity and sympathy that converted him entirely. These values strengthen our hope in goodness.

**13. Answer the following in 30-40 words : [2 × 4 = 8]**

**(a) What forced Dr. Sadao to be impatient and irritable with his patient?**

**(b) Why does Jo insist that her father should tell her the story with a different ending?**

**(c) If you were to give a different ending to the story 'On the face of it' how would you end it?**

**(d) What were the articles in the stalls and shops that fascinated Bama on her way back from school?****

**Answer :**

(a) When Dr. Sadao saw the difficulties faced by his wife in the absence of maids and servants due to the presence of an enemy in his house, he felt impatient and irritable with him. Besides when this secret was out that he was sheltering an enemy, he feared of getting into serious trouble. It caused a great deal of tension to him.

(b) Jo insists that her father should tell her the story with a different ending because she wants that other animals should play with Roger Skunk and for that he must smell like roses. According to her, the mother is wrong as she made him smell foul again and the wizard should hit his mother for that. In fact, she wants to assert her views on her father.

(c) If I were given a chance to end the story, surely it would be a happy ending. Mr. Lamb would not have died and Derry along with him would have started a new life. They would have been living happily in each other's company.

** Answer is not given due to the change in syllabus

# ENGLISH

## Delhi [Set-I] 2013

**Time allowed : 3 hours** **Maximum marks : 100**

**Note : Except for the following questions, all the remaining questions have been asked in previous sets.**

**SECTION-B**

**(WRITING SKILLS)**

**3. You are Smitha/Sunil, Secretary AVM Housing Society. You going to organise a blood donation camp. Write a notice in not more than 50 words, urging the members of your society to come in large numbers for this noble cause. Invent all the necessary details. [5]**

**OR**

**You are General Manager, Hotel Dosa, Gurgaon. You need a lady Front Office Assistant with sound knowledge of computers. She must be a graduate and good in communication skills with pleasing manners. Draft an advertisement in not more than 50 words to be published in Gurgaon Times. [10]**

**Answer :**

**AVM HOUSING SOCIETY**

**NOTICE**

26th October, 20XX

**BLOOD DONATION CAMP**

A blood donation camp is being organised by our society on 1st November, 20XX in the front park of the society. All the residents are, therefore, requested to donate their blood for a noble cause and save a life. For any details please contact the undersigned.
Smitha/Sunil
Secretary
AVM Housing Society

**OR**

**ASSISTANT REQUIRED**

A lady Front Office Assistant, having sound knowledge of computers, a graduate with good communication skills and pleasing manners. Salary negotiable. Interested may apply immediately.
Contact: General Manger
Hotel Dosa, Gurgaon

**4. Your school Sun Public School, Pune celebrated 'Environment Day' on 5th November. Write a report on the programme in 100-125 words for your school newsletter. You are Neeta/Naveen, Cultural Secretary of the school. Invent the necessary details. [10]**

**OR**

**You are Deepak/Deepika, Secretary of Ahimsa Club, Parsva Public School, Delhi. On 2nd October your school observed 'International Day for peace and Non-violence', organising various activities such as a visit to Rajghat, Charkha spinning, lectures by eminent Gandhians, etc. Write a report for the same in 100-125 words.**

**Answer :**

**ENVIRONMENT DAY CELEBRATION**

By Neeta/Naveen (Cultural Secretary)

Sun Public School, Pune, celebrated 'Environment Day' on 5th November very enthusiastically. Everybody was excited and showed eagerness towards the celebration. The students wore eco-friendly dresses. The corridors were decorated with the posters made by the children with the help of their art teacher. Many activities were organised on this occasion. A small play was presented by the students in connection with the environment friendly measures which should be adopted by the people. There was also an exhibition of the things made by the junior section of the school, which included paper bags, jute bags, and products made up of biodegradable matter. Our Principal gave an inspiring speech on how to preserve our environment. Student speakers also presented their views on the conservation of environment. After the function was over, tree plantation drive started. The first tree was planted by the Principal followed by the faculty members and then the children. In the end, students took the oath of planting a tree every year to make the environment neat and clean.

It was a great step towards the awareness about the environment.

**OR**

**INTERNATIONAL DAY FOR PEACE AND NON-VIOLENCE OBSERVED**

By Deepika/Deepak (Secretary, Ahimsa Club)

2nd October was a memorable day for our school as it observed 'International day for Peace and Non-violence.' We all know that this day is celebrated to commemorate Mahatma Gandhi, who was the greatest messenger of peace and non-violence. Various activities were organised by the school. Students were told the importance of 'Charkha Spinning' and they enjoyed doing it by themselves. At the back stage, a group of students sang Bapu's favourite bhajan 'Vaishnav jan to tinhe kahiye...' in a choir. There were speeches made on the way of living that Gandhiji advocated all his life and the means to achieve them which according to Bapu should be noble and high. After that the children were taken to the Samadhi of Gandhiji known as 'Rajghat'. The teachers as well as children paid homage to Bapu. The eminent Gandhians delivered speeches and told that he was a great example of selfless service, self-discipline, renunciation and deep love for suffering humanity.

In the evening, we returned back to the school to disperse to our homes but there was a sense of content and peace in our mind and soul.

**5. As a regular commuter by bus from Noida to Delhi, you have been witnessing rash driving by the bus drivers daily without an exception. Write a letter to the editor "The Times of India" drawing the attention of the General Manager of Delhi Transport Corporation to this problem. You are Priti/Prakash, 15, Udyog Vihar, Noida. [5]**

**OR**

**Write a letter to Lightways Sport, Amrapalli, Thane, placing an order for sports articles (Minimum 4) to be supplied to your school, ABC Matriculation School, Civil Lines, Pune, Sign as Ravi/Raveena, Sports Secretary.**

**Answer :**

15, Udyog Vihar,
Noida

24th September, 20XX

The Editor,
The Times of India

**Subject : Rash Bus Driving**

Sir,

Through the columns of your esteemed newspaper, I would like to draw the attention of the General Manager, Delhi Transport Corporation towards the increasing cases of rash bus driving. This has created havoc in the city as it has become a routine habit of the drivers. Daily one or two cases of mishaps due to the negligence of the drivers are coming into light. It's not an overheard thing, but I have experienced it myself as I am a regular commuter by bus from Noida to Delhi. The number of accidents have been increasing daily which results in the loss of many lives. This should be checked as drivers are indifferent towards the sufferings of the public and do not pay heed towards the traffic rules. The most common sufferers are the elders and the children. Immediate action is the need of the hour. Either the registration of their vehicle should be cancelled or their driving license, to bring such mishaps under control.

I expect a favorable response.

Thank you,
Yours Sincerely,
Priti/Prakash

**OR**

ABC Matriculation School,
Civil Lines, Pune

26th October, 20XX

The Manager,
Light ways Sport,
Amrapali, Thane.

**Sub : Order for Sports Articles**

Dear Sir/Madam,

We need to purchase some sports articles for our school. The list of the articles is as follows :

| Item | Quantity |
|---|---|
| 1. Basketballs | 4 |
| 2. Cricket bats | 6 |
| 3. Volleyballs | 4 |
| 4. Footballs | 3 |

Kindly, send the articles after proper checking at the school sports complex between 10. a.m. to 2. p.m. Please make the delivery as soon as possible. As we have been a regular buyer of sports goods from your shop for past many years, a good amount of discount is expected on the above mentioned items. Payment will be done at the time of delivery through demand draft.

Thank You.
Yours Sincerely,
Raveena/Ravi
(Sports Secretary)

**6. Your family has recently shifted from Kota in Rajasthan to Ernakulam in Kerala, where your house is situated in the midst of beautiful flowering plants and fruit yielding trees. Every minute and every second, you are experiencing the joy of being in the lap of nature. Write an article in 150-200 words on the diversity of nature that you have experienced. You are Latha/Lalitha of class XII. [10]**

**OR**

**Write an article in 150-200 words on the topic, 'Poverty is the cause of all evils', to be published in the Young World of 'The Hindu', Chennai.**

**Answer :**

**IN THE LAP OF NATURE**

By Latha/Lalitha Class XII

One can find huge diversity in nature while living in India. I experienced it personally when we shifted to Ernakulam in Kerala from Kota, Rajasthan. It seemed as if we have shifted not places but nature. From the brown mountains to the green canopies, the change was completely welcomed by all the family members. For the first time in my whole life, I saw such greenery. The words used for Kerala as 'God's Own Country', appeared to be true which I had only heard or read till now. Living in the lap of nature, is a completely new experience. The greenery is extremely enchanting and to add to its beauty, is the location of my house amidst the beautiful flowering plants and fruit yielding trees. The fragrance of the blooming flowers fill the house and the aroma enters the soul. The view of fruit laden trees is equally fascinating. There is no sign of pollution here. Evergreen beaches, fishing nets spread all over the sea, richness of coconut and jackfruit, all make it a desired place to live in. The place has not lacked even in the preservation of its culture. The boat race in the backwaters is done during the festival of Onam. It is famous for its spices and 'Ayurvedic Kerala Massage.' In all, I must say that I enjoy the bracing environment, relaxed life, temples, backwaters and shaded woods of Kerala.

**OR**

**POVERTY IS THE CAUSE OF ALL EVILS**

By Xyz

India is a country where most of the population live in villages and below poverty line. Poverty prevails not only in villages but also in big cities. The metro cities are full of slums, where majority of people are living in inhuman conditions. Their basic needs for food, shelter and clothing are not fulfilled. Millions and millions of people are the victims of malnutrition, hunger, poor health and illiteracy. Their condition is not better than stray dogs. The prices are rising at a very high rate and in the same way their condition is also deteriorating. The rich are becoming richer, and the poor are becoming poorer. Circumstances and negligence force them to adopt evil means and criminal activities like robbery, murder, kidnapping, looting and so on. Children are made to beg and women are compelled to become prostitutes. If it is not acceptable to anyone, then they commit mass suicide.

The society, reformatory organisations, government, and non-profit organisations should take a step forward to improvise the condition of these people. There should be a check on the fund allocated for poverty removal and also whether it is properly utilised or not.

## SECTION- C
## (LITERATURE :TEXT BOOK)

7. **Read the extract given below and answer the questions that follow :** [3]

Perhaps the Earth can teach us
As when everything seems dead
And later proves to be alive
Now I'll count up to twelve
And you keep quiet and I will go.

(i) **What does the Earth teach us?** [1]

(ii) **What does the poet mean to achieve by counting up to twelve?** [1]

(iii) **What is the significance of 'keeping quiet'?** [1]

**OR**

A flowery band to bind us to the Earth,
Spite of despondence, of the inhuman dearth
of noble natures, of the gloomy days,
Of all the unhealthy and o'er-darkened ways
made for our searching :

**OR**

(i) **What are we doing every day?** [1]

(ii) **Which evil thing do we possess and suffer from?** [1]

(iii) **What are the circumstances that contribute towards making humans unhappy and disillusioned with life?** [1]

**Answer :**

(i) The Earth teaches us to remain quiet even in the adverse situations.

(ii) The poet, asks us to keep quiet while he counts up to twelve as he wants us to self-analyse.

(iii) It is quite necessary to keep quiet for some time as it would help us introspect and understand ourselves, so that we can solve our problems.

**OR**

(i) Every day we are weaving a beautiful string of flowers that would bind us to earth, and motivate us to live our life happily instead of pains and sufferings.

(ii) We possess evils of hatred and jealousy that do not let us live freely and happily. They steal away the joys of life.

(iii) There is competition everywhere and the world is full of selfish people. There is dearth of goodness and nobility in human nature. This makes us gloomy, unhappy and disillusioned with life.

8. **Answer any three of the following questions in 30-40 words each :** [2 × 3 = 6]

(a) **Why has the poet's mother been compared to the 'late winter's moon'?**

(b) **What message does Stephen Spender convey through the poem 'An Elementary school Classroom in a Slum'?**

(c) **What was the plea of the folk who had put up the roadside stand?****

(d) **What will happen to Aunt Jennifer's tigers when she is dead?**

** Answer is not given due to the change in syllabus

**Answer :**

(a) The poet's mother has been compared to the 'Late winter's moon because she is too old and tired in the same way as the winter's moon is dull and without shine.

(b) The poet conveys the message that the life of the children living in slum is limited to their filthy dark room only. They have no future or hope and they don't know the outer world. This problem can be solved only when a school inspector, educationist or a governor comes here.

(d) After Aunt Jennifer's death, she will enter another world and her plight would remain same but her tigers would lead a fearless life and walk with grace, elegance and confidence.

**9. Answer the following questions in 30-40 words each : [2 × 3 = 6]**

**(a) How did Franz react to the declaration that it was their last French lesson?**

**(c) How did the instructor turn Douglas into a swimmer?**

**(d) Why did Sophie like her brother Geoff more than any other person?****

**Answer :**

(a) Franz was shocked to hear that there would be no French lessons from now onwards and felt extremely sorry for neglecting his lessons and regretted for not having the chance to learn its grammar rules. His French teacher M. Hamel would also be seen no more in the schoorl.

(c) The instructor taught Douglas to swim step by step. He told him to exercise and practice daily making his limbs and hands stronger. He also taught him to exhale and inhale outside and inside the water. He trained him perfectly by teaching all the strokes of swimming and thus, turned Douglas into a swimmer.

**10. Answer the following in 125-150 words : [10]**

**How did the peddler feel after robbing the crofter? What course did he adopt and how did he react to the new situation? What does his reaction reveal?**

**Answer :**

The peddler robbed thirty kronors from crofter's house who gave him shelter for the whole night in his cottage showing him kindness and hospitality. The peddler felt very happy after that and believed that it was not safe to walk along the public highway. So, he went into the woods but lost his way and when he could not get out of it, he thought the whole world with its lands and seas, its cities and villages is nothing but a big 'rattrap' and he was the victim of it. He felt that the bait in his case was the money that he had stolen from crofter's house. It was now his turn to be caught in the world's rattrap. Whenever we become the victim to temptations, we get trapped in this vicious circle.

Peddler could not help but think, and was trying to prove his act of robbery by thinking that crofter had put the money to trap him due to which, he got tempted. He was not the only one who was tempted, but almost everyone; some or the other day gets trapped.

**11. Answer the following in about 100 words : [5]**

**Exploitation is a universal phenomenon. The poor indigo farmers were exploited by the British landlords to which Gandhiji objected. Even after our independence we find exploitation of unorganised labour :**

**What values do we learn from Gandhiji's campaign to counter the present day problems of exploitation?**

**Answer :**

The conditions that prevailed during the time of Gandhiji are still existing. The weak and the poor are living in the same condition as they used to, prior to the independence, only the ways of exploitation have altered. The poor sharecroppers were exploited by the British landlords and their lands were snatched from them. Gandhiji fought for them systematically by gathering information and presenting the facts courageously and managed to get justice for the fear-stricken peasants. The striking feature of this was that he did not use any form of violence or unfair means. These methods can be adopted by us also, as we must be aware of our strengths and weaknesses and then proceed accordingly as one should not fall into the trap by any means. We should never compromise at the cost of our dignity or self-respect and thus sought out the solutions wisely.

**12. Answer the following in 125-150 words : [7]**

**What is the bond that unites the two-the old Mr. Lamb and Derry, the small boy? How does the old man inspire the small boy?**

**OR**

**Do you think Dr. Sadao's final decision was the best possible one in the circumstances? Why/ why not? Explain with reference to the story 'The Enemy'.**

**Answer :**

Derry and Mr. Lamb had the same sense of loneliness and they both lived a secluded life due to their physical inability. Derry experienced Mr. Lamb to be a peculiar person saying strange things which he was not able to understand, still he felt some kind of attraction towards him. He was differently-abled but was calm and

** Answer is not given due to the change in syllabus

maintained peace with people around him. He was not irritated by the children who called him 'Lamey-Lamb'. Before coming in contact with Mr. Lamb, Derry felt lonely due to the behaviour of people but lamb inspired him and filled him with self-confidence and determination. He taught him to live his life to the fullest without brooding over his burnt face. He told him to move ahead in life as his brain and senses were working completely. He imbibed positive thinking in him and showed him the way to live life in spite of his inability because he could do much better than others. This was the bond that united both of them.

OR

Yes, Dr. Sadao's final decision was the best possible one in the given circumstances because, he was hiding an enemy in his house and everybody was aware of it. Although he had no love for Americans and he was a true patriot still he saved him realising his 'sacred' duty to save a dying man. This brought an open defiance from his domestic staff and his wife had to do all the household work. Moreover, there was a high risk as he was sheltering an enemy. So, when he felt that the man was out of danger, he helped him to escape. He got a boat ready with food, water and extra clothing and asked the soldier to row the boat to a little island that was not too far from the coast and from there he could take a Korean fishing boat. One evening, there was no signalling and Dr. Sadao understood that the man was off safely on a Korean fishing boat and it was then that he had a sound sleep.

**13. Answer the following in 30-40 words : [2×2=4]**
**What sort of hunts did the Maharaja offer to organize for the high-ranking British officer? What trait of the officer does it reveal?****

**OR**

**What did Jo want the wizard to do when Mommy Skunk approached him?**

**Answer :**

OR

Jo wanted the wizard to hit Mommy Skunk because she wanted Roger Skunk to smell foul like before and that's why she took him to the wizard. There she hit him with the umbrella and Jo disliked it as according to her, the wizard was right.

## Delhi [Set-II]

**Note : Except for the following questions, all the remaining questions have been asked in previous sets.**

### SECTION- B
### (WRITING SKILLS)

**4. You are General Kanpur Senior Secondary School, Kanpur celebrate 'No Tobacco Day' on 4th September. Write a report in 100-125 words on the activities performed for your school newsletter. You Shobita/Sameer, cultural Secretary of your school. Invent the necessary details. [10]**

**OR**

**As a staff reporter of *The Times of India,* Delhi, you are asked to cover an incident of daylight robbery on the outskirts of Delhi, While the inmates were present in the house. Write a report in 100-125 words.**

**Answer :**

NO TOBACCO DAY

By Sameer/Shobita (Cultural Secretary)

4th June 20XX was celebrated as 'No Tobacco Day' by our school. An interesting and educative programme was organised to highlight the evil effects of tobacco on human body. The programme was inaugurated by the Director of Education. Students had exhibited pictures, posters and cartoons depicting the harms of smoking. A demonstration was presented to highlight the miserable state of tobacco addicts. The messages 'No Smoking' and 'Smoking is injurious to health' were preached quiet overtly. Then, the school doctor gave an illuminating discourse, on the harms caused by tobacco. He used slides, graphs and pictures to illuminate his talk. A short film was also screened to drive home the menace caused by tobacco smoking or chewing. The plight of the patients of cancer, caused by tobacco, made us shudder. We resolved not to smoke or chew tobacco.

OR

A DAYLIGHT ROBBERY

By Xyz (TOI, Correspondent)

**New Delhi, 2nd March, 20XX :** A daring daylight robbery occurred today in the house of Mr. William at 2.00 p.m. After lunch, the inmates were enjoying their afternoon siesta, when a group of marked young men entered the house, from the roof top. At first, they shut all the doors from inside and bolted the main gate from within. They cut off the telephone connection and then pointed their gun at the chest of Mr. William and demanded the keys of the safe, in which ornaments, cash, valuables and other important documents were kept. The head of the family was helpless and had

** Answer is not given due to the change in syllabus

to submit the keys. According to Mr. William, the robbers decamped with all the cash and jewellery they could. Apparently there was no loss of life. Personal rivalry is assumed to be one of the motives behind the crime.

**5. You are Kamini/Kamal of 10, Civil Lines Extension, Chanakyapuri, Agra. The main road leading to your locality has three open manholes causing frequent accidents at night, especially during the rainy season. Write a letter to the Editor, *The Times of India* drawing attention of the Municipal Commissioner to this problem of the residents . [10]**

**OR**

**You are Anoop/Akriti, of Baden Powel Senior Secondary School, Delhi. Write a letter to the Sales Manager, Aparna Publishing House, placing an order for the books (minimum four titles) for your school library. Invent the necessary details.**

**Answer :**[**]

10, Civil lines Extension,
Chanakyapuri, Agra.
20th June, 20XX
The Editor,
The Times of India,
**Subject : Open Manholes in the locality**
Dear Sir,
Through the column of your esteemed daily, I wish to draw the attention of the concerned authority to the problem caused by three open man-holes, on the main road of our locality. The number of accidents have increased due to these, specially during the rainy season. School students are vulnerable to such situations. All the complaint letters written to the concerned authorities till date have been in vain. The manholes need to be cemented at the earliest to avoid future mishaps.
Thank You.
Yours Sincerely,
Kamini/Kamal

**OR**

Baden Powel Sr. Secondary School,
Delhi.
20th July, 20XX
Sales Manager,
Aparna Publishing House,
Delhi.
**Subject : Order for Books**
Dear Sir,
We have been purchasing books from you for the past ten years and hence we want to place an order for the following books for our school library.

| | Name | Quantity |
|---|---|---|
| 1. | CBSE Complete Course English Core Class XIIth. | 5 |
| 2. | CBSE Political Science Textbook Class XIIth. | 20 |
| 3. | CBSE Biology Part I and II Textbook Class XIIth. | 10 |
| 4. | CBSE Home Science Textbook Class XIIth. | 25 |

Kindly, send these books and provide us the necessary discount. Payment will be done by cheque after receiving the books.
Thank You.
Yours Sincerely,
Anoop/Akriti

## SECTION- C
## (LITERATURE : TEXT BOOK)

**9. Answer any three of the following question in 30-40 words each : [6]**

**(a) How did Franz react to the declaration that it was their last French lesson?**

**(b) What made the lawyer lose his job? What was about it? (Poets and Pancakes)**[**]

**Answer :**

**(a)** Franz is overwhelmed with a feeling of remorse and regret when he heard that henceforth only German was to be taught in the schools of Alsace and Lorraine. He wished that he had spent so much time in trivial pursuit like seeking birds eggs, or going sliding in the saar. The books that seemed such a nuisance earlier appeared like an old friend now.

**13. Answer the following in 30-40 words each: [8]**

**(c) What according to the author, Tishani Doshi are the issues engaging public attention and interest during our tone?**[**]

●●

# Delhi [Set-III]

**Note: Except for the following questions, all the remaining questions have been asked in previous sets.**

## SECTION- B
## (WRITING SKILLS)

**3. You are Smitha/Sunil, Secretary AVM Housing Society. You are going to organise a yoga camp. Write a notice in not more than 50 words, urging the members of your society to come in large numbers to attend the camp. Invent all the necessary details. [5]**

** Answer is not given due to the change in syllabus

**Answer:**

AVM HOUSING SOCIETY, DELHI
NOTICE

1st March, 20XX

REJUVENATION YOGA CAMP

All residents are hereby informed that a yoga camp is being organised from March 5, 20XX to April 5, 20XX, at 6:00 am, in the Central Park of the Society. Residents from all age groups can participate in the camp. Please get yourselves registered at the earliest by contacting the undersigned.

Smitha/Sunil
Secretary
AVM Housing Society

5. **You have noticed many stray animals on the road during the busy hours of the day. These animals cause traffic jams as well as accidents. You have already written to the concerned authorities but no action has been taken so far. Write a letter to the Editor, The Hindu, drawing attention of the Muncipal Commissioner, Chennai. You are Shantha/Suresh, 12 M.G. Road, Chennai. [10]**

**OR**

**You are Nalini/Vishal, Hostel Warden, Zenith Public School, Kosikalan. Write a letter to the Sales Manager, Bharat Electronics and Domestic appliances Limited, New Delhi, placing an order for a few fans, microwave ovens and geysers that you wish to purchase for the hostel. Also ask for the discount permissible on the purchase.**

**Answer:**

12, M.G. Road,
Chennai

13th December, 20XX
The Editor,

The Hindu, Chennai.

**Subject : Havoc created by Stray Animals.**

Dear Sir,

Through the columns of your esteemed newspaper, I would like to draw the attention of the concerned authorities towards the havoc created by the stray animals in the city. These animals are left to roam during the busy hours of the day, resulting in huge traffic jams due to which the public has to face a lot of problem. Sometimes, accidents also take place as they appear suddenly in front of moving vehicles. The concerned authorities have already been notified but no action has been taken yet. The issue needs to be addressed at the earliest, in order to avoid any such inconvenience and mishaps.

Thank you.

Yours Sincerely,

Shantha/Suresh

**OR**

**Answer:**

Zenith Public School,
Kosi Kalan.
25th May, 20XX
The Sales Manager
Bharat Electronics and
Domestic Appliances Ltd.
New Delhi

**Subject : Placing an order for Electronic Appliances**

Dear Sir,

We need to purchase some electronic items for the school hostel. The list of items is as follows :

| Item | Quantity |
|---|---|
| 1. Ceiling fans | 10 |
| 2. Microwave ovens | 5 |
| 3. Geysers | 6 |

All the items should be properly checked and delivered at the school hostel between 10 a.m. to 2 p.m. Please make sure to give the necessary discount permissible on every item. The payment will be done through cheque only after the delivery of the items.

Thank You.

Yours Sincerely,
Nalini/Vishal
Hostel Warden

## SECTION- C
## (LITERATURE ; TEXT BOOK)

8. **Answer the following in 30-40 words each :**
   **(b) The poet says, 'And yet, for these children, these windows, not this map, their world'. Which world do these children belong to? Which world is inaccessible to them?**

**Answer :**

**(b)** Through these words, the poet is conveying the message that the children dwelling in the slums have no idea of the outer world. They are living a dark and gloomy life, with no future or hope. The world map hanging in their classroom is meaningless to them. They have the view of the outer world only though the windows of the classroom. So, these windows should be much bigger so that the children may be able to come out of them.

9. **Answer the following in 30-40 words each : [2]**
   **(a) What did M. Hamel tell them about the French language? What did he ask them to do and why?**

**Answer :**

**(a)** It was, the last day of M. Hamel in the school and he told them the importance of French language. He praised the language as the most beautiful and sweetest language of the world. If one keeps his language with him, he can liberate himself from any type of slavery.

●●

# PHYSICAL EDUCATION 2020

**Time allowed : 3 hours** **Maximum marks : 70**

**General Instructions :**

(i) The question paper consists of **34** questions.
(ii) All questions are compulsory.
(iii) Question **1–20** carry **1** mark each and are multiple choice questions.
(iv) Question **21–30** carry **3** marks each and should not exceed **80–100** words.
(v) Question **31–34** carry **5** marks each and should not exceed **150–200** words.

## SECTION–A

**1. The total number of matches in a knock out tournament of 34 teams are: [1]**

(a) 31 (b) 32
(c) 33 (d) 35

**Ans.** (c) 33

**2. The primary goal of Intramural competition is: [1]**

(a) To provide opportunity for mass participation of students
(b) To participate in inter-school competition
(c) To provide intra-school competition
(d) All of the above

**Ans.** (a) To provide opportunity for mass participation of students.

**3. The food component present in sugar is: [1]**

(a) fats (b) protein
(c) vitamin (d) carbohydrate

**Ans.** (d) Carbohydrates.

**4. The main source of Vitamin C is: [1]**

(a) Guave (b) Egg
(c) Milk (d) Banana

**Ans.** (a) Guava.

**5. Which asana is helpful in maintaining normal blood pressure? [1]**

(a) Shavasana (b) Padmasana
(c) Shalabhasana (d) Vakrasana

**Ans.** (a) Shavasana.

**6. Gomukhasana, Chakrasana and Matsyasana are helpful in curing which disease? [1]**

(a) Diabetes (b) Backpain
(c) Asthama (d) Obesity

**Ans.** (c) Asthama

**7. When child is not able to adjust within society or having no friends, is suffering from: [1]**

(a) ADHD (b) ASD
(c) ODD (d) OCD

**OR**

**Obsessive Compulsive Disorder is a/an:**

(a) Argumentative disorder
(b) Anxiety disorder
(c) Receiving and responding disorder
(d) All of these

**Ans.** (c) ODD

**OR**

(b) Anxiety disorder

**8. Cognitive disability may cause difficulty in which of the following activity? [1]**

(a) Reading (b) Writing
(c) Mathematics (d) All of the above

**Ans.** (d) All of the above

**9. Menarche is defined as the: [1]**

(a) Ending of menstrual period in women
(b) Beginning of menstrual period in women
(c) Time of pregnancy
(d) Beginning of pregnancy

**Ans** (b) Beginning of menstrual period in women.

**10. Scoliosis is a postural deformity related with: [1]**

(a) foot (b) leg
(c) vertebral column (d) hand

**Ans.** (c) Vertebral column.

**11. Sit and reach test is conducted for: [1]**

(a) Flexibility (b) Motor fitness
(c) Endurance (d) Speed

**Ans.** (a) Flexibility.

**12. Barrow Fitness Test does not include: [1]**

(a) Medicine Ball Put
(b) Zig-zag Run
(c) 600 metres Run
(d) Standing Broad Jump

**OR**

**Rock Port one mile test is conducted to measure:**

(a) Cardio-vascular fitness
(b) Senior citizen's fitness
(c) Vital capacity
(d) Muscular strength

**Ans.** (c) 600 metres run.

**OR**

(a) Cardio vascular fitness.

**13. Movement possible in Ball and Socket joint are: [1]**

(a) Rotation (b) Flexion
(c) Extension (d) All of the above

**OR**

**The law of acceleration is also known as: [1]**

(a) Law of inertia
(b) Law of action and reaction
(c) Law of momentum
(d) Boyle's law

**Ans.** (d) All of the above.

**OR**

(c) Law of Momentum

**14. Match List 1 and List 2, Selection the correct option: [1]**

| Sr. No. | List 1 | List 2 |
|---|---|---|
| 1. | Abrasion | Joint Injury |
| 2. | Green Stick Fracture | Soft Tissue Injury |
| 3. | Shoulder Dislocation | Cause of Sports Injury |
| 4. | Lack of Fitness | Bone Injury |

(a) 2, 4, 1, 3 (b) 3, 2, 4, 1
(c) 4, 3, 1, 2 (d) 1, 3, 2, 4

**Ans.** (a) 2,4,1,3

**15. Bio-mechanics helps in which of the following? [1]**

(a) In improving technique
(b) In improving designs of sports equipment
(c) In improving performance
(d) All of these

**Ans.** (d) All of the above.

**16. The force of friction depends upon: [1]**

(a) Nature of surface of contact
(b) Material of objects in contact
(c) Both (a) and (b)
(d) None of the above

**Ans.** (c) Both (a) and (b)

**17. Emotionally unstable, anxiety, sadness are attributes of which personality dimension? [1]**

(a) Extroversion (b) Neuroticism
(c) Agreebleness (d) Openness

**OR**

**Aggressive behaviour of a sportsperson is influenced by:**

(a) Emotional identification with the team
(b) Tactical ability
(c) Goal orientation
(d) All of the above

**Ans.** (b) Neuroticism.

**OR**

(d) All of the above

**18. The body structure of mesomorphic people is like: [1]**

(a) Fatty
(b) Large muscles and bones
(c) Solid
(d) Obese

**Ans.** (b) Large Muscles and bones

**19. Fartlek training was developed in: [1]**

(a) Sweden (b) The USA
(c) India (d) The U.K.

**Ans.** (a) Sweden

**20. If a muscle contracts and changes its length to produce force, the contraction type is: [1]**

(a) Isotonic (b) Isometric
(c) Isokinetic (d) None of these

**Ans.** (a) Isotonic

## SECTION – B

**21. Explain the procedure of Harvard step test in detail. [3]**

**OR**

**List down the test items of Rikli and Jones fitness test and explain the procedure of any one.**

**Ans.** The Harvard Step test is a test of aerobic fitness, developed by Brouha et al. (1943).

**Equipment:** A bench or box 20″ for male or 16″ for female, stop watch.

**Administration**

(i) A tester gives a demonstration of the stepping- up Style to be followed by the subjects during the test.
(ii) After the command 'Go' the stop watch is started and athlete steps – up and down for five minute at the rate of 30 steps per minute.
(iii) The stop watch is stopped after 5 min and athlete is asked to stop.
(iv) The heart – rate is measured between 1 to 1.5 min, after finishing the exercise.
(v) The same is repeated after 2 to 2.5 minutes and 3 to 3.5 minutes. The pulse of all the 3 half minute counts are recorded together and computed in the fitness index.

**OR**

The, tests designed by Rikli and Jones for senior citizen fitness are

(i) Chair stand test for lower body strength.
(ii) Arm curl test for upper body strength and endurance.
(iii) Chair sit and reach test for lower body flexibility.
(iv) Back scratch test for upper body flexibility.
(v) Eight foot up and go test for coordination and agility.
(vi) Six minute walk test for aerobic fitness and endurance.

The Arm Curl Test is a test of upper body strength. The purpose of this test is to measure upper body strength and endurance. The subject has to do as many arms curls as possible in 30 sec. This test is conducted on the dominant arm side (or stronger side), the procedure is:

**(i)** The subject sits on the chair holding the weight (8 pounds for men / 5 pounds for women) in the hand using a suitcase grip (palm facing towards the body) with the arm in a vertically down position beside the chair.

**(ii)** The upper arm is placed against the body so that only the lower arm is moving (the tester may assist to hold the upper arm steady).

**(iii)** The subject curls the arm up through a full range of motion, gradually turning the palm up (flexion with supination).

**(iv)** Then the arm is lowered through the full range of motion, gradually return to the starting position. The arm must be fully bent and then fully straightened at the elbow.

**(v)** Repeat this action as many times as possible within 30 sec.

**(vi)** Score is the total number of controlled arm curls performed in 30 sec.

**22. League tournament is a better way to judge the best team of the tournament. Comment. [3]**

**Ans.** A league tournament is also known as round-robin tournament. In this type of tournament, all teams play against each other irrespective of winning or losing. In single league tournament all participating teams compete once, with each other, where as in double league, each team plays with every other team twice, without any consideration of victory or defeat.

**Advantages of League Tournament**

**(i)** Since there is no early elimination, the better teams reach to win.

**(ii)** Every team gets maximum opportunity to show its efficiency.

**(iii)** These tournaments take long time and lot of matches are played and spectators can enjoy more.

**(iv)** Selectors get better opportunity to watch the performance of each player as all teams play equal number of matches.

**(v)** Ample number of opportunities is available to the teams/players to improve their performance.

**(vi)** Also, the ranking of team can be judged by the maximum number of matches or points won by the team.

**23. "Asanas can be used as a preventive measure." Comment. [3]**

**Ans.** Asana plays an important role to prevent the lifestyle related diseases. They are as follows:

**(i) Maintain Weight:** Nowadays, obesity has become a challenging problem for the society. By doing asanas daily, we exercise entire organs of our body. By practising asanas regularly, it burns the excess fat of the body, and hence maintains healthy weight.

**(ii) Endocrine function normalize:** Diabetes ~~in~~ is one of the most common lifestyle related disease. Diabetes occur due to either the pancreas not producing enough insulin or the cell of the body not responding properly to the insulin produced. The asana like Bhujangasana, paschimotan asana, pavanmuktasana, ardh-matsyendrasana etc., help to stimulate and regulate pancreas to produce insulin which helps to maintain glucose level in the body.

**(iii) Respiratory efficiency increases:** Asthma is a common lifestyle related disease. It occurs due to narrowing of trachea, due to which oxygen supply to lungs and body is compromised. Asana help in maintaining the trachea wide open, increase the lungs capacity and blood flow to lungs. Thus, oxygen supply is maintained and asthmatic episodes are reduced and respiratory efficiency is increased.

**(iv) Regulates Blood Flow:** Yoga or Asanas regulate blood flow in the body and help to maintain a constant blood pressure. They help the body to relax and thus reduce the high blood pressure. Yoga calm down the body and the mind and thus blood pressure is normalised.

**24. Discuss in detail any one type of disorder. [3]**

**Ans.** Disorder is usually used for mental disabilities. Disorder is any ailment that disturbs the health of an individual. Disorder creates hindrance in an individual's performance and reduces his efficiency. In the beginning disorder seems to be ordinary but they usually grow or spread in a harmful manner in an individual. Most probably, a disorder cannot be detected on time, as a result of which, a simple disorder is changed into a disability. A disorder disrupts the normal functioning of an individual.

**Types of disorders**

**(i)** Attention Deficit Hyperactivity Disorder (ADHD)

**(ii)** Sensory Processing Disorder (SPD)

**(iii)** Autism Spectrum Disorder (ASD)

**(iv)** Oppositional Defiant Disorder (ODD)

**(v)** Obsessive Compulsive Disorder (OCD)

**Attention Deficit Hyperactivity Disorder (ADHD):** It is a group of behavioural symptoms that include inattentiveness, hyperactivity and impulsiveness. It is medical conditions that affects how well can someone sit still, focus and pay attention. The individual with ADHD have some problems focusing in some activities. This type of disorder is found more common in boys than in girls.

**The various causes of ADHD are as follows:**

**(i) Genetic factors**: It is not a disorder that passed socially. Studies shows that parents, siblings, and children of people with ADHD may be up to five times more likely to have the disorder than the people who are not related to someone with ADHD.

**(ii) Brain injuries:** When a baby's brain is damaged before or after birth this could make the baby more likely to develop ADHD later on.

**(iii) Low birth weight:** It is observed that children with low birth weight are more likely to develop ADHD.

**(iv) Trauma and brain diseases:** Trauma during birth and brain diseases may lead to develop ADHD.

**(v) Diet:** There are a number of evidences which shows that taking a particular type of food or food additives play a significant role in causing ADHD.

**25. Suggest physical exercises for childhood and adulthood. [3]**

**Ans.** Regular exercise keeps the human body livelier, fitter and in better condition, thus delaying the ageing processes. Regular physical activity promotes growth and development and has multiple benefits for physical, mental, and psychosocial health that undoubtedly contribute to learning. Specifically, physical activity reduces the risk for heart disease, diabetes mellitus, osteoporosis, high blood pressure, obesity, and metabolic syndrome; improves various other aspects of health and fitness, including aerobic capacity, muscle and bone strength, flexibility, insulin sensitivity, and lipid profiles; and reduces stress, anxiety, and depression.

**Physical exercises for childhood**

**(i)** Exercise must be planned which include co-coordinative ability like catching, throwing, jumping and kicking.

**(ii)** Exercise must be structured which improve body control and strength.

**(iii)** In this stage children should be introduced to competitive sports to develop social consciousness.

**(iv)** Exercise like running, swimming, cycling, resistance exercise and fast work or rapid movement exercise must be included for whole some development of the child.

**Physical exercises for adulthood**

**(i)** Adult should do brisk walking, bike riding, dancing, and swimming for active life style.

**(ii)** They should also indulge in running, aerobic exercise, weight training, Calisthenics for muscular strength.

**(iii)** They should engage themselves in resistance training for at least two days a week.

**(iv)** They should actively engage themselves in team events or endurance activity.

**26. What do you understand by food myths? [3]**

**OR**

**What are the pitfalls of dieting?**

**Ans.** Various food myths exist not only in India but all over the World as they sound like they could be true. Individuals usually get confused with the questions like what to eat, when to eat and how often to eat. Food myths are unscientific unfounded practices followed by people regarding food items. They are more psychological in nature than actual. Wrong information regarding nutrition without any logic that we start following creates food myth.

**Various food myths are:**

**(i) Eggs increases cholesterol level so avoid them:** There is no doubt that eggs are good source of health. An egg provides the various nutrients. It is as per daily requirements of cholesterol by the body.

**(ii) Drinking while eating makes you fat:** The actual fact behind this misconception is that enzymes and their digestive juices will be diluted by drinking water while eating which slows down the digestion which may lead to excess body fat.

**(iii) Myth:** Low fat or No fat diet are good. **Fact:** Body needs fats for energy, tissue repair and to transport vitamin A, D, E.K. Just cut down on salivated fat eating unsaturated fats.

**(iv) Myth Crash:** Dieting or Fasting may lose weight.

**Fact:** It may be true in short term but ultimately it hinder weight loss. Loosing over the long term burns off fat whereas crash dieting or fasting not only removes fat but who leans muscles.

**(v) Myth:** Food eaten late night is more fattening.

**Fact:** It does not make much change.

**(vi) Myth:** Low fat milk has less calcium that full fat milk.

**Fact:** Skimmed and semi skimmed actually have more calcium because it is in watery part and not in creamy part of milk.

**(vii) Myth:** Vegetarian cannot build muscles.

**Fact:** Vegetarian can built muscles as meat eaters by getting their proteins from vegetables such as cheese, nuts, pulses etc.

**(viii) Myth:** Healthy food is expensive.

**Fact:** Tinned, stored, packed food is expensive. Whereas local and seasonal food is inexpensive.

**(ix) Other Food myths are:** Some of the other food myths are:

(a) Potatoes make you fat.
(b) Fat – free products will help you lose weight.
(c) Eggs cause heart problems.
(d) Weight gain is caused by unprocessed food.
(e) Food cravings are driven by deficiency.
(f) Do not drink milk after eating fish.
(g) Spicy food cause ulcer.
(h) Eat less if you have fever.

**OR**

Food restriction and dieting can lead to a foggy mind and lack of focus. A balanced diet includes a variety of foods, which means it has foods from all food groups, including protein, grains, dairy, and fruits and vegetables. If the body and brain are not getting enough of a certain macronutrient group, such as carbs or fat, or enough energy (calories), they will not work optimally.

**Following are pitfalls of dieting -**

**(i) Extreme Reduction of Calories:** People reduces the diet considerable which causes low level of energy thus people feel tiredness and body aches.

**(ii) Skipping meals:** People often skip meals to reduce weight where as in subsequent meal they take large amount of food.

**(iii) Low energy Diet:** The person take diet without fats and less carbohydrates by which health is affected.

**(iv) Not performing physical Activity:** People often consider that reducing diet for controlling weight but they neglect physical activities which is equally important for healthy Lifestyle.

**(v) Taking less liquid:** People often think that drinking water or liquid makes them to gain weight which is wrong.

**27. Discuss the preventive measure of sports injuries. [3]**

**Ans.** "Sports injuries" are the type of injuries that occur during participating in sports or competition, training sessions or sports activities. One of the important objectives of sports medicine is preventing injuries. It also prevents other physical, mental, social and financial harm accompanying sports injuries. General techniques that can prevent sports injuries are:

**(i) Warm-up and Cool-down:** A well-structured warm-up and cool-down is necessary to increase blood and nutrient flow and concentration. Also it helps in relaxation, improved flexibility and recovery of muscles.

**(ii) Planning a Session**: Careful planning of training and rehabilitation sessions allows gradual specific adaptations. It reduces the damage to the tissues as a result of training.

**(iii) Using Protective Equipment:** The use of protective equipment like proper footwear, helmets, goggles, gum shield, shin pads and gloves prevents many sports injuries.

**(iv) Psychological training:** Some form of mental skills training and practice could reduce injuries by reducing anxiety and improving concentration.

**(v) Adherence to the rules:** If all performers are aware of and adhere to the rules and laws of the particular sport, then injuries can be reduced to great extent.

**28. With suitable examples explain the application of Newton's law in sports. [3]**

**Ans.** Newton's laws of motion and their application in the field of sports are:

**The three laws of motion formulated by Newton are:**

(i) **Law of inertia:** According to this law a body at rest will remain at rest and a body in motion will remain in motion at the same speed and in the same direction unless acted upon by an external force. There are great examples of this law in sports such as starting in rowing, starting in sprinting, starting in throwing the hammer. Basically, if an object is in motion, it remains in motion unless something or some external force stops it. The external force may be gravitational force, the surface of playing field or a defensive player etc. For Example: Starting in sprinting, starting in rowing, starting in hammer throw.

(ii) **Law of acceleration:** According to this law, A change in motion is directly proportional to the force producing it and inversely proportional to its mass. If two unequal forces are applied to objects of equal mass, the object that has greater force applied will move faster. Conversely, if two equal forces are applied to objects of different masses, the lighter mass will travel at a faster speed. For example: In hammer throw, a thrower who is stronger will throw the hammer farther than a thrower who is less strong.

(iii) **Law of reaction:** According to this law, 'For every action there is an equal and opposite reaction.' There are so many examples in sports where this law is applied. For example: In swimming a swimmer pushes the water backwards (action) and the water pushes the swimmer forward (reaction) with the same force. In swimming, a swimmer pushes the water backwards (action). The water pushes the swimmer forward (reaction) with the same force.

**29. Discuss in detail any three techniques of motivation. [3]**

**OR**

**Define balanced Diet. Explain any four Micro Nutrients.**

**Ans.** In games and sports performance does not depends upon a single factor. The performance of an athlete directly depends upon his/her coach, the environment in which he or she took training, his personal and social dimension, and the most important factor is motivation. Following are the techniques used in games and sports by an athlete and his/her coach.

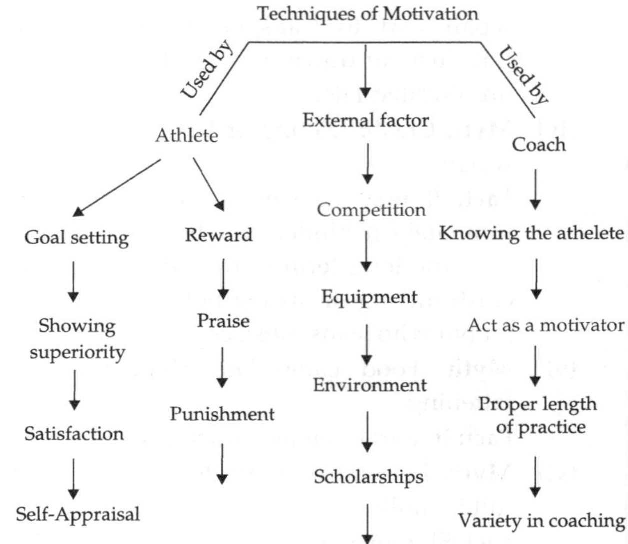


Techniques of motivation for higher achievement in sports are:

(i) **Knowing Your Athlete:** Each athlete comes from a different background with varied experiences and having different degrees of maturity.

(ii) **Level of Aspiration or Goal Setting:** As we know that achieving performance goals is a sign of competence that affects motivation positively, it is necessary to set realistic goals based on one's own abilities.

(iii) **Identifying Incentive Factors:** If an athlete perceives that any particular kind of experiences are available to him in a given sport and he feels that these will be pleasant, enjoyable or satisfying, then he will choose to participate in that game or sport and not any other.

**OR**

A diet that contains the right amount of carbohydrates, proteins, fats, minerals, salts, vitamins and water is called a balanced diet.

**Elements of balance diet are:**

(i) **Carbohydrates:** These provide essential growth as well as help in repairing muscles and other body tissues.

(ii) **Proteins:** Body building elements for example: Fish, Meat are good source of it.

(iii) **Fats:** These are the source of energy, plays an important role in many chemical processes in the body.

(iv) **Minerals:** These are essential for proper growth and functioning of the body.

(v) **Water:** Cleans Body and regulates the temperature of the body.

Micro-elements are those minerals which are also known as trace elements and are required in very less amount in the body, 0·01 g of each every day.

(i) Iron is needed for the production of haemoglobin as its deficiency would cause anaemia.

(ii) Potassium is essential to keep the nervous system and muscular system fit and active.

(iii) Phosphorus helps in the formation of bones and teeth. It keeps the muscles and nerve activities normal.

(iv) Calcium is needed for the growth and development of bones and teeth.

**30. Define flexibility and its types.** **[3]**

**Ans.** Flexibility is the range of movement of joints. Flexibility can be defined as the ability to execute movement with greater amplitude or range.

**Types of Flexibility are:**

(i) **Active Flexibility:** It is the ability to do movement for a longer distance with greater amplitude without external help e.g. Exercise without partner. Ability to do movement for a greater distance with external help.

(ii) **Passive Flexibility**: These exercise are performed with the help of an external support (Static or dynamic). It is always more than active flexibility.

Flexibility plays an important role in training the athletes by the means of increasing the movement and helping in the prevention of injuries. Different methods can be used to improve the flexibility such as Ballistic method, slow stretch and hold method.

(i) **Ballistic Method:** It is the oldest form of doing stretching exercises. This method involves jerk in movement.

(ii) **Slow Stretching Method:** In this method the muscle or joint involved is stretched to the maximum possible limit using slow movement.

(iii) **Slow Stretching and Holding Method:** It is the extension of slow stretching method.

(iv) **Post-Isometric Stretching:** This method of-flexibility development is based on the principle of proprioceptive neuromuscular facilitation.

## SECTION – C

**31. What is circuit training ? Draw a diagram of circuit training with 12 stations and explain its importance in sports.** **[5]**

OR

**Draw a knock out fixture for 25 teams with all steps involved.**

**Ans.** Circuit training is a very popular and effective method of training for the improvement of physical fitness components. In circuit training, several exercises are done one after the other. Completion of one set of each exercise in rotation is called one round. There are normally three or more rounds in circuit training which consists of 7-12 exercises generally. The exercises in a circuit are arranged in such a manner that different muscle groups are exercised in rotation. A specific circuit, however, can consist of several exercises involving the same middle groups.

**Importance of Circuit Training in Sports:**

(i) It is the best method for beginners as it develops strength and endurance. Maximum functioning of muscles can be gained in a single circuit.

(ii) It gives relief from any kind of tension. The trainee gains good result in a short period. It does not create boredom as variety of exercises can be included etc.

(iii) It is a workout routine that combines cardiovascular fitness and resistance training.

(iv) The initial routines were arranged in a circle, alternating between different muscle groups.

(v) Circuit training plays an integral role in the offseason workouts of many professional athletes.

(vi) Develops strength and endurance.

(vii) An appropriate form of training for most sports.

(viii) Can be adjusted to suit age, fitness and health of the athlete.

(ix) Exercises are simple enough to make each athlete feel a sense of achievement in completing them.

(x) A wide range of exercises to select from which will maintain the athlete's enthusiasm.

**OR**

Tournament is a large contest of many rounds among various teams. It is a competition held among various teams in a particular activity according to a fixed schedule where a winner is decided.

Total Number of teams = 25

Total Number of matches= 25 – 1 = 24

Number of teams in upper half = $\frac{N+1}{2} = \frac{25+1}{2}$

= 13 team

Number of teams in lower half = $\frac{N-1}{2} = \frac{25-1}{2}$

= 12 team

Total number of byes is = 32 – 25 = 7 byes

Number of by is in upper half = $\frac{NB-1}{2} = \frac{7-1}{2}$

= 3 byes

Number of byes in lower half = $\frac{NB+1}{2} = \frac{7+1}{2}$

= 4 byes

Number of round = 2 × 2 × 2 × 2 × 2 = 5 Round

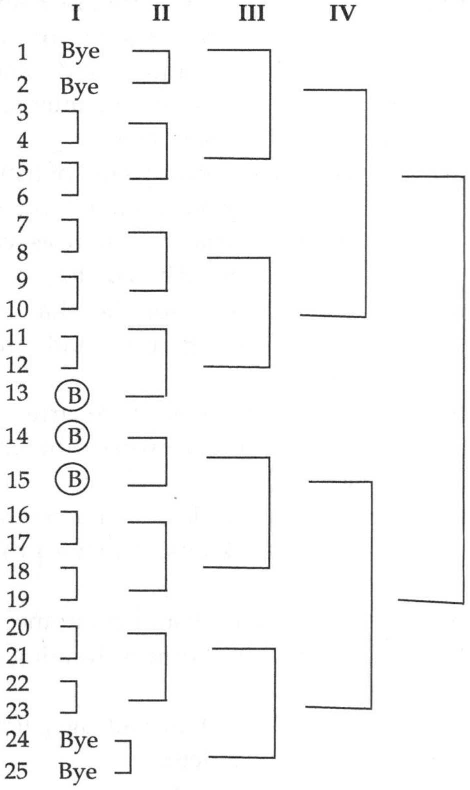

I II III IV

1 Bye
2 Bye
3
4
5
6
7
8
9
10
11
12
13 (B)
14 (B)
15 (B)
16
17
18
19
20
21
22
23
24 Bye
25 Bye

**32. How physical activities are helpful for children with special need? Explain strategies to make physical activities assessable for them. [5]**

**Ans.** Children need to be more engaged in the physical activity and exercises rather than being a spectator or counting score in the name of taking part of physical education. Physical activities not only provide health benefits but also other important social benefits as well.

**Advantages of physical activities for children with Special needs are:**

(i) **Reduced level of anxiety, stress and depression:** Physical activities may help in reducing the level of anxiety, stress and depression of children with disabilities.

(ii) **Improved social interaction:** Physical activities provide ample opportunities for improving the social interaction among children with special needs. Social relations are developed during involvement in physical activities.

(iii) **Better emotional and psychological health**: Physical activities are beneficial for children with special needs because such activities improve psychological and emotional health.

(iv) **Cognitive benefits:** Physical activities lead to cognitive skill improvement in children with disabilities. These activities allow them to discover and access strengths that cannot be challenged in the classroom setting.

**Strategies to make physical activities accessible for children with special needs**

(i) **Medical check-up:** if we want to make physical activities accessible for the children with special needs, we need to understand the type of disabilities of children and for this purpose complete medical check-up of the children is required because without complete medical check-up, the teachers of physical education cannot come to know about the type of disability child is facing.

(ii) **Activities based on interests:** Physical activities must be based on interest, aptitudes, abilities, previous experience and limitations of children with special needs. The teachers of physical education should have deep knowledge of limitations, interest and aptitudes of children.

(iii) **Different instructional strategies:** A variety of different instructional strategies such as verbal, visual and peer teaching should be used for performing various types of physical activities. By this children get opportunity to learn by their own and become independent.

**(iv) Modification of rules:** Rules can be modified according to the needs of the children. They can be provided extra time or attempt to perform a physical activity.

**(v) Specific environment:** For children with special needs the area should be limited. In case of children who have autism, they must be provided specific area because they may need some time to relax.

**33. Give your outlook on participation of Indian women in sports. [5]**

**Ans.** Over the past several decades the participation of women in sports in sports field has increased tremendously. But really, it is a matter of regret for all of us to know that sports is such a field where gender inequality in strongly evident. The general social environment has not only inhibited women from participation in sports but has also criticise them when they participate. Many people comment for women "Why don't they stay in the kitchen where they belong"? But now time has changed. Women are capable of changing society. Now the ideology suggests that women are participating in every sphere of life and proving themselves globally. There are number of reasons for the low rate of sports participation by women in India for example:

**(i) Gender equity and social attitude**

1. No parental encouragement
2. Traditional society
3. Less motivation and inspiration
4. Women constraining other women

**(ii) Lack of plans and initiatives for sports women by the Government**

1. Male dominant culture
2. Less availability of women coaches'
3. No independent games facilities for women
4. More emphasis on study
5. Less competition

(iii) Economic Factors

(iv) Social Customs and Rights

(v) Low Health Consciousness

(vi) Stress on Academics

(vii) Media Coverage

**(viii) Lack of Incentives and Career:** Attitude and prejudices of society play significant role in sports participation; some females avoid certain sports for fear of being perceived masculine. Due to such attitude and prejudices of society regarding sexuality inhibit females to participate.

**34. Which asanas will you suggest for back pain? Explain the procedure benefits and contraindications of any two asanas recommended for back pain. [5]**

**OR**

**What is the effect of exercise on cardio respiratory system and muscular system?**

**Ans.** Back pain is a hazard of modern lifestyle. It is mainly caused due to sedentary, lifestyle sitting in front of television or computers for long time, by attaining improper habits of sitting, etc.

**Yoga asanas recommended for prevention and management of back pain are:**

(i) Tadasana

(ii) Ardhamatasyendrasana

(iii) Vakrasana

(iv) Bhujangasana

(v) Shalabhasana

**(i) Vakrasana**

**(A) Procedure of Vakrasana**

1. Sit in dandasana position.
2. Keep your right leg straight and stretched.
3. Bend the left leg from knee and place the left foot beside the right knee and the left knee raised upward.
4. Inhale and raise the arms shoulder high, keeping the elbows straight.
5. Exhaling, twist to the left, place the right arm by the outer side of the left knee and hold the left ankle with the right hand.
6. Take the left hand behind the back keeping the palms on the floor.
7. Look backward towards the left side.
8. Hold on the position as long as comfortable while breathing normally.
9. Then inhale and raise the right arm shoulder high, keeping the elbows straight.
10. Exhaling, release the left twist, place the right hand by the side of right kip and left hand by the side of left hip.
11. Take a deep breath and relax.
12. Repeat the same from the other side.

**(B) Benefits of Vakrasana**

1. Increase hips and spine flexibility.
2. Removes the wastes and improves digestion.
3. Relieve fatigue, sciatica, backache and menstrual discomfort.

4. Removes extra fat from the waist hence prevents obesity.
5. Energises and stretches the backbone.

**(C) Contraindications of Vakrasana**

1. It is recommended to avoid during pregnancy and menstruation due to the strong twist in the abdomen.
2. People with Heart, abdominal or brain surgeries should not practice this asana.
3. People having severe spinal problems should avoid this asana.

**(ii) Shalabhasana**

**(A) Procedure of Shalabhasana**

1. Lie flat on the stomach with the legs and feet together.
2. Place the arms under the body. Keep the hands clenched and the palms downward.
3. Slightly stretch the chin in the forward direction. Rest the chin on the floor.
4. Gently close the eyes, relax the whole body, inhale deeply. This is the starting position.
5. Slowly start raising the legs as high as possible without straining. Do not bend the knees, keep the legs straight and together.
6. The elevated state of the legs is achieved by employing pressure with the arms against the floor while contracting the lower back muscles.
7. Without straining the body, hold the final position for as long as you find it comfortable.
8. Exhale and slowly lower the legs on the ground. Come back to the starting position. Relax the whole body and turn the head to one side, whichever deems comfortable.
9. Let the breathing and heartbeat return to normal.

**(B) Benefits of Shalabhasana**

1. Strengthens the lower back and tones the organs of pelvic region.
2. Improves the functioning of stomach, liver and bowels.
3. Relieves from back pain, slipped disc and sciatica but only if the condition is not severe.
4. Enhances the functioning of abdominal organs particularly of the lower abdomen.
5. Strengthens the hips muscles and increases appetite.

**(C) Contraindications of Shalabhasana**

1. Do not strain the body while performing this asana.
2. Do not bend the knees or raise the chin above the floor.
3. This asana is not suitable for people having high blood pressure or a weak heart.
4. People suffering from abdominal surgeries, spinal injuries or hernia should avoid practicing this asana.

**OR**

Exercise strengthens the heart and improves the circulation. The increased blood flow raises the oxygen levels in the body. This helps lower the risk of heart diseases such as high cholesterol, coronary artery disease, and heart attack. Regular exercise can also lower the blood pressure and triglyceride levels.

**The effects of exercise on the cardiovascular system are:**

(i) Cardiac output is the amount of blood pumped by the heart in 1 min. This increases directly with increasing exercise intensity.

(ii) The heart rate increases from a resting rate of 72 beats / min to 150 beats / min or even more.

(iii) The stroke volume, meaning the amount of blood pumped into the Aorta with every heartbeat, increases from a resting volume of 70-90 mL to 100-120 mL per beat.

(iv) Exercise increases the plasma volume of blood by 12 Op: but total blood volume may reduce slightly.

(v) Blood flow is redistributed with more blood going to the muscles, heart and skin, while blood in the kidneys and abdomen is reduced.

(vi) Blood pressure increases due to exercise because there is more blood flowing in the blood vessels.

(vii) Cardiac hypertrophy occurs, meaning that the heart size increases to take care of the larger requirement of blood during exercise.

**Effects of exercise on respiratory system are:**

(i) **Increase in Tidal air capacity:** By doing regular exercise it has been noted that there is an increase in the amount of tidal air capacity of an individual.

(ii) **Decrease in rate of respiration:** When a beginner starts exercising his rate of respiration increases. But when the same individual perform exercise daily, his rate of respiration decreases in comparison to the beginner at rest.

(iii) **Strong will power:** Regular exercise increases the will power of an individual. As pranayama, the specific exercise for lungs increases the will power of the doer.

(iv) **Unused alveolus becomes active:** Regular exercise activates the unused alveolus because much amount of oxygen is required in vigorous activities of daily routine. The passive alveolus become active.

(v) **Increase in vital air capacity:** The capacity of vital air varies from 3500cc to 4500cc in a normal adult. Due to regular exercise its capacity increases upto 5500cc.

**Effects of exercise on muscular system are:**

(i) **Size and shape of muscle changes:** Regular exercise changes the shape and size of the muscle. Cells of the muscles are enlarged which change the shape and size of the muscle.

(ii) **Correct body posture:** Regular exercise keeps the correct posture of the body by strengthening the muscles. The postural deformities do not occur. If there is any physical deformity, then it is removed.

(iii) **Food storage increases:** The capacity of food storage in body can be enhanced by doing regular exercises. This stored food can be utilised immediately when required.

(iv) **Toned muscles:** Regular exercise helps in keeping the muscles in toned position. Muscles become firm and maintain a slight, a steady pull on the attachments.

(v) **Efficient movement of muscles:** The movement of muscles becomes efficient and smooth. The movements during different activities become attractive.

(vi) **Change in connective tissues:** The connective tissues become powerful. These tissues can bear the stress of strenuous activity.

●●

# PHYSICAL EDUCATION 2019
## Outside Delhi

**Time allowed : 3 hours** **Maximum marks : 70**

### SECTION–A

**1. What do you understand by 'Planning in Sports'? [1]**

**Ans.** Planning is an intellectual process of thinking in advance about setting of goals and developing strategies which are required to attain the goal efficiently in the field of sports. Deciding the future course of action and developing strategies to attain the goals efficiently in the field of sports is known as planning.

**2. Write briefly about 'Micronutrients'. [1]**

**OR**

**Suggest two reasons, why our body requires food-supplements.**

**Ans.** Micronutrients are those nutrients which are needed only in minuscule or small amounts. They (Vitamins and Minerals) enable the body to produce enzymes, hormones and other substances essential for proper growth and development. Lack of it may cause a major threat to the health. Example : various micro nutrients are vitamin A, B, C, D, E and K, minerals such as iron, calcium, magnesium, iodine etc. Iodine deficiency may result in impaired cognitive development in children. Vitamin A deficiency results in night blindness, Iron deficiency results in anaemia.

**OR**

**Ans.** Two Reasons for requirement of Food Supplement in the body are:

**(i)** To balance the lack of proper Nutrition in our diet.

**(ii)** Desired intake of vitamins and minerals.

**(iii)** Helps in proper growth and development

**(iv)** To overcome deficiency diseases

**(v)** Regulate important functions of cardio-vascular, nervous, endocrine and digestive systems.

**3. Mention any two disability etiquettes. [1]**

**Ans.** Two Disability Etiquettes are:

**(i)** When you meet a person with visual impairment, always identify yourself and others who may be with you.

**(ii)** In case of introduction to a person with a disability, it is appropriate to shake hands.

**(iii)** Always avoid asking personal questions to an individual who is Differently abled.

**(iv)** Always have conversation at a normal tone of voice. Don't talk in high pitch to such individuals.

**4. What do you mean by Intellectual Disability? [1]**

**Ans.** It is a disability characterised by significant limitations both in intellectual functioning (reasoning, learning, problem solving) and in adaptive behaviour, which covers a range of everyday social and practical skills. Usually it occurs before the age of 18.

It involves problems with general mental abilities that affect functioning in two areas:

**(i)** Intellectual functioning (such as learning, problem solving, judgement).

**(ii)** Adaptive functioning (activities of daily life such as communication and independent living).

**5. Which type of deformity is "Kyphosis"? [1]**

**OR**

**What is Motor Development?**

**Ans.** Kyphosis implies an increase of a backward posterior curve or a decrease of a forward curve. It is also called round upper back. Depression of chest is common in Kyphosis.

**OR**

**Ans.** Motor Development refers to the development of movement and various motor abilities from birth till death. It is the ability to move around and manipulate his/her environment. Motor development can be divided into two sections: gross motor development and fine motor development.

**(i)** Gross motor development involves the development of the large muscles in the child's body. These muscles allow us to sit, stand, walk and run, among other activities.

**(ii)** Fine motor development involves the small muscles of the body, especially in the hand.

**6. Among females, what type of Menstrual Dysfunction is called Amenorrhea? [1]**

**Ans.** Amenorrhea is a menstrual disorder in women where girls of 18 years and above either never began menstruating or there is an absence of menstruation for three months or more than that in woman with a history of normal menstrual cycle.

**(i) Primary amenorrhea**—Menstruation never takes place. It fails to occur at puberty.

**(ii) Secondary amenorrhea**—Menstruation starts but then stops.

**7. What do you mean by "Bulimia" ? [1]**

**OR**

**Give two objectives of Extramural activities.**

**Ans.** Bulimia is a life-threatening eating disorder. It is an eating disorder in which female athlete eats excessive amount of food and then vomits it in order not to gain weight.

They try to get rid of the extra calories in an unhealthy way. There is a fear to put on weight.

**OR**

**Ans.** Two objectives of Extramural activities are :

**(i)** To provide enriching experience to students.

**(ii)** To provide the knowledge of new rules and advanced techniques.

**(iii)** To improve standard of sports.

**(iv)** To broaden the base of sports.

**8. Which test will you suggest to measure general motor ability? [1]**

**Ans.** Barrow Motor test measures the general motor ability (speed, strength, coordination, power). In this test there are three item standing broad jump, zigzag run and medicine ball throw.

**9. On the basis of physiological parameters, mention any two gender differences.** [1]**

**10. Which type of sports injury is known as "Strain" ? [1]**

**OR**

**What do you mean by soft tissue injuries?**

**Ans.** Strain is an injury to muscle or tendon. Tendons are fibrous cords of tissue that attach muscles to the bone. Strains often occur in foot, leg (typically the hamstring) or back. It is a type of soft tissue injury.

**OR**

**Ans.** Soft-tissue injury includes damage of muscles, ligaments, and tendons throughout the body. The result can be pain, swelling, bruising, and damage. Soft-tissue injuries are classified as the following : Contusions, Sprains, Tendonitis, Bursitis, Stress injuries, Strains and Abrasion.

** Answer is not given due to change in the present syllabus.

**11. Which method will you suggest to develop endurance? [1]**

**Ans.** Methods to develop endurance are Interval Training Method, Repetition Method, Continuous Method, Fartlek Method.

## SECTION—B

**12. Explain about the procedure and advantages of 'Bhujangasana'. [3]**

**Ans.** The meaning of Bhujang is cobra and asana means yoga pose. Bhujangasana is the stretching yoga exercise of the front torso and the spine. Bhujangasana is the final stage when the cobra is ready to attack its prey by raising its hood; hence the name is cobra pose.

**Procedure:**

**(i)** Lie down on the stomach by keeping legs together.

**(ii)** Put palms besides shoulder and the head should rest on the ground.

**(iii)** With inhaling raise head up to navel region and try to see the roof.

**(iv)** Maintain the position till 10 to 60 seconds with steadily inhaling and exhaling.

**(v)** Come to the original position slowly with deep exhalation.

**(vi)** For better result repeat this asana 3 to 5 times.

**Advantages:**

**(i)** It helps to strengthen up the spine and relieves backache.

**(ii)** Helps in improving the efficiency of the kidneys.

**(iii)** It is good for digestion. It gives a good massage to the organs of abdominal region such as stomach, pancreas, liver and gallbladder.

**(iv)** It is helpful in treating of many digestion related conditions such as constipation, indigestion, etc.

**(v)** It is beneficial against slipped disc or sciatica.

**(vi)** It regulates thyroid gland thus helps to maintain good health.

**(vii)** It improves lung capacity. The cobra pose helps to expand the chest thus beneficial for breathing related problems.

**(viii)** Decreases obesity.

**(ix)** Strengthens the arms and shoulders

**13. What is the role of Yoga in preventing lifestyle diseases? [3]**

**OR**

**Discuss about meal intake guidelines for pre, during and post sports event.****

**Ans.** Yoga places a great importance on a proper and healthy lifestyle whose main components are healthy activities, healthy relationships, healthy thoughts, healthy food, and healthy recreation. The holistic art and science of yoga is the best lifestyle ever designed and is effective in managing prevalent lifestyle disorders such as diabetes and hypertension.

There are various lifestyle diseases like obesity, diabetes, asthma, hypertension and backache. Through regular participation in yoga :

(i) Bones and joints become strong.

(ii) Muscles becomes stronger and flexible.

(iii) Circulation of blood becomes normal.

(iv) Respiratory organs become efficient.

(v) Efficiency of digestive system increases.

(vi) Better neuro muscular coordination.

(vii) Strengthen the immune system.

**14. Pawan got admission in class IX in a reputed school. All the students of his class welcomed him. After few days Aryaman, the class monitor observed that Pawan's behaviour is different from other students. He is aggressive, rude and non-cooperative. He finds it difficult to make friends. Even in school, nobody wants to be his friends. Aryaman had read about ODD in a science magazine in the library, which was similar to Pawan's behaviour. He discussed this situation with his classmates and suggested to interact and be friends with Pawan. This would help him to remain calm and happy.**

**Answer the following questions based on the above passage :**

**(a) What do you mean by ODD?**

**(b) What are the symptoms of ODD?**

**(c) What values are shown by Aryaman?** [3]**

**Ans. (a)** Oppositional Defiant Disorder (ODD) is a childhood disorder that is defined by a pattern of hostile, disobedient, and defiant behaviours directed at adults or other authority figures.

**(b)** Symptoms of ODD are as follows :

**(i)** Rude, aggressive and non-cooperative behaviour.

**(ii)** Difficulty to make friends.

**(iii)** Often and easily loses temper.

**(iv)** Is frequently touchy and easily annoyed by others.

**(v)** Is often angry and argumentative.

**(vi)** Often argues with adults or people in authority.

**(vii)** Often actively refuses to comply with adults' requests or rules.

**15. What are the causes of Osteoporosis? [3]**

**OR**

**Write briefly about the prevention and management of "Anorexia".**

**Ans.** Osteoporosis causes bones to become weak and brittle, so brittle that a fall or even mild stresses such as bending over or coughing can cause a fracture.

**Causes of Osteoporosis are as follows:**

**(i)** Women are much more likely to develop osteoporosis than are men. Lowered sex hormone levels tend to weaken bone. The reduction of oestrogen levels in women at menopause is one of the strongest risk factors for developing osteoporosis.

**(ii)** The old age people are at greater risk of osteoporosis.

**(iii)** Having a parent or sibling with osteoporosis puts greater risk to have osteoporosis.

**(iv)** Low calcium intake contributes to diminished bone density, early bone loss and an increased risk of fractures.

**(v)** Severely restricting food intake and being underweight weakens bone in both men and women.

**(vi)** Long-term use of oral or injected corticosteroid medications, such as prednisone and cortisone, interferes with the bone-rebuilding process.

**OR**

**Ans.** Anorexia is an eating disorder characterised by low weight, food restrictions, Fear of gaining weight and strong desire to be thin.

**Prevention of Anorexia**

Prevention is a term that can refer to either the elimination of factors causing eating disorders:

**(i)** Encourage a healthy view of the self and others means refraining from commenting on the body sizes of other children.

**(ii)** Make children aware about their genetics, body shape and size.

** Answer is not given due to change in the present syllabus.

(iii) Make them to eat healthy, nutritious food and be physically active.

(iv) Stay away from the people, places and activities which induce anorexia.

**Management of Anorexi**

(i) Face the reality.

(ii) Restoring healthy weight.

(iii) Individual Psychotherapy.

(iv) Antidepressants to aid the process of recovery.

**16. Explain the various factors affecting projectile trajectory.**** **[3]**

**17. Explain the various types of axes of rotation.**** **[3]**

**18. What do you mean by coping strategies? Write briefly.**** **[3]**

**OR**

**Explain the strategies for enhancing adherence to exercise.**

**Ans.** Getting started and continuing an exercise program can be a challenging yet, rewarding undertaking. There is no clear 'best strategy' for increasing exercise adherence, rather following interventions should be considered a tool box of strategies that could be used in order to find a successful formula for different groups or individuals for the purpose of adhering to exercise :

(i) **Work out with friends.**

It helps to have companion, social support and encouragement.

(ii) **Choose an activity you like.**

Start by identifying activities that you enjoy. One is more likely to stick to a program if it is fun and convenient. Activity of one's choice will pull one to continue practicing it.

(iii) **Learning new activity.**

The new challenge will prompt one to come back and learn challenge.

(iv) **Begin easy and slowly increase effort.**

Slowly increase the load of performing the activity reduces, the chances of sore muscles or injury. Aim for small but regular improvements.

(v) **Cross-training.**

Performing same thing every day may be boring. Try different activities. This type of training will keep one away from becoming bored with exercise.

(vi) **Goal Setting**

Goals should be set for yourself flexible, and time based. Setting realistic short-term goal will motivate one to do task and note ones progress toward goal.

(vii) **Keep a record of progress.**

It will act as a progress chart showing where one has reached and how much one has to progress. Seeing one's improvement motivates one to work towards acheivement of goals.

(viii) **Establish a routine.**

Fix a time for exercise. As one sticks to routine, exercise will become a habit.

(ix) **Combine family-time with exercise.**

Evening walk, yoga with family will help to develop bond with family members and also in exercise adherence.

(x) **Provide cues for exercises.**

Place posters and quotes to motivate while performing exercise.

**19. Write briefly about the techniques of Stress management.**** **[3]**

## SECTION C

**20. What is league tournament? Draw a fixture of nine (9) teams on the basis of league tournament using cyclic method. Explain British method to declare the winner.** **[5]**

**Ans.** In a league tournament, each team competes with every other team participating in the tournament. Winners are decided on the basis of the points acquired after all the competitions. For each win a team may be awarded 2 points, for draw 1 point and for a loss no point is awarded. Formulae to find the number of matches in a league tournament = $n(n-1)/2$,

where $n$ = number of teams participating in the tournament. The number of matches in a league tournament of 9 teams = $9(9-1)/2 = 36$ matches. In a single league tournament, every team competes with every other team participating in the tournament once. In a double league tournament, every team competes with every other team participating in the tournament twice.

** Answer is not given due to change in the present syllabus.

**League Fixture for Nine Team Using Cyclic Method**

| 1 Round | 2 Round | 3 Round | 4 Round | 5 Round | 6 Round | 7 Round | 8 Round | 9 Round |
|---|---|---|---|---|---|---|---|---|
| 9 Vs Bye | 8 Vs Bye | 7 Vs Bye | 6 Vs Bye | 5 Vs Bye | 4 Vs Bye | 3 Vs Bye | 2 Vs Bye | 1 Vs Bye |
| 8 Vs 1 | 7 Vs 9 | 6 Vs 8 | 5 Vs 7 | 4 Vs 6 | 3 Vs 5 | 2 Vs 4 | 1 Vs 3 | 9 Vs 2 |
| 7 Vs 2 | 6 Vs 1 | 5 Vs 9 | 4 Vs 8 | 3 Vs 7 | 2 Vs 6 | 1 Vs 5 | 9 Vs 4 | 8 Vs 3 |
| 6 Vs 3 | 5 Vs 2 | 4 Vs 1 | 3 Vs 9 | 2 Vs 8 | 1 Vs 7 | 9 Vs 6 | 8 Vs 5 | 7 Vs 4 |
| 5 Vs 4 | 4 Vs 3 | 3 Vs 2 | 2 Vs 1 | 1 Vs 9 | 9 Vs 8 | 8 Vs 7 | 7 Vs 6 | 6 Vs 5 |

Number of Matches = N(N-1)/2 = 9(9-1)/2= 36

N = Number of teams

**British Method to Decide the Winner****

In British method to decide winner in a league tournament, team which wins secures 2 points and for a draw 1 point and for a loss a team gets zero points.

Formulae for deciding the winner in a league tournament by British Method is number of points secured divided by maximum point which can be secured whole multiplied by 100.

Percentage scored by a team

= (Number of points secured/Maximum point which can be secured) 100

If there are 9 teams maximum points which can be scored by a team are 18 points (9×2) and the team which has secured 10 points will have a percentage score of (10/18 ) 100 is equal to 55 percentage. The team which has higher percentage will be considered winner.

**21. Explain macro-nutrients and their role in our diet. [5]**

**Ans.** Macronutrients are defined as chemical substances required in large amounts by the body for survival. The three macro-nutrients are protein, carbohydrates, fats and water. The amounts of different macro-nutrients a person needs as well as the ratio of nutrients to each other varies by age, lifestyle (sedentary, active, or very active), gender, health status, and health goals. They constitute the majority of an individual's diet.

| Types of Macronutrients | Calories per Macronutrient |
|---|---|
| Carbohydrates | 4 calories per gram of carbohydrates |
| Protein | 4 calories per gram of protein |
| Fats | 9 calories per gram of fat |

**Role of macro-nutrients in our diet :**

(i) Carbohydrates:

1. Act as major fuel for muscular contraction.
2. It provides the energy to our body.
3. It helps to maintain body weight and body temperature of the body.
4. Important for different digestive operations in our body.

**(ii) Fats:**

1. It provides heat and energy to the body.
2. It protects the body from extreme cold and hot climate.
3. Helps in regulation of body temperature.
4. It also helps to protect internal organs of the body.

**(iii) Proteins:**

1. It also plays an important role in the physical and mental development of an individual.
2. Necessary for our growth and development and for repairing the wear and tear of tissues.
3. It helps in the formation of enzymes and hormones and also act as a source of energy.
4. Transport oxygen and nutrient.
5. Regulates balance of water and acid.

(iv) Water:

1. Helps in transportation of nutrients to cells of body.
2. Regulates body temperature.
3. Vital for various chemical reactions taking place in the body.
4. Essential for body metabolism.
5. Keeps the body hydrated.

**22. Explain 'Flat Foot' and 'Knock Knees' and also suggest corrective measures for both postural deformities. [5]**

**Ans. Flat Foot**

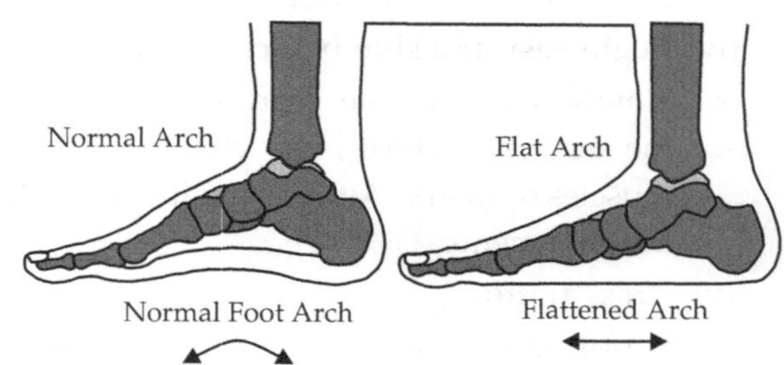


Flat foot is one physical deformity on which the medial longitudinal and transverse arches of foot

** Answer is not given due to change in the present syllabus.

are depressed and medial border of foot comes in contact with ground. In other words, in this deformity the arch of the foot collapses, with the entire sole of the foot coming into complete or near-complete contact with the ground. Flat foot deformity can be diagnosed with a very simple test called as 'Wet Foot test'. It creates problem in walking, running, standing and jumping.

**Causes of Flat Foot are:**

**(i)** Birth defect.
**(ii)** Overweight or obesity.
**(iii)** Deficiency of the vitamin D, calcium, etc. during childhood.

**Corrective Measure for Flat foot deformity:**

Following exercises can be performed as a remedy for Flat Foot deformity:

**(i)** Jumping on toes
**(ii)** Rope skipping
**(iii)** Walk on toes
**(iv)** Stand up and down on heels
**(v)** Walk on heels
**(vi)** Walking on inner and outer side of foot
**(vii)** Perform Vajrasana and Yogic exercises.

**Knock Knee**

Knock-knee is a type of postural deformity. In this deformity, both the knees touch each other in normal standing position. It creates problem in walking and running.

**Causes of Knock Knee are :**

**(i)** Birth defect.
**(ii)** Overweight or obesity.
**(iii)** Deficiency of the vitamin D, calcium, etc. during childhood.

**Corrective Measures for Knock Knee deformity:**

Following exercises can be performed as a remedy for Flat Foot deformity:

**(i)** Walking on heels.
**(ii)** Walking on inner and outer side of feet.
**(iii)** Walking on toes.
**(iv)** To perform up and down the heels.
**(v)** Jumping on toes for some time.
**(vi)** To skip on rope.
**(vii)** To perform the yogic asana.
**(viii)** Surgery
**(ix)** Standing on toes
**(x)** Picking small stone with toes and placing it a little far
**(xi)** Perform yogic asana like Tadasana (Tree posture), Vajrasana and Utkatasana (Chair posture).
**(xii)** Horse riding is one of the best exercise.
**(xiii)** Keep a pillow between the knees and stand for some time every day.
**(xiv)** Use of walking caliper may be beneficial.
**(xv)** Perform Padmasana and Gomukhasana.

** Answer is not given due to change in the present syllabus.

**23. Write in detail about AAPHER (American Alliance For Health, Physical Education and Recreation) Motor Fitness Test.**** **[5]**

**OR**

**What do you know about Harvard Step Test? Explain its procedure and administration.**

**Ans.** The Harvard Step test is a test to test aerobic fitness.

Cardiovascular fitness is the ability of the heart and lungs to supply oxygen-rich blood to the working muscle tissues and the ability of the muscles to use oxygen to produce energy for movements. Harvard Step Test is a cardiovascular fitness test. It is also called aerobic fitness test.

**Administrative procedure of Harvard Step Test:**

**Purpose:** To measure the general capacity of the heart and circulatory system for measurement of cardiovascular efficiency.

**Time Allotment:** 5 minutes

**Facilities and Equipment:** A stop watch, 20□ height bench, partners, stethoscope, metronome, score sheet.

**Procedure:** The athlete stands in front of the bench or box. On the command 'Go' the athlete steps up and down on the bench or box at a rate of 30 steps per minute (one second up one second down) for 5 minutes (150 steps). Stopwatch is also started simultaneously at the start of the stepping. After that the athlete sits down immediately after completion of the test i.e. after 5 minutes. The total number of heartbeats are counted between 1 to 1.5 minutes after completion of the last step. The heartbeats are counted for 30 seconds period. Again the heartbeats are noted for 30 seconds after the finishing of the test. After that third time the heartbeats are noted after 3 minutes of completion of the test for 30 seconds period. The same foot must start the step up each time, and an erect posture must be assumed on the bench.

**Calculation of the Score:** The athlete's fitness index score is calculated with the help of following formula:

Fitness index score = (100 x test duration in seconds) / (2 x sum of heart beats in recovery periods)

| Rating | Fitness index (long form) |
|---|---|
| Excellent | > 96 |
| Good | 83 - 96 |
| Average | 68 - 82 |
| Below average | 54 - 67 |
| Poor | < 54 |

**24. What are the effects of exercise on Respiration System ? Write in detail.**** **[5]**

**25. What do you understand by fracture? How can fracture be classified? Explain.** **[5]**

**Ans.** Broken or cracked bone is known as Fracture. It is very common injury in games and sports. A bone fracture is a medical condition where the continuity of the bone is broken. A break in the bone that does not damage surrounding tissue or tear through the skin is known as a closed fracture. On the other hand, one that damages surrounding skin and penetrates the skin is known as a compound fracture or an open fracture.

**Classification of Fracture:**

(i) **Comminuted fracture:** The bone is shattered into many pieces.

(ii) **Compression (crush) fracture :** Generally occurs in the spongy bone of the spine. For example, the front portion of a vertebra in the spine may collapse due to osteoporosis.

(iii) **Greenstick fracture:** The bone partly fractures on one side, but does not break completely because the rest of the bone can bend. This is more common among children, whose bones are softer and more elastic.

(iv) **Hairline fracture:** It is a partial fracture of the bone. Sometimes this type of fracture is harder to detect with routine X-rays.

(v) **Impacted fracture:** When the bone is fractured, one fragment of bone goes into another.

(vi) **Longitudinal fracture:** The break is along the length of the bone.

(vii) **Oblique fracture:** A fracture that is diagonal to a bone's long axis.

(viii) **Spiral fracture:** A fracture where at least one part of the bone has been twisted.

(ix) **Stress fracture:** More common among athletes. A bone breaks because of repeated stresses and strains.

(x) **Transverse fracture:** A straight break right across a bone.

(xi) **Complicated fracture:** Structures surrounding the fracture are injured.

**26. What do you understand by coordinative ability? Discuss about different types of coordinative abilities.** **[5]**

**OR**

**Write in detail about strength improving methods–Isometric, Isotonic and Isokinetic.**

**Ans.** Coordination is the ability to integrate separate motor systems with varying sensory modalities into efficient movements. These are those abilities which enable an individual to do various related activities accurately and efficiently. Coordinative abilities mainly depend on the central Nervous System. The harmonious working together of the synchrony, rhythm, and sequencing aspect of one's movements is crucial to coordinated movement. Various parts of the body may be involved, such as eye foot coordination, as in kicking a ball or walking upstairs. Eye-hand coordination is evident in fine motor activities such as bead stringing, tracing and clay modelling or in gross motor activities such as catching, striking or volleying a ball.

**Different Types of Coordinative Ability**

Following are different types of coordinative abilities:

(i) **Coupling Ability** : Coupling ability to synchronize movements and direction efficiently and to let different parts to cooperate in order to achieve a desired and smooth movement. Coupling ability of the body is considered as capacity of an individual to perform better coordination of different parts of the body. For example, while dribbling a basketball, the coordination of both the limbs is required.

(ii) **Reaction Ability** : Reaction ability is a skill to react quickly to a stimulus. For example, chaser in Kho-Kho runs quickly after Kho is given.

(iii) **Balance Ability** : Balance ability is known as capability of an individual to maintain balance while the body is in motion. Even during distributed balance regaining the lost balance is known as balance ability.

(iv) **Orientation Ability**: The orientation ability is a skill to determine and change

** Answer is not given due to change in the present syllabus.

the position in complex situations. For example, defending an opponent in football, the position is changed according to the movement of opponent.

**(v) Adaptation Ability** : Adaptation ability is the capability of a person to change the movement on the basis of predicted changes. This ability is achieved after mastering the skills.

**(vi) Rhythm Ability :** Rhythm ability is to make out motor actions on a well- defined rhythm. Rhythm ability enables to grasp and express rhythm which is externally determined or contained in the motor activity itself.

**OR**

**Ans. (i) Isometric Exercises :**

They are exercises in which muscular tension is built without any visible movement of muscle being used. Contraction during such an exercise is called isometric contraction. Work done during isometric exercise is zero. Isometric exercise is practiced by pushing or pulling an immovable object like a wall or bar anchored to the floor. Isometric exercise increases muscle tension significantly but fails to change the length of the muscles. Today, it is primarily used for rehabilitation purposes.

**Examples :**

1. Balancing on one foot.
2. Pressing against the wall.
3. Standing with heavy load on head.
4. Maintaining a sitting posture without a chair.
5. Holding a weight at arm's length.
6. Attempting to lift an immoveable object.

**Characteristic of isometric exercise**

1. The amount of heat generated and tension developed is higher than that of isotonic.
2. Increased muscular tension leads to tremor in muscles.

**(ii) Isotonic Exercises :**

Exercise where we can see movement of an object on application of force is called Isotonic exercise and contraction of the muscle during such an exercise is called isotonic contraction. Isotonic exercises are carried out against a fixed resistance. As each muscle moves through its complete range, isotonic contraction creates tension with maximum effort at the beginning and end of each exercise. Two types of muscular contraction take place while performing Isotonic exercise they are **concentric contraction** in which length of the muscle decreases from the normal size and **eccentric contraction** in which length of the muscle increases from the normal size.

**Examples** :

1. Light rhythmic exercise
2. Running downhill
3. Sit-ups and push up
4. Lifting free weights, like dumbbells and barbells
5. Hammer throw
6. Triple Jump
7. Landing on the ground after a jump

**(iii) Isokinetic Exercises :**

These are the exercises in which speed remains same throughout the range of movement. It can be also called as a combination of isometric and isotonic contraction. Isokinetic exercise uses a machine that controls the speed of contraction within the range of motion. Cybex and Biodex machines provide this kind of workout, but these machines are generally used by physical therapists and not readily available to most people. Exercise which can be found similar to isokinetic exercise in sports activity are :

1. Ice skating
2. Chin ups
3. Rope climbing
4. Rowing
5. Swimming

●●

# PHYSICAL EDUCATION

## Outside Delhi 2018

**Time allowed : 3 hours** **Maximum marks : 70**

1. **What do you mean by 'Seeding'?** **[1]**

**Ans.** Seeding is a method in which strong teams/players are selected to keep them at appropriate places in the fixture so that they may not meet in the earlier rounds. For the selection of strong teams, the organisers must be well aware regarding the previous performance of the teams.

2. **What do you mean by food intolerance?** **[1]**

**Ans.** Food intolerance is the inability of a person to digest certain food properly. Its symptoms include nausea,vomiting, acidity, stomach upset, etc.

3. **What is the main physiological cause of Asthma?** **[1]**

**Ans.** The physiological cause of asthma is narrowing of airway passago and a subsequent interference with airflow.

4. **What is Attention Deficit/Hyperactivity Disorder (ADHD)?** **[1]**

**Ans.** It is a brain disorder that includes inattentiveness, hyperactivity and impulsiveness. It is a medical condition that affects how well can someone sit still, focus and pay attention.

5. **What do you mean by Congenital Deformity?** **[1]**

**Ans.** The deformity in children that is present at birth is known as congenital deformity.

6. **What is Bulimia?** **[1]**

**Ans.** Bulimia is an eating disorder in which a person eats excessive amount of food. People with bulimia make up for the excess energy intake using a variety of techniques including vomiting, fasting, excessive exercise etc. to control body weight.

7. **Explain the meaning of cardiac output.** **[1]**

**Ans.** The amount of blood pumped by the heart in one minute is known as cardiac output. It is a product of stroke volume and heart and is measured in litres/min.

8. **What type of sports injury can be termed as 'Laceration' in sports ?** **[1]**

**Ans.** Laceration is an irregular cut on the skin with a sharp object or sharp edged sports equipment.

9. **Which field of study in sports is called 'Biomechanics'?** **[1]**

**Ans.** Biomechanics is the study of internal and external forces and their effects on living system or athlete

10. **What do you mean by the term 'Aggression' in sports?** **[1]**

**Ans.** In sports, aggresion means an intention to harm or injure player outside the rules of game so as to gain a competitive advantage or late tackling to stop an opponent from scoring.

11. **What is coordinative ability?** **[1]**

**Ans.** Coordinative ability is the ability of the body to bring together different elements of a complex activity into an efficient relationship.

12. **What are the advantages and disadvantages of food supplements for adolescents ? Write briefly.**** **[3]**

13. **Sushant was a good athlete. He used to practise regularly to achieve a position at the State level. But he could not get success. He got frustrated with his poor performance and started misbehaving with his teachers and friends in school. Due to depression and anxiety, he started taking drugs. The Principal counselled Sushant and called his parents.**

**They took him to a rehabilitation centre for treatment. After a few months, he recovered and came back home.**

**On the basis of above passage answer the following questions:**

**(a) Do you think that consuming drugs is a solution to emotion-focussed problems ?**

**(b) What values are shown by the Principal?****

**(c) What should be the attitude of the teachers and the parents after his recovery?** **[3]**

**Ans.** (a) No, consuming drug is not a solution to emotion-focussed problems or failure or poor performance.

(c) Teachers and parents should motivate Sushant for achieving his goal. They should try to make him understand that success comes after failure. So, he has to work hard and prove himself in front of everyone.

---

** Answer is not given due to change in the present syllabus.

**14. What are the types of disability? Explain briefly. [3]**

**Ans. Types of disability are:**

**(i) Cognitive disability:** Cognitive disability generally refers to anyone with lower than average intellectual functioning. A person who has a cognitive disability has trouble in performing mental tasks that the average person would be able to do.

**(ii) Intellectual disability:** Intellectual disability is a disability characterized by significant limitations in both intellectual functioning (such as learning, problem solving, judgement) and in adaptive behaviour (activities of daily life such as communication). This disability originates before the age of 18.

**(iii) Physical disability:** A physical disability is a long term loss or impairment of part of body's physical function. It can involve difficulties in walking and mobility, sitting and standing, use of hands and arms, sight, hearing, speech, muscle control etc.

**15. What are the major muscles involved in running, jumping and throwing? Explain.** [3]**

**16. Keeping in view the Indian ideology, critically analyse the sociological aspect of participation by women athletes in sports.** [3]**

**17. What do you understand by 'First Aid'? Discuss briefly about the aims and objectives of First Aid. [3]**

**Ans. First Aid:** It is the immediate care given to an injured or ill person. It does not take the place of proper medical treatment. It is a temporary assistance before actual medical care is obtained.

**Aims and Objectives:** The most obvious objective of first aid is to save one's life before the victim gets actual medical help. The objective can be simplified into the five Ps:

**(i)** Preserve life.

**(ii)** Protect the unconscious.

**(iii)** Prevent injury or illness becoming worse.

**(iv)** Promote recovery.

**(v)** Procure medical aid.

If a life-threatening situation does exist, the first aider should have knowledge of procedures to keep the casualty alive. A casualty should be isolated from all further dangers and placed in a position where their injury or illness will not become worse. The casualty's injuries or illness should be attended to in order to prevent complications arising from his or her injury or illness.

**18. Mention briefly about the common sports injuries and their prevention. [3]**

**Ans.** Common sport injuries are broadly divided in three categories:

**(i)** Soft Tissues Injuries

**(ii)** Bone Injuries

**(iii)** Joint Injuries

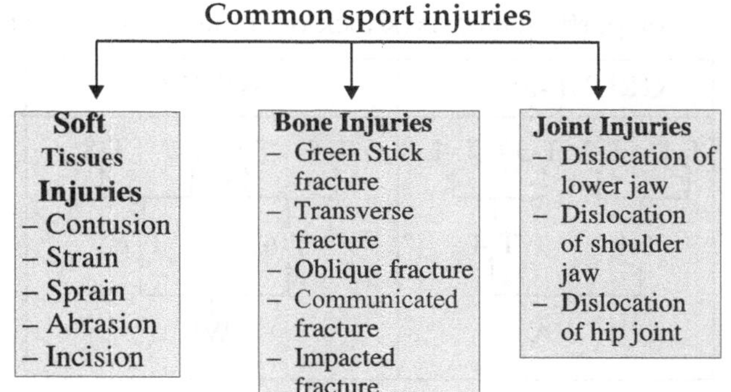


**Prevention of sports injuries :**

**(i)** Use properly fitted sports equipment and gear.

**(ii)** Hydrate adequately to maintain health and minimize cramps.

**(iii)** Perform proper warm-up and cool–down routines.

**(iv)** Have a pre-season health and wellness evaluation.

**(v)** Body conditioning.

**(vi)** Proper use of techniques.

**(vii)** Balanced diet and adequate rest.

**19. Explain the "Eight Foot Up and Go" test for measuring agility and dynamic balance. [3]**

**Ans.** The 'eight foot up and go' test is a coordination and agility test for the elderly. The purpose of this test is to measure speed, agility and balance while moving. The procedure of this test is:

**(i)** Place the chair next to a wall and mark 8 feet in front of the chair.

**(ii)** The subject starts by fully seated, hand resting on knees and feet flat on ground.

**(iii)** On the command "GO", timing started and the subject stands and walks as quickly as possible (no running) around the cone, returning to initial position and sit on the chair.

** Answer is not given due to change in the present syllabus.

**(iv)** As the person sits down, the timer is stopped.

**(v)** Perform two trials.

**(vi)** The score is better time of the two trials to the nearest 1/10th of a second.

Regular exercise increases agility and dynamic balance.

**20. Define Combination Tournament. Draw a fixture of 16 teams using Knock-out cum League Method. [5]**

**Ans.** Combination Tournament are those tournaments in which initial round of tournament is played on particular basis (knock-out or league) and rest of the tournament played on another particular basis (knock-out/league).

(i) Knock out cum knock out

(ii) League cum League

(iii) Knock out cum League

(iv) League cum Knock out

**Fixture of Knock out cum League Tournament:** Divide number of team into four groups, then they are made to compete with each other on the basis of knock out tournament, the winners from four groups are made to compete on the basis of league Tournament. This will declare one team as a winner.

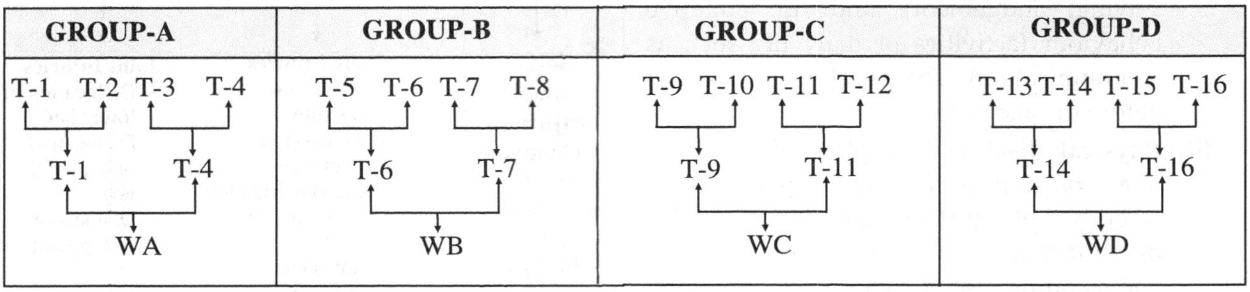

| GROUP-A | GROUP-B | GROUP-C | GROUP-D |
|---|---|---|---|
| T-1 T-2 T-3 T-4 | T-5 T-6 T-7 T-8 | T-9 T-10 T-11 T-12 | T-13 T-14 T-15 T-16 |
| T-1 T-4 | T-6 T-7 | T-9 T-11 | T-14 T-16 |
| WA | WB | WC | WD |

Henceforth winners from each group WA, WB, WC and WD move to League Tournament. And each team in League Tournament plays once with other team. Team winning most matches with maximum points is declared as winner of tournament.

**Four Teams**

WA (1) WA Vs WB — Winner WA — 2 Points

WB (2) WC Vs WD — Winner WC — 2 Points

WC (3) WA Vs WC — Winner WA — 2 Points

WD (4) WB Vs WD — Winner WD — 2 Points

(5) WA Vs WD — Winner WA — 2 Points

(6) WB Vs WC — Winner WC — 2 Points

**Final Points Tally**

| | | |
|---|---|---|
| WA $\Rightarrow 2 + 2 + 2$ | = | 6 Points |
| WB $\Rightarrow 0$ | = | 0 Points |
| WC $\Rightarrow 2 + 2$ | = | 4 Points |
| WD $\Rightarrow 2$ | = | 2 Points |

Here WA winning 3 matches with 6 points is winner.

**21. What are the causes of back pain? Explain the procedures, benefits and contraindications of two asanas recommended to cure back pain. [5]**

**Ans. Causes of Back Pain:** The human back is composed of a complex structure of muscles, ligaments, tendons, disks and bones. Problem with any of these components can lead to back pain. The most common causes of back pain are strained muscles, strained ligaments and a muscle spasm which arises due to:

(i) Incorrect body posture

(ii) Lifting something that is too heavy

Back pain can be reduced by regular excercise, correcting the posture and doing proper asanas like Tadasana and Bhujangasana.

**(a) Tadasana**

Tadasana is an asana in which the posture appears like standing in the form or shape of a mountain.

(i) **Procedure :** Stand erect and place your thighs, slightly apart, with your hands hanging alongside your body. Strengthen the inner arches of inner ankles as you lift them. Look slightly upward. Now breathe in and stretch your shoulders, arms and chest upwards. Raise your heels, making sure that your body is on your toes. Hold the pose for a few seconds. Then exhale and release.

(ii) **Benefits : (1)** It is helpful in developing physical and mental balance.

**(2)** It improves body posture.

(iii) **Contraindication :** It is best to avoid this pose if someone is having problems like severe migraine, insomnia or low blood pressure.

**(b) Bhujangasana**

Bhujangasana is a basic Hatha pose that strengthens the entire shoulder and upper back.

(i) **Procedure :** Lie flat on your stomach. Place your hands on the side and ensure that toes touch each other. Then move your hands to the front, placing your body's weight on your palms. Inhale and raise your head and trunk. Press your hips, thighs and feet to the floor. Hold the asana for 15 to 30 seconds then release the pose by slowly bringing your hands back to the sides.

(ii) **Benefits : (1)** It makes spine stronger and more flexible.

**(2)** It gives lungs, shoulder, chest and abdomen a good stretch.

(iii) **Contraindication :** People should avoid this asana if they suffer from problems like hernia and back injuries. Pregnant women should not perform this pose in any case.

**22. Describe exercise guidelines at different stages of growth in children. Give suitable examples for every stage. [5]**

**Ans. Excerise guidelines at different stages of growth**: Development is a never ending process whereas growth refers to an increase caused by physiological change in which the child becomes bigger in size and heavier in weight. As every individual is unique in himself, individual differences and specificity of activity should be considered.

(i) **Infant activity (0 to 2 yrs.) :** In this stage, the growth pattern involves control over big muscles. Clean, safe and hygienic environment is the need in this stage. In this tender age, the activities have to be done under the observation of parents and guardians. Main exercises for this tender age are practice of balancing, walking and hearing from different directions.

(ii) **Early childhood activity (3 to 8 yrs) :** During this stage of growth, the pattern involves control over small muscles and bone development. The environment needed at this stage should be clean and safe, moreover proper checks by parents and teachers are required. Suggested sports- Running, throwing etc. Parents should keep a check that he or she does not fall on ground to avoid injuries.

(iii) **Later childhood activities (8 to 12 yrs):** During this stage, flexibility and co-ordinated activity of growth occurs. During this stage of growth, good quality apparatus and safety measures should be considered. Wrong habits should be checked. Suggested exercise and sports—100 metres races, hanging from horizontal bars, Cricket, Football, Tennis, Judo, Karate etc.

**23. Write in detail about the various tests items and their administration of the American Alliance for Health, Physical Education and Recreation (AAHPER) Test.** [5]**

**24. What are the long term effects of regular exercise on the cardio-vascular system? Explain. [5]**

**Ans. Effect of exercise on the cardio-vascular system:**

(i) **Cardiac output increases:** The cardiac output tends to increase as a result of regular exercise.

(ii) **Resting heart rate decreases:** The resting heart rate decreases due to regular exercise. After a duration of 10 week training programme, the resting rate may reduce upto 10 beats/min from the normal 72 beats/min.

(iii) **Stroke volume increases at resting condition:** The stroke volume at rest remains up to 50-70 ml/beat in untrained individuals, but in trained individuals it ranges from 70-90 ml/beat.

(iv) **Blood volume increases:** The blood volume increases due to regular exercises. Actually, as the blood volume increases, there is an increase in plasma volume. Also, the body produces a greater number of red blood cells.

(v) **Blood flow increases:** The body increases its number of capillaries to the requirement of supplying more oxygen during exercise to the muscles. The existing capillaries open wider as well.

** Answer is not given due to change in the present syllabus.

**25. What is aerodynamics? Describe the various forces of aerodynamics.**** **[5]**

**26. What is circuit training? Draw a diagram of 10 stations to improve general fitness. How can load be increased in circuit training?** **[5]**

**Ans.** Circuit training is a formal type of training in which an athlete goes through a series of selected exercises that are arranged in a circuit with 5 to 10 stations. In circuit training, several exercises are done one after the another. Completion of one set of each exercise in rotation is called one round. It was designed by GT Anderson and RE Morgan in 1953. It is very effective method for strength, speed and endurance development.

**Diagram of Circuit Training With 10 Stations**

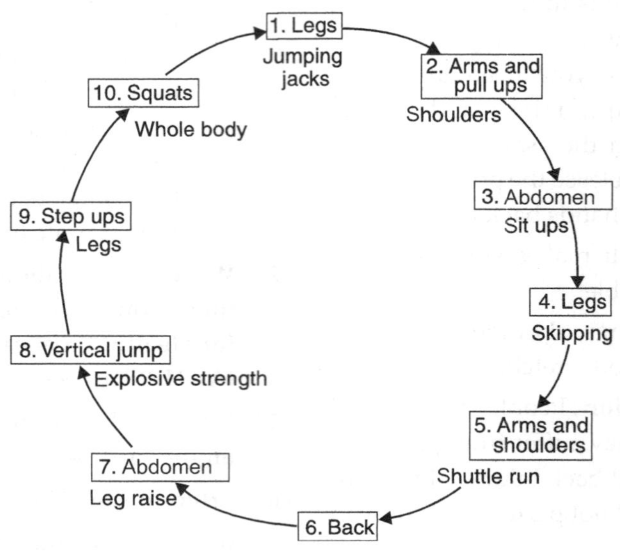


In circuit training, the load can be increased in the following two ways :

**(i)** By increasing the duration.

**(ii)** By enhancing the intensity of sportsperson.

●●

# PHYSICAL EDUCATION 2018

## Delhi

**Time allowed : 3 hours** **Maximum marks : 70**

**1. What do you mean by 'Fixture'?** **[1]**

**Ans.** The organized method in which teams participate in any tournament is a fixture.

**2. Why does the weightlifter's diet include lots of proteins?** **[1]**

**Ans.** Since weightlifter has to build muscles to keep them energetic they take high–protein diet. Protein is important for muscle growth and to repair body tissues. Protein is also used by the body for energy.

**3. What is Hypertension?** **[1]**

**Ans.** Hypertension means the increased blood pressure. It may eventually cause health problems such as heart diseases.

**4. What do you understand by 'Physical Disability'?** **[1]**

**Ans.** **Physical disability**: It is a limitation on individual's physical functioning, mobility, dexterity or stamina. Physical disability may either be motor deficiency or sensory impairment.

**5. What is the main cause of 'Scoliosis'?** **[1]**

**Ans.** Main cause of Scoliosis are:

(i) Injury of the bones and joints

(ii) Faulty posture

**6. What is 'Amenorrhoea' in female athletes triad?** **[1]**

**Ans.** It is the absence of menses for 6 months or absence of menstrual cycle for the three cycles.

**7. What do you know about the term 'Cardiac Output'?** **[1]**

**Ans.** Cardiac output is the amount of blood pumped by the heart in 1 minute. This increases exercise intensity.

**8. What type of fracture is known as Greenstick Fracture?** **[1]**

**Ans.** Greenstick is a fracture in a young, soft bone. In green stick, the bone bends and breaks. This fracture is very difficult to detect and is usually found in children.

**9. What do you mean by 'Kinesiology'?**** **[1]**

**10. Define the term 'Emotions'.** **[1]**

**Ans.** Emotions are feelings which result in physical and psychological changes that influence the behaviour of and individual.

**11. Explain the term 'Sports Training'.** **[1]**

**Ans.** **Sports Training** : Sports training is a specialised process of all round physical conditioning aimed at the preparation of sports- person for improving its performance in games and sports.

**12. Naman was a Class VI student. He used to bring junk-food in his lunch-box daily. His teacher observed that he was neither concentrating on his studies nor actively participating in physical activities. In this matter, he had a talk with his parents and came to know that he refuses to eat roti, dal, fruits and vegetables. Due to this he is facing these problems.**

**Based on the above passage, answer the following questions:**

**(a) What types of problems was Naman facing?**

**(b) Why should junk-food not be recommended?**

**(c) What values has his teacher shown in this matter ?**** **[3]**

**Ans.** (a) Naman had problem with proper balanced diet. He was eating junk food instead of roti, dal, fruits and vegetables.

(b) Junk food contains only fat and leads to obesity. Also junk food does not have protein, vitamin required for athletes, hence it is not recommended.

**13. Write briefly about some considerations of fluid intake pre, during and post competition, that affect the sportsperson's performance.**** **[3]**

**14. What are the benefits of physical activities for children with special needs? Explain.** **[3]**

**Ans.** Benefits of physical activities for children with special needs are:

(i) Physical improvement

(ii) Reduce risk of health complications

(iii) Mental improvement

** Answer is not given due to change in the present syllabus.

**(iv)** Behavioral patterns

**(v)** Better emotional and psychological health.

(i) **Reduced level of anxiety, stress and depression:** Physical activities may help in reducing the level of anxiety, stress and depression of children with special needs.

(ii) **Improved social interaction:** Physical activities provide ample opportunities for improving the social interaction among children with special needs. Some relations are developed during involvement in physical activities.

(iii) **Cognitive benefits:** Physical activities lead to the development of cognitive skill, improvement in children with disabilities. These activities allow them to discover and access strengths that cannot be challenged in classroom setting.

**15. How does participation in Games and Sports contribute to the psychological development of women athletes in India ? Explain.**** **[3]**

**16. Write briefly about menstrual dysfunctions and their effect on sports participation of female athletes.** **[3]**

**Ans. Menstrual dysfunction:** It is a disorder or irregular menstrual cycle in woman. It can also be defined as "An abnormal bleeding during the Menstrual cycle".

**Effects on sports participation:**

(i) During menstrual period, women athlete performance level is low. Women cannot perform so effectively.

(ii) During the cycle of menstrual period, women feel tension, anxiousness and uncomfortableness within body.

(iii) Often at this time, women have complaints of vomiting and pain in stomach.

(iv) During the menstrual period, woman athelete cannot take part in many competitions like, jumping, weightlifting, wrestling, boxing etc.

**17. What do you understand by 'First-Aid'? How will you manage joint injuries? Explain.** **[3]**

**Ans. First Aid:** It is the immediate care given to an injured or ill person. It does not take the place of proper medical treatment. It is a temporary assistance before actual medical care is obtained.

**Management of joint injuries:**

(i) Cold compression should be used immediately. The application of ice or cold water should not be used for more than 40 minutes persistently.

(ii) The cold compression should be performed 5 to 6 times daily.

(iii) If there is more swelling at the sight of contusion, the anti-inflammatory medicine should be given.

(iv) If the swelling persists, consult the doctor immediately.

(v) For the purpose of rehabilitation, the flexibility exercises should be performed carefully.

**18. What do you understand by Sports Medicine? Discuss briefly about the scope of Sports Medicine.**** **[3]**

**19. What are the factors affecting Self-Esteem and Body Image ? Explain.**** **[3]**

**20. Mention all calculations and steps involved to draw a Knock-out fixture of 19 teams, where 4 teams are to be seeded.** **[5]**

**Ans. Knock-out tournaments**: In knock-out tournaments, the teams which get eliminated get out automatically from the tournament. Only the winners continue in the competition. It is the fastest method to know about the winner team in the tournament.

**Seeding a Team:** The sorting of the teams and fitting them in the fixtures , so that the stronger teams do not meet each other in earlier rounds is known as seedings.

**Advantages of knock-out tournament:**

(i) These are less expensive.

(ii) Helpful in enhancing standard of sports.

(iii) It requires less time to complete the tournament.

(iv) Minimum number of officials are required

Total number of rounds played in tournament depends upon number of teams participating in tournament.

**Drawing a Knock-out fixtures of 19 Teams:**

Number of teams—19

Total number of matches = N – 1 = 19 – 1 = 18

** Answer is not given due to change in the present syllabus.

Number of teams in upper half $= \frac{N+1}{2}$

$= \frac{19+1}{2} = 10$

Number of teams in lower half $= \frac{N-1}{2}$

$= \frac{19-1}{2} = 9$

(i) Number of upper half is divided in two parts Q1 and Q2 with 5 teams each.

(ii) Number of lower half is divided in two parts Q3 and Q4 with 5 and 4 teams respectively.

(iii) Number of byes in quarter where 5 teams are participating $= \frac{5+1}{2} = \frac{6}{2} = 3$

(iv) Number of byes in quarter where 4 teams are participating (in present case)—All 4 teams.

(v) In such scenario, the match shall be as per the following :

1 Rd 2 Rd 3th Rd 4th Rd 5th Rd

UPPER HALF

Q1: 1 Seed 1 – Bye; 2 – Bye ] –1 ; 3, 4 ] –3 ; 5 Bye ] –3 ; –1, –3 ] –1

Q2: 6 Seed 2 – Bye; 7 – Bye ] –6 ; 8, 9 ] –9 ; 10 – Bye ] –10 ; –6, –10 ] 10

–1, 10 ] –1

LOWER HALF

Q3: 11 Seed 3 – Bye; 12 – Bye ] –11 ; 13, 14 ] –13 ; 15 – Bye ] –15 ; –11, –15 ] 11

Q4: 16 Seed 4 – Bye; 17 – Bye ] –16 ; 18 – Bye; 19 – Bye ] 18 ; –16, 18 ] 18

11, 18 ] –11

–1, –11 ] WINNER(11)

UPPER HALF

As per above distribution, the groups can be formulated in a way that any team gets only one bye (maximum).

**21. Briefly explain the symptoms and causes of Asthma. Explain the procedure, benefits and contraindications of any two asanas to prevent Asthma. [5]**

**Ans. Asthma:** Asthma disease is associated with the swelling of respiratory tracts which makes the tracts very sensitive and makes this process pungent with the touch of any effective things. These reactions cause contractions in the tubes, this reduces the amount of air in the lungs due to which it becomes difficult to breathe. Common symptoms of asthma are coughing, heavy breathing, chest tightness, fatigue, pain in hands, feet, shoulder and back. Reasons are dust, smoke pollen grains, animal skin, hair and feather etc. Asthma can be controlled by Matsyasana, Gomukhasana etc.

**Matsyasana :** This asana is performed in water body, hence it is called as Matsyasana.

**Pre-Stage :** Sit in Padmasana pose.

**Method :**

**(i)** Take support of your elbow and lie on your back, bend your neck with support of your hands and try to touch your head to the ground.

**(ii)** Hold toes of the feet firmly with both hands and touch ground with the elbows.

**(iii)** Stretch the stomach as up as possible.

**Benefits:** This asana is very useful for asthama patient, it provide relief from indigestion and other digestive problems, keeps the blood clean. Helps in cough and respiratory problems. It makes body and face attractive by bringing natural glow on it.

**Contraindications:**

**(i)** **Abnormal blood pressure:** Individuals suffering from high or low blood pressure should avoid this posture.

**(ii)** **Neck injury:** Neck injuries or any part of the lower back or middle back can make it difficult to practice this Fish pose and hence should be avoided.

**Gomukhasana:** This asana gets its name because while doing this asana, body resembles a cow face pose.

**Pre-stage :** Sit in Sukhasana pose.

**Method:**

**(i)** Sit in sukhasana pose.

**(ii)** Place the ankle of left leg near the right butt.

**(iii)** Place the right leg over the left thigh, so that knees should place over each other.

**(iv)** Sweep your left hand behind your back, facing palms upwards.

**(v)** Now interlock fingers of both hands behind your back.

**(vi)** Stretch both the hands in their respective direction, look straight.

**(vii)** Repeat with changing leg position.

**Benefits:** Helps in curing Asthma, reduces weight, makes body flexible.

**Contra-indications:**

**(i)** People with very stiff shoulder may have to take it slow and easy.

**(ii)** If you have sciatica, then crossing the knee may not too good as there will be pressure at the sciatic nerve.

**22. What are the types of motor development? Describe any six factors affecting motor development in children. [5]**

**Ans. Motor Development:** It refers to the development of general body control, fine motor skills and large muscle movements. There are three stages of motor development in children infancy or infancyhood, early childhood and later childhood.

It is of two types:

**(i)** **Gross Motor development :** It involves the development of large muscles in the child's body such as sitting, walking, running etc.

**(ii)** **Fine Motor development :** It involves the development of small muscles of the body, especially during the small movements of the fingers and hands, Example : holding of Javelin, discus and pole, catching a cricket ball etc.

**Factors affecting Motor Developments:**

**(i)** **Nutrition:** Nutritious food promotes good motor development. Sensory motor development is dependent upon nutrition that the child gets to a good extent. Children get stronger if they get nutritious food.

**(ii)** **Immunisation:** If mother and child both are immunised at proper time, it leads to a good sensory motor development.

**(iii)** **Environment:** Encouragement, love and security help the child to take risk to explore fearlessly and to know more about environment which leads to a better sensory development.

**Some of other factors affecting Motor development:**

**(i)** Sleep

**(ii)** Stimulation and interaction

**(iii)** Opportunities

**(iv)** Training and practice

**(v)** Recreation

**(vi)** Education : Learning and productivity

**(vii)** Gender

**(viii)** Posture deformities

**(ix)** Sensory impairment

**(x)** Obesity

**(xi)** Social skills

**(xii)** Mental health.

**23. Write in detail about 'Barrow's Motor Ability Test' to measure motor fitness components. [5]**

**Ans. General Motor Fitness:** Barrow's Motor Ability test—This test is performed by Standing Broad Jump, Zig Zag Run and Medicine Ball put. The test battery comprised of three items which are explained below:

(i) **Standing broad Jump:** To measure not only power but also agility, speed and strength.

1. **Equipment and facilities required :** One 5 × 12 feet mat marked with take off line and parallel lines 2 inches apart.
2. **Directions :** The subject is advised to stand just behind the take off line with feet together. Then by giving swing of arms and bending knees, a jump is taken, the arms are swung. Three trials are given.
3. **Scoring :** The final score is the distance covered in inches from the best jump.

| Grade T-Score | 7 | 8 | 9 | 10 | 11 | T-Score |
|---|---|---|---|---|---|---|
| 80 | 90up | 97up | 103up | 105up | 112up | 80 |
| 75 | 86-89 | 92-96 | 98-102 | 101-104 | 107-111 | 75 |
| 70 | 82-85 | 88-91 | 93-97 | 97-100 | 103-106 | 70 |
| 65 | 77-81 | 83-87 | 88-92 | 92-96 | 97-102 | 65 |
| 60 | 73-76 | 78-82 | 83-87 | 88-91 | 93-96 | 60 |
| 55 | 69-72 | 73-77 | 79-82 | 83-87 | 88-92 | 55 |
| 50 | 65-68 | 69-72 | 74-78 | 79-82 | 83-87 | 50 |
| 45 | 61-64 | 64-68 | 69-73 | 75-78 | 78-82 | 45 |
| 40 | 56-60 | 59-63 | 64-68 | 71-74 | 74-77 | 40 |
| 35 | 52-55 | 54-58 | 59-63 | 66-70 | 69-73 | 35 |
| 30 | 48-51 | 50-53 | 54-58 | 62-65 | 64-68 | 30 |
| 25 | 44-47 | 45-49 | 49-53 | 58-61 | 59-63 | 25 |
| 20 | 43 Down | 44 Down | 48 Down | 57 Down | 58 Down | 20 |

(ii) **ZigZag Run :**

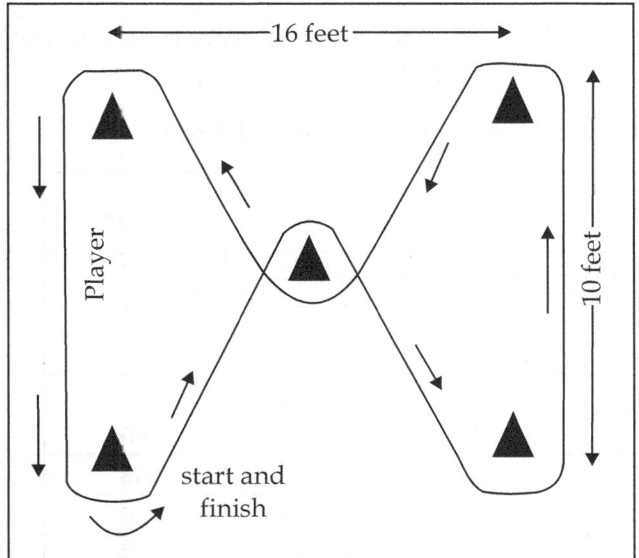


1. **Purpose:** To measure mainly agility and speed.
2. **Facilities and Equipment:** One stop watch and 5 flags or cones as shown in diagram.
3. **Direction:** First of all a demonstration is given.The subject is advised to stand just behind the starting line. Then, running in a zig-zag manner, three rounds are completed.
4. **Instructions:** The subject can start from standing position. After the start, running is done in a manner of figure of eight. Like this, three circuits are completed.
5. **Scoring:** The final score is the time taken in seconds for all the three laps.

**Table-ZigZag Run (T-score for High school and Junior High Schools Boys)**

| Grade T-Score | 7 | 8 | 9 | 10 | 11 | T-Score |
|---|---|---|---|---|---|---|
| 80 | 20.1 Down | 17.8 Down | 20.2 Down | 21.6 Down | 21.5 Down | 80 |
| 75 | 21.4-20.2 | 19.5-17.9 | 21.3-20.3 | 22.7-21.7 | 22.6-21.6 | 75 |

| 70 | 22.7-21.5 | 21.2-19.6 | 22.4-21.4 | 23.8-22.8 | 23.7-22.7 | 70 |
|---|---|---|---|---|---|---|
| 65 | 24.0-22.8 | 22.8-21.3 | 23.5-22.5 | 24.8-23.9 | 24.7-23.8 | 65 |
| 60 | 25.2-24.1 | 24.5-22.9 | 24.6-23.6 | 25.8-24.9 | 25.8-24.8 | 60 |
| 55 | 26.5-25.3 | 26.2-24.6 | 25.7-24.7 | 26.9-25.9 | 26.8-25.9 | 55 |
| 50 | 27.8-26.6 | 27.8-26.3 | 26.8-25.8 | 27.9-27.0 | 27.8-26.9 | 50 |
| 45 | 29.0-27.9 | 29.5-27.9 | 27.9-26.9 | 28.9-28.0 | 28.9-29.9 | 45 |
| 40 | 30.3-29.1 | 31.2-29.6 | 29.0-28.0 | 29.9-29.0 | 29.9-29.0 | 40 |
| 35 | 31.6-30.4 | 32.8-31.3 | 30.1-29.1 | 31.0-30.0 | 31.0-30.0 | 35 |
| 30 | 32.8-31.7 | 34.5-32.9 | 31.2-30.2 | 32.1-31.1 | 32.0-31.1 | 30 |
| 25 | 34.1-32.9 | 36.2-34.6 | 32.3-31.3 | 33.1-32.2 | 33.0-32.1 | 25 |
| 20 | 34.2 up | 36.3 up | 32.4 up | 33.2 up | 33.1 up | 20 |

**(iii) Six Pound Medicine Ball Put :**

1. **Purpose of the Test :** To measure the strength of arms and shoulder girdle. It also measures degree of arm and shoulder co-ordination, speed and balance.
2. **Facilities and Equipment required :** A space in the gymnasium hall with 90 feet × 25 feet dimensions available for use. A 15 feet restraining line marked. One medicine ball weighing six pounds and one measuring tape.
3. **Directions :** First of all the event is explained and a demonstration is given. The subject is advised to put the ball but "no throwing" is allowed. Three trials are given, but if the subject commits three faults, then he must try until he makes a fair put.
4. **Instructions :** The subject must take the position before the restraining line. No over stepping is allowed over the restraining line. The medicine ball is put, not thrown.
5. **Scoring :** The final score is the distance of the best put in feet.

**Table : Medicine Ball Put (T-scores for High School and Junior High School Boys)**

| Grade T-Score | 7 | 8 | 9 | 10 | 11 | T-Score |
|---|---|---|---|---|---|---|
| 80 | 43 up | 45 up | 49 up | 50 up | 54 up | 80 |
| 75 | 38 - 42 | 43 - 44 | 46 - 48 | 47 - 49 | 51 - 53 | 75 |
| 70 | 35 - 37 | 40 - 42 | 44 - 45 | 44 - 46 | 48 - 50 | 70 |
| 65 | 33 - 34 | 37 - 39 | 41 - 43 | 42 - 43 | 46 - 47 | 65 |
| 60 | 30 - 32 | 34 - 36 | 38 - 40 | 39 - 41 | 43 - 45 | 60 |
| 55 | 27 - 29 | 31 - 33 | 35 - 37 | 37 - 38 | 40 - 42 | 55 |
| 50 | 25 - 26 | 28 - 30 | 32 - 34 | 34 - 36 | 37 - 39 | 50 |
| 45 | 22 - 24 | 25 - 27 | 29 - 31 | 32 - 33 | 34 - 36 | 45 |
| 40 | 19 - 21 | 23 - 24 | 27 - 28 | 29 - 31 | 31 - 33 | 40 |
| 35 | 17 - 18 | 20 - 22 | 24 - 26 | 27 - 28 | 28 - 30 | 35 |
| 30 | 14 - 16 | 17 - 19 | 21 - 23 | 24 - 26 | 25 - 27 | 30 |
| 25 | 12 - 13 | 14 - 16 | 18 - 20 | 22 - 23 | 22 - 24 | 25 |
| 20 | 11 Down | 13 Down | 17 Down | 21 Down | 21 Down | 20 |

**24. Write in detail about the physiological changes taking place due to ageing.** **[5]**

**Ans.** The physiological changes, due to ageing, which take place are mentioned herein:

(i) **Changes in Nervous System:** The reaction time and movement time slows down with increase in age. The brain waits, the size of its network and its blood flow decreases with age.

(ii) **Change in Gastrointestinal System:** With an increase in age, there is a reduction in the production of Hydrochloric acid, digestive enzymes and saliva. These changes may result in delayed digestion, impaired swallowing. The breakdown and absorption of food may also be impaired. The liver becomes less efficient in metabolising drugs and repairing damage liver cells.

(iii) **Changes in Urinal system :** With an increasing age the mass of the kidney decreases, which leads to reduction in blood filtration by the kidneys. The capacity of the bladders decreases and there is an increase in residual urine. This increases the chance of urine infections.

(iv) **Changes in Senses:** With an advance in age, the senses such as vision, hearing, taste, smell and touch may become less active. Vision and hearing are affected the most by ageing. The taste buds are reduced with age. So, they loose interest in food.

(v) **Change in Respiratory System :** Pulmonary function is impaired with advancing age. The airways and lung tissues become less elastic and less efficient. There is decrease in the oxygen uptake and oxygen exchange.

(vi) **Change in Fitness :** The elasticity of tendons, ligaments and joint capsules decrease with ageing. The range of the movements is restricted and muscle mass decreases as the age increases. This leads a decrease in flexibility, endurance, strength and speed along with a shortness of breath, blood flow, enzymes etc.

**25. Mention in detail about the Aerodynamic Principles and also discuss about the application of aerodynamics in sports.**** **[5]**

**26. What do you understand by High Altitude Training in sports ? What is the impact of high altitude training on athletes ? Write in detail.**** **[5]**

●●

** Answer is not given due to change in the present syllabus.

# PHYSICAL EDUCATION 2017
## Outside Delhi

**Time allowed : 3 hours** **Maximum marks : 70**

SECTION-A

1. **Round-Robin Tournament is of two types. Name them and give one major difference between them.** [1]

**Ans.** There are two types of round-robin tournaments. In single round robin schedule, each participant plays other participant once and if each contestant plays all other twice, it is called a double round robin.

2. **Define leadership.**** [1]

3. **What do you mean by food intolerance?** [1]

**Ans.** Food intolerance is that when a person has difficulty in digesting a particular food and develops any disease like nausea, vomiting, acidity, motions, stomach upset, headache etc.

4. **State the common postural deformities.** [1]

**Ans.** Some common postural deformities are spinal curvature, bow legs, knock knees, round shoulders and flat foot etc.

5. **Name the motor development stages in children.** [1]

**Ans.** The motor development in children is categorised as follows:

(i) **Infants** : (0 to 2 years)

(ii) **Early Childhood :** The body development of child in age from 2 to 6 of years.

(iii) **Middle Childhood :** This stage is from 7 to 10 years of age.

(iv) **Late childhood :** (11 to 2 years)

6. **Your grandmother feels that she has reduced her upper body flexibility and therefore she wants to test herself. Which test would you suggest to her ?** [1]

**Ans.** Back Scratch test for upper body flexibility.

7. **What is oxygen uptake?** [1]

**Ans.** The amount of oxygen, which can be absorbed and consumed by the working muscles from the blood.

8. **What is incision?** [1]

**Ans.** An incision wound is a cut in the skin caused by sharp object, such as knife, broken glass, scissors or surgeon's scalpel. Incision wounds are 'neat' and the edges of the skin are usually smooth.

9. **An object thrown into the space either horizontally or at an acute angle under the action of gravity is called a projectile. Name the two forces which act on a projectile.**** [1]

10. **What do you mean by body image ?**** [1]

11. **What does the term 'Fartlek' mean and who developed this training method ?** [1]

**Ans.** The word 'Fartlek' is a Swedish word which means 'speed play'. Gosta Holmar developed it in 1937.

12. **What are specific sports programmes? Explain with suitable example.** [3]

**Ans.** Sports and games programmes are organised in the world as well as in our country to promote the games and sports for a specific cause. The programme motivates and creates the feeling of awareness for health, environment, unity and more and more people take part in these sports programmes. These specific sports programme are usually organized by the federations, NGO etc., to create health consciousness among the people and take part in health related sports programmes.

The various important specific programmes are:

(i) Sports day

  (1) School : Annual sports day

  (2) National sports day

(ii) Health run

(iii) Run for fun

(iv) Run for Unity

(v) Run for awareness

(vi) Run for specific cause.

13. **Write in brief about the leadership qualities that one inculcates by participating in adventure sports.**** [3]

14. **How can women's participation in sports and games be encouraged in India? Explain.** [3]

**Ans.** In Indian history of sports and games, there are only few names who participated actively in Olympics or other sports. Karnan Malleshwari became the first Indian woman to secure a medal in Olympics. Saina Nehwal and MC Mary Kom put colourful cap of India in Olympics. Though, P. T. Usha is also a famous name but she could

** Answer is not given due to change in the present syllabus.

not win a medal. Other names who have participated in other games are Sania Mirza in Tennis, Phogat Sisters in Wrestling, Anjum Chopra, Diana Edulji in Cricket.

But women are victims of gender inequality. Discrimination against woman can be seen even in the field of media.

**Reasons of less participation:**

(i) No parental encouragement

(ii) Traditional society

(iii) Less knowledge

(iv) Lack of sport infrastructure and inadequate training

(v) Concerns for personal safety and harassment or exploitation

**Methods to increase women's participation in sports:**

(i) Coaching programmes

(ii) Safety

(iii) Facilities

(iv) Grant of funds and increasing prizes at every level *i.e.*, schools to national levels.

**15. Write in brief about Osteoporosis. What are the causes of osteoporosis in women? [3]**

**Ans.** Osteoporosis is a disease of bones. When bone density decreases and the body stops producing as much bone as it did before. It is most likely to occur in women after menopause because of sudden decrease in estrogen. As the bones become weaker, there is high risk of fracture during a fall in even a minor knock. It happens when women lose too much bone, make too little bone or both. As a result, bones may break from a minor fall. Estrogen is lower in case of female athletes. Low estrogen levels and poor nutrition, especially low calcium intake, can lead to Osteoporosis.

**The cause of osteoporosis for women are:**

(i) Due to lower levels of estrogen after menopause.

(ii) Inadequate intake of calcium and vitamin-D.

(iii) Lack of weight bearing exercise.

(iv) Abnormal or absence of menstrual periods.

**16. Explain the Rockport test. [3]**

**Ans.** Rockport one mile walking test : The rockport walking test is sub-maximal field test to estimate $VO_{2\,max}$ in males and females ages 20 to 69 years.

**Equipment** : One mile track and stopwatch, weighing scale.

(i) **Procedure:** A level, one mile (1.6 KM) course is required. The inside lane of a one mile track is preferred, but any uninterrupted course of precisely one mile is suitable.

(ii) Participant with proper cloth and shoes, performs 5-10 minimum of light stretching before commencing the walk.

(iii) Instruct the participant to walk one mile as quickly as possible.

(iv) Record the participant's heart rate (HR) immediately after the completion of the mile. It is preferable to have the participant wear a heart rate monitor for this measurement but the assessment of heart rate via palpation is a suitable alternative.

(v) Estimate the participant's $VO_{2max}$ using the following formula which incorporates his/her body weight (Ib), age (yr), gender (males=1, females =0), time to complete one mile (min), and post exercise heart rate (bpm).

**Estimated $VO_{2max}$** ml-kg$^{-1}$-min$^{-1}$ = 132. – 0.0769(Weight) – 0.3877 (Age) + 6.315 (Gender) – 3.2649 (Time) – 0.1565 (HR)

**17. A famous cricket star Phillip Hughes was struck behind the ear by a ball while batting and died two days after the injury. He was wearing a helmet but the possible reason mentioned was that even when using a helmet, possibly a significant part of the neck remained exposed and the ball hit him there. And now most of the top cricketers across the would use deeper protection.**

**(a) Do you feel protective gears are important? Lay stress on your views.**

**(b) What first aid should be provided during injury at the superficial layer of the skin? [3]**

**Ans.** In a sport like Cricket, protective gears or sport gears are very vital. In recent past, it is observed that when a forceful ball hits the batsman or fielder, it may cause injuries or even death.

(a) In the game of Cricket one should protect head, arms, chest, thighs, private parts, knees, legs and feet. There are various protective gears such as helmet, batting gloves, chest guard, thigh guards, arm pads, shoulder pads, abdominal guard, leg protector, shoes, wicket keeping gloves etc. These sport gears must be of good quality. The proper protective equipments play significant role in preventing the sports injury.

(b) (i) Clean the affected part with fresh water.
(ii) Use clean sterilised qauze to wipe the affected part.
(iii) Antiseptic ointment should be used.
(iv) Refer to the doctor in case of serious injury.

**18. What is meant by 'motivation'? Explain the different techniques of motivation to achieve high goals in sports. [3]**

**Ans.** Motivation means to be inspired to do something. Indeed, motivation is the inner condition of the individual that directs his behaviour towards a goal. The force of motivation incites individuals to perform some activities. It also determines the individual's direction of action and his intensity of action.

There are two types of motivation:

(i) **Intrinsic Motivation:** Intrinsic motivation is internal. It occurs when people are compelled to do something out of pleasure, importance or desire. Motivation is always intrinsic when the force comes from within oneself. For example, when a person indulges in any sport to have mastery, to display superiority or to gain social approval, it is an intrinsic motivation.

(ii) **Eccentric motivation** : Eccentric motivation is external. It occurs when external factors compel the person to do something. Motivation is always eccentric, when external forces, positive or negative, produce a behavioural change. Reward, punishment, praise, blame or cash prize are example of eccentric motivation. It has been found that such devices motivate some persons more strongly than others. Infact eccentric motivation includes factors that motivate the individual in achieving the goal.

**19. Explain Interval training method. [3]**

**Ans. Interval training method** : It is a training of heart, through endurance training. It is also called terrace training. Dr. Woldemar and Gerschler, introduced this training method in 1930. In this method athlete use to run 400 m race 10 to 20 times daily instead of running 10 or 20 miles daily. Then, this training method came to be known as interval training method. Infact, this training method is based upon "effort and recovery" principle. During interval training, recovery time is given to the athlete after each speedy workout. Recovery period can be adjusted according to the efficiency of athlete. For an athlete of 400 m race following example can be applicable for his training.

(i) 400 m race with 80% race.
(ii) Walking or jogging until his heart rate comes down to 120 to 140 approximately.

**Advantages**

(i) More workout can be performed in short duration.
(ii) It is more beneficial for respiratory and circulatory systems and they can be trained both at same time.
(iii) The progress of the athlete can be measured easily.

**20. What are the Nutritive and Non-nutritive components of diet? Explain. [5]**

**Ans.** A diet which consists of all the essential food constituent viz. proteins, carbohydrates, fats, vitamins, minerals and water in correct proportions is called balanced diet. In balanced diet both nutritive and non-nutritive components are necessary.

(i) **Nutritive component of diet:** Nutritive component of diet are those components which contribute or provide energy or calorie. There are various nutritive components of diet, such as proteins, carbohydrates, fats, vitamins and minerals etc.

(1) **Fat:** Fat is considered as a major source of energy for athletes. Various type of sport require different proportions of fat to muscles for maximum performance. A minimum amount of fat is desirable for a distance runner, high jumper or gymnast. Distance swimmers need a certain amount of fat distributed near the skin surface to diminish the heat loss to the water.

(2) **Minerals:** The mineral deficiency can decrease the performance specially during the exercises in hot climate. Sweating reduces the amount of sodium and chloride in the body. Excess amount of salt intake can lead to potassium loss and increased water retention. This condition can be harmful for the persons with hypertension.

(3) **Vitamins:** Vitamins are chemicals which are required in very small amount to keep our body healthy. If a particular vitamin is not taken in diet, it may cause a deficiency disease. For example, if vitamin 'C' is not taken in diet, it may cause the scurvy disease.

(4) **Proteins:** The word 'Protein' is derived from the Greek word' Proteus', which means 'the best substance from foodstuffs'. The protoplasm is formed by protein which is base of living cell. Our physical growth and development will be retarded, if we take less amount of protein in our food.

(5) **Carbohydrates:** This is also a essential constituent of food. Carbohydrates act as a fuel in our body and give energy. We also need carbohydrates for the formation of fats. The individuals, who perform hard work, need maximum amount of carbohydrates. Due to lack of carbohydrates in the body, he loses weight. The skin becomes loose and wrinkles start appearing on the face. The source of carbohydrates are rice, maize, jowar, bajra, pulses, gram, dry pea, dates, grape, potato, banana etc.

(ii) **Non-nutritive component of diet** : Non-nutritive components of diet are those components which do not contribute or provide energy or calories. Roughage water, color, flavour and pesticide residue etc., are among the non-nutritive components of diet or food.

**21. What do you mean by correct posture? Explain the standing and sitting postures. What are the causes of bad posture?** [5]**

**22. Explain 'weight training' as one of the oldest methods for development of strength. Describe its advantages and disadvantages. [5]**

**Ans.** Weight training is the oldest method of exercise to develop strength. It includes those exercises, that are designed to strengthen specific muscles by causing them to overcome a fixed resistance, usually in the form of Barbells or Dumbles.

**Advantages of Weight training:**

(i) Improves postures and range of motion.

(ii) Increases muscles strength, bone density and endurances.

(iii) Improves motor performance.

(iv) Promotes healthy blood pressure and cholesterol level.

(v) Maintains healthy weight.

(vi) Develops confidence and self esteem.

(vii) Improves immune system functions.

(viii) Improves psyco-socio well being.

(ix) Promotes and develops exercise habits.

**Disadvantages of weight training:**

(i) Maturity

(ii) Introduce injury

(iii) Safety

(iv) Loss of flexibility

**23. Explain the physiological factors determining speed. [5]**

**Ans. Definition of speed:** Speed is the ability to move quickly aross the ground or move limbs rapidly to grab or throw.

**Factors affecting speed:**

(i) **Neuro-muscular coordination :** It relates to the ability of the nervous system to efficiently recruit a muscle or group of muscles in order to perform a specific task. Neuro-muscular coordination works on two levels:

(1) **Intra-Muscular Coordination:** If the fibers in the muscle all contract and relax in synchronization, then you are producing more muscle power. Basically, the better intra-muscular coordination, the more efficiently the muscle in working.

(2) **Inter-Muscular Coordination:** The coordination between different muscles or group muscles while performing a specific task or activity.

Intra and Inter-muscular coordination improves with repetition training; muscles adopt and learn to do that specific sport movement more and more efficiently, once a person develops a certain level of coordination in a movement they don't need to focus as much on it.

(ii) **Speed in the muscle contraction:** Speed of contraction depends on the characteristic and distribution of slow and fast twitch fibres each person mainly has and it is genetically determined so it does not change significantly with training. The fast

** Answer is not given due to change in the present syllabus.

muscle moves 5 times faster than the slow muscle and the super-fast moves 10 times faster than the slow muscle fibre:

**(iii) Muscle strength levels:** There is strong link between speed and strength; both contribute to the power.

**(iv) Gender and age:** Speed level increases at the same time that the nervous system and the locomotor system develop. Boys and girls have the same speed levels until the beginning of puberty; during puberty, muscular strength develops in both genders but is more pronounced in boys, therefore it provides them higher speed levels. The maximum speed development is achieved around the age of 20 years and with the appropriate training it can be maintained and even get better until 30-35 years of age.

**(v) Proper technique and sport performance:** It will allow being more efficient in movements, doing them faster and also avoiding injuries. lt will also improve the biomechanical sport movements by teaching the body and the mind the correct motor patterns and store them in your muscle memory so you can repeat them without thinking particularly when you are tired.

**(vi) Attentional states:** Influence the response time to the stimulus. Developing anticipation skill is one of the characteristics of great athletes. Anticipation is a sport specific ability to reduce the time it takes to respond to a stimulus, being able to react to an event before it has actually happened. A tennis player who anticipates the type of serve the opponent will use, detecting certain clues early in the serving sequence that predicts the potential type of server so that player can start moving towards the direction of server quicker than usual. Experience is also a very important factor in anticipation.

**24. Classify sports injuries. Explain 'P.R.I.C.E' procedure as a treatment of soft tissue injuries. [5]**

**Ans.** It is common fact that the injuries in every sport is of different type. We can classify sports injuries as below:

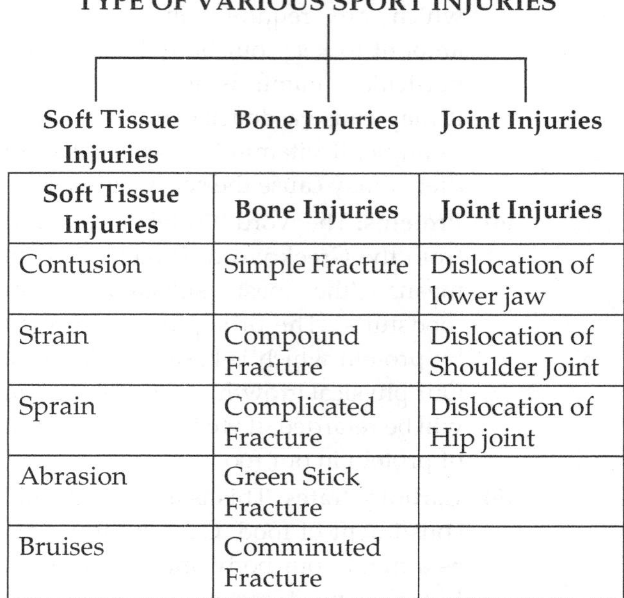


| Soft Tissue Injuries | Bone Injuries | Joint Injuries |
|---|---|---|
| Contusion | Simple Fracture | Dislocation of lower jaw |
| Strain | Compound Fracture | Dislocation of Shoulder Joint |
| Sprain | Complicated Fracture | Dislocation of Hip joint |
| Abrasion | Green Stick Fracture | |
| Bruises | Comminuted Fracture | |

**P. R. I. C. E. PROCEDURE FOR TREATMENT OF SOFT TISSUE INJURIES**

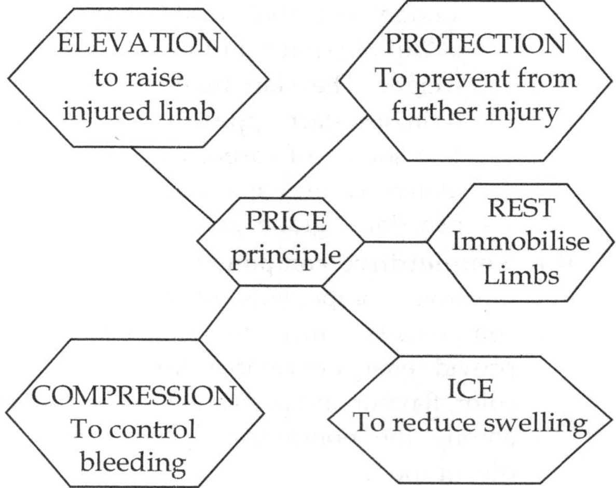


**25. What is personality? Explain its different dimensions. [5]**

**Ans. Personality** is a combination of two works *i.e.*, PERSONAL + IDENTITY = Personality.

Thus, any variable or component that makes a person unique, different from all other constitutes a part of one's a Holistic Personality.

**Holistic personality** means all round development of

**(i)** BODY

**(ii)** MIND (Knowledge, communication and emotion)

**(iii)** SOUL

**There are 11 Domains of personality**

R = Regularity in,

S = Safety,

S = Self Responsibility
M = Mental Activity,
O = Occupational Activities,
V = Vital,
E = Environmental,
S = Social
P = Physical,
E = Emotional,
S = Spiritual,
T = Temporal (Physical Growth)

**Personality Dimensions :**

The five-factor model is comprised of five personality dimensions (OCEAN): Openness to experience, Conscientiousness, Extraversion, Agreeableness and Neuroticism. The five dimensions are held to be a complete of personality.

**26. What are the various types of friction? How is friction advantageous or disadvantageous in the field of games and sports? Explain with suitable examples. [5]**

**Ans.** The force acting along two surfaces in contact which opposes the motion of one body over the other is called the force of friction. It is very important in sports. The larger the area of contact between the surfaces, the greater is the force of friction. When both the surfaces are smooth, the force of friction is reduced to almost zero.

**Three types of friction are:**

**(i) Static Friction:** The opposing force that comes into play when one body tends to move over another surface but the actual motion has not started.

**(ii) Limiting Friction:** The maximum opposing force that comes into play when one body is just on the verge of moving over the surface of other body.

**(iii) Kinetic Friction:** The opposing force that comes into play when one body is actually moving over the surface of another body.

Friction is mostly advantageous. It means that it is essential in games and sports. Without friction, we cannot give a better performance in the field of sport. Example: Spiked shoes are used by athletes for running and studs used in football boots of the players. However, friction has disadvantages also. In cycling, there should not be more friction between road surface and tyres of the cycle.

●●

# PHYSICAL EDUCATION 2016
## Outside Delhi

**Time allowed : 3 hours** **Maximum marks : 70**

### SECTION-A

**1. Write formula for giving Bye.** **[1]**

Ans. Formula for giving Bye = Next power of 2 – No. of Teams

Upper Half = nb + $\frac{1}{2}$,

Lower Half = nb – $\frac{1}{2}$

**2. What are Natural Resources in Adventure sports?**** **[1]**

**3. What are the Micro Nutrients?** **[1]**

Ans. Vitamins, minerals are the micronutrients. These are required in small proportion in our diet. They are considered as supplement food, they act as defensive food and help the proper functioning of various organs such as vitamins, minerals.

**4. Suggest any two free hand exercises for correcting round shoulder.** **[1]**

Ans. (i) Chakra Asana, Dhanur Asana , Bhujanga Asana , Ustra Asana, Backward Bending.

(ii) Pull the shoulders backward and see upward.

**5. Explain the importance of fluid intake during competition.** **[1]**

Ans. During the activity or competition sports drinks helps to extend time for exhaustion and improves the performance in the competition and lasting longer in extreme heat. Sports drinks provide hydration as well as carbohydrates.

**6. What is Oesteoporosis?** **[1]**

Ans. Osteoporosis is weakening of bone mass due to deficiency of calcium and vitamin D resulting in poor bones. This condition declines the performance and leads to bone injuries.

**7. Explain the procedure for Eight Foot Up and Go Test.** **[1]**

Ans. This test is designed to measure speed, agility (coordinative ability) and balance while moving. The equipments required are stopwatch, straight back chair (approximately 45 cm high) cone markers, measuring tape, safe and clear area.

(i) Place the chair against wall and mark a cone eight feet away in front of chair.

(ii) Clear the path between the chair and the marker.

(iii) The person sits on chair, hands resting on knees and feet flat on chair.

(iv) On the command 'GO' timing starts and person is instructed to walk (on running) around the cone as quickly as possible and sit down on the chair.

(v) Scoring is counted as time taken to perform this. Two trials are conducted and lowest time in seconds is considered for scoring.

**8. What is "Stroke Volume"?** **[1]**

Ans. Amount of blood pumped by left ventricle per beat.

At rest period–50 to 70 ml/beat.

During Exercise–110 to 130 ml/beat

Or

Amount of blood ejected by heart in one stroke.

**9. What kind of sports injury can be termed as "Abrasion"?** **[1]**

Ans. Abrasion is injury on the surface of the skin. In this injury skin is scrapped or rubbed by friction.

It causes severe pain and sometimes bleeding from the affected part. Abrasion are very common sports injuries mainly caused by fall on hard surfaces.

**10. Explain, what is "Dynamic Friction"?** **[1]**

Ans. Dynamic friction is the force of friction which comes into play when a body moves over the surface of another body.

(i) Sliding Friction

(ii) Rolling Friction.

**11. How participation in physical activities enhances self esteem among children?**** **[1]**

**12. Explain, what is strength and write the methods of improving strength?** **[3]**

Ans. Muscular strength is the highest amount of effort exerted by the muscles of the body in order to overcome the most resistance in a single effort. Strength refers to a muscle's ability to generate

** Answer is not given due to change in the present syllabus.

force against physical objects, Training methods for strength development.

(i) Isometric Exercises.

(ii) Isotonic Exercises.

(iii) Isokinetic Exercises.

**13. Write about the deformities of spinal curvature. [3]**

Ans. There are three types of spinal curvature deformities like kyphosis, lordosis and scoliosis.

**(i) Kyphosis 'round shoulders'**: It is the postural defect in which shoulders are projected forward (shoulders are bent forward, the chin is downward and head is bent forward.)

**(ii) Lordosis:** Lordosis is the problem of lumbarspine. Lordosis is an inward curvature of a portion of the vertebral column. Here the vertebral column is curved backward and the individual leans backward while standing.

**(iii) Scoliosis:** It is the problem of spine in which vertebral column bends to sideward. Scoliosis causes one shoulder down and other is raised up.

**14. What safety measures children should be taught while participating in River Rafting?** [3]**

**15. Draw a fixture of 11 Football teams participating in a Tournament on the basis of knock out. [3]**

Ans. Total number of teams = 11

$$\text{Upper half} = \frac{n+1}{2} = \frac{11+1}{2} = 6$$

$$\text{Lower half} = \frac{n-1}{2} = \frac{11-1}{2} = 5$$

Power of two's = 24 – No. of teams

= 16 – 11 = 5

No. of Byes = 5

$$\text{No. of Byes in Lower Half} = \frac{nb+1}{2} = \frac{5+1}{2} = 3$$

$$\text{No. of Byes in Upper Half} = \frac{nb-1}{2} = \frac{5-1}{2} = 2$$

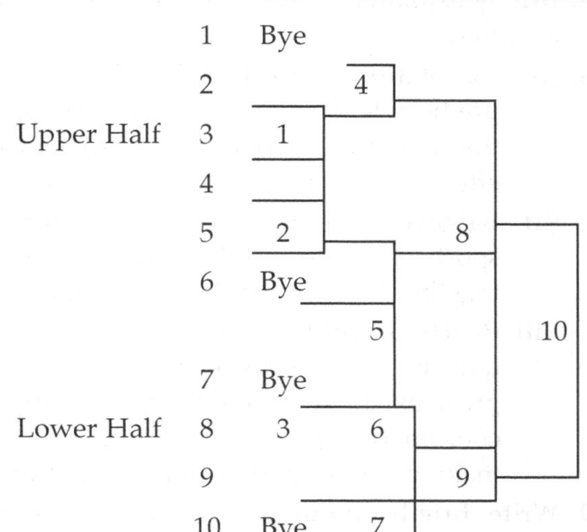


**16. How various committies are formed for Tournaments? Write briefly. [3]**

Ans. Arrangement committee, Technical committee, Discipline committee, Receptioncommittee, Boarding and Lodging committee, Ceremonial committee, Press and Media committee, Certificate Writing committee, Medical committee, Announcement committee, Recorders and Bulletin board committee.

**17. Sports are good for all age groups. Competitive sport is a showcase of power at the international arena. Supremacy over the other country is indirectly shown by standing at the top of medal tally. To achieve their target few countries are imparting very strict training to growing children. Over the years many organisations have raised their voice against the intensity with which training is scheduled for making an international athletes. [1 + 2 = 3]**

**(i) Elaborate any two physiological benefits of exercise in children.**

**(ii) What are the disadvantages of giving high intensity training to the growing children?**

**Ans.** (i) Improves cardiovascular system and health.

**(ii)** Disadvantages of giving high intensity training to the growing children are:

1. Stress or fatigue.

2. Chances of overload.

3. Early maturity.

4. Can reach too early in top form.

5. Less flexibility.

6. Chances of injury.

** Answer is not given due to change in the present syllabus.

**18. Briefly explain different types of co-ordinative abilities. [3]**

**Ans.** (i) **Adaptation ability**: According to predictable changes in circumstances, it is the capability of a sports person to result in an effectual alteration in the movement.

(ii) **Balance ability**: It is the capability of a sports person to sustain balance of the body together in static and dynamic conditions.

(iii) **Reaction ability**: It is the ability to react quickly and effectively to a signal. To a great extent it depends on sense organs, coordinative processes of CNS, decision making, concentration and anticipation.

**19. Write briefly about protein as an essential component of diet. [3]**

**Ans.** Protein is needed for growth and development of body. It helps to repair or replace the worn out tissues, it does not provide energy in normal routine whereas it acts as energy source only under extreme starvation (hunger). There are two types of proteins.

(i) Essential protein.

(ii) Non-essential protein.

**20. Write in detail about the physical and physiological advantages of physical exercises during the childhood stage. [5]**

**Ans.** (i) **Exercise helps in healthy growth and development**: Exercise is an important part of keeping children healthy. Encouraging healthy lifestyles in children and adolescents is important when they grow older. Participating in organized sports and games is not only of great fun but is very essential for healthy growth and development.

(ii) **Exercise improves self-esteem** : Exercise is necessary for your physical and mental health. Self-esteem can play a great role in how children feel about themselves and also how much they enjoy things or worry about things. Exercise reduces depressive symptoms and improves self-esteem in children.

(iii) **Exercise makes healthier weight range**: These days a lot of children not only in India but also throughout the world are getting overweight which is the root cause of various physiological problems. Regular exercise is important to control weight as it helps in burning extra calories and lowers the weight.

(iv) **Exercise makes stronger bones, muscles and joints**: Exercise is vital for strong muscles, bones and joints. Exercise may help children lower their risk of chronic pain related to muscles, bones and joints in the future. Exercise increases bone density which helps to make bones stronger. Exercise is beneficial in building strong and healthy muscles. Joints require motion to stay healthy.

(v) **Exercise strengthens the heart** : Exercise helps to improve heart health, and can even reverse some heart disease risk factors. The resting heart rate of those who exercise is also slower, because less effort is needed to pump blood. A child who exercises often has the lowest risk for heart disease, but even a small amount of exercise is beneficial.

**21. How the Cardiovascular Fitness is measured with the help of "Harvard Step Test"? Write in detail about its administrative procedure. [5]**

**Ans.** This test was developed in 1943 by Brouha to measure the cardiovascular fitness of the individual.

Fitness Index Score

$$= \frac{100 \times \text{test duration in seconds}}{2 \times \text{sum of heart beats in recovery periods}}$$

After above calculation fitness rating will be measured.

**Requirements of test:** (i) Gym Bench (45cm high) (ii) Stopwatch (iii) Assistant or Helper.

In this test, the student step-up and step-down on the gym bench for 5 minutes or until exhaustion at a rate of 30 steps per minute.

(i) Firstly, the student performs warm-up and comes near the gym bench.

(ii) At the command of 'GO' the student starts to step-up and step-down on the gym bench whereas assistant starts the stopwatch at same time.

(iii) This goes on for five minutes.

(iv) The assistant measures the heart rate (beats per minute) after one minute of test finish as pulse 1.

(v) The assistant measures the heart rate (beats per minute) after two minutes of test finish as pulse 2.

(vi) The assistant again measures the heart rate (bpm) after three minutes of test finish as pulse 3.

RESULT = 30000/(pulse 1 + pulse 2 +pulse 3)

**22. Keeping in view of the Indian Ideology, critically analyse the sociological aspect of Women Athletes in sports participation.** [5]**

**23. Participation in physical activity for a longer duration maintains functional fitness among aged population. Justify. [5]**

Ans. On the ageing process, physical activity can play very effective role, however, it should be implemented in a suitable manner. It is a well known fact that regular physical activity is the supplement to delay the ageing process. It can in some cases, overturn the most common signs of ageing viz., loss of muscles and bones, increased body fats, decline in memory, decreased metabolism, flexibility and blood flow etc. Participation in physical activity for a longer duration maintains functional fitness among aged population in following manner :

(i) **Reduces loss of muscles mass**: With advancing age the muscle mass decreases. The accumulation of fats reduces due to regular exercise. Ageing has a depressing consequence on metabolism. Regular physical activity results in dropping the metabolic rate and decreases the loss of lean body mass.

(ii) **Maintains bone density**: With advancement of age, bone density decreases as well generally leading to the rupture or osteoporosis. Physical activity assists in maintaining bone mass and prevent osteoporosis. Bone growth gets stimulated with resistance exercises.

(iii) **Improves lungs capacity**: Oxygen uptake and exchange increases and enhances the lungs capacity due to regular exercise. In maintenance of strong lungs, it plays prominent role and reduces the loss of elasticity of the lungs and chest wall.

(iv) **Lessens stress and tension**: Regular physical activity has a distinctive capability to slow down the depression process by reducing stress and tension. Actually, regular physical activity lessens the levels of body's stress hormones like adrenaline and cortisol. The body's natural pain killers and mood elevators like the endorphins are produced due to regular physical activity.

**24. Differentiate between Intrinsic and Extrinsic motivation. Explain in detail Goal setting and Reinforcement as technique of motivation. [5]**

**Ans.**

| S.No. | Intrinsic Motivation | Extrinsic Motivation |
|---|---|---|
| (i) | Natural Motivation and internal | Always form outside external factors |
| (ii) | Depends upon emot-ional needs, biological needs, personal needs and social needs. | Rewards, prizes, social recognition. |
| (iii) | Satisfaction, behaviour, happiness | Job, teachers, parents. |
| (iv) | Inner Desire, force comes from inside Interest | Appreciation |

(i) **Goal setting**: Achieving performance goals is a symbol of competency that affects motivation positively, hence it is essential to set realistic goals based on individual's own abilities. The level of the motivation gets adversely affected when goals are set up too high or too low. The goal should be realistic, precise and within reach but should at the same time be a difficult one. Goal settings have been acknowledged as an influential motivational technique as it mobilizes an athlete's hard work and extends his determination.

(ii) **Reinforcement**: Reinforcement is a vital motivation means. It refers to some kind of occurrence that increases or decreases the possibility of a similar reaction taking place in the future. Positive reinforcements enlighten the athlete at what time he is doing something accurate and support the continuance of the activity in the precise direction. Negative reinforcement is in general, of slight importance since it simply indicates that the actions are inaccurate devoid of providing information with respect to the accurate reaction or behaviour. It has been established that positive reinforcement to a large extent, has more motivational value than negative reinforcement.

**25. Write in detail about the Dislocation and Fractures among the bones and joint injuries. [5]**

** Answer is not given due to change in the present syllabus.

**Ans. Dislocation**: It is an injury of joint in which the adjoining bones are displaced from their original position. Dislocations mainly occurs in contact sports, such as football, hockey, and in sports that may involve falls, such as downhill skiing, gymnastics, volleyball and also during a motor vehicle accidents.

Dislocation takes place in different location of the joints like:

(i) Dislocation of shoulder joint.

(ii) Dislocation of lower jaw.

(iii) Dislocation of hip joint.

**Fracture**: Bone injuries occur in the bone due to some impact or minimal trauma injury as a result of certain medical conditions that weaken the bone. There are many types of bone injuries which are as follows:

**(i)** Simple fracture

**(ii)** Compound fracture

**(iii)** Greenstick fracture

**(iv)** Comminuted fracture

**(v)** Transverse fracture

**(vi)** Oblique fracture

**(vii)** Spiral fracture

**(viii)** Pathologic fracture

**(ix)** Impacted fracture

**(x)** Stress fracture.

**26. Differentiate mechanically between walking and running. Explain in detail mechanical analysis of walking. [5]**

**Ans. (i) Walking**: Walking is rolling down from heel to toe action. The first phase is heel strikes the ground followed by the transfer of weight and balance making. Afterwards the heel is rolled forward and push is given by toe and swing comes after. During walking the head should be straight without tension in shoulder. In the torso region the abdomen muscles should be slightly tightened, loose abdomen muscles cause back pain. The arm swing helps to gain speed in walking.

**(ii) Running**: Running is a toe action movement. For running, the focus is on the three major producing action-push off, knee-drive and paw-back. These three phases are done in a sequential manner to provide efficient running action, push off is the primary toe action movement needed to supply the driving force in forward direction. The force comes from the powerful contraction of calf muscles by extension of ankle joint. The slight flexion of knee joint provides horizontal force instead of vertical force. Body is also bent forward.

●●

# PHYSICAL EDUCATION 2015

## Outside Delhi

Time allowed : 3 hours  Maximum marks : 70

1. **Playgrounds are essential for creating sports environment. Justify your answer. ** [1]**

2. **What do you mean by "surfing" in adventure sports ?** [1]**

3. **Enlist two non-nutritive components of diet. [½ + ½ = 1]**

**Ans.** Non-nutritive components of diet are:
   (i) Colour compounds
   (ii) Flavour compounds

4. **What does the school intend by stating that, "only such students shall participate in the Basketball Intramurals who have not represented the school in Basketball in the past and minimum 10 substitutions shall be compulsory." [1]**

**Ans.** For promoting mass participation.
   To explore the hidden talent of the student.

5. **What is "an abnormal curvature of spine at front" termed as? [1]**

**Ans.** Lordosis is an abnormal curvature of spine at front.

6. **What type of resistances can be used for developing strength among children? [1]**

**Ans.** (i) Own body weight
   (ii) Gravitational force
   Static and dynamic resistance, own body weight, gravitational force can be used for developing strength among children.

7. **Which test would you suggest for your grandmother to test lower body flexibility? [1]**

**Ans.** Chair sit and reach test.

8. **Why does involvement in regular exercise delay the onset of fatigue? [1]**

**Ans.** If we do regular exercise, our fitness level will be increased and it also develops endurance because of the fatigue level delays.

9. **What is energy? [1]**

**Ans.** Energy is defined as an ability or capacity of a body to perform work. Energy is denoted by the letter 'E' and the 'S.I.' unit of energy is joule (J).

10. **Explain intrinsic motivation. [1]**

**Ans.** Intrinsic motivation is internal. It occurs when people are compelled to do something out of interest, pleasure, importance and desire.

** Answer is not given due to change in the present syllabus.

11. **"Pace races means, running the whole distance of a race at a constant speed". Which are the races included in pace races? [½ × ½ = 1]**

**Ans.** 800 mts and above or 800 and 1500 mts.

12. **Mention any three objectives of adventure sports.** [3]**

13. **Recently Sarita Devi refused to accept the Bronze medal during the ceremony. The international body (AIBA) which regulates boxing has taken a stringent action against Sarita Devi and the coaches. [2 + 1 = 3]**
   **(i) Do you agree with the decision of Sarita Devi? Justify your answer.**
   **(ii) What values do you think Sarita Devi has not shown by her behaviour during the medal distribution ceremony?****

**Ans.** (i) I do not agree with the decision of Sarita Devi because she did not show sportsmanship/ she did not obey the AIBA rules.

**OR**

I do agree with the decision of Sarita Devi because she raised the voice against the decision of the umpire/ injustice.

14. **What do you mean by "round shoulders"? Suggest any four physical activities for correcting round shoulders. [1 + 2 = 3]**

**Ans.** Round shoulder is a postural deformity in which the shoulders become round and sometimes they seem to be bent forward.
   Physical activities for round shoulder are:
   (i) Sit on a chair, rest the back against it, pull the shoulders backward and see upwards.
   (ii) Hold the horizontal bar for some titme.
   (iii) Perform Dhanurasana (arch formation) regularly.
   (iv) Perform Chakraasana (arch formation) regularly.

15. **Critically explain the use of dietary supplements in heavy dose for longer duration. Justify your answer with two suitable examples. [2 + 1 = 3]**

**Ans.** (i) Excess calcium in diet for longer time can cause heart diseases/formation of stones in kidney.
   (ii) Excess iron causes siderosis (vomiting and headache)
   (iii) Vitamin E can cause prostates cancer

**16. Explain in brief "The Harvard Step Test". [3]**

**Ans.** The Harvard step test is a cardiovascular fitness test. It is also called aerobic fitness test. It is used to measure the cardiovascular fitness or aerobic fitness by checking the recovery rate.

(i) **Equipment required:** A gym bench or box of 20 inches high for man and 16 inches for woman, stopwatch and cadence tape.

(ii) **Procedure:** The athlete stands in the front of the bench or box. On the command "GO" the athlete steps up and down on the bench or box at a rate of 30 steps per minute. Stopwatch is also started at the start of the stepping.

(iii) **Calculation of the scores :** Calculate with the help of following formula.

"fitness index score = (100 × test duration in seconds)/(2 × sum of heartbeat in recovery period)".

**17. "Regular physical activity can delay your ageing process" Justify your answer in light of the effect of activities on physiological changes. [3]**

**Ans.** (i) **Change in the nervous system:** After 25 years of age everyone loses nerve cells. Gradually over time, this results in a reduced ability of nerve transmission, changing response time and coordination. The brain also shrinks in size, which does not significantly effect functioning except in the most extreme cases. These changes may also affect sleeping patterns somewhat by decreasing the length of total sleep time and REM sleep.

(ii) **Maintains bone density:** With advancement of age, bone density decreases as well generally leading to the rupture or osteoporosis. Physical exercises assist in maintaining bone mass and prevent osteoporosis. Bone growth gets stimulated with resistance exercises. With the help of regular exercise, the aged persons can increase their bone density as depicted by several research studies.

(iii) **Changes in the endocrine system:** The endocrine or metabolic system is responsible for changing food into energy. After 25 years of age everyone experiences approximately a 1% decrease per year in their metabolic rate. This overall slowing results in food being absorbed and utilised less. It's consequences are reduced stamina and reserves as well as greater susceptibility to drug toxicity.

**18. How does angle of projection help as a factor for athletes in games and sports?** [3]**

**19. Dynamic strength is divided into three parts. Write in brief about each. [3]**

**Ans.** (i) **Maximum Strength :** In a single muscular contraction, it is the ability of muscle to contract over resistance of utmost intensity of stimulus. The most excellent examples are weight lifting and throwing events in track and field.

(ii) **Explosive strength :** It can be stated as the ability to prevail over resistance by means of high speed. It combines strength and speed abilities and based on the nature of the blend of strength and speed, the explosive strength can be sub-divided further into start strength, power and speed strength.

(iii) **Strength Endurance :** Similar to explosive strength, it is a result of two motor abilities as well. Under conditions of fatigue, it is the ability to work against resistance. Depending on the actuality whether the movement is static or dynamic, strength endurance can be there in form of static or dynamic strength. The strength endurance can be divided further into proper strength endurance and strength endurance depending on the kind of the blend of strength and endurance.

**20. What are the five essential elements of positive sports environment ?** [5]**

**21. Draw a knock out fixture of 21 teams mentioning all the steps involved. [5]**

**Ans.** Total No. of matches = N – 1 = 21 – 1 = 20 Matches

$$\text{No. of teams in upper half} = \frac{N+1}{2} = \frac{21+1}{2} = 11 \text{ teams}$$

$$\text{No. of teams in lower half} = \frac{N-1}{2} = \frac{21-1}{2} = 10 \text{ teams}$$

Total No. of byes = 32 – 21 = 11 byes

$$\text{No. of byes in upper half} = \frac{NB-1}{2} = \frac{11-1}{2} = 5 \text{ byes}$$

** Answer is not given due to change in the present syllabus.

$$\text{No. of byes in lower half} = \frac{NB + 1}{2}$$

$$= \frac{11 + 1}{2} = 6 \text{ byes}$$

Total rounds = 5

For quarter : In addition to it.

$$\text{No. of byes in I quarter} = \frac{NB - 1}{2}$$

$$= \frac{5 - 1}{2} = 2 \text{ byes}$$

$$\text{No. of byes in II quarter} = \frac{NB + 1}{2}$$

$$= \frac{5 + 1}{2} = 3 \text{ byes}$$

$$\text{No. of byes in III quarter} = \frac{NB}{2}$$

$$= \frac{6}{2} = 3 \text{ byes}$$

$$\text{No. of byes in IV quarter} = -\frac{NB}{2}$$

$$= \frac{6}{2} = 3 \text{ byes}$$

**22. What are the various factors affecting physiological fitness? Explain. [5]**

**Ans.** The various physiological factors which determine the strength of an individual are stated below:

**(i) Size of the muscle:** The strength of the muscle largely depends upon the size of the muscle. It is a well known fact that bigger and larger muscles can produce more force. The force produced by the same size of muscle in males and females is approximately the same but males are found to be stronger because they have larger muscles and bigger muscles in comparison to females. With the help of different methods of strength training such as weight training the size of the muscle can be increased and as a result of that strength is improved. So, the strength is determined by the size of the muscle.

**(ii) Body weight:** It is also a well known fact that the individuals who are heavier are stronger than the individuals who are lighter. There is a positive co-relation between body weight and strength among international weightlifters. The heavier weightlifters lift the heavier weight. So, body weight also determines the strength of an individual.

**(iii) Muscle composition :** Each muscle consists of basically two types of muscle fibres *i.e.,* fast twitch fibres (white fibres) and slow twitch fibres (red fibres). The fast twitch fibres are capable to contract faster and therefore, they can produce more force. On the contrary, the slow twitch fibres are not capable to contract faster but they are capable to contract for a longer duration. The muscles which have more percentage of fast twitch fibres, can produce more strength. The percentage of fast twitch fibres and slow twitch fibre is genetically determined and can not be changed through training. So it can be said that the percentage of these fibres determines the strength.

**(iv) Intensity of the nerve impulse :** A muscle is composed of a number of motor units. The total force of the muscle depends on the number of contracting motor units. Whenever, a stronger nerve impulse from central nervous system excites more number of motor units, the muscle will contract more strongly or it can be said that the muscle will produce more force or strength. So, the intensity of the nerve impulse also determines the amount of strength.

**23. Explain the cognitive aspect of stress. Suggest any three techniques briefly to overcome stress** [2 + 3 = 5]**

**24. Differentiate between 1 : 1 and 1 : 2 ratio interval training, with suitable examples. [5]**

**Ans.** 1 : 1 means load and the rest is equal e.g, 1 minute exercise followed by 1 minute of rest. Similarly 1 : 2 means that the period of rest is double of the load. e.g, 1 minute exercise followed by 2 minute of rest.

The slow and extensive interval training methods can be given by using 1: 1 and 1:2. Fast or intensive interval training methods can be given by using 1 : 1 and 1 : 2.

Most of the games and sports are based on the principle of load and rest again load and rest. The rest period during training work load should not exceed five minutes.

** Answer is not given due to change in the present syllabus.

Another example canbe of a sprinter:

100 m sprint 100 m walk 1 : 1

100 m sprint 200 m walk 1 : 2.

**25. Vitamins are very essential for working of the body and are divided into two groups. Explain about them.** **[5]**

**Ans.** **(i)** **Fat soluble vitamins:** The fat-soluble vitamins are those vitamins that are soluble in fat. These include vitamin A, D, E and K stored in the liver and in body fat.

**(1) Vitamin 'A':** Vitamin A (retinol) is a fat-soluble vitamin that acts as an antioxidant. The Vitamin A that we obtain from animal products is called retinoid and can be used by our body without any modification. The form of Vitamin A found in fruits and vegetables is known as carotenoids.

**(2) Vitamin 'D':** Vitamin D is different from other essential vitamins because our own body can manufacture it with sunlight exposure. The main function of Vitamin D is to regulate the absorption of calcium and phosphorus in our bones and aid in cell-to-cell communication throughout the body.

**(3) Vitamin 'E':** Vitamin E (tocopherol) is a powerful, fat-soluble antioxidant that helps to protect cell membranes against the damage caused by free radicals, prevents the oxidation of LDL cholesterol and is also important in the formation of red blood cells (RBC), thus essential for blood coagulation.

**(4) Vitamin 'K':** Vitamin K occurs in two forms- Vitamin K1 and Vitamin K2. Vitamin K is, known as the clotting vitamin, because without it, blood would not clot. It also helps in the prevention of haemorrhage.

**(ii)** **Water soluble vitamins :** The water-soluble vitamins are those vitamins that are soluble in water. Vitamin C and members of the Vitamin B complex are water-soluble vitamins.

**(1) Vitamin 'B1' (Thiamin) :** Vitamin B1 or Thiamine, helps to release energy from food, promotes normal appetite, and is important in maintaining proper nervous system function.

**(2) Vitamin '$B_2$' (Riboflavin) :** Riboflavin or Vitamin B2, helps to release energy from foods, promotes good vision, and healthy skin. It also helps to convert the amino acid tryptophan (which makes up protein) into niacin.

**(3) Vitamin '$B_3$' (Niacin) :** Vitamin B3, or niacin, works with other B-complex vitamins to metabolise food and provide energy for the body. Vitamin B3 is involved in energy production, normal enzyme function, digestion, promoting normal appetite, healthy skin and nerves.

**(4) Vitamin '$B_6$' (Pyridoxine) :** Vitamin B6 is a key factor in protein and glucose metabolism as well as in the formation of hemoglobin. Hemoglobin is a component of red blood cells-it carries oxygen. Vitamin B6 is also involved in keeping the lymph nodes, thymus healthy.

**(5) Vitamin '$B_{12}$' (Cobalamin) :** Vitamin B12 is also known as Cobalamin, aids in the building of genetic material, production of normal red blood cells, and maintenance of the nervous system.

**(6) Biotin :** Biotin is a coenzyme and a vitamin and is also known as Vitamin H. As a supplement, biotin is sometimes used for diabetes, brittle nails, and other conditions. Biotin helps to release energy from carbohydrates and aids in the metabolism of fats, proteins and carbohydrates from food.

**(7) Vitamin 'C':** Vitamin C is also called Ascorbic acid, which is a water-soluble vitamin and cannot be stored in the body. Most plants and animals can produce their own vitamin C but humans cannot. Vitamin C is needed for proper growth, development, and to heal wounds. It is used to make the collagen tissue for healthy teeth, gums, blood vessels and bones.

**26. Weight training is one of the oldest methods for development of strength. What are its advantages and disadvantages ?**** **[5]**

●●

** Answer is not given due to change in the present syllabus.

# PHYSICAL EDUCATION 2015
## Delhi

**Time allowed : 3 hours** **Maximum marks : 70**

**1. Suggest any four ways through which women participation in sport across age group can be enhanced.** [¼ × 4 =1]

**Ans.** (i) Modification in Legislation.

(ii) Better coverage of women sports.

(iii) Improvement in fitness and wellness movements.

(iv) Educating Women.

**2. Trekking is a long adventurous journey undertaken on foot in areas where common means of transport are generally not available. Name any four important materials required that should be carried along.**** [¼ × 4 = 1]

**3. Enlist two objectives of Intramurals.** [½ + ½ = 1]

**Ans.** (i) Develops moral and ethical values of the students.

(ii) Provide opportunity to the maximum number of students to participate in sports.

**4. Enlist two sources for calcium and iron separately.** [¼ × 4 = 1]

**Ans. Calcium sources :** Cheese, Milk, Orange Juice, Eggs, Yogurt.

**Iron sources :** Liver, Meat, Orange Juice, Egg.

**5. Explain correct sitting posture.**** [1]

**6. Calculate the physical fitness index using short formula for a 12 year old boy having completed Harvard Step Test for a duration of 3 minutes and a pulse rate of 54 beats for 1 to 1.5 minute.** [1]

**Ans.** The athlete's fitness index score is calculated with the help of following formula.

Fitness index score = (100 × test duration in seconds) divided by (2 × sum of heart beats in recovery period).

(100 × 180 Sec)/ 2 × 54 = 500/3 = 166.66

**7. Your grandmother feels she has reduced her upper body flexibility and therefore she wants to test herself. Which test would you suggest her?** [1]

**Ans.** Back scratch test for upper body flexibility.

**8. Explain the term Hypertrophy of muscles.** [1]

**Ans.** Increase in size of the muscle fiber due to regular exercises or Hypertrophy is enlargement of heart due to regular exercises which is called "Athletic heart".

**9. What do you understand by linear movement?** [1]

**Ans.** Linear movement refers to any movement along a straight line in one direction.

**10. Explain the term "realistic" in goal setting principles.** [1]

**Ans.** Realistic goals are achievable goals.

**11. Suggest any two isometric exercises for shoulder region.** [½ +½ =1]

**Ans.** (i) Pushing against the wall.

(ii) Holding push up position.

**12. What safety measures children should be taught while participating in trekking?** [3]

**Ans.** (i) Avoid trekking during bad weather conditions.

(ii) To prevent insect bite do wear full sleeves shirts and full pants.

(iii) Wear proper footwear so that you do not slip while trekking.

(iv) Do not eat local leaves, flowers etc. while trekking, they may be poisonous.

**13. Briefly explain the functions and resources of three fat soluble vitamins.** [1 × 3 = 3]

**Ans.** Fat Soluble vitamins are A, D, E, K.

**Functions :**

(i) **Vitamin 'A':** Vitamin A acts as an antioxidant which may help in reducing cancer. It also helps to prevent night blindness, keratomalalia and irregular growth of teeth.

(ii) **Vitamin 'D':** The main function of vitamin D is to regulate the absorption of calcium and phosphorus in our bones and aid in cell-to-cell communication throughout the body.

(iii) **Vitamin 'E':** Vitamin E (tocopherol) is a powerful, fat-soluble antioxidant that helps to protect cell membranes against the damage caused by free radicals, prevents the oxidation of LDL cholesterol and is also important in the formation of red blood cells (RBC), thus essential for blood coagulation.

** Answer is not given due to change in the present syllabus.

(iv) **Vitamin 'K':** Vitamin K occurs in two forms-Vitamin K1 and Vitamin K2. Vitamin K is known as the clotting vitamin, because without it, blood would not clot. It also helps in prevention of haemorrhage.

(1) **Sources of Vitamin A:** Ghee, milk, curd, egg yolk, fish, tomato, papaya, spinach, carrot, pumpkin etc.

(2) **Sources of Vitamin D:** Egg yolk, fish, sunlight. Vegetables, cod liver oil, milk, cream, butter, tomato, carrot etc.

(3) **Sources of Vitamin E:** Green vegetables, kidney, liver, heart cotton seed, sprouts seeds, coconut oil.

(4) **Sources of Vitamin K :** Cauliflower, spinach, cabbage, tomatoes, wheat, egg and meat etc.

**14. Neeti along with her father was regular at district park in early morning. She realized that most of the children are obese. She along with her few classmates wanted to help those children. She discussed with her physical education teacher and the principal of the school. School decided to organize awareness rally for the neighbourhood.**

**(i) How obesity can be prevented? Give two ways.**

**(ii) Give any two disadvantages of obesity.**

**(iii) What values are shown by Neeti and her classmates ?**** [1 × 3 = 3]

**Ans.** (i) **Obesity can be prevented by:**

(1) Avoiding fast food, fatty food and over eating

(2) Regular exercise/ physical activity

(ii) **Disadvantages of obesity:**

(1) More chances of injury

(2) More disease/ physical health problems

**15. Briefly explain the six physical benefits of exercise to children.** [½ × 6 = 3]

**Ans.** (i) **Exercise helps in healthy growth and development:** Encouraging healthy lifestyles in children and adolescents is important when they grow older. Participating in organized sports and games is not only of great fun but is very essential for healthy growth and development.

(ii) **Exercise improves self-esteem:** Self-esteem can play a great role in how children feel about themselves and also how much they enjoy things or worry about things. Exercise reduces depressive symptoms and improves self-esteem in children.

(iii) **Exercise reduces blood sugar level:** Lowering blood sugar level is not only the health issue of adults but also for children as they are also becoming trapped with high sugar level. Exercise is helpful in preventing sugar from accumulating in the blood by triggering muscles to receive more glucose from blood and utilise it for energy.

(iv) **Exercise helps in motor development:** Exercise helps in increasing the motor development of infants and children at a very rapid pace which in due course improves in building fine movements later on in life. The movements of muscles become efficient and smooth by doing exercise in premature age.

(v) **Exercise strengthens the heart :** Exercise helps in improving heart health, and can even reverse some heart disease risk factors. A child who exercises often has the lowest risk for heart disease, but any amount of exercise is beneficial.

(vi) **Exercise makes stronger bones, muscles and joints :** Exercise is vital for strong muscles, bones and joints. Exercise may help children lower their risk of chronic pain related to muscles, bones and joints in the future. Exercise increases bone density, which helps to make bones stronger.

**16. Explain the procedure for conducting Kraus-Weber test for measuring minimum muscular strength.**** [3]

**17. Maintaining physical activities for a longer period, brings desirous changes in circulatory system. Justify your answer by highlighting three benefits of exercise.** [3]

**Ans.** (i) **Heart size increases:** The size of heart and strength of the cardiac muscles increases due to regular exercises as to the maximum extent the left ventricle adapts. The walls of the heart develop into stronger and thicker as shown in recent studies and the thickness of myocardial wall increases as well.

(ii) **Resting heart rate decreases:** The resting heart rate decreases due to regular exercises. After duration of 10 week training programme, the resting heart rate may reduce up to 10 beats per minute from the normal of 72 beats per minute. The heart

** Answer is not given due to change in the present syllabus.

becomes more efficient due to regular exercises. In highly conditioned athletes the resting heart rate decreases to 30 beats/ minute.

**(iii) Stroke volume increases at resting conditions:** The stroke volume increases at resting conditions due to regular exercises. The stroke volume at rest remains up to 50-70 ml/beat in untrained individuals; in trained individuals it ranges from 70-90 ml/beat and in the elite endurance athletes it ranges from 90-110 ml/beat.

**18. What is the difference between linear and angular motion? Explain through example. [3]**

**Ans.** Linear motion is any motion that moves along a straight line in one direction. The direction can either be horizontal, vertical or inclined direction. Example, approach run.

Angular motion is rotatory motion, it occurs when all points on a body or object move in a circular path about the same fixed central line or axis. A child swing and rotations in hammer throw are the best example.

**19. What do you understand by relative strength? Explain the importance of body weight in determining relative strength. [3]**

**Ans.** Relative strength is strength in relation to your body weight. Relative strength have a determining importance in sports in which the athlete shifts his body in space without any additional external weight (H/J and L/J) as well as in sports in which he has to restrict his own weight within the framework of weight division (*e. g.* boxing, wrestling, weight lifting etc.) *e.g.* if 1 RM (repetition maximum) is 50 kgs and body wt. is 50 kg and if 1 RM is 70 kg and body wt. is 50 kg, in second category the relative strength is more.

**20. What role an individual can play in improvement of sport environment ?****

**21. Being sports captain of the school, prepare five important committees with their responsibilities to conduct one day run for health race. [5]**

**Ans. (i) Publicity committee:** The committee for publicity informs and announces the sports events, dates and venues in advance to the institutions, departments and public through newspapers, media, e-mail and website etc. Their main responsibility is to advertise the sports events.

**(ii) Boarding and lodging committee:** The main responsibility of this committee is ensuring and making needful arrangements for providing accommodation and serving food to all the officials and sportspersons.

**(iii) Transportation committee:** This committee is mainly responsible for providing the facilities regarding transportation of various teams to the venue of sports events or to the place of boarding and lodging. Their main duty is to make arrangements for transportation.

**(iv) Grounds and equipments committee:** This committee comes under the technical committee which mainly deals with the technical area of ground marking and layouts of track and field etc. This committee also makes necessary arrangements of equipments related to the game and athletic meet.

**(v) Official committee :** This committee selects various officials such as referees, judges, and various qualified officials as per requirements of the particular sport.

**22. What are the important functions of our skeletal system?** [5]**

**23. Explain Sheldon's classification of personality and explain its importance in physical education and sports. [5]**

**Ans. (i) Endomorphic:** An Endomorphic somatotype is also known as a viscerotonic. The characteristic traits of this somatotype usually includes being relaxed, tolerant, comfortable, and sociable. Psychologically, they are also fun-loving, good humored, even-tempered, and they love food and affection. The Endomorph is physically "round". They have wide hips and narrow shoulders that give a pear-shape.

They tend to have a lot of extra fat on their body and on their arms and thighs. They have skinny ankles and wrists that make the rest of their body look even bigger.

**(ii) Ectomorphic:** An ectomorph is the complete opposite of the Endomorph. Physically, they have narrow shoulders, thin legs and arms, little fat on the body, a narrow face and a narrow chest. They may eat just as much as the endomorph but never seem to gain any weight. They always stay skinny. Personality wise, they tend to be

** Answer is not given due to change in the present syllabus.

self-conscious, socially anxious, artistic, thoughtful, quiet, and private. They always keep to themselves and are to branch out.

**(iii) Mesomorphic:** The mesomorph is in between the endomorph and thin ectomorph. They have an attractive and desirable body. Physically, they tend to have a large head and broad shoulders with a narrow waist. They have a strong muscular body and strong arms and legs and little fat on the body. They work for the body they have so that they could have an attractive body. Psychologically, the mesomorph is adventurous and courageous. They are not afraid to break out and do new things with new people. They are assertive.

**24. What is movement speed? Explain the methods to develop speed endurance. [5]**

**Ans.** Movement speed is the time taken between the initiation of movement and the completion of the movement. It depends upon techniques, explosive strength, flexibility and coordinative abilities. It plays a vital role in boxing, gymnastics, swimming; throws and jumps etc. where the minimum time is taken to complete the movement.

To develop the speed endurance more work have to be done on pace races because pace races means running the whole distance at a constant speed. Generally, 800 meters and above races are included in pace races. As a matter of fact, an athlete can run a distance of 300 meters at full speed but, in longer races such as 800 meters or above races he must conserve his energy by reducing his speed. For example, if there is a runner of 800 meter race his best time is 1 minute 40 second, so, he should run first 400 meter in 49 seconds and next 400 meter in 51 seconds.

**25. Diet for sportspersons are important. What should be the aims of preparing diet for sportsperson? [1¼ × 4 = 5]**

**Ans.** Aims of preparing diet for sports person:

**(i)** Maintaining body weight and body composition desired for that specific sport.

**(ii)** Maintaining adequate pool of nutrient levels in the body.

**(iii)** Adopting healthy nutritional practices during training and competition.

**(iv)** Carrying on with healthy nutritional practices during off season as well i.e. when competition are not taking place.

**26. "Involvement in physical activities for longer period of time with moderate intensity can improve the quality of life." Justify your answer. [5]**

**Ans. (i) Exercise helps in healthy growth and development:** Exercise is an important part of keeping children healthy. Encouraging healthy lifestyles in children and adolescents is important when they grow older. Participating in organised sports and games is not only of great fun but is very essential for healthy growth and development.

**(ii) Exercises improve self-esteem:** Exercise is necessary for your physical and mental health. Self-esteem can play a great role in how children feel about themselves and also how much they enjoy things or worry about things. Exercise reduces depressive symptoms and improves self-esteem in children.

**(iii) Enhances flexibility:** The stiffness of joints decreases due to exercising, in a way improving the flexibility. The elasticity of tendons, ligaments and joint capsules improves due to regular exercise.

**(iv) Lessens stress and tension:** Regular exercise has a distinctive capability to slow down the depression process by reducing stress and tension. Actually, regular exercise lessens the levels of body's stress hormones like adrenaline and cortisol. The body's natural painkillers and mood elevators i.e., the endorphins are produced due to regular exercise. These benefits of work out facilitate in delaying the process of ageing.

**(v) Connect with others :** Spend time with positive people who enhance your life. A strong support system will buffer you from the negative effects of stress.

**(vi) Keep your sense of humor :** This includes the ability to laugh on one self. The act of laughing helps the body fight stress in a number of ways.

●●

# PHYSICS

## Outside Delhi [Set-I]

2020

**Time allowed : 3 hours** **Maximum marks : 70**

*General Instructions :*

**Read the following instructions very carefully and strictly follow them :**

(i) *This question paper comprises **four** sections – A, B, C and D.*

(ii) *There are **37** questions in the question paper. **All** questions are compulsory.*

(iii) *Section A : Q. no. **1** to **20** are very short-answer type questions carrying 1 mark each.*

(iv) *Section B : Q. no. **21** to **27** are short-answer type questions carrying **2** marks each.*

(v) *Section C : Q. no. **28** to **34** are long-answer type questions carrying **3** marks each.*

(vi) *Section D : Q. no. **35** to **37** are also long-answer type questions carrying **5** marks each.*

(vii) *There is no overall choice in the question paper. However, an internal choice has been provided in **two** questions of **one** mark, **two** questions of **two** marks, **one** question of **three** marks and all the **three** questions **five** marks. You have to attempt only one of the choices in such questions.*

(viii) *However, separate instructions are given with each section and question, wherever necessary.*

(ix) *Use of calculators and log tables is not permitted.*

(x) *You may use the following values of physical constants wherever necessary :*

$$c = 3 \times 10^8 \text{ m/s}$$
$$h = 6.63 \times 10^{-34} \text{ Js}$$
$$e = 1.6 \times 10^{-19} \text{ C}$$
$$\mu_0 = 4\pi \times 10^{-7} \text{ T m A}^{-1}$$
$$\varepsilon_0 = 8.854 \times 10^{-12} \text{ C}^2 \text{ N}^{-1} \text{ m}^{-2}$$
$$\frac{1}{4\pi\varepsilon_0} = 9 \times 10^9 \text{ Nm}^2 \text{ C}^{-2}$$

Mass of electron ($m_e$) = $9.1 \times 10^{-31}$ kg

Mass of neutron = $1.675 \times 10^{-27}$ kg

Mass of proton = $1.673 \times 10^{-27}$ kg

Avogadro's number = $6.023 \times 10^{23}$ per gram mole

Boltzmann constant = $1.38 \times 10^{-23}$ $JK^{-1}$

### SECTION-A

**Select the most appropriate option from those given below each question.**

**1. A cell of emf (E) and internal resistance *r* is connected across a variable external resistance R. The graph of terminal potential difference V as a function of R is :**

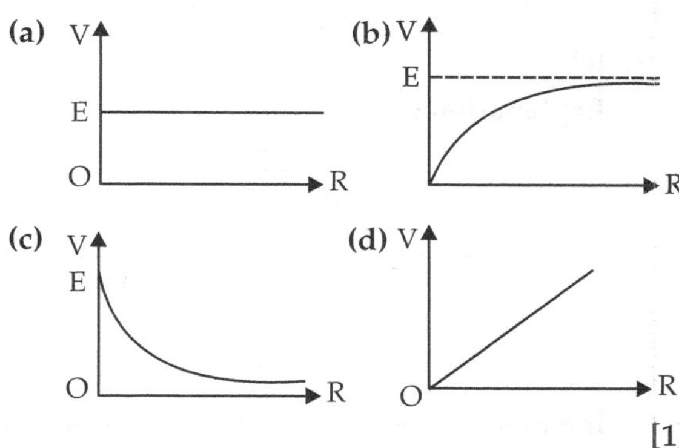


**[1]**

**Ans. (b)**

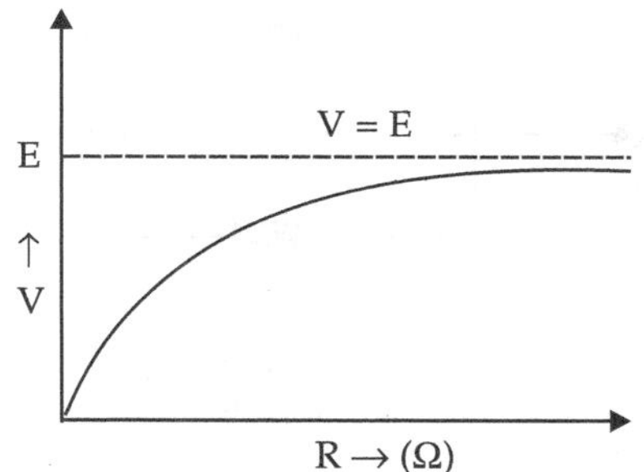


**Explanation :**

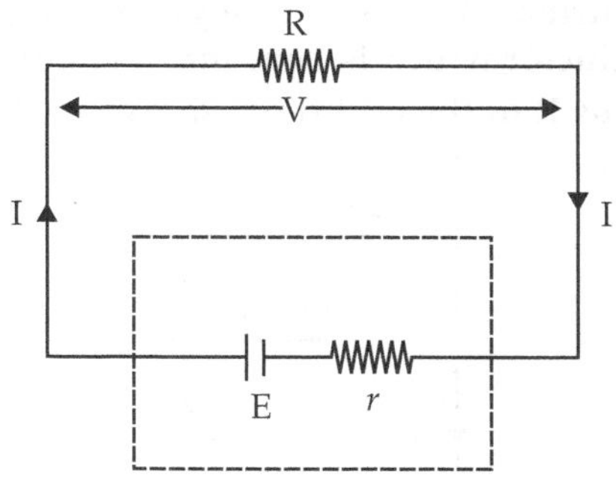


Current in circuit $I = \frac{E}{R + r}$ ...(i)

Potential difference across R will be

$$V = IR \quad \text{...(ii)}$$

Putting the value of I from (i) in (ii)

$$V = \frac{ER}{R + r}$$

$$V = \frac{E}{\left(1 + \frac{r}{R}\right)}$$

When, $R \to \infty, V = E$

$R \to 0, V = 0$

**2. A uniform wire of resistance 2R is bent in the form of a circle. The effective resistance between the ends of any diameter of the circle is :**

(a) 2R (b) R

(c) $\frac{R}{2}$ (d) $\frac{R}{4}$ [1]

**Ans.** (c)

**Explanation :**

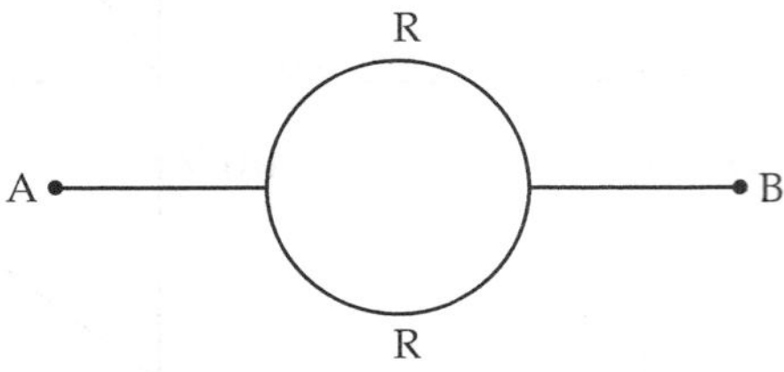


If a loop is made, then the resistances will be in parallel.

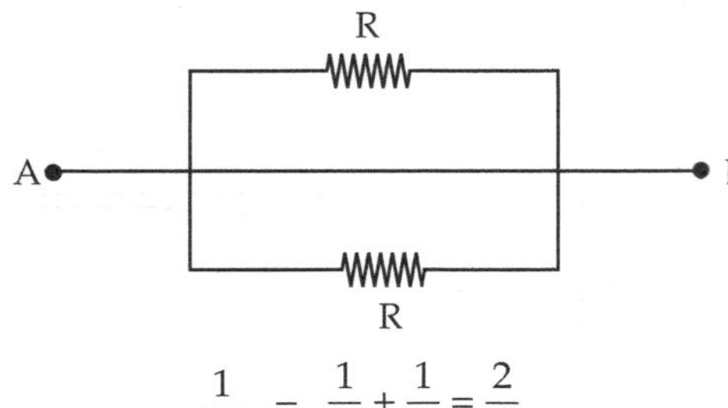


$$\frac{1}{R'} = \frac{1}{R} + \frac{1}{R} = \frac{2}{R}$$

$$R' = \frac{R}{2}$$

**3. A current I flows through a long straight conductor which is bent into a circular loop of radius R in the middle as shown in the figure.**

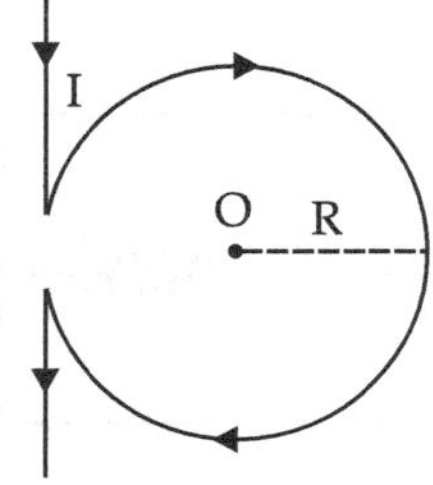


**The magnitude of the net magnetic field at point O will be :**

(a) Zero (b) $\frac{\mu_0 I}{2R}(1 + \pi)$

(c) $\frac{\mu_0 I}{4\pi R}$ (d) $\frac{\mu_0 I}{2R}\left(1 - \frac{1}{\pi}\right)$ [1]

**Ans.** (d)

**Explanation :**

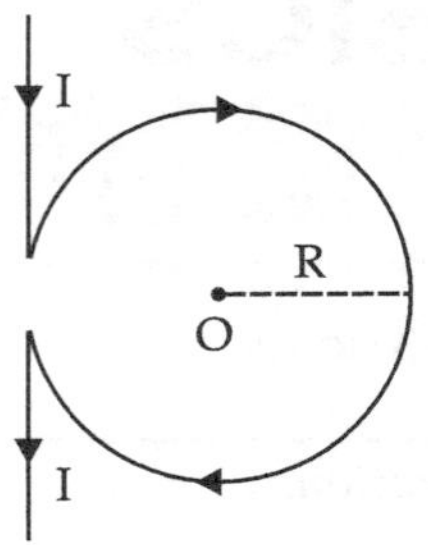


Let $\hat{n}$ be the unit vector $\perp r$ to the plane of paper. Magnetic field $\vec{B}$ at O due to current through the entire structure is :

$$\vec{B} = \frac{\mu_0}{4\pi}\frac{2\pi I}{R}(\hat{n}) + \frac{\mu_0}{4\pi}\frac{2I}{R}(-\hat{n})$$

$$\vec{B} = \frac{\mu_0 I}{2R}\left[1 - \frac{1}{\pi}\right]\hat{n}$$

$$\vec{B} = \frac{\mu_0 I}{2R}\left[1 - \frac{1}{\pi}\right] \text{ downward}$$

**4. A circular loop of radius *r*, carrying a current I lies in *y-z* plane with its centre at the origin. The net magnetic flux through the loop is :**

**(a) directly proportional to *r***

**(b) zero**

**(c) inversely proportional to *r***

**(d) directly proportional to I** [1]

**Ans.** (b)

**Explanation :**

Circular loop behaves as a magnetic dipole, whose one surface will be N-pole and another will be S-pole. Therefore, magnetic lines of force emerges from north and will meet at south. Hence, total magnetic flux through *y-z* plane is zero.

**5. The kinetic energy of a proton and that of an α-particle are 4 eV and 1 eV, respectively. The ratio of the de-Broglie wavelengths associated with them, will be :**

(a) 2 : 1 (b) 1 : 1

(c) 1 : 2 (d) 4 : 1 [1]

**Ans.** (b)

**Explanation :**

$$\lambda = \frac{h}{\sqrt{2m(K.E)}}$$

$$\frac{\lambda_{\rho^+}}{\lambda_\alpha} = \sqrt{\frac{m\alpha}{m_{\rho^+}}} \times \sqrt{\frac{K.E_\alpha}{K.E_{\rho^+}}}$$

$$\frac{\lambda_{\rho^+}}{\lambda_\alpha} = \sqrt{\frac{4m_{\rho^+}}{m_{\rho^+}}} \times \sqrt{\frac{1}{4}}$$

$$\frac{\lambda_{\rho^+}}{\lambda_\alpha} = \frac{2 \times 1}{1 \times 2} = \frac{1}{1} = 1:1$$

**6. A photocell connected in an electrical circuit is placed at a distance '*d*' from a source of light. As a result, current I flows in the circuit. What will be the current in the circuit when the distance is reduced to '*d*/2'?**

**(a) I** **(b) 2I**

**(c) 4I** **(d) 1/2** **[1]**

**Ans. (c)**

**Explanation :**

Intensity of light $\propto \frac{1}{d^2}$

current $\propto$ intensity of light

At distance $d$, current is I.

At distance $d/2$, intensity of light becomes four times of its initial value.

So, current will also become 4 times.

**7. A current of 10 A is flowing from east to west in a long straight wire kept on a horizontal table. The magnetic field developed at a distance of 10 cm due north on the table is :**

**(a) $2 \times 10^{-5}$ T, acting downwards**

**(b) $2 \times 10^{-5}$ T, acting upwards**

**(c) $4 \times 10^{-5}$ T, acting downwards**

**(d) $4 \times 10^{-5}$ T, acting upwards** **[1]**

**Ans. (a)**

**Explanation :**

$$\vec{B} = \frac{\mu_0 I}{2\pi r}$$

$$\vec{B} = \frac{4\pi \times 10^{-7} \times 10}{2\pi \times 10 \times 10^{-2}}$$

$$\vec{B} = 2 \times 10^{-5} \text{ T}$$

By using, right hand thumb rule it should be acting downwards.

**8. When a wave undergoes reflection at an interface from rarer to denser medium, adhoc change in its phases is :**

**(a)** $\frac{\pi}{2}$ **(b)** 0

**(c)** $\pi$ **(d)** $\frac{\pi}{4}$ **[1]**

**Ans. (c)**

**Explanation :**

When light is reflected from rarer to denser medium there is a phase change of $\pi$.

**9. Paschen series of atomic spectrum of hydrogen gas lies in :**

**(a) Infrared region**

**(b) Ultraviolet region**

**(c) Visible region**

**(d) Partly in ultraviolet and partly in visible region** **[1]**

**Ans. (a)**

**Explanation :**

$$\bar{v} = R\left[\frac{1}{3^2} - \frac{1}{n_2^{\,2}}\right], n_2 = 4, 5, 6$$

Lies in the infrared region.

**10. In the α-particle scattering experiment, the shape of the trajectory of the scattered α-particles depend upon :**

**(a) only on impact parameter**

**(b) only on the source of α-particles**

**(c) both impact parameter and source of α-particles.**

**(d) impact parameter and the screen material of the detector.** **[1]**

**Ans. (a)**

**Explanation :**

The shape of the trajectory of the scattered α-particles depends on the impact parameter.

$$b = \frac{kze^2 \cot \theta/2}{\frac{1}{2}mv^2}$$

**Note : Fill in the blanks with appropriate answer :**

**11. Torque acting on an electric dipole placed in an electric field is maximum when the angle between the electric field and the dipole moment is ...................... .** **[1]**

**Ans.** $\theta = 90°$

**Explanation :**

$$\vec{\tau} = PE \sin \theta$$

$$\theta = 90°$$

$$\vec{\tau}_{max} = PE$$

**12. A proton released from rest in an electric field, will start moving towards a region of ................ potential in the field.** **[1]**

**Ans.** decreasing

**Explanation :**

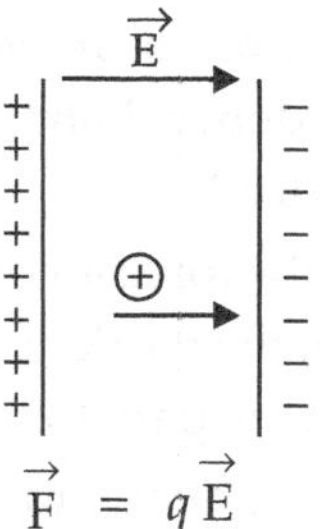

$$\vec{F} = q\vec{E}$$

Negative polarity attract positive proton.

**13. To minimize the percentage error in the determination of unknown resistance of a conductor in meter bridge experiment, the balance point is adjusted near............... of the wire. [1]**

**OR**

**In potentiometer, a long uniform wire is used to ...............potential gradient along the wire.**

**Ans.** mid-point (middle) or centre of wire.

**Explanation :**

A meter bridge is most sensitive when all four resistances are nearly equal. In that case, the null point is very close to the middle of the wire.

**OR**

Decrease

**Explanation :**

Since, $$K = \frac{V}{L}$$

When length increases, potential gradient decreases. Hence, potentiometer becomes more sensitive.

**14. Unpolarised light of intensity $I_o$ is incident on two crossed polaroids. The intensity of light transmitted by the combination will be .......... . [1]**

**Ans.** Zero

**Explanation :** As first polaroid half the light *i.e.*, $\frac{I_0}{2}$. Then the next polaroid is placed at 90° with first polaroid.

Then, $$I' = \frac{I_0}{2}\cos^2 90° = 0$$

**15. Name the particle emitted spontaneously in the following nuclear reaction :**

$${}^{32}_{15}P \longrightarrow {}^{32}_{16}S + \overline{\upsilon} + \text{———} \quad [1]$$

**Answer the following :**

**Ans.** $\beta^-$ – particle $\left({}^{\ 0}_{-1}e\right)$

Equation : $${}^{32}_{15}P \longrightarrow {}^{32}_{16}S + \overline{\upsilon} + {}^{\circ}_{-1}e$$

**16. The work done in moving a charge particle between two points in an uniform electric field, does not depend on the path followed by the particle. Why ? [1]**

**Ans.** Since electrostatic force is conservative in nature. In other words, the work done by electric field is independent of the path followed by the particle.

**17. Define 'magnetic declination' at a place earth. [1]**

**Ans.** The angle between the geographical meridian and the magnetic meridian at a place is called the magnetic declination ($\alpha$) at a place.

**18. An A.C. source with variable frequency is connected to a parallel plate capacitor. How will the displacement current be affected with the decrease in frequency of the source ? [1]**

**Ans.** When frequency of A. C. source is decreased, resistance increases so flow of current decreases. By dropping the variable frequency there will be drop or decrease in displacement current through the capacitor as it carries displacement current.

Since $$I_C = \frac{V}{X_C} = \frac{V}{\frac{1}{\omega C}} = \omega CV$$

**19. An astronomical telescope may be a refracting type or a reflecting type. Which of the two produces image of better quality ? Justify your answer. [1]**

**Ans.** The astronomical telescope of reflecting type produces better quality of image. In these telescopes, mirror of large aperture of high resolving power are used due to which image is free from chromatic and spherical aberration.

**20. Can a slab of *p*-type semi-conductor be physically joined to another *n*-type semiconductor slab to form *p-n* junction ? Justify your answer. [1]**

**OR**

**In a *p-n* juction diode the forward bias resistance is low as compared to the reverse bias resistance. Give reason.**

**Ans.** No, any slab, howsoever flat will have roughness much larger than the interatomic crystal spacing (2Å to 3Å) and hence continuous contact at the atomic junction will not be possible.The junction will behave as discontinuous for the following charge carriers.

**OR**

When the *p-n* junction diode is forward biased, the diode's depletion layer or potential barrier shrinks. This can be considered as a practical short circuit.

Hence, resistance is negligible and current flow in forward bias is maximum.

Similarly, when it is reverse bias, depletion layer width (or potential barrier) increases. Hence, resistance increases and very small amount of current flows in reverse bias.

## SECTION-B

**21. Find the total charge stored in the network of capacitors connected between A and B as shown in figure :** **[2]**

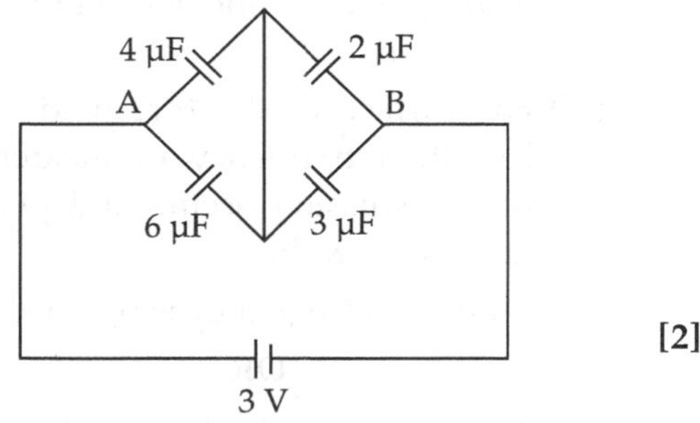


**Ans.** For capacitors 4μF and 2μF connected in series

$$\frac{1}{C_1} = \frac{1}{4} + \frac{1}{2} = \frac{1+2}{4} = \frac{3}{4}$$

$$C_1 = \frac{4}{3}\mu F$$

For capacitors 6μF and 3μF connected in series

$$\frac{1}{C_2} = \frac{1}{6} + \frac{1}{3} = \frac{1+2}{6} = \frac{3}{6} = \frac{1}{2}$$

$$C_2 = 2\ \mu F$$

Now, $C_1$ and $C_2$ are in parallel

$$\therefore \quad C_P = C_1 + C_2$$

$$C_P = \frac{4}{3} + 2 = \frac{4+6}{3} = \frac{10}{3}$$

$$C_P = \frac{10}{3}\mu F$$

$$Q = C_P V = \frac{10}{3} \times 3$$

$$Q = 10\ \mu C$$

**22. A wire of length $L_o$ has a resistance $R_o$. It is gradually stretched till its length becomes $2L_o$.**

**(a) Plot a graph showing variation of its resistance R with its length $l$ during stretching**

**(b) What will be its resistance when its length becomes $2L_o$?** **[2]**

**Ans. (a)**

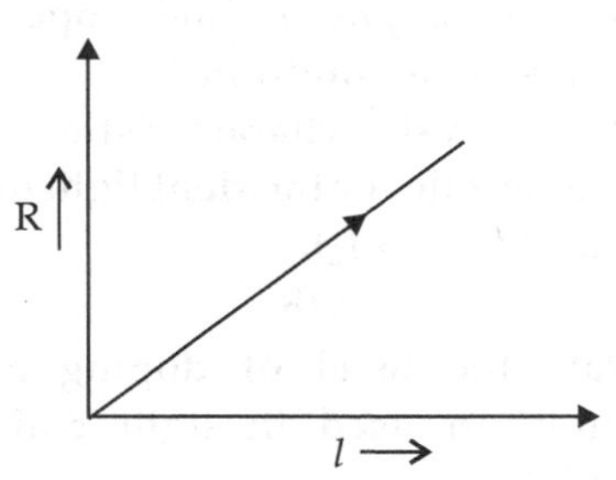


**Ans.** No

**(b)** Since, $R_o = \frac{\rho L_o}{A}$

Here, $L_{new} = 2L_o$

$$A_{new} = \frac{A}{2}$$

ρ remains same (Property of material)

$$\therefore \quad R_{new} = \frac{\rho \times 2L_o}{A/2}$$

$$R_{new} = \frac{4\rho L_o}{A} = 4R_o$$

New resistance will become 4 times of the original.

**23. A resistor R and an inductor L are connected in series to a source of voltage $V = V_o \sin \omega t$. The voltage is found to lead current in phase by $\pi/4$. If the inductor is replaced by a capacitor C, the voltage lags behind current in phase by $\pi/4$. When L, C and R are connected in series with the same source,**

**Find the :**

**(i) average power dissipated and**

**(ii) instantaneous current in the circuit.** **[2]**

**Ans. (i)**

$$P_{av} = V_{rms} \times I_{rms} \cos \phi$$

But $\cos \phi$ is 1 at resonance

$$\therefore \quad P_{av} = V_{rms} \times I_{rms} = \frac{V^2_{rms}}{R}$$

For LCR circuit

$$\tan \theta = \frac{X_L - X_C}{R} = \frac{\omega L - \frac{1}{\omega C}}{R}$$

$$\phi = \frac{\pi}{4}$$

Initially,

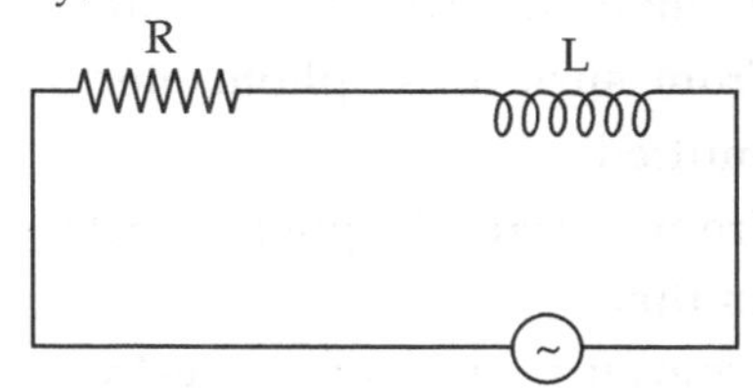


$$V = V_0 \sin \omega t$$

So, $\tan \phi = \frac{\omega L}{R}$

$$\tan \frac{\pi}{4} = \frac{\omega L}{R}$$

$$1 = \frac{\omega L}{R}$$

$$R = \omega L \quad \text{...(i)}$$

When, inductor is removed and capacitor is placed.

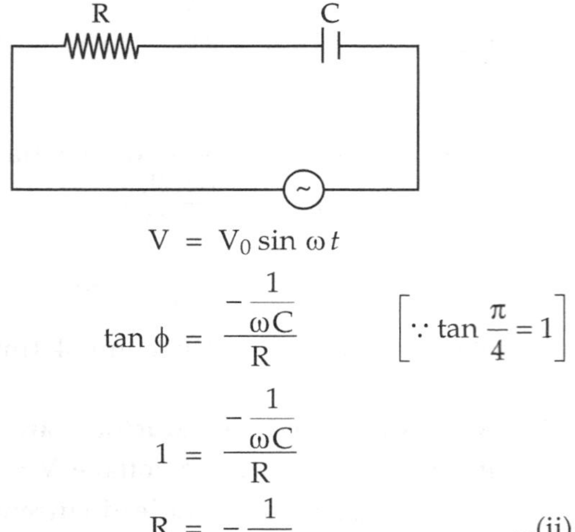


$$V = V_0 \sin \omega t$$

$$\tan \phi = \frac{-\frac{1}{\omega C}}{R} \qquad \left[\because \tan \frac{\pi}{4} = 1\right]$$

$$1 = \frac{-\frac{1}{\omega C}}{R}$$

$$R = -\frac{1}{\omega C} \qquad \text{...(ii)}$$

The – ve sign indicates that the current leads the voltage

From (i) and (ii)

$$\omega L = \frac{1}{\omega C} \text{ or } \omega = \frac{1}{\sqrt{LC}}$$

This implies that the LCR circuit is in resonance condition.

$$P_{av} = V_{rms} \times I_{rms} \cos \phi$$

But $\cos \phi$ is 1 at resonance

$$\therefore \qquad P_{av} = V_{rms} \times I_{rms} = \frac{V^2_{rms}}{R}$$

**(ii)** $$I \text{ (Instantaneous)} = \frac{V}{R} = \frac{V_0}{R} \sin \omega t$$

$$= I_0 \sin \omega t.$$

**24. Light of same wavelength is incident on three photo-sensitive surfaces A, B and C. The following observations are recorded.**

**(i) From surface A, photo electrons are not emitted.**

**(ii) From surface B, photo electrons are just emitted.**

**(iii) From surface C, photo electrons with some kinetic energy are emitted.**

**Compare the threshold frequencies of the three surfaces and justify your answer. [2]**

**OR**

**If the frequency of light incident on the cathode of a photocell is increased, how will the following be affected ? Justify your answer**

**(i) Energy of the photo electrons.**

**(ii) Photo current.**

**Ans.** From the observations made, (surface A, B and C) on the basis of photoelectric effect:

(i) For surface A, threshold frequency is more than frequency of incident light, so no emission occurs.

(ii) For surface B, threshold frequency is equal to frequency of incident light, so emission just occurs.

(iii) For surface C, the threshold frequency is less than frequency of incident light. So photo emission occurs and photo-electron have some K. E.

Threshold frequency is $\upsilon_A > \upsilon_B > \upsilon_C$.

**OR**

(i) The energy of the photoelectron also increases as $E = h\upsilon$.

(ii) No effect on photoelectric current. This is because of incident photon of increased energy cannot eject more than one electron from the metal surface and energy of electrons increases but no effect on photoelectric current.

**25. Briefly explain how a potential barrier is set up across a *p-n* junction as a result of diffusion and drift of the charge carriers. [2]**

**Ans.** Potential Barrier

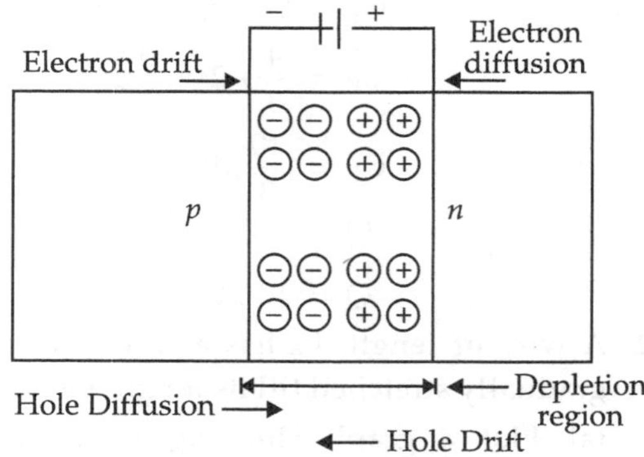


The accumulation of negative charges in the *p*-region and positive charges in the *n*-region set up a potential difference across the junction. This acts as a barrier and is called potential barrier.

**26. (a) Why is a photo diode operated under reverse bias conditon ?**

**(b) Draw V-I characteristic curves of photodiode for incident light of intensities $I_1$ and $I_2$ ($I_1 > I_2$). [2]**

**OR**

**(a) State the level of doping and biasing condition used in light emitting diode (LED).**

**(b) Write any two advantages of LED over the conventional low power lamps.**

**Ans. (a)** Photodiodes convert incident light to electric current more effectively in reverse bias condition than in forward bias because the width of the depletion region increases as we increase the applied reverse bias voltage across the diode (directly proportional) in a reverse biased *p-n* junction. Hence, by applying a larger voltage, more of the incident photons get converted to electric current. So photodiodes work better in reverse biased condition.

**(b)**

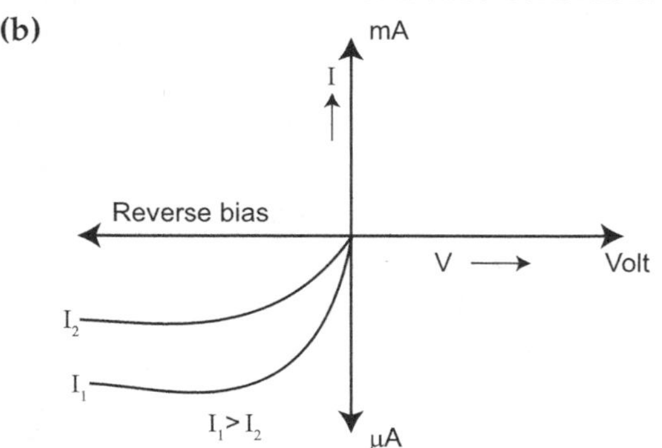


OR

**(a)** (LED) is a heavily doped *p-n* junction which under forward bias emits spontaneous radiations. When the diode is forward biased, electrons are sent from $n \rightarrow p$ and holes are sent from $p \rightarrow n$. At the junction boundary, the concentration of minority carriers increases as compared to the equilibrium conditions. Thus, at the junction boundary on either side of the junction, excess minority carriers are there, which recombine with majority carriers near the junction and on recombination, the energy is released in the form of photons. When the forward current of diode is small, the intensity of light emiting is small and increases when light reaches maximum.

Further, increase in forward current results in decrease of light intensity. LEDs are biased such that the light emiting efficiency is maximum.

**(b)** LEDs are preferred because :

(i) Low operational voltage and less power

(ii) Fast action and no warmup time required.

(iii) Long life and ruggedness.

**27. (a) Explain the formation of energy bands in crystalline solids.**

**(b) Draw the energy band diagrams of (i) a metal and (ii) a semiconductor. [2]**

**Ans. (a)** An enormously large number of energy levels closely spaced in a very small energy range constitute an energy band.

The allowed energy bands are separated by regions in which energy levels cannot exist. These forbidden regions are called band gaps or energy gaps.

The highest energy band occupied by the valence electrons is called the valence band and the next empty allowed band is called the conduction band.

**(b)** (i)

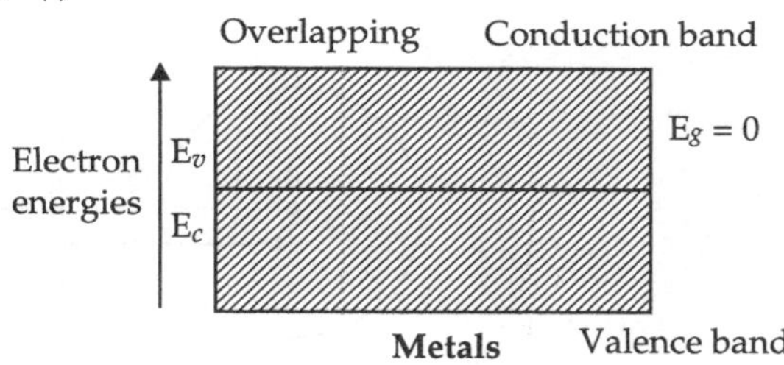


(ii)

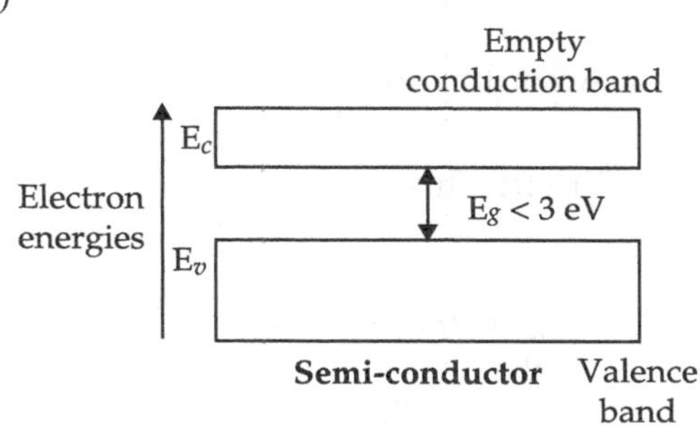


## SECTION-C

**28. A hollow conducting sphere of inner radius $r_1$ and outer radius $r_2$ has a charge Q on its surface. A point charge $-q$ is also placed at the centre of the sphere.**

**(a) What is the surface charge density on the (i) inner and (ii) outer surface of the sphere?**

**(b) Use Gauss' law of electrostatics to obtain the expression for the electric field at a point lying outside the sphere. [3]**

OR

**(a) An infinitely long thin straight wire has a uniform linear charge density $\lambda$. Obtain the expression for the electric field (E) at a point lying at a distance $x$ from the wire, using Gauss' law.**

**(b) Show graphically the variation of this electric field E as a function of distance $x$ from the wire.**

**Ans. (a)**

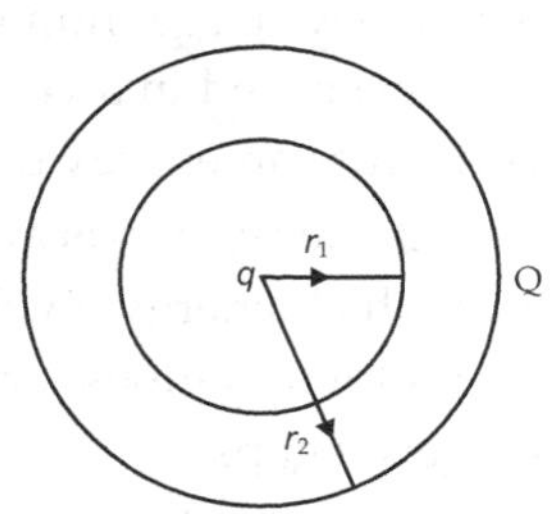


(i) $\sigma_{Inner} = \frac{-q}{4\pi r_1^2}$ (ii) $\sigma_{Outer} = \frac{Q-q}{4\pi r_2^2}$

**(b)**

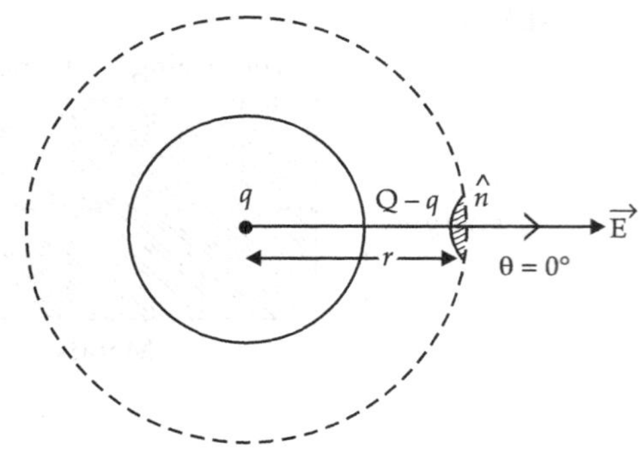


According to Gauss Law :

$$\text{Electric flux,} \quad \phi = \frac{Q-q}{\varepsilon_o}$$

$$\oint_0^{4\pi r^2} \vec{E}\, dA \cos 0° = \frac{Q-q}{\varepsilon_o}$$

$$\vec{E} \oint_0^{4\pi r^2} dA = \frac{Q-q}{\varepsilon_o}$$

$$\vec{E}\,[A]_0^{4\pi r^2} = \frac{Q-q}{\varepsilon_o}$$

$$\vec{E} = \frac{Q-q}{4\pi\varepsilon_0 r^2} = K\left(\frac{Q-q}{r^2}\right)$$

**OR**

**(a)**

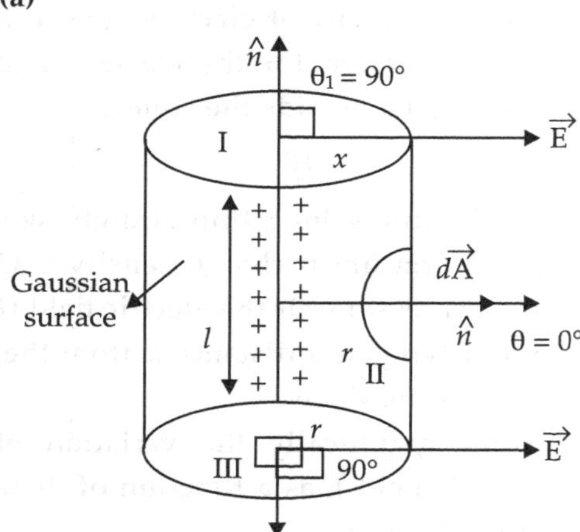


$$\lambda = \frac{q}{l} \qquad \text{...(i)}$$

Apply Gauss Law :

$$\oint_I \vec{E}\, d\vec{A} \cos 90° + \oint_{II} \vec{E}\, d\vec{A} \cos 0° + \oint_{III} \vec{E}\, d\vec{A} \cos 90°$$

$$0 + \oint_{II} \vec{E}\, d\vec{A} \cos 0° + 0 = \frac{q}{\varepsilon_0}$$

$$\vec{E} \oint_0^{2\pi x l} dA = \frac{q}{\varepsilon_0}$$

$$\vec{E}\,[A]_0^{2\pi x l} = \frac{\lambda l}{\varepsilon_0} \qquad \text{[From (i)]}$$

$$\vec{E} \times 2\pi x l = \frac{\lambda l}{\varepsilon_0}$$

$$\vec{E} = \frac{\lambda}{2\pi\varepsilon_0 x}$$

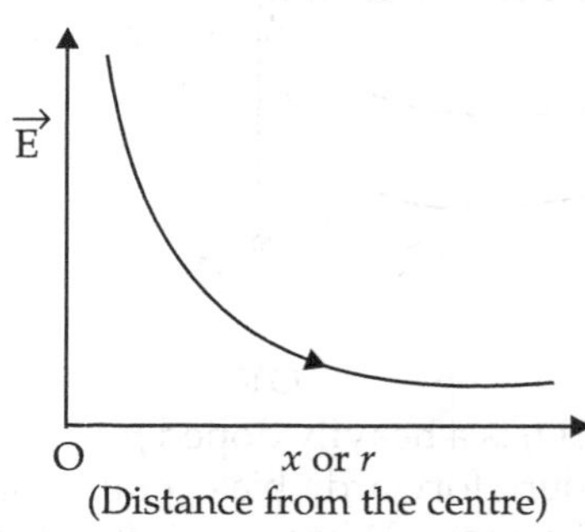


**29. (a) Explain the principle of working of a potentiometer.**

**(b) In a potentiometer, a standard source of emf 5 V and negligible internal resistance maintains a steady current through the potentiometer wire of length 10 m. Two primary cells of emf $E_1$ and $E_2$ are joined together in a series with (i) same polarity and (ii) opposite polarity. The combination is connected to the potentiometer circuit in each case. The balancing length of the wire in the two cases are found to be 700 cm and 100 cm, respectively. Find the values of emf of the two cells. [3]**

**Ans. (a) Principle of working of potentiometer wire :** The voltage drop along the wire (of uniform cross scetion) is directly proportional to the length of the wire. The potentiometer works without drawing any current from the voltage source.

**(b)** Emf, V = 5V

Length of potentiometer wire = 10 m = 1000 cm

Let the emf of two cells be $E_1$ and $E_2$

Balancing length in same polarity = 700 cm

Balancing length in different polarity = 100 cm

Potential gradient

$$= \frac{V}{l} = \frac{5V}{1000 \text{ cm}} = \frac{1}{200} \text{ V/cm}$$

$$E_1 + E_2 = Kl_1 \quad \text{(For same polarity)}$$

$$\Rightarrow \quad E_1 + E_2 = \frac{1}{200} \times 700$$

$$\Rightarrow \quad E_1 + E_2 = 3.5 \text{ cm} \quad ...(i)$$

and $$E_1 - E_2 = Kl_2 \quad \text{(For different polarity)}$$

$$\Rightarrow \quad E_1 - E_2 = \frac{1}{200} \times 100$$

$$\Rightarrow \quad E_1 - E_2 = 0.5 \text{ cm} \quad ...(ii)$$

On adding (i) and (ii), we get

$$2E_1 = 4$$

$$E_1 = 2V$$

and $$E_2 = 3.5 - 2$$

$$E_2 = 1.5 \text{ V}$$

Hence, the emf of cells are 2 V and 1.5 V.

**30. (a) Differentiate between self inductance and mutual inductance.**

**(b) The mutual inductance of two coaxial coils is 2H. The current in one coil is changed uniformly from zero to 0.5A in 100 ms. Find the :**

**(i) change in magnetic flux through the other coil.**

**(ii) emf induced in the other coil during the change.** [3]

**Ans. (a)**

| Self Inductance | Mutual Inductance |
|---|---|
| 1. When the current flowing through a coil changes then induced emf is produced in the coil itself. This pheno-menon is called self inductance. | 1. When current in one of the two closely lying coils is changed, the emf is induced in the second coil. This phenomenon is called mutual inductance. |
| 2. Phenomenon of self inductance is the characteristic of single coil. | 2. Phenomenon of mutual inductance is the characteristic of two coils. |
| 3. The induced current is produced in the same coil in which main current is flowing. Hence, the induced current affects the main current. | 3. Main current flows in one coil and induced current is produced in the second coil. Hence, main current is not directly affected. |

**(b)** Mutual inductance between coils M = 2H

Rate of change of current

$$\frac{dI}{dt} = \frac{0.5}{100 \times 10^{-3}}$$

$$= \frac{500}{100} = 5 \text{A/s}$$

(i) Change in magnetic flux is

$$\Delta\phi = e\Delta t \quad \left[\because e = \frac{Mdi}{dt}\right]$$

$$= 2 \times 0.5$$

$$= 1 \text{ Wb}$$

(ii) Change in magnetic flux is

$$e = \frac{-MdI}{dt}$$

$$= -2 \times 5$$

$$= -10 \text{ V}$$

**31. Explain with the help of a diagram, the working of a step-down transformer. Why is a laminated iron core used in a transformer ?** [3]

**Ans.**

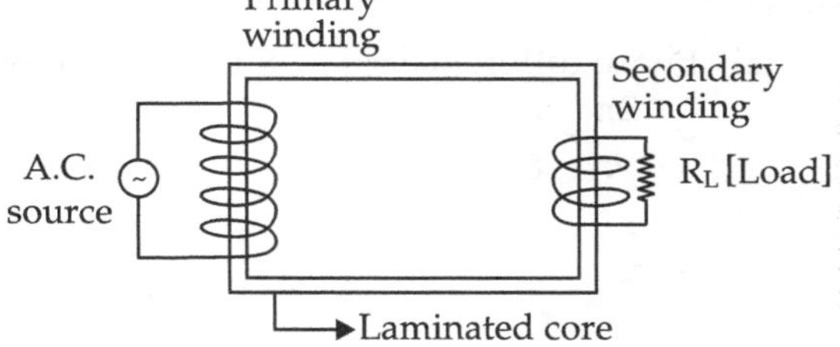


As the a.c. flows through the primary, it generates an alternating magnetic flux in the core, which also passes through the secondary. This changing flux sets up an induced emf in the secondary, also a self induced emf in the primary. If there is no leakage of magnetic flux, then flux linked with each turn of the primary will be equal to that linked with each turn of the secondary.

$$\frac{I_1}{I_2} = \frac{\varepsilon_2}{\varepsilon_1} = \frac{N_2}{N_1}$$

$$\eta = \frac{\text{Power output}}{\text{Power input}} \times 100\%$$

In step down $N_2 < N_1$

So, $\varepsilon_2 < \varepsilon_1$

Laminated iron core is used to prevent energy loss due to eddy current.

**32. Name the electro-magnetic waves with their frequency range, produced in :**

**(a) some radioactive decay**

**(b) sparks during electric welding**

**(c) TV remote [3]**

**Ans.** (a) Gamma radiations are produced during the radioactive decay of a nucleus. Frequency range is $> 3 \times 10^{17}$ Hz.

(b) Welding sparks produces ultraviolet rays. Frequency range is $7.5 \times 10^{14}$ Hz$-3 \times 10^{16}$ Hz.

(c) T.V. remote control uses infrared radiations. Frequency range is $3 \times 10^{12}$ Hz $- 4.3 \times 10^{14}$ Hz.

**33. Two coherent light waves of intensity $5 \times 10^{-2}$ Wm$^{-2}$ each super-impose and produce the interference pattern on a screen. At a point where the path difference between the waves is λ/6, λ being wavelength of the wave, find the**

**(a) phase difference between the waves.**

**(b) resultant intensity at the point.**

**(c) resultant intensity in terms of the intensity at the maximum. [3]**

**Ans.** Here,

$I_1 = I_2 = 5 \times 10^{-2}$ Wb/m$^2$

Here path difference = λ/6

**Ans.** Here,

(a) Phase difference $= \frac{2\pi}{\lambda} \times$ Path difference

$$= \frac{2\pi}{\lambda} \times \frac{\lambda}{6}$$

$$= \frac{\pi}{3}$$

(b) Resultant Intensity, $I = I_1 + I_2 + 2\sqrt{I_1 I_2} \cos\phi$

$$= 5 \times 10^{-2} + 5 \times 10^{-2} + 2\sqrt{(5 \times 10^{-2})^2} \cos\frac{\pi}{3}$$

$$= 2 \times 5 \times 10^{-2} + 2 \times 5 \times 10^{-2} \cos 60^\circ$$

$$= 10 \times 10^{-2} + 5 \times 10^{-2}$$

$$= 15 \times 10^{-2} \text{ Wb/m}^2$$

(c) The intensity is maximum when phase difference is an integral multiple of 2π

$$\therefore \quad I_{max} = I_1 + I_2 + 2\sqrt{I_1 I_2}$$

But $I_1 = I_2 = I$

$$I_{max} = 2I + 2I$$

$$= 4I$$

∴ Resultant intensity in terms of $I_{max}$ is

$$I' = I_{max} \cos^2 \phi/2$$

$$= I_{max} \cos^2\left(\frac{\pi/3}{2}\right) \left[\because \phi = \frac{\pi}{3}\right]$$

$$= I_{max} \cos^2\left(\frac{\pi}{6}\right)$$

$$= \frac{3}{4} I_{max}$$

But $I' = \frac{3}{4}[4I] = 3I$

$$= 15 \times 10^{-2} \text{ Wb/m}^2$$

**34. Two objects P and Q when placed at different positions in front of a concave mirror of focal length 20 cm, form real images of equal size. Size of object P is three times size of object Q. If the distance of P is 50 cm from the mirror, find the distance of Q from the mirror. [3]**

**Ans.** Here, focal length of concave mirror, $f = -20$ cm

Let, $I_P$ and $I_Q$ be the image of P and Q and $O_P$ and $O_Q$ be the object size of P and Q

Given,

$$I_P = I_Q$$

and $O_P = 3O_Q$

$$u_P = -50 \text{ cm}$$

$$u_Q = ?$$

Now magnification of object P, $m_P = \frac{f}{f - u_P}$

$$\Rightarrow \quad \frac{I_P}{O_P} = \frac{f}{f - u_P}$$

$$\Rightarrow \quad \frac{I_P}{O_P} = \frac{-20}{-20-(-50)} = \frac{-2}{3}$$

Now, $m_Q = \frac{I_Q}{O_Q} = \frac{I_P}{O_P/3}$

$$= \frac{3I_P}{O_P} = 3 \times \left(\frac{-2}{3}\right) = -2$$

But $m_Q = \frac{f}{f - u_Q}$

$$-2 = \frac{-20}{-20 - u_Q}$$

$$\Rightarrow \quad 40 + 2u_Q = -20$$

$$\Rightarrow \quad u_Q = -30 \text{ cm}$$

Hence, the distance of object Q from the mirror is 30 cm.

## SECTION-D

**35. (a) Show that a current carrying solenoid behaves like a small bar magnet. Obtain**

the expression for the magnetic field at an external point lying on its axis.

**(b)** **A steady current of 2A flows through a circular coil having 5 turns of radius 7 cm. The coil lies in X-Y plane with its centre at the origin. Find the magnitude and direction of the magnetic dipole moment of the coil.**

**OR**

**(a)** **Derive the expression for the force acting between two long parallel current carrying conductors. Hence, define 1 A current.**

**(b)** **A bar magnet of dipole moment 3 Am² rests with its centre on a frictionless pivot. A force F is applied at right angles to the axis of the magnet, 10 cm from the pivot. It is observed that an external magnetic field of 0.25 T is required to hold the magnet in equilibrium at an angle of 30° with the field.**

**Calculate the value of F.**

**How will the equilibrium be effected if F is withdrawn ?**

**Ans. (a)** Solenoid behaves like a magnet as it has a soft iron core with insulated copper wire over it. If the north pole of a bar magnet is brought near to the negative terminal of the battery, then the solenoid repels the bar magnet as like poles repel each other. Similarly, the other end behaves as a south pole.

Consider a circular coil of radius R with centre O having '$n$' turns. Suppose 'P' be any point on the axis at a distance of $x$ from the centre O. Consider a small element of current of length $dl$.

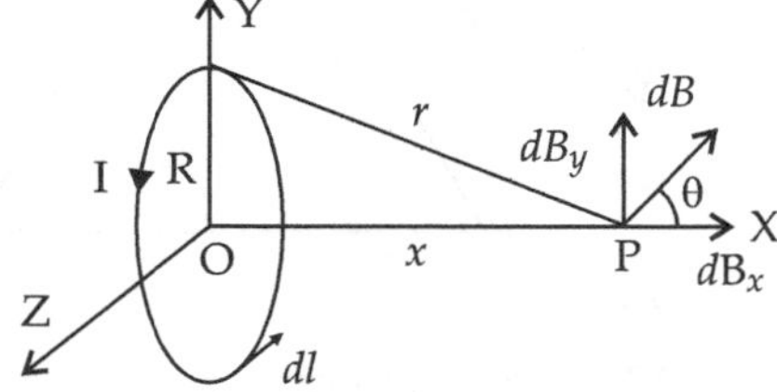


According to Biot-Savart's law, the magnetic field at point P will be

$$dB = \frac{\mu_0 I}{4\pi}\frac{|dl \times r|}{r^3}$$

Where $r^2 = x^2 + R^2$

$|dl \times r| = r\,dl$

So, $$dB = \frac{\mu_0}{4\pi} \times \frac{I\,dl}{r^2}$$

[∵ Both are perpendicular]

$$dB = \frac{\mu_0}{4\pi} \times \frac{I\,dl}{(x^2+R^2)}$$

$dB$ has two components *i.e.* $dB_x$ and $dB_y$. $dB_y$ is cancelled out and only the $x$-component remains.

$$dB_x = dB\cos\theta \quad \left[\because \cos\theta = \frac{R}{\sqrt{x^2+R^2}}\right]$$

$$dB_x = \frac{\mu_0 I\,dl}{4\pi} \cdot \frac{R}{(x^2+R^2)^{3/2}}$$

But $dl = 2\pi R$

So, $$B = \frac{\mu_0}{4\pi}\frac{IR \times 2\pi R}{(x^2+R^2)^{3/2}}$$

For $n$ turns in the circular loop,

$$B = \frac{\mu_0\, nIR^2}{2\pi\,(x^2+R^2)^{3/2}}\;\hat{i}$$

**(b)** Current in coil, $I = 2A$

Number of turns, $N = 5$

Radius of coil, $r = 7\text{cm} = 7 \times 10^{-2}$ m

Magnetic dipole moment,

$$M = NIA$$
$$= NI(\pi r^2)$$
$$= 5 \times 2 \times \pi(7 \times 10^{-2})$$
$$= 1540 \times 10^{-4}\ \text{Am}^2$$
$$= 0.154\ \text{Am}^2$$

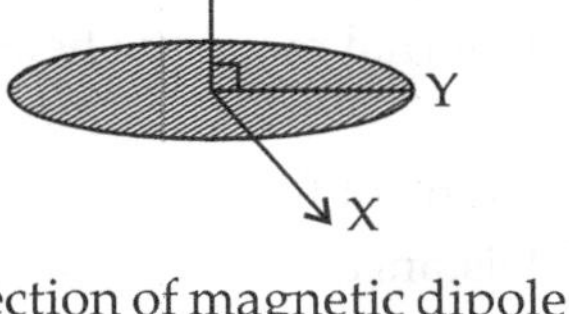


So, the direction of magnetic dipole moment is perpendicular to the plane of the coil *i.e.* along Z-axis.

**OR**

**(a)** Suppose two long thin straight long conductors (or wires) PQ and RS are placed parallel to each other, carrying currents $I_1$ and $I_2$ respectively. When the current flow in same direction they experience an attractive force and when they carry current in opposite directions, they experience a repulsive force.

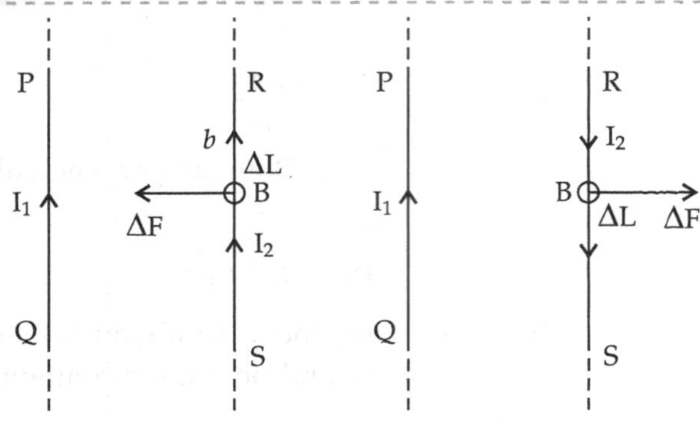


Consider a current element $\Delta L$ of wire RS. The magnetic field produced by current carrying conductor PQ at the location of other wire RS.

$$B_1 = \frac{\mu_0 I_1}{2\pi r} \quad \text{...(i)}$$

According to Maxwell's right hand rule, the direction of magnetic field will be perpendicular to the plane of the paple.

$$\Delta F = B_1 I_2 \Delta L \sin 90°$$

$$= \frac{\mu_0 I_1}{2\pi r} I_2 \Delta L$$

$\therefore$ Total force on length L

$$F = \frac{\mu_0 I_1 I_2}{2\pi r} \text{SDL} = \frac{\mu_0 I_1 I_2 L}{2\pi r}$$

$\therefore$ Force acting on per unit length of conductor

$$f = \frac{F}{L} = \frac{\mu_0 I_1 I_2}{2\pi r} \text{ N/m}$$

Thus, 1 ampere is defined as the current which when flow in each of the parallel conductors placed at separation of 1m in vacuum exerts a force of $2 \times 10^{-7}$ N on 1m length of either wire.

**(b)** Dipole moment, $M = 3 \text{ Am}^2$

External magnetic field,

$$B = .25 \text{ T}$$

Angle of field, $\theta = 30°$

Distance at which force applied from pivot,

$$r = 10 \text{ cm} = 0.10 \text{ m}$$

As magnetic bar is in equlibirium under external force.

Now, $\tau_{\text{Force}} = \tau_{\text{M.F.}}$

$\Rightarrow \quad F \times r = MB \sin\theta$

$\Rightarrow \quad F \times r \times \sin 90° = 3 \times 0.25 \times \sin 30°$

$$\Rightarrow \quad F = \frac{3 \times 0.25}{0.10} \times \frac{1}{2}$$

$$= \frac{75}{20}$$

$$= 3.75 \text{ N}$$

If F will be removed then it will not remain in equilibrium and start rotating around its pivot.

**36. (a) Draw the ray diagram showing refraction of ray of light through a glass prism. Derive the expression for the refractive index μ of the material of prism in terms of the angle of prism A and angle of minimum deviation $\delta_m$.**

**(b) A ray of light PQ enters an isosceles right angled prism ABC of refractive index 1.5 as shown in figure.**

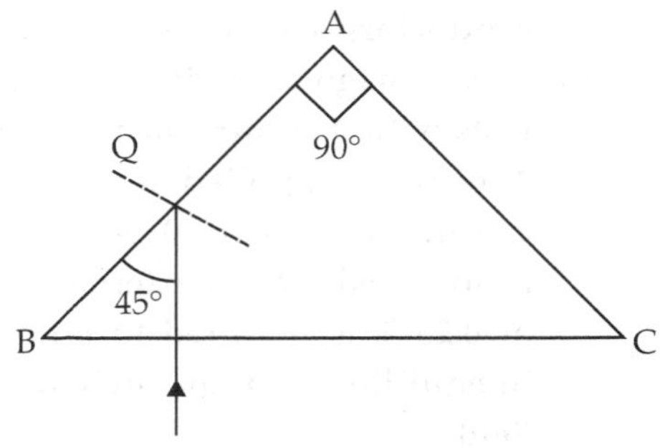


**(i) Trace the path of the ray through the prism**

**(ii) What will be the effect on the path of the ray if the refractive index of the prism is 1.4 ?**

**OR**

**(a) Two thin lenses are placed coaxially in contact. Obtain the expression for the focal length of this combination in terms of the focal lengths of the two lenses.**

**(b) A converging lens of refractive index 1.5 has a power of 10 D. When it is completely immersed in a liquid, it behaves as a diverging lens of focal length 50 cm. Find the refractive index of the liquid.**

**Ans. (a)**

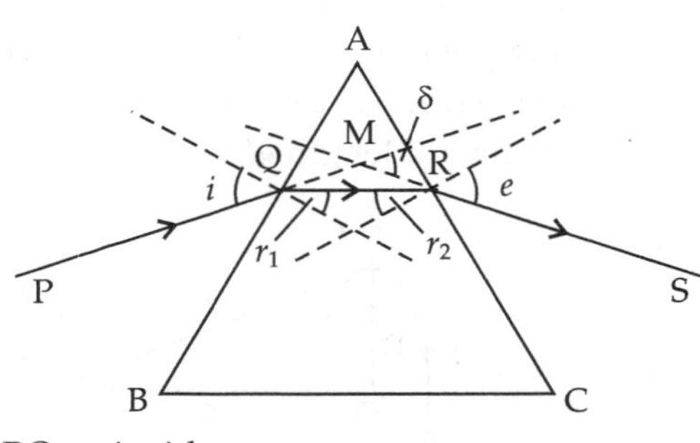


PQ = incident ray

QR = refracted ray

RS = emergent ray

$\angle A$ = angle of prism

$\angle i$ = angle of incidence

$\angle r_1$ = angle of refraction

$\angle e$ = angle of emergence

$\angle \delta$ = angle of deviation

Consider the above diagram and μ as the refractive index of the prism.

In case of minimum deviation, $\angle r_1 = \angle r_2 = \angle r$

And $\quad A = \angle r_1 + \angle r_2$

So, $\quad A = \angle r + \angle r = 2\angle r$

or $\quad \angle r = \frac{A}{2}$

Now again,

$$A + \delta = i + e$$
$$= 2i$$

or $\quad i = \frac{A+\delta}{2}$

Now from Snell's Law

$$\mu = \frac{\sin i}{\sin r} = \frac{\sin\left(\frac{A+\delta}{2}\right)}{\sin\left(\frac{A}{2}\right)}$$

$$\mu = \frac{\sin\left(\frac{A+\delta}{2}\right)}{\sin\left(\frac{A}{2}\right)}$$

is the required expression

**(b) (i)** For critical angle,

$$\sin C = \frac{1}{\mu} = \frac{1}{1.5} = \frac{2}{3}$$

$$C = \sin^{-1}\left(\frac{2}{3}\right) = 41°8'$$

So, if incident angle is greater than 41°8', Total internal reflection will occur.

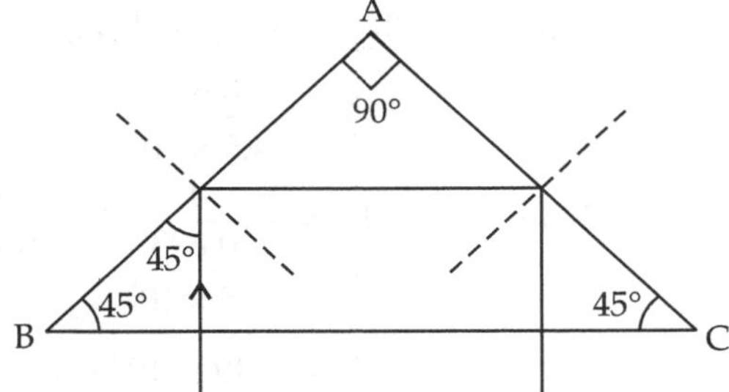


**(i)** If μ changes to 1.4

Then, $\sin C' = \frac{1}{\mu} = \frac{1}{1.4} = \frac{5}{7}$

$$C = \sin^{-1}\left(\frac{5}{7}\right) = 45°9'$$

Since, incident angle is less than the critical angle. Then total internal reflection will not occur.

**OR**

**(a)** Consider two thin lenses $L_1$ and $L_2$ of focal lenth $f_1$ and $f_2$ respectively, held coaxially in contact with each other. Let P be the point where the optical centres of the lenses coincide. (lenses being thin)

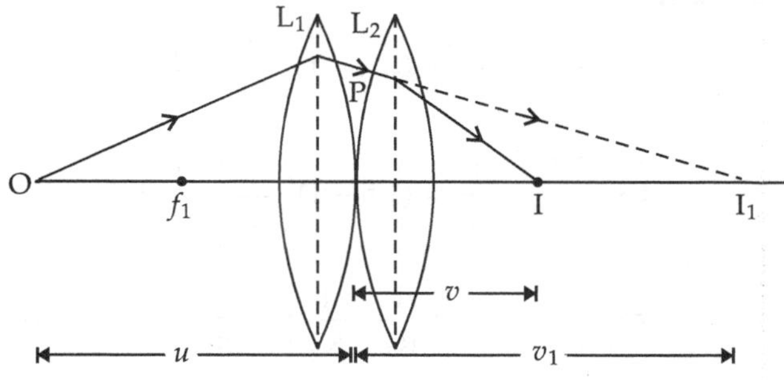


Here, OP = $u$ (object distance)

Let $L_1$ alone forms the image at $I_1$ where, $PI_1 = v_1$ (image distance). The image $I_1$ serve as a virtual object for lens $L_2$ which forms a final image I at distance PI = $v$

Now using lens formula for the lens $L_1$, we have,

$$\frac{1}{f_1} = \frac{1}{v_1} - \frac{1}{u} \quad ...(i)$$

For the image formation by second lens $L_2$, we have

$$\frac{1}{f_2} = \frac{1}{v} - \frac{1}{v_1} \quad ...(ii)$$

on adding (i) and (ii), we get

$$\frac{1}{f_1} + \frac{1}{f_2} = \frac{1}{v_1} - \frac{1}{u} + \frac{1}{v} - \frac{1}{v_1}$$

$$\Rightarrow \quad \frac{1}{f_1} + \frac{1}{f_2} = \frac{1}{v} - \frac{1}{u} \quad (iii)$$

If two lenses are considered a single lens of focal length $f$, which forms an image I at a distance $v$ with an object distance being $u$, then we get

$$\frac{1}{v} - \frac{1}{u} = \frac{1}{f} \quad (iv)$$

On comparing (iii) and (iv), we get

$$\Rightarrow \quad \frac{1}{f} = \frac{1}{f_1} + \frac{1}{f_2}$$

$$\Rightarrow \quad f = \frac{f_1 f_2}{f_1 + f_2}$$

Hence, the focal length of the combined system is given by

$$f = \frac{f_1 f_2}{f_1 + f_2}$$

**(b)** Power of lens, P = 10D

Focal length of lens = $\frac{1}{10}$ = 10cm

Refractive index of lens ${}_a\mu_g = 1.5$

Focal length of diverging lens, $f_1$ = 50cm

According to lens formula

$$\frac{1}{f} = ({}_a\mu_g - 1)\left(\frac{1}{R_1} - \frac{1}{R_2}\right)$$

$$\Rightarrow \quad \frac{1}{10} = (1.5 - 1)\left(\frac{1}{R_1} - \frac{1}{R_2}\right)$$

$$\Rightarrow \quad \left(\frac{1}{R_1} - \frac{1}{R_2}\right) = \frac{1}{5} \qquad ...(i)$$

Now, when lens is emerged in water, we get a diverging lens

$$\therefore \quad \frac{1}{f} = ({}_g\mu_l - 1)\left(\frac{1}{R_1} - \frac{1}{R_2}\right)$$

$$-\frac{1}{50} = ({}_g\mu_l - 1)\left(\frac{1}{5}\right) \qquad \text{[using (i)]}$$

$$\Rightarrow \quad ({}_g\mu_l - 1) = -\frac{1}{10}$$

$$\Rightarrow \quad {}_g\mu_l = 1 - \frac{1}{10} = \frac{9}{10}$$

$$\Rightarrow \quad \frac{\mu_g}{\mu_l} = \frac{9}{10}$$

$$\Rightarrow \quad \mu_l = \mu_g \times \frac{10}{9}$$

$$= 1.5 \times \frac{10}{9}$$

$$= 1.67$$

Hence, refractive index of liquid is 1.67.

**37. (a) Derive the law of radioactive decay**

$$\mathbf{N = N_o e^{-\lambda t}}$$

**(b) The half life of $^{238}_{92}U$ undergoing $\alpha$-decay is $4.5 \times 10^9$ years. Find its mean life.**

**(c) What fraction of the initial mass of a radio-active substance will decay in five half-life periods ? [5]**

**OR**

**(a) State the postulates of Bohr's model of hydrogen atom and derive the expression for Bohr radius.**

**(b) Find the ratio of the longest and the shortest wavelengths amongst the spectral lines of Balmer series in the spectrum of hydrogen atom.**

**Ans. (a)** Suppose, initially the number of atom in radioactive element is $N_o$ and N be the number of atoms after time $t$. After time $t$, let $dN$ be the number of atoms which disintegrate in a short interval $dt$, then rate of disintegration will be, $\frac{dN}{dt}$, this is also called activity of element.

According to Rutherford Soddy law,

$$\frac{dN}{dt} \propto N$$

$$\frac{dN}{dt} = -\lambda N \qquad ...(i)$$

Where, $\lambda$ is a constant called decay or disintegration constant.

On integrating equation (i) we get

$$\log_e N = -\lambda t + c \qquad ...(ii)$$

At $t = 0, N = N_o$

$$\therefore \quad \log_e N_o = 0 + c$$

$$\Rightarrow \quad c = \log_e N_o \qquad ...(iii)$$

Put the value of $c$ from equation (iii) in equation (ii)

$$\therefore \quad \log_e N - \log_e N_o = -\lambda t$$

$$\Rightarrow \quad \log_e\left(\frac{N}{N_o}\right) = -\lambda t$$

$$\Rightarrow \quad \frac{N}{N_o} = e^{-\lambda t}$$

$$\Rightarrow \quad N = N_o e^{-\lambda t}$$

**(b)** Mean life, $T_\alpha = \frac{1}{\lambda}$

Half life, $T_{1/2} = \frac{0.693}{\lambda}$

But $T_{1/2} = 4.5 \times 10^9$ years

$$\therefore \quad \lambda = \frac{0.693}{4.5 \times 10^9}$$

$$\text{Now} \quad T_\alpha = \frac{1}{\frac{0.693}{4.5 \times 10^9}} = \frac{4.5 \times 10^9}{0.693}$$

$$= 6.49 \times 10^9 \text{ years}$$

**(c)** Let $N_o$ be the original number of nuclei. In one half life the number reduces to (1/2) of the original number and similar goes on.

$\therefore$ Number of atoms remained after 5 half lives

$$= N_o\left(\frac{1}{2}\right)^5$$

$\therefore$ Fraction of atoms decayed

$$= N_o - \frac{N_o}{32}$$

$$= \frac{31}{32} N_o$$

OR

**(a)** The main postulates of Bohr's model of hydrogen atom are as follows :

(i) The electrons in the hydrogen atom revolve around the nucleus in a circular paths of fixed radius and energy called orbits or allowed energy states.

(ii) An electron do not radiate energy if it stayed in one orbit and therefore do not fall into the nucleus.

(iii) An electron moves from a lower energy state to a higher energy state after absorbing energy but energy is emitted when electrons move from the higher energy state to lower energy state.

Suppose $m$ be the mass of electron, $v_n$ be its speed in $n^{th}$ orbit of radius $r_n$.

$\therefore$ Centripetal force = Electrostatic force

$$\frac{mv_n^2}{r_n} = \frac{1}{4\pi\varepsilon_0}\frac{Z_e^2}{r_n^2} \quad ...(i)$$

From Bohr's postulate for quantisation of angular momentum.

$$mv_nr_n = \frac{nh}{2\pi}$$

$$\Rightarrow \quad v_n = \frac{nh}{2\pi mr_n}$$

Put this value in (i), we get

$$\frac{m}{r_n}\left(\frac{nh}{2\pi mr_n}\right)^2 = \frac{1}{4\pi\varepsilon_0}\frac{Ze^2}{r_n^2}$$

or $$r_n = \frac{\varepsilon_0 n^2h^2}{\pi mZe^2}$$

For Bohr's radius

$$n = 1$$

$$\therefore \quad r_1 = \frac{\varepsilon_0 h^2}{\pi mZe^2}$$

This is the expression for Bohr's radius.

**(b)** For Balmer series, wavelength is

$$\therefore \quad \frac{1}{\lambda} = k\left[\frac{1}{2^2} - \frac{1}{n^2}\right]$$

For shortest wavelength, $n = \infty$

$$\therefore \quad \frac{1}{\lambda_s} = 10^7\left[\frac{1}{2^2} - \frac{1}{\infty^2}\right]$$

$$\frac{1}{\lambda_s} = \frac{10^7}{4}$$

$$\lambda_s = 4 \times 10^{-7} \text{ m}$$

For longest wavelength, $n = 3$

$$\frac{1}{\lambda_e} = 10^7\left[\frac{1}{2^2} - \frac{1}{3^2}\right]$$

$$= 10^7\left[\frac{9-4}{36}\right]$$

$$\therefore \quad \lambda_e = \frac{36}{5} \times 10^{-7}$$

$$\frac{\lambda_e}{\lambda_s} = \frac{\frac{36}{5}\times 10^{-7}}{4 \times 10^{-7}} = \frac{9}{5}$$

$$\therefore \quad \lambda_e : \lambda_s = 9 : 5$$

# Outside Delhi [Set-II]

**Note : Except for the following questions, all the remaining questions have been asked in previous set.**

## SECTION-A

**Select the most appropriate option from those given below each question.**

**1. An electron and a proton are moving along the same direction with the same kinetic energy. They enter a uniform magnetic field acting perpendicular to their velocities. The dependence of radius of their paths on their masses is :**

(a) $r \propto m$ (b) $r \propto \sqrt{m}$

(c) $r \propto \frac{1}{m}$ (d) $\frac{1}{\sqrt{m}}$ [1]

**Ans. (a)** $r \propto m$

**Explanation :**

Since, radius of circular path

$$r = \frac{mv}{qB}$$

$$\therefore \quad r \propto m$$

**6. A current of 5 A is flowing from east to west in a long straight wire kept on a horizontal table. The magnetic field developed at a distance of 10 cm due south on the table is :**

**(a) $1 \times 10^{-5}$ T acting downwards**

**(b) $1 \times 10^{-5}$ T acting upwards**

**(c) $2 \times 10^{-5}$ T acting downwards**

**(d) $2 \times 10^{-5}$ T acting upwards** [1]

**Ans.** (a) $1 \times 10^{-5}$ T downwards

5A

10 cm

$$B = \frac{\mu_0 i}{2\pi d}$$

$$= \frac{4\pi \times 10^{-7} \times 5}{2\pi \times \frac{10}{100}}$$

$$= 2 \times 10^{-7} \times 5 \times 10$$

$$B = 10^{-5}\,T = 1 \times 10^{-5}\,T$$

The direction is found by right hand thumb rule.

**Note : Fill in the blanks with appropriate answer :**

**11. In β-decay, a neutron is converted into a proton, $e^-$ and ................ . [1]**

**Ans.** Electron antineutrino

$$n \longrightarrow p + e^- + \bar{v}_e$$

**12. In a linearly polarized light, the electric vector oscillates along a direction perpendicular to the aligned molecules of the polaroid, called ......... axis. [1]**

**Ans.** Pass

**19. Depict equipotential surface due to an electric dipole. [1]**

**Ans.** Equipotential surface is the region in space where every point on it is same and at same electric potential. So, an electric dipole have 2 opposite charges, then a straight line passing through centre of electric dipole will be equipotential surface as shown in figure.

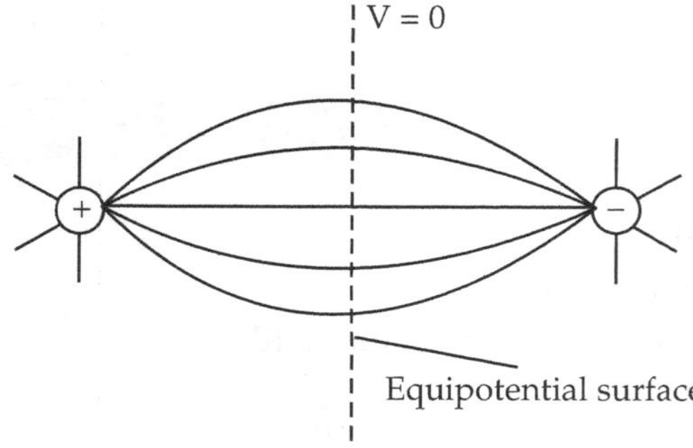


**20. Define 'angle of dip' at a place on earth. [1]**

**Ans.** The angle of dip is the angle between the direction of Earth's magnetic field and the plane of the horizon. It is also known as magnetic dip or magnetic inclination.

## SECTION-B

**21. A uniform wire is cut into three parts with their lengths in the ratio 2 : 3 : 6. The ends of each of these three part are connected across an ideal battery of 10 V. If a current of 5 A is drawn from the battery, find the initial resistance of the wire. [2]**

**Ans.** Let the resistance of wire be R, its length be $11l$ (sum of all ratios) and area of cross section 'A' and its resistivity be ρ.

$$\therefore \quad R = \frac{\rho(11l)}{A}$$

Let, the length of wire cutted be $2l$, $3l$ and $6l$

$\therefore$ Resistance of 1st part, $R_1 = \frac{\rho(2l)}{A}$

Resistance of 2nd part, $R_2 = \frac{\rho(3l)}{A}$

Resistance of 3rd part, $R_3 = \frac{\rho(6l)}{A}$

Since, 3 resistances are connected in parallel.

$$\therefore \quad \frac{1}{R_{eq}} = \frac{1}{R_1} + \frac{1}{R_2} + \frac{1}{R_3}$$

$$= \frac{A}{\rho(2l)} + \frac{A}{\rho(3l)} + \frac{A}{\rho(6l)}$$

$$= \frac{A}{\rho l}\left[\frac{1}{2} + \frac{1}{3} + \frac{1}{6}\right]$$

$$= \frac{A}{\rho l}\left[\frac{3+2+1}{6}\right] = \frac{A}{\rho l} \times \left(\frac{6}{6}\right)$$

$$\frac{1}{R_{eq}} = \frac{A}{\rho l}$$

or $$R_{eq} = \frac{\rho l}{A}$$

But, $$R_{eq} = \frac{V}{I} = \frac{10}{5} = 2\Omega$$

$$\therefore \quad \frac{\rho l}{A} = 2$$

Now, initial resistance is $R = \frac{\rho(11l)}{A}$

$$= 11\left(\frac{\rho l}{A}\right)$$

$$= 11 \times 2 = 22\Omega$$

Hence, the initial resistance of wire was 22 Ω.

**23. Find the total charge stored in the network of capacitors connected between A and B as shown in figure.**

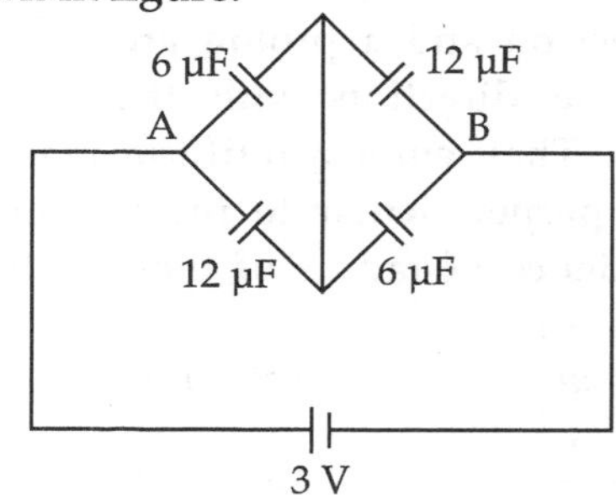


**[2]**

**Ans.**

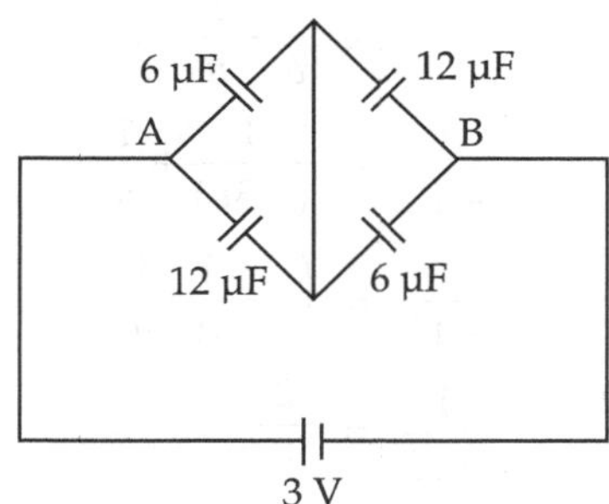


2 capacitors are connected in series between A and B.

$$\therefore \quad \frac{1}{C_1} = \frac{1}{6} + \frac{1}{12}$$

$$\frac{1}{C_1} = \frac{2+1}{12}$$

$$\Rightarrow \quad C_1 = \frac{12}{3} = 4\,\mu F$$

And 2 other capacitors are also in series between A and B.

$$\therefore \quad \frac{1}{C_2} = \frac{1}{6} + \frac{1}{12}$$

$$\frac{1}{C_2} = \frac{2+1}{12}$$

$$\Rightarrow \quad C_2 = \frac{12}{3} = 4\,\mu F$$

Now, $C_1$ and $C_2$ are in parallel

$$\therefore \quad C_{eq} = C_1 + C_2 = 4\,\mu F + 4\,\mu F = 8\,\mu F$$

$\therefore$ Total charge stored, $Q = CV = 8 \times 3\,\mu C = 24\,\mu C$

Hence, total charge stored is $2.4 \times 10^{-7}$ C.

## SECTION-C

**30. Explain the principle of the device with diagram, which is used to provide electricity at the proper voltage for household purposes. Briefly discuss loss of energy in it due to flux leakage and its minimization. [3]**

**Ans.** The device used to provide electricity at proper voltage for household purpose is stepdown transformer. Transformer is a static device which converts electrical power from one circuit to another without changing its frequency.

**Principle** : It works on the principle of mutual induction of two coils or Faraday's law of electromagnetic induction when an alternating source of voltage is introduced across a coil (which is physically coupled to another coil), the changing current through it induces an emf across the second coil.

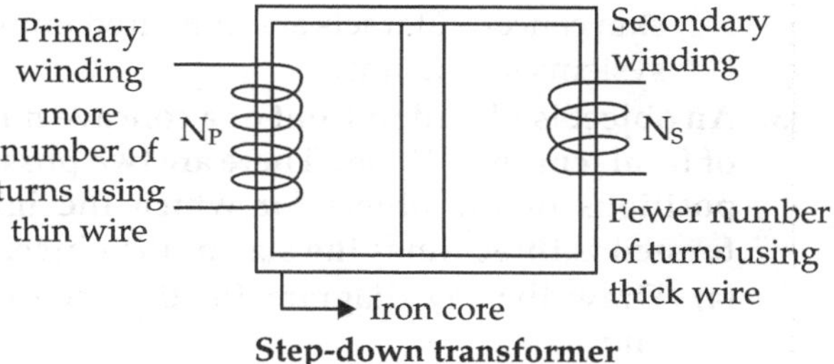


**Step-down transformer**

The possible sources of power losses in power transformer is :

**(a) Flux leakage** : Small amount of flux coupled from primary to secondary.
To avoid flux leakage, the coils are wound over each other again and again.

**(b) Resistance of windings** : The transformer coil wires cannot have absolutely zero resistance, thus small amount of loss due to heating arises in coil.

**(c) Core eddy currents** : Since, the core is a very good conductor itself, currents are induced in it due to changing magnetic fields, called eddy currents :
It can be reduced by making core of thin sheets by reducing the area.

**(d) Hysteresis** : Some part of energy is frozen in to the core permanently in the form of a residual magnetic field due to its ferromagnetic character.
It can be reduced by special core material which reached to zero or near zero flux density after removal of current.

**31. (a) We feel the warmth of the sun light but not the pressure on our hands. Explain.**

**(b) Which out of wavelength, frequency and speed of an electromagnetic wave does not change on passing from one medium to another ?**

**(c) A thin ozone layer in the upper atmosphere is crucial for human survival on earth, why ? [3]**

**Ans. (a)** Because the photons of light have only energy and do not exhibit mass.

**(b)** When electromagnetic waves passes from one medium to another, frequency remains constant. Wavelength and speed both changes when medium of the wave changes.

**(c)** The thin ozone layer around the earth's atmosphere is very crucial for human survival. It protects the Earth from harmful ultraviolet (UV) rays from the sun. Without the ozone layer in the atmosphere, life on Earth would be very difficult. Plants cannot live and grow in heavy ultraviolet radiations. Ultraviolet radiations can cause skin cancer, cataracts and impaired immune systems in humans.

**33. An object is placed in front of a concave mirror of focal length of 12 cm. There are two possible positions of the object for which the image formed is three times the size of the object.**

**(a) Draw the ray diagram for the each case, and**

**(b) Find the distance between the two positions of the object. [3]**

**Ans.** Given, focal length of concave mirror, $f = -12$ cm

Magnification, $m = \pm 3$

If $m = 3$, then

$\Rightarrow \quad 3 = \frac{v}{u}$

$\Rightarrow \quad v = 3u$

By mirror formula,

$$\frac{1}{f} = \frac{1}{u} + \frac{1}{v}$$

$$\Rightarrow \quad \frac{1}{-12} = \frac{1}{u} + \frac{1}{3u}$$

$$\Rightarrow \quad \frac{1}{-12} = \frac{3+1}{3u}$$

$$\Rightarrow \quad u = \frac{-4}{3} \times 12$$

$$\Rightarrow \quad u = -16 \text{ cm}$$

If $m = -3$, then

$$-3 = \frac{v}{u}$$

$$\Rightarrow \quad v = -3u$$

By mirror formula

$$\frac{1}{f} = \frac{1}{v} + \frac{1}{u}$$

$$\Rightarrow \quad \frac{1}{-12} = \frac{1}{(-3u)} + \frac{1}{u}$$

$$\Rightarrow \quad \frac{1}{-12} = \left(1 - \frac{1}{3}\right)\frac{1}{u}$$

$$\Rightarrow \quad \frac{1}{-12} = \frac{2}{3u}$$

$$\Rightarrow \quad u = -8 \text{ cm}$$

(a)

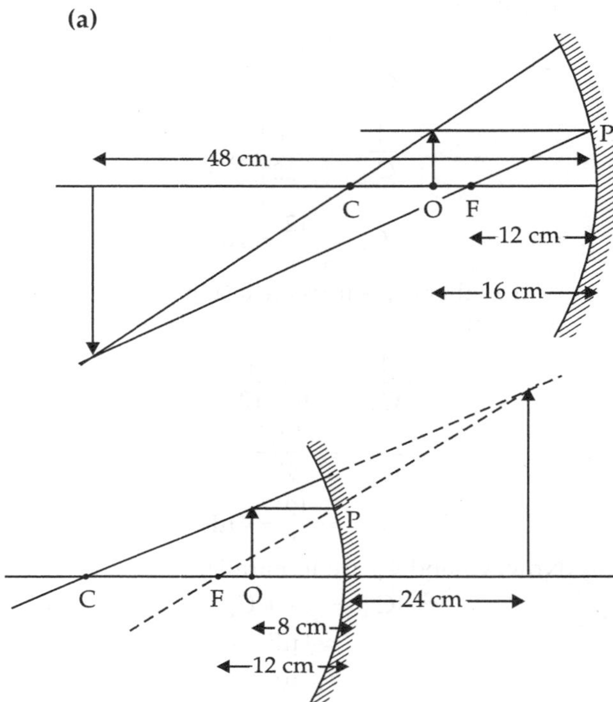


**(b)** The two positions of object is 16 cm and 8 cm.

$\therefore$ Difference = 16 – 8

= 8 cm

## Outside Delhi [Set-III]

**Note : Except for the following questions, all the remaining questions have been asked in previous sets.**

### SECTION-A

**Select the most appropriate option from those given below each question.**

**2. There are uniform electric and magnetic fields in a region pointing along X-axis. An α-particle is projected along Y-axis with a velocity $v$. The shape of the trajectory will be :**

**(a) circular in XZ plane**

**(b) circular in YZ plane**

**(c) helical with its axis parallel to X-axis**

**(d) helical with its axis parallel to Y-axis [1]**

**Ans.** (c) helical with its axis parallel to $x$-axis.

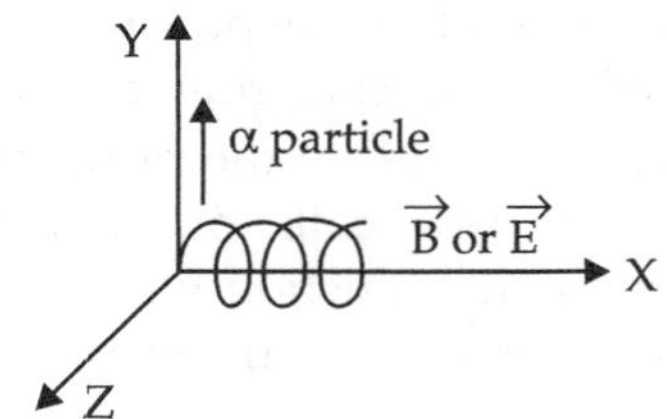

5. **A current of 10 A is following from east to west in a long straight wire kept on a horizontal table. The magnetic field developed at a distance 10 cm vertically above the wire is :**

   **(a) $1.2 \times 10^{-5}$ T, acting towards south**

   **(b) $2 \times 10^{-5}$ T, acting towards north**

   **(c) $3 \times 10^{-5}$ T, acting downwards**

   **(d) $2 \times 10^{-5}$ T, acting upwards** **[1]**

**Ans.** (b) $2 \times 10^{-5}$ T, acting towards north

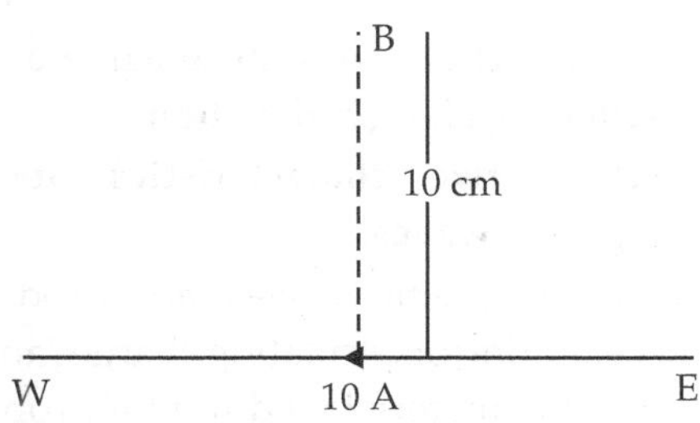


$$B = \frac{\mu_0 i}{2\pi d} = \frac{4\pi \times 10^{-7} \times 10}{2\pi \times \frac{10}{100}}$$

$$= 2 \times 10^{-5}\ T$$

Direction is calculated by using right hand thumb rule.

**Note : Fill in the blanks with appropriate answer :**

13. **A neutron is bombarded on a $^{10}_{5}B$ nucleus and an alpha particle is emitted.**

    **The nuclear reaction involved is :**

    $$^{1}_{0}n + ^{10}_{2}B \longrightarrow ^{4}_{2}He + ..........$$ **[1]**

**Ans.** $$^{10}_{5}B + ^{1}_{0}n \longrightarrow ^{4}_{2}He + ^{7}_{3}Li$$

19. **What is the change in the value of angle of dip when one goes from the equator to the north pole of earth ?** **[1]**

**Ans.** As we move from equator to the north pole, the angle of dip increases. The angle of dip is 0° at equator and 90° at the pole.

20. **A charged particle (+$q$) moves in a uniform electric field $(\vec{E})$ in the direction opposite to $\vec{E}$. What will be the effect on its electrostatic potential energy during its motion ?** **[1]**

**Ans.**

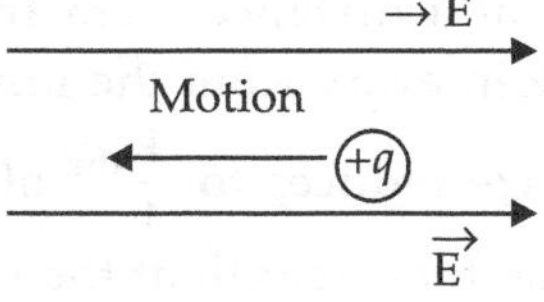


When a positive charge moves opposite to the direction of the electric field, the field does negative work on it and the potential energy increases.

## SECTION-B

21. **Briefly explain how a potential barrier is set up across a *p-n* junction as a result of diffusion and drift of the charge carriers.** **[2]**

**Ans.** During the formation of a *p-n* junction, the holes diffuse from *p*-type semiconductors to the *n*-type, and electrons diffuse from *n*-type to *p*-type. This is because of the concentration gradient across *p*-side and *n*-side.

When a hole diffuse from *p* to *n*-type, it leaves an unmovable negative charge. Similarly, when an electron diffuses from *n* to *p*-type, it leaves an unmovable positive charge. When the diffusion of holes and electrons takes place continuously across the junctions, a layer of unmovable positive and negative charges are developed on either side of the junction. This layer is called depletion layer or depletion region and the potential difference across the region is called barrier potential.

23. **A wire of length $L_o$ has a resistance $R_o$. It is gradually stretched till its length becomes 1.5 $L_o$.**

    **(a) Plot the graph showing variation of its resistance R with its length *l* during stretching.**

    **(b) What will be its resistance when its length becomes 1.5 $L_o$ ?** **[2]**

**Ans.** As the wire stretches the resistance increases.

Since, $R \propto l$

So more the distance to covered by current, more is the resistance of the wire.

(a)

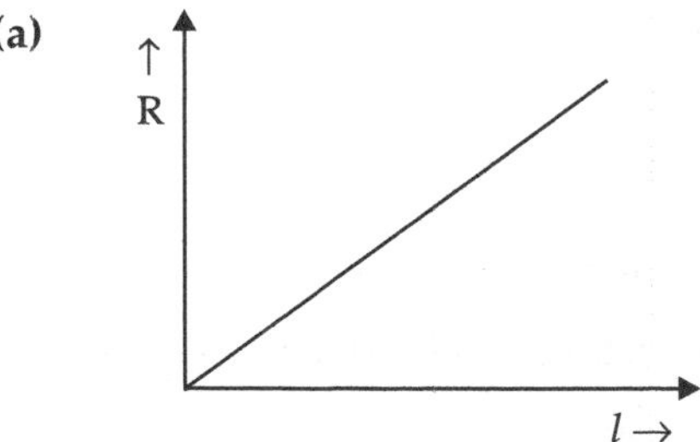


(b) Original Resistance,

$$R_0 = \frac{\rho L_0}{A_0}$$

Now, new length is 1.5 $L_0$

Volume of the wire remains constant.

Let, the new area be A′

$\therefore \quad A_0L_0 = 1.5\, L_0A'$

$\Rightarrow \quad A' = \dfrac{A_0}{1.5}$

$\therefore$ New resistance, $R = \dfrac{\rho(1.5\,L_0)}{A'}$

$= \rho\dfrac{(1.5\,L_0)}{\dfrac{A_0}{1.5}}$

$= 2.25\dfrac{\rho L_0}{A_0}$

$= 2.25\,R_0 \quad \left[\because R_0 = \dfrac{\rho L_0}{A_0}\right]$

So, new resistance becomes 2.25 times of the original resistance.

**24. You are given three capacitors of 2 μF, 3 μF and 4 μF, respectively.**

**(a) Form a combination of all these capacitors of equivalent capacitance $\frac{13}{3}$ μF.**

**(b) What is the maximum and minimum value of the equivalent capacitance that can be obtained by connecting these capacitors ? [2]**

**Ans.** Given, capacitor are 2 μF, 3 μF and 4 μF.

(a) The combination of 2 μF and 4 μF in series with 3 μF in parallel.

$$\frac{1}{C_s} = \frac{1}{2}+\frac{1}{4}$$

$$\Rightarrow \quad \frac{1}{C_s} = \frac{2+1}{4}$$

$$\Rightarrow \quad C_s = \frac{4}{3}\,\mu F$$

$$\text{Now, } C_{equivalent} = \frac{4}{3} + 3\,\mu F = \frac{13}{3}\,\mu F$$

So, we can get an equivalent of $\frac{13}{3}$ μF by connecting 2 μF and 4 μF in series and 3 μF in parallel.

(b) Maximum capacitance can be achieved by joining the capacitors in parallel

*i.e.,* $C_{max} = 2+3+4 = 9\,\mu F$

Minimum capacitance is achieved by joining the capacitors in series

*i.e.,* $\dfrac{1}{C_{min}} = \dfrac{1}{2}+\dfrac{1}{3}+\dfrac{1}{4} = \dfrac{6+4+3}{12}$

$C_{min} = \dfrac{12}{13}\,\mu F$

## SECTION-C

**29. (a) How are electro-magnetic waves produced ? Depict an electromagnetic wave propagating in Z-direction with its magnetic field $\vec{B}$ oscillating along X-direction.**

**(b) Write two characteristics of electro-magnetic waves. [3]**

**Ans.** (a) Electromagnetic waves are produced by accelerating or oscillating charges and do not need any material medium for propagation. The source of energy of these waves are energy of oscillating charged particles.

Mathematical expressions for electric and magnetic fields of an electromagnetic wave propagating along the *z*-axis are :

$$E_y = E_0 \sin(kz - \omega t)$$
$$B_x = B_0 \sin(kz - \omega t)$$

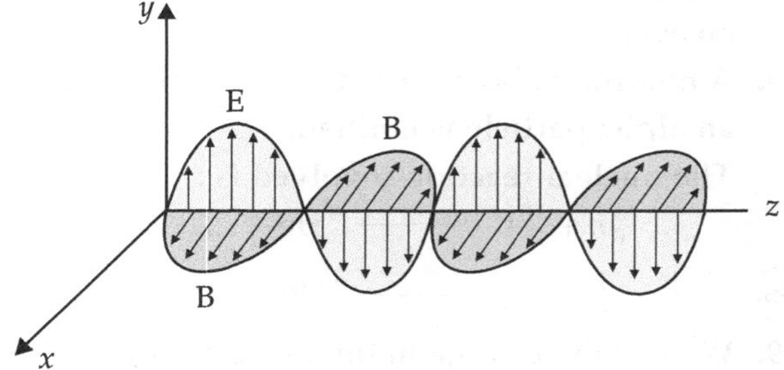


(b) Properties of electromagnetic waves are :

1. The direction of E, B and direction of propogation of waves are mutually perpendicular to one another.
2. The amplitude ratio of E and B gives velocity of light.
3. Electromagnetic waves are not deflected by electric and magnetic fields.

**30. A concave mirror forms a real image of an object kept at a distance 9 cm from it. If the object is taken away from the mirror by 6 cm, the image size reduces to $\frac{1}{4}$ the of its previous size. Find the focal length of the mirror. [3]**

**Ans.** Let, the focal length of concave mirror be $-x$.

First object distance, $u_1 = -9$ cm

Second object distance, $u_2 = -15$ cm

I case, using mirror formula

$$\frac{1}{f} = \frac{1}{u_1} + \frac{1}{v_1}$$

$$\Rightarrow \frac{1}{-x} = \frac{1}{-9} + \frac{1}{v_1}$$

$$\Rightarrow \frac{1}{v_1} = \frac{1}{9} - \frac{1}{x}$$

$$\Rightarrow \frac{1}{v_1} = \frac{x-9}{9x}$$

$$\Rightarrow v_1 = \frac{9x}{x-9} \quad ...(i)$$

II case, using mirror formula again

$$\Rightarrow \frac{1}{f} = \frac{1}{u_2} + \frac{1}{v_2}$$

$$\Rightarrow \frac{1}{-x} = \frac{1}{-15} + \frac{1}{v_2}$$

$$\Rightarrow \frac{1}{v_2} = \frac{1}{15} - \frac{1}{x} = \frac{x-15}{15x}$$

$$\Rightarrow v_2 = \frac{15x}{x-15} \quad ...(ii)$$

Now, magnification for I case,

$$m_1 = \frac{v_1}{u_1} = \frac{9x}{x-9} \times \frac{1}{(-9)} = \frac{I_1}{O}$$

Now, magnification for II case,

$$m_2 = \frac{v_2}{u_2} = \frac{15x}{x-15} \times \frac{1}{(-15)} = \frac{I_2}{O}$$

According to question,

$$I_2 = \frac{1}{4} I_1 \quad \text{[Object remains of the same size]}$$

$$\frac{x}{-(x-15)} = \frac{1}{4} \times \frac{x}{-(x-9)}$$

$$\Rightarrow \frac{1}{x-15} = \frac{1}{4(x-9)}$$

$$\Rightarrow 4x - 36 = x - 15$$

$$\Rightarrow 3x = 21$$

$$\Rightarrow x = 7 \text{ cm}$$

Hence, the focal length of mirror is 7 cm.

**32. Draw the labelled diagram of an AC generator. Briefly explain its working and obtain the expression for the emf produced in the coil. [3]**

**Ans.** Diagram of AC generator.

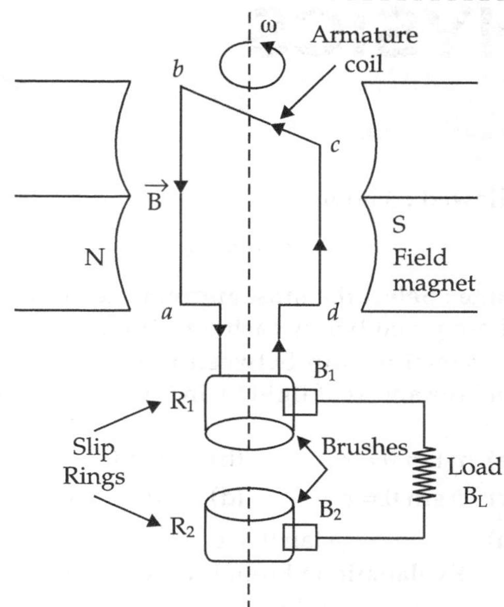


**Principle and working** : It works on the principle of electromagnetic induction *i.e.*, when a coil rotates continuously in a magnetic field, the effective area of the coil, linked (normally) with the magnetic field lines, changes continuously with time. The variation of a magnetic flux with time results in the production of a alternating emf in the coil.

When the armature rotates between the poles of the magnet upon an axis perpendicular to the magnetic field, the flux which links with the armature changes continuously. Due to this, an emf is induced in the armature this produces an electric current through galvanometer and the slip rings and brushes.

If N is the number of turns in coil, A be the area of coil and B be the magnetic induction, then flux $\phi$ is given by :

$$\phi = B.A$$

$$\phi = BA \cos \omega t$$

The emf induced in the coil is given by :

$$e = -N \frac{d\phi}{dt}$$

$$e = -N \frac{d}{dt}(BA \cos \omega t)$$

$$e = +NBA\,\omega \sin \omega t$$

$$e = e_0 \sin \omega t$$

Where, $e_0 = \sin \omega t$

# PHYSICS

## Delhi [Set-I]

# 2020

**Time allowed : 3 hours** **Maximum marks : 70**

### SECTION–A

**Note : Select the most appropriate option from those given below each question.**

1. **The relationship between Brewester angle '$\theta$' and the speed of light '$v$' in the denser medium is :** **[1]**

(a) $v \tan \theta = c$ (b) $c \tan \theta = v$

(c) $v \sin \theta = c$ (d) $c \sin \theta = v$

**Ans.** (a) $v \tan \theta = c$

**Explanation :** From Brewster's law :

Refractive index,

$$\mu = \tan \theta$$

$$\frac{c}{v} = \tan \theta \qquad \left[\because \mu = \frac{c}{v}\right]$$

$$\therefore \quad c = v \tan \theta$$

2. **Photo diodes are used to detect** **[1]**

(a) **radio waves** (b) **gamma rays**

(c) **IR rays** (d) **optical signals**

**Ans.** (d) Optical signals.

**Explanation :** Photodiodes are used to detect the visible light.

3. **The selectivity of a series LCR a.c. circuit is large, when** **[1]**

(a) **L is large and R is large**

(b) **L is small and R is small**

(c) **L is large and R is small**

(d) **L = R**

**Ans.** (c) L is large and R is small.

**Explanation :** Since, selectivity depends on the quality of resonance. The quality factor is given by $Q = \omega_0 L/R$. High value of quality factor make sure that the resonance curve is sharp. Sharper the resonance curve, more selective is the LCR circuit.

4. **The graph showing the correct variation of linear momentum (p) of a charge particle with its de-Broglie wavelength ($\lambda$) is :** **[1]**

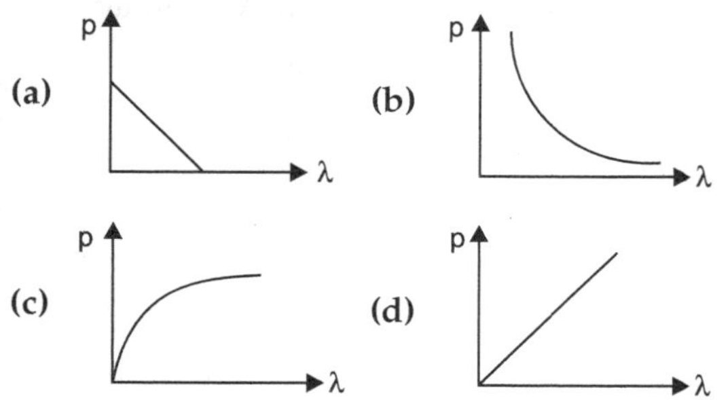


**Ans.** (b)

**Explanation :** Since $p = \frac{E}{C} = \frac{h}{\lambda}$

$$p \propto \frac{1}{\lambda},$$

By this relation we get graph between momentum and de-Broglie wavelength is a rectangular hyperbola.

5. **The wavelength and intensity of light emitted by a LED depend upon :** **[1]**

(a) **forward bias and energy gap of the semiconductor**

(b) **energy gap of the semiconductor and reverse bias**

(c) **energy gap only**

(d) **forward bias only**

**Ans.** (a) Forward bias and energy gap of the semiconductor.

**Explanation :** Since, the wavelength depends on energy gap as $E_g = hc/\lambda$.

And intensity will depends on the current in forward bias.

6. **A charge particle after being accelerated through a potential difference 'V' enters in a uniform magnetic field and moves in a circle of radius r. If V is doubled, the radius of the circle will become :** **[1]**

(a) $2r$ (b) $\sqrt{2}r$

(c) $4r$ (d) $r/\sqrt{2}$

**Ans.** (b) $\sqrt{2}r$

**Explanation :** Since, $qV = \frac{1}{2}mv^2$

$\therefore$ velocity acquired, $v = \sqrt{\frac{2qV}{m}}$

Now, radius of circular path

$$r = \frac{mv}{qB} = \frac{m}{qB}\sqrt{\frac{2qV}{m}}$$

$$= \frac{1}{B}\sqrt{\frac{2mV}{q}}$$

If V is doubled, then

$$r' = \frac{1}{B}\sqrt{\frac{2m \times (2V)}{q}}$$

$$r' = \sqrt{2}\, r$$

7. **The electric flux through a closed Gaussian surface depends upon : [1]**
   (a) **Net charge enclosed and permittivity of the medium**
   (b) **Net charge enclosed, permittivity of the medium and the size of the Gaussian surface**
   (c) **Net charge enclosed only**
   (d) **Permittivity of the medium only**

**Ans.** (a) Net charge enclosed and permittivity of medium.

**Explanation :** As per Gauss law,

$$\text{Electric flux } \phi = \frac{q}{\varepsilon_0}$$

8. **If photons of frequency $v$ are incident on the surfaces of metals. A & B of threshold frequencies $v/2$ and $v/3$ respectively, the ratio of the maximum kinetic energy of electrons emitted from A to that from B is : [1]**
   (a) **2 : 3** (b) **3 : 4**
   (c) **1 : 3** (d) **$\sqrt{3}:\sqrt{2}$**

**Ans.** (b) 3 : 4

**Explanation :** $E_k = h(v - v_0)$

$$\therefore \quad \frac{E_A}{E_B} = \frac{h(v-v/2)}{h(v-v/3)} = \frac{v/2}{2v/3} = \frac{3}{4}$$

9. **The power factor of a series LCR circuit at resonance will be : [1]**
   (a) **1** (b) **0**
   (c) **1/2** (d) **$1/\sqrt{2}$**

**Ans.** (a) 1

**Explanation :** Power Factor,

$$\cos\phi = \frac{R}{\sqrt{R^2+(X_L-X_C)^2}}$$

At resonance,

$$X_C = X_L$$

$$\therefore \quad \cos\phi = \frac{R}{\sqrt{R^2}} = 1$$

10. **A biconcave lens of power P vertically splits into two identical plano concave parts. The power of each part will be : [1]**
   (a) **2P** (b) **P/2**
   (c) **P** (d) **P/$\sqrt{2}$**

**Ans.** (b) P/2

**Explanation:** Let the focal length of biconcave lens be $f$ and that of the planoconcave lens be $f$.

By Lens' maker formula, for biconcave lens

$$\frac{1}{f} = (\mu-1)\left(\frac{1}{(-R)} - \frac{1}{R}\right) = \frac{-2(\mu-1)}{R}$$

[where R, is the radius of curvature of biconcave lens

For planoconcave lens, the focal length

$$\frac{1}{f'} = (\mu-1)\left(\frac{1}{(-R)}\right) = -\left(\frac{\mu-1}{R}\right)$$

Thus, $f' = 2f$

Thus, Power $P' = \frac{P}{2}$

**Note : Fill in the blanks with appropriate answer.**

11. **The physical quantity having SI unit $NC^{-1}$ m is .................. . [1]**

**Ans.** Electric field intensity or electric potential.

12. **A copper wire of non-uniform area of cross-section is connected to a d.c. battery. The physical quantity which remains constant along the wire is .................. . [1]**

**Ans.** Electric current

**Explanation :** Since, cross section area is nonuniform then current density electric field and drift velocity will not remain same as they vary inversely with area of cross-section.

13. **A point charge is placed at the centre of a hollow conducting sphere of internal radius '$r$' and outer radius '$2r$'. The ratio of the surface charge density of the inner surface to that of the outer surface will be .................. . [1]**

**Ans.** –4 : 1

$$\sigma_{\text{inside}} = \frac{-q}{4\pi r^2}$$

As, $$\sigma_{\text{outside}} = \frac{+q}{4\pi(2r)^2} = \frac{q}{16\pi r^2}$$

$$\therefore \quad \frac{\sigma_{\text{inside}}}{\sigma_{\text{outside}}} = \frac{\frac{-q}{4\pi r^2}}{\frac{q}{16\pi r^2}} = \frac{-4}{1}$$

$$= -4 : 1$$

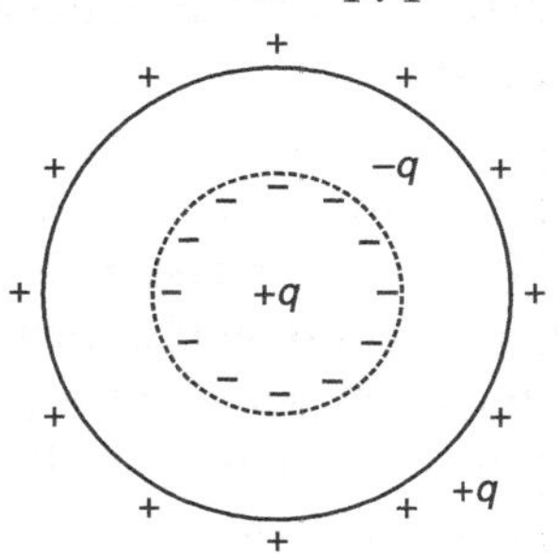

**14. The .................. , a property of materials C, Si and Ge depends upon the energy gap between their conduction and valence bands. [1]**

**Ans.** Conductivity.

**15. The ability of a junction diode to .................. an alternating voltage, is based on the fact that it allows current to pass only when it is forward biased. [1]**

**Ans.** rectify

**Note : Answer the following :**

**16. Define the term 'current sensitivity' of a moving coil galvanometer. [1]**

**Ans.** The current sensitivity of a galvano-meter is defined as the deflection produced in the galvanometer when a unit current is flowing through it.

*i.e.,* $$I_s = \frac{\alpha}{I} = \frac{NAB}{k}$$

**17. Depict the fields diagram of an electromagnetic wave propagating along positive X-axis with its electric field along Y-axis. [1]**

**Ans.**

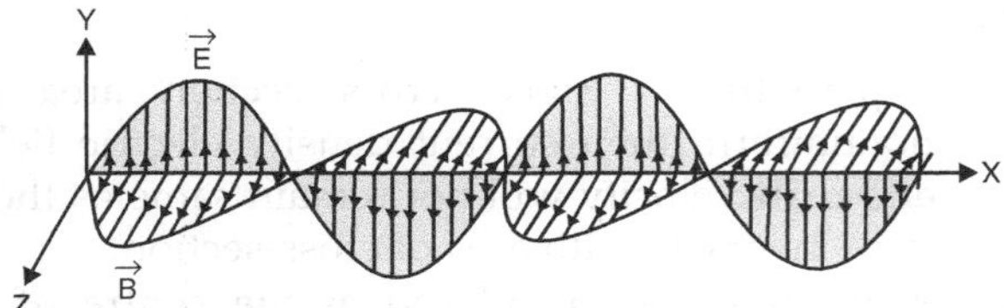


**18. Write the conditions on path difference under which (i) constructive (ii) destructive interference occur in Young's double slit experiment. [1]**

**Ans.** (i) Constructive interference :

$$\Delta x = n\lambda,$$

where, $n = 0, 1, 2$ .............

(ii) Destructive interference :
Path difference,

$$\Delta x = (2n+1)\frac{\lambda}{2}$$

where, $n = 0, 1, 2,$ ..............:

**19. Plot a graph showing variation of induced *e.m.f.*, with the rate of change of current flowing through a coil.**

**OR**

**A series combination of an inductor (L), capacitor (C) and a resistor (R) is connected across an ac source of emf of peak value $E_0$ and angular frequency (ω). Plot a graph to show variation of impedance of the circuit with angular frequency (ω).**

**Ans.**

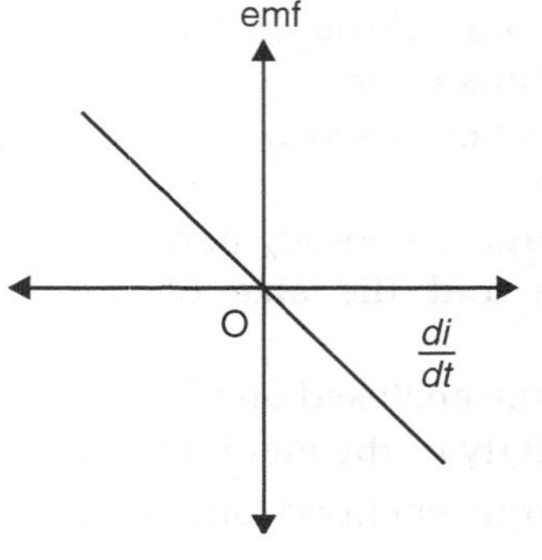


OR

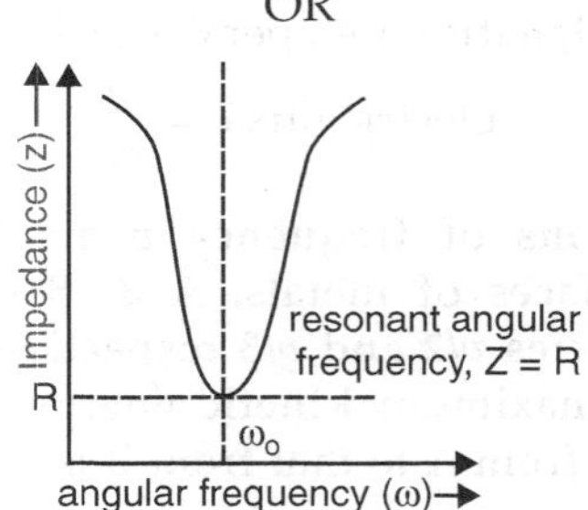


Here, $\omega_0$ represents the resonance frequency for LCR circuit.

**20. An electron moves along +*x* direction. It enters into a region of uniform magnetic field $\vec{B}$ directed along – *z* direction as shown in fig. Draw the shape of trajectory followed by the electron after entering the field. [1]**

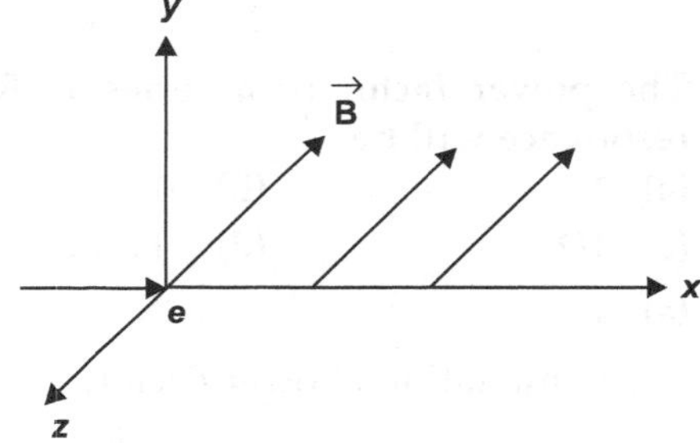


**OR**

**A square shaped current carrying loop MNOP is placed near a straight long current carrying wire AB as shown in the fig. The wire and the loop lie in the same plane. If the loop experiences a net force F towards the wire, find the magnitude of the force on the side 'NO' of the loop.**

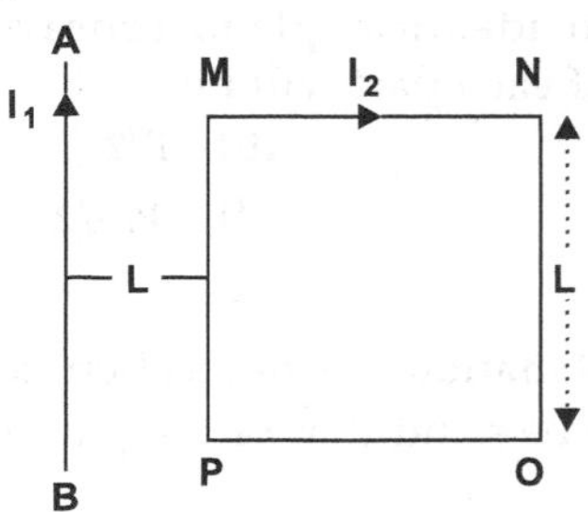

**Ans.** According to Fleming's left hand rule, the electron will move downward *i.e.*, in negative Y direction. It will follow a semi-circular path.

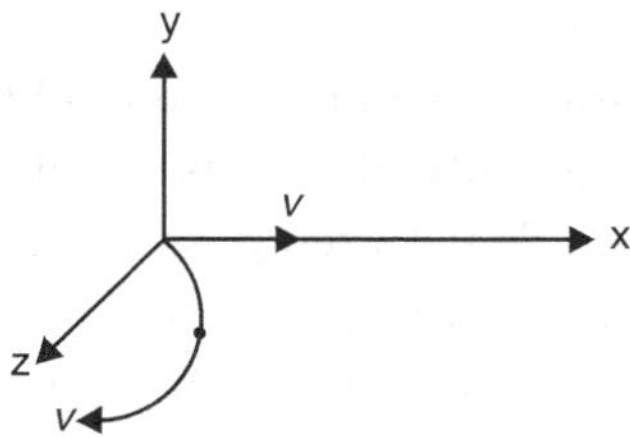


**OR**

The force on side NO will be repulsive, as current in side NO and wire AB is in opposite direction.

$$\therefore \quad F_{NO} = \frac{\mu_0 I_1 I_2 L}{2\pi r}$$

$$= \frac{\mu_0 I_1 I_2 L}{2\pi(L+L)}$$

$$= \frac{\mu_0 I_1 I_2 L}{2\pi \times 2L}$$

$$= \frac{\mu_0 I_1 I_2}{4\pi} \quad \text{(away from the wire)}$$

**SECTION–B**

**21. Derive the expression for the torque acting on an electric dipole, when it is held in a uniform electric field. Identify the orientation of the dipole in the electric field, in which it attains a stable equilibrium. [2]**

**OR**

**Obtain the expression for the energy stored in a capacitor connected across a dc battery. Hence define energy density of the capacitor.**

**Ans. Dipole in a Uniform External Field :**

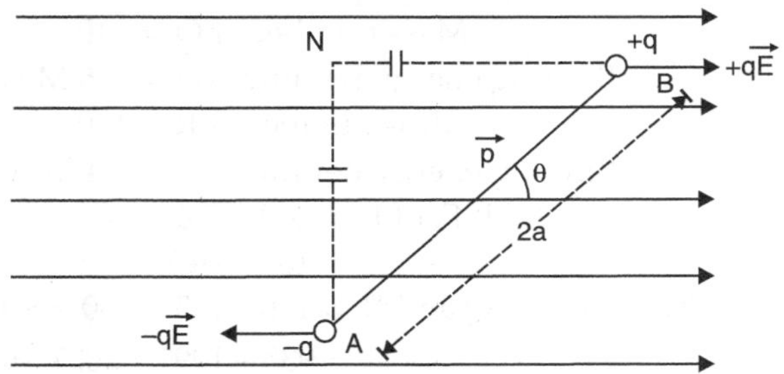


Consider an electric dipole consisting of charges $-q$ and $+q$ and of length $2a$ placed in a uniform electric field $\vec{E}$ making an angle θ with electric field.

Force on charge $-q$ at A $= -q\vec{E}$ (opposite to $\vec{E}$)

Force on charge $+q$ at B $= +q\vec{E}$ (along $\vec{E}$)

Electric dipole is under the action of two equal and unlike parallel force, which give rise to a torque on the dipole

τ = Force × Perpendicular distance between the two forces

$$\tau = qE(AN) = qE \times (2a \sin\theta)$$

$$\tau = q(2a)\, E \sin\theta$$

$$\tau = pE \sin\theta \qquad (\because p = q \times 2a)$$

$$\therefore \quad \vec{\tau} = \vec{p} \times \vec{E}$$

The dipole will be in stable equilibrium when $\theta = 0°$, then the net torque acting on it will be zero.

**OR**

The work done in charging the capacitor gets stored in it in the form of potential energy. The electric field on one plate due to charge Q and charge density σ on it $= \frac{\sigma}{2\varepsilon_0}$

The potential energy of one plate at a distance $d$ from the other plate.

$U_E$ = (electric p.d. between them) × (electric charge on the other plate)

$$= \frac{\sigma d}{2\varepsilon_0} \times Q$$

(taking reference potential on the first plate as zero).

Putting $\sigma = \frac{Q}{A}$

$$U_E = \frac{dQ}{2\varepsilon_0 A} = \frac{Q^2}{2C}$$

$$\left[\text{as } C = \frac{\varepsilon_0 A}{d} \text{ for parallel plate capacitor}\right]$$

$$U_E = \frac{Q^2}{2C} = \frac{1}{2}CV^2 = \frac{1}{2}QV$$

This is the required expression.

**Energy Density :** The energy density of a capacitor is defined as the energy stored per unit volume. It is denoted by $u$.

**22. Gamma rays and radio waves travel with the same velocity in free space. Distinguish between them in terms of their origin and the main application. [2]**

**Ans.**

| Radio waves | Gamma rays |
|---|---|
| 1. They are originated from rapid acceleration and deceleration of electrons. They are generated artificially by transmeters. | 1. They are originated from the setting process of an excited nucleus of a radionuclide after it undergoes radio-active decay. |
| 2. The wavelengths of radio waves is greater than gamma rays . | 2. The wavelength of gamma rays are smaller than radio waves. |
| 3. Radio waves are used in communication and broadcasting system. | 3. They are used in medicine (radio-therapy), industry (sterilization and disinfection) and the nuclear industry. |

**23. Light from a sodium lamp (S) passes through two polaroid sheets $P_1$ and $P_2$ as shown in fig. What will be the effect on the intensity of the light transmitted (i) by $P_1$ and (ii) by $P_2$ on rotating polaroid $P_1$ about the direction of propagation of light ? Justify your answer in both cases.** **[2]**

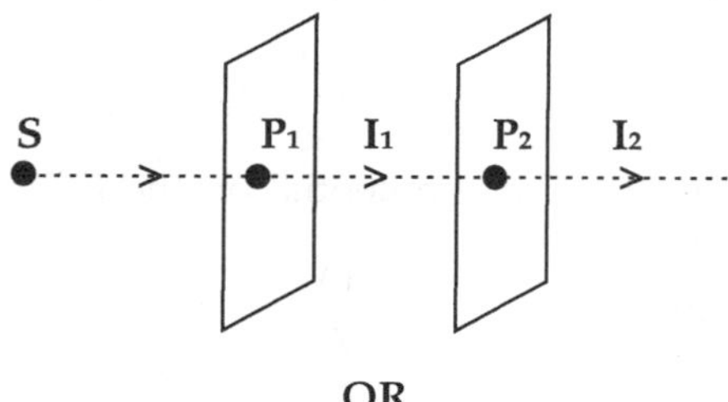


**OR**

**Define the term 'wave front of light'. A plane wave front AB propagating from denser medium (1) into a rarer medium (2) is incident on the surface $P_1P_2$ separating the two media as shown in fig.**

**Using Huygen's principle, draw the secondary wavelets and obtain the refracted wave front in the diagram.**

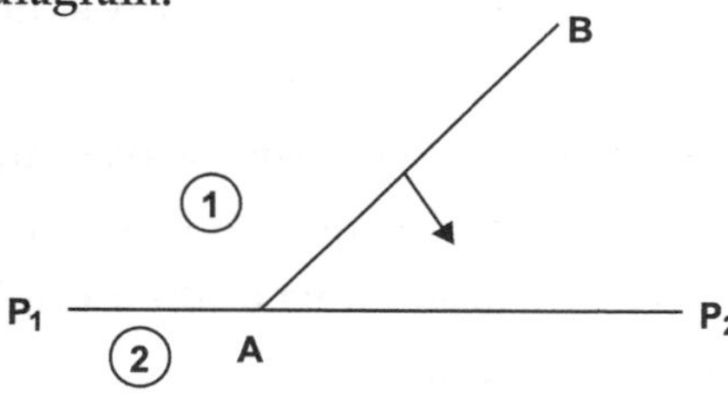


**Ans.** (i) There is no change in $I_1$ on rotation of $P_1$, because the intensity of light does not change by orientation of pass axis of the polaroid.

(ii) $I_1$ = Intensity of polarised light from $P_1$

$I_2$ = Intensity of polarised light from $P_2$

$\theta$ = Angle between pass axis of $P_1$ and $P_2$

By Malus law $I_2 = I_1 \cos^2\theta$

Thus, $I_2$ change when $P_1$ is rotated as $\theta$ changes

$I_2 = 0$ when, $\theta = 90°$

$I_2 = I_1$ when, $\theta = 0°$

**OR**

**Wavefront of light :** A wavefront is the locus of all the points in space that reach a particular distance by a propagating wave in same phase at any instant.

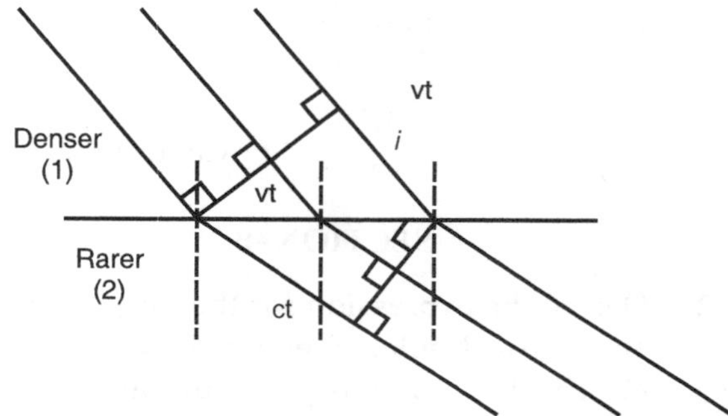


**24. A heavy nucleus P of mass number 240 and binding energy 7.6 MeV per nucleon splits in to two nuclei Q and R of mass numbers 110, 130 and binding energy per nucleon 8.5 MeV and 8.4 MeV, respectively. Calculate the energy released in the fission.** **[2]**

**Ans. Given :** Mass number of P = 240

Binding energy per nucleon = 7.6 MeV

Mass number of Q = 110

Binding energy per nucleon = 8.5 MeV

Mass number of R = 130

Binding energy per nucleon = 8.4 MeV

Total B.E. of P $= 240 \times 7.6$

$= 1824$ MeV

Total B.E. of Q and R $= 110 \times 8.5 + 130 \times 8.4$

$= 935 + 1092 = 2027$ MeV

$\therefore$ Energy released in the fission $= 2027 - 1824$

$= 203$ MeV

**25. Figure shows the stopping potential ($V_0$) for the photo electron versus ($1/\lambda$) graph, for two metals A and B, $\lambda$ being the wavelength of incident light.**

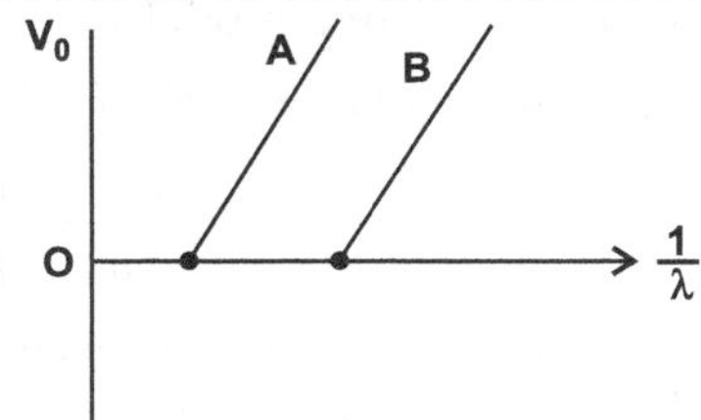


**(a) How is the value of Planck's constant determined from the graph?**

**(b) If the distance between the light source and the surface of metal A is increased, how will the stopping potential for the electrons emitted from it be effected? Justify your answer. [2]**

**Ans.** (a) We know that

$$eV_0 = \frac{hc}{\lambda} - \phi$$

or $$V_0 = \frac{hc}{e}\left(\frac{1}{\lambda}\right) - \frac{\phi}{e} \quad \text{...(i)}$$

where $e$ = Charge on electron

$h$ = Planck's constant,

$\phi$ = work function of metal surface

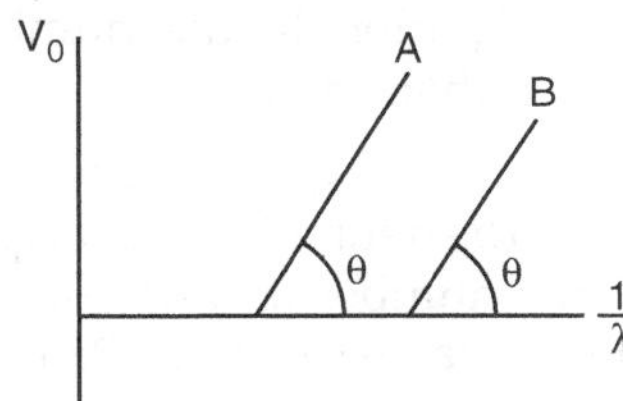


Equation (i) is a straight line as shown in the graph.

Here, slope of line = $\tan\theta = \frac{hc}{e}$

or $$h = \frac{e\tan\theta}{c} \quad \text{...(ii)}$$

Planck's constant can easily be determined by substituting the values of the slope of graph, speed of light and electronic charge in equation (ii)

(b) Stopping potential only depends on frequency of incident light. If the distance is increased, intensity will decrease but no change takes place in stopping potential.

**26. Use Bohr's model of hydrogen atom to obtain the relationship between the angular momentum and the magnetic moment of the revolving electron. [2]**

**Ans.** Circulating electron around a heavy nucleous constitutes a current given by :

$$I = \frac{e}{T} = \frac{e}{\frac{2\pi R}{v}} = \frac{ev}{2\pi R}$$

So, there will be magnetic moment associated with this circulating current is given by :

$$\mu_e = I.A$$

$$= \frac{ev}{2\pi R} \times \pi R^2$$

$$= \frac{evR}{2} \quad \text{...(i)}$$

Angular momentum of circulating electron is given by

$$l = mvR$$

so $$vR = \frac{l}{m} \quad \text{...(ii)}$$

Put the value of equation (ii) in (i), we get

$$\mu_e = \frac{el}{2m}$$

which is the required expression.

**27. In a single slit diffraction experiment, the width of the slit is increased. How will the (i) size and (ii) intensity of central bright band be affected? Justify your answer. [2]**

**Ans.** In a single slit diffraction experiment, if the width of the slit is increased then :

(a) the size of the central bright band reduces to half. As width of central maxima is $\frac{2\lambda}{a}$. If 'a' increases, then width decreases.

(b) Intensity of the central bright band increases and more brighter fringe is obtained on the screen, because more light will be allowed to pass. And the light energy is now squeezed into a smaller area on the screen because the size of the central maxima is reduced. These factors make the intensity increase manyfold.

## SECTION–C

**28. (a) Differentiate between electrical resistance and resistivity of a conductor.**

**(b) Two metallic rods, each of length L, area of cross $A_1$ and $A_2$, having resistivities $\rho_1$ and $\rho_2$ are connected in parallel across a d.c. battery. Obtain the expression for the effective resistivity of this combination. [3]**

**Ans. (a)**

| Electrical Resistance | Resistivity |
|---|---|
| 1. Resistance opposes the flow of free electrons charges in the conductor | 1. Resistivity is the property of the material which defines the resistance of the material having specific dimension. |
| 2. It is measured by the ratio of the p.d.(V) across it to the current (I) flowing through it. $R = \frac{V}{I}$ | 2. It is numerically equal to the resistance of a sample of unit length and unit area of cross-section at certain temperature. |
| 3. Its S.I. unit is Ω | 3. Its. S.I. unit is Ω mv |

**(b)** Here, length of each rod is L.

Area of cross sections are $A_1$ and $A_2$ and resistivities are $\rho_1$ and $\rho_2$ respectively

Resistance of 1st metallic rod, $R_1 = \frac{\rho_1 L}{A_1}$

Resistance of 2nd metallic rod, $R_2 = \frac{\rho_2 L}{A_2}$

∴ Effective resistance when connected in parallel.

$$R_{eq} = \frac{R_1 R_2}{R_1 + R_2}$$

$$\frac{\rho_{eq} L}{A_1 + A_2} = \frac{\rho_1 \rho_2 L^2 A_1 A_2}{A_1 A_2 L(\rho_1 A_2 + \rho_2 A_1)}$$

$$\rho_{eq} = \frac{\rho_1 \rho_2 (A_1 + A_2)}{\rho_1 A_2 + \rho_2 A_1}$$

**29. Calculate the de-Broglie wavelength associated with the electron revolving in the first excited state of hydrogen atom. The ground state energy of the hydrogen atom is – 13.6 eV. [3]**

**Ans.** De Broglie wavelength,

$$\lambda = \frac{h}{mv} = \frac{h}{p}$$

where, $p$ is momentum of electron

Kinetic energy (KE) and momentum ($p$) are related as

$$KE = \frac{p^2}{2m} \quad \text{(where, } m = \text{mass of electron)}$$

$$\Rightarrow \quad p = \sqrt{2m(KE)}$$

$$\Rightarrow \quad \lambda = \frac{h}{\sqrt{2m(KE)}}$$

According to Bohr's model,

Kinetic energy of $e^-$ = | T.E. of $e^-$ |

$$= \left| -\frac{13.6 \times Z^2}{n^2} \right| eV$$

for hydrogen, Z = 1 and first excited state implies $n = 2$.

$$K.E. = \frac{13.6 \times 1^2}{2^2} = 3.4 eV$$

$$= 3.4 \times 1.6 \times 10^{-19} J$$

$$= 5.44 \times 10^{-19} J$$

$$\therefore \quad \lambda = \frac{h}{\sqrt{2m(K.E.)}}$$

$$= \frac{6.63 \times 10^{-34}}{\sqrt{2 \times 9.1 \times 10^{-31} \times 5.44 \times 10^{-19}}}$$

$$= 6.66 \times 10^{-10} m$$

$$= 6.66 \text{Å}$$

**30. (a) Define the term decay constant of a radioactive substance. [3]**

**(b) The half life of $^{238}_{92}U$ undergoing α decay is 4.5 x 10⁹ years. Calculate the activity of 10g sample of $^{238}_{92}U$. [3]**

**Ans.** (a) Decay constant of a radioactive substance is defined as the ratio of its instantaneous rate of disintegration to the number of atoms present at that time.

*i.e.,* $N = N_0 e^{-\lambda t}$

It can also be defined as the reciprocal of time when the number of atoms in a radioactive substance decreases to 0.308 of the number present intially.

(b) We have, half life of uranium $T_{1/2} = 4.5 \times 10^9$ years

$= 4.5 \times 10^9 \times 365 \times 24 \times 60 \times 60$ seconds.

Number of atoms in 1 g of uranium,

$$N = \frac{6.023 \times 10^{23}}{238} \text{ atoms}$$

Number of atoms in 10g of uranium,

$$N_1 = \frac{6.023 \times 10^{23} \times 10}{238} \text{ atoms}$$

Activity of sample = $\lambda N$

$$= \frac{0.693}{T_{1/2}} \times N_1$$

$$= \frac{0.693 \times 6.023 \times 10^{23} \times 10}{4.5 \times 10^9 \times 3.156 \times 10^7 \times 238}$$

$= 1.235 \times 10^5$ disintegration/second.

**31. What is a solar cell? Draw its V-I characteristics. Explain the three processes involved in its working. [3]**

OR

**Draw the circuit diagram of a full wave rectifier. Explain its working showing its input and output waveforms.**

**Ans.** **Solar Cell :** It is a semiconductor device used to convert photons of solar light into electricity.

**V.I. characteristic of solar cell :**

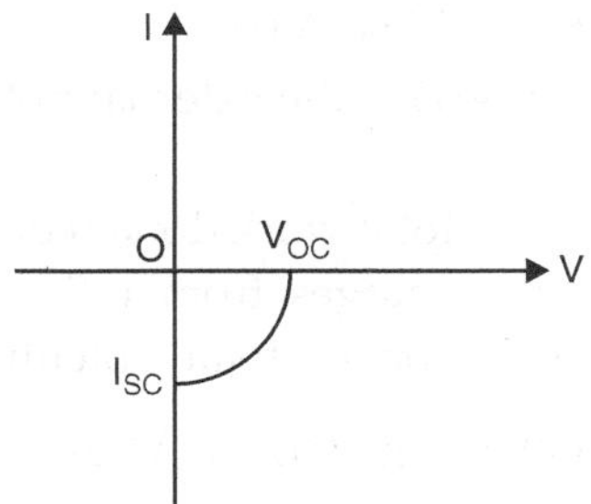


1. $I_{SC}$ is the short circuit current when the load resistance $R_L$ is zero.
2. $V_{OC}$ is the open circuit voltage when $R_L$ is infinity.

The generation of emf by a solar cell, when light falls on it, is due to the following 3 basic process : generation, separation and collection.

(i) Generation of e-h pairs due to light (with $h\upsilon > E_g$) close to the junction.

(ii) Separation of electrons and holes due to electric field of the depletion region. Electrons are swept to *n*-side and holes to *p*-side.

(iii) The electrons reaching the *n*-side are collected by the front contact and holes reaching *p*-side are collected by the back contact. Thus, *p*-side becomes positive and *n*-side becomes negative giving rise to photo voltage.

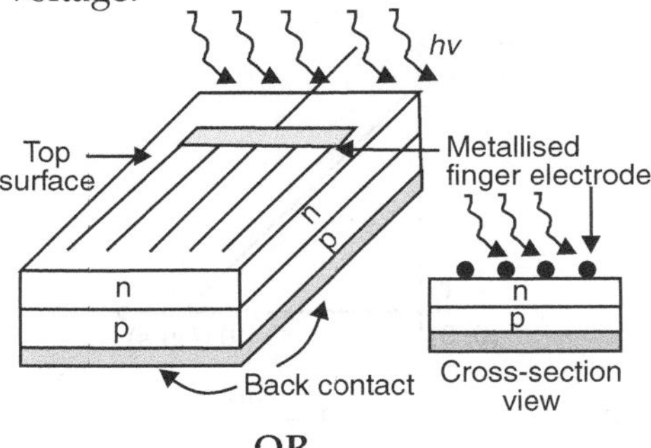


OR

**Full wave rectifier :**

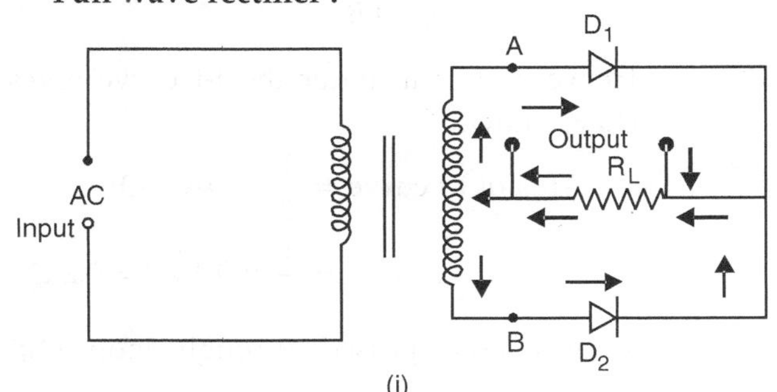


(i)

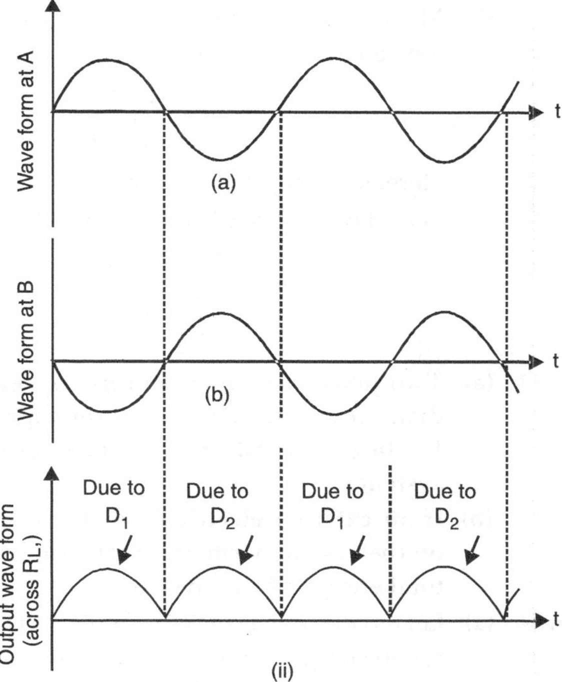


(ii)

**Explanation :** In positive half cycle of AC, end A becomes positive and $D_1$ becomes forward biased and $D_2$ is reverse biased, so $D_1$ conducts and $D_2$ doesn't. So conventional current flows through $D_1$, $R_L$ and upper half of secondary winding . Similarly, during negative half cycle of AC diode $D_2$ becomes forward biased and $D_1$ is reverse biased, current flows through $D_2$, $R_L$ and lower half of secondary winding. Thus, current flows in same direction in both half cycle input AC voltage.

**32. An optical instrument uses a lens of power 100 D for objective lens and 50 D for its eyepiece. When the tube length is kept at 25cm, the final image is formed at infinity. [3]**

**(a) Identify the optical instrument.**

**(b) Calculate the magnification produced by the instrument**

**Ans.** (a) Power of lens $= \dfrac{1}{\text{focal length}}$

So, focal length of objective lens $= \dfrac{1}{100}m$ $= 1\text{cm}$

Focal length of eye piece $= \dfrac{1}{50}m = 2\text{cm}$

Since, the objective has a smaller focal length than the eye piece, the instrument is a compound microscope.

**(b)** Magnification produced when the image is formed at infinity is given by,

$$m = \left(\frac{L}{f_o}\right)\left(\frac{D}{f_e}\right)$$

Here, tube length, L = 25cm

Least Distance of distinct vision, D = 25cm

$$\therefore \quad m = \frac{25\times25}{1\times2} = \frac{625}{2}$$

$$= 312.5$$

**33. (a) Two point charges $q_1$ and $q_2$ are kept at a distance of $r_{12}$ in air. Deduce the expression for the electrostatic potential energy of this system.** **[3]**

**(b) If an external electric field (E) is applied on the system, write the expression for the total energy of this system.**

**Ans. (a)** Let us consider a charge $q_1$ at position vector $r_1$ and charge $q_2$ which is at infinity and to be brought at point $P_2$ having position vector $r_2$ and $dW$ is the small amount of work done in moving a charge through a small distance $dx$.

$+q_1$ ←— $r_{12}$ —→ $q_2$ - - - - - - $q_2$

$P_1(r_1)$ $\quad P_2(r_2)$ $\quad$ (at ∞)

$$dW = F.dx$$

$$= -k\frac{q_1q_2}{r^2}dx$$

$$W = \int_{\infty}^{r_{12}} \frac{kq_1q_2}{r^2}dx$$

$$= -kq_1q_2\int_{\infty}^{r_{12}} \frac{1}{r^2}dx$$

$$= -kq_1q_2\left[\frac{-1}{r}\right]_{\infty}^{r_{12}}$$

$$= kq_1q_2\left[\frac{1}{r_{12}}\right]$$

$$W = \frac{kq_1q_2}{r_{12}}$$

**(b)** Let us consider a system of two charges $q_1$ and $q_2$ from the origin. Let the external electric field applied be E.

Let the work done in bringing the charge $q_1$ from infinity to $r_1$ is given as $qV(r_1)$ and the work done in bringing charge $q_2$ from infinity to $r_2$ be $q_2V(r_2)$. But the work required to be done on $q_2$ will include the field of $q_1$ along the electric field E is given as, $\frac{kq_1q_2}{r_{12}}$.

The potential energy of the charge $q$ at a distance $r$ from the origin in an external electric field is $qV(r)$.

Where, $V(r)$ is the external potential at that point.

Thus, the total work done required to bring both the charges from infinity to present configuration or total potential energy of system is $= q_1V(r_1) + q_2V(r_2) + \frac{kq_1q_2}{r_{12}}$

**34. When a conducting loop of resistance 10Ω and area 10 cm$^2$ is removed from an external magnetic field acting normally, the variation of induced current in the loop with time is shown in the figure.**

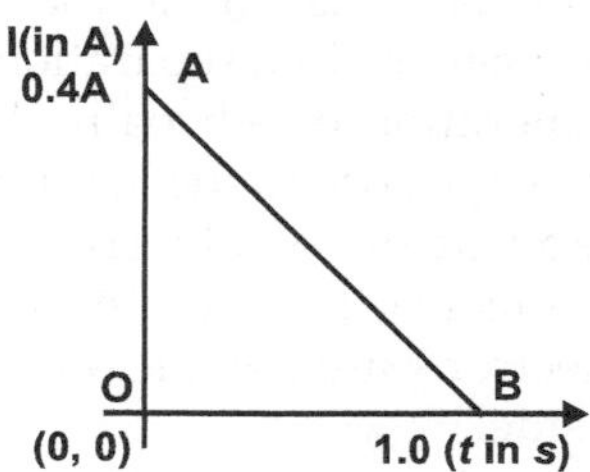


**Find the :**

**(i) total charge passed through the loop.**

**(ii) change in magnetic flux through the loop.**

**(iii) magnitude of the magnetic field applied.** **[3]**

**Ans.**

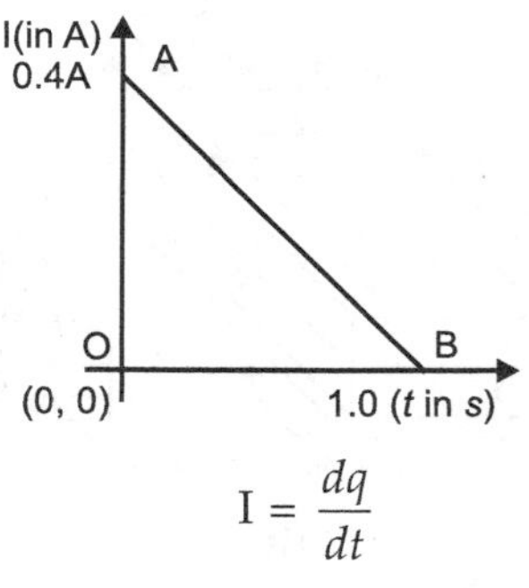


(i)
$$I = \frac{dq}{dt}$$

$$dq = Idt$$

Hence, the area under the I-t curve gives charge flow.

$$\text{Area of I-}t\text{ curve} = \frac{1}{2} \times OA \times OB$$

$$= \frac{1}{2} \times 0.4 \times 1 = 0.2\ C$$

Total charge passed through loop $(\Delta q)$ = 0.2 C

(ii) Now, we know

$$\Delta q = \frac{\Delta\phi}{R}$$

$$\therefore \quad \Delta\phi = \Delta q \times R$$

$$= 0.2 \times 10 = 2\ Wb$$

(iii) Now, change in flux, $\Delta\phi = B\,(\Delta A)$

$$2 = B\left(\frac{10}{100\times100}m\right)$$

$$\therefore \quad B = \frac{2\times100\times100}{10}$$

$$B = 2000T$$

Hence, the magnitude of field applied = 2000T

## SECTION–D

**35. (a) Define the term 'focal length of a mirror'. With the help of a ray diagram, obtain the relation between its focal length and radius of curvature.**

**(b) Calculate the angle of emergence (e) of the ray of light incident normally on the face AC of a glass prism ABC of refractive index $\sqrt{3}$. How will the angle of emergence change qualitatively, if the ray of light emerges from the prism into a liquid of refractive index 1.3 instead of air? [5]**

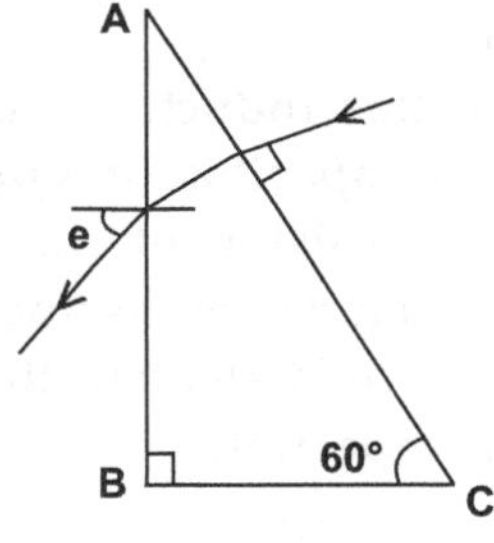


OR

**(a) Define the term 'resolving power of a telescope'. How will the resolving power be effected with the increase in :**

**(i) Wavelength of light used.**

**(ii) Diameter of the objective lens.**

**Justify your answers.**

**(b) A screen is placed 80 cm from an object. The image of the object on the screen is formed by a convex lens placed between them at two different locations separated by a distance 20 cm. Determine the focal length of the lens.**

**Ans. (a) Focal length of a mirror :** It is defined as the distance between the centre of a lens or curved mirror and its focus.

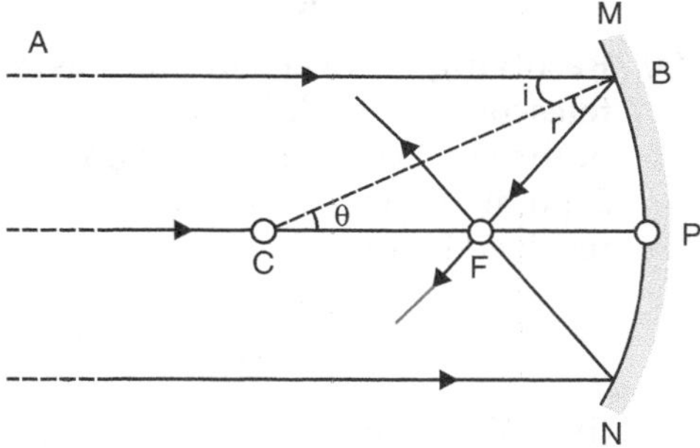


Consider a ray of light AB, parallel to the principal axis and incident on a spherical mirror at point B. The normal to the surface at point B is CB and CP = CB = R, radius of curvature. The ray AB after reflection from mirror passes through the focus F (concave mirror) or will appear to diverge from F (convex mirror) and obeys the law of reflection, $i = r$

Now, $\angle BCP = \theta = i$

In $\Delta CBF$, $\theta = r$

$\therefore \quad BF = FC$

If the aperture of mirror is small, B lies close to P and therefore BF = PF.

or $\quad FC = FP = PF$

or $\quad PC = PF + FC = PF + PF$

or $\quad R = 2PF = 2f$

or $\quad f = \frac{R}{2}$

This is the required expression.

**(b)** By Snell's law, $\mu_1 \sin i = \mu_2 \sin r$

$$\mu_1 = \mu_{prism} = \sqrt{3}, \mu_2 = 1$$

$$\therefore \quad \mu_{prism} \sin 30° = \sin e$$

$$\sqrt{3}\times\frac{1}{2} = \sin e$$

$$\Rightarrow \quad e = 60°$$

Now, when external medium is changed to liquid of $\mu_2 = 1.3$, then

$$\mu_{prism} \sin 30° = \mu_2 \sin e$$

$$\sqrt{3}\times\frac{1}{2} = 1.3 \sin e$$

$$\therefore \quad e = \sin^{-1}\left(\frac{\sqrt{3}}{2\times1.3}\right)$$

$$= 41.77°$$

Hence, the angle of emergence reduces to 41.77° from 60°.

OR

**(a) Resolving power of an astronomical telescope:** It is defined as the reciprocal of the smallest angular separation between 2 point objects whose images can be just resolved by the telescope :

$$\text{R.P.} = \frac{1.22\lambda}{D}$$

(i) With the increase in wavelength of light, the resolving power increases. (directly proportional by formula)

(ii) With the increase of diameter of objective lens the resolving power decreases. (inversely proportional by formula)

**(b) Case 1**

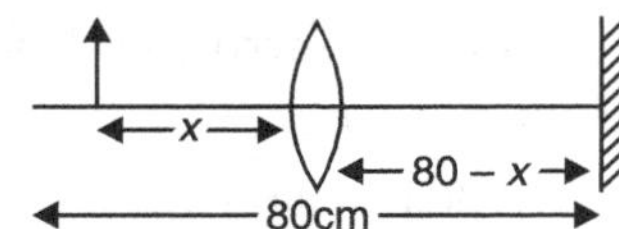


Let

Object distance, $u = x$
image distance, $v = 80 - x$
focal length $= f$

According to lens formula,

$$\frac{1}{f} = \frac{1}{v} + \frac{1}{u}$$

$$\frac{1}{f} = \frac{1}{80-x} + \frac{1}{x} \quad ...(i)$$

Similarly,

**Case 2**

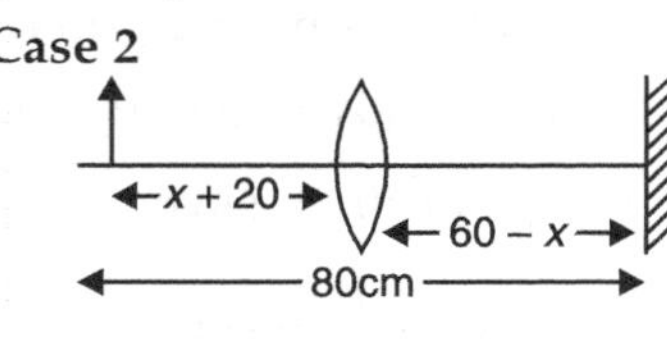


Now, $u = x + 20$

$v = 60 - x$

$$\therefore \quad \frac{1}{f} = \frac{1}{60-x} + \frac{1}{20+x} \quad ...(ii)$$

on comparing (i) and (ii) we get

$$\frac{1}{80-x} + \frac{1}{x} = \frac{1}{60-x} + \frac{1}{20+x}$$

$$\Rightarrow \quad \frac{80}{(80-x)x} = \frac{80}{(60-x)(20+x)}$$

$$80x - x^2 = 1200 + 40x - x^2$$

$$\Rightarrow \quad 40x = 1200$$

$$\Rightarrow \quad x = 30 \text{ cm}$$

Put the value of $x$ in equation (i)

$$\therefore \quad \frac{1}{f} = \frac{1}{80-30} + \frac{1}{30}$$

$$\frac{1}{f} = \frac{1}{50} + \frac{1}{30}$$

$$\Rightarrow \quad f = 18.75 \text{ cm}$$

Hence, the focal length of lens is 18.75 cm.

**36. (a) Show that an ideal inductor does not dissipate power in an a.c. circuit.**

**(b) The variation of inductive reactance ($X_L$) of an inductor with the frequency ($f$) of the ac source of 100 V and variable frequency is shown in the figure given.**

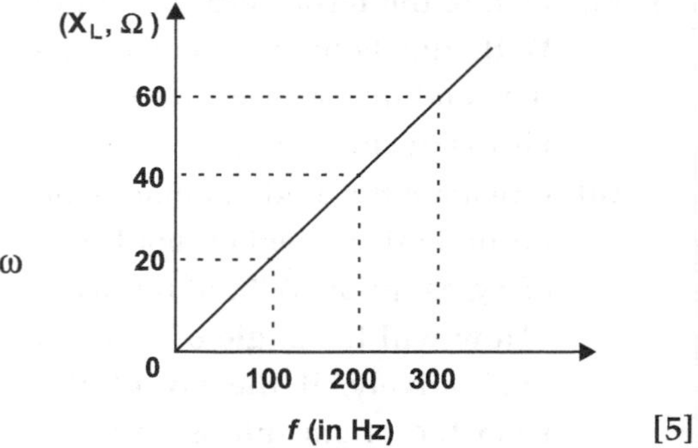


[5]

**(i) Calculate the self-inductance of the inductor.**

**(ii) When this inductor is used in series with a capacitor of unknown value and a resistor of 10 Ω at 300$s^{-1}$, maximum power dissipation occurs in the circuit. Calculate the capacitance of the capacitor.**

OR

**(a) A conductor of length '$l$' is rotated about one of its ends at a constant angular speed '$\omega$' in a plane perpendicular to a uniform magnetic field B. Plot graphs to show variations of the emf induced across the ends of the conductor with (i) angular speed $\omega$ and (ii) length of the conductor $l$.**

**(b) Two concentric circular loops of radius 1 cm and 20 cm are placed coaxially.**

**(i) Find mutual inductance of the arrangement.**

**(ii) If the current passed through the outer loop is changed at a rate of 5 A/ms, find the emf induced in the inner loop. Assume the magnetic field on the inner loop to be uniform.**

**Ans. (a)** Power in a.c., circuit,

$$P = VI \cos \phi$$

For pure inductive circuit, the phase difference between current and voltage is $\pi/2$

$$\therefore \quad \phi = \pi/2$$

$$\cos \phi = \cos \pi/2 = 0$$

Therefore, zero power is dissipated. This current is sometimes referred to as watt-less current.

**(b)** (i) Since, $X_L = \omega L$

and $\omega = 2\pi f$, $f$ is the frequency in Hz.

$$\text{So, } L = \frac{X_L}{\omega} = \frac{X_L}{2\pi f}$$

$$= \frac{20}{2\pi \times 100} = \frac{40}{2\pi \times 200} = \frac{60}{2\pi \times 300}$$

$$= 31.84 \times 10^{-3} \approx 32 \text{ mH.}$$

(ii) Since, power dissipation is maximum when

$$X_L = X_C$$

$$\text{or} \quad \omega L = \frac{1}{\omega C}$$

$$\text{or} \quad C = \frac{1}{\omega^2 L}$$

$$= \frac{1}{4\pi^2 f^2 L}$$

$$= \frac{1}{4 \times 3.14 \times 3.14 \times 300 \times 300 \times 32 \times 10^{-3}}$$

$$= 8.8 \mu F$$

OR

**(a)** (i)

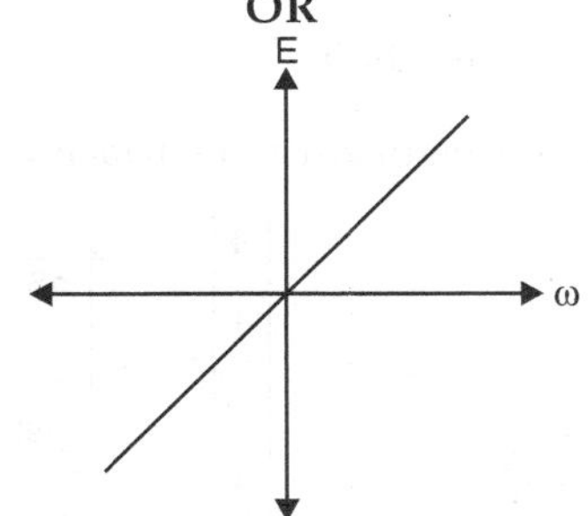


$$\text{Induced Emf} = E = \frac{B\omega L^2}{2}$$

(ii)

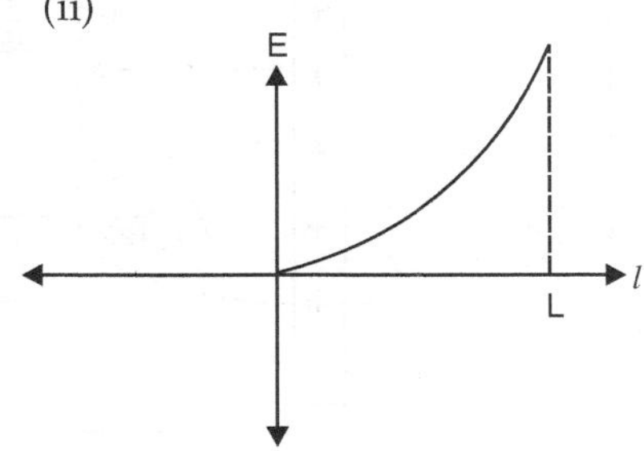


**(b)** (i) Since, magnetic flux, $\phi = MI$

And magnetic field at the centre of bigger loop, $\vec{B}$.

$$\frac{\mu_0 I}{2R} = \frac{4\pi \times 10^{-7}}{2 \times 20 \times 10^{-2}}$$

$$= \pi \times 10^{-6} I$$

Flux through the smaller loop

$$\phi = BA_s$$

$$= \frac{4\pi \times 10^{-5} I}{40} \times \pi(0.01)^2$$

$$= \pi^2 \times 10^{-10} \times I$$

$$\therefore \quad M = \frac{\phi}{I}$$

$$= \frac{\pi^2 \times 10^{-10} \times I}{I}$$

$$= 9.86 \times 10^{-10} \text{ H}$$

(ii) Emf included

$$= \frac{-d\phi}{dt}$$

$$= -9.86 \times 10^{-10} \times \frac{dI}{dt}$$

$$= -9.86 \times 10^{-10} \times 5$$

$$= -4.93 \times 10^{-9} \text{ V}$$

**37. (a) Write two important characteristics of equipotential surfaces.**

**(b) A thin circular ring of radius $r$ is charged uniformly so that its linear charge density becomes $\lambda$. Derive an expression for the electric field at a point P at a distance $x$ from it along the axis of the ring. Hence, prove that at large distances ($x >> r$), the ring behaves as a point charge. [5]**

**OR**

**(a) State Gauss's law on electrostatics and derive an expression for the electric field due to a long straight thin uniformly charged wire (linear charge density $\lambda$) at a point lying at a distance $r$ from the wire.**

**(b) The magnitude of electric field (in $NC^{-1}$) in a region varies with the distance $r$ (in $m$) as**

$$E = 10r + 5$$

**By how much does the electric potential increase in moving from point at $r$ = 1m to a point at $r$ = 10m.**

**Ans. (a)** Characteristics of equipotential surfaces are :

(i) Potential remains same at all the points on the equipotential surface.

(ii) No work is required to move a charge on an equipotential surface.

**(b)** Let us consider a thin circular ring of radius $r$ with charge density as $\lambda$.

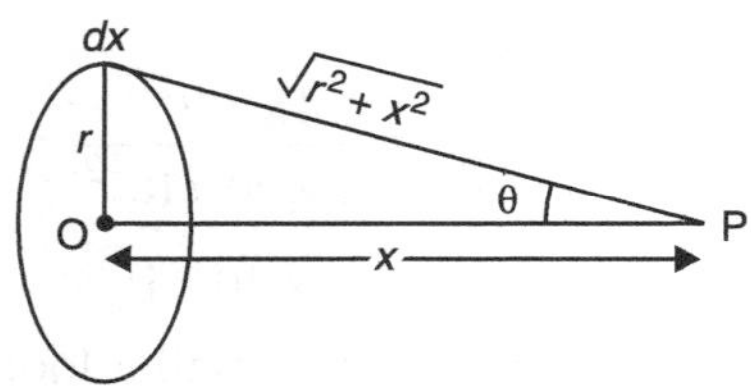


Let us consider a small charge element $(dx)$ on the ring having small charge $(dq)$

$$dq = \lambda dx$$

The electric field due to this charge element at point P is given as :

$$dE = \frac{1}{4\pi\varepsilon_0}\frac{dq}{(r^2+x^2)}$$

$$dE = \frac{1}{4\pi\varepsilon_0}\frac{\lambda dx}{(r^2+x^2)}$$

Electric field at point P will have 2 components $d\vec{E}\cos\theta$ in horizontal direction and $d\vec{E}\sin\theta$ in vertical direction. The vertical component will cancel out the effect of each other due to the presence of the diametrically opposite element and horizontal component will survive at point P.

$$\therefore \quad \cos\theta = \frac{x}{\sqrt{r^2+x^2}}$$

Now, $$dE\cos\theta = \frac{\lambda x dx}{4\pi\varepsilon_0(r^2+x^2)^{3/2}}$$

$$\therefore \quad dq = \lambda dx$$

$$\therefore \quad dE\cos\theta = \frac{xdq}{4\pi\varepsilon_0(r^2+x^2)^{3/2}}$$

On integrating, we get

$$E_x = \int dE\cos\theta$$

$$= \int\frac{xdq}{4\pi\varepsilon_0(r^2+x^2)^{3/2}}$$

$$= \frac{x.Q}{4\pi\varepsilon_0(r^2+x^2)^{3/2}}$$

Where, Q is the total charge on ring

When $x >> r$, then $r^2 + x^2 \approx x^2$

$$\therefore \quad E_x = \frac{xQ}{4\pi\varepsilon_0 x^3} = \frac{Q}{4\pi\varepsilon_0 x^2}$$

So, at large distance, the ring behaves like a point charge.

**OR**

**(a) Electric Field $(\vec{E})$ due to a long straight uniformly charged wire of charge density $\lambda$ :** Consider a cylindrical Gaussian surface of radius $r$ and length $l$ coaxial with line charge. The cylindrical Gaussian surface may be divided into three parts : (1) curved surface $S_1$ (2) flat surface $S_2$ and (3) flat surface $S_3$.

By symmetry, the electric field has the same magnitude E at each point of curved surface $S_1$ and is directed radially outward. We consider small elements of surfaces $S_1$, $S_2$ and $S_3$. The surface element vector $d\vec{S_1}$ is directed along the direction of electric field (*i.e.*, angle between $\vec{E}$ and $d\vec{S_1}$ is 0°); the elements $d\vec{S_2}$ and $d\vec{S_3}$ are directed perpendicular to field vector $\vec{E}$ (*i.e.*, angle between $d\vec{S_2}$ and $\vec{E}$ and $d\vec{S_3}$ and $\vec{E}$ is 90°).

Electric flux through the cylindrical surface.

$$\oint_S \vec{E}\, d\vec{S} = \oint_{S_1}\vec{E}\, d\vec{S_1} + \oint_{S_2}\vec{E}\, d\vec{S_2} + \oint_{S_3}\vec{E}\, d\vec{S_3}$$

$$= \oint_{S_1} E\,dS_1\cos 0° + \oint_{S_2} E\,dS_2\cos 90° + \oint_{S_3} E\,dS_3\cos 90°$$

$$= E\int dS_1 + 0 + 0$$

$= E \times$ area of curved surface $= E \times 2\pi rl$

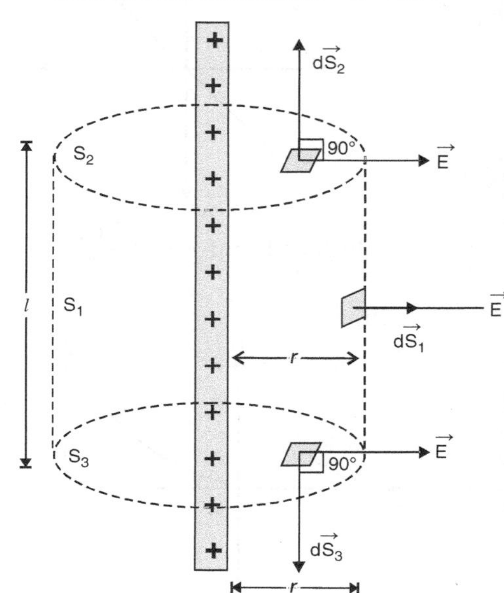

Charge enclosed,

$$q = \lambda l$$

By Gauss' theorem

$$\phi_E = \frac{q}{\varepsilon_0}$$

$$E.2\pi rl = \frac{\lambda l}{\varepsilon_0}$$

$$E = \frac{\lambda}{2\pi\varepsilon_0 r}$$

**(b)** Let point A be at $r = 1$ m and point B be given at $r = 10$m

$V_A$ = potential at A

$V_B$ = potential at B

$$\therefore \quad V_B - V_A = \int_A^B \vec{E}.d\vec{r}$$

$$= -\int_1^{10} (10r+5)dr$$

$$= -\left[\frac{10r^2}{2}+5r\right]_1^{10}$$

$$= -\left[\left(\frac{10(10)^2}{2}+5(10)\right)-\left(\frac{10(1)^2}{2}+5\times1\right)\right]$$

$$= -[550-10]$$

$$= -540 \text{ V}$$

## Delhi [Set-II]

**Note : Except for the following questions, all the remaining questions have been asked in previous set.**

### SECTION–A

**Note : Select the most appropriate option from those given below each question.**

**6. A biconvex lens of focal length $f$ is cut into two idendical plano-convex lenses. The focal length of each part will be :** **[1]**

(a) $f$ (b) $f/2$

(c) $2f$ (d) $4f$

**Ans.** (c) $2f$

**Explanation :** Let the focal length of biconvex lens be $f$ and plano-convex lens be $f'$.

Then by lens maker's formula

For biconvex lens

$$\frac{1}{f} = (\mu-1)\left(\frac{1}{R}+\frac{1}{R}\right) = \frac{2(\mu-1)}{R}$$

Where, R is the radius of curvature of biconvex lens.

For plano-convex lens

$$\frac{1}{f'} = (\mu-1)\left(\frac{1}{R}\right) = \frac{\mu-1}{R}$$

Thus $$\frac{1}{f} = \frac{2}{f'}$$

or $$f' = 2f$$

**7. The phase difference between the current and the voltage in series LCR circuit at resonance is :** **[1]**

(a) $\pi$ (b) $\pi/2$

(c) $\pi/3$ (d) zero

**Ans.** (d) zero

**Explanation :** At resonance, the circuit is purely resistive and there is no phase difference between current and voltage.

**8. Photons of frequency $v$ are incident on the surfaces of metals A and B of threshold frequencies 3/4 $v$ and 2/3 $v$, respectively. The ratio of the maximum kinetic energy of electrons emitted from A to that from B is :** **[1]**

(a) 2 : 3 (b) 4 : 3

(c) 3 : 4 (d) 3 : 2

**Ans.** (c) 3 : 4

**Explanation :** According to Einstein photoelectric equation

$$hv = hv_0 + K_{max}$$

For first metal

$$hv = h\left(\frac{3}{4}v\right)+K_1$$

$$\Rightarrow \quad K_1 = \frac{hv}{4}$$

For second metal

$$hv = h\left(\frac{2}{3}v\right)+K_2$$

$$\Rightarrow \quad K_2 = \frac{hv}{3}$$

Thus, the ratio is

$$\frac{K_1}{K_2} = \frac{hv/4}{hv/3} = \frac{3}{4}$$

## SECTION–B

**21. (a) Define one Becquerel.**

**(b) A radioactive substance disintegrates into two types of daughter nuclei, one type with disintegration constant $\lambda_1$ and the other type with disintegration constant $\lambda_2$. Determine the half-life of the radioactive substance.**

**Ans.** (a) One Becquerel (B$q$) is defined as the activity of a quantity of radioactive sample in which one nucleus decays per second.

(b) $X \xrightarrow{\lambda_1} D_1$

$X \xrightarrow{\lambda_2} D_2$

(Where, X is a radioactive nuclei and $D_1$, $D_2$ are the daughter nuclei)

$$\lambda_{effective} = \lambda_1 + \lambda_2$$

$$t_{1/2} = \frac{\log 2}{\lambda_{effective}} = \frac{0.693}{\lambda_1 + \lambda_2}$$

**23. In case of photo electric effect experiment, explain the following facts giving reasons.**

**(a) The wave theory of light could not explain the existence of the threshold frequency.**

**(b) The photo electric current increases with increase of intensity of incident light.**

**Ans.** (a) If light would only be a wave its energy would be represented by the amplitude of the light wave. A more intense light source, even if it will be of lower frequency, would be able to eject out photoelectrons from the metal surface and photoelectric current would be generated.

But actually a light below a certain threshold frequency could not generate any current, no matter how intense the light is.

(b) With the increased intensity of light the more photons will be incident on metal surface. If the frequency of incident light is more than the threshold frequency, then the more number of photons means more photoelectrons will be emitted. This result in more photoelectric current.

## SECTION–C

**30. (a) Define internal resistance of a cell.**

**(b) A cell of emf E and internal resistance $r$ is connected across a variable resistor R. Plot the shape of graphs showing variation of terminal voltage V with (i) R and (ii) circuit current I.** [3]

**Ans.** (a) Internal resistance is the resistance which is present within the battery that resists the current flow when connected to a circuit.

(b) (i) Since, $$V = \frac{ER}{R+r}$$

$\Rightarrow$ $$V = \frac{E}{1+\frac{r}{R}}$$

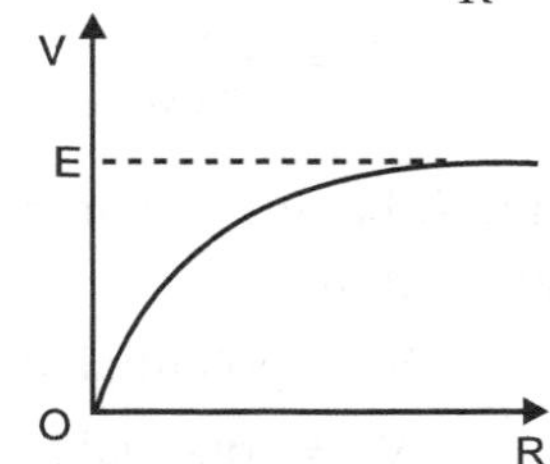


(ii) Since $V = E - IR$

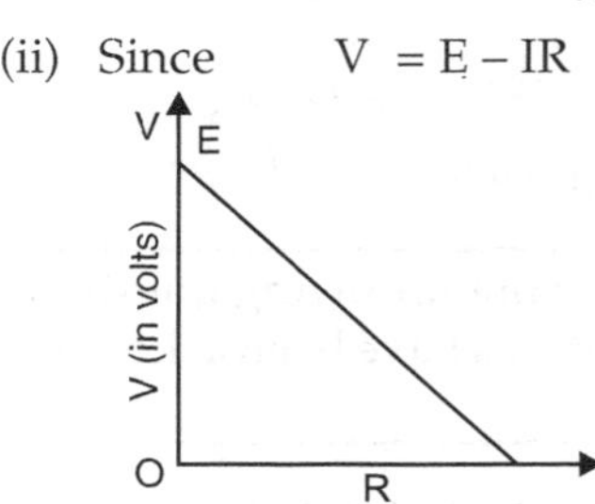


**32. When a conducting loop of resistance 10 Ω and area 10 cm² is removed from an external magnetic field acting normally, the variation of induced current I in the loop with times is as shown in the figure.**

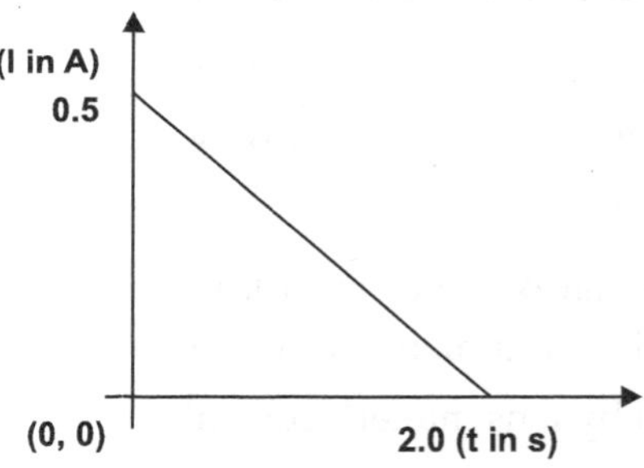


**Find the :**

**(i) total charge passed through the loop.**

**(ii) change in magnetic flux through the loop.**

**(iii) magnitude of the magnetic field applied.** [3]

**Ans.** (a) Since, $$I = \frac{dq}{dt}$$

$\Rightarrow$ $$dq = Idt$$

Area of I – $t$ curve = Total charge flows

$\therefore$ $$Q = \frac{1}{2} \times 2 \times 0.5 = 0.5C.$$

Hence, total charge that flow through loop is 0.5C

(b) Since, $\Delta Q = \frac{\Delta\phi}{k}$

$\Rightarrow \Delta\phi = \Delta Q \times k$

$= \frac{1}{2} \times 10\Omega$

$= 5Wb$

Hence, change in magnetic flux is 5Wb.

(c) Now, $\Delta\phi = B(\Delta A)$

$B = \frac{\Delta\phi}{\Delta A}$

$= \frac{5}{\frac{10}{100 \times 100}} = 5000 \text{ T}$

Hence, the magnetic field applied is 5000 T

**33. Draw the curve showing the variations of binding energy per nucleon with the mass number of nuclei. Using it explain the fusion of nuclei lying on ascending part and fission of nuclei lying on descending part of this curve.**

**Ans.**

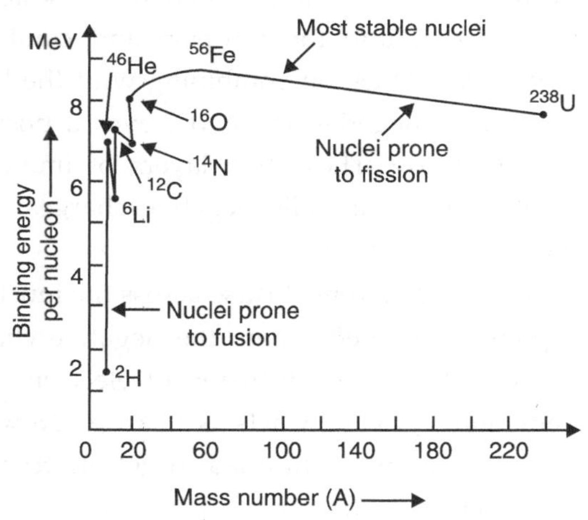


The graph above tells us that binding energy per nucleon is smaller for heavier nuclei as well as for lighter nuclei than for middle order nuclei (with mass no. between 30 to 170). Heavier nuclei are unstable and they undergo fission and lighter nuclei which are unstable undergo fusion in order to form a stable nuclei lying in the range of mass number 30 to 170.

**34. An optical instrument uses a lens of power 100D for objective lens and 50 D for its eyepiece. When the tube length is kept at 20cm, the final image is formed at infinity. [3]**

**(a) Identify the optical instrument.**

**(b) Calculate the magnification produced by the instrument.**

**Ans.** (a) Power of lens $= \frac{1}{\text{Focal length}}$

So, focal length of objective lens

$= \frac{1}{100} m = 1\text{cm}$

Focal length of eye piece $= \frac{1}{50} m = 2\text{cm}$

Since, the objective has a smaller focal length than the eye piece, the instrument is a compound microscope.

(b) Magnification produced is formed at infinity is given by,

$m = \left(\frac{L}{f_o}\right)\left(\frac{D}{fe}\right)$

where, tube length L = 20cm and least distance of distinct vision, D = 25cm

$m = \frac{20}{1} \times \frac{25}{2} = 250$

●●

# Delhi [Set-III]

**Note : Except for the following questions, all the remaining questions have been asked in previous sets.**

## SECTION–B

**21 Write shortcomings of Rutherford atomic model. Explain how these were overcome by the postulates of Bohr's atomic model. [2]**

**Ans.** According to Rutherford's model, electrons revolve around the nucleus in a circular path. But particles that are in motion on a circular path would undergo acceleration and it causes radiation of energy by charged particles. Eventually, electrons will loose energy and fall on to the nucleus. Thus, making nucleus unstable. Rutherford's model cannot explain the stability of the atom.

Bohr modified this model by proposing that the electrons move in orbits of fixed size and energy. The energy of an electrons depends on the size of the orbit and is lower for smaller orbits. Radiation occur only when electron jump from one orbit to another.

**24. Calculate for how many years will the fusion of 2.0 kg deuterium keep 800 W electric lamp glowing. Take the fusion reaction as :**

$$^2_1H + ^2_1H \longrightarrow ^3_1He + ^1_0n + 3.27 \text{ MeV}$$

**Ans.** 2 deuterium nuclei fuse to produce 3.27 MeV

$= 3.27 \times 1.6 \times 10^{-13}$ J $= 5.238 \times 10^{-13}$ J

Energy produced per nuclei

$$= \frac{5.238 \times 10^{-13}}{2} = 2.61 \times 10^{-13} \text{ J}$$

2 kg deuterium atoms will have

$$= \frac{6.023 \times 10^{23}}{2} \times 2000$$

$= 6.023 \times 10^{26}$ deuterium atom

$6.022 \times 10^{26}$ deuterium atoms will produce

$= 2.61 \times 10^{-13} \times 6.023 \times 10^{26}$ J

$= 15.77 \times 10^{13}$ J

$$\because \quad \text{Power} = \frac{E}{t}$$

$$800 = \frac{15.77 \times 10^{13}}{t}$$

$$\Rightarrow \quad t = \frac{15.77 \times 10^{13}}{800}$$

$= 1.971 \times 10^{11}$ seconds

$$t = \frac{1.971 \times 10^{11}}{365 \times 24 \times 60 \times 60}$$

$= 6.24 \times 10^{3}$ years.

Hence, electric lamp will glow for $6.24 \times 10^{3}$ years.

## SECTION–C

**28. (a) Define the term 'half-life' of a radioactive substance.**

**(b) The half life of $^{238}_{92}$U undergoing $\alpha$ decay is $4.5 \times 10^9$ years. Calculate the activity of 5g sample of $^{238}_{92}$U. [3]**

**Ans. (a)** Half life of a radioactive element is a time in which the given amount of the substance reduces to half of the original value as a consequence of decay.

**(b)** We have

$$t_{1/2} = 4.5 \times 10^9 \text{ years} = \frac{0.693}{\lambda}$$

$$\lambda = \frac{0.693}{4.5 \times 10^9 \times 365 \times 24 \times 60 \times 60}$$

Number of nuclei present in sample,

$$N_0 = \frac{5 \times 6.022 \times 10^{23}}{238}$$

$$\text{Activity} = A = \frac{dN}{dt} = \lambda N_0$$

$$= \left[\frac{0.693}{4.5 \times 10^9 \times 365 \times 24 \times 3600}\right] \times \frac{5 \times 6.022 \times 10^{23}}{238}$$

$\cong 0.61 \times 10^5$ disintegration/sec.

**29. Explain the formation of potential barrier and depletion region in a *p-n* junction diode. What is the effect of applying forward bias on the width of depletion region?**

**OR**

**What is photo diode? Briefly explain its working and draw its V-I characteristics.**

**Ans.** Two important processes involved during the formation of *p-n* junction are diffusion and hole.

**Formation of depletion region and potential barrier :** At the instant of *p-n* junction formation, the free electrons near the junction diffuse across the junction into the *p*-region and combines with holes. Thus, on combining with the hole, it makes a negative ion and leaves a positive ion on *n*-side. These two layers of immobile positive and immobile negative charges form the depletion charge.

Further, as electrons diffuse across the junction, a point is reached where the negative charge repels only further diffusion of electron. This depletion region now acts as a barrier. Now the external energy is supplied to get electrons to move across the barrier of electric field.

The potential difference required to move the electron through the electric field is called barrier potential.

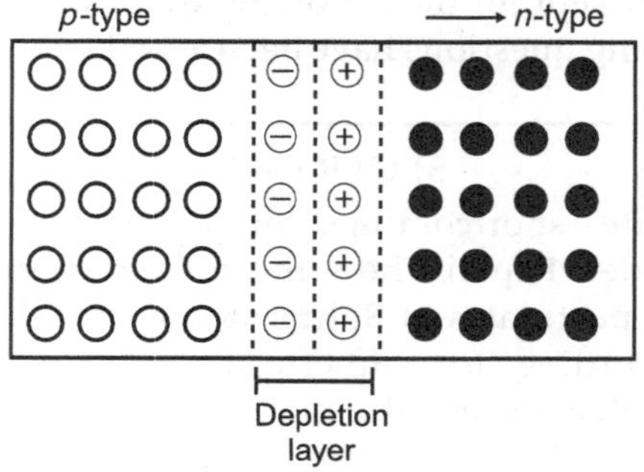


The effect of forward bias on *p-n* region is that the depletion layer will be reduced.

OR

**Photodiode** : It is an optoelectronic device in which current carriers are generated by photons through photo-excitation *i.e.*, photo conduction by light.

A photodiode is a special *p-n* junction diode made of photosensitive semiconducting material. It is operated under reverse bias below the breakdown voltage. The conductivity of *p-n* junction photodiode is modulated by the absorption of incident light in or near the depletion layer which exists at the *p-n* junction. The conductivity of *p-n* junction photodiode increases with the intensity of light falling on it.

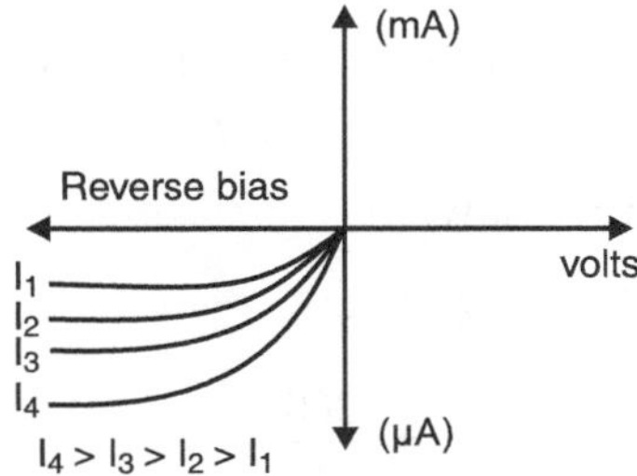


When the photodiode is reverse biased with a voltage less than its breakdown voltage and no light is incident on its junction, the reverss current is extremely small. This current is called dark current.

**30. Calculate the de-Broglie wavelength associated with the electron in the first 2nd excited state of hydrogen atom. The ground state energy of the hydrogen atom is 13.6 eV.** **[3]**

**Ans.** De broglie wavelength is, $\lambda = \frac{h}{mv} = \frac{h}{p}$

where, $p$ is the momentum of an electron K.E. and momentum are related by,

$$\text{K.E.} = \frac{p^2}{2m}$$

$$\Rightarrow \quad p = \sqrt{2m\text{K.E.}}$$

$$\Rightarrow \quad \lambda = \frac{h}{\sqrt{2m\text{K.E.}}}$$

Then, K.E of $e^-$ = |T.E of $e^-$|

$$= \left| -\frac{13.6\times Z^2}{n^2} \right| eV$$

For Hydrogen (Z = 1) and second excited state implies $n = 3$

$$\text{K.E.} = \frac{13.6\times 1^2}{3^2}$$

$$= 1.51 \text{ eV}$$

$$= 1.51 \times 1.6 \times 10^{-19} \text{ J}$$

$$= 2.41 \times 10^{-19}\text{J}$$

$$\therefore \quad \lambda = \frac{h}{\sqrt{2m(\text{KE})}}$$

$$= \frac{6.63\times 10^{-34}}{\sqrt{2\times 9.1\times 10^{-31}\times 2.41\times 10^{-19}}}$$

$$= 1.001 \times 10^{-9} \text{ m}$$

$$= 10\text{Å}$$

**32. (a) Two point charges $+Q_1$ and $-Q_2$ are placed $r$ distance apart. Obtain the expression for the amount of work done to place a third charge $Q_3$ at the midpoint of the line joining two charges.**

**(b) At what distance from charge $+Q_1$ on the line joining two charges (in terms of $Q_1$, $Q_2$ and $r$) will this work done be zero.** **[3]**

**Ans.** According to the question,

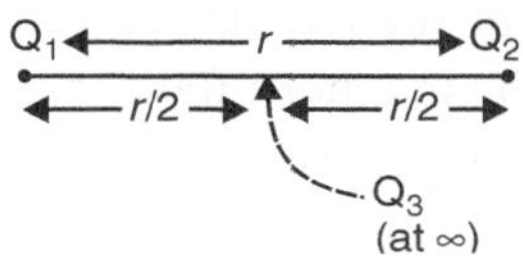


(a) Work done :

$$W = \frac{kQ_1Q_3}{r/2} - \frac{kQ_3Q_2}{r/2}$$

$$W = \frac{2rQ_1Q_3}{r} - \frac{2rQ_3Q_2}{r}$$

$$= \frac{2rQ_3(Q_1-Q_2)}{r}$$

$$= \frac{2Q_3(Q_1-Q_2)}{4\pi\varepsilon_0 r}$$

**(b) For work done to be zero :** Work done will be zero for the point where the electric potential is zero consider a point P to be lying in between the two charges, where electric potential is zero.

Let the distance of the point P from the charge $Q_1$ be $x$.

$$V = \frac{kQ_1}{x} - \frac{kQ_2}{(r-x)} = 0$$

$$\Rightarrow \quad \frac{Q_1}{x} - \frac{Q_2}{r-x} = 0$$

$$\Rightarrow \quad rQ_1 - xQ_1 - xQ_2 = 0$$

$$\Rightarrow \quad x = \frac{rQ_1}{(Q_1+Q_2)}$$

# PHYSICS 2019

## Outside Delhi [Set-I]

**Time allowed : 3 hours** **Maximum marks : 70**

### SECTION-A

**1. Draw equipotential surfaces for an electric dipole. [1]**

**Ans.**

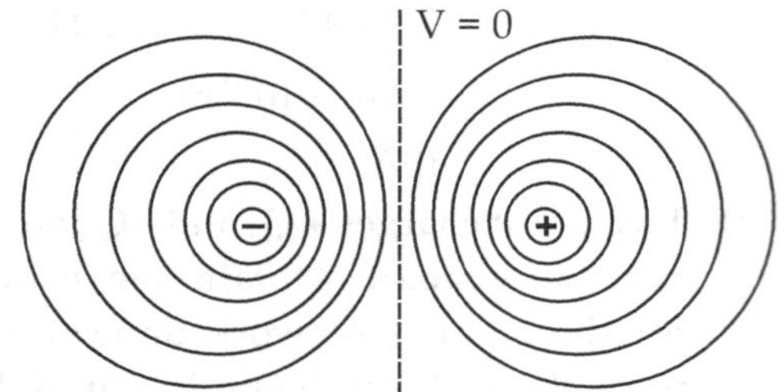


Equipotential surface for an electric dipole.

**2. A proton is accelerated through a potential difference V, subjected to a uniform magnetic field acting normal to the velocity of the proton. If the potential difference is doubled, how will the radius of the circular path described by the proton in the magnetic field change ? [1]**

**Ans.** Given, proton accelerated through potential difference V, the direction of magnetic field is normal to velocity of proton.

As we know,

$$\frac{1}{2}m_p v^2 = eV$$

$$v = \sqrt{\frac{2eV}{m_p}}$$

[during acceleration of proton, work done will be converted to kinetic energy]

When V is doubled,

$$V' = 2V,$$

$$\therefore \quad v' = \sqrt{\frac{2e.2V}{m_p}} = \sqrt{2}\,v$$

$$qvB = \frac{m_p v^2}{r}$$

$$\Rightarrow \quad r = \frac{m_p v}{qB}$$

$$\Rightarrow \quad r' = \frac{m_p}{qB}v'$$

$$\Rightarrow \quad r' = \frac{m_p}{qB}\sqrt{2}\,v$$

$$\therefore \quad r' = \sqrt{2}r$$

** Answer is not given due to change in the present Syllabus.

**3. The magnetic susceptibility χ of magnesium at 300 K is $1{\cdot}2 \times 10^5$. At what temperature will its magnetic susceptibility become $1{\cdot}44 \times 10^5$? [1]**

**OR**

**The magnetic susceptibility χ of a given material is – 0·5. Identify the magnetic material.**

**Ans.** Given,

$$\chi_{mg} \text{ at } 300\text{ K} = 1{\cdot}2 \times 10^5$$

$$\chi'_{mg} \text{ at } t \text{ temp.} = 1{\cdot}44 \times 10^5$$

$$t = ?$$

From Curies law,

$$\chi \propto \frac{1}{T}$$

$$\frac{\chi'_{mg}}{\chi_{mg}} = \frac{300}{t}$$

$$\Rightarrow \quad \frac{1{\cdot}44\times10^5}{1{\cdot}2\times10^5} = \frac{300}{t}$$

$$t = \frac{300\times1{\cdot}2}{1{\cdot}44}$$

$$= 250\text{ K}$$

**OR**

Diamagnetic as $1 < \chi < 0$.

**4. Identify the semiconductor diode whose V-I characteristics are as shown. [1]**

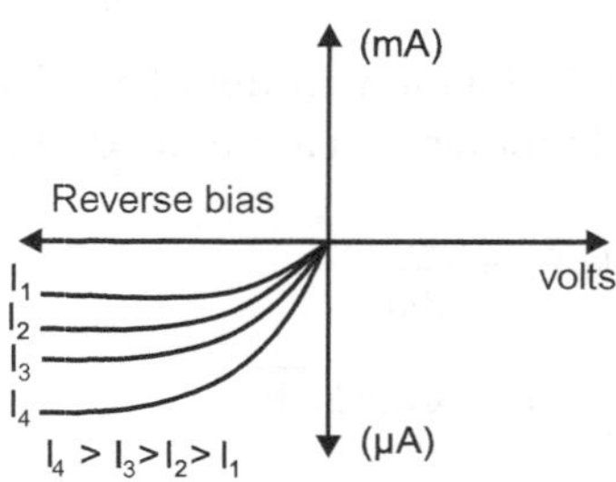


**Ans.** Photodiode.

**5. Which part of the electromagnetic spectrum is used in RADAR? Give its frequency range. [1]**

**OR**

**How are electromagnetic waves produced by accelerating charges ?****

**Ans.** Microwaves [1 GHz to 100 GHz]

## SECTION B

**6. A capacitor made of two parallel plates, each of area 'A' and separation '*d*' is charged by an external d.c.-source. Show that during charging, the displacement current inside the capacitor is the same as the current charging the capacitor. [2]**

**Ans.**

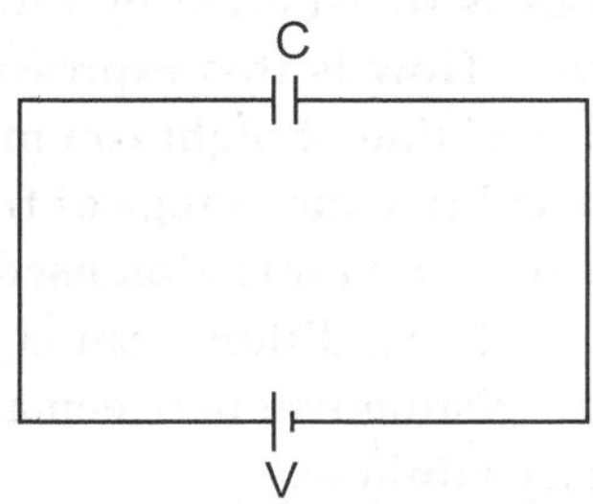


Displacement current between the plates of the capacitor is,

$$i_d = \varepsilon_0 \frac{d\phi_E}{dt}$$

where, $\phi_E$ = electric flux

$$i_d = \varepsilon_0 \frac{d(EA)}{dt}$$

$$i_d = \varepsilon_0 A \frac{d\left(\frac{v}{d}\right)}{dt}$$

$$i_d = \frac{\varepsilon_0 A}{d} \cdot \frac{dv}{dt}$$

$$i_d = C\frac{dv}{dt}$$

$i_d =$

$$i_d = \frac{d\,(q)}{dt}$$

$$i_d = \frac{dq}{dt} = i_c$$

Therefore both the displacement current and conduction currents are equal.

**7. A photon and a proton have the same de-Broglie wavelength λ. Prove that the energy of the photon is (2m λ *c*/*h*) times the kinetic energy of the proton. [2]**

**Ans.** $\lambda_{pr} = \dfrac{h}{m_p v}$ de-Broglie wavelength of proton

$$\lambda_{ph} = \frac{hc}{E}, \qquad \left(\because E_{ph} = \frac{hc}{\lambda}\right)$$

(de-Broglie wavelength of photon)

$$KE_{proton} = \frac{1}{2} m_p v^2$$

$$\Rightarrow \quad \lambda_{pr} = \lambda_{ph} = \lambda$$

Hence,

$$\frac{hc}{E} = \frac{h}{m_p v}$$

$$v = \frac{h}{m_p} \cdot \frac{E}{hc} = \frac{E}{m_p c}$$

$$K.E. = \frac{1}{2} \cdot \frac{E}{m_p c} \cdot \frac{E}{m_p c} m_p$$

$$= \frac{1}{2} \cdot \frac{E \cdot E}{m_p\, c.c}$$

$$= \frac{1}{2} \cdot \frac{E}{m_p} \cdot \frac{hc}{\lambda c.c}$$

$$K.E. = \frac{1}{2} \frac{E}{m_p} \cdot \frac{h}{c\lambda}$$

$$K.E.\left(\frac{2m_p c\lambda}{h}\right) = E$$

$$K.E.\left(\frac{2mc\lambda}{h}\right) = E$$

**8. A photon emitted during the de-excitation of electron from a state n to the first excited state in a hydrogen atom, irradiates a metallic cathode of work function 2 eV, in a photo cell, with a stopping potential of 0.55V. Obtain the value of the quantum number of the state n. [2]**

**OR**

**A hydrogen atom in the ground state is excited by an electron beam 12·5 eV energy. Find out the maximum number of lines emitted by atom from its excited state.**

**Ans.**

$$E_n = 13.6\left[\frac{1}{1^2} - \frac{1}{n^2}\right] eV$$

Work function = 2eV

∴ Maximum kinetic Energy

$$= 13.6\left[\frac{1}{4} - \frac{1}{n^2}\right] - 2eV$$

$$= 0.55 \text{ eV}$$

$$\therefore \quad \frac{13.6}{4} - \frac{13.6}{n^2} - 2 = 0.55$$

$$\frac{13.6}{4} - 2 - 0.55 = \frac{13.6}{n^2} = 0.85$$

$$n^2 = \frac{13.6}{0.85} = 16,$$

$$n = 4$$

**OR**

Given : $\Delta E = 12.5$ eV

Let the electron jump from $n = 1$ to $n = n$ level.

$$\Delta E = E_n - E_1$$

$$\therefore \quad 12.5 = -\frac{13.6}{n^2} - \left(\frac{-13.6}{1^2}\right)$$

$$12.5 = 13.6\left(1-\frac{1}{n^2}\right)$$

$$1-\frac{12.5}{13.6} = \frac{1}{n^2}$$

$$\frac{1.1}{13.6} = \frac{1}{n^2}$$

$$\frac{13.6}{1.1} = n^2$$

$$12.36 = n^2$$

$$n = 3.5$$

$$n = 3^{rd}$$

Maximum number of lines = 3

**9. Draw the ray diagram of an astronomical telescope showing image formation in the normal adjustment position. Write the expression for its magnifying power. [2]**

**OR**

**Draw a labelled ray diagram to show image formation by a compound microscope and write the expression for its resolving power.**

**Ans.**

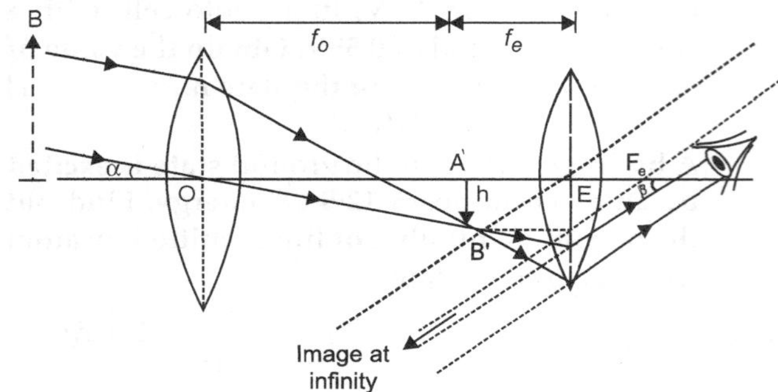


The magnifying power $m$ is the ratio of the angle β subtended at the eye by the final image to the angle α which the object subtends at the lens or the eye. Hence,

$$m = \frac{\beta}{\alpha} = \frac{h}{f_e}\cdot\frac{f_o}{h} = \frac{f_o}{f_e}$$

**OR**

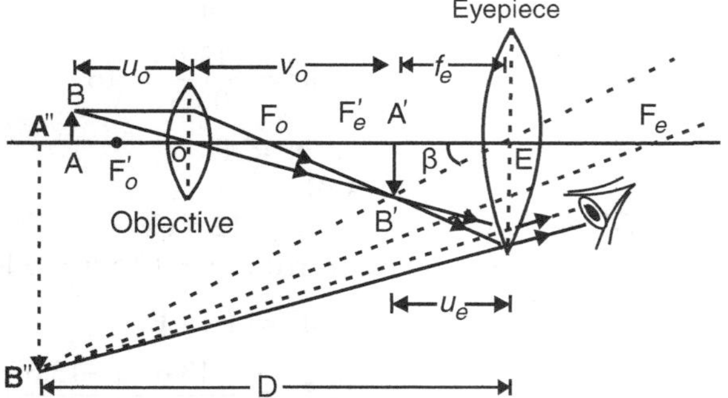


The resolving power of microscope is the reciprocal of the minimum distance.Therefore, we have

$$\text{R.P.} = \frac{1}{d_{min}} = \frac{2n\sin\beta}{\lambda}$$

**10. Write the relation between the height of a TV antenna and the maximum range up to which signals transmitted by the antenna can be received. How is this expression modified in the case of line of sight communication by space waves? In which range of frequencies, is this mode of communication used?** [2]**

**11. Under which conditions can a rainbow be observed? Distinguish between a primary and a secondary rainbow. [2]**

**Ans.** The rainbow is an example of the dispersion of sunlight by the water drops in the atmosphere. This is a phenomenon due to combined effect of dispersion, refraction and reflection of Sunlight by spherical water droplets of rain. The conditions for observing a rainbow are that the Sun should be shining in one part of the sky (say near western horizontal while it is raining in the opposite part of the sky (say eastern horizon).

**Difference between Primary and Secondary Rainbow :**

| S.N. | Primary Rainbow | Secondary Rainbow |
|---|---|---|
| 1 | Three Step process (Refraction-Reflection and Refraction). | Four Step process (Refraction Reflection reflection and Refraction) |
| 2. | Appearance intensity better than Secondary. | Appearance intensity lesser than Primary. |
| 3. | Single reflection occurs. | Double reflection occurs. |
| 4. | 2 Degree range. | 3 Degree range. |
| 5. | Fig.1 | Fig. 2 |

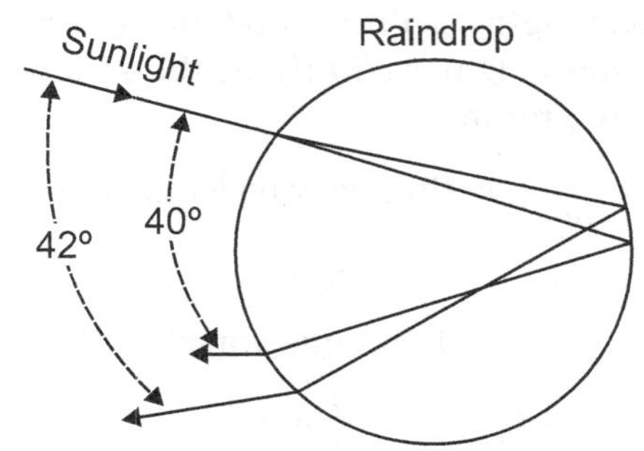


Fig. (1)

** Answer is not given due to change in the present Syllabus.

Fig. (2)

**12. Explain the following : [2]**

**(a) Sky appears blue.**

**(b) The Sun appears reddish at**

**(i) sunset, (ii) sunrise.**

**Ans.** (a) Light from the sun reaches the atmosphere that is comprised of the tiny particles of the atmosphere. These act as a prism and cause the different components of light to scatter. As blue light travels in shorter and smaller waves in comparison to the other colours of spectrum. It is scattered the most , causing the sky to appear bluish.

(b) The molecules of the atmosphere and other particles that are smaller than the longest wavelength of visible light are more effective in scattering light of shorter wavelengths than light of longer wavelengths. The amount of scattering is inversely proportional to the fourth power of the wavelength. (Rayleigh Effect) Light from the Sun near the horizon passes through a greater distance in the Earth's atmosphere than does the light received when the Sun is overhead. The correspondingly greater scattering of short wavelengths accounts for the reddish appearance of the Sun at sunset and at sunrise.

## SECTION-C

**13. A capacitor (C) and resistor (R) are connected in series with an ac source of voltage of frequency 50 Hz. The potential difference across C and R are respectively 120 V, 90 V, and the current in the circuit is 3 A. Calculate (i) the impedance of the circuit (ii) the value of the inductance, which when connected in series with C and R will make the power factor of the circuit unity. [3]**

**OR**

**The figure shows a series LCR circuit connected to a variable frequency 230 V source.**

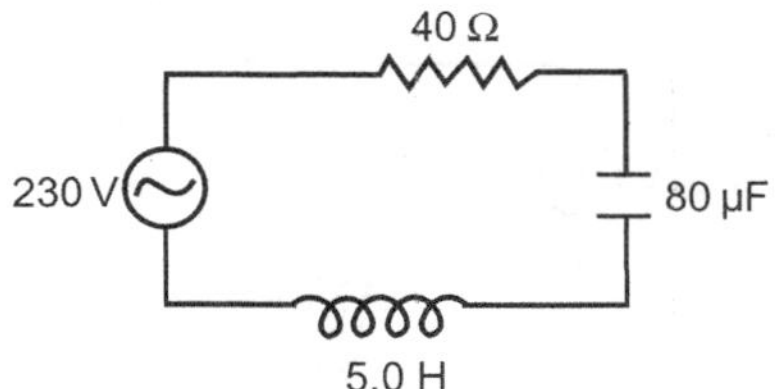


**(a) Determine the source frequency which drives the circuit in resonance.**

**(b) Calculate the impedance of the circuit and amplitude of current at resonance.**

**(c) Show that potential drop across LC combination is zero at resonating frequency.**

**Ans.**

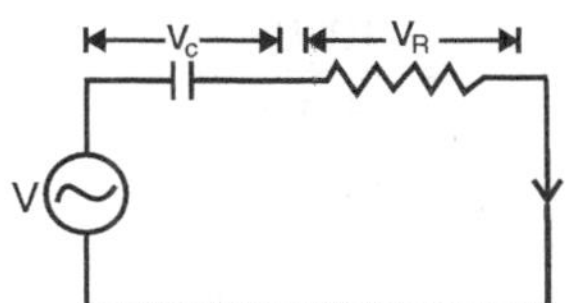


Given,

$$V_c = 120, V_R = 90\,V, I = 3A, Z = ?$$

(i)
$$\varepsilon_{rms} = \sqrt{V_R^2 + V_C^2} = \sqrt{90^2 + 120^2} = 150 \text{ Volt}$$

$$\text{Impedance, } z = \frac{\varepsilon_{rms}}{I_{rms}} = \frac{150}{3} = 50\,\Omega$$

(ii) $\because$
$$V_c = IX_c$$

$$V_c = I\frac{1}{\omega C}$$

$$\Rightarrow \quad C = \frac{I}{\omega V_c} \quad ...(1)$$

$$L = \frac{1}{\omega^2 C} \quad ...(2)$$

From eq. (1) and (2)

$$L = \frac{V_c}{\omega I} = \frac{120}{2\pi f I}$$

$$L = \frac{120}{2\pi \times 50 \times 3}$$

$$L = \frac{2}{5\pi}\text{ H}$$

OR

**(a)** Source frequency will be same as resonance frequency of LC circuit,

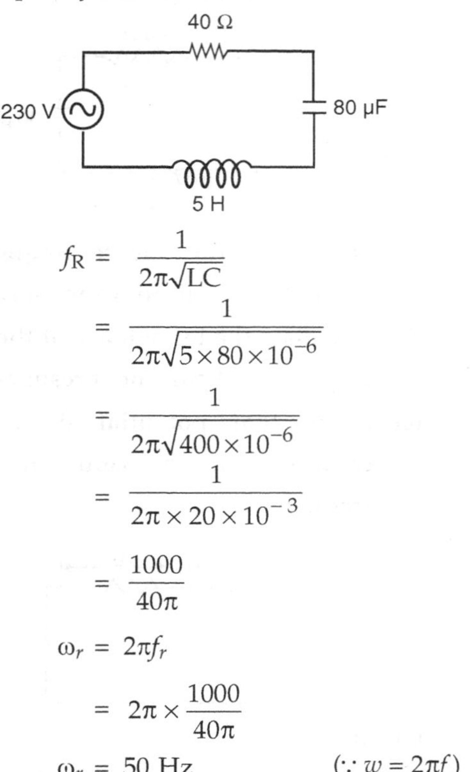


$$f_R = \frac{1}{2\pi\sqrt{LC}}$$

$$= \frac{1}{2\pi\sqrt{5 \times 80 \times 10^{-6}}}$$

$$= \frac{1}{2\pi\sqrt{400 \times 10^{-6}}}$$

$$= \frac{1}{2\pi \times 20 \times 10^{-3}}$$

$$= \frac{1000}{40\pi}$$

$$\omega_r = 2\pi f_r$$

$$= 2\pi \times \frac{1000}{40\pi}$$

$$\omega_r = 50 \text{ Hz} \qquad (\because w = 2\pi f)$$

**(b)** Impedance of circuit

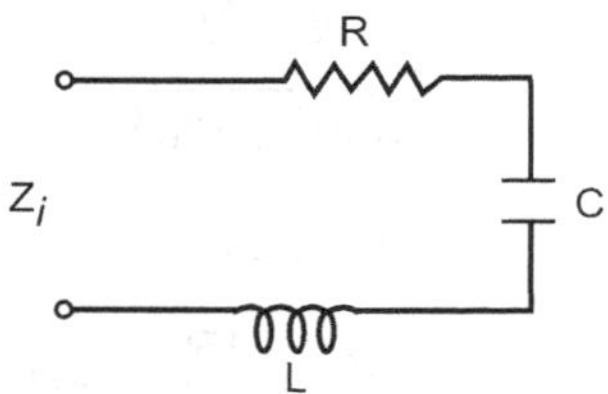


$$\text{Impedance } (z) = \sqrt{R^2 + (X_L - X_C)^2}$$

At resonance $X_L = X_C \Rightarrow Z = R = 40\Omega$

Amplitude of the current at the resonating frequency is given as :

$$I_0 = \frac{V_0}{Z} = \frac{\sqrt{2}\,V}{Z} = \frac{\sqrt{2} \times 230}{40} = 8.13\text{A}$$

**(c)** As at resonance frequency impedance of combination of L and C is 0.

Hence, voltage drop across LC combination is zero at resonating frequency.

**14. Give reason to explain why n and p regions of a Zener diode are heavily doped. Find the current through the Zener diode in the circuit given below : [3]**

**(Zener breakdown voltage is 15 V)**

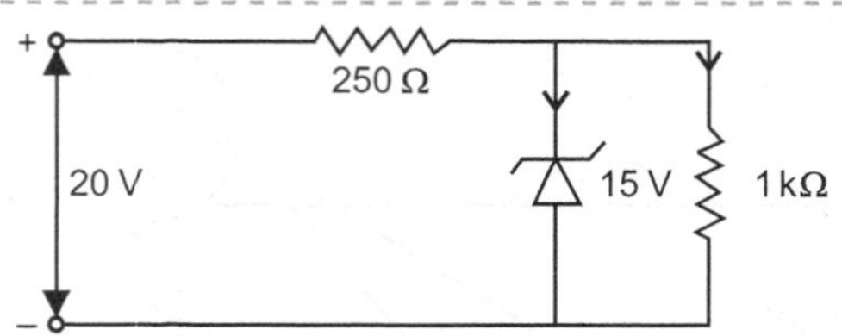


**Ans.** By heavily doping both *p* and *n* sides of the junction, depletion region formed is very thin, *i.e.*, $< 10^{-6}$ m. Hence, electric field across the junction is very high ($\sim 5 \times 10^6$ V/m) even for a small reverse bias voltage. This can lead to a break down during reverse biasing.

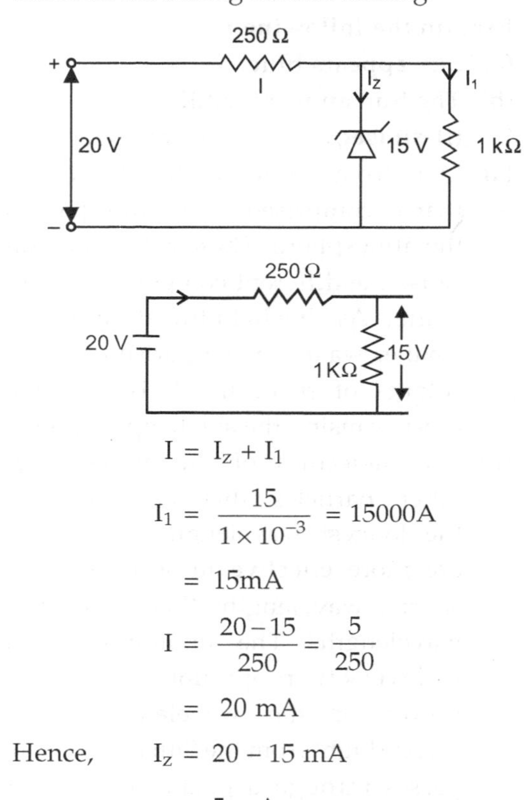


$$I = I_z + I_1$$

$$I_1 = \frac{15}{1 \times 10^{-3}} = 15000\text{A}$$

$$= 15\text{mA}$$

$$I = \frac{20 - 15}{250} = \frac{5}{250}$$

$$= 20 \text{ mA}$$

Hence, $\quad I_z = 20 - 15$ mA

$$= 5 \text{ mA}$$

**15. Draw a labelled diagram of cyclotron. Explain its working principle. Show that cyclotron frequency is independent of the speed and radius of the orbit. [3]**

OR

**(a) Derive, with the help of a diagram, the expression for the magnetic field inside a very long solenoid having n turns per unit length carrying a current I.**

**(b) How is a toroid different from a solenoid?**

**Ans. Cyclotron** : Cyclotron is a device by which the positively charged particles like protons, deutrons, etc. can be accelerated at a very high speed.

**Principle :** Cyclotron works on the principle that a positively charged particle can be acclerated by making it to cross the same electric field repeatedly with the help of a mgnetic field.

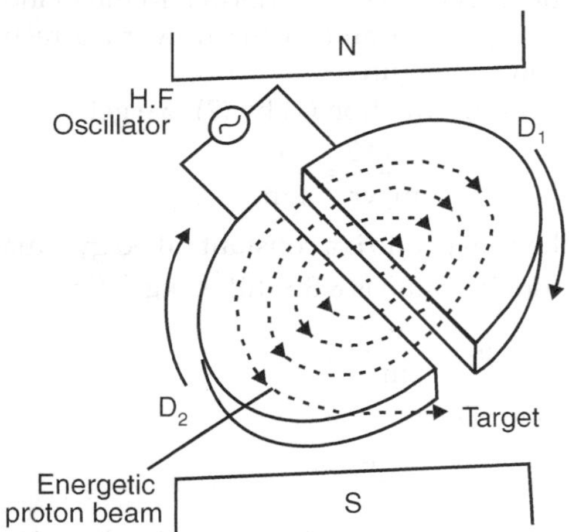


**Construction :** The construction of a simple cyclotron is shown in figure above. it consist of two-semi cylindrical boxes $D_1$ and $D_2$, which are called Dees They are enclosed in an evacuated chamber.

Chamber is kept between the poles of a powerful magnet so that uniform magnetic field acts perpendicular to the plane of the dees. An alternating voltage is applied in the gap between the two dees by the help of a high frequency oscillator. The electric field is zero inside the dees.

**Working and theory :** At a certain instant, let $D_1$ be positive and $D_2$ be negative. A proton from an ion source will be accelerated towards $D_2$, it describes a semi-circular path with a constant speed and is acted upon only by the magnetic field. The radius of the circular path is given by.

$$qvB = \frac{mv^2}{r}$$

From the above equation we get,

$$r = \frac{mv}{qB} \quad \text{...(i)}$$

The period of revolution is given by.

$$T = \frac{2\pi r}{v} = \frac{2\pi r}{v} \cdot \frac{mv}{qB}$$

$$T = \frac{2\pi m}{qB}$$

The frequency of revolution is given by,

$$f = \frac{1}{T} = \frac{qB}{2\pi m}$$

From the above equation it follows that frequency $f$ is independent of both $v$ and $r$ and is called cyclotron frequency. Also if we make the frequency of applied a.c. equal to $f$, then every time the proton reaches the gap between the dees, the direction of electric field is reversed and proton receives a push and finally it gains very high kinetic energy. The proton follows a spiral path and finally gets directed towards the target and comes out from it.

**OR**

**(a)** Magnetic field inside the solenoid

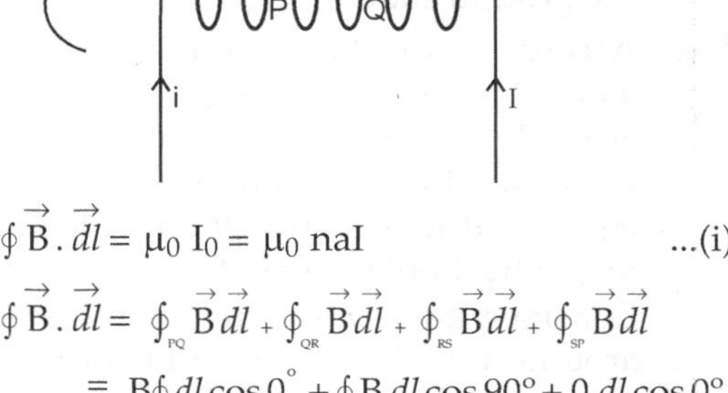


$$\oint \vec{B}.\vec{dl} = \mu_0 I_0 = \mu_0 \text{naI} \quad \text{...(i)}$$

$$\oint \vec{B}.\vec{dl} = \oint_{PQ} \vec{B}\vec{dl} + \oint_{QR} \vec{B}\vec{dl} + \oint_{RS} \vec{B}\vec{dl} + \oint_{SP} \vec{B}\vec{dl}$$

$$= B\oint dl\cos 0^\circ + \oint B.dl\cos 90^\circ + 0.dl\cos 0^\circ + \oint B.dl\cos 90^\circ$$

$$= B\oint dl = B.a \quad \text{...(ii)}$$

From equation (i) and (ii),

$\Rightarrow$ $B = \mu_0 nI$

$n$ = No. of turn per unit length.

$a$ = length of the path.

I = current passing through

**(b)** Toroid is a form in which a conductor is wound around a circular body. In this case we get magnetic field inside the core but poles are absent because circular body don't have ends. Toroid is used in toroidal inductor, toroidal transformer.

Solenoid is a form in which conductor is wound around a cylindrical body with limb. In this case magnetic field creates two poles N and S. Solenoids have some flux leakage. This is used in relay, motors, electro-magnets.

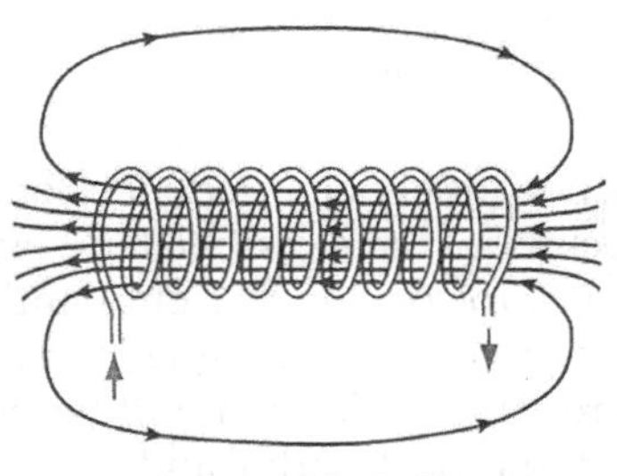

**Solenoid Coil**

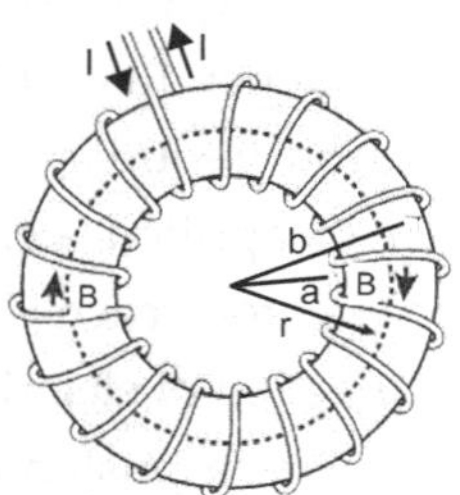


Toroidal Coil

**16. Prove that the magnetic moment of the electron revolving around a nucleus in an orbit of radius r with orbital speed v is equal to evr/2. Hence using Bohr's postulate of quantization of angular momentum, deduce the expression for the magnetic moment of hydrogen atom in the ground state. [3]**

**Ans.** According to Bohr model of hydrogen-like atoms, negatively charged electron revolves around the positively charge nucleus. This uniform circular motion of the electron is equivalent to a current loop which possesses a magnetic dipole moment.

Consider an electron revolving anticlockwise around a nucleus in an orbit of radius *r* with speed v and time period T.

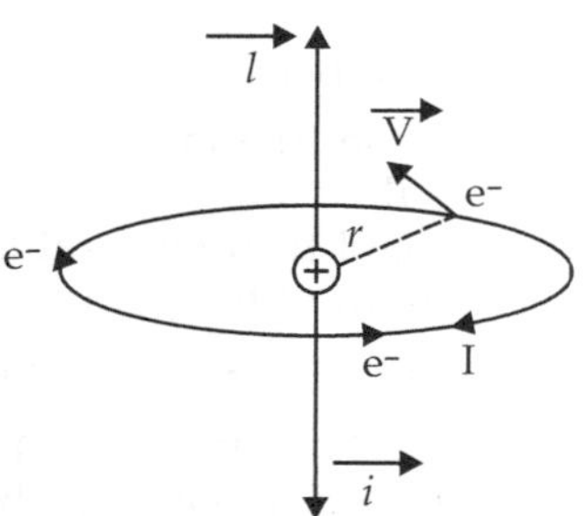


**Orbital magnetic moment of a revolving electron**

Equivanet current,

$$I = \frac{\text{Charge}}{\text{Time}} = \frac{e}{T} = \frac{e}{2\pi r / V} = \frac{eV}{2\pi r}$$

Area of the current loop, $A = \pi r^2$

Therefore, the orbital magnetic moment (magnetic moment due to orbital motion) of the electron is :

$$m_i = \frac{evr}{2}$$

As the negatively charged electron is revolving anticlockwise, the associated current flows clockwise. According to right hand thumb rule, the direction of the magnetic dipole moment of the revolving electron will be perpendicular to the plane of its orbit and in the downward direction.

Also, the angular momentum of the electron due to its orbital motion is :

$$l = m_e vr \quad ...(2)$$

The direction of $\vec{l}$ is normal to the plane of the electron orbit and in the upward direction, as shown in Figure.

Dividing equation (1) by (2), we get

$$\frac{m_i}{l} = \frac{evr/2}{m_e vr} = \frac{e}{2m_e}$$

The above ratio is a constant called gyromagnetic ratio. Its value is $8.8 \times 10^{10}$ C kg$^{-1}$. So

$$\mu_i = \frac{e}{2m_e} l$$

Vector Form,

$$\vec{\mu}_i = -\frac{e}{2m_e}\vec{l}$$

The negative sign shows hat the direction of $\vec{l}$ is opposite to that of $\mu_i$. According to Bhor's quantisation condition, the angular momentum of an electron in any permissible orbit is integral multiple of $h/2\pi$, where h is Planck's constant, *i.e.*,

$$l = \frac{nh}{2\pi}, \text{ where } n = 1, 2, 3, \text{.........}$$

$$\therefore \quad \mu_i = n\left(\frac{eh}{4\pi m_e}\right)$$

This equation gives orbital magnetic moment of an electron revolving in n$^{th}$ orbit.

**17. Two large charged plane sheets of charge densities σ and –2σ C/m² are arranged vertically with a separation of d between them. Deduce expressions for the electric field at points (i) to the left of the first sheet, (ii) to the right of the second sheet, and (iii) between the two sheets. [3]**

**OR**

**A spherical conducting shell of inner radius $r_1$ and outer radius $r_2$ has a charge Q.**

**(a) A charge q is placed at the centre of the shell. Find out the surface charge density on the inner and outer surfaces of the shell.**

**(b) Is the electric field inside a cavity (with no charge) zero; independent of the fact whether the shell is spherical or not? Explain.**

**Ans. (i)** Electric field due to plane short toward right,

$$E_1 = \frac{\sigma}{2\varepsilon_o}$$

Electic field due to plane short towards left,

$$E_2 = \frac{-\sigma}{\varepsilon_o}$$

$E_2 \leftarrow \bullet \rightarrow E_1$ (I) | $\rightarrow E_1$, $\bullet \rightarrow E_2$ (II), $\rightarrow \hat{r}$ | $E_2 \leftarrow \bullet \rightarrow E_1$ (III)

Electric field in region

$$\text{I} = \vec{E}_2 + \vec{E}_1$$

$$= \frac{\sigma}{\varepsilon_o}\hat{r} - \frac{\sigma}{2\varepsilon_o}\hat{r}$$

**(ii)** Electric field in region

$$\text{II} = \vec{E}_1 + \vec{E}_2$$

$$= \frac{-\sigma}{2\varepsilon_o}\hat{r} - \frac{\sigma}{\varepsilon_o}\hat{r}$$

**(iii)** Electric field in region

$$\text{III} = \vec{E}_1 + \vec{E}_2$$

$$= \frac{-\sigma}{2\varepsilon_o}\hat{r} - \frac{\sigma}{\varepsilon_o}\hat{r}$$

**OR**

**Ans. (a)** The charge + Q resides on the outer surface of the shell. As the charge q is placed at the centre of the shell, there is charge –v induced on the inner surface and a charge +q is induced on the outer surface of the shell. Thus the total charge on the inner surface of the shell is –q and on the outer surface of the shell is (Q + q).

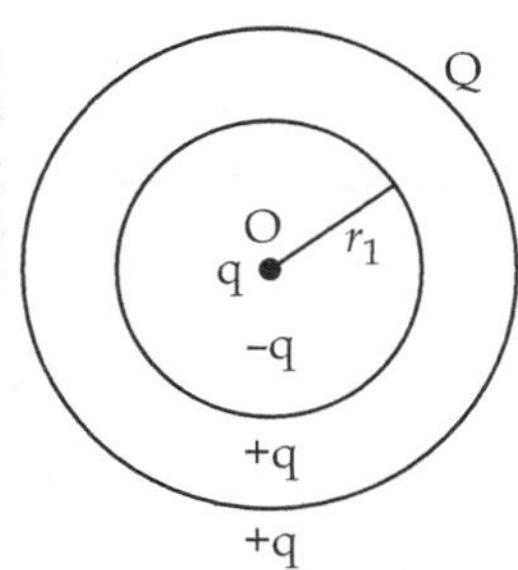


The surface charge density on the inner surface,

$$\overline{V}_1 = -\frac{q}{4\pi r_1^2}$$

The surface charge density on the outer surface,

$$\overline{V}_2 = -\frac{(Q+q)}{4\pi r_2^2}$$

**(b)** Yes, the electric field inside a cavity is zero. If the shape of the shell is not spherical. If we take a closed-loop which is inside the cavity along a field line and the rest outside it, the net work done by the field in carrying a test charge over the close-loop will not be zero. So the electric field inside a cavity with no charge is always zero.

Electric field inside a cavity with no charge will be zero as from Gauss's law

$$\phi_E = \frac{q_{enclosed}}{\varepsilon_0}$$

as $q_{enclosed} = 0$. Thus $\phi_E$ will also be zero.

**18. A signal of low frequency $f_m$ is to be transmitted using a carrier wave of frequency $f_c$. Derive the expression for the amplitude modulated wave and deduce expression for the lower and upper sidebands produced. Hence, obtain the expression for modulation index. [3]**

**Ans.** As given modulating (message) signal $m(t)$ is :

$m(t) = A_m \sin W_m t$ $\qquad W_m = 2\pi t_m$

and carrier signals

$C(t) = A_C \sin W_c t$ $\qquad W_C = 2\pi f_C$

Thus modulation signal $C_m(t)$ is

$$C_m(t) = (A_C + Am \sin W_m t) \sin W_C t$$

$$= A_C\left(1 + \frac{Am}{A_C}\sin W_m t\right)\sin W_C t$$

$$= A_C (1 + \mu \sin W_m t) \sin W_C t$$

$$\mu = \frac{A_m}{A_C} \text{ modulation index}$$

$$C_m(t) = A_C \sin W_C t + \mu A_C \sin W_m t \sin W_c t$$

$$= A_C \sin W_C t + \frac{\mu A_C}{2}\cos(W_C - W_m)t$$

$$= + \frac{\mu A_C}{2}\cos(W_C + W_m)t$$

$W_C - W_m$ & $W_C + W_m$ are the lower side and upper side band frequencies respectivel.

Amplitude of both will be $\frac{\mu A_C}{2}$

**19. Draw a plot of $\alpha$-particle scattering by a thin foil of gold to show the variation of the number of the scattered particles with scattering angle. Describe briefly how the large angle scattering explains the existence of the nucleus inside the atom, Explain with the help of impact parameter picture, how Rutherford scattering serves a powerful way to determine and upper limit on the size of the nucleus. [3]**

**Ans.**

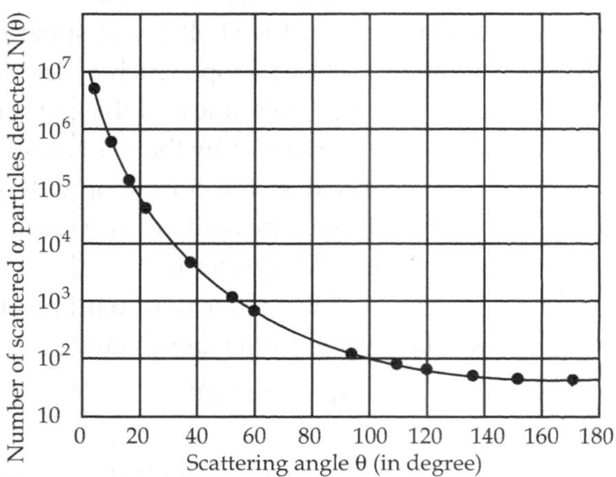


From the plot it is clear that most of the α-particles passed through the foil. Only 0.14% of the incident α particles scatter by more than 1° and about 1 in 8000 deflect by more than 90° and rest of the α-particles deflected backward due to strong repulsive force. This force will come from positive charge concentrated at the centre as most of the particles get deflected by small angles.

The α-particles trajectory depends on collision's impact parameter (b) for a given beam of α-particles, distribution of impact parameters as beam gets scattered in different directions with different probabilities.

Relation between the impact parameter b and scattering angle is given by

$$b = \frac{1}{4\pi\varepsilon_0}\frac{Ze^2 \cot e/2}{E}$$

Fig.2 shows α-particle close to nucleus suffers large scattering.

Impact parameter is minimum for head on collision α-particles rebound by 180°.

Impact parameter is high, for undeviated α-particles, with deflection angle ≅ 0°.

As these of nucleus was $10^{-14}$ m to $10^{-15}$ m w.r.t. $10^{-10}$ m size of an atom which is 10,000 to 100,000 times larger hence most of the space is empty, only small % of the incident particles rebound back indicates that number of α-particle goes head on collision. Hence most of the mass of the atom is concentrated in small volume.

Thus, Rutherford scattering is a strong tool to determine upper limit to the size of the nucleus.

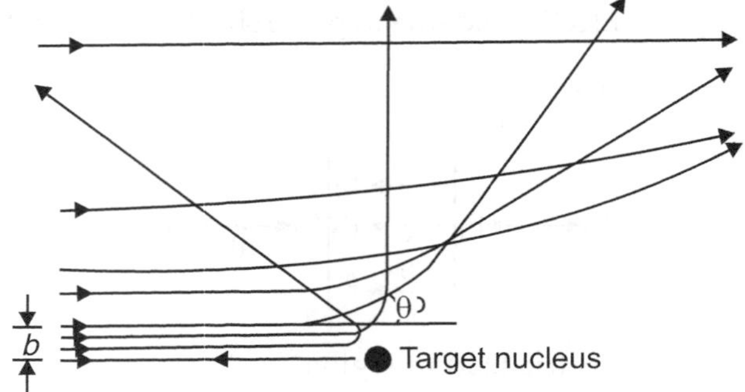


Fig. 2

**20. A 200 μF parallel plate capacitor having plate separation of 5 mm is charged by a 100 V dc source. It remains connected to the source. Using an insulated handle, the distance between the plates is doubled and a dielectric slab of thickness 5 mm and dielectric constant 10 is introduced between the plates. Explain with reason, how the (i) capacitance, (ii) electric field between the plates, (iii) energy density of the capacitor will change ? [3]**

**Ans.** Given, $C = 200 \times 10^{-6}$ F,

$d = 5 \text{ mm} = 5 \times 10^{-3}$ m  V = 100V

$Q = CV = 200 \times 10^{-6} \times 100$

$= 2 \times 10^{-2}$ coulomb.

$\varepsilon_0 A = Cd = 200 \times 10^{-6} \times 5 \times 10^{-3} = 10^{-6}$

$A = 10^{-6}/\varepsilon_0$

As dielectric of 5 mm is inserted with spacing between the dielectric doubled then it will act as following-Fig.A and Fig-B.

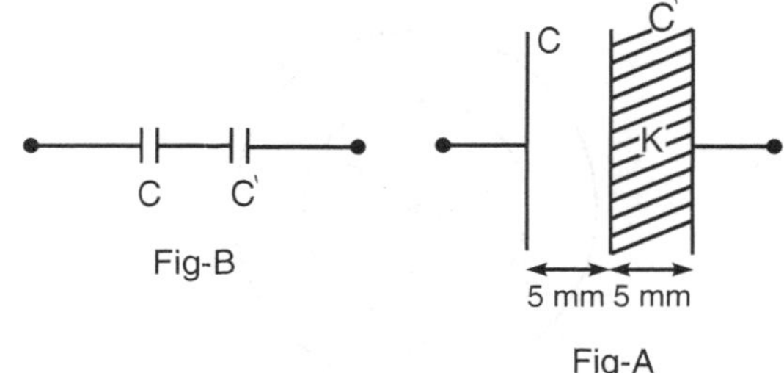


**(i)** Here, C' will be the capacitance with dielectric of 5 mm separation between plates.

$C' = KC = 10 \times 200 \times 10^{-6} = 2 \times 10^{-3}$ F

Thus, equivalent capacitance will be

$$\frac{1}{C_{eq}} = \frac{1}{C} + \frac{1}{C'}$$

$$= \frac{1}{200 \times 10^{-6}} + \frac{1}{2 \times 10^{-3}}$$

$$= \frac{10^4}{2} + \frac{10^3}{2}$$

$$= 5 \times 10^3 + 0.5 \times 10^3$$

$$\frac{1}{C_{eq}} = 5.5 \times 10^3$$

$$C_{eq} = \frac{1}{5.5 \times 10^3}$$

$$= \frac{1}{5.5} \times 10^{-3} = 0.181 \times 10^{-3}$$

$$= 181 \ \mu F$$

**(ii)** Electric field inside – dielectric will be

$$E' = \frac{E}{K} = \frac{V}{2d.K}$$

$$= \frac{100}{10 \times 10^{-3} \times 10}$$

$$= 1000 \ V/m$$

Electric field remains same, other than dielectric area,

E'

E

**(iii)**

$$U = \frac{Q^2}{2C}$$

$$U'' = U' + U$$

$$= \frac{1}{2} \cdot \frac{Q^2}{C} + \frac{1}{2} \cdot \frac{Q^2}{C'}$$

$$= \frac{1}{2} . Q^2 \left[\frac{1}{C} + \frac{1}{C'}\right]$$

$$= \frac{1}{2} \times 2 \times 10^{-2} \left[\frac{1}{0.2 \times 10^{-3}} + \frac{1}{2 \times 10^{-3}}\right]$$

$$\left[\because \frac{1}{C} + \frac{1}{C'} = \frac{1}{C_{eq}}\right]$$

$$= 10^{-2} \left[\frac{1000}{0.2} + \frac{1000}{2}\right]$$

$$= 10^{-2} \times 5 \times 10^3 \times 0.5 \times 10^3$$

$$U'' = 2.5 \times 10^4 J$$

**21. Why is it difficult to detect the presence of an anti-neutrino during β-decay ? Define the term decay constant of a radioactive nucleus and derive the expression for its mean life in terms of the decay constant.** [3]

**OR**

**(a) State two distinguishing features of nuclear force.**

**(b) Draw a plot showing the variation of potential energy of a pair of nucleons as a function of their separation. Mark the regions on the graph where the force is (i) attractive, and (ii) repulsive.**

**Ans.** The symbols $\bar{v}$ and $v$ represent antineutrino and neutrino respectively during β decay : both are neutral particles with very little or no mass. These particles are emitted from the nucleus along with the electron or positron during the decay process. Neutrions interact very weakly with matter, they can even penetrate the earth without being absorbed. It is for this reason that their detection is extremely difficult and their presence went unnoticed for long.

**Decay constant :** Decay constant of a radioactive element is the reciprocal of time during which the number of atoms left in the sample reduces to $\frac{1}{e}$ times the number of atoms in the original sample.

**Derivation of mean life :**

Let us consider, $N_0$ be the total number of radioactive atoms present initially. After time $t$, total no. of atoms present (undecayed) be N. In further $dt$ time, $d$N be the no. of atoms disintegrated. So, the life of $d$N atoms ranges lies between $t + dt$ and $dt$. Since $dt$ is very small time, the most appropriate life of $d$N atom is $t$. So the total life of N atom = $t.d$N

sum of ages of all atoms = $\int_0^{N_0} t dN$ ...(i)

$$N = N_0 e^{-\lambda t}$$

$$dN = N_0 (-\lambda) e^{-\lambda t} dt$$

Now, substituting the value of $d$N and changing the limit in equation (i) from (ii) we get

$$= \int_{\infty}^{0} t N_0 (-\lambda) e^{-\lambda t} dt$$

$$= -\int_0^{\infty} t N_0 (-\lambda) e^{-\lambda t} dt$$

$$= N_0 \lambda \int_0^{\infty} t e^{-\lambda t} dt$$

= sum of life of all atoms

$$\text{Mean life } (\tau) = \frac{N_0 \lambda \int_0^{\infty} t e^{-\lambda t} dt}{N_0}$$

$$\tau = \lambda \int_0^{\infty} t e^{-\lambda t} dt$$

$$\tau = \lambda \times \frac{1}{\lambda^2}$$

$$\tau = \frac{1}{\lambda}$$

This expression gives the relation between mean life and decay constant. Hence, mean life is reciprocal of decay constant.

**OR**

**(a)** Distinguishing features of nuclear force are :

**(i)** Nuclear forces are very strong binding forces (attractive force.)

**(ii)** It is independent of the charges on protons and neutrons (charge independent.)

**(iii)** It depends on the spin of the nucleons.

**(b)** Plot showing variation of potential energy of a pair of nucleons as a function of separation mark attractive and repulsive region.

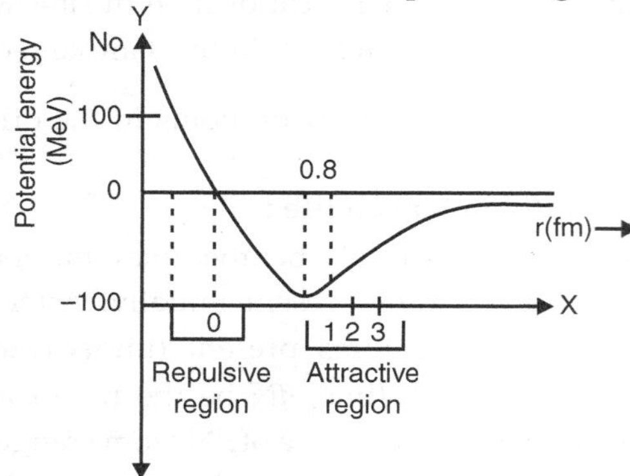


X-axis shows separation between pair of nucleons and Y-axis shows variation of potential energy with respect to separation (in $\times 10^{-15}$ m).

**22. A triangular prism of refracting angle 60° is made of a transparent material of refractive index $2/\sqrt{3}$. A ray of light is incident normally on the face KL as shown in the figure. Trace the path of the ray as it passes through the prism and calculate the angle of emergence and angle of deviation.** **[3]**

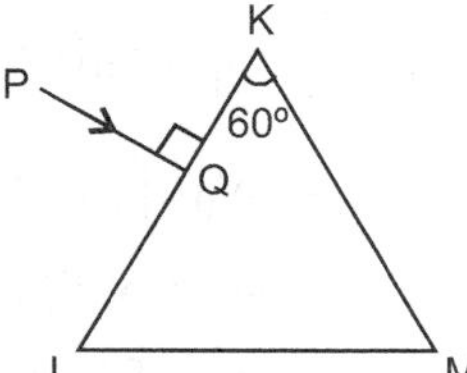


**Ans.** From diagram it is clear that incidence angle at face KM is 60°.

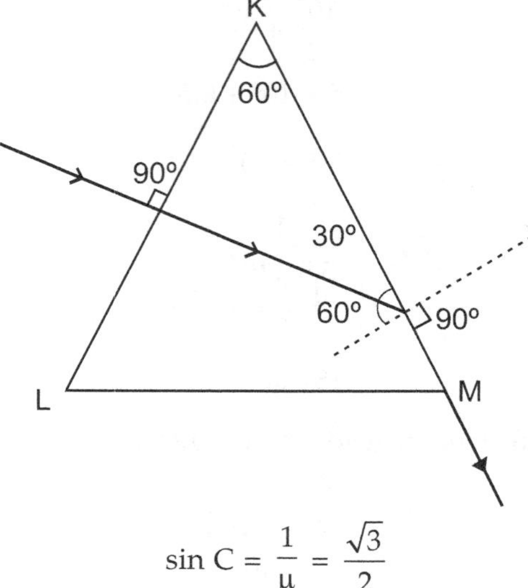


$$\sin C = \frac{1}{\mu} = \frac{\sqrt{3}}{2}$$

Hence, critical angle is also 60°.

Therefore, incident light ray will not emerge from KM face due to total internal reflection at this face. Hence, it will move along face KM.

Angle of emergence = 90°.

Hence angle of deviation = 30° (from fig.)

**23. Prove that in a common-emitter amplifier, the output and input differ in phase by 180°. In a transistor, the change of base current by 30 μA produces change of 0·02 V in the base-emitter voltage and a change of 4 mA in the collector current. Calculate the current amplification factor and the load resistance used, if the voltage gain of the amplifier is 400.** **[3]**

**Ans.**

Where, $V_i = 0$

$V_{CC} = V_{CE} + I_C R_L$

Similarly $V_{BB} = V_{BE} + I_B R_B$

If $V_i$ is not zero,

$$V_{BE} + V_i = V_{BE} + I_B R_B + \Delta I_B (R_B + R_i)$$

$$V_i = \Delta I_B (R_B + R_i)$$

$$V_i = r \Delta I_B$$

Change in $I_B$ caules a change in $I_C$

$$B_{ac} = \frac{\Delta I_C}{\Delta I_B} = I_C / I_B$$

As $\Delta V_{CC} = \Delta V_{CE} + R_L \Delta I_C = 0$

$$\Delta V_{CE} = -R_L \Delta I_C$$

$$\Delta V_{CE} = V_B$$

$$V_0 = -B_{ac} \Delta I_B R_L$$

Voltage gain = $A_V$

$V_i = \Delta I_{Br}$

$\frac{V_i}{r} = \Delta I_B$

$V_0 = -F_{ac}\frac{R_L}{r}V_i$

Negative sign shows phase reversal or Phase difference of 180°. Hence Proved.

$\Delta I_b = 30\mu A = 30 \times 10^{-6}$ A

$\Delta V_{BE} = 0.02V$

$\Delta I_C = 4$ mA $= 4 \times 10^{-3}$ A

$B_{ac} = ?$

Voltage gain = 400

$R_L = ?$

$$B_{ac} = \frac{\Delta I_c}{\Delta I_b} = \frac{4\times10^{-3}}{30\times10^{-6}} = \frac{4}{30}\times10^3$$

$400 \times 0.02$ = Voltage

Voltage = 8V

$\Delta I_C R_L = 8$

$8 = \Delta I_C R_L$

$$\frac{8}{4\times10^{-3}} = R_L$$

$R_L = 2 \times 10^3\ \Omega = 2$ kΩ.

**24. Show, on a plot, variation of resistivity of (i) a conductor, and (ii) a typical semiconductor as a function of temperature.**

**Using the expression for the resistivity in terms of number density and relaxation time between the collisions, explain how resistivity in the case of a conductor increases while it decreases in a semiconductor, with the rise of temperature. [3]**

**Ans.**

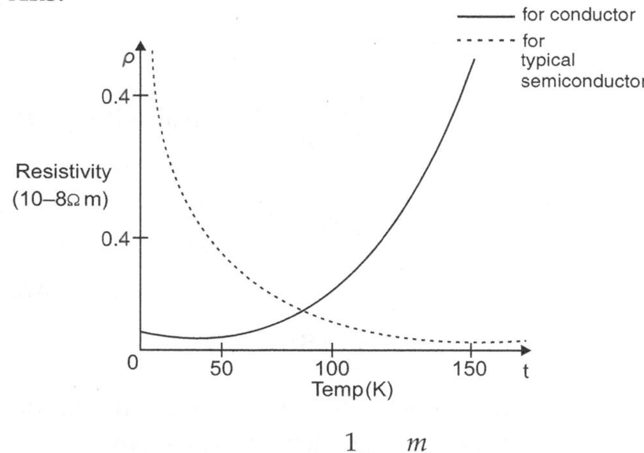


$$\rho = \frac{1}{\sigma} = \frac{m}{ne^2\tau}$$

**Ans.** $n \rightarrow$ number of free electrons

$\tau \rightarrow$ Average time between collisions.

In metals $n$ is not dependent on temperature to any appreciable extent and thus the decrease in the value of $\tau$ with rise in temperature causes $\rho$ to increase.

For semiconductors, $n$ increases with temperature. This increase more than compensates any decrease in t, so that for such materials, $\rho$ decreases with temperature.

## SECTION : D

**25. (a) Derive an expression for the induced emf developed when a coil of N turns, and area of cross-section A, is rotated at a constant angular speed ω in a uniform magnetic field B.**

**(b) A wheel with 100 metallic spokes each 0·5 m long is rotated with a speed of 120 rev/min in a plane normal to the horizontal component of the Earth's magnetic field. If the resultant magnetic field at that place is $4 \times 10^{-4}$ T and the angle of dip at the place is 30°, find the emf induced between the axle and the rim of the wheel. [5]**

**OR**

**(a) Derive the expression for the magnetic energy stored in an inductor when a current I develops in it. Hence, obtain the expression for the magnetic energy density.**

**(b) A square loop of sides 5 cm carrying a current of 0·2 A in the clockwise direction is placed at a distance of 10 cm from an inifinitely long wire carrying a current of 1 A as shown. Calculate (i) the resultant magnetic force, and (ii) the torque, if any, active on the loop.**

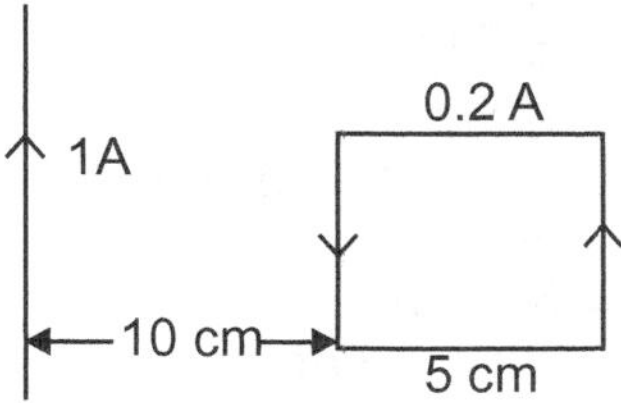


**Ans. (a)** As the armature coil is rotated in the magnetic field, angle θ between the field and normal to the coil changes continuously.Therefore, magnetic flux linked with the coil changes. An e.m.f. is induced in the coil. According to Fleming's right hand rule, current induced in AB is from A to B and it is from C to D in CD. In the external circuit, current flows from $B_2$ to $B_1$.

**To calculate the magnitude of e.m.f. induced :**

Suppose,

A → Area of each turn of the coil

N → Number of turns in the coil

B → Strength of magnetic field

θ → Angle which normal to the coil makes with $\vec{B}$ at any instant $t$

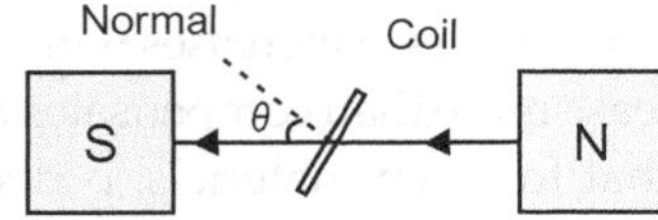


∴ Magnetic flux linked with the coil in this position.

$\phi = N(\vec{B}.\vec{A}) = NBA \cos\theta = NBA \cos \omega t$ ....(i)

Where, ω is angular velocity of the coil

As the coil rotates, angle θ changes. Therefore, magnetic flux ϕ linked with the coil changes and hence, an e.m.f. is induced in the coil. At this instant $t$, if $e$ is the e.m.f. induced in the coil, then

$$e = -\frac{d\theta}{dt} = -\frac{d}{dt}(NAB \cos \omega t)$$

$$= -NAB\frac{d}{dt}(\cos \omega t)$$

$$= -NAB(-\sin \omega t)\,\omega$$

$$\therefore \quad e = NAB\,\omega \sin \omega t$$

$$e = \varepsilon_0 \sin \omega t \text{ (Here } \varepsilon_0 = NBA\omega)$$

**(b)** We have number of spokes (N) = 100

Length of each spoke (L) = 0.5 m

Magnetic field (B) = $4 \times 10^{-4}$T

Frequency ($f$) = 120 rpm = 2 rps

$$B_H = B \cos 30°$$

$$= 4 \times 10^{-4} \times \frac{\sqrt{3}}{2} = 2\sqrt{3} \times 10^{-4}$$

Induced e.m.f. between axle and rim is given by

$$e = \frac{1}{2}B\omega^2$$

$$= \frac{1}{2} \times 2\sqrt{3} \times 10^{-4} \times 4\pi \times (0{\cdot}5)^2$$

$$= 5{\cdot}4 \times 10^{-4} \text{ V.}$$

**OR**

**Ans. (a) Energy stored in an inductor :** When a current flows through an inductor, a back e.m.f. is set up which opposes the growth of current. So, work needs to be done against back e.m.f. ($e$) in building up the current. This work done is stored as magnetic potential energy.

Let, I be the current through the inductor L at any instant $t$.

$$L = \frac{\Phi}{I}$$

Induced curret

The current rises at the rate $\frac{dI}{dt}$.

So, the induced e.m.f. is :

$$e = \frac{-L\,dI}{dt}$$

The work done against induced e.m.f.e. in moving change $dq$ is:

$$dW = e\,dq$$

$$= \frac{L\,dI}{dt} \times dq$$

$$= L\left(\frac{dq}{dt}\right)dI$$

$$= LI\,dI$$

For total work from 0 to $I_0$ current,

$$W = \int_0^{I_0} dW$$

$$= \int_0^{I_0} LI\,dI$$

$$= L\left[\frac{I^2}{2}\right]_0^{I_0}$$

$$= \frac{1}{2}LI_0^2$$

Hence, this work done is stored as the magnetic potential energy U in the inductor

$$U = \frac{1}{2}LI^2$$

The magnetic energy is given by

$$U_B = \frac{1}{2}LI^2$$

$$= \frac{1}{2}L\left(\frac{B}{\mu_0 n}\right)^2 \quad \text{(since } B = \mu_0 nI)$$

$$= \frac{1}{2}(\mu_0 n^2 Al)\left(\frac{B}{\mu_0 n}\right)^2$$

$$(\because L = \mu_0 n^2 Al)$$

$$= \frac{1}{2\mu_0}B^2 Al$$

The magnetic energy per unit volume or magnetic energy density is given by,

$$u_B = \frac{U_B}{V} \quad \text{(where, V is volume that contains flux)}$$

$$= \frac{U_B}{Al}$$

$$= \frac{B^2}{2\mu_0}$$

(b)

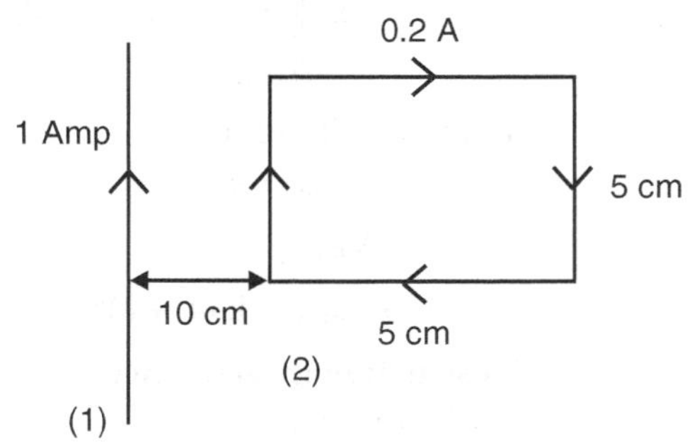


To calculate resultant magnetic force,

$$F = I(\vec{L} \times \vec{B})$$

$$F_{AB} = I_2(\vec{L} \times \vec{B_1})$$

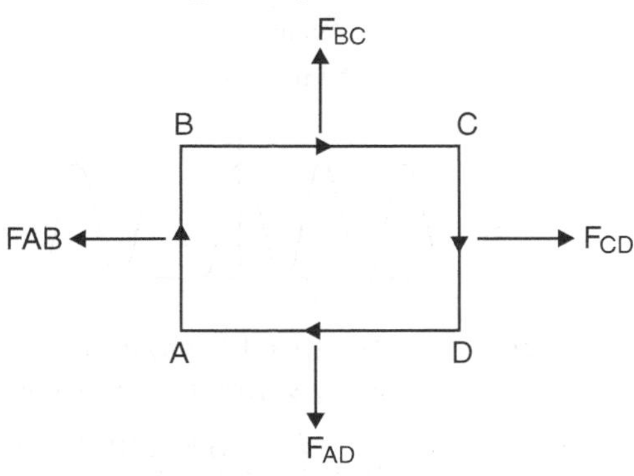


Direction of field B due to current in long wire inward ⊗

$$B = \frac{\mu_0 I_1}{2\pi r}$$

Where, $r$ = distance from long wire

$$F_{AB} = \frac{\mu_0 I_1 I_2}{2\pi r}$$

$$F_{CD} = \frac{2 \times 10^{-7} \times 1 \times 0.2 \times 5 \times 10^{-2}}{10 \times 10^{-2}}$$

$= 2 \times 10^{-8}$ Attractive

$$F_{CD} = |\vec{F}_{CB}|$$

$$= \frac{2 \times 10^{-7} \times 1 \times 0.2 \times 5 \times 10^{-2}}{15 \times 10^{-2}}$$

$= 1.3 \times 10^{-8}$ Repulsive

$$|\vec{F}_{BC}| = |\vec{F}_{AD}| \qquad \because |F_Y Net| = 0$$

Thus net magnetic force,

$$F_{net} = |\vec{F}_{AB}| - |\vec{F}_{CD}|$$

$= 2 \times 10^{-8} - 1.3 \times 10^{-8}$

$= 7 \times 10^{-9}$N.

(ii) The forces acting on all sides of the square due to current of infinite length wire are lying in the plane of coil. Thus, there is no net torque. Thus torque is zero.

**26. Explain, with the help of a diagram, how plane polarized light can be produced by scattering of light from the Sun.**

**Two polaroids $P_1$ and $P_2$ are placed with their pass axes perpendicular to each other. Unpolarised light of intensity I is incident on $P_1$. A third polaroid $P_3$ is kept between $P_1$ and $P_2$ such that its pass axis makes an angle of 45° with that of $P_1$. Calculate the intensity of light transmitted $P_1$, $P_2$ and $P_3$.** **[5]**

**OR**

**(a) Why cannot the phenomenon of interference be observed by illuminating two pin holes with two sodium lamps?**

**(b) Two monochromatic waves having displacements $y_1 = a \cos \omega t$ and $y_2 = a \cos(\omega t + \phi)$ from two coherent sources interfere to produce an interference pattern. Derive the expression for the resultant intensity and obtain the conditions for constructive and destructive interference.**

**(c) Two wavelengths of sodium light of 590 nm and 596 nm are used in turn to study the diffraction taking place at a single slit of aperture $2 \times 10^{-6}$ m. If the distance between the slit and the screen is 1·5 m, calculate the separation between the positions of the second maxima of diffraction pattern obtained in the two cases.**

**Ans.** Molecules behave like dipole radiators and scatter no energy along the dipole axis by this way plane polarized light can be produced during scattering of light.

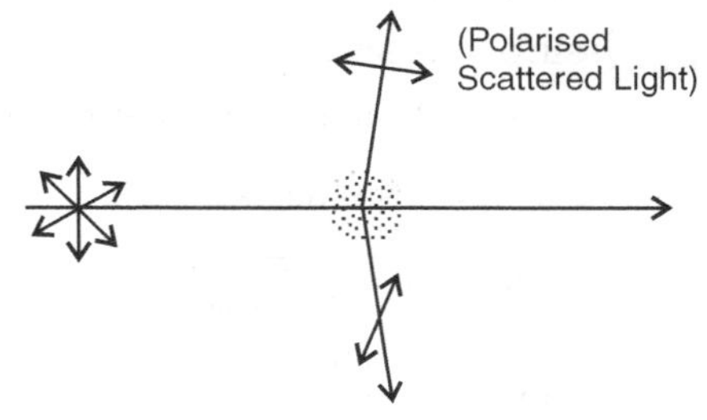


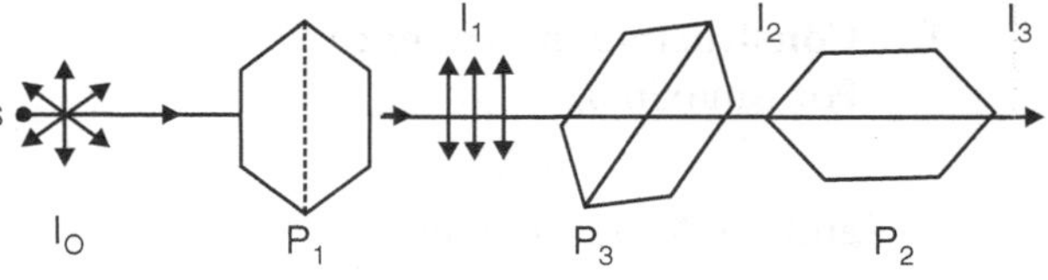

Intensity of light after passing through the polaroid $P_1(I')$

$$I_1 = I_o/2$$

Intensity of light after passing through $P_3$

$$I_2 = I_1 \cos^2 45^\circ$$

$$I_2 = \frac{I_0}{2} = \frac{I_0}{4}$$

Intensity of light after passing through $P_2$

$$I_3 = I_2 \cos^2 45^\circ$$

$$= \frac{I_0}{4} \times \frac{1}{2} = \frac{I_0}{8}$$

**OR**

**Ans (a)** Phenomenon of interference can't be observed by illuminating two pin holes with two sodium lamps because these sources are not coherent sources (it means they are not in same phase).

**(b)** Consider two monochromatic coherent sources A and B with waves $y_1 = a \cos \omega t$ and $y_2 = b \cos (\omega t + \phi)$ respectively.

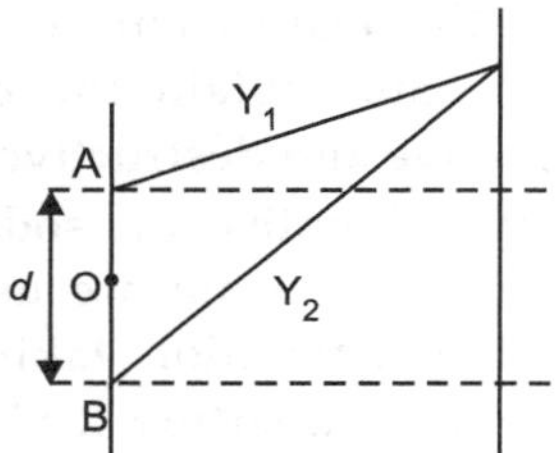


From principle of superposition,

$$y = y_1 + y_2$$
$$= a \sin \omega t + b \sin (\omega t + \phi)$$
$$= a \sin \omega t + b \sin \omega t \cos\phi + b \cos\omega t \sin\phi$$
$$= (a + b \cos \phi) \sin \omega t + b \sin\phi \cos \omega t$$

Let,

$$a + b \cos \phi = A \cos\delta$$
$$b \sin\phi = A \sin\delta$$
$$y = A \sin \omega t \cos\delta + A \cos \omega t \sin\delta$$
$$= A \sin (\omega t + \delta)$$
$$A = \sqrt{a^2 + b^2 + 2ab\cos\phi}$$
$$\tan\delta = \frac{b \sin\phi}{a + b\cos\phi}$$

1. **Constructive interference**

For I-maxima,

$$I \propto A^2$$

and for A to be maximum,

$$\cos \phi = 1$$
$$\cos \phi = \cos 2n\pi, n = 0, 1, 2, 3.....$$
$$\phi = 2n\pi$$

and path difference,

$$\Delta x = n\lambda$$
$$A_{max} = a + b$$
$$I \rightarrow I_{max} = k (a + b)^2$$

2. **Destructive Interference**

For I – minima

$$\cos\phi = -1$$
$$\Delta\phi = (2n + 1)\pi$$

Path difference,

$$\Delta x = (2n + 1) \lambda/2$$
$$A_{min} = a - b$$
$$I \rightarrow I_{min} = k (a - b)^2$$

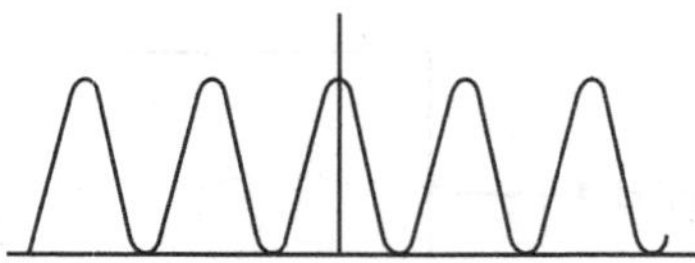

**(c)**
$$\theta = (n + 1/2) \lambda/a$$
$$a = 2 \times 10^{-6} \text{ (aperture of slit)}$$
$$\theta_1 = \frac{\lambda}{2a} = \frac{590\times10^{-9}}{4\times10^{-6}} = \frac{590}{4}\times10^{-3}$$
$$= 147.5 \times 10^{-3}$$
$$\theta_1 = \frac{\lambda'}{2a} = \frac{596\times10^{-9}}{4\times10^{-6}} = 149 \times 10^{-3}$$
$$\theta_2 - \theta_1 = 1.5 \times 10^{-3} \text{ (Angular difference)}$$
$$\lambda_1 = 596 \text{ nm}, \lambda_2 = 590 \text{ nm}$$
$$a = 2 \times 10^{-6}$$
$$D = 1.5 \text{ m}$$
$$y = \frac{3\lambda D}{2a}$$
$$y_1 - y_2 = \frac{3D}{2a} (\lambda_1 - \lambda_2)$$
$$= \frac{3\times1.5}{2\times2\times10^6} (596 - 590) \times 10^{-9} \text{ m}$$
$$= \frac{4.5}{4\times10^{-6}}\times6\times10^{-9}\text{m}$$
$$= \frac{4.5\times6}{4}\times10^{-3}\text{m}$$
$$= 6.78 \text{ nm}$$

27. (a) **Describe briefly, with the help of a circuit diagram, the method of measuring the internal resistance of a cell.**

(b) **Give reason why a potentiometer is preferred over a voltmeter for the measurement of emf of a cell.**

(c) **In the potentiometer circuit given below, calculate the balancing length $l$. Give reason, whether the circuit will work, if the driver cell of emf 5 V is replaced with a cell of 2 V, keeping all other factors constant.** **[5]**

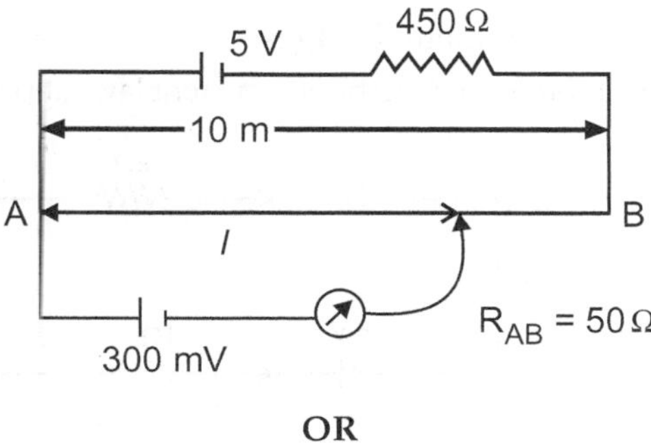


**OR**

27. (a) **State the working principle of a meter bridge used to measure an unknown resistance.**

(b) **Give reason.**

(i) **Why the connections between the resistors in a metre bridge are made of thick copper strips.**

(ii) **Why is it generally preferred to obtain the balance length near the mid-point of the bridge wire.**

(c) **Calculate the potential difference across the 4 Ω resistor in the given electrical circuit, using Kirchhoff's rules.**

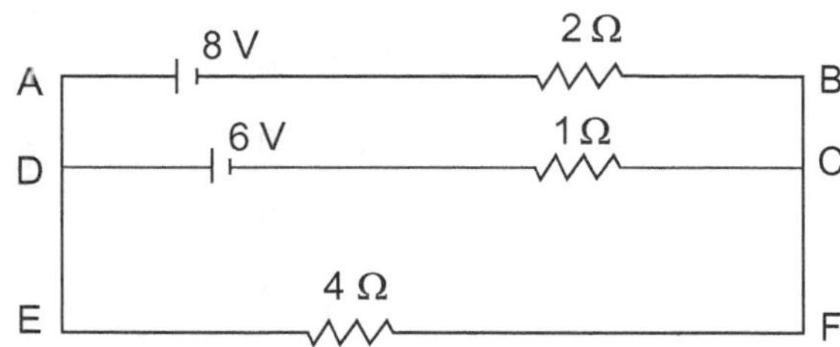


**Ans** (a) Close the key $K_1$. A current current flow through potentiometer wire. With key $K_2$ kept open, move the jockey, along AB till it balances the emf, e of the cell. Let $l_1$ be balancing length of wire. If $\phi$ is potential gradient then. When $K_2$ is open, balance is obtained at

$$l_1 = AN_1$$

$$\varepsilon = \phi l_1$$

With the help of Resistance Box (R. B.) introduce a resistance R and close key $K_2$. Find balance point for v of cell. If $l_2$ is blancing lengths.

when $K_2$ is closed

$$V = \phi l_2,$$

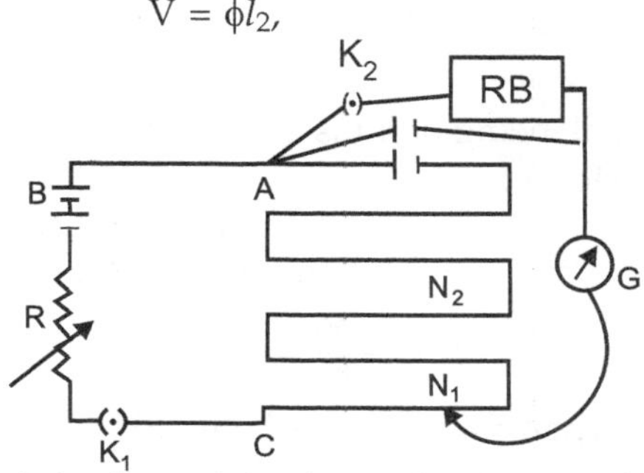


**Circuit for determining internal resistance of cell.**

$$\frac{\varepsilon}{V} = \frac{l_1}{l_2}$$

But,

$$\varepsilon = I(r + R) \text{ and } V = IR$$

Thus,

$$\frac{\varepsilon}{V} = \frac{r+R}{R} = \frac{l_1}{l_2}$$

$$r = R\left(\frac{l_1}{l_2} - 1\right)$$

where, $r$ is the internal resistance of cell.

(b) Potentiometer is preferred over voltmeter for measurement of e.m.f. of cell because a voltmeter draws some current from the cell while potentiometer draws no current. Therefore, the potentionmeter measures the actual e.m.f. of cell whereas voltmeter measures the terminal voltage.

(c)

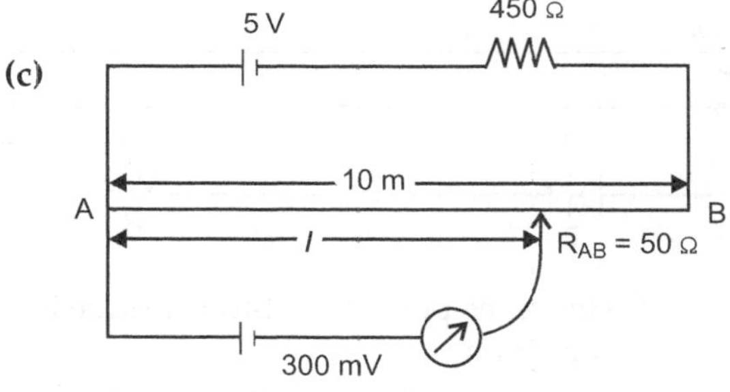


For current through AB,

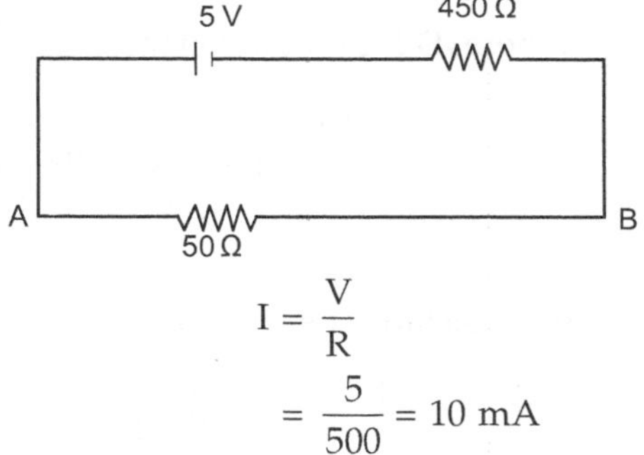


$$I = \frac{V}{R}$$

$$= \frac{5}{500} = 10 \text{ mA}$$

10 mA current passes through AB. Thus, voltage drop through AB = $V_{AB}$

$$= 10\times10^{-3} \times 50 = 500 \text{ mV}$$

Voltage drop per unit length

$$(\text{volts/m}) = \frac{500 \text{ mV}}{10}$$

$$= 50 \text{ mV/m}$$

balancing point is at 300 mV

Hence,

$$l = \frac{\text{Potential Difference}}{\text{Potential Gradiant}}$$

$$l = \frac{300}{50} = 6 \text{ m}$$

When 5V is replaced with 2V

$$I_{AB} = \frac{2}{500} = 4\text{mA}$$

$$V_{AB} = 4 \text{ mA} \times 50 = 200 \text{ mV}$$

Hence, balancing will not be possible as fall of potential along the potentiometer wire is less than emf of cell to be measured. It needs to cater to 300 mV.

**OR**

**(a)** Meter bridge is the practical apparatus which works on principle of Wheat-Stone bridge. It is used to measure unknown resistance experimentally.

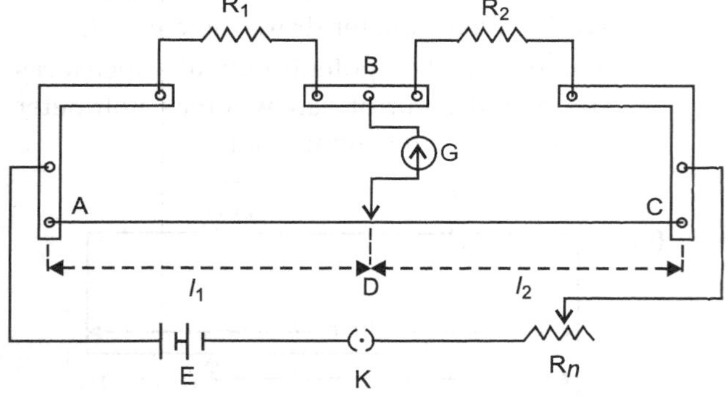


Hence, as per Wheat-Stone bridge balance condition :

$$\frac{R_1}{R_2} = \frac{\text{Resistance of wire AD}}{\text{Resistance of wire DC}}$$

Where, D is point of balance.

$$\frac{R_1}{R_2} = \frac{\rho l_1}{\rho l_2} = \frac{l_1}{l_2} = \frac{l_1}{(100 - l_1)}$$

$$(\because \; l_1 + l_2 = 100 \text{ cm})$$

$\rho$ = resistance per unit length of wire.

$$R_1 = R_2 \frac{l_1}{(100 - l_1)}$$

**(b) (i)** Connection between resistors are made of thick copper strips so that it will have maximum resistance and location of point of balance (D) will be more accurate which results in correct measurement of unknown resistance.

**(ii)** It is preferred to obtain the balance length near the mid-point of the bridge wire because it increases the sensitivity of meter bridges.

**(c)** From KCL (Kirchhoff's current law) at point D

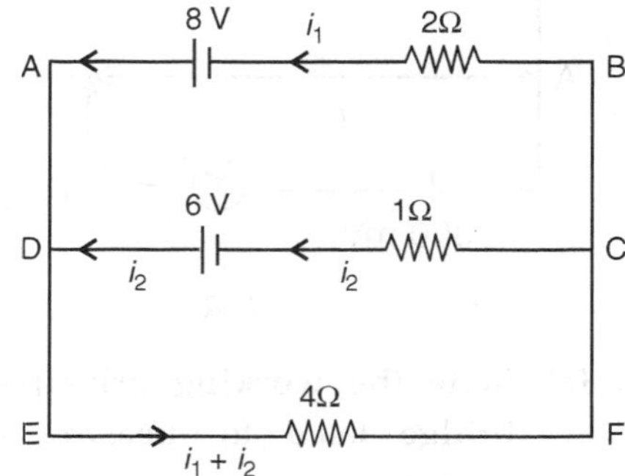


Current flowing through 4Ω resistance

$$i_1 + i_2 = i$$

KVL in loop DEFCD,

$$\Sigma V = 0$$

$$-(i_1 + i_2)4 - i_2 + 6 = 0$$

$$-4i_1 - 5i_2 + 6 = 0$$

(Voltage drop is negative and gain is positive)

$$4i_1 + 5i_2 = 6$$

KVL in loop AEFBA,

$$-(i_1 + i_2)\,4 - 2i_1 + 8 = 0$$

$$-4i_1 - 4i_2 - 2i_1 + 8 = 0$$

$$-6i_1 - 4i_2 + 8 = 0$$

$$6i_1 + 4i_2 = 8 \quad \text{....(i)}$$

$$4i_1 + 5i_2 = 6 \quad \text{....(ii)}$$

Afer solving both (i) and (ii)

$$i_1 = 8/7$$

$$i_2 = 2/7$$

$$i_1 + i_2 = \frac{10}{7}$$

So $$V = IR = \frac{10}{7} \times 4 = \frac{40}{7} \text{ V}$$

●●

# Outside Delhi [Set-II]

> **Note : Except for the following questions, all the remaining questions have been asked in previous set.**

## SECTION : A

1. **Write the relation for the force acting on a charged particle $q$ moving with velocity $\vec{v}$ in the presence of a magnetic field $\vec{B}$.** **[1]**

**Ans.** $\vec{F} = q(\vec{v} \times \vec{B})$

3. **Draw the pattern of electric field lines due to an electric dipole.** **[1]**

**Ans.**

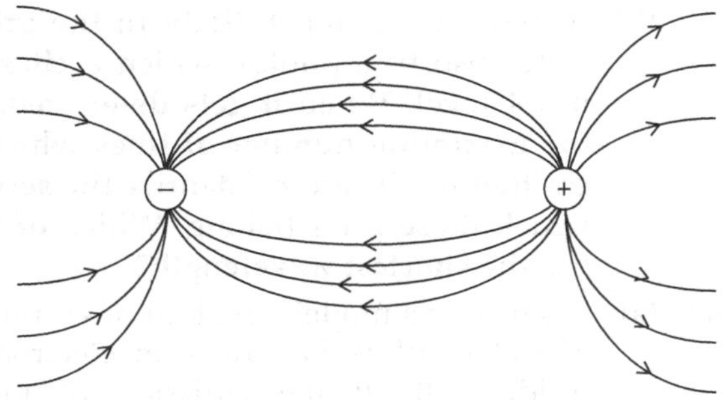


5. **Identify the semiconductor diode whose I-V characteristics are as shown.** **[1]**

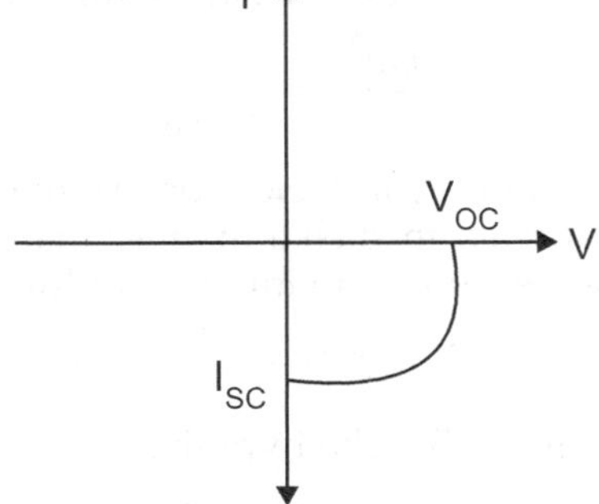


**Ans.** Photodiode/Solar cell.

## SECTION-B

7. **How is the equation for Ampere's circuital law modified in the presence of displacement current? Explain.** **[2]**

**Ans.**

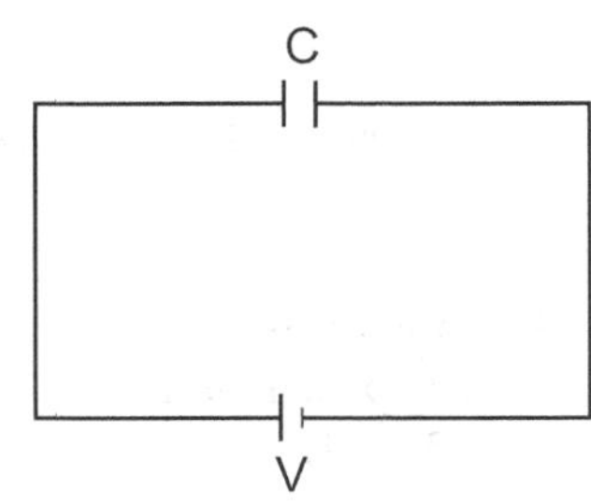


From Ampere's law,

$$\oint B \cdot dl = \mu_0 i(t)$$

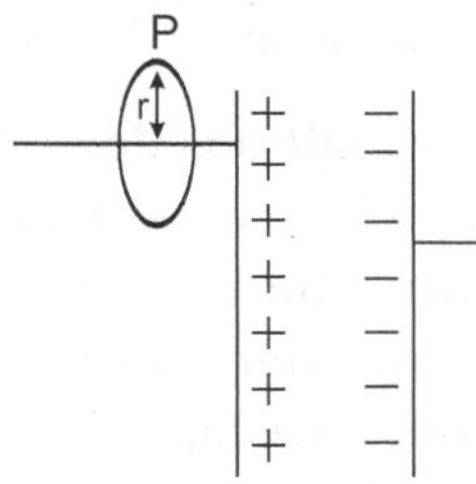


**Case-1**

Let, the case-1 where a point P is considered outside capacitor charging from Ampere's law. Magnetic field at point will be

$$B. (2\pi r) = \mu_0\, i(t)$$

$$B = \frac{\mu_0}{2\pi r}\, i(t)$$

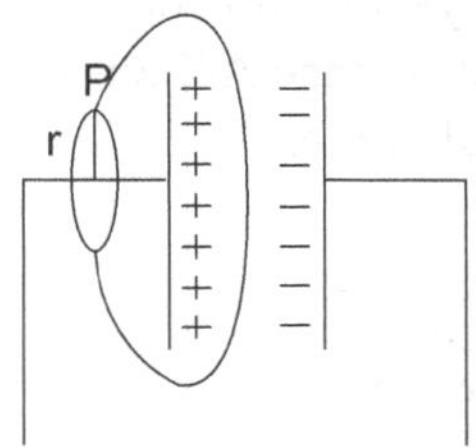


**Case-2**

Now take case-2 where shape surface under consideration covers capacitor's plate as we consider there is no current through capacitor then this value of B will be zero.

Hence, there is a contradiction.

Therefore this Ampere's law was modified with addition of displacement current inside capacitor.

$$\phi_E = |E|A = \frac{1}{\varepsilon_0} \cdot \frac{Q}{A} .A = \frac{Q}{\varepsilon_0}$$

$$\frac{d\phi_E}{dt} = \frac{1}{\varepsilon_0} \cdot \frac{dQ}{dt}$$

$$\varepsilon_0 \frac{d\phi_E}{dt} = \frac{dQ}{dt} = i_d$$

Where, $i_d$ is displacement current.

During charging of capacitor, outside the capacitor, $i_c$ (conduction current) flows and inside $i_d$ displacement current flows.

$$i = i_c + i_d = i_c + \varepsilon_0 \frac{d\phi_E}{dt}$$

Thus modified form of Amper's Circuital Law is :

$$\oint \vec{B} . \vec{dt} = \mu_0 \left(i_c + \varepsilon_0 \frac{d\phi\varepsilon}{dt}\right)$$

Outside the capacitor $i_d = 0$ hence, $i = i_c$

and inside the capacitor $i_c = 0$ hence, $i = i_d$.

**8. Under what conditions does the phenomenon of total internal reflection take place? Draw a ray diagram showing how a ray of light deviates by 90° after passing through a right-angled isosceles prism. [2]**

**Ans.** The phenomenon of total internal reflection occurs when,

1. Angle of incidence is equal or greater than critical angle.

$$i \geq C$$

2. When light travels from more denser medium to less denser medium.

In case of right angle isosceles triangle if light rays fall normally on AB then light incident on face AC with angle of incidence > critical angle.

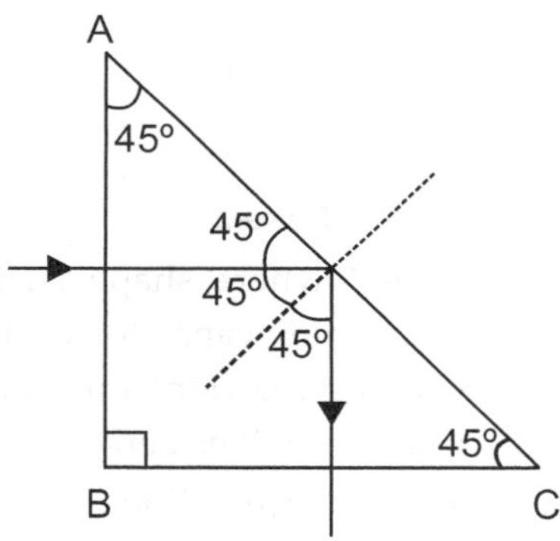


Hence, total internal reflection will occur with normal to the surface of BC.

**11. A beam of light converges at a point P. Draw ray diagrams to show where the beam will converge if (i) a convex lens, and (ii) a concave lens is kept in the path of the beam. [2]**

**Ans.** (i)

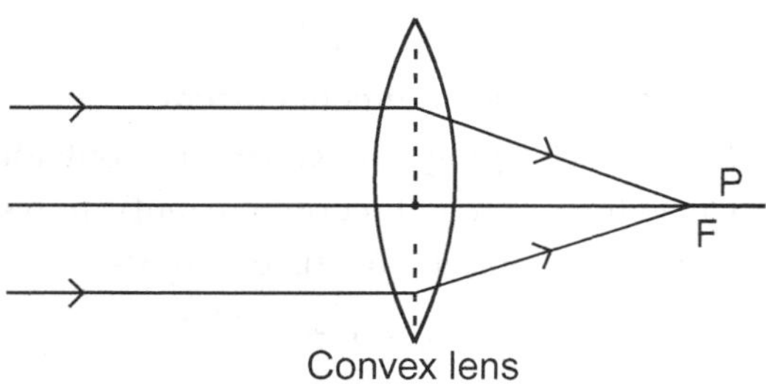


(ii)

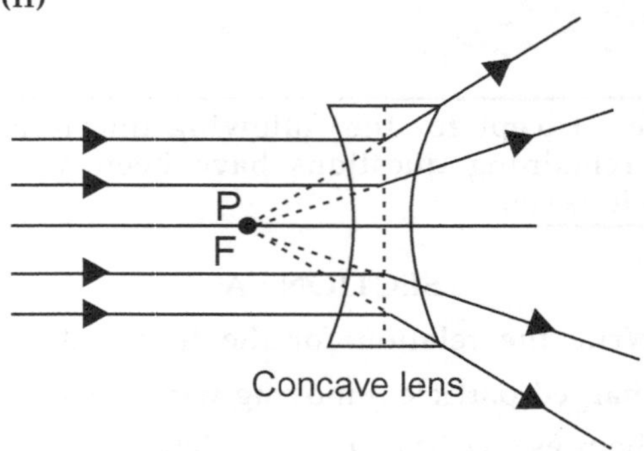


## SECTION : C

**14. (a) How is the stability of hydrogen atom in Bohr model explained by de-Broglie's hypothesis ?**

**(b) A hydrogen atom initially in the ground state absorbs a photon which excites it to n = 4 level. When it gets de-excited, find the maximum number of lines which are emitted by the atom. Identify the series to which these lines belong. Which of them has the shortest wavelength? [3]**

**Ans.** (a) From Bohr's model-An atom has a number of stable orbits in which an electron can reside without the emission of radiant energy. Each orbit corresponds to a certain energy level.

∵ Electron revolves in circular orbit so

$$\therefore \quad \frac{mv^2}{r} = \frac{e^2}{4\pi\varepsilon_0 r^2}$$

The motion of an electron in circular orbits is restricted in such a manner that its angular momentum is an integral multiple of $h/2\pi$

Thus, $\quad L = mvr = \dfrac{nh}{2\pi}$

From de-Broglie hypothesis

$$\lambda = \frac{h}{p} = \frac{h}{mv}$$

And from Bohr model

$$n\lambda = 2\pi r \text{ (for } n^{th} \text{ orbit)}$$

$$\lambda = \frac{2\pi r}{n}$$

$$\frac{h}{mv} = \frac{2\pi r}{n}$$

$$\frac{nh}{2\pi} = mvr = L$$

(b) $n_i = 1, n_f = 4$.

Possible transitions are :

$\Rightarrow$ $4 \to 3, 4 \to 2, 4 \to 1$

or $3 \to 2, 3 \to 1,$

or $2 \to 1$

Hence, Six lines are possible.

$\left.\begin{matrix}4\rightarrow 1\\ 3\rightarrow 1\\ 2\rightarrow 1\end{matrix}\right\}$ Lyman series

$\left.\begin{matrix}3\rightarrow 2\\ 4\rightarrow 2\end{matrix}\right\}$ Balmer series

$4 \rightarrow 3$ Paschen series

$4 \rightarrow 1$ has smallest wavelength.

**16. What is the reason to operate photodiodes in reverse bias?**

**A *p-n* photodiode is fabricated from a semiconductor with a band gap of range of 2·5 eV to 2·8 eV. Calculate the range of wavelengths of the radiation which can be detected by the photodiode. [3]**

**Ans.** Photodiode are reverse biased for working in photo-conductive mode. This reduces the response time because the additional reverse bias increases the width of the depletion layer, which decreases the junction capacitance. The reverse bias also increases the dark current without much change in the photo current.

Given, $E = 2.5 - 2.8$ eV

$$E = \frac{hc}{\lambda}$$

$$\lambda_1 = \frac{hc}{E_1}$$

$$= \frac{6.63\times10^{-34}\times3\times10^{8}}{2.5\times1.6\times10^{-19}}$$

$$= \frac{3\times6.63\times10^{-7}}{-2.5\times1.6}$$

$$= 497.25 \text{ nm}$$

$$\lambda_2 = \frac{hc}{E_2}$$

$$= \frac{6.63\times10^{-34}\times3\times10^{8}}{2.8\times1.6\times10^{-19}}$$

$$= 443.97 \text{ nm}$$

The range of wavelengths of radiation which can be detected by the photodiode is 443.97 nm to 497.25 nm.

**18. A ray of light incident on the face AB of an isosceles triangular prism makes an angle of incidence (i) and deviates by angle β as shown in the figure. Show that in the position of minimum deviation ∠β = ∠α. Also find out the condition when the refracted ray QR suffers total internal reflection. [3]**

**Ans.**

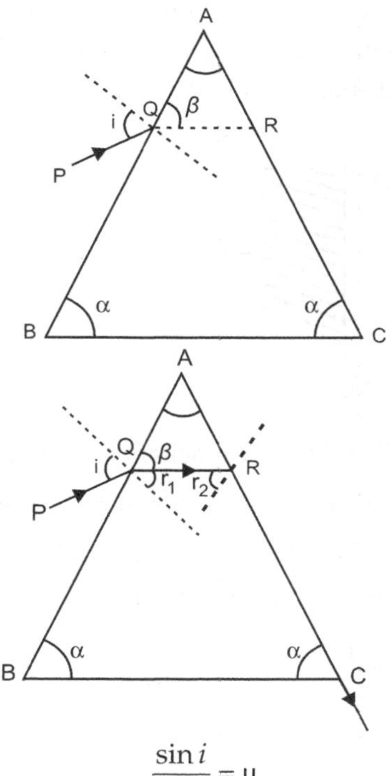


$$\frac{\sin i}{\sin r} = \mu$$

**(i)** Condition for minimum deviation :

**1.** $A = 180 - 2\alpha$

**2.** $\dfrac{\sin i}{\sin(90-\beta)} = \mu$

when, $r_1 = r_2 = r >$ critical angle

$$r_1 + r_2 = 180 - 2\alpha$$

$$2r = 180 - 2\alpha \qquad [\because r_1 = r_2]$$

$$r_1 = r = 90 - \alpha$$

$$\beta = 90 - r_1$$

$$= 90 - 90 + \alpha$$

$$\beta = \alpha$$

condition when QR have total internal reflection :

$$\angle QRC \geq \text{critical angle for the prism}$$

$$\angle 180° - \beta \geq \text{critical angle}$$

or $\angle 180° - \alpha \geq$ critical angle

**22. A 100 μF parallel plate capacitor having plate separation of 4 mm is charged by 200 V dc. The source is now disconnected. When the distance between the plates is doubled and a dielectric slab of thickness 4 mm and dielectric constant 5 is introduced between the plates, how will (i) its capacitance, (ii) the electric field between the plates, and (iii) energy density of the capacitor get affected? Justify your answer in each case. [3]**

**Ans.** Given,

$$c = 100\ \mu F$$

$$d = 4 \times 10^{-3} \text{m.}$$

$$V = 200V$$

$$k = 5$$

$$Q = CV$$

$$= 200 \times 100 \times 10^{-6}$$

$$= 2 \times 10^{-2} \text{ Coulomb}$$

As dielectric of 4 mm is inserted between the plates of capacitor and the spacing between the plates is doubled then it will acts as following Fig-A and Fig-B.

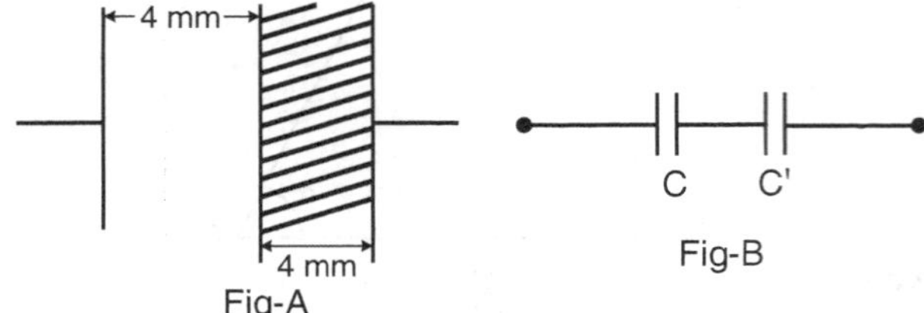

Fig-A
Fig-B

Here, C' will be capacitance with dielectric of 4 mm & 8 mm is separation between the plates.

$C' = KC = 5 \times 100 \times 10^{-6} = 0.5 \times 10^{-3}$ F

**(i)** Thus, equivalent capacitance

$$\frac{1}{C_{eq}} = \frac{1}{C} + \frac{1}{C'} = \frac{1}{100 \times 10^{-6}} + \frac{1}{0.5 \times 10^{-3}}$$

$$= 10 \times 10^3 + 2 \times 10^3$$

$$= 12 \times 10^3$$

$$\frac{1}{C_{eq}} = 12 \times 10^3$$

$$C_{eq} = \frac{1}{12 \times 10^3}$$

$$= \frac{1}{12} \times 10^{-3}$$

$$C_{eq} = 83.33\ \mu F$$

**(ii)** Charge on capacitor plates when 200V pd is applied.

$$q = C_0V_0$$

$$= 100 \times 10^{-6} \times 200$$

$$q = 2 \times 10^{-2} C$$

Even after battery is removed, the charge of $2 \times 10^{-2}$C on the capacitor plates remain the same.

So $C_0V_0 = CV = q$

$$V = \frac{C_0V_0}{C} = \frac{2 \times 10^{-2}}{83.33 \times 10^{-6}}$$

$$V = 240V$$

Now' $E' = \frac{V}{2d}$

$$= \frac{240}{2 \times 4 \times 10^{-3}}$$

$$E' = 30 \times 10^3 V/m$$

Electric field inside dielectric

$$= \frac{E'}{K}$$

$$= \frac{30 \times 10^3}{5}$$

$$= 6 \times 10^3\ V/m$$

**(iii)** Energy density of capacitor is given by :

$$U = \frac{Q^2}{2C}$$

$$= U' + U$$

$$= \frac{Q^2}{2C} + \frac{Q^2}{2C'}$$

$$\Rightarrow \frac{Q^2}{2}\left[\frac{1}{C} + \frac{1}{C'}\right] = \frac{2 \times 10^{-2} \times 2 \times 10^{-2}}{2}[12 \times 10^3]$$

$$\left(\because \frac{1}{C} + \frac{1}{C'} = \frac{1}{C_{eq}}\right)$$

$$= 2 \times 10^{-1} \times 12$$

$$= 2.4\ J$$

## Outside Delhi [Set-III]

**Note : Except for the following questions, all the remaining questions have been asked in previous sets.**

### SECTION : A

**2. Draw a pattern of electric field lines due to two positive charges placed a distance d apart. [1]**

**Ans.** Electric field lines due to two positive charge placed to at distance *d* apart :

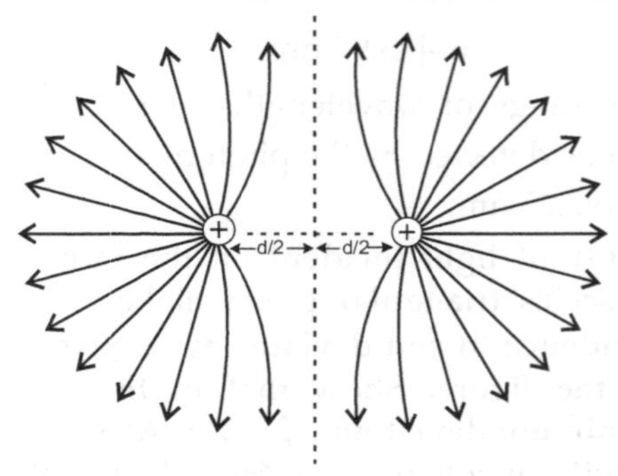


** Answer is not given due to change in the present Syllabus.

4. **When a charge $q$ is moving in the presence of electric (E) and magnetic (B) fields which are perpendicular to each other and also perpendicular to the velocity v of the particle, write the relation expressing v in terms of E and B.** **[1]**

**Ans.** $v = E/B$

5. **Draw the I-V characteristics of a Zener diode.** **[1]**

**Ans.**

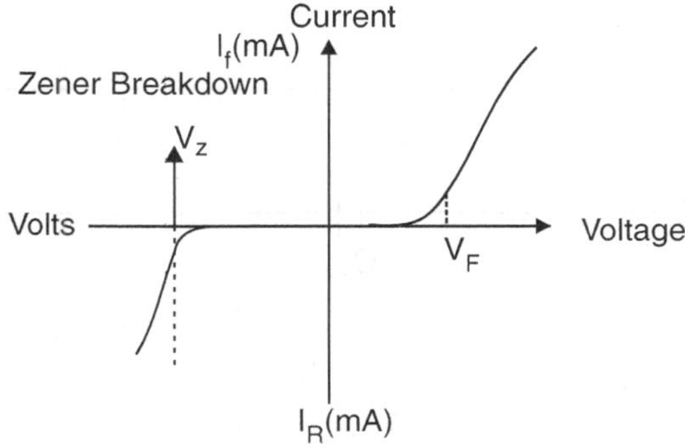


## SECTION : B

6. **State, with the help of a ray diagram, the working principle of optical fibres. Write one important use of optical fibres.** **[2]**

**Ans.**

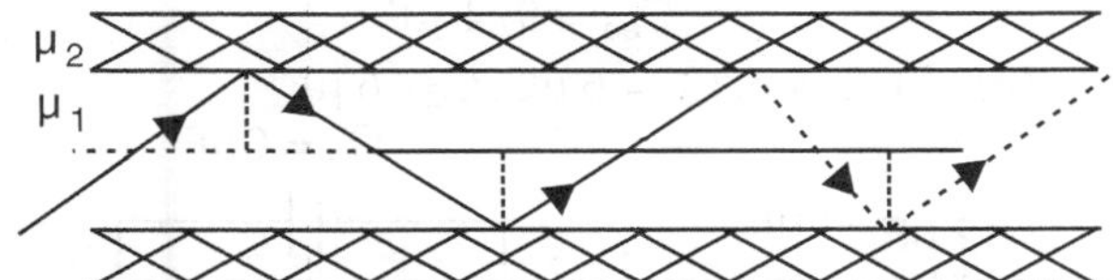


Optical fibre works on principle of total internal reflection. When angle of incidence is greater than critical angle, then incident rays are totally reflected back in same media.

When, $\theta_i > \theta_C$, Total internal reflection occurs and if $\theta_i < \theta_C$, refraction occurs.

**Application** : Optical fibre are used for communication due to very high bandwidth of media.

8. **How are electromagnetic waves produced by oscillating charges? What is the source of the energy associated with the EM waves ?**** **[2]**

9. **The wavelength of light from the spectral emission line of sodium is 590 nm. Find the kinetic energy at which the electron would have the same de-Broglie wavelength.** **[2]**

**Ans.** Given $\lambda = 590 \text{ nm} = 590 \times 10^{-9}\text{m}$

$KE = ?$

For same de-Broglie wavelength,

$$\text{De-Broglie wavelength } (\lambda) = \frac{h}{mv}$$

$$h\text{-planck constant} = 590\times10^{-9}$$

$$\frac{h}{m_e \times 590\times10^{-9}} = v$$

$$KE = \frac{1}{2} m_e v^2$$

$$= \frac{1}{2} \times m_e \times \frac{h^2}{m_e^2 \times (590\times10^{-9})^2}$$

$$= \frac{1}{2} \times \frac{h^2}{m_e \times (590 \times 10^{-9})^2}$$

$$= \frac{1}{2} \times \frac{6.62\times10^{-34} \times 6.62\times10^{-34} \times 10^9 \times 10^9}{9.1\times10^{-31} \times 590 \times 590}$$

$$KE = 6.91 \times 10^{-17} \text{ J}$$

## SECTION–C

13. **(a) Draw the energy level diagram for the line spectra representing Lyman series and Balmer series in the spectrum of hydrogen atom.**

**(b) Using the Rydberg formula for the spectrum of hydrogen atom, calculate the largest and shortest wavelengths of the emission lines of the Balmer series in the spectrum of hydrogen atom. (Use the value of Rydberg constant R = $1.1 \times 10^7$ m$^{-1}$.** **[3]**

**Ans. (a)** Energy level diagram showing Lyman and Balmer series :

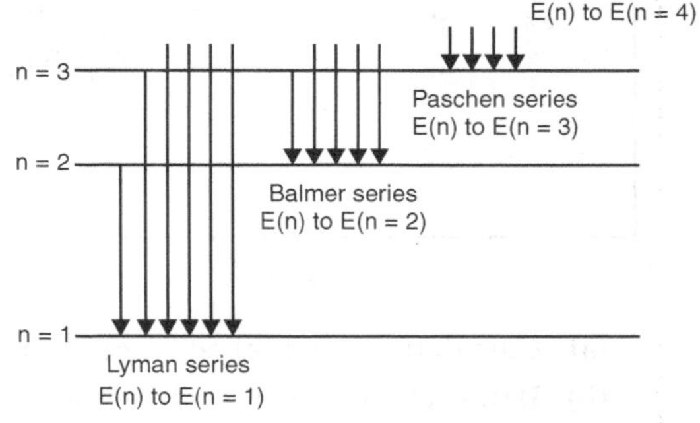


Spectrum wavelngths of both series for hydrogen atom

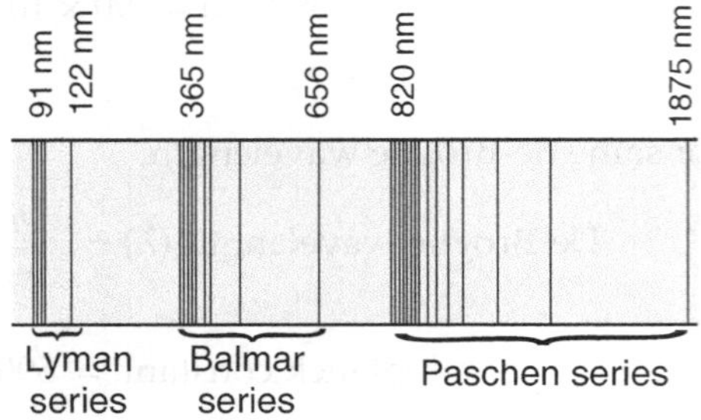


**(b)** Rydberg formula,

$$\frac{1}{\lambda} = 1.1 \times 10^7 \left[\frac{1}{n_1^2} - \frac{1}{n_2^2}\right]$$

$$\frac{1}{\lambda_1} = 1.1 \times 10^7 \left[\frac{1}{4} - \frac{1}{9}\right]$$

$$\frac{1}{\lambda_1} = 1.1 \times 10^7 \times 0.1389$$

$$\lambda_1 = \frac{1}{1.1} \times 10^7 \times 0.1389$$

$$\lambda_1 = \frac{100 \times 10^{-9}}{0.153}$$

$$\lambda_{max}\ (\lambda_1) = 653.6 \text{ nm}$$

$$\frac{1}{\lambda_2} = \frac{1.1 \times 10^7}{4}$$

$$\lambda_2 = \frac{4}{1.1 \times 10^7}$$

$$= \frac{400}{1.1} \times 10^{-9}$$

$$\lambda_{min}\ (\lambda_2) = 363.6 \text{ nm}$$

**15. In a network, four capacitors $C_1$, $C_2$, $C_3$ and $C_4$ are connected as shown in the figure. [3]**

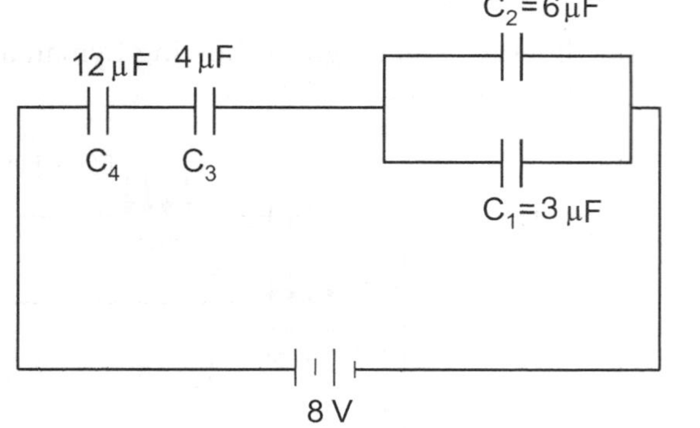


**(a) Calculate the net capacitance in the circuit.**

**(b) If the charge on the capacitor $C_1$ is 6 μC, (i) calculate the charge on the capacitors $C_3$ and $C_4$, and (ii) net energy stored in the capacitors $C_3$ and $C_4$ connected in series.**

**Ans. (a)**

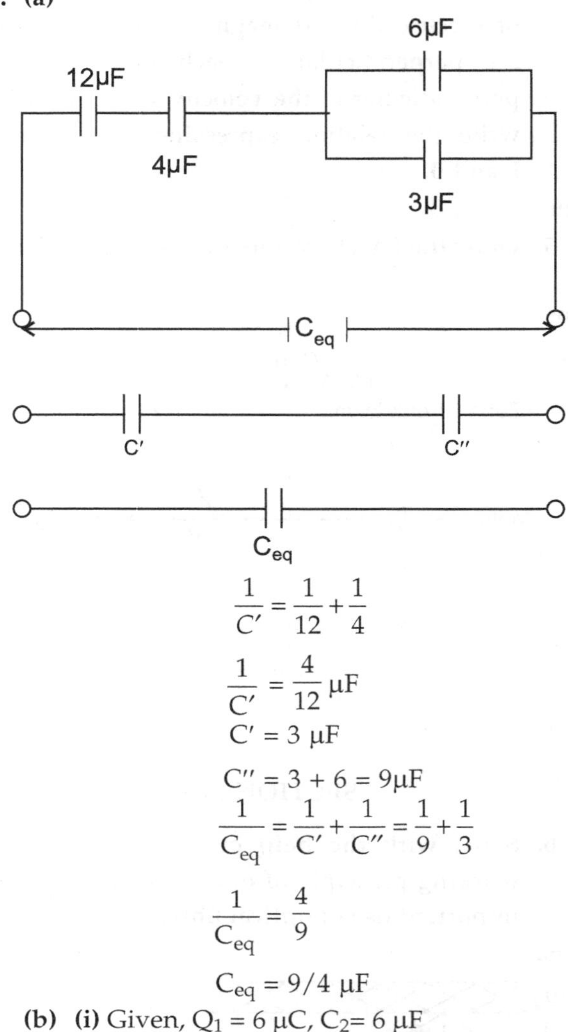


$$\frac{1}{C'} = \frac{1}{12} + \frac{1}{4}$$

$$\frac{1}{C'} = \frac{4}{12} \mu F$$

$$C' = 3\ \mu F$$

$$C'' = 3 + 6 = 9 \mu F$$

$$\frac{1}{C_{eq}} = \frac{1}{C'} + \frac{1}{C''} = \frac{1}{9} + \frac{1}{3}$$

$$\frac{1}{C_{eq}} = \frac{4}{9}$$

$$C_{eq} = 9/4\ \mu F$$

**(b) (i)** Given, $Q_1 = 6\ \mu C$, $C_2 = 6\ \mu F$

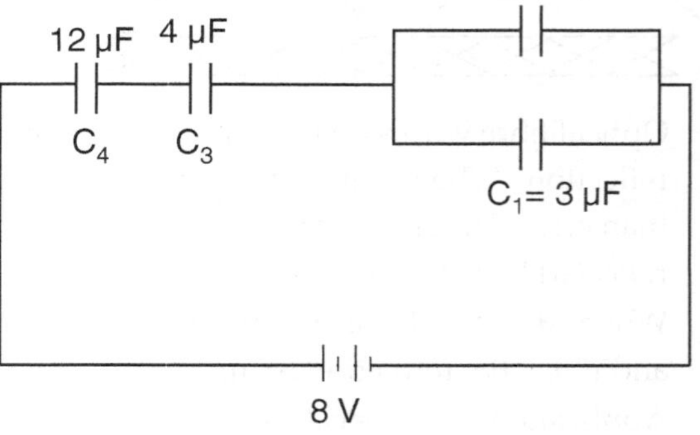


$Q_1$ is charge on $C_1$

$Q_2$ is charge on $C_2$

Q is net charge flows through $C_3$ and $C_4$

$$\frac{Q_1}{C_1} = V = \frac{Q_2}{C_2}$$

$$Q_2 = \frac{Q_1}{C_1} \times C_2 = \frac{6}{3} \times 6$$

$$Q_2 = 12\ \mu C$$

$$Q = Q_1 + Q_2 = 12 + 6 = 18\ \mu C$$

(ii) Voltage across $C_1$ and $C_2$ is 2 volts. Thus, voltage across combination of $C_4$ and $C_3$ will be

$$V = 8 - 2 = 6V$$

$$\frac{1}{C} = \frac{1}{C_3} + \frac{1}{C_4}$$

$$\frac{1}{C} = \frac{1}{4} + \frac{1}{12}$$

$$\frac{1}{C} = \frac{3+1}{12} = \frac{4}{12}$$

$$C = 3\ \mu F$$

Hence, Energy

$$E = \frac{1}{2}CV^2$$

$$= \frac{1}{2} \times 3 \times 10^{-6} \times 6^2 = \frac{36 \times 3 \times 10^{-6}}{2}$$

$$= 54 \times 10^{-6}\ J$$

**17. Draw the circuit diagram of a full wave rectifier. Explain its working principle. Show the input waveforms given to the diodes $D_1$ and $D_2$ and the corresponding output waveforms obtained at the load connected to the circuit. [3]**

**Ans.**

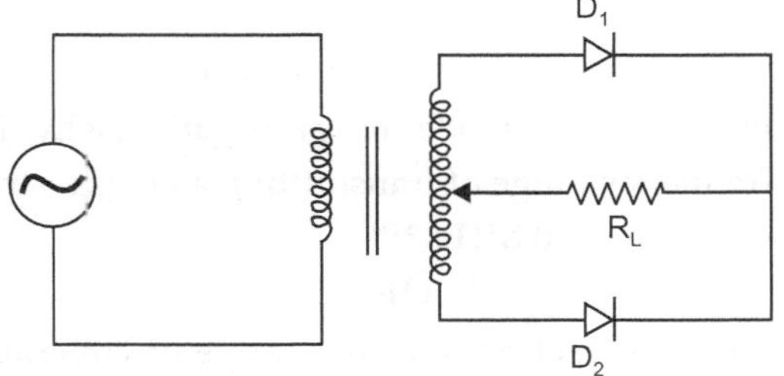


Full wave rectifier

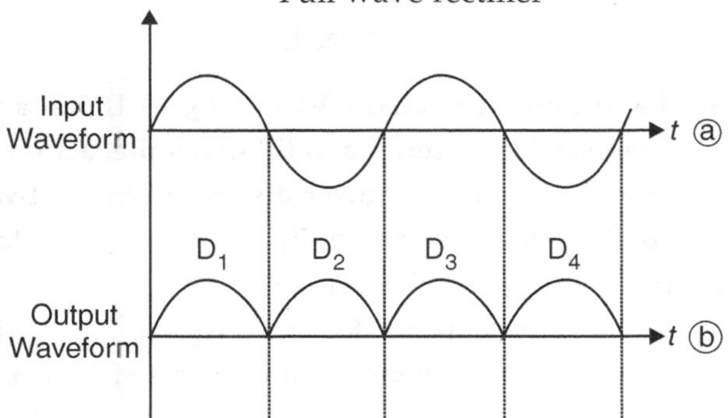


During first half of input sinusoidal ac signal, diode $D_1$ conducts as it is forward bias and during second half of input ac signal, diode $D_2$ conducts as it is forward bias now. $D_2$ and $D_1$ are inverse bias conditions during first and second half respectively and doesn't conduct.

Due to this output appears as waveform (b).

**21. (a) When a convex lens of focal length 30 cm is in contact with a concave lens of focal length 20 cm, find out if the system is converging or diverging.**

**(b) Obtain the expression for the angle of incidence of a ray of light which is incident on the face of a prism of refracting angle A so that it suffers total internal reflection at the other face. (Given the refractive index of the glass of the prism is μ) [3]**

**Ans. (a)**

$$f_1 = f_{convex} = 30 \text{ cm}$$

$$f_2 = f_{concave} = -20 \text{ cm}$$

$$\frac{1}{f} = \frac{1}{f_1} + \frac{1}{f_2} = \frac{1}{30} - \frac{1}{20} = \frac{2-3}{60}$$

$$\frac{1}{f} = -\frac{1}{60} \ [\therefore \text{ concave (diverging) lens}]$$

**(b)** Let C is the critical angle,

$$\sin C = \frac{1}{\mu}$$

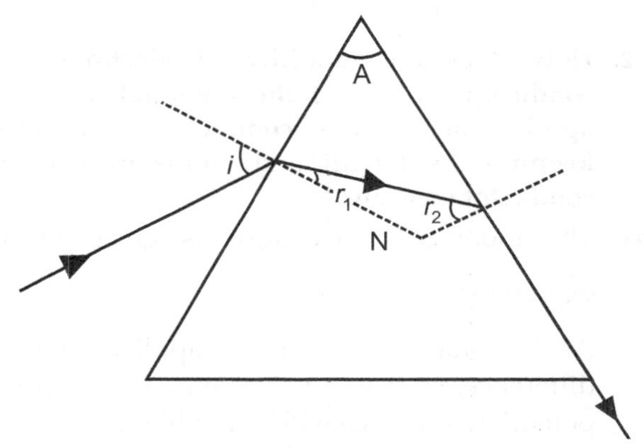


$$r_2 = C \text{ (for total internal reflection)}$$

$$r_1 + r_2 = A$$

$$r_1 + C = A$$

$$r_1 = A - C$$

$$\mu = \frac{\sin i}{\sin r_1}$$

$$\Rightarrow \sin i = \mu \sin r_1$$

$$\sin i = \mu \sin (A - C)$$

$$i = \sin^{-1} [\mu \sin (A - C)]$$

$$i = \sin^{-1}\left[\frac{\sin (A - C)}{\sin C}\right]$$

# PHYSICS

## Delhi [Set-I]

# 2019

**Time allowed : 3 hours** **Maximum marks : 70**

### SECTION-A

1. **Draw the pattern of electric field lines, when a point charge $-Q$ is kept near an uncharged conducting plate.** **[1]**

**Ans.** The positive charge will be induced on the uncharged conducting plate, kept near it. So, the lines of forces will start from metal plate and end at – Q.

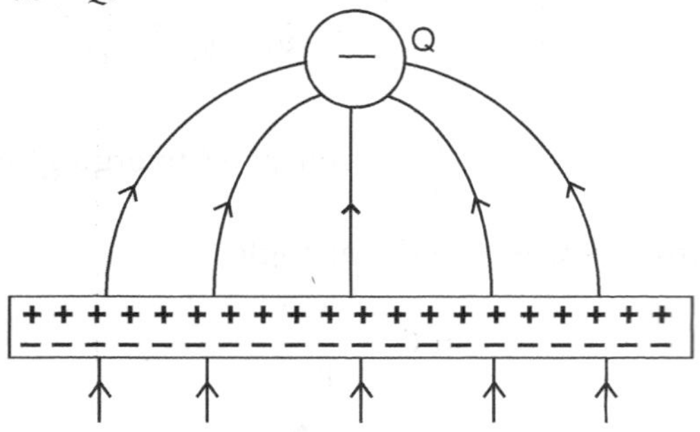


2. **How does the mobility of electrons in a conductor change, if the potential difference applied across the conductor is doubled, keeping the length and temperature of the conductor constant ?** **[1]**

**Ans.** The mobility of electrons is given by the expression, $\mu = \frac{er}{m}$

As it's independent of the applied potential difference, so it will not change if the applied potential difference will be doubled.

3. **Define the term "threshold frequency" in the context of photoelectric emission.** **[1]**

**OR**

**Define the term 'Intensity' in photon picture of electromagnetic radiation.**

**Ans.** The minimum frequency of the radiation incident on a metal surface below which there is no photoelectric emission is called threshold frequency. It is a characteristic property of a photosensitive material.

**OR**

The amount of light energy or photon energy in terms of number of energy quanta (photons) incident per meter square per second is called intensity of radiation.

If total energy emitted = $nh\nu$

Then $$I = \frac{nh\nu}{At}$$

4. **What is the speed of light in a denser medium of polarising angle 30° ?** **[1]**

**Ans.** Given : $i_p = 30^\circ$

since, $\mu = \tan i_p$ (By Brewster's law)

$\therefore$ $\mu = \tan 30^\circ$

$\Rightarrow$ $\mu = \frac{1}{\sqrt{3}}$

Since, $\mu = \frac{C}{v}$

$\therefore$ $v$ = velocity of light in medium $= \frac{C}{\mu}$

$$= \frac{3 \times 10^8}{1/\sqrt{3}}$$

$$= 3\sqrt{3} \times 10^8 \text{ m/s}$$

$$= 5.196 \times 10^8 \text{ m/s}$$

5. **In sky wave mode of propagation, why is the frequency range of transmitting signals restricted to less than 30 MHz ?**** **[1]**

**OR**

**On what factors does the range of coverage in ground wave propagation depend ?****

### SECTION-B

6. **Two bulbs are rated ($P_1$, V) and ($P_2$, V). If they are connected (i) in series and (ii) in parallel across a supply V, find the power dissipated in the two combinations in terms of $P_1$ and $P_2$.** **[2]**

**Ans.** (i) **In series combination :**

Net resistance, $R' = R_1 + R_2$ ...(i)

As heating elements are operated at same voltage V, we have

$$R = \frac{V^2}{P}, R_1 = \frac{V^2}{P_1} \text{ and } R_2 = \frac{V^2}{P_2}$$

$\therefore$ From equation (i)

$$\frac{V^2}{P} = \frac{V^2}{P_1} + \frac{V^2}{P_2}$$

$$\Rightarrow \quad \frac{1}{P} = \frac{1}{P_1} + \frac{1}{P_2}$$

** Answer is not given due to change in the present Syllabus.

$\Rightarrow$ Net power dissipated :

$$\Rightarrow \quad P = \frac{P_1 P_2}{P_1 + P_2}$$

**(ii) In parallel combination :**

Net resistance,

$$\frac{1}{R} = \frac{1}{R_1} + \frac{1}{R_2} \quad \text{...(ii)}$$

$$\Rightarrow \quad \frac{P}{V^2} = \frac{P_1}{V^2} + \frac{P_2}{V^2}$$

$$\left[\because R = \frac{V^2}{P}, R_1 = \frac{V^2}{P_1}, R_2 = \frac{V^2}{P_2}\right]$$

$$\Rightarrow \quad P = P_1 + P_2$$

Hence, net power dissipated is $P = P_1 + P_2$.

**7. Calculate the radius of curvature of an equiconcave lens of refractive index 1.5, when it is kept in a medium of refractive index 1.4, to have a power of –5 D ? [2]**

**OR**

**An equilateral glass prism has a refractive index 1.6 in air. Calculate the angle of minimum deviation of the prism, when kept in a medium of refractive index $4\sqrt{2}/5$.**

**Ans.** In an equi-concave lens, radius of curvature of both surfaces are equal

$$\therefore \quad \frac{1}{f} = (\mu - 1)\left(-\frac{1}{R} - \frac{1}{R}\right)$$

since, $\quad P = -5D$

$$\therefore \quad \frac{1}{f(\text{m})} = P = -5$$

$$\Rightarrow \quad f = \frac{-100}{5}\text{cm} = -20 \text{ cm}$$

and $\quad \mu_e = 1.5$ and $\mu_m = 1.4$

$$\therefore \quad \frac{1}{-20} = \left(\frac{1.5}{1.4} - 1\right)\left(\frac{-2}{R}\right)$$

$$\Rightarrow \quad \frac{1}{-20} = \frac{0.1}{1.4} \times \frac{-2}{R}$$

$$\Rightarrow \quad R = \frac{0.1 \times (-2)(-20)}{1.4}$$

$$= \frac{20}{7} \text{ cm } (= 2.86 \text{ cm})$$

**OR**

As we know,

$${}_1n_2 = \frac{\sin\left(\frac{A + \delta_{min}}{2}\right)}{\sin\frac{A}{2}}$$

where, $n_2 = 1.6$, refractive index of glass prism

$n_1 = \frac{4\sqrt{2}}{5}$, refractive index of medium

$A = 60°$, angle of prism

$\delta_{min}$ = angle of minimum deviation

$$\therefore \quad \frac{1.6}{4\sqrt{2}/5} = \frac{\sin\left(\frac{60° + \delta_m}{2}\right)}{\sin 30}$$

$$\Rightarrow \quad \frac{0.4 \times 5}{\sqrt{2}} = \sin\left(\frac{60° + \delta_m}{2}\right) \times 2$$

$$\left[\because \sin 30° = \frac{1}{2}\right]$$

$$\Rightarrow \quad \sin\left(\frac{60° + \delta_m}{2}\right) = \frac{1}{\sqrt{2}}$$

$$\Rightarrow \quad \frac{60° + \delta_m}{2} = \sin^{-1}\left(\frac{1}{\sqrt{2}}\right)$$

$$= \sin^{-1}(\sin 45°) = 45°$$

$$\Rightarrow \quad \delta_m = 30°$$

**8. An α-particle and a proton of the same kinetic energy are in turn allowed to pass through a magnetic field $\vec{B}$, acting normal to the direction of motion of the particles. Calculate the ratio of radii of the circular paths described by them. [2]**

**Ans.** As we know, radius of the charged particle is given by

$$r = \frac{\sqrt{2mK}}{qB}$$

Since, K. E. of particles and $\vec{B}$ are equal.

$\therefore$ For α- particle

$$r_\alpha = \frac{\sqrt{2m_\alpha K}}{q_\alpha B} \quad \text{...(i)}$$

For proton,

$$r_p = \frac{\sqrt{2m_p K}}{q_p B} \quad \text{...(ii)}$$

on comparing equation (i) and (ii)

$$\frac{r_\alpha}{r_p} = \frac{\sqrt{m_\alpha} \times q_p}{q_\alpha \times \sqrt{m_p}}$$

since, $m_\alpha = 4m_p$ and $q_\alpha = 2q_p$

$$\therefore \quad \frac{r_\alpha}{r_p} = \frac{\sqrt{4m_p} \cdot q_p}{\sqrt{m_p} \cdot 2q_p} = \frac{1}{1}$$

$$\Rightarrow \quad r_\alpha : r_p = 1 : 1$$

**9. State Bohr's quantization condition of angular momentum. Calculate the shortest wavelength of the Bracket series and state to which part of the electromagnetic spectrum does it belong. [2]**

OR

**Calculate the orbital period of the electron in the first excited state of hydrogen atom.**

**Ans.** According to Bohr's quantization of angular momentum, the stationary orbits are those in which angular momentum of electron is an integral multiple of $\frac{h}{2\pi}$ *i.e.,*

$$mvr = n\frac{h}{2\pi}$$

Where, $n = 1, 2, 3$

For Bracket series, $n_1$=2 and $n_2 = \infty$

$$\frac{1}{\lambda_{min}} = R.\left(\frac{1}{n^2}\right)$$

$$\Rightarrow \frac{1}{\lambda_{min}} = \frac{R}{16} \qquad [\because n = 4]$$

Where, R = Rydberg constant

$$= 1.096 \times 10^7 \text{ m}^{-1}$$

$$\Rightarrow \lambda_{min} = \frac{16}{1.096 \times 10^7} \text{m}$$

$$= 14.599 \times 10^{-7}$$

$$\Rightarrow = 14599 \text{ Å}$$

Bracket series is found in infrared region ($10^6$ Å to 7000 Å) of electromagnetic spectrum.

OR

$$r = 0.53\frac{n^2}{Z} \times 10^{-10} m$$

for first excited state $n = 2$ and $Z = 1$

$$r = 0.53 \times \frac{2^2}{1} \times 10^{-10}$$

$$r = 2.12 \times 10^{-10} m$$

$$v = v_0 \times \frac{z}{n} m/s$$

$$v = 2.188 \times 10^6 \times \frac{z}{n} m/s$$

$$v = 2.188 \times 10^6 \times \frac{1}{2}$$

$$[n = 2, z = 1]$$

$$v = 1.094 \times 10^6 m/s$$

$$\text{orbital period} = \frac{2\pi r}{v}$$

$$= \frac{2 \times 3.14 \times 2.12 \times 10^{-10}}{1.094 \times 10^6}$$

$$= 1.22 \times 10^{-15} \text{ sec}$$

**10. Why a signal transmitted from a TV tower cannot be received beyond a certain distance? Write the expression for the optimum separation between the receiving and the transmitting antenna.** [2]**

** Answer is not given due to change in the present Syllabus.

**11. Why is wave theory of electromagnetic radiation not able to explain photo electric effect ? How does photon picture resolve this problem ? [2]**

**Ans.** Wave theory cannot explain the photoelectric effect due to following reasons:

**(i)** The instantaneous emission of photoelectrons.

**(ii)** Existence of threshold frequency for metal surface.

**(iii)** K.E. of emitted electrons is independent of intensity of light and depends on frequency.

Photo theory can explain the photoelectric effect due to following reasons::

1. Increasing the intensity means increasing the number of photons that does not change the maximum K.E. but changes the number of affected electrons.
2. E = $h\nu$ means Energy is dependent on the frequency. Hence threshold frequency is there for photon emission.
3. As photon strikes metallic surface it is absorbed and hence electron is ejected instantaneously.

**12. Plot a graph showing variation of de Broglie wavelength (λ) associated with a charged particle of a mass *m*, versus $1/\sqrt{V}$, where V is the potential difference through which the particle is accelerated. How does this graph give us the information regarding the magnitude of the charge of the particle ? [2]**

**Ans.**

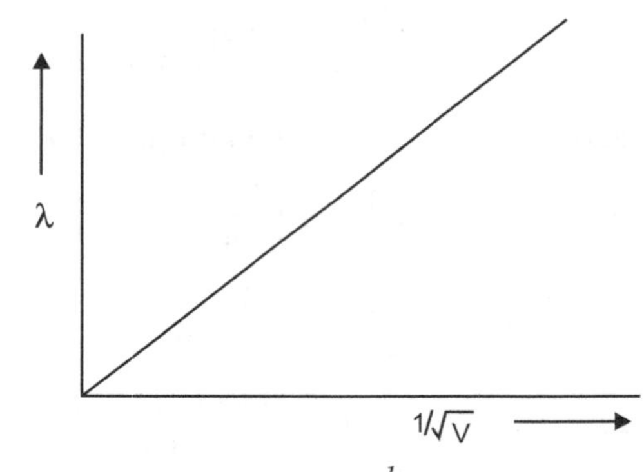


Since, $$\lambda = \frac{h}{\sqrt{2mqV}}$$

$$\Rightarrow \frac{\lambda}{\left(\frac{1}{\sqrt{V}}\right)} = \frac{h}{\sqrt{2mq}} = \text{slope}$$

*i.e.,* $$q = \frac{h^2}{2m\,(\text{slope})^2}$$

∴ By slope and mass, we can calculate the charge.

## SECTION-C

**13. (a) Draw the equipotential surfaces corresponding to a uniform electric field in the z-direction.**

**(b) Derive an expression for the electric potential at any point along the axial line of an electric dipole. [3]**

**Ans. (a)**

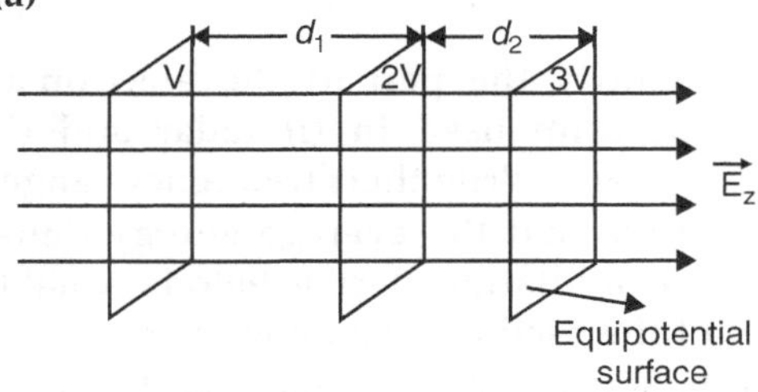


For uniform electric field

$$d_1 = d_2$$

**(b) Electric potential due to dipole along axial line :**

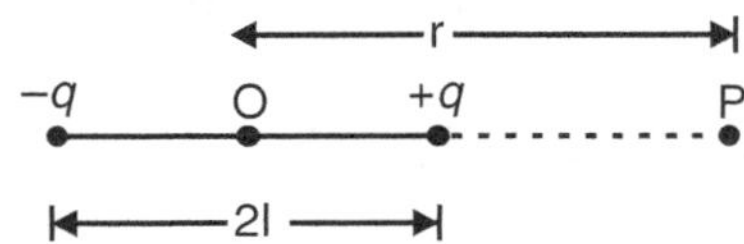


The figure shows a dipole consisting of two equal and opposite charges $+q$ and $-q$ separated by a distance $2l$. Let, P be a point on the end-on position of the dipole. OP is the axial line of dipole.

The potential due to charge $+q$ at point P $= V_1$

$$V_1 = \frac{1}{4\pi\varepsilon_0} \times \frac{q}{(r-l)}$$

The potential due to charge $-q$ at point P $= V_2$

$$V_2 = \frac{1}{4\pi\varepsilon_0} \times \frac{(-q)}{(r+l)}$$

Total potential at point P $= V = V_1 + V_2$

$$V = \frac{1}{4\pi\varepsilon_0} \times \frac{q}{(r-l)} + \frac{1}{4\pi\varepsilon_0} \times \frac{(-q)}{(r+l)}$$

$$= \frac{q}{4\pi\varepsilon_0}\left[\frac{1}{r-l} - \frac{1}{r+l}\right]$$

$$= \frac{q}{4\pi\varepsilon_0}\left[\frac{r+l-r+l}{(r-l)(r+l)}\right]$$

$$= \frac{1}{4\pi\varepsilon_0} \frac{2ql}{(r-l)(r+l)}$$

$$= \frac{1}{4\pi\varepsilon_0} \frac{2ql}{(r^2-l^2)}$$

But the dipole moment, $P = 2lq$

So, $$V = \frac{1}{4\pi\varepsilon_0} \times \frac{P}{(r^2-l^2)}$$

**14. Using Kirchhoff's rules, calculate the current through the 40 Ω and 20 Ω resistors in the following circuit : [3]**

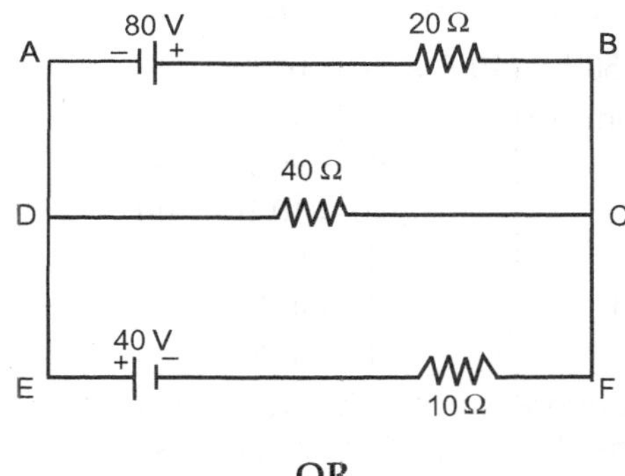


**OR**

**What is end error in a metre bridge? How is it overcome? The resistances in the two arms of the metre bridge are R = 5 Ω and S respectively When the resistance S is shunted with an equal resistance, the new balance length found to be 1.5 $l_1$, where $l_1$ is the initial balancing length. Calculate the value of S.**

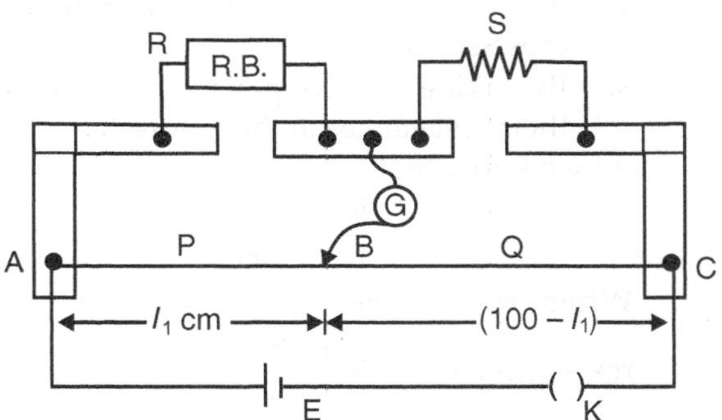


**Ans.**

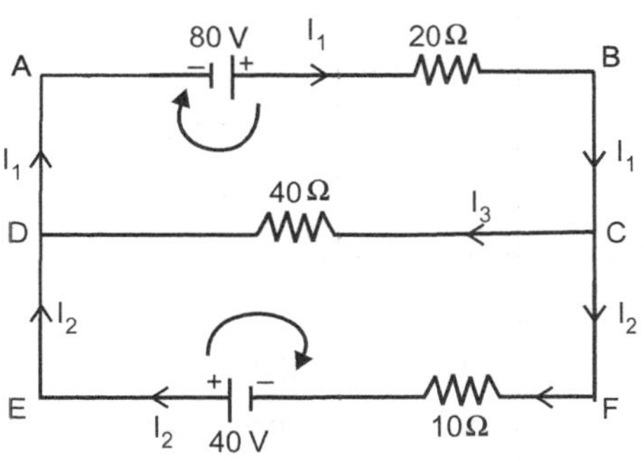


Applying Kirchhoff's first law at junction C

$$I_1 = I_2 + I_3$$

$$I_2 = I_1 - I_3 \quad \text{...(i)}$$

Applying the Kirchhoff's second law in closed loop ABCDA

$$80-20\,I_1 - 40I_3 = 0 \quad ...(ii)$$

Applying Kirchhoff's second law in closed loop DCFED

$$40I_3 - 10\,I_2 + 40 = 0 \quad ...(iii)$$

From equations (i) and (iii),

$$40\,I_3 - 10\,(I_1 - I_3) + 40 = 0$$

$$50\,I_3 - 10\,I_1 + 40 = 0 \quad ...(iv)$$

From equations (ii) and (iv)

$$80 = 20\,I_1 + 40\,I_3$$

$$40 = 10\,I_1 - 50\,I_3$$

We get, $I_3 = 0$

and $I_1 = 4$ A

$I_2 = 4$ A

Therefore, current through 40 Ω resistor is 0 and current through 20 Ω resistor is 4 A.

**OR**

The shifting of zero of the scale at different points gives rise to the end error in metre bridge wire.

This can be overcome by,

1. By trying to obtain the balance point in the middle of the metre bridge with two interchanged positions of R and S and taking the average value of '*s*' from two readings.
2. By taking metre bridge wire of uniform cross-section.
3. By taking the copper strips thick, so that their resistance can be ignored.

As we know that,

$$\frac{R}{S} = \frac{l_1}{100 - l_1} \quad ...(i)$$

When resistance S is shunted with equal resistance. Then, $S' = \frac{S}{2}$

Therefore, new balancing point is,

$$\Rightarrow \quad \frac{R}{S/2} = \frac{1.5\,l_1}{100 - 1.5\,l_1}$$

$$\Rightarrow \quad \frac{R}{S} = \frac{1.5\,l_1}{2(100 - 1.5\,l_1)} \quad ...(ii)$$

On comparing equations (i) and (ii)

$$\frac{l_1}{100 - l_1} = \frac{1.5\,l_1}{2(100 - 1.5\,l_1)}$$

$$200 - 3l_1 = 150 - 1.5l_1$$

$$1.5\,l_1 = 50$$

$$l_1 = \frac{50}{1.5}$$

From equation (i),

$$\therefore \quad \frac{5}{S} = \frac{50}{1.50} \div \left(100 - \frac{50}{1.5}\right)$$

$$(\because R = 5)$$

$$\Rightarrow \quad \frac{5}{S} = \frac{50 \times 1.5}{1.5(100)}$$

$$\Rightarrow \quad S = \frac{500}{50} = 10\ \Omega$$

**15. (a) Identify the part of the electromagnetic spectrum used in (i) radar and (ii) eye surgery. Write their frequency range.****

**(b) Prove that the average energy density of the oscillating electric field is equal to that of the oscillating magnetic field. [3]**

**Ans. (b)** Energy density in oscillating electric field is :

$$u_E = \frac{1}{2}\varepsilon_0 E^2$$

Energy density in oscillating magnetic field is

$$u_B = \frac{1}{(2\mu_0)}B^2$$

We know, $E = cB$

And $c = \frac{1}{\sqrt{\mu_0\varepsilon_0}}$

So,

$$u_E = \frac{1}{2}\varepsilon_0 (cB)^2 = \frac{1}{2}\varepsilon_0 \frac{1}{\mu_0\varepsilon_0} \times B^2$$

$$= \frac{1}{(2\mu_0)}B^2$$

So, $u_E = u_B$

**16. Define the term wavefront. Using Huygen's wave theory, verify the law of reflection. [3]**

**OR**

**Define the term, "refractive index" of a medium. Verify Snell's law of refraction when a plane wavefront is propagating from a denser to a rarer medium.**

**Ans.** A wavefront is the continuous locus of vibrating particles which are in the same state of vibration or phase at given instant of time.

**Laws of Reflection from Huygen's principle**

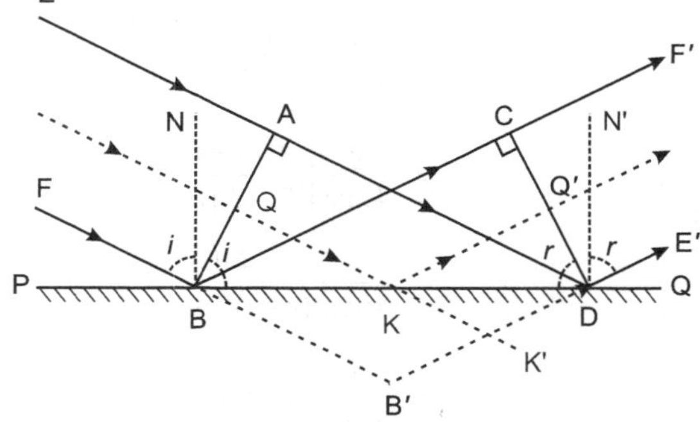


** Answer is not given due to change in the present Syllabus.

Consider a plane wavefront AB incident at an angle '$i$' on a plane reflecting surface PQ, Let $v$ be the velocity of the wave. At time $t = 0$, one end of the wavefront just touches the reflecting surface at B. Draw normal NB to PQ. When the wavefront strikes the reflecting surface, then due to the presence of it, it cannot advance further. When wavefront strikes at B, secondary wavelets starts emitting from B. The secondary wavelets will travel a distance AD = $vt$ during the time the other end A of the wavefront AB reaches the surface PQ at D.

To find the reflected wavefront, B as a centre and AD as radius draw an arc, which represent the secondary wavelets originating from B. As the incident wavefront AB advances, the secondary wavelets will touch CD simultaneously.

According to Huygen's principle, CD represents the reflected wavefront corresponding to incident wavefront AB. BD is common in triangles BAD and CBD and BC = AD = $vt$. Therefore, two triangles are congruent.

So, $\angle ABD = \angle CDB$

*i.e.*, $\angle i = \angle r$

Thus, angle of incidence is equal to the angle of reflection. This is the second law of reflection. Also, the incident wavefront AB, reflecting surface PQ and the reflected wavefront CD are perpendicular to the plane of the paper. So, the incident ray, reflected ray and the normal at the point of incidence, all lie in same plane. This is first law of reflection.

Thus, Huygen's principle explains both the laws of reflection.

**OR**

The refractive index of a medium is defined as the ratio of speed of light in vacuum to that of the speed of the light in medium.

*i.e.*, $$n = \frac{c}{v}$$

Where, $n$ = refractive index of medium when light ray passes from vaccum into a medium.

$c$ = velocity of light in vacuum

$v$ = velocity of light in the medium

**Proof of Snell's law of refraction**

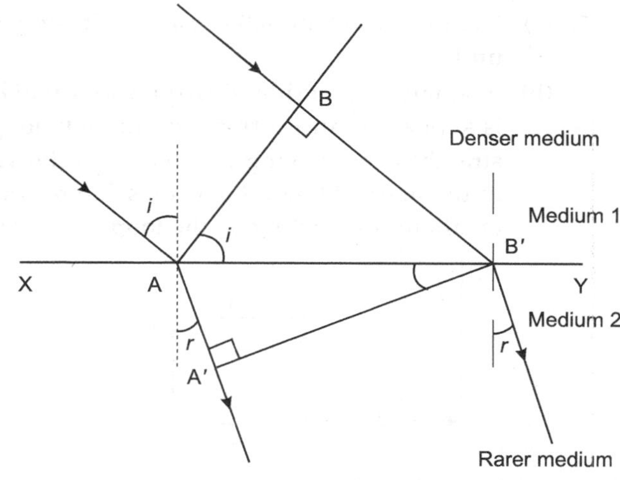


When a wavefront travels from one medium to other, it deviates from its path. In travelling from one medium to other, the frequency of wave remains same and speed and wavelength changes. Let, XY be a surface separating two media '1' and '2'. Let the speed of waves of $v_1$ and $v_2$.

Suppose, a plane wavefront AB in first medium is incident obliquely on the boundary surface XY and its end touches the surface at A at time $t = 0$, while the other end B reaches the surface at point B' after time-interval '$t$'.

Clearly, BB' = $v_1 t$.

In the same time, wavelets starts from A and reaches A' in time '$t$' with velocity $v_2$.

Therefore, $AA' = v_2 t$

According to Huygen's principle, A'B' is the new position of the wavefront in second medium. A'B' is a refracted wavefront.

Let, the incident wavefront (AB) and reflected wavefront (A'B') makes angle $i$ and $r$ with surface XY.

In $\Delta AB'B$,

$$\angle ABB' = 90^\circ$$

$$\therefore \quad \sin i = \frac{BB'}{AB'} = \frac{v_1 t}{AB'} \quad ...(i)$$

In $\Delta AA'B'$,

$$\angle AA'B' = 90^\circ$$

$$\therefore \quad \sin r = \frac{AA'}{AB'} = \frac{v_2 t}{AB'} \quad ...(ii)$$

Dividing equation (i) by (ii), we get

$$\frac{\sin i}{\sin r} = \frac{v_1}{v_2} = \text{constant}$$

Hence, ratio of sine of angle of incidence and the sine of angle of refraction for a given pair of media is constant. This is Snell's law of refraction.

**17. (a) Define mutual inductance and write its S.I. unit.**

**(b) A square loop of side '$a$' carrying a current $I_2$ is kept at distance $x$ from an infinitely long straight wire carrying a current $I_1$ as shown in the figure. Obtain the expression for the resultant force acting on the loop. [3]**

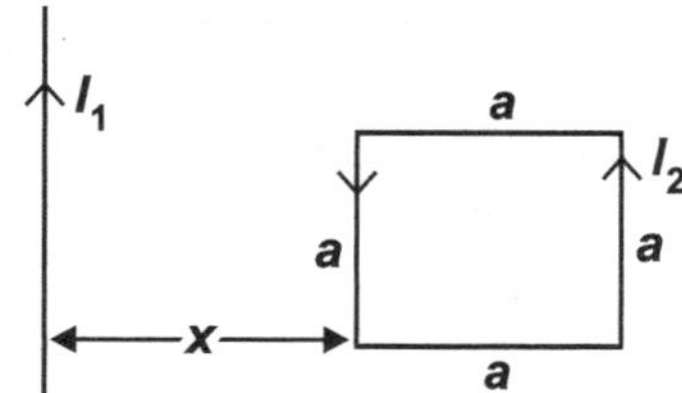


**Ans. (a) Mutual inductance :** Mutual inductance equals the magnetic flux associated with a coil when unit current flows in its neighbouring coil.

Its SI unit is henry (H).

**(b)**

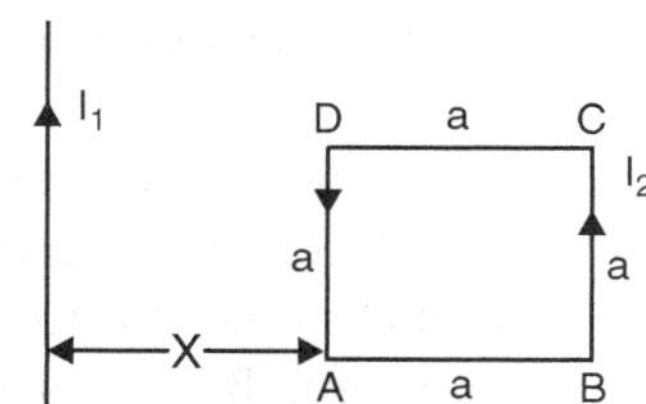


The resultant magnetic force of sides AB and DC is zero. The force on side AD is repulsive and on side BC is attractive. Since, AD is near the straight wire. So, the net force will be repulsive.

Force on side AD,

$$F_1 = \frac{\mu_0}{2\pi}\frac{I_1 I_2 a}{x}$$

(away from straight wire)

Force on side BC,

$$F_2 = \frac{\mu_0}{2\pi}\frac{I_1 I_2}{(x+a)} \times a$$

(towards the straight wire)

Therefore, resultant force,

$$F = F_1 - F_2$$

$$= \frac{\mu_0}{2\pi} I_1 I_2 a\left(\frac{1}{x} - \frac{1}{x+a}\right)$$

$$F = \frac{\mu_0}{2\pi} I_1 I_2 a\left(\frac{a}{x(x+a)}\right)$$

$$F = \frac{\mu_0}{2\pi} \times \frac{I_1 I_2 a^2}{x(x+a)},$$

(away from wire)

**18. (a) Derive the expression for the torque acting on a current carrying loop placed in a magnetic field.**

**(b) Explain the significance of a radial magnetic field when a current carrying coil is kept in it. [3]**

**Ans.** Torque on current carrying loop in magnetic field :

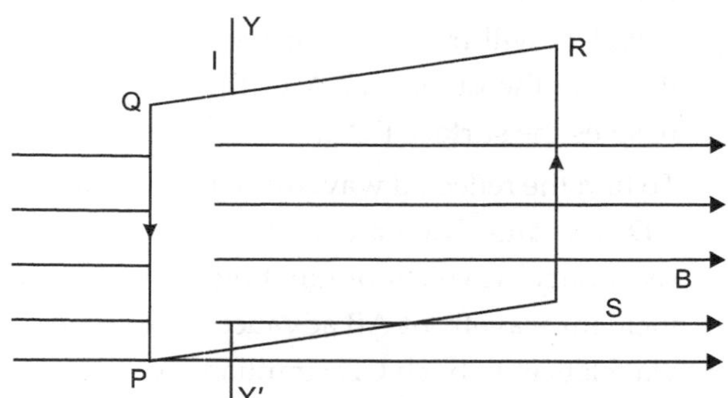


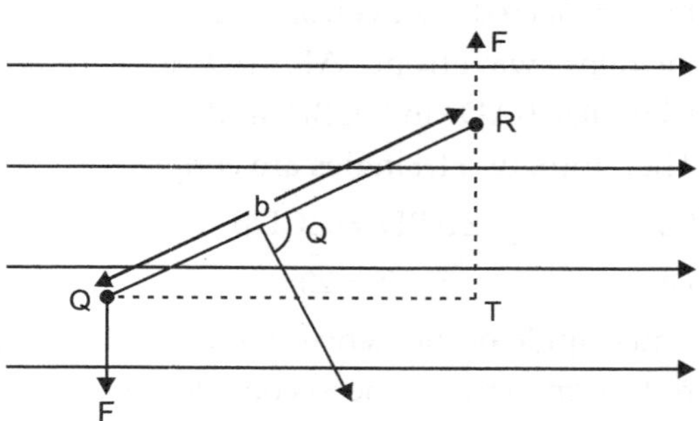


**Ans. (a)** Consider a coil PQRS placed in a magnetic field. Let θ be the angle between the plane of the coil and direction of B. When current flow through coil, each side experiences a force. The forces on vertical side will constitute a couple.

Moment of torque

= One of forces × Perpendicular distance between lines of action of force

$$\tau = F \times QT = F \times b \sin\theta$$

$$= IlB\, b \sin\theta$$

$$\tau = I\,(lb)\, B \sin\theta$$

$$\tau = I\left(\vec{A} \times \vec{B}\right) \quad (\text{Area} = lb)$$

If there are N turns, then

$$\tau = N I\left(\vec{A} \times \vec{B}\right) = \vec{M} \times \vec{B}$$

Where, $\vec{M} = NI\vec{A}$ is magnetic moment of loop.

**(b)** A magnetic field, in which the plane of the coil in all positions remains parallel to the direction of magnetic field is called radial magnetic field.

In a radial magnetic field, magnetic torque remains maximum for all the positions of the coil.

**19. Draw a labelled ray diagram of an astronomical telescope in the near point adjustment position. A giant refracting telescope at an observatory has an objective lens of focal length 15m and an eyepiece of focal length 1.0 cm. If this telescope is used to view the Moon, find the diameter of the image of the Moon formed by the objective lens. The diameter of the Moon is 3.48 × 10⁶ m and the radius of lunar orbit is 3.8 × 10⁸ m. [3]**

**Ans.** **Astronomical telescope in near points adjustment :**

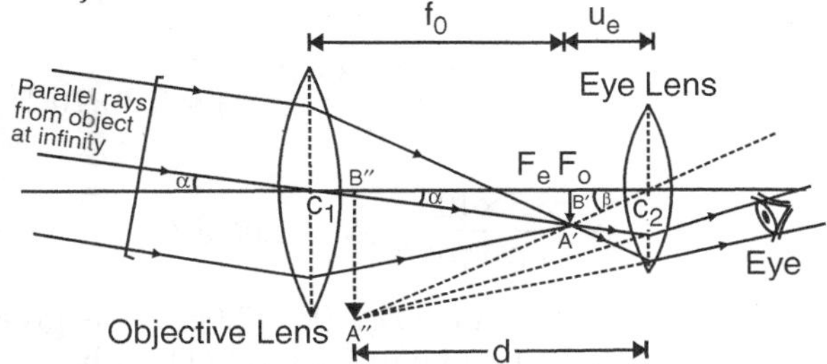


Here, $f_0 = 15$ m, $f_e = 1$ cm $= 0.01$ m

Diameter of moon, $d = 3.48 \times 10^6$ m

Radius of lunar orbit $u_0 = 3.8 \times 10^8$ m

size of the image of moon, I = ?

The angle subtended by the moon at the objective lens,

$$\tan \alpha = \frac{d}{|u_0|} \quad ...(i)$$

And the angle subtended by the image of the moon at the objective lens,

$$\tan \beta = \frac{I}{f_o} \quad ...(ii)$$

Both these angles in equations (i) and (ii) will be equal.

$$\therefore \quad \frac{d}{|u_0|} = \frac{I}{f_o}$$

$$\Rightarrow \quad I = \frac{d \times f_0}{|u_0|}$$

$$= \frac{3.48 \times 10^6 \times 15}{3.8 \times 10^8} \text{ m}$$

$$= 13.73 \times 10^{-2} \text{ m}$$

$$= 13.73 \text{ cm}$$

**20. (a) State Gauss' law for magnetism. Explain its significance.**

**(b) Write the four important properties of the magnetic field lines due to a bar magnet. [3]**

**OR**

**Write three points of differences between para-, dia- and ferro-magnetic materials, giving one example for each.**

**Ans. (a)** **Gauss law for magnetism :** If a closed surface is imagined in a magnetic field, the number of lines of force emerging from the surface must be equal to the number entering it. That is, the net magnetic flux out of any closed surface is zero.

Gauss law signifies that magnetic mono-poles does not exist.

**(b)** 1. In a bar magnet, each lines of force, starts from a north pole and reaches the south pole externally and then goes from south pole to a north pole internally. Thus, magnetic line of force forms a closed loop.

2. No two lines of force will never intersect each other.

3. In a uniform field, the lines are parallel and equidistant from each other.

4. The lines of force are crowded near the poles.

**OR**

| **Properties** | **Ferro-magnetic Materials** | **Para-magnetic Materials** | **Dia-magnetic Materials** |
|---|---|---|---|
| State | They are solid | They can be solid, liquid or gas. | They can be solid, liquid or gas. |
| Effect of Magnet | Strongly attracted by a magnet | Weakly attracted by a magnet. | Weakly repelled by a magnet. |
| Effect of Temperature | Above curie point, it becomes a paramagnetic. | With the rise of temperature, it becomes a diamagnetic. | No effect. |
| Example | iron, nickel | Lithium, molybdenum | Copper, silver |

**21. Define the term 'decay constant' of a radioactive sample. The rate of disintegration of a given radioactive nucleus is 10000 disintegrations/s and 5,000 disintegrations/s after 20 hr. and 30 hr. respectively from start. Calculate the half life and initial number of nuclei at $t = 0$. [3]**

**Ans.** 'Decay constant' of a radioactive sample is defined as the ratio of its instantaneous rate of disintegration to the number of atoms present at that time.

*i.e.,* $\frac{dN}{dt} = -\lambda n,$

$$\lambda = \frac{-dN/dt}{N}$$

Let, the initial number of nuclei be $N_o$ at $t = 0$

$$\frac{dN}{dt} = 10{,}000 \quad \text{at } t = 20 \text{ hr}$$

and $\frac{dN'}{dt} = 5{,}000 \quad \text{at } t = 30 \text{ hr}$

We have, $\frac{dN}{dt} = \lambda N,$

For first case,

$$10{,}000 = \lambda N,$$

$$\Rightarrow \quad N = \frac{10{,}000}{\lambda}$$

and $N = N_0 e^{-\lambda t},$

$$\frac{10{,}000}{\lambda} = N_0 e^{-20\lambda} \quad \text{...(i)}$$

For second case,

$$5{,}000 = \lambda N'$$

$$N' = \frac{5{,}000}{\lambda}$$

and $N' = N_0 e^{-30\lambda}$

$$\frac{5{,}000}{\lambda} = N_0 e^{-30\lambda} \quad \text{...(ii)}$$

on dividing equation (i) by equation (ii), we get

$$\Rightarrow \quad \frac{e^{-20\lambda}}{e^{-30\lambda}} = 2$$

Taking log on both sides,

$$\Rightarrow \quad -20\lambda + 30\lambda = \log_e 2$$

$$\Rightarrow \quad 10\lambda = 2.302 \times 0.3010$$

$$\Rightarrow \quad \lambda = 0.0693$$

$$\therefore \quad \text{Half life}, \tau = \frac{0.693}{0.0693} = 10 \text{ hr}$$

This means that in 20 hours (= 2 half lives), the original number of nuclei must have gone down by a factor of 4.

Hence, Rate of decay at $t = 0$,

$$R_o = \lambda N_o = 4\lambda N_{20} = 4 \times 10000$$

$$= 40{,}000 \text{ disintegrations per second.}$$

$$\Rightarrow \quad N_o = \frac{40000}{\lambda} = \frac{40000}{\ln 2} \times 10 \times 60 \times 60$$

$$\Rightarrow \quad N_o = \frac{144 \times 10^7}{0.693} = 2.08 \times 10^9 \text{ nuclei}$$

**22. (a) Three photo diodes $D_1$, $D_2$ and $D_3$ are made of semiconductors having band gaps of 2.5 eV, 2 eV and 3 eV respectively. Which of them will not be able to detect light of wavelength 600 nm ?**

**(b) Why photodiodes are required to operate in reverse bias? Explain. [3]**

**Ans. (a)** Energy corresponding to wavelength 600 nm is,

$$E = \frac{hc}{\lambda}$$

$$= \frac{6.6 \times 10^{-34} \times 3 \times 10^{8}}{600 \times 10^{-9}} \text{ J} = 3.3 \times 10^{-19} \text{ J}$$

$$= \frac{3.3 \times 10^{-19}}{1.6 \times 10^{-19}} \text{ eV} = 2.06 \text{ eV}$$

The photon energy (E = 2.06 eV) is greater than the band gap for diode $D_2$ only. Hence, diode $D_2$ will not be able to detect the given wavelength.

**(b)** A photodiode is operated in reverse bias because in reverse bias, it is easier to observe change in current with change in light intensity.

**23. (a) Describe briefly the functions of the three segments of n-p-n transistor.**

**(b) Draw the circuit arrangement for studying the output characteristics of n-p-n transistor in CE configuration. Explain how the output characteristics is obtained. [3]**

**OR**

**Draw the circuit diagram of a full wave rectifier and explain its working. Also, give the input and output waveforms.**

**Ans.**

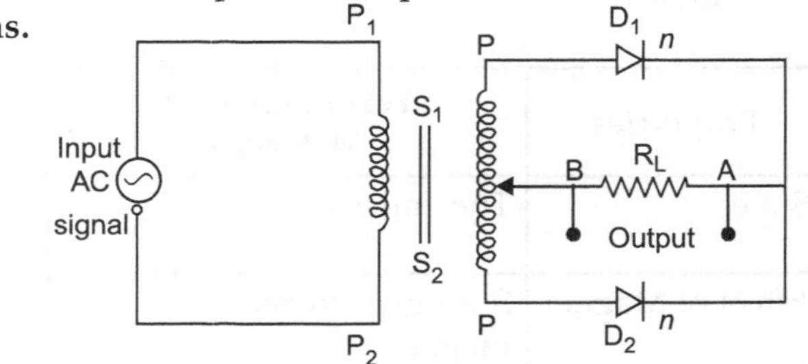


For full wave rectifier, we use two junction diodes as shown in figure.

** Answer is not given due to change in the present Syllabus.

During first half cycle of input a.c. signal, the terminal $S_1$ is positive relative to S and $S_2$ is negative, then diode $D_1$ is in forward biased and diode $D_2$ is in reverse biased. Therefore, current flows in $D_1$, not in $D_2$. In next half cycle, $S_1$ is negative and $S_2$ is positive relative to S. Then $D_1$ is in reverse biased and $D_2$ is in forward biased. Therefore, current flows in $D_2$, not in $D_1$. Thus, for input a.c. signal, the output current is a continuous series of unidirectional (rectified) pulse.

The input and output waveforms are shown in the figure.

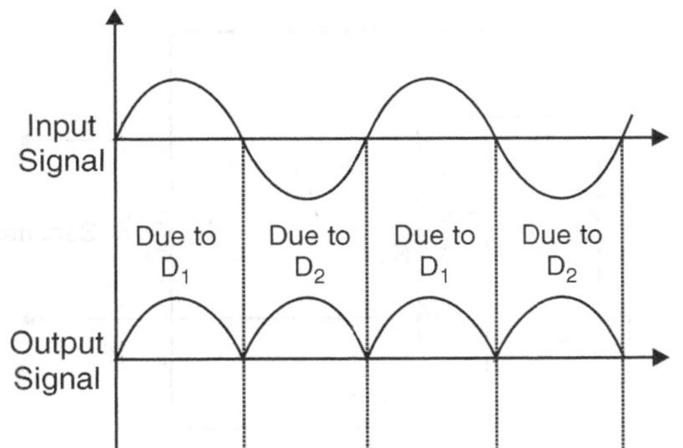


**24. (a) If A and B represent the maximum and minimum amplitudes of an amplitude modulated wave, write the expression for the modulation index in terms of A & B.**

**(b) A message signal of frequency 20 kHz and peak voltage 10 V is used to modulate a carrier of frequency 2 MHz and peak voltage of 15 V. Calculate the modulation index. Why the modulation index is generally kept less than one ?**** **[3]**

## SECTION-D

**25. (a) In a series LCR circuit connected across an ac source of variable frequency, obtain the expression for its impedance and draw a plot showing its variation with frequency of the ac source.**

**(b) What is the phase difference between the voltages across inductor and the capacitor at resonance in the LCR circuit ?**

**(c) When an inductor is connected to 200 V dc voltage, a current of 1 A flows through it. When the same inductor is connected to a 200 V, 50 Hz ac source, only 0.5 A current flows. Explain, why ? Also, calculate the self inductance of the inductor.** **[5]**

**OR**

**(a) Draw the diagram of a device which is used to decrease high ac voltage into a low ac voltage and state its working principle. Write four sources of energy loss in this device.**

**(b) A small town with a demand of 1200 kW of electric power at 220 V is situated 20 km away from an electric plant generating power at 440 V. The resistance of the two wire line carrying power is 0.5 Ω per km. The town gets the power from the line through a 4000 V - 220 V step-down transformer at a substation in the town. Estimate the line power loss in the form of heat.**

**Ans. (a)** Consider an alternating e.m.f. is connected in series with an inductor, resistance R and capacitance C. Let E and I be the instantaneous values of e.m.f. and current in the LCR circuit and $V_L$, $V_C$ and $V_R$ be the instantaneous values of voltage across inductor, capacitor and resistor respectively. Then,

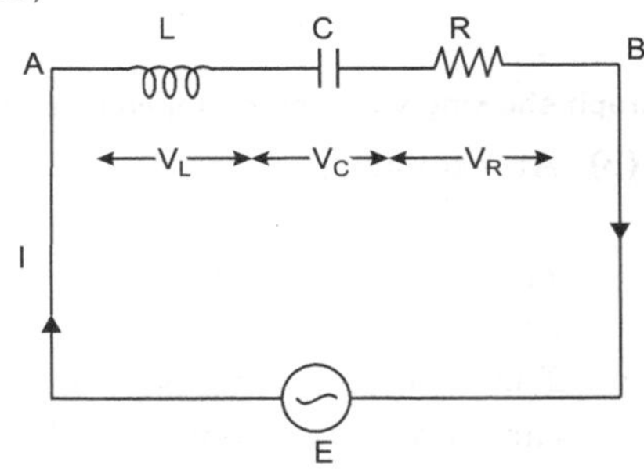


$$V_L = IX_L,\ V_C = IX_C \text{ and } V_R = IR$$

Here, $X_L = \omega L$ and $X_C = \dfrac{1}{\omega C}$

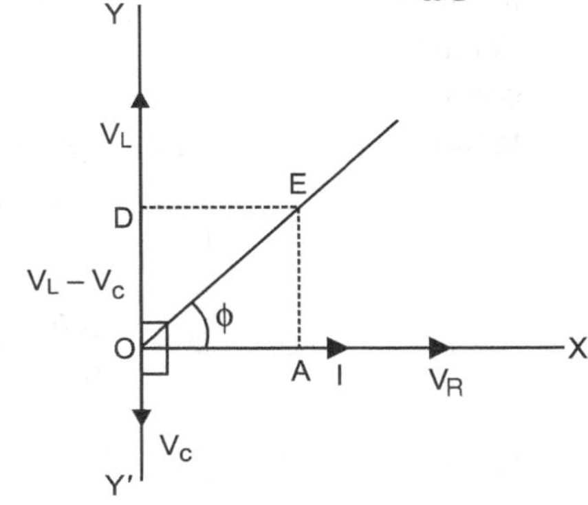


In an a.c. circuit $V_R$ and I are in same phase, $V_L$ leads I by $\pi/2$ and $V_C$ lags I by $\pi/2$.

From the graph, in right angled ΔOEA

$$OE = \sqrt{OA^2 + AE^2} = \sqrt{OA^2 + OD^2}$$

$$= \sqrt{V_R^2 + (V_L - V_c)^2}$$

$$\therefore \quad E = \sqrt{(IR)^2 + (IX_L - IX_C)^2}$$

$$E = I\sqrt{R^2 + (X_L - X_C)^2}$$

$$I = \frac{E}{\sqrt{R^2 + (X_L - X_C)^2}}$$

Therefore, effective resistance or impedance of LCR circuit is–

$$Z = \sqrt{R^2 + (X_L - X_C)^2}$$

$$= \sqrt{R^2 + \left(\omega L - \frac{1}{\omega C}\right)^2}$$

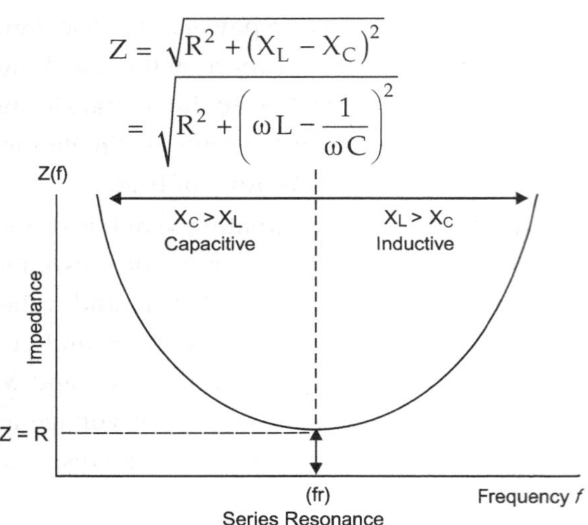


**Graph showing variation of impedance with frequency**

**(b)** At resonance,

$$X_L = X_C$$

*i.e.*, $\quad iX_L = iX_C$

or $\quad V_L = V_C$

The voltage across inductance and capacitance are equal and have a phase difference of 180° at resonance.

**(c)** Since the reactance of an inductor is zero for d.c. circuit. But the inductor offers resistance to an a.c. circuit. Therefore, the current decreases for the same inductor when it is connected with an a.c. source.

**When inductor is connected in a.c. circuit :**

$$R = \frac{V_{rms}}{I_{rms}} = \frac{200}{1} = 200\,\Omega$$

$$Z = \frac{V_{rms}}{I_{rms}} = \frac{200}{0.5} = 400\,\Omega$$

Since, $\quad Z = \sqrt{R^2 + X_L^2}$

$$\therefore (400)^2 - (200)^2 = X_L^2$$

$$X_L = \sqrt{600 \times 200} = 346.4\Omega$$

$$\text{Inductance (L)} = \frac{X_L}{\omega} = \frac{346.4}{2\pi f} = \frac{346.4}{2\pi\,(50)} = 1.1\text{H}$$

**OR**

**(a) Step-down transformer :**

It is a device used for converting high alternating voltage at low current into low alternating voltage at high current and vice-versa.

The device works on the principle of mutual induction *i.e.*, if the current or magnetic flux linked with a coil changes then an e.m.f. is induced in the other coil.

In step-down transfomer $N_p > N_s$ and transformation ratio is less than 1.

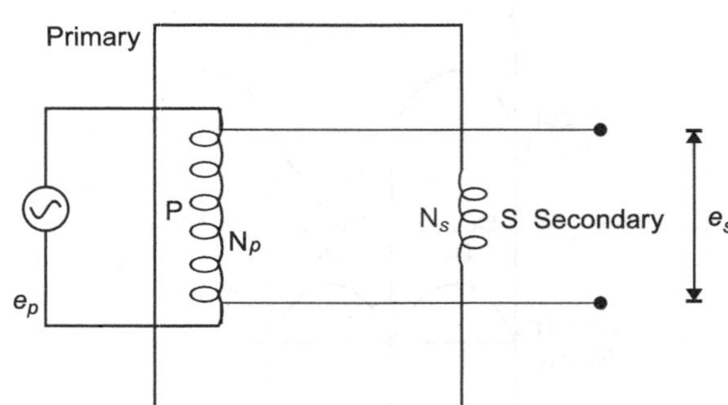


Energy losses are :

1. **Copper losses :** Due to resistance of windings in primary and secondary coils, some electrical energy is converted into heat energy.
2. **Flux losses :** Some of the flux produced in primary coil is not linked up with secondary coils.
3. **Hysteresis losses :** When the iron core is subjected to a cycle of magnetisation the core gets heated up due to hysteresis known as hysteresis loss.
4. **Iron losses :** The varying magnetic flux produces eddy current in the iron core, which leads to the wastage of energy in the form of heat.

**(b)** Length of wire line = 20 × 2 = 40 km

Resistance of wire line, $r = 40 \times 0.5 = 20\ \Omega$

Power to be supplied = 1200 kW

$$= 1200 \times 10^3\ \text{W}$$

Voltage at which power supplied = 4000 V

Since, $\quad P = VI$

$$\Rightarrow \quad I = \frac{P}{V}$$

$$\Rightarrow \quad I = \frac{1200 \times 10^3}{4000} = 300\ \text{A}$$

Therefore, line power loss
$= I^2 \times R$
$= (300)^2 \times 20$
$= 1800$ kW

$\therefore$ Line power loss in the form of heat is 1800 kW.

**26. (a) Describe any two characteristic features which distinguish between interference and diffraction phenomena. Derive the expression for the intensity at a point of the interference pattern in Young's double slit experiment.**

**(b) In the diffraction due to a single slit experiment, the aperture of the slit is 3 mm. If monochromatic light of wavelength 620 nm is incident normally on the slit, calculate the separation between the first order minima and the 3rd order maxima on one side of the screen. The distance between the slit and the screen is 1.5 m. [5]**

**OR**

**(a) Under what conditions is the phenomenon of total internal reflection of light observed? Obtain the relation between the critical angle of incidence and the refractive index of the medium.**

**(b) Three lenses of focal length + 10 cm, –10 cm and +30 cm are arranged coaxially as in the figure given (adjacent). Find the position of the final image formed by the combination.**

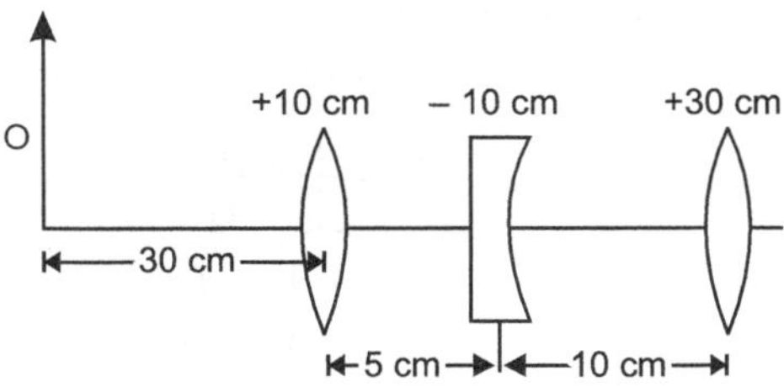


**Ans. (a)**

| | Interference | Diffraction |
|---|---|---|
| 1. | It is the result of interaction of light coming from two different wavefronts originating from two coherent sources. | 1. It is the result of interaction of light come from different parts of same wavefronts. |
| 2. | All the bright fringes are of same intensity. | 2. The bright fringes are of varying intensity (Intensity of bright fringes decreases from central bright fringe on either sides.) |

A sources of monochromatic light illuminates two narrow slits $S_1$ and $S_2$.The two illuminated slits act as the two coherent sources. The two slits are very close to each other and at equal distance from source. The wavefront $S_1$ and $S_2$ spread in all direction and superpose and produce dark and bright fringe on screen.

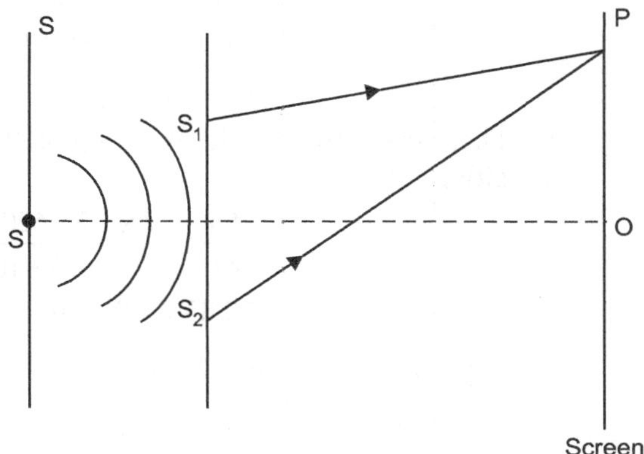


Let the displacement of waves from $S_1$ and $S_2$ at point P on screen at time $t$ is:

$$y_1 = a_1 \sin \omega t$$

$$y_2 = a_2 \sin (\omega t + \phi)$$

The resultant displacement at point P is given by

$$y = y_1 + y_2$$
$$= a_1 \sin \omega t + a_2 \sin (\omega t + \phi)$$
$$= a_1 \sin \omega t + a_2 \sin \omega t \cos \phi + a_2 \cos \omega t \sin \phi$$
$$= (a_1 + a_2 \cos \phi) \sin \omega t + a_2 \sin \phi \cos \omega t \quad ....(i)$$

Let, $a_1 + a_2 \cos \phi = A \cos \theta \quad ....(ii)$

$a_2 \sin \phi = A \sin \theta \quad ...(iii)$

Therefore, equation (i) becomes

$$y = A \cos \theta \sin \omega t + A \sin \theta \cos \omega t$$
$$= A \sin (\omega t + \theta)$$

This is the resultant displacement.

Now, squaring and adding equations (ii) and (iii)

$$A^2 \cos^2 \theta + A^2 \sin^2 \theta = (a_1 + a_2 \cos \phi)^2 + a_2^2 \sin^2 \phi$$

$$A^2 = a_1^2 + a_2^2 + 2a_1 a_2 \cos \phi$$

The is intensity of light is directly proportional to the square of the amplitude

*i.e.*, $I = a_1^2 + a_2^2 + 2a_1 a_2 \cos \phi$

This is the expression for intensity at a point of interference pattern.

or $$I = I_1 + I_2 + 2\sqrt{I_1 I_2}\cos\phi$$

**(b)** Here, $\lambda = 620 \text{ nm} = 620 \times 10^{-9}$ m

$a = 3\times10^{-3}$ m, $D = 1.5$ m

Distance of first order minima from the centre,

$$y_1 = \frac{D\lambda}{a} = \frac{1.5\times620\times10^{-9}}{3\times10^{-3}}$$

$$= 3.1 \times 10^{-4} \text{ m}$$

Distance of third order maxima on the same side,

$$y_2 = \frac{7D\lambda}{2a} = \frac{7\times1.5\times620\times10^{-9}}{2\times3\times10^{-3}}$$

$$= 10.85 \times 10^{-4} \text{ m}$$

Separation between them,

$$y = y_2 - y_1$$

$$= 10.85 \times 10^{-4} - 3.1 \times 10^{-4}$$

$$= 7.75 \times 10^{-4} \text{ m}$$

**OR**

**(a) Conditions for total internal reflection :**

1. The ray must travel from a denser medium into a rarer medium.
2. The angle of incidence in the denser medium must be greater than the critical angle for the pair of media.

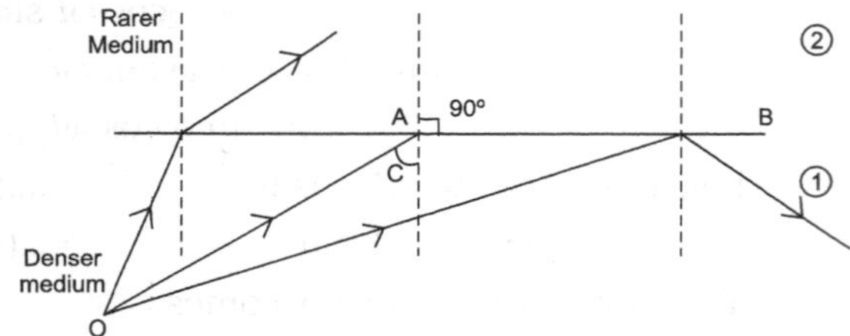


**Relation between critical angle and refractive index :**

When a ray of light travels from denser to rarer medium, the ray bends away from the normal. When the angle of incidence is equal to the critical angle then the refracted ray grazes the surface of separation represented by ray $\overrightarrow{OA}$ and $\overrightarrow{AB}$.

By Snell's law,

$$\frac{\sin i}{\sin r} = \frac{n_1}{n_2}$$

where, $i = C$, $r = 90°$ and $n_2$ = refractive index of denser medium.

$n_1$ = refractive index of Rarer medium

$$\therefore \quad \frac{\sin C}{\sin 90°} = \frac{n_1}{n_2}$$

$$^1n_2 = \frac{1}{\sin C}$$

**(b)** For first lens, $u_1 = -30$ cm, $f_1 = 10$ cm

$\therefore$ From lens formula,

$$\frac{1}{f_1} = \frac{1}{v_1} - \frac{1}{u_1}$$

$$\Rightarrow \quad \frac{1}{v_1} = \frac{1}{f_1} + \frac{1}{u_1}$$

$$= \frac{1}{10} - \frac{1}{30}$$

$$\Rightarrow \quad v_1 = 15 \text{ cm}$$

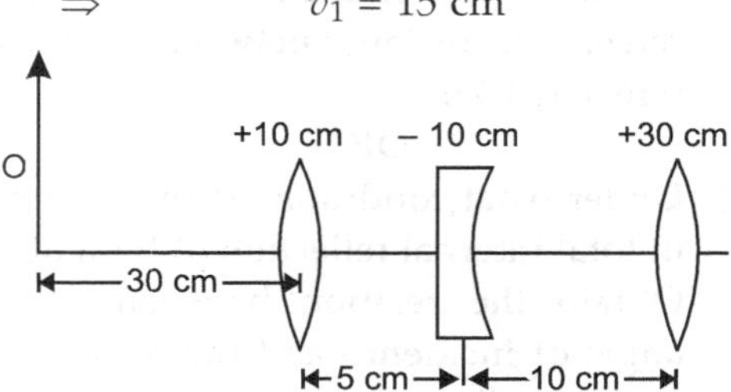


This means that image formed by first lens is at a distance of 15 cm to the right of the first lens. This image serves as the virtual object for second lens.

For second lens,

$f_2 = -10$ cm, $u = 15 - 5 = +10$ cm

$$\therefore \quad \frac{1}{v_2} = \frac{1}{f_2} + \frac{1}{u_2}$$

$$= -\frac{1}{10} + \frac{1}{10}$$

$$\Rightarrow \quad v_2 = \infty$$

This means that the real image is formed by second lens at infinite distance. This acts as an object for third lens.

for third lens,

$$f_3 = +30 \text{ cm}, u_3 = \infty$$

$$\therefore \quad \frac{1}{v_3} = \frac{1}{f_3} + \frac{1}{u_3}$$

$$\therefore \quad = \frac{1}{30} + \frac{1}{\infty}$$

$$v_3 = 30 \text{ cm}$$

*i.e.*, final image is formed at a distance of 30 cm to the right of third lens.

**27. (a) Describe briefly the process of transferring the charge between the two plates of a parallel plate capacitor when connected to a battery. Derive an expression for the energy stored in a capacitor.**

**(b) A parallel plate capacitor is charged by a battery to a potential difference V. It is disconnected from battery and then connected to another uncharged capacitor of the same capacitance. Calculate the ratio of the energy stored in the combination to the initial energy on the single capacitor. [5]**

**OR**

**(a) Derive an expression for the electric field at any point on the equatorial line of an electric dipole.**

**(b) The identical point charges, $q$ each, are kept 2 m apart in air. A third point charge Q of unknown magnitude and sign is placed on the line joining the charges such that the system remains in equilibrium. Find the position and nature of Q.**

**Ans. (a)** When the plates of the parallel plate capacitor is connected to a battery. Then the first insulated metal plate gets the positive charge till its potential become maximum. Then, the charge will leak to surroundings. So, the negative charge will be induced on the nearer face of the second plate and the positive charge will be induced on its farther plate. Process contnues till the potential difference between two plates equals the potential of the battery.

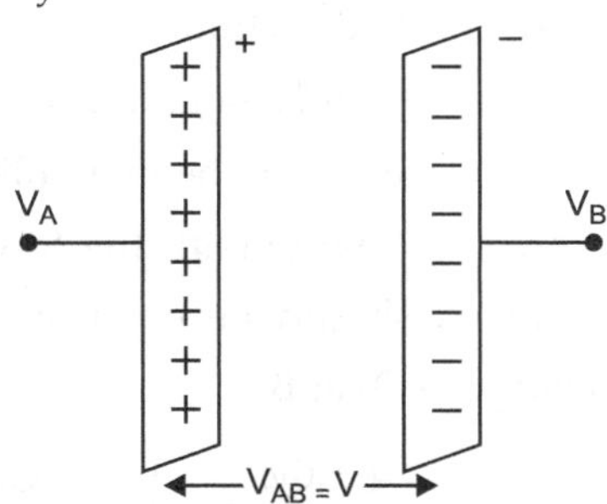


Consider a capacitor of capacitance C. Initial charge and potential difference be zero. Let, a charge Q be given in small steps. Let at any instant when charge on capacitor be $q$, the potential difference between its plates,

$$V = \frac{q}{C}$$

Now work done in giving an additional charge $dq$ is,

$$dW = V\,dq = \frac{q}{C}\,dq$$

Total work done in giving charge from 0 to Q is,

$$W = \int_0^Q V dq = \int_0^Q \frac{q}{C} dq$$

$$= \frac{1}{C}\left[\frac{q^2}{2}\right]_0^Q = \frac{1}{C}\left[\frac{Q^2}{2} - \frac{0}{2}\right] = \frac{Q^2}{2C}$$

$$W = \frac{(CV)^2}{2C} = \frac{1}{2}CV^2 = \frac{1}{2}QV \quad [\because Q = CV]$$

∴ Electrostatic potential energy,

$$U = \frac{Q^2}{2C} = \frac{1}{2}CV^2 = \frac{1}{2}QV$$

**(b)** Here, $C_1 = C$, $C_2 = C$

$$V_1 = V, V_2 = 0$$

Now initially, energy stored in first capacitor, as second capacitor is uncharged.

$$U_1 = \frac{1}{2}C\,V_1^2$$

$$= \frac{1}{2}CV^2 \quad ...(i)$$

Now, when $C_1$ and $C_2$ are connected, the two capacitors form a parallel combination Equivalent capacitance,

$$C' = C_1 + C_2 = 2C$$

$$\text{Final potential} = \frac{\text{Total charge}}{\text{Total capacitance}}$$

$$V' = -\frac{q}{2C}$$

[Total charge will be of first capacitor, which is distributed]

$$= \frac{CV}{2C} = \frac{V}{2}$$

Final energy stored in combination,

$$U_2 = \frac{1}{2}C'V'^2 = \frac{1}{2} \times 2C \times \left(\frac{V}{2}\right)^2 = \frac{CV^2}{4} \quad ...(ii)$$

On dividing equations (i) with (ii), we get

$$\frac{U_2}{U_1} = \frac{\frac{1}{4}CV^2}{\frac{1}{2}CV^2} = \frac{1}{2}$$

$$U_2 : U_1 = 1 : 2$$

**OR**

**(a)** Consider an electric dipole of charges $-q$ and $+q$ separated by a distance $2a$ and placed in a free space. Let P be a point on equatorial line of dipole at a distance $r$ from the centre of a dipole.

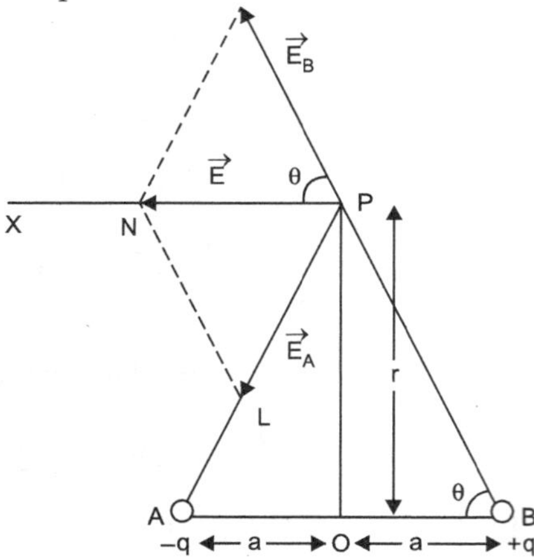


Let, $\vec{E}_A$ and $\vec{E}_B$ be the electric field at point P due to charges $-q$ and $+q$ respectively. Then resultant electric field at point P is

$$\vec{E} = \vec{E}_A + \vec{E}_B$$

Now,

$$|\vec{E}_A| = \frac{1}{4\pi\varepsilon_0} \cdot \frac{q}{AP^2} = \frac{1}{4\pi\varepsilon_0} \frac{q}{(r^2 + a^2)}$$

(along PA)

$$|\vec{E}_B| = \frac{1}{4\pi\varepsilon_0} \cdot \frac{q}{BP^2} = \frac{1}{4\pi\varepsilon_0} \frac{q}{(r^2 + a^2)}$$

(along BP)

The resultant intensity is the vector sum of $E_A$ and $E_B$.

$E_A$ and $E_B$ can be resolved into two components. The Y-components cancel out each other. And X-component will add up to give the resultant field.

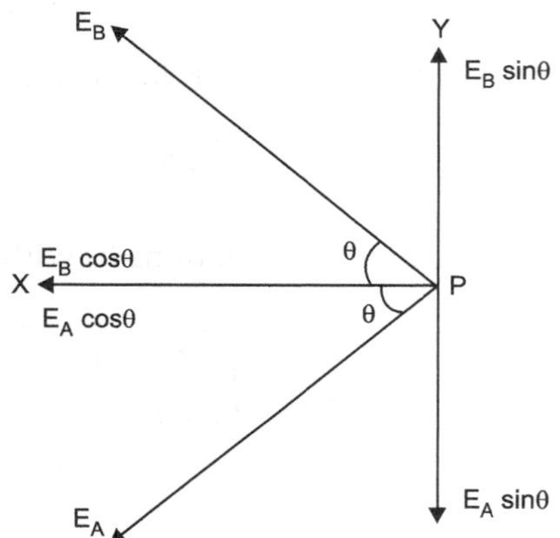


$$\therefore |\vec{E}| = E_A \cos\theta + E_B \cos\theta$$

Now in right triangle ORB

$$\cos\theta = \frac{OB}{BP} = \frac{a}{\sqrt{r^2 + a^2}}$$

$$\therefore \quad E = 2 \times \frac{1}{4\pi\varepsilon_0} \times \frac{q}{r^2 + a^2} \frac{a}{(r^2 + a^2)^{1/2}}$$

$$= \frac{2qa}{4\pi\varepsilon_0 (r^2 + a^2)^{3/2}}$$

$$= \frac{1}{4\pi\varepsilon_0} \frac{p}{(r^2 + a^2)^{3/2}} \qquad [\because 2qa = p]$$

This is the required expression.

**(b)** Let the two charges of + $q$ each placed at point A and B at a distance 2 m apart in air.

$$\overset{q}{A} \bullet \text{———} \overset{Q}{+} \text{———} \bullet \overset{q}{B}$$

$$\longleftarrow x \longrightarrow O \longleftarrow (2 - x) \longrightarrow$$

Suppose, the third charge Q (unknown magnitude and charge) is placed at a point O on the line joining the other two charges, such that OA= $x$ and OB = $2-x$.

For the system to be in equilibrium, net force on each of the 3 charges must be zero. If we assume that charge Q placed at O is positive, the force on it at O may be zero. But the force on charge $q$ at point A or B will not be zero. It is because the forces on a charge $q$ due to the other two charges will act in same direction. If charge Q is negative, then the forces on $q$ due to other two charges will act in opposite direction.

Hence, Q will be negative in nature.

For charge (–Q) to be in equilibrium.

Force on charge ($-q$) due to charge ($+q$) at point A should be equal and opposite to charge (+Q) at B

$$\frac{1}{4\pi\varepsilon_0} \frac{Qq}{x^2} = \frac{1}{4\pi\varepsilon_0} \cdot \frac{Qq}{(2-x)^2}$$

$$\Rightarrow \quad (2-x)^2 = x^2$$

$$\Rightarrow \quad x = (2-x) \Rightarrow x = 1 \text{ m}$$

Therefore, for the system to be in equilibrium a charge $-$Q is placed at a mid point between the two charges of + $q$ each.

●●

# Delhi [Set-II]

**Note : Except for the following questions, all the remaining questions have been asked in previous set.**

## SECTION-A

**2. When unpolarised light is incident on the interface separating the rarer medium and the denser medium, Brewster angle is found to be 60°. Determine the refractive index of the denser medium.** **[1]**

**Ans.** According to Brewster's law,

$$\tan i_p = n$$

Since, $i_p = 60°$

$\therefore$ Refractive index of medium,

$$n = \tan 60°, = \sqrt{3}$$

$$n = 1.732$$

**4. When a potential difference is applied across the ends of a conductor, how is the drift velocity of the electrons related to the relaxation time?** **[1]**

**Ans.** $$\vec{v}_d = -\frac{e\vec{E}}{m}\tau$$

where, $m$ = mass of electron

$e$ = charge of electron

E = potential difference applied

$\tau$ = relaxation time

$\therefore v_d \propto \tau$ *i.e.* drift velocity is directly proportional to relaxation time.

**5. Draw the equipotential surfaces due to an isolated point charge.** **[1]**

**Ans.**

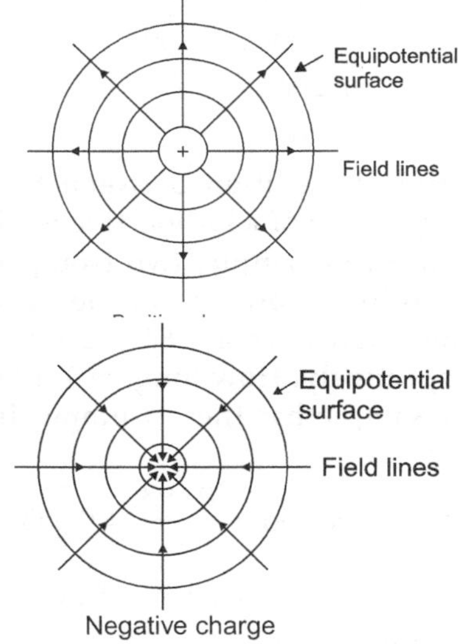


## SECTION-B

**4. Explain with the help of Einstein's photoelectric equation any two observed features in photoelectric effect which cannot be explained by wave theory.** **[2]**

**Ans.** Features of photoelectric equation which can not be explained by wave theory :

**(a)** The wave theory could not explain the instantaneous process of photoelectric effect.

**(b)** 'Maximum kinetic energy' of the emitted photoelectrons is independent of intensity of incident light.

**7. A deuteron and an alpha particle having same momentum are in turn allowed to pass through a magnetic field $\vec{B}$, acting normal to the direction of motion of the particles. Calculate the ratio of the radii of the circular paths described by them.** **[2]**

**Ans.** Radius of circular path,

$$r = \frac{mv}{qB}$$

*i.e.*, $$r = \frac{p}{qB} \qquad [\because p = mv]$$

Momentum of deuteron and alpha-particle are same and in same magnetic field

$$\therefore \quad r_d = \frac{p}{q_d B} = \frac{p}{eB} \qquad [\because q_d = e]$$

$$\text{Now } r_\alpha = \frac{p}{q_\alpha B} = \frac{p}{2eB} \qquad [\because q_\alpha = 2e]$$

Therefore, $r_d : r_\alpha = 2 : 1$

**11. (a) Plot a graph showing variation of de Broglie wavelength (λ) associated with a charged particle of mass *m*, versus $\sqrt{V}$, where V is the accelerating potential.**

**(b) An electron, a proton and an alpha particle have the same kinetic energy. Which one has the shortest wavelength ?** **[2]**

**Ans. (a)** We know,

$$\lambda = \frac{1.22}{\sqrt{V}} \text{Å}$$

$$\lambda\sqrt{V} = \text{constant}$$

The nature of the graph between $\lambda$ and $\sqrt{V}$ is hyperbola.

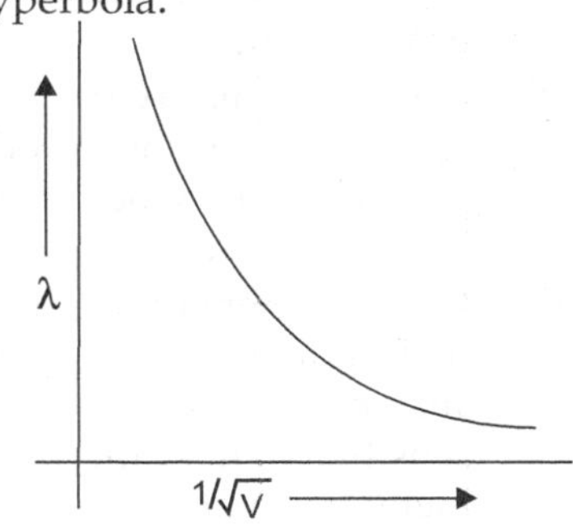

**(b)** According to de-Broglie wavelength,

$$\lambda = \frac{h}{\sqrt{2mE_k}}$$

For same kinetic energy,

$$\lambda_\alpha = \frac{1}{\sqrt{m}}$$

We know that

$$m_\alpha > m_p > m_e$$

Since alpha-particle has the maximum mass, therefore, alpha-particle will have minimum wavelength, therefore

$$\lambda_e > \lambda_p > \lambda_\alpha.$$

## SECTION-C

**13. (a) State the underlying principle of a moving coil galvanometer.**

**(b) Give two reasons to explain why a galvanometer cannot as such be used to measure the value of the current in a given circuit.**

**(c) Define the terms : (i) voltage sensitivity and (ii) current sensitivity of a galvanometer.** [3]

**Ans. (a)** **The Principle :** When a current flows through the conductor coil, a torque acts on it due to the external radial magnetic field. Under the action of this torque the coil rotates and the deflection in a moving coil is directly proportional to the current flowing throgh coil.

**(b)** A galvanometer can be used as such to measure current due to following two reasons:

**(i)** A galvanometer has a finite large resistance and is connected in series in the circuit, so it will increase the resistance of circuit and hence change the value of current in the circuit.

**(ii)** A galvanometer is a very sensitive device, it gives a full scale deflection for the current of the order of microampere, hence if connected as such, it will not measure current of the order of ampere.

**(c) (i)** **Voltage sensitivity** : It is defined as the deflection produced in the galvanometer when a unit voltage is applied across it.

**(ii)** **Current Sensitivity :** The ratio of deflection produced by the coil $\phi$ to the current in the coil is called the current sensitivity. It is the deflection of the meter per unit current.

**15. (a) Draw equipotential surfaces corresponding to the electric field that uniformly increases in magnitude along with the *z*-directions.**

**(b) Two charges $-q$ and $+q$ are located at point $(0, 0, -a)$ and $(0, 0, a)$. What is the electrostatic potential at the points $(0, 0, \pm z)$ and $(x, y, 0)$ ?** [3]

**Ans. (a)**

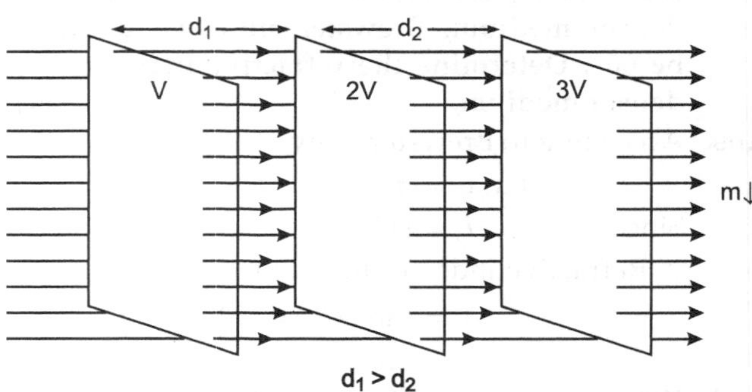


**(b)** Since, two charges are on Z-axis. Therefore, the electric potential on an axial line. *i.e.*, at $(0, 0, \pm Z)$ is given by

$$V_1 = \frac{1}{4\pi\varepsilon_0} \times \frac{2lq}{r^2 - l^2}$$

$$= \frac{1}{4\pi\varepsilon_0} \times \frac{2aq}{(z^2 - a^2)} \qquad [\because l = a \because r = \pm z]$$

Now electric potential at the position $(x, y, 0)$ (given position will be equatorial plane) is given by

$$V = \frac{1}{4\pi\varepsilon_0} \cdot \frac{2aq\cos\theta}{r^2}$$

But $\theta = 90°$

$\therefore \quad V_2 = 0$

**17. (a) Write the relation between half life and average life of a radioactive nucleus.**

**(b) In a given sample, two isotopes A and B are initially present in the ratio of 1 : 2. Their half lives are 60 years and 30 years respectively. How long will it take so that the sample has these isotopes in the ratio of 2 : 1 ?** [3]

**Ans. (a)** Half life period $= \frac{0.693}{\lambda} = 0.693\,\tau$

$= 69.3\%$ of average life

**(b)** We have, $\quad N = N_0 e^{-\lambda t}$

For two isotopes, we can write

$$N_A = N_0 e^{-\lambda_A t_A} \quad \text{...(i)}$$

$$N_B = 2N_0 e^{-\lambda_B t_B} \quad \text{...(ii)}$$

Let the time be $t$ after which,

$$\frac{N_A}{N_B} = \frac{2}{1}$$

*i.e.,* $t_A = t_B = t$

$\therefore \quad N_0 e^{-\lambda_A t} = 4N_0 e^{-\lambda_B t}$

$\Rightarrow \quad e^{-\lambda_A t + \lambda_B t} = 4$

$\Rightarrow \quad e^{\lambda_B t - \lambda_A t} = 4$

$\Rightarrow \quad (\lambda_B t - \lambda_A t) = \log_e 2^2$

$\Rightarrow \quad (\lambda_B - \lambda_A)t = 2\log_e 2$

$$\Rightarrow \left[\frac{\log_e 2}{T_B} - \frac{\log_e 2}{T_A}\right] t = 2\log_e 2 \left[\because \lambda = \frac{\log_e 2}{T}\right]$$

$$\Rightarrow \left[\frac{1}{30} - \frac{1}{60}\right] t = 2$$

$$\Rightarrow \quad t = \frac{30 \times 60 \times 2}{30} = 120 \text{ years}$$

**19. (a) Define the term 'self inductance' of a coil. Write its S.I. unit.**

**(b) A rectangular loop of sides *a* and *b* carrying current $I_2$ is kept at a distance '*a*' from an infinitely long straight wire carrying current $I_1$ as shown in the figure. Obtain an expression for the resultant force acting on the loop.** **[3]**

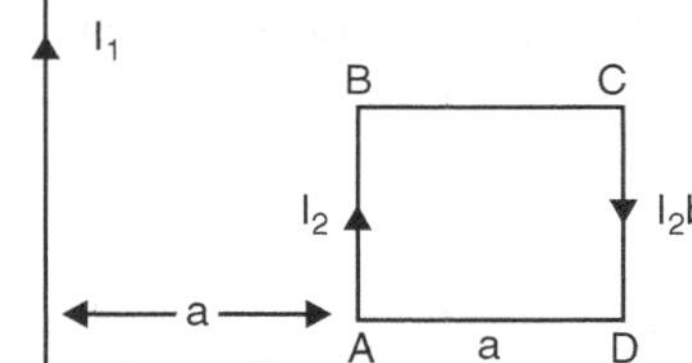


**Ans. (a) Self-inductance :** Inductance of the coil is defined as the property of coil due to which it opposes the change of current flowing through it. Inductance is attained by coil due to self induced emf produced in coil itself by changing current through it.

The S.I. unit of Self-inductance is henry (H) or weber per Ampere, $1 \text{ H} = 1\ wb/A$

**(b)** The force on the side AB of rectangle is attractive as current is flowing in the same direction and on side CD will be repulsive as the current is flowing in opposite direction with respect to straight conductor. The resultant magnetic force on sides AD and BC is zero. The side AB is near to the straight wire. So, the net force will be attractive and rectangular loop will move towards the straight wire.

Now, force between AB and straight wire

$$F_1 = \frac{\mu_0}{2\pi}\frac{I_1 I_2}{a} b \qquad ...(i)$$

Force between CD and straight wire

$$F_2 = \frac{\mu_0}{2\pi} \times \frac{I_1 I_2 b}{(a+a)} \qquad ...(ii)$$

$$\therefore \quad F_{net} = F_1 - F_2$$

$$= \frac{\mu_0}{2\pi} \cdot \frac{I_1 I_2 b}{a} - \frac{\mu_0}{2\pi}\frac{I_1 I_2 b}{(a+a)}$$

$$= \frac{\mu_0}{2\pi} I_1 I_2 b \left[\frac{1}{a} - \frac{1}{2.a}\right]$$

$$= \frac{\mu_0}{2\pi} \times I_1 I_2 \frac{b}{2a}$$

$$F_{net} = \frac{\mu_0}{4\pi} I_1 I_2 \times \frac{b}{a}$$

[Towards the wire]

●●

# Delhi [Set-III]

**Note : Except for the following questions, all the remaining questions have been asked in previous set.**

## SECTION-A

**1. Distinguish between unpolarized and linearly polarized light.** **[1]**

**Ans. Unpolarized light :** The light having vibration of electric field vector in all possible directions perpendicular to the direction of wave propagation the light is known a unpolarized light.

**Linearly polarized light :** The light having vibrations of electric field vector in only one direction perpendicular to the direction of propagation of light is as plane or linearly polarized light.

**2. How is the drift velocity in a conductor affected with the rise in temperature ?** **[1]**

**Ans.** With the rise in temperature, the collision of electrons occurs more frequently, so average relaxation time decreases and hence drift velocity also decreases.

$$v_d = \frac{eE}{m}\tau$$

$$v_d \propto \tau$$

**3. Draw the pattern of electric field lines when a point charge $+q$ is kept near an uncharged conducting plate.** **[1]**

**Ans.** The lines of force start from + Q and terminates at metal plate inducing negative charge on it.

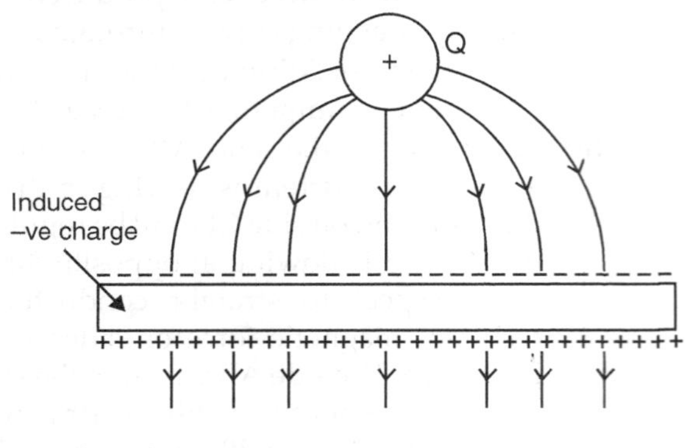


## SECTION - B

**6. (a) Define the terms, (i) threshold frequency and (ii) stopping potential in photoelectric effect. [2]**

**Ans. (a) (i) Threshold frequency** : The minimum frequency of incident light which is just capable of ejecting electrons from a metal is called the threshold frequency. It is denoted by $\nu_0$.

**(ii) Stopping potential** : The minimum retarding potential applied to anode of a photoelectric tube which is just capable of stopping photoelectric current is called the stopping potential. It is denoted by $V_0$ (or $V_s$).

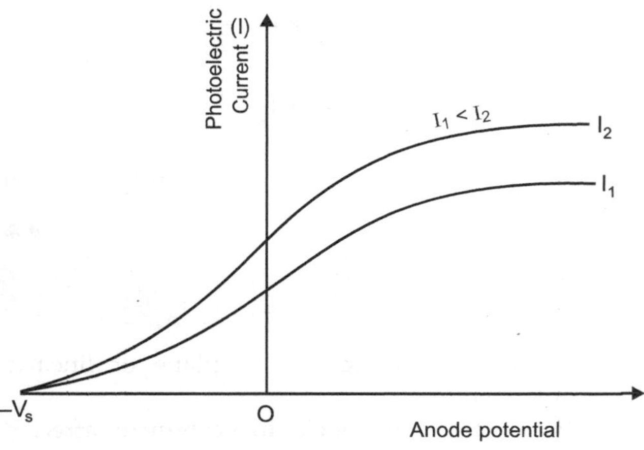


**11. Obtain the expression for the ratio of the de-Broglie wavelengths associated with the electron orbiting in the second and third excited states of hydrogen atom. [2]**

**Ans.** According to Bohr's postulate,

$$mvr = \frac{nh}{2\pi}$$

*i.e.,*
$$2\pi r = \frac{nh}{mv}$$

or
$$\frac{h}{mv} = \frac{2\pi r}{n}$$

since, $\frac{h}{mv} = \frac{h}{p} = \lambda$, by de -Broglie hypothesis

Therefore, $\frac{2\pi r}{n} = \lambda$

Now for second excited state

$$\lambda_2 = \frac{2\pi r_3}{n} \quad \text{...(i)}$$

and for third excited state

$$\lambda_3 = \frac{2\pi r_4}{n} \quad \text{...(ii)}$$

on dividing equations (i) by (ii),

$$\frac{\lambda_2}{\lambda_3} = \frac{r_3}{r_4} = \frac{9}{16}$$

This is the required expression.

**12. A charged particle $q$ is moving in the presence of a magnetic field B which is inclined to an angle 30° with the direction of the motion of the particle. Draw the trajectory followed by the particle in the presence of the field and explain how the particle describes this path. [2]**

**Ans.**

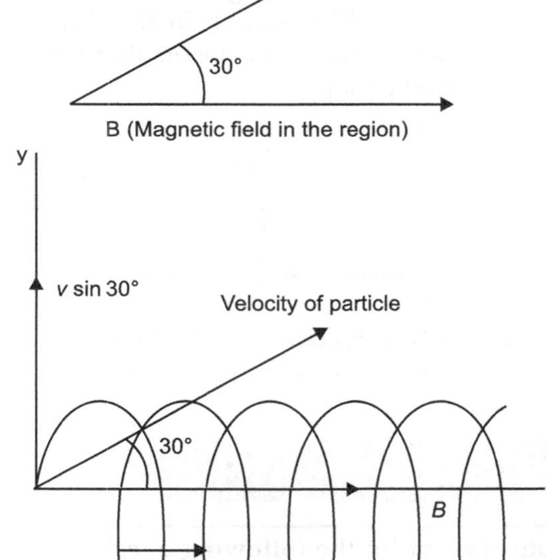


**Ans.** When a charged particle enter in a magnetic field making angle at 30°, then velocity component is resolved into 2 components: $v \cos \theta$ (along the magnetic field) and $v \sin \theta$ (normal to the magnetic field).

As the charged particle moves along XY-plane due to velocity component $v \sin\theta$, it also advances linearly due to the velocity component $v \cos \theta$. As a result, the charged particle will move in a helical path as shown in figure.

## SECTION-C

**13. (a) Explain briefly how Rutherford scattering of α–particle by a target nucleus can provide information on the size of the nucleus.**

**(b) Show that density of nucleus is independent of its mass number A.** **[3]**

**Ans. (a)** In Rutherford's scattering experiment of α-particle, it was observed that the fast and heavy α-particles could be deflected through 180°. But only very small number of particles *i.e.,* 1 in about 8,000 α-particles are deflected through 180°, that too from centre only.

So by this, it was assumed that the size of central part *i.e.,* nucleus is about $\frac{1}{10,000}$th of the size of the atom and whole positive charge is concentrated in it.

**(b)** Consider an atom whose mass number is A and R be the radius of the nucleus. If we neglect the mass of orbital electrons, then mass of the nucleus of the atom of mass number A = A a.m.u.

Mass of nucleus, $m = A \times 1.66 \times 10^{-27}$ kg

Volume of nucleus,

$$V = \frac{4}{3}\pi R^3$$

$$= \frac{4}{3}\pi (R_0 A^{1/3})^3 \quad \left[\because R = R_0 A^{1/3}\right]$$

where, $R_0 = 1.1 \times 10^{-15}$ m,

Therefore, $V = \frac{4}{3}\pi(1.1\times10^{-15})^3 \times A$ m$^3$

$$\text{Density of nucleus, } \rho = \frac{\text{Mass of nucleus}}{\text{Volume of nucleus}}$$

$$= \frac{A\times1.66\times10^{-27}}{\frac{4}{3}\pi(1.1\times10^{-15})^3\times A}$$

$$= 2.97 \times 10^{17} \text{ kg/m}^3$$

Therefore, density of nucleus is independent of mass number.

**14. State the underlying principle of a cyclotron. Explain its working with the help of a schematic diagram. Obtain the expression for cyclotron frequency.** **[3]**

**Ans. (a) Principle** : It is based on a principle that a positive ion can acquire sufficiently large energy with a comparatively smaller alternating potential difference by making it to cross the same electric field again and again by making use of a strong magnetic field.

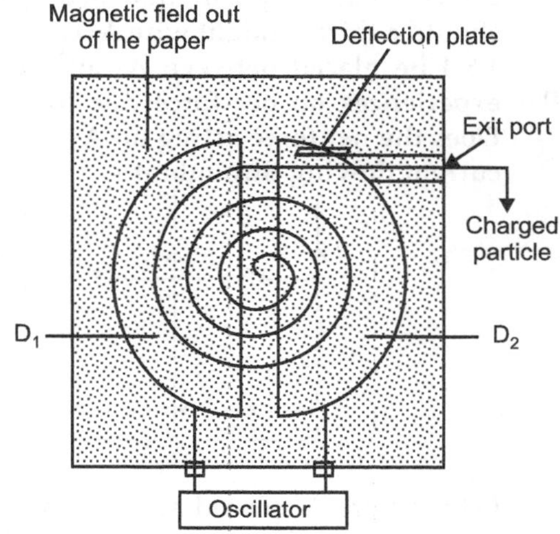


**Working** : The positive ions are produced from the source at the centre are accelerated by a dee which is at negative potential at that moment. Due to the presence of perpendicular magnetic field, the ion will move in a circular path inside the dees. The magnetic field and the frequency of a.c source are so chosen that as the ion comes out of a dee, it changes its polarity and the ion is further accelerated and moves with higher velocity along a circular path of greater radius. This phenomenon is continued till the ion reaches the periphery of the dees where a deflecting plate deflects the accelerated ion on the target to be bombarded.

**Expression for cyclotron frequency**

Suppose a position ion with charge $q$ moving with a velocity $v$, then

$$qvB = \frac{mv^2}{r}$$

or

$$r = \frac{mv}{qB}$$

Therefore, angular velocity ω is

$$\omega = \frac{v}{r} = \frac{qB}{m}$$

The time taken by ion in describing a semi circle is (*i.e.,* an angle π)

$$t = \frac{\pi}{\omega} = \frac{\pi m}{qB}$$

This is a semi-periodic time (T/2)

$$\therefore \quad \frac{T}{2} = t = \frac{\pi m}{qB}$$

*i.e.,* $T = \frac{2\pi m}{qB}$ time period of revolution

∴ Frequency of revolution is

$$f = \frac{1}{T} = \frac{qB}{2\pi m}$$

This frequency is called the cyclotron frequency.

**15. Two infinitely long straight wire $A_1$ and $A_2$ carrying currents I and 2 I flowing in the same direction are kept '$d$' distance apart. Where should a third straight wire $A_3$ carrying current 1.5 I be placed between $A_1$ and $A_2$ so that it experiences no net force due to $A_1$ and $A_2$ ? Does the net force acting on $A_3$ depend on the current flowing throught it ?** [3]

**Ans.**

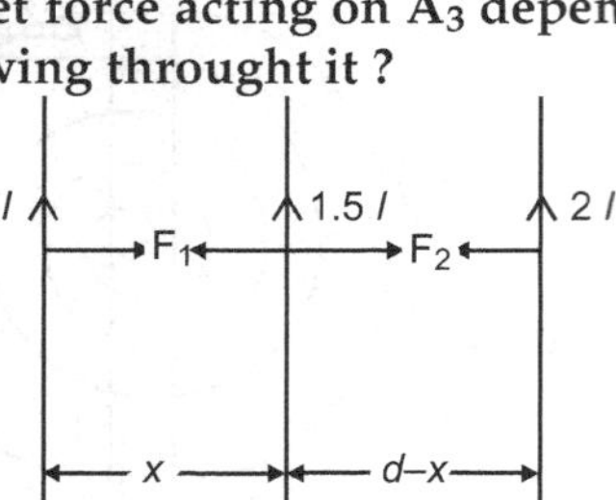


Let the wire $A_3$ placed at a distance $x$ from the wire $A_1$. And the distance of wire $A_3$ from $A_2$ be $(d - x)$. The current flowing through them is I, 1.5 I and 2I in $A_1$, $A_3$ and $A_2$ respectively. Let the current flowing in the wire $A_3$ be in the same direction as $A_1$ and $A_2$.

Therefore, force between $A_1$ and $A_3$ is,

$$F_1 = \frac{\mu_0}{2\pi}\cdot\frac{I \times 1.5\, I}{x} \qquad ...(i)$$

Force between $A_3$ and $A_2$ is,

$$F_2 = \frac{\mu_0}{2\pi}\cdot\frac{2\, I \times 1.5\, I}{(d-x)} \qquad ...(ii)$$

Since, the net force on $A_3$ is zero,

$$\therefore \quad F_1 = F_2$$

$$\Rightarrow \quad \frac{\mu_0}{2\pi}\times\frac{I \times 1.5\, I}{(x)} = \frac{\mu_0}{2\pi}\times\frac{2\, I \times 1.5\, I}{(d-x)}$$

$$\Rightarrow \quad \frac{1.5}{x} = \frac{3}{d-x}$$

$$\Rightarrow \quad 1.5\, d - 1.5x = 3x$$

$$\Rightarrow \quad 4.5\, x = 1.5\, d$$

Therefore, $x = \frac{d}{3}$

The wire $A_3$ is placed at a distance of $\frac{d}{3}$ from $A_1$ and $\frac{2d}{3}$ from $A_2$.

Yes, Net force $F_{net} = (F_1 - F_2)$

$$= \frac{\mu_0}{4\pi}\left[\frac{I \times IA_3}{x} - \frac{2IIA_3}{(d-x)}\right]$$

It depends on value of current flow through $A_3$.

**16. (a) Draw the equipotential surfaces due to an electric dipole.**

**(b) Derive an expression for the electric field due to a dipole of dipole moment $\vec{p}$ if at a point on its perpendicular bisector.** [3]

**Ans. (a)**

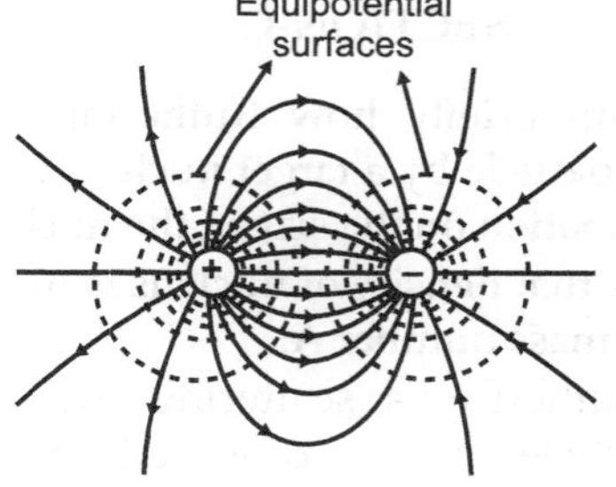


**(b)**

The Magnitudes of the electric field due to the two charges $+q$ and $-q$ given by.

$$E_{+q} = \frac{q}{4\pi\varepsilon_0}\frac{1}{r^2 + a^2} \qquad ...(i)$$

$$E_{-q} = \frac{q}{4\pi\varepsilon_0}\frac{1}{r^2 + a^2} \qquad ...(ii)$$

$$\therefore \quad E_{+q} = E_{-q}$$

The directions of $E_{+q}$ and $E_{-q}$ are as shown in the figure. The components normal to the dipole axis cancel away. The components along the dipole axis add up.

$\therefore$ Total electric field,

$$E = -(E_{+q} + E_{-q})\cos\theta\,\hat{p}$$

[Negative sign shows that field is opposite to $\hat{p}$]

$$E = -\frac{2qa}{4\pi\varepsilon_0 (r^2 + a^2)^{\frac{3}{2}}}\hat{p} \qquad ...(iii)$$

At large distances ($r >> a$), this reduces to

$$E = -\frac{2qa}{4\pi\varepsilon_0 r^3}\hat{p} \qquad ...(iv)$$

$$\because \quad \vec{p} = q \times 2\vec{a}\,\hat{p}$$

$$\therefore \quad E = \frac{-\vec{p}}{4\pi\varepsilon_0 r^3} \quad (r >> a)$$

# PHYSICS

## Delhi [Set-I] 2018

**Time allowed : 3 hours** **Maximum marks : 70**

### SECTION–A

**1. A proton and an electron travelling along parallel paths enter a region of uniform magnetic field, acting perpendicular to their paths. Which of them will move in a circular path with higher frequency? [1]**

**Ans.** Frequency of revolution of a particle,

$$f = \frac{qB}{2\pi m}$$

or $$f \propto \frac{1}{m}$$

Since mass of electron is less than that of proton, therefore, its frequency of revolution will be higher than that of proton.

**2. Name the electromagnetic radiations used for (a) water purification, and (b) eye surgery. [1]**

**Ans.** **(a) Water purification :** Ultraviolet radiation.

**(b) Eye surgery :** Ultraviolet radiation/laser.

**3. Draw graphs showing variation of photoelectric current with applied voltage for two incident radiations of equal frequency and different intensities. Mark the graph for the radiation of higher intensity. [1]**

**Ans.** Graph for photoelectric current (I) versus applied potential V for radiations of same frequency and varying intensity.

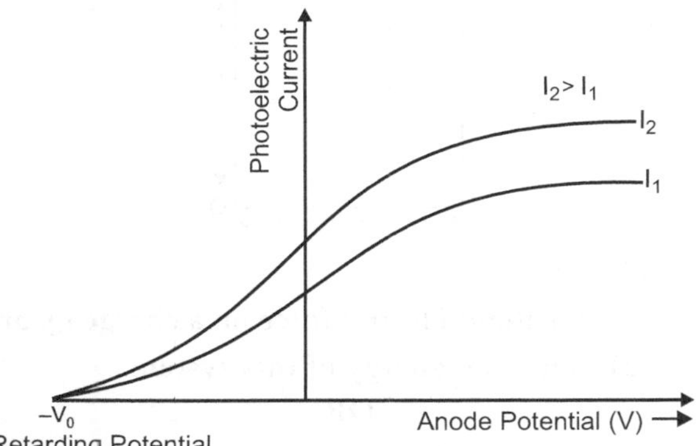


**4. Four nuclei of an element undergo fusion to form a heavier nucleus, with release of energy. Which of the two— the parent or the daughter nucleus—would have higher binding energy per nucleon? [1]**

**Ans.** When lighter nuclei combine to form a heavier nucleus, binding energy per nucleon increases and energy is released. Thus, the daughter nucleus would have higher binding energy per nucleon.

**5. Which mode of propagation is used by short wave broadcast services? [1]**

**Ans.** Sky wave propagation is used for short wave broadcast for frequency range upto 30 MHz.

### SECTION–B

**6. Two electric bulbs P and Q have their resistances in the ratio of 1 : 2. They are connected in series across a battery. Find the ratio of the power dissipation in these bulbs. [2]**

**Ans.** Let resistances of bulbs P and Q be R and 2R respectively.

As they are connected in series, so current through each bulb is same. Let the current be I.

$$\therefore \qquad \frac{P_1}{P_2} = \frac{I^2R_1}{I^2R_2} = \frac{R_1}{R_2} = \frac{1}{2}$$

*i.e.* $$P_1 : P_2 = 1 : 2$$

**7. A 10 V cell of negligible internal resistance is connected in parallel across a battery of emf 200 V and internal resistance 38 Ω as shown in the figure. Find the value of current in the circuit. [2]**

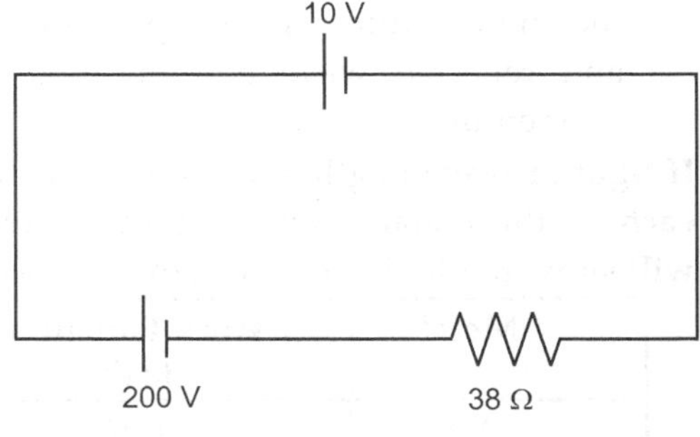


**OR**

**In a potentiometer arrangement for determining the emf of a cell, the balance point of the cell in open circuit is 350 cm. When a resistance of 9 Ω is used in the external circuit of the cell, the balance point shifts to 300 cm. Determine the internal resistance of the cell.**

**Ans.** Given : $E_1$ = 10 V; $E_2$ = 200 V, $r$ = 38 Ω, I = ?

Net emf, $E = E_2 - E_1$ (∵ two cells being in opposition)

$= 200\text{ V} - 10\text{ V} = 190\text{ V}$

$$\therefore \quad \text{Current, } I = \frac{\text{Net emf}}{\text{Resistance}} = \frac{190\text{ V}}{38\Omega} = 5\text{ A.}$$

**OR**

Balance point in open circuit, $l_1$= 350 cm. R = 9 Ω,

New balance point $l_2$ = 300 cm.

Internal resistance of a cell is given by,

$$r = R\left(\frac{l_1 - l_2}{l_2}\right) = 9\left(\frac{350-300}{300}\right) = 1.5\ \Omega.$$

**8. (a) Why are infra-red waves often called heat waves ? Explain.**

**(b) What do you understand by the statement, "Electromagnetic waves transport momentum" ? [1]**

**Ans. (a)** Infra-red waves are called heat waves because they raise the temperature of the object on which they fall and hence increase their thermal motion. They also affect the photographic plate and are readily absorbed by most of the materials.

**(b)** Electromagnetic waves transport momentum. This means that when an electromagnetic wave travels through space with energy U and speed $c$, then it transports linear momentum $p = \frac{U}{c}$.

If a surface absorbs the waves completely, then momentum '$p$' is delivered to the surface. If the surface reflects the wave, then momentum delivered by both incident and reflected wave add on to give '$2p$' momentum.

**9. If light of wavelength 412.5 nm is incident on each of the metals given below, which ones will show photoelectric emission and why? [2]**

| Metal | Work Function (eV) |
|---|---|
| Na | 1.92 |
| K | 2.15 |
| Ca | 3.20 |
| Mo | 4.17 |

**Ans.** Wavelength of incident light,

$$\lambda = 412.5\text{ nm} = 412.5 \times 10^{-9}\text{m}$$

∴ Energy of incident light

$$E = \frac{hc}{\lambda} = \frac{6.63\times10^{-34}\times3\times10^{8}}{412.5\times10^{-9}}$$

$$= 4.82 \times 10^{-19}\text{ J}$$

$$= \frac{4.82\times10^{-19}}{1.6\times10^{-19}}\text{ eV}$$

$$= 3.01\text{ eV}$$

Since the energy of incident radiation is greater than the work function of sodium and potassium, but less than that of calcium and molybdenum, therefore, photoelectric emission will take place in sodium and potassium.

**10. A carrier wave of peak voltage 15 V is used to transmit a message signal. Find the peak voltage of the modulating signal in order to have a modulation index of 60%. [2]**

**Ans.** Peak voltage of carrier wave,

$$E_c = 15\text{ V.}$$

Modulation index, $m = 60\% = 0.6$

Peak voltage of modulating signal,

$$E_m = ?$$

We know, $m = \frac{E_m}{E_c}$

or $E_m = mE_c = 0.6 \times 15 = 9\text{ V}$

## SECTION–C

**11. Four point charges Q, $q$, Q and $q$ are placed at the corners of a square of side '$a$' as shown in the figure.**

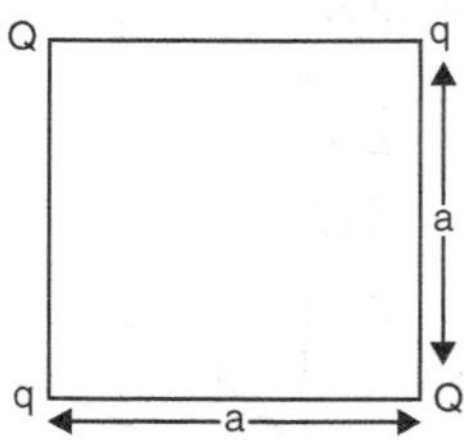


**Find the :**

**(a) resultant electric force on a charge Q, and**

**(b) potential energy of this system. [3]**

**OR**

**(a) Three point charges $q$, – $4q$ and $2q$ are placed at the vertices of an equilateral triangle ABC of side '$l$' as shown in the figure. Obtain the expression for the magnitude of the resultant electric force acting on the charge $q$.**

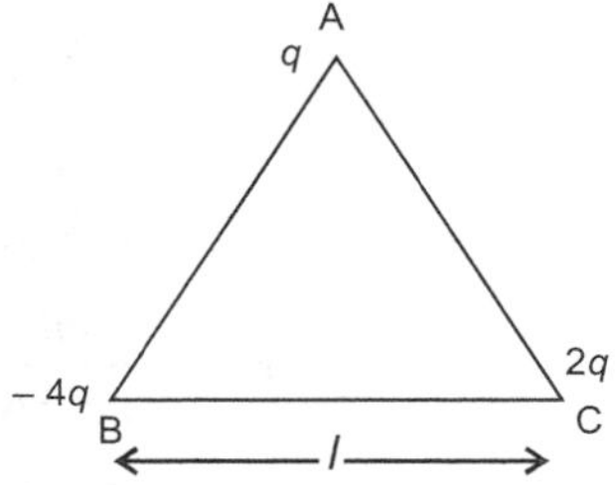


**(b) Find out the amount of the work done to separate the charges at infinite distance.**

**Ans.** Let us find the force on charge Q placed at point B.

**(a)** Force on charge Q at B due to charge $q$ at A

$$F_1 = \frac{1}{4\pi\varepsilon_0}\frac{qQ}{a^2} \text{ along } \overrightarrow{BY}$$

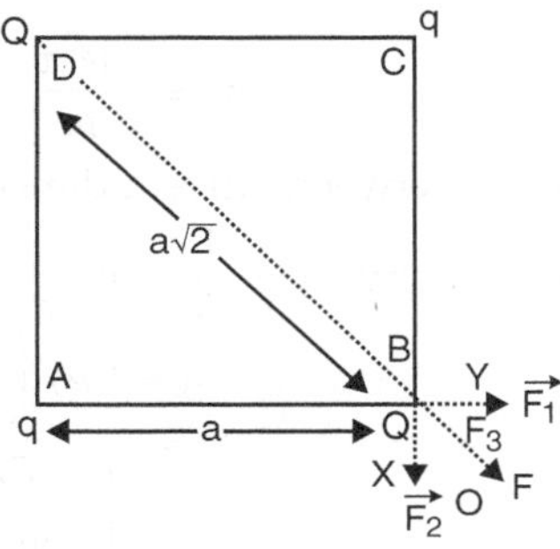


Force on charge Q at B due to charge q at c.

$$F_2 = \frac{1}{4\pi\varepsilon_0}\frac{qQ}{a^2} \text{ along } \overrightarrow{BX}$$

Resultant force, $F = \sqrt{F_1^2 + F_2^2}$ along $\overrightarrow{BO}$

$$= \sqrt{2\left(\frac{1}{4\pi\varepsilon_0}\frac{qQ}{a^2}\right)^2}$$

$$= \frac{qQ\sqrt{2}}{4\pi\varepsilon_0 a^2}$$

Force due to charge Q = $F_3$

$$= \frac{1}{4\pi\varepsilon_0}\frac{Q.Q}{(\sqrt{2}a)^2} \text{ along } \overrightarrow{BO}$$

Total resultant force = F + $F_3$ along $\overrightarrow{BO}$

$$= \frac{1}{4\pi\varepsilon_0}\frac{\sqrt{2}\,Qq}{a^2} + \frac{1}{4\pi\varepsilon_0}\frac{Q.Q}{2a^2}$$

$$= \left[\frac{1}{4\pi\varepsilon_0}\frac{Q}{a^2}\left(\sqrt{2}q + \frac{Q}{2}\right)\right] \text{ along } \overrightarrow{BO}$$

**(b)** Total potential energy of the system,

$$U = \frac{1}{4\pi\varepsilon_0}\left[\frac{Qq}{a} + \frac{Qq}{a} + \frac{Qq}{a} + \frac{Qq}{a} + \frac{Q.Q}{\sqrt{2}a} + \frac{q.q}{\sqrt{2}a}\right]$$

$$= \frac{1}{4\pi\varepsilon_0}\left[\frac{4Qq}{a} + \frac{Q^2}{\sqrt{2}a} + \frac{q^2}{\sqrt{2}a}\right]$$

$$= \frac{1}{4\pi\varepsilon_0} \times \frac{1}{a}\left[4Qq + \frac{Q^2 + q^2}{\sqrt{2}}\right]$$

**OR**

**(a)** Electric force at A due to charge $2q$

$$F_C = \frac{1}{4\pi\varepsilon_0}\frac{q \times 2q}{l^2} \text{ along } \overrightarrow{CA}$$

Electric force at A due to charge $(-4q)$,

$$F_B = \frac{1}{4\pi\varepsilon_0}\frac{q \times (-4q)}{l^2} \text{ along } \overrightarrow{AB}$$

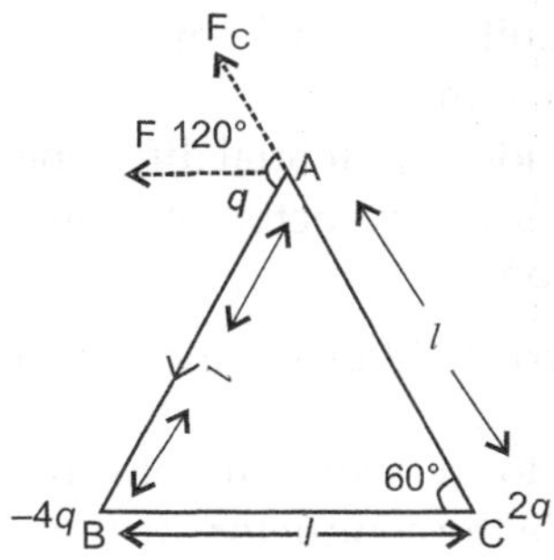


Resultant force,

$$F = \sqrt{F_B^2 + F_C^2 + 2F_BF_C \cos 120°}$$

$$= \frac{1}{4\pi\varepsilon_0}\frac{q^2}{l^2}\sqrt{(-4)^2 + (2)^2 + 2(-4)(2) \times \left(-\frac{1}{2}\right)}$$

$$= \frac{1}{4\pi\varepsilon_0}\frac{2\sqrt{7}q^2}{l^2}$$

**(b)** Work done to separate the charges to infinity :

Initial potential energy,

$$U_i = \frac{1}{4\pi\varepsilon_0}\left[\frac{(-4q)q}{l} + \frac{(-4q)(2q)}{l} + \frac{(q)(2q)}{l}\right]$$

$$= \frac{1}{4\pi\varepsilon_0}\frac{q^2}{l}[-4 - 8 + 2]$$

$$= \frac{1}{4\pi\varepsilon_0}\left(\frac{-10q^2}{l}\right)$$

Final potential energy, $U_f = 0$

Thus, work done

$$= U_f - U_i = 0 - \left(\frac{-10q^2}{4\pi\varepsilon_0 l}\right) = \frac{10q^2}{4\pi\varepsilon_0 l} = \frac{5q^2}{2\pi\varepsilon_0 l}$$

**12. (a) Define the term 'conductivity' of a metallic wire. Write its SI unit.**

**(b) Using the concept of free electrons in a conductor, derive the expression for the conductivity of a wire in terms of number density and relaxation time. Hence obtain the relation between current density and the applied electric field E.** **[3]**

**Ans. (a)** Conductivity of a metallic wire is defined as its ability to allow electric charges or heat to pass through it.

Numerically, conductivity of a material is reciprocal of its resistivity.

SI unit : $ohm^{-1}$ $m^{-1}$ or mho $m^{-1}$ or Siemen $m^{-1}$.

**(b)** Consider a potential difference V be applied across a conductor of length $l$ and cross-section A.

Electric field inside the conductor, $E = \frac{V}{l}$.

Due to the external field the free electrons inside the conductor drift with velocity $v_d$.

Let, number of electrons per unit volume = $n$,

charge on an electron = $e$

$\therefore$ Total electrons in length $l$ = $nAl$

And, total charge, $q = neAl$

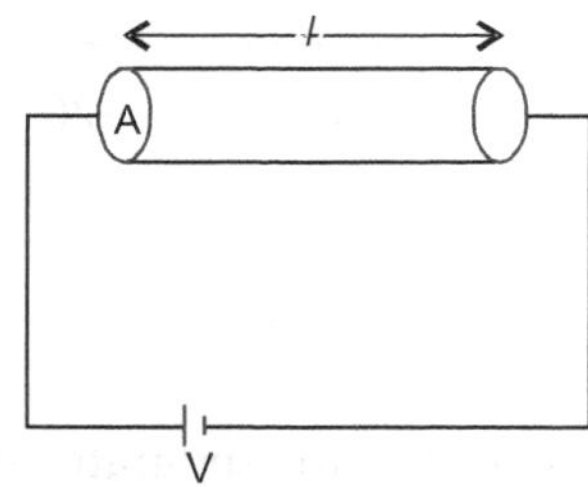

Time taken by electrons to enter and leave the conductor,

$$t = \frac{l}{v_d}$$

$$\text{Current, } I = \frac{q}{t} = \frac{neAl}{l/v_d}$$

$$= ne\,A\,v_d$$

$$\text{Current density, } J = \frac{I}{A} = nev_d \quad \text{...(i)}$$

We know, $$v_d = \frac{eE\tau}{m} = \frac{eV\tau}{ml}$$

$$\therefore \quad I = neAv_d = \frac{neAVe\tau}{ml}$$

$$\Rightarrow \quad \frac{V}{I} = \frac{ml}{ne^2\tau A}$$

$$\Rightarrow \quad R = \frac{ml}{ne^2\tau A}$$

$$\left[\because \frac{V}{I} = R; \text{ Ohm's law}\right]$$

$$\text{Resistivity, } \rho = \frac{RA}{l} = \frac{mlA}{ne^2\tau A \times l}$$

or $$\rho = m/ne^2\tau$$

$$\text{Since conductivity, } \sigma = \frac{1}{\rho}$$

$$\therefore \quad \sigma = \frac{ne^2\tau}{m} \quad \text{...(ii)}$$

Relation between current density and field :

$$\text{current density, } J = \frac{I}{A} = nev_d \quad \text{[from (i)]}$$

$$J = (ne)\left(\frac{eE\tau}{m}\right)$$

$$= \left(\frac{ne^2\tau}{m}\right)E$$

$$\Rightarrow \quad J = \sigma E \quad \text{[from (ii)]}$$

which is the required relation.

**13. A bar magnet of magnetic moment 6 J/T is aligned at 60° with a uniform external magnetic field of 0.44 T. Calculate (a) the work done in turning the magnet to align its magnetic moment (i) normal to the magnetic field, (ii) opposite to the magnetic field, and (b) the torque on the magnet in the final orientation in case (ii).** **[3]**

**Ans.** Given, magnetic momentum, $m = 6\ JT^{-1}$

External magnetic field, B = 0.44 T,

$$\theta_1 = 60° \Rightarrow \cos\theta_1 = \cos 60° = \frac{1}{2}.$$

**(a)** Work done in turning the magnet normal to the field,

$$W = -mB(\cos\theta_2 - \cos\theta_1)$$

**(i)** Here, $\theta_2 = 90°$

$$\therefore \quad W = +mB\cos\theta_1$$

$$= 6 \times 0.44 \times \frac{1}{2} = 1.32\ J$$

**(ii)** Here, $\theta_2 = 180°$

$$\therefore \quad W = -mB(\cos\theta_2 - \cos\theta_1)$$

$$W = -6 \times 0.44\left(-1 - \frac{1}{2}\right)$$

$$= 3.96\ J$$

**(b)** Torque on magnet when moment is aligned opposite to the field,

$$\text{Torque} = |\vec{m} \times \vec{B}|$$
$$\tau = mB \sin\theta$$
$$= 6 \times 0.44 \times \sin 180^\circ$$
$$= 0 \qquad (\because \sin 180^\circ = 0)$$

**14. (a) An iron ring of relative permeability $\mu_r$ has windings of insulated copper wire of $n$ turns per metre. When the current in the windings is I, find the expression for the magnetic field in the ring.**

**(b) The susceptibility of a magnetic material is 0.9853. Identify the type of magnetic material. Draw the modification of the field pattern on keeping a piece of this material in a uniform magnetic field. [3]**

**Ans. (a)** Consider a ring of radius $r$ having $n$ turns per metre. Then total number of turns in the ring $= 2\pi rn = N$

Current enclosed $= NI = 2\pi rnI$.

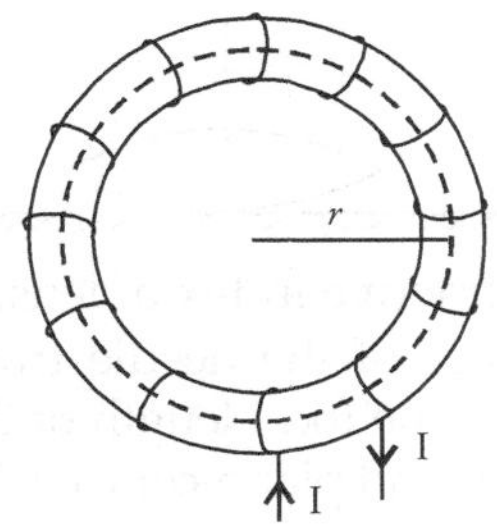


By Ampere's circuital law, $\oint \vec{B}.\vec{dt} = \mu_0\mu_r I$

$$B \times 2\pi r = \mu_0\mu_r 2\pi rnI$$

or $$B = \mu_0\mu_r nI$$

**(b)** Given, susceptibility, $\chi = 0.9853$.

The material is paramagnetic in nature.

If a piece of this material is kept in uniform magnetic field, then field pattern gets modified as follows :

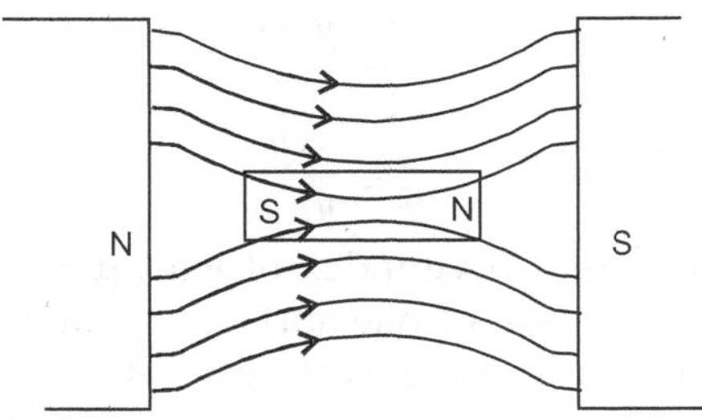


The lines of force tend to pass through the material rather than the surrounding air.

**15. (a) Show using a proper diagram how unpolarised light can be linearly polarised by reflection from a transparent glass surface.**

**(b) The figure shows a ray of light falling normally on the face AB of an equilateral glass prism having refractive index $\frac{3}{2}$, placed in water of refractive index $\frac{4}{3}$. Will this ray suffer total internal reflection on striking the face AC ? Justify your answer. [3]**

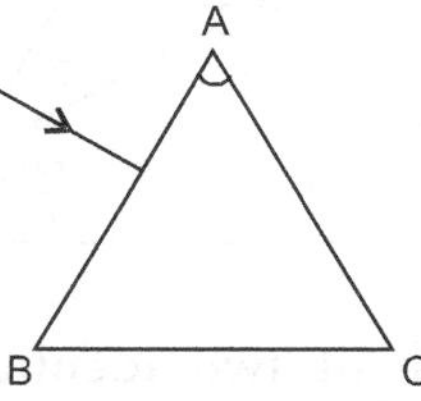


**Ans. (a)** Polarisation of light by reflection from a transparent glass surface :

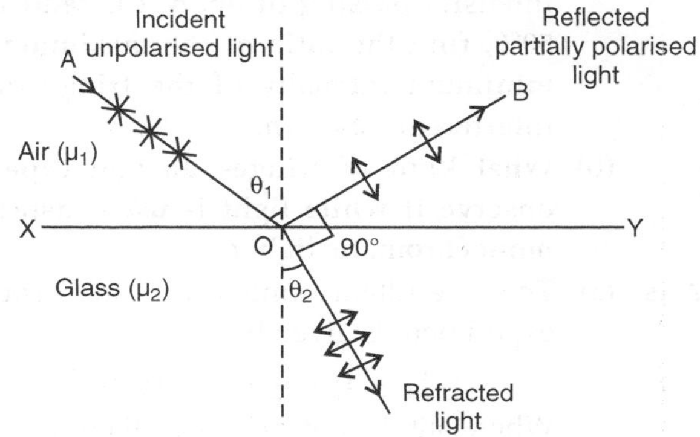


$$\text{or } \theta_2 = \frac{\pi}{2} - \theta_1$$

$$\text{or } \mu = \frac{\sin\theta_1}{\sin\theta_2} = \frac{\sin\theta_1}{\sin(90-\theta_1)} = \frac{\sin\theta_1}{\cos\theta_1} = \tan\theta_1$$

Light gets totally polarised by reflection when it is incident at an angle $\theta_1$ (Brewster Angle),

$$\theta_1 = \tan^{-1}\mu$$

**(b)** Given,

$$_a\mu_g = \frac{3}{2} \text{ and } _a\mu_g = \frac{4}{3}$$

$$_w\mu_g = \frac{_a\mu_g}{_a\mu_w} = \frac{3/2}{4/3} = \frac{9}{8} = 1.125$$

We know at critical angle C,

$$_w\mu_g = \frac{1}{\sin C}$$

$$\text{or } \sin C = \frac{1}{_w\mu_g} = \frac{1}{1.125} = 0.88$$

*i.e.* $C = 62.73^\circ$

But from the figure, light will incident on face AC making an angle 60° with the normal *i.e.*, $i = 60°$

So, $i < C$

$\therefore$ Light will not suffer total internal reflection.

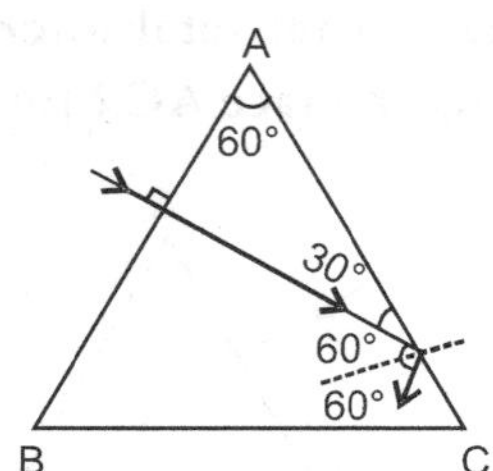


**16. (a) If one of two identical slits producing interference in Young's experiment is covered with glass, so that the light intensity passing through it is reduced to 50%, find the ratio of the maximum and minimum intensity of the fringe in the interference pattern.**

**(b) What kind of fringes do you expect to observe if white light is used instead of monochromatic light? [3]**

**Ans. (a)** The resultant intensity in Young's experiment is given by

$$I_R = I_1 + I_2 + 2\sqrt{I_1 I_2}\cos\phi$$

When slit is not covered, then $I_0$ is the intensity from each slit.

Maximum intensity ($I_{max}$) occurs when $\phi = 0°$.

Minimum intensity ($I_{min}$) occurs when $\phi = 180°$.

If one slit is covered with glass to reduce its intensity by 50%, then

$$I_{max} = I_0 + \frac{I_0}{2} + 2\sqrt{I_0 \times \frac{I_0}{2}}\cos 0°$$

$$= 1.5\, I_0 + 2 \times 0.707\, I_0 = 2.914\, I_0$$

$$I_{min} = I_0 + \frac{I_0}{2} + 2\sqrt{I_0 \times \frac{I_0}{2}}\cos 180°$$

$$= 1.5\, I_0 - 2 \times 0.707\, I_0 = 0.086\, I_0$$

$$\text{Ratio} = \frac{I_{max}}{I_{min}} = \frac{2.914\, I_0}{0.086\, I_0} = 33.884 \approx 34$$

**(b)** If instead of monochromatic light, white light is used, then the central fringe will be white and the fringes on either side will be coloured. Blue colour will be nearer to central fringe and red will be farther away. The path difference at the centre on perpendicular bisector of slits will be zero for all colours and each colour produces a bright fringe thus resulting in white fringe. Further, the shortest visible wave, blue, produces a bright fringe first.

**17. A symmetric biconvex lens of radius of curvature R and made of glass of refractive index 1.5, is placed on a layer of liquid placed on top of a plane mirror as shown in the figure. An optical needle with its tip on the principal axis of the lens is moved along the axis until its real, inverted image coincides with the needle itself. The distance of the needle from the lens is measured to be *x*. On removing the liquid layer and repeating the experiment, the distance is found to be *y*. Obtain the expression for the refractive index of the liquid in terms of *x* and *y*. [3]**

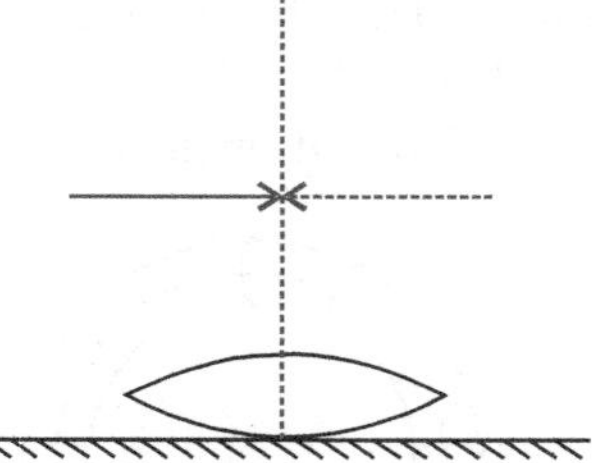

**Ans.** Given, refractive index of lens, $\mu_g = 1.5$

The distance of the needle from the lens in the first case = The focal length of the combination of convex lens and plano-concave lens formed by the liquid, $f = x$

And, the distance measured in the second case = Focal length of the convex lens, $f_1 = y$

If the focal length of plano-concave lens formed by the liquid be $f_2$, then

$$\frac{1}{f} = \frac{1}{f_1} + \frac{1}{f_2}$$

$$\Rightarrow \quad \frac{1}{f_2} = \frac{1}{f} - \frac{1}{f_1} = \frac{1}{x} - \frac{1}{y}$$

$$= \frac{y-x}{xy}$$

or $$f_2 = \frac{xy}{y-x}$$

Now, refractive index of lens, $\mu = 1.5$, radius of curvature of one surface = R and radius of curvature of other surface = – R

$$\therefore \quad \frac{1}{f_1} = (\mu_g - 1)\left(\frac{1}{R} - \frac{1}{-R}\right)$$

$$\frac{1}{y} = (1.5 - 1)\left(\frac{2}{R}\right) \Rightarrow R = y$$

And, if refractive index of liquid is $\mu_l$,

Radius of curvature on the side of plane mirror = $\infty$

Radius of curvature on the side of lens = – R

$$\because \quad \frac{y-x}{xy} = (\mu_l - 1)\left(\frac{1}{-R} - \frac{1}{\infty}\right)$$

$$= (\mu_l - 1)\left(-\frac{1}{y}\right)$$

$$\Rightarrow \quad (\mu_l - 1) = -\left(\frac{y-x}{x}\right)$$

or

$$\mu_l = \frac{x-y}{x} + 1$$

$$= \frac{2x-y}{x}$$

**18. (a) State Bohr's postulate to define stable orbits in hydrogen atom. How does de Broglie's hypothesis explain the stability of these orbits ?**

**(b) A hydrogen atom initially in the ground state absorbs a photon which excites it to the $n$ = 4 level. Estimate the frequency of the photon. [3]**

**Ans. (a) Bohr's postulate for stable orbits in hydrogen atom :** An electron can revolve only in those circular orbits in which its angular momentum is an integral multiple of $h/2\pi$, where $h$ is planck's constant.

If $n$ is the principal quantum number of orbit, then an electron can revolve only in certain orbits or definite radii. These are called stable orbits.

**de Broglie explanation of stability of orbits :** According to de Broglie, orbiting electron around the nucleus is associated with a stationary wave. Electron wave is a circular standing wave. Since destructive interference will occur if a standing wave does not close upon itself, only those de Broglie waves exist for which the circumference of circular orbit contains a whole number of wavelengths *i.e.,* for orbit circumference of $n^{th}$ orbit as $2\pi r_n$,

$$2\pi r_n = n\lambda$$

$$2\pi r_n = n\left(\frac{h}{mv}\right)$$

or

$$mvr_n = n\left(\frac{h}{2\pi}\right)$$

which is quantum condition proposed by Bohr.

**(b)** In ground state, $n = 1$

In excited state, $n = 4$

$$\frac{1}{\lambda} = R\left[\frac{1}{(1)^2} - \frac{1}{(4)^2}\right], \text{ where R is Rydberg constant}$$

$$\frac{1}{\lambda} = 1.097 \times 10^7 \times \frac{15}{16}$$

$$= 10284375$$

$$= 1.028 \times 10^7 \text{ m}^{-1}$$

Frequency,

$$\nu = \frac{c}{\lambda} = 3 \times 10^8 \times 1.028 \times 10^7$$

$$= 3.09 \times 10^{15} \text{ Hz}$$

**19. (a) Explain the processes of nuclear fission and nuclear fusion by using the plot of binding energy per nucleon (BE/A) versus the mass number A.**

**(b) A radioactive isotope has a half-life of 10 years. How long will it take for the activity to reduce to 3.125% ? [3]**

**Ans. (a)** Plot of binding energy per nucleon mass number :

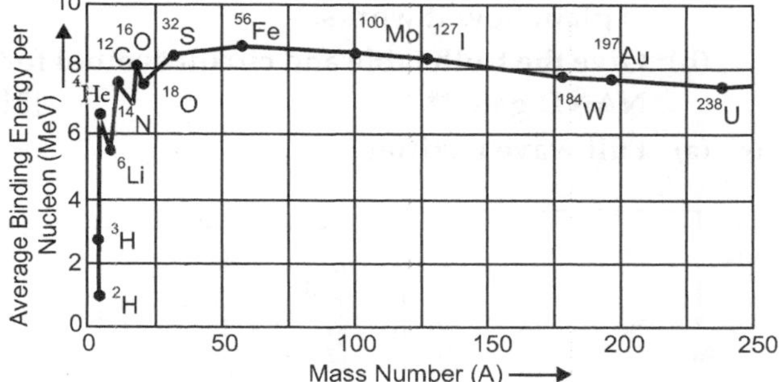


**Ans. (a)** Plot

1. When we move from the heavy nuclei region to the middle region of the plot, we find that there will be a gain in the overall binding energy and hence results in release of energy. This indicates that energy can be released when a heavy nucleus (A ≈ 240) breaks into two roughly equal fragments. This process is called nuclear fission.
2. Similarly, when we move from lighter nuclei to heavier nuclei, we again find that there will be gain in the overall binding energy and hence release of energy takes place. This indicates that energy can be released when two or more lighter nuclei fuse together to form a heavy nucleus. This process is called nuclear fusion.

**(b)** We know that,

$$\frac{R}{R_0} = \left(\frac{1}{2}\right)^n$$

Given,

$$\frac{R}{R_0} = 3.125\% = \frac{3.125}{100}$$

$$\therefore \quad \frac{3.125}{100} = \left(\frac{1}{2}\right)^n$$

or $$\frac{1}{32} = \left(\frac{1}{2}\right)^n$$

$$\Rightarrow \quad \left(\frac{1}{2}\right)^5 = \left(\frac{1}{2}\right)^n$$

$$\Rightarrow \quad n = 5$$

Given, T = 10 years

As $$n = \frac{t}{T}$$

$$\therefore \quad \frac{t}{T} = 5 \quad [\because T = 10 \text{ year}]$$

or $$t = 5 \times 10 = 50 \text{ years}$$

**20. (a) A student wants to use two *p-n* junction diodes to convert alternating current into direct current. Draw the labelled circuit diagram she would use and explain how it works.**

**(b) Give the truth table and circuit symbol for NAND gate.** [3]**

**Ans. (a) Full wave rectifier :**

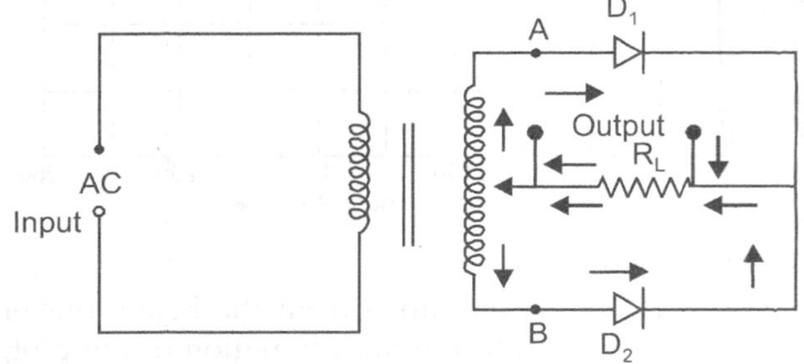


centre tap transformer

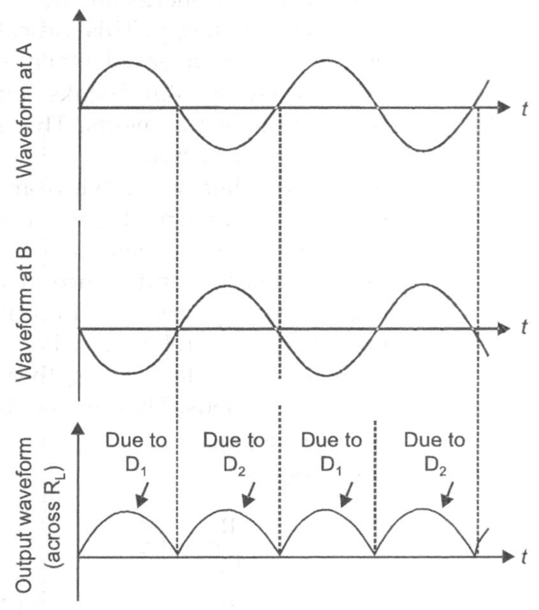


**Explanation :** In positive half cycle of AC, end A becomes positive and $D_1$ becomes forward biased and $D_2$ is reverse biased, so $D_1$ conducts and $D_2$ doesn't. So conventional current flows through $D_1$, $R_L$ and upper half of secondary winding. Similarly, during negative half cycle of AC, diode $D_2$ becomes forward biased and $D_1$ is reverse biased, current flows through $D_2$, $R_L$ and lower half of secondary winding. Thus, current flows in same direction in both half cycles of input AC voltage.

**21. Draw the typical input and output characteristics of an *n-p-n* transistor in CE configuration. Show how these characteristics can be used to determine (a) the input resistance ($r_i$), and (b) current amplification factor (β).** [3]**

**22. (a) Give three reasons why modulation of a message signal is necessary for long distance transmission.****

**(b) Show graphically an audio signal, a carrier wave and an amplitude modulated wave.** [3]**

## SECTION–D

**23. The teachers of Geeta's school took the students on a study trip to a power generating station, located nearly 200 km away from the city. The teacher explained that electrical energy is transmitted over such a long distance to their city, in the form of alternating current (ac) raised to a high voltage. At the receiving end in the city, the voltage is reduced to operate the devices. As a result, the power loss is reduced. Geeta listened to the teacher and asked questions about how the ac is converted to a higher or lower voltage. [4]**

**(a) Name the device used to change the alternating voltage to a higher or lower value. State one cause for power dissipation in this device.**

**(b) Explain with an example, how power loss is reduced if the energy is transmitted over long distances as an alternating current rather than a direct current.**

**(c) Write two values each shown by the teachers and Geeta.**

**Ans.** (a) The device used to change alternating voltage to a higher or lower value is a transformer.

Causes of power dissipation in this device are :

1. Core losses due to eddy currents and hysteresis loop due to alternating flux.

**Answer is not given due to change in the present Syllabus.

2. Copper losses due to resistance of winding in primary and secondary coils.
3. Loss of power due to leakage of magnetic flux in coil.

**(b)** The loss of power in the transmission lines is $I^2R$, where I is the strength of current and R is the resistance of the wires. To reduce the power loss, a.c. is transmitted over long distances at extremely high voltages. This reduces I in the same ratio. Therefore, $I^2R$ becomes negligibly low.
For the same reason, at the generating stations, the voltage is stepped up to transmit it over long distances to minimize power loss. Therefore a.c. is used because stepping up is not possible for direct current.

## SECTION–E

**24. (a) Define electric flux. Is it a scalar or a vector quantity ? A point charge $q$ is at a distance of $d/2$ directly above the centre of a square of side $d$, as shown in the figure. Use Gauss' law to obtain the expression for the electric flux through the square.**

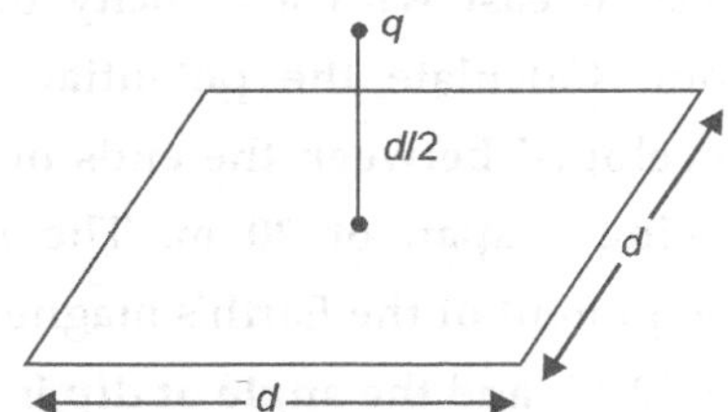


**(b) If the point charge is now moved to a distance '$d$' from the centre of the square and the side of the square is doubled, explain how the electric flux will be affected.** [5]

**OR**

**(a) Use Gauss' law to derive the expression for the electric field (E) due to a straight uniformly charged infinite line of charge density λ C/m.**

**(b) Draw a graph to show the variation of E with perpendicular distance $r$ from the line of charge.**

**(c) Find the work done in bringing a charge $q$ from perpendicular distance $r_1$ to $r_2$ $(r_2 > r_1)$.**

**Ans. (a) Electric flux :** Electric flux through an area is defined as the product of electric field strength E and area $dS$ perpendicular to the field. It represents the field lines crossing the area. It is a scalar quantity.

Imagine a cube of edge $d$, enclosing the charge. The square surface is one of the six faces of this cube. According to Gauss' theorem in electrostatics,

Total electric flux through the cube = $\frac{q}{\varepsilon_0}$

This is the total flux through all six surface S

∴ Electric flux through the square surface

$= \frac{q}{6\varepsilon_0}$

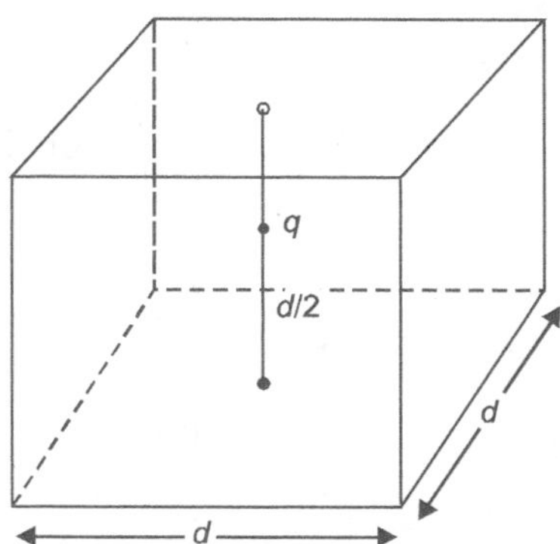


**(b)** On moving the charge to distance $d$ from the centre of square and making side of square $2d$, does not change the flux at all because flux is independent of side of square or distance of charge in this case.

**OR**

**(a) Electric field ($\vec{E}$) due to a straight uniformly charged infinite line of charge density λ :** Consider a cylindrical Gaussian surface of radius $r$ and length $l$ coaxial with line charge. The cylindrical Gaussian surface may be divided into three parts : (i) curved surface $S_1$ (ii) flat surface $S_2$ and (iii) flat surface $S_3$.
By symmetry, the electric field has the same magnitude E at each point of curved surface $S_1$ and is directed radially outward.

We consider small elements of surfaces $S_1$, $S_2$ and $S_3$. The surface element vector $d\vec{S}$ is directed along the direction of electric field (*i.e.*, angle between $\vec{E}$ and $d\vec{S_1}$ is 0°); the elements $d\vec{S_2}$ and $d\vec{S_3}$ are directed perpendicular to field vector $\vec{E}$ (*i.e.*, angle between $d\vec{S_2}$ and $\vec{E}$, and $d\vec{S_3}$ and $\vec{E}$ is 90°).

Electric flux through the cylindrical surface

$$\oint_S \vec{E}.\vec{dS} = \oint_{S_1} \vec{E}.\vec{dS_1} + \oint_{S_2} \vec{E}.\vec{dS_2} + \oint_{S_3} \vec{E}.\vec{dS_3}$$

$$= \oint_{S_1} E dS_1 \cos 0° + \oint_{S_2} E dS_2 \cos 90° + \oint_{S_3} E dS_3 \cos 90°$$

$$= E\int dS_1 + 0 + 0$$

$= E \times$ area of curved surface $= E \times 2\pi rl$

Total Charge enclosed, $q = \lambda l$

and flux, by Gauss' theorem,

$$\oint E = \frac{q}{\varepsilon_0} = \frac{\lambda l}{\varepsilon_0}$$

$$E.\ 2\pi rl = \frac{\lambda l}{\varepsilon_0} \quad \text{or } E = \frac{\lambda}{2\pi\varepsilon_0 r}$$

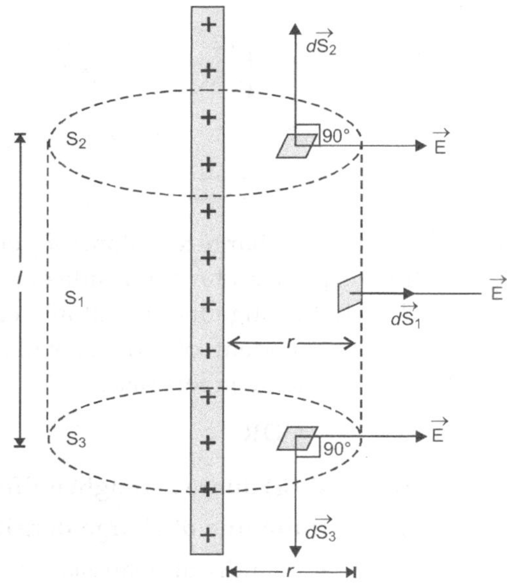


**(b) Graph showing variation of E with perpendicular distance from line of charge** : The electric field is inversely proportional to distance '$r$' from line of charge.

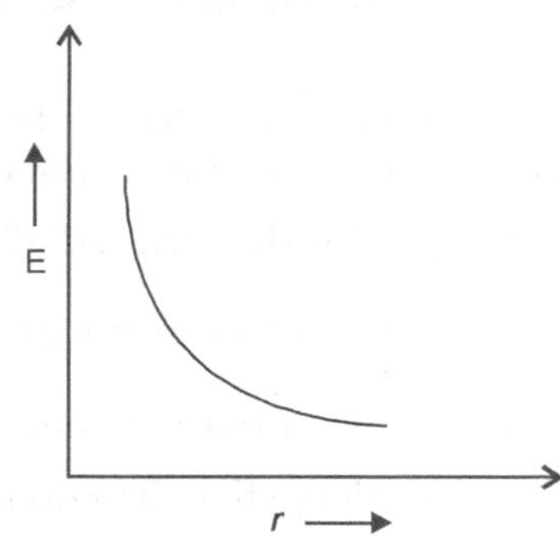


(c) Work done in bringing a charge $q$ through a small displacement '$dr$'.

$$dW = \vec{F}.\vec{dr}$$

$$dW = q\vec{E}.\vec{dr}$$

$$= qEdr \cos 0°$$

$$dW = q \times \frac{\lambda}{2\pi\varepsilon_0 r} dr$$

Work done in moving the given charge from $r_1$ to $r_2$ $(r_2 > r_1)$

$$W = \int_{r_1}^{r_2} dW = \int_{r_1}^{r_2} \frac{\lambda q dr}{2\pi\varepsilon_0 r}$$

$$W = \frac{\lambda q}{2\pi\varepsilon_0}[\log_e r_2 - \log_e r_1]$$

$$W = \frac{\lambda q}{2\pi\varepsilon_0} . \log_e\left(\frac{r_2}{r_1}\right)$$

**25. (a) State the principle of an ac generator and explain its working with the help of a labelled diagram. Obtain the expression for the emf induced in a coil having N turns each of cross-sectional area A, rotating with a constant angular speed '$\omega$' in a magnetic field $\vec{B}$, directed perpendicular to the axis of rotation.**

**(b) An aeroplane is flying horizontally from west to east with a velocity of 900 km/hour. Calculate the potential difference developed between the ends of its wings having a span of 20 m. The horizontal component of the Earth's magnetic field is $5 \times 10^{-4}$ T and the angle of dip is 30°. [5]**

**OR**

**A device X is connected across an ac source of voltage $V = V_0 \sin \omega t$. The current through X is given as $I = I_0 \sin\left(\omega t + \frac{\pi}{2}\right)$**

**(a) Identify the device X and write the expression for its reactance.**

**(b) Draw graphs showing variation of voltage and current with time over one cycle of ac, for X.**

**(c) How does the reactance of the device X vary with frequency of the ac ? Show this variation graphically.**

**(d) Draw the phasor diagram for the device X.**

**Ans. (a) Principle of ac generator :** The ac generator is based on the principle of electromagnetic induction. When closed coil is rotated in a uniform field with its axis perpendicular to field, then magnetic flux changes and emf is induced.

**Working :** When the armature coil rotates, the magnetic flux linked with it changes and produces induced current. If initially, coil PQRS is in vertical position and rotated clockwise, then PQ moves down and SR moves up. By Fleming's right hand rule, induced current flows from Q to P and S to R which is the first half rotation of coil. Brush $B_1$ is positive terminal and $B_2$ is negative. In second half rotation, PQ moves up and SR moves down. So induced current reverses and the alternating current is produced in this manner by the generator.

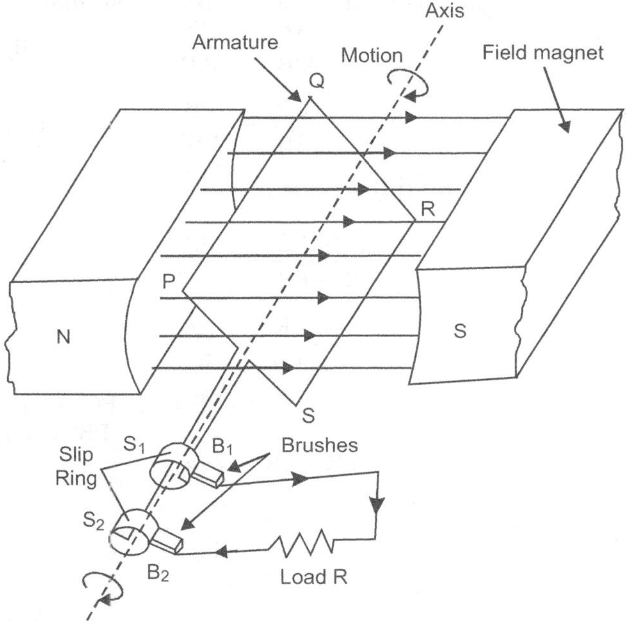


Expression for emf induced : If number of turns in coil = N, cross-section = A, angular speed of rotation = ω, magnetic field = $\vec{B}$, then to find emf induced,

Flux through the coil when its normal makes angle θ with the field,

$$\phi = BA \cos \theta$$

When coil rotates with angular velocity ω and turns through θ in time 't' then $\theta = \omega t$

$$\Rightarrow \quad \phi = BA \cos \omega t$$

When coil rotates, ϕ changes to set an induced emf.

$$\varepsilon = \frac{-d\phi}{dt} = -\frac{d}{dt}(BA \cos \omega t)$$
$$= BA\,\omega \sin \omega t$$

For N turns, total induced emf,

$$\varepsilon = NBA\,\omega \sin \omega t$$

**(b)** Given :

Velocity ($v$) = 900 km/h

$$= 900 \times \frac{5}{18} \text{ m/s}$$
$$= 250 \text{ ms}^{-1}$$

Wing span ($l$) = 20 m

Horizontal component of earth's field

$$(B_H) = 5 \times 10^{-4} \text{ T}$$

Angle of dip (δ) = 30°

$$B_V = B_H \tan \delta$$
$$= 5 \times 10^{-4} \times \tan 30°$$
$$= 5 \times 10^{-4} \times \frac{1}{1.732}$$

$$\text{emf induced} = B_V lv$$
$$= \frac{5 \times 10^{-4}}{1.732} \times 20 \times 250$$
$$= 1.44\text{V}$$

**OR**

$$V = V_0 \sin \omega t \,;\, I = I_0 \sin\left(\omega t + \frac{\pi}{2}\right).$$

**(a)** Since current leads the voltage by $\frac{\pi}{2}$ radians, X is a capacitor.

Capacitive reactance, $X_C = \frac{1}{\omega C} = \frac{1}{2\pi f C}$

where, ω = angular velocity, C = capacitance.

**(b)**

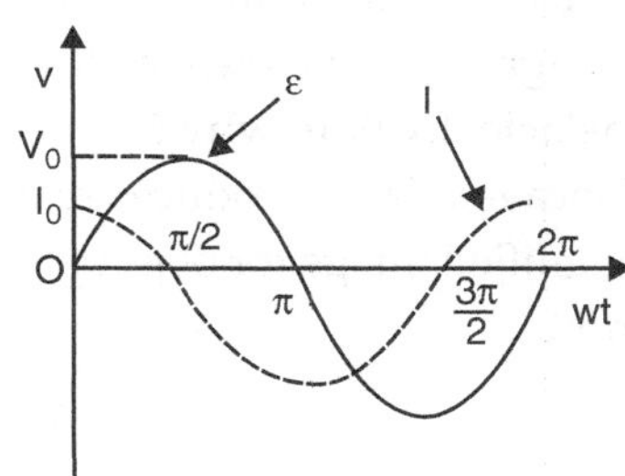


**(c)** Capacitive reactance varies inversely with frequency as $X_C = \frac{1}{\omega C} = \frac{1}{2\pi f C}$

$$X_C \propto \frac{1}{f}$$

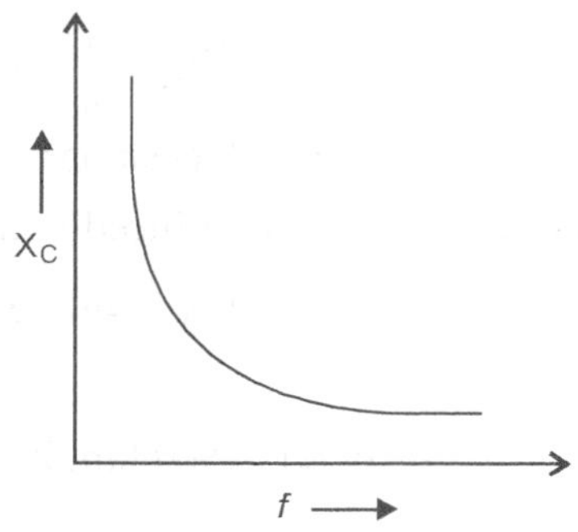


(d)

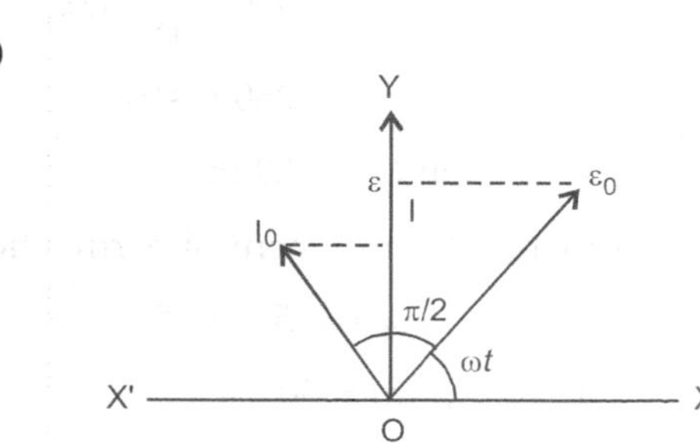


**26. (a) Draw a ray diagram to show image formation when the concave mirror produces a real, inverted and magnified image of the object.**

**(b) Obtain the mirror formula and write the expression for the linear magnification.**

**(c) Explain two advantages of a reflecting telescope over a refracting telescope. [5]**

**OR**

**(a) Define a wavefront. Using Huygens' principle, verify the laws of reflection at a plane surface.**

**(b) In a single slit diffraction experiment, the width of the slit is made double the original width. How does this affect the size and intensity of the central diffraction band ? Explain.**

**(c) When a tiny circular obstacle is placed in the path of light from a distant source, a bright spot is seen at the centre of the obstacle. Explain why ?**

**Ans. (a)** Concave mirror produces real, inverted and magnified image for object placed between F and C :

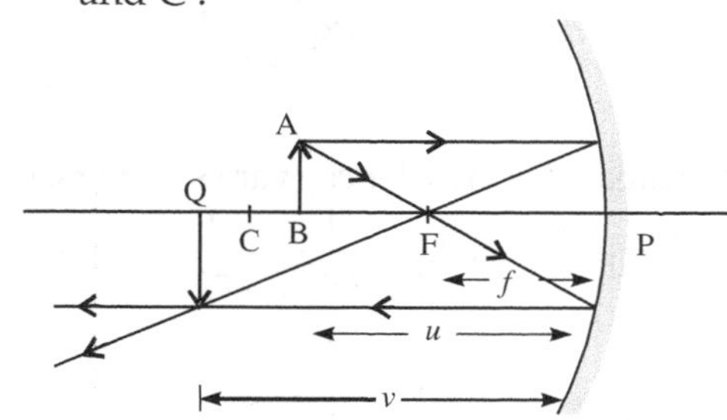


**(b) Derivation for mirror formula and magnification:** Consider an object AB be placed in front of a concave mirror beyond centre of curvature C.

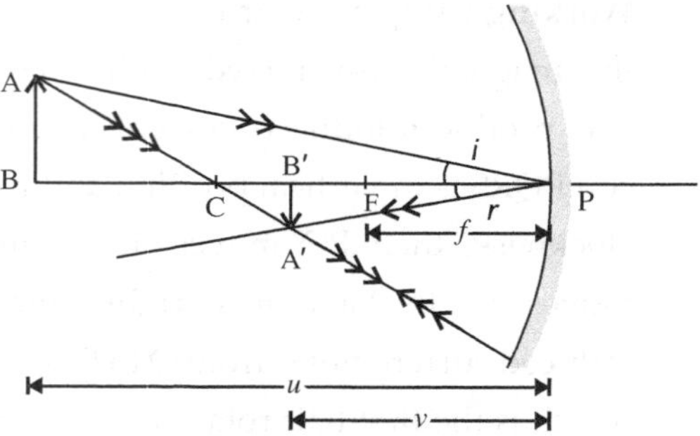


If F is the focus, focal length = $f$; object distance = $u$; image distance = $v$

As Δ ABC and ΔA'B'C are similar,

$$\frac{AB}{A'B'} = \frac{CB}{CB'} \quad \text{... (i)}$$

Again, as Δ ABP and ΔA'B'P are similar

$$\frac{AB}{A'B'} = \frac{PB}{B'P} \quad \text{... (ii)}$$

$$\Rightarrow \quad \frac{CB}{CB'} = \frac{PB}{PB'} \quad \text{... (iii)}$$

∴ From (iii),

$$\frac{PB-PC}{PC-PB'} = \frac{PB}{PB'}$$

$$\Rightarrow \quad \frac{-u+R}{-R+v} = \frac{-u}{v}$$

$[\because PB = -u, PC = -R, PB' = -v]$

$$\Rightarrow \quad uR + vR = 2uv$$

$$\Rightarrow \quad \frac{1}{v}+\frac{1}{u}=\frac{2}{R}$$

$$\Rightarrow \quad \boxed{\frac{1}{v}+\frac{1}{u}=\frac{1}{f}} \quad (\because R = 2f)$$

Linear magnification,

$$m = \frac{\text{Image height } (h_i)}{\text{Object height } (h_0)}$$

From equation (ii), $\dfrac{A'B'}{AB} = \dfrac{PB'}{PB}$

Using new Cartesian sign conventions, $A'B' = -h_2$, $AB = +h_1$, $PB' = -v$, $PB = -u$

$$\therefore \quad \frac{-h_2}{h_1} = \frac{-v}{u} = \frac{v}{u}$$

or $$m = \frac{h_2}{h_1} = \frac{-v}{u}$$

**(c)** Advantages of reflecting telescope over a refracting telescope are :

1. Reflecting telescope is free from chromatic and spherical aberrations unlike refracting telescope. Thus image formed is sharp and bright.
2. It has a larger light gathering power so that a bright image of even far off object is obtained.
3. Resolving power of reflecting telescope is large.

**OR**

**(a)** Wavefront is defined as the continuous locus of all the particles of a medium which are vibrating in the same phase.

**Verification of laws of reflection using Huygen's principle :** Let XY be a reflecting surface at which a wavefront is being incident obliquely. Let $v$ be the speed of the wavefront and at time $t = 0$, the wavefront touches the surface XY at A. After time $t$, point B of wavefront reaches the point B′ of the surface.

According to Huygen's principle, each point of wavefront acts as a source of secondary waves. When the point A of wavefront strikes the reflecting surface, then due to presence of reflecting surface, it cannot advance further; but the secondary wavelet originating from point A begins to spread in all directions in the first medium with speed $v$. As the wavefront AB advances further, its points $A_1$, $A_2$, $A_3$ .......... etc. strike the reflecting surface successively and send spherical secondary wavelets in the first medium.

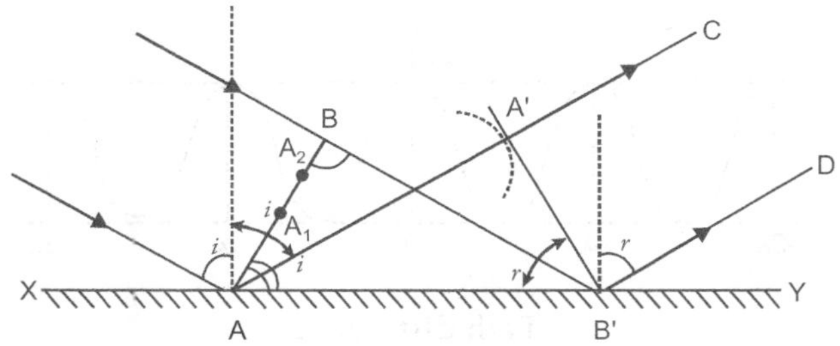


First of all the secondary wavelet starts from point A and traverses distance AA′ (= $vt$) in first medium in time $t$. In the same time $t$, the point B of wavefront, after travelling a distance BB′, reaches point B′ (of the surface), from where the secondary wavelet now starts. Now taking A as centre we draw a spherical arc of radius AA′(= $vt$) and draw tangent A′B′ on this arc from point B′. As the incident wavefront AB advances, the secondary wavelets starting from points between A and B′, one after the other and will touch A′B′ simultaneously. According to Huygen's principle wavefront A′B′ represents the new position of AB, *i.e.*, A′B′ is the reflected wavefront corresponding to incident wavefront AB.

Now in right-angled triangles ABB' and AA′B′

$$\angle ABB' = \angle AA'B' = 90^\circ$$

$$BB' = AA' = vt$$

Now

and AB′ is common

*i.e.*, both triangles are congruent.

$$\therefore \qquad \angle BAB' = \angle AB'A'$$

*i.e*, incident wavefront AB and reflected wavefront A′B′ make equal angles with the reflecting surface XY. As the rays are always normal to the wavefront, therefore the incident and the reflected rays make equal angles with the normal drawn on the surface XY, *i.e.*,

angle of incidence $i$ = angle of reflection $r$

This is the second law of reflection.

Since AB, A'B' and XY are all in the plane of paper, therefore the perpendiculars dropped on them will also be in the same plane. Therefore, we conclude that the incident ray, reflected ray and the normal at the point of incidence, all lie in the same plane. This is the first law of reflection. Thus, Huygen's principle explains both the laws of reflection.

**(b)** Width of central diffraction band = $2D.\frac{\lambda}{a}$

where a is the width of the slit.

So, on doubling the width of the slit, the size of the central diffraction band reduces to half value. But, the light amplitude becomes double, which increases the intensity four fold.

**(c)** If a tiny circular obstacle is kept in the path of light, a bright spot is seen at the centre of the obstacle. This is because the waves which get diffracted from the edge of the circular obstacle interfere constuctively at the centre of the shadow producing a bright spot.

**Note : All the sets of outside Delhi and Delhi are same.**

●●

# PHYSICS

## Outside Delhi [Set-I] 2017

Time allowed : 3 hours | Maximum marks : 70

### SECTION – A

**1. Nichrome and copper wires of same length and same radius are connected in series. Current I is passed through them. Which wire gets heated up more ? Justify your answer. [1]**

**Ans.** We know that,

$$R = \rho\frac{l}{A}$$

$$\frac{R_{Cu}}{R_{Ni}} = \frac{\rho_{Cu}}{\rho_{Ni}}$$

$\because$ $\rho_{Cu} < \rho_{Ni}$

$\therefore$ $R_{Cu} < R_{Ni}$

$\because$ $H = I^2Rt$

$$H \propto R$$

$\therefore$ Nichrome wire will get heated up more.

**2. Do electromagnetic waves carry energy and momentum ? [1]**

**Ans.** Yes, the electromagnetic waves carry energy and momentum because,

$$\text{Energy, } E = h\upsilon$$

$$\text{Momentum, } p = \frac{h}{\lambda}$$

**3. How does the angle of minimum deviation of a glass prism vary, if the incident violet light is replaced by red light ? Give reason. [1]**

**Ans.** We have,

$$\mu = \frac{\sin\left(\frac{A+\delta_m}{2}\right)}{\sin(A/2)}$$

$\therefore$ $\mu \propto \delta_m$

And $\mu_{red} < \mu_{violet}$

$\therefore$ Angle of minimum deviation $\delta_m$ will decrease.

**4. Name the phenomenon which shows the quantum nature of electromagnetic radiation. [1]**

**Ans.** Photoelectric effect.

**5. Predict the polarity of the capacitor in the situation described below : [1]**

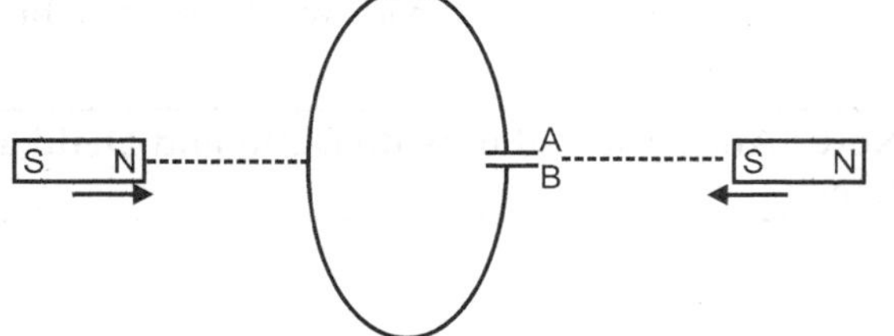


**Ans.** A will be positive and B will be negative.

### SECTION – B

**6. Draw the intensity pattern for single slit diffraction and double slit interference. Hence, state two differences between interference and diffraction patterns. [2]**

**OR**

**Unpolarised light is passed through a polaroid $P_1$. When this polarised beam passes through another polaroid $P_2$ and if the pass axis of $P_2$ makes angle $\theta$ with the pass axis of $P_1$, then write the expression for the polarised beam passing through $P_2$. Draw a plot showing the variation of intensity when $\theta$ varies from 0 to $2\pi$.**

**Ans. Intensity pattern for single slit diffraction :**

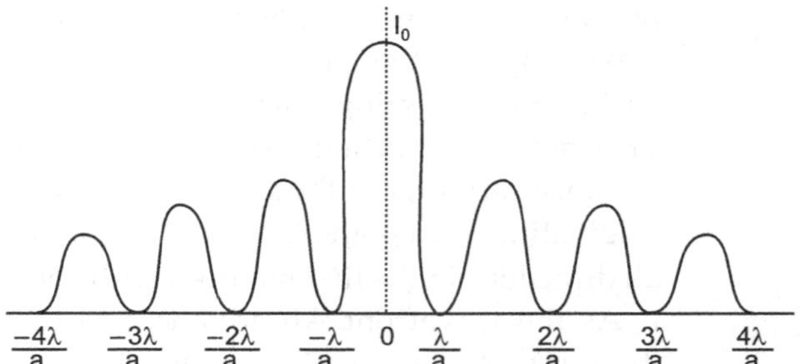


**Path difference**

**Intensity pattern for double slit interference :**

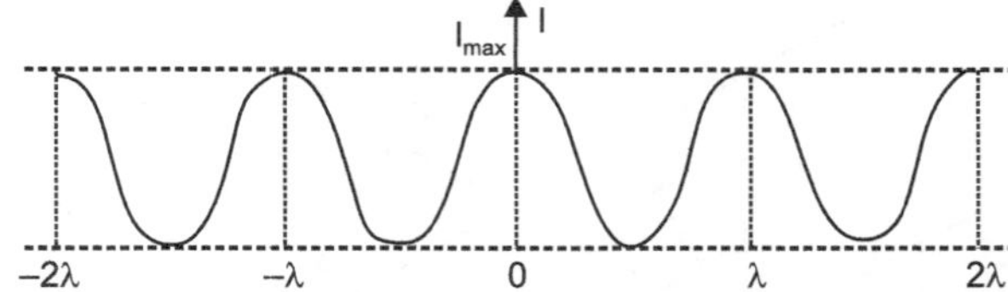


**Path difference**

**Difference between interference and diffraction patterns :**

**(a)** Interference fringes are of the same width while diffraction fringes are not of the same width.

**(b)** In interference pattern, all bright bands are of same intensity while in diffraction pattern, all bright bands are not of same intensity.

OR

$$I = \frac{I_0}{2}\cos^2\theta$$

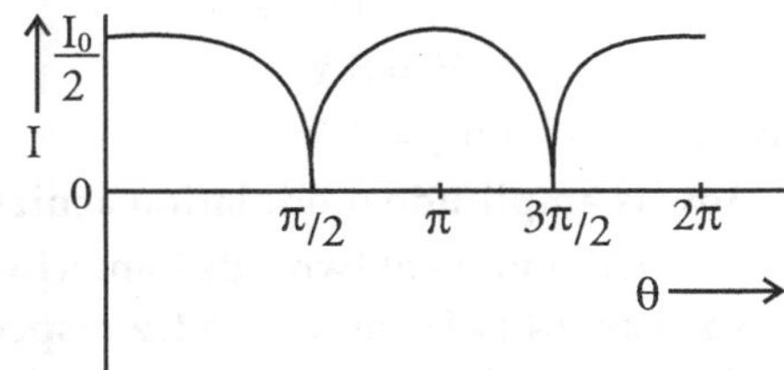


**7. Identify the electromagnetic waves whose wavelengths vary as :**
**(a) $10^{-12}$ m < λ < $10^{-8}$ m**
**(b) $10^{-3}$ m < λ < $10^{-1}$ m**
**Write one use for each. [2]**

**Ans. (a)** X-rays → To detect fractures in the human body.

**(b)** Microwaves → For aircraft navigation in RADAR systems.

**8. Find the condition under which the charged particles moving with different speeds in the presence of electric and magnetic field vectors can be used to select charged particles of a particular speed. [2]**

**Ans. (a)** The velocity $\vec{v}$, of the charged particles, and the $\vec{E}$ and $\vec{B}$ vectors, should be mutually perpendicular.

Also the forces on $q$, due to $\vec{E}$ and $\vec{B}$, must be oppositely directed.

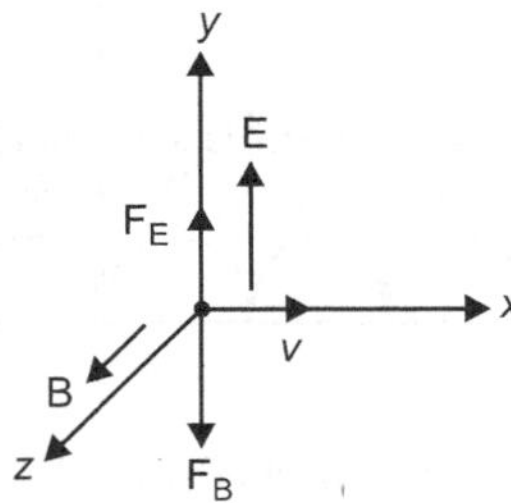


**(b)** If magnetic force = electrostatic force

$$F_m = F_e$$

$$qvB = qE$$

$$v = \frac{E}{B}$$

**9. A 12.5 eV electron beam is used to excite a gaseous hydrogen atom at room temperature. Determine the wavelengths and the corresponding series of the lines emitted. [2]**

**Ans.** Given :

$$\Delta E = 12.5 \text{ eV}$$

Let the electron jump from $n = 1$ to $n = n$ level.

$$\Delta E = E_n - E_1$$

$$\therefore \quad 12.5 = -\frac{13.6}{n^2} - \left[\frac{-13.6}{1^2}\right]$$

$$12.5 = 13.6\left(1 - \frac{1}{n^2}\right)$$

$$1 - \frac{12.5}{13.6} = \frac{1}{n^2}$$

$$\frac{1.1}{13.6} = \frac{1}{n^2}$$

$$\frac{13.6}{1.1} = n^2$$

$$12.36 = n^2$$

$$n = 3.5$$

$$\therefore \quad n = 3^{rd}$$

For $n = 3$ to $n = 1$,

$$\frac{1}{\lambda} = R\left(\frac{1}{1^2} - \frac{1}{n^2}\right)$$

$$\frac{1}{\lambda} = R\left(\frac{1}{1^2} - \frac{1}{3^2}\right)$$

$$\frac{1}{\lambda} = 1.097 \times 10^7\left(1 - \frac{1}{9}\right)$$

$$\frac{1}{\lambda} = 1.097 \times 10^7 \times \frac{8}{9}$$

$$\lambda = 102.55 \text{ nm}$$

For $n = 2$ to $n = 1$,

$$\frac{1}{\lambda} = 1.097 \times 10^7\left(\frac{1}{1^2} - \frac{1}{2^2}\right)$$

$$\frac{1}{\lambda} = 1.097 \times 10^7\left(\frac{3}{4}\right)$$

$$\lambda = 121.54 \text{ nm}$$

For $n = 3$ to $n = 2$,

$$\frac{1}{\lambda} = 1.097 \times 10^7\left(\frac{1}{2^2} - \frac{1}{3^2}\right)$$

$$\frac{1}{\lambda} = 1.097 \times 10^7\left(\frac{1}{4} - \frac{1}{9}\right)$$

$$\frac{1}{\lambda} = 1.097 \times 10^7 \times \frac{5}{36}$$

$$\lambda = 656.33 \text{ nm}$$

Hence, λ = 102.5 nm and 121.5 nm → Lyman series and λ = 656.33 nm → Balmer series.

**10. Write two properties of a material suitable for making (a) a permanent magnet, and (b) an electromagnet. [2]**

**Ans.** **Properties of a material suitable for making permanent magnet :**
1. High retentivity.
2. High coercivity.

**Properties of a material suitable for making electromagnet :**
1. High permeability.
2. Low retentivity.

## SECTION – C

**11. (a) The potential difference applied across a given resistor is altered so that the heat produced per second increases by a factor of 9. By what factor does the applied potential difference change ?**

**(b) In the figure shown, an ammeter A and a resistor of 4 Ω are connected to the terminals of the source. The emf of the source is 12V having the internal resistance of 2Ω. Calculate the voltmeter and ammeter readings. [3]**

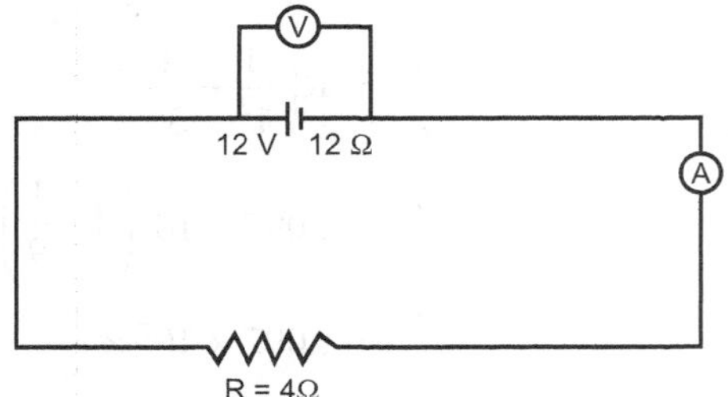


**Ans. (a)** Heat produced,

$$H = \frac{V^2}{R}t$$

$$\Rightarrow \quad H \propto V^2$$

$$\frac{H'}{H} = \frac{V'^2}{V^2}$$

$$\frac{9H}{H} = \frac{V'^2}{V^2} \qquad [\because H' = 9H]$$

$$V'^2 = 9V^2$$

$$V' = 3V$$

So, potential difference is increased by a factor of 3.

**(b)**
$$\text{Current, } I = \frac{E}{R+r}$$

$$I = \frac{12}{4+2} = \frac{12}{6} = 2\text{ A}$$

Ammeter reading = 2A

Potential difference,

$$V = E - Ir$$

$$V = 12 - 2 \times 2$$

$$= 12 - 4$$

$$V = 8\text{ V}$$

Voltmeter reading = 8 V.

**12. (a) How is amplitude modulation achieved ?****

**(b) The frequencies of two side bands in an AM wave are 640 kHz and 660 kHz respectively. Find the frequencies of carrier and modulating signal. What is the bandwidth required for amplitude modulation ?** [3]**

**13. (a) In the following diagram, is the junction diode forward biased or reverse biased ?**

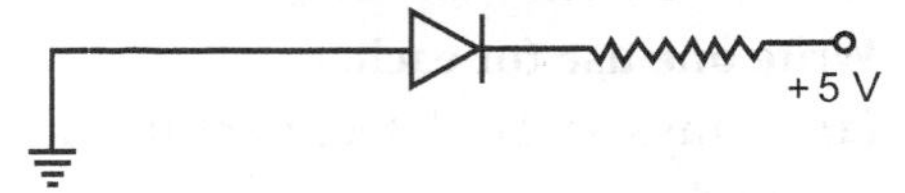


**(b) Draw the circuit diagram of a full wave rectifier and state how it works. [3]**

**Ans. (a)** Reverse biased.

**(b) Full wave Rectifier :**

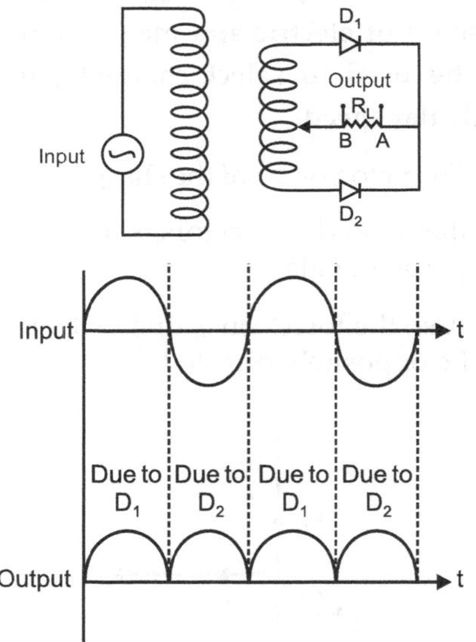


Diode $D_1$ conducts only when the junction is forward biased. Hence during first half cycle of input A.C., $D_1$ will conduct while $D_2$ will not and current in $R_L$ will flow from A to B. Diode $D_2$ is reverse biased. During second half cycle of input A.C., diode $D_2$ will conduct while $D_1$ will not conduct and current will again flow from A to B in $R_L$. Hence complete cycle will become unidirectional.

**14. Using photon picture of light, show how Einstein's photoelectric equation can be established. Write two features of photoelectric effect which cannot be explained by wave theory. [3]**

** Answer is not given due to change in the present Syllabus.

**Ans. Einstein's photoelectric equation :** Einstein explained the various laws of photo-electric emission on the basis of Planck's quantum theory. According to Planck's quantum theory, light radiations consist of tiny packets of energy called quanta. One quantum of light radiation is called a photon which travels with the speed of light.

The energy of a photon is given by,

$$E = h\nu$$

where $h$ is Planck's constant and $\nu$ is the frequency of light radiation.

Einstein assumed that one photoelectron is ejected from a metal surface if one photon of suitable light radiation falls on it.

Consider a photon of light of frequency $\nu$, incident on a photosensitive metal surface. The energy of the photon $(= h\nu)$ is spent in two ways :

**(a)** A part of the energy of the photon is used in liberating the electron from the metal surface which is equal to the work function $\phi_0$ of the metal.

**(b)** The rest of the energy of the photon is used in imparting the kinetic energy to the emitted photo electron.

If $v_{max}$ is the maximum velocity of the emitted photoelectron and $m$ is its mass, then

Max. K.E. of the photoelectron,

$$K_{max} = \frac{1}{2}mv_{max}^2$$

$$\therefore \quad h\nu = \phi_0 + \frac{1}{2}mv_{max}^2$$

This equation is called Einstein's photoelectric equation.

**Features of photoelectric effect which can not be explained by wave theory :**

1. The wave theory could not explain the instantaneous process of photoelectric effect.
2. Maximum kinetic energy of the emitted photoelectrons is independent of intensity of incident light.

**15. (a) Monochromatic light of wavelength 589 nm is incident from air on a water surface. If μ for water is 1.33, find the wavelength, frequency and speed of the refracted light.**

**(b) A double convex lens is made of a glass of refractive index 1.55, with both faces of the same radius of curvature. Find the radius of curvature required, if the focal length is 20 cm.** [3]

**Ans. (a)** Given :

$$\mu = 1.33$$

$$\lambda_a = 589 \text{ nm}$$

We know that,

$$\mu = \frac{c}{v}$$

$$v = \frac{c}{\mu} = \frac{3\times10^8}{1.33}$$

$$\text{Speed, } v = 2.26 \times 10^8 \text{ m/s}$$

Frequency remains same.

$$\therefore \quad \nu = \frac{c}{\lambda_a} = \frac{3\times10^8}{589\times10^{-9}}$$

$$\text{Frequency, } \nu = 5.09 \times 10^{14} \text{ Hz}$$

$$\text{Wavelength, } \lambda_w = \frac{\lambda_a}{\mu} = \frac{589}{1.33} = 442.8 \text{ nm}$$

**(b)** Given :

$$\mu = 1.55, f = 20 \text{ cm}$$

We know that,

$$\frac{1}{f} = (\mu - 1)\left(\frac{1}{R_1} - \frac{1}{R_2}\right)$$

$$\frac{1}{20} = (1.55 - 1)\left(\frac{1}{R} + \frac{1}{R}\right)$$

$$\frac{1}{20} = 0.55 \times \frac{2}{R}$$

$$R = 20 \times 0.55 \times 2$$

$$R = 22 \text{ cm}$$

**16. Define mutual inductance between a pair of coils. Derive an expression for the mutual inductance of two long coaxial solenoids of same length wound one over the other.** [3]

**OR**

**Define self-inductance of a coil. Obtain the expression for the energy stored in an inductor L connected across a source of emf.**

**Ans.** Mutual inductance is numerically equal to the induced e.m.f in the secondary coil when the current in the primary coil changes by unity.

Suppose two long co-axial solenoids each of length $l$. We denote the area of the inner solenoid $S_1$ by $A_1$ and the number of turns per unit length by $n_1$. The corresponding quantities for the outer solenoid $S_2$ are $A_2$ and $n_2$ respectively. Let $N_1$ and $N_2$ be the total number of turns of coils $S_1$ and $S_2$ respectively.

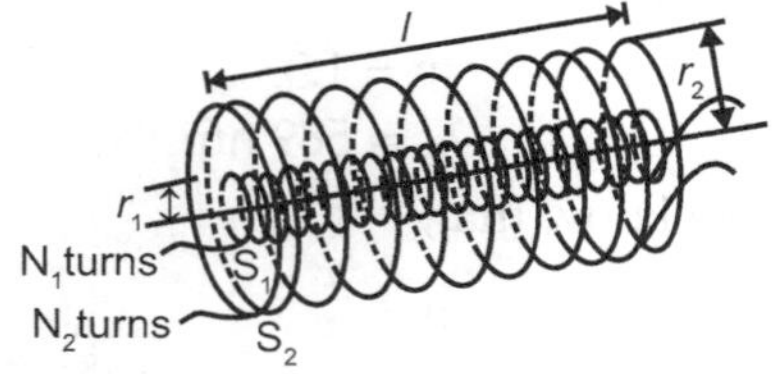


Let a curent $i_2$, flow in the secondary coil

$$\therefore \quad B_2 = \frac{\mu_0 N_2 i_2}{l}$$

$\therefore$ Flux linked with the primary coil

$$= N_1 A_1 B_2 = \frac{\mu_0 N_2 N_1 A_1 i_2}{l} = M_{12} i_2$$

Hence, $$M_{12} = \frac{\mu_0 N_2 N_1 A_1}{l}$$

$$= \mu_0 n_2 n_1 A_1 l \left( n_1 = \frac{N_1}{l}; n_2 = \frac{N_2}{l} \right)$$

**OR**

**Self inductance :**

Self inductance of a coil is equal to the magnitude of induced emf produced in the coil when rate of change of current through the coil is unity.

now, $$\phi = LI$$

$\therefore$ Induced emf (e) $$e = -\frac{d\phi}{dt}$$

$$e = -L\frac{dI}{dt}$$

**Energy stored in an inductor :** For the current I at an instant in a circuit, the rate of work done is

$$\frac{dW}{dt} = |e| I$$

$$\frac{dW}{dt} = LI\frac{dI}{dt}$$

$$dW = L I dI$$

Total work done in increasing the current from 0 to I is given by

$$W = \int_0^I LIdI$$

$$W = L\left[\frac{I^2}{2}\right]_0^I$$

$$W = \frac{1}{2}LI^2$$

This work done is stored as magnetic potential energy.

$$\therefore \quad U = \frac{1}{2}LI^2$$

**17. (a) Write the principle of working of a metre bridge.**

**(b) In a metre bridge, the balance point is found at a distance $l_1$ with resistances R and S as shown in the figure.**

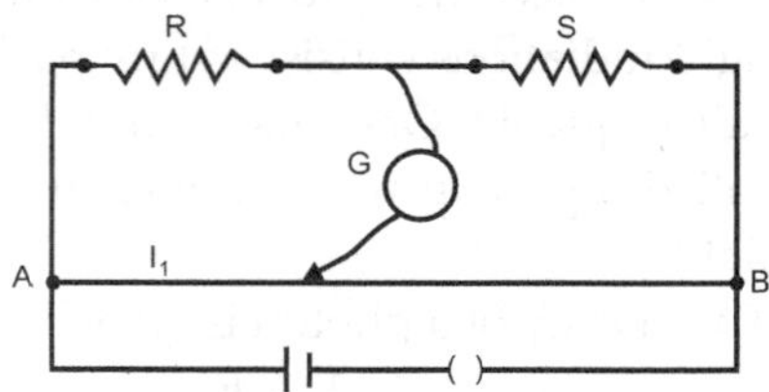


**An unknown resistance X is now connected in parallel to the resistance S and the balance point is found at a distance $l_2$. Obtain a formula for X in terms of $l_1$, $l_2$ and S. [3]**

**Ans. (a)** Meter bridge works on the principle of balanced Wheatstone bridge *i.e.,* when the bridge is balanced ($i_g = 0$),

$$\frac{R_1}{R_2} = \frac{R_3}{R_4}$$

where $R_1$, $R_2$, $R_3$ and $R_4$ are resistances connected in four arms of Wheatstone's bridge.

**(b)** In first case,

$$\frac{R}{S} = \frac{l_1}{100 - l_1} \qquad ...(i)$$

In second case,

$$\frac{R}{\frac{XS}{(X+S)}} = \frac{l_2}{100 - l_2} \qquad ...(ii)$$

Dividing (ii) by (i),

$$\frac{X+S}{X} = \frac{l_2}{l_1}\left(\frac{100-l_1}{100-l_2}\right)$$

$$1 + \frac{S}{X} = \frac{l_2}{l_1}\left(\frac{100-l_1}{100-l_2}\right)$$

$$X = \frac{S}{\frac{l_2}{l_1}\left(\frac{100-l_1}{100-l_2}\right) - 1}$$

$$X = \frac{l_1(100-l_2)}{100(l_2-l_1)}S$$

**18. Draw a block diagram of a generalized communication system. Write the functions of each of the following :** [3]**

**(a) Transmitter**

**(b) Channel**

**(c) Receiver**

**19. (a) Write the functions of the three segments of a transistor.****

** Answer is not given due to change in the present Syllabus.

**(b) The figure shows the input waveforms A and B for 'AND' gate. Draw the output waveform and write the truth table for this logic gate.**** **[3]**

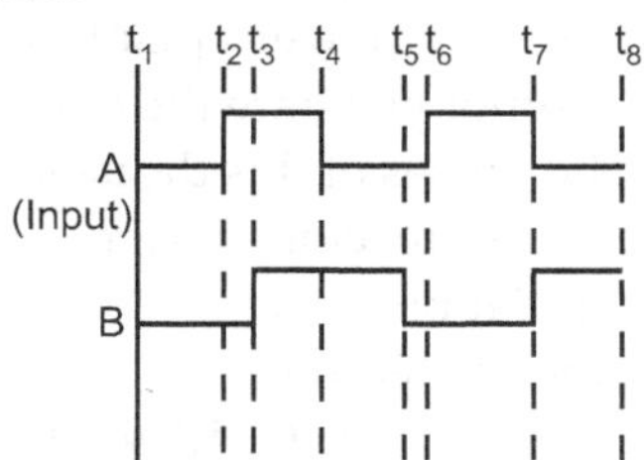


**20. (a) Draw a ray diagram depicting the formation of the image by an astronomical telescope in normal adjustment.**

**(b) You are given the following three lenses. Which two lenses will you use as an eyepiece and as an objective to construct an astronomical telescope ? Give reason.** **[3]**

| Lenses | Power (D) | Aperture (cm) |
|---|---|---|
| $L_1$ | 3 | 8 |
| $L_2$ | 6 | 1 |
| $L_3$ | 10 | 1 |

**Ans. (a)**

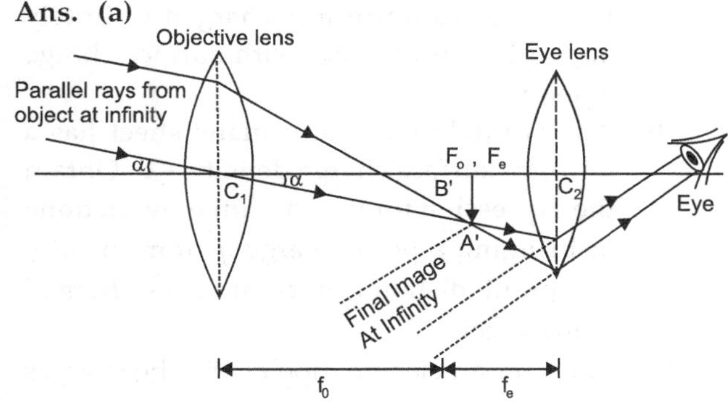


**(b)** An astronomical telescope should have an objective of large aperture and large focal length while an eyepiece of small aperture and small focal length.

Therefore, we will use $L_1$ as an objective and $L_3$ as an eyepiece.

**21. (a) State Biot–Savart law and express this law in the vector form.**

**(b) Two identical circular coils, P and Q each of radius R, carrying currents 1 A and $\sqrt{3}$ A respectively, are placed concentrically and perpendicular to each other lying in the XY and YZ planes. Find the magnitude and direction of the net magnetic field at the centre of the coils.** **[3]**

**Ans. (a) Biot-Savart Law :**

It states that the magnetic field strength $(\vec{dB})$ produced due to a current element I and length $dl$ at a point having position vector $\vec{r}$ relative to current element is

(i) directly proportional to the current I *i.e.*, $dB \propto I$.

(ii) directly proportional to the length $dl$ of the element *i.e.*, $dB \propto dl$.

(iii) directly proportional to sin θ, where θ is the angle between $dl$ and $r$, *i.e.*, $dB \propto \sin\theta$.

(iv) inversely proportional to the square of the distance $r$ from the current element *i.e.*, $dB \propto \frac{1}{r^2}$.

$$\therefore \quad dB \propto \frac{Idl\sin\theta}{r^2} = \frac{\mu_0}{4\pi}\frac{Idl\sin\theta}{r^2}$$

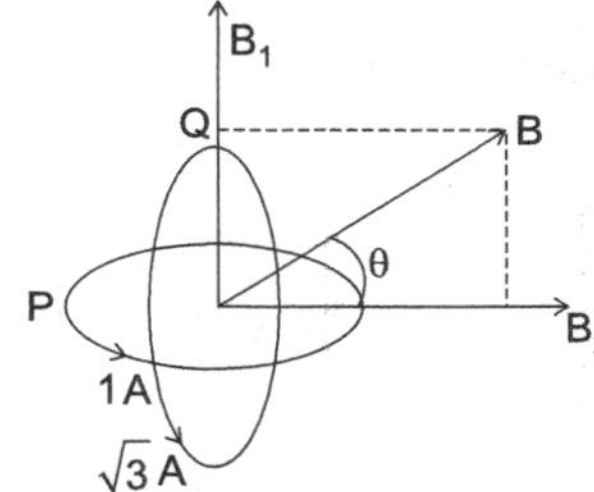


In vector form,

$$\vec{dB} = \frac{\mu_0}{4\pi}\cdot\frac{I\,\vec{dl}\times\vec{r}}{|\vec{r}|^3}$$

**(b)**

We know that,

$$B = \frac{\mu_0}{2R}I$$

$$B_1 = \frac{\mu_0}{2R}.1 \text{ (along Z - direction)}$$

$$B_2 = \frac{\mu_0}{2R}.\sqrt{3} \text{ (along X - direction)}$$

$$B = \sqrt{B_1^2 + B_2^2}$$

$$= \sqrt{\left(\frac{\mu_0}{2R}\right)^2 + \left(\frac{\mu_0}{2R}.\sqrt{3}\right)^2}$$

$$B = \frac{\mu_0}{2R}\sqrt{1+3} = \frac{\mu_0}{2R}.\sqrt{4}$$

$$B = \frac{\mu_0}{R}$$

$$\tan\theta = \frac{B_1}{B_2} = \frac{1}{\sqrt{3}}$$

$$\Rightarrow \quad \theta = 30^\circ \text{ in XZ - plane}$$

**22. Two identical parallel plate capacitors A and B are connected to battery of V volts with the switch S closed. The switch is now opened and the free space between the plates of the capacitors is filled with a dielectric of dielectric constant K. Find the ratio of the total electrostatic energy stored in both capacitors before and after the introduction of the dielectric.** **[3]**

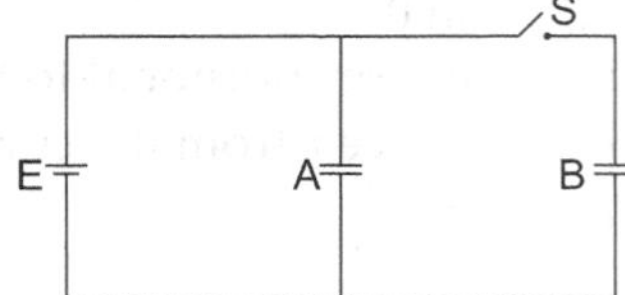


**Ans.** Energy stored $= \frac{1}{2}CV^2 = \frac{1}{2}\frac{Q^2}{C}$

Net capacitance with switch S closed $= C + C = 2C$

$\therefore$ Energy stored $= \frac{1}{2} \times 2C \times V^2 = CV^2$ ...(i)

After the switch S is opened, capacitance of each capacitor $= KC$

$\therefore$ Energy stored in capacitor A $= \frac{1}{2}KCV^2$

For capacitor B, Energy stored

$$= \frac{1}{2}\frac{Q^2}{KC} = \frac{1}{2}\frac{C^2V^2}{KC} = \frac{1}{2}\frac{CV^2}{K}$$

$\therefore$ Total energy stored

$$= \frac{1}{2}KCV^2 + \frac{1}{2}\frac{CV^2}{K}$$

$$= \frac{1}{2}CV^2\left(K + \frac{1}{K}\right)$$

$$= \frac{1}{2}CV^2\left(\frac{K^2+1}{K}\right) \quad \text{(ii)}$$

on dividing (i) by (ii)

$\therefore$ Required ratio $= \frac{2CV^2K}{CV^2(K^2+1)} = \frac{2K}{(K^2+1)}$

## SECTION – D

**23. Asha's mother read an article in the newspaper about a disaster that took place at Chernobyl. She could not understand much from the article and asked a few questions from Asha regarding the article. Asha tried to answer her mother's questions based on what she learnt in Class XII Physics.** **[4]**

**(a) What was the installation at Chernobyl where the disaster took place ? What, according to you, was the cause of this disaster ?**

**(b) Explain the process of release of energy in the installation at chernobyl.**

**(c) What, according to you, were the values displayed by Asha and her mother ?****

**Ans. (a)** Nuclear power plant was installed at Chernobyl where the disaster took place. The causes of this disaster were deficiencies in the reactor design and in operating regulations.

**(b)** Nuclear fission.

## SECTION – E

**24. (a) Derive an expression for the electric field E due to a dipole of length '2*a*' at a point distant *r* from the centre of the dipole on the axial line.**

**(b) Draw a graph of E versus *r* for $r >> a$.**

**(c) If this dipole were kept in a uniform external electric field $E_0$, diagrammatically represent the position of the dipole in stable and unstable equilibrium and write the expressions for the torque acting on the dipole in both the cases.** **[5]**

**OR**

**(a) Use Gauss's theorem to find the electric field due to a uniformly charged infinitely large plane thin sheet with surface charge density σ.**

**(b) An infinitely large thin plane sheet has a uniform surface charge density + σ. Obtain the expression for the amount of work done in bringing a point charge *q* from infinity to a point, distant *r*, in front of the charged plane sheet.**

**Ans. (a)** Consider an electric dipole AB. The charges $-q$ and $+q$ of dipole are situated at A and B respectively, as shown in the figure. The separation between the charges is $2a$. Electric dipole moment is given by

$$p = q \cdot 2a \quad \text{...(i)}$$

Consider a point P on the axis of dipole at a distance *r* from mid point O of electric dipole. The distance of point P from charge $+q$ at B is, $BP = r - a$ and distance of point P from charge $-q$ at A is, $AP = r + a$ Let $E_1$ and $E_2$ be the electric field strengths at point P due to charges $+q$ and $-q$ respectively.

2a
A B P
$-q$ O $+q$ $E_1$ $E_2$
r
$r - a$
$r + a$

---

** Answer is not given due to change in the present Syllabus.

The resultant electric field due to electric dipole is given by

$$E = E_1 - E_2$$

$$= \frac{1}{4\pi\epsilon_0}\cdot\frac{q}{(r-a)^2} - \frac{1}{4\pi\epsilon_0}\cdot\frac{q}{(r+a)^2}$$

$$E = \frac{1}{4\pi\epsilon_0} q\cdot\left[\frac{1}{(r-a)^2} - \frac{1}{(r+a)^2}\right]$$

$$E = \frac{1}{4\pi\epsilon_0}\cdot q\left[\frac{(r+a)^2-(r-a)^2}{(r-a)^2(r+a)^2}\right]$$

$$E = \frac{1}{4\pi\epsilon_0}\cdot q\left[\frac{4ra}{(r^2 - a^2)^2}\right]$$

$$E = \frac{1}{4\pi\epsilon_0}\cdot\frac{2pr}{(r^2-a^2)^2} \quad \text{[From (i)]}$$

**(b)** If $r >> a$,

$$E = \frac{1}{4\pi\epsilon_0}\cdot\frac{2p}{r^3}$$

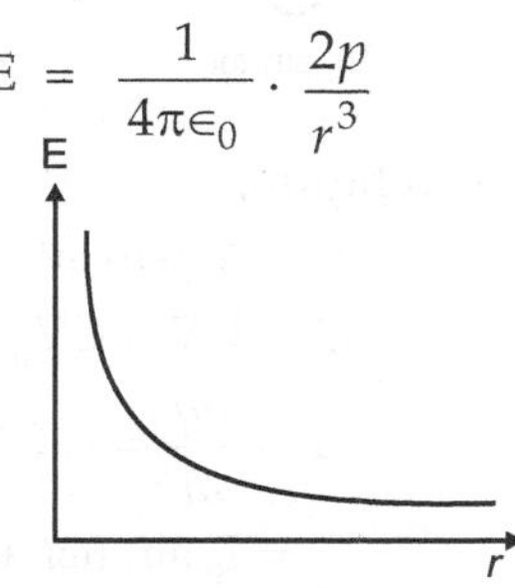


**(c) Position of dipole in stable equilibrium :**

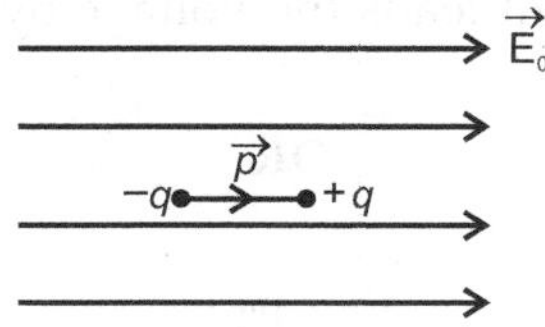


In stable equilibrium, $\theta = 0°$ ($\vec{P}$ is parallel to $\vec{E}_0$)

$$\vec{\tau} = \vec{p} \times \vec{E} = pE\sin\theta = pE\sin 0°$$

$$\vec{\tau} = 0$$

**Position of dipole in unstable equilibrium :**

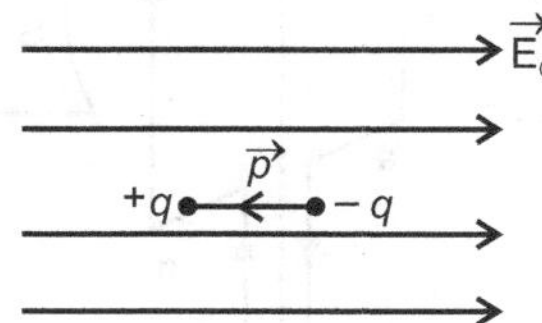


In unstable equilibrium, $\theta = 180°$ ($\vec{P}$ is antiparallel to $\vec{E}_0$)

$$\vec{\tau} = pE\sin 180°$$

$$\vec{\tau} = 0$$

**OR**

**(a)**

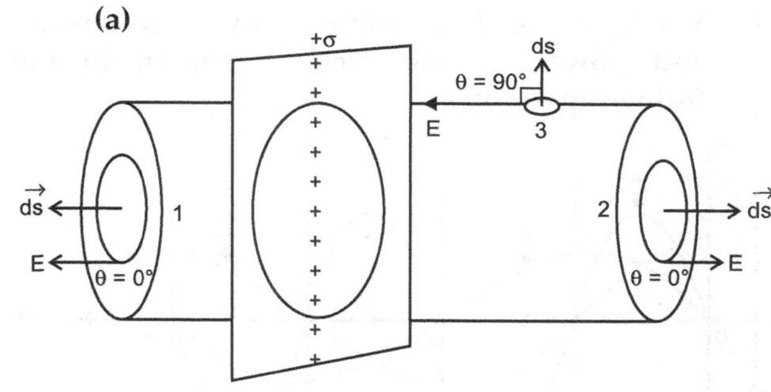


Let electric charge be uniformly distributed over the surface of a thin non-conducting infinite sheet. Let the surface charge density be σ.

According to Gauss' theorem :

$$\oint_1 \vec{E}.\vec{ds} + \oint_2 \vec{E}.\vec{ds} + \oint_3 \vec{E}.\vec{ds} = \frac{q}{\epsilon_0}$$

$$\oint_1 Eds\cos 0° + \oint_2 Eds\cos 0° + \oint_3 Eds\cos 90° = \frac{q}{\epsilon_0}$$

$$\oint_1 Eds + \oint_2 Eds = \frac{q}{\epsilon_0}$$

$$2\oint Eds = \frac{q}{\epsilon_0}$$

$$2E.S = \frac{\sigma S}{\epsilon_0}$$

$$E = \frac{\sigma}{2\epsilon_0}$$

**(b)** The electric field due to a uniformly charged infinitely large thin sheet with surface charge density σ is,

$$E = \frac{\sigma}{2\epsilon_0} \quad ...(i)$$

The amount of work done in bringing a point charge $q$ from infinity to a point, at a distance $r$ is given by

$$W = \int_\infty^r F.dr$$

where F is the force experienced by charge $q$.

Now, $$W = q\int_\infty^r -E.dr \quad [\because F = -Eq]$$

$$W = -q\frac{\sigma}{2\epsilon_0}\int_\infty^r dr \quad \text{[From (i)]}$$

$$W = q\frac{\sigma}{2\epsilon_0}[\infty - r] = \infty$$

**25. A device 'X' is connected to an ac source $V = V_0 \sin \omega t$. The variation of voltage, current and power in one cycle is shown in the following graph :** **[5]**

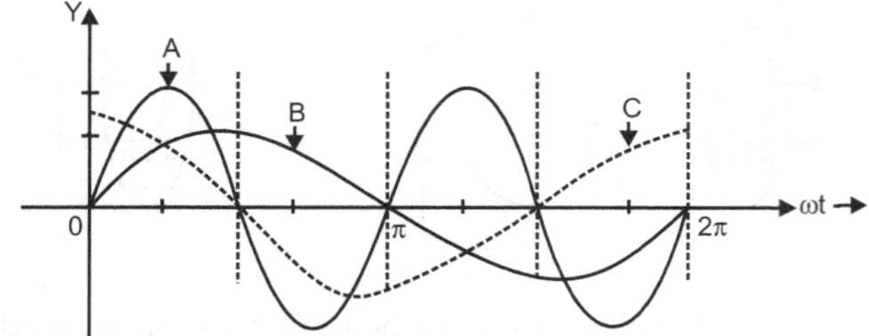


**(a) Identify the device 'X'.**

**(b) Which of the curves A, B and C represent the voltage, current and the power consumed in the circuit ? Justify your answer.**

**(c) How does its impedance vary with frequency of the ac source ? Show graphically.**

**(d) Obtain an expression for the current in the circuit and its phase relation with ac voltage.**

**OR**

**(a) Draw a labelled diagram of an ac generator. Obtain the expression for the emf induced in the rotating coil of N turns each of cross-sectional area A, in the presence of magnetic field $\vec{B}$.**

**(b) A horizontal conducting rod 10 m long extending from east to west is falling with a speed 5.0 ms$^{-1}$ at right angles to the horizontal component of the Earth's magnetic field, 0.3 × 10$^{-4}$ Wbm$^{-2}$. Find the instantaneous value of the emf induced in the rod.**

**Ans. (a)** The device X is a capacitor.

**(b)** The curves A,B and C represents power consumption, voltage and current respectively.

Since, it is given $V = V_0 \sin \omega t$; this sinusoidal variation is represented by the curve B. In case of capacitor, current (I) leads the voltage by 90° which is being represented by curve C. Now, we know power is given by

$$P = V \times I$$

So, the power would be positive for those cycles where both V and I are either positive or negative. Power would be negative when one of the two voltage or current, is negative. This illustration is followed by the curve A.

**(c)** Capacitive reactance, $X_C = \frac{1}{\omega C} = \frac{1}{2\pi f C}$

This shows that capacitive reactance vary inversely with the frequency.

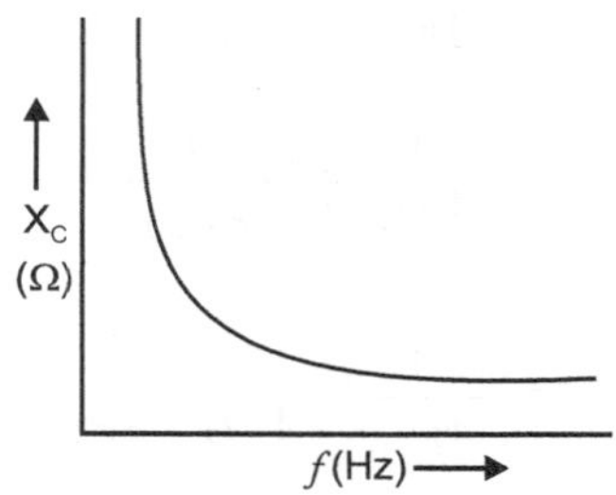


**(d)**

C
$V_0 \sin \omega t$

In the given figure,

$$V = V_0 \sin \omega t$$

$$Q = CV = CV_0 \sin \omega t$$

$$I = \frac{dq}{dt} = \omega CV_0 \cos \omega t$$

$$= I_0 \sin (\omega t + \pi/2)$$

Where, $I_0 = \omega CV_0$

Current leads the voltage by a phase angle of 90°.

**OR**

**(a)**

ω
Armature coil
b
c
N
$\vec{B}$
S
Field magnet
a
d
$B_1$
$R_1$
Slip rings
Brushes
Load $R_L$
$R_2$
$B_2$

If N is the number of turns in coil, A is the area of coil and B the magnetic induction, then flux $\phi$ is given by

$$\phi = B.A$$

$$\phi = BA \cos \omega t$$

The e.m.f induced in the coil is given by

$$e = -N\frac{d\phi}{dt}$$

$$e = -N\frac{d}{dt}(BA \cos \omega t)$$

$$e = +NBA\omega \sin \omega t$$

$$e = e_0 \sin \omega t$$

where, $e_0 = NAB\omega$

**(b)** $e = B_H l v$

$$= 0.3 \times 10^{-4} \times 10 \times 5$$

$$e = 1.5 \times 10^{-3}\ V = 1.5\ mV$$

**26. (a) Define wavefront. Use Huygens' principle to verify the laws of refraction.**

**(b) How is linearly polarised light obtained by the process of scattering of light ? Find the Brewster angle for air-glass interface, when the refractive index of glass = 1.5. [5]**

**OR**

**(a) Draw a ray diagram to show the image formation by a combination of two thin convex lenses in contact. Obtain the expression for the power of this combination in terms of the focal lengths of the lenses.**

**(b) A ray of light passing from air through an equilateral glass prism undergoes minimum deviation when the angle of incidence is $\frac{3}{4}$th of the angle of prism. Calculate the speed of light in the prism.**

**Ans. (a) Wavefront :** A wavefront is a continuous locus of all the particles of a medium which are vibrating in the same phase.

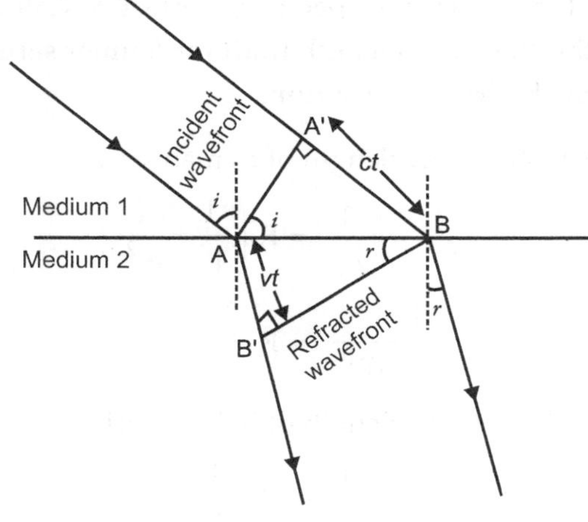


In $\Delta AA'B$,

$$\sin i = \frac{A'B}{AB} \quad ...(i)$$

In $\Delta AB'B$,

$$\sin r = \frac{AB'}{AB} \quad ...(ii)$$

Dividing equation (i) by (ii),

$$\frac{\sin i}{\sin r} = \frac{A'B/AB}{AB'/AB}$$

$$\frac{\sin i}{\sin r} = \frac{A'B}{AB'}$$

$$= \frac{ct}{vt}$$

$$\frac{\sin i}{\sin r} = \frac{c}{v} = \mu$$

This is the Snell's law of refraction.

**(b)** When a beam of white light is passed through a medium containing particles whose size is of the order of wavelength of light, then the beam gets scattered. When the scattered light is viewed through an analyser in a direction at the right angle to the direction of incidence, it is found to be plane polarised. This is called polarisation by scattering.

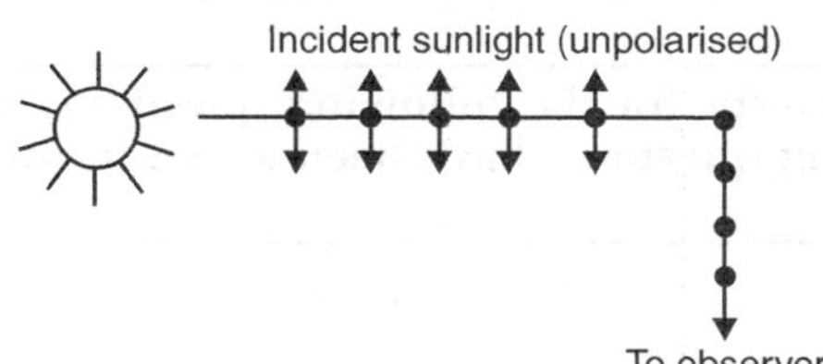


We know that,

$$\tan i_p = \mu \text{ (Brewster law)}$$

$$\Rightarrow \tan i_p = 1.5$$

$$\Rightarrow i_p = \tan^{-1}(1.5) = 56.31°$$

**OR**

**(a)**

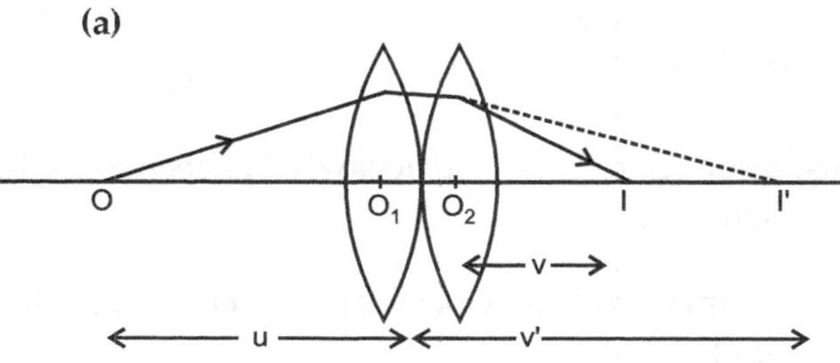


For first lens, object is at O, and image is at I'.

$$\frac{1}{f_1} = \frac{1}{v'} - \frac{1}{u} \quad ...(i)$$

For second lens, object is at I and image is at I.

$$\frac{1}{f_2} = \frac{1}{v} - \frac{1}{v'} \quad ...(ii)$$

Adding (i) and (ii),

$$\frac{1}{f_1} + \frac{1}{f_2} = \frac{1}{v} - \frac{1}{u} \quad ...(iii)$$

If $f$ is the combined focal length then

$$\frac{1}{f} = \frac{1}{v} - \frac{1}{u}$$

From equation (iii),

$$\frac{1}{f} = \frac{1}{f_1} + \frac{1}{f_2}$$

$$P = \frac{1}{f_1} + \frac{1}{f_2} \quad [\because \frac{1}{f} = P]$$

$$P = \frac{f_1 + f_2}{f_1 f_2}$$

**(b)** Given : $A = 60°$

$$i = \frac{3}{4} A$$

$$i = \frac{3}{4} \times 60°$$

$$i = 45°$$

For angle of minimum deviation,

$$r = A/2 = \frac{60°}{2}$$

$$r = 30°$$

$$\because \quad \mu = \frac{\sin i}{\sin r}$$

$$\frac{c}{v} = \frac{\sin 45°}{\sin 30°} \quad [\because \mu = \frac{c}{v}]$$

$$\frac{3 \times 10^8}{v} = \frac{\frac{1}{\sqrt{2}}}{\frac{1}{2}}$$

$$\frac{3 \times 10^8}{v} = \sqrt{2}$$

$$v = \frac{3 \times 10^8}{1.414} \text{ m/s}$$

∴ Speed of light in the prism

$$= 2.12 \times 10^8 \text{ m/s}$$

●●

# Outside Delhi [Set-II]

**Note : Except for the following questions, all the remaining questions have been asked in previous set.**

## SECTION-A

**1. A bar magnet is moved in the direction indicated by the arrow between two coils PQ and CD. Predict the direction of the induced current in each coil. [1]**

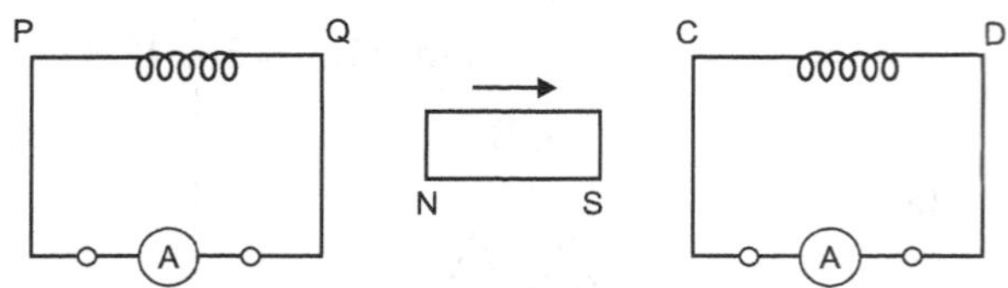


**Ans.** From Q to P (*i.e.*, anticlockwise) as seen from the left end.

From C to D (*i.e.*, clockwise) as seen from the left end.

**2. Write the relation for the speed of electromagnetic waves in terms of the amplitudes of electric and magnetic fields. [1]**

**Ans.** $c = \frac{E_0}{B_0}$

**7. Identify the electromagnetic waves whose wavelengths lie in the range**

**(a) $10^{-11} < \lambda < 10^{-14}$ m**

**(b) $10^{-4}$ m $< \lambda < 10^{-6}$ m**

**Write one use of each. [2]**

**Ans.** (a) $\gamma$-rays

**Use :** For treatment of cancer.

(b) Infrared rays

**Use :** In remote control of T.V., V.C.R, etc.

**9. The short wavelength limit for the Lyman series of the hydrogen spectrum is 913.4 Å. Calculate the short wavelength limit for Balmer series of the hydrogen spectrum. [2]**

**Ans.** For short wavelength of Lyman series,

$$\frac{1}{\lambda_L} = R\left(\frac{1}{1^2} - \frac{1}{\infty^2}\right)$$

$$\frac{1}{913.4 \times 10^{-10}} = R$$

For short wavelength of Balmer series,

$$\frac{1}{\lambda_L} = R\left(\frac{1}{2^2} - \frac{1}{\infty^2}\right)$$

$$\frac{1}{\lambda_L} = \frac{1}{913.4\times10^{-10}} \times \frac{1}{4}$$

$$\lambda_B = 4 \times 913.4 \times 10^{-10}$$

$$\lambda_B = 3653.6 \times 10^{-10}\,m$$

$$= 3653.6\ Å$$

**12. (a) Draw a ray diagram showing the formation of image by a reflecting telescope.**

**(b) Write two advantages of a reflecting telescope over a refracting telescope. [3]**

**Ans. (a)** Ray diagram

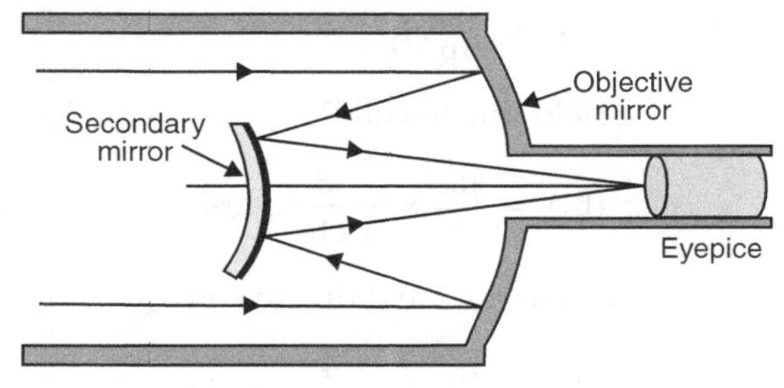


**(b) Advantages**

(i) Spherical and chromatic aberration is absent.

(ii) Light gathering power is more.

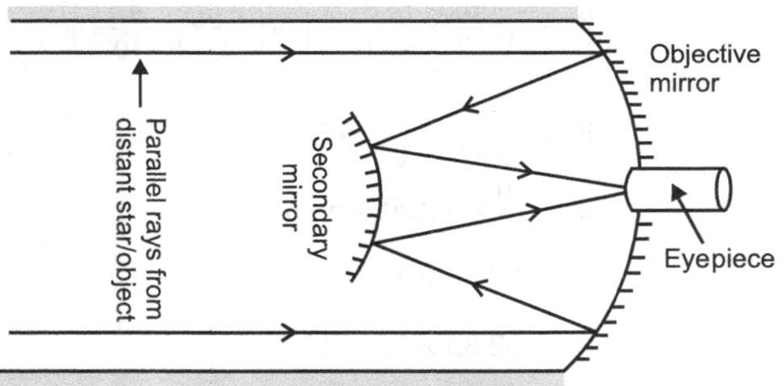


**15. Explain giving reasons for the following : [3]**

**(a) Photoelectric current in a photocell increases with the increase in the intensity of the incident radiation.**

**(b) The stopping potential ($V_0$) varies linearly with the frequency ($\nu$) of the incident radiation for a given photosensitive surface with the slope remaining the same for different surfaces.**

**(c) Maximum kinetic energy of the photoelectrons is independent of the intensity of incident radiation.**

**Ans. (a)** Since number of photoelectrons emitted is directly proportional to the intensity of incident radiation, therefore, as intensity increases, the number of electron-hole pairs also increases.

**(b)**

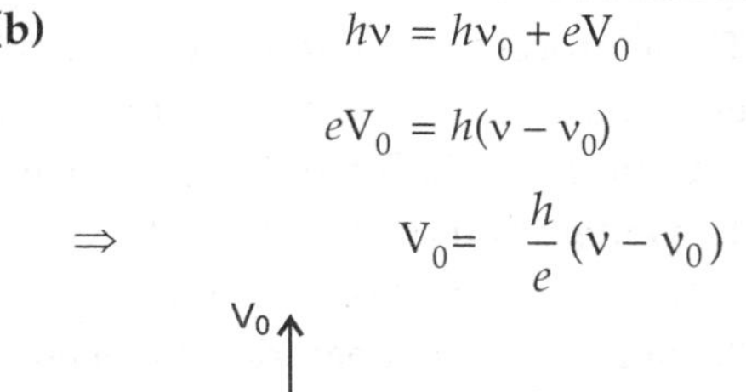

$$h\nu = h\nu_0 + eV_0$$

$$eV_0 = h(\nu - \nu_0)$$

$$\Rightarrow \quad V_0 = \frac{h}{e}(\nu - \nu_0)$$

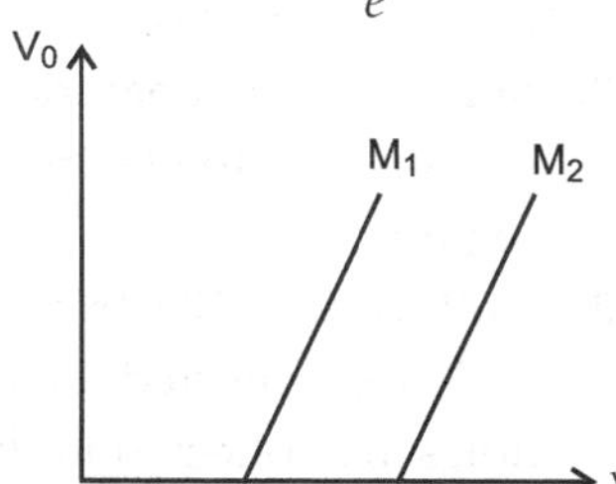


Further from graph, slope is $h/e$ which is a constant and it does not depend on $\nu$.

**(c)** $K.E_{max} = h(\nu - \nu_0)$

Hence, it depends on the frequency and not on the intensity of the incident radition.

As intensity increases, the number of photons increases but the energy remains same.

**16. (a) In the following diagram which bulb out of $B_1$ and $B_2$ will glow and why ?**

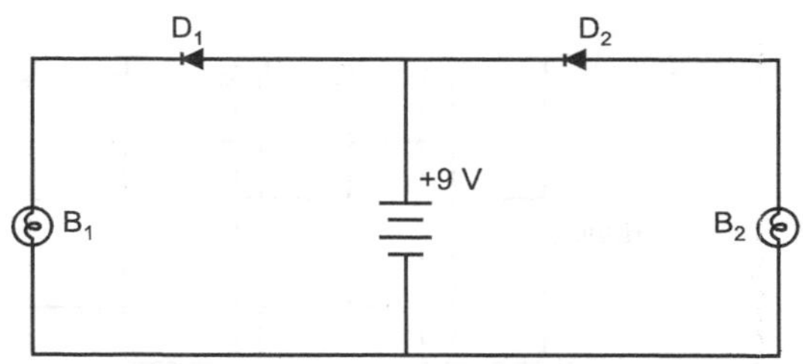


**(b) Draw a diagram of an illuminated *p-n* junction solar cell.**

**(c) Explain briefly the three processes due to which generation of emf takes place in a solar cell. [3]**

**Ans. (a)** $B_1$ will glow because only diode $D_1$ is forward biased.

**(b) Diagram**

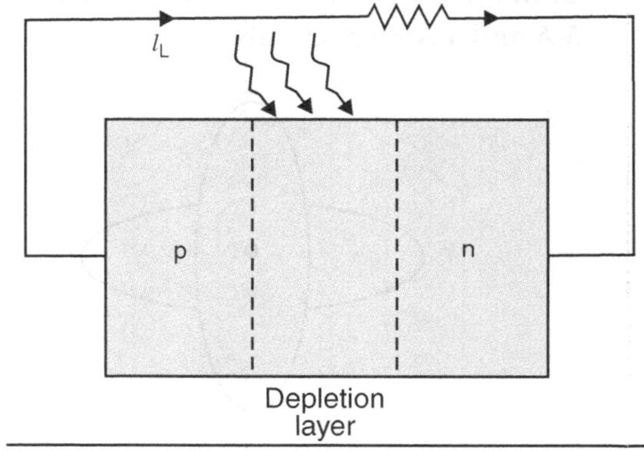


** Answer is not given due to change in the present Syllabus.

(c) Three processes due to which generation of emf takes place in solar cell are :

1. Generation, 2. Separation, 3. Collection.

1. **Generation :** Generation of electron-hole pairs take place due to light.
2. **Separation :** Separation of electron-hole pairs is due to electric field of depletion region.
3. **Collection :** Electrons reach the $n$-side and are collected in front contact and holes are collected in the back contact.

**19. (a) Draw the circuit diagram for studying the characteristics of a transistor in common emitter configuration. Explain briefly and show how input and output characteristics are drawn.****

**(b) The figure shows input waveforms A and B to a logic gate. Draw the output waveform for an OR gate. Write the truth table for this logic gate and draw its logic symbol.**** [3]

**20. Two identical loops P and Q each of radius 5 cm are lying in perpendicular planes such that they have a common centre as shown in the figure. Find the magnitude and direction of the net magnetic field at the common centre of the two coils, if they carry currents equal to 3 A and 4 A respectively.** [3]

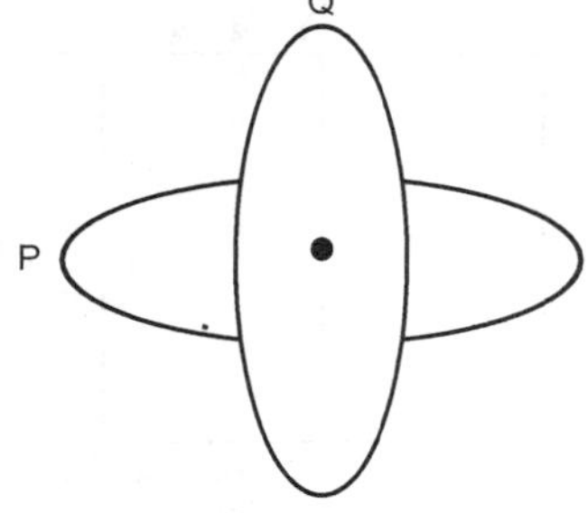


** Answer is not given due to change in the present Syllabus.

**Ans.**

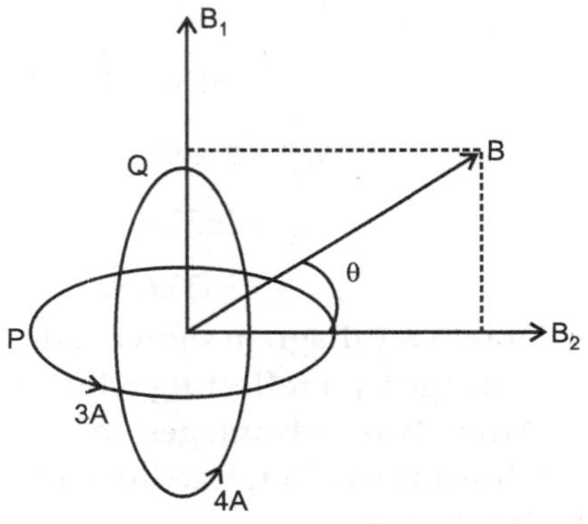


$$B_1 = \frac{\mu_0}{2R} . I_1$$

Field due to coil P,

$$(B_1) = \frac{\mu_0}{2} \times \frac{3}{5\times10^{-2}} \text{ tesla}$$

Similarly, Field due to coil Q,

$$(B_2) = \frac{\mu_0}{2} \times \frac{4}{5\times10^{-2}} \text{ tesla}$$

Resultant magnetic field is,

$$B = \sqrt{B_1^2 + B_2^2}$$

$$B = \sqrt{\left(\frac{\mu_0}{2}\times\frac{3}{5\times10^{-2}}\right)^2 + \left(\frac{\mu_0}{2}\times\frac{4}{5\times10^{-2}}\right)^2}$$

$$B = \frac{\mu_0}{2\times5\times10^{-2}} \times 5$$

$$B = \frac{\mu_0}{2} \times 100 = 50\,\mu_0$$

$$= 50 \times 4\pi \times 10^{-7}$$

$$= 62.83 \times 10^{-6}\,\text{T}$$

Let the field make an angle θ with the vertical

$$\tan\theta = \frac{B_1}{B_2}$$

$$\tan\theta = \frac{3}{4}$$

$$\theta = \tan^{-1}\left(\frac{3}{4}\right)$$

Thus direction of magnetic field makes an angle $\tan^{-1}\left(\frac{3}{4}\right)$ with the horizontal.

# Outside Delhi [Set-III]

**Note : Except for the following questions, all the remaining questions have been asked in the previous sets.**

**SECTION-A**

**3. What is the direction of induced currents in metal rings 1 and 2 when current I in the wire is increasing steadily ? [1]**

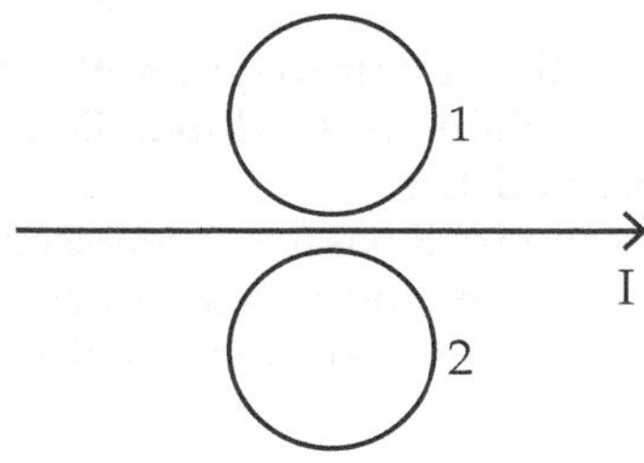


**Ans.** Ring - 1 → Clockwise
Ring - 2 → Anticlockwise

**4. In which directions do the electric and magnetic field vectors oscillate in an electromagnetic wave propagating along the x-axis ? [1]**

**Ans.** Electric component → Y - axis
Magnetic component → Z - axis

**8. Why does current in a steady state not flow in a capacitor connected across a battery ? However momentary current does flow during charging or discharging of the capacitor. Explain. [2]**

**Ans.** During charging/discharging, when there is change in the electric flux, there will be a displacement current and in steady state, when flux is constant, the displacement as well as conduction current will be zero.

$$I_D = \epsilon_0 \frac{d\phi_E}{dt}$$

**9. The ground state energy of hydrogen atom is – 13.6 eV. If an electron makes a transition from an energy level – 1.51 eV to – 3.4 eV, calculate the wavelength of the spectral line emitted and name the series of hydrogen spectrum to which it belongs. [2]**

**Ans.** When energy is – 1.51 eV then $n = 3$.
When energy is – 3.4 eV then $n = 2$.

$$\therefore \quad \frac{1}{\lambda} = R\left(\frac{1}{2^2} - \frac{1}{3^2}\right)$$

$$\frac{1}{\lambda} = 1.1 \times 10^7 \left(\frac{1}{4} - \frac{1}{9}\right)$$

$$= 1.1 \times 10^7 \times \frac{5}{36}$$

$$\lambda = \frac{36}{5 \times 1.1} \times 10^{-7}\, m$$

$$\lambda = 6.545 \times 10^{-7}\, m$$

$$\lambda = 6545\, Å$$

It belongs to the Balmer series.

**14. (a) Draw the circuit diagram of an *n-p-n* transistor amplifier in common emitter configuration.****

**(b) Derive an expression for voltage gain of the amplifier and hence show that the output voltage is in opposite phase with the input voltage.** [3]**

**17. (a) Draw a ray diagram for the formation of image by a compound microscope.**

**(b) You are given the following three lenses. Which two lenses will you use as an eyepiece and as an objective to construct a compound microscope ?**

| Lenses | Power (D) | Aperture (cm) |
|---|---|---|
| $L_1$ | 3 | 8 |
| $L_2$ | 6 | 1 |
| $L_3$ | 10 | 1 |

**(c) Define resolving power of a microscope and write one factor on which it depends. [3]**

**Ans.** (a) Ray diagram for compound microscope :

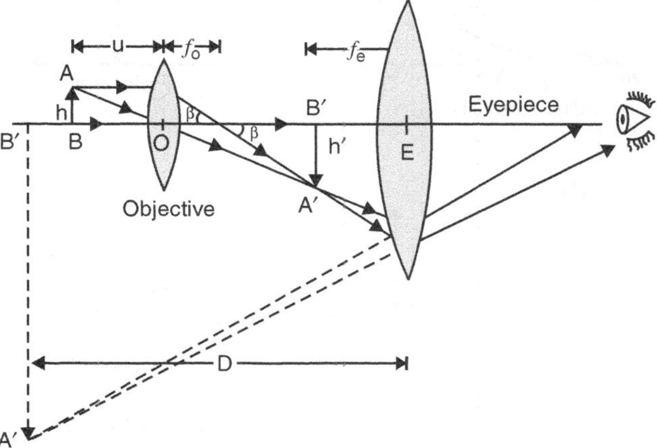


(b) Objective lens → $L_3$
Eye lens → $L_2$

(c) The resolving power of a microscope is its ability to form distinctly separate images of two neighbouring objects.

$$R.P. = \frac{2\mu \sin\theta}{\lambda}$$

** Answer is not given due to change in the present Syllabus.

It depends on the wavelength ($\lambda$) of the light used.

**18. The following graph shows the variation of photocurrent for a photosensitive metal : [3]**

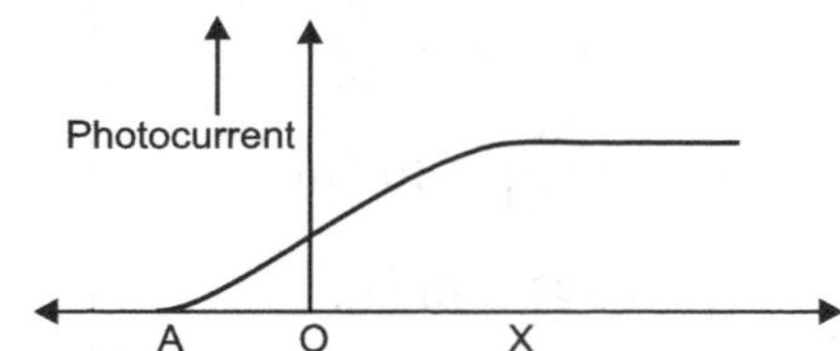


**(a) Identify the variable X on the horizontal axis.**

**(b) What does the point A on the horizontal axis represent ?**

**(c) Draw this graph for three different values of frequencies of incident radiation $\nu_1$, $\nu_2$, and $\nu_3$ ($\nu_1 > \nu_2 > \nu_3$) for same intensity.**

**(d) Draw this graph for three different values of intensities of incident radiation $I_1$, $I_2$ and $I_3$ ($I_1 > I_2 > I_3$) having same frequency.**

**Ans.** (a) X is collector plate potential.

(b) A is stopping potential.

(c)

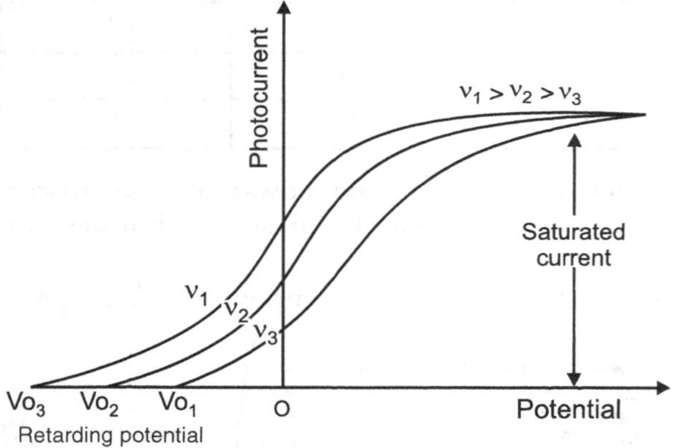


(d)

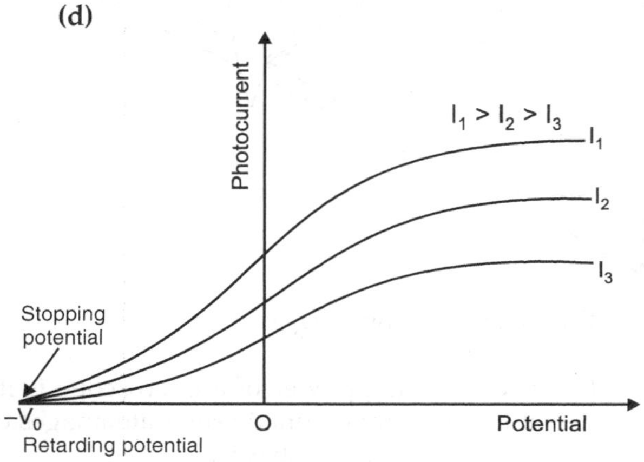


**21. (a) In the following diagram 'S' is a semiconductor. Would you increase or decrease the value of R to keep the reading of the ammeter A constant when S is heated ? Give reason for your answer.**

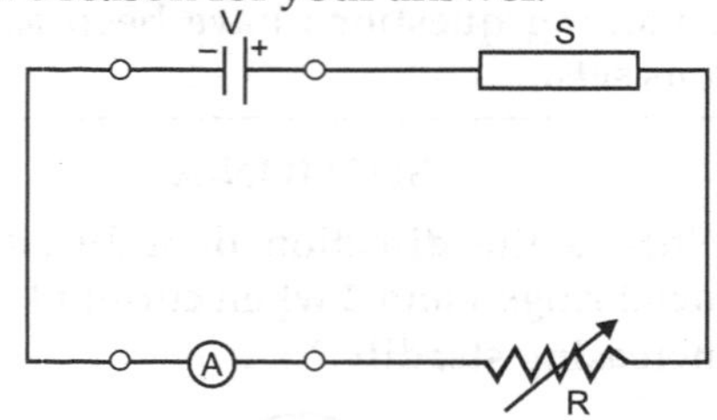


**(b) Draw the circuit diagram of a photodiode and explain its working. Draw its I – V characteristics. [3]**

**Ans.** (a) The value of R has to be increased because on heating, the conductivity of a semiconductor increases *i.e.* resistance of S decreases on heating.

(b)

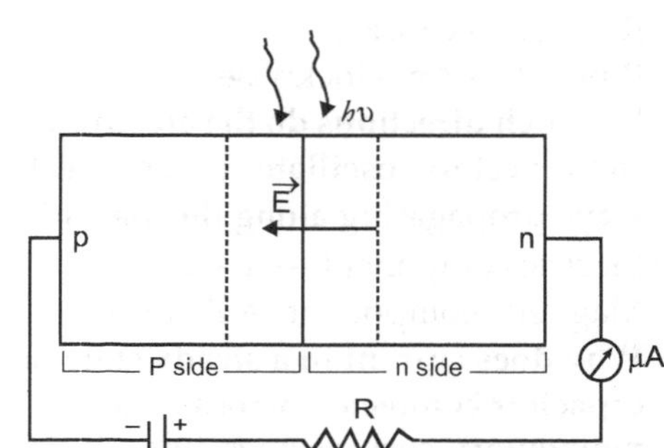


**Working** : In photodiode, an electric field exists across the junction from n-side to p-side. When visible light with energy hν greater than energy gap ($E_g$) illuminates the junction, then electron-hole pairs are generated in the depletion layer. Due to electric field, electron moves towards *n* side and holes towards *p*-side giving rise to an emf. When an external load is connected, current flows.

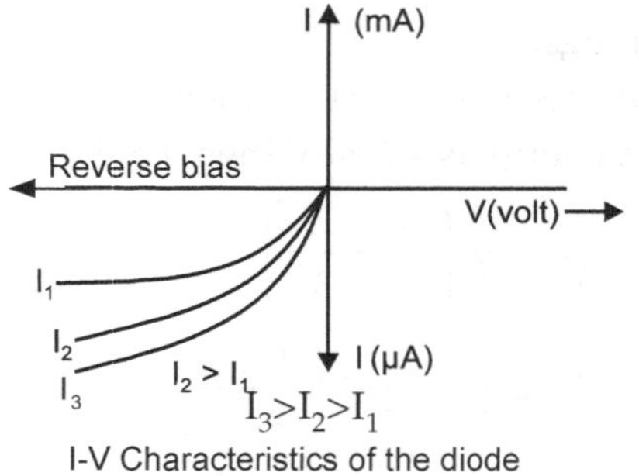


I-V Characteristics of the diode

# PHYSICS

## Delhi [Set-I]

## 2017

Time allowed : 3 hours

Maximum marks : 70

### SECTION – A

**1. Does the charge given to a metallic sphere depend on whether it is hollow or solid ? Give reason for your answer. [1]**

**Ans.** No, because all the charge resides on the surface of the sphere only.

**2. A long straight current carrying wire passes normally through the centre of circular loop. If the current through the wire increases, will there be an induced emf in the loop ? Justify. [1]**

**Ans.** No, the emf will not be induced as the magnetic field lines are parallel to the plane of the circular loop. So magnetic flux will remain zero.

$$\phi = BA\cos\theta = BA\cos 90°$$
$$= 0$$

Hence, induced emf, $\varepsilon = -\frac{d\phi}{dt} = 0$

**3. At a place, the horizontal component of earth's magnetic field is B and angle of dip is 60°. What is the value of horizontal component of the earth's magnetic field at equator ? [1]**

**Ans.**

$$B_H = B' \cos 60°$$
$$B = B' \cos 60° \quad (\text{given } B_H = B)$$
$$B = B'\left(\frac{1}{2}\right)$$
$$B' = 2B$$

At equator, $\theta = 0°$

$$B_H = B' \cos\theta$$
$$= B' \cos 0°$$
$$B_H = 2B$$

**4. Name the junction diode whose I-V characteristics are drawn below : [1]**

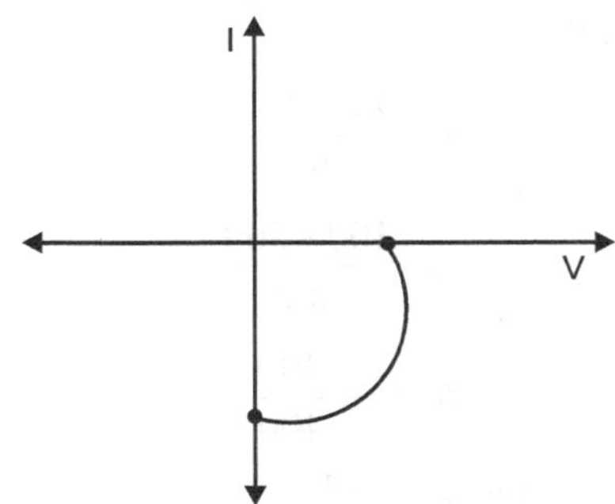


**Ans.** This is the characteristic of solar cell.

**5. How is the speed of em-waves in vacuum determined by the electric and magnetic fields? [1]**

**Ans.** The speed of em-waves are determined by the ratio of the peak values of electric and magnetic field vectors.

$$c = \frac{E_0}{B_0}$$

### SECTION–B

**6. How does Ampere-Maxwell law explain the flow of current through a capacitor when it is being charged by a battery ? Write the expression for the displacement current in terms of the rate of change of electric flux. [2]**

**Ans.** During the charging of capacitor, electric flux between the plates of capacitor keeps on changing, due to which displacement current between the plates is produced. Hence circuit becomes complete and current flows through the circuit.

$$I_d = \epsilon_0 \frac{d\phi_E}{dt}$$

**7. Define the distance of closest approach. An α-particle of kinetic energy 'K' is bombarded on a thin gold foil. The distance of the closest approach is 'r'. What will be the distance of closest approach for an α-particle of double the kinetic energy ? [2]**

**OR**

**Write two important limitations of Rutherford nuclear model of the atom.**

**Ans.** It is defined as the minimum distance of the charged particle from the nucleus at which initial kinetic energy of the particle is equal to the potential energy of the system.

At this distance,

$$K.E = P.E$$
$$K = \frac{1}{4\pi\epsilon_0}\frac{Ze(2e)}{r}$$
$$r = \frac{1}{4\pi\epsilon_0}\frac{2Ze^2}{K} \quad ...(i)$$

If K is doubled,

$$r' = \frac{1}{4\pi\epsilon_0}\frac{2Ze^2}{2K} \quad ...(ii)$$

Dividing (ii) by (i), we get

$$\frac{r'}{r} = \frac{1}{2}$$

$$\Rightarrow \qquad r' = \frac{r}{2}$$

**OR**

**Limitations of Rutherford Model : (a)** It is not in accordance with the Maxwell's theory and could not explain the stability of an atom.

**(b)** It did not say anything about the arrangement of electrons in an atom.

**8. Find out the wavelength of the electron orbiting in the ground state of hydrogen atom. [2]**

**Ans.** We know that, according to de Broglie relation,

$$2\pi r = n\lambda$$

For first orbit,

$$n = 1$$

$$2\pi r = n\lambda$$

$$\lambda = 2\pi \times 0.53 \text{ Å} \qquad [\because r = 0.53 \text{ Å}]$$

$$\lambda = 3.331 \text{ Å}$$

**9. Define the magnifying power of a compound microscope when the final image is formed at infinity. Why must both the objective and the eyepiece of a compound microscope have short focal lengths ? Explain. [2]**

**Ans. Magnifying Power :** The magnifying power of a compound microscope is defined as the ratio of the angle subtended at the eye by the final image to the angle subtended at the eye by the object, when both the final image and the object are situated at the least distance of distinct vision from the eye.

Magnifying power of the compound microscope when the final image is at infinity,

$$m = \frac{L}{f_0} \times \frac{D}{f_e}$$

From the above equation, we can see that to achieve a large magnification, the objective and eyepiece should have small focal lengths.

**10. Which basic mode of communication is used in satellite communication ? What type of wave propagation is used in this mode ? Write, giving reason, the frequency range used in this mode of propagation.** [2]**

** Answer is not given due to change in the present Syllabus.

## SECTION – C

**11. (a) Find the value of the phase difference between the current and the voltage in the series LCR circuit shown below. Which one leads in phase : current or voltage ?**

**(b) Without making any other change, find the value of the additional capacitor $C_1$, to be connected in parallel with capacitor C, in order to make the power factor of the circuit unity. [3]**

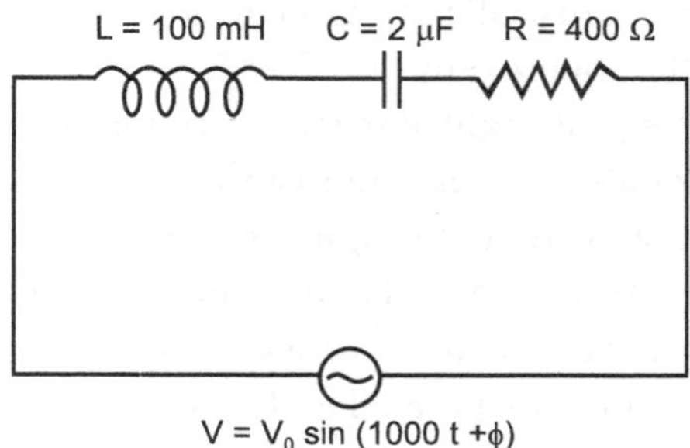


**Ans. (a)** Given :

$$L = 100 \times 10^{-3} \text{ H}$$

$$C = 2 \times 10^{-6} \text{ F}$$

$$R = 400 \, \Omega$$

$$\omega = 1000$$

Now, $X_L = \omega L$

$$X_L = (1000 \times 100 \times 10^{-3}) \, \Omega$$

$$X_L = 100 \, \Omega$$

And $X_C = \frac{1}{\omega C}$

$$= \frac{1}{1000 \times 2 \times 10^{-6}}$$

$$= \frac{10^6}{2 \times 10^3}$$

$$X_C = \frac{1000}{2} = 500 \, \Omega$$

Here $X_C > X_L$ so, the current will lead the voltage.

$$\tan \phi = \frac{X_L - X_C}{R}$$

$$= \frac{100 - 500}{400}$$

$$\tan \phi = -\frac{400}{400}$$

$$\tan \phi = -1$$

$\Rightarrow$ phase angle $\phi = -45°$

**(b)** For unity power factor,

$$X_L = X_C$$

$$\omega L = \frac{1}{\omega C_{eq}}$$

$$C_{eq} = \frac{L}{L\omega^2}$$

$$C_{eq} = \frac{1}{(1000)^2 \times 100 \times 10^{-3}} = 10\,\mu F$$

Now, $C_{eq} = C + C_1$

$$10 = 2 + C_1$$

$$10 - 2 = C_1$$

$$C_1 = 8\,\mu F$$

**12. Write the two processes that take place in the formation of a *p-n* junction. Explain with the help of a diagram, the formation of depletion region and barrier potential in a *p-n* junction. [3]**

**Ans.** Two important processes involved during the formation of p-n junction are diffusion and drift.

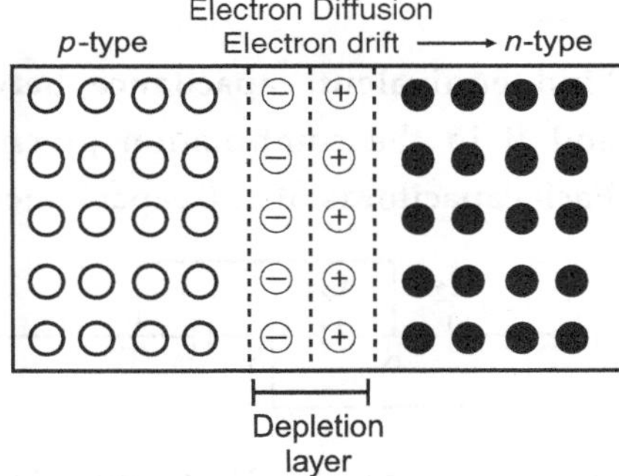


**Formation of depletion region and barrier potential :** At the instant of *p-n* junction formation, the free electrons near the junction diffuse across the junction into the *p* region and combine with holes. Thus, on combining with the hole, it makes a negative ion and leaves a positive ion on n-side. These two layers of immobile positive and immobile negative charges form the depletion region.

Further, as electrons diffuse across the junction, a point is reached where the negative charge repels any further diffusion of electron. This depletion region now acts as a barrier. Now the external energy is supplied to get the electrons to move across the barrier of electric field. The potential difference required to move the electrons through the electric field is called barrier potential.

**13. (a) Obtain the expression for the cyclotron frequency.**

**(b) A deuteron and a proton are accelerated by the cyclotron. Can both be accelerated with the same oscillator frequency ? Give reason to justify your answer. [3]**

**Ans. (a)** Suppose the positive ion with charge $q$ moves in a dee with a velocity $v$, then

$$\frac{mv^2}{r} = qvB$$

$$r = \frac{mv}{qB} \quad \ldots(i)$$

where $m$ is the mass and $r$ is the radius of the path of ion in the dee and B is the strength of the magnetic field.

The time taken by the ion,

$$T = \frac{2\pi r}{v}$$

$$\frac{1}{f} = \frac{2\pi r}{v}$$

$$f = \frac{v}{2\pi r}$$

$$\therefore \quad f = \frac{qB}{2\pi \times m} \quad \text{[Using (i)]}$$

This frequency is called the cyclotron frequency.

**(b)** No, Deuteron and proton have different masses and cyclotron frequency depends inversely on mass. Hence, they cannot be accelerated with the same oscillator frequency.

**14. (a) How does one explain the emission of electrons from a photosensitive surface with the help of Einstein's photoelectric equation?**

**(b) The work function of the following metals is given : Na = 2.75 eV, K = 2.3 eV, Mo = 4.17 eV and Ni = 5.15 eV. Which of these metals will not cause photoelectric emission for radiation of wavelength 3300 Å from a laser source placed 1 m away from these metals ? What happens if the laser source is brought nearer and placed 50 cm away ? [3]**

**Ans. (a)** According to Einstein's photoelectric equation,

$$h\nu = h\nu_0 + E_{max}$$

$$E_{max} = h\nu - h\nu_0$$

$$E_{max} = h(\nu - \nu_0)$$

If $v < v_0$ then $E_{max}$ will be negative so, no emission takes place.

If $v > v_0$ then $E_{max}$ will be positive and directly proportional to $u$ so, emission takes place.

**(b)**

$$E = \frac{hc}{\lambda}$$

$$E = \frac{6.63\times10^{-34}\times3\times10^{8}}{3300\times10^{-10}\times1.6\times10^{-19}}$$

$$E = 3.76 \text{ eV}$$

The work function of Na and K is less than the energy of incident radiation.

Therefore, Na and K will cause photoelectric emission while Mo and Ni will not cause photoelectric emission.

There will be no effect on photoelectric emission if the source is brought nearer or farther.

**15. A resistance of R draws current from a potentiometer. The potentiometer wire, AB, has a total resistance of $R_0$. A voltage V is supplied to the potentiometer. Derive an expression for the voltage across R when the sliding contact is in the middle of potentiometer wire. [3]**

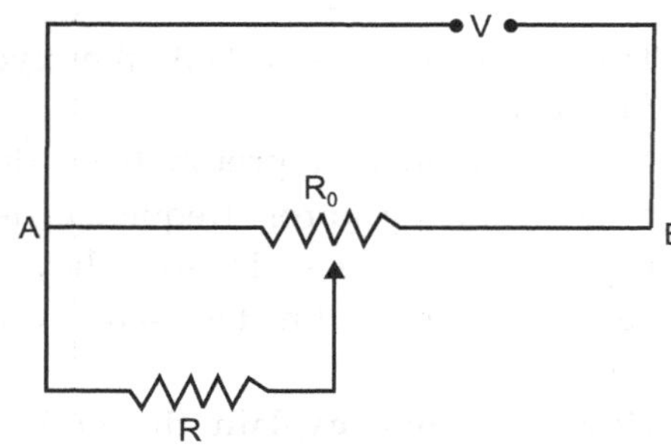


**Ans.**

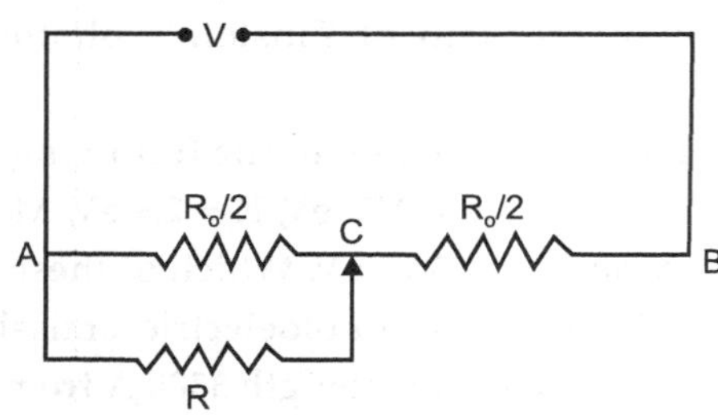


The equivalent resistance (R′) between A and C,

$$\frac{1}{R'} = \frac{1}{R} + \frac{1}{R_0/2}$$

$$R' = \frac{R\left(\frac{R_0}{2}\right)}{R+\frac{R_0}{2}} = \frac{RR_0}{2R+R_0} \quad ...(i)$$

Equivalent resistance ($R_{eq}$) between A and B,

$$R_{eq} = R' + \frac{R_0}{2}$$

$$R_{eq} = \frac{RR_0}{2R+R_0} + \frac{R_0}{2}$$

$$R_{eq} = \frac{R_0(4R+R_0)}{2(2R+R_0)} \quad ...(ii)$$

Current in the circuit,

$$I = \frac{V}{R_{eq}} = \frac{V.2(2R+R_0)}{R_0(4R+R_0)}$$

$\therefore$ Voltage across R,

$$V_R = I.R'$$

$$V_R = \frac{V.2(2R+R_0)}{R_0(4R+R_0)} \times \frac{RR_0}{(2R+R_0)}$$

$$V_R = \frac{2VR}{4R+R_0}$$

**16. Define the term 'amplitude modulation.' Explain any two factors which justify the need for modulating a low frequency base-band signal.** [3]**

**17. (a) Find equivalent capacitance between A and B in the combination given below. Each capacitor is of 2μF capacitance.**

A —$C_1$— P —$C_2$— R —$C_3$— S —$C_4$— T —$C_5$— B

**(b) If a dc source of 7 V is connected across AB, how much charge is drawn from the source and what is the energy stored in the network? [3]**

**Ans. (a)** In the given figure, $C_2$, $C_3$ and $C_4$ are in parallel, so the equivalent capacitance of parallel capacitors is given by C'.

$$C' = C_2 + C_3 + C_4 = (2 + 2 + 2)\ \mu F$$

$$C' = 6\ \mu F$$

Again, $C_1$, C' and $C_5$ are in series. Therefore, the equivalent capacitance $C_{eq}$ is given by

$$\frac{1}{C_{eq}} = \frac{1}{2}+\frac{1}{6}+\frac{1}{2}$$

$$\frac{1}{C_{eq}} = \frac{3+1+3}{6} = \frac{7}{6}$$

$$C_{eq} = \frac{6}{7}\mu F$$

** Answer is not given due to change in the present Syllabus.

**(b)** The charge remains same in series circuit.

$\therefore \quad Q = C_{eq}V$

$= \frac{6}{7} \times 7$

$Q = 6\mu C$

Now, $U = \frac{1}{2} C_{eq} V^2$

$U = \frac{1}{2} \times \frac{6}{7} \times (7)^2 \times 10^{-6}$

$U = 21 \times 10^{-6}$ Joule $= 21\ \mu J$

**18. (a) Derive the expression for electric field at a point on the equatorial line of an electric dipole.**

**(b) Depict the orientation of the dipole in (i) stable, (ii) unstable equilibrium in a uniform electric field. [3]**

**Ans. (a)** Consider a point P on broad side position of dipole formed of charges $+q$ and $-q$ at separation $2l$. The distance of point P from mid point (O) of electric dipole is r. Let $\vec{E_1}$ and $\vec{E_2}$ be the electric field strengths due to charges $+q$ and $-q$ of electric dipole.

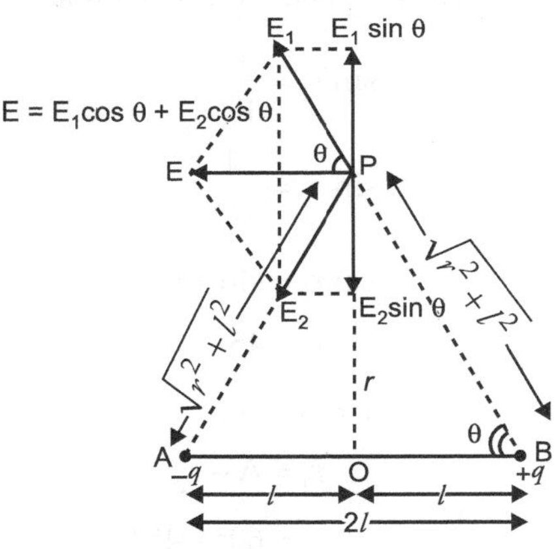


From the figure,

$AP = BP = \sqrt{r^2 + l^2}$

$\therefore \quad \vec{E_1} = \frac{1}{4\pi\varepsilon_0} \frac{q}{r^2 + l^2}$ along B to P

$\vec{E_2} = \frac{1}{4\pi\varepsilon_0} \frac{q}{r^2 + l^2}$ along P to A

The components of $\vec{E_1}$ and $\vec{E_2}$ perpendicular to AB, $E_1 \sin\theta$ and $E_2 \sin\theta$, being equal and opposite cancel each other, while the cosine components of $\vec{E_1}$ and $\vec{E_2}$ parallel to AB, $E_1 \cos\theta$ and $E_2 \cos\theta$ being in the same direction add up and give the resultant electric field whose direction is parallel to $\vec{BA}$.

$\therefore$ Resultant electric field at P is E

$= E_1 \cos\theta + E_2 \cos\theta$

But $\quad E_1 = E_2 = \frac{1}{4\pi\varepsilon_0} \frac{q}{(r^2 + l^2)}$

From the figure,

$$\cos\theta = \frac{OB}{PB} = \frac{l}{\sqrt{r^2 + l^2}}$$

$$= \frac{l}{(r^2 + l^2)^{1/2}}$$

$$E = 2E_1 \cos\theta$$

$$= 2 \times \frac{1}{4\pi\varepsilon_0} \frac{q}{(r^2 + l^2)} \cdot \frac{l}{(r^2 + l^2)^{1/2}}$$

$$= \frac{1}{4\pi\varepsilon_0} \frac{2ql}{(r^2 + l^2)^{3/2}}$$

But $q.2l = p$ = electric dipole moment

$$E = \frac{1}{4\pi\varepsilon_0} \frac{p}{(r^2 + l^2)^{3/2}}$$

**(b)** (i) $\vec{E}$

$\vec{p}$

$-q \quad +q$

Stable equilibrium

*i.e.*, $\theta = 0°$

(ii)

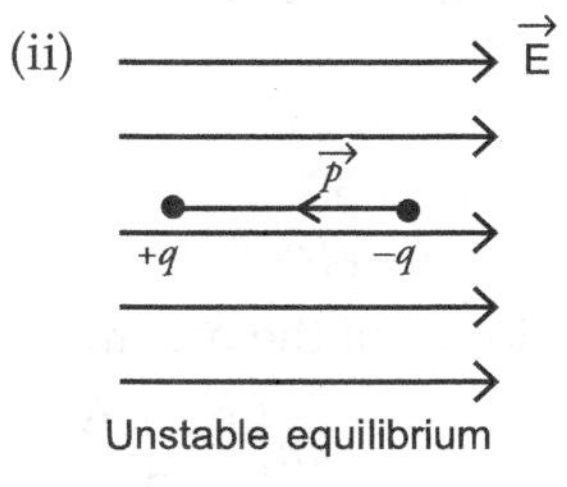


Unstable equilibrium

*i.e.*, $\theta = 180°$

**19. (a) A radioactive nucleus 'A' undergoes a series of decays as given below :**

$$A \xrightarrow{\alpha} A_1 \xrightarrow{\beta} A_2 \xrightarrow{\alpha} A_3 \xrightarrow{\gamma} A_4$$

**The mass number and atomic number of $A_2$ are 176 and 71 respectively. Determine the mass and atomic numbers of $A_4$ and A.**

**(b) Write the basic nuclear process underlying $\beta^+$ and $\beta^-$ decays.** **[3]**

**Ans. (a)**

$$A \xrightarrow{\alpha} A_1 \xrightarrow{\beta} {}^{176}_{71}A_2 \xrightarrow{\alpha} A_3 \xrightarrow{\gamma} A_4$$

$${}^{176}_{71}A_2 \xrightarrow{\alpha} {}^{172}_{69}A_3 \xrightarrow{\gamma} {}^{172}_{69}A_4$$

$$A_4 \longrightarrow \begin{cases} 172 = \text{Mass number} \\ 69 = \text{Atomic number} \end{cases}$$

$${}^{180}_{72}A \xrightarrow{\alpha} {}^{176}_{70}A_1 \xrightarrow{\beta} {}^{176}_{71}A_2$$

$$A \longrightarrow \begin{cases} 180 = \text{Mass number} \\ 72 = \text{Atomic number} \end{cases}$$

**(b)** The process of spontaneous emission of a positron ($e^+$) from nucleus is called $\beta^+$ decay.

$$p \rightarrow n + {}^{0}_{1}e + \overline{\nu}$$

The process of spontaneous emission of an electron ($e^-$) from nucleus is called $\beta^-$ decay.

$$n \rightarrow p + {}^{0}_{-1}e + \overline{\nu}$$

**20. (a) A ray of light incident on face AB of an equilateral glass prism, shows minimum deviation of 30°. Calculate the speed of light through the prism.**

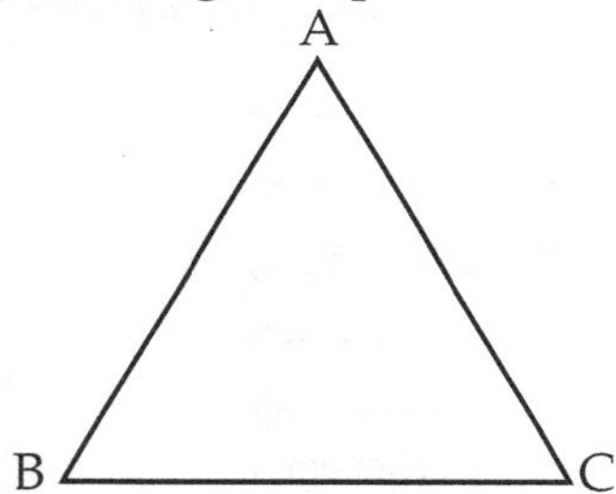


**(b) Find the angle of incidence at face AB so that the emergent ray grazes along the face AC.** **[3]**

**Ans. (a)** Given :

$$\delta_m = 30^\circ$$
$$A = 60^\circ$$

Refractive index of the prism,

$$\mu = \frac{\sin\left(\frac{\delta_m + A}{2}\right)}{\sin\left(A/2\right)}$$

$$\mu = \frac{\sin\left(\frac{30^\circ + 60^\circ}{2}\right)}{\sin\left(\frac{60^\circ}{2}\right)}$$

$$\mu = \frac{\sin(45^\circ)}{\sin(30^\circ)}$$

$$\mu = \frac{1/\sqrt{2}}{1/2} = \frac{2}{\sqrt{2}} = \sqrt{2}$$

Now, $\mu = \frac{c}{v}$

$$v = \frac{c}{\mu}$$

$$v = \frac{3 \times 10^8}{\sqrt{2}}$$

$$v = \frac{3 \times 10^8}{1.414}$$

$$v = 2.12 \times 10^8 \text{ m/s}$$

**(b)**

Since the light grazes along the face AC. Therefore, $r_2$ is the critical angle.

$$\sin r_2 = \frac{1}{\mu}$$

$$\sin r_2 = \frac{1}{\sqrt{2}}$$

$$\sin r_2 = \sin 45^\circ$$

$$r_2 = 45^\circ$$

$\therefore$ $r_1 + r_2 = A$

$$r_1 = A - r_2$$

$$= 60^\circ - 45^\circ$$

$$r_1 = 15^\circ$$

Now, $\mu = \frac{\sin i}{\sin r_1}$

By snell's law

$$\sin i = \mu \sin r_1$$

$$\sin i = \sqrt{2}\sin(15^\circ)$$

$$i = \sin^{-1}(\sqrt{2}\sin 15^\circ)$$

$$\sin i = 0.366$$

$$i = 21^\circ \text{ (approx)}$$

**21. For a CE-transistor amplifier, the audio signal voltage across the collector resistance of 2 kΩ is 2 V. Given the current amplification factor of the transistor is 100, find the input signal voltage and base current, if the base resistance is 1 kΩ.**** **[3]**

**22. Describe the working principle of a moving coil galvanometer. Why is it necessary to use (i) a radial magnetic field and (ii) a cylindrical soft iron core in a galvanometer ? Write the expression for current sensitivity of the galvanometer.**

**Can a galvanometer as such be used for measuring the current ? Explain.** **[3]**

**OR**

**(a) Define the term 'self-inductance' and write its S.I. unit.**

**(b) Obtain the expression for the mutual inductance of two long co-axial solenoids $S_1$ and $S_2$ wound one over the other, each of length L and radii $r_1$ and $r_2$ and $n_1$ and $n_2$ number of turns per unit length, when a current I is set up in the outer solenoid $S_2$.**

**Ans. Moving coil galvanometer :**

**Principle** : It works on the principle that a current carrying coil placed in a magnetic field experiences a torque.

(i) **Radial magnetic field :** It is necessary to eliminate the effect of θ on torque and hence enable us to make the scale of galvanometer linear (deflection proportional to current).

(ii) **Cylindrical soft iron core :** It makes the magnetic lines of force pointing along the radii of the circle and also due to high permeability, it intensifies the magnetic field and hence increases the sensitivity of the galvanometer.

**Current sensitivity :**

$$I_s = \frac{NBA}{K} \text{ or } \frac{\theta}{I}$$

where θ is the deflection of coil.

No, it can not be used to measure electric current because it can only defect current. It is a highly current sensitive device which would show large deflection even with the passage of small current. And the galvanometer coil is likely to be damaged by currents in (mA/A) range.

**OR**

(a) **Self-inductance :** The self inductance of a coil may be defined as the induced emf set up in the coil due to a unit rate of change of current through it.

Its S.I. unit is Henry (H).

(b) Magnetic field due to current $I_2$ sin $S_2$ is

$$B_2 = \mu_0 n_2 I_2$$

Magnetic flux through $S_1$ is

$$\phi_1 = B_2 A_1 N_1$$

[where $N_1 = n_1 l$]

$$\phi_1 = (\mu_0 n_2 I_2)(\pi r_1^2)(n_1 l) \quad ...(1)$$

By defination of mutual inductance

$$\phi_1 = M_{12}(I_2) ...(2)$$

By equation (1) and equation (2)

$$M_{12} = \mu_0 n_1 n_2 \pi r_1^2 l$$

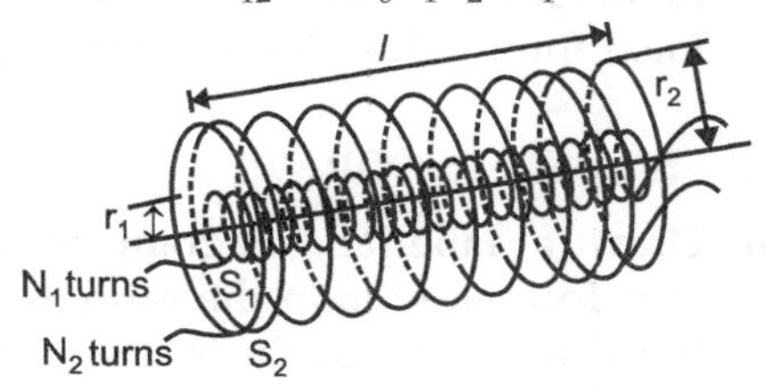


## SECTION–D

**23. Mrs. Rashmi Singh broke her reading glasses. When she went to the shopkeeper to order new spects, he suggested that she should get spectacles with plastic lenses instead of glass lenses. On getting the new spectacles, she found that the new ones were thicker than the earlier ones, she asked this question to the shopkeeper but he could not offer satisfactory explanation for this. At home, Mrs. Singh raised the same question to her daughter Anuja who explained why plastic lenses were thicker.**

**(a) Write two qualities displayed each by Anuja and her mother.****

**(b) How do you explain this fact using lens maker's formula ?** **[4]**

**Ans.** (b) According to lens maker's formula,

$$\frac{1}{f} = (n-1)\left(\frac{1}{R_1} - \frac{1}{R_2}\right)$$

where $f$ is the focal length, n is the refractive index and $R_1$ and $R_2$ is the radius of curvature of the lens.

Since $n_g > n_p$

where, g stands for glass and p stands for plastic.

Therefore, we get $(n_g - 1) > (n_p - 1)$

** Answer is not given due to change in the present Syllabus.

Now, using the lens maker's formula, we see that focal length is inversely proportional to $(n - 1)$.

Hence, $f_p > f_g$

Thus, in the case of plastic lens the thickness of the lens should be increased to keep the same focal length as that of the glass lens to give the same power.

## SECTION–E

**24. (a) Draw a labelled diagram of AC generator. Derive the expression for the instantaneous value of the emf induced in the coil.**

**(b) A circular coil of cross-sectional area 200 cm$^2$ and 20 turns is rotated about the vertical diameter with angular speed of 50 rad s$^{-1}$ in a uniform magnetic field of magnitude 3.0 × 10$^{-2}$ T. Calculate the maximum value of the current in the coil.** **[5]**

**OR**

**(a) Draw a labelled diagram of a step-up transformer. Obtain the ratio of secondary to primary voltage in terms of number of turns and currents in the two coils.**

**(b) A power transmission line feeds input power at 2200 V to a step-down transformer with its primary windings having 3000 turns. Find the number of turns in the secondary to get the power output at 220 V.**

**Ans. (a)**

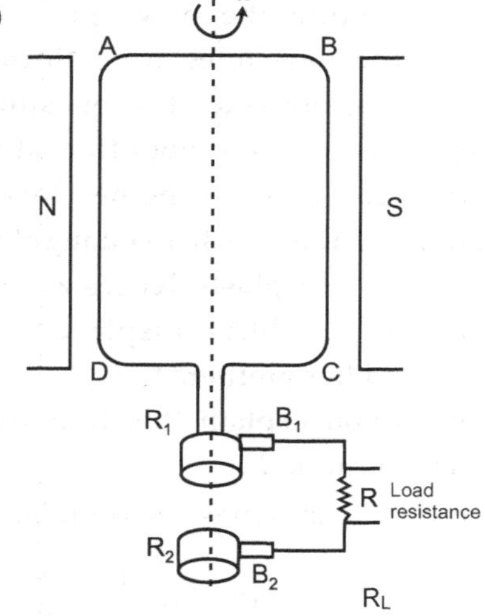


where, $\left.\begin{matrix}B_1\\B_2\end{matrix}\right\}$ are brushes.

$\left.\begin{matrix}R_1\\R_2\end{matrix}\right\}$ are slip rings.

$R_L$ is load resistance.

ABCD is armature.

If N is the number of turns in coil, A area of coil and B the magnetic induction, then flux $\phi$ is given by

$$\phi = B.A$$

$$\phi = BA \cos \omega t$$

When the coil is rotated with constant angular speed 'ω' in time 't'

The emf induced in the coil is given by,

$$e = -N\frac{d\phi}{dt} = -N\frac{d}{dt}(BA \cos \omega t)$$

$$= -NBA(-\omega \sin \omega t)$$

$$= NBA\omega \sin \omega t$$

$$e = e_0 \sin \omega t \qquad (\text{where } e_0 = NBA\omega)$$

**(b)** We know that,

$$e_0 = NBA\omega$$

$$e_0 = 20 \times 3 \times 10^{-2} \times 200 \times 10^{-4} \times 50$$

$$e_0 = 0.6 \text{ volt} = 600 \text{ mV}$$

$$I_0 = \frac{e_0}{R}$$

$$I_0 = \frac{0.6}{R} \text{ A} = \frac{600}{R} mA$$

**OR**

**(a)**

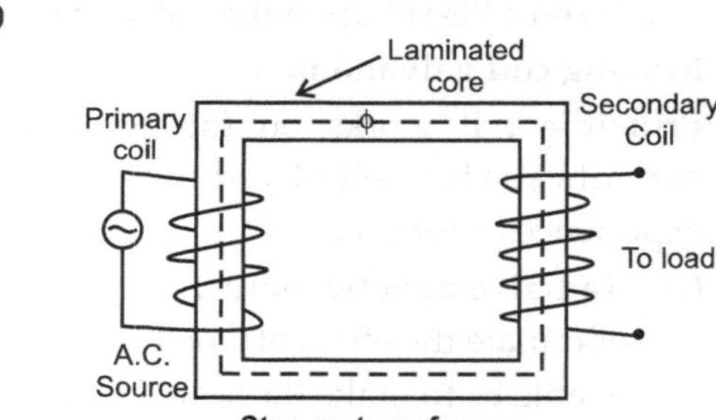


**Step up transformer**

The emf induced in primary and secondary coil,

$$e_s = -N_s \frac{d\phi}{dt} \qquad ...(i)$$

$$e_p = -N_p \frac{d\phi}{dt} \qquad ...(ii)$$

Dividing (i) by (ii), we get

$$\frac{e_s}{e_p} = \frac{N_s}{N_p} \qquad ...(iii)$$

According to the law of conservation of energy in an ideal transformer,

$$e_s I_s = e_p I_p$$

$$\frac{e_s}{e_p} = \frac{I_p}{I_s} \qquad ...(iv)$$

From equation (iii) and (iv), we get

$$\frac{V_s}{V_p} = \frac{e_s}{e_p} = \frac{N_s}{N_p} = \frac{I_p}{I_s}$$

**(b)** Given :

$$e_p = 2200 \text{ V}$$
$$N_p = 3000$$
$$N_s = ?$$
$$e_s = 220 \text{ V}$$

$$\therefore \quad \frac{e_s}{e_p} = \frac{N_s}{N_p}$$

$$\frac{220}{2200} = \frac{N_s}{3000}$$

$$\Rightarrow \quad N_s = 300$$

**25. (a) Distinguish between unpolarised light and linearly polarised light. How does one get linearly polarised light with the help of a polaroid ?**

**(b) A narrow beam of unpolarised light of intensity $I_0$ is incident on a polaroid $P_1$. The light transmitted by it is then incident on a second polaroid $P_2$ with its pass axis making angle of 60° relative to the pass axis of $P_1$. Find the intenstity of the light transmitted by $P_2$.** **[5]**

**OR**

**(a) Explain two features to distinguish between the interference pattern in Young's double slit experiment with the diffraction pattern obtained due to a single slit.**

**(b) A monochromatic light of wavelength 500 nm is incident normally on a single slit of width 0.2 mm to produce a diffraction pattern. Find the angular width of the central maximum obtained on the screen.**

**Estimate the number of fringes obtained in Young's double slit experiment with fringe width 0.5 mm, which can be accommodated with the region of total angular spread of the central maximum due to single slit.** **[5]**

**Ans. (a) Unpolarised light :** If the electric component of light vibrates in all the direction in the plane perpendicular to the direction of propagation, the light is known as unpolarised light.

**Linearly polarised light :** If the vibration of electric components of light are restricted in one direction only in the plane perpendicular to the propagation of light then light is known as linearly polarised light.

A polaroid consists of long chain molecules aligned in a particular direction. The electric vectors along the direction of the aligned molecules get absorbed. So, when an unpolarised light falls on a polaroid, it lets only those of its electric vectors that are oscillating along a direction perpendicular to its aligned molecules to pass through it. The incident light thus gets linearly polarised.

**(b)**

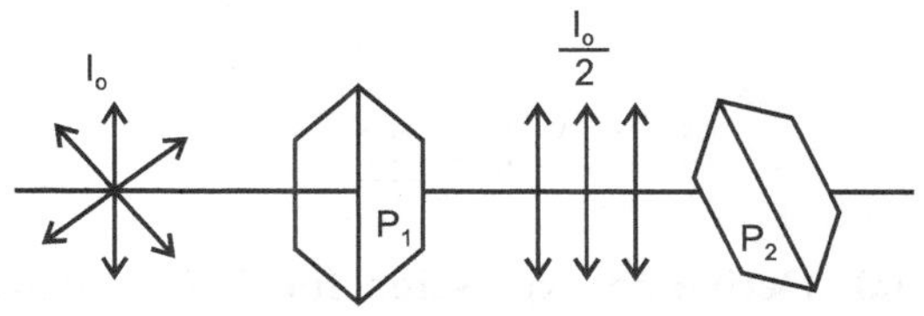


Intensity of light after passing through the polaroid $P_1$ (I') $= \frac{I_0}{2}$

Intensity of light after passing through the polaroid $P_2$, according to Malus law

$$I = I_0 \cos^2\theta$$

$$I = I' \cos^2 60° \qquad [\therefore \theta = 60°]$$

$$I = \frac{I_0}{2}\left(\frac{1}{2}\right)^2$$

$$I = \frac{I_0}{2} \times \frac{1}{4}$$

$$I = \frac{I_0}{8}$$

**OR**

**(a)**

| | Interference | Diffraction |
|---|---|---|
| **1.** | All bright and dark fringes are of equal width. | The width of central bright fringe is twice the width of any secondary maxima. |
| **2.** | All the bright fringes are of same intensity. | Intensity of bright fringes decreases as we move away from central bright fringe. |

**(b)** Angular width of central maxima, $\omega = \frac{2\lambda}{a}$

$$= \frac{2 \times 500 \times 10^{-9}}{0.2 \times 10^{-3}}$$

$$= 5 \times 10^{-3} \text{ rad.}$$

Now, fringe width,

$$\beta = \frac{\lambda D}{d}$$

Linear width of central maxima in the diffraction pattern 0.02

$$w' = \frac{2\lambda D}{a}$$

Let 'n' be the number of interference fringes which can be accommodated in the central maxima.

$$n \times \beta = w'$$

$$n = \frac{2\lambda D}{a} \times \frac{d}{\lambda D}$$

$$n = \frac{2d}{a}$$

**26. (a) Derive an expression for drift velocity of electrons in a conductor. Hence deduce Ohm's law.**

**(b) A wire whose cross-sectional area is increasing linearly from its one end to the other, is connected across a battery of V volts. Which of the following quantities remain constant in the wire ?**

**(i) drift speed (ii) current density**

**(iii) electric current (iv) electric field**

**Justify your answer. [5]**

**OR**

**(a) State the two Kirchhoff's laws. Explain briefly how these rules are justified.**

**(b) The current is drawn from a cell of emf E and internal resistance r connected to the network of resistors each of resistance r as shown in the figure. Obtain the expression for (i) the current drawn from the cell and (ii) the power consumed in the network.**

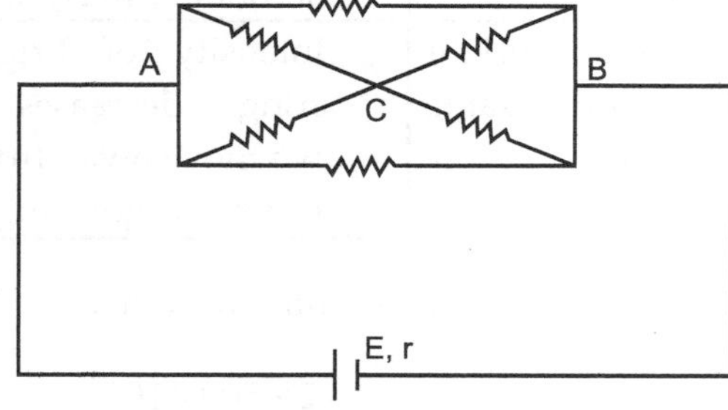


**Ans. (a) Expression of Drift velocity :** It is defined as the average velocity gained by the free electrons of a conductor in the opposite direction of the externally applied electric field.

If an electron have initial velocity $u_1$ and accelerated for time $\tau_1$ then velocity attain by it will be

$$\vec{v}_1 = \vec{u}_1 + \vec{a}\tau_1$$

Similarly

$$\vec{v}_2 = \vec{u}_2 + \vec{a}\tau_2$$

$$\vec{v}_3 = \vec{u}_3 + \vec{a}\tau_3$$

$$\vec{v}_n = \vec{u}_n + \vec{a}\tau_n$$

Then average velocity,

$$\vec{v}_d = \frac{\vec{v}_1 + \vec{v}_2 + \vec{v}_3 + ....\vec{v}_n}{n}$$

$$\vec{v}_d = \frac{(\vec{u}_1 + \vec{a}\tau_1) + (\vec{u}_2 + \vec{a}\tau_2) + ......(\vec{u}_n + \vec{a}\tau_n)}{n}$$

$$\vec{v}_d = \frac{(\vec{u}_1 + \vec{u}_2 + .....\vec{u}_n)}{n} + \frac{\vec{a}(\tau_1 + \tau_2 + .....\tau_n)}{n}$$

$$\vec{v}_d = 0 + \vec{a}\tau$$

$$\vec{v}_d = \vec{a}.\tau$$

where $\tau = \frac{\tau_1 + \tau_2 + ... + \tau_n}{n}$ = average relaxation time.

$$\vec{v}_d = \frac{-e\vec{E}\tau}{m} \qquad \left[\because \vec{a} = \frac{-e\vec{E}}{m}\right]$$

$$|\vec{v}_d| = \left|\frac{-e\vec{E}\tau}{m}\right|$$

Now,

$$v_d = \frac{eE\tau}{m}$$

$$v_d = \frac{eV\tau}{ml} \qquad \left[\because E = \frac{V}{l}\right]$$

Now charge flowing through conductor at an instant is

and $\quad I = neAv_d$

$$I = neA\frac{eV\tau}{ml}$$

$$\frac{V}{I} = \frac{ml}{ne^2\tau A}$$

If physical condition remain unchanged then right hand side will be constant.

$$\therefore \quad \frac{V}{I} = \text{Constant}$$

$$\frac{V}{I} = R \text{ or } V = IR$$

Hence, Ohm's law is derived.

**(b)** (iii) The electric current will remain constant in the wire as it does not depend upon the cross-sectional area whereas the drift speed, current density and electric field depend upon the increasing area of cross-section with the following relations :

Drift speed, $v_d = \frac{I}{neA}$

Current density, $J = \frac{I}{A}$

Electric field, $E = \frac{J}{\sigma}$

**OR**

**(a) Kirchhoff's laws :**

1. **Kirchhoff's first law or junction rule :** In an electric circuit, the algebraic sum of currents at any junction is zero.

$$\Sigma I = 0$$

2. **Kirchhoff's second law or loop rule :** Around any closed loop of a network, the algebraic sum of changes in the potential must be zero.

$$\Sigma \Delta V = 0$$

**Justification :** First law is based on the law of conservation of charge and second law is based on the law of conservation of energy.

**(b)** (i)

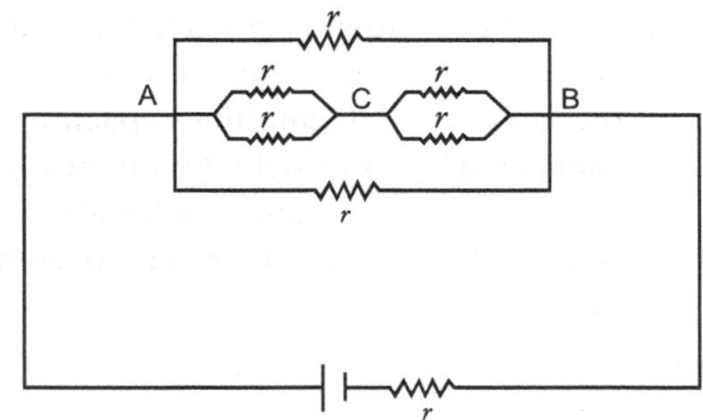


The equivalent resistance between AC and CB is given by

$$\frac{1}{r'} = \frac{1}{r} + \frac{1}{r}$$

$$\frac{1}{r'} = \frac{2}{r}$$

$$r' = \frac{r}{2}$$

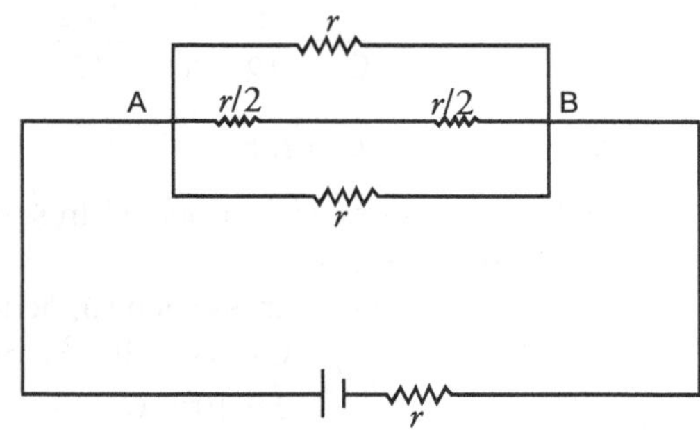


The equivalent resistance of middle branch is given by,

$$r'' = \frac{r}{2} + \frac{r}{2} = r$$

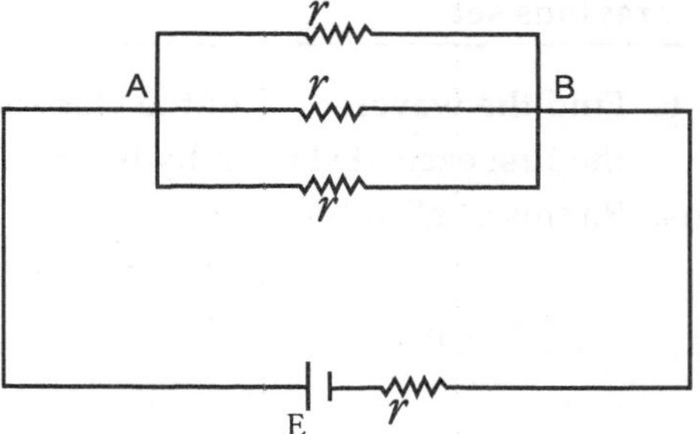


The equivalent resistance between A and B is given by

$$\frac{1}{r'''} = \frac{1}{r} + \frac{1}{r} + \frac{1}{r}$$

$$r''' = \frac{r}{3}$$

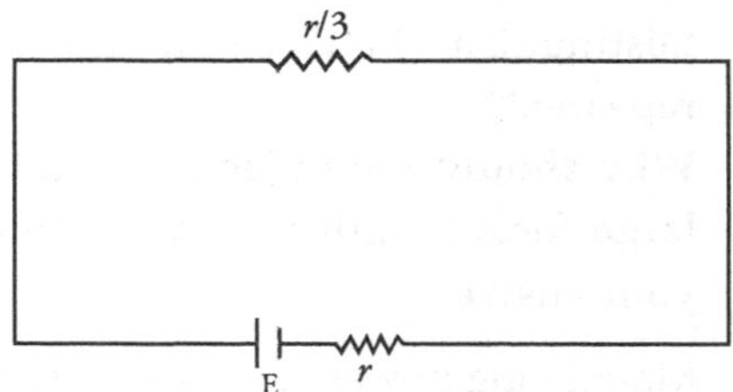


Total resistance of the circuit is given by,

$$R = \frac{r}{3} + r$$

$$R = \frac{4r}{3}$$

Now current through the circuit is given by

$$I = \frac{E}{\frac{4r}{3}}$$

$$I = \frac{3E}{4r}$$

(ii) Power consumption

$$= I^2R$$

$$= \left(\frac{3E}{4r}\right)^2 \frac{4r}{3}$$

$$P = \frac{3E^2}{4r}$$

●●

# Delhi [Set-II]

**Note : Except for the following questions, all the remaining questions have been asked in previous set.**

**6. Find the wavelength of the electron orbiting in the first excited state in hydrogen atom. [2]**

**Ans.** Radius of n[th] orbit,

$$r_n = r_0 n^2 = 0.53\, n^2 \text{ Å}$$

For 1[st] excited state $n = 2$.

$$\therefore \quad r_2 = 0.53 \times 4 \text{ Å} = 2.12 \text{ Å}$$

For an electron revolving in n[th] orbit, according to de Broglie relation,

$$2\pi r_n = n\lambda,$$

$$2 \times 3.14 \times 2.12 \times 10^{-10} = 2\lambda$$

$$\lambda = 3.14 \times 2.12 \times 10^{-10} \text{ m} = 6.67 \text{ Å}$$

**7. Distinguish between a transducer and a repeater.** [2]**

**10. Why should the objective of a telescope have large focal length and large aperture ? Justify your answer. [2]**

**Ans.** Magnifying power of telescope, $m = -\dfrac{f_0}{f_e}$

$\therefore$ For large magnifying power $f_0$ should be large.

Also, Resolving proper of telescope $= \dfrac{A}{1.22\lambda}$

$\therefore$ Aperture should also be large to increase resolving power.

**12. In the study of a photoelectric effect the graph between the stopping potential V and frequency ν of the incident radiation on two different metals P and Q is shown below : [3]**

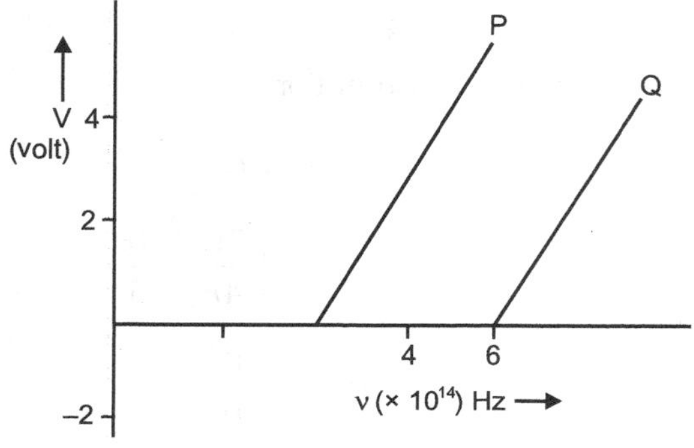


**(i) Which one of the two metals has higher threshold frequency ?**

**(ii) Determine the work function of the metal which has greater value.**

**(iii) Find the maximum kinetic energy of electron emitted by light of frequency $8 \times 10^{14}$ Hz for this metal.**

**Ans.** (i) Q has higher threshold frequency.

(ii) Work function of Q

$$\phi_0 = h\nu_0$$

$$h\nu_0 = (6.6 \times 10^{-34}) \times \frac{6\times10^{14}}{1.6\times10^{-19}} \text{ eV} = 2.5 \text{ eV}$$

(iii) $K_{max} = h(\nu - \nu_0)$

$$= \frac{6.6\times10^{-34}(8\times10^{14} - 6\times10^{14})}{1.6\times10^{-19}} \text{ eV}$$

$$= \frac{6.6\times10^{-34} \times 2\times10^{14}}{1.6\times10^{-19}} \text{ eV}$$

$$K_{max} = 0.83 \text{ eV}$$

**13. A 12 pF capacitor is connected to a 50 V battery. How much electrostatic energy is stored in the capacitor ? If another capacitor of 6 pF is connected in series with it with the same battery connected across the combination, find the charge stored and potential difference across each capacitor. [3]**

**Ans.** Energy stored in the capacitor of capacitance 12 pF,

$$U = \frac{1}{2}CV^2$$

$$= \frac{1}{2} \times 12 \times 10^{-12} \times 50 \times 50 \text{ J}$$

$$= 1.5 \times 10^{-8} \text{ J}$$

Let C be the equivalent capacitance of 12 pF and 6 pF capacitor connected in series and it is given by

$$\frac{1}{C} = \frac{1}{12} + \frac{1}{6} = \frac{1+2}{12}$$

$$\therefore \quad C = 4pF$$

As the capacitors are connected in series, so the charge remains same

$\therefore$ Charge stored across each capacitor,

$$q = CV = 4 \times 10^{-12} \times 50 = 2 \times 10^{-10} \text{ C}$$

** Answer is not given due to change in the present Syllabus.

Potential difference across capacitor $C_1$,

$$V_1 = \frac{2\times10^{-10}}{12\times10^{-12}} \text{ volt} = \frac{50}{3} \text{ V}$$

Potential difference across capacitor $C_2$,

$$V_2 = \frac{2\times10^{-10}}{6\times10^{-12}} \text{ volt} = \frac{100}{3} \text{ V}$$

**18. A zener diode is fabricated by heavily doping both *p* and *n*- sides of the junction. Explain, why ? Briefly explain the use of zener diode as a dc voltage regulator with the help of a circuit diagram. [3]**

**Ans.** By heavily doping both p and n sides of the junction, depletion region formed is very thin, *i.e.* $< 10^{-6}$ m. Hence, electric field across the junction is very high ($\sim 5 \times 10^6$ V/m) even for a small reverse bias voltage. This can lead to a break down during reverse biasing.

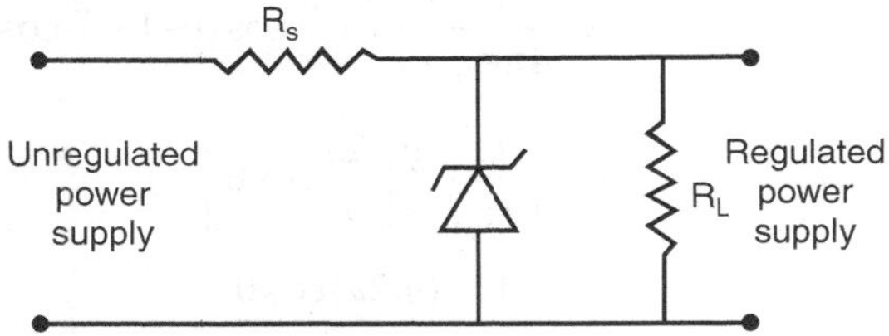


If the input voltage increases/decreases then current through resistor $R_S$ and Zener diode also increases/decreases. This increases/decreases the voltage drop across $R_S$ without any change in voltage across the zener diode.

This is because in the breakdown region, zener voltage remains constant even though the current through the zener diode changes.

**21. A electron of mass $m_e$ revolves around a nucleus of charge +Ze. Show that it behaves like a tiny magnetic dipole. Hence prove that the magnetic moment associated with it is expressed as**

$$\vec{\mu} = -\frac{e}{2m_e}\vec{L}$$

**, where $\vec{L}$ is the orbital angular momentum of the electron. Give the significance of negative sign. [3]**

**Ans.** Electron in circular motion around the nucleus, constitutes a current loop which behaves like a magnetic dipole.

Current associated with the revolving electron,

$$I = \frac{e}{T}$$

and $$T = \frac{2\pi r}{v}$$

$\therefore$ $$I = \frac{e}{2\pi r}v$$

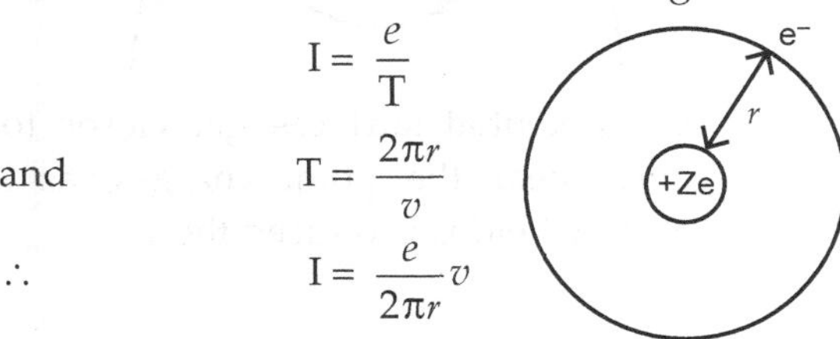


Magnetic moment of the loop,

$$\mu = IA$$

$$\mu = IA = \frac{ev}{2\pi r}\pi r^2$$

$$= \frac{evr}{2} = \frac{e.m_e vr}{2m_e} \quad \text{(i)}$$

Orbital angular momentum of the electron,

$$L = m_e vr$$

Put the value of L in equation (i) (ii)

$$\vec{\mu} = \frac{-e}{2m_e}\vec{L}$$

The negative sign signifies that the angular momentum of the revolving electron is opposite in direction to the magnetic moment associated with it.

**22. (a) Derive the expression for the electric potential due to an electric dipole at a point on its axial line.**

**(b) Depict the equipotential surfaces due to an electric dipole. [3]**

**Ans. (a)**

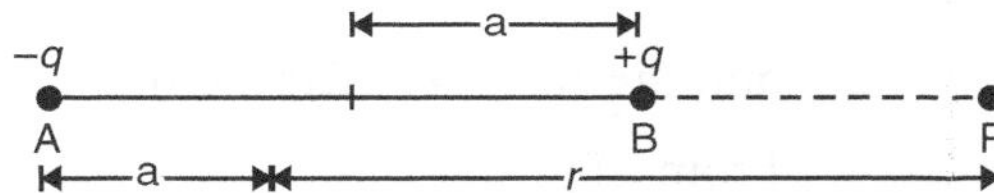


Potential due to charge at A, $V_A = \frac{1}{4\pi\varepsilon_0}\frac{-q}{(r+a)}$

Potential due to charge at B, $V_B = \frac{1}{4\pi\varepsilon_0}\frac{-q}{(r+a)}$

$\therefore$ Potential at point P.

$$V = V_B + V_A$$

Net potential at P $= \frac{q}{4\pi\varepsilon_0}\left[\frac{-1}{(r+a)} + \frac{1}{(r-a)}\right]$

$$V = \frac{q\times 2a}{4\pi\varepsilon_0(r^2 - a^2)}$$

Consider an electric dipole having charges $-q$ and $+q$ at separation '$2a$'. The dipole moment of dipole is $\vec{p} = q(2\vec{a})$, directed from $-q$ to $+q$.

The electric potential due to dipole is the algebraic sum of potentials due to charges $+q$ and $-q$.

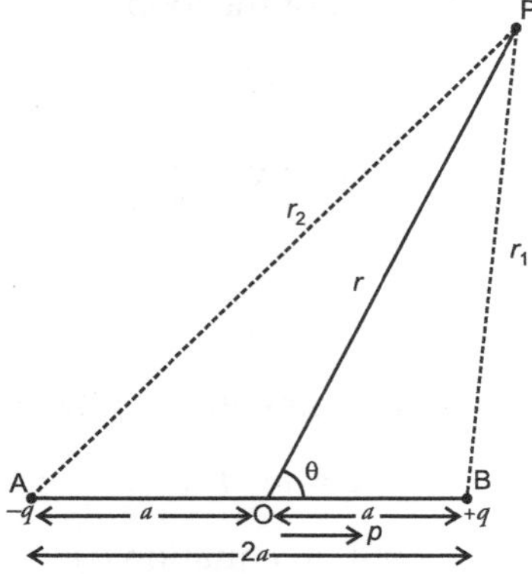


If $r_1$ and $r_2$ are distances of any point P from charge $+q$ and $-q$ respectively as shown in the figure, then the potential due to electric dipole at point P, is

$$V = \frac{1}{4\pi\epsilon_0}\frac{q}{r_1} - \frac{1}{4\pi\epsilon_0}\frac{q}{r_2}$$

$$= \frac{q}{4\pi\epsilon_0}\left[\frac{1}{r_1} - \frac{1}{r_2}\right] \quad ...(i)$$

If $(r, \theta)$ are polar coordinates of point P with respect to mid-point O of dipole, then

By geometry,

$$r_1^2 = r^2 + a^2 - 2ar\cos\theta \quad ...(ii)$$

and $$r_2^2 = r^2 + a^2 + 2ar\cos\theta \quad ...(iii)$$

From (ii),

$$r_1^2 = r^2\left[1 - \frac{2a\cos\theta}{r} + \frac{a^2}{r^2}\right]$$

If $r >> a$ *i.e.*, $\frac{a}{r} << 1$, then it is sufficient to retain terms only upto first order in $\left(\frac{a}{r}\right)$.

$$\therefore \quad r_1^2 = r^2\left[1 - \frac{2a\cos\theta}{r}\right]$$

$$\Rightarrow \quad r_1 = r\left[1 - \frac{2a\cos\theta}{r}\right]^{1/2} \quad ...(iv)$$

Similarly from (iii),

$$r_2^2 = r^2\left[1 + \frac{2a\cos\theta}{r}\right]$$

$$\Rightarrow \quad r_2 = r\left[1 + \frac{2a\cos\theta}{r}\right]^{1/2} \quad ...(v)$$

From (iv) and (v),

$$\frac{1}{r_1} = \frac{1}{r}\left[1 - \frac{2a\cos\theta}{r}\right]^{-1/2}$$

and $$\frac{1}{r_2} = \frac{1}{r}\left[1 + \frac{2a\cos\theta}{r}\right]^{-1/2}$$

Using binomial theorem and retaining terms upto first order in $\frac{a}{r}$ only, we have

$$\frac{1}{r_1} = \frac{1}{r}\left[1 - \left(-\frac{1}{2}\right)\frac{2a\cos\theta}{r}\right]$$

$$= \frac{1}{r}\left[1 + \frac{a}{r}\cos\theta\right] \quad ...(vi)$$

and $$\frac{1}{r_2} = \frac{1}{r}\left[1 - \frac{a}{r}\cos\theta\right] \quad ...(vii)$$

Substituting these values in (i), we get

$$V = \frac{q}{4\pi\epsilon_0}\left[\frac{1}{r}\left(1 + \frac{a}{r}\cos\theta\right) - \frac{1}{r}\left(1 - \frac{a}{r}\cos\theta\right)\right]$$

$$= \frac{1}{4\pi\epsilon_0}\frac{q}{r}\left[1 + \frac{a}{r}\cos\theta - 1 + \frac{a}{r}\cos\theta\right]$$

$$= \frac{1}{4\pi\epsilon_0}\frac{q}{r}\left[\frac{2a}{r}\cos\theta\right]$$

$$= \frac{1}{4\pi\epsilon_0}\frac{(q.2a)\cos\theta}{r^2}$$

or $$V = \frac{1}{4\pi\epsilon_0}\frac{p\cos\theta}{r^2} \quad (\because p = q.2a) \quad ...(viii)$$

When point P lies on the axis of dipole, then $\theta = 0°$

$$\therefore \quad \cos\theta = \cos 0° = 1$$

$$\therefore \quad V = \frac{1}{4\pi\varepsilon_0}.\frac{p}{r^2}$$

**(b)**

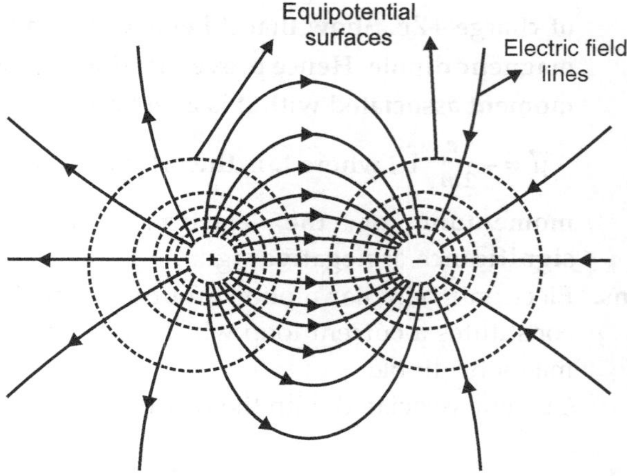


Equipotential surfaces get closer to each other near the point charges, as strong electric field is produced there.

# Delhi [Set-III]

**Note : Except for the following questions, all the remaining questions have been asked in previous sets.**

**7. When are two objects just resolved ? Explain. How can the resolving power of a compound microscope be increased ? Use relevent formula to support your answer. [2]**

**Ans.** When the central maxima of diffraction pattern from one object coincide with the first minima of second object then they are said to be just resolved.

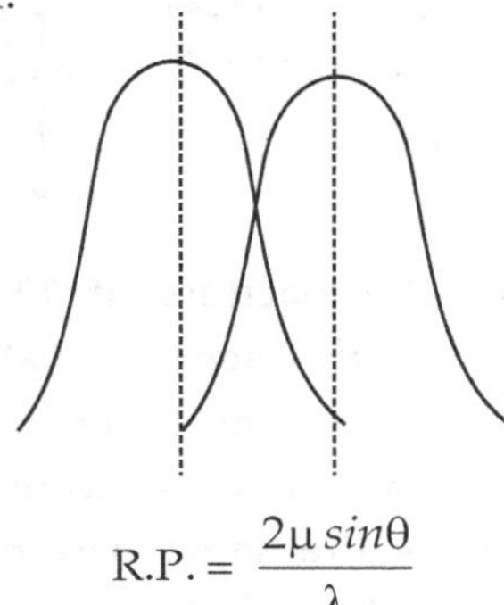

$$\text{R.P.} = \frac{2\mu \sin\theta}{\lambda}$$

The resolving power of a compound microscope can be increased by increasing $\mu$ and by decreasing $\lambda$.

**8. (a) What is the line of sight communication ?****

**(b) Why is it not possible to use sky waves for transmission of T.V. signals ? Upto what distance can a signal be transmitted using an antenna of height '$h$' ?** [2]**

**9. An $\alpha$-particle and a proton are accelerated through the same potential difference. Find the ratio of their de Broglie wavelengths. [2]**

**Ans.** de Broglie wavelength,

$$\lambda = \frac{h}{P} = \frac{h}{\sqrt{2mK}} = \frac{h}{\sqrt{2mqV}}$$

$$\therefore \quad \frac{\lambda_\alpha}{\lambda_p} = \sqrt{\frac{m_p q_p}{m_\alpha q_\alpha}}$$

$$= \sqrt{\frac{m_p\, q_p}{4m_p\, 2q_p}} \qquad \left[\begin{matrix}\because m_\alpha\ 4m_p \\ q_\alpha = 2q_p\end{matrix}\right]$$

$$\frac{\lambda_\alpha}{\lambda_p} = \sqrt{\frac{1}{8}} = \frac{1}{2\sqrt{2}}$$

$$\therefore \quad \lambda_a : \lambda_p = 1 : 2\sqrt{2}$$

**14. (a) State two important features of Einstein's photoelectric equation.**

**(b) Radiation of frequency $10^{15}$ Hz is incident on two photosensitive surfaces P and Q. There is no photoemission from surface P. Photoemission occurs from surface Q but photoelectrons have zero kinetic energy. Explain these observations and find the value of work function for surface Q. [3]**

**Ans. (a)** 1. By Einstein's photoelectric equation, we can explain the laws of photoelectric emission.

2. By this, we can find the value of Planck's constant and work function.

**(b)** Given : $\nu = 10^{15}$ Hz

$$\text{Wavelength, } \lambda = \frac{c}{\nu} = \frac{3\times10^8}{10^{15}} = 300 \text{ nm}$$

$$\text{Energy, E} = \frac{hc}{\lambda} = \frac{6.6\times10^{-34}\times3\times10^8}{300\times10^{-9}}$$

$$= \frac{6.6\times10^{-19}}{1.6\times10^{-19}} \text{ eV}$$

$$= 4.125 \text{ eV}$$

The work function of P is more than the 4.125eV. Therefore, there is no photoemission from surface P.

For surface Q, the kinetic energy is zero,

$$\therefore \quad \text{K.E}_{max} = \text{E} - \phi_0$$

$$0 = \text{E} - \phi_0$$

$\Rightarrow$ Work function, $\phi_0 = \text{E} = 4.15$ eV

**16. (a) Obtain the expression for the torque $\vec{\tau}$ experienced by an electric dipole of dipole moment $\vec{p}$ in a uniform electric field, $\vec{E}$.**

**(b) What will happen if the field were not uniform ? [3]**

**Ans. (a)**

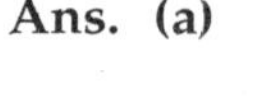

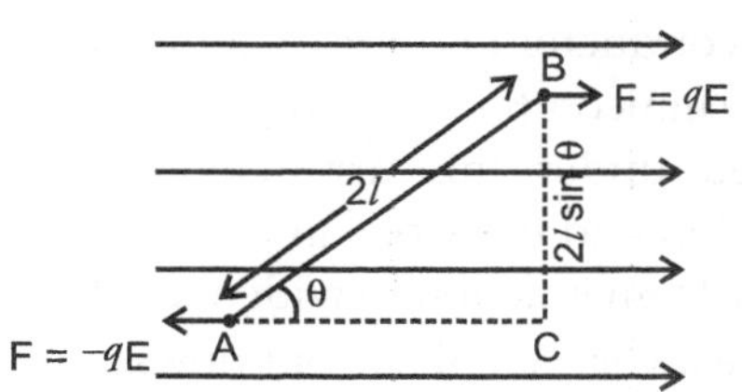


** Answer is not given due to change in the present Syllabus.

$\tau$ = Force × Perpendicular distance

$\tau = F \times BC$

$\tau = qE \times 2l \sin\theta$

$\tau = pE \sin\theta \qquad (\because p = q \times 2l)$

$\vec{\tau} = \vec{p} \times \vec{E}$

**(b)** If the field is non-uniform then there will be a net force acting on the dipole. Also, a net torque acting on the dipole which depends on the location of the dipole.

**17. Explain briefly with the help of necessary diagrams, the forward and the reverse biasing of a p-n junction diode. Also draw their characteristic curves in the two cases. [3]**

**Ans. Forward biasing :** If the positive terminal of the battery is connected to the p type semiconductor and negative terminal to the n type semiconductor then it is said to be in forward biased under this arrangement holes move from *p*-region to *n*-region and electrons moves opposite and thus curent is setup.

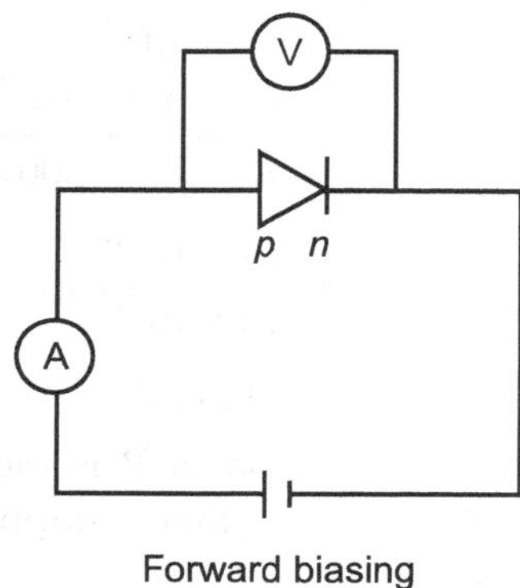

Forward biasing

**Characteristics :**

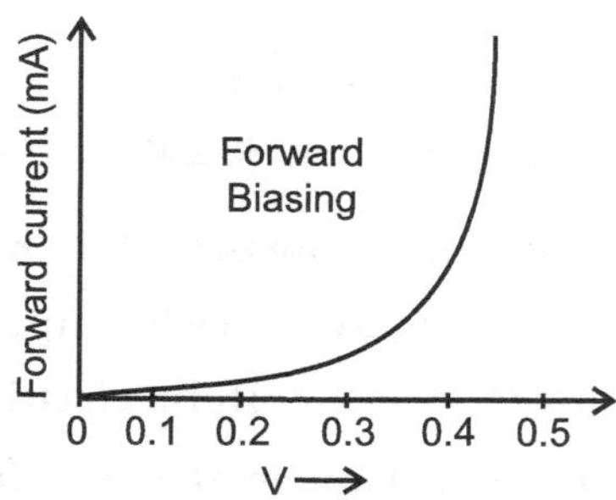


**Reverse Biasing :** If the positive terminal of the battery is connected to the n type semiconductor and the negative termal to the p type semiconductor then it is said to be reverse biased under this arrangement the holes in *p*-region and electrons in *n* – region are pushed away from junction to widen the depletion layer hence a very small current flows.

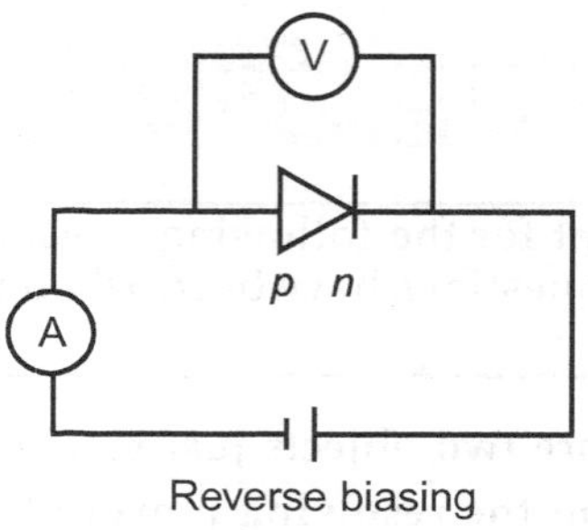

Reverse biasing

**Characteristics :**

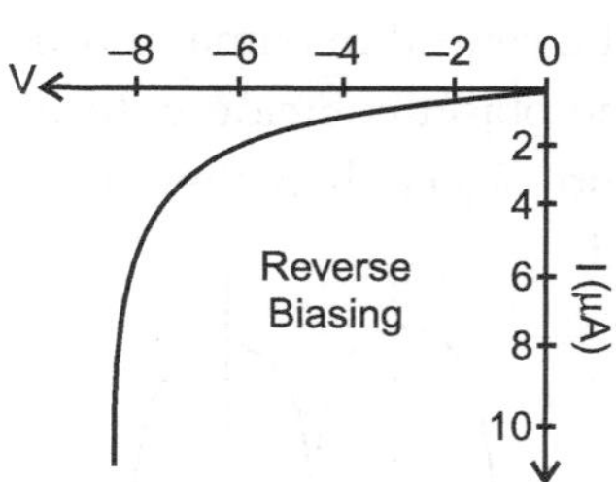


**20. Two identical capacitors of 12 pF each are connected in series across a battery of 50 V. How much electrostatic energy is stored in the combination ? If these were connected in parallel across the same battery, how much energy will be stored in the combination now ? Also find the charge drawn from the battery in each case. [3]**

**Ans. Case I :** When the capacitors are connected in series.

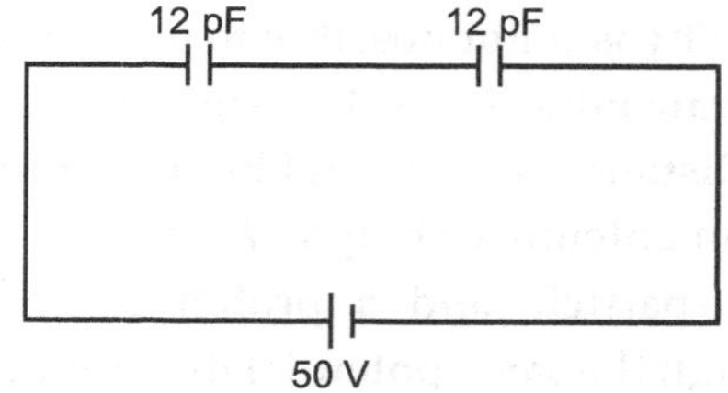


The equivalent capacitance is given by

$$C_{eq} = \frac{12 \times 12}{12 + 12} = \frac{12 \times 12}{24}$$

$$C_{eq} = 6 \text{ pF} = 6 \times 10^{-12} \text{F}$$

Electrostatic energy,

$$U = \frac{1}{2} CV^2$$

$$= \frac{1}{2} \times 6 \times 10^{-12} \times 50 \times 50$$

$$U = 75 \times 10^{-10} \text{ J.}$$

As the capacitors are connected in series so, the charge remains the same.

$\therefore \quad Q = C_{eq}V$

$Q = 6 \times 10^{-12} \times 50$

$Q = 300 \text{ pF} = 300 \times 10^{-12}\text{C}$

**Case II :** When the capacitors are connected is parallel.

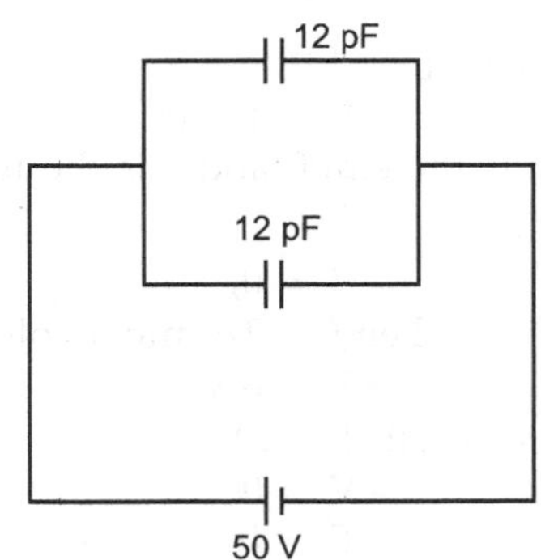


The equivalent capacitance is given by,

$C_{eq} = (12 + 12) \text{ pF}$

$= 24 \text{ pF} = 24 \times 10^{-12}\text{F}$

Electrostatic energy,

$$U = \frac{1}{2} \times 24 \times 10^{-12} \times (50)^2$$

$U = 3 \times 10^{-8} \text{ J}$

As the capacitors are connected in parallel so, the charge on each capacitor is different but potential difference remains the same.

Charge on $C_1$,

$Q_1 = C_1V$

$Q_1 = 12 \times 10^{-12} \times 50$

$= 600 \times 10^{-12} \text{ F}$

Similarly, $Q_2 = 600 \times 10^{-12} \text{ F}$

Total charge, $Q = Q_1 + Q_2$

$= (600 + 600) \times 10^{-12}$

$= 1200 \times 10^{-12}$

$= 12 \times 10^{-10} \text{ F}$

**21. (a) Write the expression for the force $\vec{F}$ acting on a particle of mass m and charge q moving with velocity $\vec{v}$ in a magnetic field $\vec{B}$. Under what conditions will it move in (i) a circular path and (ii) a helical path ?**

**(b) Show that the kinetic energy of the particle moving in magnetic field remains constant.** **[3]**

**Ans. (a)** $\vec{F} = \theta(\vec{v} \times \vec{B}) = qvB \sin\theta$

**(i)** If the angle between v and B is 90°, then it will move in circular path.

**(ii)** If the angle is other than 0°, 90° and 180°, the path will be helical.

**(b)** Since the work done on the charged particle moving in the magnetic field is zero because force experienced is perpendicular which cannot bring any change in speed of charged particles. Hence the change in K.E. is zero.

●●

# PHYSICS

## Outside Delhi [Set-I] 2016

**Time allowed : 3 hours** **Maximum marks : 70**

### SECTION — A

1. **A charge '$q$' is moved from a point A above a dipole of dipole moment '$p$' to a point B below the dipole in equatorial plane without acceleration. Find the work done in the process. [1]**

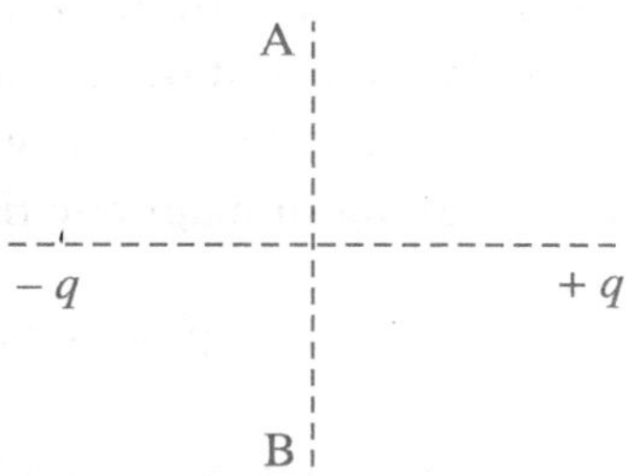


**Ans.** Work done, $W = q(V_B - V_A) = q \times 0 = 0$

2. **In what way is the behaviour of a diamagnetic material different from that of a paramagnetic, when kept in an external magnetic field ? [1]**

**Ans.** The magnetic field lines bends towards the paramagnetic material while the magnetic field lines move away from the diamagnetic material when placed in external magnetic field or paramagnetic material get aligned along B and diamagnetic aligned perpendicular to B.

3. **Name the essential components of a communication system.** [1]**

4. **Why does sun appear red at sunrise and sunset? [1]**

**Ans.** Sun appears red at sunrise and sunset due to the least scattering of red light as it has the longest wavelength.

5. **The plot of the variation of potential difference across a combination of three identical cells in series, versus current is shown below. What is the emf and internal resistance of each cell ? [1]**

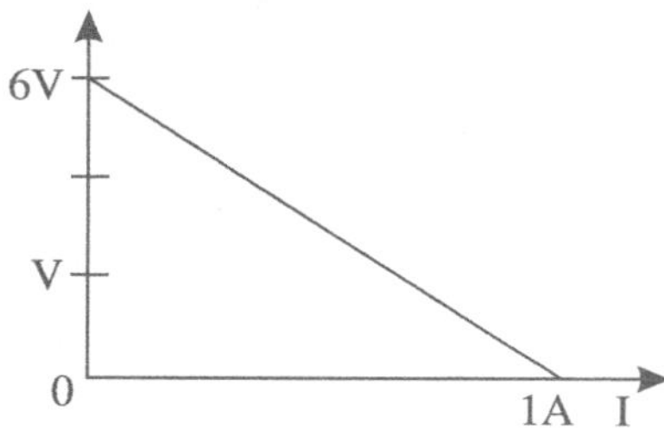


**Ans.** We know that,

$$V = E - Ir$$

where E is the e.m.f. and $r$ is the total internal resistance.

When $I = 0$,

Total emf = Terminal voltage

$3E = 6$ V

emf. of each cell $E = 2$ V

When $V = 0$

$E = Ir$

$$r = \frac{\text{Total E.m.f}}{I} = \frac{6}{1} = 6\ \Omega$$

As the cells are connected in series. So, the internal resistance of each cell = 2Ω.

### SECTION–B

6. **Define modulation index. Why is it kept low ? What is the role of a bandpass filter ? [2]**

**Ans.** **Modulation index :** It is the ratio of the amplitude of the modulating wave to that of the carrier wave.

So, $\mu = E_m / E_c$.

The modulation index is kept low because when $E_m$ is greater than $E_c$, the carrier wave is overmodulated, *i.e.*, $\mu > 1$, which causes distortion during the reception.

The main function of a bandpass filter is to limit the bandwidth of the output signal to the band allocated for the transmission.

7. **A ray PQ incident normally on the refracting face BA is refracted in the prism BAC made of material of refractive index 1.5. Complete the path of ray through the prism. From which face will the ray emerge ? Justify your answer. [2]**

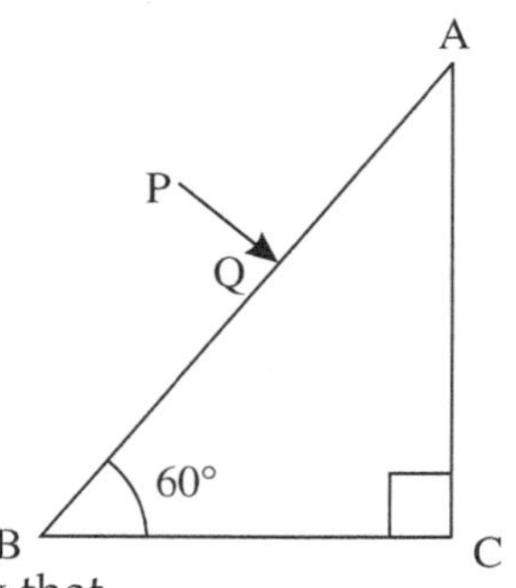


**Ans.** We know that,

$$n = \frac{1}{\sin C}$$

**Answer is not given due to change in the present Syllabus.

where, $n$ = Refractive index
C = Critical angle.

So, $\sin C = 1/n = 1/1.5 = 0.667$
$\sin C = \sin 41° 49'$
$C = 41° 49'$
$C > 30°$

Critical Angle on face AC is greater than incident angle 30°

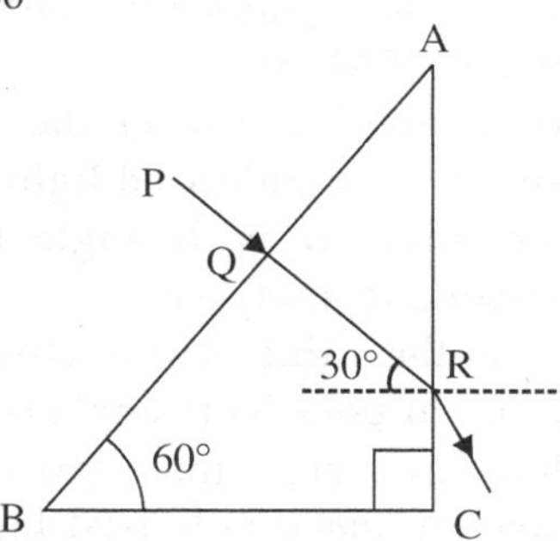


Thus, the light ray PQ will emerge out from face AC.

**8. Calculate the de-Broglie wavelength of the electron orbitting in the $n = 2$ state of hydrogen atom. [2]**

**Ans.** The velocity of an orbitting electron,

$$v = \frac{Zc}{137n}$$

Where, $c$ = Speed of light in vacuum.

So, $$v = \frac{1 \times 3 \times 10^8}{137 \times 2} = 1.09 \times 10^6 \text{ m/s.}$$

The de-Broglie wavelength, $\lambda = h/mv$

Where $h$ is Planck's constant $= 6.6 \times 10^{-34}$ Js

and mass of electron $= 9.1 \times 10^{-31}$ kg

So, $\lambda = 6.6 \times 10^{-34}/(9.1 \times 10^{-31} \times 1.09 \times 10^6)$
$= 6.65 \times 10\text{m}^{-10}$

**9. Define ionization energy.**

**How would the ionization energy change when electron in hydrogen atom replaced by a particle of mass 200 times that of the electron but having the same charge? [2]**

**OR**

**Calculate the shortest wavelength of the spectral lines emitted in Balmer series. [Given Rydberg constant, R= $10^7$ m$^{-1}$]**

**Ans.** Ionization energy is defined as the amount of energy needed to remove the valence electron of an isolated gaseous atom.

The ionization energy of hydrogen atom is

$$E_H \propto m_e \quad \text{...(i)}$$

When mass of electron is replaced by a particle having mass = 200 $m_e$, then

$$E'_H \propto 200\, m_e \quad \text{...(ii)}$$

From (i) and (ii), we have

$$E'_H = 200\, E_H = 200 \times (-13.6) = -2720 \text{ eV}$$

$(\because E_H = -13.6 \text{ eV})$

**OR**

For Balmer series, wavelength is given by :

$$\frac{1}{\lambda} = R[1/2^2 - 1/n^2]$$

Where, $n = 3, 4, 5$

For shortest wavelength, $n = \infty$

So, $$\frac{1}{\lambda} = R\left[\frac{1}{2^2} - \frac{1}{\infty^2}\right]$$

$$\frac{1}{\lambda} = 10^7\left[\frac{1}{4}\right]$$

$$\lambda = 4 \times 10^{-7} \text{ m.}$$

**10. A battery of emf 12 V and internal resistance 2 Ω is connected to a 4 Ω resistor as shown in the figure. [2]**

**(a) Show that a voltmeter when placed across the cell and across the resistor, in turn, gives the same reading.**

**(b) To record the voltage and the current in the circuit, why is voltmeter placed in parallel and ammeter in series in the circuit ?**

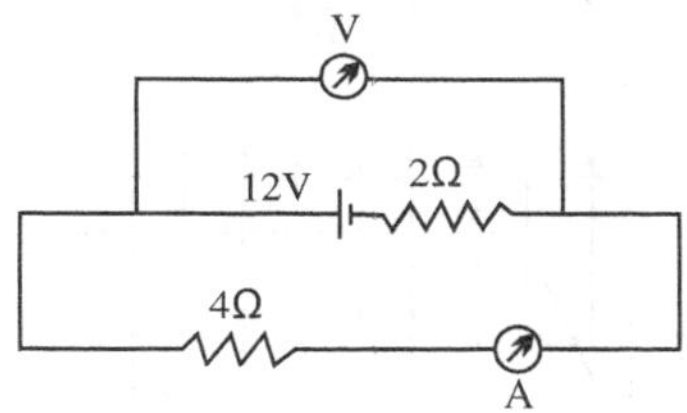


**Ans.** We know that,

**(a)** Effective resistance of the circuit $R_E = 6\ \Omega$

$$\therefore \quad I = \frac{12}{6} = 2\text{A}$$

Terminal potential difference across the cell can be calculated as,

$$V = E - Ir = 12 - 2 \times 2 = 12 - 4 = 8 \text{ V}$$

Also, Potential difference across 4Ω resistor can be calculated as

$$V = IR = 2 \times 4 = 8 \text{ V.}$$

So, a voltmeter when placed across the cell and across the resistor, gives the same reading.

**(b)** An ammeter is connected in series because it has very low resistance. So, when, an ammeter is connected in series, then there is not much increase in the resistance of the circuit and hence the current through the circuit unchanged.

A voltmeter is connected in parallel because it has very high resistance. So, it draws a very small current from the circuit.

## SECTION–C

**11. Define an equipotential surface. Draw equipotential surfaces : [3]**

**(i) in the case of a single point charge and**

**(ii) in a constant electric field in Z-direction. Why the equipotential surfaces about a single charge are not equidistant ?**

**(iii) Can electric field exist tangential to an equipotential surface ? Give reason.**

**Ans.** An equipotential surface is the surface which has same potential at its every point.

(i)

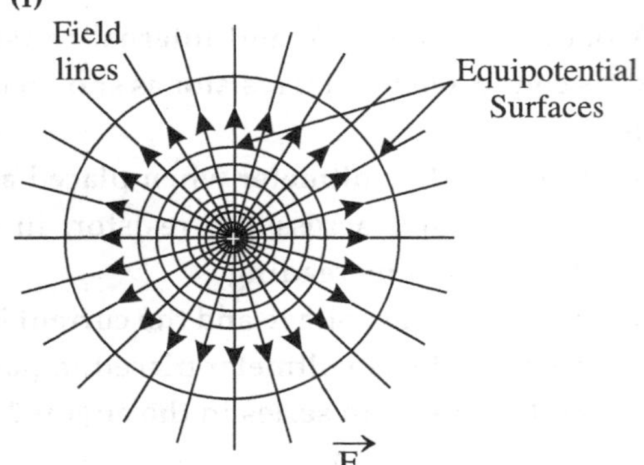


(ii)

The electric field due to single charge is not constant, this is the reason why the equipotential surfaces about a single charge are not equidistant and potential vary inversely with radius i.e.,

$$V \propto \frac{1}{r}$$

**(iii)** No, electric field cannot exist tangential to an equipotential surface. If it happen then a charged particle will experience a force along the tangential line and can move along it. As a charged particle can move only due to the potential difference *i.e.,* along the direction of change of potential, this contradicts the concept of an equipotential surface.

**12. (i) State law of Malus.**

**(ii) Draw a graph showing the variation of intensity (I) of polarised light transmitted by an analyser with angle (θ) between polariser and analyser.**

**(iii) What is the value of refractive index of a medium of polarising angle 60°? [3]**

**Ans. (i)** Malus discovered that when a beam of completely plane polarized light is passed through the analyser, the intensity 'I' of transmitted light changes directly as the square of the cosine of the angle θ between the transmission axes of the plane of polarizer and analyzer.

This is known as the law of Malus.

$$I \propto \cos^2 \theta$$

or $$I = I_0 \cos^2 \theta$$

Where, $I_0$ is the maximum intensity of the transmitted light.

(ii)

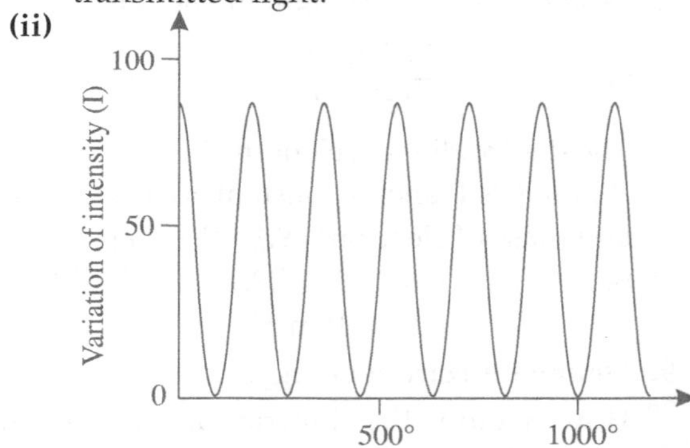


**(iii)** From the Brewster's law of polarization :

$$\mu = \tan i_p$$
$$= \tan 60^\circ$$
$$= \sqrt{3}$$
$$= 1.7320$$

Thus, the refractive index of the material is 1.73.

**13. Sketch the graphs showing variation of stopping potential with frequency of incident radiations for two photosensitive materials A and B having threshold frequencies $\nu_A > \nu_B$. [3]**

**(i) In which case is the stopping potential more and why ?**

**(ii) Does the slope of the graph depend on the nature of the material used ? Explain.**

**Ans.**

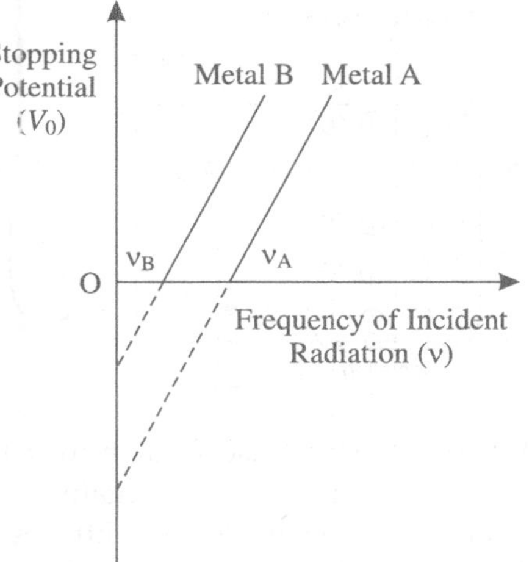


**(i)** We know that,

$$V_0 = \frac{h}{e}(\nu - \nu_0)$$

For the same value of ν, stopping potential is more for which threshold frequency ($\nu_0$) is less.

Given, $\nu_A > \nu_B$

Therefore, threshold frequency for metal B is less than the threshold frequency for metal A. Hence, stopping potential is more for metal B.

**(ii)** No, the slope of the graph tells us the value of $h/e$ which is same constent for both the materials. So, it does not depend on the nature of the materials.

**14. (a) Write the basic nuclear process involved in the emission of $\beta^+$ in a symbolic form, by a radioactive nucleus.**

**(b) In the reactions given below :**

**(i)** $^{11}_{6}\text{C} \longrightarrow {}^{z}_{y}\text{B} + x + \nu$

**(ii)** $^{12}_{6}\text{C} + {}^{12}_{6}\text{C} \longrightarrow {}^{20}_{a}\text{Ne} + {}^{c}_{b}\text{He}$

**Find the value of $x, y, z$ and $a, b, c$.** **[3]**

**Ans. (a)** In $\beta^+$-decay, the atomic number of the radioactive nucleus decreases by one and its mass number remains same. In this process, a positron ($e^+$) and a new particle neutrino ($\nu$) are emitted from the nucleus.

Generally, $^{A}_{Z}\text{X} \rightarrow {}_{Z-1}^{\;\;A}\text{Y} + e^+ + \nu$

**(b) (i)** $^{11}_{6}\text{C} \rightarrow {}^{z}_{y}\text{B} + x + \nu$

$^{11}_{6}\text{C} \rightarrow {}^{11}_{5}\text{B} + {}^{0}_{1}\beta + \nu$

The corresponding $y$ and $z$ are 5 and 11, respectively. The $x$ is the positron.

**(ii)** $^{12}_{6}\text{C} + {}^{12}_{6}\text{C} \rightarrow {}^{20}_{a}\text{Ne} + {}^{c}_{b}\text{He}$

$\Rightarrow {}^{12}_{6}\text{C} + {}^{12}_{6}\text{C} \rightarrow {}^{20}_{10}\text{Ne} + {}^{4}_{2}\text{He}$

The corresponding values of $a$, $b$, $c$ are 10, 2 and 4, respectively.

**15. (i) Derive an expression for drift velocity of free electrons.**

**(ii) How does drift velocity of electrons in a metallic conductor vary with increase in temperature ? Explain.** **[3]**

**Ans. (i)** Consider a conductor in which an electric field E is produced. Let a free electron experience a force ($-e$E) in this electric field. So, the acceleration of free electron is

$$a = \text{F}/m = -e\text{E}/m \qquad \text{...(i)}$$

E

$\vec{F} = -e\vec{E}$

$l$

Here, $e$ = Charge on electron.

$m$ = Mass of an electron.

So, the final velocity of the free electron in time interval $t_1$ is,

$$v_1 = u_1 + at_1$$

For $n$ free electrons, the final velocities be $v_2$, $v_3, ..., v_n$.

So, the average velocity of the free electrons or the drift velocity

$$v_d = \frac{(v_1 + v_2 + v_3 + ... + v_n)}{n}$$

$$\text{or } v_d = \frac{(u_1 + at_1 + u_2 + at_2 + ... + u_n + at_n)}{n}$$

$$\text{or } v_d = \frac{[(u_1 + u_2 + ... + u_n) + (at_1 + at_2 + ... + at_n)]}{n}$$

$$\text{or } v_d = \frac{[(u_1 + u_2 + ... + u_n) + a(t_1 + t_2 + ... + t_n)]}{n}$$

But, $\frac{(u_1 + u_2 + ... + u_n)}{n}$ = average initial velocity of free electrons = 0.

and $\frac{(t_1 + t_2 + ... + t_n)}{n}$ = average time taken between two consecutive collision = $\tau$

where $\tau$ is relaxation time.

So, $\quad v_d = a\tau$

or, $\quad v_d = \frac{-e\vec{E}}{m}\tau \qquad$ [ from (i)]

This is the required relation.

**(ii)** The drift velocity of free electrons in a metallic conductor decreases with increase in temperature. because, if we increase the temperature of the metallic conductor the collision between the electrons and ions increases, which decreases relaxation time. Hence, drift velocity decreases.

**16. (i) When an AC source is connected to an ideal inductor show that the average power supplied by the source over a complete cycle is zero.**

**(ii) A lamp is connected in series with an inductor and an AC source. What happens to the brightness of the lamp when the key is plugged in and an iron rod is inserted inside the inductor ? Explain.** **[3]**

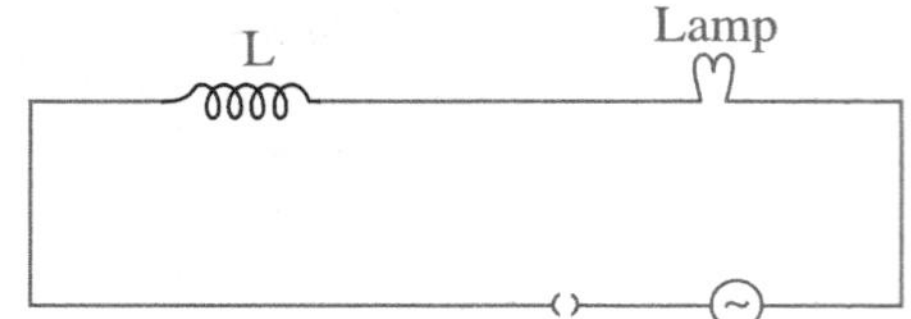


**Ans. (i)** The average power supplied by the source over a complete cycle is

$$P_{av} = E_{rms}.I_{rms}.\cos\phi$$

When the circuit contains an ideal inductor, then the phase difference between the current and voltage is $\pi/2$.

So, $\phi = \pi/2$. So, $\cos\phi = \cos\pi/2 = 0$.

Hence $P_{av} = 0$.

So, when an ac source is connected to an ideal inductor, the average power supplied by the source over a complete cycle is zero.

**(ii)** The brightness of the lamp will decrease. When the key is plugged in and the iron rod is inserted inside the inductor, it increases the inductance. Hence, the reactance of the inductor ($X_L = \omega L$) increases. So, the impedance of the circuit ($Z = R + j\omega L$) increases, which decreases the current in the circuit.

**17. (i) Explain with the help of a diagram the formation of depletion region and barrier potential in a *p-n* junction.**

**(ii) Draw the circuit diagram of a half wave rectifier and explain its working.** **[3]**

**Ans. (i)** During the formation of *p-n* junction, the holes diffuse from *p*-type semiconductor to the *n*-type, and electrons diffuse from *n*-type to *p*-type. This is because of the concentration gradient across *p*-side and *n*-side.

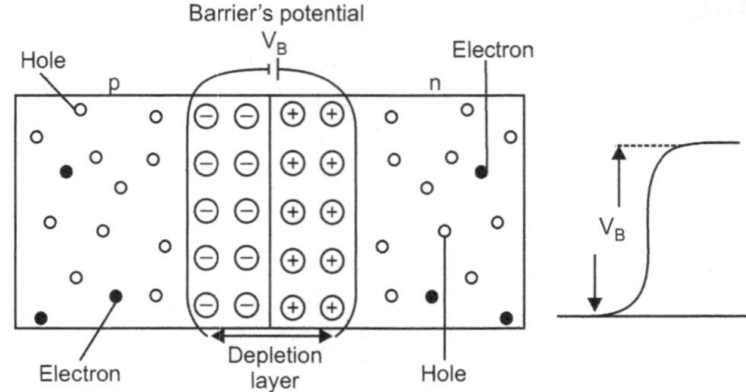


When a hole diffuse from *p* to *n* type, it leaves an unmovable negative charge. Similarly, when an electron diffuses from *n* to *p* type, it leaves an unmovable positive charge. When the diffusion of holes and electrons takes place continuously across the junction, a layer of unmovable positive and negative charges are developed on either side of the junction. This layer is called the depletion layer or the depletion region and the potential difference across the region is called barrier potential.

**(ii) Half wave rectifier :**

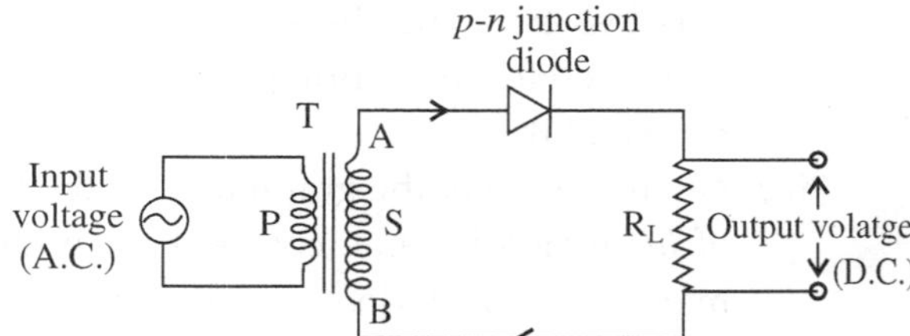


**Working :** When an input ac voltage is applied across the primary coil, a potential difference is developed across the ends of the secondary coil. Consider that in half cycle of input ac signal, the end A acts as the positive end and B acts as the negative end of the battery. So, the diode is in forward bias and we get output across the ends of the load resistance $R_L$.

In the second half cycle, ends A and B reverse in polarity. Now, A acts as the negative end and B acts as the positive end. So, the diode D is in reverse bias and no output is obtained due to the high resistance offered by the diode.

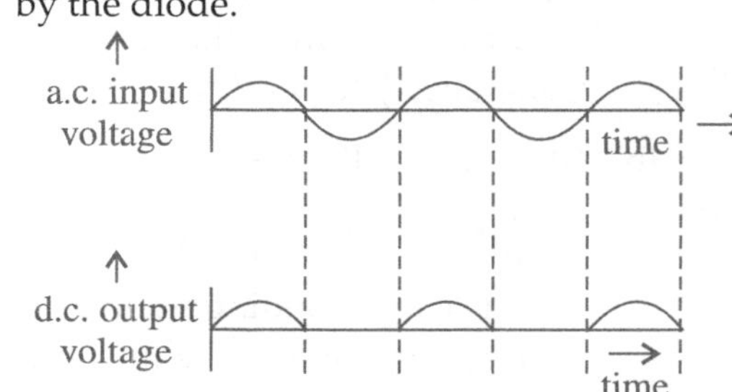

So, in this process, we get output alternately, and hence the diode is called the half wave rectifier.

**18. (i) Which mode of propagation is used by short wave broadcast service having frequency range from a few MHz upto 30 MHz ? Explain diagrammatically how long distance communication can be achieved, by this mode.**

**(ii) Why is there an upper limit to frequency of waves used in this mode ?**** [3]

**19. (i) Identify the part of the electromagnetic spectrum which is :**

**(a) suitable for radar system used in aircraft navigation,**

**(b) produced by bombarding a metal target by high speed electron.**

**(ii) Why does a galvanometer show a momentary deflection at the time of charging or discharging a capacitor ? Write the necessary expression to explain this observation.** [3]

**Ans. (i) (a)** Microwaves, **(b)** X-rays

**(ii)** During the charging and discharging of a capacitor, a flow of charges take place from the battery to the plates of the capacitor. This produces a conduction current in the circuit and a displacement current between plates. Hence the galvanometer shows a momentary deflection.

$$\int B.dl = \mu_0 I + I_d$$

Where, $I_d = \dfrac{\varepsilon_0 d\phi_E}{dt}$

**20. For a CE-transistor amplifier, the audio signal voltage across the collector resistance of 2 kΩ is 2 V. Suppose the current amplification factor of the transistor is 100, find the input signal voltage and base current, if the base resistance is 1 kΩ.**** [3]

**21. Define the term wave front. State Huygen's principle.**

**Consider a plane wave front incident on a thin convex lens. Draw a proper diagram to show how the incident wave front traverses through the lens and after refraction focusses on the focal point of the lens, giving the shape of the emergent wave front.** [3]

**OR**

**Explain the following, giving reasons :**

**(i) When monochromatic light is incident on a surface separating two media, the reflected and refracted light both have the same frequency as the incident frequency.**

**(ii) When light travels from a rarer to a denser medium, the speed decreases. Does this decrease in speed imply a reduction in the energy carried by the wave ?**

**(iii) In the wave picture of light, intensity of light is determined by the square of the amplitude of the wave. What determines the intensity in the photon picture of light.**

**Ans. Wave front :** A wave front is the locus of all the points in space that reach a particular distance by a propagating wave in same phase at any instant.

**Huygen's principle :** It is based on two assumptions :

**(a)** Each point of the wavefront behaves like a source of secondary disturbances and secondary wavelets from there points spread out in all directions with the same speed as that of the original wave front.

**(b)** When we draw an envelope in the forward direction of the secondary disturbances at any instant, And this envelope tells the new position of the wavefront at that instant.

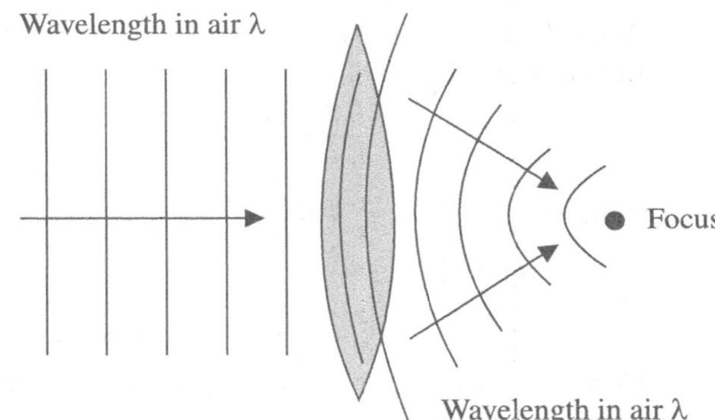


**OR**

**(i)** Both the reflection and refraction takes place due to the interaction of light with the atoms at the surface of the separation. Light incident on these atoms, force them to vibrate with the frequency of light. But, the light emitted by these charged atoms is equal to their own frequency of oscillation. So, both the reflected and refracted lights have same frequency as the frequency of incident light hence frequency remains changed.

**(ii)** The energy carried by a wave depends on the amplitude of the wave. It does not depend on the speed of the wave propagation. Hence the energy of the wave remains same and does not decrease.

**(iii)** The intensity of light is determined by the number of photons incident per unit area per unit time around the point at which intensity is to be determined.

**22. Use Biot-Savart law to derive the expression for the magnetic field on the axis of a current carrying circular loop of radius R.**
**Draw the magnetic field lines due to a circular wire carrying current I.** **[3]**

**Ans.** Imagine a circular coil of radius R with centre O. Let the current flowing through the circular loop be I. Suppose P is any point on the axis at a distance of $r$ from the centre O. Let the circular coil be made up of a large number of small elements of current, each having a length of $dl$.

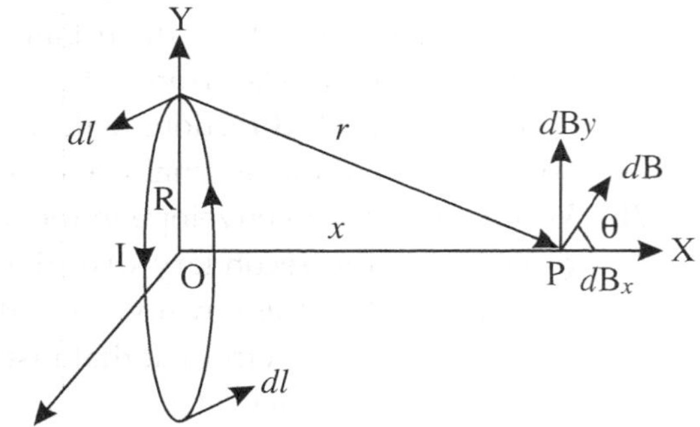


According to Biot-Savart's law, the magnetic field at Point P will be

$$dB = \frac{\mu_0 I}{4\pi} \times \frac{|dl \times r|}{r^3}$$

where, $r^2 = x^2 + R^2$

$|dl \times r| = rdl$

[ ∵ Both are perpendicular]

Here, $r$ is the position vector of point O from the current element.

So, $$dB = \frac{\mu_0}{4\pi} \times \frac{I.\,dl}{r^2}$$

$$dB = \frac{\mu_0}{4\pi} \times \frac{I.dl}{(x^2 + R^2)}$$

$dB$ has two components *i.e.*, $dB_x$ and $dB_y$. $dB_y$ is cancelled out and only the $x$– component remains.

∴ $$dB_x = dB \cos\theta$$

$$\cos\theta = \frac{R}{\sqrt{x^2 + R^2}}$$

$$dB_x = \frac{\mu_0 Idl}{4\pi} \cdot \frac{R}{(x^2 + R^2)^{3/2}}$$

But, $$\int dl = 2\pi R$$

So, $$B = \frac{\mu_0 IR \times 2\pi R}{4\pi(x^2 + R^2)^{3/2}}$$

For $n$ turns in the circular loop,

$$B = \frac{\mu_0 nIR^2}{2(x^2 + R^2)^{3/2}} \cdot \hat{i}$$

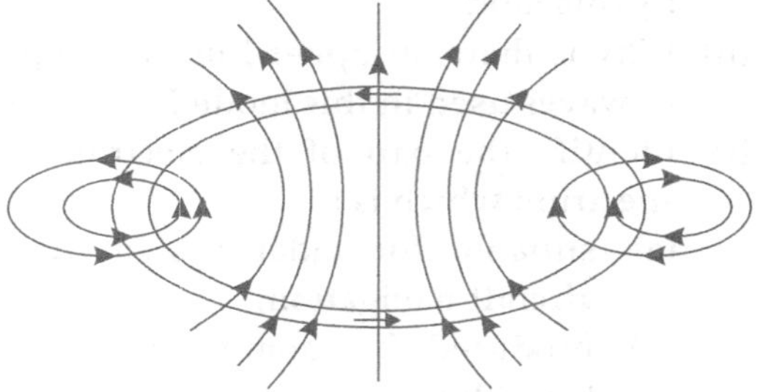

## SECTION–D

**23. Ram is a student of class X in a village school. His uncle gifted him a bicycle with a dynamo fitted in it. He was very excited to get it. While cycling during night, he could light the bulb and see the objects on the road. He, however, did not know how this device works. He asked this question to his teacher. The teacher considered it an opportunity to explain the working to the whole class.**
**Answer the following questions :** **[4]**
**(a) State the principle and working of a dynamo.**
**(b) Write two values each displayed by Ram and his school teacher.****

**Ans. (a) Principle :** A dynamo works on the principle of electro-magnetic induction.

A working dynamo includes a coil attached to a small turbine fitted with a plastic cap. The coil is placed in a magnetic field. When the plastic cap comes in contact with moving tyres of the bicycle, the coil placed between the poles of a magnet rotates, thus, the flux through the coil changes continuously. This induces a sinosoidal verying emf in the coil which is connected to a bulb which lights up according to $e = e_0 \sin \omega t$.

As long as the bicycle is moving, the coil keeps on rotating, and hence, the flux keeps on changing. At a steady rate, we get a steady current and hence a light of steady intensity is obtained.

**Answer is not given due to change in the present Syllabus.**

## SECTION–E

**24. (i) Draw a labelled diagram of a step-down trans former. State the principle of its working.**

**(ii) Express the turns ratio in terms of voltages.**

**(iii) Find the ratio of primary and secondary currents in terms of turns ratio in an ideal transformer.**

**(iv) How much current is drawn by the primary of a transformer connected to 220 V supply when it delivers power to a 110 V— 550 W refrigerator ? [5]**

**OR**

**(a) Explain the meaning of the term mutual inductance. Consider two concentric circular coils, one of radius $r_1$ and the other of radius $r_2$ ($r_1 < r_2$) placed coaxially with centres coinciding with each other. Obtain the expression for the mutual inductance of the arrangement.**

**(b) A rectangular coil of area A, having number of turns N is rotated at '$f$' revolutions per second in a uniform magnetic field B, the field being perpendicular to the coil. Prove that the maximum emf induced in the coil is $2\pi f$ NBA.**

**Ans. (i)**

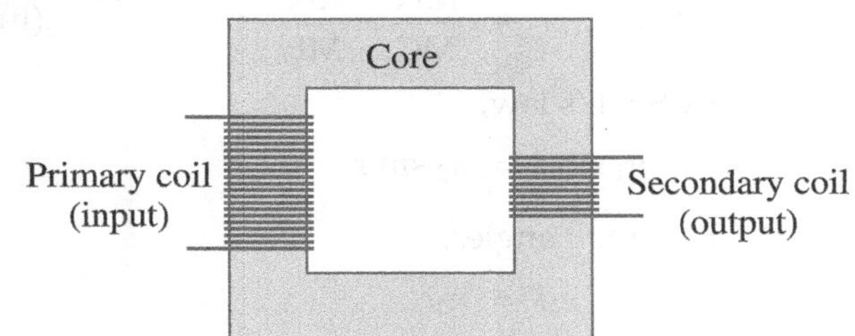


**Principle :** A transformer works on the principle of mutual induction. Whenever the amount of magnetic flux linked with a coil changes, an emf is induced in the neighbouring coil.

**Working :** When an alternating current source is connected to the ends of primary coil, the current changes continuously in the primary coil, due to which magnetic flux linked with the secondary coil changes continuously. Therefore, the alternating emf of same frequency is developed across the secondary terminals. According to Faraday's laws the e.m.f. induced in the primary coil,

$$E_P = -N_P \frac{\Delta\phi}{\Delta t} \quad ...(i)$$

and emf induced in the secondary coil

$$E_S = -N_S \frac{\Delta\phi}{\Delta t} \quad ...(ii)$$

From (i) and (ii)

$$\frac{E_S}{E_P} = \frac{N_S}{N_P} = K \quad ...(iii)$$

For step-down transformer, $K < 1$.

$$\therefore \quad E_S < E_P$$

**(ii)** The induced emf in primary coil,

$$E_P = -N_P\,(d\phi/dt)$$

The induced emf in secondary coil,

$$E_S = -N_S\,(d\phi/dt)$$

$$E_S/E_P = N_S/N_P = K$$

Where K is the turns ratio or the transformation ratio.

**(iii)** If the transformer is ideal, then Input electrical power = Output electrical power

$$E_P I_P = E_S I_S$$

$$E_S/E_P = I_P/I_S$$

$$\frac{I_P}{I_S} = \frac{E_S}{E_P} = \frac{N_S}{N_P} = K$$

**(iv)** Given, Power ,

$$P = 550 \text{ W}$$

Supply voltage,

$$V_S = 220 \text{ V}$$

$$\text{Power} = V_P I_P$$

$$550 = 220 \times I_P$$

$$\Rightarrow \quad I_P = 5/2 = 2.5 \text{ A}$$

**OR**

**(a) Mutual inductance :** It is the property of a pair of coils due to which an e.m.f. is induced in one coil due to the change in the flux or current in the other coil.

Let a current $I_2$ flow through the outer circular coil. The magnetic field at the centre of the coil is

$$B_2 = \frac{\mu_0 I_2}{2r_2} \quad ...(i)$$

As the inner coil placed co-axially has very small radius, therefore, $B_2$ may be taken as constant over its cross-sectional area. Hence, flux associated with inner coil is

$$\phi_1 = \pi r_1^2 B_2$$

$$= \pi r_1^2 \frac{\mu_0 I_2}{2r_2} \quad \text{[From (i)]}$$

$$= \left(\frac{\mu_0 \pi r_1^2}{2r_2}\right) I_2$$

$$= M_{12} I_2$$

$$\therefore \quad M_{12} = \frac{\mu_0 \pi r_1^2}{2r_2}$$

Now, $M_{21} = M_{12} = \dfrac{\mu_0 \pi r_1^2}{2r_2}$

**(b)** Let N be the number of turns of the rectangular coil and A be its cross-sectional area placed in a magnetic field B, then, the magnetic flux linked with the coil,

$$\phi = NBA \cos \theta$$

The induced emf,

$$e = -d\phi/dt$$

$$e = -\frac{d\phi}{dt} = \left(-NBA(-\sin\theta)\frac{d\theta}{dt}\right)$$

$$= NBA.\sin\theta\,(2\pi f) \qquad \left[\because \; \theta = \omega t, \; \frac{d\theta}{dt} = \omega = 2\pi f\right]$$

For maximum induced e.m.f.

$$\sin\theta = 1$$

$$\therefore \quad e = NBA\,(2\pi f)$$

**25. (a) Derive the mathematical relation between refractive indices $n_1$ and $n_2$ of two media and radius of curvature R for refraction at a convex spherical surface. Consider the object to be a point since lying on the principle axis in rarer medium of refractive index $n_1$ and a real image formed in the denser medium of refractive index $n_2$. Hence, derive Lens Maker's formula.**

**(b) Light from a point source in air falls on a convex spherical glass surface of refractive index 1.5 and radius of curvature 20 cm. The distance of light source from the glass surface is 100 cm. At what position is the image formed ? [5]**

**OR**

**(a) Draw a labelled ray diagram to obtain the real image formed by an astronomical telescope in normal adjustment position. Define its magnifying power.**

**(b) You are given three lenses of power 0.5 D, 4 D and 10 D to design a telescope.**

**(i) Which lenses should be used as objective and eyepiece ? Justify your answer.**

**(ii) Why is the aperture of the objective preferred to be large ?**

**Ans. (a)**

$$\tan\angle NOM = \frac{MN}{OM};$$

$$\tan\angle NCM = \frac{MN}{MC};$$

$$\tan\angle NIM = \frac{MN}{MI}$$

For $\Delta$NOC, $i$ is the exterior angle.

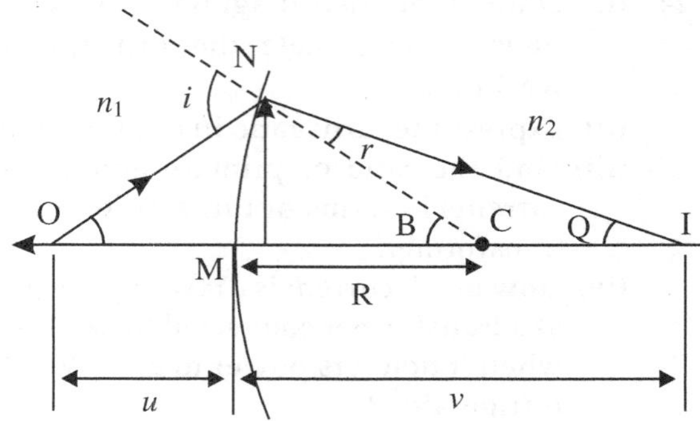


Assuming the incident ray is very close to the principal axis, all the angles are very small. Hence, for very small angles,

$$\tan x = x = \sin x$$

$$\therefore \quad i = \angle NOM + \angle NCM$$

$$i = \frac{MN}{OM} + \frac{MN}{MC} \quad \text{...(i)}$$

Similarly,

$$r = \angle NCM - \angle NIM$$

*i.e.,*

$$r = \frac{MN}{MC} - \frac{MN}{MI} \quad \text{...(ii)}$$

By Snell's law,

$$n_1 \sin i = n_2 \sin r$$

For small angles,

$$n_1 i = n_2 r$$

On substituting the values of $i$ and $r$ in equations, we get

$$n_1\left(\frac{MN}{OM} + \frac{MN}{MC}\right) = n_2\left(\frac{MN}{MC} - \frac{MN}{MI}\right)$$

$$\frac{n_1}{OM} + \frac{n_2}{MI} = \frac{n_2 - n_1}{MC} \quad \text{...(iii)}$$

On applying new Cartesian sign conventions,

$$OM = -u, \; MI = +v, \; MC = +R$$

Substituting these values in equation (iii), we get

$$\frac{n_2}{v} - \frac{n_1}{u} = \frac{n_2 - n_1}{R} \quad \text{...(iv)}$$

In deriving Lens maker's formula, we adopt the coordinate geometry sign convention and make the assumptions :

**(i)** The lens is thin so that the distances measured from the poles of its two surfaces can be taken as equal to the distances from its optical centre.

**(ii)** The aperture of the lens is small.

**(iii)** The object is a point-object placed on the principal axis of the lens.

**(iv)** The incident and the refracted rays make small angles with the principal axis.

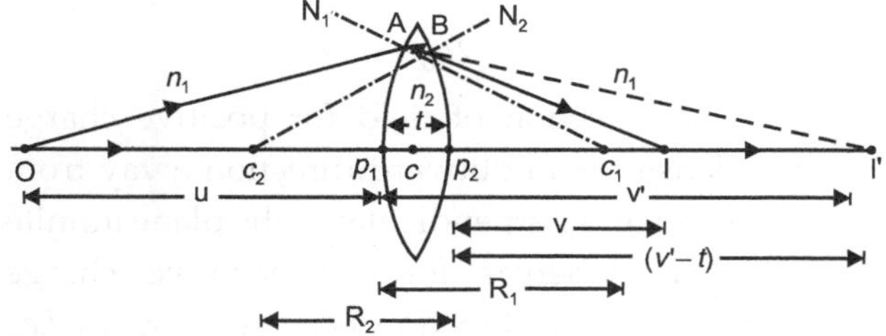


Refraction at the first surface,

$$\frac{n_2}{v'} - \frac{n_1}{u} = \frac{n_2 - n_1}{R_1}$$

Refraction at the second surface,

$$\frac{n_1}{v} - \frac{n_2}{v'-t} = \frac{n_1 - n_2}{R_2}$$

The lens is 'thin', hence $t << v'$ and can be ignored. Then, we have

$$\frac{n_1}{v} - \frac{n_2}{v'} = \frac{n_1 - n_2}{R_2} \quad ....(ii)$$

Adding equation (i) and (ii), we get

$$\frac{n_1}{v} - \frac{n_1}{u} = (n_2 - n_1)\left(\frac{1}{R_1} - \frac{1}{R_2}\right)$$

Putting $n_2/n_1 = n$, the refractive index of the material of the lens with respect to the surrounding medium,

we have

$$\frac{1}{v} - \frac{1}{u} = (n-1)\left(\frac{1}{R_1} - \frac{1}{R_2}\right) \quad ....(iii)$$

When the object is at infinity, the image will be formed at the principal focus of the lens, *i.e.*, when $u = \infty$,

$$\frac{1}{f} - \frac{1}{\infty} = (n-1)\left(\frac{1}{R_1} - \frac{1}{R_2}\right)$$

$$\frac{1}{f} = (n-1)\left(\frac{1}{R_1} - \frac{1}{R_2}\right)$$

This is the Lens Maker's formula.

**(b)** Given, refractive index, $n_2 = 1.5$, $n_1 = 1$ (air)

Radius of curvature, R = 20 cm

Object distance, $u = -100$ cm

To find , Image distance, $v$

$$\frac{n_2}{v} - \frac{n_1}{u} = \frac{n_2 - n_1}{R}$$

$$\frac{1.5}{v} + \frac{1}{(100)} = \frac{1.5-1}{20} = \frac{1}{40}$$

$$\frac{1.5}{v} = \frac{1}{40} - \frac{1}{100} = \frac{5-2}{200} = \frac{3}{200}$$

$$v = \frac{200}{3} \times 1.5 = 100 \text{ cm}$$

The image is formed at 100 cm away from surface.

**OR**

**(a)**

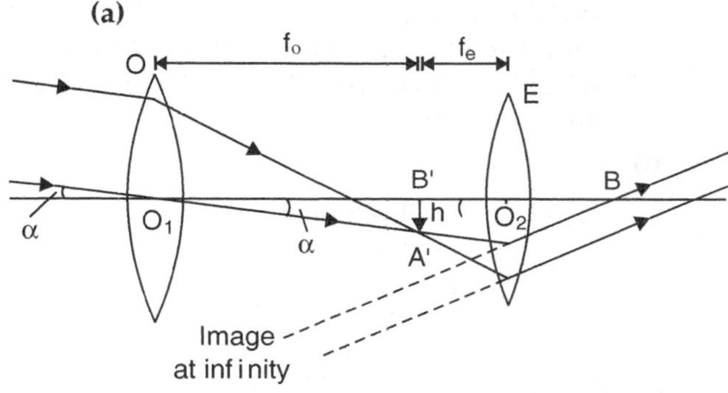


**Magnifying power :** The magnifying power of a refracting type astronomical telescope is defined as the ratio of angle subtended by the final image at eye to the angle subtended by the object at eye.

**(b) (i)** We know that,

$$\text{Magnification, } m = \frac{f_o}{f_e} = \frac{P_e}{P_o}$$

Therefore, the lens of 0.5 D should be used as objective and the lens of 10 D should be used as eye-piece in order to achieve higher magnification.

**(ii)** The aperture of the objective lens is made larger so, that it receives as much light as coming from the distant object and the resolving power of the telescope increases.

**26. (i) Use Gauss's law to find the electric field due to a uniformly charged infinite plane sheet. What is the direction of field for positive and negative charge densities ?**

**(ii) Find the ratio of the potential differences that must be applied across the parallel and series combination of two capacitors $C_1$ and $C_2$ with their capacitances in the ratio 1 : 2 so that the energy stored in the two cases becomes the same. [5]**

**OR**

**(i) If two similar large plates, each of area A having surface charge densities + σ and – σ are separated by a distance in air, find the expressions for**

**(a) field at points between the two plates and on outer side of the plates. Specify the direction of the field in each case.**

**(b) the potential difference between the plates.**

**(c) the capacitance of the capacitor so formed.**

**(ii) Two metallic spheres of radii R and 2R are charged so that both of these have same surface charge density σ. If they are connected to each other with a conducting wire, in which direction will the charge flow and why ?**

**Ans. (i)**

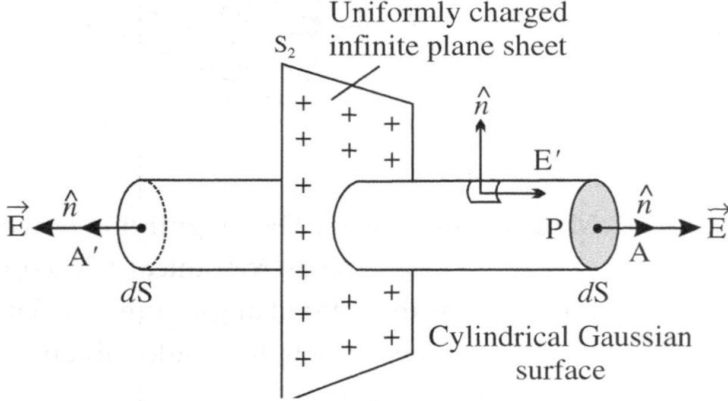


Consider a thin infinite uniformly charged plane sheet having the surface charge density of σ. The electric field is normally outward to the plane sheet and is same in magnitude but opposite in direction.

Now, draw a Gaussian surface in the form of cylinder around an axis. Let its cross-sectional area be A. The cylinder is made from three surfaces A, $S_2$, and A′ and the electric flux linked with $S_2$ is 0. So, the total electric flux linked through the Gaussian surface is

$\phi_E$ = electric flux through A + electric flux through $S_2$ + electric flux through A′

$\phi_E = EA\cos 0^0 + 0 + EA\cos 0^0.$

$$\phi = 2EA \quad ...(i)$$

According to Gauss theorem,

$$\phi = \frac{q}{\varepsilon_0}$$

$$\phi = \frac{\sigma A}{\varepsilon_0} \quad (\because q = \sigma A) \;...(ii)$$

From equations (i) and (ii)

$$2EA = \frac{\sigma A}{\varepsilon_0}$$

$$E = \frac{\sigma}{2\varepsilon_0}$$

The direction of field for positive charge density is in outward direction away from sheet and perpendicular to the plane infinite sheet whereas for the negative charge density the direction becomes inward *i.e.*, towards the sheet and perpendicular to the sheet.

**(ii)** Given

$$C_1 : C_2 = 1 : 2$$

$$\Rightarrow \quad C_2 = 2C_1$$

For parallel combination of capacitor,

$$C_p = C_1 + C_2 = C_1 + 2C_1 = 3C_1$$

The energy stored in capacitor

$$E = \frac{1}{2}C_pV_p^2$$

$$= \frac{1}{2}3C_1V_p^2 = \frac{3}{2}C_1V_p^2 \quad ...(i)$$

For series combination of capacitor,

$$\frac{1}{C_s} = \frac{1}{C_1} + \frac{1}{C_2}$$

$$C_s = \frac{2}{3}C_1$$

The energy stored in capacitor

$$E = \frac{1}{2}C_sV_s^2$$

$$E = \frac{C_1V_s^2}{3} \quad ...(ii)$$

Equating equation (i) and (ii), since energy stored in both cases are same we get,

$$\frac{3}{2}C_1V_p^2 = \frac{C_1V_s^2}{3}$$

$$\frac{V_p}{V_s} = \frac{\sqrt{2}}{3}$$

OR

**(i) (a)** Consider a parallel plate capacitor with two identical plates X and Y, each having an area of A, and separated by a distance $d$. Let the space between the plates be filled by a dielectric medium with its dielectric constant as K and $\sigma$ be the surface charge density on each of the plates.

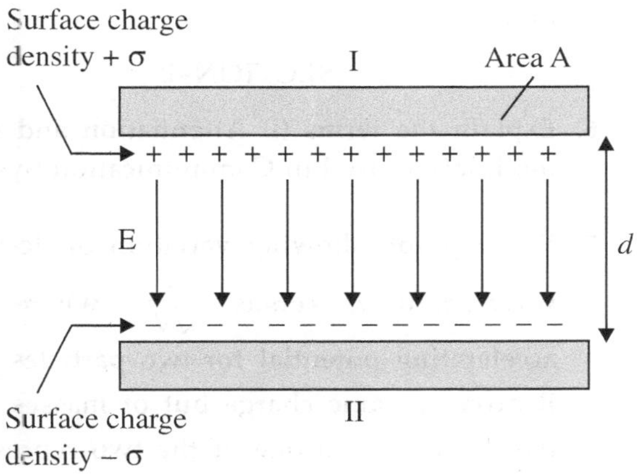

Surface charge density $+\sigma$ I Area A

E $d$

Surface charge density $-\sigma$ II

Surface charge density of plate I

$$\sigma = Q/A$$

and that of plate II is $-\sigma$.

Electric field in outer region I,

$$E = \frac{\sigma}{2\varepsilon_0} - \frac{\sigma}{2\varepsilon_0} = 0$$

Electric field in outer region II,

$$E = \frac{\sigma}{2\varepsilon_0} - \frac{\sigma}{2\varepsilon_0} = 0$$

In the inner region between plates 1 and 2, the electric fields due to the two charged plates add up. So

$$E = \frac{\sigma}{2\varepsilon_0} + \frac{\sigma}{2\varepsilon_0} = \frac{\sigma}{\varepsilon_0}$$

**(b)** For uniform electric field, potential difference is simply the electric field multiplied by the distance between the plates, *i.e.*,

$$V = Ed = \frac{1}{\varepsilon_0}\frac{Qd}{A}$$

**(c)** Now, the capacitance of the parallel plate capacitor,

$$C = \frac{Q}{V} = \frac{Q.\varepsilon_0 A}{Qd} = \frac{\varepsilon_0 A}{d}$$

**(ii)** We know that the potential difference of the metallic sphere is given by,

$$V = \frac{Q}{4\pi\varepsilon_0 r}$$

where $r$ is the radius of the sphere.

Now, the potential of the metallic sphere of radius R is given by,

$$V_R = \frac{Q}{4\pi\varepsilon_0 r}$$

$$V_R = \frac{\sigma(4\pi R^2)}{4\pi\varepsilon_0 R}$$

$$V_R = \frac{\sigma R}{\varepsilon_0} \quad \text{...(i)}$$

Similarly, the potential of the metallic sphere of radius $V_{2R}$ is given by

$$V_{2R} = \frac{Q}{4\pi\varepsilon_0 2R}$$

$$V_{2R} = \frac{\sigma(4\pi (2R)^2)}{4\pi\varepsilon_0 2R}$$

$$V_{2R} = \frac{\sigma 2R}{\varepsilon_0} \quad \text{...(ii)}$$

From the relation (i) and (ii) we find that $V_{2R} > V_R$.

The charge will flow from the sphere of radius of 2R to the sphere of radius R, if the spheres are connected.

●●

# PHYSICS

## Delhi [Set-I]

# 2016

Time allowed : 3 hours

Maximum marks : 70

### SECTION – A

1. **A point charge +Q is placed at point O as shown in the figure. Is the potential difference $V_A - V_B$ positive, negative or zero? [1]**

+Q •---------• A ----• B (O)

**Ans.** Potential at a distance $r$ from a given point charge Q is given by

$$V = \frac{1}{4\pi\varepsilon_0}\frac{Q}{r}$$

+Q at O, points A and B; $r_A$, $r_B$

$$V_A = \frac{1}{4\pi\varepsilon_0 r_A}$$

$$V_B = \frac{Q}{4\pi\varepsilon_0 r_B}$$

Since $r_A < r_B$

$\Rightarrow$ $V_A > V_B$

Hence, $V_A - V_B$ is positive.

2. **How does the electric flux due to a point charge enclosed by a spherical Gaussian surface get affected when its radius is increased? [1]**

**Ans.** According to Gauss's law

$$\phi = \oint \vec{E}.\vec{ds} = \frac{q_{en}}{\varepsilon_0}$$

Flux depends only on the charge enclosed and not on the radius.

Hence, the electric flux remains constant.

3. **Write the underlying principle of a moving coil galvanometer. [1]**

**Ans.** When a current carrying coil is placed in magnetic field then it experiences a torque.

This is the underlying principle of a moving coil galvenometer.

4. **Why are microwaves considered suitable for radar systems used in aircraft navigation? [1]**

**Ans.** Microwaves are considered suitable because they have a short wavelenght range they are suitable for radar system used in aircraft navigation.

5. **Define 'quality factor' of resonance in series LCR circuit. What is its SI unit? [1]**

**Ans.** The Q–factor of series resonance circuit is defined as the ratio of the voltage developed across the inductor or capacitor at resonance to the impressed voltage, which is the voltage across R.

$$Q = \frac{1}{R}\sqrt{\frac{L}{C}}$$

It is dimensionless quantity. Hence, it has no units.

### SECTION–B

6. **Explain the terms (i) Attenuation and (ii De-modulation used in Communication System.** [2]**

7. **Plot a graph showing variation of de-Broglie wavelength $\lambda$ versus $\frac{1}{\sqrt{V}}$, where V is accelerating potential for two particles A and B carrying same charge but of masses $m_1$, $m_2$ ($m_1 > m_2$). Which one of the two represents a particle of smaller mass and why? [2]**

**Ans.** We know that,

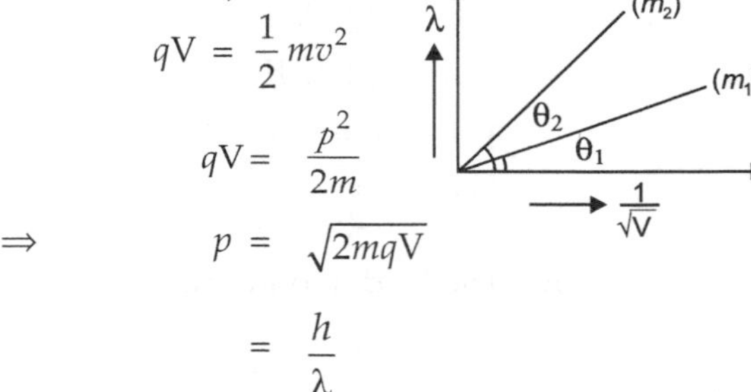


$$qV = \frac{1}{2}mv^2$$

$$qV = \frac{p^2}{2m}$$

$$\Rightarrow \quad p = \sqrt{2mqV}$$

$$= \frac{h}{\lambda}$$

$$\Rightarrow \quad \lambda = \frac{h}{\sqrt{2mqV}}$$

$$\Rightarrow \quad \text{Slope} \propto \frac{1}{\sqrt{m}}$$

Hence, the particle with lower mass ($m_2$) will have greater slope.

8. **A nucleus with mass number A = 240 and BE/A = 7.6 MeV breaks into two fragments each of A = 120 with BE/A = 8.5 MeV. Calculate the released energy. [2]**

**OR**

**Calculate the energy in fusion reaction :**

**$^2_1H + ^2_1H \longrightarrow ^3_2He + n$, where BE of $^2_1H$ = 2.23 MeV and of $^3_2He$ = 7.73 MeV.**

**Ans.** Binding energy of the nucleus,

$$B_1 = 7.6 \times 240 = 1824 \text{ MeV}.$$

Binding energy of each product nucleus,

$$B_2 = 8.5 \times 120 = 1020 \text{ MeV}$$

Then, energy released as the nucleus breaks,

$$E = 2\,B_2 - B_1 = 2 \times 1020 - 1824 = 216 \text{ MeV.}$$

**OR**

${}^2_1H + {}^2_1H \longrightarrow {}^3_2He + n$, where BE of ${}^2_1H = 2.23$ MeV and of ${}^3_2He = 7.73$ MeV.

The energy released in the fusion reaction is

$$\Delta E = (7.73) - 2\,(2.23)$$
$$= 7.73 - 4.46$$
$$= 3.27 \text{ MeV}$$

**9. Two cells of emfs 1.5 V and 2.0 V having internal resistances 0.2 Ω and 0.3 Ω respectively are connected in parallel. Calculate the emf and internal resistance of the equivalent cell. [2]**

**Ans.** Given, $E_1 = 1.5 \text{ V}, r_1 = 0.2\ \Omega,$

$$E_2 = 2 \text{ V}, r_2 = 0.3\ \Omega$$

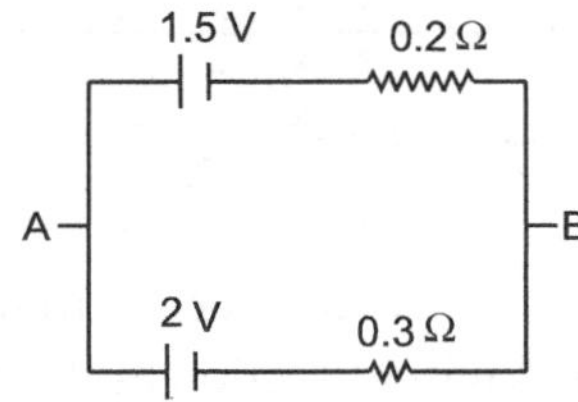


Equivalent emf,

$$E = \frac{\frac{E_1}{r_1} + \frac{E_2}{r_2}}{\frac{1}{r_1} + \frac{1}{r_2}} = \frac{E_1 r_2 + E_2 r_1}{r_1 + r_2}$$

$$= \frac{(1.5 \times 0.3) + (2 \times 0.2)}{0.2 + 0.3}$$

$$= \frac{0.45 + 0.4}{0.5}$$

$$= \frac{0.85}{0.5} = 1.7 \text{ volt}$$

Equivalent internal resistance

$$= \frac{r_1 r_2}{r_1 + r_2}$$

$$= \frac{0.2 \times 0.3}{0.2 + 0.3}$$

$$= \frac{0.06}{0.5}$$

$$r_{eq} = 0.12\ \Omega$$

**10. State Brewster's law.**

**The value of Brewster angle for a transparent medium is different for light of different colours. Give reason. [2]**

**Ans. Brewster's law** : The law states that the tangent of the polarising angle of incidence for a given medium is equal to the refractive index of the medium. The light incident at this angle when reflects back is perfectly polarised.

*i.e.* $\mu = \tan i_p$

The refractive index of a material depends on the colour or wavelength of light. As the polarising angle depends on refractive index ($\mu = \tan i_p$), so it also depends on wavelength of light.

## SECTION–C

**11. A charge is distributed uniformly over a ring of radius 'a'. Obtain an expression for the electric intensity E at a point on the axis of the ring. Hence show that for points at large distances from the ring, it behaves like a point charge. [3]**

**Ans.** Suppose we have a ring of radius $a$ that carries a uniformly distributed positive charge $q$.

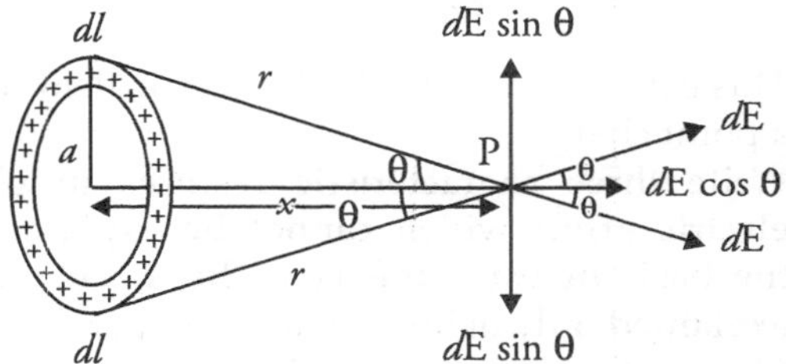


**Ans.** As the total charge $q$ is uniformly distributed, the charge $dq$ on the element $dl$ is

$$dq = \frac{q}{2\pi a} \cdot dl$$

**Ans.** ∴ The magnitude of the electric field produced by the element $dl$ at the axial point P is

$$dE = k \cdot \frac{dq}{r^2} = \frac{kq}{2\pi a} \cdot \frac{dl}{r^2}$$

The electric field $dE$ has two components.

**(i)** The axial components $dE \cos\theta$ and

**(ii)** The perpendicular component $dE \sin\theta$.

Since the perpendicular component of any two diametrically opposite elements are equal and opposite, they cancel out in pairs. Only the axial components will add up to produce the resultant field.

E at point P is given by

$$E = \int_0^{2\pi a} dE \cos\theta$$

[∵ Only the axial components contribute towards E]

$$E = \int_0^{2\pi a} \frac{kq}{2\pi a} \cdot \frac{dl}{r^2} \cdot \frac{x}{r} \qquad \left[\because \cos\theta = \frac{x}{r}\right]$$

$$= \frac{kqx}{2\pi a} \cdot \frac{1}{r^3} \int_0^{2\pi a} dl$$

$$= \frac{kqx}{2\neq a} \cdot \frac{1}{r^3} [l]_0^{2\neq a}$$

$$= \frac{kqx}{2\pi a} \cdot \frac{1}{(x^2 + a^2)^{3/2}} \cdot 2\pi a$$

$$[\because r^2 = x^2 + a^2]$$

$$E = \frac{kqx}{(x^2 + a^2)^{3/2}}$$

$$\left[\text{where } k = \frac{1}{4\pi\varepsilon_0} \quad a = \text{constant}\right]$$

$$= \frac{1}{4\pi\varepsilon_0} \cdot \frac{qx}{(x^2 + a^2)^{3/2}}$$

If $x >> a$, then $x^2 + a^2 \approx x^2$

$$E = \frac{1}{4\pi\varepsilon_0} \frac{qx}{(x^2)^{3/2}}$$

$$E = \frac{1}{4\pi\varepsilon_0} \frac{q}{x^2}$$

This expression is similar to electric field due to a point charge.

**12. Write three characteristic features in photo-electric effect which cannot be explained on the basis of wave theory of light, but can be explained only using Einstein's equation. [3]**

**Ans.** (i) **Existence of threshold frequency** : According to wave theory, there should not exist any threshold frequency but Einstein's theory explains the existence of threshold frequency.

(ii) **Dependence of kinetic energy on frequency of incident light** : According to wave theory, the maximum kinetic energy of emitted electrons should depend on intensity of incident light and not on frequency whereas Einstein's equation explains that it's dependence on frequency and not on intensity of the incident light.

(iii) **Instantaneous emission of electrons** : According to wave theory there should be time lag between emission of electrons and incident of light whereas Einstein's equation explains why there is no time lag between incident of light and emission of electrons.

**13. (a) Write the expression for the magnetic force acting on a charged particle moving with velocity $v$ in the presence of magnetic field B.**

**(b) A neutron, an electron and an alpha particle moving with equal velocities, enter a uniform magnetic field going into the plane of the paper as shown. Trace their paths in the field and justify your answer. [3]**

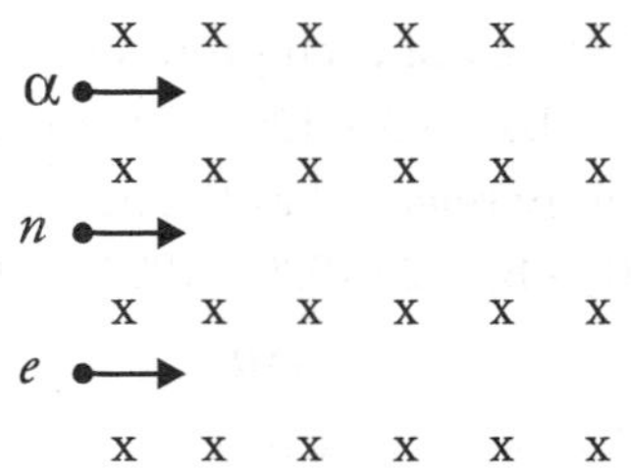

x x x x x x
$\alpha$ •→
x x x x x x
$n$ •→
x x x x x x
$e$ •→
x x x x x x

**Ans.** (a) A charge particle having charge $q$ is moving with velocity '$v$' in a magnetic field of field strength 'B' then the force acting on it is given by the formula $F = q(\vec{v} \times \vec{B})$ and $F = qvB \sin\theta$ (where $\theta$ is the angle between velocity vector and magnetic field).

Direction of force is given by the cross product of velocity and magnetic field.

(b)

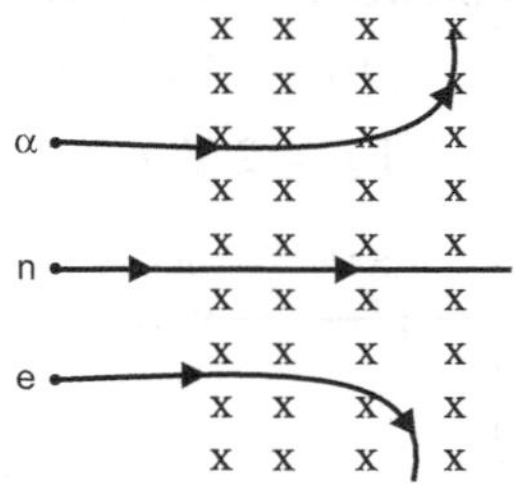

α
n
e

$\alpha$ particle will trace circular path in anticlockwise direction as it's deviation will be in the direction of $(\vec{v} \times \vec{B})$.

Neutron will pass without any deviation as magnetic field does not exert any force on neutral particle.

Electron will trace circular path in clockwise direction as its deviation will be in the direction opposite to $(\vec{v} \times \vec{B})$ with a smaller radius due to negative charge and large charge/mass ratio as $r = \frac{mv}{qB}$.

**14. (a) Define mutual inductance.**

**(b) A pair of adjacent coils has a mutual inductance of 1.5 H. If the current in one coil changes from 0 to 20 A in 0.5 s, what is the change of flux linkage with the other coil ? [3]**

**Ans.** (a) Mutual induction is the phenomenon of production of induced emf in one coil due to change of current or flux in the neighbouring coil. The coil in which the current changes is called primary coil and the coil in which emf is induced is called the secondary coil.

**(b)**

$$M = 1.5\,H$$
$$I_i = 0\,A$$
$$I_f = 20\,A$$
$$dI = 20\,A, \quad \Delta t = 0.5\,s$$

$$e = \frac{-MdI}{dt}$$
$$= -1.5 \times \frac{20}{0.5}$$
$$= -60\,V$$

So the flux linked with the other coil is given by

$$\Delta\phi = -e\Delta t = 60 \times 0.5$$
$$= 30\text{ Wb.}$$

**15. Two parallel plate capacitors X and Y have the same area of plates and same separation between them. X has air between the plates while Y contains a dielectric medium of $\varepsilon_r = 4$.** **[3]**

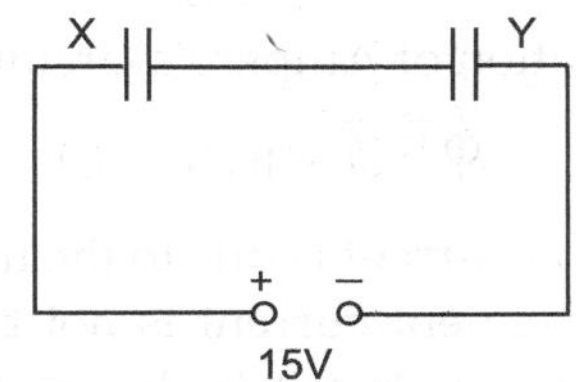


**(i) Calculate capacitance of each capacitor if equivalent capacitance of the combination is 4 μF.**

**(ii) Calculate the potential difference between the plates of X and Y.**

**(iii) Estimate the ratio of electrostatic energy stored in X and Y.**

**Ans. (i)** Let capacitance of X be $C_1$ and capacitance of Y be $C_2$.

$$C_1 = \frac{\varepsilon_0 A}{d}$$
$$C_2 = \frac{\varepsilon_r \varepsilon_0 A}{d}$$

Taking the ratio of $C_1$ and $C_2$

$$\frac{C_1}{C_2} = \frac{1}{\varepsilon_r}$$
$$\Rightarrow \quad C_2 = \varepsilon_r C_1$$

Let $C_1 = C$

Then $C_2 = 4C \qquad \{\because \varepsilon_r = 4\}$

Since two capacitance are connected in series so, equivalent capacitance will be

$$\frac{1}{C_{eq}} = \frac{1}{C_1} + \frac{1}{C_2}$$
$$C_{eq} = \frac{C_1 C_2}{C_1 + C_2}$$
$$4\,\mu F = \frac{C \times 4C}{C + 4C}$$
$$\Rightarrow \quad C = 5\,\mu F$$

So, $C_1 = 5\,\mu$ F and $C_2 = 20\,\mu F$

**(ii)**

$$C_{eq} V_{net} = Q_{Total}$$
$$4\mu F \times 15V = Q_{Total}$$
$$Q_{Total} = 60\,\mu C$$

Since in series configuration charge on each capacitor is equal.

Hence, $Q_1 = Q_2 = Q_{Total}$

$$= 60\,\mu C$$

Using, $Q = CV$

$$V_1 = \frac{Q_1}{C_1} = \frac{60\,\mu C}{5\,\mu F} = 12\,V$$
$$V_2 = \frac{Q_2}{C_2} = \frac{60\,\mu C}{20\,\mu F} = 3\,V$$

**(iii)**

$$U_1 = \frac{1}{2}\frac{Q_1^2}{C_1} = \frac{1}{2}\frac{(60\,\mu C)^2}{5\,\mu F} = 360\,\mu J$$
$$U_2 = \frac{1}{2}\frac{Q_2^2}{C_2} = \frac{1}{2}\frac{(60\,\mu C)^2}{20\,\mu F} = 90\,\mu J$$
$$\Rightarrow \quad \frac{U_1}{U_2} = \frac{4}{1}$$
$$\Rightarrow U_1 : U_2 = 4 : 1$$

**16. Two long straight parallel conductors carry steady current $I_1$ and $I_2$ separated by a distance $d$. If the currents are flowing in the same direction, show how the magnetic field set up if one produces an attractive force on the other. Obtain the expression for this force. Hence define one ampere.** **[3]**

**Ans.** Magnetic field produced on the wire (carrying current $I_2$) due to $I_1$ will be the same force towards wire 2. Thus both wines attracted each other with same force F.

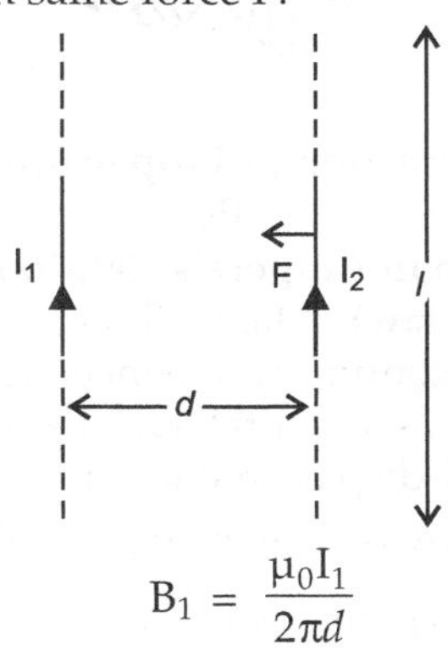


$$B_1 = \frac{\mu_0 I_1}{2\pi d}$$

Force acting at $l$ length is

$$F = I_2 l B_1$$

$$F = \frac{\mu_0 I_1 I_2 l}{2\pi d} \text{ towards } I_1$$

According to fleming Left-hand rule, the force on wire 2 acts in the plane of paper perpendicular to wire 2 directed towards wire 1. Similarly wire 1 also experience the same force towards wires. Thus both wires attract each other with same force F.

If $l = 1$ m, $d = 1$m, $I_1 = I_2 = I$ and $F = 2 \times 10^{-7}$ N

$\Rightarrow$ $I = 1$ A

So one ampere is defined as the current, which when maintained in two parallel infinite length conductors, held at a separation of one metre will produce a force of $2 \times 10^{-7}$ N per metre on each conductor.

**17. How are e.m. waves produced by oscillating charges ?**

**Draw a sketch of linearly polarized e.m. waves propagating in the Z-direction. Indicate the directions of the oscillating electric and magnetic fields. [3]**

**OR**

**Write Maxwell's generalization of Ampere's circuital law. Show that in the process of charging a capacitor, the current produced within the plates of the capacitor is $i = \varepsilon_0 \frac{d\phi_E}{dt}$**

**where $\phi_E$ is the electric flux produced during charging of the capacitor plates.**

**Ans.** A charge oscillating with some frequency, produces an oscillating electric field in space, which in turn produces an oscillating magnetic field perpendicular to the electric field. This process goes on repeating, producing e.m. waves in space perpendicular to both the fields.

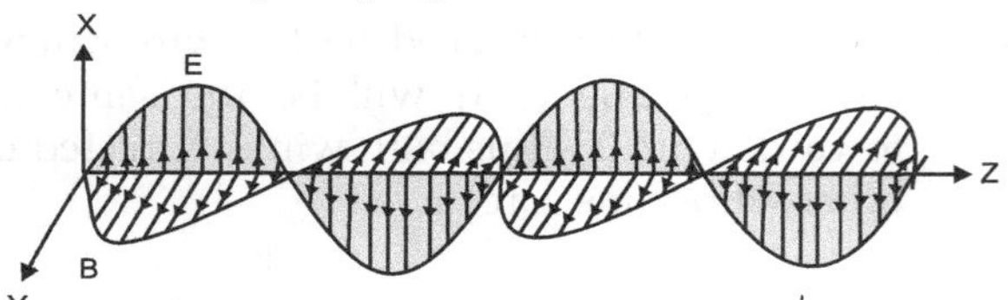


Direction of Propagation

**OR**

**Correction in Ampere's circuital law (Modified Ampere's law) :** Maxwell removed the problem of current continuity and inconsistency observed in Ampere's circuital law by introducing the concept of displacement current. Displacement current arises due to change in electric flux with time and is given by $i_d = \varepsilon_0 \frac{d\phi_E}{dt}$

Consider that a parallel capacitor is charging in a circuit.

Electric flux through the loop

$$\phi_E = EA$$

$$= \frac{\sigma}{\varepsilon_0} A = \frac{Q}{A}\frac{A}{\varepsilon_0} = \frac{Q}{\varepsilon_0}$$

(where, Q = charge on either plates)

$$\phi_E = \frac{Q}{\varepsilon_0}$$

$$\frac{d\phi_E}{dt} = \frac{1}{\varepsilon_0}\frac{dQ}{dt}$$

$$\varepsilon_0 \frac{d\phi_E}{dt} = \frac{dQ}{dt}$$

$\frac{dQ}{dt}$ is called conduction current which is equal to $\varepsilon_0 \frac{d\phi_E}{dt}$ which is displacement current.

Hence, $i_c = i_d$

Generalization of Ampere's circuital law is :

$$\oint \vec{B}.\vec{dt} = \mu_0 (i_c + i_d)$$

Conduction current is due to the flow of charges but displacement current is not because of the flow of charges but it is due to the change in electric flux.

**18. (a) Explain any two factors which justify the need of modulating a low frequency signal.****

**(b) Write two advantages of frequency modulation over amplitude modulation.** [3]**

**19. (a) Write the functions of three segments of a transistor.**

**(b) Draw the circuit diagram for studying the input and output characteristics of *n-p-n* transistor in common emitter configuration. Using the circuit, explain how input, output characteristics are obtained.** [3]**

**20. (a) Calculate the distance of an object of height $h$ from a concave mirror of radius of curvature 20 cm, so as to obtain a real image of magnification 2. Find the location of image also.**

**(b) Using mirror formula, explain why does a convex mirror always produce a virtual image. [3]**

**Ans. (a)** Given, Height of object = $h$

Radius of curvature = – 20 cm

Magnification, $m$ = – 2

Object distance, $u$ = ?

Image distance, $v$ = ?

$$\text{Magnification, M} = \frac{-v}{u} = \frac{h_i}{h_O}$$

$$-2 = -\frac{v}{u}$$

$$v = 2u \quad ...(i)$$

Using mirror formula,

$$\frac{1}{v}+\frac{1}{u} = \frac{1}{f}$$

$$\frac{1}{2u}+\frac{1}{u} = \frac{-2}{20}$$

$$\frac{+3}{2u} = \frac{-2}{20}$$

$$u = -\frac{60}{4} = -15 \text{ cm}$$

**(b)** For convex mirror,

$$f = +\text{ve (always)}$$

Mirror formula,

$$\frac{1}{v}+\frac{1}{u} = \frac{1}{f}$$

$$\frac{1}{v} = \frac{1}{f}-\frac{1}{u}$$

As, $u = -ve$ (for real object)

$$\frac{1}{v} = \frac{1}{f}-\left(\frac{1}{-u}\right)$$

$$\frac{1}{v} = \frac{1}{f}+\frac{1}{u}$$

$$v = +ve$$

Hence, it will form virtual and erect image.

**21. (a) State Bohr's quantization condition for defining stationary orbits. How does de-Broglie hypothesis explain the stationary orbits ?**

**(b) Find the relation between the three wavelengths $\lambda_1$, $\lambda_2$ and $\lambda_3$ from the energy level diagram shown below. [3]**

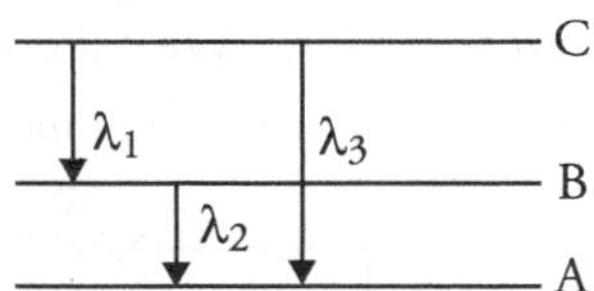


**Ans. (a) Quantization condition** : Of all possible circular orbits allowed by the classical theory, the electrons are permitted to circulate only in those orbits in which the angular momentum of an electron is an integral multiple of $\frac{h}{2\pi}$; $h$ being Planck's constant.

Therefore, for any permitted orbit,

$$L = mvr = \frac{nh}{2\pi};\ n = 1, 2, 3, .......$$

Where L, $m$, and $v$ are the angular momentum, mass and speed of the electron respectively, $r$ is the radius of the permitted orbit and $n$ is positive integer called principle quantum number.

The above equation is Bohr's famous quantum condition. When an electron of mass $m$ is confined to move in a line of length $l$ with velocity $v$, the de-Broglie wavelength $\lambda$ associated with electron is :

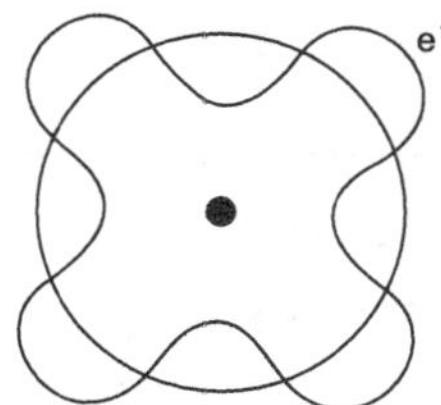


$$\lambda = \frac{h}{mv} = \frac{h}{p}$$

or $p$ = Linear momentum

$$\Rightarrow \quad p = \frac{h}{\lambda} = \frac{h}{2l/n} = \frac{nh}{2l}$$

When electron revolves in a circular orbit of radius '$r$' then $2l = 2\pi r$.

$$\therefore \quad p = \frac{nh}{2\pi r} \text{ or } p \times r = \frac{nh}{2\pi}$$

or angular momentum $|\vec{L}| = p \times r$ is an integral multiple of $h/2\pi$, which is Bohr's quantisation of angular momentum.

**(b)**

$$E_{CB} = \frac{hc}{\lambda_1}$$

$$E_{BA} = \frac{hc}{\lambda_2}$$

$$E_{CA} = \frac{hc}{\lambda_3}$$

Now, $E_{CA} = E_{CB} + E_{BA}$

where,

$E_{CB}$ = Energy gap between level B and C,

$E_{BA}$ = Energy gap between level A and B,

$E_{CA}$ = Energy gap between level A and C.

$$\frac{hc}{\lambda_3} = \frac{hc}{\lambda_1} + \frac{hc}{\lambda_2}$$

$$\frac{1}{\lambda_3} = \frac{1}{\lambda_1} + \frac{1}{\lambda_2}$$

$$\lambda_3 = \frac{\lambda_1\lambda_2}{\lambda_2 + \lambda_1}$$

**22. Draw a schematic ray diagram of reflecting telescope showing how rays coming from a distant object are received at the eye-piece. Write its two important advantages over a refracting telescope.** [3]

**Ans.**

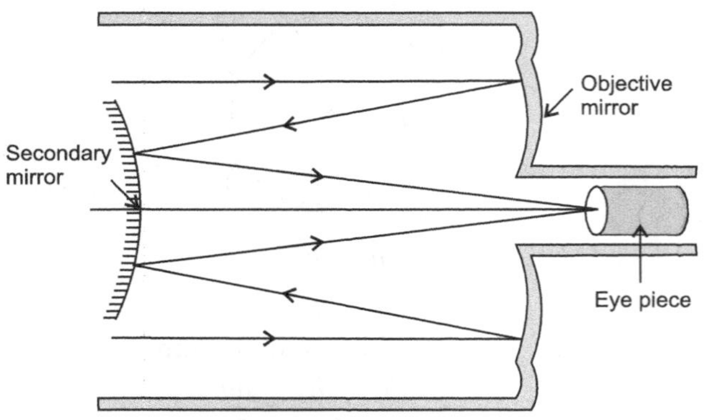


**Ans. Advantages :**

(i) The resolving power (the ability to observe two object distinctly) is high, due to the large diameter of the objective.

(ii) There is no chromatic aberration as the object is a mirror.

## SECTION – D

**23. Meeta's father was driving her to the school. At the traffic signal she noticed that each traffic light was made of many tiny lights instead of a single bulb. When Meeta asked this question to her father, he explained the reason for this.**

**Answer the following questions based on above information:**

**(i) What were the values displayed by Meeta and her father?****

**(ii) What answer did Meeta's father give ?**

**(iii) What are the tiny lights in traffic signals called and how do these operate ?** [4]

**Ans. (ii)** Meeta's father said that these are LED light which consume less power and have high reliability.

(iii) The tiny lights in traffic signals are Light Emitting Diode. These are operated by connecting the *p-n* junction diode in forward biased condition and work on the principle of de-excitation.

## SECTION–E

**24. (a) An a.c. source of voltage $V = V_0 \sin \omega t$ is connected to a series combination of L, C and R. Use the phasor diagram to obtain expressions for impedance of the circuit and phase angle between voltage and current. Find the condition when current will be in phase with the voltage. What is the circuit in this condition called ?**

**(b) In a series LR circuit $X_L = R$ and power factor of the circuit is $P_1$. When capacitor with capacitance C such that $X_L = X_C$ is put in series, the power factor becomes $P_2$. Calculate $P_1/P_2$.** [5]

**OR**

**(a) Write the function of a transformer. State its principle of working with the help of a diagram. Mention various energy losses in this device.**

**(b) The primary coil of an ideal step up transformer has 100 turns and transformation ratio is also 100. The input voltage and power are respectively 220 V and 1100 W. Calculate :**

**(i) number of turns in secondary.**

**(ii) current in primary.**

**(iii) voltage across secondary.**

**(iv) current in secondary.**

**(v) power in secondary.**

**Ans. (a)** Let a series LCR circuit is connected to an a.c. source V (Fig). We take the voltage of the source to be $V = V_0 \sin \omega t$

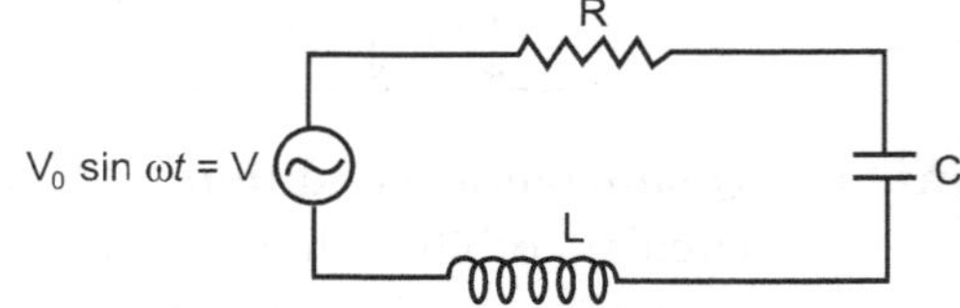


The a.c. current in each element is the same at any time, having the same amplitude and phase. It is given by,

$$I = I_0 \sin (\omega t + \phi)$$

**Answer is not given due to change in the present Syllabus.

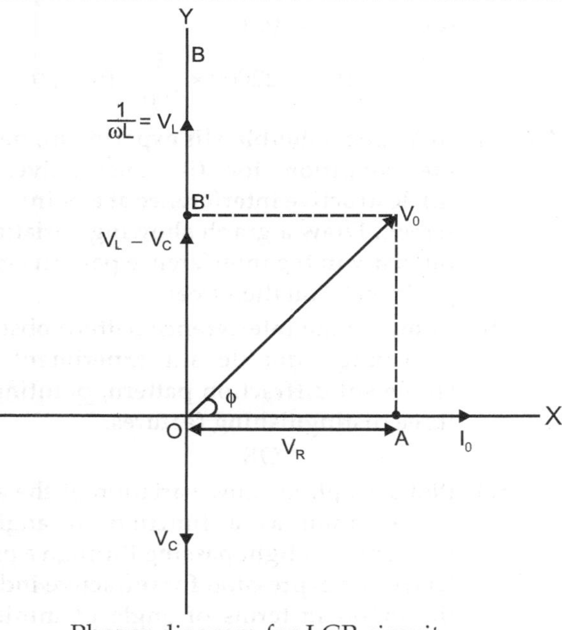


Phasor diagram for LCR circuit

Let $V_L$, $V_R$, $V_C$ and V represent the voltage across the inductor, resistor, capacitor and the source respectively. But $V_L = I_0X_L$, $V_R = I_0R$ and $V_c = I_0X_c$

Let $V_L > V_C$

$\vec{V}_0$ is the resultant of $\vec{V}_R$ and $(\vec{V}_L - \vec{V}_C)$

$$\Rightarrow \vec{V}_0 = \vec{V}_R + (\vec{V}_L + \vec{V}_C)$$

$$\therefore \quad V_0^2 = V_R^2 + (V_L - V_C)^2$$

$$V_0^2 = (I_0R)^2 + (I_0X_L - I_0X_C)^2$$

$$V_0^2 = I_0^2\left[R^2 + (X_L - Xc)^2\right]$$

and
$$I_0 = \frac{V_0}{\sqrt{R^2 + (X_L - X_C)^2}}$$

$$\Rightarrow \quad I_0 = \frac{V_0}{Z}$$

Where $Z = \sqrt{R^2 + (X_L - X_C)^2}$

It is called the impedance in an a. c. circuit.

From the figure,

$$\tan\phi = \frac{V_L - V_C}{V_R}$$

$$= \frac{I_0 X_L - I_0 X_C}{I_0 R}$$

$$\tan\phi = \frac{X_L - X_C}{R}$$

$$\Rightarrow \quad \phi = \tan^{-1}\left(\frac{X_L - X_C}{R}\right)$$

**Condition** : The current will be in phase with the voltage at resonance condition.

At resonance condition,

$$X_L = X_C$$

$$\omega L = \frac{1}{\omega C}$$

$$\omega = \frac{1}{\sqrt{LC}}$$

$$2\pi f = \frac{1}{\sqrt{LC}}$$

$$f = \frac{1}{2\pi\sqrt{LC}}$$

The circuit is known as non-inductive circuit.

**(b)** As $\cos\phi = \frac{R}{Z}$

In LR circuit,

$$P_1 = \cos\phi$$

$$P_1 = \frac{R}{\sqrt{R^2 + X_L^2}} = \frac{R}{\sqrt{2R^2}} \quad [X_L = R]$$

$$P_1 = \frac{1}{\sqrt{2}}$$

In LCR circuit when, $X_L = X_C$

$$P_2 = \frac{R}{\sqrt{R^2 + (X_L - X_C)^2}}$$

$$= \frac{R}{R} = 1 \quad [X_L = X_C]$$

$$\therefore \quad \frac{P_1}{P_2} = \frac{1}{\sqrt{2}}$$

**OR**

**(a)** A transformer is an electrical device for converting an alternating current at low voltage into high voltage or vice-versa.

1. If it increases the input a.c. voltage and decreases the, it is called step up tranformer.
2. If it decreases the input a.c. voltage and in ccreases the current, it is called step down transformer.

**Principle** : It works on the principle of mutual induction *i.e.,* when a changing current or flux is passed through one of the two inductively coupled coils, an induced emf is set up in the other coil.

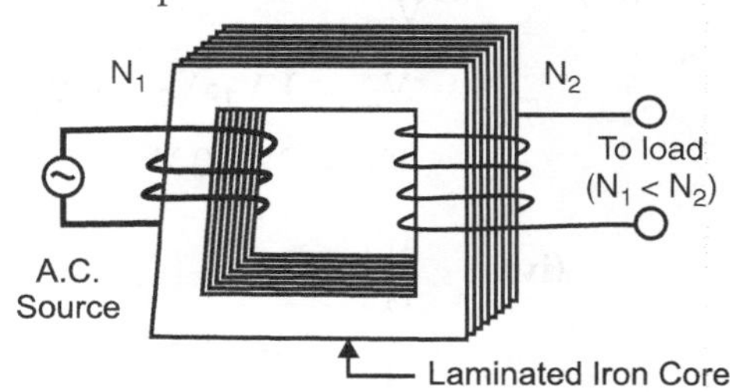


Step-up transformer

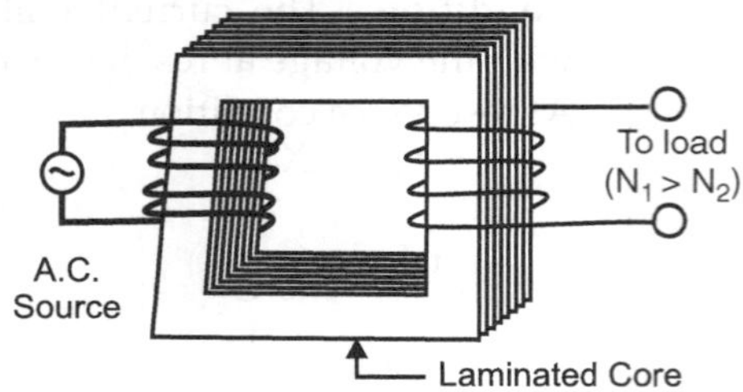

Step-down transformer

Energy losses in transformer :

1. **Copper loss** : Some energy is lost due to the heating of copper wires used in the primary and secondary windings. This power loss ($P = I^2R$) can be minimised by using thick copper wires of low resistance.
2. **Eddy current loss or Iron loss** : The alternating magnetic flux induces eddy current in the iron core which leads to some energy loss in the form of heat. This loss can be reduced by using laminated iron core.
3. **Hysteresis loss** : When the iron core is subjected to a cycle of magnetisation the core gets heated up due to hysteresis, having low hysteresis loop.
4. **Flux leakage** : The magnetic flux produced by the primary may not fully pass through the secondary. Some of the flux may leak into air. This loss can be minimised by winding the primary and secondary coils over one another.

**(b)** Given, $N_1 = 100$

$K = 100$

$V_1 = 220$ V

$P_1 = 1100$ W

**(i)** As, $K = \frac{N_2}{N_1}$

$N_2 = KN_1 = 100 \times 100$

$N_2 = 10000$

**(ii)** $P_1 = V_1I_1$

$I_1 = \frac{P_1}{V_1} = \frac{1100}{220} = 5$ A

**(iii)** $\frac{V_2}{V_1} = K$

$V_2 = KV_1; V_2 = 100 \times 220$

$V_2 = 22000$ V

**(iv)** $\frac{I_1}{I_2} = K$

$I_2 = \frac{I_1}{K} = \frac{5}{100}$

$I_2 = 0.05$ A

**(v)** $P_2 = V_2I_2$

$P_2 = 22000 \times \frac{5}{100}; P_2 = 1100$ W

**25. (a) In Young's double slit experiment, deduce the condition for (i) constructive, and (ii) destructive interference at a point on the screen. Draw a graph showing variation of intensity in the interference pattern against position '$x$' on the screen.**

**(b) Compare the interference pattern observed in Young's double slit experiment with single slit diffraction pattern, pointing out three distinguishing features. [5]**

**OR**

**(a) Plot a graph to show variation of the angle of deviation as a function of angle of incidence for light passing through a prism. Derive an expression for refractive index of the prism in terms of angle of minimum deviation and angle of prism.**

**(b) What is dispersion of light ? What is its cause ?**

**(c) A ray of light incident normally on one face of a right isosceles prism is totally reflected as shown in fig. What must be the minimum value of refractive index of glass? Give relevant calculations.**

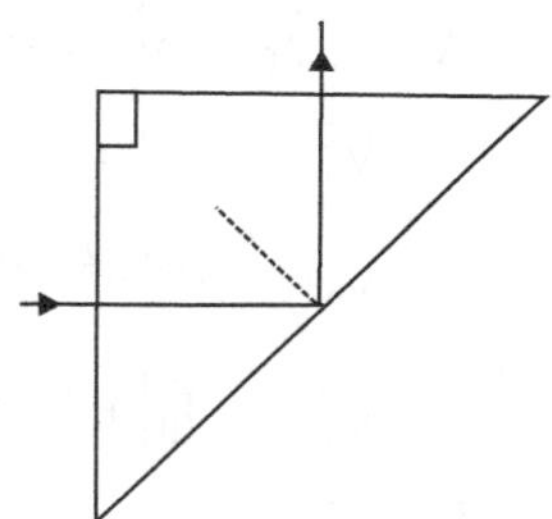

**Ans. (a)** Let the two waves arising from the slits A and B have the amplitudes $a$ and $b$ and the phase difference $\phi$. Such that $y_1 = a \sin \omega t$ and $y_2 = b \sin (\omega t + \phi)$.

The resultant displacement is given as :

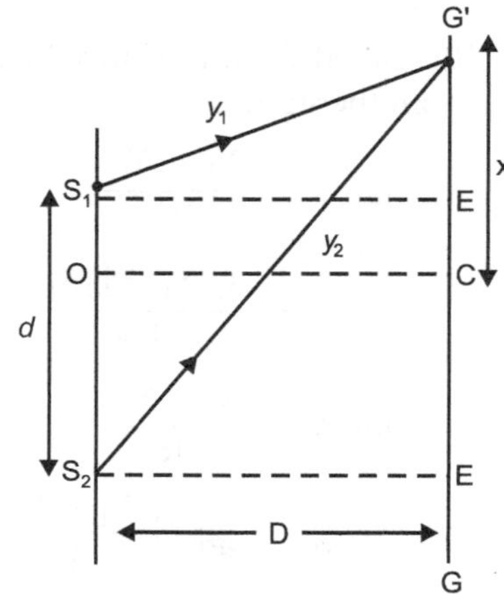

$$y = y_1 + y_2$$
$$y = a \sin \omega t + b \sin (\omega t + \phi)$$
$$y = a \sin \omega t + b \sin \omega t \cos \phi + b \cos \omega t \sin \phi$$
$$y = (a + b \cos \phi) \sin \omega t + b \sin \phi \cos \omega t \quad ...(i)$$

Let $a + b \cos \phi = A \cos \delta$ ...(ii)

and $b \sin \phi = A \sin \delta$ ...(iii)

Hence, $y = A \sin \omega t \cos \delta + A \cos \omega t \sin \delta$

$$y = A \sin (\omega t + \delta) \quad ...(iv)$$

Where the amplitude A of the resultant wave can be given as:

$$A = \sqrt{a^2 + b^2 + 2ab \cos \phi} \quad ...(v)$$

and $\tan \delta = \dfrac{b \sin \phi}{a + b \cos \phi}$ ...(vi)

**(i) Constructive interference** : Intensity $I \propto A^2$ and for A to be maximum

$$\cos \phi = 1$$

or $\cos \phi = \cos 2n\pi, n = 0, 1, 2, 3, .....$

$$\phi = 2n\pi \quad ...(i)$$

and path difference

$$\Delta x = n\lambda \quad ...(ii)$$
$$A_{max} = a + b$$
$$I \rightarrow I_{max} = k(a + b)^2$$

**(ii) Destructive interference** : For I to be minimum

$$\cos \phi = -1$$

Phase difference :

$$\Delta\phi = (2n + 1)\pi$$

and path difference :

$$\Delta x = (2n + 1)\frac{\lambda}{2}$$
$$A_{min} = a - b$$
$$I \rightarrow I_{min} = k(a - b)^2$$

Graph showing interference pattern against position 'x' on the screen.

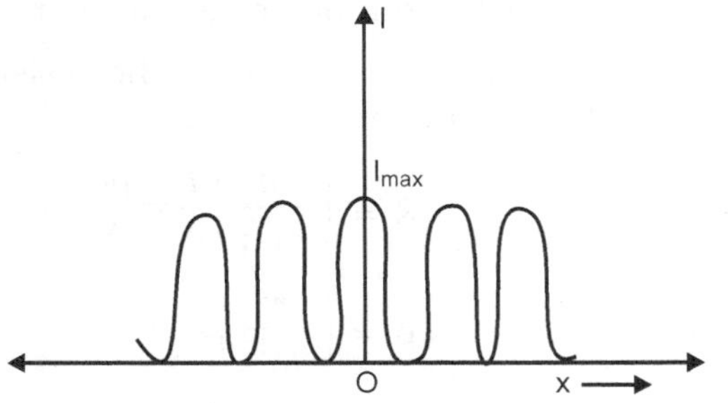


**(b)** Comparison of interference pattern observed in Young's double slits and the single slit diffraction :

| S. No. | Interference | Diffraction |
|---|---|---|
| 1. | Interference is the result of superposition of secon dary waves starting from two different wave fronts originating from two coherent sources. | Diffraction is the result of superposition of secondary waves starting from different part of same wavefront. |
| 2. | All bright and dark fringes are of equal width. | The width of central bright fringe is twice the width of any secondary maxima. |
| 3. | All bright fringes are of same intensity | Intensity of bright fringes decreases as we move away from central bright fringes on either side. |

**OR**

**(a)**

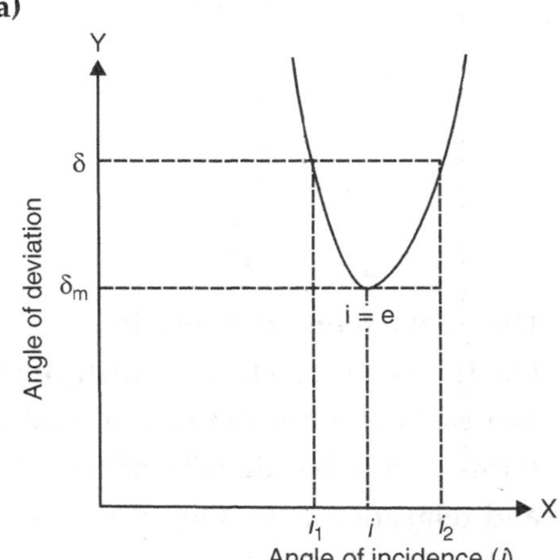


From figure, $\delta = \delta_m, i = e$ which implies $r_1 = r_2$

$$2r = A, \text{ or } r = A/2$$

Using $\delta = i + e - A$

$$\delta_m = 2i - A$$
$$i = \frac{A + \delta_m}{2}$$
$$\mu = \frac{\sin i}{\sin r} = \frac{\sin\left(\frac{A + \delta_m}{2}\right)}{\sin A/2}$$

(b) **Dispersion of light** : Dispersion is often observed as light passes through a triangular prism. Upon passing through the prism, the white light is separated into its component colours : red, orange, yellow, green, blue, and voilet. The separation of visible light into its different colours is known as dispersion. Dispersion occurs because for different colour of light a transparent medium will have different refractive indices (μ). as different colours have different speed in transparent medium.

(c) **For total internal reflection :**

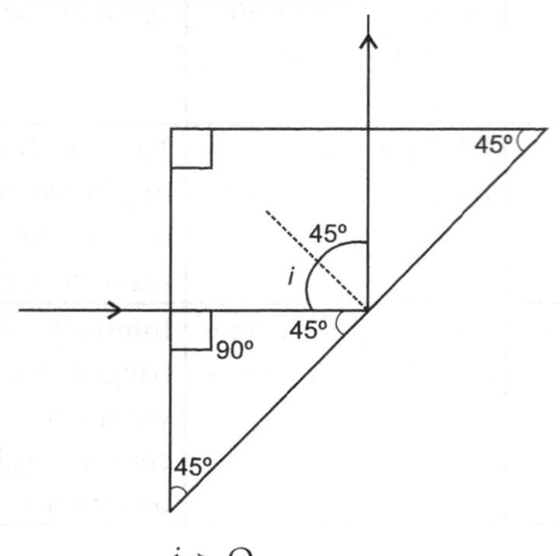


$$i \geq Q_c$$
$$\sin i \geq \sin Q_c$$
$$\sin 45° \geq \frac{1}{\mu}$$
$$\frac{1}{\sqrt{2}} \geq \frac{1}{\mu}$$
$$\mu \geq \sqrt{2}$$
$$\mu_{min} = \sqrt{2}$$

**26. (a) Define the term drift velocity.**

**(b) On the basis of electron drift, derive an expression for resistivity of a conductor in terms of number density of free electrons and relaxation time. On what factors does resistivity of a conductor depend ?**

**(c) Why alloys like constantan and manganin are used for making standard resistors ?** [5]

OR

**(a) State the principle of working of a potentiometer.**

**(b) In the following potentiometer circuit AB is a uniform wire of length 1 m and resistance 10 Ω. Calculate the potential gradient along the wire and balance length AO (= $l$).**

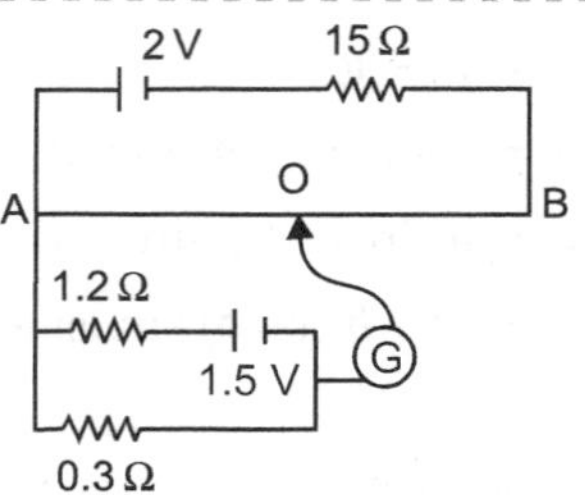


**Ans. (a)** Drift velocity is defined as the average velocity with which the free electrons are drifted towards the positive terminal under the effect of applied electric field. Thermal velocities are randomly distributed and average thermal velocity is zero.

$$\frac{\vec{u}_1 + \vec{u}_2 + \dots\dots\dots + \vec{u}_N}{N} = 0$$

$$i.e.\ v_d = -\frac{eE\tau}{m}$$

(b) We know that the current flowing through the conductor is :

$$I = nAev_d$$

$$\therefore \quad I = neA\left(-\frac{eE\tau}{m}\right)$$

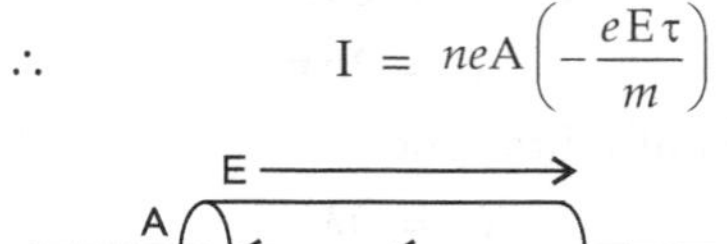


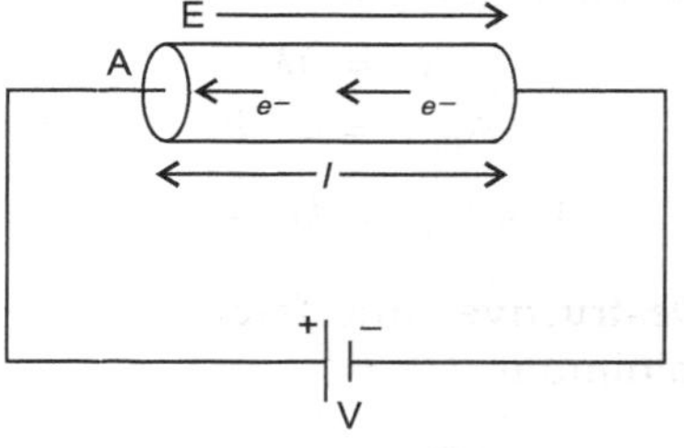


Using $\quad E = -\frac{V}{l}$

$$I = neA\left(\frac{eV}{ml}\right)\tau$$

$$= \left(\frac{ne^2A\tau}{ml}\right)V = \frac{1}{R}V$$

$I \propto V \rightarrow$ by Ohm's law

Where, $R = \dfrac{ml}{nAe^2\tau}$ is a constant for a particular conductor at a particular temperature and is called the resistance of the conductor.

$$R = \left(\frac{m}{ne^2\tau}\right)\frac{l}{A} = \frac{\rho l}{A}$$

$$\rho = \left(\frac{m}{ne^2\tau}\right)$$

Where $\rho$ is the specific resistance or resistivity of the material of the wire. It depends on number of free electron per unit volume and temperature.

**(c)** They are used to make standard resistors because :

1. They have high value of resistivity.
2. Temperature coefficient of resistance is less.
3. They are least affected by temperature.

**OR**

**(a) Principle of potentiometer** : The basic principle of potentiometer is that when a constant current flows through a wire of uniform cross-section area then the potential drop across any length of the wire is directly proportional to that length. A potentiometer is a device used to measure an unknown emf or potential difference and internal resistance of a cell accurately.

**(b)** Total resistance of the primary circuit $= 15 + 10 = 25\ \Omega$,

$$\text{emf} = 2\ \text{V}$$

$\therefore$ Current in the wire AB

$$I = \frac{2}{25} = 0.08\text{A}$$

P. D. across the wire AB = Current × Resistance of wire AB

$$= 0.08 \times 10 = 0.8\ \text{V}$$

$$\text{Potential gradient} = \frac{\text{P. D.}}{\text{Length}} = \frac{0.8}{100}$$

$$= 0.008\ \text{V cm}^{-1}$$

Resistance of secondary circuit

$$= 1.2 + 0.3 = 1.5\ \Omega$$

$$\text{emf} = 1.5\ \text{V}$$

Current in the secondary circuit $= \frac{1.5}{1.5} = 1.0\ \text{A}$

The same is the current in 0.3 Ω resistor.

P. D. between points A and O,

P. D. across 0.3 Ω resistor in the zero-deflection condition.

$$= \text{Current} \times \text{Resistance}$$

$$= 1.0 \times 0.3 = 0.3\text{V}$$

$$\text{Length AO} = \frac{\text{Potential difference}}{\text{Potential gradient}}$$

$$= \frac{0.3\ \text{V}}{0.008\ \text{Vcm}^{-1}} = 37.5\ \text{cm}$$

●●

# PHYSICS

## Outside Delhi [Set-I] 2015

**Time allowed : 3 hours** **Maximum marks : 70**

### SECTION–A

1. **Define the term 'self-inductance' of a coil. Write its S.I. unit.** **[1]**

**Ans.** Self-inductance of a coil is numerically equal to the amount of magnetic flux linked with the coil when unit current flows through the coil.

The S.I. unit of self inductance is henry (H) or weber per ampere.

$$1 \text{ H} = 1 \text{ Wb/A}$$

2. **Why does bluish colour predominate in a clear sky ?** **[1]**

**Ans.** While light from the sun reaches the atmosphere that is comprised of the tiny particles of the atmosphere. These act as a prism and cause the different components to scatter. As blue light travels in shorter and smaller waves in comparison to the other colours of the spectrum, it is scattered the most, causing the sky to appear bluish.

3. **I-V graph for a metallic wire at two different temperatures, $T_1$ and $T_2$ is as shown in the figure.**
**Which of the two temperature is lower and why ?** **[1]**

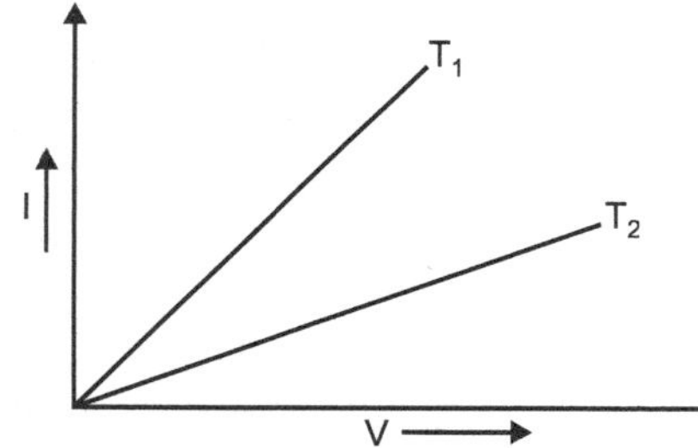


**Ans.** The slope of a V-I graph is given by the formula $\frac{I}{V} = \frac{1}{R}$. Thus, the smaller the slope larger is the resistance. As the resistance of a metal increases with the increase in temperature, so resistance at $T_2$ is higher and $T_1$ is lower.

4. **Which basic mode of communication is used for telephonic communication ?**** **[1]**

5. **Why do the electrostatic field lines not form closed loops ?** **[1]**

**Ans.** Electrostatic field lines never form loops because they do not converge at only one point as in the case with magnetic field lines. They depend on +ve and –ve charges that can extend to infinity in any particular direction.

### SECTION–B

6. **When an electron in hydrogen atom jumps from the third excited state to the ground state, how would the de-Broglie wavelength associated with the electron change ? Justify your answer.** **[2]**

**Ans.** de-Broglie wavelength associated with a moving charge particle having a K.E. 'K' can be given as

$$\lambda = \frac{h}{p} = \frac{h}{\sqrt{2mK}} \left[K = \frac{1}{2}mv^2 = \frac{p^2}{2m}\right] \quad \text{... (i)}$$

The kinetic energy of the electron in any orbit of hydrogen atom can be given as

$$K = -E = -\left(\frac{13.6}{n^2} eV\right) = \frac{13.6}{n^2} eV \quad \text{... (ii)}$$

Let $K_1$ and $K_4$ be the K.E. of the electron in ground state and third excited state, where $n_1 = 1$ shows ground state and $n_2 = 4$ shows third excited state. Using the concept of equations (i) and (ii), we have

$$\frac{\lambda_1}{\lambda_4} = \sqrt{\frac{K_4}{K_1}} = \sqrt{\frac{n_1^2}{n_2^2}}$$

$$\frac{\lambda_1}{\lambda_4} = \sqrt{\frac{1^2}{4^2}} = \frac{1}{4}$$

$$\Rightarrow \quad \lambda_1 = \frac{\lambda_4}{4}$$

*i.e.*, the wavelength in the ground state will decrease.

7. **Write two factors which justify the need of modulating a low frequency signal into high frequencies before transmission.**** **[2]**

8. **Use Kirchhoff's rules to determine the potential difference between the points A and D when no current flows in the arm BE of the electric network shown in the figure.** **[2]**

---

** Answer is not given due to change in the present Syllabus.

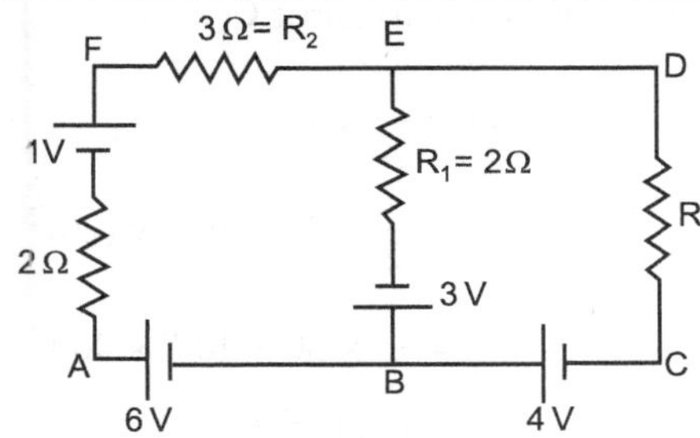


**Ans.**

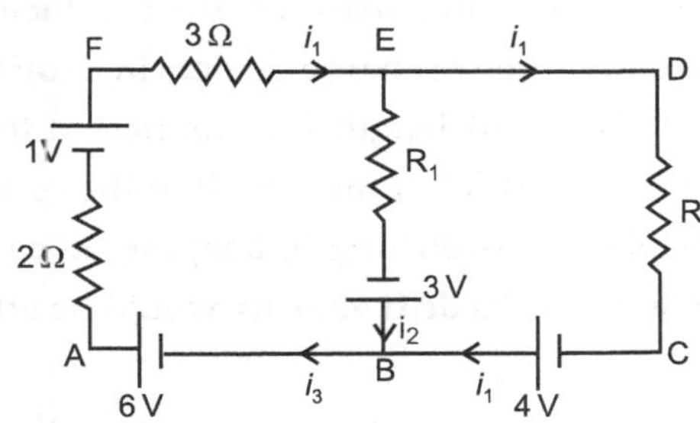


According Kirchhoff's junction law at B,

$$i_3 = i_1 + i_2$$

As $\quad i_2 = 0$ (given)

$\therefore \quad i_3 = i_1$

Applying kirchhoff's second law to loop AFEB,

$$\therefore \quad i_3 \times 2 + i_1 \times 3 + i_2 R_1 = 1 + 3 + 6$$

$$2i_3 + 3i_3 = 10$$

$$5i_3 = 10$$

$$i_3 = i_1 = 2 \text{ A}$$

The potential difference between A to D along the branch AFD,

$$\therefore \quad V_{AD} = 2i_3 - 1 + 3 \times i_1$$

$$= (4 - 1 + 6)\text{V}$$

$$= 9 \text{ V}$$

**9. You are given two converging lenses of focal lengths 1.25 cm and 5 cm to design a compound microscope. If it is desired to have a magnification of 30, find out the separation between the objective and the eyepiece. [2]**

**OR**

**A small telescope has an objective lens of focal length 150 cm and eye-piece of focal length 5 cm. What is the magnifying power of the telescope for viewing distant objects in normal adjustment ?**

**If this telescope is used to view a 100 m tall tower 3 km away, what is the height of the image of the tower formed by the objective lens ?**

**Ans.** Given, $f_e = 5$ cm; $f_o = 1.25$ cm and M = – 30.

Let L be the tube length (distance between the objective and the eyepiece)

$$\therefore \quad M = \frac{-L}{f_o}\left(1 + \frac{d}{f_e}\right)$$

[$d$ is a constant and equals to the normal distance of clear vision of the human eye]

Hence, $$-30 = -\frac{L}{1.25}\left(1 + \frac{25}{5}\right)$$

$$\therefore \quad L = \frac{30 \times 1.25}{6}$$

$$= 6.25 \text{ cm}$$

Hence, the tube length = 6.25 cm.

**OR**

If the telescope is in normal adjustment, *i.e.*, the final image is at infinity.

$$M = \frac{f_o}{f_e}$$

Since $f_o = 150$ cm, $f_e = 5$ cm

$$M = \frac{150}{5} = 30$$

If tall tower is at distance 3 km from the objective lens of focal length 150 cm. It will form its image at distance $v_0$. So,

$$\frac{1}{f_o} = \frac{1}{v_o} - \frac{1}{u_o}$$

$$\frac{1}{150 \text{ cm}} = \frac{1}{v_o} - \frac{1}{(-3\ km)}$$

$$\frac{1}{v_o} = \frac{1}{1.5\ m} - \frac{1}{3000\ m}$$

$$v_0 = \frac{3000 \times 1.5}{3000 - 1.5} = \frac{4500}{2998.5} = 1.5 \text{ m}$$

$$\text{Magnification, } m_0 = \frac{I}{O} = \frac{h_i}{h_o} = \frac{v_o}{u_o}$$

$$\frac{h_i}{100\ m} = \frac{1.5\ m}{3\ km} = \frac{1.5\ m}{3000 m}$$

$$h_i = \frac{1.5 \times 100}{3000} = \frac{1}{20} m$$

$$h_i = 0.05 \text{ m}$$

**10. Calculate the shortest wavelength in the Balmer series of hydrogen atom. In which region (infrared, visible, ultraviolet) of hydrogen spectrum does this wavelength lie ? [2]**

**Ans.** The formula for wavelength (λ) in Balmer series is :

$$\frac{1}{\lambda} = R\left(\frac{1}{2^2} - \frac{1}{n^2}\right) \text{ where } n = 3, 4, 5 \text{ and}$$

$R = 1.097 \times 10^7 m^{-1}$ is the Rydberg constant.

Now, the shortest wavelength in Balmer series when $n = \infty$ is

$$\frac{1}{\lambda} = 1.097 \times 10^7 \text{ m}^{-1} \times \left(\frac{1}{2^2} - \frac{1}{\infty^2}\right)$$

$$\frac{1}{\lambda} = 1.097 \times 10^7 \times (0.25 - 0)$$

$$\frac{1}{\lambda} = 0.2743 \times 10^7$$

$$\lambda = 3.6456 \times 10^{-7}$$

$$= 364.56 \times 10^{-9} \text{ m}$$

$$= 364.56 \text{ nm.}$$

This wavelength lies in the visible region of the hydrogen spectrum.

## SECTION–C

**11. Calculate the potential difference and the energy stored in the capacitor $C_2$ in the circuit shown in the figure. Given potential at A is 90 V, $C_1 = 20\ \mu F$, $C_2 = 30\ \mu F$ and $C_3 = 15\ \mu F$. [3]**

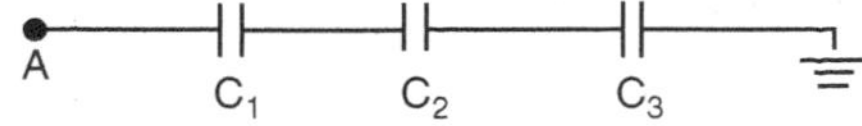


**Ans.** The given capacitors are connected in series,

$$\therefore \quad \frac{1}{C} = \frac{1}{C_1} + \frac{1}{C_2} + \frac{1}{C_3}$$

$$= \frac{1}{20} + \frac{1}{30} + \frac{1}{15}$$

$$= \frac{3+2+4}{60} = \frac{9}{60} = \frac{3}{20}$$

$$C = \frac{20}{3} \mu F,$$

As the capacitors are connected in series, so the charge remains same.

$$\therefore \quad Q = CV = \frac{20}{3} \times 90 = 600\ \mu C$$

Potential difference across $C_2$,

$$V_2 = \frac{Q}{C_2}$$

$$= \frac{600}{30} = 20 \text{ V}$$

Energy stored in capacitor $C_2$,

$$U = \frac{1}{2} C_2 (V_2)^2$$

$$= \frac{1}{2} \times 30 \times 10^{-6} \times (20)^2$$

$$= \frac{1}{2} \times 30 \times 10^{-6} \times 400$$

$$= 6 \times 10^{-3} \text{ J}$$

**12. Find the relation between drift velocity and relaxation time of charge carriers in a conductor. A conductor of length L is connected to a d.c. source of emf 'E'. If the length of the conductor is tripled by stretching it, keeping 'E' constant, explain how its drift velocity would be affected. [3]**

**Ans.** All free electrons suffer collisions with the heavy fixed ions inside the conductor. After collisions, these electrons again emerge with the same speed, but in random directions. So, at given time, net velocity of the electrons is zero *i.e.,*

$$\vec{u}_{avg} = \frac{\vec{u}_1 + \vec{u}_2 + .... + \vec{u}_n}{n} = 0$$

If the electric field established inside the conductor, electrons get accelerated, so

$$a = -\frac{eE}{m}$$

Now, the average velocity of all electrons is given by

$$\vec{v}_d = \vec{u}_{avg} - \frac{eE}{m}\tau$$

$$\vec{v}_d = 0 - \frac{eE}{m}\tau$$

$$\vec{v}_d = \frac{-eE}{m}\tau$$

where τ is the average relaxation time.

But $\quad E = \frac{V}{l}$

$$\vec{v}_d = -\frac{eV\tau}{ml}$$

$$v_d \propto \frac{1}{l}$$

As the drift velocity is inversely proportional to length of conductor, so the drift velocity would be reduced by one third if the length of the conductor is tripled.

**13. State clearly how an unpolarised light gets linearly polarised when passed through a polaroid. [3]**

**(i) Unpolarised light of intensity $I_0$ is incident on a polaroid $P_1$ which is kept near another polaroid $P_2$ whose pass axis is parallel to that of $P_1$. How will the intensities of light, $I_1$ and $I_2$, transmitted by the polaroids $P_1$ and $P_2$ respectively, change on rotating $P_1$ without disturbing $P_2$ ?**

**(ii) Write the relation between the intensities $I_1$ and $I_2$.**

**Ans.** Polaroid filters are made of a special material that is capable of blocking of the two planes of vibration of an electromagnetic wave. In this sense, a polaroid acts as a device that filter out half of the vibrations on transmission of the light through the filter. When unpolarised light is transmitted through a polaroid filter, it emerge with one-half the intensity and with vibrations in a single plane; it emerges as polarised light.

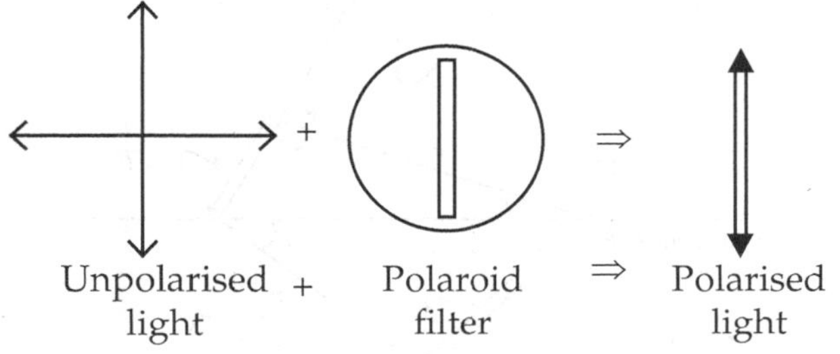


Unpolarised light + Polaroid filter ⇒ Polarised light

(i) When an unpolarised light of intensity $I_0$ is incident on a polaroid $P_1$ of the light intensity $I_1$ will become half of the incident intensity.

$$I_1 = \frac{I_0}{2}$$

and when it will pass through $P_2$ polaroid kept near P. whose pass axis is parallel to that of P, then, there is no change in the intensity but the angle will change by $\cos^2 \theta$

$$I_2 = I_1 \cos^2 \theta = \frac{I_0}{2} \cos^2 \theta$$

(ii) The relation between $I_1$ and $I_2$ is given by

$$\frac{I_2}{I_1} = \cos^2 \theta$$

**14. Define modulation index. Why is its value kept, in practice, less than one ?**

**A carrier wave of frequency 1.5 MHz and amplitude 50 V is modulated by a sinusoidal wave of frequency 10 kHz producing 50% amplitude modulation. Calculate the amplitude of the AM wave and frequencies of the side bands produced.** [3]**

** Answer is not given due to change in the present Syllabus.

**15. A uniform magnetic field B is set up along the positive X-axis. A particle of charge 'q' and mass 'm' moving with a velocity $v$ enters the field at the origin in X-Y plane such that it has velocity components both along and perpendicular to the magnetic field B. Trace, giving reason, the trajectory followed by the particle. Find out the expression for the distance moved by the particle along the magnetic field in one rotation. [3]**

**Ans.** If component $v_x$ of the velocity vector is along the magnetic field, and remain constant, the charge particle will follow a helical trajectory; as shown in figure.

If the velocity component $v_y$ is perpendicular to the magnetic field B, the magnetic force acts like a centripetal force $qv_y$ B.

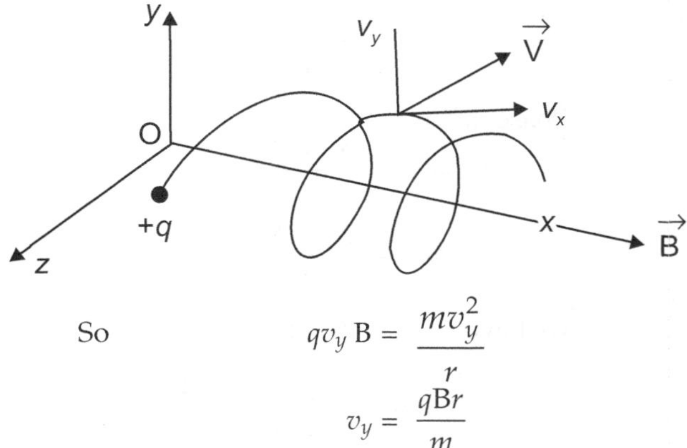


So

$$qv_y B = \frac{mv_y^2}{r}$$

$$v_y = \frac{qBr}{m}$$

Since tangent velocity $v_y = r\omega$

$$\Rightarrow \quad r\omega = \frac{qBr}{m}$$

$$\Rightarrow \quad \omega = \frac{qB}{m}$$

Time taken for one revolution,

$$T = \frac{2\pi}{\omega} = \frac{2\pi m}{qB} \quad [\because W = 2\pi f]$$

and the distance moved along the magnetic field in the helical path is

$$x = v_x.T$$

$$= v_x.\frac{2\pi m}{qB}$$

**16. (a) Determine the value of phase difference between the current and the voltage in the given series LCR circuit.**

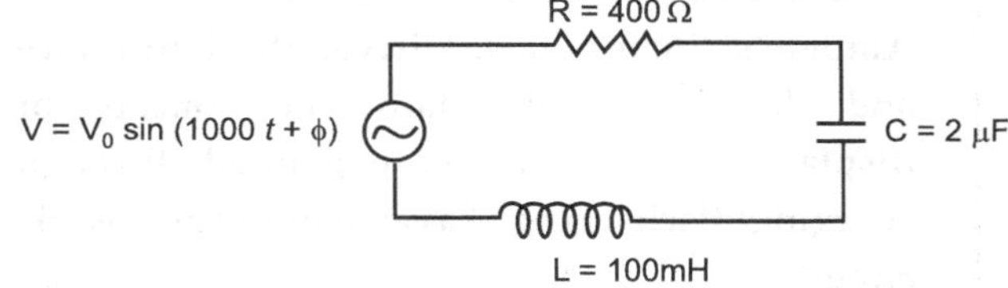

**(b) Calculate the value of the additional capacitor which may be joined suitably to the capacitor C that would make the power factor of the circuit unity.** **[3]**

**Ans. (a)** $V = V_0 \sin(1000t + \phi)$

Comparing this with $V = V_0 \sin(\omega t + \phi)$

we have,

$$\omega = 1000$$

Given : $L = 100 \times 10^{-3}$ H, $R = 400\ \Omega$, $C = 2\ \mu F$

$$X_L = \omega L$$

$$X_L = 1000 \times 100 \times 10^{-3} = 100\ \Omega$$

$$X_C = \frac{1}{\omega C} = \frac{1}{1000 \times 2 \times 10^{-6}} =$$

$$= \frac{1}{2} \times 10^3$$

$$= 500\ \Omega$$

Phase difference,

$$\tan\phi = \frac{X_L - X_C}{R}$$

$$\tan\phi = \frac{X_C - X_L}{R}, X_C > X_L$$

we have, $R = 400\ \Omega$

$$\tan\phi = \frac{500 - 100}{400} = \frac{400}{400} = 1$$

$$\tan\phi = 1$$

$$\therefore \quad \tan\phi = \tan 45^\circ$$

$$\phi = 45^\circ$$

**(b)** The power factor of the circuit is unity. It means that the given circuit is in resonance. It is possible, if another capacitor C′ is used in the circuit.

$$X_C' = X_L$$

$$\frac{1}{\omega C'} = \omega L$$

$$C' = \frac{1}{\omega^2 L} = \frac{1}{(1000)^2 \times 100 \times 10^{-3}}$$

$$C' = 10\ \mu F$$

Since C′ > C, so an additional capacitor of 8 μF would be connected in parallel to the capacitor of C = 2μF.

**17. Write the expression for the generalized form of Ampere's circuital law. Discuss its significance and describe briefly how the concept of displacement current is explained through charging/discharging of a capacitor in an electric circuit.** **[3]**

**Ans. Generalized form of Ampere circuital law :**

$$\oint \vec{B}.\vec{dt} = \mu_0 \left( I_c + \varepsilon_o \frac{d\phi}{dt} \right)$$

It signifies that the source of magnetic field is not just due to the conduction electric current due to flow of charge but also due to the time rate of change of electric field called displacement current. During charging and discharging of a capacitor the electric field between the plates will change, because of which there will be a change of electric flux (displacement current) between the plates.

**18. Use Huygen's principle to show how a plane wavelength propagates from a denser to rarer medium. Hence verify Snell's law of refraction.** **[3]**

**Ans.**

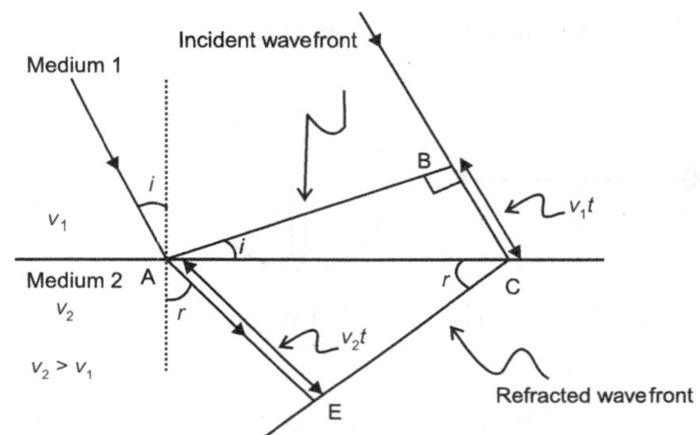


In Δ ABC,

$$\sin i = \frac{BC}{AC} = \frac{v_1 t}{AC} \quad \ldots(i)$$

In Δ CEA,

$$\sin r = \frac{AE}{AC} = \frac{v_2 t}{AC} \quad \ldots(ii)$$

From equetion (i) and (ii)

$$\therefore \quad \frac{\sin i}{\sin r} = \frac{BC}{AE} = \frac{v_1 t}{v_2 t} = \frac{v_1}{v_2} \quad \ldots(iii)$$

∴ But $\mu_1 = \frac{c}{v_1}$

$$\mu_2 = \frac{c}{v_2}$$

$$\therefore \quad \frac{\mu_2}{\mu_1} = \frac{v_1}{v_2} \quad \ldots(iv)$$

on comparing equation (iii) and (iv)

$$\therefore \quad \frac{\sin i}{\sin r} = \frac{\mu_2}{\mu_1}$$

or $\mu_2 \sin r = \mu_1 \sin i,$

It is Snell's law.

**19. Identify the gates P and Q shown in the figure. Write the truth table for the combination of the gates shown.**

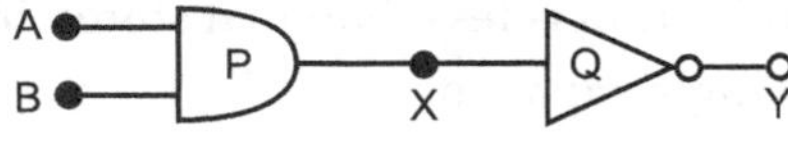


**Name the equivalent gate representing this circuit and write its logic symbol. [3]**

**20. Draw a circuit diagram of a C.E. transistor amplifier. Briefly explain its working and write the expression for (i) current gain, (ii) voltage gain of the amplifier. [3]**

**21. (a) Write three characteristic properties of nuclear force.****

**(b) Draw a plot of potential energy of a pair of nucleons as a function of their separation. Write two important conclusions that can be drawn from the graph. [3]**

**Ans.** **(a)** 1. Nuclear forces are short range forces ;

2. Nuclear forces are primarily attractive and extremly strong;

3. Nuclear forces are charge independent.

**(b)**

Potential energy (MeV) →, 100, 0, −100, Repulsion, Attraction, 1, 2, 3, r (fm) →

**22. (a) Describe briefly three experimentally observed features in the phenomenon of photoelectric effect.**

**(b) Discuss briefly how wave theory of light cannot explain these features. [3]**

**OR**

**(a) Write the important properties of photons which are used to establish Einstein's photoelectric equation.**

**(b) Use this equation to explain the concept of (i) threshold frequency and (ii) stopping potential.**

**19. (a)** 1. The photoelectric effect will not occur when the frequency of the incident light is less than the threshold frequency. Different materials have different threshold frequencies and most elements have threshold frequencies in the ultraviolet region of the electromagnetic spectrum.

2. The maximum KE of a stream of photoelectrons increases linearly with the frequency of the incident light above the threshold frequency.

3. The rate at which photoelectrons are emitted from a photosensitive surface is directly proportional to the intensity of incident light when the frequency is constant.

**(b)** Classical wave theory cannot explain :

1. The existence of threshold frequency because it predicts that electrons would absorb enough energy to escape and there would not be any threshold frequency.

2. The almost immediate emission of photoelectrons as, according to this theory, electrons require a period of time before sufficient energy is absorbed by it to escape from the metal; however, such a thing does not happen practically.

3. It cannot explain why maximum KE is dependent on the frequency and independent of intensity.

**OR**

(a) The important property of photons that is useful in establishing Einstein's photoelectric equation is their ability to hold on to the electrons of an atom by their forces of attraction.

(b) Einstein's photoelectric equation states that :

$$E_{max} = h\nu - \phi$$

If $E_{max} > 0$ then $h\nu - \phi > 0$

or $$h\nu - h\nu_0 > 0$$

Here, $\nu_0$ is the threshold frequency.

When $V < V_0$ there will be no emission of electrons.

Hence, V must greeder than $V_0$.

Again, loss in KE = gain in electromagnetic PE.

At $V = V_0$, $E_{max} = eV_s = 0$

Where $V_s$ is called stopping potential.

or $$E_{max} = eV_s$$

Where $V_s$ represents the stopping potential.

** Answer is not given due to change in the present Syllabus.

## SECTION–D

**23. One morning an old man walked bare-foot to replace the fuse wire in kit kat fitted with power supply mains for his house. Suddenly he screamed and collapsed on the floor. His wife cried loudly for help. His neighbour's son Anil heard the cries and rushed to the place with shoes on. He took a wooden baton and used it to switch off the main supply.** **[4]**

**Answer the following questions :**

**(i) What is the voltage and frequency of mains supply in India ?**

**(ii) These days most of the electrical devices we use require a.c. voltage. Why ?**

**(iii) Can a transformer be used to step up d.c. voltage ?**

**(iv) Write two qualities displayed by Anil by his action.****

**Ans.** (i) The voltage and frequency of mains supply in India are 240 V and frequency is 50 Hz.

(ii) Most electrical devices require a.c. voltage because a.c. is available by default through the mains supply and also because d.c., is actually a one-way current and is available only through batteries.

(iii) No, a transformer cannot be used to step-up d.c. voltage because induced emf in the secondary coil of the transformer is only due to of change of magnetic flux in primary coil.

## SECTION–E

**24. (a) Define electric flux. Write its S.I. unit.**

**"Gauss's law in electrostatics is true for any closed surface, no matter what its shape or size is". Justify this statement with the help of a suitable example.**

**(b) Use Gauss's law to prove that electric field inside a uniformly charged spherical shell is zero.** **[5]**

**OR**

**(a) Derive the expression for the energy stored in a parallel plate capacitor. Hence obtain the expression for the energy density of the electric field.**

**(b) A fully charged parallel plate capacitor is connected across an uncharged identical capacitor. Show that the energy stored in the combination is less than that stored initially in the single capacitor.**

**Ans.** (a) The electric flux through an area is defined as the electric field multiplied by the area of the surface projected on a plane, perpendicular to the field. Its S.I. unit is voltmeter (Vm) or Newton metre square per coulomb (Nm$^2$ C$^{-1}$).

The given statement is justified because while measuring the flux, the surface area is more important than its volume or its size.

(b) To prove that the electric field inside a uniformly charged spherical shell is zero, we place a single positive point charge '$q$' at the centre of an imaginary spherical surface with radius R. The field lines of this point radiate outside equally in all directions. The magnitude E of the electric field at every point on the surface is given by E = $\frac{1}{4\pi\varepsilon_0} \times \frac{q}{R^2}$.

At each point on the surface, $\vec{E}$ is $\perp$ to the surface and its magnitude is the same.

Thus, the total electric flux ($\phi_E$) is the product of their field magnitude E and the A.

Hence,
$$\phi_E = EA$$
$$= \frac{1}{4\pi\varepsilon_0} \frac{q}{R^2}(4\pi R^2)$$
$$= \frac{q}{\varepsilon_0}$$

If the sphere is uniformly charged, then there is zero charge inside the sphere, according to Gauss's law

When $q = 0$, $\phi_E = \frac{0}{\varepsilon_0} = 0$

**OR**

(a) Consider a capacitor of capacitance C. Initial charge on capacitor is zero. Initial potential difference between capacitor plates is zero. Let a charge Q be given to it in small steps. When charge is given to capacitor, the potential difference between its plates increases. Let at any instant when charge on capacitor be $q$, the potential difference between its plates $V = \frac{q}{C}$.

** Answer is not given due to change in the present Syllabus.

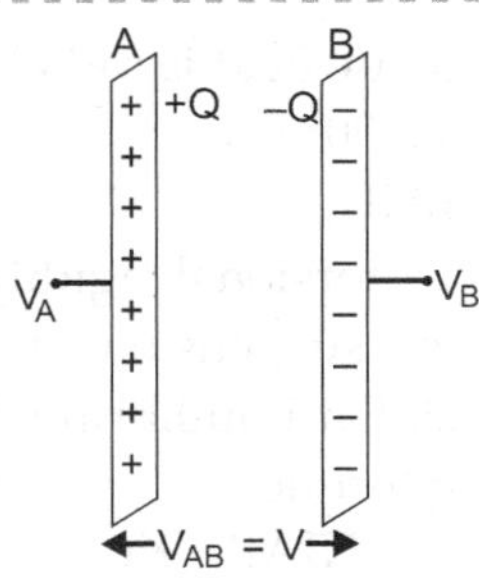


Now work done in giving an additional infinitesimal charge $dq$ to capacitor

$$dW = V\,dq = \frac{q}{C}\,dq$$

The total work done in giving charge from 0 to Q will be equal to the sum of all such infinitesimal works, which may be obtained by integration. Therefore, total work done.

$$W = \int_0^Q V\,dq = \int_0^Q \frac{q}{C}\,dq$$

$$= \frac{1}{C}\left[\frac{q^2}{2}\right]_0^Q = \frac{1}{C}\left(\frac{Q^2}{2} - \frac{0}{2}\right) = \frac{Q^2}{2C}$$

If $V$ is the final potential difference between capacitor plates, then Q = CV

$$\therefore \quad W = \frac{(CV)^2}{2C} = \frac{1}{2}CV^2 = \frac{1}{2}QV$$

This work is stored as electrostatic potential energy of capacitor *i.e.,*

Electrostatic potential energy,

$$U = \frac{Q^2}{2C} = \frac{1}{2}CV^2 = \frac{1}{2}QV$$

**Energy density** : Consider a parallel plate capacitor consisting of plates, each of area A, separated by a distance $d$. If space between the plates is filled with a medium of dielectric constant K, then

Capacitance of capacitor, $C = \frac{K\varepsilon_0 A}{d}$

If σ is the surface charge density of plates, then electric field strength between the plates

$$E = \frac{\sigma}{K\varepsilon_0} \Rightarrow \sigma = K\,\varepsilon_0\,E$$

Charge on each plate of capacitor, $Q = \sigma A = K\,\varepsilon_0\,E\,A$

∴ Energy stored by capacitor,

$$U = \frac{Q^2}{2C} = \frac{(K\,\varepsilon_0 EA)^2}{2\,(K\varepsilon_0\,A/d)}$$

$$= \frac{1}{2}K\,\varepsilon_0\,E^2\,A\,d$$

But $Ad$ = Volume of space between capacitor plates

∴ Energy stored, $U = \frac{1}{2}K\,\varepsilon_0\,E^2\,Ad$

Electrostatic energy stored per unit volume,

$$u_e = \frac{U}{Ad} = \frac{1}{2}K\varepsilon_0\,E^2$$

This is expression for electrostatic energy density in medium of dielectric constant K.

**(b)** Initially, if we consider a charged capacitor, then its charge would be

$$Q = CV$$



and energy stored is

$$U_1 = \frac{1}{2}CV^2 \qquad \text{...(i)}$$

Then, this charged capacitor is connected to uncharged capacitor.

Let the common potential be V. The charge flows from first capacitor to the other capacitor unless both the capacitors attain common potential.

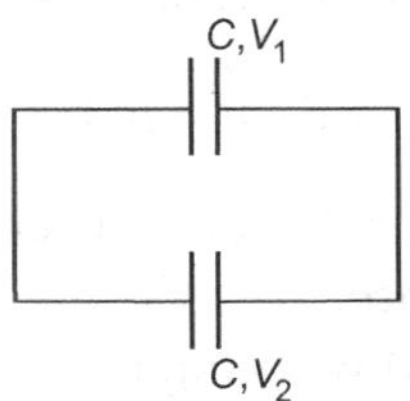


$$\Rightarrow \quad Q_1 = CV_1 \text{ and } Q_2 = CV_2$$

Applying conservation of charge,

$$Q = Q_1 + Q_2 \Rightarrow CV = CV_1 + CV_2$$

$$\Rightarrow \quad V = V_1 + V_2 \Rightarrow V_1 = \frac{V}{2}$$

$$[\because V_1 = V_2]$$

Total energy stored,

$$U_2 = \frac{1}{2}CV_1^2 + \frac{1}{2}CV_2^2$$

$$U_2 = \frac{1}{2}C\left(\frac{V}{2}\right)^2 + \frac{1}{2}C\left(\frac{V}{2}\right)^2$$

$$\Rightarrow \quad U_2 = \frac{1}{4}CV^2 \qquad \text{...(ii)}$$

From Equations (i) and (ii), we get

$$U_2 < U_1$$

⇒ Energy stored in the combination is less than that stored initially in single capacitor.

**25. Explain, using a labelled diagram, the principle and working of a moving coil galvanometer. What is the function of (i) uniform radial magnetic field, (ii) soft iron core ?**

**Define the terms (i) current sensitivity and (ii) voltage sensitivity of a galvanometer. Why does increasing the current sensitivity not necessarily increase voltage sensitivity ?** [5]

OR

**(a) Write using Biot-Savart law, the expression for the magnetic field $\vec{B}$ due to an element $\vec{dl}$ carrying current I at a distance $\vec{r}$ from it in a vector form.**

**Hence derive the expression for the magnetic field due to a current carrying loop of radius R at a point P distant X from its centre along the axis of the loop.**

**(b) Explain how Biot-Savart law enables one to express the Ampere's circuital law in the integral form,** viz.,

$$\oint \vec{B}.\vec{dl} = \mu_0 I$$

**Where I is the total current passing through the surface.**

**Ans. Principle and working :**

The basic principle of a moving coil galvanometer is that when a current carrying coil is placed in a magnetic field, it experiences a torque.

When the current I is passed through the coil, the torque experienced is given by $\tau$ = NIAB sin $\theta$

Where N = No. of turns of the coil,

A = Area of the coil

B = Magnetic field and

$\theta$ = Angle between normal of coil and magnetic field

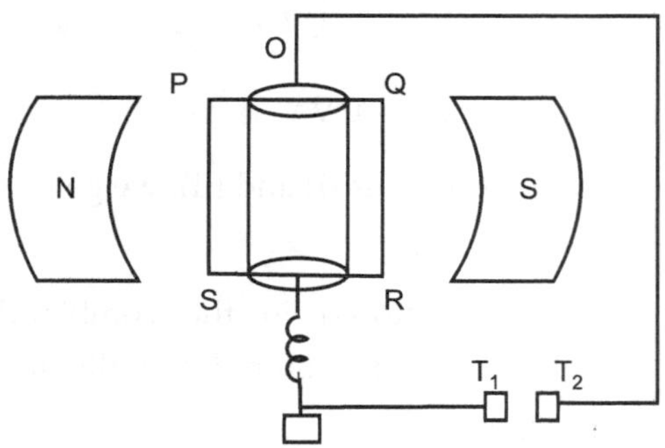


When magnetic field is radial than

$\theta = 90° \Rightarrow \sin 90° = 1$

then $\tau$ = NIAB.

Now, If K is torsional rigidity of wire and $\theta$ is the twist of suspension strip then restoring torque = K$\theta$. For Equilibrium Deflecting Torque = Restoring torque.

$$NIAB = K\theta$$

$$\theta = \frac{NAB}{K} I$$

$$\theta \propto I.$$

(i) The uniform radial magnetic field is used to make the scale linear.

(ii) The soft iron core increases the strength of the magnetic field.

The current sensitivity is defined as the deflection produced in the galvanometer, while passing a current of 1 ampere. (1 amp). through it.

Thus, current sensitivity $\left(\frac{\alpha}{I}\right) = \frac{NBA}{K}$.

The voltage sensitivity is defined as the deflection produced in the galvanometer when a potential difference of 1V is applied to the coil.

Thus, voltage sensitivity $\left(\frac{\alpha}{V}\right) = \frac{NBA}{KR}$.

Where R is the resistance.

Increasing the current sensitivity does not necessarily increase the voltage sensitivity as there is an increase in the resistance as well.

OR

(a) Suppose we have a conductor of length $l$ in which current $i$ is flowing. We need to calculate the magnetic field at a point P in vacuum. If $i\vec{dl}$ is one of the infinitely small current element, the magnetic field $\vec{dB}$ at point P is given by

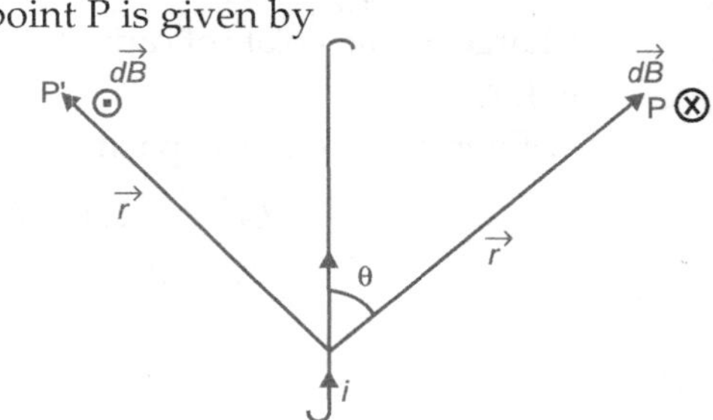

$$d\vec{B} \propto \frac{i\,\vec{dl} \times \vec{r}}{r^3}$$

$$d\vec{B} = \frac{\mu_0}{4\pi} \frac{i\,\vec{dl} \times \vec{r}}{r^3}$$

Where $\frac{\mu_0}{4\pi}$ is a proportionality constant.

Suppose there is a circular coil of radius R, carrying a current *i*. Let P be a point at the axis of the coil at a distance *x* from the centre, at which the field in required.

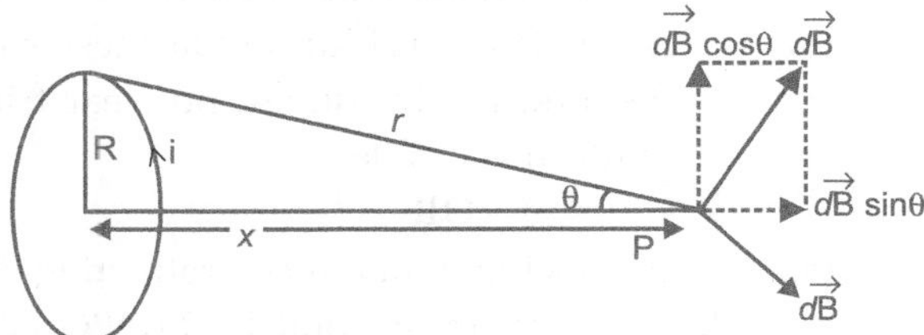


Consider a conducting element *dl* of the loop. The magnetic field due to *dl* is given by the Biot-Savart law,

$$dB = \frac{\mu_0}{4\pi} \frac{i\,|\vec{dl} \times \vec{r}\,|}{r^3}$$

$$dB = \frac{\mu_0}{4\pi} \frac{idl}{(R^2 + x^2)}$$

The direction of dB is perpendicular to the plane formed by *dl* and *r*. It has an X-component $dB_x$ and a component perpendicular to X-axis, $dB_{\perp y}$. When the components perpendicular to the X-axis are summed over, they cancel out and we obtain null result. Thus, only the X-component survives.

So the resultant field $\vec{B}$ at P in given by

$$B = \int dB \sin\theta$$

$$B = \frac{\mu_0}{4\pi} \frac{i}{r^2} \int dl \sin\theta$$

$$B = \frac{\mu_0}{4\pi} \frac{iR}{r^3} \int dl \qquad \left[\because \sin\theta = \frac{R}{r}\right]$$

But $\int dl = 2\pi R$ and $r = (R^2 + x^2)^{1/2}$

$$\therefore \quad B = \frac{\mu_0}{4\pi} \frac{2\pi i R^2}{(R^2 + r^2)^{3/2}}$$

If the coil has N turns, then each turn will contribute equally to B. Then,

$$B = \frac{\mu_0 N i R^2}{2(x^2 + R^2)^{3/2}}$$

(b) Bio-Savart law can be expressed as Ampere's circuital law by considering the surface to be made up a large number of loops. The sum of the tangential components of magnetic field multiplied by the length of all scuh elements give.

**26. (a) Consider two coherent sources $S_1$ and $S_2$ producing monochromatic waves to produce interference pattern. Let the displacement of the wave produced by $S_1$ be given by $Y_1 = a \cos \omega t$ and the displacement by $S_2$ be $Y_2 = a \cos (\omega t + \phi)$. Find out the expression for the amplitude of the resultant displacement at a point and show that the intensity at that point will be $I = 4a^2 \cos^2 \phi/2$. Hence establish the conditions for constructive and destructive interference.**

**(b) What is the effect on the interference fringes in Young's double slit experiment when (i) the width of the source slit is increased; (ii) the monochromatic source is replaced by a source of white light. [5]**

**OR**

**(a) A ray 'PQ' of light is incident on the face AB of a glass prism ABC (as shown in the figure) and emerges out of the face AC. Trace the path of the ray. Show that $\angle i + \angle e = \angle A + \angle \delta$. When δ and *e* denote the angle of deviation and angle of emergence respectively.**

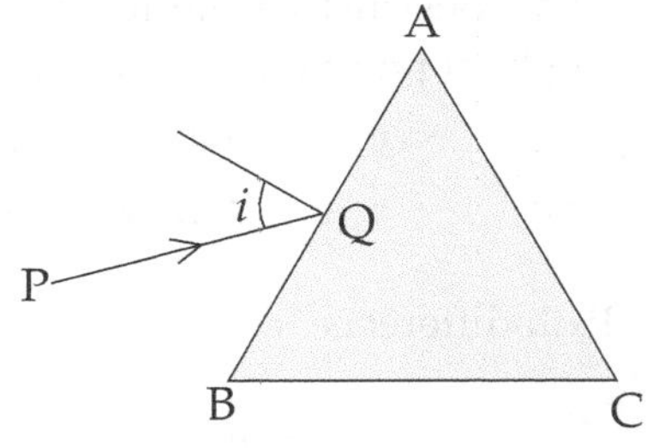


**Plot a graph showing the variation of the angle of deviation as a function of angle of incidence. State the condition under which $\angle\delta$ is minimum.**

**(b) Find out the relation between the refractive index (μ) of the glass prism and $\angle$A for the case when the angle of prism (A) is equal to the angle of minimum deviation ($\delta_m$). Hence obtain the value of the refractive index for angle of prism A = 60°.**

**Ans. (a)** Let S be a narrow slit illuminated by a monochromatic source of light and $S_1$ and $S_2$ two similar parallel slits very close together and equidistant from S.

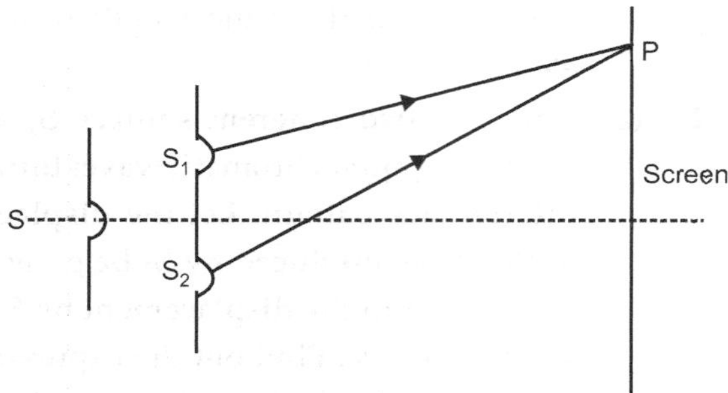


Displacement of the wave produced by $S_1$ is given by

$$y_1 = a \cos \omega t$$

and the displacement of the wave produced by $S_2$ is given by

$$y_2 = a \cos(\omega t + \phi)$$

The resultant displacement is given by

$$y = y_1 + y_2$$

$$y = a \cos\omega t + a \cos(\omega t + \phi)$$

$$y = 2a \cos\phi/2 \cos(\omega t + \phi/2)$$

The amplitude of the resultant displacement is $2a \cos\phi/2$. The intensity of light is directly proportional to the square of amplitude of the wave.

The resultant intensity is given by

$$I = 4a^2\cos^2 \phi/2$$

For constructive interference the intensity of light at point P is maximum, if

$$\cos \phi = \max = +1 : \phi = 0, 2\pi, 4\pi.....$$

*i.e.,* $\phi = 2n\pi$ where $n = 0, 1, 2, ........$

$$\text{Path difference} = \frac{\lambda}{2\pi} \times 2n\pi$$

$$= n\lambda$$

For destructive interference the intensity of light at point P is minimum, if

$$\cos \phi = -1$$

$\therefore$ $\phi = \pi, 3\pi, 5\pi, ......$

or $\phi = (2n - 1)\pi$ where $n = 1, 2, 3......$

The corresponding path difference between the two waves

$$x = \frac{\lambda}{2\pi}\phi = \frac{\lambda}{2\pi}(2n-1)\pi = (2n-1)\frac{\lambda}{2}$$

**(b) (i)** As the width of the slits is increased, the fringe width decreases. because,

$$\beta \propto \frac{1}{d}$$

**(ii)** The different colours of white light will produce different interference patterns but the central bright fringes due to all colours are at the same positions. Therefore, the central bright fringe is white in colour. Since the wavelength of the blue light is smallest, the fringe closed on the either side of the central white fringe is blue and farthest is red. Beyond a few fringes, no clear fringe pattern is visible.

**OR**

**(a)** Let the incident ray meet refracting face AB of the prism at point P. Ray PQ is the refracted ray inside the prism and $\delta_2$ and $r_1$ are the angle of the deviation and refraction at interface AB. At interface AC the ray goes out of the prism. Let $e$ be the angle of emergence. The angle of deviation at point Q is $\delta_2$ as shown in figure.

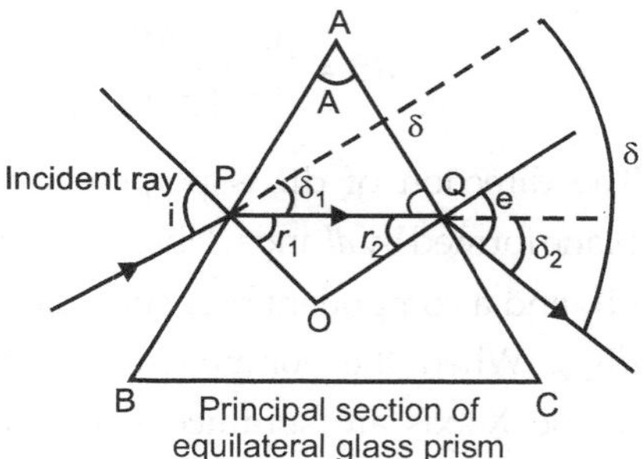


Principal section of equilateral glass prism

Using geometry, we see that at point P,

$$i = \delta_1 + r_1 \qquad \therefore \; \delta_1 = i - r_1$$

and at point Q,

$$e = \delta_2 + r_2$$

$$\delta_2 = e - r_2$$

The total deviation $\delta$, suffered by the incident ray is equal to $\delta_1 + \delta_2$.

or
$$\delta = \delta_1 + \delta_2$$

$$= (i - r_1) + (e - r_2)$$

$$= (i + e) - (r_1 + r_2) \qquad ...(i)$$

In quadrilateral POQA, the sum of all four angle is 360°.

$$\angle P + \angle O + \angle Q + \angle A = 360°$$

as P and Q both are right angles

$$\angle P + \angle Q = 180°$$

$$\angle O + \angle A = 180° \qquad ...(ii)$$

In triangle POQ

$\oint \vec{B}.\vec{dt} = \mu_0 I.$

$$O + r_1 + r_2 = 180° \quad ...(iii)$$

Comparing equations (ii) and (iii), we have

$$A = r_1 + r_2$$

Substituting this value in equation (i)

$$\delta = i + e - A$$

$$\delta + A = i + e \quad ...(iv)$$

So angle of deviation produced by a prism depends upon the angle of incidence, refracting angle of prism, and the material of the prism.

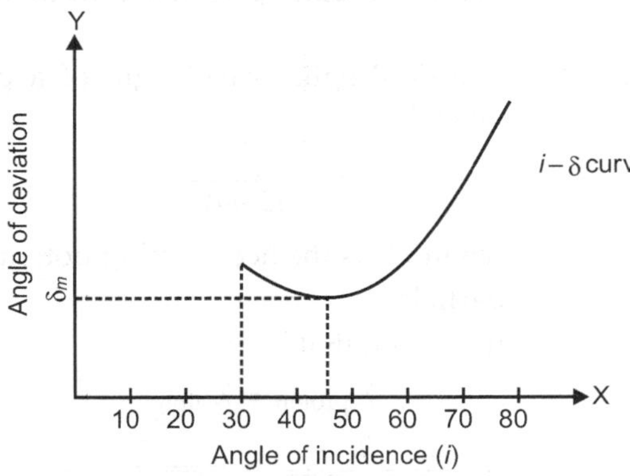


When prism is in the position of minimum deviation,

then $i = e$ and $r_1 = r_2$

From equation (iv),

$$\delta_m + A = i + i$$

$$i = \frac{(\delta_m + A)}{2}$$

**(b)** by Snell's law, $\mu = \dfrac{\sin i}{\sin r}$ putting the value of $i$ and $r$

$$\mu = \frac{\sin\left(\frac{A + \delta_{min}}{2}\right)}{\sin \frac{A}{2}}$$

But $A = \delta_m$

$$\therefore \quad \mu = \frac{\sin\left(\frac{A + A}{2}\right)}{\sin \frac{A}{2}}$$

$$= \frac{\sin A}{\sin \frac{A}{2}} = \frac{2 \sin \frac{A}{2} . \cos \frac{A}{2}}{\sin \frac{A}{2}} = 2 \cos \frac{A}{2}$$

If $A = 60°$,

$$\mu = 2 \cos \frac{60°}{2} = 2 \cos 30°$$

$$= 2 \times \frac{\sqrt{3}}{2} = \sqrt{3}$$

**Note : All questions of Outside Delhi Set II are from Outside Delhi Set I and Outside Delhi Set III are from Set I and Set II.**

●●

# PHYSICS

## Delhi [Set-I] 2015

**Time allowed : 3 hours** **Maximum marks : 70**

### SECTION–A

1. **Define capacitor reactance. Write its S.I. units.** [1]

**Ans.** Capacitor reactance is the resistance offered by a capacitor to the flow of a.c. It is given by

$$X_C = \frac{1}{\omega C}$$

Where $\omega = 2\pi f$, $f$ = frequency of the source

$$X_C = \frac{1}{2\pi f C}$$

$\omega$ = Angular frequency of the source

C = Capacitance of the capacitor.

The SI unit of capacitor reactance is ohm ($\Omega$).

2. **What is the electric flux through a cube of side 1 cm which encloses an electric dipole ?** [1]

**Ans.** The electric flux through a cube of side 1 cm which encloses an electric dipole will be zero, as net charge enclosed by a cube is zero.

3. **A concave lens of refractive index 1.5 is immersed in a medium of refractive index 1.65. What is the nature of the lens ?** [1]

**Ans.** Since $\mu_g$ lens < $\mu_m$ surroundings.

It behaves like a converging lens.

4. **How are side bands produced ?**** [1]

5. **Graph showing the variation of current versus voltage for a material Ga As is shown in the figure. Identify the region of :** [1]

   **(i) negative resistance**

   **(ii) where Ohm's law is obeyed.**

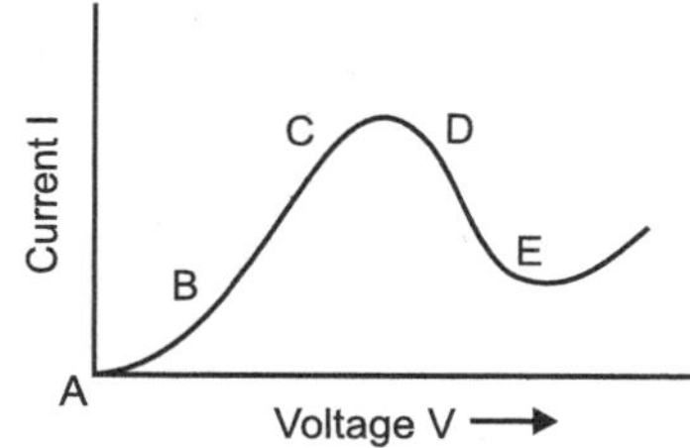


**Ans. (i)** DE is the region of negative resistance because the slope of curve in this part is negative.

**(ii)** BC is the region where Ohm's law is obeyed because in this part, the current varies linearly with the voltage.

### SECTION–B

6. **A proton and an $\alpha$-particle have the same de-Broglie wavelength. Determine the ratio of (i) their accelerating potentials (ii) their speeds.** [2]

**Ans. (i)** The de-Broglie wavelength of a particle is given by

$$\lambda = \frac{h}{\sqrt{2mqV}}$$

where, V is the accelerating potential of the particle.

It is given that

$$\lambda_{proton} = \lambda_{alpha}$$

$$\Rightarrow \frac{h}{\sqrt{2m_p q_p V_p}} = \frac{h}{\sqrt{2m_\alpha q_\alpha V_\alpha}}$$

$$\sqrt{2m_\alpha q_\alpha V_\alpha} = \sqrt{2m_p q_p V_p}$$

On squaring both sides, we get

$$2m_\alpha q_\alpha V_\alpha = 2\, m_p q_p V_p$$

$$\Rightarrow \frac{V_p}{V_\alpha} = \frac{m_\alpha q_\alpha}{m_p q_p} = \frac{4m_p \times 2q_p}{m_p q_p}$$

$$\left[\because m_\alpha = 4m_p \text{ and } q_\alpha = 2q_p\right]$$

$$= \frac{8}{1}$$

**(ii)** We can also write de-Broglie wavelength as

$$\lambda = \frac{\hbar}{mv}$$

where $h$ is Planck's constant, $m$ is mass of the particle and $v$ is speed of the particle.

We know

$$m_{alpha} = 4m_{proton}$$

$$\therefore \quad \lambda_{alpha} = \frac{h}{4m_{proton}\, v_{alpha}}$$

It is given that

$$\lambda_{proton} = \lambda_{alpha}$$

** Answer is not given due to change in the present Syllabus.

$$\Rightarrow \frac{h}{m_{proton}v_{proton}} = \frac{h}{4m_{proton}v_{alpha}}$$

$$\frac{v_{proton}}{v_{alpha}} = 4$$

**7. Show that the radius of the orbit in hydrogen atom varies as $n^2$. Where $n$ is the principal quantum number of the atom. [2]**

**Ans.** According to the Bohr's theory of hydrogen atom, the angular momentum of a revolving electron is given by

$$mvr = \frac{nh}{2\pi} \quad ...(i)$$

Here,

$m$ = Mass of the electron
$v$ = Velocity of the electron
$r$ = Radius of the orbit
$h$ = Planck's constant
$n$ = Principal quantum number of the atom

If an electron of mass $m$ and velocity $v$ is moving in a circular orbit of radius $r$, then the centripetal force required is given by

$$F = \frac{mv^2}{r}$$

Also, if the charge on the nucleus is $Ze$, then the force of electrostatic attraction between the nucleus and the electron will provide the necessary centripetal force.

$$F = \frac{1}{4\pi\varepsilon_0}\frac{(Ze)(e)}{r^2} = \frac{KZe^2}{r^2}$$

where $$K = \frac{1}{4\pi\varepsilon_0}$$

$$\therefore \quad \frac{mv^2}{r} = \frac{KZe^2}{r^2} \quad ...(ii)$$

From (i), we get

$$v = \frac{nh}{2\pi mr}$$

Putting this value in (ii), we get

$$\frac{m}{r}\frac{n^2h^2}{4\pi^2m^2r^2} = \frac{KZe^2}{r^2}$$

$$\Rightarrow \quad r = \frac{n^2h^2}{4\pi^2 mKZe^2}$$

$$\Rightarrow \quad r \propto n^2.$$

**8. Distinguish between 'intrinsic' and 'extrinsic' semiconductors. [2]**

**Ans.**

| Intrinsic semiconductor | Extrinsic semiconductor |
|---|---|
| 1. It is pure semiconducting material with no impurity atoms added to it. | It is prepared by doping a small quantity of impurity atoms to the pure semiconductor. |
| 2. The number of free electrons in the conduction band and the number of holes in valence band are exactly equal. | The number of free electrons and holes are never equal. There is an excess of electrons in $n$-type semiconductors and an excess of holes in $p$-type semiconductors. |
| 3. Its electrical conductivity is a function of temperature alone. | Its electrical conductivity depends upon the temperature and the amount of impurity added in it. |

**9. Use the mirror equation to show that an object placed between $f$ and $2f$ of a concave mirror produces a real image beyond $2f$. [2]**

**OR**

**Find an expression for intensity of transmitted light when a polaroid sheet is rotated between two crossed polaroids. In which position of the polaroid sheet will the transmitted intensity be maximum ?**

**Ans.** Mirror equation is

$$\frac{1}{f} = \frac{1}{u} + \frac{1}{v}$$

Where $u$ is the distance of object from the mirror, $v$ is the distance of image from the mirror and $f$ is the focal length of the mirror.

For a concave mirror $f$ is negative *i.e.* $f < 0$, $u < 0$

For a real object (on the left of mirror)

$$\therefore \quad 2f < u < f$$

$$\Rightarrow \quad \frac{1}{2f} > \frac{1}{u} > \frac{1}{f}$$

$$\frac{1}{2f} - \frac{1}{f} > \frac{1}{u} - \frac{1}{f} > \frac{1}{f} - \frac{1}{f}$$

$$\frac{1}{f} - \frac{1}{2f} < \frac{1}{f} - \frac{1}{u} < \frac{1}{f} - \frac{1}{f}$$

$$\Rightarrow \quad \frac{1}{2f} < \frac{1}{v} < 0$$

$$\Rightarrow \quad v < 0. \text{ Also } v > 2f$$

$\therefore$ Image is real.

This implies that $v$ is negative and greater than $2f$. This means that the image lies beyond $2f$.

**OR**

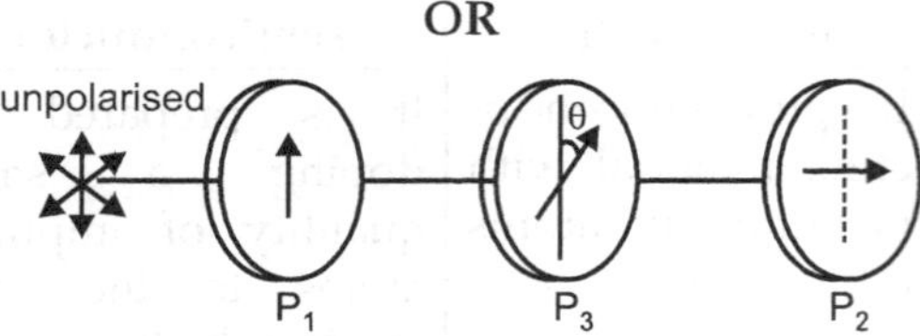


Let us consider two crossed polarisers $P_1$ and $P_2$ with a polaroid sheet $P_3$ placed between them.

Let $I_0$ be the intensity of polarised light after passing through the first polarizer $P_1$. If θ is the angle between the axes of $P_1$ and $P_3$, then the intensity of the polarized light after passing through $P_3$ will be $I = I_0 \cos^2 \theta$.

As $P_1$ and $P_2$ are crossed. The angle between the axes of $P_1$ and $P_2 = 90°$.

∴ Angle between the axes of $P_2$ and $P_3 = (90° - \theta)$

The intensity of light emerging from $P_2$ will be given by

$$I' = I \cos^2 (90 - \theta)$$

$$I' = [I_0 \cos^2 \theta] \cos^2 (90° - \theta)$$

$$\Rightarrow \quad I' = [I_0 \cos^2 \theta] \sin^2 \theta$$

$$\Rightarrow \quad I' = \frac{I_0}{4}(4 \cos^2 \theta \sin^2 \theta)$$

$$\Rightarrow \quad I' = \frac{I_0}{4}(2 \sin \theta \cos \theta)^2$$

$$\Rightarrow \quad I' = \frac{I_0}{4}(\sin^2 2\theta)$$

The intensity of polarized light transmitted from $P_2$ will be maximum when

$$\sin 2\theta = \text{maximum} = 1$$

$$\Rightarrow \quad \sin 2\theta = \sin 90°$$

$$\Rightarrow \quad 2\theta = 90°$$

$$\Rightarrow \quad \theta = 45°$$

The maximum transmitted intensity of polarised light is

$$I' = \frac{I_0}{4}$$

**10. Use Kirchhoff's rules to obtain conditions for the balance condition in a Wheatstone bridge.** **[2]**

**Ans.** Let us consider a Wheatstone bridge arrangement as shown below :

Wheatstone bridge is a special bridge type circuit which consists of four resistances, a galvanometer and a battery. It is used to determine unknown resistance.

In figure four resistance P, Q, R and S are connected in the form of four arms of a quadrilateral. Let the current given by battery in the balanced position be I. This current on reaching point A is divided into two parts $I_1$ and $I_2$. As there is no current in galvanometer in balanced state, therefore, current in resistances P and Q is $I_1$ and in resistances R and S it is $I_2$.

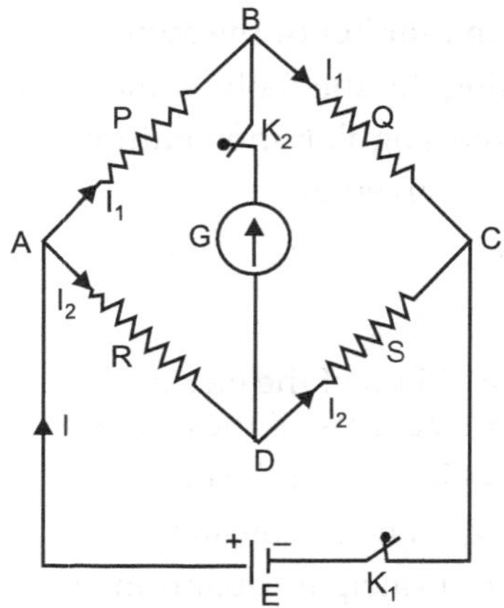


Applying Kirchhoff's first law i.e. junction law at point A

$$I - I_1 - I_2 = 0 \text{ or } I = I_1 + I_2 \quad \text{...(i)}$$

Applying Kirchhoff's second law to closed mesh ABDA

$$-I_1P + I_2R = 0 \text{ or } I_1P = I_2R \quad \text{...(ii)}$$

Applying Kirchhoff's second law to mesh BCDB

$$-I_1Q + I_2S = 0 \text{ or } I_1Q = I_2S \quad \text{...(iii)}$$

Dividing equation (ii) by (iii), we get

$$\frac{I_1P}{I_1Q} = \frac{I_2R}{I_2S} \quad \text{or} \quad \frac{P}{Q} = \frac{R}{S}$$

This is condition of balance Wheatstone's bridge.

## SECTION–C

**11. Name the parts of the electromagnetic spectrum which is**

**(a) suitable for radar systems used in aircraft navigation**

**(b) used to treat muscular strain**

**(c) used as a diagnostic tool in medicine**

**Write in brief, how these waves can be produced.**

**Ans.** (a) Microwaves are suitable for radar systems that are used in aircraft navigation.

These rays are produced by special vacuum tubes, namely klystrons, magnetrons and Gunn diodes.

(b) Infrared waves are used to treat muscular strain.

These rays are produced by hot bodies and molecules.

**(c)** X-rays are used as a diagnostic tool in medical.

These rays are produced when high energy electrons are stopped suddenly on a metal of high atomic number.

**12. (i) A giant refracting telescope has an objective lens of focal length 15 m. If an eye-piece of focal length 1.0 cm is used. What is the angular magnification of the telescope ?**

**(ii) If this telescope is used to view the moon. What is the diameter of the image of the moon formed by the objective lens ? The diameter of the moon is $3.48 \times 10^6$ m and the radius of lunar orbit is $3.8 \times 10^8$ m. [3]**

**Ans. (i)** Let

$f_o$ = Focal length of the objective lens = 15 m = 1500 cm

$f_e$ = Focal length of the eye lens = 1.0 cm

Angular magnification of the giant refracting telescope is given by

$$m_0 = \left|\frac{f_o}{f_e}\right|$$

$$\therefore \quad m_0 = \left|\frac{1500}{1}\right| = 1500$$

**(ii)** Diameter of the image of the moon formed by the objective lens, $d = \alpha f_o$

$$\Rightarrow \quad d = \frac{\text{Diameter of the moon}}{\text{Radius of the lunar orbit}} \times f_o$$

$$\Rightarrow \quad d = \frac{3.48 \times 10^6}{3.8 \times 10^8} \times 15$$

$$\Rightarrow \quad d = 0.1374 \text{ m.}$$

$$d = 13.74 \text{ cm}$$

**13. Write Einstein's photoelectric equation and mention which important features in photoelectric effect can be explained with the help of this equation.**

**The maximum kinetic energy of the photoelectrons gets doubled when the wavelength of light incident on the surface changes from $\lambda_1$ to $\lambda_2$. Derive the expressions for the threshold wavelength $\lambda_0$ and work function for the metal surface. [3]**

**Ans.** Einstein's photoelectric equation is given by

$$K_{max} = \frac{1}{2} mv^2_{max}$$

$$K_{max} = h\nu - \phi_0$$

or

$$h\nu = h\nu_0 + \frac{1}{2} mv^2_{max}$$

where

$K_{max}$ = Maximum kinetic energy of the photo-electron

$v_{max}$ = Maximum velocity of the emitted photoelectron

$m$ = Mass of the photoelectron

$\nu$ = Frequency of the light radiation

$\phi_0$ = Work function

$h$ = Planck's constant

If $\nu_0$ is the threshold frequency, then the work function can be written as

$$W = \phi_0 = h\nu_0$$

$$\Rightarrow \quad K_{max} = \frac{1}{2} mv^2_{max} = h\nu - h\nu_0 = h(\nu - \nu_0)$$

The above equations explains the following results :

1. If $\nu < \nu_0$, then the maximum kinetic energy is negative, which is impossible. Hence, photoelectric emission does not take place for the incident radiation below the threshold frequency. Thus, the photoelectric emission can take place if $\nu > \nu_0$.
2. The maximum kinetic energy of emitted photoelectrons is directly proportional to the frequency of the incident radiation. This means that maximum kinetic energy of photoelectron depends only on the frequency of incident light not on the intesity

According to the photoelectric equation,

$$K_{max} = \frac{1}{2} mv^2_{max} = h\nu - \phi_0$$

$$K_1 = \frac{hc}{\lambda_1} - \phi_0 \quad \text{...(i)}$$

Let the maximum kinetic energy for the wavelength $\lambda_2$ be $K_2$.

$$K_2 = \frac{hc}{\lambda_2} - \phi_0 \quad \text{...(ii)}$$

$$K_2 = 2K_1 \quad \text{(given)}$$

From (i) and (ii), we have

$$\frac{hc}{\lambda_2} - \phi_0 = 2\left(\frac{hc}{\lambda_1} - \phi_0\right)$$

$$\Rightarrow \quad \phi_0 = hc\left(\frac{2}{\lambda_1} - \frac{1}{\lambda_2}\right)$$

$$\Rightarrow \quad h\nu_0 = hc\left(\frac{2}{\lambda_1} - \frac{1}{\lambda_2}\right)$$

$$\Rightarrow \quad \frac{c}{\lambda_0} = c\left(\frac{2}{\lambda_1} - \frac{1}{\lambda_2}\right)$$

$$\Rightarrow \quad \frac{1}{\lambda_0} = \left(\frac{2}{\lambda_1} - \frac{1}{\lambda_2}\right)$$

$$\Rightarrow \quad \lambda_0 = \left(\frac{\lambda_1\lambda_2}{2\lambda_2 - \lambda_1}\right)$$

Work function is the energy required to eject a photoelectron from the metal.

$$W = \frac{hc}{\lambda_0}$$

$$\therefore \quad W = \frac{hc\,(2\lambda_2 - \lambda_1)}{\lambda_1\lambda_2}$$

**14. In the study of Geiger-Marsdon experiment on scattering of $\alpha$-particles by a thin foil of gold, draw the trajectory of $\alpha$-particles in the Coulomb field of target nucleus. Explain briefly how one gets the information on the size of the nucleus from this study.**

**From the relation $R = R_0 A^{1/3}$, where $R_0$ is constant and A is the mass number of the nucleus, show that nuclear matter density is independent of A.** **[3]**

**OR**

**Distinguish between nuclear fission and fusion. Show how in both these processes energy is released. Calculate the energy release in MeV in the deuterium-tritium fusion reaction :**

$${}^{2}_{1}H + {}^{3}_{1}H \rightarrow {}^{4}_{2}He + {}_{0}n^{1}$$

**Using the data :**

$$m\left({}^{2}_{1}H\right) = 2.014102\ u$$

$$m\left({}^{3}_{1}H\right) = 3.016049u$$

$$m\left({}^{4}_{2}He\right) = 4.002603u$$

$$mn = 1.008665u$$

$$1u = 931.5\ MeV/c^2$$

**Ans.**

Trajectory of a Particles in Coulomb Field of Target Nucleus

Incident $\alpha$ particles; 3, 2, 1, 2, 3; $\theta \le 90°$; $\theta > 180°$; Atomic nucleus; $\alpha$; $r_0$

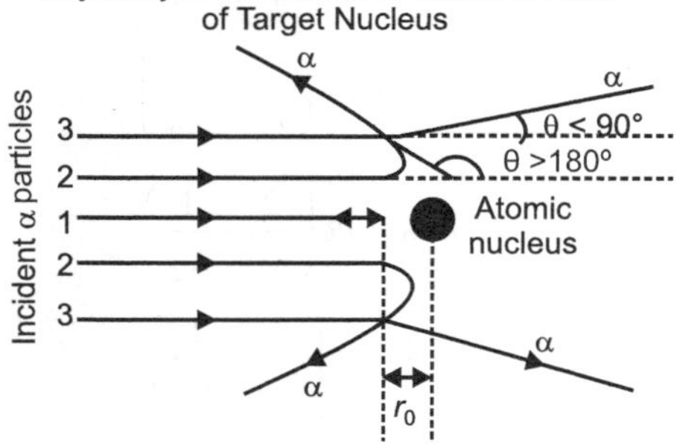

From this experiment, the following is observed :

1. Most of the alpha particles pass straight through the gold foil. It means that they do not suffer any collision with gold atoms.
2. About one alpha particle in every 8000 alpha particles deflects by more than 90°.

   As most of the alpha particles go undeflected and only a few get deflected, this shows that most of the space in an atom is empty and at the centre of the atom, there exists a nucleus. By the number of the alpha particles deflected, the information regarding size of the nucleus can be known.

   If $m$ is the average mass of a nucleon and R is the nuclear radius, then mass of nucleus = $m$A, where A is the mass number of the element.

   Volume of the nucleus,

$$V = \frac{4}{3}\pi R^3$$

Given $R = R_0 A^{1/3}$

$$\Rightarrow \quad V = \frac{4}{3}\pi\left(R_0 A^{1/3}\right)^3$$

$$\Rightarrow \quad V = \frac{4}{3}\pi R_0^3 A$$

Density of nuclear matter, $\rho = \frac{mA}{V}$

$$\Rightarrow \quad \rho = \frac{mA}{\frac{4}{3}\pi R_0^3 A}$$

$$\Rightarrow \quad \rho = \frac{3m}{4\pi R_0^3}$$

This shows that the nuclear density is independent of A.

**OR**

**Nuclear fission :** Nuclear fission is a disintegration process, in which a heavier nucleus gets split up into two lighter nuclei, with the release of a large amount of energy due to difference in Binding Energy.

$${}_{92}U^{235} + {}_{0}n^{1} \rightarrow {}_{56}Ba^{141} + {}_{36}Kr^{92} + 3{}_{0}n^{1} + Q$$

Here, the energy released per fission of ${}_{92}U^{235}$ is 200.4 MeV.

**Nuclear fusion :** When two or more light nuclei combine to form a heavy stable nuclide, part of mass disappears in the process is converted into energy. This phenomenon is called nuclear fusion.

Energy released in this process

$${}^{2}_{1}H + {}^{3}_{1}H \rightarrow {}^{4}_{2}He + {}_{0}n^{1}$$

$\therefore \Delta m = (2.014102 + 3.016049) - (4.002603 + 1.008665)$
$= 0.018883$

Energy released, $Q = 0.018883 \times 931.5$ MeV
$= 17.589$ MeV

**15. Draw a block diagram of a detector for AM signal and show, using necessary processes and the waveforms, how the original message signal is detected from the input AM wave.**** **[3]**

**16. A cell of emf 'E' and internal resistance 'r' is connected across a variable load resistor R. Draw the plots of the terminal voltage V versus (i) R and (ii) the current i.**

**It is found that when R = 4 Ω, the current is 1 A when R is increased to 9 Ω, the current reduces to 0.5 A. Find the values of the emf E and internal resistance *r*.** **[3]**

**Ans. (i)** Graph between terminal voltage (V) and resistance (R) :

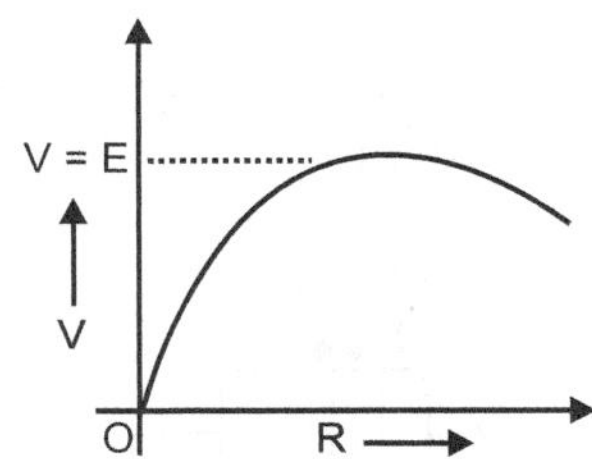


**(ii)** Graph between terminal voltage (V) and current (I) :

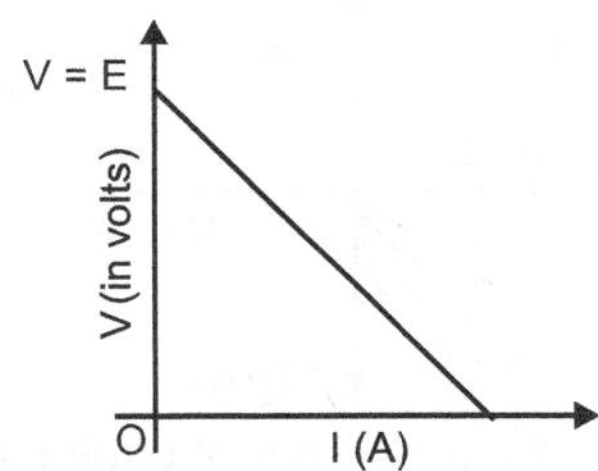


**(iii)** When R = 4 Ω and I = 1 A

We know,

Terminal voltage, $V = E - Ir$

So, we have

$$V = IR = 4 = E - Ir$$

$$E - r = 4 \quad \text{... (i)}$$

When R = 9 Ω and I = 0.5 A

$$V = IR = 0.5 \times 9 = E - 0.5r$$

$$E - 0.5r = 4.5 \quad \text{...(ii)}$$

Subtracting (i) from (ii), we get

$$E - 0.5r - E + r = 4.5 - 4$$

$$0.5r = 0.5$$

$$r = 1\,\Omega$$

Substituting value of $r$ in (i)

$$E - 1 = 4$$

$$E = 5\text{ V}$$

Thus $r = 1\,\Omega$ and $E = 5$ V.

**17. Two capacitors of unknown capacitances $C_1$ and $C_2$ are connected first in series and then in parallel across a battery of 100 V. If the energy stored in the two combinations is 0.045 J and 0.25 J respectively, determine the value of $C_1$ and $C_2$. Also calculate the charge on each capacitor in parallel combination.** **[3]**

**Ans.** When the capacitors are connected in parallel.

Equivalent capacitance, $C_p = C_1 + C_2$

The energy stored in the combination of the capacitors,

$$E_p = \frac{1}{2}C_pV^2$$

$$\Rightarrow \quad E_p = \frac{1}{2}(C_1 + C_2)(100^2) = 0.25\text{J}$$

$$\Rightarrow \quad (C_1 + C_2) = 5 \times 10^{-5} \quad \text{...(i)}$$

When the capacitors are connected in series.

Equivalent capacitance,

$$C_s = \frac{C_1C_2}{C_1 + C_2}$$

The energy stored in the combination of the capacitors,

$$E_s = \frac{1}{2}C_sV^2$$

$$\Rightarrow E_s = \frac{1}{2}\frac{C_1C_2}{C_1 + C_2}(100)^2 = 0.045\text{J}$$

$$\Rightarrow \frac{1}{2}\frac{C_1C_2}{5 \times 10^{-5}}(100)^2 = 0.045\text{J} \quad [\because c_1 + c_2 = 5 \times 10^{-5}]$$

$$\Rightarrow C_1C_2 = 0.045 \times 10^{-4} \times 5 \times 10^{-5} \times 2 = 4.5 \times 10^{-10} \quad \text{...(ii)}$$

Now, $(C_1 - C_2)^2 = (C_1 + C_2)^2 - 4C_1C_2$

$$\Rightarrow (C_1 - C_2) = \sqrt{7 \times 10^{-10}} = 2.64 \times 10^{-5}$$

[using (i) & (ii)]

Now, $C_1 - C_2 = 2.64 \times 10^{-5}$ ...(iii)

Solving (i) and (iii), we get

$$C_1 = 38.2\ \mu\text{F and } C_2 = 11.8\ \mu\text{F}$$

** Answer is not given due to change in the present Syllabus.

When the capacitors are connected in parallel, the charge on each of them can be obtained as follows :

$$Q_1 = C_1V = 38.2 \times 10^{-6} \times 100 = 38.2 \times 10^{-4}\ C$$

$$Q_2 = C_2V = 11.8 \times 10^{-6} \times 100 = 11.8 \times 10^{-4}\ C$$

**18. State the principle of working of a galvanometer. A galvanometer of resistance G is converted into a voltmeter to measure upto V volts by connecting a resistance $R_1$ in series with the coil. If a resistance $R_2$ is connected in series with it, then it can measure upto V/2 volts. Find the resistance, in terms of $R_1$ and $R_2$, required to be connected to convert it into a voltmeter that can read upto 2 V. Also find the resistance G of the galvanometer in terms of $R_1$ and $R_2$.** **[3]**

**Ans.** **Principle :** When a current-carrying coil is placed in a magnetic field, it experiences a torque. From the measurement of the deflection of the coil, the strength of the current can be computed. A high resistance is connected in series with the galvanometer to convert it into voltmeter. The value of the resistance is given by

$$R = \frac{V}{I_g} - G$$

Here,

V = Potential difference across the terminals of the voltmeter

$I_g$ = Current through the galvanometer

G = Resistance of the galvanometer

When the resistance $R_1$ is connected in series with the galvanometer.

$$V = I_g (G + R_1)$$

When the resistance $R_2$ is connected in series with the galvanometer

$$\frac{V}{2} = I_g (G + R_2) \qquad \text{...(ii)}$$

Divide eq. (i) by eq. (ii)

$$\Rightarrow \qquad 2 = \frac{G+R_1}{G+R_2}$$

$$G = R_1 - 2R_2$$

Let $R_3$ be the resistance required for conversion of galvanometer into voltmeter of range 2V

$$\therefore \qquad 2V = I_g (G + R_3)$$

$$\text{Also} \qquad V = I_g (G + R_1)$$

$$\therefore \qquad 2 = \frac{G+R_3}{G+R_1}$$

$$\therefore R_3 = G + 2R_1 = R_1 - 2R_2 + 2R_1 = 3R_1 - 2R_2$$

**19. With what considerations in view, a photodiode is fabricated ? State its working with the help of a suitable diagram.**

**Even though the current in the forward bias is known to be more than in the reverse bias, yet the photodiode works in reverse bias. What is the reason ?** **[3]**

**Ans.** A photodiode is used to observe the change in current with change in the light intensity under reverse bias condition.

In fabrication of photodiode, material chosen should have band gap –1.5 eV or lower so that solar conversion efficiency is better. This is the reason to choose Si or GaAs material.

**Working :** It is a *p-n* junction fabricated with a transparent window to allow light photons to fall on it. These photons generate electron hole pairs upon absorption. If the junction is reverse biased using an electrical circuit, these electron hole pair move in opposite directions so as to produce current in the circuit. This current is very small and is detected by the microammeter placed in the circuit.

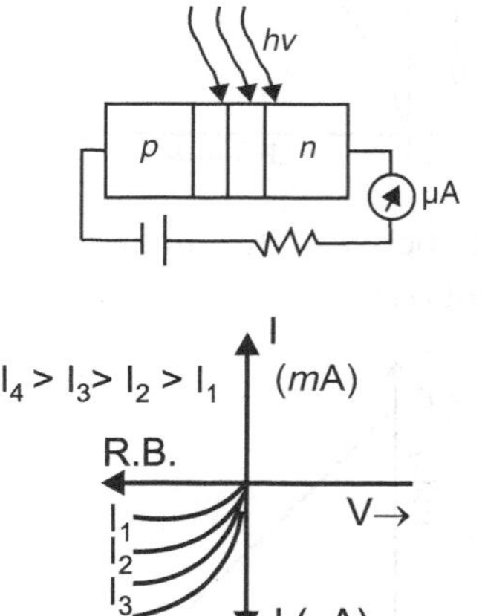


I-V characteristics of photodiode for different illumination intensities

A photodiode is preferably operated in reverse bias condition. Consider an *n*-type semiconductor. Its majority carrier (electron) density is much larger than the minority hole density *i.e.* $n >> p$. When illuminated with light, both types of carriers increase equally in number

$$n' = n + \Delta n; \ p' = p + \Delta p$$

Now $n >> p$ and $\Delta n = \Delta p$

$$\frac{\Delta n}{n} << \frac{\Delta p}{p}$$

That is, the fractional increase in majority carriers is much less than the fractional increase in minority carriers. Consequently, the fractional change due to the photo-effects on the minority carrier dominated reverse bias current is more easily measurable than the fractional change in the majority carrier dominated forward bias current. Hence, photodiodes are preferably used in the reverse bias condition for measuring light intensity.

**20. Draw a circuit diagram of a transistor amplifier in CE configuration.**
**Define the terms :**
**(i) Input resistance and (ii) Current amplification factor. How are these determined using typical input and output characteristics?** [3]**

**21. Answer the following questions : [3]**

**(a) In a double slit experiment using light of wavelength 600 nm, the angular width of the fringe formed on a distant screen is 0.1°. Find the spacing between the two slits.**

**(b) Light of wavelength 5000 Å propagating in air gets partly reflected from the surface of water. How will the wavelengths and frequencies of the reflected and refracted light be affected ?**

**Ans. (a)** Angular width ($\theta$) of fringe in double-slit experiment is given by $\theta = \frac{\lambda}{d}$

Where $d$ = Spacing between the slits.

Given : Wavelength of light, $\lambda$ = 600 nm

Angular width of fringe,

$$\theta = 0.1° = 0.1 \times \frac{\pi}{180} \text{ rad} = 0.0018 \text{ rad}$$

$$\therefore \quad d = \frac{\lambda}{\theta}$$

$$d = \frac{600 \times 10^{-9}}{18 \times 10^{-4}} = 33.33 \times 10^{-5}$$

$$\Rightarrow \quad d = 0.33 \times 10^{-3} \text{m}$$

**(b)** The frequency and wavelength of reflected wave will not change.

The refracted wave will have same frequency, only wavelength will change.

The velocity of light in water is given by

$$v = \lambda f$$

where, $v$ = Velocity of light

$f$ = Frequency of light

$\lambda$ = Wavelength of light

As light ray is travelling from rarer (air) medium to denser medium, its speed will decrease. Hence wavelength ($\lambda$) will also decrease.

**22. An inductor L of inductance $X_L$ is connected in series with a bulb B and an ac source. How would brightness of the bulb change when (i) number of turn in the inductor is reduced. (ii) an iron rod is inserted in the inductor and (iii) a capacitor of reactance $X_C = X_L$ is inserted in series in the circuit. Justify your answer in each case. [3]**

**Ans.**

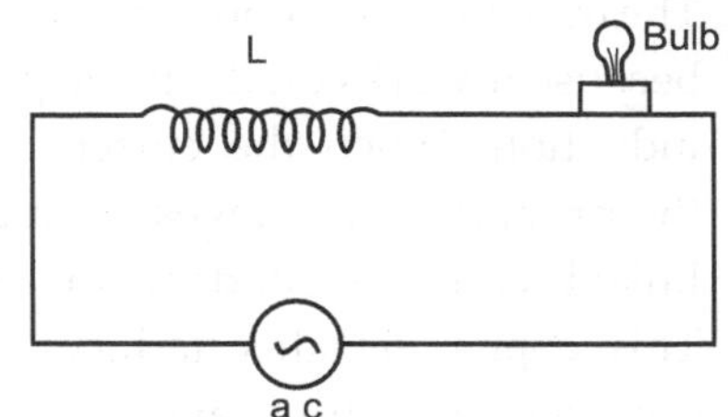


(i) When the number of turns in the inductor is reduced, its reactance $X_L$ decreases. The current in the circuit increases and hence brightness of the bulb increases.

(ii) When an iron rod is inserted in the inductor, the self inductance increases. Consequently, the inductive reactance $X_L = \omega L$ increases. This decreases the current in the circuit and the bulb glows dimmer.

(iii) With capacitor of reactance $X_C = X_L$, the impedance

$$Z = \sqrt{R^2 + (X_L - X_C)^2} = R$$

becomes minimum, the current in circuit becomes maximum. Hence, the bulb glows with maximum brightness.

## SECTION–D

**23. A group of students while coming from the school noticed a box marked "Danger H.T. 2200 V" at a substation in the main street. They did not understand the utility of such a high voltage, while they argued the supply was only 220 V. They asked their teacher this question the next day. The teacher thought it to be an important question and therefore, explained to the whole class. [4]**

** Answer is not given due to change in the present Syllabus.

**Answer the following questions :**

**(i) What device is used to bring the high voltage down to low voltage of a.c. current and what is the principle of its working ?**

**(ii) Is it possible to use this device for bringing down the high dc voltage to the low voltage? Explain.**

**(iii) Write the values displayed by the students and the teacher.****

**Ans.** (i) The device that is used to bring high voltage down to low voltage of an a.c. current is a transformer. It works on the principle of mutual induction of two windings or circuits. When current in one circuit changes, emf is induced in the neighbouring circuit.

(ii) The transformer cannot convert d.c. voltages because it works on the principle of mutual induction. When the current linked with the primary coil changes the magnetic flux linked with the secondary coil also changes. This change in flux induces emf in the secondary coil. If we apply a direct current to the primary coil the current will remain constant. Thus, there is no mutual induction and hence no emf is induced.

## SECTION–E

**24. (a) State Ampere's circuital law. Use this law to obtain the expression for the magnetic field inside an air cored toroid of average radius, having 'n' turns per unit length and carrying a steady current I.**

**(b) An observer to the left of a solenoid of N turns each of cross section area A observes that a steady current I in it flows in the clockwise direction. Depict the magnetic field lines due to the solenoid specifying its polarity and show that it acts as a bar magnet of magnetic momentum M = NIA.** **[5]**

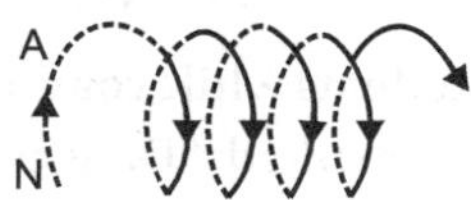


**OR**

**(a) Define mutual inductance and write its S.I. units.**

**(b) Derive an expression for the mutual inductance of two long co-axial solenoids of same length wound one over the other.**

**(c) In an experiment two coils $c_1$ and $c_2$ are placed close to each other. Find out the expression for the emf induced in the coil $c_1$ due to a change in the current through the coil $c_2$.**

**Ans.** (a) Ampere's circuital law in electromagnetism is analogous to Gauss' law in electrostatics. This law states that "The line integral of resultant magnetic field along a closed plane curve is equal to $\mu_0$ time the total current crossing the area bounded by the closed curve provided the electric field inside the loop remains constant. Thus $\oint B.dl = \mu_0 I_{enc}$ where $\mu_0$ is permeability of free space and $I_{enc.}$ is the net current enclosed by the loop.

A toroid is a hollow circular ring on which a large number of turns of a wire are closely wound. Consider an air-cored toroid (as shown below) with centre O.

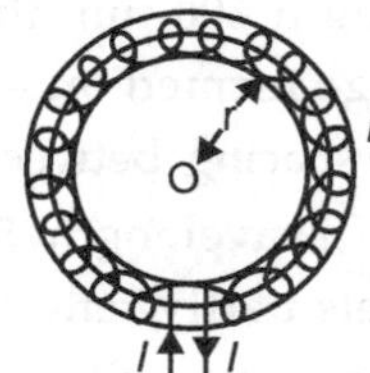


Given : $r$ = Average radius of the toroid

I = Current through the solenoid

$n$ = Number of turns per unit length

To determine the magnetic field inside the toroid, we consider three amperian loops (loop 1, loop 2 and loop 3) as shown in the figure below.

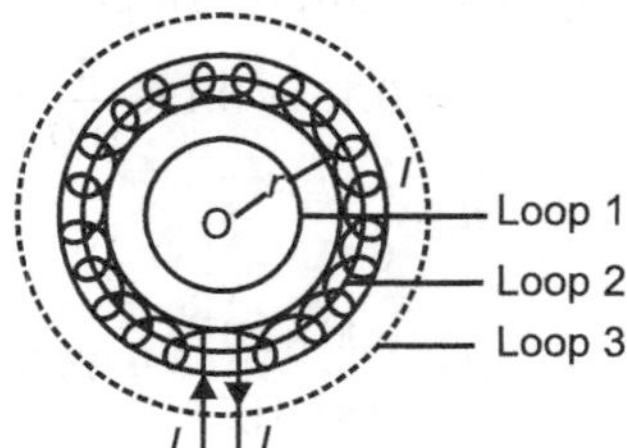


According to Ampere's circuital law, we have

$$\oint \vec{B}.\vec{dl} = \mu_0 I_{enc.} \text{ (Total current)}$$

Total current for loop 1 is zero because no current is passing through this loop

So, for loop 1

$$\oint \vec{B}.\vec{dl} = \mu_0 \text{ (Total current)} = 0$$

For loop 3

According to Ampere's circuital law, we have

$\oint \vec{B}.\vec{dl} = \mu_0$ (Total current) = 0

Total current for loop 3 is zero because net current coming out of this loop is equal to the net current going inside the loop.

For loop 2 :

The total current flowing through the toroid is NI, where N is the total number of turns

$$\oint \vec{B}.\vec{dl} = 0 = \mu_0 (NI) \quad ...(i)$$

Now, $\vec{B}$ and $\vec{dl}$ are in the same direction

$$\oint \vec{B}.\vec{dl} = B \int dl$$

$$\Rightarrow \quad \oint \vec{B}.\vec{dl} = B(2\pi r) \quad ...(ii)$$

Comparing (i) and (ii), we get

$$B(2\pi r) = \mu_0 NI$$

$$\Rightarrow \quad B = \frac{\mu_0 NI}{2\pi r}$$

Number of turns per unit length is given by

$$n = \frac{N}{2\pi r}$$

$$\therefore \quad B = \mu_0 nI$$

This is the expression for magnetic field inside air-cored toroid.

**(b)** Given that the current flows in the clockwise direction for an observer on the left side of the solenoid. This means that left face of the solenoid acts as south pole and right face acts as north pole. Inside a bar magnet the magnetic field lines are directed from south to north. Therefore, the magnetic field lines are directed from left to right in the solenoid. Magnetic moment of single current carrying loop is given by $m' = IA$

where I = Current flowing through the loop

A = Area of the loop

So, Magnetic moment of the whole solenoid is given by

$$M = Nm' = N(IA)$$

**OR**

**(a)** Mutual inductance is the property of two coils by the virtue of which each opposes any change in the value of current flowing through the other by developing an induced emf. The SI unit of mutual inductance is henry and its symbol is H.

**(b)** Consider two long solenoids $S_1$ and $S_2$ of same length $l$ such that solenoid $S_2$ surrounds solenoid $S_1$ completely.

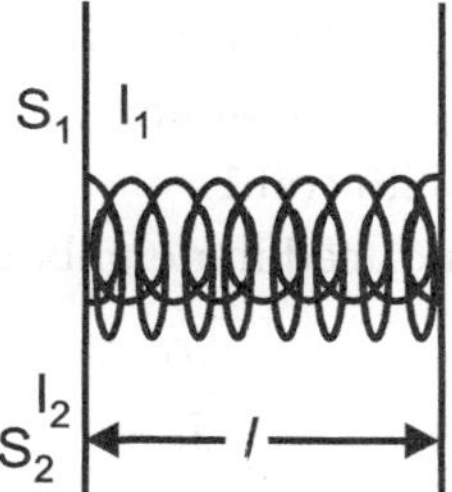


Let :

$n_1$ = Number of turns per unit length of $S_1$

$n_2$ = Number of turns per unit length of $S_2$

$I_1$ = Current passed through solenoid $S_1$

$\phi_{21}$ = Flux linked with $S_2$ due to current flowing through $S_1$

$\phi_{21} \propto I_1$

$\phi_{21} = M_{21}I_1$

where $M_{21}$ = Coefficient of mutual induction of the two solenoids

When current is passed through solenoid $S_1$, an emf is induced in solenoid $S_2$.

Magnetic field produced inside solenoid $S_1$ on passing current through it is given by

$$B_1 = \mu_0 n_1 I_1$$

Magnetic flux linked with each turn of solenoid $S_2$ will be equal to $B_1$ times the area of cross-section of solenoid $S_2$.

Magnetic flux linked with each turn of the solenoid $\phi_{21} = B_1 A$

Therefore, total magnetic flux linked with the solenoid $S_2$ is given by

$$\phi_{21} = B_1 A = \mu_0 n_1 I_1 A$$

$$M_{21} = \frac{n_2 \phi_{21}}{I_1}$$

$$M_{21} = \frac{n_2 \mu_0 n_1 I_1 A}{I_1} = \mu_0 n_1 n_2 A$$

where $N_2$ is total number of turns wound over the secondary coil.

$$\therefore \quad M_{21} = \mu_0 n_1 n_2 A$$

Similarly the mutual inductance between the two solenoids when current is passed through solenoid $S_2$ and induced emf is produced in solenoid $S_1$ is given by

$$M_{12} = \mu_0 n_1 n_2 A$$

where $n_1$ total numbers of turns wound over primary coil.

(c) It is found that,

$$\phi \propto I$$

$$\phi = MI$$

Where, I is the strength of current in coil 2, and $\phi$ is the total amount of magnetic flux linked with coil 1.

E.m.f. induced in neighbouring coil $C_1$ is,

$$e = -\frac{d\phi}{dt}$$

$$= -\frac{d(MI)}{dt} = -\frac{MdI}{dt}$$

**25. (a) A point object 'O' is kept in a medium of refractive index $n$ in front of a convex spherical surface of radius of curvature R which separate the second medium of refractive index $n_2$ from the first one as shown in the figure.**

**Draw the ray diagram showing the image formation and deduce the relationship between the object distance and the image distance in terms of $n_1$, $n_2$ and R.**

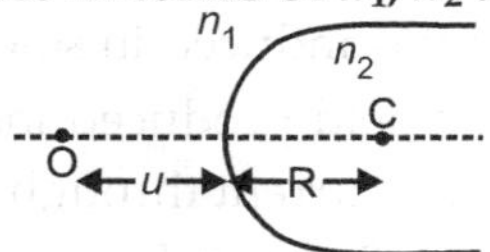


**(b) When the image formed above acts as a virtual object for a concave spherical surface separating the medium $n_2$ from $n_1$ ($n_2 > n_1$) draw this ray diagram and write the similar (similar to (a)) relation. Hence obtain the expression for the Lens Maker's formula.** **[5]**

**Ans.** (a) Let a spherical surface separate a rarer medium of refractive index $n_1$ from second medium of refractive index $n_2$. Let C be the centre of curvature and R = MC be the radius of the surface.

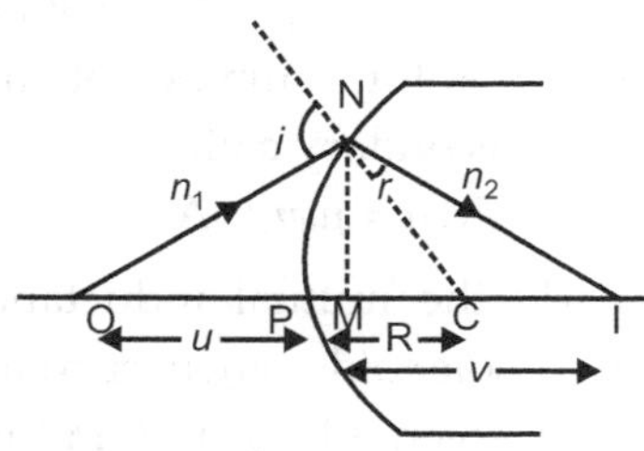


Consider a point object O lying on the principal axis of the surface. Let a ray starting from O incident normally on the surface along OM and pass straight. Let another ray of light incident on NM along ON and refract along NI.

From M, draw MN perpendicular to OI.

The above figure shows the geometry of formation of image I of an object O and the principal axis of a spherical surface with centre of curvature C and radius of curvature R.

Let us make the following assumptions :

**(i)** The aperture of the surface is small as compared to the other distance involved.

**(ii)** The object consists only of a point lying on the principal axis of the spherical refracting surface.

$$\tan \angle NOM = \frac{MN}{OM}$$

$$\tan \angle NCM \frac{MN}{MC}$$

$$\tan \angle NIM = \frac{MN}{MI}$$

For $\Delta NOC$, $i$ is the exterior angle.

$$\therefore \quad i = \angle NOM + \angle NCM$$

For small angles $\tan i \approx i$,

$$i = \frac{MN}{OM} + \frac{MN}{MC} \quad \text{...(i)}$$

Similarly,

$$r = \angle NCM - \angle NIM$$

$$r = \frac{MN}{MC} - \frac{MN}{MI} \quad \text{...(ii)}$$

By Snell's law,

$$n_1 \sin i = n_2 \sin r$$

For small angles,

$$n_1 i = n_2 r$$

Substituting the values of $i$ and $r$ from (i) and (ii), we obtain

$$n_1\left(\frac{MN}{OM} + \frac{MN}{MC}\right) = n_2\left(\frac{MN}{MC} - \frac{MN}{MI}\right)$$

or, $$\frac{n_1}{OM} + \frac{n_2}{MI} = \frac{n_2 - n_1}{MC} \quad \text{...(iii)}$$

Applying new Cartesian sign conventions, we get

$$OM = -u, MI = +v, MC = +R$$

Substituting these values in equation (iii), we obtain

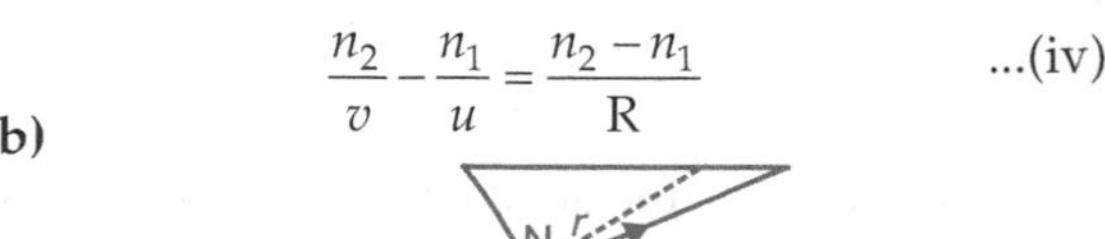

$$\frac{n_2}{v} - \frac{n_1}{u} = \frac{n_2 - n_1}{R} \quad \text{...(iv)}$$

(b)

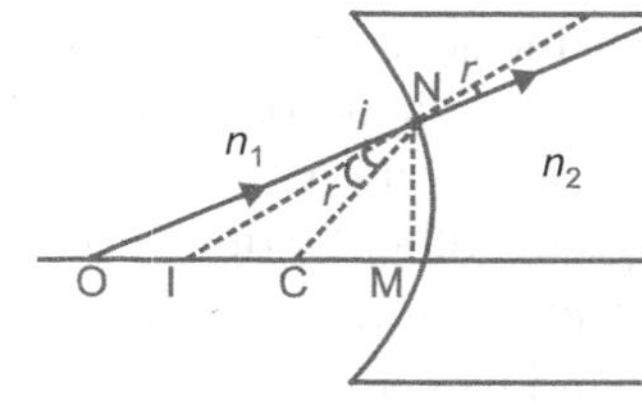

Now the image I acts as virtual object for the second surface that will form real image at $l$. As refraction takes place from denser to rarer medium

$$\therefore \quad \frac{-n_2}{v} + \frac{n_1}{v'} = \frac{n_1 - n_2}{R} \quad ...(v)$$

Adding (iv) and (v) we get

$$\frac{n_1}{v'} - \frac{n_1}{u} = (n_2 - n_1)\left[\frac{1}{R} - \frac{1}{R'}\right]$$

$$\frac{1}{f} = (n_{21} - 1)\left[\frac{1}{R} - \frac{1}{R'}\right]$$

$$\left[\because n_{21} = \frac{n_2}{n_1}, \frac{1}{f} = \frac{1}{v'} - \frac{1}{u}\right]$$

**26. (a) An electric dipole of dipole moment $\vec{p}$ consists of point charges $+q$ and $-q$ separated by a distance $2a$ apart. Deduce the expression for the electric field $\vec{E}$ due to the dipole at a distance x from the centre of the dipole on its axial line in terms of the dipole moment $\vec{p}$. Hence show that in the limit**

$$x >> a, \vec{E} \rightarrow \frac{2\vec{p}}{4\pi\varepsilon_0 x^3}$$

**(b) Given the electric field in the region $\vec{E} = 2\hat{x}l$, find the net electric flux through the cube and the charge enclosed by it. [5]**

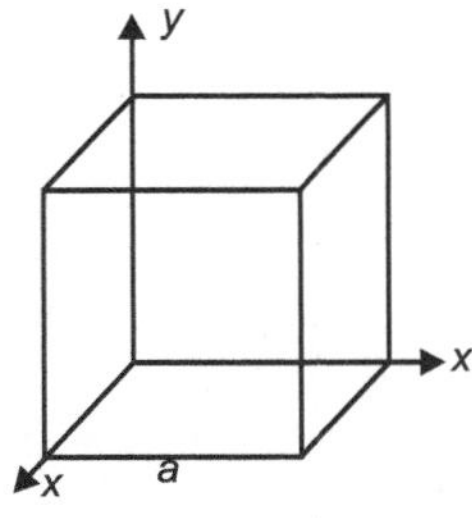


**OR**

**(a) Explain, using suitable diagrams, the difference in the behaviour of a (i) conductor and (ii) dielectric in the presence of external electric field. Define the terms polarization of a dielectric and write its relation with susceptibility.**

**(b) A thin metallic spherical shell of radius carries a charge Q on its surface. A point charge $\frac{Q}{2}$ is placed at its centre C and another charge +2Q is placed outside the shell at a distance $x$ from the centre as shown in the figure. Find (i) the force on the charge at the centre of shell and at the point A, (ii) the electric flux through the shell.**

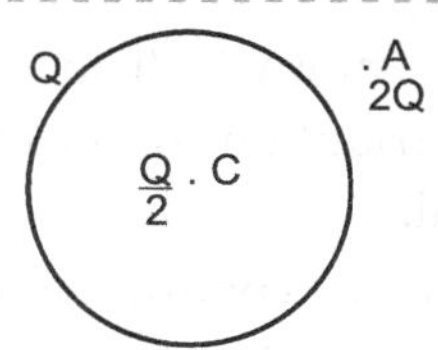


**Ans. (a)** Electric field at a point on the axial line

$$|\vec{E}_{+q}| = \frac{kq}{(x-a)^2}$$

$$|\vec{E}_{-q}| = \frac{kq}{(x+a)^2}$$

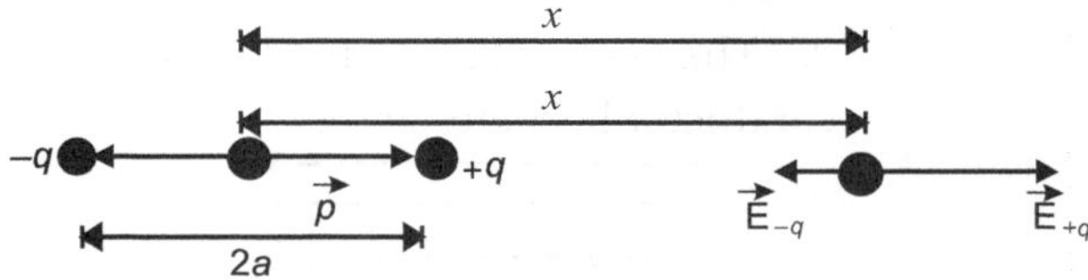


$$\vec{E} = \vec{E}_{+q} - \vec{E}_{-q} = \frac{kq}{(x-a)^2} - \frac{kq}{(x+a)^2}$$

$$\Rightarrow \quad |\vec{E}| = \frac{kq4ax}{(x^2 - a^2)^2}$$

$$\vec{E} = \frac{2k\vec{p}x}{(x^2-a^2)^2} \text{ (Parallel to } \vec{p}) \quad [\because \vec{p} = 2aq]$$

If $x >> a$, $\vec{E} = \frac{2\vec{p}}{4\pi\varepsilon_0 x^3}$.

In vector form, $\vec{E} = \frac{2\vec{p}}{4\pi\varepsilon_0 x^3}$ $\quad \left[\because k = \frac{1}{4\pi\varepsilon_o}\right]$

**(b)** Since, the electric field is parallel to the faces and parallel to $xy$ and $xz$ planes, the electric flux through them is zero.

Electric flux through the left face

$$\phi_L = (E_L)(a^2)\cos 180^\circ$$
$$= (0)(a^2)\cos 180^\circ = 0$$

Electric flux through the right face,

$$\phi_R = (E_R)(a^2)\cos 0^\circ$$
$$= (2a)(a^2) \times 1$$
$$= 2a^3$$

$$\text{Total flux}(\phi) = 2a^3 = \frac{q_{enclosed}}{\varepsilon_0}$$

$$\therefore q_{enclosed} = 2a^3\varepsilon_0$$

**OR**

**(a) (i) Conductor :**

Let $E_0$ = external field, and

$E_m$= internal field created by the redistribution of electrons inside the metal.

When a conductor like a metal is subjected to external electric field, the electrons experience a force in the opposite direction collecting on the left hand side.

A positive charge is therefore induced on the right hand side. This creates an opposite electric field ($E_m$) that balances out ($E_0$).

∴ The net electric field inside the conductor becomes zero.

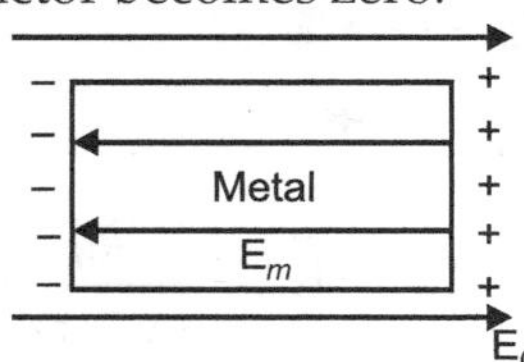


**(ii) Dielectric :**

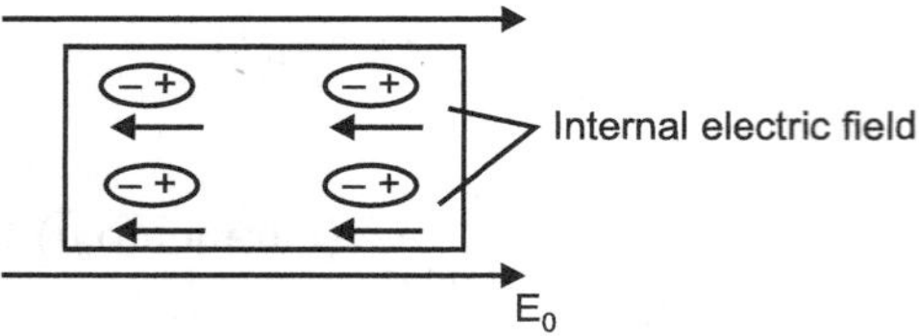


When an external electric field is applied, dipoles are created (in case of non-polar dielectrics). The placement of dipoles is as shown in the given figure. An internal electric field is created which reduces the external electric field.

Polarization of dielectric (P) is defined as the Induced dipole moment per unit volume of the polarized dielectric.

$$P = \chi_e \varepsilon_0 E$$

Where $\chi_e$ = Susceptibility

$E$ = Electric field

**(b)** **(i)** Net force on the charge $\frac{Q}{2}$, placed at the centre of the shell is zero.

Net Force on charge 2Q kept at a point A

$$F = E \times 2Q = \frac{1.\left(\frac{3Q}{2}\right)2Q}{4\pi\varepsilon_o r^2} = \frac{k3Q^2}{r^2}$$

**(ii)** Electric flux through the shell,

$$\phi = \frac{Q}{2\varepsilon_o} \qquad \left[\because \phi = \frac{q_{enclosed}}{\varepsilon_0}\right]$$

**Note : All questions of Delhi Set II are from Delhi Set I and Delhi Set III are from Set I and Set II.**

# PHYSICS

## Outside Delhi [Set-I] 2014

**Time allowed : 3 hours** **Maximum marks : 70**

1. **Using the concept of force between two infinitely long parallel current carrying conductors, define one ampere of current.** **[1]**

**Ans.** One ampere is the current which when flowing through each of the two parallel uniform long linear conductors placed in free space at a distance of one metre from each other will attract or repel each other with a force of $2 \times 10^{-7}$ N per metre of their length.

2. **To which part of the electromagnetic spectrum does a wave of frequency $5 \times 10^{19}$ Hz belong ?** **[1]**

**Ans.** The frequency $5 \times 10^{19}$ Hz lies in the gamma region of the electromagnetic spectrum.

3. **What is the force between two small charges of $2 \times 10^{-7}$C and $3 \times 10^{-7}$C placed 30 cm apart in air?** **[1]**

**Ans.** We know that,

$$F = \frac{1}{4\pi\varepsilon_0}\frac{q_1q_2}{r^2}$$

$$= 9 \times 10^9 \times \frac{(2\times10^{-7})(3\times10^{-7})}{(0.30)^2}$$

$$= 6 \times 10^{-3}\text{ N}$$

4. **Define intensity of radiation on the basis of photon picture of light. Write its S.I. unit.** **[1]**

**Ans.** The intensity of radiation can be defined as the energy associated with photons emitted from a unit surface area in unit time. Its S.I. unit is joule/metre$^2$second (J/m$^2$s).

5. **The electric current flowing in a wire in the direction from B to A is decreasing. Find out the direction of the induced current in the metallic loop kept above the wire as shown.** **[1]**

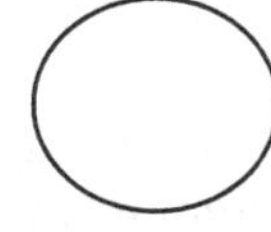

A B

**Ans.** The direction of the current in the loop will be such as to oppose the decrease of this field (Lenz's Law).

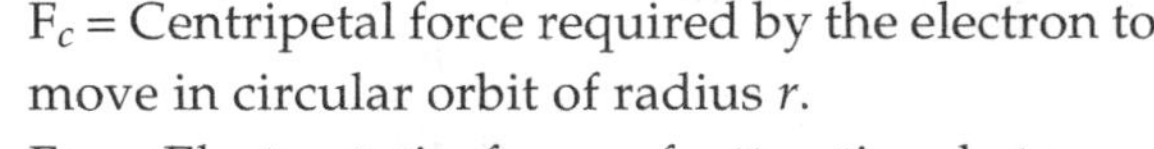

6. **Why is it found experimentally difficult to detect neutrinos in nuclear $\beta$-decay ?** **[1]**

**Ans.** Neutrinos are difficult to detect experimentally in $\beta$-decay because these are uncharged particles with almost zero mass and also they interact weakly with matter.

7. **Why is the use of A.C. voltage preferred over D.C. voltage ? Give two reasons.** **[1]**

**Ans.** The use of A.C. voltage is preferred over the use of D.C. voltage because of the following reasons :

(i) The energy losses while transmission of A.C. voltage are very less as compared to D.C. voltage.

(ii) A.C. voltage can be controlled as required by using a transformer (*i.e.*, stepped up or stepped down).

8. **A biconvex lens made of a transparent material of refractive index 1.25 is immersed in water of refractive index 1.33. Will the lens behave as a converging or a diverging lens ? Give reason.** **[1]**

**Ans.** In this case the biconvex lens will behave as a diverging lens because the refractive index of water (1.33) is more than that of the material (1.25) of the lens.

9. **Using Rutherford model of the atom, derive the expression for the total energy of the electron in hydrogen atom. What is the significance of total negative energy possessed by the electron ?** **[2]**

**OR**

**Using Bohr's postulates of the atomic model, derive the expression for radius of $n^{th}$ electron orbit. Hence obtain the expression for Bohr's radius.**

**Ans.** In a hydrogen atom, an electron having charge $-e$ revolves around the nucleus having charge $+e$ in a circular orbit of radius $r$.

Let,

$F_c$ = Centripetal force required by the electron to move in circular orbit of radius $r$.

$F_e$ = Electrostatic force of attraction between revolving electron and nucleus.

The electrostatic force of attraction ($F_e$) provides the necessary centripetal force,

$$F_c = F_e$$

$$\frac{mv^2}{r} = \frac{(e)(e)}{4\pi\varepsilon_0 r^2} \quad ...(i)$$

K.E. of electron in the orbit,

$$K.E. = \frac{1}{2}mv^2$$

From equation (i),

$$K.E. = \frac{e^2}{8\pi\varepsilon_0 r}$$

Potential energy of electron in orbit,

$$P.E. = \frac{(e)(-e)}{4\pi\varepsilon_0 r} = \frac{-e^2}{4\pi\varepsilon_0 r}$$

∴ Total energy of electron in hydrogen atom

$$E = E.K. + P.E. = \frac{e^2}{8\pi\varepsilon_0 r} - \frac{e^2}{4\pi\varepsilon_0 r}$$

$$E.K. = -\frac{e^2}{8\pi\varepsilon_0 r}$$

Here, negative sign indicates that the revolving electron is bound to the positive nucleus.

**OR**

Suppose $m$ be the mass of an electron and $v_n$ be its speed in $n^{th}$ orbit of radius $r_n$. From Rutherford model, the centripetal force for revolution is produced by electrostatic attraction between electron and nucleus.

$$\frac{mv_n^2}{r_n} = \frac{1}{4\pi\varepsilon_0}\frac{Ze^2}{r_n^2} \quad ...(i)$$

From Bohr's postulate for quantization of angular momentum of $n^{th}$ orbit.

$$mv_n r_n = \frac{nh}{2\pi} \Rightarrow v_n = \frac{nh}{2\pi m r_n}$$

Substituting this value in equation (i), we get

$$\frac{m}{r_n}\left[\frac{nh}{4\pi m r_n}\right]^2 = \frac{1}{4\pi\varepsilon_0}\frac{Ze^2}{r_n^2}$$

or $$r_n = \frac{\varepsilon_0 n^2 h^2}{\pi m Z e^2}$$

For Bohr's radius $n = 1$,

$$r_1 = \frac{\varepsilon_0 h^2}{\pi m Z e^2}$$

This is the expression for Bohr's radius.

**10. A parallel plate capacitor of capacitance C is charged to a potential V. It is then connected to another uncharged capacitor having the same capacitance. Find out the ratio of the energy stored in the combined system to that stored initially in the single capacitor. [2]**

**Ans.** Let '$q$' be the charge on the charged capacitor. Energy stored in it is,

$$U = \frac{q^2}{2C}$$

When another similar uncharged capacitor is connected, the net capacitance of the system is,

$$C' = 2C$$

The charge on the system is constant. So, the energy stored in the system now is,

$$U' = \frac{q^2}{2(C')^2}$$

$$\Rightarrow \quad U' = \frac{q^2}{2(2C)}$$

$$\Rightarrow \quad U' = \frac{q^2}{4C}$$

Thus, the required ratio is,

$$\frac{U'}{U} = \frac{\frac{q^2}{4C}}{\frac{q^2}{2C}}$$

$$\Rightarrow \quad \frac{U'}{U} = \frac{1}{2}$$

**11. Considering the case of a parallel plate capacitor being charged, show how one is required to generalize Ampere's circuital law to include the term due to displacement current. [2]**

**Ans.** Gauss' law states that the electric flux $\phi_E$ of a parallel plate capacitor having an area A, and a total charge Q is given by

$$\phi_E = EA = \frac{Q}{\varepsilon_0 A} \times A$$

$$= \frac{Q}{\varepsilon_0}$$

As the charge Q on the capacitor plates change with time, so current is given by

$$i = \frac{dQ}{dt}$$

$$\therefore \quad \frac{d\phi_E}{dt} = \frac{d}{dt}\left(\frac{Q}{\varepsilon_0}\right) = \frac{1}{\varepsilon_0}\frac{dQ}{dt}$$

$$\Rightarrow \quad \varepsilon_0 \frac{d\phi_E}{dt} = \frac{dQ}{dt} = i$$

This is the missing term in Ampere's Circuital law.

So, the total current through the conductor is

$i$ = Conduction current ($i_c$) + Displacement current ($i_d$)

$$i = i_c + \varepsilon_0 \frac{d\phi_E}{dt} \quad ...(i)$$

Ampere's circuital law states that,

$$\oint \vec{B}.\vec{dl} = \mu_0 i \quad ...(ii)$$

Putting equation (i) in (ii), we get

$$\oint \vec{B}.\vec{dl} = \mu_0 i_c + \mu_0 \varepsilon_0 \frac{d\phi_E}{dt}$$

This is the required generalized form of Ampere's circuital law.

**12. A cell of emf 'E' and internal resistance 'r' is connected across a variable resistor 'R'. Plot a graph showing variation of terminal voltage 'V' of the cell versus the current 'I'. Using the plot, show how the emf of the cell and its internal resistance can be determined. [2]**

**Ans.** The terminal voltage 'V' of the cell is given by

$$V = E - Ir$$

where E is the emf of the cell,

$r$ is the internal resistance of the cell

and, I is the current through the circuit.

Comparing with the equation of a straight line $y = mx + c$, we get

$$y = V; x = I; m = -r; c = E$$

Graph showing variation of terminal voltage 'V' of the cell versus the current 'I'.

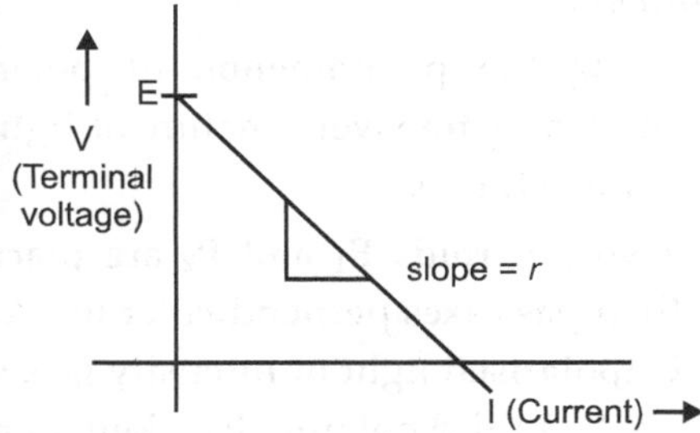


Emf of the cell = Intercept on V axis.

Internal resistance = slope of the line.

**13. Explain, with the help of a circuit diagram, the working of a $p$-$n$ junction diode as a half-wave rectifier. [2]**

**Ans.** The circuit diagram for a $p$-$n$ junction diode as a half wave rectifier is shown below :

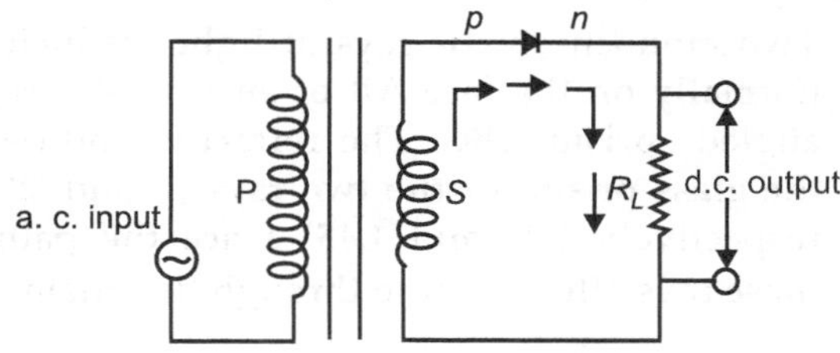


**Working :** During the positive half cycle of the input a.c., the $p$-$n$ junction is forward biased *i.e.*, the forward current flows from $p$ to $n$. In the forward biasing, the diode provides a very low resistance and allows the current to flow. Thus, we get output across load.

During the negative half cycle of the input a.c., the $p$-$n$ junction is reversed biased. In the reverse biasing, the diode provides a high resistance and hence a very small amount of current will flow through the diode which is of negligible amount. Thus no output is obtained across the load. During the next half cycle, output is again obtained as the junction diode gets forward biased. Thus, a half wave rectifier gives discontinuous and pulsating d.c. output across the load resistance.

The waveform of input and output is shown below :

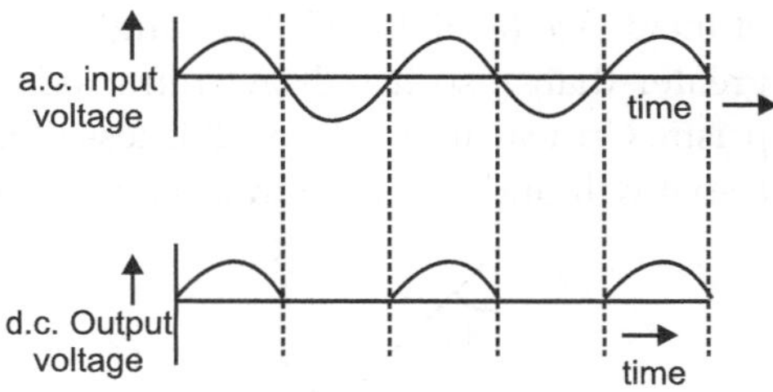


**14. Estimate the average drift speed of conduction electrons in a copper wire of cross-sectional area $1.0\times10^{-7}m^2$ carrying a current of 1.5 A. Assume the density of conduction electrons to be $9 \times 10^{28} m^{-3}$. [2]**

**Ans.** Since, drift velocity, $v_d = \frac{I}{nAq}$

where,

I is the current,

$n$ is the charge density,

$q$ is charge of the electron, and

A is cross-sectional area.

$$v_d = \frac{1.5}{9\times10^{28}\times1.0\times10^{-7}\times1.6\times10^{-19}}$$

$$v_d = 10.4 \times 10^{-4} \text{ m/s}$$

**15. Two monochromatic rays of light are incident normally on the face AB of an isosceles right-angled prism ABC. The refractive indices of the glass prism for the two rays '1' and '2' are respectively 1.35 and 1.45. Trace the path of these rays after entering through the prism. [2]**

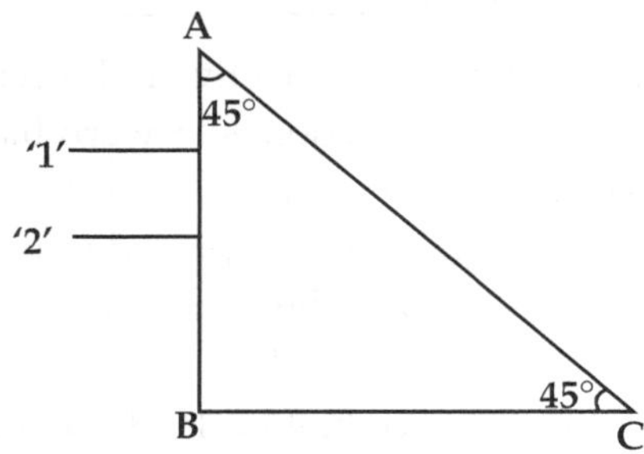


**Ans.** Critical angle of ray '1':

$$\sin(c_1) = \frac{1}{\mu_1} = \frac{1}{1.35}$$

$$\Rightarrow \quad c_1 = \sin^{-1}\left(\frac{1}{1.35}\right) = 47.79°$$

Similarly, critical angle of ray '2':

$$\sin(c_2) = \frac{1}{\mu_2} = \frac{1}{1.45}$$

$$\Rightarrow \quad c_1 = \sin^{-1}\left(\frac{1}{1.45}\right) = 43.6°$$

Ray '1' and '2' will fall on the side AC at an angle of incidence ($i$) of 45°. Critical angle of ray '1' is greater than $i$, so it will get refracted from the prism. Critical angle of ray '2' is less than that of $i$, so it will undergo total internal reflection.

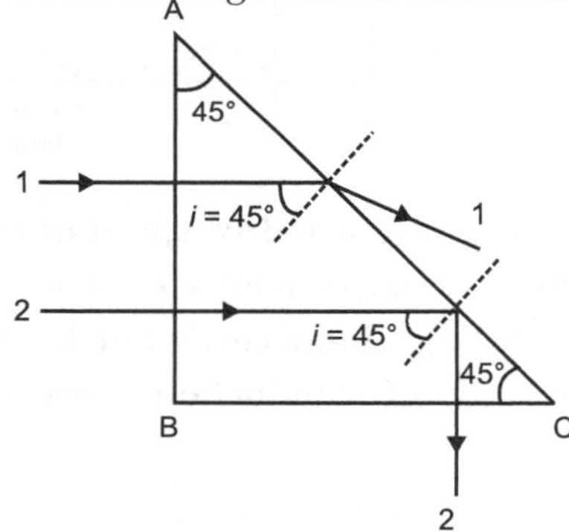


**16. Write the functions of the following in communication systems : [2]**

**(i) Transducer**

**(ii) Repeater**

**17. Show diagrammatically the behaviour of magnetic field lines in the presence of (i) paramagnetic and (ii) diamagnetic substances. How does one explain this distinguishing feature ? [2]**

** Answer is not given due to change in the present Syllabus.

**Ans.** (i) The behaviour of magnetic field lines in the presence of a paramagnetic substance is shown below :

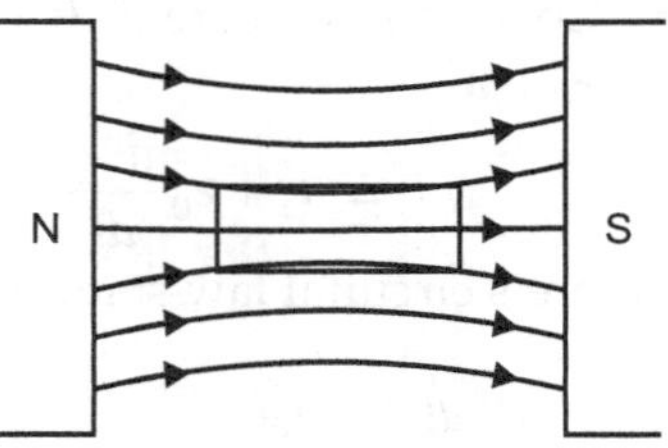


(ii) The behaviour of magnetic field lines in the presence of a diamagnetic substance is shown below :

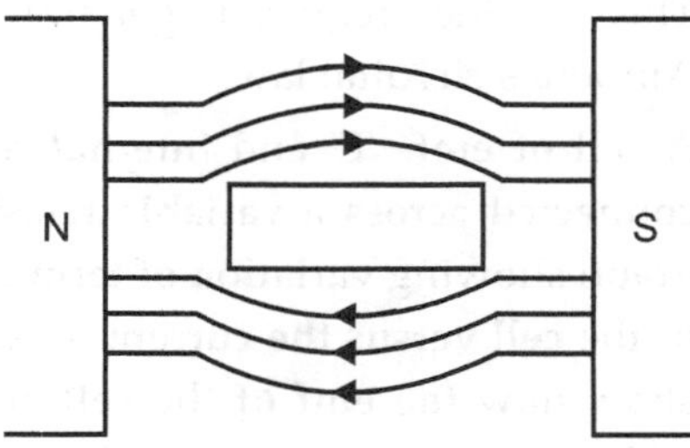


This distinguishing feature is because of the difference in their relative permeability. The relative permeability of diamagnetic substance is less than 1, so the magnetic lines of force do not prefer passing through the substance whereas the relative permeability of a paramagnetic substance is greater than 1, so the magnetic lines of force prefer passing through the substance.

**18. Draw a circuit diagram of *n-p-n* transistor amplifier in CE configuration. Under what condition does the transistor act as an amplifier?** [2]**

**19. (a) Using the phenomenon of polarization, show how transverse nature of light can be demonstrated.**

**(b) Two polaroids $P_1$ and $P_2$ are placed with their pass axes perpendicular to each other. Unpolarised light of intensity $I_0$ is incident on $P_1$. A third polaroid $P_3$ is kept in between $P_1$ and $P_2$ such that its pass axis makes an angle of 30° with that of $P_1$. Determine the intensity of light transmitted through $P_1$, $P_2$ and $P_3$. [3]**

**Ans.** (a) Suppose that an ordinary light is incident normally on a pair of crystals $C_1$ and $C_2$. When the incident ray of light passes through crystal $C_1$, it gets plane polarised in the direction perpendicular to the length of crystal. Now, we see that when the axis of two crystals are parallel, the intensity of the emerging light will be maximum. When the second crystal is placed perpendicular with respect to the first crystal, the intensity of light observed is zero. This is due to the electric field of the plane polarised light obtained from $C_1$ can vibrate only in one direction. Hence, when the axis of the crystal $C_2$ is perpendicular to its direction of vibration of electric field, it gets blocked. This shows the transverse nature of light.

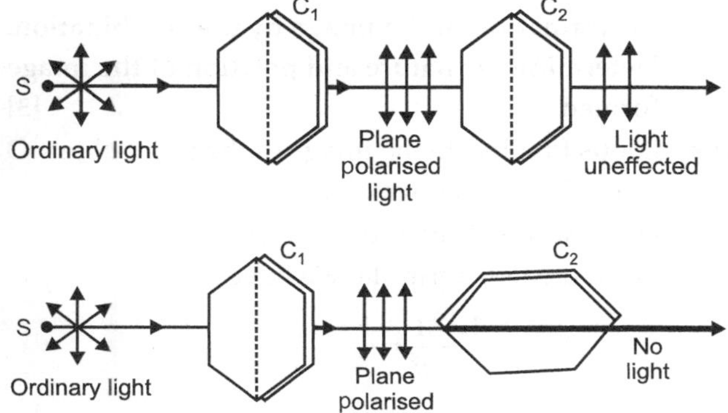


**(b)** Intensity of light after falling on $P_1$,

$$I = \frac{I_0}{2}$$

Intensity of light after falling on $P_3$,

$$I' = I\cos^2(\theta) = \frac{I_0}{2}\cos^2(30°) = \frac{3I_0}{8}$$

$$\left[\because \cos 30° = \frac{\sqrt{3}}{2}\right]$$

Therefore, a light of intensity $\frac{3I_0}{8}$ will pass through the $P_2$, and the angle between $P_3$ and $P_2$ will be 60° because of the condition given in the question. Intensity of light after falling on $P_2$,

$$I'' = I'\cos^2 60°$$

$$I'' = \frac{3I_0}{8}\cos^2(60°) = \frac{3I_0}{32} \quad \left[\because \cos 60° = \frac{1}{2}\right]$$

**20. Define the term 'mutual inductance' between the two coils.**

**Obtain the expression for mutual inductance of a pair of long coaxial solenoids each of length $l$ and radii $r_1$ and $r_2$ ($r_2 >> r_1$). Total number of turns in the two solenoids are $N_1$ and $N_2$ respectively. [3]**

**Ans.** Mutual inductance of two coils is equal to the e.m.f. induced in one coil when rate of change of current through the other coil is unity.

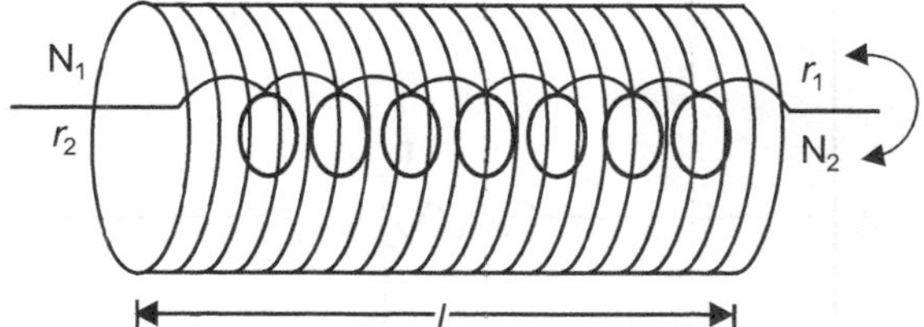


**Mutual inductance of two co-axial solenoids :**

Consider two long co-axial solenoid each of length $l$ with number of turns $N_1$ and $N_2$ wound one over the other. Number of turns per unit length in solenoid, $n = \frac{N_1}{l}$. If $I_1$ is the current flowing in primary solenoid, the magnetic field produced within this solenoid.

$$B_1 = \frac{\mu_0 N_1 I_1}{l}$$

The flux linked with each turn of inner solenoid coil is $\phi_2 = B_1 A_2$, where $A_2$ is the cross-sectional area of inner solenoid. The total flux linkage with inner coil of $N_2$ turns.

$$\phi_2 = N_2\phi_2$$

$$= N_2 B_1 A_2$$

$$= N_2\left(\frac{\mu_0 N_1 I_1}{l}\right)A_2$$

$$= \frac{\mu_0 N_1 N_2 A_2 I_1}{l}$$

$$\text{Mutual Inductance, } M_{21} = \frac{\phi_2}{I_1} = \frac{\mu_0 N_1 N_2 A_2}{l}$$

If $n_1$ is number of turns per unit length of outer solenoid and $r_2$ is radius of inner solenoid, then

$$M = \mu_0 n_1 N_2 \pi r_2^2.$$

**21. Answer the following : [3]**

**(a) Why are the connections between the resistors in a meter bridge made of thick copper strips ?**

**(b) Why is it generally preferred to obtain the balance point in the middle of the meter bridge wire ?**

**(c) Which material is used for the meter bridge wire and why ?**

**OR**

**A resistance of R Ω draws current from a potentiometer as shown in the figure. The potentio meter has a total resistance $R_0$ Ω. A voltage V is supplied to the potentiometer.**

**Derive an expression for the voltage across R when the sliding contact is in the middle of the potentiometer.**

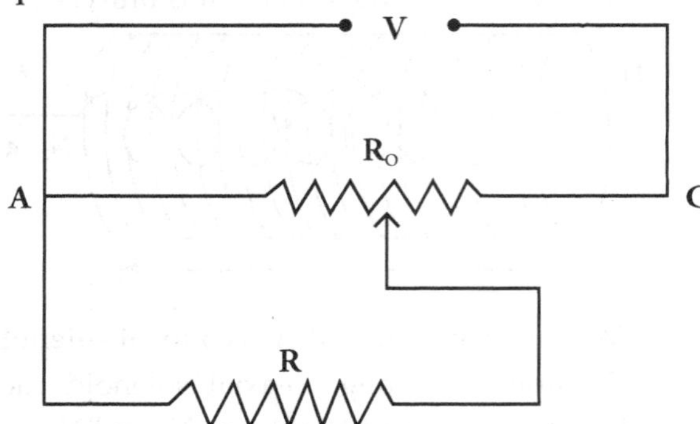


**Ans.** **(a)** The connection between the resistors in a meter bridge is made of thick copper strips because the resistivity of a copper wire is very low. As, the connections are thick, so the area becomes large and the resistance of the wires becomes almost negligible.

**(b)** It is preferred to obtain the balance point in the middle of the meter bridge wire because it improves the sensitivity of the meter bridge.

**(c)** Constantan is used for meter bridge wire because its temperature coefficient of resistance is almost negligible due to which the resistance of the wire does not get affected on increasing temperature of the wire due to flow of current.

**OR**

As the slide is in the middle of the potentiometer so, only half of its resistance ($R_0/2$) will be in parallel with the resistance R. Hence, the total resistance ($R_1$) will be given by the following expression :

$$\frac{1}{R_1} = \frac{1}{R} + \frac{1}{(R_0/2)}$$

$$R_1 = \frac{R_0 R}{R_0 + 2R}$$

The total resistance between A and C is $R_1 + R_0/2$.

$\therefore$ The current flowing through the potentiometer will be

$$I = \frac{V}{R_1 + R_0/2} = \frac{2V}{2R_1 + R_0}$$

The voltage $V_1$ taken from the potentiometer will be the product of current I and resistance $R_1$.

$$V_1 = IR_1 = \left(\frac{2V}{2R_1 + R_0}\right) \times R_1$$

Substituting for $R_1$, we have

$$V_1 = \frac{2V}{2\left(\frac{R_0 \times R}{R_0 + 2R}\right) + R_0} \times \frac{R_0 R}{(R_0 + 2R)}$$

$$V_1 = \frac{2VR}{2R + R_0 + 2R}$$

or $$V_1 = \frac{2VR}{R_0 + 4R}$$

**22. A convex lens of focal length 20 cm is placed coaxially with a convex mirror of radius of curvature 20 cm. The two are kept at 15 cm apart from each other. A point object lies 60 cm in front of the convex lens. Draw a ray diagram to show the formation of the image by the combination. Determine the nature and position of the image formed.** **[3]**

**Ans.** Let us first locate the image of the point object O formed by the convex lens.

Here : $u = -60$ cm and $f = 20$ cm

From the lens formula, we have :

$$\frac{1}{v} - \frac{1}{u} = \frac{1}{f}$$

$$\frac{1}{v} = \frac{1}{f} + \frac{1}{u}$$

$$= \frac{1}{20} + \frac{1}{-60} = \frac{3-1}{60} = \frac{2}{60} = \frac{1}{30}$$

$v = +30$ cm

The positive sign indicates that the image is formed at the right of the lens.

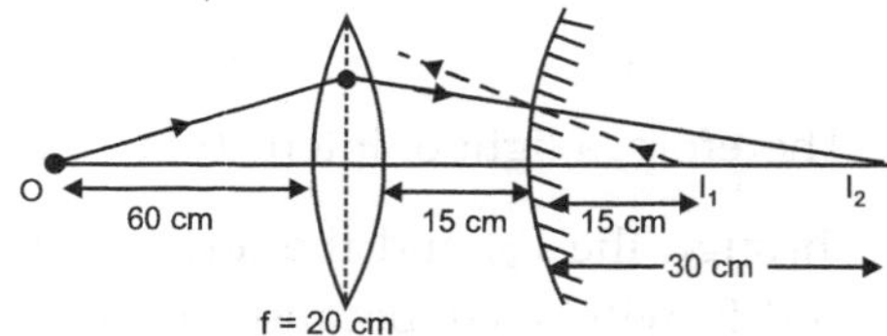


The image $I_1$ is formed behind the mirror and acts as a virtual oject for the mirror. The convex mirror forms the image $I_2$, whose distance from the mirror can be determined as :

$$\frac{1}{v} + \frac{1}{u} = \frac{1}{f}$$

Here, $$u = 15 \text{ cm}$$

and, $$f = \frac{R}{2} = 10 \text{ cm}$$

$$\Rightarrow \quad \frac{1}{v} = \frac{1}{f} - \frac{1}{u}$$

$$\Rightarrow \quad \frac{1}{v} = \frac{1}{10} - \frac{1}{15}$$

$$\Rightarrow \quad \frac{1}{v} = \frac{3-2}{30} = \frac{1}{30}$$

$$\Rightarrow \quad v = 30 \text{ cm}$$

Hence, the final virtual image is formed at a distance of 30 cm from the convex mirror.

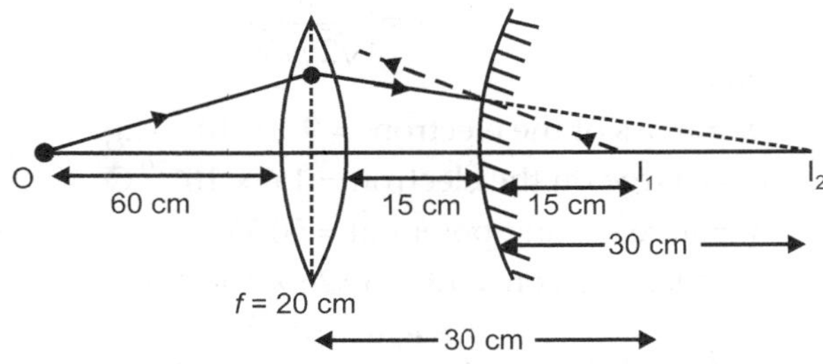


**23. A voltage V = $V_0$ sin ω*t* is applied to a series LCR circuit. Derive the expression for the average power dissipated over a cycle. Under what condition is (i) no power dissipated even though the current flows through the circuit, (ii) maximum power dissipated in the circuit ? [3]**

**Ans.** Voltage $V = V_0 \sin \omega t$ is applied to a series LCR circuit.

Current is

$$I = I_0 \sin(\omega t + \phi)$$

$$I_0 = \frac{V_0}{Z}$$

$$\phi = \tan^{-1}\left(\frac{X_C - X_L}{R}\right)$$

Instantaneous power supplied by the source is

$$P = VI = (V_0 \sin \omega t) \times (I_0 \sin(\omega t + \phi))$$

$$= \frac{V_0 I_0}{2}[\cos\phi - \cos(2\omega t + \phi)]$$

The average power over a cycle is average of the two terms on the R.H.S. of the above equation. The second term is time dependent, so, its average is zero.

So,
$$P = \frac{V_0 I_0}{2}\cos\phi$$

$$= \frac{V_0 I_0}{\sqrt{2}\sqrt{2}}\cos\phi$$

$$= VI\cos\phi$$

$$P = I^2 Z\cos\phi$$

where, cos ϕ is called the power factor.

**Case 1.**

For pure inductive circuit or pure capacitive circuit, the phase difference between current and voltage *i.e.*, ϕ is $\frac{\pi}{2}$.

$$\therefore \quad \phi = \frac{\pi}{2}$$

So
$$\cos\phi = 0$$

Therefore, no power is dissipated.

**Case 2.**

For power dissipated at resonance in an LCR circuit,

$$X_C - X_L = 0, \phi = 0$$

$$\therefore \quad \cos\phi = 1$$

So, maximum power is dissipated.

**24. Write any two distinguishing features between conductors, semiconductors and insulators on the basis of energy band diagrams. [3]**

**Ans. Conductors :**

(i) In case of conductors, the valence band is completely filled and the conduction band can have two cases-either it is partially filled with an extremely small energy gap between the valence and conduction bands or it is empty, with the two bands overlapping each other as shown below :

| Case 1 | Case 2 |
|---|---|
| Conduction Band | Conduction Band |
| Valence Band | Valence Band |

(ii) Even when a small current is applied, conductors can conduct electricity.

**Insulators :**

(i) In case of insulators, the energy gap between the conduction and valence bands is very large and the conduction band is practically empty.

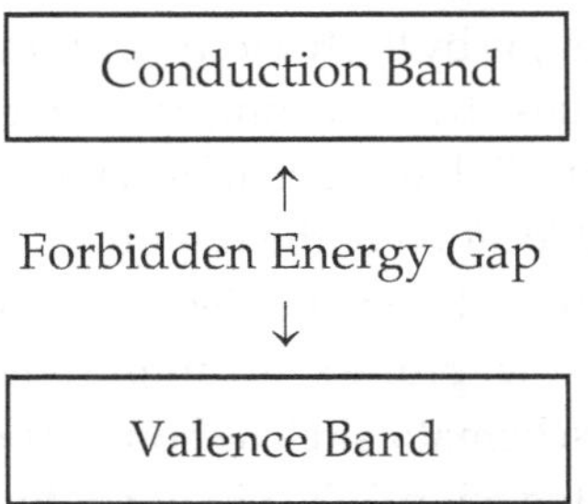


(ii) When an electric field is applied to such kind of material, the electrons find hard to receive such a large amount of energy to reach the conduction band. Thus, the conduction band remains empty. That is why no current flows through insulators.

**Semiconductors :**

(i) In case of semiconductor, the energy band structure of semiconductors is similar to insulators, But in this case, the size of forbidden energy gap is quite smaller than that of the insulators.

| Conduction Band |
|---|
| ↑ Forbidden Energy Gap ↓ |
| Valence Band |

(ii) When an electric field is applied to a semiconductor, the electrons in the valence band find it relatively easier to jump to the conduction band. So, the conductivity of semiconductors lies between the conductivity of conductors and insulators.

**25. For the past some time, Aarti had been observing some erratic body movement, unsteadiness and lack of coordination in the activities of her sister Radha, who also used to complain of severe headache occasionally. Aarti suggested to her parents to get a medical check-up of Radha. The doctor thoroughly examined Radha and diagnosed that she has a brain tumour.** [3]

**(a) What, according to you, are the values displayed by Aarti?**

**(b) How can radioisotopes help a doctor to diagnose brain tumour ?**

**Ans.** (a) awareness & responsibility towards her sister.

(b) A little amount of radioisotope like radioiodine is inserted into the body along with organic dyes which are absorbed strongly by the tumour tissue than the normal tissues. By detecting the emitted radiation, the radiologist get information about the size and location of the tumour.

**26. Write two basic modes of communication. Explain the process of amplitude modulation. Draw a schematic sketch showing how amplitude modulated signal is obtained by superposing a modulating signal over a sinusoidal carrier wave.**** [3]

**27. An electron microscope uses electrons accelerated by a voltage of 50 kV. Determine the de-Broglie wavelength associated with the electrons. Taking other factors, such as numerical aperture etc., to be same, how does the resolving power of an electron microscope compare with that of an optical microscope which used yellow light ?** [3]

**Ans.** The de-Broglie wavelength of the electrons is given by :

$$\lambda = \frac{h}{\sqrt{2meV}}$$

Here,

$m$ = mass of the electron $= 9.1 \times 10^{-31}$ kg

$e$ = charge on the electron $= 1.6 \times 10^{-19}$ C

V = accelerating potential = 50 kV

$h$ = Planck's constant $= 6.626 \times 10^{-34}$ Js

$$\Rightarrow \lambda = \frac{6.626 \times 10^{-34}}{\sqrt{2(9.1 \times 10^{-31})(1.6 \times 10^{-19})(50 \times 10^{3})}}$$

$$\Rightarrow \lambda = 0.0549 \text{Å}$$

Resolving power of a microscope, $R = \frac{2\mu \sin \theta}{\lambda}$

This formula shows that to enhance resolution, we have to use shorter wavelength and media with large indices of refraction.

For an electron microscope, $\mu$ is equal to 1 (vacuum).

For an electron microscope, the electrons are accelerated through a 60,000 V potential difference.

Thus, the wavelength of electrons is given by,

$$\lambda = \frac{12.27}{\sqrt{V}} = \frac{12.27}{\sqrt{60000}} = 0.05 \text{ Å}$$

As, $\lambda$ is very little (roughly $10^{-5}$ times smaller) for electron microscope than an optical microscope which uses yellow light of wavelength (5700 Å to 5900 Å). So, the resolving power of electron microscope is about $10^5$ greater than that of optical microscope.

**28. Draw a labelled diagram of Van de Graaff generator. State its working principle to show how by introducing a small charged sphere into a larger sphere, a large amount of charge can be transferred to the outer sphere. State the use of this machine and also point out its limitations.** [5]

**OR**

**(a) Deduce the expression for the torque acting on a dipole of dipole moment $\vec{p}$ in the presence of a uniform electric field $\vec{E}$.**

** Answer is not given due to change in the present Syllabus.

**(b) Consider two hollow concentric spheres, $S_1$ and $S_2$, enclosing charges 2Q and 4Q respectively as shown in the figure. (i) Find out the ratio of the electric flux through them. (ii) How will the electric flux through the sphere $S_1$ change if a medium of dielectric constant '$\varepsilon_r$' , '$\varepsilon_r$' is introduced in the space inside $S_1$ in place of air?**

**Deduce the necessary expression.**

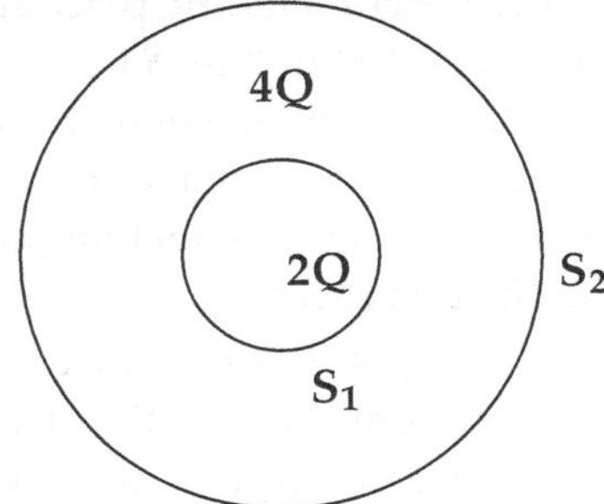


**Ans.** Van de Graaff generator is a device used for building up high potential differences of the order of a few million volts.

**Principle :** The working of Van de Graaff generator is based on the following two electrostatic phenomena.

**(i)** Discharging action at sharp points (corona discharge) *i.e.,* electric discharge takes place in air or gases readily at the pointed ends of conductors.

**(ii)** If a charged conductor is brought into internal contact with a hollow conductor, all of its charge transfers to the hollow conductor, howsoever high the potential of the latter may be.

**Construction :** It has a big spherical conducting shell (S) kept over insulating pillars. A long narrow insulating belt is wound around two pulleys $P_1$ and $P_2$. $B_1$ and $B_2$ are two metal combs with sharp points. $B_1$ is known as spray comb and $B_2$ collecting comb.

**Working :** The spray comb provides positive potential by high tension source. The positive charge is sprayed on belt.

As belt moves and touches the sphere, a negative charge is induced on the sharp ends of collecting comb $B_2$ and similar positive charge is induced on the further end of $B_2$. This positive charge moves immediately to the outer surface of S, because of discharging action of sharp points of $B_2$, the positive charge on the belt is neutralized. The uncharged belt moves downwards and collects the positive charge from $B_1$, which is then collected by $B_2$. This process is repeated and the positive charge on S goes on accumulating. In this way, voltage differences of as much as 6 or 8 million volts (with respect to the ground) can be created.

**Use :** Van de Graaff generator is used to create high potential differences that are used to accelerate charged particles such as electrons, protons, ions, etc., used for nuclear reactions

**Limitations :**

1. It is a series of combination that allows only one way for moving charge.
2. It can accelerate only the charged particles and not the uncharged particles.

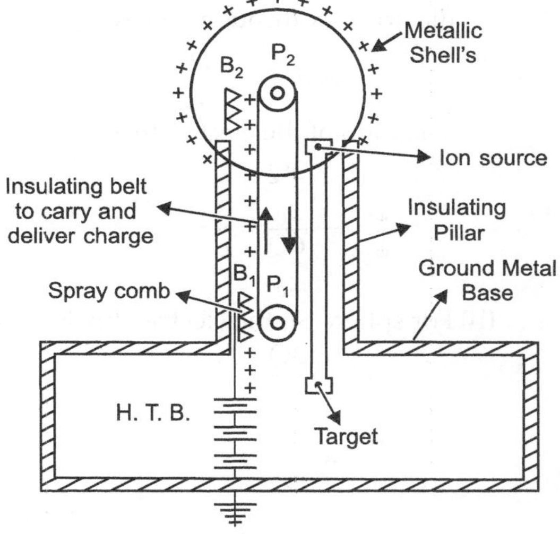


**OR**

**(a) Dipole in a Uniform External Field**

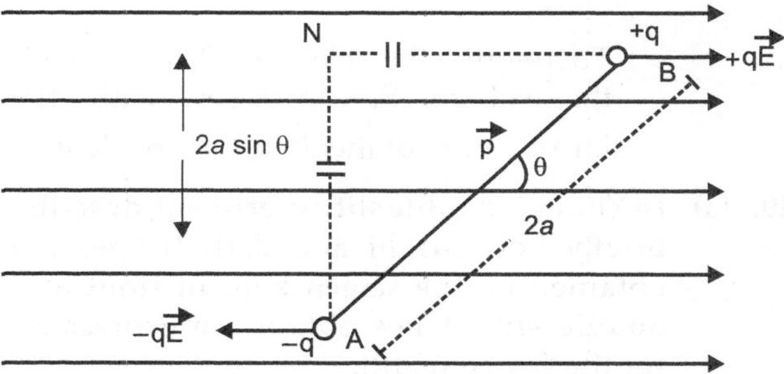


Consider an electric dipole consisting of charges $-q$ and $+q$ and of length $2a$ placed in a uniform electric field $\vec{E}$ making an angle $\theta$ with electric field.

Force on charge $-q$ at A = $-q\vec{E}$ (opposite to $\vec{E}$)

Force on charge $+q$ at B = $+q\vec{E}$ (along $\vec{E}$)

Electric dipole is under the action of two equal and unlike parallel force, which give rise to a torque on the dipole.

$\tau$ = Force × Perpendicular distance between the two forces

$$\tau = qE(AN) = qE(2a \sin\theta)$$

$$\tau = pE \sin\theta \qquad [\because 2qa = P]$$

$$\vec{\tau} = \vec{p} \times \vec{E}$$

**(b)** **(i)** Charge enclosed by sphere $S_1 = 2Q$

By Gauss' law, electric flux through sphere $S_1$ is

$$\phi_1 = \frac{2Q}{\varepsilon_0}$$

Charge enclosed by sphere $S_2$ is

$$Q' = 2Q + 4Q = 6Q$$

Electric flux through sphere $S_2$ is

$$\therefore \quad \phi_2 = \frac{6Q}{\varepsilon_0}$$

The ratio of the electric flux is

$$\frac{\phi_1}{\phi_2} = \frac{\frac{2Q}{\varepsilon_0}}{\frac{6Q}{\varepsilon_0}} = \frac{2}{6} = \frac{1}{3}$$

**(ii)** For sphere $S_1$, the electric flux is

$$\phi' = \frac{2Q}{\varepsilon_r}$$

$$\therefore \quad \frac{\phi'}{\phi_1} = \frac{\varepsilon_0}{\varepsilon_r}$$

$$\Rightarrow \quad \phi' = \phi_1 \times \frac{\varepsilon_0}{\varepsilon_r}$$

$$\because \quad \varepsilon_r > \varepsilon_0$$

$$\therefore \quad \phi' < \phi_1$$

Therefore, the electric flux through the sphere $S_1$ decreases with the introduction of the dielectric inside it.

**29. (a) In Young's double slit experiment, describe briefly how bright and dark fringes are obtained on the screen kept in front of a double slit. Hence obtain the expression for the fringe width.**

**(b) The ratio of the intensities at minima to the maxima in the Young's double slit experiment is 9 : 25. Find the ratio of the widths of the two slits.** [5]

OR

**(a) Describe briefly how a diffraction pattern is obtained on a screen due to a single narrow slit illuminated by a monochromatic source of light. Hence obtain the conditions for the angular width of secondary maxima and secondary minima.**

**(b) Two wavelengths of sodium light of 590 nm and 596 nm are used in turn to study the diffraction taking place at a single slit of aperture $2 \times 10^{-6}$ m. The distance between the slit and the screen is 1.5 m. Calculate the separation between the positions of first maxima of the diffraction pattern obtained in the two cases.**

**Ans. (a)** In Young's double slit experiment, the wave fronts from the two illuminated slits supere mpose on the screen. This results in formation of alternate bright and dark fringes because of constructive and destructive interference, respectively. The intensity of light is maximum at the centre C of the screen and it is called central maxima.

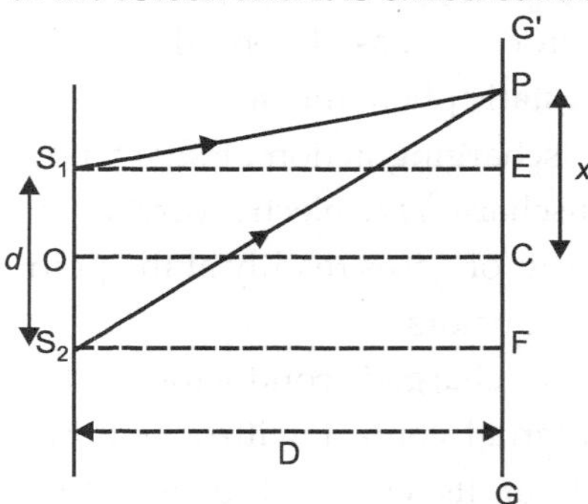


Let $S_1$ and $S_2$ be two slits separated by a distance $d$. GG' is the screen at a distance D from the slits $S_1$ and $S_2$. Both the slits are equidistant from point C. The intensity of light will be maximum at this point due to the path difference of the waves reaching this point will be zero.

At point P, the path difference between the rays coming from the slits $S_1$ and $S_2$ is $S_2P - S_1P$.

Now, $S_1S_2 = d$, $EF = d$, and $S_2F = D$

$\therefore$ In $\Delta S_2PF$,

$$S_2P = [S_2F^2 + PF^2]^{\frac{1}{2}}$$

$$S_2P = \left[D^2 + \left(x + \frac{d}{2}\right)^2\right]^{\frac{1}{2}}$$

$$= D\left[1+\frac{\left(x+\frac{d}{2}\right)^2}{D^2}\right]^{\frac{1}{2}}$$

Similarly, in $\Delta S_1PE$

$$S_1P = D\left[1+\frac{\left(x-\frac{d}{2}\right)^2}{D^2}\right]^{\frac{1}{2}}$$

$$S_2P - S_1P = D\left[1+\frac{\left(x+\frac{d}{2}\right)^2}{D^2}\right]^{\frac{1}{2}} - D\left[1+\frac{\left(x-\frac{d}{2}\right)^2}{D^2}\right]^{\frac{1}{2}}$$

On expanding it binomially,

$$S_2P - S_1P = \frac{1}{2D}\left[4x \times \frac{d}{2}\right] = \frac{xd}{D}$$

For bright fringes (constructive interference), the path difference is an integral multiple of wavelengths, *i.e.*, path difference is $n\lambda$.

$$\therefore \quad n\lambda = \frac{xd}{D}$$

$$x = \frac{n\lambda D}{d}, \text{ where } n = 0, 1, 2, 3, 4, \ldots\ldots$$

For $n = 0, \quad x_0 = 0$

$n = 1, \quad x_1 = \frac{\lambda D}{d}$

$n = 2, \quad x_2 = \frac{2\lambda D}{d}$

$n = 3, \quad x_3 = \frac{3\lambda D}{d}$

$\vdots$

$n = n, \quad x_n = \frac{n\lambda D}{d}$

The separation between the centres of two consecutive bright interference fringes is the width of a dark fringe.

$$\therefore \quad \boxed{\beta_1 = x_n - x_{n-1} = \frac{\lambda D}{d}}$$

Similarly, for dark fringes,

$$x_n = (2n-1)\frac{\lambda D}{2d}$$

For $n = 1, \quad x_1 = \frac{\lambda D}{2d}$

For $n = 2, \quad x_2 = \frac{3\lambda D}{2d}$

The separation between the centres of two consecutive dark interference fringes is the width of a bright fringe.

$$\therefore \quad \boxed{\beta_2 = x_n - x_{n-1} = \frac{\lambda D}{d}}$$

$$\therefore \quad \beta_1 = \beta_2$$

All the bright and dark fringes are of equal width as $\beta_1 = \beta_2$

**(b)** Let $w$, $a$ and I represent the slit width, amplitude and intensity respectively.

$$\frac{I_{min}}{I_{max}} = \frac{(a_1-a_2)^2}{(a_1+a_2)^2} = \frac{9}{25}$$

$$\frac{(a_1-a_2)}{(a_1+a_2)} = \frac{3}{5}$$

$$\frac{(a_1-a_2)+(a_l+a_2)}{(a_1+a_2)-(a_1-a_2)} = \frac{3+5}{5-3}$$

(By Componendo and dividendo)

$$\frac{2a_1}{2a_2} = \frac{8}{2}$$

or $$\frac{a_1}{a_2} = \frac{4}{1}$$

and $$\frac{w_1}{w_2} = \frac{(a_1)^2}{(a_2)^2} = \frac{16}{1}$$

**OR**

**(a)** The phenomenon of bending of light round the sharp corners of an obstacle and spreading into the regions of the geometrical shadow is called diffraction.

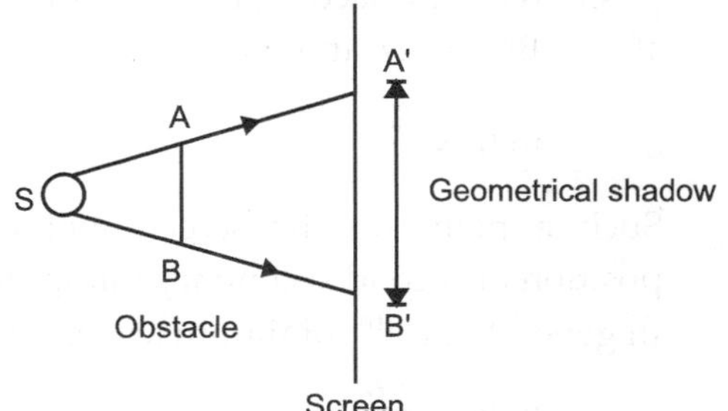


**Expression for Fringe width**

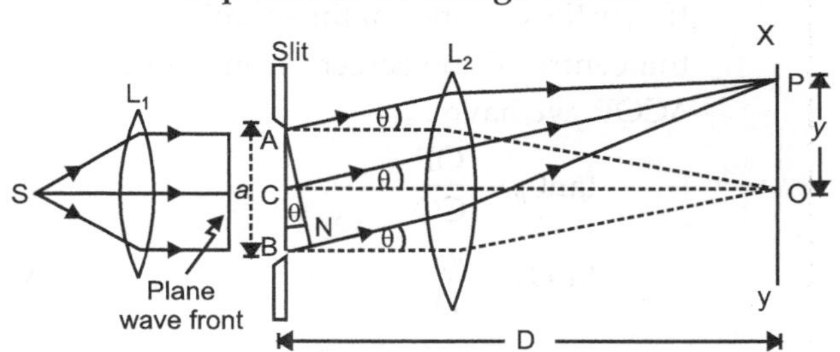

Consider a parallel light beam from a lens is incident on slit AB. As diffraction happens, the pattern is focussed on screen XY with the help of lens $L_2$. We will get a diffraction pattern that is a central maximum at the centre O flanked by a number of dark and bright fringes known as secondary maxima and minima.

**Central Maximum :** Each point on the plane wave front AB sends secondary wavelets in all directions. The waves from points equidistant from the centre C kept on the upper and lower half reach point O with zero path difference and so, reinforce each other, making maximum intensity at point O.

**PositionsandWidthsofSecondaryMaximaand Minima**

Consider a point P on screen at which wavelets moving in a direction making angle $\theta$ with CO are brought to focus by the lens. The wavelets from points A and B will have a path difference similar to BN :

From the right-angled $\Delta$ANB, we have :

$$BN = AB \sin\theta$$

$$BN = a \sin\theta \quad \text{...(i)}$$

Suppose BN = $\lambda$ and $\theta = \theta_1$

Then, the above equation becomes

$$\lambda = a \sin\theta_1,$$

$$\sin\theta_1 = \frac{\lambda}{a} \quad \text{...(ii)}$$

Such a point on the screen will be the position of first secondary minimum,

If BN = $2\lambda$ and $\theta = \theta_2$, then

$$\sin\theta_2 = \frac{2\lambda}{a} \quad \text{...(iii)}$$

Such a point on the screen will be the position of second secondary minimum.

In general, for $n^{th}$ minimum at point P,

$$\sin\theta_n = \frac{n\lambda}{a} \quad \text{...(iv)}$$

If $y_n$ is the distance of the $n^{th}$ minimum from the centre of the screen, from right-angled $\Delta$COP, we have :

$$\tan\theta_n = \frac{OP}{CO}$$

$$\tan\theta_n = \frac{y_n}{D} \quad \text{...(v)}$$

In case $\theta_n$ is small, $\tan\theta_n \approx \sin\theta_n$

$\therefore$ Equations (iv) and (v) give

$$\frac{y_n}{D} = \frac{n\lambda}{a}$$

$$y_n = \frac{nD\lambda}{a}$$

Width of the secondary maximum,

$$\beta = y_n - y_{n-1} = \frac{nD\lambda}{a} - \frac{(n-1)D\lambda}{a}$$

$$\beta = \frac{D\lambda}{a} \quad \text{...(vi)}$$

$\therefore$ $\beta$ is independent of $n$, all the secondary maxima are of the same width $\beta$.

If BN = $\frac{3\lambda}{2}$ and $\theta = \theta'$, from equation (i), we have :

$$\frac{3\lambda}{2} = a\sin\theta'_1$$

$$\sin\theta'_1 = \frac{3\lambda}{2}$$

In general, for the $n^{th}$ maximum at point P,

$$\sin\theta'_n = \frac{(2n+1)\lambda}{2a} \quad \text{...(vii)}$$

If $y'_n$ is the distance of $n^{th}$ maximum from the centre of the screen, then the angular position of the maximum is given by

$$\tan\theta'_n = \frac{y'_n}{D} \quad \text{...(viii)}$$

In case $\theta'_n$ is small,

$$\sin\theta'_n \approx \tan\theta'_n$$

$$\therefore \quad y'_n = \frac{(2n+1)D\lambda}{2a} \quad \text{..(viii)}$$

Width of the secondary minimum,

$$\beta' = y'_n - y'_{n-1} = \frac{(2n+1)D\lambda}{2a} - \frac{(2n-1)D\lambda}{2a}$$

$$\beta' = \frac{D\lambda}{a} \quad \text{...(ix)}$$

Since $\beta'$ is independent of $n$, all the secondary minima are of the same width $\beta'$.

**(b)** For first maxima of the diffraction pattern we know

$$\sin\theta = \frac{3\lambda}{2a}$$

where $a$ is aperture of slit.

For small value of $\theta$, $\sin\theta \approx \tan\theta = \frac{y}{D}$

Where $y$ is the distance of first minima from central line and D is the distance between the slit and the screen.

So, $$y = \frac{3\lambda}{2a}D$$

When $\lambda = 590$ nm, $a = 2 \times 10^{-6}$m, D = 1.5m

$$y_1 = \frac{3 \times 590 \times 10^{-9}}{2 \times 2 \times 10^{-6}} \times 1.5$$

$$y_1 = 0.66375 \text{ m}$$

When $\lambda = 596$ nm, $a = 2 \times 10^{-6,}$ D = 1.5m

$$y_2 = \frac{3 \times 596 \times 10^{-9}}{2 \times 2 \times 10^{-6}} \times 1.5$$

$$y_2 = 0.6705 \text{ m}$$

Separation between the positions of first maxima = $y_2 - y_1 = 0.00675$ m or 6.75 mm

**30. (a) Deduce an expression for the frequency of revolution of a charged particle in a magnetic field and show that it is independent of velocity or energy of the particle.**

**(b) Draw a schematic sketch of a cyclotron. Explain, giving the essential details of its construction, how it is used to accelerate the charged particles. [5]**

**OR**

**(a) Draw a labelled diagram of a moving coil galvanometer. Describe briefly its principle and working.**

**(b) Answer the following :**

**(i) Why is it necessary to introduce a cylind-rical soft iron core inside the coil of a galvanometer.**

**(ii) Increasing the current sensitivity of a galvanometer may not necessarily increase its voltage sensitivity. Explain, giving reason.**

**Ans. (a)** When a charged particle having charge $q$ moves inside a magnetic field $\vec{B}$ having velocity $v$, it experiences a force, which is given by :

$$\vec{F} = q(\vec{v} \times \vec{B})$$

Here, $\vec{v}$ is perpendicular to $\vec{B}$,

$\vec{F}$ is the force on the charged particle which behaves as the centripetal force and make it to move in a circular path.

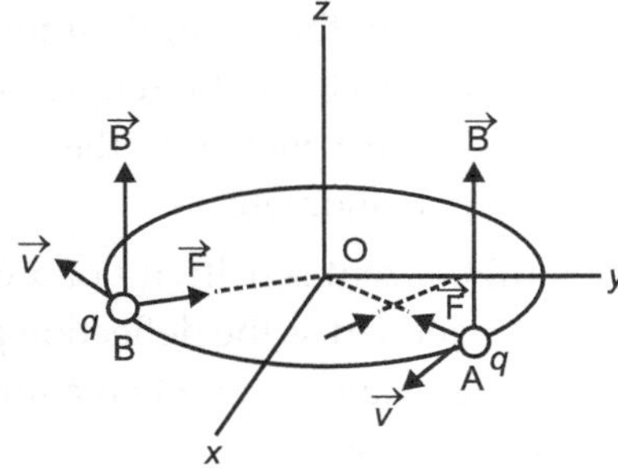


Let $m$ be the mass of the charged particle and $r$ be the radius of the circular path.

$$\therefore \quad q(\vec{v} \times \vec{B}) = \frac{mv^2}{r}$$

$v$ and B are at right angles :

$$\therefore \quad qvB = \frac{mv^2}{r}$$

$$r = \frac{mv}{Bq}$$

Time period of circular motion of the charged particle can be calculated by,

$$T = \frac{2\pi r}{v}$$

$$= \frac{2\neq}{v}\frac{mv}{Bq}$$

$$T = \frac{2\neq m}{Bq}$$

$\therefore$ Angular frequency is

$$\omega = \frac{2\neq}{T}$$

$$\therefore \quad \omega = \frac{Bq}{m}$$

Therefore, the frequency of the revolution of the charged particle is independent of the velocity or the energy of the particle.

**(b)** The working principle of a cyclotron is that a charge particle can be accelerated to high energy by an oscillating electric field.

A cyclotron uses an electric field to accelerate charge particles across the gap between the two D-shaped magnetic field regions. The magnetic field is perpendicular to the paths of the charged particles that makes them follow in circular paths within the two Ds. Each time the charged particles cross the Ds, it is accelerated by an alternating voltage. As its speed increases the radius of path of each particle also increases. So, the accelerated particles move in a spiral path to the other wall of the cyclotron.

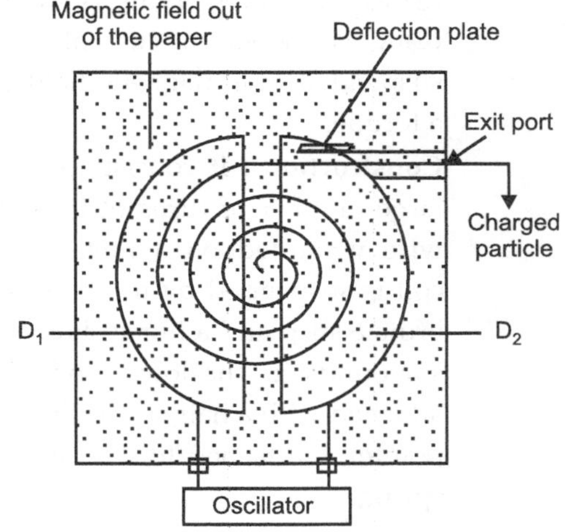

Square wave electric fields are used to accelerate the charged particles in a cyclotron.

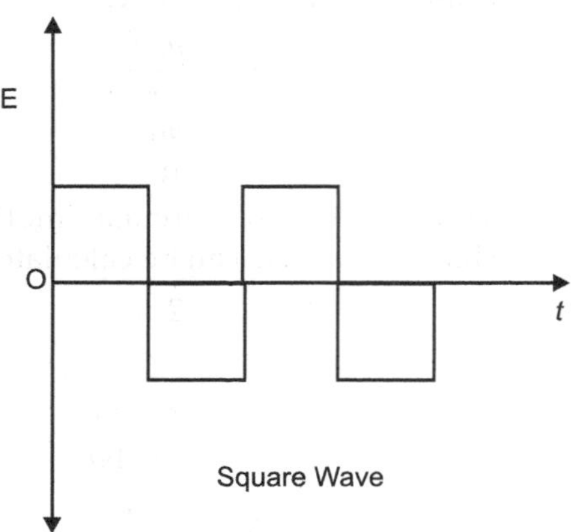


At the time the charge particle finishes its half circle, the accelerating electric field reverse so that it gets accelerated across the gap between the Ds.

The particle gets accelerated again and again, and its velocity increases. Therefore, a high kinetic energy is achieved.

**OR**

**(a) Moving coil galvanometer :** It is a device used for detecting and measuring small electric current.

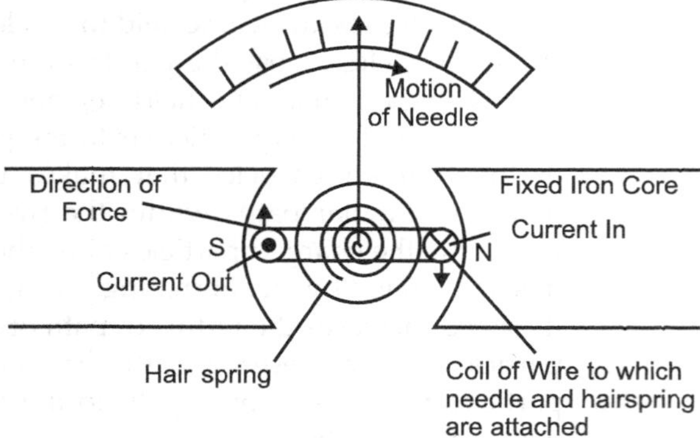


**Principle :** The working is based upon the principle, when a current carrying coil suspended in a magnetic field experiences a torque.

**Construction :** It consists of a coil with a large number of turns of insulated copper wire wounded on a metallic frame. The coil is suspended by means of a phosphor bronze strip and a horse shoe magnet's NS surrounds it. The lower end of the coil is attached with a hair spring. The scale of the pointer is attached to the other end of the spring.

**Working :** When current is passed, the couple acts on it. Since the plane remains parallel to the magnetic field in all position of the coil, the force on the vertical arms always remains perpendicular to the place of the coil.

Let I = the current flowing through coil,

B = magnetic field supposed to be uniformed and always parallel to the coil, and

A = area of the coil.

Deflection acting on the coil is

$\tau = \text{NIBA} \sin 90° = \text{NIBA}$ $[\because \sin 90° = 1]$

Due to the deflection torque, the coil rotates and suspended wire gets twisted. The suspension fiber experiences a restoring torque. If $\phi$ is angle through the coil rotates and $k$ is the restoring torque per unit angular twist, then

Restoring torque, $\tau = k\phi$

In equilibrium, Deflection torque = Restoring torque

$$\text{NIBA} = k\phi$$

$$\phi = \left(\frac{\text{NBA}}{k}\right)\text{I}$$

$$\Rightarrow \qquad \phi = \left(\frac{\text{I}}{\text{G}}\right)\cdot\text{I}$$

$$\text{I} = \text{G}\phi$$

where $\text{G} = \dfrac{k}{\text{NBA}}$

known as galvanometer constant.

$\therefore \qquad \phi \propto \text{I}$

This provides a linear scale for the galvanometer.

**(b) (i)** When a soft iron core is used the magnetic field lines tend to crowd through the core. It is because, soft iron core is ferromagnetic in nature. As a result, the strength of the magnetic field due to the field magnet increases, which in turn increases the sensitivity of the galvanometer.

**(ii)** Current sensitivity of a galvanometer is defined as the deflection produced in the galvanometer when a unit current flows through it.

$$\text{I}_s = \frac{\theta}{\text{I}} = \frac{n\text{BA}}{\text{K}} \qquad \text{...(i)}$$

Voltage sensitivity of a galvanometer is defined as the deflection produced in the galvanometer when a unit voltage is applied across two terminals.

$$V_s = \frac{\theta}{V} = \frac{\theta}{IR} = \frac{nBA}{KR} = \frac{I_s}{R} \quad \text{...(ii)}$$

Where

$n$ = number of turns in the coil of galvanometer

B = Magnetic field around coil

A = Area of coil

K = restoring torque per unit twist

From equation (i) we can say that current sensitivity increases by increasing $n$, B, A and decreases by decreasing K.

From equation (ii) we can say that voltage sensitivity increases by increasing $n$, B, A and decreases by de reasing K, R.

In case of current sensitivity if we increases the no. of turns $n$ its current sensitivity increases. Since resistance of galvano- meter R also increases, voltage sensitivity remains same or unchanged. Therefore increase in current sensitivity of galvanometer may not necessarily increase the voltage.

## Outside Delhi [Set-II]

**Note : Except for the following questions, all the remaining questions have been asked in previous Set.**

**1. A conducting loop is held above a current carrying wire 'PQ' as shown in the figure. Depict the direction of the current induced in the loop when the current in the wire PQ is constantly increasing. [1]**

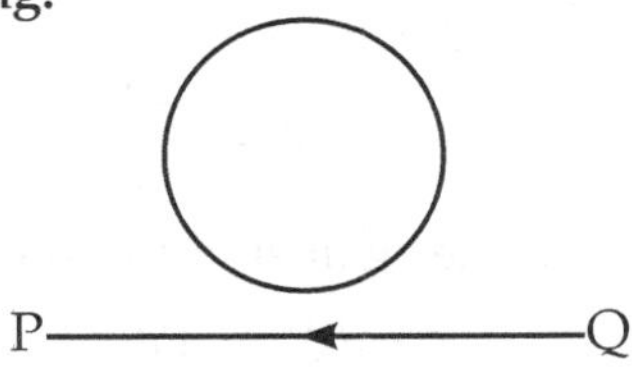


**Ans.** Anticlockwise

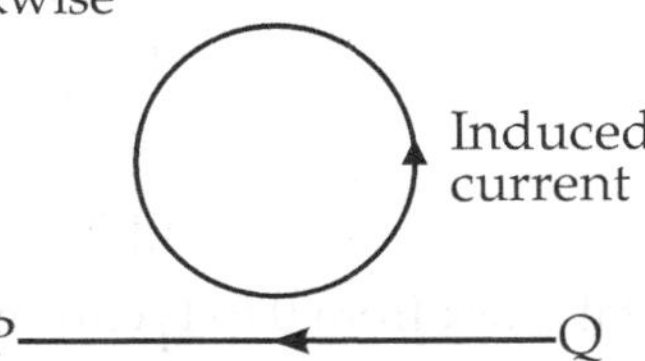


**4. Why do the electrostatic field lines not form closed loops ? [1]**

**Ans.** The electrostatic field lines originate from positive charge and end at the negative charge. As the isolated positive and negative charge do exist, the electrostatic field lines do not form closed loops.

**5. A biconvex lens made of a transparent material of refractive index 1.5 is immersed in water of refractive index 1.33. Will the lens behave as a converging or a diverging lens ? Give reason. [1]**

**Ans.** The refractive index of material of lens (1.5) is greater than the refractive index of water (1.33). So, it will behave as a converging lens.

**7. To which part of the electromagnetic spectrum does a wave of frequency $3 \times 10^{13}$ Hz belong ? [2]**

**Ans.** The frequency $3 \times 10^{13}$ Hz belongs to infrared region of electromagnetic spectrum.

**9. Estimate the average drift speed of conduction electrons in a copper wire of cross-sectional area $2.5 \times 10^{-7} m^2$ carrying a current of 1.8 A. Assume the density of conduction electrons to be $9 \times 10^{28}$ $m^{-3}$. [2]**

**Ans.** Given : Current (I) = 1.8 A

Charge density ($n$) = $9 \times 10^{28}$ $m^{-3}$

Cross-section area (A) = $2.5 \times 10^{-7}$ $m^2$

Charge of electron ($q$) = $1.6 \times 10^{-19}$ C

$$v_d = \frac{I}{nAq}.$$

$$= \frac{1.8}{9 \times 10^{28} \times 2.5 \times 10^{-7} \times 1.6 \times 10^{-19}}$$

$$= 0.0005 \text{ m/s} = 0.5 \text{ mm s}^{-1}$$

**13. Write the functions of the following in communication systems : [3]**

**(i) Transmitter**

**(ii) Modulator****

**21. (a) Show with the help of a diagram, how unpolarised sunlight gets polarised due to scattering.**

** Answer is not given due to change in the present Syllabus.

**(b) Two polaroids $P_1$ and $P_2$ are placed with their pass axes perpendicular to each other. Unpolarised light of intensity $I_0$ is incident on $P_1$. A third polaroid $P_3$ is kept in between $P_1$ and $P_2$ such that its pass axis makes an angle of 45° with that of $P_1$. Determine the intensity of light transmitted through $P_1$, $P_2$ and $P_3$.** **[3]**

**Ans. (a)** The given figure shows, the incident sunlight is unpolarised. The dots stand for polarisation perpendicular to

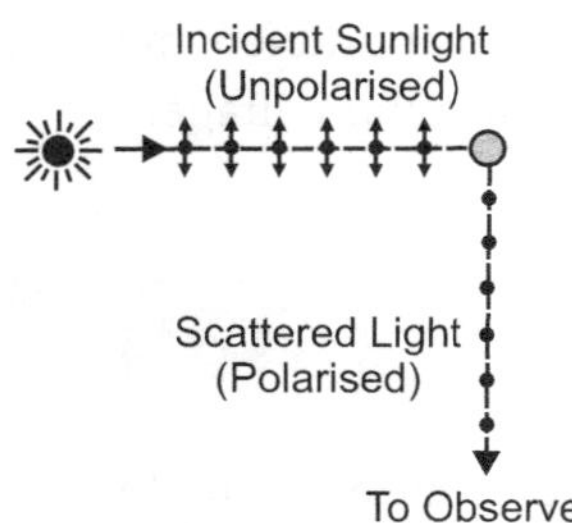


the plane of the figure. The double arrows show polarisation in the plane of the figure. Under the influence of the electric field of the incident wave the electrons in the molecules acquire components of motion in both these directions. We have drawn an observer looking at 90° to the direction of the sun. Clearly, charges accelerating parallel to the double arrows do not radiate energy towards this observer since their acceleration has no transverse component. The radiation scattered by the molecule is therefore represented by dots. It is polarised perpendicular to the plane of the figure. This explains the polarisation of scattered light from the sky.

**(b)** As given in the question, the polaroids $P_1$ and $P_2$ are placed with their pass axes perpendicular to each other. Also, $P_3$ is placed at an angle of 45° with respect to $P_1$.

Now, we have :

Intensity of light after falling on $P_1$, $I = I_0/2$

Intensity of light after falling on $P_3$,

$$I' = I\cos^2\theta = \frac{I_0}{2}\cos^2 45° = \frac{I_0}{4}$$

Therefore, a light of intensity $I_0/4$ will pass through $P_3$ and the angle between $P_3$ and $P_2$ will be 45° because of the condition given in the question.

Intensity of light after falling on $P_2$,

$$I'' = I'\cos^2(\theta) = \frac{I_0}{4}\cos^2 45° = \frac{I_0}{8}$$

**22. Define the term self-inductance of a solenoid. Obtain the expression for the magnetic energy stored in an inductor of self-inductance L to build up a current I through it.** **[3]**

**Ans.** The ratio of magnetic flux through the solenoid to the current passing through it is called self-inductance of a solenoid. It is given by

$$L = \frac{\phi}{I}$$

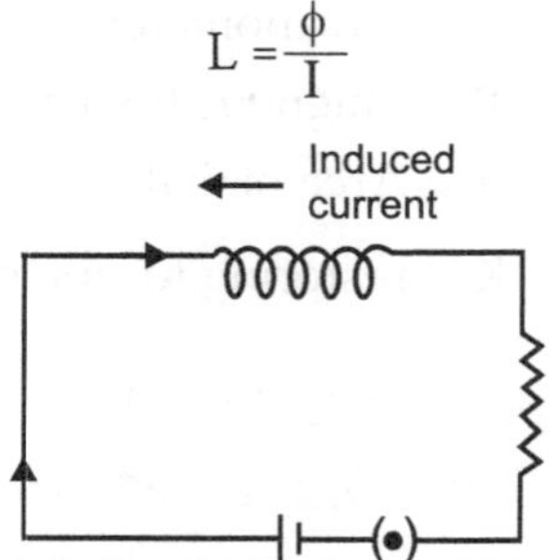


**Energy stored in an inductor :** When a current grows through an inductor, a back e.m.f. is set up which opposes the growth of current. So work needs to be done against back e.m.f. ($e$) in building up the current. This work done is stored as magnetic potential energy.

Let I be the current through the inductor L at any instant $t$. The current rises at the rate $dI/dt$.

So the induced e.m.f. is

$$e = \frac{-L\,dI}{dt}$$

The work done against induced e.m.f. in $dt$ is

$$dW = P\,dt$$

$$= -e\,I\,dt \qquad [P = VI]$$

$$= \frac{L\,dI}{dt} I\,dt$$

$$= LI\,dI$$

For total work from 0 to $I_0$ current

$$W = \int dW$$

$$= \int_0^{I_0} LI\,dI$$

$$= L\left[\frac{I^2}{2}\right]_0^{I_0}$$

$$= \frac{1}{2}LI_0^2$$

Hence, this work done is stored as the magnetic potential energy U in the inductor

$$U = \frac{1}{2}LI^2$$

**24. A convex lens of focal length 20 cm is placed coaxially with a concave mirror of focal length 10 cm at a distance of 50 cm apart from each other. A beam of light coming parallel to the principal axis is incident on the convex lens. Find the position of the final image formed by this combination. Draw the ray diagram showing the formation of the image. [5]**

**Ans.** The beam incident on lens L is parallel to principal axis. Hence the lens forms an image $I_1$ at its focus. *i.e.*, at a distance $OI_1$ (= 20 cm) from the lens.

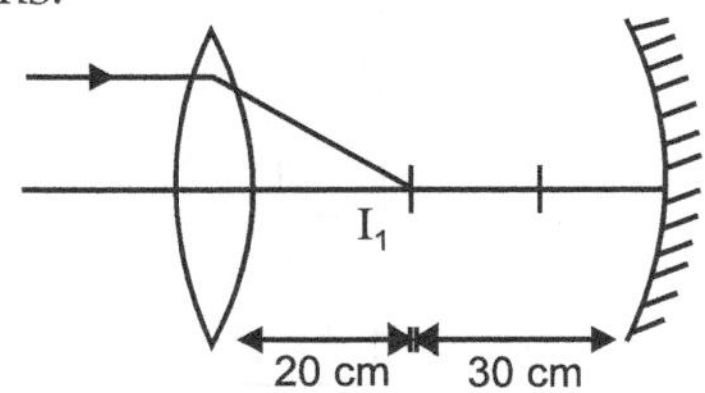


The image $I_1$ is formed in front of the mirror and hence, acts as a real source for the mirror. The concave mirror forms the image $I_2$, whose distance from the mirror can be calculated as;

$$\frac{1}{v}+\frac{1}{u}=\frac{1}{f}$$

Here : $u = -30$ cm, and $f = -10$ cm

$$\Rightarrow \quad \frac{1}{v}=\frac{1}{f}-\frac{1}{u}$$

$$\Rightarrow \quad \frac{1}{v}=-\frac{1}{10}+\frac{1}{30}$$

$$\Rightarrow \quad \frac{1}{v}=\frac{1-3}{30}=-\frac{2}{30}$$

$$\Rightarrow \quad v=-15 \text{ cm}$$

Hence, the final image is formed at a distance of 15 cm from the concave mirror, as shown in the following figure.

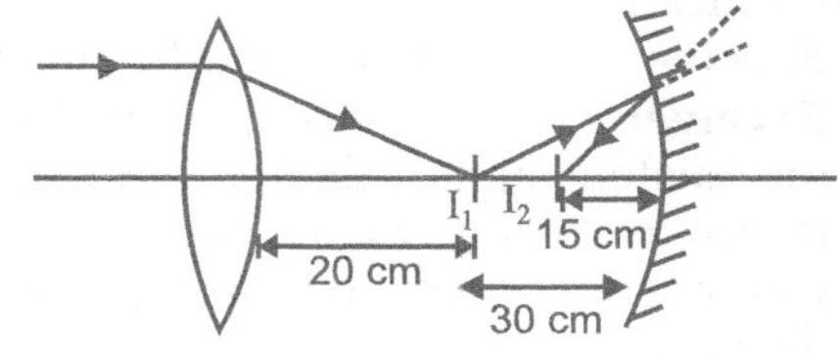


# Outside Delhi [Set-III]

**Note : Except for the following questions, all the remaining questions have been asked in previous Sets.**

**1. A conducting loop is held below a current carrying wire PQ as shown. Predict the direction of the induced current in the loop when the current in the wire is constantly increasing. [1]**

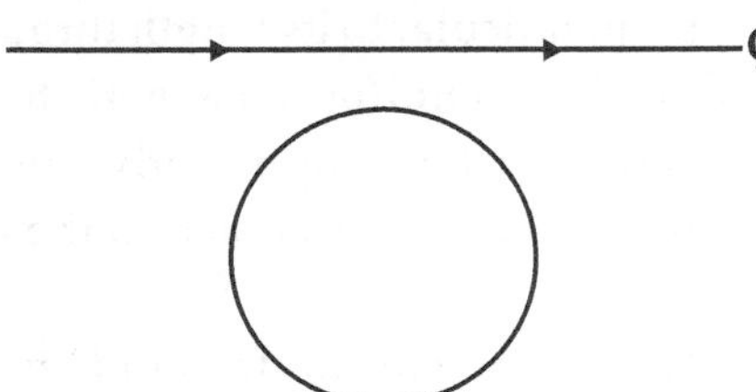


**Ans.** Anticlockwise direction

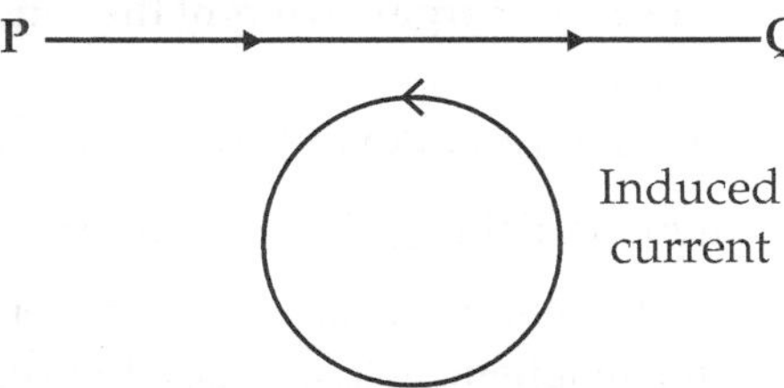


**2. The graph shows variation of stopping potential $V_0$ versus frequency of incident radiation ν for two photosensitive metals A and B. Which of the two metals has higher threshold frequency and why ? [1]**

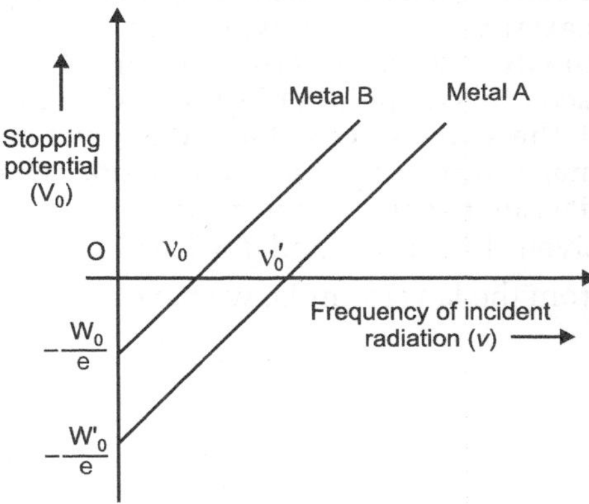


**Ans.** Metal A has higher threshold frequency because from the graph it is clear that the minimum frequency required to start photoemission is more in A than that of B.

**5. Why do the electric field lines never cross each other ? [1]**

**Ans.** At any point, if electric field lines cross each other than two tangents can be drawn, it means at that point there are two directions of electric field, which is impossible.

**6. To which part of the electromagnetic spectrum does a wave of frequency $5 \times 10^{11}$ Hz belong ? [2]**

**Ans.** A wave of frequency $5 \times 10^{11}$ Hz will belong to the microwaves of electromagnetic spectrum.

**10. Estimate the average drift speed of conduction electrons in a copper wire of cross-sectional area $2.5\times10^{-7}m^2$ carrying a current of 2.7 A. Assume the density of conduction electrons to be $9\times10^{28}$ $m^{-3}$.** **[2]**

**Ans.** We know that drift velocity, $v_d = \frac{I}{nAq}$ Where I is the current, $n$ is charge density, $q$ is charge of electron and A is cross-sectional area.

$$v_d = \frac{2.7}{9\times10^{28}\times2.5\times10^{-7}\times1.6\times10^{-19}}$$

$$v_d = 7.5\times10^{-4}\ m/s$$

or $\quad v_d = 0.75\ mm\ s^{-1}$

This is the required average drift velocity.

**18. Write the functions of the following in communication systems :** **[3]**

**(i) Receiver (ii) Demodulator****

**Ans.** (i) **Receiver :** The device that picks up the modulated signal, demodulates it, boosts it up by using an amplifier and finally converts the recovered message signal into the original form.

(ii) **Demodulator :** Demodulator is a device which recovers the original information signal from the modulated wave at the receiver end.

**19. A convex lens of focal length 20 cm is placed coaxially with a convex mirror of radius of curvature 20 cm. The two are kept at 15 cm from each other. A point object placed 40 cm in front of the convex lens. Find the position of the image formed by this combination. Draw a ray diagram to show the formation.** **[3]**

**Ans.** Given, 4 = –40 cm and, $f$ = 20 cm

From the lens formula, we have :

$$\frac{1}{v}-\frac{1}{u}=\frac{1}{f}$$

$$\Rightarrow \quad \frac{1}{v}=\frac{1}{f}+\frac{1}{u}$$

$$\Rightarrow \quad \frac{1}{v}=\frac{1}{20}+\frac{1}{(-40)}$$

$$\Rightarrow \quad \frac{1}{v}=\frac{2-1}{40}=\frac{1}{40}$$

$\Rightarrow v = 40$ cm

The positive sign describes that the image is formed to the right of the lens.

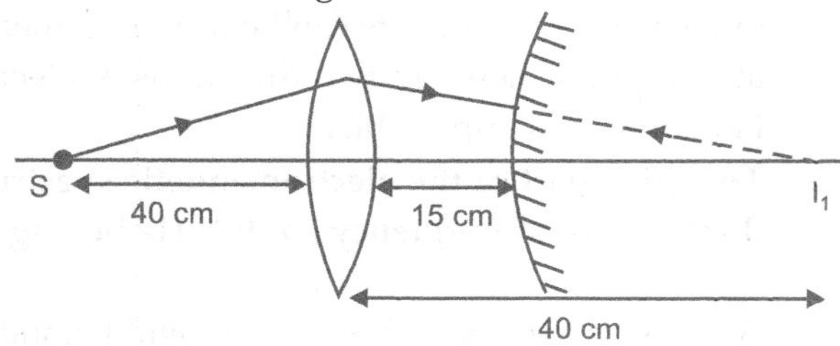


The image $I_1$ is formed behind the mirror and thus acts as a virtual source for the mirror. The convex mirror forms the image $I_2$, whose distance from the mirror is given by :

$$\frac{1}{v}+\frac{1}{v} = \frac{1}{f}$$

Here : $u$ = 25 cm

$$f = \frac{R}{2} = 10\ cm$$

$$\frac{1}{v}+\frac{1}{v} = \frac{1}{f}$$

$$\Rightarrow \quad \frac{1}{v} = \frac{1}{f}-\frac{1}{u}$$

$$\Rightarrow \quad \frac{1}{v} = \frac{1}{10}-\frac{1}{25}$$

$$\Rightarrow \quad \frac{1}{v} = \frac{5-2}{50}=\frac{3}{50}$$

$$\Rightarrow \quad v = +16.67\ cm$$

Hence, the final image is formed at a distance of 16.67 cm behind the convex mirror.

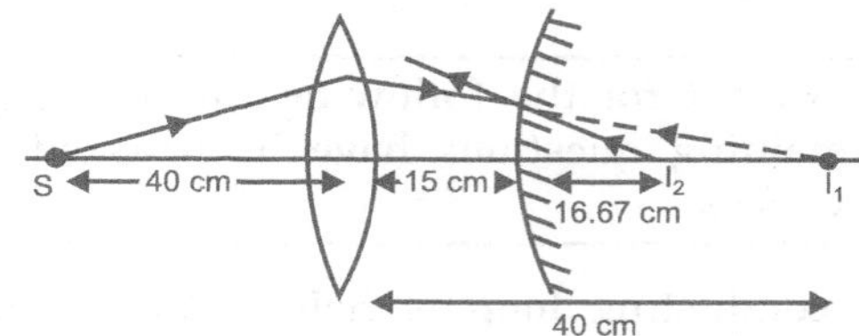


**25. (a) A rod of length '$l$' is moved horizontally with a uniform velocity '$v$' in a direction perpendicular to its length through a region in which a uniform magnetic field is acting vertically downward. Derive the expression for the emf induced across the ends of the rod.**

**(b) How does one understand this motional emf by involving the Lorentz force acting on the free charge carriers of the conductor? Explain.** **[5]**

**Ans.** (a) Consider a rod PQ of length $l$, moving in a magnetic field $\vec{B}$ with a constant velocity $\vec{v}$. The length of the rod is perpendicular to the magnetic field and also the velocity is perpendicular to both the rod and field. The free electrons of the rod also move at this velocity $\vec{v}$ because of which it experiences a magnetic force.

** Answer is not given due to change in the present Syllabus.

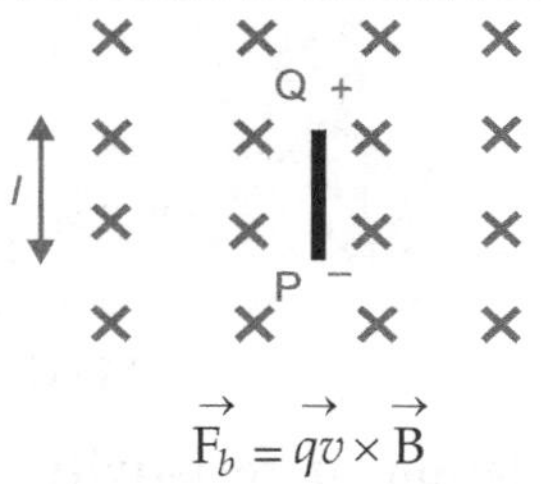


$$\vec{F_b} = q\vec{v} \times \vec{B}$$

This force is towards Q to P.

Thus, the free electrons will move towards P and positive charge will appears at Q. An electrostatic field E is developed within the wire from Q to P. This field exerts a force.

$$\vec{F_e} = q\vec{E}$$

on each free electron. The charge keeps on gathering until

$$\vec{F_b} = \vec{F_e}$$

$$\Rightarrow \quad |q\vec{v} \times \vec{B}| = |q\vec{E}|$$

$$vB = E$$

After this, resultant force on the free electrons of the wire PQ becomes zero. The potential difference between the ends Q and P is given by,

$$V = El = vBl$$

Thus, the potential difference is maintained by the magnetic force on the moving free electron and hence, produces an emf, $e = Bvl$

**(b)** Lorentz force acting on a charge $q$ which is moving with a speed $v$ in a (normal) uniform magnetic field B, is $Bqv$.

All the charges will experience the same force.

Work done to move the charge from P to Q,

$$W = Bqv \times l$$

$$\therefore \quad e = \frac{W}{q} = \frac{Bqvl}{q} = Blv$$

**26. (a) Show, giving via suitable diagram, how unpolarized light can be polarised by reflection.**

**(b) Two polaroids $P_1$ and $P_2$ are placed with their pass axes perpendicular to each other. Unpolarised light of intensity $I_0$ is incident on $P_1$. A third polaroid $P_3$ is kept in between $P_1$ and $P_2$ such that its pass axis makes an angle of 60° with that of $P_1$. Determine the intensity of light transmitted through $P_1$, $P_2$ and $P_3$.** **[5]**

**Ans. (a)** On reflection from a transparent medium a normal light beam becomes partially polarised. As the angle of incidence gets higher, the degree of polarization gets higher.

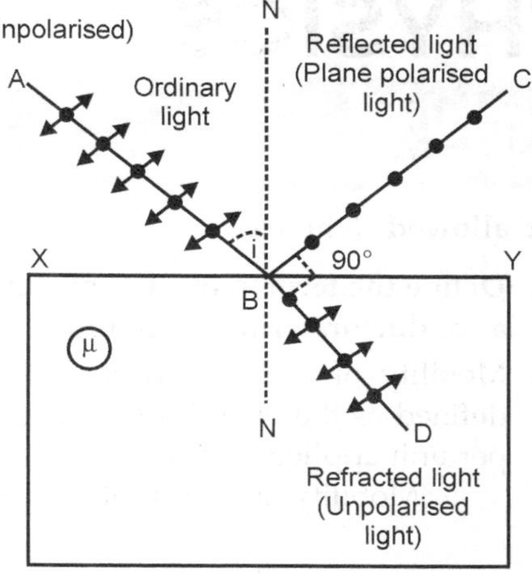


The reflected light beam becomes fully polarised at a certain angle. This angle of incidence is known as polarizing angle ($p$). At the interface of a refracting medium when light is incident at polarizing angle, the refractive index of the medium is similar to the tangent of the polarizing angle.

$$\mu = \tan p$$

where $\mu \rightarrow$ Refractive index of the refracting medium

$p \rightarrow$ Polarising angle

**(b)** As given, the polaroid $P_1$ and $P_2$ are placed with their axes perpendicular to each other and polaroid $P_3$ placed at 60° with respect to $P_1$.

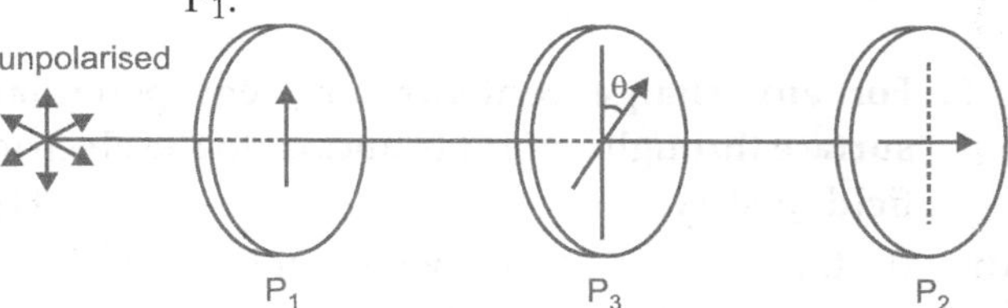


Therefore, Intensity of light after falling on first polaroid $P_1$,

$$I = \frac{I_0}{2}$$

Intensity of light after falling on third polaroid $P_3$,

$$I' = I\cos^2(\theta) = \frac{I_0}{2}\cos^2(60°) = \frac{I_0}{8}$$

Therefore, the intensity $\frac{I_0}{8}$ will pass through the $P_3$ and angle between $P_3$ and $P_2$ is 30°. Because of the condition given in the question,

Intensity of light after falling on second polaroid $P_2$,

$$I'' = I'\cos^2\theta$$

$$= I''\cos^2 30° = \frac{3I_0}{32}$$

# PHYSICS

## Delhi [Set-I]

# 2014

**Time allowed : 3 hours** **Maximum marks : 70**

1. **Define the term 'mobility' of charge carriers in a conductor. write its S.I. unit.** **[1]**

**Ans.** Mobility of charge carriers in a conductor is defined as the magnitude of their drift velocity per unit applied electric field.

Mobility, μ = Drift of electric field

$$\mu = \frac{v_d}{E}$$

S.I. unit of mobility is $m^2V^{-1}s^{-1}$.

2. **The carrier wave is given by $C(t) = 2 \sin(8\pi t)$ volt.**

**The modulating signal is a square wave as shown. Find modulation index.**** **[1]**

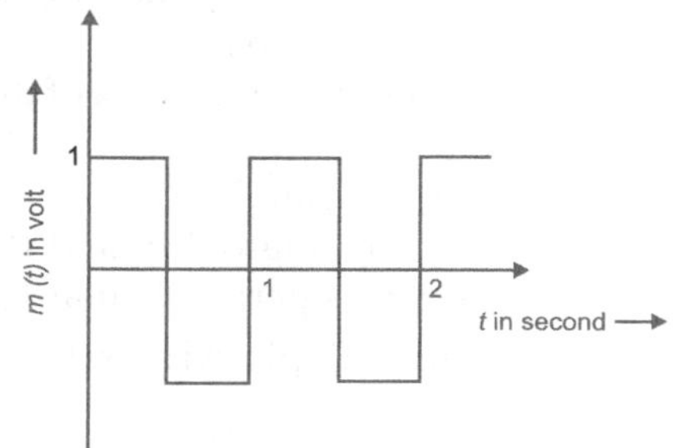


3. **For any charge configuration, equipotential surface through a point is normal to the electric field. Justify.** **[1]**

**Ans.** If the electric field were not normal to equipotential surface, it would have non-zero component along the surface. To move a charge against this component, work would have to be done. But no work is needed to move a test charge on an equipotential surface. Hence electric field must be normal to the equipotential surface at every point.

4. **Two spherical bobs, one metallic and the other of glass, of the same size are allowed to fall freely from the same height above the ground. Which of the two would reach earlier and why ?** **[1]**

**Ans.** Glass bob will reach the ground earlier than the metallic bob. As the metallic bob falls, it intercepts earth's magnetic field and induced currents are set up in it which oppose its downward motion. But no such currents are induced in the glass bob.

** Answer is not given due to change in the present Syllabus.

5. **Show variation of resistivity of copper as a function of temperature in a graph.** **[1]**

**Ans.** The variation of resistivity of copper with temperature is parabolic in nature. This is shown in the following graph :

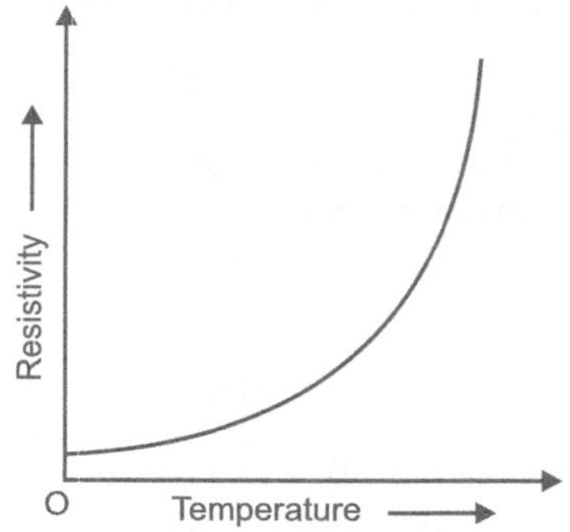


6. **A convex lens is placed in contact with a plane mirror. A point object at a distance of 20 cm on the axis of this combination has its image coinciding with itself. What is the focal length of the lens ?** **[2]**

**Ans.** The figure shows a convex lens L placed in contact with a plane mirror M. P is the point object, kept in front of

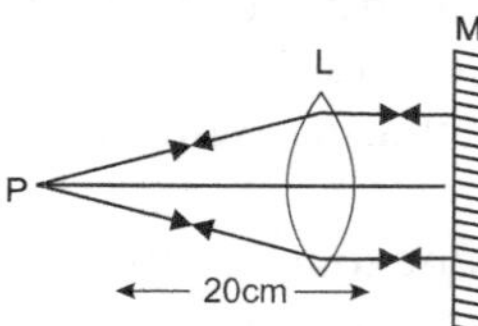


this combination at a distance of 20 cm, from it. As the image coincides with itself, the rays from the object, after refraction from lens, should fall normally on the mirror M, so that they retrace their path. For this, the rays from P, after refraction from the lens must form a parallel beam perpendicular to M. For clarity, M has been shown at a small distance from L (in diagram). As the rays from P, form a parallel beam after refraction, P must be at the focus of the lens. Hence, the focal length of the lens is 20 cm.

7. **Write the expression, in a vector form, for the Lorentz magnetic force $\vec{F}$ due to a charge moving with velocity $\vec{v}$ in a magnetic field $\vec{B}$. What is the direction of the magnetic force ?** **[2]**

**Ans.** The Lorentz magnetic force is given by the following relation :

$$\vec{F} = q(\vec{v} \times \vec{B})$$

Here $q$, is the magnitude of the moving charge. The direction of the magnetic force is perpendicular to the plane containing the velocity vector $\vec{v}$ and the magnetic field vector $\vec{B}$.

**8. The figure given below shows the block diagram of a generalized communication system. Identify the element labelled 'X' and write its function.** [2]**

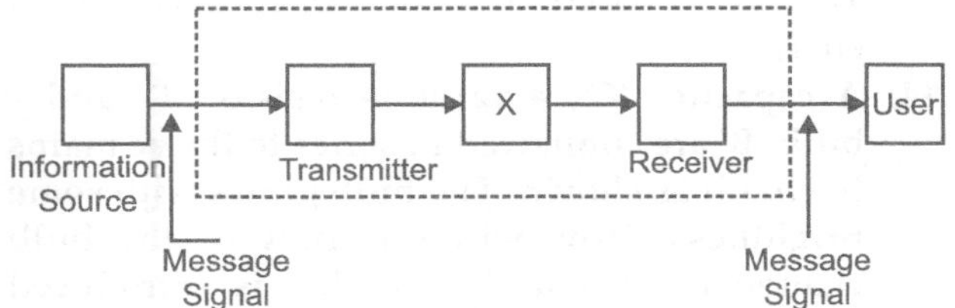


**9. Out of the two magnetic materials, 'A' has relative permeability slightly greater than unity while 'B' has less than unity. Identify the nature of the materials 'A' and 'B'. Will their susceptibilities be positive or negative? [2]**

**Ans.** For a paramagnetic material, the relative permeability lies between $1 < \mu r < 1 + \varepsilon$ and its susceptibility lies between $0 < \chi < \varepsilon$.

Hence, 'A' is a paramagnetic material and its susceptibility is positive. This is because its relative permeability is slightly greater than unity. For a diamagnetic material, the relative permeability lies between $0 \leq \mu_r < 1$ and its susceptibility lies between $-1 < \chi < 0$. Hence, 'B' is a diamagnetic material and its susceptibility is negative. This is because its relative permeability is less than unity.

Here $\mu_r$ and $\chi$ refer to the relative permeability and susceptibility.

**10. Given a uniform electric field $\vec{E} = 5\times10^3\,\hat{i}$ N/C find the flux of this field through a square of 10 cm on a side whose plane is parallel to the YZ plane. What would be the flux through the same square if the plane makes a 30° angle with the X-axis ? [2]**

**Ans.** When the plane is parallel to the Y-Z plane :

Electric flux, $\phi = EA$

Here, $E = 5\times10^3\,\hat{i}$ N/C

$A = 10 \text{ cm}^2\,\hat{i} = 10^{-2}\,\hat{i}\text{ m}^2 = 10^{-2}\,\hat{i}\text{ m}^2$

$\phi = 5 \times 10^3\,\hat{i} \cdot 10^{-2}\,\hat{i}$

$\Rightarrow \quad \phi = 50$ weber or $Nm^2C^{-1}$

When the plane makes a 30° angle with the X-axis, the area vector makes 60° with the X-axis.

$$\phi = \vec{E} \cdot \vec{A}$$

$$\Rightarrow \quad \phi = EA \cos\theta$$

$$\Rightarrow \quad \phi = 5 \times 10^3 \cdot 10^{-2} \cos 60°$$

$$\Rightarrow \quad \phi = \frac{50}{2}$$

$$\Rightarrow \quad \phi = 25 \text{ weber or } Nm^2C^{-1}$$

**11. For a single slit of width $a$, the first minimum of the interference pattern of a monochromatic light of wavelength $\lambda$ occurs at an angle of $\frac{\lambda}{a}$. At the same angle of $\frac{\lambda}{a}$, we get a maximum for two narrow slits separated by a distance $a$. Explain. [3]**

**Ans.** When a single slit is used, the interference pattern is due to the diffraction phenomenon. In case of diffraction from a single slit of width $a$ using monochromatic light of wavelength $\lambda$, the first minimum of the interference pattern occurs at an angle $\theta_1$, which is given by

$$\sin\theta_1 = \frac{\lambda}{a}$$

Since $\lambda$ is very small, $\theta_1$ will also be very small.

$$\therefore \quad \theta_1 \approx \sin\theta_1 = \frac{\lambda}{a} \quad \text{...(i)}$$

In case of interference from two narrow slits separated by a distance $a$ using monochromatic light of wavelength $\lambda$, the first maximum occurs at a distance $y_1$ from the centre of the screen, which is given by.

$$y_1 = \frac{D\lambda}{a}$$

Here, D is the distance of the screen from the centre of the slits.

If the first maximum occurs at an angle $\theta_1'$, then

$$\tan\theta_1' = \frac{y_1}{D} = \frac{\lambda}{a}$$

Again as $\lambda$ is very small, $\theta_1'$ will also be very small.

$$\therefore \quad \tan\theta_1' \approx = \theta_1' = \frac{\lambda}{a} \quad \text{...(ii)}$$

** Answer is not given due to change in the present Syllabus.

From the equations (i) and (ii), we have

$$\theta_1 = \theta_1' = \frac{\lambda}{a}$$

Hence, it proves the result.

**12. Write the truth table for the combination of the gates shown. Name the gates used. [3]**

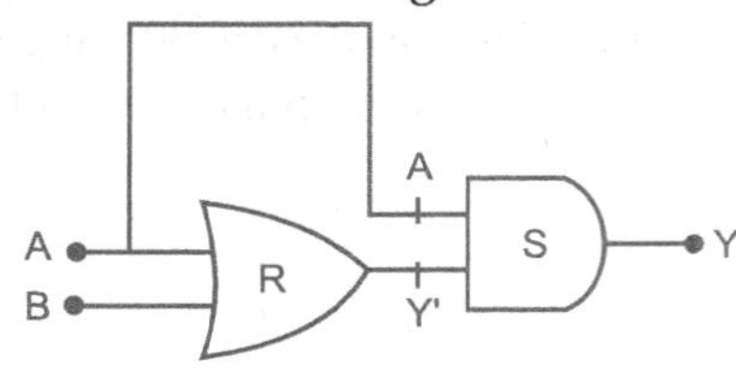


**OR**

**Identify the logic gates marked 'P' and 'Q' in the given circuit. Write the truth table for the combination.****

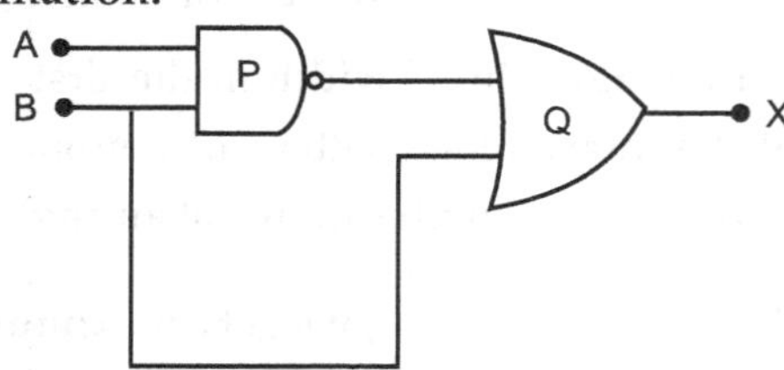


**13. State Kirchhoff's rules. Explain briefly how these rules are justified. [3]**

**Ans. Kirchhoff's first Law–Junction Rule**

In an electrical circuit, the algebraic sum of the currents meeting at a junction is always zero.

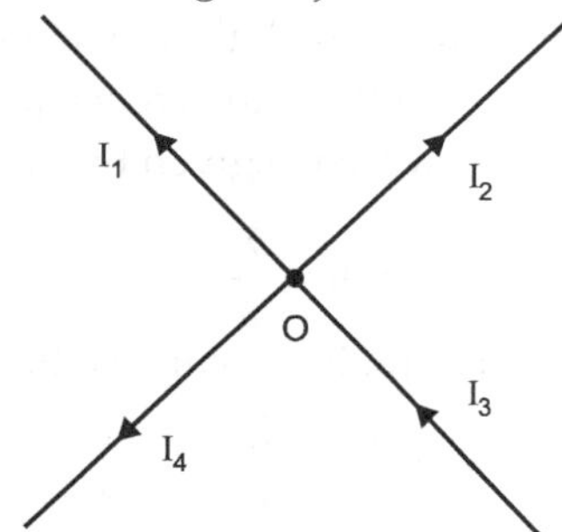


$I_1$, $I_2$, $I_3$ and $I_4$ are the currents flowing through the respective wires.

**Convention :** The current flowing towards the junction is taken as positive.

The current flowing away from the junction is taken as negative.

$$I_3 + (-I_1) + (-I_2) + (-I_4) = 0$$

This law is based on the law of conservation of charge.

**Kirchhoff's Second Law – Loop rule**

In a closed loop, the algebraic sum of the emf's is equal to the algebraic sum of the products of the resistances and the currents flowing through them.

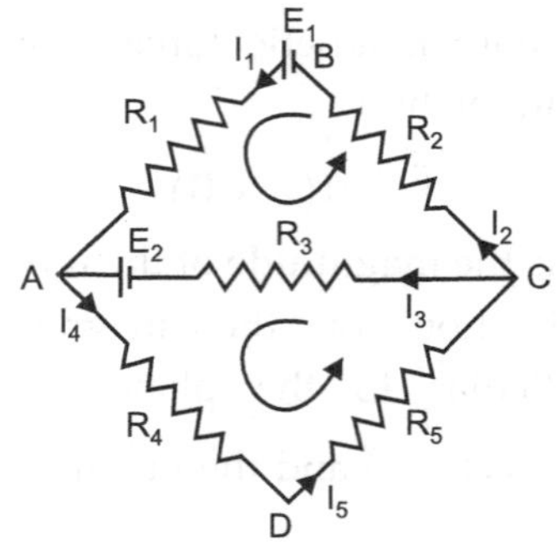


For the closed loop BACB :

$$E_1 - E_2 = I_1R_1 + I_2R_2 - I_3R_3$$

For the closed loop CADC :

$$E_2 = I_3R_3 + I_4R_4 + I_5R_5$$

This law is based on the law of conservation of energy.

**14. A capacitor 'C', a variable resistor 'R' and a bulb 'B' are connected in series to the ac mains in circuit as shown. The bulb glows with some brightness. How will the glow of the bulb change if (i) a dielectric slab is introduced between the plates of the capacitor, keeping resistance R to be the same; (ii) the resistance R is increased keeping the same capacitance ? [3]**

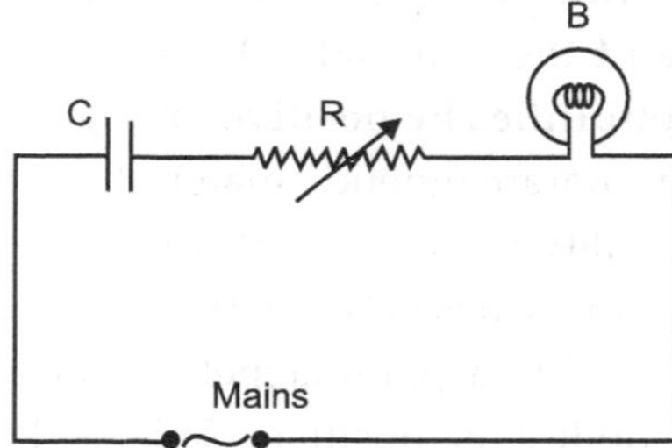


**Ans. (i)** As the dielectric slab is introduced between the plates of the capacitor, its capacitance will increase. Hence, the potential drop across the capacitor will decrease (V= Q/C). As a result, the potential drop across the bulb will increase (since both are connected in series). So, its brightness will increase.

**(ii)** As the resistance (R) is increased, the potential drop across the resistor will increase. As a result, the potential drop across the bulb will decrease (since both are connected in series). So, its brightness will decrease.

**15. State the underlying principle of a cyclotron. Write briefly how this machine is used to accelerate charged particles to high energies. [3]**

**Ans.** The underlying principle of a cyclotron is that an oscillating electric field can be used to accelerate a charged particle to high energy.

** **Answer is not given due to change in the present Syllabus.**

A cyclotron involves the use of an electric field to accelerate charge particles across the gap between the two D-shaped magnetic field regions. The magnetic field is perpendicular to the paths of the charged particles that makes them follow in circular paths within the two Ds. An alternating voltage accelerates the charged particles each time they cross the Ds. The radius of each particle's path increases with its speed. So, the accelerated particles spiral toward the outer wall of the cyclotron.

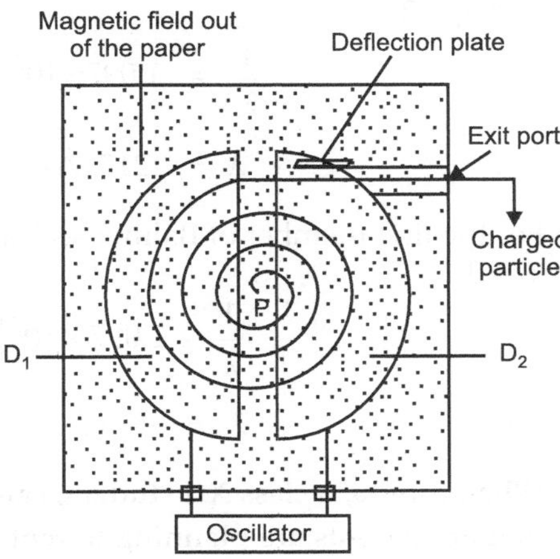


The Ds are the semi-circular structures ($D_1$ and $D_2$) between which the charges move. The accelerating voltage is maintained across the opposite halves of the Ds.

Square wave electric fields are used to accelerate the charged particles in a cyclotron.

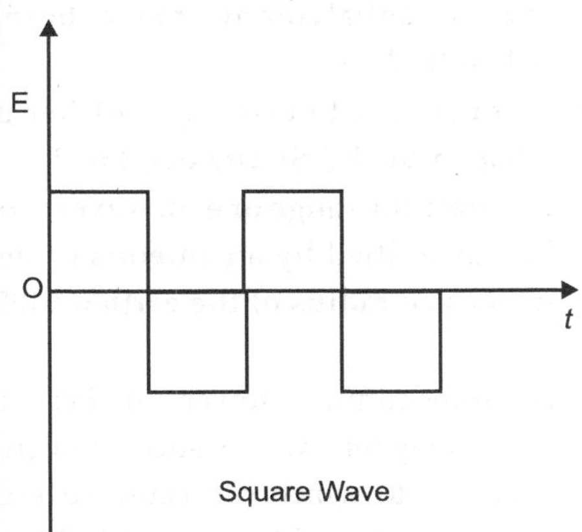


The accelerating electric field reverses just at the time the change particle finishes its half circle. So that it gets accelerated across the gap between the Ds.

The particle gets accelerated again and again, and its velocity increases. Therefore, it attains high kinetic energy.

The positively charged ion adopts a circular path with a constant speed $v$, under the action of magnetic field B, which is perpendicular to the planes of D's of radius $r$.

$$r = \frac{mv}{qB}$$

**16. An electric dipole of length 4 cm, when placed with its axis making an angle of 60° with a uniform electric field, experiences a torque of $4\sqrt{3}$ Nm. Calculate the potential energy of the dipole, if it has charge ± 8 nC. [3]**

**Ans.** As $\tau = pE \sin\theta$

$\therefore$ $4\sqrt{3} = pE \sin\theta$

$\Rightarrow$ $pE \times \frac{\sqrt{3}}{2} = 4\sqrt{3}$

$\Rightarrow$ $pE = 8$

Potential energy of dipole,

$$U = -pE\cos\theta$$
$$= -pE\cos 60°$$
$$= -8 \cdot \frac{1}{2}$$
$$= -4 \text{ J.}$$

**17. A proton and a deuteron are accelerated through the same accelerating potential. Which one of the two has**

**(a) greater value of de-Broglie wavelength associated with it, and**

**(b) less momentum ?**

**Give reasons to justify your Answer. [3]**

**Ans. (a)** de-Broglie wavelength,

$\lambda \propto 1/\text{mass}$ (for same accelerating potential)

Mass of a proton is less as compared to a deuteron. So, proton will have greater value of de-Broglie wavelength associated with it.

**(b)** Momentum, $p \propto$ mass (for same accelerating potential)

Mass of deuteron is more as compared to a proton. So, it will have a greater value of momentum.

**18. (i) Monochromatic light of frequency $6.0 \times 10^{14}$ Hz is produced by a laser. The power emitted is $2.0 \times 10^{-3}$ W. Estimate the number of photons emitted per second on an average by the source.**

**(ii) Draw a plot showing the variation of photoelectric current versus the intensity of incident radiation on a given photosensitive surface. [3]**

**Ans. (i)** The energy of a photon of frequency ν is $E = h\nu = (6.63 \times 10^{-34} \text{ Js}) \times (6 \times 10^{14} \text{ s}^{-1}) = 3.98 \times 10^{-19}$ J. $\simeq 4\ 10^{-19}$ J

If $n$ be the number of photons emitted by the source per second, then the power P transmitted in the beam is given by

$$P = nE$$

$$\therefore \quad n = \frac{P}{E}$$

$$\Rightarrow \quad n = \frac{2\times10^{-3}}{4\times10^{-19}} = 5 \times 10^{15} \text{ photons/sec.}$$

**(ii)**

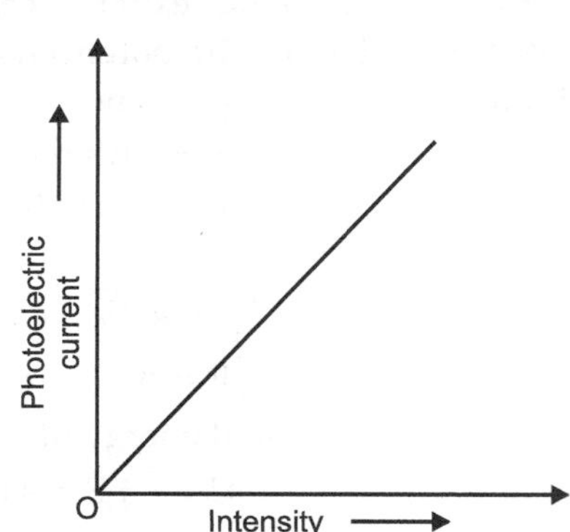


**19. A 12.5 eV electron beam is used to bombard gaseous hydrogen at room temperature. Upto which energy level the hydrogen atoms would be excited ? Calculate the wavelength of the first member of Lyman and first member of Balmer series.** **[3]**

**Ans.** Energy of the electron in the $n^{th}$ state of an atom

$$= \frac{-13.6}{n^2} \times Z^2 eV$$

Here, Z is the atomic number of the atom. For hydrogen atom, Z is equal to 1. Energy required to excite an atom from the initial state ($n_i$) to the final state ($n_f$).

$$E_f = \frac{-13.6}{n_f^2} + \frac{13.6}{n_i^2} eV$$

This energy must be equal to or less than the energy of the incident electron beam.

$$\Rightarrow -\frac{-13.6}{n_f^2} + \frac{13.6}{n_i^2} = 12.5$$

Energy of the electron in the ground state = 13.612 = –13.6 eV

$$\Rightarrow \frac{-13.6}{n_f^2} + 13.6 = 12.5 \qquad [\because n_i = 1]$$

$$\Rightarrow 13.6 - 12.5 = \frac{13.6}{n_f^2}$$

$$\Rightarrow n_f^2 = \frac{13.6}{1.1}$$

$$\Rightarrow n_f^2 = 12.36$$

$$\Rightarrow n_f = 3.5$$

State cannot be a fraction number $n_f = 3$

Hence, hydrogen atom would be excited up to 3rd energy level.

Rydberg formula for the spectrum of the hydrogen atom is given below :

$$\frac{1}{\lambda} = R\left(\frac{1}{n_f^2} - \frac{1}{n_i^2}\right)$$

Here, $\lambda$ is the wavelength and R is the Rydberg constant.

$$R = 1.097 \times 10^7 \text{ m}^{-1}$$

For the first member of the Lyman series : $n_f = 1$ and $n_i = 2$

$$\frac{1}{\lambda} = 1.097\times10^7\left(\frac{1}{1^2} - \frac{1}{2^2}\right)$$

$$\lambda = 1215 \text{ (Å)}$$

For the first member of Balmer series : $n_f = 2$ and $n_i = 3$

$$\frac{1}{\lambda} = 1.097\times10^7\left(\frac{1}{2^2} - \frac{1}{3^2}\right)$$

$$\Rightarrow \quad \lambda = 6563 \text{ Å}$$

**20. When Sunita, a class XII student, came to know that her parents are planning to rent out the top floor of their house to a mobile company she protested. She tried hard to convince her parents that this move would be a health hazard.**

**Ultimately her parents agreed :** **[3]**

**(1) In what way can the setting up of transmission tower by a mobile company in a residential colony prove to be injurious to health ?**

**(2) By objecting to this move of her parents, what value did Sunita display ?**

**(3) Estimate the range of e.m. waves which can be transmitted by an antenna of height 20 m. (Given radius of the earth = 6400 km)**** **[3]**

**Ans.** (1) A transmitting tower makes use of electromagnetic waves such as microwaves, exposure to which can cause severe health hazards like, giddiness, headache, tumour and cancer. Also, the transmitting antenna operates on a very high power, so the risk of someone getting severely burnt in a residential area increases.

(2) By objecting to this move of her parents, sunita has displayed awareness towards the health and environment of her society.

**21. A potentiometer wire of length 1 m has a resistance of 10 Ω. It is connected to a 6 V battery in series with a resistance of 5 Ω. Determine the emf of the primary cell which gives a balance point at 40 cm.** [3]

**Ans.** From the figure below :

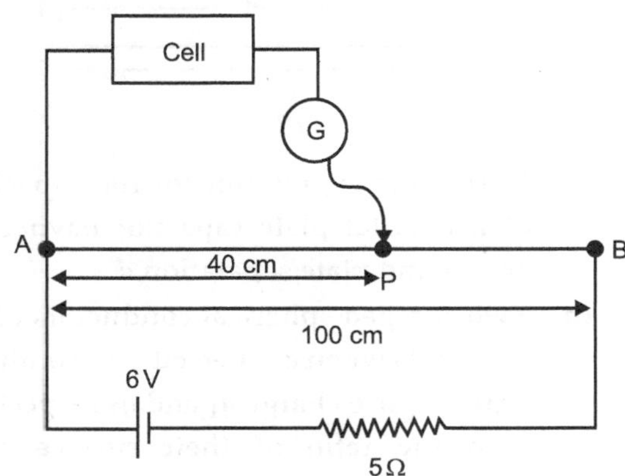


Total resistance of the circuit,

$$R = (R_{AB} + 5) = (10 + 5)\ \Omega = 15\ \Omega$$

Current in the circuit, $I = \frac{V}{R} = \frac{6}{15} A$

$$V_{AB} = 6 - \frac{6}{15} \times 5$$

$$= 6 - 2 = 4\ V$$

∴ Voltage across AB,

Emf of the cell, $e = \left(\frac{l}{L}\right) V_{AB}$

Here : $l$ = 40 m (balance point)

AB = L = 1 m = 100 cm (total length of the wire)

∴ $$e = \left(\frac{40}{100}\right) 4 = 1.6 V$$

**22. (a) Draw a labelled ray diagram showing the formation of a final image by a compound microscope at least distance of distinct vision.**

**(b) The total magnification produced by a compound microscope is 20. The magnification produced by the eye piece is 5. The microscope is focussed on a certain object. The distance between the objective and eye-piece is observed to be 14 cm. If least distance of distinct vision is 20 cm, calculate the focal length of the objective and the eye-piece.** [3]

**Ans. (a)**

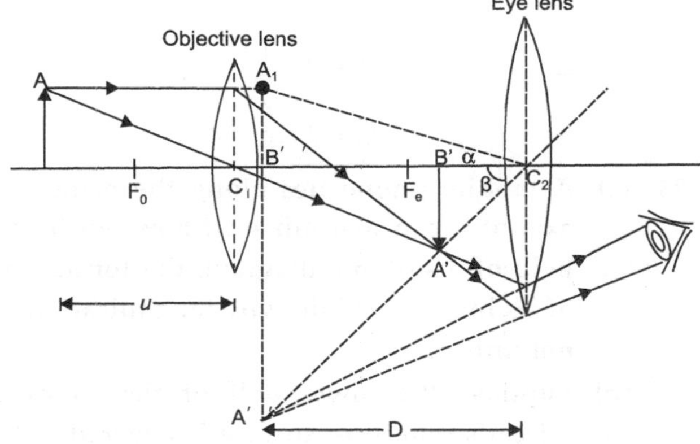


**(b)** For the least distance of clear vision, the total magnification is given by :

$$m = \frac{L}{f_o}\left(1 + \frac{D}{fe}\right) = m_o m_e \quad ...(i)$$

where, L is the separation between the eyepiece and the objective

$f_o$ is the focal length of the objective

$f_e$ is the focal length of the eyepiece.

D is the least distance for clear vision.

Also, the given magnification for the eyepiece :

$$m_e = 5\ m_e = 1 + \frac{D}{f_e}$$

$$\Rightarrow \quad 5 = 1 + \frac{20}{f_e}$$

$$\Rightarrow \quad f_e = 5 \text{ cm}$$

Substituting the value of $m$ and $m_e$ in equation (i), we get :

$$m = m_o\, m_e$$

$$20 = m_o \times 5$$

$$m_o = 4$$

For the eyepiece $v_e$ = – 20cm, fe = 5cm

$$\therefore u_e = \frac{v_e f_e}{f_e - v_e} = \frac{20 \times 5}{5 + 20} = -4\text{cm}$$

Now, we have

$$|u_e| + |v_o| = 14$$

(*i. e.*, distance between objective and eye piece)

$$u_e + v_o = 14$$

$$v_o = (14 - 4) \text{ cm}$$

$$\Rightarrow \quad v_o = 10 \text{ cm}$$

Now, $m_o = 1 - \frac{v_o}{f_o}$

$\Rightarrow \quad -4 = 1 - \frac{10}{f_o}$

$f_o = 2$ cm

**23. (a) A mobile phone lies along the principal axis of a concave mirror. Show, with the help of a suitable diagram, the formation of its image. Explain why magnification is not uniform.**

**(b) Suppose the lower half of the concave mirror's reflecting surface is covered with an opaque material. What effect this will have on the image of the object ? Explain.**

**[4]**

**Ans. (a)** The image of the mobile phone formed by the convave mirror will be as shown in fig. given below :

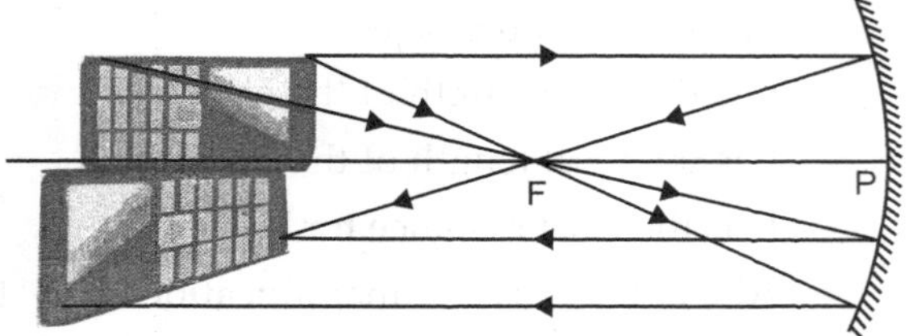


The magnification produced by a concave mirror,

$$m = \frac{I}{O} = \frac{f}{f-u}$$

Since the object distance for the different parts of the mobile phone along its length are different, the different parts will be magnified differently. Hence, the magnification is not uniform.

**(b)** As the laws of reflection are true for all points of the mirror, the height of the whole image will be produced. However, as the area of the reflecting surface has been reduced, the image intensity will be reduced. In other words, the image produced will be less bright.

**24. (a) Obtain the expression for the energy stored per unit volume in a charged parallel plate capacitor.**

**(b) The electric field inside a parallel plate capacitor is E. Find the amount of work done in moving a charge $q$ over a closed rectangular loop *abcda*.** **[5]**

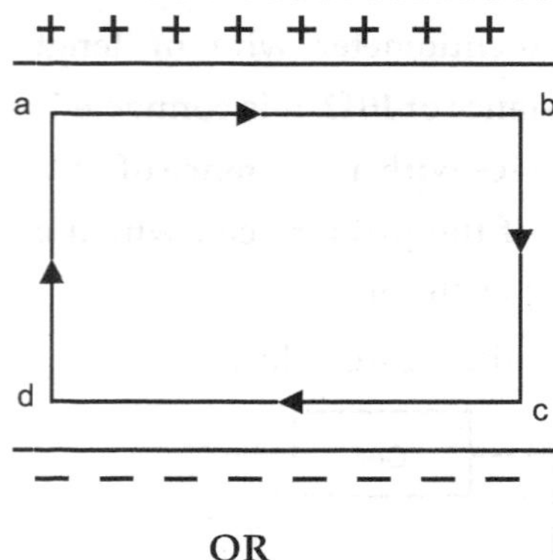


**OR**

**(a) Derive the expression for the capacitance of a parallel plate capacitor having plate area A and plate separation *d*.**

**(b) Two charged spherical conductors of radii $R_1$ and $R_2$ when connected by a conducting wire acquire charge $q_1$ and $q_2$ respectively. Find the ratio of their surface charge densities in terms of their radii.**

**Ans. (a)** Let us consider a parallel-plate capacitor of plate area A. If separation between plates is $d$ metre, capacitance C is given by

$$C = \frac{\varepsilon_0 A}{d}$$

We know that, the magnitude of the electric field between the charged plates of the capacitor is

$$E = \frac{\sigma}{\varepsilon_0}$$

Where σ in surface density of either plate. Therefore, the plate charge is

$$Q = \sigma A = \varepsilon_0 EA$$

Now, the energy stored in the capacitor is

$$U = \frac{1}{2}\frac{Q^2}{C} = \frac{1}{2}\frac{(\varepsilon_0 EA)^2}{\varepsilon_0 A/d}$$

$$U = \frac{1}{2}\varepsilon_0 E^2 (Ad) \text{ J}$$

The volume between the plates is A$d$ metre$^3$. Therefore, the energy per unit volume is given by

$$U' = \frac{U}{Ad} = \frac{1}{2}\varepsilon_0 E^2 \text{ J/m}^3$$

+Q –Q

F F

a b

**(b)** Work done, W = F.$d$

Here, F is the force exerted on the charge ($q$) due to electric field (E) and is given by,

$$F = qE$$

Net displacement, $d = 0$

$$\therefore \quad W = 0$$

**OR**

**(a) Derivation for the capacitance of parallel plate capacitor:**

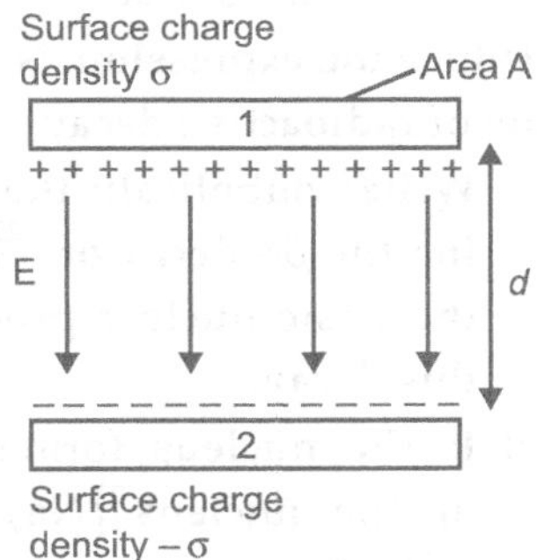


A parallel plate capacitor consists of two large plane parallel conducting plates separated by a small distance. The two plates have charges $q$ and $-q$ and distance between them is $d$.

Plate 1 has charge density, $\sigma = \frac{q}{A}$

Plate 2 has charge density, $\sigma = -\frac{q}{A}$

In the inner region between the plates 1 and 2, the electric fields due to the two charged plates add up.

$$E = \frac{\sigma}{2\varepsilon_0} + \frac{\sigma}{2\varepsilon_0} = \frac{\sigma}{\varepsilon_0} = \frac{q}{A\varepsilon_0}$$

The direction of electric field is from the positive to the negative plate.

For this electric field, potential difference between the plates is given by,

$$V = Ed = \frac{1}{\varepsilon_0}\frac{qd}{A}$$

The capacitance C of the parallel plate capacitor is then

$$C = \frac{Q}{V} = \frac{\varepsilon_0 A}{d}$$

**(b)** The surface charge density for a spherical conductor of radius $R_1$ is given by :

$$\sigma_1 = \frac{q_1}{4\pi R_1^2}$$

Similarly, for spherical conductor $R_2$, the surface charge density is given by :

$$\sigma_2 = \frac{q_1}{4\pi R_1^2}$$

$$\therefore \quad \frac{\sigma_1}{\sigma_2} = \frac{q_1 R_2^2}{q_2 R_1^2} \quad ...(i)$$

As the spheres are connected so the charges will flow between the spherical conductors till their potential become equal.

$$i.e., \quad \frac{kq_1}{R_1} = \frac{kq_2}{R_2}$$

$$\Rightarrow \quad \frac{q_1}{q_2} = \frac{R_1}{R_2} \quad ...(ii)$$

Using (ii) in (i)

$$\text{We have,} \quad \frac{\sigma_1}{\sigma_2} = \frac{R_1}{R_2}\cdot\frac{R_2^2}{R_1^2} = \frac{R_2}{R_1}$$

$$\Rightarrow \quad \frac{\sigma_1}{\sigma_2} = \frac{R_2}{R_1}$$

**25. (a) State Ampere's circuital law, expressing it in the integral form.**

**(b) Two long coaxial insulated solenoids, $S_1$ and $S_2$ of equal lengths are wound one over the other as shown in the figure. A steady current "I" flow through the inner solenoid $S_1$ to the other end B, which is connected to the outer solenoid $S_2$ through which the same current "I" flows in the opposite direction so as to come out at end A. If $n_1$ and $n_2$ are the number of turns per unit length, find the magnitude and direction of the net magnetic field at a point (i) inside on the axis and (ii) outside the combined system.** **[5]**

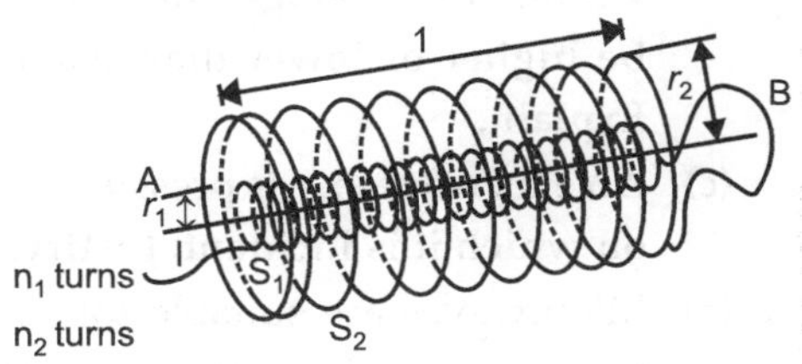


**Ans. (a)** Ampere's circuital law states that the circulation of the resultant magnetic field along a closed, plane curve is equal to $\mu_0$ times the total current crossing the area bounded by the closed curve, provided the electric field inside the loop remains constant.

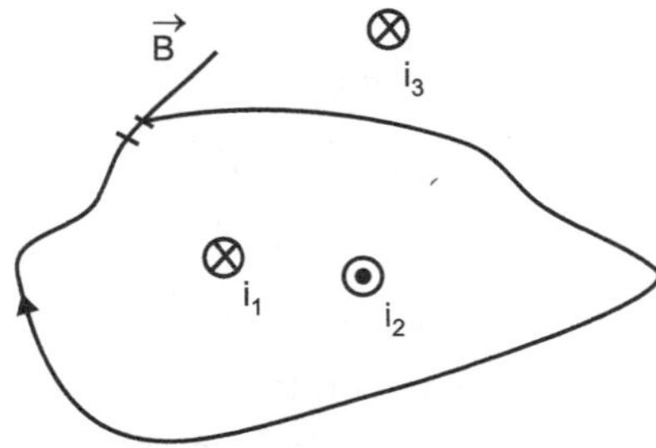


In the above illustration, the Ampere's circuital law can be written as follows :

$$\oint \vec{B}.\vec{dl} = \mu_0 i$$

where $i = i_1 - i_2$

**(b)** (i) The magnetic field due to a current carrying solenoid :

$$B = \mu_0 ni$$

where, $n$ = number of turns per unit length, $i$ = current through the solenoid.

Now, the magnetic field due to solenoid $S_1$ will be in the upward direction and the magnetic field due to $S_2$ will be in the downward direction (by right-hand screw rule).

$$B_{net} = B_1 - B_2$$

$$B_{net} = \mu_0 n_1 I - \mu_0 n_2 I$$

$$= \mu_0 I(n_1 - n_2)$$

(ii) The magnetic field is zero outside a solenoid.

**26. Answer the following :** **[5]**

**(a) Name the e.m. waves which are suitable for radar systems used in aircraft navigation. Write the range of frequency of these waves.**

**(b) If the earth did not have atmosphere, would its average surface temperature be higher or lower than what it is now ? Explain.**

**(c) An e.m. wave exerts pressure on the surface on which it is incident. Justify.**

**Ans. (a)** Microwaves are suitable for radar systems used in aircrafts navigation. The range of frequency for these waves is $10^9$ Hz to $10^{12}$ Hz.

**(b)** In the absence of atmosphere, there would be no greenhouse effect on the surface of the Earth. As a result, the temperature of the Earth would decrease rapidly, making it difficult for human survival.

**(c)** The momentum transported by electromagnetic waves is given by

$$p = \frac{U}{c}$$

where U is the energy transported by electroma-gnetic waves in a given time and $c$ is speed of electromagnetic waves in free space. As a result, when these waves strike a surface, pressure and hence force is exerted by them on the surface.

**27. (a) Deduce the expression, $N = N_0 e^{-\lambda t}$, for the law of radioactive decay.**

**(b) (i) Write symbolically the process expressing the $\beta$+ decay of $^{22}_{11}$Na . Also write the basic nuclear process underlying this decay.**

**(ii) Is the nucleus formed in the decay of the nucleus $^{22}_{11}$Na, an isotope or isobar?** **[5]**

**Ans. (a)** Suppose initially the number of atoms in radioactive element is $N_0$ and $N$ the number of atoms after time $t$. After time $t$, let $dN$ be the number of atoms which disintegrate in a short interval $dt$, then rate of disintegration will be , $\frac{dN}{dt}$ , this is also called the activity of the element.

According to Rutherford-Soddy law,

$$\frac{dN}{dt} \propto N$$

or $$\frac{dN}{dt} = -\lambda N \quad \text{...(i)}$$

where λ is a constant, called decay constant or disintegration constant of the element. Its unit is $s^{-1}$. Negative sign shows that the rate of disintegration decreases with increase of time. For a given element λ is a constant and is different for different elements. Equation (i) may be rewritten as

$$\frac{dN}{N} = -\lambda dt$$

Integrating

$$\log_e N = -\lambda t + C \quad \text{...(ii)}$$

where C is a constant of integration.

At $t = 0, N = N_0$

$\therefore$ $\log_e N_0 = 0 + C \Rightarrow C = \log_e N_0$

$\therefore$ Equation (ii) gives

$$\log_e N = -\lambda t + \log_e N_0$$

or $\log_e N - \log_e N_0 = -\lambda t$

or $\log_e \frac{N}{N_0} = -\lambda t$

$$\frac{N}{N_0} = e^{-\lambda t}$$

$$N = N_0 e^{-\lambda t}$$

(b) (i) The $\beta^+$ decay of ${}_{11}N^{22}$ is expressed as below :

${}_{11}N^{22} \rightarrow {}_{10}Ne^{22} + 1e^0 + \nu$

The neutron-proton ratio is also an important factor that determines the stability of a nucleus. A proton-rich nucleus undergoes positron emission ($\beta^+$ decay), so as to improve the neutron-proton ratio. The positron emission decreases proton number relative to neutron number and the product nucleus becomes relatively more stable. In $\beta^+$ decay, neutrino ($\nu$) is emitted so as to conserve spin.

(ii) The nucleus so formed is an isobar of $11N^{22}_{a}$ because the mass number is same, but the atomic numbers are different.

**28. (a) (i) 'Two independent mono-chromatic sources of light cannot produce a sustained interference pattern'. Give reason.**

**(ii) Light wave each of amplitude '$a$' and frequency '$\omega$', emanating from two coherent light sources superpose at a point. If the displacements due to these waves is given by $y_1 = a \cos \omega t$ and $y_2 = a \cos (\omega t + \phi)$ where $\phi$ is the phase difference between the two, obtain the expression for the resultant intensity at the point.**

**(b) In Young's double slit experiment, using monochromatic light of wavelength $\lambda$, the intensity of light at a point on the screen where path difference is $\lambda$, is K units. Find out the intensity of light at a point where path difference is $\lambda/3$.** [5]

OR

**(a) How does one demonstrate, using a suitable diagram, that unpolarised light when passed through a polaroid gets polarized ?**

**(b) A beam of unpolarised light is incident on a glass-air interface. Show, using a suitable ray diagram, that light reflected from the interface is totally polarised, when $\mu = \tan i_B$, where $\mu$ is the refractive index of glass with respect to air and $i_B$ is the Brewster's angle.**

**Ans. (a)** (i) The condition for the sustained interference is that both the sources must be coherent (*i.e.*, they must have the same wavelength and the same frequency, and they must have the same phase or constant phase difference).

Two sources are monochromatic if they have the same frequency and wavelength. Since they are independent, *i.e.*, they have different phases with irregular difference, they are not coherent sources

(ii)

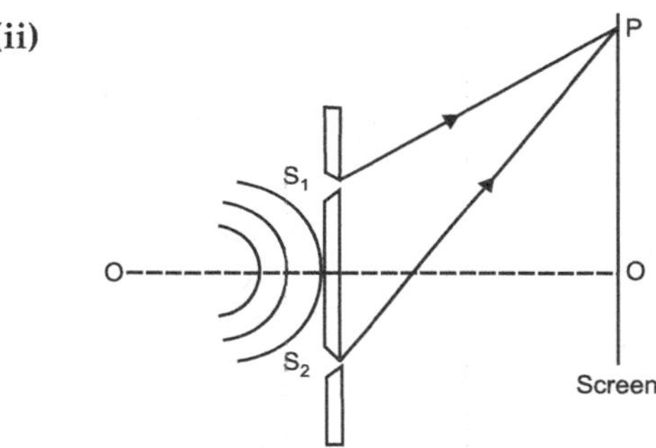


Let the displacement of the waves from the sources $S_1$ and $S_2$ at point P on the screen at any time $t$ be given by :

$$y_1 = a \cos \omega t$$

and $$y_2 = a \cos(\omega t + \phi)$$

Where, $\phi$ is the constant phase difference between the two waves.

By the superposition principle, the resultant displacement at point P is given by :

$$y = y_1 + y_2$$

$$y = a \cos \omega t + a \cos (\omega t + \phi)$$

$$y = 2a\left[\cos\left(\frac{\omega t + \omega t + \phi}{2}\right)\cos\left(\frac{\omega t - \omega t - \phi}{2}\right)\right]$$

$$y = 2a \cos\left(\omega t + \frac{\phi}{2}\right)\cos\left(\frac{\phi}{2}\right) \quad \text{...(i)}$$

Let $2a \cos (\phi/2) = A$ ...(ii)

Then, equation (i) becomes

$$y = A \cos (\omega t + \phi/2)$$

Now, we have :

$$A^2 = 4a^2 \cos^2(\phi/2) \qquad ...(iii)$$

The intensity of light is directly proportional to the square of the amplitude of the wave. The intensity of light at point on the screen is given by

$$I = 4a^2\cos^2 \frac{\phi}{2}$$

**(b)** Intensity $I = 4I_0 \cos^2 \frac{\phi}{2}$

When path difference is $\lambda$, phase difference is $2\pi$

$$I = 4I_0 \cos^2 \pi$$
$$= 4I_0 = k \text{ (given)}$$

When path difference, $\Delta = \frac{\lambda}{3}$, the phase difference

$$\phi_1 = \frac{2\pi}{\lambda}\Delta$$
$$= \frac{2\pi}{\lambda} \times \frac{\lambda}{3} = \frac{2\pi}{3}$$

$$I_1 = 4I_0.\cos^2 \frac{2\neq}{3} \qquad \text{(since } k = 4I_0)$$
$$= k\cos^2 \frac{2\neq}{3}$$
$$= k \times \left(-\frac{1}{2}\right)^2$$
$$= \frac{1}{4}k$$

**OR**

**(a)**

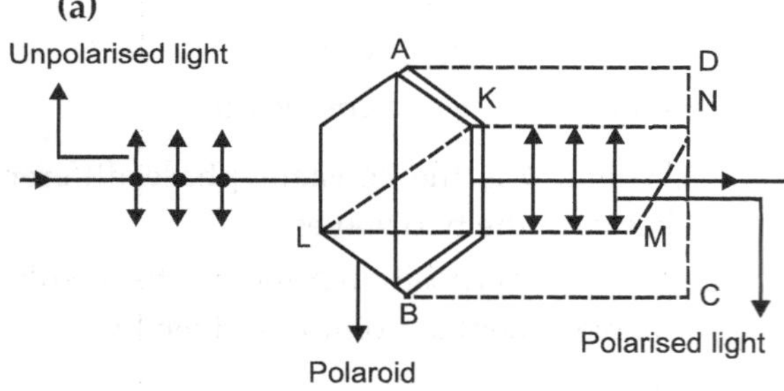


The phenomenon of restricting the vibration of light (electric vector) in a particular direction perpendicular to the direction of the wave propagation is called polarization of light.

When unpolarised light is passed through a polaroid, only those vibrations of light pass through the crystal, which are parallel to the axis of the crystal (AB). All other vibrations are absorbed and that is why intensity of the emerging light is reduced.

The plane ABCD in which the vibrations of the polarised light are confined is called the plane of vibration. The plane KLMN that is perpendicular to the plane of vibration is defined as the plane of polarization.

**(b)** When unpolarised light is incident on the glass-air interference at Brewster angle $i_B$, then reflected light is totally polarised. This is called Brewster's Law.

When light is incident at Brewster angle, the reflected component OB and the refracted component OC are mutually perpendicular to each other.

From the figure,

We have :

$$\angle BOY + \angle YOC = 90^\circ$$
$$(90^\circ - i_B) + (90^\circ - r) = 90^\circ$$
$$i_B + \mu = 90^\circ$$

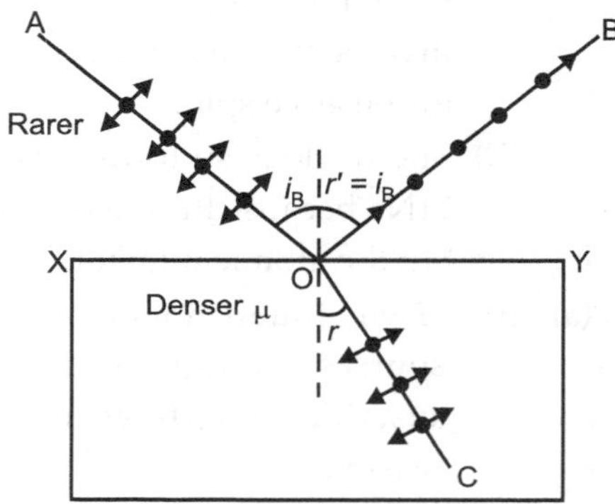


Where, $r$ is angle of refraction

According to the Snell's law :

$$\mu = \frac{\sin i_B}{\sin r} = \frac{\sin i_B}{\sin(90^\circ - i_B)}$$
$$\mu = \frac{\sin i_B}{\cos i_B}$$
$$\mu = \tan i_B$$

**Hence proved.**

**29. (a) Describe a simple experiment (or activity) to show that the polarity of emf induced in a coil is always such that it tends to produce a current which opposes the change of magnetic flux that produces it.**

**(b) The current flowing through an inductor of self inductance L is continuously increasing. Plot a graph showing the variation.**

**(i) Magnetic flux versus the current**

**(ii) Induced emf versus $dI/dt$**

**(iii) Magnetic potential energy stored versus the current. [5]**

OR

**(a) Draw a schematic sketch of an ac generator describing its basic elements. State briefly its working principle. Show a plot of variation of**

**(i) Magnetic flux and**

**(ii) Alternating emf versus time generated by a loop of wire rotating in a magnetic field.**

**(b) Why is choke coil needed in the use of fluorescent tubes with ac mains ?**

**Ans. (a) Lenz's law :** According to Lenz's law, the polarity of the induced emf is such that it opposes a change in magnetic flux responsible for its production.

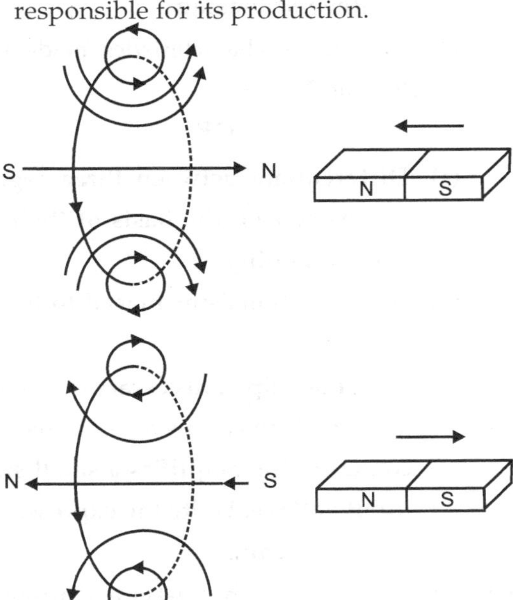


When the north pole of a bar magnet is pushed towards the coil, the amount of magnetic flux linked with the coil increase. Current is reduced in the coil from a direction such that 'it opposes the increase in magnetic flux. This is possible only when the current induced in the coil is in anti-clockwise direction, with respect to an observer. The magnetic moment M associated with this induced emf has north polarity, towards the north pole of the approaching bar magnet. Similarly, when the north pole of the bar magnet is moved away from the coil, the magnetic flux linked with the coil decreases. To counter this decrease in magnetic flux, current is induced in the coil in clockwise direction so that its south pole faces the receding north pole of the bar magnet. This would result in an attractive force which opposes the motion of the magnet and the corresponding decrease in magnetic flux.

**(b) (i)** Since $\phi = LI$

where,

I = strength of current through the coil at any time

$\phi$ = Amount of magnetic flux linked with all turns of the coil at that time and,

L = Constant of proportionally called coefficient of self induction.

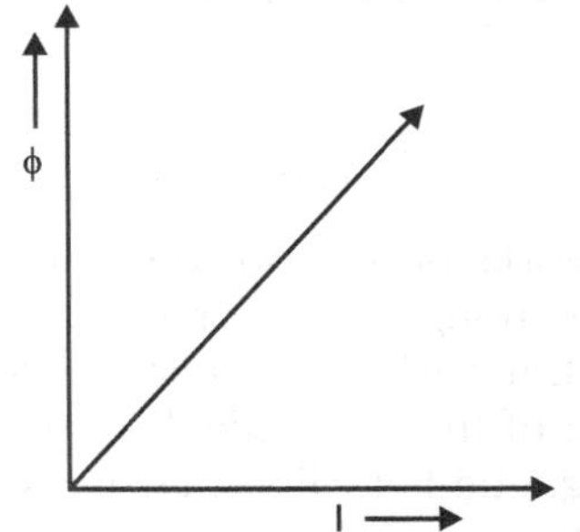


**(ii)** Induced emf,

$$e = \frac{-d\phi}{dt} = -\frac{d(LI)}{dt}$$

*i.e.* $$e = -L\left(\frac{dI}{dt}\right)$$

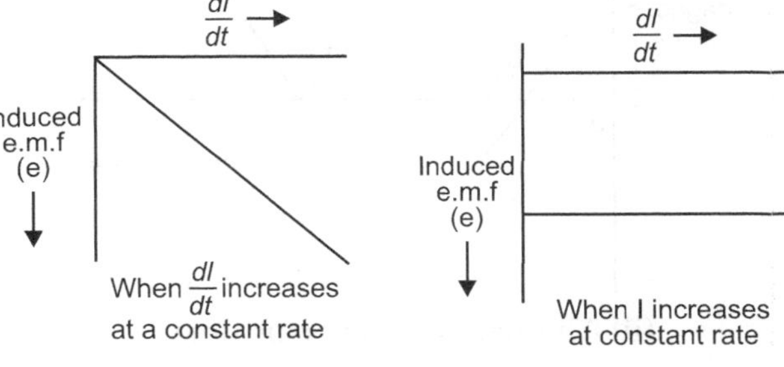


**(iii)** Since magnetic potential energy is given by

$$U = \frac{1}{2}LI^2$$

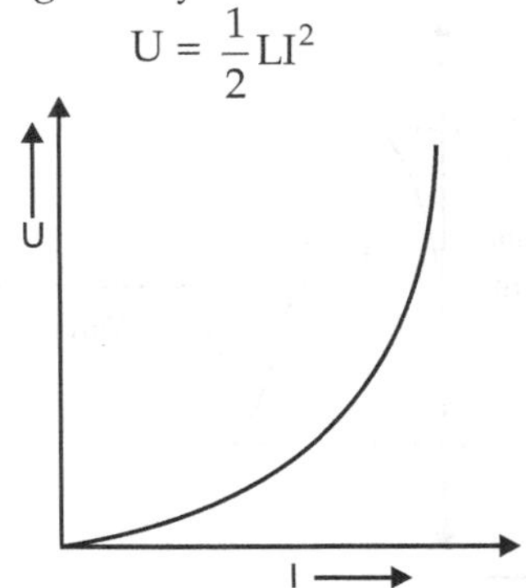

OR

(a)

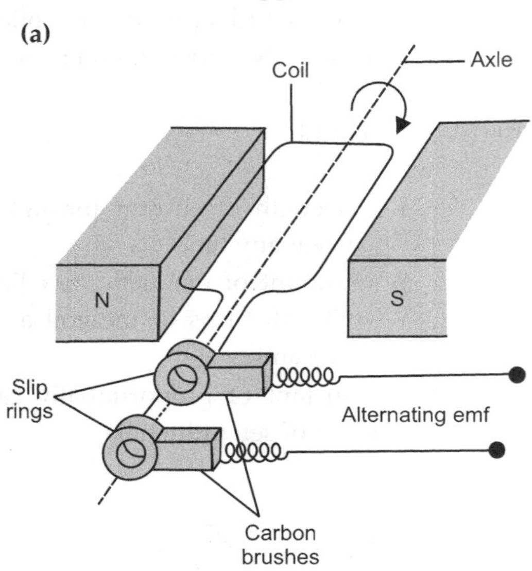


It works on the process of electromagnetic induction, *i.e.*, when a coil rotates continuously in a magnetic field, the effective area of the coil, linked (normally) with the magnetic field lines, changes continuously with time. This variation of magnetic flux with time results in the production of a (alternating) emf in the coil.

**(i)** Magnetic flux versus time

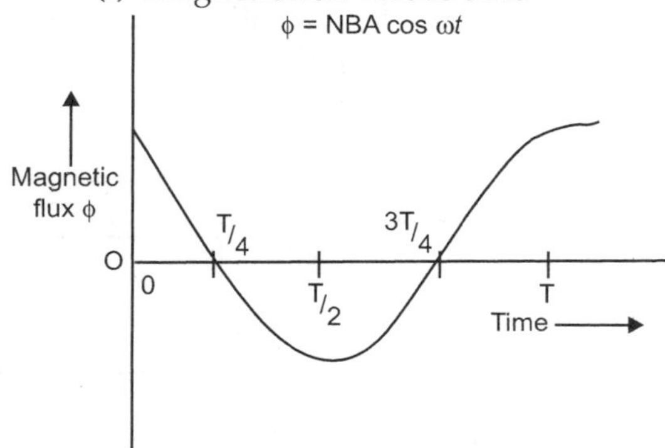


**(ii)** Alternating emf versus time

$$e = NAB\omega \sin \omega t = e_0 \sin \omega t$$

The graph between alternating emf versus time is shown below :

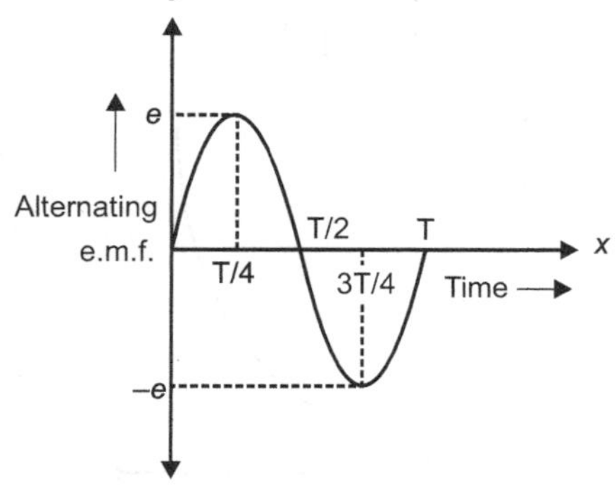


**(b)** A choke coil is an electrical appliance used for controlling current in an a.c. circuit. Therefore, if we use a resistance R for the same purpose, a lot of energy would be wasted in the form of heat etc.

**30. (a) State briefly the processes involved in the formation of *p-n* junction explaining clearly how the depletion region is formed.**

**(b) Using the necessary circuit diagram, show how the V-I characteristics of a *p-n* junction are obtained in**

**(i) Forward biasing**

**(ii) Reverse biasing**

**How are these characteristics made use of in rectification ?** [5]

OR

**(a) Differentiate between three segments of a transistor on the basis of their size and level of doping.**

**(b) How is a transistor biased to be in active state ?****

**(c) With the help of necessary circuit diagram, describe briefly how *n-p-n* transistor in CE configuration amplifies a small sinusoidal input voltage. Write the expression for the ac current gain.**

**Ans. (a)** As we know that *n*-type semi-conductor has more concentration of electrons than that of a hole and *p*-type semi-conductor has more concentration of holes than an electron. Due to the difference in concentration of charge carriers in the two regions of *p-n* junction, the holes diffuse from *p*-side to *n*-side and electrons diffuse from *n*-side to *p*-side.

When an electron diffuses from *n* to *p*, it leaves behind an ionized donor on *n*-side. The ionised donor (+*ve* charge) is immobile as it is bounded by the surrounding atoms. Therefore, a layer of positive charge is developed on the *n*-side of the junction. Similarly, a layer of negative charge is developed on the *p*-side.

** **Answer is not given due to change in the present Syllabus.**

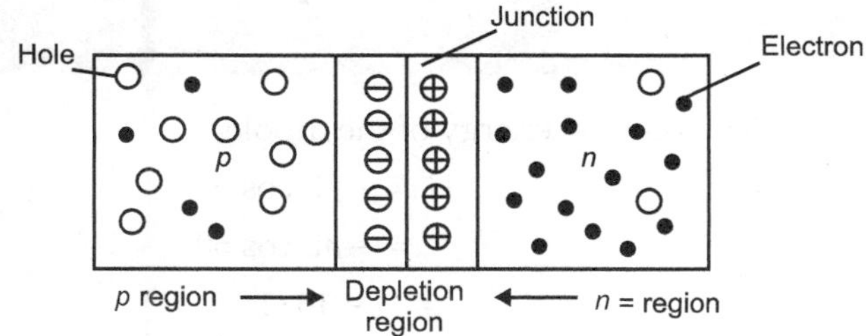


Hence, a space-charge region is formed on both side of the junction, which has immobile ions and is devoid of any charge carrier, called as depletion layer or depletion region.

**(b)**

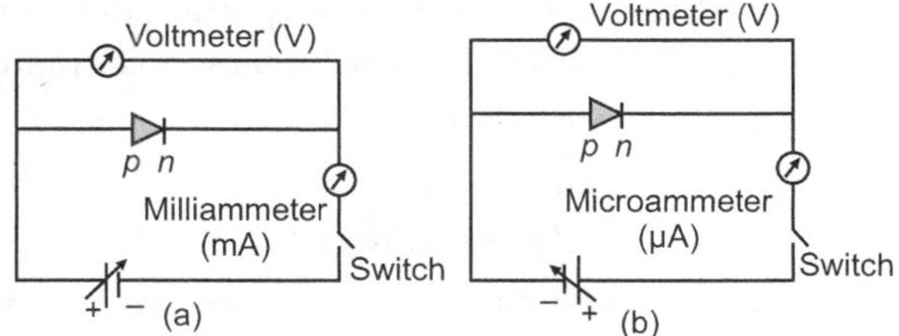


Using the circuit arrangements shown in fig (a) and fig (b), we study the variation of current with applied voltage to obtain the V-I characteristics shown below.

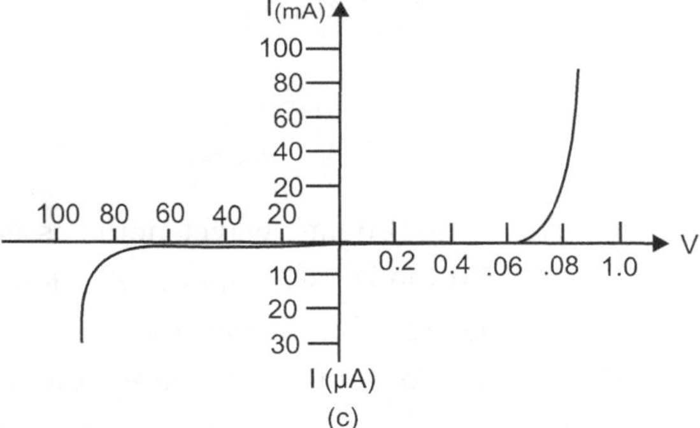


From the V-I characteristics of a junction diode, it is clear that it allows the current to pass only when it is forward biased. So when an alternating voltage is applied across the diode, current flows only during that part of the cycle when it is forward biased.

**OR**

**(a)** **Emitter (E) :** It is the left hand side thick layer of the transistor, which is heavily doped.

**Base (B) :** It is the central thin layer of the transistor, which is lightly doped.

**Collector (C) :** It is the right hand side thick layer of the transistor, which is moderately doped.

**(b)** There are two conditions for a transistor to be into an active region.

(1) The input circuit should be forward biased by using a low voltage battery.

(2) The output circuit should be reverse biased by using a high voltage battery.

**(c)** ***n-p-n* transistor as an amplifier :**

The operating point is fixed in the middle of its active region.

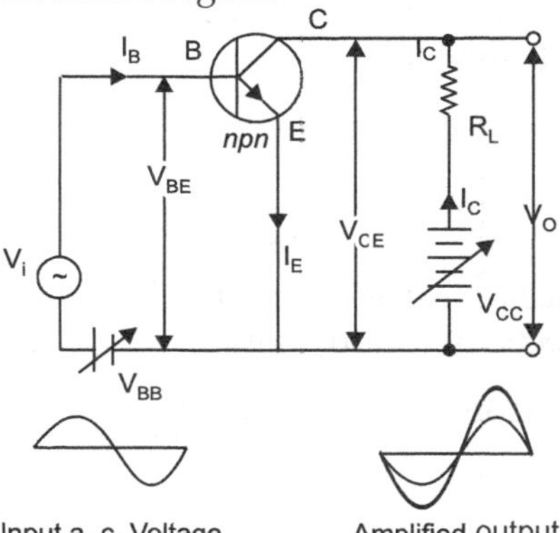


An a. c. input signal $V_i$ is superimposed on base bias $V_{BB}$(dc). The output is taken between the collector and the ground.

Applying Kirchhoff's law to the output loop,

$$I_E = I_B + I_C$$

If $V_{CE}$ is the collector voltage then,

$$V_{CE} = V_{CC} - I_C R_L \quad ...(i)$$

When the input signal voltage is fed to the emitter base circuit, it will change the emitter base voltage and hence to the emitter current, which in turn will change the collector current. Due to this the collector voltage $V_{CE}$ will vary in accordance with relation (i). This variation in collector voltage appears as amplified output.

Let $R_i$ be the input resistance of emitter-base circuit, $V_i$ be the input voltage applied to the emitter-base circuit, and $\Delta I_B$, $\Delta I_C$ be the change in base current and collector current respectively.

Then $\quad V_i = \Delta I_B \times R_i$

The a.c. current gain of transistor is,

$$\beta_{ac} = \frac{\Delta I_C}{\Delta I_B}$$

# Delhi [Set-II]

**Note : Except for the following questions, all the remaining questions have been asked in previous Set.**

1. **Define the term 'electrical conductivity' of a metallic wire. Write its S.I. unit.** **[1]**

**Ans.** The electrical conductivity of a metallic wire is defined as the ratio of the current density to the electric field it creates.

Electrical conductivity, $\sigma = \frac{J}{E}$

S.I. unit = $(\text{ohm m})^{-1}$

**OR**

The reciprocal of resistivity of a material is called its electrical conductivity.

$$\text{Conductivity} = \frac{1}{\text{Resistivity}}$$

or $$\sigma = \frac{1}{\rho}$$

2. **The carrier wave is represented by C (*t*) = 5 sin (10 $\pi t$) volt A modulating signal is a square wave as shown. Determine modulation index.**** **[1]**

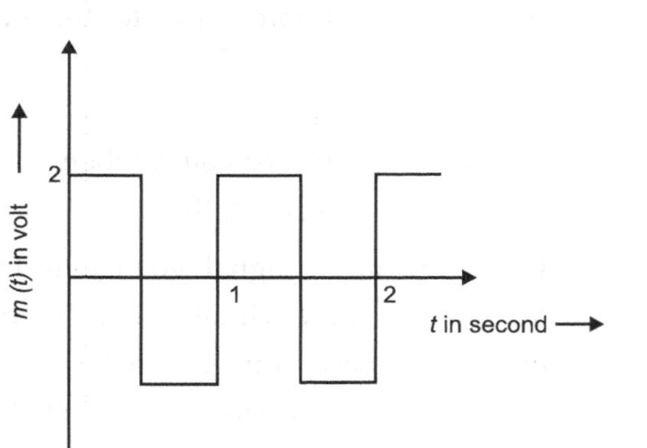


10. **An electric dipole of length 2 cm, when placed with its axis making an angle of 60° with a uniform electric field, experiences a torque of $8\sqrt{3}$ Nm. Calculate the potential energy of the dipole, if it has a charge of ±4 nC.** **[2]**

**Ans.** As

$$\tau = pE \sin\theta$$

$$\therefore \quad 8\sqrt{3} = pE \sin\theta$$

$$= pE \sin 60°$$

$$= \frac{pE\sqrt{3}}{2}$$

$$\Rightarrow \quad pE = 16$$

Also potential energy of the dipole,

$$\Rightarrow \quad U = -pE\cos\theta$$

$$= -pE \cos 60°$$

$$= \frac{-16 \cdot 1}{2} = -8 \text{ J}.$$

15. **A proton and an alpha particle are accelerated through the same potential ? Which one of the two has (i) greater value of de-Broglie wavelength associated with it and (ii) less kinetic energy. Give reasons to justify your answer.** **[3]**

**Ans.** (i) de-Broglie wavelength of a particle depends upon its mass and charge for same accelerating potential such that

$$\lambda \propto \frac{1}{\sqrt{(\text{Mass})(\text{Charge})}}$$

If mass and charge of a proton are $m_p$ and $e$ respectively, then, mass and charge of an alpha particle are $4m_p$ and $2e$ respectively. Where, $e$ is the charge of an electron

$$\frac{\lambda_p}{\lambda_\alpha} = \sqrt{\frac{m_\alpha q_\alpha}{m_p\, q_p}}$$

$$= \sqrt{\frac{(4m_p)(2e)}{(m_p)(e)}} = 2\sqrt{2}$$

Thus, de-Broglie wavelength associated with proton is $2\sqrt{2}$ times of the de-Broglie wavelength of alpha particle.

(ii) K.E. $\propto q$ (for same accelerating potential).

Charge of an alpha particle is more as compared to a proton. So, it will have a greater value of K.E. Hence, proton will have lesser kinetic energy.

16. **Given a uniform electric field $\vec{E} = 2 \times 10^3 \hat{i}$ N/C. Find the flux of this field through a square of side 20 cm, whose plane is parallel to the Y-Z plane. What would be the flux through the same square, if the plane makes an angle of 30° with the X-axis ?** **[3]**

**Ans.** When the plane is parallel to the Y-Z plane

$$\phi = \vec{E} \cdot \vec{A}$$

Here,

$$\vec{E} = 2 \times 10^3 \hat{i}$$

** Answer is not given due to change in the present Syllabus.

$$A = (20\text{ cm})^2 i = 0.04\text{ m}^2 i.$$

$$\therefore \quad \phi = (2 \times 10^3 i).(0.04 i).$$

$$\Rightarrow \quad \phi = 80 \text{ weber or Nm}^2\text{C}^{-1}.$$

When the plane makes a 30° angle with the X-axis, the area vector makes a 60° angle with the X-axis.

$$\phi = \vec{E}\cdot\vec{A}$$

$$\Rightarrow \quad \phi = EA\cos\theta$$

$$\Rightarrow \quad \phi = 2 \times 10^3 \times 0.04 \times \cos 60°$$

$$\Rightarrow \quad \phi = 2 \times 10^3 \times 0.04 \times \frac{1}{2}$$

$$\Rightarrow \quad \phi = 40 \text{ weber or Nm}^2\text{C}^{-1}.$$

**20. A 12.9 eV beam of electronic is used to bombard gaseous hydrogen at room temperature. Upto which energy level the hydrogen atoms would be excited ? Calculate the wavelength of the first member of Paschen series and first member of Balmer series. [3]**

**Ans.** Energy of the electron in the $n^{th}$ state of an atom

$$= \frac{-13.6Z^2}{n^2}$$

Here, Z is the atomic number of the atom.

For hydrogen atom, Z = 1

Energy required to excite an atom from initial state ($n_i$) to final state ($n_f$),

$$E = -13.6\left(\frac{1}{n_f^2} - \frac{1}{n_i^2}\right)\text{eV}$$

$$\Rightarrow \quad \frac{-13.6}{n_f^2} + \frac{13.6}{n_i^2} = 12.9$$

This energy must be equal to or less than the energy of the incident electron beam.

$$\Rightarrow \quad 13.6 - 12.9 = \frac{13.6}{n_f^2}. \qquad [\because n' = 1]$$

$$n_f^2 = \frac{13.60}{0.7} = 19.43.$$

$$\Rightarrow \quad n_f = 4.4.$$

State cannot be a fraction number.

$$\Rightarrow \quad n_f = 4.$$

Hence, the hydrogen atom would be excited up to 4th energy level.

Rydberg's formula for the spectrum of the hydrogen atom is given by :

$$\frac{1}{\lambda} = R\left(\frac{1}{n_2^1} - \frac{1}{n_2^2}\right)$$

Here, λ is the wavelength.

Rydberg's constant, $R = 1.097 \times 10^7 \text{ m}^{-1}$

For the first member of the Paschen series

$$n_1 = 3, n_2 = 4$$

$$\frac{1}{\lambda} = 1.097 \times 10^7\left(\frac{1}{3^2} - \frac{1}{4^2}\right)$$

$$\lambda = 18752.4 \text{ Å}$$

For the first member of Balmer series

$$n_1 = 2, n_2 = 3$$

$$\frac{1}{\lambda} = 1.097 \times 10^7\left(\frac{1}{2^2} - \frac{1}{3^2}\right)$$

$$\lambda = 6563.3 \text{ Å}$$

**22. Answer the following : [3]**

**(a) Name the e.m. waves which are used for the treatment of certain forms of cancer. Write their frequency range.**

**(b) Thin ozone layer on top of stratosphere is crucial for human survival. Why ?**

**(c) Why is the amount of the momentum transferred by the e.m. waves incident on the surface so small ?**

**Ans.** (a) Gamma rays are used for the treatment of certain forms of cancer. Their frequency range is $3 \times 10^{19}$ $H_3$ to $5 \times 10^{20}$ Hz.

(b) The thin ozone layer on top of stratosphere absorb most of the harmful ultraviolet rays coming from the Sun towards the Earth. They include UVA, UVB and UVC radiations, which can destroy the life system on the Earth. Hence, this layer is crucial for human survival.

(c) The momentum transported by electroma-gnetic waves is given by

$$p = \frac{U}{c} = \frac{hv}{c}$$

where U is the energy transported by electroma-gnetic waves in a given time and $c$ is speed of electromagnetic waves in free space.

Now, $h = 6.62 \times 10^{-34}$ J s, $c = 3 \times 10^8$ m s$^{-1}$

Therefore, even for γ-rays ($\nu \approx 10^{20}$ Hz),

$$p = \frac{6.62 \times 10^{-34} \times 10^{20}}{3 \times 10^{8}}$$

$$= 2.2 \times 10^{-22} \text{ kg m s}^{-1}$$

Thus, the amount of the momentum transferred by the e.m. waves incident on a surface is very small.

**24. A potentiometer wire of length 1.0 m has a resistance of 15 Ω. It is connected to a 5 V battery in series with a resistance of 5 Ω. Determine the emf of the primary cell which gives a balance point at 60 cm.** **[5]**

**Ans.** Current through potentiometer wire,

$$I = \frac{V}{R_{AB} + R}$$

$$= \frac{5}{15 + 5} 0.25 \text{ A}$$

P.D. across the potentiometer wire,

$$V = IR_{AB}$$

$$= 0.25 \times 15 = 3.75 \text{ V}$$

Potential gradient,

$$k = V/l$$

$$= 3.75/100$$

$$= 0.0375 \text{ V/cm}$$

So, unknown e.m.f. balanced against 60 cm of the wire,

$$e = kl' = 0.0375 \times 60$$

$$= 2.25 \text{ V}$$

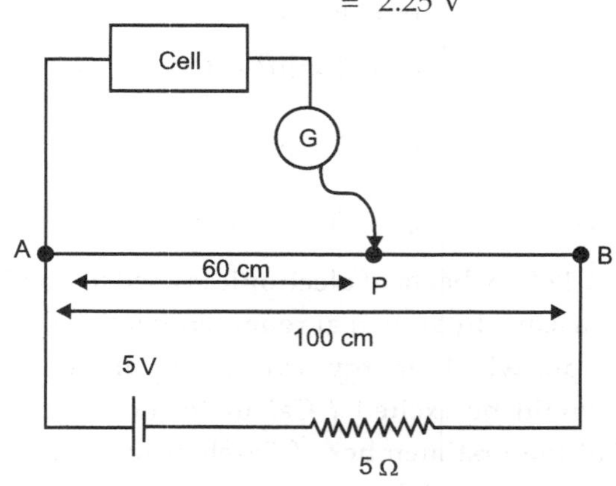


## Delhi [Set-III]

> **Note : Except for the following questions, all the remaining questions have been asked in previous Sets.**

**1. Define the term 'drift velocity' of charge carriers in a conductor and write its relationship with the current flowing through it.** **[1]**

**Ans.** The net speed achieved by an electron in a current carrying conductor is called as drift velocity.

The average velocity acquired by the free electrons along the length of a metallic conductor under a potential difference applied across the conductor is called drift velocity of the electrons.

$$v_d = \frac{I}{ne\text{A}}$$

Here :

$I$ is the current flowing through the conductor.

$n$ is the number density of an electron.

$A$ is the area of the conductor,

$e$ is the charge of the electron.

**2. The carrier wave of a signal is given by C($t$) = 3 sin (8π$t$) volt. The modulating signal is a square wave as shown. Find its modulation index.**** **[1]**

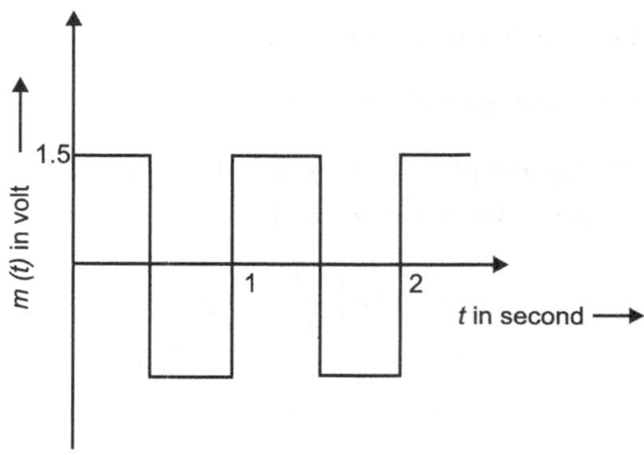


**4. Plot a graph showing variation of current versus voltage for the material Ga.** **[1]**

**Ans.** Current–Voltage characteristics graph for Ga :

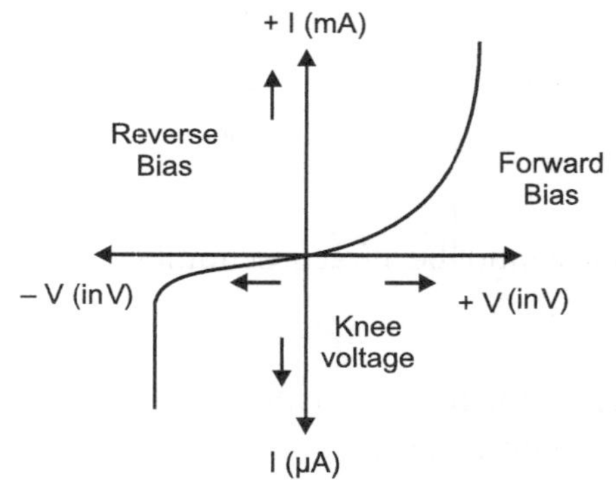


** Answer is not given due to change in the present Syllabus.

**9. An electric dipole of length 2 cm, when placed with its axis making an angle of 60° with a uniform electric field, experiences a torque of $6\sqrt{3}$ Nm. Calculate the potential energy of the dipole, if it has a charge of ±2 nC.** **[1]**

**Ans.** As

$$\tau = pE\sin\theta$$

$$6\sqrt{3} = pE\sin 60°$$

$$= pE \times \frac{\sqrt{3}}{2}$$

or $$pE = 12$$

Potential energy of a dipole,

$$U = -pE\cos\theta$$

$$= -12 \times \cos 60°$$

$$= -12 \times \frac{1}{2} = -6\text{ J}$$

**12. A deuteron and an alpha particle are accelerated with the same accelerating potential.** **[3]**
**Which one of the two has**

**(1) greater value of de-Broglie wavelength, associated with it, and**

**(2) less kinetic energy ? Explain.**

**Ans. (1)** de-Broglie wavelength of a particle is dependent on its mass and charge for same accelerating potential, such that

$$\lambda \propto \frac{1}{\sqrt{(\text{Mass})(\text{Charge})}}$$

Mass and charge of a deuteron are $2m_p$ and $e$ respectively and, mass and charge of an alpha particle are $4m_p$ and $2e$ respectively.

Where,

$m_p$ = the mass of a proton.

and $e$ = the charge of an electron.

$$\frac{\lambda_D}{\lambda_\alpha} = \sqrt{\frac{m_\alpha q_\alpha}{m_D q_D}} = \sqrt{\frac{(4m_p)(2e)}{(2m_p)(e)}} = \frac{2}{1}.$$

Thus, de-Broglie wavelength related with deuteron is twice of the de-Broglie wavelength of alpha particle.

**(2)** $$\text{K.E.} \propto q$$

(for same accelerating potential)

Charge of a deuteron is less as compared to an alpha particle. So, deuteron will have less kinetic energy.

**15. Given a uniform electric field $\vec{E} = 4 \times 10^3 \hat{i}$ N/C. Find the flux of this field through a square of 5 cm on a side whose plane is parallel to the Y-Z plane. What would be the flux through the same square, if the plane makes an angle of 30° with the X-axis ?** **[3]**

**Ans.** When the plane is parallel to the Y-Z plane :

Electric flux, $\phi = \vec{E}\cdot\vec{A}$

Here,

$$\vec{E} = 4 \times 10^3 \hat{i}\text{ N/C}$$

$$\vec{A} = (5\text{ cm})^2\hat{i} = 0.25 \times 10^{-2}\hat{i}\,m^2$$

$\therefore$ $$\phi = (4 \times 10^3\hat{i})\cdot(25 \times 10^{-4}\hat{i})$$

$\Rightarrow$ $$\phi = 10\text{ weber or Nm}^2\text{C}^{-1}$$

When the plane makes a 30° angle with X-axis, the area vector makes a 60° angle with the X-axis.

$$\phi = \vec{E}\cdot\vec{A}$$

$\Rightarrow$ $$\phi = EA\cos\theta$$

$\Rightarrow$ $$\phi = (4 \times 10^3)(25 \times 10^{-4})\cos 60°$$

$\Rightarrow$ $$\phi = 10/2$$

$\Rightarrow$ $$\phi = 5\text{ weber or Nm}^2\text{C}^{-1}$$

**20. A 12.3 eV electron beam is used to bombard gaseous hydrogen at room temperature. Upto which energy level the hydrogen atoms would be excited ? Calculate the wavelength of the second member of Lyman series and second member of Balmer series.** **[3]**

**Ans.** Let the hydrogen atoms be excited to $n^{th}$ energy level.

$$E = -13.6\left(\frac{1}{n_f^2} - \frac{1}{n_i^2}\right) \quad [\because n_i = 1]$$

$\Rightarrow$ $$12.3 = \left(\frac{1}{1^2} - \frac{1}{n^2}\right)$$

$\Rightarrow$ $$12.3 = 13.6 - \frac{13.6}{n^2}$$

$\Rightarrow$ $$\frac{13.6}{n^2} = 13.6 - 12.3 = 1.3$$

$\Rightarrow$ $$n^2 = \frac{13.6}{1.3}$$

$\Rightarrow$ $$n \oplus 3$$

The formula for calculating the wavelength of Lyman series is given below :

$$\frac{1}{\lambda} = R\left(\frac{1}{1^2} - \frac{1}{n^2}\right)$$

For second member of Lyman series, $n = 3$

$$\therefore \quad \frac{1}{\lambda} = R\left(1 - \frac{1}{3^2}\right)$$

$$\Rightarrow \quad \frac{1}{\lambda} = (1.09737 \times 10^7)\left(\frac{8}{9}\right)$$

$$\Rightarrow \quad \lambda = 1025.1 \text{ Å}$$

The formula for calculating the wavelength of Balmer series is given below :

$$\therefore \quad \frac{1}{\lambda} = R\left(\frac{1}{4} - \frac{1}{n^2}\right)$$

For second member of Balmer series :

$$n = 4$$

$$\therefore \quad \frac{1}{\lambda} = R\left(\frac{1}{4} - \frac{1}{4^2}\right)$$

$$\Rightarrow \quad \frac{1}{\lambda} = (1.09737 \times 10^7)\left(\frac{3}{16}\right)$$

$$\Rightarrow \quad \lambda = 4861 \text{ Å}$$

**24. Answer the following : [5]**

**(a) Name the em waves which are used for the treatment of certain forms of cancer. Write their frequency range.**

**(b) Welders wear special glass goggles while working. Why ? Explain.**

**(c) Why are infrared waves often called as heat waves ? Give their one application.**

**Ans.** (a) Gamma rays are used for the treatment of certain forms of cancer. The frequency range of Gamma rays is $3 \times 10^{19}$ to $5 \times 10^{20}$ Hz.

(b) Welders wear special glass goggles while working so that they can protect their eyes from harmful electromagnetic radiation.

(c) Infrared waves are often called as heat waves because they induce resonance in molecules and increase internal energy in a substance.

Infrared waves are used in burglar alarms, security lights and remote controls for television and DVD players.

●●

# PHYSICS

## Outside Delhi [Set-I]

# 2013

**Time allowed : 3 hours** **Maximum marks : 70**

1. **Two charges of magnitudes –2Q and +Q are located at point (a, 0) and (4a, 0) respectively. What is the electric flux due to these charges through a sphere of radius '3a' with its centre at the origin ?** **[1]**

**Ans.** Gauss's theorem states that the electric flux through a closed surface enclosing a charge is equal to $\frac{1}{\varepsilon_0}$ times the magnitude of the charge enclosed. The sphere encloses a charge of –2Q thus, $\phi = \frac{2Q}{\varepsilon_0}$

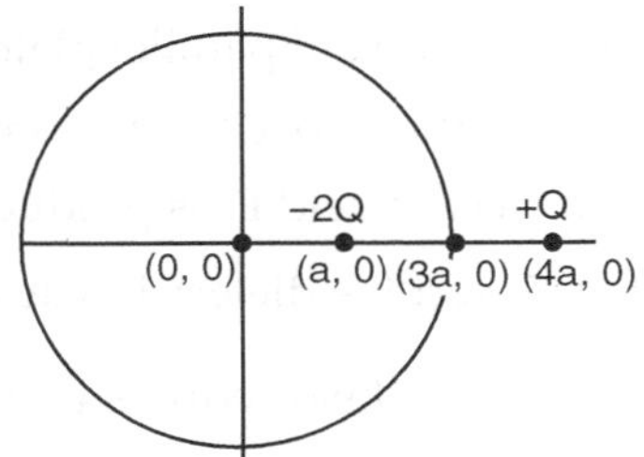


2. **How does the mutual inductance of a pairs of coils change when**
   **(i) distance between the coils is increased and**
   **(ii) number of turns in the coils is increased ?** **[1]**

**Ans.** (i) As $\phi$ = MI, with the increase in the distance between the coils, the magnetic flux linked with the stationary coil decreases and hence the mutual inductance of the two coils will decrease.

(ii) Mutual inductance of two coils can be found out by

$$M = \mu_0 n_1 n_2 AI$$

*i.e.*, $M \propto n_1 n_2$ so with the increase in number of turns, mutual inductance increases.

3. **The graph shown in the figure represents a plot of current versus voltage for given semiconductor. Identify the region, if any, over which the semiconductor has a negative resistance.** **[1]**

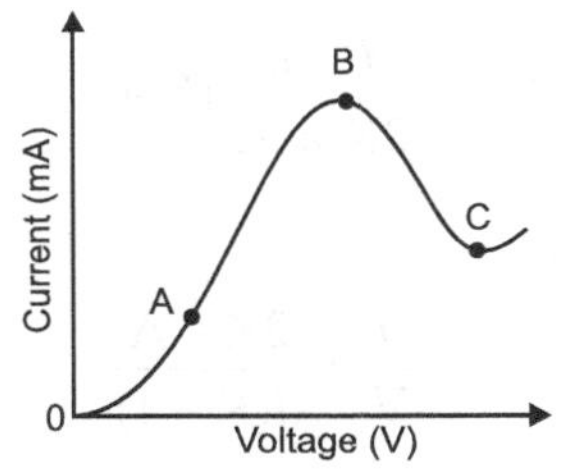


**Ans.** Resistance of a material can be found out by the slope of the V-I curve. Part BC of the curve shows the negative resistance as with the increase in voltage, current decreases.

4. **Two identical cells, each of emf E , having negligible internal resistance, are connected in parallel with each other across an external resistance R. What is the current through this resistance ?** **[1]**

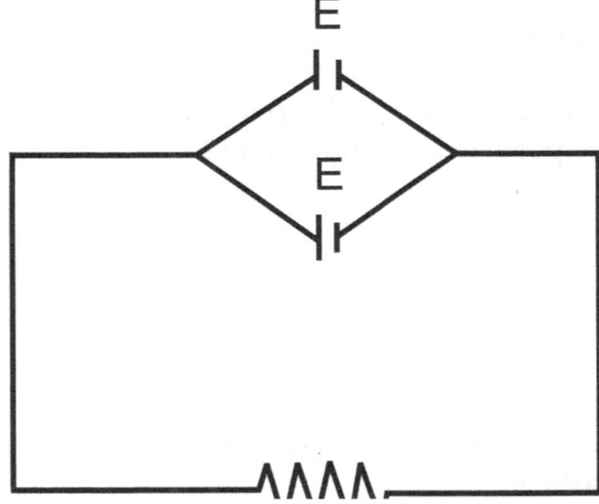


**Ans.** The cells are arranged as shown in the circuit diagram. As the internal resistance is negligible, so total resistance of the circuit = R.

So, current through the resistance, $I = \frac{E}{R}$.

5. **The motion of copper plate is damped when it is allowed to oscillate between the two poles of a magnet. What is the cause of this damping ?** **[1]**

**Ans.** As the copper plates oscillate in the magnetic field between the two poles of the magnet, there is a continuous change of magnetic flux linked with the copper plate. Due to this, eddy currents are set up in the copper plate which try to oppose the motion of the plate according to the Lenz's law and finally bring it to rest.

6. **Define the activity of a given radioactive substance. Write its S.I. unit.** **[1]**

**Ans.** Activity of a radioactive substance is defined as number of radioactive disintegrations taking place in one second in the sample.

S. I. unit of activity is Becquerel (Bq).

1 Becquerel = 1 Bq = 1 decay per second.

**7. Welders wear special goggles or face masks with glass windows to protect their eyes from electromagnetic radiations. Name the radiations and write the range of their frequency. [1]**

**Ans.** Welders wear special goggles with glass windows to protect the eyes from ultraviolet rays (UV rays). The range of frequency of UV rays is $8 \times 10^{14}$Hz to $3 \times 10^{16}$Hz.

**8. Write the expression for the de-Broglie wavelength associated with a charged particle having charge 'q' and mass 'm', when it is accelerated by a potential V. [1]**

**Ans.** Let $v$ be the velocity gained by the given charge particle when it is accelerated through a potential difference of 'V' volts.

Kinetic energy of the particle $= \frac{1}{2}mv^2$.

Kinetic energy of the particle = Work done on the particle by electric field.

$$\therefore \quad \frac{1}{2}mv^2 = qV$$

$$\text{or} \quad \frac{p^2}{2m} = qV$$

$$p = \sqrt{2mqV}$$

de-Broglie wavelength ($\lambda$) associated with the particle,

$$\lambda = \frac{h}{p} = \frac{h}{\sqrt{2mqV}}$$

**9. Draw a typical output characteristics of an *n-p-n* transistor in CE configuration. Show how these characteristics can be used to determine output resistance.** [1]**

**10. A parallel beam of light of 500 nm falls on a narrow slit and the resulting diffraction pattern is observed on a screen 1 m away. It is observed that the first minimum is at a distance of 2.5 mm from the centre of the screen. Calculate the width of the slit. [1]**

**Ans.** The distance of the $n^{th}$ minimum from the centre of the screen is,

$$x_n = \frac{nD\lambda}{a}$$

Where, D = distance of slit from screen = 1 m

$\lambda$ = wavelength of the light = $500 \times 10^{-9}$m

$a$ = width of the slit

For first minimum, $n = 1$

Thus, $2.5 \times 10^{-3} = \frac{1(1)(500 \times 10^{-9})}{a}$,

$$a = 2 \times 10^{-4} \text{ m}$$

$$a = 0.2 \text{ mm}$$

**11. A slab of material of dielectric constant K has the same area as that of the plates of a parallel plate capacitor but has the thickness $d/2$, where $d$ is the separation between the plates. Find out the expression for its capacitance when the slab is inserted between the plates of the capacitor. [2]**

**Ans.** Initially when there is vacuum between the two plates, the capacitance of the two parallel plate is

$$C_0 = \frac{\varepsilon_0 A}{d}$$

Where, A is the area of parallel plates.

Suppose that the capacitor is connected to a battery, an electric field $E_0$ is produced.

Now, if we insert the dielectric slab of thickness $t = \frac{d}{2}$, the electric field reduces to E.

Now, the gap between plates is divided in two parts, for distance $t$ there is electric field E and for the remaining distance $(d - t)$ the electric field is $E_0$.

If V is the potential difference between the plates of the capacitor, then $V = Et + E_0 (d-t)$.

$$V = \frac{Ed}{2} + \frac{E_0 d}{2} = \frac{d}{2}(E + E_0) \quad \left[\because t = \frac{d}{2}\right]$$

$$V = \frac{d}{2}E_0\left(\frac{E}{E_0} + 1\right) \quad \left[\because \frac{E_0}{E} = K\right]$$

$$= \frac{dE_0}{2K}(K+1)$$

Now $\quad E_0 = \frac{\sigma}{\varepsilon_0} = \frac{q}{\varepsilon_0 A}$

$$\Rightarrow \quad V = \frac{d}{2K}\frac{q}{\varepsilon_0 A}(K+1)$$

We know,

$$C = \frac{q}{V} = \frac{2K\varepsilon_0 A}{d(K+1)}$$

** Answer is not given due to change in the present Syllabus.

**12. A capacitor, made of two parallel plates each of plate area A and separation d, being charged by an external ac source. Show that the displacement current inside the capacitor is same as the current charging the capacitor. [2]**

**Ans.** Let the alternating emf charging the plates of capacitor be $E = E_0 \sin \omega t$.

Charge on the capacitor,

$$q = CE = CE_0 \sin \omega t$$

Instantaneous current is I.

$$I = \frac{dq}{dt} = \frac{d}{dt}(CE_0 \sin\omega t)$$

$$= \omega CE_0 \cos \omega t$$

$$= I_0 \cos \omega t \quad ...(i)$$

where, $I_0 = \omega CE_0$

Displacement current,

$$I_D = \varepsilon_0 \frac{d\phi_E}{dt}$$

$$= \varepsilon_0 A \frac{d(E)}{dt} \quad [\phi = E \cdot A]$$

$$= \varepsilon_0 A \frac{d}{dt}\left(\frac{q}{\varepsilon_0 A}\right) \quad \left[\because E = \frac{q}{\varepsilon_0 A}\right]$$

$$= \varepsilon_0 A \frac{d}{dt}\left(\frac{CE_0 \sin \omega t}{\varepsilon_0 A}\right)$$

$$= \frac{d}{dt}(CE_0 \sin \omega t)$$

$$= \omega CE_0 \cos \omega t$$

$$= I_0 \cos \omega t \quad ...(ii)$$

From equations (i) and (ii), the displacement current inside the capacitor is same as the current charging the capacitor.

**13. Explain the term 'drift velocity' of electrons in a conductor. Hence obtain the expression for the current through a conductor in terms of 'drift velocity'. [3]**

**OR**

**Describe briefly, with the help of a circuit diagram, how a potentiometer is used to determine the internal resistance of a cell.**

**Ans. Drift velocity of electrons in a conductor :** Metals contain a large number of free electrons. These electrons are in continuous random motion. Due to the random motion, the free electrons collide with positive metal ions with high frequency and undergo change in direction at each collision. So the average velocity for the electrons in a conductor is zero.

Now, when this conductor is connected to a source of emf, an electric field is established in the conductor, such that

$$E = \frac{V}{L}$$

Where, V = potential difference across the conductor and,

L = length of the conductor.

The electric field exerts an electrostatic force '$-Ee$' on each free electron in the conductor.

The acceleration of each electron is given by

$$\vec{a} = -\frac{e\vec{E}}{m}$$

Where, $e$ = electric charge on the electron and

$m$ = mass of electron.

The negative sign indicates the force and hence the acceleration is in a direction opposite to the direction of the electric field. Due to this acceleration, the electrons attain a velocity in addition to thermal velocity in the direction opposite to that of electric field. The average velocity of all the free electrons in the conductor is called the drift velocity of free electrons of the conductor.

Thus, the expression for the drift velocity in electric field (E)

$$\vec{v} = \frac{e\vec{E}}{m}\tau \quad ...(i)$$

Where $\tau$ = relaxation time between two successive collision,

Let $n$ = number density of electrons in the conductor.

No. of free electrons in the conductor = $nAL$

Total charge on the conductor, $q = nALe$

Time taken by this charge to cover the length L of the conductor,

$$t = \frac{L}{v_d}$$

Current $$I = \frac{q}{t}$$

$$= \frac{n\,ALe}{L} \times v_d$$

$$= nAev_d$$

**OR**

Measurement of internal resistance of a cell using potentiometer is shown in figure. The cell of emf, E is connected across a resistance box (R) through key $K_2$.

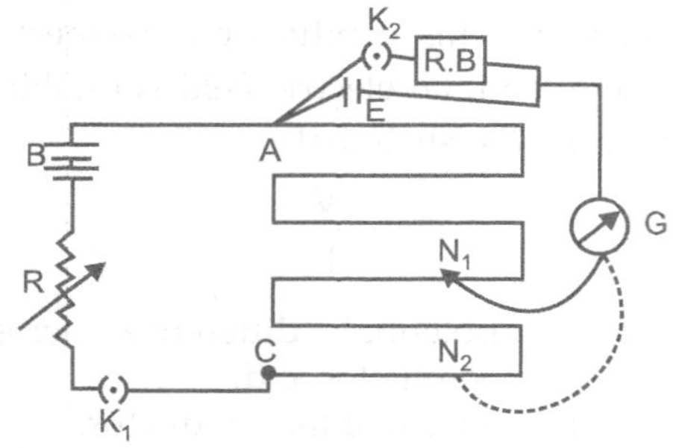


When $K_2$ is open, balance length is obtained at length $AN_1 = l_1$

$$E \propto l_1$$

$$E = \phi\, l_1 \qquad ...(i)$$

When $K_2$ is closed :

Let V be the terminal potential difference of cell and the balance is obtained at $AN_2 = l_2$

$$V \propto l_2$$

$$V = \phi\, l_2 \qquad ...(ii)$$

From equations (i) and (ii), we get

$$\frac{E}{V} = \frac{l_1}{l_2} \qquad ...(iii)$$

But $E = I(r + R)$

$$V = IR$$

$$\frac{E}{V} = \frac{r+R}{R} \qquad ...(iv)$$

From equations (iii) and (iv), we get

$$\frac{R+r}{R} = \frac{l_1}{l_2}$$

$$\therefore \quad r = R\left(\frac{l_1}{l_2} - 1\right)$$

We know $l_1$, $l_2$ and R, so we can calculate $r$.

**14. A convex lens of focal length $f_1$ is kept in contact with a concave lens of focal length $f_2$. Find the focal length of the combination. [2]**

**Ans.** For convex lens, focal length, $f = f_1$ and for concave lens, the focal length, $f = -f_2$

The equivalent focal length of a combination of convex lens and concave lens is given as :

$$\frac{1}{F} = \frac{1}{f_1} + \frac{1}{-f_2}$$

$$F = \frac{f_1 f_2}{f_1 - f_2}$$

** Answer is not given due to change in the present Syllabus.

**15. In the block diagram of a simple modulator for obtaining an AM signal, shown in the figure, identify the boxes A and B. Write their functions.** [2]**

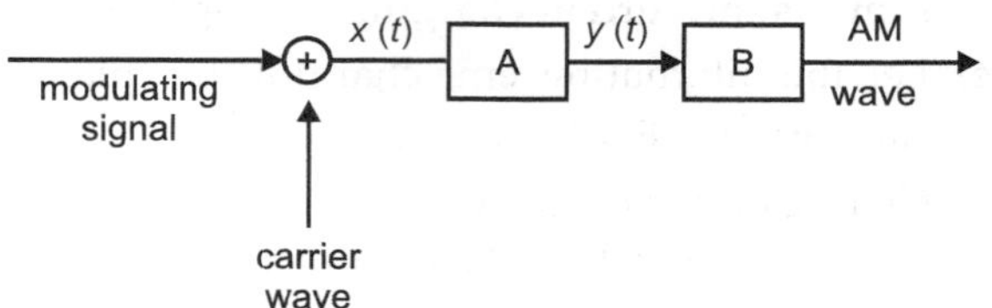


**16. In the circuit shown in the figure, identify the equivalent gate of the circuit and make its truth table.** [2]**

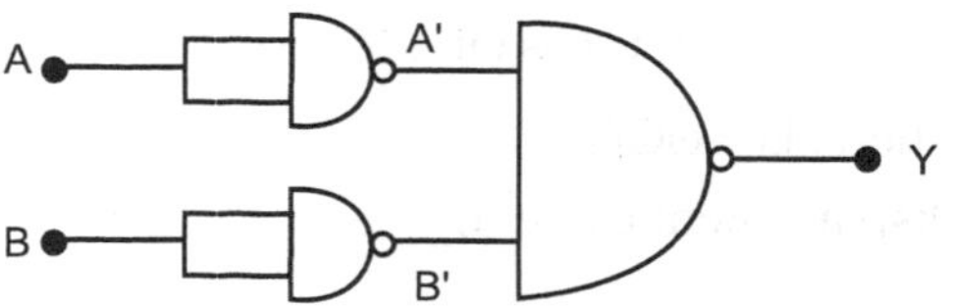


**17. (a) For a given a.c., $i = i_m \sin \omega t$, show that the average power dissipated in a resistor R over a complete cycle is $\frac{1}{2} i_m^2 R$.**

**(b) A light bulb is rated at 100 W for a 220 V a.c. supply. Calculate the resistance of the bulb. [2]**

**Ans. (a)** The average power dissipated

$$P = |i^2 R| = |i_m^2 R \sin^2 \omega t|$$

$$= i_m^2 R |\sin^2 \omega t|$$

$$\sin^2 \omega t = \frac{1}{2}(1 - \cos 2\omega t)$$

$$\Rightarrow |\sin^2 \omega t| = \frac{1}{2}(1 - |\cos 2\omega t|) = \frac{1}{2}$$

$$[\because (|\cos 2\omega t|) = 0]$$

$$\therefore \quad \bar{P} = \frac{1}{2} i_m^2 R$$

**(b)** Power of the bulb, P = 100 W and voltage, V= 220 V.

The resistance of the bulb is given as

$$R = \frac{V^2}{P}$$

$$= \frac{(220)^2}{100}$$

$$= 484\ \Omega$$

**18. A rectangular conductor LMNO is placed in a uniform magnetic field of 0.5 T. The field is directed perpendicular to the plane of the conductor. When the arm MN of length of 20 cm is moved towards left with a velocity of 10 $ms^{-1}$, calculate the emf induced in the arm. Given the resistance of the arm to be 5 Ω (assuming that other arms are of negligible resistance) find the value of the current in the arm.** **[2]**

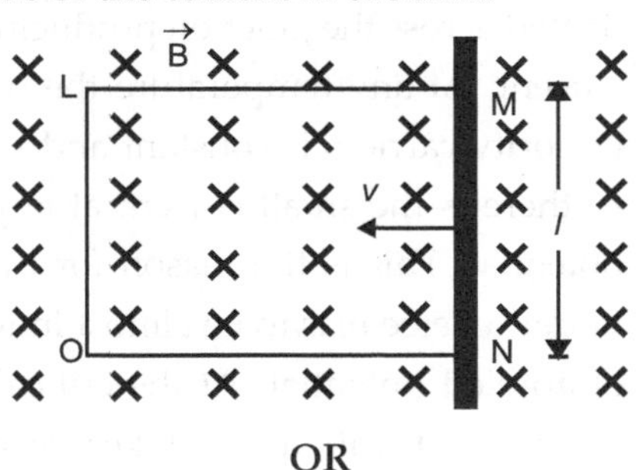


**OR**

**A wheel with 8 metallic spokes each 50 cm long is rotated with a speed of 120 rev/min in a plane normal to the horizontal component of the Earth's magnetic field. The Earth's magnetic field at the place is 0.4 G and the angle of dip is 60°. Calculate the emf induced between the axle and the rim of the wheel. How will the value of emf be affected if the number of spokes were increased ?**

**Ans.** Let ON be $x$ at some instant.

The emf induced in the loop = $e$.

$$e = \frac{d\phi}{dt} = \frac{-d(Blx)}{dt}$$

$$e = Bl\left(-\frac{dx}{dt}\right)$$

$$= Blv = 0.5 \times 0.2 \times 10 \quad \left[\because v = \frac{dx}{dt}\right]$$

$$= 1 \text{ V}$$

Current in the arm, $I = \frac{e}{R} = \frac{1}{5} = 0.2\text{ A}$

**OR**

$H = B\cos\theta = 0.4\cos 60° = 0.2\text{ G} = 0.2 \times 10^{-4}\text{T}$.

This component is parallel to the plane of the wheel. Thus, the emf induced is given as,

$$E = \frac{1}{2}Bl^2\omega$$

Where,

$$\omega = \frac{2\pi N}{t}$$

$$E = \frac{1}{2} \times 0.2 \times 10^{-4} \times (0.5)^2 \times \frac{2 \times 3.14 \times 120}{60}$$

$$= 3.14 \times 10^{-5}\text{V}$$

The value of emf induced is independent of the number of spokes as the emf across the spokes are in parallel. So, the emf will be unaffected with the increase in spokes.

**19. Define the current sensitivity of a galvanometer. Write S.I. unit. Figure shows two circuits each having a galvanometer and a battery of 3 V. When the galvanometers in each arrangement do not show any deflection, obtain the ratio $R_1/R_2$.** **[2]**

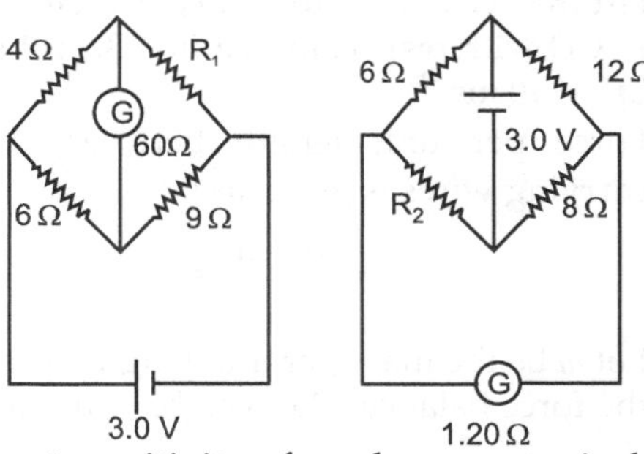


**Ans.** Current sensitivity of a galvanometer is defined as the deflection in galvanometer per unit current. Its SI unit is radians/ampere.

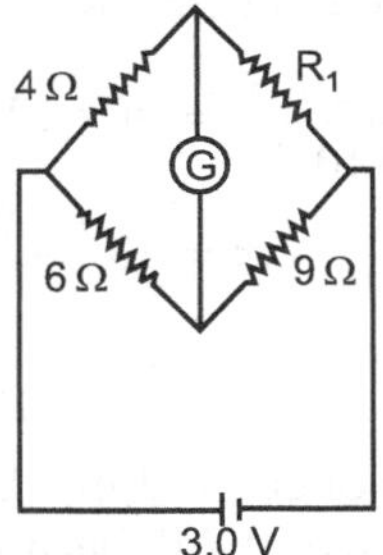


For balanced Wheatstone bridge, there will be no deflection in the galvanometer.

$$\frac{4}{R_1} = \frac{6}{9}$$

$$R_1 = \frac{4 \times 9}{6} = 6\ \Omega$$

For the equivalent circuit, when the Wheatstone bridge is balanced, there will be no deflection in the galvanometer.

$$\frac{12}{8} = \frac{6}{R_2}$$

$$R_2 = \frac{6 \times 8}{12}$$

$$= 4\ \Omega$$

$$\therefore \quad \frac{R_1}{R_2} = \frac{6}{4} = \frac{3}{2}$$

**20. A wire AB is carrying a steady current 12 A and is lying on the table. Another wire CD carrying 5 A is held directly above AB at a height of 1 mm. Find the mass per unit length of the wire CD so that it remains suspended at its position when left free. Give the direction of the current flowing in CD with respect to that in AB. [Take the value of $g = 10\ ms^{-2}$] [2]**

**Ans.** Force per unit length between the current carrying wires is given as :

$$F = \frac{\mu_0}{4\pi}\frac{2I_1I_2}{r}$$

Let $m$ be the mass per unit length of wire CD. As the force balances the weight of the wire,

$$\therefore \quad \frac{\mu_0}{4\pi}\frac{2I_1I_2}{r} = mg$$

Here, $m$ is mass per unit length.

$$10^{-7} \times \frac{2\times 12\times 5}{1\times 10^{-3}} = m \times 10$$

$$m = 10^{-7}\times\frac{2\times 12\times 5}{1\times 10^{-3}}\times\frac{1}{10}$$

$$= 1.2 \times 10^{-3}\ kg\ m^{-1}$$

The direction of current in CD is opposite to the direction of current AB. because in this they will repel each other and CD will remain suspended.

**21. Draw V-I characteristics of a *p-n* junction diode. Answer the following questions, giving reasons :**

**(i) Why is the current under reverse bias almost independent of the applied potential upto a critical voltage ?**

**(ii) Why does the reverse current show a sudden increase at the critical voltage ?**

**Name any semiconductor device which operates under the reverse bias in the breakdown region. [3]**

**Ans.** V-I characteristics of *p-n* junction diode :

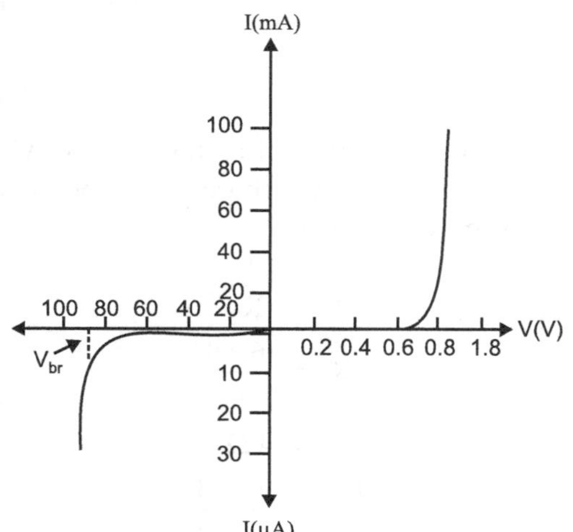


(i) Under the reverse bias condition, the holes of *p*-side are attracted towards the negative terminal of the battery and the electrons of the *n*-side are attracted towards the positive terminal of the battery. This increases the depletion layer and the potential barrier. However the minority charge carriers are drifted across the junction producing a small current. At any temperature the number of minority carriers is constant and very small so there is the small current at any applied potential. This is the reason for the current under reverse bias to be almost independent of applied potential. At the critical voltage, avalanche break down takes place which results in a sudden flow of large current.

(ii) At the critical voltage, the holes in the *n*-side and conduction electrons in the *p*-side are accelerated due to the reverse-bias voltage. These minority carriers acquire sufficient kinetic energy from the electric field and collide with valence electrons. Thus, the bond is finally broken and the valence electrons move into the conduction band resulting in enormous flow of electrons and thus result in formation of hole-electron pairs. Thus, there is a sudden increase in the current at the critical voltage.

Zener diode is a semiconductor device which operates under the reverse bias condition in the breakdown region.

**22. Draw a labelled ray diagram of a refracting telescope. Define its magnifying power and write the expression for it.**

**Write two important limitations of a refracting telescope over a reflecting type telescope. [3]**

**Ans.** Refracting telescope :

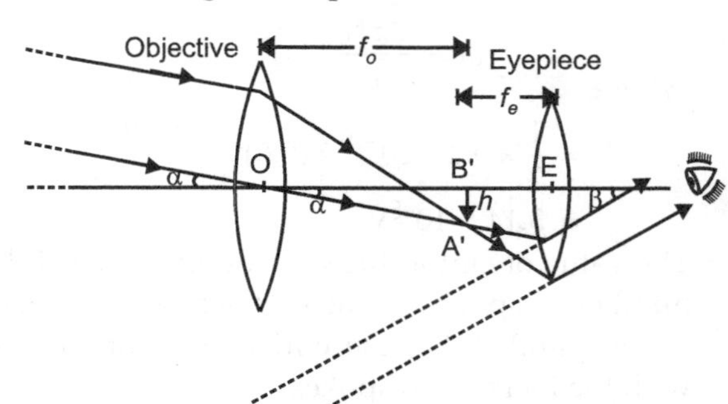

**Magnifying power :** The magnifying power is the ratio of the angle β subtended at the eye by the final image to the angle α which the object subtends at the lens or the eye.

$$m = -\frac{\beta}{\alpha} = -\frac{h}{f_e}.\frac{f_o}{h} = -\frac{f_o}{f_e}$$

**Limitations of refracting telescope over reflecting type telescope :**

**(i)** Refracting telescope suffers from chromatic aberration as it uses large sized lenses.

**(ii)** The requirement of big lenses tend to be very heavy and therefore, difficult to make and support by their edges.

**23. Write Einstein's photoelectric equation and point out any two characteristic properties of photons on which this equation is based.**
**Briefly explain the three observed features which can be explained by this equation. [4]**

**Ans.** Einstein's photoelectric effect equation :

$$K_{max} = \frac{1}{2}mv^2_{max} = (h\nu - \phi_o)$$

or

$$K_{max} = h(\nu - \nu_o)$$

The two characteristics properties of the photon on which this equation is based :

**(i)** The light travel in the form of photons *i.e.* small packet of enery and each photon has energy $h\nu$.

**(ii)** Each photon interact with one electron.

Three observed features that can be explained by this equation.

**(i)** When $\nu < \nu_o$, $K_{max}$ becomes negative. The negative value of K./E. has no physical significance. Hence, no photoelectric effect is possible below there shold frequency.

**(ii)** When $\nu > \nu_o$, $K_{max} \propto \nu$, *i.e.* max. K.E. of the emitted photo electrons depends linearly on the frequency of incident radiations.

**(iii)** It is clear from the photoelectric equation that the maximum K.E. of photoelectrons is independent of intensity of incident reactions.

**24. Name the type of waves which are used for line of sight (LOS) communication. What is the range of their frequencies ?**
**A transmitting antenna at the top of a tower has a height of 20 m and the height of the receiving antenna is 45 m. Calculate the maximum distance between them for satisfactory communication in LOS mode. (Radius of the Earth = 6.4 × $10^6$m)** [5]**

** Answer is not given due to change in the present Syllabus.

**25. (a) What is linearly polarised light. Describe briefly using a diagram how sunlight is polarised.**
**(b) Unpolarised light is incident on a Polaroid. How would the intensity of transmitted light change when the Polaroid is rotated ? [5]**

**Ans. (a)** The light in which the vibrations of electric vector are restricted to a particular plane, is called plane polarised light.

The incident sunlight is unpolarised. The dot and double arrows show the polarisation in the perpendicular and in the plane of the figure respectively. Under the influence of the electric field of the incident wave, the electrons in the molecules of the atmosphere acquire components of motion in both these directions. For an observer looking at 90° to the direction of the sun, the charges accelerating parallel to the double arrows do not radiate energy towards this observer since their acceleration has no transverse component. The radiation scattered by the molecule is therefore, represented by dots. It is linearly polarised light perpendicular to the plane of the figure.

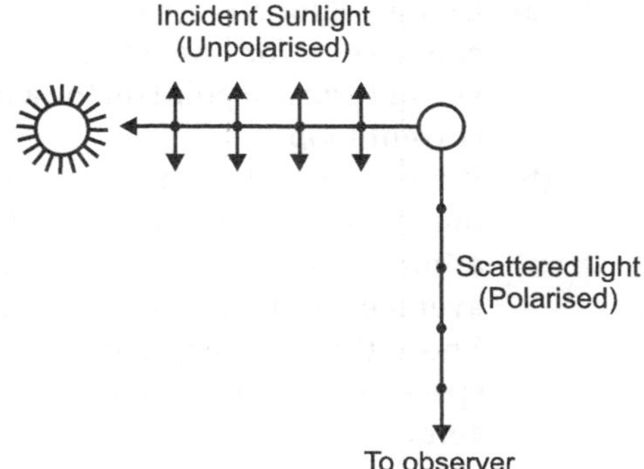


**(b)** If the unpolarised light is incident on a polaroid, the intensity is reduced by half. Even if the polaroid is rotated by angle θ the average over $\cos^2\theta = \frac{1}{2}$. Thus, from Malus' law : $I = I_0 \cos^2\theta$

$$|I| = |I_0 \cos^2\theta|$$

$$= I_0|\cos^2\theta| = \frac{I_0}{2}$$

Thus, the intensity of the transmitted light remains unchanged when the polaroid is rotated.

**26. One day Chetan's mother developed a severe stomach ache all of a sudden. She rushed to the doctor who suggested for an immediate endoscopy test and gave an estimate of expenditure for the same. Chetan immediately contacted his class teacher and shared the information with her. The class teacher arranged for the money and rushed to the hospital. On realizing that Chetan belonged to a below average income group family, even the doctor offered concession for the test fee. The test was conducted successfully.**

**Answer the following questions based on the above information : [5]**

**(a) Which principle in optics is made use of in endoscopy ?**

**(b) Briefly explain the values reflected in the action taken by the teacher.**

**(c) In what way do you appreciate the response of the doctor on the given situation ?**

**Ans.** (a) Endoscopy is based on total internal reflection principle. It has tubes which are made up of optical fibres and are used for transmitting and receiving electrical signals, which are converted into light by suitable transducer.

(b) The values reflected from the tracher are helping attitude and caring nature. Being kind towards other makes us a better human being.

(c) Doctor gave monetary help to Chetan by understanding his poor financial condition.

**27. (a) Using Biot-Savart's law, derive the expression for the magnetic field in the vector form at a point on the axis of a circular current loop.**

**(b) What does a toroid consist of ? Find out the expression for the magnetic field inside a toroid for N turns of the coil having the average radius r and carrying a current I. Show that the magnetic field in the open space inside and exterior to the toroid is zero. [5]**

**OR**

**(a) Draw a schematic sketch of a cyclotron. Explain clearly the role of crossed electric and magnetic field in accelerating the charge. Hence derive the expression for the kinetic energy acquired by the particles.**

**(b) An α-particle and a photon are released from the centre of the cyclotron and made to accelerate.**

**(i) Can both be accelerated at the same cyclotron frequency ? Give reason to justify your answer.**

**(ii) When they are accelerated in turn, which of the two will have higher velocity at the exit slit of the dees ?**

**Ans.** (a) Magnetic field on the axis of a circular loop

I → Current

R → Radii

X → Axis

$x$ → Distance of OP

$dl$ → Conducting element of the loop

According to Biot-Savart's law, the magnetic field at P is

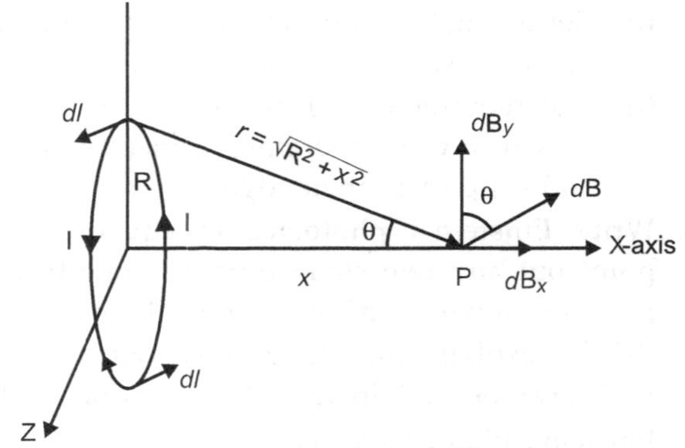


$$dB = \frac{\mu_0}{4\pi}\frac{I\vec{dl} \times \vec{r}}{r^2}$$

$$dB = \frac{\mu_0}{4\pi}\frac{Idl \sin\theta}{r^2}$$

$$dB = \frac{\mu_0}{4\pi}\frac{Idl}{r^2} \qquad \left[\because \vec{dl} \perp \hat{r}\right]$$

$dB$ has two components $dB_x$ and $dB_y$. $dB_y$ components being in opposite direction cancel out each other and $dB_x$ component being in same direction added up.

$$\therefore \quad dB_x = dB \sin\theta$$

$$= \frac{\mu_0 Idl}{4\pi r^2} \times \frac{R}{\left(x^2 + R^2\right)^{1/2}}$$

$$dB_x = \frac{\mu_0 Idl}{4\pi} \times \frac{R}{\left(x^2 + R^2\right)^{\frac{3}{2}}}$$

$$[\because r^2 = x^2 + R^2]$$

Total magnetic field at P

$$B = \int dB_x$$

$$= \int \frac{\mu_0 IRdl}{4\pi (x^2 + R^2)^{3/2}}$$

$$\frac{\mu_0 IR}{4\pi (x^2 + R^2)^{3/2}} \int dl = \frac{\mu_0 IR}{4\pi \left(x^2 + R^2\right)^{3/2}} \times 2\pi R$$

$$B = \frac{\mu_0 IR^2}{2\left(x^2 + R^2\right)^{3/2}}$$

$$\vec{B} = \frac{\mu_0 IR^2}{2\left(x^2 + R^2\right)^{3/2}}\hat{i}$$

**(b)** Toroid is a hollow circular ring on which a large number of turns of a wire are closely wound.

Figure shows a sectional view of the toroid. The direction of the magnetic field inside is clockwise as per the right-hand thumb rule for circular loops. Three circular Amperian loops 1, 2 and 3 are shown by dashed lines.

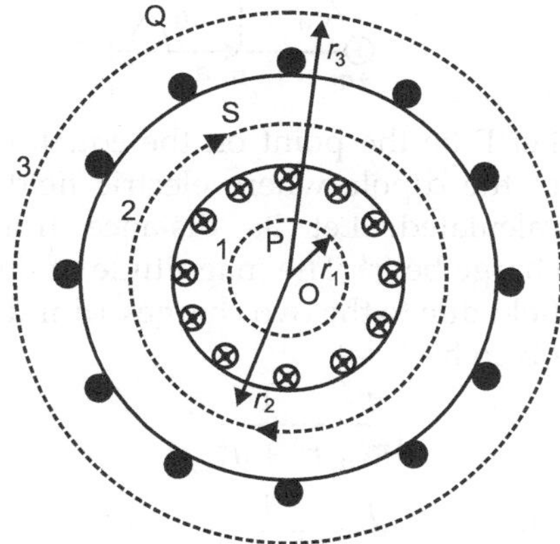


By symmetry, the magnetic field should be tangential to each of them and constant in magnitude for a given loop.

Let the magnetic field inside the toroid be B.

Then by Ampere's circuital law,

$$\int \vec{B}\cdot\vec{dl} = \mu_0 I \Rightarrow BL = \mu_0 NI$$

Where, L is the length of the loop for which B is tangential, I is the current enclosed by one loop and N is the number of turns.

We find, $L = 2\pi r$

The current threads the ring as many times are there are turns in the solenoid, therefore total current in the solenoid is NI.

$$B(2\pi r) = \mu_0 NI$$

or $$B = \frac{\mu_0 NI}{2\pi r}$$

Open space inside the toroid encloses no current thus, I = 0.

Hence, B = 0

Open space exterior to the toroid :

Each turn of current carrying wire is cut twice by the loop 3. Thus, the current coming out of the plane of the paper is cancelled exactly by the current going into it. Thus, I= 0, and B = 0

**OR**

**(a) Schematic sketch of cyclotron :** The combination of crossed electric and magnetic field is used to increase the energy of the charged particle. Cyclotron uses the fact that the frequency of revolution of the charged particle in a magnetic field is independent of its energy. Inside the dees particle is shielded from the electric field and magnetic field acts on the particle and makes it to go round in a circular path. Every time the particle moves from one dee to the other it comes under the influence of electric field which ensures to increase the energy of the particles as the sign of the electric field changed alternately. The increase in energy increases the radius of the circular path so the accelerated particle moves in a spiral path.

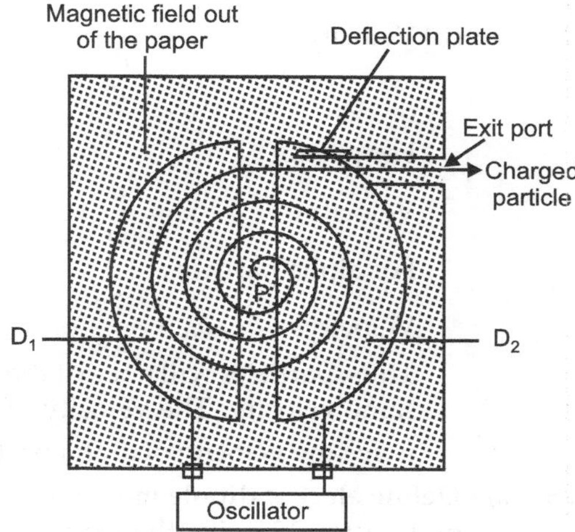


Since, magnetic force provides required centripetal force

$$\therefore \quad qvB = \frac{mv^2}{r}$$

or $$r = \frac{mv}{qB}$$

Here, $r$ is the radius of semi-circular path which a particle covers inside a dee.

Hence, the kinetic energy of ions,

$$K.E. = \frac{1}{2}mv^2$$

$$K.E. = \frac{1}{2}m\frac{r^2q^2B^2}{m^2} \quad \left[\because v = \frac{rqB}{m}\right]$$

$$K.E. = \frac{1}{2}\frac{r^2q^2B^2}{m^2}$$

**(b) (i)** Let us consider :

$m \rightarrow$ mass of proton

$q \rightarrow$ charge of proton

$\therefore$ Mass of alpha particle = 4 $m$

Charge of alpha particle = 2 $q$

Cyclotron frequency is given by,

$$\nu = \frac{Bq}{2\pi m}$$

or $$\nu \propto \frac{q}{m}$$

For proton :

Frequency, $\nu_p \propto \frac{q}{m}$

For alpha particle,

Frequency, $\nu_\alpha \propto \frac{2q}{4m}$

or $\nu_\alpha \propto \frac{q}{2m}$

Thus, particles will not accelerate with same cyclotron frequency. The frequency of proton is twice than the frequency of alpha particle.

**(ii)** Velocity, $v = \frac{Bqr}{m}$

or $v \propto \frac{q}{m}$

For proton :

Velocity, $v \propto \frac{q}{m}$

For alpha particle :

Velocity, $v_\alpha \propto \frac{2q}{4m}$ or $v_\alpha \propto \frac{q}{2m}$

Thus, particles will not exit the dees with same velocity. The velocity of proton is twice the velocity of alpha particle.

**28. (a) Define electric dipole moment. Is it a scalar or a vector ? Derive the expression for the electric field of a dipole at a point on the equatorial plane of the dipole.**

**(b) Draw the equipotential surfaces due to an electric dipole. Locate the points where the potential due to the dipole is zero. [5]**

**OR**

**Using Gauss' law deduce the expression for the electric field due to a uniformly charged spherical conducting shell of radius R at a point (i) outside and (ii) inside the shell. Plot a graph showing variation of electric field as a function of $r > R$ and $r < R$. ($r$ being the distance from the centre of the shell)**

**Ans. (a) Electric dipole moment :** Electric dipole moment is defined as the product of magnitude of either charges and the distance between the positive and negative charge of the electric dipole. The strength of any electric dipole is measured by the magnitude of its electric dipole moment.

q ⊕ ———— ⊖ q
←—— 2a ——→

Electric dipole moment, $p = q \times 2a$

It is a vector quantity.

In vector form it is written as $\vec{p} = q \times \vec{2a}$

where the direction of $\vec{a}$ is from negative charge to positive charge.

**Electric Field of dipole at points on the equatorial plane :**

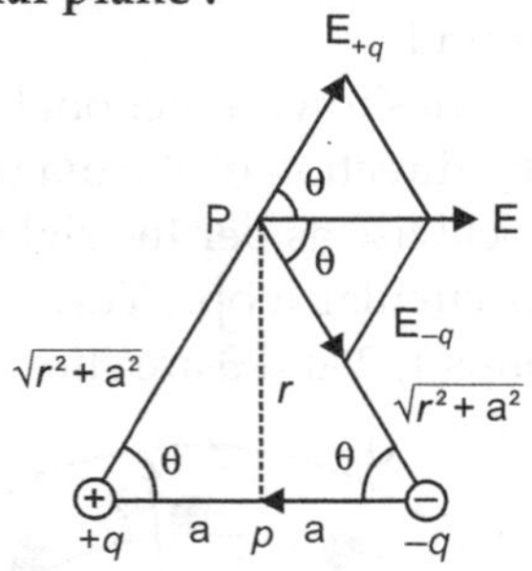


Let P be the point on the equatorial plane of the dipole where electric field is to be calculated. Let its distance from either charge be '$r$'. The magnitude of the electric field due to the two charges $+q$ and $-q$ at P is given by

$$E_{+q} = \frac{q}{4\pi\varepsilon_0} \frac{1}{r^2 + a^2} \quad ...(i)$$

$$E_{-q} = \frac{q}{4\pi\varepsilon_0} \frac{1}{r^2 + a^2} \quad ...(ii)$$

From equations (i) and (ii),

$E_{+q} = E_{-q} = E(\text{say})$

The direction of $E_{+q}$ and $E_{-q}$ are shown in figure.

Total electric field at P,

$$\vec{E}_p = \vec{E}_{+q} + \vec{E}_{-q}$$

Magnitude of E (using | | gm law of vector addition),

$$E_p = \sqrt{E_{+q}^2 + E_{-q}^2 + 2E_{+q}E_{-q}\cos 2\theta}$$

$$= \sqrt{E^2 + E^2 + 2E^2 \cos 2\theta}$$

$$= \sqrt{2E^2 + 2E^2 \cos 2\theta}$$

$$= \sqrt{2E^2 (1 + \cos 2\theta)}$$

$$= \sqrt{2E^2\, 2\cos^2\theta}$$

$$= \sqrt{4E^2 \cos^2\theta}$$

$$E_p = 2E\cos\theta$$

$$= 2 \times \frac{1}{4\pi\varepsilon_0} \frac{q}{(r^2 + a^2)} \times \frac{a}{(r^2 + a^2)^{1/2}}$$

$$E_p = \frac{2aq}{4\pi\varepsilon_0 (r^2 + a^2)^{3/2}}$$

$$E_p = \frac{p}{4\pi\varepsilon_0 (r^2 + a^2)^{3/2}} \quad [\because p = 2aq]$$

$$\vec{E}_p = \frac{\vec{p}}{4\pi\varepsilon_0 (r^2 + a^2)^{3/2}}$$

**(b) Equipotential surface due to electric dipole :**

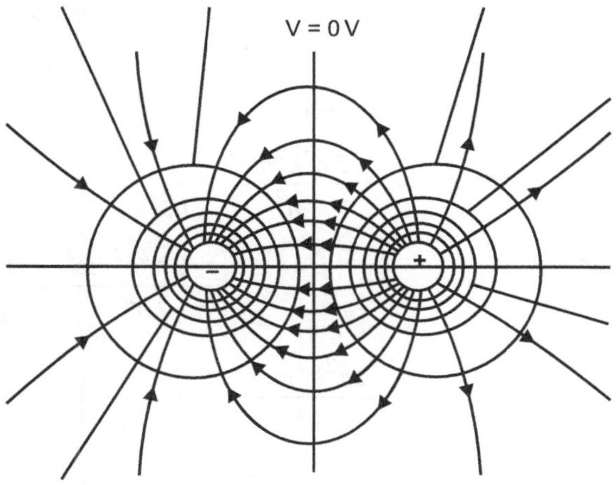


The potential due to the dipole is zero at the line bisecting the dipole length.

**OR**

Electric field due to a uniformly charged thin spherical shell :

**(i) When point P lies outside the spherical shell :** Consider a spherical shell of radius R and centre O. Let $q$ be the charge on the spherical shell. Suppose that we have to calculate electric field at the point P at a distance $r$ ($r$ > R) from its centre. Draw the Gaussian surface through point P so as to enclose the charged spherical shell. The Gaussian surface is a spherical shell of radius $r$ and centre O.

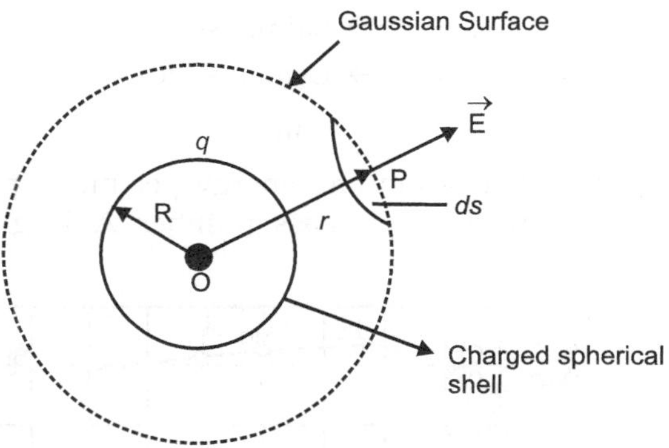


Let $\vec{E}$ be the electric field at point P. Then, the electric flux through area element $\vec{ds}$ is given by

$$d\phi = \vec{E} \cdot \vec{ds}$$

Since $\vec{ds}$ is also along normal to the surface,

$$\therefore \quad d\phi = E ds$$

$\therefore$ Total electric flux through the Gaussian surface is given by

$$\phi = \oint_s E.ds$$

$$= E\oint_s ds$$

Now,

$$\oint ds = 4\pi r^2$$

$$\therefore \quad \phi = E \times 4\pi r^2 \quad ...(i)$$

Since the charge enclosed by the Gaussian surface is $q$, according to Gauss theorem,

$$\phi = \frac{q}{\varepsilon_0} \quad ...(ii)$$

From equations (i) and (ii), we get,

$$E \times 4\pi r^2 = \frac{q}{\varepsilon_0}$$

$$E = \frac{1}{4\pi\varepsilon_0} \cdot \frac{q}{r^2} \quad (\text{for } r > R)$$

**(ii) When point P lies inside the spherical shell :** In such a case the Gaussian surface encloses no charge,

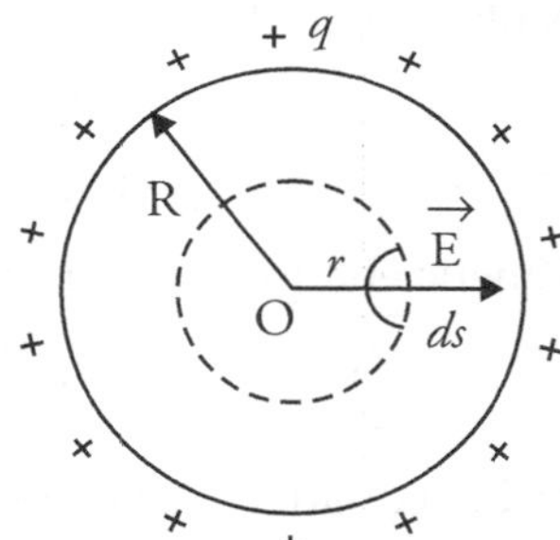


According to Gauss law,

$$E \times 4\pi r^2 = 0$$

*i.e.,* $\quad E = 0 \; (r < R)$

Graph showing the variation of electric field as a function of $r$.

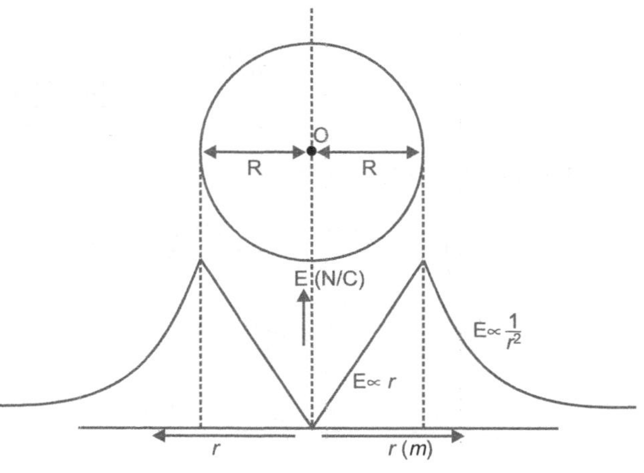

**29. Using Bohr's postulates, derive the expression for the frequency of radiation emitted when electron in hydrogen atom undergoes transition from higher energy state (quantum number $n_i$) to the lower state, ($n_f$).**

**When electron in hydrogen atom jumps from energy state $n_i$ = 4 to $n_f$ = 3, 2, 1, identify the spectral series to which the emission lines belong.** **[5]**

**OR**

**(a) Draw the plot of binding energy per nucleon (BE/A) as a function of mass number. Write two important conclusions that can be drawn regarding the nature of nuclear source.**

**(b) Use this graph to explain the release of energy in both the processes of nuclear fusion and fission.**

**(c) Write the basic nuclear process of neutron undergoing β-decay. Why is the detection of neutrinos found very difficult ?**

**Ans.** In the hydrogen atom,

Radius of electron orbit,

$$r = \frac{n^2h^2}{4\pi^2kme^2} \quad ...(i)$$

Kinetic energy of electron,

$$E_k = \frac{1}{2}mv^2 = \frac{ke^2}{2r}$$

Using equation (i), we get

$$E_k = \frac{ke^2}{2}\frac{4\pi^2kme^2}{n^2h^2} = \frac{2\pi^2k^2me^4}{n^2h^2}$$

Potential energy,

$$E_p = \frac{-k(e)\times(e)}{r} = \frac{-ke^2}{r}$$

Using equation (i), we get

$$E_p = -ke^2 \times \frac{4\pi^2k\,kme^2}{n^2h^2}$$

$$= -\frac{4\pi^2k^2me^4}{n^2\,h^2}$$

Total energy of electron,

$$E = \frac{2\pi^2k^2me^4}{n^2h^2} - \frac{4\pi^2k^2me^4}{n^2h^2}$$

$$= -\frac{2\pi^2k^2me^4}{n^2h^2}$$

$$= -\frac{2\pi^2k^2me^4}{h^2}\times\left(\frac{1}{n^2}\right)$$

Now, according to Bohr's frequency condition when electron in hydrogen atom undergoes transition from higher energy state to the lower energy state ($n_f$) is,

$$h\nu = E_{ni} - E_{nf}$$

or
$$h\nu = -\frac{2\pi^2k^2me^4}{h^2}\times\frac{1}{n_i^2} - \left(-\frac{2\pi^2k^2me^4}{h^2}\times\frac{1}{n_f^2}\right)$$

or
$$h\nu = \frac{2\pi^2k^2me^4}{h^2}\times\left(\frac{1}{n_f^2}-\frac{1}{n_i^2}\right)$$

$$\nu = \frac{2\pi^2k^2me^4}{h^3}\times\left(\frac{1}{n_f^2}-\frac{1}{n_i^2}\right)$$

$$\nu = \frac{c2\pi^2k^2me^4}{ch^3}\times\left(\frac{1}{n_f^2}-\frac{1}{n_i^2}\right)$$

$$\frac{2\pi^2k^2me^4}{ch^3} = R \text{ (Rydberg constant)}$$

$$R = 1.097\times10^7 m^{-1}$$

Thus,

$$\nu = Rc\times\left(\frac{1}{n_f^2}\ \frac{1}{n_i^2}\right)$$

Now, higher state,

$$n_i = 4$$

Lower state,

$$n_f = 3, 2, 1$$

For the transition

$n_i = 4$ to $n_f = 3$ → Paschen series

$n_i = 4$ to $n_f = 2$ → Balmer series

$n_i = 4$ to $n_f = 1$ → Lyman series

**OR**

**(a)** Plot of binding energy per nucleon as the function of mass number A is given as below :

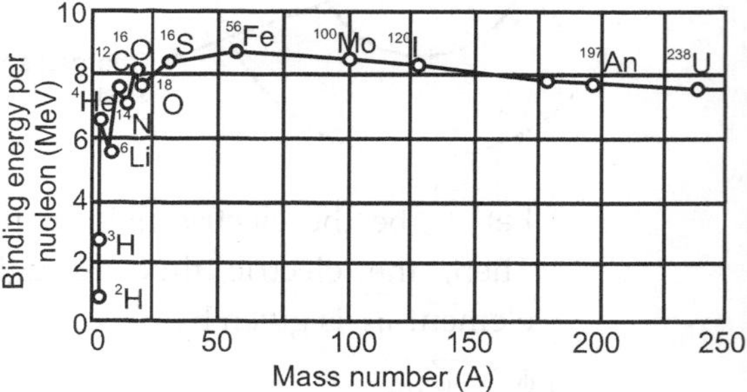


Following are the two conclusion that can be drawn regarding the nature of the nuclear force :

(i) The force is attractive and strong enough to produce a binding energy of few MeV per nucleon.

(ii) The constancy of the binding energy in the range 30 <A<170 is a consequence of the fact that the nuclear force is short range force.

(b) **Nuclear fission :** A very heavy nucleus (say, A = 240) has lower binding energy per nucleon as compared to the nucleus with A = 120. Thus, if the heavier nucleus breaks to the lighter nucleus with high binding energy per nucleon, nucleons are tightly bound.
This implies that energy will be released in the process which justifies the energy release in fission reaction.

**Nuclear fusion :** When two light nuclei (A<10) are combined to form a heavier nuclei, the binding energy of the fused heavier nuclei is more than the binding energy per nucleon of the lighter nuclei. Thus the final system is more tightly bound than the initial system. Again the energy will be released in fusion reaction.

(c) The basic nuclear process of neutron undergoing β-decay is given as :

$$n \rightarrow p + e^{-} + \overline{\nu}$$

Neutrinos are massless and chargeless particles. Neutrinos interact very weakly with matter that it becomes very difficult to detect them. That's why the detection of neutrinos is found very difficult.

●●

# Outside Delhi [Set-II]

> **Note : Except for the following questions, all the remaining questions have been asked in previous set.**

**4. Two charges of magnitudes –3Q and +2Q are located at points (*a*, 0) and (4*a*, 0) respectively. What is the electric flux due to these charges through a sphere of radius '5*a*' with its centre at the origin ? [1]**

**Ans.** According to Gauss' theorem, the electric flux through a closed surface enclosing a charge is equal to $\frac{1}{\varepsilon_0}$ times the magnitude of the charge enclosed.

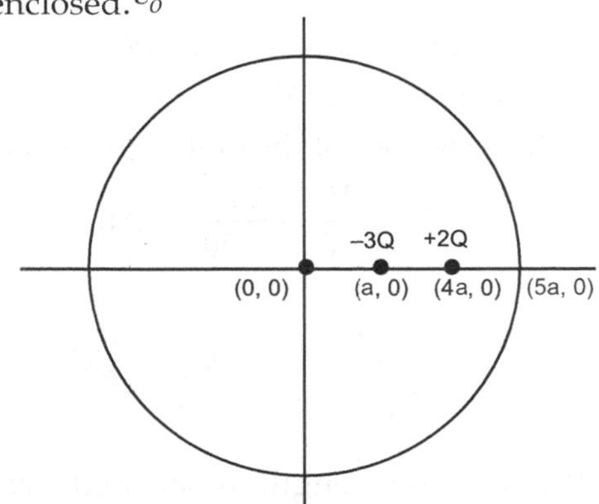


The sphere enclose charge = –3Q + 2Q = –Q

$\therefore$ Total electric flux, $\phi = \frac{Q}{\varepsilon_0}$

**7. A light metal disc on the top of an electromagnet is thrown up as the current is switched on. Why ? Give reason. [2]**

**Ans.** A light metal disc on the top of an electromagnet is thrown up as the current is switched on because when the current flows through the electromagnet, the magnetic flux through the disc increases which leads in setting up an eddy current in the disc in the same direction of the electromagnetic current. So the upper surface of electromagnet and the lower surface of the disc acquire same polarity. Since body with same polarity repels, therefore, the disc is thrown up.

**9. In the circuit shown in the figure, identify the equivalent gate of the circuit and make its truth table.** [2]**

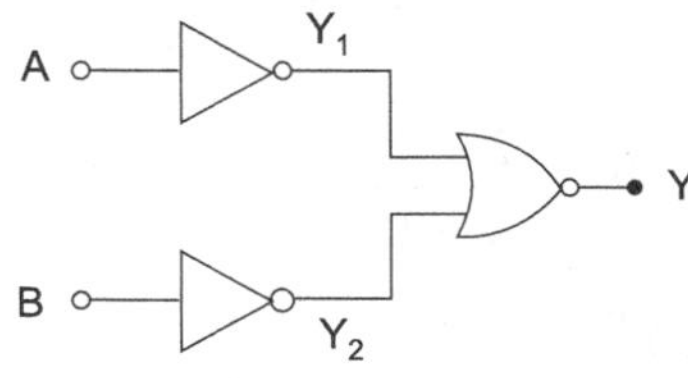


**13. A parallel beam of light of 600 nm falls on a narrow slit and the resulting diffraction pattern is observed on a screen 1.2 m away. It is observed that the first minimum is at a distance of 3 mm from the centre of the screen. Calculate the width of the slit. [3]**

**Ans.** The distance of the $n^{th}$ minimum from the centre of the screen is,

$$x_n = \frac{nD\lambda}{a}$$

where, D = distance of slit from screen
$\lambda$ = wavelength of the light
$a$ = width of the slit

** Answer is not given due to change in the present Syllabus.

For first minimum, $n = 1$

$$x_n = 3 \times 10^{-3} \text{ m}$$
$$\lambda = 600 \times 10^{-9} \text{ m}$$
$$D = 1.2 \text{ m}$$
$$3 \times 10^{-3} = \frac{1 \times 1.2 \times 600 \times 10^{-9}}{a}$$
$$a = \frac{1 \times 1.2 \times 600 \times 10^{-9}}{3 \times 10^{-3}}$$
$$= \frac{1.2 \times 6 \times 10^{-4}}{3}$$
$$= 2.4 \times 10^{-4} \text{ m}$$
$$= 0.24 \times 10^{-3} \text{ m}$$
$$a = 0.24 \text{ mm}$$

**19. A wire AB is carrying a steady current of 10 A and is lying on the table. Another wire CD carrying 6 A is held directly above AB at a height of 2 mm. Find the mass per unit length of the wire CD so that it remains suspended at its position when left free. Give the direction of the current flowing in CD with respect to that in AB. [Take the value of $g = 10 \text{ ms}^{-2}$] [3]**

**Ans.** Force per unit length between the current carrying wires is given as :

$$F = \frac{\mu_0}{4\pi} \frac{2I_1I_2}{r}$$

where,

The current in wire AB, $I_1 = 10$ A

The current in wire CD, $I_2 = 6$ A

The distance between wires, $r = 2 \text{ mm} = 2 \times 10^{-3}\text{m}$

Let $m$ be the mass per unit length of wire CD.

As this force balances the weight of the wire CD.

$$\therefore \quad \frac{\mu_0}{4\pi} \frac{2I_1I_2}{r} = mg$$

$$\Rightarrow \quad 10^{-7} \times \frac{2 \times 10 \times 6}{2 \times 10^{-3}} = m \times 10$$

$$\therefore m = 10^{-7} \times \frac{2 \times 10 \times 6}{2 \times 10^{-3}} \times \frac{1}{10} = 6 \times 10^{-4} \text{ kg/m}$$

The direction of current in CD w.r.t. AB is opposite (antiparallel) because in this condition both the wires will repel each other.

**23. Name the type of waves which are used for line of sight (LOS) communication. What is the range of their frequencies ?**

**A transmitting antenna at the top of a tower has a height of 45 m and the height of the receiving antenna is 80 m. Calculate the maximum distance between them for satisfactory communication in LOS mode. (Radius of the Earth = 6.4 × $10^6$ m)** [4]**

**24. (a) For a given a.c., $i = i_m \sin\omega t$, show that the average power dissipated in a resistor R over a complete cycle is $\frac{1}{2} i_m^2 R$R**

**(b) A light bulb is rated at 120 W for a 240 V a.c. supply. Calculate the resistance of the bulb. [5]**

**Ans.** (a) The average power dissipated

$$P_{av} = (i^2R) = (i_m^2 R \sin^2\omega t)$$
$$= i_m^2 R (\sin^2\omega t)$$

since,

$$\sin^2 \omega t = \frac{1}{2}(1 - \cos 2\omega t)$$
$$\therefore (\sin^2 \omega t) = \frac{1}{2}[1 - (\cos 2\omega t)] = \frac{1}{2}$$
$$[\because \cos 2\omega t = 0]$$
$$\therefore \quad P_{av} = \frac{1}{2} i_m^2 R$$

(b) Power of the bulb, P = 120 W

Voltage, V = 240 V

The resistance of the bulb is given as

$$R = \frac{V^2}{P} = \frac{240^2}{120} = 480 \ \Omega$$

## Outside Delhi [Set-III]

Note : Except for the following questions, all the remaining questions have been asked in previous set.

**1. Write the expression for the de-Broglie wavelength associated with a charged particle having charge '$q$' and mass '$m$', when it is accelerated by a potential V. [1]**

**Ans.** de-Broglie wavelength associated with the particle,

$$\lambda = \frac{h}{p} = \frac{h}{\sqrt{2mqV}}$$

**3. Two charges of magnitudes + 4Q and –Q are located at points ($a$, 0) and (–3$a$, 0) respectively. What is the electric flux due to these charges through a sphere of radius '2$a$' with its centre at the origin ? [1]**

** Answer is not given due to change in the present Syllabus.

**Ans.** Gauss's theorem states that the electric flux through a closed surface enclosing a charge is equal to $\frac{1}{\varepsilon_0}$ times the magnitude of the charge enclosed.

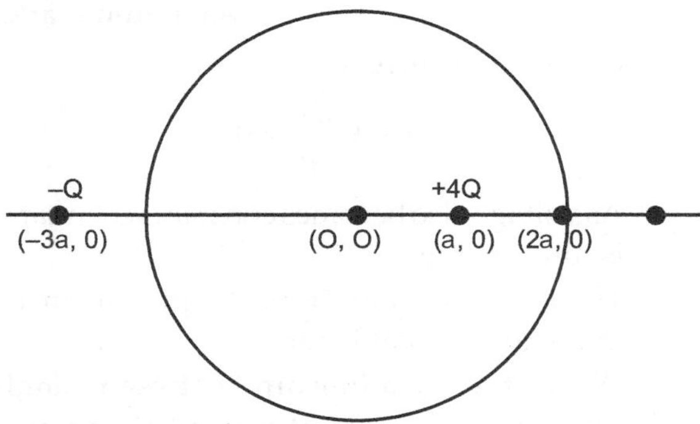


The sphere enclose charge = + 4Q

Thus, total electric flux $\phi = \frac{4Q}{\varepsilon_0}$

**9. In the circuit shown in the figure, identify the equivalent gate of the circuit and make its truth table.**** **[1]**

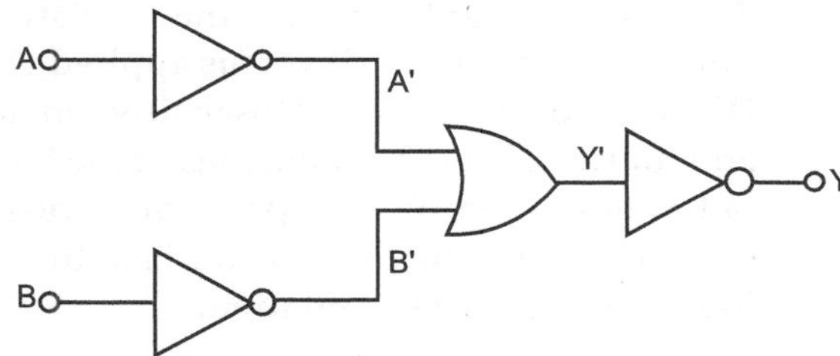


**11. A parallel beam of light of 450 nm falls on a narrow slit and the resulting diffraction pattern is observed on a screen 1.5 m away. It is observed that the first minimum is at a distance of 3 mm from the centre of the screen. Calculate the width of the slit.** **[2]**

**Ans.** The distance of the $n^{th}$ minimum from the centre of the screen,

$$x_n = \frac{nD\lambda}{a}$$

Where, D → distance of slit from screen.

λ → wavelength of the light.

$a$ → width of the slit

For first minimum, $n = 1$,

$$x_n = 3 \times 10^{-3} \text{ m}$$

$$D = 1.5 \text{ m}$$

$$\lambda = 450 \times 10^{-9} \text{ m}$$

$$3\times10^{-3} = \frac{1\times(1.5)\times(450\times10^{-9})}{a} \frac{nD\lambda}{a}$$

$$a = \frac{1\times(1.5)\times(450\times10^{-9})}{3\times10^{-3}}$$

$$a = 0.255\text{mm}$$

**17. A wire AB is carrying a steady current of 6 A and is lying on the table. Another wire CD carrying 4 A is held directly above AB at a height of 1 mm. Find the mass per unit length of the wire CD so that it remains suspended at its position when left free. Give the direction of the current flowing in CD with respect to that in AB. [Take the value of $g = 10 \text{ ms}^{-2}$]** **[3]**

**Ans.** Force per unit length between the current carrying wire is given as :

$$F = \frac{\mu_0}{4\pi}\frac{2I_1I_2}{r}$$

Let $m$ be the mass per unit length of wire CD. As the force balance the weight of the wire

$$\therefore \quad \frac{\mu_0}{4\pi}\frac{2I_1I_2}{r} = mg$$

Here $m$ → Mass per unit length

$$\frac{10^{-7}\times2\times6\times4}{1\times10^{-3}} = m \times 10$$

$$m = 4.8 \times 10^{-4} \text{ kg m}^{-1}$$

The current in CD is in opposite direction to that in AB than they will repel each other and CD wire remain suspended in air.

**18. (b) A light bulb is rated at 125 W for a 250 V a.c. supply. Calculate the resistance of the bulb.** **[3]**

**Ans. (b)** Power of the bulb, P = 125 W

Voltage, V = 250 V

The resistance of the bulb is given as

$$R = \frac{V^2}{P} = \frac{(250)^2}{125}$$

$$= 500\ \Omega$$

**25. Name the type of waves which are used for line of sight (LOS) communication. What is the range of their frequencies ?**

**A transmitting antenna at the top of a power has a height of 45 m and the receiving antenna is on the ground. Calculate the maximum distance between them for satisfactory communication in LOS mode. (Radius of the Earth = $6.4 \times 10^6$m)**** **[5]**

** Answer is not given due to change in the present Syllabus.

# PHYSICS

## Delhi [Set-I]

# 2013

**Time allowed : 3 hours** **Maximum marks : 70**

1. **What are permanent magnets ? Give one example.** [1]

**Ans.** The magnets which have high retentivity and high coercivity are known as permanent magnets.
For example : Steel, cobalt

2. **What is the geometrical shape of equipotential surface due to a single isolated charge ?** [1]

**Ans.** The equipotential surfaces of an isolated charge are concentric spherical shells and the distance between the shells increase with the decrease in electric field and vice-versa.

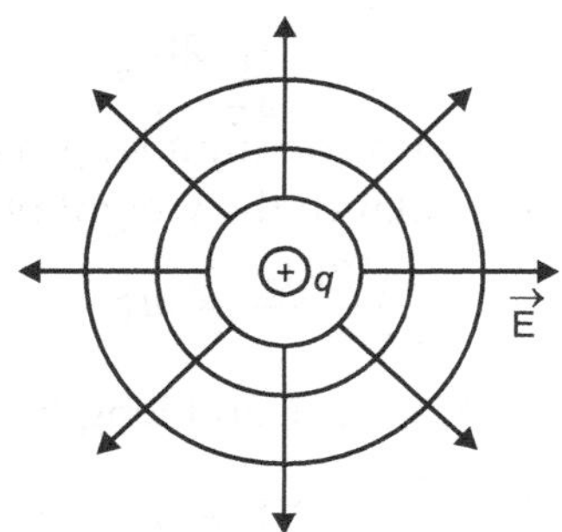


3. **Which of the following waves can be polarized (i) Heat waves (ii) Sound waves ? Give reason to support your answer.** [1]

**Ans.** Heat waves can be polarized as they are transverse waves whereas sound waves cannot be polarized as they are longitudinal waves.
Transverse waves can oscillate in the direction perpendicular to the direction of its transmission but longitudinal waves oscillate only along the direction of its transmission. So, longitudinal waves cannot be polarized.

4. **A capacitor has been charged by a dc source. What are the magnitude of conduction and displacement current, when it is fully charged ?** [1]

**Ans.** Electric flux through plates of capacitor,

$$\phi = \frac{q}{\varepsilon_0}$$

Here charge $q$ is constant as the capacitor is fully charged.
Displacement current,

$$I_D = \varepsilon_0 \frac{d\phi}{dt} = \varepsilon_0 \frac{d\left(\frac{q}{\varepsilon_0}\right)}{dt} = 0$$

[since $q$ is constant]

Conduction current,

$$I = C\frac{dV}{dt} = 0$$

As voltage also becomes constant when capacitor is fully charged.
Therefore, magnitude of di splacement to and induction current is zero.

5. **Write the relationship between angle of incidence '$i$', angle of prism 'A' and angle of minimum deviations $\delta_m$ for a triangular prism.** [1]

**Ans.** The relation between the angle of incidence $i$, angle of prism A, and the angle of minimum deviation $\delta_m$, for a triangular prism is given by

$$i = \frac{A + \delta_m}{2}$$

6. **The given graph shows the variation of photoelectric current (I) versus applied voltage (V) for two different photosensitive materials and for two different intensities of the incident radiations. Identify the pairs of curves that correspond to different materials but same intensity of incident radiation.** [2]

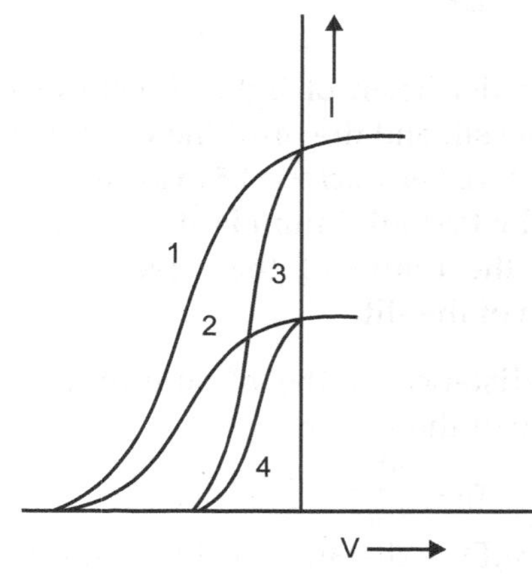


**Ans.** Curves 1 and 2 correspond to similar materials while curves 3 and 4 represent different materials, since the value of stopping potential for 1, 2 and 3, 4 are the same. For the given frequency of the incident radiation, the stopping potential is independent of its intensity.
So, the pairs of curves (1 and 3) and (2 and 4) correspond to different materials but same intensity of incident radiation.

7. **A 10 V battery of negligible internal resistance is connected across a 200 V battery and a resistance of 38 Ω as shown in the figure. Find the value of the current in circuit.** **[2]**

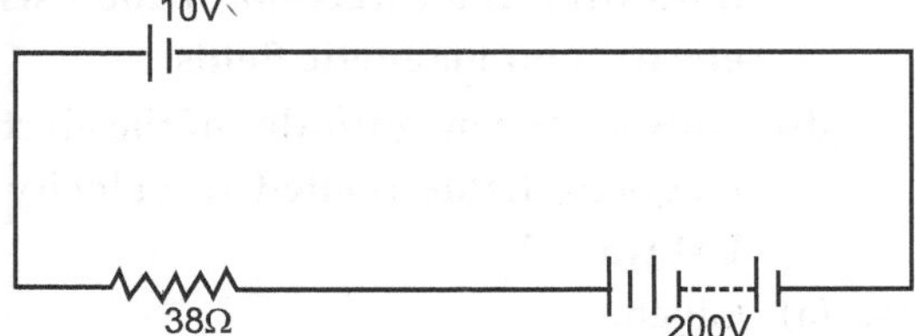


**Ans.** Since, the positive terminal of the batteries are connected together (So they oppose each other), so the equivalent emf of the batteries is given by

$$E = 200 - 10 = 190 \text{ V}$$

Hence, the current in the circuit is given by

$$I = \frac{E}{R} = \frac{190}{38} = 5A$$

8. **The emf of a cell is always greater than its terminal voltage. Why ? Give reason.** **[2]**

**Ans.** The emf of a cell is greater than its terminal voltage because there is some potential drop across the cell due to its small internal resistance.

9. **(a) Write the necessary conditions for the phenomenon of total internal reflection to occur.**
   **(b) Write the relation between the refractive index and critical angle for a given pair of optical media.** **[2]**

**Ans.** (a) Necessary conditions for total internal reflection to occur are :

1. The incident ray on the interface should travel in optically denser medium.
2. The angle of incidence should be greater than the critical angle for the given pair of optical media.

(b) $${}_a\mu_b = \frac{1}{\sin C}$$

Where *a* and *b* are the rarer and denser media respectively. C is the critical angle for the given pair of optical media.

10. **State Lenz's law.**
    **A metallic rod held horizontally along east-west direction, is allowed to fall under gravity. Will there be an emf induced at its ends ? Justify your answer.** **[2]**

**Ans.** Lenz's law states that the polarity of induced emf is such that it produces a current which opposes the change in magnetic flux that produces it.

Yes, emf will be induced in the rod as there is change in magnetic flux.

When a metallic rod held horizontally along east-west direction, is allowed to fall freely under gravity *i.e.*, fall from north to south, the magnetic flux changes due to vertical component of Earth's magnetic field, which keeps on changing and the emf is induced in it.

11. **A convex lens of focal length 25 m is placed coaxially in contact with a concave lens of focal length 20 cm. Determine the power of the combination. Will the system be converging or diverging in nature ?** **[2]**

**Ans.** We have, focal length of convex lens,

$$f_1 = +25 \text{ cm} = +0.25 \text{ m}$$

Focal length of the concave lens,

$$f_2 = -20 \text{ cm} = -0.20 \text{ m}$$

Equivalent focal length,

$$\frac{1}{F} = \frac{1}{f_1} + \frac{1}{f_2} = \frac{1}{25} + \frac{1}{-20} = -\frac{1}{100}$$

$$\therefore \quad F = -100 \text{ cm}$$

Power of convex lens,

$$P_1 = \frac{1}{f_1} = \frac{1}{0.25}$$

Power of concave lens,

$$P_2 = \frac{1}{f_2} = \frac{1}{-0.20}$$

Power of the combination,

$$P = P_1 + P_2$$

$$P = \frac{1}{0.25} + \frac{1}{-0.20} = \frac{100}{25} - \frac{100}{20} = \frac{400 - 500}{100}$$

$$= -\frac{100}{100} = -1$$

$$\therefore \quad P = -1 \text{ D}$$

The focal length of the combination

$$= -1\text{m} = -100 \text{ cm.}$$

As the focal length is negative, the system will be diverging in nature.

12. **An ammeter of resistance 0.80 Ω can measure current up to 1.0 A.** **[3]**
    **(i) What must be the value of shunt resistance to enable the ammeter to measure current up to 5.0 A ?**
    **(ii) What is the combined resistance of the ammeter and the shunt ?**

**Ans.** We have, resistance of ammeter, $R_A = 0.80\ \Omega$

Maximum current through ammeter, $I_A = 1.0$ A

So, voltage across ammeter, $V = IR = 1 \times 0.80 = 0.8V$

Let the value of shunt be x.

(i) Resistance of ammeter with shunt,

$$R = \frac{R_A x}{R_A + x} = \frac{0.8x}{0.8 + x}$$

Current through ammeter, I = 5 A

$$\therefore \left(\frac{0.8x}{0.8+x}\right) \times 5 = 0.8$$

$$\Rightarrow \quad 0.8x \times 5 = 0.8(0.8 + x)$$

$$4x = 0.64 + 0.8x$$

$$x = \frac{0.64}{3.2}$$

$$x = 0.2$$

Thus, the shunt resistance is 0.2 Ω

(ii) Combined resistance of the ammeter and the shunt,

$$R = \frac{0.8x}{0.8 + x} = \frac{0.8 \times 0.2}{0.8 + 0.2} = \frac{0.16}{1} = 0.16\ \Omega$$

**13. In the given circuit diagram a voltmeter 'V' is connected across a lamp 'L'. How would (i) the brightness of the lamp and (ii) voltmeter reading 'V' be affected, if the value of resistance 'R' is decreased ? Justify your answer. [3]**

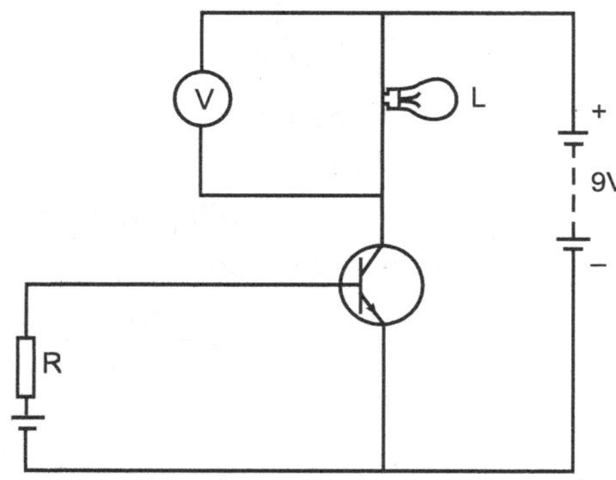


**Ans.** The given figure is Common Emitter (CE) configuration of an *n-p-n* transistor. The input circuit is forward biased and collector circuit is reverse biased.

If resistance R decreases, forward biased in the input circuit will increase, thus the base current ($I_B$) will decrease and the emitter current ($I_E$) will increase. This will increase the collector current ($I_C$) as $I_E = I_B + I_C$.

When $I_C$ increases which flows through the lamp, the voltage across the bulb will also increase making the lamp brighter and since the voltmeter is connected in parallel with the lamp, the reading in the voltmeter will also increase.

**14. (a) An EM wave is travelling in a medium with a velocity $v = v\hat{i}$. Draw a sketch showing the propagation of the EM wave, indicating the direction of the oscillating electric and magnetic fields.**

**(b) How are the magnitudes of the electric and magnetic fields related to velocity of the EM wave ? [3]**

**Ans. (a)** Given.

Velocity, $v = v\hat{i}$ and electric field, E along Y-axis and magnetic field, B along Z-axis.

The propagation of EM wave is shown below :

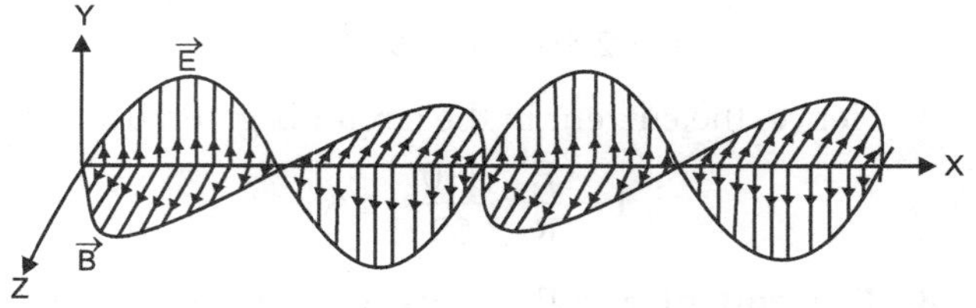


**(b)** Speed of EM wave can be given as the ratio of magnitude of electric field ($E_0$) to the magnitude of magnetic field ($B_0$),

$$c = \frac{E_0}{B_0}$$

**15. Block diagram of a receiver is shown in the figure :**

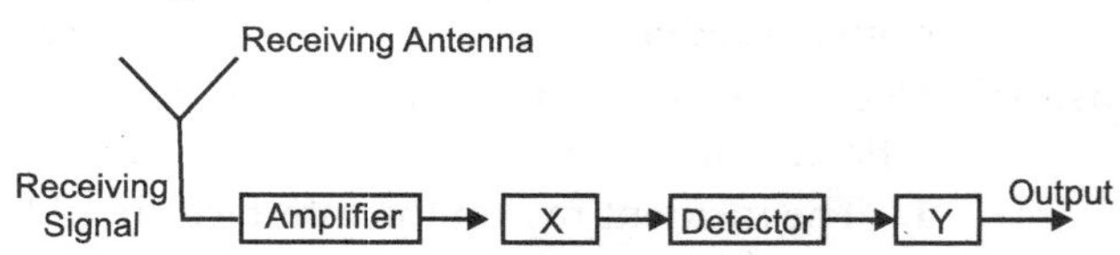


**(a) Identify 'X' and 'Y'.**

**(b) Write their functions.****

**16. Explain, with the help of a circuit diagram, the working of a photodiode. Write briefly how it is used to detect the optical signals. [3]**

**OR**

**Mention the important consideration required while fabricating a *p-n* junction diode to be used as a Light Emitting Diode (LED). What should be the order of band gap of an LED if it is required to emit light in the visible range ?**

**Ans.** A junction diode made from light sensitive semiconductor is called a photodiode.

An electrical device that is used to detect and convert light into an energy signal with the use of a photo detector is known as a photodiode. The light that falls on it controls the function of

** Answer is not given due to change in the present Syllabus.

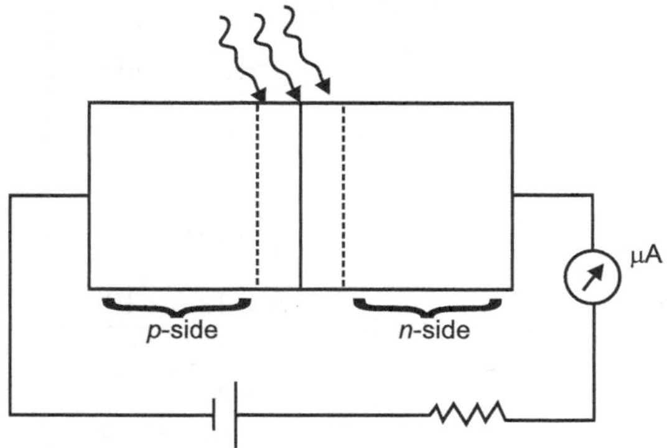


pn-junction. Suppose, the wavelength is such that the energy of a photon $\frac{hc}{\lambda}$ is enough to break a valence bond. There is an increase in number of charge carriers and hence the conductivity of the junction also increases. New hole-electron pairs are created when such light falls on the junction. If the junction is connected in a circuit, the intensity of the incident light controls the current in the circuit.

OR

(i) The reverse breakdown voltage of LEDs are very low, which is around 5V. So enough care is to be taken while fabricating a *pn*-junction diode such that the high reverse voltages do not occur across them.

(ii) There exist very small resistance to limit the current in LED. So, a resistor must be placed in series with the LED such that no damage is occurred to the LED.

The semiconductor used for fabrication of visible LEDs must at least have a band gap of 1.8 eV.

**17. Write three important factors which justify the need of modulating a message signal. Show diagrammatically how an amplitude modulated wave is obtained when a modulating signal is superimposed on a carrier wave. [3]**

**18. A capacitor of unknown capacitance is connected across a battery of V volts. The charge stored in it is 360 μC. When potential across the capacitor is reduced by 120 V, the charge stored in it becomes 120 μC.**

**Calculate :**

**(i) The potential V and the unknown capacitance C.**

**(ii) What will be the charge stored in the capacitor, if the voltage applied had increased by 120 V ? [3]**

**OR**

**A hollow cylindrical box of length 1 m and area of cross-section 25 cm² is placed in a three-dimensional coordinate system as shown in the figure. The electric field in the region is given by $\vec{E} = 50x\hat{i}$ where E is NC$^{-1}$ and $x$ is in metres. Find :**

**(i) Net flux through the cylinder.**

**(ii) Charge enclosed by the cylinder.**

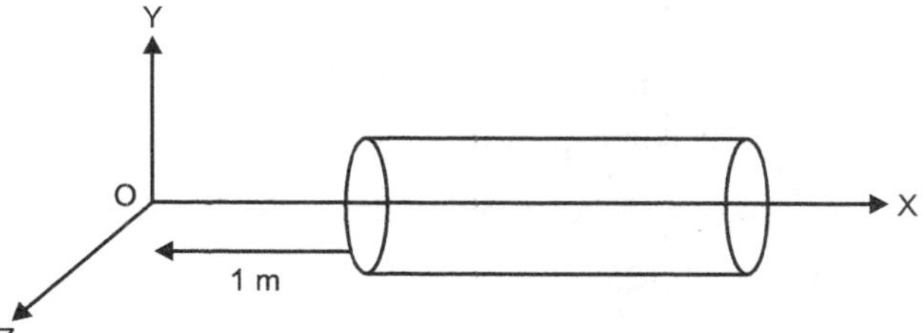


**Ans.** (i) Initial voltage, $V_1$ = V volts and charge stored,

$$Q_1 = 360 \mu C$$

$$Q_1 = CV_1 \quad ...(i)$$

Changed potential,

$$V_2 = V - 120$$

$$Q_2 = 120 \mu C$$

$$Q_2 = CV_2 \quad ...(ii)$$

By dividing (ii) from (i), we get

$$\frac{Q_1}{Q_2} = \frac{CV_1}{CV_2}$$

$$\Rightarrow \quad \frac{360}{120} = \frac{V}{V-120}$$

$$V = 180 \text{ volts}$$

$$\therefore \quad C = \frac{Q_1}{V_1} = \frac{360 \times 10^{-6}}{180}$$

$$= 2 \times 10^{-6} \text{ F} = 2 \mu F$$

(ii) If the voltage applied had increased by 120 V, then

$$V_3 = 180 + 120 = 300 \text{ V.}$$

Hence, charge stored in the capacitor,

$$Q_3 = CV_3 = 2 \times 10^{-6} \times 300 = 600 \mu C$$

OR

(i) Given,

$$\vec{E} = 50xi, \text{ and } \Delta s = 25 \text{ cm}^2 = 25 \times 10^{-4} \text{ m}^2$$

** Answer is not given due to change in the present Syllabus.

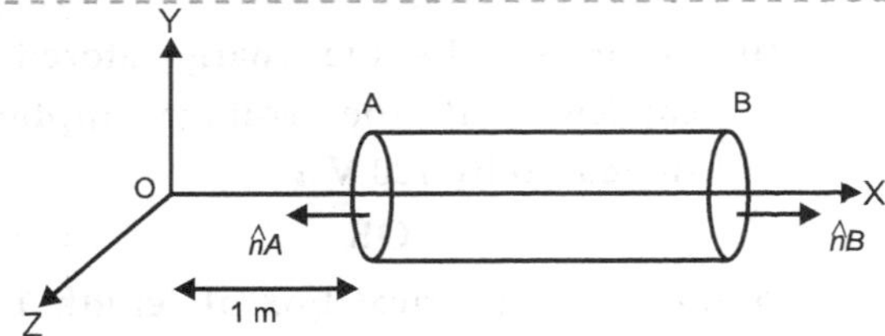


As the electric field is only along the X-axis, so, flux will pass only through the cross-section of cylinder.

Magnitude of electric field at cross-section A,

$$E_A = 50 \times 1 = 50 \text{ NC}^{-1}$$

Magnitude of electric field at cross-section B,

$$E_B = 50 \times 2 = 100 \text{ NC}^{-1}$$

The corresponding electric fluxes are :

$\phi_A = \vec{E}\cdot\vec{\Delta s} = 50 \times 25 \times 10^{-4} \times \cos 180°$

$= -0.125 \text{ Nm}^2/\text{C}$

$\phi_B = \vec{E}\cdot\vec{\Delta s} = 100 \times 25 \times 10^{-4} \times \cos 0°$

$= 0.25 \text{ Nm}^2/\text{C}$

So, the net flux through the cylinder,

$$\phi = \phi_A + \phi_B = -0.125 + 0.25 = 1.25\text{Nm}^2/\text{C}$$

**(ii)** Using Gauss's Law :

$$\oint \vec{E}.\vec{dS} = \frac{q}{\varepsilon_0}$$

$$\Rightarrow \quad 0.125 = \frac{q}{8.85 \times 10^{-12}}$$

$$q = 8.85 \times 0.125 \times ^{-12} = 1.1 \times 10^{-12} \text{ C}$$

**19. (a) In a typical nuclear reaction, *e.g***

$${}^2_1\text{H} + {}^2_1\text{H} \rightarrow {}^3_2\text{He} + {}^1_0n + 3.27 \text{ MeV},$$

**although number of nucleons is conserved, yet energy is released. How ? Explain.**

**(b) Show that nuclear density in a given nucleus is independent of mass number A.** **[3]**

**Ans. (a)** In a nuclear reaction, the aggregate of the masses of the target nucleus (${}^2_1$H) and the bombarding particle may be greater or less than the aggregate of the masses of the product nucleus (${}^3_2$He) and the outgoing particle (${}^1_0n$). So from the law of conservation of mass-energy some energy (3.27 MeV) is evolved or involved in a nuclear reaction. This energy is called Q-value of the nuclear reaction.

**(b)** Density of the nucleus $= \dfrac{\text{Mass of nucleus}}{\text{Volume of nucleus}}$

Mass of the nucleus = A amu

$= \text{A} \times 1.666 \times 10^{-27}$ kg

Volume of the nucleus

$$V = \frac{4}{3}\pi R^3 = \frac{4}{3}\pi\left(R_0 A^{1/3}\right)^3$$

Where, $R = R_0 A^{1/3}$

$$V = \frac{4}{3}\pi R_0^3 A$$

Thus, density $= \dfrac{A \times 1.66 \times 10^{-27}}{\left(\frac{4}{3}\pi R_0^3\right)A}$

$$= \frac{1.66 \times 10^{-27}}{\left(\frac{4}{3}\pi R_0^3\right)} \quad [\because f = \frac{M}{V}]$$

which shows that the density is independent of mass number A.

**20. (a) Why photoelectric effect cannot be explained on the basis of wave nature of light ? Give reasons.**

**(b) Write the basic features of photon picture of electro-magnetic radiation on which Einstein's photoelectric equation is based.** **[3]**

**Ans. (a)** Wave nature of radiation cannot explain the following :

1. The immediate ejection of photoelectrons.
2. The presence of threshold frequency for a metal surface.
3. The fact that kinetic energy of the emitted electrons is independent of the intensity of light and depends upon its frequency.

Thus, the photoelectric effect cannot be explained on the basis of wave nature of light.

**(b)** Photon picture of electromagnetic radiation on which Einstein's photoelectric equation is based on particle nature of light.

Its basic features are :

1. In interaction with matter, radiation behaves as if it is made up of particles called photons.
2. Each photon has energy ($E = h\nu$), momentum $\left(p = \dfrac{h\nu}{c}\right)$, where $c$ is the speed of light
3. All photons of light of a particular frequency $\nu$, or wavelength $\lambda$, have the same energy $\left(E = h\nu = \dfrac{hc}{\lambda}\right)$ and momentum $\left(p = \dfrac{h\nu}{c}\right)$

4. By increasing the intensity of light of given wavelength, there is only an increase in the number of photons emitted per second crossing a given area, with each photon having the same energy. Thus, photon energy is independent of intensity of radiation.
5. Photons are electrically neutral and are not deflected by electric and magnetic fields.
6. In a photon–particle collision (such as photon-electron collision), the total energy and total momentum are conserved. However, number of photons may not be observed.

**21. A metallic rod of length '$l$' is rotated with a frequency $\nu$ with one end hinged at the centre and the other end at the circumference of a circular metallic ring of radius $r$, about an axis passing through the centre and perpendicular to the plane of the ring. A constant uniform magnetic field B parallel to the axis is present everywhere. Using Lorentz force, explain how emf is induced between the centre and the metallic ring and hence obtained the expression for it. [3]**

**Ans.** (a) Suppose the length of the rod is greater than the radius of the circle and rod rotates anticlockwise and suppose the direction of electrons in the rod at any instant be along $+y$ direction.

Suppose the direction of the magnetic field is along $+z$ direction.

Then, using Lorentz law, we get the following :

$$\vec{F} = e(\vec{v} \times \vec{B})$$
$$\vec{F} = -(n\hat{j} \times B\hat{k})$$
$$\vec{F} = -eB\hat{i}$$

Thus, the direction of force on the electrons is along $-x$ axis.

So, the electrons will move towards the centre *i.e.*, the fixed end of the rod. This movement of electrons will effect in current and thus it will generate an emf in the rod between the fixed end and the point touching the ring.

Let $\theta$ be the angle between the rod and radius of the circle at any time $t$.

Then, area swept by the rod inside the circle

$$= \frac{1}{2}\pi r^2\theta$$

Induced emf

$$= B \times \frac{d}{dt}\left(\frac{1}{2}\pi r^2\theta\right) = \frac{1}{2}\pi r^2 B\frac{d\theta}{dt} = \frac{1}{2}\pi r^2 B\omega$$
$$= \frac{1}{2}\pi r^2 B(2\pi v) = \pi^2 r^2 Bv$$

**22. Output characteristics of an $n$-$p$-$n$ transistor in CE configuration is shown in the figure. [3]**

**Determine :**

**(i) dynamic output resistance**

**(ii) dc current gain and**

**(iii) ac current gain at an operating point $V_{CE}$ = 10 V, when $I_B$ = 30 μA.****

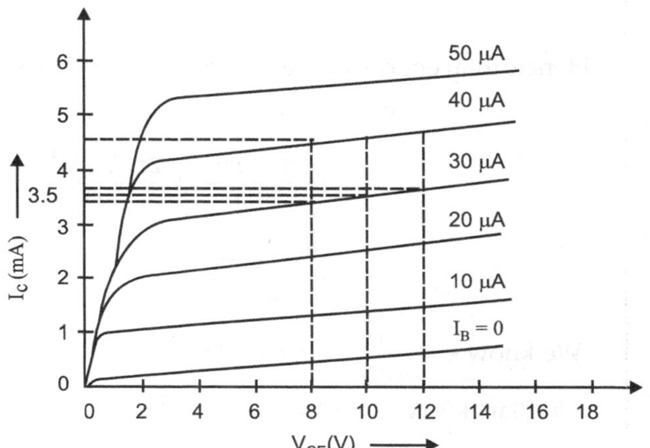


**23. Using Bohr's postulates, obtain the expression for the total energy of the electron in the stationary states of the hydrogen atom. Hence draw the energy level diagram showing how the line spectra corresponding to Balmer series occur due to transition between energy levels. [4]**

**Ans.** In a hydrogen atom,

Radius of electron orbit,

$$r = \frac{n^2h^2}{4\pi^2 k\, me^2} \quad ...(i)$$

and, angular momentum

$$mvr = \frac{nh}{2\pi}$$

or

$$v = \frac{nh}{2\pi\, mr}$$

$$v = \frac{nh}{2\pi m\left(\frac{n^2h^2}{4\pi^2\, k\, ne^2}\right)}$$

$$= \frac{2\pi ke^2}{nh}$$

** Answer is not given due to change in the present Syllabus.

So, kinetic energy,

$$E_k = \frac{1}{2}mv^2 = \frac{1}{2}m\left(\frac{2\pi ke^2}{nh}\right)^2$$

$$= \frac{4\pi^2k^2e^4m}{2n^2h^2} = \frac{2\pi^2k^2e^4m}{n^2h^2}$$

Potential energy

$$E_p = \frac{-k(e)\times(e)}{r} = -\frac{ke^2}{r}$$

Using equation (i), we get

$$E_p = -ke^2 \times \frac{4\pi^2kme^2}{n^2h^2}$$

$$= -\frac{4\pi^2k^2me^4}{n^2h^2}$$

Hence, total energy of the electron in the $n^{th}$ orbit

$$E = E_p + E_k$$

$$= -\frac{4\pi^2k^2me^4}{n^2h^2} + \frac{2\pi^2k^2me^4}{n^2h^2}$$

$$= -\frac{2\pi^2k^2me^4}{n^2h^2}$$

We know $k = \frac{1}{4\pi\varepsilon_0} = 9 \times 10^9 \text{ Nm}^2\text{C}^{-2}$

$h$ (Planck's constant) $= 6.6 \times 10^{-34}$ Js

$m$ for H-atom $= 1.67 \times 10^{-27}$ kg

$e = 1.6 \times 10^{-19}$ C

Substituting these values, we get

$$E = \frac{-13.6}{n^2} \text{ eV}$$

When the electron in a hydrogen atom jumps from higher energy level to the lower energy level, the difference of energies of the two energy levels is emitted as a radiation of particular wavelength. It is called a spectral line.

In H-atom, when an electron jumps from the orbit $n_i$ to orbit $n_f$, the wavelength of the emitted radiation is given by,

$$\frac{1}{\lambda} = R\left(\frac{1}{n_f^2} - \frac{1}{n_i^2}\right)$$

Where,

$R \rightarrow$ Rydberg's constant $= 1.09678 \times 10^7 \text{ m}^{-1}$

For Balmer series, $n_f = 2$ and $n_i = 3, 4, 5, ....$

$$\frac{1}{\lambda} = R\left(\frac{1}{2^2} - \frac{1}{n_i^2}\right)$$

Where, $n_i = 3, 4, 5, ...$

These spectral lines lie in the visible region.

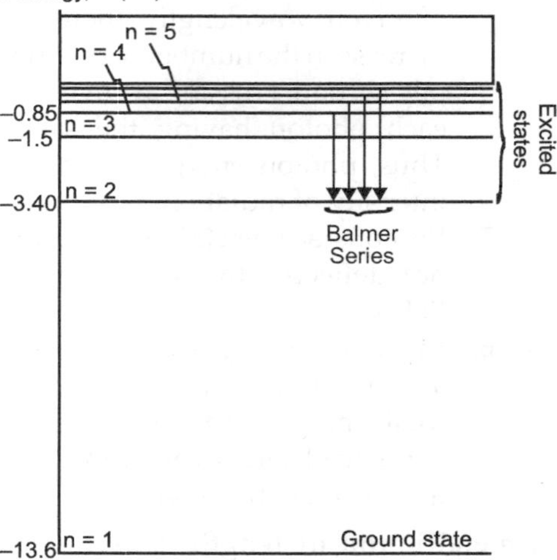


**24. (a) In what way is diffraction from each slit related to the interference pattern in a double slit experiment.**

**(b) Two wavelengths of sodium light 590 nm and 596 nm are used, in turn to study the diffraction taking place at a single slit of aperture 2 × 10⁻⁴ m. The distance between the slit and the screen is 1.5 m. Calculate the separation between the positions of the first maxima of the diffraction pattern obtained in the two cases. [5]**

**Ans. (a)** If the width of each slit is comparable to the wavelength of light used, the interference pattern thus obtained in the double-slit experiment is modified by diffraction from each of the two slits.

**(b)** Given that : Wavelength of the light beam,

$\lambda_1 = 590 \text{ nm} = 5.9 \times 10^{-7}\text{m}$

Wavelength of another light beam,

$\lambda_2 = 596 \text{ nm} = 5.96 \times 10^{-7}\text{m}$

Distance of the slits from the screen = D = 1.5 m

Distance between the two slits, $a = 2 \times 10^{-4}$m

For the first secondary maxima, $\sin\theta = \frac{3\lambda_1}{2a} = \frac{x_1}{D}$

$$x_1 = \frac{3\lambda_1 D}{2a} \text{ and } x_2 = \frac{3\lambda_2 D}{2a}$$

$$x_1 = \frac{3\times590\times10^{-9}\times15}{2\times2\times10^{-4}},$$

$$x_2 = \frac{3\times596\times10^{-9}\times1.5}{2\times2\times10^{-4}}$$

$x_1 = 663.75 \times 10^{-5}$, $x_2 = 670.50 \times 10^{-5}$

∴ Spacing between the positions of first secondary maxima of two sodium lines

$x_2 - x_1 = 6.75 \times 10^{-5}$ m = 0.0675 mm.

**25. In a series LCR circuit connected to an ac source of variable frequency and voltage $v = v_m \sin \omega t$, draw a plot showing the variation of current (I) with angular frequency ($\omega$) for two different values of resistance $R_1$ and $R_2$ ($R_1 > R_2$). Write the condition under which the phenomenon of resonance occurs. For which value of the resistance out of the two curves, a sharper resonance is produced ? Define Q-factor of the circuit and give its significance. [5]**

**Ans.** Figure shows the variation of $i_m$ with $\omega$ in a LCR series circuit for two values of Resistance $R_1$ and $R_2$ ($R_1 > R_2$),

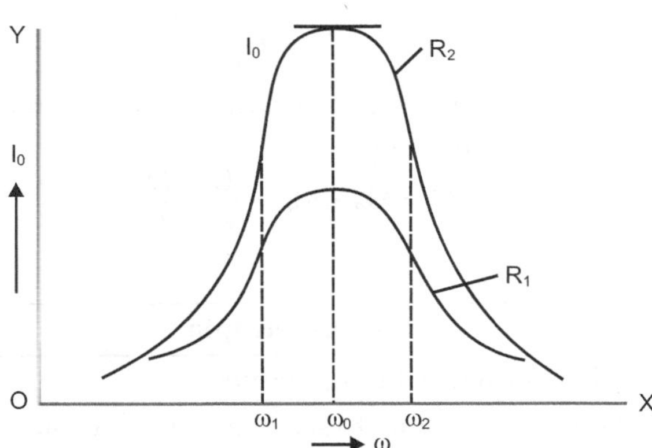


The condition for resonance in the LCR circuit is,

$$\omega_0 = \frac{1}{\sqrt{LC}}$$

We can observe that the current amplitude is maximum at the resonant frequency $\omega_0$. Since $i_m = V_m/R$ at resonance, the current amplitude for case $R_2$ is sharper to that for case $R_1$.

Quality factor or simply the Q-factor of a resonant LCR circuit is defined as the ratio of voltage drop across the capacitor (or inductor) to that of applied voltage.

It is given by $Q = \frac{1}{R}\sqrt{\frac{L}{C}}$

The Q factor determines the sharpness of the resonance curve and if the resonance is less sharp, the maximum current decreases and also the circuit is close to the resonance for a larger range $\Delta\omega$ of frequencies and the regulation of the circuit will not be good. So, less sharp the resonance, less is the selectivity of the circuit while higher is the Q, sharper is the resonance curve and lesser will be the loss in energy of the circuit and circuit will be more selective.

**26. While travelling back to his residence in the car, Dr. Pathak was caught up in a thunderstorm. It became very dark. He stopped driving the car and waited for thunderstorm to stop. Suddenly he noticed a child walking alone on the road. He asked the boy to come inside the car till the thunderstorm stopped. Dr. Pathak dropped the boy at his residence. The boy insisted that Dr. Pathak should meet his parents. The parents expressed their gratitude to Dr. Pathak for his concern for safety of the child.**

**Answer the following questions based on the above information : [5]**

**(a) Why is it safer to sit inside a car during a thunderstorm ?**

**(b) Give an example of similar action on your part in the part from everyday life.**

**Ans.** (a) It is safer to be inside a car during thunderstorm because the car acts like a Faraday cage.

(b) Once I came across to a situation where a puppy was struck in the middle of a busy road during rain and was not able to cross due to heavy flow, So I quickly rushed and helped him.

**27. (a) Draw a ray diagram showing the image formation by a compound microscope. Hence obtain expression for total magnification when the image is formed at infinity.**

**(b) Distinguish between myopia and hypermetropia. Show diagrammatically how these defects can be corrected. [5]**

**OR**

**(a) State Huygen's principle. Using this principle draw a diagram to show how a plane wave front incident at the interface of the two media gets refracted when it propagates from a rarer to a denser medium. Hence verify Snell's law of refraction.**

**(b) When monochromatic light travels from a rarer to a denser medium, explain the following, giving reasons :**

**(i) Is the frequency of reflected and refracted light same as the frequency of incident light ?**

**(ii) Does the decrease in speed imply a reduction in the energy carried by light wave ?**

**Ans.** (a) A compound microscope consists of two convex lenses parallel separated by some distance. The lens nearer to the object is called the objective. The lens through which the final image is viewed is called the eyepiece.

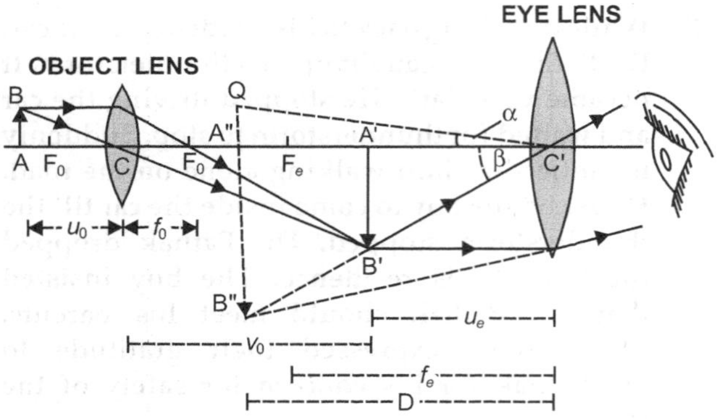

**Magnifying power when final image is at infinity :** The magnification produced by the compound microscope is the product of the magnifications produced by the eyepiece and objective.

$\therefore$ $$M = M_e \times M_o \quad ...(i)$$

Where, $M_e$ and $M_o$ are the magnifying powers of the eyepiece and objective respectively.

If $u_0$ is the distance of the object from the objective and $v_0$ is the distance of the image from the objective, then the magnifying power of the objective is

$$M_o = \frac{h'}{h} = \frac{L}{f_o}$$

Where, $h$, $h'$ are object and image heights respectively and $f_o$ is the focal length of the objective and L is the tube length *i.e.*, the distance between the second focal point of the objective and the first focal point of the eyepiece.

When the final image is at infinity,

$$M_e = \frac{D}{f_e}$$

Magnifying power of compound microscope,

$$M = M_o \times M_e = \frac{L}{f_o} \times \frac{D}{f_e}$$

| | Myopia | Hypermetropia |
|---|---|---|
| 1. | Also known as nearsightedness. | Also known as farsightedness. |
| 2. | Person is not able to see far objects clearly. | Person is not able to see near objects clearly. |
| 3. | For myopic eye, far point is less than infinity. | For hypermetropic eye, near point is greater than D *i.e.*, 25 cm (least distance of distinct vision). |
| 4. | To correct myopia, concave lens of appropriate focal length is used.<br>**Myopic eye :**<br>From ∞<br>**Correction :** | To correct hypermetropia convex lens of appropriate focal length is used.<br>**Hypermetropic eye :**<br>**Correction :**<br>R |

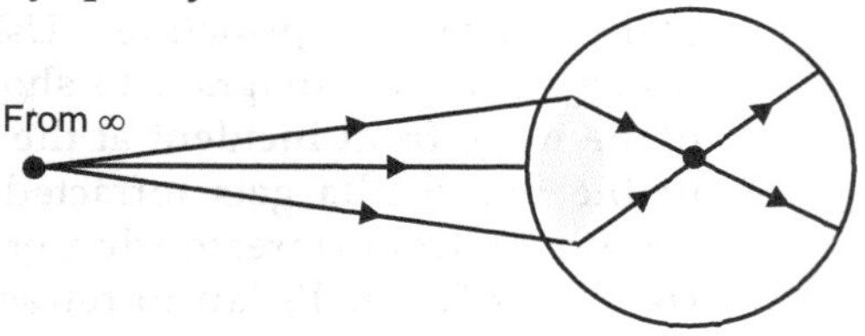

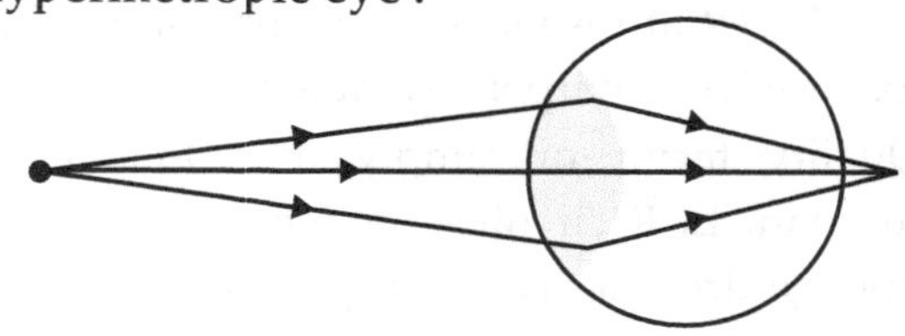

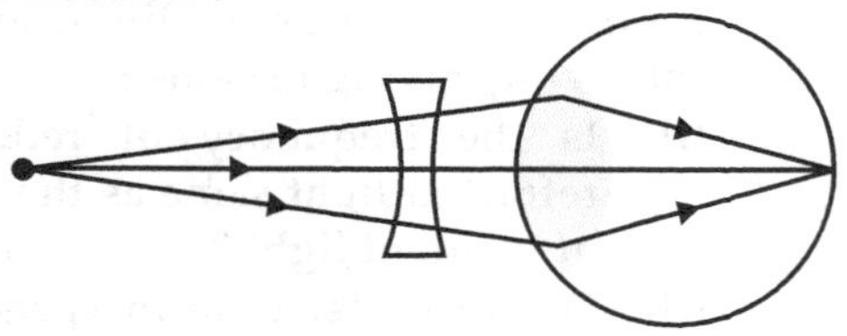

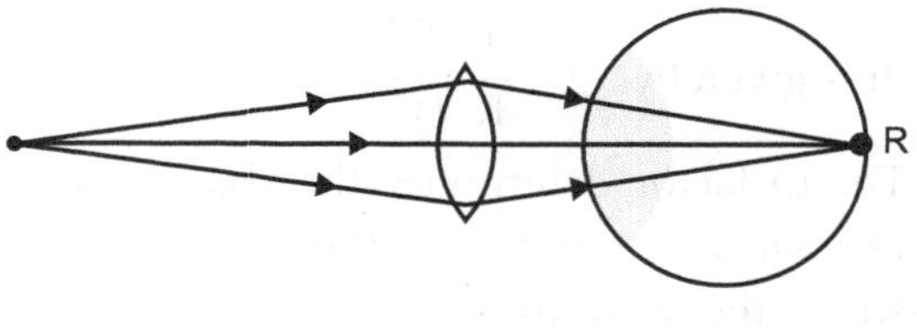

OR

(a) **Huygen's Principle :**

1. Each point on the primary wave front acts as a source of secondary wavelets, transferring out disturbance in all directions in the same way as the original source of light does.
2. The new position of the wave front at any instant is the envelope of the secondary wavelets at that instant.

**Refraction on the basis of wave theory**

1. Consider any point Q on the incident wave front.

2. Suppose when disturbance from point P on incident wave front reaches point P' on the refracted wave front, the disturbance from point Q reaches Q' on the refracting surface XY.

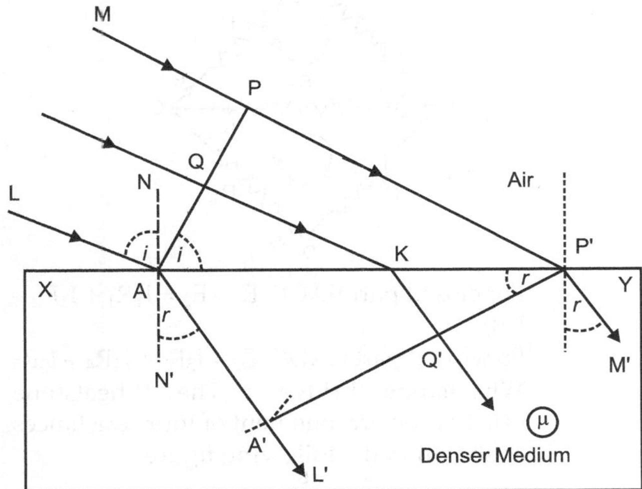


3. Since P'A' represents the refracted wave front, the time taken by light to travel from a point on incident wave front to the corresponding point on refracted wave front should always be the same. Now, time taken by light to go from Q to Q′ will be

$$t = \frac{QK}{c} + \frac{QK}{v} \quad ...(i)$$

In right-angled $\Delta AQK$, $\angle QAK = i$

$$\therefore \quad QK = AK \sin i \quad ...(ii)$$

In right-angled $\Delta P'Q'K$, $\angle Q'P'K = r$ and $Q'K = KP' \sin r$ ...(iii)

Substituting (ii) and (iii) in equation (i),

$$t = \frac{AK \sin i}{c} + \frac{KP' \sin r}{v}$$

$$t = \frac{AK \sin i}{c} + \frac{(AP' - AK \sin r)}{v}$$

$$[\because KP' = AP' - AK]$$

$$t = \frac{AP'}{v} \sin r + AK\left(\frac{\sin i}{c} - \frac{\sin r}{v}\right) \quad ...(iv)$$

The rays from different points on the incident wave front will take the same time to reach the corresponding points on the refracted wave front *i.e.*, given equation (iv) is independent of AK. It will happen so, if

$$\frac{\sin i}{c} - \frac{\sin r}{v} = 0$$

$$\Rightarrow \quad \frac{\sin i}{\sin r} = \frac{c}{v}$$

$$\Rightarrow \quad n = \frac{\sin i}{\sin r}$$

This is the Snell's law for refraction of light.

(b) (i) The frequency of reflected and refracted light remains constant as the frequency of incident light because frequency only depends on the source of light.

(ii) Since the frequency remains same, hence there is no reduction in energy.

28. **(a) State the working principle of a potentiometer. With the help of the circuit diagram, explain how a potentiometer is used to compare the emf's of two primary cells. Obtain the required expression used for comparing the emfs.**

**(b) Write two possible causes for one sided deflection in a potentiometer experiment.** **[5]**

**OR**

**(a) State Kirchhoff's rules for an electric network. Using Kirchhoff's rules, obtain the balance condition in terms of the resistance of four arms of Wheatstone bridge.**

**(b) In the meter bridge experimental set-up, shown in the figure, the null point 'D' is obtained at a distance of 40 cm from end A of the meter bridge wire. If a resistance of 10 Ω is connected in series with $R_1$, null point is obtained at AD = 60 cm. Calculate the values of $R_1$ and $R_2$.**

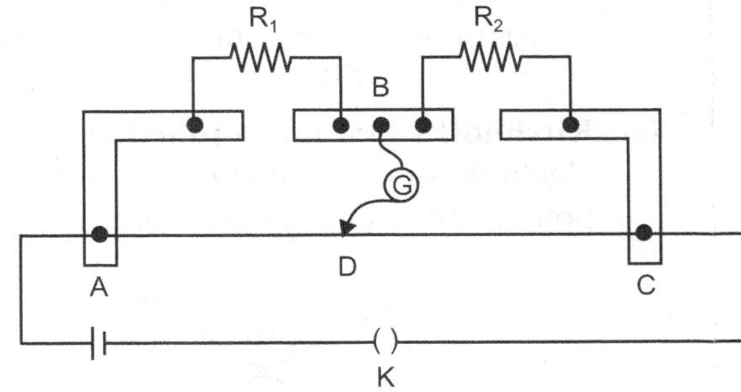


**Ans. (a) Working principle of Potentiometer :** When a constant current is passed through a wire of uniform area of cross-section, the potential drop across any portion of the wire is directly proportional to the length of that portion.

**Applications of Potentiometer for comparing emf's of two cells :** The following figure shows an application of the potentiometer to compare the emf of two cells of emf $E_1$ and $E_2$.

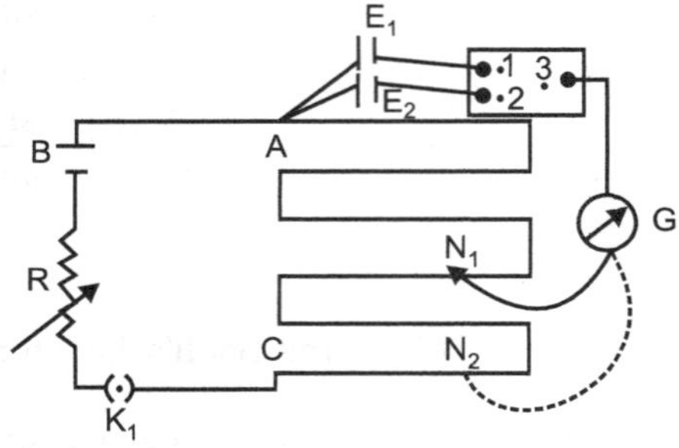


$E_1$, $E_2$ are the emf of the two cells.

1, 2, 3 form a two way key.

When 1 and 3 are connected, $E_1$ is connected to the galvanometer (G).

Jockey is moved to $N_1$, which is at a distance $L_1$ from A, of cell E the balancing length.

By the potentiometer principle

$E_1 = kl_1$ where $k$ is the potential gradient

Similarly for $E_2$ is balanced at $l_2$ ($AN_2$)

$E_2 = kl_2$

By (i) and (ii)

$$\frac{E_1}{E_2} = \frac{l_1}{l_2} \quad ...(iii)$$

Thus, we can compare the emf's of any two sources. Generally, one of the cells is chosen as a standard cell whose emf is known to a high degree of accuracy. The emf of the other cell is then calculated from equation (iii).

**(b)** **(i)** The emf of the cell connected in main circuit may not be more than the emf of the primary cells whose emfs are to be compared.

**(ii)** The positive ends of all cells are not connected to the same end of the wire.

**OR**

**(a)** **Kirchhoff's first law – Junction rule :** The algebraic sum of the currents meeting at a point in an electrical circuit is always zero.

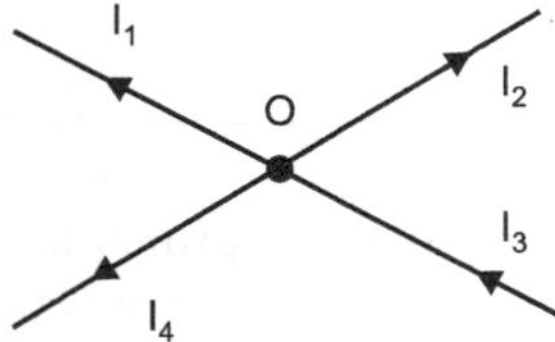


Let the currents be $I_1$, $I_2$, $I_3$, and $I_4$.

**Convention :** Current towards the junction positive

Current away from the junction negative

$$I_3 + (-I_1) + (-I_2) + (-I_4) = 0$$

This law is in accordance with conservation of charge

**Kirchhoff's second law–Loop rule :** In a closed loop, the algebraic sum of the emfs is equal to the algebraic sum of the products of the resistance and current flowing through them.

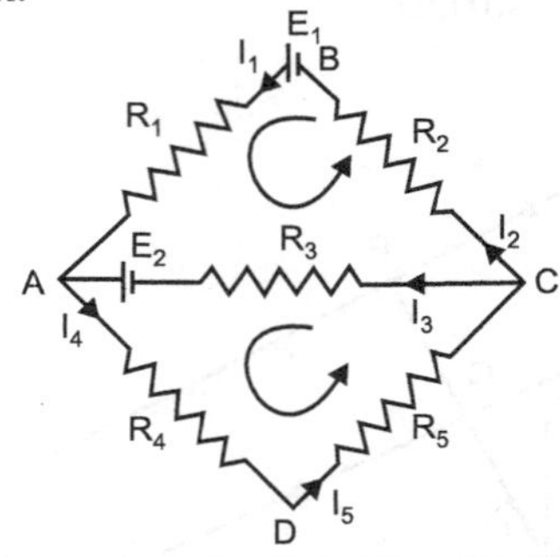


For closed part BACB, $E_1 - E_2 = I_1R_1 + I_1R_2 - I_3R_3$

For closed part CADC, $E_2 = I_3R_3 + I_4R_4 + I_5R_5$

**Wheatstone bridge :** The Wheatstone bridge is an arrangement of four resistances as shown in the following figure.

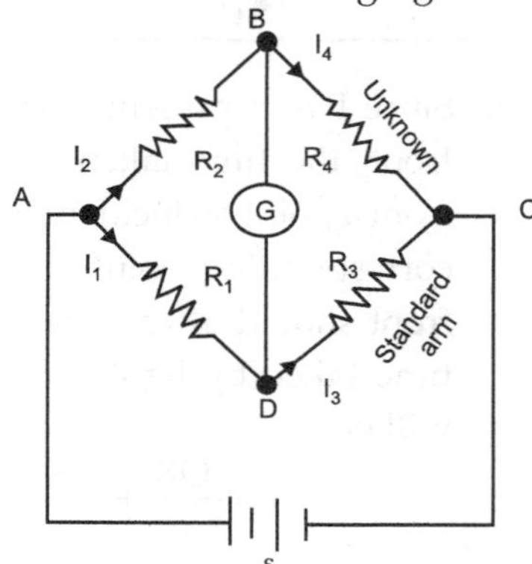


$R_1$, $R_2$, $R_3$, and $R_4$ are the four resistances.

Galvanometer (G) has a current $I_g$ flowing through it.

At balanced condition, $I_g = 0$

Applying junction rule at B,

$\therefore \quad I_2 = I_4$

Applying loop rule to closed loop ADBA,

$-I_1 R_1 + 0 + I_2 R_2 = 0$

$$\therefore \quad \frac{I_1}{I_2} = \frac{R_2}{R_1} \quad ...(i)$$

Applying loop rule to closed loop CBDC,

$I_2R_4 + 0 - I_1R_3 = 0 \quad (\because I_3 = I_1, I_4 = I_2)$

$$\therefore \quad \frac{I_1}{I_2} = \frac{R_4}{R_3} \quad ...(ii)$$

From equations (i) and (ii),

$$\frac{R_2}{R_1} = \frac{R_4}{R_3}$$

This is the required balanced condition of Wheatstone bridge.

**(b)** Considering both the situations and writing them in the form of equations Let R′ be the resistance per unit length of the potentiometer wire,

$$\frac{R_1}{R_2}=\frac{l_1 R'}{(100-l_1)R'}$$

$$\Rightarrow \frac{R_1}{R_2}=\frac{R'\times 40}{R'(100-40)}=\frac{40}{60}=\frac{2}{3} \quad \ldots(i)$$

$$\Rightarrow \frac{R_1+10}{R_2}=\frac{R'\times 60}{R'(100-60)}=\frac{60}{40}=\frac{3}{2} \quad \ldots(ii)$$

Putting the value of $R_1$ from equation (i) and substituting in equation (ii)

$$\frac{2}{3}+\frac{10}{R_2}=\frac{3}{2}$$

$$R_2 = 12\ \Omega$$

Recalling equation (i) again

$$\frac{R_1}{12}=\frac{2}{3}$$

$$R_1 = 8\ \Omega$$

**29. (a) Derive the expression for the torque on a rectangular current carrying loop suspended in a uniform magnetic field.**

**(b) A proton and a deuteron having equal momenta enter in a region of a uniform magnetic field at right angle to the direction of the field. Depict their trajectories in the field.** [5]

OR

**(a) A small compass needle of magnetic moment '*m*' is free to turn about an axis perpendicular to the direction of uniform magnetic field 'B'. The moment of inertia of the needle about the axis is 'I'. The needle is slightly disturbed from its stable position and then released. Prove that it executes simple harmonic motion. Hence deduce the expression for its time period.**

**(b) A compass needle, free to turn in a vertical plane orients itself with its axis vertical at a certain place on the earth. Find out the values of (i) horizontal component of earth's magnetic field and (ii) angle of dip at the place.**

**Ans. (a)** Consider a rectangular loop PQRS of length *l*, breadth *b* suspended in a uniform magnetic field $\vec{B}$. The length of loop = PQ = RS = *l* and breadth = QR = SP = *b*. Let at any instant the normal to the plane of loop make an angle θ with the direction of magnetic field $\vec{B}$ and *I* be the current in the loop. We know that a force acts on a current carrying wire placed in a magnetic field. Therefore, each side of the loop will experience a force. The net force and torque acting on the loop will be determined by the forces acting on all sides of the loop. Suppose that the forces on sides PQ, QR, RS and SP are $\vec{F}_1, \vec{F}_2, \vec{F}_3$ and $\vec{F}_4$ respectively. The sides QR and SP make angle (90° – θ) with the direction of magnetic field. Therefore each of the forces $\vec{F}_2$ and $\vec{F}_4$ acting on these sides has same magnitude F = B*Ib* sin (90° – θ) = B*Ib* cos θ. According to Fleming's left hand rule the forces $\vec{F}_2$ and $\vec{F}_4$ are equal and opposite but their line of action is same. Therefore, these forces cancel each other *i.e.*, the resultant of $\vec{F}_2$ and $\vec{F}_4$ is zero.

The sides PQ and RS of current loop are perpendicular to the magnetic field, therefore, the magnitude of each of forces $\vec{F}_1$ and $\vec{F}_3$ is zero.

$$F = IlB \sin 90° = IlB.$$

According to Fleming's left hand rule the forces $\vec{F}_1$ and $\vec{F}_3$ acting on side PQ and RS are equal and opposite, but their lines of action are different; therefore, the resultant force of $\vec{F}_1$ and $\vec{F}_3$ is zero, but they form a couple called the deflecting couple. When the normal to plane of loop makes and angle θ with the direction of magnetic field B, the perpendicular distance between $F_1$ and $F_3$ is *b* sin θ.

∴ Moment of couple or Torque,

τ = (Magnitude of one force F) × Perpendicular distance

= (BI*l* ). (*b* sin θ) = I (*lb*) B sin θ

But *lb* = area of loop = A (say)

∴ Torque, τ = IAB sin θ

If the loop contains N-turns, then τ = NI AB sin θ

In vector form $\vec{\tau} = NI\,\vec{A}\times\vec{B}$.

The magnetic dipole moment of rectangular current loop

= M = NIA

$$\therefore \quad \vec{\tau} = \vec{M}\times\vec{B}$$

Direction of torque is perpendicular to direction of area of loop as well as the direction of magnetic field *i.e.,* along $(\vec{A} \times \vec{B})$

**(b)** We know, Lorentz force, F = B$qv$ sin θ

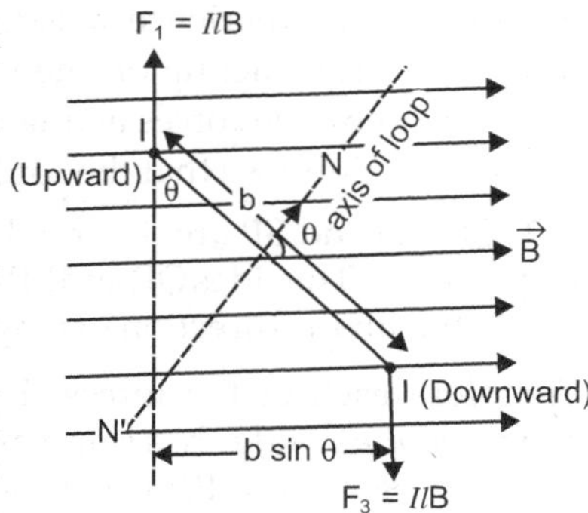


Where θ = angle between velocity of particle and magnetic field = 90°

So, Lorentz force, F = B$qv$

Thus, the particles will move in circular path and required centripetal force is provided by this Lorentz force.

$$Bqv = \frac{mv}{r} \Rightarrow r = \frac{mv}{Bq}$$

Let $m_p$ = mass of proton, $m_d$ = mass of deuteron, $v_p$ = velocity of proton and $v_d$ = velocity of deuteron.

$$r_p = \frac{m_p v_p}{Bq} \quad \text{...(i)}$$

$$r_d = \frac{m_d v_d}{Bq} \quad \text{...(ii)}$$

The charge of proton and deuteron are equal.

Given that, $m_p v_p = m_d v_d$

Since, RHS of equation (i) and (ii) are equal, so $r_p = r_d = r$.

Thus, the trajectory of both the particles will be same.

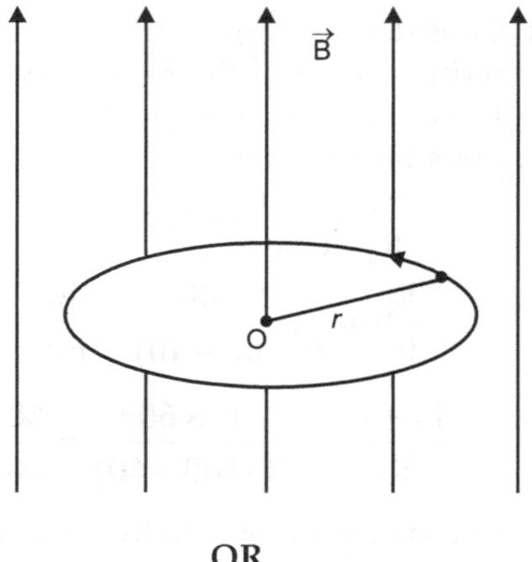


**OR**

**(a)** The torque on the needle is τ = $m$ × B

In magnitude τ = $m$B sin θ

Here τ is restoring torque and θ is the angle between $m$ and B.

Therefore, in equilibrium $I\frac{d^2\theta}{dt^2} = mB \sin\theta$

Negative sign with $m$B sin θ implies that restoring torque is in a position to deflecting torque. For small values of θ in radians, we approximate sin θ ≈ θ and get

$$I\frac{d^2\theta}{dt^2} = -mB\sin\theta$$

or $$\frac{d^2\theta}{dt^2} = -\frac{mB}{I}\theta$$

This represents a simple harmonic motion. The square of the angular frequency $\omega^2 = mB/I$ and the time period is,

$$T = 2\pi\sqrt{\frac{1}{mB}} \quad [\because = \frac{2\pi}{T}]$$

**(b)** Since the needle rotates in vertical plane, so the place here is magnetic pole of earth.

Therefore, $B_H = 0$

and, angle of dip at magnetic pole, δ = 90°

●●

# Delhi [Set-II]

**Note : Except for the following questions, all the remaining questions have been asked in previous Set.**

**1. A cell of emf 'E' and internal distance '$r$' draws a current 'I'. Write the relation between terminal voltage 'V' in terms of E, I, $r$. [1]**

**Ans.** When the current I draws from a cell of emf E and internal resistance $r$, then the terminal voltage is V = E – I$r$

**2. Which of the following substances are diamagnetic ? [1]**

**Bi, Al, Na, Cu, Ca and Ni**

**Ans.** Bi and Cu are diamagnetic substances.

**3. A heating element is marked 210 V, 630 W. What is the value of current drawn by the element when connected to a 210 V dc source ? [1]**

**Ans.** In dc source, P = VI

Given that P = 630 W and V = 210 V.

So, $$I = \frac{P}{V} = \frac{630}{210} = 3A$$

**10. An ammeter of resistance 1 Ω can measure current upto 1.0 A. (i) What must be the value of the shunt resistance to enable the ammeter to measure upto 5.0 A ? (ii) What is the combination resistance of the ammeter and the shunt ? [1]**

**Ans.** We have, Resistance of ammeter, $R_A = 1\ \Omega$

Maximum current across ammeter, $I_A = 1.0$ A

So, Voltage drop across ammeter, V = IR = 1.0 × 1.0 = 1V

**(i)** Resistance of ammeter with shunt,

$$R = \frac{R_A x}{R_A + x} = \frac{x}{1+x}$$

Current through ammeter,

$$I = 5A$$

Now, V = IR

$$\left(\frac{x}{1+x}\right) \times 5 = 1.0$$

$$x \times 5 = 1 + x$$

$$4x = 1$$

$$x = 0.25$$

Thus the shunt resistance is 0.25 Ω

**(ii)** Combined resistance of the ammeter and the shunt

$$R = \frac{x}{1+x}$$

$$R = \frac{0.25}{1+0.25}$$

$$R = \frac{0.25}{1.25}$$

$$R = 0.2\ \Omega$$

**14. A convex lens of focal length 20 cm is placed coaxially in contact with a concave lens of focal length 25 cm. Determine the power of the combination. Will the system be converging or diverging in nature ? [2]**

**Ans.** We have, Focal length of concave lens, $f_1$ = +20 cm = +0.20 m

Focal length of concave lens, $f_2$ = −25 cm = −0.25 m

Power of convex lens,

$$P_1 = \frac{1}{f_1} = \frac{1}{0.20}$$

Power of concave lens,

$$P_2 = \frac{1}{f_2} = \frac{1}{-0.25}$$

Power of the combination lens,

$$P = P_1 + P_2$$

$$= \frac{1}{0.20} + \frac{1}{-0.25}$$

$$= \frac{100}{20} + \frac{100}{-25}$$

$$= \frac{500-400}{100}$$

$$= \frac{100}{100}$$

$$= 1D$$

As the power is positive, the system will be converging in nature.

**15. Using Bohr's postulates, obtain the expressions for (i) kinetic energy and (ii) potential energy of the electron in stationary state of hydrogen atom.**

**Draw the energy level diagram showing how the transitions between energy levels result in the appearance of Lyman series. [2]**

**Ans.** According to Bohr's postulates, in a hydrogen atom, a single electron revolves around a nucleus of charge + *e*.

And the radius of orbit in which the electron revolves, is given by

$$r = \frac{n^2h^2}{4\pi^2kme^2} \quad ...(i)$$

**(i)** Kinetic energy of electron,

$$E_K = \frac{1}{2}mv^2 = \frac{ke^2}{2r}$$

Using equation (i), we get

$$E_K = \frac{ke^2}{2}\frac{4\pi^2kme^2}{n^2h^2}$$

**(ii)** Potential energy

$$E_P = \frac{-k(e)\times(e)}{r} = \frac{-ke^2}{r}$$

Using equation (i), we get

$$E_P = -ke^2 \times \frac{4\pi^2kme^2}{n^2h^2}$$

$$= -\frac{4\pi^2k^2me^4}{n^2h^2}$$

Energy level diagram showing the transitions between energy levels result in the appearance of Lyman series :

For Lyman series, $n_f = 1$ and $n_i = 2, 3, 4, 5, \ldots..$

$$\frac{1}{\lambda} = R_H\left(\frac{1}{1^2} - \frac{1}{n^2}\right)$$

Where, $n_i = 2, 3, 4, 5, \ldots...$

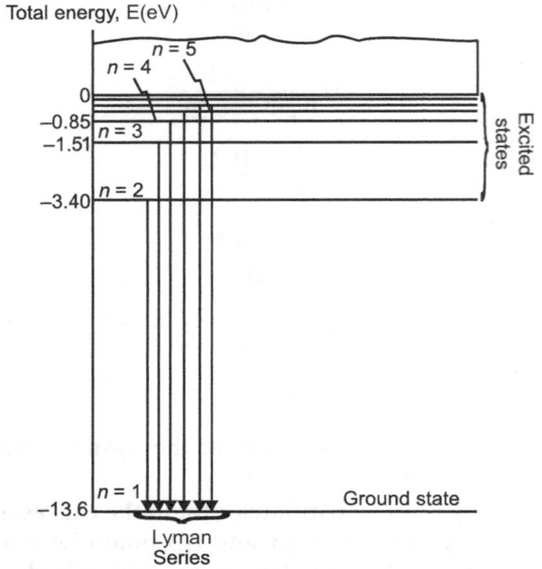

Total energy, E(eV)
n = 5
n = 4
0
−0.85 n = 3
−1.51
−3.40 n = 2
Excited states
−13.6 n = 1
Ground state
Lyman Series

**22. Figure shows a rectangular loop conducting PQRS in which the arm PQ is free to move. A uniform magnetic field acts in the direction perpendicular to the plane of the loop. Arm PQ is moved with a velocity $v$ towards the arm RS Assuming that the arms QR, RS and SP have negligible resistances and the moving arm PQ has the resistance $r$, obtain the expression for (i) the current in the loop (ii) the force and (iii) the power required to move the arm PQ.**

S P v Q R

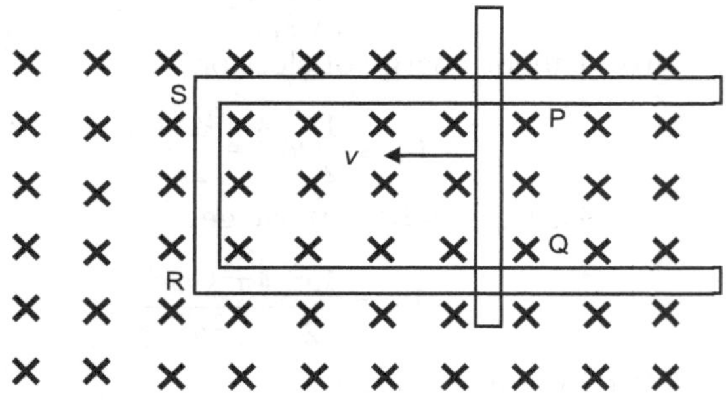

[3]

**Ans.** Let the length RQ = $x$ and RS = $l$.

Let the magnitude of the uniform magnetic field be B.

**(i)** The magnetic flux $\phi$ enclosed by the loop PQRS is given by, $\phi = Blx$

As, $x$ is changing with time, the rate of change of flux $\frac{d\phi}{dt}$ will induce an emf given by :

$$E = -\frac{d\phi}{dt} = -\frac{d(Blx)}{dt} = -Bl\frac{dx}{dt} = Blv \quad \left[as, \frac{dx}{dt} = -v\right]$$

Current in the loop is given by,

$$I = \frac{E}{r} = \frac{Blv}{r}$$

**(ii)** The magnetic force on the PQ is,

$$F = BIl = B\left(\frac{Blv}{r}\right) \times l = \frac{B^2l^2v}{r}$$

**(iii)** Power required to move the arm PQ is,

$$P = F \times v = \frac{B^2l^2v}{r} \times v = \frac{B^2l^2v^2}{r}$$

**23. Distinguish between 'sky waves' and 'space waves' modes of propagation in communication system.**** **[4]**

**(a) Why is sky wave mode propagation restricted to frequencies upto 40 MHz ?**

**(b) Give two examples where space wave mode of propagation is used.**

**Ans. Sky wave :** Sky waves are the AM radio waves which are received after being reflected from the ionosphere. The propagation of radio wave signals from one point to another via reflection from ionosphere is known as sky wave propagation. The sky wave propagation is an important consequence of the total internal reflection of radio waves. As we go higher in the ionosphere, there is an increase in the free electron density. Consequently there is a decrease of refractive index. Thus, as a radio wave travels up in the ionosphere, it finds itself travelling from denser to rarer medium. It continuously bends away from its path till it suffers total internal reflection to reach back the Earth.

**Space waves :** Space waves are the waves which are used for satellite communication and line of sight path. The waves have frequencies up to 40 MHz provides essential communication and limited the line of sight paths.

**(a)** The em waves of frequencies greater than 40 MHz penetrate the ionosphere and escape so, the sky wave propagation is restricted to the frequencies upto 40 MHz.

**(b)** In television broadcast and satellite communication, the space wave mode of propagation is used.

** Answer is not given due to change in the present Syllabus.

# Delhi [Set-III]

**Note : Except for the following questions, all the remaining questions have been asked in previous Set.**

**6. A 5 V battery of negligible internal resistance is connected across a 200 V battery and a resistance of 39 Ω as shown in the figure. Find the value of the current. [2]**

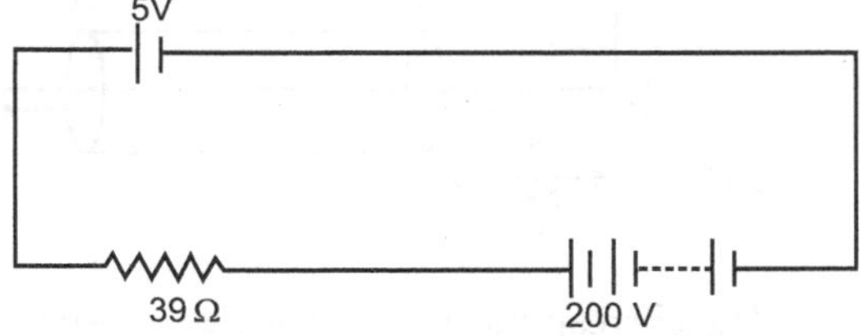


**Ans.** Let I be the current flowing in the circuit.

Using Kirchhoff's law,

$$39\,I = 200 - 5$$

$$I = \frac{195}{39}$$

$$I = 5\text{ A}.$$

**7. Which of the following substances are paramagnetic ? [2]**

**Bi, Al, Cu, Ca, Pb, Ni**

**Ans.** Paramagnetic substances are Aluminium (Al) and Calcium (Ca).

**9. An ammeter of resistance 0.6 Ω can measure current upto 1.0 A. Calculate (i) The shunt resistance required to enable the ammeter to measure current upto 5.0 A. (ii) The combined resistance of the ammeter and the shunt. [2]**

**Ans.** We have, resistance of ammeter, $R_A = 0.60$ ohm.

Maximum current across ammeter, $I_A = 1.0$ A

So, voltage across ammeter,

$$V = IR$$

$$= 1.0 \times 0.60$$

$$V = 0.6\text{ V}$$

Let the value of shunt be $x$.

**(i)** Resistance of ammeter with shunt

$$R = \frac{R_A x}{R_A + x}$$

$$= \frac{0.6x}{0.6 + x}$$

Current through ammeter I = 5A

$$\because \quad V = IR$$

$$\text{Now, } \left(\frac{0.6x}{0.6+x}\right) \times 5 = 0.6$$

$$0.6x \times 5 = 0.6(0.6 + x)$$

$$3x = 0.36 + 0.6x$$

$$3x - 0.6x = 0.36$$

$$x = \frac{0.36}{2.4}$$

$$x = 0.15\,\Omega$$

Therefore, the shunt resistance is 0.15 ohm.

**(ii)** Combined resistance of the ammeter and the shunt is given by

$$R = \frac{0.6x}{0.6 + x}$$

$$R = \frac{0.6 \times 0.15}{0.6 + 0.15}$$

$$R = 0.12\,\Omega$$

**15. A convex lens of focal length 30 cm is placed coaxially in contact with a concave lens of focal length 40 cm. Determine the power of the combination. Will the system be converging or diverging in nature ? [3]**

**Ans.** We have, focal length of convex lens,

$$f_1 = +30\text{ cm}$$

$$= 0.30\text{ m}$$

Focal length of concave lens

$$f_2 = -40\text{ cm}$$

$$= -0.40\text{ m}$$

Equivalent focal length

$$\frac{1}{F} = \frac{1}{f_1} + \frac{1}{f_2}$$

$$= \frac{1}{30} + \frac{1}{-40}$$

$$= \frac{40 - 30}{1200} = \frac{1}{120}$$

$$\therefore \quad F = 120\text{ cm}$$

$$= 1.2\text{ m}$$

$\therefore$ Power of the combination,

$$P = \frac{1}{F} = \frac{1}{1.2} = 0.83\,D$$

The focal length of combination = 1.2 m = 120 cm. As the focal length is positive the system will be converging in nature.

**18. A capacitor of unknown capacitance is connected across a battery of V volts. The charge stored in it is 300 μC. When potential across the capacitor is reduced by 100 V, the charge stored in it becomes 100 μC. Calculate the potential V and the unknown capacitance. What will be the charge stored in the capacitor if the voltage applied had increased by 100 V ? [3]**

OR

**A hollow cylindrical box of length 0.5 m and area of cross-section 20 cm is placed in a three-dimensional coordinate system as shown in the figure. The electric field in the region is given by $E = 20x\hat{i}$, where E is $NC^{-1}$ and $x$ is in meters. Find**

**(i) Net flux through the cylinder**

**(ii) Charge enclosed in the cylinder.**

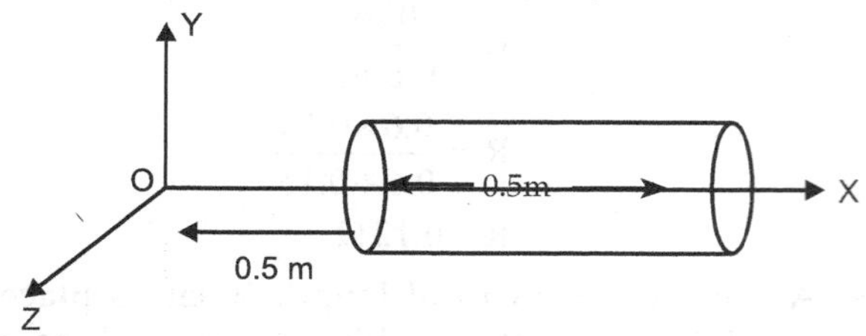


**Ans. (i)** Initial voltage

$V_1 = V$ volts

Charge stored,

$Q_1 = 300\ \mu C$

$Q_1 = CV_1$ ...(i)

Changed potential, $V_2 = V - 100$ V

$Q_2 = 100\ \mu C$

$Q_2 = CV_2$ ...(ii)

By dividing (ii) from (i), we get

$$\frac{Q_1}{Q_2} = \frac{CV_1}{CV_2}$$

$$\frac{Q_1}{Q_2} = \frac{V_1}{V_2}$$

$$\frac{300}{100} = \frac{V}{V-100}$$

$V = 150$ volts

$$C = \frac{Q_1}{V_1} = \frac{300 \times 10^{-6}}{150}$$

$= 2 \times 10^{-6}$F

$C = 2\mu$F

If the voltage applied had increased by 100 V, then

$V_3 = 150 + 100$

$= 250$ V

Hence, charge stored in the capacitor,

$Q_3 = CV_3 = 2 \times 10^{-6} \times 250$

$= 500\ \mu C$

OR

**Ans. (i)** Given, $\vec{E} = 20x\,\hat{i}$

and $\Delta s = 20\ cm^2$

$= 20 \times 10^{-4} m^2$

As the electric field is only along the X-axis, so flux will pass only through the cross-section of cylinder.

Magnitude of electric field at A,

$E_A = 20 \times 0.5 = 10$ N/C

Magnitude of electric field at cross-section B.

$E_B = 20 \times 1 = 20$ N/C

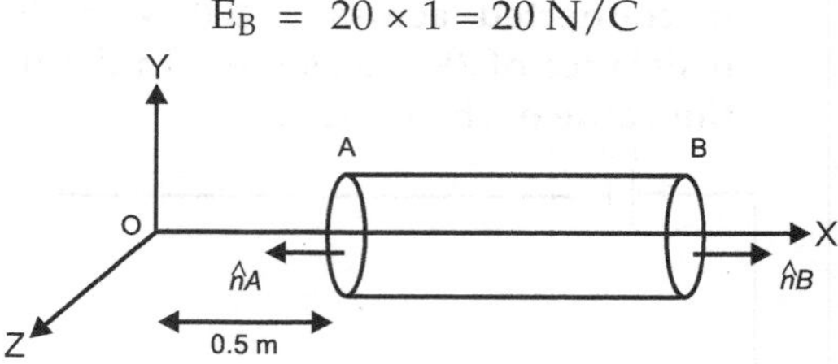


The corresponding electric fluxes are :

$\phi_A = \vec{E} \cdot \Delta\vec{s}$

$= 10 \times 20 \times 10^{-4} \times \cos 180°$

$= -0.02\ Nm^2C^{-1}$

$\phi_A = \vec{E} \cdot \Delta\vec{s}$

$= 20 \times 20 \times 10^{-4} \times \cos 0°$

$= 0.04\ Nm^2C^{-1}$

So, the net flux through the cylinder,

$\phi = \phi_A + \phi_B$

$= -0.02 + 0.04$

$= 0.02\ Nm^2C^{-1}$

**(ii)** Using Gauss's law :

$$\int \vec{E} \cdot \vec{ds} = \frac{q}{\varepsilon_0}$$

$$0.02 = \frac{q}{8.85 \times 10^{-12}}$$

$q = 8.85 \times 0.02 \times 10^{-12}$

$q = 0.177 \times 10^{-12}$C

**19. (a) Write two characteristic features distinguishing the diffraction pattern from the interference fringes obtained in Young's double slit experiment.**

**(b) Two wavelength of sodium light 590 nm and 596 nm are used, in turn, to study the diffraction taking place due to a single slit of aperture $1 \times 10^{-4}$ m. The distance between the slit and the screen is 1.8 m. Calculate the separation between the positions of the first maxima of the diffraction pattern obtained in the two cases. [3]**

**Ans. (a)** Two characteristic features distinguishing the diffraction pattern from the interference fringes obtained in Young's double slit experiment are :

**(i)** The interference fringes may or may not be of the same width whereas the fringes of diffraction pattern are always of varying width.

**(ii)** In interference the bright fringes are of same intensity whereas in diffraction pattern the intensity falls as we go to successive maxima away from the centre, on either side.

**(b)** Wavelength of the light beam, $\lambda_1 = 590$ nm $= 5.9 \times 10^{-7}$ m

Wavelength of another light beam, $\lambda_2 = 596$ nm $= 5.96 \times 10^{-7}$ m

Distance of the slits from the screen = D = 1.8 m

Distance between the two slits $= 1 \times 10^{-4}$ m

For the first secondary maxima,

$$\sin\theta = \frac{3\lambda_1}{2a} = \frac{x_1}{D}$$

$$x_1 = \frac{3\lambda_1 D}{2a}$$

and $$x_2 = \frac{3\lambda_2 D}{2a}$$

$\therefore$ Spacing between the positions of first secondary maxima of two sodium lines

$$= x_2 - x_1$$

$$= \frac{3D}{2a}(\lambda_2 - \lambda_1)$$

$$= \frac{3\times1.8\times6\times10^{-9}}{2\times10^{-4}} = 1.62 \times 10^{-4}\text{m}$$

**21. (a) In a nuclear reaction**

$${}^3_2\text{He} + {}^3_2\text{He} \rightarrow {}^4_2\text{He} + {}^1_1\text{H} + {}^1_1\text{H} + 12.86\,\text{MeV},$$

**though the number of nucleons is conserved on both sides of the reaction, yet the energy is released. How ? Explain.**

**(b) Draw a plot of potential energy between a pair of nucleons as a function of their separation. Mark the regions where potential energy is (i) positive and (ii) negative. [3]**

**Ans. (a)** In a nuclear reaction, the sum of the masses of the target nucleus ${}^3_2$He may be greater or less the sum of the masses of the product nucleus ${}^4_2$He and the ${}^1_1$H. So from the law of conservation of mass energy, some energy, (12.86 MeV) is evolved in nuclear reaction. This energy is called Q-value of the nuclear reaction. The binding energy of the nucleus on the left side is not equal to the right side. The difference in the binding energies on two sides appears as energy released or absorbed in the nuclear reaction.

**(b)** The potential energy is minimum at $r_0$. For distance larger than $r_0$ the negative potential energy goes on decreasing and for the distances less than $r_0$ the negative potential energy decrease to zero and then becomes positive and increases abruptly. Thus, A to B is the positive potential energy region and B to C is the negative potential energy region.

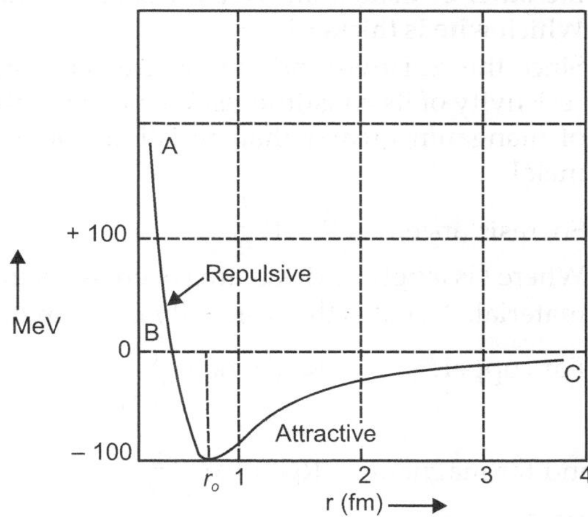


**25. (b) What is the significance of negative sign in the expression for the energy ?**

**(c) Draw the energy level diagram showing how the line spectra corresponding to Paschen series occur due to transition between energy levels. [5]**

**Ans. (b)** Negative sign indicates that revolving electron is bound to the positive nucleus.

**(c)** For Paschen series, $n = 3$ and $n_i = 4, 5, \ldots$

$$\frac{1}{\lambda} = R\left(\frac{1}{3^2} - \frac{1}{n_i^2}\right)$$

where, $n_i = 4, 5, \ldots..$

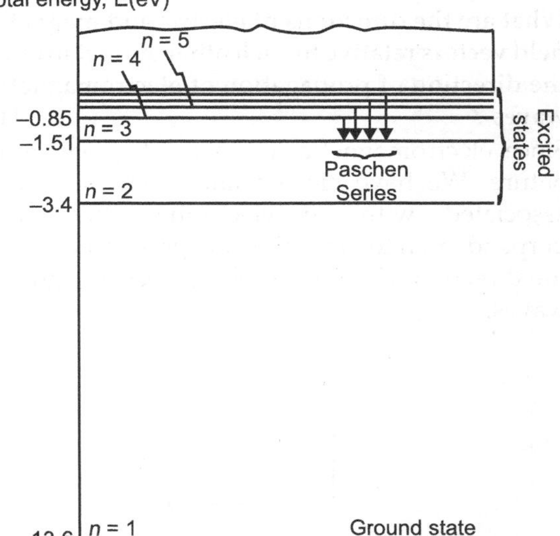

# PHYSICS

## Outside Delhi [Set-I] 2012

**Time allowed : 3 hours** **Maximum marks : 70**

1. **Two wires of equal length, one of copper and the other of manganin have the same resistance. Which wire is thicker ?** **[1]**

**Ans.** Since the resistivity of alloy is greater than the resistivity of its constituents. We have resistivity of manganin greater than resistivity of copper metal.

So, resistance, $R = \rho \frac{l}{A}$

Where $l$ is length and A is area of cross-section of material. Also ρ is the resistivity of the material.

For copper, $R_{Cu} = \rho_{cu} \frac{l_{Cu}}{A_{Cu}}$

and for magnanin, $R_M = \rho_M \frac{l_M}{A_M}$

We have $\rho_M > \rho_{Cu}$

or $\frac{\rho_M}{\rho_{Cu}} > 1$

As $l_{Cu} = l_M$

On Dividing $\frac{R_{Cu}}{R_M} = \frac{\rho_{Cu} A_M}{\rho_M A_{Cu}}$

Again $R_{Cu} = R_M$

∴ $\rho_{Cu} A_M = \rho_M A_{Cu}$

$$\frac{\rho_M}{\rho_{Cu}} = \frac{A_M}{A_{Cu}} > 1$$

Hence, $A_M > A_{Cu}$.

⇒ Manganin wire is thicker.

2. **What are the directions of electric and magnetic field vectors relative to each other and relative to the direction of propagation of electromagnetic waves ?** **[1]**

**Ans.** Since electromagnetic waves are transverse in nature. We have electric and magnetic fields associated with an electromagnetic wave perpendicular to each other and perpendicular to the direction of propagation of electromagnetic waves.

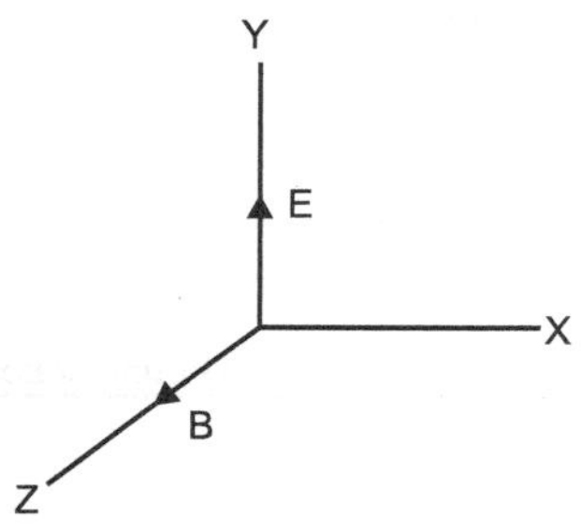


Let the direction of electric field and magnetic field is along Y and Z-axis then the direction of propagation of EM waves will be along positive X-axis.

3. **How does the angular separation between fringes in single-slit diffraction experiment change when the distance of separation between the slit and screen is doubled ?** **[1]**

**Ans.** We know angular separation is given as

$$\theta = \frac{\beta}{D} = \frac{\frac{D\lambda}{d}}{D} = \frac{\lambda}{d}$$

Since θ is independent of D *i.e.*, the distance of separation between the screen and the slit, so when D is doubled, angular separation would remain same.

4. **A bar magnet is moved in the direction indicated by the arrow between two coils PQ and CD. Predict the directions of induced current in each coil.** **[1]**

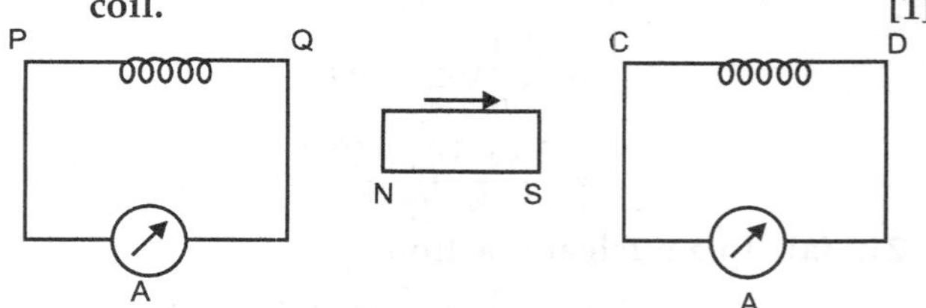


**Ans.** According to Lenz's, law the polarity of the induced emf is such that it opposes the change in magnetic flux responsible for its production. Since, North pole of bar magnet is receding away from the coil so the right end of the coil will develop South pole *i.e.*, induced current as seen from the left end will be anticlockwise.

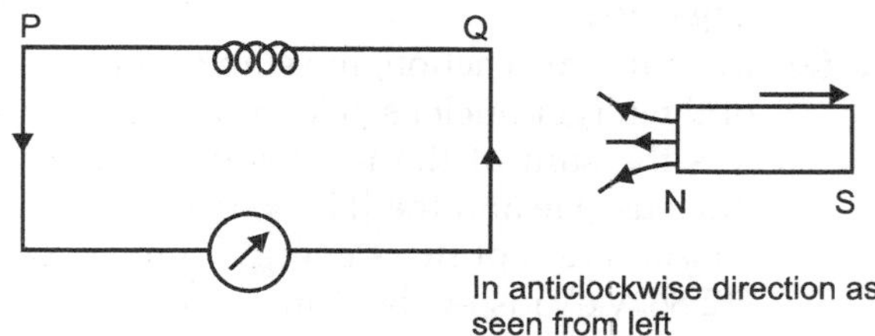

In anticlockwise direction as seen from left

Again, since South pole is pushing towards the second coil so the left end of the coil will develop South pole in order to repel it and decrease the flux *i.e.*, induced current as seen from the left end will be clockwise.

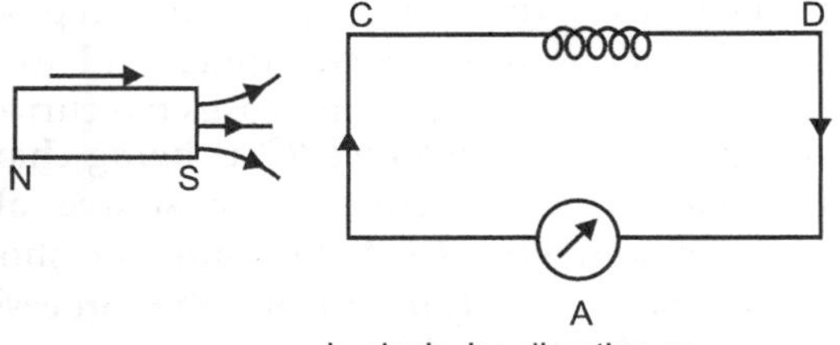

In clockwise direction as seen from left

5. **For the same value of angle of incidence, the angles of refraction in three media A, B and C are 15°, 25° and 35° respectively. In which medium would the velocity of light be minimum ? [1]**
**Answer :** As light travels from a rarer to denser medium it bends towards the normal as its speed decreases. So, if the bending is more, the speed of the light would be less in that medium, compared to other media. As the angle of refraction is measured with respect to the normal, the ray making the least angle of refraction would bend more and the speed of light would be minimum in that case. So, the correct option is medium A where refracting angle is 15°.

6. **A proton and an electron have same kinetic energy. Which one has greater de-Broglie wavelength and why ? [1]**

**Ans.** Since de-Broglie wavelength λ in terms of kinetic energy is given as

$$\lambda = \frac{h}{\sqrt{2mE_k}}$$

Where, $E_k$ is kinetic energy, $m$ is mass of electron and $h$ is the Planck's constant.
Thus, for electron and proton with same kinetic energy, de-Broglie wavelength would depend on mass.

Since $\qquad \lambda \propto \frac{1}{\sqrt{m}}$

As $\qquad m_p > m_e$

$\Rightarrow \qquad \lambda_e > \lambda_p.$

Hence, wavelength of electron is greater than wavelength of proton.

7. **Mention the two characteristic properties of the material suitable for making core of a transformer. [1]**

**Ans.** Two characteristic properties of material :
**(i)** Low hysteresis loss.
**(ii)** Low coercivity.

8. **A charge '$q$' is placed at the centre of a cube of side $l$. What is the electric flux passing through each face of the cube ? [1]**

**Ans.** By using Gauss's law,

$$\phi = \oint \vec{E} \cdot d\vec{S} = \frac{q}{\varepsilon_0}$$

Now, the flux passing through all the six faces would be

$$\phi = 6\phi = \frac{q}{\varepsilon_0}$$

And the flux passing through each face would be

$$\phi = \frac{q}{6\varepsilon_0}$$

9. **A test charge 'q' is moved without acceleration from A to C along the path from A to B and then from B to C in electric field E as shown in the figure. (i) Calculate the potential difference between A and C. (ii) At which point (of the two) is the electric potential more and why ? [2]**

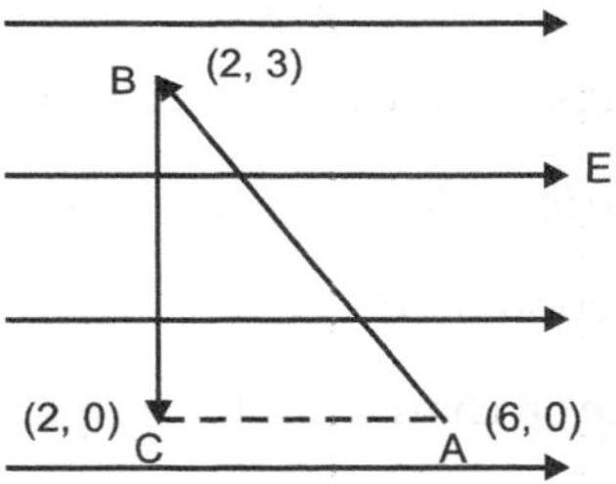


**Ans. (i)** Since work done is independent of the path therefore, we may directly move from A to C.
Potential difference between A and C is given by,

$$V_C - V_A = -\int_A^C \vec{E} \cdot \vec{dl}$$

negative sign is due to the work done against electric field

$$= -\int_A^C E \cdot dl \cos 180°$$
$$= -E(-1)\int_A^C dl$$
$$= E \times 4$$
$$= 4E$$

So, $\qquad V_C - V_A = 4E$

**(ii)** Electric potential will be more at point C as direction of electric field is in decreasing potential.
Hence,

$$V_C > V_A.$$

**10. An electric dipole is held in a uniform electric field. [2]**
**(i) Show that the net force acting on it is zero.**
**(ii) The dipole is aligned parallel to the field. Find the work done in rotating it through the angle of 180°.**

**Ans. (i)** Consider an electric dipole consisting of two equal and opposite point charges, $-q$ at A and $+q$ at B, separated by a small distance $2a$.

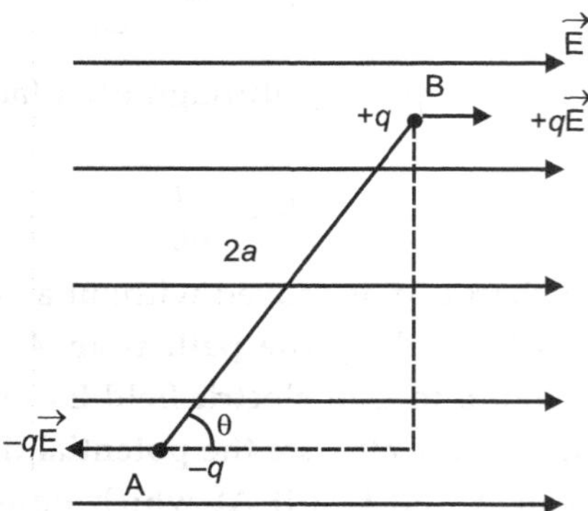


Dipole moment of a dipole is given by,

$$|\vec{p}| = q(2a)$$

Let this dipole be held in a uniform external electric field $\vec{E}$ at an angle θ with the direction of $\vec{E}$.

Force on charge $-q$ at A $= -q\vec{E}$, in a direction opposite to $\vec{E}$. Force on charge $+q$ at B $= +q\vec{E}$, along the direction to $\vec{E}$.

Net force on the dipole $= q\vec{E} - q\vec{E} = 0$

**(ii)** Work done on dipole,

$$W = \Delta U = pE(\cos\theta_1 - \cos\theta_2)$$
$$W = pE(\cos 0^\circ - \cos 180^\circ)$$
$$W = 2\,pE.$$

**11. State the underlying principle of a transformer. How is the large scale transmission of electric energy over long distances done with the use of transformers ? [2]**

**Ans.** Transformer is a device which converts high voltage AC into low voltage AC and vice-versa. It is based on the principle of mutual induction. When alternating current is passed through a coil, an induced emf is set up in the neighbouring coil.

Transformers are used for transmission of electrical energy over long distances.

It step up the output voltage of power plant using step up transformer which reduces the current and inereases the Voltage and hence reduces resistive power loss. Then a step down transformer is used at consumer end to step down the voltage and to increases the current.

**12. A capacitor of capacitance 'C' is being charged by connecting it across a dc source along with an ammeter. Will the ammeter show a momentary deflection during the process of charging ? If so, how would you explain this momentary deflection and the resulting continuity of current in the circuit ? Write the expression for the current inside the capacitor. [2]**

**Ans.** Yes, the ammeter shows a momentary deflection during the process of charging because during charging, conduction current flows through the wires which leads to deposition of charges on the plates of capacitor. This produce an electric field of increasing magnitude, which in turn, produces a displacement currents between the plates and this displacement current maintains the continuity of current in the circuit.

Current inside the capacitor,

$$I_D = \varepsilon_0 \frac{d\phi_E}{dt}$$

**13. An object AB is kept in front of a concave mirror as shown in the figure. [2]**

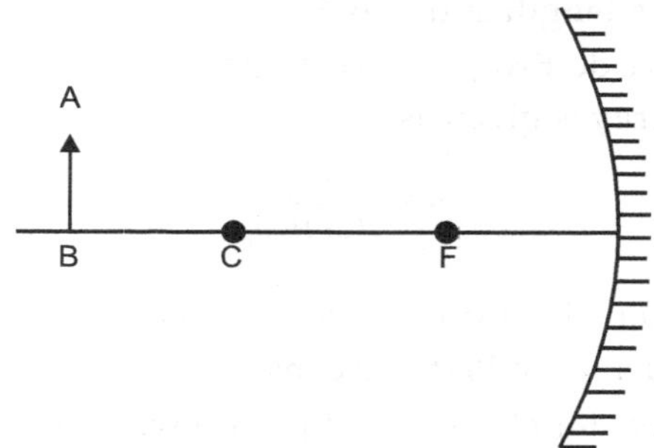


**(i) Complete the ray diagram showing the image formation of the object.**
**(ii) How will the position and intensity of the image be affected if the lower half of the mirror's reflecting surface is painted black ?**

**Ans. (i)**

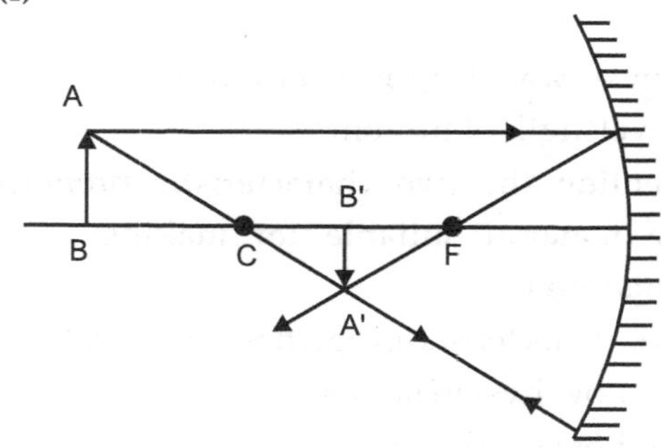

Image formed will real and inverted, between focus and center of curvature and small in size.

**(ii)** If the lower half of the mirror's reflecting surface is painted black, the position of image will be same but its intensity gets reduced.

**14. Draw a labelled ray diagram of a reflecting telescope. Mention its two advantages over the refracting telescope. [2]**

**Ans. Reflecting Telescope :**

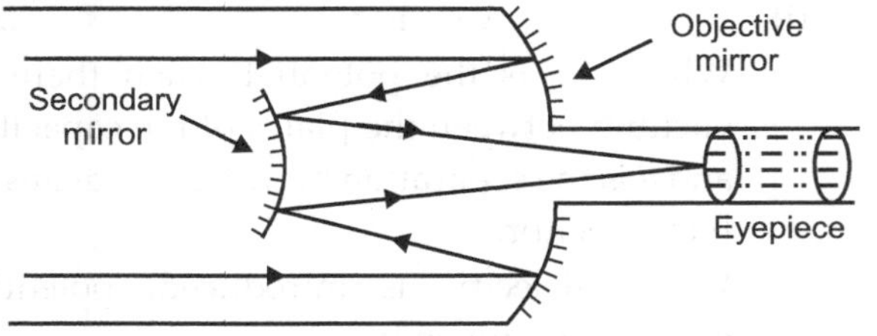


Advantages over refracting telescope :

1. There is no chromatic aberration in reflecting telescope.
2. The resolving power of a large aperture mirror is high and hence minute details of distant stars can be obtained.

**15. Describe briefly with the help of a circuit diagram, how the flow of current carriers in a *p-n-p* transistor is regulated with emitter-base junction forward biased and base-collector junction reverse biased. [2]**

**Ans.** In *p-n-p* transistor, the emitter base junction is always forward biased with voltage $V_{EE}$ and the collector base junction is always reverse biased with voltage $V_{CC}$.

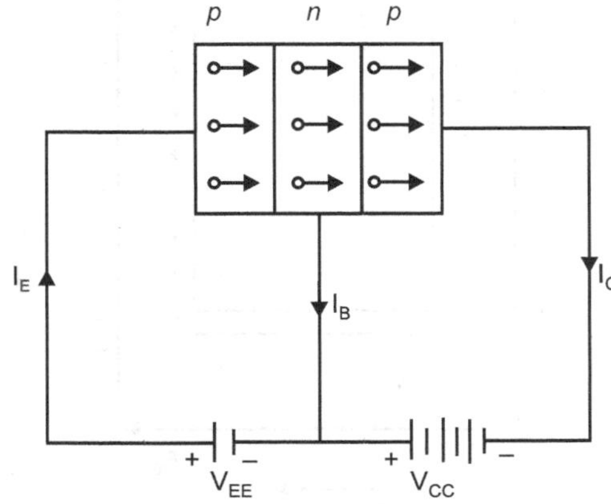


The holes in the emitter are pushed into the base by the positive terminal of battery of voltage $V_{EE}$. Since, base is thin and lightly doped so only few holes combine with electrons in the base. Thus, base current $I_B$ is small. Since $V_{CC}$ is quite large, almost 99% of holes coming from emitter are collected by collector. For each hole reaching the collector, an electron is released from the negative terminal of the collector base battery to neutralize the hole. For each hole consumed in collector, a bond breaks in emitter and electron is released that enters positive terminal of emitter-base battery. Thus, we can say the current is carried by holes inside the transistor and by electrons in external circuit.

**16. In the given block diagram of a receiver, identify the boxes labelled as X and Y and write their functions.** [2]**

**Ans.**

**17. A light bulb is rated 100 W for 220 V ac supply of 50 Hz. Calculate : [2]**

**(i) The resistance of the bulb;**

**(ii) The rms current through the bulb.**

**OR**

**An alternative voltage given by V = 140 sin 314*t* is connected across a pure resistor of 50 Ω. Find**

**(i) the frequency of the source.**

**(ii) the rms current through the resistor.**

**Ans.** P = 100 W and $V_{rms}$ = 220 V, f = 50 Hz

(i)Resistance, $R = \frac{V_{rms}^2}{P} = \frac{220 \times 220}{100} = 848$

**(ii)** $P = I_{rms} \times V_{rms}$

$$I_{rms} = \frac{P}{V_{rms}} = \frac{100}{220} = 0.45 \text{ A}$$

**OR**

Given, V = 140 sin 314*t*, R = 50 Ω

(i) Comparing with $V = V_0 \sin \omega t$

Thus, $V_0 = 140$ V

And, $\omega = 314$

$2\pi v = 314$

$$v = \frac{314}{2 \times 3.14} = 50 \text{ Hz}$$

(ii) $V_{rms} = \frac{V_0}{\sqrt{2}} = \frac{140}{\sqrt{2}} = 98.99$ V

$$I_{rms} = \frac{V_{rms}}{R} = \frac{98.99}{50} = 1.97 \text{ A}$$

**18. A circular coil of 'N' turns and radius 'R' carries a current 'I'. It is unwound and rewound to make another coil of radius 'R/2', current 'I' remaining the same. Calculate the ratio of the magnetic moments of the new coil and original coil. [2]**

**Ans.** The magnetic moment *m* of a current, carrying loop,

$$m = NIA$$

** Answer is not given due to change in the present Syllabus.

Where, N = number of turns.

Let $m_1$ and $m_2$ be the magnetic moments of circular original coil of radius 'R' and new coil of radius 'R/2'.

Length of wire remains same. Thus,

$$N(2\pi r) = N'\left(2\pi\left(\frac{R}{2}\right)\right)$$

$$N' = 2N$$

Now, $m_1 = NIA_1 = NI\pi R^2$

$$m_2 = N'IA_2 = 2NI\left[\pi\left(\frac{R}{2}\right)^2\right]$$

$$= 2NI\left[\pi\left(\frac{R}{2}\right)^2\right] = \frac{1}{2}NI\pi R^2$$

$$\frac{m_2}{m_1} = \frac{\frac{1}{2}}{1} = \frac{1}{2}$$

**19. Deduce the expression for the electrostatic energy stored in a capacitor of capacitance 'C' and having charge 'Q'.**

**How will the (i) energy stored and (ii) the electric field inside the capacitor be affected when it is completely filled with a dielectric material of dielectric constant 'K' ?** [3]

**Ans. Energy stored in a charged capacitor :** The energy of a charged capacitor is measured by the total work done in charging the capacitor to a given potential. We know that, Capacitance,

$$C = \frac{q}{V}$$

Where $q$ is the charge on the plates and V is potential difference.

When an additional amount of charge $dq$ is transferred from negative to positive plate, the small work done is given by

$$dW = Vdq = \frac{q}{C}dq$$

The total work done in transferring total charge Q is given by

$$W = \int_0^Q \frac{q}{C}dq = \frac{1}{C}\int_0^Q q\,dq = \frac{1}{C}\left[\frac{q^2}{2}\right]_0^Q = \frac{1}{C}\left[\frac{Q^2}{2} - 0\right]$$

$$= \frac{Q^2}{2C}$$

This work is stored as electrostatic potential energy U in the capacitor.

$$\therefore \quad U = \frac{Q^2}{2C}$$

or $$U = \frac{(CV)^2}{2C} \qquad [\because Q = CV]$$

or $$U = \frac{1}{2}CV^2$$

or $$U = \frac{1}{2}QV$$

When dielectric material of dielectric constant 'K' is introduced inside the capacitor :

**Ans.** (i) $$V_0 = E_0 d \qquad ...(i)$$

Where $V_0$ is the potential when there is vacuum between the plates of the capacitor and $d$ is the separation between the plates of the capacitor.

When dielectric is introduced, potential difference is given by

$$V = Ed \qquad ...(2)$$

Dividing equations (i) and (ii)

$$\frac{V_0}{V} = \frac{E_0}{E} = K$$

But $K > 1$

$\therefore$ $V_0 > V$

Thus, potential difference also decreases.

We have energy stored as $U = \frac{1}{2}QV$

Since V decreases, U also decreases.

**(ii)**

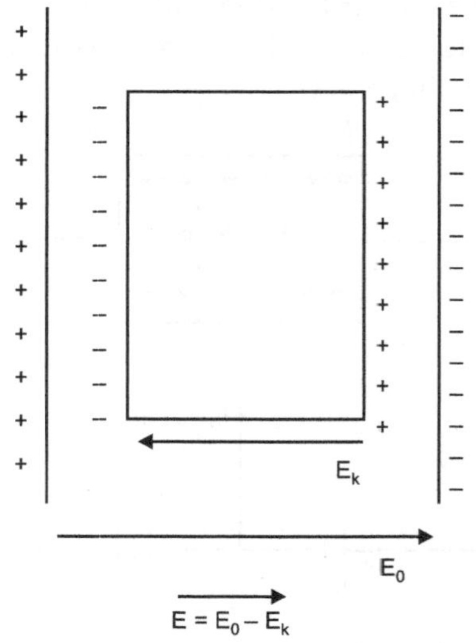

$E_k$

$E_0$

$\vec{E} = E_0 - E_k$

But $K > 1 \; \frac{E_0}{E} = K$

So, $\frac{E_0}{E} > 1$ or $E_0 > E$

Thus, electric field is reduced.

**20. Calculate the value of the resistance R in the circuit shown in the figure so that the current in the circuit is 0.2 A. What would be the potential difference between points B and E ? [3]**

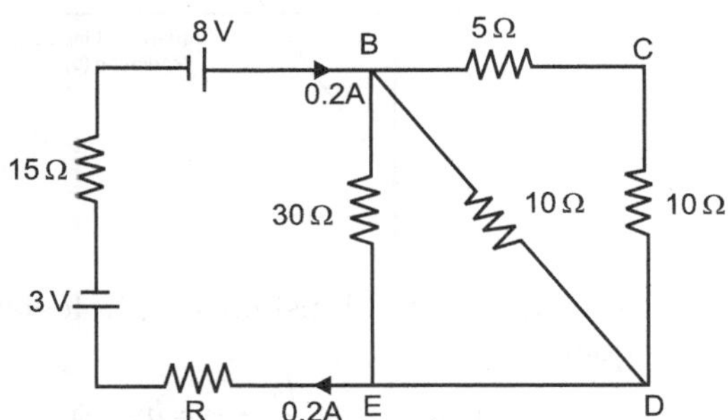


**Ans.** 

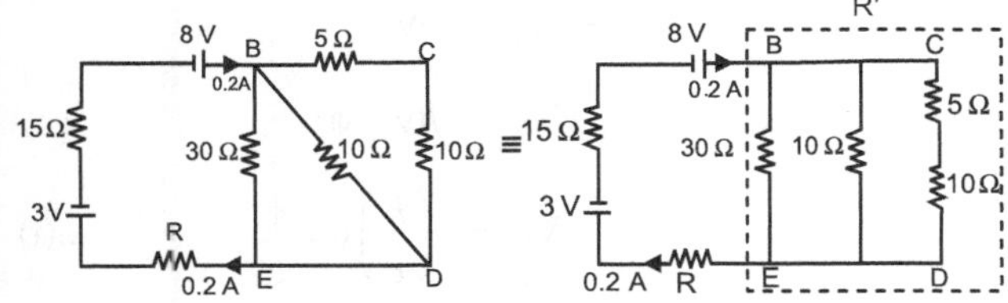


Equivalent resistance R' is given by

$$\frac{1}{R'} = \frac{1}{30} + \frac{1}{10} + \frac{1}{10+5}$$

$$= \frac{1}{30} + \frac{1}{10} + \frac{1}{15} = \frac{6}{30}$$

$$\Rightarrow \quad R' = 5\ \Omega$$

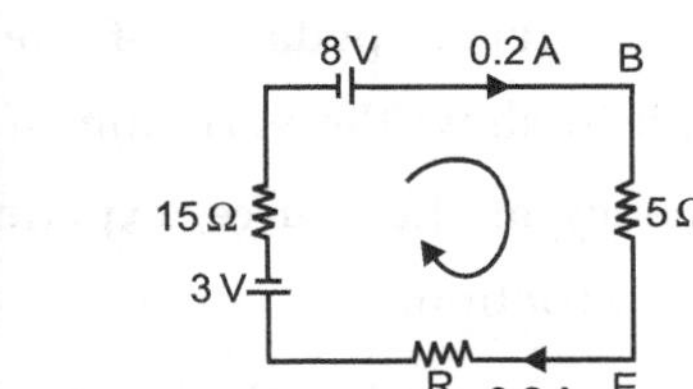


On applying Kirchhoff's second law, we get

$\Rightarrow 5(0.2) + R(0.2) + 15(0.2) = 8 - 3$

$\Rightarrow$ and $R = 5\ \Omega$

$$V_{BE} = 5(0.2) = 1\ V$$

$[\because R_{BE} = R' = 5\ \Omega]$

**21. You are given three lenses $L_1$, $L_2$ and $L_3$ each of focal length 20 cm. An object is kept at 40 cm in front of $L_1$, as shown. The final real image is formed at the focus 'I' of $L_3$. Find the separations between $L_1$, $L_2$ and $L_3$. [3]**

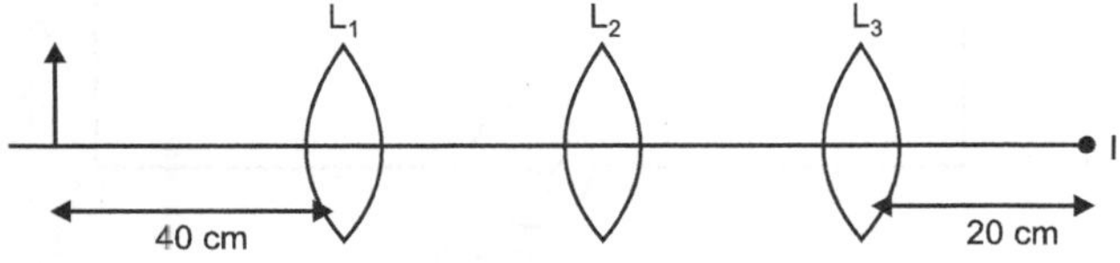


**Ans.**

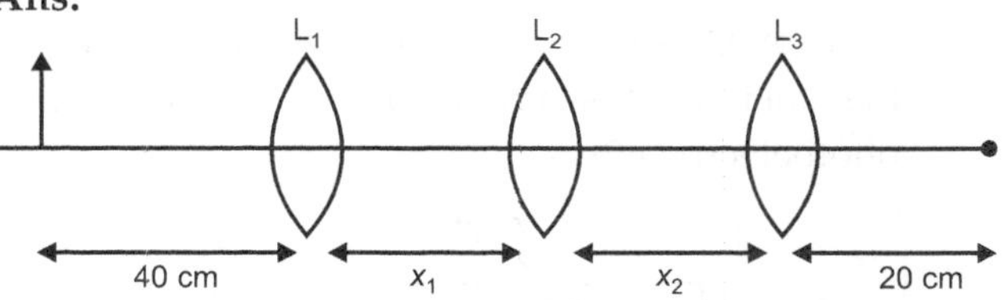


Here, $f_1 = f_2 = f_3 = 20$ cm

Now, $u_1 = -40$ cm

For lens $L_1$, from Lens Makers formula,

$$\frac{1}{v_1} = \frac{1}{u_1} = \frac{1}{f_1}$$

$$\frac{1}{v_1} = \frac{1}{f_1} + \frac{1}{u_1}$$

$$= \frac{1}{20} + \frac{1}{40}$$

$$= \frac{2-1}{40} = \frac{1}{40}$$

$$v_1 = 40 \text{ cm}$$

For lens $L_3$

$f_3 = 20$ cm, $v_3 = 20$ cm, $u_3 = ?$

By lens formula,

$$\frac{1}{v_3} - \frac{1}{u_3} = \frac{1}{f_3}$$

$$\Rightarrow \quad \frac{1}{20} - \frac{1}{u_3} = \frac{1}{20}$$

$$\Rightarrow \quad \frac{1}{u_3} = 0 \Rightarrow u_3 = \infty$$

Thus lens $L_2$ should produce image at infinity. Hence, for $L_2$, its object should be at focus.

But we have seen above that image by $L_1$ is formed at 40 cm on the right of $L_1$ which is at 20 cm left of $L_2$ (focus of $L_2$).

So, $X_1$ = distance between $L_1$ and $L_2$ = (40 + 20) cm = 60 cm

Again distance between $L_2$ and $L_3$ does not matter as the image by $L_2$ is formed at infinity. Hence, the distance between $L_2$ and $L_3$ can have any value.

**22. Define the terms (i) 'cut-off voltage' and (ii) 'threshold frequency' in relation to the phenomenon of photoelectric effect.**

**Using Einstein's photoelectric equation show how the cut-off voltage and threshold frequency for a given photosensitive material can be determined with the help of a suitable plot/graph. [3]**

**Ans.** When light of suitable frequency is incident on a metal surface, electrons are ejected from the metal. This phenomenon is called the photoelectric effect.

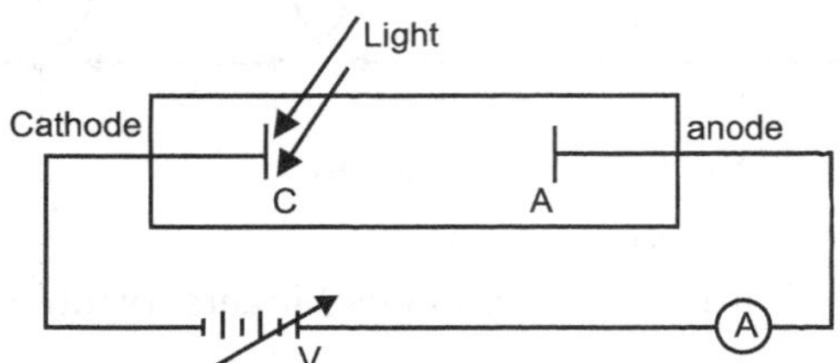


**(i)** The cathode is illuminated with light of some fixed frequency $v$ and fixed intensity $I_1$. A small photoelectric current is observed due to few electrons that reach anode just because they have sufficiently large velocity of emission. If we make the potential of the anode negative with respect to cathode then the electrons emitted by cathode are repelled. Some electrons even go back to the cathode so that the current decreases. At a certain value of this negative potential, the current is completely stopped. The least value of this anode potential which just stops the photocurrent is called cut off potential or stopping potential.

**(ii)** For a given material, there is a certain minimum frequency that if the incident radiation has a frequency below this threshold, no photoelectric emission will take place, howsoever intense the radiation may be falling. This minimum frequency is called threshold frequency.

According to Einstein's photoelectric equation, maximum K.E. is given as

$$K.E_{max} = \frac{hc}{\lambda} - \phi = hv - \phi$$

Where $\lambda$ is wavelength of light and $\nu$ is corresponding frequency and $\phi$ is work function.

We expose a material to lights of various frequencies and thus photoelectric current is observed and cut off potential needed to reduce this current to zero is noted. A graph is plotted and that is straight line.

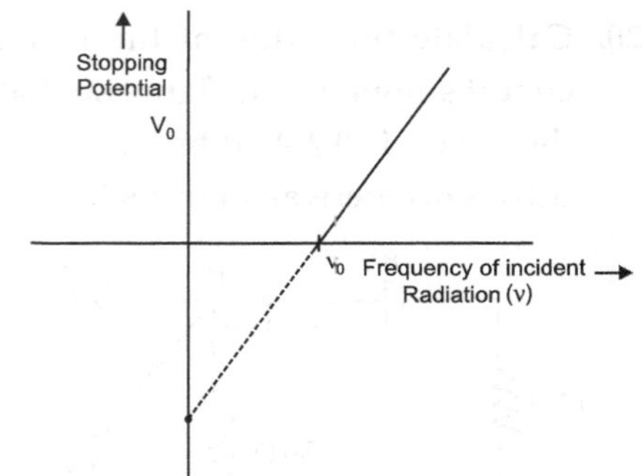


According to Einstein's photoelectric equation

$$K.E_{max} = \frac{hc}{\lambda} - \phi = h\nu - \phi$$

$$K.E_{max} = eV_0$$

$$\therefore \quad eV_0 = h\nu - \phi$$

$$V_0 = \left(\frac{h}{e}\right)\nu - \frac{\phi}{e} \quad \text{...(i)}$$

We can read the value of threshold frequency from graph.

From equation (i), we can find the value of stopping potential ($V_0$).

**23. A series LCR circuit is connected to an ac source. Using the phasor diagram, derive the expression for the impedance of the circuit. Plot a graph to show the variation of current with frequency of the source, explaining the nature of its variation.** **[3]**

**Ans.** Let an alternating emf $E = E_0 \sin \omega t$ is applied to a series combination of inductor L, capacitor C and resistance R. Since all three of them are connected in series the current through them is same. But the voltage across each element has a different phase relation with current.

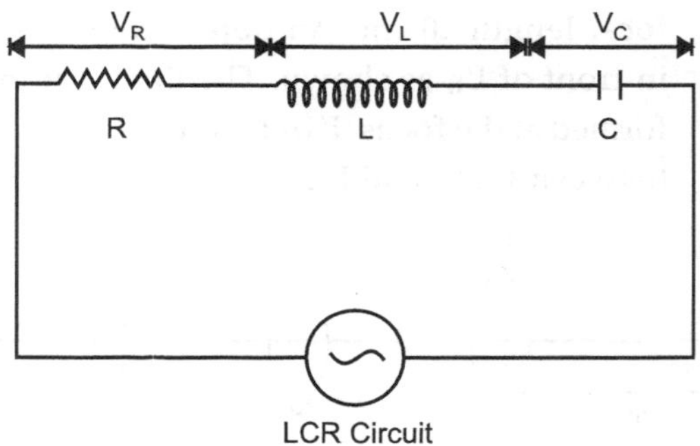


LCR Circuit

The potential difference $V_L$, $V_C$ and $V_R$ across L, C and R at any instant is given by

$$V_L = IX_L,\ V_C = IX_C \text{ and } V_R = IR$$

Where I is the current at that instant.

$X_L$ is inductive reactance and

$X_C$ is capacitive reactance.

$V_R$ is in phase with I, $V_L$ leads I by 90° and $V_C$ lags behind I by 90°.

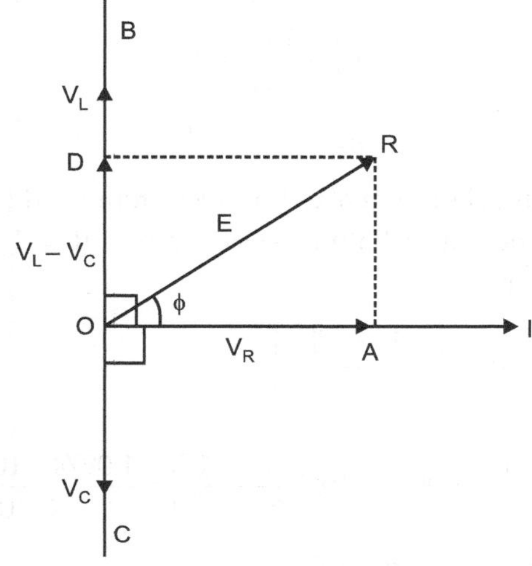


In the phasor diagram,

$V_L$ and $V_C$ are opposite to each other. If $V_L > V_C$ then resultant $(V_L - V_C)$ is represented by OD. OR represent the resultant of $V_R$ and $(V_L - V_C)$. It is equal to the applied emf E.

$$E^2 = V_R^2 + (V_L - V_C)^2$$
$$E^2 = I^2\left[R^2 + (X_L - X_C)^2\right]$$

or
$$I = \frac{E}{\sqrt{R^2 + (X_L - X_C)^2}}$$

The term $\sqrt{R^2 + (X_L - X_C)^2}$ is called impedance Z of the LCR circuit.

$$Z = \sqrt{R^2 + (X_L - X_C)^2} = \sqrt{R^2 + \left(L\omega - \frac{1}{C\omega}\right)^2}$$

Emf leads current by a phase angle φ, which is given by $r_o$

$$\tan\phi = \frac{V_L - V_C}{V_R} = \frac{X_L - X_C}{R} = \frac{L\omega - \frac{1}{C\omega}}{R}$$

When resonance takes place, $X_L = X_C$

or
$$\omega L = \frac{1}{\omega C}$$

At resonance, impedance of circuit becomes equal to R. Current becomes maximum and is equal to $\frac{E}{R}$

The graph of variation of peak current $I_{max}$ with frequency is shown below :

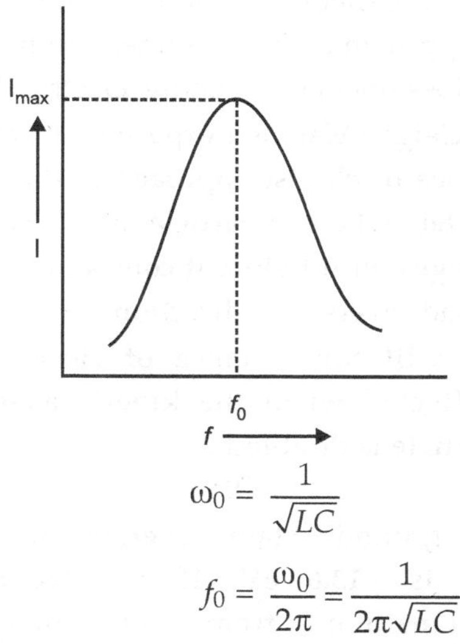


$$\omega_0 = \frac{1}{\sqrt{LC}}$$

$$f_0 = \frac{\omega_0}{2\pi} = \frac{1}{2\pi\sqrt{LC}}$$

With increase in frequency, current first increases and then decreases. At resonant frequency, the current amplitude is maximum.

**24. Mention three different modes of propagation used in communication system. Explain with the help of a diagram how long distance communication can be achieved by ionospheric reflection of radiowaves.**** **[3]**

**25. Draw a plot of potential energy of a pair of nucleons as a functions of their separations. Mark the regions where the nuclear force is (i) attractive and (ii) repulsive. Write any two characteristic features of nuclear forces.** **[3]**

**Ans.** Potential energy of a pair of nucleons as a function of their separation :

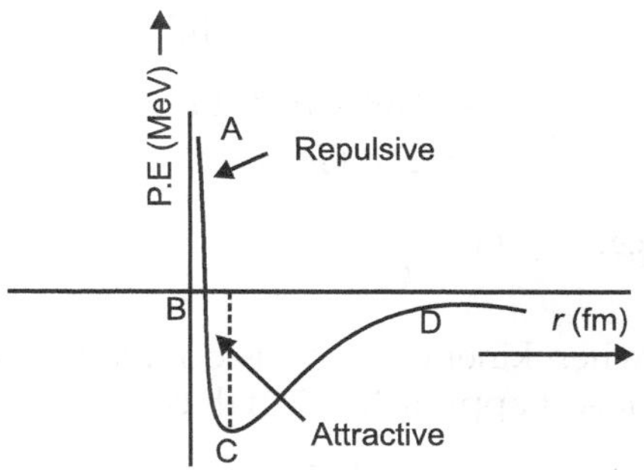


Here, Part AB represents repulsive force and part BCD represents attractive force.

$r_0$ is the distance at which potential energy is minimum.

** **Answer is not given due to change in the present Syllabus.**

Characteristic features of nuclear forces are :

**(i)** Nuclear forces are much stronger then Coulomb forces acting between charges or the gravitational forces between masses.

**(ii)** The nuclear force between neutron-neutron, proton-neutron and proton-proton is approximately the same. The nuclear force does not depend on the electric charge.

**26. In a Geiger-Marsden experiment, calculate the distance of closest approach to the nucleus of Z = 80, when α-particle of 8 MeV energy impinges on it before it comes momentarily to rest and reverses its direction.**

**How will the distance of closest approach be affected when the kinetic energy of the α-particle is doubled ? [3]**

**OR**

**The ground state energy of hydrogen atom is –13.6 eV. If an electron makes a transition from an energy level –0.85 eV to –3.4 eV, calculate the wavelength of the spectral line emitted. To which series of hydrogen spectrum does the wavelength belong?**

**Ans.** Let $r_0$ be the centre to centre distance between the alpha-particle and nucleus.

Given, $Z = 80, E_k = 8$ MeV

Now, $$E_k = \frac{1}{4\pi\varepsilon_0}\frac{(Ze)(2e)}{r_0}$$

or $$r_0 = \frac{1}{4\pi\varepsilon_0}\frac{(2e)(Ze)}{E_k}$$

$$= \frac{9\times10^9\times80\times2(1.6\times10^{-19})^2}{8\times10^6\times1.6\times10^{-19}}$$

$$= \frac{18\times10\times10^9\times1.6\times10^{-19}}{8\times10^6}$$

$$= 2.88\times10^{-14}\text{ m}$$

$$= 28.8\text{ fm}$$

Since, $$r_0 \propto \frac{1}{E_k}$$

So when kinetic energy is doubled the distance of closest approach $r_0$ is halved.

**OR**

We know that, $E_n = -\frac{13.6}{n^2}$ eV

Here, ground state energy for $n = 1$,

$$E_1 = -13.6\text{ eV}$$

Now, electron transist from $Ep = -0.85$ eV to $Eq = -3.4$ eV

$$-0.85 = \frac{-13.6}{n_P^2}$$

$$n_P^2 = \frac{13.6}{0.85} = 16$$

Thus, $$n_p = 4$$

Again, $$-3.4 = \frac{-13.6}{n_q^2}$$

$$n_q^2 = \frac{13.6}{3.4} = 4$$

$\Rightarrow$ $$n_q = 2$$

Thus, electron makes transition from $n = 4$ to $n = 2$. Hence, it is Balmer series. Now, $R = 1.0974 \times 10^7 \text{m}^{-1}$

$$\frac{1}{\lambda} = R\left(\frac{1}{n_q^2} - \frac{1}{n_p^2}\right)$$

$$\frac{1}{\lambda} = 1.0974\times10^7\left(\frac{1}{2^2} - \frac{1}{4^2}\right) = \frac{1.0974\times10^7\times12}{4\times16}$$

$$\frac{1}{\lambda} = 0.2057\times10^7$$

$$\lambda = 4.861\times10^{-7}\text{ m}$$

$$\lambda = 4861\text{ Å}.$$

**27. Define relaxation time of the free electrons drifting in a conductor. How is it related to the drift velocity of free electrons ? Use this relation to deduce the expression for the electrical resistivity of the material. [3]**

**Ans.** Relaxation time (τ) is the time for which a free electron accelerates before it undergoes a collision with the positive ion in the conductor. Or, we can say it is the average time elapsed between two successive collisions. It is of the order $10^{-14}$ seconds. It decreases with increase of temperature and is given as

$$\vec{v}_d = \vec{a}\,\tau$$

or $$\vec{v}_d = -\frac{eE}{m}\tau$$

$$\left[\because \vec{a} = \frac{-e\vec{E}}{m}\right]$$

Where $\vec{v}_d$ is the drift velocity, E is the applied electric field. $e$ and $m$ are the charge and mass of electron respectively.

Again consider a conductor with length $l$ and A as area of cross-section. Let $n$ be the number of electrons per unit volume in the conductor.

We know, $v_d = -\frac{eE}{m}\tau$

Magnitude of drift velocity

And, the current flowing through the conductor due to drift of electron is given by,

$$I = neAv_d$$

Substituting value of $v_d$

$$I = nA\left(\frac{eE\tau}{m}\right)e$$

$$I = \frac{nAe^2E\tau}{m}$$

If V is potential difference applied across the two ends then

$$E = \frac{V}{l}$$

So, $$I = \frac{nAe^2V\tau}{ml}$$

$$\frac{V}{I} = \frac{m}{ne^2\tau}\cdot\frac{l}{A}$$

Now, according to Ohm's law $\frac{V}{I}$ = R(Resistance of conductor)

Thus,

$$R = \frac{ml}{ne^2\tau A} = \frac{m}{me^2\tau}\frac{l}{A}$$

Comparing this with formula of resistance,

$$R = \rho\cdot\frac{l}{A}$$

Where $\rho$ is the resistivity of the material.

We get, $\rho = \frac{m}{ne^2\tau}$

**28. (a) In Young's double slit experiment, derive the condition for**

**(i) constructive interference and**

**(ii) destructive interference at a point on the screen.**

**(b) A beam of light consisting of two wavelengths, 800 nm and 600 nm is used to obtain the interference fringes in a Young's double slit experiment on a screen placed 1.4 m away. If the two slits are separated by 0.28 mm, calculate the least distance from the central bright maximum where the bright fringes of the two wavelengths coincide. [5]**

**OR**

**(a) How does an unpolarized light incident on a polaroid get polarized ? Describe briefly, with the help of a necessary diagram, the polarization of light by reflection from a transparent medium.**

**(b) Two polaroids 'A' and 'B' are kept in crossed position. How should a third polaroid 'C' be placed between them so that the intensity of polarized light transmitted by polaroid B reduces to 1/8th of the intensity of unpolarized light incident on A ?**

**Ans. (a) Condition of constructive and destructive interference :**

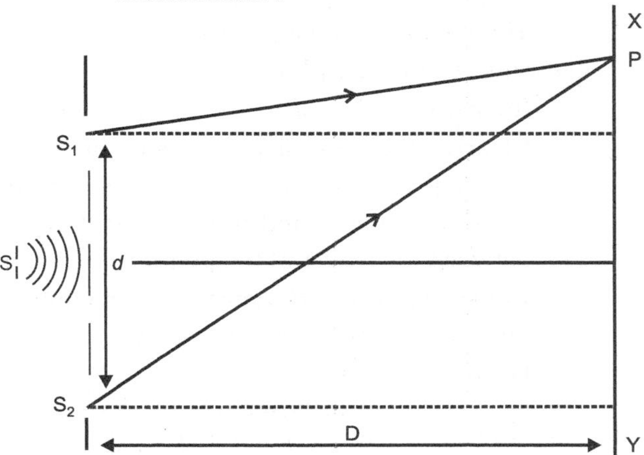


In the given figure, S is a monochromatic source of light. $S_1$ and $S_2$ are two narrow pin holes equidistant from S and they act as coherent sources. Consider a point P on the screen XY placed parallel to $S_1$ and $S_2$.

Let $a_1$ be the amplitude of the waves from $S_1$ and $a_2$ is from $S_2$. Let $\phi$ be the phase difference between the two waves reaching the point P. Let $y_1$ and $y_2$ be the displacements of the two waves, arriving at P.

$$y_1 = a_1 \sin \omega t \quad ...(1)$$

$$y_2 = a_2 \sin(\omega t + \phi) \quad ...(2)$$

By the principle of superposition, the resultant displacement of the two waves at P is

$$y = y_1 + y_2 = a_1 \sin\omega t + a_2 \sin(\omega t + \phi)$$

$$y = a_1 \sin \omega t + a_2 \sin \omega t \cos \phi + a \cos \omega t \sin\phi$$

$$= \sin \omega t\,(a_1 + a_2 \cos\phi) + \cos \omega t\,(a_2 \sin\phi) \quad ...(3)$$

$$a_1 + a_2 \cos\phi = A\cos\theta \quad ...(4)$$

$$a_2 \sin\phi = A\sin\theta \quad ...(5)$$

Substituting in equation (3),

$$y = \sin \omega t \, A \cos\theta + \cos \omega t \, A \sin\theta$$
$$= A \sin(\omega t + \theta) \quad ...(6)$$

Thus, the resultant vibration is S.H.M. and the resultant amplitude is A.

Squaring and adding equation (4) and equation (5),

$$A^2 \cos^2\theta + A^2 \sin^2\theta = (a_1 + a_2 \cos\phi)^2 + a_2^2 \sin^2\phi$$
$$A^2 = a_1^2 + 2a_1a_2 \cos\phi + a_2^2 \cos^2\phi + a_2^2 \sin^2\phi$$
$$= a^2{}_1 + a_2{}^2 + 2a_1a_2 \cos\phi$$

The amplitude of the resultant wave is A

$$A = \sqrt{a_1^2 + a_2^2 + 2a_1a_2 \cos\phi}$$

The resultant intensity at P is given by the square of the amplitude

Intensity, $I = A^2 = a^2{}_1 + a_2{}^2 + 2a_1a_2 \cos\phi$

Clearly, the intensity of resultant wave at any point depends on the amplitude of individual waves and the phase difference between the waves at the point.

**Constructive interference :** For maximum intensity at any point, $\cos\phi = +1$

For phase difference $\phi$

$$= 0, 2\pi, 4\pi, 6\pi \, .......$$
$$= 2n\pi \, (n = 0, 1, 2, .......)$$

The maximum intensity,

$$I_{max} = a_1^2 + a_2^2 + 2a_1a_2 = (a_1 + a_2)^2$$

Path difference,

$$\Delta = \frac{\lambda}{2\pi} \times \text{Phase difference}$$
$$= \frac{\lambda}{2\pi} \times 2n\pi$$
$$= n\lambda$$

Clearly, the maximum intensity is obtained in the region of superposition at those points where waves meet in the same phase or the phase difference between the waves is even multiple of $\pi$ or path difference between them is the integral multiple of $\lambda$ and maximum intensity is $(a_1 + a_2)^2$ which is greater than the sum intensities of individual waves by an amount $2a_1a_2$.

**Destructive interference :** For minimum intensity at any point, $\cos\phi = -1$.

for phase difference $\phi = \pi, 3\pi, 5\pi, 7\pi......$

$$= (2n - 1)\pi, \, n = 1, 2, 3......$$

In this case the minimum intensity,

$$I_{min} = a_1^2 + a_2^2 - 2a_1a_2 = (a_1 - a_2)^2$$

Path difference,

$$\Delta = \frac{\lambda}{2\pi} \times \text{Phase difference}$$
$$= \frac{\lambda}{2\pi} \times (2n-1)\,\pi = (2n-1)\frac{\lambda}{2}$$

Clearly, the minimum intensity is obtained in the region of superposition at those points where waves meet in opposite phase or the phase difference between the waves is odd multiple of $\pi$ or path difference between the waves is odd multiple of $\frac{\lambda}{2}$ and minimum intensity $= (a_1 - a_2)^2$ which is less than the sum of intensities of the individual waves by an amount $2a_1a_2$.

**(b)** Given : $\lambda_1 = 800 \text{ nm} = 800 \times 10^{-9}\text{m}$

$\lambda_2 = 600 \text{ nm} = 600 \times 10^{-9}\text{m}$

$D = 1.4 \text{ m}$

$d = 0.28 \text{ mm} = 0.28 \times 10^{-3}\text{m}$

Suppose $n_1{}^{th}$, maximum corresponds to wavelength $\lambda_1$ and it coincides with $n_2{}^{th}$ maximum corresponding to wavelength $\lambda_2$.

$$\therefore \quad n_1 \frac{\lambda_1 D}{d} = n_2 \frac{\lambda_2 D}{d}$$

$$\text{or} \quad \frac{n_1}{n_2} = \frac{\lambda_2}{\lambda_1} = \frac{600}{800} = \frac{3}{4}$$

Thus, 3rd maximum corresponding to wavelength 800 nm coincides with 4th maximum corresponding to wavelength 600 nm.

And the minimum distance is given by,

$$x_{min} = n_1 \frac{\lambda_1 D}{d} = \frac{3 \times 800 \times 10^{-9} \times 1.4}{0.28 \times 10^{-3}}$$
$$x_{min} = 12 \text{ mm}$$

**OR**

**(a)** Polaroid is made up of a special material which blocks one of the two planes of vibration of an electromagnetic wave. Because of its chemical composition, it allows only those vibrations of the electromagnetic wave which are parallel to its crystallographic axis.

** **Answer is not given due to change in the present Syllabus.**

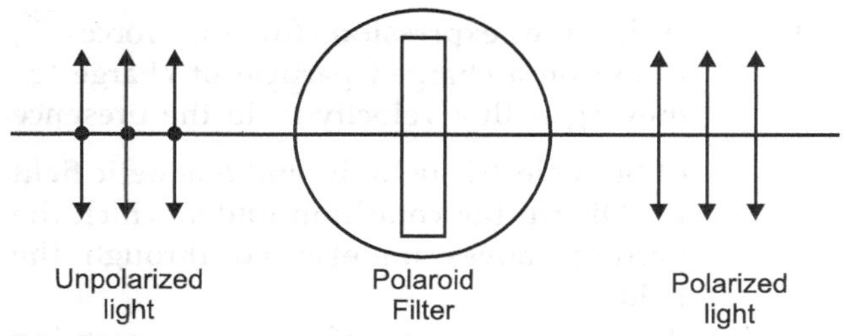


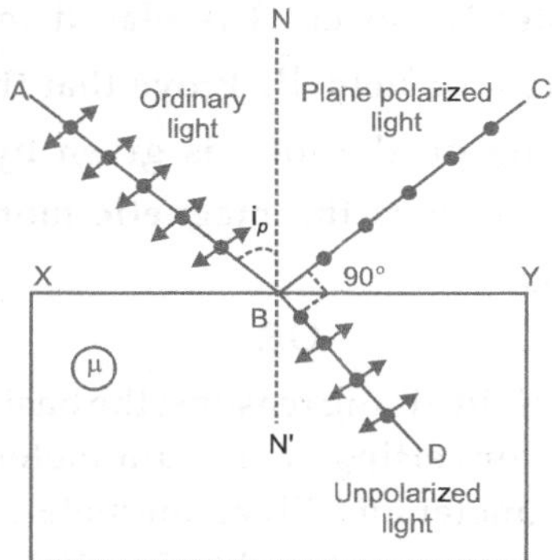


An ordinary beam of light on reflection from a transparent medium becomes partially polarized. The degree of polarization increases as the angle of incidence is increased. At a particular value of angle of incidence, the reflected beam becomes completely polarized. This angle of incidence is called the polarizing angle ($i_p$).

**(b)** By Malus law, the intensity of light emerging from the middle polaroid 'C' is given by

$$I_1 = I_0 \cos^2 \theta$$

This light (intensity $I_1$) falls on the polaroid 'B' whose polarization axis makes an angle of (90° – θ) with the polarization axis of the polaroid 'C'. Therefore, the intensity of light emerging from 'B' is given by

$$I_2 = I_1 \cos^2 (90° - \theta) = I_0 \cos^2 \theta \cos^2 (90° - \theta)$$

$$= I_0 \cos^2\theta \sin^2\theta = \frac{2I_0}{8}$$

Where $2I_0$ is intensity of unpolarized light incident on A

$$4 \cos^2 \theta \sin^2 \theta = 1$$

$$(2 \sin \theta \cos \theta)^2 = 1$$

$$(\sin 2\theta)^2 = 1$$

$$\sin 2\theta = 1 = \sin 90°$$

$$\theta = \frac{90°}{2} = 45°$$

Thus, polaroid 'C' must be placed at an angle 45° with polaroid 'B'.

**29. (a) Describe briefly, with the help of a diagram, the role of the two important processes involved in the formation of a *p-n* junction.**

**(b) Name the device which is used as a voltage regulator. Draw the necessary circuit diagram and explain its working. [5]**

**OR**

**(a) Explain briefly the principle on which a transistor-amplifier works as an oscillator. Draw the necessary circuit diagram and explain its working.**

**(b) Identify the equivalent gate for the following circuit and write its truth table.****

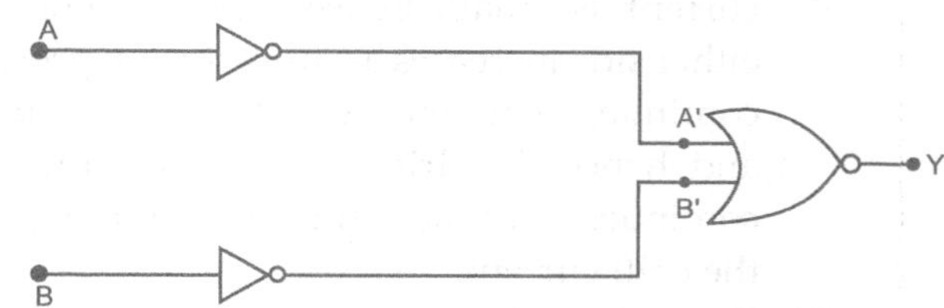


**Ans. (a)** Two important process involved in the formation of a *p-n* junction are (i) diffusion and (ii) drift.

In *n*-type semiconductor, electrons are the majority carriers and holes are minority carriers. In the same way, in *p*-type semiconductor holes are majority and electrons are minority carriers. During the formation of *p-n* junction, due to concentration gradient, the holes diffuse from *p* side to *n* side and electrons diffuse from *n* side to *p* side. This motion gives rise to **diffusion current** across the junction.

When an electron diffuses from *n* to *p* side, it leaves behind a positive charge. In such a manner a positively charged layer forms on *n*-side of the junction.

Similarly, when a hole diffuses from *p* to *n* side, it leaves behind a negative charge and a negatively charged layer forms on *p* side of the junction.

This space charge region is known as **depletion region**. An electric field directed from *n*-region to *p*-region develops across the junction. Due to this field, electrons on *p* side of the junction move to *n*-side and holes on *n* side of the junction move to *p* side. This motion of charge carriers due to the electric field is called **drift**.

Drift current is opposite in direction to the diffusion current.

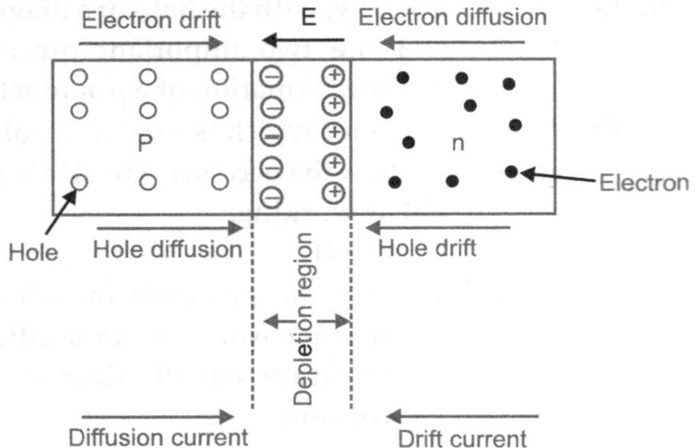


Initially, diffusion current is large and drift current is small. Space-charge region on either side increases as the diffusion process continues. This increases the electric field and hence the drift current. This process continues until the diffusion current equals the drift current.

**(b)** Zener diode is used as a voltage regulator.

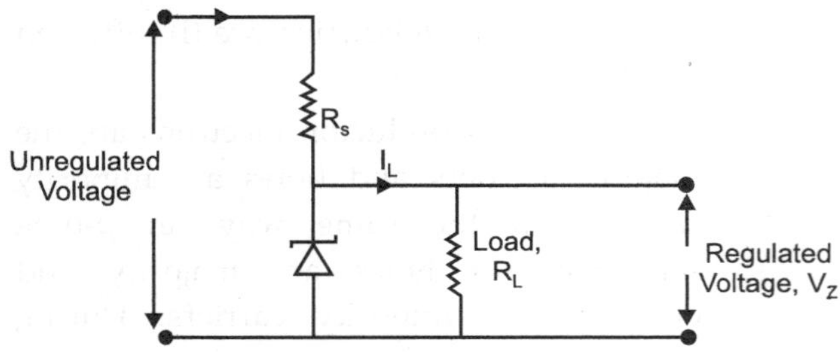


Voltage regulator converts an unregulated dc voltage into a constant regulated dc voltage using zener diode. The unregulated voltage is connected to the zener diode through a series resistance $R_S$ such that the zener diode is reverse biased. If the input voltage increases, the current through $R_S$ and zener diode also increases. This increases the voltage drop across $R_S$ without any change in voltage drop across zener diode. This is because in the breakdown region, zener voltage remain constant even though the current through zener diode changes.

Similarly, if the input voltage decreases, the current through $R_S$ and zener diode decrease. The voltage drop across $R_S$ decrease without any change in the voltage across the zener diode. Thus any change in input voltage results the change in voltage drop across $R_S$ without any change in voltage across the zener diode.

**30. (a) Write the expression for the force $\vec{F}$, acting on a charged particle of charge 'q', moving with a velocity $\vec{v}$ in the presence of both electric field $\vec{E}$ and magnetic field $\vec{B}$. Obtain the condition under which the particle moves undeflected through the fields.**

**(b) A rectangular loop of size l × b carrying a steady current I is placed in a uniform magnetic field $\vec{B}$. Prove that the torque $\vec{\tau}$ acting on the loop is given by, $\vec{\tau} = \vec{m} \times \vec{B}$ where $\vec{m}$ is the magnetic moment of the loop.** **[5]**

**OR**

**(a) Explain giving reasons, the basic difference in converting a galvanometer into (i) a voltmeter and (ii) an ammeter.**

**(b) Two long straight parallel conductors carrying steady currents $I_1$ and $I_2$ are separated by a distance 'd'. Explain briefly, with the help of a suitable diagram, how the magnetic field due to one conductor acts on the other. Hence deduce the expression for the force acting between the two conductors. Mention the nature of this force.**

**Ans. (a)** Force acting on a charge '$q$' moving with velocity $\vec{v}$ in the presence of both electric field $\vec{E}$ and magnetic field $\vec{B}$ is given by,

$$\vec{F} = q\vec{E} + q(\vec{v} \times \vec{B})$$

Consider a region in which magnetic field, electric field and velocity of charge particle are perpendicular to each other.

To move charge particle undeflected, the net force acting on the particle must be zero *i.e.*, the electric force must be equal and opposite to the magnetic force.

$$q\vec{E} = -q(\vec{v} \times \vec{B})$$

$$\vec{E} = -(\vec{v} \times \vec{B})$$

$$\vec{E} = \vec{B} \times \vec{v}$$

$$\vec{E} = Bv\sin\theta = Bv$$

$$v = \frac{E}{B}$$

The direction of electric and magnetic forces are in opposite direction. Their magnitudes are in such a way that they cancel out each other to give net force zero and so the charge particle does not deflect.

**(b)** When an electric current flows in closed loop of wire, placed in a uniform magnetic field, the magnetic forces produce a torque which

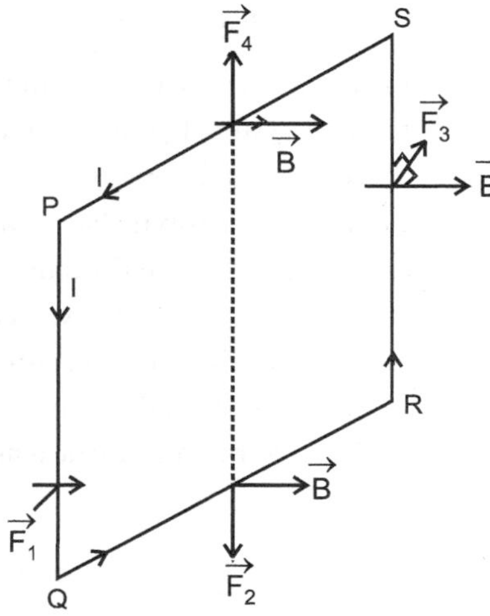


**Fig. (a)**

tends to rotate the loop so that area of the loop is perpendicular to the direction of the magnetic field.

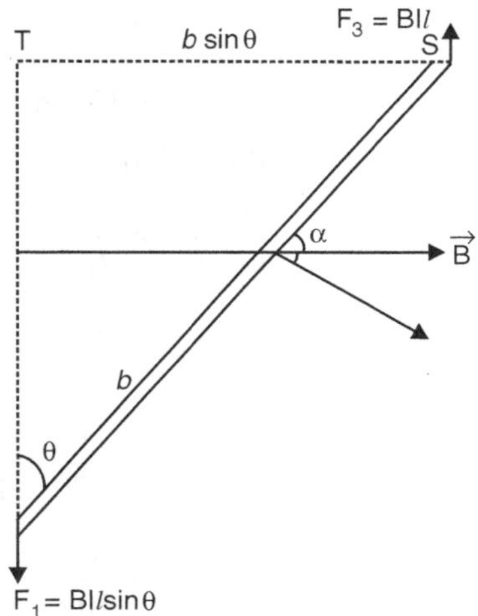


**Fig. (b)**

Consider a rectangular coil PQRS placed in an external magnetic field as shown in Fig (a). Let 'I' be the current flowing through the coil. Each part of the coil experiences a Lorentz force. Forces of each part is $\vec{F_1}$, $\vec{F_2}$, $\vec{F_3}$ and $\vec{F_4}$ as shown. The $\vec{F_4}$ and $\vec{F_2}$ are equal in magnitude but act in opposite directions along the same straight line. Hence, they cancel out each other.

The force $\quad \vec{F_1} = I(\vec{PQ} \times \vec{B})$

$F_1 = IlB \qquad (\because \theta = 90°)$

($\vec{F_1}$ acts in direction perpendicular to the plane of paper)

Similarly, $\quad \vec{F_3} = I(\vec{RS} \times \vec{B})$

$F_3 = IlB$

These two forces constitute a couple and so rotates the coil in the anticlockwise direction. The torque

$$\tau = \text{force} \times \text{arm of couple}$$
$$\tau = Fb\cos\theta$$
$$\tau = IlBb\cos\theta$$
$$\tau = IAB\cos\theta \qquad [\because l \times b = A]$$

If the coil has N turns then

$$\tau = NIAB\cos\theta$$

The area vector A makes an angle α with $\vec{B}$ so

$$\theta + \alpha = 90°$$
$$\cos\theta = \cos(90 - \alpha) = \sin\alpha$$
$$\therefore \quad \tau = NIAB\sin\alpha$$
$$\tau = mB\sin\alpha$$

or $\quad \vec{\tau} = \vec{m} \times \vec{B}$

where $m$ = NIA is called the magnetic dipole moment of the loop.

**OR**

**(a)** **(i)** In converting a galvanometer into a voltmeter, a very high suitable resistance is connected in series to its coil, so, that galvanometer gives full scale deflection

**(ii)** In converting a galvanometer into an ammeter, a very small suitable resistance is connected in parallel to its coil. The remaining pair of the current *i.e.*, $(I - I_g)$ flows through the resistance.

**(b)** **Assumption** : Current flowing in the conductors are in the same direction.

Using right hand thumb rule, the direction of the magnetic field at point P due to current $I_2$ is perpendicular to the plane of paper and inwards.

Similarly, at point Q on $X_2Y_2$, the direction of magnetic field due to current $I_1$ is perpendicularly outward.

Using Fleming's left hand rule we can find the direction of forces $F_{12}$ and $F_{21}$ which are in opposite directions,

Thus, by Ampere's circuital law, we have,

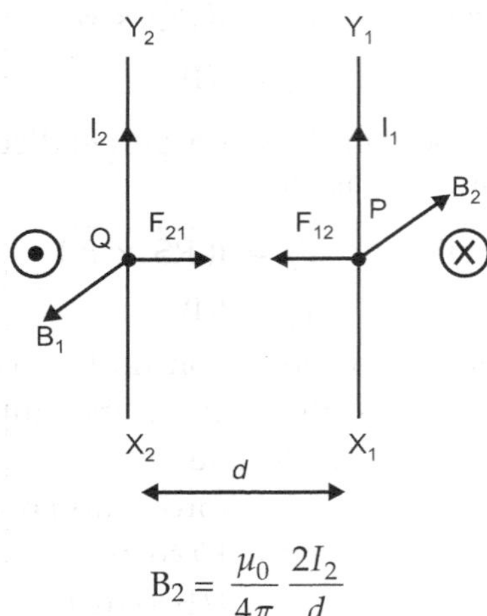


$$B_2 = \frac{\mu_0}{4\pi}\frac{2I_2}{d}$$

Now, $F_{12} = I_1LB_2$ (Where L = length of the conductors)

$$F_{12} = \frac{\mu_0}{4\pi}\frac{2I_1I_2L}{d} = \frac{\mu_0}{2\pi}\frac{I_1I_2L}{d}$$

In similar manner, we get

$$F_{21} = \frac{\mu_0}{2\pi}\frac{I_1I_2L}{d} \quad ...(i)$$

From above we get the magnitude of forces $F_{12}$ and $F_{21}$ are equal but in opposite direction. So, $F_{12} = -F_{21}$

Therefore, two parallel straight conducting carrying current in the same direction attract each other. Similarly, we can prove if two parallel straight conductors carry currents in opposite direction, they repel each other with the same magnitude as equation (i).

## Outside Delhi [Set-II]

**Note : Except for the following questions, all the remaining questions have been asked in previous Set.**

**8. In a single-slit diffraction experiment, the width of the slit is made double the original width. How does this affect the size and intensity of the central diffraction band ? [1]**

**Ans.** In single-slit diffraction experiment, fringe width is given as

$$\beta = \frac{2\pi D}{d}$$

Where, $\beta$ = fringe width

$\lambda$ = wavelength of light used

D = the distance between screen and slit

$d$ = the width of the slit.

If $d$ is doubled, the width of the central maximum is halved. Thus there is a reduction in the size of the central diffraction band. Intensity of central band of the diffraction pattern varies square of the slit width so as the slit gets double, intensity will get four times.

**19. You are given three lens $L_1$, $L_2$, $L_3$ each of focal length 15 cm. An object is kept at 20 cm in front of $L_1$, as shown. The final real image is formed at the focus 'I' of $L_3$. Find the separations between $L_1$, $L_2$ and $L_3$. [3]**

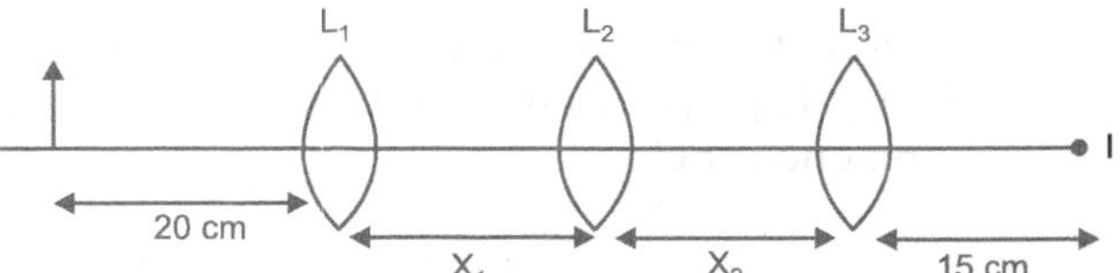


**Ans.** We have, $f_1 = f_2 = f_3 = 15$ cm

By lens formula for lens $L_1$,

$$\frac{1}{v_1} - \frac{1}{u_1} = \frac{1}{f_1}$$

$$\frac{1}{v_1} - \frac{1}{-20} = \frac{1}{15}$$

$$\frac{1}{v_1} = \frac{1}{-20} + \frac{1}{15}$$

$$\frac{1}{v_1} = \frac{1}{60}$$

$$\Rightarrow \quad v_1 = 60 \text{ cm}$$

For lens $L_3$, $v_3 = f_3 = 15$ cm, it means image by $L_3$ is formed at focus. By using lens formula we find that the object lie at infinity of $L_3$. Hence, $L_2$ will produce image at infinity. So, we can conclude that object for $L_2$ should be at its focus.

But, we have seen above the image by $L_1$ is formed at 60 cm right of $L_1$ which is at 15 cm left of $L_2$ (focus of $L_2$).

So, $X_1$ = distance between $L_1$ and $L_2$ = (60 + 15) cm = 75 cm

Again distance between $L_2$ and $L_3$ does not matter as the image by $L_2$ is formed at infinity so $X_2$ can take any value.

**27. In a Geiger Marsden experiment, calculate the distance of closest approach to the nucleus of Z = 75, when an particle of 5 MeV energy impinges on it before it comes momentarily to rest and reverse its direction.**

**How will the distance of closest approach be affected when the kinetic energy of the-particle is doubled ?** [3]

**OR**

**The ground state energy of hydrogen atom is –13.6 eV. If an electron make a transition from an energy level –0.85 eV to –1.51 eV, calculate the wavelength of the spectral line emitted. To which series of hydrogen spectrum does this wavelength belong ?**

**Ans.** Let $r_0$ be the centre to centre distance between the alpha-particle and nucleus when the α-particle is at its stopping point.

$$\text{Now, } E_k = \frac{1}{4\pi\varepsilon_0}\frac{(2e)(Ze)}{r_0}$$

$$r_0 = \frac{1}{4\pi\varepsilon_0}\frac{(2e)(Ze)}{E_K}$$

Given, $E_K = 5 \times 10^6$ eV

$= 5 \times 10^6 \times 1.6 \times 10^{-19}$ V

Z = 75

$$r_0 = \frac{9\times10^9 \times 75\times 2\times(1.6\times10^{-19})^2}{5\times10^6\times1.6\times10^{-19}}$$

$$= \frac{3456\times10^9\times10^{-38}}{8\times10^{-13}}$$

$= 432 \times 10^{-16}$m

$= 43.2 \times 10^{-15}$m

= 43.2 fm

Since distance of closest approach ($r_0$) is given as

$$r_0 = \frac{1}{4\pi\varepsilon_0}\frac{(2e)(Ze)}{E_K}$$

$$\Rightarrow \quad r_0 \propto \frac{1}{E_K}$$

So, when kinetic energy of the α-particle is doubled the distance between closest approach $r_0$ is halved.

**OR**

Since energy of hydrogen atom is given as

$$E_n = \frac{-13.6}{n^2}\,eV$$

Ground state energy, $E_1 = -13.6\,eV$

Now electron makes a transition from an energy level

$E_p = -0.85$ eV to $E_q = -1.51$ eV

Now,

$$E_p = \frac{-13.6}{n_p^2}eV$$

$$n_p^2 = \frac{-13.6}{-0.85} = 16$$

$$n_p = 4$$

Again,

$$E_q = \frac{-13.6}{n_p^2}eV$$

$$n_p^2 = \frac{-13.6}{-1.51}$$

$$n_q = 3$$

Thus, we have transition from $n = 4$ to $n = 3$. Since transition corresponds to the transition of an electron from $n = 4$ to an orbit having $n = 3$. It is Paschen series of the hydrogen spectrum.

Wavelength is given as

$$\frac{1}{\lambda} = R\left(\frac{1}{n_q^2} - \frac{1}{n_p^2}\right)$$

Where R is Rydberg's constant

$$\Rightarrow \quad \frac{1}{\lambda} = 1.0974 \times 10^7\left(\frac{1}{9} - \frac{1}{16}\right)$$

$$\Rightarrow \quad \frac{1}{\lambda} = 1.0974 \times 10^7 \times \frac{7}{9\times16}$$

$$\Rightarrow \quad \lambda = \frac{9\times16}{1.097\times10^7\times7}$$

$= 18.752 \times 10^{-7}$m

= 1875 mm.

# Outside Delhi [Set-III]

**Note : Except for the following questions, all the remaining questions have been asked in previous Sets.**

1. **How does the fringe width, in Young's double-slit experiment change when the distance of separation between the slits and screen is doubled ? [1]**

**Ans.** We have, frings width

$$\beta = \frac{\lambda D}{d}$$

$$\Rightarrow \quad \beta \propto D$$

The fringe width becomes double when the distance of separation between the slits and screen is doubled.

4. **The speed of an electromagnetic wave in a material medium is given by $v = \frac{1}{\sqrt{\mu\varepsilon}}$, $\mu$ being the permeability of the medium and $\varepsilon$ is its permittivity. How does its frequency change ? [1]**

**Ans.** The frequencies of electromagnetic waves have its inherent characteristics. When an electromagnetic wave travels from one medium to another, its wavelength changes but frequency remains unchanged.

7. **A proton and an electron have same kinetic energy. Which one has smaller de-Broglie wavelength and why ? [1]**

**Ans.** In terms of kinetic energy, wavelength is given by

$$\lambda = \frac{h}{\sqrt{2m E_k}}$$

$$\Rightarrow \quad \lambda \propto \frac{1}{\sqrt{m}}$$

So wavelength is inversely proportional to $\sqrt{m}$, *i.e.*, more the mass, less the wavelength and vice-versa.

So, for same kinetic energy, as a proton has a larger mass than an electron, thus a proton has smaller de-Broglie wavelength than a electron.

9. **A circular coil of closely wound N turns and radius $r$ carries a current I. Write the expression for the following : [2]**
   **(i) the magnetic field at its centre**
   **(ii) the magnetic moment of this coil**

**Ans.** (i) Magnetic field at its centre is given by,

$$B = \frac{\mu_0 NI}{2r}$$

**(ii)** Magnetic moment of this coil is

$$m = NIA = NI\,\pi r^2$$

12. **A light bulb is rated 150 W for 220 V a.c supply of 60 Hz. Calculate [2]**
    **(i) the resistance of the bulb**
    **(ii) the rms current through the bulb**

**OR**

**An alternating voltage given by $V = 70 \sin 100\pi t$ is connected across a pure resistor of 25 Ω. Find**
**(i) the frequency of the source.**
**(ii) the rms current through the resistor.**

**Ans.** (i) Resistance of the bulb,

$$R = \frac{V^2}{P} = \frac{(220)^2}{150} = \frac{48400}{150}$$

$$R = 322.6\ \Omega$$

**(ii)** The rms current through the bulb

$$I = \frac{P}{V}$$

$$\therefore \quad I = \frac{150}{220}$$

$$I = 0.68\ A$$

**OR**

An ac voltage is given by

$$V = V_0 \sin \omega t$$

Given, $V = 70 \sin 100\,\pi t$

$\therefore \quad \omega = 100\pi$ and $V_0 = 70$ V

**(i)** Frequency of the source,

$$f = \frac{\omega}{2\pi} = \frac{100\pi}{2\pi} = 50\ \text{Hz}$$

**(ii)** $$V_{rms} = \frac{V_0}{\sqrt{2}} = \frac{70}{\sqrt{2}} = 49.5\ V$$

$$I_{rms} = \frac{V_{rms}}{R} = \frac{49.5}{25} = 1.98\ A$$

**20. Explain briefly the following terms used in communication system :**** **[3]**

**(i) Transducer**

**(ii) Repeater**

**(iii Amplification**

**22. You are given three lens $L_1$, $L_2$ and $L_3$ each of focal length 10 cm. An object is kept at 15 cm in front of $L_1$, as shown. The final real image is formed at the focus 'I' of $L_3$. Find the separations between $L_1$, $L_2$ and $L_3$.** **[3]**

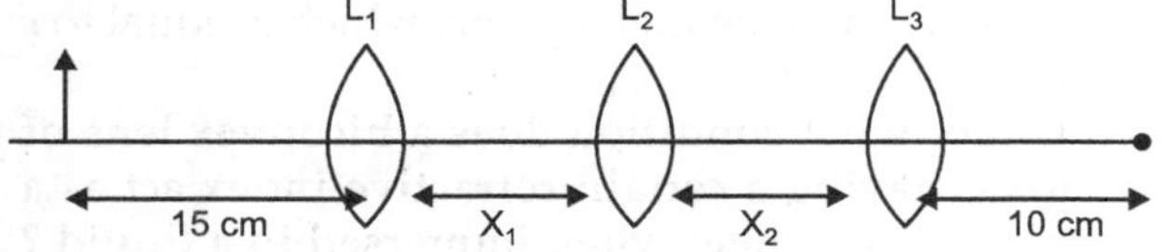


**Ans.** We have,

$$f_1 = f_2 = f_3 = 10 \text{ cm}$$

For lens $L_1$, from lens makers formula,

$$u_1 = -15 \text{ cm}$$

$$f_1 = 10 \text{ cm}$$

By lens formula,

$$\frac{1}{v_1} \quad \frac{1}{u_1} = \frac{1}{f_1}$$

$$\frac{1}{v_1} - \frac{1}{-15} = \frac{1}{10}$$

$$\frac{1}{v_1} = \frac{1}{10} - \frac{1}{15}$$

$$\frac{1}{v_1} = \frac{3-2}{30} = \frac{1}{30}$$

$$\frac{1}{v_1} = \frac{1}{30}$$

$$\Rightarrow \qquad v_1 = 30 \text{ cm}$$

For lens $L_3$, $v_3 = f_3 = 10$ cm, it means image by $L_3$ is formed at focus. So by using lens formula, we get that object should lie at infinity for $L_3$. Hence, $L_2$ will produce image at infinity. So, we can conclude that object for $L_2$ should be at its focus.

But, we have seen above that image by $L_1$ is formed at 30 cm right of $L_1$ which is at 10 cm left of $L_2$ (focus of $L_2$).

So, $X_1$ = distance between $L_1$ and $L_2$ = (30 + 10) cm = 40 cm

Again distance between $L_2$ and $L_3$ does not matter as the image by $L_2$ is formed at infinity so $X_2$ can take any value.

** Answer is not given due to change in the present Syllabus.

# PHYSICS

## Delhi [Set-I]

# 2012

**Time allowed : 3 hours** **Maximum marks : 70**

1. **When electrons drift in a metal from lower to higher potential, does it mean that all the free electrons of the metal are moving in the same direction ?** [1]

**Ans.** No, when electric field is applied, the electrons will have net drift from lower to higher field but locally electrons may collide with ions and may change their direction of motion.

2. **The horizontal component of the earth's magnetic field at a place is B and angle of dip is 60°. What is the value of vertical component of earth's magnetic field at equator ?** [1]

**Ans.** On the equator, the value of both angle of dip ($\delta$) and vertical component of earth's magnetic field is zero. So, in this case, $B_v = 0$.

3. **Show on a graph, the variation of resistivity with temperature for a typical semiconductor.** [1]

**Ans.** The following curve shows the variation of resistivity with temperature for a typical semiconductor.

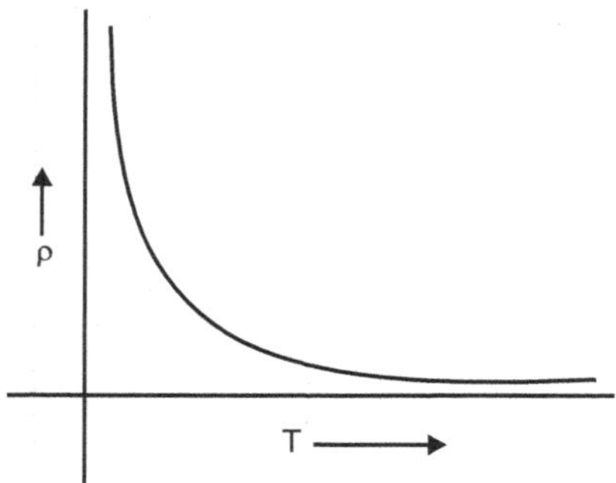


This is because, for a semiconductor, resistivity decreases rapidly with increasing temperature.

4. **Why should electrostatic field be zero inside a conductor ?** [1]

**Ans.** Charge on conductor resides on its surface. So if we consider a Gaussian surface inside the conductor to find the electrostatic field,

$$\phi = \frac{q}{\varepsilon_0}$$

Where, $q$ = charge enclosed in Gaussian surface. $q = 0$, inside the conductor, Hence the electrostatic field inside the conductor is zero.

5. **Name the physical quantity which remains same for microwaves of wavelength 1 mm and UV radiations of 1600 Å in vacuum.** [1]

**Ans.** Both microwaves and UV rays are a part of the electromagnetic spectrum. Thus, the physical quantity that remains same for both types of radiation will be their speeds which is equal to $c$ ($3 \times 10^8$ m/s).

6. **Under what condition does a biconvex lens of glass having a certain refractive index act as a plane glass sheet when immersed in a liquid ?** [1]

**Ans.** A biconvex lens will act like a plane sheet of glass if it is immersed in a liquid having the same index of refraction as itself.

7. **Predict the directions of induced currents in metal rings 1 and 2 lying in the same plane where current I in the wire is increasing steadily.** [1]

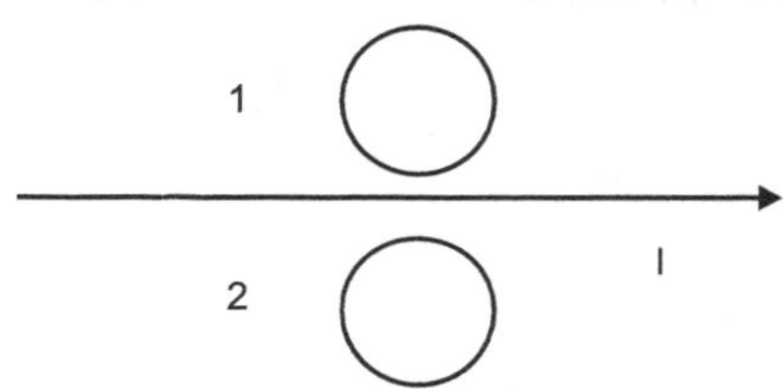


**Ans.** Using Lenz's law we can predict the direction of induced current in both the rings. Induced current oppose the cause of increase of magnetic flux. So,

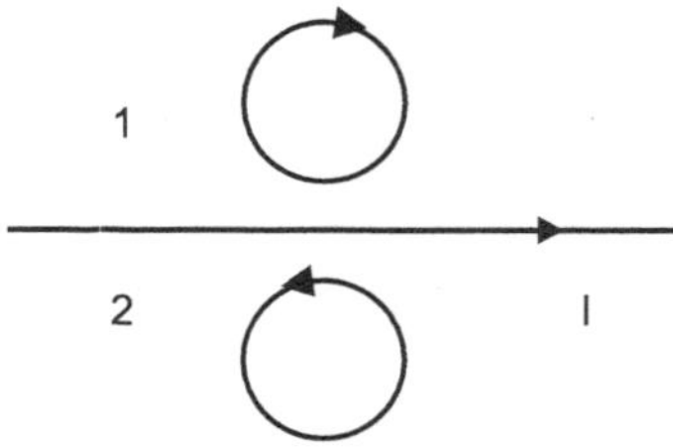


It will be clockwise in ring 1 and anticlockwise in ring 2.

8. **State de-Broglie hypothesis.** [1]

**Ans.** de Broglie postulated that the material particles may exhibit wave aspect. Accordingly a moving material particle behaves as wave and the wavelength associated with material particle is

$$\lambda = \frac{h}{mv}$$

where $h$ = Planck's constant
$m$ = mass of the object
$v$ = velocity of the object

**9. A ray of light incident on an equilateral prism ($\mu_g = \sqrt{3}$) moves parallel to the base line of the prism inside it. Find the angle of incidence for this ray. [2]**

**Ans.** It is given that the prism is equilateral in shape. So, all the angles are equal to 60°. Thus, the angle of prism, A = 60°.

The angle of refraction in case of a prism,

$$r = \frac{A}{2} = 30°$$

We can now apply Snell's law

$$\mu_a \sin i = \mu_g \sin r$$

Here, $\mu_a$ = refractive index of air = 1

$\mu_g$ = refractive index of glass = $\sqrt{3}$

$i$ = angle of incidence

$$\sin i = \left(\frac{\mu_g}{\mu_a}\right)\sin r = \left(\frac{\mu_g}{\mu_a}\right)\sin 30°$$

$$\sin i = \frac{\sqrt{3}}{2}$$

So, the angle of incidence is $i = 60°$.

**10. Distinguish between 'Analog and Digital signals'. [2]**

**OR**

**Mention the functions of any two of the following used in communication system :**

**(i) Transducer**

**(ii) Repeater**

**(iii) Transmitter**

**(iv) Bandpass Filter**

**11. A cell of emf E and internal resistance r is connected to two external resistance $R_1$ and $R_2$ and a perfect ammeter. The current in the circuit is measured in four different situations**

**(i) without any external resistance in the circuit**

**(ii) with resistance $R_1$ only**

**(iii) with $R_1$ and $R_2$ in series combination**

**(iv) with $R_1$ and $R_2$ in parallel combination**

**The currents measured in the four cases are 0.42 A, 1.05 A, 1.4 A and 4.2 A, but not necessarily in the order. Identify the currents corresponding to the four cases mentioned above. [2]**

**Ans.** The current relating to corresponding situations is as follows :

(i) Without any external resistance in the circuit :

$$I_1 = \frac{E}{r}$$

The current in this case would be maximum.

So, $I_1 = 4.2$ A

** Answer is not given due to change in the present Syllabus.

(ii) With resistance $R_1$ only :

$$I_2 = \frac{E}{r + R_1}$$

The current in this case will be second smallest value,

So, $I_2 = 1.05$ A

(iii) With $R_1$ and $R_2$ in series combination

$$I_3 = \frac{E}{r + (R_1 + R_2)}$$

The current in this case will be minimum as the resistance will be maximum.

So, $I_3 = 0.42$ A.

(iv) With $R_1$ and $R_2$ in parallel combination

$$I_4 = \frac{E}{r + \left(\frac{R_1 R_2}{R_1 + R_2}\right)}$$

The current in this case would be the second largest value.

So, $I_4 = 1.4$ A.

**12. The susceptibility of a magnetic material is –2.6 × 10$^{-5}$. Identify the type of magnetic material and state its two properties. [2]**

**Ans.** Diamagnetic materials have negative susceptibility. So the given magnetic material is diamagnetic.

Two properties of diamagnetic material :

(i) They do not obey Curie's law.

(ii) They are feebly repelled by a magnet.

**13. Two identical circular wires P and Q each of radius R and carrying current 'I' are kept in perpendicular planes such that they have a common centre as shown in the figure. Find the magnitude and direction of the net magnetic field at the common centre of the two coils. [2]**

**Ans.** Magnetic field produced by the two coils at their common centre having currents $I_1$ and $I_2$, radius $a_1$ and $a_2$, number of turns $N_1$ and $N_2$ respectively are given by :

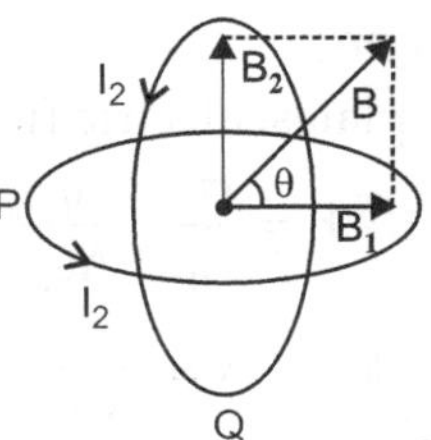


$$B_1 = \frac{\mu_0 N_1 I_1}{2a_1}$$

$$B_2 = \frac{\mu_0 N_2 I_2}{2a_2}$$

The resultant field at the common centre is :

$$B = \sqrt{B_1^2 + B_2^2}$$

$$= \sqrt{\left(\frac{\mu_0 N_1 I_1}{2a_1}\right)^2 + \left(\frac{\mu_0 N_2 I_2}{2a_2}\right)^2}$$

Here $N_1 = N_2 = 1$, $I_1 = I_2 = I$, $a_1 = a_2 = R$

$$B = \sqrt{\left(\frac{\mu_0 \times 1 \times I}{2R}\right)^2 + \left(\frac{\mu_0 \times 1 \times I}{2R}\right)^2}$$

$$= \frac{\mu_0 I}{2R}\sqrt{1+1} = \sqrt{2}\left(\frac{\mu_0 I}{2R}\right)$$

Direction of B, $\tan\theta = \dfrac{B_2}{B_1} = 1$

$\therefore \quad \theta = 45°$

Hence, the net magnetic field is directed at an angle of 45° with either of the fields.

**14. When an ideal capacitor is charged by a d.c. battery, no current flows. However, when an a.c. source is used, the current flows continuously. How does one explain this based on the concept of displacement current ? [2]**

**Ans.** When an ideal capacitor is charged by d.c. battery, charge flows till the capacitor gets fully charged.

When an a.c. source is connected then conduction current $i_c = \dfrac{dq}{dt}$ flows in the connecting wire.

Due to changing current, charge deposited on the plates of the capacitor changes with time. Changing charge causes electric field between the plates of capacitor to be varying, giving rise to displacement current $i_d = \varepsilon_0 \dfrac{d\phi_c}{dt}$ [As displacement current is proportional to the rate of flux variation].

Between the plates, electric field

$$E = \frac{\sigma}{\varepsilon_0} = \frac{q}{A\varepsilon_0}$$

Electric flux, $\phi_c = EA = \dfrac{q}{A\varepsilon_0}A$

$$\text{So, } i_d = \frac{\varepsilon_0 d\phi_c}{dt} = \frac{\varepsilon_0 d}{dt}\left(\frac{qA}{A\varepsilon_0}\right) = \frac{\varepsilon_0 \times A}{\varepsilon_0 \times A}\frac{dq}{dt} = \frac{dq}{dt} = i_c$$

Displacement current brings continuity in the flow of current between the plates of the capacitor.

**15. Draw a plot showing the variation of (i) electric field (E) and (ii) electric potential (V) with distance r due to a point charge Q. [2]**

**Ans.** We know that for a point charge Q.

Electric potential, $V = \dfrac{Q}{4\pi\varepsilon_0 r}$

or $\quad V \propto \dfrac{1}{r}$

Electric field, $\quad E = \dfrac{Q}{4\pi\varepsilon_0 r^2}$

or $\quad E \propto \dfrac{1}{r^2}$

Thus, electric potential shows an inverse relationship with distance $r$ while electric field shows a inverse square relationship with $r$. So, their corresponding plots would be

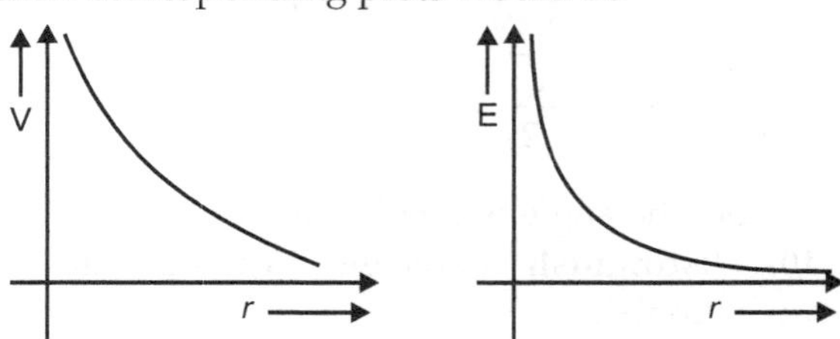


**16. Define self-inductance of a coil. Show that magnetic energy required to build up the current I in a coil of self inductance L is given by $\frac{1}{2}LI^2$. [2]**

**Ans.** Self inductance is the inherent inductance of a circuit, given by the ratio of the electromotive force produced in the circuit by self-induction to the rate of change of current producing it. It is also called coefficient of self-induction.

Suppose I = Current flowing in the coil at any time.

$\phi$ = Amount of magnetic flux linked.

It is found that $\phi \propto I$

$$\phi = LI$$

where, L is the constant of proportionality and is called coefficient of self induction.

SI unit of self-inductance is henry.

Let at $t = 0$, the current in the inductor is zero. So at any instant $t$, the current in the inductor is I and rate of growth of I is $dI/dt$.

Then, the induced emf is $E = L \times dI/dt$

If the source is sending a constant current I through the inductor for a small time $dt$, then small amount of work done by the source is given by

$$dW = EI\,dt = (L\,dI/dt)I\,dt = LI\,dI$$

The total amount of work done by the source of emf, till the current increases from its initial value I = 0 to its final value I is given by

$$W = \int_0^1 LIdI = L\int_0^1 IdI = L\left[\frac{I^2}{2}\right] = \frac{1}{2}LI^2$$

This work done by the source of emf used in building up current from zero to I is stored in the inductor in energy form. Therefore, energy stored in the inductor is

$$U = \frac{1}{2}LI^2$$

**17. The current in the forward bias is known to be more (~ mA) than the current in the reverse bias (~ μA). What is the reason, then, to operate the photodiode in reverse bias ? [2]**

**Ans.** The current in the forward bias is due to majority carriers whereas current in the reverse bias is due to minority carriers. So current in forward bias is more (~mA) than current in reverse bias (~μA).

On illumination of photodiodes with light, the fractional change in the majority carriers would be much less than that in minority carriers. It implies that fractional change due to light on minority carriers dominated reverse bias current is more easily measurable than fractional change in forward bias current. So, photodiodes are operated in reverse bias condition.

**18. A metallic rod of 'L' length is rotated with angular frequency of 'ω' with one end hinged at the centre and the other end at the circumference of a circular metallic ring of radius L about an axis passing through the centre and perpendicular to the plane of the ring. A constant and uniform magnetic field B parallel to the axis is present everywhere. Deduce the expression for the emf between the centre and the metallic ring. [2]**

**Ans.** The induced emf,

$$e = \frac{d\phi_B}{dt}$$

$$e = \frac{d}{dt}(BA\cos\phi) \quad [\because \phi_B = BA\cos\phi]$$

$$= B\frac{dA}{dt} \quad [\because \phi = 0°]$$

where $\frac{dA}{dt}$ = Rate of change of area of loop formed by the sector OPQ. Let θ be the angle between the rod and the radius of the circle at P at time $t$.

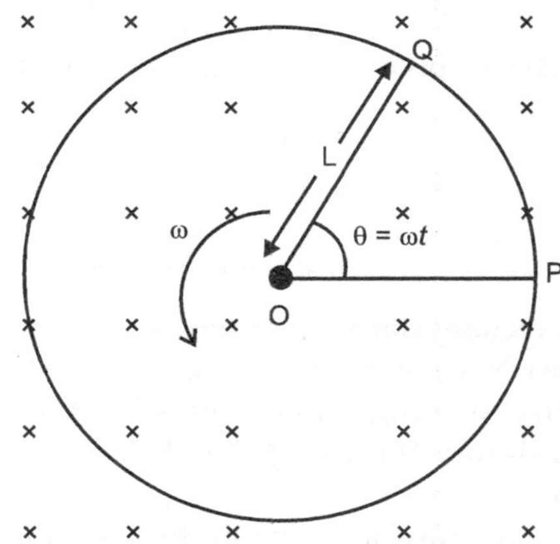


The area of the sector OPQ = $\pi L^2 \times \frac{\theta}{2\pi} = \frac{1}{2}L^2\theta$

where L = Radius of the circle.

Hence $e = B \times \frac{d}{dt}\left(\frac{1}{2}L^2\theta\right) = \frac{1}{2}BL^2\frac{d\theta}{dt} = \frac{B\omega L^2}{2}$

**19. The figure shows a series LCR circuit with L = 5.0 H, C = 80 μF, R = 40 Ω connected to a variable frequency 240 V source, calculate [3]**

**(i) the angular frequency of the source which drives the circuit at resonance,**

**(ii) the current at the resonating frequency,**

**(iii) the rms potential drop across the inductor at resonance.**

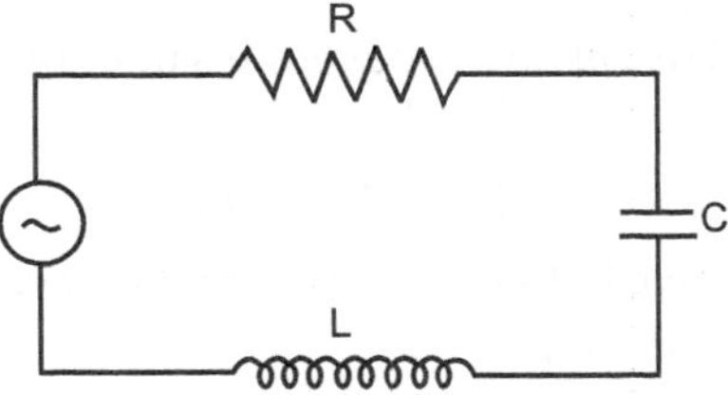


**Ans.** Given, L = 5.0 H, C = 80 μF, R = 40 Ω, V = 240 V

(i) Resonant angular frequency

$$\omega = \frac{1}{\sqrt{LC}} = \frac{1}{\sqrt{5\times 80\times 10^{-6}}}$$

$$= \frac{1}{\sqrt{400\times 10^{-6}}}$$

$$= \frac{1}{20\times 10^{-3}}$$

$$= \frac{1000}{20}$$

$$\omega = 50 \text{ rad s}^{-1}$$

**(ii)** At resonant frequency, we know that the inductive reactance cancels out the capacitive reactance. i.e, $X_L = X_\omega$

$\therefore$ Impedance, $Z = R = 40\ \Omega$

The current at resonant frequency

$$I_{rms} = \frac{V_{rms}}{R} = \frac{240}{40} = 6\text{ A}$$

**(iii)** rms potential drop across inductor

$$V_L = I_{rms} \times X_L$$
$$= I_{rms} \times \omega L$$
$$= 6 \times 50 \times 5 = 1500\text{ V}$$

**20. A rectangular loop of wire of size 4 cm × 10 cm carries a steady current of 2 A. A straight long wire carrying 5 A current is kept near the loop as shown. If the loop and the wire are coplanar, find** [3

**(i) the torque acting on the loop and**

**(ii) the magnitude and direction of the force on the loop due to the current carrying wire.**

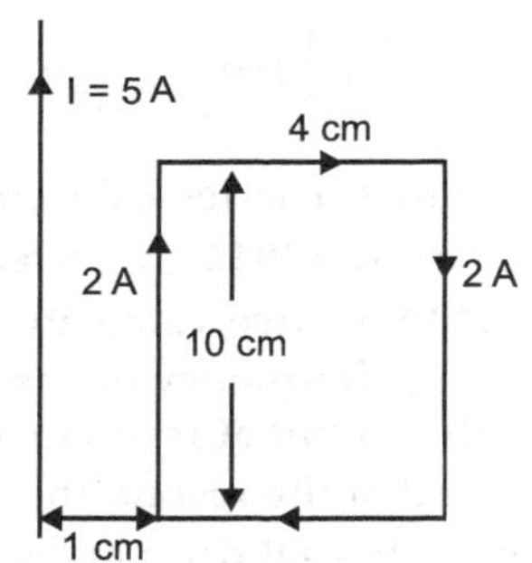


**Ans. (i)** $\vec{\tau} = \vec{m} \times \vec{B} = \vec{m}\, B \sin\theta$

Here, $m$ and B have the same direction

$\therefore$ $\theta = 0°$

$|\vec{\tau}| = mB \sin\theta = 0$

**(ii)** Force between two current carrying wires is given by,

$$F = \frac{\mu_0}{2\pi} \cdot \frac{I_1 I_2}{r} \times l$$

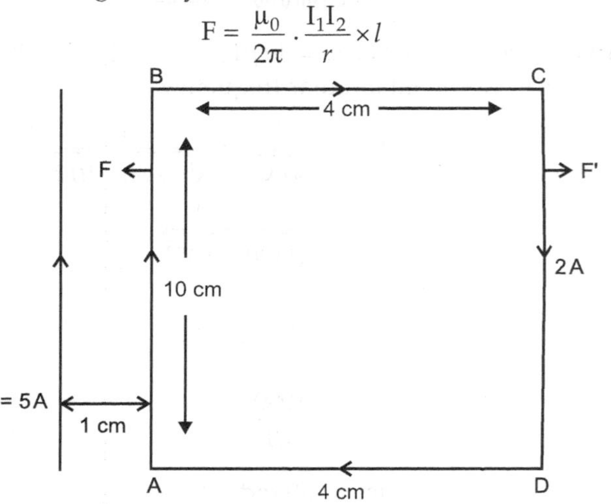


Force on wires BC and AD being equal and opposite will cancel out each other. Now, force on wire AB

$$F = \frac{\mu_0}{2\pi} \times \frac{5 \times 2}{1 \times 10^{-2}} \times (0.10)$$

$$F = \frac{\mu_0}{2\pi} \times 100$$

force will be attractive as current is flowing in same direction

And force on wire CD

$$F' = \frac{\mu_0}{2\pi} \times \frac{5 \times 2}{5 \times 10^{-2}} \times (0.10)$$

$$F' = \frac{\mu_0}{2\pi} \times 20$$

force will be repulsive as current is in opposit direction

Now, resultant force on loop,

$$F_{net} = F - F'$$
$$= \frac{\mu_0}{2\pi}[100 - 20]$$
$$= 2 \times 10^{-7} \times 80 = 16 \times 10^{-6}\text{ N}$$

The direction of net force is towards the straight wire *i.e.*, attractive.

**21. (a) Using Bohr's second postulate of quantization of orbital angular momentum show that the circumference of the electron in the $n^{th}$ orbital state in hydrogen atom is $n$ times the de Broglie wavelength associated with it.**

**(b) The electron in hydrogen atom is initially in the third excited state. What is the maximum number of spectral lines which can be emitted when it finally moves to the ground state ?** **[3]**

**Ans. (a)** According to Bohr's second postulate of quantization, the electron can revolve around the nucleus only in those circular orbits in which the angular momentum of the electron is integral multiple of $\frac{h}{2\pi}$ where $h$ is Planck's constant ($= 6.62 \times 10^{-34}$ Js).

So, if $m$ is the mass of electron and $v$ is the velocity of electron in permitted quantized orbit with radius $r$ then

$$mvr = n\frac{h}{2\pi} \quad ...(1)$$

Where $n$ is the principal quantum number and can take integral values like

$n = 1, 2, 3, .....$

This is the Bohr's quantization condition.

Now, de-Broglie wavelength is given as

$$\lambda = \frac{h}{mv}$$

Where $\lambda$ = Wavelength of wave associated with electron

$m$ = Mass of the electron

$h$ = Planck's constant

$$mv = \frac{h}{\lambda}$$

$$\text{or } mvr = \frac{hr}{\lambda} \quad ...(2)$$

From equations (i) and (ii)

$$\therefore \quad \frac{h}{\lambda}r = \frac{nh}{2\neq}$$

$$\frac{r}{\lambda} = \frac{n}{2\neq}$$

$$2\pi r = n\lambda$$

Now, circumference of the electron in the $n^{th}$ orbital state of Hydrogen atom with radius $r$ is $n\lambda$.

**(b)** If $n$ is quantum number of the highest energy level involved in the transitions, then the total number of possible spectral lines emitted is

$$N = \frac{n(n-1)}{2}$$

Third excited state means fourth energy level *i.e.* $n = 4$. Here, electron makes transition from $n = 4$ to $n = 1$. So, highest $n$ is $n = 4$.

Thus, possible spectral lines

$$N = \frac{4(4-1)}{2}$$

$$= \frac{4 \times 3}{2} = 6$$

The maximum possible number of spectral lines is 6.

**22. In the figure a long uniform potentiometer wire AB is having a constant potential gradient along its length. The null points for the two primary cells of emfs $\varepsilon_1$ and $\varepsilon_2$ connected in the manner shown are obtained at a distance of 120 cm and 300 cm from the end A. Find (i) $\varepsilon_1/\varepsilon_2$ and (ii) position of null point for the cell $\varepsilon_1$.**

**How is the sensitivity of a potentiometer increased ?** [3]

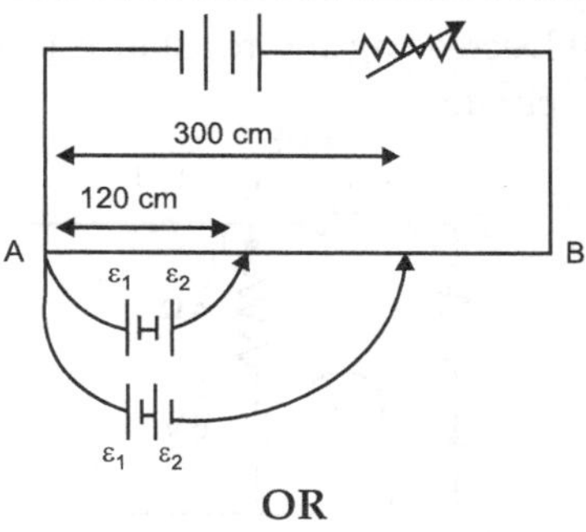


**OR**

**Using Kirchhoff's rules determine the value of unknown resistance R in the circuit so that no current flows through 4 Ω resistance. Also find the potential between A and D.**

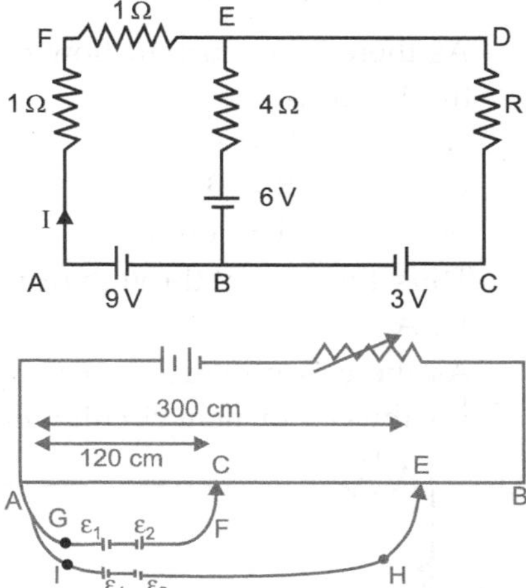


**Ans.**

**(i)** Apply Kirchhoff's law in loop ACFGA :

$$k(120) = \varepsilon_1 - \varepsilon_2$$

$k$ = potential drop per unit length

$$\text{or} \quad \varepsilon_1 = \varepsilon_2 + k(120) \quad ...(1)$$

For loop AEHIA :

$$k(300) = \varepsilon_2 + \varepsilon_1$$

By substituting value of $\varepsilon_1$ from equation (i),

$$\varepsilon_2 + [\varepsilon_2 + k(120)] = k(300)$$

$$2\varepsilon_2 = k(300 - 120)$$

$$\text{or,} \quad \varepsilon_2 = 90k \quad ...(ii)$$

$$\text{Thus,} \quad \varepsilon_1 = 90k + 120k$$

$$\varepsilon_1 = 210k \quad ...(iii)$$

$$\text{Hence,} \quad \frac{\varepsilon_1}{\varepsilon_2} = \frac{210}{90} = \frac{7}{3}$$

**(ii)** As we know, $\varepsilon = kl$

Thus, from equations (ii) and (iii),

Null point for cell $\varepsilon_2$ is 90 cm

And for cell $\varepsilon_1$, it is 210 cm.

Sensitivity of the potentiometer can be increased by :

**(a)** Increasing the length of the potentiometer wire.

(b) Decreasing the resistance in the primary circuit.

**OR**

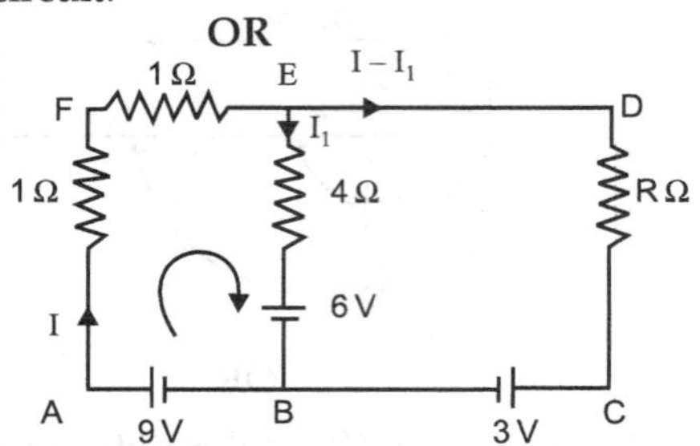


Apply Kirchhoff's law in loop AFEBA :

$$I + I + 4I_1 = 9 - 6$$

$$2I + 4I_1 = 3 \quad \text{...(i)}$$

As there is no current flowing through the 4 Ω resistance,

$\therefore \quad I_1 = 0$

or $\quad 2I = 3$

or $\quad I = 1.5$ A

Thus, the current through resistance R is 1.5 A.

As there is no current through branch EB, thus equivalent circuit will be,

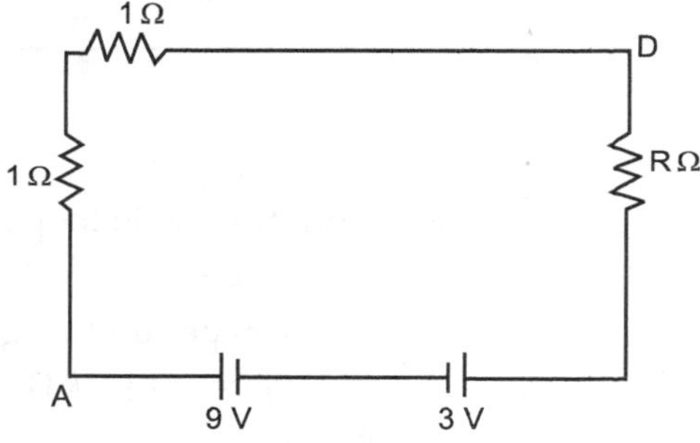


By applying Kirchhoff's loop law, we get

$$1.5 + 1.5 + R(1.5) = 9 - 3$$

$$R = 2\ \Omega$$

Potential difference between A and D = $2 \times 1.5 = 3$ V.

**23. (i) What characteristic property of nuclear force explains the constancy of binding energy per nucleon (BE/A) in the range of mass number 'A' lying 30 < A < 170 ?**

**(ii) Show that the density of nucleus over a wide range of nuclei is constant and independent of mass number A.** **[3]**

**Ans. (i)** The constancy of BE/A over most of the range is due to saturation property of nuclear force.

In heavy nuclei : nuclear size > range of nuclear force.

So, nucleons experiences nearly constant interaction.

**(ii)** To find the density of nucleus of an atom, we have an atom with mass number let say A and let mass of the nucleus of the atom of the mass number A be $m$ A.

Where, $m$ is mass of one nucleon.

Let radius of nucleus be R.

Then, volume of nucleus $= \frac{4}{3}\pi R^3$

$$= \frac{4}{3}\pi(R_0 A^{1/3})^3 \quad [\because R = R_0 A^{1/3}]$$

$$= \frac{4}{3}\pi R_0^3 A$$

Density of the nucleus,

$$\rho = \frac{\text{Mass of nucleus}}{\text{Volume of nucleus}}$$

$$\rho = \frac{mA}{\frac{4}{3}\pi R_0^3 A} = \frac{3m}{4\pi R_0^3}$$

This expression is independent of mass number A and since $m$, π and $R_0$ have same values for any atom, therefore, density is constant also.

**24. Write any two factors which justify the need for modulating a signal.**

**Draw a diagram showing an amplitude modulated wave by superposing a modulating signal over a sinusoidal carrier wave.**** **[3]**

**25. Write Einstein's photoelectric equation. State clearly how this equation is obtained using the photon picture of electromagnetic radiation.**

**Write the three salient features observed in photoelectric effect which can be explained using this equation.** **[3]**

**Ans.** Einstein's photoelectric equation,

$$h\nu = K_{max} + \phi$$

Where, $h$ = Planck's constant

$\nu$ = frequency of radiation

$\phi$ = Work function

According to Planck's quantum theory, light radiations consist of small packets of energy.

Einstein postulated that a photon of energy $h\nu$ is absorbed by the electron of the metal surface, then the energy equal to $\phi$ is used to liberate electron from the surface and rest of the energy $h\nu - \phi$ becomes the kinetic energy of the electron.

** **Answer is not given due to change in the present Syllabus.**

$\therefore$ Energy of photon is,

$$E = h\nu$$

The minimum energy required by the electron of a material to escape out of its work function '$\phi$'.

The additional energy acquired by the electron appears as the maximum kinetic energy '$K_{max}$' of the electron.

*i.e.,* $\quad K_{max} = h\nu - \phi$

or $\quad h\nu = K_{max} + \phi$

where $\phi = eV_0$.

Sailent features observed in photoelectric effect :

1. The stopping potential and hence the maximum kinetic energy of emitted electrons varies linearly with the frequency of incident radiation.
2. There exists a minimum cut-off frequency $\nu_0$, for which the stopping potential is zero.
3. Photoelectric emission is instantaneous.

**26. (a) Why are coherent sources necessary to produce a sustained interference pattern ?**

**(b) In Young's double slit experiment using monochromatic light of wavelength λ, the intensity of light at a point on the screen where path difference is λ, is K units. Find out the intensity of light at a point where path difference is λ/3.** **[3]**

**Ans.** (a) Coherent sources have constant phase difference between them *i.e.,* phase difference does not change with time. Hence, the intensity distribution on the screen remains constant and sustained.

(b) We know

Phase difference $= \left(\frac{2\pi}{\lambda}\right) \times$ Path difference

At path difference λ

Phase difference, $\phi = \frac{2\pi}{\lambda} \times \lambda = 2\pi$

Intensity, $I = 4I_0 = \cos^2 \frac{\phi}{2}$

But I = K at path difference λ

$$\therefore \quad K = 4I_0 \cos^2 \frac{2\pi}{2}$$

or $\quad K = 4I_0$

or $\quad I_0 = \frac{K}{4}$

Now, at path difference $= \frac{\lambda}{3}$

$$\phi' = \frac{2\pi}{\lambda} \times \frac{\lambda}{3} = \frac{2\pi}{3}$$

Intensity, $\quad I' = 4I_0 \cos^2 \frac{1}{2}\left(\frac{2\pi}{3}\right)$

$$I' = 4 \times \frac{K}{4} \cos^2 \frac{\pi}{3}$$

$$= K \times \left(\frac{1}{2}\right)^2$$

$$= \frac{1}{4}K$$

**27. Use Huygen's principle to explain the formation of diffraction pattern due to a single slit illuminated by a monochromatic source of light.**

**When the width of the slit is made double the original width, how would this affect the size and intensity of the central diffraction band ?** **[3]**

**Ans.** Consider a parallel beam of monochromatic light is incident normally on a slit of width *b* as shown in figure. According to Huygen's principle, every point of slit acts as a source of secondary wavelets spreading in all directions. Screen is placed at a larger distance.

Consider a particular point P on the screen that receives waves from all the secondary sources. All these waves start from different point of the slit and interfere at point P to give resultant intensity.

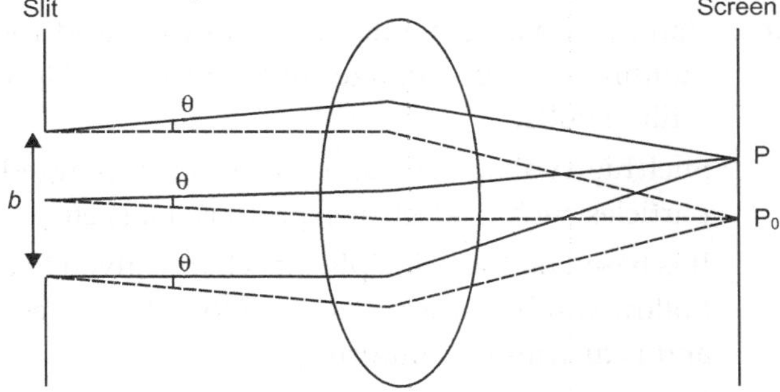


Point $P_0$ is a bisector plane of the slit. At $P_0$, all waves are travelling equal optical path. So all wavelets are in phase thus interfere constructively with each other and maximum intensity is observed. As we move from $P_0$, the wave arrives with different phases and intensity is changed. Intensity at point P is given by

$$I = I_0 \frac{\sin^2 \alpha}{\alpha}$$

Where $\quad \alpha = \frac{\pi}{\lambda} b \sin\theta$

For central maxima, $\alpha = 0$ thus,

$$I = I_0$$

When the width of slit is made double the original width intensity will get four times of its original value.

Width of central maximum is given by,

$$\beta = \frac{2D\lambda}{b}$$

Where, D = Distance between screen and slit,

$\lambda$ = Wavelength of the light,

$b$ = size of slit

So with the increase in size of slit the width of central maxima decrease. Hence, double the size of the slit would result in half the width of the central maxima.

**28. Explain the principle of a device that can build up high voltage of the order of a few million volts.**

**Draw a schematic diagram and explain the working of this device.**

**Is there any restriction on the upper limit of the high voltage set up in this machine ? Explain. [5]**

**OR**

**(a) Define electric flux. Write its S.I. units.**

**(b) Using Gauss's law, prove that the electric field at a point due to a uniformly charged infinite plane sheet is independent of the distance from it.**

**(c) How is the field directed if (i) the sheet is positively charged, (ii) negatively charged ?**

**Ans.** Van de Graaff generator is the device used for building up high voltages of the order of a few million volts.

Such high voltages are used to accelerate charged particles such as electrons, protons, ions, etc.

It is based on the principle that charge given to a hollow conductor is transferred to outer surface and is distributed uniformly over it.

**Construction :**

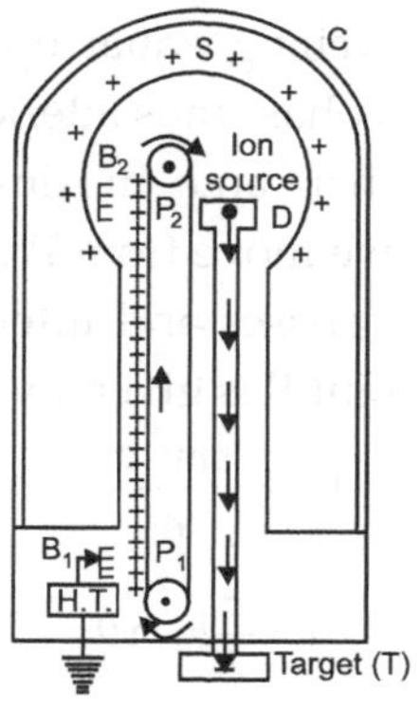


It consists of large spherical conducting shell (S) supported over the insulating pillars. A long narrow belt of insulating material is wound around two pulleys $P_1$ and $P_2$, $B_1$ and $B_2$ are two sharply pointed metal combs. $B_1$ is called the spray comb and $B_2$ is called the collecting comb.

**Working :** The spray comb is given a positive potential by high tension source. The positive charge gets sprayed on the belt.

As the belt moves and reaches the sphere, a negative charge is induced on the sharp ends of collecting comb $B_2$ and an equal positive charge is induced on the farther end of $B_2$.

This positive charge shifts immediately to the outer surface of S. Due to discharging action of sharp points of $B_2$, the positive charge on the belt is neutralized. The uncharged belt returns down and collects the positive charge from $B_1$, which in turn is collected by $B_2$. This is repeated. Thus, the positive charge of S goes on accumulating. In this way, potential differences of as much as 6 or 8 million volts (with respect to the ground) can be built up.

The main limiting factor on the value of high potential is the radii of sphere.

If the electric field just outside the sphere is sufficient for dielectric breakdown of air, no more charge can be transferred to it.

For a conducting sphere,

Electric field just outside sphere

$$E = \frac{Q}{4\pi\varepsilon_0 R^2}$$

and electric potential

$$V = \frac{Q}{4\pi\varepsilon_0 R}$$

Thus, $E = V/R$

Now, for $E = 3 \times 10^6$ V/m (dielectric breakdown)

Radius of sphere should be 1 m.

Thus, the maximum potential of a sphere of radius 1 m would be $3 \times 10^6$ V.

**OR**

**(a) Electric Flux :** It is the number of electric field lines passing through a surface normally.

$$\phi = \vec{E} \cdot \vec{A}$$

where E = Electric field, A = Area

S.I. unit of flux is $Nm^2C^{-1}$

**(b)** Consider a uniformly charged infinite plane sheet of charge density σ.

We have to find electric field E at point P as shown in figure.

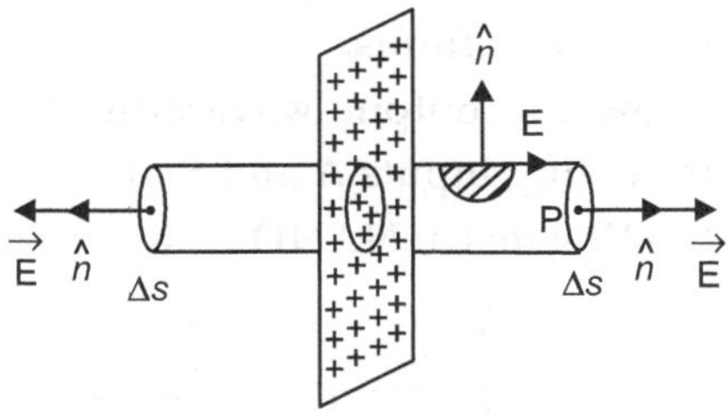


Now, we construct a Gaussian surface as shown in figure in the form of cylinder.

Applying Gauss's law,

$$\phi = \oint \vec{E} \cdot \vec{ds} = \frac{\sigma \Delta s}{\varepsilon_0}$$

$$\Rightarrow \quad E\Delta s + E\Delta s + 0 = \frac{\sigma \Delta s}{\varepsilon_0}$$

$$= 2E\Delta s = \frac{\sigma \Delta s}{\varepsilon_0}$$

$$\Rightarrow \quad E = \frac{\sigma}{2\varepsilon_0}$$

It shows that electric field is uniform due to charged infinite plane sheet. Also, we can say that E is independent from distance from the sheet.

**(c)** $$E = \frac{\sigma}{2\varepsilon_0}$$

**(i)** Direction of field will be away from the sheet if sheet is positively charged.

**(ii)** $$E = \frac{-\sigma}{2\varepsilon_0}$$

Direction of field will be towards the sheet if sheet is negatively charged.

**29. Define magnifying power of a telescope. Write its expression.**

**A small telescope has an objective lens of focal length 150 cm and an eye-piece of focal length 5 cm. If this telescope is used to view a 100 m high tower 3 km away, find the height of the final image when it is formed 25 cm away from the eye-piece. [5]**

**OR**

**How is the working of a telescope different from that of a microscope ?**

**The focal lengths of the objective and eyepiece of a microscope are 1.25 cm and 5 cm respectively. Find the position of the object relative to the objective in order to obtain an angular magnification of 30 in normal adjustment.**

**Ans.** Magnifying power of a telescope is defined as the ratio of the angle subtended at the eye by the image formed at the least distance of distinct vision to the angle subtended at the eye by the object lying at infinity, when seen directly.

**The formula for magnifying power is,**

$$\text{Magnifying power, } M = -\frac{f_o}{f_e}\left(1 + \frac{f_e}{D}\right)$$

Where, $f_o$ = Focal length of the objective = 150 cm

$f_e$ = Focal length of the eye-piece = 5 cm

D = Least distance of distinct vision = 25 cm

$$M = \frac{-150}{5} \times \left(1 + \frac{5}{25}\right) = -36$$

Also, by definition, $M = \frac{\beta}{\alpha}$

or $\quad M = \frac{\tan \beta}{\tan \alpha}$

(As angles *a* and *b* are small)

$$\tan \alpha = \frac{\text{Height of object}}{\text{Distance of object from objective}}$$

$$= \frac{H}{u} = \frac{100}{300} = \frac{1}{30}$$

$$M = \frac{\tan \beta}{\left(\frac{1}{30}\right)}$$

$$\tan \beta = \frac{-36}{30}$$

$$\tan \beta = \frac{\text{Height of image}}{\text{Distance of image from eye-piece}} = \frac{H'}{D}$$

Thus,

$$H = \frac{-36 \times 25}{30} = -30 \text{ cm}$$

Negative sign indicates that we get an inverted image.

**OR**

A microscope is used to look into smaller objects like structure of cells etc. On the other hand, a telescope is used to see larger objects that are very far away like stars, planets etc.

Telescope mainly focuses on collecting the light into the objective lens, which should thus be large, where the microscope already has a focus and the rest is blurred around it.

There is a big difference in their magnification factors.

For telescope, the angular magnification is given by

$$M = \frac{f_o}{f_e}$$

Where $f_o$ is the focal length of the objective lens and $f_e$ is the focal length of the eyepiece.

For microscope, the angular magnification is given by

$M = 1 + \frac{D}{f_o}$ when image is formed at distance of least distinct vision.

$M = \frac{D}{f_o}$ when image is formed infinity.

Where D is the distance of least distinct vision.

Given : $f_o = 1.25$ cm

$f_e = 5$ cm

M = –30 (Magnifying power is negative)

We know,

$$M = \frac{v_0}{u_0}\left(1 + \frac{D}{f_o}\right)$$

Where, $v_0$ = Distance of image from objective

$u_0$ = Distance of object from objective

$$-30 = \frac{v_0}{u_0}\left(1 + \frac{25}{5}\right)$$

$$v_0 = -5\mu_0$$

Using Lens formula,

$$\frac{1}{f_o} = \frac{1}{u_0} + \frac{1}{v_0}$$

$$\frac{1}{1.25} = -\frac{1}{u_0} - \frac{1}{5u_0}$$

$$u_0 = -1.5 \text{ cm}$$

Thus the distance of object from objective is 1.5 cm.

**30. Draw a simple circuit of a CE transistor amplifier. Explain its working. Show that the voltage gain, $A_V$, of the amplifier is given by**

**$A_V = \frac{-\beta_{ac} RL}{r_i}$, where $\beta_{ac}$ is the current gain; $R_L$ is the load resistance and $r_i$ is the input resistance of the transistor. What is the significance of the negative sign in the expression for the voltage gain ?** [5]**

**OR**

**(a) Draw the circuit diagram of a full wave rectifier using *p-n* junction diode. Explain its working and show the output, input waveforms.**

**(b) Show the output waveforms (Y) for the following inputs A and B of (i) OR gate (ii) NAND gate**

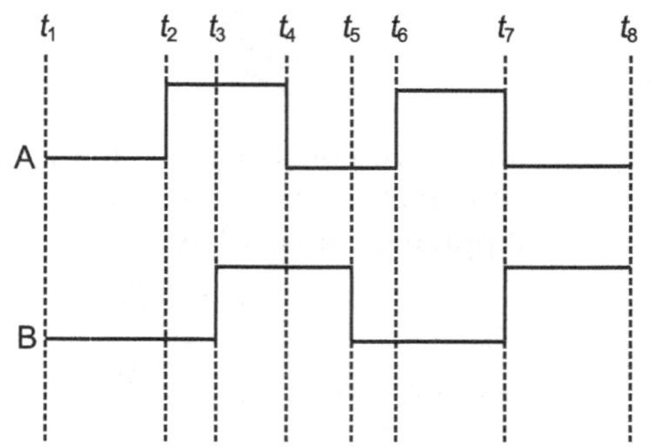


**Ans.**

**OR**

**(a) Full wave rectifier :**

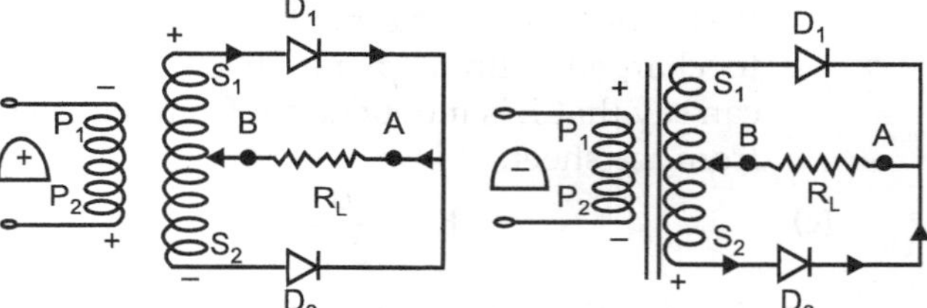


**Working :** When the diode rectifies the whole of the ac wave, it is called full wave rectifier.

The figure shows the arrangement for using diode as full wave rectifier. The alternating input signal is fed to the primary $P_1P_2$ of a transformer. The output signal appears across the load resistance $R_L$.

During the positive half of the input signal, suppose $P_1$ and $P_2$ are negative and positive respectively. This would mean that $S_1$ and $S_2$ are positive and negative respectively. Therefore, the diode $D_1$ is forward biased and $D_2$ is reverse biased. The flow of current in the load resistance $R_L$ is from A to B.

During the negative half of the input signal, $S_1$ and $S_2$ are negative and positive respectively.

** Answer is not given due to change in the present Syllabus.

Therefore, the diode $D_1$ is reverse biased and $D_2$ is forward biased. The flow of current in the load resistance $R_L$ is from A to B.

**Output, input waveforms :**

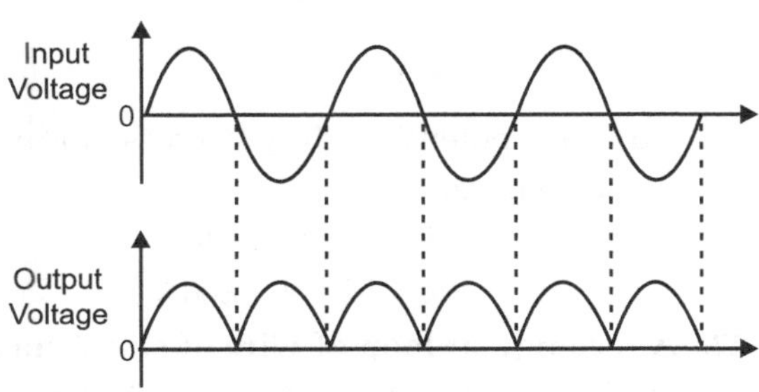


**(b) Output waveforms (Y) :**

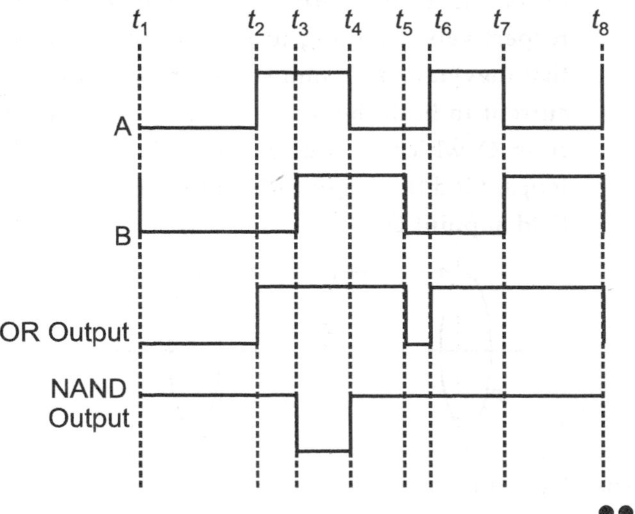


## Delhi [Set-II]

**Note : Except for the following questions, all the remaining questions have been asked in previous set.**

1. **Why must electrostatic field be normal to the surface at every point of a charged conductor ? [1]**

**Ans.** The electrostatic field must be normal to the surface of the conductor at every point because if the electric field is not normal to the surface of the charged conductor, there will be a component of the electric field along the surface of the conductor, which would exert a force on the charges at the surface. Due to this, charge starts flowing which is not possible.

6. **Predict the direction of induced current in a metal ring when the ring is moved towards a straight conductor with constant speed *v*. The conductor is carrying current I in the direction shown in the figure. [1]**

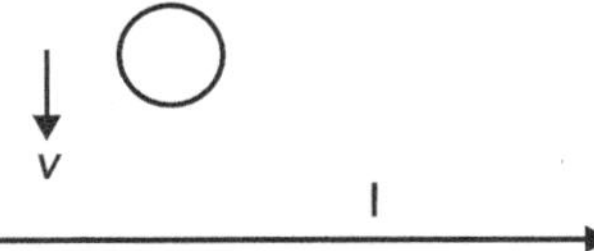


**Ans.** Clockwise

10. **Derive the expression for the self inductance of a long solenoid of cross-sectional area A and length *l*, having *n* turns per unit length. [2]**

**Ans. Self-inductance of a long Solenoid :**

The magnetic field B at any point inside a solenoid is constant and given by,

$$B = \frac{\mu_0 NI}{l} \quad ...(i)$$

Where, $\mu_0$ = Absolute magnetic permeability of free space

N = Total number of turns in the solenoid

∴ Magnetic flux through each turn of the solenoid coil = B × Area of each turn

$$B = \frac{\mu_0 NI}{l}.A$$

Where, A = Area of each turn of the solenoid.

Total magnetic flux linked with the solenoid

$$\phi = \frac{\mu_0 NIA}{l} \times N \quad ...(ii)$$

If L is coefficient of self-inductance of the solenoid, then

$$\phi = LI \quad ...(iii)$$

Equating (ii) and (iii),

$$LI = \frac{\mu_0 NIA}{l} \times N$$

$$L = \frac{\mu_0 N^2 A}{l}$$

If core is of any other magnetic material, $\mu_0$ is replaced by $\mu = \mu_0\mu_r$ then,

$$L = \mu_r\mu_0 N^2 A/l.$$

If *n* is the number of turns per unit length then

$$L = \mu_r \mu_0 n^2 Al$$

This is self inductance of a long solenoid.

**16. Two identical circular loops, P and Q, each of radius r and carrying currents I and 2I respectively are lying in parallel planes such that they have a common axis. The direction of current in both the loops is clockwise as seen from O which is equidistance from the both loops. Find the magnitude of the net magnetic field at point O.** **[2]**

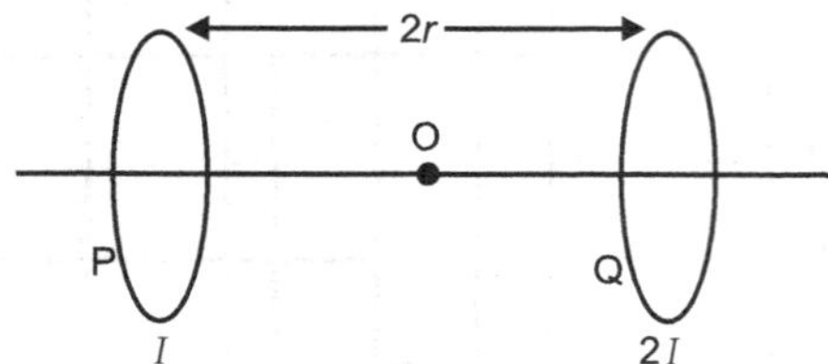


**Ans.** Field due to loop P,

$$\vec{B}_P = \frac{\mu_0 r^2 I}{2(r^2+r^2)^{3/2}} = \frac{\mu_0 I}{4\sqrt{2}r} \text{ towards P.}$$

Field due to loop Q,

$$\vec{B}_Q = \frac{\mu_0 r^2 (2I)}{2(r^2+r^2)^{3/2}} = \frac{\mu_0 (2I)}{4\sqrt{2}r} \text{ towards Q.}$$

So, net field at O

$$\vec{B} = \vec{B_Q} - \vec{B_P}$$

$$= \frac{\mu_0(2I)}{4\sqrt{2}r} - \frac{\mu_0 I}{4\sqrt{2}r} = \frac{\mu_0 I}{4\sqrt{2}r} \text{ towards Q.}$$

**20. A series LCR circuit with L = 4.0 H, C = 100 μF, R = 60 Ω connected to a variable frequency 240 V source as shown in figure calculate :** **[3]**

**(i) the angular frequency of the source which drives the circuit at resonance,**

**(ii) the current at the resonating frequency,**

**(iii) the rms potential drop across the inductor at resonance.**

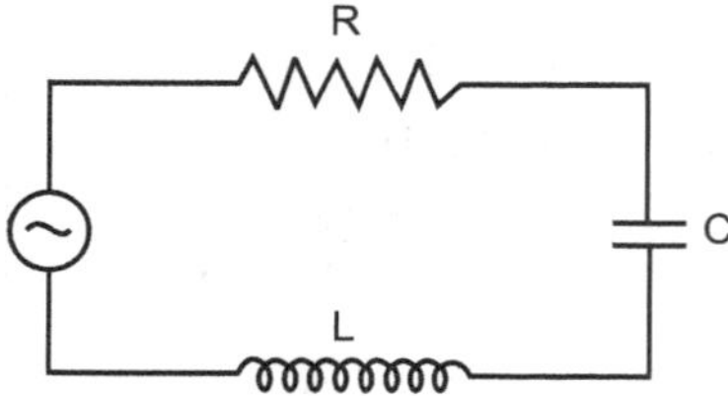


**Ans. (i)** Angular frequency at resonance

$$\omega = \frac{1}{\sqrt{LC}} = \frac{1}{\sqrt{4\times 100\times 10^{-6}}}$$

$$= \frac{1}{2\times 10^{-2}}$$

$$= 50 \text{ rad s}^{-1}$$

**(ii)** At Resonance, we know the inductive reactance cancels out the capacitive reactance.

$$Z = R = 60\ \Omega$$

∴ The current at resonant frequency

$$I_{rms} = \frac{V_{rms}}{Z} = \frac{240}{60}$$

$$= 4 \text{ A}$$

**(iii)** rms potential drop across inductor at resonance

$$V_L = I_{rms}\cdot X_L = I_{rms}\cdot \omega L$$

$$= 4\times 50\times 4 = 800 \text{ V}$$

**22. A rectangular loop of wire of size 2 cm × 5 cm carries a steady current of 1 A. A straight long wire carrying 4 A current is kept near the loop as shown in the figure. If the loop and the wire are coplanar, find (i) the torque acting on the loop and (ii) the magnitude and direction of the force on the loop due to the current carrying wire.** **[3]**

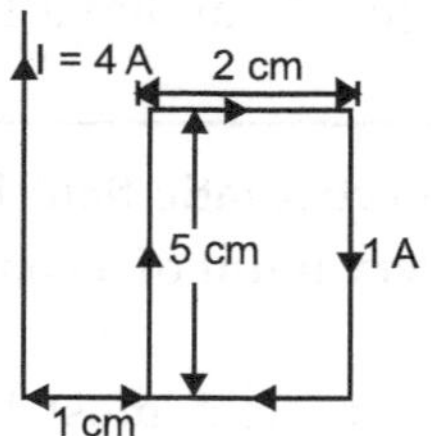


**Ans. (i)** $|\vec{\tau}| = \vec{m}\times\vec{B} = mB\sin\theta$

Here $m$ and B have the same direction

$$\theta = 0$$

$$\Rightarrow \quad |\vec{\tau}| = mB\sin\theta = 0$$

**(ii)** Force between two current carrying wires is given by,

$$F = \frac{\mu_0}{2\pi}\cdot\frac{I_1 I_2}{r}\times l$$

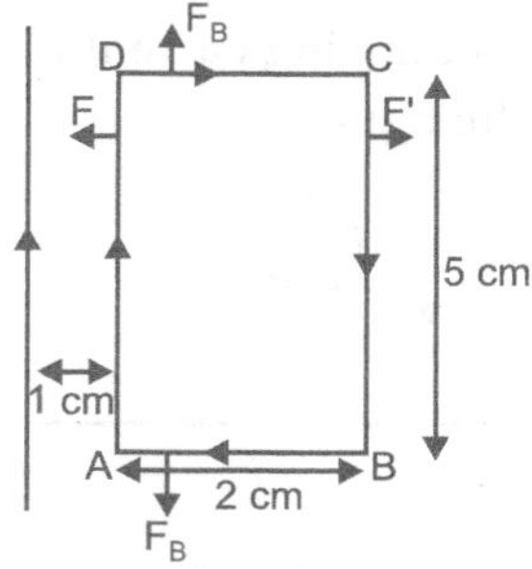


On wire AB and CD, the magnetic forces are equal and opposite, hence they cancel out each other.

Magnetic force on wire AD,

$$F = \frac{\mu_0}{2\pi}\frac{4\times1}{(0.01)}(0.05)$$

(Attractive)

$$F = \frac{\mu_0}{2\pi}\times 20$$

Magnetic force on line CB,

$$F' = \frac{\mu_0}{2\pi}\frac{4\times1}{0.03}(0.05)$$

(Repulsive)

$$= \frac{\mu_0}{2\pi}\times\frac{20}{3}$$

Now, resultant force on loop

$$F_{net} = F - F'$$

$$= \frac{\mu_0}{2\pi}20\left(1-\frac{1}{3}\right)$$

$$= 2\times10^{-7}\times20\left(\frac{2}{3}\right)$$

$$= 26.67\times10^{-7}\text{ N}$$

The direction of net force is towards the wire *i.e.*, attractive.

**27. Name the three different modes of propagation of electromagnetic waves. Explain, using a proper diagram the mode of propagation used in the frequency range above 40 MHz.** [3]**

## Delhi [Set-III]

**Note : Except for the following questions, all the remaining questions have been asked in previous set.**

**6. Why is electrostatic potential constant throughout the volume of the conductor and has the same value (as inside) on its surface ? [1]**

**Ans.** We know that the electric field inside a conductor is zero. This is the reason why electrostatic potential is constant, as

$$E = -\frac{dV}{dr}.$$

So, if $E = 0$

$$\frac{dV}{dr} = 0$$

Thus, V = constant.

**8. Predict the direction of induced current in metal rings 1 and 2 when current I in the wire is steadily decreasing ? [1]**

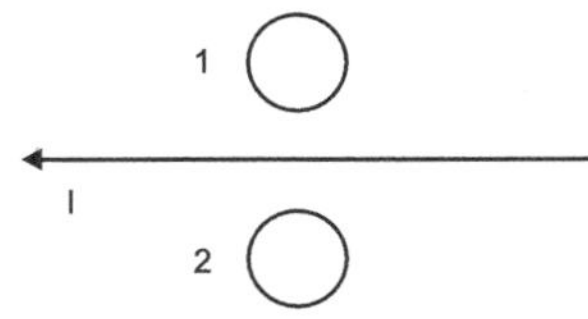


**Ans.** Using Lenz's law, we can predict the direction of induced current in the ring. Induced current oppose the cause of change of magnetic flux in moving towards the conductor.

In metal Ring 1, induced current will be in is clockwise.

In metal Ring 2, induced current will be in is anticlockwise.

** Answer is not given due to change in the present Syllabus.

**9. The relative magnetic permeability of a magnetic material is 800. Identify the nature of magnetic material and state its two properties. [2]**

**Ans.** A magnetic material having relative permeability of 800 would be classified as a ferromagnet. A few examples of such materials include iron and nickel.

Its two properties are :

(i) All ferromagnetic materials become paramagnetic when heated to a temperature above the Curie temperatue ($T_C$).

(ii) These materials show a strong attraction towards magnetic fields and have a tendency to become magnets themselves.

**16. Two identical circular loops, P and Q, each of radius r and carrying equal currents are kept in the parallel planes having a common axis passing through O. The direction of current in P is clockwise and in Q is anti-clockwise as seen from O which is equidistant from the loops P and Q. Find the magnitude of the net magnetic field at O. [2]**

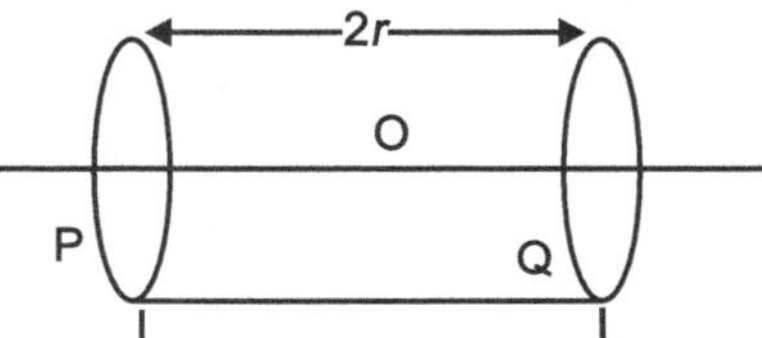


**Ans.** The standard formula for field at an axial point is given as

$$\frac{\mu_0 Ia^2}{2(a^2+x^2)^{3/2}}$$

Field at O due to loop P

$$B_p = \frac{\mu_0 I(r)^2}{2\left[r^2 + \left(\frac{2r}{2}\right)^2\right]^{3/2}} = \frac{\mu_0 I}{2\pi}$$

And, field at O due to loop Q

$$B_Q = \frac{\mu_0 I (r)^2}{2\left[r^2 + \left(\frac{2r}{2}\right)^2\right]^{3/2}} = \frac{\mu_0 I}{2\pi}$$

Now, as the current flowing in loop P is clockwise, by using right hand thumb's rule, the direction of the magnetic field will be towards left and as the current in loop Q is clockwise then the direction of magnetic field is towards left. So the net magnetic field at point O will be the sum of the magnetic fields due to loops P and Q.

So, net field

$$B = B_P + B_Q = \frac{2\mu_0 I}{2r} = \frac{\mu_0 I}{r}$$

**23. A rectangular loop of wire of size 2.5 cm × 4 cm carries a steady current of 1 A. A straight wire carrying 2 A current is kept near the loop as shown. If the loop and the wire are co-planar, find the (i) torque acting on the loop and (ii) the magnitude and direction of the force on the loop due to the current carrying wire.** **[3]**

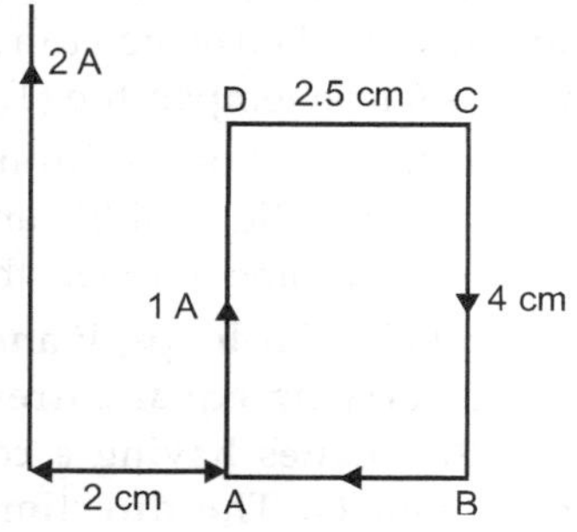


**Ans. (i)** $|\vec{\tau}| = \vec{m} \times \vec{B} = mB \sin\theta$

Here $m$ and B have the same direction

$$\therefore \quad \theta = 0°$$

$$\Rightarrow \quad \vec{\tau} = mB \sin\theta = 0$$

**(ii)** We know that force between two current carrying wires is given by,

$$F = \frac{\mu_0}{2\pi} \cdot \frac{I_1 I_2}{r} l$$

On line AB and CD, the magnetic forces are equal and opposite. So they cancel out each other.

Magnetic force on line AD,

$$F = \frac{\mu_0}{2\pi} \cdot \frac{2 \times 1 \times 0.04}{0.02} \quad \text{(Attractive)}$$

$$= \frac{\mu_0}{2\pi} \times 4$$

Magnetic force on line CB,

$$F' = \frac{\mu_0}{2\pi} \cdot \frac{2 \times 1 \times 0.04}{0.045}$$

$$= \frac{\mu_0}{2\pi} \times \frac{16}{9}$$

So, net force

$$F_{net} = F - F'$$

$$= \frac{\mu_0}{2\pi}\left[4 - \frac{16}{9}\right]$$

$$= 2 \times 10^{-7} \times \frac{20}{9}$$

$$= 4.44 \times 10^{-7} \text{ N}$$

The direction of the force on the loop will be towards the wire *i.e.*, attractive.

●●

# CHEMISTRY

## Outside Delhi [Set-I]

# 2020

**Time allowed : 3 hours** **Maximum marks : 70**

**General Instructions :**

Read the following instructions very carefully and strictly follow them :

(i) Question paper comprises four sections – **A, B, C** and **D.**

(ii) There are **37** questions in the question paper. **All** questions are compulsory.

(iii) **Section – A :** Q. No. **1** to **20** are very short answer type questions, carrying **1** mark each. Answer these questions in one word or one sentence.

(iv) **Section – B :** Q. No. **21** to **27** are short answer type questions, carrying **2** marks each.

(v) **Section – C :** Q. No. **28** to **34** are long answer type-I questions, carrying **3** marks each.

(vi) **Section – D :** Q. No. **35** to **37** are long answer type-II questions, carrying **5** marks each.

(vii) There is no overall choice in the question paper. However, an internal choice has been provided in **2** questions of **two** marks, **2** questions of **three** marks and all the **3** questions of **five** marks. You have to **attempt only one of the choices** in such questions.

(viii) In addition to this, separate instructions are given with each section and question, wherever necessary.

(ix) Use of calculators and log tables is **not** permitted.

### SECTION – A

**Read the given passage and answer the questions 1 to 5 that follow :** 20

Colloidal particles always carry and electric charge which may be either positive or negative. For example, when $AgNO_3$ solution is added to KI solution, a negatively charged colloidal sol is obtained. The presence of equal and similar charges on colloidal particles provide stability to the colloidal sol and if, somehow, charge is removed, coagulation of sol occurs. Lyophobic sols are readily coagulated as compare to lyophilic sols.

**1. What is the reason for the charge on sol particles?**

**Ans.** Preferential adsorption of either positive or negative ions on surface of sol particles is the reason of the charge on them.

**2. Why the presence of equal and similar charges on colloidal particles provide stability?**

**Ans.** Due to the presence of similar and equal charges, the colloidal particles repel one another and are thus unable to combine together to form larger particles and coagulate. Hence, presence of equal and similar charge on colloidal particles provide stability to colloidal solution.

**3. Why a negatively charged sol is obtained on adding $AgNO_3$ solution to KI solution ?**

**Ans.** When $AgNO_3$ solution is added to KI solution then negatively charged sol is formed due to the adsorption of iodide ions from KI which is disperse medium onto the precipitated AgI particles.

**4. Name one method by which coagulation of lyophobic sol can be carried out.**

**Ans** The coagulation of the lyophobic sols can be carried out by electrophoresis.

**5. Out of KI or $K_2SO_4$, which electrolyte is better in the coagulation of positive sol ?**

**Ans.** $K_2SO_4$ would be better in coagulation of positive sol due to higher flocculating power of $SO_4^{2-}$.

**Questions 6 to 10 are one word answer :**

**6. Name the method applied for the concentration of Bauxite ore in the extraction of Aluminium.**

**Ans.** Leaching or Baeyer's process

**7. Out of $C_6H_{11}$–Cl and $C_6H_{11}$–$CH_2$–Cl, which one is more reactive towards $S_N1$ reaction ?**

**Ans.** $C_6H_{11}$–Cl.

**8. Write an isomer of $C_3H_9N$ which gives foul smell of isocyanide when treated with chloroform and ethanolic NaOH.**

**Ans.** 1-Amino propane; $CH_3CH_2CH_2NH_2$

**9. Which one of the following is an antidepressant drug?**

**Chloramphenicol, Luminal, Bithional.**

**Ans.** Luminal

**10. Write the name of component of starch which is water soluble.**

**Ans.** Amylose

**Questions 11 to 15 are Multiple Choice Questions :**

**11. How many ions are produced from the complex $[Co(NH_3)_5Cl]Cl_2$ in solution?**

**(a)** 4 **(c)** 3

**(b)** 2 **(d)** 5

**Ans.** (c) 3

**12. In a lead storage battery**

**(a) $PbO_2$ is reduced to $PbSO_4$ at the cathode.**

**(b) Pb is oxidised to $PbSO_4$ at the anode.**

**(c) Both electrodes are immersed in the same aqueous solution of $H_2SO_4$.**

**(d) All the above are true.**

**Ans. (d)** All the above are true.

**13. The slope in the plot of *ln* [R] Vs. time gives**

(a) +k (c) – k

(b) $\frac{+k}{2.303}$ (d) $\frac{-k}{2.303}$

**(where [R] is the final concentration of reactant.)**

**Ans.** (c) – k

**14. The pair $[Co(NH_3)_4 Cl_2] Br_2$ and $[Co(NH_3)_4Br_2] Cl_2$ will show**

**(a) Linkage isomerism**

**(b) Hydrate isomerism**

**(c) Ionization isomerism**

**(d) Coordinate isomerism**

**Ans.** (c) Ionization isomerism

**15. An α-helix is a structural feature of**

**(a) Sucrose** **(c) Nucleotides**

**(b) Polypeptides** **(d) Starch**

**Ans. (b)** Polypeptides

**Questions 16 to 20 :**

**(A) Both Assertion (A) and Reason (R) are correct statements, and Reason (R) is the correct explanation of the Assertion (A).**

**(B) Both Assertion (A) and Reason (R) are correct statements, but Reason (R) is not the correct explanation of the Assertion (A).**

**(C) Assertion (A) is correct, but Reason (R) is wrong statement.**

**(D) Assertion (A) is wrong, but Reason (R) is correct statement.**

**16. Assertion (A)** : $F_2$ is a strong oxidising agent.

**Reason (R)** : Electron gain enthalpy of fluorine is less negative.

**Ans. (B)** As $F_2$ has low bond dissociation energy and high hydration energy than $Cl_2$ but less electron gain enthalpy due to its smaller size. Due to these factors $F_2$ wins in getting reduced fastly. Thus, both Assertion (A) and Reason (R) are correct statements but Reason (R) is not the correct explanation of the Assertion (A).

**17. Assertion (A)** : $(CH_3)_3 C—O—CH_3$ gives $(CH_3)_3 C—I$ and $CH_3OH$ on treatment with HI.

**Reason (R)** : The reaction occurs by $S_N1$ mechanism.

**Ans. (A)** $SN^1$ reaction occurs in compound that has steric hindrance and $(CH_3)_3—C—OCH_3$ has very much hindrance to attack by reagent. Thus both Assertion (A) and Reason (R) are correct statements, and Reason (R) is the correct explanation of the Assertion (A).

**18. Assertion (A)** : Transition metals have low melting points.

**Reason (R)** : The involvement of greater number of $(n - 1)d$ and ns electrons in the interatomic metallic bonding.

**Ans. (D)** Transition metals have high melting point. Thus, Assertion (A) is wrong, but Reason (R) is correct statement.

**19. Assertion (A)** : Hydrolysis of an ester follows first order kinetics.

**Reason (R)** : Concentration of water remains nearly constant during the course of the reaction.

**Ans. (A)** Hydrolysis of an ester follows first order kinetics, because concentration of water remains constant during the course of the reaction. Both Assertion (A) and Reason (R) are correct statements, and Reason (R) is the correct explanation of the Assertion (A).

**20. Assertion (A)** : Benzoic acid does not undergo Friedal-Crafts reaction.

**Reason (R)** : The carboxyl group is activating and undergo electrophilic substitution reaction.

**Ans. (C)** Benzoic acid does not undergo Friedal-crafts reaction due to deactivation of benzene ring by electron withdrawing effect of COOH group. Thus, Assertion (A) is correct, but Reason (R) is wrong statement.

## SECTION-B

**21. What happens when**

**(i) a pressure greater than osmotic pressure is applied on the solution side separated from solvent by a semipermeable membrane ?**

**(ii) acetone is added to pure ethanol ? [1 + 1 = 2]**

**Ans. (i)** When a pressure greater than osmotic pressure is applied on the solution side, the solvent flows out to the solution through the semipermeable membrane. This is known as reverse osmosis.

**(ii)** Mixture of ethanol and acetone shows a positive deviation from Raoult's law and the vapour pressure over the solution increases. The acetone molecules weaken the hydrogen bonding between molecules of ethanol hence making escape of ethanol molecules easy.

**22. Write the principle of the following refining methods :**

**(a) vapour phase refining**

**(b) chromatography** **[1 + 1 = 2]**

**OR**

**Write chemical equations involved to obtain :**

**(i) Cu from $Cu_2S$**

**(ii) Ag from $[Ag (CN)_2]^-$ complex**

**Ans.(a)** In vapour phase refining the metal is converted into its volatile compound with a chemical reagent and is collected elsewhere. Then it is decomposed to give pure metal.

**(b)** Chromatography is a physical method of separation of different components of a mixture, that are partitioned differentially between the stationary and mobile phase, when mobile phase is run over the stationary phase loaded with the mixture at one point, they are adsorbed to different extent.

OR

**(i)** The sulphide ore is roasted to give oxide:

$$2Cu_2S + 3O_2 \rightarrow 2Cu_2O + 2SO_2.$$

The oxide is then reduced to metallic copper using coke :

$$Cu_2O + C \rightarrow 2Cu + CO$$

**(ii)** The silver cyanide complex is treated with zinc:

$$2[Ag(CN)_2]^-(aq) + Zn(s) \rightarrow 2Ag(s) + [Zn(CN)_4]^{2-}(aq).$$

**23. Write the balanced chemical equations involved in the preparation of $KMnO_4$ from pyrolusite ore ($MnO_2$).** **[2]**

OR

**Write the balanced ionic equations showing the oxidising action of acidified dichromate ($Cr_2O_7^{2-}$) solution with (i) Iron (II) Ion and (ii) tin (II) ion.**

**Ans.** Potassium permanganate ($KMnO_4$) is prepared from pyrolusite ore ($MnO_2$). The finely powdered pyrolusite ore ($MnO_2$) is fused with an alkali metal hydroxide like KOH in the presence of air or an oxidising agent like $KNO_3$ to give the dark green potassium manganate ($K_2MnO_4$). Potassium manganate disproportionates in a neutral or acidic solution to give permanganate.

1. $2MnO_2 + 4KOH + O_2 \rightarrow 2K_2MnO_4 + 2H_2O$
2. $3MnO_4^{2-} + 4H^+ \rightarrow 2MnO_4^- + MnO_2 + 2H_2O$

OR

**(i)** With Iron (II) ion

$$Cr_2O_7^{2-} + 14H^+ + 6Fe^{2+} \rightarrow 2Cr^{3+} + 6Fe^{3+} + 7H_2O$$

**(ii)** With tin (II) ion

$$Cr_2O_7^{2-} + 14H^+ + 3Sn^{2+} \rightarrow 2Cr^{3+} + 3Sn^{4+} + 7H_2O$$

**24. Write the IUPAC names and hybridisation of the following complexes :**

**(i) $[Ni(CN)_4]^{2-}$** **(ii) $[Fe(H_2O)_6]^{2+}$**

**(Given : Atomic number Ni = 28, Fe = 26)** **[1 + 1 = 2]**

**Ans. (i)** $[Ni(CN)_4]^{2-}$

**IUPAC name :** Tetracyanonickelate (II).

**Hybridisation :** $dsp^2$

**(ii)** $[Fe(H_2O)_6]^{2-}$

**IUPAC name :** Hexaaquairon(II) ion.

**Hybridisation :** $sp^3d^2$

**25. Define the following terms with a suitable example in each :** **[1 + 1 = 2]**

**(i) Antibiotics** **(ii) Antiseptics**

**Ans. (i) Antibiotics:** They are naturally occurring, semisynthetic, or synthetic anti-infective agents that destroys or inhibits the growth of microorganisms when ingested in low concentration. Example: Penicillin, Tetracycline etc.

**(ii) Antiseptics:** They are antimicrobial agents applied to the living tissues such as wounds, cuts, ulcers and other disease skin surfaces to destroy microbes or inhibit growth of microbes. These are not ingested as antibiotics. Example : Iodine solution, Dettol etc.

**26. Write the reactions showing the presence of the following in the open structure of glucose :**

**(i) a carbonyl group** **[1 + 1 =2]**

**(ii) Straight chain with six carbon atoms**

**Ans. (i)** Glucose reacts with hydroxylamine to form an oxime and adds hydrogen cyanide to give cyanohydrin. These reactions confirms the presence of carbonyl group in the glucose molecule.

$$\underset{CH_2OH}{\overset{CHO}{(CH_2OH)_4}} + NH_2OH \longrightarrow \underset{CH_2OH}{\overset{CH=N-OH}{(CHOH)_4}}$$

Glucose oxime

$$\underset{CH_2OH}{\overset{CHO}{(CHOH)_4}} + HCN \longrightarrow \underset{CH_2OH}{\overset{CH(OH)CN}{(CHOH)_4}}$$

Glucose cyanohydrin

**(ii)** Glucose forms n-hexane on prolonged heating with HI, this shows that it has a linear structure.

$$\underset{CH_2OH}{\overset{CHO}{(CHOH)_4}} \xrightarrow[\text{Reduction}]{HI,\ \Delta} CH_3-(CH_2)_4-CH_3$$

D-Glucose      n-hexane

**27. State Henry's law. Calculate the solubility of $CO_2$ in water at 298 K under 760 mm Hg. ($K_H$ for $CO_2$ in water at 298 K is 1.25 $10^6$ mm Hg)** **[2]**

**Ans.** Henry's law states that "at a constant temperature, The partial pressure of the gas in vapour phase (p) is directly proportional to the mole fraction of gas(x) in the solution".

Mathematically $p = K_Hx$, where $K_H$ is constant.

To calculate the solubility of $CO_2$ in water :

Given : T = 298 K; $K_H = 1.25 \times 10^6$ mm Hg;

$p = 760$ mm Hg

We know by Henry's Law; $p = K_Hx$, where $x$ is mole fraction of the gas in the solution.

Substituting the given values:

$$760 = 1.25 \times 10^6 \times x$$

$$x = 760 / 1.25 \times 10^6$$

$$x = 608 \times 10^{-6}$$

Now, we know that mole fraction of gas is given

$$x = \frac{\text{moles of gas}}{(\text{moles of gas} + \text{moles of water})}$$

$$608 \times 10^{-6} = \frac{\text{moles of gas}}{\text{moles of gas} + 55.5}$$

(since number of moles of water in 1 L = 55.5)

As, moles of gas <<<< moles of water

Moles of gas = $608 \times 10^{-6} \times 55.5$ mol

= $33744 \times 10^{-6} = 33.74 \times 10^{-3}$ mol.

Therefore, solubility of $CO_2$ in water is $33.74 \times 10^{-3}$ mol/L.

## SECTION C

**28. The freezing point of a solution containing 5g of benzoic acid (M = 122 g $mol^{-1}$) in 35g of benzene is depressed by 2.94 K. What is the percentage association of benzoic acid if it forms a dimer in solution ?**

**($K_f$ for benzene = 4.9 K kg $mol^{-1}$)** [3]

**Ans.** Given :

Weight of benzoic acid (solute) =5 g

Weight of benzene (solvent) =35 g

$K_f$ =4.9 K Kg $mol^{-1}$

$\Delta T_f$ = 2.94 K

We know that molecular weight of solute can be calculated as-

$$M_2 = \frac{(K_f \times W_2 \times 1000)}{\Delta T_f \times W_1}$$

Substituting the values –

$$M_2 = \frac{(K_f \times W_2 \times 1000)}{\Delta T_f \times W_1}$$

$$= \frac{(4.9 \times 5 \times 1000)}{2.94 \times 35}$$

$$= \frac{24500}{102.9}$$

$$= 238.09 \text{ g}$$

Now let us consider the following equilibrium for the benzoic acid :

$$2\ C_6H_5COOH \rightleftharpoons (C_6H_5COOH)_2$$

If $x$ represents the degree of association of the solute then it would have (1-$x$) mol of benzoic acid left in unassociated form and correspondingly $x/2$ as associated moles of benzoic acid at equilibrium. Therefore, total number of moles of particles at equilibrium is:

$$(1 - x) + x/2 = 1 - (x/2)$$

Thus, total number of moles of particles at equilibrium equals to van't Hoff factor i

But, $$i = \frac{\text{normal molar mass}}{\text{abnormal molar mass}}$$

So, van't Hoff factor, $i = \frac{122}{238.09} = 0.512$

Hence,

$$1-(x/2) = 0.512$$
$$(x/2) = 0.512$$
$$= 0.488$$
$$x = 0.976$$

Hence, the percentage association is 97.6% for benzoic acid.

**29. The rate constant for the first order decomposition of $N_2O_5$ is given by the following equation :**

**$k = (2.5\ \ 10^{14}\ s^{-1})\ e^{(-25000K)/T}$**

**Calculate $E_a$ for this reaction and rate constant if its half-life period be 300 minutes.** [3]

**Ans.** Arrhenius equation is given by:

$$k = Ae^{-Ea/RT}$$
$$k = (2.5 \times 10^{14}\ s^{-1})\ e^{(-25000K/T)}$$

Comparing the two equations :

$$E_a/RT = \frac{25000\ K}{T}$$

$$E_a = 25000\ K \times 8.314\ J\ K^{-1}mol^{-1}$$
$$= 207850\ J\ mol^{-1}$$
$$E_a = 207.9\ kJ\ mol^{-1}$$

Now, for first order reactions, rate constant is given by :

$$k = \frac{0.693}{t_{1/2}}$$

Substituting the value of $t_{1/2}$:

$$k = 0.693 / 300$$
$$k = 2.3 \times 10^{-3} s^{-1}$$

**30. Write the name and structures of monomer(s) in the following polymers :** **[1 + 1 + 1 = 3]**

**(i) Nylon-6 (ii) PVC (iii) Neoprene**

**Ans. (i)** Caprolactum is monomer of Nylon–6.

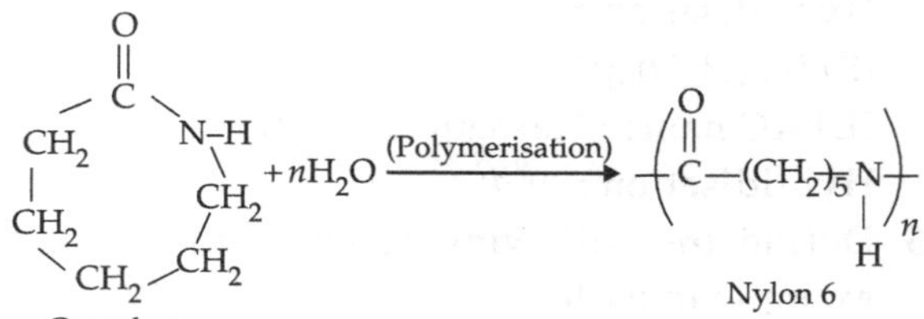

$$\text{Caprolactum (cyclic: C=O, N–H, } CH_2 \times 5) + nH_2O \xrightarrow{\text{(Polymerisation)}} \left( \overset{O}{\overset{\|}{C}} - (CH_2)_5 - \underset{H}{\underset{|}{N}} \right)_n$$

Caprolactum　　　　Nylon 6

**(ii)** PVC or Polyvinyl chloride is made up of Vinyl chloride (chloroethene) monomers.

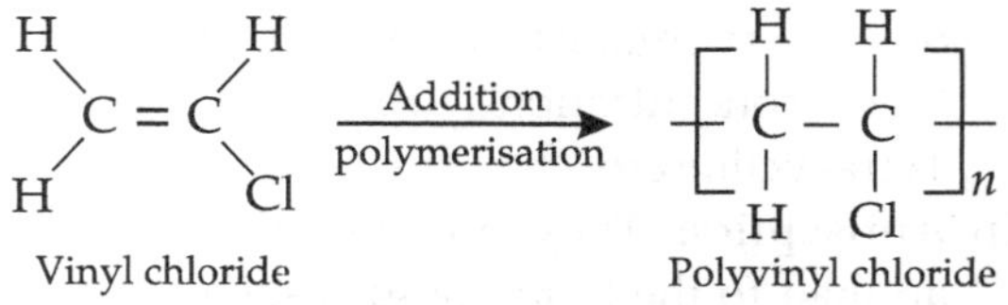

$$\underset{\text{Vinyl chloride}}{(H)(H)C = C(H)(Cl)} \xrightarrow[\text{polymerisation}]{\text{Addition}} \underset{\text{Polyvinyl chloride}}{\left[ \underset{H}{\overset{H}{C}} - \underset{Cl}{\overset{H}{C}} \right]_n}$$

**(iii)** Chloroprene is the monomer of Neoprene.

$$n\,CH_2=\underset{\underset{Cl}{|}}{C}-CH=CH_2 \xrightarrow{\text{Polymerisation}} \left(-CH_2-\underset{\underset{Cl}{|}}{C}=CH-CH_2-\right)_n$$

Chloroprene                Neoprene

**31. Following ions are given :**

**$Cr^{2+}$, $Cu^{2+}$, $Cu^{+}$, $Fe^{2+}$, $Fe^{3+}$, $Mn^{3+}$**

**Identify the ion which is**

**(i) a strong reducing agent.**

**(ii) unstable in aqueous solution.**

**(iii) a strong oxidising agent.**

**Give suitable reason in each.** [1 + 1 + 1 = 3]

**Ans.** (i) $Cr^{2+}$ is a strong reducing agent as it is very easily oxidised to more stable $Cr^{3+}$ ($d^3$) where it attains a stable half-filled $t_{2g}^3$ configuration.

(ii) $Cu^+$ is unstable in aqueous solution because it disproportionates in water to form $Cu^{2+}$ and Cu.

(iii) $Mn^{3+}$ is a strong oxidising agent because it gets reduced to $Mn^{2+}$ ($d^5$) in the process and attains an extra stable half-filled *d*-orbital configuration.

**32. (i) Write the structure of major alkene formed by β-elimination of 2, 2, 3–trimethyl–3–bromopentane with sodium ethoxide in ethanol.**

**(ii) Which one of the compounds in the following pairs is chiral ?**

Br  Br

**(iii) Identify (A) and (B) in the following :**

$$(A) \xleftarrow{\text{Na/dry ether}} C_6H_5Br \xrightarrow{\text{Mg/dry ether}} (B)$$

[1 + 1 + 1 = 3]

**OR**

**How can you convert the following ?**

**(i) But-1-ene to 1-iodobutane**

**(ii) Benzene to acetophenone**

**(iii) Ethanol to propanenitrile**

**Ans.** (i) 2,2,3–trimethyl-3-bromopentane will undergo elimination reaction in presence of sodium ethoxide and alcohol and will provide 2,2, 3–Trimethylpent-3-ene as major production following Saytzeff's rule.

$$H_3C-CH_2-\underset{\underset{CH_3}{|}}{\overset{\overset{Br}{|}}{C}}-\underset{\underset{CH_3}{|}}{\overset{\overset{CH_3}{|}}{C}}-CH_3 \xrightarrow[\text{Ethanol}]{\text{Sodium ethoxide}} H_3C-CH=\underset{\underset{CH_3}{|}}{C}-\underset{\underset{CH_3}{|}}{\overset{\overset{CH_3}{|}}{C}}-CH_3$$

2, 2, 3 – trimethyl - 3 - bromo pentane                3,4,4–trimethylpent-2-ene

(ii) The following compound is chiral compound.

Br

(iii) Compound A is biphenyl compound.

$$2\,C_6H_5Br \xrightarrow{\text{Na, Dry ether}} C_6H_5-C_6H_5 + 2\,NaBr$$

Bromobenzene                Biphenyl (A)

Compound B is Phenyl magnesium bromide

$$C_6H_5Br + Mg \xrightarrow{\text{dry ether}} C_6H_5MgBr$$

Bromobenzene                Phenylmagensium bromide

**OR**

(i) But-1-ene to 1-Iodobutane conversion

$$CH_3CH_2CH=CH_2 \xrightarrow{\text{HBr/Peroxide}} CH_3CH_2CH_2CH_2-Br \xrightarrow{\text{NaI/dry acetone}} CH_3CH_2CH_2CH_2-I$$

(ii) Benzene to Acetophenone conversion

$$C_6H_6 + H_3C-COCl \xrightarrow{\text{anhydrous } AlCl_3} C_6H_5COCH_3 + HCl$$

Benzene    Acetyl chloride    Acetophenone

(iii) Ethanol to propane nitrile conversion :

This can be carried out in two steps, first, conversion of ethanol to ethyl chloride by reacting with zinc chloride and then reacting ethyl chloride with KCN in aq. ethanol.

$$\underset{\text{Ethanol}}{C_2H_5OH} \xrightarrow{ZnCl_2} \underset{\text{Ethyl chloride}}{C_2H_5Cl}$$

$$\underset{\text{Ethyl Chloride}}{CH_3-CH_2-Cl} \xrightarrow[-KCl]{\text{KCN/aq.ethanol}} \underset{\text{Propane nitrile}}{CH_3-CH_2-CN}$$

**33. Arrange the following compounds as directed:**

**(i) In increasing order of solubility in water :**
**$(CH_3)_2NH$, $CH_3NH_2$, $C_6H_5NH_2$**

**(ii) In decreasing order of basic strength in aqueous solution :**
**$(CH_3)_3N$, $(CH_3)_2NH$, $CH_3NH_2$**

**(iii) In increasing order of boiling point**
**$(C_2H_5)_2NH$, $(C_2H_5)_3N$, $C_2H_5NH_2$**

**[1 + 1 + 1 =3]**

**Ans.** **(i)** Increasing order of solubility in water :

$C_6H_5NH_2 < (CH_3)_2NH < CH_3NH_2$

**(ii)** Increasing order of basic strength in aqueous solution :

$(CH_3)_3N < CH_3NH_2 < (CH_3)_2NH$

**(iii)** Increasing order of boiling point :

$(C_2H_5)_3N < (C_2H_5)_2NH < C_2H_5NH_2$

**34. Write the product(s) of the following reactions:**

**[1 + 1 + 1 = 3]**

**(i)** Cyclohexenol (OH) $\xrightarrow{\text{PCC}}$

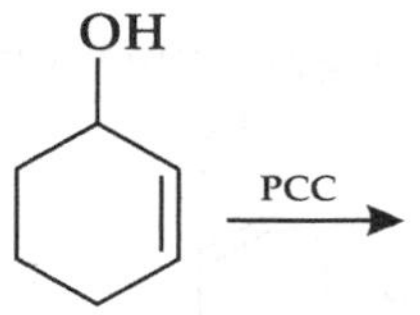

**(ii)** Salicylic acid (OH, COOH) $\xrightarrow[CH_3COOH]{(CH_3CO)_2O}$

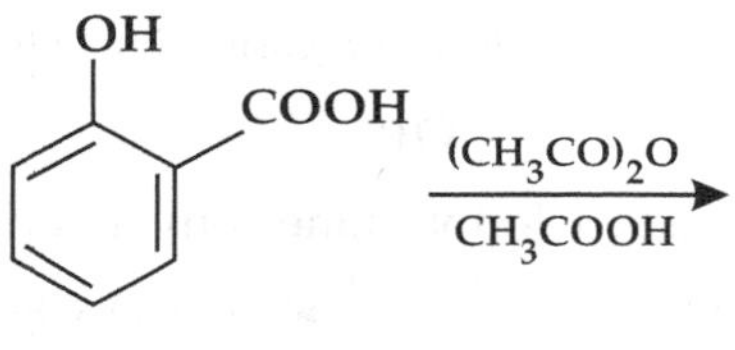

**(iii)** Cyclohexanone (O) $+ CH_3MgBr \xrightarrow{H_3O^+}$

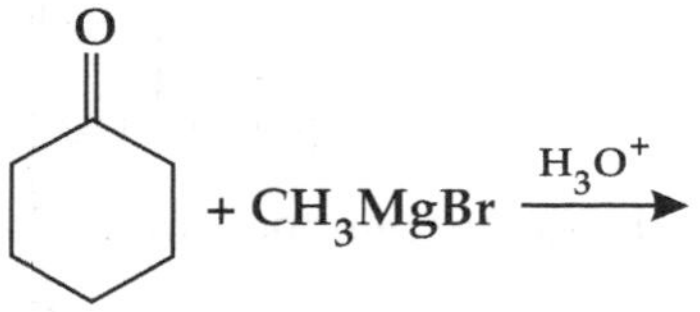

**OR**

**(a) Write the mechanism of the following $S_N1$ reaction:**

$$(CH_3)_3C-Br \xrightarrow{\text{Aq.NaOH}} (CH_3)_3C-OH + NaBr$$

**(b) Write the equation for the preparation of 2-methyl-2-methoxypropane by Williamson synthesis.** **[2 + 1 = 3]**

**Ans.**

**(i)** OH (cyclohexenol) $\xrightarrow{\text{PCC}}$ O (cyclohexenone)

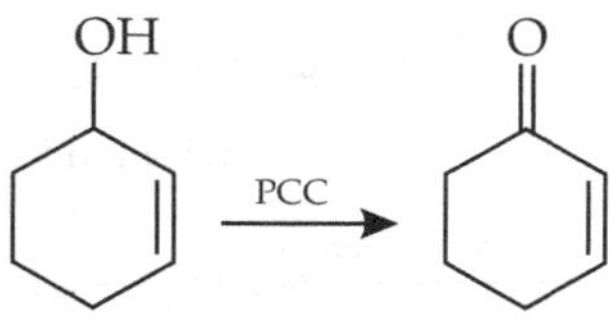

Cyclohexenol   Cyclohexenone

**(ii)**

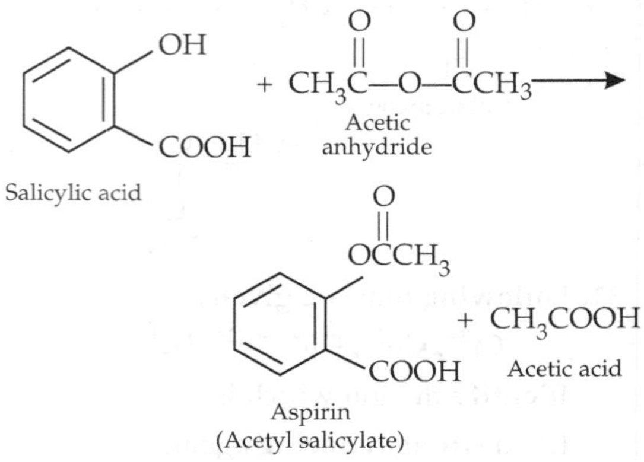

Salicylic acid (OH, COOH) $+ CH_3\overset{O}{\overset{\|}{C}}-O-\overset{O}{\overset{\|}{C}}CH_3 \longrightarrow$ (Acetic anhydride)

$\longrightarrow$ Aspirin (Acetyl salicylate) ($O\overset{O}{\overset{\|}{C}}CH_3$, COOH) $+ CH_3COOH$ (Acetic acid)

**(iii)**

Cyclohexanone $+ CH_3MgBr \xrightarrow{H_3O^+}$ 1-methyl cyclohexanol (HO, $CH_3$)

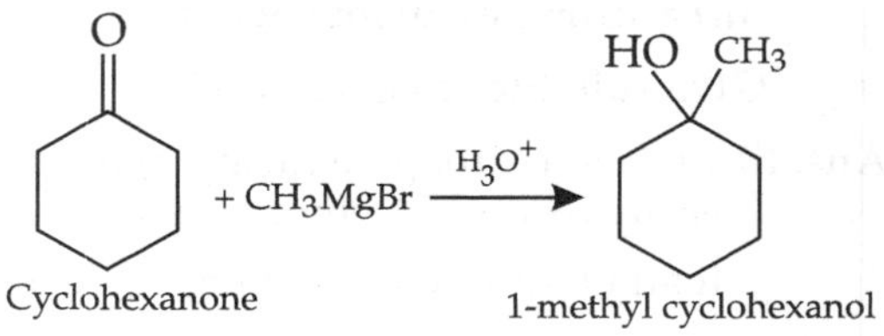

**OR**

**(a)** $(CH_3)_3Br \xrightarrow{\text{Aq. NaOH}} (CH_3)_3OH + NaBr$

The above $S_N1$ reaction happens in two stages:

Firstly a small proportion of halogenoalkane ionises to give a carbocation (carbonium ion) and a bromide ion.

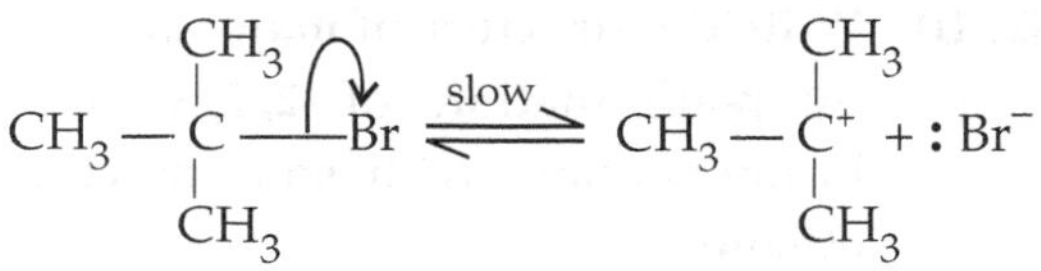

$$CH_3-\underset{CH_3}{\overset{CH_3}{\underset{|}{\overset{|}{C}}}}-Br \underset{}{\overset{\text{slow}}{\rightleftharpoons}} CH_3-\underset{CH_3}{\overset{CH_3}{\underset{|}{\overset{|}{C^+}}}} + :Br^-$$

Once the carbocation is formed it would react immediately with an $OH^-$ ion in the second stage of reaction. The lone pair on the nucleophile is strongly attracted towards the positive carbon, and moves towards it to create a new bond.

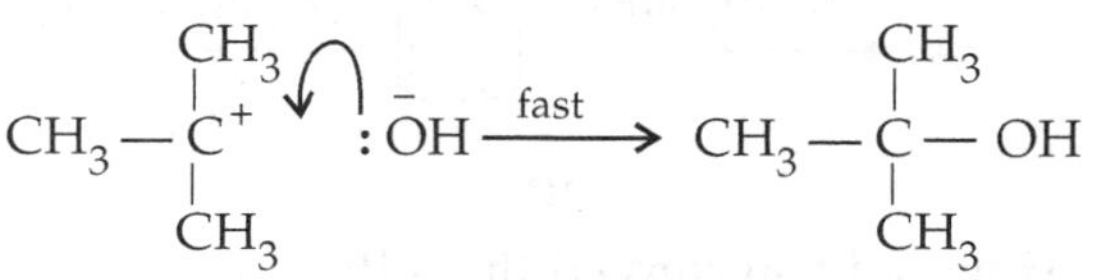

$$CH_3-\underset{CH_3}{\overset{CH_3}{\underset{|}{\overset{|}{C^+}}}} \;\; :\bar{O}H \xrightarrow{\text{fast}} CH_3-\underset{CH_3}{\overset{CH_3}{\underset{|}{\overset{|}{C}}}}-OH$$

**(b)** Preparation of 2-methyl-2-methoxypropane can be done using sodium *tert*-butoxide and methyl chloride as starting materials :

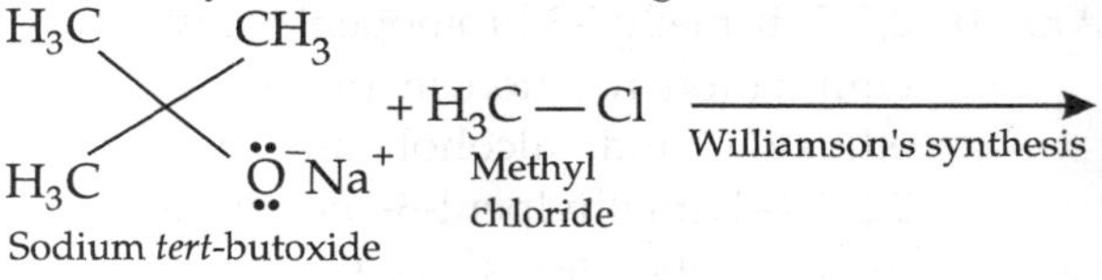

$(H_3C)(CH_3)(H_3C)C-\ddot{O}^-Na^+$ (Sodium *tert*-butoxide) $+ H_3C-Cl$ (Methyl chloride) $\xrightarrow{\text{Williamson's synthesis}}$

$$\begin{array}{c} CH_3 \\ | \\ H_3C \quad \ddot{O} \\ \diagdown \diagup \\ C \\ \diagup \diagdown \\ H_3C \quad CH_3 \end{array}$$

2-methyl-2-methoxypropane

## SECTION – D

**35. (a) The electrical resistance of a column of 0.05 M KOH solution of length 50 cm and area of cross-section 0.625 cm$^2$ is 5 × 10$^3$ ohm. Calculate its resistivity, conductivity and molar conductivity.**

**(b) Predict the products of electrolysis of an aqueous solution of $CuCl_2$ with platinum electrodes.**

**(Given : $E_{Cu^{2+}/Cu}$ = 0.34 V, $E_{(½ Cl_2/Cl^-)}$ = + 1.36 V**

**$E_{H^+/H_2 (g), Pt}$ = 0.00V, $E_{(½ O_2/H_2O)}$ = + 1.23 V)**

**[3 + 2 = 5]**

**OR**

**(a) Calculate e.m.f. of the following cell :**

**Zn(s)/$Zn^{2+}$ (0.1 M) || (0.01 M) $Ag^+$/Ag(s)**

**Given : $E_{Zn^{2+}/Zn}$ = – 0.76 V, $E_{Ag^+/Ag}$ = + 0.80V**

**[Given : log 10 = 1]**

**(b) X and Y are two electrolytes. On dilution molar conductivity of 'X' increases 2.5 times while that Y increases 25 times. Which of the two is a weak electrolyte and why ?**

**Ans. (a)** Resistivity, $\rho = \frac{RA}{l}$ where, R is the electrical resistance; $l$ is the length and A is the area of cross-section of the column.

$$\rho = \frac{RA}{l}$$

Substituting the given values :

$$\rho = \frac{RA}{l} = \frac{(5\times10^3 \text{ohm} \times 0.625 \text{ cm}^2)}{50 \text{ cm}}$$

$$= 62.5 \text{ ohm cm}$$

The inverse of resistivity is called conductivity. Hence,

$$\kappa = \frac{1}{\rho} = \frac{1}{62.5 \text{ ohm}^{-1}\text{cm}^{-1}}$$

$$= 0.016 \text{ S cm}^{-1}$$

Now, the molar conductivity is given by :

$$\Lambda_m = \frac{\kappa}{c} \times 1000$$

Substituting the values :

$$\Lambda_m = \frac{0.016 \times 1000}{0.05}$$

$$= 320 \text{ S cm}^2\text{mol}^{-1}$$

**(b)** Reactions which will occur in the solution are:

$$CuCl_2(s) \leftrightarrow Cu^{2+} + 2Cl^-$$

$$H_2O \leftrightarrow H^+ + OH^-$$

Reaction at cathode :

$$Cu^{2+} + 2e^- \rightarrow Cu(s)$$

Because $E^\circ_{Cu^{2+}/Cu} > E^\circ_{H^+/H^2}$

Reaction at anode :

$$2Cl^- \rightarrow Cl_2 + 2e^-$$

This reaction occurs at the anode due to over potential of $O_2$ oxidation of $Cl^-$ is preferred. Hence, Cu will deposit at cathode and $Cl_2$ gas will be liberated at anode.

**OR**

**(a)** The half-cell reactions can be written as :

$Zn^{2+} + 2e^- \rightarrow Zn$; $E_{el} = -0.76 + (0.0591/2)(\log[Zn^{2+}])$

Substituting the concentration value and calculating :

$$E_{el} = -0.76 + (0.0591/2)(\log 0.1)$$

$$= -0.76 - 0.0295$$

and $= -0.789$ V ...(i)

$$Ag^+ + e^- \rightarrow Ag;$$

$$E_{el} = 0.8 \text{ V} + 0.0591 (\log [Ag^+])$$

Substituting the concentration value and calculating:

$$E_{el} = 0.8 + 0.0591 (\log 0.01)$$

$$= 0.8 - 0.1182$$

$= 0.6818$ ...(ii)

Here, $Ag^+$ will reduce and Zn will oxidise—

$2Ag^+ (aq) + Zn(s) \rightarrow Zn^{2+} (aq) + 2Ag(s)$

will be the cell reaction

So, $E_{cell}$ will be :

$$E_{cell} = 0.6818 - (-0.789) \text{ V}$$

$$= 1.471 \text{ V}$$

**(b)** Y is weak electrolyte. On dilution weak electrolytes undergo complete dissociation and thus, there is a steep increase in molar conductivity observed. Whereas, in case of strong electrolytes, as they are already dissociated completely, dilution does not affect the conductivity very much. So, in this case X is a strong electrolyte.

**36. (a) An organic compound (A) having molecular formula $C_4H_8O$ gives orange red precipitate with 2, 4-DNP reagent. It does not reduce Tollens' reagent but gives yellow precipitate of iodoform on heating with NaOH and $I_2$. Compound (A) on reduction with $NaBH_4$ gives compound (B) which undergoes**

dehydration reaction on heating with conc. $H_2SO_4$ to form compound (C). Compound (C) on Ozonolysis gives two molecules of ethanal.

Identify (A), (B) and (C) and write their structures. Write the reactions of compound (A) with (i) $NaOH/I_2$ and (ii) $NaBH_4$.

**(b)** Give reasons :

(i) Oxidation of propanal is easier than propanone.

(ii) α-hydrogen of aldehydes and ketones is acidic in nature. [3 + 2 = 5]

OR

**(a)** Draw structures of the following derivatives.

(i) Cyanohydrin of cyclobutanone

(ii) Hemiacetal of ethanal

**(b)** Write the major product(s) in the following :

(i) $CH_3-CH=CH-CH_2-CN \xrightarrow[\text{(ii) } H_3O^+]{\text{(i) DIBAL-H}}$

(ii) $CH_3-CH_2-OH \xrightarrow{CrO_3}$

**(c)** How can you distinguish between propanal and propanone ? [2 + 2 + 1 = 5]

**Ans (a)** $C_4H_8O$ (A) gives precipitate with 2, 4-DNP, it indicates that compound A contains a carbonyl ($>C=O$) group. It does not reduce Tollen's reagent so it is a ketone compound. It gives Iodoform test as it contains a methyl ketone group ($-COCH_3$). Let us assume tentative structure of A as $X-COCH_3$. Now, A gives compound B on $NaBH_4$ reduction, so the structure of compound B becomes $X-CHOH-CH_3$. Compound B further is subjected to dehydration which means the new compound C will contain a double bond. Compound C gives two molecules of ethanal on Ozonolysis, this tells us that compound C is symmetrical and the double bond divides the molecule in two equal parts.

Collating all the above given information, the reaction sequence can be formulated as :

$$CH_3CH_2COCH_3 \text{ (Butanone (A))} \xrightarrow[\text{Reduction}]{NaBH_4} CH_3CH_2CH(OH)CH_3 \text{ (Butan-2-ol (B))} \xrightarrow{\text{Conc.}H_2SO_4}$$

$$CH_3CH=CHCH_3 \text{ (But-2-ene (C))} \xrightarrow{\text{Ozonolysis}} 2\,CH_3CHO \text{ (Ethanal (D))}$$

$$CH_3CH_2COCH_3 \text{ (Butanone (A))} \xrightarrow{NaOH/I_2} CH_3CH_2COO^-Na^+ \text{ (Sodium propanoate)} + CHI_3 \text{ (Iodomethane)}$$

$$CH_3CH_2COCH_3 \text{ (Butanone (A))} \xrightarrow[\text{Reduction}]{NaBH_4} CH_3CH_2CH(OH)CH_3 \text{ (Butan-2-ol)}$$

**(b) (i)** Oxidation of aldehyde is easier than ketones as in propanal there is one oxidisable -H present attached to the carbonyl carbon whereas propanone being ketone does not have the oxidisable -H atom. Also, as the carbonyl carbon contains two alkyl groups attached to it in ketone, it becomes less reactive responsible for less reactivity of propanone compared to propanal

**(ii)** In aldehydes and ketone, α-hydrogen is the hydrogen atom attached to the alpha (α) carbon atom, which in turn is attached to the carbonyl carbon. After removal of α-hydrogen atom, conjugated base so obtained is resonance stabilised. Hence, it becomes easy to lose the α-hydrogen atoms in basic medium and thus aldehydes and ketones are acidic in nature. The phenomenon is shown below :

[carbonyl (with :B abstracting α-H) ⇌ enolate ⟷ resonance form]

carbonyl enolate

OR

**(a) (i)** Cyanohydrin of cyclobutanone

$$\text{Cyclobutanone} \xrightarrow{HCN} \text{Cyclobutane cyanohydrin (C(OH)(C}\equiv\text{N) ring)}$$

Cyclobutanone Cyclobutane cyanohydrin

**(ii)** Hemiacetal of Ethanal

$$CH_3CHO \text{ (Ethanal)} \underset{H^+}{\overset{ROH}{\rightleftharpoons}} CH_3CH(OH)OR \text{ (Hemiacetal of ethanal)}$$

**(b) (i)**

$$H_3C-CH=CH-CH_2-CN \text{ (Pent-3-enenitrile)} \xrightarrow[\text{(ii) } H_3O^+]{\text{(i) DIBAL-H}} H_3C-CH=CH-CH_2-CHO \text{ (Pent-3-enal)}$$

(ii) $H_3C\diagdown\diagup OH \xrightarrow[\text{Oxidation}]{CrO_3} H_3C-CHO$

Ethanol → Ethanal

(c) Propanal and propanone can be distinguished by Iodoform reaction. When both are treated with NaOH and $I_2$, propanone (Acetone) gives a yellow precipitate of Iodoform whereas propanal does not react in these conditions.

$$\underset{\text{Acetone}}{CH_3-\overset{\overset{O}{\|}}{C}-CH_3} + 3I_2 + 4NaOH \longrightarrow \underset{\text{Iodoform}}{CHI_3} + CH_3COONa + 3NaI + 3H_2O$$

$$\text{Propanal} + I_2 + NaOH \longrightarrow \text{No reaction}$$

**37. (a) Account for the following :**

**(i) Tendency to show –2 oxidation state decreases from oxygen to tellurium.**

**(ii) Acidic character increases from HF to HI.**

**(iii) Moist $SO_2$ gas acts an a reducing agent.**

**(b) Draw the structure of an oxoacid of sulphur containing S—O—S linkage.**

**(c) Complete the following equation :**

**$XeF_2 + H_2O \rightarrow$ [3 + 1 + 1 = 5]**

**OR**

**(a) Among the hydrides of group 16, write the hydride**

**(i) Which is a strong reducing agent.**

**(ii) Which has maximum bond angle.**

**(iii) Which is most thermally stable.**

**Give situable reason in each.**

**(b) Complete the following equations :**

**(i) $S + H_2SO_4 \longrightarrow$**
**(Conc.)**

**(ii) $Cl_2 + NaOH \longrightarrow$**
**(Cold and dilute)**

**Ans. (a)** (i) Tendency to show -2 oxidation states decreases from oxygen to tellurium due to increase in atomic size and decrease in electronegativity.

(ii) Acidic nature is determined by the ability to donate $H^+$ ion in solution. HI being the least stable due to its large size and its inability to form strong hydrogen bonds lose $H^+$ easily. Also, its bond enthalpy is least among the hydrogen halides, breakage of H-I bond becomes easy thus making it lose $H^+$ quickly and hence most acidic followed by HBr, HCl, HF due to their increasing order of stability.Thus the acidic character increases from HF to HI.

(iii) In $SO_2$, sulphur is in +4 oxidation state and can further get reduced to achieve +3 oxidation state. So, moist sulphur dioxide behaves as a reducing agent as shown in following reaction :

$$2Fe^{3+} + SO_2 + 2H_2O \rightarrow 2Fe^{2+} SO_4^{2-} + 4H^+$$

(b) Pyrosulphuric acid contains -S-O-S- linkage

$$HO-\overset{\overset{O}{\|}}{\underset{\underset{O}{\|}}{S}}-O-\overset{\overset{O}{\|}}{\underset{\underset{O}{\|}}{S}}-OH$$

(c) $2XeF_2(s) + 2H_2O(I) \rightarrow 2Xe(g) + 4HF(aq) + O_2(g)$

**OR**

(a) (i) $H_2Te$ is the strongest reducing agent among all the hydrides of group 16 due to the low bond dissociation enthalpy for H-Te bond.

(ii) $H_2O$ has maximum bond angle of 104.45°. This is because oxygen forms 2 covalent bonds with 2 hydrogen atoms, hence, 2 pairs of electrons are bonding pairs and 2 pairs of electrons are lone pairs (non-bonding pairs) around the central oxygen atom. The repulsion between lone pairs of electrons is greater than the repulsion between lone pairs and the bonding pairs, which is greater than the repulsion between the bonding pairs of electrons, so the bonding pairs of electrons are pushed even closer together. Hence, a tetrahedral distribution of electron pairs is there in water molecule.

(iii) Thermal stability of the hydrides of group 16 decrease on the decrease in enthalpy for the dissociation of H-E bond down the group. Hence, $H_2O$ is thermally most stable.

(b) (i) Sulphur is oxidised to sulphur dioxide by conc. $H_2SO_4$.

$$S + 2H_2SO_4(\text{Conc.}) \rightarrow 3SO_2 + 2H_2O$$

(ii) Chlorine reacts with cold and dilute sodium hydroxide solution and forms the sodium chloride, sodium hypochlorite along with the water molecule in the solution.

$$Cl_2 + 2NaOH\,(\text{cold and dilute}) \rightarrow NaCl + NaOCl + H_2O$$

●●

# Outside Delhi [Set-II]

**Note : Except for the following questions all the remaining questions have been asked in previous set.**

## SECTION-A

**6. Name the depressant which is used to separate PbS and ZnS containing ore in forth floatation process.**

**Ans.** NaCN

**7. Out of —$CH_2-Cl$ and —$CH_2-Cl$ which will react faster in $S_N1$ reaction with $OH^-$?**

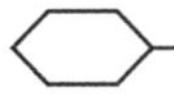

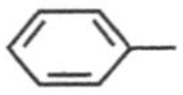

**Ans.** 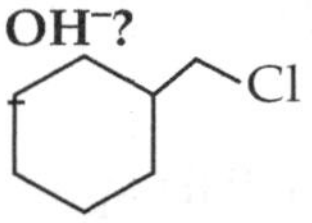 Cl

**8. Out of $CH_3NH_2$ and $CH_3OH$, which has higher boiling point ?**

**Ans.** $CH_3OH$

**9. Which one of the following is a narcotic analgesic ?**
**Penicillin, Codeine, Ranitidine.**

**Ans.** Codeine

**10. Write the name of linkage joining two monosaccharides.**

**Ans.** Glycosidic linkage

**11. The coordination number of 'Co' in the complex $[Co(en)_3]^{3+}$ is**

(a) 3 (c) 4

(b) 6 (d) 5

**Ans. (b)** 6

**12. An electrochemical cell behaves like an electrolytic cell when**

(a) $E_{cell} = E_{external}$ (c) $E_{external} > E_{cell}$

(b) $E_{cell} = 0$ (d) $E_{external} < E_{cell}$

**Ans. (c)** $E_{external} > E_{cell}$

**13. The half-life period for a zero order reaction is equal to**

(a) $\frac{0.693}{k}$ (c) $\frac{2.303}{k}$

(b) $\frac{2k}{[R]_0}$ (d) $\frac{[R]_0}{2k}$

**(Where $[R]_0$ is initial concentration of reactant and k is rate constant)**

**Ans. (d)** $\frac{[R]_0}{2K}$

**14. The crystal field splitting energy for octahedral ($\Delta_0$) and tetrahedral ($\Delta_t$) complexes is related as**

(a) $\Delta_t = \Delta_0$

(b) $\Delta_t = \frac{5}{9}\Delta_0$

(c) $\Delta_t = \frac{4}{9}\Delta_0$

(d) $\Delta_t = 2\Delta_0$

**Ans. (c)** $\Delta_t = \frac{4}{9}\Delta_0$

**15. α–D(+) glucose and (β–D(+) glucose are .**

(a) Geometrical isomers

(b) Enantiomers

(c) Anomers

(d) Optical isomers

**Ans. (c)** Anomers

**Questions 16 to 20 :**

**(A) Both Assertion (A) and Reason (R) are correct statements, and Reason (R) is the correct explanation of the Assertion (A).**

**(B) Both Assertion (A) and Reason (R) are correct statements, but Reason (R) is not the correct explanation of the Assertion (A).**

**(C) Assertion (A) is correct, but Reason (R) is wrong statement.**

**(D) Assertion (A) is wrong, but Reason (R) is correct statement.**

**16. Assertion (A) :** $F_2$ has lower bond dissociation enthalpy than $Cl_2$.
**Reason (R) :** Fluorine is more electronegative than chlorine.

**Ans. (B)** Fourine has low bond dissoiation enthalpy due to small atomic size number of electrons create large repulsion in bonded electron. Thus, Both Assertion (A) and Reason (R) are correct statements, but Reason (R) is not the correct explanation of the Assertion (A).

**18. Assertion (A) :** Transition metals have high melting point.
**Reason (R) :** Transition metals have completely filled *d*-orbitals.

**Ans. (C)** Transition metals have high melting point due to involvement of *d* and *S* orbitals in bonding. Thus, assertion (A) is correct, but Reason (R) is wrong statement.

## SECTION-B

**23. Write IUPAC name and hybridization of the following complexes :**

(i) $[Ni(CO)_4]$ (ii) $[CoF_6]^{3-}$

**(Atomic number Ni = 28, Co = 27) [1 + 1 = 2]**

**Ans. (i) IUPAC name :** Tetracarbonylnickel (O)
**Hybridisation :** $sp^3$

**(ii) IUPAC name :** Hexa fluoridocobaltate (III) ion
**Hybridisation :** $sp^3d^2$

**25. Define the following terms with a suitable example in each :** **[1 + 1 = 2]**

**(i) Tranquilizers (ii) Anionic detergent**

**Ans.** (i) **Tranquilizers:** They are a class of neurologically active chemical compounds used for treatment of stress and mild or even severe mental diseases. These relive anxiety by affecting the message transfer mechanism from nerve to receptor. Example: Chlordiazepoxide, Equanil etc.

(ii) **Anionic detergent :** They are sodium salts of sulphonated long chain alcohols or hydrocarbons. Alkyl hydrogen sulphates formed by treating long chain alcohols with concentrated sulphuric acid are neutralised with alkali to form anionic detergents. Example : Sodium lauryl sulphate.

**26. Write the reactions showing the presence of following in the open structure of glucose.**

**(i) an aldehyde group (ii) a primary alcohol** **[1 + 1 = 2]**

**Ans.** (i) Glucose gets oxidised to six carbon carboxylic acid (gluconic acid) on reaction with mild oxidising agent like bromine water. This confirms the presence of an aldehyde group in glucose molecule.

$$\underset{\text{Glucose}}{\begin{matrix} CHO \\ | \\ (CHOH)_4 \\ | \\ CH_2OH \end{matrix}} + [O] \xrightarrow{Br_2/Water} \underset{\text{Gluconic acid}}{\begin{matrix} COOH \\ | \\ (CHOH)_4 \\ | \\ CH_2OH \end{matrix}}$$

(ii) On oxidation with nitric acid glucose yields a dicarboxylic acid, saccharic acid.

It indicates the presence of a primary alcoholic group (–OH) in glucose .

$$\underset{\text{Glucose}}{\begin{matrix} CHO \\ | \\ (CHOH)_4 \\ | \\ CH_2OH \end{matrix}} + [O] \xrightarrow{conc.HNO_3} \underset{\text{(Saccharic acid)}}{\begin{matrix} COOH \\ | \\ (CHOH)_4 \\ | \\ CH_2OH \end{matrix}}$$

SECTION - C

**31. Write the name and structures of monomers in the following polymers :** **[1 + 1 + 1 = 3]**

**(i) Nylon 6, 6 (ii) Terylene (iii) PHBV**

**Ans.** (i) Adipic acid and Hexamethylene dichloride are the monomers of Nylon 6,6 polymer.

$$\underset{\text{Adipic acid}}{HOOC-(CH_2)_4-COOH} + \underset{\text{Hexamethylene diamine}}{H_2N-(CH_2)_6-NH_2} \longrightarrow \underset{\text{Nylon 6,6}}{\left[ \overset{O}{\overset{\|}{C}}-(CH_2)_4-\overset{O}{\overset{\|}{C}}-\underset{H}{\underset{|}{N}}-(CH_2)_6-\overset{H}{\overset{|}{N}} \right]}$$

(ii) Ethylene glycol and Terephthalic acid are the monomers of Terylene polymer.

$$\underset{\text{Ethylene glycol}}{HO-CH_2-CH_2-OH} + \underset{\text{Terephthalic acid}}{HOOC-C_6H_4-COOH} \xrightarrow{-H_2O} \underset{\text{Polyethylene terephthalate or (Terylene)}}{\left[ O-CH_2-CH_2-OOC-C_6H_4-\overset{O}{\overset{\|}{C}}- \right]_n}$$

(iii) 3-Hydroxybutanoic acid and 3-Hydroxy pentanoic acid are the monomers of PHBV polymer :

$$\underset{\text{3-Hydroxy butanoic acid}}{CH_3-\overset{OH}{\overset{|}{CH}}-CH_2-COOH} + n\,\underset{\text{3-Hydroxy pentanoic acid}}{CH_3-CH_2-\overset{OH}{\overset{|}{CH}}-CH_2-COOH} \xrightarrow{-H_2O} \underset{\text{PHBV}}{\left[ O-\underset{CH_3}{\underset{|}{CH}}-CH_2-\underset{O}{\underset{\|}{C}}-O-\underset{CH_2-CH_3}{\underset{|}{CH}}-CH_2-\underset{O}{\underset{\|}{C}}-O \right]}$$

## Outside Delhi [Set-III]

**Note : Except for the following questions all the remaining questions have been asked in previous sets.**

SECTION-A

**6. Name the method used for the refining of Zinc.**

**Ans.** Electrolytic refining or Distillation

**7. Out of $CH_3CH_2CH_2Cl$ and $CH_2 = CH - CH_2 - Cl$, which one is more reactive towards $S_N1$ reaction?**

**Ans.** $CH_2 = CH - CH_2Cl$

**8. Write an isomer of $C_3H_9N$ which does not react with Hinsberg reagent.**

**Ans.** $(CH_3)_3N$; Trimethylamine

**9. What type of protein is present in keratin ?**

**Ans.** Fibrous protein

**10. Name the compound which is added to soap to provide antiseptic properties.**

**Ans.** Bithional

**11. Which of the following is the most stable complex?**

**(a)** $[Fe(CO)_5]$ **(c)** $[Fe(C_2O_4)_3]^{3-}$

**(b)** $[Fe(H_2O)_6]^{3+}$ **(d)** $[Fe(CN)_6]^{3-}$

**Ans. (c)** $[Fe(C_2O_4)_3]^{3-}$

**12. Which of the following is correct for spontaneity of a cell ?**

**(a)** $\Delta G = -ve\ E = +ve$

**(b)** $\Delta G = +ve\ E = 0$

**(c)** $\Delta G = -ve\ E = 0$

**(d)** $\Delta G = +ve E = -ve$

**Ans. (a)** $\Delta G = -ve\ E° = +ve$

**13. For a zero order reaction, the slope in the plot of [R] Vs. time is**

**(a)** $\frac{-k}{2.303}$ **(c)** $\frac{+k}{2.303}$

**(b)** $-k$ **(d)** $+k$

**(where [R] is the final concentration of reactant)**

**Ans. (b)** $-k$

**14. What type of isomerism is shown by the pair $[Cr(H_2O)_6]Cl_3$ and $[Cr(H_2O)_5Cl]Cl_2.H_2O$ ?**

**(a) Ionization isomerism**

**(b) Coordination isomerism**

**(c) Solvate isomerism**

**(d) Linkage isomerism**

**Ans. (c)** Solvate isomerism

**15. Which one is the complementary base of cytosine in one strand to that in other strand of DNA ?**

**(a) Adenine** **(c) Thymine**

**(b) Guanine** **(d) Uracil**

**Ans. (b)** Guanine

**Questions 16 to 20 :**

**(A) Both Assertion (A) and Reason (R) are correct statements, and Reason (R) is the correct explanation of the Assertion (A).**

**(B) Both Assertion (A) and Reason (R) are correct statements, but Reason (R) is not the correct explanation of the Assertion (A).**

**(C) Assertion (A) is correct, but Reason (R) is wrong statement.**

**(D) Assertion (A) is wrong, but Reason (R) is correct statement.**

**16. Assertion (A): $F_2$ has low reactivity.**

**Reason (R): F-F bond has low $\Delta_{bond}H$**

**Ans. (D)** $F_2$ is more reactive than other halogens because its valence electrons are more closer to nucleus and its more electronegative so, bonded electrons repel each other causing low bond dissociation enthalpy. Thus, assertion is wrong , but Reason is correct statement.

## SECTION-B

**23. Write the IUPAC name and hybridisation of the following complexes :**

**(i)** $[Co(NH_3)_6]^{3+}$ **(ii)** $[NiCl_4]^{2-}$

**(Given : Atomic number : Ni = 28, Co = 27)**

**[1 + 1 =2]**

**Ans. (i) IUPAC Name:** Hexamminecobalt (III) ion

**Hybridisation:** $d^2sp^3$.

**(ii) IUPAC Name :** Tetrachloridonickelate (II ion

**Hybridisation :** $sp^3$.

**25. Write the reactions showing the presence of the following in the open structure of glucose :**

**(i) five – OH groups**

**(ii) a carbonyl group** **[1 + 1 = 2]**

**Ans. (i)** On acetylation with acetic anhydride, glucose gives glucose pentaacetate. This confirms presence of 5 –OH groups in glucose molecule.

$$\underset{\text{Glucose}}{CHO-(CHOH)_4-CH_2OH} + \underset{\text{Acetic anhydride}}{5(CH_3CH)_2O} \longrightarrow \underset{\text{Glucose penta acetate}}{CHO-(CHOCOCH_3)_4-CH_2O-\overset{O}{\overset{\|}{C}}-CH_3} + 5CH_3COOH$$

**(ii)** On reaction with hydroxylamine, glucose gives glucose oxime. This confirms presence of a $>C=O$ group in glucose molecule.

$$CHO-(CH_2OH)_4-CH_2OH + NH_2OH \longrightarrow \underset{\text{Glucose oxime}}{CH=N-OH-(CHOH)_4-CH_2OH}$$

**27. Define the following terms with a suitable example in each :** **[1 + 1 = 2]**

**(i) Antacids**

**(ii) Artificial Sweetener**

**Ans. (i) Antacids–**They are used to relieve the symptoms of hyperacidity, heartburn or indigestion. Over production of acid in the stomach causes irritation and pain. By neutralising the acid in the stomach, they relieve the symptoms of hyperacidity.

**Example:** A mixture of aluminium and magnesium hydroxide.

**(ii) Artificial Sweetener :** They are substances which have no nutritive value and are used to provide sweetness to food with low calories. These do not increase the calorie intake values. **Example :** Aspartame.

## SECTION-C

**31. Write the names and structures of monomers in the following polymers :**

**(a) Buna-S**

**(b) Glyptal**

**(c) Bakelite** [1 + 1 + 1 = 3]

**Ans. (a)** Butadiene and Styrene copolymerise to provide Buna-S polymer.

$$nCH_2=CH-CH=CH_2 + nCH_2=CH(C_6H_5) \xrightarrow{\text{Copolymerisation}} \left[-CH_2-CH=CH-CH_2-CH_2-CH(C_6H_5)-\right]_n$$

1, 3-butadiene; Styrene

Butadiene-Styrene copolymer (Buna -S)

**(b)** Ethylene glycol and Phthalic acid are the monomers of Glyptal polymer.

$$HOCH_2CH_2OH + C_6H_4(COOH)_2 \longrightarrow \left[-OCH_2-CH_2OOC-C_6H_4-CO-\right]_n$$

Ethylene glycol; Phthalic acid; Glyptal

**(c)** Phenol and formaldehyde are the monomers of Bakelite-

$$C_6H_5OH + H_2C=O \xrightarrow{\text{(many steps)}} \text{Bakelite}$$

Phenol; Formaldehyde; Bakelite

# CHEMISTRY

## Delhi [Set-I]

# 2020

**Time allowed : 3 hours** **Maximum marks : 70**

### SECTION-A

**Read the given passage and answer the questions 1 to 5 that follow :**

The halogens have the smallest atomic radii in their respective periods. The atomic radius of fluorine is extremely small. All halogens exhibit –1 oxidation state. They are strong oxidising agents and have maximum negative electron gain enthalpy. Among halogens, fluorine shows anomalous behaviour in many properties. For example electronegativity and ionisation enthalpy are higher for fluorine than expected whereas bond dissociation enthalpy, m.p. and b.p. and electron gain enthalpy are quite lower than expected. Halogens react with hydrogen to give hydrogen halides (HX) and combine amongst themselves to form a number of compounds of the type XX', $XX'_3$, $XX'_5$ and $XX'_7$ called inter-halogens.

**1. Why halogens have maximum negative electron gain enthalpy ?**

**Ans.** The outer electronic configuration of halogens is $ns^2\ np^5$, hence it completes its octet and gets stabilised by accepting one electron. So, halogens have maximum electron gain enthalpy.

**2. Why fluorine shows anomalous behaviour as compared to other halogens ?**

**Ans.** Fluorine is the first element in halogen group, its atomic number is 9 and electronic configuration is $[He]2s^2\ 2p^5$. The anomalous behaviour of fluorine is due to its small size, highest electronegativity, low F-F bond dissociation enthalpy, and non-availability of *d* orbitals in valence shell.

**3. Arrange the hydrogen halides (HF to HI) in the decreasing order of their reducing character.**

**Ans.** HF < HCl < HBr < HI

**4. Why fluorine is a stronger oxidizing agent than chlorine ?**

**Ans.** Fluorine has high electronegativity low heat of dissociation, high heat of hydration and low electron affinity than chlorine which makes attraction of electrons towards fluorine more easy than chlorine, hence it is a better oxidising agent compared to oxygen.

**5. What are the sizes of X and X' in the interhalogen compounds ?**

**Ans.** X is halogen of larger size and X' is halogen of smaller size.

**Question 6 to 10 are one word answers.**

**6. Name the cell used in hearing aids and watches.**

**Ans.** Mercury cell

**7. How much charge in terms of Faraday is required to reduce one mol of $MnO_4^-$ to $Mn^{2+}$ ?**

**Ans.** $5 \times 96500 = 482500F$

**8. Write the slope value obtained in the plot of log $[R_0]/[R]$ Vs. time for a first order reaction.**

**Ans.** Slope = $k/2.303$

**9. Name the sweetening agent used in the cooking of sweets for a diabetic patient.**

**Ans.** Sucralose

**10. Name the polymer which is used for making electrical switches and combs.**

**Ans.** Bakelite

**Question 11 to 15 are multiple choice questions.**

**11. In the Mond's process the gas used for the refining of a metal is**

**(a)** $H_2$ **(c)** CO
**(b)** $CO_2$ **(d)** $N_2$

**Ans. (c)** CO

**12. The conversion of an alkyl halide into an alcohol by aqueous NaOH is classified as**

**(a) a dehydrohalogenation reaction**
**(b) a substitution reaction**
**(c) an addition reaction**
**(d) a dehydration reaction**

**Ans. (b)** a substitution reaction

**13. $CH_3CONH_2$ on reaction with NaOH and $Br_2$ in alcoholic medium gives**

**(a)** $CH_3CH_2NH_2$ **(c)** $CH_3NH_2$
**(b)** $CH_3CH_2Br$ **(d)** $CH_3COONa$

**Ans. (c)** $CH_3NH_2$

**14. The oxidation state of Ni in $[Ni(CO)_4]$ is**

**(a)** 0 **(c)** 3
**(b)** 2 **(d)** 4

**Ans. (a)** 0

**15. Amino acids are**

**(a)** acidic **(c)** amphoteric
**(b)** basic **(d)** neutral

**Ans. (c)** amphoteric

Questions 16 to 20.

**(A) Both Assertion (A) and Reason (R) are correct statements, and Reason (R) is the correct explanation of the Assertion (A).**

(B) Both Assertion (A) and Reason (R) are correct statements, but Reason (R) is not the correct explanation of the Assertion (A).

(C) Assertion (A) is correct, but Reason (R) is wrong statement.

(D) Assertion (A) is wrong, but Reason (R) is correct statement.

**16. Assertion (A) :** Conductivity of an electrolyte increases with decrease in concentration.

**Reason (R) :** Number of ions per unit volume decreases on dilution.

**Ans. (D)** With increase in dilution, the number of ions in solution deceases. So, conductivity of electrolytic solution decease on dilution increasing. Thus, Assertion (A) is wrong, but Reason (R) is correct statement.

**17. Assertion (A) :** The C—O—C bond angle in ethers is slightly less than tetrahedral angle.

**Reason (R) :** Due to the repulsive interaction between the two alkyl groups in ethers.

**Ans. (D)** In ethers, bond angle around oxygen is not exactly 109° 28'. There is deviation in angle caused due to repulsive interactions between bulkier alkyl groups. Thus, Assertion (A) is wrong, but Reason (R) is correct statement.

**18. Assertion (A) :** Low spin tetrahedral complexes are rarely observed.

**Reason (R) :** Crystal field splitting energy is less than pairing energy for tetrahedral complexes.

**Ans. (A)** In tetrahedral complexes, none of the orbitals is pointing directly towards ligands due to which splitting is less such that the magnitude of crystal field spliting in field is quite small and is always less than pairing energy and consequently paring of electrons will never be energetically favourable. Thus, tetrahedral complexes are high spin complexes So, Both Assertion (A) and Reason (R) are correct statements, and Reason (R) is the correct explanation of the Assertion (A).

**19. Assertion (A) :** Elevation in boiling point is a colligative property.

**Reason (R) :** Elevation in boiling point is directly proportional to molarity.

**Ans. (C)** Elevation in boiling point is directly proportional to molarity. Thus, Assertion (A) is correct, but Reason (R) is wrong statement.

**20. Assertion (A) :** Oxidation of ketones is easier than aldehydes.

**Reason (R) :** C—C bond of ketones is stronger than C—H bond of aldehydes.

**Ans. (D)** Oxidation of aldehydes is easier than ketones. Thus, Assertion (A) is wrong, but Reason (R) is correct statement.

## SECTION – B

**21. State Raoult's law for a solution containing volatile components. What is the similarity between Raoult's law and Henry's law ? [2]**

**Ans.** Raoult's law states

For a solution of volatile liquids, the partial vapour pressure of each component of the solution is directly proportional to its mole fraction present in solution.

$$p_1 = p^\circ_1 x_1$$

According to Henry's law, the partial vapour pressure of a gas (the component is volatile and exists as a gas) in a liquid is

$$p_1 = K_H x_1$$

If we compare the equations for Raoult's law and Henry's law, it can be seen that the partial pressure of the volatile component or gas is directly proportional to its mole fraction in solution. Thus, Raoult's law becomes a special case of Henry's law when $p^\circ_1 = K_H$.

**22. Write the role of**

**(a) Dilute NaCN in the extraction of Gold.**

**(b) CO in the extraction of Iron. [1 + 1 = 2]**

**OR**

**How is leaching carried out in the case of low grade copper ores ? Name the method used for refining of copper metal.**

**Ans. (a)** Dilute NaCN is used to leach gold metal from its ore in presence of air –

$$4Au(s) + 8CN^-(aq) + 2H_2O(aq) + O_2(g) \rightarrow 4[Au(CN)_2]^-(aq) + 4OH^-(aq)$$

**(b)** CO acts as a reducing agent due to which iron in the ore reduces to iron metal.

$$3CO(g) + Fe_2O_3(s) \rightarrow 3CO_2(g) + 2Fe(l)$$

**OR**

Copper is extracted by hydrometallurgy from low grade ores. It is leached out using acid or bacteria. The solution containing $Cu^{2+}$ is treated with scrap iron or $H_2$

$$Cu^{2+}(aq) + H_2(g) \rightarrow Cu(s) + 2H^+(aq)$$

Copper is refined by electrolytic refining method. Anodes are made up of impure copper and pure copper strips are taken as cathode.

**23. Define adsorption with an example. What is the role of adsorption in heterogeneous catalysis ? [2]**

**OR**

**Define Brownian movement. What is the cause of Brownian movement in colloidal particles?**

**How is it responsible for the stability of Colloidal Sol?**

**Ans.** The accumulation of molecular species at the surface rather than in the bulk of a solid or a liquid is known as adsorption. It is essentially a surface phenomenon and the substance which concentrates at the surface is known as adsorbate and the material on the surface of which adsorption takes place is known as adsorbate. For example : Aqueous solution of raw sugar become colourless when passed through a bed of charcoal. Adsorption plays an important role in heterogenous catalysis. The catalyst adsorbs the reactants, forms an intermediate complex known as activated complex hence increases the concentration of reactant molecules. This increases the chances of reactant molecule coming close to each other for effective reaction. Heat of adsorption released is used in formation of activated complex and hence increases the rate of reaction.

**OR**

The continuous irregular and random zigzag motion of colloidal particles in the colloidal solution, when observed under an ultramicroscope, is called Brownian movement. Brownian movement is caused due to the collision between the molecules of dispersion medium and dispersed particles because of the unbalanced bombardment of the particles by the molecules of the dispersion medium. The Brownian movement gives a stirring effect to the colloidal solution, which does not permit the particles to settle and thus, is responsible for the stability of sols.

**24. (a) Write the IUPAC name and hybridisation of the complex $[Fe(CN)_6]^{3-}$.**
**(Given : Atomic number of Fe = 26)**

**(b) What is the difference between an ambidentate ligand and a chelating ligand?** **[1 + 1 = 2]**

**Ans. (a)** $[Fe(CN)_6]^{3-}$

**IUPAC name -** Hexacyanoferrate(III)

**Hybridisation-** $d^2sp^3$

**(b)** Ambidentate ligands are the ligand which can ligate through two different atoms such as $-NO_2$ ligand which can ligate through N atom as well as O atom as shown below :

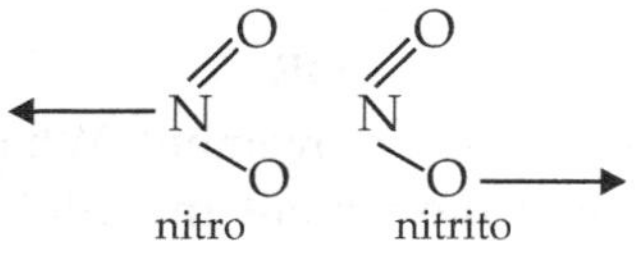

$\leftarrow N(=O)-O$ (nitro) $\quad N(=O)-O \rightarrow$ (nitrito)

Whereas, ligands those bind *via* more than one atom are termed as chelating ligands. These are commonly formed by linking donor groups via organic linkers. The below given example explains the same:

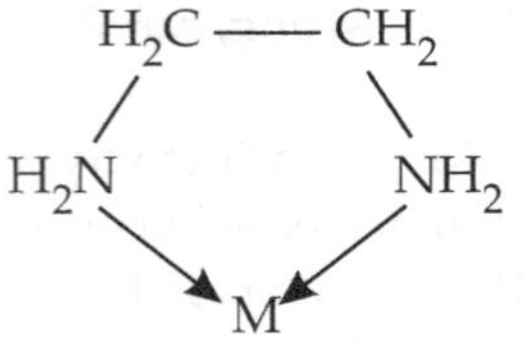

$H_2C-CH_2$ with $H_2N$ and $NH_2$ both coordinating to M ($H_2N \rightarrow M \leftarrow NH_2$)

A chelating ligand forms a more stable complex as compared to an ambidentate ligand. Chelating ligand forms a cyclic complex while ambidentate ligand forms a non-cyclic complex.

**25. How do antiseptics differ from disinfectants? Name a substance which can be used as a disinfectant as well as an antiseptic.** **[2]**

**Ans.** Antiseptics are applied to the living tissues such as wounds, cuts, ulcers and other disease skin surfaces. These are not ingested as antibiotics. Disinfectants are applied to inanimate objects such as floors, drainage systems, instruments to sterilize. Same chemical can be used as antiseptic and disinfectant by varying concentration. Phenol is a substance which when used in 0.2 % concentration works as antiseptic whereas its 1 % solution is disinfectant.

**26. Identify the monomers in the following polymers:** **[1 + 1 = 2]**

(i) 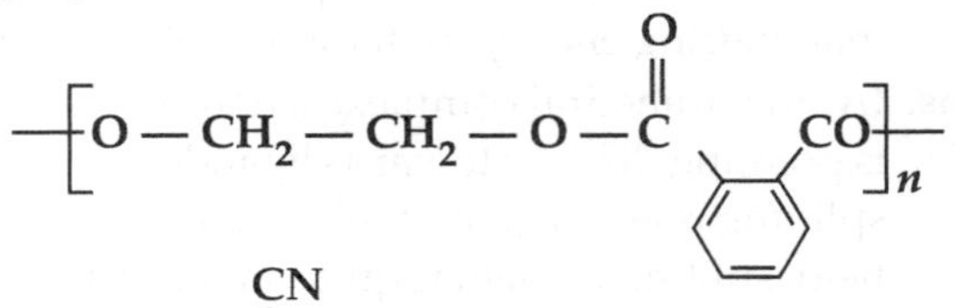 $\left[ O-CH_2-CH_2-O-\overset{O}{\overset{\|}{C}}-C_6H_4-CO \right]_n$

(ii) 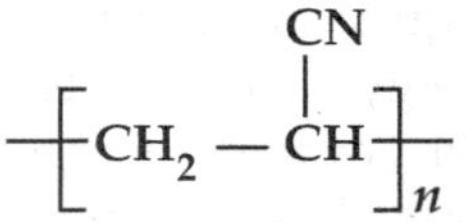 $\left[ CH_2-\underset{}{\overset{CN}{\overset{|}{CH}}} \right]_n$

**Ans. (i)** The given polymer is Glyptal and it is prepared from ethylene glycol and phthalic acid monomers.

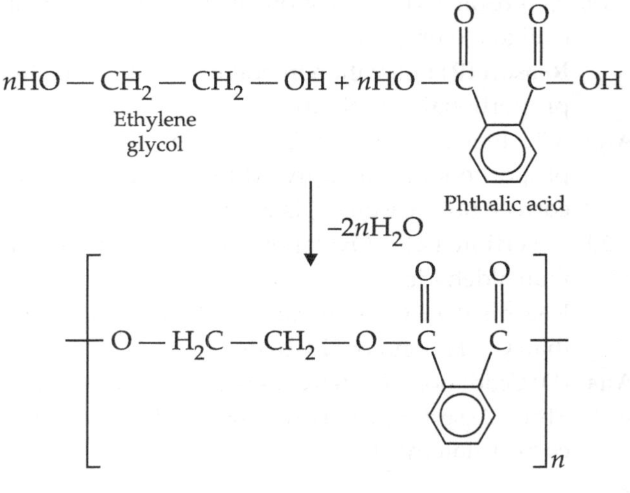

$$nHO-CH_2-CH_2-OH + nHO-\overset{O}{\overset{\|}{C}}-C_6H_4-\overset{O}{\overset{\|}{C}}-OH$$

Ethylene glycol; Phthalic acid

$$\downarrow -2nH_2O$$

$$\left[ O-H_2C-CH_2-O-\overset{O}{\overset{\|}{C}}-C_6H_4-\overset{O}{\overset{\|}{C}} \right]_n$$

**(ii)** The polymer is polyacrylonitrile and it is prepared through heating monomer acrylonitrile.

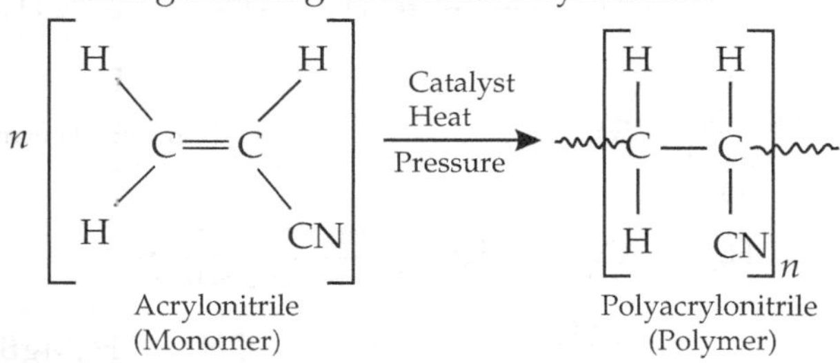


**27. Draw the structures of the following :**

**(i) $H_2S_2O_8$ (ii) $XeF_6$ [1 + 1 = 2]**

**Ans. (i)** Structure of $H_2S_2O_8$ (Peroxydisulphuric acid)

$$HO-\overset{O}{\underset{O}{\overset{\|}{\underset{\|}{S}}}}-O-O-\overset{O}{\underset{O}{\overset{\|}{\underset{\|}{S}}}}-OH$$

**(ii)** Structure of $XeF_6$ (Xenon hexafluoride)

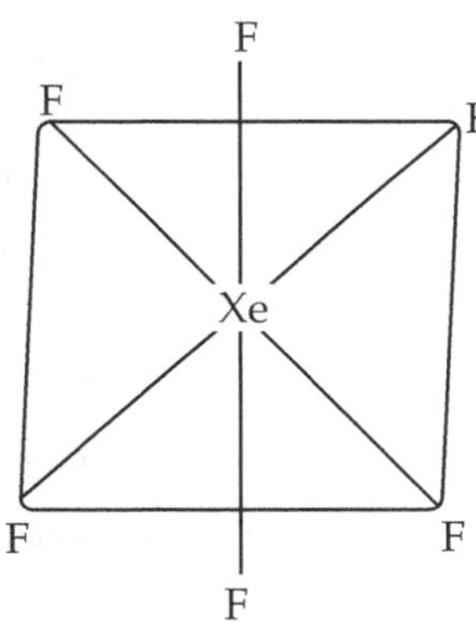


## SECTION – C

**28. A 0.01 m aqueous solution of $AlCl_3$ freezes at –0.068 C. Calculate the percentage of dissociation. [Given : $K_f$ for Water = 1.86 K kg $mol^{-1}$] [3]**

**Ans.** Given : K kg. $mol^{-1}$

$$\Delta T_f = 0.068, K_f = 1{\cdot}86$$

Sustituting the values

$$\Delta T_f = iK_f m$$

$$0{\cdot}068 = i \times 1{\cdot}86 \times 0{\cdot}01$$

$$i = 3{\cdot}65$$

degree of dissociation = $\alpha$

$$\alpha = \frac{i-1}{n-1}$$

$$\alpha = 0{\cdot}883$$

or 88·3% is the percentage of dissociation of $AlCl_3$ in water.

**29. When a steady current of 2A was passed through two electrolytic cells A and B containing electrolytes $ZnSO_4$ and $CuSO_4$ connected in series, 2 g of Cu were deposited at the cathode of cell B. How long did the current flow ? What mass of Zn was deposited at cathode of cell A ?**

**[Atomic mass : Cu = 63.5 g $mol^{-1}$, Zn = 65 g $mol^{-1}$; 1F = 96500 C $mol^{-1}$] [3]**

**Ans.** For deposition of Cu, reaction is :

$$Cu^{2+} + 2e^- \rightarrow Cu$$

Thus, 1 mole, i.e., 63.5 g Cu is deposited by 2F = 2 × 96500 C

$\therefore$ 2 g of Cu will be deposited by (96500 / 63.5) × 2 C = 3039.37 ≈ 3039·4S

Now, Charge Q = I × t

$$\frac{m_1}{m_2} = \frac{\text{eq wt}_1}{\text{eq wt}_2}$$

$$\frac{2}{m_2} = \frac{63.5/2}{65/2}$$

$$m_2 = 2{\cdot}05 \text{ g}$$

2·05 g of Zn was deposited at cathode of cell A.

**30. Differentiate between following :**

**(i) Amylose and Amylopectin**

**(ii) Globular protein and Fibrous protein**

**(iii) Nucleotide and Nucleoside [1 + 1 + 1 = 3]**

**Ans. (i)** Difference between Amylose and amylopectin

| Amylose | Amylopectin |
|---|---|
| 1. Straight chain polymer of D-glucose. | Branched chain polymer of D-glucose. |
| 2. Constitutes about 20% of starch. | Constitutes about 80% of starch. |
| 3. Amylose is water soluble component of starch. | Amylopectin is insoluble in water. |
| 4. Gives a dark blue black colour with iodine. | Gives orange colour with iodine. |

**(ii)** Difference between Globular protein and Fibrous protein :

| Globular protein | Fibrous protein |
|---|---|
| 1. Consists of coiled and folded polypeptide chains forming spherical shape. The structure is unstable. | Consists of long, parallel polypeptide chains forming helical structures or pleated sheets. The structure is linear and stable. |
| 2. Soluble in water. | Insoluble in water. |

| | |
|---|---|
| 3. Takes part in metabolite and chemical processes. Example : Enzymes and Haemoglobin. | Plays a major role in mechanical and structural functions. Example : Keratin and Collagen. |

**(iii)** Difference between Nucleotide and Nucleoside:

| Nucleotide | Nucleoside |
|---|---|
| 1. Nucleotide is a compound formed by the union of a nitrogen base, a pentose sugar and phosphate. | Nucleoside is a compound formed by the union of a nitrogen base with a pentose sugar. |
| 2. Nucleotide is formed through phosphorylation of nucleoside. | It is a component of nucleotide. |
| 3. A nucleotide is acidic in nature. Example : Thymidylic acid | It is slightly basic in nature. Example : Adenosine |

**31. Identify A, B, C, D, E and F in the following :**

$E \xleftarrow{H_2O} D \xleftarrow[\text{dry ether}]{Mg} CH_3-CH(CH_3)-CH_2-Br$

$CH_3-CH(CH_3)-CH_2-Br \xrightarrow{NaOC_2H_5} F$

$CH_3-CH(CH_3)-CH_2-Br \xrightarrow{\text{alcoholic KOH}} A$

$A \xrightarrow{HBr} B \xrightarrow{\text{Na/dry ether}} C$

[6 × ½ = 3]

**Ans.** $H_3C-CH(CH_3)-CH_2Br \xrightarrow[\text{Elimination}]{\text{Alc. KOH}} H_3C-C(CH_3)=CH_2$

2-Methyl-1-bromopropane → 2-Methyl prop-1-ene (A)

$H_3C-C(CH_3)=CH_2 \xrightarrow{HBr} H_3C-C(Br)(CH_3)-CH_3$

2-Methyl prop-1-ene (A) → 2-Bromo-2-methyl propane (B)

$H_3C-C(Br)(CH_3)-CH_3 \xrightarrow{\text{Na, dry ether}} H_3C-C(CH_3)_2-C(CH_3)_2-CH_3$

2-Bromo-2-methyl propane (B) → 2,2,3,3-Tetramethyl butane (C)

$H_3C-CH(CH_3)-CH_2Br \xrightarrow[\text{Grignard's reaction}]{\text{Mg, dry ether}} H_3C-CH(CH_3)-CH_2MgBr$

2-Methylpropyl magnesium bromide (D)

$H_3C-CH(CH_3)-CH_2MgBr \xrightarrow{H_2O} H_3C-CH(CH_3)-CH_3$

2-Methylpropyl magnesium bromide (D) → 2-Methylpropane-1 ol (E)

$H_3C-CH(CH_3)-CH_2Br \xrightarrow[\text{Substitution}]{NaOC_2H_5} H_3C-CH(CH_3)-CH_2OC_2H_5$

1-Ethoxy-2-methylpropane (F)

**32. Give the structures of final products expected from the following reactions :**

**(i) Hydroboration of propene followed by oxidation with $H_2O_2$ in alkaline medium.**

**(ii) Dehydration of $(CH_3)_3C-OH$ by heating it with 20% $H_3PO_4$ at 358 K.**

**(iii) Heating of $C_6H_5-CH_2-O-C_6H_5$ with HI.**

[3 1 = 3]

**OR**

**How can you convert the following ?**

**(i) Phenol to o-hydroxybenzaldehyde.**

**(ii) Methanal to ethanol**

**(iii) Phenol to phenyl ethanoate.**

**Ans. (i)** Propanol is obtained.

$H_3C-CH=CH_2 \xrightarrow[\text{Hydroboration Oxidation}]{\text{1. } BH_3, THF \quad \text{2. } H_2O_3, OH} H_3C-CH_2-CH_2-OH$

Propene → Propanol

(ii) 2-Methyl propene is obtained.

$$(CH_3)_2C(CH_3)OH \xrightarrow[\text{Dehydration}]{20\%\ H_3PO_3} (CH_3)_2C{=}CH_2$$

2-Methyl-propan-2-ol → 2-Methyl propene

(iii) Benzyl iodide is obtained along with phenol.

$$C_6H_5CH_2{-}O{-}C_6H_5 \xrightarrow{HI} C_6H_5CH_2I + C_6H_5OH$$

Benzyl iodide Phenol

**OR**

(i) Reimer-Tiemann reaction will occur by electrophilic substitution reaction by reacting phenol with Chloroform in presence of aqueous alkali at 340K.

$$C_6H_5OH \xrightarrow[3\ KOH]{CHCl_3} o\text{-}HOC_6H_4CHO$$

Phenol o-Hydroxy benzaldehyde

(ii) Methanal reacts with Grignard's reagent to form products which on hydrolysis give alcohols.

$$H{-}\underset{\underset{O}{\|}}{C}{-}H + H_3C{-}Mg{-}Br \xrightarrow{\text{dry ether}}$$

Methanal Methyl magnesium bromide

$$H_3C{-}\underset{H}{\overset{H}{C}}{-}O{-}Mg{-}Br$$

$$\downarrow H_2O$$

$$H_3C{-}CH_2{-}OH + Mg\begin{matrix}Br\\OH\end{matrix}$$

Ethanol

(iii) Phenol with acetic anhydrides in presence of $H_2SO_4$ react to give esters. This is called acylation.

$$C_6H_5OH + (CH_3CO)_2O \xrightarrow{H^+} C_6H_5OCO{-}CH_3 + CH_3COOH$$

Phenol Acetic anhydride Phenyl ethanoate

**33. Give reasons :**

**(i) Aniline does not undergo Friedal-Crafts reaction.**

**(ii) Aromatic primary amines cannot be prepared by Gabriel's phthalimide synthesis.**

**(iii) Aliphatic amines are stronger bases than ammonia.** **[3 × 1 = 3]**

**Ans. (i)** Aniline is basic in nature and AlCl3, the catalyst used in Friedal-Crafts reaction is a lewis acid and forms a salt with aniline.

$$C_6H_5NH_2 + AlCl_3 \longrightarrow C_6H_5\overset{+}{N}H_2\overset{-}{A}lCl_3 + HCl$$

Aniline Salt

In this salt the N atom possesses a positive charge due to which the benzene ring is deactivated for electrophilic substitution and hence aniline does not undergo Friedal-Crafts reaction.

(ii) Gabriel phthalimide synthesis involves nucleophilic substitution ($S_N2$) of alkyl halides by the anion formed by the phthalimide as follows :

$$C_6H_4(CO)_2N{-}H \xrightarrow{KOH} C_6H_4(CO)_2\overset{-}{N}\overset{+}{K} \xrightarrow{R-X} C_6H_4(CO)_2N{-}R$$

Phthalimide N-Alkylphthalimide

$$C_6H_4(CO)_2N{-}R \xrightarrow[(aq)]{NaOH} C_6H_4(CO\overset{-}{O}\overset{+}{Na})_2 + R{-}NH_2$$

1° amine

Since, aryl halides do not undergo nucleophilic substitution with the anion formed by the phthalimide, aromatic primary amines cannot be prepared by this process.

(iii) Aliphatic amines are stronger bases than ammonia due to +I effect of the aliphatic group attached to $-NH_2$ group. The alkyl group increases electron density on N and thus renders the N atom as a more electron rich center, hence increasing the basicity.

$$H_3C{\rightarrow}\ddot{N}H_2$$

increase in electron density of nitrogen lone pair

**34. Write three differences between lyophobic sol and lyophilic sol.** [3]

**OR**

**Define the following terms :**

**(i) Protective colloid**

**(ii) Zeta potential**

**(iii) Emulsifying agent**

**Ans.**

| Lyophobic Sol | Lyophilic Sol |
|---|---|
| 1. Interaction between dispersed phase and dispersion medium are weak. | Interaction between dispersed phase and dispersion medium are strong. |
| 2. Irreversible | Reversible |
| 3. Can be easily coagulated. | Cannot be easily coagulated |

**OR**

(i) Lyophilic colloids have a unique property of protecting lyophobic colloids. When a lyophilic sol is added to the lyophobic sol, the lyophilic particles form a layer around lyophobic particles and thus protect the latter from electrolytes. Lyophilic colloids used for this purpose are called protective colloids.

(ii) A colloidal particle acquires charge by adsorbing a selective charge (either positive or negative). This potential difference between the fixed layer and the diffused layer of opposite charges is called the zeta potential by a colloid.

(iii) Emulsifying agent is a substance that help an emulsion to become more stable. They are added to an emulsion to prevent the coalescence of the globules of the dispersed phase.

## SECTION-D

**35. (a) Give reasons :**

**(i) Transition metals and their compounds show catalytic activities.**

**(ii) Separation of a mixture of Lanthanoid elements is difficult.**

**(iii) Zn, Cd and Hg are soft and have low melting point.**

**(b) Write the preparation of the following :**

**(i) $Na_2Cr_2O_7$ from $Na_2CrO_4$**

**(ii) $K_2MnO_4$ from $MnO_2$** [3 + 2 = 5]

**OR**

**(a) Account for the following :**

**(i) $Ti^{3+}$ is coloured whereas $Sc^{3+}$ is colourless in aqueous solution.**

**(ii) $Cr^{2+}$ is a strong reducing agent.**

**(b) Write two similarities between chemistry of lanthanoids and actinoids.**

**(c) Complete the following ionic equation :**

**$3\,MnO_4^{2-} + 4H^+ \longrightarrow$**

**Ans. (a)** (i) Transition metals and their derivatives show catalytic properties because they have partially filled *d*-orbitals which enables them to show variable oxidation states and form complexes with other chemical entities. Hence, they can adsorb other substances on their surface and activate them for reaction process.

(ii) In lanthanoid series, there is a regular decrease in the atomic and ionic radii with increase in atomic number. This steady but small decrease in the sizes of the atoms or ions of the Lanthanides is called lanthanide contraction. Lanthanides are difficult to separate because of their almost similar sizes which is a result of lanthanide contraction.

(iii) In Zn, Cd and Hg, all the electrons in outer *d*-subshell are paired. They have no unpaired electron. Hence the metallic bonds present in them are weak. That is why they are soft metals and have low melting and boiling points.

**(b)** (i) Preparation of $Na_2Cr_2O_7$ from $Na_2CrO_4$ is done by acidifying the solution of latter with sulphuric acid-

$$2Na_2\,CrO_4 + 2H^+ \rightarrow Na_2Cr_2O_7 + 2Na^+ + H_2O$$

(ii) Preparation of $K_2MnO_4$ from $MnO_2$ is done by using KOH solution-

$$2MnO_2 + 4KOH + O_2 \rightarrow 2K_2MnO_4 + 2H_2O$$

**OR**

**(a)** (i) Electronic configuration of $Ti^{3+}$ ion is [Ar] $4s^0\,3d^1$ and $Sc^{3+}$ [Ar] $4s^0\,3d^0$. As there is no unpaired electron in d orbitals of $Sc^{3+}$ it does not show colour, while $Ti^{3+}$ is coloured due to presence of one unpaired electron in 3*d* orbital.

(ii) $Cr^{2+}$ is a strong reducing agent as by losing one more electron while reducing the other reacting species, it gains an oxidation state of +3 which is favourable and more stable.

**(b)** 1. Both lanthanoids and actinoids constitute the *f*-block of the periodic table, with the successive filling of the inner orbitals (4*f* and 5*f* respectively). They both show *f-f* transitions.

2. Both lanthanoids and actinoids elements show a gradual decrease in atomic and ionic sizes with increase in atomic number. This effects the properties and reactivities of these elements to a large extent. They both show contration in atomic radii.

(c) $3MnO_4^{2-} + 4H^+ \rightarrow 2MnO_4^- + MnO_2 + 2H_2O$

**36. (a) Write the products formed when benzaldehyde reacts with the following reagents :**

**(i) $CH_3CHO$ in presence of dilute NaOH**

**(ii) $H_2N - NH - C_6H_5$**

**(iii) Conc. NaOH**

**(b) Distinguish between following :**

**(i) $CH_3 - CH = CH - CO–CH_3$ and $CH_3 - CH_2 –CO - CH = CH_2$**

**(ii) Benzaldehyde and Benzoic acid.**

[3 + (1 + 1) = 5]

OR

**(a) Write the final products in the following :**

**(i)** $(CH_3)_2C = O \xrightarrow[\text{Conc. HCl}]{\text{Zn/Hg}}$

**(ii)** $C_6H_5-COONa \xrightarrow[\Delta]{\text{NaOH/CaO}}$

**(iii)** $CH_2 = CH - CH_2 - CN \xrightarrow[\text{(b) } H_3O^+]{\text{(a) DIBAL-H}}$

**(b) Arrange the following in the increasing order of their reactivity towards nucleophilic addition reaction :**

**$CH_3COCH_3$, HCHO, $CH_3CHO$, $C_6H_5-COCH_3$**

**(c) Draw the structure of 2, 4 DNP derivative of acetaldehyde.**

**Ans. (a) (i)** Cinnamaldehyde is formed when benzaldehyde reacts with acetaldehyde in presence of dil NaOH by Aldol condensation.

$$\underset{\substack{\text{Benzaldehyde} \\ (\text{no } \alpha\text{-hydrogen})}}{C_6H_5CHO} + CH_3CHO \xrightarrow[\text{Aldol condensation}]{\text{dil NaOH}} \underset{\text{Aldol.}}{C_6H_5CH(OH)CH_2CHO}$$

$$\xrightarrow{H^+, \text{Heat}, -H_2O} \underset{\text{Cinnamaldehyde}}{C_6H_5CH = CHCHO}$$

(ii) Benzaldehyde reacts with phenyl-hydrazine to give phenylhydrazone.

$$\underset{\text{Phenylhydrazine}}{C_6H_5-NH-NH_2} + \underset{\text{Benzaldehyde}}{C_6H_5CHO} \longrightarrow \underset{\text{Phenyl hydrazone}}{C_6H_5-NH-N=CH-C_6H_5}$$

(iii) Benzaldehyde do not have α-hydrogen so on heating with concentrated KOH solution a mixture of alcohol and salt is formed this is Cannizzaro's reaction.

$$\underset{\text{Benzaldehyde}}{C_6H_5CHO} + \text{Conc.KOH} \longrightarrow \underset{\text{Potassium benzoate}}{C_6H_5COOK} + \underset{\text{Benzyl alcohol}}{C_6H_5CH_2OH}$$

**(b) (i)** Among the two compounds given, $CH_3CH = CH–CO–CH_3$, has a terminal–$CH_3$ group attached to carbonyl carbon ($–CO–CH_3$) is present which will undergo Iodform reaction to give yellow coloured precipitate of triiodomethane.

$$CH_3CH = CH-CO-CH_3 + 4NaOH + 3I_2 \rightarrow CH_3CH = CH - COO^-Na^+ + 3NaI + 3H_2O + \underset{\text{(Yellow solid)}}{CHI_3}$$

Whereas the other compound $CH_3 - CH_2 - CO - CH = CH_2$ will not give any such reaction.

**(ii)** Benzaldehyde when treated with ammoniacal silver nitrate gets oxidized and reduces $Ag^+$ ions to Ag (elemental) hence, a deposition of silver is obtained which is known as silver mirror, whereas benzoic acid does not undergo this reaction hence can be distinguished with the help of Silver mirror test from benzaldehyde.

$$\underset{\text{Benzaldehyde}}{C_6H_5CHO} \xrightarrow[NH_3/H_2O]{AgO_2} C_6H_5COOH + Ag$$

Metallic silver is deposited in a thin mirror coating

OR

(a) (i) Clemmenson reduction :

$$CH_3-CO-CH_3 \xrightarrow[\text{Conc HCl}]{\text{Zn/Hg}} CH_3-CH_2-CH_3 + H_2O$$

Acetone → Propane

(ii)

Sodium benzoate (COONa on benzene ring) $\xrightarrow[\Delta]{NaOH + CaO}$ Benzene $+ Na_2CO_3 + H_2O$

(iii)

$H_2C{=}CH{-}CH_2{-}CN$ $\xrightarrow[\text{(b) } H_3O^+]{\text{(a) DIBAL H}}$ $H_2C{=}CH{-}CH_2{-}CHO$

But-3-enenitrile → But-3-enal

(b) Reactivity in increasing order is–

$C_6H_5COCH_3 < CH_3COCH_3 < CH_3CHO < HCHO$

(c)

$CH_3CH{=}O$ + (2, 4-dinitrophenyl) hydrazine ($H_2N{-}NH{-}C_6H_3(NO_2)_2$) → $CH_3CH{=}N{-}NH{-}C_6H_3(NO_2)_2$

(1E)-acetaldehyde (2-methyl-4dinitrophenyl) hydrazone

**37. (a) A first order reaction is 25% complete in 40 minutes. Calculate the value of rate constant. In what time will the reaction be 80% completed ?**

**(b) Define order of reaction. Write the condition under which a bimolecular reaction follows first order kinetics.** **[3 + 2 = 5]**

**OR**

**(a) A first order reaction is 50% complete in 30 minutes at 300 K and in 10 minutes at 320 K. Calculate activation energy ($E_a$) for the reaction. (R = 8.314 $JK^{-1}$ $mol^{-1}$)**

**(b) Write the two conditions for collisions to be effective collisions.**

**(c) How order of reaction and molecularity differ towards a complex reaction ?**

**[Given : log 2 = 0.3010, log 3 = 0.4771, log 4 = 0.6021, log 5 = 0.6991]**

**Ans. (a)** For a first order reaction–

$$k = \left(\frac{2.303}{t}\right)\log\left(\frac{a}{a}-x\right)$$

When $x = \left(\frac{25}{100}\right) a = 0.25$ a

$t = 40$ minutes (given)

$$\text{Therefore, } k = \left(\frac{2.303}{40}\right)\log\left(\frac{a}{a}-0.25a\right)$$

$$k = \left(\frac{2.303}{40}\right)\log\left(\frac{1}{0.75}\right)$$

$$k = 0.00719 \text{ min}^{-1}$$

Hence the value of the rate constant is 0.00719 $min^{-1}$

Now, we need to find time for rest 50% reaction t = ?, when $x = 0.50$ a

From above, k = 0.00719 $min^{-1}$

$$\text{Therefore, } t = \left(\frac{2.303}{0.00719}\right)\log\left(\frac{a}{a}-0.50a\right)$$

$$t = \left(\frac{2.303}{0.00719}\right)\log\left(\frac{1}{0.50}\right)$$

$$t = 230 \text{ min}$$

The time at which the reaction will be 50% complete is 230 min.

**(b)** Order of reaction for elementary reaction is the sum of the powers to which the reactant concentrations are raised in the rate law equation. It relates the rate of a chemical reaction with the concentrations of the reacting substances.

When concentration of one of the reactant is in excess and practically unchanged during the reaction for example when solvent used for reaction is one of the reactants, then the reaction follows first order kinetics as it depends on the concentration of only one reactant in practical sense.

**OR**

**(a)** The problem can be solved by using half-life equation–

by using half life equation, we get

$$K = \frac{0.693}{t_{1/2}}$$

$$K = \frac{0.693}{30\,\text{min}} = 0.0231 \text{ at } 300 \text{ K}$$

$$K = \frac{0.693}{10\,\text{min}} = 0.0693 \text{ at } 320 \text{ K}$$

From the Arrhenius equation

$$\log\frac{k_2}{k_1} = \frac{E_a}{2.303R}\left[\frac{T_2-T_1}{T_1T_2}\right]$$

$$E_a = \frac{2.303\times R\times T_1\times T_2}{T_2-T_1}\log\frac{k_2}{k_1}$$

$$= \frac{2.303 \times 8.314 JK^{-1}mol^{-1} \times 300 \times 320K}{320K - 300\ K} \times i$$

$$\left(\log \frac{0.06935}{0.2315}\right) = 43.848\ kJ\ mol^{-1}$$

(b) Two conditions for collisions to be effective collisions are :

1. Collisions must be sufficiently energetic to break chemical bonds; and it should be more than the threshold energy.

2. The reacting particle must be properly oriented during collision, so as to confirm maximum interaction between the reacting particles for successful reaction.

(c) 1. Order of reaction is an experimental quantity. It can be zero and even in fraction but molecularity cannot be zero or a non integer.

2. Order is applicable to elementary as well as complex reaction where as molecularity is applicable only for elementary reactions.

3. For complex reaction, order is given by the slowest step and molecularity of the slowest step is same as the order of overall reaction.

●●

# Delhi [Set-II]

**Note : Except for the following questions all the remaining questions have been asked in previous set.**

## SECTION-A

**6. Out of** chlorobenzene ($C_6H_5Cl$) **and** benzyl chloride ($C_6H_5CH_2$—Cl)**, which will undergo $S_N1$ reaction faster with $OH^-$?**

**Ans.** $C_6H_5CH_2Cl$.

**7. Write the IUPAC name of $CH_3$—N($CH_3$)—$C_6H_5$.**

**Ans.** N, N-Dimethyl benzenamine.

**8. What type of linkage is present in polysaccharides?**

**Ans.** Glycosidic linkages.

**9. Name an artificial sweetener whose use is limited to cold drinks.**

**Ans.** Aspartame

**10. Name the polymer which is used for making non-stick utensils.**

**Ans.** Teflon or PTFE

**11. Kohlrausch given the following relation for strong electrolytes :**

$$\wedge = \wedge_0 - A\sqrt{C}$$

**Which of the following equality holds ?**

**(a)** $\Lambda = \Lambda_0$ **as** $C \longrightarrow \sqrt{A}$

**(b)** $\Lambda = \Lambda_0$ **as** $C \longrightarrow \infty$

**(c)** $\Lambda = \Lambda_0$ **as** $C \longrightarrow 0$

**(d)** $\Lambda = \Lambda_0$ **as** $C \longrightarrow 1$

**Ans.** (c) $\Lambda = \Lambda_0$ as $C \longrightarrow 0$

**12. In an electrochemical process, a salt bridge is used.**

**(a) as a reducing agent.**

**(b) as an oxidising agent.**

**(c) to complete the circuit so that current can flow.**

**(d) None of these**

**Ans.** (c) to complete the circuit so that current can flow.

**13. In a chemical reaction $X \rightarrow Y$, it is found that the rate of reaction doubles when the concentration of X is increased four times. The order of the reaction with respect to X is**

**(a) 1** **(c) 2**

**(b) 0** **(d) 1/2**

**Ans.** (d) 1/2

**14. Which of the following will give a white precipitate upon reacting with $AgNO_3$ ?**

**(a)** $K_2[Pt(en)_2Cl_2]$ **(c)** $[Cr(H_2O)_6]Cl_3$

**(b)** $[Co(NH_3)_3Cl_3]$ **(d)** $[Fe(H_2O)_3Cl_3]$

**Ans.** (c) $[Cr(H_2O)_6]Cl_3$

**15. Copper matte contains**

**(a)** $Cu_2S$, $Cu_2O$ **and silica**

**(b)** $Cu_2S$, $CuO$ **and silica**

**(c)** $Cu_2S$, $FeO$ **and silica**

**(d)** $Cu_2S$, $FeS$ **and silica**

**Ans.** (d) $Cu_2S$, FeS and silica

**Questions 16 to 20 :**

**(A) Both Assertion (A) and Reason (R) are correct statements, and Reason (R) is the correct explanation of the Assertion (A).**

**(B) Both Assertion (A) and Reason (R) are correct statements, but Reason (R) is not the correct explanation of the Assertion (A).**

**(C) Assertion (A) is correct, but Reason (R) is wrong statement.**

**(D) Assertion (A) is wrong, but Reason (R) is correct statement.**

**16. Assertion (A) :** 0.1 M solution of KCl has greater osmotic pressure than 0.1 M solution of glucose at same temperature.

**Reason (R) :** In solution, KCl dissociates to produce more number of particles.

**Ans. (A)** KCl can dissociates in water but glucose does not due to dissociation of ions, solution exhibit higher colligative property. Thus, Both Assertion (A) and Reason (R) are correct statements, and Reason (R) is the correct explanation of the Assertion (A).

**18. Assertion (A) :** Ortho and para-nitrophenols can be separated by steam distillation.

**Reason (R) :** Ortho isomer associates through intermolecular hydrogen bonding while Para isomer associates through intramolecular hydrogen bonding.

**Ans. (C)** Ortho and para isomers of nitro- phenol can be separated by steam distillation because of nearby same boiling point of both and Ortho isomers associate by intramolecular hydrogen bonding and Para isomers associate by hydrogen bonding. Thus, Assertion (A) is correct, but Reason (R) is wrong statement.

## SECTION - B

**23. Draw the structures of the following :**

**(i) $H_2S_2O_7$**

**(ii) $BrF_5$** **[2]**

**Ans. (i)** $H_2S_2O_7$ -Pyrosulphuric acid

HO–S(=O)(=O)–O–S(=O)(=O)–OH

**(ii)** $BrF_5$ - Bromine pentafluoride

[Square pyramidal structure: Br with five F atoms and one lone pair]

**25. Identify the monomers in the following polymers:** **[2]**

(i) $-[C_6H_3(OH)-CH_2-C_6H_3(OH)-CH_2]_n-$

(ii) $-[NH-(CH_2)_6-NH-\overset{O}{\overset{\|}{C}}-(CH_2)_4-\overset{O}{\overset{\|}{C}}]_n-$

**Ans. (i)** The given polymer is Bakelite and the monomers are phenol and formaldehyde.

Phenol + HCHO (Formaldehyde) $\xrightarrow{NH_4OH}$ Bakelite polymer

Bakelite polymer

**(ii)** The given polymer is Nylon 6,6. It's monomers are hexamethylene diamine and adipic acid-

$$nHO-\overset{O}{\overset{\|}{C}}-(CH_2)_4-\overset{O}{\overset{\|}{C}}-OH + nH_2N-(CH_2)_6-NH_2 \longrightarrow$$

Adipic acid Hexamethylene diamine

$$\left[-\overset{O}{\overset{\|}{C}}-(CH_2)_4-\overset{O}{\overset{\|}{C}}-NH(CH_2)_6-NH-\right]_n + 2nH_2O$$

Nylon 6,6

**26. Discuss the nature of bonding in metal carbonyls.** **[2]**

**Ans. Nature of Bonding in Metal Carbonyls:** CO ligands donate electron pair to metal atom to generate a σ bond. Also, they accept electron density from the metal into their π* orbitals which is known as back bonding. This back bonding stabilizes the complex further. The lower the formal charge on metal ion, more it is willing to donate electrons to π orbitals of the CO ligands. Hence, zero valent metal atoms easily form CO complexes compared to the positively charged metal ions.

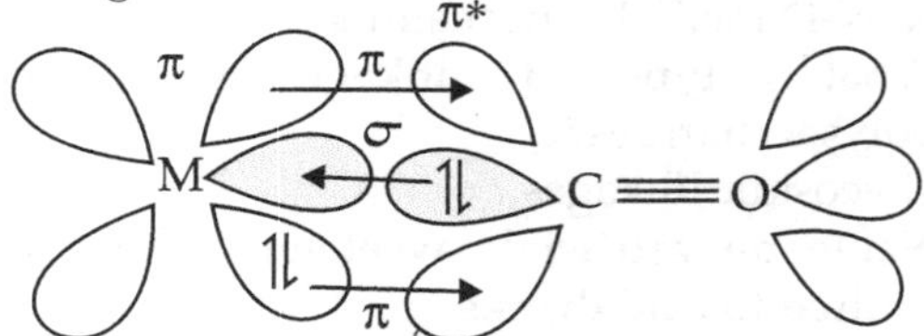


Synergic bonding in metal carbonyls

## SECTION - C

**30. Define the following terms with a suitable example in each :**

**(a) Polysaccharides**

**(b) Denatured protein**

**(c) Fibrous protein** **[1 + 1 + 1 = 3]**

**Ans. (i) Polysaccharides**–They contain large number of simple carbohydrate units joined together by glycosidic linkages. For example : Starch

**(ii) Denatured proteins**–When the protein molecules in their native form are subjected to physical changes like change in temperature, pH etc. the hydrogen bonds are disturbed within the structure of the proteins and they lose their tertiary structure resulting in loss of biological activity. This is known as denaturation of proteins. For Example: Hardening of egg upon boiling.

**(iii) Fibrous protein–**They are a fiber-like structure formed by the polypeptide chains. These proteins are held together by strong hydrogen and disulphide bonds and they run parallel to each other. It is usually insoluble in water. Fibrous proteins are usually used for structural purposes. For Example : Keratin present in nails and hair.

# Delhi [Set-III]

**Note : Except for the following questions all the remaining questions have been asked in previous sets.**

## SECTION-A

**6. A hydrocarbon $C_5H_{12}$ gives only one monochloride on photochemical chlorination. Identify the compound.**

**Ans.** The compound is 2, 2,-Dimethyl propane

$$H_3C-\underset{CH_3}{\overset{CH_3}{C}}-CH_3$$

2,2-Dimethyl propane

**7. Out of $(CH_3)_3N$ and $(CH_3)_2NH$, which one is more basic in aqueous solution ?**

**Ans.** $(CH_3)_2NH$

**8. Out of Cis – $[Pt(en)_2Cl_2]^{2+}$ and Trans $[Pt(en)_2Cl_2]^{2+}$, which one is optically active ?**

**Ans.** Cis-$[Pt(en)_2Cl_2]^{2+}$

**9. Name the method of refining used to obtain semiconductor of very high purity.**

**Ans.** Zone refining

**10. Is** $\left[ CH_2-CH=CH-CH_2-CH_2-\underset{CN}{\underset{|}{CH}} \right]_n$ **a homopolymer or copolymer ?**

**Ans.** Copolymer

**11. The amount of electricity required to produce one mole of Zn from $ZnSO_4$ solution will be :**

(a) 3F
(b) 2F
(c) 1F
(d) 4F

**Ans.** (b) 2F

**12. Zinc is coated over iron to prevent rusting of iron because**

(a) $E_{Zn^{2+}/Zn} = E_{Fe^{2+}/Fe}$
(b) $E_{Zn^{2+}/Zn} < E_{Fe^{2+}/Fe}$
(c) $E_{Zn^{2+}/Zn} > E_{Fe^{2+}/Fe}$
(d) None of these

**Ans.** (b) $E^\circ_{Zn^{2+}/Zn} < E^\circ_{Fe^{2+}/Fe}$

**13. The unit of rate constant depends upon the**

(a) molecularity of the reaction.
(b) activation energy of the reaction.
(c) order of the reaction
(d) temperature of the reaction

**Ans.** (c) Order of the reaction

**14. The formula of the complex triamminetri (nitrito-O) Cobalt (III) is**

(a) $[Co(ONO)_3 (NH_3)_3]$
(b) $[Co(NO_2)_3 (NH_3)_3]$
(c) $[Co(ONO_2)_3 (NH_3)_3]$
(d) $[Co(NO_2)(NH_3)_3]$

**Ans.** (a) $[Co(ONO)_3 (NH_3)_3]$

**15. Which of the following is a disaccharide ?**

(a) Glucose
(b) Starch
(c) Cellulose
(d) Lactose

**Ans.** (d) Lactose

**Questions 16 to 20 :**

**(A) Both Assertion (A) and Reason (R) are correct statements, and Reason (R) is the correct explanation of the Assertion (A).**

**(B) Both Assertion (A) and Reason (R) are correct statements, but Reason (R) is not the correct explanation of the Assertion (A).**

**(C) Assertion (A) is correct, but Reason (R) is wrong statement.**

**(D) Assertion (A) is wrong, but Reason (R) is correct statement.**

**16. Assertion (A) :** An ideal solution obeys Henry's law.

**Reason (R) :** In an ideal solution, solute-solute as well as solvent solvent interactions are similar to solute-solvent interaction.

**Ans.** (D) An ideal solution obeys Raoults' law. Thus, Assertion (A) is wrong, but Reason (R) is correct statement.

**18. Assertion (A) :** Benzaldehyde is less reactive than ethanal towards nucleophilic addition reactions.

**Reason (R) :** Ethanal is more sterically hindered.

**Ans.** (C) Benzaldehyde is less electrophillic than ethanal due to resonance stability of benzaldehyde. Thus, Assertion (A) is correct, but Reason (R) is wrong statement.

## SECTION - B

**22. Draw the structures of the following :**

(i) $HClO_4$
(ii) $XeOF_4$ [1 + 1 = 2]

**Ans.** (i) $HClO_4$ - Perchloric acid

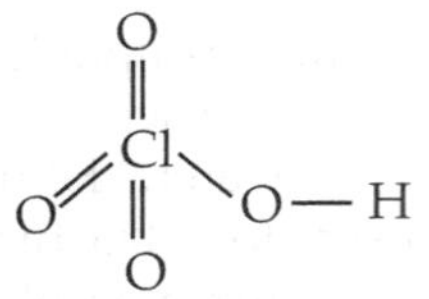

(Structure: $HClO_4$ — Cl bonded to three O by double bonds and to O—H)

**(ii)** $XeOF_4$ - Xenon oxytetrafluoride

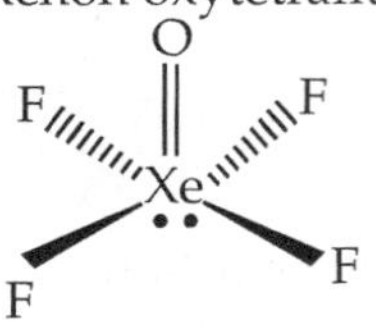

(Structure: Xe with one O, four F and one lone pair)

**24. Identify the monomers in the following polymers :**

(i) 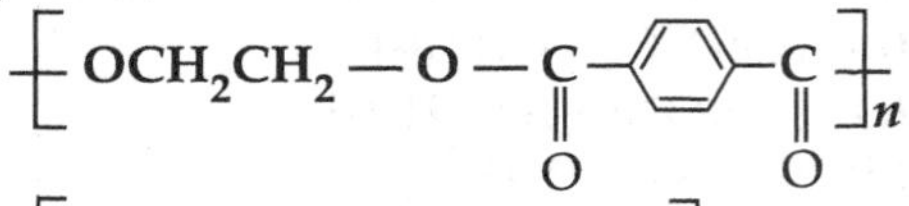 $\left[ OCH_2CH_2-O-\overset{\overset{O}{\|}}{C}-C_6H_4-\overset{\overset{O}{\|}}{C} \right]_n$

(ii) 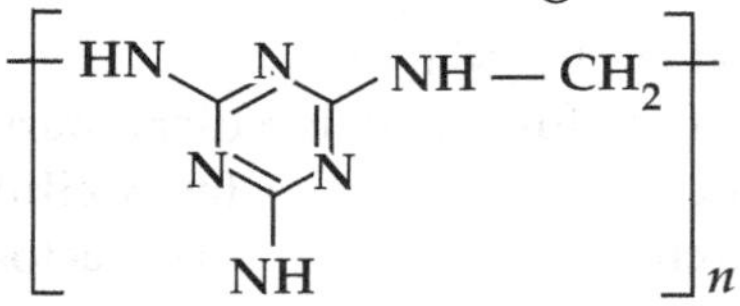 $\left[ HN-C_3N_3(NH)-NH-CH_2 \right]_n$ **[1+ 1 = 2]**

**Ans. (i)** The given polymer is Glyptal and it is prepared from ethylene glycol and phthalic acid monomers.

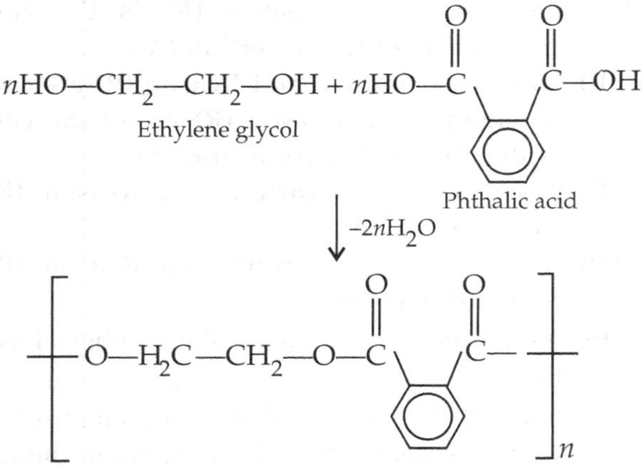

$$nHO-CH_2-CH_2-OH + nHO-\overset{\overset{O}{\|}}{C}-C_6H_4-\overset{\overset{O}{\|}}{C}-OH \xrightarrow{-2nH_2O} \left[ O-H_2C-CH_2-O-\overset{\overset{O}{\|}}{C}-C_6H_4-\overset{\overset{O}{\|}}{C} \right]_n$$

Ethylene glycol  Phthalic acid

**(ii)** The given polymer is melamine polymer and it is prepared from melamine and formaldehyde monomers.

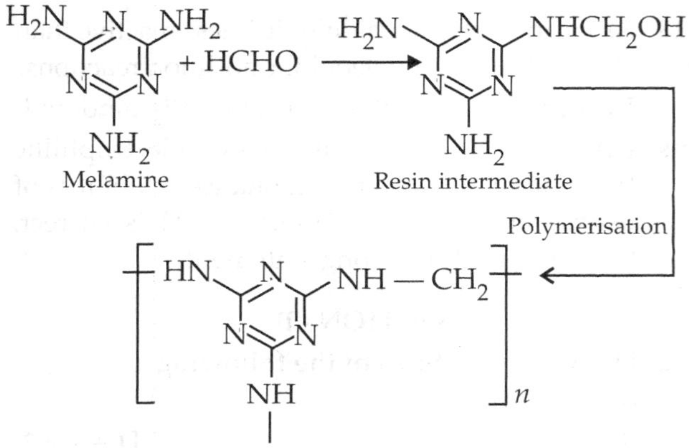

$$\underset{\text{Melamine}}{H_2N-C_3N_3(NH_2)-NH_2} + HCHO \longrightarrow \underset{\text{Resin intermediate}}{H_2N-C_3N_3(NH_2)-NHCH_2OH} \xrightarrow{\text{Polymerisation}} \left[ HN-C_3N_3(NH)-NH-CH_2 \right]_n$$

Melamine-formaldehyde polymer

**27. Define the following terms with a suitable example in each :**

**(i) Bacteriocidal antibiotics**

**(ii) Food preservatives.** **[1 + 1 = 2]**

**Ans. (i) Bacteriocidal antibiotics**–They are the antibiotics which kill the bacteria. Example : Penicillin.

**(ii) Food preservatives**–They are added to food items to prevent spoilage of food due to microbial growth. Example : Sodium benzoate.

## SECTION - C

**31. (i) What are the hydrolysis products of DNA ?**

**(ii) What happens when D-glucose is treated with Bromine water ?**

**(iii) What is the effect of denaturation on the structure of proteins ?** **[1 + 1 + 1 =3]**

**Ans. (i)** Complete hydrolysis of DNA yields a pentose sugar, phosphoric acid and nitrogen contaning heterocycles (nitrogenous bases). For example–

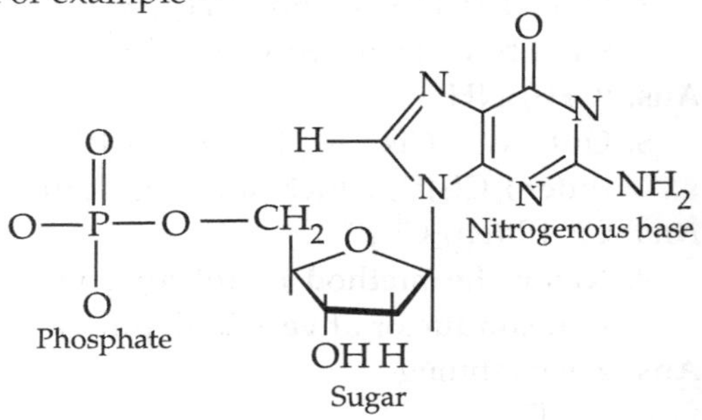

(Structure: Phosphate — O—P(=O)(O)—O—$CH_2$ attached to Sugar (OH H), attached to Nitrogenous base with $NH_2$)

Phosphate  Sugar  Nitrogenous base

**(ii)** When D-Glucose is treated with bromine water, gluconic acid is obtained:

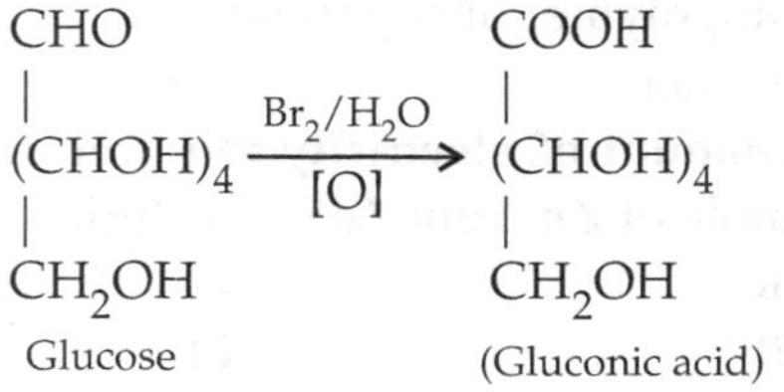

$$\underset{\text{Glucose}}{CHO-(CHOH)_4-CH_2OH} \xrightarrow[{[O]}]{Br_2/H_2O} \underset{\text{(Gluconic acid)}}{COOH-(CHOH)_4-CH_2OH}$$

**(iii)** When the protein molecules in their native form are subjected to physical changes like change in temperature, pH etc. the hydrogen bonds are disturbed within the structure of the proteins, this brings about denaturation of proteins due to which they lose their tertiary structure resulting in loss of biological activity. Example : Hardening of egg upon boiling.

●●

# CHEMISTRY

## Outside Delhi [Set-I]

2019

Time allowed : 3 hours

Maximum marks : 70

### SECTION-A

**1. Out of KCl and AgCl, which one shows Schottky defect and why?** [1]

OR

**Why does ZnO appear yellow on heating?**

**Ans.** KCl will show the Schottky defect because of comparable sizes of $K^+$ and $Cl^-$ ions.

OR

On heating, excess $Zn^{2+}$ ions move to interstitial sites and the electrons to neighbouring interstitial sites because of metal excess defect due to presence of extra $Zn^{2+}$ cations at interstitial sites it undergoes transitions and thus appears yellow on heating.

**2. Arrange the following in decreasing order of basic character :** [1]

**$C_6H_5NH_2$, $(CH_3)_3N$, $C_2H_5NH_2$**

**Ans.** Decreasing order of basic character :

$$(CH_3)_3N > C_2H_5NH_2 > C_6H_5NH_2$$

**3. What type of colloid is formed when a solid is dispersed in a liquid ? Give an example.** [1]

**Ans.** Sols are formed when a solid is dispersed in liquid. Example – Paints.

**4. Out of Chlorobenzene and Cyclohexyl chloride, which one is more reactive towards nucelophilic substitution reaction and why ?** [1]

**Ans.** Cyclohexyl chloride is more reactive towards nucleophilic substitution reaction, because the carbon bearing the chlorine atom is deficient in electron and seeks a nucleophile. In Chlorobenzene the carbon bearing the halogen is a part of aromatic ring and is electron rich due to the electron density in the ring.

**5. What is the basic structural difference between starch and cellulose ?** [1]

OR

**Write the products obtained after hydrolysis of DNA.**

**Ans.** Starch is a polymer of $\alpha$-glucose whereas cellulose is a polymer of $\beta$-glucose.

OR

Hydrolysis of DNA yields a pentose sugar ($\beta$-D-2deoxyribose), phosphoric acid and nitrogen containing heterocyclic compounds called bases (Adenine, Guanine, Cytosine and Thymine).

### SECTION-B

**6. Write balanced chemical equations for the following processes :**

**(a) $Cl_2$ is passed through slaked lime.**

**(b) $SO_2$ gas is passed through an aqueous solution of Fe (III) salt.** [2]

OR

**(a) Write two poisonous gases prepared from chlorine gas.**

**(b) Why does $Cu^{2+}$ solution give blue colour on reaction with ammonia ?**

**Ans. (a)** $Cl_2$ is passed through slaked lime to give bleaching powder $[Ca(OCl)_2]$

$$2Ca(OH)_2 + 2Cl_2 \rightarrow Ca(OCl)_2 + CaCl_2 + 2H_2O$$

**(b)** When $SO_2$ gas is passed through an Fe(III) aqueous solution, Fe(III) is reduced to Fe(II) ion :

$$2Fe^{3+} + SO_2 + 2H_2O \rightarrow 2Fe^{2+} + SO_4^{2-} + 4H^+$$

OR

**(a)** Two poisonous gases prepared from chlorine – Phosgene ($COCl_2$) and tear gas ($CCl_3NO_2$).

**(b)** Nitrogen in ammonia has a lone pair of electrons, which makes it a Lewis base. It donates the electron pair and forms linkage with metal ions.

$$\underset{\text{(Blue)}}{Cu^{2+}(aq)} + 4NH_3(aq) \rightleftharpoons \underset{\text{(Deep blue)}}{[Cu(NH_3)_4]^{2+}(aq)}$$

**7. Give reasons :**

**(a) Cooking is faster in pressure cooker than in cooking pan.**

**(b) Red Blood Cells (RBC) shrink when placed in saline water but swell in distilled water.** [2]

**Ans. (a)** Boiling points increase on increasing the pressure in case of liquids. Water used for cooking attains higher temperature than usual boiling temperature inside the pressure cooker due to the existing high pressure inside the pressure cooker vessel. This leads to faster flow of water inside the vegetables or grains etc. resulting in faster cooking of food in a pressure cooker than in the cooking pan.

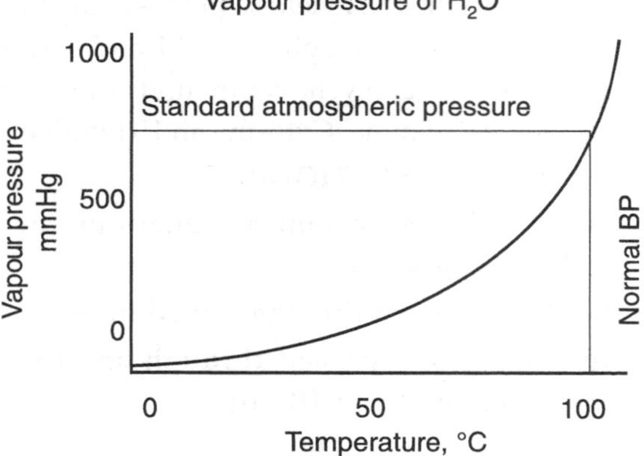


**(b)** Red blood cells shrink when placed in saline water because of exosmosis, *i.e.*, water comes out from the cell to surrounding (more concentrated) to equate the concentration. Whereas, when placed in distilled water concentration within the cell becomes more than the surrounding, hence water comes inside and endosmosis takes place to equate the concentrations.

**8. Define order of reaction. Predict the order of reaction in the given graphs:**

**(a)**

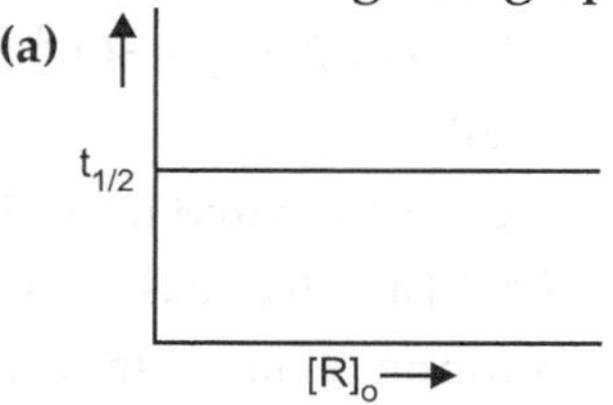


**(b)**

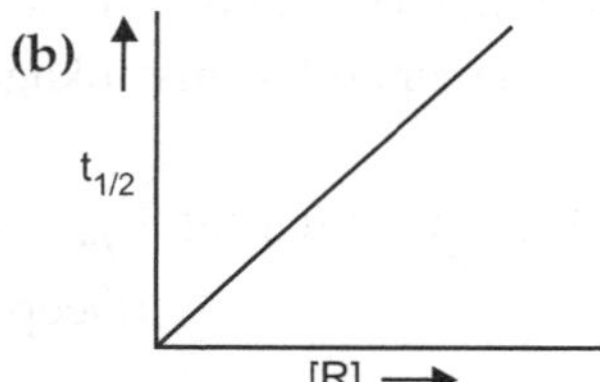


**where $[R]_0$ is the initial concentration of reactant and $t_{1/2}$ is half-life.** **[2]**

**Ans.** Order of reaction is defined as the sum of powers to which the concentration terms are raised in the rate law equation.

**(a)** In this graph as $t_{1/2}$ is independent of initial reactant concentration, it is a first order reaction.

**(b)** In this graph as $t_{1/2}$ is directly proportional to initial concentration of reactant hence, it is a zero order reaction.

**9. When $FeCr_2O_4$ is fused with $Na_2CO_3$ in the presence of air it gives a yellow solution of compound (A). Compound (A) on acidification gives compound (B). Compound (B) on reaction with KCl forms an orange coloured compound (C). An acidified solution of compound (C) oxidises $Na_2SO_3$ to (D). Identify (A), (B), (C) and (D).** **[2]**

**Ans.** $4FeCr_2O_4 + 8Na_2CO_3 + 7O_2 \rightarrow 8Na_2CrO_4 + 2Fe_2O_3 + 8CO_2$

$8Na_2CrO_4$ — Sodium chromate (yellow) A

$Na_2Cr_2O_4 + H^+ \rightarrow Na_2Cr_2O_7 + 2Na^+ + H_2O$

$Na_2Cr_2O_7$ — Sodium dichromate (B)

$Na_2Cr_2O_7 + 2KCl \rightarrow K_2Cr_2O_7 + 2NaCl$

$K_2Cr_2O_7$ — Potassium dichromate (C)

$K_2Cr_2O_7 + 3Na_2SO_3 + 4H_2SO_4 \rightarrow Cr_2(SO_4) + 3Na_2SO_4 + K_2SO_4 + 4H_2O$

$Cr_2(SO_4)$ — Chromium sulphate (D)

**10. Write IUPAC name of the complex $[Co(en)_2(NO_2)Cl]^+$. What type of structural isomerism is shown by this complex ?** **[2]**

**OR**

**Using IUPAC norms, write the formulae for the following complexes :**

**(a) Hexaaquachromium (III) chloride**

**(b) Sodium trioxalatoferrate (III)**

**Ans.** IUPAC name of $[Co(en)_2(NO_2)Cl]^+$ is chloridobis (ethane-1,2-diamine) nitrito-N-cobalt (III) ion.

This compound shows linkage isomerism.

**OR**

**(a)** Hexaaquachromium(III) chloride–

$[Cr(H_2O)_6]Cl_3$

**(b)** Sodium trioxalatoferrate(III) –

$Na_3[Fe(C_2O_4)_3]$

**11. (a) Although both $[NiCl_4]^{2-}$ and $[Ni(CO)_4]$ have $sp^3$ hybridisation yet $[NiCl_4]^{2-}$ is paramagnetic and $[Ni(CO)_4]$ is diamagnetic. Give reason. (Atomic no. of Ni = 28).**

**(b) Write the electronic configuration of $d^5$ on the basis of crystal field theory when**

**(i) $\Delta_0 < P$ and**

**(ii) $\Delta_0 > P$** **[2]**

**Ans. (a)** $[NiCl_4]^{2-}$ is a high spin complex and there are two unpaired electrons with $3d^8$ electronic configuration of central metal atom, hence it is paramagnetic. Whereas in $[Ni(CO)_4]$ Ni is in zero oxidation state and contains no unpaired electrons, hence it is diamagnetic in nature.

**(b) (i)** Electronic configuration of $d^5$ when $\Delta_0 < P$ is given as $t_{2g}{}^3\, e_g{}^2$

**(ii)** Electronic configuration of $d^5$ when $\Delta_0 > P$ is given as $t_{2g}{}^5\, e_g{}^0$

**12. Write structures of main compounds A and B in each of the following reactions :** **[2]**

**(a)** $C_6H_5COOH \xrightarrow{PCl_5} A \xrightarrow{H_2\text{-Pd-BaSO}_4} B$

**(b)** $CH_3CN \xrightarrow[\text{(ii) } H_3O^+]{\text{(i) } CH_3MgBr} A \xrightarrow{\text{Zn(Hg)/conc. HCl}} B$

**Ans. (a)**

COOH $\xrightarrow{PCl_5}$ COCl $\xrightarrow{H_2\text{-Pd, BaSO}_4}$ CHO

Benzoic acid — Benzoyl chloride (A) — Benzaldehyde (B)

**(b)**

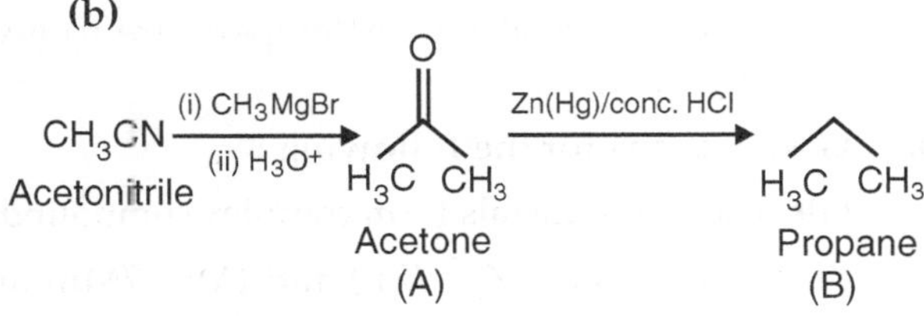


## SECTION–C

**13. The following data were obtained for the reaction :**

$$A + 2B \rightarrow C$$

| Experiment | [A]/M | [B]/M | Initial rate of formation of C/M min$^{-1}$ |
|---|---|---|---|
| 1 | 0·2 | 0·3 | $4{\cdot}2 \times 10^{-2}$ |
| 2 | 0·1 | 0·1 | $6{\cdot}0 \times 10^{-3}$ |
| 3 | 0·4 | 0·3 | $1{\cdot}68 \times 10^{-1}$ |
| 4 | 0·1 | 0·4 | $2{\cdot}40 \times 10^{-2}$ |

**(a) Find the order of reaction with respect to A and B.**

**(b) Write the rate law and overall order of reaction.**

**(c) Calculate the rate constant ($k$).** **[3]**

**Ans.** The reaction is-

$$A + 2B \rightarrow C$$

**(a)** It can be seen that when concentration of A is doubled keeping B constant, then the rate increases by a factor of 4 (from $4.2 \times 10^{-2}$ to $1.68 \times 10^{-1}$). This indicates that the rate depends on the square of the concentration of the reactant A. Also, when concentration of reactant B is made four times, keeping the concentration of reactant A constant, the reaction rate also becomes 4 times ($2.4 \times 10^{-2}$ to $6.0 \times 10^{-3}$). This indicates that the rate depends on concentration of reactant B to the first power.

Order with respect to A = 2, B = 1.

**(b)** So, the rate equation will be :

$$\text{Rate} = k[A]^2[B]$$

Overall order of reaction will be 2+1 = 3.

**(c)** Rate constant can be calculated by putting the values given.

$$4.2 \times 10^{-2}\ \text{M min}^{-1} = k\,(0.2)^2(0.3)\ \text{M}$$

$$k = \frac{0.042}{0.012}$$

$$= 3.5\ \text{min}^{-1}$$

**14. (a) Write the dispersed phase and dispersion medium of dust.**

**(b) Why is physisorption reversible whereas chemisorption is irreversible ?**

**(c) A colloidal sol is prepared by the method given in the figure. What is the charge of AgI colloidal particles formed in the test tube ?**

**How is this sol represented ?** **[3]**

KI Solution

$AgNO_3$

**Ans. (a)** In dust the dispersed phase is solid particles and dispersion medium is air (gas).

**(b)** Physisorption occurs only because of physical attractive forces, like van der Waals forces between molecules of adsorbate and adsorbent, hence that can be reversed on application of bigger forces but chemisorption occurs due to chemical reaction between molecules of adsorbate and adsorbent, and hence cannot be reversed.

(c) When KI solution is added to $AgNO_3$ a positively charged sol results due to absorption of $Ag^+$ ions from dispersion medium–$AgI/Ag^+$(positively charged)

**15. An element X with an atomic mass of 81 u has density 10·2 g cm$^{-3}$ . If the volume of unit cell is 2·7 × 10$^{-23}$ cm$^3$, identify the type of cubic unit cell. (Given : $N_A$ = 6·022 × 10$^{23}$ mol$^{-1}$) [3]**

**Ans.** Density of element is given by,

$$d = \frac{zM}{N_A \times a^3}$$

Or

$$Z = \frac{d \times N_A \times a^3}{M}$$

$$= \frac{10.2 \times 6.022 \times 10^{23} \times 2.7 \times 10^{-23}}{81}$$

$$= 2$$

Since Z =2 thus the lattice is Body Centred Cubic (BCC)

**16. A solution containing 1·9 g per 100 mL of KCl (M = 74·5 g mol$^{-1}$) is isotonic with a solution containing 3 g per 100 mL of urea (M = 60 g mol$^{-1}$). Calculate the degree of dissociation of KCl solution. Assume that both the solutions have same temperature. [3]**

**Ans.** Two solutions having same osmotic pressure at a given temperature are called isotonic solution. Now in the given problem, the KCl and urea solutions are given to be isotonic.

Osmotic pressure π is given by the equation –

$\pi = (n_2/V)RT$, where, $n_2$= moles of solute, V = volume of a solution in litre.

Also, $n_2 = w_2/M_2$, where $w_2$ = grams of solute and $M_2$ = molar mass of solute.

The other given information are–

Molar mass of KCl = 74.5 g mol$^{-1}$

Weight of KCl, $w_2$ = 1.9 g, V = 100 mL

So, for KCl–

$$\pi = (w_2/M_2 \times V)RT$$

$$\pi\ RT_{KCl} = 1.9/(74.5 \times 100)$$

$$= 2.55 \times 10^{-4}$$

Now as the solutions are isotonic at same temperature :

$$\pi RT_{KCl} = \pi RT_{Urea}$$

Hence, substituting the values for urea :

$$2.55 \times 10^{-4} = 3/M_2 \times 100$$

$$M_2 = 117.6$$

So, the experimentally determined molecular weight of urea is found to be as 117.6, so the degree of dissociation can be given as :

Osmotic pressure (π) =

$$\frac{\text{Experimentally determined molecular weight}}{\text{Actual molecular weight}}$$

$$= 117.6 / 60$$

$$= 1.96 \approx 2$$

$$\alpha = \frac{i-1}{n-1} = \frac{1.96-1}{2-1} = 96\%$$

So, Urea diamerised in the given experimental solution.

**17. Write the name and principle of the method used for refining of (a) Zinc, (b) Germanium, (c) Titanium. [3]**

**Ans. (a)** Distillation is used for refining zinc. As zinc is a low boiling metal, the impure metal is evaporated and the pure metal is obtained as a distillate.

**(b)** Zone refining is used for refining Germanium. This method is based on principle that the impurities are more soluble in the melt than in the solid state of the metal.

**(c)** Titanium is refined by van Arkel method. This method is used for removal of oxygen and nitrogen present as impurity. In this method the, metal forms a volatile compound which decomposes at higher temperature to give pure metal.

**18. Give reasons for the following :**

**(a) Transition metals form complex compounds.**

**(b) E° values of ($Zn^{2+}$/Zn) and ($Mn^{2+}$/Mn) are more negative than expected.**

**(c) Actinoids show wide range of oxidation states. [3]**

**Ans. (a)** Transition elements have partly filled d-orbitals due to which they have variable oxidation states which enables them to bind with a variety of ligands because of their small size and high ionic charge. Thus, they form complex compounds.

**(b)** Oxidation of Zn to $Zn^{2+}$ leads to a completely filled $d^{10}$ configuration in $Zn^{2+}$, making it more stable. Also, $Mn/Mn^{2+}$ conversion leads to a half filled stable $d^5$ configuration of $Mn^{2+}$ ion. Hence, E° value for $Zn/Zn^2$and $Mn/Mn^{2+}$conversion have negative values.

**(c)** Actinoids show wide range of oxidation states due to their partially filled *f*-orbitals and they have comparable energies of 5*f*, 6*d* and 7*s* orbitals.

**19. Write the structures of monomers used for getting the following polymers :**
**(a) Nylon-6** **(b) Terylene**
**(c) Buna-N** [3]

**OR**

**(a) Is $\left[ CH_2–CH(C_6H_5) \right]_n$ homopolymer or copolymer ? Give reason.**
**(b) Write the monomers of the following polymer :**

$$\left[ O-\underset{CH_3}{\underset{|}{CH}}-CH_2-\underset{O}{\underset{||}{C}}-O-\underset{C_2H_5}{\underset{|}{CH}}-CH_2-\underset{O}{\underset{||}{C}} \right]_n$$

**(c) Write the role of benzoyl peroxide in polymerisation of ethene.**

**Ans. Structures of monomers—**

(a) Caprolactum is monomer of Nylon-6

Caprolactum (ring: N–H, C=O, $CH_2$, $H_2C$–$CH_2$, $H_2C$, $H_2C$) $\xrightarrow[H_2O]{533\text{-}543\ K}$ $\left[ \overset{O}{\overset{||}{C}}-(CH_2)_5-\overset{H}{\overset{|}{N}} \right]_n$ Nylon 6

(b) Ethylene glycol and terephthalic acid polymerise to give Terylene –

$n\,HOH_2C-CH_2OH + n\,HOOC-C_6H_4-COOH$
Ethylene glycol    Terephthalic acid

$$\downarrow$$

$$\left[ OCH_2-CH_2-\overset{O}{\overset{||}{C}}-C_6H_4-\overset{O}{\overset{||}{C}} \right]_n$$

Terylene

(c) 1,3-Butadiene and Acrylonitrile polymerise to give Buna-N—

$nCH_2=CH–CH=CH_2$
1, 3-Butadiene
+ $\xrightarrow{\text{Copolymerisation}}$ $\left[ CH_2–CH=CH–CH_2–CH_2–\overset{CN}{\overset{|}{CH}} \right]_n$
$n\overset{CH}{\overset{|}{CH_2=CH}}$
Acrylonitrile

Buna–N

**OR**

(a) $\left[ CH_2\text{-}CH(C_6H_5) \right]$ is a homopolymer as it is formed by addition polymerization of monomer $CH_2 = CHC_6H_5$(Styrene).

(b) The monomers are 3-Hydroxybutanoic acid and 3-Hydroxypentanoic acid—

$CH_3–\overset{OH}{\overset{|}{CH}}–CH_2–COOH$
3-Hydroxybutanoic acid
+ $\xrightarrow{\text{Copolymerisation}}$
$CH_3–CH_2–\overset{OH}{\overset{|}{CH}}–CH_2–COOH$

$$\left( O–\underset{CH_3}{\underset{|}{CH}}–CH_2–\underset{O}{\underset{||}{C}}–O–\underset{CH_2CH_3}{\underset{|}{CH}}–CH_2–\underset{O}{\underset{||}{C}} \right)_n$$

(c) Benzoyl peroxide acts as free radical initiator in polymerization of ethene.

Chain initiation step

$$C_6H_5–\overset{O}{\overset{||}{C}}–O–O–\overset{O}{\overset{||}{C}}–C_6H_5 \longrightarrow 2\left[ C_6H_5–\overset{O}{\overset{||}{C}}–\dot{O} \right]$$

Benzoyl peroxide    Benzoyl free radical

$$\longrightarrow \dot{C}_6H_5 + CO_2$$

Phenyl free radical

$$\dot{C}_6H_5 + CH_2=CH_2 \longrightarrow C_6H_5-CH_2-\dot{C}H_2$$

Ethene (Monomer)    Big radical (I)

Chain Propagating step

$$C_6H_5-CH_2-\dot{C}H_2 + CH_2=CH_2 \longrightarrow$$

Ethene

$$C_6H_5-CH_2-CH_2-CH_2-\dot{C}H_2$$

$$\downarrow$$

$$n\,C_6H_5-(CH_2-CH_2)_n\,CH_2-\dot{C}H_2$$

Bigger free radical (II)

**20. (a) Pick out the odd one from the following on the basis of their medicinal properties :**
**Equanil, Seconal, Bithional, Luminal**
**(b) What types of detergents are used in diswashing liquids ?**
**(c) Why is the use of aspartame limited to cold foods ?** [3]

**OR**

**Define the following terms with suitable example of each :**
**(a) Antibiotics** **(b) Antiseptics**
**(c) Anionic detergents**

**Ans. (a)** 'Bithionol' is the odd one here, as it is an antiseptic whereas others are tranquilizers.

**(b)** Liquid dishwashing detergents are non-ionic type.

(c) Aspartame is an artificial sweetener which is unstable at cooking temperature hence its use is limited to cold foods.

OR

(a) **Antibiotics** –These are the compounds (produced by microorganisms or synthetically) which either inhibit the growth of microorganism kill them. Example: Penicillin.

(b) **Antiseptics** – These are the chemicals used to kill or prevent the growth of microorganisms when applied to the living tissues. Example – Soframicin, Dettol.

(c) **Anionic detergents** – These are sodium salts of sulphonated long chain alcohols or hydrocarbons. In these the anionic part of the molecule is involved in the cleansing action. Example – Sodium lauryl sulphate.

**21. Among all the isomers of molecular formula $C_4H_8Br$, identify :**

**(a) the one isomer which is optically active.**

**(b) the one isomer which is highly reactive towards $S_N2$.**

**(c) the two isomers which give same product on dehydrohalgenation with alcoholic KOH.** [3]

**Ans.** The possible isomers are :

1. $H_3C—CH_2—CH_2—CH_2—Br$

1-Bromobutane

2. $H_3C—CH_2—CH(Br)—CH_3$

2-Bromobutane

3. $H_3C—CH(CH_3)—CH_2Br$

1-Bromobutane-2-methylpropane

4. $H_3C—C(CH_3)_2—Br$

2-Bromo-2-methylpropane

(a) 2- Bromobutane is optically active as C-2 is a chiral carbon here having all the four different groups attached to it.

(b) 1-Bromobutane being primary alkyl halide is highly reactive towards $S_N2$ reaction.

(c) 2- Bromo-2-methylpropane and 1-Bromo-2-methylpropane would give same product after dehydrohalogenation.

$$H_3C-C(CH_3)_2-Br \xrightarrow[\text{Dehydrohalogenation}]{KOH} H_3C-C(CH_3)=CH_2$$

2-Bromo-2-methyl propane → 2-Methylpropene

$$(CH_3)_2CH-CH_2Br \xrightarrow[\text{Dehydrohalogenation}]{KOH} H_3C-C(CH_3)=CH_2$$

2-Methylpropene

**22. Complete the following reactions :** **[3]**

(a) Cyclobutyl–CN $\xrightarrow{H_2/Ni}$

(b) 2-Bromo-4-methyl... (ring with $CH_3$, Br and $N_2^+Cl^-$) $\xrightarrow{H_3PO_2 + H_2O}$

(c) $C_6H_5CH_2-NH_2 + CHCl_3 \xrightarrow{\text{Ethanolic KOH}}$

OR

**How do you convert the following :**

**(a) N-phenylethanamide to *p*-bromaniline**

**(b) Benzene diazonium chloride to nitrobenzene**

**(c) Benzoic acid to aniline**

**Ans.**

(a)

$$\text{Cyclobutyl–CN} \xrightarrow{H_2/Ni} \text{Cyclobutyl–}CH_2NH_2$$

Cyclobutylnitrile → Cyclobutylmethylamine

(b)

$$\text{(ring: } CH_3\text{, Br, } N_2^+Cl^-) \xrightarrow{H_3PO_2,\ H_2O} \text{(ring: } CH_3\text{, Br)}$$

2-Bromo-4-methylbenzene diazonium chloride → 2-Bromotoluene

(c)

$$C_6H_5CH_2NH_2 + CHCl_3 \xrightarrow{\text{Eth. KOH}} C_6H_5CH_2NC$$

Ethyl amine → Benzyl isocyanide

OR

(a)

$$CH_3CONHC_6H_5 \xrightarrow[CH_3COOH]{Br_2} p\text{-}BrC_6H_4NHCOCH_3 \xrightarrow{H_2} p\text{-}BrC_6H_4NH_2$$

N-Phenylethanamide → p-Bromoacetanilide → p-Bromoaniline

(b)

$$C_6H_5N_2^+Cl^- \xrightarrow{HBF_4} C_6H_5N_2^+BF_4^- \xrightarrow[Cu,\Delta]{NaNO_2} C_6H_5NO_2 + N_2 + NaBF_4$$

Benzene diazonium chloride → Nitrobenzene

(c)

$$C_6H_5COOH \xrightarrow[(ii)\ \Delta]{(i)\ NH_3} C_6H_5CONH_2 \xrightarrow[\text{Hoffmann bromamide reaction}]{Br_2,\ NaOH} C_6H_5NH_2$$

Benzoic acid → Benzamide → Aniline

**23. (a) Give reasons :**

**(i) Benzoic acid is a stronger acid than acetic acid.**

**(ii) Methanal is more reactive towards nucleophilic addition reaction than ethanal.**

**(b) Give a simple chemical test to distinguish between propanal and propanone. [3]**

**Ans. (a) (i)** Benzoic acid is a stronger acid than acetic acid because the benzoate anion (conjugate base of benzoic acid) formed after loss of $H^+$ is stabilized by resonance, whereas acetate ion ($CH_3COO^-$) has no such extra stability. Hence, Benzoic acid has more tendency of losing proton compared to acetic acid hence more acidic.

(Resonance structures of benzoate ion)

**(ii)** Methanal is more reactive towards nucleophilic addition reaction than ethanal because in ethanal there is a methyl group attached to the carbonyl carbon (centre for nucleophile attack) and +I effect of the methyl group decreases the nucleophilicity of carbonyl carbon by increasing the electron density at carbonyl carbon.

$HCHO$ (Methanal) $CH_3CHO$ (Ethanal)

**(b)** Propanal and propanone can be distinguished using Tollen's reagent by silver mirror test. Propanal being an aldehyde reacts with Tollen's reagent to give silver deposition whereas propanone being a ketone does not give the reaction.

$$CH_3CH_2CHO + 2[Ag(NH_3)_2]^+ + 3OH^- \longrightarrow CH_3COO^- + 2Ag + 2H_2O + 4NH_3$$

Ethanal, Tollen's reagent, Silver mirror

$$CH_3COCH_3 + 2[Ag(NH_3)_2]^+ + 3OH^- \longrightarrow \text{No reaction}$$

Propanone

**24. (a) What are product of hydrolysis of maltose ?**

**(b) What type of bonding provides stability to α-helix structure of protein ?**

**(c) Name the vitamin whose deficiency causes pernicious anaemia. [3]**

**OR**

**Define the following terms :**

**(a) Invert sugar**

**(b) Native protein**

**(c) Nucleotide**

**Ans. (a)** On hydrolysis maltose gives two molecules of α-D-glucose.

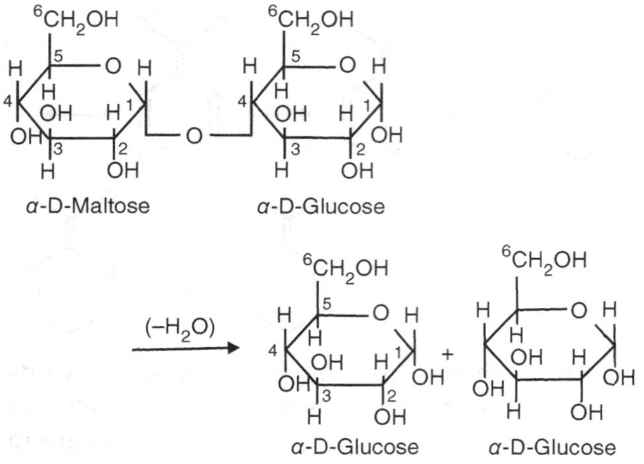

α-D-Maltose α-D-Glucose

$\xrightarrow{(-H_2O)}$

α-D-Glucose + α-D-Glucose

**(b)** α-Helix structure of proteins is stabilized by hydrogen bonds between –NH group of each amino acid and –COOH group of amino acid at adjacent turn.

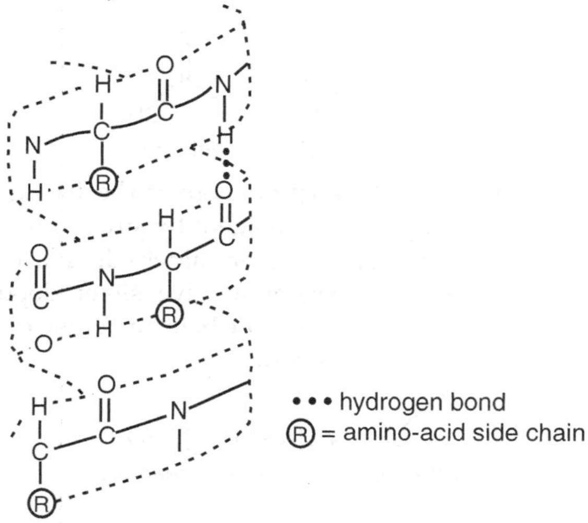

••• hydrogen bond
Ⓡ = amino-acid side chain

**(c)** Deficiency of Vitamin $B_{12}$ causes pernicious anaemia.

**OR**

**(a) Invert sugar** – It is a mixture of glucose and fructose obtained after hydrolysis of sucrose. Sucrose is dextrorotatory, but after hydrolysis gives a mixture of dextrorotatory glucose and levorotatory fructose which outweighs in magnitude and thus the whole mixture becomes levorotatory hence the mixture obtained is called invert sugar.

**(b) Native protein** – Protein found in a biological system with a unique three-dimensional structure and biological activity is called a native protein.

**(c) Nucleotide** –They are building blocks of DNA or RNA. These consist of a pentose sugar moiety attached to a nitrogenous base at 1′ position and a phosphoric acid molecule at 5′ position.

Example :

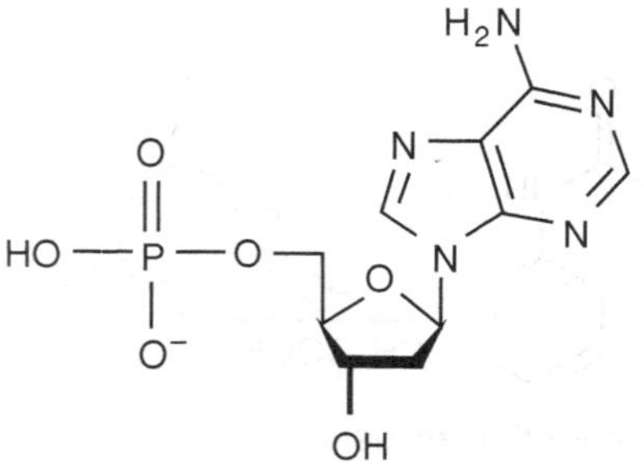

## SECTION-D

**25. (a) The conductivity of 0·001 mol $L^{-1}$ acetic acid is 4·95 × $10^{-5}$ S $cm^{-1}$. Calculate the dissociation constant if $\Lambda_m^0$ for acetic acid is 390·5 S $cm^2$ $mol^{-1}$.**

**(b) Write Nernst equation for the reaction at 25°C :**

**$2\ Al\ (s) + 3\ Cu^{2+}\ (aq) \rightarrow 2\ Al^{3+}\ (aq) + 3\ Cu\ (s)$**

**(c) What are secondary batteries ? Give an example. [5]**

**OR**

**(a) Represent the cell in which the following reaction takes place :**

**$2Al\ (s) + 3\ Ni^{2+}\ (0{\cdot}1M) \rightarrow 2Al^{3+}\ (0{\cdot}01\ M) + 3\ Ni\ (s)$**

**Calculate its emf if $E^0_{cell}$ = 1·41 V.**

**(b) How does molar conductivity vary with increase in concentration for strong electrolyte and weak electrolyte ? How can you obtain limiting molar conductivity ($\Lambda_m^0$) for weak electrolyte ?**

**Ans. (a)** Conductivity $\Lambda_m$ of solution is given by the following equation :

$\Lambda_m = \frac{k}{c}$, where $k$ is dissociation constant and $c$ is the concentration of solution.

Here, given.

Conductivity, $k = 4.95 \times 10^{-5}$ S $cm^{-1}$

Limiting molar conductivity,

$$\Lambda_m^0 = 390.5\ S\ cm^2\ mol^{-1}$$

Concentration,

$$c = 0.001\ mol\ L^{-1}$$
$$= 1 \times 10^{-3}\ mol\ L^{-1}$$

Substituting the given values in above equation

Molar conductivity,

$$\Lambda_m = \left(\frac{49.5 \times 10^{-5}\ S\ cm^{-1}}{1 \times 10^{-3}\ mol\ L^{-1}}\right) 10^3 cm^3 L^{-1}$$
$$= 49.5\ S\ cm^2\ mol^{-1}$$

Now, degree of dissociation α, is given by

$$\alpha = \frac{\Lambda_m}{\Lambda^\circ_m}$$

Putting the values,

$$\alpha = \frac{49.5}{390.5} \text{ S cm}^2 \text{ mol}^{-1}$$

$$= 0.1267$$

Dissociation constant, $k$, for acetic acid, can be given as

$$k = \frac{c\alpha^2}{(1-\alpha)}$$

$$= \frac{[1\times10^{-3}\text{mol L}^{-1}\times(0.1267)^2]}{(1-0.1267)}$$

$$= \frac{0.016\times10^{-3}\text{mol L}^{-1}}{0.8733}$$

$$= 1.8 \times 10^{-5} \text{ mol L}^{-1}$$

**(b)** Nernst equation for the given reaction can be written as

$$E_{cell} = \frac{E^\circ_{cell} - RT}{2F \ln\left(\frac{[Al_3{}^{+}]^2}{[Cu^{2+}]^3}\right)}$$

**Or**

$$\text{Ecell} = \text{E}^\circ\text{cell} - \frac{0.059}{6} \log\frac{[Ai^{3+}]^2}{[Cu^{2+}]^3}$$

**(c)** A secondary battery can be recharged after use, by passing current through it in opposite direction so that it can be used again. Example: The most important secondary cell is lead storage cell. It consists of lead anode and a grid of lead packed with lead dioxide as cathode. A 38% solution of sulphuric acid is used as an electrolyte.

OR

**(a)** The cell can be represented as

Al | $Al^{3+}$(0.01 M) | | $Ni^{2+}$ (0.1) | Ni

Cell potential is given by the following equation in this case

$$E_{cell} = E^{o}_{(cell)} - \frac{0.059}{6}\log\frac{[Al^3]^2}{[NI^{2+}]^3}$$

$$E_{cell} = 1.41V - \frac{0.059}{6}\log\frac{[0.01]^2}{[0.1]^3}$$

$$E_{cell} = 1.4198V$$

or $E_{cell} = 1.42V$

**(b)** For strong electrolytes the molar conductivity is increased only slightly on dilution. A strong electrolyte is completely dissociated in solution and thus, furnishes all ions for conductance. However, at higher concentrations, the dissociated ions are close to each other and thus, the interionic attractions are greater. These forces retard the motion of the ions and thus, conductivity is low. With decrease in concentration (dilution), the ions move away from each other thereby feeling less attractive forces from the counter ions. This results in an increase in molar conductivity with dilution. The molar conductivity approaches a maximum limiting value at infinite dilution designated as $\Lambda_m{}^{0.}$

In case of weak electrolytes as the solution of a weak electrolyte is diluted, its ionisation is increased. This results in more number of ions in solution and thus, there is an increase in molar conductivity, also there is a large increase in the value of molar conductivity with dilution, especially near infinite dilution. However, the conductance of a weak electrolyte never approaches a limiting value. Or in other words it is not possible to find conductance at infinite dilution (zero concentration).

So, limiting molar conductivity for weak electrolytes is obtained by using Kohlrausch law, from the limiting molar conductivities of individual ions ($\lambda^\circ$).

Kohlrausch law of independent migration of ions states that limiting molar conductivity of an electrolyte can be represented as the sum of the individual contributions of the anion and the cation of the electrolyte.

$$\Lambda^\circ_m = V_+ \lambda^\circ_+ + V_- \lambda^\circ_-$$

**26. (a) Give equation of the following reactions :**

**(i) Phenol is treated with conc. $HNO_3$.**

**(ii) Propene is treated with $B_2H_6$ followed by $H_2O_2/OH^-$.**

**(iii) Sodium t-butoxide is treated with $CH_3Cl$.**

**(b) How will you distinguish between butan–1–ol and butan–2–ol ?**

**(c) Arrange the following in increasing order of acidity :**

**Phenol, ethanol, water** [5]

OR

**(a) How can your obtain Phenol from (i) Cumene, (ii) Benzene sulphonic acid, (iii) Benzene diazonium chloride ?**

**(b) Write the structure of the major product obtained from dinitration of 3-methylphenol.**

**(c) Write the reaction involved in Kolbe's reaction.**

**Ans. (a) (i)** Phenol is treated with conc. $HNO_3$ to obtain 2,4,6-trinitrophenol picric acid.

OH
conc. $HNO_3$ → $O_2N$, OH, $NO_2$, $NO_2$
Phenol
2,4,6-Trinitrophenol

**(ii)** Propene undergoes hydroboration-oxidation when treated with $B_2H_6$ followed by hydrogen peroxide in basic medium to give propan-1-ol.

$H_3C$–CH=$CH_2$ —(i) $B_2H_6$, (ii) $H_2O/OH$→ $H_3C$–$CH_2$–$CH_2$–OH
Propene Propan-1-ol

**(iii)** Methyl tert-butyl ether is produced when sodium tert-butoxide is treated with methyl chloride.

$H_3C$—Cl —t $BuO^-Na^+$→ $(H_3C)_3C$–O–$CH_3$
Methyl chloride Methyl tert-butyl ether

**(b)** Butan-1-ol and Butan-2-ol can be distinguished using Lucas reagent ($ZnCl_2$+HCl), where butan-2-ol would react with Lucas reagent in around 5 minutes to give a white precipitate of 2-chlorobutane, whereas butan-1-ol would not give any reaction at room temperature.

**(c)** Increasing order of acidity can be given as Ethanol < water < phenol

**OR**

**(a) (i)** Phenol from cumene

$H_3C$–CH($CH_3$)–$C_6H_5$ —$O_2$→ $H_3C$–C(O—OH)($CH_3$)–$C_6H_5$
Cumene Cumene hydroperoxide

—$H^+$, $H_2O$→ $C_6H_5OH$ + $CH_3COCH_3$
Phenol

**(ii)** Phenol from benzene sulphonic acid

$C_6H_5SO_3H$ —(i) NaOH, (ii) $H^+$→ $C_6H_5OH$
Benzene sulphonic acid Phenol

**(iii)** Phenol from benzene diazonium chloride

$C_6H_5N_2^+Cl^-$ —(i) $H_2O$, warm→ $C_6H_5OH$
Benzene diazonium chloride Phenol

**(b)** The combined influence of –OH and –$CH_3$ groups determine the position of the entering groups, also the sterically hindered positions are not substituted.

OH, $CH_3$ —(i) $HNO_3$→ OH, $O_2N$, $CH_3$, $NO_2$
3-Methyl phenol
3-Methyl-4, 6 dinitrophenol

**(c)** In Kolbe's reaction phenol is reacted with $CO_2$ in presence of sodium hydroxide, followed by acidification, to give a carboxylic acid group on 2-position of phenol-

$C_6H_5OH$ —NaOH→ $C_6H_5ONa$ —(i) $CO_2$, (ii) $H^+$→ ONa, COOH
Phenol Sodium phenoxide 2-Hydroxybenzoic acid (Salicylic acid)

**27. (a) Account for the following :**

**(i) Tendency to show – 3 oxidation state decreases from N to Bi in group 15.**

**(ii) Acidic character increases from $H_2O$ to $H_2Te$.**

**(iii) $F_2$ is more reactive than $ClF_3$, whereas $ClF_3$ is more reactive than $Cl_2$.**

**(b) Draw the structure of (i) $XeF_2$, (ii) $H_4P_2O_7$.** **[5]**

**OR**

**(a) Give one example to show the anomalous reaction of fluorine.**

**(b) What is the structural difference between white phosphorous and red phosphorous ?**

**(c) What happens when $XeF_6$ reacts with NaF ?**

**(d) Why is $H_2S$ a better reducing agent than $H_2O$ ?**

**(e) Arrange the following acids in the increasing order of their acidic character :**

**HF, HCl, HBr and HI**

**Ans. (a) (i)** Tendency to show-3 oxidation state decreases from N to Bi in group 15 because of increase in size and metallic character from N to Bi in group 15.

**(ii)** Acidic character increases from $H_2O$ to $H_2Te$ due to decrease in E—H bond dissociation enthalpy down the group. Thus it becomes easy to lose proton going down the group.

**(iii)** $F_2$ is more reactive than $ClF_3$ because of small size of fluorine atom F—F bond, bond dissociation enthalpy is low (thus is reactive).

Whereas $ClF_3$ is more reactive than $Cl_2$ because $ClF_3$ is an interhalogen compound with weak Cl—F bond (compared to Cl—Cl bond) due to difference in atomic sizes (hence ineffective overlap of orbitals).

**(b) (i)** Structure of $XeF_2$ is linear.

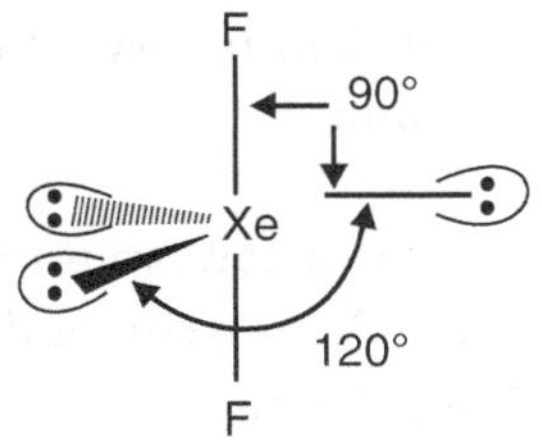


**(ii)** Structure of $H_4P_2O_7$ :

**OR**

**(a)** Fluorine reacts with cold sodium hydroxide solution to give $OF_2$. $2F_2\ (g) + 2NaOH\ (aq) \rightarrow 2NaF\ (aq) + OF_2\ (g) + H_2O\ (l)$

**(b)** White phosphorus is discrete tetrahedral whereas red phosphorus is polymeric.

**(c)** $XeF_6$ reacts with NaF as follows :

$XeF_6 + NaF \rightarrow Na^+[XeF_7]^-$

**(d)** Ability to reduce is judged by ease with which an atom can donate its electrons to the species which is getting reduced. Now, the size of oxygen atom in $H_2O$ is smaller than that of Sulphur atom in $H_2S$, due to which the lone pair of electrons on oxygen are more attracted by the oxygen nucleus, making it difficult to donate the electrons (by oxygen compared to Sulphur while in $H_2S$ the influence of nucleus is less on lone pair of electrons of sulphur) i.e.; the bond dissociation enthalpy of H-S bond is lower than H-O bond hence, it can give away its electrons, easily compared to oxygen, and thus acts as a better reducing agent.

**(e)** The increasing order of acidic character can be written as:

$HF < HCl < HBr < HI$

## Outside Delhi [Set-II]

**Note: Except for the following questions all the remaining questions have been asked in the previous set.**

### SECTION-A

**2. Arrange the following in increasing order of $pK_b$ values :**

**$C_6H_5CH_2NH_2$, $C_6H_5NHCH_3$, $C_6H_6NH_2$ [1]**

**Ans.** These can be arranged in increasing order of $pK_b$ values as follows :

$C_6H_5CH_2NH_2 < C_6H_5NHCH_3\ C_6H_5NH_2$

**3. What type of colloid is formed when a liquid is dispersed in a solid ? Give an example. [1]**

**Ans.** When a liquid is dispersed in a solid, a 'gel' is formed. Example : Butter.

**4. Out of chlorobenzene and *p*-nitrochlorobenzene, which one is more reactive towards nucleophilic substitution reaction and why ? [1]**

**Ans.** *p*-Nitro chlorobenzene would be more reactive towards nucleophilic substitution reaction compared to chlorobenzene. In chlorobenzene the carbon bearing the halogen is a part of aromatic ring and is electron rich due to the electron density in the ring so, it does not attract the nucleophile. The $-NO_2$ substitution lessens the electron density on the benzene ring due to its electron withdrawing nature, making the electron density on ring less compared to chlorobenzene, hence *p*-nitro chlorobenzene attracts nucleophiles better.

## SECTION-B

**7. Give reasons :**

**(a) A decrease in temperature is observed on mixing ethanol and acetone.**

**(b) Potassium chloride solution freezes at a lower temperature than water. [2]**

**Ans. (a)** Upon mixing molecules of ethanol and acetone have strong intermolecular attractions due to which heat is evolved from reaction system and hence cooling of mixture is observed.

**(b)** Potassium chloride solution is a solution of non -volatile solute KCl and water solution. We know that, at the freezing point of a substance, the solid phase (here ice) is in dynamic equilibrium with the liquid phase. A solution freezes when its vapour pressure equals the vapour pressure of pure solid solvent. Now, according to Raoult's law when a non -volatile solid is added to the solvent (in this case it is KCl), its vapour pressure decreases and now it would become equal to that of solid solvent at lower temperature. Thus, the freezing point of solvent decreases.

**10. Define the following terms with a suitable example of each : [2]**

**(a) Chelate complex**

**(b) Ambidentate ligand**

**OR**

**Using IUPAC norms, write the formulae for the following complexes :**

**(a) Tetraamminediaquacobalt (III) chloride**

**(b) Dibromidobis (ethane-1, 2-diamine) platinum (IV) nitrate**

**Ans. (a) Chelate complex** –Chelate complexes are coordination or complex compound consisting of a central metal atom attached to a large molecule, called a ligand, in a cyclic or ring structure. The ligands are bi- or polydentate *i.e.*, they can attach to metal atom through two or more than two binding sites. An example of a chelate ring occurs in the ethylenediamine-nickel complex.

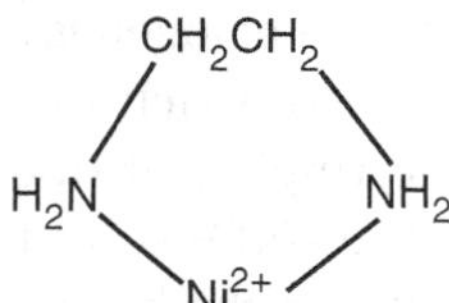


A chelate complex

**(b) Ambidentate ligand :** Ligands which can ligate (attach to metal atom) through two different atoms is called ambidentate ligand. One example of such ligand is $NO_2^-$, this can bind through both the atoms, nitrogen and oxygen.

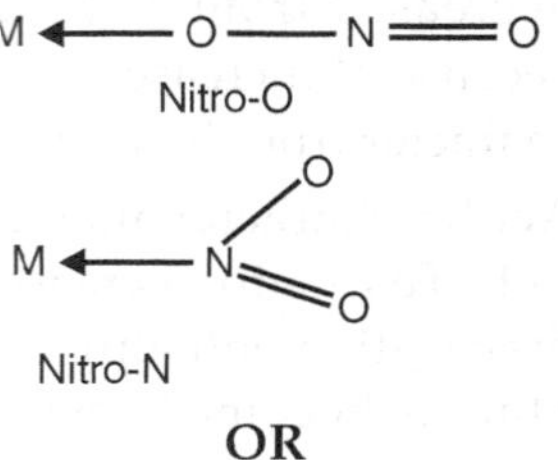


**OR**

IUPAC names

**(a)** Tetramminediaquacobalt(III)chloride

$[Co(NH_3)_4(H_2O)_2]Cl_3$

**(b)** Dibromidobis(ethane-1,2-diamine platinum(IV) nitrate –$[PtBr_2(en)_2](NO_3)_2$

**11. (a) Using valence bond theory, write the hybridisation and magnetic character of the complex $[Fe(CN)_6]^{4-}$. (Atomic no. of Fe = 26)**

**(b) Write the electronic configuration of $d^6$ on the basis of field theory when**

**(i) $\Delta_0 < P$ and**

**(ii) $\Delta_0 > P$ [2]**

**Ans. (a)** $[Fe(CN)_6]^{4-}$ is a low spin or inner orbital complex as $CN^-$ is a strong field ligand.

**Hybridisation Scheme**

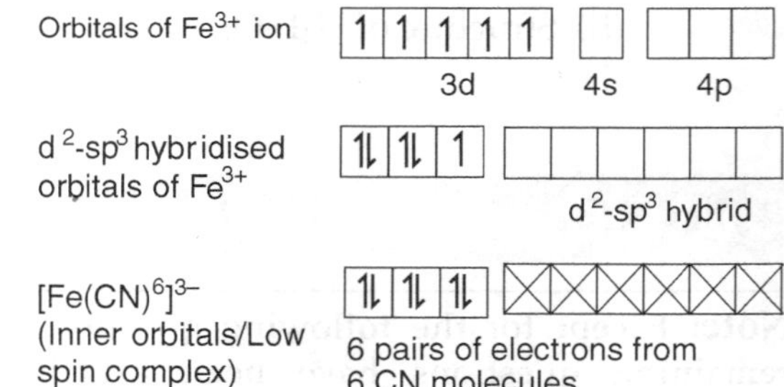


$[Fe(CN)_6]^{4-}$ is diamagnetic as there is no unpaired electron left.

**(b) (i)** Electronic configuration of $d^6$ when $\Delta_0 < P$, that is in case of weak field ligand

$t_{2g}^4 e_g^2$

$dxy^2, dxz^1, d_{yz}{}^1, d_{x^2y^2}{}^1, d_{z^2}{}^1$

**(ii)** Electronic configuration of $d^6$ when $\Delta_0 > P$, that is in case of strong field ligand

$t_{2g}^6 e_g^0$

$dxy^2, dxz^2, d_{yz}{}^2, d_{x^2y^2}{}^0, d_{z^2}{}^0$

## SECTION C

**13. (a) Write the dispersed phase and dispersion medium of milk.**

**(b) Why is adsorption exothermic in nature ?**

**(c) Write Freundlich adsoption isortherm for gases at high pressure. [3]**

**Ans. (a)** Dispersed phase of milk is liquid and dispersion medium of milk is liquid.

**(b)** During the process of adsorption molecules of adsorbate and adsorbent come closer to form physical or chemical bonds hence getting stabilized, in this process heat is evolved leading the overall process to be exothermic.

**(c)** Freundlich adsorption isotherm for gases at high pressure.

It is an empirical relationship between the quantity of gas adsorbed by unit mass of solid adsorbent and pressure at a particular temperature :

$$x/m = k\,p^{1/n}(n > 1)$$

Where, x is the mass of the gas adsorbed on mass $m$ of the adsorbent at pressure $p$, $k$ and $n$ are constants which depend on the nature of the adsorbent and the gas at a particular temperature.

**15. Write the name and principle of the method used for refining of (a) Tin, (b) Copper, (c) Nickel. [3]**

**Ans. (a) Tin**: It is refined through liquation. In this method a low melting metal like tin is made to flow on a sloping surface, where the higher melting impurities are left behind and the lower melting metal is collected at the slope end.

**(b) Copper :** It is refined through electrolytic refining. Anode is made of impure copper and pure copper stripes are taken as cathode. They are dipped in acidified solution of copper sulphate as electrolyte. The net result of electrolysis is the transfer of copper in pure form from the anode to the cathode and the impurities gets deposited as anode mud.

**(c) Nickel :** It is refined through Mond's process. In this process, Nickel is heated in a stream of carbon monoxide forming a volatile complex, nickel tetracarbonyl.

$$Ni + 4\,CO \xrightarrow{330-350\text{ K}} Ni(CO)_4$$

The carbonyl is subjected to higher temperature so that it is decomposed giving the pure metal.

$$Ni(CO)_4 \xrightarrow{450-470\text{ K}} Ni + 4CO$$

**16. Give reason for the following :**

**(a) Transition metals show variable oxidation states.**

**(b) E° value of $(Zn^{2+}/Zn)$ is negative while that of $(Cu^{2+}/Cu)$ is positive.**

**(c) Higher oxidation state of Mn with fluorine is +4 whereas with oxygen is +7. [3]**

**Ans. (a)** Transition metals show variable oxidation states because their *d*-orbitals are incompletely filled and different arrangements of electrons are possible according to the chemical environment of metal ion hence, the ions can occupy variable oxidation states.

**(b)** E° value for $Zn/Zn^{2+}$ is negative because conversion of Zn to $Zn^{2+}$ gives it a completely filled $d^{10}$ configuration and extra stability gained by $Zn^{2+}$. Whereas, conversion of Cu to $Cu^{2+}$ does not give any extra stability, hence it has a positive E° value.

**(c)** Mn has the highest oxidation state of +4 with fluorine but with oxygen it is +7. This is due to the ability of oxygen to form multiple bonds with the metal ion, whereas fluorine being of small size and devoid of *d*-orbitals cannot form multiple bonds.

**19. Write the structures of monomers used for getting the following polymers :**

**(a) Nylon-6, 6 (b) Bakelite**

**(c) Buna-S [3]**

**OR**

**(a) Write one example each of :**

**(i) Thermoplastic polymer**

**(ii) Elastomers**

**(b) Arrange the following polymers in the increasing order of their intermolecular forces:**

**Polythene, Nylon-6, 6, Buna-S**

**(c) Which factor provides crystalline nature to a polymer like Nylon ?**

**Ans. (a)** Monomers of Nylon-6, 6 are adipic acid and hexamethylene diamine.

HO–C(=O)–(CH2)4–C(=O)–OH + H2N–(CH2)6–NH2

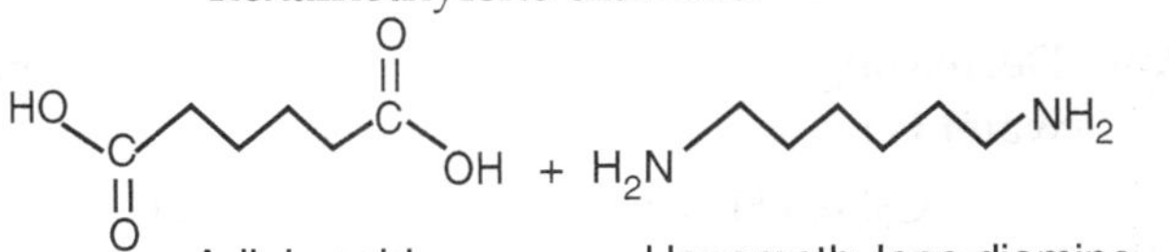

Adipic acid    Hexamethylene diamine

→ –[C(=O)–(CH2)4–C(=O)–NH–(CH2)6–NH]–

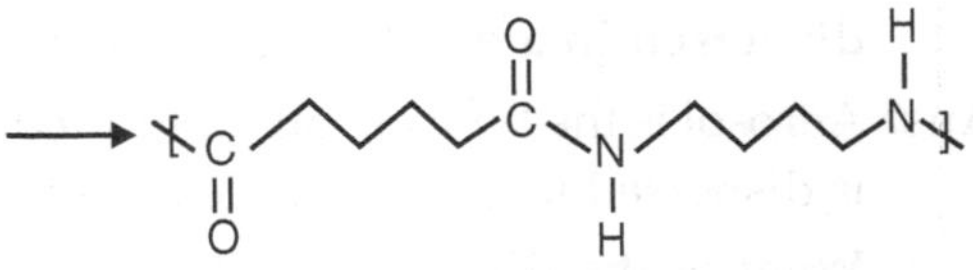

Nylon 6, 6

**(b)** Monomers of bakelite are phenol and formaldehyde :

OH + H–C(=O)–H + OH → Bakelite + $H_2O$

Phenol   Formaldehyde   Bakelite

**(c)** Monomers of Buna-S are 1, 3-Butadiene and Styrene

$$nCH_2{=}CH{-}CH{=}CH_2 + n\,C_6H_5CH{=}CH_2 \xrightarrow[\text{Polymerisation}]{\text{Na, Heat}} {-}\!(CH_2{-}CH{=}CH{-}CH_2{-}CH_2{-}CH(C_6H_5))\!{-}_n$$

1, 3-Butadiene   Styrene

(Butadiene-Styrene Copolymer)
(SBR or BUNA-S)

**OR**

**(a) (i)** Example of thermoplastic polymer – polythene, polystyrene.

**(ii)** Example of elastomer – Neoprene, Buna S.

**(b)** In increasing order of their intermolecular force, they can be arranged as :

Buna-S < Polythene < Nylon-6, 6

**(c)** Strong intermolecular forces between the polymer molecules, such as hydrogen bonding leads to closed packed structure, thus imparting crystalline nature to the polymers.

## Outside Delhi [Set-III]

**Note: Except for the following questions all the remaining questions have been asked in the previous sets.**

### SECTION-A

**1. Out of Chlorobenzene and Cyclohexyl chloride, which one is more reactive towards nucleophilic substitution reaction and why ? [1]**

**Ans.** Cyclohexyl chloride is more reactive towards nucleophilic substitution reaction, because the carbon bearing the chlorine atom is deficient in electron and seeks a nucleophile. In Chlorobenzene the carbon bearing the halogen is a part of aromatic ring and is electron rich due to the electron density in the ring.

**2. Arrange the following in decreasing order of solubility in water :**

**$(C_2H_5)_2NH$, $C_2H_5NH_2$, $C_6H_5NH_2$ [1]**

**Ans.** Decreasing order of solubility in water is :

$$C_2H_5NH_2 > (C_2H_5)_2NH > C_6H_5NH_2$$

**3. What type of colloid is formed when a solid is dispersed in a gas ? Give an example. [1]**

**Ans.** Aerosol is the type of colloid formed when solid is dispersed in gas. Example : smoke and dust.

**5. What is the difference between amylose and amylopectin ? [1]**

**OR**

**Write the products obtained after hydrolysis of lactose.**

**Ans.** Amylose is a linear polymer of α-glucose and is water soluble whereas amylopectin is a branched polymer of α-glucose and is water insoluble.

**OR**

Galactose and Glucose are the products obtained after hydrolysis of lactose.

Lactose —β-Galactosidase→ Galactose + Glucose

(β-linkage)

### SECTION-B

**7. Give reasons :**

**(a) An increase in temperature is observed on mixing chloroform and acetone.**

**(b) Aquatic animals are more comfortable in cold water than in warm water. [2]**

**Ans. (a)** A mixture of chloroform and acetone forms a solution with negative deviation from Raoult's law. This is because chloroform molecule is able to form hydrogen bond with acetone molecule as shown by the following figure :

$$\overset{\delta-}{Cl}-\underset{\underset{\delta-}{Cl}}{\overset{\overset{\delta-}{Cl}}{C}}{}^{\delta-}-\overset{\delta+}{H}\cdots\cdots\overset{\delta-}{O}=C(CH_3)_2$$

Hydrogen bonding between chloroform and acetone

This decreases the escaping tendency of molecules for each component, and consequently the vapour pressure decreases and the temperature of the solution is increased because of stability attained by the molecule by associating and releasing energy due to the stronger interaction between chloroform and acetone than pure chloroform or acetone interactions.

**(b)** Solubility of gases in liquid increases on decreasing temperature, hence cold water has more dissolved oxygen because of which aquatic species find themselves more comfortable in cold water as compared to hot water.

**10. Define the following terms with a suitable example of each : [2]**

**(a) Polydentate ligand**

**(b) Homoleptic complex**

**OR**

**Using IUPAC norms, write the formulae for the following complexes :**

**(a) Potassium tri (oxalato) chromate (III)**

**(b) Hexaaquamanganese (II) sulphate.**

**Ans. (a) Polydentate ligands :** Ligands with several donor atoms are called polydentate ligands. These can bond with metal ion in a complex with the different donor atoms present in them. Example : $N(CH_2CH_2NH_2)_3$.

**(b) Homoleptic complex :** Complexes in which a metal atom is bound to only one kind of donor groups, *e.g.*, $[Co(NH_3)_6]^{2+}$are known as homoleptic complex.

**OR**

**(a)** $K_3[Cr(C_2O_4)_3]$

**(b)** $[Mn(H_2O)_6]SO_4$

**12. Write structures of main compounds A and B in each of the following reactions : [2]**

**(a)** $CH_3CH_2OH \xrightarrow{PCC} A \xrightarrow{CH_3OH/dry\ HCl(g)} B$

**(b)** $C_6H_5COCH_3 \xrightarrow{NaOI} A + B$

**Ans. (a)**

$$\underset{\text{Ethanol}}{CH_3CH_2OH} \xrightarrow{PCC} \underset{\text{Ethanal (A)}}{CH_3CHO} \xrightarrow[\text{HCl (gas) [dry]}]{CH_3CO} \underset{\text{Hemiacetal}}{\left[CH_3\text{-}CH(O\text{-}CH_3)(OH)\right]}$$

$$\xrightarrow[\text{HCl (gas) [dry]}]{CH_3OH} \underset{\text{Acetal (B)}}{H_3C\text{-}CH(O\text{-}CH_3)(OH)}$$

**(b)** $\underset{\text{Acetophenone}}{C_6H_5COCH_3} \xrightarrow{NaOI} \underset{\text{Iodoform (A)}}{CHI_3} + \underset{\text{Sodium benzoate (B)}}{C_6H_5COONa}$

## SECTION-C

**14. (a) Write the dispersed phase and dispersion medium of butter.**

**(b) Why does physiosorption decrease with increase in temperature ?**

**(c) A colloidal sol is prepared by the method given in the figure. What is the charge on AgI colloidal particles formed in the test tube ? How is this sol represented ? [3]**

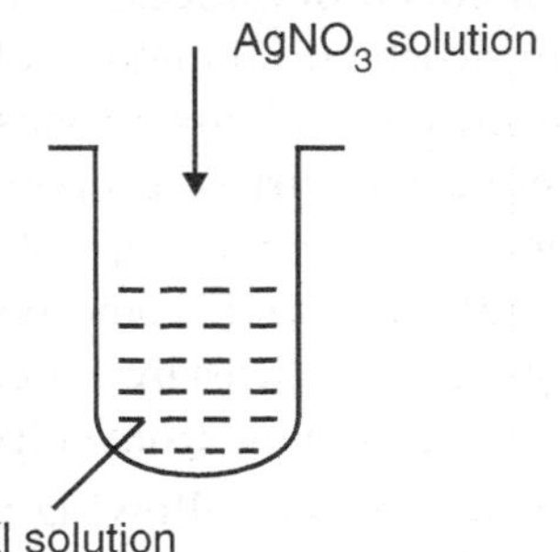


**Ans. (a)** Butter is an example of 'Gel' type of colloid. Here the dispersed phase is liquid and dispersion medium is solid.

**(b)** Physisorption occurs because of physical attractive forces, like van der Waals forces between molecules of adsorbate and adsorbent, hence that can be reversed on application of bigger forces. Hence, when temperature is increased, the movement of adsorbed molecules increases, resulting in disturbed attractive forces, detachment of adsorbed molecules from adsorbent surface hence physisorption decreases.

(c) When $AgNO_3$ solution is added to KI, silver iodide, AgI, is precipitated. The precipitated silver iodide adsorbs iodide ions from dispersion medium and negatively charged colloidal sol results. It can be shown as $AgI/I^-$ (negatively charged).

**17. Write the principle of the following :**

**(a) Hydraulic washing**

**(b) Chromatography**

**(c) Froth-floatation process** [3]

**Ans.** (a) **Hydraulic washing :** This method of concentration of ores is based on the differences in gravities of the ore and the gangue particles. It is a type of gravity separation. An upward stream of running water is used to wash the powdered ore. The lighter gangue particles are washed away and the heavier ores are left behind.

(b) **Chromatography :** Chromatography is a physical method of separation of a mixture in which the components to be separated are distributed between two phases, stationary and mobile phase. Different compounds of a mixture are differently adsorbed on an adsorbent. The stationary phase may be a solid or a liquid supported on a solid or a gel. The mobile phase may be either a liquid or a gas.

(c) **Froth floatation process :** Froth floatation is a physicochemical method of concentrating fine minerals. This process utilises the difference in surface properties of valuable mineral and gangue (impurity) particles. The mineral particles become wet by oil while the gangue particles by water. For example, removal of gangue from sulphide ores.

**18. Give reasons for the following :**

**(a) Transition metals have high enthalpies of atomization.**

**(b) Manganese has lower melting point even though it has a higher number of unpaired electrons for bonding.**

**(c) $Ce^{4+}$ is a strong oxidising agent.** [3]

**Ans.** (a) Transition element have high effective nuclear charge and a large number of valence electrons (($n$–1) $d$ electrons). So, as a result of greater number of electrons participating, very strong metallic bonds are formed. As a result of the strong inter-atomic metallic bonding, the transition metals have high enthalpies of atomization.

(b) Manganese has lower melting point even though it has a higher number of unpaired electrons for bonding. Melting point depends on the intermolecular or interatomic forces. Stronger the forces, higher the melting point. In Mn there is half filled 3$d$ sub-shell (3$d^5$ configuration) which makes it stable and hence, it does not make additional covalent bonds with nearby atoms hence, it has less melting point.

(c) $Ce^{4+}$ is a strong oxidising agent because $Ce^{4+}$ oxidises others and itself gets reduced to the common and preferred 3+ oxidation state of lanthanoid elements and Ce is more stable in +3 oxidation state.

**19. Write the structures of monomers used for getting the following polymers :**

**(a) Novolac** **(b) Neoprene**

**(c) Buna-S** [3]

**OR**

**(a) Write on example each of**

**(i) Cross-linked polymer**

**(ii) Natural polymer**

**(b) Arrange the following in the increasing order of their intermolecular forces : Terylene, Buna-N, Polystyrene**

**(c) Define biodegradable polymers with an example.**

**Ans.** (a)

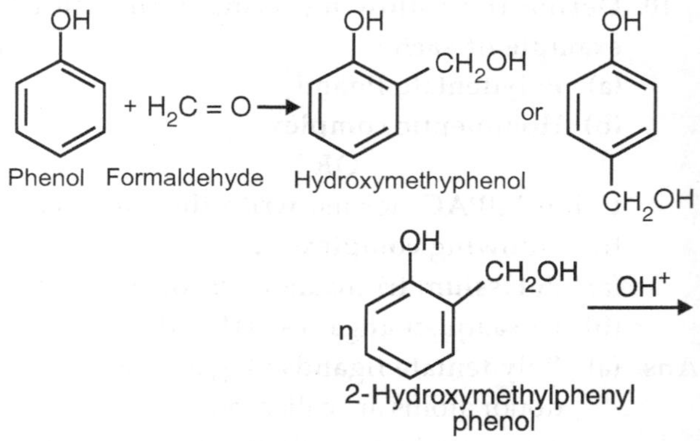

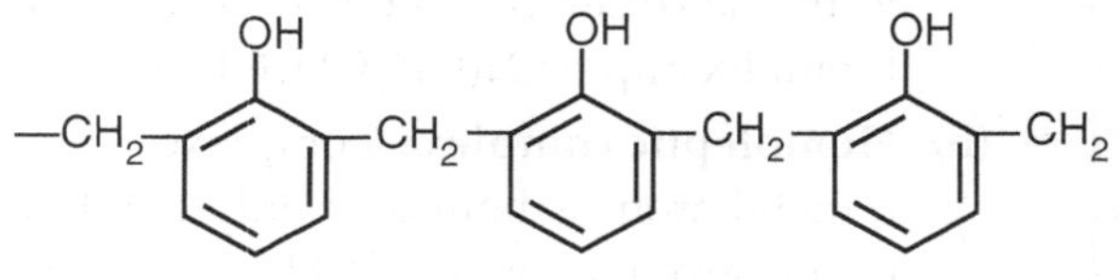

OH (Phenol) + $H_2C = O$ (Formaldehyde) → Hydroxymethyphenol (o-$CH_2OH$) or (p-$CH_2OH$)

n 2-Hydroxymethylphenyl phenol $\xrightarrow{OH^+}$

$-CH_2-C_6H_3(OH)-CH_2-C_6H_3(OH)-CH_2-C_6H_3(OH)-CH_2$

Novalac

(b) Neoprene is polymer of chloroprene or 2-chloro-1,3-butadiene—

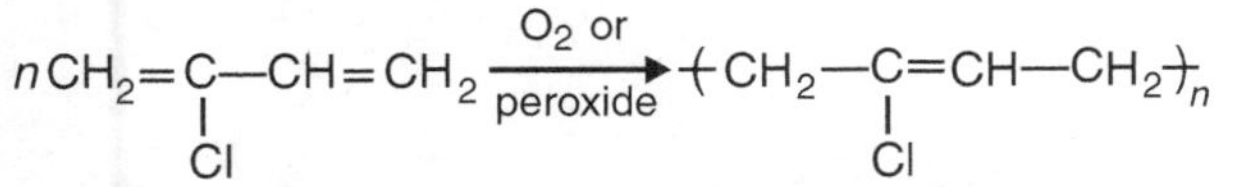

$$nCH_2=\underset{Cl}{\underset{|}{C}}-CH=CH_2 \xrightarrow[\text{peroxide}]{O_2 \text{ or}} \left( CH_2-\underset{Cl}{\underset{|}{C}}=CH-CH_2 \right)_n$$

(Chloroprene) Neoprene

**(c)** Buna-S is a polymer of 1, 3-Butadiene and Styrene monomers.

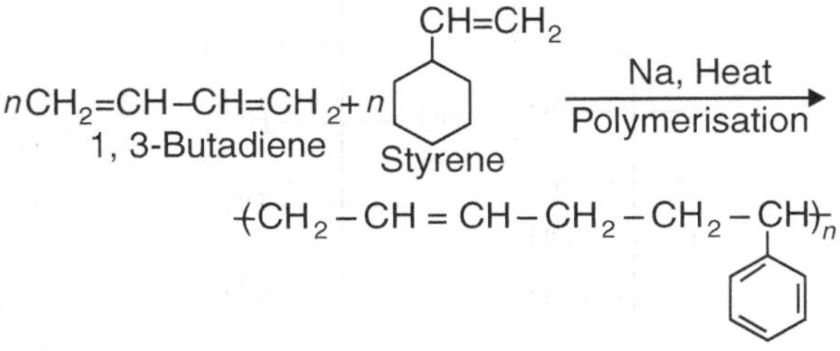

$$nCH_2=CH-CH=CH_2 + n\,C_6H_5CH=CH_2 \xrightarrow[\text{Polymerisation}]{\text{Na, Heat}}$$

1, 3-Butadiene Styrene

$$\left( CH_2-CH=CH-CH_2-CH_2-\underset{C_6H_5}{\underset{|}{CH}} \right)_n$$

(Butadiene-Styrene Copolymer)
(SBR or BUNA-S)

**OR**

**(a) (i)** Bakelite is an example of cross-linked polymer—

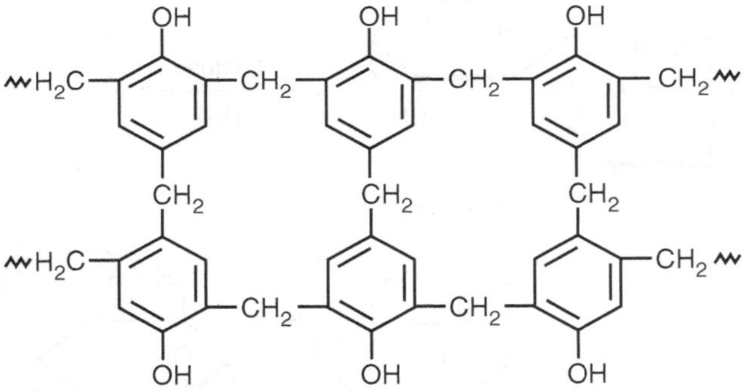

Bakelite

**(ii)** Rubber is an example of natural polymer—

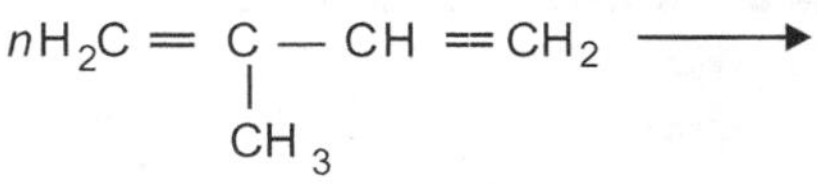

$$nH_2C=\underset{CH_3}{\underset{|}{C}}-CH=CH_2 \longrightarrow$$

Isoprence
[2-methyl butadiene]

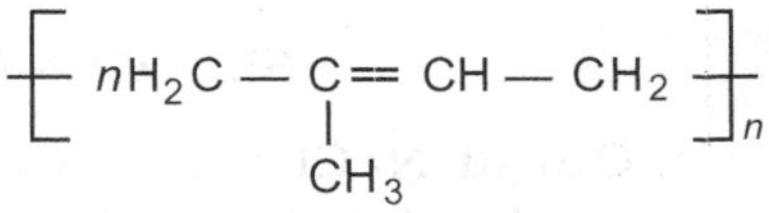

$$\left[ nH_2C-\underset{CH_3}{\underset{|}{C}}=CH-CH_2 \right]_n$$

Natural rubber
[Poly isoprene]

**(b)** Increasing order of their molecular forces :

Buna-N < Polystyrene < Terylene

**(c) Biodegradable polymer :** These are synthetic polymers designed so as to contain functional groups similar to ones present in biopolymers. These are thus easily degraded by environmental degradation process hence, known as Biodegradable polymers.

Example : Poly β-hydroxybutyrate-co-β-hydroxy valerate (PHBV).

**23. (a) Write the product when D-glucose reacts with $Br_2$ (aq).** **[3]**

**Ans. (a)** When D-glucose reacts with Bromine (aq.) gluconic acid is formed.

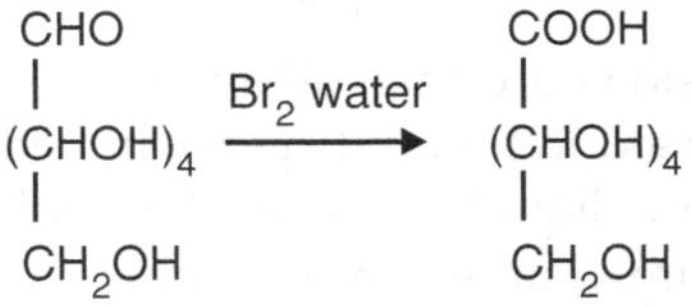

$$\begin{array}{c} CHO \\ | \\ (CHOH)_4 \\ | \\ CH_2OH \end{array} \xrightarrow{Br_2 \text{ water}} \begin{array}{c} COOH \\ | \\ (CHOH)_4 \\ | \\ CH_2OH \end{array}$$

Glucose Gluconic acid

●●

# CHEMISTRY

## Delhi [Set-I] 2019

**Time allowed : 3 hours** **Maximum marks : 70**

### SECTION–A

1. **Out of NaCl and AgCl, which one shows Frenkel defect and why ?** **[1]**

**Ans.** AgCl, will show the Frenkel defect because of small size of $Ag^+$ ion.

2. **Arrange the following in increasing order of boiling points :**

$$(CH_3)_3N, C_2H_5OH, C_2H_5NH_2 \quad [1]$$

**Ans.** Increasing order of boiling point would be:

$$(CH_3)_3N < C_2H_5NH_2 < C_2H_5OH$$

3. **Why are medicines more effective in colloidal state ?** **[1]**

**OR**

**What is difference between an emulsion and a gel ?**

**Ans.** Medicines are effective in colloidal state because in colloidal state they have large surface area and are easily assimilated in body.

**OR**

Emulsions are the colloids made up of liquids dispersed in liquid dispersion medium whereas gels are liquids dispersed in solid dispersion medium. For example : milk is an emulsion whereas butter is a gel.

4. **Define ambidient nucleophile with an example.** **[1]**

**Ans** Ambidient nucleophiles are the ones which can attack through two different atoms present in the same nucleophilic species, such as $NO_2^-$.This can attack the electrophilic center either through nitrogen or the oxygen. For example : $CN^-$, $SCN^-$ etc.

5. **What is the basic structural difference between glucose and fructose ?** **[1]**

**OR**

**Write the products obtained after hydrolysis of lactose.**

**Ans.** Both glucose and fructose have the molecular formula $C_6H_{12}O_6$ but Glucose has an aldehydic functional group at C-1 (in its open chain structure) and Fructose has a ketonic functional group at C-2. Glucose is an aldohexose whereas fructose is a ketohexose.

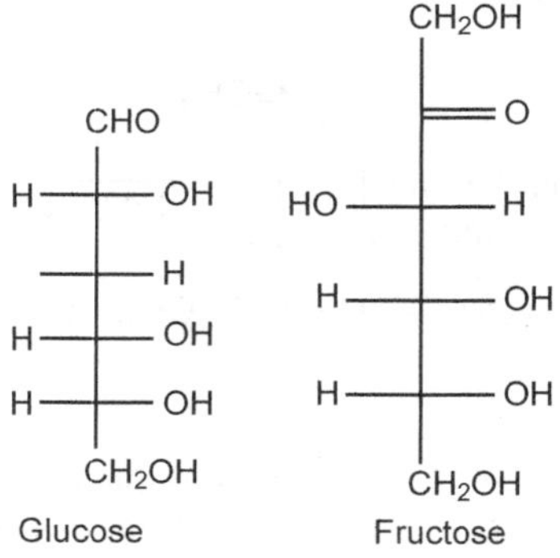


Glucose Fructose

**OR**

The products obtained after hydrolysis of lactose are β-D-glucose and β-D-galactose.

Lactose $\xrightarrow[H_2O]{Lactose}$

D-Galactose + D-Glucose

### SECTION–B

6. **Write balanced chemical equations for the following processes :**
   **(i) $XeF_2$ undergoes hydrolysis.**
   **(ii) $MnO_2$ is heated with conc. HCl.** **[2]**

**OR**

**Arrange the following in order of property indicated for each set :**
**(i) $H_2O$, $H_2S$, $H_2Se$, $H_2Te$ – increasing acidic character**
**(ii) HF, HCl, HBr, HI – decreasing bond enthalpy**

**Ans. (i)** $XeF_2$ undergoes hydrolysis to give Xe, HF and $O_2$

$$2XeF_2(s) + 2H_2O\ (l) \rightarrow 2Xe(g) + 4HF(aq) + O_2(g)$$

**(ii)** $MnO_2$ is heated with conc. HCl to give $MnCl_2$, $Cl_2$ and $H_2O^-$

$$MnO_2 + 4HCl \rightarrow MnCl_2 + Cl_2 + 2H_2O$$

OR

(i) Increasing order of acidity–

$$H_2O < H_2S < H_2Se < H_2Te$$

(ii) Decreasing bond enthalpy–

$$HF > HCl > HBr > HI$$

**7. State Raoult's law for a solution containing volatile components. Write two characteristics of the solution which obeys Raoult's law at all concentrations. [2]**

**Ans.** Raoult's law states that for any solution the partial vapour pressure of each volatile component in the solution is directly proportional to its mole fraction.

The solutions which obey Raoult's law at all concentrations are known as ideal solutions. The two important properties of ideal solutions are as follows :

1. Enthalpy of mixing of the pure components to form the solution is zero

$$\Delta H_{mix} = 0$$

2. Volume of mixing of the pure components to form the solution is zero

$$\Delta V_{mix} = 0$$

**8. For a reaction**

$$2H_2O_2 \xrightarrow[\text{Alkaline medium}]{I^-} 2H_2O + O_2$$

**the proposed mechanism is as given below :**

**(1) $H_2O_2 + I^- \rightarrow H_2O + IO^-$ (slow)**

**(2) $H_2O_2 + IO^- \rightarrow H_2O + I^- + O_2$ (fast)**

**(i) Write rate law for the reaction.**

**(ii) Write the overall order of reaction.**

**(iii) Out of steps (1) and (2), which one is rate determining step ? [2]**

**Ans. (i)** Rate law of the reaction is given by :

$$\text{Rate} = -d[H_2O]/dt = k[H_2O_2][I^-]$$

**(ii)** As the rate law shows that the reaction is first order with $H_2O_2$ and $I^-$ both, hence the overall order of the reaction becomes 1+1 = 2.

**(iii)** Step no. 1 is the slowest among two steps, hence this is the rate determining step.

**9. When $MnO_2$ is fused with KOH in the presence of $KNO_3$ as an oxidizing agent, it gives a dark green compound (A). Compound (A) disproportionates in acidic solution to give purple compound (B). An alkaline solution of compound (B) oxidises KI to compound (C) whereas an acidified solution of compound (B) oxidises KI to (D). Identify (A), (B), (C) and (D). [2]**

**Ans.** The reaction sequence can be written as:

$$2MnO_2 + 4KOH + O_2 \xrightarrow{KNO_3} \underset{\substack{\text{Potassium} \\ \text{manganate (A)} \\ \text{Green}}}{2K_2MnO_4} + 2H_2O$$

$$\underset{(A)}{3MnO_4^{2-}} + 4H^+ + \longrightarrow \underset{\substack{\text{Potassium} \\ \text{permanganate (B)} \\ \text{Purple}}}{2MnO_4^-} + MnO_2 + 2H_2O$$

$$\underset{(B)}{2KMnO_4} + H_2O + KI \xrightarrow{\text{Alkaline}} 2MnO_2 + 2KOH + \underset{\substack{\text{Potassium} \\ \text{iodate (C)}}}{KIO_3}$$

$$10KI + \underset{(B)}{2KMnO_4} + 16HCl + \longrightarrow 2MnCl_2 + 8H_2O + \underset{\text{Iodine (D)}}{5I_2} + 12KCl$$

**10. Write IUPAC name of the complex $[Pt(en)_2Cl_2]$. Draw structures of geometrical isomers for this complex. [2]**

**OR**

**Using IUPAC norms write the formulae for the following :**

**(i) Hexaamminecobalt (III) sulphate**

**(ii) Potassium trioxalatochromate (III)**

**Ans.** Dichlorido bis(ethane-1, 2-diammine platinum(II)

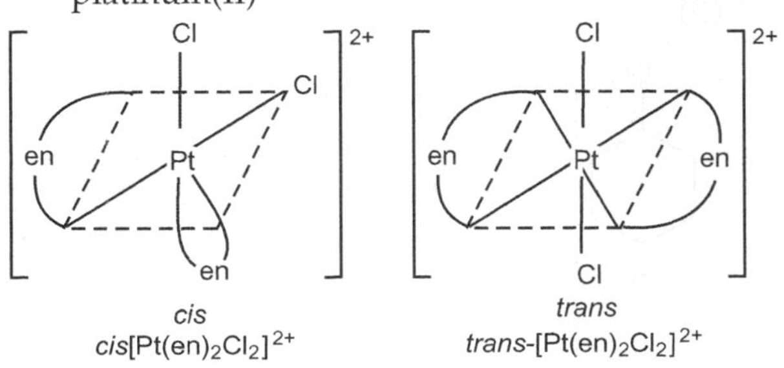


*cis*      *trans*

*cis*$[Pt(en)_2Cl_2]^{2+}$      *trans*-$[Pt(en)_2Cl_2]^{2+}$

OR

**(i)** Hexaamminecobalt(III) sulphate

$[Co(NH_3)_6]_2(SO_4)_3$

**(ii)** Potassium trioxalatochromate(III)

$K_3[Cr(C_2O_4)_3]$

**11. Out of $[CoF_6]^{3-}$ and $\{Co(en)_3]^{3+}$, which one complex is :**

**(i) paramagnetic**

**(ii) more stable**

**(iii) inner orbital complex and**

**(iv) high spin complex**

**(Atomic no. of Co = 27) [2]**

**Ans. (i)** $[CoF_6]^{3-}$ is paramagnetic as it has four unpaired electrons.

**(ii)** $[Co(en)_3]^{3+}$ is more stable.

**(iii)** $[Co(en_3)]^{3+}$ is inner orbital complex.

**(iv)** $[CoF_6]^{3-}$ is high spin complex.

**12. Write structures of compound A and B in each of the following reactions : [2]**

(i) $CH_2CH_3$ (benzene ring) $\xrightarrow{KMnO_4 - KOH}$ (A) $\xrightarrow{H_3O^+}$ (B)

(ii) OH (cyclohexane ring) $\xrightarrow{CrO_3}$ (A) $\xrightarrow{H_2N\text{-}NH\text{-}CONH_2}$ (B)

**Ans. (i)**

$CH_2CH_3$ $\xrightarrow[\text{Oxidation}]{KMnO_4 - KOH}$ $COO^-K^+$ $\xrightarrow{H_3O^+}$

Ethyl benzene — Potassium benzoate (A)

COOH

Benzoic acid (B)

**(ii)**

OH $\xrightarrow{CrO_3}$ O $\xrightarrow{H_2N\text{-}NH\text{-}CONH_2}$

Phenol — Cyclohexanone (A)

O, N, NH, $NH_2$

Cyclohexyl semicarbazide (B)

## SECTION–C

**13. The decomposition of $NH_3$ on platinum surface is zero order reaction. If rate constant (k) is 4, × $10^{-3}$ $Ms^{-1}$, how long will it take to reduce the initial concentration of $NH_3$ from 0.1 M to 0.064 M. [3]**

**Ans.** Rate law for a zero order reaction can be given as–

$k = [R]^0 - [R] / t$

Substituting the given values–

$4 \times 10^{-3} Ms^{-1} = (0.1 - 0.064)\ M / t$

$t = 0.036\ M / 4 \times 10^{-3} Ms^{-1}$

$= 9\ s$

So, the time taken would be 9 seconds.

**14. (i) What is the role of activated charcoal in gas mask ?**

**(ii) A colloidal sol is prepared by the given method in figure. What is the charge on hydrated ferric oxide colloidal particles formed in the test tube ? How is the sol represented ? [3]**

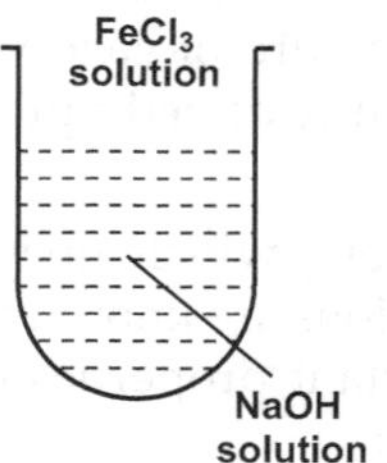


**(iii) How does chemisorption vary with temperature ?**

**Ans. (i)** Activated charcoal present in gas mask adsorbs the harmful suffocating gases which surround the user and protects them from the particulate matter and ashes present in coal mines.

**(ii)** In the given method, $FeCl_3$ is being added to the NaOH solution to give a sol of hydrated ferric oxide, $Fe_2O_3.xH_2O$. The sol formed preferentially adsorbs the negatively charged $OH^-$ present in the solution and hence acquire a negative charge on sol particles. The negatively charged sol formed can be represented as

$Fe_2O_3.\ x\ H_2O/OH^-$ (negatively charged)

**(iii)** Chemisorption increases with temperature upto a certain extent as then after it starts decreasing, chemisorption occurs due to a chemical reaction between adsorbate and adsorbent, and chemical reactions increase with increase in temperature.

**15. An element crystallizes in fcc lattice and with a cell edge of 300 pm. The density of the element is 10.8 g $cm^{-3}$. Calculate the number of atoms in 108 g of the element. [3]**

**Ans.** We know that :

$$M = \frac{a^3 \times N_a \times d}{Z}$$

$$= \frac{27 \times 10^{-24} \times 6.022 \times 10^{23} \times 10.8}{4}$$

$$= 43.88\ g\ mol^{-1}$$

43.88 g $mol^{-1}$ contains $6.02 \times 10^{23}$ atoms

So, 108 g contains

$$= \frac{6.02 \times 10^{23} \times 108}{43.88} = 1.48 \times 10^{24} \text{ atoms}$$

Thus 108 g of the element will have $1.48 \times 10^{24}$ atoms.

**16. A 4% solution (w/w) of sucrose (M = 342 g mol$^{-1}$) in water has a freezing point of 271.15 K. Calculate the freezing point of 5% glucose (M = 180 g mol$^{-1}$) in water. (Given : Freezing point of pure water = 273.15 K)** **[3]**

**Ans.** Depression in freezing point can be shown by the given formula–

$$\Delta Tf = k_f m$$

The difference in freezing point,

$$\Delta T_f = 273.15 - 271.15 = 2 \text{ K}$$

and $$m = \frac{(w_2 \times 1000)}{(m_2 \times w_1)}$$

For 4% sucrose solution in water–

$$2\text{K} = \frac{k_f \times 4 \times 1000}{342 \times 96}$$

$$k_f = 16.42 \text{ K kg/mol}$$

Now for 5% glucose solution–

$w_2 = 5$ g; $m_2 = 180$ g mol$^{-1}$; $w_1 = 100 - 5$ g $= 95$ g

$$273.15 - T_{f\text{glucose}} = \frac{16.42 \times 5 \times 1000}{180 \times 95}$$

$$273.15 - T_{f\text{glucose}} = 1.877$$

$$T_{f\text{glucose}} = 273.15 - 4.8$$

$$T_{f\text{glucose}} = 268.35\text{K}$$

So, freezing point of 5% solution of glucose is 268.35 K.

**17. (a) Name the method of refining which is**

**(i) used to obtain semiconductor of high purity.**

**(ii) used to obtain low boiling metal.**

**(b) Write chemical reactions taking place in the extraction of copper from $Cu_2S$.** **[3]**

**Ans. (a) (i)** Zone refining is being used for production of semiconductors like germanium, silicon, boron, gallium etc. of high purity.

**(ii)** Distillation method is used to obtain low boiling metals such as zinc and mercury.

**(b)** Chemical reactions involved in extraction of Cu from $Cu_2S$ are :

$Cu_2S$ ore is heated in a reverberatory furnace after mixing with silica. In the furnace, iron oxide 'slags of' as iron silicate and copper is produced in the form of copper matte. This contains $Cu_2S$ and FeS.

The reactions undergo as given below.

$$2FeS + 3O_2 \rightarrow 2FeO + 2SO_2$$

$$FeO + SiO_2 \rightarrow FeSiO_3$$

$$2Cu_2S + 3O_2 \rightarrow 2Cu_2O + 2SO_2$$

$$2Cu_2O + Cu_2S \rightarrow 6Cu + SO_2$$

**18. Give reasons for the following :**

**(i) Transition elements and their compounds act as catalysts.**

**(ii) E° value for ($Mn^{2+}$ / Mn) is negative whereas for ($Cu^{2+}$ / Cu) is positive.**

**(iii) Actinoids show irregularities in their electronic configuration.** **[3]**

**Ans. (i)** Transition elements and their compounds act as catalyst due to their ability to adopt variable oxidation states and to form complexes.

**(ii)** $Mn^{2+}$ has $d^5$ configuration (stable half-filled configuration) and it prefers to stay in +2 oxidation state, so, E° value for $Mn^{2+}$/Mn is negative. Whereas, the high energy to transform Cu(s) to $Cu^{2+}$ (aq) is not balanced by its hydration enthalpy, hence $Cu^{2+}$ is not favoured over Cu° state and it has positive E° value.

**(iii)** Actinoids show irregularities in their oxidation states due to extra stability of empty, half filled and fully filled *f* subshells and due to comparable energies of 5*f*, 6*d* and 7*s* orbitals.

**19. Write the structures of monomers used for getting the following polymers :**

**(i) Nylon–6,6** **(ii) Glyptal**

**(iii) Buna-S** **[3]**

**OR**

**(i) Is** $\left[ CH_2 - \underset{\displaystyle |}{\overset{CH_3}{CH}} \right]_n$ **a homopolymer or copolymer ? Give reason.**

**(ii) Write the monomers of the following polymer :**

$$\left( -HN-C_3N_3(NH-)-NH-CH_2- \right)_n$$

**(iii) What is the role of Sulphur in vulcanization of rubber ?**

**Ans. (i)** Monomers of Nylon-6,6 are adipic acid and hexamethylene diamine.

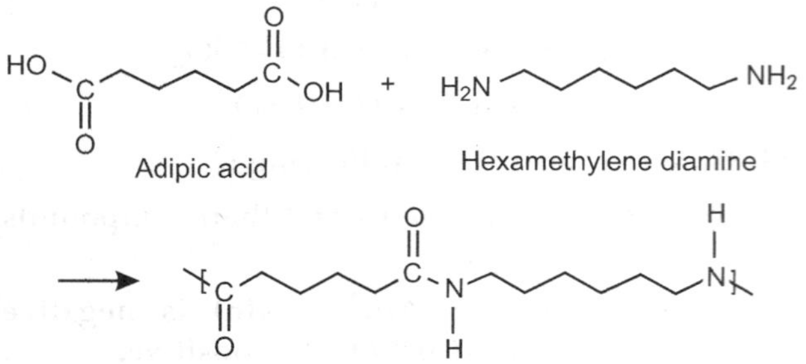

Adipic acid + Hexamethylene diamine → Nylon 6,6

**(ii)** Monomers of glyptal are ethylene glycol and phthalic acid.

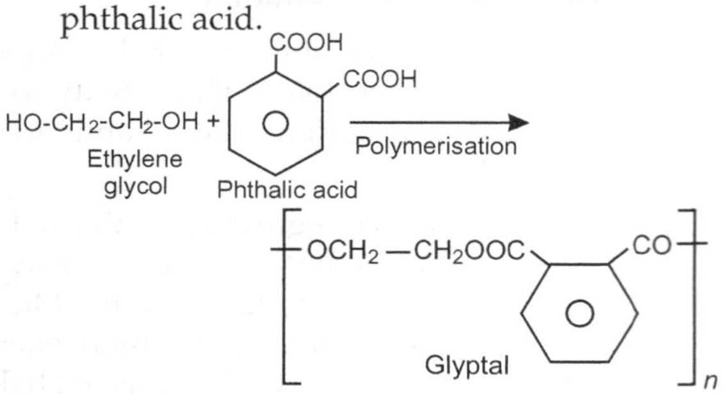

HO-$CH_2$-$CH_2$-OH (Ethylene glycol) + Phthalic acid $\xrightarrow{\text{Polymerisation}}$ $-[OCH_2-CH_2OOC-C_6H_4-CO]_n-$ Glyptal

**(iii)** Monomers of Buna-S are Butadiene and Styrene.

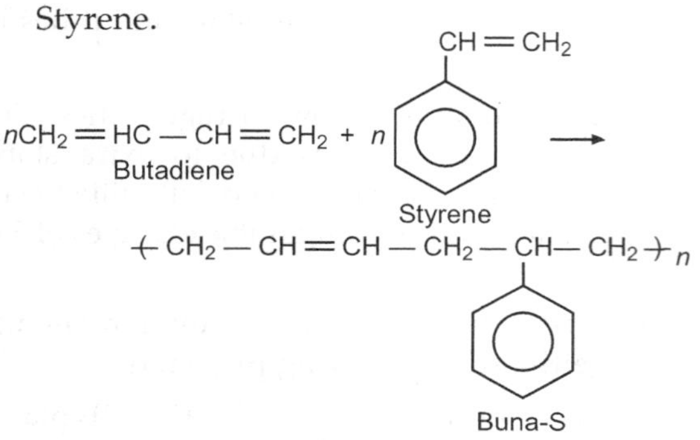

$nCH_2=HC-CH=CH_2$ (Butadiene) + $n$ Styrene ($C_6H_5CH=CH_2$) → $-[CH_2-CH=CH-CH_2-CH(C_6H_5)-CH_2]_n-$ Buna-S

**OR**

**(i)** $-[CH_2\text{-}CH_2(CH_3)]-$ is a homopolymer and the monomer from which it is obtained is $CH_2=CH(CH_3)$.

**(ii)** The given polymer is melamine polymer and the monomers are Melamine and Formaldehyde.

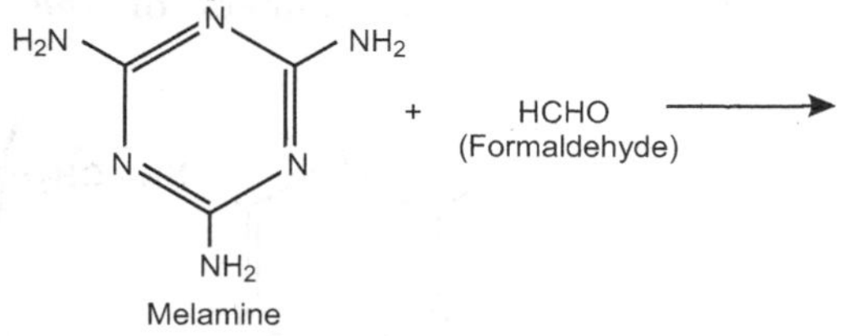

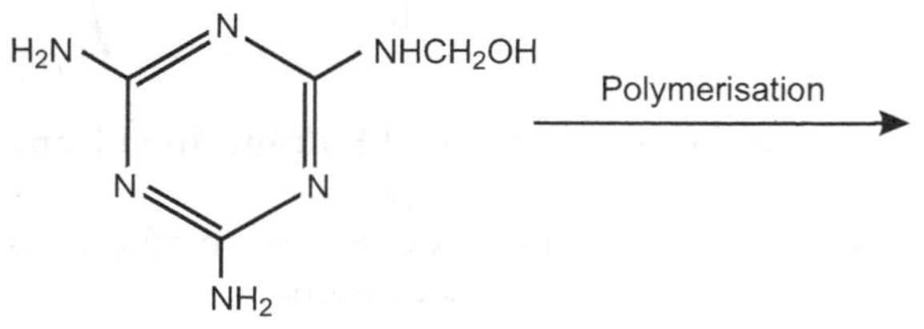

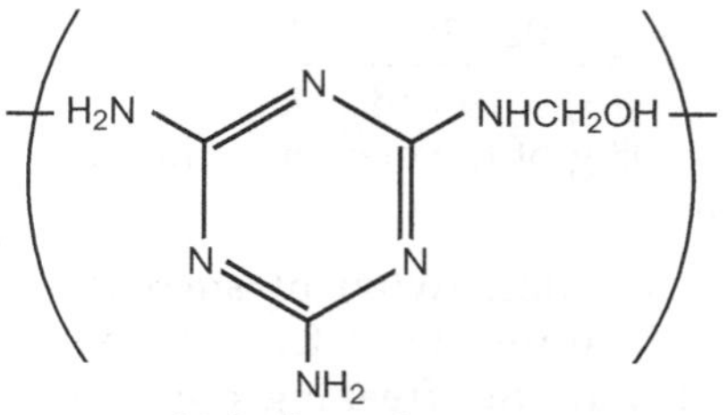

Melamine + HCHO (Formaldehyde) → melamine–$NHCH_2OH$ $\xrightarrow{\text{Polymerisation}}$ Melamine polymer

**(iii)** Natural rubber becomes soft at high temperatures. In the vulcanisation process, sulphur forms cross links at the reactive sites of double bonds in rubber monomers and thus the rubber gets stiffened. This improves the physical properties of rubber.

**20. (i) What type of drug is used in sleeping pills?**

**(ii) What type of detergents are used in toothpastes ?**

**(iii) Why the use of alitame as artificial sweetener is not recommended ? [3]**

**OR**

**Define the following terms with a suitable example in each :**

**(i) Broad-spectrum antibiotics**

**(ii) Disinfectants**

**(iii) Cationic detergents**

**Ans. (i)** Tranquilizers are used in sleeping pills.

**(ii)** Anionic detergents are used in toothpastes.

**(iii)** It is difficult to control the sweetness of food by using alitame hence its use is not recommended.

**OR**

**(i) Broad spectrum antibiotics :** Antibiotics which kill or inhibit a wide range of Gram-positive and Gram-negative bacteria are said to be broad spectrum antibiotics. Ampicillin is an example.

**(ii) Disinfectants :** These are chemicals which are applied to inanimate objects such as floors, drainage system, instruments etc. Example is 1% solution of phenol.

**(iii) Cationic detergents :** These are quarter-nary ammonium salts of amines with acetates, chlorides or bromides as anions. Example is Cetyltrimethyl ammonium bromide.

**21. (i) Out of $(CH_3)_3C$–Br and $(CH_3)_3C$–I, which one is more reactive towards $S_N1$ and why?**

**(ii) Write the product formed when *p*-nitro-chlorobenzene is heated with aqueous NaOH at 443 K followed by acidification.**

**(iii) Why *dextro* and *laevo* – rotatory isomers of Butan-2-ol are difficult to separate by fractional distillation ? [3]**

**Ans. (i)** $(CH_3)_3C–I$ is more reactive than $(CH_3)_3C$—Br towards $S_N1$ reaction because, C–I bond being weaker than C–Br bond (due to larger size of $I^-$ compared to $Br^-$) forms the tertiary carbocation easily. Due to its large size iodine is a better leaving group, than Br.

**(ii)** 4-Nitrophenol is formed as product when *p*-nitro chlorobenzene is heated with NaOH at 443 K and acidified later.

4-Nitrochlorobenzene $\xrightarrow[\text{(ii) } H^+]{\text{(i) Aq. NaOH, 443 K}}$ 4-Nitrophenol

**(iii)** *Dextro-* and *laevo-* rotatory isomers of Butan-2-ol are stereoisomers of each other and have same physical properties. As they have same boiling points, thus it is difficult to isolate them through fractional distillation.

**22. An aromatic compound 'A' on heating with $Br_2$ and KOH forms a compound 'B' of molecular formula $C_6H_7N$ which on reacting with $CHCl_3$ and alcoholic KOH produces a foul smelling compound 'C'. Write the structures and IUPAC names of compounds A, B and C. [3]**

**Ans.** As the chemical combination $Br_2$ and KOH are used for Hofmann Bromamide reaction where an amide is reduced to amine, the compound B, $C_6H_7N$ seems to be Aniline. The reaction can be suggested as follows :

Benzamide (A) $\xrightarrow[\text{KOH}]{Br_2}$ Aniline (B)

Then aniline undergoes carbylamine reaction to give the foul smelling compound C.

Aniline (B) $\xrightarrow[\text{Alc. KOH}]{CHCl_3}$ Phenyl isocyanide (C)

IUPAC names are as follows :

Compound A – Benzamide

Compound B – Aniline or Benzenamine

Compound C – Isocyanobenzene or phenylisocyanide.

**23. Complete the following reactions: [3]**

**(i)** $C_6H_5CHO \xrightarrow{\text{NaCN/HCl}}$

**(ii)** $(C_6H_5CH_2)_2Cd + 2CH_3COCl$

**(ii)** $CH_3—CH(CH_3)—COOH \xrightarrow[\text{(ii) } H_2O]{\text{(i) } Br_2\text{/Red } P_4}$

**OR**

**Write chemical equations for the following reactions :**

**(i) Propanone is treated with dilute $Ba(OH)_2$.**

**(ii) Acetophenone is treated with Zn(Hg)/conc. HCl.**

**(iii) Benzoyl chloride is hydrogenated in presence of Pd/$BaSO_4$.**

**Ans. (i)**

Benzaldehyde $\xrightarrow{\text{NaCN/HCl}}$ $C_6H_5CH(OH)CN$ (2-Phenyl-2-hydroxy ethane nitrile)

**(ii)** $(C_6H_5CH_2)_2Cd + 2CH_3COCl \longrightarrow 2\,CH_3COCH_2C_6H_5 + CdCl_2$

Methyl benzyl ketone

**(iii)** $H_3C–CH(CH_3)–COOH \xrightarrow[\text{(ii) } H_2O]{\text{(i) } Br_2\text{/Red } P_4} H_3C–C(CH_3)(Br)–COOH$

2-Bromo-2-methyl propanoic acid

**OR**

**(i)** $H_3C–CO–CH_3 \xrightarrow[\text{Aldol condensation}]{\text{dil. } Ba(OH)_2} (CH_3)_2C(OH)–CH_2–CO–CH_3$

4-Hydroxy-4-methyl pentane 2-one

$\xrightarrow{-H_2O}$ $H_3C-C(CH_3)=CH-C(=O)-CH_3$

4-Methylpent-3-ene-2-one

**(ii)**

$$C_6H_5COCH_3 \xrightarrow[\text{Clemmenson reduction}]{\text{Zn/Hg, Conc. HCl}} C_6H_5CH_2CH_3$$

Acetophenone → Ethyl benzene

**(iii)**

$$C_6H_5COCl \xrightarrow[\text{Rosenmund's Reduction}]{\text{Pd/BaSO}_4,\ H_2} C_6H_5CHO$$

Benzoyl chloride → Benzaldehyde

**24. Differentiate between the following :**

**(i) Amylose and Amylopectin**

**(ii) Peptide linkage and Glycosidic linkage**

**(iii) Fibrous proteins and Globular proteins.**

[3]

**OR**

**Write chemical reactions to show that open structure of D-glucose contains the following :**

**(i) Straight chain**

**(ii) Five alcohol groups**

**(iii) Aldehyde as carbonyl group.**

**Ans. (i)** Difference between Amylose and Amylopectin:

| S.No. | Amylose | Amylopectin |
|---|---|---|
| 1. | Water soluble. | Water insoluble. |
| 2. | Constitutes about 15-20% of starch. | Constitutes about 80-85% of starch. |
| 3. | It has un-branched chain. | It has branched chain. |

**(ii)** Difference between Peptide linkage and Glycosidic linkage :

| S.No. | Peptide linkage | Glycosidic linkage |
|---|---|---|
| 1. | It is an amide formed between –COOH group of one amine and $-NH_2$ group of second amino acid molecule. | It is an oxide linkage, between two monosaccharides are joined through an oxygen atom. |
| 2. | It is found in protein molecules. | It is found in carbohydrate molecules. |

**(iii)** Difference between Fibrous proteins and Globular Proteins :

| S.No. | Fibrous protein | Globular protein |
|---|---|---|
| 1. | In these polypeptide chains run parallel and are held together by hydrogen and disulphide bonds. | In these chains of polypeptide coil around to give a spherical shape. |
| 2. | Usually insoluble in water. Examples are Keratin and myosin. | Usually soluble in water. Examples are albumin and insulin. |

**OR**

**(i)** Open chain structure of Glucose is straight chain can be shown as follows:

On prolonged heating with HI, it forms *n*-hexane, this shows that glucose has straight open chain structure.

$$\underset{\text{Glucose}}{CHO-(CHOH)_4-CH_2OH} \xrightarrow{HI} \underset{\text{n-hexane}}{H_3C-CH_2-CH_2-CH_2-CH_2-CH_3}$$

**(ii)** On acetylation with acetic anhydride, glucose gives glucose pentaacetate, this shows that glucose has five alcoholic groups.

$$\underset{\text{Glucose}}{CHO-(CHOH)_4-CH_2OH} \xrightarrow{\text{Acetic anhydride}} \underset{\text{Glucose pentaacetate}}{CHO-(CH-O-CO-CH_3)_4-CH_2-O-CO-CH_3}$$

**(iii)** Glucose gets oxidized to six carbon carboxylic acid (Gluconic acid) on reaction with a mild oxidising agent like bromine water. This indicates that the carbonyl group is present as an aldehydic group.

$$\underset{\text{Glucose}}{\begin{matrix} CHO \\ | \\ (CHOH)_4 \\ | \\ CH_2OH \end{matrix}} \xrightarrow[\text{Oxidation}]{Br_2 \text{ water}} \underset{\text{Gluconic acid}}{\begin{matrix} COOH \\ | \\ (CHOH)_4 \\ | \\ CH_2OH \end{matrix}}$$

## SECTION–D

**25. $E^o_{cell}$ for the given redox reaction is 2.71 V $Mg_{(s)} + Cu^{2+}{}_{(0.01\ M)} \rightarrow Mg^{2+}{}_{(0.001\ M)} + Cu_{(s)}$ Calculate $E_{cell}$ for the reaction. Write the direction of flow of current when an external opposite potential applied is**

**(i) less than 2.71 V and**

**(ii) greater than 2.71 V. [5]**

**OR**

**(a) A steady current of 2 amperes was passed through two electrolytic cells X and Y connected in series containing electrolytes $FeSO_4$ and $ZnSO_4$ until 2.8g of Fe deposited at the cathode of cell X. How long did the current flow ? Calculate the mass of Zn deposited at the cathode of cell Y.**

**(Molar mass : Fe = 56 g $mol^{-1}$, Zn = 65.3 g $mol^{-1}$, 1F = 96500 C $mol^{-1}$)**

**(b) In the plot of molar conductivity ($\Lambda_m$) vs square root of concentration ($c^{1/2}$) following curves are obtained for two electrolytes A and B :**

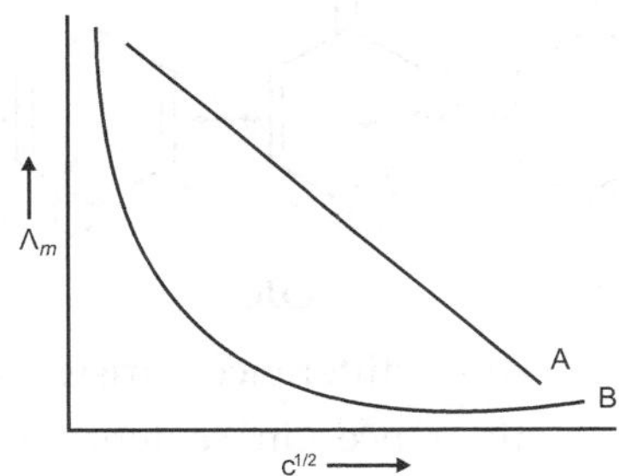


**Answer the following :**

**(i) Predict the nature of electrolytes A and B.**

**(ii) What happens on extrapolation of $\Lambda_m$ to concentration approaching zero for electrolytes A and B ?**

**Ans.** The cell can be represented as :

$Mg\,|\,Mg^{2+}\ (0.001\ M)\ ||\ Cu^{2+}\ (0.01\ M)\,|\,Cu$

$$E_{cell} = E^o_{Cell} - \frac{0.059}{n} \log kc$$

Putting the given values,

$$E_{cell} = E^o_{Cell} - \frac{0.059}{2} \log \frac{10^{-3}}{10^{-2}}$$

$$= 2.71 + 0.0295$$

$$E_{cell} = 2.7395V$$

(i) When external opposite potential is less than 2.71 V then electron flows from Mg rod to Cu rod hence current flows from Cu to Mg.

(ii) When external opposite potential is greater than 2.71 V then electron flows from Cu rod to Mg rod and current flows from Mg to Cu.

**OR**

(a) Charge = Current × Time

Now, in the given experiment, 2.8 g of iron was deposited or

$\frac{2.8}{56}$ = 0.05 moles of iron were deposited.

Now, as it is a 2 electron transfer process :
1 mol of iron is deposited by 2 × 96500 C of charge

Hence, 0.05 mol of iron will need 0.05 × 2 × 96500 C of charge

= 9650 C

So,

9650 C = 2 A × $t$

Time = 4825 seconds = 80.41 minutes

Similarly,

So, same amount of charge will flow to deposit Zn as well, keeping that in mind 2 × 96500 C of charge can deposit 1 mol of Zn

Hence, 9650 C of charge would deposit

$$= \left(\frac{1}{2} \times 96500\ C\right) \times 9650 = 0.05 \text{ mol}$$

Weight of Zinc deposited

$$= 0.05 \text{ mol} \times 65.3 \text{ g mol}^{-1} = 3.26 \text{ g}$$

(b) (i) The electrolyte A is a strong electrolyte, and the electrolyte B is a weak electrolyte.

(ii) On extrapolation for electrolyte A limiting value of conductance is obtained that is conductance at zero concentration or infinite dilution.

The curve obtained for a strong electrolyte shows that there is a small decrease in molar conductivity with increase in concentration. In other words, the molar conductivity is increased only slightly on dilution (for observing dilution effects, go towards zero on X-axis). A strong electrolyte is completely dissociated in solution and thus, furnishes all ions for conductance. However, at higher concentrations, the dissociated ions are close to each other and thus, the inter-ionic attractions are greater. These forces retard the motion of the ions and thus, conductivity is low. With decrease in concentration (dilution), the ions move away from each other thereby feeling less attractive forces from the counter ions. This results in an increase in molar conductivity with dilution. The molar conductivity approaches a maximum limiting value at infinite dilution designated as $\Lambda_m^0$.

For electrolyte B :

The curve obtained for B shows that there is a large increase in the value of molar conductivity with dilution, especially near infinite dilution. This is because as the solution of a weak electrolyte is diluted, its ionization is increased. This results in more number of ions in solution and thus, there is an increase in molar conductivity. However, the conductance of a weak electrolyte never approaches a limiting value. Or in other words it is not possible to find conductance at infinite dilution (zero concentration).

**26. (a) How do you convert the following :**

**(i) Phenol to Anisole**

**(ii) Ethanol to Propan–2–ol**

**(b) Write mechanism of the following reaction :**

$$CH_3OH \xrightarrow[443\,K]{H_2SO_4} CH_2 = CH_2 + H_2O$$

**(c) Why phenol undergoes electrophilic substitution more easily than benzene ?[5]**

**OR**

**(a) Account for the following :**

**(i) *o*–nitrophenol is more steam volatile than *p*-nitrophenol.**

**(ii) *t*-butyl chloride on heating with sodium methoxide gives 2-methylpropene instead of *t*-butyl methylether.**

**(b) Write the reaction involved in the following:**

**(i) Riemer Tiemann reaction**

**(ii) Friedal Crafts Alkylation of Phenol**

**(c) Give simple chemical test to distinguish between Ethanol and Phenol.**

**Ans. (a) (i)** Conversion of phenol to anisole can be done as follows.

$$\underset{\text{Phenol}}{C_6H_5OH} + NaOH \longrightarrow \underset{\text{Phenol}}{C_6H_5ONa} \xrightarrow[AlCl_3]{CH_3Cl} \underset{\text{Anisole}}{C_6H_5OCH_3}$$

Friedel Crafts reaction

**(ii)** Ethanol can be converted to propan-2-ol through the following sequence :

$$\underset{\text{Ethanol}}{C_2H_5OH} \xrightarrow[\text{Oxidation}]{CrO_3} \underset{\text{Ethanal}}{CH_3CHO} \xrightarrow[\text{(ii) } H^+/H_2O]{\text{(i) } CH_3MgI} \underset{\text{Propan-2-ol}}{H_3C-CH(OH)-CH_3}$$

Grignard reaction

**(b)** The mechanism of dehydration of ethanol to give ethene can be written as follows:

$$CH_3-CH_2-\ddot{O}-H + H^+ \xrightarrow{\text{fast}} CH_3-CH_2-\overset{+}{O}H_2$$

$$CH_3-CH_2-\overset{+}{O}H_2 \xrightarrow{\text{slow}} CH_3-\overset{+}{C}H_2 + H_2O$$

$$H-CH_2-\overset{+}{C}H_2 \rightleftharpoons CH_2=CH_2 + H^+$$

**(c)** The –OH group in phenol increases the electron density on the benzene ring hence electrophilic substitution reaction is more prominent in phenol compared to benzene. The lone pair of oxygen on phenolic–OH group takes part in the resonance and makes the ring electron rich, hence activating the ring for incoming electrophiles.

$:\ddot{O}H$ ⟷ $^{+}\ddot{O}H$ ⟷ $^{+}\ddot{O}H$ ⟷ $^{+}\ddot{O}H$

**OR**

**(a) (i)** This difference arises due to the difference in extent of association of molecules with each other. In *o*-nitrophenol the nitro and hydroxyl group present in same molecule forms a hydrogen bond (intra-molecular) and thus have least association with nearby molecules, whereas in *p*-nitrophenol the nitro and hydroxyl groups of adjacent molecules form hydrogen bonds (intermolecular) which result in long range association of molecules. Hence, large amount of energy is required to break the intermolecular hydrogen bonds hence, high boiling point than the 2-nitrophenol molecules.

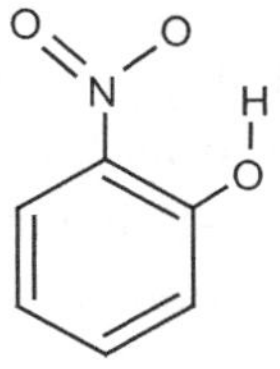

2-Nitrophenol

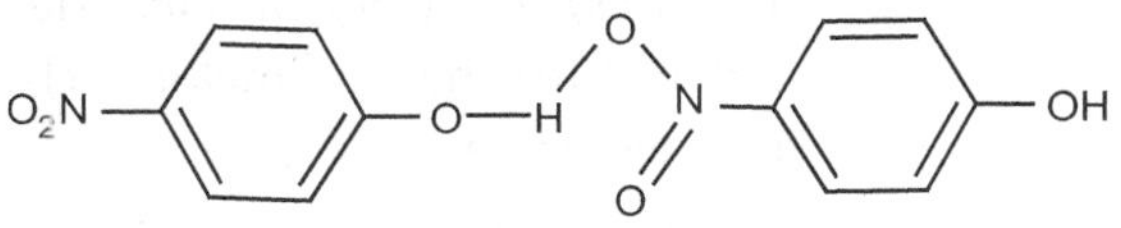

4-Nitrophenol

**(ii)** Sodium methoxide acts as strong base and extracts a proton from one of the methyl groups of *t*-butyl chloride giving rise to a primary carbanion which quickly loses Cl - to give a double bond, hence the product formed is 2-methyl propene.

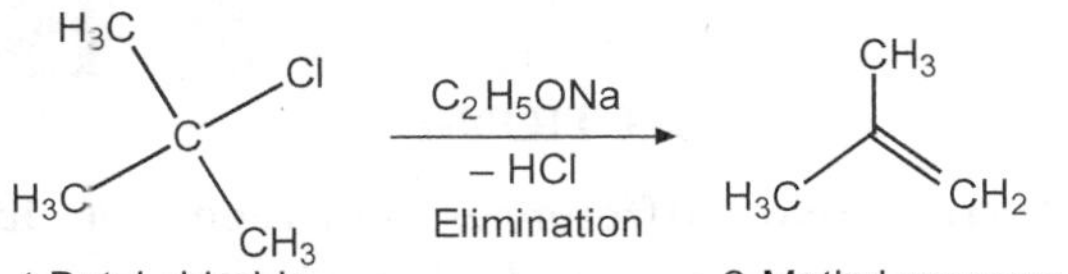

$$(CH_3)_3C\text{-}Cl \xrightarrow[\text{Elimination}]{C_2H_5ONa,\ -HCl} (CH_3)_2C{=}CH_2$$

t-Butyl chloride → 2-Methyl propene

**(b) (i)** Reimer-Tiemann reaction:

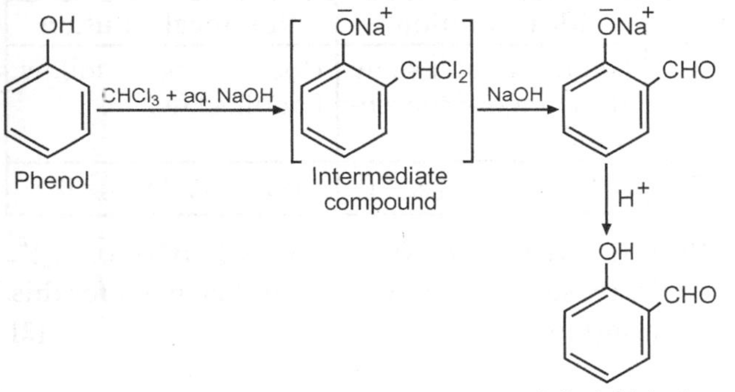

Phenol $\xrightarrow{CHCl_3 + aq.\ NaOH}$ [Intermediate compound ($\bar{O}\overset{+}{Na}$, $CHCl_2$)] $\xrightarrow{NaOH}$ ($\bar{O}\overset{+}{Na}$, CHO) $\xrightarrow{H^+}$ Salicylaldehyde (OH, CHO)

**(ii)** Friedal Crafts alkylation of phenol can be shown as following :

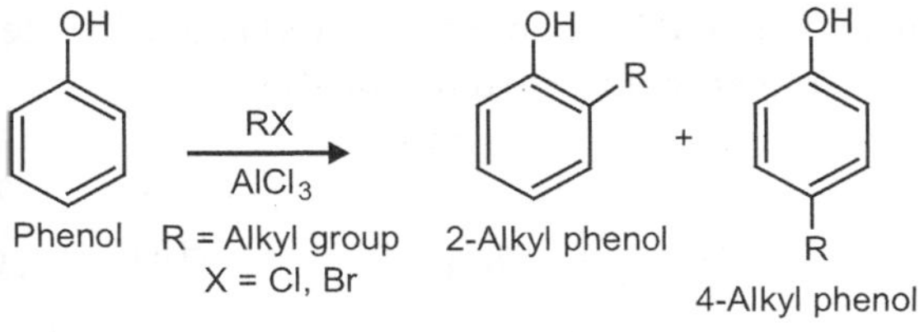

Phenol $\xrightarrow[AlCl_3]{RX}$ 2-Alkyl phenol + 4-Alkyl phenol

R = Alkyl group
X = Cl, Br

**(c)** Chemical test to distinguish between ethanol and phenol :

On adding neutral $FeCl_3$, phenol gives a violet coloured solution but ethanol does not.

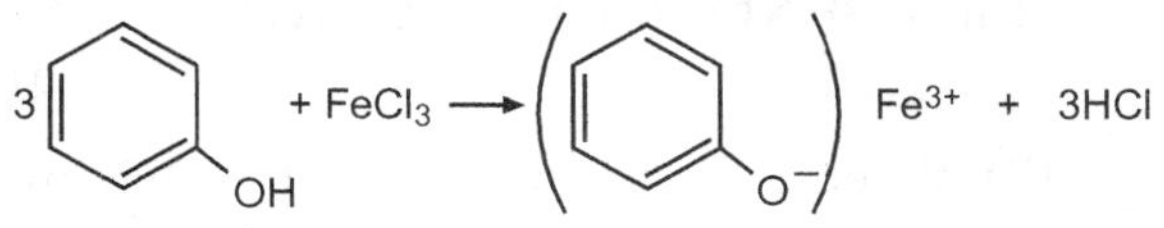

$$3\,C_6H_5OH + FeCl_3 \longrightarrow (C_6H_5O^-)_3Fe^{3+} + 3HCl$$

Violet colour

**27. Give reasons for the following :**

**(i) Sulphur in vapour state shows paramagnetic behaviour.**

**(ii) N-N bond is weaker than P—P bond.**

**(iii) Ozone is thermodynamically less stable than oxygen.**

**(b) Write the name of gas released when Cu is added to**

**(i) dilute $HNO_3$ and**

**(ii) conc. $HNO_3$** **[5]**

**OR**

**(a) (i) Write the disproportionation reaction of $H_3PO_3$.**

**(ii) Draw the structure of $XeF_4$.**

**(b) Account for the following :**

**(i) Although Fluorine has less negative electron gain enthalpy yet $F_2$ is strong oxidizing agent.**

**(ii) Acidic character decreases from $N_2O_3$ to $Bi_2O_3$ in group 15.**

**(c) Write a chemical reaction to test sulphur dioxide gas. Write chemical equation involved.**

**Ans. (a) (i)** In vapour phase sulphur partly exists as $S_2$ molecule and $S_2$ molecule like $O_2$ molecule has two unpaired electrons in anti-bonding $\pi$ orbital, hence it shows paramagnetic behaviour.

**(ii)** N-N bond is weaker than P-P bond due to greater inter-electronic repulsion in N-N bond in comparison to the P-P bond.

**(iii)** Ozone ($O_3$) is thermodynamically less stable than dioxygen ($O_2$) because decomposition of ozone into dioxygen results in the liberation of heat ($\Delta H$ is negative) and increase in entropy ($\Delta S$ is positive) resulting in large negative Gibbs energy change ($\Delta G$) for its conversion into oxygen. These two effects reinforce each other.

**(b) (i)** When Cu is added to dil. $HNO_3$ Nitrogen monoxide or nitric oxide (NO) is released:

$3Cu + 8HNO_3 \text{ (dil.)} \rightarrow 3Cu(NO_3)_2 + 2NO + 4H_2O$

**(ii)** When Cu is added to conc. $HNO_3$ Nitrogen dioxide ($NO_2$) is released :

$$Cu + 4HNO_3\text{(conc.)} \rightarrow Cu(NO_3)_2 + 2NO_2 + 2H_2O$$

**OR**

**(a) (i)** $4H_3PO_3 \longrightarrow 3H_3PO_4 + PH_3$

**(ii)** Structure of $XeF_4$

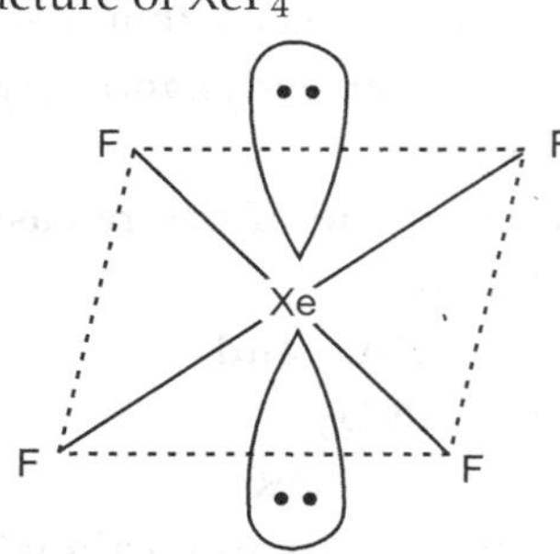


**(b) (i)** Although Fluorine has less electron gain enthalpy yet it is a strong oxidising agent due to its low enthalpy of dissociation of F—F bond and high enthalpy of hydration of $F^-$ ion.

**(ii)** Acidic character decreases from $N_2O_3$ to $Bi_2O_3$ in group 15 because as the size increases, electronegativity decreases and in turn the non-metallic character also decreases.

**(c)** Sulphur dioxide behaves as a reducing agent when moist, this property is being used for its test in laboratory. It decolorises the purple coloured potassium permanganate(VII) solution :

$$5SO_2 + 2KMnO_4 + 2H_2O \rightarrow 2H_2SO_4 + 2MnSO_4 + K_2SO_4$$

Purple colour — light brown colour solution

●●

# Delhi [Set-II]

**Note : Except for the following questions, all the remaining questions have been asked in previous set.**

### SECTION–A

**2. Arrange the following in increasing order of base strength in gas phase : [1]**

**$(CH_2H_5)_3N$, $C_2H_5NH_2$, $(C_2H_3)_2NH$**

**Ans.** In gas phase the basicity order will be –
$(C_2H_5)_3N > (C_2H_5)_2NH > C_2H_5NH_2$

**3. Why conductivity of silicon increases on doping with phosphorus ? [1]**

**Ans.** The conductivity of silicon increases on doping with Phosphorous due to the formation of n-type semiconductor providing free electrons for the conduction.

**5. Write IUPAC name of the given compound : [1]**

Cl

$SO_3H$

**Ans.**

Cl

$SO_3H$

4-Chlorobenzene-1-sulphonic acid

### SECTION-B

**8. Write two differences between an ideal solution and a non-ideal solution. [2]**

**Amswer :** Differences between Ideal and non-ideal solutions–

| S.No. | Ideal solution | Non-ideal solution |
|---|---|---|
| 1. | Follows Raoult's law at all range of concentration. | Does not follow Raoult's law. |
| 2. | $\Delta H_{mix} = 0, \Delta V_{mix} = 0$ | $\Delta H_{mix} \neq 0, \Delta V_{mix} \neq 0$ |

**10. Write IUPAC of the complex $[Cr(NH_3)_4Cl_2]^+$. Draw structures of geometrical isomers for this complex. [2]**

**OR**

**Using IUPAC norms write the formulae for the following :**

**(i) Pentamminenitrito-O-cobalt(III) chloride**

**(ii) Potassium tetracyanidonickelate (II)**

**Ans.** The IUPAC name for $[Cr(NH_3)_4Cl_2]^+$ is tetraamminedichlorochromium(III) :

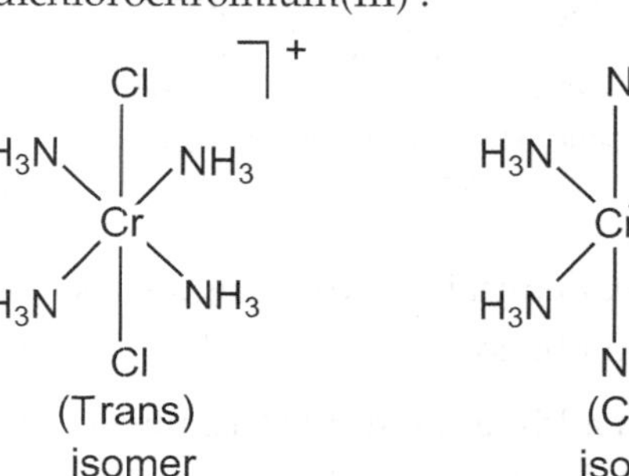

OR

(i) Pentamminenitritocobalt(III)chloride :

$[Co(NH_3)_5ONO]Cl_2$

(ii) Potassium tetracyanidonickelate (II) :

$K_2[Ni(CN)_4]$

**11. Out of $[CoF_6]^{3-}$ and $[Co(C_2O_4]^{3-}$, which one complex is :**

**(i) diamagnetic (ii) more stable**

**(iii) outer orbital compelx and**

**(iv) low spin complex ?**

**(Atomic no. of (Co = 27) [2]**

**Ans.** $[Co(C_2O_4)_3]^{3-}$has $d^2sp^3$ hybridisation (low spin complex) and $[CoF_6]^{3-}$has $sp^3d^2$hybridisation (high spin complex).

(i) $[Co(C_2O_4)_3]^{3-}$ is diamagnetic

(ii) $[Co(C_2O_4)_3]^{3-}$ is more stable.

(iii) $[CoF_6]^{3-}$ is outer orbital complex

(iv) $[Co(C_2O_4)_3]^{3-}$is low spin complex.

## SECTION-B

**17. (i) Write the role of 'CO' in the purification of nickel.**

**(ii) What is the role of silica in the extraction of copper ?**

**(iii) What type of metals are generally extracted by electrolytic method ? [3]**

**Ans. (i) Role of CO in purification of Nickel:** Mond process is the technique used to purify nickel. The impure nickel reacts with carbon monoxide at 50-60° C to form the gas nickel carbonyl, leaving the impurities as solids. Nickel carbonyl is subjected to higher temperature so that it is decomposed giving the pure metal.

$$Ni + 4CO \xrightarrow{330\text{-}350\ K} Ni(CO)_4$$

$$Ni(CO)_4 \xrightarrow{450\text{-}470\ K} Ni + 4CO$$

**(ii) Role of silica in extraction of Copper :** The role of silica in the metallurgy of copper is to remove the iron oxide obtained during the process of roasting. If the sulphide ore of copper contains iron, then silica ($SiO_2$) is added as flux before roasting.

$$FeO + SiO_2 \rightarrow FeSiO_3(slag)$$

**(iii)** Metals whose ions easily get reduced in solution or molten state are generally extracted by electrolytic method due to their large negative electrode potential because they are more reactive.

**18. Give reasons for the following :**

**(i) Transition metals form alloys.**

**(ii) $Mn_2O_3$ is basic whereas $Mn_2O_7$ is acidic.**

**(iii) $Eu^{2+}$ is a strong reducing agent. [3]**

**Ans. (i)** Transition metal forms alloys because the atomic sizes of transition metals are very similar to each other. As the atomic sizes are very similar, one metal can replace the other metal from its lattice and form a solid solution. This solid solution is known as alloy.

**(ii)** In $Mn_2O_3$ manganese has +3 oxidation state, it has lone pairs of electrons which can be donated (Lewis base), hence it is basic in nature. Whereas in $Mn_2O_7$ manganese has +7 oxidation state. Higher oxidation states are short of electrons, meaning they can accept electrons and thus function as Lewis acids. So, $Mn_2O_7$ is acidic.

**(iii)** Reducing agent is that which can reduce other species and itself gets oxidised. $Eu^{2+}$ readily changes to the common +3 oxidation state shown by the lanthanides by losing one more electron. So, $Eu^{2+}$is regarded as a strong reducing agent.

**20. (i) Why bithional is added in soap ?**

**(ii) Why magnesium hydroxide is a better antacid than sodium bicarbonate ?**

**(ii) Why soaps are biodegradable whereas detergents are non-biodegradable ? [3]**

OR

**Define the following terms with a suitable example in each :**

**(i) Antibiotics**

**(ii) Artificial sweeteners**

**(iii) Analgesics**

**Ans. (i)** Bithional is added in soap to impart antiseptic properties to soap.

**(ii)** Sodium bicarbonate if taken in excess can make the stomach alkaline in turn stimulating more acid release, hence magnesium hydroxide is better antacid than sodium bicarbonate because being insoluble, it does not increase the pH above neutrality.

**(iii)** Soaps are sodium or potassium salts of long chain fatty acids whereas the hydrocarbon portion of synthetic detergents contain highly branched hydrocarbon chain which is not easy for the bacteria to degrade. Hence, soaps are biodegradable but synthetic detergents are not, because in soaps hydrocarbon chains are not branched.

OR

(i) **Antibiotics:** They are the compounds (produced by microorganisms or synthetically) which either inhibit the growth of bacteria or kill bacteria. Example: Penicillin.

(ii) **Artificial sweeteners:** They are the compounds which make the food sweet in taste without adding calories to the food. Example – Aspartame, saccharin etc.

(iii) **Analgesics :** These are the compounds which reduce or abolish pain without causing impairment of consciousness, mental confusion or any other disturbances to central nervous system. They are of two types, Narcotic (Morphine) and non-narcotic (Example : Paracetamol).

**21. Write the structures of main products when benzene diazonium chloride reacts with the following reagents.**
**(i) CuCN (ii) $CH_3CH_2OH$**
**(iii) KI [3]**

**Ans.** The products will be as follows :

(i)

$C_6H_5N_2^+Cl^-$ (Benzene diazonium chloride) + CuCN $\xrightarrow{KCN}$ $C_6H_5CN$ (Cyanobenzene) + $N_2$

(ii)

$C_6H_5N_2^+Cl^-$ (Benzene diazonium chloride) + $CH_3CH_2OH$ $\longrightarrow$ $C_6H_6$ (Benzene) + $N_2$ + $CH_3CHO$ + HCl

(iii)

$C_6H_5N_2^+Cl^-$ (Benzene diazonium chloride) + KI $\longrightarrow$ $C_6H_5I$ (Iodobenzene) + $N_2$ + KCl

# Delhi [Set-III]

**Note : Except for the following questions, all the remaining questions have been asked in previous sets.**

## SECTION–A

**1. Arrange the following in decreasing order of solubility in water :**
**$(CH_3)_3N$, $(CH_3)_2NH$, $CH_3NH_2$ [1]**

**Ans.** Decreasing order of solubility in water is–

$$CH_3NH_2 > (CH_3)_2NH > (CH_3)_3N$$

**2. What type of stoichiometric defect is shown by ZnS and why ? [1]**

**Ans.** Frenkel defect is shown by ZnS due to large difference in size of Zn and S ions.

**3. Write one sterochemical difference between $S_N1$ and $S_N2$ reactions. [1]**

**Ans.** Products of $S_N1$ reactions are usually racemic in nature, whereas products of $S_N2$ reactions have inverted configuration compared to the starting reactant.

## SECTION-B

**7. State Henry's law and write its two applications. [2]**

**Ans.** Henry's law states that the solubility of a gas in a liquid is directly proportional to the partial pressure of the gas present above the surface of

The most commonly used form of above law can be put as–

'The partial pressure of the gas in vapour phase ($p$) is proportional to the mole fraction of the gas ($x$) in the solution:.

$$p = K_H x$$

**Applications :**

1. To increase the solubility of $CO_2$ in soft drinks and soda water, the sealing is done under high pressure.
2. Tanks used by scuba divers are filled with air diluted with helium gas in order to avoid accumulation of nitrogen in bubbles in their blood. As increased pressure underwater increases solubility of nitrogen in the blood.

**11. Write the hybridization and magnetic character of following complexes :**
**(a) $[Fe(H_2O)_6]^{2+}$ (b) $[Fe(CO)_5]$**
**(Atomic no. of Fe = 26) [2]**

**Ans.** (a) The hybridisation in $[Fe(H_2O)_6]^{3+}$ is $sp^3d^2$. As there are five unpaired electrons, it is strongly paramagnetic in nature.

$[Fe(H_2O)_6]^{3+}$ :

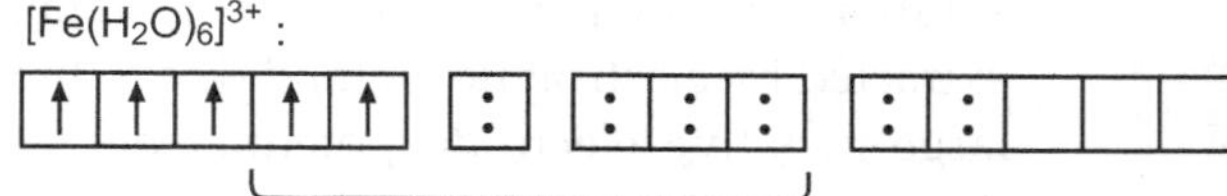

(b) $[Fe(CO)_5]$ has $dsp^3$ hybridisation and no unpaired electron, hence it is diamagnetic in nature.

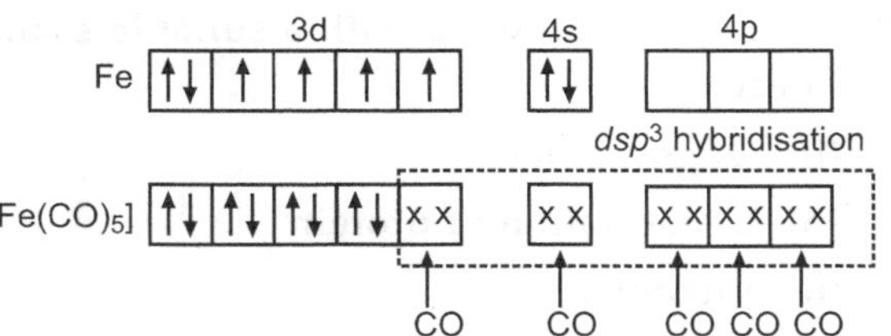


**12. Write structures of main compounds A and B in each of the following reactions : [2]**

(i) $CH_3CH_2CN \xrightarrow{CH_3MgBr\ /\ H_2O^+} A \xrightarrow{LiAlH_4} B$

(ii) Toluene ($C_6H_5CH_3$) $\xrightarrow[(ii)\ H_3O^+\ /\ \Delta]{(i)\ CrO_3\ /\ (CH_3CO)_2O} A \xrightarrow{H_2N-NH_2} B$

**Ans. (i)**

$CH_3CH_2CN$ (Propanenitrile) $\xrightarrow{CH_3MgBr/H_3O^+}$ $CH_3CH_2-C(=O)-CH_3$ Butan-2-one (A) $\xrightarrow{LiAlH_4}$ $CH_3CH_2-CH(OH)-CH_3$ Butan-2-ol (B)

(ii)

Toluene $\xrightarrow[(ii)\ H_3O^+/\ \Delta]{(i)\ CrO_3/(CH_3CO_2)O}$ Benzaldehyde ($C_6H_5CHO$) (A)

$\xrightarrow{H_2N-NH_2}$ $C_6H_5-HC=N-NH_2$ Benzaldehyde hydrazone (B)

## SECTION-C

**17. How will you convert the following :**

**(i) Impure nickel to pure nickel**

**(ii) Zinc blende to zinc metal**

**(iii) $[Ag(CN)_2]^-$ to Ag [3]**

**Ans. (i)** Impure nickel can be converted to pure nickel by Mond's process. In this process, Nickel is heated in a stream of carbon monoxide forming a volatile complex, nickel tetracarbonyl

$$Ni + 4CO \xrightarrow{330\text{-}350\ K} Ni(CO)_4$$

The carbonyl is subjected to higher temperature so that it is decomposed giving the pure metal.

**(ii)** Zinc can be obtained from zinc blende by roasting ZnS to give ZnO which is heated with reducing agent coke to give Zn.

$$\underset{\text{Zinc Blende}}{2ZnS} + \underset{\text{From air}}{O_2} \xrightarrow{\text{Roasting}} 2\,ZnO + 2SO_2$$

$$\underset{\text{Zinc oxide}}{ZnO} + C \xrightarrow{\text{Reduction}} Zn + CO_2$$

**(iii)** $[Ag(CN)_2]^{2-}$ can be converted into Ag by treating with zinc (a more electropositive element than silver). Hence, on treatment with zinc metal, Zn oxidizes to furnish $Zn^{2+}$ ions, which go into solution replacing the $Ag^+$ ions, which are in turn reduced to metallic silver and deposited in pure form.

$$2[Ag(CN)_2]^{2-}(aq) + Zn(s) \rightarrow [Zn(CN)_4]^{2-}(aq) + 2Ag(s)$$

**18. Give reasons for the following :**

**(i) The transition metals generally form coloured compounds.**

**(ii) E° value for $(Mn^{3+}/Mn^{2+})$ is highly positive than that for $(Cr^{3+}/Cr^{2+})$ couple.**

**(iii) The chemistry of actinoids elements is not so smooth as that of the lanthanoids. [3]**

**Ans. (i)** Transition metals have partly filled *d*-orbitals. So, the single electrons available in *d*-orbitals absorb energy and go to higher unoccupied electronic energy levels due to *d–d* transitions. When they come back, they emit energy in visible range and hence impart colour.

(ii) E° value for $Mn^{3+}/Mn^{2+}$ is highly positive because $Mn^{2+}$ has a stable $d^5$ configuration and it is reluctant to lose one electron to achieve the 3+ state. $Cr^{2+}$ has $3d^4$ configuration and losing another electron to achieve $3d^3$ configuration is not that difficult, hence the E° value is not more positive compared to $Mn^{3+}/Mn^{2+}$.

(iii) Chemistry of actinoid elements is not so smooth in view of their ability to exist in different oxidation states. Also, many of the actinoid elements are radioactive which makes the study of these elements difficult and the energy difference between 5*f*, 6*d* and 7*s* orbitals is very less as compared to lanthanoids.

**22. Write equations of the following reactions :**

**(i) Acetylation of aniline**

**(ii) Coupling reaction**

**(iii) Carbyl amine reaction** [3]

**Ans. (i)** Acetylation of aniline :

$$C_6H_5NH_2 \xrightarrow{(CH_3CO_2)_2} C_6H_5NHCOCH_3$$

Aniline → Acetanilide

**(ii) Coupling reaction :** Benzene diazonium chloride reacts with other suitable aromatic compounds to give azo compounds. This reaction is known as coupling reaction.

$$C_6H_5N^+\equiv NCl^- + H-C_6H_4-OH \xrightarrow{OH^-} C_6H_5-N=N-C_6H_4-OH + Cl^- + H_2O$$

Benzene diazonium chloride + Phenol → *p*-Hydroxyazobenzene (orange dye)

**(iii) Carbylamine reaction :** Aliphatic and aromatic primary amines on heating with chloroform and ethanolic potassium hydroxide form isocyanides or carbylamines.

$$C_2H_5NH_2 + CHCl_3 + 3KOH \longrightarrow C_2H_5NC + 3KCl + 3H_2O$$

Ethylamine, Chloroform, Potassium hydroxide → Ethyl isonitrile

**24. Define the following with a suitable example in each :**

**(i) Oligosaccharides**

**(ii) Denaturation of protein**

**(iii) Vitamins** [3]

**OR**

**Write the reactions involved when D-glucose is treated with the following reagents :**

**(i) $Br_2$ water**

**(ii) $H_2N–OH$**

**(iii) $(CH_3CO)_2O$**

**Ans. (i) Oligosaccharide :** Carbohydrates that yield two to ten monosaccharide units on hydrolysis are known as oligosaccharides. They are further classified as disaccharides, trisaccharides, tertrasaccharides etc., depending upon the number of monosaccharides they provide upon hydrolysis. Disaccharides are the most common, *e.g.*, maltose, sucrose.

**(ii) Denaturation of protein :** Protein found in a biological system with a unique three-dimensional structure and biological activity is called a native protein. When a native protein in its native form is subjected to physical change like temperature, chemical or pH change, the hydrogen bonds within the protein structure are disturbed causing globules to unfold and α-helix to uncoiled. The secondary and tertiary structures of protein molecule is destroyed, in turn losing their biological activity, this is known as denaturation of protein. Example is coagulation of egg white on boiling.

**(iii) Vitamins :** These are the organic compounds required in the diet in small amounts to perform specific biological functions for normal maintenance of optimum growth and health of organisms. Example – Vitamins A, B, C, D etc.

**OR**

(i) Reaction of glucose with $Br_2$ water (a mild oxidising agent) gives gluconic acid, this reaction indicates that carbonyl group of glucose is an aldehydic group :

$$\underset{\text{Glucose}}{\begin{array}{c} CHO \\ | \\ (CHOH)_4 \\ | \\ CH_2OH \end{array}} \xrightarrow{Br_2 \text{ water}} \underset{\text{Gluconic acid}}{\begin{array}{c} COOH \\ | \\ (CHOH)_4 \\ | \\ CH_2OH \end{array}}$$

**(ii)** Reaction of glucose with $H_2N—OH$ (hydroxylamine) gives oxime, this reaction confirms that there is a carbonyl group present in glucose :

$$\underset{\text{Glucose}}{\begin{array}{c} CHO \\ | \\ (CHOH)_4 \\ | \\ CH_2OH \end{array}} \xrightarrow{H_2N–OH} \underset{\text{Glucose oxime}}{\begin{array}{c} CH{=}NOH \\ | \\ (CHOH)_4 \\ | \\ CH_2OH \end{array}}$$

**(iii)** Reaction of glucose with $(CH_3CO)_2O$ (acetic anhydride) gives glucose pentaacetate, this reaction confirms that there are five hydroxyl groups present in glucose :

$$\underset{\text{Glucose}}{\begin{array}{c} CHO \\ | \\ (CHOH)_4 \\ | \\ CH_2OH \end{array}} \xrightarrow{(CH_3CO)_2O} \underset{\text{Glucose pentaacetate}}{\begin{array}{l} CHO \\ | \\ (CH—O—\overset{O}{\overset{||}{C}}—CH_3)_4 \\ | \\ CH_2—O—\overset{O}{\overset{||}{C}}—CH_3 \end{array}}$$

●●

# CHEMISTRY

# 2018

**Time allowed : 3 hours** **Maximum marks : 70**

1. **Analysis shows that FeO has a non-stoichiometric composition with formula $Fe_{0.95}O$. Give reason.** **[1]**

**Ans.** FeO has a non-stoichiometric composition with formula $Fe_{0.95}O$ because it shows metal deficiency defect and it is a mixture of $Fe^{2+}$ and $Fe^{3+}$ ions and some $Fe^{2+}$ ions are oxidised to $Fe^{3+}$ ions.

2. **CO(g) and $H_2$(g) react to give different products in the presence of different catalysts. Which ability of the catalyst is shown by these reactions?** **[1]**

**Ans.** CO(g) and $H_2$(g) react in presence of different catalysts to give different products, this shows that action of a catalyst is highly selective in nature.

3. **Write the coordination number and oxidation state of Platinum in the complex $[Pt(en)_2 Cl_2]$.** **[1]**

**Ans.** Coordination number : 6; Oxidation state : +2

4. **Out of chlorobenzene and benzyl chloride, which one gets easily hydrolysed by aqueous NaOH and why ?** **[1]**

**Ans.** Benzyl chloride would be easily hydrolysed compared to chlorobenzene. In the given reaction condition, hydrolysis proceeds by nucleophilic substitution mechanism and the benzylic carbonium ion formed after losing the leaving group (—Cl) is better stabilized (through resonating structures) hence reacts easily.

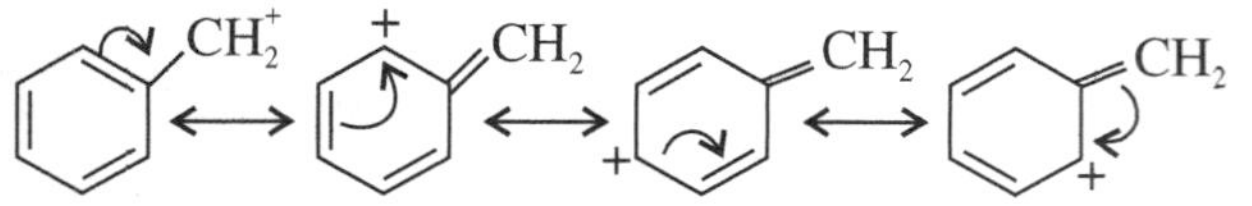

$C_6H_5\overset{+}{C}H_2 \longleftrightarrow \longleftrightarrow \longleftrightarrow$ (resonance structures of benzyl carbocation, with $CH_2$ and + charge on ring)

5. **Write the IUPAC name of the following :** **[1]**

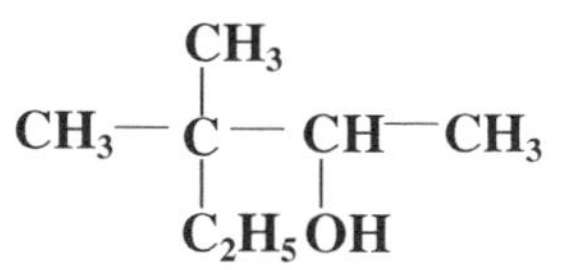

$$CH_3-\underset{C_2H_5}{\overset{CH_3}{\overset{|}{\underset{|}{C}}}}-\underset{OH}{\underset{|}{CH}}-CH_3$$

**Ans.** The IUPAC name would be 3, 3- Dimethy1-pentan-2-o1.

6. **Calculate the freezing point of a solution containing 60 g of glucose (Molar mass = 180 g $mol^{-1}$) in 250 g of water.**
**($K_f$ of water = 1.86 K kg $mol^{-1}$)** **[2]**

**Ans.** Molality ($m$) of given solution of Glucose :

$$m = [(60/180) \text{ g mol}^{-1}/250 \text{ g}] \times 1000$$
$$= 1.33 \text{ mol kg}^{-1}$$

Now, depression in freezing point is given by, $\Delta T_f = K_f m$

Putting the given values,

$$\Delta T_f = K_f m$$
$$= 1.86 \times 1.33$$
$$= 2.5$$

So, freezing point of the solution would be 273.15 K – 2.5 K = 270.65 K.

7. **For the reaction** **[2]**

$$2N_2O_5(g) \rightarrow 4NO_2(g)+O_2(g)$$

**the rate of formation of $NO_2$(g) is $2.8 \times 10^{-3}$ $Ms^{-1}$. Calculate the rate of disappearance of $N_2O_5$(g).**

**Ans.** Rate of reaction for the given reaction can be given as,

$$\text{Rate} = 1/2\ \{-\Delta[N_2O_5]/\Delta t\}$$

or $\{-\Delta[N_2O_5]/\Delta t\} = 1/2\ \{[NO_2/\ \Delta t]\}$

So, rate of disappearance of $N_2O_5$ would be half of rate of production of $NO_2$ (given $2.8 \times 10^{-3} Ms^{-1}$). So, the rate of disappearance of $N_2O_5$ is $1.4 \times 10^{-3}$ $Ms^{-1}$.

8. **Among the hydrides of Group-15 elements, which have the**
   (a) **lowest boiling point ?** **[2]**
   (b) **maximum basic character ?**
   (c) **highest bond angle ?**
   (d) **maximum reducing character ?**

**Ans.** Hydrides of group 15 element are : $NH_3$, $PH_3$, $AsH_3$, $SbH_3$, $BiH_3$

(a) Lowest boiling point is of $PH_3$ (185.5 K)
(b) Maximum basic character is shown by $NH_3$
(c) Highest bond angle is for $NH_3$.
(d) $BiH_3$ has the maximum reducing character.

9. **How do you convert the following ?** **[2]**
   (a) **Ethanal to Propanone**
   (b) **Toluene to Benzoic acid**

**OR**

**Account for the following :**
(a) **Aromatic carboxylic acids do not undergo Friedel- Crafts reaction.**

**(b) $pK_a$ value of 4-nitrobenzoic acid is lower than that of benzoic acid.**

**Ans. (a)** Conversion of ethanal to Propanone :

$$H_3C-CHO \xrightarrow[\text{Grignard reaction}]{\text{(i) } CH_3MgBr \text{ (ii) } H_2O} H_3C-CH(OH)-CH_3$$

Ethanal → Propan-2-ol

$$\xrightarrow[\text{Oxidation}]{CrO_3} H_3C-CO-CH_3$$

Propanone

**(b) Conversion of toluene to benzoic acid :**

$$C_6H_5CH_3 \xrightarrow[\text{Heat}]{KMnO_4,\ KOH} C_6H_5COOK \xrightarrow{H_3O^+} C_6H_5COOH$$

Toluene → Benzoic acid

OR

**(a)** Aromatic carboxylic acids do not undergo Friedel-Crafts reaction because the carboxyl group is deactivating for electrophilic substitution reaction, secondarily, the catalyst aluminium chloride gets bonded to the carboxyl group.

**(b)** $pK_a$ value of 4-Nitrobenzoic acid is lower than benzoic acid, which means 4-Nitrobenzoic acid is more acidic than the benzoic acid. Being an electron withdrawing group, the—$NO_2$ group withdraws electrons towards itself resulting in ease of carboxylic proton release, hence increasing the acidity.

$$O_2N-C_6H_4-COOH$$

**10. Complete and balance the following chemical equations :**

**(a) $Fe^{2+} + MnO_4^- + H^+ \rightarrow$**

**(b) $MnO_4^- + H_2O + I^- \rightarrow$** [2]

**Ans.**

**(a)** $5\,Fe^{2+} + MnO_4^- + 8H^+ \rightarrow Mn^{2+} + 4H_2O + 5Fe^{3+}$

**(b)** $2MnO_4^- + H_2O + I^- \rightarrow 2MnO_2 + 2\,OH^- + IO_3^-$

**11. Give reasons for the following :** [3]

**(a) Measurement of osmotic pressure method is preferred for the determination of molar masses of macromolecules such as proteins and polymers.**

**(b) Aquatic animals are more comfortable in cold water than in warm water.**

**(c) Elevation of boiling point of 1 M KCl solution is nearly double than that of 1 M sugar solution.**

**Ans. (a)** Molar masses of macromolecules like polymers and proteins are measured through osmotic pressure method. The osmotic pressure method uses 'molarity' of solution (instead of molality) which has a large magnitude even for dilute solutions, given that polymers have poor solubility, osmotic pressure measurement is used for determination of their molar masses. Macromolecules such as proteins are not stable at high temperatures and because measurement of osmotic pressure is done at around room temperature, it is useful for determination of molar masses of proteins.

**(b)** Solubility of gases in liquid decreases on increasing the temperature. Hence, the availability of dissolved oxygen in water is more at lower temperatures hence, the aquatic animals feel more comfortable at lower temperatures than at the higher temperatures because oxygen is more soluble in cold water or at low temperature.

**(c)** Elevation of boiling point is a colligative property and hence depends on the number of solute particles in the solution. Now, 1 M KCl would have twice the number of solute particles, as KCl dissociates into K+ and Cl–, compared to sugar solution (as sugar does not undergo any dissociation). So, elevation of boiling point is nearly double for 1 M KCl solution compared to 1 M sugar solution.

**12. An element 'X' (At. mass = 40 g $mol^{-1}$) having f.c.c. structure, has unit cell edge length of 400 pm. Calculate the density of 'X' and the number of unit cells in 4 g of 'X'. ($N_A = 6.022 \times 10^{23}\ mol^{-1}$)** [3]

**Ans.** Given: Atomic mass of element (M) = 40 g $mol^{-1}$ Length of unit cell ($a$) = 400 pm $= 4 \times 10^{-8}$ cm; Z = 4 (*fcc*).

Now density is given by formula,

Density ($d$) = $(Z \times M)/N_A \times V$

Volume of the unit cell, $V = a^3 = (4 \times 10^{-8})^3$ cm $= 64 \times 10^{-24}$

Putting the values-

$$\text{Density } (\mathbf{d}) = \frac{(4\times40)}{\left[(6.022\times10^{23})\times64\times10^{-24}\right]}$$
$$= 160 / 38.5$$
$$= 4.1 \text{ g cm}^{-3}$$

$$\text{Number of unit cells} = \frac{\text{Total Number of atoms}}{4}$$

$$= \frac{\frac{4}{40}\times6.022\times10^{23}}{4}$$

Number of unit cells = $1.5 \times 10^{22}$

**13. A first order reaction is 50% completed in 40 minutes at 300 K and in 20 minutes at 320 K. Calculate the activation energy of the reaction. (Given : log 2 = 0.3010, log 4 = 0.6021, R = 8.314 JK$^{-1}$ mol$^{-1}$)** **[3]**

**Ans.** Rate constant for a first order reaction is given by,

Number of unit cells = $1.5 \times 10^{22}$

$$k = \frac{2.303}{t}\log\frac{[R_0]}{[R_1]}$$

So, at 300 K,

$$k_{300} = \frac{2.303}{40}\log\left(\frac{100}{50}\right)$$
$$= 0.058 \times \log 2$$
$$= 0.058 \times 0.301$$
$$= 0.017$$

$$k_{320} = \frac{2.303}{20}\log\left(\frac{100}{50}\right)$$
$$= 0.11 \times \log 2$$
$$= 0.11 \times 0.3010 = 0.034$$

$$\text{Now, } \log\frac{k_{320}}{k_{300}} = \left(\frac{E_a}{2.303R}\right)\left[\frac{T_2 - T_1}{T_1 T_2}\right]$$

Putting the values,

$$\log\frac{0.034}{0.017} = \left(\frac{E_a}{2.303\times8.314\text{JK}^{-1}\text{mol}^{-1}}\right)\left[\frac{320-300}{320\times200}\right]\text{K}$$

$$0.3010 = \frac{Ea}{19.14}\times\left[\frac{20}{96000}\right]$$

$$\frac{0.3010\times19.14\times96000}{20} = E_a$$

$$E_a = 27663.8 \text{ J/mol or}$$
$$E_a = 27.66 \text{ KJ/mol}$$

**14. What happens when** **[3]**

**(a) a freshly prepared precipitate of $Fe(OH)_3$ is shaken with a small amount of $FeCl_3$ solution ?**

**(b) persistent dialysis of a colloidal solution is carried out ?**

**(c) an emulsion is centrifuged ?**

**Ans. (a)** When $FeCl_3$ is added to a freshly prepared precipitate of $Fe(OH)^3$, a positively charged sol of hydrated ferric oxide is formed due to adsorption of $Fe^{3+}$ ions.

**(b)** When persistent dialysis of colloidal solution is carried out, traces of electrolytes present in the sol are removed almost completely leaving the colloids unstable and finally coagulation takes place.

**(c)** Emulsions are centrifuged to carry out the process of Demulsification; to separate them into constituent liquids.

**15. Write the chemical reactions involved in the process of extraction of Gold. Explain the role of dilute NaCN and Zn in this process.** **[3]**

**Ans.** Extraction of gold involves leaching the metal with dilute solution of NaCN or KCN in the presence of air (for $O_2$) from which the metal is obtained later by replacement method (using Zinc).

The reactions involved are :

$$4Au(s) + 8CN^-(aq) + 2\,H_2O(aq) + O_2(g) \rightarrow 4[Au(CN)_2]^-(aq) + 4OH^-(aq)$$

$$2\,[Au(CN)_2]^-(aq) + Zn(s) \rightarrow 2Au\,(s) + [Zn(CN)_4]^{2-}(aq)$$

NaCN leaches gold and acts as the leaching agent in the process while Zn acts as reducing agent it displaces gold in the process of extraction of gold.

**16. Give reasons :**

**(a) E$^0$ value for $Mn^{3+}/Mn^{2+}$ couple is much more positive than that for Fe $^{3+}/Fe^{2+}$.**

**(b) Iron has higher enthalpy of atomization than that of copper .**

**(c) $Sc^{3+}$ is colourless in aqueous solution whereas $Ti^{3+}$ is coloured.** **[3]**

**Ans. (a)** $Mn^{2+}$ has a $d^5$ configuration, and the extra stability of half filled *d*-orbitals is compromised when another electron is taken out to give $Mn^{3+}$, On the contrary $Fe^{3+}$

attains a half filled orbital configuration when $Fe^{2+}$ gets oxidized to $Fe^{3+}$. Hence, the E° value for $Mn^{3+}/Mn^{2+}$ couple has more positive E° value.

**(b)** Fe has a $3d^6 4s^2$ outer electronic configuration whereas Cu has $3d^{10}\ 4s^1$ configuration. Now, more the number of unpaired electrons in *d*-orbital, more favourable are interatomic attractions and thus higher atomization enthalpies. Hence, Fe having 4 unpaired *d*-electrons has more enthalpy of atomization than copper having no unpaired d-electron.

**(c)** $Sc^{3+}$ has a $3d^0$ configuration whereas $Ti^{3+}$ has a $3d^1$ configuration there is 1 unpaired electron and it undergoes *d-d* transitions. As there are no electrons in *d* orbital for $Sc^{3+}$ ion, there is no transition of electrons by absorption of energy and hence no emission in visible range imparting colour to the $Sc^{3+}$ ion.

**17. (a) Identify the chiral molecule in the following pair : [3]**

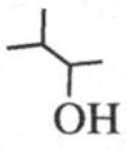

OH (i) and OH (ii)

**(b) Write the structure of the product when chlorobenzene is treated with methyl chloride in the presence of sodium metal and dry ether.**

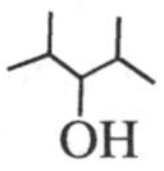

**(c) Write the structure of the alkene formed by dehydrohalogenation of 1-bromo-1 methylcyclohexane with alcoholic KOH.**

**Ans. (a)** The molecule **(i)** is a chiral molecule.

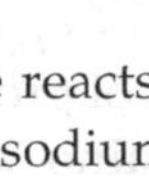

OH

**(b)** Chlorobenzene reacts with methyl chloride in presence of sodium metal and dry ether to give toluene. This reaction is known as Wurtz-Fittig reaction.

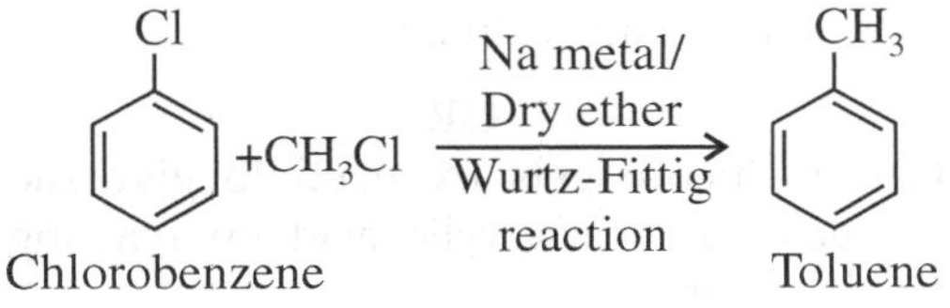

Cl (Chlorobenzene) $+CH_3Cl$ —Na metal/ Dry ether, Wurtz-Fittig reaction→ $CH_3$ (Toluene)

**(c)** In the 1-bromo-1-methylcyclohexane, all β-hydrogen atoms are equivalent. Thus dehydrohalogenation takes place, in the reaction of this compound with KOH.

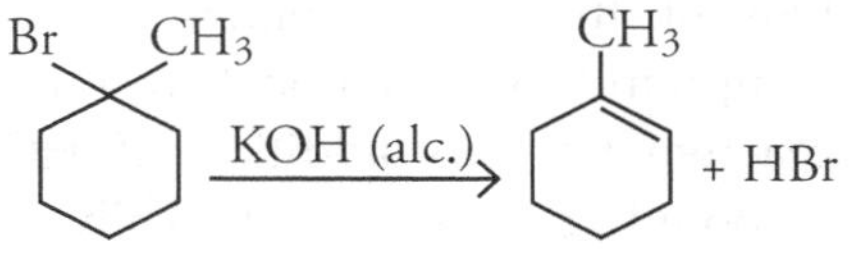

Br $CH_3$ —KOH (alc.)→ $CH_3$ + HBr

**18. (A), (B) and (C) are three non-cyclic functional isomers of a carbonyl compound with molecular formula $C_4H_8O$. Isomers (A) and (C) give positive Tollen's test whereas isomer (B) does not give Tollen's test but gives positive Iodoform test. Isomers (A) and (B) on reduction with Zn (Hg)/conc. HCl give the same product (D).**

**(a) Write the structures of (A), (B), (C) and (D).**

**(b) Out of (A), (B) and (C) isomers, which one is least reactive towards addition of HCN ? [3]**

**Ans. (a)** Compound A and C give positive Tollen's test which indicates that they are aldehydes. Compound C gives Iodoform test which means it contains a carbonyl group with a methyl group attached to the carbonyl carbon so, with formula $C_4H_8O$ the structure of compound would be $CH_3COCH_2CH_3$ (Butanone). Now upon reduction with Zn(Hg)/conc. HCl, the corresponding alkanes are obtained, so reduction of B gives Butane (D), so the isomer A have to be a linear chain aldehyde (Butanal), giving Butane (compound D) on reduction. So, the last isomer possible is compound C, 2-Methyl propanaldehyde. The reactions involved are shown below with the structures of compounds:

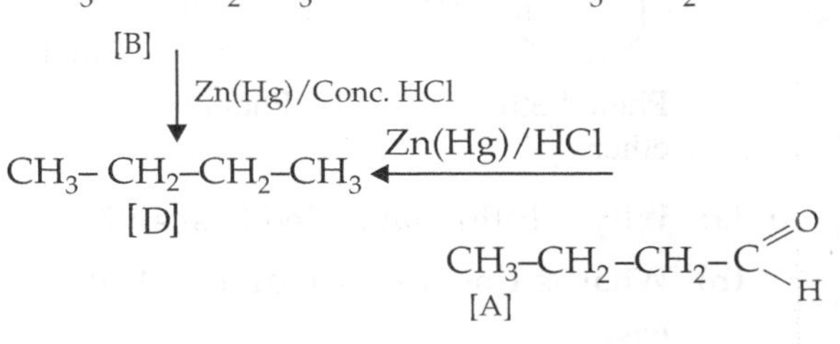

$CH_3-\overset{O}{\overset{\|}{C}}-CH_2CH_3$ [B] $\xrightarrow{NaOH, I_2}$ $CH_3-CH_2COONa$

[B] $\xrightarrow{Zn(Hg)/Conc.\ HCl}$ $CH_3-CH_2-CH_2-CH_3$ [D]

$CH_3-CH_2-CH_2-CH_3$ [D] $\xleftarrow{Zn(Hg)/HCl}$ $CH_3-CH_2-CH_2-C(=O)H$ [A]

isomer of [A] is compound [C] whose structure is

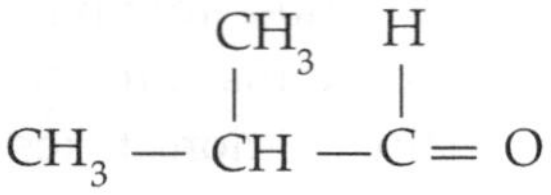

$$\begin{array}{c} CH_3 \quad H \\ | \qquad | \\ CH_3 - CH - C = O \end{array}$$

[C]

(b) Out of the three isomers A, B and C, compound B (Butanone) would be least reactive towards addition of HCl as the carbonyl carbon is sterically hindered and most reactive would be compound A (Butanal) towards addition of HCN.

**19. Write the structures of the main products in the following reactions : [3]**

(i) (cyclohexanone ring)–$CH_2-\underset{\|}{\overset{}{C}}(O)-OCH_3$ $\xrightarrow{NaBH_4}$

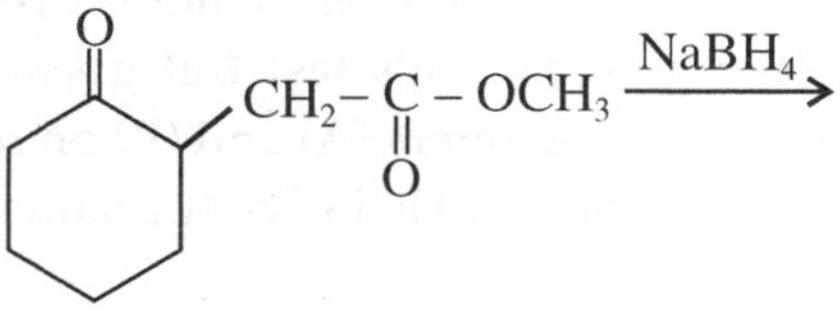

(ii) (benzene ring)–$CH=CH_2$ $+ H_2O \xrightarrow{H^+}$

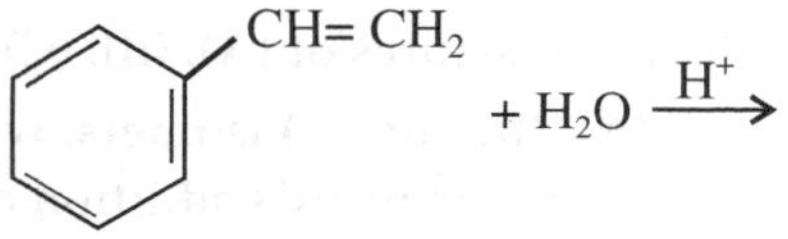

(iii) (benzene ring)–$OC_2H_5$ $+ HI \longrightarrow$

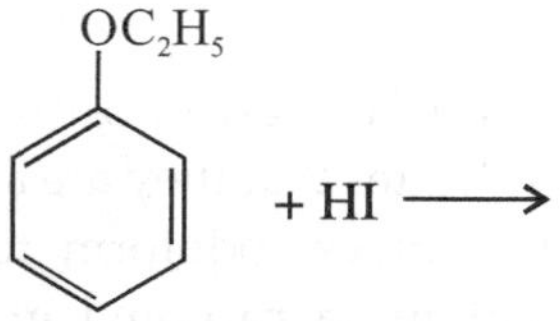

**Ans. (i)** Sodium borohydride doesn't reduce esters, so product would be,

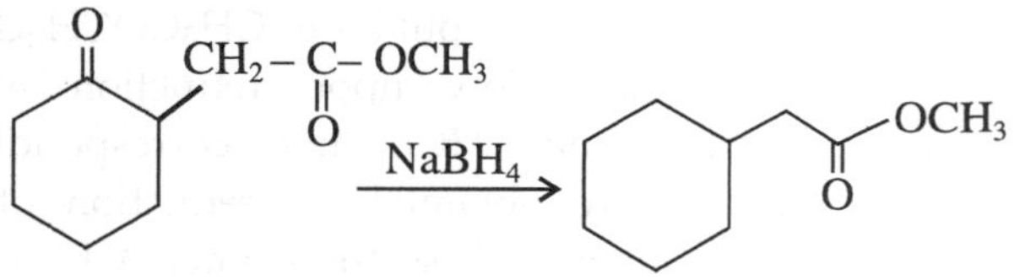

(cyclohexanone ring)–$CH_2-C(=O)-OCH_3$ $\xrightarrow{NaBH_4}$ (cyclohexane ring)–$CH_2-C(=O)-OCH_3$

(ii) (benzene ring)–$CH = CH_2$ $+H_2O \xrightarrow{H^+}$ (benzene ring)–$CH(OH)-CH_3$

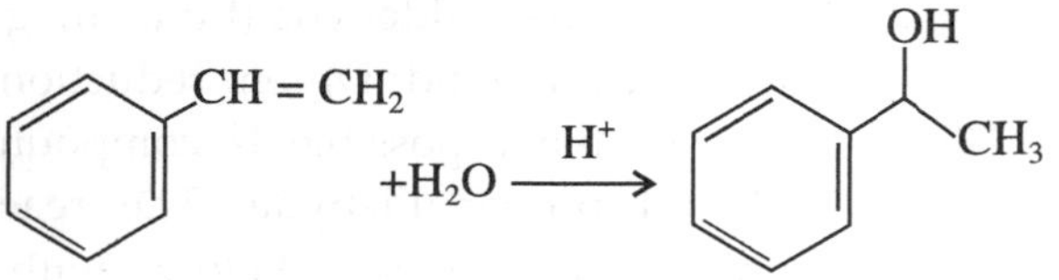

(iii) (benzene ring)–$O-C_2H_5$ $+ HI \longrightarrow$ (benzene ring)–$OH$ $+ C_2H_5I$

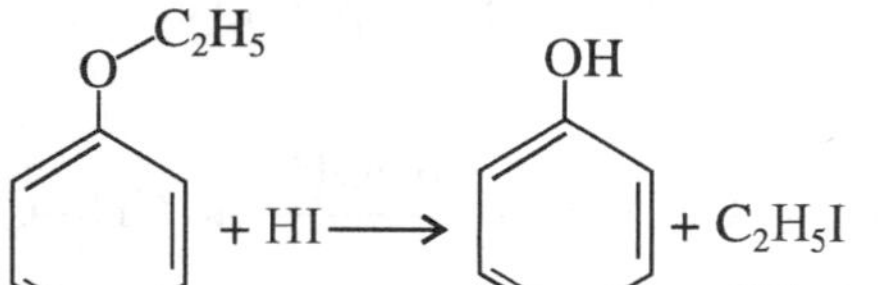

Phenyl ethyl ether — Phenol — Ethyl iodide

**20. (a) Why is bithional added to soap ? [3]**

**(b) What is tincture of iodine ? Write its one use.**

**(c) Among the following, which one acts as a food preservative ?**

**Aspartame, Aspirin, Sodium Benzoate, Paracetamol**

**Ans. (a)** Bithional is added to soaps to impart antiseptic properties to soap.

**(b)** Tincture of iodine is 2-3 percent mixture of iodine in alcohol water mixture. It is used as an antiseptic.

**(c)** Sodium benzoate is used as a food preservative.

**21. Define the following with an example of each : [3]**

**(a) Polysaccharides**

**(b) Denatured protein**

**(c) Essential amino acids**

**OR**

**(a) Write the product when D-glucose reacts with conc. $HNO_3$.**

**(b) Amino acids show amphoteric behaviour. Why ?**

**(c) Write one difference between $\alpha$–helix and $\beta$-pleated structures of proteins.**

**Ans. (a) Polysaccharides :** Polysaccharides are food storage materials and most commonly found carbohydrates in nature. These are the compound which are formed of large number of monosaccharide units joined together by glycosidic linkages. Example. Starch, main storage polysaccharide of plants.

**(b) Denatured protein :** Proteins have an unique three dimensional structure in their native form. If the native form of protein is subjected to any physical change (such as temperature change) or any chemical change (such as change in pH), the hydrogen bonds are disturbed. Due to this, globules unfold and helix get uncoiled due to which protein loses its biological activity. This is called denaturation of protein. During denaturation 2° and 3° structures of proteins are destroyed but 1° structure remains intact. Coagulation of egg white is an example of denaturation of protein.

**(c) Essential amino acids :** The amino acids which are not synthesized in our body and have to be obtained through diet are known as essential amino acids. Example : Tryptophan, Valine etc.

**OR**

**(a)** D-Glucose gets oxidized to give saccharic acid, a dicarboxylic acid on reacting with nitric acid.

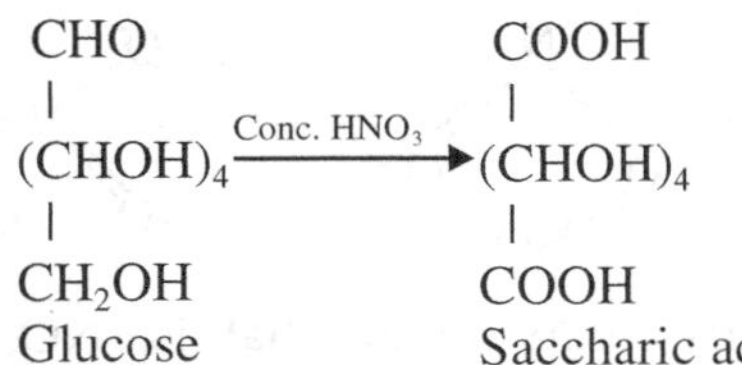


**(b)** Amino acids show amphoteric behaviour due to the presence of both acidic (carboxylic group) and basic (amino group) in the same molecule. So, in basic medium the carboxyl group can lose a proton and in acidic medium amino group can accept a proton.

**(c)** In $\alpha$-helix structure the polypeptide chain forms all possible hydrogen bonds by twisting into a right handed screw (helix) with the —NH group of each amino acid residue gets hydrogen bonded to the —C = O of an adjacent turn of the helix (intramolecular bonding), whereas in $\beta$-structure all peptide chains are stretched out to nearly maximum extension and then laid side by side which are held together by intermolecular hydrogen bonds intramolecular bonding).

**22. (a) Write the formula of the following coordination compound:**

**Iron (III) hexacyanoferrate (II)**

**(b) What type of isomerism is exhibited by the complex $[Co(NH_3)_5 Cl]SO_4$ ?**

**(c) Write the hybridisation and number of unpaired electrons in the complex $[CoF_6]^3$. (Atomic number of Co =27)** **[3]**

**Ans. (a)** Molecular formula of Iron(III) $\alpha-$cyanoferrate(II) is $Fe_4[Fe(CN)_6]_3$

**(b)** $[Co(NH_3)_5Cl]SO_4$ will show Ionisation isomerism and the possible isomers are $[Co(NH_3)_5Cl]SO_4$ and $[Co\ (NH_3)_5SO_4]Cl$

**(c)** Electronic configuration of $Co^{3+}$ ion is,

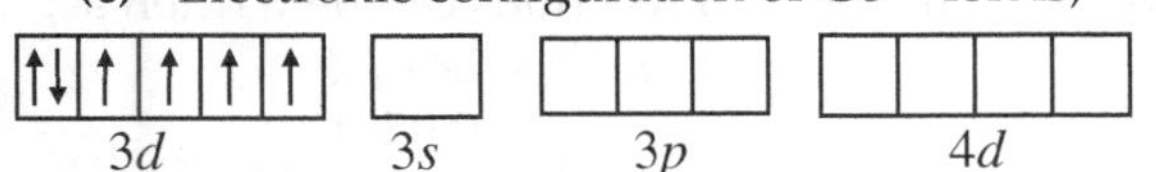


Electronic configuration of $sp^3d^2$ hybridized (as $F^-$ is a weak field ligand) orbitals of $Co^{3+}$, with six pairs of electrons from six $F^-$ ions.

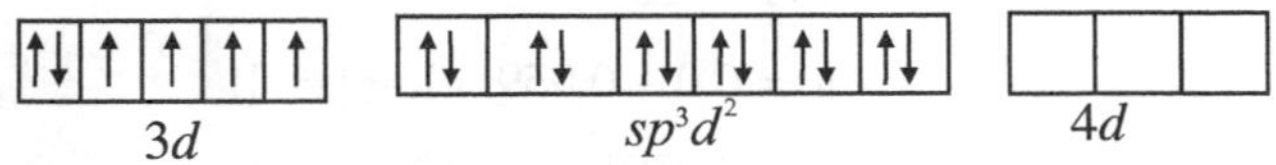


There are 4 unpaired electrons in $[CoF_6]^3$.

** Answer is not given due to the change in present syllabus.

**23. Shyam went to a grocery shop to purchase some food items. The shopkeeper packed all the items in polythene bags and gave them to Shyam. But Shyam refused to accept the polythene bags and asked the shopkeeper to pack the items in paper bags. He informed the shopkeeper about the heavy penalty imposed by the government for using polythene bags. The shopkeeper promised that he would use paper bags in future in place of polythene bags.** **[4]**

**Answer the following :**

**(a) Write the values (at least two) shown by Shyam.****

**(b) Write one structural difference between low-density polythene and high-density polythene.**

**(c) Why did Shyam refuse to accept the items in polythene bags ?**

**(d) What is a biodegradable polymer ? Give an example.**

**Ans. (b)** Low density polythene has branched chain structure, whereas the high density polythene has linear chain structure.

**(c)** Shyam refused to take the items in polythene bags as polythene is non-biodegradable neither recyclable,

**(d)** Biodegradable polymers contain functional groups similar to functional groups present in biopolymers, so they get degraded in environment by certain microorganisms and thus are environment friendly. For example: Poly $\beta$-hydroxybutyrate-co-$\beta$-hydroxyvalerate (PHBV).

**24. (a) Give reasons :** **[5]**

**(i) $H_3PO_3$ undergoes disproportionation reaction but $H_3PO_4$ does not.**

**(ii) When $Cl_2$ reacts with excess of $F_2$, $ClF_3$ is formed and not $FCl_3$.**

**(iii) Dioxygen is a gas while Sulphur is a solid at room temperature.**

**(b) Draw the structures of the following :**

**(i) $XeF_4$ (ii) $HClO_3$**

**OR**

**(a) When concentrated sulphuric acid was added to an unknown salt present in a test tube a brown gas (A) was evolved. This gas intensified when copper turnings were added to this test tube. On cooling, the gas (A) changed into a colourless solid (B).**

(i) **Identify (A) and (B).**

(ii) **Write the structures of (A) and (B).**

(iii) **Why does gas (A) change to solid on cooling ?**

(b) **Arrange the following in the decreasing order of their reducing character :**

**HF, HCl, HBr, HI**

(c) **Complete the following reaction :**

$$XeF_4 + SbF_5 \rightarrow$$

**Ans.** (a) (i) In $H_3PO_3$ (orthophosphorus acid) oxidation state of phosphorus is + 3 and it contains one P-H bond in addition to P=O and P–OH bonds. These type of oxoacids tend to undergo disproportionation to give orthophosphoric acid (P has +5 state) and phosphine (P has +3 state). Whereas in $H_3PO_4$ (orthophosphoric acid), Phosphorus is in +5 state hence no disproportionation takes place in $H_3PO_4$.

(ii) When $Cl_2$ reacts with excess of $F_2$, $ClF_3$ is formed and not $FCl_3$ because Fluorine can't expand its valency and can show only –1 oxidation state, whereas Cl can expand its valency due to the availability of *d*-orbitals.

(iii) Dioxygen is a gas while sulphur is a solid at room temperature this is because sulphur have $S_8$ molecules and these are packed to give different crystal structure, whereas dioxygen is a diatomic molecule ($O_2$) and it does not have enough intermolecular attraction and thus exists in gaseous form.

(b)

| (i) $XeF_4$ is square planar in structure | (ii) $HClO_3$ or chloric acid |
|---|---|
| F, F, Xe, F, F | H, O, Cl, O, O |

OR

(a) (i) The brown gas A is $NO_2$ or nitrogen dioxide. On cooling it dimerises to $N_2O_4$ and solidifies as a colourless solid.

$$2NO_2 \underset{\text{Heat}}{\overset{\text{Cool}}{\rightleftharpoons}} N_2O_4$$

(ii) 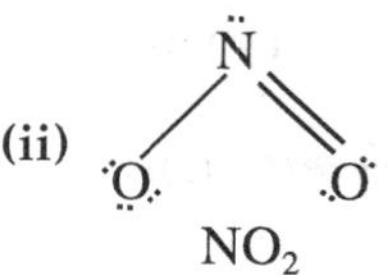

$NO_2$

O, N=N, O, O, O

$N_2O_4$

(iii) Compoud A, that is, $NO_2$ contains odd number of valence electrons. It behaves as a typical odd molecule. On dimerization, it is converted to stable $N_2O_4$ molecule with even number of electrons (thus colourless) and have better intermolecular forces to get solidified. Thus, it changes to solid on cooling.

(b) Decreasing order of reducing character-

HI > HBr > HCl > HF

(c) $XeF_4 + SbF_5 \rightarrow [XeF_3]^+ + [SbF_6]^-$

**25. (a) Write the cell reaction and calculate the e.m.f. of the following cell at 298 K : [5]**

**Sn(s) | $Sn^{2+}$ (0.004 M) | | $H^+$ (0.020 M) | $H_2(g)$ (1 bar) | Pt (s)**

**(Given : $E°_{Sn^{2+}/Sn} = -0.14V$)**

**(b) Give reasons :**

**(i) On the basis of E° values, $O_2$ gas should be liberated at anode but it is $Cl_2$ gas which is liberated in the electrolysis of aqueous NaCl.**

**(ii) Conductivity of $CH_3COOH$ decreases on dilution.**

**OR**

**(a) For the reaction**

**$2AgCl(s) + H_2(g)$ (1 atm) $\rightarrow 2Ag(s) + 2H^+$ (0.1M) $+ 2Cl^-$ (0.1M),**

**$\Delta G° = -43600$ J at 25°C.**

**Calculate the e.m.f. of the cell.**

**[ $\log 10^{-n} = -n$]**

**(b) Define fuel cell and write its two advantages.**

**Ans.** (a) The half cell reactions can be written as;

$$Sn^{2+} + 2e^- \rightarrow Sn$$

$$E_{cell} = -0.14 + \frac{0.0591}{2} (\log [Sn^{2+}])$$

$$= -0.11 \times (\log 0.004) \text{ V}$$

$$= 0.26 \text{ V} \quad .....(i)$$

$$2H^+ + 2e^- \rightarrow H_2$$

$$E_{cell} = 0.0 + 0.0591 (\log [H^+])$$

$$= 0.0591 \times (-1.7)$$

$$= -0.10V \quad .....(ii)$$

Considering,

$2H^+(aq) + Sn(s) \rightarrow Sn^{2+}(aq) + H_2(g)$, as the cell reaction

So, $E_{cell}$ will be,

$$E_{cell} = 0.26 \text{ V} - (-0.10) \text{ V} = 0.36 \text{ V}$$

**(b) (i)** During the electrolysis of aqueous NaCl, there are two possible reactions at anode,

$$Cl^- (aq) \rightarrow \frac{1}{2} Cl_2(g) + e^-; E_{cell} = 1.36 \text{ V}$$

$$2 H_2O(I) \rightarrow O_2 (g) + 4H^+ (aq) + 4e^-; E_{cell} = 1.23 \text{ V}$$

The reaction at anode with lower value of $E_{cell}$ is preferred and therefore, water should get oxidized to give $O_2$ but on account of over potential of oxygen, $Cl^-$ gets oxidized preferably, liberating $Cl_2$ gas.

**(ii)** Conductivity of $CH_3COOH$ decreases on dilution because the number of ions per unit volume that carry the current in a solution decreases on dilution.

**OR**

**(a)** $E^\circ_{cell}$ can be obtained from the formula,

$$\Delta G^\circ = -nE^\circ_{cell}$$

$$E^\circ_{cell} = \frac{\Delta G^\circ}{nF}$$

$$= \frac{-43600 \text{ J}}{2 \times 96487 \text{ C mol}^{-1}}$$

$$= 0.226 \text{ V}$$

Now, let us consider the given reaction equation,

$2AgCl(s) + H_2(g) (1 \text{ atm}) \rightarrow 2Ag(s) + 2H^+ (0.1 \text{ M}) + 2 Cl^- (0.1 \text{ M})$

or $AgCl(s) + ½ H_2 (g) (1 \text{ atm}) \rightarrow Ag (s) + H^+ (0.1 \text{ M}) + Cl^- (0.1 \text{ M})$

According to Nernst equation :

$$E_{cell} = E^\circ_{cell} - \frac{2.303 \text{ RT}}{nF} \log \frac{[\text{product}]}{[\text{reactant}]}$$

$$E_{cell} = E^\circ_{cell} - \frac{2.303 \text{ RT}}{nF} \log \frac{[H^+]^2[Cl^-]^2[Ag]}{[AgCl][H_2]^{1/2}}$$

As the activity of solid and $H_2$ gas at 1 atm is taken unity, the equation becomes,

$$E_{cell} = E^\circ_{cell} - \frac{2.303 \text{ RT}}{nF} \log ([H^+] [Cl^-]) \quad .......(i)$$

Now, putting the values in equation (i) above,

$$E_{cell} = 0.026 - (0.059)/2 \log ((0.1)^2 \times (0.1)^2)$$

$$= 0.026 - \frac{0.059}{2} \log(10^{-4})$$

$$\Rightarrow 0.026 + 0.118$$

$$E_{cell} = 0.344 \text{V}$$

So, EMF of the given cell is 0.344 V

**(b)** Galvanic cells that are designed to convert the energy of combustion of fuels like hydrogen, methane, methanol etc. directly into electrical energy are called fuel cells.

**Advantages of fuel cells are :**

1. Fuel cells produce electricity with an efficiency of about 70% compared to thermal plants whose efficiency is about 40%.
2. Fuel cells are pollution free.

**26. (a) Write the reactions involved in the following : [5]**

**(i) Hofmann bromamide degradation reaction**

**(ii) Diazotisation**

**(iii) Gabriel phthalimide synthesis**

**(b) Give reasons :**

**(i) $(CH_3)_2NH$ is more basic than $(CH_3)_3N$ in an aqueous solution.**

**(ii) Aromatic diazonium salts are more stable than aliphatic diazonium salts.**

**OR**

**(a) Write the structures of the main products of the following reactions :**

(i) $C_6H_5NH_2 \xrightarrow[\text{Pyridine}]{(CH_3CO)_2O}$

(ii) $C_6H_5SO_2Cl \xrightarrow{(CH_3)_2NH}$

(iii) $C_6H_5N_2^+Cl^- \xrightarrow{CH_3CH_2OH}$

**(b) Give a simple chemical test to distinguish between Aniline and N,N-dimethylaniline.**

**(c) Arrange the following in the increasing order of their $pK_b$ values :**

**$C_6H_5NH_2$, $C_2H_5NH_2$, $C_6H_5NHCH_3$**

**Ans.**

**(a) (i) Hofmann bromamide degradation reaction:** Acetamide can be considered for example. In this reaction, Acetamide ($CH_3CONH_2$) undergoes Hofmann degradation in presence of Bromine and NaOH to give Methanamine.

$CH_3CONH_2 + Br_2 + 4NaOH \rightarrow CH_3NH_2$

$Na_2CO_3 + 2NaBr + 2H_2O$

**(ii) Diazotisation:** The conversion of primary aromatic amines into diazonium salts is known as diasotisation.

$NH_2$

+$NaNO_2$ + 2 HCl $\xrightarrow{273–278\ K}$

Aniline

$N_2^+Cl^-$

+NaCl + 2 $H_2O$

Benzene diazonium chloride

**(iii) Gabriel phthalimide synthesis :** This reaction is used for the preparation of primary amines. Phthalimide on treatment with ethanolic potassium hydroxide forms potassium salt of phthalimide which on heating with alkyl halide followed by alkaline hydrolysis produces the corresponding primary amine.

Phthalimide —KOH→ $NK^+$ —RX→

N–R

N-Alkylophthalimide

NaOH (aq)

$RNH_2$ + ONa, ONa

Primary amine

**(b) (i)** $(CH_3)_2$ NH is more basic than $(CH_3)_3N$ in aqueous solutions because, in $(CH_3)_3N$ the lone pair of electrons on nitrogen atom is responsible for its basicity are quite hindered by the three methyl groups, hence are less available. Due to which it is less basic as compared to $(CH_3)_2NH$.

**(ii)** Aromatic diazonium salts are more stable than aliphatic diazonium salts because the positive charge on nitrogen atom is stabilized by the resonance with attached phenyl group.

**OR**

**(a)**

**(i)** $NH_2$ $\xrightarrow[\text{Pyridine}]{(CH_3CO)_2O}$ $NHCOCH_3$

Aniline N-Acetylanilne

**(ii)** $SO_2Cl$ $\xrightarrow{(CH_3)_2NH}$ $SO_2N(CH_3)_2$

Benzene sulphonyl chloride N-Dimethylbenzene sulphonamide

**(iii)** $N_2^+Cl^-$ $\xrightarrow{CH_3CH_2OH}$ Benzene $+N_2+CH_3CHO$ + HCl

Benzene diazonium chloride Benzene

**(b)** Aniline can be distinguished from N, N-dimethyl aniline by diazo coupling reaction. Aniline would react with benzene diazonium chloride to give a yellow dye, whereas N, N-dimethyl aniline won't undergo this reaction.

$N_2^+Cl^-$ + $NH_2$ $\xrightarrow{H^+}$

Benzene diazonium chloride Aniline

–N = N– –$NH_2$ + HO

*p*-Aminoazobenzen (yellow dye)

$N_2Cl^-$ + $H_3C$ $CH_3$ N ⟶ No reaction

Benzene diazonium chloride N, N-dimethylaniline

**(c)** Increasing order of $pK_b$ values is

$C_2H_5NH_2 < C_6H_5NHCH_3 < C_6H_5NH_2$

# CHEMISTRY

## Outside Delhi [Set-I] 2017

**Time allowed : 3 hours** **Maximum marks : 70**

1. **Write the formula of the compound of phosphorus which is obtained when conc. $HNO_3$ oxidises $P_4$.** **[1]**

**Ans.** $P_4 + 20HNO_3 \longrightarrow 4H_3PO_4 + 20NO_2 + 4H_2O$

2. **Write the IUPAC name of the following compound:** **[1]**

$$H_3C—\underset{CH_3}{\underset{|}{C}} = \underset{Br}{\underset{|}{C}}—CH_2—OH$$

**Ans.** $H_3\overset{4}{C}—\underset{CH_3}{\underset{|}{\overset{3}{C}}} = \underset{Br}{\underset{|}{\overset{2}{C}}}—\overset{1}{C}H_2—OH$

2-Bromo 3-methyl-but. 2-en-1-ol

3. **What is the effect of adding a catalyst on** **[1]**
   **(a) Activation energy ($E_a$), and**
   **(b) Gibbs energy ($\Delta G$) of a reaction ?**

**Ans.** On adding a catalyst
(a) Activation energy of the reaction decreases.
(b) Gibbs energy doesn't change.

4. **Out of [cyclohexene with X on carbon adjacent to double bond] and [cyclohexene with X on double-bond carbon], which is an example of allylic halide ?** **[1]**

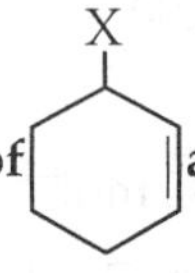

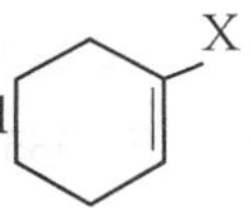

**Ans.** 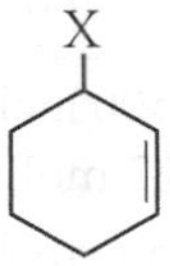 [cyclohexene with X on carbon adjacent to double bond] is an example of allylic halide

5. **What type of colloid is formed when a liquid is dispersed in a solid ? Give an example.** **[1]**

**Ans.** When a liquid is dispersed in solid, 'gel' colloid is formed. **Examples** Jelly, butter, cheese, curd etc.

6. **(a) Arrange the following compounds in the increasing order of their acid strength :** **[2]**
   ***p*-cresol, *p*-nitrophenol, phenol**
   **(b) Write the mechanism (using curved arrow notation) of the following reaction :**

$$CH_2 = CH_2 \xrightarrow{H_3O^+} CH_3 — CH_2^+ + H_2O$$

**OR**

**Write the structures of the products when Butan-2-ol reacts with the following :**
**(a) $CrO_3$** **(b) $SOCl_2$**

**Ans.**

(a) [p-cresol: benzene ring with $CH_3$ and OH] < [Phenol: benzene ring with OH] < [p-nitrophenol: benzene ring with $NO_2$ and OH]

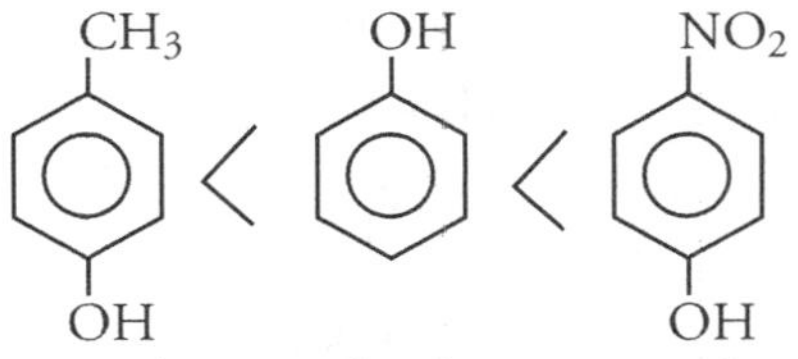

*p*-cresol < Phenol < *p*-nitrophenol

(b)

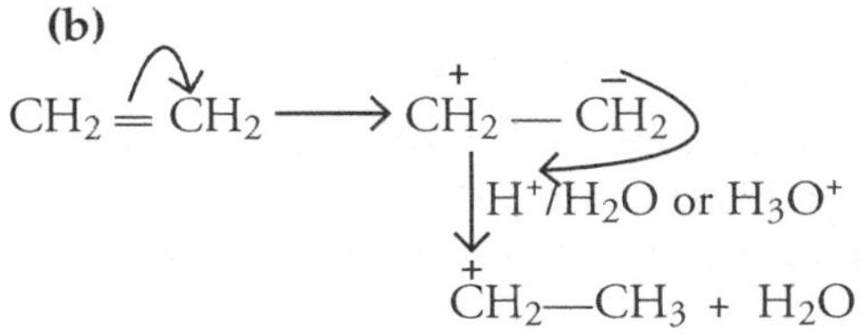

$$CH_2 = CH_2 \longrightarrow \overset{+}{C}H_2 — \overset{-}{C}H_2$$

$\downarrow$ $H^+/H_2O$ or $H_3O^+$

$$\overset{+}{C}H_2—CH_3 + H_2O$$

**OR**

(a)

$$CH_3—CH_2—\overset{OH}{\overset{|}{C}H}—CH_3 \xrightarrow{CrO_3}$$

Butane–2–ol

$$CH_3—CH_2—\overset{O}{\overset{||}{C}H}—CH_3$$

Butane-2-one

(b)

$$H_3C—CH_2—\overset{OH}{\overset{|}{C}H}—CH_3 \xrightarrow{SOCl_2}$$

Butane-2-ol

$$H_3C—CH_2—\overset{Cl}{\overset{|}{C}H}—CH_2$$

2-Chloro-butane

7. **Calculate the number of unit cells in 8.1g of aluminium if it crystallises in a face-centred cubic (f.c.c.) structure. (Atomic mass of Al= 27 g $mol^{-1}$).** **[2]**

**Ans.** 1 mol of Al = 27g $mol^{-1}$ = $6.022 \times 10^{23}$ atoms

No. of atoms present in 8.1g of Al

$$= \frac{6.02 \times 10^{23}}{27} \times 8.1$$

$$= 1.8069 \times 10^{23} \text{ atoms}$$

An fcc unit cell contains 4 atoms

$$\text{No. of unit cells present} = \frac{1.8069 \times 10^{23}}{4}$$

$= 0.4517 \times 10^{23}$

No. of unit cells $= 4.5 \times 10^{22}$

**8. Draw the structures of the following :**

**(a) $H_2SO_3$**

**(b) $HClO_3$** **[2]**

**Ans. (a)**

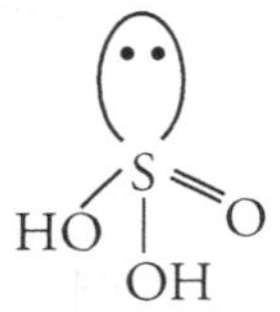

(Structure: S with lone pair, bonded to HO, OH and =O)

$H_2SO_3$

**(b)**

(Structure: Cl with lone pair, bonded to HO, =O and =O)

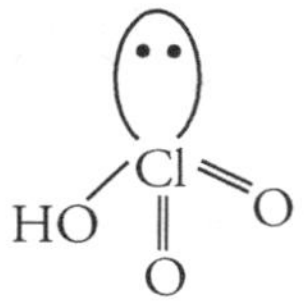

$HClO_3$

**9. Write the name of the cell which is generally used in hearing aids. Write the reactions taking place at the anode and the cathode of this cell.** **[2]**

**Ans.** Electrolytic cells are generally used is hearing aids. At cathode, reduction of metal takes place and at anode, oxidation of metal takes place.

Cathode : $M + e^- \longrightarrow M^+$

Anode : $M^+ \longrightarrow M + e^-$

**10. Using IUPAC norms write the formulae for the following :**

**(a) Sodium dicyanidoaurate (I)**

**(b) Tetraamminechloridonitrito-N-platinum (IV) sulphate** **[2]**

**Ans. (a)** Sodium dicyanoaurate (I)

$Na[Au(CN)_2]$

**(b)** Tetraammine chloridonitrito–N–platinum (IV) Sulphate $[Pt(NH_3)_4(Cl)(NO_2)]SO_4$

**11. (a) Based on the nature of intermolecular forces, classify the following solids :**

**Silicon carbide, Argon**

**(b) ZnO turns yellow on heating. Why ?**

**(c) What is meant by groups 12-16 compounds ? Give an example.** **[3]**

**Ans. (a) Silicon carbide** : covalent crystal or Network solid

**Argon.** Non-polar molecular solid

**(b)** When ZnO is heated, it loses oxygen as :

$$ZnO \xrightarrow{heat} Zn^{2+} + \frac{1}{2} O_2 + 2e^-$$

$Zn^{2+}$ are entrapped in the interstitial sites and electrons are entrapped in the neighbouring interstitial sites to maintain electrical neutrality. This results in metal excess defects and F-centres are created. Due to the presence of electrons in the interstitial voids, the colour is yellow.

**(c)** When a compound is formed between group-12 and group-16 elements, then it is known as group 12-16 compounds.

**For example :** $CO_2$, ZnS

$C \longrightarrow$ Group–12 element

$O \longrightarrow$ Group –16 element

**12. (a) The cell in which the following reaction occurs :**

$$2Fe^{3+}(aq) + 2I^-(aq) \longrightarrow 2Fe^{2+}(aq) + I_2(s)$$

**has $E^o_{cell} = 0.236$ V at 298 K. Calculate the standard Gibbs energy of the cell reaction. (Given : 1F = 96,500 C $mol^{-1}$)**

**(b) How many electrons flow through a metallic wire if a current of 0.5 A is passed for 2 hours? (Given : 1 F = 96,500 C $mol^{-1}$)** **[3]**

**Ans. (a)**
$$\Delta G^o = -nF E^o cell$$
$$= -2 \times 96500 \times 0.236$$
$$= -45.548 \text{ kJ/mol}$$

**(b)** According to Faraday's first law the amount of metal deposited (W).

$$W = i \times t$$
$$= 0.5 \times 7200$$
$$= 3600C$$

$\because$ $1F = 96500 \text{ C mol}^{-1}$

That is $e^-$ flows from 96500 C = 1 mol

$\therefore$ $e^-$ flows from 3600 C $= \frac{1 \times 3600}{96500 \text{ mol}}$

$= 0.037$ mol.

No. of electrons $= 0.037 \times 6.023 \times 10^{23}$

$= 0.2246 \times 10^{23}$

$= 22.46 \times 10^{21}$ electrons

**13. (a) What type of isomerism is shown by the complex $[Co(NH_3)_5(SCN)]^{2+}$ ?**

**(b) Why is $[NiCl_4]^{2-}$ paramagnetic while $[Ni(CN)_4]^{2-}$ is diamagnetic ? (Atomic number of Ni = 28)**

**(c) Why are low spin tetrahedral complexes rarely observed ?** **[3]**

**Ans. (a)** Linkage isomerism

**(b)** $[NiCl_4]^{2-}$, $Ni^{2+} = 1s^2 2s^2 2p^6 3s^2 3p^6 3d^8$

$Cl^-$ is a weak field ligand.

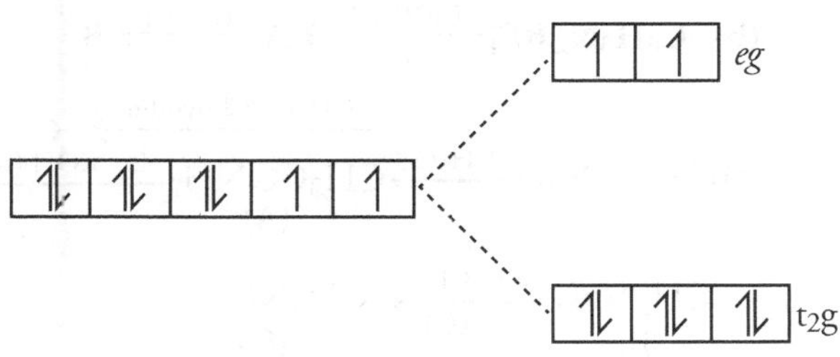


2 electrons are unpaired in $[NiCl_4]^{2-}$ which provides paramagnetism to the complex. $[Ni(CN)_4]^{2-}$

$Ni^{2+} = 1s^2\ 2s^2\ 2p^6\ 3s^2\ 3p^6\ 3d^8$

$CN^-$ is a strong field ligand

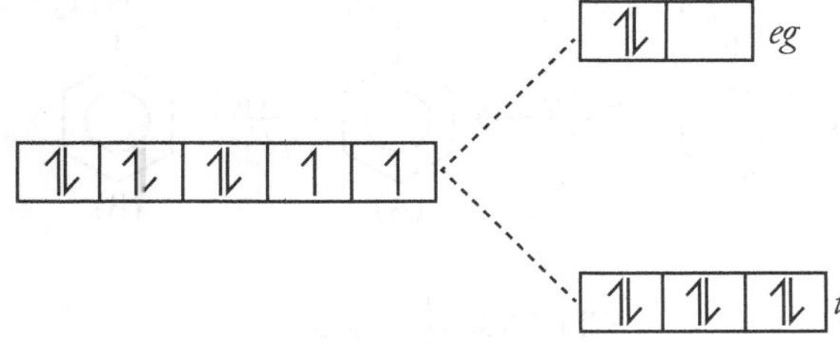


no electron is unpaired in $[Ni(CN)_4]^{2-}$ That's why the complex is diamagnetic.

**(c)** In tetrahedral complex, CFSE is very low and it is difficult for the tetrahedral complexes to exceed the pairing energy. Usually electrons prefer to move to higher energy orbitals for pairing. Thus they usually forms high spin complexes.

$$(CFSE)_{tetrahedral} = \frac{4}{9}(CFSE)_{octahedral}$$

**14. Write one difference in each of the following :**

**(a) Multimolecular colloid and Associated colloid**

**(b) Coagulation and Peptization**

**(c) Homogeneous catalysis and Heterogeneous catalysis** **[3]**

**OR**

**(a) Write the dispersed phase and dispersion medium of milk.**

**(b) Write one similarity between physisorption and chemisorption.**

**(c) Write the chemical method by which $Fe(OH)_3$ sol is prepared from $FeCl_3$.**

**Ans. (a)** Multimolecular colloids are the colloids in which the dispersed phase consists of aggregates of atoms or molecules with molecular size less than 1nm whereas associated colloids are the substances that are dissolved in a medium, behave as normal electrolytes at low concentration but as colloids at higher concentration.

**(b)** Coagulation is the process of precipitation of a colloidal solution by the addition of excess of an electrolyte whereas peptization is the process responsible for the formation of stable dispersion of colloidal particles in dispersion medium.

**(c)** Homogeneous catalysis is the one in which the phases of the reactants and the catalysts are the same whereas in heterogeneous catalysis the phases of the reactants and the catalysts are not the same.

**OR**

**(a)** Milk

Dispersed phase — Liquid
Dispersion medium — Liquid

**(b)** Both physisorption and chemisorption depends on the surface area. Both increases with an increase in the surface area. They both are surface phenomenons.

**(c)** $Fe(OH)_3$ sol is prepared from $FeCl_3$ by hydrolysis method.

$$FeCl_3 + H_2O \xrightarrow{hydrolysis} \underset{(Sol)}{Fe(OH)_3} + 3HCl$$

**15. A first order reaction takes 20 minutes for 25% decomposition. Calculate the time when 75% of the reaction will be completed.**

**Given : log 2 = 0.3010, log 3 = 0.4771, log 4 = 0.6021)** **[3]**

**Ans.** For first order reaction.

$$K = \frac{2.303}{t}\log\frac{a}{a-x}$$

$a \rightarrow$ initial amount

$a - x \rightarrow$ amount left after time

for 25% decomposition :

$$K = \frac{2.303}{20}\log\frac{100}{75} \quad .......(1)$$

for 75% decomposition :

$$K = \frac{2.303}{t}\log\frac{100}{25} \quad .......(2)$$

K is constant throughout the process eq. (1) = eq. (2)

Thus on comparing eq. (1) and eq. (2) we have

$$\frac{2.303}{20}\log\frac{100}{75} = \frac{2.303}{t}\log\frac{100}{25}$$

$$\frac{1}{20}(\log 100 - \log 75) = \frac{1}{t}(\log 100 - \log 25)$$

$$\frac{1}{20}[2 - 1.875] = \frac{1}{t}[2 - 1.398]$$

$$\frac{0.125}{20} = \frac{0.602}{t}$$

$$t = \frac{12.04}{0.125}$$

$t = 96.32$ minutes

**16. The following compounds are given to you :**
**2-Bromopentane, 2-Bromo-2-methylbutane, 1–Bromopentane**

**(a) Write the compound which is most reactive towards $S_N^2$ reaction.**

**(b) Write the compound which is optically active.**

**(c) Write the compound which is most reactive towards β-elimination reaction. [3]**

**Ans. (a)** 1-bromopentane > 2-bromopentane > 2-bromo -2-methylpentane (Reactivity towards, $S_N^2$ reaction). Thus 1-bromopentane will be more reactive towards $S_N^2$ reaction.

**(b)** 2-bromopentane is optically active

$CH_3CH_2CH_2CHBrCH_3$

**(c)** $H_3C-CH_2-CH_2-C(Br)(CH_3)-CH_3$

2-bromo-2-methylpentane

This compound is most reactive towards β-elimination.

**17. Write the principle of the following :**

**(a) Zone refining**

**(b) Froth floatation process**

**(c) Chromatography [3]**

**Ans. (a)** It is based on the principle that Impurities are more soluble in the melt than in the solid metal.

**(b)** It is based on the principle that Mineral particles are wetted by oils forming froth while gangue particles are wetted by water and settle down.

**(c)** It is based on the principle that different components of a mixture are differently adsorbed on an adsorbent.

**18. Write the structures of compounds A, B and C in the following reactions :**

**(a)** $CH_3COOH \xrightarrow{NH_3/\Delta} A \xrightarrow{Br_2/KOH(aq)} B \xrightarrow{CHCl_3 + alc.\ KOH} C$

**(b)** $C_6H_5N_2^+BF_4^- \xrightarrow[\Delta]{NaNO_2/Cu} A \xrightarrow{Fe/HCl} B \xrightarrow{CH_3COCl/pyridine} C$ **[3]**

**Ans. (a)** $CH_3COOH \xrightarrow{NH_3/\Delta} \underset{(A)}{CH_3CONH_2} \xrightarrow{Br_2/KOH\ (aq)}$

$\underset{(B)}{CH_3NH_2} \xrightarrow[alc.\ KOH]{CH_3Cl_3 +} \underset{(C)}{CH_3NC}$

(A) $CH_3CONH_2 \rightarrow$ Acetamide

(B) $CH_3NH_2 \rightarrow$ Methylamine

(C) $CH_3NC$—Methylisocyanide

**(b)** $C_6H_5N_2^+BF_4^- \xrightarrow[\Delta]{NaNO_2/Cu} \underset{(A)}{C_6H_5NO_2} \xrightarrow{Fe/HCl} \underset{(B)}{C_6H_5NH_2}$

$\downarrow CH_3COCl$ | Pyridine

$\underset{(C)}{C_6H_5NH-C(=O)-CH_3} + HCl$

(A) Nitrobenzene — $C_6H_5NO_2$

(B) Aniline — $C_6H_5NH_2$

(C) Acetanilide — $C_6H_5NHCOCH_3$

**19. Write the structures of the monomers used for getting the following polymers :**

**(a) Nylon-6,6**

**(b) Melamine-formaldehyde polymer**

**(c) Buna-S [3]**

**Ans. (a) Monomers of Nylon-6,6**

$H_2N-(CH_2)_6-NH_2$ and $HOOC-(CH_2)_4-COOH$

Hexamethylenediamine    Adipic acid

**(b) Monomers of Melamine-formaldehyde polymer**

Melamine (2,4,6-triamino-1,3,5-triazine: triazine ring bearing three $NH_2$ groups) and HCHO (Formaldehyde)

**(c) Buna-S (monomer)**

$CH_2=CH-CH=CH_2$ (1,3-butadiene) and $C_6H_5-CH=CH_2$ (Styrene)

**20. Define the following :** **[3]**

**(a) Anionic detergents**

**(b) Limited spectrum antibiotics**

**(c) Antiseptics**

**Ans. (a) Anionic detergents:** They are sodium salts of sulphonated long chain alcohols or hydrocarbons whose anionic part is involved in cleansing action.

**(b) Limited Spectrum Antibiotics :** The antibiotics which are effective against single organism or disease are called limited spectrum antibiotics, example–streptomycin.

**(c) Antiseptics :** The chemical substances that are used to either kill or prevent the growth of microorganisms are called antiseptics. These are not harmful to living tissues and can be safely applied on wounds, cuts, ulcers etc., example Soframycin.

**21. Give reasons for the following :**

**(a) Red phosphorus is less reactive than white phosphorus.**

**(b) Electron gain enthalpies of halogens are largely negative.**

**(c) $N_2O_5$ is more acidic than $N_2O_3$.** **[3]**

**Ans. (a)** White phosphorus has a discrete tetrahedral basic unit where four phosphorus atoms are arranged at the four corners of a tetrahedron.

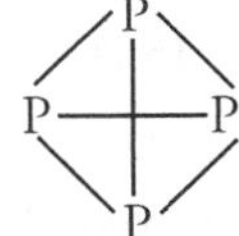

White phosphorous

This creates a very strained system which is unstable and very reactive. On the other hand red phosphorus has an polymeric structure and is therefore a lot more stable and less reactive.

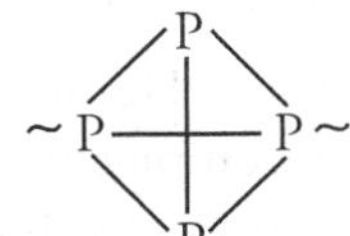

Red phosphorus

**(b)** Electron gain enthalpies of halogens are largely negative in their respective periods. This is due to the fact that the atoms of these elements have only one electron less than the stable noble gas ($ns^2\ np^6$) configuration. Therefore, they have maximum tendency to accept an additional electron.

**(c)** The oxidation state of N in $N_2O_5$ is +5 and that of N in $N_2O_3$ is +3 thus more number of oxygen atoms are attached with N in $N_2O_5$ as compared to $N_2O_3$. Thus, the acidic strength of $N_2O_5$ is higher than $N_2O_3$.

**22. Give reasons for the following :**

**(a) Acetylation of aniline reduces its activation effect.**

**(b) $CH_3NH_2$ is more basic than $C_6H_5NH_2$.**

**(c) Although-$NH_2$ is *o/p* directing group, yet aniline on nitration gives a significant amount of m-nitroaniline.** **[3]**

**Ans. (a)** Aniline ($C_6H_5NH_2$) + $Cl-\overset{O}{\overset{\|}{C}}-CH_3$ ⟶ Acetanilide ($C_6H_5NHCOCH_3$)

Aniline Acetyl chloride Acetanilide

In acetanilide, the oxygen atom of the group withdraws electrons from $\ddot{N}H_2$ group as shown below :

$C_6H_5-\ddot{N}H-C(=\ddot{O}:)-CH_3 \longleftrightarrow C_6H_5-\overset{+}{N}H=C(-O^-)-CH_3$

As a result, the electron pair on nitrogen gets displaced to the carboxyl group. Therefore, the unshared pair of electron on nitrogen is less available for donation of the electron to aromatic ring.

**(b)** In aniline, lone pair of $e^-$ present on 'N' is in conjugation with the benzene ring and become less available for protonation because of resonance.

$\ddot{N}H_2$ ⟶ $\overset{+}{N}H_2$ ⟷ $\overset{+}{N}H_2$ ⟷ $\overset{+}{N}H_2$ (resonance structures of aniline with negative charge at ortho and para positions)

This conjugation of lone pair of $e^-$ is not present in case of methyl amine and lone pair of $e^-$ of 'N' are fully available for protonation. That's why the basicity order of aniline and methyl amine is:

(c) $CH_3NH_2 > C_6H_5NH_2$

$NH_2$ Aniline $\xrightarrow[\Delta]{HNO_3 \text{ Conc. } H_2SO_4}$ o-nitro-aniline (2%) + m-nitro-aniline (47%) + p-nitro-aniline (51%)

The reason for the formation of large amount of *m*-nitroaniline is that under strongly acidic conditions, aniline gets protonated to anilinium ion ($-\overset{+}{N}H_3$ group). This is a deactivating group and is meta-directing in nature.

**23. After watching a programme on TV about the presence of carcinogens (cancer causing agents) Potassium bromate and potassium iodate in bread and other bakery products, Rupali a Class XII student decided to make others aware about the adverse effects of these carcinogens in foods. She consulted the school principal and requested him to instruct the canteen contractor to stop selling sandwiches, pizzas, burgers and other bakery products to the students. The principal took an immediate action and instructed the canteen contractor to replace the bakery products with some protein and vitamin rich food like fruits, salads, sprouts, etc. The decision was welcomed by the parents and the students.**

**After reading the above passage, answer the following questions :**

**(a) What are the values (at least two) displayed by Rupali ?****

**(b) Which polysaccharide component of carbohydrates is commonly present in bread?**

**(c) Write the two types of secondary structures of proteins.**

**(d) Give two examples of water soluble vitamins. [4]**

** Answer is not given due to the change in present syllabus.

**Ans. (b)** Starch

**(c)** 1. α-helix structure.

2. β-pleated sheet structure.

**(d)** Vitamin B and Vitamin C

**24. (a) Account for the following :**

**(i) Transition metals show variable oxidation states.**

**(ii) Zn, Cd and Hg are soft metals.**

**(iii) E° value for the $Mn^{3+}/Mn^{2+}$ couple is highly positive (+1.57 V) as compared to $Cr^{3+}/Cr^{2+}$.**

**(b) Write one similarity and one difference between the chemistry of lanthanoid and actinoid elements. [5]**

**OR**

**(a) Following are the transition metal ions of 3*d* series :**

**$Ti^{4+}$, $V^{2+}$, $Mn^{3+}$, $Cr^{3+}$**

**(Atomic numbers : Ti = 22, V = 23, Mn = 25, Cr = 24)**

**Answer the following :**

**(i) Which ion is most stable in an aqueous solution and why ?**

**(ii) Which ion is a strong oxidising agent and why ?**

**(iii) Which ion is colourless and why ?**

**(b) Complete the following equation :**

**(i) $2MnO_4^- + 16\,H^+ + 5S^{2-} \longrightarrow$**

**(ii) $KMnO_4 \xrightarrow{Heat}$**

**Ans. (a) (i)** Transition metal ions shows variable oxidation states due to the participation of (n-1) *d* electrons in addition to outer *ns*–electrons because the energies of ns and $(n-1)d$ subshells are almost equal. As a result of which the electrons of $(n-1)$ *d* and ns subshell both part in bond formation.

**(ii)** Zn, Cd and Hg are soft metals because of their completely filled 3*d*, 4*d* and 5*d* orbitals respectively Due to completely filled d-orbitals these metals are reluctant to form Zn–Zn, Cd–Cd and Hg–Hg bonds.

**(iii)** Highly positive value of E° for $Mn^{3+}/Mn^{2+}$ shows that $Mn^{2+}$ ($d^5$) is particularly stable. While low value of E° for $Cr^{3+}/Cr^{2+}$ shows that $Cr^{2+}$ ($d^4$) is less stable than $Cr^{3+}$ ($d^3$)

(b) **Similarity :** In lanthanoids and actinoids both the added electron enters the antipenultimate shell 4*f* and 5*f* respectively.

**Difference :** Lanthanoids show a common oxidation state of +3 while actinoids show different oxidation states other than +3.

**OR**

(a) $Ti^{4+} = 1s^2\, 2s^2\, 2p^6\, 3s^2\, 3p^6$
$V^{2+} = 1s^2\, 2s^2\, 2p^6\, 3s^2\, 3p^6\, 3d^3$
$Mn^{3+} = 1s^2\, 2s^2\, 2p^6\, 3s^2\, 3p^6\, 3d^4$
$Cr^{3+} = 1s^2\, 2s^2\, 2p^6\, 3s^2\, 3d^3$

(i) $Ti^{4+}$ is most stable in an aqueous solution because of full filled valence shell ($3s^2\, 3p^6$) onfiguration (noble gas configuration).

(ii) $Mn^{3+}$ is the strong agent as it oxidises other species it will reduce itself by taking an electron and will stabilise its configuration to $3d^5$.

(iii) $Ti^{4+}$ is colourless due to absence of unpaired electrons ($3s^2\, 3p^6$)

(b) (i) $2MnO_4^- + 16H^+ + 5S^{2-} \rightarrow 2Mn^{2+} + 5S + 8H_2O$

(ii) $2KMnO_4 \xrightarrow{\text{Heat}} K_2MnO_4 + MnO_2 + O_2$

**25. (a) A 10% solution (by mass) of sucrose in water has a freezing point of 269.15 K. Calculate the freezing point of 10% glucose in water if the freezing point of pure water is 273.15 K.**

**Given :**

**Molar mass of sucrose = 342 g mol⁻¹**

**Molar mass of glucose = 180 g mol⁻¹**

**(b) Define the following terms :**

**(i) Molality (*m*)**

**(ii) Abnormal molar mass** [5]

**OR**

**(a) 30 g of urea (M=60 g mol⁻¹) is dissolved in 846 g of water. Calculate the vapour pressure of water for this solution if vapour pressure of pure water at 298 K is 23.8 mm Hg.**

**(b) Write two differences between ideal solutions and non-ideal solutions.**

**Ans.** (a) $T_o$ (freezing point of water) = 273.15K

$T_s$ (freezing point of sucrose solution) = 269.15K

Weight of sucrose in solution = 10 g

Weight of glucose in solution = 10 g

Molar mass of sucrose = 342 g mol⁻¹

Molar mass of glucose = 180 g mol⁻¹

Freezing point of glucose = $x$

Depression in freezing point

$$\Delta T_f = \frac{K_f \times W_B \times 1000}{W_A \times 1000}$$

$$K_f = \frac{\Delta T_f \times W_A \times M_B}{W_B \times 1000}$$

In case of sucrose solution

$$K_f = \frac{(273.15 - 269.15) \times 90 \times 342}{10 \times 1000} \quad ...(1)$$

In case of glucose solution

$$K_f = \frac{(273.15 - x) \times 90 \times 180}{10 \times 1000} \quad ....(2)$$

∵ $K_f$ is constant

thus equation (1) = equation (2)

$$\frac{(273.15-269.15) \times 90 \times 342}{10 \times 1000} = \frac{273.15 - x \times 90 \times 180}{10 \times 1000}$$

$$4 \times 342 = (273.15 - x) \times 180$$

$$(273.15 - x) = \frac{40 \times 342}{180} = 7.6$$

$$x = 265.55 \text{ K}$$

**So, freezing point of glucose solution = 265.55 K.**

(b) (i) **Molality :** It is the number of moles of the solute dissolved per 1000 g of the solvent. It is denoted by *m*.

$$\text{Molality} = \frac{\text{Moles of solute}}{\text{Mass of solvent (in gram)}} \times 1000$$

(ii) **Abnormal molar mass :** Those solute that dissociate or associate in solution, show abnormal molar mass in solution.

for example, Molar mass of ethanoic acid is greater than normal molar mass.

$$\underset{\text{Ethanoic acid}}{2CH_3COOH} \rightleftharpoons (CH_3COOH)_2$$

Molar mass of KCl in solution is reduced than normal molar mass.

$$KCl \rightleftharpoons K^+ + Cl^-$$

**OR**

(a) $W_B = 30\text{ g}$ $\quad M_B = 60 \text{ g mol}^{-1}$

$W_A = 846\,g \quad M_A = 18\,g\,mol^{-1}$

$P^o = 23.8mm\ Hg$

$P_s = x$

Relative lowering of vapour pressure

$$\frac{P^o - Ps}{P^o} = \frac{W_B \times M_A}{M_B \times W_A}$$

$$\frac{23.8 - x}{23.8} = \frac{30 \times 18}{60 \times 846}$$

$$23.8 - x = \frac{23.8}{94}$$

$$23.8 - x = 0.253$$

$$x = 23.8 - 0.253 = 23.547$$

So, vapour pressure of water for this solution = 23.597 mmHg

(b)

| Ideal Solutions | Non-ideal Solutions |
|---|---|
| 1. The interactions between the components are similar to those in the pure components. It obeys Raoults' law. | The interactions between the components are different from those of the pure components. It does not obey Raoults' law. |
| 2. There is no enthalpy change on mixing $\Delta H_{mix} = 0$ | There is enthalpy change on mixing $\Delta H_{mix} \neq o$ |

**26. (a) Write the product(s) in the following reactions:** **[5]**

(i) Cyclohexanone + **HCN** $\longrightarrow$ **?**

(ii) $C_6H_5COONa$ + **NaOH** $\xrightarrow[\Delta]{CaO}$ **?**

(iii) $CH_3 - CH = CH - CN \xrightarrow[(b)\ H_2O]{(a)\ DIBAL-H}$ **?**

**(b) Give simple chemical tests to distinguish between the following pairs of compounds:**

**(i) Butanal and Butan-2-one**

**(ii) Benzoic acid and Phenol**

**OR**

**(a) Write the reactions involved in the following:**

**(i) Etard reaction**

**(ii) Stephen reduction**

**(b) How will you convert the following in not more than two steps :**

**(i) Benzoic acid to Benzaldehyde**

**(ii) Acetophenone to Benzoic acid**

**(iii) Ethanoic acid to 2-Hydroxyethanoi acid**

**Ans. (a)**

(i) Cyclohexanone (=O) + HCN $\longrightarrow$ Cyclohexane with OH and CN (cyanohydrin)

(ii) $C_6H_5COONa + NaOH \xrightarrow[\Delta]{CaO}$ Benzene $+ Na_2CO_3$

Benzene

(iii) $CH_3 - CH = CH - CN \xrightarrow[(b)\ H_2O]{(a)\ DIBAL-H} CH_3-CH=CH-CHO$

**(b) (i) Butanal and Butan-2-one**

1. $CH_3CH_2CH_2CHO + 2[Ag(NH_3)_2]^+ + 3OH^-$ (Butanal) $\downarrow$ Tollen's reagent

$CH_3CH_2CH_2COO^- + 2Ag + 2H_2O + 4NH_3$

This reaction is known as silver mirror test—

2. $CH_3 - CH_2 - \overset{O}{\overset{||}{C}} - CH_3 + 2[Ag(NH_3)_2]^+ + 3OH^-$ (Butan-2-one) $\downarrow$ Tollen's reagent

No reaction

Thus Butanal gives silver mirror test with Tollen's reagent whereas Butan-2-one does not.

**(ii) Benzoic acid and phenol**

$C_6H_5COOH + NaHCO_3 \longrightarrow C_6H_5COONa + H_2O + CO_2$

(Benzoic acid) (Sodium bicarbonate) (Sodium benzoate)

$C_6H_5OH + NaHCO_3 \longrightarrow$ No reaction

(Phenol) (Sodium bicarbonate)

Thus, Benzoic acid gives sodium benzoate on reaction with sodium bicarbonate whereas phenol gives no reaction with sodium bicarbonate.

OR

(a) (i) **Etard reaction :** The oxidation of toluene to benzaldehyde with chromyl chloride ($CrO_2Cl_2$) dissolved in $CCl_4$ or $CS_2$.

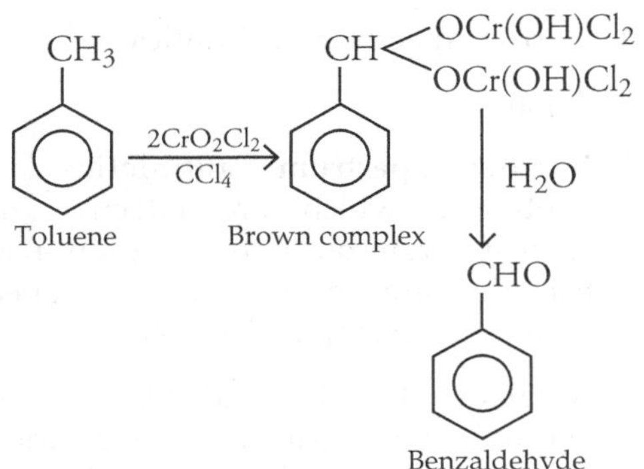


(ii) **Stephen reaction :** The partial reduction of alkyl or aryl cyanides to the corresponding aldehydes with a suspension of anhydrous $SnCl_2$ in ether saturated with HCl at room temperature followed by hydrolysis.

$$SnCl_2 + 2HCl \longrightarrow SnCl_4 + 2[H]$$

$$CH_3C \equiv N + 2[H] + 2HCl \longrightarrow \underset{\text{Acetaldoxime hydrochloride}}{CH_3\,CH \equiv NHHCl} \xrightarrow[\text{boil}]{H_2O} \underset{\text{Acetaldehyde}}{CH_3CHO + NH_4Cl}$$

$$C_6H_5 - C \equiv N + 2[H] + 2HCl \longrightarrow \underset{\text{Benzaldoxime hydrochloride}}{C_6H_5CH = NH.HCl} \xrightarrow{H_2O \,|\, \text{Boil}} \underset{\text{Benzaldehyde}}{C_6H_5CHO + NH_4Cl}$$

**(b) (i) Benzoic acid to Benzaldehyde**

$$\underset{\text{Benzoic acid}}{C_6H_5COOH} \xrightarrow{\text{Soda lime}} \underset{\text{Benzene}}{C_6H_6} \xrightarrow[\text{Anhyd.}AlCl_3]{CO,\ HCl} \underset{\text{Benzaldehyde}}{C_6H_5CHO}$$

(ii) Acetophenone to Benzoic acid

$$\underset{\text{Acetophenone}}{C_6H_5COCH_3} \xrightarrow{Zn/Hg\text{–}HCl} \underset{\text{Ethyl Benzene}}{C_6H_5CH_2CH_3} \xrightarrow{[O]} \underset{\text{Benzoic acid}}{C_6H_5COOH}$$

**(iii) Ethanoic acid to 2-hydroxyethanoic acid**

$$\underset{\text{Acetic acid or Ethanoic acid}}{CH_3COOH} \xrightarrow{P,\ Cl_2} \underset{|\ Cl}{CH_2COOH} \xrightarrow{\text{aq. KOH}} \underset{\text{2-Hydroxyethanoic acid}}{CH_2(OH) - COOH}$$

●●

# Outside Delhi [Set-II]

**Note : Except for the following questions, all the remaining questions have been asked in previous sets.**

**3. Write the IUPAC name of the following compound.** **[1]**

$C_6H_5-CH = CH - CH_2 - OH$

**Ans.** $C_6H_5-CH = CH - CH_2 - OH$

3-Phenyl-prop-2-en-1-ol

**5. Out of** (cyclohexene with X on the $sp^3$ carbon adjacent to the double bond) **and** (cyclohexene with X on the double-bonded carbon)**, which is an example of vinylic halide ?** **[1]**

**Ans.** Cyclohexene with X on the double-bonded carbon is an example of vinylic halide.

**6. Using IUPAC norms write the formulae for the following :**

**(a) Tris (ethane-1, 2-diamine) chromium (III) chloride.**

**(b) Potassium tetrahydroxozincate (II).** **[2]**

**Ans.** (a) $[Cr(en)_3]Cl_3$

(b) $K_2[Zn(OH)_4]$

**7. Draw the structures of the following :** **[2]**

(a) $H_2S_2O_8$

(b) $ClF_3$

**Ans. (a)**

$$HO-S(=O)_2-O-O-S(=O)_2-OH$$

$H_2S_2O_8$

**(b)**

$$ClF_3$$

**8. Write the name of the cell which is generally used in inverters. Write the reactions taking place at the anode and the cathode of this cell.** **[2]**

**Ans.** Lead storage battery is commonly used in inverters.

**Reactions taking place at anode**

$$Pb(s) \longrightarrow Pb^{2+}(aq) + 2e^-$$

$$Pb^{2+}_{(aq.)} + SO_{4\,(aq)}^{2-} \longrightarrow PbSO_4(s)$$

The overall reaction at anode is

$$Pb(s) + SO_4^{2-}(aq) \longrightarrow PbSO_4(s) + 2e^-$$

**Reactions taking place at cathode.**

$$PbO_2(s) + 4H^+(aq) + SO_4^{2-}(aq) + 2e^- \longrightarrow PbSO_4(s) + 2H_2O(l)$$

**11. (a) Write the principle of vapour phase refining.**

**(b) Write the role of dilute NaCN in the extraction of silver.**

**(c) What is the role of collectors in the froth floatation process ? Give an example of a collector.** **[2]**

**Ans. (a) Vapour phase refining :** This method is based on the principle that certain metals are converted to their volatile compounds which on strong heating is decomposed to give pure metal. While the impurities are not affected during compound formation.

**(b)** NaCN is used to leach the silver ore in the presence of air. Pure silver is obtained by replacement in the process of extraction of silver.

$$Ag_2S + 4NaCN \longrightarrow \underset{\text{Sodium dicyano-argentate}}{2Na[Ag(CN)_2]} + Na_2S$$

$$2Na[Ag(CN)_2] + Zn \longrightarrow 2Ag(s) + \underset{\text{Sodium tetracyanozincate (II)}}{Na\,[Zn(CN_4)](aq)}$$

**(c)** In the froth floatation process, collectors enhances the non-wettability of the mineral particles. Example of collectors are pine oil, eucalyptus oil, fatty acids etc.

**16. Define the following :**

**(b) Narrow spectrum antibiotics.**

**(c) Antacids** **[3]**

**Ans (b) Narrow spectrum antibiotics :** The antibiotics which are effective mainly against gram-positive or gram-negative bacteria are called narrow spectrum antibiotics Example: Penicillin.

**(c) Antacids :** The chemical substances which neutralizes excess acids in the gastric juices and gives relief from acid indigestion, acidity, heart burns and gastric ulcers are called antacids. Example Sodium hydrogen-carbonate (baking soda) in water.

**17. Write the structures of the monomers used for getting the following polymers :**

**(a) Polyvinyl chloride (PVC)**

**(c) Buna-N** **[3]**

**Ans. (a) Monomer of polyvinyl chloride (PVC)**

Vinyl chloride $CH_2 = CH - Cl$

**(c) Monomer of Buna-N**

$$\underset{\text{1, 3-butadiene}}{H_2C = CH - CH = CH_2} \text{ and } \underset{\text{Acrylonitrile}}{CH_2 = CH - CN}$$

**22. (a) Based on the nature of intermolecular forces, classify the following solids :**

**Benzene, Silver**

**(b) AgCl shows Frenkel defect while NaCl does not. Give reason.**

**(c) What type of semiconductor is formed when Ge is doped with Al ?** **[3]**

**Ans. (a) Benzene:** Non-polar molecular solid

Silver–Metallic solid.

**(b)** Frenkel defect is not found in NaCl because $Na^+$ ions cannot get into the interstitial sites due to larger size. While, $Ag^+$ ions in AgCl has smaller size and easily fit into the interstitial sites.

**(c)** When Ge is doped with Al, p-type semi-conductor is formed.

●●

# Outside Delhi [Set-III]

Note : Except for the following questions, all the remaining questions have been asked in previous set.

**1. Out of** (benzene ring with $CHCl_2$) **and** (benzene ring with $CH_2CH_2Cl$), **which is an example of a benzylic halide ? [1]**

**Ans.** 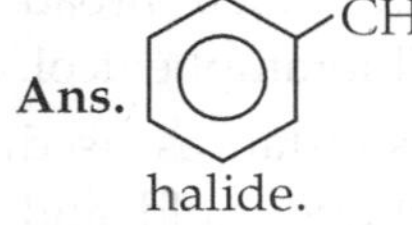

is an example of benzylic halide.

**3. Write the formula of the compound of iodine which is obtained when conc. $HNO_3$ oxidises $I_2$. [1]**

**Ans.** Iodic acid, $HIO_3$ is obtained on the oxidation of $I_2$ by $HNO_3$.

$$I_2 + 10HNO_3 \longrightarrow \underset{\text{Iodic acid}}{2HIO_3} + 10NO_2 + 4H_2O$$

**4. What type of colloid is formed when a gas is dispersed in a liquid ? Give an example. [1]**

**Ans.** 'Foam' colloid is formed when a gas is dispersed in a liquid. For example whipped cream or soda water.

**5. Write the IUPAC name of the following compound : [1]**

$$CH_3-O-\underset{CH_3}{\overset{CH_3}{\underset{|}{\overset{|}{C}}}}-CH_3$$

**Ans.** $$H_3C-O-\underset{CH_3}{\overset{CH_3}{\underset{|}{\overset{|}{C}}}}-CH_3$$

2-Methoxy-2methyl-propane

**6. Draw the structures of the following :**

**(a) $XeF_4$** **(b) $BrF_5$** **[2]**

**Ans. (a) $XeF_4$**

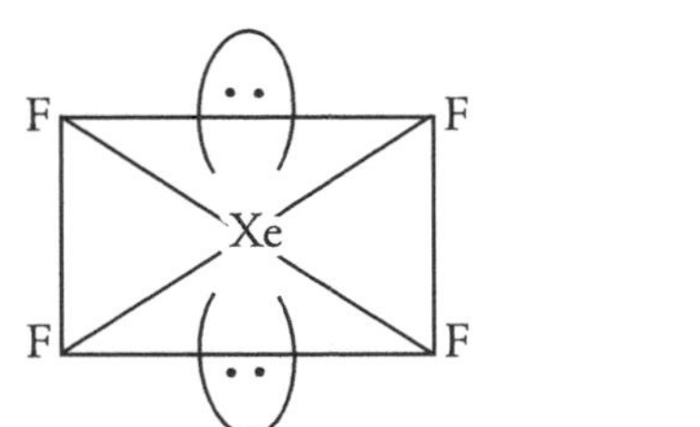


**(b) $BrF_5$**

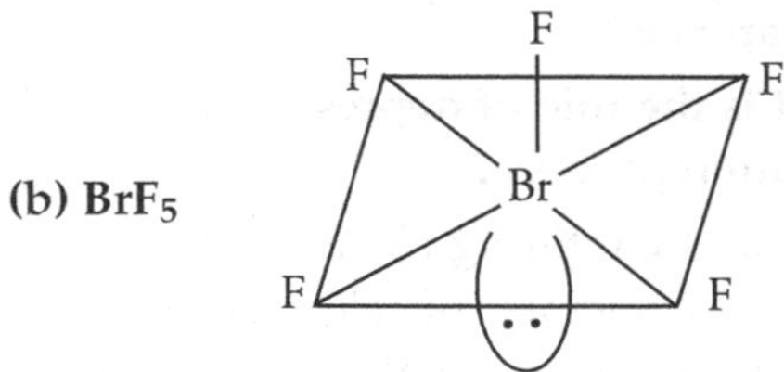


**7. Write the name of the cell which is generally used in transistors. Write the reactions taking place at the anode and the cathode of this cell. [2]**

**Ans.** Dry cells are used in transistors.

At anode

$$Zn \longrightarrow Zn^{2+} + 2e^-$$

At cathode

$$2NH_4^+(aq) + 2MnO_2(s) + 2e^- \longrightarrow 2MnO(OH) + 2NH_3$$

**9. Using IUPAC norms write the formulae for the following :**

**(a) Potassium trioxalatoaluminate (III).**

**(b) Dichloridobis (ethane-1, 2-diamine) cobalt (III) [2]**

**Ans.** (a) $K_3[Al(Ox)_3]$

(b) $[CoCl_2(en)_2]^+$

**14. (a) Based on the nature of intermolecular forces, classify the following solids : Sodium sulphate, Hydrogen**

**(b) What happens when $CdCl_2$ is doped with AgCl?**

**(c) Why do ferrimagnetic substances show better magnetism than antiferromagnetic substances? [3]**

**Ans.** (a) **Sodium sulphate** : Ionic solid

**Hydrogen** : Non-polar molecular solid

(b) When $CdCl_2$ is doped with AgCl, electrical conductivity of AgCl solid is increased due to schottky defect.

(c) In ferrimagnetic substances, magnetic moments of domains are aligned in parallel and antiparallel directions in unequal amount so that resulting magnetic moment is not zero. While in case of antiferromagnetic substances, due to equal number of opposite domains, net magnetic moment is zero. That's why ferrimagnetic substances show better magnetism than antifer-romagnetic substances.

**15. (a) Write the principle of electrolytic refining.**

**(b) Why does copper obtained in the extraction from copper pyrites have a blistered appearance ?**

**(c) What is the role of depressants in the froth floatation process ? [3]**

**Ans. (a) Electrolytic refining :** This method is based on the principle of electrolysis. In this method impure metal is made to act as anode and a strip of same metal in pure form is used as cathode. Both anode and cathode are placed in a suitable electrolytic bath containing soluble salt of same metals.

**(b)** In the extraction of copper from $CuFeS_2$, $SO_2$, $N_2$ and $O_2$ escape from the metal. As the metal solidifies, the dissolved gases escape producing blisters on the metal surface, which provides blister appearance to copper.

**(c)** Depressents are used to prevent certain types of particles from forming the froth with air bubbles. For example : NaCN can be used as a depressent in the separation of ZnS and PbS.

**19. Define the following :**

**(a) Cationic detergents**

**(b) Broad spectrum antibiotics**

**(c) Tranquilizers [3]**

**Ans. (a) Cationic detergents :** These are the quaternary ammonium salts of amines with acetates, chlorides or bromides as anions. The cationic part possesses a long hydrocarbon chain with a positive charge on nitrogen atom. Example. Cetyltrimethyl ammonium chloride.

$$\left[ H_3C-(CH_2)_{15}-\underset{\substack{|\\CH_3}}{\overset{\substack{CH_3\\|}}{N}}-CH_3 \right]^+ Br^-$$

**(b) Broad spectrum antibiotics :** Antibiotics which kills or inhibit a wide range of gram-positive and gram-negative bacteria are called broad spectrum antibiotics. Example Chloramphenicol.

**(c) Tranquilizers :** The chemical substances used for the treatment of stress, fatigue, mild and severe mental diseases are called tranquilizers. Example : Phenelzine (Nardil).

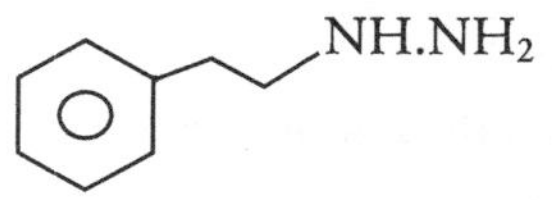

$$C_6H_5-CH_2-CH_2-NH.NH_2$$

**20. Write the structures of the monomers used for getting the following polymers :**

**(a) Teflon**

**(c) Neoprene [3]**

**Ans. (a) Monomer of Teflon :**

$$F_2C=CF_2$$

Tetrafluoroethene

**(c) Neoprene (monomer) :**

$$CH_2=\underset{\substack{|\\Cl}}{C}-CH=CH_2$$

Chloroprene or 2-chlorobuta-1,3-diene

●●

# CHEMISTRY

## Delhi [Set-I]

## 2017

**Time allowed : 3 hours** **Maximum marks : 70**

**1. Write the formula of an oxo-anion of Manganese (Mn) in which it shows the oxidation state equal to its group number. [1]**

**Ans.** Manganese belongs to group number 7 and its oxidation state in $KMnO_4$ is +7 *i.e.,*

$$KMnO_4$$

$$1 + x + 4(-2) = 0$$

$$1 + x - 8 = 0$$

$$x = 7$$

Thus, the formula of the oxo-anion is $KMnO_4$.

**2. Write IUPAC name of the following compound :**

**$(CH_3CH_2)_2NCH_3$ [1]**

**Ans.** N-Ethyl-N-methylethanamine.

**3. For a reaction $R \longrightarrow P$, half-life ($t_{1/2}$) is observed to be independent of the initial concentration of reactants. What is the order of reaction ? [1]**

**Ans.** Since half life is independent of the initial concentration of the reactants. Thus it is a first order Reaction. Formula for half-life of the first order reaction.

$$\boxed{t_{1/2} = \frac{0.693}{K}}$$

**4. Write the structure of 1-bromo-4-chlorobut-2-ene. [1]**

**Ans.** $Br\,CH_2CH = CHCH_2Cl$

**5. Write one similarity between physisorption and Chemisorption. [1]**

**Ans.** Physisorption and chemisorption both are the surface phenomenon and both increases the surface area during the process of adsorption.

**6. Complete the following reactions :**

**(i) $NH_3 + 3Cl_2(\text{excess}) \longrightarrow$**

**(ii) $XeF_6 + 2H_2O \longrightarrow$**

**OR**

**What happens when**

**(i) $(NH_4)_2Cr_2O_7$ is heated ?**

**(ii) $H_3PO_3$ is heated ?**

**Write the equations. [2]**

**Ans. (i)** $NH_3 + 3Cl_2 \text{ (excess)} \longrightarrow NCl_3 + 3HCl$.

**(ii)** $XeF_6 + 2H_2O \longrightarrow XeO_2F_2 + 4HF$

**OR**

**(i)** $(NH_4)_2Cr_2O_7(s) \longrightarrow Cr_2O_3(s) + N_2(g) + 4H_2O(g)$

**(ii)** $4H_3PO_3 \longrightarrow 3H_3PO_4 + PH_3$.

**7. Define the following terms :**

**(i) Colligative properties**

**(ii) Molality ($m$) [2]**

**Ans. (i)** Colligative properties are those which depends on number of moles of solute irrespective of their nature. They are independent of nature of solute.

**(ii)** Molality is defined as the number of moles of solute dissolved per kg of the solvent. It is independent of temperature.

$$\text{Molality} = \frac{\text{Moles of solute}}{\text{Mass of solvent in kilograms}}$$

**8. Draw the structures of the following :**

**(i) $H_2S_2O_7$**

**(ii) $XeF_6$ [2]**

**Ans. (i)** $H_2S_2O_7$

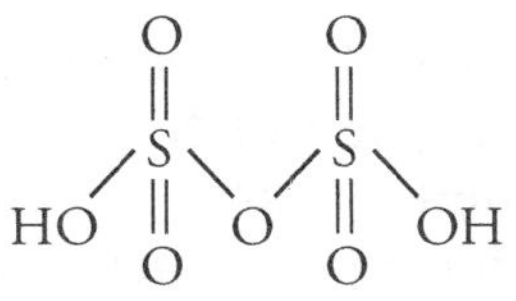

HO–S(=O)(=O)–O–S(=O)(=O)–OH

**(ii)** $XeF_6$

Xe with one lone pair and six F atoms

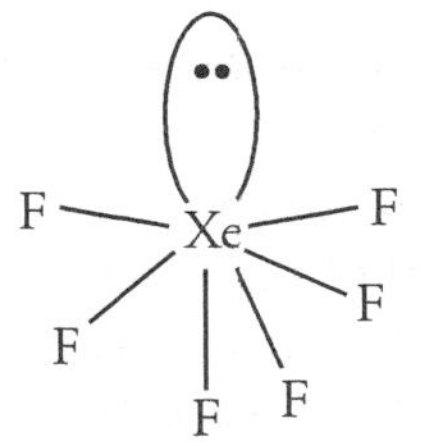

**9. Calculate the degree of dissociation ($\alpha$) of acetic acid if its molar conductivity ($\wedge_m$) is 39.05 S cm$^2$ mol$^{-1}$.**

**Given $\lambda^\circ (H^+) = 349.6$ S cm$^2$ mol$^{-1}$ and $\lambda^\circ (CH_3COO^-) = 40.9$ S cm$^2$ mol$^{-1}$. [2]**

**Ans.** Given : Molar conductivity ($\wedge_m$) for acetic acid = 39.05 S cm$^2$ mol$^{-1}$.

$$\lambda^\circ (H^+) = 349.6 \text{ S cm}^2 \text{ mol}^{-1}$$

$$\lambda^\circ (CH_3COO^-) = 40.95 \text{ S cm}^2 \text{ mol}^{-1}$$

We know that :

$$\lambda_m (CH_3COOH) = \lambda^\circ(H^+)+\lambda^\circ(CH_3COO^-)$$
$$39.05 = 349.6 + 40.9$$
$$39.05 = 390.5 \text{ S cm}^2/\text{mol}$$

also; $$\alpha = \frac{\wedge_m}{\wedge^\circ_m}$$
$$\alpha = \frac{39.05}{390.5}$$
$$\alpha = 0.1$$

Thus, the degree of dissociation of acetic acid is 0.1.

**10. Write the equations involved in the following reactions :**
**(i) Wolff–Kishner reduction**
**(ii) Etard reaction.** **[2]**

**Ans. (i) Wolff-Kishner reduction :**

$$\underset{\text{Acetaldehyde}}{CH_3 - \underset{|}{\overset{}{C}}(H) = O} + \underset{\text{Hydrazine}}{NH_2NH_2} \xrightarrow{-H_2O} \underset{\text{Hydrazone}}{CH_3 - \underset{|}{C}(H) = NNH_2}$$

$$\xrightarrow[\substack{453\text{-}473\text{ K} \\ \text{Glycol}}]{KOH} CH_3 - CH_3 + N_2$$

**(ii) Etard Reaction :**

$$\underset{\text{Toluene}}{C_6H_5CH_3} + CrO_2Cl_2 \xrightarrow{CS_2} \underset{\text{Chromium complex}}{C_6H_5CH(OCrOHCl_2)_2} \xrightarrow{H_3O^+} \underset{\text{Benzaldehyde}}{C_6H_5CHO}$$

**11. A 10% solution (by mass) of sucrose in water has freezing point of 269.15 K. Calculate the freezing point of 10% glucose in water, if freezing point of pure water is 273.15K.**
**[Given : (Molar mass of sucrose = 342 g mol$^{-1}$)**
**(Molar mass of glucose = 180 g mol$^{-1}$]** **[3]**

**Ans.** Given : Freezing point of = 269.15 K
10% solution of glucose
Freezing point of pure water = 273.15 K
Molar mass of sucrose = 342 g mol$^{-1}$.
Molar mass of glucose = 180 g mol$^{-1}$.
We know that :

$$\Delta T_f = K_f \times m$$

Also; $$m = \frac{W_2 \times 1000}{M_2 \times M_1}$$

**For the sucrose solution :**

$$273.15 - 269.15 = \frac{K_f \times 10 \times 1000}{342 \times 90}$$
$$4 \times 242 \times 90 = K_f \times 10 \times 1000$$
$$\frac{4 \times 342 \times 90}{10 \times 1000} = K_f$$
$$K_f = 12.3 \text{ k kg/mol}$$

**For the glucose solution :**

$$\Delta T_f = K_f \times m$$
$$= \frac{12.3 \times 1\cancel{0} \times 100\cancel{0}}{18\cancel{0} \times 9\cancel{0}}$$
$$\Delta T_f = 7.6 \text{ K}$$

Thus $$T_f = 273.15 - 7.6$$
$$T_f = 265.5 \text{ K}$$

The freezing point of 10% glucose in water is 265.5 K.

**12. (a) Calculate the mass of Ag deposited at cathode when a current of 2 amperes was passed through a solution of $AgNO_3$ for 15 minutes.**
**Given : Molar mass of Ag = 108 g mol$^{-1}$, 1F = 96500 C mol$^{-1}$)**
**(b) Define fuel cell** **[3]**

**Ans. (a)** Given :

Current = 2 amperes
Time = 15 minutes
Molar mass of Ag = 108 g mol$^{-1}$
1F = 96500 C mol$^{-1}$

Amount of metal deposited ($m$) = ZQ

$$Q = It$$
$$= 2 \times 15 \times 60 = 1800 \text{ C}$$

Silver deposited

$$Ag^+ + 2e^- \rightarrow Ag(s)$$

1 mole of electron or 1×96500 C of current deposit silver = 108 g

1800 C of current will deposit

$$= \frac{108 \times 1800}{96500}$$

Amount of Ag deposited = 2.01g

**(b)** Fuel cell is the cell which converts the energy of combustion of fuels directly into electrical energy.

**13. (i) What type of isomerism is shown by the complex $[Co(NH_3)_6]Cr(CN)_6]$ ?**
**(ii) Why a solution of $[Ni(H_2O)_6]^{2+}$ is green while a solution of $[Ni(CN)_4]^{2-}$ is colourless ? (At no. of Ni= 28)**
**(iii) Write the IUPAC name of the following complex :** **[3]**

**$[Co(NH_3)_5(CO_3)]Cl$.**

**Ans. (i)** Both shows coordination isomerism because both cationic and anionic entities and isomers differ in the distribution of ligands in the coordination entity of cationic and anionic part.

**(ii)** In $[Ni(H_2O)_6]^{2+}$ Ni is in +2 oxidation state with electronic configuration $3d^8$. In the presence of weak ligand $H_2O$ the two unpaired electrons do not pair up and hence the complex has two unpaired electrons. Therefore, it is coloured and shows *d-d* transitions which absorbs red light and emits green complimentary light.

In case of $[Ni(CN)_4]^{2-}$ Ni also shows +2 oxidation state but CN ligand is strong ligand and two unpaired electrons undergo pairing, to no *d-d* transitions takes place and it shows no colour.

**(iii)** Pentaamminecarbonatocobalt(III) chloride.

**14. Write one difference in each of the following :**

**(i) Lyophobic sol and Lyophilic sol.**

**(ii) Solution and Colloid**

**(iii) Homogeneous catalysis and Heterogeneous catalysis.** **[3]**

**Ans. (i)** Lyophobic colloidal sols are not hydrated and have weak affinity with the dispersion medium whereas lyophilic colloidal sols are heavily hydrated and have strong affinity with the dispersion medium.

**(ii)** Solution is a homogeneous mixture of solute and solvent whereas colloid is the heterogeneous mixture of dispersed phase and dispersion medium.

**(iii)** Homogeneous catalysis is the catalysis in which the reactants and the catalysts are in the same phase whereas in the heterogeneous catalysis the reactants and the catalysts are in the different phases.

**15. Following data are obtained for reaction :**

$$N_2O_5 \rightarrow 2NO_2 + 1/2O_2$$

| *t*/s | 0 | 300 | 600 |
|---|---|---|---|
| $[N_2O_5]$/mol $L^{-1}$ | $1.6\times10^{-2}$ | $0.8\times10^{-2}$ | $0.4\times10^{-2}$ |

**(a) Shows that it follows first order reaction.**

**(b) Calculate the half-life.**

**(Given log 2=0.3010, log 4= 0.6021)** **[3]**

**Ans. (a)** $K = 2.303/t \log [A_0]/[A]$

$= 2.303/300 \log 1.6\times10^{-2}/0.8\times10^{-2}$

$= 2.303/300 \log 2 = 2.31\times10^{-3}\ s^{-1}$

$=$ At 600 s, $K = 2.303/t \log [A_0]/[A]$

$= 2.303/300 \log 1.6\times10^{-2}/0.4\times10^{-2}$

$= 2.303/600 \log 4 = 2.31\times10^{-3}\ s^{-1}$

Since K is constant when using first order equation therefore, it follows first order kinetics.

**(b)** $t_{1/2} = 0.693/k$

$= 0.693/2.31\times10^{-3} = 300$ s

Thus, the half life of the reaction is 300 s.

**16. Following compounds are given to you :**

**2-Bromopentane, 2-Bromo-2-methylbutane, 1-Bromopentane**

**(i) Write the compound which is most reactive towards $S_N2$ reaction.**

**(ii) Write the compound which is optically active.**

**(iii) Write the compound which is most reactive towards β-elimination reaction.** **[3]**

**Ans. (i)** 1-Bromopentane is most reactive towards $S_N2$ reaction as it follows the order $1° > 2° > 3°$.

**(ii)** 2-Bromopentane is optically active.

**(iii)** 2-Bromo-2-methylbutane is most reactive towards β-elimination reaction.

**17. (a) Write the principle of method used for the refining of germanium.**

**(b) Out of PbS and $PbCO_3$ (ores of lead), which one is concentrated by froth floatation process preferably ?**

**(c) What is the significance of leaching in the extraction of aluminium ?** **[3]**

**Ans (a)** Zone refining method is used for the refining of germanium and it is based on the principle that the impurities are more soluble in the melt than in the solid state of the metal.

**(b)** PbS, Sulphide ore has more tendency to stick to the oil which comes on the surface being lighter and easily skimmed off so PbS is concentrated by froth floatation method.

**(c)** Leaching of alumina is done to remove the impurities like $SiO_2$ by using NaOH solution and pure alumina is obtained.

**18. Write structures of compounds A,B and C in each of the following reactions :** **[3]**

**(i)** $C_6H_5Br \xrightarrow{\text{Mg/dry ether}} A \xrightarrow[\text{(b) } H_3O^+]{\text{(a) } CO_{2(g)}} B \xrightarrow{PCl_5} C$

**(ii)** $CH_3CN \xrightarrow[\text{(b) } H_3O^+]{\text{(a) } SnCl_2/HCl} A \xrightarrow{\text{dil. NaOH}} B \xrightarrow{\Delta} C$

**OR**

**Do the following conversions in not more than two steps :**

**(a) Benzoic acid to Benzaldehyde**

**(b) Ethyl benzene to Benzoic acid**

**(c) Propanone to Propene**

**Ans. (i)** $$C_6H_5Br \xrightarrow[\text{dry ether}]{Mg/} \underset{(A)}{C_6H_5MgBr} \xrightarrow[\text{(b) } H_3O^+]{\text{(a) } CO_2(g)}$$

$$\underset{(B)}{C_6H_5COOH} \xrightarrow{PCl_5} \underset{(C)}{C_6H_5COCl}$$

**(ii)** $$CH_3CN \xrightarrow[\text{(b) } H_3O^+]{\text{(a) } SnCl_2/HCl} \underset{(A)}{CH_3CHO} \xrightarrow{\text{dil. NaOH}}$$

$$\underset{(B)}{CH_3CH(OH)CH_2CHO} \xrightarrow{\Delta} \underset{(C)}{CH_3CH{=}CHCHO}$$

**OR**

**(a)**

$$\underset{\text{Benzoic acid}}{C_6H_5COOH} \xrightarrow{SOCl_2} C_6H_5COCl \xrightarrow[BaSO_4]{H_2/Pd} \underset{\text{Benzaldehyde}}{C_6H_5CHO}$$

**(b)** $$\underset{\text{Ethylbenzene}}{C_6H_5C_2H_5} \xrightarrow[\Delta]{KMnO_4\text{-}KOH} C_6H_5COOK \xrightarrow{H_3O^+} \underset{\text{Benzoic acid}}{C_6H_5COOH}$$

**(c)**

$$\underset{\text{Propanone}}{CH_3COCH_3} \xrightarrow{NaBH_4} CH_3CH(OH)CH_3 \xrightarrow[\text{dehydration}]{\text{Conc. } H_2SO_4} \underset{\text{Propene}}{CH_3CH{=}CH_2}$$

**19. Write the structure of the monomers used for getting the following polymers :**

**(i) Dacron** **[3]**

**Ans (i) Monomers of Dacron :**

$$\underset{\text{Ethylene glycol}}{HOH_2C-CH_2OH} \text{ and } \underset{\text{Terephthalic acid}}{HOOC-C_6H_4-COOH}$$

**21. Give reasons :**

**(i) Thermal stability decreases from $H_2O$ to $H_2Te$.**

**(ii) Fluoride ion has higher hydration enthalpy than chloride ion.**

**(iii) Nitrogen does not form pentahalide.** **[3]**

**Ans. (i)** As we move down in a group atomic radius increases as a result bond length increases. Larger the bond length lesser will be the bond dissociation enthalpy. So thermal stability decreases from O to Te.

**(ii)** Fluoride ion is the smallest ion in the group and it has high charge density and charge size ratio. That is why it has high hydration enthalpy.

**(iii)** Nitrogen belongs to 2nd period and there is absence of 3*d* orbital so it can not extend its covalency beyond 3. Thus it does not forms pentahalides.

**24. (a) Account for the following :**

**(i) Transition metals form large number of complex compounds.**

**(ii) The lowest oxide of transition metal is basic whereas the highest oxide is amphoteric or acidic.**

**(iii) E° value for the $Mn^{3+}/Mn^{2+}$ couple is highly positive (+1.57 V) as compare to $Cr^{3+}/Cr^{2+}$.**

**(b) Write one similarity and one difference between the chemistry of lanthanoid and actinoid elements.** **[5]**

**OR**

**(a) (i) How is the variability in oxidation states of transition metals different from that of the *p*-block elements ?**

**(ii) Out of $Cu^+$ and $Cu^{2+}$, which ion is unstable in aqueous solution and why ?**

**(iii) Orange colour of $Cr_2O_7^{2-}$ ion changes to yellow when treated with an alkali. Why ?**

**(b) Chemistry of actinoids is complicated as compared to lanthanoids. Give two reasons.**

**Ans. (a) (i)** Transition metals forms large number of complexes due to :

1. Small size of atoms and ions of transition metals.
2. High nuclear charge.
3. Presence of incompletely filled *d*-orbitals.

**(ii)** As the oxidation state increases the size of ion goes on decreasing thus the covalent character increases as a result of this amphoteric and acidic strength increases. While in case of lower oxides of transition metals ionic size increases and thus basic character increases. Therefore higher the oxidation state, higher is the acidic character.

**(iii)** E° value for the $Mn^{3+}/Mn^{2+}$ couple is highly positive (+1.57V) as compare to $Cr^{3+}$ $Cr^{2+}$ because $Mn^{2+}$ has $3d^5$ as a stable oxidation state which is half filled and stable. Mn has very high third ionization energy for change from $d^5$ to $d^4$ but in case of $Cr^{3+}$, $3d^3$ is more stable due to completely half filled $t_{2g}^3$ orbitals (crystal field spitting theory) and that is why $Mn^{3+}/Mn^{2+}$ is highly positive as compared to $Cr^{3+}/Cr^{2+}$.

**(b)** Both Lanthanoids and Actinoids have the +3 oxidation state and both show contraction or irregular electronic configuration while the major difference between the lanthanoids and actinoids is actinoids are radioactive while lanthanoids are not radioactive in nature.

OR

**(a) (i)** In $p$ block elements the difference in oxidation state is 2 and in transition elements the difference is 1.

**(ii)** $Cu^+$ is unstable in aq. solution because it undergoes disproportionation reaction and has low hydration enthalpy.

**(iii)** In alkaline medium dichromate ions $Cr_2O_7^{2-}$ changes to chromate ion $CrO_4^{2-}$ , which is yellow in colour due to which the colour changes when treated with an alkali.

**(b)** Chemistry of actinoids is complicated as compared to lanthanoids due to the following reasons :

1. They show multiple oxidation states namely +5, +6 and +7 oxidation states respectively which permits the formation of higher oxidation states through the removal of the periphery electrons.
2. They are radioactive and have a strong propensity to form complex reactions because of its unstable isotopes, some actinoids are formed naturally by radioactive decay.

**25. (a) An element has atomic mass 93g $mol^{-1}$ and density 11.5 g $cm^{-3}$. If the edge length of its unit cell is 300 pm, identify the type of unit cell.**

**(b) Write any two differences between amorphous solids and crystalline solids. [5]**

OR

**(a) Calculate the number of unit cells in 8.1 g of aluminium if it crystallizes in a f.c.c. structure. (Atomic mass of Al = 27 g $mol^{-1}$)**

**(b) Give reasons :**

**(i) In stoichiometric defects, NaCl exhibits Schottky defect and not Frenkel defect.**

**(ii) Silicon on doping with phosphorous forms $n$-type semiconductor.**

**(iii) Ferrimagnetic substances show better magnetism than antiferromagnetic substances.**

**Ans. (a)** Given :

Atomic mass = 93 g $mol^{-1}$

Density = 11.5 g $cm^{-3}$

Edge length = 300 pm

We know that

$$\rho = \frac{Z \times m}{a^3 \times N_A}$$

$$11.5 = \frac{Z \times 93}{\left[(300 \times 10^{-10})3 \times 6.022 \times 10^{23}\right]}$$

$$Z = 2$$

$\therefore$ the unit cell is Body centered structure (BCC) since Z = 2.

**(b) Difference between Crystalline and Amorphous solids :**

| Crystalline solids | Amorphous solids |
|---|---|
| 1. The internal arrangement of particles is regular. | The internal arrangement of particles is irregular. |
| 2. They have sharp melting point. | They do not have sharp meting point. |

OR

**(a)** Given: Aluminium crystallizes in F.C.C. structure.

Atomic mass of Al = 27 g $mol^{-1}$

we know that :

$$n = \frac{\text{Given mass}}{\text{Molar mass} \times \text{Avogadro's number}}$$

$$= \frac{8.1}{27 \times 6.022 \times 10^{23}}$$

Number of atoms in one unit cell = 4 (fcc)

$$\text{Number of unit cells} = \frac{8.1}{27 \times 6.022 \times 10^{23} / 4}$$

$$= 4.5 \times 10^{22}.$$

Thus, 8.1 g of aluminium has $4.5 \times 10^{22}$ unit cells.

**(b) (i)** In stoichiometric defects NaCl exhibits schottky defect due to the comparable size of $Na^+$ (cation) and $Cl^-$ (anion). So it can't occupy the interstitial sites and does not shows Frenkel defect.

**(ii)** P has 5 valence electrons, when it is doped with silicon which has 4 valence electrons it leaves one electron. This extra electron helps in conducting current and results in the formation of *n*-type semiconductor.

**(iii)** Ferrimagnetic substances shows better magnetism than antiferromagnetic substances because in ferrimagnetic substances domains or magnetic moments are aligned in opposite directions in unequal numbers whereas in the antiferromagnetic substances domains are aligned in opposite direction in equal numbers so they cancel the magnetic moments completely and thus, the net magnetism is zero

**26. (a) Write the product(s) in the following reactions:**

(i) [Salicylic acid: benzene ring with OH and COOH] $\xrightarrow[H^+]{(CH_3CO)_2O}$ ?

(ii)

$$CH_3-\underset{\displaystyle CH_3\,|}{CH}-O-CH_2-CH_3 \xrightarrow{HI} ? + ?$$

(iii)

$$CH_3-CH=CH-CH_2-OH \xrightarrow{PCC} ?$$

**(b) Give simple chemical tests to distinguish between the following pairs of compounds:**

**(i) Ethanol and Phenol**

**(ii) Propanol and 2-methylpropan-2-ol [5]**

**OR**

**(a) Write the formula of reagents used in the following reactions :**

**(i) Bromination of phenol to 2, 4, 6-tribromophenol**

**(ii) Hydroboration of propene and then oxidation to propanol.**

**(b) Arrange the following compound groups in the increasing order of their property indicated:**

**(i) *p*-nitrophenol, ethanol, phenol (acidic character)**

**(ii) Propanol, propane, Propanal (boiling point)**

**(c) Write the mechanism (using curved arrow notation) of the following reaction :**

$$CH_3-CH_2-\overset{+}{O}H_2 \xrightarrow{CH_3CH_2OH}$$

$$CH_3-CH_2-\underset{\displaystyle |\atop H}{\overset{+}{O}}-CH_2-CH_3 + H_2O$$

**Ans. (a) (i)**

[Salicylic acid: benzene ring with OH and COOH] $\xrightarrow[H^+]{(CH_3CO)_2O}$ [Benzene ring with COOH and $OCOCH_3$]

(ii)

$$CH_3-\underset{\displaystyle CH_3\,|}{CH}-O-CH_2-CH_3 \xrightarrow{HI} (CH_3)_2\,CHOH + CH_3CH_2I$$

(iii) $CH_3-CH=CH-CH_2-OH \xrightarrow{PCC} CH_3CH=CHCHO$

**(b) (i) Ethanol and phenol :** When neutral ferric chloride is added to both the compounds phenol gives violet coloured complex whereas ethanol does not gives this complex when treated with ferric chloride solution.

**(ii) Propanol and 2-methyl propan-2-ol :** When both the solutions were treated with anhydrous $ZnCl_2$ and conc. HCl (Luca's test) the 2-methylpropan-2-ol gives the turbidity immediately whereas propanol does not gives the turbidity immediately.

**OR**

(a) (i) Aq-$Br_2$

(ii) $B_2H_6$ and then $H_2O_2$ and $OH^-$

(b) (i) Ethanol < Phenol < *p*-Nitrophenol.

(ii) Propane < Propanal < Propanol

(c) $CH_3-CH_2-OH_2^+ \xrightarrow{CH_3CH_2OH}$

$$CH_3-CH_2-\underset{\displaystyle |\atop H}{\overset{+}{O}}-CH_2-CH_3 + H_2O$$

Mechanism for this above reaction is :

$$CH_3CH_2-\overset{+}{O}\begin{matrix}H\\H\end{matrix} + CH_3-CH_2-\underset{\displaystyle |\atop H}{\ddot{O}}: \longrightarrow$$

$$CH_3-CH_2-\underset{\displaystyle |\atop H}{\overset{+}{O}}-CH_2-CH_3 + H_2O$$

## Delhi [Set-II]

**Note : Except for the following questions, all the remaining questions have been asked in previous set.**

**2. Write the structure of 2, 4-dinitrochlorobenzene.** **[1]**

**Ans.**

(Structure: benzene ring with Cl at position 1, $NO_2$ at positions 2 and 4)

(2, 4-dinitrochlorobenzene)

**4. Write IUPAC name of the following compound: $CH_3NHCH(CH_3)_2$** **[1]**

**Ans.** N-methylpropan-2-amine.

**5. Write the formula of an oxo-anion of chromium (Cr) in which it shows the oxidation state equal to its group number.** **[1]**

**Ans.** Chromium belongs to group number 6 and its oxidation state in $K_2Cr_2O_7$ is +6 *i.e.*, $K_2Cr_2O_7$

$$1 \times 2 + 2x + (-2 \times 7) = 0$$

$$2 + 2x - 14 = 0$$

$$2x - 12 = 0$$

$$2x = 12$$

$$x = 6$$

Thus, the formula of the oxo-anion is $K_2Cr_2O_7$

**7. Draw the structures of the following :**

**(i) $H_3PO_2$** **[2]**

**Ans.**

(Structure: P with =O, two H and one OH)

**8. Define the following terms :**

**(i) Ideal solution**

**(ii) Molarity (M)** **[2]**

**Ans. (i)** Ideal solutions are those solutions that obeys Raoult's law over entire range of concentration. Example : Benzene and toluene, *n*-heptane and *n*-hexane.

**(ii)** Molarity is defined as the number of moles of solute dissolved per liter of solution.

$$M = \frac{W_b \times 1000}{M_b \times V}$$

**9. Complete the following reactions :** **[2]**

**(i) $Cl_2 + H_2O \rightarrow$**

**(ii) $XeF_6 + 3H_2O \rightarrow$**

**OR**

**What happens when**

**(i) conc. $H_2SO_4$ is added to Cu ?**

**(ii) $SO_3$ is passed through water ?**

**Write the equations.**

**Ans. (i)** $Cl_2 + H_2O \rightarrow 2HCl + [O]$

**(ii)** $XeF_6 + 3H_2O \rightarrow XeO_3 + 6HF$

**OR**

**(i)** $Cu + 2H_2SO_4 \rightarrow CuSO_4 + SO_2 + 2H_2O$

**(ii)** $SO_3 + H_2O \rightarrow H_2SO_4$

**10. Write the reactions involved in the following :**

**(i) Hell-Volhard-Zelinsky reaction**

**(ii) Decarboxylation reaction** **[2]**

**Ans. (i) Hell-Volhard Zelinsky reaction :**

$$\underset{\text{Acetic acid}}{CH_3COOH} \xrightarrow[-HCl]{+Cl_2,\ Red\ P} \underset{\text{Monochloro acetic acid}}{CH_2Cl.COOH} \xrightarrow[-HCl]{+Cl_2,\ Red\ P}$$

$$\underset{\text{Dichloro acetic acid}}{CHCl_2.COOH} \xrightarrow[-HCl]{+Cl_2,\ Red\ P} \underset{\text{Trichloro acetic acid}}{CCl_3.COOH}$$

**(ii) Decarboxylation reaction :**

$$\underset{\text{Sodium acetate}}{CH_3COONa} \xrightarrow[\Delta]{NaOH\ and\ CaO} \underset{\text{Methane}}{CH_4} + Na_2CO_3$$

**13. Write the principles of the following methods :**

**(i) Vapour phase refining** **[3]**

**Ans. (i) Vapour phase refining :** It is based on the principle that the metal is converted into its volatile compound and collected elsewhere. It is then decomposed at high temperature to give pure metal.

**15. Define the following :**

**(iii) Disinfectants** **[3]**

**Ans. (iii) Disinfectants :** These are the substances that are applied to non-living objects to destroy microorganisms that are present on the objects. Example : 1-2% Phenol solution.

# Delhi [Set-III]

**Note : Except for the following questions, all the remaining questions have been asked in previous sets.**

**4. Write the structure of 3-Bromo-2-methylprop-1-ene.** **[1]**

**Ans.** $BrCH_2(CH_3)C = CH_2$

**5. Write IUPAC name of the following compound: $(CH_3)_2N–CH_2CH_3)$** **[1]**

**Ans.** N,N-dimethylethanamine

**6. Write the reactions involved in the following reactions :**

**(i) Clemmensen reduction**

**(ii) Cannizzaro reaction** **[2]**

**Ans. (i) Clemmensen reduction :**

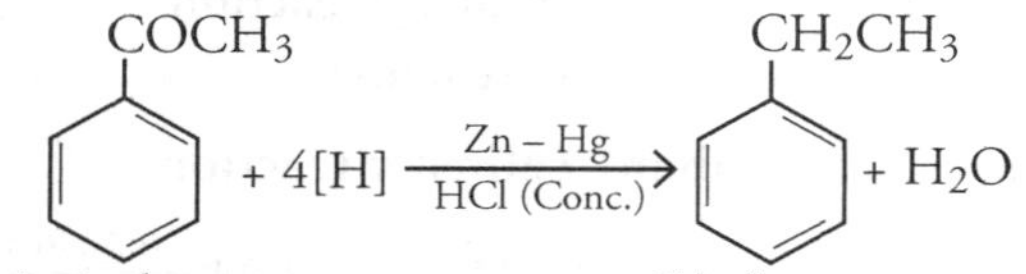

$$C_6H_5COCH_3 \text{ (Acetophenone)} + 4[H] \xrightarrow[\text{HCl (Conc.)}]{\text{Zn – Hg}} C_6H_5CH_2CH_3 \text{ (Ethylbenzene)} + H_2O$$

**(ii) Cannizzaro reaction :**

$$2C_6H_5CHO \text{ (Benzaldehyde)} \xrightarrow{\text{Conc. KOH}} C_6H_5CH_2OH \text{ (Benzylalcohol)} + C_6H_5COO^-Na^+ \text{ (Sodium benzoate)}$$

**7. Draw the structures of the following :**

**(i) $H_4P_2O_7$**

**(ii) $XeOF_4$** **[2]**

**Ans (i)** $H_4P_2O_7$

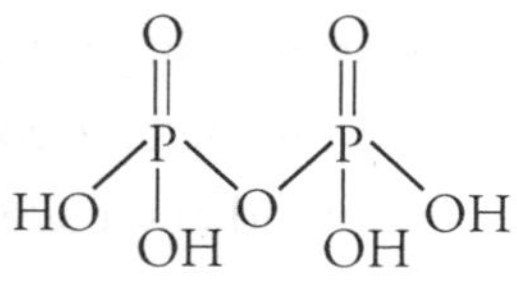

(Structure: HO–P(=O)(OH)–O–P(=O)(OH)–OH)

**(ii)** $XeOF_4$

(Structure: Xe with one double-bonded O, four F atoms and one lone pair)

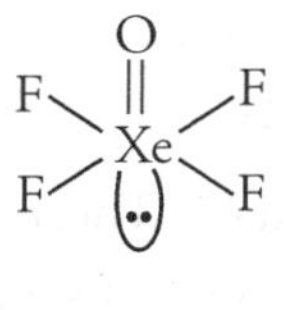

**8. Define the following terms :**

**(ii) van't Hoff factor (*i*)** **[2]**

**Ans (ii) van't Hoff factor (i) :** It is defined as the extent of dissociation or association or the ratio of the observed colligative property to the calculated colligative property.

**10. Complete the following chemical equations :[2]**

**(i) $F_2 + 2Cl^- \rightarrow$**

**(ii) $2XeF_2 + 2H_2O \rightarrow$**

**OR**

**What happens when**

**(i) HCl is added to $MnO_2$ ?**

**(ii) $PCl_5$ is heated ?**

**Write the equations involved.**

**Ans.** (i) $F_2 + 2Cl^- \rightarrow 2F^- + Cl_2$

(ii) $2XeF_2 + 2H_2O \rightarrow 2Xe + 4HF + O_2$

**OR**

(i) $MnO_2 + 4HCl \rightarrow MnCl_2 + Cl_2 + 2H_2O$

(ii) $PCl_5 \xrightarrow{\Delta} PCl_3 + Cl_2$

## SECTION-C

**14. Write the structures of the monomers used for getting the following polymers :**

**(i) Nylon-6** **[3]**

**Ans. (i) Monomers of Nylon-6 :**

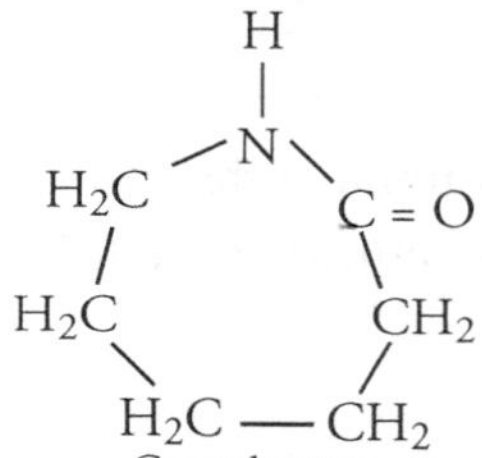

(Structure: seven-membered ring N(H)–C(=O)–$CH_2$–$CH_2$–$H_2C$–$H_2C$–$H_2C$–)

Caprolactum

**19. Write one difference between each of the following :**

**(i) Multimolecular colloid and Macromolecular colloid**

**(ii) Sol and Gel**

**(iii) O/W emulsion and W/O emulsion** **[3]**

**Ans. (i)** In multimolecular colloids a large number of atoms or smaller molecules of a substance aggregates together to form species having size in the colloidal range. Example : Sulphur sol whereas in macro-molecular colloids the colloidal particles are large molecules having colloidal dimensions. Example: Starch.

(ii) In sol the dispersing phase is solid and dispersing medium is liquid; Example: paint, gold sol etc., whereas in Gel the dispersing phase is liquid and dispersing medium is solid; Example : Jelly, butter etc.

(iii) In O/W emulsion, oil is the dispersed phase while water is the dispersion medium. Example : milk, vanishing cream etc whereas in W/O emulsion water is the dispersed phase while oil is the dispersion medium. Example: Cold cream, butter etc.

**20. (i) What type of isomerism is shown by the complex $[Co(en)_3]Cl_3$ ?**

**(ii) Write the hybridisation and magnetic character of $[Co(C_2O_4)_3]^{3-}$**

**(At. no. of Co = 27)**

**(iii) Write IUPAC name of the following Complex $[Cr(NH_3)_3Cl_3]$.** [3]

**Ans. (i)** Since the given coordinate compound does not have a plane of symmetry and the ligand attached is bidentate ligand so it will show optical isomerism.

**(ii)** $[Co(C_2O_4)_3]^{3-}$ Co is in + 3 oxidation state with electronic configuration of $3d^6$ Oxalate is a strong field ligand so pairing of electrons take place.

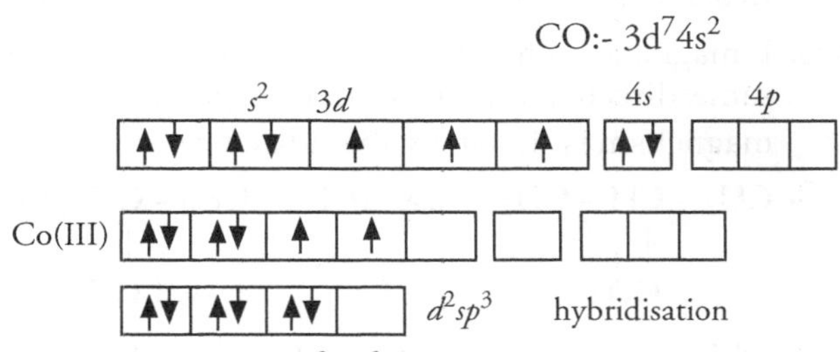


It shows $d^2sp^3$ hybridisation and is diamagnetic in nature.

**(iii)** Triamminetrichloridochromium(III).

# CHEMISTRY

## Outside Delhi [Set-I]

## 2016

Time allowed : 3 hours

Maximum marks : 70

**1. What type of magnetism is shown by a substance if moments of domains are arranged in same direction ? [1]**

**Ans.** If magnetic moments of domains are arranged in same directions then the ferromagnetic type of magnetism is shown by the substance.

**2.** $CH_3-\underset{\underset{CH_3}{|}}{CH}-CH_2-Cl$ and $CH_3-CH_2-\underset{\underset{CH_3}{|}}{CH}-Cl$,

**which is more reactive towards $S_N{}^1$ reaction and why ? [1]**

**Ans.** $CH_3-CH_2-\underset{\underset{CH_3}{|}}{CH}-Cl$, is more reactive towards $SN^1$ reaction as secondary (2°) carbonation formed as the reaction intermediate $\left(CH_3-CH_2-\underset{\underset{CH_3}{|}}{\overset{\oplus}{C}}\right)$ is more stable than primary (1°) carbocation $\left(CH_3\ \underset{\underset{CH_3}{|}}{CH}-\overset{\oplus}{CH_2}\right)$.

**3. On adding NaOH to ammonium sulphate a colourless gas with pungent odour is evolved which forms a blue coloured complex with $Cu^{2+}$ ion. Identify the gas. [1]**

**Ans.** The gas evolved is ammonia ($NH_3$).

$$(NH_4)_2SO_4(aq) + 2NaOH \rightarrow Na_2SO_4\,(aq) + 2H_2O(l) + 2NH_3(g)$$

Ammonia reacts with solution of $Cu^{2+}$ ion to form a deep blue coloured complex, $\left[Cu\left(NH_3\right)_4\right]^{2+}$.

**4. Write the main reason for the stability of colloidal solutions. [1]**

Ans. All the colloidal particles in a given solution carry the same charge and the dispersion medium has an opposite and equal charge; the system as a whole being electrically neutral. This is the main reason for the stability of the colloidal solution.

**5. Write the IUPAC name of the given compound. [1]**

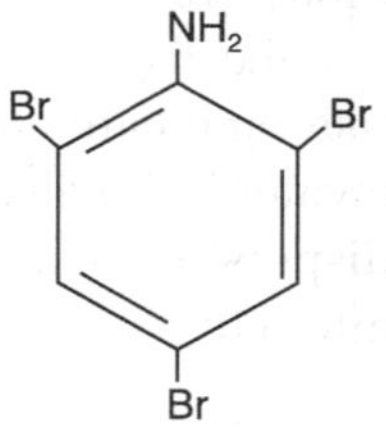

NH₂ at C-1; Br at positions 2, 4, 6 (benzene ring structure)

**Ans.** (Structure with ring numbered 1–6: $NH_2$ at 1, Br at 2, 4, 6)

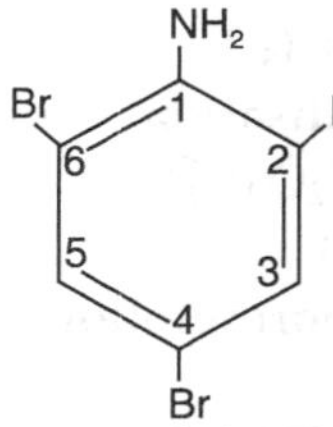

2,4,6-Tribromoaniline.

**6. When a coordination compound $CrCl_3.6H_2O$ is mixed with $AgNO_3$, 2 mole of AgCl are precipitated per mole of the compound. Write [2]**

**(i) Structural formula of the complex.**

**(ii) IUPAC name of the complex**

**Ans. (i)** Structural formula of the complex is $[Cr\,(H_2O)_5\,Cl]\,Cl_2.\,H_2O$, because two moles of chlorine are outside coordination entity to form two moles of AgCl from per mole compound.

**(ii)** Pentaaquachlorido Chromium(III) chloride monohydrate.

**7. From the given cells : [2]**

**Lead storage cell, Mercury cell, Fuel cell and Dry cell**

**Answer the following :**

**(i) Which cell is used in hearing aids ?**

**(ii) Which cell was used in Apollo Space Programme ?**

**(iii) Which cell is used in automobiles and inverters ?**

**(iv) Which cell does not have long life ?**

**Ans. (i)** Hearing aid–Mercury cell.

**(ii)** Apollo Space Programme–Fuel cell.

**(iii)** Automobile and inverters–Lead storage cell.

**(iv)** Cell does not have long life–Dry cell.

**8. When chromite ore $FeCr_2O_4$ is fused with NaOH in presence of air, a yellow coloured compound**

**(A) is obtained which on acidification with dilute sulphuric acid gives a compound (B), compound (B) on reaction with KCl forms a orange coloured crystalline compound (C). [2]**

**(i) Write the formulae of the compounds (A), (B) and (C).**

**(ii) Write one use of compound (C).**

**OR**

**Complete the following chemical equations :**

**(i) $8MnO_4^- + 3S_2O_3^{2-} + H_2O \rightarrow$**

**(ii) $Cr_2O_3^{2-} + 3Sn^{2+} + 14 H^+ \rightarrow$**

**Ans.** (i) On fusing chromite ore with Sodium hydroxide in presence of air, yellow coloured "sodium chromate" (A) is formed.

$$FeCr_2O_4 + 16 NaOH + 7O_2 \rightarrow \underset{(A)}{8Na_2CrO_4} + 2 Fe_2O_3 + 8H_2O$$

On acidification with dil. $H_2SO_4$ it forms sodium dichromate (B).

$$\underset{(A)}{2Na_2CrO_4} + H_2SO_4 \rightarrow \underset{(B)}{Na_2Cr_2O_7} + Na_2SO_4 + H_2O.$$

Compound (B) *i.e.* sodium dichromate forms potassium dichromate, orange coloured crystals (C) on treating with KCl.

$$\underset{(B)}{Na_2Cr_2O_7} + 2KCl \rightarrow \underset{(C)}{K_2Cr_2O_7} + 2NaCl.$$

The formula of compounds are :

(A) Sodium chromate —$Na_2CrO_4$

(B) Sodium dichromate —$Na_2Cr_2O_7$

(C) Potassium dichromate —$K_2Cr_2O_7$

(ii) Potassium dichromate, $K_2Cr_2O_7$ is most commonly used as on oxidizing agent in various laboratory and industrial applications and as a primary stand and in volumetric analysis.

**OR**

(i) $8MnO_4^- + 3S_2O_3^{2-} + H_2O \rightarrow 8MnO_2 + 6SO_4^{2-} + 2OH^-$

(ii) $Cr_2O_7^{2-} + 3Sn^{2+} + 14H^+ \rightarrow 3Sn^{4+} + 2Cr^{3+} + 7H_2O$

**9. Write the mechanism of the following reaction: [2]**

$$2CH_3CH_2OH \xrightarrow[413K]{Conc.H_2SO_4} CH_3CH_2 - O - CH_2 - CH_3$$

**Ans.** Formation of ether from alcohol is a nucleophilic bimolecular reaction ($S_N2$). A protonated alcohol is attacked by another alcohol molecule.

**Reaction Steps :**

$$CH_3CH_2—\ddot{O}—H + H^+ \longrightarrow CH_3—CH_2—\overset{+}{\ddot{O}}(H)—H$$

$$CH_3—CH_2—\ddot{O}(H): + CH_3—CH_2—\overset{+}{O}(H)—H \xrightarrow{-H_2O} CH_3CH_2—\overset{+}{O}(H)CH_2CH_3 + H_2O$$

$$CH_3CH_2—\overset{+}{O}(H)—CH_2CH_3 \xrightarrow{-H^+} CH_3—CH_2—O—CH_2CH_3 + H^+$$

**10. For a reaction: $2NH_3 (g) \xrightarrow[Rate = k]{Pt} N_2 (g) + 3H_2 (g)$**

**(i) Write the order and molecularity of this reaction.**

**(ii) Write the unit of *k*. [2]**

**Ans.** (i) For any reaction ; Rate = K [A] order

[A] = concentration of reactant.

Hence it is a zero order reaction and its molecularity is two.

(ii) Unit of K for a zero order reaction is mol $L^{-1}$ $sec^{-1}$.

**11. The rate constant for the first order decomposition of $H_2O_2$ is given by the following equation :**

$$\log k = 14.2 - \frac{1.0 \times 10^4}{T} K$$

**Calculate $E_a$ for this reaction and rate constant K if its half-life period be 200 minute. (Given : R = 8.314 J $K^{-1}$ $mol^{-1}$) [3]**

**Ans.** According to Arrhenius equation

$$\log k = \log A - \frac{1.0 \times 10^4}{T} K \quad ...(i)$$

So equating similar terms in the equations

$$\log K = 14.2 - \frac{1.0 \times 10^4}{T} K \quad ...(ii)$$

On comparing the equation no. **(i)** and **(ii)**

$$\frac{E_a}{2.303 RT} = \frac{1.0 \times 10^4}{T} K ;$$

$$E_a = \frac{2.30 \times R \times T \times 1.0 \times 10^4}{T} K$$

$$\Rightarrow E_a = 2.303 \times 8.314 K^{-1} Mol^{-1} \times 1.0 \times 10^4 K$$

$$= 19.14 \times 10^4 J Mol^{-1} = 191.4 kJ mol^{-1}$$

For a first order reaction :

$$t_{1/2} = \frac{0.693}{200} \quad ...(iii)$$

$$t_{1/2} = 200 \text{ min}$$

$$K = \frac{0.693}{200} = 0.0034\ min^{-1}$$

**12. (i) Differentiate between adsorption and absorption.**

**(ii) Out of $MgCl_2$ and $AlCl_3$, which one is more effective in causing coagulation of negatively charged sol and why ?**

**(iii) Out of sulphur sol and proteins, which one forms multimolecular colloids ? [3]**

**Ans. (i)**

| | Adsorption | Absorption |
|---|---|---|
| **1.** | It is a surface phenomenon, Adsorbate molecules are held at the surface of adsorbent. | Absorption occurs in the bulk of absorbing substance. It is a Bulk phenomena. |
| **2.** | Initially, rate of adsorption is rapid. | Absorption occurs at uniform rate. |
| **3.** | The concentration of the adsorbent surface is much more than that in the bulk. E.g., water vapours on silica gel. | Absorbed material is uniformly distributed throughout the bulk means concentration is same throughout and it is essentially a bulk phenomenon. E.g., Water vapours are absorbed by anhydrous $CaCl_2$. |

**(ii)** According to Hardy-Schulze law the ions carrying opposite charge to that on sol are responsible for coagulation of the sol. These are called active ions. Hence as the sol is negative, $Mg^{2+}$ and $Al^{3+}$ ions will cause coagulation.

As coagulation power of electrolyte is proportional to the valency of oppositely charged ion, so $AlCl_3$ will be more effective than $MgCl_2$.

**(iii)** Sulphur sol will form the multimolecular colloid. A sol of sulphur consists of colloidal particles which are aggregates of $S_8$ molecules.

13. **Give reasons : [3]**

**(i) C–Cl bond length in chlorobenzene is shorter than C–Cl bond length in $CH_3$–Cl.**

**(ii) The dipole moment of chlorobenzene is lower than that of cyclohexyl chloride.**

**(iii) $S_N1$ reactions are accompanied by racemization in optically active alkyl halides.**

**Ans. (i)** C–Cl bond length in chloro-benzene is shorter than $CH_3$–Cl, C–Cl bond as in chlorobenzene due to resonance C–Cl bond has partial double bond character which reduces the bond length.

**(ii)** In cyclohexyl chloride, carbon in C–Cl bond is $sp^3$ hybridised whereas in chlorobenzene C–Cl bond carbon is $sp^2$ hybridised, $sp^2$ is more electronegative than $sp^3$ carbon. Hence C–Cl bond of chlorobenzene is less polar.

**(iii)** In $S_N1$ reaction a planar carbocation intermediate is formed. Carbon in carbocation formed is $sp^2$ hybridised. In case of optically active alkyl halide the attack of nucleophile in the next step to carbocation can occur from both the faces of the trigonal planar species in equal probability. Thus 50 : 50 racemic mixture is obtained.

$Nu-C(R_1)(R_2)(R_3)$ ⟵ $\ddot{N}u$ → $R_1R_2\overset{\oplus}{C}R_3$ ← $\ddot{N}u$ ⟶ $R_2-C(R_1)(R_3)-Nu$

carbocation

**14. An element crystallizes in a f.c.c. lattice with cell edge of 250 pm. Calculate the density if 300 g of this element contain $2 \times 10^{24}$ atoms. [3]**

**Ans.** Density of an element $= \frac{Z \times M\ (\text{molar mass})}{a^3 \times N_A}$

$a$ (edge length) = 250 pm = $250 \times 10^{-10}$ cm

$N_A$ (Avogadro's no.) = $6.022 \times 10^{23}$

Z (no. of elements in unit cell) = 4 in f.c.c.

M (molar mass) = mass of $6.022 \times 10^{23}$ atoms of element.

given that $2 \times 10^{24}$ atoms are in 300 g of element.

Hence $6.022 \times 10^{23}$ atoms are in.

$$\frac{300 \times 6.022 \times 10^{23}}{2 \times 10^{24}}\ g = M$$

Substitute the value in density formula.

$$d = \frac{4 \times 300 \times 6.022 \times 10^{23}}{(2.50 \times 10^{-10}\ cm)^3 \times 6.022 \times 10^{23} \times 2 \times 10^{24}}$$

$$d = 38.4 g/cm^3$$

**15. Give reasons :** **[3]**

**(i) Mn shows the highest oxidation state of + 7 with oxygen but with flourine it shows the highest oxidation state of + 4**

**(ii) Transition metals show variable oxidation states.**

**(iii) Actinoids show irregularities in their electronic configurations.**

**Ans. (i)** Mn shows the highest oxidation state of +7 with oxygen because it can form $p\pi$-$d\pi$ multiple bonds. On the other hand Mn shows highest oxidation state of +4 with flourine because it can form only single bond.

**(ii)** Transition metals show variable oxidation state because of use of ns and ($n$–1) d shell electrons while bonding as the shells have similar energy.

**(iii)** Actinoids show irregularities in their electronic configurations because 6$d$, 7$s$ and 5$f$ electrons or shells have less energy difference and electrons can be accommodated in any of them.

**16. Write the main product(s) in each of the following reactions :** **[3]**

(i) $CH_3-C(CH_3)_2-O-CH_3 + HI \rightarrow$

(ii) $CH_3-CH=CH_2 \xrightarrow[\text{(ii) } 3H_2O_2/OH^-]{\text{(i) } B_2H_6}$

(iii) $C_6H_5-OH \xrightarrow[\text{(ii) } CO_2, H^+]{\text{(i) aq. NaOH}}$

**Ans. (i)** $CH_3-C(CH_3)_2-O-CH_3 + HI \longrightarrow CH_3-C(CH_3)_2-I + CH_3OH$

(ii) $3CH_3-CH=CH_2 \xrightarrow[\text{(ii) } 3H_2O_2/OH^-]{\text{(i) } B_2H_6}$

$3CH_3-CH_2-CH_2-OH + B(OH)_3$

Antimarkovnikov addition.

(iii) $C_6H_5OH \xrightarrow[\text{(ii) } CO_2, H^+]{\text{(i) aq. NaOH}}$ (o-HO–$C_6H_4$–COOH)

Salicylic acid

**17. (i) Name the method of refining of metals such as Germanium.**

**(ii) In the extraction of Al, impure $Al_2O_3$ is dissolved in conc. NaOH to form sodium aluminate and leaving impurities behind. What is the name of this process ?**

**(iii) What is the role of coke in the extraction of iron from its oxides ?** **[3]**

**Ans. (i)** Zone refining method is used for refining of metals such as germanium which is based on the principle that the impurities are more soluble in the molten state (melt) than in the solid state of the metal.

**(ii) Leaching :** This method consists of treating the powdered ore with a suitable reagent which can selectively dissolve the ore but not the impurities.

**(iii)** Coke act as a reducing agent and it reduces the iron ore hematite.

**18. Calculate e.m.f. of the following cell at 298 K.** **[3]**

**$2Cr(s) + 3Fe^{2+} (0.1M) \rightarrow 2Cr^{3+} (0.01M) + 3\,Fe(s)$**

**Given : $E^\circ_{(Cr^{3+}|Cr)} = -0.74$ V $E^\circ_{(Fe^{2+}|Fe)} = -0.44$ V**

**Ans.** $Cr|Cr^{3+}(0.01M)\,||\,Fe^{2+}(0.1M)|Fe_{(s)}$.

The half cell reactions for they given cell :

$Cr \rightarrow Cr^{3+} + 3e^-] \times 2$

$Fe^{2+} + 2e^-\ Fe\ ] \times 3$

Hence $n = 6$, T = 298 K, R = 8.314 J $K^{-1}$ $mol^{-1}$.

Nernst equation for the cell ;

$$E_{cell} = E^\circ_{cell} - \frac{2.303RT}{nF}\log\frac{[Cr^{3+}]^2}{[Fe^{2+}]^3}$$

$$E^\circ cell = [E^\circ_R - E^\circ_L]V$$

$$[-0.44 - (-0.74)]V = +0.30\text{ V.}$$

$$E_{cell} = 0.30\text{ V} - \frac{2.303\times8.314\times298}{6\times96500}\log\frac{[0.01]^2}{[0.1]^3}$$

$$= 0.30V - \frac{0.059}{6}\log\frac{[10^{-2}]^2}{[0.1]^3}$$

$$= 0.30 - \frac{0.059}{6}\log 10^{-1}\text{ V} = 0.30\text{ V} - \frac{0.059}{6}(-1)\text{ V}$$

$= 0.30 \text{ V} + \frac{0.059}{6} \text{ V} = 0.30 + 0.0098$

$= 0.3098 \text{ V}$

**19. (i) Write the name of two monosaccharides obtained on hydrolysis of lactose sugar.**

**(ii) Why Vitamin C cannot be stored in our body?**

**(iii) What is the difference between a nucleoside and nucleotide ? [3]**

**Ans. (i)** The two monosaccharides are β-D-galactose and β-D-glucose.

**(ii)** Vitamin C is, a water soluble vitamin and hence get excreted by urine. So it cannot be stored in body and needs to be supplemented regularly.

**(iii)** When a base (purine or pyriamidine) get attached to 1′ position of a pentose sugar a nucleoside is formed. When a nucleoside is further linked to phosphoric acid at 5′ position of sugar moeity, we get a nucleotide.

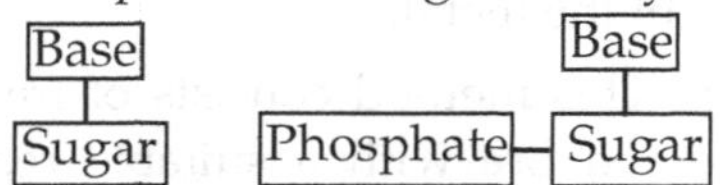


**20. (a) For the complex $[Fe(CN)_6]^{3-}$, write the hybridization type, magnetic character and spin nature of the complex. (At number Fe = 26).**

**(b) Draw one of the geometrical isomers of the complex $[Pt(en)_2Cl_2]^{2+}$, which is optically active ? [3]**

**Ans. (a)** $[Fe(CN)_6]^{3-}$, Fe, Z = 66 $[Ar] 4s^2 3d^6$

Fe has +3 oxidation state : Electronic configuration; $[Ar] 4s^0 3d^5$

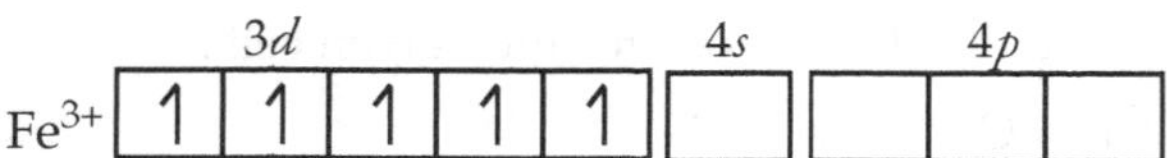


$CN^-$ is a strong field ligand hence pairing will take place to accommodate 6 pair of ligand electrons.

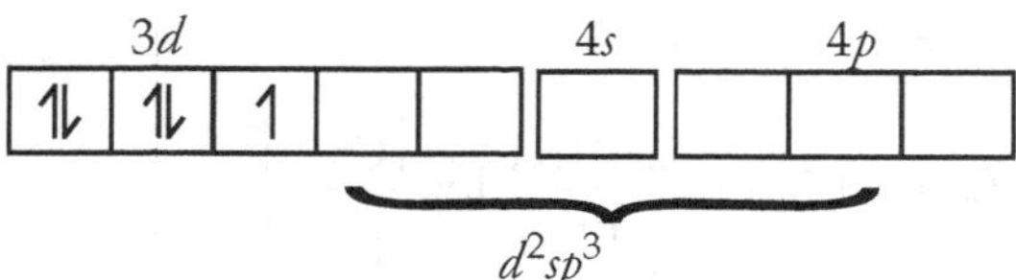


Hence the complex has $d^2sp^3$ hybridization.

Type : Octahedral complex, inner orbital complex.

Magnetic character : One unpaired electron hence paramagnetic.

It will have a total electron spin moment of 1 electron :

$(\mu) = \sqrt{1(1+2)} = \sqrt{3}$ BM, ∴ Low spin complex

**(b)** $[Pt(en)_2Cl_2]^{2+}$, that is isomer is optically active.

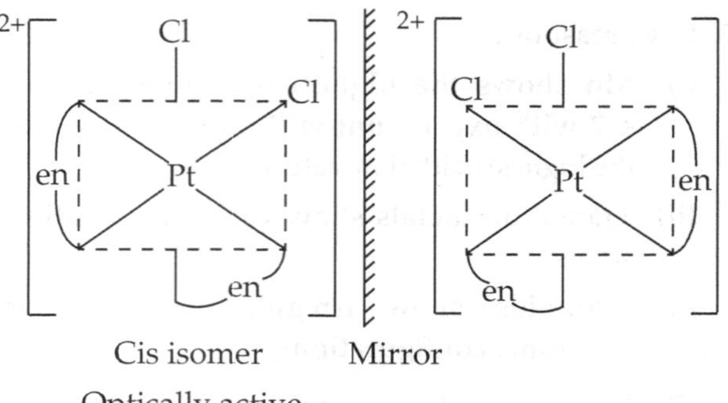


Cis isomer Mirror

Optically active.

**21. Write the structure of A, B and C in the following :**

**(i)**

$C_6H_5—CO—NH_2 \xrightarrow{aq./KOH} A \xrightarrow[0–5° C]{NaNO_2 + HCl} B \xrightarrow{KI} C$

**(ii)**

$CH_3—Cl \xrightarrow{KCN} A \xrightarrow{LiAlH_4} B \xrightarrow{CHCl_3 + alc KOH} C$ **[3]**

**Ans. (i)** $C_6H_5—CO—NH_2 \xrightarrow{Br_2/aq. KOH} C_6H_5NH_2$ (A) Aniline

$\xrightarrow[0–5°C]{NaNO_2 + HCl} C_6H_5N^+_2Cl^-$ (B) Benzene diazonium chloride $\xrightarrow{KI} C_6H_5I$ (C) Iodobenzene

**(ii)** $CH_3Cl \xrightarrow{KCN} CH_3CN$ (A) Ethanenitrile $\xrightarrow{LiAlH_4} CH_3CH_2NH_2$ (B) Ethanamine

$\xrightarrow[\Delta]{CH_3Cl + alc. KOH} CH_3CH_2NC$ (C) (Ethyl isocyanide)

**22. (i) What is the role of *t*-butyl peroxide in the polymerization of ethane ?**

**(ii) Identify the monomers in the following polymer :**

$—[NH—(CH_2)_6—NH—CO—(CH_2)_4—CO—]_n$

**(iii) Arrange the following polymers in the increasing order of their intermolecular forces.**

**Polystyrene, Terylene, Buna-S. [3]**

**OR**

**Write the mechanism of free radical polymerisation of ethene.**

**Ans. (i)** Polymerisation of ethene to low density polyethene (L.D.P.) needs presence of a free radical generating initiator (catalyst). *t*-butyl peroxide helps in starting the chain of radical formations. It acts as the initiator of free radical.

**(ii)** $\left[-NH-[CH_2]_6NH-\underset{\underset{O}{\|}}{C}-(CH_2)_4-\underset{\underset{O}{\|}}{C}-\right]_n$

This is Nylon–6, 6 and its monomers are :

Hexamethylene diamine

$H_2N-(CH_2)_6-NH_2$ and Adipic acid

$HOOC-(CH_2)_4-COOH$.

**(iii)** Buna–S <Polystyrene <Terylene

**OR**

**Mechanism of Polymerisation of ethene.**

1. **Initiation :** The process starts with the formation of a free radical by addition of catalyst free radical like phenyl or benzoyl etc., generating a new and larger free radical.

$$\underset{\text{Benzoylperoxide}}{C_6H_5-\overset{\overset{O}{\|}}{C}-O-O-\overset{\overset{O}{\|}}{C}-C_6H_5} \longrightarrow 2C_6H_5-\overset{\overset{O}{\|}}{C}-\dot{O}$$

$$\longrightarrow \underset{\text{Phenyl Free radical}}{2\dot{C}_6H_5}$$

$$\underset{\text{Free radical}}{\dot{C}_6H_5} + CH_2 = CH_2 \longrightarrow C_6H_5-CH_2-\dot{C}H_2$$

2. **Propagation :** The radical reacts with another molecule of ethene thus forming a bigger radical molecule. The process continues till the required length of chain we need.

$$C_6H_5-CH_2-\dot{C}H_2 + CH_2 = CH_2 \longrightarrow C_6H_5-CH_2-CH_2-CH_2-\dot{C}H_2$$

$$\downarrow$$

$$C_6H_5-(CH_2-CH_2)_n-CH_2-\dot{C}H_2$$

3. **Chain Termination:** When free radical combine with each other the chain terminates resulting in formation of a polymer.

$$\begin{matrix} C_6H_5(CH_2-CH_2)_n CH_2-\dot{C}H_2 \\ C_6H_5(CH_2-CH_2)_n CH_2-\dot{C}H_2 \end{matrix} \longrightarrow$$

$$C_6H_5 + (CH_2-CH_2)_n CH_2-CH_2-CH_2(CH_2-CH_2)_n C_6H_5$$

**23. Due to hectic and busy schedule Mr. Angad made his life full of tensions and anxiety. He started taking sleeping pills to overcome the depression without consulting the doctor. Mr. Deepak a close friend of Mr. Angad advised him to stop taking sleeping pills and suggested to change his lifestyle by doing Yoga, meditation and some physical exercise. Mr. Angad followed his friend's advice and after few days he started feeling better. After reading the above passage answer the following :** **[4]**

**(i) What are the values (at least two) displayed by Mr. Deepak?****

**(ii) Why is it not advisable to take sleeping pills without consulting doctor?**

**(iii) What are tranquilizers ? Give two examples.**

**Ans. (ii)** Sleeping pills are tranquilizers and may cause harmful side effects as they slow down the working of the brain and nervous system. Hence a doctor must be consulted to regularise the doses of such drugs.

**(iii)** Tranquilizers are neurologically active of drugs or chemicals which are used to treat stress and mental disease. Example : Iproniazid, valium and Equanil.

24. **(a) Write the structures of A, B, C, D and E in the following reactions :** **[5]**

$$C_6H_6 \xrightarrow[\text{Anhyd. }AlCl_3]{CH_3COCl} A \xrightarrow{\text{Zn–Hg/conc. HCl}} B$$

$$A \xrightarrow{NaOI} D + E \xrightarrow[\text{(ii) }H_3O^+]{\text{(i) }KMnO_4\text{–}KOH_2\,\Delta} C$$

**OR**

**(a) Write the chemical equation for the reaction involved in Cannizzaro reaction.**

**(b) Draw the structure of the semicarbazone of ethanal.**

**(c) Why pKa of F–$CH_2$–COOH is lower than that of Cl–$CH_2$–COOH ?**

**(d) Write the product in the following reaction;**

$$CH_3-CH = CH-CH_2CN \xrightarrow[\text{(ii) }H_2O]{\text{(i) DIBAL–H}}$$

**(e) How can you distinguish between propanal and proanone ?**

**Ans. (a)** $$\underset{\text{Benzene}}{C_6H_6} \xrightarrow[\text{AnhyAlCl}_3]{\text{(i) }CH_3COCl} \underset{\underset{(A)}{\text{Acetophenone}}}{C_6H_5-\overset{\overset{O}{\|}}{C}-CH_3}$$

** Answer is not given due to the change in present syllabus.

$$C_6H_5-\overset{\overset{O}{||}}{C}-CH_3 \xrightarrow[\text{conc. HCl}]{\text{Zn-Hg/}} C_6H_5-CH_2-CH_3$$

(A) (B) Ethyl Benzene

(A) $\xrightarrow{\text{NaOI}}$ $C_6H_5COONa + CHI_3$

(D) Sodium Benzoate (E) Iodoform

(B) $\xrightarrow[\Delta]{\text{(i) } KMnO_4/KOH_3 \quad \text{(a) } H_3O^+}$ $C_6H_5COOH$

(C) Benzoic acid

**OR**

**(a)** For aldehydes which do not have α-hydrogen atom self oxidation and reduction takes place in presence of concentrated alkali. This produces one mole of alcohol and one mole of salt of carboxylic acid. This is called Cannizzaro's reaction.

$$\underset{\text{Formaldehyde}}{HCHO + HCHO} + \underset{\text{Sodium hydroxide}}{NaOH(conc.)} \longrightarrow \underset{\text{Methylalcohol}}{CH_3OH} + \underset{\text{(Sodium formate)}}{HCOO^-Na^+}$$

**(b)** $$\underset{\text{Ethanal}}{CH_3CHO} + \underset{\text{Semicarbazide}}{H_2N-NH-\overset{\overset{O}{||}}{C}-NH_2} \longrightarrow \underset{\text{Ethanal Semicarbozone}}{CH_3-CH=N-NH-\overset{\overset{O}{||}}{C}-NH_2}$$

**(c)** pKa of $F-CH_2-COOH$ is lower than that of pKa of $Cl-CH_2-COOH$ as $F-CH_2-COOH$ is a stronger acid. This is because of higher electronegativity of F atom than Cl atom.

**(d)** $$\underset{\text{Nitrile}}{CH_3-CH=CH-CH_2CN} \xrightarrow[\text{(ii) } H_2O]{\text{(i) DIBAL-H}} \underset{\text{aldehyde}}{CH_3-CH=CHCH_2CHO}$$

**(e)** Tollen's reagent will give a positive test of silver mirror formation with propanal, while propane does not gives this test since aldehydes can oxidise Tollens' reagent to metallic silver but ketones cannot.

$$R-CHO + \underset{\text{Tollen's reagent}}{2[Ag(NH_3)_2]^+} + 3OH^- \longrightarrow \underset{\text{Silver colour deposit}}{2Ag\downarrow} + R-COO^- + H_2O$$

**25. (a) Calculate the freezing point of solution when 1.9 g (of $MgCl_2$ (M = 95 g/mol) was dissolved in 50 g of water, assuming $MgCl_2$. Undergoes complete ionization. [5]**

**($K_f$ for water = 1.86 K kg mol$^{-1}$)**

**(b) (i) Out of 1 M glucose and 2 M glucose, which one has a higher boilling point and why ?**

**(ii) What happens when the external pressure applied becomes more than the osmotic pressure of solution ?**

**OR**

**(a) When 2.56 g of sulphur was dissolved in 100 g of $CS_2$, the freezing point lowered by 0.383 K. Calculate the formula of sulphur ($S_X$).[5]**

**($K_f$ for $CS_2$ = 3.83 K kg mole$^{-1}$, Atomic mass of sulphur = 32 g/mol$^{-1}$).**

**(b) Blood cells are isotonic with 0.9% sodium chloride solution what happens if we place blood cells in a solution containing.**

**(i) 1.2% sodium chloride solution ?**

**(ii) 0.4% sodium chloride solution ?**

**Ans. (a)** $MgCl_2$ on ionisation gives 3 ions each mole.

$$MgCl_2(s) \longrightarrow Mg^{2+}(aq) + 2Cl^-(q)$$

Hence Vant's Hoff factor, $i = 3$

$$\Delta T_f = T_{f(water)} - T_{f(MgCl_2)} = i \times K_f \times m$$

$$m(\text{molality}) = \frac{1.9g \times 1000g}{95g\ mol^{-1} \times 50g} = 0.4 \text{ mole}$$

$\therefore$ $T_g(\text{water}) = 273$ K

Hence $\Delta T_f = 273\ K - T_{f(MgCl_2)}$

$= 3 \times 1.86\ K.kg\ mol^{-1} \times 0.4\ molkg^{-1}$

$= 2.23$ K.

$$T_{f(MgCl_2)} = (273 - 2.23)\ K = 270.77\ K$$

**(b) (i)** 2M glucose will have higher boiling point because boiling point of a solution of a non-volatile liquid increases with increase in concentration

**(ii)** When the external pressure exerted on the solution is higher than the osmotic pressure, pure solvent starts flowing out of the solution through the semi permeable membrane. This process is known as reverse-osmosis.

**OR**

**(a)** Weight of sulphur ($W_2$) = 2.5 g, $W_1(CS_2)$ = 100 g,

$$\Delta T_f = 0.383\ K$$

$$M_2 = 32 \text{ g/mol}$$
$$K_f = 3.83 \text{ K kg/ mol}$$

Applying the formula

$$\Delta T_f = \frac{k_f \times W_2 \times 1000}{W_1 \times M}$$

$$0.383 = \frac{3.83 \times 2.56 \times 1000}{100 \times M}$$

$$M = 256 \text{ g/Mol}$$

**Formula of Sulphur :** One atom of S = 32 g $mol^{-1}$

atoms of S in molecule

$$= \frac{256 \text{ gmol}^{-1}}{32 \text{ gmol}^{-1}} = 8 .$$

Formula = $S_8$

**(b) (i)** 1.2 % Sodium chloride is hypertonic than blood cells, hence cells will shrink. Plasmolyis will take place.

**(ii)** 0.4% Sodium chloride solution is hypotonic than blood cell, so cells will swell. Endoosmosis will take place.

**26. (a) Account for the following :** [5]

**(i) Ozone is thermodynamically unstable.**

**(ii) Solid $PCl_5$ is ionic in nature.**

**(iii) Fluorine forms only one oxoacid HOF.**

**(b) Draw the structure of : (i) $BrF_5$, (ii) $XeF_4$**

**OR**

**(i) Compare the oxidising action of $F_2$ and $Cl_2$ by considering parameters such as bond dissociation enthalpy, electron gain enthalpy and hydration enthalpy.**

**(ii) Write the conditions to maximize the yield of $H_2SO_4$ by contact process.**

**(iii) Arrange the following in the increasing order of property mentioned.**

**(a) $H_3PO_3$, $H_3PO_4$, $H_3PO_2$ (Reducing Characters)**

**(b) $NH_3$, $PH_3$, $AsH_3$, $SbH_3$, $BiH_3$ (Base strength)**

**Ans. (a) (i)** Ozone easily decompose to give nascent oxygen :

$$O_3 \rightarrow O_2 + [O]$$

because the reaction is exothermic, ($\Delta H$ = negative). and results in the increases in entropy ($\Delta S$ = positive). Overall Gibb's energy change is quite high and negative.

**(ii)** In solid state $PCl_5$ exist as $[PCl_4]^+ [PCl_6]^-$. The cation is tetrahedral and is formed by breaking of weaker P–Cl axial bonds of trigonal bipyramidal structure to form $[PCl_4]^+$ and it bonds ionically with octahedral $[PCl_6]^-$.

**(iii)** Due to high electronegativity , small size, -1 oxidation state and absence of d-orbitals fluorine forms only one oxoacid, HOF.

**(b) (i)** $BrF_5$

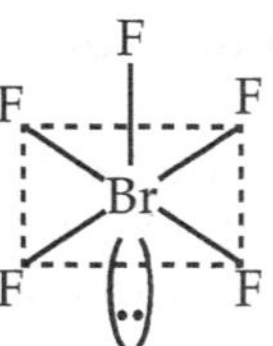


**(ii)** $XeF_4$

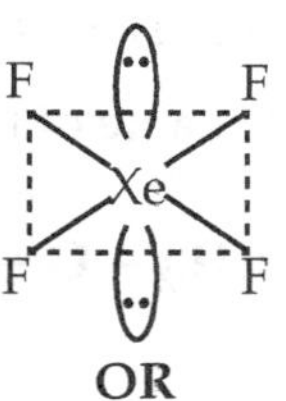


**OR**

**(i)** Fluorine is a much stronger oxidizing agent than chlorine. The oxidizing power depends on three factors.

1. Low enthalpy of dissociation of F-F bond.
2. Less negative electron gain enthalpy of F.
3. High hydration enthalpy of $F^-$ ion.

The electron gain enthalpy of chlorine is more negative than that of Fluorine. However, the bond dissociation energy of fluorine is much lesser than that of chlorine. Also, because of its small size, the hydration energy of fluorine is much higher than that of chlorine. Therefore, the latter two factors compensate more than for the less negative electron gain enthalpy of fluorine. Thus, fluorine is a much stronger oxidizing agent than chlorine.

**(ii)** The condition necessary to maximize the yield of $H_2SO_4$ by contact process are.

1. A moderately low temperature of about 720 K and high pressure of about 2 bar yields maximum $H_2SO_4$ acid.
2. Its an exothermic reaction and forward reaction causes decrease in pressure. The presence of catalyst also increases the yield.

(iii) **(a)** $H_3PO_4 < H_3PO_3 < H_3PO_2$.

**(b)** $BiH_3 < SbH_3 < AsH_3 < PH_3 < NH_3$.

**Note : All questions are same in Outside Delhi Set II and III and Delhi Set-I, II and III**

# CHEMISTRY

## Outside Delhi [Set-I]

# 2015

**Time allowed : 3 hours** **Maximum Marks : 70**

1. **Write the formulae of any two oxoacids of sulphur.** **[1]**

**Ans.** Oxoacids of sulphur are chemical compounds that contain sulphur, oxygen and hydrogen. Two oxoacids of sulphur are :

1. Peroxodisulphuric acid : $H_2S_2O_8$
2. Dithionic acid : $H_2S_2O_6$.

2. **Write the IUPAC name of the given compound :** **[1]**

$$CH_3-\underset{\displaystyle CH_3}{\underset{|}{CH}}-CH_2-O-CH_2-CH_3$$

**Ans.** Ethoxy-2-methyl propane.

3. **A delta is formed at the meeting point of sea water and river water. Why ?** **[1]**

**Ans.** River water is the negatively charged colloidal solution whereas sea water contains a number of electrolytes the meeting point of sea water and river water, the electrolytes present in sea water coagulate the colloidal solution of clay resulting in its deposition with the formation of delta.

4. **Which would undergo $S_N^1$ reactions faster in the following pair :** **[1]**

$$CH_3-CH_2-CH_2-Br \text{ and } CH_3-\underset{\displaystyle Br}{\underset{|}{CH}}-CH_3$$

**Ans.**

$$CH_3-CH_2-CH_2-Br \xrightarrow{-Br^-} CH_3-CH_2-\overset{+}{C}H_2 \quad \text{...(i)}$$

$$CH_3-\underset{\displaystyle Br}{\underset{|}{CH}}-CH_3 \xrightarrow{-Br^-} CH_3-\overset{+}{C}H-CH_3 \quad \text{...... (ii)}$$

During $S_N1$ reaction carbocation **(i)** and **(ii)** are formed. Out of **(i)** and **(ii)**, **(ii)** is more stable due to +I effect of two methyl group.

Thus $CH_3-\underset{\displaystyle Br}{\underset{|}{CH}}-CH_3$ gives faster $S_N^1$ reaction than $CH_3-CH_2-CH_2-Br$.

5. **What is the formula of a compound in which the element Y forms ccp lattice and atoms of X occupy 2/3rd of tetrahedral voids ?** **[1]**

**Ans.** The formula of the compound is $X_4Y_3$.

6. **Write one similarity and one difference between the chemistry of lanthanoids and that of actinoids.** **[2]**

**Ans.** Similarity

1. Both involve the filling of *f*-orbital (*i.e.* 4*f* and 5*f* )
2. Both show contraction *i.e.* Lanthanoide contraction and actinoide contraction.

**Difference between lanthanoids and Actinoids :**

| S. No. | Lanthanoids | Actinoids |
|---|---|---|
| 1. | Except Pm (promethium) all lanthanoids are non-radioactive. | All Actinoids are radioactive. |
| 2. | Lathanoids do not show wide range of oxidation state. | Actinoids shows the wide range of oxidation state. |
| 3. | Lanthanoid ions are generally coloured. | Actinoid ions are colourless |

7. (i) **Write down the IUPAC name of the following complex : $[Co(NH_3)_5Cl]^{2+}$**

(ii) **Write the formula for the following complex: Potassium tetrachloridonickelate (II).** **[2]**

**Ans. (i)** $[Co(NH_3)_5Cl]^{2+}$:
Pentaamminechlorocobalt (III) ion

**(ii)** Potassium tetrachloridonickelate(II): $K_2[NiCl_4]$

8. **Write the reagents required in the following reactions :**

(i) $CH_2=CH-CH_2OH \xrightarrow{?} CH_2=CH-CHO$

(ii) $CH_3-COOH \xrightarrow{?} CH_3CONH_2$ **[2]**

**OR**

**Arrange the following compounds in increasing order of their property as indicated :**

(i) $CH_3COCH_3$, $C_6H_5COCH_3$, $CH_3CHO$ **(reactivity towards nucleophilic addition reaction)**

(ii) $Cl-CH_2-COOH$, $F-CH_2-COOH$, $CH_3-COOH$ **(acidic character).**

**Ans. (i)** $CH_2 = CH - CH_2OH \xrightarrow[\text{in isopropanol } (C_5H_5NHCrO_3Cl) \text{ or Cu at 573 K}]{\text{Pyridinium Chloro chromate}} CH_2 = CH - CHO$

**(ii)** $CH_3 - COOH \xrightarrow{NH_4OH \text{ or liq } NH_3/\Delta} CH_3 - CO{-}NH_2$

**OR**

**(i)**

$$\underset{(I)}{CH_3-\overset{\overset{O}{||}}{C}-CH_3} \quad \underset{(II)}{C_6H_5-\overset{\overset{O}{||}}{C}-CH_3} \quad \underset{(III)}{CH_3-\overset{\overset{O}{||}}{C}-H}$$

Nucleophiles are negatively charged entity, positivity of carbon in carbonyl group facilitates nucleophilic addition. Thus positivity of carbon in carbonyl group is in order III > I > II. Due to +I effect of – $CH_3$ group and benzene ring π-cloud, positivity of carbonyl carbon is decreased so much than (I) and (II). Thus nucleophilic addition order III > I > II.

**(ii)** $Cl - CH_2 - COOH \xrightarrow{-H^+} Cl - CH_2 - \overset{\overset{O}{||}}{C} - O^-$ ....(i)

$F - CH_2 - COOH \xrightarrow{-H^+} F - CH_2 - \overset{\overset{O}{||}}{C} - O^-$ ....(ii)

$CH_3 - COOH \xrightarrow{-H^+} CH_3 - \overset{\overset{O}{||}}{C} - O^-$ ....(iii)

The stability order II > I > III. Due to +I effect of $-CH_3$ group electron density of oxygen is more. But in (I) and (II), –I effect of –F is more than –Cl. Thus II is more stabilized than I. Thus acidity order is II > I > III.

**9. (i) On mixing liquid X and liquid Y, volume of the resulting solution decreases. What type of deviation from Raoult's law is shown by the resulting solution ? What change in temperature would you observe after mixing liquids X and Y ?**

**(ii) What happens when we place the blood cell in water (hypotonic solution) ? Give reason.** **[2]**

**Ans. (i)** Negative deviation of Raoult's law occurred. There is a elevation of boiling point *i.e.*, temperature of the solution increases.

**(ii)** When blood cell is placed in water (hypotonic solution), water penetrate to blood cell and blood cell gets bulged and then disrupt.

**10. Calculate the time to deposit 1.27 g of copper at cathode when a current of 2A was passed through the solution of $CuSO_4$.**

**(Molar mass of Cu = 63.5 g mol$^{-1}$, 1 F = 96500 C mol$^{-1}$).** **[2]**

**Ans.** $Cu^{2+} + 2e^- \rightarrow Cu$

63.5 g of copper deposited by 2 × 96500 C

∴ 1.27 g of copper will be deposited by

$$\frac{2 \times 96500 \times 1.27}{63.5} \text{ C} = 3860 \text{ C}$$

We know $Q = It$

HereI = 2A and Q = 3860 C

$$t = \frac{Q}{I} = \frac{3860}{2} = 1930 \text{ s.}$$

**11. A solution is prepared by dissolving 10 g of non-volatile solute in 200 g of water. It has a vapour pressure of 31.84 mm Hg at 308 K. Calculate the molar mass of the solute.**

**(Vapour pressure of pure water at 308 K = 32 mm Hg).** **[3]**

**Ans.** According to Raoult's law,

$$\frac{P^o - P}{P^o} = \frac{w \times M}{m \times W}$$

Where

$P^o \rightarrow$ Vapour pressure of pure water

$P \rightarrow$ Vapour pressure of solution

$w \rightarrow$ Weight of solute

$m \rightarrow$ Molecular weight of solute

$W \rightarrow$ Weight of solvent

$M \rightarrow$ Molecular weight of solvent

Thus, $\dfrac{32 - 31.84}{32} = \dfrac{10 \times 18}{m \times 200}$

or, 0.005 m = 0.9 or, $m$ = 180

Thus molar mass of the solute 180 g./mol

**12. (i) Name the method of refining to obtain silicon of high purity.**

**(ii) What is the role of $SiO_2$ in the extraction of copper ?**

**(iii) What is the role of depressants in froth floatation process ?** **[3]**

**Ans. (i)** Zone refining is the method of refining to obtain silicon of high purity.

**(ii)** $SiO_2$ combined with the iron in the copper ore to form iron(II) silicate slag which is easily removed. Thus it ($SiO_2$) acts as a flux to remove the impurity of Iron oxide.

$$\underset{\text{gauge}}{FeO} + \underset{\text{flux}}{SiO_2} \longrightarrow \underset{\text{slag}}{FeSiO_3}$$

(iii) In froth flotation process, depressant prevents the formation of froth. It is used to separate two sulphide ore by preventing the formation of froth of one sulphide ore and allowing the other to form the froth *e.g.*, NaCN a depressant selectively prevents ZnS from coming in froth but allows PbS to come with the froth.

**13. (i) Which one of the following is a polysaccharide :**
**Starch, maltose, fructose, glucose.**
**(ii) Write one difference between α-helix and β-pleated sheet structures of protein.**
**(iii) Write the name of the disease caused by the deficiency of vitamin $B_{12}$.** [3]

**Ans. (i)** Starch is a polysaccharide.

**(ii)**

| | α-helix | β-pleated |
|---|---|---|
| 1. | It is rod like structure. | It is sheet like structure. |
| 2. | It is stabilized by intramolecular hydrogen bonding. | It is stabilized by intermolecular hydrogen bonding. |

**(iii)** Pernicious Anaemia is caused by the deficiency of Vitamin $B_{12}$.

**14. (i) What type of isomerism is shown by the complex $[Cr(H_2O)_6]Cl_3$ ?**
**(ii) On the basis of crystal field theory, write the electronic configuration for $d^4$ ion if $\Delta_o > P$.**
**(iii) Write the hybridization and shape of $[CoF_6]^{3-}$.**
**(Atomic number of Co = 27).** [3]

**Ans. (i)** Hydration isomerism is shown by the complex $[Cr(H_2O)_6]Cl_3$.

$[Cr(H_2O)_5Cl]Cl_2(H_2O)$;

$[Cr(H_2O)_3Cl_3]3(H_2O)$.

**(ii)** Since $\Delta_o > P$ it is on splitting $t_{2g}^4 e_g^0$

**(iii)** $[CoF_6]^{3-}$ : $sp^3d^2$ hybridisation and the shape is octahedral.

$[CoF_6]^{3-}$ : [⇅|↑|↑|↑|↑] (3d) [××] (4s) [××|××|××] (4p) [××|××| | | ] (4d)

$[CoF_6]^{3-}$ has four unpaired electron. Thus it is paramagnetic.

**15. How can the following conversion be carried out :**
**(i) Aniline to bromobenzene**
**(ii) Chlorobenzene to 2-chloroacetophenone**
**(iii) Chloroethane to butane.** [3]

**OR**

**What happens when**
**(i) Chlorobenzene is treated with $Cl_2/FeCl_3$**
**(ii) Ethyl chloride is treated with $AgNO_2$**
**(iii) 2-bromopentane is treated with alcoholic KOH ?**

**Write the chemical equations in support of your answer.**

**Ans. (i)**

$$\underset{\text{Aniline}}{C_6H_5NH_2} \xrightarrow[\substack{(NaNO_2 + HCl)\\ 0-5°C}]{HCl + HNO_2} \underset{\text{Benzene diazonium chloride}}{C_6H_5N_2^+Cl^-} \xrightarrow[HBr]{CuBr} \underset{\text{Bromo Benzene}}{C_6H_5Br} + N_2$$

**(ii)**

$$\underset{\text{Chlorobenzene}}{C_6H_5Cl} \xrightarrow[\text{Anhydrous } AlCl_3]{(CH_3CO)_2O} \underset{\text{2-Chloroacetophenone}}{C_6H_4(Cl)COCH_3} + HCl$$

**(iii)** $$\underset{\text{Chloroethane}}{2CH_3CH_2Cl} \xrightarrow[Na]{\text{dry ether}} \underset{\text{butane}}{CH_3CH_2CH_2CH_3} + 2NaCl$$

**OR**

**(i)**

$$\underset{\text{Chlorobenzene}}{C_6H_5Cl} \xrightarrow{Cl_2, FeCl_3} \underset{\text{1, 4-dichloro-benzene (major)}}{p\text{-}C_6H_4Cl_2} + \underset{\text{1, 2-dicholoro Benzene (minor)}}{o\text{-}C_6H_4Cl_2}$$

**(ii)** $$\underset{\text{Chloroethane}}{CH_3CH_2Cl} \xrightarrow{AgNO_2} \underset{\text{Nitroethane}}{CH_3CH_2NO_2} + AgCl$$

**(iii)**

$$CH_3-\underset{Br}{\underset{|}{CH}}-CH_2-CH_2-CH_3 \xrightarrow{\text{Alc. KOH}} \underset{\text{pent-1-ene (minor product)}}{CH_2=CH-CH_2-CH_2-CH_3} + \underset{\text{pent-2-ene (major product)}}{CH_3-CH_2-CH=CH-CH_3}$$

**16. Examine the given defective crystal :**

$X^+$ $Y^-$ $X^+$ $Y^-$ $X^+$

$Y^-$ O $Y^-$ $X^+$ $Y^-$

$X^+$ $Y^-$ $X^+$ O $X^+$

$Y^-$ $X^+$ $Y^-$ $X^+$ $Y^-$

**Answer the following questions :**

**(i) Is the above defect stoichiometric or non-stoichiometric ?**

**(ii) Write the term used for this type of defect. Give an example of the compound which shows this type of defect.**

**(iii) How does this defect affect the density of the crystal ?** [3]

**Ans.** (i) Stoichiometric defect

(ii) Schottky defects *e.g.* NaCl, CsCl.

(iii) The presence of a large number of Schottky defect in a crystal will lower its density.

**17. Conductivity of $2.5 \times 10^{-4}$ M methanoic acid is $5.25 \times 10^{-5}$ S cm$^{-1}$.**

**Calculate its molar conductivity and degree of dissociation.**

**Given : $\lambda_{(H^+)} = 349.5$ S cm$^2$ mol$^{-1}$ and $\lambda^0_{(HCOO^-)} = 50.5$ S cm$^2$ mol$^{-1}$.** [3]

**Ans.** We know molar conductivity

$$\therefore \quad (\lambda_m) = \frac{1000 \times \text{conductivity (k)}}{\text{concentration (c)}}$$

$$\therefore \quad \lambda_m = \frac{1000 \times 5.25 \times 10^{-5}}{2.5 \times 10^{-4}} = 210 \text{ cm}^2 \text{ Mol}^{-1}$$

$\lambda^0 HCOOH = \lambda^0_{H^+} + \lambda^0_{(HCOO^-)} = (349.5 + 50.5) = 400$ S cm$^2$mol$^{-1}$

$$\therefore \quad \alpha = \frac{\lambda_m}{\lambda_0} = \frac{210}{400} = 0.52$$

or, $\alpha = 52.5$ %

**18. Write any three differences between Physisorption and Chemisorption.** [3]

**Ans.**

| | Physisorption | Chemisorption |
|---|---|---|
| 1. | It forms multimolecular layers | It forms unimolecular layers |
| 2. | It is reversible | It is irreversible |
| 3. | Its does not require activation energy | It requires activation energy |
| 4. | It is not very specific | It is very specific |
| 5. | Force of attraction are Vander Waals forces | Force of attraction are chemical bond |

**19. Give reasons for the following :**

**(i) Phenol is more acidic than methanol.**

**(ii) The C – O – H bond angle in alcohols is slightly less than the tetrahedral angle (109°28').**

**(iii) $(CH_3)_3C$ – O – $CH_3$ on reaction with HI gives $(CH_3)_3C$ – I and $CH_3$ – OH as the main products and not $(CH_3)_3C$ – OH and $CH_3$ – I.** [3]

**Ans.** (i) Phenoxide ion is more stabilized than $CH_3 - O^-$ ion because of resonance :

Thus phenol is more acidic than $CH_3$ – OH.

(ii) Due to presence of lone pair on oxygen which causes repulsion,the bond angle in alcohol is slightly less than tetrahedral angle (109.28').

(iii) Since $(CH_3)_3 C^{\oplus}$ is more stabilized through +I effect of three methyl group thus $(CH_3)_3$ C – I and $CH_3$ – OH formed on treatment of HI with – $(CH_3)_3C$ – O – $CH_3$.

$$Me_3C - O - Me \rightarrow Me_3C^{\oplus} + Me - O^- \xrightarrow{H^+I^-} Me_3C - I + MeOH.$$

**20. Predict the products of the following reactions :** [3]

**(i)** $CH_3 - \underset{\underset{CH_3}{|}}{C} = O \xrightarrow{H_2N-NH_2} ?$

**(ii)** $C_6H_5 - CH_3 \xrightarrow[\text{(b) } H^+]{\text{(a) } KMnO_4/KOH} ?$

**(iii)** $C_6H_5COOH \xrightarrow{Br_2/FeBr_3}$

**Ans.** (i) $CH_3 - \underset{\underset{CH_3}{|}}{C} = \xrightarrow{H_2N-NH_2} H_3C - \underset{\underset{CH_3}{|}}{C} = N - NH_2$

Propanone     Propanone hydrozone

(ii)

$$C_6H_5 - CH_3 = \xrightarrow[\text{(b) } H^+]{\text{(a) } KMnO_4/KOH} C_6H_5 - COOH + H_2O$$

(iii)

$$C_6H_5COOH + HBr \xrightarrow{Br_2/FeBr_3} m\text{-}BrC_6H_4COOH$$

Benzoic acid → (3-Bromobenzoic acid)

**21. (a) Account for the following :**

**(i) $Cu^+$ is unstable in an aqueous solution.**

**(ii) Transition metals form complex compounds.**

**(b) Complete the following equation :**

**$Cr_2O_7^{2-} + 8H^+ + 3NO_2^- \rightarrow$** **[3]**

**Ans. (a)** (i) $2Cu^+(aq) \rightarrow Cu^{2+}_{(aq)} + Cu(s)$

The higher stability of $Cu^{2+}$ ion in aqueous solution is due to its greater negative charge than Cu. It compensates the second ionisation enthalpy of Cu involved in the formation of $Cu^{2+}$ ions.

(ii) Since transition metals have unfilled or partially filled *d*-orbital thus to satisfy its octate forms complex compounds.

**(b)** $\overset{+6}{Cr_2}O_7^{2-} + 14H^+ + 6e^- \rightarrow 2Cr^{3+} + 7H_2O$

$3NO_2^- + H_2O \rightarrow \overset{+5}{N}O_3^- + 2H^+ + 2e^- ] \times 3$

$Cr_2O_7^{2-} + 3NO_2^- + 8H^+ \rightarrow 2Cr^{3+} + 3NO_3^- + 4H_2O$

**22. Write the names and structures of the monomers of the following polymers :**

**(i) Terylene**

**(ii) Buna-S**

**(iii) Neoprene** **[3]**

**Ans.** (i) Monomers are Ethylene Glycol $(HO{-}CH_2{-}CH_2{-}OH)$ and Terephthalic acid $(HO_2C{-}C_6H_4{-}COOH)$.

(ii) Monomers are Styrene $(C_6H_5{-}CH = CH_2)$ and 1,3-Butadiene $(H_2C = CH - CH = CH_2)$.

(iii) Monomer is Chloroprene

$$\left(H_2C = \overset{\overset{\displaystyle Cl}{|}}{C} - CH = CH_2\right).$$

**23. Seeing the growing cases of diabetes and depression among young children, Mr. Chopra, the principal of one reputed school organized a seminar in which he invited parents and principals. They all resolved this issue by strictly banning junk food in schools and introducing healthy snacks and drinks like soup, lassi, milk, etc. in school canteens. They also decided to make compulsory half an hour of daily physical activities for the students in the morning assembly. After six months, Mr. Chopra conducted the health survey in most of the schools and discovered a tremendous improvement in the health of the students.**

**After reading the above passage, answer the following questions :**

**(i) What are the values (at least two) displayed by Mr. Chopra ?****

**(ii) As a student, how can you spread awareness about this issue ?**

**(iii) Why should antidepressant drugs not be taken without consulting a doctor ?**

**(iv) Give two examples of artificial sweeteners.** **[4]**

**Ans.** (ii) Awareness can be spread by performing nukkad natak in community, displaying posters, cartoons and slogans and by conducting seminars.

(iii) Antidepressant drugs have lots of side effects like indigestion, headache, stomach aches, drowsiness, weight gain. That is why it should not be taken without consulting with doctors.

(iv) Example of artificial sweeteners are Aspartame, Saccharin, Sucralose etc.

**24. (a) Account for the following :**

**(i) Acidic character increases from HF to HI.**

**(ii) There is a large difference between the melting and boiling points of oxygen and sulphur.**

**(iii) Nitrogen does not form pentahalide.**

**(b) Draw the structures of the following :**

**(i) $ClF_3$**

**(ii) $XeF_4$** **[5]**

**OR**

**(i) Which allotrope of phosphorus is reactive and why ?**

**(ii) How are the supersonic jet aeroplanes responsible for the depletion of ozone layer ?**

**(iii) $F_2$ has lower bond dissociation enthalpy than $Cl_2$. Why ?**

**(iv) Which noble gas is used in filling balloons for meteorological observations ?**

**(v) Complete the following equation :**

**$XeF_2 + PF_5 \rightarrow$** **[5]**

**Ans. (a)** (i) Size of halide ions is of the order $F^- < Cl^- < Br^- < I^-$ with increase in size negative charge is dispersed throughout ions and

** Answer is not given due to the change in present syllabus.

ions gets stabilized. Thus acidity order

HF < HCl < HBr < HI.

**(ii)** Due to presence of vacant *d*-orbital and combining forms of sulphur ($S_8$) which is not present in oxygen, the cohesive energy of sulphur is more than oxygen leading to large melting point and boiling point.

**(iii)** Due to non-availability of *d*-orbitals nitrogen can not extend its octate to form pentahalide.

**(b) (i)** $F-Cl(F)(F)$ with two lone pairs Shape→T-shaped

**(ii)** $XeF_4$ with two lone pairs Shape→Square planar

OR

**(i)** White phosphorus is most reactive of all the allotropes of phosphours. It is because it exist as $P_4$ discrete tetrahedral units with 60° angle, which results in angular strain and makes it highly reactive.

**(ii)** The exhaust emitted from supersonic jet aeroplane contains $CO_2$, NO and other particles which are the killers of stratospheric ozone layer along with the supersonic sound produced by these aeroplanes are destroy ozone layer.

**(iii)** Due to smaller size and high electro-negativity of fluorine, more energy is required to break the bond of $F_2$ than $Cl_2$. Thus $F_2$ has lower bond dissociation energy than $Cl_2$.

**(iv)** Helium (He) gas is used in filling balloons for meteorological observations because it is non-inflammable and light gas.

**(v)** $XeF_2 + PF_5 \rightarrow [XeF]^+ [PF_6]^-$.

**25. An aromatic compound 'A' of molecular formula $C_7H_6O_2$ undergoes a series of reactions as shown below. Write the structures of A, B, C, D and E in the following reactions : [5]**

$$(C_7H_6O_2)A \xrightarrow{NH_3/heat} C_6H_5CONH_2 \xrightarrow{Br_2 + NaOH} B \xrightarrow{(CH_3CO)_2O} C$$

$C_6H_5CONH_2 \xrightarrow{LiAlH_4/ether} D$; $B \xrightarrow{Br_2(aq)} E$

OR

**(a) Write the structures of main products when benzene diazonium chloride reacts with the following reagents :**

**(i) $H_3PO_2 + H_2O$**

**(ii) CuCN/KCN**

**(iii) $H_2O$**

**(b) Arrange the following in the increasing order of their basic character in an aqueous solution :**

**$C_2H_5NH_2$, $(C_2H_5)_2NH$, $(C_2H_5)_3N$**

**(c) Give a simple chemical test to distinguish between the following pair of compounds :**

**$C_6H_5-NH_2$ and $C_6H_5-NH-CH_3$ [5]**

**Ans.**

$$\underset{\text{(Benzoic acid)}}{\underset{(A)}{C_7H_6O_2}} \xrightarrow[\text{heat}]{NH_3/} C_6H_5CONH_2 \xrightarrow[NaOH]{Br_2+} \underset{\text{Aniline}}{\underset{(B)}{C_6H_5NH_2}}$$

$C_6H_5CONH_2 \xrightarrow{LiAlH_4/ether} \underset{(D)}{C_6H_5CH_2NH_2}$ Benzylamine

$C_6H_5NH_2 \xrightarrow{(CH_3CO)_2O} \underset{(C)}{C_6H_5NHCOCH_3}$ (N-phenyl ethanamide)

$$\underset{\text{Aniline}}{\underset{(B)}{C_6H_5NH_2}} \xrightarrow{Br_2(aq)} C_6H_5(Br)_3NH_2$$

2, 4, 6-Tribromo aniline

OR

**(a) (i)**

$C_6H_5N_2^+Cl^-$ (Benzene Diazonium Chloride) $\xrightarrow{H_3PO_2 + H_2O}$ $C_6H_6$ (Benzene)

**(ii)**

$C_6H_5N_2^+Cl^-$ (Benzene Diazonium Chloride) $\xrightarrow{CuCN/KCN}$ $C_6H_5CN$ (Benzonitrile) $+ N_2$

**(iii)**

$C_6H_5N_2^+Cl^-$ (Benzene diazonium chloride) $\xrightarrow{H_2O}$ $C_6H_5OH$ (Phenol) $+ N_2 + HCl$

**(b)** $C_2H_5NH_2 < (C_2H_5)_3N < (C_2H_5)_2NH$

+I effect of three $C_2H_5$ group increases the enormous availability of lone pair of nitrogen atom than +I effect of 2 ethyl group. Thus the order of basicity is I < III < II.

**(c)** $C_6H_5 - NH_2$ (Primary amine) and

$C_6H_5 - NH - CH_3$ (Secondary amine) can be distinguished by Hinsberg's test. In this test amines are allowed to react with Hinsberg's reagent, benzenesulphonyl chloride ($C_6H_5SO_2Cl$). Primary amines reacts with this reagent to form N-alkylbenzene sulphonyl amide which is soluble in alkali but secondary amines gives sulphonamide which is insoluble in alkali.

**26. For the hydrolysis of methyl acetate in aqueous solution, the following results were obtained :**

| t/s | 0 | 10 | 20 |
|---|---|---|---|
| $[CH_3COOCH_3]$/mol $L^{-1}$ | 0.10 | 0.05 | 0.025 |

**(a) Show that it follows pseudo first order reaction, as the concentration of water remains constant.**

**(b) Calculate the average rate of reaction between the time interval 10 to 20 seconds.**

**(Given : log 2 = 0.3010, log 4 = 0.6021). [5]**

**OR**

**(a) For a reaction A + B → P, the rate is given by**

**Rate = $k[A][B]^2$**

**(i) How is the rate of reaction affected if the concentration of B is doubled ?**

**(ii) What is the overall order of reaction if A is present in large excess ?**

**(b) A first order reaction takes 30 minutes for 50% completion. Calculate the time required for 90% completion of this reaction.**

**Ans. (a)** $[A]_0 = 0.01$ Mol/L

$[A] = 0.05$ Mol/L at time $t = 10$ s.

$$k = \frac{2.303}{10}\log\frac{0.10}{0.05}$$

$$k = \frac{2.303}{10}\log\frac{0.10}{0.05}$$

$$k = 0.0693\ s^{-1}$$

$$t = 20\,s$$

$$k = \frac{2.303}{t}\log\frac{[A]_0}{[A]} = \frac{2.303}{10}\log\frac{0.10}{0.05}$$

$k = 0.0693\,s^{-1}$, Thus it is pseudo first order reaction.

**(b)** The average rate constant is

$$\frac{-\Delta[R]}{\Delta t} = \frac{\Delta[CH_3COOCH_3]}{\Delta t}$$

$$= \frac{[0.025\text{mol L}^{-1} - 0.05\text{mol L}^{-1}]}{20s - 10s} = \frac{0.025}{10s}\ \text{mol L}^{-1}$$

$$= 0.0025\ \text{molL}^{-1}\text{s}^{-1}$$

**OR**

**(a) (i)** Since the given reaction has order two with respect to reactant B, thus if the concentration of B is doubled in the given reaction, then the rate of reaction will become four times.

**(ii)** It the concentration of B is doubled *i.e.*; $[B]^2$ the overall reaction will be two, because if A is present in large excess, then the reaction will be independent of the concentration of A and will be dependent only on the concentration of B. Order of reaction = 2.

**(b)** $t_{1/2} = 30$ min

$$[R] = [R]_0 - 90\% \text{ of } [R]_0$$

$$= [R]_0 - \frac{90[R]_0}{100}$$

$$[R] = \frac{[R]_0}{10}$$

$$K = \frac{0.693}{t_{1/2}} = \frac{0.693}{30} = 0.0231\ \text{min}^{-1}$$

$$t = \frac{2.303}{K}\log\frac{[R]_0}{[R]}$$

$$t = \frac{2.303}{0.0231}\log\frac{[R]_0}{[R]} - \frac{2.303}{0.0231}\log 10$$

$$t = 99.7\ \text{min}$$

**Note : All questions are same in Outside Delhi Set II and III.**

# CHEMISTRY

## Delhi [Set-I] 2015

**Time allowed : 3 hours** **Maximum Marks : 70**

1. **What is the basicity of $H_3PO_4$ ?** **[1]**

**Ans.** $H_3PO_4$ has three ionisable hydrogen atoms. Hence, its basicity is 3.

O
||
P—OH
HO / \ OH

2. **Write the IUPAC name of the given compound :** **[1]**

$NO_2$ OH
$NO_2$

**Ans.**

$NO_2$ 5 6 OH
1
4 2 $NO_2$
3

2,5-Dinitrophenol

3. **Which would undergo $S_N2$ reaction faster in the following pair and why ?** **[1]**

$$CH_3—CH_2—Br \text{ and } CH_3—\underset{\underset{Br}{|}}{\overset{\overset{CH_3}{|}}{C}}—CH_3$$

**Ans.** Primary alkyl halides will undergo $S_N2$ reactions faster than tertiary alkyl halides because of less steric hindrance experienced by the approaching nucleophile. Hence, out of the given pair, primary alkyl halide ($CH_3CH_2Br$) would undergo $S_N2$ reaction faster.

4. **Out of $BaCl_2$ and KCl, which one is more effective in causing coagulation of a negatively charged colloidal Sol ? Give reason.** **[1]**

**Ans.** According to the Hardy-Schulze rule, greater the valency of a flocculating ion, the greater is its power to cause precipitation. Between $Ba^{2+}$ (from $BaCl_2$) and $K^+$ (from KCl), $Ba^{2+}$ has greater valency. Therefore, $BaCl_2$ will be more effective in causing the coagulation of a negatively charged colloidal sol.

5. **What is the formula of a compound in which the element Y forms ccp lattice and atoms of X occupy 1/3 rd of tetrahedral voids ?** **[1]**

**Ans.** Given : Y form CCP lattice

No. of Y atom in CCP lattice = 1

No. of tetrahedral void = 2

No. of X atom $= 2 \times \frac{1}{3} = \frac{2}{3}$

Therefore ratio of X : Y $= \frac{2}{3} : 1$

Formula of compound : $X_2Y_3$

6. **What are the transition elements ? Write two characteristics of the transition elements.** **[2]**

**Ans.** Elements which in their ground state or in any of their oxidation state have partially filled *d*-orbital are called transition elements. The name 'transition' given to the elements of *d*-block is only because of their position between *s*-block and *p*-block elements.

The two characteristics of transition elements are:

1. They show variable oxidation states.
2. They generally form coloured compounds.

7. **(i) Write down the IUPAC name of the following complex :**

**$[Cr(NH_3)_2Cl_2(en)]Cl$ (en = ethylenediamine)**

**(ii) Write the formula for the following complex: Pentaamminenitrito-o-Cobalt(III).** **[2]**

**Ans. (i)** Diamminedichloridoethylenediaminechromium(III) chloride.

**(ii)** $[Co(NH_3)_5(ONO)]^{2+}$.

8. **Name the reagents used in the following reactions :**

(i) $CH_3-CO-CH_3 \xrightarrow{?} CH_3-\underset{OH}{CH}-CH_3$

(ii) $C_6H_5-CH_2-CH_3 \xrightarrow{?} C_6H_5-COO^-K^+$ **[2]**

**Ans. (i)** Sodium borohdride ($NaBH_4$)/$LiAlH_4$/$H_2$, *Pt*.

**(ii)** Alkaline potassium permanganate ($KMnO_4$, KOH).

9. **What is meant by positive deviations from Raoult's law ? Give an example. What is the sign of $\Delta_{mix}H$ for positive deviation ?** **[2]**

**OR**

**Define azeotropes. What type of azeotrope is formed by positive deviation from Raoult's law ? Give an example.**

**Ans.** Positive deviation from Rauolt's law means that the observed vapour pressure is greater than expected, and it occurs when the A – B attractions are weaker than the average of the attractions in the pure component of the mixture. For example : A mixture of ethanol and acetone shows a positive deviation from Raoult's law.

In case of solutions showing positive deviations, absorption of heat takes palce; *i.e.*, $\Delta_{mix}H$ has a positive (+) sign.

OR

Azeotropes are the binary mixtures which have the same composition in liquid and vapour phases and boil at a constant temperature.

A minimum-boiling azeotrope is formed by solutions showing a large positive deviation from Raoult's law at a specific composition.

**Example :** $C_2H_5OH + H_2O$ (An ethanol-water mixture)

**10. (a) Following reactions occur at cathode during the electrolysis of aqueous silver chloride solution :**

$$Ag^+(aq) + e^- \longrightarrow Ag(s); E° = +.80 \text{ V}$$

$$H^+(aq) + e^- \longrightarrow \frac{1}{2}H_2(g); E° = 0.00 \text{ V}$$

**On the basis of their standard reduction electrode potential (E°) values, which reaction is feasible at the cathode and why ?**

**(b) Define limiting molar conductivity. Why conductivity of an electrolyte solution decreases with decrease in concentration ?** **[2]**

**Ans. (a)** The relationship between the standard free energy change and emf of a cell reaction is given by :

$$\Delta G° = -nFE°$$

Thus, more positive the standard reduction potential of a reaction, the more negative is the standard free energy change associated with the process and consequently, the higher is the feasibility of the reaction.

Since $E°_{Ag^+/Ag}$ has a greater positive value than $E°_{H^+/H_2}$ the reaction which is feasible at the cathode is given by

$$Ag^+(aq) + e^- \longrightarrow Ag(s)$$

**(b)** The limiting molar conductivity of an electrolyte is defined as its molar conductivity when the concentration of the electrolyte in the solution approaches zero.

The conductivity of an electrolyte solution is the conductance of ions present in a unit volume of the solution. The number of ions (responsible for carrying current) decreases when the solution is diluted or the concentration is decreased. As a result, the conductivity of an electrolyte solution decreases with the decrease in concentration.

**11. 3.9 g of benzoic acid dissolved in 49 g of benzene shows a depression in freezing point of 1.62 K. Calculate the Van't Hoff factor and predict the nature of solute (associated or dissociated).**

**(Given : Molar mass of benzoic acid = 122 g $mol^{-1}$, $K_f$ for benzene = 4.9 K kg $mol^{-1}$)** **[3]**

**Ans.** We know that the depression in freezing point by

Given : Mass of solute ($W_B$) *i.e.*, $C_6H_5COOH$ = 3.9 g

$$\text{Mass of solvent } (W_A) = 49 \text{ g} = \frac{49}{1000} \text{ kg}$$

$$\text{Molar mass of } C_6H_5COOH(M_S) = 122 \text{ g/mol}$$

$$K_f = 4.9 \text{ K kg/mol}$$

$$\Delta T_f = 1.62 \text{ K}$$

To find : Van't Hoff fact ($i$ = ?)

We know that the depression in freezing point is given by :

$$\Delta T_f = i \times K_f \times m$$

$$\Delta T_f = i \times K_f \times \frac{W_B}{M_S} \times \frac{1}{W_A(kg)}$$

$$i = \frac{\Delta T_f \times M_S \times W_A(kg)}{K_f \times W_B}$$

$$i = \frac{1.62 \times 122 \times 49}{4.9 \times 3.9 \times 1000}$$

Since, $i < 1$, hence solute benzoic acid ($C_6H_5COOH$) will undergo association in benzene.

**12. (i) Indicate the principle behind the method used for the refining of zinc.**

**(ii) What is the role of silica in the extraction of copper ?**

**(iii) Which form of the iron is the purest form of commercial iron ?** **[3]**

**Ans. (i)** Zinc is a metal having low boiling point and it is refined by distillation method leaving behind impurities. This method is used for metals having low boiling points.

**(ii)** During extraction of Copper, FeS or FeO is present as impurity hence $SiO_2$ is added as flux to form silicate (slag) which can be removed easily as it floats over molten copper (Cu)

$$\underset{\text{Impurity}}{FeS(s)} + O_2(g) \longrightarrow FeO(s) + SO_2(g)$$

$$\underset{\text{Impurity}}{FeO(s)} + \underset{\text{Flux}}{SiO_2(s)} \longrightarrow \underset{\text{Iron silicate (slag)}}{FeSiO_3(g)}$$

**(iii)** Wrought iron is purest form of commercial iron.

**13. An element with molar mass 27 g mol$^{-1}$ forms a cubic unit cell with edge length 4.05 × 10$^{-8}$cm. If its density is 2.7 g cm$^{-3}$, what is the nature of the cubic unit cell ?** **[3]**

**Ans.** Given,

Molar mass of the given element, M = 27 g mol$^{-1}$ = 0.027 kg mol$^{-1}$

Edge length, $a = 4.05 \times 10^{-8}$cm $= 4.05 \times 10^{-10}$m

Density, $d = 2.7$g cm$^{-3}$ $= 2.7 \times 10^{-3}$ kg m$^3$

We know that

$$d = \frac{Z \times M}{a^3 \times N_A}$$

Where, Z is the number of atoms in the unit cell and $N_A$ is the Avogadro number.

Thus,

$$Z = \frac{d \times a^3 \times N_A}{M}$$

$$Z = \frac{2.7 \times 10^3 (4.05 \times 10^{-10})^3 \times 6.022 \times 10^{23}}{0.027} \quad (N_A = 6.022 \times 10^{23})$$

Z = 4 or Z ≈ 4 (f.c.c.)

Since the number of atoms in the unit cell is 4, the given cubic unit cell has a face-centred cubic (fcc) or cubic-closed packed (ccp) structure.

**14. (a) How would you account for the following :**

**(i) Actinoid contraction is greater than lanthanoid contraction.**

**(ii) Transition metals form coloured compounds.**

**(b) Complete the following equation :**

**$2MnO_4^- + 6H^+ + 5NO_2^- \longrightarrow$** **[3]**

**Ans. (a) (i)** The 5*f*-orbitals (in case of actinoids) have a poorer shielding effect than 4*f*-orbitals (in lanthanoids). Thus, the effective nuclear charge experienced by electrons in valence shells in case of actinoids is much more than that experienced by electrons in valence shells in case of lanthanoids. Hence, the contraction in size in actinoids is greater than that in lanthanoids.

**(ii)** In the presence of ligands, the *d*-orbitals of transition metal ions split up into two sets of orbitals having different energies. Thus, the transition of electrons takes place from one set to another. The energy required for these transitions is quite less and falls in the visible region of radiation. The ions of transition metals absorb the radiation of a particular wavelength and the rest is reflected, imparting colour to the solution.

**(b)** $2MnO_4^- + 6H^+ + 5NO_2^- \longrightarrow 2Mn^{2+} + 5NO_3^- + 3H_2O$

**15. (i) Draw the geometrical isomers of complex $[Pt(NH_3)_2Cl_2]$.**

**(ii) On the basis of crystal field theory, write the electronic configuration for $d^4$ ion if $\Delta_0$ < P.**

**(iii) Write the hybridization and magnetic behaviour of the complex $[Ni(CO)_4]$.**

**(At. no. of Ni = 28).** **[3]**

**Ans. (i)** Geometrical isomers of complex $[Pt(NH_3)_2Cl_2]$.

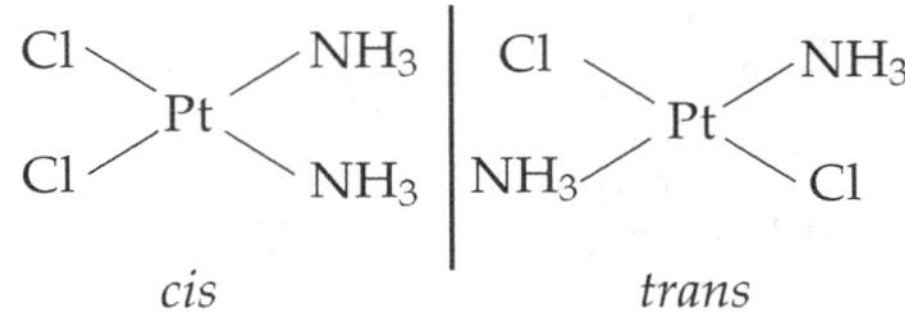


**(ii)** On the basis of crystal field theory, for a $d^4$ ion if, $\Delta_0$ < P, then complex is a high spin complex formed by association of weak field ligands with the metal ion. As a result, the fourth electron enters one of the $e_g$ orbitals, thereby, exhibiting the electronic configuration $t_{2g}3$ eg'.

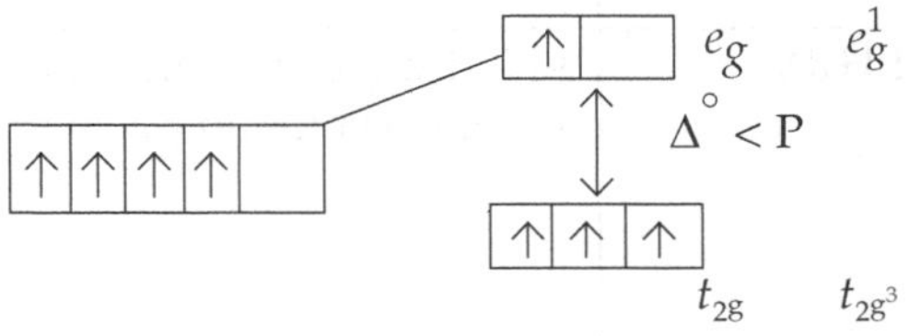


**(iii)** Outer electronic configuration of Ni atom in ground state.

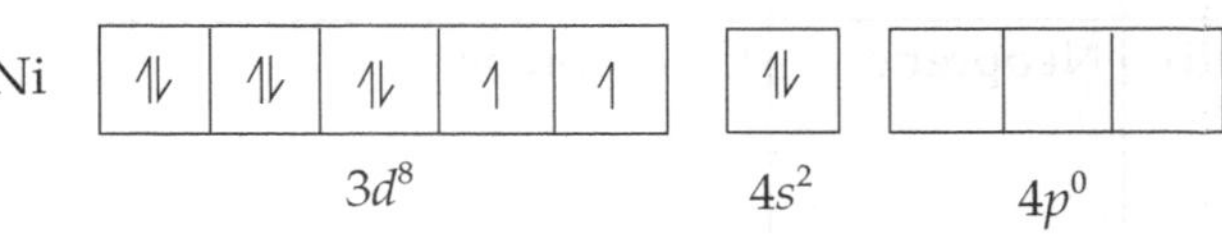


Outer electronic configuration of Ni atom in $Ni(CO)_4$

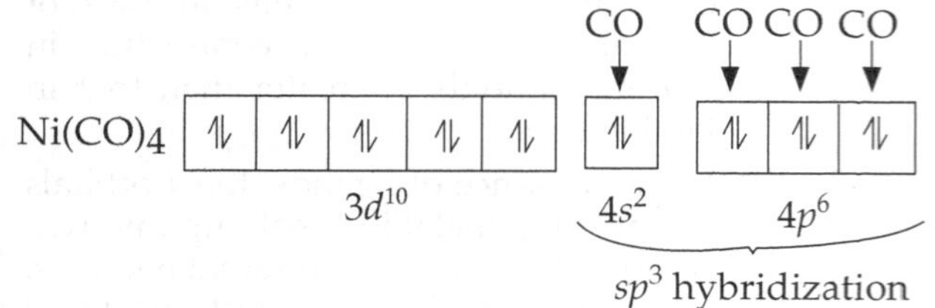


Carbonyl, CO being a strong field ligand causes the pairing of valence electrons in the Ni atom against the Hund's Rule of Maximum Multiplicity . This results in the formation of an inner orbital complex, $[Ni(CO)_4]$ . Since the complex $[Ni(CO)_4]$ has no unpaired electron, it is diamagnetic in nature and posses tetrahedral shape.

**16. Calculate emf of the following cell at 25°C :**

**$Fe \mid Fe^{2+}(0.001M) \mid H^+(0.01M) \mid \mid H_2(g)$ (1bar) Pt(s) $E°(Fe^{2+} \mid Fe) = -0.44$ V$E°(H^+/H_2) = 0.00$ V.** [3]

**Ans.** For the given cell representation, the cell reaction will be

$$Fe(s) + 2H^+(aq) \rightarrow Fe^{2+}(aq) + H_2(g)$$

The standard emf of the cell will be

$E°_{cell} = E°_{H^+/H_2} - E°_{Fe^{2+}/Fe}$

$E°_{cell} = 0 - (-0.44) = 0.44$ V

The Nernst equation for the cell reaction at 25°C will be

$$E_{cell} = E°_{cell} \frac{0.0591}{n} \log \frac{[Fe^{2+}]}{[H^+]^2}$$

$$= 0.44 - \frac{0.0591}{2} \log \frac{0.001}{(0.01)^2}$$

$$= 0.44 - \frac{0.0591}{2} \log 10$$

$$= 0.44 - \frac{0.0591}{2}$$

$$E_{cell} = 0.4105 \text{ V} \approx 0.41 \text{ V}$$

**17. Give reasons for the following observations :**

**(i) Leather gets hardened after tanning.**

**(ii) Lyophilic sol is more stable than lyophobic sol.**

**(iii) It is necessary to remove CO when ammonia is prepared by Haber's process.** [3]

**Ans. (i)** Animal skin (hide) is colloidal in nature and has postively charged colloidal particles. When a hide is soaked in tanning, mutual coagulation takes place and as a result, leather gets hardened.

**(ii)** The stability of lyophillic solution depends on the two factors, the presence of a charge and the solvation of colloid particles. On the other hand, the stability of lyophobic solutions is only because of the presence of a charge. Thus, lyophilic solution is more stable than lyophobic solution due to the extensive solvation.

**(iii)** It is necessary to remove CO when ammonia is prepared by Haber's process because in this process the CO act as a poison and adversely affects the activity of iron catalyst, used in the process.

**18. Write the names and structures of the monomers of the following polymers :**

**(i) Nylon-6, 6** **(ii) PHBV**

**(iii) Neoprene.** [3]

**Ans.**

| | Polymer | Name of Monomer(s) | Structure of Monomer(s) |
|---|---|---|---|
| (i) | Nylon-6, 6 | Hexamethylenediamine and adipic acid | $H_2N-CH_2-CH_2-CH_2-CH_2-CH_2-CH_2-NH_2$ Hexamethylenediamine $HOOC-CH_2-CH_2-CH_2-CH_2-COOH$ Adipic acid |
| (ii) | PHBV | 3-Hydroxypentanoic acid and 3-Hydroxybutanoic acid. | $CH_3-CH_2-CH(OH)-CH_2-COOH$ 3-Hydroxypentanoic acid $CH_3-CH(OH)-CH_2-COOH$ 3-Hydroxybutanoic acid |
| (iii) | Neoprene | Chloroprene | $H_2C=C(Cl)-C(H)=CH_2$ Chloroprene |

**19. Predict the products of the following reactions :**

(i) $CH_3-\underset{\displaystyle CH_3}{\underset{|}{C}}=O \xrightarrow[\text{(ii) KOH/Glycol},\Delta]{\text{(i) } H_2N-NH_2}$

(ii) $C_6H_5-CO-CH_3 \xrightarrow{NaOH/I_2} ?+?$

(iii) $CH_3COONa \xrightarrow[\Delta]{NaOH/CaO} ?$ [3]

**Ans.** (i)

$$\underset{\text{Acetone}}{(H_3C)_2C=O} \xrightarrow{\text{(i) } H_2N-NH_2} (H_3C)_2C=N-NH_2$$

$$\xrightarrow{\text{(ii) KOH/Glycol } \Delta} (H_3C)_2CH_2 + N_2$$

Propane

(Wolff Kishner Reduction)

(ii) $\underset{\text{Acetophenone}}{C_6H_5-CO-CH_3} \xrightarrow[\Delta]{NaOH/I_2} CHI_3\downarrow + C_6H_5COONa$

Iodoform

(Yellow)

(Haloform Reaction)

(iii) $\underset{\text{Sodium acetate}}{CH_3-COONa} \xrightarrow[\Delta]{NaOH/CaO} \underset{\text{Methane}}{CH_4} + Na_2CO_3$

(Decarboxylation Reaction)

**20. How do you convert the following :**

**(i) Phenol to anisole**

**(ii) Propan 2-ol to 2-methylpropan-2-ol**

**(iii) Aniline to phenol** [3]

OR

**(a) Write the mechanism of the following reaction :**

$$2CH_3CH_2OH \xrightarrow{H^+} CH_3CH_2-O-CH_2CH_3$$

**(b) Write the equation involved in the acetylation of Salicyclic acid.** [3]

**Ans.** (i)

$$\underset{\text{Phenol}}{C_6H_5OH} + NaOH \longrightarrow C_6H_5O^-Na^+ \xrightarrow[-NaBr]{CH_3Br} \underset{\text{Anisole}}{C_6H_5OCH_3}$$

(ii)

$$\underset{\text{Propan-2-ol}}{CH_3-\underset{OH}{\underset{|}{\overset{H}{\overset{|}{C}}}}-CH_3} \xrightarrow[\text{(Oxidation)}]{CrO_3 \text{ or } Cu/573K} \underset{\text{Propanone}}{CH_3-\underset{O}{\underset{\|}{C}}-CH_3}$$

$$\xrightarrow{CH_3-MgBr} \underset{\text{Adduct}}{\left[CH_3-\underset{CH_3}{\underset{|}{\overset{CH_3}{\overset{|}{C}}}}-\overset{-\;+}{O}MgBr\right]}$$

$$\xrightarrow{H_2O} \underset{\text{2-methylpropan-2-ol}}{CH_3-\underset{CH_3}{\underset{|}{\overset{OH}{\overset{|}{C}}}}-CH_3} + Mg(OH)Br$$

(iii)

$$\underset{\text{Aniline}}{C_6H_5NH_2} \xrightarrow[0^\circ-5^\circ C]{NaNO_2+HCl} \underset{\text{Benzene Diazonium Chloride}}{C_6H_5N_2Cl} \xrightarrow[\text{Warm}]{H_2O} \underset{\text{Phenol}}{C_6H_5OH} + N_2 + HCl$$

OR

(a) The given reaction follows $S_N^2$ mechanism as shown below :

**Step 1 :** Protonation

$$CH_3-CH_2-\ddot{\underset{..}{O}}-H + H^+ \rightarrow \underset{\text{Protonated alcohol}}{CH_3-CH_2-\overset{+}{\ddot{O}}-H} \quad (\text{with } -H \text{ on O})$$

**Step 2 :** Nucleophilic attack ($S_N2$)

$$CH_3-CH_2-\underset{H}{\underset{|}{\ddot{O}}}: + CH_3-CH_2-\overset{+}{O}\begin{matrix}H\\H\end{matrix}$$

$$\longrightarrow \underset{\text{Protonated ether}}{CH_3CH_2-\underset{H}{\underset{|}{O^+}}-CH_2-CH_3} + H_2O$$

**Step 3 :**

$$CH_3CH_2-\overset{+}{\underset{H}{O}}-CH_2-CH_3 \rightarrow \underset{\text{ether}}{CH_3-CH_2-O-CH_2-CH_3} + H^+$$

**(b)** Acetylation of Salicyclic acids.

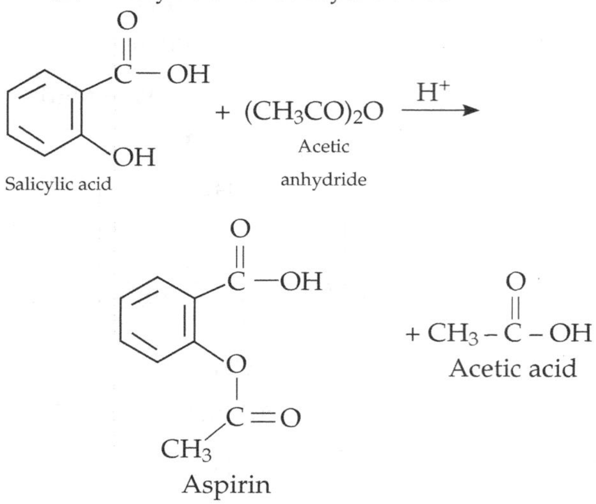

$$\text{Salicylic acid} + \underset{\text{Acetic anhydride}}{(CH_3CO)_2O} \xrightarrow{H^+} \text{Aspirin} + \underset{\text{Acetic acid}}{CH_3-\overset{O}{\overset{\|}{C}}-OH}$$

**21. (i) Which one of the following is a disaccharide: Starch, Maltose, Fructose, Glucose ?**

**(ii) What is the difference between fibrous protein and globular protein ?**

**(iii) Write the name of vitamin whose deficiency causes bone deformities in children. [3]**

**Ans. (i)** Maltose is a disaccharide, as it contains α-D-glucose units.

**(ii)**

| | Fibrous Proteins | Globular Proteins |
|---|---|---|
| 1. | In fibrous proteins, the polypeptide chains, run parallel and are held together by strong hydrogen and disulphide bonds. | In globular proteins, the polypeptide chains are folded looped or tursted around themselves, giving these proteins a spherical structure. |
| 2. | They are usually insoluble in water. | They are usually soluble in water. |
| 3. | They are structural proteins. For example, keratin is present in nails and hair, collagen is present in tendons and myosin is present in muscles. | They are functional proteins. For example, most enzymes and some hormones like insulin. |

**(iii)** Deficiency of Vitamin D causes the bone deformities in children.

**22. Give reasons :**

**(a) *n*-Butyl bromide has higher boiling point than *t*-butyl bromide.**

**(b) Racemic mixture is optically inactive.**

**(c) The presence of nitro group ($-NO_2$) at o/p positions increases the reactivity of haloarenes towards nucleophilic substitution reactions. [3]**

**Ans. (a)** *n*-butyl bromide is a straight chain molecule having larger surface area and therefore, has stronger intermolecular forces. On the other hand, *t*-butyl bromide $(CH_3)_3CBr$ is branched molecule, so it has a smaller surface area. Hence, it has weaker intermolecular force.

**(b)** Racemic mixture is an equimolar solution of two enantiomers (*d* and *l* forms) and thus, the rotation due to one isomer is cancelled by the rotation due to another. Therefore, it has zero optical rotation and hence, it is optically inactive.

**(c)** This is because nitro group ($-NO_2$) at o/p positions withdraw the electrons from the benzene ring. It acts as electron withdrawing group. The reduced electron density at o/p positions for haloarenes facilitate the attack of nucleophile. The negative charge in the Carbanion formed, at o/p positions with respect to halogen atom is stabilised through resonance and due to the presence of nitro groups ($NO_2$) respectively.

**23. Mr. Roy, the principal of one reputed school organized a seminar in which he invited parents and principals to discuss the serious issue of diabetes and depression in students. They all resolved this issue by strictly banning the junk food in school and to introduce healthy snacks and drinks like soup, lassi, milk etc. in school canteens. They also decided to make compulsory half an hour physical activities for the students in the morning assembly daily. After six months, Mr. Roy conducted the health survey in most of the schools and discovered a tremendous improvement in the health of students.**

**After reading the above passage, answer the following :**

**(i) What are the values (at least two, displayed by Mr. Roy ?****

**(ii) As a student, how can you spread awareness about this issue ?**

**(iii) What are tranquilizers ? Give an example.**

**(iv) Why is use of aspartame limited to cold foods and drinks ? [4]**

** Answer is not given due to the change in present syllabus.

**Ans. (ii)** As a student, I can spread awareness regarding diabetes and depression among students by conducting seminars, health camps, debates, distribution of pemplates, and workshops by doctors. So as to highlight the needs to follow healthy eating.

**(iii)** Tranquilizers are neurologically active drugs that induce a sense of well being and are used to treat stress, anxiety and mild or severe mental disease. They perform their function by inhibiting the message transfer mechanism from nerve to receptor *e.g.*, equanil, meprobamate and iproniazid etc.

**(iv)** The use of aspartame is limited to foods and cold drinks because aspartame is unstable at cooking temperature.

**24. (a) Account for the the following :**

**(i) Acidic character increases from HF to HI.**

**(ii) There is large difference between melting and boiling points of oxygen and sulphur.**

**(iii) Nitrogen does not form pentahalide.**

**(b) Draw the structures of the following :**

**(i) $ClF_3$ (ii) $XeF_4$** [5]

**OR**

**(i) Which allotrope of phosphorus is more reactive and why ?**

**(ii) How the supersonic jet aeroplanes are responsible for the depletion of ozone layers ?**

**(iii) $F_2$ has lower bond dissociation enthalpy than $Cl_2$. Why ?**

**(iv) Which noble gas is used in filling balloons for meteorological observations ?**

**(v) Complete the equation :**

**$XeF_2 + PF_5 \longrightarrow$**

**Ans. (a) (i)** The acidic strength of the hydrohalic acids increases from HF to HI because the stability of the acids decreases from HF to HI on account of decrease in bond dissociation enthalpy of H – X bond from HF to HI.

**(ii)** The difference in melting point and boiling point of oxygen and sulphur is due to the difference in their atomicities oxygen exists as a diatomic ($O_2$) molecule, while sulphur exists as a polyatomic ($S_8$) molecule and also oxygen is small in size and have high electronegativities.

**(iii)** Being an element of second period, Nitrogen has no *'d'* orbitals and its maximum covalency is restricted to four. Hence, due to the non-availability of *d*-orbitals, it can't form pentahalides.

**(b) (i)** Structure of $ClF_3$

Bent-T-Shaped

**(ii)** Structure of $XeF_4$

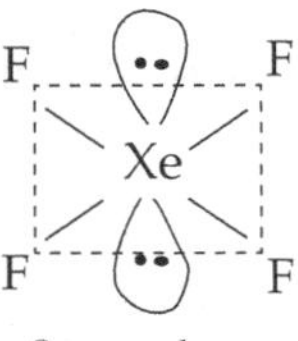


Square planar

**OR**

**(i)** White phosphorus is most reactive of all the allotropes of phosphorus. It is because it exists as $P_4$ discrete tetrahedral units with 60° angle, which results in angular strain and makes it highly reactive.

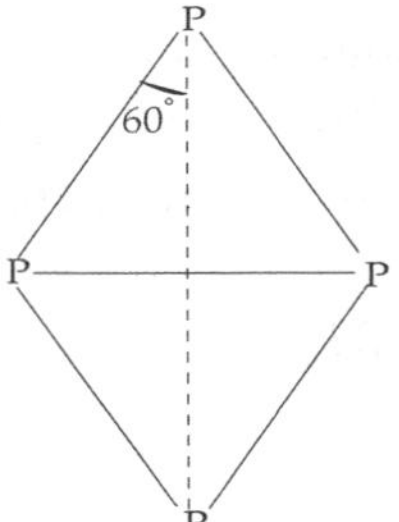


White phosphorus

**(ii)** Since supersonic jets fly in the stratosphere near the ozone layer, they are responsible for the depletion of ozone layer. The oxides of nitrogen emitted from the exhausts of supersonic jet aeroplanes readily combine with ozone to form nitrogen dioxide and diatomic oxygen.

$$NO(g) + O_3(g) \longrightarrow NO_2(g) + O_2(g)$$

(iii) The size of a fluorine atom is very small as compared to a chlorine atom. Therefore, the repulsion between electrons in the outer most shell of the two atoms in a fluorine molecule is much greater than that in a chlorine molecule. Hence, it requires less energy to break up the fluorine molecule, making its bond dissociation energy lesser than that of chlorine molecule.

(iv) Helium, gas is used for filling of balloons for meteorological observations.

(v) $XeF_2 + PF_5 \longrightarrow [XeF]^+ [PF_6]^-$

**25. An aromatic compound 'A' of molecular formula $C_7H_7ON$ undergoes a series of reactions as shown below. Write the structures of A, B, C, D and E in the following reactions.**

$(C_7H_7ON)\ A \xrightarrow{Br_2/KOH} C_6H_5NH_2 \xrightarrow[273\ K]{NaNO_2 + HCl} B \xrightarrow{CH_3CH_2OH} C$

$C_6H_5NH_2 \xrightarrow{CHCl_3 + NaOH} D$

$B \xrightarrow{KI} E$

[5]

OR

**(a) Write the structures of main products when aniline reacts with the following reagents :**

**(i) $Br_2$ water**

**(ii) HCl**

**(iii) $(CH_3CO)_2O$/pyridine.**

**(b) Arrange the following in the increasing order of their boiling point :**

**$C_2H_5NH_2$, $C_2H_5OH$, $(CH_3)_3N$**

**(c) Give a simple chemical test to distinguish between the following pair of compounds :**

**$(CH_3)_2NH$ and $(CH_3)_3N$.**

**Ans. (A)** $C_6H_5-CONH_2$

Benzamide

**(B)** $C_6H_5-\overset{+}{N_2}\overset{-}{Cl}$

Benzene diazonium chloride

**(C)** $C_6H_6$

Benzene

**(D)** $C_6H_5-NC$

Phenylisocyanide

**(E)** $C_6H_5-I$

Iodobenzene

OR

(a) (i) $C_6H_2Br_3NH_2$ (NH at position 1, Br at 2, 4, 6)

2, 4, 6-Tribromo aniline

(ii) $C_6H_5-\overset{+}{N}H_3\overset{-}{Cl}$

Anilinium Chloride

(iii) $C_6H_5-NHCOCH_3$

Acetanilide

(b) Ethanol has high boiling point then ethylamine because oxygen, being more electronegative forms strong extensive hydrogen bond as compared to that of nitrogen. In trimethylamine there is no hydrogen and hence has the lowest boiling point.

Increasing order of their boiling point.

$(CH_3)_3N < C_2H_5NH_2 < C_2H_5OH.$

(c) $(CH_3)_2NH$ reacts with benzene sulphonyl chloride (Hinsberg reagent) as follows :

$$C_6H_5-\underset{\underset{O}{\|}}{\overset{\overset{O}{\|}}{S}}-Cl + H-\underset{\underset{CH_3}{|}}{N}-CH_3 \xrightarrow{\text{(Hinsberg Reagent)}}$$

Benzene Sulphonyl Chloride

$$C_6H_5-\overset{\overset{O}{||}}{\underset{\underset{O}{||}}{S}}-\underset{\underset{CH_3}{|}}{N}-CH_3 + HCl$$

N, N - Dimethyl benzenesulphonamide.

$(CH_3)_3N$ does not react with benzene sulphonyl chloride.

**26. For the hydrolysis of methyl acetate in aqueous solution, the following results were obtained :**

| t/s | 0 | 30 | 60 |
|---|---|---|---|
| $[CH_3COOCH_3]$ mol $L^{-1}$ | 0.60 | 0.30 | 0.15 |

**(i) Show that it follows pseudo first order reaction, as the concentration of water remains constant.**

**(ii) Calculate the average rate of reaction between the time interval 30 to 60 seconds.**

**(Given log 2 = 0.3010, log 4 = 0.6021)** [5]

**OR**

**(a) For a reaction A + B →P, the rate is given by**

**Rate = $k$[A] $[B]^2$**

**(i) How is the rate of reaction affected if the concentration of B is doubled ?**

**(ii) What is the overall order of reaction if A is present in large excess ?**

**(b) A first order reaction takes 30 minutes for 50% completion. Calculate the time required for 90% completion of this reaction.**

**(log 2 = 0.3010)**

**Ans.** For the hydrolysis of methyl acetate in aqueous solution :

(i) $$K = \frac{2.303}{t} \log \frac{[A]_0}{[A]}$$

where $[A]_0$ = Initial concentration of reactant

[A] = Final concentration of reactant.

At $t_1$ = 30 sec,

$$K = \frac{2.303}{30} \log \frac{0.60}{0.30}$$

$$K = 0.07677 \log 2$$

$$K = 0.0231\ s^{-1}$$

for $t$ = 60 sec

$$K = \frac{2.303}{60} \log \frac{0.60}{0.15}$$

$$K = 0.07677 \log 2$$

$$K = 0.0231\ s^{-1}$$

∴ K is same for both the cases hence it is pseudo first order reaction.

(ii) Average rate of reaction between the time interval of 30 – 60 seconds is given by

$$K = \frac{-\Delta[CH_3COOCH_3]}{\Delta t}$$

$$K = -\left(\frac{0.15-0.30}{60-30}\right)$$

$$= \frac{0.15}{30} = 0.005\ mol\ L^{-1}s^{-1}$$

**OR**

(a) (i) Since the given reaction has order two with respect to reactant B, thus if the concentration of B is doubled in the given reaction, then the rate of reaction will become four times.

(ii) If the concentration of B is doubled *i.e.*; $[B]^2$ the overall reaction will be two, because if A is present in large excess, then the reaction will be independent of the concentration of A and will be dependent only on the concentration of B. Order of reaction = 2.

(b) $t_{½}$ = 30 min

$$[R] = [R]_0 - 90\% \text{ of } [R]_0$$

$$= [R]_0 - \frac{90[R]_0}{100}$$

$$[R] = \frac{[R]_0}{10}$$

$$K = \frac{0.693}{t_{½}} = \frac{0.693}{30} = 0.0231\ min^{-1}$$

$$t = \frac{2.303}{K} \log \frac{[R]_0}{[R]}$$

$$t = \frac{2.303}{0.0231} \log \frac{[R]_0}{\frac{[R]_0}{10}}$$

$$= \frac{2.303}{0.0231} \log 10$$

$$t = 99.7\ min$$

**Note : All questions of Delhi Set-II are from Delhi Set I and Delhi Set-III are from Set I and II.**

# CHEMISTRY

## Outside Delhi [Set-I] 2014

**Time allowed : 3 hours** **Maximum marks : 70**

1. **What is the effect of temperature on chemisorption ?** **[1]**

**Ans.** Chemisorption initially increases then decreases with rise in temperature. The initial increase is due to the high energy of activation and the decrease afterwards is due to the exothermic nature of adsorption equilibrium.

2. **What is the role of zinc metal in the extraction of silver ?** **[1]**

**Ans.** Zinc is used as a reducing agent to recover silver from its cyanide complex. It reduces $Ag^+$ to Ag and itself get oxidised to $Zn^{2+}$.

3. **What is the basicity of $H_3PO_3$ ?** **[1]**

**Ans.** The $H_3PO_3$ ionises to give two $H^+$ ions. Its basicity is two.

4. **Identify the chiral molecule in the following pair :** **[1]**

Cl Cl

**Ans.** $H_3C-CH_2-CH(Cl)-CH_3$ = (Cl) is chiral due to pairing with four different groups.

5. **Which of the following is a natural polymer ? Buna-S, Proteins, PVC Economics type** **[1]**

**Ans.** Protein, is a natural polymer having amino acid as a monomer.

6. **The conversion of primary aromatic amines into diazonium salts is known as ______.** **[1]**

**Ans.** Diazotisation reaction.

7. **What are the products of hydrolysis of sucrose ?** **[1]**

**Ans.** The products of Hydrolysis of sucrose are Glucose and Fructose.

8. **Write the structure of *p*-methylbenzaldehyde.** **[1]**

**Ans.** $CH_3-C_6H_4-CHO$

9. **An element with density 2.8 g $cm^{-3}$ forms a *f.c.c.* unit cell with edge length $4 \times 10^{-8}$cm. Calculate the molar mass of the element. (Given $N_A$ = $6.022 \times 10^{23}$ $mol^{-1}$)** **[2]**

**Ans.** $d = \frac{ZM}{a^3N_A}$

$d = 2.8\ g/cm^3$

$a = 4 \times 10^{-8}$ cm

$N_A = 6.022 \times 10^{23}\ mol^{-1}$

$$M = \frac{da^3N_A}{Z} = \frac{2.8 \times (400 \times 10^{-3})^3 \times 6.022 \times 10^{23}}{4}$$

$M = 26.98\ g\ mol^{-1} \approx 27\ g/mol.$

Therefore, molar mass of the element is approximately 27 g/mol.

10. (i) **What type of non-stoichiometric point defect is responsible for the pink colour of LiCl ?****

(ii) **What type of stoichiometric defect is shown by NaCl ?** **[2]**

**OR**

**How will you distinguish between the following pairs of terms ?**

(i) **Tetrahedral and octahedral voids**

(ii) **Crystal lattice and unit cell**

**Ans.** (i) Metal excess defect is due to anionic vacancies which is responsible for pink colour of LiCl

(ii) Schottky defect is shown by NaCl.

**OR**

(i)

| Tetrahedral Voids | Octahedral Voids |
|---|---|
| 1. A void surrounded by four spheres occupying the corners of tetrahedron is called tetrahedral voids. | A void surrounded by six spheres along the corners of an octahedral void. |
| 2. Radius ratio $(r^+/r^-)$ for tetrahedral void is 0.225. | Radius ratio $(r^+/r^-)$ for Octahedral void is 0.414. |

(ii)

| Crystal Lattice | Unit Cell |
|---|---|
| The 3-dimensional arrangement of constiuent particles in a crystal in which each particle is depicted as a point is called the crystal lattice. | Unit cell is the smallest portion of a crystal lattice which, when repeated in different directions, generates the entire lattice. |

11. **State the Kohlrausch law of independent migration of ions. Why does the conductivity of a solution decreases with dilution ?** **[2]**

** Answer is not given due to the change in present syllabus.

**Ans.** It states that the limiting molar Conductivity of an electrolyte can be expressed as sum of individual contribution of anion and cation of the electrolyte.

With dilution, the number of ions per unit volume of electrolytes decreases and hence, conductivity decreases.

**12. For a chemical reaction R→P, the variation in the concentration (R) Vs. time (*t*) plot is given as [2]**

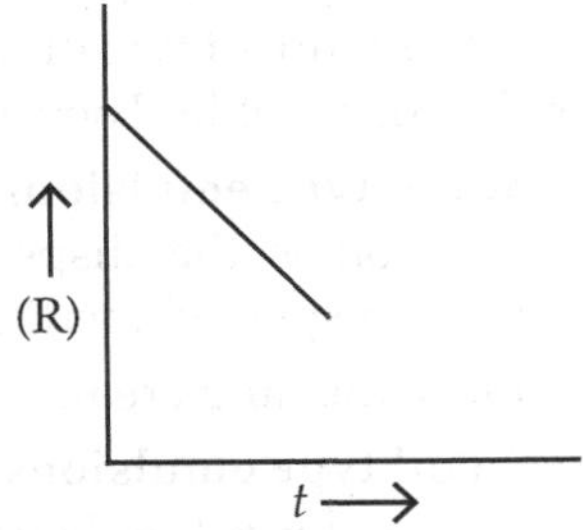


**(i) Predict the order of the reaction.**

**(ii) What is the slope of the curve ?**

**Ans.** (i) Zero order reaction

(ii) Slope of the curve is (–K). *i.e.*, negative slope.

**13. Explain the principle of the method of electrolytic refining of metals. Give one example. [2]**

**Ans.** In electrolytic method, the impure metal is made anode. A strip of the same metal in pure form is used as cathode. They are put in a suitable electrolytic bath containing soluble salt of the same metal. The more basic metal remains in the solution and the less basic ones go to the anode mud.

For *e.g.* : Copper is refined using an electrolytic method, the net result of electrolysis is the transfer of copper in pure form from the anode to the cathode.

Anode : $Cu \longrightarrow Cu^{2+} + 2e^-$

Cathode : $Cu^{2+} + 2e^- \longrightarrow Cu$

**14. Complete the following equations : [2]**

**(i) $P_4 + H_2O \longrightarrow$**

**(ii) $XeF_4 + O_2F_2 \longrightarrow$**

**Ans.** (i) $P_4 + H_2O \longrightarrow$ No reaction

(ii) $XeF_4 + O_2F_2 \longrightarrow XeF_6 + O_2$

**15. Draw the structures of the following :**

**(i) $XeF_2$ (ii) $BrF_3$ [2]**

**Ans.**

F
Xe
F

(i)

F
Br —— F
F

(ii)

**16. Write the equations involved in the following reactions : [2]**

**(i) Reimer-Tiemann reaction**

**(ii) Williamson synthesis**

**Ans. (i) Reimer Tiemann reaction :**

Phenol (OH) $\xrightarrow[\text{aq. NaOH}]{CHCl_3}$ [ONa, $CHCl_2$] $\xrightarrow{NaOH}$

ONa, CHO $\xrightarrow{H^+}$ OH, CHO

Salicylaldehyde

**(ii) Williamson synthesis :**

$CH_3CH_2I + NaOCH_2CH_3 \longrightarrow CH_3\text{–}CH_2\text{–}O\text{–}CH_2\text{–}CH_3$

Ethyl Iodide Sodium ethoxide Diethyl ether

**17. Write the mechanism of the following reaction :**

$$CH_3CH_2OH \xrightarrow{HBr} CH_3CH_2Br + H_2O \quad [2]$$

**Ans.** $S_N2$ Mechanism

1. $H\text{–}Br \rightleftharpoons H^+ + :Br^-$

$CH_3\text{–}CH_2\ddot{O}H + H^+ \rightleftharpoons CH_3\text{–}CH_2O^+H_2$

Ethyl oxonium ion

2.

$:Br^- + CH_2(CH_3)\text{–}O^+H_2 \longrightarrow \left[\overset{\delta^-}{Br} \text{----} CH_2(CH_3) \text{----} \overset{\delta^+}{OH_2}\right]$

Transition state

↓

$Br\text{–}CH_2\text{–}CH_3 + H_2O$

Ethyl bromide

**18. Write the name of monomers used for getting the following polymers : [2]**

**(i) Bakelite (ii) Neoprene**

**Ans. (i)** Phenol and formaldehyde are the monomers used for the formation of Bakelite.

**(ii)** Chloroprene is the monomer used to prepare Neoprene.

**19. (a) Calculate $\Delta_r G^0$ for the reaction**

**$Mg(s) + Cu^{2+}(aq) \rightarrow Mg^{2+}(aq) + Cu(s)$**

**Given : $E^0_{cell} = 2.71$ V, 1 F = 96500 C $mol^{-1}$**

**(b) Name the type of cell that was used in Apollo space program for providing electrical power.** **[3]**

**Ans. (a)** $Mg(s) + Cu^{2+}(aq) \rightarrow Mg^{2+}(aq) + Cu(s)$

$$\Delta_r G^\circ = -nFE^\circ_{cell}$$
$$= -2 \times 96500 \times 2.71$$
$$= -523.03 \text{ kJ mol}^{-1}$$

**(b)** Hydrogen-oxygen fuel cells and solar cells.

**20. The following data were obtained during the first order thermal decomposition of $SO_2Cl_2$ at a constant volume :**

**$SO_2Cl(g) \longrightarrow SO_2(g) + Cl_2(g)$**

| Experiment | Time ($s^{-1}$) | Total pressure (atm) |
|---|---|---|
| 1 | 0 | 0.4 |
| 2 | 100 | 0.7 |

**Calculate the rate constant. (Given: log 4 = 0.6021, log 2 = 0.3010)** **[3]**

**Ans.** $SO_2Cl_2(g) \rightarrow SO_2(g) + Cl_2(g)$

$dt_1$ $t = 0$ $P_0$ ................ 0

At $t = t$ $P_0 - P$ ............. P

The total pressure of the thermal decomposition of $SO_2Cl_2$ time $t$.

$$P_t = (P_0 - P) + P + P$$
$$P_t = P_0 + P$$

Hence, $P = P_t - P_0$

$\therefore$ $P_0 - P = P_0 - (P_1 - P_0)$

$$= 2P_0 - P_t$$

We know,

$$SO_2Cl_2 \rightarrow SO_2 + Cl_2$$

At $t = 0$ s 0.4 atm 0 atm 0 atm

At $t = 100$ s$(0.4 - x)$ atm $x$ atm $x$ atm

$$P_t = 0.4 - x + x + x = 0.4 + x$$
$$0.7 = 0.4 + x$$
$$x = 0.3$$

for first order reaction.

$$K = \frac{2.303}{t} \log \frac{P_0}{2P_0 - P_t}$$
$$= \frac{2.303}{100} \log \frac{0.4}{2 \times 0.4 - 0.7}$$
$$= \frac{2.303}{100} \log \frac{0.4}{0.1}$$
$$K = \frac{2.303 \times 0.602}{100}$$
$$K = 1.386 \times 10^{-2} \text{ s}^{-1}$$

**21. What are emulsions ? What are their different types ? Give one example of each type.** **[3]**

**Ans.** Emulsions are colloids in which both the dispersed phase and dispersing medium are liquid. It is a mixture of two or more liquids that are normally immiscible. They are of two types:

1. **Oil in water type emulsions (O/W) :** In such emulsions, oil is the dispersed phase and water is the dispersed medium.

   *e.g.,* Milk, Vanishing cream.

2. **Water in oil type emulsions (W/O) :** In such emulsions, water is the dispersed phase and oil is the dispersed medium.

   *e.g.,* Butter, Cod liver oil.

**22. Given reasons for the following :**

**(i) $(CH_3)_3$ P = O exists but $(CH_3)_3N$ = O does not.**

**(ii) Oxygen has less electron gain enthalpy with negative sign than sulphur.**

**(iii) $H_3PO_2$ is a stronger reducing agent than $H_3PO_3$.** **[3]**

**Ans. (i)** Due to the absence of *d*-orbitals, nitrogen cannot form $p\pi$-$d\pi$ multiple bonds. Thus, N cannot expand its covalency beyond four. So, the compound $(CH_3)_3N$= O does not exist. On the other hand, P due to the presence of *d*-orbitals forms. $p\pi$-$d\pi$ multiple bonds and hence forms $(CH_3)_3P$ = O in which covalency of P is 5.

**(ii)** Due to small size and high electronegativity of oxygen compared to sulphur, oxygen has less electron gain enthalpy than sulphur.

**(iii)** Acids which contain P-H bonds have reducing character, Since $H_3PO_2$ contains two P-H bonds while $H_3PO_3$ contains only one P–H bond therefore, $H_3PO_2$ is a stronger reducing agent than $H_3PO_3$.

**23. (i) Write the IUPAC name of the complex $[Cr(NH_3)_4Cl_2]Cl$.**

**(ii) What type of isomerism is exhibited by the complex $[Co(en)_3]^{3+}$ ?**

**(en = ethane-1, 2-diamine)**

**(iii) Why is $[NiCl_4]^{2-}$ paramagnetic but $[Ni(Co)_4]$ is diamagnetic** **[3]**

**(At. nos. : Cr = 24, Co = 27, Ni = 28)**

**Ans.** (i) Tetraammine dichlorido chromium(III) chloride

(ii) Optical isomerism is shown by the complex $[Co(en)_3]^{3+}$

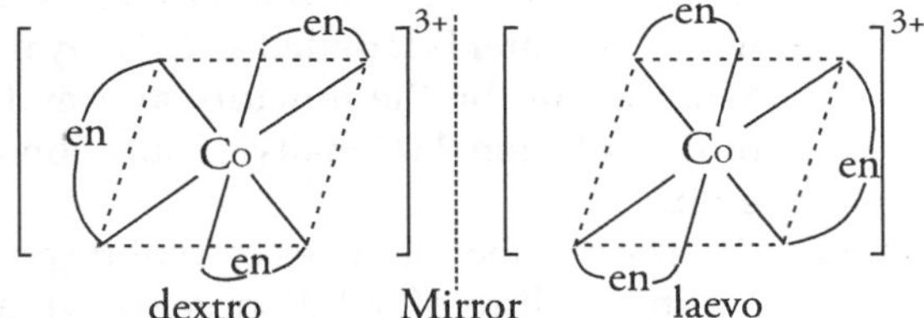


(iii) $[Ni(Cl)_4]^{2-}$, Ni is in +2 oxidation state with the electronic configuration $3d^8 4s^0$. As $Cl^-$ is a weak ligand it cannot pair up the electrons in 3*d* orbitals therefore, $[NiCl_4]^{2-}$ is paramagnetic. In $[Ni(Co)_4]$, Ni is in 0 oxidation state with the electronic configuration $3d^8 4s^2$. Co is a strong ligand it causes 4*s* electrons to shift to 3*d* and pair up 3*d* electrons. Therefore, $[Ni(Co)_4]$ is diamagnetic.

**24. (a) Draw the structure of major monohalo products in each of the following reactions:**

(i) $C_6H_{11}-CH_2OH \xrightarrow{PCl_5}$

(ii) $C_6H_5-CH_2-CH=CH_2 + HBr \rightarrow$

**(b) Which halogen compound in each of the following pairs will react faster in $S_N2$ reaction :**

**(i) $CH_3Br$ or $CH_3I$**

**(ii) $(CH_3)_3C-Cl$ or $CH_3-Cl$** [3]

**Ans. (a)** (i) $C_6H_5-CH_2OH \xrightarrow{PCl_5} C_6H_5-CH_2Cl + POCl_3 + HCl$

(ii) $C_6H_5-CH_2-CH=CH_2 + HBr \longrightarrow C_6H_5-CH_2-\underset{\underset{Br}{|}}{CH}-CH_3$

**(b)** (i) $CH_3-I$ will react faster because Iodine (I)-is a better leaving group.

(ii) $CH_3-Cl$ will react faster because it is a primary halide and it undergoes $S_N^2$ reaction faster.

**25. Account for the following :**

**(i) Primary amines ($R-NH_2$) have higher boiling point than tertiary amines ($R_3N$).**

**(ii) Aniline does not undergo Friedel-Crafts reaction.**

**(iii) $(CH_3)_2NH$ is more basic than $(CH_3)_3N$ in an aqueous solution.** **[3]**

**OR**

**Give the structures of A, B and C in the following reactions :**

**(i) $C_6H_5NO_2 \xrightarrow{Sn+HCl} A \xrightarrow{NaNO_2+HCl} B \xrightarrow{H_2O} C$**

**(ii) $CH_3CN \xrightarrow{H_2O/H^+} A \xrightarrow[\Delta]{NH_3} B \xrightarrow{Br_2+KOH} C$**

**Ans.** (i) Due to intermolecular hydrogen bonding in primary amines (presence of more number of H-atoms). They have high boiling point in comparison to tertiary amines.

(ii) Aniline does not undergo Friedel-Crafts reaction because Aniline is basic in nature and reacts with $AlCl_3$ to form a salt. The positive charge on the N-atom, electrophilic substitution in the benzene ring is deactivated.

$$\underset{\text{Aniline}}{C_6H_5-NH_2} + AlCl_3 \longrightarrow \underset{\text{Salt}}{C_6H_5-\overset{+}{N}H_2\overset{-}{A}lCl_3} + HCl$$

(iii) Due to steric hindrance and solvation effect in case of $(CH_3)_3N$, it is less basic than $(CH_3)_2NH$ in aqueous solution.

**OR**

(i) $C_6H_5NO_2 \xrightarrow{Sn+HCl} \underset{\text{(A) Benzylamine}}{C_6H_5NH_2}$

$\downarrow NaNO_2 + HCl$

$\underset{\text{(C) (Phenol)}}{C_6H_5OH} \xleftarrow{H_2O} \underset{\text{(B) Benzene diazonium chloride}}{C_6H_5N_2^+Cl^-}$

(ii) $CH_3CN \xrightarrow{H_2O/H^+} \underset{\text{(A) Acetic acid}}{CH_3COOH} \xrightarrow[\Delta]{NH_3} \underset{\text{Ethanamide (B)}}{CH_3CONH_2}$

$\downarrow Br_2+KOH$

$\underset{\text{Methanamine}}{CH_3NH_2(C)}$

**26. Define the following terms related to proteins :** **[3]**

**(i) Peptide linkage**

**(ii) Primary structure**

**(iii) Denaturation**

**Ans. (i)** Peptide linkage is the amide linkage formed by —COOH group of one α-amino acid and —$NH_2$ group of other α-amino acid by lose of a water molecule.

**(ii)** The sequence in which various amino acids are arranged in a linear structure with no intermediate bonding is called primary structure of a protein.

**(iii)** When a protein in its native form is subjected to a change, such as temperature, pressure etc. Due to this protein loses its biological activity and this is called denaturation of a protein *e.g.*, curdling of milk.

**27. On the occasion of World Health Day.**

**Dr. Satpal organized a 'health camp' for the poor farmers living in a nearby village. After check-up, he was shocked to see that most of the farmers suffered from cancer due to regular exposure to pesticides and many were diabetic. They distributed free medicines to them. Dr. Satpal immediately reported the matter to the National Human Rights Commission (NHRC). On the suggestions of NHRC, the government decided to provide medical care, financial assistance, setting up of super-speciality hospitals for treatment and prevention of the deadly disease in the affected villages all over India.**

**(i) Write the values shown by (a) Dr. Satpal (b) NHRC ?****

**(ii) What type of analgesics are chiefly used for the relief of pains of terminal cancer ?**

**(iii) Give an example of artificial sweetener that could have been recommended to diabetic patients.** **[3]**

**Ans. (ii)** Narcotic analgesics like morphine and heroin.

**(iii)** Aspartame.

**28.(a) Define the following terms :**

**(i) Molarity (ii) Molal elevation constant ($k_b$)**

**(b) A solution containing 15 g urea (molar mass = 60 g mol$^{-1}$) per litre of solution in water has the same osmotic pressure (isotonic) as a solution of glucose (molar mass = 190 g mol$^{-1}$) in water. Calculate the mass of glucose present in one litre of its solution.** **[2,3]**

** Answer is not given due to the change in present syllabus.

**OR**

**(a) What type of deviation is shown by a mixture of ethanol and acetone ? Give reason.**

**(b) A solution of glucose (molar mass = 180 g mol$^{-1}$) in water is labelled as 10% (by mass). What would be the molality and molarity of the solution ? (Density of solution = 1.2 g mL$^{-1}$)**

**Ans. (a) (i)** The number of moles of a solute present in one litre of solution is known as its molarity.

**(ii)** The elevation in boiling point of a solution when one mole of non-volatile solute is dissolved in one kilogram of a volatile solvent is known as molal elevation constant ($K_b$).

$$\Delta T_b = \frac{K_b \times W_B \times 1000}{M_B \times W_A}$$

$W_B$ = Mass of solute

$M_B$ = Molar mass of solute

$W_A$ = Mass of solvent.

**(b)** For isotonic solution,

$\pi_1 = \pi_2$

$C_1 = C_2$ (at same temp.)

or $n_1 = n_2$ (at same Vol.)

$$\therefore \frac{15}{60} = \frac{x}{180}$$

$x$ = 45 g, mass of glucose per litre of solution.

**OR**

**(a)** The mixture of ethanol and acetone shows positive deviation from Raoults law. In pure ethanol hydrogen bond exist between the molecules. On adding acetone to ethanol, acetone molecules get in between the molecules of ethanol thus breaking some of the hydrogen bonds and weakening the molecular interactions this leads to an increase in vapour pressure resulting in positive deviation from Raoult's law.

**(b)** Mass of glucose = 10 g

Mass of solution = 100 g

Mass of water = 100–10 = 90 g

$$\text{Volume of solution} = \frac{100\text{g}}{1.2\text{g ml}} = 83{\cdot}33 \text{ ml}$$

$$= 0.083 \text{ L}$$

$$\text{Molarity} = \left(\frac{10}{180}\right)\cdot\frac{1000}{83.33}$$

$$= 0.67 \text{ mol L}^{-1}$$

$$\text{Molality} = \frac{10}{180}\cdot\frac{1000}{90} = 0.617 \text{ mol kg}^{-1}$$

**29. (a) Complete the following equations :**

**(i) $Cr_2O_7^{2-}+2OH^- \rightarrow$**

**(ii) $MnO_4^-+4H^++3e^- \rightarrow$**

**(b) Account for the following :**

**(i) Zn is not considered as a transition element.**

**(ii) Transition metals form a larger number of complexes.**

**(iii) The E value for the $Mn^{3+}/Mn^{2+}$ couple is much more positive than that for $Cr^{3+}/Cr^{2+}$ couple.** [2,3]

**OR**

**(i) With reference to structural variability and chemical reactivity, write the difference between lanthanoids and actionoids.**

**(ii) Name a member of the lanthanoid series which is well known to exhibit +4 oxidation state.**

**(iii) Complete the following equation :**

$$MnO_4^-+8H^++5e^- \rightarrow$$

**(iv) Out of $Mn^{3+}$ and $Cr^{3+}$, which is more paramagnetic and why ? (atomic nos : Mn=25, Cr=24)** [5]

**Ans. (a)** (i) $Cr_2O_7^{2-} + 2OH^- \rightarrow 2CrO_4^{2-} + H_2O$

(ii) $MnO_4^- + 4H^+ + 3e^- \rightarrow MnO_2 + 2H_2O$

**(b)** (i) Zinc has $3d^{10}4s^2$ configuration with no unpaired *d*-orbital electron and hence it is not considered as a transition element.

(ii) Transition metals have vacant orbitals to accommodate lone pairs of electrons for bond formation and have high charge density, therefore, they form complexes.

(iii) Due to high stability of $Mn^{2+}$ (due to its half-filled 3d sub shell) than $Mn^{3+}$ while $Cr^{+3}$ is more stable than $Cr^{+2}$.

OR

(i)

| | Actinoids | Lanthanoids |
|---|---|---|
| 1 | Actinoids have stronger tendency to form complexes. | Lathanoids form less complexes. |
| 2. | They show more number of oxidation states. | They show less number of oxidation states. |
| 3. | They are radioactive. | Lanthanoids except promethium are not radioactive. |

**(ii)** Cerium (Ce) is lanthanoid element, which is well known to exhibit +4 oxidation state.

**(iii)** $MnO_4^- + 8H^+ + 5e^- \rightarrow Mn^{2+} + 4H_2O$

**(iv)** $Mn^{3+}$ is more paramagnetic as it has four unpaired electrons while $Cr^{3+}$ has only three.

$Mn^{+3} = [Ar]\ 4s^o3d^4$

$Cr^{+3} = [Ar]\ 4s^o3d^3$

**30. (a) Write the products formed when $CH_3CHO$ reacts with the following reagents :**

**(i) HCN**

**(ii) $H_2N-OH$**

**(iii) $CH_3CHO$ in the presence of dilute NaOH**

**(b) Give simple chemical tests to distinguish between the following pairs of compounds.**

**(i) Benzoic acid and Phenol**

**(ii) Propanal and Propanone.** [3,2]

**OR**

**(a) Account for the following :** (2, 2, 1)

**(i) $Cl^-CH_2COOH$ is a stronger acid than $CH_3COOH$.**

**(ii) Carboxylic acids do not give reactions of carbonyl group.**

**(b) Write the chemical equations to illustrate the following name reactions :**

**(i) Rosenmund reduction**

**(ii) Cannizzaro's reaction**

**(c) Out of $CH_3CH_2-CO-CH_3$ and $CH_3CH_2-CH_2-CO-CH_3$, which gives iodoform test ?**

**Ans. (a)** (i) On reaction of acetaldehyele with hydrogen cyanide it gives 2-Hydroxypropanenitrile as product.

$$\underset{\text{Acetaldehyde}}{CH_3-CHO} + \underset{\text{Hydrogen cyanide}}{HCN} \longrightarrow \underset{\text{2-Hydroxypropanenitrile.}}{H_3C-\overset{\overset{H}{|}}{\underset{\underset{CN}{|}}{C}}-OH}$$

(ii) On reaction of acetaldehyde with Hydroxylamine ($NH_2$-OH) it gives acetaldoxime.

$$\underset{\text{Acetaldehyde}}{CH_3\text{-}CHO} + \underset{\text{Hydroxylamine}}{NH_2\text{-}OH} \longrightarrow \underset{\text{Acetaldoxime}}{CH_3-CH=NOH} + H_2O$$

(iii) When 2 molecules of acetaldehyde reacts with each other in presence of dil NaOH, 3-Hydroxybutanal is obtained. This is a aldol reaction and further proceeds the reaction when heated.

$$CH_3-CHO + CH_3-CHO \xrightleftharpoons{\text{dil.NaOH}} \underset{\substack{|\\OH}}{CH_3-CH}-CH_2-CHO$$

3–Hydroxybutanal (aldol reaction)

**(b) (i)** Phenol reacts with $FeCl_3$ to give violet coloured precipitate while benzoic acid gives buff coloured precipitate.

$$\underset{\text{Phenol}}{3C_6H_5OH} + FeCl_3 \longrightarrow \underset{\text{Violet Colour}}{(C_6H_5O)_3Fe} + 3HCl$$

$$\underset{\text{Benzoic acid}}{3C_6H_5COOH} + FeCl_3 \longrightarrow \underset{\text{Buff colour}}{(C_6H_5COO)_3Fe} + 3HCl$$

**(ii)** Propanone gives iodoform test but propanal does not.

$$CH_3-\underset{\substack{||\\O}}{C}-CH_3 \xrightarrow{I_2+NaOH} CHI_3 \text{(Iodoform)}$$

$$CH_3-CH_2-CHO \xrightarrow{I_2+NaOH} \text{No reaction}$$

**OR**

**(a) (i)** $Cl^-CH_2COOH$ is a stronger acid than $CH_3COOH$ because of –I effect of Cl atom in $ClCH_2COOH$ and +I effect of $CH_3$ group in $CH_3COOH$ the electron density in the O–H bond in $ClCH_2COOH$ is much lower than $CH_3COOH$. Thus $Cl^-CH_2COOH$ acid is stronger acid than $CH_3COOH$.

**(ii)** In carboxylic acids, the carboxyl group is not free as it is involved in resonance

**(b) (i)** $$\underset{\text{Acetyl chloride}}{CH_3COCl} \xrightarrow[\text{Pd-BaSO}_4]{H_2} \underset{\text{Acetaldehyde}}{CH_3CHO} + HCl$$

**(ii)** $$\underset{\text{Formaldehyde}}{2HCHO} \xrightarrow{NaOH} \underset{\text{Methyl alcohol}}{CH_3OH} + HCOONa$$

**(c)** $CH_3CH_2CH_2COCH_3$, being a methyl ketone gives iodoform test.

# Outside Delhi [Set-II]

> **Note : Except for the following questions, all the remaining questions have been asked in previous set.**

**1. Why is adsorption always exothermic ? [1]**

**Ans.** Adsorption being a surface phenomenon leads to decrease in surface energy and hence is always exothermic in nature.

**2. Name the method used for refining of Nickel. [1]**

**Ans.** Mond's process is used for refining of Nickel.

**3. Why does $NO_2$ dimerise ? [1]**

**Ans.** $NO_2$ is an odd-electron molecule and hence dimerises to form the stable $N_2O_4$ molecule.

**4. Based on molecular forces, what type of polymer is neoprene ? [1]**

**Ans.** Neoprene is an Elastomer based on molecular forces.

**5. What are the products of hydrolysis of maltose? [1]**

**Ans.** Two molecules of α-D-glucose.

**6. Write the structure of 4-chloropentan-2 one. [1]**

**Ans.** $$CH_3-\underset{\substack{|\\Cl}}{CH}-CH_2-\underset{\substack{||\\O}}{C}-CH_3$$

**9. Write the name of monomers used for getting the following polymers : [2]**

**(i) Terelyne (ii) Nylon-6, 6**

**Ans. (i)** Ethylene glycol and Terephthalic acid.

**(ii)** Adipic acid and Hexamethylene diamine.

**10. Describe the role of the following : [2]**

**(i) $SiO_2$ in the extraction of copper from copper matte.**

**(ii) NaCN in froth floatation process.**

**Ans. (i)** $SiO_2$ (Silica) acts as a flux in the extraction of copper from copper matte to remove ferrous oxide as ferrous silicate (slag).

$$FeO + SiO_2 \rightarrow \underset{\text{(Slag)}}{FeSiO_3}$$

**(ii)** NaCN is used as a depressant as it forms zinc complex, $Na_2[Zn(CN)_4]$ on the surface of ZnS thereby preventing it from forming froth.

**11. Complete the following equations : [2]**

**(i) $Ag + PCl_5$·**

**(ii) $CaF_2 + H_2SO_4$·**

**Ans. (i)** $2Ag + PCl_5 \rightarrow 2AgCl + PCl_3$

**(ii)** $CaF_2 + H_2SO_4 \rightarrow CaSO_4 + 2HF$

**12. Draw the structures of the following : [2]**

**(i) $XeF_4$** **(ii) $HClO_4$**

**Ans. (i)**

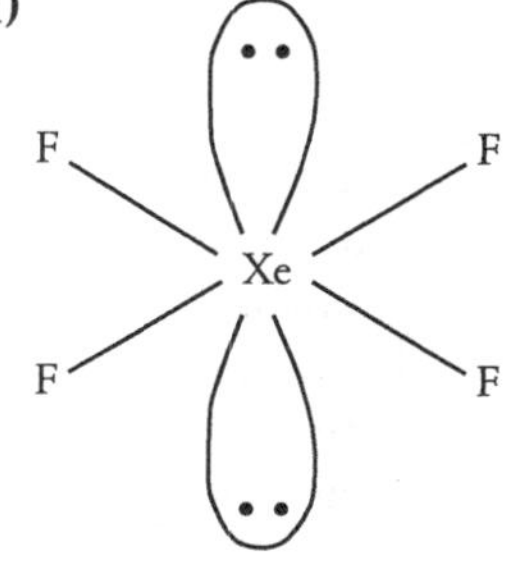

(ii) 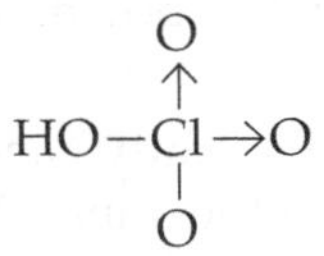 HO–Cl(→O)(=O)→O

$$HO-\underset{\underset{O}{|}}{\overset{\overset{O}{\uparrow}}{Cl}}\rightarrow O$$

**13. (i) Write the type of magnetism observed when the magnetic moment are oppositely aligned and cancel out each other.**

**(ii) Which stoichiometric defect does not change the density of the crystal ? [2]**

**Ans.** (i) Antiferromagnetism is observed when the magnetic moment are oppositely aligned and cancel out each other.

(ii) Frenkel defect does not change the density of the crystal.

**14. Define the following terms : [2]**

**(i) Fuel cell**

**(ii) Limiting molar conductivity ($\Lambda°m$)**

**Ans.** (i) Fuel cell is a device that converts chemical energy from combustion of a fuel into electric energy through a chemical reaction.

(ii) Molar conductivity of electrolyte at infinite dilution or when concentration approaches zero is called limiting molar conductivity. It is expressed as $\Lambda°m$

**19. Define the following terms: [3]**

**(i) Glycosidic linkage**

**(ii) Invert sugar**

**(iii) Oligo saccharides**

**Ans.** (i) The linkage between two monosac-charide units through oxygen atom is known as glycosidic linkage.

(ii) Sucrose is called invert sugar as it produces equimolar quantities of glucose and fructose onhydrolysisitgivesanequimolarmixture of D – (+)-glucose, and D – (–)–fructose, which is laevo rotatory. This change of specific rotation from dextrorotation to laevo-rotation is called inversion of sugar and the mixture so obtained is called invert sugar.

(iii) Carbohydrate which on hydrolysis give two to ten molecules of monosaccharides are called oligosaccharide *e.g.* sucrose.

## Outside Delhi [Set-III]

**Note : Except for the following questions, all the remaining questions have been asked in previous sets.**

**1. What are the dispersed phase and dispersion medium in milk ? [1]**

**Ans.** Dispersed phase is oil or fat and dispersion medium is water.

**2. Name the method used for refining of copper metal. [1]**

**Ans.** Electrolytic refining.

**3. Why does $NH_3$ act as a Lewis base ? [1]**

**Ans.** Due to the presence of lone pair on nitrogen in $NH_3$, It can donate its lone pair of electrons and it forms coordinate bonds with Lewis acids and acts as Lewis base.

**5. Which of the following is a fibre ? [1]**

**Nylon, Neoprene, PVC**

**Ans.** Nylon

**6. Write the products of hydrolysis of lactose. [1]**

**Ans.** $$\underset{\text{Lactose}}{C_{12}H_{22}O_{11}} + H_2O \longrightarrow \underset{\substack{\text{D-(+)}\\ \text{Glucose}}}{C_6H_{12}O_6} + \underset{\substack{\text{D-(+)}\\ \text{Galactose}}}{C_6H_{12}O_6}$$

**8. Write the structure of 2-hydroxybenzoic acid. [1]**

**Ans.** 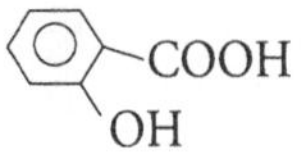 (benzene ring with –COOH and ortho –OH)

COOH

OH

**9. Complete the following equations : [2]**

**(i) $Cu + 2H_2SO_4$ (Conc.) $\rightarrow$**

**(ii) $XeF_2 + H_2O \rightarrow$**

**Ans.** (i) $Cu + \text{conc. } 2H_2SO_4 \rightarrow CuSO_4 + SO_2 + 2H_2O$

(ii) $2XeF_2 + 2H_2O \longrightarrow 2Xe + 4HF + O_2$

**10. Draw the structure of the following : [2]**

**(i) $XeO_3$ (ii) $H_2SO_4$**

**Ans.**

(i) 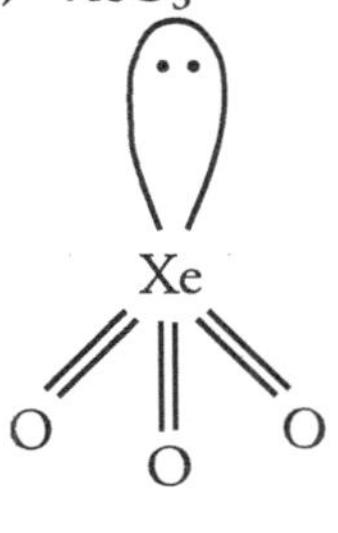 Xe with a lone pair, double bonded to three O atoms

(ii) 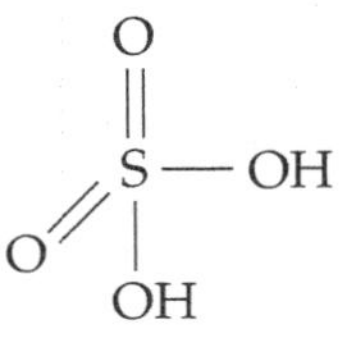 S double bonded to two O atoms, with two –OH groups

**11. Write the name of monomers used for getting the following polymers : [2]**

**(i) Teflon (ii) Buna-N**

**Ans.** (i) Tetrafluoroethene

(ii) 1,3-Butadiene and Acrylonitrile

**13. (i) Write the type of magnetism observed when the magnetic moment are aligned in parallel and anti-parallel directions in unequal numbers. [2]**

**(ii) Which stoichiometric defect decreases the density of the crystal ?**

**Ans. (i)** Ferrimagnetism.

**(ii)** Schottky defect decreases the density of the crystal.

**14. Define the following terms :**

**(i) Molar conductivity (Λ*m*)**

**(ii) Secondary batteries** **[2]**

**Ans. (i)** Molar conductivity is defined as the conducting power of all the ions produce by dissolving one mole of an electrolyte in solution.

$$\Lambda_m = \frac{K}{C}$$

**(ii)** Secondary batteries are those batteries which can be recharged by passing electric current through them and hence can be used over again *e.g.* Lead storage battery.

**17. Write the principle behind the froth floatation process. What is the role of collectors in this process ?** **[2]**

**Ans.** Froth floatation method has been in use for removing gangue from sulphide ores. In this process, a suspension of the powdered ore is made with water.

Collectors (*e.g.* pine oils, fatty acids, etc.) enhance non-wettability of the mineral particles and used to skim the froth off the surface.

**23. Define the following terms :** **[3]**

**(i) Nucleotide**

**(ii) Anomers**

**(iii) Essential amino acids**

**Ans. (i)** A Nucleotide contain all the three basic components of nucleic acid *i.e.,* pentose sugar, a nitrogeneous base and a phosphoric acid. When nucleoside is linked to phosphoric acid at 5' position of sugar moiety, we get a nucleotide.

**(ii)** The carbohydrate which differ in configuration at the glycosidic carbon (*i.e.,* $C_1$ in aldoses and $C_2$ in ketoses) are called anomers. *e.g.* α-D-(+) glucose and β-D-(+) glucose.

**(iii)** Essential amino acids are those amino acids which cannot be synthesised by the body and need to be consumed through diet. *eg.* Valine.

●●

# CHEMISTRY

## Delhi [Set-I] 2014

**Time allowed : 3 hours** **Maximum marks : 70**

**1. Give one example each of 'oil in water' and 'water in oil' emulsion.** [1]

**Ans.** Oil in water – Milk
Water in oil – Butter

**2. Which reducing agent is employed to get copper from the leached low grade copper ore ?** [1]

**Ans.** Scrap iron is used as reducing agent to obtain copper metal from the solution containing copper.

$$Cu^{2+}(aq) + Fe(s) \rightarrow Cu(s) + Fe^{2+}(aq)$$

**3. Which of the following is more stable complex and why ? $[Co(NH_3)_6]^{3+}$ and $[Co(en)_3]^{3+}$** [1]

**Ans.** $NH_3$isaunidentateligandand$H_2NCH_2CH_2NH_2$(en) is a bidentate ligand Chelating ligands form more stable complexes compared to non-chelating ligands. Thus, $[Co(en)_3]^{3+}$ is more stable.

**4. Write the IUPAC name of the compound.** [1]

$$\underset{\displaystyle \overset{|}{OH}}{CH_3-CH}-CH_2-COOH$$

*d*

**Ans.** 3-hydroxybutan-1-oic acid

**5. Which of the following isomers is more volatile: *o*-nitrophenol or *p*-nitrophenol ?** [1]

**Ans.** *o*-nitrophenol is more volatile because of the presence of intra-molecular hydrogen bonding.

**6. What are isotonic solutions ?** [1]

**Ans.** Two or more solutions having same osmotic pressure are called isotonic solutions. *e.g.* 0.5 M NaCl, 0.5 M KCl and 1M glucose are isotonic.

**7. Arrange the following compounds in increasing order of solubility in water :** [1]

$$C_6H_5NH_2, (C_2H_5)_2NH, C_2H_5NH_2$$

**Ans.** $C_6H_5NH_2 < (C_2H_5)_2NH < C_2H_5NH_2$

**8. Which of the two components of starch is water soluble ?** [1]

**Ans.** Amylose is the water soluble content of starch between Amylose and Amylopectin.

**9. An element with density 11.2 gm cm$^{-3}$ forms f.c.c. lattice with edge length 4 × 10$^{-8}$cm. Calculate the atomic mass of the element. (Given $N_A = 6.022 \times 10^{23}mol^{-1}$)** [2]

**Ans.** Given,

$d = 11.2$ g cm$^{-1}$ $N_A = 6.022 \times 10^{23}$ mol$^{-1}$

$a = 4 \times 10^{-8}$cm for fcc, number of atoms present per unit cells (Z) = 4

From the formula, $d = \frac{Z \times M}{N_A \times a^3}$

$$11.2 \text{ g cm}^{-3} = \frac{4 \times M}{6.022 \times 10^{23} \text{mol}^{-1} \times (4 \times 10^{-8}\text{cm})^3}$$

$$M = \frac{11.2 \times 6.022 \times 10^{23} \times 64 \times 10^{-24}}{4} \text{g mol}^{-1}$$

$$= 107.91 \text{ g mol}^{-1}$$

∴ Atomic mass of the element is 107.91 g mol$^{-1}$

**10. Examine the given defective crystal** [2]

| | | | | |
|---|---|---|---|---|
| $A^+$ | $B^-$ | $A^+$ | $B^-$ | $A^+$ |
| $B^-$ | O | $B^-$ | $A^+$ | $B^-$ |
| $A^+$ | $B^-$ | $A^+$ | O | $A^+$ |
| $B^-$ | $A^+$ | $B^-$ | $A^+$ | $B^-$ |

**Answer the following questions :**

**(i) What type of stoichiometric defect is shown by crystal ?**

**(ii) How is the density of the crystal affected by this defect ?**

**(iii) What type of ionic substances show such defect ?**

**Ans. (i)** Schottky defect, as equal number of cations and anions are missing.

**(ii)** The density of the crystal decreases as ions are missing from crystal lattice.

**(iii)** Highly ionic substances with almost comparable size of cation and anion show this defect, *e.g.* KCl.

**11. Calculate the mass of compound (molar mass = 256 g mol$^{-1}$) to be dissolved in 75 g of benzene to lower its freezing point by 0.48 K ($K_f$ = 5.12 K kg mol$^{-1}$).** [2]

**Ans.** Given, $M_B$ = 256 g mol$^{-1}$, $W_A$ = 75 g

$K_f$ = 5.12 kg mol$^{-1}$, $\Delta T_f$ = 0.48K

From the formula,

$$\Delta T_f = K_f \times \frac{W_B}{M_B} \times \frac{1000}{W_A}$$

Mass of solute, $W_B = \frac{\Delta T_f \times M_B \times W_A}{K_f \times 1000}$

$$= \frac{0.48 \times 256 \times 75}{5.12 \times 1000}$$

$$= \frac{(0.48K)(256\ g\,mol^{-1})(0.075\ kg)}{(5.12\,K\,kg\,mol^{-1})}$$

$$= 1.8\ g$$

**12. Define an ideal solution and write one of its characteristics. [2]**

**Ans.** A solution which obeys Raoult's law at all temperatures and concentrations is called an ideal solution.

For an ideal solution,

$$P = P_A + P_B$$

$$P_A = P_A^0 \times A$$

$$P_B = P_B^0 \times B$$

$$\therefore \quad P = P_A^0 \times A + P_B^0 \times B$$

Where, P = Total pressure and $P_A$, $P_B$, $P^0_A$, $P^0_B$ and $X_A$, $X_B$ are partial pressure, pressure of pure component and mole fraction of component A and B respectively.

**Characteristics :** In ideal solution, no volume or enthalpy change takes place while mixing to form such a solution, *i.e.*, $\Delta H_{mix} = 0$ and $\Delta V_{mix} = 0$.

**13. Write two differences between 'order of reaction' and 'molecularity of reaction'. [2]**

**Ans.**

| Molecularity | Order of Reaction |
|---|---|
| 1. It is the number of reacting species taking part in an elementary reaction. | It is equal to the sum of all the exponents of all the reactants present in the rate law expression. |
| 2. It is a theoretical concept and cannot be zero or fractional. | It is determined experimentally and can be equal to zero or can have fractional value. |

**14. Outline the principles behind the refining of metals by the following methods : [2]**

**(i) Zone refining method**

**(ii) Chromatographic method**

**Ans. (i) Zone refining Method :** This method is employed when impurities are more soluble in the melt than in solid form of the metal. It is used to obtain metals such as Germanium (Ge), Silicon (Si), Gallium (Ga), etc in their purest form.

**(ii) Chromatographic Method :** This method is used when the impurities are not very different in chemical properties from the element to be purified. The components of the mixture show different mobility on the stationary phase, *i.e.*, the components are adsorbed differently on the adsorbent.

**15. Complete the following chemical equations: [2]**

**(i) $Ca_3P_2 + H_2O \rightarrow$**

**(ii) $Cu + H_2SO_4$ (conc.) $\rightarrow$**

**OR**

**Arrange the following in order of property indicated against each set :**

**(i) HF, HCl, HBr, HI : increasing bond dissociation enthalpy**

**(ii) $H_2O$, $H_2S$, $H_2Se$, $H_2Te$, : increasing acidic character.**

**Ans. (i)** $Ca_3P_2 + 6H_2O \rightarrow 3Ca(OH)_2 + 2PH_3$
Calcium hydroxide, Phosphine

**(ii)**

$Cu + 2H_2SO_4$ (conc.) $\rightarrow CuSO_4 + SO_2 + 2H_2O$
Copper (II) Suphate, Suphur dioxide

**OR**

(i)

$$HI < HBr < HCl < HF \xrightarrow{\text{Increasing order of Bond Dissociation Energy.}}$$

(ii) $$H_2O < H_2S < H_2Se < H_2Te \xrightarrow{\text{Increasing acidic character}}$$

**16. Write the IUPAC name of the complex $[Cr(NH_3)_4Cl_2]^+$. What type of isomerism does it exhibit ? [2]**

**Ans.** Tetraamminedichloridochromium(III) ion. The complex exhibits geometric (cis-trans isomerism.

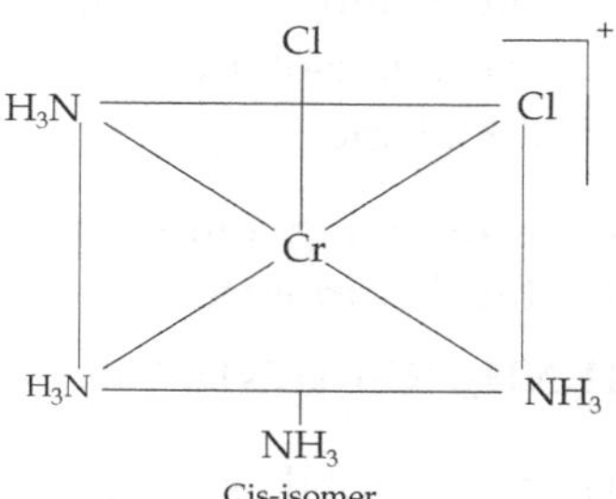

Cis-isomer

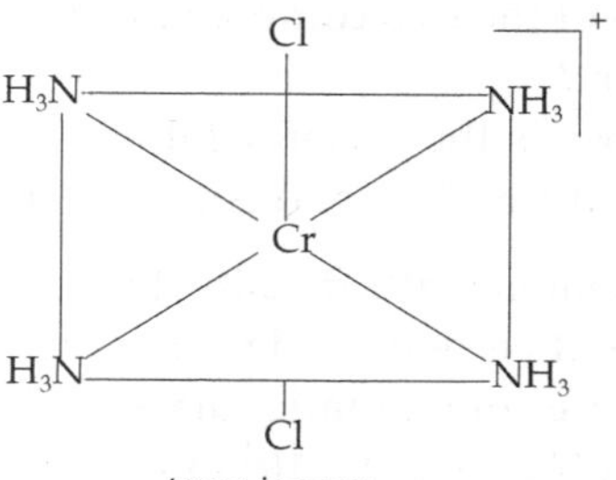

trans-isomer

17. (i) **Which alkyl halide from the following pair is chiral and undergoes faster $S_N2$ reaction?** **[2]**

(a) (b)

(ii) **Out of $S_N1$ and $S_N2$, which reaction occurs with**

(a) **Inversion of configuration**

(b) **Racemisation**

**Ans.** (i) Compound **(b)**, *i.e.*, $CH_3CHBrCH_2CH_3$ possesses chiral centre and undergoes faster $S_N2$ reaction.

(ii) (a) Inversion of configuration results in $S_N2$ reaction as there is formation of intermediate transition state in which there is simultaneous attack and migration of leaving group.

(b) Racemisation in $S_N1$ reaction is due to attack of nucleophile on both sides of the planar carbocation.

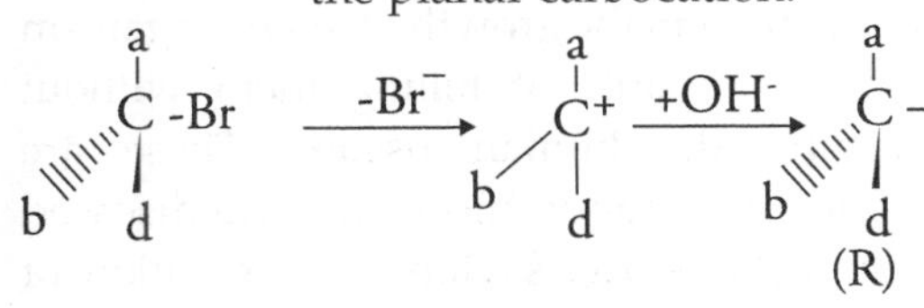


Racemic forms (Optically inactive mixture)

18. **Draw the structure of major monohalo product in each of the following reactions :** **[2]**

(i) Cyclohexanol-OH $\xrightarrow{SOCl_2}$

(ii) $C_6H_5$—$CH_2$—CH—$CH_2$ + HBr $\xrightarrow{\text{Peroxide}}$

**Ans.** (i) Cyclohexyl—OH $\xrightarrow{SOCl_2}$ Cyclohexyl—Cl

(ii) $C_6H_5$—$CH_2$—CH—$CH_2$ + HBr $\xrightarrow{\text{Peroxide}}$ $C_6H_5$—$CH_2$—$CH_2CH_2Br$

19. (a) **In reference to Freundlich adsorption isotherm, write the expression for adsorption of gases on solids in the form of an equation.** **[3]**

(b) **Write an important characteristic of lyophilic sols.**

(c) **Based on type of particles of dispersed phase, give one example each of associated colloid and multimolecular colloid.**

**Ans.** (a) Expression for Freundlich adsorption isotherm :

$$\frac{x}{m} = kp^{\frac{1}{n}} (\text{where } n > 1) \text{ Or}$$

$$\log (x/m) = \log k + \frac{1}{n} \log p$$

Where, $x$ = mass of the gas adsorbed (adsorbate)

$m$ = mass of the adsorbent (solid)

$p$ = pressure of the gas

$n$ and $k$ are constants, which depend on the nature of the adsorbate and adsorbent.

(b) Lyophilic sols are stable and reversible in nature

(c) **Associated colloid** : Soap solution or detergent solution.

**Multimolecular colloid:** Sulphur sol or Gold sol.

20. (a) **Draw the structures of the following molecules :** **[3]**

(i) $XeOF_4$ (ii) $H_2SO_4$

(b) **Write the structural difference between white phosphorus and red phosphorus**

**Ans.** (a)

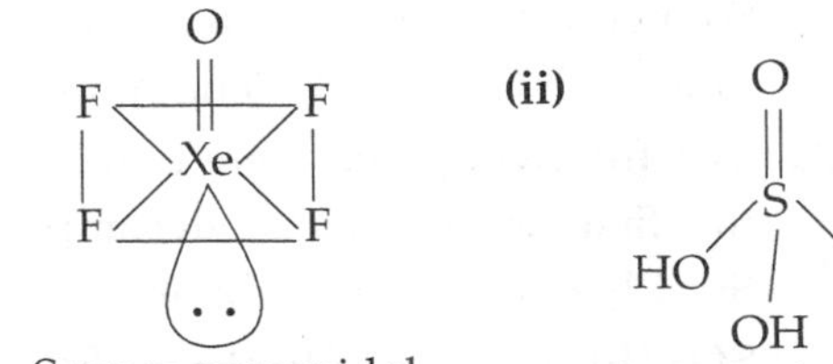


Square pyramidal Tetrahedral

(b) **Structural Difference : White Phosphorus:** In white phosphorus, $P_4$ molecules are held by weak Vander Waal's forces and exist as a discrete tetrahedral.

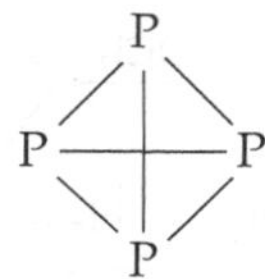

**Red Phosphorus** : In red phosphorus, $P_4$ molecules are held by covalent bonds in polymeric structure.

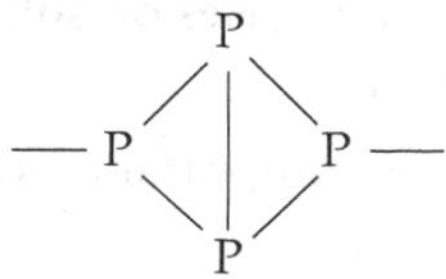

**21. Account for the following :** **[3]**

**(i) $PCl_5$ is more covalent than $PCl_3$.**

**(ii) Iron on reaction with HCl forms $FeCl_2$ and not $FeCl_3$.**

**(iii) The two O-O bond lengths in the ozone molecule are equal.**

**Ans. (i)** According to Fajan's rule higher the oxidation state of the central atom, more is its polarising power, thus more is the covalent character of the bond formed. In the $PCl_5$ molecule the oxidation state of P is +5 while in $PCl_3$ it is +3. Thus, $PCl_5$ is more covalent in nature than $PCl_3$.

**(ii)** As HCl is a mild oxidising agent and thus the hydrogen gas produced in the reaction prevents further oxidation of $Fe^{2+}$ to $Fe^{3+}$.

$$Fe + 2HCl \rightarrow FeCl_2 + H_2.$$

**(iii)** An ozone molecule is a resonance hybrid of the molecule in which there is single bond with one terminal oxygen and double bond with other terminal oxygen of the central oxygen atom.

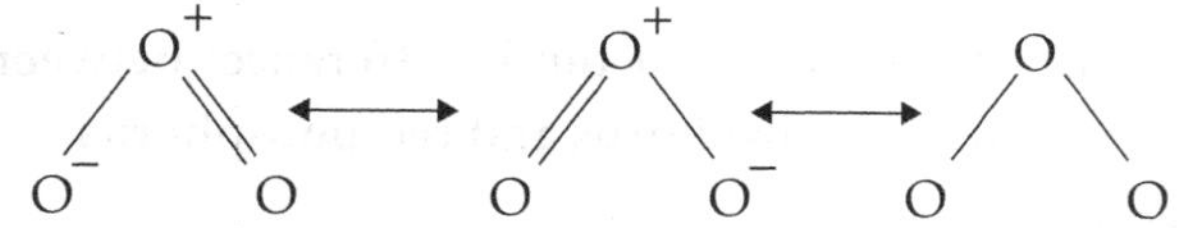

So, neither single or double bond is pure. Thus both O – O bond lengths are equal.

**22. The following data were obtained during the first order thermal decomposition of $SO_2Cl_2$ at a constant volume :** **[3]**

$$SO_2Cl_2(g) \longrightarrow SO_2(g) + Cl_2(g)$$

| Experiment | Time/$s^{-1}$ | Total pressure/atm |
|---|---|---|
| 1 | 0 | 0.4 |
| 2 | 100 | 0.7 |

**Calculate the rate constant.**

**(Given : log 4 = 0.6021, log 2 = 0.3010)**

**Ans.** $SO_2Cl_2 \rightarrow SO_2 + Cl_2$

At $t = 0$ 0.4 0 0

At $t = 100$ $0.4 - x$ $x$ $x$

$$P_t = 0.4 - x + x + x$$

$$0.7 = 0.4 + x$$

$$x = 0.3$$

$$K = \frac{2.303}{t} \log \frac{P_0}{2P_0 - P_t}$$

$$= \frac{2.303}{100} \log \frac{0.4}{2 \times 0.4 - 7}$$

$$= \frac{2.303}{1000} \log \frac{0.4}{0.1}$$

$$K = \frac{2.303 \times 0.602}{100}$$

$$= 1.38 \times 10^{-2}\ s^{-1}$$

**23. (i) Give two examples of macromolecules that are chosen as drug targets.** **[3]**

**(ii) What are antiseptics ? Give an example.**

**(iii) Why is use of aspartame limited to cold foods and soft drinks ?**

**Ans. (i)** Proteins, enzymes, nucleic acids and lipids are called drug targets as drug interacts with these macromolecules.

**(ii)** Antiseptics are the chemical substances which prevent the growth of micro-organism and are capable of killing them without harming the human tissues. These are applied on wounds, ulcer, cuts and diseased skin surfaces, *e.g.* savlon, 0.2% solution of phenol, dettol, iodine tincture etc.

**(iii)** Aspartame decomposes on heating hence it is used as an artificial sweetner for foods and soft drinks at low temperatures.

**24. (i) Deficiency of which vitamin causes night-blindness ?**

**(ii) Name the base that is found in nucleotide of RNA only.**

**(iii) Glucose on reaction with HI gives n-hexane. What does it suggest about the structure of glucose ?** **[3]**

**Ans. (i)** Vitamin A.

**(ii)** Uracil

**(iii)** Glucose exists in acyclic straight six membered carbon chain *i.e.;* it has open chain structure.

$$CHO-(CHOH)_4-CH_2OH$$

open chain structure of Glucose

**25. After the ban on plastic bags, students of a school decided to make the people aware of the harmful effects of plastic bags on the environment and Yamuna River. To make the awareness more impactful, they organized rally by joining hands with other schools and distributed paper bags to vegetable vendors, shopkeepers and departmental stores. All the students pledged not to use polythene bags in the future to save the Yamuna River. [3]**

**After reading the above passage, answer the following questions :**

**(i) What values are shown by the students ?****

**(ii) What are bio-degradable polymers ? Give one example.**

**(iii) Is polyethene a condensation or the addition polymer ?**

**Ans. (ii)** Polymers that are decomposed over a period of time either by itself or by the action of micro-organisms are called biodegradable polymers. PHBV (Poly β-hydroxy butrate Co-β-hydroxy valerate)

**(iii)** Polythene is an addition polymer.

**26. (a) Write the mechanism of the following reaction : [3]**

$$CH_3CH_2OH \xrightarrow{HBr} CH_3CH_2Br + H_2O$$

**(b) Write the equation involved in Reimer-Tiemann reaction.**

**Ans. (a)** Mechanism of the reaction

$$HBr \longrightarrow H^+ + Br$$

$$CH_3-CH_2-\ddot{O}-H \xrightarrow{H^+} CH_3-CH_2-\overset{+}{O}(H)-H$$

$$CH_3-CH_2-\overset{+}{O}(H)-H \longrightarrow CH_3-\overset{+}{C}H_2 + H_2O$$

$$CH_3-\overset{+}{C}H_2 \xrightarrow{Br^-} CH_3-CH_2-Br$$

This is an $S_N2$ reaction mechanism.

**(b)**

Phenol (OH) $+ CHCl_3$ → [o-$\overset{-}{O}\overset{+}{Na}$, $CHCl_2$ benzene] $\xrightarrow{NaOH}$ o-$\overset{-}{O}\overset{+}{Na}$ benzaldehyde (CHO) $\xrightarrow{H^+}$ o-hydroxybenzaldehyde (OH, CHO)

Salicylaldehyde

Reimer-Tiemann reaction

**27. Give the structures of A, B and C in the following reactions : [3]**

**(i)** $CH_3Br \longrightarrow A \xrightarrow{LiAlH_4} B \xrightarrow[273\ K]{HNO_2} C$

**(ii)** $CH_3COOH \longrightarrow A \longrightarrow B \xrightarrow{CHCl_3+NaOH} C$

**OR**

**How will you convert the following ?**

**(i) Nitrobezene into aniline.**

**(ii) Ethanoic acid into methanamine.**

**(iii) Aniline into N-phenylethanamide.**

**(Write the chemical equations involved.)**

**Ans. (i)** $CH_3Br \xrightarrow{KCN} \underset{\text{(A) Ethanenitrile}}{CH_3CN} \xrightarrow{LiAlH_4} \underset{\text{(B) Ethaneamine}}{CH_3CH_2NH_2}$

$\xrightarrow[273K]{HNO_2} \underset{\text{(C) Ethanol}}{C_2H_5OH} + N_2 + HCl$

**(ii)** $CH_3COOH \xrightarrow{NH_3} \underset{\text{(A) Ethanamide}}{CH_3CONH_2} \xrightarrow[\text{Hoffman's Bromide reaction}]{Br + KOH}$

$\underset{\text{(B) Methanamine}}{CH_3NH_2} \xrightarrow[\text{Carbylamine reaction}]{CHCl_3 + NaOH} \underset{\text{(C) Methyl isonitrite (Methyl Carbylamine)}}{CH_3NC}$

**OR**

**(i)** Nitrobenzene to Aniline

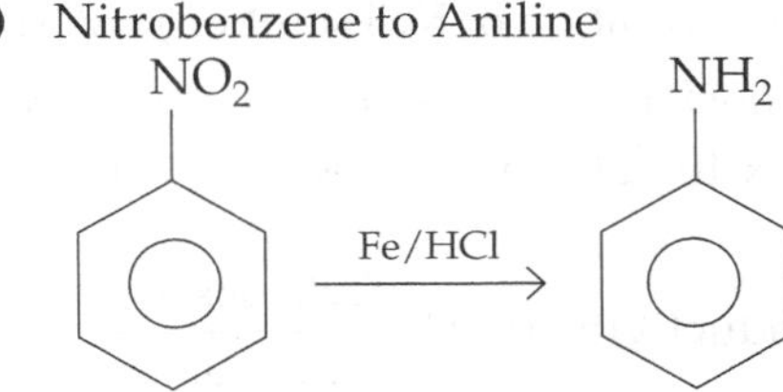


** Answer is not given due to the change in present syllabus.

**(ii)** Ethanoic acid into methanamine

$$CH_3COOH \xrightarrow[\Delta]{NH_3} CH_3CONH_2 \xrightarrow[NaOH]{Br_2} CH_3NH_2$$

**(iii)** Aniline into N-Phenylethanamide

$$C_6H_5NH_2 \xrightarrow[\text{Pyridine}]{(CH_3(CO)_2O} C_6H_5NH-\overset{O}{\overset{||}{C}}-CH_3$$

**28. (a) Define the following terms :** **[2,3]**

**(i) Limiting molar conductivity**

**(ii) Fuel cell**

**(b) Resistance of a conductivity cell filled with 0.1 mol $L^1$ KCl solution is 100 Ω. If the resistance of the same cell when filled with 0.02 mol $L^{-1}$ KCl solution is 520 Ω, calculate the conductivity and molar conductivity of 0.02 mol $L^{-1}$ KCl solution. The conductivity of 0.1 mol $L^{-1}$ KCl solution is $1.29 \times 10^{-2} \Omega^{-1} cm^{-1}$.**

**OR**

**(a) State Faraday's first law of electrolysis. How much charge in terms of Faraday is required for the reduction of 1 mol of $Cu^{2+}$ to Cu.**

**(b) Calculate *emf* of the following cell at 298 K :**

**Mg(s) | $Mg^{2+}$(0.1M) | | $Cu^{2+}$ (0.01 M) | Cu(s)**

**[Given $E^0_{cell}$ = +2.71 V, 1 F = 96500 C $mol^{-1}$** **[5]**

**Ans. (a) (i)** The limiting molar conductivity is the molar conductivity of a solution when the concentration approaches zero (infinite dilution, $C \rightarrow 0$).

**(ii)** A fuel cell is a galvanic cell in which the heat energy of the fuel is directly converted into electrical energy.

**(b)** Given, $R_1 = 100\Omega$

$C_1$ (Concentration of KCl solution) = 0.1 mol $L^{-1}$

$K_1 = 1.29 \times 10^{-2} \Omega^{-1} cm^{-1}$

$R_2 = 520\Omega$

$C_2$ (Concentration of KCl solution) = 0.02 mol $L^{-1}$

Cell constant (G) = conductivity × resistance $= 1.29 \times 10^{-2} \Omega^{-1} cm^{-1} \times 100\Omega = 1.29\ cm^{-1}$

$$\therefore \text{Conductivity } (K_2) = \frac{\text{Cell constant}}{\text{Resistance}}$$

$$= \frac{1.29\ cm^{-1}}{520\Omega} = 2.48 \times 10^{-3} \Omega^{-1} cm^{-1}$$

Molar Conductivity ($\Lambda_m$)

$$= \frac{K_2 \times 1000}{C_2}$$

$$= \frac{2.48 \times 10^{-3} \times 1000\ Scm^{-1}}{0.02\ mol\ cm^{-3}}$$

$$= 1.24 \times 10^2 \text{ or } 124\ Scm^2\ mol^{-1}$$

**OR**

**(a)** The mass of substance deposited or liberated at any eletrode is proportional to the quantity of electric charge passed through an electrolysis cell.

Reduction of 1mol of $Cu^{2+}$ to 2 moles Cu

$$\underset{1\text{ mole}}{Cu^{2+}} + \underset{2\text{ moles}}{2e^-} \rightarrow Cu$$

One mole of electrons carry 1F charge. Since 2 moles of electrons are required for reduction of 1 mol of $Cu^{2+}$, So charge required is 2F.

**(b)** $Mg + Cu^{2+} \rightarrow Mg^{2+} + Cu$

$$E_{cell} = E^\circ_{cell} - \frac{0.059}{n} \log \frac{[Mg^{2+}]}{[Cu^{2+}]}$$

$$= 2.71 - \frac{0.059}{2} \log \frac{[0.1]}{[0.01]}$$

$$= 2.71 - 0.0295 \log 10$$

$$= 2.71 - 0.0295$$

$$= 2.6805\ V$$

**29. (a) How do you prepare :** **[2,3]**

**(i) $K_2MnO_4$ from $MnO_2$ (ii) $Na_2Cr_2O_7$ from $Na_2CrO_4$**

**(b) Account for the following :**

**(i) $Mn^{2+}$ is more stable than $Fe^{2+}$ towards oxidation to +3 state.**

**(ii) The enthalpy of atomization is lowest for Zn in 3*d* series of the transition elements.**

**(iii) Actinoid elements show wide range of oxidation states.**

**OR**

**(i) Name the elements of 3*d* transition series which shows maximum number of oxidation states. Why does it show so ?**

**(ii) Which transition metal of 3*d* series has positive $E^\circ_{(M^{2+}/M)}$ value and why ?**

**(iii) Out of $Cr^{3+}$ and $Mn^{3+}$, which is a stronger oxidizing agent and why ?**

**(iv) Name a member of the Lanthanoid series which is well known to exhibit +2 oxidation state.**

**(v) Complete the following equation**

$$MnO_4^- + 8H^+ + 5e^- \longrightarrow$$

**Ans. (a)** (i) Pyrolusite is fused with KOH in presence of atmospheric oxygen or an oxidizing agent ($KNO_3$ or $KClO_3$) to give potassium permanganate.

$$2MnO_2 + 4KOH + O_2 \xrightarrow{\Delta} 2K_2MnO_4 + 2H_2O$$

(ii) Acidification of yellow solution of sodium chromate with green core $H_2SO_4$ gives orange solution of $Na_2Cr_2O_7$.

$$\underset{\text{Yellow}}{2Na_2CrO_4} + H_2SO_4 \longrightarrow \underset{\text{Orange}}{Na_2Cr_2O_7} + NaSO_4 + H_2O$$

**(b)** (i) The electronic configuration of $Mn^{2+}$ is [Ar] $3d^5$ and $Fe^{2+}$ is [Ar] $3d^6$. As it is clear from the electronic configuration, $Mn^{2+}$ is already in half-filled stable $d^5$ configuration whereas $Fe^{2+}$ ion by losing one electron it can attain stable $d^5$ configuration. Thus, $Mn^{2+}$ shows stability while $Fe^{+2}$ can be easily oxidized to $Fe^{3+}$ state.

(ii) Unpaired electrons account for stronger metallic bond. Zinc lacks unpaired electrons as it has electronic configuration of [Ar] $3d^{10}\ 4s^2$, thus metallic bonding is the weakest. So, Zn has the lowest enthalpy of atomization.

(iii) Due to comparable energies of 5*f*, 6*d* and 7*s* orbitals, actinoids show wide range of oxidation states.

**OR**

(i) Manganese shows maximum number of oxidation states in 3*d* transition series. This is because all the five *d*-orbital electrons are unpaired *i.e* [Ar] $3d^5 4s^2$. Thus, Mn shows oxidation states from +2 to +7.

(ii) Cu has positive $E^\circ_{(Cu^{2+}/Cu)}$ value *i.e.* +0.34. This is because Cu has high value of $\Delta_aH$ and low value of $\Delta_{hyd}H$ $\Delta_iH$ required is not compensated by the energy released. Therefore, $E^\circ(Cu^{2+}/Cu)$ is positive.

(iii) $E^\circ_{(Cr^{3+}/Cr^{2+})} = -0.4V$ and $E^\circ_{(M^{3+}/Mn^{2+})} = +1.5V$ The negative value of E° shows the stability of $Cr^{3+}$ and high positive value of E° in case of manganese shows that $Mn^{3+}$ is easily converted to $Mn^{2+}$. Thus, $Mn^{2+}$ is stronger oxidizing agent than $Cr^{3+}$.

(iv) Europium shows +2 oxidation state owing to its half-filled electronic configuration.

(v) $MnO_4^- + 8H^+ + 5e^- \rightarrow Mn^{2+} + 4H_2O$

**30. (a) Write the products of the following reactions: [2,3]**

**(i)** $C_6H_{10}{=}O + H_2N{-}OH \xrightarrow{H^+}$

**(ii)** $2C_6H_5CHO$ **+ Conc. NaOH** $\longrightarrow$

**(iii)** $CH_3COOH \xrightarrow{Cl_2/P}$

**(b) Give simple chemical tests to distinguish between the following pairs of compound:**

**(i) Benzaldehyde and Benzoic acid**

**(ii) Propanal and Propanone**

**OR**

**(a) Account for the following :**

**(i)** $CH_3CHO$ **is more reactive than** $CH_3COCH_3$ **towards reaction with HCN.**

**(ii) Carboxylic acid is a stronger acid than phenol.**

**(b) Write the chemical equations to illustrate the following name reactions :**

**(i) Wolff-Kishner reduction (ii) Aldol condesnation**

**(iii) Cannizzaro reaction**

**Ans. (a)** (i)

$$C_6H_{10}{=}O + H_2N{-}OH \longrightarrow \underset{\text{Oxime}}{C_6H_{10}{=}N{-}OH} + H_2O$$

(ii) $2C_6H_5CHO + NaOH\text{ (conc.)} \rightarrow$

$$\underset{\text{Sodium benzoate}}{C_6H_5COONa} + \underset{\text{Benzylalcohol}}{C_6H_5CH_2OH}$$

(iii)

$$CH_3COOH \xrightarrow{Cl_2/\text{red P}} \underset{\alpha\text{-chloroacetic acid}}{CH_2(Cl){-}COOH}$$

**(b) (i)** Benzoic acid reacts with $NaHCO_3$ to produce effervescence of $CO_2$ gas but benzaldehyde does not

$$C_6H_5COOH + NaHCO_3 \rightarrow C_6H_5COONa + H_2O + CO_2$$

**(ii)** Propanone gives iodoform test but propanal does not because propanone has $CH_3CO$- group

$$CH_3COCH_3 + 3NaOH \xrightarrow[\Delta]{NaOH, I_2} CH_3COONa + CHI_3$$

**OR**

**(a) (i)** $CH_3CHO$ is more reactive than $CH_3COCH_3$ towards reaction with HCN because the reactivity of compound depends on the steric hindrance due to the groups present around the carbonyl group. More the steric hindrance, less will be the reactivity of the compound. Therefore, due to the presence of more electrophilic carbonyl carbon in $CH_3CHO$, it is more reactive than $CH_3COCH_3$.

**(ii)** Carboxylic acid is stronger than phenol because of resonance, stabilisation of more electronegative oxygen atom in carboxylate ion than carbon atom in phenoxide ion.

$$R-C(=O)-O-H \longrightarrow R-C(=O)-O^-$$

carboxylate ion

$$R-C(=O)-O^- \longleftrightarrow R-C(-O^-)=O$$

I II

$$C_6H_5OH \xrightarrow{-H^+} C_6H_5O^-$$

phenoxide ion

I ⟷ II ⟷ III ⟷ IV

The reason for greater stability of carboxylate ion is that it involves equivalent resonance in which negative charge is present on oxygen atom all the time whereas, in phenoxide ion, out of total four resonating structures, in three structures, negative charge is present on carbon atom which makes it less stable.

**(b) (i)** Reduction of aldehyde or ketone to respective hydrocarbon.

$$R-\overset{O}{\overset{\|}{C}}-R' + NH_2NH_2 \xrightarrow[\Delta]{KOH} R-CH_2-R'$$

R = any alkyl group.

R' = –H or any alkyl group

**(ii)** Formation of β-hydroxyaldehyde or β-hydroxy ketone from aldehydes and ketones respectively.

$$2R-CH_2-\overset{O}{\overset{\|}{C}}-R' \xrightarrow{\text{dil KOH}} R-CH_2-\underset{R'}{\underset{|}{\overset{OH}{\overset{|}{C}}}}-\underset{R}{\underset{|}{CH}}-\overset{O}{\overset{\|}{C}}-R'$$

aldehyde/ketone β-hydroxybenzaldehyde/ketone

**(iii)** $$2H-\overset{O}{\overset{\|}{C}}-H \xrightarrow{\text{Conc. KOH}} H-\underset{O^-K^+}{\underset{|}{\overset{O}{\overset{\|}{C}}}} + H-\underset{O}{\underset{|}{\overset{OH}{\overset{\|}{C}}}}-H$$

$$2C_6H_5\overset{O}{\overset{\|}{C}}-H \xrightarrow{\text{Conc. KOH}} C_6H_5\overset{O}{\overset{\|}{C}}-O^-K^+ + C_6H_5CH_2OH$$

# Delhi [Set-II]

**Note : Except for the following questions, all the remaining questions have been asked in previous set.**

**1. Give one example each of sol and gel.** **[1]**

**Ans.** Sol-paint, gel-butter

**3. Write the IUPAC name of the compound** **[1]**

$$CH_3-\underset{NH_2}{\underset{|}{CH}}-CH_2-CHO$$

**Ans.** 3-aminebutan-1-al

**5. Some liquids on mixing form 'azeotropes'. What are 'azeotropes' ? [1]**

**Ans.** Azeotropes are mixtures with fixed concentrations of components such that they boil at constant temperatures. They are also called constant boiling point mixtures *e.g.* 95% ethanol and 5% water by mass mixture.

**7. Which component of starch is a branched polymer of α-glucose and insoluble in water ? [1]**

**Ans.** Amylopectin is a branched polymer of α-glucose and insoluble in water.

**9. State Henry's law. What is the effect of temperature on the solublity of a gas in a liquid. [2]**

**Ans.** Henry's law states that "the partial pressure of the gas in vapour phase ($p$) is directly proportional to the mole fraction of gas ($x$) in the solution."

$$p = K_H x \text{ (}K_H\text{–Henry's law constant)}$$

The solubility of a gas in liquid decreases with rise in temperature as dissolution of a gas in a liquid is an exothermic process.

**10. Define the following terms : [2]**

**(i) Pseudo first-order reaction**

**(ii) Half-life period of reaction ($t_{1/2}$).**

**Ans. (i)** Reactions which are actually not first order but behave as first order under certain conditions like excess of one of the reactants, is a psuedo first order reaction. *e.g.* Acid hydrolysis of ethyl acetate.

**(ii)** The time taken for the concentration of reactants to be reduced to half of its initial concentration is called the half life of a reaction.

**11. Write the principle behind the following methods of refining : [2]**

**(i) Hydraulic washing**

**(ii) Vapour-phase refining**

**Ans. (i)** Hydraulic washing is based on the differences in densities or gravities of the ore and the gangue particles. The lighter gangue particles are washed away and the heavier ores are left behind.

**(ii)** In Vapour phase refining method the metal is converted into its volatile compound and is collected elsewhere. It is then thermally decomposed to get the pure metal. *e.g.* Mond's process.

**22. (a) Draw the structures of the following : [3]**

**(i) $XeF_2$ (ii) $BrF_3$**

**(b) Write the structural difference between white phosphorus and red phosphorus.**

**Ans. (a) (i)**

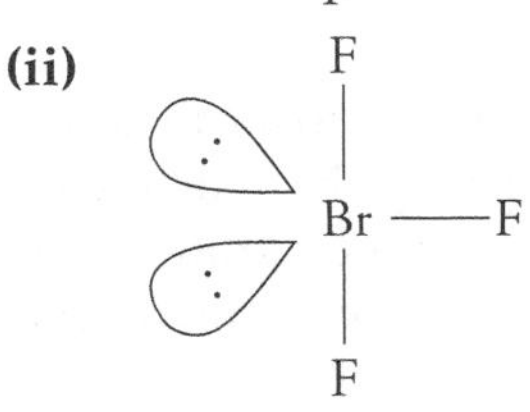


**(ii)**

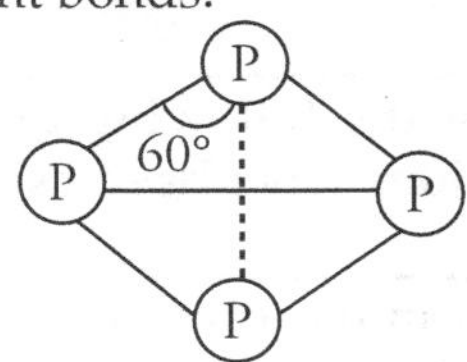


**(b)** White phosphorus consist of discrete tetrahedral $P_4$ molecule with six P-P covalent bonds.



Red Phosphorus has polymeric structure in which $P_4$ tetrahedral are linked together through P-P covalent bond to form chain.

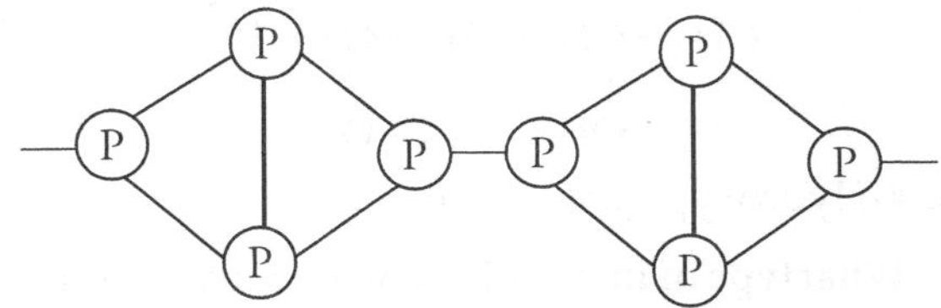


**23. Account for the following : [3]**

**(i) Bi(V) is a stronger oxidizing agent than Sb(V).**

**(ii) N – N single bond is weaker than P – P single bond.**

**(iii) Noble gases have very low boiling points.**

**Ans. (i)** Due to inert pair effect +3 oxidation state of Bi is more stable than its +5 oxidation state while oxidation state of Sb is more stable than its +3 oxidation state. Therefore, Bi (V) can accept a pair of electrons to form more stable Bi (III) more easily than Sb (V).

**(ii)** N has small size so lone pair is more concentrated over N hence repulsion takes place and bond becomes weak but P has large size therefore, no repulsion takes place and Hence, bond becomes strong.

(iii) Noble gas are monoatomic with weak Van der Waals forces of attraction, Hence, noble gases have very low boiling point.

**24. (i) Name the sweetening agents used in the preparation of sweets for a diabetic patient. [3]**

**(ii) What are antibiotics ? Give an example.**

**(iii) Give two example of macromolecules that are chosen as drug targets.**

**Ans (i)** Sucralose.

**(ii)** An antibiotic refers to a substance produced wholly or partly by chemical synthesis, which is low concentrations inhibits the growth or destroys microorganisms by intervening in their metabolic processes. Penicillin.

**(iii)** Carbohydrate, lipid, proteins, enzymes, nucleic acid.

**27. (i) Deficiency of which vitamin causes rickets? [3]**

**(ii) Give an example for each of fibrous protein and globular protein.**

**(iii) Write the product formed on reaction of D-Glucose with $Br_2$ water.**

**Ans. (i)** Vitamin D.

**(ii)** Fibrous protein : Keratin or myosin
Globular protein : Insulin or albumin.

**(iii)** Gluconic acid *i.e.*
$HOOC\,(CH_2OH)_4.CH_2OH$.

# Delhi [Set-III]

**Note : Except for the following questions, all the remaining questions have been asked in previous sets.**

**1. Give one example each of lyophobic sol and lyophillic sol. [1]**

**Ans.** Lyophobic sol : Metal sol or metal sulphide.
**Lyophilic sol :** Gum, Starch, gelatin.

**2. Write the IUPAC name of the compound. [1]**

$$CH_3-\underset{\displaystyle OH}{\underset{|}{CH}}-CH_2-\underset{\displaystyle O}{\underset{||}{C}}-CH_3$$

**Ans.** 4-Hydroxypentan-2-one.

**3. What type of intermolecular attractive interaction exists in the pair of methanol and acetone ? [1]**

**Ans.** Hydrogen bonding (intermolecular)

**6. Name the products of hydrolysis of sucrose. [1]**

**Ans.** Glucose and fructose.

**9. State Raoult's law for the solution containing volatile components. What is the similarity between Raoult's law and Henry's law ? [2]**

**Ans.** Raoult's law states that the partial pressure of the vapour of a volatile component in a solution is directly proportional to its mole fraction in the solution. Raoult's law is a special case of Henry's law.

**10. Explain the following terms : [2]**

**(i) Rate constant (*k*)**

**(ii) Half life period of reaction ($t_{½}$).**

**Ans. (i)** Rate constant (*k*) is the rate of reaction when the concentration of its reactants is unity.

**(ii)** Half life period ($t_{½}$) of reaction is the time in which the concentration of reactant is reduced to half of its initial concentration.

**11. Write the principles of the following methods : [2]**

**(i) Froth floatation method**

**(ii) Electrolytic refining**

**Ans. (i)** It is based on the difference in wetting qualities of gangue and the sulphide ore particles with water and oil. Whereas the ore particles are wetted by oil, the gangue or the earthy particles are wetted by water.

**(ii)** In this process a slab of impure copper is used as anode and a thin sheet of pure copper as cathode. The copper sulphate is used as electrolytic solution. By passing electricity through the cell copper is dissolved from the anode and deposited on cathode. The impurities either remain in solution or collect as an insoluble gangue.

**20. (a) Draw structure of the following compounds:**

**(i) $XeF_4$ (ii) $N_2O_5$**

**(b) Write the structural difference between white phos phorus and red phosphorus. [3]**

**Ans. (a) (i)** F F Xe F F

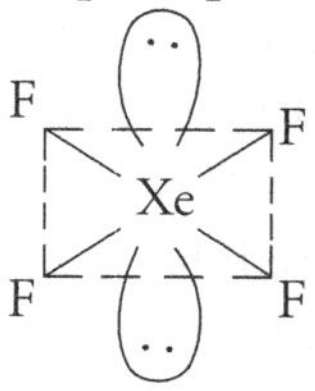

**(ii)** :Ö N :Ö: N Ö: :Ö Ö

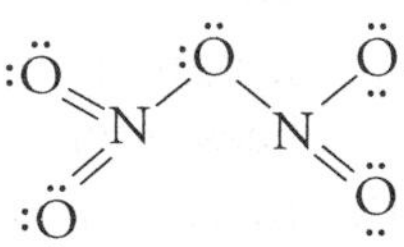

**(b)** White phosphorus consist of discrete tetrahedral $P_4$ molecule with six P-P covalent bonds.

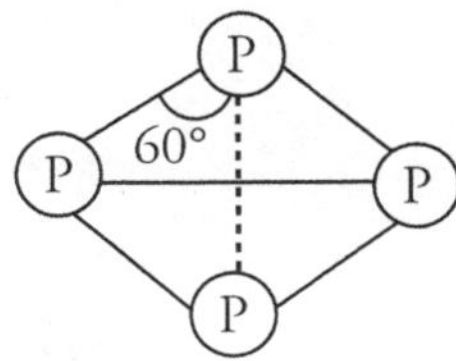


Red Phosphorus has polymeric structure in which $P_4$ tetrahedral are linked together through P-P covalent bond to form chain.

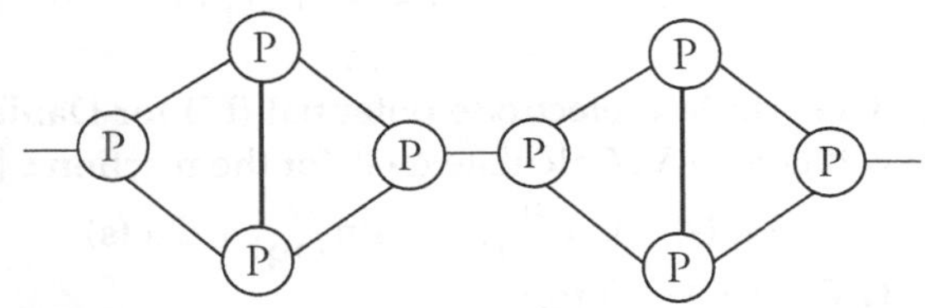


**22. Account for the following :**

**(i) Sulphur in vapour form exhibit paramagnetic behaviour.**

**(ii) $SnCl_4$ is more covalent than $SnCl_2$.**

**(iii) $H_3PO_2$ is stronger reducing agent than $H_3PO_3$.** **[2]**

**Ans. (i)** In vapour form sulphur partly exits as $S_2$ molecule which have two unpaired electrons in the antibonding *p* molecular orbitals like $O_2$ molecule and hence exhibits paramagnetism.

**(ii)** It is due to higher oxidation state (+4) of Sn in $SnCl_4$ or because of its high polarising power which increases the covalent character of bond formation between the central atom and the atoms around it.

**(iii)** It is because of two P–H bonds in $H_3PO_2$ whereas there is only one P–H bond in $H_3PO_3$.

**23. (i) What are disinfectants ? Give an example.** **[3]**

**(ii) Give two examples of macromolecules that are chosen as drug targets.**

**(iii) What are anionic detergents ? Give an example.**

**Ans. (i)** Chemicals which are used to kill micro-organisms and applied on non-living objects like floors and drains are called disinfectants *e.g.* 1% phenol solution.

**(ii)** Proteins, amino acids and enzymes.

**(iii)** Detergents in which the anionic part of the molecule is responsible for cleansing action are called anionic detergents *eg.* Sodium laurylsulphate.

**24. (i) Deficiency of which vitamin causes scurvy?**

**(ii) What type of linkage is responsible for the formation of proteins ?**

**(iii) Write the product formed when glucose is treated with HI.** **[3]**

**Ans. (i)** Vitamin-C

**(ii)** Peptide linkage

**(iii)** *n*-hexane : $CH_3CH_2CH_2CH_2CH_2CH_3$

●●

# CHEMISTRY

## Outside Delhi [Set-I] 2013

Time allowed : 3 hours Maximum marks : 70

**1. Of physisorption or chemisorption, which has a higher enthalpy of adsorption? [1]**

**Ans.** Chemisorption.

**2. Name the method used for refining of copper metal. [1]**

**Ans.** Electrolytic refining of copper metal.

**3. Name two poisonous gases which can be prepared from chlorine gas. [1]**

**Ans.** Chloropicrin or tear gas ($CCl_3NO_2$) and phosgene gas ($COCl_2$).

**4. Write the IUPAC name of the following compound: [1]**

$$CH_3-\underset{\underset{CH_3}{|}}{\overset{\overset{CH_3}{|}}{C}}-\underset{\underset{Cl}{|}}{CH}-CH_3$$

**Ans.** 2-chloro-3, 3-dimethyl butane.

**5. Rearrange the following compounds in the increasing order of their boiling points: [1]**

**$CH_3–CHO$, $CH_3–CH_2–OH$, $CH_3 – CH_2 – CH_3$**

**Ans.** $CH_3CH_2CH_3 < CH_3CHO < CH_3CH_2OH$

**6. Write the structure of N-methylethanmine. [1]**

**Ans.** $CH_3-\underset{\underset{H}{|}}{N}-CH_2-CH_3$

**7. What are the products of hydrolysis of sugar? [1]**

**Ans.** Glucose and fructose.

**8. Is $(CH_2-\underset{\underset{Cl}{|}}{CH})$ a homopolymer or a copolymer? [1]**

**Ans.** Homopolymer.

**9. Account for the following: [2]**

**(i) Schottky defects lower the density of related solids.**

**(ii) Conductivity of silicon increases on doping it with phosphorus.**

**Ans. (i)** Due to the missing of equal number of cations and anions from lattice, the density of lattice solid decreases in Schottky defect.

**(ii)** Presence of one unshared electron in phosphorous increases its conductivity and make silicon an *n*-type semiconductor.

**10. Aluminium crystallizes in an FCC structure. Atomic radius of the metal is 125 pm, what is the length of the side of the unit cell of the metal? [2]**

**Ans.**

$$r = \frac{a}{2\sqrt{2}}$$

For fcc,

$$a = 2\sqrt{2}\, rr$$
$$r = 125 \text{ pm}$$
$$= 2 \times 1.414 \times 125 \text{ pm}$$
$$= 353.5 \text{ pm}$$

**11. The standard electrode potential (E°) for Daniell cell is + 1.1 V. Calculate $\Delta G°$ for the reaction : [2]**

$$Zn\,(s) + Cu^{2+}_{(aq)} \rightarrow Zn^{2+}_{(aq)} + Cu\,(s)$$

**(1 F = 96500 C/mol)**

**Ans.** $n = 2$, $E°_{cell} = 1.1$ volt, $1F = 96500\ C\ mol^{-1}$

$$\Delta_r G° = -nFE°_{cell}$$
$$= -2 \times 96500 \times 1.1$$
$$= -212.3\ kJ\ mol^{-1}$$

**12. For a reaction $A + B \rightarrow P$, the rate law is given by [2]**

$$r = k\,[A]^{1/2}\,[B]^2$$

**(a) What is the order of this reaction?**

**(b) A first order reaction is found to have a rate constant $k = 5.5 \times 10^{-14}\ s^{-1}$. Find the half life of the reaction.**

**Ans. (a)** Order of reaction $= \frac{1}{2} + 2 = 5/2$

**(b)** For first order reaction,

$$\text{Half life } (t_{1/2}) = \frac{0.693}{k}$$

$$= \frac{0.693}{5.5 \times 10^{-14}} = 1.26 \times 10^{13}\ S$$

**13. (a) Name the method used for removing gangue from sulphide ores. [2]**

**(b) How is wrought iron different from steel?**

**Ans. (a)** Froth floatation method.

**(b)** Wrought iron is the purest form of iron with 0.2-0.5% carbon and steel is an alloy of iron. It has carbon content of 0.1-1.5%.

**14. Draw the structures of the following molecules:**

**(i) $XeOF_4$ (ii) $H_3PO_3$ [2]**

**Ans. (i)**

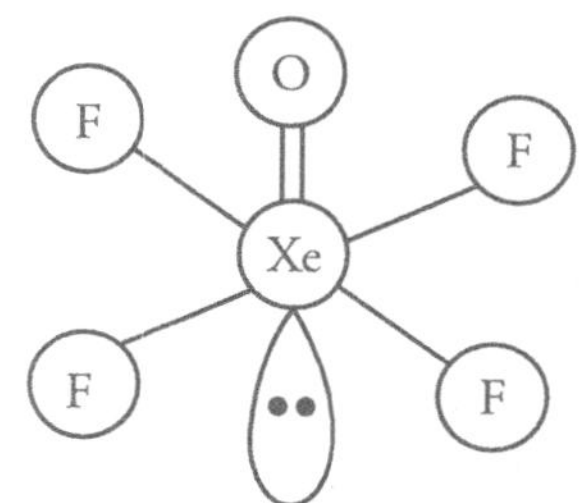

(ii) Structure: P with =O, H, OH, OH

**15. How are interhalogen compounds formed? What general compositions can be assigned to them? [2]**

**Ans.** Interhalogen compounds are formed by direct combination of halogens or by action of halogen on lower interhalogen compounds. The general composition of interhalogen compounds are :

$$XX', XX'_3, XX'_5, XX'_7$$

**16. Explain the mechanism of the following reaction: [2]**

$$CH_3-CH_2-OH \xrightarrow[443\,K]{H^+} CH_2=CH_2+H_2O$$

**Ans. Mechanism :** It involves three steps :

1. Formation of protonated alcohol

$$H-\underset{H}{\overset{H}{C}}-\underset{H}{\overset{H}{C}}-\ddot{O}-H+H^+ \xrightleftharpoons{\text{Fast}} H-\underset{H}{\overset{H}{C}}-\underset{H}{\overset{H}{C}}-\overset{H}{\ddot{O}^+}-H$$

Ethanol

2. Formation of carbocation

$$H-\underset{H}{\overset{H}{C}}-\underset{H}{\overset{H}{C}}-\overset{H}{\ddot{O}^+}-H \xrightleftharpoons{\text{Slow}} H-\underset{H}{\overset{H}{C}}-\underset{H}{\overset{H}{C^+}}+H_2O$$

3. Formation of ethene by elimination of a proton

$$H-\underset{H}{\overset{H}{C}}-\underset{H}{\overset{H}{C^+}} \rightleftharpoons \underset{H}{\overset{H}{>}}C=C\underset{H}{\overset{H}{<}} + H^+$$

Ethene

**17. Write the equations involved in the following reactions: [2]**

**(i) Reimer-Tiemann reaction**

**(ii) Williamson ether synthesis**

**Ans. (i)**

Phenol $\xrightarrow[\text{aq. NaOH}]{CHCl_3+}$ [Intermediate: $C_6H_4(O^-Na^+)(CHCl_2)$] $\xrightarrow{NaOH}$ $C_6H_4(O^-Na^+)(CHO)$ $\xrightarrow{H^+}$ $C_6H_4(OH)(CHO)$

Salicylaldehyde

**(ii)** $C_6H_5-ONa \xrightarrow{CH_3I} C_6H_5-OCH_3 + NaI$

Sodium Phenoxide — Methyoxy Benzene

**18. Define thermoplastic and thermosetting polymers. Give one example of each. [2]**

**OR**

**What is a biodegradable polymer? Give an example of biodegradable aliphatic polyester.**

**Ans. Thermoplastic Polymers:** They are plastics which becomes soft on heating and can be remoulded and hardened on cooling. because they become infusible on heating and form insoluble hard mass. They have less forces e.g. Polythene, Teflon etc.

**Thermosetting Polymers:** They do not become soft on heating and cannot be remoulded because they become infusible on heating and form insoluble hard mass. They have strong force of attraction *e.g.* bakelite.

**OR**

The polymers which can be degraded by the microorganisms are called Biodegradable polymers.

Example of biodegradable aliphatic polyester is PHBV (Poly-β-hydroxybutyrate-co-β-hydroxy Valerate).

**19. The rate of a reaction becomes four times when the temperature changes from 293 K to 313 K. Calculate the energy of activation ($E_a$) of the reaction assuming that it does not change with temperature. [3]**

**[R = 8.314 J/K mol$^{-1}$, log 4 = 0.6021**

**Ans.** Given,

$T_1 = 293K, T_2 = 313$ K.

$R = 8.314\ J\ K^{-1}\ Mol^{-1}$

log 4 = 0.6021

$$Ea = \frac{2.303RT_1T_2}{T_2-T_1}\log\frac{K_2}{K_1}$$

$$Ea = \frac{2.303\times8.314\times293\times313}{20}\times\log4$$

$$Ea = 52.86\,k\,J\,mol^{-1}$$

**20. What are the characteristics of the following colloids? Give one example of each. [3]**

**(i) Multimolecular colloids**

**(ii) Lyophobic sols**

**(iii) Emulsions.**

**Ans. (i)** On dissolution, a large number of atoms or molecules of a substance aggregate to form colloidal particles. This colloid is called a multimolecular colloid. *e.g.* Sulphur sol.

**(ii)** The colloids in which there is no affinity between the particles of dispersed phase and dispersion medium are called lyophobic sols. They are not stable, that is easily coagulated and irreversible. They are produced only by special methods. eg. $As_2S_3$ sol, $Fe(OH)_3$ sol.

**(iii)** Emulsions are colloids in which both dispersed phase and dispersing medium are liquid and immiscible with each other *eg.* Milk, cod liver oil, etc.

**21. Give reasons for the following : [3]**

**(i) When R is an alkyl group $R_3 P = 0$ exist but $R_3 N = 0$ doesn't.**

**(ii) $PbCl_4$ is more covalent than $PbCl_2$.**

**(iii) $N_2$ is much less reactive at room temperature.**

**Ans. (i)** Due to absence of d-orbitals, N cannot form $p\pi-d\pi$ multiple bonds. Thus, it cannot expand its covalency beyond 4. In $R_3 N = 0$, N has covalency 5 so it does not exist. On the other hand, due to presence of d-orbitals, P forms $p\pi-d\pi$ multiple bonds and expand its covalency beyond 4. In $R_3 P=0$ covalency of P is 5 hence it exists.

**(ii)** Because 'Pb' is in +4 oxidation state in $PbCl_4$ and has high charge/size ratio than $Pb^{2+}$. According to Fajan's rule, a higher charge on cation or anion makes compound more covalent, +4 state is more stable than +2 state. Hence $PbCl_4$ is more covalent than $PbCl_2$

**(iii)** $N_2$ has strong $p\pi-d\pi$ overlap resulting in triple bond ($N \equiv N$). Due to short bond length of ($N \equiv N$) triple bond, $N_2$ has high bond energy and hence it is less reactive at room temperature.

**22. For the complex $[NiCl_4]^{2-}$, write [3]**

**(i) the IUPAC name**

**(ii) the hybridization type**

**(iii) the shape of the complex**

**(Atomic no. of Ni = 28)**

**OR**

**'What is meant by crystal field splitting energy? On the basis of crystal field theory, write the electronic configuration of $d^4$ in terms of $t_{2g}$ and $e_g$ in an octahedral field when**

**(i) $\Delta_o > P$** **(ii) $\Delta_o < P$**

**Ans. (i)** Tetrachloridonickelate (II) ion

**(ii)** $sp^3$ hybridisation

**(iii)** Tetrahedral shape.

**OR**

The energy involved in splitting the degenerate d-orbitals into two sets t2g and eg is called crystal field splitting energy.

(i) When $\Delta_0 > P$, electronic configuration is $t_{2g}^4 e_g^0$.

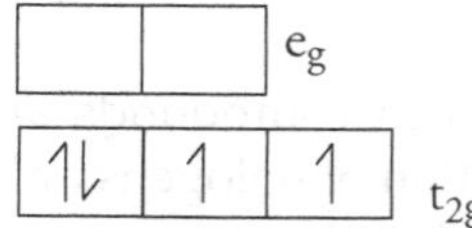


(ii) When $\Delta_0 < P$, electronic configuration is $t_{2g}^3 e_g^1$

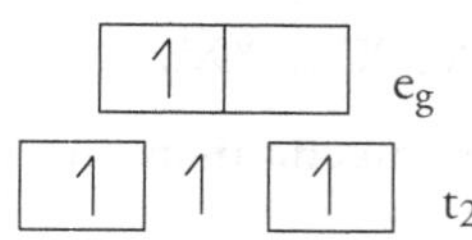


**23. Give reasons for the following: [3]**

**(i) Ethyl iodide undergoes $S_N2$ reaction faster than ethyl bromide.**

**(ii) (±) 2-Butanol is optically inactive.**

**(iii) C–X bond length in halobenzene is smaller than C–X bond length in $CH_3$–X.**

**Ans. (i)** $I^-$ ion is better leaving group than $Br^-$ion, therefore, ethyl iodide reacts faster than ethyl bromide in $S_N2$ reaction.

**(ii)** (±) 2-butanol is optically inactive because it is racemic mixture *i.e.* it has enantiomers in equal amount and hence they cancel each other's rotation of plane polarised light.

**(iii)** In halobenzene halogen atom is attached to $sp^3$ hybridized carbon while in $CH_3$-X $sp^2$ hybridized carbon which is smaller in size than $sp^3$ hybridized carbon attached to the halogen. Therefore C–X bond in halobenzene is shorter than in $CH_3$–X.

**24. Complete the following reactions: [3]**

**(i) $CH_3 CH_2 NH_2 + CHCl_3 + Alc.KOH \rightarrow$**

**(ii) $C_6H_5 N_2^+ Cl^- \xrightarrow[\text{Room Temp.}]{\text{Water}}$**

**(iii)** $C_6H_5NH_2$ (aniline, ring with $NH_2$) **+ HCl (aq.) $\longrightarrow$**

**Ans. (i)** $CH_3CH_2NH_2 + CHCl_3 + 3KOH(aq) \rightarrow CH_3CH_2-NC + 3KCl + 3H_2O$

**(ii)** $C_6H_5N_2^+ Cl^- \xrightarrow[\text{(room temp.)}]{H_2O} C_6H_5OH + N_2 + HCl$

**(iii)** $C_6H_5NH_2 + HCl(aq) \rightarrow C_6H_5NH_3Cl$

**25. (i) What class of drug is Ranitidine?** **[3]**

**(ii) If water contains dissolved $Ca^{2+}$ ions, out of soaps and synthetic detergents, which will you use for cleaning clothes?**

**(iii) Which of the following is an antiseptic?**

**0.2% phenol, 1% phenol**

**Ans.** (i) Antacid.

(ii) Soap gets precipitated in hard water hence it can't be used to wash clothes. On the other hand, synthetic detergents do not precipitate in hard water because its calcium salt is soluble in water. Therefore, it can be used to wash clothes in hard water.

(iii) 0.2% phenol.

**26. Calculate the emf of the following cell at 25°C:** **[3]**

**$Ag(s)|Ag^+(10^{-3}M)||Cu^{2+}(10^{-1}M)|Cu(s)$**

**Given $E^0_{cell} = +0.46V$ & $\log 10^n = n$.**

***Given cell notation is incorrect correct cell formula is***

***$Cu^{2+}(10^{-1}M)|Cu(s)||Ag^+(10^{-3}M)|Ag(s)$***

**Ans.** According to Nernst equation,

$$E_{cell} = E^\circ_{cell} - \frac{0.0591}{n}\log\frac{[Cu^{2+}]}{[Ag^-]^2}$$

$$= 0.46 - \frac{0.0591}{2}\log\frac{10^{-1}}{[10^{-3}]^2}$$

$$= 0.46 - \frac{0.591}{2}\log 10^5$$

$$= 0.46 - \frac{0.0591}{2} \times 5$$

$$= 0.46 - 0.14775$$

$$E_{cell} = 0.31V$$

**OR**

$$E_{cell} = E^\circ_{cell} - \frac{0.059}{2}\log\frac{[Ag^+]^2}{[Cu^{2+}]}$$

$$= 0.46\ V - \frac{0.059}{2}\log\frac{[10^{-3}]^2}{[0.1]}$$

$$= 0.46\ V + 0.0295 \times 5$$

$$= 0.6075\ V$$

**27. Shanti, a domestic helper of Mrs. Anuradha, fainted while mopping the floor. Mrs. Anuradha immediately took her to the nearby hospital where she was diagnosed to be severely 'anaemic'. The doctor prescribed an iron rich diet and multivitamins supplement to her. Mrs. Anuradha supported her financially to get the medicines. After a month, Shanti was diagnosed to be normal.**

**(i) What values are displayed by Mrs. Anuradha?****

**(ii) Name the vitamin whose deficiency causes 'pernicious anaemia'.** **[3]**

**(iii) Give an example of a water soluble vitamin.**

**Ans.** (ii) Pernicious anaemia is caused due to deficiency of Vitamin $B_{12}$.

(iii) Vitamin C and Vitamin B.

**28. (a) State Raoult's law for a solution containing volatile components. How does Raoult's law become a special case of Henry's law?**

**(b) 1.00 g of a non-electrolyte solute dissolved in 50 g of benzene lowered the freezing point of benzene by 0.40 K. Find the molar mass of the solute. ($K_f$ for benzene = 5.12 K kg/ mol)** **[5]**

**OR**

**(a) Define the following terms :**

**(i) Ideal solution**

**(ii) Azeotrope**

**(iii) Osmotic pressure.**

**(b) A solution of glucose ($C_6H_{12}O_6$) in water is labelled as 10% by weight. What would be the molality of the solution?**

**(Molar mass of glucose = 180 g/mol)**

**Ans.** (a) Raoult's law states that the partial pressure of a vapour of a component in the solution is directly proportional to its mole fraction in the solution.

Raoult's law becomes a special case of Henry's law as it states that the partial pressure of gas in vapour phase is directly proportional to the mole fraction of the gas in the solution.

(b) $$M = \frac{1000.k_f.w_2}{w_1.\Delta T_f} = \frac{1000 \times 5.12 \times 1}{50 \times 0.4}$$

$$M = 256 g/mol$$

**OR**

(a) (i) A solution that obeys Raoult's law over all ranges of temperature and concentration and shows no attractive forces between components, is called as ideal solution.

(ii) A liquid mixture which distill at constant temperature without undergoing any change in its composition is called Azeotropes.

**Answer is not given due to the change in present syllabus.

**(iii)** The minimum external pressure required to prevent osmosis is known as osmotic pressure.

**(b)** Given:

Molecular Mass of Glucose ($M_B$) = 180

% by wt. = ($W_B$) = 10

Molality ($m$) =?

We know

$$m = \frac{W_B \times 1000}{M_B \times W_A}$$

$$W_A = 100 - 10 = 90$$

$$m = \frac{10 \times 1000}{90 \times 180}$$

$$m = 0.617 \text{ m}$$

**29. (a) Give reasons for the following:** **[5]**

**(i) $Mn^{3+}$ is a good oxidising agent.**

**(ii) $E^o_{M^{2+}/M}$ values are not regular for first row transition metals (3d series).**

**(iii) Although 'F' is more electronegative than 'O', the highest Mn fluoride is $MnF_4$, whereas the highest oxide is $Mn_2O_7$.**

**(b) Complete the following equations:**

**(i) $2CrO_4^{2-} + 2H^+ \rightarrow$ (ii) $KMnO_4 \xrightarrow{\Delta}$**

**OR**

**(a) Why do transition elements show variable oxidation states?**

**(i) Name the element showing maximum number of oxidation states among the first series of transition metals from Sc (Z = 21) to Zn (Z = 30).**

**(ii) Name the element which shows only +3 oxidation state.**

**(b) What is lanthanoid contraction? Name an important alloy which contains some of the lanthanoid metals.**

**Ans. (a) (i)** $Mn^{+3}$ has electronic configuration $3d^4$. It gains one electron on reduction and become $3d^5$. which is half filled stable configuration. Hence it is a good oxidizing agent.

**(ii)** Due to extra stability of half and fully filled d-orbitals and variations in ionization energies the $E^o_{M^{2+}/M}$ values are not regular.

**(iii)** Due to the ability of oxygen to form multiple bonds with $Mn_2O_7$ and because oxygen stabilizes the highest oxidation state even more than fluorine.

**(b) (i)** $2CrO_4^{2-} + 2H^+ \longrightarrow Cr_2O_7^{2-} + H_2O$

**(ii)** $2KMnO_4 \xrightarrow{\Delta} K_2MnO_4 + O_2 + MnO_2$

**OR**

**(a)** Due to the presence of incomplete d-orbitals, transition elements shows variable oxidation states.

**(i)** Manganese shows oxidation states from +2 to +7.

**(ii)** Scandium (Sc).

**(b)** Lanthanide contraction refers to the steady and regular decrease in atomic size along the period from $La^{3+}$ to $Lu^{3+}$ eg. Misch metal alloy which contains 95% lanthanoids and 5% iron.

**30. (a) How will you convert the following:** **[5]**

**(i) Propanone to Propan-2-ol**

**(ii) Ethanal to 2-hydroxy propanoic acid**

**(iii) Toluene to benzoic acid**

**(b) Give simple chemical test to distinguish between :**

**(i) Pentan-2 -one and Pentan-3 -one**

**(ii) Ethanal and Propanal**

**OR**

**(a) Write the products of the following reactions:**

**(i)** $CH_3-CO-CH_3 \xrightarrow[\text{Conc.HCl}]{\text{Zn - Hg}}$ **?**

**(ii)** $CH_3-CO-Cl \xrightarrow{\text{Pd - BaSO}_4}$ **?**

**(iii)** $C_6H_5COOH$ (benzoic acid) $\xrightarrow{Br_2/FeBr_3}$

**(b) Which acid of each pair shown here would you expect to be stronger?**

**(i)** $F-CH_2-COOH$ **or** $Cl-CH_2-COOH$

**(ii)** $C_6H_5OH$ (phenol) **or** $CH_3COOH$

**Ans. (a) (i)** $\underset{\text{Propanone}}{CH_3COCH_3} \xrightarrow[\text{NaBH}_4]{\text{LiAlH}_4 \text{ or}} \underset{\text{Propan-2-ol}}{CH_3-CH(OH)-CH_3}$

(ii)

$$CH_3-\underset{\text{Ethanal}}{\overset{H}{\overset{|}{C}}}=O \xrightarrow{H^+CN^-} CH_3-\overset{H}{\underset{CN}{\overset{|}{\underset{|}{C}}}}-OH \xrightarrow[H_2O]{H^+}$$

$$CH_3-\overset{OH}{\underset{H}{\overset{|}{\underset{|}{C}}}}-COOH$$

2-Hydroxy propanoic acid

(iii) $C_6H_5^+-CH_3 \xrightarrow[\text{Co/Mn acetate}]{O_2} C_6H_5-COOH$

(b) (i) Pentan-2-one gives iodoform test but pentan-3-one does not

$$\underset{\text{Pentan-2-one}}{CH_3-\overset{O}{\overset{||}{C}}-CH_2-CH_3} + \underset{\text{sodium hypoidite}}{3NaOI} \xrightarrow[\text{heat}]{NaOH/I_2} \underset{\text{Iodoform (yellow ppt.)}}{CHI_3\downarrow} +$$

$$2NaOH + CH_3CH_2CH_2COONa$$

$$\underset{\text{Pentan-3-one}}{CH_3-CH_2-\overset{O}{\overset{||}{C}}-CH_2-CH_3} \xrightarrow[NaOH/I_2]{3NaOI} \rightarrow \text{No reaction}$$

(ii) Ethanal gives iodoform test but propanal does not.

$$CH_3-\overset{O}{\overset{||}{C}}-H + 3I_2 + 4NaOH \xrightarrow{\text{heat}}$$

$$\underset{\text{Sodium Formate}}{H-\overset{O}{\overset{||}{C}}-ONa} + \underset{\text{Idoform}}{CHI_3} + 3NaOI + 3H_2O$$

$$\underset{\text{Propanal}}{CH_3-CH_2-\overset{O}{\overset{||}{C}}-H} + I_2 + NaOH \rightarrow \text{No Reaction}$$

**OR**

(a) (i)

$$\underset{\text{Acetone}}{CH_3-CO-CH_3} \xrightarrow[\text{Conc. HCl}]{\text{Zn-Hg}} \underset{\text{Propane}}{CH_3CH_2CH_3} + H_2O$$

(ii)

$$\underset{\text{Acetyl chloride}}{CH_3-CO-Cl} + H_2 \xrightarrow[BaSO_4]{Pd} \underset{\text{Acetaldehyde}}{CH_3-CHO} + HCl$$

(iii)

$$\underset{\text{Benzoic acid}}{C_6H_5^-COOH} \xrightarrow{Br_2/FeBr_3} \text{(Br)}C_6H_4-COOH + HBr$$

m-bromo-benzoic acid

(b) (i) $F-CH_2-COOH$ is a stronger acid than $Cl-CH_2-COOH$, because F is more electronegative than Cl, so it will favour release of $H^+$ ion faster by dragging electron density towards itself more as compared to Cl.

(ii) Acetic acid $CH_3COOH$ is stronger acid than phenol. Acetic acid forms carboxylate ion and phenol forms phenoxide ion. Carboxylate ion is more stable than phenoxide ion due to resonance.

## Outside Delhi [Set-II]

**Note: Except for the following questions, all the remaining questions have been asked in previous set.**

**1. Write the structure of 2-aminotoluene.** [1]

**Ans.** Benzene ring with $CH_3$ and, on the adjacent carbon, $NH_2$ (2-aminotoluene).

**2. Which aerosol depletes ozone layer?** [1]

**Ans.** Chloroflurocarbon (CFCs) present in aerosols depletes ozone layer.

**4. Ethanal is soluble in water. Why?** [1]

**Ans.** Ethanal is soluble in water due to Hydrogen bonding.

**5. Write the IUPAC name of the following compound :** [1]

$$CH_3-\underset{Br}{\underset{|}{CH}}-CH_2-\underset{Cl}{\underset{|}{CH}}-CH_3$$

**Ans.** 2-bromo-4-chloropentane.

**7. Write the name of linkage joining two amino acids.** [1]

**Ans.** Peptide linkage.

**8. Give one example of a condensation polymer.** [1]

**Ans.** Dacron or Nylon–6, 6.

**9. (a) Why does presence of excess of lithium makes LiCl crystals pink?** [2]

**(b) A solid with cubic crystal is made of two elements P and Q Atoms of Q are at the**

**corners of the cube and P at the body-centre. What is the formula of the compound?**

**Ans.** (a) Excess of Li in LiCl causes metal excess defect due to anionic vacancy in the crystal. Pink colour results due to excitation of electrons on absorbing visible light.

(b) Atoms of Q are present at the corners of the cube. Therefore, member of atoms of Q in one unit cell $= 8 \times \frac{1}{8} = 1$. Atoms of P are present at the body centre. Therefore, number of atoms of P in one unit cell = 1. Hence, the formula of compound is PQ.

**14. Draw the structures of the following molecules:** **[2]**

**(i) $XeF_6$** **(ii) $H_2S_2O_7$**

**Ans.** (i)

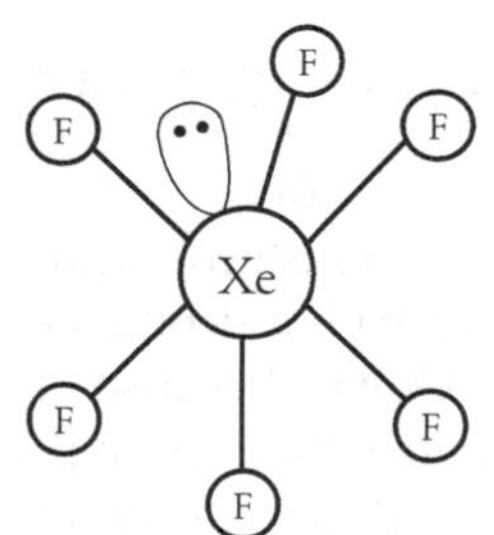


(ii)

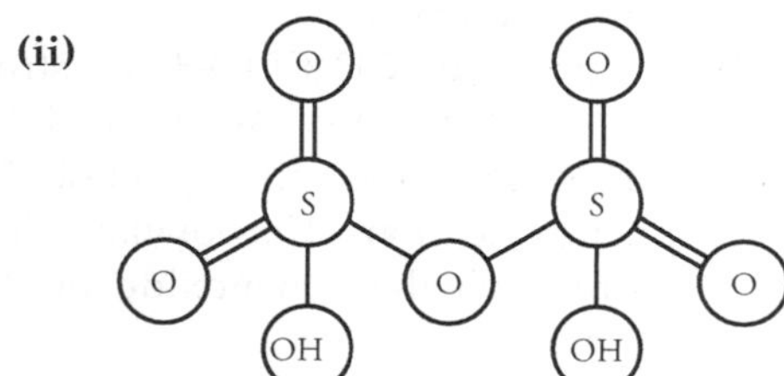


**18. Outline the principles of refining of metals by the following methods:** **[2]**

**(i) Zone refining**

**(ii) Vapour phase refining**

**Ans.** (i) This method is based on the principle that the impurities are more soluble in the molten state than in the solid state of metal.

(ii) In this process the metal is converted to its volatile compound, collected elsewhere and then decomposed to give pure metal.

**19. Define the following terms giving an example of each:** **[3]**

**(i) Associated colloids** **(ii) Lyophilic sol**

**(iii) Adsorption.**

**Ans.** (i) Associated colloids are colloidal substances which at low concentrations behave as normal electrolytes but at higher concentration they aggregate to form colloids. eg. Soaps and detergents.

(ii) Lyophilic sols are the colloids in which the particles of dispersed phase have a strong affinity for the dispersion medium. They are reversible in nature because on precipitation they can be easily converted back to colloidal form by adding dispersion medium. *eg.* Starch sol.

(iii) The aggregation of a substance on the surface of liquid or solid is known as adsorption eg. Adsorption of poisonous gases on charcol.

**22. Write the main products of the following reactions:** **[2]**

(i) $C_6H_5N_2^+Cl^- \xrightarrow{H_3PO_2 + H_2O} ?$

(ii) $NH_2$-substituted benzene (aniline) $\xrightarrow{Br_2(aq)}$

(iii) $CH_3-\underset{\underset{O}{\|}}{C}-NH_2 \xrightarrow{Br_2 + NaOH} ?$

**Ans.** (i)

$$\underset{\text{Benzenediazonium chloride}}{C_6H_5N_2^+Cl^-} \xrightarrow{H_3PO_2 + HCl} \underset{\text{Benzene}}{C_6H_6} + N_2 + HCl + H_3PO_3$$

(ii)

$$\underset{\text{Aniline}}{C_6H_5NH_2} \xrightarrow[\text{Bromination}]{Br_2(aq)} \text{2,4,6-}Br_3C_6H_2NH_2$$

2,4,6-Tribromoaniline

(iii)

$$\underset{\text{Acetamide}}{CH_3-\underset{\underset{O}{\|}}{C}-NH_2} \xrightarrow{Br_2 + 4NaOH} \underset{\text{Methylamine}}{CH_3NH_2} + 2NaBr + Na_2CO_3 + 2H_2O$$

**27. Give reasons for the following :** **[3]**

**(i) Oxygen is a gas but sulphur is a solid.**

**(ii) $O_3$ acts as a powerful oxidising agent.**

**(iii) $BiH_3$ is the strongest reducing agent amongst all the hydrides of Group 15 elements.**

**Ans.** (i) Oxygen is smaller in size than sulphur. Due to small size, it can effectively form $p\pi - p\pi$ bonds, and forms diatomic $O_2$ molecule. The intermolecular forces in oxygen are weak van der Waals forces, which causes it to exists as gas. On the other hand, sulphur does not form strong S=S double bonds and exists as puckered structure held together by covalent bonds and exists as polyatomic molecule. So, it exists as solid.

(ii) $O_3$ act as powerful oxidising agent because it decompose to give nascent oxygen.

(iii) On moving down the group, the bond length, increases and its strength decreases. Bi-H bond is the weakest. It can break easily and evolves $H_2$ gas which is a strong reducing agent. Hence, $BiH_3$ is the strongest reducing agent among group 15 elements.

●●

# Outside Delhi [Set-III]

**Note: Except for the following questions, all the remaining questions have been asked in previous sets.**

1. **What is especially observed when a beam of light is passed through a colloidal solution? [1]**

**Ans.** Tyndall effect is observed due to scattering of light.

2. **What is the basicity of $H_3PO_3$ and why? [1]**

**Ans.** The basicity of $H_3PO_3$ is 2.

$$\begin{array}{c} O \\ \| \\ HO - P - OH \\ | \\ H \end{array}$$

Because the third hydrogen is directly attached to phosphorous is not acidic.

3. **Write the IUPAC name of the following compound: [1]**

(Benzene ring with $CH_3$ at position 1, Cl at position 2 and Cl at position 4)

**Ans.** 2, 4–dichlorotoluene

8. **Write the structure of prop-2-en-1-amine. [1]**

**Ans.** $H_2C = CH - CH_2 - NH_2$

12. **Draw the structures of the following molecules:**

(i) $N_2O_5$

(ii) $XeF_2$ **[2]**

**Ans.**

(i) $:\ddot{O}=N(\rightarrow\ddot{O}:)-\ddot{O}-N(\rightarrow\ddot{O}:)=\ddot{O}:$

(ii)

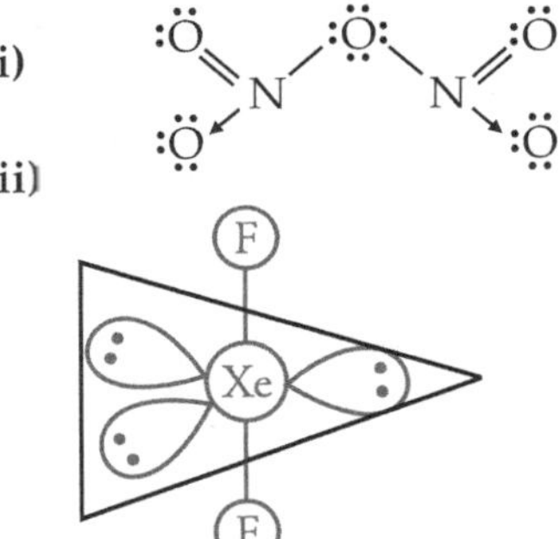


13. **(a) What change occurs when AgCl is doped with $CdCl_2$?**

**(b) What type of semiconductor is produced when silicon is doped with boron? [2]**

**Ans. (a)** When AgCl is doped with $CdCl_2$ impurity defect is produced one $Cd^{2+}$ ion will replace two $Ag^+$ ions to maintain electrical conductivity. One position is occupied by $Cd^{2+}$ ion and other will be left as a hole similar to schottky defect.

**(b)** *p*-type semiconductor

18. **Name the principal ore of aluminium. Explain the significance of leaching in extraction as aluminium. [2]**

**Ans.** Bauxite ($Al_2O_3.2H_2O$) is the principal ore of aluminium. The significance of leaching in extraction of aluminium is to prepare pure alumina from the bauxite ore in the following steps:

$$\underset{(s)}{Al_2O_3} + \underset{(aq)}{2NaOH} + \underset{(Cl)}{3H_2O} \xrightarrow[\Delta]{473\text{-}523\text{ K}} \underset{\substack{\text{Sod.metaaluminate} \\ (aq)}}{2Na[Al(OH)_4]}$$

$$SiO_2 + 2NaOH \xrightarrow[\Delta]{473\text{–}523K} \underset{\substack{\text{Sodium} \\ \text{Silicate}}}{Na_2SiO_3} + H_2O$$

After filtration of impurities sodium meta aluminate is neutralized by passing $CO_2$.

$$\underset{(aq)}{2Na[Al(OH)_4]} + \underset{(g)}{2CO_2} \rightarrow \underset{(s)}{Al_2O_3.xH_2O} \downarrow + \underset{(aq)}{2NaHCO_3}$$

$$\underset{(s)}{Al_2O_3xH_2O} \xrightarrow{1473K} \underset{\substack{\text{Alumina} \\ (s)}}{Al_2O_3} + \underset{(g)}{xH_2O}$$

19. **Define the following terms with an example in each case : [3]**

**(i) Macromolecular Sol**

**(ii) Peptization**

**(ii) Emulsion.**

**Ans. (i) Macromolecular sol :** They are molecules of large size having high molecular masses. Due to long chain, the van der Waals forces are stronger. Eg. rubber, nylon, etc.

**(ii) Peptization :** It is the process of converting a precipitate into colloidal solution by shaking it with dispersion medium in the presence of small amount of electrolyte. E.g. A preciptate of AgI can be peptized by shaking with a dilute solution of silver nitrate.

**(iii) Emulsion:** A type of colloidal solution in which both the dispersed phase and dispersion medium are liquid and are immiscible with each other is called emulsion. Ex. milk.

**21. Give reasons for the following:** **[3]**

**(i) Though nitrogen exhibits +5 oxidation state, it does not form pentahalide.**

**(ii) Electron gain enthalpy with negative sign of flourine is less than that of chlorine.**

**(iii) The two oxygen-oxygen bond lengths in ozone molecules are identical.**

**Ans.** (i) Due to non-availability of empty d-orbitals in nitrogen, it does not form pentahalide.

(ii) Due to small size, fluorine atom has high electro negativity and strong electron – electron repulsions in its compact 2p orbitals, its electron gain enthalpy is less than that of chlorine.

(iii) The two oxygen oxygen–bond lengths in ozone molecules are identical due to resonance.

●●

# CHEMISTRY

## Delhi [Set-I] 2013

**Time allowed : 3 hours** **Maximum marks : 70**

1. **How many atoms constitute one unit cell of a face-centered cubic crystal? [1]**

**Ans.** 4 atoms constitute one unit cell of a face centered cubic crystal. It can be determined from the number of atoms contributed from the faces and the corners of the unit cell as :

$$(8\text{ corners} \times \frac{1}{8}\text{ atom per corner} = 1\text{ atom}) + (6\text{ faces} \times \frac{1}{2}\text{ atom per unit face} = 3\text{ atoms})$$

$\therefore$ Total number of atoms per unit cell = 1 + 3 = 4 atoms.

2. **Name the method used for refining of Nickel metal. [1]**

**Ans.** Method used for refining of nickel metal is Mond's process.

3. **What is the covalency of nitrogen in $N_2O_5$? [1]**

**Ans.** In $N_2O_5$, the covalency of N is restricted to '4' due to $sp^2$ hybridisation of nitrogen atom involving one 2s and three 2p orbitals.

4. **Write the IUPAC name of:**

$$CH_3-\underset{\displaystyle Cl}{\underset{|}{CH}}-CH_2-CH=CH_2$$

**[1]**

**Ans.** 4-chloropent-1-ene.

5. **What happens when $CH_3$-Br is treated with KCN? [1]**

**Ans.** $\underset{\text{Methyl Bromide}}{CH_3Br} + \underset{\text{Potassium cyanide}}{KCN} \rightarrow \underset{\text{Methyl Cyanide}}{CH_3CN} + \underset{\text{Potassium Bromide}}{KBr}$

It is a nucleophilic substitution reaction.

6. **Write the structure of 3-methyl butanal. [1]**

**Ans.** $H_3C-\underset{\displaystyle CH_3}{\underset{|}{CH}}-CH_2-CHO$

7. **Arrange the following in increasing order of their basic strength in aqueous solution: $CH_3NH_2$, $(CH_3)_3N$, $(CH_3)_2NH$. [1]**

**Ans.** $\underset{3^\circ\text{amine}}{(CH_3)_3N} < \underset{1^\circ\text{amine}}{CH_3NH_2} < \underset{2^\circ\text{amine}}{(CH_3)_2NH}$

8. **What are the types of RNA molecule which perform different functions? [1]**

**Ans.** Three types of RNA molecules which perform different functions are:

(i) Messenger RNA (m-RNA)

(ii) Transfer RNA (t-RNA)

(iii) Ribosomal RNA (r-RNA)

9. **18 g of glucose, $C_6H_{12}O_6$ (Molar Mass = 180g/mol) is dissolved in 1 kg of water in a sauce pan. At what temperature will this solution boil? [2] ($K_b$ for water = 0.52 K kg mol$^{-1}$, boiling point of pure water = 373.15 K)**

**Ans.** Given,

Weight of solvent ($H_2O$) ($W_1$) = 1 kg

Weight of solute ($C_6H_{12}O_6$) ($W_2$) = 18 gm

Molar mass of solute ($M_2$) = 180 g/mol

$K_b = 0.52\text{ K kg mol}^{-1}$

$T_b^\circ = 373.15\text{ K}$

We know that

$$\therefore \Delta T_b = \frac{K_b \times 1000 \times W_2}{M_2 \times W_1}$$

$$= \frac{0.52 \times 1000 \times 18}{180 \times 1000}$$

$$= 0.052\text{ K}$$

$\therefore \Delta T_b = T_b - T_b^\circ = 0.052$

$0.052 = T_b - 373.15$

$T_b = 373.202\text{ K}$

Hence, boiling point of solution is 373.202 K

10. **The conductivity of 0.20 M solution of KCl at 298 K is 0.025 S cm$^{-1}$. Calculate its molar conductivity. [2]**

**Ans.** Given $k = 0.025\text{ S cm}^{-1}$

Molarity, M = 0.20 M

Molar conductivity

$$(\Lambda m) = \frac{\kappa \times 1000}{M}$$

$$= \frac{0.025 \times 1000}{0.20}$$

$\therefore \Lambda_m = 125\text{ S cm}^2\text{mol}^{-1}$

11. **Write the dispersed phase and dispersion medium of the following colloidal system : [2]**

**(i) Smoke (ii) Milk.**

**OR**

**What are lyophilic and lyophobic colloids? Which of these sols can be easily coagulated on addition of small amounts of electrolytes?**

**Ans.** (i) Dispersed Phase in Smoke : Solid, dispersion medium in smoke: Gas

(ii) Dispersed Phase in Milk: Fats (liquid), dispersion medium in milk: Water (liquid)

**OR**

**Ans.** **Lyophilic Colloids:** These are the colloidal solutions in which dispersed phase has great affinity for dispersion medium. Such solutions are quite stable and are reversible in nature. e.g. starch, proteins, etc.

**Lyophobic Colloids (Liquid Hating):** These are the colloidal solutions in which dispersed phase has very little affinity for the dispersion medium. Such solutions are unstable and are irreversible in nature. *eg.*, ($As_2S_3$ Sol).

Lyophobic colloids can be easily coagulated because on addition of small amount of electrolyte, the charge on colloidal particles is removed, as a result the particles will come closer to each other and then aggregate to form a cluster which settle down under the force of gravity.

**12. Write the differences between physisorption and chemisorption with respect to the following: [2]**

**(i) Specificity (ii) Temperature dependence**

**(iii) Reversibility and (iv) Enthalpy change**

**Ans.**

| S. No. | Point of difference | Physisorption | Chemisorption |
|---|---|---|---|
| (i) | Specificity | It is not specific in nature *i.e.,* all gases are adsorbed on all solids to some extent. | This is highly specific in nature. *i.e.,* it occurs only when there is some possibility of compound formation between the gas being adsorbed and the solid being adsorbent. |
| (ii) | Temperature dependence | Low temperature is favourable for physisorption. It decreases with increase in temperature | High temperature is favourable for chemisorptions. It increases with the increase in temperature. |
| (iii) | Reversibility | It is reversible in nature. | It is irreversible in nature. |
| (iv) | Enthalpy change | Enthalpy of adsorption is low *i.e.,* 20-40 $kJmol^{-1}$ | Enthalpy of adsorption is high *i.e.,* 40-4000 $kJ\ mol^{-1}$ |

**13. (a) Which solution is used for the leaching of silver metal in the presence of air in the metallurgy of silver?**

**(b) Out of 'C' and 'CO', which is a better reducing agent at the lower temperature range in the blast furnace to extract iron from the oxide ore? [2]**

**Ans.** (a) Dilute solution *i.e.,* 0.5% NaCN and KCN is used for leaching of silver metal in the presence of air.

$$4Ag + 8NaCN + 2H_2O + O_2 \rightarrow \underset{\text{Sodium. dicyanoargentate (I)}}{4Na[Ag(CN)_2]} + 4NaOH$$

(b) Out of C and CO, CO is a better reducing agent at lower temperature range in the blast furnace to extract iron from the oxide ore because in Ellingham diagram $\Delta G_{(CO, CO_2)} < \Delta G_{(Fe, FeO)}$ SO, CO will reduce FeO to Fe and will oxidized to $CO_2$.

$$FeO + CO \xrightarrow{1123\,K} Fe + CO_2$$

**14. What happens when [2]**

**(i) $PCl_5$ is heated?**

**(ii) $H_3PO_3$ is heated?**

**Write the reaction involved.**

**Ans.** (i) $PCl_5$ on heating gives $PCl_3$ and $Cl_2$.

$$PCl_5 \xrightarrow{Heat} PCl_3 + Cl_2$$

(ii) $H_3PO_3$ on heating gives orthophosphoric acid and phosphine.

$$4H_3PO_3 \xrightarrow{\Delta} \underset{\text{orthophosphoric acid}}{3H_3PO_4} + \underset{\text{Phosphine}}{PH_3}$$

**15. (a) Which metal in the first transition series (3*d* series) exhibits + 1 oxidation state most frequently and why?**

**(b) Which of the following cations are coloured in aqueous solutions and why? [2]**

**$Sc^{3+}$, $V^{3+}$, $Ti^{4+}$, $Mn^{2+}$**

**(Atomic Nos. Sc = 21, V = 23, Ti = 22, Mn = 25)**

**Ans.** (a) Cu is the only metal in the first transition series (3*d* series) which exhibits +1 oxidation state more frequently. This is because the electronic configuration of Cu is $3d^{10}\ 4s^1$ and after losing one s electron it acquires a stable $3d^{10}$ configuration.

(b) The colour of cations depend upon the number of unpaired electrons present in

$d$-orbital. The electronic configuration of the following cations is as follows:

Sc (Atomic number 21) = $3d^1 4s^2$ and $Sc^{3+}$ = $3d^0 4s^0$. As d-orbital is empty, it is colourless.

V (atomic number 23) = $3d^3 4s^2$ and $V^{3+}$ = $3d^2 4s^0$. As $d$-orbital is having 2 unpaired electrons, it undergoes $d$-$d$ transition and depicts green colour.

Ti (Atomic number 22) = $3d^2 4s^2$ and $Ti^{+4}$ = $3d^0 4s^0$. As '$d$' orbital is empty, it is colourless.

Mn(Atomic number 25) = $3d^5 4s^2$ and $Mn^{+2}$=$3d^5 4s^0$. As '$d$' orbital has 5 unpaired electrons, it depicts pink colour.

Thus, $V^{3+}$ and $Mn^{2+}$ ions are coloured in their aqueous solution due to the presence of unpaired electron.

**16. Chlorobenzene is extremely less reactive towards a nucleophilic substitution reaction. Give two reasons for the same. [2]**

**Ans.** Chlorobenzene is extremely less reactive towards a nucleophilic substitution reaction because of the following reasons:

1. **Resonance Effect:** The electron pair on chlorine atom is in conjugation with the benzene-electrons of the benzene ring which results in the following resonating structures:

:Cl: ⟷ :Cl+ ⟷ :Cl+ ⟷ :Cl+ ⟷ :Cl:

This results in delocalization of electrons of C-Cl bond and a partial double bond character develops in the bond, which makes it difficult for the nucleophile to cleave the C-Cl bond.

2. The nucleophile suffers repulsion from the increased electron density on the benzene ring as a result the nucleophile is unable to make a close approach for the attack on the molecule.

**17. Explain the mechanism of the following reaction: [2]**

$$2CH_3—CH_2—OH \xrightarrow[413\ K]{H^+} CH_3CH_2—O—CH_2—CH_3 + H_2O$$

**Ans.** The mechanism of the reaction is given below:

**Step 1**: Protonation of alcohol

$$CH_3-CH_2-\ddot{O}-H + H^+ \rightleftharpoons CH_3-CH_2-\overset{+}{O}\begin{matrix}H\\H\end{matrix}$$

First molecule of ethanol — Ethyl oxonium ion (Protonated Salt)

**Step 2**: Attack by nucleophile on protonated alcohol molecule

$$CH_3-CH_2-\ddot{O}H + CH_3-CH-\overset{+}{O}(H)-H \xrightarrow[Slow]{S_N2}$$

Second molecule of ethanol

$$CH_3-CH_2-\overset{+}{O}(H)-CH_2CH_3 + H_2O$$

Oxonium ion (Protonated ethoxy ethane)

**Step 3**: Loss of proton : to form ethoxy ethane

$$CH_3-CH_2-\overset{+}{O}\begin{matrix}CH_2CH_3\\H\end{matrix} + H-\ddot{O}-CH_2-CH_3$$

$$\xrightarrow{fast} CH_3CH_2-O-CH_2CH_3 + CH_3CH_2-\overset{+}{O}\begin{matrix}H\\H\end{matrix}$$

Ethoxy ethane — Protonated ethanol

**18. How will you convert: [2]**

**(i) Propene to Propan-2-ol?**

**(ii) Phenol to 2, 4, 6 – trinitrophenol?**

**Ans. (i)** When $H_2SO_4$ is added to propene, propan-2-ol is formed. The addition of $H_2SO_4$ takes place in accordance with Markovnikov's rule.

$$CH_3CH=CH_2 \xrightarrow[\text{Markonikov's addition}]{HBr} CH_3-\underset{|}{\overset{Br}{CH}}-CH_3$$

Isopropyl bromide

$$\xrightarrow[\substack{-KBr\\ \text{hydodysis}}]{Aq\ KOH;\Delta} CH_3-\overset{OH}{CH}-CH_3$$

Propan – 2 – ol

**(ii)** When concentrated nitric acid is added to phenol in the presence of sulphuric acid it gives 2, 4, 6 – trinitrophenol.

$$C_6H_5OH + 3HNO_3 \xrightarrow[\Delta]{H_2SO_4\ Conc.} C_6H_2(OH)(NO_2)_3 + 3H_2O$$

Phenol — 2, 4, 6 - Trinitrophenol (Picric acid)

**19. (a) What type of semiconductor is obtained when silicon is doped with boron? [3]**

**(b) What type of magnetism is shown in the following alignment of magnetic moments?**

**(c) What type of point defect is produced when AgCl is doped with $CdCl_2$?**

**Ans.** (a) When silicon is doped with boron, *p*-type semiconductor is obtained.

(b) Ferromagnetism is shown in this alignment of magnetic moments.

(c) Impurity defect of ionic solids is produced when AgCl is doped with $CdCl_2$.

**20. Determine the osmotic pressure of a solution prepared by dissolving $2.5 \times 10^{-2}$ g of $K_2SO_4$ in 2 L of water at 25°C, assuming that it is completely dissociated. [3]**

**(R = 0.0821 L atm $K^{-1}$ $mol^{-1}$, Molar mass of $K_2SO_4$ = 174 g $mol^{-1}$)**

**Ans.** We know

$$\pi = iCRT \Rightarrow \pi = \frac{inRT}{V} \Rightarrow \pi = i \times \frac{w}{M} \times \frac{1}{V} RT$$

Given : $w = 2.5 \times 10^{-2}$, g = 0.025 g

V = 2 L, T = 25°C = 298 K

M = $K_2SO_4$ = 2 × 39 + 32 + 4 × 16 = 174 g $mol^{-1}$

$K_2SO_4$ dissociates completely as

$K_2SO_4 \rightarrow 2K^+ + SO_4^{2-}$

Ions produced = 3 *i.e.*, *i* = 3

$$\text{Hence, } \pi = \frac{3 \times 0.025\text{g}}{174\text{ g mol}^{-1}} \times \frac{1}{2\text{L}} \times 0.0821 \times 298\text{ K}$$

$$\pi = 5.27 \times 10^{-3}\text{atm}$$

**21. Calculate the emf of the following cell at 298 K: [3]**

**Fe(s) | $Fe^{2+}$ (0.001 M) | | $H^+$ (1M) | $H_2$(g) (1 bar), Pt(s)**

**(Given $E^\circ_{cell}$ = +0.44 V)**

**Ans.** At anode : Fe → $Fe^{2+} + 2e^-$

At cathode: $2H^+ + 2e^- \rightarrow H_2$

So, total number of electrons (n) transferred = 2

Given that : $E^\circ_{cell}$ = + 0.44 V

Temperature (T) = 298 K

We know,

$$E_{cell} = E^\circ_{cell} - \left(\frac{2.303RT}{nF}\right) \log \frac{a_{oxi}}{a_{red}}$$

$$E_{cell} = E^\circ_{cell} - \left(\frac{0.05916\text{ V}}{n}\right) \log \frac{[Fe^{2+}]}{[H^+]^2}$$

$$= 0.44 - \frac{0.5916}{2} \log \frac{0.001}{1}$$

$$= 0.44 - 0.02955 \times (-3)$$

$$= 0.44 + 0.08865$$

$\therefore$ $E_{cell}$ = 0.53 V

**22. How would you account for the following? [3]**

**(i) Transition metals exhibit variable oxidation states.**

**(ii) Zr (Z = 40) and Hf (Z = 72) have almost identical radii.**

**(iii) Transition metals and their compounds act as catalyst.**

OR

**Complete the following chemical equations:**

**(i) $Cr_2O_7^{2-} + 6Fe^{2+} + 14H^+ \rightarrow$**

**(ii) $2CrO_4^{2-} + 2H^+ \rightarrow$**

**(iii) $2MnO_4^- + 5C_2O_4^{2-} + 16H^+ \rightarrow$**

**Ans.** (i) The variable oxidation states of transition elements is due to the participation of ns and (*n*-1)*d*-electrons in bonding. Lower oxidation state is exhibited when ns-electrons take part in bonding. Higher oxidation states are exhibited when (*n*-1) *d*-electrons take part in bonding.

(ii) This is because atomic radii of 4*d* and 5*d* transition elements are nearly same. This similarity in size is a consequence of lanthanide contraction which is due to weak shielding of *d*-electrons. As a result, the radii of Hf becomes nearly equal to that of Zr.

(iii) The catalytic activity of transition elements and their compound is due to the following reasons :

1. Due to their tendency to show variable oxidation state transition metal form instable intermediate compounds and provides a new path for the reaction with lower activation energy.

2. In some cases, the transition metals provide a suitable large surface area with free valencies or ion which reactant can adsorbed.

OR

(i) $Cr_2O_7^{2-} + 6Fe^{2+} + 14H \rightarrow 2Cr^{3+} + 6Fe^{3+} + 7H_2O$

(ii) $2CrO_4^{2-} + 2H^+ \rightarrow Cr_2O_7^{2-} + H_2O$

(iii) $2MnO_4^- + 5C_2O_4^{2-} + 16H^+ \rightarrow 2Mn^{2+} + 10CO_2 + 8H_2O$

**23. Write the IUPAC names of the following coordination compounds : [3]**

**(i) $[Cr(NH_3)_3Cl_3]$**

**(ii) $K_3[Fe(CN)_6]$**

**(iii) $[CoBr_2(en)_2]^+$, (en = ethylenediamine)**

**Ans.** (i) Triamminetrichloridochromium(III)

(ii) Potassium hexacyanoferrate(III)

(iii) Dibromidobis-(ethylene-diammine) cobalt (III) ion

**24. Give the structures of A, B and C in the following reactions: [3]**

**(i)**

$$C_6H_5N_2^+Cl^- \xrightarrow{CuCN} A \xrightarrow{H_2O/H^+} B \xrightarrow[\Delta]{NH_3} C$$

**(ii)**

$$C_6H_5NO_2 \xrightarrow{Sn + HCl} A \xrightarrow[273K]{NaNO_2+HCl} B \xrightarrow[\Delta]{H_2O/H^+} C$$

**Ans.** (i)

$$\underset{\text{Benzene diazonium chloride}}{C_6H_5N_2^+Cl^-} \xrightarrow{CuCN} \underset{\text{Benzonitrile}}{\underset{(A)}{C_6H_5CN}} \xrightarrow{H_2O/H^+}$$

$$\underset{\text{Benzoic acid}}{\underset{(B)}{C_6H_5COOH}} \xrightarrow[\Delta]{NH_3} \underset{\text{Ammonium benzoate}}{C_6H_5COO^-NH_4^+} \xrightarrow{\Delta} \underset{\text{Benzamide}}{\underset{(C)}{C_6H_5CONH_2}}$$

(ii)

$$\underset{\text{Nitrobenzene}}{C_6H_5NO_2} \xrightarrow{Sn+HCl} \underset{\text{Aniline}}{\underset{(A)}{C_6H_5NH_2}}$$

$$\xrightarrow[273K]{NaNO_2+HCl} \underset{\text{Benzene diazonium Chloride}}{\underset{(B)}{C_6H_5N^+\equiv NCl^-}} \xrightarrow[\Delta]{H_2O/H^+} \underset{\text{Phenol}}{\underset{(C)}{C_6H_5OH}}$$

**25. Write the names and structures of the monomers of the following polymers: [3]**

**(i) Buna–S**

**(ii) Neoprene**

**(iii) Nylon-6, 6**

**Ans.** (i) Buna - S:

$\underset{\text{stryene}}{CH_2 = CH(C_6H_5)}$ and $\underset{\text{Buta-1,3-diene}}{CH_2 = CH-CH = CH_2}$

(ii) Neoprene:

$$\underset{\text{Chloroprene}}{CH_2 = CH-C(Cl) = CH_2}$$

(iii) Nylon-6, 6:

$$\underset{\text{Adipic acid}}{HO-\overset{O}{\overset{\|}{C}}-(CH_2)_4-\overset{O}{\overset{\|}{C}}-OH}\text{ and }$$

$$\underset{\text{Hexamethylenediamine}}{H_2N-(CH_2)_6-NH_2}$$

**26. After watching a programme on TV about the adverse effects of junk food and soft drinks on the health of school children, Sonali, a student of Class XII, discussed the issue with the principal. Principal immediately instructed the canteen contractor to replace the fast food with the fibre and vitamins rich food like sprouts, salad, fruits etc. This decision was welcomed by the parents and the students. [3]**

**After reading the above passage, answer the following questions :**

**(a) What values are expressed by Sonali and the Principal of the school?****

**(b) Give two examples of water-soluble vitamins.**

**Ans.** (b) The two water soluble vitamins are vitamin B and Vitamin C (Ascorbic acid).

**27. (a) Which one of the following is a food preservative? [3]**

**Equanil, Morphine, Sodium benzoate**

**(b) Why is bithional added to soap?**

**(c) Which class of drugs is used in sleeping pills?**

**Ans.** (a) Sodium benzoate is used as a food preservative whereas equanil is a tranquilizer and morphine is a narcotic analgesic.

(b) Bithional is an antiseptic so it is added to soaps to reduce the odours produced by bacterial decomposition of organic matter on the skin.

(c) Tranquilizers relieve stress, fatigue by inducing sense of well being, so they are used in sleeping pills.

**28. (a) A reaction is second order in A and first order in B. [5]**

**(i) Write the differential rate equation.**

**(ii) How is the rate affected on increasing the concentration of A three times?**

**(iii) How is the rate affected when the concentrations of both A and B are doubled?**

**(b)** **A first order reaction takes 40 minutes for 30% decomposition. Calculate $t_{1/2}$ for this reaction.**

**(Given : log 1.428 = 0.1548)**

**OR**

**(a)** **For a first order reaction, show that time required for 99% completion is twice the time required for the completion of 90% of reaction.**

**(b)** **Rate constant 'k' of a reaction varies with temperature 'T' according to the equation :**

$$\log k = \log A - \frac{E_a}{2.303R}\left(\frac{1}{T}\right)$$

**Where $E_a$ is the activation energy. When a graph is plotted for log k Vs. $\frac{1}{T}$, a straight line with a slope of –4250 K is obtained. Calculate '$E_a$' for the reaction. (R = 8.314 $JK^{-1}$ $mol^{-1}$)**

**Ans. (a)** (i) A reaction is second order in A and first order in B.

$$\text{Rate} = \frac{-d[R]}{dt} = K[A]^2[B]$$

(ii) On increasing the concentration of 'A' three times *i.e.* 3A; the rate of reaction becomes 9 times of the initial rate.

$$\text{Rate} = K[3A]^2[B] = 9K[A]^2[B] = 9 \text{ times rate.}$$

(iii) On increasing the concentration of A and B as 2A and 2B. The rate of reaction becomes 8 times of the initial rate.

$$\text{Rate} = K[2A]^2[2B] = 8K[A]^2[B] = 8 \text{ times rate}$$

**(b)**

$$A \rightarrow P$$
$$t = \quad 0 \quad 0$$
$$t = t \quad (a-x) \quad x$$

Now, it takes 40 min for 30% decomposition *i.e.* reactant left after 40 min is 70% of its initial concentration.

$$\text{So, } (a-x) = \frac{70}{100} \times a = \frac{7}{10}a$$

Using the formula,

$$= \frac{2.303}{t}\log\frac{a}{a-x}$$

$$= \frac{2.303}{40}\log\frac{a}{(7/10)\,a}$$

$$= 0.00891\,\text{min}^{-1}$$

$$t_{1/2} = \frac{0.693}{k} = \frac{0.693}{0.008913}$$

$$\therefore \quad t_{1/2} = 77.78 \text{ min}$$

**OR**

**(a)** For a first order reaction,

$$A \rightarrow P$$
$$\text{At, } t=0 \quad a \quad 0$$
$$\text{At, } t=t \quad (a-x) \quad x$$

$$t_{99\%} = \frac{2.303}{K}\log\frac{100}{1} \qquad \text{eq. (i)}$$

$$t_{90\%} = \frac{2.303}{K} = \frac{100}{10} \qquad \text{eq. (ii)}$$

on Comparing eq (i) and (ii)

$$\frac{t_{99\%}}{t_{90\%}} = \frac{\log 100}{\log 1} \quad \therefore t_{99\%} = 2t_{90\%}$$

∴ the time required for 99% completion of 1st order reaction is twice the time required for 90% completion.

**(b)** We know that, $\log k = \log A - \frac{E_a}{2.303R}\left(\frac{1}{T}\right)$

The above equation is like $y = mx + c$, where if we plot '$y$' vs '$x$' we get a straight line with slope '$m$' and intercept '$c$'.

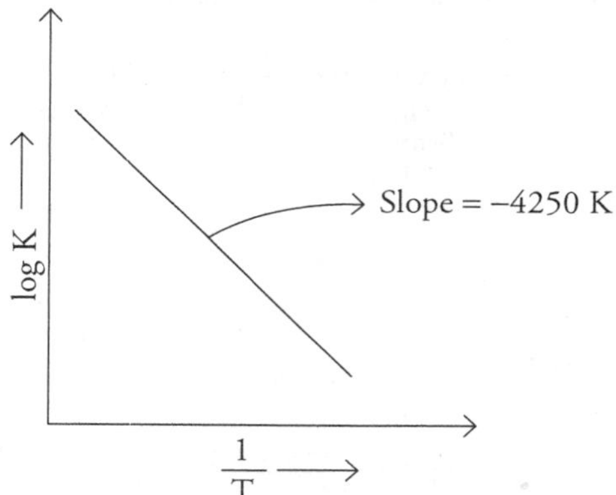


Hence, slope is equal to $= \frac{-E_a}{2.303\ R}$

$$\frac{-E_a}{2.303R} = -4250 \text{ K}$$

$$\therefore Ea = 4250 \times 2.303 \times 8.314\ (JK^{-1}\ mol^{-1})$$
$$= 81{,}375.3535\ Jmol^{-1}$$
$$Ea = 81.3733\ kJ\ mol^{-1}$$

**29. (a)** **Give reasons for the following :** **[5]**

**(i)** **Bond enthalpy of $F_2$ is lower than that of $Cl_2$.**

**(ii)** **$PH_3$ has lower boiling point than $NH_3$.**

**(b)** **Draw the structures of the following molecules :**

**(i)** **$BrF_3$**

**(ii)** **$(HPO_3)_3$**

**(iii)** **$XeF_4$**

OR

**(a) Account for the following :**

**(i) Helium is used in diving apparatus.**

**(ii) Flourine does not exhibit positive oxidation state.**

**(iii) Oxygen shows catenation behaviour less than sulphur.**

**(b) Draw the structures of the following molecules.**

**(i) $XeF_2$**

**(ii) $H_2S_2O_8$**

**Ans. (a)** (i) Bond enthalpy of $F_2$ is lower than that of $Cl_2$ because 'F' atom is small in size and due to this the electron-electron repulsions between the lone pairs of F–F electrons are very large. Thus, the bond dissociation energy of $F_2$ is lower than that of $Cl_2$.

(ii) $PH_3$ has lower boiling point than $NH_3$ because $NH_3$ molecule possess intermolecular hydrogen bonding which binds them strongly, whereas, $PH_3$ has weaker van der Waals forces. Thus $PH_3$ has a lower boiling point than $NH_3$.

**(b)** (i) $BrF_3$, Bent T-shape

(ii) $(HPO_3)_3$, Cyclic structure

(iii) $XeF_4$, square planar

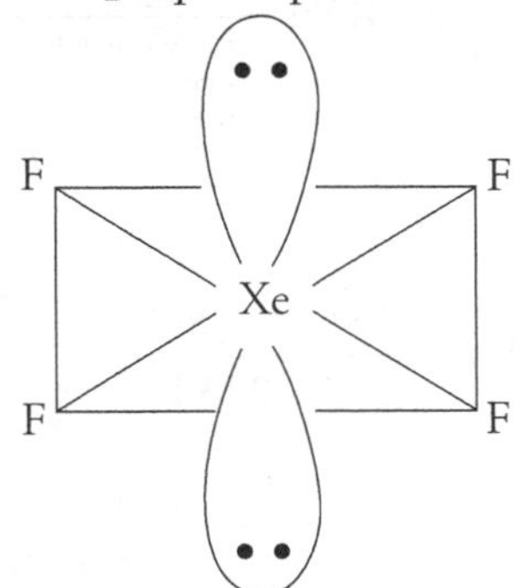


OR

**(a)** (i) Helium mixed with oxygen under pressure is given to sea-divers for artificial respiration because of its very low solubility in blood. Air is not given because nitrogen present in air being soluble in blood will give a painful sensation called bends by bubbling out blood on moving from high pressure to the atmospheric pressure. Thus, oxygen-helium mixture is used.

(ii) Fluorine being the most electronegative atom does not exhibit positive oxidation state because, it does not have d-orbitals for octet expansion and therefore, it shows only a negative oxidation state of –1.

(iii) Oxygen shows catenation behaviour less than sulphur because the oxygen atom is smaller in size as compared to sulphur due to this the lone pair of electron in O–O bonds in oxygen experiences more repulsion as compared to the S–S bonds and thus, S–S forms strong bond.

**(b)** (i) $XeF_2$,

$XeF_2$, Shape : Linear

(ii) $H_2S_2O_8$

$$HO-\overset{\overset{O}{\|}}{\underset{\underset{O}{\|}}{S}}-O-O-\overset{\overset{O}{\|}}{\underset{\underset{O}{\|}}{S}}-OH$$

**30. (a) Although phenoxide ion has more number of resonating structures than carboxylate ion, carboxylic acid is a stronger acid than phenol. Give two reasons. [5]**

**(b) How will you bring about the following conversions?**

**(i) Propanone to propane**

**(ii) Benzoyl chloride to benzaldehyde**

**(iii) Ethanal to but-2-enal.**

**OR**

**(a) Complete the following reactions :**

**(i)** $2H-\underset{\underset{O}{||}}{C}-H \xrightarrow{\text{Conc·KOH}}$

**(ii)** $CH_3COOH \xrightarrow{Br_2/P}$

**(iii)** $C_6H_5CHO \xrightarrow[273-283\ K]{HNO_3/H_2SO_4}$

**(b) Give simple chemical tests to distinguish between the following pairs of compounds :**

**(i) Ethanal and Propanal**

**(ii) Benzoic acid and Phenol.**

**Ans. (a)** On losing a proton, carboxylic acids forms carboxylate ion and phenol forms phenoxide ion as follows:

$RCOO^-$

Carboxylate ion

$C_6H_5O^-$

Phenoxide ion

Now, the negative charge is delocalized in both molecules as follows :

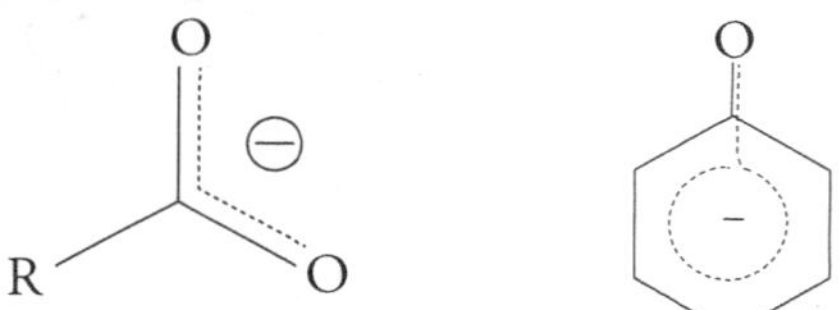

The conjugate base of carboxylic acid has two resonance structures in which negative charge is declocalized over two oxygen atoms which stabilizes the carboxylate ion.

On the other hand, in phenoxide ion the charge is delocalized over entire molecule on the less electronegative atom, thus resonance of phenoxides is not important in comparison to resonance in carboxylate ion.

Further, in carboxylate ion the negative charge is effectively delocalized over two oxygen atoms whereas it is less effectively delocalized over one oxygen atom and less electronegative carbon atom.

Thus, phenol is less acidic than carboxylic acids. In other words, carboxylic acids are stronger acids than phenol.

**(b) (i)** Conversion of propanone to propane:

$$CH_3-\underset{\underset{O}{||}}{C}-CH_3 \xrightarrow[\text{Reduction}]{NaBH_4,\ CH_3OH} \underset{\text{2–Propanol}}{CH_3-\overset{\overset{OH}{|}}{CH}-CH_3}$$

$$\xrightarrow[443K]{Conc.H_2SO_4} \underset{\text{Propene}}{CH_3-CH=CH_2} \xrightarrow{+H_2,\ Ni} \underset{\text{Propane}}{CH_3-CH_2-CH_3}$$

**(ii)** Conversion of benzoyl chloride to benzaldehyde :

$$\underset{\text{Benzoyl chloride}}{C_6H_5COCl} \xrightarrow[+H_2]{Pd\text{-}BaSO_4} \underset{\text{Benzaldehyde}}{C_6H_5CHO}$$

**(iii)** On treatment with dilute alkali, ethanol produces 3-hydroxybutanal gives But-2-enal on heating

$$\underset{\text{Ethanal}}{CH_3CHO} + H-CH_2-\overset{\overset{O}{||}}{C}-H \xrightarrow{\text{dil. NaOH}} \underset{\text{3-hydroxybutanal}}{CH_3-\overset{\overset{OH}{|}}{CH}-CH_2-CHO}$$

$$\xrightarrow[-H_2O]{\Delta} \underset{\text{But–2–enal}}{CH_3-CH=CH-CHO}$$

**OR**

**(a) (i)**

$$2\ \underset{\text{Formaldehyde}}{H_2C=O} + \xrightarrow[\text{Cannizzaro reaction}]{\text{Conc. KOH}} \underset{\text{Methanol}}{H-\overset{\overset{H}{|}}{\underset{\underset{H}{|}}{C}}-OH} + \underset{\text{Potassium formate}}{HCOOK}$$

**(ii)** $$\underset{\text{Acetic acid}}{CH_3COOH} \xrightarrow[\text{Hell–Volhard Zelinsky reaction}]{Br_2/P} \underset{\alpha\text{-Bromoacetic acid}}{\underset{\underset{Br}{|}}{CH_2}-COO}$$

**(iii)**

$$\underset{\text{Benzaldehyde}}{C_6H_5CHO} \xrightarrow[273-283\ K]{HNO_3/H_2SO_4} \underset{m-\text{Nitrobenzaldehyde}}{m\text{-}O_2N-C_6H_4-CHO} + H_2O$$

**(b) (i)** Ethanol and Propanal:

**Iodoform Test:** When ethanol is treated with sodium hydroxide and ammonia, iodoform is obtained.

$$CH_3CHO + 4NaOH + 3I_2 \longrightarrow \underset{\substack{\text{Iodoform}\\\text{(yellow ppt.)}}}{CHI_3} + \underset{\text{Sod. formate}}{HCOONa} + 4NaI + 3H_2O$$

Propanal does not give this test

$$CH_3CH_2CHO + 4NaOH + 3I_2 \rightarrow \text{No reaction}$$

**(ii)** Benzoic acid and phenol :

**$NaHCO_3$ Test**: When benzoic acid is treated with $NaHCO_3$ brisk efferversence of $CO_2$ gas were evolved.

$$C_6H_5COOH + NaHCO_3 \longrightarrow C_6H_5COONa + H_2O + CO_2\uparrow \text{ (Brisk effervescence)}$$

Phenol does not give this test:

$$C_6H_5OH + NaHCO_3 \rightarrow \text{No reaction}$$

## Delhi [Set-II]

**Note: Except for the following questions, all the remaining questions have been asked in previous set.**

**1. What type of stoichiometric defect is shown by AgCl?** **[1]**

**Ans.** Frenkel defect is shown by AgCl.

**2. Write the IUPAC name of:** **[1]**

$$CH_3CH = CH - \overset{\displaystyle CH_3}{\underset{\displaystyle Br}{C}} - CH_3$$

**Ans.** 4-Bromo-4-methylpent-2-ene.

**4. What type of bonding helps in stabilizing the α-helix structure of proteins?** **[1]**

**Ans.** Hydrogen bonding between the NH group of each amino acid residue and the >C = O of an adjacent turn of the helix helps in stabilizing the α-helix structure of proteins.

**6. What inspired N. Bartlett for carrying out reaction between Xe and $PtF_6$?** **[1]**

**Ans.** N. Bartlett observed that the first ionization enthalpy of molecular oxygen is almost identical with that of xenon. So after preparing red coloured compound $O_2^+PtF_6^-$ he made efforts to prepare $Xe^+PtF_6^-$ by mixing $PtF_6$ and Xe.

**7. What happens when ethyl chloride is treated with aqueous KOH?** **[1]**

**Ans.** When $C_2H_5Cl$ reacts with aq. KOH, substitution Nucleophilic bimolecular ($S_N2$) reaction takes place and Ethanol is formed.

$$OH^- + CH_3CH_2Cl \longrightarrow [HO^{\delta+}\cdots CH_2(CH_3)\cdots Cl^{\delta-}] \longrightarrow \underset{\text{Ethanol}}{HO-CH_2CH_3} + Cl^-$$

And inversion of configuration also takes place.

**8. Write the structure of 4-chloropentan-2-one. [1]**

**Ans.** $CH_3 - \underset{\displaystyle Cl}{CH} - CH_2 - \overset{\displaystyle O}{\overset{\|}{C}} - CH_3$

**9. How will you convert the following?** **[2]**

**(i) Propan-2-ol to propanone.**

**(iii) Phenol to 2, 4, 6 – tribromophenol?**

**Ans. (i)** Propan-2-ol to propanone:

$$\underset{\text{Propane}-2-\text{ol}}{CH_3 - \underset{\displaystyle OH}{CH} - CH_3} \xrightarrow{K_2Cr_2O_7 + H_2SO_4} \underset{\text{Propanone}}{CH_3 - \underset{\displaystyle O}{\underset{\|}{C}} - CH_3}$$

**(ii)** Phenol to 2, 4, 6-tribromophenol:

$$\underset{\text{Phenol}}{C_6H_5OH} \xrightarrow{Br_2 \text{ in } H_2O} \underset{2,4,6-\text{Tribromophenol}}{C_6H_2Br_3OH}$$

**11. What is the difference between oil/water (O/W) type and water/oil (W/O) type emulsions? Give an example of each type. [2]**

**Ans.** Emulsion of oil-in-water has oil as dispersed phase and water as dispersion medium. For example, Milk etc.

Emulsion of water-in-oil has water as dispersed phase and oil as dispersion medium. For example, Cod liver oil etc.

**17. (a) Which of the following ores can be concentrated by froth floatation method and why?**

**$Fe_2O_3$, ZnS, $Al_2O_3$**

**(b) What is the role of silica in the metallurgy of Copper? [2]**

**Ans. (a)** Only sulphide ores are concentrated by this process because pine oil selectively wets the sulphide ore and hence bring it to the froth.

**(b)** Silica is added in the reverberatory furnance during the extraction of copper to remove iron oxide present in the ore. Iron oxide reacts with silica and is removed as slag of iron silicate.

$$FeO + \underset{\text{Flux}}{SiO_2} \rightarrow \underset{\text{Slag}}{FeSiO_3}$$

**18. (a) Why does *p*-dichlorobenzene has a higher m.p. than its *o*- and *m*- isomers? [2]**

**(b) Why is (–) – Butan-2-ol optically inactive?**

**Ans. (a)** p-dichlorobenzene has higher melting point than ortho and meta isomer. This is because the para isomer is having a symmetrical structure and therefore, its packing is more efficient as compared to the ortho and meta isomer, therefore, it shows higher melting point.

*p*-dichlorobenzene 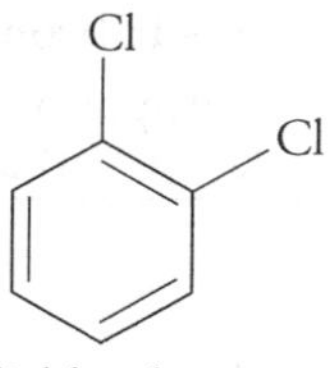 *o*-dichlorobenzene 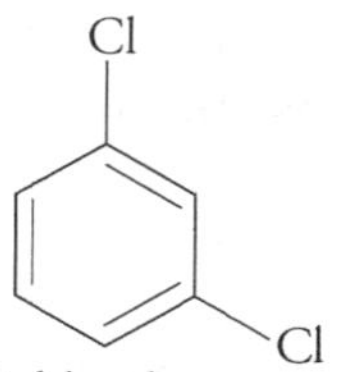 *m*-dichlorobenzene

**(b)** The (±) – Butan-2-ol is optically inactive because it is racemic mixture and exists in two enantiomeric forms which are non-superimposable mirror images of each other. Both the isomers are present in equal amounts therefore, it does not rotate the plane of polarized light and is optically inactive.

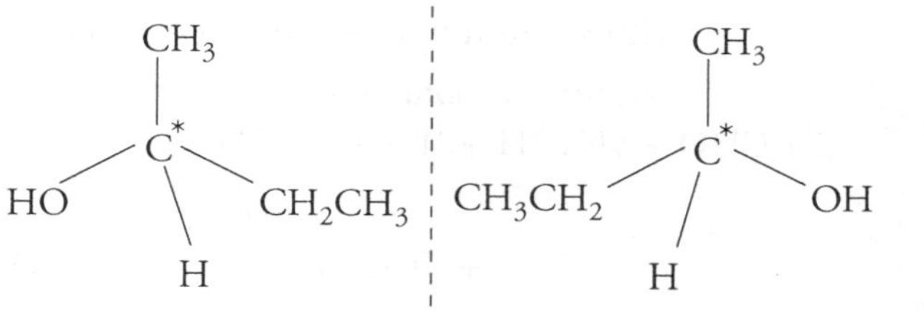

50% – (+) Butan -2-ol | 50% – (–) Butan -2-ol

**23. Write the names and structures of the monomers of the following polymers: [3]**

**(i) Polystyrene**

**(ii) Dacron**

**(iii) Teflon**

**Ans.** Polymers with their monomers and their structures:

| Polymer | Monomer | Structure |
|---|---|---|
| (i) Polystyrene | Styrene (Vinyl benzene) | 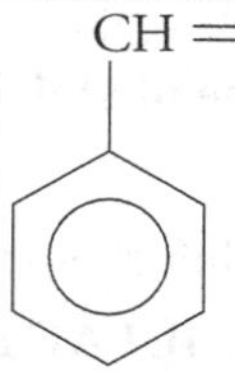 $C_6H_5$–CH = $CH_2$ |
| (ii) Dacron | 1. Ethylene glycol<br>2. Terephthalic acid | 1. $HOH_2C - CH_2OH$<br>2.  HOOC–$C_6H_4$–COOH |
| (iii) Teflon | Tetrafluoroethene | $CF_2 = CF_2$ |

**27. Write the types of isomerism exhibited by the following complexes: [3]**

**(i) $[Co(NH_3)_5Cl]SO_4$**

**(ii) $[Co(en)_3]^{3+}$**

**(iii) $[Co(NH_3)_6][Cr(CN)_6]$**

**Ans. (i)** Ionisation isomerism

**(ii)** Optical isomerism

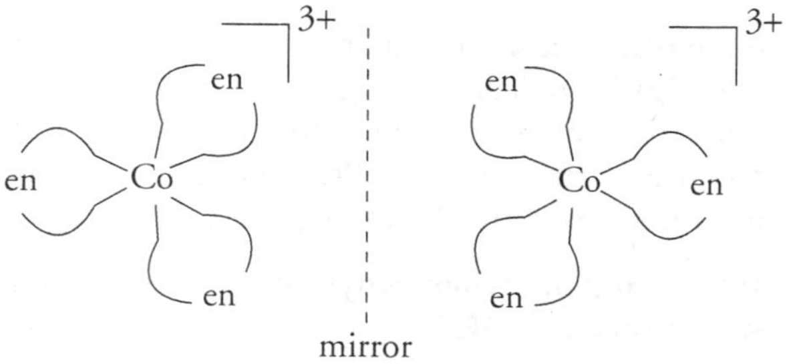

mirror

*dextro* *laevo*

**(iii)** Coordination isomerism

# Delhi [Set-III]

**Note: Except for the following questions, all the remaining questions have been asked in previous sets.**

1. **What type of substances would make better Permanent Magnets, Ferromagnetic or Ferrimagnetic?** [1]

**Ans.** Ferromagnetic substance would make better permanent magnets because when the ferromagnetic substance is placed in a magnetic field, all domains get oriented in the direction of magnetic field and a strong magnetic effect is produced.

3. **What is the composition of 'Copper matte'?** [1

**Ans.** Composition of 'Copper matte' is $Cu_2S$ and FeS.

5. **What is a glycosidic linkage?** [1]

**Ans.** The linkage between the two monosaccharide units through oxygen atom accompanied by the loss of a water molecule is called glycosidic linkage.

6. **Write the IUPAC name of $(CH_3)_2CH.CH(Cl)CH_3$** [1]

**Ans.**

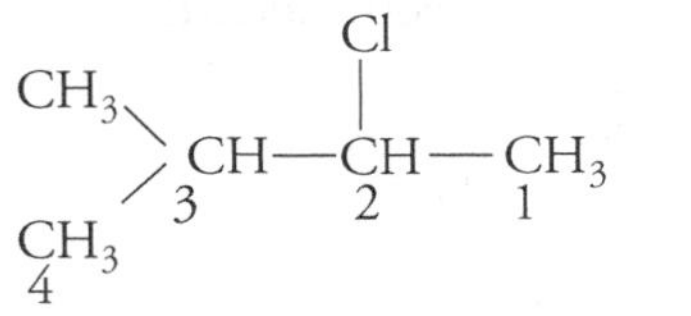

$$\overset{4}{CH_3}-\underset{\underset{CH_3}{|}}{\overset{3}{CH}}-\overset{Cl}{\overset{|}{\underset{2}{CH}}}-\underset{1}{CH_3}$$

2-Chloro-3-methylbutane

7. **Which compound in the following pair undergoes faster $S_N1$ reaction?** [1]

Cl and Cl

**Ans.**

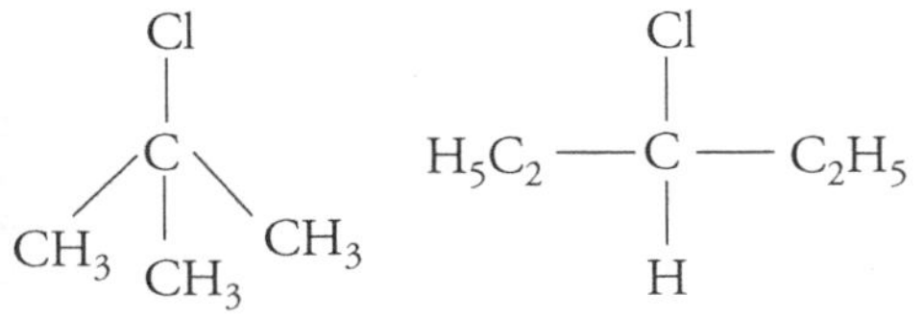

$$\underset{\text{I}}{(CH_3)_3C-Cl} \qquad \underset{\text{II}}{H_5C_2-\underset{\underset{H}{|}}{\overset{\overset{Cl}{|}}{C}}-C_2H_5}$$

The carbocation formed when compounds I and II undergo $S_N1$ reaction are shown below:

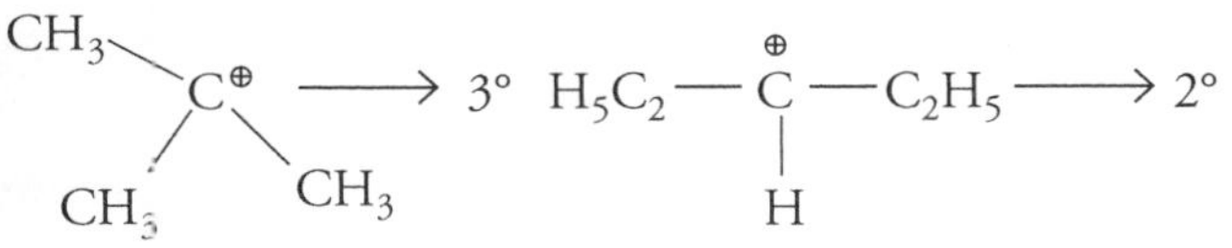

$$(CH_3)_3C^{\oplus} \longrightarrow 3° \quad H_5C_2-\underset{\underset{H}{|}}{\overset{\oplus}{C}}-C_2H_5 \longrightarrow 2°$$

As 3° carbocation is more stable than 2° carbocation, hence compound I undergoes faster $S_N1$ reaction.

8. **Write the structure of *p*-Methylbenzaldehyde molecule.**

**Ans.** [1]

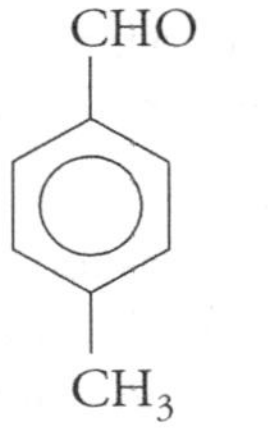

CHO

(benzene ring)

$CH_3$

9. **What is the difference between multi-molecular and macromolecular colloids? Give one example of each.** [2]

**Ans.** Difference between Multi-molecular and Macromolecular colloids:

| Point of Difference | Multi-molecular Colloid | Macro-molecular Colloid |
|---|---|---|
| **Definition** | When a large number of atoms or small molecules (having diameters of less than 1 nm) of a substance combine together in a dispersion medium to form aggregates having size in the colloidal range, the colloidal solutions thus formed are called multi-molecular colloid. Example : Gold sol, sulphur sol. | When substances which have very high molecular masses are dispersed in suitable dispersion medium, the resulting colloidal solutions are known as macromolecular colloids. Example: Starch, cellulose. |

14. (a) **Give an example of zone refining of metals.**

(b) **What is the role of cryolite in the metallurgy of aluminum?** [2]

**Ans.** (a) **Zone Refining of Metals:** This method is used for production of semiconductor and other metals of very high purity like germanium, silicon, boron, gallium and indium.

(b) **Role of Cryolite in Metallurgy of Aluminium :** Cryolite is added to lower the melting point of mixture and to increase the conductivity of electrolyte.

**17. Account for the following: [2]**

**(i) The C—Cl bond length in chlorobenzene is shorter than that in $CH_3$—Cl.**

**(ii) Chloroform is stored in closed dark brown bottles.**

**Ans.** (i) This is due to partial double bond character to C-Cl bond (due to resonance in $C_6H_5Cl$).

(ii) Chloroform in the presence of air gets oxidized to phosgene. Phosgene is carbonyl chloride and is represented as $COCl_2$. To prevent the formation of phosgene, chloroform is stored in dark coloured bottles. The reaction represented as $CHCl_3 + \frac{1}{2}O_2 \rightarrow COCl_2 + HCl$

**18. How will you convert: [2]**

**(i) Propene to Propan-1-ol?**

**(ii) Ethanal to Propan-2-ol?**

**Ans.** (i) $\underset{\text{Propene}}{CH_3CH=CH_2} + HBr \xrightarrow{\text{Peroxide}}$

$CH_3CH_2CH_2Br + \underset{\text{(aq.)}}{KOH}$

$\downarrow \Delta$

$\underset{\text{Propan-1-ol}}{CH_3CH_2CH_2OH}$

(ii) $\underset{\text{Ethanal}}{CH_3(H)C=O} \xrightarrow{CH_3MgI} CH_3(H)C(OMgI)CH_3 \xrightarrow{H^+, H_2O} \underset{\text{2-propanol}}{CH_3(H)C(OH)CH_3}$

**23. Give the structures of products A, B and C in the following reactions : [3]**

**(i)**

$$CH_3CH_2Br \xrightarrow{KCN} A \xrightarrow{LiAlH_4} B \xrightarrow[0°C]{HNO_3} C$$

**(ii)** $$CH_3COOH \xrightarrow[\Delta]{NH_3} A \xrightarrow{NaOH+Br_2} B \xrightarrow{CHCl_3+Alc.KOH} C$$

**Ans.** (i)

$$CH_3CH_2Br \xrightarrow{KCN} \underset{A}{CH_3CH_2CN} \xrightarrow{LiAlH_4} \underset{B}{CH_3CH_2CH_2NH_2} \xrightarrow[0°C]{HNO_2} \underset{C}{CH_3CH_2CH_2OH}$$

(ii)

$$CH_3COOH \xrightarrow[\Delta]{NH_3} \underset{A}{CH_3CONH_2} \xrightarrow{NaOH+Br_2} \underset{B}{CH_3NH_2} \xrightarrow{CHCl_3+Alc.KOH} \underset{C}{CH_3NC}$$

**27. Write the names and structures of the monomers of the following polymers : [3]**

**(i) Bakelite**

**(ii) Nylon-6**

**(iii) Polythene**

**Ans.** (i) OH (benzene ring) and HCHO

Phenol and Formaldehyde

(Caprolactam ring: C=O, N–H)

(ii) Caprolactam

(iii) $CH_2=CH_2$

Ethene

# CHEMISTRY

## Outside Delhi [Set-I] 2012

**Time allowed : 3 hours** **Maximum marks : 70**

1. **How may the conductivity of an intrinsic semiconductor be increased?** **[1]**

**Ans.** By doping the intrinsic semiconductor with elements like arsenic and phosphorous, the conductivity can be increased.

2. **Define 'peptization'.** **[1]**

**Ans.** The process of conversion of freshly prepared precipitate into a colloidal solution by adding suitable electrolyte is called peptization.

3. **How is copper extracted from a low grade ore of it?** **[1**

**Ans.** Copper from its low grade ore is leached out using acid or bacteria and then $Cu^{2+}$ ions in the solution are reduced to copper by treating it with Hydrogen or Iron. This method is called hydrometallurgy.

4. **Which is a stronger reducing agent, $SbH_3$ or $BiH_3$, and why?** **[1]**

**Ans.** $BiH_3$ is stronger reducing agent because of its tendency to liberate maximum $H_2$ atom thus reducing the element.

5. **What happens when bromine attacks $CH_2 = CH - CH_2 - C \equiv CH$?** **[1]**

**Ans.**

$$CH_2 = CH - CH_2 - C \equiv CH + Br_2 \rightarrow CH_2 = CH - CH_2 - \underset{\substack{|\\ Br}}{C} = CH - Br$$

And if $Br_2$ is in excess then

$$\underset{\substack{|\\ Br}}{CH_2} - \underset{\substack{|\\ Br}}{CH} - CH_2 - \overset{\substack{Br\\ |}}{\underset{\substack{|\\ Br}}{C}} - \overset{\substack{H\\ |}}{\underset{\substack{|\\ Br}}{C}} - Br$$

6. **Write the IUPAC name of the following:** **[1]**

$$CH_3 - CH_2 - CH = CH - \overset{\substack{O\\ \|}}{C} - H$$

**Ans.** Pent-2-enal

7. **Write the structure of the product obtained when glucose is oxidised with nitric acid.** **[1]**

**Ans.**

$$\underset{\text{Glucose}}{\begin{matrix} CHO \\ | \\ (CHOH)_4 \\ | \\ CH_2OH \end{matrix}} + HNO_3 \rightarrow \underset{\text{Saccharic acid}}{\begin{matrix} COOH \\ | \\ (CHOH)_4 \\ | \\ COOH \end{matrix}}$$

8. **Differentiate between disinfectants and antiseptics.** **[1]**

**Ans.** Antiseptics are used on living tissues to kill or prevent the growth of microorganisms. It does not cause any harm to the living tissues *e.g.* 0.2% solution of phenol.

Disinfectants are used on non-living objects to kill the micro-organisms. They are harmful to the living tissues and hence, cannot be applied to the skin *e.g.* 1% solution of phenol.

9. **Express the relation among cell constant, resistance of the solution in the cell and conductivity of the solution. How is molar conductivity of a solution related to its conductivity.** **[2]**

**OR**

**The molar conductivity of a 1.5 M solution of an electrolyte is found to be 138.9 S cm$^2$ mol$^{-1}$. Calculate the conductivity of this solution.**

**Ans.** $CG^* = k$

Where, C = conductance, $G^*$ = cell constant, k = conductivity

$$G^* \times \frac{1}{R} k \Rightarrow G^* = Rk$$

$$\therefore \Lambda_m = \frac{k \times 1000}{M} Scm^2mol^{-1}$$

**OR**

Given : M = 1.5 M

$\Lambda_m = 138.9\ S\ cm^2\ mol^{-1}$

$$\Lambda_m = \frac{1000K}{M}$$

$$M = \frac{k.\Lambda_m}{1000} = \frac{1.5 \times 138.9}{1000}$$

$M = 0.2085\ S\ cm^{-1}$

10. **A reaction is of second order with respect to a reactant. How is its rate affected if the concentration of the reactant is (i) doubled (ii) reduced to half?** **[2]**

**Ans.** **(i)** Rate increases by four times,

$R' = 2R \Rightarrow r = k[2R]^2 = 4kR^2$

**(ii)** Rate is reduced by ¼ times,

$$R' = \frac{1}{2}R \Rightarrow r = k\left[\frac{R}{2}\right]^2 = \frac{k}{4}R^2$$

**11. Which methods are usually employed for purifying the following metals : [2]**

**(i) Nickel, (ii) Germanium.**

**Mention the principle behind each one of them.**

**Ans. (i) Mond's Process :** In this method Nickel forms volatile complex with Co which decomposes on heating to give pure nickel.

**(ii) Zone Refining :** It is based on the principle that the impurities are more soluble in the molten state than in the solid state of metal.

**12. Explain the following facts giving appropriate reason in each case : [2]**

**(i) $NF_3$ is an exothermic compound whereas $NCl_3$ is not.**

**(ii) All the bond in $SF_4$ are not equivalent.**

**Ans. (i)** The bond energy of F-F bond is lower that of N-F bond. So $NF_3$ is an exothermic compound whereas bond energy of Cl-Cl bond is higher than N-Cl bond so $NCl_3$ is an endothermic compound.

**(ii)** $SF_4$ has see-saw structure with bond pairs at two equatorial and two axial positions. Hence, all the bonds in $SF_4$ are not equivalent.

**13. Complete the following chemical reaction equations : [2]**

**(i) $Cr_2O_7^{2-} + I^- + H^+ \longrightarrow$**

**(ii) $MnO_4^- + NO_2^- + H^+ \longrightarrow$**

**Ans.**

(i) $Cr_2O_7^{2-} + 14H^+ + 6I^- \rightarrow 2Cr^{3+} + 7H_2O + 3I_2$

(ii) $2MnO_4^- + 5NO_2^- + 6H^+ \rightarrow 2Mn^{2+} + 3H_2O + 5NO_3^-$

**14. Explain the mechanism of acid catalysed hydration of an alkene to form corresponding alcohol. [2]**

**Ans. Step1:** Carbocation is formed due to electrophilic attack by $H_3O^+$.

$$>C=C< + H-\overset{+}{O}(H)-H \longrightarrow -\underset{|}{\overset{H}{\overset{|}{C}}}-\overset{+}{C}< + H_2O$$

**Step 2:** Nucleophilic water attacks the carbocation.

$$>\overset{H}{\overset{|}{C}}-\overset{+}{C}< + H_2\ddot{O} \longrightarrow >\overset{H}{\overset{|}{C}}-\underset{H}{\underset{|}{\overset{H}{\overset{|}{C}}}}-\overset{H}{\overset{|}{\overset{+}{O}}}-H$$

**Step 3:** Deprotonation to form an alcohol

$$-\underset{H}{\underset{|}{\overset{H}{\overset{|}{C}}}}-\underset{H}{\underset{|}{\overset{O}{\overset{|}{C}}}}-\overset{+}{O}-H + H_2\ddot{O} \longrightarrow >\overset{H}{\overset{|}{C}}-\overset{:\ddot{O}H}{\overset{|}{C}}< + H_3O^+$$

**15. Explain the following behaviours: [2]**

**(i) Alcohols are more soluble in water than the hydrocarbons of comparable molecular masses.**

**(ii) Ortho-nitrophenol is more acidic than ortho-methoxyphenol.**

**Ans. (i)** Due to the formation of hydrogen bonds by alcohols with water molecules. They are more soluble in water than the hydrocarbons of comparable molecular masses.

**(ii)** The phenoxide ion is more stable for *o*-nitrophenol as nitro group is electron withdrawing due to resonance while methoxy group is electron donating via resonance. Thus orthonitro phenol is more acidic than orthomethoxy phenol.

**16. Describe the following giving the relevant chemical equation in each case: [2]**

**(i) Carbylamine reaction**

**(ii) Hofmann's bromamide reaction.**

**Ans. (i) Carbylamine Reaction :** When aliphatic and aromatic primary amines are heated with chloroform and ethanolic KOH solution, they form isocyanides or carbylamines which are foul smelling substances.

$$R-NH_2 + CHCl_3 + 3KOH \xrightarrow{\Delta} \underset{\text{Isocyanide}}{R-NC} + 3KCl + 3H_2O$$

**(ii) Hofmann's Bromamide Reaction:** In this reaction, primary amines are prepared by treating an amide with $Br_2$ in an aqueous or alcoholic solution of NaOH

$$R-\overset{O}{\overset{||}{C}}-NH_2 + Br_2 + 4NaOH \rightarrow R-NH_2 + Na_2CO_3 + 2NaBr + 2H_2O$$

**17. Complete the following reaction equations: [2]**

**(i) $C_6H_5N_2Cl + H_3PO_2 + H_2O \longrightarrow$**

**(ii) $C_6H_5NH_2 + Br_2(aq.) \longrightarrow$**

**Ans. (i)** $C_6H_5N_2Cl + H_3PO_2 + H_2O \rightarrow C_6H_6 + N_2 + H_2PO_3 + HCl$

**(ii)** $C_6H_5NH_2 + 3Br_{2(aq)} \rightarrow$ 2,4,6-tribromoaniline (ring with $NH_2$ at top, Br at 2, 4, 6) $+ 3HBr$

**18. What are food preservatives? Name two such substances.** **[2]**

**Ans.** Chemicals used to prevent spoilage of food by preventing growth of microorganisms like bacteria, fungus etc are called food preservatives. Sodium benzoate, nitrogen gas are two such substances which are used as food preservatives.

**19. Copper crystallizes with face centred cubic unit cell. If the radius of copper atom is 127.8 pm, Calculate the density of copper metal.** **[3]**

**(Atomic mass of Cu = 63.55 u and Avogadro's number $N_A = 6.02 \times 10^{23}$ mol$^{-1}$)**

**OR**

**Iron has a body centred cubic unit cell with the cell dimension of 286.65 pm. Density of iron is 7.87 g cm$^{-3}$. Use this information to calculate Avogadro's number. (Atomic mass of Fe = 56.0 u)**

**Ans.** Given:

$$r = 127.8 \text{ pm} = 127.8 \times 10^{-10} \text{ cm} = 1.278 \times 10^{-8} \text{ cm}$$

$$N_A = 6.02 \times 10^{23} \text{ mol}^{-1}$$

$$M = 63.55 \text{ u}$$

From the formula, $d = \dfrac{Z \times M}{a^3 \times N_A}$

In fcc lattice, Z = 4 atoms

For fcc lattice for copper $a = 2\sqrt{2}r$

$$a^3 = (2\sqrt{2}) \Rightarrow a^3 = 8 \times 2\sqrt{2}\ (1.278 \times 10^{-8} \text{ cm})^3$$

$$\Rightarrow a^3 = 4.723 \times 10^{-23} \text{ cm}^3$$

$$d = \frac{4 \times 63.55 \text{ g mol}^{-1}}{4.723 \times 10^{-23} \text{ cm}^3 \times 6.02 \times 10^{23} \text{ mol}^{-1}}$$

$$\Rightarrow d = \frac{254.2 \text{ g mol}^{-1}}{28.43} = 8.94 \text{ g cm}^{-3}$$

**OR**

From the formula, $d = \dfrac{Z \times M}{a^3 \times N_A}$

Z for bcc = 2

$$NA = \frac{Z \times M}{a^3 \times d} = \frac{2 \times 56 \text{ g mol}^{-1}}{(286.65 \times 10^{-10}) \times 7.87 \text{ cm}^{-3}}$$

$$= 6.043 \times 10^{23} \text{ mol}^{-1}$$

**20. The electrical resistance of a column of 0.05 M NaOH solution of diameter 1 cm and length 50 cm is 5.55 × 10$^3$ ohm. Calculate its resistivity, conductivity and molar conductivity.** **[3]**

**Ans.** $A = \pi r^2 = 3.14 \times (0.5)^2 = 0.785 \text{ cm}^2, l = 50 \text{ cm}$

$$R = \frac{\rho l}{A} \quad \rho = \frac{AR}{l} = \frac{5.55 \times 10^{-3} \times 0.785}{50}$$

$$\rho = 87.18 \ \Omega \text{ cm}$$

$$\text{Conductivity, } K = \frac{1}{\rho} = 0.01147 \text{ S cm}^{-1}$$

$$\text{Molar Conductivity} = \frac{1000 \times K}{C} = \frac{0.01147 \times 1000}{0.05}$$

$$\Lambda_m = 229.4 \text{ S cm}^2 \text{mol}^{-1}$$

**21. The reaction $N_2(g) + O_2(g) \rightleftharpoons 2NO(g)$ contributes to air pollution whenever a fuel is burnt in air at a high temperature. At 1500 K, equilibrium constant K for it is 1.0 × 10$^{-5}$. Suppose in a case $[N_2]$ = 0.80 mol L$^{-1}$ and $[O_2]$ = 0.20 mol L$^{-1}$ before any reaction occurs. Calculate the equilibrium concentrations of the reactants and the product after the mixture has been heated to 1500 K.** **[3]**

**Ans.**

| | $N_2(g)$ + | $O_2(g)$ $\rightleftharpoons$ | $2NO(g)$ |
|---|---|---|---|
| Initial conc. | 0.80 | 0.20 | 0 |
| at equilibrium | $0.80 - x$ | $0.20 - x$ | $2x$ |
| Initial conc. at : equilibrium | 0.80-$x$ | 0.80 | 0.20 $2x$ |

$$K = \frac{[NO]^2}{[O_2][N_2]} = \frac{(2x)^2}{(0.8 - x)(0.2 - x)} = 10^{-5}$$

$$\Rightarrow \frac{(4x)^2}{0.16 - 1x + x^2} = 10^5$$

$$\Rightarrow 400000x^3 = x^2 - x + 0.16$$

$$\Rightarrow 399999\, x^2 + x - 0.16 = 0$$

$$b^2 - 4ac = (1)^2 - 4(399999)(-0.16)$$

$$= 1 + 255999.36 = 256000.36$$

$$\Rightarrow \sqrt{b^2 - 4ac} = \sqrt{256000.36} = 505.96$$

$$x = \frac{-1 \pm 505.96}{2 \times 399999} = \frac{504.96}{2 \times 399999}$$

$$= 0.63 \times 10^{-3}$$

$$[NO] = 2x = 2 \times 0.63 \times 10^{-3} = 1.26 \times 10^{-3} \text{ mol/litre}$$

**22. Explain the following terms giving a suitable example for each:** **[3]**

**(i) Aerosol, (ii) Emulsion, (iii) Micelle.**

**Ans. (i)** An aerosol is a colloid in which dispersed phase is a solid and dispersion medium is a gas *eg.*, Dust, smoke.

**(ii)** Emulsion is a colloid solution in which both the dispersed phase and dispersion medium are in liquid state. *eg.*, Milk and cod liver oil.

(iii) They are associated colloids showing colloidal behaviour at high concentration and strong electrolytes at low concentration. *e.g.*, Soap.

**23. How would you account for the following: [3]**

**(i) Among lanthanoids, Ln (III) compounds are predominant. However, occasionally in solution or in solid compounds, +2 and +4 ions are also obtained.**

**(ii) The $E^\circ M^{2+}/M$ for copper is positive (0.34 V). Copper is the only metal in the first series of transition elements showing this behaviour.**

**(iii) The metallic radii of the third (5*d*) series of transition metals are nearly the same as those of the corresponding members of the second series.**

**Ans. (i)** Some lanthanoid show +2 and +4 oxidation states in ionic solutions or solid components due to the extra stability that arises because of empty half-filled or fully filled 4*f*-subshell.

**(ii)** This is because copper has high enthalpy of atomization and low enthalpy of hydration. Hence, $E^\circ_{Cu^{2+}/Cu}$ is positive.

**(iii)** This is because of lanthanoid contraction the metallic radii of the third (5*d*) series of transition metals are nearly the same as those of the corresponding members of the second series.

**24. Name the following coordination entities and draw the structures of their steroisomers: [3]**

**(i) $[Co(en)_2Cl_2]^+$ (en = ethan-1, 2-diamine)**

**(ii) $[Cr(C_2O_4)_3]^{3-}$**

**(iii) $[Co(NH_3)_3Cl_3]$**

**(Atomic numbers Cr = 24, Co = 27)**

**Ans. (i)** $[Co(en)_2Cl_2]^+$ dichloridobis (ethan-1,2 diamine) cobalt **(III)** ion

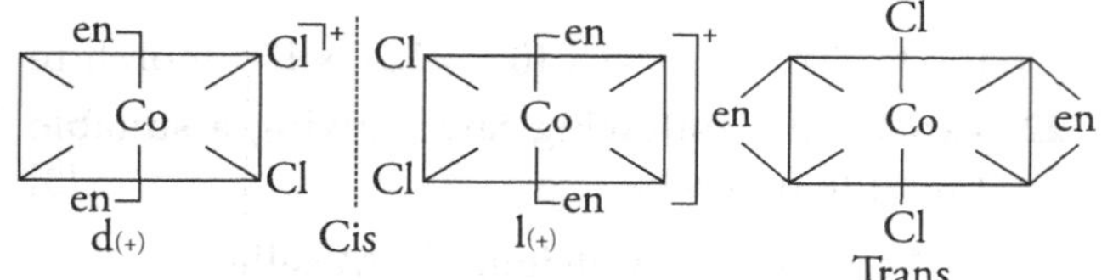


**(ii)** $[Cr(C_2O_4)_3]^{3-}$: Trioxalatochromate **(III)** ion

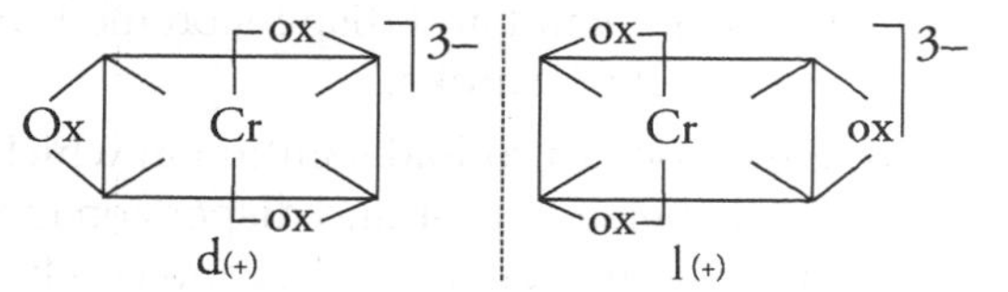


**(iii)** $[Co(NH_3)_3Cl_3]$ triamminetrichloridocobalt **(III)**

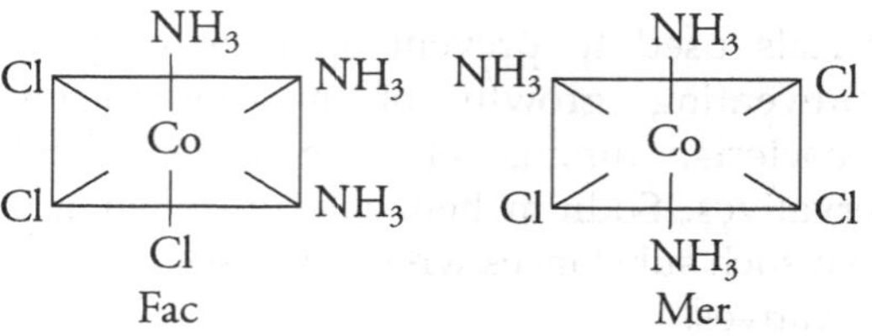


**25. Answer the following questions: [3]**

**(i) What is meant by chirality of a compound? Give an example.**

**(ii) Which one of the following compounds is more easily hydrolyzed by KOH and why?**

$CH_3CHClCH_2CH_3$ or $CH_3CH_2CH_2Cl$

**(iii) Which one undergoes $S_N2$ substitution reaction faster and why ?**

**Ans. (i)** Chirality is the property of a molecule to have non-superimposable mirror images. These molecules contain one asymmetric carbon atom. e.g., Butan - 2- ol

**(ii)** $CH_3CHClCH_2CH_3$ is more easily hydrolyzed due to the formation of more stable secondary carbocation.

**(iii)** $CH_3CH_2CH_2Cl$ undergoes $S_N2$ substitution reaction faster because it is a better leaving group due to its large size and less electronegativity.

**26. What is essentially the difference between α-glucose and β-glucose? What is meant by pyranose structure of glucose? [3]**

**Ans.** α-glucose and β-glucose are two cyclic hemiacetal forms of glucose which differ only in the configuration of hydroxyl group (–OH) at anomeric carbon. Such isomers are called anomers. The six-membered cyclic structure of glucose is called pyranose structure.

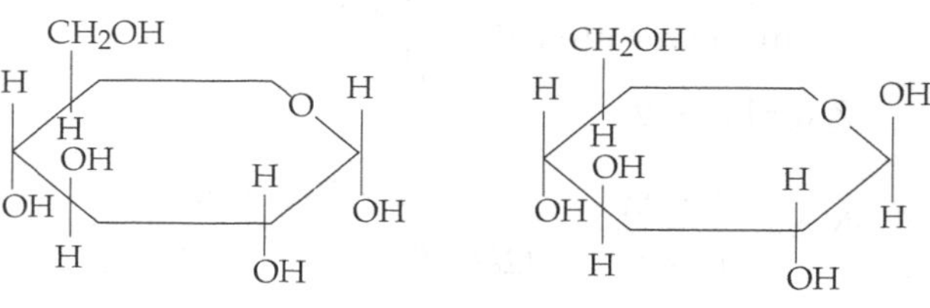


**Pyranose Structure of Glucose :** The six membered ring contains oxygen atom because of its resemblance with pyran it is called pyranose form.

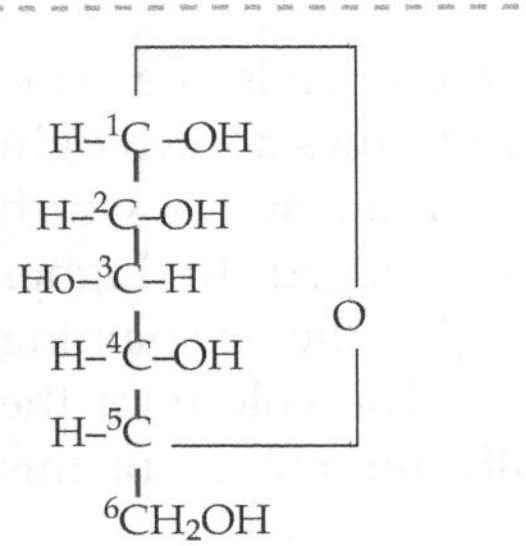

α–D–(+)– Glucopyranose β –D–(+)– Glucopyranose

**27. Differentiate between thermoplastic and thermosetting polymers. Give one example of each.** **[3]**

**Ans. Thermoplastic Polymers:** These polymers do not have cross-links between their chains and hence can be reshaped upon heating eg. Polyethylene, Polypropene etc.

**Thermosetting Polymers:** These polymers have cross-links between their chains and hence cannot be reshaped upon heating. *eg*. Bakelite, Melamine etc.

**28. (a) Define the following terms :** **[5]**

**(i) Mole fraction**

**(ii) Ideal solution.**

**(b) 15.0 g of an unknown molecular material was dissolved in 450 g of water. The resulting solution was found to freeze at –0.34 °C. What is the molar mass of this material ? ($K_f$ for water = 1.86 K kg mol$^{-1}$)**

**OR**

**(a) Explain the following :**

**(i) Henry's law about dissolution of a gas in a liquid.**

**(ii) Boiling point elevation constant for a solvent.**

**(b) A solution of glycerol ($C_3H_8O_3$) in water was prepared by dissolving some glycerol in 500 g of water. This solution has a boiling point of 100.42°C. What mass of glycerol was dissolved to make this solution ? ($K_b$ for water = 0.512 K kg mol$^{-1}$).**

**Ans. (a) (i)** The ratio of number of moles of a solute (components of a mixture) to the total number of moles in the mixture is called mole fraction $x_A = \frac{n_A}{n_A + n_B}$ or $x_B = \frac{n_B}{n_A + n_B}$

**(ii)** A solution that obey's Raoult's law at all temperature and concentration conditions is called an ideal solution.

**(b)** Given: $W_2 = 15.0$ g, $\Delta T_f = 0.34°C$

$W_1 = 450$ g, $K_f = 1.86$ K kg mol$^{-1}$

from the formula, $\Delta T_f = \frac{1000.K_f \times W_2}{W_1 \times M_2}$

$$M_2 = \frac{1000\ K_f \times W_2}{\Delta T_f \times W_1} = \frac{1000 \times 1.86 \times 15}{0.34 \times 450}$$

$M_2 = 182.35$ g/mol

**OR**

**(a) (i)** Henry's law states that at a constant temperature, the solubility of a gas in a liquid is directly proportional to the pressure of the gas. *i.e.*, $p = K_H.x$

**(ii)** The boiling point elevation constant for a solvent is defined as the elevation in boiling point when the molality of the solution is unity.

**(b)** Given:

$W_1 = 500$ g

$K_b = 0.512$ K kg mol$^{-1}$

$\Delta T_b = (100.42 - 100)°C$

$= 0.42°C$

$$\Delta T_b = \frac{1000 \times K_b \times W_2}{W_1 \times W_2}$$

$$W_2 = \frac{\Delta T_b \times W_1 \times M_2}{100 \times Kb} = \frac{0.42 \times 500 \times 92}{1000 \times 0.512}$$

$W_2 = 37.7$ g

**29. (a) Draw the molecular structures of the following compounds:** **[2,3]**

**(i) $N_2O_5$**

**(ii) $XeOF_4$.**

**(b) Explain the following observations :**

**(i) Sulphur has a greater tendency for catenation than oxygen.**

**(ii) I-Cl is more reactive than $I_2$.**

**(iii) Despite lower value of its electron gain enthalpy with negative sign, fluorine ($F_2$) is a stronger oxidizing agent than $Cl_2$.**

**OR**

**(a) Complete the following chemical equations:**

**(i) $Cu + HNO_3$ (dilute) $\longrightarrow$**

**(ii) $XeF_4 + O_2F_2 \longrightarrow$**

**(b) Explain the following observations:**

**(i) Phosphorus has greater tendency for catenation than nitrogen.**

**(ii) Oxygen is a gas but sulphur a solid.**

**(iii) The halogens are coloured. Why ?**

**Ans. (a) (i)**

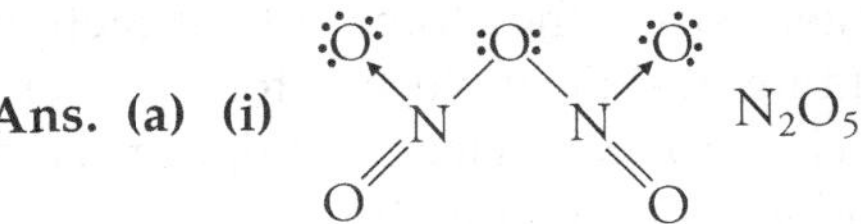

(ii) [Structure: Xe with double-bonded O on top, four F atoms, and a lone pair] $XeOF_4$

**(b) (i)** Due to strong S—S bond and less interelectronic repulsion, sulphur has greater tendency for catenation.

**(ii)** I—Cl bond is polar and hence more reactive compound to $I_2$ in which I—I bond is non-polar.

**(iii)** Due to high electronegativity and small size of fluorine, it acts as a stronger oxidizing agent.

**OR**

**(a) (i)** $3Cu + 8HNO_3\,(dil.) \longrightarrow 3Cu\,(NO_3)_2 + 2NO + 4H_2O$

**(ii)** $XeF_4 + O_2F_2 \longrightarrow XeF_6 + O_2$

**(b) (i)** Catenation (*i.e.* linking of atoms of the same kind with one another) is related to the atom-atom bond energy. Greater the atom-atom bond energy, greater is the catenation. Because of low N—N bond energy (163.8 kJ $mol^{-1}$) nitrogen shows little tendency for catenation. P—P bond energy ($201\times 10^{-6}$ kJ/mol) is quite high, hence, it shows more tendency for catenation than nitrogen.

**(ii)** Oxygen forms $p\pi - p\pi$ multiple bonds. Due to small size and high electronegativity oxygen exists as diatomic ($O_2$) molecule. These molecules are held together by weak van der Waals' forces. Hence $O_2$ is a gas at room temperature.

Sulphur because of its bigger size and lower electronegativity, prefer to form S—S single bonds. Further because of stronger S—S than O—O single bonds, sulphur has a much greater tendency for catenation than oxygen. Thus, sulphur because of its higher tendency for catenation and lower tendency for $p\pi - p\pi$ multiple bonds, forms octa atomic ($S_8$) molecules, having eight-membered puckered ring structure because of bigger size, the force of attraction holding the $S_8$ molecule together are much stronger. Hence, sulphur is a solid at room temperature.

**(iii)** All halogens are coloured. It is due to the reason that their molecules absorb light in the visible region as a result of which their electrons get excited to higher energy levels while the remaining light is transmitted. The colour of the halogens is actually the colour of this transmitted light.

**30. (a) Write a suitable chemical equation to complete each of the following transformations:** **[2, 3]**

**(i) Butan-1-ol to butanoic acid**

**(ii) 4-Methylacetophenone to benzene-1, 4-dicarboxylic acid**

**(b) An organic compound with molecular formula $C_9H_{10}O$ forms 2,4-DNP derivative, reduces Tollen's reagent and undergoes Cannizzaro's reaction. On vigorous oxidation it gives 1, 2-benzenedicarboxylic acid. Identify the compound.**

**OR**

**(a) Give chemical tests to distinguish between:**

**(i) Propanol and propanone**

**(ii) Benzaldehyde and acetophenone**

**(b) Arrange the following compounds in an increasing order of their property as indicated :**

**(i) Acetaldehyde, Acetone, Methyl tert-butyl ketone (reactivity towards HCN)**

**(ii) Benzoic acid, 3, 4-dinitrobenzoic acid, 4-Methoxy-benzoic acid (acid strength)**

**(iii) $CH_3CH_3CH(Br)COOH$, $CH_3CH(Br)CH_2COOH$, $(CH_3)_2CHCOOH$(acid strength)**

**Ans. (a) (i)**

$$\underset{\text{Butan-1-ol}}{CH_3CH_2CH_2CH_2OH} \xrightarrow{KMnO_4} \underset{\text{Butanoic acid}}{CH_3CH_2CH_2COOH}$$

**(ii)**

$$\underset{\text{4-Methyl acetophenone}}{CH_3C_6H_4COCH_3} \xrightarrow[\Delta]{KMnO_4/KOH} \overset{-\,+}{KOOC}C_6H_4\overset{-\,+}{COOK} \xrightarrow[-K_2SO_4]{\text{dil. } H_2SO_4} \underset{\text{Benzene-1,4-dicarboxylic acid}}{HOOCC_6H_4COOH}$$

**(b) 1.** The given compound with molecular formula $C_9H_{10}O$ forms a 2, 4-DNP derivatitve and reduces Tollen's reagent it must be an aldehyde.

**2.** As the compond undergoes cannizzaro reaction, therefore CHO group is directly attached to the benzene ring.

3. On vigrous oxidation, it gives 1, 2-benzene dicarboxylic acid, therefore, it must be an ortho-substituted benzaldehyde.

$Ag\downarrow +$ 2-Ethylbenzoate ($COO^-$, $C_2H_5$) ⟵ O-Ethyl benzaldehyde ($COOH$, $C_2H_5$) $\xrightarrow{[O]}$ 1, 2-Benzene dicarboxylic acid ($COOH$, $COOH$)

O-Ethyl benzaldehyde $\xrightarrow{2,4,\text{-DNP}}$ $CH=NNH$–($NO_2$)–$NO_2 + H_2O$ ($C_2H_5$)

2, 4-DNP derivative

OR

**Ans. (a) (i) Iodoform Test:** This test is given by propanone and not by propanol. Propanone on reacting with hot $NaOH/I_2$ gives a yellow precipitate of $CHI_3$ while propanol does not.

$$2NaOH + I_2 \rightarrow NaI + NaOI + H_2O$$

$$CH_3-\overset{O}{\overset{\|}{C}}-CH_3 + \underset{\text{Sodium hypo-Iodite}}{3NaOI} \rightarrow \underset{\text{yellow ppt.}}{CHI_3} + \underset{\text{Sodium acetate}}{CH_3COONa} + 2NaOH$$

**(ii) Silver Mirror Test:** Benzaldehyde being an aldehyde reduces Tollens' reagent to give silver mirror test but acetophenone being a ketone does not give this test.

$$\underset{\text{Benzaldehyde}}{C_6H_5CHO} + \underset{\text{Tollens' reagent}}{2[Ag(NH_3)_2]^+} + 3OH^- \xrightarrow{\text{Heat}}$$

$$C_6H_5COO^- + 2H_2O + \underset{\text{Silver mirror}}{2Ag\downarrow} + 4NH_3$$

**(b) (i)** Methyl tert - butyl ketone < Acetone < Acetaldehyde.

**(ii)** 4-methoxybenzoic acid < benzoic acid < 3,4 dinitrobenzoic acid

**(iii)** $(CH_3)_2CHCOOH < CH_3CH(Br)CH_2COOH < CH_3CH_2CH(Br)COOH$

# Outside Delhi [Set-II]

**Note : Except for the following questions, all the remaining questions have been asked in previous set.**

**1. Which stoichiometric defect increases the density of a solid? [1]**

**Ans.** Interstitial defect.

**2. What is meant by 'shape selective catalysis'? [1]**

**Ans.** Zeolites are shape selective catalysts and shape selective catalysis depends upon the structure of the pores present in the catalyst and size of the reactant and product molecules. ZSM-5 is an example of shape selective catalyst, which is used in converting alcohol directly into gasoline.

**3. What is the role of collectors in Froth Floatation process? [1]**

**Ans.** Collectors help in attachment of ore particle to an air bubble in froth *e.g.*, Sodium xanthates.

**6. Write the IUPAC name of Ph – CH = CH – CHO. [1]**

**Ans.** 3-phenyl prop-2-enal

**17. Explain the cleaning action of soap. Why do soaps not work in hard water ? [2]**

**Ans.** When soap is rubbed on dirty cloth in water, concentration of soap becomes greater than (CMC) critical micelle concentration, micelle formation takes place. These micelles get adsorb at the dirt or grease

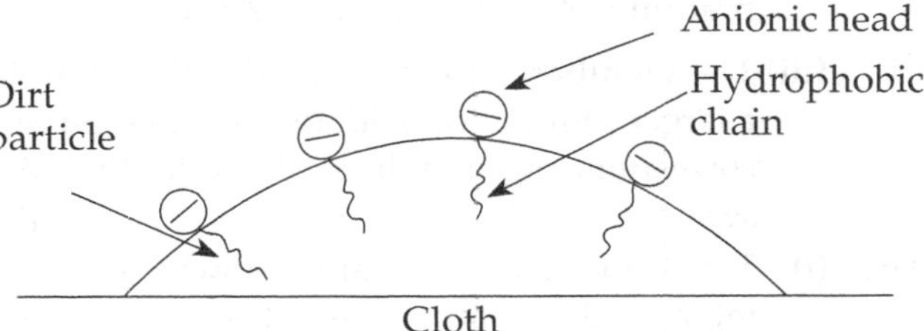


In such a manner that hydrophobic end get adsorbs and anionic head remains at the surface. Therefore dirt particle become negatively charged and when rinsing is done, this particle moves along with water and clothes become free from dirt or grease.

In soap sodium or potassium salts of fatty acids are present *e.g.*, $C_{17}H_{35}COO^-Na^+$ (sodium stearate) when it is added in hard water, presence of calcium or magnesium salts makes insoluble salts of Ca or Mg with carboxylate ion called scum which stick to cloth as gummy mass.

**20. A voltaic cell is set up at 25°C with the following half cells :**

**$Al/Al^{3+}$ (0.001 M) and $Ni/Ni^{2+}$ (0.50 M)**

**Write an equation for the reaction that occurs when the cell generates an electric current and determine the cell potential. [3]**

**$E_{Ni^{2+}/Ni} = -0.25$ V and $E^\circ_{Al^{3+}/Al} = -1.66$ V. ($\log 8 \times 10^{-6} = -0.54$)**

**Ans.**

$Al \rightarrow Al^{3+} + 3e^-$

$Ni^{2+} + 2e^- \rightarrow Ni$

$2Al + 3Ni^{2+} \rightarrow 2Al^{3+} + 3Ni$

$$E^\circ_{cell} = E^\circ_{red} - E^\circ_{ox}$$

$$E^\circ cell = (-0.25) - (-1.66)$$

$$= 1.41\ V$$

$$E^\circ_{Mn+/M} = E^\circ_{cell} = \frac{0.0591}{6} \log \frac{[Al^{+3}]^2}{[Ni^{+2}]^3}$$

$$E^\circ_{Mn+/M} = 1.41 = \frac{0.0591}{6} \log \frac{(1 \times 10^{-3})^2}{(5 \times 10^{-1})^3}$$

$$= 1.41 - 0.00985 \log \frac{1000 \times 10^{-3}}{125 \times 10^{-3}}$$

$$= 1.41 - 0.00985 \log 2 \times 10^{-6}$$

$$= 1.41 - (0.00985 \times - 0.54)$$

$$= 1.41 - 0.005319 = 1.415\ V$$

**23. Explain the following observations :**

**(i) Many of the transition elements are known to form interstitial compounds.**

**(ii) There is a general increase in density from titanium (Z = 22) to copper (Z = 29).**

**(iii) The members of the actinoid series exhibit a larger number of oxidation states than the corresponding members of the lanthanoid series. [3]**

**Ans. (i)** Transition elements are known to form interstitial compounds because these elements are capable of entrapping smaller atoms of other elements such as H,C and N in the interstitial sites in their crystal lattice.

**(ii)** It is due to increase in atomic mass whereas atomic size decreases from Ti to Cu. Therfore, density goes on increasing.

**(iii)** The members of the actinoid series exhibit a larger number of oxidation states than the corresponding members of the lanthanoid series because of the comparable energies of 5*f*, 6*d* and 7*s* orbitals in actinoids, they exhibit larger number of oxidation states than the corresponding members of lanthanoid series.

**27. Explain the following terms giving a suitable example for each : [3]**

**(i) Elastomers**

**(ii) Condensation polymers**

**(iii) Addition polymers.**

**Ans. (i) Elastomers:** Their polymer chain are held together by weakest intermolecular forces so that they can be stretched. That is why they exhibit elastic properties and are rubber-like solids. *eg.*, : Buna–S, Buna–N etc.

**(ii) Condensation Polymers :** They are formed by repeated condensation reaction between two different bi or tri-functional monomeric units with the elimination of smaller molecules such as water, alcohol, HCl etc. *e.g.*, : Nylon 6, 6.

**(iii) Addition Polymers :** They are formed by the repeated addition of monomer molecules having double or triple bonds. *eg*: Polyethylene.

**30. (a) Draw the structures of the following molecules: [5]**

**(i) $H_3PO_2$, (ii) $ClF_3$.**

**(b) Explain the following observations :**

**(i) Nitrogen is much less reactive than phosphorus.**

**(ii) Despite having greater polarity, hydrogen fluoride boils at a lower temperature than water.**

**(iii) Sulphur has a greater tendency for catenation than oxygen in the same group.**

**OR**

**(a) Draw the structures of the following molecules:**

**(i) $N_2O_5$, (ii) $HClO_4$.**

**(b) Explain the following observations:**

**(i) $H_2S$ is more acidic than $H_2O$.**

**(ii) Fluorine does not exhibit any positive oxidation state.**

**(iii) Helium forms no real chemical compound.**

**Ans. (a) (i)**

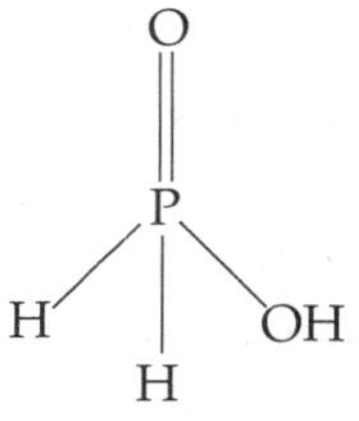

O
‖
P (bonded to H, OH, H)

(Hypophosphorous acid)

$H_3PO_2$

**(ii)**

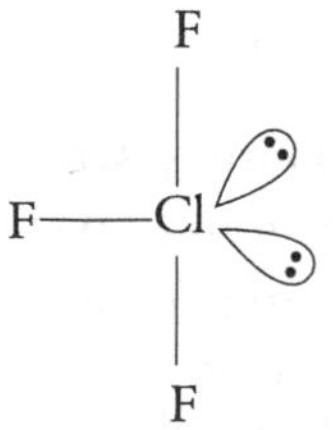

F
|
F—Cl (two lone pairs)
|
F

T shape, $ClF_3$

$sp^3d$ hybridisation

**(b) (i)** Due to N≡N triple bond in $N_2$ molecule, it is inert and hence is less reactive as compared to phosphorous which has P-P single bond,which makes it more reactive.

**(ii)** In case of HF, average two intermolecular hydrogen bonds are present. As a result, Vander Waals forces of attraction increases in water molecule and hence, boiling point increases.

**(iii)** Sulphur has a greater tendency for catenation than oxygen in the same group because S–S bond is stronger than O–O bonds as there is more interelectronic repulsion in O–O than in S–S.

**OR**

**(a) (i)**

Planar Structure ($N_2O_5$)

**(ii)**

Perchloric acid ($HClO_4$)

**(b) (i)** $H_2S$ is more acidic than $H_2O$ because bond dissociation enthalpy of H–S bond in $H_2S$ is less than that of H–O bond in $H_2O$ .

**(ii)** Flourine is the most electronegative element in nature and it has no *d*-orbitals and therefore, there is no scope for electron promotion. Hence, it can show only –1 oxidation state in its compounds.

**(iii)** Helium forms no real chemical compound because it is an inert gas and has very high ionization enthalpy, therefore no real chemical compound of helium is known.

## Outside Delhi [Set-III]

**Note : Except for the following questions, all the remaining questions have been asked in previous sets.**

**1. What are *n*-type semiconductors? [1]**

**Ans.** Semiconductors which are doped with Group 15 elements like phosphorous are *n*-type semiconductors.

**4. What is the basicity of $H_3PO_2$ acid and why? [1]**

**Ans.** $H_3PO_2$ is monobasic as it has one replaceable hydrogen atom.

**5. Write the IUPAC name of the following: [1]**

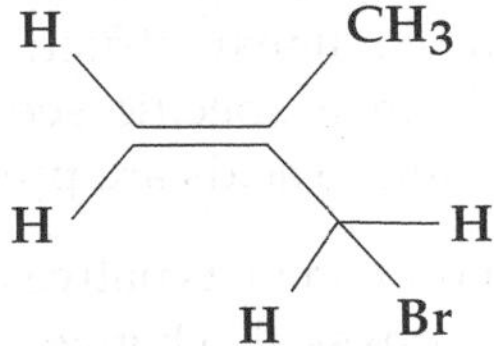


**Ans.** 3 Bromo-2-methyl propene.

**7. Write a reaction which shows that all the carbon atoms in glucose are linked in a straight chain. [1]**

**Ans.**

$$\underset{\text{Glucose}}{\begin{array}{l}CHO\\ |\\ (CHOH)_4\\ |\\ CH_2OH\end{array}} + HI \xrightarrow{\Delta} \underset{\text{n–hexane}}{CH_3-CH_2-CH_2-CH_2-CH_2-CH_3}$$

**8. What is the cause of a feeling of depression in human beings? Name a drug which can be useful in treating this depression. [1]**

**Ans.** The inability to achieve one's goal and extra work may cause the level of noradrenaline low, as a result the signal-sending activity becomes low and the person suffers from depression. Tranquilizers such as Equanil can help in treating depression.

**11. Explain the role of each of the following: [2]**

**(i) NaCN in the extraction of silver.**

**(ii) $SiO_2$ in the extraction of copper.**

**Ans. (i)** Dil. NaCN solution is used to leach silver (Ag) from silver ore in the presence of air. The silver metal is obtained by replacement.

$$4Ag + 8CN^- + 2H_2O + O_2 \rightarrow 4[Ag(CN)_2]^- + 4OH^-$$

$$2[Ag(CN)_2] + Zn \rightarrow [Zn(CN)_4]^2 + 2Ag$$

**(ii)** $SiO_2$ is used to remove impurity and it reacts with FeO to form easily removable feasible slag

$$FeO + SiO_2 \rightarrow \underset{\text{slag}}{FeSiO_3}$$

**22. Write three distinct features of chemisorption which are not found in physisorption. [3]**

**Ans.** The three distinct features of chemisorption are :

**(i)** It is irreversible.

**(ii)** It occurs by chemical bond formation and hence requires activation energy.

**(iii)** It is highly specific in nature.

**23. Explain each of the following observations: [3]**

**(i) With the same *d*-orbital configuration ($d^4$), $Cr^{2+}$ is a reducing agent while $Mn^{3+}$ is an oxidising agent.**

**(ii) Actinoids exhibit a much larger number of oxidation states than the lanthanoids.**

**(iii) There is hardly any increase in atomic size with increasing atomic numbers in a series of transition metals.**

**Ans. (i)** $Cr^{2+}$ is reducing because when it loses one electron to form $Cr^{3+}$: $[Ar]3d^3$, it has three unpaired electrons in lower energy *d*-orbitals which are more stable whereas $M^{3+}$ is oxidizing because after gaining one electron it becomes $Mn^{2+}$ which has more stable electronic configuration due to half-filled *d*-orbitals $Mn^{2+}$:$[Ar]3d^5$

**(ii)** Due to comparable energies of 5*f*, 6*d* and 7*s* orbitals and unpaired electrons in these orbitals, actinoids exhibit much larger number of oxidation states than the lanthanoids.

**(iii)** There is hardly any increase in atomic size with increasing atomic numbers in a series of transition metals because along transition series, nuclear charge increases which tends to decrease the size but the addition of electrons in the penultimate *d*-subshell increases the screening effect which counter balances the effect of increased nuclear charge. Thus, atomic radii does not change.

**24. Name the following coordination entities and describe their structures : [3]**

**(i) $[Fe(CN)_6]^{4-}$ (ii) $[Cr(NH_3)_4Cl_2]^+$ (iii) $[Ni(CN)_4]^{2-}$**

**(Atomic Numbers Fe = 26, Cr = 24, Ni = 28)**

**Ans. (i)** $[Fe(CN)_6]^{4-}$: Hexacyanoferrate (II) ion.

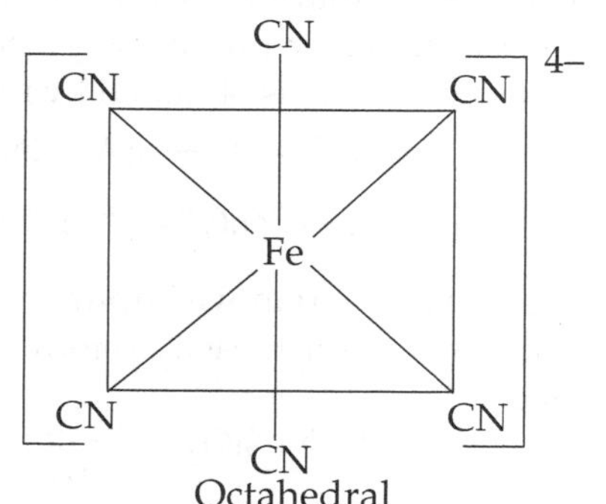

Octahedral

**(ii)** $[Cr(NH_3)_4Cl_2]^+$ : Tetraamminnedichlorido Chromium (III) ion\

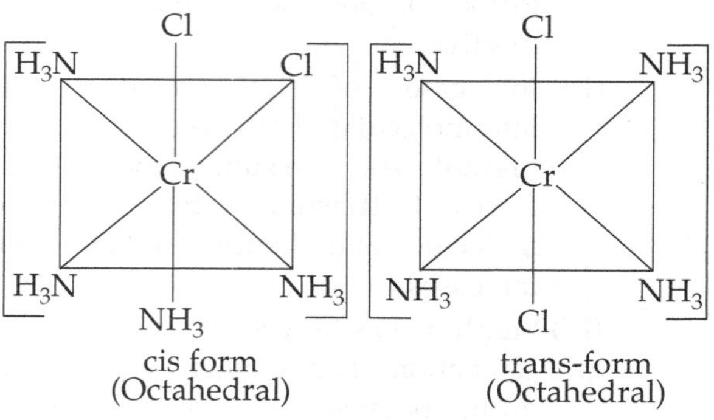

cis form (Octahedral) trans-form (Octahedral)

**(iii)** $[Ni(CN)_4]^{2-}$: Tetracyanonickelate (II) ion

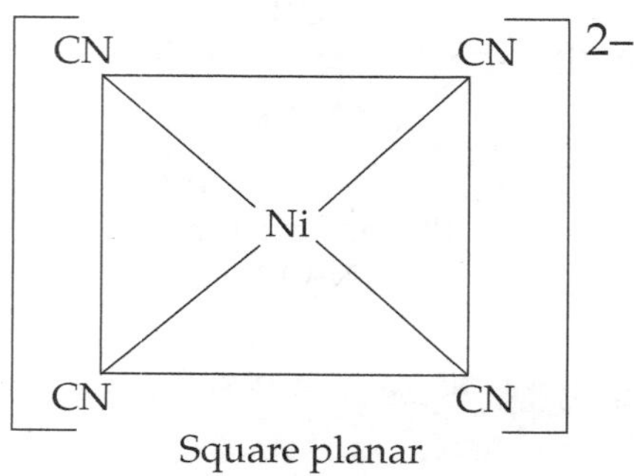

Square planar

**26. Define the following as related to proteins: [3]**

**(i) Peptide linkage**

**(ii) Primary structure**

**(iii) Denaturation.**

**Ans. (i)** Peptide linkage is the amide linkage between two amino acids to form proteins and polypeptides. CO-NH is peptide linkage.

**(ii)** The simple linear structure of a protein molecule in a specific sequence in which various amino acids are present.

**(iii)** When a protein in its native form is subjected to any physical change like change in temperature or chemical change like change in pH, the H-bonds gets disturbed. Due to this globules unfold and helix get uncoiled and protein loses its biological activity. This is called denaturation of protein.

●●

# CHEMISTRY

## Delhi [Set-I]

## 2012

**Time allowed : 3 hours** **Maximum marks : 70**

1. **What is meant by 'doping' in a semiconductor ?** [1]

**Ans.** Addition of impurities (like phosphorous and arsenic) to a semiconductor (like silicon) to improve the conductivity is called doping.

2. **What is the role of graphite in the electrometallurgy of aluminium ?** [1]

**Ans.** Graphite is used as anode and useful for the reduction of $Al_2O_3$ into Al.

3. **Which one of $PCl_4^+$ and $PCl_4^-$ is not likely to exist and why ?** [1]

**Ans.** The oxidation state of P in $PCl_4^+$ is +5 while that in $PCl_4^-$ is +3. As we move down the group, the stability of +5 oxidation state decreases and +3 oxidation state increases. Chlorine belongs to third period. Hence, $PCl_4^+$ compound is more likely to exist.

4. **Give the IUPAC name of the following compound.**

$$CH_2=\underset{\underset{CH_3}{|}}{C}-CH_2Br$$

**Ans.** 3-bromo-2-methylpropene.

5. **Draw the structural formula of 2-methyl-propan-2ol molecule.** [1]

**Ans.** $$H_3C-\underset{\underset{OH}{|}}{\overset{\overset{CH_3}{|}}{C}}-CH_3$$

6. **Arrange the following compounds in an increasing order of their reactivity in nucleophilic addition reactions: ethanol, propanal, propanone, butanone.**

**Ans.** Butanone < Propanone < Proponal < Ethanol

7. **Arrange the following in the decreasing order of their basic strength in aqueous solutions:**

**$CH_3NH_2$, $(CH_3)_2NH$, $(CH_3)_3N$ and $NH_3$** [1]

**Ans.** $(CH_3)_2NH > CH_3NH_2 > (CH_3)_3N > NH_3$

8. **Define the term, 'homopolymerisation' giving an example.** [1]

**Ans.** Polymerisation involving the presence of only one monomer is called homopolymerisation. *e.g.*, polyethene is a homopolymer

$$\underset{\text{Ethene}}{n(CH_2=CH_2)} \rightarrow \underset{\text{Polyethene}}{\text{+}CH_2-CH_2\text{+}_n}$$

9. **A 1.00 molal aqueous solution of trichloroacetic acid ($CCl_3COOH$) is heated to its boiling point. The solution has the boiling point of 100.18 °C. Determine the Van't Hoff factor for trichloroacetic acid. ($K_b$ for water = 0.512 K kg mol$^{-1}$)** [2]

**OR**

**Define the following terms :**

**(i) Mole fraction** **(ii) Isotonic solutions**
**(iii) Van't Hoff factor** **(iv) Ideal solution**

**Ans.** Given,

$\Delta T_b = 373.18-373. = 0.18$ K.

$K_b = 0.512$ K kg mol$^{-1}$

$m = 1$

$\Delta T_b = i K_b m$

$$i = \frac{\Delta T_b}{K_b m} = \frac{0.18}{0.512 \times 1} = 0.35$$

**OR**

(i) Ratio of the number of moles of a component in a mixture to the total number of moles in the mixture is called the mole fraction of that component. It is denoted by '$x$'.

(ii) Two solutions having the same molar concentration are said to be isotonic solutions. *e.g.*, All intravenous injections must be isotonic with body fluids.

(iii) The ratio of observed colligative property to the calculated colligative property is called the Van't Hoff factor. It is denoted by '$i$'.

(iv) Solutions that follow Raoult's law at all temperatures and concentrations are called ideal solutions.

10. **What do you understand by the 'order of a reaction'? Identify the reaction order from each of the following units of reaction rate constant:** [2]

**(i) L$^{-1}$ mol s$^{-1}$** **(ii) L mol$^{-1}$ s$^{-1}$**

**Ans.** The sum of the powers to which the concentration of reactants are raised in the rate law expression is called the order of a reaction.

(i) Zero order reaction

(ii) Second order reaction.

**11. Name the two groups into which phenomenon of catalysis can be divided. Give an example of each group with the chemical equation involved. [2]**

**Ans.** Catalysis can be positive, that is, it increases the rate of the reaction or negative *i.e.*, decreases the rate of reaction. Depending on the phase of the reactants and the catalyst, catalysis can be:

1. **Homogenous Catalysis:** The reactants and catalyst are in the same phase.

   *e.g.*, $2SO_2(g) + O_2(g) \xrightarrow{NO(g)} 2SO_3(g)$

2. **Heterogeneous catalysis :** The reactants and catalyst are in different phase.

$$4NH_3\ (g) + 5SO_2\ (g) \xrightarrow{Pt(s)} 4NO + 5S + 6H_2O$$

**12. What is meant by coagulation of a colloidal solution? Describe briefly any three methods by which coagulation of lyophobic sols can be carried out. [2]**

**Ans.** The process of setting of colloidal particles is called coagulation of sol. Methods of coagulation are :

1. **Electrophoresis:** In this process, the colloidal particles move towards opposite charged electrodes and get discharged and precipitated.
2. **Mixing two Oppositely Charged Sols:** Equal proportions of oppositely charged sols are mixed, they get neutralized and get precipitated.
3. **Dialysis:** Electrolytes are removed from the sol and colloid becomes unstable and gets coagulated.

**13. Describe the principle involved in each of the following processes. [2]**

**(i) Mond process for refining of Nickel.**

**(ii) Column chromatography for purification of rare elements.**

**Ans. (i)** Nickel combined with carbon monoxide to form volatile complex which is further decomposed to get back pure nickel.

$$Ni(g) + 4CO(g) \xrightarrow{330\text{-}350K} \underset{\text{Nickel tetracarbonyl}}{Ni(CO)_4(g)}$$

$$Ni(CO)_4(g) \xrightarrow[450-470K]{\Delta} \underset{\text{Nickel}}{Ni} + 4CO$$

**(ii)** The basic principle involved in column chromatography is that different elements present in a mixture are adsorbed on adsorbent at different extents.

**14. Explain the following giving an appropriate reason in each case. [2]**

**(i) $O_2$ and $F_2$ both stabilize higher oxidation states of metals but $O_2$ exceeds $F_2$ in doing so.**

**(ii) Structures of xenon fluorides cannot be explained by Valence Bond approach.**

**Ans. (i)** Due to the difference in atomic size of oxygen and fluorine and the property of oxygen to form multiple bonds with metals, $O_2$ exceeds $F_2$, stabilize higher oxidation states.

**(ii)** For explaining the structures of xenon fluorides, we need to use VSEPR and hybridization theories because in VBT, covalent bonds are formed by overlapping of half filled atomic orbital. But xenon has fully filled electronic configuration.

**15. Complete the following chemical equations: [2]**

**(i) $Cr_2O_7^{2-} + H^+ + I^-$**

**(ii) $MnO_4^- + NO_2^- + H^+ \rightarrow$**

**Ans. (i)** $Cr_2O_7^{2-} + 14H^+ + 6I^- \rightarrow 2Cr^{3+} + 7H_2O + 3I_2$

**(ii)** $2MnO_4^- + 5NO_2^- + 6H^+ \rightarrow 2Mn^{2+} + 3H_2O + 5O_3^-$

**16. What is meant by (i) peptide linkage (ii) biocatalysts ? [2]**

**Ans. (i)** Peptides linkage is the amide bond that helps to connect amino acids to form proteins. It is formed between –COOH and $-NH_2$ group of two amino acids with the loss of water molecule.

**(ii)** Biocatalysts are enzymes that catalyses the biochemical reactions in the bodies of living organisms. *e.g.*, Amylase.

**17. Write any two reactions of glucose which cannot be explained by the open chain structure of glucose molecule. [2]**

**Ans.** Two reactions which can't be explained by open chain structure of glucose are :

(i) Despite having the aldehyde group, glucose does not gives 2, 4 –DNP test.

(ii) The pentaacetate of glucose does not react with hydroxylamine indicating the absence of free –CHO group.

**18. Draw the structure of the monomer for each of the following polymers : [2]**

**(i) Nylon 6,(ii) Polypropene.**

**Ans.**

(i) Caprolactum: ring with $H_2C$–N(H)–C=O, $H_2C$, $CH_2$, $H_2C$–$CH_2$

(ii) $CH_3$—CH=$CH_2$
Propene

**19. Tungsten crystallizes in body centered cubic unit cell. If the edge of the unit cell is 316.5 pm, what is the radius of tungsten atom? [3]**

**OR**

**Iron has a body centered cubic unit cell with a cell dimension of 286.65 pm. The density of iron is 7.874 g cm$^{-3}$. Use this information to calculate Avogadro's number. (At mass of Fe = 55.845 u)**

**Ans.** Given, $a$ = 316.5 pm.

For bcc lattice,

$$r = \frac{\sqrt{3}}{4}a$$

$$r = \frac{\sqrt{3}}{4} \times 316.5 \text{ pm} = 137.04 \text{ pm}$$

**OR**

Given,

$$a = 286.65 \text{ pm} = 286.65 \times 10^{-10}\text{c}$$
$$d = 7.874 \text{ g cm}^{-3}$$

Atomic mass of Fe = 55.8454

$$N_A = ?$$

$$N_A = \frac{Zm}{a^3 d}$$

$$N_A = \frac{2 \times 55.845}{(286.65 \times 10^{-10})^3 \times 7.875}$$

$$N_A = 6.022 \times 10^{23} \text{ mol}^{-1}$$

**20. Calculate the amount of KCl which must be added to 1 kg of water so that the freezing point is depressed by 2K. ($K_f$ for water = 1.86 K kg mol$^{-1}$) [3]**

**Ans.** Given,

$$K_f = 1.86 \text{ K kg mol}^{-1}$$
$$i = 2, \Delta T_f = 2\text{K}$$
$$M = 74.5$$
$$\Delta T_f = iMK_f$$
$$2 = 2 \times 1.86, \frac{\text{Mass of KCl}}{74.5}$$
$$\text{Mass of KCl} = \frac{74.5}{1.86} = 40.05 \text{ gm}$$

**21. For the reaction [3]**

$$2NO(g) + Cl_2(g) \longrightarrow 2NOCl(g)$$

**the following data were collected. All the measurements were taken at 263 K:**

| Experiment No. | Initial [NO] (M) | Initial [$C_2$] (M) | Initial rate of disappearance of $Cl_2$ (M/min) |
|---|---|---|---|
| 1 | 0.15 | 0.15 | 0.60 |
| 2 | 0.15 | 0.30 | 1.20 |
| 3 | 0.30 | 0.15 | 2.40 |
| 4 | 0.25 | 0.25 | ? |

**(a) Write the expression for rate law.**

**(b) Calculate the value of rate constant and specify its units.**

**(c) What is the initial rate of disappearance of $Cl_2$ in exp. 4 ?**

**Ans.** (a) Rate law = $k[NO]^2[Cl_2]^1$

(b)
$$R = k[NO]^2[Cl_2]^1$$
$$k = \frac{R}{[NO]^2} = \frac{0.60}{(0.15)^2(0.15)^1}$$
$$= \frac{0.60}{(0.15)^3} = 177.75 \text{ mol}^{-2}\text{L}^2 \text{ min}^{-1}$$

(c)
$$R_4 = k[NO]^2[Cl_2]^1$$
$$= 177.75 \times (0.25)^2 (0.25)$$
$$= 2.78 \text{ M/min.}$$

**22. How would you account for the following? [3]**

**(i) Many of the transition elements are known to form interstitial compounds.**

**(ii) The metallic radii of the third (5*d*) series of transition metals are virtually the same as those of the corresponding group members of the second (4*d*) series.**

**(iii) Lanthanoids form primarily +3 ions, while the actinoids usually have higher oxidation states in their compounds, +4 or even +6 being typical.**

**Ans.** (i) Transition metal lattice have voids and hence these voids can trap small atoms like H, C and N to form interstitial compounds.

(ii) It is because electrons first fill the 4*f* orbitals and than the 5*d*-orbitals causing decline in radii of third (5*d*) series.

(iii) This is because of the comparable energies of 5*f*, 6*d* and 7*s* orbitals so all of them can participate

**23. Give the formula of each of the following coordination entities: [3]**

**(i) $Co^{3+}$ ion is bound to one $Cl^-$, one $NH_3$ molecule and two bidentate ethylene diamine (en) molecules.**

**(ii) $Ni^{2+}$ ion is bound to two water molecules and two oxalate ions.**

**Write the name and magnetic behavior of each of the above coordination entities.**

**(At. Nos. Co = 27, Ni = 28)**

**Ans.** (i) $[Co(NH_3)(en)_2Cl]^{2+}$ Amminebis (ethylenediamine) chlorido cobalt (III) ion, diamagnetic.

(ii) $[Ni(H_2O)_2(C_2O_4)_2]^{2-}$

Diaquadioxalatonickelate (II) ion, paramagnetic.

**24. Although chlorine is an electron withdrawing group, yet it is ortho, para-directing in electrophilic aromatic substitution reactions. Explain why it is so?** **[3]**

**Ans.** Due to resonance, electron density is increased on ortho and para positions for electron donating (via resonance) groups like chlorine. Thus chlorine is an electron withdrawing group being electrophilic it is ortho, para-direction in aromatic substitution reaction.

**25. Draw the structure and name the product formed if the following alcohols are oxidized. Assume that an excess of oxidizing agent is used.** **[3]**

**(i) $CH_3CH_2CH_2CH_2OH$, (ii) 2-butanol**

**(iii) 2-methyl-l-propanol**

**Ans.** (i)

$$\underset{\text{Butan-1-ol}}{CH_3\text{–}CH_2\text{–}CH_2\text{–}CH_2\text{–}OH} \xrightarrow[CrO_3\text{–}H_2SO_4]{[O]} \underset{\text{Butanoic acid}}{CH_3\text{–}CH_2\text{–}CH_2\text{–}COOH}$$

(ii)

$$\underset{\text{But-2-enol}}{CH_3\text{–}CH{=}CH_2\text{–}CH_2\text{–}OH} \xrightarrow[PCC]{[O]} \underset{\text{Butanoic acid}}{CH_3\text{–}CH{=}CH\text{–}CHO}$$

(iii)

$$\underset{\text{2-methyl propan-1-ol}}{CH_3\text{–}\overset{\overset{CH_3}{|}}{C}H\text{–}CH_2\text{–}OH} \xrightarrow{[O]} \underset{\text{2-MethylPropanoic acid}}{CH_3\text{–}\overset{\overset{CH_3}{|}}{C}H\text{–}CH\text{–}COOH}$$

**26. Write chemical equations for the following conversions :** **[3]**

**(i) Nitrobenzene to benzoic acid.**

**(ii) Benzyl chloride to 2-phenylethanamine.**

**(iii) Aniline to benzyl alcohol.**

**Ans.** (i)

$$\underset{\text{(Nitrobenzene)}}{(C_6H_5NO_2} \xrightarrow[BaSO_4]{H_2/Pd} \underset{\text{Aniline}}{C_6H_5NH_2} \xrightarrow[HCl]{NaNO_2}$$

$$\underset{\text{Benzenediazonium chloride}}{C_6H_5N_2^+Cl^-)} \longrightarrow CuCN$$

$$\underset{\text{Benzoic acid}}{C_6H_5COOH} \xleftarrow[H^+]{H_2O} \underset{\text{Benzonitrile}}{C_6H_5 - CN}$$

(ii)

$$\underset{\text{Benzylcholride}}{C_6H_5-CH_2Cl} + KCN \longrightarrow \underset{\text{Phenylethanenitrile}}{C_6H_5-CH_2-CN}$$

$$\downarrow LiAlH_4$$

$$\underset{\text{2-phyenylethanamine}}{C_6H_5CH_2-\underset{\underset{NH_2}{|}}{CH_2}}$$

(iii)

$$\underset{\text{Aniline}}{C_6H_5NH_2} \xrightarrow[HCl]{NaNO_2} \underset{\text{Benzenediazonium}}{C_6H_5N_2Cl^-} \xrightarrow{CuCN}$$

CN-substituted benzene ring: Benzonitrile

Benzonitrile → Benzoic acid (COOH on benzene ring) $\xrightarrow{LiAlH_4}$ Benzylalcohol ($CH_4ON$ on ring)

**27. What are the following substances? Give one example of each one of them.** **[3]**

**(i) Tranquilizers**

**(ii) Food preservatives**

**(iii) Synthetic detergents.**

**Ans.** (i) Tranquilizers are drugs that act on the central nervous system to get relief from anxiety, stress etc. They are used in treatment of stress related mental disorders *eg.*,Equanil.

(ii) These are chemicals used to preserve food by protecting it against microbial growth. *e.g.*, Sodium benzoate.

(iii) Synthetic detergents have all the properties of soap but they do not precipitate in hard water. *e.g.*, Sodium *p*-dodecyl benzenesulphonate.

**28. (a) What type of a battery is the lead storage battery? Write the anode and the cathode reactions and the overall reactions occurring in a lead storage battery when current is drawn from it.** **[2],[3]**

**(b) In the button cell, widely used in watches, the following reaction takes place.**

$$Zn(s) + Ag_2O(s) + H_2O(l) \rightarrow Zn^{2+}(aq) + 2Ag(s) + 2OH^-(aq)$$

**Determine E° and ΔG° for the reaction**

**(Given : $E^\circ_{Ag^+/Ag} = +0.80V, E^\circ_{Zn^{2+}/Zn} = -0.76\ V$)**

OR

**(a) Define molar conductivity of a solution and explain how molar conductivity changes with change in concentration of solution for a weak and a strong electrolyte.**

**(b) The resistance of a conductivity cell containing 0.001 M KCl solution at 298 K is 1500 Ω. What is the cell constant if the conductivity of 0.001 M KCl solution at 298 K is $0.146 \times 10^{-3}$ S cm$^{-1}$?**

**Ans. (a)** Lead storage battery is secondary cell and can be recharged by passing current through it.

**At anode :** $Pb_{(s)} + SO_4^{2-}{}_{(aq)} \longrightarrow PbSO_{4(s)} + 2e^-$

**At cathode :** $PbO_{2(s)} + SO_4^{2-}{}_{(aq)} + 4H^+ + 2e^- \longrightarrow PbSO_{4(s)} + 2H_2O$

**Overall cell reaction :**

$Pb + PbO_2 + H_2SO_4\,(aq) \rightarrow 2PbSO_4 + 2H_2O$

**(b)** $E°cell = 0.8 - (-0.76) = 1.56$ V

$\Delta G° = -nFE°_{Cell}$

$= -2 \times 96500 \times 1.56$

$= -301$ kJ mol$^{-1}$

OR

**(a)** Molar conductivity of a solution is the conductance of the solution with one mole of electrolyte placed between electrodes 1 cm apart.

With dilution, molar conductivity of weak electrolytes increases sharply and for strong electrolytes also, it increases.

**(b)** $k = 0.146 \times 10^{-3}$ cm$^{-1}\Omega^{-1}$

$R = 1500\Omega$

$$\text{Specific conductance} = \frac{\text{Cell constant}}{\text{Resistance}}$$

Cell constant $= \kappa R$

$= 0.146 \times 10^{-3} \times 1500$

$= 0.219$ cm$^{-1}$

**29. (a) Complete the following chemical reactions. equations :** **[2],[3]**

**(i) $P_4 + SO_2Cl_2 \rightarrow$**

**(ii) $XeF_6 + H_2O \rightarrow$**

**(b) Predict the shape and asked the angle (90° or more or less) in each of the following cases :**

**(i) $SO_3^{2-}$ and the angle O–S–O**

**(ii) $ClF_3$ and the angle F–Cl–F**

**(iii) $XeF_2$ and the angle F–Xe–F**

OR

**(a) Complete the following chemical equations:**

**(i) $NaOH + Cl_2 \rightarrow$**

**(ii) $XeF_4 + O_2F_2 \rightarrow$**

**(b) Draw the structures of the following molecules :**

**(i) $H_3PO_2$**

**(ii) $H_2S_2O_7$**

**(iii) $XeOF_4$**

**Ans. (a) (i)** $P_4 + 10SO_2Cl_2 \rightarrow 4PCl_5 + 10SO_2$

**(ii)** $XeF_6 + 3H_2O \rightarrow XeO_3 + 6HF$

**(b) (i)** **(ii)**

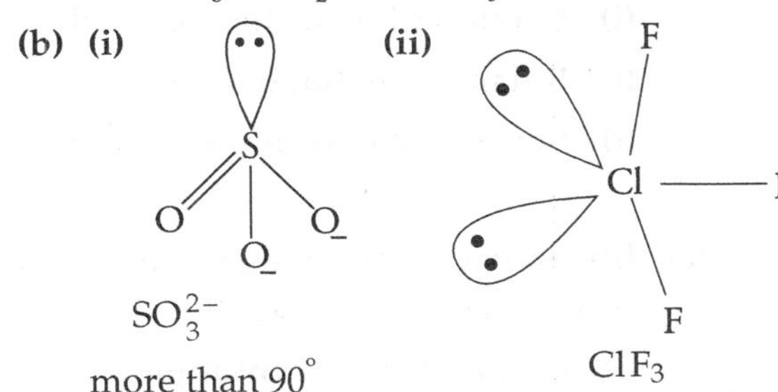


$SO_3^{2-}$
more than 90°

$ClF_3$
less than 90°

**(iii)**

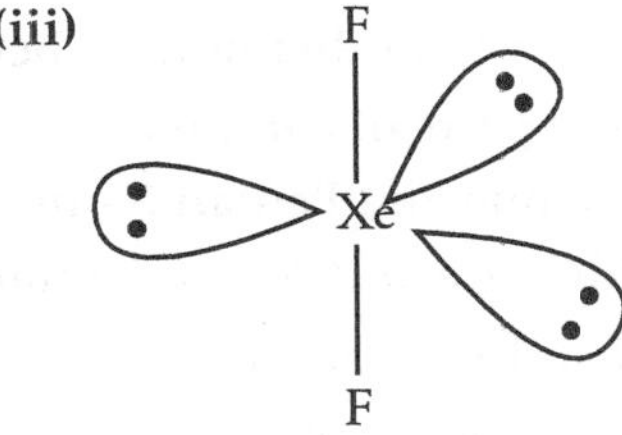


$XeF_2$
more than 90°

OR

**(a) (i)** $6NaOH + 3Cl_2 \rightarrow 5NaCl + NaClO_3 + 3H_2O$

**(ii)** $XeF_4 + O_2F_2 \rightarrow XeF_6 + O_2$

**(b) (i)**

O
‖
P
H / | \ OH
H

$H_3PO_2$

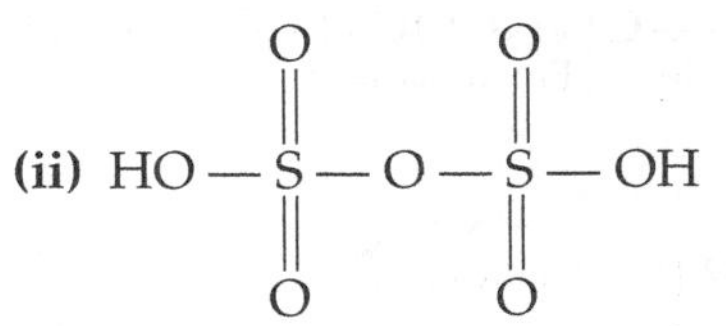


Disulphuric acid
$H_2S_2O_7$

(iii)

$XeOF_4$ (square pyramidal structure: O double-bonded to Xe at top, four F atoms around Xe, lone pair at bottom)

**30. (a) Illustrate the following name reactions giving suitable example in each case: [2],[3]**

**(i) Clemmensen reduction**

**(ii) Hell-Volhard-Zelinsky reaction**

**(b) How are the following conversions carried out ?**

**(i) Ethylcyanide to ethanoic acid**

**(ii) Butan-l-ol to butanoic acid**

**(iii) Benzoic acid to *m*-bromobenzoic acid**

**OR**

**(a) Illustrate the following reactions giving a suitable example for each:**

**(i) Cross aldol condensation,**

**(ii) Decarboxylation**

**(b) Give simple tests to distinguish between the following pairs of compounds:**

**(i) Pentan-2-one and Pentan-3-one**

**(ii) Benzaldehyde and Acetophenone**

**(iii) Phenol and Benzoic acid**

**Ans. (a)** (i) Aldehydes and ketones are reduced to $CH_2$ group on treatment with zinc-amalgam and conc. HCl.

$$(H_3C)_2C{=}O \xrightarrow[HCl]{Zn-Hg} CH_3CH_2CH_3 + H_2O$$

(ii) Carboxylic acids containing α-hydrogen atom gives halo carboxylic acids on treatment with halogens in the presence of red phosphorous.

$$CH_3COOH + Cl_2 \xrightarrow{\text{Red P}} ClCH_2COOH + HCl$$

**(b) (i)**

$$\underset{\text{Ethylcyanide}}{C_2H_5CN} \xrightarrow{H^+} \underset{\text{Propanoic acid}}{C_2H_5COOH} \xrightarrow{NH_3/\Delta^+} C_2H_5CONH_2 \xrightarrow[KOH]{Br_2}$$

$$\underset{\text{Ethylamine}}{C_2H_5NH_2} \xrightarrow[\text{Ethyamine}]{HNO_2} \underset{\text{Ethylalcohol}}{C_2H_5OH} \xrightarrow[KMnO_4]{\text{alkaline}} \underset{\text{Ethanoic acid}}{CH_3COOH}$$

(ii)

$$\underset{\text{Butanol}}{CH_3CH_2CH_2CH_2OH} \xrightarrow[K_2Cr_2O_7]{[O]} \underset{\text{Butanoic acid}}{CH_3CH_2CH_2COOH}$$

(iii)

$$\underset{\text{Benzoic acid}}{C_6H_5COOH} \xrightarrow{Br_2/FeBr_3} \underset{m\text{-Bromobenzoic acid}}{m\text{-}BrC_6H_4COOH}$$

**OR**

**(a)** (i) When aldol condensation is carried out between two different aldehydes or ketones it is called cross aldol condensation.

$$CH_3CHO + CH_3CH_2OH \xrightarrow[2]{1.NaOH}$$

$$\underset{\text{But-2-enal}}{CH_3-CH=CH-CHO} + \underset{\text{2 Methyl pent-2enal}}{CH_3CH_2-CH=C(CH_3)-CHO}$$

$$+ \underset{\text{2-Methyl but-2 enal}}{CH_3\ CH\ C(CH_3)\ CHO} + \underset{\text{Pent-2-enal}}{CH_3CH_2CH=CHCHO}$$

(Cross Aldol Product)

(ii) Carboxylic acid lose carbon dioxide to form hydrocarbon, when their salts are heated with sodium. The reaction is known as decarboxylation reaction

$$R\text{—}COONa \xrightarrow[\text{Heat}]{NaOH\ \&\ CaO} R\text{—}H + Na_2CO_3$$

**(b)** (i) Pentan-2-one gives iodoform test but pentan-3-one does not.

$$\underset{\text{Pentan–2–one}}{CH_3CH_2CH_2\overset{O}{\overset{\|}{C}}CH_3} + 3NaOI \rightarrow CH_3CH_2CH_2COONa + \underset{\substack{\text{Iodoform}\\\text{(yellow ppt)}}}{CHI_3\downarrow} + 2NaOH$$

$$\underset{\text{Pentan–3–one}}{CH_3CH_2\overset{O}{\overset{\|}{C}}CH_2CH_3} + NaOI \rightarrow \text{No yellow ppt. of Iodoform.}$$

(ii) Benzaldehyde does not gives iodoform test but Acetophenone gives iodoform test.

$$\underset{\text{Acetophenone}}{C_6H_5COCH_3} + 3NaOI \rightarrow \underset{\substack{\text{Iodoform}\\\text{(yellow ppt.)}}}{CHI_3\downarrow} + C_6H_5COONa + 2NaOH$$

$$\underset{\text{Benzaldehyde}}{C_6H_5CHO} + NaOI \rightarrow \text{No yellow ppt. of Iodoform}$$

(iii) Phenol gives violet colour with neutral $FeCl_3$ solution but benzoic acid does not.

$$\underset{\text{Phenol}}{6C_6H_5OH} + FeCl_3 \rightarrow \underset{\text{Violet colour}}{[Fe(OC_6H_5)_6]^{3-}} + 3H^+ + 3HCl$$

$$\underset{\text{Benzoic-acid}}{3C_6H_5COOH} + FeCl_3 \rightarrow \underset{\text{Buff colour}}{(C_6H_5COO)_3Fe} + 3HCl$$

●●

## Delhi [Set-II]

**Note : Except for the following questions, all the remaining questions have been asked in previous set.**

1. **Write a point of distinction between a metallic solid and an ionic solid other than metallic luster.** **[1]**

**Ans.** Metallic solids do not ionize in aqueous state but ionic solids ionize immediately.

11. **Describe a conspicuous change observed when:** **[2]**
   (i) **a solution of NaCl is added to a sol of hydrated ferric oxide.**
   (ii) **a beam of light is passed through a solution of NaCl and then through a sol.**

**Ans.** (i) Coagulation of ferric hydroxide sol. would take place.

(ii) NaCl solution is transparent so when beam of light is passed, no tyndall effect is produced. But on passing through solution the path of light becomes visible due to Tyndall effect.

13. **Describe the following:** **[2]**
   (i) **The role of cryolite in electro-metallurgy of aluminium.**
   (ii) **The role of carbon monoxide in the refining of crude nickel.**

**Ans.** (i) Cryolite lowers the melting point of the mixture and brings conductivity. Therefore, it is mixed with alumina during metallurgy of aluminium.

(ii) Carbon monoxide forms a volatile complex with nickel. The volatile complex is then subjected to high temperature to get pure metal through decomposition.

$$\underset{\text{Impure}}{Ni} + 4CO \xrightarrow{300-350K} \underset{\text{(Nickel tetracarbonyl)}}{Ni(CO)_4}$$

$$Ni(CO_4) \xrightarrow{400-450K} \underset{\text{Pure}}{Ni} + 4CO$$

14. **What is meant by (i) peptide linkage, (ii) biocatalysts?** **[2]**

**Ans.** (i) **Peptide Linkage:** It is the linkage between amino acids, formed due to loss of a water molecule and amide linkage is formed.

$$\left(-\overset{\overset{\displaystyle O}{\|}}{C}-NH-\right)$$

(ii) **Biocatalysts:** Biocatalysts are enzymes which are used to perform chemical reactions on organic compounds.

18. **Write the main structural difference between DNA and RNA. Of the two bases, thymine and uracil, which one is present in DNA?** **[2]**

**Ans.**

| DNA | RNA |
|---|---|
| 1. DNA has deoxyribose sugar. | RNA has ribose sugar. |
| 2. It has double helical structure. | It is single stranded. |
| 3. It is less reactive due to presence of C-H bonds. | It has larger groo-ves so it is more reactive. |

DNA contains thymine base and uracil base is present in RNA.

23. **How would you account for the following?** **[3]**
   (i) **With the same d-orbital configuration ($d^4$) $Cr^{2+}$ is a reducing agent while $Mn^{3+}$ is an oxidizing agent.**
   (ii) **The actinoids exhibit a larger number of oxidation states than the corresponding members in the lanthanoid series.**
   (iii) **Most of the transition metal ions exhibit characteristic in colours in aqueous solutions.**

**Ans.** (i) $Cr^{2+}$ has $d^4$ configuration. It gets oxidized to $Cr^{3+}$ with electronic configuration $d^3$ which is more stable. Therefore, $Cr^{2+}$ is a reducing agent. $Mn^{3+}$ has $d^4$ configuration. It gets reduced to $Mn^{+2}$ with $d^5$ configuration. This is half-filled d-orbital and is stable. Therefore $Mn^{3+}$ is an oxidizing agent.

(ii) The actinoids exhibit a larger number of oxidation states than the corresponding members in the lanthanoid series because 5*f*, 6*d* and 7*s* energy levels has small energy gap in the actinoid series. Due to these orbitals actinoids exhibit large number of oxidation states.

**(iii)** Due to partial absorption of visible light the electron from one orbital gets promoted to another orbital of the *d* subshell. Due to the presence of unpaired electrons transition metals are coloured.

**30. (a) Give a possible explanation for each one of the following: [2],[3]**

**(i) There are two $-NH_2$ groups in semicarbazide. However, only one such group is involved in the formation of semicarbazones.**

**(ii) Cyclohexanone forms cyanohydrin in good yield but 2, 4, 6-trimethylcyclohexanone does not.**

**(b) An organic compound with molecular formula $C_9H_{10}O$ forms 2, 4, -DNP derivative, reduces Tollens' reagent and undergoes Cannizzaro's reaction. On vigorous oxidation it gives 1, 2-benzene-di-carboxylic acid. Identify the compound.**

**OR**

**(a) Give chemical tests to distinguish between:**

**(i) Phenol and Benzoic acid**

**(ii) Acetophenone and Benzophenone**

**(b) Write the structures of the main products of following reactions:**

**(i)** $C_6H_6 + C_6H_5COCl \xrightarrow[CS_2]{\text{Anhydrous}AlCl_3}$

**(ii)** $H_3C-C\equiv C-H \xrightarrow{Hg^{2+}, H_2SO_4}$

**(iii)** p-$CH_3C_6H_4NO_2$ $\xrightarrow[2.\ H_3O^+]{1.\ CrO_2Cl_2}$

**Ans. (a) (i)** Semicarbazide show resonance involving one of the two $-NH_2$ groups, which is attached to the carboxyl carbon atom. Due to which electron density on $-NH_2$ group involved in resonance decreases. So it cannot act as nucleophile. Other $-NH_2$ group can act as nucleophile to produce semicarbazones with aldehydes and ketones.

**(ii)** In cyclohexanone $CN^-$ can easily attack without any steric hindrance. But in 2, 4, 6 – Trimethylcyclohexanone due to presence of methyl groups steric hindrance is produced and $CN^-$ cannot attack effectively.

**(b)** $C_9H_{10}O$ is aldehyde because it reduces Tollen's reagent. It undergoes Cannizaro's reaction therefore it is substituted benzaldehyde. It gives 1, 2, Benezene-di carboxylic acid. The compound is 2-Ethylbenzaldehyde.

$$C_6H_4(CHO)(C_2H_5) \xrightarrow{[O]} C_6H_4(COOH)_2$$

2Ethylbenzaldehyde → 1,2-Benzene-dicarboxylic acid

**OR**

**(a) (i)** Phenol gives violet colour with neutral $FeCl_3$ solution but benzoic acid does not.

$$\underset{\text{Phenol}}{6C_6H_5OH} + FeCl_3 \rightarrow \underset{\text{Violet colour}}{[Fe(OC_6H_5)]^{3-}} + 3H^+ + HCl$$

$$\underset{\text{Benzoic acid}}{3C_6H_5COOH} + FeCl_3 \rightarrow \underset{\text{Buff Colour}}{(C_6H_5COO)_3Fe} + 3HCl$$

**(ii)** Acetophenone gives iodoform test but benzophenone does not

$$\underset{\text{Acetophenone}}{C_6H_5COCH_3} + 3NaOI \xrightarrow{\Delta} C_6H_5COONa + \underset{\substack{\text{Iodoform} \\ \text{(yellow ppt)}}}{CHI_3} + 2NaOI$$

$$\underset{\text{Benzophenone}}{C_6H_5COC_6H_5} \xrightarrow{NaOI} \text{No yellow ppt of Iodoform}$$

**(b) (i)**

$$C_6H_6 + C_6H_5COCl \xrightarrow[CS_{32}]{\text{Anhyd.}AlCl_3} \underset{\text{Benzophenone}}{C_6H_5-\overset{O}{\overset{\|}{C}}-C_6H_5} + HCl$$

**(ii)**

$$CH_3-C\equiv C-H \xrightarrow[H_2SO_4]{Hg^{2+}} \underset{\text{Acetone}}{CH_3-\underset{\underset{O}{\|}}{C}-CH_3}$$

**(iii)**

$$\text{p-}CH_3C_6H_4NO_2 \xrightarrow[H_3O^+]{CrO_2Cl_2} \text{p-}OHC\text{-}C_6H_4NO_2$$

Para nitrobenzaldehyde

## Delhi [Set-III]

**Note : Except for the following questions, all the remaining questions have been asked in previous sets.**

**3. Out of $PH_3$ and $H_2S$, which is more acidic and why? [1]**

**Ans.** $H_2S$ is more acidic than $PH_3$ due to smaller size and higher electronegativity of sulphur. Therefore S-H bond is polar than P-H bond and easy to remove.

**5. Draw the structure of hex-1-en-3-ol compound. [1]**

**Ans.** $CH_3CH_2CH_2\underset{\displaystyle OH}{\underset{|}{C}H}-CH=CH_2$

**12. Explain the following terms giving one example for each: [2]**

**(i) Micelles**

**(ii) Aerosol.**

**Ans.** (i) Micelles are aggregates which exhibit colloidal behaviour at higher concentration, with the hydrophilic part outside and the hydrophobic part towards the oil and dirt particle. *e.g.*, soap.

(ii) An aerosol is a colloid of fine solid particle or liquid drops in air or another gas. It can be natural or artificial. e.g. dust particle and smoke.

**20. 15.0 g of an unknown molecular material was dissolved in 450 g of water. The resulting solution was found to freeze at –0.34 °C. What is the molar mass of this material ? ($K_f$ for water = 1.86 K kg mol$^{-1}$) [3]**

**Ans.** Given,

$W_1 = 15$ g, $W_2 = 450$ g

$K_f = 1.86$ K kg/mol

$\Delta T_f = -0.34$°C

$$\Delta T_f = \frac{K_f \times W_2 \times 1000}{W_1 \times M_2}$$

$$M_2 = \frac{K_f \times W_2 \times 1000}{W_1 \times \Delta T_f} = \frac{1.86 \times 15 \times 1000}{450 \times 0.34}$$

$M_2 = 182.35$ g / mol

**22. Explain the following observations giving an appropriate reason for each: [3]**

**(i) The enthalpies of atomization of transition elements are quite high.**

**(ii) There occurs much more frequent metal-metal bonding in compounds of heavy transition metals (*i.e.*, 3rd series).**

**(iii) $Mn^{2+}$ is much more resistant than $Fe^{2+}$ towards oxidation.**

**Ans.** (i) Due to the presence of metallic bonds as a result of large number of valence electrons. The enthalpies of atomization of transition elements are quite high.

(ii) The presence of valence electrons and unpaired *d*-orbital electrons help heavy transition metals to form metallic bonds.

(iii) Due to stability of $Mn^{2+}$ because of half filled-subshell ($3d^5$) it does not gets oxidized. But $Fe^{2+}$ has $3d^6$ configuration and it can lose one electron to become $3d^5$ which is stable. Therefore it is easily oxidized.

**23. Write the name, the structure and the magnetic behaviour of each of the following complexes: [3]**

**(i) $(Pt(NH_3)Cl(NO_2)]$, (ii) $(Co(NH_3)_4Cl_2]Cl$**

**(iii) $Ni(CO)_4$**

**(At. Nos. Co = 27, Ni = 28, Pt = 78)**

**Ans.** (i) Amminechloridonitrito–N-platinum (II) trigonal planar, diamagnetic.

(ii) Tetra-ammine dichlorido cobalt **(III)** chloride, octahedral, diamagnetic

(iii) Tetracarbonyl nickel (0), Tetrahedral, diamagnetic

**27. Explain the following terms giving one example of each type: [3]**

**(i) Antacids,**

**(ii) Disinfectants,**

**(iii) Enzymes.**

**Ans.** (i) Substances consumed to reduce acidity in the stomach by neutralizing excess HCl produced by the stomach. *e.g.* Milk of magnesia.

(ii) Disinfectants are chemicals used to kill microorganism, applied only to non-living objects like floors and drains. *e.g.* 1% phenol solution.

(iii) Enzymes are structurally globular proteins that catalyse biochemical reactions in living organisms *e.g.* Trypsin.

**30. (a) Draw the molecular structures of following compounds: [2],[3]**

**(i) $XeF_6$ (ii) $H_2S_2O_8$**

**(b) Explain the following observations:**

**(i) The molecules $NH_3$ and $NF_3$ have dipole moments which are of opposite direction.**

**(ii) All the bonds of $PCl_5$ molecule are not equivalent.**

**(iii) Sulphur in vapour state exhibits parama-gnetism.**

**OR**

**(a) Complete the following chemical equations:**

**(i) $XeF_4 + SbF_5 \rightarrow$**

**(ii) $Cl_2 + F_2$ (excess) $\rightarrow$**

**(b) Explain each of the following:**

**(i) Nitrogen is much less reactive than phosphorus.**

**(ii) The stability of +5 oxidation state decreases down group 15.**

**(iii) The bond angles (O-N-O) are not of the same value in $NO_2^-$ and $NO_2^+$.**

**Ans. (a) (i)**

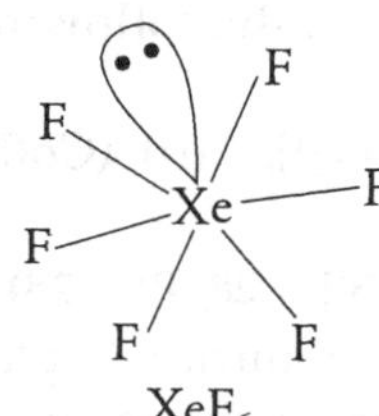


$XeF_6$

**(ii)**

(Structure: two S atoms each with two S=O, one OH/HO, linked by O—O)

$(H_2S_2O_8)$

**(b) (i)** Fluorine is more electronegative than nitrogen while hydrogen is less electronegative than nitrogen resulting in opposite dipole moments of $NH_3$ and $NF_3$. Dipole points towards N in $NH_3$ and towards F in $NF_3$.

**(ii)** $PCl_5$ has trigonal bipyramidal shape. The two axial bonds feel more repulsion from the three equatorial bond pairs and are hence longer. Therefore all bonds in $PCl_5$ are not equivalent.

**(iii)** Sulphur in vapour state exists as $S_2$ and hence like $O_2$ has two unpaired electrons in antibonding $\pi^*$ orbitals leading to paramagnetism.

**OR**

**(a) (i)** $XeF_4 + SbF_5 \rightarrow [XeF_3]^+[SbF_6]^-$

**(ii)** $Cl_2 + \underset{\text{(excess)}}{3F_2} \rightarrow 2ClF_3$

**(b) (i)** Due to the small size, absence of *d*-orbitals and presence of triple bond in nitrogen, it is less reactive than phosphorous.

**(ii)** Due to the increase in size and inert pair effect, +5 oxidation state decreases down the group.

**(iii)** Bond angles in $NO_2^-$ and $NO_2^+$ are not of same value because $NO_2^+$ is linear and $NO_2^-$ has bent shape.

●●

# MATHEMATICS

## Outside Delhi [Set-I] 2020

**Time allowed : 3 hours** **Maximum marks : 80**

**General Instructions :**

Read the following instructions very carefully and strictly follow them :

(i) This question paper comprises **four** sections : A, B, C and D.
This question paper carries **36** questions. **All** questions are compulsory.

(ii) **Section A** – Question no. **1** to **20** comprises of **20** questions of **one** mark each.

(iii) **Section B** – Question no. **21** to **26** comprises of **6** questions of **two** marks each.

(iv) **Section C** – Question no. **27** to **32** comprises of **6** questions of **four** marks each.

(v) **Section D** – Question no. **33** to **36** comprises of **4** questions of **six** marks each.

(vi) There is no overall choice in the question paper. However, an internal choice has been provided in 3 questions of one mark, 2 questions of two marks, 2 questions of four marks and 2 questions of six marks. Only one of the choices in such questions have to be attempted.

(vii) In addition to this, separate instructions are given with each section and question, wherever necessary.

(viii) Use of calculators is not permitted.

### SECTION–A

**Question numbers 1 to 10 are multiple choice questions of 1 mark each. Select the correct option :**

**1. The value of $\sin^{-1}\left(\cos\frac{3\pi}{5}\right)$ is**

(a) $\frac{\pi}{10}$ (b) $\frac{3\pi}{5}$

(c) $\frac{-\pi}{10}$ (d) $\frac{-3\pi}{5}$ [1]

**Sol.** $\sin^{-1}\left(\cos\frac{3\pi}{5}\right) = \frac{\pi}{2} - \cos^{-1}\left(\cos\frac{3\pi}{5}\right)$

$= \frac{\pi}{2} - \frac{3\pi}{5}$

$= \frac{5\pi - 6\pi}{10}$

$= \frac{-\pi}{10}$

$\therefore$ Option (c) is correct.

**2. If A = [2 – 3 4], B = $\begin{bmatrix}3\\2\\2\end{bmatrix}$, X= [1 2 3] and Y = $\begin{bmatrix}2\\3\\4\end{bmatrix}$, then AB + XY equals.**

**(a) [28]** **(b) [24]**

**(c) 28** **(d) 24** [1]

**Sol.** $AB + XY = [2\ \ -3\ \ 4]\begin{bmatrix}3\\2\\2\end{bmatrix} + [1\ 2\ 3]\begin{bmatrix}2\\3\\4\end{bmatrix}$

$= [8] + [20]$

$= [28]$

$\therefore$ Option (a) is correct.

**3. If $\begin{vmatrix}2 & 3 & 2\\ x & x & x\\ 4 & 9 & 1\end{vmatrix} + 3 = 0$, then the value of $x$ is**

**(a) 3** **(b) 0**

**(c) –1** **(d) 1** [1]

**Sol.** Given : $\begin{vmatrix}2 & 3 & 2\\ x & x & x\\ 4 & 9 & 1\end{vmatrix} + 3 = 0$

$\Rightarrow \quad 2(x - 9x) - 3(x - 4x) + 2(9x - 4x) + 3 = 0$

$\Rightarrow \quad -16x + 9x + 10x + 3 = 0$

$\Rightarrow \quad 3x = -3$

$\Rightarrow \quad x = -1$

$\therefore$ Option (c) is correct.

**4. $\int_0^{\pi/8} \tan^2(2x)\,dx$ is equal to**

(a) $\frac{4-\pi}{8}$ (b) $\frac{4+\pi}{8}$

(c) $\frac{4-\pi}{4}$ (d) $\frac{4-\pi}{2}$ [1]

**Sol.** $\int_0^{\pi/8} \tan^2(2x)dx = \int_0^{\pi/8} \left(\sec^2(2x) - 1\right)dx$

$$= \left[\frac{\tan(2x)}{2} - x\right]_0^{\pi/8}$$

$$= \frac{1}{2} - \frac{\pi}{8} - 0$$

$$= \frac{4-\pi}{8}$$

∴ Option (a) is correct.

**5. If $\vec{a}\cdot\vec{b} = \frac{1}{2}|\vec{a}||\vec{b}|$, then the angle between $\vec{a}$ and $\vec{b}$ is**

**(a) 0°** **(b) 30°**

**(c) 60°** **(d) 90°** **[1]**

**Sol.** $\vec{a}.\vec{b} = \frac{1}{2}|\vec{a}||\vec{b}| = |\vec{a}||\vec{b}|\cos\theta$

$\Rightarrow \quad \cos\theta = \frac{1}{2}$

$\Rightarrow \quad \theta = 60°$

∴ Option (c) is correct.

**6. The two lines $x = ay + b$, $z = cy + d$; and $x = a'y + b'$, $z = c'y + d'$ are perpendicular to each other, if**

**(a)** $\frac{a}{a'} + \frac{c}{c'} = 1$ **(b)** $\frac{a}{a'} + \frac{c}{c'} = -1$

**(c)** $aa' + cc' = 1$ **(d)** $aa' + cc' = -1$ **[1]**

**Sol.** Option (d) is correct.

**7. The two planes $x - 2y + 4z = 10$ and $18x + 17y + kz = 50$ are perpendicular, if $k$ is equal to**

**(a) –4** **(b) 4**

**(c) 2** **(d) –2** **[1]**

**Sol.** $18 - 34 + 4k = 0$

{using $a_1a_2 + b_1b_2 + c_1c_2 = 0$}

$\Rightarrow \quad 4k = 16$

$\Rightarrow \quad k = 4$

∴ Option (b) is correct.

**8. In an LPP, if the objective function $z = ax + by$ has the same maximum value on two corner points of the feasible region, then the number of points at which $z_{max}$ occurs is**

**(a) 0** **(b) 2**

**(c) finite** **(d) infinite** **[1]**

**Sol.** Option (d) is correct.

**9. From the set {1, 2, 3, 4, 5}, two numbers $a$ and $b$ ($a \neq b$) are chosen at random. The probability that $\frac{a}{b}$ is an integer is** **[1]**

**(a)** $\frac{1}{3}$ **(b)** $\frac{1}{4}$

**(c)** $\frac{1}{2}$ **(d)** $\frac{3}{5}$

**Sol.** Total possible outcomes of $(a, b)$ = {(1, 2), (1, 3), (1, 4), (1, 5), (2, 1), (2, 3), (2, 4), (2, 5), (3, 1), (3, 2), (3, 4), (3, 5), (4, 1), (4, 2), (4, 3), (4, 5), (5, 1), (5, 2), (5, 3), (5, 4)}

⇒ Number of possible outcomes = 20

Also, favourable outcomes of $(a, b)$ = {(2, 1), (3, 1), (4, 1), (4, 2), (5, 1)}

⇒ Number of possible outcomes = 5

∴ $P\left(\frac{a}{b} \text{ is integer}\right) = \frac{5}{20} = \frac{1}{4}$

∴ Option (b) is correct.

**10. A bag contains 3 white, 4 black and 2 red balls. If 2 balls are drawn at random (without replacement), then the probability that both the balls are white is**

**(a)** $\frac{1}{18}$ **(b)** $\frac{1}{36}$

**(c)** $\frac{1}{12}$ **(d)** $\frac{1}{24}$ **[1]**

**Sol.** Required probability = $\frac{^3C_2}{^9C_2} = \frac{1}{12}$

∴ Option (c) is correct.

**In Q. numbers 11 to 15, fill in the blanks with correct word/sentence :**

**11. If $f : R \rightarrow R$ be given by $f(x) = (3 - x^3)^{1/3}$, then $fof(x)$ = ________** **[1]**

**Sol.** $x$

**Explanation:** $(fof)(x) = f(f(x))$

$= f((3 - x^3)^{1/3})$

$= \left[3 - \left[(3 - x^3)^{1/3}\right]^3\right]^{1/3}$

$= [3 - 3 + x^3]^{1/3}$

$= x$

**12. If $\begin{bmatrix} x+y & 7 \\ 9 & x-y \end{bmatrix} = \begin{bmatrix} 2 & 7 \\ 9 & 4 \end{bmatrix}$, then $x \cdot y$ = ______** **[1]**

**Sol.** – 3

**Explanation:** Comparing elements of both matrices, we get

$$x + y = 2 \quad ...(i)$$

and $$x - y = 4 \quad ...(ii)$$

Adding (i) and (ii), we get

$$2x = 6$$

$$\Rightarrow \quad x = 3$$

From (i), $$y = 2 - x = 2 - 3 = -1$$

$$\therefore \quad x.y = 3(-1) = -3$$

**13. The number of points of discontinuity of $f$ defined by $f(x) = |x| - |x+1|$ is ______. [1]**

**Sol.** 0

**Explanation:** Here $f(x) = |x| - |x+1|$

Let $h(x) = |x|$

and $g(x) = |x+1|$

Then $h(x)$ and $g(x)$ are both continuous functions as modulus function and polynomial function are continuous everywhere.

Also $(hog)(x) = h(g(x)) = h(x+1) = |x+1|$

Since composite of two continuous functions is also continuous in its domain

$\therefore (hog)(x) = |x+1|$ is also continuous

Also difference of two continuous functions is continuous everywhere

$\therefore f(x) = |x| - |x+1| = h(x) - (hog)(x)$

is also continuous everywhere.

Hence, the number of points of discontinuity of $f(x) = |x| - |x+1|$ is zero.

**14. The slope of the tangent to the curve $y = x^3 - x$ at the point (2, 6) is ______. [1]**

**OR**

**The rate of change of the area of a circle with respect to its radius $r$, when $r = 3$ cm, is ______.**

**Sol.** 11

**Explanation:** $$y = x^3 - x$$

$$\Rightarrow \quad \frac{dy}{dx} = 3x^2 - 1$$

$\therefore$ Slope of tangent at (2, 6) $= 3(2)^2 - 1 = 11$

**OR**

$6\pi$ cm$^2$/cm

**Explanation:** Area of circle, $A = \pi r^2$

$$\therefore \quad \frac{dA}{dr} = 2\pi r$$

$$\left[\frac{dA}{dr}\right]_{r = 3\text{cm}} = 2 \times \pi \times 3 = 6\pi \text{cm}^2/\text{cm}$$

**15. If $\vec{a}$ is a non-zero vector, then $(\vec{a}.\hat{i})\hat{i} + (\vec{a}.\hat{j})\hat{j} + (\vec{a}.\hat{k})\hat{k}$ equals ______. [1]**

**OR**

**The projection of the vector $\hat{i} - \hat{j}$ on the vector $i + j$ is______.**

**Sol.** $\vec{a}$

**Explanation:** Let $\vec{a} = a_1\hat{i} + b_1\hat{j} + c_1\hat{k}$

$$\therefore (\vec{a}.\hat{i})\hat{i} + (\vec{a}.\hat{j})\hat{j} + (\vec{a}.\hat{k})\hat{k} = a_1\hat{i} + b_1\hat{j} + c_1\hat{k} = \vec{a}$$

**OR**

0

**Explanation:** Projection of $\hat{i} - \hat{j}$ on $\hat{i} + \hat{j}$

$$= \frac{(\hat{i}-\hat{j}).(\hat{i}+\hat{j})}{|\hat{i}+\hat{j}|} = 0$$

**Q. 16 to 20 are very short answer questions.**

**16. Find adj A, if A = $\begin{bmatrix} 2 & -1 \\ 4 & 3 \end{bmatrix}$ [1]**

**Sol.** Given : $A = \begin{bmatrix} 2 & -1 \\ 4 & 3 \end{bmatrix}$

For a square matrix of order 2, the adj A can be obtained by interchanging $a_{11}$ and $a_{22}$ and by changing sign of $a_{12}$ and $a_{21}$.

$$\therefore \quad \text{adj } A = \begin{bmatrix} 3 & 1 \\ -4 & 2 \end{bmatrix} \quad \textbf{Ans.}$$

**17. Find $\int \frac{2^{x+1} - 5^{x-1}}{10^x} dx$ [1]**

**Sol.** $$\int \frac{2^{x+1} - 5^{x-1}}{10^x} dx = \int \left(\frac{2.2^x}{10^x}\right)dx - \int \left(\frac{1}{5}.\frac{5^x}{10^x}\right)dx$$

$$= 2\int \frac{1}{5^x}dx - \frac{1}{5}\int \frac{1}{2^x}dx$$

$$= 2\int 5^{-x}dx - \frac{1}{5}\int 2^{-x}dx$$

$$= \frac{-2(5^{-x})}{\log 5} + \frac{1}{5}\frac{(2^{-x})}{\log 2} + C$$

$$= \frac{-2}{5^x \log 5} + \frac{1}{5.2^x \log 2} + C \quad \textbf{Ans.}$$

**18. Evaluate $\int_0^{2\pi} |\sin x|\, dx$** **[1]**

**Sol.** Let $I = \int_0^{2\pi} |\sin x|\, dx$

$$= \int_0^{\pi} \sin x\, dx - \int_{\pi}^{2\pi} \sin x\, dx$$

$$= -[\cos x]_0^{\pi} + [\cos x]_{\pi}^{2\pi}$$

$$= -[\cos \pi - \cos 0] + [\cos 2\pi - \cos \pi]$$

$$= -[-1-1] + [1+1] = 4$$ **Ans.**

**19. If $\int_0^a \frac{dx}{1+4x^2} = \frac{\pi}{8}$, then find the value of $a$.** **[1]**

**OR**

**Find** $\int \frac{dx}{\sqrt{x}+x}$

**Sol.** Given,

$$\int_0^a \frac{dx}{1+4x^2} = \frac{\pi}{8}$$

$$\Rightarrow \int_0^a \frac{d}{1^2+(2x)^2} = \frac{\pi}{8}$$

$$\Rightarrow \frac{1}{2}[\tan^{-1} 2x]_0^a = \frac{\pi}{8}$$

$$\Rightarrow \tan^{-1} 2a - 0 = \frac{\pi}{4}$$

$$\Rightarrow \tan^{-1} 2a = \frac{\pi}{4}$$

$$\Rightarrow 2a = \tan\frac{\pi}{4}$$

$$\Rightarrow 2a = 1$$

$$\Rightarrow a = \frac{1}{2}$$ **Ans.**

**OR**

Let $I = \int \frac{dx}{\sqrt{x}+x} = \int \frac{1}{\sqrt{x}(1+\sqrt{x})}\, dx$

Let $\sqrt{x} = t$

$$\Rightarrow \frac{1}{2\sqrt{x}} dx = dt$$

$$\Rightarrow \frac{dx}{\sqrt{x}} = 2dt$$

$$\therefore I = 2\int \frac{1}{(1+t)}\, dt$$

$$= 2 \log |(1+t)| + C$$

$$= 2\log|1+\sqrt{x}| + C$$ **Ans.**

**20. Show that the function $y = ax + 2a^2$ is a solution of the differential equation $2\left(\frac{dy}{dx}\right)^2 + x\left(\frac{dy}{dx}\right) - y = 0$** **[1]**

**Sol.** Given, $y = ax + 2a^2$

$$\Rightarrow \frac{dy}{dx} = a$$

Now, L.H.S. $= 2\left(\frac{dy}{dx}\right)^2 + x\left(\frac{dy}{dx}\right) - y$

$$= 2a^2 + x(a) - (ax + 2a^2)$$

$$= 2a^2 + ax - ax - 2a^2$$

$$\Rightarrow = 0 = \text{R.H.S.}$$

**Hence Proved.**

## SECTION – B

**Q. Nos. 21 to 26 carry 2 marks each.**

**21. Check if the relation R on the set A = {1, 2, 3, 4, 5, 6} defined as R = {(x, y) : y is divisible by x} is (i) symmetric, (ii) transitive.** **[2]**

**OR**

**Prove that :**

$$\frac{9\pi}{8} - \frac{9}{4}\sin^{-1}\left(\frac{1}{3}\right) = \frac{9}{4}\sin^{-1}\left(\frac{2\sqrt{2}}{3}\right)$$

**Sol.** Given, A = {1, 2, 3, 4, 5, 6}

R = {(x, y) : y is divisible by x}

**(i)** $(2, 4) \in R$ $\{\because$ 4 is divisible by 2$\}$

But $(4, 2) \notin R$ $\{\because$ 2 is not divisible by 4$\}$

$\therefore$ R is not symmetric. **Ans.**

**(ii)** Let $(a, b) \in R$ & $(b, c) \in R$

$\Rightarrow$ $b = \lambda a$ and $c = \mu b$

Now, $c = \mu b = \mu(\lambda a) \Rightarrow (a, c) \in R$

$\Rightarrow$ $c$ is divisible by $a$

$\therefore$ R is transitive. **Ans.**

**OR**

$$\text{L.H.S.} = \frac{9\pi}{8} - \frac{9}{4}\sin^{-1}\frac{1}{3}$$

$$= \frac{9}{4}\left(\frac{\pi}{2} - \sin^{-1}\frac{1}{3}\right)$$

$$= \frac{9}{4}\cos^{-1}\frac{1}{3} \quad \left\{\because \frac{\pi}{2} - \sin^{-1} x = \cos^{-1} x\right\}$$

$$= \frac{9}{4}\sin^{-1}\sqrt{1-\left(\frac{1}{3}\right)^2}$$

$$\left\{\because \cos^{-1}x = \sin^{-1}\sqrt{1-x^2}\right\}$$

$$= \frac{9}{4}\sin^{-1}\sqrt{\frac{8}{9}}$$

$$= \frac{9}{4}\sin^{-1}\frac{2\sqrt{2}}{3} = \text{R.H.S.}$$ **Hence Proved.**

**22. Find the value of $\frac{dy}{dx}$ at $\theta = \frac{\pi}{3}$, if**

**$x = \cos\theta - \cos 2\theta$, $y = \sin\theta - \sin 2\theta$.** **[2]**

**Sol.** Given, $x = \cos\theta - \cos 2\theta$

$\Rightarrow \frac{dx}{d\theta} = -\sin\theta + 2\sin 2\theta$

Also, $y = \sin\theta - \sin 2\theta$

$\Rightarrow \frac{dy}{d\theta} = \cos\theta - 2\cos 2\theta$

Now, $\frac{dy}{dx} = \frac{dy}{d\theta}\times\frac{d\theta}{dx} = \frac{\cos\theta - 2\cos 2\theta}{2\sin 2\theta - \sin\theta}$

$$\Rightarrow \left[\frac{dy}{dx}\right]_{\theta=\frac{\pi}{3}} = \frac{\cos\frac{\pi}{3} - 2\cos\frac{2\pi}{3}}{2\sin\frac{2\pi}{3} - \sin\frac{\pi}{3}} = \sqrt{3}$$

**Ans.**

**23. Show that the function $f$ defined by $f(x) = (x-1)e^x + 1$ is an increasing function for all $x > 0$.** **[2]**

**Sol.** Given, $f(x) = (x-1)e^x + 1, x > 0$

$\therefore f'(x) = (x-1)e^x + e^x(1) + 0$

$= e^x(x-1+1) = xe^x > 0 \ \forall \ x > 0$

$\therefore f'(x)$ is an increasing function for all $x > 0$.

**Hence Proved.**

**24. Find $|\vec{a}|$ and $|\vec{b}|$, if $|\vec{a}| = 2|\vec{b}|$ and $(\vec{a}+\vec{b}).(\vec{a}-\vec{b}) = 12$.** **[2]**

**OR**

**Find the unit vector perpendicular to each of the vectors $\vec{a} = 4\hat{i} + 3\hat{j} + \hat{k}$ and $\vec{b} = 2\hat{i} - \hat{j} + 2\hat{k}$.**

**Sol.** Given, $|\vec{a}| = 2|\vec{b}|$ ...(i)

And, $(\vec{a}+\vec{b}).(\vec{a}-\vec{b}) = 12$

$\Rightarrow |\vec{a}|^2 - |\vec{b}|^2 = 12$

$\Rightarrow (2|\vec{b}|)^2 - |\vec{b}|^2 = 12$ [Using (i)]

$\Rightarrow 3|\vec{b}|^2 = 12$

$\Rightarrow |\vec{b}|^2 = 4$

$\Rightarrow |\vec{b}| = 2$

From (i), $|\vec{a}| = 2(2) = 4$

$\therefore |\vec{a}| = 4$ and $|\vec{b}| = 2$ **Ans.**

**OR**

Given, $\vec{a} = 4\hat{i} + 3\hat{j} + \hat{k}$

and $\vec{b} = 2\hat{i} - \hat{j} + 2\hat{k}$

$$\therefore \vec{a}\times\vec{b} = \begin{vmatrix} \hat{i} & \hat{j} & \hat{k} \\ 4 & 3 & 1 \\ 2 & -1 & 2 \end{vmatrix}$$

$$= 7\hat{i} - 6\hat{j} - 10\hat{k}$$

$\therefore$ Unit vector perpendicular to both $\vec{a}$ and $\vec{b}$

$$= \pm\frac{(\vec{a}\times\vec{b})}{|\vec{a}\times\vec{b}|}$$

$$= \pm\frac{(7\hat{i} - 6\hat{j} - 10\hat{k})}{\sqrt{49+36+100}}$$

$$= \pm\frac{(7\hat{i} - 6\hat{j} - 10\hat{k})}{\sqrt{185}}$$ **Ans.**

**25. Find the equation of the plane with intercept 3 on the $y$-axis and parallel to $xz$-plane.** **[2]**

**Sol.** Equation of any plane parallel to $xz$-plane is $y = b$ (where $b$ is a constant)

Since its $y$-intercept is 3,

$\therefore$ Equation of required plane is $y = 3$. **Ans.**

**26. Find [P(B/A) + P(A/B)], if P(A) = $\frac{3}{10}$, P(B) = $\frac{2}{5}$ and P (A ∪ B) = $\frac{3}{5}$.** **[2]**

**Sol.** We have,

$P(A \cup B) = P(A) + P(B) - P(A \cap B)$

$$\Rightarrow \quad \frac{3}{5} = \frac{3}{10}+\frac{2}{5}-P(A\cap B)$$

$$\Rightarrow \quad P(A\cap B) = \frac{3}{10}+\frac{2}{5}-\frac{3}{5}$$

$$\Rightarrow \quad P(A\cap B) = \frac{1}{10}$$

$$\therefore \quad P(B/A)+P(A/B) = \frac{P(A\cap B)}{P(A)}+\frac{P(A\cap B)}{P(B)}$$

$$= \frac{1/10}{3/10}+\frac{1/10}{2/5}$$

$$= \frac{1}{3}+\frac{1}{4} = \frac{7}{12}$$ **Ans.**

## SECTION–C

**Q. Nos. 27 to 32 carry 4 marks each.**

**27. Prove that the relation R on Z, defined by R $\{(x, y) : (x-y)$ is divisible by 5$\}$ is an equivalence relation.** **[4]**

**Sol.** R on Z is defined as :

$R = \{(x, y) : (x-y)$ is divisible by 5$\}$

**For reflexive :** Let $x \in Z$

$\Rightarrow x - x = 0$, which is divisible by 5

$\Rightarrow \quad (x, x) \in R \ \forall x \in Z$

Hence, R is a reflexive relation.

**For symmetric :** Let $(x, y) \in R$

$\Rightarrow x - y$ is divisible by 5

$\Rightarrow -(x-y)$ is divisible by 5

$\Rightarrow y - x$ is divisible by 5

$\Rightarrow \quad (y, x) \in R$

Hence, R is a symmetric relation.

**For transitive :** Let $(x, y) \in R$ and $(y, z) \in R$

$\Rightarrow x - y$ and $y - z$ are divisible by 5

$\Rightarrow x - y + y - z$ is divisible by 5

$\Rightarrow x - z$ is divisible by 5

$\Rightarrow (x, z) \in R$

Hence, R is a transitive relation.

Hence, R is an equivalence relation.

**Hence Proved.**

**28. If $y = \sin^{-1}\left(\frac{\sqrt{1+x}+\sqrt{1-x}}{2}\right)$, then show that $\frac{dy}{dx} = \frac{-1}{2\sqrt{1-x^2}}$.** **[4]**

**OR**

**Verify the Rolle's Theorem for the function $f(x) = e^x \cos x$ in $\left[-\frac{\pi}{2}, \frac{\pi}{2}\right]$**

**Sol.** Let $x = \cos\theta$

$$\Rightarrow \quad y = \sin^{-1}\left(\frac{\sqrt{1+\cos\theta}+\sqrt{1-\cos\theta}}{2}\right)$$

$$= \sin^{-1}\left(\frac{\sqrt{2}\cos\frac{\theta}{2}+\sqrt{2}\sin\frac{\theta}{2}}{2}\right)$$

$$= \sin^{-1}\left(\frac{1}{\sqrt{2}}\cos\frac{\theta}{2}+\frac{1}{\sqrt{2}}\sin\frac{\theta}{2}\right)$$

$$= \sin^{-1}\left(\sin\frac{\pi}{4}\cos\frac{\theta}{2}+\cos\frac{\theta}{4}\sin\frac{\theta}{2}\right)$$

$$= \sin^{-1}\left[\sin\left(\frac{\theta}{2}+\frac{\pi}{4}\right)\right] = \frac{\theta}{2}+\frac{\pi}{4}$$

$$\therefore \quad y = \frac{1}{2}\cos^{-1}x+\frac{\pi}{4} \qquad [\because x = \cos\theta]$$

$$\Rightarrow \quad \frac{dy}{dx} = -\frac{1}{2\sqrt{1-x^2}}$$ **Hence Proved.**

**OR**

Given, $\quad f(x) = e^x \cos x$ in $\left[-\frac{\pi}{2}, \frac{\pi}{2}\right]$

We know that $e^x$ and cosine function are continuous everywhere and product of any two continuous function is also continuous in its domain.

Hence, $f(x) = e^x \cos x$ is continuous in $\left[-\frac{\pi}{2}, \frac{\pi}{2}\right]$

Now, $f'(x) = e^x(-\sin x) + \cos x . e^x$

$= e^x (\cos x - \sin x)$

It is defined $\forall x \in \left(-\frac{\pi}{2}, \frac{\pi}{2}\right)$

$\therefore$ $f$ is differentiable in $\left(-\frac{\pi}{2}, \frac{\pi}{2}\right)$

And, $\quad f\left(-\frac{\pi}{2}\right) = e^{-\pi/2}\cos\left(-\frac{\pi}{2}\right) = 0$

$$f\left(\frac{\pi}{2}\right) = e^{\pi/2}\cos\left(\frac{\pi}{2}\right) = 0$$

$$\therefore \quad f\left(-\frac{\pi}{2}\right) = f\left(\frac{\pi}{2}\right)$$

$\therefore$ All the three postulates of Rolle's theorem are satisfied. Hence there exists a real number,

$c \in \left[-\frac{\pi}{2}, \frac{\pi}{2}\right]$ such that

$$f'(c) = 0$$

$$\Rightarrow \quad e^c(\cos c - \sin c) = 0$$

$$\Rightarrow \quad \cos c - \sin c = 0 \qquad \{\because e^c \neq 0\}$$

$$\Rightarrow \quad \cos c = \sin c$$

$$\Rightarrow \quad \tan c = 1 = \tan\frac{\pi}{4}$$

$$\Rightarrow \quad c = \frac{\pi}{4} \in \left(-\frac{\pi}{2}, \frac{\pi}{2}\right)$$

**Hence Verified.**

**29. Evaluate : $\int_0^{\pi} \frac{x \sin x}{1+\cos^2 x} dx$. [4]**

**Sol.** Let $\mathrm{I} = \int_0^{\pi} \frac{x\sin x}{1+\cos^2 x}dx$ ...(i)

Using $\int_0^{\pi} f(x)dx = \int_0^{a} f(a-x)dx$

$$\mathrm{I} = \int_0^{\pi} \frac{(\pi - x)\sin(\pi - x)}{1+\cos^2(\pi - x)}dx$$

$$\Rightarrow \quad \mathrm{I} = \int_0^{\pi} \frac{(\pi - x)\sin x}{1+\cos^2 x}dx \quad ...(ii)$$

On adding equations (i) and (ii), we get

$$\Rightarrow \quad 2\mathrm{I} = \pi\int_0^{\pi} \frac{\sin x}{1+\cos^2 x}dx$$

Let $\cos x = t$

$\Rightarrow$ $\sin x\, dx = -dt$

When $x = 0$, $t = 1$

When $x = \pi$, $t = -1$

$$\therefore \quad \mathrm{I} = \frac{\pi}{2}\int_1^{-1} \frac{-dt}{1+t^2}$$

$$\Rightarrow \quad \mathrm{I} = -\frac{\pi}{2}(\tan^{-1} t)_1^{-1}$$

$$= -\frac{\pi}{2}[\tan^{-1}(-1) - \tan^{-1}(1)]$$

$$= -\frac{\pi}{2}\left(-\frac{\pi}{4} - \frac{\pi}{4}\right)$$

$$= -\frac{\pi}{2}\left(-\frac{\pi}{2}\right)$$

$$\therefore \quad \mathrm{I} = \frac{\pi^2}{4} \qquad \textbf{Ans.}$$

**30. For the differential equation given below, find a particular solution satisfying the given condition $(x + 1)\frac{dy}{dx} = 2e^{-y} + 1$; $y = 0$ when $x = 0$. [4]**

**Sol.** Given,

$$(x+1)\frac{dy}{dx} = \frac{2}{e^y} + 1,\ y = 0,\ x = 0$$

$$\Rightarrow \quad (x+1)\frac{dy}{dx} = \frac{2+e^y}{e^y}$$

$$\Rightarrow \quad \int \frac{e^y}{e^y+2}dy = \int \frac{dx}{x+1}$$

Let $e^y + 2 = t$

$\Rightarrow$ $e^y dy = dt$

$$\Rightarrow \quad \int \frac{dt}{t} = \int \frac{dx}{x+1}$$

$$\Rightarrow \quad \log|t| = \log|x+1| + \log|c|$$

$$\Rightarrow \quad \log|e^y + 2| = \log|c(x+1)|$$

$$\Rightarrow \quad e^y + 2 = c(x+1)$$

Put $x = 0, y = 0$

$$\Rightarrow \quad e^0 + 2 = c(0+1)$$

$$\Rightarrow \quad 1 + 2 = c$$

$$\Rightarrow \quad c = 3$$

$\therefore$ Solution is,

$$e^y + 2 = 3(x+1)$$

$$\Rightarrow \quad e^y = 3x + 1 \qquad \textbf{Ans.}$$

**31. A manufacturer has three machines I, II and III installed in his factory. Machine I and II are capable of being operated for atmost 12 hours wherease machine III must be operated for atleast 5 hours a day. He produces only two items M and N each requiring the use of all the three machines.**

**The number of hours required for producing 1 unit of M and N on three machines are given in the following table :**

| Items | Number of hours required on machines | | |
|---|---|---|---|
| | I | II | III |
| M | 1 | 2 | 1 |
| N | 2 | 1 | 1.25 |

**He makes a profit of ₹ 600 and ₹ 400 on one unit of items M and N respectively. How many units of each item should he produce so as to maximize his profit assuming that he can sell all the items that he produced. What will be the maximum profit ? [4]**

**Sol.** Let number of units of item M produced be $x$ and that of item N be $y$.

$$\text{Profit (in ₹)} = 600x + 400y$$

$$Z = 600x + 400y$$

$\therefore$ LPP is maximized.

Subject to constraints :

$$x + 2y \leq 12 \quad \text{...(i)}$$

$$2x + y \leq 12 \quad \text{...(ii)}$$

$$x + 1.25y \geq 5 \quad \text{...(iii)}$$

$$x \geq 0$$

$$y \geq 0$$

To solve this LPP, we draw the lines

$$x + 2y = 12$$

$$2x + y = 12$$

$$x + 1.25y = 5$$

$$x = 0$$

and $$y = 0$$

$$x + 2y = 12$$

| $x$ | 0 | 12 |
|---|---|---|
| $y$ | 6 | 0 |

$$2x + y = 12$$

| $x$ | 0 | 6 |
|---|---|---|
| $y$ | 12 | 0 |

$$x + 1.25y = 5$$

| $x$ | 0 | 5 |
|---|---|---|
| $y$ | 4 | 0 |

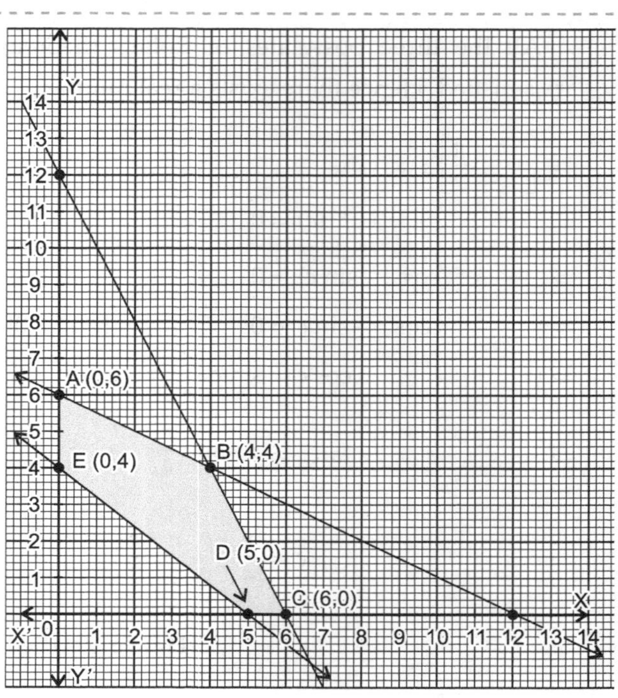


| Corner point | $Z = 600x + 400y$ |
|---|---|
| A (0, 6) | Z = 2400 |
| B (4, 4) | Z = 4000 Maximum |
| C (6, 0) | Z = 3600 |
| D (5, 0) | Z = 3000 |
| E (0, 4) | Z = 1600 |

$\therefore$ Maximum profit is ₹ 4000 at B (4, 4), *i.e.* when 4 units of item M and 4 units of item N are produced. **Ans.**

**32. A coin is biased so that the head is three times as likely to occur as tail. If the coin is tossed twice, find the probability distribution of number of tails. Hence find the mean of the number of tails. [4]**

**OR**

**Suppose that 5 men out of 100 and 25 women out of 1000 are good orators. Assuming that there are equal number of men and women, find the probability of choosing a good orator.**

**Ans.** According to question :

$$P(H) = 3P(T)$$

But $$P(H) + P(T) = 1$$

$$\Rightarrow \quad 3P(T) + P(T) = 1$$

$$\Rightarrow \quad 4P(T) = 1$$

$$\Rightarrow \quad P(T) = \frac{1}{4}$$

So, $$P(H) = \frac{3}{4}$$

Let X denotes number of tails obtained when the coin is tossed twice. Then, X can be 0, 1 or 2.

$$P(X = 0) = P(HH) = \left(\frac{3}{4}\right)^2 = \frac{9}{16}$$

$$P(X = 1) = P(HT \text{ or } TH) = \frac{3}{4}\times\frac{1}{4}+\frac{1}{4}\times\frac{3}{4} = \frac{6}{16}$$

$$P(X = 2) = P(TT) = \frac{1}{4}\times\frac{1}{4}=\frac{1}{16}$$

∴ Probability distribution table is

| $X_i$ | 0 | 1 | 2 |
|---|---|---|---|
| $P(X_i)$ | $\frac{9}{16}$ | $\frac{6}{16}$ | $\frac{1}{16}$ |

Now, $$\text{Mean} = E(X) = \sum_{i=1}^{n} X_i P(X_i)$$

$$= 0\times\frac{9}{16}+1\times\frac{6}{16}+2\times\frac{1}{16}$$

$$= \frac{8}{16} = \frac{1}{2}$$

∴ $\text{Mean} = \frac{1}{2}$ **Ans.**

**OR**

Let $E_1$ be the event of selecting a man and $E_2$ be the event of selecting a woman.
Let A be the event of selecting a good orator.

Then $$P(E_1) = P(E_2) = \frac{1}{2}$$

{∵ No. of men and women are equal}

$$P(A/E_1) = \frac{5}{100} = \frac{1}{20}$$

$$P(A/E_2) = \frac{25}{1000} = \frac{1}{40}$$

∴By total theorem of probability :

$$P(A) = P(E_1)\,P(A/E_1) + P(E_2)\,P(A/E_2)$$

$$= \frac{1}{2}\times\frac{1}{20}+\frac{1}{2}\times\frac{1}{40}$$

$$= \frac{1}{40}+\frac{1}{80} = \frac{3}{80}$$ **Ans.**

## SECTION – D

**Q. Nos. 33 to 36 carry 6 marks each.**

**33. Using properties of determinants prove that :**

$$\begin{vmatrix} a-b & b+c & a \\ b-c & c+a & b \\ c-a & a+b & c \end{vmatrix} = a^3 + b^3 + c^3 - 3abc.$$ **[6]**

**OR**

**If** $A = \begin{bmatrix} 1 & 3 & 2 \\ 2 & 0 & -1 \\ 1 & 2 & 3 \end{bmatrix}$, **then show that** $A^3 - 4A^2 - 3A + 11I = 0$. **Hence find** $A^{-1}$.

**Sol.** $$\text{L.H.S.} = \begin{vmatrix} a-b & b+c & a \\ b-c & c+a & b \\ c-a & a+b & c \end{vmatrix}$$

Using $C_2 \to C_2 + C_3$ and then taking $(a + b + c)$ common from $C_2$,

$$= (a+b+c)\begin{vmatrix} a-b & 1 & a \\ b-c & 1 & b \\ c-a & 1 & c \end{vmatrix}$$

Applying $C_1 \to C_1 - C_3$,

$$= (a+b+c)\begin{vmatrix} -b & 1 & a \\ -c & 1 & b \\ -a & 1 & c \end{vmatrix}$$

Applying $R_3 \to R_3 - R_1$ and $R_2 \to R_2 - R_1$,

$$(a+b+c)\begin{vmatrix} -b & 1 & a \\ b-c & 0 & b-a \\ b-a & 0 & c-a \end{vmatrix}$$

Expanding along $C_2$,

$= (a + b + c)(-1)[(b - c)(c - a) - (b - a)^2]$

$= (a + b + c)(-1)[(bc - ab - c^2 + ca - b^2 - a^2 + 2ab)]$

$= (a + b + c)(a^2 + b^2 + c^2 - ac - ab - ca)$

$= a^3 + b^3 + c^3 - 3abc = \text{R.H.S.}$ **Hence Proved.**

**OR**

Given, $$A = \begin{bmatrix} 1 & 3 & 2 \\ 2 & 0 & -1 \\ 1 & 2 & 3 \end{bmatrix}$$

∴ $$A^2 = \begin{bmatrix} 1 & 3 & 2 \\ 2 & 0 & -1 \\ 1 & 2 & 3 \end{bmatrix}\begin{bmatrix} 1 & 3 & 2 \\ 2 & 0 & -1 \\ 1 & 2 & 3 \end{bmatrix}$$

$$= \begin{bmatrix} 9 & 7 & 5 \\ 1 & 4 & 1 \\ 8 & 9 & 9 \end{bmatrix}$$

Also, $A^3 = AA^2 = \begin{bmatrix} 1 & 3 & 2 \\ 2 & 0 & -1 \\ 1 & 2 & 3 \end{bmatrix} \begin{bmatrix} 9 & 7 & 5 \\ 1 & 4 & 1 \\ 8 & 9 & 9 \end{bmatrix}$

$$= \begin{bmatrix} 28 & 37 & 26 \\ 10 & 5 & 1 \\ 35 & 42 & 34 \end{bmatrix}$$

Now, L.H.S. $= A^3 - 4A^2 - 3A + 11I$

$$= \begin{bmatrix} 28 & 37 & 26 \\ 10 & 5 & 1 \\ 35 & 42 & 34 \end{bmatrix} - 4\begin{bmatrix} 9 & 7 & 5 \\ 1 & 4 & 1 \\ 8 & 9 & 9 \end{bmatrix} - 3\begin{bmatrix} 1 & 3 & 2 \\ 2 & 0 & -1 \\ 1 & 2 & 3 \end{bmatrix} + 11\begin{bmatrix} 1 & 0 & 0 \\ 0 & 1 & 0 \\ 0 & 0 & 1 \end{bmatrix}$$

$$= \begin{bmatrix} 28-36-3+11 & 37-28-9 & 26-20-6 \\ 10-4-6 & 5-16+11 & 1-4+3 \\ 35-32-3 & 42-36-6 & 34-36-9+11 \end{bmatrix}$$

$$= \begin{bmatrix} 0 & 0 & 0 \\ 0 & 0 & 0 \\ 0 & 0 & 0 \end{bmatrix} = 0 = \text{R.H.S.}$$ **Hence Proved.**

Now, $A^3 - 4A^2 - 3A + 11I = 0$

Pre-multiplying by $A^{-1}$

$(A^{-1}A)A^2 - 4(A^{-1}A)A - 3A^{-1}(A) + 11A^{-1}I = A^{-1}0$

$\Rightarrow \quad IA^2 - 4IA - 3I + 11A^{-1} = 0$

$\Rightarrow \quad A^{-1} = \frac{1}{11}\{4A + 3I - A^2\}$

$$= \frac{1}{11}\left\{4\begin{bmatrix} 1 & 3 & 2 \\ 2 & 0 & -1 \\ 1 & 2 & 3 \end{bmatrix} + 3\begin{bmatrix} 1 & 0 & 0 \\ 0 & 1 & 0 \\ 0 & 0 & 1 \end{bmatrix} - \begin{bmatrix} 9 & 7 & 5 \\ 1 & 4 & 1 \\ 8 & 9 & 9 \end{bmatrix}\right\}$$

$$= \frac{1}{11}\begin{bmatrix} 4+3-9 & 12-7 & 8-5 \\ 8-1 & 3-4 & -4-1 \\ 4-8 & 8-9 & 12+3-9 \end{bmatrix}$$

$\Rightarrow \quad A^{-1} = \frac{1}{11}\begin{bmatrix} -2 & 5 & 3 \\ 7 & -1 & -5 \\ -4 & -1 & 6 \end{bmatrix}$ **Ans.**

**34. Find the intervals on which the function $f(x) = (x-1)^3 (x-2)^2$ is**

**(a) strictly increasing,**

**(b) strictly decreasing.** **[6]**

**OR**

**Find the dimensions of the rectangle of perimeter 36 cm which will sweep out a volume as large as possible, when revolved about one of its side. Also, find the maximum volume.**

**Sol.** Given, $f(x) = (x-1)^3 (x-2)^2$

$\therefore f'(x) = (x-1)^3 2(x-2) + (x-2)^2 3(x-1)^2$

$\Rightarrow f'(x) = (x-1)^2 (x-2)[2(x-1) + 3(x-2)]$

$= (x-1)^2 (x-2)(5x-8)$

Putting $f'(x) = 0$

$\Rightarrow \quad x = 1, 2, \frac{8}{5}$

| Intervals | Sign of $f'(x) = (x-1)^2 (x-2)(5x-8)$ | Strictly increasing/ decreasing |
|---|---|---|
| $(-\infty, 1)$ | + | Strictly increasing |
| $\left(1, \frac{8}{5}\right)$ | + | Strictly increasing |
| $\left(\frac{8}{5}, 2\right)$ | – | Strictly decreasing |
| $(2, \infty)$ | + | Strictly increasing |

**(a)** $f$ is strictly increasing in $(-\infty, 1) \cup \left(1, \frac{8}{5}\right) \cup (2, \infty)$ **Ans.**

**(b)** $f$ is strictly decreasing in $\left(\frac{8}{5}, 2\right)$ **Ans.**

**OR**

Let the length of rectangle be $x$ cm and the breadth of rectangle be $y$ cm.

$\therefore \quad \text{Perimeter} = 2(l + b)$

$\Rightarrow \quad 2(x + y) = 36$

$\Rightarrow \quad x + y = 18$

$\Rightarrow \quad x = 18 - y$ ...(i)

When the rectangle is revolved about its length then the dimensions of cylinder so formed are :

$h = x$ cm

$r = y$ cm

$\therefore$ Volume of cylinder is

$V = \pi r^2 h$

$\Rightarrow \quad V = \pi y^2 x$

$\Rightarrow \quad V = \pi y^2(18 - y)$ [Using (i)]

$\Rightarrow \quad V = \pi(18y^2 - y^3)$

$\Rightarrow \quad \dfrac{dV}{dy} = \pi(36y - 3y^2)$

Put $\dfrac{dV}{dy} = 0$

$\Rightarrow \quad 36y - 3y^2 = 0$

$\Rightarrow \quad 3y(12 - y) = 0$

$\Rightarrow \quad y = 0$ (Rejected), $y = 12$

Now, $\quad \dfrac{d^2V}{dy^2} = \pi(36 - 6y)$

$= 6\pi(6 - y)$

When $y = 12$, $\quad \dfrac{d^2V}{dy^2} = 6\pi(6 - 12)$

$= 6\pi(-6) < 0$

$\therefore$ Volume is maximum when $y = 12$ cm.

From (i),

$x = 18 - y = 18 - 12 = 6$ cm

$\therefore$ For maximum volume, sides of rectangle are 6 cm and 12 cm.

Maximum volume , $V = \pi y^2 x$

$= \pi(12)^2 6$

$= \pi \times 144 \times 6$

$= 864\pi \text{ cm}^3$ **Ans.**

**35. Find the area of the region lying in the first quadrant and enclosed by the x-axis, the line $y = x$ and the circle $x^2 + y^2 = 32$.** **[6]**

**Sol.**

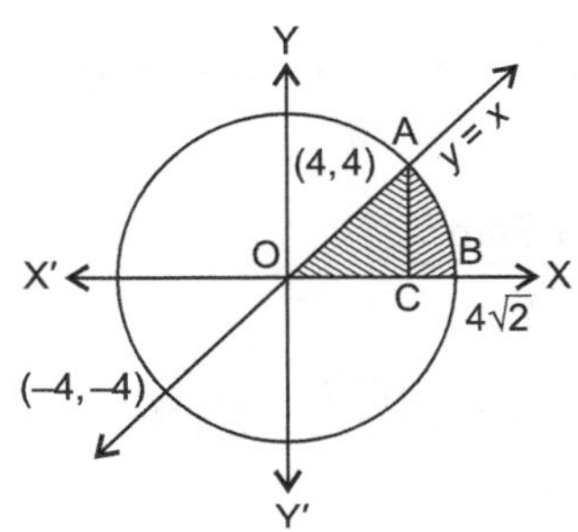


Given, $\quad x^2 + y^2 = 32$

$\Rightarrow \quad x^2 + y^2 = (4\sqrt{2})^2$ ...(i)

$\therefore$ Centre of this circle is (0, 0)

and $\quad$ Radius $= 4\sqrt{2}$

Given line is $\quad y = x$ ...(ii)

From (i) and (ii),

$x^2 + x^2 = 32$

$\Rightarrow \quad 2x^2 = 32$

$\Rightarrow \quad x^2 = 16$

$\Rightarrow \quad x = \pm 4$

$\therefore$ From (ii), $\quad y = \pm 4$

Hence points of intersection of circle and line are (4, 4) and (– 4, – 4).

Required area = $ar(\text{AOC}) + ar(\text{ABC})$

$$= \int_0^4 x\,dx + \int_4^{4\sqrt{2}} \sqrt{(4\sqrt{2})^2 - x^2 dx}$$

$$= \left[\frac{x^2}{2}\right]_0^4 + \left[\frac{x}{2}\sqrt{32 - x^2} + \frac{32}{2}\sin^{-1}\left(\frac{x}{4\sqrt{2}}\right)\right]_4^{4\sqrt{2}}$$

$$= \left(\frac{16}{2} - 0\right) + \left[(0 + 16\sin^{-1}(1)) - \left(2\sqrt{16} + 16\sin^{-1}\left(\frac{1}{\sqrt{2}}\right)\right)\right]$$

$$= 8 + \left[16\left(\frac{\pi}{2}\right) - \left(8 + 16\left(\frac{\pi}{4}\right)\right)\right]$$

$= 8 + (8\pi - 8 - 4\pi)$

$= 8 + 4\pi - 8$

$= 4\pi$ square units **Ans.**

**36. Show that the lines $\vec{r} = \vec{a} + \lambda\vec{b}$ and $\vec{r} = \vec{b} + \mu\vec{a}$ are coplanar and the plane containing them is given by $\vec{r}\cdot(\vec{a} \times \vec{b}) = 0$.** **[6]**

**Sol.** Given lines are :

$\vec{r} = \vec{a} + \lambda\vec{b}$ ...(i)

$\vec{r} = \vec{b} + \mu\vec{a}$ ...(ii)

We know that the lines $\vec{r} = \vec{a_1} + \lambda\vec{b_1}$, $\vec{r} = \vec{a_2} + \mu\vec{b_2}$ are coplanar only if

$$\left(\vec{a_2} - \vec{a_1}\right)\cdot\left(\vec{b_1} \times \vec{b_2}\right) = 0$$

Here, $\vec{a_1} = \vec{a}$, $\vec{b_1} = b$, $\vec{a_2} = b$, $\vec{b_2} = \vec{a}$.

$$\therefore \left(\vec{a_2} - \vec{a_1}\right)\cdot\left(\vec{b_1} \times \vec{b_2}\right) = \left(\vec{b} - \vec{a}\right)\cdot\left(\vec{b} \times \vec{a}\right)$$

$$= \vec{b}\cdot\left(\vec{b} \times \vec{a}\right) - \vec{a}\cdot\left(\vec{b} \times \vec{a}\right)$$

$$= \left[\vec{b}\ \vec{b}\ \vec{a}\right] - \left[\vec{a}\ \vec{b}\ \vec{a}\right]$$

$= 0 - 0$

{$\because$ When two of three vectors are identical, then their scalar triple product is 0}

$= 0$

Hence given lines are coplanar.

Also equation of plane containing the given lines is :

$$\left(\vec{r}-\vec{a}\right)\cdot\left(\vec{b_1}\times\vec{b_2}\right)=0\left[\because\left(\vec{r}-\vec{a}\right).\vec{n}=0\right]$$

$$\Rightarrow \quad \left(\vec{r}-\vec{a}\right)\cdot\left(\vec{b}\times\vec{a}\right)=0$$

$$\Rightarrow \vec{r}.\left(\vec{b}\times\vec{a}\right)-\vec{a}.\left(\vec{b}\times\vec{a}\right)=0$$

$$\Rightarrow \quad \vec{r}.\left(\vec{b}\times\vec{a}\right)=0$$

$$\Rightarrow \quad -\vec{r}.\left(\vec{a}\times\vec{b}\right)=0 \quad \left[\because\left[\vec{a}\,\vec{b}\,\vec{a}\right]=0\right]$$

$$\Rightarrow \quad \vec{r}.\left(\vec{a}\times\vec{b}\right)=0$$

**Hence Proved.**

## Outside Delhi [Set-II]

**Note : Except for the following questions, all the remaining questions have been asked in the previous set.**

### SECTION – A

**8. Let A = $\begin{bmatrix}200 & 50\\ 10 & 2\end{bmatrix}$ and B = $\begin{bmatrix}50 & 40\\ 2 & 3\end{bmatrix}$, then | AB | is equal to**

**(a) 460** **(b) 2000**

**(c) 3000** **(d) –7000** **[1]**

**Sol.**

$$|AB| = |A|\,|B|$$
$$= (400-500)\,(150-80)$$
$$= (-100)\,(70)$$
$$= -7000$$

$\therefore$ Option (d) is correct.

**9. Let $\vec{a}=\hat{i}-2\hat{j}+3\hat{k}$. If $\vec{b}$ is a vector such that $\vec{a}.\vec{b}=|\vec{b}|^2$ and $|\vec{a}-\vec{b}|=\sqrt{7}$, then $|\vec{b}|$ equals**

**(a) 7** **(b) 14**

**(c) $\sqrt{7}$** **(d) 21** **[1]**

**Sol.** $\left|\vec{a}-\vec{b}\right|^2=\left|\vec{a}\right|^2+\left|\vec{b}\right|^2-2\,\vec{a}.\vec{b}$

$$\Rightarrow \quad (\sqrt{7})^2=(1+4+9)+\left|\vec{b}\right|^2-2\left|\vec{b}\right|^2$$

$$\Rightarrow \quad 7-14=-\left|\vec{b}\right|^2$$

$$\Rightarrow \quad \left|\vec{b}\right|^2=7$$

$$\therefore \quad \left|\vec{b}\right|=\sqrt{7}$$

$\therefore$ Option (c) is correct.

**10. Three dice are thrown simultaneously. The probability of obtaining a total score of 5 is**

**(a) $\frac{5}{216}$** **(b) $\frac{1}{6}$**

**(c) $\frac{1}{36}$** **(d) $\frac{1}{49}$** **[1]**

**Sol.** When three dice are thrown :

$$n(S)=6\times6\times6=216$$

Let A be the event of getting total score 5.

Then A = {(1, 1, 3), (1, 3, 1), (3, 1, 1), (2, 2, 1), (2, 1, 2), (1, 2, 2)}

$$\therefore \quad n(A)=6$$

$$\therefore \quad P(A)=\frac{n(A)}{n(S)}$$
$$=\frac{6}{216}=\frac{1}{36}$$

$\therefore$ Option (c) is correct.

**15. If $f(x)=2|x|+3|\sin x|+6$, then the right hand derivative of $f(x)$ at $x=0$ is __________ .** **[1]**

**Sol.** 5

**Explanation:** $f(x)=2|x|+3|\sin x|+6$

Right hand derivative of $f$ at $x=0$ is :

$Rf'(0)=\lim_{h\to0}\frac{f(0+h)-f(0)}{h}$, where $h$ is a very small positive real number.

$$=\lim_{h\to0}\frac{(2|h|+3|\sin h|+6)-(2|0|+3\sin|0|+6)}{h}$$

$$=\lim_{h\to0}\frac{(2h+3\sin h+6)-(0+0+6)}{h}$$

$$=\lim_{h\to0}\frac{2h+3\sin h}{h}$$

$$=\lim_{h\to0}\frac{2h}{h}+3\lim_{h\to0}\frac{\sin h}{h}=2+3(1)=5$$

**Ans.**

**19. Find $\int\sin^5\left(\frac{x}{2}\right).\cos\left(\frac{x}{2}\right)dx$** **[1]**

**Sol.** Let, $I=\int\sin^5\left(\frac{x}{2}\right).\cos\left(\frac{x}{2}\right)dx$

Let $\sin\frac{x}{2}=t$

$$\Rightarrow \quad \cos\left(\frac{x}{2}\right)dx=2dt$$

$$\therefore \quad I = 2\int t^5 dt$$

$$\Rightarrow \quad I = \frac{2t^6}{6} + c$$

$$= \frac{1}{3}\sin^6\left(\frac{x}{2}\right) + c \quad \textbf{Ans.}$$

**20. If A = $\begin{bmatrix} 1 & 0 \\ 1 & 1 \end{bmatrix}$, then find A$^3$.** **[1]**

**Sol.** Given,

$$A = \begin{bmatrix} 1 & 0 \\ 1 & 1 \end{bmatrix}$$

$$A^2 = \begin{bmatrix} 1 & 0 \\ 1 & 1 \end{bmatrix}\begin{bmatrix} 1 & 0 \\ 1 & 1 \end{bmatrix}$$

$$= \begin{bmatrix} 1 & 0 \\ 2 & 1 \end{bmatrix}$$

And,

$$A^3 = A.A^2$$

$$= \begin{bmatrix} 1 & 0 \\ 1 & 1 \end{bmatrix}\begin{bmatrix} 1 & 0 \\ 2 & 1 \end{bmatrix}$$

$$\Rightarrow \quad A^3 = \begin{bmatrix} 1 & 0 \\ 3 & 1 \end{bmatrix} \quad \textbf{Ans.}$$

## SECTION – B

**25. Find the derivative of $x^{\log x}$ w.r.t. $\log x$.** **[2]**

**Sol.** Let

$$u = x^{\log x}$$

Taking log on both sides, we get

$$\log u = \log x^{\log x}$$

$$\Rightarrow \quad \log u = \log x . \log x$$

$$\Rightarrow \quad \log u = (\log x)^2$$

Diff. both sides w.r.t $x$,

$$\Rightarrow \quad \frac{1}{u}\frac{du}{dx} = \frac{2\log x}{x}$$

$$\Rightarrow \quad \frac{du}{dx} = \frac{2u\log x}{x}$$

$$\Rightarrow \quad \frac{du}{dx} = \frac{2x^{\log x}\log x}{x}$$

Also, let

$$v = \log x$$

$$\therefore \quad \frac{dv}{dx} = \frac{1}{x}$$

$$\therefore \quad \frac{du}{dv} = \frac{du}{dx} \times \frac{dx}{dv} = \frac{2x^{\log x}\log x}{x} \times \frac{x}{1}$$

$$\Rightarrow \quad \frac{du}{dv} = 2x^{\log x}.\log x \quad \textbf{Ans.}$$

**26. Find the distance between the parallel planes $2x + y + 2z = 8$ and $4x + 2y + 4z + 5 = 0$** **[2]**

**Sol.** Given lines are :

$$2x + y + 2z = 8 \quad \text{...(i)}$$

$$4x + 2y + 4z + 5 = 0 \quad \text{...(ii)}$$

Multiplying equation (i) by 2,

$$4x + 2y + 4z - 16 = 0$$

Distance between given parallel lines

$$= \frac{|D_1 - D_2|}{\sqrt{A^2 + B^2 + C^2}}$$

$$= \frac{|-16-5|}{\sqrt{4^2 + 2^2 + 4^2}} = \frac{21}{\sqrt{36}}$$

$$= \frac{21}{6} = \frac{7}{2}\text{units} \quad \textbf{Ans.}$$

## SECTION – C

**31. Show that the function $f$ : R → R defined by $f(x) = \dfrac{x}{x^2+1}$ ∀ $x \in$ R is neither one-one nor onto.** **[4]**

**Sol.** Given, $f(x) = \dfrac{x}{x^2+1}$

$$\because \quad f(2) = \frac{2}{2^2+1} = \frac{2}{5}$$

and

$$f\left(\frac{1}{2}\right) = \frac{\frac{1}{2}}{\left(\frac{1}{2}\right)^2 + 1} = \frac{\frac{1}{2}}{\frac{1}{4}+1} = \frac{\frac{1}{2}}{\frac{5}{4}}$$

$$= \frac{1}{2} \times \frac{4}{5} = \frac{2}{5}$$

$$\therefore \quad f(2) = f\left(\frac{1}{2}\right)$$

But

$$2 \neq \frac{1}{2}$$

∴ $f$ is not one-one.

**For onto :**

Let $y \in$ R

Such that

$$f(x) = y$$

$$\Rightarrow \quad \frac{x}{x^2+1} = y$$

$$\Rightarrow \quad x^2y + y = x$$

$$\Rightarrow \quad x^2y - x + y = 0$$

$$\Rightarrow \quad x = \frac{-(-1) \pm \sqrt{1-4y^2}}{2(y)}$$

$$\Rightarrow \quad x = \frac{1 \pm \sqrt{1-4y^2}}{2y}$$

Here $x$ is not real when $1 - 4y^2 < 0$

$\Rightarrow \quad 1 < 4y^2$

$\Rightarrow \quad y > \left(\frac{1}{2}\right)^2$

$\Rightarrow \quad y < \frac{-1}{2}$

or $\quad y > \frac{1}{2}$

$\therefore$ for $y \in \left(-\infty, \frac{-1}{2}\right) \cup \left(\frac{1}{2}, \infty\right)$

There exists no $x \in$ R such that $f(x) = y$.

Hence $f$ is not an onto function. **Hence Proved.**

**32. Evaluate :** $\int_{-1}^{2} |x^3 - x|\, dx$ **[4]**

**Sol.** Let, $\quad I = \int_{-1}^{2} |x^3 - x|\, dx$

Putting $\quad x^3 - x = 0$

$\Rightarrow \quad x(x^2 - 1) = 0$

$\Rightarrow \quad x(x - 1)(x + 1) = 0$

$\Rightarrow \quad x = 0, x = 1, x = -1$

$$\therefore I = \int_{-1}^{0} |x^3 - x|\, dx + \int_{0}^{1} |x^3 - x| + \int_{1}^{2} |x^3 - x|\, dx$$

$\therefore \quad I = I_1 + I_2 + I_3$

$$I_1 = \int_{-1}^{0} |x^3 - x|\, dx = \int_{-1}^{0} [x^3 - x]\, dx$$

$[\because x^3 - x \geq 0 \ \forall \ x \leq 0]$

$$= \left[\frac{x^4}{4} - \frac{x^2}{2}\right]_{-1}^{0} = 0 - \left(\frac{1}{4} - \frac{1}{2}\right) = \frac{1}{4}$$

$$I_2 = \int_{0}^{1} |x^3 - x| dx = -\int_{0}^{1} (x^3 - x)\, dx$$

$$= -\left[\frac{x^4}{4} - \frac{x^2}{2}\right]_{0}^{1}$$

$$= -\left[\left(\frac{1}{4} - \frac{1}{2}\right) - 0\right] = \frac{1}{4}$$

$$I_3 = \int_{1}^{2} |x^3 - x| dx$$

$$= \int_{1}^{2} (x^3 - x)\, dx = \left[\frac{x^4}{4} - \frac{x^2}{2}\right]_{1}^{2}$$

$$= \left[(4 - 2) - \left(\frac{1}{4} - \frac{1}{2}\right)\right]$$

$$= \left[2 + \frac{1}{4}\right] = \frac{9}{4}$$

Now, $\quad I = I_1 + I_2 + I_3$

$$= \frac{1}{4} + \frac{1}{4} + \frac{9}{4} = \frac{11}{4}$$ **Ans.**

## SECTION – D

**36. Using integration find the area of the region :**

**$\{(x, y) : 0 \leq y \leq x^2, 0 \leq y \leq x, 0 \leq x \leq 2\}$ [6]**

**Sol.** Given,

$\{(x, y) : 0 \leq y \leq x^2, 0 \leq y \leq x, 0 \leq x \leq 2\}$

$\Rightarrow \quad y \leq x^2 \quad$ ...(i)

$y \leq x \quad$ ...(ii)

$x \leq 2 \quad$ ...(iii)

$x \geq 0 \quad$ ...(iv)

$y \geq 0 \quad$ ...(v)

Considering inequalities as equation :

$y = x^2, y = x, x = 2, x = 0, y = 0$

Solving $y = x^2$ and $y = x$

$\Rightarrow \quad x^2 = x$

$\Rightarrow \quad x(x - 1) = 0$

$\Rightarrow \quad x = 0, x = 1$

$\therefore \quad y = 0, y = 1$

$\therefore$ Points of intersection of curve (i) and the (ii) are (0, 0) and (1, 1)

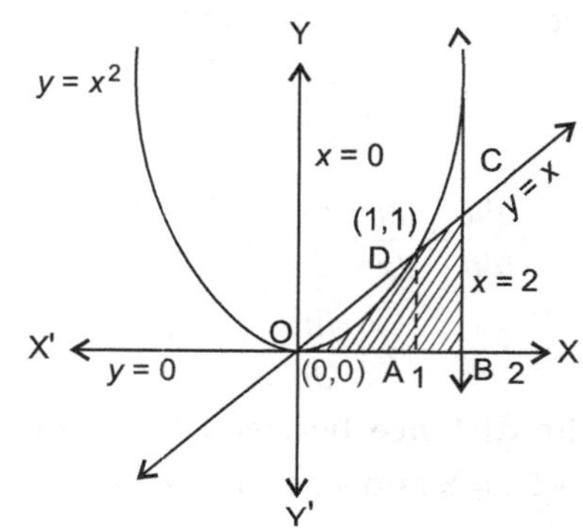

$\therefore$ Required area $= \int_0^1 (y \text{ of the parabola})\, dx$

$+ \int_1^2 (y \text{ of the line}) dx$

$\therefore$ Required area $= \int_0^1 x^2 dx + \int_1^2 x\, dx$

$= \left[\frac{x^3}{3}\right]_0^1 + \left[\frac{x^2}{2}\right]_1^2$

$= \left(\frac{1}{3} - 0\right) + \left(\frac{4}{2} - \frac{1}{2}\right)$

$= \frac{1}{3} + \left(2 - \frac{1}{2}\right)$

$= \frac{1}{3} + \frac{3}{2} = \frac{2+9}{6}$

$= \frac{11}{6}$ square units. **Ans.**

# Outside Delhi [Set-III]

**Note : Except for the following questions, all the remaining questions have been asked in the previous set.**

## SECTION – A

**8. The value of $\tan^{-1}\left[\frac{1}{2}\cos^{-1}\left(\frac{\sqrt{5}}{3}\right)\right]$ is**

(a) $\frac{3+\sqrt{5}}{2}$ (b) $\frac{3-\sqrt{5}}{2}$

(c) $\frac{-3+\sqrt{5}}{2}$ (d) $\frac{-3-\sqrt{5}}{2}$ [1]

**Sol.** $\tan^{-1}\left[\frac{1}{2}\cos^{-1}\left(\frac{\sqrt{5}}{3}\right)\right]$

This question is wrong.

It should be $\tan\left[\frac{1}{2}\cos^{-1}\left(\frac{\sqrt{5}}{3}\right)\right]$

Now, let $y = \tan\left[\frac{1}{2}\cos^{-1}\left(\frac{\sqrt{5}}{3}\right)\right]$

Putting, $x = \cos^{-1}\left(\frac{\sqrt{5}}{3}\right)$

$\Rightarrow$ $\cos x = \frac{\sqrt{5}}{3}$ ...(i)

Now, $y = \tan\left(\frac{1}{2} \times x\right)$

$y = \tan\frac{x}{2}$

$\therefore$ $y = \sqrt{\frac{1-\cos x}{1+\cos x}}$ (using property)

$\therefore$ [From (i)] $y = \sqrt{\frac{1-\sqrt{5}/3}{1+\sqrt{5}/3}}$

$\therefore$ $y = \sqrt{\frac{3-\sqrt{5}}{3+\sqrt{5}}} = \sqrt{\frac{(3-\sqrt{5})(3-\sqrt{5})}{(3+\sqrt{5})(3-\sqrt{5})}}$

$\therefore$ $y = \sqrt{\frac{(3-\sqrt{5})^2}{9-5}} = \frac{3-\sqrt{5}}{\sqrt{4}} = \frac{3-\sqrt{5}}{2}$

$\therefore$ $y = \frac{3-\sqrt{5}}{2}$

*i.e.,* $\tan\left[\frac{1}{2}\cos^{-1}\left(\frac{\sqrt{5}}{3}\right)\right] = \frac{3-\sqrt{5}}{2}$

$\therefore$ Option (b) is correct.

**9. If A = $\begin{bmatrix} a & 0 & 0 \\ 0 & a & 0 \\ 0 & 0 & a \end{bmatrix}$, then det (adj A) equals**

**(a)** $a^{27}$ **(b)** $a^9$

**(c)** $a^6$ **(d)** $a^2$ [1]

**Sol.** $|A| = a^3$

$\therefore$ $|\text{adj } A| = |A|^2$

$= (a^3)^2 = a^6$

$\therefore$ Option (c) is correct.

**10. The line $\frac{x-2}{3}=\frac{y-3}{4}=\frac{z-4}{5}$ is parallel to the plane**

**(a) $2x + 3y + 4z = 0$ (b) $3x + 4y - 5z = 7$**

**(c) $2x + y - 2z = 0$ (d) $x - y + z = 2$ [1]**

**Sol.** Option (b) or (c) is correct.

As for (b) → $3.3 + 4 \times 4 - 5 \times 5 = 0$

or (c) → $2.3 + 4 \times 1 - 2 \times 5 = 0$

**15. If $f(x) = x\,|x|$, then $f'(x) =$ ________. [1]**

**Sol.** $2x, x \geq 0, -2x, x < 0$

**Explanation:** $f(x) = x\,|x|$

$$= \begin{cases} x^2, & \text{if } x \geq 0 \\ -x^2, & \text{if } x < 0 \end{cases}$$

$$\therefore \quad f'(x) = \begin{cases} 2x, & \text{if } x \geq 0 \\ -2x, & \text{if } x < 0 \end{cases}$$

**19. Find $\int \frac{1}{x(1+x^2)} dx$**

**Sol.** Let, $I = \int \frac{1}{x(1+x^2)} dx$

$$= \int \frac{x}{x^2(x^2+1)} dx$$

Let $x^2 = t$

$$\Rightarrow x\,dx = \frac{dt}{2}$$

$$\therefore \quad I = \frac{1}{2}\int \frac{dt}{t(t+1)} = \frac{1}{2}\int \frac{(t+1)-t}{t(t+1)} dt$$

$$= \frac{1}{2}\int \left(\frac{1}{t} - \frac{1}{t+1}\right) dt$$

$$= \frac{1}{2}[\log|t| - \log|t+1|] + c$$

$$= \frac{1}{2}\log\left|\frac{t}{t+1}\right| + c = \frac{1}{2}\log\left|\frac{x^2}{x^2+1}\right| + c$$ **Ans.**

**20. If $[x]$ denotes the greatest integer function, then find $\int_0^{3/2} [x^2]\,dx$. [1]**

**Sol.** $$\int_0^{3/2} [x^2]dx = \int_0^1 [x^2]dx + \int_1^{\sqrt{2}} [x^2]dx + \int_{\sqrt{2}}^{3/2} [x^2]dx$$

$$= \int_0^1 0\,dx + \int_1^{\sqrt{2}} 1\,dx + \int_{\sqrt{2}}^{3/2} 2\,dx = 0 + [x]_1^{\sqrt{2}} + 2[x]_{\sqrt{2}}^{3/2}$$

$$= \sqrt{2} - 1 + 2\left(\frac{3}{2} - \sqrt{2}\right) = \sqrt{2} - 1 + 3 - 2\sqrt{2}$$

$$= 2 - \sqrt{2}$$ **Ans.**

## SECTION–B

**25. Show that the function $f(x) = \frac{x}{3} + \frac{3}{x}$ decreases in the intervals $(-3, 0) \cup (0, 3)$. [2]**

**Sol.** Given, $f(x) = \frac{x}{3} + \frac{3}{x}$

$$\Rightarrow \quad f'(x) = \frac{1}{3} - \frac{3}{x^2}$$

$$\Rightarrow \quad f'(x) = \frac{x^2 - 9}{3x^2}$$

Now, $f$ is decreasing when $f'(x) < 0$

$$\Rightarrow \quad \frac{x^2-9}{3x^2} < 0$$

$$\Rightarrow \quad x^2 < 9, x \neq 0$$

$$\Rightarrow \quad x^2 < 3^2, x \neq 0$$

$$\Rightarrow \quad -3 < x < 3, x \neq 0$$

$\Rightarrow f(x)$ decreases in $(-3, 0) \cup (0, 3)$

**Hence Proved.**

**26. Three distinct numbers are chosen randomly from the first 50 natural numbers. Find the probability that all the three numbers are divisible by both 2 and 3. [2]**

**Sol.** Here, $n(S) = {}^{50}C_3$

Numbers divisible by both 2 and 3 from first 50 natural numbers are 6, 12, 18, 24, 30, 36, 42, 48

*i.e.*, 8 numbers

$$\therefore \text{ Required Probability } = \frac{{}^8C_3}{{}^{50}C_3}$$

$$= \frac{8 \times 7 \times 6}{50 \times 49 \times 48} = \frac{1}{350}$$ **Ans.**

## SECTION–C

**31. Evaluate $\int_0^1 \sqrt{3 - 2x - x^2}\,dx$ [4]**

**Sol.** $\sqrt{3 - 2x - x^2}dx = \int_0^1 \sqrt{-[x^2 + 2x - 3]}\,dx$

$$= \int_0^1 \sqrt{-[(x+1)^2 - 2^2]}\,dx = \int_0^1 \sqrt{2^2-(x+1)^2}\,dx$$

$$= \left[\frac{x+1}{2}\sqrt{3-2x-x^2} + \frac{2^2}{2}\sin^{-1}\left(\frac{x+1}{2}\right)\right]_0^1$$

$$= \left[\frac{2}{2}\sqrt{0}+2\sin^{-1}(1)\right] - \left[\frac{1}{2}\sqrt{3}+2\sin^{-1}\left(\frac{1}{2}\right)\right]$$

$$= 2\left(\frac{\pi}{2}\right) - \frac{\sqrt{3}}{2} - 2\left(\frac{\pi}{6}\right) = \pi - \frac{\sqrt{3}}{2} - \frac{\pi}{3} = \frac{2\pi}{3} - \frac{\sqrt{3}}{2}$$

**Ans.**

**32. Find the general solution of the differential equation $\frac{dy}{dx} + \frac{1}{x} = \frac{e^y}{x}$.** **[4]**

**Sol.** Given,

$$\frac{dy}{dx} + \frac{1}{x} = \frac{e^y}{x}$$

$$\Rightarrow \quad \frac{dy}{dx} = \frac{e^y - 1}{x}$$

$$\Rightarrow \quad \int \frac{dy}{e^y - 1} = \int \frac{dx}{x}$$

$$\Rightarrow \quad \int \frac{e^y}{e^y(e^y-1)}dy = \int \frac{dx}{x}$$

Let $\quad e^y = t$

$$\Rightarrow \quad e^y dy = dt$$

$$\Rightarrow \quad \int \frac{dt}{t(t-1)} = \int \frac{dx}{x}$$

$$\Rightarrow \quad \int \frac{t-(t-1)}{t(t-1)}dt = \int \frac{dx}{x}$$

$$\Rightarrow \quad \int \frac{1}{t-1}dt - \int \frac{1}{t}dt = \int \frac{dx}{x}$$

$$\Rightarrow \quad \log|t-1| - \log|t| = \log|x| + \log|C|$$

$$\Rightarrow \quad \log\left|\frac{t-1}{t}\right| = \log|Cx|$$

$$\Rightarrow \quad \frac{t-1}{t} = Cx$$

$$\Rightarrow \quad e^y - 1 = Cxe^y$$

$$\Rightarrow \quad 1 - e^{-y} = Cx$$

**Ans.**

## SECTION – D

**36. Find the image of the point (–1, 3, 4) in the plane $x - 2y = 0$.** **[6]**

**Sol.**

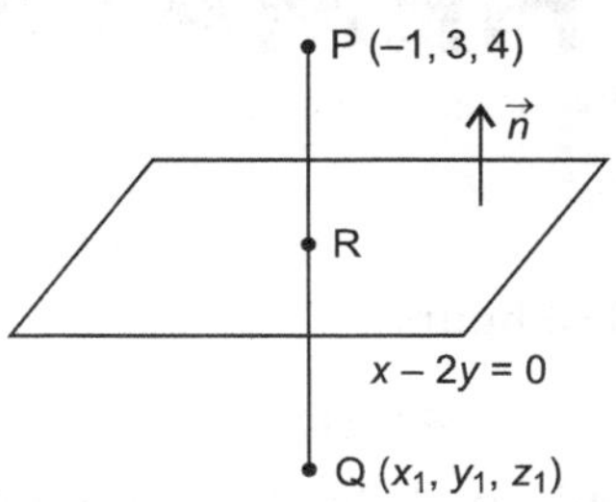


Let the image of the point P(–1, 3, 4) in the plane $x - 2y = 0$ be Q $(x_1, y_1, z_1)$.

$\therefore$ Direction ratios of $\overrightarrow{PQ}$ are

$a_1 = x_1 + 1,\ b_1 = y_1 - 3,\ c_1 = z_1 - 4$

Also, direction ratios of $\vec{n}$ are

$a_2 = 1,\ b_2 = -2,\ c_2 = 0$

Since $\overrightarrow{PQ} \parallel \vec{n}$

$$\Rightarrow \quad \frac{a_1}{a_2} = \frac{b_1}{b_2} = \frac{c_1}{c_2}$$

$$\Rightarrow \quad \frac{x_1+1}{1} = \frac{y_1-3}{-2} = \frac{z_1-4}{0} = \lambda \text{ (say)}$$

$$\Rightarrow \quad x_1 = \lambda - 1,\ y_1 = 3 - 2\lambda,\ z_1 = 4$$

$$\therefore \quad \text{Point Q} = (\lambda - 1, 3 - 2\lambda, 4)$$

$$\text{Mid-point of PQ} = R\left(\frac{\lambda-1-1}{2}, \frac{3-2\lambda+3}{2}, \frac{4+4}{2}\right)$$

$$= R\left(\frac{\lambda-2}{2}, 3-\lambda, 4\right)$$

$\because$ Point R lies on the plane $x - 2y = 0$.

$$\therefore \quad \frac{\lambda-2}{2} - 2(3-\lambda) = 0$$

$$\Rightarrow \quad \frac{\lambda-2}{2} = 6 - 2\lambda$$

$$\Rightarrow \quad \lambda - 2 = 12 - 4\lambda$$

$$\Rightarrow \quad 5\lambda = 14$$

$$\Rightarrow \quad \lambda = \frac{14}{5}$$

$$\therefore \quad Q = \left(\frac{14}{5} - 1,\ 3 - \frac{28}{5},\ 4\right)$$

$$= \left(\frac{9}{5}, \frac{-13}{5}, 4\right)$$

Hence, the image of the point (–1, 3, 4) in the plane $x - 2y = 0$ is $\left(\frac{9}{5}, -\frac{13}{4}, 4\right)$ **Ans.**

● ●

# MATHEMATICS

## Delhi [Set-I] 2020

Time allowed : 3 hours Maximum marks : 80

### SECTION–A

**Question numbers 1 to 10 are multiple choice questions of 1 mark each. Select the correct option :**

**1. If A is a square matrix of order 3, such that A (adj A) = 10 I, then |adj A| is equal to :**

**(a) 1** **(b) 10**

**(c) 100** **(d) 101** **[1]**

**Sol.** Given : $A\,(\text{adj } A) = 10I$

We know, $A.\text{adj } A = |A|.I$

$\therefore$ $|A| = 10$

$\because$ $|\text{adj } A| = |A|^{n-1}$

$\therefore$ $|\text{adj } A| = |A|^{3-1} = 10^2 = 100$

$\therefore$ Option (c) is correct.

**2. If A is a 3 × 3 matrix such that |A| = 8, then |3A| equals:**

**(a) 8** **(b) 24**

**(c) 72** **(d) 216** **[1]**

**Sol.** Given : $|A| = 8$

We know, $|kA| = k^3|A|$, where $k$ is constant

$\therefore$ $|3A| = 3^3|A|$

$= 27 \times 8 = 216$

$\therefore$ Option (d) is correct.

**3. If $y = Ae^{5x} + Be^{-5x}$, then $\frac{d^2y}{dx^2}$ is equal to:**

**(a) 25$y$** **(b) 5$y$**

**(c) – 25$y$** **(d) 15$y$** **[1]**

**Sol.** Given : $y = Ae^{5x} + Be^{-5x}$

$\therefore$ $\frac{dy}{dx} = 5Ae^{5x} - 5Be^{-5x}$

$\Rightarrow$ $\frac{d^2y}{dx^2} = 25Ae^{5x} + 25Be^{-5x}$

$\Rightarrow$ $\frac{d^2y}{dx^2} = 25y$

$\therefore$ Option (a) is correct.

**4. $\int x^2 e^{x^3}\, dx$ equals :**

**(a) $\frac{1}{3}e^{x^3} + C$** **(b) $\frac{1}{3}e^{x^4} + C$**

**(c) $\frac{1}{2}e^{x^3} + C$** **(d) $\frac{1}{2}e^{x^2} + C$** **[1]**

**Sol.** Let $I = \int x^2 e^{x^3}\, dx$

Put $x^3 = t$

$\Rightarrow$ $3x^2dx = dt$

$\Rightarrow$ $x^2dx = \frac{1}{3}dt$

$\therefore$ $I = \frac{1}{3}\int e^t dt$

$I = \frac{1}{3}e^t + C$

$I = \frac{1}{3}e^{x^3} + C$

$\therefore$ Option (a) is correct.

**5. If $\hat{i}, \hat{j}, \hat{k}$ are unit vectors along three mutually perpendicular directions, then**

**(a) $\hat{i}.\hat{j} = 1$** **(b) $\hat{i} \times \hat{j} = 1$**

**(c) $\hat{i}.\hat{k} = 0$** **(d) $\hat{i} \times \hat{k} = 0$** **[1]**

**Sol.** Option (c) is correct.

**6. ABCD is a rhombus whose diagonals intersect at E. Then $\vec{EA} + \vec{EB} + \vec{EC} + \vec{ED}$ equals :**

**(a) $\vec{0}$** **(b) $\vec{AD}$**

**(c) $2\vec{BC}$** **(d) $2\vec{AD}$** **[1]**

**Sol.** Option (a) is correct.

**7. The lines $\frac{x-2}{1} = \frac{y-3}{1} = \frac{4-z}{k}$ and $\frac{x-1}{k} = \frac{y-4}{2} = \frac{z-5}{-2}$ are mutually perpendicular if the value of $k$ is :**

**(a) $-\frac{2}{3}$** **(b) $\frac{2}{3}$**

**(c) – 2** **(d) 2** **[1]**

**Sol.** First line is,

$$\frac{x-2}{1} = \frac{y-3}{1} = \frac{z-4}{-k}$$

$\therefore$ Direction ratios of first line = $< 1, 1, -k >$

Second line is,

$$\frac{x-1}{k} = \frac{y-4}{2} = \frac{z-5}{-2}$$

$\therefore$ Direction ratios of second line = $< k, 2, -2 >$

We know, two lines are perpendicular to each other, if the dot product of their direction ratios is zero.

*i.e.,* $1 \times k + 1 \times 2 + (-k) \times (-2) = 0$

$$\Rightarrow \quad k + 2 + 2k = 0$$

$$\Rightarrow \quad 3k = -2$$

$$\Rightarrow \quad k = -\frac{2}{3}$$

$\therefore$ Option (a) is correct.

**8. The graph of the inequality $2x + 3y > 6$ is**

**(a) half plane that contains the origin**

**(b) half plane that neither contains the origin nor the points of the line $2x + 3y = 6$**

**(c) whole XOY–plane excluding the points on the line $2x + 3y = 6$**

**(d) entire XOY plane. [1]**

**Sol.** Option (b) is correct.

**9. A card is picked at random from a pack of 52 playing cards. Given that the picked card is a queen, the probability of this card to be a card of spade is :**

**(a)** $\frac{1}{3}$ **(b)** $\frac{4}{13}$

**(c)** $\frac{1}{4}$ **(d)** $\frac{1}{2}$ **[1]**

**Sol.** Let event A = Picked card is a queen. and event B = card of spade is picked.

$$\therefore \quad P(A) = \frac{4}{52}, \; P(B) = \frac{13}{52}$$

and, $P(A \cap B) = \frac{1}{52}$ [$\because A \cap B$ = Queen of spade]

Now, required probability,

$$P(B/A) = \frac{P(A \cap B)}{P(A)} = \frac{1/52}{4/52} = \frac{1}{4}$$

$\therefore$ Option (c) is correct.

**10. A die is thrown once. Let A be the event that the number obtained is greater than 3. Let B be the event that the number obtained is less than 5. Then P(A∪B) is :**

**(a)** $\frac{2}{5}$ **(b)** $\frac{3}{5}$

**(c) 0** **(d) 1** **[1]**

**Sol.**

$$P(A) = \frac{3}{6}$$

$$P(B) = \frac{4}{6}$$

and $$P(A \cap B) = \frac{1}{6}$$

We know, $$P(A \cup B) = P(A) + P(B) - P(A \cap B)$$

$$= \frac{3}{6} + \frac{4}{6} - \frac{1}{6}$$

$$= \frac{6}{6} = 1$$

$\therefore$ Option (d) is correct.

**Fill in the blanks in Questions from 11 to 15.**

**11. A relation in a set A is called ________ relation, if each element of A is related to itself. [1]**

**Sol.** reflexive

**12. If $A + B = \begin{bmatrix} 1 & 0 \\ 1 & 1 \end{bmatrix}$ and $A - 2B = \begin{bmatrix} -1 & 1 \\ 0 & -1 \end{bmatrix}$, then $A =$ ________ . [1]**

**Sol.** $\begin{bmatrix} \frac{1}{3} & \frac{1}{3} \\ \frac{2}{3} & \frac{1}{3} \end{bmatrix}$

**Explanation:**

Given : $$A + B = \begin{bmatrix} 1 & 0 \\ 1 & 1 \end{bmatrix} \quad ...(i)$$

and $$A - 2B = \begin{bmatrix} -1 & 1 \\ 0 & -1 \end{bmatrix} \quad ...(ii)$$

$$\Rightarrow \quad A - 2\left\{\begin{bmatrix} 1 & 0 \\ 1 & 1 \end{bmatrix} - A\right\} = \begin{bmatrix} -1 & 1 \\ 0 & -1 \end{bmatrix} \quad \text{[From (i)]}$$

$$\Rightarrow \quad A - \begin{bmatrix} 2 & 0 \\ 2 & 2 \end{bmatrix} + 2A = \begin{bmatrix} -1 & 1 \\ 0 & -1 \end{bmatrix}$$

$$3A = \begin{bmatrix} -1 & 1 \\ 0 & -1 \end{bmatrix} + \begin{bmatrix} 2 & 0 \\ 2 & 2 \end{bmatrix}$$

$$3A = \begin{bmatrix} 1 & 1 \\ 2 & 1 \end{bmatrix}$$

$$A = \frac{1}{3}\begin{bmatrix} 1 & 1 \\ 2 & 1 \end{bmatrix} = \begin{bmatrix} 1/3 & 1/3 \\ 2/3 & 1/3 \end{bmatrix}$$

**13. The least value of the function $f(x) = ax + \frac{b}{x}$ $(a > 0, b > 0, x > 0)$ is ________ . [1]**

**Sol.** $2\sqrt{ab}$

**Explanation:**

$$\because \quad f(x) = ax + \frac{b}{x}$$

$$\therefore \quad f'(x) = a - \frac{b}{x^2}$$

For maximum and minimum function,

Put $f'(x) = 0$

$\Rightarrow \quad a - \dfrac{b}{x^2} = 0$

$\Rightarrow \quad ax^2 - b = 0$

$\Rightarrow \quad ax^2 = b$

$\Rightarrow \quad x = \sqrt{\dfrac{b}{a}} \quad [\because x > 0]$

Now, $f''(x) = 0 + \dfrac{2b}{x^3}$

$= \dfrac{2b}{\left(\sqrt{b/a}\right)^3} > 0$

$\therefore f(x)$ is minimum at $x = \sqrt{\dfrac{b}{a}}$

$\therefore$ The least value of $f(x)$ is,

$$f\left(\sqrt{\frac{b}{a}}\right) = a\sqrt{\frac{b}{a}} + \frac{b}{\sqrt{\frac{b}{a}}}$$

$$= \sqrt{ab} + \sqrt{ab} = 2\sqrt{ab}$$

**14. The integrating factor of the differential equation $x\dfrac{dy}{dx} + 2y = x^2$ is ______.** **[1]**

**OR**

**The degree of the differential equation $1 + \left(\dfrac{dy}{dx}\right)^2 = x$ is ______.**

**Sol.** $x^2$

**Explanation:**

$\because \quad x\dfrac{dy}{dx} + 2y = x^2$

$\Rightarrow \quad \dfrac{dy}{dx} + \dfrac{2y}{x} = x$

Here, $P = \dfrac{2}{x}, \; Q = x$

$\therefore \quad \text{I.F.} = e^{\int P dx} = e^{\int \frac{2}{x} dx}$

$= e^{2\log x} = e^{\log x^2}$

$= x^2$

**OR**

2

**15. The vector equation of a line which passes through the points (3, 4, – 7) and (1, –1, 6) is ______.** **[1]**

**OR**

**The line of shortest distance between two skew lines is ______ to both the lines.**

**Sol.** $(3\hat{i} + 4\hat{j} - 7\hat{k}) + \lambda(-2\hat{i} - 5\hat{j} + 13\hat{k})$

**Explanation:**

$\because$ A = (3, 4, – 7) and B = (1, – 1, 6)

$\therefore$ Direction ratios of AB = < – 2, –5, 13 >

$\therefore$ Equation of required line is

$\vec{r} = (3\hat{i} + 4\hat{j} - 7\hat{k}) + \lambda(-2\hat{i} - 5\hat{j} + 13\hat{k})$

$\vec{r} = (3\hat{i} + 4\hat{j} - 7\hat{k}) - \lambda(2\hat{i} + 5\hat{j} - 13\hat{k})$

**OR**

Perpendicular

**Q. Nos. 16 to 20 are of very short answer type questions.**

**16. Find the value of $\sin^{-1}\left[\sin\left(-\dfrac{17\pi}{8}\right)\right]$.** **[1]**

**Sol.** We know that

$\sin^{-1}(\sin x) = x$

$$\text{So, } \sin^{-1}\left[\sin\left(\frac{-17\pi}{8}\right)\right] = \sin^{-1}\left[-\sin\frac{17\pi}{8}\right]$$

$$= \sin^{-1}\left[-\sin\left(2\pi + \frac{\pi}{8}\right)\right]$$

$$= \sin^{-1}\left(-\sin\frac{\pi}{8}\right)$$

$$= \sin^{-1}\left[\sin\left(-\frac{\pi}{8}\right)\right]$$

$$= -\frac{\pi}{8}$$ **Ans.**

**17. For A = $\begin{bmatrix} 3 & -4 \\ 1 & -1 \end{bmatrix}$ write $A^{-1}$.** **[1]**

**Sol.** Given : $A = \begin{bmatrix} 3 & -4 \\ 1 & -1 \end{bmatrix}$

$\therefore \quad |A| = 3 \times (-1) - (-4) \times 1$

$= -3 + 4 = 1$

and $\quad \text{adj } A = \begin{bmatrix} -1 & 4 \\ -1 & 3 \end{bmatrix}$

$\because \quad A^{-1} = \dfrac{1}{|A|}\text{adj } A$

$\therefore \quad A^{-1} = \dfrac{1}{1}\begin{bmatrix} -1 & 4 \\ -1 & 3 \end{bmatrix} = \begin{bmatrix} -1 & 4 \\ -1 & 3 \end{bmatrix}$ **Ans.**

**18. If the function $f$ defined as :**

$$f(x) = \begin{cases} \dfrac{x^2-9}{x-3} & , x \neq 3 \\ k & , x = 3 \end{cases}$$

**is continuous at $x = 3$, find the value of $k$. [1]**

**Sol.** Clearly, $f(x) = \dfrac{x^2-9}{x-3} = x+3$

and $f(3) = k$

Now, for R.H.L.,

$$\lim_{x \to 3^+} f(x) = \lim_{h \to 0}(3+h)+3 = 6$$

and, for L.H.L.,

$$\lim_{x \to 3^-} f(x) = \lim_{h \to 0}(3-h)+3 = 6$$

$\because f(x)$ is continuous at $x = 3$

$\therefore$ RHL = LHL = $f(3)$

$\therefore$ $6 = k$

$\therefore$ $k = 6$ **Ans.**

**19. If $f(x) = x^4 - 10$, then find the approximate value of $f(2.1)$. [1]**

**OR**

**Find the slope of the tangent to the curve $y = 2\sin^2(3x)$ at $x = \dfrac{\pi}{6}$.**

**Sol.** Given : $f(x) = x^4 - 10$

$\therefore$ $f'(x) = 4x^3$

Let $x = 2$ and $h = 0.1$

We know that

$$f'(x) = \frac{f(x+h)-f(x)}{h}$$

or $h.f'(x) = f(x+h) - f(x)$

$\Rightarrow$ $0.1 \times \{4(2)^3\} = f(2.1) - (2^4 - 10)$

$\Rightarrow$ $3.2 = f(2.1) - 6$

$\Rightarrow$ $f(2.1) = 9.2$ **Ans.**

**OR**

Given equation of curve is,

$$y = 2\sin^2(3x)$$

$\therefore$ $\dfrac{dy}{dx} = 4\sin 3x \cos 3x.3$

$\Rightarrow$ $\dfrac{dy}{dx} = 12\sin 3x.\cos 3x$

$\therefore$ $\left[\dfrac{dy}{dx}\right]_{x=\frac{\pi}{6}} = 12\sin\dfrac{\pi}{2}\cos\dfrac{\pi}{2} = 0$

$\therefore$ Slope of tangent = 0 **Ans.**

**20. Find the value of $\displaystyle\int_1^4 |x-5|\,dx$ [1]**

**Sol.** Let $I = \displaystyle\int_1^4 |x-5|\,dx$

$\Rightarrow$ $I = \displaystyle\int_1^4 -(x-5)\,dx$

$\Rightarrow$ $I = \dfrac{-1}{2}[x^2]_1^4 + 5[x]_1^4$

$\Rightarrow$ $I = \dfrac{-1}{2}(16-1)+5(4-1)$

$\Rightarrow$ $I = \dfrac{-15}{2}+15 = \dfrac{15}{2}$ **Ans.**

## SECTION-B

**Q. Nos. 21 to 26 carry 2 marks each.**

**21. If $f(x) = \dfrac{4x+3}{6x-4}$, $x \neq \dfrac{2}{3}$, then show that $(fof)(x) = x$, for all $x \neq \dfrac{2}{3}$ Also, write inverse of $f$. [2]**

**OR**

**Check if the relation R in the set R of real numbers defined as R = {$(a, b) : a < b$} is (i) symmetric, (ii) transitive**

**Sol.** Given: $f(x) = \dfrac{4x+3}{6x-4}$

Now, $(fof)x = f(f(x))$

$$= \frac{4\left(\dfrac{4x+3}{6x-4}\right)+3}{6\left(\dfrac{4x+3}{6x-4}\right)-4}$$

$$= \frac{16x+12+18x-12}{24x+18-24x+16}$$

$$= \frac{34x}{34} = x$$

$\therefore$ $(fof)x = x$ **Hence Proved.**

To find inverse of $f(x)$,

Let $f(x) = y$

$\therefore$ $y = \dfrac{4x+3}{6x-4}$

$\Rightarrow$ $6xy - 4y = 4x + 3$

$\Rightarrow$ $6xy - 4x = 4y + 3$

$\Rightarrow \quad x(6y-4) = 4y+3$

$\Rightarrow \quad x = \dfrac{4y+3}{6y-4}$

$= f^{-1}(y) \quad \begin{bmatrix} \because f(x) = y \\ \therefore x = f^{-1}(y) \end{bmatrix}$

$\therefore \quad f^{-1}(x) = \dfrac{4x+3}{6x-4}, x \neq \dfrac{2}{3}$ **Ans.**

**OR**

Given : $R = \{(a, b) : a < b\}$

Let A be the set of real numbers.

**(i)** For symmetric : Let $a, b \in A$

Then, $(a, b) \in R$ such that $a < b$

$\Rightarrow \quad b < a$ which is false

$\Rightarrow \quad (b, a) \notin R$

$\therefore$ R is not symmetric. **Ans.**

**(ii)** For transitive : Let $a, b, c \in A$

Then, $(a, b) \in R$ and $(b, c) \in R$

$\Rightarrow \quad a < b$ and $b < c$

$\Rightarrow \quad a < c$

$\Rightarrow \quad (a, c) \in R$

$\therefore$ R is transitive. **Ans.**

**22. Find $\int \dfrac{x}{x^2+3x+2} dx$** **[2]**

**Sol.** Let $\quad I = \int \dfrac{x}{x^2+3x+2} dx$

$= \int \dfrac{x}{(x+1)(x+2)} dx$

Let $\quad \dfrac{x}{(x+1)(x+2)} = \dfrac{A}{x+1} + \dfrac{B}{x+2}$

$\Rightarrow \quad x = A(x+2) + B(x+1)$

$\Rightarrow \quad x = (A+B)x + (2A+B)$

Comparing coefficients of $x$ both sides, we get

$1 = A + B \quad$ ...(i)

Comparing constant function both sides, we get

$0 = 2A + B \quad$ ...(ii)

Solving equations (i) and (ii), we get

$A = -1$ and $B = 2$

$\therefore \quad I = -\int \dfrac{dx}{x+1} + 2\int \dfrac{dx}{x+2}$

$\Rightarrow \quad I = -\log|x+1| + 2\log|x+2| + C$

$\Rightarrow \quad I = -\log|x+1| + \log|x+2|^2 + C$

$\Rightarrow \quad I = \log\left|\dfrac{(x+2)^2}{x+1}\right| + C$ **Ans.**

**23. If $x = a\cos\theta$; $y = b\sin\theta$, then find $\dfrac{d^2y}{dx^2}$.** **[2]**

**OR**

**Find the differential of $\sin^2 x$ w.r.t. $e^{\cos x}$.**

**Sol.** Given: $x = a\cos\theta$ and $y = b\sin\theta$

$\therefore \quad \dfrac{dx}{d\theta} = a(-\sin\theta) = -a\sin\theta$

and $\quad \dfrac{dy}{d\theta} = b(\cos\theta) = b\cos\theta$

$\therefore \quad \dfrac{dy}{dx} = \dfrac{dy}{d\theta} \times \dfrac{d\theta}{dx} = \dfrac{dy}{d\theta} \times \dfrac{1}{dx/d\theta}$

$= b\cos\theta \times \dfrac{1}{(-a\sin\theta)}$

$= -\dfrac{b}{a}\cot\theta$

Now, $\quad \dfrac{d^2y}{dx^2} = -\dfrac{b}{a}(-\operatorname{cosec}^2\theta)\dfrac{d\theta}{dx}$

$= \dfrac{b}{a}\operatorname{cosec}^2\theta \times \dfrac{1}{(-a\sin\theta)}$

$= -\dfrac{b}{a^2}\operatorname{cosec}^3\theta$ **Ans.**

**OR**

Let $\quad u = \sin^2 x$

$\therefore \quad \dfrac{du}{dx} = 2\sin x\cos x$

Also, $\quad v = e^{\cos x}$

$\therefore \quad \dfrac{dv}{dx} = -\sin x\, e^{\cos x}$

Now, $\quad \dfrac{du}{dv} = \dfrac{du}{dx} \times \dfrac{dx}{dv}$

$\dfrac{du}{dv} = \dfrac{2\sin x\cos x}{-\sin x\, e^{\cos x}}$

$= -2\cos x\, e^{-\cos x}$ **Ans.**

**24. Evaluate $\int_1^2 \left[\dfrac{1}{x} - \dfrac{1}{2x^2}\right] e^{2x} dx$** **[2]**

**Sol.** Let $I = \int_1^2 \left[\dfrac{1}{x} - \dfrac{1}{2x^2}\right] e^{2x} dx$

$$\Rightarrow I = \int_1^2 \frac{1}{x}e^{2x}dx - \int_1^2 \frac{1}{2x^2}e^{2x}dx$$

$$\Rightarrow I = \left[\frac{1}{x}\frac{e^{2x}}{2} - \int\left(\frac{-1}{x^2}\right)\frac{e^{2x}}{2}dx\right]_1^2 - \int_1^2 \frac{1}{2x^2}e^{2x}dx$$

[Using by parts]

$$\Rightarrow I = \frac{1}{2}\left[\frac{1}{x}e^{2x}\right]_1^2 + \int_1^2 \frac{1}{2x^2}e^{2x}dx - \int_1^2 \frac{1}{2x^2}e^{2x}dx$$

$$\Rightarrow I = \frac{1}{2}\left[\frac{1}{2}e^4 - 1e^2\right]$$

$$\Rightarrow I = \frac{e^4}{4} - e^2/2$$ **Ans.**

**25. Find the value of $\int_0^1 x(1-x)^n dx$.** **[2]**

**Sol.** Let $I = \int_0^1 x(1-x)^n dx$

$$\Rightarrow I = \int_0^1 (1-x)x^n dx \quad \left[\because \int_0^a f(x)dx = \int_0^a f(a-x)dx\right]$$

$$\Rightarrow I = \int_0^1 x^n dx - \int_0^1 x^{n+1}dx$$

$$\Rightarrow I = \frac{1}{n+1}[x^{n+1}]_0^1 - \frac{1}{n+2}[x^{n+2}]_0^1$$

$$\Rightarrow I = \frac{1}{n+1}[1-0] - \frac{1}{n+2}[1-0]$$

$$\Rightarrow I = \frac{1}{n+1} - \frac{1}{n+2}$$

$$\Rightarrow I = \frac{n+2-n-1}{(n+1)(n+2)} = \frac{1}{(n+1)(n+2)}$$ **Ans.**

**26. Given two independent events A and B such that P(A) = 0.3 and P(B) = 0.6, find P(A′∩ B′).** **[2]**

**Sol.** Given : P(A) = 0.3 and P(B) = 0.6

∵ A and B are independent events

Then,

$$\begin{aligned} P(A' \cap B') &= P(A')\,P(B') \\ &= (1 - P(A))\,(1 - P(B)) \\ &= (1 - 0.3)\,(1 - 0.6) \\ &= 0.7 \times 0.4 \\ &= 0.28 \end{aligned}$$ **Ans.**

## SECTION-C

**Q. Nos. 27 to 32 carry 4 marks each.**

**27. Solve for $x$ : $\sin^{-1}(1-x) - 2\sin^{-1}(x) = \frac{\pi}{2}$** **[4]**

**Sol.** Given :

$$\sin^{-1}(1-x) - 2\sin^{-1}(x) = \frac{\pi}{2}$$

$$\Rightarrow \sin^{-1}(1-x) = \frac{\pi}{2} + 2\sin^{-1}(x)$$

$$\Rightarrow 1 - x = \sin\left(\frac{\pi}{2} + 2\sin^{-1}x\right)$$

$$\Rightarrow 1 - x = \cos(2\sin^{-1}x)$$

$$\left[\because \sin\left(\frac{\pi}{2}+\theta\right) = \cos\theta\right]$$

Let $\sin^{-1}x = \theta$

$\Rightarrow x = \sin\theta$

$$\therefore \cos(2\sin^{-1}x) = \cos(2\theta)$$

$$\Rightarrow \quad = 1 - 2\sin^2\theta = 1 - 2(x)^2 = 1 - 2x^2$$

$\therefore 1 - x = 1 - 2x^2$

$\Rightarrow 2x^2 - x = 0$

$\Rightarrow x(2x - 1) = 0$

$\Rightarrow x = 0 \text{ or } \frac{1}{2}$

For $x = \frac{1}{2}$,

$$\sin^{-1}(1-x) - 2\sin^{-1}x = \sin^{-1}\frac{1}{2} - 2\sin^{-1}\frac{1}{2}$$

$$= -\sin^{-1}\frac{1}{2}$$

$$= -\frac{\pi}{2} \neq \frac{\pi}{2}$$

$\therefore x = 0$ **Ans.**

**28. If $y = (\log x)^x + x^{\log x}$, then find $\frac{dy}{dx}$.** **[4]**

**Sol.** Given : $y = (\log x)^x + x^{\log x}$

Let $(\log x)^x = u$

and $x^{\log x} = v$

$\therefore y = u + v$

and $\frac{dy}{dx} = \frac{du}{dx} + \frac{dv}{dx}$

Now, $u = (\log x)^x$

Taking log on both sides, we get

$$\log u = \log(\log x)^x$$

$$\Rightarrow \quad \log u = x \log(\log x)$$

Differentiate both sides w.r.t. $x$, we get

$$\frac{1}{u}\times\frac{du}{dx} = x\times\frac{1}{\log x}\times\frac{1}{x}+\log(\log x)$$

$$\Rightarrow \quad \frac{du}{dx} = u\left[\frac{1}{\log x}+\log(\log x)\right]$$

$$\Rightarrow \quad \frac{du}{dx} = (\log x)^x\left[\frac{1}{\log x}+\log(\log x)\right]$$

and $\quad v = x^{\log x}$

Taking log on both sides, we get

$$\log v = \log(x)^{\log x}$$

$$\Rightarrow \quad \log v = \log x . \log x$$

Differentiate both sides w.r.t. $x$ we get

$$\frac{1}{v}\frac{dv}{dx} = \frac{\log x}{x}+\frac{\log x}{x}$$

$$\Rightarrow \quad \frac{dv}{dx} = v\left(\frac{2\log x}{x}\right)$$

$$\Rightarrow \quad \frac{dv}{dx} = x^{\log x}\left(\frac{2\log x}{x}\right)$$

Now, $\quad \frac{dy}{dx} = \frac{du}{dx}+\frac{dv}{dx}$

$$\Rightarrow \quad \frac{dy}{dx} = (\log x)^x\left[\frac{1}{\log x}+\log(\log x)\right] + x^{\log x}\left[\frac{2\log x}{x}\right]$$

**Ans.**

**29. Solve the differential equation :**

$$x \sin\left(\frac{y}{x}\right)\frac{dy}{dx}+x-y \sin\left(\frac{y}{x}\right)=0$$

**Given that $x = 1$ when $y = \frac{\pi}{2}$.** **[4]**

**Sol.** Given differential equation is,

$$x\sin\left(\frac{y}{x}\right)\frac{dy}{dx}+x-y\sin\left(\frac{y}{x}\right) = 0$$

$$\Rightarrow \quad x\sin\left(\frac{y}{x}\right)\frac{dy}{dx} = y\sin\left(\frac{y}{x}\right)-x$$

or $\quad \frac{dy}{dx} = \frac{y\sin\left(\frac{y}{x}\right)-x}{x\sin\left(\frac{y}{x}\right)}$

Put $\quad y = vx$

$$\therefore \quad \frac{dy}{dx} = v+x\frac{dv}{dx}$$

$$\Rightarrow \quad v+x\frac{dv}{dx} = \frac{(vx)\sin\left(\frac{vx}{x}\right)-x}{x\sin\left(\frac{vx}{x}\right)}$$

$$\Rightarrow \quad v+x\frac{dv}{dx} = \frac{v\sin v-1}{\sin v}$$

$$\Rightarrow \quad \frac{xdv}{dx} = \frac{v\sin v-1}{\sin v}-v$$

$$= \frac{v\sin v-1-v\sin v}{\sin v}$$

$$\Rightarrow \quad x\frac{dv}{dx} = \frac{-1}{\sin v}$$

$$\Rightarrow \quad \sin v\, dv = -\frac{dx}{x}$$

Integrating both sides, we get

$$\int \sin v\, dv = -\int\frac{dx}{x}$$

$$\Rightarrow \quad -\cos v = -\log x + C$$

$$\Rightarrow \quad -\cos\left(\frac{y}{x}\right) = -\log x + C \quad (\because y = vx)$$

It is given that,

$$x = 1 \text{ when } y = \frac{\pi}{2}$$

$$\therefore \quad -\cos\left(\frac{\pi}{2}\right) = -\log 1 + C$$

$$\Rightarrow \quad C = 0$$

$$\therefore \quad -\cos\left(\frac{y}{x}\right) = -\log x$$

$$\Rightarrow \quad \cos\left(\frac{y}{x}\right) = \log x$$

or $\quad \log x - \cos\left(\frac{y}{x}\right) = 0$ **Ans.**

**30. If $\vec{a} = \hat{i}+2\hat{j}+3\hat{k}$ and $\vec{b} = 2\hat{i}+4\hat{j}-5\hat{k}$ represent two adjacent sides of a parallelogram, find unit vectors parallel to the diagonals of the parallelogram.** **[4]**

**OR**

**Using vectors, find the area of the triangle ABC with vertices A (1, 2, 3), B (2, –1, 4) and C (4, 5, – 1).**

**Sol.** Given : $\vec{a} = \hat{i}+2\hat{j}+3\hat{k}$

and $\vec{b} = 2\hat{i}+4\hat{j}-5\hat{k}$

Let $\vec{d_1}$ and $\vec{d_2}$ be the diagonal vectors of the parallelogram.

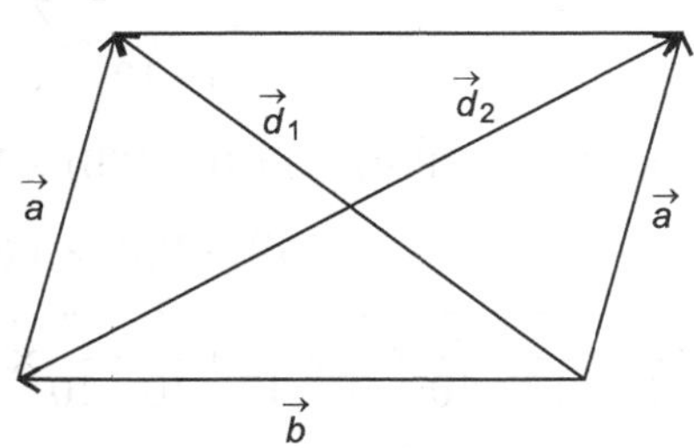


$\therefore \quad \vec{d_1} = \vec{a}+\vec{b}$

and $\quad \vec{d_2} = \vec{a}-\vec{b}$

So, $\vec{d_1} = (\vec{i}+2\vec{j}+3\vec{k})+(2\vec{i}+4\vec{j}-5\vec{k})$

$= 3\vec{i}+6\vec{j}-2\vec{k}$

and $\quad \vec{d_2} = (\vec{i}+2\vec{j}+3\vec{k})-(2\vec{i}+4\vec{j}-5\vec{k})$

$= -\vec{i}-2\vec{j}+8\vec{k}$

$\therefore$ Unit vector parallel to $\vec{d_1} = \dfrac{\vec{d_1}}{|\vec{d_1}|}$

$$= \frac{3\hat{i}+6\hat{j}-2\hat{k}}{\sqrt{3^2+6^2+(-2)^2}}$$

$$= \frac{3}{7}\hat{i}+\frac{6}{7}\hat{j}-\frac{2}{7}\hat{k}$$

and, unit vector parallel to $\vec{d_2} = \dfrac{\vec{d_2}}{|\vec{d_2}|}$

$$= \frac{-\hat{i}-2\hat{j}+8\hat{k}}{\sqrt{(-1)^2+(-2)^2+8^2}}$$

$$= -\frac{1}{\sqrt{69}}\hat{i}-\frac{2}{\sqrt{69}}\hat{j}+\frac{8}{\sqrt{69}}\hat{k}$$

Hence, the unit vectors are $\frac{3}{7}\hat{i}+\frac{6}{7}\hat{j}-\frac{2}{7}\hat{k}$ and $-\frac{1}{\sqrt{69}}\hat{i}-\frac{2}{\sqrt{69}}\hat{j}+\frac{8}{\sqrt{69}}\hat{k}$. **Ans.**

**OR**

Given : Coordinates of vertices of $\Delta$ABC are A(1, 2, 3), B (2, – 1, 4) and C(4, 5, – 1)

$\therefore \quad \vec{AB} = \hat{i}-3\hat{j}+\hat{k}$

and $\quad \vec{AC} = 3\hat{i}+3\hat{j}-4\hat{k}$

We know that,

$$ar(\Delta ABC) = \frac{1}{2}\left|\vec{AB}\times\vec{AC}\right|$$

Now, $\vec{AB}\times\vec{AC} = \begin{vmatrix} \vec{i} & \vec{j} & \vec{k} \\ 1 & -3 & 1 \\ 3 & 3 & -4 \end{vmatrix}$

$= \vec{i}(12-3)-\vec{j}(-4-3)+\vec{k}(3+9)$

$= 9\vec{i}+7\vec{j}+12\vec{k}$

$\therefore \quad \left|\vec{AB}\times\vec{AC}\right| = \sqrt{9^2+7^2+(12)^2} = \sqrt{274}$

So, $ar(\Delta ABC) = \frac{1}{2}\sqrt{274}$ sq. units **Ans.**

**31. A company manufactures two types of novelty souvenirs made of plywood. Souvenirs of type A requires 5 minutes each for cutting and 10 minutes each for assembling. Souvenirs of type B require 8 minutes each for cutting and 8 minutes each for assembling. Given that total time for cutting is 3 hours 20 minutes and for assembling 4 hours. The profit for type A souvenir is ₹ 100 each and for type B souvenir, profit is ₹ 120 each. How many souvenirs of each type should the company manufacture in order to maximize the profit ? Formulate the problem as an LPP and solve it graphically. [4]**

**Sol.** Let the company manufactures $x$ souvenirs of type A and $y$ souvenirs of type B.

Then the objective function is,

Maximize Z = $100x + 120y$

Subject to constraints

$5x + 8y \le 3$ h 20 min

$\Rightarrow \quad 5x + 8y \le 200$ ...(i)

and $\quad 10x + 8y \le 4$ h

$\Rightarrow \quad 10x + 8y \leq 240$

$\Rightarrow \quad 5x + 4y \leq 120 \quad$ ...(ii)

Now, $\quad 5x + 8y = 200$

| $x$ | 40 | 0 |
|---|---|---|
| $y$ | 0 | 25 |

$5x + 4y = 120$

| $x$ | 24 | 0 |
|---|---|---|
| $y$ | 0 | 30 |

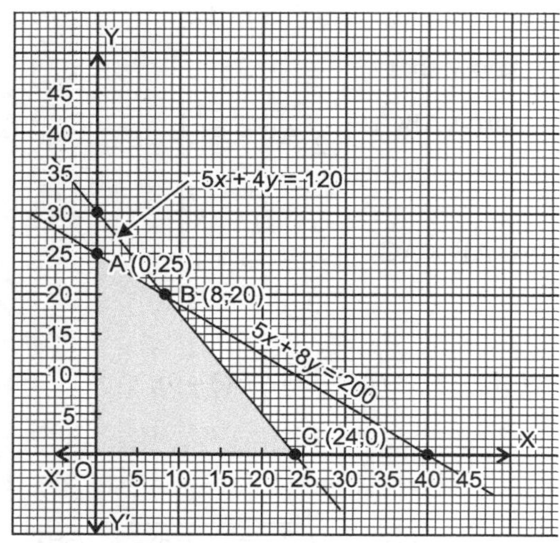


The feasible region is the shaded region OABC .

| Points | Objective function Z = 100$x$ + 120$y$ |
|---|---|
| O (0, 0) | Z = 0 |
| A (0, 25) | Z = 100(0) + 120(25) = 3000 |
| B (8, 20) | Z = 100 (8) + 120 (20) = 3200 (Maximum) |
| C (24, 0) | Z = 100 (24) + 120 (0) = 2400 |

Hence, the profit will be maximum if 8 souvenirs of type A and 20 souvenirs of type B are manufactured. **Ans.**

**32. Three rotten apples are mixed with seven fresh apples. Find the probability distribution of the number of rotten apples, if three apples are drawn one by one with replacement. Find the mean of the number of rotten apples. [4]**

**OR**

**In a shop X, 30 tins of ghee of type A and 40 tins of ghee of type B which look alike, are kept for sale. While in shop Y, similar 50 tins of ghee of type A and 60 tins of ghee of type B are there. One tin of ghee is purchased from one of the randomly selected shop and is found to be of type B. Find the probability that it is purchased from shop Y.**

**Sol.** Let A be the event of taking out a rotten apple and B be the event of taking out a fresh apple.

$\therefore \quad P(A) = \frac{3}{10}$

and $\quad P(B) = \frac{7}{10}$

Let $x_i$ be the number of rotten apples taken out in the three chances.

So, $\quad i = 0, 1, 2, 3$

$\therefore \quad P(x=0) = P(B) \times P(B) \times P(B)$

$= \frac{7}{10} \times \frac{7}{10} \times \frac{7}{10} = \frac{343}{1000}$

$P(x = 1) = P(A) \times P(B) \times P(B) + P(B) \times P(A) \times P(B) + P(B) \times P(B) \times P(A)$

$= \frac{3}{10} \times \frac{7}{10} \times \frac{7}{10} + \frac{7}{10} \times \frac{3}{10} \times \frac{7}{10} + \frac{7}{10} \times \frac{7}{10} \times \frac{3}{10}$

$= \frac{147}{1000} + \frac{147}{1000} + \frac{147}{1000} = \frac{441}{1000}$

$P(x = 2) = P(A) \times P(A) \times P(B) + P(A) \times P(B) \times P(A) + P(B) \times P(A) \times P(A)$

$= \frac{3}{10} \times \frac{3}{10} \times \frac{7}{10} + \frac{3}{10} \times \frac{7}{10} \times \frac{3}{10} + \frac{7}{10} \times \frac{3}{10} \times \frac{3}{10}$

$= \frac{63}{1000} + \frac{63}{1000} + \frac{63}{1000} = \frac{189}{1000}$

$P(x = 3) = P(A) \times P(A) \times P(A)$

$= \frac{3}{10} \times \frac{3}{10} \times \frac{3}{10} = \frac{27}{1000}$

$\therefore$ The probability distribution is,

| $x_i$ | 0 | 1 | 2 | 3 |
|---|---|---|---|---|
| $P(x_i)$ | $\frac{343}{1000}$ | $\frac{441}{1000}$ | $\frac{189}{1000}$ | $\frac{27}{1000}$ |

Now, $\quad \text{Mean} = \sum_{i=0}^{3} x_i P(x_i)$

$= 0 \times \frac{343}{1000} + 1 \times \frac{441}{1000} + 2 \times \frac{189}{1000} + 3 \times \frac{27}{1000}$

$= 0 + \frac{441}{1000} + \frac{378}{1000} + \frac{81}{1000}$

$= \frac{900}{1000} = 0.9$

Hence, the mean of number of rotten apples is 0.9. **Ans.**

**OR**

Let $H_1$ and $H_2$ be the events of purchasing tins of ghee from shop X and Y respectively.

$\therefore \quad P(H_1) = P(H_2) = \frac{1}{2}$

Also, let E be the event that ghee of type B is purchased.

$\therefore \quad P\left(\frac{E}{H_1}\right) = \frac{40}{70} = \frac{4}{7}$

and $\quad P\left(\frac{E}{H_2}\right) = \frac{60}{110} = \frac{6}{11}$

By Bayes' theorem, required probability is,

$$P\left(\frac{H_2}{E}\right) = \frac{P\left(\frac{E}{H_2}\right)P(H_2)}{P\left(\frac{E}{H_1}\right)P(H_1)+P\left(\frac{E}{H_2}\right)P(H_2)}$$

$$= \frac{\frac{6}{11}\times\frac{1}{2}}{\frac{4}{7}\times\frac{1}{2}+\frac{6}{11}\times\frac{1}{2}}$$

$$= \frac{\frac{6}{11}}{\frac{44+42}{77}} = \frac{6}{11}\times\frac{77}{86} = \frac{21}{43}$$

**Ans.**

## SECTION-D

**Q. 33 to 36 carry 6 marks each.**

**33. Find the vector and cartesian equations of the line which is perpendicular to the lines with equations**

$$\frac{x+2}{1} = \frac{y-3}{2} = \frac{z+1}{4} \text{ and } \frac{x-1}{2} = \frac{y-2}{3} = \frac{z-3}{4}$$

**and passes through the point (1, 1, 1). Also find the angle between the given lines. [6]**

**Sol.** Given : Equation of given lines are,

$$\frac{x+2}{1} = \frac{y-3}{2} = \frac{z+1}{4}$$

and $$\frac{x-1}{2} = \frac{y-2}{3} = \frac{z-3}{4}$$

$\therefore$ Direction ratios of first line = < 1, 2, 4 >

Direction ratios of second line = < 2, 3, 4 >

Let the direction ratios of required line be $< a, b, c >$.

$\because$ The required line is perpendicular to the given lines,

$\therefore \quad a + 2b + 4c = 0 \quad$ ...(i)

and $\quad 2a + 3b + 4c = 0 \quad$ ...(ii)

Solving equations (i) and (ii) by cross multiplication method, we get

$$\frac{a}{8-12} = \frac{-b}{4-8} = \frac{c}{3-4} = k \text{ (say)}$$

$\Rightarrow a = -4k$, $b = 4k$ and $c = -k$

$\therefore$ Direction ratios of required line = <–4, 4, –1>

Now, vector equation of line passing through $(x, y, z)$ and having direction ratios $< a, b, c >$ is given as,

$$(x\hat{i}+y\hat{j}+z\hat{k})+\lambda(a\hat{i}+b\hat{j}+c\hat{k}) = 0$$

$$\Rightarrow \quad (\hat{i}+\hat{j}+\hat{k})+\lambda(-4\hat{i}+4\hat{j}-\hat{k}) = 0$$ **Ans.**

And its cartesian form is,

$$\frac{x-1}{-4} = \frac{y-1}{4} = \frac{z-1}{-1}$$ **Ans.**

Now angle between two lines is given as,

$$\cos\theta = \frac{a_1a_2+b_1b_2+c_1c_2}{\sqrt{a_1^2+b_1^2+c_1^2}\sqrt{a_2^2+b_2^2+c_2^2}}$$

Here, $\quad < a_1, b_1, c_1 > = < 1, 2, 4 >$

and $\quad < a_2, b_2, c_2 > = < 2, 3, 4 >$

$$\Rightarrow \quad \cos\theta = \frac{1\times2+2\times3+4\times4}{\sqrt{1^2+2^2+4^2}\sqrt{2^2+3^2+4^2}}$$

$$= \frac{2+6+16}{\sqrt{21}\sqrt{29}} = \frac{24}{\sqrt{609}}$$

$$\Rightarrow \quad \theta = \cos^{-1}\left(\frac{24}{\sqrt{609}}\right)$$

Hence, the angle between the given lines is $\cos^{-1}\left(\frac{24}{\sqrt{609}}\right)$. **Ans.**

**34. Using integration find the area of the region bounded between the two circles $x^2 + y^2 = 9$ and $(x - 3)^2 + y^2 = 9$. [6]**

**OR**

**Evaluate the following integral as the limit of sums $\int_1^4 (x^2 - x)\,dx$.**

**Sol.** Given :

Equation of first circle is,

$$x^2 + y^2 = 9 \quad \text{...(i)}$$

$\therefore \quad$ Its centre = (0, 0)

and $\quad$ Radius = 3 units

Also, equation of second circle is,

$$(x - 3)^2 + y^2 = 9 \quad \text{...(ii)}$$

$\therefore \quad$ Its centre = (3, 0)

and $\quad$ Radius = 3 units

From equations (i) and (ii), we get

$$(x-3)^2 + 9 - x^2 = 9$$

$$\Rightarrow \quad x^2 - 6x + 9 - x^2 = 0$$

$$\Rightarrow \quad -6x + 9 = 0$$

$$\Rightarrow \quad x = \frac{3}{2}$$

Putting $x = \frac{3}{2}$ in equation (i), we get

$$\left(\frac{3}{2}\right)^2 + y^2 = 9$$

$$\Rightarrow \quad y^2 = 9 - \frac{9}{4} = \frac{36-9}{4} = \frac{27}{4}$$

$$\Rightarrow \quad y = \pm\frac{3\sqrt{3}}{2}$$

∴ The points of intersection of two circles are $\left(\frac{3}{2}, \frac{3\sqrt{3}}{2}\right)$ and $\left(\frac{3}{2}, \frac{-3\sqrt{3}}{2}\right)$.

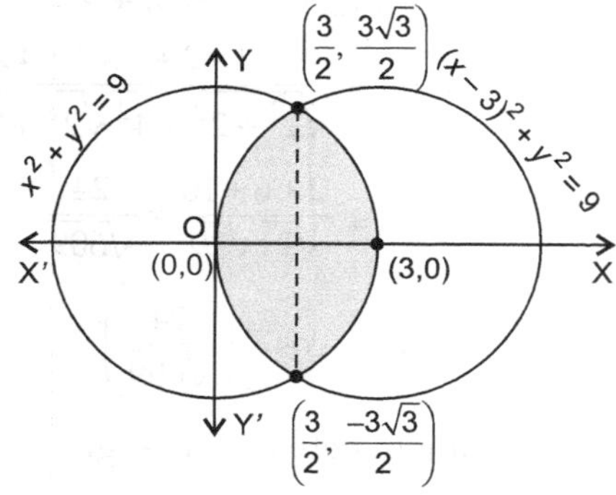


Now, area of bounded region

$$= 2\left[\int_0^{3/2} (y \text{ of } 2^{nd} \text{ circle})\,dx + \int_{3/2}^{3} (y \text{ of } 1^{st} \text{ circle})\,dx\right]$$

$$\Rightarrow I = 2\left[\int_0^{3/2} \sqrt{3^2-(x-3)^2}\,dx + \int_{3/2}^{3} \sqrt{3^2-x^2}\,dx\right]$$

$$\Rightarrow I = 2\,[I_1 + I_2]$$

Now, $$I_1 = \int_0^{3/2} \sqrt{3^2-(x-3)^2}\,dx$$

$$= \left[\frac{x-3}{2}\sqrt{3^2-(x-3)^2} + \frac{9}{2}\sin^{-1}\left(\frac{x-3}{3}\right)\right]_0^{3/2}$$

$$= \frac{1}{2}\left(\frac{3}{2}-3\right)\sqrt{3^2-\left(\frac{3}{2}-3\right)^2} + \frac{9}{2}\sin^{-1}\left\{\frac{1}{3}\left(\frac{3}{2}-3\right)\right\}$$

$$- \left\{\left(\frac{0-3}{2}\right)\sqrt{3^2-(0-3)^2} + \frac{9}{2}\sin^{-1}\left(\frac{0-3}{3}\right)\right\}$$

$$= \frac{1}{2}\left(\frac{-3}{2}\right)\sqrt{9-\frac{9}{4}} + \frac{9}{2}\sin^{-1}\left(\frac{-1}{2}\right) - \left(\frac{-3}{2}\right)\sqrt{9-9} - \frac{9}{2}\sin^{-1}(-1)$$

$$= -\frac{3}{4}\cdot\frac{3\sqrt{3}}{2} + \frac{9}{2}\left(-\frac{\pi}{6}\right) - \frac{9}{2}\left(-\frac{\pi}{2}\right)$$

$$= \frac{-9\sqrt{3}}{8} - \frac{3\pi}{4} + \frac{9\pi}{4}$$

$$= \frac{3\pi}{2} - \frac{9\sqrt{3}}{8} = \frac{1}{2}\left[3\pi - \frac{9\sqrt{3}}{4}\right]$$

Similarly,

$$I_2 = \int_{3/2}^{3} \sqrt{3^2-x^2}\,dx$$

$$\Rightarrow I_2 = \left[\frac{x}{2}\sqrt{9-x^2} + \frac{9}{2}\sin^{-1}\left(\frac{x}{3}\right)\right]_{3/2}^{3}$$

$$= \frac{1}{2}\left[\{0 + 9\sin^{-1}(1)\} - \left\{\frac{3}{2}\sqrt{9-\frac{9}{4}} + 9\sin^{-1}\left(\frac{\frac{3}{2}}{3}\right)\right\}\right]$$

$$\Rightarrow \quad I_2 = \frac{1}{2}\left[\frac{9\pi}{2} - \left\{\frac{3}{2}\sqrt{\frac{27}{4}} + 9\sin^{-1}\frac{1}{2}\right\}\right]$$

$$\Rightarrow \quad I_2 = \frac{1}{2}\left[\frac{9\pi}{2} - \frac{9\sqrt{3}}{4} - \frac{9\pi}{6}\right]$$

$$\Rightarrow \quad I_2 = \frac{1}{2}\left[\frac{9\pi}{2} - \frac{9\sqrt{3}}{4} - \frac{3\pi}{2}\right]$$

$$\Rightarrow \quad I_2 = \frac{1}{2}\left[\frac{6\pi}{2} - \frac{9\sqrt{3}}{4}\right]$$

$$\Rightarrow \quad I_2 = \frac{1}{2}\left[3\pi - \frac{9\sqrt{3}}{4}\right]$$

$$\because \quad I = 2(I_1 + I_2)$$

$$\Rightarrow \quad I = 2\times\frac{1}{2}\left[3\pi - \frac{9\sqrt{3}}{4} + 3\pi - \frac{9\sqrt{3}}{4}\right]$$

$$\Rightarrow \quad I = \left(6\pi - \frac{18}{4}\sqrt{3}\right)$$

$$= \left(6\pi - \frac{9\sqrt{3}}{2}\right) \text{ sq. units}$$ **Ans.**

**OR**

Let $$I = \int_1^4 (x^2 - x)\,dx$$

Here, $a = 1$, $b = 4$ and $nh = 4 - 1 = 3$

Let $f(x) = x^2 - x$

Clearly,

$$f(1) = (1)^2 - 1$$
$$f(1+h) = (1+h)^2 - (1+h)$$
$$f(1+2h) = (1+2h)^2 - (1+2h)$$
$$f[1+(n-1)h] = [1+(n-1)h]^2 - [1+(n-1)h]$$

$\therefore$ Required sum

$$= \lim_{x\to 0} h\left[f(1)+f(1+h)+...+f(1+(n-1)h)\right]$$

$$= \lim_{h\to 0} h\,[(1^2-1) + \{(1+h)^2-(1+h)\} + \{(1+2h)^2 - (1+2h)\} + .... + \{(1+(n-1)h)^2 - (1+(n-1)h)\}]$$

$$= \lim_{h\to 0} h[(1-1) + (1+2h+h^2-1-h) + (1+4h + 4h^2 - 1 - 2h) + .... + \{1+2(n-1)h + \{(n-1)h\}^2 - 1 - (n-1)h\}]$$

$$= \lim_{h\to 0} h[0 + (h^2+h) + (4h^2+2h) + ...\{(n-1)^2h^2 + (n-1)h\}]$$

$$= \lim_{h\to 0} h\,[(h^2+4h^2+9h^2+ ... + (n-1)^2h^2) + (h+2h+3h+...(n-1)h)]$$

$$= \lim_{h\to 0} h\,[h^2(1+2^2+3^2+ ... (n-1)^2) + h\,(1+2+3+ ... (n-1))]$$

$$= \lim_{h\to 0} h\left[h^2\frac{n(n-1)(2n-1)}{6} + h\frac{n(n-1)}{2}\right]$$

$$= \lim_{h\to 0}\left[\frac{h^3n(n-1)(2n-1)}{6} + \frac{h^2n(n-1)}{2}\right]$$

$$= \lim_{h\to 0}\left[\frac{nh(nh-h)(2nh-h)}{6} + \frac{nh(nh-h)}{2}\right]$$

$$= \frac{3(3-0)(6-0)}{6} + \frac{3(3-0)}{2}$$

$$= 9 + \frac{9}{2} = \frac{27}{2}$$ **Ans.**

**35. Find the minimum value of $(ax + by)$, where $xy = c^2$.** **[6]**

**Sol.** Given : $xy = c^2$

$$\Rightarrow \quad y = \frac{c^2}{x} \quad ...(i)$$

Let $t = ax + by$

$$\Rightarrow \quad t = ax + \frac{bc^2}{x} \quad \text{[Using (i)]}$$

Now, $$\frac{dt}{dx} = a - \frac{bc^2}{x^2}$$

For minimum value of $t$,

$$\frac{dt}{dx} = 0$$

$$\Rightarrow \quad a - \frac{bc^2}{x^2} = 0$$

$$\Rightarrow \quad ax^2 - bc^2 = 0$$

$$\Rightarrow \quad x = \pm c\sqrt{\frac{b}{a}}$$

Now,

$$\frac{d^2t}{dx^2} = 0 - (bc^2)\left(\frac{-2}{x^3}\right) = \frac{2bc^2}{x^3}$$

$$\Rightarrow \quad \left[\frac{d^2t}{dx^2}\right]_{x=c\sqrt{\frac{b}{a}}} = \frac{2bc^2}{c^3\frac{b\sqrt{b}}{a\sqrt{a}}} = \frac{2a}{c}\sqrt{\frac{a}{b}}$$

and $$\left[\frac{d^2t}{dx^2}\right]_{x=-c\sqrt{\frac{b}{a}}} = \frac{2bc^2}{-c^3\frac{b\sqrt{b}}{a\sqrt{a}}} = -\frac{2a}{c}\sqrt{\frac{a}{b}}$$

$\therefore$ Minimum value of $t$ occurs at $x = c\sqrt{\frac{b}{a}}$

$$\therefore \text{ Minimum value of } t = a\left(c\sqrt{\frac{b}{a}}\right) + b\left(\frac{c^2}{c\sqrt{\frac{b}{a}}}\right)$$

$$= c\sqrt{ab} + c\sqrt{ab}$$

$$= 2c\sqrt{ab}$$ **Ans.**

**36. If $a, b, c$ are $p^{th}$, $q^{th}$ and $r^{th}$ terms respectively of a G.P., then prove that $\begin{vmatrix} \log a & p & 1 \\ \log b & q & 1 \\ \log c & r & 1 \end{vmatrix} = 0$** **[6]**

**OR**

**If $A = \begin{bmatrix} 2 & -3 & 5 \\ 3 & 2 & -4 \\ 1 & 1 & -2 \end{bmatrix}$, then find $A^{-1}$.**

**Using $A^{-1}$, solve the following system of equations :**

$$2x - 3y + 5z = 11$$
$$3x + 2y - 4x = -5$$
$$x + y - 2z = -3$$

**Sol.** Given : $a, b, c$ are $p^{th}$, $q^{th}$ and $r^{th}$ terms of G.P., respectively.

Let A be the first term and R be the common ratio of the G.P.

$$\therefore \quad a_p = a$$
$$\Rightarrow \quad AR^{p-1} = a$$
$$\Rightarrow \quad \log a = \log(AR^{p-1})$$
$$= \log A + \log R^{p-1}$$

Thus, $\log a = \log A + (p-1)\log R$

Similarly, $\log b = \log A + (q-1)\log R$

And, $\log c = \log A + (r-1)\log R$

Now,

$$\text{L.H.S.} = \begin{vmatrix} \log a & p & 1 \\ \log b & q & 1 \\ \log c & r & 1 \end{vmatrix}$$
$$= \begin{vmatrix} \log A+(p-1)\log R & p & 1 \\ \log A+(q-1)\log R & q & 1 \\ \log A+(r-1)\log R & r & 1 \end{vmatrix}$$

Applying $C_1 \rightarrow C_1 - \log A\ C_3$, we get

$$\text{L.H.S.} = \begin{vmatrix} (p-1)\log R & p & 1 \\ (q-1)\log R & q & 1 \\ (r-1)\log R & r & 1 \end{vmatrix}$$

Taking out log R common from $C_1$,

$$\text{L.H.S.} = \log R \begin{vmatrix} p-1 & p & 1 \\ q-1 & q & 1 \\ r-1 & r & 1 \end{vmatrix}$$

Applying $C_1 \rightarrow C_1 + C_3$, we get

$$\text{L.H.S.} = \log R \begin{vmatrix} p & p & 1 \\ q & q & 1 \\ r & r & 1 \end{vmatrix}$$

$\because C_1$ and $C_2$ are identical.

$$\therefore \quad \begin{vmatrix} p & p & 1 \\ q & q & 1 \\ r & r & 1 \end{vmatrix} = 0$$

$\therefore$ L.H.S. = log R × 0 = 0 = R.H.S. **Hence Proved.**

**OR**

Given : $A = \begin{bmatrix} 2 & -3 & 5 \\ 3 & 2 & -4 \\ 1 & 1 & -2 \end{bmatrix}$

$$\therefore \quad |A| = 2(-4+4) + 3(-6+4) + 5(3-2)$$
$$= 0 + (-6) + 5$$
$$= -1 \neq 0$$

$\therefore$ Matrix A is invertible, *i.e.* $A^{-1}$ exists.

Now, cofactors of matrix A are

$$C_{11} = 0, C_{21} = -1, C_{31} = 2$$
$$C_{12} = 2, C_{22} = -9, C_{32} = 23$$
$$C_{13} = 1, C_{23} = -5, C_{33} = 13$$

$$\therefore \text{Cofactor matrix} = \begin{bmatrix} 0 & 2 & 1 \\ -1 & -9 & -5 \\ 2 & 23 & 13 \end{bmatrix}$$

$$\because \quad \text{adj A} = \begin{bmatrix} 0 & 2 & 1 \\ -1 & -9 & -5 \\ 2 & 23 & 13 \end{bmatrix}^T$$
$$= \begin{bmatrix} 0 & -1 & 2 \\ 2 & -9 & 23 \\ 1 & -5 & 13 \end{bmatrix}$$

Now,
$$A^{-1} = \frac{1}{|A|}\text{adj A}$$
$$A^{-1} = \frac{1}{(-1)}\begin{bmatrix} 0 & -1 & 2 \\ 2 & -9 & 23 \\ 1 & -5 & 13 \end{bmatrix}$$
$$A^{-1} = \begin{bmatrix} 0 & 1 & -2 \\ -2 & 9 & -23 \\ -1 & 5 & -13 \end{bmatrix}$$ **Ans.**

Now, given equations are

$$2x - 3y + 5z = 11$$
$$3x + 2y - 4z = -5$$
$$x + y - 2z = -3$$

Let
$$A = \begin{bmatrix} 2 & -3 & 5 \\ 3 & 2 & -4 \\ 1 & 1 & -2 \end{bmatrix}$$
$$X = \begin{bmatrix} x \\ y \\ z \end{bmatrix}$$

and
$$B = \begin{bmatrix} 11 \\ -5 \\ -3 \end{bmatrix}$$

$$\therefore \quad AX = B$$

or
$$X = A^{-1}B \quad \text{...(i)}$$

$$\Rightarrow \begin{bmatrix} x \\ y \\ z \end{bmatrix} = \begin{bmatrix} 0 & 1 & -2 \\ -2 & 9 & -23 \\ -1 & 5 & -13 \end{bmatrix} \begin{bmatrix} 11 \\ -5 \\ -3 \end{bmatrix}$$

$$\Rightarrow \begin{bmatrix} x \\ y \\ z \end{bmatrix} = \begin{bmatrix} 0-5+6 \\ -22-45+69 \\ -11-25+39 \end{bmatrix}$$

$$\begin{bmatrix} x \\ y \\ z \end{bmatrix} = \begin{bmatrix} 1 \\ 2 \\ 3 \end{bmatrix}$$

$\therefore$ $x = 1, y = 2$

and $z = 3$ **Ans.**

## Delhi [Set-II]

**Note : Except for the following questions, all the remaining questions have been asked in the previous set.**

### SECTION-A

**1. If $[x \quad 1]\begin{bmatrix} 1 & 0 \\ -2 & 0 \end{bmatrix} = 0$, then $x$ equals**

**(a) 0** **(b) −2**

**(c) −1** **(d) 2** [1]

**Sol.** Given :

$$[x \quad 1]\begin{bmatrix} 1 & 0 \\ -2 & 0 \end{bmatrix} = 0$$

$$\Rightarrow [x-2 \quad 0] = [0 \quad 0]$$

$$\Rightarrow x - 2 = 0$$

$$\Rightarrow x = 2$$

$\therefore$ Option (d) is correct.

**2. $\int 4^x 3^x dx$ equals**

**(a)** $\dfrac{12^x}{\log 12} + C$ **(b)** $\dfrac{4^x}{\log 4} + C$

**(c)** $\left(\dfrac{4^x . 3^x}{\log 4 . \log 3}\right) + C$ **(d)** $\dfrac{3^x}{\log 3} + C$ [1]

**Sol.** Let, $I = \int 4^x 3^x dx$

$$\Rightarrow I = \int 12^x dx$$

$$\Rightarrow I = \frac{12^x}{\log 12} + C$$

$$\left[\because \int a^x \, dx = \frac{a^x}{\log x} + C\right]$$

$\therefore$ Option (a) is correct.

**3. A number is chosen randomly from numbers 1 to 60. The probability that the chosen number is multiple of 2 or 5 is**

**(a)** $\dfrac{2}{5}$ **(b)** $\dfrac{3}{5}$

**(c)** $\dfrac{7}{10}$ **(d)** $\dfrac{9}{10}$ [1]

**Sol.** Let A = Numbers multiple of 2

and B = Numbers multiple of 5

$$\therefore P(A \cup B) = P(A) + P(B) - P(A \cap B)$$

$$= \frac{30}{60} + \frac{12}{60} - \frac{6}{60}$$

$$= \frac{36}{60} = \frac{3}{5}$$

$\therefore$ Option (b) is correct.

**11. A relation R on a set A is called ________, if $(a_1, a_2) \in R$ and $(a_2, a_3) \in R$ implies that $(a_1, a_3) \in R$ for $a_1, a_2, a_3 \in A$.** [1]

**Sol.** Transitive

**16. Evaluate : $\sin\left[\dfrac{\pi}{3} - \sin^{-1}\left(\dfrac{-1}{2}\right)\right]$** [1]

**Sol.** $\sin\left[\dfrac{\pi}{3} - \sin^{-1}\left(\dfrac{-1}{2}\right)\right]$

$$= \sin\left[\frac{\pi}{3} + \sin^{-1}\frac{1}{2}\right] \quad [\because \sin^{-1}(-x) = -\sin x]$$

$$= \sin\left(\frac{\pi}{3} + \frac{\pi}{6}\right) = \sin\frac{\pi}{2} = 1$$ **Ans.**

**17. Using differential, find the approximate value of $\sqrt{36.6}$ upto 2 decimal places.** [1]

**OR**

**Find the slope of tangent to the curvey $y = 2\cos^2(3x)$ at $x = \dfrac{\pi}{6}$.**

**Sol.** Let $y = \sqrt{x}$, $x = 36$ and $\Delta x = 0.6$

Now, $y = \sqrt{x}$

$$\Rightarrow \quad \frac{dy}{dx} = \frac{1}{2\sqrt{x}}$$

$$\because \quad \Delta y = \frac{dy}{dx}\Delta x$$

$$= \left(\frac{1}{2\sqrt{x}}\right)_{x=36} \times 0.6$$

$$= \frac{1}{2\sqrt{36}} \times 0.6$$

$$= \frac{1}{2\times 6} \times 0.6 = 0.05$$

Now, $\Delta y = f(x + \Delta x) - f(x)$

$$\Rightarrow \quad 0.05 = \sqrt{36+0.6} - \sqrt{36}$$

$$\Rightarrow \quad 0.05 = \sqrt{36.6} - 6$$

$$\Rightarrow \quad \sqrt{36.6} = 6.05 \quad \textbf{Ans.}$$

**OR**

Given : $y = 2\cos^2(3x)$

$$\Rightarrow \quad \frac{dy}{dx} = 2[2\cos 3x.(-\sin 3x).3]$$

$$= -12\cos 3x \sin 3x$$

$$= -6\times 2\cos 3x \sin 3x$$

$$= -6\sin 6x$$

$$\Rightarrow \quad \left(\frac{dy}{dx}\right)_{x=\frac{\pi}{6}} = -6\sin\pi = 0$$

$\because$ Slope of tangent $= \frac{dy}{dx}$

$\therefore$ Slope of tangent $= 0$ **Ans.**

## SECTION–B

**21. Find $\int \frac{x+1}{x(1-2x)}dx$.** **[2]**

**Sol.** Let $I = \int \frac{x+1}{x(1-2x)}dx$

Also, let $\frac{x+1}{x(1-2x)} = \frac{A}{x} + \frac{B}{1-2x}$

$$\Rightarrow \quad \frac{x+1}{x(1-2x)} = \frac{A(1-2x)+Bx}{x(1-2x)}$$

$$\Rightarrow \quad x + 1 = x(-2A + B) + A$$

Comparing coefficient of $x$ both sides, we get

$$1 = -2A + B \quad \text{...(i)}$$

Comparing constant term both sides, we get

$$A = 1$$

$\therefore$ From equation (i),

$$1 = -2(1) + B$$

$$\Rightarrow \quad B = 3$$

$$\therefore \quad \int \frac{x+1}{x(1-2x)}dx = \int \frac{1}{x}dx + \int \frac{3}{1-2x}dx$$

$$= \log|x| + \frac{3\log|1-2x|}{-2} + C$$

$$= \log|x| - \frac{3}{2}\log|1-2x| + C \quad \textbf{Ans.}$$

**22. Evaluate $\int \frac{x\sin^{-1}(x^2)}{\sqrt{1-x^4}}dx$.** **[2]**

**Sol.** Let $I = \int \frac{x\sin^{-1}(x^2)}{\sqrt{1-x^4}}dx$

Let $\sin^{-1}(x^2) = t$

$$\therefore \quad \frac{1}{\sqrt{1-x^4}}(2x)dx = dt$$

$$\Rightarrow \quad \frac{x}{\sqrt{1-x^4}}dx = \frac{dt}{2}$$

$$\therefore \quad I = \int \frac{tdt}{2} = \frac{1}{2}\int tdt$$

$$I = \frac{1}{2}\cdot\frac{t^2}{2} + c$$

$$I = \frac{1}{4}(\sin^{-1}x^2)^2 + C \quad \textbf{Ans.}$$

## SECTION–C

**27. Prove that $\tan\left[2\tan^{-1}\frac{1}{2} - \cot^{-1}3\right] = \frac{9}{13}$** **[4]**

**Sol.** L.H.S. $= \tan\left[2\tan^{-1}\frac{1}{2} - \cot^{-1}3\right]$

$$= \tan\left[\tan^{-1}\left(\frac{2\times\frac{1}{2}}{1-\left(\frac{1}{2}\right)^2}\right) - \tan^{-1}\frac{1}{3}\right]$$

$$\left[\because 2\tan^{-1}x = \tan^{-1}\left(\frac{2x}{1-x^2}\right) \text{ and } \cot^{-1}x = \tan^{-1}\frac{1}{x}\right]$$

$$= \tan\left[\tan^{-1}\left(\frac{1}{1-\frac{1}{4}}\right) - \tan^{-1}\frac{1}{3}\right]$$

$$= \tan\left[\tan^{-1}\frac{4}{3} - \tan^{-1}\frac{1}{3}\right]$$

$$= \tan\left[\tan^{-1}\left(\frac{\frac{4}{3}-\frac{1}{3}}{1+\frac{4}{3}\times\frac{1}{3}}\right)\right] = \tan\left[\tan^{-1}\left(\frac{1}{\frac{9+4}{9}}\right)\right]$$

$$= \tan\left[\tan^{-1}\left(\frac{9}{13}\right)\right] = \frac{9}{13} = \text{R.H.S.}$$ **Hence Proved.**

**28. If $y = (\cos x)^x + \tan^{-1}\sqrt{x}$, find $\frac{dy}{dx}$.** **[4]**

**Sol.** Given : $y = (\cos x)^x + \tan^{-1}\sqrt{x}$

Let $u = (\cos x)^x$

and $v = \tan^{-1}\sqrt{x}$

$\therefore$ $y = u + v$

$$\Rightarrow \frac{dy}{dx} = \frac{du}{dx} + \frac{dv}{dx} \quad ...(i)$$

Now, $u = (\cos x)^x$

Taking log on both sides, we get

$$\log u = \log (\cos x)^x$$

$$\Rightarrow \log u = x \log (\cos x)$$

Differentiating both sides w.r.t. $x$, we get

$$\frac{1}{u}\frac{du}{dx} = \frac{x(-\sin x)}{\cos x} + \log(\cos x) \times 11$$

$$\Rightarrow \frac{du}{dx} = u\,[-x \tan x + \log(\cos x)]$$

$$\Rightarrow \frac{du}{dx} = (\cos x)^x\,[-x \tan x + \log(\cos x)]$$

Similarly, $v = \tan^{-1}\sqrt{x}$

Differentiating both sides w.r.t. $x$, we get

$$\frac{dv}{dx} = \frac{1}{\left(1+(\sqrt{x})^2\right)} \times \frac{1}{2\sqrt{x}}$$

$$\Rightarrow \frac{dv}{dx} = \frac{1}{2\sqrt{x}(1+x)}$$

Putting the values of $\frac{du}{dx}$ and $\frac{dv}{dx}$ in (i), we get

$$\frac{dy}{dx} = (\cos x)^x\,[-x \tan x + \log \cos x] + \frac{1}{2\sqrt{x}(1+x)}$$

**Ans.**

## SECTION–D

**36. Find the point on the curve $y^2 = 4x$ which is nearest to the point (2, 1).** **[6]**

**Sol.** Let A($x$, $y$) be the required point on the curve $y^2 = 4x$.

$$\therefore \quad \text{Distance} = \sqrt{(2-x)^2 + (1-y)^2}$$

$$\Rightarrow D = \sqrt{(2-x)^2 + (1-y)^2}$$

$$\Rightarrow D(y) = \sqrt{\left(2-\frac{y^2}{4}\right)^2 + (1-y)^2}$$

$$\left[\because y^2 = 4x \quad \therefore x = \frac{y^2}{4}\right]$$

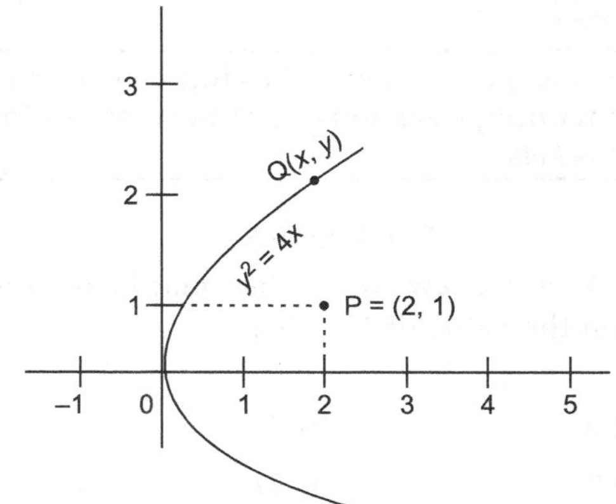


Squaring both sides, we get

$$D^2(y) = \left(2-\frac{y^2}{4}\right)^2 + (1-y)^2$$

Let $Z = D^2(y)$

$$\therefore \quad Z = \left(2-\frac{y^2}{4}\right)^2 + (1-y)^2$$

Now, $$\frac{dZ}{dy} = 2\left(2-\frac{y^2}{4}\right)\left(\frac{-2y}{4}\right) + 2(1-y)(-1)$$

$$\Rightarrow \frac{dZ}{dy} = -y\left(2-\frac{y^2}{4}\right) - 2(1-y)$$

$$\Rightarrow \frac{dZ}{dy} = -2y + \frac{y^3}{4} - 2 + 2y$$

$$\Rightarrow \frac{dZ}{dy} = \frac{y^3}{4} - 2$$

For maximum and minimum value,

Put $\frac{dZ}{dy} = 0$

$$\Rightarrow \frac{y^3}{4} - 2 = 0$$

$$\Rightarrow \frac{y^3}{4} = 2$$

$$\Rightarrow y^3 = 8$$

$$\Rightarrow y = 2$$

Now, $\frac{d^2 Z}{dy^2} = \frac{3y^2}{4}$

$$\left[\frac{d^2Z}{dy^2}\right]_{y=2} = \frac{3(2)^2}{4} = \frac{12}{4} = 3 > 0$$

So, Z is minimum at $y = 2$.

$\because \quad y^2 = 4x$

$\Rightarrow \quad (2)^2 = 4x$

$\Rightarrow \quad x = 1$

$\therefore$ Required point on the curve is (1, 2). **Ans.**

# Delhi [Set-III]

**Note : Except for the following questions, all the remaining questions have been asked in the previous sets.**

## SECTION-A

**1. If A is a skew symmetric matrix of order 3, then the value of $|A|$ is :**

**(a) 3** **(b) 0**

**(c) 9** **(d) 27** **[1]**

**Sol.** Since A is a skew symmetric matrix,

$\therefore \quad A^T = -A$

$\Rightarrow \quad |A^T| = |-A|$

$\Rightarrow \quad |A| = |-1|^3|A|$

$\Rightarrow \quad |A| = -|A|$

$\Rightarrow \quad 2|A| = 0$

$\Rightarrow \quad |A| = 0$

$\therefore$ Option (b) is correct.

**6. If $y = \log_e\left(\frac{x^2}{e^2}\right)$, then $\frac{d^2y}{dx^2}$ equals :**

**(a) $-\frac{1}{x}$** **(b) $-\frac{1}{x^2}$**

**(c) $\frac{2}{x^2}$** **(d) $-\frac{2}{x^2}$** **[1]**

**Sol.** Given : $y = \log\left(\frac{x^2}{e^2}\right)$

$$\therefore \quad \frac{dy}{dx} = \frac{1}{\frac{x^2}{e^2}} \times \frac{1}{e^2}(2x)$$

$$\Rightarrow \quad \frac{dy}{dx} = \frac{2}{x}$$

$$\Rightarrow \quad \frac{d^2y}{dx^2} = \frac{-2}{x^2}$$

$\therefore$ Option (d) is correct.

**9. The distance of the origin (0, 0, 0) from the plane $-2x + 6y - 3z = -7$ is :**

**(a) 1 unit** **(b) $\sqrt{2}$ units**

**(c) $2\sqrt{2}$ units** **(d) 3 units** **[1]**

**Sol.** Given : Equation of plane is,

$$-2x + 6y - 3z = -7$$

$$\Rightarrow \quad -2x + 6y - 3z + 7 = 0$$

We know distance of point $(x_1, y_1, z_1)$ from a plane $Ax + By + Cz + D = 0$ is given as

$$\text{Distance} = \left|\frac{Ax_1 + By_1 + Cz_1 + D}{\sqrt{A^2+B^2+C^2}}\right|$$

$$\therefore \quad \text{Distance} = \left|\frac{-2(0)+6(0)+(-3)(0)+7}{\sqrt{4+36+9}}\right|$$

$$= \left|\frac{7}{\sqrt{49}}\right| = \left|\frac{7}{7}\right| = 1 \text{ unit}$$

$\therefore$ Option (a) is correct.

**11. If A and B are square matrices each of order 3 and $|A| = 5$, $|B| = 3$, then the value of $|3AB|$ is ________ .** **[1]**

**Sol.** 405

**Explanation :** $\because \quad |A| = 5, |B| = 3$

$\therefore \quad |3\,AB| = 3^3|A||B|$

$= 27 \times 5 \times 3 = 405$

**16. Find the cofactors of all the elements of $\begin{bmatrix} 1 & -2 \\ 4 & 3 \end{bmatrix}$.** **[1]**

**Sol.** Let $A = \begin{bmatrix} 1 & -2 \\ 4 & 3 \end{bmatrix}$

$\therefore$ Co-factors of A are

$C_{11} = 3, C_{21} = 2, C_{12} = -4$ and $C_{22} = 1$ **Ans.**

**17. Let $f(x) = x|x|$, for all $x \in R$ check its differentiability at $x = 0$.** **[1]**

**Sol.** Given,

$$f(x) = x|x|$$

$$\therefore \quad f(x) = \begin{cases} +x(x), & x \geq 0 \\ -x(x), & x < 0 \end{cases}$$

$$\Rightarrow \quad f(x) = \begin{cases} x^2, & x \geq 0 \\ -x^2, & x < 0 \end{cases}$$

Now,

$$\text{R.H.D.} = \lim_{h \to 0} \frac{f(0+h) - f(0)}{h}$$

$$= \lim_{h \to 0} \frac{(0+h)^2 - (0)^2}{h}$$

$$= \lim_{h \to 0} \frac{h^2}{h} = 0$$

$$\text{L.H.D.} = \lim_{h \to 0} \frac{f(0-h) - f(0)}{-h}$$

$$= \lim_{h \to 0} \frac{-(0-h)^2 - (0)^2}{-h}$$

$$= \lim_{h \to 0} \frac{-h^2}{-h} = 0$$

$\because$ L.H.D. = R.H.D

So, $f(x)$ is differentiable at $x = 0$. **Ans.**

## SECTION – B

**21. Find $\int \frac{x+1}{(x+2)(x+3)} dx$.** **[2]**

**Sol.** Let $I = \int \frac{x+1}{(x+2)(x+3)}$

Let $\frac{x+1}{(x+2)(x+3)} = \frac{A}{x+2} + \frac{B}{x+3}$

$\Rightarrow \quad x + 1 = A(x + 3) + B(x + 2)$

$\Rightarrow \quad x + 1 = (A + B)x + (3A + 2B)$

Comparing coefficients of $x$ both sides, we get

$1 = A + B$ ...(i)

Comparing constant terms both sides, we get

$1 = 3A + 2B$ ...(ii)

Solving equations (i) and (ii),

$3A + 2(1 - A) = 0$

$\Rightarrow \quad 3A + 2 - 2A = 0$

$\Rightarrow \quad A + 2 = 0$

$\Rightarrow \quad A = -2$

Putting $A = -2$ in equation (i), we get

$B = 3$

$$\therefore \quad I = -2\int \frac{dx}{x+2} + 3\int \frac{dx}{x+3}$$

$$\Rightarrow \quad I = -2 \log |x + 2| + 3 \log|x + 3| + C$$

$$= \log |x + 2|^{-2} + 3 \log|x + 3|^3 + C$$

$$= \log\left|\frac{1}{x+2}\right|^2 + \log|x+3|^3 + C$$

$$\Rightarrow \quad I = \log\left|\frac{(x+3)^3}{(x+2)^2}\right| + C$$ **Ans.**

**26. Find the value of $\int_0^1 \tan^{-1}\left(\frac{1-2x}{1+x-x^2}\right) dx$.** **[2]**

**Sol.** Let $I = \int_0^1 \tan^{-1}\left(\frac{1-2x}{1+x-x^2}\right) dx$

$$\Rightarrow \quad I = \int_0^1 \tan^{-1}\left(\frac{(1-x)-x}{1+x(1-x)}\right) dx$$

$$\Rightarrow \quad I = \int_0^1 \left[\tan^{-1}(1-x) - \tan^{-1} x\right] dx$$

$$\left[\because \tan^{-1} A - \tan^{-1} B = \tan^{-1}\left(\frac{A-B}{1+AB}\right)\right]$$

$$\Rightarrow \quad I = \int_0^1 \tan^{-1}(1-x)dx - \int_0^1 \tan^{-1} x\, dx$$

$$\Rightarrow \quad I = \int_0^1 \tan^{-1}[1-(1-x)]dx - \int_0^1 \tan^{-1} x\, dx$$

$$\left[\because \int_0^a f(x)dx = \int_0^a f(a-x)dx\right]$$

$$\Rightarrow \quad I = \int_0^1 \tan^{-1} x\, dx - \int_0^1 \tan^{-1} x\, dx$$

$\Rightarrow \quad I = 0.$ **Ans.**

## SECTION - C

**27. Solve the equation for $x$ :**

**$\sin^{-1}\left(\frac{5}{x}\right) + \sin^{-1}\left(\frac{12}{x}\right) = \frac{\pi}{2} (x \neq 0)$** **[4]**

**Sol.** Given : $\sin^{-1}\left(\frac{5}{x}\right) + \sin^{-1}\left(\frac{12}{x}\right) = \frac{\pi}{2}$

$$\Rightarrow \sin^{-1}\left\{\frac{5}{x}\sqrt{1-\left(\frac{12}{x}\right)^2}+\frac{12}{x}\sqrt{1-\left(\frac{5}{x}\right)^2}\right\} = \frac{\pi}{2}$$

$$\left[\because \sin^{-1}x+\sin^{-1}y=\sin^{-1}\left(x\sqrt{1-y^2}+y\sqrt{1-x^2}\right)\right]$$

$$\Rightarrow \frac{5}{x}\sqrt{\frac{x^2-144}{x^2}}+\frac{12}{x}\sqrt{\frac{x^2-25}{x^2}} = \sin\frac{\pi}{2}$$

$$\Rightarrow \frac{5}{x}\frac{\sqrt{x^2-144}}{x}+\frac{12}{x}\frac{\sqrt{x^2-25}}{x} = 1$$

$$\Rightarrow \frac{5\sqrt{x^2-144}}{x^2}+\frac{12\sqrt{x^2-25}}{x^2} = 1$$

$$\Rightarrow 5\sqrt{x^2-144}+12\sqrt{x^2-25} = x^2$$

$$\Rightarrow 5\sqrt{x^2-144} = x^2 - 12\sqrt{x^2-25}$$

Squaring both sides, we get

$$\Rightarrow 25(x^2-144) = x^4 + 144(x^2-25) - 2\times 12x^2\sqrt{x^2-25}$$

$$\Rightarrow 25x^2 - 3600 = x^4 + 144x^2 - 3600 - 24x^2\sqrt{x^2-25}$$

$$\Rightarrow 25x^2 - 144x^2 - x^4 = -24x^2\sqrt{x^2-25}$$

$$\Rightarrow -119x^2 - x^4 = -24x^2\sqrt{x^2-25}$$

$$\Rightarrow -x^2(119+x^2) = -24x^2\sqrt{x^2-25}$$

$\Rightarrow x = 0$ (Rejected, as $x \neq 0$)

or $\quad 119 + x^2 = -24\sqrt{x^2-25}$

Again squaring both sides, we get

$$14161 + x^4 + 238x^2 = 576(x^2-25)$$

$$\Rightarrow x^4 + 238x^2 + 14161 = 576x^2 - 14400$$

$$\Rightarrow x^4 - 338x^2 + 28561 = 0$$

$$\Rightarrow x^2 = \frac{338\pm\sqrt{114244-114244}}{2}$$

$$\Rightarrow x^2 = \frac{338}{2}$$

$$\Rightarrow x^2 = 169$$

$$\Rightarrow x = \pm 13$$

$\therefore \quad x = 13$ **Ans.**

**28. Find the general solution of the differential equation :**

**$ye^{x/y}\,dx = (xe^{x/y} + y^2)\,dy,\ y \neq 0$** **[4]**

**Sol.** Given differential equation is,

$$ye^{\frac{x}{y}}dx = \left(xe^{\frac{x}{y}}+y^2\right)dy$$

$$\Rightarrow \frac{dx}{dy} = \frac{xe^{\frac{x}{y}}+y^2}{ye^{\frac{x}{y}}}$$

Put $\quad x = vy$

$$\Rightarrow \frac{dx}{dy} = v+y\frac{dv}{dy}$$

Now, $$v+y\frac{dv}{dy} = \frac{(vy)e^{vy/y}+y^2}{ye^{vy/y}}$$

$$\Rightarrow v+y\frac{dv}{dy} = \frac{vye^v+y^2}{ye^v}$$

$$\Rightarrow v+y\frac{dv}{dy} = \frac{ve^v+y}{e^v}$$

$$\Rightarrow y\frac{dv}{dy} = \frac{ve^v+y}{e^v}-v$$

$$\Rightarrow y\frac{dv}{dy} = \frac{ve^v+y-ve^v}{e^v}$$

$$\Rightarrow y\frac{dv}{dy} = \frac{y}{e^v}$$

$$\Rightarrow \frac{dv}{dy} = \frac{1}{e^v}$$

$$\Rightarrow e^v dv = dy$$

Integrating both sides,

$$\Rightarrow \int e^v dv = \int dy$$

$$\Rightarrow e^v = y + C$$

$$\Rightarrow e^{x/y} = y + C$$

or $\quad y = e^{x/y} + K$

which is the required solution of given differential equation. **Ans.**

## SECTION - D

**33. Find the distance of the point P (3, 4, 4) from the point, where the line joining the points A(3, – 4, – 5) and B (2, – 3, 1) intersects the plane 2$x$ + $y$ + $z$ = 7. [6]**

**Sol.** Equation of line passing through points A(3, –4, – 5) and B(2, –3, 1) is,

$$\frac{x-3}{2-3}=\frac{y+4}{-3-(-4)}=\frac{z+5}{1-(-5)}$$

$$\Rightarrow \quad \frac{x-3}{-1}=\frac{y+4}{1}=\frac{z+5}{6}$$

Let $$\frac{x-3}{-1}=\frac{y+4}{1}=\frac{z+5}{6}=k$$

$\therefore \quad x = 3 - k,\ y = k - 4$ and $z = 6k - 5$

$\therefore$ Coordinates of any point on the line are M($3 - k$, $k - 4$, $6k - 5$)

Since, point M lies on the plane $2x + y + z - 7 = 0$

$\therefore 2(3 - k) + (k - 4) + (6k - 5) - 7 = 0$

$\Rightarrow \quad 6 - 2k + k - 4 + 6k - 5 - 7 = 0$

$\Rightarrow \quad 5k - 10 = 0$

$\Rightarrow \quad k = 2$

$\therefore$ Coordinates of M = (3 – 2, 2 – 4, 6(2) – 5)
= (1, – 2, 7)

Now, distance between points P(3, 4, 4) and M(1, – 2, 7) is,

$$PM = \sqrt{(1-3)^2+(-2-4)^2+(7-4)^2}$$

$$= \sqrt{(-2)^2+(-6)^2+(3)^2}$$

$$= \sqrt{4+36+9} = \sqrt{49}$$

= 7 units **Ans.**

# MATHEMATICS
## Outside Delhi [Set-I] 2019

**Time allowed : 3 hours** **Maximum marks : 100**

### SECTION–A

**1. If A is a square matrix satisfying A′ A = I, write the value of | A |.** **[1]**

**Sol.** Given, $A'A = I$

Then $|A'A| = |I|$

$\Rightarrow |A'||A| = |I|$

$\Rightarrow |A||A| = |I| \quad [\because |A'| = |A|]$

$\Rightarrow |A|^2 = 1 \quad [\because |I| = 1]$

$\Rightarrow |A| = \pm 1$ **Ans.**

**2. If $y = x\,|x|$, find $\frac{dy}{dx}$ for $x < 0$.** **[1]**

**Sol.** If $y = x\,|x|$

Then,

$$y = \begin{cases} -x^2 & , x < 0 \\ x^2 & , x > 0 \end{cases}$$

$\Rightarrow \frac{dy}{dx} = -2x$ when $x < 0$ **Ans.**

**3. Find the order and degree (if defined) of the differential equation**

$$\frac{d^2y}{dx^2} + x\left(\frac{dy}{dx}\right)^2 = 2x^2 \log\left(\frac{d^2y}{dx^2}\right).$$ **[1]**

**Sol.**

Order of this equation is 2.

Degree of this equation is not defined. **Ans.**

**4. Find the direction consines of a line which makes equal angles with the coordinate axes. [1]**

**Sol.** Let the direction cosines of the line make an angle $\alpha$ with each of the coordinate axes and direction cosines be $< l, m, n >$.

$\therefore l = \cos\alpha, \quad m = \cos\alpha$ and $n = \cos\alpha$

$l^2 + m^2 + n^2 = 1$

$\therefore \quad \cos^2\alpha + \cos^2\alpha + \cos^2\alpha = 1$

$\Rightarrow 3\cos^2\alpha = 1$

$\Rightarrow \cos^2\alpha = \frac{1}{3}$

$\Rightarrow \cos\alpha = \pm\frac{1}{\sqrt{3}}$

The direction cosines are $\left(\pm\frac{1}{\sqrt{3}}, \pm\frac{1}{\sqrt{3}}, \pm\frac{1}{\sqrt{3}}\right)$

**Ans.**

**OR**

**A line passes through the point with position vector $2\hat{i} - \hat{j} + 4\hat{k}$ and is in the direction of the vector $\hat{i} + \hat{j} - 2\hat{k}$. Find the equation of the line in cartesian form.**

**Sol.** The line passes through a point (2, – 1, 4) and has direction ratios proportional to (1, 1, – 2).

Cartesian equation of the line,

$$\Rightarrow \frac{x - x_1}{a} = \frac{y - y_1}{b} = \frac{z - z_1}{c}$$

$$\Rightarrow \frac{x - 2}{1} = \frac{y + 1}{1} = \frac{z - 4}{-2}$$ **Ans.**

### SECTION–B

**5. Examine whether the operation * defined on R, the set of all real numbers, by $a * b = \sqrt{a^2 + b^2}$ is a binary operation or not, and if it is a binary operation, find whether it is associative or not.**** **[2]**

**6. If $A = \begin{bmatrix} 4 & 2 \\ -1 & 1 \end{bmatrix}$, show that (A – 2I) (A – 3I) = 0. [2]**

**Sol.** Given,

$$A = \begin{bmatrix} 4 & 2 \\ -1 & 1 \end{bmatrix} \text{ and } I = \begin{bmatrix} 1 & 0 \\ 0 & 1 \end{bmatrix}$$

$$A - 2I = \begin{bmatrix} 4 & 2 \\ -1 & 1 \end{bmatrix} - 2\begin{bmatrix} 1 & 0 \\ 0 & 1 \end{bmatrix}$$

$$= \begin{bmatrix} 4 & 2 \\ -1 & 1 \end{bmatrix} - \begin{bmatrix} 2 & 0 \\ 0 & 2 \end{bmatrix}$$

$$= \begin{bmatrix} 2 & 2 \\ -1 & -1 \end{bmatrix}$$

$$A - 3I = \begin{bmatrix} 4 & 2 \\ -1 & 1 \end{bmatrix} - 3\begin{bmatrix} 1 & 0 \\ 0 & 1 \end{bmatrix}$$

$$= \begin{bmatrix} 4 & 2 \\ -1 & 1 \end{bmatrix} - \begin{bmatrix} 3 & 0 \\ 0 & 3 \end{bmatrix}$$

** Answer is not given due to change in the present syllabus.

$$= \begin{bmatrix} 1 & 2 \\ -1 & -2 \end{bmatrix}$$

Then,

$$(A-2I)(A-3I) = \begin{bmatrix} 2 & 2 \\ -1 & -1 \end{bmatrix}\begin{bmatrix} 1 & 2 \\ -1 & -2 \end{bmatrix}$$

$$= \begin{bmatrix} 0 & 0 \\ 0 & 0 \end{bmatrix} = 0$$ **Hence Proved**

**7. Find : $\int \sqrt{3-2x-x^2}\,dx$.** **[2]**

**Sol.** Let $I = \int \sqrt{3-2x-x^2}\,dx$

$$= \int \sqrt{-(x^2+2x-3)}\,dx$$

$$= \int \sqrt{-(x^2+1+2x-3-1)}\,dx$$

$$= \int \sqrt{-[(x+1)^2-4]}\,dx$$

$$= \int \sqrt{4-(x+1)^2}\,dx$$

$$= \int \sqrt{2^2-(x+1)^2}\,dx$$

$$= \frac{1}{2}(x+1)\sqrt{4-(x+1)^2} + \frac{4}{2}\sin^{-1}\left(\frac{x+1}{2}\right) + c$$

$$\left[\because \int \sqrt{a^2-x^2}\,dx = x\frac{a^2-x^2}{2} + \frac{a^2}{2}\sin^{-1}\left(\frac{x}{a}\right) + c\right]$$

$$\therefore \quad I = \frac{1}{2}(x+1)\sqrt{3-2x-x^2} + 2\sin^{-1}\frac{(x+1)}{2} + c$$ **Ans.**

**8. Find :**

$$\int \frac{\sin^3 x + \cos^3 x}{\sin^2 x \cos^2 x}dx.$$ **[2]**

**Sol.** Let $I = \int \frac{\sin^3 x + \cos^3 x}{\sin^2 x.\cos^2 x}dx$

$$= \int \frac{\sin^3 x\,dx}{\sin^2 x.\cos^2 x} + \int \frac{\cos^3 x}{\sin^2 x.\cos^2 x}dx$$

$$= \int \frac{\sin x}{\cos^2 x}dx + \int \frac{\cos x}{\sin^2 x}dx$$

$$= \int \tan x.\sec x\,dx + \int \cot x.\text{cosec}\,x\,dx$$

$$= \sec x - \text{cosec}\,x + c$$ **Ans.**

**OR**

**Find : $\int \frac{x-3}{(x-1)^3}e^x dx$**

**Sol.**

Let, $I = \int \frac{x-3}{(x-1)^3}e^x\,dx$

$$= \int \frac{x-2-1}{(x-1)^3}e^x\,dx$$

$$= \int \frac{x-1-2}{(x-1)^3}e^x\,dx$$

$$= \int \left[\frac{x-1}{(x-1)^3} - \frac{2}{(x-1)^3}\right]e^x\,dx$$

$$= \int \left[\frac{1}{(x-1)^2} - \frac{2}{(x-1)^3}\right]e^x\,dx \quad ...(i)$$

Here, let $f(x) = \left[\frac{1}{(x-1)^2}\right]$

Then, $f'(x) = \frac{-2}{(x-1)^3}$

We know that,

$$\int [f(x)+f'(x)]\,e^x dx = e^x f(x) + c$$

Then from equation (i), we have

$$\int \left[\frac{1}{(x-1)^2} - \frac{2}{(x-1)^3}\right]e^x dx = \frac{e^x}{(x-1)^2} + c$$ **Ans.**

**9. Find the differential equation of the family of curves $y = Ae^{2x} + Be^{-2x}$, where A and B are arbitrary constants.** **[2]**

**Sol.** Given, $y = Ae^{2x} + Be^{-2x}$ ...(i)

On differentiating equation (i) w.r.t. $x$, we get

$$\frac{dy}{dx} = 2Ae^{2x} - 2Be^{-2x} \quad ...(ii)$$

Again differentiating equation (ii) w.r.t. $x$, we get

$$\frac{d^2y}{dx^2} = 4Ae^{2x} + 4Be^{-2x}$$

$$= 4(Ae^{2x} + Be^{-2x})$$

$$= 4y$$

$\therefore \quad \frac{d^2y}{dx^2} - 4y = 0$ is the required differential equation. **Ans.**

**10. If $|\vec{a}| = 2$, $|\vec{b}| = 7$ and $\vec{a} \times \vec{b} = 3\hat{i} + 2\hat{j} + 6\hat{k}$, find the angle between $\vec{a}$ and $\vec{b}$.** **[2]**

**Sol.** Given,

$$|\vec{a}| = 2, |\vec{b}| = 7 \text{ and } \left(\vec{a} \times \vec{b}\right) = 3\hat{i} + 2\hat{j} + 6\hat{k}$$

$\therefore$ Angle between $\vec{a}$ and $\vec{b}$ is given by

$$\sin\theta = \frac{|\vec{a} \times \vec{b}|}{|\vec{a}|\,|\vec{b}|} \quad ...(i)$$

Then, $$|\vec{a} \times \vec{b}| = \sqrt{(3)^2 + (2)^2 + (6)^2} = \sqrt{49} = 7$$

$\therefore$ $$\sin\theta = \frac{7}{2 \times 7} = \frac{1}{2}$$

$\Rightarrow$ $$\theta = \frac{\pi}{6} \text{ or } \frac{5\pi}{6}$$ **Ans.**

**OR**

**Find the volume of a cuboid whose edges are given by $-3\hat{i} + 7\hat{j} + 5\hat{k}$, $-5\hat{i} + 7\hat{j} - 3\hat{k}$ and $7\hat{i} - 5\hat{j} - 3\hat{k}$.**

**Sol.** If $a, b, c$ are edges of a cuboid.

Then, volume of cuboid $= \vec{a}.(\vec{b} \times \vec{c})$

Here,

$$\vec{a} = -3\hat{i} + 7\hat{j} + 5\hat{k}$$

$$\vec{b} = -5\hat{i} + 7\hat{j} - 3\hat{k}$$

$$\vec{c} = 7\hat{i} - 5\hat{j} - 3\hat{k}$$

Then,

$$\vec{a}.(\vec{b} \times \vec{c}) = \begin{vmatrix} \hat{i} & \hat{j} & \hat{k} \\ -3 & 7 & 5 \\ -5 & 7 & -3 \\ 7 & -5 & -3 \end{vmatrix}$$

$= -3(-21 - 15) - 7(15 + 21) + 5(25 - 49)$

$= -3 \times (-36) - 7 \times 36 + 5 \times (-24)$

$= 108 - 252 - 120$

$= -264$

But volume cannot be negative.

$\therefore$ Volume of cuboid = 264 cubic units **Ans.**

**11. If P(not A) = 0·7, P(B) = 0·7 and P(B/A) = 0·5, then find P(A/B).** **[2]**

**Sol.**

Given, P(not A) = 0.7

P (B) = 0.7

and P (B/A) = 0.5

We know that

$$P(B/A) = \frac{P(A \cap B)}{P(A)}$$

$$0.5 = \frac{P(A \cap B)}{0.3}$$

[$\because$ P(not A) = 0.7 then P(A) = 1 – 0.7 = 0.3]

$\Rightarrow$ $$P(A \cap B) = 0.15$$

Also we know,

$$P(A/B) = \frac{P(A \cap B)}{P(B)} = \frac{0.15}{0.70} = \frac{15}{70} = \frac{3}{14}$$ **Ans.**

**12. A coin is tossed 5 times. What is the probability of getting (i) 3 heads, (ii) at most 3 heads?** **[2]**

**Sol.**

Here, $n = 5$, $p = \frac{1}{2}$ and $q = \frac{1}{2}$

We know that,

$$p(x) = {}^nC_x\, p^x\, q^{n-x} = {}^5C_x \left(\frac{1}{2}\right)^x \left(\frac{1}{2}\right)^{5-x}$$

(i) For 3 heads, $x = 3$

Then, the probability of getting 3 heads is

$$P(3) = {}^5C_3 \left(\frac{1}{2}\right)^3 \left(\frac{1}{3}\right)^{5-3}$$

$$= {}^5C_3 \left(\frac{1}{2}\right)^3 \times \left(\frac{1}{2}\right)^2$$

$$= {}^5C_3 \times \left(\frac{1}{2}\right)^5$$

$$= \frac{5 \times 4 \times 3!}{2 \times 3!} \left(\frac{1}{2}\right)^5$$

$$= 10\left(\frac{1}{2}\right)^5 = \frac{5}{16}$$ **Ans.**

(ii) Probability of getting at most 3 heads is

$P(x \leq 3) = P(0) + P(1) + P(2) + P(3)$

Then, $$P(0) = {}^5C_0 \left(\frac{1}{2}\right)^5 = \frac{1}{32}$$

$$P(1) = {}^5C_1\left(\frac{1}{2}\right)^5 = \frac{5}{32}$$

$$P(2) = {}^5C_2\left(\frac{1}{2}\right)^5 = \frac{5}{16}$$

$$P(3) = {}^5C_3\left(\frac{1}{2}\right)^5 = \frac{5}{16}$$

$$\therefore \quad P(x \le 3) = \frac{1}{32}+\frac{5}{32}+\frac{5}{16}+\frac{5}{16}$$

$$= \frac{26}{32} = \frac{13}{16} \quad \textbf{Ans.}$$

**OR**

**Find the probabilty distribution of X, the number of heads in a simultaneous toss of two coins.**

**Sol.**

If we toss two coins simultaneously then sample space is given by (HH, HT, TH, TT)

Then probability distribution is,

| X(No. of heads) | 0 | 1 | 2 |
|---|---|---|---|
| P($x$) | $\frac{1}{4}$ | $\frac{1}{2}$ | $\frac{1}{4}$ |

**Ans.**

## SECTION–C

**13. Check whether the relation R defined on the set A = {1, 2, 3, 4, 5, 6} as R = {($a$, $b$) : $b$ = $a$ + 1} is reflexive, symmetric or transitive. [4]**

**Sol.** Here, R = {($a$, $b$) : $b = a + 1$}

$\therefore \quad R = \{(a, a+1) : a, a+1 \in (1, 2, 3, 4, 5, 6)\}$

$\Rightarrow \quad R = \{(1,2,)\ (2,3), (3,4), (4,5), (5,6)\}$

(i) R is not reflexive as $(a, a) \notin R\, \forall a$

(ii) R is not symmetric as $(1, 2) \in R$ but $(2, 1) \notin R$

(iii) R is not transitive as $(1, 2) \in R$, $(2, 3) \in R$ but $(1, 3) \notin R$ **Ans.**

**OR**

**Let $f$ : N → Y be a function defined as**

**$f(x) = 4x + 3$,**

**where Y = {$y \in$ N : $y = 4x + 3$, for some $x \in$ N}. Show that $f$ is invertible. Find its inverse.**

**Sol.** Consider an arbitrary element of Y. By the definition of $y$, $y = 4x + 3$, for some $x$ in the domain N.

This shows that $x = \frac{y-3}{4}$

Define $g : Y \to N$ by $g(y) = \frac{y-3}{4}$

Now, $gof(x) = g(f(x)) = g(4x+3) = \frac{4x+3-3}{4} = x$

and $fog(y) = f(g(y)) = f\left(\frac{y-3}{4}\right) = \frac{4(y-3)}{4} + 3$

$= y - 3 + 3 = y$

This shows that $gof = I_N$ and $fog = I_y$ which implies that $f$ is invertible and g is the inverse of $f$.

**Hence Proved.**

$\therefore$ Inverse of $f = g(y) = \frac{y-3}{4}$ **Ans.**

**14. Find the value of $\sin\left(\cos^{-1}\frac{4}{5}+\tan^{-1}\frac{2}{3}\right)$. [4]**

**Sol.**

$$\sin\left(\cos^{-1}\frac{4}{5}+\tan^{-1}\frac{2}{3}\right) = \sin\left(\cos^{-1}\frac{4}{5}+\tan^{-1}\frac{2}{3}\right)$$

$$= \sin\left(\tan^{-1}\frac{3}{4}+\tan^{-1}\frac{2}{3}\right)$$

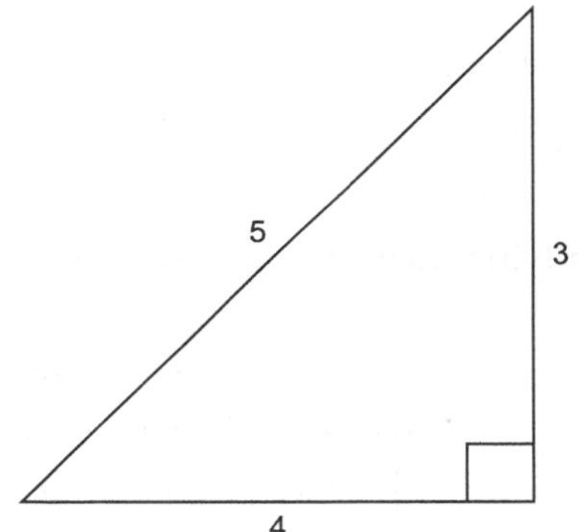


$$= \sin\left(\tan^{-1}\frac{\frac{3}{4}+\frac{2}{3}}{1-\frac{3}{4}\times\frac{2}{3}}\right)$$

$$\left[\because \tan^{-1}x + \tan^{-1}y = \tan^{-1}\left(\frac{x+y}{1-xy}\right)\right]$$

$$= \sin\left(\tan^{-1}\frac{\frac{9+8}{12}}{1-\frac{6}{12}}\right)$$

$$= \sin\left(\tan^{-1}\frac{\frac{17}{12}}{\frac{6}{12}}\right)$$

$$= \sin\left(\tan^{-1}\frac{17}{6}\right)$$

$$= \sin\left(\sin^{-1}\frac{17}{\sqrt{325}}\right)$$

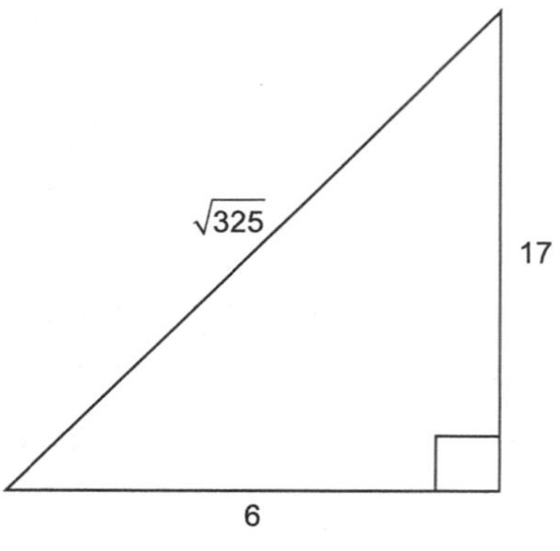


$$= \frac{17}{\sqrt{325}}$$ **Ans.**

**15. Using properties of determinants, show that**

$$\begin{vmatrix} 3a & -a+b & -a+c \\ -b+a & 3b & -b+c \\ -c+a & -c+b & 3c \end{vmatrix} = 3(a+b+c)(ab+bc+ca)$$ **[4]**

**Sol.**

$$\text{L.H.S.} = \begin{vmatrix} 3a & -a+b & -a+c \\ -b+a & 3b & -b+c \\ -c+a & -c+b & 3c \end{vmatrix}$$

On applying operation $C_1 \to C_1 + C_2 + C_3$, we have

$$= \begin{vmatrix} 3a-a+b-a+c & -a+b & -a+c \\ -b+a+3b-b+c & 3b & -b+c \\ -c+a-c+b+3c & -c+b & 3c \end{vmatrix}$$

$$= \begin{vmatrix} a+b+c & -a+b & -a+c \\ a+b+c & 3b & -b+c \\ a+b+c & -c+b & 3c \end{vmatrix}$$

$$= (a+b+c)\begin{vmatrix} 1 & -a+b & -a+c \\ 1 & 3b & -b+c \\ 1 & -c+b & 3c \end{vmatrix}$$

Again, applying operations $R_1 \to R_1 - R_3$ and $R_2 \to R_2 - R_3$, we get

$$= (a+b+c)\begin{vmatrix} 0 & -a+c & -a-2c \\ 0 & 2b+c & -b-2c \\ 1 & -c+b & 3c \end{vmatrix}$$

Now applying $R_2 \to R_2 - R_1$, we get

$$= (a+b+c)\begin{vmatrix} 0 & -a+c & -a-2c \\ 0 & 2b+a & -b+a \\ 1 & -c+b & 3c \end{vmatrix}$$

Expanding about $C_1$, we get

$= (a+b+c)\,[(c-a)(a-b) + (a+2c)(2b+a)]$

$= (a+b+c)\,[ac - bc - a^2 + ab + 2ab + a^2 + 4bc + 2ac]$

$= (a+b+c)\,[3ac + 3bc + 3ab]$

$= (a+b+c) \times 3\,(ac + bc + ab)$

$= 3(a+b+c)\,(ab + bc + ca) =$ R. H. S.

**Hence Proved.**

**16. If $x\sqrt{1+y} + y\sqrt{1+x} = 0$ and $x \neq y$, prove that $\frac{dy}{dx} = -\frac{1}{(x+1)^2}$** **[4]**

**Sol.** Given, $x\sqrt{1+y} + y\sqrt{1+x} = 0$

$$\Rightarrow \quad x\sqrt{1+y} = -y\sqrt{1+x}$$

On squaring both sides, we get

$$\Rightarrow \quad x^2(1+y) = y^2(1+x)$$

$$\Rightarrow \quad x^2 + x^2y = y^2 + y^2x$$

$$\Rightarrow \quad x^2 + x^2y - y^2 - y^2x = 0$$

$$\Rightarrow \quad x^2 - y^2 + x^2y - y^2x = 0$$

$$\Rightarrow (x-y)(x+y) + xy(x-y) = 0$$

$$\Rightarrow \quad (x-y)[x+y+xy] = 0$$

$$\Rightarrow \quad x + y + xy = 0 \qquad [\because x \neq y]$$

$$\Rightarrow \quad y = -\frac{x}{1+x} \qquad \text{...(i)}$$

On differentiating equation (i) w.r.t. $x$, we get

$$\frac{dy}{dx} = \frac{(1+x)(-1) + x(1)}{(1+x)^2}$$

$$= \frac{-1-x+x}{(1+x)^2}$$

$$= \frac{-1}{(1+x^2)}$$

$\therefore$ L.H.S = R.H.S. **Hence Proved.**

**OR**

**If $(\cos x)^y = (\sin y)^x$, find $\frac{dy}{dx}$.**

**Sol.** Given, $(\cos x)^y = (\sin y)^x$

On taking log on both sides, we get

$$y \log(\cos x) = x \log(\sin y) \qquad \text{..(i)}$$

On differentiating equation (i) w.r.t. $x$,

$$\Rightarrow y \times \frac{d}{dx}(\log \cos x) + \log \cos x \frac{dy}{dx}$$

$$= \log \sin y \frac{dx}{dx} + x\frac{d}{dx}(\log \sin y)$$

$$\Rightarrow y \times \frac{1}{\cos x}(-\sin x) + \log \cos x \frac{dy}{dx}$$

$$= \log \sin y + x \cdot \frac{\cos y}{\sin y} \cdot \frac{dy}{dx}$$

$$\Rightarrow -y\tan x + \frac{dy}{dx}\log\cos x = \log\sin y + x\cot y\frac{dy}{dx}$$

$$\Rightarrow \frac{dy}{dx}(\log\cos x - x\cot y) = \log \sin y + y\tan x$$

$$\therefore \qquad \frac{dy}{dx} = \frac{\log\sin y + y\tan x}{\log\cos x - x\cot y} \qquad \textbf{Ans.}$$

**17. If $(x - a)^2 + (y - b)^2 = c^2$, for some $c > 0$, prove that $\dfrac{\left[1+\left(\dfrac{dy}{dx}\right)^2\right]^{3/2}}{\dfrac{d^2y}{dx^2}}$ is a constant independent of $a$ and $b$.** **[4]**

**Sol.** If $\quad (x-a)^2 + (y-b)^2 = c^2, c > 0 \qquad$ ...(i)

On differentiating equation (i) w.r.t. $x$, we get

$$2(x-a) + 2(y-b)\frac{dy}{dx} = 0$$

$$\Rightarrow \quad x - a + (y-b)\frac{dy}{dx} = 0$$

$$\Rightarrow \quad \frac{dy}{dx} = -\frac{(x-a)}{(y-b)} \qquad \text{...(ii)}$$

Again, differentiating equation (ii) w.r.t. $x$, we get

$$\frac{d^2y}{dx^2} = -\left[\frac{(y-b)-(x-a)\,dy/dx}{(y-b)^2}\right]$$

$$= -\left[\frac{(y-b)-(x-a)\,dy/dx}{(y-b)^2}\right]$$

$$= -\left[\frac{(y-b) + \dfrac{(x-a)(x-a)}{(y-b)}}{(y-b)^2}\right] \qquad \text{[from equation (ii)]}$$

$$= -\left[\frac{\dfrac{(y-b)^2+(x-a)^2}{(y-b)}}{(y-b)^2}\right]$$

$$= -\left[\frac{(y-b)^2+(x-a)^2}{(y-b)^3}\right]$$

$$\frac{d^2y}{dx^2} = -\left[\frac{c^2}{(y-b)^3}\right] \qquad \text{reference (iii)}$$

[$\because$ From equation (i)]

$$\text{Now,} \quad \frac{\left[1+\left(\dfrac{dy}{dx}\right)^2\right]^{3/2}}{\dfrac{d^2y}{dx^2}} = \frac{\left[1+\dfrac{(x-a)^2}{(y-b)^2}\right]^{3/2}}{\left[\dfrac{c^2}{(y-b)^3}\right]}$$

[from equation (ii) and (iii)]

$$= -\frac{\left[(y-b)^2+(x-a)^2\right]^{3/2}\times(y-b)^3}{(y-b)^{2\times 3/2}\times c^2}$$

$$= -\frac{\left[(y-b)^2+(x-a)^2\right]^{3/2}\times(y-b)^3}{(y-b)^3\times c^2}$$

$$= -\frac{\left[(y-b)^2+(x-a)^2\right]^{3/2}}{c^2} = -\frac{-c^{2\times 3/2}}{c^2}$$

[from equation (i)]

$$= \frac{-c^3}{c^2} = -c = \text{constant}$$

It shows that $\dfrac{\left[1+\left(\dfrac{dy}{dx}\right)^2\right]^{3/2}}{\dfrac{d^2y}{dx^2}}$ is independent of $a$ and $b$ **Hence Proved.**

**18. Find the equation of the normal to the curve $x^2 = 4y$ which passes through the point $(-1, 4)$. [4]**

**Sol.** Suppose the normal at $P(x_1, y_1)$ on the parabola $x^2 = 4y$ passes through $(-1, 4)$

Since $P(x_1, y_1)$ lies on $x^2 = 4y$

$$\therefore \qquad x_1^2 = 4y_1 \qquad \text{...(i)}$$

The equation of curve is $x^2 = 4y$

Differentiating with respect to $x$, we have

$$2x = 4\frac{dy}{dx}$$

$$\Rightarrow \quad \frac{dy}{dx} = \frac{x}{2}$$

$$\Rightarrow \quad \left(\frac{dy}{dx}\right)_{(x_1, y_1)} = \frac{x_1}{2}$$

The equation $(x_1, y_1)$ of normal at $P(x_1, y_1)$ is

$$y - y_1 = \frac{-1}{\dfrac{dy}{dx}}(x - x_1)$$

$$y - y_1 = \frac{-2}{x_1}(x - x_1) \qquad \text{...(ii)}$$

$\because$ It passes through $(-1, 4)$,

$\therefore$ Putting $x = -1$ and $y = 4$, we get

$$4 - y_1 = \frac{-2}{x_1}(-1 - x_1)$$

$$\Rightarrow \quad 4 - y_1 = \frac{2}{x_1}(1 + x_1)$$

$$\Rightarrow \quad 4x_1 - x_1y_1 = 2 + 2x_1$$

$$\Rightarrow \quad 2x_1 = 2 + x_1y_1$$

$$\Rightarrow \quad \frac{2x_1 - 2}{x_1} = y_1 \quad \text{....(iii)}$$

Putting value of y from (iii) into (i), we have

$$x_1^2 = 4\left(\frac{2x_1 - 2}{x_1}\right)$$

$$x_1^3 = 8x_1 - 8$$

$$\Rightarrow \quad x_1^3 - 8x_1 + 8 = 0$$

$$(x_1 - 2)(x_1^2 + 2x_1 - 4) = 0$$

$$\Rightarrow \quad x_1 = 2 \text{ or } \pm\sqrt{5} - 1;$$

Putting $x_1 = 2$ in (iii), we get $y_1 = 1$

Putting values of $x_1, y_1$ in (ii), we get equation of normal is:

$$y - 1 = -1(x - 2)$$

$$x + y - 3 = 0$$

**Ans.**

**19. Find :** $\int \frac{x^2 + x + 1}{(x + 2)(x^2 + 1)} dx.$ **[4]**

**Sol.** Let,

$$I = \int \frac{x^2 + x + 1}{(x + 2)(x^2 + 1)} dx$$

By partial fractions

$$\frac{x^2 + x + 1}{(x + 2)(x^2 + 1)} = \frac{A}{x + 2} + \frac{Bx + C}{x^2 + 1}$$

$$x^2 + x + 1 = A(x^2 + 1) + Bx(x + 2) + C(x + 2)$$

$$= Ax^2 + A + Bx^2 + 2Bx + Cx + 2C$$

$$x^2 + x + 1 = x^2(A + B) + x(2B + C) + A + 2C \quad \text{...(i)}$$

On comparing coefficients of equation (i), we get

$$1 = A + B \quad \text{reference (ii)}$$

and $1 = 2B + C$ ...(iii)

and $1 = A + 2C$ ...(iv)

Solving (ii), (iii) and (iv), we get

$$A = \frac{3}{5}, B = \frac{2}{5} \text{ and } C = \frac{1}{5}$$

Hence, $\frac{x^2 + x + 1}{(x + 2)(x^2 + 1)} = \frac{3}{5(x + 2)} + \frac{\frac{2}{5}x + \frac{1}{5}}{x^2 + 1}$

Then,

$$\int \frac{(x^2 + x + 1)dx}{(x + 2)(x^2 + 1)} = \frac{3}{5}\int \frac{1}{x + 2} dx + \frac{2}{5}\int \frac{x}{x^2 + 1} dx + \frac{1}{5}\int \frac{1}{(x^2 + 1)} dx$$

$$= \frac{3}{5}\log|x + 2| + \frac{1}{5}\int \frac{2x}{x^2 + 1} dx + \frac{1}{5}\int \frac{dx}{x^2 + 1}$$

$$= \frac{3}{5}\log|x + 2| + \frac{1}{5}\log|x^2 + 1| + \frac{1}{5}\tan^{-1} x + c$$

**Ans.**

**20. Prove that :** $\int_0^a f(x)\, dx = \int_0^a f(a - x)\, dx$ **and hence evaluate** $\int_0^{\pi/2} \frac{x}{\sin x + \cos x} dx.$ **[4]**

**Sol.**

$$\text{R.H.S.} = \int_0^a f(a - x)\, dx$$

Let $a - x = v$ ...(i)

Differentiating (i) w.r.t. $x$, we get

$$-1 = \frac{dv}{dx}$$

for $x = 0, v = a$

$x = a, v = 0$

Then,

$$\text{R.H.S.} = \int_a^0 f(v)(-dv)$$

$$= -\int_a^0 f(v)(dv) = \int_0^a f(v)(dv)$$

$$\left[\because \int_a^b f(x)dx = -\int_b^a f(x)dx\right]$$

Now, replacing $v$ by $x$,

$$= \int_0^a f(x)\, dx$$

Hence, $\int_0^a f(x)\, dx = \int_0^a f(a - x)dx$ **Hence Proved.**

Now, Let

$$I = \int_0^{\pi/2} \frac{x}{\sin x + \cos x} dx \quad \text{...(ii)}$$

$$I = \int_0^{\pi/2} \frac{\frac{\pi}{2} - x}{\sin\left(\frac{\pi}{2} - x\right) + \cos\left(\frac{\pi}{2} - x\right)} dx$$

$$\left[\because \int_0^a f(x)dx = \int_0^a f(a - x)dx\right]$$

$$I = \int_0^{\pi/2} \frac{\frac{\pi}{2} - x}{\cos x + \sin x} dx \quad \text{(iii)}$$

Adding equations (ii) and (iii), we get

$$2I = \int_0^{\pi/2} \frac{x + \frac{\pi}{2} - x}{\sin x + \cos x} dx$$

$$2I = \int_0^{\pi/2} \frac{\frac{\pi}{2}}{\sin x + \cos x} dx$$

$$2I = \frac{\pi}{2}\int_0^{\pi/2} \frac{1}{\sin x + \cos x} dx$$

$$2I = \frac{\pi}{2}\int_0^{\pi/2} \frac{1}{\frac{2\tan x/2}{1+\tan^2 x/2} + \frac{1-\tan^2 x/2}{1+\tan^2 x/2}} dx$$

$$2I = \frac{\pi}{2}\int_0^{\pi/2} \frac{1+\tan^2\frac{x}{2}}{2\tan\frac{x}{2} + 1 - \tan^2\frac{x}{2}} dx$$

$$2I = \frac{\pi}{2}\int_0^{\pi/2} \frac{\sec^2 x/2}{-\tan^2 x/2 + 2\tan x/2 + 1} dx$$

Let, $\tan\frac{x}{2} = t$

Then, $\frac{d}{dx}\left(\tan\frac{x}{2}\right) = \frac{d}{dx}(t)$

$$\frac{1}{2}\sec^2\frac{x}{2}\,dx = dt$$

$$\sec^2\frac{x}{2}dx = 2dt$$

For $x = 0, t = 0$

$x = \pi/2, t = 1$

$$\therefore \quad 2I = \frac{\pi}{2}\int_0^1 \frac{2dt}{-t^2+2t+1}$$

$$= \pi\int_0^1 \frac{dt}{-t^2+2t+1}$$

$$= \pi\int_0^1 \frac{dt}{-(t^2-2t-1)}$$

$$= \pi\int_0^1 \frac{dt}{-\left[(t-1)^2-2\right]}$$

$$= \pi\int_0^1 \frac{dt}{(\sqrt{2})^2-(t-1)^2}$$

$$= \pi \times \frac{1}{2\sqrt{2}}\left[\log\left|\frac{\sqrt{2}+t-1}{\sqrt{2}-t+1}\right|\right]_0^1$$

$$= \pi \times \frac{1}{2\sqrt{2}}\left[\log 1 - \log\left(\frac{\sqrt{2}-1}{\sqrt{2}+1}\right)\right]$$

$$= \frac{-\pi}{2\sqrt{2}}\log\left(\frac{\sqrt{2}-1}{\sqrt{2}+1}\right) = \frac{\pi}{2\sqrt{2}}\log\left(\frac{\sqrt{2}+1}{\sqrt{2}-1}\right)$$

$$\Rightarrow \quad I = \frac{\pi}{4\sqrt{2}}\log\left[\frac{\sqrt{2}+1}{\sqrt{2}-1}\right]$$

$$\Rightarrow \quad I = \frac{\pi}{4\sqrt{2}}\log\left\{\frac{(\sqrt{2}+1)^2}{(\sqrt{2}-1)(\sqrt{2}+1)}\right\}$$

$$\Rightarrow \quad I = \frac{\pi}{4\sqrt{2}}\log(\sqrt{2}+1)^2$$

$$\Rightarrow \quad I = \frac{2\pi}{4\sqrt{2}}\log(\sqrt{2}+1)$$

$$\Rightarrow \quad I = \frac{\pi}{2\sqrt{2}}\log(\sqrt{2}+1)$$

**Ans.**

**21. Solve the differential equation :**

$$x\frac{dy}{dx} = y - x\tan\left(\frac{y}{x}\right)$$ **[4]**

**Sol.**

Given, $x\frac{dy}{dx} = y - x\tan\left(\frac{y}{x}\right)$

Let, $y = vx$

Then, $\frac{dy}{dx} = v + x\frac{dv}{dx}$

$\therefore \quad x\frac{dy}{dx} = y - x\tan\left(\frac{y}{x}\right)$

$\Rightarrow \quad x\left(v + x\frac{dv}{dx}\right) = vx - x\tan\left(\frac{v}{x}\right)$

$\Rightarrow \quad x\left(v + x\frac{dv}{dx}\right) = x(v - \tan v)$

$\Rightarrow \quad xv + x^2\frac{dv}{dx} = xv - x\tan v$

$\Rightarrow \quad x^2\frac{dv}{dx} = -x\tan v$

$\Rightarrow \quad x\frac{dv}{dx} = -\tan v$

$\Rightarrow \quad \frac{dv}{\tan v} = -\frac{dx}{x}$

Then $\int \cot v\, dv = -\int\frac{dx}{x}$

$\log\sin v = -\log x + \log c$

$\Rightarrow \quad \log\sin v = \log\frac{c}{x}$

$\Rightarrow \quad \sin\frac{y}{x} = \frac{c}{x}$

$$\Rightarrow \quad x \sin \frac{y}{x} = c$$ **Ans.**

**OR**

**Solve the differential equation :**

$$\frac{dy}{dx} = -\left[\frac{x + y\cos x}{1 + \sin x}\right]$$

**Sol.**

$$\frac{dy}{dx} = -\left[\frac{x + y\cos x}{1 + \sin x}\right]$$

$$\Rightarrow \quad \frac{dy}{dx} = -\frac{x}{1+\sin x} - \frac{y\cos x}{1+\sin x}$$

$$\Rightarrow \quad \frac{dy}{dx} + \frac{y\cos x}{1+\sin x} = \frac{-x}{1+\sin x}$$

Here , $P = \frac{\cos x}{1+\sin x}$ and $Q = \frac{-x}{1+\sin x}$

Then,

$$\text{I.F.} = e^{\int P\,dx} = e^{\int \frac{\cos x}{1+\sin x}dx} = e^{\log|1+\sin x|} = 1 + \sin x$$

Then,

$$y \times \text{I.F.} = \int Q \times \text{I.F.}\, dx + c$$

$$y\,(1 + \sin x) = \int \frac{-x}{1+\sin x} \times (1+\sin x)dx = \int -x\,dx$$

$$y\,(1+\sin x) = \frac{-x^2}{2} + c$$ **Ans.**

**22. The scalar product of the vector $\vec{a} = \hat{i} + \hat{j} + \hat{k}$ with a unit vector along the sum of the vectors $\vec{b} = 2\hat{i} + 4\hat{j} - 5\hat{k}$ and $\vec{c} = \lambda\hat{i} + 2\hat{j} + 3\hat{k}$ is equal to 1. Find the value of $\lambda$ and hence find the unit vector along $\vec{b} + \vec{c}$.** **[4]**

**Sol.** Here, $\vec{a} = \hat{i} + \hat{j} + \hat{k}, \hat{n}$ is unit vector

$$\vec{b} = 2\hat{i} + 4\hat{j} - 5\hat{k}$$

$$\vec{c} = \lambda\hat{i} + 2\hat{j} + 3\hat{k}$$

$$\vec{b} + \vec{c} = (2+\lambda)\hat{i} + 6\hat{j} - 2\hat{k}$$

Then,

$$\hat{n} = \frac{(2+\lambda)\hat{i} + 6\hat{j} - 2\hat{k}}{\sqrt{(2+\lambda)^2 + 36 + 4}}$$

Given, $\vec{a}.\hat{n} = 1$

$$\left(\hat{i} + \hat{j} + \hat{k}\right) \cdot \left(\frac{(2+\lambda)\hat{i} + 6\hat{j} - 2\hat{k}}{\sqrt{(2+\lambda)^2 + 40}}\right) = 1$$

$$\Rightarrow \quad (2+\lambda) + 6 - 2 = \sqrt{(2+\lambda)^2 + 40}$$

$$\Rightarrow \quad (2+\lambda) + 4 = \sqrt{(2+\lambda)^2 + 40}$$

$$\Rightarrow \quad \lambda + 6 = \sqrt{(2+\lambda)^2 + 40}$$

On squaring both sides, we get

$$(6+\lambda)^2 = (2+\lambda)^2 + 40$$

$$\Rightarrow \quad 36 + \lambda^2 + 12\lambda = 4 + \lambda^2 + 4\lambda + 40$$

$$\Rightarrow \quad 8\lambda - 8 = 0$$

$$\Rightarrow \quad \lambda = 1$$

Then,

$$\vec{b} + \vec{c} = (2+\lambda)\hat{i} + 6\hat{j} - 2\hat{k} = 3\hat{i} + 6\hat{j} - 2\hat{k}$$

$$\text{unit vector along } \left(\vec{b} + \vec{c}\right) = \frac{3\hat{i} + 6\hat{j} - 2\hat{k}}{\sqrt{9 + 36 + 4}} = \frac{3\hat{i} + 6\hat{j} - 2\hat{k}}{\sqrt{49}} = \frac{3\hat{i} + 6\hat{j} - 2\hat{k}}{7} = \frac{3}{7}\hat{i} + \frac{6}{7}\hat{j} - \frac{2}{7}\hat{k}$$ **Ans.**

**23. If the lines $\frac{x-1}{-3} = \frac{y-2}{2\lambda} = \frac{z-3}{2}$ and $\frac{x-1}{3\lambda} = \frac{y-1}{2} = \frac{z-6}{-5}$ are perpendicular, find the value of $\lambda$. Hence find whether the lines are intersecting or not.** **[4]**

**Sol.** The equation of the given lines are,

$$\frac{x-1}{-3} = \frac{y-2}{2\lambda} = \frac{z-3}{2} \text{ and } \frac{x-1}{3\lambda} = \frac{y-1}{2} = \frac{z-6}{-5}$$

If these lines are perpendicular, then

$$a_1a_2 + b_1b_2 + c_1c_2 = 0$$

$$\Rightarrow \quad -3 \times 3\lambda + 2\lambda \times 2 + 2 \times (-5) = 0$$

$$\Rightarrow \quad -9\lambda + 4\lambda - 10 = 0$$

$$\Rightarrow \quad -5\lambda = 10$$

$$\Rightarrow \quad \lambda = -2$$

Now, the lines are

$$\frac{x-1}{-3} = \frac{y-2}{-4} = \frac{z-3}{2} \text{ and } \frac{x-1}{-6} = \frac{y-1}{2} = \frac{z-6}{-5}$$

The co-ordinates of any point on first line are given by :

$$\frac{x-1}{-3} = \frac{y-2}{-4} = \frac{z-3}{2} = \alpha$$

or $x - 1 = -3\alpha \Rightarrow x = -3\alpha + 1$

$y - 2 = -4\alpha \Rightarrow y = -4\alpha + 2$

$z - 3 = 2\alpha \Rightarrow z = 2\alpha + 3$

So, coordinates of any point on this line are, $(-3\alpha + 1, -4\alpha + 2, 2\alpha + 3)$.

The coordinates of any point on second line are given by :

$$\frac{x-1}{-6} = \frac{y-1}{2} = \frac{z-6}{-5} = \beta$$

or $x - 1 = -6\beta \Rightarrow x = -6\beta + 1$

$y - 1 = 2\beta \Rightarrow y = 2\beta + 1$

$z - 6 = -5\beta \Rightarrow z = -5\beta + 6$

So, co-ordinates of any point on second line are $(-6\beta + 1, 2\beta + 1, -5\beta + 6)$.

If lines intersect then they have a common point. So, for some value of $\alpha$ and $\beta$, we have

$-3\alpha + 1 = -6\beta + 1$

$\Rightarrow$ $-3\alpha = -6\beta$

$\Rightarrow$ $\alpha = 2\beta$ ...(i)

and $-4\alpha + 2 = 2\beta + 1$

$\Rightarrow$ $-4\alpha + 1 = 2\beta$ ...(ii)

On solving, (i) and (ii) we have

$$\alpha = \frac{1}{5}, \text{ and } \beta = \frac{1}{10}$$

The values of $\alpha$ and $\beta$ do not satisfy the third equation. Hence, lines do not intersect each other. **Ans.**

## SECTION–D

**24. If** $A = \begin{bmatrix} 1 & 3 & 4 \\ 2 & 1 & 2 \\ 5 & 1 & 1 \end{bmatrix}$, **find** $A^{-1}$ **[6]**

**Hence solve the system of equations**

$x + 3y + 4z = 8$

$2x + y + 2z = 5$

**and** $5x + y + z = 7$

**Sol.**

If $$A = \begin{bmatrix} 1 & 3 & 4 \\ 2 & 1 & 2 \\ 5 & 1 & 1 \end{bmatrix}$$

$$|A| = 1(1-2) - 3(2-10) + 4(2-5)$$
$$= 11 \neq 0, A^{-1} \text{ exists}$$

Then cofactors of A are

$A_{11} = -1, A_{12} = 8, A_{13} = -3, A_{21} = 1, A_{22} = -19,$
$A_{23} = 14, A_{31} = 2, A_{32} = 6$ and $A_{33} = -5$

Then, $$\text{adj } A = \begin{bmatrix} A_{11} & A_{12} & A_{13} \\ A_{21} & A_{22} & A_{23} \\ A_{31} & A_{32} & A_{33} \end{bmatrix}^T$$

$$= \begin{bmatrix} -1 & 8 & -3 \\ 1 & -19 & 14 \\ 2 & 6 & -5 \end{bmatrix}^T$$

$$\text{adj } A = \begin{bmatrix} -1 & 1 & 2 \\ 8 & -19 & 6 \\ -3 & 14 & -5 \end{bmatrix}$$

Now, $$A^{-1} = \frac{1}{|A|}\text{adj}A$$

$$A^{-1} = \frac{1}{11}\begin{bmatrix} -1 & 1 & 2 \\ 8 & -19 & 6 \\ -3 & 14 & -5 \end{bmatrix} \quad ...(ii)$$

**Ans.**

Given system of equations are

$x + 3y + 4z = 8$

$2x + y + 2z = 5$

$5x + y + z = 7$

Let, $A = \begin{bmatrix} 1 & 3 & 4 \\ 2 & 1 & 2 \\ 5 & 1 & 1 \end{bmatrix}$ $B = \begin{bmatrix} 8 \\ 5 \\ 7 \end{bmatrix}$ and $X = \begin{bmatrix} x \\ y \\ z \end{bmatrix}$

Then, $AX = B$

$$\begin{bmatrix} 1 & 3 & 4 \\ 2 & 1 & 2 \\ 5 & 1 & 1 \end{bmatrix} X = \begin{bmatrix} 8 \\ 5 \\ 7 \end{bmatrix}$$

$$X = \begin{bmatrix} 1 & 3 & 4 \\ 2 & 1 & 2 \\ 5 & 1 & 1 \end{bmatrix}^{-1} \begin{bmatrix} 8 \\ 5 \\ 7 \end{bmatrix}$$

$$= \frac{1}{11}\begin{bmatrix} -1 & 1 & 2 \\ 8 & -19 & 6 \\ -3 & 14 & -5 \end{bmatrix}\begin{bmatrix} 8 \\ 5 \\ 7 \end{bmatrix}$$

[Using (ii)]

$$= \frac{1}{11}\begin{bmatrix} -8+5+14 \\ 64-95+42 \\ -24+70-35 \end{bmatrix}$$

$$X = \frac{1}{11}\begin{bmatrix} 11 \\ 11 \\ 11 \end{bmatrix}$$

$$\begin{bmatrix} x \\ y \\ z \end{bmatrix} = \begin{bmatrix} 1 \\ 1 \\ 1 \end{bmatrix}$$

Hence, $x = 1, y = 1, z = 1$ **Ans.**

**OR**

**Find the inverse of the following matrix, using elementary transformation :**

$$\mathbf{A} = \begin{bmatrix} 2 & 0 & -1 \\ 5 & 1 & 0 \\ 0 & 1 & 3 \end{bmatrix}$$

**Sol.**

Given, $$A = \begin{bmatrix} 2 & 0 & -1 \\ 5 & 1 & 0 \\ 0 & 1 & 3 \end{bmatrix}$$

We know that,

$$AA^{-1} = I$$

$\therefore$ $$A = IA$$

$$\begin{bmatrix} 2 & 0 & -1 \\ 5 & 1 & 0 \\ 0 & 1 & 3 \end{bmatrix} = \begin{bmatrix} 1 & 0 & 0 \\ 0 & 1 & 0 \\ 0 & 0 & 1 \end{bmatrix} A$$

On applying $R_3 \to R_3 + 3R_1$, we get

$$\begin{bmatrix} 2 & 0 & -1 \\ 5 & 1 & 0 \\ 6 & 1 & 0 \end{bmatrix} = \begin{bmatrix} 1 & 0 & 0 \\ 0 & 1 & 0 \\ 3 & 0 & 1 \end{bmatrix} A$$

Applying $R_3 \to R_3 - R_2$, we get

$$\begin{bmatrix} 2 & 0 & -1 \\ 5 & 1 & 0 \\ 1 & 0 & 0 \end{bmatrix} = \begin{bmatrix} 1 & 0 & 0 \\ 0 & 1 & 0 \\ 3 & -1 & 1 \end{bmatrix} A$$

Interchanging $R_1 \leftrightarrow R_3$, we get

$$\begin{bmatrix} 1 & 0 & 0 \\ 5 & 1 & 0 \\ 2 & 0 & -1 \end{bmatrix} = \begin{bmatrix} 3 & -1 & 1 \\ 0 & 1 & 0 \\ 1 & 0 & 0 \end{bmatrix} A$$

Applying $R_2 \to R_2 - 5R_1$ and $R_3 \to R_3 - 2R_1$, we get

$$\begin{bmatrix} 1 & 0 & 0 \\ 0 & 1 & 0 \\ 0 & 0 & -1 \end{bmatrix} = \begin{bmatrix} 3 & -1 & 1 \\ -15 & 6 & -5 \\ -5 & 2 & -2 \end{bmatrix} A$$

Applying $R_3 \to (-1) R_3$, we get

$$\begin{bmatrix} 1 & 0 & 0 \\ 0 & 1 & 0 \\ 0 & 0 & 1 \end{bmatrix} = \begin{bmatrix} 3 & -1 & 1 \\ -15 & 6 & -5 \\ 5 & -2 & 2 \end{bmatrix} A$$

Hence, the required inverse of the matrix is

$$\begin{bmatrix} 3 & -1 & 1 \\ -15 & 6 & 5 \\ 5 & -2 & 2 \end{bmatrix}$$ **Ans.**

**25. Show that the height of the cylinder of maximum volume that can be inscribed in a sphere of radius R is $\frac{2R}{\sqrt{3}}$. Also find the maximum volume.** **[6]**

**Sol.** Let, '$x$' be the diameter of the base of the cylinder and let '$h$' be height of the cylinder.

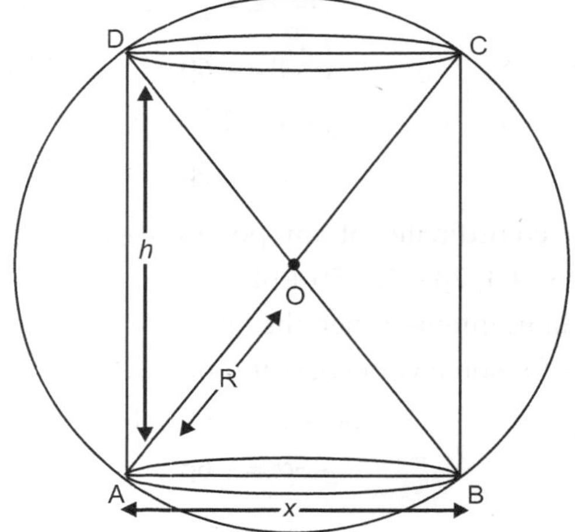


In $\Delta ABC$, we have

$$(BC)^2 + (AB)^2 = (AC)^2$$

$$h^2 + x^2 = (2R)^2$$

$$x^2 = 4R^2 - h^2 \quad \text{...(i)}$$

Volume of cylinder, $V = \pi r^2 h$

$\Rightarrow$ $$V = \pi \times \left(\frac{x}{2}\right)^2 \times h$$

$\Rightarrow$ $$V = \pi \times \frac{x^2}{4} \times h$$

$\Rightarrow$ $$V = \frac{\pi(4R^2 - h^2)}{4} \times h$$ [Using (i)]

$\Rightarrow$ $$V = \frac{4\pi R^2 \times h}{4} - \frac{\pi h^3}{4}$$

$\Rightarrow$ $$V = \pi h R^2 - \pi \frac{h^3}{4} \quad \text{...(ii)}$$

On differentiating equation (ii) w.r.t. $h$, we get

$$\frac{dV}{dh} = \pi R^2 - \frac{\pi}{4}(3h^2)$$

$\Rightarrow$ $$\frac{dV}{dh} = \pi R^2 - \frac{3\pi h^2}{4} \quad \text{...(iii)}$$

For maximum volume, $\frac{dV}{dh} = 0$

$$\Rightarrow \quad 0 = \pi R^2 - \frac{3\pi h^2}{4}$$

$$\Rightarrow \quad h = \frac{2R}{\sqrt{3}}$$

Again, differentiating equation (iii) w.r.t. $h$, we get

$$\Rightarrow \quad \frac{d^2V}{dh^2} = 0 - \frac{3}{4}\pi \times 2h$$

$$\Rightarrow \quad \frac{d^2V}{dh^2} = -\frac{3\pi h}{2}$$

At $h = \frac{2R}{\sqrt{3}}$, we have

$$\frac{d^2V}{dh^2} = \frac{-3\pi}{2}\left(\frac{2R}{\sqrt{3}}\right)$$

$$= -\sqrt{3}\,\pi R$$

$$\frac{d^2V}{dh^2} < 0$$

Hence, $h = \frac{2R}{\sqrt{3}}$ is a point of maxima.

So, V is maximum when $h = \frac{2R}{\sqrt{3}}$

Hence, the height of the cylinder of maximum volume that can be inscribed in a sphere of radius R is $\frac{2R}{\sqrt{3}}$. **Hence Proved.**

From (i), we have

$$x^2 = 4R^2 - h^2$$

$$\Rightarrow \quad x^2 = 4R^2 - \left(\frac{2R}{\sqrt{3}}\right)^2$$

$$\Rightarrow \quad x^2 = \frac{8}{3}R^2$$

$\therefore$ Maximum Volume of cylinder

$$= \pi\left(\frac{x}{2}\right)^2 \times h$$

$$= \pi \times \frac{x^2}{4} \times h$$

$$= \frac{\pi}{4}x^2 h$$

$$= \frac{\pi}{4} \times \frac{8R^2}{3} \times \frac{2R}{\sqrt{3}}$$

$$= \frac{4\pi R^3}{3\sqrt{3}}$$ **Ans.**

**26. Using method of integration, find the area of the triangle whose vertices are (1, 0), (2, 2) and (3, 1).** **[6]**

**Sol.** A(1, 0), B(2, 2) and C(3, 1)

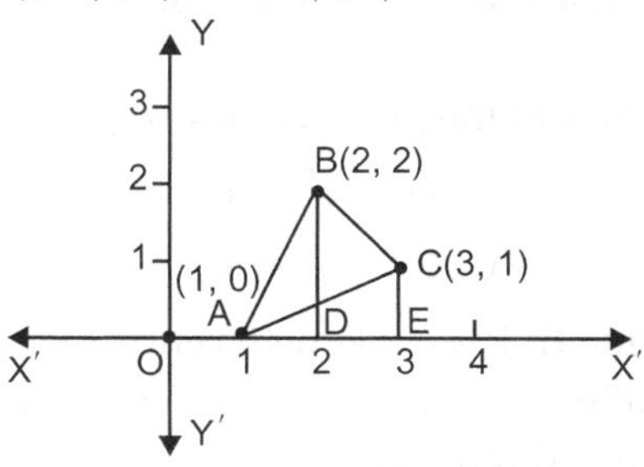


Let A(1, 0), B(2, 2) and C(3, 1) be the vertices of a triangle ABC.

Area of $\Delta$ABC = Area of $\Delta$ABD + Area of trapezium BDEC – Area of $\Delta$AEC

Now, Equation of line AB,

$$y - y_1 = \frac{y_2 - y_1}{x_2 - x_1}(x - x_1)$$

$$y - 0 = 2(x - 1)$$

$$y = 2(x - 1) \quad \text{...(i)}$$

Equation of line BC,

$$y - 2 = \frac{1-2}{3-2}(x - 2)$$

$$y - 2 = \frac{-1}{1}(x - 2)$$

$$y - 2 = -(x - 2)$$

$$y = 4 - x \quad \text{...(ii)}$$

Equation of line AC,

$$y - 0 = \frac{1-0}{3-1}(x - 1)$$

$$y - 0 = \frac{1}{2}(x - 1)$$

$$y = \frac{1}{2}(x - 1) \quad \text{...(iii)}$$

Hence, area of $\Delta$ABC

$$= \int_1^2 2(x-1)dx + \int_2^3 (4-x)dx - \int_1^3 \frac{x-1}{2}dx$$

$$= 2\left[\frac{x^2}{2} - x\right]_1^2 + \left[4x - \frac{x^2}{2}\right]_2^3 - \frac{1}{2}\left[\frac{x^2}{2} - x\right]_1^3$$

$$= 2\left[\left(\frac{2^2}{2} - 2\right) - \left(\frac{1}{2} - 1\right)\right] + \left[\left(4 \times 3 - \frac{3^2}{2}\right) - \left(4 \times 2 - \frac{2^2}{2}\right)\right] - \frac{1}{2}\left[\left(\frac{3^2}{2} - 3\right) - \left(\frac{1}{2} - 1\right)\right]$$

$$= 2\left(\frac{1}{2}\right) + \frac{3}{2} - 1 = \frac{3}{2} \text{ sq. units.}$$ **Ans.**

**OR**

**Using method of integration, find the area of the region enclosed between two circles $x^2 + y^2 = 4$ and $(x - 2)^2 + y^2 = 4$.**

**Sol.**

Equations of the given circles are,

$$x^2 + y^2 = 4 \quad \text{...(i)}$$

$$(x - 2)^2 + y^2 = 4 \quad \text{...(ii)}$$

Equation (i) is a circle with centre O at the origin and radius 2. Equation (ii) is a circle with centre C (2, 0) and radius 2.

Solving equation (i) and (ii) we have

$$(x - 2)^2 + y^2 = x^2 + y^2$$

or $$x^2 - 4x + 4 + y^2 = x^2 + y^2$$

or $x = 1$ which gives $y = \pm\sqrt{3}$

Thus, the points of intersection of the given circles are A $(1, \sqrt{3})$ and A′$(1, -\sqrt{3})$

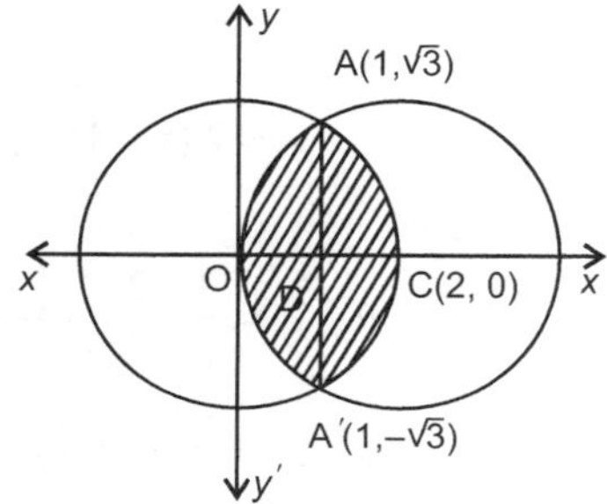


Required area of the enclosed region OACA′O between circles

= 2 [area of region ODCAO]

= 2 [area of region ODAO + area of region DCAD]

$$= 2\left[\int_0^1 y\,dx + \int_1^2 y\,dx\right]$$

$$= 2\left[\int_0^1 \sqrt{4 - (x-2)^2}\,dx + \int_1^2 \sqrt{4 - x^2}\,dx\right]$$

[from (i) and (ii)]

$$= 2\left[\frac{1}{2}(x-2)\sqrt{4-(x-2)^2} + \frac{1}{2}\times 4\sin^{-1}\left(\frac{x-2}{2}\right)\right]_0^1$$

$$+ 2\left[\frac{1}{2}x\sqrt{4-x^2} + \frac{1}{2}\times 4\sin^{-1}\frac{x}{2}\right]_1^2$$

$$= \left[(x-2)\sqrt{4-(x-2)^2} + 4\sin^{-1}\left(\frac{x-2}{2}\right)\right]_0^1$$

$$+ \left[x\sqrt{4-x^2} + 4\sin^{-1}\frac{x}{2}\right]_1^2$$

$$= \left[-\sqrt{3} + 4\sin^{-1}\left(\frac{-1}{2}\right) - 4\sin^{1}(-1)\right]$$

$$+ \left[4\sin^{-1}1 - \sqrt{3} - 4\sin^{-1}\frac{1}{2}\right]$$

$$= \left[-\sqrt{3} - 4\times\frac{\pi}{6} + 4\times\frac{\pi}{2}\right] + \left[4\times\frac{\pi}{2} - \sqrt{3} - 4\times\frac{\pi}{6}\right]$$

$$= \left(\frac{8\pi}{3} - 2\sqrt{3}\right) \text{ sq. units}$$ **Ans.**

**27. Find the vector and cartesian equations of the plane passing through the points having position vectors $\hat{i} + \hat{j} - 2\hat{k}$, $2\hat{i} - \hat{j} + \hat{k}$ and $\hat{i} + 2\hat{j} + \hat{k}$. Write the equation of a plane passing through a point (2, 3, 7) and parallel to the plane obtained above. Hence, find the distance between the two parallel planes. [6]**

**Sol.** Let A, B,C be the points with position vectors $\hat{i} + \hat{j} - 2\hat{k}$, $2\hat{i} - \hat{j} + \hat{k}$ and $\hat{i} + 2\hat{j} + \hat{k}$ respectively.

Then, $\overrightarrow{AB}$ = P. V. of B – P. V. of A

$$= \left(2\hat{i} - \hat{j} + \hat{k}\right) - \left(\hat{i} + \hat{j} - 2\hat{k}\right)$$

$$= \hat{i} - 2\hat{j} + 3\hat{k}$$

and $\overrightarrow{BC}$ = P. V. of C – P.V. of B

$$= \left(\hat{i} + 2\hat{j} + \hat{k}\right) - \left(2\hat{i} - \hat{j} + \hat{k}\right)$$

$$= -\hat{i} + 3\hat{j} + 0\hat{k}$$

A vector normal to the plane containing points A, B & C is

$$\vec{n} = \overrightarrow{AB} \times \overrightarrow{AC}$$

$$= \begin{vmatrix} \hat{i} & \hat{j} & \hat{k} \\ 1 & -2 & 3 \\ -1 & 3 & 0 \end{vmatrix}$$

$$= -9\hat{i} - 3\hat{j} + \hat{k}$$

The required plane passes through the point having position vector $\vec{a} = \hat{i} + \hat{j} - 2\hat{k}$ and is normal to the vector $-9\hat{i} - 3\hat{j} + 2\hat{k}$. So, its vector equation is,

$$\left(\vec{r} - \vec{a}\right) \cdot \vec{n} = 0$$

$$\Rightarrow \quad \vec{r} \cdot \vec{n} - \vec{a} \cdot \vec{n} = 0$$

$$\Rightarrow \quad \vec{r} \cdot \vec{n} = \vec{a} \cdot \vec{n}$$

$$\Rightarrow \quad \vec{r} \cdot (-9\hat{i} - 3\hat{j} + \hat{k}) = (\hat{i} + \hat{j} - 2\hat{k}) \cdot \left(-9\hat{i} - 3\hat{j} + \hat{k}\right)$$

$\Rightarrow \vec{r}.(-9\hat{i}-3\hat{j}+\hat{k}) = -9-3-2$

$\Rightarrow \vec{r}.(-9\hat{i}-3\hat{j}+\hat{k}) = -14$

This is the required vector equation of the plane
The cartesian equation of plane is given by

$$(x\hat{i}+y\hat{j}+z\hat{k}).(-9\hat{i}-3\hat{j}+\hat{k}) = -14$$

$$-9x-3y+z = -14$$

$$9x+3y-z-14 = 0$$

Direction ratios of this plane are (9, 3, – 1)

Then the equation of plane parallel to the above plane and passing through (2,3,7) is

$\Rightarrow \quad a(x-x_1)+b(y-y_1)+c(z-z_1) = 0$

$\Rightarrow \quad 9(x-2)+3(y-3)-1(z-7) = 0$

$\Rightarrow \quad 9x+3y-z-20 = 0$

This is the required parallel plane.
Let $P(x_1, y_1, z_1)$ be any point on $9x+3y-z-14=0$
Then,

$$9x_1+3y_1-z_1-14 = 0 \quad \text{...(i)}$$

Let $d$ be the distance between planes. Then,

$d$ = length of perpendicular from $P(x_1, y_1, z_1)$ to $9x+3y-z-20=0$

$$d = \left|\frac{9x_1+3y_1-z_1-20}{\sqrt{(9)^2+(3)^2+(-1)^2}}\right|$$

[from (i)]

$$= \left|\frac{+14-20}{\sqrt{91}}\right|$$

$$= \frac{6}{\sqrt{91}} \text{ units.}$$ **Ans.**

**OR**

**Find the equation of the line pasing through (2, –1, 2) and (5, 3, 4) and of the plane passing through (2, 0, 3), (1, 1, 5) and (3, 2, 4). Also, find their point of intersection.**

**Sol.** We know that the equation of line passing through points $(x_1, y_1, z_1)$ and $(x_2, y_2, z_2)$ is given by

$$\frac{x-x_1}{x_2-x_1} = \frac{y-y_1}{y_2-y_1} = \frac{z-z_1}{z_2-z_1}$$

So equation of line is,

$$\frac{x-2}{3} = \frac{y+1}{4} = \frac{z-2}{2}$$

Now, equation of plane passing through points (2, 0,3), (1,1,5) and (3,2,4) is given by

$$\begin{vmatrix} x-x_1 & y-y_1 & z-z_1 \\ x_2-x_1 & y_2-y_1 & z_2-z_1 \\ x_3-x_1 & y_3-y_1 & z_3-z_1 \end{vmatrix} = 0$$

$$\begin{vmatrix} x-2 & y-0 & z-3 \\ -1 & 1 & 2 \\ 1 & 2 & 1 \end{vmatrix} = 0$$

$\Rightarrow (x-2)(-3)-y(-1-2)+(z-3)(-2-1) = 0$

$\Rightarrow \quad -3x+6+3y-3z+9 = 0$

$\Rightarrow \quad -x+y-z+5 = 0 \quad \text{...(i)}$

This is the required equation of plane.

Now,

Let $\frac{x-2}{3} = \frac{y+1}{4} = \frac{z-2}{2} = k$

$\therefore \quad x = 3k+2$

$y = 4k-1$

$z = 2k+2$

On putting these values in equation (i), we have

$-(3k+2)+4k-1-(2k+2)+5 = 0$

$\Rightarrow \quad -3k-2+4k-1-2k-2+5 = 0$

$\Rightarrow \quad -k = 0$

$\Rightarrow \quad k = 0$

Then coordinates of intersection point are

$x = 3k+2 = 2,$

$y = 4k-1 = -1,$

$z = 2k+2 = 2$

$\therefore$ Point of intersection is (2, – 1, 2) **Ans.**

**28. There are three coins. One is a two-headed coin, another is a biased coin that comes up heads 75% of the time and the third is an unbiased coin. One of the three coins is chosen at random and tossed. If it shows heads, what is the probability that it is the two-headed coin? [6]**

**Sol.** Given, there are three coins.

Let, $E_1$ = coin is two headed
$E_2$ = biased coin
$E_3$ = unbiased coin
A = shows only head

Here, $P(E_1) = P(E_2) = P(E_3) = \frac{1}{3}$

Then $P\left(\frac{A}{E_1}\right) = 1$

$P\left(\frac{A}{E_2}\right) = \frac{75}{100} = \frac{3}{4}$ (given)

$$P\left(\frac{A}{E_3}\right) = \frac{1}{2}$$

Now, Probability of two headed coin

$P(E_1/A)$

$$= \frac{P(E_1)P(A/E_1)}{P(E_1)\times P(A/E_1)+P(E_2)\times P(A/E_2)+P(E_3)\times P(A/E_3)}$$

$$= \frac{\frac{1}{3}\times 1}{\frac{1}{3}\times 1+\frac{1}{3}\times\frac{3}{4}+\frac{1}{3}\times\frac{1}{2}} = \frac{1}{\frac{4+3+2}{4}} = \frac{4}{9}$$ **Ans.**

**29. A company produces two types of goods, A and B, that require gold and silver. Each unit of type A requries 3 g of silver and 1 g of gold while that of type B requires 1 g of silver and 2 g of gold. The company can use at the most of 9 g of silver and 8 of gold. If each unit of type A brings a profit of ₹ 40 and that of type B ₹ 50, find the number of units of each type that the company should produce to maximize profit. Formulate the above LPP and solve it graphically and also find the maximum profit. [6]**

**Sol.** There are two types of goods, A and B and let units of type A be $x$ and units of type B be $y$.

| | A | B |
|---|---|---|
| Gold | 1 | 2 |
| Silver | 3 | 1 |
| Profit | 40 | 50 |

Then, Total profit of goods

$$P = 40x + 50y, x \geq 0, y \geq 0$$

Hence, the mathematical formulation of the problem is as follows :

Maximise $\quad P = 40x + 50y \quad$ ...(i)

Subject to the constraints :

$x + 2y \leq 8 \quad$ ...(ii)

$3x + y \leq 9 \quad$ ...(iii)

$x \geq 0, y \geq 0$

To solve this LPP, we draw the lines

$x + 2y = 8$

$3x + y = 9$

$x = 0$ and $y = 0$

$x + 2y = 8$

| $x$ | 0 | 8 |
|---|---|---|
| $y$ | 4 | 0 |

and $3x + y = 9$

| $x$ | 0 | 3 |
|---|---|---|
| $y$ | 9 | 0 |

Plotting these points on the graph

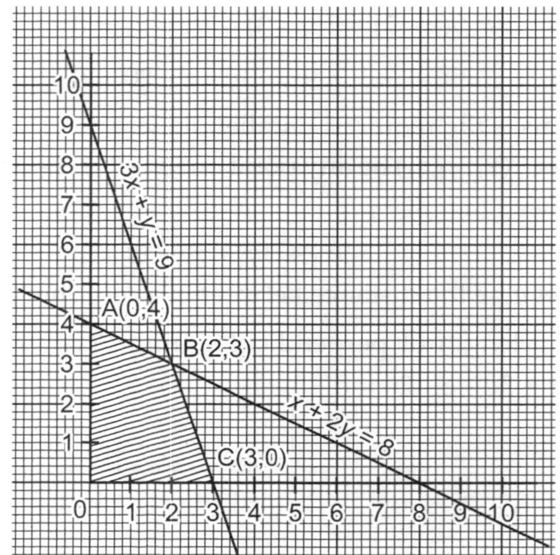


The shaded region is the required feasible region.

| Corner Points | Maximum P = $40x + 50y$ |
|---|---|
| A(0,4) | $0 + 50 \times 4 = 200$ |
| B(2,3) | $2 \times 40 + 3 \times 50 = 230$ |
| C(3,0) | $40 \times 3 + 0 = 120$ |
| O (0,0) | $0 + 0 = 0$ |

Clearly, P is maximum at B (2,3) and the maximum profit is ₹ 230. **Ans.**

## Outside Delhi [Set-II]

**Note : Except for the following questions, all the remaining questions have been asked in previous set.**

### SECTION–A

**1. Find $|AB|$, if $A = \begin{bmatrix} 0 & -1 \\ 0 & 2 \end{bmatrix}$ and $B = \begin{bmatrix} 3 & 5 \\ 0 & 0 \end{bmatrix}$. [1]**

**Sol.**

Given, $A = \begin{bmatrix} 0 & -1 \\ 0 & 2 \end{bmatrix}$ and $B = \begin{bmatrix} 3 & 5 \\ 0 & 0 \end{bmatrix}$

Then, $AB = \begin{bmatrix} 0 & -1 \\ 0 & 2 \end{bmatrix}\begin{bmatrix} 3 & 5 \\ 0 & 0 \end{bmatrix}$

$= \begin{bmatrix} 0 & 0 \\ 0 & 0 \end{bmatrix}$

$\therefore \quad |AB| = 0$ **Ans.**

**2. Differentiate $e^{\sqrt{3x}}$ with respect to $x$. [1]**

**Sol.**

Let $y = e^{\sqrt{3x}}$ ...(i)

On differentiating equation (i) w.r.t $x$, we get

$$\frac{dy}{dx} = e^{\sqrt{3x}} \times \sqrt{3} \times \frac{1}{2} x^{\frac{-1}{2}}$$

$$= \frac{\sqrt{3}}{2} \frac{e^{\sqrt{3x}}}{\sqrt{x}}$$ **Ans.**

## SECTION–B

**6. If $A = \begin{bmatrix} p & 2 \\ 2 & p \end{bmatrix}$ and $|A^3| = 125$, find the value of $p$. [2]**

**Sol.**

Given, $A = \begin{bmatrix} p & 2 \\ 2 & p \end{bmatrix}$ and $|A^3| = 125$

Now, $|A|^3 = 125$

$\Rightarrow \quad |A|^3 = 5^3$

$\Rightarrow \quad |A| = 5$

Also we have $|A| = \begin{vmatrix} p & 2 \\ 2 & p \end{vmatrix}$

$\Rightarrow \quad p^2 - 4 = 5$

$\Rightarrow \quad p^2 = 9$

$\Rightarrow \quad p = \pm 3$ **Ans.**

**12. Find the general solution of the differential equation $\frac{dy}{dx} = e^{x+y}$ [2]**

**Sol.**

Given, $\frac{dy}{dx} = e^{x+y}$

$\Rightarrow \quad \frac{dy}{dx} = e^x . e^y$

$\Rightarrow \quad \frac{dy}{e^y} = e^x .dx$

$\Rightarrow \quad e^{-y}\, dy = e^x .dx$

On integrating both sides, we get

$$\int e^{-y} dy = \int e^x dx$$

$\Rightarrow -e^{-y} = e^x + c$, which is the required solution.

**Ans.**

## SECTION–C

**21. If $(a + bx)\, e^{y/x} = x$, then prove that**

$$x^3 \frac{d^2y}{dx^2} = \left( x\frac{dy}{dx} - y \right)^2$$ **[4]**

**Sol.** We have

$$(a + bx)\, e^{y/x} = x$$

$\Rightarrow \quad e^{y/x} = \frac{x}{a+bx}$

$\Rightarrow \quad \frac{y}{x} = \log \frac{x}{a+bx}$

$\Rightarrow \quad \frac{y}{x} = \log x - \log (a + bx)$

On differentiating w.r.t. $x$, we get

$$\frac{x\frac{dy}{dx} - y}{x^2} = \frac{1}{x} - \frac{1}{a+bx} \times b$$

$\Rightarrow \quad \frac{1}{x^2}\left( x\frac{dy}{dx} - y \right) = \frac{1}{x} - \frac{b}{a+bx}$

$\Rightarrow \quad x\frac{dy}{dx} - y = x^2 \left( \frac{1}{x} - \frac{b}{a+bx} \right)$

$\Rightarrow \quad x\frac{dy}{dx} - y = \frac{ax}{a+bx}$ ...(i)

Again, differentiating both sides w.r.t. $x$, we get

$$x\frac{d^2y}{dx^2} + \frac{dy}{dx} - \frac{dy}{dx} = \frac{(a+bx)a - ax \times b}{(a+bx)^2}$$

$\Rightarrow \quad x\frac{d^2y}{dx^2} = \frac{a^2 + abx - abx}{(a+bx)^2}$

$\Rightarrow \quad x\frac{d^2y}{dx^2} = \frac{a^2}{(a+bx)^2}$

On multiplying both sides by $x^2$, we get

$$x^3 \frac{d^2y}{dx^2} = \frac{a^2x^2}{(a+bx)^2}$$

$$x^3 \frac{d^2y}{dx^2} = \left( \frac{ax}{a+bx} \right)^2$$ ...(ii)

From (i) and (ii), we get

$$x^3 \frac{d^2y}{dx^2} = \left(x\frac{dy}{dx} - y\right)^2$$

**Hence Proved.**

**22. The volume of a cube is increasing at the rate of 8 cm³/s. How fast is the surface area increasing when the length of its edge is 12 cm? [4]**

**Sol.** Let $x$ be the length of side, V be the volume and S be the surface area of cube.

Then, $V = x^3$ and $S = 6x^2$, where $x$ is a function of time $t$

Now, $\frac{dV}{dt} = 8\text{cm}^3/\text{s}$ (given)

$\therefore \quad 8 = \frac{dV}{dt} = \frac{d}{dt}(x^3) = 3x^2\frac{dx}{dt}$

$\Rightarrow \quad 8 = 3x^2\frac{dx}{dt}$

$\Rightarrow \quad \frac{8}{3x^2} = \frac{dx}{dt}$ ...(i)

Now, $\frac{dS}{dt} = \frac{d(6x^2)}{dt}$

$= 12x\frac{dx}{dt}$

$= 12x \times \frac{8}{3x^2}$ [Using (i)]

$\Rightarrow \quad \frac{dS}{dt} = \frac{32}{x}$

Hence, when $x = 12$ cm

Then, $\frac{dS}{dt} = \frac{32}{12} = \frac{8}{3}\text{cm}^2/\text{s}$ **Ans.**

**23. Find the cartesian and vector equations of the plane passing through the point A(2, 5, – 3), B(– 2, – 3, 5) and C(5, 3, – 3). [4]**

**Sol.** We know that the general equation of the plane passing through three points $(x_1, y_1, z_1)$ $(x_2, y_2, z_2)$, $(x_3, y_3, z_3)$

$$\begin{vmatrix} x - x_1 & y - y_1 & z - z_1 \\ x_2 - x_1 & y_2 - y_1 & z_2 - z_1 \\ x_3 - x_1 & y_3 - y_1 & z_3 - z_1 \end{vmatrix} = 0$$

Then the plane passing through A (2, 5, –3), B (– 2, – 3, 5,), C (5, 3, –3)

$$\begin{vmatrix} x-2 & y-5 & z+3 \\ -2-2 & -3-5 & 5+3 \\ 5-2 & 3-5 & -3+3 \end{vmatrix} = 0$$

$$\Rightarrow \begin{vmatrix} x-2 & y-5 & z+3 \\ -4 & -8 & 8 \\ 3 & -2 & 0 \end{vmatrix} = 0$$

$\Rightarrow \quad (x-2)16 - (y-5)\,24 + (z+3)(8+24) = 0$

$\Rightarrow 16(x-2) + 24(y-5) + 32\,(z+3) = 0$

$\Rightarrow \quad 8\,[2x - 4 + 3y - 15 + 4z + 12] = 0$

$\Rightarrow \quad 2x + 3y + 4z - 7 = 0$

$\Rightarrow \quad 2x + 3y + 4z = 7$

This is the required cartesian equation of the plane.

Now,

The required plane passes through the point A(2, 5, –3) whose position vector is $a = 2\hat{i} + 5\hat{j} - 3\hat{k}$ and is normal to the vector $\vec{n}$ given by $\vec{n} = \vec{AB} \times \vec{AC}$

$\therefore \quad \vec{AB} = -2\hat{i} - 3\hat{j} + 5\hat{k} - (2\hat{i} + 5\hat{j} - 3\hat{k})$

$\vec{AB} = -4\hat{i} - 8\hat{j} + 8\hat{k}$

$\vec{AC} = 5\hat{i} + 3\hat{j} - 3\hat{k} - \left(2\hat{i} + 5\hat{j} - 3\hat{k}\right)$

$= 3\hat{i} - 2\hat{j}$

Now, $\vec{n} = \vec{AB} \times \vec{AC} = \begin{vmatrix} \hat{i} & \hat{j} & \hat{k} \\ -4 & -8 & 8 \\ 3 & -2 & 0 \end{vmatrix}$

$= 16\hat{i} - (-24)\hat{j} + (8+24)\hat{k}$

$= 16\hat{i} + 24\hat{j} + 32\hat{k}$

The vector equation of the plane is given by

$\vec{r}\,.\,\vec{n} = \vec{a}\,.\,\vec{n}$

$\Rightarrow \vec{r}\,.(16\hat{i} + 24\hat{j} + 32\hat{k})$

$= (2\hat{i} + 5\hat{j} - 3\hat{k})(16\hat{i} + 24\hat{j} + 32\hat{k})$

$= 32 + 120 - 96$

$\Rightarrow \vec{r}\,.\,(16\hat{i} + 24\hat{j} + 32\hat{k}) = 56$

$\Rightarrow \quad \vec{r}\,.\,(2\hat{i} + 3\hat{j} + 4\hat{k}) = 7$

This is the required vector equation. **Ans.**

## SECTION–D

**24. Find the point on the curve $y^2 = 4x$, which is nearest to the point (2, – 8). [6]**

**Sol.** Given curve is of the form, $y^2 = 4x$ and let $p(x,y)$ is a point on the curve which is nearest to the point (2, – 8).

$\because \quad y^2 = 4x$

then $\quad \frac{y^2}{4} = x$

$\therefore$ P(x, y) will be $P\left(\frac{y^2}{4}, y\right)$

$$y^2 = 4x$$

$$= P(x, y)$$

$$= A(2, -8)$$

Now, the distance between point A and P is given by

$$AP = \sqrt{(x-2)^2 + (y+8)^2}$$

$$= \sqrt{\left(\frac{y^2}{4} - 2\right)^2 + (y+8)^2}$$

$$= \sqrt{\frac{y^4}{16} - y^2 + 4 + y^2 + 16y + 64}$$

$$= \sqrt{\frac{y^4}{16} + 16y + 68}$$

and let $Z = AP^2 = \frac{y^4}{16} + 16y + 68$ ...(i)

Now, differentiate equation (i) w.r.t. $y$, we get

$$\frac{dZ}{dy} = \frac{1}{16} \times 4y^3 + 16$$

$$\Rightarrow \quad \frac{dZ}{dy} = \frac{y^3}{4} + 16 \quad \text{...(ii)}$$

For maximum or minumim value of Z, we have

$$\frac{dZ}{dy} = 0$$

$$\Rightarrow \quad \frac{y^3}{4} + 16 = 0$$

$$\Rightarrow \quad y^3 + 64 = 0$$

$$\Rightarrow \quad (y + 4)(y^2 - 4y + 16) = 0$$

($\because y^2 - 4y + 16 = 0$ gives imaginary value of $y$)

$$\Rightarrow \quad y = -4$$

Again, differentiating equation (ii) w.r.t. x,

$$\frac{d^2Z}{dy^2} = \frac{1}{4} \times 3y^2 = \frac{3}{4}y^2$$

At $y = -4$,

$$\frac{d^2Z}{dy^2} = \frac{3}{4}(-4)^2$$

$$= 12 > 0$$

Thus, Z is minimum when $y = -4$

Substituting $y = -4$ in equation of curve $y^2 = 4x$, we have $x = 4$

Hence, the point (4, – 4) on the curve $y^2 = 4x$ is nearest to the point (2, – 8). **Ans.**

**25. Find $\int_1^3 (x^2 + 2 + e^{2x})\, dx$ as the limit of sums. [6]**

**Sol.** We have

$$\int_a^b f(x)dx = \lim_{h \to 0} h\, [f(a) + f(a+h) + f(a+2h) + \ldots f(a + (n-1)h]$$

where $\quad h = \frac{b-a}{n}$

Here, $\quad a = 1$ and $b = 3$

$$f(x) = x^2 + 2 + e^{2x}$$

$$\therefore \quad h = \frac{3-1}{n} = \frac{2}{n}$$

$$\Rightarrow \quad nh = 2$$

Now, $I = \int_1^3 (x^2 + 2 + e^{2x})dx$

$$= \lim_{h \to 0} h[f(1) + f(1+h) + f(1+2h) \ldots f(1 + (n-1)h]$$

$$I = \lim_{h \to 0} h\, [(1 + 2 + e^2) + \{(1+h)^2 + 2 + e^{2(1+h)}\} + \{(1+2h)^2 + 2 + e^{2(1+2h)}\} + \ldots \{(1 + (n-1)h)^2 + 2 + e^{2(1+(n-1)h)}\}]$$

$$I = \lim_{h \to 0} h\Big[1 + 2 + e^2 + 1 + h^2 + 2h + 2 + e^{2(1+h)} + 1 + 4h^2 + 4h + 2 + e^{2(1+2h)} + \ldots\{1 + (n-1)^2 h^2 + 2(n-1)h + 2 + e^{2(1+(n-1)h)}\}]$$

$$I = \lim_{h \to 0} h[1 + 1 + 1 + \ldots + 2 + 2 + \ldots 2h + 4h + \ldots 2(n-1)h + h^2 + 4h^2 + \ldots (n-1)^2 h^2 + e^2 + e^{2(1+h)} + e^{2(1+2h)} + \ldots + e^{2(1+(n-1)h)}]$$

$$I = \lim_{h\to 0} h[1+1+1+......+2+2+.....+2h+4h + ..... 2(n-1)h + h^2 + 4h^2 + ..... (n-1)^2h^2 + e^2(1+e^{2h}+e^{4h}+.....e^{2(n-1)h}]$$

$$I = \lim_{h\to 0} h\Big[n + 2n + 2h(1+2+3+..... n-1) + h^2(1^2+2^2+3^2+...... (n-1)^2 + \frac{e^2 \times (1-e^{2hn})}{1-e^{2h}}\Big]$$

$$\left[\because \begin{array}{l} 1+1+1+.....n = n \\ 2+2+2+.....n = 2n \end{array}\right]$$

$$I = \lim_{h\to 0} h\left[3n + 2h\frac{n(n-1)}{2} + \frac{h^2(n-1)n(2n-1)}{6} + e^2\frac{(1-e^{2hn})}{1-e^{2h}}\right]$$

$$I = \lim_{h\to 0}\left[3nh + nh^2(n-1) + \frac{h^3(n-1)\,n(2n-1)}{6} + \frac{e^2h(1-e^{2hn})}{1-e^{2h}}\right]$$

$$I = \lim_{h\to 0}\left[nh \times 3 + nh \times h(n-1) + \frac{nh\,(nh-h)(2nh-h)}{6} + \frac{e^2h(1-e^{2hn})}{1-e^{2h}}\right]$$

Put $nh = 2$

$$I = \lim_{h\to 0}\left[6 + 2\,(2-h) + \frac{(2-h)2(4-h)}{6} + \frac{e^2}{2} \times \frac{2h(1-e^4)}{(1-e^{2h})}\right]$$

Now,

$$I = 6 + 4 + \frac{2\times 2\times 4}{6} - \frac{e^2(1-e^4)}{2}$$

$$\Rightarrow \quad I = 10 + \frac{8}{3} - \frac{e^2}{2}(1-e^4)$$

$$\Rightarrow \quad I = \frac{38}{3} - \frac{e^2}{2}(1-e^4)$$

$$\Rightarrow \quad I = \frac{38}{3} + \frac{e^2}{2}(e^4-1)$$ **Ans.**

**OR**

**Using integration, find the area of the triangular region whose sides have the equation $y = 2x + 1$, $y = 3x + 1$ and $x = 4$.**

**Sol.** The equations of sides of triangle are

$y = 2x + 1,$ ...(i)

$y = 3x + 1$ ...(ii)

and $x = 4$ ...(iii)

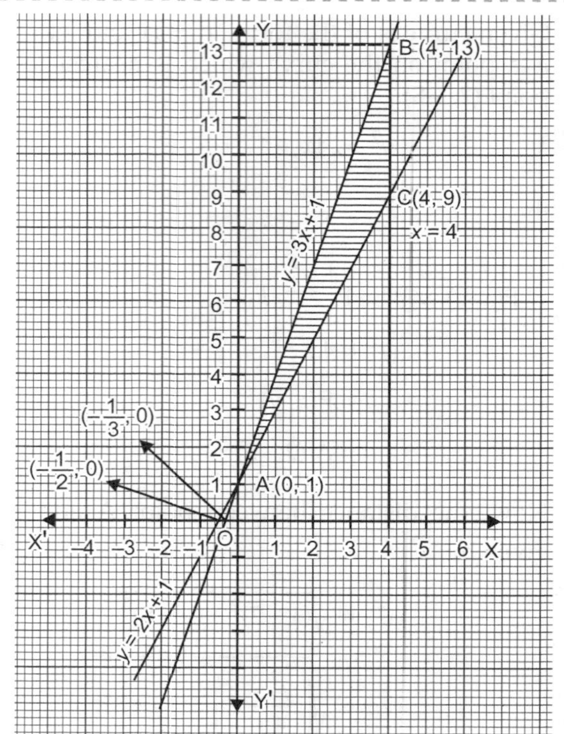


The equation $y = 2x + 1$ meets $x$ and $y$ axes at $\left(-\frac{1}{2}, 0\right)$ and (0,1). By joining these two points we obtain the graph of $y = 2x + 1$. Similarly, graphs of other equations are drawn.

Solving equation (i), (ii) and (iii) in pairs, we obtain the coordinates of verticies of $\Delta$ABC are A(0,1), B (4,13) and C (4,9).

Then, area of $\Delta$ABC = Area (OLBAO) – Area (OLCAO)

$$= \int_0^4 (3x+1)\,dx - \int_0^4 (2x+1)\,dx$$

$$= \int_0^4 (3x+1-2x-1)\,dx$$

$$= \int_0^4 x\,dx$$

$$= \left[\frac{x^2}{2}\right]_0^4$$

$$= \left[\frac{x^2}{2}\right]_0^4$$

$$= \frac{1}{2} \times 4 \times 4 = 8 \text{ square units.}$$ **Ans.**

●●

# Outside Delhi [Set-III]

**Note : Except for the following questions, all the remaining questions have been asked in previous sets.**

## SECTION–A

**1. Find the differenital equation representing the family of curves $y = ae^{2x} + 5$, where $a$ is an arbitrary constant. [1]**

**Sol.** Given, $y = ae^{2x} + 5$ ...(i)

On differentiating w.r.t. $x$, we get

$$\frac{dy}{dx} = a.e^{2x}.(2)+0$$

$$\Rightarrow \quad \frac{dy}{dx} = 2a.e^{2x}$$

$$\frac{dy}{dx} = 2(y-5) \quad \text{[from equation (i)]}$$

Again, differentiating w.r.t. $x$, we get

$$\frac{d^2y}{dx^2} = 2\frac{dy}{dx}$$

$$\Rightarrow \quad \frac{d^2y}{dx^2} - 2\frac{dy}{dx} = 0$$

This is the required differential equation. **Ans.**

**2. If $y = \cos\sqrt{3x}$, then find $\frac{dy}{dx}$. [1]**

**Sol.**

Given, $y = \cos\left(\sqrt{3x}\right)$

On differentiating w.r.t.$x$, we get

$$\frac{dy}{dx} = -\sin\left(\sqrt{3x}\right) \times \sqrt{3} \times \frac{1}{2\sqrt{x}}$$

$$\Rightarrow \quad \frac{dy}{dx} = -\frac{\sqrt{3}\sin\left(\sqrt{3x}\right)}{2\sqrt{x}}$$ **Ans.**

## SECTION–B

**5. Show that the points A$(-2\hat{i} + 3\hat{j} + 5\hat{k})$, B$(\hat{i} + 2\hat{j} + 3\hat{k})$ and C$(7\hat{i} - \hat{k})$ are collinear. [2]**

**Sol.** Given,

$$\vec{A} = -2\hat{i} + 3\hat{j} + 5\hat{k}$$

$$\vec{B} = \hat{i} + 2\hat{j} + 3\hat{k}$$

$$\vec{C} = 7\hat{i} + 0\hat{j} - \hat{k}$$

Now, $\vec{AB} = \hat{i} + 2\hat{j} + 3\hat{k} - \left(-2\hat{i} + 3\hat{j} + 5\hat{k}\right)$

$$\vec{AB} = \hat{i} + 2\hat{j} + 3\hat{k} + 2\hat{i} - 3\hat{j} - 5\hat{k}$$

$$\vec{AB} = 3\hat{i} - \hat{j} - 2\hat{k}$$

$$\vec{BC} = 7\hat{i} + 0\hat{j} - \hat{k} - (\hat{i} + 2\hat{j} + 3\hat{k})$$

$$\vec{BC} = 7\hat{i} + 0\hat{j} - \hat{k} - \hat{i} - 2\hat{j} - 3\hat{k}$$

$$\vec{BC} = 6\hat{i} - 2\hat{j} - 4\hat{k}$$

$$\vec{BC} = 2(3\hat{i} - \hat{j} - 2\hat{k})$$

Clearly, $\vec{BC} = 2\vec{AB} \Rightarrow AB || BC$

But B is a common point to $\vec{AB}$ and $\vec{BC}$.

$\therefore$ $\vec{AB}$ and $\vec{BC}$ are collinear vectors.

Hence, points A, B and C are collinear.

**Hence Proved.**

**OR**

**Find $|\vec{a} \times \vec{b}|$ if $\vec{a} = 2\hat{i} + \hat{j} + 3\hat{k}$ and $\vec{b} = 3\hat{i} + 5\hat{j} - 2\hat{k}$.**

**Sol.**

Given, $\vec{a} = 2\hat{i} + \hat{j} + 3\hat{k}$

and $b = 3\hat{i} + 5\hat{j} - 2\hat{k}$

Then, $$\vec{a} \times \vec{b} = \begin{vmatrix} \hat{i} & \hat{j} & \hat{k} \\ 2 & 1 & 3 \\ 3 & 5 & -2 \end{vmatrix}$$

$$= \hat{i}(-2-15) - \hat{j}(-4-9) + \hat{k}(10-3)$$

$$= -17\hat{i} + 13\hat{j} + 7\hat{k}$$

$$\left|\vec{a} \times \vec{b}\right| = \sqrt{(-17)^2 + (13)^2 + (7)^2}$$

$$= \sqrt{289+169+49} = \sqrt{507}$$ **Ans.**

**6. Find : $\int \frac{x-5}{(x-3)^3} e^x dx$. [2]**

**Sol.** Let $I = \int \frac{x-5}{(x-3)^3} e^x dx$

$$I = \int \frac{(x-3)-2}{(x-3)^3} . e^x dx$$

$$= \int \left[\frac{(x-3)}{(x-3)^3} - \frac{2}{(x-3)^3}\right] e^x dx$$

$$= \int \left[ \frac{1}{(x-3)^2} - \frac{2}{(x-3)^3} \right] e^x dx \quad \text{...(i)}$$

Now,

$$\frac{d}{dx}(x-3)^{-2} = -2(x-3)^{-3} = \frac{-2}{(x-3)^3}$$

and we know that

$$\int [f(x) + f'(x)]e^x dx = e^x f(x) + c$$

Here, $f(x) = \frac{1}{(x-3)^2}$ and $f'(x) = \frac{-2}{(x-3)^3}$

Then,

$$\int \frac{x-5}{(x-3)^3} e^x dx = e^x \times \frac{1}{(x-3)^2} + C = \frac{e^x}{(x-3)^2} + C$$

**Ans.**

## SECTION–C

**13. Solve for $x$ :**

**$\tan^{-1}(x+1) + \tan^{-1}(x-1) = \tan^{-1}\left(\frac{8}{31}\right)$.** **[4]**

**Sol.**

Given,

$$\tan^{-1}(x+1) + \tan^{-1}(x-1) = \tan^{-1}\frac{8}{31}$$

We know that

$$\tan^{-1} x + \tan^{-1} y = \tan^{-1}\left[\frac{x+y}{1-xy}\right]$$

$\therefore \tan^{-1}(x+1) + \tan^{-1}(x-1)$

$$= \tan^{-1}\left[\frac{x+1+x-1}{1-[(x+1)(x-1)]}\right]$$

$$= \tan^{-1}\left[\frac{2x}{1-(x^2-1)}\right]$$

$$= \tan^{-1}\left[\frac{2x}{2-x^2}\right]$$

Now, $\tan^{-1}\frac{2x}{2-x^2} = \tan^{-1}\frac{8}{31}$

$\Rightarrow \quad \frac{2x}{2-x^2} = \frac{8}{31}$

$\Rightarrow \quad 62x = 16 - 8x^2$

$\Rightarrow \quad 4x^2 + 31x - 8 = 0$

$\Rightarrow \quad 4x^2 + 32x - x - 8 = 0$

$\Rightarrow \quad 4x(x+8) - 1(x+8) = 0$

$\Rightarrow \quad (x+8)(4x-1) = 0$

$\Rightarrow \quad x = -8$ and $x = \frac{1}{4}$ **Ans.**

**14. If $x = ae^t(\sin t + \cos t)$ and $y = ae^t(\sin t - \cos t)$, then prove that $\frac{dy}{dx} = \frac{x+y}{x-y}$.** **[4]**

**Sol.** Given, $x = ae^t(\sin t + \cos t)$ ...(i)

and $y = ae^t(\sin t - \cos t)$ ...(ii)

On differentiating equation (i) with w.r.t. $t$, we get

$$\frac{dx}{dt} = a\frac{d}{dt}\left[e^t(\sin t + \cos t)\right] = a\frac{d}{dt}\left[e^t \sin t + e^t \cos t\right]$$

$$= a\left[e^t \cos t + \sin t\, e^t + e^t(-\sin t) + e^t \cos t\right]$$

$$= a\left[e^t \cos t + e^t \sin t - \sin t\, e^t + e^t \cos t\right]$$

$$\frac{dx}{dt} = 2ae^t \cos t \quad \text{...(iii)}$$

On differentiating equation (ii) w.r.t. $t$, we get

$$\frac{dy}{dt} = a\frac{d}{dt}\left[e^t(\sin t - \cos t)\right]$$

$$= a\frac{d}{dt}\left[e^t \sin t - e^t \cos t\right]$$

$$= a\left[e^t \cos t + \sin t\, e^t - \left(-e^t \sin t + \cos e^t\right)\right]$$

$$= a\left[e^t \cos t + e^t \sin t + \sin t\, e^t - e^t \cos t\right]$$

$$\frac{dy}{dt} = 2ae^t \sin t \quad \text{...(iv)}$$

On dividing equation (iv) by (iii), we get

$$\frac{dy/dt}{dx/dt} = \frac{2ae^t \sin t}{2ae^t \cos t}$$

$$\frac{dy}{dx} = \frac{\sin t}{\cos t}$$

$$\text{L.H.S.} = \frac{dy}{dx} = \tan t \quad \text{...(v)}$$

$$\text{Now, R.H.S.} = \frac{x+y}{x-y}$$

$$= \frac{ae^t(\sin t + \cos t) + ae^t(\sin t - \cos t)}{ae^t(\sin t + \cos t) - ae^t(\sin t - \cos t)}$$

$$= \frac{2ae^t \sin t}{2ae^t \cos t} = \tan t \quad \text{...(vi)}$$

From equations (v) and (vi), we have

$$\frac{dy}{dx} = \frac{x+y}{x-y}$$

**Hence Proved**

**OR**

**Differentiate $x^{\sin x} + (\sin x)^{\cos x}$ with respect to $x$.**

**Sol.** Let $y = x^{\sin x} + (\sin x)^{\cos x}$

Let $x^{\sin x} = u$

and $(\sin x)^{\cos x} = v$

Then, $y = u + v$

$$\frac{dy}{dx} = \frac{du}{dx} + \frac{dv}{dx} \quad ...(i)$$

Now, $u = x^{\sin x}$

$$\log u = \sin x \log x$$

On differentiating w.r.t. $x$, we get

$$\frac{1}{u}\frac{du}{dx} = \sin x \times \frac{1}{x} + \log x \cos x$$

$$\frac{du}{dx} = u\left(\frac{\sin x}{x} + \log x \cos x\right)$$

$$\frac{du}{dx} = x^{\sin x}\left(\frac{\sin x}{x} + \log x \cos x\right) \quad ...(ii)$$

Now, $v = (\sin x)^{\cos x}$

$$\log v = \cos x \log \sin x$$

On differentiating w.r.t. $x$, we get

$$\frac{1}{v}\frac{dv}{dx} = \cos x \frac{1}{\sin x} \times \cos x + \log \sin x\,(-\sin x)$$

$$\frac{1}{v}\frac{dv}{dx} = \cot x \times \cos x - \sin x \log(\sin x)$$

$$\frac{dv}{dx} = v(\cot x \cos x - \sin x \log \sin x)$$

$$\frac{dv}{dx} = \left((\sin x)^{\cos x} (\cot x \cos x - \sin x \log \sin x)\right) \quad ...(iii)$$

From (i), (ii) and (iii),

$$\frac{dy}{dx} = x^{\sin x}\left(\frac{\sin x}{x} + \log x \cos x\right) + (\sin x)^{\cos x}(\cot x \cos x - \sin x \log \sin x)$$

**Ans.**

**15. Find :**

$$\int \frac{2\cos x}{(1 - \sin x)(2 - \cos^2 x)} dx$$ **[4]**

**Sol.**

Let $$I = \int \frac{2\cos x\, dx}{(1 - \sin x)(2 - \cos^2 x)}$$

$$I = \int \frac{2\cos x\, dx}{(1 - \sin x)(1 + \sin^2 x)}$$

Now, let $\sin x = t$

$\Rightarrow$ $\cos x\, dx = dt$

Then, $$I = \int \frac{2dt}{(1 - t)(1 + t^2)}$$

Now, solving it by partial fraction,

$$\Rightarrow \frac{2}{(1-t)(1+t^2)} = \frac{A}{(1-t)} + \frac{Bt + C}{(1+t^2)}$$

$$\Rightarrow 2 = A(1 + t^2) + (Bt + C)(1 - t)$$

$$\Rightarrow 2 = A + At^2 + Bt - Bt^2 + C - Ct$$

$$\Rightarrow 2 = t^2(A - B) + t(B - C) + A + C$$

Equating the coefficient of $t^2$, $t$ and of constant terms of both sides, we get

$$A - B = 0$$
$$B - C = 0$$
and $$A + C = 2$$

On solving the above equations we get

$$A = C = 1 \text{ and } B = 1$$

$$\therefore \int \frac{2dt}{(1-t)(1+t^2)} = \int \left[\frac{1}{(1-t)} + \left(\frac{t+1}{1+t^2}\right)\right] dt$$

$$I = \int \frac{1}{1-t} dt + \int \frac{t\, dt}{(1+t^2)} + \int \frac{1}{(1+t^2)} dt$$

$$= -\log|1 - t| + \frac{1}{2}\log|1 + t^2| + \tan^{-1} t$$

$$= -\log|1 - \sin x| + \frac{1}{2}\log|1 + \sin^2 x| + \tan^{-1}(\sin x) + c$$ **Ans.**

## SECTION–D

**24. Show that for the matrix A = $\begin{bmatrix} 1 & 1 & 1 \\ 1 & 2 & -3 \\ 2 & -1 & 3 \end{bmatrix}$, $A^3 - 6A^2 + 5A + 11I = 0$. Hence, find $A^{-1}$.** **[6]**

**Sol.** Given, $$A = \begin{bmatrix} 1 & 1 & 1 \\ 1 & 2 & -3 \\ 2 & -1 & 3 \end{bmatrix}$$

$$A^2 = \begin{bmatrix} 1 & 1 & 1 \\ 1 & 2 & -3 \\ 2 & -1 & 3 \end{bmatrix}\begin{bmatrix} 1 & 1 & 1 \\ 1 & 2 & -3 \\ 2 & -1 & 3 \end{bmatrix}$$

$$= \begin{bmatrix} 1+1+2 & 1+2-1 & 1-3+3 \\ 1+2-6 & 1+4+3 & 1-6-9 \\ 2-1+6 & 2-2-3 & 2+3+9 \end{bmatrix}$$

$$A^2 = \begin{bmatrix} 4 & 2 & 1 \\ -3 & 8 & -14 \\ 7 & -3 & 14 \end{bmatrix}$$

Now, $A^3 = \begin{bmatrix} 4 & 2 & 1 \\ -3 & 8 & -14 \\ 7 & -3 & 14 \end{bmatrix}\begin{bmatrix} 1 & 1 & 1 \\ 1 & 2 & -3 \\ 2 & -1 & 3 \end{bmatrix}$

$$= \begin{bmatrix} 4+2+2 & 4+4-1 & 4-6+3 \\ -3+8-28 & -3+16+14 & -3-24-42 \\ 7-3+28 & 7-6-14 & 7+9+42 \end{bmatrix}$$

$$A^3 = \begin{bmatrix} 8 & 7 & 1 \\ -23 & 27 & -69 \\ 32 & -13 & 58 \end{bmatrix}$$

Now, putting $A^3$, $A^2$ in $A^3 - 6A^2 + 5A + 11I$

$$= \begin{bmatrix} 8 & 7 & 1 \\ -23 & 27 & -69 \\ 32 & -13 & 58 \end{bmatrix} - 6\begin{bmatrix} 4 & 2 & 1 \\ -3 & 8 & -14 \\ 7 & -3 & 14 \end{bmatrix} + 5\begin{bmatrix} 1 & 1 & 1 \\ 1 & 2 & -3 \\ 2 & -1 & 3 \end{bmatrix} + 11\begin{bmatrix} 1 & 0 & 0 \\ 0 & 1 & 0 \\ 0 & 0 & 1 \end{bmatrix}$$

$$= \begin{bmatrix} 8 & 7 & 1 \\ -23 & 27 & -69 \\ 32 & -13 & 58 \end{bmatrix} - \begin{bmatrix} 24 & 12 & 6 \\ -18 & 48 & -84 \\ 42 & -18 & 84 \end{bmatrix} + \begin{bmatrix} 5 & 5 & 5 \\ 5 & 10 & -15 \\ 10 & -5 & 15 \end{bmatrix} + \begin{bmatrix} 11 & 0 & 0 \\ 0 & 11 & 0 \\ 0 & 0 & 11 \end{bmatrix}$$

$$= \begin{bmatrix} 8-24 & 7-12 & 1-6 \\ -23+18 & 27-48 & -69+84 \\ 32-42 & -13+18 & 58-84 \end{bmatrix} + \begin{bmatrix} 5+11 & 5+0 & 5+0 \\ 5+0 & 10+11 & -15+0 \\ 10+0 & -5+0 & 15+11 \end{bmatrix}$$

$$= \begin{bmatrix} -16 & -5 & -5 \\ -5 & -21 & 15 \\ -10 & 5 & -26 \end{bmatrix} + \begin{bmatrix} 16 & 5 & 5 \\ 5 & 21 & -15 \\ 10 & -5 & 26 \end{bmatrix}$$

$$= \begin{bmatrix} 0 & 0 & 0 \\ 0 & 0 & 0 \\ 0 & 0 & 0 \end{bmatrix} = 0 = \text{R.H.S.}$$

**Hence Proved.**

We have, $A^3 - 6A^2 + 5A + 11I = 0$

Multiplying by $A^{-1}$ on both sides, we get

$$A^{-1}(A^3 - 6A^2 + 5A + 11I) = 0.A^{-1}$$

$\Rightarrow \quad A^3A^{-1} - 6A^2.A^{-1} + 5A.A^{-1} + 11I.A^{-1} = 0$

$\Rightarrow \quad A^2(AA^{-1}) - 6A.(AA^{-1}) + 5(AA^{-1}) + 11IA^{-1} = 0$

$\Rightarrow \quad A^2I - 6AI + 5I + 11A^{-1} = 0$

$\left[\because AA^{-1} = I \text{ and } 11IA^{-1} = 11A^{-1}\right]$

$\Rightarrow A^2 - 6A + 5I + 11A^{-1} = 0$

$\Rightarrow \quad 11A^{-1} = -A^2 + 6A - 5I$

$\Rightarrow \quad A^{-1} = \frac{1}{11}(-A^2 + 6A - 5I)$

$$A^{-1} = \frac{-(A^2 - 6A + 5I)}{11}$$

$$A^{-1} = \frac{-1}{11}\left(\begin{bmatrix} 4 & 2 & 1 \\ -3 & 8 & -14 \\ 7 & -3 & 14 \end{bmatrix} - 6\begin{bmatrix} 1 & 1 & 1 \\ 1 & 2 & -3 \\ 2 & -1 & 3 \end{bmatrix} + 5\begin{bmatrix} 1 & 0 & 0 \\ 0 & 1 & 0 \\ 0 & 0 & 1 \end{bmatrix}\right)$$

$$A^{-1} = \frac{-1}{11}\left(\begin{bmatrix} 4 & 2 & 1 \\ -3 & 8 & -14 \\ 7 & -3 & 14 \end{bmatrix} - \begin{bmatrix} 6 & 6 & 6 \\ 6 & 12 & -18 \\ 12 & -6 & 18 \end{bmatrix} + \begin{bmatrix} 5 & 0 & 0 \\ 0 & 5 & 0 \\ 0 & 0 & 5 \end{bmatrix}\right)$$

$$A^{-1} = \frac{-1}{11}\left(\begin{bmatrix} 4-6 & 2-6 & 1-6 \\ -3-6 & 8-12 & -14+18 \\ 7-12 & -3+6 & 14-18 \end{bmatrix} + \begin{bmatrix} 5 & 0 & 0 \\ 0 & 5 & 0 \\ 0 & 0 & 5 \end{bmatrix}\right)$$

$$A^{-1} = \frac{-1}{11}\left(\begin{bmatrix} -2 & -4 & -5 \\ -9 & -4 & 4 \\ -5 & 3 & -4 \end{bmatrix} + \begin{bmatrix} 5 & 0 & 0 \\ 0 & 5 & 0 \\ 0 & 0 & 5 \end{bmatrix}\right)$$

$$A^{-1} = \frac{-1}{11}\begin{bmatrix} -2+5 & -4+0 & -5+0 \\ -9+0 & -4+5 & 4+0 \\ -5+0 & 3+0 & -4+5 \end{bmatrix}$$

$$A^{-1} = \frac{1}{11}\begin{bmatrix} 3 & -4 & -5 \\ -9 & 1 & 4 \\ -5 & 3 & 1 \end{bmatrix}$$

$$A^{-1} = \frac{1}{11}\begin{bmatrix} -3 & 4 & 5 \\ 9 & -1 & -4 \\ 5 & -3 & -1 \end{bmatrix}$$

**Ans.**

OR

**Using matrix method, solve the following system of equations :**

$$3x - 2y + 3z = 8$$
$$2x + y - z = 1$$
$$4x - 3y + 2z = 4$$

**Sol.** The given equations are

$$3x - 2y + 3z = 8$$
$$2x + y - z = 1$$
$$4x - 3y + 2z = 4$$

These equations can be written in the form AX = B, where

$$A = \begin{bmatrix} 3 & -2 & 3 \\ 2 & 1 & -1 \\ 4 & -3 & 2 \end{bmatrix}, X = \begin{bmatrix} x \\ y \\ z \end{bmatrix} \text{ and } B = \begin{bmatrix} 8 \\ 1 \\ 4 \end{bmatrix}$$

Now, $|A| = 3(2 - 3) + 2(4 + 4) + 3(-6 - 4)$
$= -17 \neq 0$

Hence, A is non-singular and so its inverse exists.
Confactors of A are

$A_{11} = -1,\ A_{12} = -8,\ A_{13} = -10$
$A_{21} = -5,\ A_{22} = -6,\ A_{23} = 1$
$A_{31} = -1,\ A_{32} = 9,\ A_{33} = 7$

$$\therefore \quad \text{adj } A = \begin{bmatrix} -1 & -8 & -10 \\ -5 & -6 & 1 \\ -1 & 9 & 7 \end{bmatrix}^T$$

$$= \begin{bmatrix} -1 & -5 & -1 \\ -8 & -6 & 9 \\ 10 & 1 & 7 \end{bmatrix}$$

Now, $A^{-1} = \frac{1}{|A|} \text{adj } A$

$$= \frac{-1}{17}\begin{bmatrix} -1 & -5 & -1 \\ -8 & -6 & 9 \\ -10 & 1 & 7 \end{bmatrix}$$

$\Rightarrow \quad X = A^{-1}B$

$$= \frac{-1}{17}\begin{bmatrix} -1 & -5 & -1 \\ -8 & -6 & 9 \\ -10 & 1 & 7 \end{bmatrix}\begin{bmatrix} 8 \\ 1 \\ 4 \end{bmatrix}$$

$$\begin{bmatrix} x \\ y \\ z \end{bmatrix} = \frac{-1}{17}\begin{bmatrix} -17 \\ -34 \\ -51 \end{bmatrix} = \begin{bmatrix} 1 \\ 2 \\ 3 \end{bmatrix}$$

Hence, $x = 1, y = 2$ and $z = 3$ **Ans.**

**26. A bag contains 5 red and 4 black balls, a second bag contains 3 red and 6 black balls. One of the two bags is selected at random and two balls are drawn at random (without replacement) both of which are found to be red. Find the probability that the balls are drawn from the second bag. [6]**

**Sol.** Let $E_1$ be the event of choosing the bag I, $E_2$ be the event of choosing the bag II and A be the event of drawing a red ball.

Then,

$$P(E_1) = P(E_2) = \frac{1}{2}$$

Now, $P\left(\frac{A}{E_1}\right) = \frac{{}^5C_2}{{}^9C_2} = \frac{20}{72}$

and $P\left(\frac{A}{E_2}\right) = \frac{{}^3C_2}{{}^9C_2} = \frac{6}{72}$

Now, the probability of drawing a ball from Bag II, if it is given that it is red, is $P\left(\frac{E_2}{A}\right)$.

Now, by Bayes' theorem, we have

$$P\left(\frac{E_2}{A}\right) = \frac{P(E_2).P(A/E_2)}{P(E_1).P\left(\frac{A}{E_1}\right) + P(E_2)\,P\left(\frac{A}{E_1}\right)}$$

$$= \frac{\frac{1}{2} \times \frac{6}{72}}{\frac{1}{2} \times \frac{20}{72} + \frac{1}{2} \times \frac{6}{72}}$$

$$= \frac{\frac{6}{2 \times 72}}{\frac{20}{2 \times 72} + \frac{6}{2 \times 72}}$$

$$= \frac{6}{20 + 6}$$

$$= \frac{6}{26} = \frac{3}{13}$$ **Ans.**

●●

# MATHEMATICS

## Delhi [Set-I] 2019

Time allowed : 3 hours | Maximum marks : 100

### SECTION–A

**1. If A and B are square matrices of the same order 3, such that | A | = 2 and AB = 2I, write the value of | B |.** **[1]**

**Sol.** Given, $|A| = 2$ and $AB = 2I$

$\therefore \quad |AB| = |2I|$

$\Rightarrow \quad |A||B| = 2^3|I|$

$\Rightarrow \quad 2|B| = 8I$

$\Rightarrow \quad |B| = 4$ **Ans.**

**2. If $f(x) = x + 1$, find $\frac{d}{dx}$ $(fof)(x)$.** **[1]**

**Sol.** Given, $f(x) = x + 1$

Now, $f o f(x) = f(f(x))$

$= f(x + 1)$

$= x + 1 + 1$

$= x + 2$

$\therefore \quad \frac{d}{dx}(fof)(x) = 1$ **Ans.**

**3. Find the order and the degree of the differential equation $x^2 \frac{d^2y}{dx^2} = \left\{1 + \left(\frac{dy}{dx}\right)^2\right\}^4$.** **[1]**

**Sol.** We have,

$$x^2 \frac{d^2y}{dx^2} = \left\{1 + \left(\frac{dy}{dx}\right)^2\right\}^4$$

$\therefore$ Order = 2 and degree = 1 **Ans.**

**4. If a line makes angles 90°, 135°, 45° with $x$, $y$ and $z$ axes respectively, find its direction cosines.** **[1]**

**Sol.** Given, $\alpha = 90^\circ, \beta = 135^\circ, \gamma = 45^\circ$

So, $l = \cos 90^\circ = 0$

$m = \cos 135^\circ = \cos(180^\circ - 45^\circ)$

$= -\cos 45^\circ = -\frac{1}{\sqrt{2}}$

and $n = \cos 45^\circ = \frac{1}{\sqrt{2}}$

$\therefore$ The required direction cosines are $0, -\frac{1}{\sqrt{2}}, \frac{1}{\sqrt{2}}$.

**Ans.**

OR

**Find the vector equation of the line which passes through the point (3, 4, 5) and is parallel to the vector $2\hat{i} + 2\hat{j} - 3\hat{k}$.**

**Sol.** Given, the line passes through the point (3, 4, 5). and parallel to the vector $2\hat{i} + 2\hat{j} - 3\hat{k}$.

D. R. s of the given vector are $< 2, 2, -3 >$.

$\therefore$ Vector equation of line,

$\vec{r} = (3\hat{i} + 4\hat{j} + 5\hat{k}) + \lambda(2\hat{i} + 2\hat{j} - 3\hat{k})$ **Ans.**

### SECTION–B

**5. Examine whether the operation* defined on R by $a * b = ab + 1$ is (i) a binary or not. (ii) if a binary operation, is it associative or not?**** **[2]**

**6. Find a matrix A such that 2A – 3B + 5C = 0, where $B = \begin{bmatrix} -2 & 2 & 0 \\ 3 & 1 & 4 \end{bmatrix}$ and $C = \begin{bmatrix} 2 & 0 & -2 \\ 7 & 1 & 6 \end{bmatrix}$.** **[2]**

**Sol.**

Given, $B = \begin{bmatrix} -2 & 2 & 0 \\ 3 & 1 & 4 \end{bmatrix}, C = \begin{bmatrix} 2 & 0 & -2 \\ 7 & 1 & 6 \end{bmatrix}$

and $2A - 3B + 5C = 0$

$\Rightarrow \quad 2A = 3B - 5C$

$\Rightarrow \quad A = \frac{1}{2}(3B - 5C)$

$\Rightarrow \quad A = \frac{1}{2}\left(3\begin{bmatrix} -2 & 2 & 0 \\ 3 & 1 & 4 \end{bmatrix} - 5\begin{bmatrix} 2 & 0 & -2 \\ 7 & 1 & 6 \end{bmatrix}\right)$

$\Rightarrow \quad A = \frac{1}{2}\left(\begin{bmatrix} -6 & 6 & 0 \\ 9 & 3 & 12 \end{bmatrix} - \begin{bmatrix} 10 & 0 & -10 \\ 35 & 5 & 30 \end{bmatrix}\right)$

$\Rightarrow \quad A = \frac{1}{2}\begin{bmatrix} -16 & 6 & 10 \\ -26 & -2 & -18 \end{bmatrix}$

$\Rightarrow \quad A = \begin{bmatrix} -8 & 3 & 5 \\ -13 & -1 & -9 \end{bmatrix}$ **Ans.**

**7. Find : $\int \frac{\sec^2 x}{\sqrt{\tan^2 x + 4}} dx$.** **[2]**

**Sol.** Let $I = \int \frac{\sec^2 x \, dx}{\sqrt{\tan^2 x + 4}}$

Let $\tan x = t$

$\sec^2 x \, dx = dt$

---

** Answer is not given due to change in the present syllabus.

$$\therefore \quad I = \int \frac{dt}{\sqrt{t^2+4}}$$

$$\Rightarrow \quad I = \int \frac{dt}{\sqrt{t^2+2^2}}$$

$$\Rightarrow \quad I = \log |t + \sqrt{t^2+4}\ | + C$$

$$\Rightarrow \quad I = \log |\tan x + \sqrt{\tan^2 x+4}\ | + C$$

**Ans.**

**8. Find : $\int \sqrt{1-\sin 2x}\, dx,\ \frac{\pi}{4} < x < \frac{\pi}{2}$.** **[2]**

**Sol.** Let $I = \int \sqrt{1-\sin 2x}\, dx$

$$\Rightarrow \quad I = \int \sqrt{(\cos x - \sin x)^2}\, dx$$

$$\Rightarrow \quad I = \int (\cos x - \sin x) dx$$

$$\Rightarrow \quad I = \sin x + \cos x + C$$ **Ans.**

**OR**

**Find : $\int \sin^{-1}(2x)\, dx$.**

**Sol.** Let $I = \int \sin^{-1}(2x)\, dx$

Let $\sin^{-1} 2x = t$

$2x = \sin t$

$2dx = \cos t\, dt$

$dx = \frac{1}{2} \cos t\, dt$

$$\therefore \quad I = \frac{1}{2} \int t \cos t\, dt$$

$$\Rightarrow \quad I = \frac{1}{2} \left[t \sin t - \int 1 \cdot \sin t\, dt\right]$$

[integrating by parts]

$$\Rightarrow \quad I = \frac{1}{2} [t \sin t + \cos t] + C$$

$$\Rightarrow \quad I = \frac{1}{2} [2x \sin^{-1} 2x + \cos (\sin^{-1} 2x)] + C$$

$$\Rightarrow \quad I = x \sin^{-1} 2x + \frac{1}{2} \sqrt{1-4x^2} + C$$

**Ans.**

**9. Form the differential equation representing the family of curves $y = e^{2x}(a + bx)$, where '$a$' and '$b$' are arbitrary constants.** **[2]**

**Sol.** Given, $y = e^{2x}(a + bx)$ ...(i)

Differentiating w.r.t. $x$, we get

$$\frac{dy}{dx} = e^{2x}(b) + 2(a + bx)e^{2x}$$

$$\frac{dy}{dx} = be^{2x} + 2y \quad \text{[From (i)]}$$

$$\frac{dy}{dx} - 2y = be^{2x} \quad \text{...(ii)}$$

Again, differentiating both sides w.r.t. $x$, we get

$$\frac{d^2y}{dx^2} - 2\frac{dy}{dx} = 2be^{2x}$$

$$\frac{d^2y}{dx^2} - 2\frac{dy}{dx} = 2\left(\frac{dy}{dx} - 2y\right) \quad \text{[From (ii)]}$$

$$\frac{d^2y}{dx^2} - \frac{2dy}{dx} = \frac{2dy}{dx} - 4y$$

$$\frac{d^2y}{dx^2} - \frac{4dy}{dx} + 4y = 0$$

This is the required differential equation. **Ans.**

**10. If the sum of two unit vectors is a unit vector, prove that the magnitude of their difference is $\sqrt{3}$.** **[2]**

**Sol.** Let $\vec{a}$ and $\vec{b}$ are two unit vectors.

Given, $|\vec{a}| = 1, |\vec{b}| = 1, |\vec{a} + \vec{b}| = 1$

We know that,

$$|\vec{a} + \vec{b}|^2 + |\vec{a} - \vec{b}|^2 = 2(|\vec{a}|^2 + |\vec{b}|^2)$$

$$\Rightarrow \quad (1)^2 + |\vec{a} - \vec{b}|^2 = 2(1^2 + 1^2)$$

$$\Rightarrow \quad (1)^2 + |\vec{a} - \vec{b}|^2 = 4$$

$$\Rightarrow \quad |\vec{a} - \vec{b}|^2 = 3$$

$$\Rightarrow \quad |\vec{a} - \vec{b}| = \sqrt{3}$$

**Ans.**

**OR**

**If $\vec{a} = 2\hat{i} + 3\hat{j} + \hat{k}$, $\vec{b} = \hat{i} - 2\hat{j} + \hat{k}$ and $\vec{c} = -3\hat{i} + \hat{j} + 2\hat{k}$, find $\left[\vec{a}\ \vec{b}\ \vec{c}\right]$.**

**Sol.** Given, $\vec{a} = 2\hat{i} + 3\hat{j} + \hat{k}$

$$\vec{b} = \hat{i} - 2\hat{j} + \hat{k}$$

$$\vec{a} = -3\hat{i} + \hat{j} + 2\hat{k}$$

Now, $\left[\vec{a}\ \vec{b}\ \vec{c}\right] = |\vec{a} . \vec{b} \times \vec{c}|$

$$= \begin{vmatrix} 2 & 3 & 1 \\ 1 & -2 & 1 \\ -3 & 1 & 2 \end{vmatrix}$$

$$= 2(-4-1) - 3(2+3) + 1(1-6)$$

$$= -10 - 15 - 5 = -30$$ **Ans.**

**11. A die marked 1, 2, 3 in red and 4, 5, 6 in green is tossed. Let A be the event "number is even" and B be the event "number is marked red". Find**

**whether the events A and B are independent or not.** **[2]**

**Sol.** Given, S = { 1, 2, 3, 4, 5, 6 }

Let the two events be

A : The number is even

B : The number is red

$$P(A) = \frac{3}{6} = \frac{1}{2}, P(B) = \frac{3}{6} = \frac{1}{2}$$

and $$P(A \cap B) = \frac{1}{6}$$

$\therefore$ $$P(A \cap B) \neq P(A) . P(B)$$

Hence, A and B are not independent. **Ans.**

**12. A die is thrown 6 times. If "getting an odd number" is a "success", what is the probability of (i) 5 successes? (ii) atmost 5 success ?** **[2]**

**Sol.** Throwing a die is a Bernouli trial

So, X has a binomial distribution

$$P ( X = x) = {}^nC_x q^{n-x} p^x \quad ...(i)$$

$n$ = number of throws = 6

$p$ = Probability of getting an odd number

$$= \frac{3}{6} = \frac{1}{2}$$

$$q = 1 - p = 1 - \frac{1}{2} = \frac{1}{2}$$

(i) Probability of 5 successes

$$= P ( X = 5)$$

$$= {}^6C_5 \left(\frac{1}{2}\right)^{6-5} \left(\frac{1}{2}\right)^5 \quad \text{[Using (i)]}$$

$$= 6 \times \left(\frac{1}{2}\right)^6$$

$$= 6 \times \frac{1}{64} = \frac{3}{32}$$

(ii) Probability at most 5 successes i.e. P (X $\leq$ 5)

$$= P (X = 0) + P (X = 1) + P (X = 2) + P (X = 3) + P (X = 4) + P (X = 5)$$

$$= 1 - P (X = 6)$$

$$= 1 - \frac{1}{64}$$

$$= \frac{63}{64}$$

OR

**The random variable X has a probability distribution P(X) of the following form where '$k$' is some number**

$$P(X = x) = \begin{cases} k, & \text{if } x = 0 \\ 2k, & \text{if } x = 1 \\ 3k, & \text{if } x = 2 \\ 0, & \text{otherwise} \end{cases}$$

**Determine the value of '$k$'.**

**Sol.** We know that,

$\Sigma P(X) = 1$

$P(x = 0) + P(x = 1) + P(x = 2) + P(x = \text{other}) = 1$

$$\Rightarrow \quad k + 2k + 3k + 0 = 1$$

$$\Rightarrow \quad 6k = 1$$

$$\Rightarrow \quad k = \frac{1}{6}$$ **Ans.**

## SECTION–C

**13. Show that the relation R on R defined as R = [($a$, $b$) : $a \leq b$], is reflexive, and transitive but not symmetric.** **[4]**

**Sol. Reflexive :**

Let $a \in R$

$\therefore$ $a \leq a$

So, $(a, a) \in R$

Hence, R is reflexive.

**Symmetric :**

Let $(a, b) \in R$

Then $(b, a) \in R$

Then, $a \leq b$

$\Rightarrow$ $b \leq a$ (which is not true)

$\therefore$ $(b, a) \notin R$

Hence, R is not symmetric.

**Transitive :**

Let, $a, b, c \in R$, such that $(a, b) \in R$ and $(b, c) \in R$

Then, $a \leq b$

and $b \leq c$

$\Rightarrow$ $a \leq c$

$\Rightarrow$ $(a, c) \in R$

Hence, R is transitive.

Hence, R is reflexive and transitive but not symmetric. **Hence Proved.**

OR

**Prove that the function $f$ : N $\to$ N, defined by $f(x) = x^2 + x + 1$ is one-one but not onto.**

**Find inverse of $f$ : N $\to$ S, where S is range of $f$.**

**Sol.**

Given, $f : N \to N, f(x) = x^2 + x + 1$

Let A be the set of natural number (domain).

and B be the set of natural number (co-domain).

For One-One :

Let $x_1, x_2 \in$ A such that $f(x_1) = f(x_2)$

$$\Rightarrow \quad x_1^2 + x_1 + 1 = x_2^2 + x_2 + 1$$

$$\Rightarrow \quad x_1^2 - x_2^2 + x_1 - x_2 = 0$$

$$\Rightarrow \quad (x_1 - x_2)[x_1 + x_2 + 1] = 0$$

$$\Rightarrow \quad x_1 - x_2 = 0 \qquad [\because x_1 + x_2 + 1 \neq 0]$$

$$\Rightarrow \quad x_1 = x_2$$

$\therefore f$ is one-one.

For Onto :

Let $\quad y = 1 \in$ N (Co domain)

$$\therefore \quad x^2 + x + 1 = 1$$

$$\Rightarrow \quad x^2 + x = 0$$

$$\Rightarrow \quad x(x+1) = 0$$

$$\therefore \quad x = 0, -1 \notin N$$

So, $f$ is not onto.

Hence, f is one-one but not onto. **Hence Proved**

Now, $f: \mathrm{N} \rightarrow \mathrm{S} : f(x) = x^2 + x + 1$

where S = range (Given)

$f: \mathrm{N} \rightarrow \mathrm{S}$ is onto as co-domain = range.

Hence, $f$ is invertible.

Let $\quad y = f(x) \Rightarrow y = x^2 + x + 1$

$$\Rightarrow \quad y = x^2 + 2x \times \frac{1}{2} + \left(\frac{1}{2}\right)^2 - \left(\frac{1}{2}\right)^2 + 1$$

$$\Rightarrow \quad y = \left(x + \frac{1}{2}\right)^2 + \frac{3}{4}$$

$$\Rightarrow \quad y - \frac{3}{4} = \left(x + \frac{1}{2}\right)^2$$

$$\Rightarrow \quad \frac{4y-3}{4} = \left(x + \frac{1}{2}\right)^2$$

$$\Rightarrow \quad \pm\frac{\sqrt{4y-3}}{2} = x + \frac{1}{2} = \frac{2x+1}{2}$$

$$\therefore \quad x = \frac{-1 \pm \sqrt{4y-3}}{2}$$

$$\Rightarrow \quad x = \frac{-1 + \sqrt{4y-3}}{2} \qquad (\because X \in N)$$

$$\therefore f^{-1}(y) = \frac{-1+\sqrt{4y-3}}{2} \text{ or } f^{-1}(x) = \frac{-1+\sqrt{4x-3}}{2}$$

**Ans.**

**14. Solve : $\tan^{-1} 4x + \tan^{-1} 6x = \frac{\pi}{4}$** **[4]**

**Sol.** We have

$$\tan^{-1} 4x + \tan^{-1} 6x = \pi/4$$

$$\Rightarrow \quad \tan^{-1}\left(\frac{4x + 6x}{1 - 4x \,.\, 6x}\right) = \frac{\pi}{4}$$

$$\Rightarrow \quad \frac{10x}{1 - 24x^2} = \tan\frac{\pi}{4}$$

$$\Rightarrow \quad \frac{10x}{1 - 24x^2} = 1$$

$$\Rightarrow \quad 10x = 1 - 24x^2$$

$$\Rightarrow \quad 24x^2 + 10x - 1 = 0$$

$$\Rightarrow \quad x = \frac{-10 \pm \sqrt{100 + 96}}{48}$$

$$\Rightarrow \quad x = \frac{-10 \pm 14}{48}$$

$$\Rightarrow \quad x = \frac{-1}{2}, \frac{1}{12}$$

As $x = -1/2$ does not satisfy the given equation,

$$\therefore \quad x = \frac{1}{12}$$ **Ans.**

**15. Using properties of determinants, prove that**

$$\begin{vmatrix} a^2 + 2a & 2a+1 & 1 \\ 2a+1 & a+2 & 1 \\ 3 & 3 & 1 \end{vmatrix} = (a-1)^3.$$ **[4]**

**Sol.**

$$\text{L.H.S} = \begin{vmatrix} a^2 + 2a & 2a+1 & 1 \\ 2a+1 & a+2 & 1 \\ 3 & 3 & 1 \end{vmatrix}$$

Applying $R_2 \rightarrow R_2 - R_1$ and we get $R_3 \rightarrow R_3 - R_1$, we get

$$= \begin{vmatrix} a^2 + 2a & 2a+1 & 1 \\ 1 - a^2 & -a+1 & 0 \\ 3 - a^2 - 2a & -2a+2 & 0 \end{vmatrix}$$

Taking $(1-a)$ common from $R_2$ & $R_3$, we get

$$= (1-a)^2 \begin{vmatrix} a^2 + 2a & 2a+1 & 1 \\ (1+a) & 1 & 0 \\ a+3 & 2 & 0 \end{vmatrix}$$

Expanding along $C_3$, we get

$$= (1-a)^2 [1(2+2a) - (a+3)]$$

$$= (1-a)^2 [2 + 2a - a - 3]$$

$$= (1-a)^2 (a-1)$$

$= (a-1)^2 (a-1)$
$= (a-1)^3 = \text{R.H.S.}$

**Hence Proved.**

**16. If $\log (x^2 + y^2) = 2 \tan^{-1}\left(\frac{y}{x}\right)$, show that $\frac{dy}{dx} = \frac{x+y}{x-y}$.** **[4]**

**Sol.**

Given, $\log\left(x^2+y^2\right) = 2\tan^1\left(\frac{y}{x}\right)$

Differentiating both sides w.r.t. $x$, we get

$$\frac{1}{x^2+y^2}\left(2x+2y\frac{dy}{dx}\right) = \frac{2}{1+\frac{y^2}{x^2}}\left(\frac{x\frac{dy}{dx}-y}{x^2}\right)$$

$$\Rightarrow \frac{x}{x^2+y^2} + \frac{y}{x^2+y^2}\cdot\frac{dy}{dx} = \frac{1}{x^2+y^2}\left(x\frac{dy}{dx}-y\right)$$

$$\Rightarrow \frac{x}{x^2+y^2} + \frac{y}{x^2+y^2}\cdot\frac{dy}{dx} = \frac{x}{x^2+y^2}\cdot\frac{dy}{dx} - \frac{y}{x^2+y^2}$$

$$\Rightarrow \frac{x}{x^2+y^2} + \frac{y}{x^2+y^2} = \frac{x}{x^2+y^2}\cdot\frac{dy}{dx} - \frac{y}{x^2+y^2}\cdot\frac{dy}{dx}$$

$$\Rightarrow \frac{x+y}{x^2+y^2} = \left(\frac{x-y}{x^2+y^2}\right)\frac{dy}{dx}$$

$$\Rightarrow \frac{dy}{dx} = \frac{x+y}{x-y}.$$

**Ans.**

**OR**

**If $x^y - y^x = a^b$, find $\frac{dy}{dx}$**

**Sol.**

$$x^y - y^x = a^b$$

Let $u = x^y$ and $v = y^x$

Then, $u - v = a^b$

$$\frac{du}{dx} - \frac{dv}{dx} = 0 \quad ...(i)$$

Now, $u = x^y$

Taking log on both sides, we get

$$\log u = y \log x$$

Differentiating both sides w.r.t. $x$, we get.

$$\frac{1}{u}\cdot\frac{du}{dx} = \frac{y}{x} + \log x\cdot\frac{dy}{dx}$$

$$\Rightarrow \frac{du}{dx} = x^y\left[\frac{y}{x} + \log x.\frac{dy}{dx}\right]$$

and $v = y^x$

Taking log on both sides, we get

$$\log v = x \log y$$

Differentiating both sides w.r.t. $x$, we get

$$\frac{1}{v}\cdot\frac{dv}{dx} = \frac{x}{y}\cdot\frac{dy}{dx} + \log y$$

$$\Rightarrow \frac{dv}{dx} = y^x\left[\frac{x}{y}\frac{dy}{dx} + \log y\right]$$

From equation (i),

$$x^y\left[\frac{y}{x} + \log x.\frac{dy}{dx}\right] - y^x\left[\frac{x}{y}\frac{dy}{dx} + \log y\right] = 0$$

$$\Rightarrow y.x^{y-1} + x^y \log x.\frac{dy}{dx} - xy^{x-1}.\frac{dy}{dx} - y^x \log y = 0$$

$$\Rightarrow \frac{dy}{dx}\left[x^y \log x - xy^{x-1}\right] = y^x \log y - yx^{y-1}$$

$$\Rightarrow \frac{dy}{dx} = \left[\frac{y^x \log y - y.x^{y-1}}{x^y \log x - xy^{x-1}}\right]$$

**Ans.**

**17. If $y = (\sin^{-1} x)^2$, prove that $(1-x^2)\frac{d^2y}{dx^2} - x\frac{dy}{dx} - 2 = 0$** **[4]**

**Sol.** Given, $y = (\sin^{-1}x)^2$ ...(i)

Differentiating both sides w.r.t. $x$, we get

$$\frac{dy}{dx} = \frac{2\sin^{-1}x}{\sqrt{1-x^2}}$$

$$\Rightarrow \sqrt{1-x^2}\frac{dy}{dx} = 2\sin^{-1}x \quad ...(ii)$$

Again differentiating both sides w.r.t. $x$, we get

$$\sqrt{1-x^2}\frac{d^2y}{dx^2} - \frac{2x}{2\sqrt{1-x^2}}\frac{dy}{dx} = \frac{2}{\sqrt{1-x^2}}$$

$$\Rightarrow (1-x^2)\frac{d^2y}{dx^2} - x\frac{dy}{dx} = 2$$

$$\Rightarrow (1-x^2)\frac{d^2y}{dx^2} - x\frac{dy}{dx} - 2 = 0$$

**Hence Proved.**

**18. Find the equation of tangent to the curve $y = \sqrt{3x-2}$ which is parallel to the line $4x - 2y + 5 = 0$. Also, write the equation of normal to the curve at the point of contact.** **[4]**

**Sol.** Given, $y = \sqrt{3x-2}$ ...(i)

Differentiating both sides w.r.t. $x$, we get

$$\frac{dy}{dx} = \frac{3}{2\sqrt{3x-2}} = m_1$$

and equation of line, $4x - 2y + 5 = 0$

$$\text{Slope} = \frac{-4}{-2} = 2 = m_2$$

Apply condition of parallel line,

*i.e.* $m_1 = m_2$

$$\frac{3}{2\sqrt{3x-2}} = 2$$

$\Rightarrow$ $3 = 4\sqrt{3x-2}$

$\Rightarrow$ $9 = 16(3x-2)$

$\Rightarrow$ $9 = 48x - 32$

$\Rightarrow$ $41 = 48x$

$\Rightarrow$ $x = \frac{41}{48}$

Putting the value of $x$ in (i), we get

$$y = \sqrt{3\left(\frac{41}{48}\right) - 2} = \frac{3}{4}$$

$\therefore$ Point of contact $\left(\frac{41}{48}, \frac{3}{4}\right)$

Slope of curve = 2

So, equation of tangent is

$$\left(y - \frac{3}{4}\right) = 2\left(x - \frac{41}{48}\right)$$

$\Rightarrow$ $\left(\frac{4y-3}{4}\right) = 2\left(\frac{48x-41}{48}\right)$

$\Rightarrow$ $\frac{4y-3}{4} = \frac{48x-41}{24}$

$\Rightarrow$ $4y - 3 = \frac{48x-41}{6}$

$\Rightarrow$ $24y - 18 = 48x - 41$

$\Rightarrow$ $48x - 24y - 23 = 0$ **Ans.**

And equation of normal is

$$\left(y - \frac{3}{4}\right) = \frac{-1}{2}\left(x - \frac{41}{48}\right)$$

$\Rightarrow$ $\frac{4y-3}{4} = \frac{-1}{2}\left(\frac{48x-41}{48}\right)$

$\Rightarrow$ $24(4y-3) = -48x + 41$

$\Rightarrow$ $96y - 72 = -48x + 41$

$\Rightarrow$ $48x + 96y - 113 = 0$ **Ans.**

**19. Find : $\int \frac{3x+5}{x^2+3x-18} dx.$** **[4]**

**Sol.**

Let $I = \int \frac{3x+5}{x^2+3x-18} dx$

$(3x+5) = A(2x+3) + B$

Comparing coefficients of $x$,

$3 = 2A$

$\therefore$ $\frac{3}{2} = A$

Comparing constant terms,

$5 = 3A + B$

$\Rightarrow$ $5 = \frac{9}{2} + B$

$\Rightarrow$ $B = \frac{1}{2}$

$\therefore$ $I = \frac{3}{2}\int \frac{2x+3}{x^2+3x-18} dx + \frac{1}{2}\int \frac{dx}{x^2+3x-18}$

Let $I_1 = \frac{3}{2}\int \frac{2x+3}{x^2+3x-18} dx$

Put $x^2 + 3x - 18 = t$

$(2x+3)\, dx = dt$

$\Rightarrow$ $I_1 = \frac{3}{2}\int \frac{dt}{t}$

$\Rightarrow$ $I_1 = \frac{3}{2} \log |t| + C_1$

$\Rightarrow$ $I_1 = \frac{3}{2} \log |x^2 + 3x - 18| + C_1$

Now, $I_2 = \frac{1}{2}\int \frac{dx}{x^2+3x-18}$

$\Rightarrow$ $I_2 = \frac{1}{2}\int \frac{dx}{x^2 + 2x.\frac{3}{2} + \left(\frac{3}{2}\right)^2 - \left(\frac{3}{2}\right)^2 - 18}$

$\Rightarrow$ $I_2 = \frac{1}{2}\left(\frac{dx}{\left(x+\frac{3}{2}\right)^2 - \left(\frac{9}{2}\right)^2}\right)$

$\Rightarrow$ $I_2 = \frac{1}{2} \times \frac{2}{2 \times 9} \log \left|\frac{x + \frac{3}{2} - \frac{9}{2}}{x + \frac{3}{2} + \frac{9}{2}}\right| + C_2$

$\Rightarrow$ $I_2 = \frac{1}{18} \log \left|\frac{2x-6}{2x+12}\right| + C_2$

$\Rightarrow$ $I_2 = \frac{1}{18} \log \left|\frac{x-3}{x+6}\right| + C_2$

$\therefore \quad I = I_1 + I_2$

$\Rightarrow \quad I = \frac{3}{2}\log\left|x^2 + 3x - 18\right| + \frac{1}{18}\log\left|\frac{x-3}{x+6}\right| + C$

**Ans.**

**20. Prove that $\int_0^a f(x)dx = \int_0^a f(a-x)dx$, hence evaluate $\int_0^{\pi} \frac{x\sin x}{1+\cos^2 x}dx$.** **[4]**

**Sol.** To prove : $\int_0^a f(x)dx = \int_0^a f(a-x)dx$

Proof : Let $x = a - t$

$dx = -dt$

Also, $x = 0 \Rightarrow t = a$

and $x = a \Rightarrow t = 0$

$\therefore \quad \int_0^a f(x)dx = -\int_a^0 f(a-t)\,dt$

$\Rightarrow \quad \int_0^a f(x)dx = \int_0^a f(a-t)\,dt \quad \left[\because \int_b^a f(x)dx = -\int_a^b f(x)dx\right]$

$\Rightarrow \int_0^a f(x)\,dx = \int_0^a f(a-x)\,dx$

$\left[\because \int_a^b f(x)dx = \int_a^b f(t)dt\right]$

**Hence Proved.**

Let $\quad I = \int_0^{\pi} \frac{x\sin x\,dx}{1+\cos^2 x} \quad$ ...(i)

$I = \int_0^{\pi} \frac{(\pi - x)\sin(\pi - x)}{1+\cos^2(\pi - x)}dx$

[Applying the property used above]

$I = \int_0^{\pi} \frac{(\pi - x)\sin x\,dx}{1+\cos^2 x} \quad$ ...(ii)

Adding equation (i) and (ii), we get

$2I = \pi\int_0^{\pi} \frac{\sin x}{1+\cos^2 x}dx$

Put $\quad \cos x = t$

$-\sin x\,dx = dt$

when $\quad x = 0, t = 1$

and $\quad x = \pi, t = -1$

$\Rightarrow \quad 2I = -\pi\int_1^{-1} \frac{dt}{1+t^2}$

$\Rightarrow \quad 2I = -\pi\left[\tan^{-1}(t)\right]_1^{-1}$

$\Rightarrow \quad 2I = -\pi\left[\tan^{-1}(-1) - \tan^{-1}(1)\right]$

$\Rightarrow \quad 2I = -\pi[-\tan^{-1}(1) - \tan^{-1}(1)]$

$\Rightarrow \quad 2I = -\pi\left[-2\tan^{-1}(1)\right]$

$\Rightarrow \quad 2I = -\pi\left(-\frac{2\pi}{4}\right)$

$\Rightarrow \quad 2I = \frac{2\pi^2}{4}$

$\Rightarrow \quad I = \frac{\pi^2}{4}$ **Ans.**

**21. Solve the differential equation $x\,dy - y\,dx = \sqrt{x^2+y^2}\,dx$, given that $y = 0$ when $x = 1$.** **[4]**

**Sol.** Given, $\quad x\,dy - y\,dx = \sqrt{x^2+y^2}\,dx$

$\Rightarrow \quad x\,dy = (\sqrt{x^2+y^2} + y)dx$

$\Rightarrow \quad \frac{dy}{dx} = \frac{\sqrt{x^2+y^2}+y}{x}$

Put $y = vx$

$\frac{dy}{dx} = v + x\frac{dv}{dx}$

So, $\quad v + x.\frac{dv}{dx} = \frac{\sqrt{x^2+v^2x^2}+vx}{x}$

$\Rightarrow \quad v + x\frac{dv}{dx} = \sqrt{1+v^2} + v$

$\Rightarrow \quad x.\frac{dv}{dx} = \sqrt{1+v^2}$

Integrating both sides, we get

$\Rightarrow \quad \int \frac{dv}{\sqrt{1+v^2}} = \int \frac{dx}{x}$

$\log\left|v + \sqrt{1+v^2}\right| = \log x + \log c$

$\Rightarrow \quad \log\left|v + \sqrt{1+v^2}\right| = \log(xc)$

$\Rightarrow \quad v + \sqrt{1+v^2} = xc$

$\Rightarrow \quad \frac{y}{x} + \sqrt{1+\frac{y^2}{x^2}} = xc$

$\Rightarrow \quad \frac{y+\sqrt{x^2+y^2}}{x} = xc$

$\Rightarrow \quad y + \sqrt{x^2+y^2} = x^2c \quad$ ...(i)

Now, given $y = 0$ when $x = 1$

$\therefore$ From equation (i),

$$0 + 1 = c$$

$$c = 1$$

Substitute the value of $c$ in (i), we get

$$y + \sqrt{x^2 + y^2} = x^2$$ **Ans.**

**OR**

**Solve the differential equation $(1 + x^2)\frac{dy}{dx} + 2xy - 4x^2 = 0$, subject to the initial condition $y(0) = 0$.**

**Sol.** Given, $(1 + x^2)\frac{dy}{dx} + 2xy - 4x^2 = 0$

$$\Rightarrow \frac{dy}{dx} + \frac{2xy}{1+x^2} - \frac{4x^2}{1+x^2} = 0$$

$$\Rightarrow \frac{dy}{dx} + \frac{2xy}{1+x^2} = \frac{4x^2}{1+x^2}$$

Here, $$P = \frac{2xy}{1+x^2} \text{ and } Q = \frac{4x^2}{1+x^2}$$

$$\text{I.F} = e^{\int P dx} = e^{\int \frac{2x}{1+x^2} dx}$$

$$\text{I.F.} = e^{\log|1+x^2|}$$

$$\text{I.F.} = 1 + x^2$$

Solution of given differential equation is

$$y\,(\text{I.F.}) = \int (\text{I.F} \times \text{Q})\, dx$$

$$\Rightarrow y\,(1 + x^2) = \int (1 + x^2)\left(\frac{4x^2}{1+x^2}\right) dx$$

$$\Rightarrow y\,(1 + x^2) = \frac{4x^3}{3} + C \quad \text{...(i)}$$

Also given $y(0) = 0$

Now, puting $x = 0, y = 0$ in (i), we get

$$0 = 0 + C$$

$$\therefore \quad C = 0$$

Substitute the value of C in (i), we get

$$y(1 + x^2) = \frac{4x^3}{3}$$

$$y = \frac{4x^3}{3(1+x^2)}$$ **Ans.**

**22. If $\hat{i} + \hat{j} + \hat{k}$, $2\hat{i} + 5\hat{j}$, $3\hat{i} + 2\hat{j} - 3\hat{k}$ and $\hat{i} - 6\hat{j} - \hat{k}$ respectively are the position vectors of points A, B, C and D, then find the angle between the straight lines AB and CD. Find whether $\overrightarrow{AB}$ and $\overrightarrow{CD}$ are collinear or not. [4]**

**Sol.**

$A\left(\hat{i} + \hat{j} + \hat{k}\right), B\left(2\hat{i} + 5\hat{j}\right), C\left(3\hat{i} + 2\hat{j} - 3\hat{k}\right)$ and $D\left(\hat{i} - 6\hat{j} - \hat{k}\right)$.

Now, $$\overrightarrow{AB} = \hat{i} + 4\hat{j} - \hat{k}$$

$$\overrightarrow{CD} = -2\hat{i} - 8\hat{j} + 2\hat{k}$$

Let $\theta$ be the angle between $\overrightarrow{AB}$ & $\overrightarrow{CD}$

So, $$\cos\theta = \frac{\overrightarrow{AB}.\overrightarrow{CD}}{|\overrightarrow{AB}||\overrightarrow{CD}|}$$

$$\cos\theta = \frac{\left(\hat{i} + 4\hat{j} - \hat{k}\right).\left(-2\hat{i} - 8\hat{j} + 2\hat{k}\right)}{\sqrt{1+16+1}\sqrt{4+64+4}}$$

$$\cos\theta = \frac{-2-32-2}{\sqrt{18}\sqrt{72}}$$

$$\cos\theta = \frac{-36}{3\sqrt{2}.6\sqrt{2}}$$

$$\cos\theta = -1$$

$$\cos\theta = \cos\pi$$

$$\theta = \pi$$

so $\overrightarrow{AB}$ and $\overrightarrow{CD}$ are collinear. **Ans.**

**23. Find the value of $\lambda$ so that the line $\frac{1-x}{3} = \frac{7y-14}{\lambda} = \frac{z-3}{2}$ and $\frac{7-7x}{3\lambda} = \frac{y-5}{1} = \frac{6-z}{5}$ are at right angles. Also, find whether the lines are intersecting or not. [4]**

**Sol.** Equation of 1st line,

$$\frac{1-x}{3} = \frac{7y-14}{\lambda} = \frac{z-3}{2}$$

$$\Rightarrow \frac{x-1}{-3} = \frac{y-2}{\lambda/7} = \frac{z-3}{2}$$

DR's of 1st line $<-3, \lambda/7, 2>$ and equation of 2nd line,

$$\frac{7-7x}{3\lambda} = \frac{y-5}{1} = \frac{6-z}{5}$$

$$\Rightarrow \frac{x-1}{-3\lambda/7} = \frac{y-5}{1} = \frac{z-6}{-5}$$

DR's of 2nd line $< \frac{-3\lambda}{7}, 1, -5 >$

Since given lines are at right angles,

$$-3\left(\frac{-3\lambda}{7}\right)+\frac{\lambda}{7}(1)+2(-5)=0$$

$$\Rightarrow \quad \frac{9\lambda}{7}+\frac{\lambda}{7}-10=0$$

$$\Rightarrow \quad 10\lambda-70=0$$

$$\Rightarrow \quad 10\lambda=70$$

$$\Rightarrow \quad \lambda=7$$

Now, equation of 1st line,

$$\frac{x-1}{-3}=\frac{y-2}{1}=\frac{z-3}{2}$$

and equation of 2nd line,

$$\frac{x-1}{-3}=\frac{y-5}{1}=\frac{z-6}{-5}$$

Here, $a_1=-3, b_1=1, c_1=2$

$a_2=-3, b_2=2, c_2=-5$

and $x_1=1, y_1=2, z_1=3$

$x_2=1, y_2=5, z_2=6$

$$\therefore \quad =\begin{vmatrix} x_2-x_1 & y_2-y_1 & z_2-z_1 \\ a_1 & b_1 & c_1 \\ a_2 & b_2 & c_2 \end{vmatrix}$$

$$=\begin{vmatrix} 0 & 3 & 3 \\ -3 & 1 & 2 \\ -3 & 1 & -5 \end{vmatrix}$$

$$=-63\neq 0$$

Hence, the given two lines do not intersect each other. **Ans.**

## SECTION-D

**24. If $A=\begin{bmatrix} 1 & 1 & 1 \\ 1 & 0 & 2 \\ 3 & 1 & 1 \end{bmatrix}$, find $A^{-1}$. Hence, solve the system of equations $x+y+z=6$, $x+2z=7$, $3x+y+z=12$** **[6]**

**Sol.** Given,

$$A=\begin{vmatrix} 1 & 1 & 1 \\ 1 & 0 & 2 \\ 3 & 1 & 1 \end{vmatrix}$$

$$|A|=1(0-2)-1(1-6)+1(1-0)$$

$$|A|=-2+5+1$$

$$|A|=4\neq 0$$

$\therefore$ $A^{-1}$ exists.

Now, cofactors of A are,

$$a_{11}=-2, a_{21}=0, a_{31}=2$$

$$a_{12}=5, a_{22}=-2, a_{32}=-1$$

$$a_{13}=1, a_{23}=2, a_{33}=-1$$

$$\text{adj } A=\begin{vmatrix} -2 & 5 & 1 \\ 0 & -2 & 2 \\ 2 & -1 & -1 \end{vmatrix}^T$$

$$\text{adj } A=\begin{vmatrix} -2 & 0 & 2 \\ 5 & -2 & -1 \\ 1 & 2 & -1 \end{vmatrix}$$

So, $$A^{-1}=\frac{1}{|A|}(\text{adj } A)$$

$$A^{-1}=\frac{1}{4}\begin{bmatrix} -2 & 0 & 2 \\ 5 & -2 & -1 \\ 1 & 2 & -1 \end{bmatrix} \quad ...(i)$$

**Ans.**

The given equations are :

$x+y+z=6,$

$x+2z=7,$

and $3x+y+z=12$

Here, $A=\begin{bmatrix} 1 & 1 & 1 \\ 1 & 0 & 2 \\ 3 & 1 & 1 \end{bmatrix}, X=\begin{bmatrix} x \\ y \\ z \end{bmatrix} B=\begin{bmatrix} 6 \\ 7 \\ 12 \end{bmatrix}$

$$X=A^{-1}B$$

$$\begin{bmatrix} x \\ y \\ z \end{bmatrix}=\frac{1}{4}\begin{bmatrix} -2 & 0 & 2 \\ 5 & -2 & -1 \\ 1 & 2 & -1 \end{bmatrix}\begin{bmatrix} 6 \\ 7 \\ 12 \end{bmatrix} \quad \text{[Using (i)]}$$

$$\begin{bmatrix} x \\ y \\ z \end{bmatrix}=\frac{1}{4}\begin{bmatrix} -12+0+24 \\ 30-14-12 \\ 6+14-12 \end{bmatrix}$$

$$\begin{bmatrix} x \\ y \\ z \end{bmatrix}=\frac{1}{4}\begin{bmatrix} 12 \\ 4 \\ 8 \end{bmatrix}$$

$$\begin{bmatrix} x \\ y \\ z \end{bmatrix}=\begin{bmatrix} 3 \\ 1 \\ 2 \end{bmatrix}$$

$\therefore$ $x=3, y=1$ and $z=2$ **Ans.**

**OR**

**Find the inverse of the following matrix using elementary operations.**

$$\mathbf{A}=\begin{bmatrix} 1 & 2 & -2 \\ -1 & 3 & 0 \\ 0 & -2 & 1 \end{bmatrix}$$

**Sol.** Given, $$A=\begin{bmatrix} 1 & 2 & -2 \\ -1 & 3 & 0 \\ 0 & -2 & 1 \end{bmatrix}$$

$$A = IA$$

$$\begin{bmatrix} 1 & 2 & -2 \\ -1 & 3 & 0 \\ 0 & -2 & 1 \end{bmatrix} = \begin{bmatrix} 1 & 0 & 0 \\ 0 & 1 & 0 \\ 0 & 0 & 1 \end{bmatrix} A$$

Applying $R_2 \to R_2 + R_1$, we get

$$\begin{bmatrix} 1 & 2 & -2 \\ 0 & 5 & -2 \\ 0 & -2 & 1 \end{bmatrix} = \begin{bmatrix} 1 & 0 & 0 \\ 1 & 1 & 0 \\ 0 & 0 & 1 \end{bmatrix} A$$

Now, Applying $R_1 \to R_1 + R_3$, we get

$$\begin{bmatrix} 1 & 0 & -1 \\ 0 & 5 & -2 \\ 0 & -2 & 1 \end{bmatrix} = \begin{bmatrix} 1 & 0 & 1 \\ 1 & 1 & 0 \\ 0 & 0 & 1 \end{bmatrix} A$$

Applying $R_2 \to \frac{1}{5} R_2$, we get

$$\begin{bmatrix} 1 & 0 & -1 \\ 0 & 1 & -\frac{2}{5} \\ 0 & -2 & 1 \end{bmatrix} = \begin{bmatrix} 1 & 0 & 1 \\ \frac{1}{5} & \frac{1}{5} & 0 \\ 0 & 0 & 1 \end{bmatrix} A$$

Applying $R_3 \to R_3 + 2R_2$, we get

$$\begin{bmatrix} 1 & 0 & -1 \\ 0 & 1 & -\frac{2}{5} \\ 0 & 0 & \frac{1}{5} \end{bmatrix} = \begin{bmatrix} 1 & 0 & 1 \\ \frac{1}{5} & \frac{1}{5} & 0 \\ \frac{2}{5} & \frac{2}{5} & 1 \end{bmatrix} A$$

Applying $R_2 \to R_2 + 2R_3$, we get

$$\begin{bmatrix} 1 & 0 & -1 \\ 0 & 1 & 0 \\ 0 & 0 & \frac{1}{5} \end{bmatrix} = \begin{bmatrix} 1 & 0 & 1 \\ 1 & 1 & 2 \\ \frac{2}{5} & \frac{2}{5} & 1 \end{bmatrix} A$$

Now, Applying $R_3 \to 5R_3$, we get

$$\begin{bmatrix} 1 & 0 & -1 \\ 0 & 1 & 0 \\ 0 & 0 & 1 \end{bmatrix} = \begin{bmatrix} 1 & 0 & 1 \\ 1 & 1 & 2 \\ 2 & 2 & 5 \end{bmatrix} A$$

Applying $R_1 \to R_1 + R_3$, we get

$$\begin{bmatrix} 1 & 0 & 0 \\ 0 & 1 & 0 \\ 0 & 0 & 1 \end{bmatrix} = \begin{bmatrix} 3 & 2 & 6 \\ 1 & 1 & 2 \\ 2 & 2 & 5 \end{bmatrix} A$$

Hence, $A^{-1} = \begin{bmatrix} 3 & 2 & 6 \\ 1 & 1 & 2 \\ 2 & 2 & 5 \end{bmatrix}$ **Ans.**

**25. A tank with rectangular base and rectangular sides, open at the top is to be constructed so that its depth is 2 m and volume is 8 m³. If building of tank costs ₹ 70 per square metre for the base and ₹ 45 per square metre for the sides, what is the cost of least expresive tank? [6]**

**Sol.** Let the length and breadth of the tank be $x$ and $y$ metres, respectively.

Given, Volume $= 8 \text{ m}^3$

$$\Rightarrow \quad 2xy = 8$$

[∵ depth of tank = 2 m]

$$\Rightarrow \quad y = \frac{8}{2x} = \frac{4}{x}$$

Let C be the cost of tank. Then

$$C = 70\,xy + 45\,(2 \times 2x + 2 \times 2y)$$

$$\Rightarrow \quad C = 70\,xy + 180x + 180\,y$$

$$\Rightarrow \quad C = 70\,x \times \frac{4}{x} + 180\,x + 180 \times \frac{4}{x}$$

$$\Rightarrow \quad C = 280 + 180x + \frac{720}{x} \quad \text{...(i)}$$

Now, differentiating both sides w.r.t. $x$, we get

$$\frac{dC}{dx} = 180 - \frac{720}{x^2}$$

For maxima and minima, $\frac{dC}{dx} = 0$

$$180 - \frac{720}{x^2} = 0$$

$$\Rightarrow \quad 180 = \frac{720}{x^2}$$

$$\Rightarrow \quad x^2 = 4$$

$$\Rightarrow \quad x = \pm 2$$

$$\Rightarrow \quad x = 2$$

Again, differentiating both sides w.r.t. $x$, we get

$$\frac{d^2C}{dx^2} = \frac{1440}{x^3}$$

$$\left.\frac{d^2C}{dx^2}\right|_{x=2} = \frac{1440}{8} = 180 > 0$$

∴ when $x = 2$, cost of tank is minimum

Substituting the value of $x$ in equation (i), we get

$$C = 280 + 180 \times 2 + \frac{720}{2}$$

$$\Rightarrow \quad C = 280 + 360 + 360$$

$$\Rightarrow \quad C = 1000$$

Hence, the cost of least expensive tank is ₹ 1000. **Ans.**

**26. Using integration, find the area of a triangle ABC, whose vertices are A(2, 5), B(4, 7) and C(6, 2). [6]**

**Sol.** Given, A (2,5), B (4,7) and C (6,2) be the vertices of a triangle.

The equation of side AB.

$$(y-5) = \frac{7-5}{4-2}(x-2)$$

$$\Rightarrow \quad y-5 = x-2$$

$$\Rightarrow \quad y = x+3$$

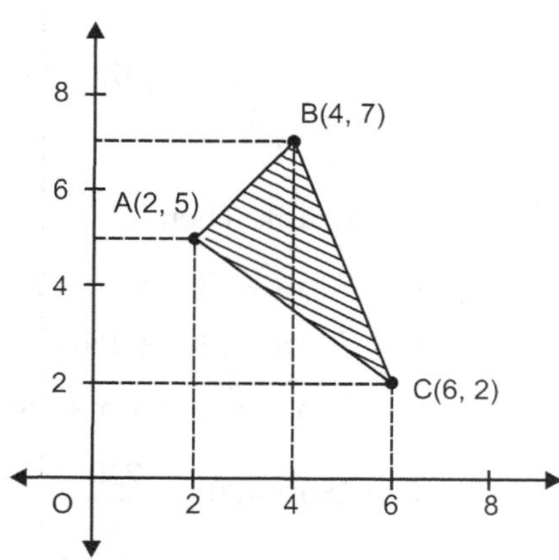


The equation of side BC,

$$(y-7) = \frac{2-7}{6-4}(x-4)$$

$$\Rightarrow \quad y-7 = \frac{-5}{2}(x-4)$$

$$\Rightarrow \quad 2y-14 = -5x+20$$

$$\Rightarrow \quad 2y = -5x+34$$

$$\Rightarrow \quad y = \frac{-1}{2}(5x-34)$$

The equation of side AC,

$$(y-5) = \frac{2-5}{6-2}(x-2)$$

$$\Rightarrow \quad y-5 = \frac{-3}{4}(x-2)$$

$$\Rightarrow \quad 4y-20 = -3x+6$$

$$\Rightarrow \quad 4y = -3x+26$$

$$y = -\frac{1}{4}(3x-26)$$

$\therefore$ Area of $\Delta ABC$

$$= \int_2^4 y_{AB}\,dx + \int_4^6 y_{BC}\,dx - \int_2^6 y_{AC}\,dx$$

$$= \int_2^4 (x+3)\,dx + \int_4^6 -\frac{1}{2}(5x-34)\,dx - \int_2^6 -\frac{1}{4}(3x-26)\,dx$$

$$= \left[\frac{x^2}{2}+3x\right]_2^4 - \frac{1}{2}\left[\frac{5x^2}{2}-34x\right]_4^6 + \frac{1}{4}\left[\frac{3x^2}{2}-26x\right]_2^6$$

$$= \left[\left(\frac{16}{2}+12\right)-\left(\frac{4}{2}+6\right)\right] - \frac{1}{2}\left[\left(\frac{180}{2}-204\right) - \left(\frac{80}{2}-136\right)\right] + \frac{1}{4}\left[\left(\frac{108}{2}-156\right)-\left(\frac{12}{2}-52\right)\right]$$

$$= [(8+12)-(2+6)] - \frac{1}{2}[(90-204)-(40-136)] + \frac{1}{4}[(54-156)-(6-52)]$$

$$= 12 + \frac{1}{2}(18) - \frac{1}{4}(56) = 12 + 9 - 14 = 7 \text{ sq. units.}$$

**Ans.**

**OR**

**Find the area of a region lying above $x$-axis and included between the circle $x^2 + y^2 = 8x$ and inside of the parabola $y^2 = 4x$.**

**Sol.** Given, equation of circle is $x^2 + y^2 = 8x$ can be expressed as

$$(x-4)^2 + y^2 = 16 \quad \text{...(i)}$$

Centre is (4,0) and radius is 4 units

and equation of parabola is $y^2 = 4x$ ...(ii)

Let required Area = I

$$\therefore I = \int_0^4 y \text{ of the parabola } dx + \int_4^8 y \text{ of the circle } dx$$

$$= \underset{I_1}{\int_0^4 2\sqrt{x}\,dx} + \underset{I_2}{\int_4^8 \sqrt{4^2-(x-4)^2}\,dx}$$

So, $$I_1 = 2\int_0^4 \sqrt{x}\,dx$$

$$\Rightarrow \quad I_1 = 2\left[\frac{x^{3/2}}{3/2}\right]_0^4$$

$$\Rightarrow \quad I_1 = \frac{4}{3}[4^{\frac{3}{2}} - 0]$$

$$\Rightarrow \quad I_1 = \frac{4}{3}(8) = \frac{32}{3}$$

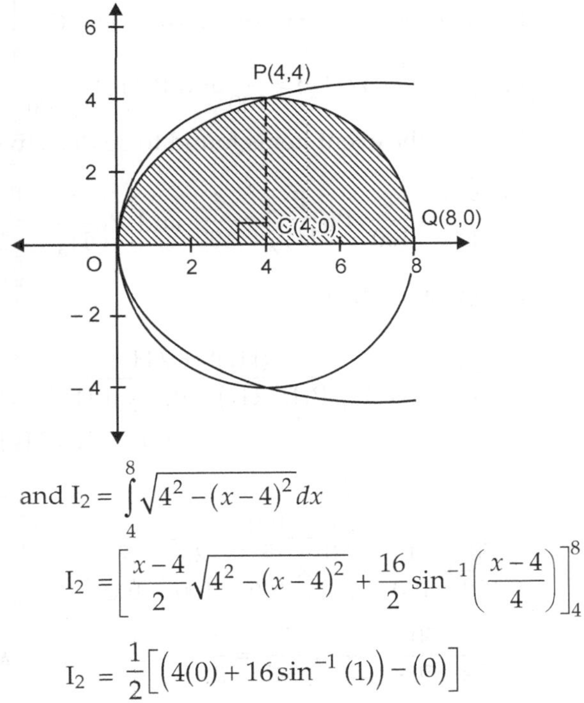


and $I_2 = \int_4^8 \sqrt{4^2-(x-4)^2}\,dx$

$$I_2 = \left[\frac{x-4}{2}\sqrt{4^2-(x-4)^2} + \frac{16}{2}\sin^{-1}\left(\frac{x-4}{4}\right)\right]_4^8$$

$$I_2 = \frac{1}{2}\left[\left(4(0)+16\sin^{-1}(1)\right)-(0)\right]$$

$$I_2 = \frac{1}{2}\times 16\frac{\pi}{2} = 4\pi$$

$\therefore \quad I = I_1 + I_2$

$$= \left(\frac{32}{3}+4\pi\right) = \frac{4}{3}(8+3\pi) \text{ sq. units.}$$ **Ans.**

**27. Find the vector and cartesian equation of the plane passing through the points (2, 2, –1), (3, 4, 2) and (7, 0, 6). Also find the vector equation of a plane passing through (4, 3, 1) and parallel to the plane obtained above.** **[6]**

**Sol.** We know that the general equation of the plane passing through three points $(x_1, y_1, z_1)$ $(x_2, y_2, z_2)$

$$\begin{vmatrix} x-x_1 & y-y_1 & z-z_1 \\ x_2-x_1 & y_2-y_1 & z_2-z_1 \\ x_3-x_1 & y_3-y_1 & z_3-z_1 \end{vmatrix} = 0$$

The the plane passing through A (2, 2, –1), B (3, 4, 2) and C (7, 0, 6)

$$\begin{vmatrix} x-2 & y-2 & z+1 \\ 1 & 2 & 3 \\ 5 & -2 & 7 \end{vmatrix} = 0$$

$\Rightarrow (x-2)(14+6)-(y-2)(7-15)+(z+1)(-2-10) = 0$

$\Rightarrow \quad 20(x-2)-(-8)(y-2)+(-12)(z+1) = 0$

$\Rightarrow \quad 4[5x-10+2y-4-3z-3] = 0$

$\Rightarrow \quad 5x+2y-3z = 17$

This is the required cartesian equation of the plane.

Now,

The required plane passes through the point A (2, 2, –1) whose position vector is $a = 2\hat{i}+2\hat{j}-\hat{k}$ and is normal to vertor $\vec{n}$ given by $\vec{n} = \vec{AB}\times\vec{AC}$

$\therefore \quad \vec{AB} = 3\hat{i}+4\hat{i}+2\hat{k}-(2\hat{i}+2\hat{j}-\hat{k})$

$= \hat{i}+2\hat{i}+3\hat{k}$

and, $\quad \vec{AC} = 7\hat{i}+0\hat{j}+6\hat{k}-(2\hat{i}+2\hat{j}-\hat{k})$

$= 5\hat{i}-2\hat{j}+7\hat{k}$

Now, $\quad \vec{n} = \vec{AB}\times\vec{AC} = \begin{vmatrix} \hat{i} & \hat{j} & \hat{k} \\ 1 & 2 & 3 \\ 5 & -2 & 7 \end{vmatrix}$

$= \hat{i}(14+6)-\hat{j}(7-15)+\hat{k}(-2-10)$

$= 20\hat{i}+8\hat{j}-12\hat{k}$

The vector equation of the plane is given by

$$\vec{r}.\vec{n} = \vec{a}.\vec{n}$$

$\Rightarrow \vec{r}.(20\hat{i}+8\hat{j}-12\hat{k})$

$= (2\hat{i}+2\hat{j}-\hat{k})(20\hat{i}+8\hat{j}-12\hat{k})$

$= 40+16+12$

$\Rightarrow \vec{r}.(20\hat{i}+8\hat{j}-12\hat{k}) = 68$

$\Rightarrow \vec{r}.(5\hat{i}+2\hat{j}-3\hat{k}) = 17$

This is the required vector equation of the plane.

Equation of required parallel plane is

$$\vec{r}.(5\hat{i}+2\hat{j}-3\hat{k}) = (4\hat{i}+3\hat{j}+\hat{k}).(5\hat{i}+2\hat{j}-3\hat{k})$$

$\Rightarrow \vec{r}.(5\hat{i}+2\hat{j}-3\hat{k}) = 23$ **Ans.**

**OR**

**Find the vector equation of the plane that contains the lines $\vec{r} = (\hat{i}+\hat{j})+\lambda(\hat{i}+2\hat{j}-\hat{k})$ and the point (– 1, 3, – 4). Also, find the length of the perpendicular drawn from the point (2, 1, 4) to the plane, thus obtained.**

**Sol.** Given, equation of the given line

$$\vec{r} = (\hat{i}+\hat{j})+\lambda(\hat{i}+2\hat{j}+\hat{k})$$

The plane passes through the point (– 1, 3, – 4). Then the equation of the plane,

$$a(x+1)+b(y-3)+c(z+4)=0 \quad ...(i)$$

Since (1, 1, 0) lies on the plane,

$$\therefore \quad 2a-2b+4c=0 \quad ...(ii)$$

Also, (1, 2, –1) lies on the plane

$$\therefore \quad 2a-b+3c=0 \quad ...(iii)$$

Solving equations (ii) and (iii), we get

$$\frac{a}{(-6+4)}=\frac{-b}{(6-8)}-\frac{c}{(-2+4)}=k \text{ (say)}$$

$$\frac{a}{-2}=\frac{-b}{-2}=\frac{c}{2}=k$$

$\Rightarrow a=-2k,\ b=2k$ and $c=2k$

Putting the values of $a$, $b$, and $c$ in (i), we get

$$-2k(x+1)+2k(y-3)+2k(z+4)=0$$

$$\Rightarrow \quad -(x+1)+(y-3)+(z+4)=0$$

$$\Rightarrow \quad -x-1+y-3+z+4=0$$

$$\Rightarrow \quad -x+y+z=0$$

$\therefore$ Vector equation of plane is,

$$\vec{r}.(-\vec{i}+\vec{j}+\vec{k})=0 \quad \textbf{Ans.}$$

$$\text{Perpendicular distance} = \left|\frac{\vec{a}.\vec{n}-d}{|\vec{n}|}\right|$$

Perpendicular distance

$$=\left|\frac{\left(2\vec{i}+\vec{j}+4\vec{k}\right).\left(-\vec{i}+\vec{j}+\vec{k}\right)-0}{\sqrt{1+1+1}}\right|$$

$$=\left|\frac{-2+1+4}{\sqrt{3}}\right|=\frac{3}{\sqrt{3}}=\sqrt{3} \text{ units.} \quad \textbf{Ans.}$$

**28. A manufacturer has three machine operators A, B and C. The first operator A produces 1% of defective items, whereas the other two operators B and C produces 5% and 7% defective items respectively. A is on the job for 50% of the time, B on the job 30% of the time and C on the job for 20% of the time. All the items are put into one stockpile and then one item is chosen at random from his and is found to be defective. What is the probability that it was produced by A? [6]**

**Sol.** Let $H_1$ be the event items are produced by A

$H_2$ be the event items are produced by B

$H_3$ be the event items are produced by C

$$P(H_1)=\frac{50}{100},\ P(H_2)=\frac{30}{100} \text{ and } P(H_3)=\frac{20}{100}$$

Let E be the event items found to be defective.

$$P\left(\frac{E}{H_1}\right)=\frac{1}{100},\ P\left(\frac{E}{H_2}\right)=\frac{5}{100} \text{ and } P\left(\frac{E}{H_3}\right)=\frac{7}{100}$$

Using Bayes' theorem,

$$P\left(\frac{H_1}{E}\right)=\frac{P(H_1)P(E/H_1)}{P(H_1)P(E/H_1)+P(H_2)P(E/H_2)+P(H_3)P(E/H_3)}$$

$$=\frac{\frac{50}{100}\times\frac{1}{100}}{\frac{50}{100}\times\frac{1}{100}+\frac{30}{100}\times\frac{5}{100}+\frac{20}{100}\times\frac{7}{100}}$$

$$=\frac{50}{50+150+140}=\frac{50}{340}=\frac{5}{34} \quad \textbf{Ans.}$$

**29. A manufacturer has employed 5 skilled men and 10 semi-skilled men and makes two models A and B of an article. The making two one item of model A requires 2 hours work by a skilled man and 2 hours work by a semi-skilled man. One item of model B requires 1 hour by a skilled man and 3 hours by a semi-skilled man. No man is expected to work more than 8 hours per day. The manufacturer's profit on an item of model A is ₹15 and on an item of model B is ₹ 10. How many of items of each model should be made per day in order to maximize daily profit? Formulate the above LPP and solve it graphically and find the maximum profit. [6]**

**Sol.** Let, $x$ be the number of items of model A and $y$ be the number of items of model B

Let Z be the required profit,

Subject to constraints :

$$2x+y\le 8\times 5$$

$$\Rightarrow \quad 2x+y\le 40$$

$$2x+3y\le 8\times 10$$

$$\Rightarrow \quad 2x+3y\le 80$$

$$x\ge 0,\ y\ge 0$$

$$\text{Maximise } Z=15x+10y$$

Changing the above inequalities into equations,

$2x + y = 40$

| $x$ | 0 | 20 |
|---|---|---|
| $y$ | 40 | 0 |

$2x + 3y = 80$

| $x$ | 0 | 40 |
|---|---|---|
| $y$ | 80/3 | 0 |

Now, ploting these points on the graph.

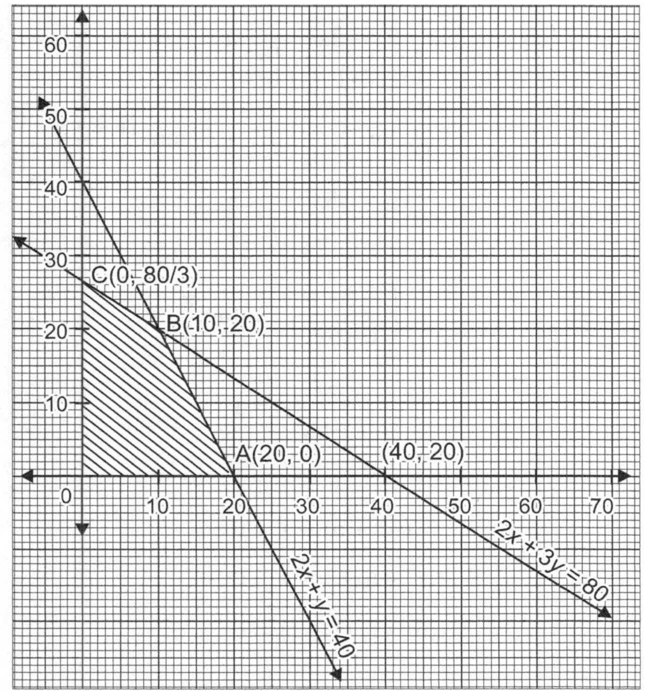


The shaded region is the required feasible region.

| Vertices | Maximum $Z = 15x + 10y$ |
|---|---|
| O (0,0) | $15 \times 0 + 10 \times 0 = 0$ |
| A (20, 0) | $15 \times 20 + 10 \times 0 = 300$ |
| B (10, 20) | $15 \times 10 + 10 \times 20 = 350$ |
| C (0, 80/3) | $15 \times 0 + 10 \times 80/3 = 266.6$ |

Thus, the maximum profit is obtained when the manufacture produces 10 items of model A and 20 items of model B and the maximum profit is ₹ 350. **Ans.**

# Delhi [Set-II]

**Note : Except for the following questions, all the remaining questions have been asked in previous set.**

## SECTION–A

**2. If $f(x) = x + 7$ and $g(x) = x - 7$, $x \in R$, then find $\frac{d}{dx}(fog)(x)$.** **[1]**

**Sol.** Given, $f(x) = x + 7$

and $g(x) = x - 7$

Given,
$$(fog)(x) = f(g(x)) = f(x-7) = (x-7) + 7 = x$$

Now, Differentiating w.r.t, $x$, we get

$$\frac{d(fog)(x)}{dx} = 1$$ **Ans.**

**3. Find the value of $x - y$, if**
$$2\begin{bmatrix} 1 & 3 \\ 0 & x \end{bmatrix} + \begin{bmatrix} y & 0 \\ 1 & 2 \end{bmatrix} = \begin{bmatrix} 5 & 6 \\ 1 & 8 \end{bmatrix}.$$ **[1]**

**Sol.** Given, $2\begin{bmatrix} 1 & 3 \\ 0 & x \end{bmatrix} + \begin{bmatrix} y & 0 \\ 1 & 2 \end{bmatrix} = \begin{bmatrix} 5 & 6 \\ 1 & 8 \end{bmatrix}$

$$\begin{bmatrix} 2 & 6 \\ 0 & 2x \end{bmatrix} + \begin{bmatrix} y & 0 \\ 1 & 2 \end{bmatrix} = \begin{bmatrix} 5 & 6 \\ 1 & 8 \end{bmatrix}$$

$$\begin{bmatrix} 2+y & 6 \\ 1 & 2x+2 \end{bmatrix} = \begin{bmatrix} 5 & 6 \\ 1 & 8 \end{bmatrix}$$

By definition of equality, we have

$$2 + y = 5$$
$$2x + 2 = 8$$
$$y = 3$$
$$2x = 6$$
$$x = 3$$

$\therefore$ $x - y = 3 - 3$

$\Rightarrow$ $x - y = 0$ **Ans.**

## SECTION–B

**6. If A = $\begin{bmatrix} 2 & 0 & 1 \\ 2 & 1 & 3 \\ 1 & -1 & 0 \end{bmatrix}$, then find ($A^2 - 5A$).** **[2]**

**Sol.** Given, $A = \begin{bmatrix} 2 & 0 & 1 \\ 2 & 1 & 3 \\ 1 & -1 & 0 \end{bmatrix}$

Now, $A^2 = \begin{bmatrix} 2 & 0 & 1 \\ 2 & 1 & 3 \\ 1 & -1 & 0 \end{bmatrix}\begin{bmatrix} 2 & 0 & 1 \\ 2 & 1 & 3 \\ 1 & -1 & 0 \end{bmatrix}$

$$A^2 = \begin{bmatrix} 4+0+1 & 0+0-1 & 2+0+0 \\ 4+2+3 & 0+1-3 & 2+3+0 \\ 2-2+0 & 0-1-0 & 1-3+0 \end{bmatrix}$$

$$A^2 = \begin{bmatrix} 5 & -1 & 2 \\ 9 & -2 & 5 \\ 0 & -1 & -2 \end{bmatrix}$$

and $5A = 5\begin{bmatrix} 2 & 0 & 1 \\ 2 & 1 & 3 \\ 1 & -1 & 0 \end{bmatrix} = \begin{bmatrix} 10 & 0 & 5 \\ 10 & 5 & 15 \\ 5 & -5 & 0 \end{bmatrix}$

$$\therefore \; A^2 - 5A = \begin{bmatrix} 5 & -1 & 2 \\ 9 & -2 & 5 \\ 0 & -1 & -2 \end{bmatrix} - \begin{bmatrix} 10 & 0 & 5 \\ 10 & 5 & 15 \\ 5 & -5 & 0 \end{bmatrix}$$

$$\Rightarrow A^2 - 5A = \begin{bmatrix} -5 & -1 & -3 \\ -1 & -7 & -10 \\ -5 & 4 & -2 \end{bmatrix}$$ **Ans.**

**12. Find : $\int \frac{\tan^2 x \sec^2 x}{1 - \tan^6 x} dx$.** **[2]**

**Sol.**

Let $I = \int \frac{\tan^2 x \sec^2 x}{1 - \tan^6 x} dx$

$\Rightarrow \quad I = \int \frac{\tan^2 x \sec^2 x}{1 - (\tan^3 x)^2} dx$

Put $\tan^3 x = t$

$3 \tan^2 x . \sec^2 x \, dx = dt$

$\tan^2 x . \sec^2 dx = \frac{dt}{3}$

$\Rightarrow \quad I = \frac{1}{3}\int \frac{dt}{1 - t^2}$

$\Rightarrow \quad I = \frac{1}{3} \times \frac{1}{2 \times 1} \log\left|\frac{1+t}{1-t}\right| + C$

$\Rightarrow \quad I = \frac{1}{6} \log\left|\frac{1+\tan^3 x}{1-\tan^3 x}\right| + C$ **Ans.**

## SECTION–C

**13. Solve for $x$ : $\tan^{-1}(2x) + \tan^{-1}(3x) = \frac{\pi}{4}$.** **[4]**

**Sol.** Given, $\tan^{-1} 2x + \tan^{-1} 3x = \frac{\pi}{4}$

$\Rightarrow \quad \tan^{-1}\left[\frac{2x + 3x}{1 - (2x)(3x)}\right] = \frac{\pi}{4}$

$\left[\because \tan^{-1} x + \tan^{-1} y = \tan^{-1}\left(\frac{x+y}{1-xy}\right)\right]$

$\Rightarrow \quad \tan^{-1}\left(\frac{5x}{1-6x^2}\right) = \frac{\pi}{4}$

$\Rightarrow \quad \left(\frac{5x}{1-6x^2}\right) = \tan\frac{\pi}{4}$

$\Rightarrow \quad \frac{5x}{1-6x^2} = 1$

$\Rightarrow \quad 5x = 1 - 6x^2$

$\Rightarrow \quad 6x^2 + 5x - 1 = 0$

$\Rightarrow \quad 6x^2 + 6x - x - 1 = 0$

$\Rightarrow \quad (6x - 1)(x + 1) = 0$

$\Rightarrow \quad x = \frac{1}{6}$ or $x = -1$

$x = -1$ doesn't satisfy the given equation

$\therefore \quad x = \frac{1}{6}$ **Ans.**

**18. Using properties of determinants, prove the following :**

$$\begin{vmatrix} a+b+c & -c & -b \\ -c & a+b+c & -a \\ -b & -a & a+b+c \end{vmatrix}$$

**= 2 (a + b) (b + c) (c + a). [4]**

**Sol.** Taking L. H. S.

$$= \begin{vmatrix} a+b+c & -c & -b \\ -c & a+b+c & -a \\ -b & -a & a+b+c \end{vmatrix}$$

Applying $R_2 \to R_2 + R_1$ and $R_3 \to R_3 + R_1$, we get

$$= \begin{vmatrix} a+b+c & -c & -b \\ a+b & a+b & -(a+b) \\ a+c & -(a+c) & a+c \end{vmatrix}$$

Taking $(a + b)$ common from $R_2$ and $(a + c)$ common from $R_3$, we get

$$= (a+b)(a+c)\begin{vmatrix} a+b+c & -c & -b \\ 1 & 1 & -1 \\ 1 & -1 & 1 \end{vmatrix}$$

Now, Applying $C_1 \to C_1 + C_2$, we get

$$= (a+b)(a+c)\begin{vmatrix} (a+b) & -c & -b \\ 2 & 1 & -1 \\ 0 & -1 & 1 \end{vmatrix}$$

Applying $C_2 \to C_2 + C_3$, we get

$$= (a+b)(a+c)\begin{vmatrix} (a+b) & -(b+c) & -b \\ 2 & 0 & -1 \\ 0 & 0 & 1 \end{vmatrix}$$

Now, expanding along $R_3$, we get

$= (a+b)(a+c)\,[1.\{0 + 2(b+c)\}]$

$= (a+b)(a+c)\,[2.(b+c)]$

$= 2(a+b)(b+c)(c+a) =$ R.H.S. **Hence Proved.**

**19. If $x = \cos t + \log \tan\left(\frac{t}{2}\right)$, $y = \sin t$, then find the values of $\frac{d^2y}{dt^2}$ and $\frac{d^2y}{dx^2}$ at $t = \pi/4$.** **[4]**

**Sol.** Given,

$$x = \cos t + \log \tan\left(\frac{t}{2}\right)$$

Differentiating w.r.t. $t$, we get

$$\frac{dx}{dt} = -\sin t + \frac{1}{\tan\left(\frac{t}{2}\right)} \cdot \frac{1}{2}\sec^2\frac{t}{2}$$

$$\Rightarrow \quad \frac{dx}{dt} = -\sin t + \frac{\cos\left(\frac{t}{2}\right)}{2.\sin\left(\frac{t}{2}\right)} \cdot \frac{1}{\cos^2\left(\frac{t}{2}\right)}$$

$$\Rightarrow \quad \frac{dx}{dt} = -\sin t + \frac{1}{2\sin\left(\frac{t}{2}\right)\cos\left(\frac{t}{2}\right)}$$

$$\Rightarrow \quad \frac{dx}{dt} = -\sin t + \frac{1}{\sin t}$$

$$\Rightarrow \quad \frac{dx}{dt} = \frac{-\sin^2 t + 1}{\sin t}$$

$$\Rightarrow \quad \frac{dx}{dt} = \frac{\cos^2 t}{\sin t} \qquad ...(i)$$

and $\quad y = \sin t \quad$ (Given)

Differentiating w.r.t. $t$, we get

$$\Rightarrow \quad \frac{dy}{dt} = \cos t \qquad ...(ii)$$

Now, $\quad \frac{dy}{dx} = \frac{dy}{dt} \times \frac{dt}{dx}$

$$\Rightarrow \quad \frac{dy}{dx} = \cos t \times \frac{\sin t}{\cos^2 t} \quad \text{[from (i) and (ii)]}$$

$$\Rightarrow \quad \frac{dy}{dx} = \tan t$$

Again, differentiating both sides w.r.t., $x$, we get

$$\frac{d^2y}{dx^2} = \sec^2 t \frac{dt}{dx}$$

$$\Rightarrow \quad \frac{d^2y}{dx^2} = \sec^2 t \times \frac{\sin t}{\cos^2 t} \qquad \text{[using (i)]}$$

$$\Rightarrow \quad \left.\frac{d^2y}{dx^2}\right|_{t=\frac{\pi}{4}} = \sec^2 \pi/4 . \frac{\sin \pi/4}{\cos^2 \pi/4}$$

$$= \left(\sqrt{2}\right)^2 . \frac{\frac{1}{\sqrt{2}}}{\left(\frac{1}{\sqrt{2}}\right)^2}$$

$$= \frac{2 \times \frac{1}{\sqrt{2}}}{\frac{1}{2}} = \frac{4}{\sqrt{2}} = 2\sqrt{2} \qquad \textbf{Ans.}$$

## SECTION–D

**24. Show that the altitude of the right circular cone of maximum volume that can be inscribed in a sphere of a radius $r$ is $\frac{4r}{3}$. Also find the maximum volume of cone.** **[6]**

**Sol.** Let R be the radius of cone.

Let OA = OB = $r$ (radius of sphere)

(height of cone) AC $= r + x$ ...(i)

Let V be the volume of cone.

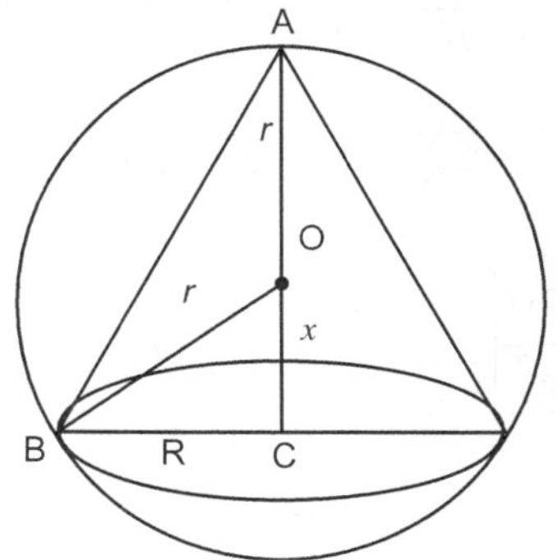


To prove : $\quad AC = \frac{4r}{3}$

Now, $\quad V = \frac{\pi}{3}(BC)^2(AC) \qquad ...(ii)$

(from the figure)

$$V = \frac{\pi}{3}(BC)^2(r+x)$$

Now in $\Delta OBC$,

$$(OB)^2 = (OC)^2 + (BC)^2$$

$$r^2 = x^2 + R^2$$

$$r^2 - x^2 = (BC)^2 \quad ...(iii)$$

$$\therefore \quad V = \frac{\pi}{3}(r^2 - x^2)(r + x)$$

$$V = \frac{\pi}{3}(r + x)^2(r - x)$$

Differentiating both sides w. r. t. $x$, we get

$$\frac{dV}{dx} = \frac{\pi}{3}[(r + x)^2(-1) + (r - x).2(r + x)]$$

$$\Rightarrow \frac{dV}{dx} = \frac{\pi}{3}(r + x)[-r - x + 2r - 2x]$$

$$\Rightarrow \frac{dV}{dx} = \frac{\pi}{3}(r + x)(r - 3x)$$

For maximum or minimum volume,

$$\frac{dV}{dx} = 0$$

$$\Rightarrow \frac{\pi}{3}(r + x)(r - 3x) = 0$$

$$\Rightarrow x = -r \text{ or } x = \frac{r}{3}$$

$x = -r$ is not possible.

so, $x = \frac{r}{3}$

Now, again differentiating w. r. t. $x$, we get

$$\frac{d^2V}{dx^2} = \frac{\pi}{3}[(r+x)(-3)+(r-3x)(1)]$$

$$\Rightarrow \frac{d^2V}{dx^2} = \frac{\pi}{3}[-3r - 3x + r - 3x]$$

$$\Rightarrow \frac{d^2V}{dx^2} = \frac{\pi}{3}(-2r - 6x)$$

$$\Rightarrow \left.\frac{d^2V}{dx^2}\right|_{x=\frac{r}{3}} = \frac{\pi}{3}\left(-2r - \frac{6r}{3}\right)$$

$$= \frac{-4\pi r}{3} < 0$$

So, volume of cone is maximum when $x = \frac{r}{3}$.

Now, put $x = \frac{r}{3}$ in (i)

$$\therefore \quad AC = r + \frac{r}{3}$$

$$\Rightarrow \quad AC = \frac{4r}{3}$$

The altitude of cone is $\frac{4r}{3}$ when its volume is maximum. **Hence Proved.**

From equation (iii),

$$(BC)^2 = r^2 - \left(\frac{r}{3}\right)^2 = r^2 - \frac{r^2}{9}$$

$$\Rightarrow \quad (BC)^2 = \frac{8r^2}{9}$$

Maximum volume of cone $= \frac{\pi}{3}\left(\frac{8r^2}{9}\right)\left(\frac{4r}{3}\right)$

[Using (ii)]

$$= \frac{32\pi r^3}{81}$$ **Ans.**

**25. If $A = \begin{bmatrix} 2 & -3 & 5 \\ 3 & 2 & -4 \\ 1 & 1 & -2 \end{bmatrix}$, then find $A^{-1}$. Hence solve the following system of equations :**

**$2x - 3y + 5z = 11, 3x + 2y - 4z = -5, x + y - 2z = -3.$**

**[6]**

**Sol.** Given, $A = \begin{bmatrix} 2 & -3 & 5 \\ 3 & 2 & -4 \\ 1 & 1 & -2 \end{bmatrix}$

$$|A| = 2(-4 + 4) + 3(-6 + 4) + 5(3 - 2)$$

$$|A| = 3(-2) + 5(1)$$

$$|A| = -1 \neq 0$$

$\therefore A^{-1}$ exists.

Cofactors of A,

$a_{11} = 0, a_{21} = -1, a_{31} = 2,$

$a_{12} = 2, a_{22} = -9, a_{32} = 23,$

$a_{13} = 1, a_{23} = -5, a_{33} = 13,$

$$\therefore \quad \text{adj } A = \begin{bmatrix} 0 & 2 & 1 \\ -1 & -9 & -5 \\ 2 & 23 & 13 \end{bmatrix}^T$$

$$\Rightarrow \quad \text{adj } A = \begin{bmatrix} 0 & -1 & 2 \\ 2 & -9 & 23 \\ 1 & -5 & 13 \end{bmatrix}$$

$$\Rightarrow \quad A^{-1} = \frac{1}{|A|}\text{adj } A$$

$$\Rightarrow \quad A^{-1} = \frac{1}{-1}\begin{bmatrix} 0 & -1 & 2 \\ 2 & -9 & 23 \\ 1 & -5 & 13 \end{bmatrix}$$

$$\Rightarrow \qquad A^{-1} = \begin{bmatrix} 0 & 1 & -2 \\ -2 & 9 & -23 \\ -1 & 5 & -13 \end{bmatrix} \qquad ...(i)$$

**Ans.**

Now, given equations are

$$2x - 3y + 5z = 11$$
$$3x + 2y - 4z = -5$$
$$x + y - 2z = -3$$

Here, $A = \begin{bmatrix} 2 & -3 & 5 \\ 3 & 2 & -4 \\ 1 & 1 & -2 \end{bmatrix}, X = \begin{bmatrix} x \\ y \\ z \end{bmatrix} B = \begin{bmatrix} 11 \\ -5 \\ -3 \end{bmatrix}$

and $AX = B$

$\therefore \quad X = A^{-1} B$

$$\Rightarrow \begin{bmatrix} x \\ y \\ z \end{bmatrix} = \begin{bmatrix} 0 & 1 & -2 \\ -2 & 9 & -23 \\ -1 & 5 & -13 \end{bmatrix} \begin{bmatrix} 11 \\ -5 \\ -3 \end{bmatrix} \quad \text{[From (i)}$$

$$\Rightarrow \begin{bmatrix} x \\ y \\ z \end{bmatrix} = \begin{bmatrix} 0 & -5 & +6 \\ -22 & -45 & +69 \\ -11 & -25 & +39 \end{bmatrix}$$

$$\Rightarrow \begin{bmatrix} x \\ y \\ z \end{bmatrix} = \begin{bmatrix} 1 \\ 2 \\ 3 \end{bmatrix}$$

Hence, $x = 1$, $y = 2$ and $z = 3$ **Ans.**

**OR**

**Obtain the inverse of the following matrix using elementary operations :**

$$A = \begin{bmatrix} -1 & 1 & 2 \\ 1 & 2 & 3 \\ 3 & 1 & 1 \end{bmatrix}$$

**Sol.**

Given, $A = \begin{bmatrix} -1 & 1 & 2 \\ 1 & 2 & 3 \\ 3 & 1 & 1 \end{bmatrix}$

$$A = IA$$

$$\begin{bmatrix} -1 & 1 & 2 \\ 1 & 2 & 3 \\ 3 & 1 & 1 \end{bmatrix} = \begin{bmatrix} 1 & 0 & 0 \\ 0 & 1 & 0 \\ 0 & 0 & 1 \end{bmatrix} A$$

Applying $R_1 \leftrightarrow R_2$, we get

$$\begin{bmatrix} 1 & 2 & 3 \\ -1 & 1 & 2 \\ 3 & 1 & 1 \end{bmatrix} = \begin{bmatrix} 0 & 1 & 0 \\ 1 & 0 & 0 \\ 0 & 0 & 1 \end{bmatrix} A$$

Applying $R_2 \rightarrow R_2 + R_1$, we get

$$\begin{bmatrix} 1 & 2 & 3 \\ 0 & 3 & 5 \\ 3 & 1 & 1 \end{bmatrix} = \begin{bmatrix} 0 & 1 & 0 \\ 1 & 1 & 0 \\ 0 & 0 & 1 \end{bmatrix} A$$

Applying $R_3 \rightarrow R_3 - 3R_1$, we get

$$\begin{bmatrix} 1 & 2 & 3 \\ 0 & 3 & 5 \\ 0 & -5 & -8 \end{bmatrix} = \begin{bmatrix} 0 & 1 & 0 \\ 1 & 1 & 0 \\ 0 & -3 & 1 \end{bmatrix} A$$

Applying $R_2 \rightarrow \frac{1}{3} R_2$, we get

$$\begin{bmatrix} 1 & 2 & 3 \\ 0 & 1 & \frac{5}{3} \\ 0 & -5 & -8 \end{bmatrix} = \begin{bmatrix} 0 & 1 & 0 \\ \frac{1}{3} & \frac{1}{3} & 0 \\ 0 & -3 & 1 \end{bmatrix} A$$

Applying $R_1 \rightarrow R_1 - 2R_2$, we get

$$\begin{bmatrix} 1 & 0 & \frac{-1}{3} \\ 0 & 1 & \frac{5}{3} \\ 0 & -5 & -8 \end{bmatrix} = \begin{bmatrix} \frac{-2}{3} & \frac{1}{3} & 0 \\ \frac{1}{3} & \frac{1}{3} & 0 \\ 0 & -3 & 1 \end{bmatrix} A$$

Applying $R_3 \rightarrow R_3 + 5R_2$, we get

$$\begin{bmatrix} 1 & 0 & \frac{-1}{3} \\ 0 & 1 & \frac{5}{3} \\ 0 & 0 & \frac{1}{3} \end{bmatrix} = \begin{bmatrix} \frac{-2}{3} & \frac{1}{3} & 0 \\ \frac{1}{3} & \frac{1}{3} & 0 \\ \frac{5}{3} & \frac{-4}{3} & 1 \end{bmatrix} A$$

Applying $R_1 \rightarrow R_1 + R_3$, we get

$$\begin{bmatrix} 1 & 0 & 0 \\ 0 & 1 & \frac{5}{3} \\ 0 & 0 & \frac{1}{3} \end{bmatrix} = \begin{bmatrix} 1 & -1 & 1 \\ \frac{1}{3} & \frac{1}{3} & 0 \\ \frac{5}{3} & \frac{-4}{3} & 1 \end{bmatrix} A$$

Applying $R_2 \rightarrow R_2 - 5R_3$, we get

$$\begin{bmatrix} 1 & 0 & 0 \\ 0 & 1 & 0 \\ 0 & 0 & \frac{1}{3} \end{bmatrix} = \begin{bmatrix} 1 & -1 & 1 \\ -8 & 7 & -5 \\ \frac{5}{3} & \frac{-4}{3} & 1 \end{bmatrix} A$$

Now, Applying $R_3 \rightarrow 3R_3$, we get

$$\begin{bmatrix} 1 & 0 & 0 \\ 0 & 1 & 0 \\ 0 & 0 & 1 \end{bmatrix} = \begin{bmatrix} 1 & -1 & 1 \\ -8 & 7 & -5 \\ 5 & -4 & 3 \end{bmatrix} A$$

Hence, $A^{-1} = \begin{bmatrix} 1 & -1 & 1 \\ -8 & 7 & -5 \\ 5 & -4 & 3 \end{bmatrix}$ **Ans.**

## Delhi [Set-III]

**Note : Except for the following questions, all the remaining questions have been asked in previous sets.**

### SECTION-A

**1. If $3A - B = \begin{bmatrix} 5 & 0 \\ 1 & 1 \end{bmatrix}$ and $B = \begin{bmatrix} 4 & 3 \\ 2 & 5 \end{bmatrix}$, then find the matrix A.** **[1]**

**Sol.** Given, $B = \begin{bmatrix} 4 & 3 \\ 2 & 5 \end{bmatrix}$

Let $C = \begin{bmatrix} 5 & 0 \\ 1 & 1 \end{bmatrix}$

$\therefore$ $3A - B = C$

$\Rightarrow$ $A = \frac{1}{3}(B + C)$

$\Rightarrow$ $A = \frac{1}{3}\left(\begin{bmatrix} 4 & 3 \\ 2 & 5 \end{bmatrix} + \begin{bmatrix} 5 & 0 \\ 1 & 1 \end{bmatrix}\right)$

$\Rightarrow$ $A = \frac{1}{3}\begin{bmatrix} 9 & 3 \\ 3 & 6 \end{bmatrix}$

$\therefore$ $A = \begin{bmatrix} 3 & 1 \\ 1 & 2 \end{bmatrix}$ **Ans.**

**2. Write the order and the degree of the following differential equation :**

$$x^3\left(\frac{d^2y}{dx^2}\right)^2 + x\left(\frac{dy}{dx}\right)^4 = 0$$ **[1]**

**Sol.** Given,

$$x^3\left(\frac{d^2y}{dx^2}\right)^2 + x\left(\frac{dy}{dx}\right)^4 = 0$$

$\therefore$ Order of the given differential equation is 2 and degree is 2. **Ans.**

### SECTION-B

**5. Find : $\int \sin x . \log \cos x \, dx$.** **[2]**

**Sol.** Let $I = \int \sin x \log \cos x \, dx$

Put $\cos x = t$

$\sin x \, dx = -dt$

$I = \int \underset{\text{I}}{\log t} . \underset{\text{II}}{1} \, dt$

$\Rightarrow$ $I = -\left\{t \log t - \int \frac{1}{t}.t dt\right\}$

[Integrating by parts]

$\Rightarrow$ $I = -t \log t + t + c$

$\Rightarrow$ $I = -\cos x (\log \cos x - 1) + C$

**Ans.**

**6. Evaluate : $\int_{-\pi}^{\pi}(1 - x^2)\sin x \cos^2 x \, dx$.** **[2]**

**Sol.**

Let $I = \int_{-\pi}^{\pi}(1-x^2)\sin x \cos^2 x \, dx$

Let $f(x) = (1-x^2) \sin x \cos^2 x$

Then $f(-x) = -(1-x^2) \sin x \cos^2 x$

$\left[\because \begin{array}{l} \cos(-\theta) = \cos\theta \\ \sin(-\theta) = -\sin\theta \end{array}\right]$

$\therefore$ $f(x) = -f(x)$

So, $f(x)$ is odd.

$\therefore$ $\int_{-a}^{a} f(x)dx = 0$

$\therefore$ $I = 0$ **Ans.**

**OR**

**Evaluate : $\int_{-1}^{2} \frac{|x|}{x} dx$.**

**Sol.** Let $I = \int_{-1}^{2} \frac{|x|}{x} dx$

$I = \int_{-1}^{0} \frac{-x}{x} dx + \int_{0}^{2} \frac{x}{x} dx$

$\Rightarrow$ $I = \int_{-1}^{0} -1 \, dx + \int_{0}^{2} 1 \, dx$

$$\Rightarrow \quad I = -[x]_{-1}^{0} + 1[x]_0^2$$

$$\Rightarrow \quad I = -[0+1] + [2-0]$$

$$\Rightarrow \quad I = -1+2$$

$$\Rightarrow \quad I = 1 \qquad \textbf{Ans.}$$

**8. Find a matrix A such that 2A – 3B + 5C = 0, where** $B = \begin{bmatrix} -2 & 2 & 0 \\ 3 & 1 & 4 \end{bmatrix}$ **and** $C = \begin{bmatrix} 2 & 0 & -2 \\ 7 & 1 & 6 \end{bmatrix}$. **[2]**

**Sol.** Given, $B = \begin{bmatrix} -2 & 2 & 0 \\ 3 & 1 & 1 \end{bmatrix}$ and $C = \begin{bmatrix} 2 & 0 & -2 \\ 7 & 1 & 6 \end{bmatrix}$

and $2A - 3B + 5C = 0$

$$\Rightarrow \quad 2A = -5C + 3B$$

$$\Rightarrow \quad A = \frac{1}{2}[3B - 5C]$$

$$\Rightarrow \quad A = \frac{1}{2}\left(3\begin{bmatrix} -2 & 2 & 0 \\ 3 & 1 & 4 \end{bmatrix} - 5\begin{bmatrix} 2 & 0 & -2 \\ 7 & 1 & 6 \end{bmatrix}\right)$$

$$\Rightarrow \quad A = \frac{1}{2}\left(\begin{bmatrix} -6 & 6 & 0 \\ 9 & 3 & 12 \end{bmatrix} - \begin{bmatrix} 10 & 0 & -10 \\ 35 & 5 & 30 \end{bmatrix}\right)$$

$$\Rightarrow \quad A = \frac{1}{2}\begin{bmatrix} -16 & 6 & 10 \\ -26 & -2 & -18 \end{bmatrix}$$

$$\Rightarrow \quad A = \begin{bmatrix} -8 & 3 & 5 \\ -13 & -1 & -9 \end{bmatrix} \qquad \textbf{Ans.}$$

## SECTION–C

**13. Using properties of determinants, prove the following:**

$$\begin{vmatrix} a & b & c \\ a-b & b-c & c-a \\ b+c & c+a & a+b \end{vmatrix} = a^3 + b^3 + c^3 - 3abc.$$

**[4]**

**Sol.** $$\text{L.H.S.} = \begin{vmatrix} a & b & c \\ a-b & b-c & c-a \\ b+c & c+a & a+b \end{vmatrix}$$

Applying $R_1 \to R_1 + R_3$, we get

$$= \begin{vmatrix} a+b+c & a+b+c & a+b+c \\ a-b & b-c & c-a \\ b+c & c+a & a+b \end{vmatrix}$$

Taking $(a+b+c)$ from $R_1$, we get

$$= (a+b+c)\begin{vmatrix} 1 & 1 & 1 \\ a-b & b-c & c-a \\ b+c & c+a & a+b \end{vmatrix}$$

Applying $C_2 \to C_2 - C_1$ and $C_3 \to C_3 - C_1$, we get

$$= (a+b+c)\begin{vmatrix} 1 & 0 & 0 \\ a-b & 2b-c-a & c-2a+b \\ b+c & a-b & a-c \end{vmatrix}$$

Now, Expanding along $R_1$, we get

$$= (a+b+c)\,[(1)\,\{(2b-c-a)(a-c) - (a-b)(c-2a+b)\}]$$

$$= (a+b+c)\,[(2ba - 2bc - ac + c^2 - a^2 + ac) - (ac - 2a^2 + ab - bc + 2ab - b^2)]$$

$$= (a+b+c)\,[2ba - 2bc + c^2 - a^2 - ac + 2a^2 - ab + bc - 2ab + b^2]$$

$$= (a+b+c)\,[a^2 + b^2 + c^2 - ab - bc - ca]$$

$$= a^3 + b^3 + c^3 - 3abc = \text{R.H.S.} \qquad \textbf{Hence Proved.}$$

**20. Find :** $\displaystyle\int \frac{\cos x}{(1+\sin x)(2+\sin x)}dx$ **[4]**

**Sol.** Let $$I = \int \frac{\cos x}{(1+\sin x)(2+\sin x)}dx$$

Put $\sin x = t$

$\cos x\, dx = dt$

$$\therefore \quad I = \int \frac{dt}{(1+t)(2+t)}$$

Let $$\frac{1}{(1+t)(2+t)} = \frac{A}{1+t} + \frac{B}{2+t}$$

$$1 = A(2+t) + B(1+t)$$

Comparing coefficient of $t$,

$$0 = A + B$$

$$\Rightarrow \quad B = -A \qquad ...(i)$$

Comparing constant terms,

$$1 = 2A + B$$

$$\Rightarrow \quad 1 = -2B + B \qquad [\text{from (i)}]$$

$$\Rightarrow \quad -1 = B$$

$$\therefore \quad A = 1$$

$$\frac{1}{(1+t)(2+t)} = \frac{1}{(1+t)} - \frac{1}{(2+t)}$$

Now, $$I = \int \frac{dt}{1+t} - \int \frac{dt}{2+t}$$

$$\Rightarrow \quad I = \log|1+t| - \log|2+t| + c$$

$$\Rightarrow \quad I = \log\left|\frac{1+t}{2+t}\right| + C$$

$$\Rightarrow \quad I = \log\left|\frac{1+\sin x}{2+\sin x}\right| + C \qquad \textbf{Ans.}$$

**21. Solve the differential equation :**

$$\frac{dy}{dx} - \frac{2x}{1+x^2}y = x^2 + 2.$$ **[4]**

**Sol.**

Given, $\frac{dy}{dx} - \frac{2x}{1+x^2} y = x^2 + 2$

Here, $P = \frac{-2x}{1+x^2}$ and $Q = x^2 + 2$

$$\text{I.F.} = e^{\int P dx} = e^{-\int \frac{2x}{1+x^2} dx}$$

$$\text{I.F.} = e^{-\log|1+x^2|}$$

$$= \frac{1}{(1+x^2)} \quad [\because e^{\log x} = x]$$

$\therefore$ The solution of given differential equation is

$$y(\text{I.F}) = \int (\text{I.F.} \times \text{Q})\, dx$$

$$\Rightarrow \quad y\left(\frac{1}{1+x^2}\right) = \int \frac{x^2+2}{1+x^2} dx$$

$$\Rightarrow \quad y\left(\frac{1}{1+x^2}\right) = \int \left(1 + \frac{1}{1+x^2}\right) dx$$

$$\Rightarrow \quad y \frac{1}{1+x^2} = x + \tan^{-1} x + C$$

or $\quad y = (1 + x^2)(x + \tan^{-1} x + C)$ **Ans.**

**OR**

**Solve the differential equations :**

$$(x + 1)\frac{dy}{dx} = 2e^{-y} - 1; y(0) = 0.$$

**Sol.** Given,

$$(x+1)\frac{dy}{dx} = 2e^{-y} - 1$$

$$\Rightarrow \quad \frac{dy}{2e^{-y} - 1} = \frac{dx}{x+1}$$

$$\Rightarrow \quad \int \frac{e^y dy}{2 - e^y} = \int \frac{dx}{x+1}$$

Put $2 - e^y dy = t$

$$-e^y dy = dt$$

$$e^y dy = -dt$$

$$\therefore \quad \int -\frac{dt}{t} = \int \frac{dx}{x+1}$$

$$\Rightarrow \quad -\log t = \log|x+1| + \log C$$

$$\Rightarrow \quad \log t^{-1} = \log[C(x+1)]$$

$$\Rightarrow \quad t^{-1} = C(x+1)$$

$$\Rightarrow \quad \frac{1}{2 - e^y} = C(x+1) \quad \text{...(i)}$$

Put $x = 0$ and $y = 0$ in (i), we get

$$\Rightarrow \quad C = 1$$

$$\therefore \quad \frac{1}{2 - e^y} = (x+1)$$

or $\quad (x + 1)(2 - e^y) = 1$ **Ans.**

## SECTION–D

**26. Prove that the curves $y^2 = 4x$ and $x^2 = 4y$ divide the the square bounded by $x = 0$, $x = 4$, $y = 4$ and $y = 0$ into three equal parts. [6]**

**Sol.** Given, $\quad y^2 = 4x \quad$ ...(i)

$$x^2 = 4y \quad \text{...(ii)}$$

$$x = 0, x = y, y = 0 \text{ and } y = 4$$

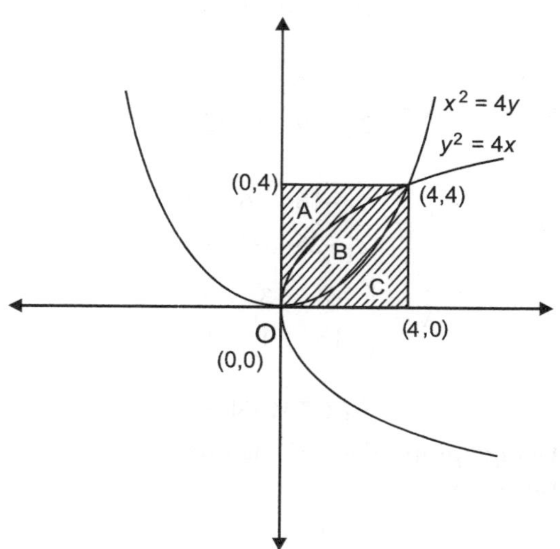


For point of interaction put $y = \frac{x^2}{4}$ in (i)

$$\Rightarrow \quad \frac{x^4}{16} = 4x$$

$$\Rightarrow \quad x^4 = 64x$$

$$\Rightarrow \quad x^4 - 64x = 0$$

$$\Rightarrow \quad x(x^3 - 64) = 0$$

$$\Rightarrow \quad x = 0, x = 4$$

Area of part A,

$$\text{I} = \int_0^4 x \text{ of the curve I}^{\text{st}}\, dy$$

$$\text{I} = \int_0^4 \frac{y^2}{4} dy$$

$$\text{I} = \frac{1}{4 \times 3}\left[y^3\right]_0^4$$

$$I = \frac{1}{12}[4^3]$$

$$I = \frac{64}{12}$$

$$I = \frac{16}{3} \text{ sq. units}$$

Area of Part B,

$$I = \int_0^4 y \text{ of the I}^{st} \text{ curve } dx - \int_0^4 y \text{ of the 2}^{nd} \text{ curve } dx$$

$$I = \int_0^4 2\sqrt{x}dx - \int_0^4 \frac{x^2}{4} dx$$

$$I = 2 \times \frac{2}{3}[x^{3/2}]_0^4 - \frac{1}{12}[x^3]_0^4$$

$$I = \frac{4}{3}[4^{3/2}] - \frac{1}{12}[4^3]$$

$$I = \frac{4}{3} \times 8 - \frac{1}{12} \times 64$$

$$I = \frac{32}{3} - \frac{16}{3}$$

$$I = \frac{16}{3} \text{ sq units}$$

Area of part C,

$$I = \int_0^4 y \text{ of the 2}^{nd} \text{ curve } dx$$

$$I = \int_0^4 \frac{x^2}{4} dx$$

$$I = \frac{1}{4}\int_0^4 x^2 dx$$

$$I = \frac{1}{12}[x^3]_0^4$$

$$I = \frac{1}{12} \times 64$$

$$\Rightarrow \quad I = \frac{16}{3} \text{ sq. units}$$

$\therefore$ The curves $y^2 = 4x$ and $x^2 = 4y$ divide the area of the square into three equal parts. **Hence Proved.**

**OR**

**Using integration, find the area of the triangle whose vertices are (2, 3), (3, 5) and (4, 4).**

**Sol.** Given, the vertices of $\Delta$ABC, A(2, 3), B (3, 5) and C (4, 4).

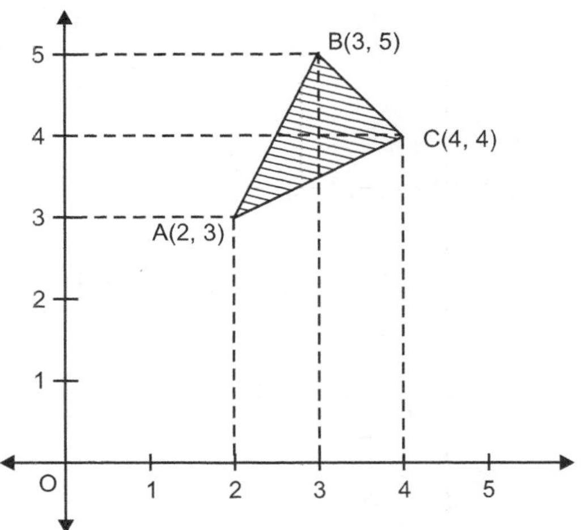


Equation of line AB,

$$(y-3) = \frac{5-3}{3-2}(x-2)$$

$$\Rightarrow \quad (y-3) = 2(x-2)$$

$$\Rightarrow \quad y = 2x-4+3$$

$$\Rightarrow \quad y = 2x-1$$

Equation of line BC,

$$(y-5) = \frac{4-5}{4-3}(x-3)$$

$$\Rightarrow \quad (y-5) = -(x-3)$$

$$\Rightarrow \quad y-5 = -x+3$$

$$\Rightarrow \quad y = -x+8$$

Equation of line AC,

$$(y-3) = \frac{4-3}{4-2}(x-2)$$

$$\Rightarrow \quad (y-3) = \frac{1}{2}(x-2)$$

$$\Rightarrow \quad y = \frac{x-2}{2}+3$$

$$\Rightarrow \quad y = \frac{x+4}{2}$$

Area of $\Delta$ ABC,

$$I = \int_2^3 y_{AB}\, dx + \int_3^4 y_{BC}\, dx - \int_2^3 y_{AC}\, dx$$

$$= \int_2^3 (2x-1)\, dx + \int_3^4 (-x+8)\, dx - \int_2^4 \frac{x+4}{2} dx$$

$$= [x^2]_2^3 - [x]_2^3 - \frac{1}{2}[x^2]_3^4 + 8[x]_3^4$$

$$-\frac{1}{2}\left[\frac{x^2}{2}\right]_2^4 - 2[x]_2^4$$

$$= [9-4]-[3-2]-\frac{1}{2}[16-9]$$

$$+8[4-3]-\frac{1}{4}[16-4]-2[4-2]$$

$$= 5-1-\frac{7}{2}+8-\frac{12}{4}-4$$

$$= 4-\frac{7}{2}+8-3-4$$

$$= 5-\frac{7}{2}$$

$$\Rightarrow \quad I = \frac{3}{2} \text{ Sq. units}$$ **Ans.**

**29. Two cards are drawn simultaneously (or successively without replacement) from a well shuffled pack of 52 cards. Find the mean and variance of the number of kings. [6]**

**Sol.** Let S = Number of kings.

and F = number of non kings

$\therefore$ $P(S) = \frac{4}{52}$ and $P(F) = \frac{48}{52}$

Let $x_i$ be the event that king is drawn.

| $x_i$ | $P(x_i)$ | $x_iP(x_i)$ | $x_i^2$ | $P(x_i)\ x_i^2$ |
|---|---|---|---|---|
| 0 | $\frac{48}{52}\times\frac{47}{51}\times 1=\frac{2256}{2652}$ | 0 | 0 | 0 |
| 1 | $\frac{4}{52}\times\frac{48}{51}\times 2=\frac{384}{2652}$ | $\frac{384}{2652}$ | 1 | $\frac{384}{2652}$ |
| 2 | $\frac{4}{52}\times\frac{3}{51}\times 1=\frac{12}{2652}$ | $\frac{24}{2652}$ | 4 | $\frac{48}{2652}$ |

| $P(x_i)$ | $2256/2652$ | $384/2652$ | $12/2652$ |
|---|---|---|---|
| $x_i$ | 0 | 1 | 2 |

$$\text{Mean } (\overline{X}) = \Sigma x_i p(x_i)$$

$$= \frac{384+24}{2652}$$

$$= \frac{408}{2652} \text{ or } \frac{2}{13}$$

$$\text{Variance } (\sigma^2) = \Sigma x_i^2 p(x_i) - (\overline{X})^2$$

$$= \frac{432}{2652} - \left(\frac{408}{2652}\right)^2$$

$$= \frac{36}{221} - \left(\frac{34}{221}\right)^2$$

$$= \frac{(36\times 221)-(34\times 34)}{(221\times 221)}$$

$$= \frac{7956-1156}{48841}$$

$$= \frac{6800}{48841} \text{ or } \frac{400}{2873}$$

**Ans.**

●●

# MATHEMATICS 2018

**Time allowed : 3 hours** **Maximum marks : 100**

## SECTION – A

**1. Find the value of $\tan^{-1}\sqrt{3} - \cot^{-1}(-\sqrt{3})$** **[1]**

**Sol.** We have,

$$\tan^{-1}\sqrt{3} - \cot^{-1}(-\sqrt{3}) = \tan^{-1}\left(\tan\frac{\pi}{3}\right) - \cot^{-1}\left[\cot\left(\pi - \frac{\pi}{6}\right)\right]$$

$$= \frac{\pi}{3} - \left(\pi - \frac{\pi}{6}\right)$$

$$= \frac{\pi}{3} - \frac{5\pi}{6} = \frac{2\pi - 5\pi}{6}$$

$$= \frac{-3\pi}{6} = \frac{-\pi}{2}$$ **Ans.**

**2. If the matrix $A = \begin{bmatrix} 0 & a & -3 \\ 2 & 0 & -1 \\ b & 1 & 0 \end{bmatrix}$ is skew-symmetric, find the values of 'a' and 'b'.** **[1]**

**Sol.** Given, $A = \begin{bmatrix} 0 & a & -3 \\ 2 & 0 & -1 \\ b & 1 & 0 \end{bmatrix}$

A is given to be skew-symmetric matrix

$\therefore$ $A^T = -A$

$\therefore$ $\begin{bmatrix} 0 & 2 & b \\ a & 0 & 1 \\ -3 & -1 & 0 \end{bmatrix} = -\begin{bmatrix} 0 & a & -3 \\ 2 & 0 & -1 \\ b & 1 & 0 \end{bmatrix}$

$$\begin{bmatrix} 0 & 2 & b \\ a & 0 & 1 \\ -3 & -1 & 0 \end{bmatrix} = \begin{bmatrix} 0 & -a & 3 \\ -2 & 0 & 1 \\ -b & -1 & 0 \end{bmatrix}$$

On comparing both sides we get

$$-a = 2 \text{ and } b = 3$$

$\Rightarrow$ $a = -2$ and $b = 3$ **Ans.**

**3. Find the magnitude of each of the two vectors $\vec{a}$ and $\vec{b}$, having the same magnitude such that the angle between them is 60° and their scalar product is $\frac{9}{2}$.** **[1]**

**Sol.** Let $\vec{a}$ and $\vec{b}$ be two such vectors.

Now, $\vec{a}\cdot\vec{b} = |\vec{a}||\vec{b}|\cos\theta$ ...(i)

It is given that $|\vec{a}| = |\vec{b}|, \theta = 60°$ and $\vec{a}\cdot\vec{b} = \frac{9}{2}$

$\therefore$ Putting these values in equation (i)

$$\frac{9}{2} = |\vec{a}||\vec{a}|\cos 60°$$

$\Rightarrow$ $\frac{9}{2} = |\vec{a}|^2 \cdot \frac{1}{2}$

$\Rightarrow$ $|\vec{a}|^2 = 9$

$\Rightarrow$ $|\vec{a}| = 3$

$\therefore$ $|\vec{a}| = |\vec{b}| = 3$ **Ans.**

**4. If $a * b$ denotes the larger of 'a' and 'b' and if $a\ o\ b = (a * b) + 3$, then write the value of (5) $o$ (10), where $*$ and $o$ are binary operations.**** **[1]**

## SECTION – B

**5. Prove that :**

**$3\sin^{-1}x = \sin^{-1}(3x - 4x^3), x \in \left[-\frac{1}{2}, \frac{1}{2}\right]$** **[2]**

**Sol.** R.H.S. $= \sin^{-1}(3x - 4x^3)$

Putting $x = \sin\theta$ in R.H.S, we get

R.H.S. $= \sin^{-1}(3\sin\theta - 4\sin^3\theta)$

$= \sin^{-1}(\sin 3\theta)$

Now, $-\frac{1}{2} \le x \le \frac{1}{2}$

$\Rightarrow$ $-\frac{1}{2} \le \sin\theta \le \frac{1}{2}$

$\Rightarrow$ $-\frac{\pi}{6} \le \theta \le \frac{\pi}{6}$

$\Rightarrow$ $-\frac{\pi}{2} \le 3\theta \le \frac{\pi}{2}$

Hence, $\sin^{-1}(\sin 3\theta) = 3\theta$

$\Rightarrow$ R.H.S. $= 3\theta$

$= 3\sin^{-1}x$

$[\because x = \sin\theta \Rightarrow \theta = \sin^{-1}x]$

= L.H.S. **Hence Proved.**

** Answer is not given due to change in the present syllabus.

**6. Given $A = \begin{bmatrix} 2 & -3 \\ -4 & 7 \end{bmatrix}$, compute $A^{-1}$ and show that $2A^{-1} = 9I - A$.** **[2]**

**Sol.** Given, $A = \begin{bmatrix} 2 & -3 \\ -4 & 7 \end{bmatrix}$

$\Rightarrow$ $|A| = 14 - 12 = 2$

$\therefore$ $A^{-1} = \frac{1}{2}\begin{bmatrix} 7 & 3 \\ 4 & 2 \end{bmatrix}$

Given, L.H.S. $= 2A^{-1} = 2\left(\frac{1}{2}\right)\begin{bmatrix} 7 & 3 \\ 4 & 2 \end{bmatrix}$

$\Rightarrow$ L.H.S. $= \begin{bmatrix} 7 & 3 \\ 4 & 2 \end{bmatrix}$ ...(i)

and R.H.S. $= 9I - A = 9\begin{bmatrix} 1 & 0 \\ 0 & 1 \end{bmatrix} - \begin{bmatrix} 2 & -3 \\ -4 & 7 \end{bmatrix}$

$= \begin{bmatrix} 9 & 0 \\ 0 & 9 \end{bmatrix} - \begin{bmatrix} 2 & -3 \\ -4 & 7 \end{bmatrix}$

$\Rightarrow$ R.H.S. $= \begin{bmatrix} 7 & 3 \\ 4 & 2 \end{bmatrix}$ ...(ii)

From equations (i) and (ii),

L.H.S. = R.H.S. **Hence Proved.**

**7. Differentiate $\tan^{-1}\left(\frac{1+\cos x}{\sin x}\right)$ with respect to $x$.** **[2]**

**Sol.** Let $y = \tan^{-1}\left(\frac{1+\cos x}{\sin x}\right)$

$$= \tan^{-1}\left(\frac{2\cos^2\frac{x}{2}}{2\sin\frac{x}{2}\cos\frac{x}{2}}\right)$$

$$= \tan^{-1}\left(\cot\frac{x}{2}\right)$$

$$= \tan^{-1}\left[\tan\left(\frac{\pi}{2} - \frac{x}{2}\right)\right]$$

$$= \frac{\pi}{2} - \frac{x}{2}$$

Differentiating w.r.t. $x$, we get

$$\frac{dy}{dx} = \frac{d}{dx}\left(\frac{\pi}{2} - \frac{x}{2}\right) = -\frac{1}{2}$$ **Ans.**

**8. The total cost $C(x)$ associated with the production of $x$ units of an item is given by $C(x) = 0.005x^3 - 0.02x^2 + 30x + 5000$. Find the marginal cost when 3 units are produced, where by marginal cost we mean the instantaneous rate of change of total cost at any level of output.** **[2]**

**Sol.** Cost function is given as

$$C(x) = 0.005x^3 - 0.02x^2 + 30x + 5000$$

Marginal cost (MC) $= \frac{d}{dx}(C(x))$

$= 0.005(3x^2) - 0.02\,(2x) + 30$

$= 0.015x^2 - 0.04x + 30$

When $x = 3$, MC $= 0.015(3)^2 - 0.04(3) + 30$

$= 0.135 - 0.12 + 30$

$= 30.015$ **Ans.**

**9. Evaluate : $\int \frac{\cos 2x + 2\sin^2 x}{\cos^2 x} dx$** **[2]**

**Sol.** We have,

$$\int \frac{\cos 2x + 2\sin^2 x}{\cos^2 x} dx = \int \frac{\cos^2 x - \sin^2 x + 2\sin^2 x}{\cos^2 x} dx$$

$$= \int \frac{\cos^2 x + \sin^2 x}{\cos^2 x} dx$$

$$= \int \frac{1}{\cos^2 x} dx$$

$$= \int \sec^2 x\, dx$$

$$= \tan x + C$$ **Ans.**

**10. Find the differential equation representing the family of curves $y = ae^{bx+5}$, where $a$ and $b$ are arbitrary constants.** **[2]**

**Sol.** Given curve is

$y = ae^{bx+5}$ ...(i)

Differentiating (i) w.r.t. $x$, we get

$$\frac{dy}{dx} = ae^{bx+5}\cdot\frac{d}{dx}(bx+5)$$

$\Rightarrow$ $\frac{dy}{dx} = ae^{bx+5}\cdot b$ ...(ii)

$\Rightarrow$ $\frac{dy}{dx} = y\cdot b$ [From (i)]

Differentiating again w.r.t. $x$, we get

$$\frac{d^2y}{dx^2} = \frac{dy}{dx}\cdot b$$

$\Rightarrow$ $\frac{d^2y}{dx^2} = \frac{dy}{dx}\left(\frac{dy}{dx}\cdot\frac{1}{y}\right)$

$\left[\because b = \frac{1}{y}\frac{dy}{dx}\right]$

$\Rightarrow$ $y\frac{d^2y}{dx^2} = \left(\frac{dy}{dx}\right)^2$ **Ans.**

**11. If θ is the angle between two vectors $\hat{i}-2\hat{j}+3\hat{k}$ and $3\hat{i}-2\hat{j}+\hat{k}$, find sin θ.** **[2]**

**Sol.** Let

$$\vec{a} = \hat{i}-2\hat{j}+3\hat{k}$$

$$\vec{b} = 3\hat{i}-2\hat{j}+\hat{k}$$

We know, $\vec{a}\cdot\vec{b} = |\vec{a}||\vec{b}|\cos\theta$

$$\Rightarrow \cos\theta = \frac{\vec{a}\cdot\vec{b}}{|\vec{a}||\vec{b}|}$$

$$= \frac{(\hat{i}-2\hat{j}+3\hat{k})\cdot(3\hat{i}-2\hat{j}+\hat{k})}{|\hat{i}-2\hat{j}+3\hat{k}||3\hat{i}-2\hat{j}+\hat{k}|}$$

$$= \frac{1(3)+(-2)(-2)+3(1)}{\sqrt{1+4+9}\sqrt{9+4+1}}$$

$$= \frac{3+4+3}{14} = \frac{10}{14} = \frac{5}{7}$$

Now, $\sin\theta = \sqrt{1-\cos^2\theta}$

$$= \sqrt{1-\left(\frac{5}{7}\right)^2} = \sqrt{1-\frac{25}{49}}$$

$$= \sqrt{\frac{49-25}{49}}$$

$$\Rightarrow \sin\theta = \sqrt{\frac{24}{49}} = \frac{2\sqrt{6}}{7}$$ **Ans.**

**12. A black and a red die are rolled together. Find the conditional probability of obtaining the sum 8, given that the red die resulted in a number less than 4.** **[2]**

**Sol.** The sample space has 36 Outcomes. Let A be event that the sum of observations is 8.

$$\therefore A = \{(2,6),(3,5),(5,3),(4,4),(6,2)\}$$

$$\Rightarrow n(A) = 5$$

$$\Rightarrow P(A) = \frac{5}{36}$$

Let B be event that observation on red die is less than 4.

$\therefore$ B = {(1, 1), (2, 1), (3, 1), (4, 1), (5, 1), (6, 1), (1, 2), (2, 2), (3, 2), (4, 2), (5, 2), (6, 2), (1, 3), (2, 3), (3, 3), (4, 3), (5, 3), (6, 3)}

$$\therefore n(B) = 18$$

$$\Rightarrow P(B) = \frac{18}{36} = \frac{1}{2}$$

Clearly, $A \cap B = \{(5,3),(6,2)\}$

$$\therefore n(A\cap B) = 2$$

$$\therefore P(A\cap B) = \frac{2}{36} = \frac{1}{18}$$

$$P\left(\frac{A}{B}\right) = \frac{P(A\cap B)}{P(B)} = \frac{\frac{1}{18}}{\frac{1}{2}}$$

$$= \frac{2}{18} = \frac{1}{9}$$ **Ans.**

## SECTION–C

**13. Using properties of determinants, prove that :**

$$\begin{vmatrix} 1 & 1 & 1+3x \\ 1+3y & 1 & 1 \\ 1 & 1+3z & 1 \end{vmatrix} = 9(3xyz + xy + yz + zx)$$

**[4]**

**Sol.**

$$\text{L.H.S.} = \begin{vmatrix} 1 & 1 & 1+3x \\ 1+3y & 1 & 1 \\ 1 & 1+3z & 1 \end{vmatrix}$$

Taking $x$, $y$ and $z$ common from $R_1$, $R_2$ and $R_3$ respectively, we get

$$= xyz\begin{vmatrix} \frac{1}{x} & \frac{1}{x} & \frac{1}{x}+3 \\ \frac{1}{y}+3 & \frac{1}{y} & \frac{1}{y} \\ \frac{1}{z} & \frac{1}{z}+3 & \frac{1}{z} \end{vmatrix}$$

On applying $R_1 \to R_1 + R_2 + R_3$, we get

$$= xyz\begin{vmatrix} \frac{1}{x}+\frac{1}{y}+\frac{1}{z}+3 & \frac{1}{x}+\frac{1}{y}+\frac{1}{z}+3 & \frac{1}{x}+\frac{1}{y}+\frac{1}{z}+3 \\ \frac{1}{y}+3 & \frac{1}{y} & \frac{1}{y} \\ \frac{1}{z} & \frac{1}{z}+3 & \frac{1}{z} \end{vmatrix}$$

Taking common $\left(\frac{1}{x}+\frac{1}{y}+\frac{1}{z}+3\right)$ from $R_1$, we get

$$= xyz\left(\frac{1}{x}+\frac{1}{y}+\frac{1}{z}+3\right)\begin{vmatrix} 1 & 1 & 1 \\ \frac{1}{y}+3 & \frac{1}{y} & \frac{1}{y} \\ \frac{1}{z} & \frac{1}{z}+3 & \frac{1}{z} \end{vmatrix}$$

On applying $C_2 \to C_2 - C_1$ and $C_3 \to C_3 - C_1$, we get

$$= xyz\left(\frac{1}{x}+\frac{1}{y}+\frac{1}{z}+3\right)\begin{vmatrix} 1 & 0 & 0 \\ \frac{1}{y}+3 & -3 & -3 \\ \frac{1}{z} & 3 & 0 \end{vmatrix}$$

On expanding along $R_1$, we get

$$= xyz\left(\frac{1}{x}+\frac{1}{y}+\frac{1}{z}+3\right) 1\,(0+9)$$

$$= xyz\left(\frac{yz+xz+xy+3xyz}{xyz}\right)(9)$$

$$= 9\,(3xyz + xy + yz + zx)$$

$=$ R.H.S. **Hence Proved.**

**14. If $(x^2 + y^2)^2 = xy$, find $\frac{dy}{dx}$.** **[4]**

**Sol.** We have,

$$(x^2 + y^2)^2 = xy \quad \text{...(i)}$$

Differentiating (i) w.r.t. $x$, we get

$$2(x^2+y^2)\left(2x+2y.\frac{dy}{dx}\right) = x\frac{dy}{dx}+y$$

$$\Rightarrow 4x\,(x^2+y^2) + 4y\frac{dy}{dx}(x^2+y^2) = x\frac{dy}{dx}+y$$

$$\Rightarrow \frac{dy}{dx}\,[4y\,(x^2+y^2) - x] = y - 4x\,(x^2+y^2)$$

$$\Rightarrow \frac{dy}{dx} = \frac{y-4x(x^2+y^2)}{4y(x^2+y^2)-x}$$

$$\Rightarrow \frac{dy}{dx} = \frac{y-4x^3-4xy^2}{4x^2y+4y^3-x}$$

**Ans.**

**OR**

**If $x = a\,(2\theta - \sin 2\theta)$ and $y = a\,(1 - \cos 2\theta)$, find $\frac{dy}{dx}$ when $\theta = \frac{\pi}{3}$.**

**Sol.** Given, $x = a\,(2\theta - \sin 2\theta)$
and $y = a\,(1 - \cos 2\theta)$
Differentiating $x$ and $y$ w.r.t. $\theta$, we get

$$\frac{dx}{d\theta} = a(2 - \cos 2\theta\cdot 2)$$

$$= 2a(1 - \cos 2\theta)$$

and

$$\frac{dy}{d\theta} = a\,(\sin 2\theta\cdot 2)$$

$$= 2a\sin 2\theta$$

$$\therefore \quad \frac{dy}{dx} = \frac{\frac{dy}{d\theta}}{\frac{dx}{d\theta}} = \frac{2a\sin 2\theta}{2a(1-\cos 2\theta)}$$

$$= \frac{\sin 2\theta}{1-\cos 2\theta}$$

$$= \frac{2\sin\theta\cos\theta}{2\sin^2\theta}$$

$$= \cot\theta$$

$$\therefore \quad \left[\frac{dy}{dx}\right]_{\theta=\frac{\pi}{3}} = \cot\frac{\pi}{3} = \frac{1}{\sqrt{3}}$$ **Ans.**

**15. If $y = \sin(\sin x)$, prove that :**

$$\frac{d^2y}{dx^2} + \tan x\frac{dy}{dx} + y\cos^2 x = 0$$ **[4]**

**Sol.** Given, $y = \sin(\sin x)$
Differentiating $y$ w.r.t. $x$, we get

$$\frac{dy}{dx} = \cos(\sin x)\cdot\cos x$$

Differentiating again w.r.t. $x$, we get

$$\frac{d^2y}{dx^2} = \cos(\sin x)(-\sin x) + \cos x\,[-\sin(\sin x)\cdot\cos x]$$

$$= -\sin x.\cos(\sin x) - \cos^2 x\cdot\sin(\sin x)$$

Now, L.H.S. $= \frac{d^2y}{dx^2} + \tan x\frac{dy}{dx} + y\cos^2 x$

$$= -\sin x\cdot\cos(\sin x) - \cos^2 x\cdot\sin(\sin x) + \tan x\,[\cos(\sin x)\cdot\cos x] + \cos^2 x\cdot\sin(\sin x)$$

$$= -\sin x\cos(\sin x) + \frac{\sin x}{\cos x}[\cos(\sin x)\cdot\cos x]$$

$$= -\sin x\cdot\cos(\sin x) + \sin x\cdot\cos(\sin x)$$

$= 0 =$ R.H.S. **Hence Proved.**

**16. Find the equations of the tangent and the normal to the curve $16x^2 + 9y^2 = 145$ at the point $(x_1, y_1)$, where $x_1 = 2$ and $y_1 > 0$.** **[4]**

**Sol.** Given curve is

$$16x^2 + 9y^2 = 145 \quad \text{....(i)}$$

Since $(x_1, y_1)$ lies on equation (i),

$$\therefore \quad 16x_1^2 + 9y_1^2 = 145$$

$$\Rightarrow \quad 16\,(2)^2 + 9y_1^2 = 145 \quad [\because x_1 = 2 \text{ (given)}]$$

$$\Rightarrow \quad 9y_1^2 = 145 - 64 = 81$$

$$\Rightarrow \quad y_1 = 3 \quad [\because y_1 > 0 \text{ (given)}]$$

$\therefore$ Point of contact is (2, 3).

Differentiating equation (i) w.r.t. $x$,

$$32x + 18y\frac{dy}{dx} = 0$$

$$\Rightarrow \quad \frac{dy}{dx} = -\frac{32x}{18y}$$

Slope of tangent at (2, 3) $= \left[\frac{dy}{dx}\right]_{(2,3)}$

$$= -\frac{32(2)}{18(3)} = -\frac{64}{54}$$

$$= -\frac{32}{27}$$

$\therefore$ Equation of tangent is,

$$y - 3 = m(x - 2)$$

$$\Rightarrow \quad y-3 = -\frac{32}{27}(x-2)$$

$$\Rightarrow \quad 27y - 81 = -32x + 64$$

$$\Rightarrow \quad 32x + 27y = 145$$

The slope of the normal $= \dfrac{-1}{\text{Slope of tangent}}$

$$= \frac{27}{32}$$

$\therefore$ Equation of normal is

$$\Rightarrow \quad y-3 = \frac{27}{32}(x-2)$$

$$\Rightarrow \quad 32y - 96 = 27x - 54$$

$$\Rightarrow \quad 27x - 32y = -42$$

**Ans.**

**OR**

**Find the intervals in which the function $f(x) = \frac{x^4}{4} - x^3 - 5x^2 + 24x + 12$ is**

**(a) strictly increasing,**

**(b) strictly decreasing.**

**Sol.** Given function is

$$f(x) = \frac{x^4}{4} - x^3 - 5x^2 + 24x + 12$$

$$\Rightarrow \quad f'(x) = \frac{4x^3}{4} - 3x^2 - 10x + 24$$

$$\Rightarrow \quad f'(x) = x^3 - 3x^2 - 10x + 24$$

For critical points, put $f'(x) = 0$

$$\therefore \quad x^3 - 3x^2 - 10x + 24 = 0$$

$$\Rightarrow \quad (x-2)(x^2 - x - 12) = 0$$

$$\Rightarrow \quad (x-2)(x-4)(x+3) = 0$$

$$\Rightarrow \quad x = 2, 4, -3$$

Therefore, we have the intervals $(-\infty, -3)$, $(-3, 2)$, $(2, 4)$ and $(4, \infty)$

$-\infty$ ← − (−3) + (2) − (4) + → $\infty$

Since $f'(x) > 0$ in $(-3, 2) \cup (4, \infty)$.

$\therefore$ $f(x)$ is strictly increasing in $(-3, 2) \cup (4, \infty)$

and $f'(x) < 0$ in $(-\infty, -3) \cup (2, 4)$

$\therefore$ $f(x)$ is strictly decreasing in $(-\infty, -3) \cup (2, 4)$.

**Ans.**

**17. An open tank with a square base and vertical sides is to be constructed from a metal sheet so as to hold a given quantity of water. Show that the cost of material will be least when depth of the tank is half of its width. If the cost is to be borne by nearby settled lower income families, for whom water will be provided, what kind of value is hidden in this question ? [4]**

**Sol.** Let the length, breadth and height of the open tank be $x$, $x$ and $y$ units respectively.

Then, Volume (V) $= x^2y$ ...(i)

Total surface area (S) $= x^2 + 4xy$

$$S = x^2 + 4x \cdot \frac{V}{x^2} \quad \text{[using (i)]}$$

$$\Rightarrow \quad S = x^2 + \frac{4V}{x}$$

Differentiating w.r.t. x,

$$\Rightarrow \quad \frac{dS}{dx} = 2x - \frac{4V}{x^2} \quad \text{...(ii)}$$

For critical points, put $\frac{dS}{dx} = 0$

$$\Rightarrow \quad 2x - \frac{4V}{x^2} = 0$$

$$\Rightarrow \quad 2x^3 = 4V$$

$$\Rightarrow \quad 2x^3 = 4x^2y \quad \text{[using (i)]}$$

$$\Rightarrow \quad x = 2y \quad \text{...(iii)}$$

Now, differentiating (ii) w.r.t. $x$,

$$\frac{d^2S}{dx^2} = 2 + \frac{8V}{x^3}$$

$$= 2 + \frac{8V}{8y^3} \quad \text{[using (iii)]}$$

$$= 2 + \frac{V}{y^3} > 0$$

Area is minimum, thus cost is minimum when $x = 2y$, *i.e.*, depth of tank is half of the width.

Value : Any relevant value. **Ans.**

**18. Find $\int \frac{2\cos x}{(1-\sin x)(1+\sin^2 x)} dx$ [4]**

**Sol.** Let $I = \int \frac{2\cos x}{(1-\sin x)(1+\sin^2 x)} dx$

Put $\sin x = t$

$$\Rightarrow \quad \cos x \, dx = dt$$

$$\therefore \quad I = \int \frac{2dt}{(1-t)(1+t^2)}$$

Let $\dfrac{2}{(1-t)(1+t^2)} = \dfrac{A}{1-t} + \dfrac{Bt+C}{1+t^2}$

$$2 = A(1+t^2) + (Bt + C)(1-t)$$

$$\Rightarrow \quad 2 = (A-B)t^2 + (B-C)t + A + C$$

Put $t = 1$, we get

$$2 = 2A$$

$$\Rightarrow \quad A = 1$$

Comparing coefficients of $t^2$ and $t$ on both sides,

$$A - B = 0$$

$$\Rightarrow \quad B = A \Rightarrow B = 1$$

Also, $B - C = 0$

$\Rightarrow \quad B = C = 1$

$$\therefore \quad I = \int\left(\frac{1}{1-t}+\frac{t+1}{t^2+1}\right)dt$$

$$= \int\frac{1}{1-t}dt+\frac{1}{2}\int\frac{2t}{t^2+1}dt+\int\frac{1}{t^2+1}dt$$

$$= -\log|1-t|+\frac{1}{2}\log|t^2+1|+\tan^{-1}t+C$$

$$= -\log(1-\sin x)+\frac{1}{2}\log(\sin^2 x+1) + \tan^{-1}(\sin x)+C$$

**Ans.**

**19. Find the particular solution of the differential equation $e^x \tan y\, dx + (2 - e^x) \sec^2 y\, dy = 0$, given that $y = \frac{\pi}{4}$ when $x = 0$.** **[4]**

**Sol.** Given differential equation is,

$$e^x \tan y\, dx + (2 - e^x) \sec^2 y\, dy = 0$$

$$\Rightarrow \quad e^x \tan y\, dx = (e^x - 2) \sec^2 y\, dy$$

$$\Rightarrow \quad \frac{e^x}{e^x-2}dx = \frac{\sec^2 y}{\tan y}dy$$

Integrating both sides, we get

$$\int\frac{e^x}{e^x-2}dx = \int\frac{\sec^2 y}{\tan y}dy$$

$$\Rightarrow \quad \log|e^x-2| = \log|\tan y| + \log C$$

$$\Rightarrow \quad \log|e^x-2| = \log|C\tan y|$$

$$\Rightarrow \quad e^x - 2 = C\tan y \qquad \text{...(i)}$$

Given, $y = \frac{\pi}{4}$ when $x = 0$

$$\therefore \quad e^0 - 2 = C\left(\tan\frac{\pi}{4}\right)$$

$$\Rightarrow \quad -1 = C$$

$$\Rightarrow \quad C = -1$$

$\therefore$ From equation (i),

$$e^x - 2 = -\tan y \Rightarrow y = \tan^{-1}(2 - e^x)$$

which is the required solution. **Ans.**

**OR**

**Find the particular solution of the differential equation $\frac{dy}{dx} + 2y \tan x = \sin x$, given than $y = 0$ when $x = \frac{\pi}{3}$.**

**Sol.** Given differential equation is,

$$\frac{dy}{dx}+(2\tan x)y = \sin x \qquad \text{....(i)}$$

which is linear differential equation of the form

$$\frac{dy}{dx}+Py = Q$$

where $P = 2\tan x$, $Q = \sin x$

Integrating factor,

$$\text{I.F.} = e^{\int P.dx} = e^{\int 2\tan x.dx}$$

$$= e^{2\log\sec x} = \sec^2 x$$

Solution of equation (i) is given by

$$y.\text{I.F.} = \int \text{Q. I. F.}\,dx + C$$

$$\Rightarrow \quad y\sec^2 x = \int \sin x\cdot\sec^2 x\cdot dx + C$$

$$\Rightarrow \quad y\sec^2 x = \int \sec x\cdot\tan x\cdot dx + C$$

$$\Rightarrow \quad y\sec^2 x = \sec x + C$$

Put $y = 0$ when $x = \frac{\pi}{3}$ (given)

$$= \sec\frac{\pi}{3} + C$$

$$\Rightarrow \quad 0 = 2 + C$$

$$\Rightarrow \quad C = -2$$

Hence, particular solution is

$$y\sec^2 x = \sec x - 2$$

or $\quad y = \cos x - 2\cos^2 x$ **Ans.**

**20. Let $\vec{a} = 4\hat{i}+5\hat{j}-\hat{k}$, $\vec{b} = \hat{i}-4\hat{j}+5\hat{k}$ and $\vec{c} = 3\hat{i}+\hat{j}-\hat{k}$. Find a vector $\vec{d}$ which is perpendicular to both $\vec{c}$ & $\vec{b}$ and $\vec{d}\cdot\vec{a} = 21$.** **[4]**

**Sol.** Given,

$$\vec{a} = 4\hat{i}+5\hat{j}-\hat{k},\ \vec{b} = \hat{i}-4\hat{j}+5\hat{k},\ \vec{c} = 3\hat{i}+\hat{j}-\hat{k}$$

Let, $\quad \vec{d} = x\hat{i}+y\hat{j}+z\hat{k}$

$\because \quad \vec{d}$ is perpendicular to $\vec{c}$ and $\vec{b}$

$$\therefore \quad \vec{d}.\vec{b} = 0$$

$$\Rightarrow \quad x - 4y + 5z = 0 \qquad \text{...(i)}$$

and $\quad \vec{d}\cdot\vec{c} = 0$

$$\Rightarrow \quad 3x + y - z = 0 \qquad \text{...(ii)}$$

Also $\quad \vec{d}.\vec{a} = 21 \qquad$ (given)

$$\Rightarrow \quad 4x + 5y - z = 21 \qquad \text{...(iii)}$$

On subtracting equation (ii) from equation (iii), we get

$$x + 4y = 21 \qquad \text{...(iv)}$$

On multiplying equation (iii) by 5 and adding equation (i), we get

$$21x + 21y = 105$$

$$\Rightarrow \quad x + y = 5 \qquad \text{....(v)}$$

On subtracting equation (v) from equation (iv), we get

$$3y = 16$$

$$\Rightarrow \quad y = \frac{16}{3}$$

From equation (v),

$$x = 5 - \frac{16}{3} = -\frac{1}{3}$$

From equation (ii),

$$z = 3x + y$$

$$= 3\left(-\frac{1}{3}\right) + \frac{16}{3} = \frac{13}{3}$$

$$\therefore \quad \vec{d} = \frac{-1}{3}\hat{i} + \frac{16}{3}\hat{j} + \frac{13}{3}\hat{k} \quad \textbf{Ans.}$$

**21. Find the shortest distance between the lines $\vec{r} = (4\hat{i} - \hat{j}) + \lambda(\hat{i} + 2\hat{j} - 3\hat{k})$ and $\vec{r} = (\hat{i} - \hat{j} + 2\hat{k}) + \mu(2\hat{i} + 4\hat{j} - 5\hat{k})$. [4]**

**Sol.** Given lines are

$$\vec{r} = (4\hat{i} - \hat{j}) + \lambda(\hat{i} + 2\hat{j} - 3\hat{k}) \quad \text{...(i)}$$

and $\vec{r} = (\hat{i} - \hat{j} + 2\hat{k}) + \mu(2\hat{i} + 4\hat{j} - 5\hat{k})$, ...(ii)

Comparing equation (i) with $\vec{r} = \vec{a_1} + \lambda\vec{b_1}$ and equation (ii) with $\vec{r} = \vec{a_2} + \mu\vec{b_2}$, we get

$$\vec{a_1} = 4\hat{i} - \hat{j}$$

$$\vec{a_2} = \hat{i} - \hat{j} + 2\hat{k}$$

$$\vec{b_1} = \hat{i} + 2\hat{j} - 3\hat{k}$$

$$\vec{b_2} = 2\hat{i} + 4\hat{j} - 5\hat{k}$$

Shortest distance between equation (i) and (ii) is given by

$$\text{S.D.} = \left|\frac{(\vec{a_2} - \vec{a_1}).(\vec{b_1} \times \vec{b_2})}{|\vec{b_1} \times \vec{b_2}|}\right|$$

$$\vec{b_1} \times \vec{b_2} = \begin{vmatrix} \hat{i} & \hat{j} & \hat{k} \\ 1 & 2 & -3 \\ 2 & 4 & -5 \end{vmatrix}$$

$$\Rightarrow \quad \vec{b_1} \times \vec{b_2} = \hat{i}(-10+12) - \hat{j}(-5+6) + \hat{k}(4-4)$$

$$= 2\hat{i} - \hat{j}.$$

$$\Rightarrow |\vec{b_1} \times \vec{b_2}| = \sqrt{4+1} = \sqrt{5}$$

$$\Rightarrow \quad \text{S.D.} = \left|\frac{(-3\hat{i} + 0\hat{j} + 2\hat{k})\cdot(2\hat{i} - \hat{j} + 0\hat{k})}{\sqrt{4+1}}\right|$$

$$= \frac{6}{\sqrt{5}} \text{ or } \frac{6\sqrt{5}}{5} \text{ units} \quad \textbf{Ans.}$$

**22. Suppose a girl throws a die. If she gets 1 or 2, she tosses a coin three times and notes the number of tails. If she gets 3, 4, 5 or 6, she tosses a coin once and notes whether a 'head' or 'tail' is obtained. If she obtained exactly one 'tail', what is the probability that she threw 3, 4, 5 or 6 with the dice? [4]**

**Sol.** Let $E_1$ be the event that girl gets 1 or 2 on the roll and $E_2$ be the event that girl gets 3, 4, 5, or 6 on the roll of a die.

$$\therefore \quad P(E_1) = \frac{2}{6} = \frac{1}{3}$$

$$P(E_2) = \frac{4}{6} = \frac{2}{3}$$

Let A be the event that she gets exactly one tail.

If she tossed coin 3 times and gets exactly one tail then possible outcomes are HTH, HHT, THH

$$\therefore \quad P(A/E_1) = \frac{3}{8}$$

If she tossed coin only once and exactly one tail shows

Then, $P(A/E_2) = \dfrac{1}{2}$

$$\therefore \quad P(E_2/A) = \frac{P(E_2)\cdot P(A/E_2)}{P(E_1)\cdot P(A/E_1) + P(E_2)\cdot P(A/E_2)}$$

$$= \frac{\frac{2}{3} \times \frac{1}{2}}{\frac{1}{3} \times \frac{3}{8} + \frac{2}{3} \times \frac{1}{2}} = \frac{\frac{1}{3}}{\frac{1}{8} + \frac{1}{3}}$$

$$= \frac{\frac{1}{3}}{\frac{11}{24}} = \frac{8}{11} \quad \textbf{Ans.}$$

**23. Two numbers are selected at random (without replacement) from the first five positive integers. Let X denote the larger of the two numbers obtained. Find the mean and variance of X. [4]**

**Sol.** First five positive integers are 1, 2, 3, 4, 5. We select two positive numbers in 5 × 4 = 20 ways. Out of these, two numbers are selected at random. Let X denote larger of the two selected numbers. Then, X can have values 2, 3, 4 or 5.

$$P(X = 2) = P(\text{larger no. is 2}) = \{(1, 2) \text{ and } (2, 1)\}$$

$$= \frac{2}{20}$$

$$P(X = 3) = \frac{4}{20}$$

$$P(X = 4) = \frac{6}{20}$$

$$P(X = 5) = \frac{8}{20}$$

Thus, the probability distribution of X is

| X | 2 | 3 | 4 | 5 |
|---|---|---|---|---|
| P(X) | $\frac{2}{20}$ | $\frac{4}{20}$ | $\frac{6}{20}$ | $\frac{8}{20}$ |

$$\therefore \quad \text{Mean} = E(X) = \sum_{i=1}^{n} x_i P(x_i)$$

$$= 2\times\frac{2}{20}+3\times\frac{4}{20}+4\times\frac{6}{20}+5\times\frac{8}{20}$$

$$= \frac{4+12+24+40}{20} = \frac{80}{20} = 4$$

$$E(X^2) = \sum_{i=1}^{n} x_i^2 P(x_i)$$

$$= 2^2\times\frac{2}{20}+3^2\times\frac{4}{20}+4^2\times\frac{6}{20}+5^2\times\frac{8}{20}$$

$$= \frac{8}{20}+\frac{36}{20}+\frac{96}{20}+\frac{200}{20}$$

$$= \frac{8+36+96+200}{20} = \frac{340}{20} = \frac{34}{2} = 17$$

$$\therefore \text{Variance} = E(X^2) - [E(X)]^2$$

$$= 17 - (4)^2 = 1$$

Therefore, mean and variance are 4 and 1 respectively. **Ans.**

## SECTION–D

**24. Let A = $\{x \in Z : 0 \le x \le 12\}$. Show that R = $\{(a, b) : a, b \in A, |a - b|$ is divisible by 4$\}$ is an equivalence relation. Find the set of all elements related to 1. Also write the equivalence class [2].** **[6]**

**Sol.** Given, R = $\{(a, b) : a, b \in A, |a - b|$ is divisible by 4$\}$

**Reflexivity :** For any $a \in A$

$|a - a| = 0$, which is divisible by 4

$(a, a) \in R$

So, $R$ is reflexive.

**Symmetry :** Let $(a, b) \in R$

$\Rightarrow |a - b|$ is divisible by 4

$\Rightarrow |b - a|$ is divisible by 4 $[\because |a - b| = |b - a|]$

$\Rightarrow (b, a) \in R$

So, R is symmetric.

**Transitive :** Let $(a, b) \in R$ and $(b, c) \in R$

$\Rightarrow |a - b|$ is divisible by 4

$\Rightarrow \quad |a - b = 4k$

$\therefore \quad a - b = \pm 4k, k \in Z \quad$ ...(i)

Also, $|b - c|$ is divisible by 4

$\Rightarrow \quad |b - c = 4m$

$\therefore \quad b - c = \pm 4m, m \in Z \quad$ ...(ii)

Adding equations (i) and (ii)

$$a - b + b - c = \pm 4(k + m)$$

$\Rightarrow \quad a - c = \pm 4(k + m)$

$|a - c|$ is divisible by 4,

$\Rightarrow \quad (a, c) \in R$

So, R is transitive.

$\Rightarrow$ R is reflexive, symmetric and transitive.

$\therefore$ R is an equivalence relation.

Let $x$ be an element of R such that $(x, 1) \in R$

Then $|x - 1|$ is divisible by 4

$$x - 1 = 0, 4, 8, 12, .....$$

$\Rightarrow \quad x = 1, 5, 9 \quad (\because x \le 12)$

$\therefore$ Set of all elements of A which are related to 1 are $\{1, 5, 9\}$.

Equivalence class of 2, *i.e.*

$[2] = \{(a, 2) : a \in A, |a - 2|$ is divisible by 4$\}$

$\Rightarrow \quad |a - 2| = 4k$

($k$ is whole number, $k \le 3$)

$\Rightarrow \quad a = 2, 6, 10$

Therefore, equivalence class [2] is $\{2, 6, 10\}$. **Ans.**

**OR**

**Show that the function $f : R \to R$ defined by $f(x) = \frac{x}{x^2+1}, \forall x \in R$ is neither one-one nor onto.**

**Also, if $g : R \to R$ is defined as $g(x) = 2x - 1$, find $f o g(x)$.**

**Sol.** Given, $\quad f(x) = \frac{x}{x^2+1}, \forall x \in R$

For one–one, $\quad f(x) = f(y)$

$$\frac{x}{x^2+1} = \frac{y}{y^2+1}$$

$\Rightarrow \quad xy^2 + x = yx^2 + y$

$\Rightarrow \quad xy^2 - yx^2 = y - x$

$\Rightarrow \quad xy\,(y - x) = y - x$

$\Rightarrow \quad xy = 1$

$\Rightarrow \quad x = \frac{1}{y}$

Since $x \ne y$, therefore, $f(x)$ is not one-one.

For onto, $\quad f(x) = y$

$\Rightarrow \quad \frac{x}{x^2+1} = y$

$\Rightarrow \quad x = yx^2 + y$

$\Rightarrow \quad x^2y + y - x = 0$

$x$ cannot be expressed in terms of $y$

$\Rightarrow f(x)$ is not onto.

As $\quad g(x) = 2x - 1$

$$\therefore \quad fog\,(x) = f\,[g\,(x)] = f(2x-1)$$

$$= \frac{2x-1}{(2x-1)^2+1}$$

$$= \frac{2x-1}{4x^2-4x+2} \quad \textbf{Ans.}$$

**25. If** $A = \begin{bmatrix} 2 & -3 & 5 \\ 3 & 2 & -4 \\ 1 & 1 & -2 \end{bmatrix}$**, find** $A^{-1}$**. Use it to solve the system of equations :** **[6]**

$$2x - 3y + 5z = 11$$
$$3x + 2y - 4z = -5$$
$$x + y - 2z = -3$$

**Sol.** $$|A| = \begin{vmatrix} 2 & -3 & 5 \\ 3 & 2 & -4 \\ 1 & 1 & -2 \end{vmatrix}$$

$$= 2(-4+4) + 3(-6+4) + 5(3-2)$$

$$|A| = 0 - 6 + 5 = -1 \neq 0$$

$\therefore A^{-1}$ exists.

Let $A_{ij}$ be the cofactors of elements $a_{ij}$ in $A = [a_{ij}]$, then

$$A_{11} = (-1)^{1+1}\begin{vmatrix} 2 & -4 \\ 1 & -2 \end{vmatrix} = -4+4 = 0$$

$$A_{12} = (-1)^{1+2}\begin{vmatrix} 3 & -4 \\ 1 & -2 \end{vmatrix} = -(-6+4) = 2$$

$$A_{13} = (-1)^{1+3}\begin{vmatrix} 3 & 2 \\ 1 & 1 \end{vmatrix} = 3-2 = 1$$

$$A_{21} = (-1)^{2+1}\begin{vmatrix} -3 & 5 \\ 1 & -2 \end{vmatrix} = -(6-5) = -1$$

$$A_{22} = (-1)^{2+2}\begin{vmatrix} 2 & 5 \\ 1 & -2 \end{vmatrix} = -4-5 = -9$$

$$A_{23} = (-1)^{2+3}\begin{vmatrix} 2 & -3 \\ 1 & 1 \end{vmatrix} = -(2+3) = -5$$

$$A_{31} = (-1)^{3+1}\begin{vmatrix} -3 & 5 \\ 2 & -4 \end{vmatrix} = (12-10) = 2$$

$$A_{32} = (-1)^{3+2}\begin{vmatrix} 2 & 5 \\ 3 & -4 \end{vmatrix} = -(-8-15) = 23$$

$$A_{33} = (-1)^{3+3}\begin{vmatrix} 2 & -3 \\ 3 & 2 \end{vmatrix} = 4+9 = 13$$

$$\therefore \quad A_{ij} = \begin{bmatrix} 0 & 2 & 1 \\ -1 & -9 & -5 \\ 2 & 23 & 13 \end{bmatrix}$$

$$\text{adj}\,(A) = [A_{ij}]^T = \begin{bmatrix} 0 & -1 & 2 \\ 2 & -9 & 23 \\ 1 & -5 & 13 \end{bmatrix}$$

$$A^{-1} = \frac{\text{adj}\,(A)}{|A|} = \frac{1}{(-1)}\begin{bmatrix} 0 & -1 & 2 \\ 2 & -9 & 23 \\ 1 & -5 & 13 \end{bmatrix}$$

$$= \begin{bmatrix} 0 & 1 & -2 \\ -2 & 9 & -23 \\ -1 & 5 & -13 \end{bmatrix}$$

Now, given system of equations can be written as

$$\begin{bmatrix} 2 & -3 & 5 \\ 3 & 2 & -4 \\ 1 & 1 & -2 \end{bmatrix}\begin{bmatrix} x \\ y \\ z \end{bmatrix} = \begin{bmatrix} 11 \\ -5 \\ -3 \end{bmatrix}$$

$$\Rightarrow \quad AX = B$$

$$\therefore \quad X = A^{-1}B$$

$$\Rightarrow \quad X = \begin{bmatrix} 0 & 1 & -2 \\ -2 & 9 & -23 \\ -1 & 5 & -13 \end{bmatrix}\begin{bmatrix} 11 \\ -5 \\ -3 \end{bmatrix}$$

$$\Rightarrow \quad \begin{bmatrix} x \\ y \\ z \end{bmatrix} = \begin{bmatrix} 0-5+6 \\ -22-45+69 \\ -11-25+39 \end{bmatrix} = \begin{bmatrix} 1 \\ 2 \\ 3 \end{bmatrix}$$

$$\therefore \quad x = 1,\ y = 2,\ z = 3 \quad \textbf{Ans.}$$

**OR**

**Using elementary row transformations, find the inverse of the matrix:**

$$A = \begin{bmatrix} 1 & 2 & 3 \\ 2 & 5 & 7 \\ -2 & -4 & -5 \end{bmatrix}.$$

**Sol.** We know that, $A = I\,A$

$$\begin{bmatrix} 1 & 2 & 3 \\ 2 & 5 & 7 \\ -2 & -4 & -5 \end{bmatrix} = \begin{bmatrix} 1 & 0 & 0 \\ 0 & 1 & 0 \\ 0 & 0 & 1 \end{bmatrix}A$$

On applying, $R_2 \to R_2 - 2R_1$ and $R_3 \to R_3 + 2R_1$, we get

$$\begin{bmatrix} 1 & 2 & 3 \\ 0 & 1 & 1 \\ 0 & 0 & 1 \end{bmatrix} = \begin{bmatrix} 1 & 0 & 0 \\ -2 & 1 & 0 \\ 2 & 0 & 1 \end{bmatrix}A$$

On applying $R_1 \to R_1 - 3R_3$ and $R_2 \to R_2 - R_3$, we get

$$\begin{bmatrix} 1 & 2 & 0 \\ 0 & 1 & 0 \\ 0 & 0 & 1 \end{bmatrix} = \begin{bmatrix} -5 & 0 & -3 \\ -4 & 1 & -1 \\ 2 & 0 & 1 \end{bmatrix} A$$

On applying $R_1 \to R_1 - 2R_2$, we get

$$\begin{bmatrix} 1 & 0 & 0 \\ 0 & 1 & 0 \\ 0 & 0 & 1 \end{bmatrix} = \begin{bmatrix} 3 & -2 & -1 \\ -4 & 1 & -1 \\ 2 & 0 & 1 \end{bmatrix} A$$

$$\therefore \quad A^{-1} = \begin{bmatrix} 3 & -2 & -1 \\ -4 & 1 & -1 \\ 2 & 0 & 1 \end{bmatrix}$$ **Ans.**

**26. Using integration, find the area of the region in the first quadrant enclosed by the X-axis, the line $y = x$ and the circle $x^2 + y^2 = 32$.** **[6]**

**Sol.** Given curve is

$$x^2 + y^2 = 32 \quad \text{... (i)}$$

$$x^2 + y^2 = \left(\sqrt{32}\right)^2 = \left(4\sqrt{2}\right)^2$$

It is a circle with centre (0, 0) and radius $4\sqrt{2}$.

Given line is $y = x$ ....(ii)

Solving equations (i) and (ii) for points of intersections,

$$x^2 + x^2 = 32 \quad \text{[using (ii) in (i)]}$$

$$2x^2 = 32$$

$$\Rightarrow \quad x^2 = 16$$

$$\Rightarrow \quad x = \pm 4$$

$$\Rightarrow \quad x = 4 \quad \text{(first quadrant)}$$

$\therefore$ Point of intersection is (4, 4).

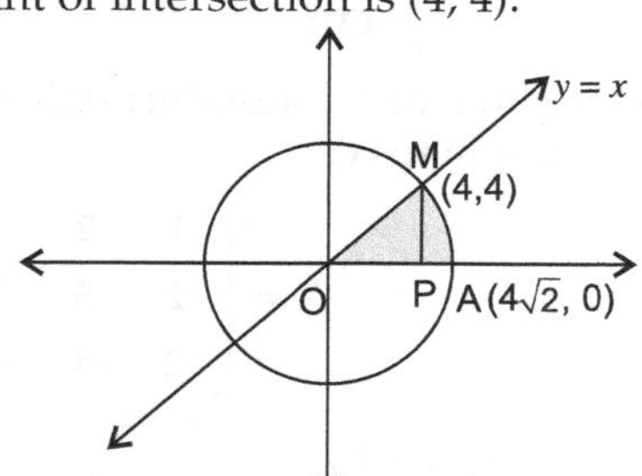


Required area = Area OMA

= Area OMP + Area MPA

$$= \int_0^4 x \cdot dx + \int_4^{4\sqrt{2}} \sqrt{(4\sqrt{2})^2 - x^2} \cdot dx$$

$$= \left[\frac{x^2}{2}\right]_0^4 + \left[\frac{x}{2}\sqrt{(4\sqrt{2})^2 - x^2} + \frac{(4\sqrt{2})^2}{2}\sin^{-1}\left(\frac{x}{4\sqrt{2}}\right)\right]_4^{4\sqrt{2}}$$

$$= \frac{16}{2} + \left[\left\{\frac{4\sqrt{2}}{2}\sqrt{(4\sqrt{2})^2 - (4\sqrt{2})^2} + \frac{32}{2}\sin^{-1}1\right\} - \left\{\frac{4}{2}\sqrt{(4\sqrt{2})^2 - (4)^2} + \frac{32}{2}\sin^{-1}\frac{1}{\sqrt{2}}\right\}\right]$$

$$= 8 + \left(2\sqrt{2}(0) + 16 \times \frac{\pi}{2}\right) - \left(2 \times 4 + 16 \times \frac{\pi}{4}\right)$$

$$= 8 + 8\pi - 8 - 4\pi$$

$$= 4\pi \text{ sq. units}$$ **Ans.**

**27. Evaluate :**

$$\int_0^{\pi/4} \frac{\sin x + \cos x}{16 + 9\sin 2x} dx$$ **[6]**

**Sol.**

Let $$I = \int_0^{\pi/4} \frac{\sin x + \cos x}{16 + 9\sin 2x} . dx$$

$$= \int_0^{\pi/4} \frac{\sin x + \cos x}{16 + 9[1 - (\sin x - \cos x)^2]} \cdot dx$$

$$\Rightarrow \quad = \int_0^{\pi/4} \frac{\sin x + \cos x}{25 - 9(\sin x - \cos x)^2} \cdot dx$$

Put $\sin x - \cos x = t$

$(\cos x + \sin x)\, dx = dt$

When $x = 0, t = -1$

and $x = \frac{\pi}{4}, t = 0$

$$I = \int_{-1}^0 \frac{dt}{25 - 9t^2} = \frac{1}{9}\int_{-1}^0 \frac{dt}{\frac{25}{9} - t^2}$$

$$= \frac{1}{9}\int_{-1}^0 \frac{dt}{\left(\frac{5}{3}\right)^2 - t^2}$$

$$= \frac{1}{9}\left[\frac{1}{2 \times \frac{5}{3}} \log\left|\frac{\frac{5}{3} + t}{\frac{5}{3} - t}\right|\right]_{-1}^0$$

$$= \frac{1}{9}\left[\frac{1}{\frac{10}{3}}\left(\log\left|\frac{\frac{5}{3}}{\frac{5}{3}}\right| - \log\left|\frac{\frac{2}{3}}{\frac{8}{3}}\right|\right)\right]$$

$$= \frac{1}{30}\left[\log 1 - \log\frac{1}{4}\right]$$

$$= \frac{1}{30}[\log 1 - (\log 1 - \log 4)]$$

$$= \frac{1}{30}\log 4 = \frac{1}{30}\log (2)^2 = \frac{2}{30}\log 2$$

$$= \frac{1}{15}\log 2$$ **Ans.**

OR

**Evaluate $\int_1^3 (x^2+3x+e^x)dx$ as the limit of the sum.**

**Sol.** We have, $\int_1^3 (x^2+3x+e^x).dx$

Here, $f(x) = x^2 + 3x + e^x$, $a = 1$, $b = 3$,

$nh = b - a = 3 - 1 = 2$

By limit of sum, we have

$$\int_a^b f(x)\,dx = \lim_{h\to 0} h[f(a)+f(a+h)+f(a+2h)+\ldots + f(a+(n-1)h)]$$

where $nh = b - a$

$$\therefore \int_1^3 (x^2+3x+e^x).dx$$

$$= \lim_{h\to 0} h[f(1)+f(1+h)+f(2+h)\ldots + f(1+\overline{n-1}h)]$$

$$= \lim_{h\to 0} h[(1+3+e)+\{(1+h)^2+3(1+h) + e^{1+h}\}+\{(1+2h)^2+3(1+2h)+e^{1+2h}\}+\ldots]$$

$$= \lim_{h\to 0} h(4+e+(1+h^2+2h+3+3h+e^{1+h}) + (1+4h^2+4h+3+6h+e^{1+2h})+\ldots]$$

$$= \lim_{h\to 0} h\left[4+e+(4+h^2+5h+e^{1+h}) + (4+4h^2+10h+e^{1+2h})+\ldots\right]$$

$$= \lim_{h\to 0} h[4n+e(1+e^h+e^{2h}+\ldots) + h^2(1^2+2^2+\ldots)+5h(1+2+\ldots)]$$

$$= \lim_{h\to 0} h\left[4n+e\left(\frac{e^{nh}-1}{e^h-1}\right)+h^2\frac{n(n-1)(2n-1)}{6} + \frac{5hn(n-1)}{2}\right]$$

$$= \lim_{h\to 0}\left[4nh+e\cdot\frac{h}{e^h-1}(e^{nh}-1) + \frac{nh(nh-h)(2nh-h)}{6}+\frac{5nh(nh-h)}{2}\right]$$

$$= 4(2)+e\cdot(e^2-1)+\frac{2(2-0)(4-0)}{6}+\frac{5(2)(2-0)}{2}$$

$$= 8+(e^3-e)+\frac{8}{3}+10$$

$$= \frac{24+8+30}{3}+e^3-e = \frac{62}{3}+e^3-e \qquad \textbf{Ans.}$$

**28. Find the distance of the point (–1, –5, –10) from the point of intersection of the line $\vec{r} = 2\hat{i}-\hat{j}+2\hat{k}+\lambda(3\hat{i}+4\hat{j}+2\hat{k})$ and the plane $\vec{r}.(\hat{i}-\hat{j}+\hat{k}) = 5$. [6]**

**Sol.** Equation of line is

$$\vec{r} = 2\hat{i}-\hat{j}+2\hat{k}+\lambda(3\hat{i}+4\hat{j}+2\hat{k}) \qquad \ldots(i)$$

Coordinates of any point on this line are

$$(2+3\lambda)\hat{i}+(-1+4\lambda)\hat{j}+(2+2\lambda)\hat{k}$$

Equation of plane is

$$\vec{r}.(\hat{i}-\hat{j}+\hat{k}) = 5 \qquad \ldots(ii)$$

Since, the point on line lies on the plane,

$$\vec{r}.(\hat{i}-\hat{j}+\hat{k}) = 5$$

$$[(2+3\lambda)\hat{i}+(-1+4\lambda)\hat{j}+(2+2\lambda)](\hat{i}-\hat{j}+\hat{k}) = 5$$

$$\Rightarrow (2+3\lambda)-(-1+4\lambda)+(2+2\lambda) = 5$$

$$\Rightarrow \lambda+5 = 5$$

$$\Rightarrow \lambda = 0$$

So equation of line is

$$\vec{r} = (2\hat{i}-\hat{j}+2\hat{k})+0(3\hat{i}+4\hat{j}+2\hat{k})$$

$$\Rightarrow \vec{r} = 2\hat{i}-\hat{j}+2\hat{k} \qquad \ldots(iii)$$

Let, point of intersection be $(x, y, z)$

$$\therefore \vec{r} = x\hat{i}+y\hat{j}+z\hat{k} \qquad \ldots(iv)$$

From equations (iii) and (iv)

$$x = 2, y = -1, z = 2$$

$\therefore$ Point of intersection is (2, –1, 2).

Distance between points (2, –1, 2) and (–1, –5, –10) is

$$= \sqrt{(-1-2)^2+(-5+1)^2+(-10-2)^2}$$

$$= \sqrt{(-3)^2+(-4)^2+(-12)^2}$$

$$= \sqrt{9+16+144} = \sqrt{169} = 13 \text{ units} \qquad \textbf{Ans.}$$

**29. A factory manufactures two types of screws A and B, each type requiring the use of two machines, an automatic and a hand-operated. It takes 4 minutes on the automatic and 6 minutes on the hand-operated machines to manufacture a packet of screws 'A' while it takes 6 minutes on the automatic and 3 minutes on the hand-operated machine to manufacture a packet of screws 'B'. Each machine is available for at most 4 hours on any day. The manufacturer can sell a packet of screws 'A' at a profit of 70 paise and screws 'B' at a profit of ₹1. Assuming that he can sell all the screws he manufactures, how many packets of each type should the factory owner produce in a day in order to maximize his profit ? Formulate the above LPP and solve it graphically and find the maximum profit. [6]**

**Sol.** Let the number of packets of screw 'A' manufactured in a day be $x$ and that of screw B be $y$.

Therefore, $x \geq 0$, $y \geq 0$

| Item | Number | Machine A | Machine B | Profit |
|---|---|---|---|---|
| Screw A | $x$ | 4 minutes | 6 minutes | ₹ 0.7 |
| Screw B | $y$ | 6 minutes | 3 minutes | ₹ 1 |
| Max. time available | | 4 hrs. = 240 min. | 4 hrs. = 240 min. | |

Then, the constraints are :

$$4x + 6y \leq 240 \text{ or } 2x + 3y \leq 120$$

$$6x + 3y \leq 240 \text{ or } 2x + y \leq 80$$

and total profit,

$$Z = 0.7\,x + y$$

So our LPP will be

$$\text{Max. } Z = 0.7\,x + y$$

Subject to the constraints :

$$2x + 3y \leq 120,$$

$$2x + y \leq 80,$$

and $x, y \geq 0$

Now, $2x + 3y \leq 120$ and $2x + y \leq 80$

| $x$ | 0 | 60 |
|---|---|---|
| $y$ | 40 | 0 |

| $x$ | 0 | 40 |
|---|---|---|
| $y$ | 80 | 0 |

Plotting the points on the graph, we get the feasible region OABC as shown (Shaded).

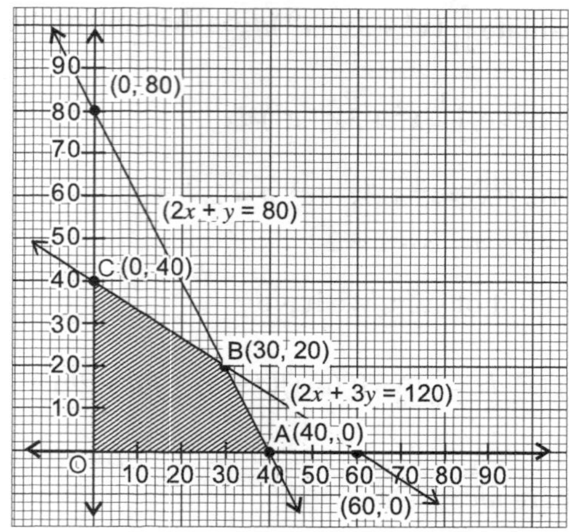


| Corner points | Value of Z = 0.7x + y |
|---|---|
| C(0, 40) | 0.7(0) + 40 = 40 |
| B (30, 20) | 0.7 (30) + 20 = 41 Maximum |
| A(40, 0) | 0.7(40) + 0 = 28 |
| O (0, 0) | 0.7 (0) + 0 = 0 |

Hence, profit will be maximum if company produces 30 packets of screw A and 20 packets of screw B and maximum profit = ₹ 41. **Ans.**

# MATHEMATICS

## Outside Delhi [Set-I]

## 2017

**Time allowed : 3 hours** **Maximum marks : 100**

### SECTION — A

**1. If for any $2 \times 2$ square matrix A,**

**$A(\text{adj } A) = \begin{bmatrix} 8 & 0 \\ 0 & 8 \end{bmatrix}$,**

**then write the value of $|A|$.** **[1]**

**Sol.** We have,

$$A(\text{adj } A) = \begin{bmatrix} 8 & 0 \\ 0 & 8 \end{bmatrix}$$

As, $$A(\text{adj } A) = |A|\, I$$

$$\begin{bmatrix} 8 & 0 \\ 0 & 8 \end{bmatrix} = 8\begin{bmatrix} 1 & 0 \\ 0 & 1 \end{bmatrix}$$

On comparing, we get

$$|A| = 8.$$ **Ans.**

**2. Determine the value of '$k$' for which the following function is continuous at $x = 3$.**

$$f(x) = \begin{cases} \dfrac{(x+3)^2 - 36}{x-3}, & x \neq 3 \\ k, & x = 3 \end{cases}$$ **[1]**

**Sol.** Given,

$$f(x) = \begin{cases} \dfrac{(x+3)^2 - 36}{x-3}, & x \neq 3 \\ k, & x = 3 \end{cases}$$

Since $f(x)$ is continuous at $x = 3$

$\therefore \quad \lim_{x \to 3} f(x) = f(3)$

$\Rightarrow \quad \lim_{x \to 3} f(x) = k$

$\Rightarrow \quad \lim_{x \to 3} \dfrac{(x+3)^2 - 36}{x-3} = k$

$\Rightarrow \quad \lim_{x+3 \to 6} \dfrac{(x+3)^2 - 6^2}{(x+3)-6} = k$

$\Rightarrow \quad \lim_{x \to 3} x + 3 + 6 = k$

$\Rightarrow \quad 12 = k$

Thus, $f(x)$ is continuous at $x = 3$; if $k = 12$. **Ans.**

**3. Find : $\int \dfrac{\sin^2 x - \cos^2 x}{\sin x \cos x} dx$.** **[1]**

**Sol.** We have,

$$\int \frac{\sin^2 x - \cos^2 x}{\sin x \cos x} dx = -2\int \frac{\cos^2 x - \sin^2 x}{2 \sin x \cos x} dx$$

$$= -2\int \frac{\cos 2x}{\sin 2x} dx$$

$$= -2\int \cot 2x \, dx$$

$$= -\log |\sin 2x| + C$$

**Ans.**

**4. Find the distance between the planes $2x - y + 2z = 5$ and $5x - 2.5y + 5z = 20$.** **[1]**

**Sol.** Since, the direction ratios of the normal to the given planes are proportional.

*i.e.,* $$\frac{2}{5} = \frac{-1}{-2.5} = \frac{2}{5}$$

Thus, the given planes are parallel.

Now, let $P(x_1, y_1, z_1)$ be any point on $2x - y + 2z - 5 = 0$

Then, $2x_1 - y_1 + 2z_1 - 5 = 0$

or$5x_1 - 2.5y_1 + 5z_1 - 12.5 = 0$ ...(i)

The length of the perpendicular from $P(x_1, y_1, z_1)$ to $5x - 2.5y + 5z - 20 = 0$,

$$= \left|\frac{5x_1 - 2.5y_1 + 5z_1 - 20}{\sqrt{(5)^2 + (-2.5)^2 + (5)^2}}\right|$$

$$= \left|\frac{12.5 - 20}{\sqrt{25 + 6.25 + 25}}\right| \quad \text{[From eq. (i)]}$$

$$= \left|\frac{-7.5}{7.5}\right| = 1 \text{ unit}$$

Therefore, the distance between the given planes is 1 unit. **Ans.**

### SECTION — B

**5. If A is a skew-symmetric matrix of order 3, then prove that det A = 0.** **[2]**

**Sol.** Given, A is a skew-symmetric matrix of order 3.

So, $$A^T = -A$$

Now, $|A^T| = |-A|$

$|A^T| = (-1)^3|A|$

[$\because |kA| = k^n|A|$ where $n$ is order of A]

$|A| = -|A|$ [$|A^T| = |A|$]

$\Rightarrow$ $|A| + |A| = 0$

$\therefore$ $2|A| = 0$ or $|A| = 0$.

*i.e.*, det A = 0 **Hence Proved.**

**6. Find the value of $c$ in Rolle's theorem for the function $f(x) = x^3 - 3x$ in $[-\sqrt{3}, 0]$. [2]**

**Sol.** We know that the polynomial function $f(x) = x^3 - 3x$ is everywhere continuous and differentiable.

So, $f(x)$ is continuous on $[-\sqrt{3}, 0]$.

Also, $f(x)$ is differentiable on $(-\sqrt{3}, 0)$

Now, $f(-\sqrt{3}) = (-\sqrt{3})^3 - 3(-\sqrt{3})$

$= -3\sqrt{3} + 3\sqrt{3} = 0$

and $f(0) = (0)^3 - 3(0) = 0$

$\therefore$ $f(-\sqrt{3}) = f(0)$

Thus, all the three conditions of Rolle's theorem are satisfied.

Now, there must exist some $c \in (-\sqrt{3}, 0)$ such that $f'(c) = 0$

Now, $f'(x) = 3x^2 - 3$

Then, $f'(c) = 0$

$\Rightarrow$ $3c^2 - 3 = 0$

$\Rightarrow$ $3(c^2 - 1) = 0$

$\Rightarrow$ $c = \pm 1$

But $c \neq 1$ as $1 \notin (-\sqrt{3}, 0)$

$\therefore$ $c = -1 \in (-\sqrt{3}, 0)$

Thus, required value of $c$ is $-1$. **Ans.**

**7. The volume of a cube is increasing at the rate of 9 cm$^3$/s. How fast is its surface area increasing when the length of an edge is 10 cm ? [2]**

**Sol.** Let, the side of the cube be $a$ cm

then, volume of cube (V) = $a^3$

Differentiating V w.r.t. $t$, we get

$$\frac{dV}{dt} = 3a^2 \frac{da}{dt}$$

$$\Rightarrow \quad 3a^2 \frac{da}{dt} = 9 \text{ cm}^3/\text{s}$$

$$\left[\text{Given, } \frac{dV}{dt} = 9 \text{ cm}^3/\text{s}\right]$$

$$\Rightarrow \quad \frac{da}{dt} = \frac{3}{a^2} \text{ cm/s}$$

and surface area of cube (S) = $6a^2$

$$\Rightarrow \quad \frac{dS}{dt} = 12a \frac{da}{dt}$$

$$\Rightarrow \quad \frac{dS}{dt} = 12a \times \frac{3}{a^2} = \frac{36}{a} \text{ cm}^2/\text{s}$$

When $a = 10$,

$$\left[\frac{dS}{dt}\right]_{a=10} = \frac{36}{10} = 3.6 \text{ cm}^2/\text{s}.$$ **Ans.**

**8. Show that the function $f(x) = x^3 - 3x^2 + 6x - 100$ is increasing on R. [2]**

**Sol.** We have,

$f(x) = x^3 - 3x^2 + 6x - 100$

then, $f'(x) = 3x^2 - 6x + 6$

$= 3(x^2 - 2x + 1) + 3$

$= 3(x - 1)^2 + 3 > 0$ for all $x \in$ R.

Hence, the function $f(x)$ is increasing on R.

**Hence Proved.**

**9. The $x$-coordinate of a point on the line joining the points P( 2, 2, 1) and Q (5, 1, – 2) is 4. Find its $z$-coordinate. [2]**

**Sol.** Given, the points P(2, 2, 1) and Q (5, 1, – 2) of a line.

Then, equation of line PQ,

$$\frac{x-2}{5-2} = \frac{y-2}{1-2} = \frac{z-1}{-2-1}$$

$$\Rightarrow \quad \frac{x-2}{3} = \frac{y-2}{-1} = \frac{z-1}{-3} = \lambda$$

$\therefore$ Coordinates of any point of line PQ is

$(3\lambda + 2, -\lambda + 2, -3\lambda + 1)$

Given, the $x$-coordinate is 4,

$\Rightarrow$ $3\lambda + 2 = 4$

$\Rightarrow$ $3\lambda = 2$

$\therefore$ $\lambda = \frac{2}{3}$

$\therefore$ $z$ coordinate is $-3\lambda + 1 = -1$. **Ans.**

**10. A die, whose faces are marked 1, 2, 3 in red and 4, 5, 6 in green, is tossed. Let A be the event "number obtained is even" and B be the event "number obtained is red". Find if A and B are independent events. [2]**

**Sol.** Since, A be the event of number obtained is even then,

A = {2, 4, 6}

and B be the event of number obtained is red then,

$$B = \{1, 2, 3\}$$

$$\therefore \quad A \cap B = \{2\}$$

So, $P(A) = \frac{3}{6} = \frac{1}{2}$; $P(B) = \frac{3}{6} = \frac{1}{2}$; $P(A \cap B) = \frac{1}{6}$

Now, $P(A \cap B) \neq P(A) \cdot P(B)$

$$\frac{1}{6} \neq \frac{1}{4}$$

Hence, the events A and B are not independent events. **Ans.**

**11. Two tailors, A and B earn ₹ 300 and ₹ 400 per day respectively. A can stitch 6 shirts and 4 pairs of trousers while B can stitch 10 shirts and 4 pairs of trousers per day. To find how many days should each of them work if it is desired to produce at least 60 shirts and 32 pairs of trousers at a minimum labour cost, formulate this as an LPP. [2]**

**Sol.** Let $x$ and $y$ be the number of days for which the tailors A and B work respectively.

Total cost per day = ₹ $(300x + 400y)$

Let Z denote the total cost in rupees, then,

$$Z = 300x + 400y$$

Since in one day 6 shirts are stitched by tailor A and 10 shirts are stitched by tailor B and it is desired to produce atleast 60 shirts.

$$\therefore \quad 6x + 10y \geq 60$$

It is given that 4 pairs of trousers are stitched by each tailor A and B per day to produce atleast 32 pairs of trousers.

$$\therefore \quad 4x + 4y \geq 32$$

Finally, the no. of shirts and pair of trousers cannot be negative.

$$\therefore \quad x \geq 0, \; y \geq 0$$

Thus, mathematical formulation of the given LPP is as follows :

Minimize $\quad Z = 300x + 400y$

Subject to constraints :

$$6x + 10y \geq 60$$

$$4x + 4y \geq 32$$

$$x \geq 0, y \geq 0$$ **Ans.**

**12. Find : $\int \frac{dx}{5 - 8x - x^2}$. [2]**

**Sol.** We have,

$$\int \frac{dx}{5 - 8x - x^2} = \int \frac{dx}{5 - 8x - x^2 + (4)^2 - (4)^2}$$

$$= \int \frac{dx}{21 - [(4)^2 + 8x + (x)^2]}$$

$$= \int \frac{dx}{(\sqrt{21})^2 - (x+4)^2}$$

$$= \frac{1}{2\sqrt{21}} \log \left| \frac{\sqrt{21} + x + 4}{\sqrt{21} - x - 4} \right| + C$$ **Ans.**

## SECTION — C

**13. If $\tan^{-1} \frac{x-3}{x-4} + \tan^{-1} \frac{x+3}{x+4} = \frac{\pi}{4}$, then find the value of $x$. [4]**

**Sol.** We have,

$$\tan^{-1}\left(\frac{x-3}{x-4}\right) + \tan^{-1}\left(\frac{x+3}{x+4}\right) = \frac{\pi}{4}$$

$$\Rightarrow \tan^{-1}\left(\frac{x-3}{x-4}\right) + \tan^{-1}\left(\frac{x+3}{x+4}\right) = \tan^{-1} 1$$

$$\Rightarrow \tan^{-1}\left(\frac{x-3}{x-4}\right) = \tan^{-1} 1 - \tan^{-1}\left(\frac{x+3}{x+4}\right)$$

$$\Rightarrow \tan^{-1}\left(\frac{x-3}{x-4}\right) = \tan^{-1}\left(\frac{1 - \frac{x+3}{x+4}}{1 + 1 \times \frac{x+3}{x+4}}\right)$$

$$\Rightarrow \tan^{-1}\left(\frac{x-3}{x-4}\right) = \tan^{-1}\left(\frac{x+4-x-3}{x+4+x+3}\right)$$

$$\Rightarrow \frac{x-3}{x-4} = \frac{1}{2x+7}$$

$$\Rightarrow (2x+7)(x-3) = x-4$$

$$\Rightarrow 2x^2 - 6x + 7x - 21 = x - 4$$

$$\Rightarrow 2x^2 + x - 21 = x - 4$$

$$\Rightarrow 2x^2 = 17$$

$$\Rightarrow x = \pm\sqrt{\frac{17}{2}}$$ **Ans.**

**14. Using properties of determinants, prove that**

$$\begin{vmatrix} a^2+2a & 2a+1 & 1 \\ 2a+1 & a+2 & 1 \\ 3 & 3 & 1 \end{vmatrix} = (a-1)^3.$$ **[4]**

**Sol.** Let $$\Delta = \begin{vmatrix} a^2+2a & 2a+1 & 1 \\ 2a+1 & a+2 & 1 \\ 3 & 3 & 1 \end{vmatrix} = \text{L. H. S.}$$

Applying, $R_1 \to R_1 - R_2, R_2 \to R_2 - R_3$,

$$\Delta = \begin{vmatrix} a^2-1 & a-1 & 0 \\ 2a-1 & a-1 & 0 \\ 3 & 3 & 1 \end{vmatrix}$$

Taking $(a-1)$ common from $R_1$ and $R_2$,

$$= (a-1)^2 \begin{vmatrix} a+1 & 1 & 0 \\ 2 & 1 & 0 \\ 3 & 3 & 1 \end{vmatrix}$$

Now, expanding along $C_3$,

$$\Delta = (a-1)^2 \{(a+1)1 - (2)1\}$$
$$= (a-1)^2 (a-1) = (a-1)^3$$
$$= \text{R. H. S.}$$ **Hence Proved.**

**OR**

**Find matrix A such that**

$$\begin{bmatrix} 2 & -1 \\ 1 & 0 \\ -3 & 4 \end{bmatrix} A = \begin{bmatrix} -1 & -8 \\ 1 & -2 \\ 9 & 22 \end{bmatrix}$$

**Sol.** We have,

$$\begin{bmatrix} 2 & -1 \\ 1 & 0 \\ -3 & 4 \end{bmatrix}_{3\times 2} A = \begin{bmatrix} -1 & -8 \\ 1 & -2 \\ 9 & 22 \end{bmatrix}_{3\times 2}$$

$\therefore$ A is of order $2 \times 2$

Let $$A = \begin{bmatrix} a & b \\ c & d \end{bmatrix}$$

Then, $$\begin{bmatrix} 2 & -1 \\ 1 & 0 \\ -3 & 4 \end{bmatrix}\begin{bmatrix} a & b \\ c & d \end{bmatrix} = \begin{bmatrix} -1 & -8 \\ 1 & -2 \\ 9 & 22 \end{bmatrix}$$

$$\begin{bmatrix} 2a-c & 2b-d \\ a & b \\ -3a+4c & -3b+4d \end{bmatrix} = \begin{bmatrix} -1 & -8 \\ 1 & -2 \\ 9 & 22 \end{bmatrix}$$

By equality of matrices, on comparing, we get

$$2a - c = -1$$
$$2b - d = -8$$
$$a = 1$$
$$b = -2$$
$$-3a + 4c = 9$$
$$-3b + 4d = 22$$

On solving the equations, we get

$a = 1, b = -2, c = 3, d = 4$

Hence, $$A = \begin{bmatrix} 1 & -2 \\ 3 & 4 \end{bmatrix}$$ **Ans.**

**15. If $x^y + y^x = a^b$, then find $\frac{dy}{dx}$.** **[4]**

**Sol.** We have, $$x^y + y^x = a^b$$

$$\Rightarrow \quad e^{\log x^y} + e^{\log y^x} = a^b$$

$$\Rightarrow \quad e^{y\log x} + e^{x\log y} = a^b$$

On differentiating both sides w.r.t. $x$, we get

$$e^{y\log x}\left(y\cdot\frac{1}{x} + \log x\cdot\frac{dy}{dx}\right) + e^{x\log y}\left(x\cdot\frac{1}{y}\frac{dy}{dx} + \log y\right) = 0$$

$$\Rightarrow \quad x^y\left(\frac{y}{x} + \log x\frac{dy}{dx}\right) + y^x\left(\frac{x}{y}\frac{dy}{dx} + \log y\right) = 0$$

$$\Rightarrow \quad yx^{y-1} + x^y\log x\frac{dy}{dx} + xy^{x-1}\frac{dy}{dx} + y^x\log y = 0$$

$$\Rightarrow \quad \frac{dy}{dx}(x^y\log x + xy^{x-1}) = -(y^x\log y + yx^{y-1})$$

$$\therefore \quad \frac{dy}{dx} = -\left(\frac{y^x\log y + y\cdot x^{y-1}}{x^y\log x + x\cdot y^{x-1}}\right)$$ **Ans.**

**OR**

**If $e^y(x+1) = 1$, then show that $\frac{d^2y}{dx^2} = \left(\frac{dy}{dx}\right)^2$.**

**Sol.** We have,

$$e^y(x+1) = 1$$

$$\Rightarrow \quad e^y = \frac{1}{x+1}$$

$$\Rightarrow \quad \log e^y = \log_e\left(\frac{1}{x+1}\right)$$

[Taking log on both sides]

$$\Rightarrow \quad y = -\log(x+1)$$

On differentiating $y$ w.r.t. $x$, we get

$$\frac{dy}{dx} = -\frac{1}{x+1} \quad ...(i)$$

Again, differentiating w.r.t. $x$, we get

$$\frac{d^2y}{dx^2} = \frac{1}{(x+1)^2}$$

$$\Rightarrow \quad \frac{d^2y}{dx^2} = \left(-\frac{1}{(x+1)}\right)^2$$

$$\Rightarrow \quad \frac{d^2y}{dx^2} = \left(\frac{dy}{dx}\right)^2 \quad \text{[using (i)]}$$

**Hence Proved.**

**16. Find : $\int \frac{\cos\theta}{(4+\sin^2\theta)(5-4\cos^2\theta)}d\theta$** **[4]**

**Sol.**

Let $$I = \int \frac{\cos\theta}{(4+\sin^2\theta)(5-4\cos^2\theta)}\,d\theta$$

$$= \int \frac{\cos\theta}{(4+\sin^2\theta)(5-4(-\sin^2\theta+1))}\,d\theta$$

$$= \int \frac{\cos\theta}{(4+\sin^2\theta)(1+4\sin^2\theta)}\,d\theta$$

Substituting $\sin\theta = y \Rightarrow \cos\theta\, d\theta = dy$

$$I = \int \frac{1}{(4+y^2)(1+4y^2)}\,dy$$

Let $$\frac{1}{(4+y^2)(1+4y^2)} = \frac{A}{(4+y^2)} + \frac{B}{(1+4y^2)}$$

$$1 = A(1+4y^2) + B(4+y^2)$$

Putting $y = 0$, we get

$$1 = A + 4B \quad \text{...(i)}$$

Putting $y = 1$, we get

$$1 = 5A + 5B \quad \text{...(ii)}$$

Solving (i) and (ii), we get

$$A = \frac{-1}{15} \text{ and } B = \frac{4}{15}$$

$$\therefore \quad I = \int\left(\frac{-1}{15(4+y^2)} + \frac{4}{15(1+4y^2)}\right)dy$$

$$\Rightarrow \quad I = \frac{-1}{15}\int\frac{dy}{((2)^2+y^2)} + \frac{4}{15}\int\frac{dy}{((1)^2+(2y)^2)}$$

$$\Rightarrow \quad I = \frac{-1}{15}\times\frac{1}{2}\tan^{-1}\frac{y}{2} + \frac{4}{15}\times\frac{1}{2}\tan^{-1}2y + C$$

$$\Rightarrow \quad I = \frac{-1}{30}\tan^{-1}\left(\frac{\sin\theta}{2}\right) + \frac{2}{15}\tan^{-1}(2\sin\theta) + C$$

**Ans.**

**17. Evaluate :** $\displaystyle\int_0^\pi \frac{x\tan x}{\sec x + \tan x}\,dx.$ **[4]**

**Sol.** Let $$I = \int_0^\pi \frac{x\tan x}{\sec x+\tan x}\,dx. \quad \text{...(i)}$$

Now, $$I = \int_0^\pi \frac{(\pi - x)\tan(\pi - x)}{\sec(\pi - x)+\tan(\pi - x)}$$

$$\Rightarrow \quad I = \int_0^\pi \frac{(\pi - x)\tan x}{\sec x+\tan x}\,dx \quad \text{...(ii)}$$

Adding (i) and (ii), we get

$$2I = \int_0^\pi \frac{x\tan x}{\sec x+\tan x}\,dx + \int_0^\pi \frac{(\pi-x)\tan x}{\sec x+\tan x}\,dx$$

$$\Rightarrow \quad 2I = \pi\int_0^\pi \frac{\tan x}{\sec x+\tan x}\,dx$$

$$\Rightarrow \quad I = \frac{\pi}{2}\int_0^\pi \frac{\tan x}{\sec x+\tan x}\times\frac{\sec x-\tan x}{\sec x-\tan x}\,dx$$

$$\Rightarrow \quad I = \frac{\pi}{2}\int_0^\pi \frac{(\sec x\tan x - \tan^2 x)}{\sec^2 x - \tan^2 x}\,dx$$

$$\Rightarrow \quad I = \frac{\pi}{2}\int_0^\pi (\sec x\tan x - \sec^2 x + 1)\,dx$$

$$\Rightarrow \quad I = \frac{\pi}{2}\left[\sec x - \tan x + x\right]_0^\pi$$

$$\Rightarrow \quad I = \frac{\pi}{2}[(\sec\pi - \tan\pi + \pi) - (\sec 0 - \tan 0 + 0]$$

$$\Rightarrow \quad I = \frac{\pi}{2}[(-1+\pi) - (1)]$$

$$\therefore \quad I = \frac{\pi(\pi-2)}{2}$$

**Ans.**

**OR**

**Evaluate :** $\displaystyle\int_1^4 \{|x-1|+|x-2|+|x-4|\}\,dx$

**Sol.** Let $$I = \int_1^4 \{|x-1|+|x-2|+|x-4|\}\,dx$$

$$\Rightarrow \quad I = \int_1^4 |x-1|\,dx + \int_1^4 |x-2|\,dx + \int_1^4 |x-4|\,dx$$

$$\Rightarrow \quad I = \int_1^4 |x-1|\,dx + \int_1^2 |x-2|\,dx + \int_2^4 |x-2|\,dx + \int_1^4 |x-4|\,dx$$

$$\Rightarrow \quad I = \int_1^4 (x-1)\,dx - \int_1^2 (x-2)\,dx + \int_2^4 (x-2)dx - \int_1^4 (x-4)\,dx$$

$$\Rightarrow \quad I = \frac{1}{2}\left[(x-1)^2\right]_1^4 - \frac{1}{2}\left[(x-2)^2\right]_1^2 + \frac{1}{2}\left[(x-2)^2\right]_2^4 - \frac{1}{2}\left[(x-4)^2\right]_1^4$$

$$= \frac{1}{2}(9-0) - \frac{1}{2}(0-1) + \frac{1}{2}(4-0) - \frac{1}{2}(0-9)$$

$$\Rightarrow \quad I = \frac{9}{2}+\frac{1}{2}+\frac{4}{2}+\frac{9}{2}$$

$$\therefore \quad I = \frac{23}{2}$$

**Ans.**

**18. Solve the differential equation**

**$(\tan^{-1}x - y)\,dx = (1+x^2)\,dy.$** **[4]**

**Sol.** We have,

$$(\tan^{-1}x - y)\,dx = (1+x^2)\,dy$$

*i.e.,* $$\frac{dy}{dx} = \frac{\tan^{-1}x - y}{1+x^2}$$

$$\Rightarrow \quad \frac{dy}{dx}+\left(\frac{1}{1+x^2}\right)y = \frac{\tan^{-1}x}{1+x^2}$$

which is a linear differential equation of the form,

$$\frac{dy}{dx}+Py = Q$$

where, $P = \frac{1}{1+x^2}$ and $Q = \frac{\tan^{-1}x}{1+x^2}$

Now, I. F. $= e^{\int P dx} = e^{\int \frac{1}{1+x^2}dx} = e^{\tan^{-1}x}$

Then, required solution is :

$$ye^{\tan^{-1}x} = \int \left(e^{\tan^{-1}x}\right)\frac{\tan^{-1}x}{1+x^2}dx + C$$

Putting, $\tan^{-1}x = t$

$$\Rightarrow \quad \frac{1}{1+x^2}dx = dt$$

$$\Rightarrow \quad ye^t = \int e^t\, t dt + C$$

$$\Rightarrow \quad ye^t = t\int e^t dt - \int\left(\frac{d}{dt}(t)\int e^t\, dt\right)dt + C$$

$$\Rightarrow \quad ye^t = te^t - e^t + C$$

$$\Rightarrow \quad ye^t = (t-1)e^t + C$$

$$\Rightarrow \quad ye^{\tan^{-1}x} = (\tan^{-1}x - 1)\,e^{\tan^{-1}x} + C$$

$[\because t = \tan^{-1}x]$

$$\therefore \quad y = \tan^{-1}x - 1 + Ce^{-\tan^{-1}x}$$

which is the required solution. **Ans.**

**19. Show that the points A, B, C with position vectors $2\hat{i}-\hat{j}+\hat{k}$, $\hat{i}-3\hat{j}-5\hat{k}$ and $3\hat{i}-4\hat{j}-4\hat{k}$ respectively, are the vertices of a right-angled triangle. Hence, find the area of the triangle. [4]**

**Sol.** Given, the position vectors of the points A, B and C are $2\hat{i}-\hat{j}+\hat{k}$, $\hat{i}-3\hat{j}-5\hat{k}$ and $3\hat{i}-4\hat{j}-4\hat{k}$, respectively.

Then, $\vec{OA} = 2\hat{i}-\hat{j}+\hat{k}$,

$\vec{OB} = \hat{i}-3\hat{j}-5\hat{k}$

and $\vec{OC} = 3\hat{i}-4\hat{j}-4\hat{k}$

Now, $\vec{AB} = \vec{OB}-\vec{OA}$

$= (\hat{i}-3\hat{j}-5\hat{k})-(2\hat{i}-\hat{j}+\hat{k})$

$= -\hat{i}-2\hat{j}-6\hat{k}$

$\vec{BC} = \vec{OC}-\vec{OB}$

$= (3\hat{i}-4\hat{j}-4\hat{k})-(\hat{i}-3\hat{j}-5\hat{k})$

$= 2\hat{i}-\hat{j}+\hat{k}$

and $\vec{CA} = \vec{OA}-\vec{OC}$

$= (2\hat{i}-\hat{j}+\hat{k})-(3\hat{i}-4\hat{j}-4\hat{k})$

$= -\hat{i}+3\hat{j}+5\hat{k}$

$|\vec{AB}|^2 = (-1)^2+(-2)^2+(-6)^2 = 1+4+36 = 41$

$|\vec{BC}|^2 = (2)^2+(-1)^2+(1)^2 = 4+1+1 = 6$

and $|\vec{CA}|^2 = (-1)^2+(3)^2+(5)^2 = 1+9+25 = 35$

$\therefore \quad |\vec{AB}|^2 = |\vec{BC}|^2+|\vec{CA}|^2$

Hence, A, B, C are the vertices of a right angled triangle.

Now, area of $\Delta ABC = \frac{1}{2}\left|\vec{BC}\times\vec{CA}\right|$

and $\vec{BC}\times\vec{CA} = \begin{vmatrix}\hat{i} & \hat{j} & \hat{k}\\ 2 & -1 & 1\\ -1 & 3 & 5\end{vmatrix}$

$= -8\hat{i}-11\hat{j}+5\hat{k}$

$\Rightarrow \quad |\vec{BC}\times\vec{CA}| = \sqrt{(-8)^2+(-11)^2+(5)^2}$

$= \sqrt{64+121+25} = \sqrt{210}$

Hence, Area of $\Delta ABC = \frac{\sqrt{210}}{2}$ square units. **Ans.**

**20. Find the value of $\lambda$, if four points with position vectors $3\hat{i}+6\hat{j}+9\hat{k}$, $\hat{i}+2\hat{j}+3\hat{k}$, $2\hat{i}+3\hat{j}+\hat{k}$ and $4\hat{i}+6\hat{j}+\lambda\hat{k}$ are coplanar. [4]**

**Sol.** Let, the four points be A, B, C and D with position vectors $3\hat{i}+6\hat{j}+9\hat{k}$, $\hat{i}+2\hat{j}+3\hat{k}$, $2\hat{i}+3\hat{j}+\hat{k}$ and $4\hat{i}+6\hat{j}+\lambda\hat{k}$, respectively.

Then, $\vec{OA} = 3\hat{i}+6\hat{j}+9\hat{k}$

$\vec{OB} = \hat{i}+2\hat{j}+3\hat{k}$

$\vec{OC} = 2\hat{i}+3\hat{j}+\hat{k}$

and $\vec{OD} = 4\hat{i}+6\hat{j}+\lambda\hat{k}$

Then, $\overrightarrow{AB} = \overrightarrow{OB} - \overrightarrow{OA}$

$= (\hat{i} + 2\hat{j} + 3\hat{k}) - (3\hat{i} + 6\hat{j} + 9\hat{k})$

$= -2\hat{i} - 4\hat{j} - 6\hat{k}$

$\overrightarrow{AC} = \overrightarrow{OC} - \overrightarrow{OA}$

$= (2\hat{i} + 3\hat{j} + \hat{k}) - (3\hat{i} + 6\hat{j} + 9\hat{k})$

$= -\hat{i} - 3\hat{j} - 8\hat{k}$

and $\overrightarrow{AD} = \overrightarrow{OD} - \overrightarrow{OA}$

$= (4\hat{i} + 6\hat{j} + \lambda\hat{k}) - (3\hat{i} + 6\hat{j} + 9\hat{k})$

$= \hat{i} + (\lambda - 9)\hat{k}$

Since, the points are coplanar, then

$$[\overrightarrow{AB}, \overrightarrow{AC}, \overrightarrow{AD}] = 0$$

*i.e.*
$$\begin{bmatrix} -2 & -4 & -6 \\ -1 & -3 & -8 \\ 1 & 0 & \lambda - 9 \end{bmatrix} = 0$$

$\Rightarrow -2(-3\lambda + 27) + 4(-\lambda + 9 + 8) - 6(3) = 0$

$\Rightarrow \quad 6\lambda - 54 - 4\lambda + 68 - 18 = 0$

$\Rightarrow \quad 2\lambda - 4 = 0$

$\Rightarrow \quad 2\lambda = 4$

$\therefore \quad \lambda = 2$ **Ans.**

**21. There are 4 cards numbered 1, 3, 5 and 7, one number on one card. Two cards are drawn at random without replacement. Let X denote the sum of the numbers on the two drawn cards. Find the mean and variance of X. [4]**

**Sol.** Given, X denote the sum of the numbers on the two drawn cards.

Then, X can take values 4, 6, 8, 10, 12

and sample space (S) = {(1, 3), (1, 5), (1, 7), (3, 1), (3, 5), (3, 7), (5, 1), (5, 3), (5, 7), (7, 1), (7, 3), (7, 5)}

So, the probability distribution of X is as given below :

| X | 4 | 6 | 8 | 10 | 12 |
|---|---|---|---|---|---|
| P(X) | $\frac{2}{12}$ | $\frac{2}{12}$ | $\frac{4}{12}$ | $\frac{2}{12}$ | $\frac{2}{12}$ |

Computation of Mean and Variance :

| $x_i$ | $P(X = x_i) = p_i$ | $p_i x_i$ | $p_i x_i^2$ |
|---|---|---|---|
| 4 | $\frac{2}{12}$ | $\frac{8}{12}$ | $\frac{32}{12}$ |
| 6 | $\frac{2}{12}$ | $\frac{12}{12}$ | $\frac{72}{12}$ |
| 8 | $\frac{4}{12}$ | $\frac{32}{12}$ | $\frac{256}{12}$ |
| 10 | $\frac{2}{12}$ | $\frac{20}{12}$ | $\frac{200}{12}$ |
| 12 | $\frac{2}{12}$ | $\frac{24}{12}$ | $\frac{288}{12}$ |
| | | $\Sigma p_i x_i = \frac{96}{12}$ | $\Sigma p_i x_i^2 = \frac{848}{12}$ |

We have, $\Sigma p_i x_i = \frac{96}{12} = 8$

$\therefore$ Mean, $\bar{X} = \Sigma p_i x_i = 8$

Now, $\text{Var}(X) = \Sigma p_i x_i^2 - (\Sigma p_i x_i)^2$

$= \frac{848}{12} - (8)^2$

$= \frac{848}{12} - 64$

$= \frac{80}{12} = \frac{20}{3}$

Hence, Mean = 8 and Variance $= \frac{20}{3}$ **Ans.**

**22. Of the students in a school, it is known that 30% have 100% attendance and 70% students are irregular. Previous year results report that 70% of all students who have 100% attendance attain A grade and 10% irregular students attain A grade in their annual examination. At the end of the year, one student is chosen at random from the school and he was found to have an A grade. What is the probability that the student has 100% attendance ? Is regularity required only in school ? Justify your answer. [4]**

**Sol.** Consider the following events :

A : the student has grade A.

$E_1$ : the student has 100% attendance.

$E_2$ : the student is irregular.

Then, probability of the students having 100% attendance :

$$P(E_1) = 30\% = 0.3$$

Similarly, $P(E_2) = 70\% = 0.7$

Now, by previous year report, the probability of the students having grade A who have 100% attendance :

$$P(A/E_1) = 70\% = 0.7$$

and the probability of the students having grade A who are irregular :

$$P(A/E_2) = 10\% = 0.1$$

Then, the probability of the student having 100% attendance who already has attained A grade

$$= P(E_1/A)$$

By Bayes' theorem,

$$P(E_1/A) = \frac{P(A/E_1)\,P(E_1)}{P(A/E_1)\,P(E_1) + P(A/E_2)\,P(E_2)}$$

$$= \frac{0.7 \times 0.3}{0.7 \times 0.3 + 0.1 \times 0.7}$$

$$= \frac{21}{21+7} = \frac{21}{28} = \frac{3}{4}$$

No, regularity is required in school as well as in life.

It helps to be disciplined in every aspect of life.

And, when you work regularly, inspiration strikes regularly. **Ans.**

**23. Maximise Z = $x + 2y$**

**subject to the constraints :**

$$x + 2y \geq 100$$
$$2x - y \leq 0$$
$$2x + y \leq 200$$
$$x, y \geq 0$$

**Solve the above LPP graphically.** **[4]**

**Sol.** Given,

Maximise $Z = x + 2y$

Subject to the constraints :

$$x + 2y \geq 100$$
$$2x - y \leq 0$$
$$2x + y \leq 200$$
$$x, y \geq 0$$

Converting the inequations into equations we obtain the lines

$$x + 2y = 100,$$
$$2x - y = 0,$$
$$2x + y = 200.$$

Then, $x + 2y = 100$

| $x$ | 0 | 100 |
|---|---|---|
| $y$ | 50 | 0 |

$2x - y = 0$

| $x$ | 10 | 20 |
|---|---|---|
| $y$ | 20 | 40 |

and $2x + y = 200$

| $x$ | 0 | 100 |
|---|---|---|
| $y$ | 200 | 0 |

Plotting these points on the graph, we get the shaded feasible region *i.e.*, ADFEA.

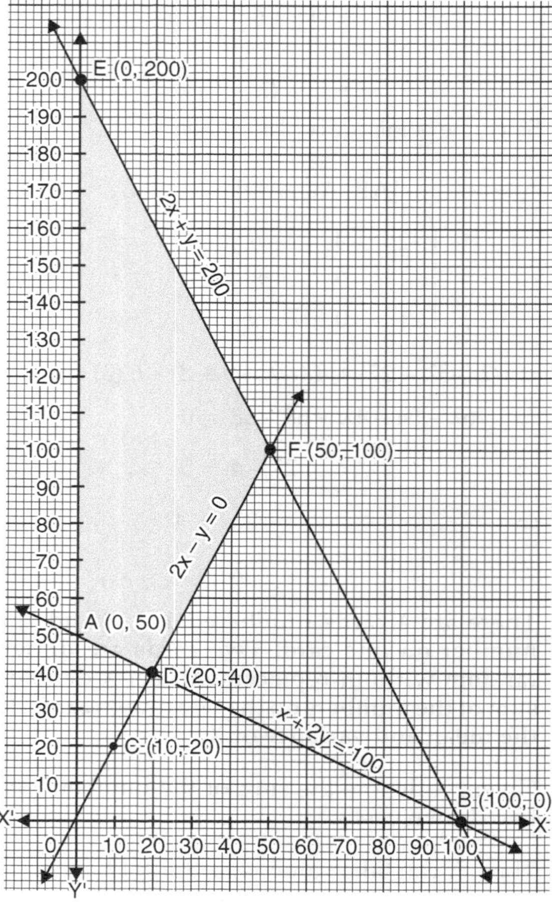


| Corner points | Value of Z = $x + 2y$ |
|---|---|
| A (0, 50) | (0) + 2(50) = 100 |
| D (20, 40) | 20 + 2(40) = 100 |
| F (50, 100) | 50 + 2(100) = 250 |
| E (0, 200) | 0 + 2(200) = 400 |

Clearly, the maximum value of Z is 400 at (0, 200). **Ans.**

## SECTION — D

**24. Determine the product :**

$$\begin{bmatrix} -4 & 4 & 4 \\ -7 & 1 & 3 \\ 5 & -3 & -1 \end{bmatrix} \begin{bmatrix} 1 & -1 & 1 \\ 1 & -2 & -2 \\ 2 & 1 & 3 \end{bmatrix}$$ **and use it to solve the system of equations $x - y + z = 4$, $x - 2y - 2z = 9$, $2x + y + 3z = 1$.** **[6]**

**Sol.**

Let $A = \begin{bmatrix} -4 & 4 & 4 \\ -7 & 1 & 3 \\ 5 & -3 & -1 \end{bmatrix}$

and $B = \begin{bmatrix} 1 & -1 & 1 \\ 1 & -2 & -2 \\ 2 & 1 & 3 \end{bmatrix}$

Then, $AB = \begin{bmatrix} -4 & 4 & 4 \\ -7 & 1 & 3 \\ 5 & -3 & -1 \end{bmatrix} \begin{bmatrix} 1 & -1 & 1 \\ 1 & -2 & -2 \\ 2 & 1 & 3 \end{bmatrix}$

$$= \begin{bmatrix} -4+4+8 & 4-8+4 & -4-8+12 \\ -7+1+6 & 7-2+3 & -7-2+9 \\ 5-3-2 & -5+6-1 & 5+6-3 \end{bmatrix}$$

$$\Rightarrow \quad AB = \begin{bmatrix} 8 & 0 & 0 \\ 0 & 8 & 0 \\ 0 & 0 & 8 \end{bmatrix} = 8I$$

$$\Rightarrow \quad AB = 8I$$

$$\Rightarrow \quad ABB^{-1} = 8IB^{-1}$$

$$\Rightarrow \quad \frac{1}{8}A = B^{-1} \qquad ...(i)$$

Now, consider the given system of equations :

$$x - y + z = 4$$
$$x - 2y - 2z = 9$$
$$2x + y + 3z = 1$$

which can be expressed as BX = C,

where, $B = \begin{bmatrix} 1 & -1 & 1 \\ 1 & -2 & -2 \\ 2 & 1 & 3 \end{bmatrix}$

$X = \begin{bmatrix} x \\ y \\ z \end{bmatrix}$ and $C = \begin{bmatrix} 4 \\ 9 \\ 1 \end{bmatrix}$

$$X = B^{-1}C$$

$$X = \frac{1}{8}AC \qquad \text{[using (i)]}$$

$$\therefore \quad X = \frac{1}{8}\begin{bmatrix} -4 & 4 & 4 \\ -7 & 1 & 3 \\ 5 & -3 & -1 \end{bmatrix}\begin{bmatrix} 4 \\ 9 \\ 1 \end{bmatrix}$$

$$= \frac{1}{8}\begin{bmatrix} -16+36+4 \\ -28+9+3 \\ 20-27-1 \end{bmatrix}$$

$$= \frac{1}{8}\begin{bmatrix} 24 \\ -16 \\ -8 \end{bmatrix} = \begin{bmatrix} 3 \\ -2 \\ -1 \end{bmatrix}$$

$$X = \begin{bmatrix} x \\ y \\ z \end{bmatrix} = \begin{bmatrix} 3 \\ -2 \\ -1 \end{bmatrix}$$

$\therefore \quad x = 3,\ y = -2,\ z = -1$ **Ans.**

**25. Consider $f : R - \left\{-\frac{4}{3}\right\} \to R - \left\{\frac{4}{3}\right\}$ given by $f(x) = \frac{4x+3}{3x+4}$. Show that $f$ is bijective. Find the inverse of $f$ and hence find $f^{-1}(0)$ and $x$ such that $f^{-1}(x) = 2$.** **[6]**

**Sol.** For one-one :

Let $x, y \in R - \left\{-\frac{4}{3}\right\}$

and $f(x) = f(y)$

$$\frac{4x+3}{3x+4} = \frac{4y+3}{3y+4}$$

$$\Rightarrow \quad (4x+3)(3y+4) = (3x+4)(4y+3)$$

$$\Rightarrow \quad 12xy + 16x + 9y + 12 = 12xy + 9x + 16y + 12$$

$$\Rightarrow \quad 16x + 9y - 9x - 16y = 0$$

$$\Rightarrow \quad 7x - 7y = 0$$

$$\Rightarrow \quad 7(x - y) = 0$$

$$\Rightarrow \quad x - y = 0$$

$$\Rightarrow \quad x = y$$

Hence, $f$ is one-one function.

For onto :

Let $y \in R - \left\{\frac{4}{3}\right\}$, then

$$f(x) = y$$

$$\Rightarrow \quad \frac{4x+3}{3x+4} = y$$

$$\Rightarrow \quad 4x + 3 = 3xy + 4y$$

$$\Rightarrow \quad (4 - 3y)x = 4y - 3$$

$$\Rightarrow \quad x = \frac{4y-3}{4-3y} \qquad ...(i)$$

As $y \in R - \left\{\frac{4}{3}\right\}$, $\quad x = \frac{4y-3}{4-3y} \in R$

Also $\dfrac{4y-3}{4-3y} \neq -\dfrac{4}{3}$

Since if, $\dfrac{4y-3}{4-3y} = -\dfrac{4}{3}$

$\Rightarrow$ $12y - 9 = -16 + 12y$

$\Rightarrow$ $9 = 16$ which is not possible.

Thus, $x = \dfrac{4y-3}{4-3y} \in R - \left\{-\dfrac{4}{3}\right\}$ such that

$$f(x) = f\left(\frac{4y-3}{4-3y}\right)$$

$$\Rightarrow \quad f(x) = \frac{4\left(\frac{4y-3}{4-3y}\right)+3}{3\left(\frac{4y-3}{4-3y}\right)+4}$$

$$= \frac{16y - 12 + 12 - 9y}{12y - 9 + 16 - 12y}$$

$$= \frac{7y}{7}$$

$\Rightarrow$ $f(x) = y$

Hence, every element $y \in R - \left\{-\frac{4}{3}\right\}$ has its preimage $x \in R - \left\{-\frac{4}{3}\right\}$

Hence, $f$ is onto and so bijective

$\Rightarrow f$ is one-one and onto, so $f$ is invertible.

Now, $f(x) = y$

$\Rightarrow$ $f^{-1}(y) = x$

$\Rightarrow$ $f^{-1}(y) = \dfrac{4y-3}{4-3y}$ [from eq. (i)]

$$f^{-1}(0) = \frac{4 \times 0 - 3}{4 - 3 \times 0}$$

$$= \frac{-3}{4}$$

Also given, $f^{-1}(x) = 2$

$\Rightarrow$ $\dfrac{4x-3}{4-3x} = 2$

$\Rightarrow$ $4x - 3 = 8 - 6x$

$\Rightarrow$ $10x = 11$

$\Rightarrow$ $x = \dfrac{11}{10}$ **Ans.**

**OR**

**Let A = Q × Q and let * be a binary operation on A defined by $(a, b) * (c, d) = (ac, b + ad)$ for $(a, b)$, $(c, d) \in$ A. Determine, whether * is commutative and associative. Then, with respect to * on A.**

**(i) Find the identity element in A**

**(ii) Find the invertible elements of A.****

**26. Show that the surface area of a closed cuboid with square base and given volume is minimum, when it is a cube. [6]**

**Sol.** Let the length and breadth of the cuboid of square base be $x$ and height be $y$.

Then, volume of the cuboid (V) $= x^2y$

$$\Rightarrow \quad y = \frac{V}{x^2} \quad \text{...(i)}$$

Now, surface area of cuboid,

$$S = 2(x^2 + yx + yx)$$

$$\Rightarrow \quad S = 2(x^2 + 2xy)$$

$$\Rightarrow \quad S = 2\left(x^2 + 2x\left(\frac{V}{x^2}\right)\right) \quad \text{[using (i)]}$$

$$\Rightarrow \quad S = 2\left(x^2 + \frac{2V}{x}\right)$$

Now, differentiating S w.r.t. $x$, we get

$$\frac{dS}{dx} = 2\left(2x - \frac{2V}{x^2}\right)$$

For maxima and minima, we have

$$\frac{dS}{dx} = 0$$

$$\Rightarrow \quad 2\left(2x - \frac{2V}{x^2}\right) = 0$$

$$\Rightarrow \quad 4\left(x - \frac{V}{x^2}\right) = 0$$

$$x = \frac{V}{x^2}$$

$$\Rightarrow \quad x = \sqrt[3]{V}$$

Now, $\dfrac{dS}{dx} = 2\left(2x - \dfrac{2V}{x^2}\right) = 4\left(x - \dfrac{V}{x^2}\right)$

Again differentiating w.r.t. $x$, we get

$$\frac{d^2S}{dx^2} = 4\left(1 + \frac{2V}{x^3}\right)$$

$$\Rightarrow \quad \left[\frac{d^2S}{dx^2}\right]_{x=\sqrt[3]{V}} = 4\left(1 + \frac{2V}{V}\right) = 12 > 0$$

Thus, S is minimum when $x = \sqrt[3]{V}$.

Putting $x = \sqrt[3]{V}$ in (i), we get

$$y = \frac{V}{x^2} = \frac{x^3}{x^2} = x$$

** Answer is not given due to change in the present syllabus.

$\Rightarrow \quad y = x$

Hence, it is a cube since the length, breadth and height of a cube are equal. **Hence Proved.**

**27. Using the method of integration, find the area of the triangle ABC, coordinates of whose vertices are A (4, 1), B (6, 6) and C (8, 4). [6]**

**Sol.** We have, A(4, 1), B (6, 6) and C(8, 4) as the vertices of a triangle ABC.

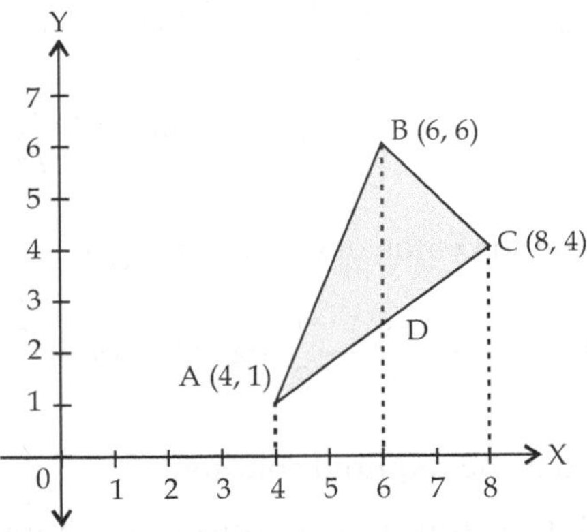


Then, equation of AB is

$$\Rightarrow \frac{x-4}{6-4} = \frac{y-6}{4-6}$$

$$\Rightarrow \frac{x-4}{2} = \frac{y-1}{5}$$

$$\Rightarrow 5x - 20 = 2y - 2$$

$$\Rightarrow 5x - 2y - 18 = 0$$

$$\Rightarrow y = \frac{5x-18}{2} \quad \text{...(i)}$$

Equation of BC is,

$$\frac{x-6}{8-6} = \frac{y-6}{4-6}$$

$$\Rightarrow \frac{x-6}{2} = \frac{y-6}{-2}$$

$$\Rightarrow -2x + 12 = 2y - 12$$

$$\Rightarrow -2x - 2y + 24 = 0$$

$$\Rightarrow x + y - 12 = 0$$

$$\Rightarrow y = 12 - x \quad \text{...(ii)}$$

and equation of CA is

$$\frac{x-8}{4-8} = \frac{y-4}{1-4}$$

$$\Rightarrow \frac{x-8}{-4} = \frac{y-4}{-3}$$

$$\Rightarrow -3x + 24 = -4y + 16$$

$$-3x + 4y + 8 = 0 \quad \text{...(iii)}$$

$$\Rightarrow y = \frac{3x-8}{4}$$

Hence, area of $\Delta$ABC

$$= \int_4^6 y_{AB}dx + \int_6^8 y_{AB}dx - \int_4^8 y_{AC}dx$$

$$= \int_4^6 \left(\frac{5x-18}{2}\right)dx + \int_6^8 (12-x)dx - \int_4^8 \left(\frac{3x-8}{4}\right)dx$$

$$= \left[\frac{5x^2}{4} - 9x\right]_4^6 + \left[12x - \frac{x^2}{2}\right]_6^8 - \left[\frac{3x^2}{8} - 2x\right]_4^8$$

$$= \left(\frac{5(6)^2}{4} - 9(6)\right) - \left(\frac{5(4)^2}{4} - 9(4)\right) + \left(12(8) - \frac{(8)^2}{2}\right) - \left(12(6) - \frac{(6)^2}{2}\right) - \left(\frac{3(8)^2}{8} - 2(8)\right) + \left(\frac{3(4)^2}{8} - 2(4)\right)$$

$$= (45 - 54) - (20 - 36) + (96 - 32) - (72 - 18) - (24 - 16) + (6 - 8)$$

$$= -9 + 16 + 64 - 54 - 8 - 2$$

$= 7$ sq. units. **Ans.**

**OR**

**Find the area enclosed between the parabola $4y = 3x^2$ and the straight line $3x - 2y + 12 = 0$.**

**Sol.** Given, the equation $4y = 3x^2$ and $3x - 2y + 12 = 0$.

*i.e.,* $$y = \frac{3x^2}{4} \quad \text{...(i)}$$

and $$y = \frac{3x}{2} + 6 \quad \text{...(ii)}$$

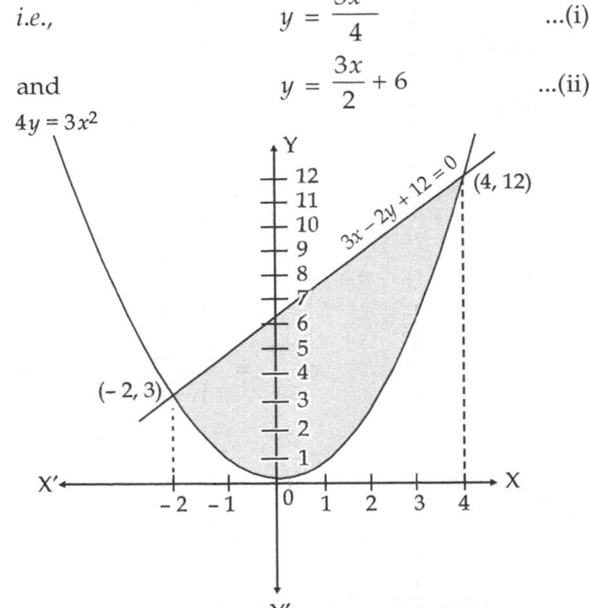


Solving these equations, we get

$$\frac{3x^2}{4} = \frac{3x}{2} + 6$$

$$\Rightarrow 3x^2 - 6x - 24 = 0$$

$$\Rightarrow x^2 - 2x - 8 = 0$$

$$\Rightarrow x^2 - 4x + 2x - 8 = 0$$

$$\Rightarrow x(x-4) + 2(x-4) = 0$$

$$\Rightarrow (x+2)(x-4) = 0$$

$$\Rightarrow x = -2, 4$$

At $x = -2$, $y = 3$

and at $x = 4,\ y = 12$

The points of intersection are (4, 12) and (– 2, 3).

$$\therefore \quad \text{Required area} = \int_{-2}^{4}\left(\frac{3x}{2}+6\right)dx - \int_{-2}^{4}\frac{3x^2}{4}\,dx$$

$$= \left[\frac{3x^2}{4}+6x\right]_{-2}^{4} - \frac{\left[x^3\right]_{-2}^{4}}{4}$$

$$= (12 + 24 - 3 + 12) - \frac{(64+8)}{4}$$

$$= 48 - 3 - 18$$

$$= 27 \text{ sq. units.}$$ **Ans.**

**28. Find the particular solution of the differential equation $(x - y)\dfrac{dy}{dx} = (x + 2y)$, given that $y = 0$ when $x = 1$.** **[6]**

**Sol.** We have,

$$(x-y)\frac{dy}{dx} = (x+2y)$$

$$\Rightarrow \quad \frac{dy}{dx} = \frac{x+2y}{x-y} \quad \text{...(i)}$$

Putting $\quad y = vx$

and $\quad \dfrac{dy}{dx} = v + x\dfrac{dv}{dx}$

$$\Rightarrow \quad v + x\frac{dv}{dx} = \frac{x+2vx}{x-vx}$$

$$\Rightarrow \quad v + x\frac{dv}{dx} = \frac{1+2v}{1-v}$$

$$\Rightarrow \quad x\frac{dv}{dx} = \frac{1+2v}{1-v} - v$$

$$\Rightarrow \quad x\frac{dv}{dx} = \frac{1+2v-v+v^2}{1-v}$$

$$\frac{1-v}{1+v+v^2}dv = \frac{dx}{x}$$

On integrating both sides, we get

$$\int\frac{1-v}{1+v+v^2}dv = \int\frac{dx}{x}$$

$$\Rightarrow \quad \frac{1}{2}\int\frac{3-1-2v}{1+v+v^2}dv = \int\frac{dx}{x}$$

$$\Rightarrow \frac{3}{2}\int\frac{1}{1+v+v^2}dv - \frac{1}{2}\int\frac{1+2v}{1+v+v^2}dv = \int\frac{dx}{x}$$

$$\Rightarrow \frac{3}{2}\int\frac{dv}{\left(v+\frac{1}{2}\right)^2+\left(\frac{\sqrt{3}}{2}\right)^2} - \frac{1}{2}\int\frac{1+2v}{1+v+v^2}dv = \int\frac{dx}{x}$$

$$\Rightarrow \frac{3}{2}\cdot\frac{2}{\sqrt{3}}\tan^{-1}\left(\frac{2v+1}{\sqrt{3}}\right) - \frac{1}{2}\log|1+v+v^2|$$

$$= \log|x| + C$$

$$\Rightarrow \sqrt{3}\tan^{-1}\left(\frac{2y+x}{\sqrt{3}x}\right) - \frac{1}{2}\log|x^2+xy+y^2| = C \quad \text{...(ii)}$$

Now, given $y = 0$, when $x = 1$

So, $\sqrt{3}\tan^{-1}\left(\dfrac{2(0)+1}{\sqrt{3}}\right) - \dfrac{1}{2}\log|1+0+0| = C$

$$\Rightarrow \quad C = \sqrt{3}\tan^{-1}\left(\frac{1}{\sqrt{3}}\right) = \frac{\sqrt{3}\pi}{6}$$

Putting the value of C in (ii), we get

$$\sqrt{3}\tan^{-1}\left(\frac{2y+x}{\sqrt{3}x}\right) - \frac{1}{2}\log|x^2+xy+y^2| = \frac{\sqrt{3}\pi}{6}$$

which is the required solution. **Ans.**

**29. Find the coordinates of the point where the line through the points (3, – 4, – 5) and (2, – 3, 1), crosses the plane determined by the points (1,2,3), (4,2,–3) and (0,4,3).** **[6]**

**Sol.** Equation of the plane determined by the points (1, 2, 3), (4, 2, – 3) and (0, 4, 3) is

$$\begin{bmatrix} x-1 & y-2 & z-3 \\ 4-1 & 2-2 & -3-3 \\ 0-1 & 4-2 & 3-3 \end{bmatrix} = 0$$

$$\begin{bmatrix} x-1 & y-2 & z-3 \\ 3 & 0 & -6 \\ -1 & 2 & 0 \end{bmatrix} = 0$$

$$(x-1)(0+12) - (y-2)(0-6) + (z-3)(6-0) = 0$$

$$\Rightarrow \quad 12x - 12 + 6y - 12 + 6z - 18 = 0$$

$$\Rightarrow \quad 12x + 6y + 6z - 42 = 0$$

$$\Rightarrow \quad 2x + y + z - 7 = 0$$

$$\Rightarrow \quad 2x + y + z = 7 \quad \text{...(i)}$$

Now, equation of line through (3, – 4, – 5) and (2, – 3, 1) is

$$\frac{x-3}{2-3} = \frac{y+4}{-3+4} = \frac{z+5}{1+5} = \lambda$$

*i.e.,* $\quad \dfrac{x-3}{-1} = \dfrac{y+4}{1} = \dfrac{z+5}{6} = \lambda$

$\therefore$ Coordinates of any point on line is :

$$P(-\lambda+3, \lambda-4, 6\lambda-5)$$

Now, the point P crosses the plane.

$\therefore$ It satisfies the equation (i) of plane.

$$2(-\lambda+3)+(\lambda-4)+(6\lambda-5) = 7$$

$$\Rightarrow \quad -2\lambda+6+\lambda-4+6\lambda-5 = 7$$

$$\Rightarrow \quad 5\lambda-3 = 7$$

$$\Rightarrow \quad 5\lambda = 10$$

$$\therefore \quad \lambda = 2$$

Hence, the point of intersection is $(1, -2, 7)$. **Ans.**

**OR**

**A variable plane which remains at a constant distance $3p$ from the origin cuts the coordinate axes at A, B, C. Show that the locus of the centroid of triangle ABC is**

$$\frac{1}{x^2}+\frac{1}{y^2}+\frac{1}{z^2} = \frac{1}{p^2}$$

**Sol.** Let the coordinates of A, B, C are $(a, 0, 0)$, $(0, b, 0)$ and $(0, 0, c)$

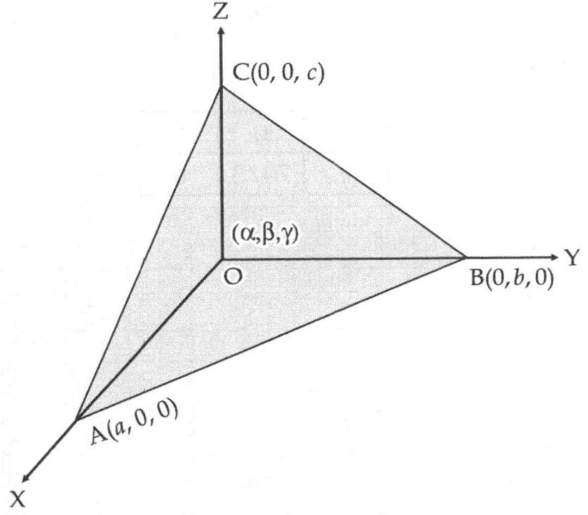


$\therefore$ The equation of plane is :

$$\frac{x}{a}+\frac{y}{b}+\frac{z}{c} = 1 \quad ...(i)$$

Since, the distance of plane is equal to $3p$ from the origin.

Then,
$$3p = \frac{\left|\frac{0}{a}+\frac{0}{b}+\frac{0}{c}-1\right|}{\sqrt{\frac{1}{a^2}+\frac{1}{b^2}+\frac{1}{c^2}}}$$

$$\Rightarrow \quad \frac{1}{a^2}+\frac{1}{b^2}+\frac{1}{c^2} = \frac{1}{9p^2} \quad ...(ii)$$

Let the centroid of $\Delta$ABC be $(x, y, z)$

$$= \left(\frac{a+0+0}{3}, \frac{0+b+0}{3}, \frac{0+0+c}{3}\right)$$

$$= \left(\frac{a}{3}, \frac{b}{3}, \frac{c}{3}\right)$$

$$\Rightarrow \quad a = 3x, \quad b = 3y, \quad c = 3z$$

Putting the value of $a$, $b$ and $c$ in (ii), we get

$$\frac{1}{(3x)^2}+\frac{1}{(3y)^2}+\frac{1}{(3z)^2} = \frac{1}{9p^2}$$

$$\frac{1}{9x^2}+\frac{1}{9y^2}+\frac{1}{9z^2} = \frac{1}{9p^2}$$

$$\frac{1}{x^2}+\frac{1}{y^2}+\frac{1}{z^2} = \frac{1}{p^2}$$

Hence, the required locus of the centroid is :

$$\frac{1}{x^2}+\frac{1}{y^2}+\frac{1}{z^2} = \frac{1}{p^2}.$$

**Hence Proved.**

## Outside Delhi [Set-II]

**Note : Except for the following questions, all the remaining questions have been asked in previous set.**

### SECTION — B

**12. The length $x$, of a rectangle is decreasing at the rate of 5 cm/minute and the width $y$, is increasing at the rate of 4 cm/minute. When $x = 8$ cm and $y = 6$ cm, find the rate of change of the area of the rectangle. [2]**

**Sol.** We have,

$$\frac{dx}{dt} = -5 \text{ cm/min} \quad ...(i)$$

and
$$\frac{dy}{dt} = 4 \text{ cm/min} \quad ...(ii)$$

Now, area of the rectangle, A = $xy$

$$\Rightarrow \quad \frac{dA}{dt} = x\frac{dy}{dt}+y\frac{dx}{dt}$$

$$\Rightarrow \quad \frac{dA}{dt} = x(4)+y(-5)$$

[using (i) and (ii)]

$$\therefore \quad \frac{dA}{dt} = 4x - 5y$$

When $x = 8$ cm and $y = 6$ cm,

$$\therefore \quad \left[\frac{dA}{dt}\right]_{x=8,\,y=6} = 4(8) - 5(6)$$

$$= 32 - 30 = 2 \text{ cm}^2/\text{min}$$

Hence, the rate of change of the area of the rectangle is 2 cm$^2$/min. **Ans.**

## SECTION — C

**20. Find :** $\int \frac{\sin\theta\, d\theta}{(4+\cos^2\theta)(2-\sin^2\theta)}$. **[4]**

**Sol.** Let

$$I = \int \frac{\sin\theta}{(4+\cos^2\theta)(2-\sin^2\theta)}\, d\theta$$

$$= \int \frac{\sin\theta}{(4+\cos^2\theta)[(2-(1-\cos^2\theta)]}\, d\theta$$

$$= \int \frac{\sin\theta}{(4+\cos^2\theta)(1+\cos^2\theta)}\, d\theta$$

Putting $\cos\theta = t$

$\Rightarrow \quad -\sin\theta\, d\theta = dt$

$$\therefore \quad I = \int \frac{-dt}{(4+t^2)(1+t^2)}$$

Let $\quad \frac{-1}{(4+t^2)(1+t^2)} = \frac{A}{4+t^2} + \frac{B}{1+t^2}$

$$-1 = A(1+t^2) + B(4+t^2)$$

Putting $t = 0$, we get

$$-1 = A + 4B \quad \text{...(i)}$$

Putting $t = 1$, we get

$$-1 = 2A + 5B \quad \text{...(ii)}$$

Solving (i) and (ii), we get

$$A = \frac{1}{3} \text{ and } B = \frac{-1}{3}$$

$$\therefore \quad I = \frac{1}{3}\int \frac{1}{4+t^2}\, dt - \frac{1}{3}\int \frac{1}{1+t^2}\, dt$$

$$\Rightarrow \quad I = \frac{1}{3} \times \frac{1}{2}\tan^{-1}\left(\frac{t}{2}\right) - \frac{1}{3} \times \tan^{-1} t + C$$

$$\Rightarrow \quad I = \frac{1}{6}\tan^{-1}\left(\frac{t}{2}\right) - \frac{1}{3} \times \tan^{-1} t + C$$

$$\Rightarrow \quad I = \frac{1}{6}\tan^{-1}\left(\frac{\cos\theta}{2}\right) - \frac{1}{3} \times \tan^{-1}(\cos\theta) + C$$

**Ans.**

**21. Solve the following linear programming problem graphically :**

**Maximise $Z = 34x + 45y$**
**under the following constraints**

$$x + y \le 300$$
$$2x + 3y \le 70$$
$$x \ge 0,\ y \ge 0$$

**[4]**

**Sol.** We have,

Maximise $Z = 34x + 45y$

Subject to the constraints :

$$x + y \le 300$$
$$2x + 3y \le 70$$
$$x \ge 0,\ y \ge 0$$

Converting the given inequalities into equations, we obtain the following equations :

$$x + y = 300$$
$$2x + 3y = 70$$

Then, $\quad x + y = 300$

| $x$ | 0 | 300 |
|---|---|---|
| $y$ | 300 | 0 |

and $\quad 2x + 3y = 70$

| $x$ | 0 | 35 |
|---|---|---|
| $y$ | 70/3 | 0 |

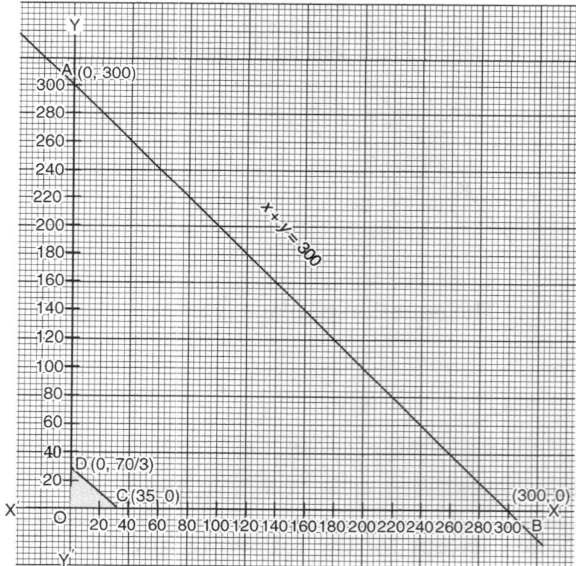


Plotting these points on the graph, we get the shaded feasible region *i.e.*, OCDO.

| Corner point | Value of $Z = 34x + 45y$ |
|---|---|
| O (0, 0) | $34(0) + 45(0) = 0$ |
| C (35, 0) | $34(35) + 45(0) = 1190$ |
| D (0, 70/3) | $34(0) + 45\,(70/3) = 1050$ |

Clearly, the maximum value of Z is 1190 at (35, 0). **Ans.**

**22. Find the value of $x$ such that the points A (3, 2, 1), B (4, $x$, 5), C(4, 2, – 2) and D (6, 5, – 1) are coplanar.** **[4]**

**Sol.** Given, the points A(3, 2, 1), B (4, $x$, 5), C (4, 2, – 2) and D (6, 5, – 1).

$$\therefore \quad \vec{AB} = 4\hat{i} + x\hat{j} + 5\hat{k} - (3\hat{i} + 2\hat{j} + \hat{k})$$

$$= \hat{i} + (x-2)\hat{j} + 4\hat{k}$$

and
$$\vec{AC} = 4\hat{i} + 2\hat{j} - 2\hat{k} - (3\hat{i} + 2\hat{j} + \hat{k})$$

$$= \hat{i} - 3\hat{k}$$

$$\vec{AD} = 6\hat{i} + 5\hat{j} - \hat{k} - (3\hat{i} + 2\hat{j} + \hat{k})$$
$$= 3\hat{i} + 3\hat{j} - 2\hat{k}$$

Since, the points are coplanar, then

$$\left[\vec{AB}, \vec{AC}, \vec{AD}\right] = 0$$

$$\begin{vmatrix} 1 & x-2 & 4 \\ 1 & 0 & -3 \\ 3 & 3 & -2 \end{vmatrix} = 0$$

$$\Rightarrow \quad 1(0+9) - (x-2)(-2+9) + 4(3-0) = 0$$

$$\Rightarrow \quad 9 - 7x + 14 + 12 = 0$$

$$\Rightarrow \quad 7x = 35$$

$$\therefore \quad x = 5$$

**Ans.**

**23. Find the general solution of the differential equation :**

$$\boldsymbol{y\,dx - (x + 2y^2)\,dy = 0}$$ **[4]**

**Sol.** We have,

$$ydx - (x + 2y^2)dy = 0$$

$$\Rightarrow \quad ydx = (x + 2y^2)dy$$

$$\Rightarrow \quad \frac{dx}{dy} = \frac{x + 2y^2}{y}$$

$$\Rightarrow \quad \frac{dx}{dy} + \left(\frac{-1}{y}\right)x = 2y$$

which is a linear differential equation of the form,

$$\frac{dx}{dy} + Px = Q$$

where $P = \frac{-1}{y}$

and $Q = 2y$

Now, $$\text{I. F.} = e^{\int P dy} = e^{\int \frac{-1}{y} dy} = e^{\log(1/y)} = \frac{1}{y}$$

So, the required solution is,

$$x \cdot \frac{1}{y} = \int 2y \cdot \frac{1}{y} dy + C$$

$$\therefore \quad \frac{x}{y} = 2y + C \text{ is the required solution.}$$

**Ans.**

## SECTION — D

**28. AB is the diameter of a circle and C is any point on the circle. Show that the area of triangle ABC is maximum, when it is an isosceles triangle.[6]**

**Sol.** Let $r$ be the radius of the circle then,

$$AB = 2r \quad \text{(AB is diameter)}$$

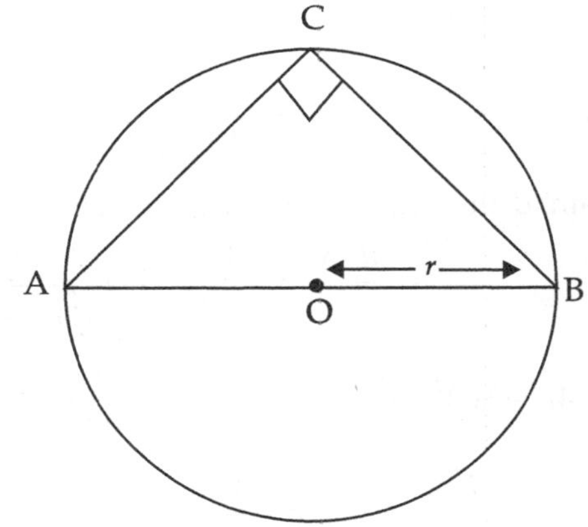


Let, $BC = x$ units

We know that angle subtended by diameter in a circle is right angle

$$\therefore \quad \angle C = 90°$$

Then, $$AC = \sqrt{(AB)^2 - (BC)^2}$$

$$AC = \sqrt{(2r)^2 - (x)^2}$$

$$= \sqrt{4r^2 - x^2} \quad \text{...(i)}$$

Now, area of $\Delta$ABC

$$A = \frac{1}{2}(AC)(BC)$$

$$A = \frac{1}{2}\sqrt{4r^2 - x^2}\,(x)$$

Differentiating A w.r.t. $x$, we get

$$\Rightarrow \quad \frac{dA}{dx} = \frac{1}{2}\left\{\sqrt{4r^2 - x^2} + x\frac{1}{2\sqrt{4r^2 - x^2}} \cdot \frac{d}{dx}(4r^2 - x^2)\right\}$$

$$\Rightarrow \quad \frac{dA}{dx} = \frac{1}{2}\left\{\sqrt{4r^2 - x^2} - \frac{x^2}{\sqrt{4r^2 - x^2}}\right\}$$

$$= \frac{1}{2}\left\{\frac{4r^2 - x^2 - x^2}{\sqrt{4r^2 - x^2}}\right\}$$

$$\Rightarrow \quad \frac{dA}{dx} = \frac{1}{2}\left\{\frac{4r^2 - 2x^2}{\sqrt{4r^2 - x^2}}\right\}$$

The critical numbers of $x$ are given by $\frac{dA}{dx} = 0$

$$\Rightarrow \quad \frac{1}{2}\left\{\frac{4r^2 - 2x^2}{\sqrt{4r^2 - x^2}}\right\} = 0$$

$$\Rightarrow \quad 4r^2 - 2x^2 = 0$$

$$\Rightarrow \quad 4r^2 = 2x^2$$

$$\therefore \quad x = \sqrt{2}r$$

Now, $$\frac{dA}{dx} = \frac{1}{2}\left\{\frac{4r^2 - 2x^2}{\sqrt{4r^2 - x^2}}\right\}$$

Again, differentiating w.r.t. $x$, we get

$$\frac{d^2A}{dx^2} = \frac{1}{2}\left\{(-4x)\frac{1}{\sqrt{4r^2 - x^2}}\right.$$

$$\left. + (4r^2 - 2x^2)\left(\frac{-1}{2}\right)(4r^2 - x^2)^{-3/2}\frac{d}{dx}(4r^2 - x^2)\right\}$$

$$\Rightarrow \quad \frac{d^2A}{dx^2} = \frac{1}{2}\left\{\frac{-4x}{\sqrt{4r^2 - x^2}} + \frac{x(4r^2 - 2x^2)}{(4r^2 - x^2)^{3/2}}\right\}$$

$$\Rightarrow \quad \left(\frac{d^2A}{dx^2}\right)_{x=\sqrt{2}r} = \frac{1}{2}\left\{\frac{-4(\sqrt{2}r)}{\sqrt{4r^2 - 2r^2}}\right.$$

$$\left. + \frac{\sqrt{2}r(4r^2 - 4r^2)}{(4r^2 - 2r^2)^{3/2}}\right\}$$

$$= \frac{-2\sqrt{2}r}{\sqrt{2}r} = -2 < 0$$

Thus, A is maximum when $x = \sqrt{2}r$.

Putting $x = \sqrt{2}r$ in (i),

$$AC = \sqrt{4r^2 - (\sqrt{2}r)^2}$$

$$\therefore \quad AC = \sqrt{2}r$$

$$\therefore \quad BC = AC = \sqrt{2}r$$

Hence, A is maximum when the triangle is isosceles. **Hence Proved.**

**29. If A = $\begin{bmatrix} 2 & -3 & 5 \\ 3 & 2 & -4 \\ 1 & 1 & -2 \end{bmatrix}$, find $A^{-1}$. Hence using $A^{-1}$ solve the system of equations $2x - 3y + 5z = 11$, $3x + 2y - 4z = -5$, $x + y - 2z = -3$.** **[6]**

**Sol.** We have,

$$A = \begin{bmatrix} 2 & -3 & 5 \\ 3 & 2 & -4 \\ 1 & 1 & -2 \end{bmatrix}$$

$$\therefore \quad |A| = \begin{vmatrix} 2 & -3 & 5 \\ 3 & 2 & -4 \\ 1 & 1 & -2 \end{vmatrix}$$

$$= 2(-4 + 4) + 3(-6 + 4) + 5(3 - 2)$$

$$= 2(0) + 3(-2) + 5(1) = -1 \neq 0$$

Hence, A is invertible and $A^{-1}$ exists.

Let $A_{ij}$ be the co-factors of elements $a_{ij}$ in $A = [a_{ij}]$. Then,

$$A_{11} = (-1)^{1+1}\begin{vmatrix} 2 & -4 \\ 1 & -2 \end{vmatrix} = 0,$$

$$A_{12} = (-1)^{1+2}\begin{vmatrix} 3 & -4 \\ 1 & -2 \end{vmatrix} = 2$$

$$A_{13} = (-1)^{1+3}\begin{vmatrix} 3 & 2 \\ 1 & 1 \end{vmatrix} = 1,$$

$$A_{21} = (-1)^{2+1}\begin{vmatrix} -3 & 5 \\ 1 & -2 \end{vmatrix} = -1$$

$$A_{22} = (-1)^{2+2}\begin{vmatrix} 2 & 5 \\ 1 & -2 \end{vmatrix} = -9,$$

$$A_{23} = (-1)^{2+3}\begin{vmatrix} 2 & -3 \\ 1 & 1 \end{vmatrix} = -5$$

$$A_{31} = (-1)^{3+1}\begin{vmatrix} -3 & 5 \\ 2 & -4 \end{vmatrix} = 2,$$

$$A_{32} = (-1)^{3+2}\begin{vmatrix} 2 & 5 \\ 3 & -4 \end{vmatrix} = 23$$

$$A_{33} = (-1)^{3+3}\begin{vmatrix} 2 & -3 \\ 3 & 2 \end{vmatrix} = 13$$

$$\therefore \quad \text{adj } A = \begin{bmatrix} 0 & 2 & 1 \\ -1 & -9 & -5 \\ 2 & 23 & 13 \end{bmatrix}^T = \begin{bmatrix} 0 & -1 & 2 \\ 2 & -9 & 23 \\ 1 & -5 & 13 \end{bmatrix}$$

$$\therefore \quad A^{-1} = \frac{1}{|A|}\text{adj } A = \frac{1}{-1}\begin{bmatrix} 0 & -1 & 2 \\ 2 & -9 & 23 \\ 1 & -5 & 13 \end{bmatrix}$$

$$= \begin{bmatrix} 0 & 1 & -2 \\ -2 & 9 & -23 \\ -1 & 5 & -13 \end{bmatrix}$$

Now, the given system of equations is expressible as

$$\begin{bmatrix} 2 & -3 & 5 \\ 3 & 2 & -4 \\ 1 & 1 & -2 \end{bmatrix}\begin{bmatrix} x \\ y \\ z \end{bmatrix} = \begin{bmatrix} 11 \\ -5 \\ -3 \end{bmatrix} \text{ or } AX = B$$

where, $A=\begin{bmatrix}2 & -3 & 5\\3 & 2 & -4\\1 & 1 & -2\end{bmatrix}$, $X=\begin{bmatrix}x\\y\\z\end{bmatrix}$ and $B=\begin{bmatrix}11\\-5\\-3\end{bmatrix}$

Now, $AX = B \Rightarrow A^{-1}AX = A^{-1}B$

$\therefore$ $X = A^{-1}B$

$\Rightarrow$ $X = \begin{bmatrix}0 & 1 & -2\\-2 & 9 & -23\\-1 & 5 & -13\end{bmatrix}\begin{bmatrix}11\\-5\\-3\end{bmatrix}$

$\Rightarrow$ $X = \begin{bmatrix}x\\y\\z\end{bmatrix} = \begin{bmatrix}1\\2\\3\end{bmatrix}$

$\therefore$ $x = 1,\ y = 2,\ z = 3.$ **Ans.**

## Outside Delhi [Set-III]

**Note : Except for the following questions, all the remaining questions have been asked in previous sets.**

### SECTION — B

**12. The volume of a sphere is increasing at the rate of 8 cm³/s. Find the rate at which its surface area is increasing when the radius of the sphere is 12 cm. [2]**

**Sol.** We have,

Volume of sphere (V) $= \frac{4}{3}\pi r^3$

where, $r$ is the radius of sphere.
Now, differentiating V w.r.t. $t$, we get

$$\frac{dV}{dt} = \frac{4}{3}\pi \times 3r^2\frac{dr}{dt} = 8\text{cm}^3/\text{s}$$

$\Rightarrow$ $$4\pi r^2\frac{dr}{dt} = 8\text{ cm}^3/\text{s}$$

$\Rightarrow$ $$\frac{dr}{dt} = \frac{2}{\pi r^2}\text{cm}^3/\text{s} \quad ....(i)$$

and surface area of sphere (S) $= 4\pi r^2$.

Then, differentiating S w.r.t. $t$, we get

$$\frac{dS}{dt} = 4\pi\cdot 2r\frac{dr}{dt}$$

$\Rightarrow$ $$\frac{dS}{dt} = 8\pi r\cdot\frac{2}{\pi r^2} \quad \text{[using (i)]}$$

$\Rightarrow$ $$\frac{dS}{dt} = \frac{16}{r}\text{ cm}^2/\text{s}.$$

When $r = 12$ cm,

$$\left[\frac{dS}{dt}\right]_{r=12} = \frac{16}{12} = \frac{4}{3}\text{ cm}^2/\text{s}.$$

Hence, the surface area is increasing at the rate of $\frac{4}{3}$ cm²/s when radius of sphere is 12 cm. **Ans.**

### SECTION — C

**20. Solve the following linear programming problem graphically : [4]**

**Maximise Z = 7x + 10y**

**subject to the constraints**

$$4x + 6y \le 240$$
$$6x + 3y \le 240$$
$$x \ge 10$$
$$x \ge 0, y \ge 0$$

**Sol.** We have,

Maximise $Z = 7x + 10y$

Subject to the constraints :

$$4x + 6y \le 240$$
$$6x + 3y \le 240$$
$$x \ge 10$$
$$x \ge 0, y \ge 0$$

Converting the given inequalities into equations, we obtain the following equations :

$$4x + 6y = 240$$
$$6x + 3y = 240$$
$$x = 10$$

Then, $4x + 6y = 240$

| $x$ | 0 | 60 |
|---|---|---|
| $y$ | 40 | 0 |

$6x + 3y = 240$

| $x$ | 0 | 40 |
|---|---|---|
| $y$ | 80 | 0 |

and $x = 10$ is a line parallel to Y-axis.

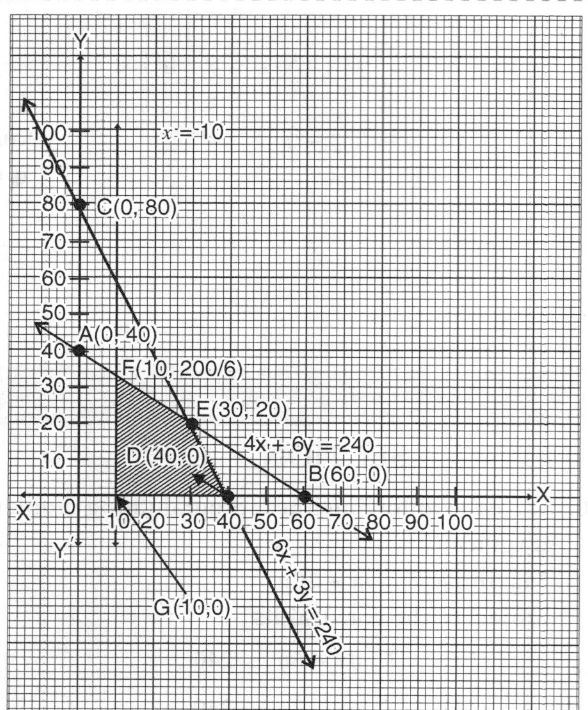


Plotting these points on the graph, we get the shaded feasible region *i.e.*, DEFGD.

| Corner points | Value of Z= $7x + 10y$ |
|---|---|
| D(40, 0) | 7(40) + 10(0) = 280 |
| E (30, 20) | 7(30) + 10(20) = 410 |
| F (10, 200/6) | 7(10) + 10(200/6) = 403.33 |
| G (10, 0) | 7(10) + 10(0) = 70 |

Clearly, the maximum value of Z is 410 at (30, 20).

**Ans.**

**21. Find :** $\int \frac{e^x dx}{(e^x - 1)^2 (e^x + 2)}$. **[4]**

**Sol.** Let $$I = \int \frac{e^x dx}{(e^x - 1)^2 (e^x + 2)}$$

Putting, $e^x = t$ and $e^x dx = dt$

$$I = \int \frac{dt}{(t-1)^2 (t+2)}$$

Let $$\frac{1}{(t-1)^2 (t+2)} = \frac{A}{t-1} + \frac{B}{(t-1)^2} + \frac{C}{t+2}$$

$$1 = A(t-1)(t+2) + B(t+2) + C(t-1)^2$$

Putting $t = 1$, we get

$$1 = 3B \Rightarrow B = 1/3$$

Putting $t = -2$, we get

$$1 = 9C \Rightarrow C = 1/9$$

Putting $t = 0$, we get

$$1 = -2A + 2B + C$$

$$\Rightarrow \quad 1 = -2A + \frac{2}{3} + \frac{1}{9}$$

$$\Rightarrow \quad A = -1/9$$

$$\therefore \quad I = \frac{-1}{9} \int \frac{1}{(t-1)} dt + \frac{1}{3} \int \frac{1}{(t-1)^2} dt + \frac{1}{9} \int \frac{1}{(t+2)} dt$$

$$\Rightarrow I = \frac{-1}{9} \log|t-1| - \frac{1}{3(t-1)} + \frac{1}{9} \log|t+2| + C$$

$$\therefore \ I = \frac{1}{9} \log\left|\frac{e^x + 2}{e^x - 1}\right| - \frac{1}{3(e^x - 1)} + C.$$ **Ans.**

**22. If $\vec{a} = 2\hat{i} - \hat{j} - 2\hat{k}$ and $\vec{b} = 7\hat{i} + 2\hat{j} - 3\hat{k}$, then express $\vec{b}$ in the form of $\vec{b} = \vec{b_1} + \vec{b_2}$, where $\vec{b_1}$ is parallel to $\vec{a}$ and $\vec{b_2}$ is perpendicular to $\vec{a}$.** **[4]**

**Sol.** We have,

$\vec{a} = 2\hat{i} - \hat{j} - 2\hat{k}$ and $\vec{b} = 7\hat{i} + 2\hat{j} - 3\hat{k}$,

Since, $\vec{b_1}$ is parallel to $\vec{a}$

$$\vec{b_1} = \lambda \vec{a}$$

$$\therefore \quad \vec{b_1} = 2\lambda\hat{i} - \lambda\hat{j} - 2\lambda\hat{k}$$

So, $$\vec{b_2} = \vec{b} - \vec{b_1}$$

$$= 7\hat{i} + 2\hat{j} - 3\hat{k} - (2\lambda\hat{i} - \lambda\hat{j} - 2\lambda\hat{k})$$

$$= (7 - 2\lambda)\hat{i} + (2 + \lambda)\hat{j} + (-3 + 2\lambda)\hat{k}$$

Since, $\vec{b_2}$ is perpendicular to $\vec{a}$,

$$\vec{a} \cdot \vec{b_2} = 0$$

$$\therefore (2\hat{i} - \hat{j} - 2\hat{k})[(7 - 2\lambda)\hat{i} + (2 + \lambda)\hat{j} + (-3 + 2\lambda)\hat{k}] = 0$$

$$\Rightarrow \quad 2(7 - 2\lambda) - 1(2 + \lambda) - 2(-3 + 2\lambda) = 0$$

$$\Rightarrow \quad 14 - 4\lambda - 2 - \lambda + 6 - 4\lambda = 0$$

$$\Rightarrow \quad -9\lambda + 18 = 0$$

$$\Rightarrow \quad 9\lambda = 18$$

$$\therefore \quad \lambda = 2$$

Hence, $$\vec{b} = (4\hat{i} - 2\hat{j} - 4\hat{k}) + (3\hat{i} + 4\hat{j} + \hat{k})$$

**Ans.**

**23. Find the general solution of the differential equation $\frac{dy}{dx} - y = \sin x$.** **[4]**

**Sol.** We have,

$$\frac{dy}{dx} - y = \sin x$$

which is a linear differential equation of the form

$$\frac{dy}{dx} + Py = Q$$

where, $P = -1$ and $Q = \sin x$

Now, $\text{I. F.} = e^{\int Pdx} = e^{\int -1dx} = e^{-x}$

So, the required solution is

$$ye^{-x} = \int e^{-x} \sin x\, dx + C_1 \quad ...(i)$$

Let $I = \int e^{-x} \sin x\, dx \quad ...(ii)$

$$I = \sin x \int e^{-x}dx - \int\left(\frac{d}{dx}(\sin x) \int e^{-x}\, dx\right) dx + C_2$$

$$\Rightarrow I = -\sin x\, e^{-x} + \int \cos x\, e^{-x}\, dx + C_2$$

$$\Rightarrow I = -\sin x\, e^{-x} + \cos x \int e^{-x}dx - \int\left(\frac{d}{dx}(\cos x)\int e^{-x}dx\right) dx + C_2$$

$$\Rightarrow I = -\sin x\, e^{-x} - \cos x\, e^{-x} - \int \sin x\, e^{-x}dx + C_2$$

$$\Rightarrow I = -\sin x\, e^{-x} - \cos x\, e^{-x} - I + C_2 \quad \text{[using (ii)]}$$

$$\Rightarrow 2I = -e^{-x}(\sin x + \cos x) + C_2$$

$$\Rightarrow I = \frac{-1}{2}e^{-x}(\sin x + \cos x) + C_2$$

Putting I in equation (i),

$$ye^{-x} = \frac{-1}{2}e^{-x}(\sin x + \cos x) + C_1 + C_2$$

$$\Rightarrow 2y = -(\sin x + \cos x) + 2Ce^{x} \quad (\because C_1 + C_2 = C)$$

$\therefore 2y = 2Ce^{x} - \sin x - \cos x$ is the required solution.

**Ans.**

## SECTION — D

**29. A window is in the form of a rectangle surmounted by a semicircular opening. The total perimeter of the window is 10 m. Find the dimensions of the window to admit maximum light through the whole opening. [6]**

**Sol.** Let ABCD be a window of rectangular form surmounted by a semicircle with diameter AB. Given, perimeter of the window (P) = 10 m

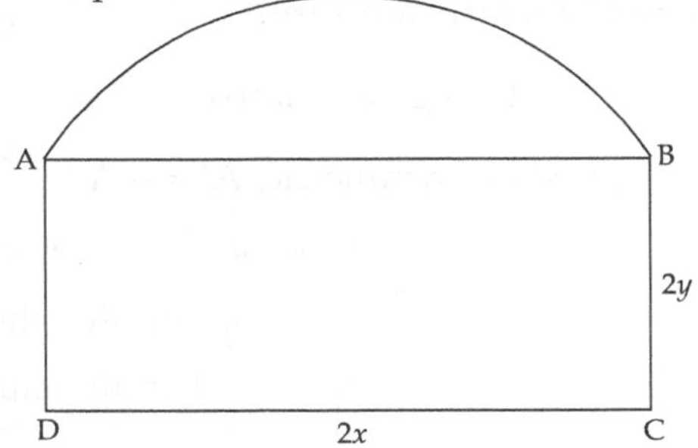


Let the length and breadth of the rectangle be $2x$ and $2y$ respectively.

Since, $P = 10$ m

*i.e.,* $2x + 4y + \pi x = 10$

$$\Rightarrow 4y = 10 - 2x - \pi x$$

$$\Rightarrow 4xy = 10x - 2x^2 - \pi x^2 \quad ...(i)$$

Now, area of the window ,

$$A = (2x)(2y) + \frac{1}{2}\pi x^2$$

$$\Rightarrow A = 4xy + \frac{1}{2}\pi x^2$$

$$\Rightarrow A = 10x - 2x^2 - \pi x^2 + \frac{1}{2}\pi x^2 \quad \text{[using (i)]}$$

$$\Rightarrow A = 10x - 2x^2 - \frac{1}{2}\pi x^2$$

On differentiating A w. r. t. $x$, we get

$$\frac{dA}{dx} = 10 - 4x - \pi x$$

The critical values of $x$ are given by $\frac{dA}{dx} = 0$

$$\Rightarrow 10 - 4x - \pi x = 0$$

$$\Rightarrow -x(4 + \pi) = -10$$

$$\Rightarrow x = \frac{10}{4 + \pi}$$

Now, $\frac{dA}{dx} = 10 - 4x - \pi x$

differentiating again, w.r.t. $x$, we get

$$\frac{d^2A}{dx^2} = -4 - \pi = -(4 + \pi) < 0$$

Thus, A is maximum when $x = \frac{10}{4 + \pi}$ m

Now, length of the window,

$$2x = \frac{20}{4 + \pi}\text{ m}$$

and width of the window,

$$2y = \frac{10 - 2x - \pi x}{2} \quad \text{[using (i)]}$$

$$= 5 - \frac{10}{4 + \pi} - \frac{\pi}{2} \cdot \frac{10}{4 + \pi}$$

$$= \frac{10(4 + \pi) - 20 - 10\pi}{2(4 + \pi)}$$

$$= \frac{40 + 10\pi - 20 - 10\pi}{2(4 + \pi)}$$

$$= \frac{10}{4 + \pi}\text{ m}$$

Also, radius of the semi-circle is $\left(\frac{10}{4 + \pi}\right)$ m

Hence, the dimensions of the rectangular part of the window are $\frac{20}{4 + \pi}$ m and $\frac{10}{4 + \pi}$ m. **Ans.**

●●

# MATHEMATICS

## Delhi [Set-I] 2017

**Time allowed : 3 hours** **Maximum marks : 100**

### SECTION — A

**1. If A is a 3 × 3 invertible matrix, then what will be the value of $k$ if det ($A^{-1}$) = (det A)$^k$. [1]**

**Sol.** Given, A is an invertible matrix.

$\therefore \quad A.A^{-1} = I$

$\Rightarrow \quad \det(A.A^{-1}) = \det(I)$

$\Rightarrow \quad \det(A).\det(A^{-1}) = 1 \quad [\because \det(I) = 1]$

$\Rightarrow \quad \det(A).(\det A)^k = 1 \quad$ (Given)

$$(\det A)^k = \frac{1}{\det(A)}$$

$\therefore \quad (\det A)^k = (\det A)^{-1}$

on comparing both sides, we get

$k = -1$ **Ans.**

**2. Determine the value of the constant '$k$' so that the function $f(x) = \begin{cases} \dfrac{kx}{|x|}, & \text{if } x < 0 \\ 3, & \text{if } x \geq 0 \end{cases}$ is continuous at $x = 0$. [1]**

**Sol.** Given, that the function is continuous at $x = 0$

$\therefore$

$$\lim_{x \to 0^-} f(x) = \lim_{x \to 0^+} f(x) = f(0) \quad ...(i)$$

$$\lim_{x \to 0^-} f(x) = \lim_{x \to 0^-} \frac{kx}{|x|}$$

$$= k \lim_{x \to 0^-} \frac{x}{-x}$$

$$= k(-1) = -k$$

$$\lim_{x \to 0^+} f(x) = \lim_{x \to 0^+} 3 = 3$$

From equation (i),

$-k = 3$

or $\quad k = -3$ **Ans.**

**3. Evaluate : $\int_2^3 3^x\, dx$. [1]**

**Sol.** Let $\quad I = \int_2^3 3^x\, dx = \left(\dfrac{3^x}{\log 3}\right)_2^3 + C$

where C is constant of integration

$$I = \frac{1}{\log 3}\left[3^3 - 3^2\right] + C$$

$$= \frac{1}{\log 3}(27 - 9) + C$$

$$= \frac{18}{\log 3} + C$$ **Ans.**

**4. If a line makes angles 90° and 60° respectively with the positive directions of X and Y-axes, find the angle which it makes with the positive direction of Z-axis. [1]**

**Sol.** We know that :

$$l^2 + m^2 + n^2 = 1 \quad ...(i)$$

and $\quad l = \cos\alpha,\ m = \cos\beta,\ n = \cos\gamma$

Given, $\alpha = 90°,\ \beta = 60°$

$\therefore \cos\alpha = \cos 90° = 0$ and $\cos\beta = \cos 60° = \dfrac{1}{2}$

From equation (i),

$$0^2 + \left(\frac{1}{2}\right)^2 + n^2 = 1$$

$\Rightarrow \quad n^2 = 1 - \dfrac{1}{4} = \dfrac{3}{4}$

$\Rightarrow \quad \cos^2\gamma = \dfrac{3}{4}$

$\Rightarrow \quad \cos\gamma = \pm\dfrac{\sqrt{3}}{2}$

$\Rightarrow \quad \gamma = \cos^{-1}\left(\dfrac{\sqrt{3}}{2}\right)$ or $\cos^{-1}\left(\dfrac{-\sqrt{3}}{2}\right)$

$\Rightarrow \quad \gamma = \dfrac{\pi}{6}$ or $\dfrac{5\pi}{6}$ **Ans.**

### SECTION — B

**5. Show that all the diagonal elements of a skew symmetric matrix are zero. [2]**

**Sol.** Let $A = [a_{ij}]$ be a given matrix

Since, it is skew symmetric, $A' = -A$

$\therefore \quad a_{ji} = -a_{ij} \quad$ For all $i, j$

$\Rightarrow \quad a_{ii} = -a_{ii}$ For all values of $i$

$\Rightarrow \quad 2a_{ii} = 0 \quad$ For all values of $i$

$$\Rightarrow \quad a_{ii} = 0 \quad \text{For all values of } i$$

$$\Rightarrow \quad a_{11} = a_{22} = a_{33} = \ldots\ldots = a_{nn} = 0$$

Hence, all the diagonal elements of a skew. symmetric matrix are zero.

**Hence Proved.**

**6. Find $\frac{dy}{dx}$ at $x = 1, y = \frac{\pi}{4}$ if $\sin^2 y + \cos xy = K$.** **[2]**

**Sol.** Given, $\sin^2 y + \cos xy = K$

Differentiating both sides w.r. t. $x$, we get

$$2 \sin y \cos y \cdot \frac{dy}{dx} + \left[-\sin xy\left(x\frac{dy}{dx} + y\right)\right] = 0$$

$$\Rightarrow \sin 2y \cdot \frac{dy}{dx} - x \sin xy \frac{dy}{dx} - y \sin xy = 0$$

$$\Rightarrow \quad \frac{dy}{dx}(\sin 2y - x \sin xy) = y \sin xy$$

$$\Rightarrow \quad \frac{dy}{dx} = \frac{y \sin xy}{\sin 2y - x \sin xy} \quad \ldots\text{(i)}$$

Given, at $x = 1, y = \frac{\pi}{4}$

From equation (i),

$$\frac{dy}{dx} = \frac{\frac{\pi}{4}\sin\frac{\pi}{4}}{\sin\frac{2\pi}{4} - 1.\sin\frac{\pi}{4}}$$

$$= \frac{\frac{\pi}{4} \times \frac{1}{\sqrt{2}}}{1 - \frac{1}{\sqrt{2}}}$$

$$= \frac{\frac{\pi}{4\sqrt{2}}}{\frac{\sqrt{2}-1}{\sqrt{2}}} = \frac{\pi}{4(\sqrt{2}-1)}$$

$$= \frac{\pi\left(\sqrt{2}+1\right)}{4\left(\left(\sqrt{2}\right)^2 - 1^2\right)}$$

$$= \frac{\pi\left(\sqrt{2}+1\right)}{4} \quad \textbf{Ans.}$$

**7. The volume of a sphere is increasing at the rate of 3 cubic centimetre per second. Find the rate of increase of its surface area, when the radius is 2 cm.** **[2]**

**Sol.** Let V be the volume and $r$ be the radius of sphere at any time $t$.

Then, $$V = \frac{4}{3}\pi r^3$$

Given, $$\frac{dV}{dt} = 3 \text{ cm}^3/\text{s}$$

Differentiating V w.r.t. $t$, we get

$$\frac{dV}{dt} = \frac{4}{3}\pi \times 3r^2 \cdot \frac{dr}{dt}$$

$$\Rightarrow \quad 3 = 4\pi r^2 \cdot \frac{dr}{dt}$$

$$\Rightarrow \quad \frac{dr}{dt} = \frac{3}{4\pi r^2} \quad \ldots\text{(i)}$$

Now, let S be the surface area of the sphere, then

$$S = 4\pi r^2$$

Differentiating S w.r.t. $t$, we get

$$\frac{dS}{dt} = 8\pi r\frac{dr}{dt}$$

$$= 8\pi r\frac{3}{4\pi r^2} \quad \text{[From eq. (i)]}$$

$$= \frac{6}{r}$$

$$\Rightarrow \quad \left(\frac{dS}{dt}\right)_{r=2} = \frac{6}{2}$$

$$= 3 \text{ cm}^2/\text{sec}$$

$\therefore$ Rate of increase of surface area of the sphere is 3 square centimetre per second. **Ans.**

**8. Show that the function $f(x) = 4x^3 - 18x^2 + 27x - 7$ is always increasing on R.** **[2]**

**Sol.** Given, $f(x) = 4x^3 - 18x^2 + 27x - 7$

Differentiating $f(x)$ w.r.t. $x$, we get

$$f'(x) = 12x^2 - 36x + 27 = 3(4x^2 - 12x + 9)$$

$$= 3(2x - 3)^2$$

$$(2x - 3)^2 \geq 0 \quad \text{for any } x \in R$$

$$\therefore \quad f'(x) \geq 0$$

$\Rightarrow$ The function is always increasing on R.

**Hence Proved.**

**9. Find the vector equation of the line passing through the point A(1, 2, – 1) and parallel to the line $5x - 25 = 14 - 7y = 35z$.** **[2]**

**Sol.** Given line is $5x - 25 = 14 - 7y = 35z$

$$\Rightarrow \quad 5(x - 5) = -7(y - 2) = 35z$$

$$\Rightarrow \quad \frac{x-5}{1/5} = \frac{y-2}{-1/7} = \frac{z-0}{1/35}$$

$$\Rightarrow \quad \frac{x-5}{7} = \frac{y-2}{-5} = \frac{z-0}{1}$$

Direction ratios of this line are 7, – 5, 1.

∴ Vector equation of the line which passes through the point A (1, 2, – 1) and its direction ratio are proportional to 7, – 5, 1 is

$$\vec{r} = \hat{i} + 2\hat{j} - \hat{k} + \lambda\left(7\hat{i} - 5\hat{j} + \hat{k}\right)$$ **Ans.**

**10. Prove that if E and F are independent events, then the events E and F′ are also independent. [2]**

**Sol.** Since E and F are independent events :

$$P(E \cap F) = P(E) \,.\, P(F) \quad ...(i)$$

$$P(E \cap F') = P(E) - P(E \cap F)$$
$$= P(E) - P(E)\,P(F) \quad \text{[From (i)]}$$
$$= P(E)\,(1 - P(F))$$
$$= P(E)\,P(F')$$

∴ E and F′ are also independent. **Hence Proved.**

**11. A small firm manufactures necklaces and bracelets. The total number of necklaces and bracelets that it can handle per day is at most 24. It takes one hour to make a bracelet and half an hour to make a necklace. The maximum number of hours available per day is 16. If the profit on a necklace is ₹ 100 and that on a bracelet is ₹ 300. Formulate an L.P.P. for finding how many of each should be produced daily to maximize the profit ? It is being given that at least one of each must be produced. [2]**

**Sol.** Let the manufacturer produces $x$ pieces of necklaces and $y$ pieces of bracelets.

Since total number of necklaces and bracelets that can be handle per day are 24.

so, $$x + y \le 24 \quad ...(i)$$

To make bracelet one needs one hour and half an hour is need to make necklace and maximum time available is 16 hours

so, $$\frac{1}{2}x + y \le 16 \quad ...(ii)$$

Now, let Z be the profit and we have to maximize it, so our LPP will be

Maximize Z = $100x + 300y$

Subject to constraints :

$$x + y \le 24$$
$$\frac{1}{2}x + y \le 16$$

or $$x + 2y \le 32$$

and $$x, y \ge 1$$ **Ans.**

**12. Find $\int \frac{dx}{x^2 + 4x + 8}$. [2]**

**Sol.** Let $I = \int \frac{dx}{x^2 + 4x + 8}$

$$= \int \frac{dx}{x^2 + 4x + 4 - 4 + 8}$$

$$= \int \frac{dx}{(x+2)^2 + (2)^2}$$

We know that,

$$\int \frac{dx}{x^2 + a^2} = \frac{1}{a}\tan^{-1}\frac{x}{a} + C,$$

where C is constant of integration

∴ $$I = \frac{1}{2}\tan^{-1}\left(\frac{x+2}{2}\right) + C$$ **Ans.**

## SECTION — C

**13. Prove that $\tan\left\{\frac{\pi}{4} + \frac{1}{2}\cos^{-1}\frac{a}{b}\right\} + \tan\left\{\frac{\pi}{4} - \frac{1}{2}\cos^{-1}\frac{a}{b}\right\} = \frac{2b}{a}$ [4]**

**Sol.** Let $\frac{1}{2}\cos^{-1}\frac{a}{b} = A$

$$\text{L.H.S.} = \tan\left(\frac{\pi}{4} + A\right) + \tan\left(\frac{\pi}{4} - A\right)$$

$$= \frac{\tan\frac{\pi}{4} + \tan A}{1 - \tan\frac{\pi}{4}\tan A} + \frac{\tan\frac{\pi}{4} - \tan A}{1 + \tan\frac{\pi}{4}\tan A}$$

$$= \frac{1 + \tan A}{1 - \tan A} + \frac{1 - \tan A}{1 + \tan A}$$

$$= \frac{(1 + \tan A)^2 + (1 - \tan A)^2}{1 - \tan^2 A}$$

$$= \frac{2 + 2\tan^2 A}{1 - \tan^2 A}$$

$$= \frac{2(1 + \tan^2 A)}{1 - \tan^2 A}$$

$$= 2 \times \frac{1}{\cos 2A}$$

$$= 2 \times \frac{1}{\cos\left(2 \times \frac{1}{2}\cos^{-1}\frac{a}{b}\right)}$$

$$= 2 \times \frac{1}{\cos\left(\cos^{-1}\frac{a}{b}\right)} = \frac{2}{a/b}$$

$$= \frac{2b}{a} = \text{R.H.S}$$ **Hence Proved.**

**14. Using properties of determinants, prove that :**

$$\begin{vmatrix} x & x+y & x+2y \\ x+2y & x & x+y \\ x+y & x+2y & x \end{vmatrix} = 9y^2(x+y).$$ **[4]**

**Sol.** Consider, $\begin{vmatrix} x & x+y & x+2y \\ x+2y & x & x+y \\ x+y & x+2y & x \end{vmatrix}$

Applying $C_1 \to C_1 + C_2 + C_3$, we get

$$= \begin{vmatrix} 3x+3y & x+y & x+2y \\ 3x+3y & x & x+y \\ 3x+3y & x+2y & x \end{vmatrix}$$

$$= 3(x+y)\begin{vmatrix} 1 & x+y & x+2y \\ 1 & x & x+y \\ 1 & x+2y & x \end{vmatrix}$$

Applying $R_1 \to R_1 - R_2$ and $R_3 \to R_3 - R_2$, we get

$$= 3(x+y)\begin{vmatrix} 0 & y & y \\ 1 & x & x+y \\ 0 & 2y & -y \end{vmatrix}$$

$= 3(x+y)[-(-y^2-2y^2)]$ (expanded along $C_1$)

$= 3(x+y)\cdot 3y^2$

$= 9y^2(x+y)$ **Hence Proved.**

**OR**

**Let A =** $\begin{pmatrix} 2 & -1 \\ 3 & 4 \end{pmatrix}$, **B =** $\begin{pmatrix} 5 & 2 \\ 7 & 4 \end{pmatrix}$, **C =** $\begin{pmatrix} 2 & 5 \\ 3 & 8 \end{pmatrix}$, **find a matrix D such that CD – AB = 0.**

**Sol.** Let D be the matrix of order 2 × 2,

$$D = \begin{bmatrix} a & b \\ c & d \end{bmatrix}$$

Given, $CD - AB = 0$

$\therefore$ $CD = AB$

$$\begin{bmatrix} 2 & 5 \\ 3 & 8 \end{bmatrix}\begin{bmatrix} a & b \\ c & d \end{bmatrix} = \begin{bmatrix} 2 & -1 \\ 3 & 4 \end{bmatrix}\begin{bmatrix} 5 & 2 \\ 7 & 4 \end{bmatrix}$$

$$\Rightarrow \begin{bmatrix} 2a+5c & 2b+5d \\ 3a+8c & 3b+8d \end{bmatrix} = \begin{bmatrix} 3 & 0 \\ 43 & 22 \end{bmatrix}$$

On comparing both sides, we get

$2a + 5c = 3$ ...(i)

$2b + 5d = 0$

$\Rightarrow b = \frac{-5}{2}d$

$3a + 8c = 43$ ...(ii)

$3b + 8d = 22$ ...(iii)

Substituting $b = \frac{-5d}{2}$ in equation (iii), we get

$$3\left(\frac{-5d}{2}\right) + 8d = 22$$

$$\Rightarrow \frac{-15d}{2} + 8d = 22$$

$$\Rightarrow -15d + 16d = 44$$

$$\Rightarrow d = 44$$

Also, $b = \frac{-5}{2}d$

$= \frac{-5}{2} \times 44 = -110$

From equation (i),

$$a = \frac{3-5c}{2}$$

Substituting in equation (ii), we get

$$3\left(\frac{3-5c}{2}\right) + 8c = 43$$

$$\Rightarrow 9 - 15c + 16c = 86$$

$$\Rightarrow c = 77$$

and $a = \frac{3 - 5\times 77}{2} = \frac{3-385}{2}$

$= \frac{-382}{2} = -191$

$\therefore$ $D = \begin{bmatrix} -191 & -110 \\ 77 & 44 \end{bmatrix}$ **Ans.**

**15. Differentiate the function $(\sin x)^x + \sin^{-1}\sqrt{x}$ with respect to $x$.** **[4]**

**Sol.** Let $y = (\sin x)^x + \sin^{-1}\sqrt{x}$

Let,

$u = (\sin x)^x$ and $v = \sin^{-1}\sqrt{x}$

$\therefore$ $y = u + v$

Differentiating $y$ w.r.t. $x$, we get

$$\frac{dy}{dx} = \frac{du}{dx} + \frac{dv}{dx} \quad ...(i)$$

Now, $u = (\sin x)^x$

Taking log on both sides,

$\log u = \log(\sin x)^x = x \log \sin x$

Differentiating w.r.t. $x$, we get

$$\frac{1}{u}\frac{du}{dx} = \frac{x}{\sin x}\cos x + \log \sin x$$

$$\Rightarrow \frac{du}{dx} = u[x\cot x + \log \sin x]$$

$$= (\sin x)^x[x\cot x + \log\sin x] \quad ...(ii)$$

Also, $v = \sin^{-1}\sqrt{x}$

Differentiating $v$ w.r.t. $x$, we get

$$\frac{dv}{dx} = \frac{1}{\sqrt{1-x}} \times \frac{1}{2\sqrt{x}} = \frac{1}{2\sqrt{x}\sqrt{1-x}} \quad ...(iii)$$

From equations (i), (ii), and (iii)

$$\frac{dy}{dx} = (\sin x)^x\ [x \cot x + \log \sin x] + \frac{1}{2\sqrt{x}\ \sqrt{1-x}}$$

$$\frac{dy}{dx} = (\sin x)^x [x \cot x + \log \sin x] + \frac{1}{2\sqrt{x - x^2}}$$ **Ans.**

**OR**

**If $x^m y^n = (x + y)^{m+n}$, prove that $\dfrac{d^2y}{dx^2} = 0$.**

**Sol.** Given, $x^m y^n = (x + y)^{m+n}$

Taking log on both sides, we get

$$\log x^m y^n = \log (x + y)^{m+n}$$

$$\Rightarrow \quad m \log x + n \log y = (m + n) \log (x + y)$$

Differentiating both sides w.r.t. $x$, we get

$$\frac{m}{x} + \frac{n}{y} \cdot \frac{dy}{dx} = \frac{m+n}{(x+y)} \cdot \left[1 + \frac{dy}{dx}\right]$$

$$\Rightarrow \quad \frac{m}{x} + \frac{n}{y} \cdot \frac{dy}{dx} = \frac{m+n}{x+y} + \frac{m+n}{x+y} \frac{dy}{dx}$$

$$\Rightarrow \quad \frac{dy}{dx}\left(\frac{n}{y} - \frac{m+n}{x+y}\right) = \frac{m+n}{x+y} - \frac{m}{x}$$

$$\Rightarrow \frac{dy}{dx}\left(\frac{nx + ny - my - ny}{y(x+y)}\right) = \frac{mx + nx - mx - my}{x(x+y)}$$

$$\Rightarrow \quad \frac{dy}{dx} = \frac{y}{x}\left(\frac{nx - my}{nx - my}\right)$$

$$\Rightarrow \quad \frac{dy}{dx} = \frac{y}{x}$$

Again, differentiating w.r.t. $x$, we get

$$\frac{d^2y}{dx^2} = \frac{x\dfrac{dy}{dx} - y}{x^2}$$

$$= \frac{x \times \dfrac{y}{x} - y}{x^2}$$

$$= \frac{y - y}{x^2} = 0$$ **Hence Proved.**

**16. Find $\displaystyle\int \frac{2x}{(x^2+1)(x^2+2)^2} dx$** **[4]**

**Sol.** Let
$$I = \int \frac{2x}{(x^2+1)(x^2+2)^2} dx$$

Put $\quad x^2 = t$

$\Rightarrow\ 2xdx = dt$

$$\therefore \qquad I = \int \frac{dt}{(t+1)(t+2)^2}$$

Let

$$\frac{1}{(t+1)(t+2)^2} = \frac{A}{t+1} + \frac{B}{t+2} + \frac{C}{(t+2)^2} \qquad \text{...(A)}$$

$$\Rightarrow \quad \frac{1}{(t+1)(t+2)^2} = \frac{A(t+2)^2 + B(t+1)(t+2) + C(t+1)}{(t+1)(t+2)^2}$$

$$1 = A(t+2)^2 + B(t+1)(t+2) + C(t+1)$$

Equating coefficient of $t^2$, $t$ and constant terms on both sides, we get

$$A + B = 0 \qquad \text{...(i)}$$

$$4A + 3B + C = 0 \qquad \text{...(ii)}$$

$$4A + 2B + C = 1 \qquad \text{...(iii)}$$

Subtracting equation (iii) from (ii), we get

$$B = -1$$

Substituting $B = -1$ in equation (i),

$$A = 1$$

Substituting the values of A and B in (ii), we get

$$4 - 3 + C = 0$$

$$\Rightarrow \qquad C = -1$$

From equation (A),

$$\frac{1}{(t+1)(t+2)^2} = \frac{1}{t+1} + \left(\frac{-1}{t+2}\right) + \left(\frac{-1}{(t+2)^2}\right)$$

$$\Rightarrow \int \frac{1}{(t+1)(t+2)^2} dt$$

$$= \int \frac{1}{t-1} dt - \int \frac{1}{t+2} dt - \int \frac{1}{(t+2)^2} dt$$

$$= \log(t+1) - \log(t+2) + \frac{1}{t+2} + C$$

where C is constant of integration.

$$\therefore \int \frac{2x\,dx}{(x^2+1)(x^2+2)^2} = \log\left|\frac{x^2+1}{x^2+2}\right| + \frac{1}{x^2+2} + C$$

**Ans.**

**17. Evaluate : $\displaystyle\int_0^{\pi} \frac{x \sin x}{1 + \cos^2 x} dx$** **[4]**

**Sol.** Let,
$$I = \int_0^{\pi} \frac{x \sin x}{1 + \cos^2 x} dx \qquad \text{...(i)}$$

$$= \int_0^{\pi} \frac{(\pi - x)\sin(\pi - x)}{1 + \cos^2(\pi - x)}$$

$$\left[\because \int_0^a f(x)\,dx = \int_0^a f(a-x)\,dx\right]$$

$$I = \int_0^{\pi} \frac{(\pi - x).\sin x}{1 + \cos^2 x} \qquad \text{...(ii)}$$

Adding equations (i) and (ii), we get

$$2I = \pi\int_0^{\pi}\frac{\sin x}{1+\cos^2 x}dx$$

Put $\cos x = t$

$\Rightarrow -\sin x\, dx = dt$

when $x = 0, t = 1$

and when $x = \pi, t = -1$

$$2I = \pi\int_1^{-1}\frac{dt}{1+t^2}$$

$$= -\pi\int_{-1}^{1}\frac{dt}{1+t^2}$$

$$\left[\because \int_a^b f(x)dx = -\int_b^a f(x)dx\right]$$

$$\Rightarrow \quad 2I = \pi.\left(\tan^{-1}t\right)_{-1}^{1}$$

$$= \pi[\tan^{-1}1 - \tan^{-1}(-1)]$$

$$\Rightarrow \quad 2I = \pi\left[\frac{\pi}{4}-\left(-\frac{\pi}{4}\right)\right] = \frac{\pi^2}{2}$$

$$\Rightarrow \quad I = \frac{\pi^2}{4}$$ **Ans.**

**OR**

**Evaluate :** $\int_0^{3/2}|x\sin\pi x|\,dx.$

**Sol.** Let $I = \int_0^{3/2}|x\sin\pi x|\,dx.$

$$|x\sin\pi x| = \begin{cases} x\sin\pi x, & 0\le x\le 1 \\ -x\sin\pi x, & 1\le x\le \frac{3}{2}\end{cases}$$

$$I = \int_0^1(x\sin\pi x)dx + \int_1^{3/2}(x\sin\pi x)dx$$

$$= \int_0^1(x\sin\pi x)dx - \int_1^{3/2}(x\sin\pi x)\,dx$$

Applying by parts on both the integrals, we get

$$I = \left(\frac{-x\cos\pi x}{\pi}\right)_0^1 - \int_0^1 -\frac{\cos\pi x}{\pi}dx$$

$$-\left[\left(\frac{-x\cos\pi x}{\pi}\right)_1^{3/2} - \int_1^{3/2} -\frac{\cos\pi x}{\pi}dx\right]$$

$$I = \left[\left(\frac{-\cos\pi+0}{\pi}\right) + \frac{1}{\pi^2}(\sin\pi x)\right]_0^1$$

$$-\left[\left(\frac{\frac{-3}{2}\cos\frac{3\pi}{2}+\cos\pi}{\pi}\right) + \frac{1}{\pi^2}(\sin\pi x)_1^{3/2}\right]$$

$$= \frac{1}{\pi} + \frac{1}{\pi^2}(\sin\pi - \sin 0)$$

$$-\left[\frac{0-1}{\pi} + \frac{1}{\pi^2}\left(\sin\frac{3\pi}{2} - \sin\pi\right)\right]$$

$$= \frac{1}{\pi} + \frac{1}{\pi^2}(0-0) + \frac{1}{\pi} - \frac{1}{\pi^2}(-1-0)$$

$$= \frac{2}{\pi} + \frac{1}{\pi^2}$$

$$\Rightarrow \quad I = \frac{2\pi+1}{\pi^2}$$ **Ans.**

**18. Prove that $x^2-y^2 = c(x^2+y^2)^2$ is the general solution of the differential equation $(x^3-3xy^2)\,dx = (y^3-3x^2y)\,dy$, where $c$ is a parameter. [4]**

**Sol.** Given differential equation is,

$$(x^3-3xy^2)\,dx = (y^3-3x^2y)\,dy$$

$$\Rightarrow \quad \frac{dy}{dx} = \frac{x^3-3xy^2}{y^3-3x^2y} \quad \text{...(i)}$$

Clearly, it is a homogeneous differential equation.

Put $y = vx$

$$\Rightarrow \quad \frac{dy}{dx} = v + x\frac{dv}{dx}$$

From equation (i),

$$v + x\frac{dv}{dx} = \frac{x^3-3x.(vx)^2}{(vx)^3-3x^2.(vx)}$$

$$= \frac{x^3(1-3v^2)}{x^3(v^3-3v)}$$

$$\Rightarrow \quad x\frac{dv}{dx} = \frac{1-3v^2}{v^3-3v} - v$$

$$= \frac{1-3v^2-v^4+3v^2}{v^3-3v}$$

$$= \frac{1-v^4}{v^3-3v}$$

$$\Rightarrow \quad \frac{v^3-3v}{1-v^4}dv = \frac{dx}{x}$$

On integrating both sides, we get

$$\int\frac{v^3-3v}{1-v^4}dv = \int\frac{dx}{x}$$

$$\int\frac{v^3}{1-v^4}dv - 3\int\frac{v}{1-v^4}dv = \int\frac{dx}{x} \quad \text{...(ii)}$$

In the first integral (For L.H.S.)

Put $1-v^4 = t$

$$\Rightarrow \quad -4v^3 dv = dt,$$

$$\Rightarrow \quad v^3\,dv = \frac{dt}{-4}$$

In the second integral, put $v^2 = z$

$$2v dv = dz \Rightarrow v dv = \frac{dz}{2}$$

From equation (ii), we get

$$\frac{-1}{4}\int\frac{dt}{t} - \frac{3}{2}\int\frac{dz}{1-z^2} = \int\frac{dx}{x}$$

$$\Rightarrow \frac{-1}{4}\log t - \frac{3}{2}\times\frac{1}{2}\log\frac{1+z}{1-z} = \log x + \log A$$

$$\left[\because \int\frac{1}{a^2-x^2} = \frac{1}{2a}\log\left|\frac{a+x}{a-x}\right| + C\right]$$

where log A is constant of integration.

$$\Rightarrow \frac{3}{4}\log\frac{1-z}{1+z} - \frac{1}{4}\log t = \log Ax$$

$$\Rightarrow \frac{1}{4}\log\left|\left(\frac{1-z}{1+z}\right)^3 \times \frac{1}{t}\right| = \log Ax$$

$$\Rightarrow \log\left(\frac{\left(1-\frac{y^2}{x^2}\right)^3}{\left(1+\frac{y^2}{x^2}\right)^3\left(1-\frac{y^4}{x^4}\right)}\right)^{1/4} = \log Ax$$

$$\Rightarrow \log\left(\frac{\frac{(x^2-y^2)^3}{(x^2)^3}}{\frac{(x^2+y^2)^3}{(x^2)^3}\;\frac{(x^4-y^4)}{x^4}}\right)^{1/4} = \log Ax$$

$$\Rightarrow \left(\frac{(x^2-y^2)^3\,x^4}{(x^2+y^2)^3.(x^2-y^2)(x^2+y^2)}\right)^{1/4} = Ax$$

$$\Rightarrow \left(\frac{(x^2-y^2)^2\,x^4}{(x^2+y^2)^4}\right)^{1/4} = Ax$$

$$\Rightarrow \frac{(x^2-y^2)^{1/2}\,x}{(x^2+y^2)} = Ax$$

$$\Rightarrow \quad (x^2-y^2)^{1/2} = A(x^2+y^2)$$

On squaring both sides, we get

$$x^2 - y^2 = A^2\,(x^2+y^2)^2$$

or $\quad x^2 - y^2 = c\,(x^2+y^2)^2,$

[where $A^2 = c$]

$x^2 - y^2 = c\,(x^2+y^2)^2$ is the solution of given differential equation **Ans.**

**19. Let $\vec{a} = \hat{i}+\hat{j}+\hat{k}$, $\vec{b} = \hat{i}$ and $\vec{c} = c_1\hat{i}+c_2\hat{j}+c_3\hat{k}$, then :**

**(a) Let $c_1 = 1$ and $c_2 = 2$, find $c_3$ which makes $\vec{a}$, $\vec{b}$ and $\vec{c}$ coplanar.**

**(b) If $c_2 = -1$ and $c_3 = 1$, show that no value of $c_1$ can make $\vec{a}$, $\vec{b}$ and $\vec{c}$ coplanar. [4]**

**Sol.** Given, $\vec{a} = \hat{i}+\hat{j}+\hat{k}$, $\vec{b} = \hat{i}$ and $\vec{c} = c_1\hat{i} + c_2\hat{j} + c_3\hat{k}$

**(a)** When $c_1 = 1$ and $c_2 = 2$

$$\vec{c} = \hat{i} + 2\hat{j} + c_3\hat{k}$$

We know that, $\vec{a}, \vec{b}, \vec{c}$ are coplanar if

$$[\vec{a}\;\vec{b}\;\vec{c}] = \vec{a}.(\vec{b}\times\vec{c}) = 0 \qquad ...(i)$$

$$\vec{b}\times\vec{c} = \begin{vmatrix}\hat{i} & \hat{j} & \hat{k}\\ 1 & 0 & 0\\ 1 & 2 & c_3\end{vmatrix} = -c_3\,\vec{j} + 2\hat{k}$$

From equation (i),

$$\vec{a}.(\vec{b}\times\vec{c}) = (\vec{i}+\vec{j}+\vec{k}).(-c_3\hat{j}+2\hat{k}) = 0$$

$$\Rightarrow \quad -c_3 + 2 = 0$$

$$\Rightarrow \quad c_3 = 2.$$ **Ans.**

**(b)** When $c_2 = -1$ and $c_3 = 1$

$$\vec{c} = c_1\hat{i} - \hat{j} + \hat{k}$$

$$\vec{b}\times\vec{c} = \begin{vmatrix}\hat{i} & \hat{j} & \hat{k}\\ 1 & 0 & 0\\ c_1 & -1 & 1\end{vmatrix} = -\hat{j} - \hat{k}$$

From equation (i),

$$(\hat{i}+\hat{j}+\hat{k}).(-\hat{j}-\hat{k}) = 0$$

$\Rightarrow -1 - 1 = 0$, which is not possible

$\Rightarrow$ No value of $c_1$ can make $\vec{a}, \vec{b}$ and $\vec{c}$ coplanar. **Hence Proved.**

**20. If $\vec{a}, \vec{b}, \vec{c}$ are mutually perpendicular vectors of equal magnitudes, show that the vector $\vec{a}+\vec{b}+\vec{c}$ is equally inclined to $\vec{a}, \vec{b}$ and $\vec{c}$. Also, find the angle which $\vec{a}+\vec{b}+\vec{c}$ makes with $\vec{a}$ or $\vec{b}$ or $\vec{c}$. [4]**

**Sol.** Let $|\vec{a}| = |\vec{b}| = |\vec{c}| = \lambda$ ...(i)

Given, $\vec{a}, \vec{b}, \vec{c}$ are mutually perpendicular

$\Rightarrow \quad \vec{a}.\vec{b} = \vec{b}.\vec{c} = \vec{c}.\vec{a} = 0$...(ii)

$\therefore \quad (\vec{a}+\vec{b}+\vec{c})^2 = (\vec{a}+\vec{b}+\vec{c}).(\vec{a}+\vec{b}+\vec{c})$

$= \vec{a}.\vec{a} + \vec{a}.\vec{b} + \vec{a}.\vec{c} + \vec{b}.\vec{a} + \vec{b}.\vec{b} + \vec{b}.\vec{c} + \vec{c}.\vec{a} + \vec{c}.\vec{b} + \vec{c}.\vec{c}$

$\Rightarrow |\vec{a}|^2 + |\vec{b}|^2 + |\vec{c}|^2 = 3\lambda^2$ [using (i) and (ii)]

$\Rightarrow \quad |\vec{a}+\vec{b}+\vec{c}| = \sqrt{3}\lambda$ ...(iii)

Suppose $\vec{a}+\vec{b}+\vec{c}$ makes angles $\theta_1, \theta_2, \theta_3$ with $\vec{a}, \vec{b}$ and $\vec{c}$ respectively, then

$(\vec{a}+\vec{b}+\vec{c}).\vec{a} = |\vec{a}+\vec{b}+\vec{c}||\vec{a}|\cos\theta_1$

$\Rightarrow \vec{a}.\vec{a} + \vec{b}.\vec{a} + \vec{c}.\vec{a} = \sqrt{3}\lambda.\lambda\cos\theta_1$

...[using (i) & (iii)]

$\Rightarrow \quad |\vec{a}|^2 = \sqrt{3}\lambda^2\cos\theta_1$

[using (ii)]

$\Rightarrow \quad \lambda^2 = \sqrt{3}\lambda^2\cos\theta_1$

$\Rightarrow \quad \cos\theta_1 = \dfrac{1}{\sqrt{3}}$

$\Rightarrow \quad \theta_1 = \cos^{-1}\left(\dfrac{1}{\sqrt{3}}\right)$

Similarly, $\theta_2 = \cos^{-1}\left(\dfrac{1}{\sqrt{3}}\right)$

and $\theta_3 = \cos^{-1}\left(\dfrac{1}{\sqrt{3}}\right)$

$\therefore \quad \theta_1 = \theta_2 = \theta_3$

Hence, $\vec{a}+\vec{b}+\vec{c}$ is equally inclined with $\vec{a}, \vec{b}$ and $\vec{c}$. **Hence Proved.**

**21. The random variable $X$ can take only the values 0, 1, 2, 3. Given that P (X = 0) = P (X = 1) = $p$ and P(X = 2) = P(X = 3) such that $\Sigma p_i x_i^2 = 2\Sigma p_i x_i$, find the value of $p$. [4]**

**Sol.** Given, P(X = 0) = P (X = 1) = $p$ and P (X = 2) = P (X = 3)

Let P (X = 2) = P(X = 3) = $k$

| X | 0 | 1 | 2 | 3 |
|---|---|---|---|---|
| P($x$) | $p$ | $p$ | $k$ | $k$ |

also given that $\Sigma p_i x_i^2 = 2\Sigma p_i x_i$

$\Rightarrow \quad 0 + p + 4k + 9k = 2(0 + p + 2k + 3k)$

$\Rightarrow \quad p + 13k = 2(p + 5k)$

$\Rightarrow \quad 3k = p$ ...(i)

also we know that $\Sigma p_i = 1$

$\Rightarrow \quad p + p + k + k = 1$

$\Rightarrow \quad 2p + 2k = 1$

$\Rightarrow \quad 6k + 2k = 1$ [using (i)]

$\Rightarrow \quad 8k = 1$

or $\quad k = \dfrac{1}{8}$

From equation (i), $\quad p = \dfrac{3}{8}$ **Ans.**

**22. Often it is taken that a truthful person commands, more respect in the society. A man is known to speak the truth 4 out of 5 times. He throws a die and reports that it is a six. Find the probability that it is actually a six.**

**Do you also agree that the value of truthfulness leads to more respect in the society ? [4]**

**Sol.** Let $E_1$, $E_2$ and A be the events defined as follows:

$E_1$ = Six appears on throwing a die.

$E_2$ = Six does not appear on throwing a die.

and A = the man reports that it is a six

We have,

$$P(E_1) = \frac{1}{6}$$

$$P(E_2) = \frac{5}{6}$$

Now P(A/$E_1$) = Probability that the man reports that there is a six on the die given that six has occurred on the die = 4/5 (probability that the man speaks truth)

and P(A/$E_2$) = Probability that the man reports that there is a six on the die given that six has not occurred on the die (probability that the man does not speak truth).

$$= 1 - \frac{4}{5} = \frac{1}{5}$$

By Bayes' theorem, we have

$$P(E_1/A) = \frac{P(E_1)\,P(A/E_1)}{P(E_1)\,P(A/E_1) + P(E_2)\,P(A/E_2)}$$

$$= \frac{\frac{1}{6}\times\frac{4}{5}}{\frac{1}{6}\times\frac{4}{5}+\frac{5}{6}\times\frac{1}{5}}$$

$$= \frac{4}{4+5} = \frac{4}{9}$$

Yes, truthfulness always leads to more respect in the society as truth always wins. **Ans.**

**23. Solve the following L.P.P. graphically :**

**Minimise** $Z = 5x + 10y$

**Subject to constraints**

$$x + 2y \le 120$$
$$x + y \ge 60$$
$$x - 2y \ge 0$$

**and** $x, y \ge 0$ **[4]**

**Sol.** We have,

Minimise $Z = 5x + 10y$

Subject to the constraints :

$$x + 2y \le 120$$
$$x + y \ge 60$$
$$x - 2y \ge 0$$

and $x, y \ge 0$

Converting the given inequalities into equations, we obtain the following equations :

$$x + 2y = 120$$
$$x + y = 60$$
$$x - 2y = 0$$

Then, $x + 2y = 120$

| $x$ | 0 | 120 |
|---|---|---|
| $y$ | 60 | 0 |

$x + y = 60$

| $x$ | 0 | 60 |
|---|---|---|
| $y$ | 60 | 0 |

$x - 2y = 0$

| $x$ | 20 | 60 |
|---|---|---|
| $y$ | 10 | 30 |

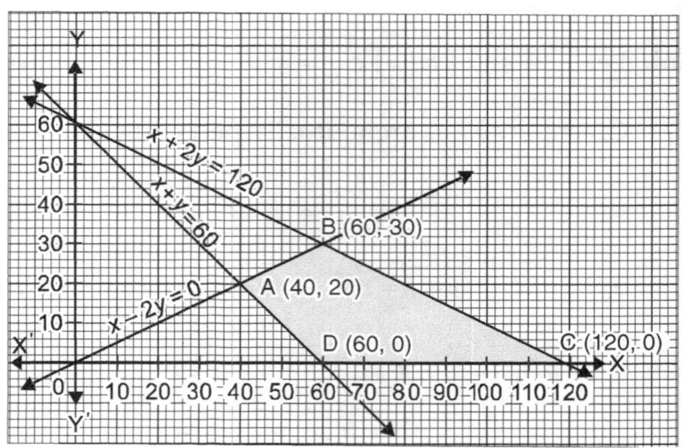


The shaded region ABCD represented by the given constraints is the feasible region. Corner points of the common shaded region are A (40, 20), B (60, 30), C (120, 0) and D (60, 0).

Value of Z at each corner point is given as :

| Corner Point | $Z = 5x + 10y$ |
|---|---|
| A (40, 20) | 200 + 200 = 400 |
| B (60, 30) | 300 + 300 = 600 |
| C (120, 0) | 600 + 0 = 600 |
| D (60, 0) | 300 + 0 = 300 ← Minimum |

Hence, minimum value of Z is 300 at (60, 0). **Ans.**

## SECTION — D

**24. Use product** $\begin{bmatrix}1 & -1 & 2\\ 0 & 2 & -3\\ 3 & -2 & 4\end{bmatrix}\begin{bmatrix}-2 & 0 & 1\\ 9 & 2 & -3\\ 6 & 1 & -2\end{bmatrix}$ **to solve the system of equations** $x + 3z = 9, -x + 2y - 2z = 4, 2x - 3y + 4z = -3.$ **[6]**

**Sol.** Consider,

$$\begin{bmatrix}1 & -1 & 2\\ 0 & 2 & -3\\ 3 & -2 & 4\end{bmatrix}\begin{bmatrix}-2 & 0 & 1\\ 9 & 2 & -3\\ 6 & 1 & -2\end{bmatrix}$$

$$= \begin{bmatrix}-2-9+12 & 0-2+2 & 1+3-4\\ 0+18-18 & 0+4-3 & 0-6+6\\ -6-18+24 & 0-4+4 & 3+6-8\end{bmatrix}$$

$$= \begin{bmatrix}1 & 0 & 0\\ 0 & 1 & 0\\ 0 & 0 & 1\end{bmatrix} = I$$

Hence, $\begin{bmatrix}1 & -1 & 2\\ 0 & 2 & -3\\ 3 & -2 & 4\end{bmatrix}^{-1} = \begin{bmatrix}-2 & 0 & 1\\ 9 & 2 & -3\\ 6 & 1 & -2\end{bmatrix}$

Now, given system of equations can be written in matrix form as follows :

$$\begin{bmatrix} 1 & 0 & 3 \\ -1 & 2 & -2 \\ 2 & -3 & 4 \end{bmatrix}\begin{bmatrix} x \\ y \\ z \end{bmatrix} = \begin{bmatrix} 9 \\ 4 \\ -3 \end{bmatrix}$$

$$AX = B$$

$$\Rightarrow \quad X = A^{-1}B$$

$$\Rightarrow \quad \begin{bmatrix} x \\ y \\ z \end{bmatrix} = \begin{bmatrix} 1 & 0 & 3 \\ -1 & 2 & -2 \\ 2 & -3 & 4 \end{bmatrix}^{-1}\begin{bmatrix} 9 \\ 4 \\ -3 \end{bmatrix}$$

$$= \left[\begin{bmatrix} 1 & -1 & 2 \\ 0 & 2 & -3 \\ 3 & -2 & 4 \end{bmatrix}^{T}\right]^{-1}\begin{bmatrix} 9 \\ 4 \\ -3 \end{bmatrix}$$

$$= \left[\begin{bmatrix} 1 & -1 & 2 \\ 0 & 2 & -3 \\ 3 & -2 & 4 \end{bmatrix}^{-1}\right]^{T}\begin{bmatrix} 9 \\ 4 \\ -3 \end{bmatrix}$$

$$= \begin{bmatrix} -2 & 0 & 1 \\ 9 & 2 & -3 \\ 6 & 1 & -2 \end{bmatrix}^{T}\begin{bmatrix} 9 \\ 4 \\ -3 \end{bmatrix}$$

$$= \begin{bmatrix} -2 & 9 & 6 \\ 0 & 2 & 1 \\ 1 & -3 & -2 \end{bmatrix}\begin{bmatrix} 9 \\ 4 \\ -3 \end{bmatrix}$$

$$\Rightarrow \quad \begin{bmatrix} x \\ y \\ z \end{bmatrix} = \begin{bmatrix} -18+36-18 \\ 0+8-3 \\ 9-12+6 \end{bmatrix} = \begin{bmatrix} 0 \\ 5 \\ 3 \end{bmatrix}$$

$\Rightarrow x = 0, \ y = 5, z = 3$ **Ans.**

**25. Consider $f : R_+ \to [-5, \infty)$, given by $f(x) = 9x^2 + 6x - 5$. Show that $f$ is invertible with $f^{-1}(y) = \left(\dfrac{\sqrt{y+6} - 1}{3}\right)$.** **[6]**

**Hence find :**

**(i) $f^{-1}(10)$**

**(ii) $y$ if $f^{-1}(y) = \dfrac{4}{3}$,**

**where $R_+$ is the set of all non-negative real numbers.**

**Sol.** For one-one :

Let $x_1, x_2 \in R_+$

$$f(x_1) = f(x_2)$$

$$\Rightarrow \quad 9x_1^2 + 6x_1 - 5 = 9x_2^2 + 6x_2 - 5$$

$$\Rightarrow \quad 9(x_1^2 - x_2^2) + 6(x_1 - x_2) = 0$$

$$\Rightarrow (x_1 - x_2)(9(x_1 + x_2) + 6) = 0$$

Since $\quad 9(x_1 + x_2) + 6 > 0 \quad [\text{as } x_1, x_2 \in R_+]$

$$\Rightarrow \quad x_1 - x_2 = 0$$

or $\quad x_1 = x_2$

∴ Function is one-one

For onto :

For every $y \in [-5, \infty)$ such that $f(x) = y$

$$\Rightarrow \quad 9x^2 + 6x - 5 = y$$

$$\Rightarrow \quad (3x)^2 + 2.(3x).1 + 1^2 - 1^2 - 5 = y$$

$$\Rightarrow \quad (3x+1)^2 - 6 = y$$

$$\Rightarrow \quad (3x+1)^2 = y + 6$$

$$\Rightarrow \quad 3x + 1 = \sqrt{y+6}$$

$$\Rightarrow \quad x = \frac{-1+\sqrt{y+6}}{3} \in R_+$$

∴ Function is onto.

Since $f$ is both one-one and onto.

∴ Function is invertible.

$$f(x) = y$$

$$\Rightarrow \quad x = f^{-1}(y) = \frac{-1+\sqrt{y+6}}{3}$$

(i) $$f^{-1}(10) = \frac{-1+\sqrt{16}}{3} = \frac{-1+4}{3} = 1$$ **Ans.**

(ii) $$f^{-1}(y) = \frac{4}{3}$$

$$\Rightarrow \quad \frac{4}{3} = \frac{-1+\sqrt{y+6}}{3},$$

$$\Rightarrow \quad 4 = -1 + \sqrt{y+6}$$

$$\Rightarrow \quad 5 = \sqrt{y+6}$$

Squaring on both sides,

$$25 = y + 6$$

$$\Rightarrow \quad y = 19$$ **Ans.**

**OR**

**Discuss the commutativity and associativity of binary operation '*' defined on A = Q – {1} by the rule $a * b = a - b + ab$ for all, $a, b \in$ A. Also find the identity element of * in A and hence find the invertible elements of A.****

**26. If the sum of lengths of the hypotenuse and a side of a right angled triangle is given, show that the area of the triangle is maximum, when the angle between them is $\dfrac{\pi}{3}$.** **[6]**

**Sol.** Let $h$, $x$ and $y$ be the length of hypotenuse and sides of the right triangle ABC.

** Answer is not given due to change in the present syllabus.

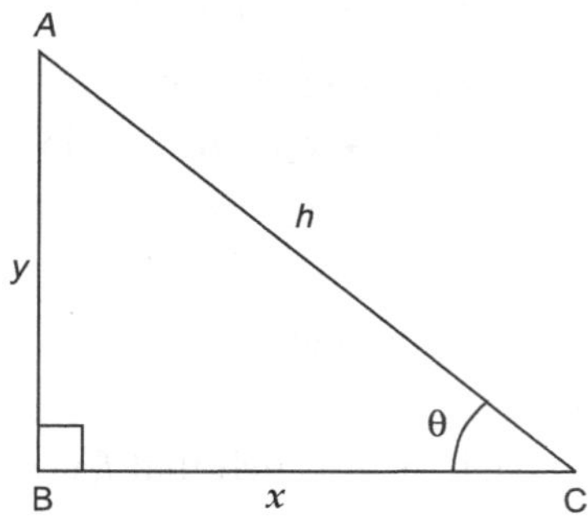


From $\Delta$ABC

$$h^2 = x^2 + y^2$$

$$\Rightarrow \quad y^2 = h^2 - x^2$$

If A be the area of the triangle ABC, then

$$A = \frac{1}{2}xy = \frac{1}{2}x\sqrt{h^2 - x^2}$$

$$\Rightarrow \quad A^2 = \frac{x^2}{4}(h^2 - x^2)$$

Let $\quad A^2 = z$

[*also, given*

$$h + x = k \text{ (constant)}$$

$$\Rightarrow \quad h = k - x]$$

$$\Rightarrow \quad z = \frac{x^2}{4}((k - x)^2 - x^2)$$

$$= \frac{x^2}{4}(k^2 - 2kx)$$

$$= \frac{k^2x^2 - 2kx^3}{4}$$

Differentiating $z$ w.r. t. $x$, we get

$$\frac{dz}{dx} = \frac{2k^2x - 6kx^2}{4}$$

$$= \frac{k^2x - 3kx^2}{2}$$

Again, differentiating w.r.t. $x$, we get

$$\frac{d^2z}{dx^2} = \frac{k^2 - 6kx}{2}$$

For maxima or minima

Put $\quad \frac{dz}{dx} = 0$

$$\Rightarrow \quad \frac{k^2x - 3kx^2}{2} = 0$$

$$\Rightarrow \quad kx.\,(k - 3x) = 0$$

$$\Rightarrow \quad k - 3x = 0, \text{ as } x \neq 0$$

or $\quad x = \frac{k}{3}$

$$h = k - x = k - \frac{k}{3} = \frac{2k}{3}$$

$$y^2 = h^2 - x^2$$

$$= \frac{4k^2}{9} - \frac{k^2}{9}$$

$$= \frac{3k^2}{9}$$

$$= \frac{k^2}{3}$$

$$\Rightarrow \quad y = \frac{k}{\sqrt{3}}$$

when $x = \frac{k}{3}$

$$\left(\frac{d^2z}{dx^2}\right)_{x=\frac{k}{3}} = \frac{1}{2}\left(k^2 - 6k \times \frac{k}{3}\right)$$

$$= \frac{-k^2}{2} < 0$$

$\therefore$ Area of the triangle is maximum.

From $\Delta$ABC

$$\cos\theta = \frac{x}{h} = \frac{k/3}{2k/3} = \frac{1}{2}$$

$$\Rightarrow \quad \theta = \frac{\pi}{3}$$

**Hence Proved.**

**27. Using integration, find the area of region bounded by the triangle whose vertices are (– 2, 1), (0, 4) and (2, 3). [6]**

**Sol.** The vertices of the $\Delta$ABC are A(– 2, 1), B(0, 4) and C(2, 3).

Equation of the side AB is

$$y - y_1 = \frac{y_2 - y_1}{x_2 - x_1}(x - x_1)$$

$$\Rightarrow \quad y - 1 = \frac{4-1}{0-(-2)}\{x - (-2)\} = \frac{3}{2}(x+2)$$

$$\Rightarrow \quad y = \frac{3}{2}x + 4$$

Equation of the side BC is,

$$y - 4 = \frac{3-4}{2-0}(x - 0) = \frac{-1}{2}x$$

$$\Rightarrow \quad y = \frac{-1}{2}x + 4$$

Equation of the side AC is,

$$y - 1 = \frac{3-1}{2-(-2)}\{x - (-2)\} = \frac{1}{2}(x+2)$$

$$\Rightarrow \quad y = \frac{1}{2}x + 2$$

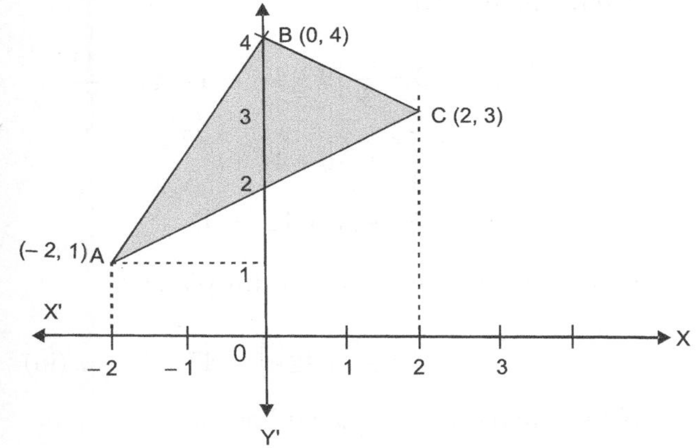


Required area = Shaded area

$$= \int_{-2}^{0}\left(\frac{3}{2}x+4\right)dx + \int_{0}^{2}\left(\frac{-1}{2}x+4\right)dx - \int_{-2}^{2}\left(\frac{1}{2}x+2\right)dx$$

$$= \left[\frac{3}{2}.\frac{x^2}{2}+4x\right]_{-2}^{0} + \left[\frac{-1}{2}.\frac{x^2}{2}+4x\right]_{0}^{2} - \left[\frac{1}{2}.\frac{x^2}{2}+2x\right]_{-2}^{2}$$

$$= (0+0)-(3-8)+(-1+8)-(0+0)-(1+4)+(1-4)$$

$$= 0+5+7-0-5-3$$

$= 4$ sq. units. **Ans.**

**OR**

**Find the area bounded by the circle $x^2 + y^2 = 16$ and the line $\sqrt{3}y = x$ in the first quadrant, using integration.**

**Sol.** Given, $x = \sqrt{3}y$

and $x^2 + y^2 = 16,$

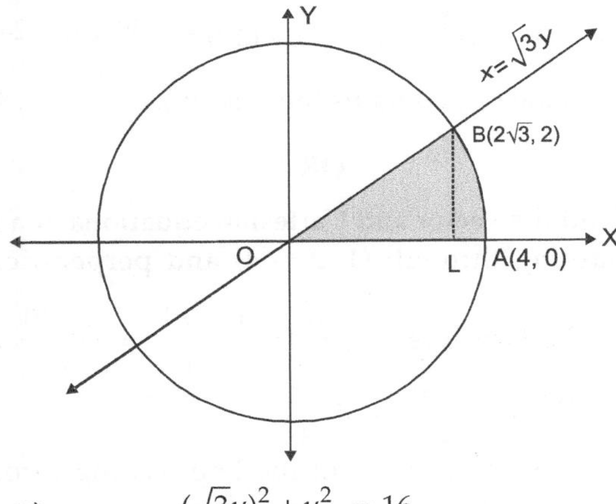


$\Rightarrow \quad (\sqrt{3}y)^2 + y^2 = 16$

$\Rightarrow \quad 4y^2 = 16$

$\Rightarrow \quad y^2 = 4$

$\Rightarrow \quad y = 2$

$\therefore \quad x = \sqrt{3}y = 2\sqrt{3}$

$\therefore$ B$(2\sqrt{3}, 2)$ is the point of intersection in first quadrant.

Required area = Area under OBL +Area under LBA

$$= \int_{0}^{2\sqrt{3}} \frac{x}{\sqrt{3}}dx + \int_{2\sqrt{3}}^{4}\sqrt{16-x^2}\,dx$$

$$= \frac{1}{\sqrt{3}}\left(\frac{x^2}{2}\right)_0^{2\sqrt{3}} + \left[\frac{x}{2}\sqrt{16-x^2}+\frac{16}{2}\sin^{-1}\frac{x}{4}\right]_{2\sqrt{3}}^{4}$$

$$= \frac{1}{2\sqrt{3}}(12-0) + \left(0+8\sin^{-1}1\right) - \left(\frac{2\sqrt{3}}{2}\sqrt{16-12}+8\sin^{-1}\frac{2\sqrt{3}}{4}\right)$$

$$= \frac{6}{\sqrt{3}} + 8\times\frac{\pi}{2} - 2\sqrt{3} - 8\sin^{-1}\frac{\sqrt{3}}{2}$$

$$= \frac{6\sqrt{3}}{3} + 4\pi - 2\sqrt{3} - 8\times\frac{\pi}{3}$$

$= 2\sqrt{3} + \frac{4\pi}{3} - 2\sqrt{3} = \frac{4\pi}{3}$ sq. units. **Ans.**

**28. Solve the differential equation $x\frac{dy}{dx} + y = x\cos x + \sin x$, given that $y = 1$ when $x = \frac{\pi}{2}$.** **[6]**

**Sol.** Given differential equation is :

$$x\frac{dy}{dx} + y = x\cos x + \sin x$$

$$\Rightarrow \quad \frac{dy}{dx} + \frac{y}{x} = \cos x + \frac{\sin x}{x}$$

which is of the form $\frac{dy}{dx} + Py = Q$,

where $P = \frac{1}{x}$, $Q = \cos x + \frac{\sin x}{x}$

$$\text{I.F.} = e^{\int P dx} = e^{\int \frac{1}{x}dx} = e^{\log x} = x$$

Required solution is

$$y\cdot\text{I}\cdot\text{F}\cdot = \int \text{Q.I.F.} + \text{C}$$

$$y.\,x = \int\left(\cos x + \frac{\sin x}{x}\right)x\,dx + C$$

$$= \int x\cos x\,dx + \int \sin x\,dx + C$$

$$= x.\int\cos x\,dx - \int\left[\frac{d}{dx}x.\int\cos x\,dx\right]dx - \cos x + C$$

$$\Rightarrow \quad xy = x\sin x - \int\sin x\,dx - \cos x + C$$

$\Rightarrow \quad xy = x \sin x + \cos x - \cos x + C$

$\Rightarrow \quad xy = x \sin x + C \qquad ...(i)$

Given, $y = 1$ when $x = \frac{\pi}{2}$

From equation (i),

$$1 \times \frac{\pi}{2} = \frac{\pi}{2} \sin \frac{\pi}{2} + C$$

$$\Rightarrow \quad \frac{\pi}{2} = \frac{\pi}{2} + C$$

$$\Rightarrow \quad C = 0$$

Substitute the value of C = 0 in (i), we get

$$xy = x \sin x$$

$\Rightarrow \quad y = \sin x$, which is the required solution.

**Ans.**

**29. Find the equation of the plane through the line of intersection of $\vec{r} \cdot (2\hat{i} - 3\hat{j} + 4\hat{k}) = 1$ and $\vec{r} \cdot (\hat{i} - \hat{j}) + 4 = 0$ and perpendicular to the plane $\vec{r} \cdot (2\hat{i} - \hat{j} + \hat{k}) + 8 = 0$. Hence find whether the plane thus obtained contains the line $x - 1 = 2y - 4 = 3z - 12$. [6]**

**Sol.** The equation of the plane passing through the line of intersection of the given planes is :

$$r[(2\hat{i} - 3\hat{j} + 4\hat{k}) - 1 + \lambda(\hat{i} - \hat{j})] + 4\lambda - 1 = 0$$

$$\Rightarrow \quad \vec{r} \cdot [(2 + \lambda)\hat{i} - (3 + \lambda)\hat{j} + 4\hat{k}] = 1 - 4\lambda$$

Taking $\vec{r} = x\hat{i} + y\hat{j} + z\hat{k}$, we get

$$(2 + \lambda)x - (3 + \lambda)y + 4z = 1 - 4\lambda \qquad ...(i)$$

$\because$ It is perpendicular to the plane,

$$\therefore \quad \vec{r} \cdot (2\hat{i} - \hat{j} + \hat{k}) + 8 = 0$$

Cartesian equation of this plane is :

$$2x - y + z + 8 = 0 \qquad ...(ii)$$

$\because$ Equations (i) and (ii) are perpendicular

$$\therefore \quad (2 + \lambda)2 + (3 + \lambda) + 4 = 0$$

$$\Rightarrow \quad 4 + 2\lambda + 3 + \lambda + 4 = 0$$

$$\Rightarrow \quad 11 + 3\lambda = 0$$

$$\text{or} \quad \lambda = \frac{-11}{3}$$

From equation (i),

$$\left(2 - \frac{11}{3}\right)x - \left(3 - \frac{11}{3}\right)y + 4z = 1 - 4 \times \frac{-11}{3}$$

$$\Rightarrow \quad \frac{-5x}{3} + \frac{2}{3}y + 4z = \frac{47}{3}$$

$$\Rightarrow \quad -5x + 2y + 12z = 47$$

Required vector equation of this plane is :

$$\vec{r} \cdot \left(-5\hat{i} + 2\hat{j} + 12\hat{k}\right) = 47 \qquad ...(iii)$$

Now, equation of the given line is,

$$x - 1 = 2y - 4 = 3z - 12$$

$$\Rightarrow \quad \frac{x-1}{1} = 2(y - 2) = 3(z - 4)$$

$$\Rightarrow \quad \frac{x-1}{1} = \frac{y-2}{1/2} = \frac{z-4}{1/3}$$

$$\Rightarrow \quad \frac{x-1}{6} = \frac{y-2}{3} = \frac{z-4}{2}$$

Vector equation of this line is :

$$\vec{r} = (\hat{i} + 2\hat{j} + 4\hat{k}) + \lambda(6\hat{i} + 3\hat{j} + 2\hat{k}) \qquad ...(iv)$$

Obviously plane (iii) contains the line (iv) since the point $\hat{i} + 2\hat{j} + 4\hat{k}$ satisfy the equation of plane (iii) $\left[\text{as } (\hat{i} + 2\hat{j} + 4\hat{k}) \cdot (-5\hat{i} + 2\hat{j} + 12\hat{k}) = -5 + 4 + 48 = 47\right]$ and vector $-5\hat{i} + 2\hat{j} + 12\hat{k}$ is perpendicular to $6\hat{i} + 3\hat{j} + 2\hat{k}$ as $\left(6\hat{i} + 3\hat{j} + 2\hat{k}\right) \cdot \left(-5\hat{i} + 2\hat{j} + 12\hat{k}\right) = -30 + 6 + 24 = 0$

$\therefore$ Plane (iii) contains the line (iv). **Ans.**

**OR**

**Find the vector and Cartesian equations of a line passing through (1, 2, – 4) and perpendicular to the two lines $\frac{x-8}{3} = \frac{y+19}{-16} = \frac{z-10}{7}$ and $\frac{x-15}{3} = \frac{y-29}{8} = \frac{z-5}{-5}$.**

**Sol.** Cartesian equation of the line passing through (1, 2, – 4) is :

$$\frac{x-1}{a} = \frac{y-2}{b} = \frac{z+4}{c} \qquad ...(i)$$

Given lines are :

$$\frac{x-8}{3} = \frac{y+19}{-16} = \frac{z-10}{7} \qquad ...(ii)$$

and $\frac{x-15}{3} = \frac{y-29}{8} = \frac{z-5}{-5}$ ...(iii)

Let $\vec{b_1}, \vec{b_2}, \vec{b_3}$ are parallel vectors of (i), (ii) and (iii) respectively :

$$\vec{b_1} = a\hat{i} + b\hat{j} + c\hat{k}$$

$$\vec{b_2} = 3\hat{i} - 16\hat{j} + 7\hat{k}$$

$$\vec{b_3} = 3\hat{i} + 8\hat{j} - 5\hat{k}$$

Given that (i) is perpendicular to both (ii) and (iii)

$\Rightarrow \quad \vec{b_1} \cdot \vec{b_2} = 0$

$3a - 16b + 7c = 0$ ...(iv)

and $\quad \vec{b_1} \cdot \vec{b_3} = 0$

$3a + 8b - 5c = 0$ ...(v)

From equations (iv) and (v),

$$\frac{a}{80-56} = \frac{b}{21+15} = \frac{c}{24+48}$$

$\Rightarrow \quad \frac{a}{24} = \frac{b}{36} = \frac{c}{72}$

$\Rightarrow \quad \frac{a}{2} = \frac{b}{3} = \frac{c}{6} = \lambda$ (say)

$\Rightarrow \quad a = 2\lambda, b = 3\lambda, c = 6\lambda$

Putting in (i),

$$\frac{x-1}{2\lambda} = \frac{y-2}{3\lambda} = \frac{z+4}{6\lambda}$$

$\Rightarrow \quad \frac{x-1}{2} = \frac{y-2}{3} = \frac{z+4}{6}$

which is the cartesian equation of the line and vector equation of this line is

$\vec{r} = \left(\hat{i} + 2\hat{j} - 4\hat{k}\right) + \lambda\left(2\hat{i} + 3\hat{j} + 6\hat{k}\right)$ **Ans.**

●●

# Delhi [Set-II]

> **Note : Except for the following questions, all the remaining questions have been asked in previous set.**

## SECTION — B

**12. For the curve $y = 5x - 2x^3$, if $x$ increases at the rate of 2 units/sec, then find the rate of change of the slope of the curve when $x = 3$. [2]**

**Sol.** Given curve is,

$$y = 5x - 2x^3$$

and $\quad \frac{dx}{dt} = 2$ units/sec.

Differentiating $y$ w.r.t. $x$, we get

$$\frac{dy}{dx} = 5 - 6x^2.$$

Differentiating both sides w.r.t. $t$, we get

$$\frac{d}{dt}\left(\frac{dy}{dx}\right) = -12x\frac{dx}{dt}$$

when $x = 3$,

$$\frac{d}{dt}\left(\frac{dy}{dx}\right)\bigg|_{x=3} = -12 \times 3 \times 2 = -72 \text{ units/sec.}$$

Thus, the slope of the curve is decreasing at the rate of 72 units/sec when $x = 3$. **Ans.**

## SECTION — C

**20. The random variable X can take only the values 0, 1, 2, 3. Given that P(2) = P(3) = $p$ and P(0) = 2P(1). If $\Sigma p_i x_i^2 = 2\Sigma p_i x_i$, find the value of $p$. [4]**

**Sol.** Given, $\quad P(2) = P(3) = p$

and $\quad P(0) = 2P(1)$

Let $\quad P(1) = k$

| X | 0 | 1 | 2 | 3 |
|---|---|---|---|---|
| P(x) | 2k | k | p | p |

also given that $\Sigma p_i x_i^2 = 2\,\Sigma p_i x_i$

$\Rightarrow \quad 0 + k + 4p + 9p = 2(0 + k + 2p + 3p)$

$\Rightarrow \quad k + 13p = 2k + 10p$

or $\quad 3p = k$ ...(i)

also we know that

$$\Sigma p_i = 1$$

$\Rightarrow \quad 2k + k + p + p = 1$

$\Rightarrow \quad 3k + 2p = 1$

$\Rightarrow \quad 9p + 2p = 1$ [using (i)]

$\Rightarrow \quad 11p = 1$

$\Rightarrow \quad p = \frac{1}{11}.$ **Ans.**

**21. Using vectors find the area of triangle ABC with vertices A(1, 2, 3), B(2, – 1, 4) and C (4, 5, – 1). [4]**

**Sol.** Vertices of the given ΔABC are A(1, 2, 3), B(2, – 1, 4) and C(4, 5, – 1)

$$\vec{AB} = \vec{OB} - \vec{OA}$$

$$= (2\hat{i} - \hat{j} + 4\hat{k}) - (\hat{i} + 2\hat{j} + 3\hat{k})$$

$$= \hat{i} - 3\hat{j} + \hat{k}$$

$$\vec{AC} = \vec{OC} - \vec{OA}$$

$$= (4\hat{i} + 5\hat{j} - \hat{k}) - (\hat{i} + 2\hat{j} + 3\hat{k})$$

$$= 3\hat{i} + 3\hat{j} - 4\hat{k}$$

$$\text{Required area} = \frac{1}{2}|\vec{AB} \times \vec{AC}|$$

$$\vec{AB} \times \vec{AC} = \begin{vmatrix} \hat{i} & \hat{j} & \hat{k} \\ 1 & -3 & 1 \\ 3 & 3 & -4 \end{vmatrix} = 9\hat{i} + 7\hat{j} + 12\hat{k}$$

$$|\vec{AB} \times \vec{AC}| = \sqrt{9^2 + 7^2 + 12^2}$$

$$= \sqrt{81 + 49 + 144} = \sqrt{274}$$

$\therefore$ Area of $\Delta ABC = \frac{1}{2}\sqrt{274}$ sq. units **Ans.**

**22. Solve the following L. P. P. graphically**
**Maximise $Z = 4x + y$**

**Subject to following constraints:**

$$x + y \leq 50,$$
$$3x + y \leq 90,$$
$$x \geq 10$$
$$x, y \geq 0$$ **[4]**

**Sol.** We have,

Maximise $Z = 4x + y$

Subject to the constraints :

$$x + y \leq 50$$
$$3x + y \leq 90$$
$$x \geq 10$$
$$x, y \geq 0$$

Converting the given inequalities into equations, we obtain the following equations :

$$x + y = 50$$
$$3x + y = 90$$
$$x = 10$$

Then, $x + y = 50$ $\quad$ $3x + y = 90$

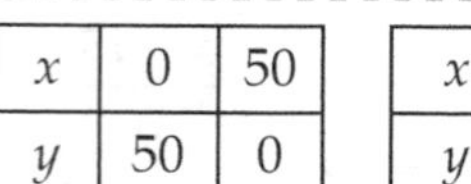

| $x$ | 0 | 50 |
|---|---|---|
| $y$ | 50 | 0 |

| $x$ | 0 | 30 |
|---|---|---|
| $y$ | 90 | 0 |

$x = 10$ is a line which is parallel to Y-axis

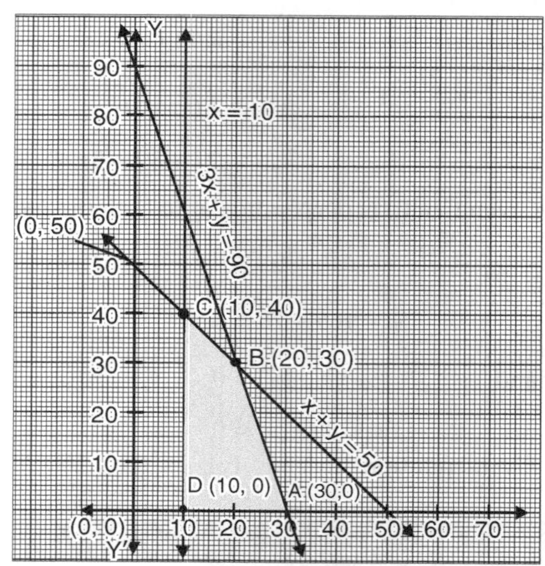


Plotting these points on the graph, we get the shaded feasible region *i.e.*, ABCD.

| Corner point | Value of $Z = 4x + y$ |
|---|---|
| A (30, 0) | $Z = 4 \times 30 + 0 = 120 \leftarrow$ Maximum |
| B (20, 30) | $Z = 4 \times 20 + 30 = 110$ |
| C (10, 40) | $Z = 10 \times 4 + 40 = 80$ |
| D (10, 0) | $Z = 10 \times 4 + 0 = 40$ |

$\therefore$ Maximum value of Z is 120 at (30, 0) **Ans.**

**23. Find : $\int \frac{2x}{(x^2 + 1)(x^4 + 4)} dx.$** **[4]**

**Sol.** Let, $$I = \int \frac{2x}{(x^2 + 1)(x^4 + 4)} dx$$

$$= \int \frac{2x}{(x^2 + 1)((x^2)^2 + 4)} dx$$

Put $x^2 = t$,

$$\Rightarrow \quad 2xdx = dt$$

$$I = \int \frac{dt}{(t + 1)(t^2 + 4)}$$

Let $$\frac{1}{(t + 1)(t^2 + 4)} = \frac{A}{t + 1} + \frac{Bt + C}{t^2 + 4} \quad \text{...(a)}$$

$$\frac{1}{(t + 1)(t^2 + 4)} = \frac{A(t^2 + 4) + (Bt + C)(t + 1)}{(t + 1)(t^2 + 4)}$$

$$1 = A(t^2 + 4) + (Bt + C)(t + 1)$$

Equating coefficients of $t^2$, $t$ and constant terms on both sides, we get

$$A + B = 0$$
$$\Rightarrow \quad A = -B \quad \text{...(i)}$$
$$B + C = 0$$
$$\Rightarrow \quad B = -C \quad \text{...(ii)}$$
$$4A + C = 1 \quad \text{...(iii)}$$

From equations (i) and (ii) we get A = C.

From equation (iii)

$$4C + C = 1$$
$$5C = 1$$
$$\Rightarrow \quad C = \frac{1}{5}$$
$$\Rightarrow \quad A = C = \frac{1}{5}$$
$$\Rightarrow \quad B = -C = \frac{-1}{5}$$

From equation (a),

$$\Rightarrow \frac{1}{(t+1)(t^2+4)} = \frac{1/5}{t+1} + \frac{\frac{-1}{5}t + \frac{1}{5}}{t^2+4}$$

$$\Rightarrow \int \frac{1}{(t+1)(t^2+4)}dt = \frac{1}{5}\int \frac{1}{t+1}dt - \frac{1}{5}\int \frac{t}{t^2+4}dt + \frac{1}{5}\int \frac{1}{t^2+4}dt + C$$

where C is constant of integration.

$$I = \frac{1}{5}\int \frac{1}{t+1}dt - \frac{1}{10}\int \frac{2t}{t^2+4}dt + \frac{1}{5}\int \frac{1}{t^2+2^2}dt + C$$

$$= \frac{1}{5}\log|t+1| - \frac{1}{10}\log|t^2+4| + \frac{1}{5}\times\frac{1}{2}\tan^{-1}\frac{t}{2} + C$$

$$\Rightarrow I = \frac{1}{5}\log|x^2+1| - \frac{1}{10}\log|x^4+4| + \frac{1}{10}\tan^{-1}\frac{x^2}{2} + C \quad \textbf{Ans.}$$

## SECTION — D

**28. A metal box with a square base and vertical sides is to contain 1024 cm$^3$. The material for the top and bottom costs ₹ 5 per cm$^2$ and the material for the sides costs ₹2.50 per cm$^2$. Find the least cost of the box. [6]**

**Sol.** Let the length, breadth and height of the metal box be $x$ cm, $x$ cm and $y$ cm respectively. It is given that,

$$x^2y = 1024$$
$$\Rightarrow \quad y = \frac{1024}{x^2} \quad \text{...(i)}$$

Let $C$ be the total cost (in rupees) of material used to construct the box.

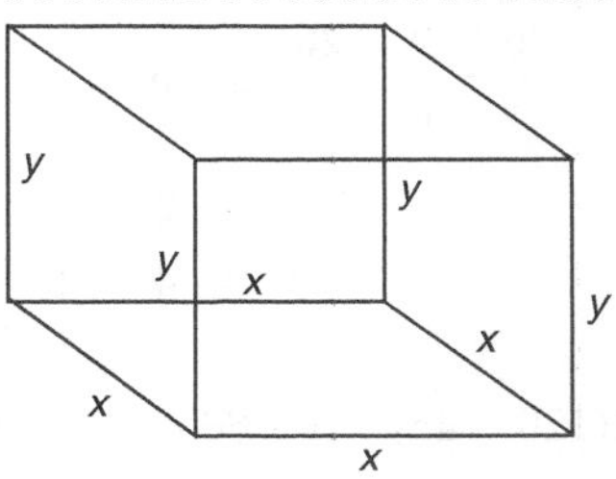


Then,
$$C = 5x^2 + 5x^2 + \frac{5}{2}\times 4xy$$
$$= 10x^2 + 10xy$$
$$= 10x^2 + 10x\times\frac{1024}{x^2}$$
$$= 10x^2 + \frac{10240}{x}$$

Now, differentiating '$C$' w.r.t. $x$, we get

$$\frac{dC}{dx} = 20x - \frac{10240}{x^2}$$

Again, differentiating w.r.t. $x$, we get

$$\frac{d^2C}{dx^2} = 20 + \frac{20480}{x^3}$$

For maxima or minima,

Put $\frac{dC}{dx} = 0$

$$\Rightarrow \quad 20x - \frac{10240}{x^2} = 0$$
$$\Rightarrow \quad x^3 = 512$$
$$\Rightarrow \quad x^3 = 8^3 \Rightarrow x = 8$$

when $x = 8$,

$$\left(\frac{d^2C}{dx^2}\right)_{x=8} = 20 + \frac{20480}{8^3} > 0$$

Thus, the cost of the box is least when $x = 8$, putting $x = 8$ in equation (i) we obtain $y = 16$, so the dimensions of the box are $8 \times 8 \times 16$

Putting, $x = 8$ and $y = 16$ in $C = 10x^2 + 10xy$, we get

$$C = 10 \times 64 + 10 \times 8 \times 16 = 1920$$

Hence, the least cost of the box is ₹ 1920. **Ans.**

**29. If A = $\begin{pmatrix} 2 & 3 & 10 \\ 4 & -6 & 5 \\ 6 & 9 & -20 \end{pmatrix}$, find A$^{-1}$. Using A$^{-1}$ solve the system of equations :**

$$\frac{2}{x} + \frac{3}{y} + \frac{10}{z} = 2;$$
$$\frac{4}{x} - \frac{6}{y} + \frac{5}{z} = 5;$$

$$\frac{6}{x}+\frac{9}{y}-\frac{20}{z}=-4 \qquad [6]$$

**Sol.** Given,
$$A=\begin{bmatrix}2 & 3 & 10\\ 4 & -6 & 5\\ 6 & 9 & -20\end{bmatrix}$$

$$|A| = 2(120-45)-3(-80-30)+10(36+36)$$
$$= 2\times 75-3\times(-110)+10\times 72$$
$$= 150+330+720=1200\neq 0$$

$\therefore$ $A^{-1}$ exists.

Let $A_{ij}$ be the cofactors of elements $a_{ij}$ in $A=[a_{ij}]$. Then,

$$A_{11}=(-1)^{1+1}\begin{bmatrix}-6 & 5\\ 9 & -20\end{bmatrix}=120-45=75$$
$$A_{12}=(-1)^{1+2}\begin{bmatrix}4 & 5\\ 6 & -20\end{bmatrix}=-(-80-30)=110$$
$$A_{13}=(-1)^{1+3}\begin{bmatrix}4 & -6\\ 6 & 9\end{bmatrix}=36+36=72$$
$$A_{21}=(-1)^{2+1}\begin{bmatrix}3 & 10\\ 9 & -20\end{bmatrix}=-(-60-90)=150$$
$$A_{22}=(-1)^{2+2}\begin{bmatrix}2 & 10\\ 6 & -20\end{bmatrix}=-40-60=-100$$
$$A_{23}=(-1)^{2+3}\begin{bmatrix}2 & 3\\ 6 & 9\end{bmatrix}=-(18-18)=0$$
$$A_{31}=(-1)^{3+1}\begin{bmatrix}3 & 10\\ -6 & 5\end{bmatrix}=15+60=75$$
$$A_{32}=(-1)^{3+2}\begin{bmatrix}2 & 10\\ 4 & 5\end{bmatrix}=-(10-40)=30$$
$$A_{33}=(-1)^{3+3}\begin{bmatrix}2 & 3\\ 4 & -6\end{bmatrix}=-12-12=-24$$

$$\text{Cofactor matrix of A}=\begin{bmatrix}75 & 110 & 72\\ 150 & -100 & 0\\ 75 & 30 & -24\end{bmatrix}$$

$$\text{adj A}=\begin{bmatrix}75 & 150 & 75\\ 110 & -100 & 30\\ 72 & 0 & -24\end{bmatrix}$$

$$A^{-1}=\frac{\text{adj A}}{|A|}$$
$$=\frac{1}{1200}\begin{bmatrix}75 & 150 & 75\\ 110 & -100 & 30\\ 72 & 0 & -24\end{bmatrix}$$

Given system of equations is,

$$\frac{2}{x}+\frac{3}{y}+\frac{10}{z}=2$$
$$\frac{4}{x}-\frac{6}{y}+\frac{5}{z}=5$$
$$\frac{6}{x}+\frac{9}{y}-\frac{20}{z}=-4$$

Let $\frac{1}{x}=u,\ \frac{1}{y}=v,\ \frac{1}{z}=w$

$\therefore$ The given linear equation becomes

$$2u+3v+10w=2$$
$$4u-6v+5w=5$$
$$6u+9v-20w=-4$$

$$A=\begin{bmatrix}2 & 3 & 10\\ 4 & -6 & 5\\ 6 & 9 & -20\end{bmatrix},\ X=\begin{bmatrix}u\\ v\\ w\end{bmatrix},\ B=\begin{bmatrix}2\\ 5\\ -4\end{bmatrix}$$

$$X=A^{-1}B$$
$$\begin{bmatrix}u\\ v\\ w\end{bmatrix}=\begin{bmatrix}2 & 3 & 10\\ 4 & -6 & 5\\ 6 & 9 & -20\end{bmatrix}^{-1}\begin{bmatrix}2\\ 5\\ -4\end{bmatrix}$$
$$=\frac{1}{1200}\begin{bmatrix}75 & 150 & 75\\ 110 & -100 & 30\\ 72 & 0 & -24\end{bmatrix}\begin{bmatrix}2\\ 5\\ -4\end{bmatrix}$$
$$=\frac{1}{1200}\begin{bmatrix}150+750-300\\ 220-500-120\\ 144+0+96\end{bmatrix}$$
$$=\frac{1}{1200}\begin{bmatrix}600\\ -400\\ 240\end{bmatrix}=\begin{bmatrix}1/2\\ -1/3\\ 1/5\end{bmatrix}$$

when $u=1/2 \Rightarrow x=2$
$v=-1/3 \Rightarrow y=-3$
and $w=1/5 \Rightarrow z=5$

$\Rightarrow$ $x=2$, $y=-3$ and $z=5$ **Ans.**

# Delhi [Set-III]

**Note :Except for the following questions, all the remaining questions have been asked in previous sets.**

## SECTION — B

**12. If $y = \sin^{-1}\left(6x\sqrt{1-9x^2}\right)$, $\frac{-1}{3\sqrt{2}} < x < \frac{1}{3\sqrt{2}}$, then find $\frac{dy}{dx}$.** **[2]**

**Sol.** Given, $$y = \sin^{-1}\left(6x\sqrt{1-9x^2}\right)$$

Put $x = \frac{\sin\theta}{3} \Rightarrow \theta = \sin^{-1}3x$.

$$y = \sin^{-1}\left(6.\frac{\sin\theta}{3}\sqrt{1-\frac{9\sin^2\theta}{9}}\right)$$

$$\Rightarrow \quad y = \sin^{-1}\left(2\sin\theta\sqrt{1-\sin^2\theta}\right)$$

$$\Rightarrow \quad y = \sin^{-1}\left(2\sin\theta\sqrt{\cos^2\theta}\right)$$

$$\Rightarrow \quad y = \sin^{-1}(2\sin\theta\cos\theta)$$

$$\Rightarrow \quad y = \sin^{-1}(\sin 2\theta) = 2\theta = 2\sin^{-1}3x$$

$$\Rightarrow \quad \frac{dy}{dx} = \frac{2}{\sqrt{1-9x^2}} \times 3 = \frac{6}{\sqrt{1-9x^2}}$$ **Ans.**

## SECTION — C

**20. Solve the following L.P.P. graphically :**

**Maximise $Z = 20x + 10y$**

**Subject to the following constraints :**

$$x + 2y \le 28,$$
$$3x + y \le 24,$$
$$x \ge 2$$
$$x, y \ge 0$$ **[4]**

**Sol.** We have

Maximise $Z = 20x + 10y$

Subject to the constraints :

$$x + 2y \le 28$$
$$3x + y \le 24$$
$$x \ge 2$$
$$x, y \ge 0$$

Converting the given inequalities into equations, we obtain the following equations :

$$x + 2y = 28$$
$$3x + y = 24$$
$$x = 2$$

Then

$x + 2y = 28$

| $x$ | 0 | 28 |
|---|---|---|
| $y$ | 14 | 0 |

$3x + y = 24$

| $x$ | 0 | 8 |
|---|---|---|
| $y$ | 24 | 0 |

$x = 2$ is a line which is parallel to y-axis.

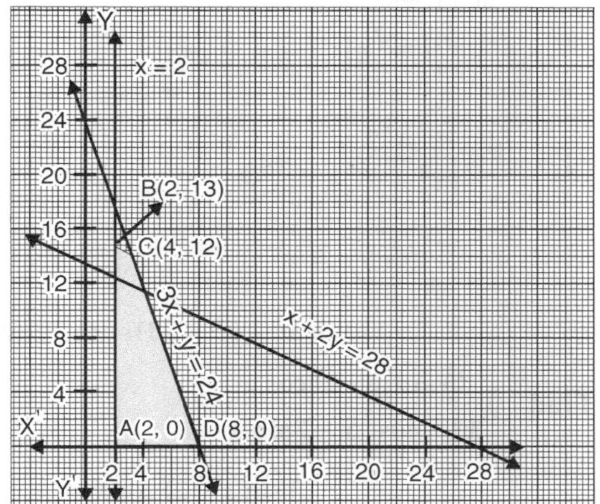


Plotting these points on the graph, we get the shaded feasible region *i.e.*, ABCD

| Corner point | Value of $Z = 20x + 10y$ |
|---|---|
| A (2, 0) | $Z = 40 + 0 = 40$ |
| B (2, 13) | $Z = 40 + 130 = 170$ |
| C (4, 12) | $Z = 80 + 120 = 200 \leftarrow$ Maximum |
| D (8, 0) | $Z = 160 + 0 = 160$ |

$\therefore$ Maximum value of Z is 200 at (4, 12). **Ans.**

**21. Show that the family of curves for which $\frac{dy}{dx} = \frac{x^2+y^2}{2xy}$, is given by $x^2 - y^2 = cx$.** **[4]**

**Sol.** Given family of curve,

$$\frac{dy}{dx} = \frac{x^2+y^2}{2xy} \quad ...(i)$$

Put $y = vx$

$$\Rightarrow \quad \frac{dy}{dx} = v + x\frac{dv}{dx}$$

From equation (i),

$$v + x\frac{dv}{dx} = \frac{x^2+(vx)^2}{2x(vx)} = \frac{x^2(1+v^2)}{x^2(2v)}$$

$$\Rightarrow \quad x\frac{dv}{dx} = \frac{1+v^2}{2v} - v$$

$$= \frac{1+v^2-2v^2}{2v} = \frac{1-v^2}{2v}$$

$$\Rightarrow \quad \frac{2v}{1-v^2}dv = \frac{dx}{x}$$

$$\Rightarrow \quad \int\frac{2v}{1-v^2}dv = \int\frac{dx}{x}$$

$$\Rightarrow -\log|1-v^2| + \log c = \log x,$$

where $\log c$ is constant of integration.

$$\log\left|\frac{c}{1-\frac{y^2}{x^2}}\right| = \log x$$

$$\Rightarrow \quad \log\left|\frac{cx^2}{x^2-y^2}\right| = \log x$$

$$\Rightarrow \quad \frac{cx^2}{x^2-y^2} = x$$

$$\Rightarrow \quad cx = x^2 - y^2$$ **Hence Proved.**

**22. Find : $\int\frac{(3\sin x - 2)\cos x}{13-\cos^2 x - 7\sin x}dx$ [4]**

**Sol.** Let
$$I = \int\frac{(3\sin x - 2)\cos x}{13-\cos^2 x - 7\sin x}dx$$

$$= \int\frac{(3\sin x - 2)\cos x}{\sin^2 x - 7\sin x + 12}dx$$

$$(\because \cos^2 x = 1 - \sin^2 x)$$

Put $\sin x = t$

$\Rightarrow \cos x\, dx = dt$

$$I = \int\frac{(3t-2)}{t^2-7t+12}dt$$
$$= \int\frac{3t-2}{(t-3)(t-4)}dt$$

Let, $$\frac{3t-2}{(t-3)(t-4)} = \frac{A}{t-3}+\frac{B}{t-4} \quad ...(A)$$

$$\Rightarrow \quad \frac{3t-2}{(t-3)(t-4)} = \frac{A(t-4)+B(t-3)}{(t-3)(t-4)}$$

$$\Rightarrow \quad 3t-2 = A(t-4)+B(t-3)$$

Equating coefficient of $t$ and constant term on both sides, we get

$$A + B = 3 \quad ...(i)$$

$$-4A - 3B = -2 \quad ...(ii)$$

Solving the above two equations, we get

$$A = -7, B = 10$$

From equation (A),

$$\frac{3t-2}{(t-3)(t-4)} = \frac{-7}{t-3}+\frac{10}{t-4}$$

$$\int\frac{3t-2}{(t-3)(t-4)}dt = -7\int\frac{1}{t-3}dt + 10\int\frac{1}{t-4}dt$$

$$I = -7\log|t-3| + 10\log|t-4| + C,$$

where C is constant of integration.

$$I = 10\log|\sin x - 4| - 7\log|\sin x - 3| + C.$$

**Ans.**

**23. Solve the following equation for $x$ :**

**$\cos(\tan^{-1}x) = \sin\left(\cot^{-1}\frac{3}{4}\right)$ [4]**

**Sol.** Given equation is,

$$\cos(\tan^{-1}x) = \sin\left(\cot^{-1}\frac{3}{4}\right) \quad ...(i)$$

$$\Rightarrow \quad \cos(\tan^{-1}x) = \cos\left[\frac{\pi}{2}-\cot^{-1}\frac{3}{4}\right]$$

$$\Rightarrow \quad \tan^{-1}x = \frac{\pi}{2}-\cot^{-1}\frac{3}{4}$$

$$\Rightarrow \quad \frac{\pi}{2}-\cot^{-1}x = \frac{\pi}{2}-\cot^{-1}\frac{3}{4}$$

$$\left(\because \tan^{-1}x + \cot^{-1}x = \frac{\pi}{2}\right)$$

$$\Rightarrow \quad \cot^{-1}x = \cot^{-1}\frac{3}{4}$$

or $$x = \frac{3}{4}$$ **Ans.**

## SECTION — D

**28. If $A = \begin{bmatrix} 2 & 3 & 1 \\ 1 & 2 & 2 \\ -3 & 1 & -1 \end{bmatrix}$, find $A^{-1}$ and hence solve the system of equations $2x + y - 3z = 13$, $3x + 2y + z = 4$, $x + 2y - z = 8$ [6]**

**Sol.** Given,

$$A = \begin{bmatrix} 2 & 3 & 1 \\ 1 & 2 & 2 \\ -3 & 1 & -1 \end{bmatrix}$$

$$|A| = 2(-2-2) - 3(-1+6) + 1(1+6)$$

$$= -8 - 15 + 7 = -16 \neq 0$$

$\therefore A^{-1}$ exists

Let $A_{ij}$ be the cofactors of elements $a_{ij}$ in $A = [a_{ij}]$. Then,

$$A_{11} = (-1)^{1+1}\begin{bmatrix} 2 & 2 \\ 1 & -1 \end{bmatrix} = -2-2 = -4$$

$$A_{12} = (-1)^{1+2}\begin{bmatrix} 1 & 2 \\ -3 & -1 \end{bmatrix} = -(-1+6) = -5$$

$$A_{13} = (-1)^{1+3}\begin{bmatrix} 1 & 2 \\ -3 & -1 \end{bmatrix} = 1 + 6 = 7$$

$$A_{21} = (-1)^{2+1}\begin{bmatrix} 3 & 1 \\ 1 & -1 \end{bmatrix} = -(-3-1) = 4$$

$$A_{22} = (-1)^{2+2}\begin{bmatrix} 2 & 10 \\ 6 & -20 \end{bmatrix} = -2 + 3 = 1$$

$$A_{23} = (-1)^{2+3}\begin{bmatrix} 2 & 3 \\ -3 & 1 \end{bmatrix} = -(2+9) = -11$$

$$A_{31} = (-1)^{3+1}\begin{bmatrix} 3 & 1 \\ 2 & 2 \end{bmatrix} = 6 - 2 = 4$$

$$A_{32} = (-1)^{3+2}\begin{bmatrix} 2 & 1 \\ 1 & 2 \end{bmatrix} = -(4-1) = -3$$

$$A_{33} = (-1)^{3+3}\begin{bmatrix} 2 & 3 \\ 1 & 2 \end{bmatrix} = 4 - 3 = 1$$

$$\therefore \quad a_{ij} = \begin{bmatrix} -4 & -5 & 7 \\ 4 & 1 & -11 \\ 4 & -3 & 1 \end{bmatrix}$$

$$\text{adj. A} = [A_{ij}]^T = \begin{bmatrix} -4 & 4 & 4 \\ -5 & 1 & -3 \\ 7 & -11 & 1 \end{bmatrix}$$

$$A^{-1} = \frac{1}{|A|}\text{adj. A} = \frac{-1}{16}\begin{bmatrix} -4 & 4 & 4 \\ -5 & 1 & -3 \\ 7 & -11 & 1 \end{bmatrix}$$

From the given linear equations, we have

Let $C = \begin{bmatrix} 2 & 1 & -3 \\ 3 & 2 & 1 \\ 1 & 2 & -1 \end{bmatrix}$, $X = \begin{bmatrix} x \\ y \\ z \end{bmatrix}$, $D = \begin{bmatrix} 13 \\ 4 \\ 8 \end{bmatrix}$

by matrix method,

$$X = C^{-1}D = (A^T)^{-1}D = (A^{-1})^T D$$

$$\begin{bmatrix} x \\ y \\ z \end{bmatrix} = \frac{-1}{16}\begin{bmatrix} -4 & 4 & 4 \\ -5 & 1 & -3 \\ 7 & -11 & 1 \end{bmatrix}^T \begin{bmatrix} 13 \\ 4 \\ 8 \end{bmatrix}$$

$$= \frac{-1}{16}\begin{bmatrix} -4 & -5 & 7 \\ 4 & 1 & -11 \\ 4 & -3 & 1 \end{bmatrix}\begin{bmatrix} 13 \\ 4 \\ 8 \end{bmatrix}$$

$$\begin{bmatrix} x \\ y \\ z \end{bmatrix} = \frac{-1}{16}\begin{bmatrix} -52-20+56 \\ 52+4-88 \\ 52-12+8 \end{bmatrix}$$

$$= \frac{-1}{16}\begin{bmatrix} -16 \\ -32 \\ 48 \end{bmatrix} = \begin{bmatrix} 1 \\ 2 \\ -3 \end{bmatrix}$$

$\Rightarrow \quad x = 1, y = 2, z = -3.$ **Ans.**

**29. Find the particular solution of the differential equation. $\tan x \cdot \frac{dy}{dx} = 2x \tan x + x^2 - y$; $(\tan x \neq 0)$ given that $y = 0$ when $x = \frac{\pi}{2}$.** **[6]**

**Sol.** Given differential equation is,

$$\tan x . \frac{dy}{dx} = 2x \tan x + x^2 - y$$

$$\Rightarrow \quad \frac{dy}{dx} = 2x + x^2 \cot x - y \cot x$$

$$\Rightarrow \quad \frac{dy}{dx} + y \cot x = 2x + x^2 \cot x,$$

which is of the form

$$\frac{dy}{dx} + Py = Q,$$

where $P = \cot x$, $Q = 2x + x^2 \cot x$

$$\text{I. F.} = e^{\int P\,dx} = e^{\int \cot x\,dx} = e^{\log|\sin x|} = \sin x$$

$\therefore$ Required solution is

$$y.\sin x = \int(2x + x^2 \cot x)\sin x\,dx + C$$

$$= \int 2x \sin x\,dx + \int x^2 \cot x \sin x\,dx + C$$

$$= 2\int x \sin x\,dx + \int x^2 \cos x\,dx + C$$

$$= 2\left[x.(-\cos x) - \int(-\cos x)dx\right] + x^2 . \sin x - \int 2x . \sin x\,dx + C$$

$$= -2x \cos x + 2 \sin x + x^2 \sin x - 2\left[x.(-\cos x) - \int -\cos x\,dx\right] + C$$

$$= -2x \cos x + 2 \sin x + x^2 \sin x + 2x \cos x - 2\sin x + C$$

$$\therefore \quad y \sin x = x^2 \sin x + C \quad \text{...(i)}$$

Given that, $y = 0$ when $x = \frac{\pi}{2}$

$$\Rightarrow \quad 0 = \frac{\pi^2}{4}\sin\frac{\pi}{2} + C$$

$$\Rightarrow \quad C = -\frac{\pi^2}{4} =$$

From equation (i),

$$y \sin x = x^2 \sin x - \frac{\pi^2}{4}$$

$$\Rightarrow \quad (x^2 - y)\sin x = \frac{\pi^2}{4}$$

**Ans.**

# MATHEMATICS

## Outside Delhi [Set-I] 2016

Time allowed : 3 hours

Maximum marks : 100

### SECTION — A

**1. If $x \in N$ and $\begin{vmatrix} x+3 & -2 \\ -3x & 2x \end{vmatrix} = 8$, then find the value of $x$.** **[1]**

**Sol.** We have, $\begin{vmatrix} x+3 & -2 \\ -3x & 2x \end{vmatrix} = 8$

$$\Rightarrow \quad (x+3) \times 2x - (-2) \times (-3x) = 8$$

$$\Rightarrow \quad 2x^2 + 6x - 6x = 8$$

$$\Rightarrow \quad 2x^2 = 8$$

$$\Rightarrow \quad x^2 = 4$$

But $x \neq -2$ as $x \in N$

$$\Rightarrow \quad x = 2$$ **Ans.**

**2. Use elementary column operation $C_2 \to C_2 + 2C_1$ in the following matrix equation :**

$$\begin{bmatrix} 2 & 1 \\ 2 & 0 \end{bmatrix} = \begin{bmatrix} 3 & 1 \\ 2 & 0 \end{bmatrix} \begin{bmatrix} 1 & 0 \\ -1 & 1 \end{bmatrix}$$ **[1]**

**Sol.** We have

$$\begin{bmatrix} 2 & 1 \\ 2 & 0 \end{bmatrix} = \begin{bmatrix} 3 & 1 \\ 2 & 0 \end{bmatrix} \begin{bmatrix} 1 & 0 \\ -1 & 1 \end{bmatrix}$$

$$P = A.B \text{ (say)}$$

Applying $C_2 \to C_2 + 2C_1$ on P, we get

$$\begin{bmatrix} 2 & 5 \\ 2 & 4 \end{bmatrix} = Q \text{ (say)}$$

Applying $C_2 \to C_2 + 2C_1$ on B, we get

$$\begin{bmatrix} 1 & 2 \\ -1 & -1 \end{bmatrix} = C \text{ (say)}$$

Now, $$AC = \begin{bmatrix} 3 & 1 \\ 2 & 0 \end{bmatrix} \begin{bmatrix} 1 & 2 \\ -1 & -1 \end{bmatrix} = Q$$

$$\therefore \quad \begin{bmatrix} 2 & 5 \\ 2 & 4 \end{bmatrix} = \begin{bmatrix} 3 & 1 \\ 2 & 0 \end{bmatrix} \begin{bmatrix} 1 & 2 \\ -1 & -1 \end{bmatrix}$$ **Ans.**

**3. Write the number of all possible matrices of order $2 \times 2$ with each entry 1, 2 or 3.** **[1]**

**Sol.** Total number of all possible matrices of order $2 \times 2$ with each entry 1, 2 or 3 are $3^4$ *i.e.*, 81. **Ans.**

**4. Write the position vector of the point which divides the join of points with position vector $3\vec{a} - 2\vec{b}$ and $2\vec{a} + 3\vec{b}$ in the ratio 2 : 1.** **[1]**

**Sol.** Let A and B be the given points with position vectors $3\vec{a} - 2\vec{b}$ and $2\vec{a} + 3\vec{b}$ respectively.

Let P and Q be the points dividing AB in the ratio 2 : 1 internally and externally respectively. Then,

$$\text{Position vector of P} = \frac{1(3\vec{a} - 2\vec{b}) + 2(2\vec{a} + 3\vec{b})}{1+2}$$

$$= \frac{7\vec{a}}{3} + \frac{4\vec{b}}{3}$$

$$\text{Position vector of Q} = \frac{1(3\vec{a} - 2\vec{b}) - 2(2\vec{a} + 3\vec{b})}{1-2}$$

$$= \vec{a} + 8\vec{b}$$ **Ans.**

**5. Write the number of vectors of unit length perpendicular to both the vectors $\vec{a} = 2\hat{i} + \hat{j} + 2\hat{k}$ and $\vec{b} = \hat{j} + \hat{k}$.** **[1]**

**Sol.** We know that the unit vectors perpendicular to the plane of $\vec{a}$ and $\vec{b}$ are $\pm \dfrac{\vec{a} \times \vec{b}}{|\vec{a} \times \vec{b}|}$

So, $$\vec{a} \times \vec{b} = \begin{vmatrix} \hat{i} & \hat{j} & \hat{k} \\ 2 & 1 & 2 \\ 0 & 1 & 1 \end{vmatrix}$$

$$= (1-2)\hat{i} - (2-0)\hat{j} + (2-0)\hat{k}$$

$$= -\hat{i} - 2\hat{j} + 2\hat{k}$$

$$\Rightarrow \quad |\vec{a} \times \vec{b}| = \sqrt{(-1)^2 + (-2)^2 + (2)^2} = \sqrt{9} = 3$$

Hence, required vectors

$$= \pm\frac{1}{3}(-\hat{i} - 2\hat{j} + 2\hat{k})$$

and number of vectors are 2. **Ans.**

**6. Find the vector equation of the plane with intercepts 3, – 4 and 2 on $x$, $y$ and $z$-axis respectively.** **[1]**

**Sol.** The equation of the required plane is,

$$\frac{x}{3} + \frac{y}{-4} + \frac{z}{2} = 1$$

$\Rightarrow \quad 4x - 3y + 6z = 12$

**In vector form :** $(x\hat{i}+y\hat{j}+z\hat{k}).(4\hat{i}-3\hat{j}+6\hat{k})=12$

*i.e.,* $\quad \vec{r}.(4\hat{i}-3\hat{j}+6\hat{k})=12$

This is the vector equation of the plane with intercept 3, – 4 and 2 on coordinate axis. **Ans.**

## SECTION–B

**7. Find the coordinates of the point where the line through the points A(3, 4, 1) and B(5, 1, 6) crosses the XZ-plane. Also find the angle which this line makes with the XZ-plane. [4]**

**Sol.** The equation of the line passing through A and B is,

$$\frac{x-3}{5-3}=\frac{y-4}{1-4}=\frac{z-1}{6-1} \text{ or } \frac{x-3}{2}=\frac{y-4}{-3}=\frac{z-1}{5} \quad ...(i)$$

The coordinates of any point on this line are given by

$$\frac{x-3}{2}=\frac{y-4}{-3}=\frac{z-1}{5}=\lambda$$

$\Rightarrow \quad x = 2\lambda + 3, y = -3\lambda + 4, z = 5\lambda + 1$

So $(2\lambda + 3, -3\lambda + 4, 5\lambda + 1)$ are the coordinates of any point on the line passing through A and B. If it lies on XZ-plane then $y = 0$.

$\therefore \quad -3\lambda + 4 = 0 \Rightarrow \lambda = \frac{4}{3}$

So, the coordinates of required point are

$$\left(2\times\frac{4}{3}+3,-3\times\frac{4}{3}+4,5\times\frac{4}{3}+1\right) i.e., \left(\frac{17}{3},0,\frac{23}{3}\right)$$

Now, the line in equation (i) is parallel to the vector $\vec{b}=2\hat{i}-3\hat{j}+5\hat{k}$ and the XZ-plane is normal to the vector $\vec{n}=\hat{j}$. Therefore, the angle $\theta$ between them is given by

$$\sin\theta = \left|\frac{\vec{b}\cdot\vec{n}}{|\vec{b}|\cdot|\vec{n}|}\right|$$

$$= \left|\frac{(2\hat{i}-3\hat{j}+5\hat{k})(\hat{j})}{\sqrt{(2)^2+(-3)^2+(5)^2}\sqrt{(1)^2}}\right|$$

$\Rightarrow \quad \sin\theta = \frac{3}{\sqrt{38}}$

$\Rightarrow \quad \theta = \sin^{-1}\left(\frac{3}{\sqrt{38}}\right)$

$\Rightarrow \quad \theta = \sin^{-1}(0.4866)$ **Ans.**

**8. The two adjacent sides of a parallelogram are $2\hat{i}-4\hat{j}-5\hat{k}$ and $2\hat{i}+2\hat{j}+3\hat{k}$. Find the two unit vectors parallel to its diagonals. Using the diagonal vectors, find the area of the parallelogram. [4]**

**Sol.** Let ABCD be a parallelogram such that

$$\vec{AB} = \vec{a} = 2\hat{i}-4\hat{j}-5\hat{k}$$

and $$\vec{BC} = \vec{b} = 2\hat{i}+2\hat{j}+3\hat{k}$$

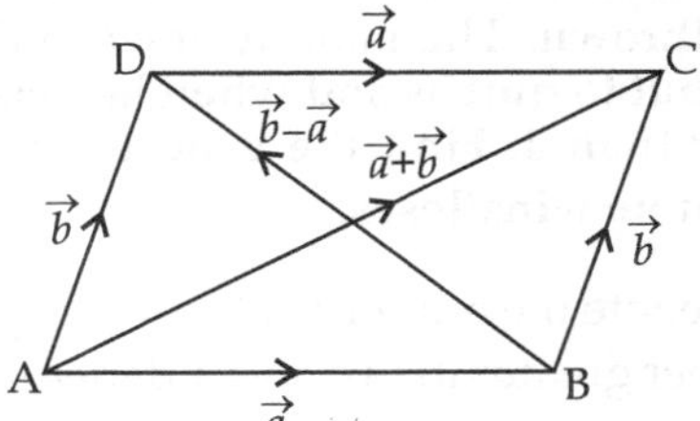


Then,

$$\vec{AB}+\vec{BC} = \vec{AC}$$

$\Rightarrow \quad \vec{AC} = \vec{a}+\vec{b} = 4\hat{i}-2\hat{j}-2\hat{k}$

and $\vec{AB}+\vec{BD} = \vec{AD}$

$\Rightarrow \quad \vec{BD} = \vec{AD}-\vec{AB}$

$\Rightarrow \quad \vec{BD} = \vec{b}-\vec{a} = 0\hat{i}+6\hat{j}+8\hat{k}$

Now, $\quad \vec{AC} = 4\hat{i}-2\hat{j}-2\hat{k}$

$$|\vec{AC}| = \sqrt{(4)^2+(-2)^2+(-2)^2} = \sqrt{24} = 2\sqrt{6}$$

and $\quad |\vec{BD}| = 0\hat{i}+6\hat{j}+8\hat{k}$

$$|\vec{BD}| = \sqrt{(0)^2+(6)^2+(8)^2} = \sqrt{100} = 10$$

Unit vector along $\vec{AC}$

$$= \frac{\vec{AC}}{|\vec{AC}|} = \frac{1}{2\sqrt{6}}(4\hat{i}-2\hat{j}-2\hat{k})$$

$$= \frac{1}{\sqrt{6}}(2\hat{i}-\hat{j}-\hat{k})$$

Unit vector along $\vec{BD}$

$$= \frac{\vec{BD}}{|\vec{BD}|} = \frac{1}{10}(6\hat{j}+8\hat{k}) = \frac{1}{5}(3\hat{j}+4\hat{k})$$

Now, area of parallelogram

$$= \frac{1}{2}|\vec{AC}\times\vec{BD}|$$

$$\Rightarrow \quad \vec{AC}\times\vec{BD} = \begin{vmatrix} \hat{i} & \hat{j} & \hat{k} \\ 4 & -2 & -2 \\ 0 & 6 & 8 \end{vmatrix}$$

$$= -4\hat{i} - 32\hat{j} + 24\hat{k}$$

Area of parallelogram

$$= \frac{1}{2}\left|\vec{AC} \times \vec{BD}\right|$$

$$= \sqrt{404} \text{ or } 2\sqrt{101} \text{ sq. units} \quad \textbf{Ans.}$$

**9. In a game, a man wins ₹ 5 for getting a number greater than 4 and loses ₹ 1 otherwise, when a fair die is thrown. The man decided to throw a die thrice but to quit as and when he gets a number greater than 4. Find the expected value of the amount he wins/loses. [4]**

**Sol.** Let $n$ denote the number of throws required to get a number greater than 4 and X denote the amount won/lost.

The man may get a number greater than 4 in the very first throw of the die or in second throw or in the third throw.

Let $p$ = Probability of getting a number greater than 4

$$p = \frac{2}{6}$$

$$q = 1 - p = \frac{4}{6}$$

Thus, we have the following probability distribution for X.

| Number of throws ($n$) | 1 | 2 | 3 | 3 |
|---|---|---|---|---|
| Amount won (X) | 5 | 4 | 3 | – 3 |
| Probability (P(X)) | $\frac{2}{6}$ | $\frac{4}{6}\times\frac{2}{6}$ | $\frac{4}{6}\times\frac{4}{6}\times\frac{2}{6}$ | $\frac{4}{6}\times\frac{4}{6}\times\frac{4}{6}$ |

$$E(X) = 5\times\frac{2}{6}+4\times\frac{4}{6}\times\frac{2}{6}+3\times\frac{4}{6}\times\frac{4}{6}\times\frac{2}{6}+(-3)\times\frac{4}{6}\times\frac{4}{6}\times\frac{4}{6}$$

$$= \frac{19}{9}$$

Expected amount he could win is ₹ $\frac{19}{9}$ **Ans.**

**OR**

**A bag contains 4 balls. Two balls are drawn at random (without replacement) and are found to be white. What is the probability that all balls in the bag are white ?**

**Sol.** We know that the number of white balls can't be less than 2.

Now, there are different cases, for the number of white balls in the bag. The total cases are:

$$^2C_2 + {}^3C_2 + {}^4C_2 = \frac{2!}{2!\times 0!}+\frac{3!}{2!\times 1!}+\frac{4!}{2!\times 2!}$$

$$= 1 + 3 + 6 = 10$$

∴ Probability of the case that there are 4 white balls is :

$$^4C_2 = 6$$

Hence, the probability that all balls in the bag are white is :

$$\frac{6}{10} \text{ or } \frac{3}{5}. \quad \textbf{Ans.}$$

**10. Differentiate $x^{\sin x} + (\sin x)^{\cos x}$ with respect to $x$. [4]**

**Sol.** We have,

$$x^{\sin x} + (\sin x)^{\cos x}$$

Let $y = x^{\sin x} + (\sin x)^{\cos x}$

Taking log on both sides,

$$\log y = \sin x \log x + \cos x \log (\sin x)$$

$$\Rightarrow \quad y = e^{\sin x \log x} + e^{\cos x \log (\sin x)}$$

On differentiating both sides w.r.t. $x$, we get

$$\frac{dy}{dx} = e^{\sin x \log x}\frac{d}{dx}(\sin x \log x) + e^{\cos x \log(\sin x)}\frac{d}{dx}(\cos x \log(\sin x))$$

$$\Rightarrow \frac{dy}{dx} = x^{\sin x}\left\{\cos x \log x + \frac{\sin x}{x}\right\} + (\sin x)^{\cos x}\left\{-\sin x \log(\sin x) + \cos x . \frac{1}{\sin x} . \cos x\right\}$$

$[\because e^{\sin x \log x} = x^{\sin x}$ and $e^{\cos x \log (\sin x)} = (\sin x)^{\cos x}]$

$$= x^{\sin x}\left\{\cos x \log x + \frac{\sin x}{x}\right\} + (\sin x)^{\cos x}\left\{-\sin x \log(\sin x) + \frac{\cos^2 x}{\sin x}\right\} \quad \textbf{Ans.}$$

**OR**

**If $y = 2 \cos (\log x) + 3 \sin (\log x)$, prove that $x^2\frac{d^2y}{dx^2} + x\frac{dy}{dx} + y = 0$.**

**Sol.** Given, $y = 2 \cos (\log x) + 3 \sin (\log x)$

On differentiating both sides w.r.t. $x$, we get

$$\frac{dy}{dx} = -2\sin(\log x).\frac{1}{x} + 3\cos(\log x).\frac{1}{x}$$

$$\Rightarrow \quad x\frac{dy}{dx} = -3\sin(\log x) - 2\cos(\log x)$$

Again differentiating both sides w.r.t. $x$, we get

$$x\frac{d^2y}{dx^2} + \frac{dy}{dx} = -2\cos(\log x).\frac{1}{x} - 3\sin(\log x).\frac{1}{x}$$

$$\Rightarrow \quad x^2\frac{d^2y}{dx^2} + x.\frac{dy}{dx} = -3\sin(\log x) - 2\cos(\log x)$$

$$\Rightarrow \quad x^2\frac{d^2y}{dx^2}+x\frac{dy}{dx} = -y$$

$$\Rightarrow \quad x^2\frac{d^2y}{dx^2}+x\frac{dy}{dx}+y = 0$$

**Hence Proved.**

**11. If $x = a \sin 2t\ (1 + \cos 2t)$ and $y = b \cos 2t\ (1 - \cos 2t)$, find $\frac{dy}{dx}$ at $t=\frac{\pi}{4}$.** **[4]**

**Sol.** We have, $x = a \sin 2t\ (1 + \cos 2t)$

$$\Rightarrow \quad \frac{dx}{dt} = 2a \cos 2t\ (1 + \cos 2t) - 2a \sin^2 2t$$

and $\quad y = b \cos 2t\ (1 - \cos 2t)$

$$\Rightarrow \quad \frac{dy}{dt} = 2b \cos 2t \sin 2t - 2b \sin 2t\ (1 - \cos 2t)$$

$$\therefore \quad \frac{dy}{dx}=\frac{dy}{dt}\times\frac{dt}{dx}$$

$$= \frac{2b\cos 2t\sin 2t-2b\sin 2t\ (1-\cos 2t)}{2a\cos 2t(1+\cos 2t)-2a\sin^2 2t}$$

$$= \frac{b\,[\cos 2t\sin 2t-\sin 2t(1-\cos 2t)]}{a\,[\cos 2t(1+\cos 2t)-\sin^2 2t]}$$

Since, we know that at $t=\frac{\pi}{4}, \sin 2t = 1$ and $\cos 2t = 0$

So, $\left(\frac{dy}{dx}\right)_{t=\frac{\pi}{4}} = \frac{b\,[0.1-1(1-0)]}{a\,[0(1+0)-1^2]} = \frac{b}{a}$ **Ans.**

**12. The equation of tangent at (2, 3) on the curve $y^2 = ax^3 + b$ is $y = 4x - 5$. Find the value of $a$ and $b$.** **[4]**

**Sol.** Since the point (2, 3) lies on the curve $y^2 = ax^3 + b$

$$\Rightarrow \quad (3)^2 = a(2)^3 + b$$

$$\Rightarrow \quad 9 = 8a + b \qquad \text{...(i)}$$

and $\quad y^2 = ax^3 + b$

On differentiating w.r.t. $x$,

$$2y\frac{dy}{dx} = 3ax^2$$

$$\Rightarrow \quad \frac{dy}{dx} = \frac{3ax^2}{2y}$$

$$\Rightarrow \quad \left(\frac{dy}{dx}\right)_{(2,3)} = \frac{3a(2)^2}{2(3)} = 2a$$

The required equation of the tangent at (2, 3) is

$$(y - 3) = 2a\ (x - 2)$$

$$\Rightarrow \quad y - 3 = 2ax - 4a$$

$$\Rightarrow \quad y = 2ax + 3 - 4a$$

Comparing the equation with $y = 4x - 5$, we get

$$4 = 2a \Rightarrow a = 2$$

From equation (i), we get

$$9 = 8(2) + b \Rightarrow b = -7$$ **Ans.**

**13. Find : $\int\frac{x^2}{x^4+x^2-2}dx$** **[4]**

**Sol.** Let,

$$I = \int\frac{x^2}{x^4+x^2-2}dx = \int\frac{x^2}{(x^2-1)(x^2+2)}dx$$

Let $x^2 = y$

Then, $$\frac{y}{(y-1)(y+2)} = \frac{A}{(y-1)}+\frac{B}{(y+2)} \qquad \text{...(i)}$$

$$y = A\,(y+2)+B\,(y-1) \qquad \text{...(ii)}$$

Putting $y = 1$ and $y = -2$ successively in (ii), we get

$$A = \frac{1}{3} \text{ and } B = \frac{2}{3}$$

Substituting the values of A and B in (i), we obtain

$$\frac{y}{(y-1)(y+2)} = \frac{1}{3(y-1)}+\frac{2}{3(y+2)}$$

Replacing $y$ by $x^2$, we obtain

$$\frac{x^2}{(x^2-1)(x^2+2)} = \frac{1}{3(x^2-1)}+\frac{2}{3(x^2+2)}$$

$$\therefore I = \int\frac{x^2}{x^4+x^2-2}dx = \frac{1}{3}\int\frac{1}{(x^2-1)}dx + \frac{2}{3}\int\frac{1}{(x^2+2)}dx$$

$$= \frac{1}{3}\cdot\frac{1}{2}\log\left|\frac{x-1}{x+1}\right| + \frac{2}{3}\cdot\frac{1}{\sqrt{2}}\tan^{-1}\left(\frac{x}{\sqrt{2}}\right)+C$$

$$= \frac{1}{6}\log\left|\frac{x-1}{x+1}\right| + \frac{\sqrt{2}}{3}\tan^{-1}\left(\frac{x}{\sqrt{2}}\right)+C$$ **Ans.**

**14. Evaluate : $\int_0^{\pi/2}\frac{\sin^2 x}{\sin x+\cos x}dx$** **[4]**

**Sol.** We have,

$$I = \int_0^{\pi/2}\frac{\sin^2 x}{\sin x+\cos x}dx \qquad \text{...(i)}$$

$$\Rightarrow \quad I = \int_0^{\pi/2}\frac{\sin^2\left(\frac{\pi}{2}-x\right)}{\sin\left(\frac{\pi}{2}-x\right)+\cos\left(\frac{\pi}{2}-x\right)}dx$$

$$\left[\text{Using} : \int_0^a f(x)dx = \int_0^a f(a-x)dx\right]$$

$$\Rightarrow \quad I = \int_0^{\pi/2}\frac{\cos^2 x}{\cos x+\sin x}dx \qquad \text{...(ii)}$$

Adding equation (i) and (ii), we get

$$2I = \int_0^{\pi/2}\frac{\sin^2 x}{\sin x+\cos x}+\frac{\cos^2 x}{\sin x+\cos x}dx$$

$$= \int_0^{\pi/2}\frac{1}{\sin x+\cos x}dx$$

$$\Rightarrow \quad 2I = \int_0^{\pi/2}\frac{1}{\frac{2\tan x/2}{1+\tan^2 x/2}+\frac{1-\tan^2 x/2}{1+\tan^2 x/2}}dx$$

$$\Rightarrow \quad 2I = \int_0^{\pi/2} \frac{1+\tan^2 x/2}{2\tan x/2+1-\tan^2 x/2}dx$$

$$= \int_0^{\pi/2} \frac{\sec^2 x/2}{2\tan x/2+1-\tan^2 x/2}dx$$

Let $\tan\frac{x}{2} = t.$

Then, $\sec^2\frac{x}{2}\cdot\frac{1}{2}dx = dt \Rightarrow \sec^2\frac{x}{2}dx = 2dt$

Also, $x = 0 \Rightarrow t = \tan 0 = 0$ and $x = \frac{\pi}{2} \Rightarrow t = \tan\frac{\pi}{4} = 1$

$$\therefore \quad 2I = \int_0^1 \frac{2dt}{2t+1-t^2} = 2\int_0^1 \frac{1}{(\sqrt{2})^2-(t-1)^2}dt$$

$$\Rightarrow \quad 2I = 2\times\frac{1}{2\sqrt{2}}\left[\log\left|\frac{\sqrt{2}+t-1}{\sqrt{2}-t+1}\right|\right]_0^1$$

$$\Rightarrow \quad 2I = \frac{1}{\sqrt{2}}\left\{\log\left(\frac{\sqrt{2}}{\sqrt{2}}\right)-\log\left(\frac{\sqrt{2}-1}{\sqrt{2}+1}\right)\right\}$$

$$\Rightarrow \quad 2I = \frac{1}{\sqrt{2}}\left\{0-\log\left(\frac{\sqrt{2}-1}{\sqrt{2}+1}\right)\right\}$$

$$\Rightarrow \quad 2I = \frac{1}{\sqrt{2}}\left\{\log\left(\frac{\sqrt{2}-1}{\sqrt{2}+1}\right)^{-1}\right\}$$

$$\Rightarrow \quad 2I = \frac{1}{\sqrt{2}}\left\{\log\left(\frac{\sqrt{2}+1}{\sqrt{2}-1}\right)\right\}$$

$$\Rightarrow \quad I = \frac{1}{2\sqrt{2}}\log\left(\frac{\sqrt{2}+1}{\sqrt{2}-1}\right)$$ **Ans.**

**OR**

**Evaluate :** $\int_0^{3/2}|x\cos\pi x|dx$

**Sol.** We have, $0 < x < \frac{3}{2} \Rightarrow 0 < \pi x < \frac{3\pi}{2}$

Now, $0 < x < \frac{1}{2} \Rightarrow 0 < \pi x < \frac{\pi}{2}$

$\Rightarrow \quad \cos\pi x > 0$

$\Rightarrow \quad x\cos\pi x > 0$

$\Rightarrow |x\cos\pi x| = x\cos\pi x$ for $0 < \pi x < \frac{\pi}{2}$

and $\quad \frac{1}{2} < x < \frac{3}{2}$

$\Rightarrow \quad \frac{\pi}{2} < \pi x < \frac{3\pi}{2}$

$\Rightarrow \quad \cos\pi x < 0$

$\Rightarrow \quad x\cos\pi x < 0$

$\Rightarrow |x\cos\pi x| = -x\cos\pi x$ for $\pi < \pi x < \frac{3\pi}{2}$

$\therefore \int_0^{3/2}|x\cos\pi x|dx$

$$= \int_0^{1/2}|x\cos\pi x|dx + \int_{1/2}^{3/2}|x\cos\pi x|dx$$

$$= \int_0^{1/2}(x\cos\pi x)dx + \int_{1/2}^{3/2}(-x\cos\pi x)dx$$

$$= \int_0^{1/2}(x\cos\pi x)dx - \int_{1/2}^{3/2}(x\cos\pi x)dx$$

$$= \left[\left(\left|\frac{x\sin\pi x}{\pi}\right|_0^{1/2} - \int_0^{1/2}\frac{\sin\pi x}{\pi}dx\right) - \left(\left|\frac{x\sin\pi x}{\pi}\right|_{1/2}^{3/2} - \int_{1/2}^{3/2}\frac{\sin\pi x}{\pi}dx\right)\right]$$

$$= \left[\left\{\frac{1}{\pi}\left(\frac{1}{2}\sin\frac{\pi}{2}-0\right)+\left|\frac{\cos\pi x}{\pi^2}\right|_0^{1/2}\right\} - \left\{\frac{1}{\pi}\left(\frac{3}{2}\sin\frac{3\pi}{2}-\frac{1}{2}\sin\frac{\pi}{2}\right)+\left|\frac{\cos\pi x}{\pi^2}\right|_{1/2}^{3/2}\right\}\right]$$

$$= \frac{1}{2\pi}+\frac{1}{\pi^2}\left(\cos\frac{\pi}{2}-\cos 0\right)-\frac{1}{\pi}\left(\frac{3}{2}(-1)-\frac{1}{2}(1)\right) + \frac{1}{\pi^2}\left(\cos\frac{3\pi}{2}-\cos\frac{\pi}{2}\right)$$

$$= \frac{1}{2\pi}+\frac{1}{\pi^2}(0-1)-\frac{1}{\pi}\left(-\frac{3}{2}-\frac{1}{2}\right)+\frac{1}{\pi^2}(0-0)$$

$$= \frac{1}{2\pi}-\frac{1}{\pi^2}+\frac{2}{\pi}$$

$$= \frac{1}{2\pi}+\frac{2}{\pi}-\frac{1}{\pi^2}$$

$$= \frac{\pi+4\pi-2}{2\pi^2} = \frac{5\pi-2}{2\pi^2}$$ **Ans.**

**15. Find :** $\int(3x+1)\sqrt{4-3x-2x^2}\,dx$ **[4]**

**Sol.** Let $3x + 1 = \lambda\frac{d}{dx}(4-3x-2x^2)+\mu.$

Then, $\quad 3x + 1 = \lambda(-3-4x) + \mu$

$\Rightarrow \quad 3x + 1 = -4\lambda x + (-3\lambda + \mu)$

Comparing the coefficients of like powers of $x$, we get

$$-4\lambda = 3 \text{ and } -3\lambda + \mu = 1$$

$$\Rightarrow \quad \lambda = \frac{-3}{4} \text{ and } \mu = \frac{-5}{4}$$

Let $\quad I = \int(3x+1)\sqrt{4-3x-2x^2}\,dx$

$$I = \int\left\{-\frac{3}{4}(-3-4x)-\frac{5}{4}\right\}\sqrt{4-3x-2x^2}\,dx$$

$$= -\frac{3}{4}\int(-3-4x)\sqrt{4-3x-2x^2}\,dx$$

$$-\frac{5}{4}\int\sqrt{4-3x-2x^2}\,dx$$

$$= -\frac{3}{4}\int\sqrt{t}\,dt - \frac{5}{4}\int\sqrt{-2\left(x^2+\frac{3}{2}x+\frac{9}{16}-\frac{9}{16}-2\right)}dx$$

where, $t = 4 - 3x - 2x^2$

$\Rightarrow \quad dt = (-3-4x)\,dx$

$$= -\frac{3}{4}\left(\frac{t^{3/2}}{3/2}\right)+c_1-\frac{5\sqrt{2}}{4}\int\sqrt{-\left(x+\frac{3}{4}\right)^2+\left(\frac{\sqrt{41}}{4}\right)^2}\,dx$$

$$= -\frac{1}{2}(4-3x-2x^2)^{3/2}+c_1$$

$$-\frac{5}{2\sqrt{2}}\left\{\frac{1}{2}\left(x+\frac{3}{4}\right)\sqrt{\left(\frac{\sqrt{41}}{4}\right)^2-\left(x+\frac{3}{4}\right)^2}\right\}$$

$$-\frac{5}{2\sqrt{2}}\times\frac{1}{2}\left(\frac{\sqrt{41}}{4}\right)^2\sin^{-1}\left(\frac{x+\frac{3}{4}}{\sqrt{41}/4}\right)+c_2$$

$$= -\frac{1}{2}(4-3x-2x^2)^{3/2}-\frac{5}{4\sqrt{2}}$$

$$\left\{\left(x+\frac{3}{4}\right)\sqrt{\frac{4-3x-2x^2}{2}}\right\}-\frac{5}{4\sqrt{2}}\times\frac{41}{16}\sin^{-1}\left(\frac{4x+3}{\sqrt{41}}\right)+c$$

(where $c = c_1 + c_2$)

$$= -\frac{1}{2}(4-3x-2x^2)^{3/2}-\frac{5}{8}\left(x+\frac{3}{4}\right)\sqrt{4-3x-2x^2}$$

$$+\frac{5\times 41}{4\sqrt{2}\times 16}\sin^{-1}\left(\frac{4x+3}{\sqrt{41}}\right)+c$$

$$= -\frac{1}{2}\left(4-3x-2x^2\right)^{3/2}-\frac{5}{8}\left(\frac{4x+3}{4}\right)\sqrt{4-3x-2x^2}$$

$$+\frac{5\times 41\sqrt{2}}{8\times 16}\sin^{-1}\left(\frac{4x+3}{\sqrt{41}}\right)+c$$

$$= -\frac{1}{2}\left(4-3x-2x^2\right)^{3/2}-\frac{5}{8}\left\{\left(\frac{4x+3}{4}\right)\sqrt{4-3x-2x^2}\right.$$

$$\left.+\frac{41\sqrt{2}}{16}\sin^{-1}\left(\frac{4x+3}{\sqrt{41}}\right)+c\right\}$$ **Ans.**

**16. Solve the differential equation :**

$$y + x\frac{dy}{dx} = x - y\frac{dy}{dx}$$ **[4]**

**Sol.** We have, $y + x\dfrac{dy}{dx} = x - y\dfrac{dy}{dx}$

$$\Rightarrow \quad x\frac{dy}{dx}+y\frac{dy}{dx} = x-y$$

$$\Rightarrow \quad \frac{dy}{dx} = \frac{x-y}{x+y} \quad \text{...(i)}$$

Which is a homogeneous differential equation.

Putting $y = Vx \Rightarrow \dfrac{dy}{dx} = V + x\dfrac{dV}{dx}$ in (i), we get

$$V + x\frac{dV}{dx} = \frac{x-Vx}{x+Vx}$$

$$\Rightarrow \quad V + x\frac{dV}{dx} = \frac{1-V}{1+V}$$

$$\Rightarrow \quad x\frac{dV}{dx} = \frac{1-V}{1+V}-V$$

$$\Rightarrow \quad x\frac{dV}{dx} = \frac{1-V-V-V^2}{1+V}$$

$$\Rightarrow \quad x\frac{dV}{dx} = \frac{1-2V-V^2}{1+V}$$

$$\Rightarrow \quad \frac{1+V}{1-2V-V^2}dV = \frac{dx}{x}, \quad x \neq 0$$

$$\Rightarrow \quad \frac{1+V}{V^2+2V-1}dV = \frac{-dx}{x} \quad \text{...(ii)}$$

Putting $t = V^2 + 2V - 1$

$\Rightarrow \quad dt = (2V + 2)\,dV$

$\Rightarrow \quad \dfrac{1}{2}dt = (V + 1)\,dV$

Now, equation (ii) becomes

$$\frac{1}{2t}dt = \frac{-dx}{x}$$

On integrating above equation

we get,

$$\frac{1}{2}\int\frac{1}{t}dt = -\int\frac{dx}{x}$$

$$\Rightarrow \quad \frac{1}{2}\log|t| = -\log|x| + \log C$$

$$\Rightarrow \quad \log|V^2+2V-1|^{1/2} = \log\left|\frac{C}{x}\right|$$

$$\Rightarrow \quad V^2+2V-1 = \left(\frac{C}{x}\right)^2$$

$$\Rightarrow \quad \left(\frac{y}{x}\right)^2+2\left(\frac{y}{x}\right)-1 = \frac{C^2}{x^2} \quad \left(\because V = \frac{y}{x}\right)$$

$$\Rightarrow \quad \frac{y^2}{x^2}+\frac{2y}{x}-1 = \frac{C^2}{x^2}$$

$$\Rightarrow \quad y^2+2xy-x^2 = C^2$$ **Ans.**

**17. Form the differential equation of the family of circles in the second quadrant and touching the coordinate axes. [4]**

**Sol.** The equation of circles in the second quadrant which touch the coordinate axes is

$$(x + a)^2 + (y - a)^2 = a^2,\ a \in R \quad ...(i)$$

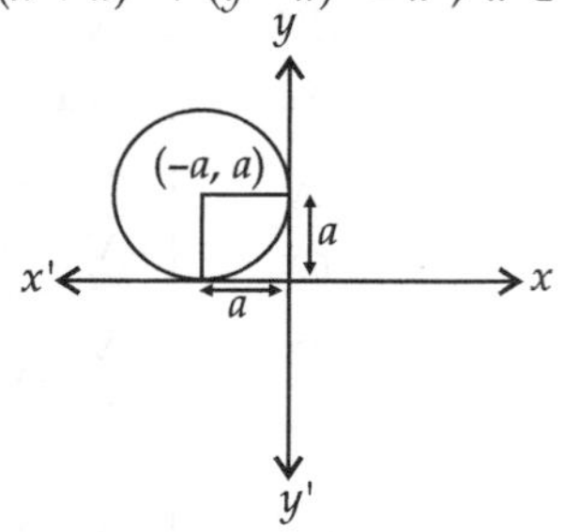

where $a$ is a parameter. This equation contains one arbitrary constant. So we shall differentiate it once only and we shall get a differential equation of first order.

Differentiating (i) w.r.t. $x$, we get

$$2(x+a)+2(y-a)\frac{dy}{dx} = 0$$

$$\Rightarrow \quad x+a+(y-a)\frac{dy}{dx} = 0$$

$$\Rightarrow \quad a = -\left(\frac{x+y\frac{dy}{dx}}{1-\frac{dy}{dx}}\right)$$

$$\Rightarrow \quad a = \frac{x+Py}{P-1}, \text{ where } P = \frac{dy}{dx}$$

Substituting the value of $a$ in (i), we get

$$\left(x+\frac{x+Py}{P-1}\right)^2 + \left(y-\frac{x+Py}{P-1}\right)^2 = \left(\frac{x+Py}{P-1}\right)^2$$

$$\Rightarrow (xP - x + x + yP)^2 + (yP - y - x - yP)^2 = (x + yP)^2$$

$$\Rightarrow \quad (x + y)^2P^2 + (x + y)^2 = (x + yP)^2$$

$$\Rightarrow \quad (x + y)^2 (P^2 + 1) = (x + yP)^2$$

$$\Rightarrow \quad (x + y)^2 \left[\left(\frac{dy}{dx}\right)^2 + 1\right] = \left(x + y\frac{dy}{dx}\right)^2$$

This is the required differential equation representing the given family of circles. **Ans.**

**18. Solve the equation for $x$ :**

**$\sin^{-1} x + \sin^{-1} (1 - x) = \cos^{-1} x.$ [4]**

**Sol.** We have, $\sin^{-1} x + \sin^{-1} (1 - x) = \cos^{-1} x$

$$\Rightarrow \quad \sin^{-1}(x\sqrt{1-(1-x)^2} + (1-x)\sqrt{1-x^2}) = \cos^{-1} x$$

$$\Rightarrow \sin^{-1}(x\sqrt{1-(1+x^2-2x)} + (1-x)\sqrt{1-x^2}) = \cos^{-1} x$$

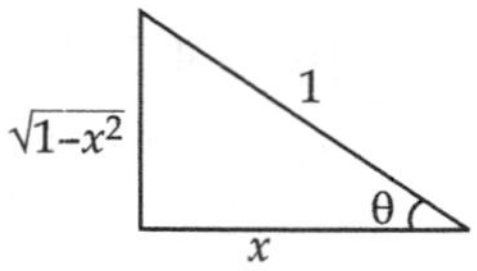

Let $\cos^{-1} x = \theta$

$\Rightarrow \quad x = \cos\theta$

and $\quad \sin\theta = \sqrt{1-x^2}$

$$\Rightarrow \quad \theta = \sin^{-1}(\sqrt{1-x^2}) = \cos^{-1}x$$

$$\Rightarrow \sin^{-1}(x\sqrt{2x-x^2} + (1-x)\sqrt{1-x^2}) = \sin^{-1}(\sqrt{1-x^2})$$

$$\Rightarrow x\sqrt{2x-x^2} + (1-x)\sqrt{1-x^2} = \sqrt{1-x^2}$$

$$\Rightarrow \quad x\sqrt{2x-x^2} = \sqrt{1-x^2}.(1-1+x)$$

$$\Rightarrow \quad x\left[\sqrt{2x-x^2} - \sqrt{1-x^2}\right] = 0$$

$$\Rightarrow \quad x = 0$$

or

$$\sqrt{2x-x^2} = \sqrt{1-x^2}$$

$$\Rightarrow \quad x = 0$$

or

$$2x - x^2 = 1 - x^2$$

$$\Rightarrow \quad x = 0 \text{ or } x = \frac{1}{2}$$ **Ans.**

**OR**

**If $\cos^{-1}\frac{x}{a} + \cos^{-1}\frac{y}{b} = \alpha$, prove that :**

$$\frac{x^2}{a^2} - 2\frac{xy}{ab}\cos\alpha + \frac{y^2}{b^2} = \sin^2\alpha$$

**Sol.** We have, $\cos^{-1}\frac{x}{a} + \cos^{-1}\frac{y}{b} = \alpha$

$$\Rightarrow \quad \cos^{-1}\left(\frac{x}{a}.\frac{y}{b} + \sqrt{1-\frac{x^2}{a^2}}\sqrt{1-\frac{y^2}{b^2}}\right) = \alpha$$

$$\Rightarrow \quad \frac{xy}{ab} + \sqrt{1-\frac{x^2}{a^2}}\sqrt{1-\frac{y^2}{b^2}} = \cos\alpha$$

$$\Rightarrow \quad \sqrt{1-\frac{x^2}{a^2}}\sqrt{1-\frac{y^2}{b^2}} = \cos\alpha - \frac{xy}{ab}$$

On squaring both sides, we get

$$\left(1-\frac{x^2}{a^2}\right)\left(1-\frac{y^2}{b^2}\right) = \left(\cos\alpha - \frac{xy}{ab}\right)^2$$

$$\Rightarrow \quad 1-\frac{x^2}{a^2}-\frac{y^2}{b^2}+\frac{x^2y^2}{a^2b^2} = \cos^2\alpha + \frac{x^2y^2}{a^2b^2} - \frac{2xy}{ab}\cos\alpha$$

$$\Rightarrow \quad 1-\frac{x^2}{a^2}-\frac{y^2}{b^2} = 1-\sin^2\alpha - \frac{2xy}{ab}\cos\alpha$$

$$\Rightarrow \quad \frac{x^2}{a^2} - \frac{2xy}{ab}\cos\alpha + \frac{y^2}{b^2} = \sin^2\alpha$$ **Hence Proved.**

**19. A trust invested some money in two types of bond. The first bond pays 10% interest and**

**second bond pays 12% interest. The trust received ₹ 2,800 as interest. However, if trust had interchanged money in bonds, they would have got ₹ 100 less as interest. Using matrix method, find the amount invested by the trust. Interest received on this amount will be given to Helpage India as donation. Which value is reflected in this question ? [4]**

**Sol.** Let the amount invested by the trust in first and second bond be $x$ and $y$ respectively.

$$\text{Interest from first bond} = \frac{10 \times x \times 1}{100} = \frac{10x}{100}$$

$$\text{Interest from second bond} = \frac{12 \times y \times 1}{100} = \frac{12y}{100}$$

Interest received by trust = ₹ 2,800

According to the question,

$$\frac{10x}{100} + \frac{12y}{100} = 2,800$$

$$\Rightarrow \quad 10x + 12y = 2,80,000 \quad \text{...(i)}$$

and

$$\frac{12x}{100} + \frac{10y}{100} = 2,700$$

$$\Rightarrow \quad 12x + 10y = 2,70,000 \quad \text{...(ii)}$$

This system of equations can be written in matrix form as follows :

$$\begin{bmatrix} 10 & 12 \\ 12 & 10 \end{bmatrix}\begin{bmatrix} x \\ y \end{bmatrix} = \begin{bmatrix} 2,80,000 \\ 2,70,000 \end{bmatrix}$$

or AX = B,

where $A = \begin{bmatrix} 10 & 12 \\ 12 & 10 \end{bmatrix}$, $X = \begin{bmatrix} x \\ y \end{bmatrix}$ and $B = \begin{bmatrix} 2,80,000 \\ 2,70,000 \end{bmatrix}$

Now, $|A| = \begin{vmatrix} 10 & 12 \\ 12 & 10 \end{vmatrix}$

$$= 100 - 144 = -44 \neq 0$$

So, $A^{-1}$ exists and the solution of the given system of equations is given by

$$X = A^{-1}B$$

Let $c_{ij}$ be the cofactor of $a_{ij}$ in $A = [a_{ij}]$. Then,

$c_{11} = 10, c_{12} = -12, c_{21} = -12, c_{22} = 10$

$$\therefore \text{adj } A = \begin{bmatrix} 10 & -12 \\ -12 & 10 \end{bmatrix}^T = \begin{bmatrix} 10 & -12 \\ -12 & 10 \end{bmatrix}$$

So, $A^{-1} = \frac{1}{|A|}(\text{adj } A) = -\frac{1}{44}\begin{bmatrix} 10 & -12 \\ -12 & 10 \end{bmatrix}$

Hence, the solution is given by

$$X = A^{-1}B = -\frac{1}{44}\begin{bmatrix} 10 & -12 \\ -12 & 10 \end{bmatrix}\begin{bmatrix} 2,80,000 \\ 2,70,000 \end{bmatrix}$$

$$\begin{bmatrix} x \\ y \end{bmatrix} = -\frac{1}{44}\begin{bmatrix} 28,00,000 - 32,40,000 \\ -33,60,000 + 27,00,000 \end{bmatrix}$$

$$\begin{bmatrix} x \\ y \end{bmatrix} = -\frac{1}{44}\begin{bmatrix} -4,40,000 \\ -6,60,000 \end{bmatrix} = \begin{bmatrix} 10,000 \\ 15,000 \end{bmatrix}$$

$\Rightarrow x = 10,000$ and $y = 15,000$

$\Rightarrow A = x + y = 10,000 + 15,000 =$ ₹ 25,000

Hence, the amount invested by the trust is ₹ 25,000.

**Value :** Giving help to those in need is a humanitarian act. **Ans.**

## SECTION–C

**20. There are two types of fertilisers 'A' and 'B'. 'A' consists of 12% nitrogen and 5% phosphoric acid whereas 'B' consists of 4% nitrogen and 5% phosphoric acid. After testing the soil conditions, farmer finds that he needs at least 12 kg of nitrogen and 12 kg of phosphoric acid for his crops. If 'A' costs ₹ 10 per kg and 'B' cost ₹ 8 per kg, then graphically determine how much of each type of fertiliser should be used so that nutrient requirements are met at a minimum cost. [6]**

**Sol.** Let the quantity of fertiliser A and B be $x$ kg and $y$ kg respectively.

To minimise : Z = ₹ $(10x + 8y)$

Subject to the constraints :

$$\frac{12}{100}x + \frac{4}{100}y \geq 12$$

or $\quad 12x + 4y \geq 1200$

and
$$\frac{5x}{100} + \frac{5y}{100} \geq 12$$

or $\quad 5x + 5y \geq 1200$

and $\quad x \geq 0, y \geq 0$

*i.e.,* $\quad 3x + y \geq 300,$

$x + y \geq 240,$

$x \geq 0, y \geq 0$

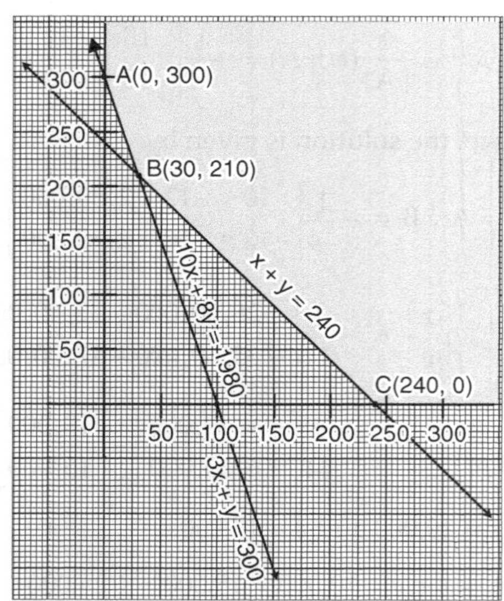


| Corner Points | Z = $10x + 8y$ |
|---|---|
| A (0,300) | $Z = 10 \times 0 + 8 \times 300 =$ ₹ 2400 |
| B (30,210) | $Z = 10 \times 30 + 8 \times 210 =$ ₹ 1980 |
| C (240,0) | $Z = 10 \times 240 + 8 \times 0 =$ ₹ 2400 |

The region of $10x + 8y < 1980$ has no point in common to the feasible region.

So, Z is minimum for $x = 30$ and $y = 210$ and the minimum value of Z is ₹ 1980.

Hence, the quantity of fertilizer A is 30 kg and of fertilizer B is 210 kg. **Ans.**

**21. Five bad oranges are accidently mixed with 20 good ones. If four oranges are drawn one by one successively with replacement, then find the probability distribution of number of bad oranges drawn. Hence find the mean and variance of the distribution. [6]**

**Sol.** Let X denotes the number of bad oranges in a draw of 4 oranges from a group of 20 good oranges and 5 bad oranges. Since there are 5 bad oranges in the group, therefore X can take values, 0, 1, 2, 3, 4.

Now, P(X = 0) = Probability of getting no bad orange.

P(X = 0) = Probability of getting 4 good oranges

$$= \left(\frac{20}{25}\right)^4 . {}^4C_0$$

P(X = 1) = Probability of getting one bad orange

$$= \frac{5}{25} \times \left(\frac{20}{25}\right)^3 . {}^4C_1$$

P(X = 2) = Probability of getting two bad oranges

$$= \left(\frac{5}{25}\right)^2 \times \left(\frac{20}{25}\right)^2 . {}^4C_2$$

P(X = 3) = Probability of getting three bad oranges

$$= \left(\frac{5}{25}\right)^3 \times \frac{20}{25} . {}^4C_3$$

P(X = 4) = Probability of getting four bad oranges

$$= \left(\frac{5}{25}\right)^4 . {}^4C_4$$

**Computation of Mean and Variance**

| $x_i$ | $p_i = P(X = x_i)$ | $p_i x_i$ | $p_i x_i^2$ |
|---|---|---|---|
| 0 | $\frac{256}{625}$ | 0 | 0 |
| 1 | $\frac{256}{625}$ | $\frac{256}{625}$ | $\frac{256}{625}$ |
| 2 | $\frac{96}{625}$ | $\frac{192}{625}$ | $\frac{384}{625}$ |
| 3 | $\frac{16}{625}$ | $\frac{48}{625}$ | $\frac{144}{625}$ |
| 4 | $\frac{1}{625}$ | $\frac{4}{625}$ | $\frac{16}{625}$ |
| | | $\Sigma p_i x_i = \frac{500}{625}$ | $\Sigma p_i x_i^2 = \frac{800}{625}$ |

We have, $\Sigma p_i x_i = \frac{500}{625} = \frac{4}{5}$ and $\Sigma p_i x_i^2 = \frac{800}{625} = \frac{32}{25}$

$$\therefore \qquad \bar{X} = \text{Mean} = \Sigma p_i x_i = \frac{4}{5}$$

and $\text{Var}(X) =$

$$\Sigma p_i x_i^2 - (\Sigma p_i x_i)^2 = \frac{32}{25} - \frac{16}{25} = \frac{16}{25}$$

Hence, mean $= \frac{4}{5}$ and variance $= \frac{16}{25}$ **Ans.**

**22. Find the position vector of the foot of perpendicular and the perpendicular distance from the point P with position vector $2\hat{i} + 3\hat{j} + 4\hat{k}$ to the plane $\vec{r}.(2\hat{i} + \hat{j} + 3\hat{k}) - 26 = 0$. Also find image of P in the plane. [6]**

**Sol.** Let L be the foot of the perpendicular drawn from $P(2\hat{i} + 3\hat{j} + 4\hat{k})$ on the plane $\vec{r}.(2\hat{i} + \hat{j} + 3\hat{k}) - 26 = 0$ We know that the position vector of the foot of perpendicular from the point P with position vector $\vec{a}$ from the plane $\vec{r}.\vec{n} = d$ is given by

$$\vec{PL} = \frac{[d - (\vec{a}.\vec{n})]\vec{n}}{|\vec{n}|^2}$$

$$= \frac{[26-(2\hat{i}+3\hat{j}+4\hat{k})(2\hat{i}+\hat{j}+3\hat{k})](2\hat{i}+\hat{j}+3\hat{k})]}{(\sqrt{4+1+9})^2}$$

$$= \frac{[26-(4+3+12)](2\hat{i}+\hat{j}+3\hat{k})}{14} = \frac{7}{14}(2\hat{i}+\hat{j}+3\hat{k})$$

$$= \hat{i}+\frac{1}{2}\cdot\hat{j}+\frac{3}{2}\hat{k}$$

and the perpendicular distance is given by $\frac{|\vec{a}.\vec{n}-d|}{|\vec{n}|}$

So, required distance

$$= \frac{|(2\hat{i}+3\hat{j}+4\hat{k})(2\hat{i}+\hat{j}+3\hat{k})-26|}{\sqrt{14}}$$

$$= \frac{7}{\sqrt{14}}$$

Let Q be the image of the point $P(2\hat{i}+3\hat{j}+4\hat{k})$ to the plane $\vec{r}\cdot(2\hat{i}+\hat{j}+3\hat{k})-26=0$. Then PQ is normal to the plane.

Therefore, equation of line PQ is

$$\vec{r} = (2\hat{i}+3\hat{j}+4\hat{k})+\lambda(2\hat{i}+\hat{j}+3\hat{k})$$

Since Q lies on line PQ. So, let the position vector of Q be $(2\hat{i}+3\hat{j}+4\hat{k})+\lambda(2\hat{i}+\hat{j}+3\hat{k})$

$$=(2+2\lambda)\hat{i}+(3+\lambda)\hat{j}+(4+3\lambda)\hat{k}$$

Since, R is the mid-point of PQ. Therefore, position vector of R is

$$\frac{[(2+2\lambda)\hat{i}+(3+\lambda)\hat{j}+(4+3\lambda)\hat{k}]+(2\hat{i}+3\hat{j}+4\hat{k})}{2}$$

$$= (2+\lambda)\hat{i}+(3+\lambda/2)\hat{j}+(4+3\lambda/2)\hat{k}$$

Since R lies on the plane $\vec{r}.[2\hat{i}+\hat{j}+3\hat{k}]-26=0$

$$\therefore \left\{(2+\lambda)\hat{i}+(3+\lambda/2)\hat{j}+(4+3\lambda/2)\hat{k}\right\}\cdot(2\hat{i}+\hat{j}+3\hat{k})-26=0$$

$$\Rightarrow \quad 4+2\lambda+3+\frac{\lambda}{2}+12+\frac{9\lambda}{2}-26=0$$

$$\Rightarrow \quad \lambda = \frac{7}{7}=1$$

Thus, the position vector of Q is $4\hat{i}+4\hat{j}+7\hat{k}$

**Ans.**

**23. Show that the binary operation * on A = R – {– 1} defined as $a * b = a + b + ab$ for all $a, b \in$ A is commutative and associative on A. Also find the identity element of * in A and prove that every element of A is invertible.**** **[6]**

** Answer is not given due to change in the present syllabus.

**24. Prove that the least perimeter of an isosceles triangle in which a circle of radius $r$ can be inscribed is $6\sqrt{3}r$.** **[6]**

**Sol.** Let ABC be an isosceles triangle with AB = AC and a circle with centre O and radius $r$, touching sides AB, BC,CA at D, F, E respectively.

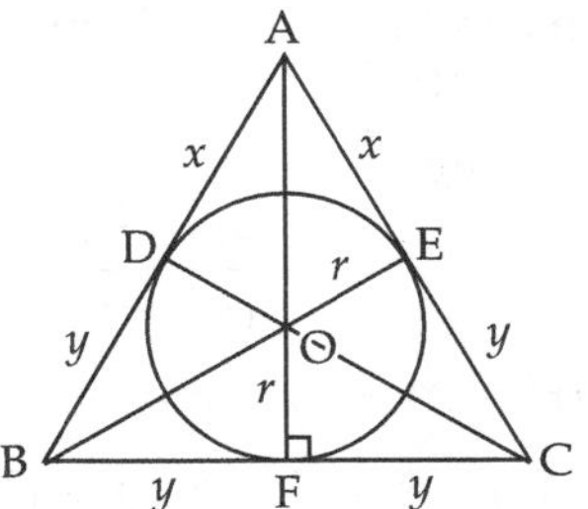


In ΔABC,

Let AD = AE = $x$, BD = BF = $y$ and CF = CE = $y$

(∵ Tangents drawn from an external point are equal)

Now, ar (ΔABC) = ar (ΔAOB) + ar (ΔAOC) + ar (ΔBOC)

$$\Rightarrow \frac{1}{2}\times 2y\left(r+\sqrt{r^2+x^2}\right) = \frac{1}{2}\{2yr+(x+y)r+(x+y)r\}$$

$$\Rightarrow \quad y\left(r+\sqrt{r^2+x^2}\right) = \frac{1}{2}\{2yr+2(x+y)r\}$$

$$\Rightarrow \quad y\left(r+\sqrt{r^2+x^2}\right) = r(y+x+y)$$

$$\Rightarrow \quad yr+y\left(\sqrt{r^2+x^2}\right) = 2yr+rx$$

$$\Rightarrow \quad \left(\sqrt{r^2+x^2}\right)y = rx+yr$$

Squaring both sides, we get

$$y^2(r^2+x^2) = r^2x^2+y^2r^2+2r^2xy$$

$$\Rightarrow \quad y^2r^2+y^2x^2 = r^2x^2+y^2r^2+2r^2xy$$

$$\Rightarrow \quad y^2x = r^2x+2r^2y$$

$$\Rightarrow \quad x = \frac{2r^2y}{y^2-r^2}$$

Now, P (Perimeter of ΔABC) = $2x + 4y$

$$\Rightarrow \quad P = \frac{4r^2y}{y^2-r^2}+4y$$

Differentiate above equation w.r.t. $y$, we get

$$\frac{dP}{dy} = \frac{4r^2(y^2-r^2)-4r^2y(2y)}{(y^2-r^2)^2}+4$$

$$= \frac{4r^2\left[y^2 - r^2 - 2y^2\right]}{(y^2 - r^2)^2} + 4$$

$$\Rightarrow \quad \frac{dP}{dy} = \frac{-4r^2(r^2 + y^2)}{(y^2 - r^2)^2} + 4$$

For maxima and minima of P, put

$$\Rightarrow \quad \frac{dP}{dy} = 0$$

$$\Rightarrow \quad \frac{-4r^2(r^2 + y^2)}{(y^2 - r^2)^2} + 4 = 0$$

$$\Rightarrow \quad r^2(r^2 + y^2) = (y^2 - r^2)^2$$

$$\Rightarrow \quad r^4 + r^2y^2 = y^4 + r^4 - 2y^2r^2$$

$$\Rightarrow \quad 3y^2r^2 = y^4$$

$$\Rightarrow \quad y^2 = 3r^2$$

$$\Rightarrow \quad y = \sqrt{3}\,r$$

Now, again differentiate w.r. to $y$, we get $\frac{d^2P}{dy^2}$

$$= \frac{-4r^2(2y)(y^2 - r^2)^2 + 4r^2(r^2 + y^2)2(y^2 - r^2) \times 2y}{(y^2 - r^2)^4}$$

$$= \frac{4r^2(y^2 - r^2)\,[-2y(y^2 - r^2) + 4y(r^2 + y^2)]}{(y^2 - r^2)^4}$$

$$= \frac{4r^2(y^2 - r^2)\,y[-2y^2 + 2r^2 + 4r^2 + 4y^2]}{(y^2 - r^2)^4}$$

$$\Rightarrow \quad \frac{d^2P}{dy^2} = \frac{4r^2y[2y^2 + 6r^2]}{(y^2 - r^2)^3}$$

$$\left.\frac{d^2P}{dy^2}\right|_{y=\sqrt{3}r} = \frac{6\sqrt{3}}{r} => 0$$

Hence, perimeter P of $\Delta$ABC is least for $y = \sqrt{3}\,r$ and least perimeter is

$$P = 4y + \frac{4r^2y}{y^2 - r^2}$$

$$= 4\sqrt{3}r + \frac{4r^2\sqrt{3}r}{2r^2} = 6\sqrt{3}r \text{ Hence Proved.}$$

**OR**

**If the sum of lengths of hypotenuse and a side of a right angled triangle is given, show that area of triangle is maximum, when the angle between them is $\frac{\pi}{3}$.**

**Sol.** Let ABC be a right angled triangle with base BC = $x$, AB = $y$ such that $x + y = k$ (constant). Let $\theta$ be the angle between base and hypotenuse. Let A be the area of the tringle. Then,

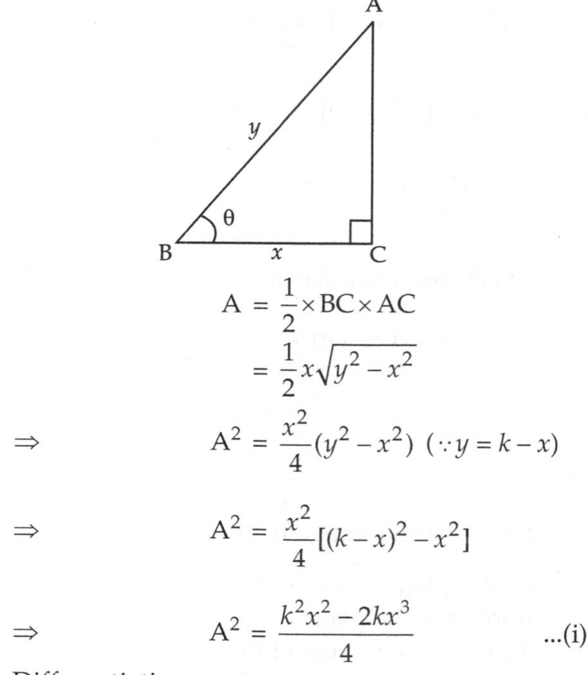


$$A = \frac{1}{2} \times BC \times AC$$

$$= \frac{1}{2}x\sqrt{y^2 - x^2}$$

$$\Rightarrow \quad A^2 = \frac{x^2}{4}(y^2 - x^2) \quad (\because y = k - x)$$

$$\Rightarrow \quad A^2 = \frac{x^2}{4}[(k - x)^2 - x^2]$$

$$\Rightarrow \quad A^2 = \frac{k^2x^2 - 2kx^3}{4} \quad \text{...(i)}$$

Differentiating w.r.t. $x$, we get

$$2A\frac{dA}{dx} = \frac{2k^2x - 6kx^2}{4}$$

$$\Rightarrow \quad \frac{dA}{dx} = \frac{k^2x - 3kx^2}{4A}$$

For maximum or minimum, we have

$$\frac{dA}{dx} = 0$$

$$\Rightarrow \quad \frac{k^2x - 3kx^2}{4A} = 0$$

$$\Rightarrow \quad x = \frac{k}{3}$$

Again differentiating (ii) w.r.t. $x$, we get

$$2\left(\frac{dA}{dx}\right)^2 + 2A\frac{d^2A}{dx^2} = \frac{2k^2 - 12kx}{4} \quad \text{....(iii)}$$

Putting $\frac{dA}{dx} = 0$ and $x = \frac{k}{3}$ in (iii), we get

$$\frac{d^2A}{dx^2} = \frac{-k^2}{4A} < 0$$

Thus, A is maximum when $= \frac{k}{3}$

Now, and $x = \frac{k}{3}$ and $y = k - x = k - \frac{k}{3} = \frac{2k}{3}$

$$\therefore \quad \cos\theta = \frac{x}{y} = \frac{k/3}{2k/3} = \frac{1}{2}$$

$$\theta = \cos^{-1}\left(\frac{1}{2}\right) = \frac{\pi}{3} \text{ Hence Proved.}$$

**25. Prove that the curves $y^2 = 4x$ and $x^2 = 4y$ divide the area of square bounded by $x = 0, x = 4, y = 4$ and $y = 0$ into three equal parts.** **[6]**

**Sol.** Let $A_1$, $A_2$ and $A_3$ denote areas OSPQO, OSPTO and OTPRO respectively.

**To prove :** $A_1 = A_2 = A_3$,

Now, $A_1 = \int_0^4 \frac{x^2}{4} dx$

$= \frac{1}{4}\int_0^4 x^2 dx$

$= \frac{1}{4}\left[\frac{x^3}{3}\right]_0^4 = \frac{1}{4}\times\frac{64}{3} = \frac{16}{3}$ sq. units

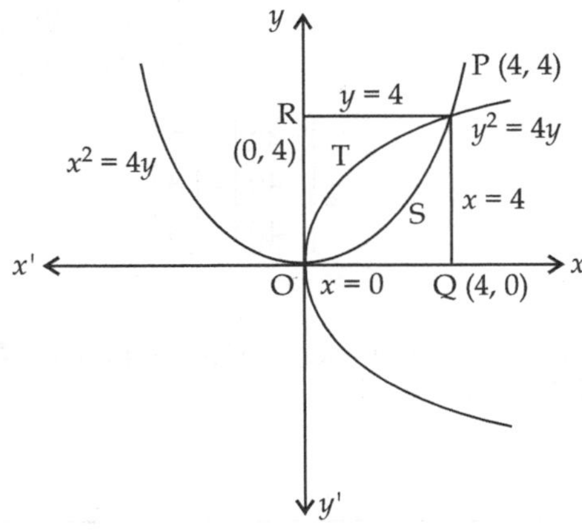


$A_2$ = (Area bounded by $y^2 = 4x$) – (Area bounded by $x^2 = 4y$)

$$A_2 = \int_0^4\left(\sqrt{4x} - \frac{x^2}{4}\right)dx$$

$$= \int_0^4\left(2\sqrt{x} - \frac{x^2}{4}\right)dx$$

$$= \left[\frac{4}{3}x^{3/2} - \frac{x^3}{12}\right]_0^4$$

$$= \left(\frac{4}{3}\times 8 - \frac{64}{12}\right) = \frac{16}{3} \text{ sq. units}$$

and $A_3 = \int_0^4 \frac{y^2}{4} dy$

$= \frac{1}{4}\left[\frac{y^3}{3}\right]_0^4 = \frac{1}{4}\times\frac{64}{3} = \frac{16}{3}$ sq. units.

Hence, $A_1 = A_2 = A_3$. **Hence Proved.**

**26. Using properties of determinants, show that $\Delta$ABC is isosceles if :** **[6]**

$$\begin{vmatrix} 1 & 1 & 1 \\ 1+\cos A & 1+\cos B & 1+\cos C \\ \cos^2 A+\cos A & \cos^2 B+\cos B & \cos^2 C+\cos C \end{vmatrix} = 0$$

**Sol.** We have,

$$\begin{vmatrix} 1 & 1 & 1 \\ 1+\cos A & 1+\cos B & 1+\cos C \\ \cos^2 A+\cos A & \cos^2 B+\cos B & \cos^2 C+\cos C \end{vmatrix} = 0$$

Applying $C_2 \to C_2 - C_1$ and $C_3$– $C_3 \to C_1$

$$\begin{vmatrix} 1 & 0 & 0 \\ 1+\cos A & \cos B - \cos A & \cos C - \cos A \\ \cos^2 A+\cos A & (\cos B - \cos A)(1+\cos A+\cos B) & (\cos C - \cos A)(1+\cos A+\cos C) \end{vmatrix} = 0$$

Taking (cos B – cos A) and (cos C – cos A) as common from $C_2$ and $C_3$ respectively.

$\Rightarrow$ (cos B – cos A) (cos C – cos A)

$$\begin{vmatrix} 1 & 0 & 0 \\ 1+\cos A & 1 & 1 \\ \cos^2 A+\cos A & 1+\cos A+\cos B & 1+\cos A+\cos C \end{vmatrix} = 0$$

$\Rightarrow$ Expanding along $R_1$

(cos B – cos A) (cos C – cos A)

[1 + cos A + cos C – 1 – cos A – cos B] = 0

$\Rightarrow$ (cos B – cos A) (cos C – cos A) (cos C – cos B) = 0

Either cos B = cos A or cos C = cos A or cos C = cos B

*i.e.*, either BC = AC or BC = AB or AC = AB

Hence, $\Delta$ ABC is isosceles. **Hence Proved.**

**OR**

**A shopkeeper has 3 varieties of pens 'A', 'B' and 'C'. Meenu purchased 1 pen of each variety for a total of ₹ 21. Jeevan purchased 4 pens of 'A' variety, 3 pens of 'B' variety and 2 pens of 'C' variety for ₹ 60. While Shikha purchased 6 pens of 'A' variety, 2 pens of 'B' variety and 3 pens of 'C' variety for ₹ 70. Using matrix method, find cost of each variety of pen.**

**Sol.** Let the cost of each variety of pen be ₹ $x$, ₹ $y$ and ₹ $z$ respectively. Then,

$$x + y + z = 21$$

$$4x + 3y + 2z = 60$$

and $6x + 2y + 3z = 70$

This system of equations can be written in matrix form as follows :

$$\begin{bmatrix} 1 & 1 & 1 \\ 4 & 3 & 2 \\ 6 & 2 & 3 \end{bmatrix}\begin{bmatrix} x \\ y \\ z \end{bmatrix} = \begin{bmatrix} 21 \\ 60 \\ 70 \end{bmatrix}$$

or $$AX = B$$

where $A = \begin{bmatrix} 1 & 1 & 1 \\ 4 & 3 & 2 \\ 6 & 2 & 3 \end{bmatrix}, X = \begin{bmatrix} x \\ y \\ z \end{bmatrix}$ and $B = \begin{bmatrix} 21 \\ 60 \\ 70 \end{bmatrix}$

Now, $$|A| = \begin{vmatrix} 1 & 1 & 1 \\ 4 & 3 & 2 \\ 6 & 2 & 3 \end{vmatrix}$$

$$= 1\,(9-4) - 1\,(12-12) + 1\,(8-18)$$

$$= -5 \neq 0$$

So, $A^{-1}$ exists and the solution of the given system of equation is given by

$$X = A^{-1}B$$

Let $c_{ij}$ be the cofactor of $a_{ij}$ in $A = [a_{ij}]$. Then,

$c_{11} = 5, c_{12} = 0, c_{13} = -10, c_{21} = -1, c_{22} = -3, c_{23} = 4,$

$c_{31} = -1, c_{32} = 2, c_{33} = -1$

$$\therefore \text{adj } A = \begin{bmatrix} 5 & 0 & -10 \\ -1 & -3 & 4 \\ -1 & 2 & -1 \end{bmatrix}^T = \begin{bmatrix} 5 & -1 & -1 \\ 0 & -3 & 2 \\ -10 & 4 & -1 \end{bmatrix}$$

$$\text{So, } A^{-1} = \frac{1}{|A|}(\text{adj } A) = -\frac{1}{5}\begin{bmatrix} 5 & -1 & -1 \\ 0 & -3 & 2 \\ -10 & 4 & -1 \end{bmatrix}$$

Hence, the solution is given by

$$X = A^{-1}B = -\frac{1}{5}\begin{bmatrix} 5 & -1 & -1 \\ 0 & -3 & 2 \\ -10 & 4 & -1 \end{bmatrix}\begin{bmatrix} 21 \\ 60 \\ 70 \end{bmatrix}$$

$$= -\frac{1}{5}\begin{bmatrix} 105-60-70 \\ 0-180+140 \\ -210+240-70 \end{bmatrix}$$

$$= -\frac{1}{5}\begin{bmatrix} -25 \\ -40 \\ -40 \end{bmatrix} = \begin{bmatrix} 5 \\ 8 \\ 8 \end{bmatrix}$$

Hence, the cost of each variety of pen are ₹ 5, ₹ 8 and ₹ 8 respectively.

**Ans.**

All questions are same in Outside Delhi Set II and Set III

# MATHEMATICS

## Delhi [Set-I] 2016

**Time allowed : 3 hours** **Maximum marks : 100**

### SECTION–A

**1. Find the maximum value of** $\begin{vmatrix} 1 & 1 & 1 \\ 1 & 1+\sin\theta & 1 \\ 1 & 1 & 1+\cos\theta \end{vmatrix}$. **[1]**

**Sol.** Let $\Delta = \begin{vmatrix} 1 & 1 & 1 \\ 1 & 1+\sin\theta & 1 \\ 1 & 1 & 1+\cos\theta \end{vmatrix}$

On applying $R_1 \to R_1 - R_2$ and $R_2 \to R_2 - R_3$, we get

$$\Delta = \begin{vmatrix} 0 & -\sin\theta & 0 \\ 0 & \sin\theta & -\cos\theta \\ 1 & 1 & 1+\cos\theta \end{vmatrix}$$

On expanding along $R_1$, we get

$$\Delta = 0 + \sin\theta\,(0 + \cos\theta) + 0$$
$$= \sin\cos\theta$$
$$= \frac{1}{2}\sin 2\theta$$

Now as $-1 \le \sin 2\theta \le 1$ for all $\theta \in R$

$\Rightarrow -\frac{1}{2} \le \frac{1}{2}\sin 2\theta \le \frac{1}{2}$ for all $\theta \in R$

Clearly, maximum value of $\Delta$ is $\frac{1}{2}$. **Ans.**

**2. If A is a square matrix such that $A^2 = I$, then find the simplified value of $(A - I)^3 + (A + I)^3 - 7A$.** **[1]**

**Sol.** Given, $(A - I)^3 + (A + I)^3 - 7A$

$$= A^3 - I^3 - 3A^2I + 3AI^2 + A^3 + I^3 + 3A^2I + 3AI^2 - 7A$$
$$= 2A^3 + 6AI^2 - 7A$$
$$= 2A.A^2 + 6AI^2 - 7A$$
$$= 2AI + 6AI - 7A$$
$$= 8A - 7A$$
$$= A$$ **Ans.**

**3. Matrix** $A = \begin{bmatrix} 0 & 2b & -2 \\ 3 & 1 & 3 \\ 3a & 3 & -1 \end{bmatrix}$ **is given to be symmetric, find values of *a* and *b*.** **[1]**

**Sol.** We have, $A = \begin{bmatrix} 0 & 2b & -2 \\ 3 & 1 & 3 \\ 3a & 3 & -1 \end{bmatrix}$

It is given that the matrix is symmetric.

$\therefore$ $A = A'$

$$\Rightarrow \begin{bmatrix} 0 & 2b & -2 \\ 3 & 1 & 3 \\ 3a & 3 & -1 \end{bmatrix} = \begin{bmatrix} 0 & 3 & 3a \\ 2b & 1 & 3 \\ -2 & 3 & -1 \end{bmatrix}$$

Now, by equality of matrices, we get

$$2b = 3$$
$$\Rightarrow b = \frac{3}{2}$$
and $$3a = -2$$
$$\Rightarrow a = \frac{-2}{3}$$

Therefore, $a = \frac{-2}{3}$ and $b = \frac{3}{2}$ **Ans.**

**4. Find the position vector of a point which divides the join of points with position vectors $\vec{a} - 2\vec{b}$ and $2\vec{a} + \vec{b}$ externally in the ratio 2 : 1.** **[1]**

**Sol.** Let A and B be the given points with position vectors $\vec{a} - 2\vec{b}$ and $2\vec{a} + \vec{b}$ respectively. Let P be the point dividing AB in the ratio 2 : 1 externally.

$A(\vec{a} - 2\vec{b})$ —————— P —————— $B(2\vec{a} + \vec{b})$

$$\therefore \text{Position vector of} = \frac{2 \times (2\vec{a} + \vec{b}) - 1 \times (\vec{a} - 2\vec{b})}{2 - 1}$$
$$= 3\vec{a} + 4\vec{b}$$ **Ans.**

**5. The two vectors $\hat{j} + \hat{k}$ and $3\hat{i} - \hat{j} + 4\hat{k}$ represent the two sides AB and AC, respectively of a $\Delta$ABC. Find the length of the median through A.** **[1]**

**Sol.** In $\Delta$ABC,

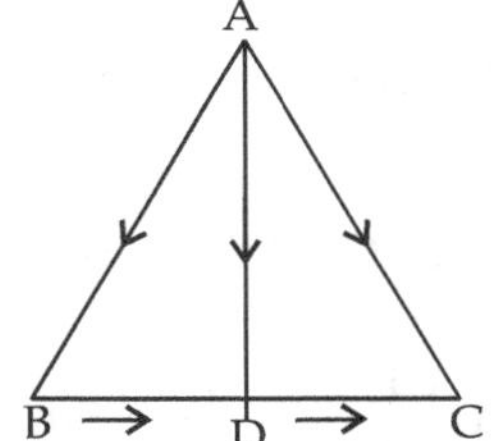

Using the triangle law of vector addition, we have

$$\vec{BC} = \vec{AC} - \vec{AB}$$

$$= (3\hat{i} - \hat{j} + 4\hat{k}) - (\hat{j} + \hat{k})$$

$$= 3\hat{i} - 2\hat{j} + 3\hat{k}$$

$$\therefore \quad \vec{BD} = \frac{1}{2}\vec{BC} = \frac{3}{2}\hat{i} - \hat{j} + \frac{3}{2}\hat{k}$$

(since AD is the median)

In $\Delta ABD$, using the triangle law of vector addition, we have

$$\vec{AD} = \vec{AB} + \vec{BD}$$

$$= (\hat{j} + \hat{k}) + \left(\frac{3}{2}\hat{i} - \hat{j} + \frac{3}{2}\hat{k}\right)$$

$$= \frac{3}{2}\hat{i} + 0\hat{j} + \frac{5}{2}\hat{k}$$

$$\therefore \quad AD = \sqrt{\left(\frac{3}{2}\right)^2 + 0^2 + \left(\frac{5}{2}\right)^2} = \frac{1}{2}\sqrt{34}$$

Hence, the length of the median through A is $\frac{1}{2}\sqrt{34}$ units. **Ans.**

**6. Find the vector equation of a plane which is at a distance of 5 units from the origin and its normal vector is $2\hat{i} - 3\hat{j} + 6\hat{k}$. [1]**

**Sol.** Here, $d = 5$ units and $\vec{n} = 2\hat{i} - 3\hat{j} + 6\hat{k}$

$$\therefore \quad \hat{n} = \frac{\vec{n}}{|\vec{n}|} = \frac{2\hat{i} - 3\hat{j} + 6\hat{k}}{\sqrt{4+9+36}} = \frac{2\hat{i} - 3\hat{j} + 6\hat{k}}{\sqrt{49}}$$

$$\Rightarrow \quad \hat{n} = \frac{2}{7}\hat{i} - \frac{3}{7}\hat{j} + \frac{6}{7}\hat{k}$$

Hence, the required equation of the plane is

$$\vec{r} \cdot \left(\frac{2}{7}\hat{i} - \frac{3}{7}\hat{j} + \frac{6}{7}\hat{k}\right) = 5 \qquad [\because \vec{r} \cdot \hat{n} = d]$$

or $\quad \vec{r} \cdot (2\hat{i} - 3\hat{j} + 6\hat{k}) = 35$ **Ans.**

## SECTION — B

**7. Prove that :**

$$\tan^{-1}\frac{1}{5} + \tan^{-1}\frac{1}{7} + \tan^{-1}\frac{1}{3} + \tan^{-1}\frac{1}{8} = \frac{\pi}{4} \quad [4]$$

**Sol.** $\text{LHS} = \left(\tan^{-1}\frac{1}{5} + \tan^{-1}\frac{1}{7}\right) + \left(\tan^{-1}\frac{1}{3} + \tan^{-1}\frac{1}{8}\right)$

$$= \tan^{-1}\left(\frac{\frac{1}{5} + \frac{1}{7}}{1 - \frac{1}{5} \times \frac{1}{7}}\right) + \tan^{-1}\left(\frac{\frac{1}{3} + \frac{1}{8}}{1 - \frac{1}{3} \times \frac{1}{8}}\right)$$

$$\left[\because \tan^{-1}A + \tan^{-1}B = \tan^{-1}\left(\frac{A+B}{1-AB}\right)\right]$$

$$= \tan^{-1}\frac{6}{17} + \tan^{-1}\frac{11}{23}$$

$$= \tan^{-1}\left(\frac{\frac{6}{17} + \frac{11}{23}}{1 - \frac{6}{17} \times \frac{11}{23}}\right)$$

$$= \tan^{-1}\left(\frac{325}{325}\right)$$

$$= \tan^{-1}(1)$$

$$= \frac{\pi}{4}$$

**Hence Proved.**

**OR**

**Solve for $x$:**

$$\mathbf{2\tan^{-1}(\cos x) = \tan^{-1}(2\,\text{cosec}\, x)}$$

**Sol.** Given, $2\tan^{-1}(\cos x) = \tan^{-1}(2\text{cosec}\, x)$

$$\Rightarrow \quad \tan^{-1}\left(\frac{2\cos x}{1 - \cos^2 x}\right) = \tan^{-1}(2\,\text{cosec}\, x)$$

$$\left[\because 2\tan^{-1}x = \tan^{-1}\left(\frac{2x}{1-x^2}\right)\right]$$

$$\Rightarrow \quad \frac{2\cos x}{\sin^2 x} = 2\,\text{cosec}\, x = \frac{2}{\sin x}$$

$$\Rightarrow \quad \cos x = \sin x$$

$$\Rightarrow \quad \tan x = 1$$

$$\Rightarrow \quad x = \frac{\pi}{4}$$

**Ans.**

**8. The monthly incomes of Aryan and Babban are in the ratio 3 : 4 and their monthly expenditures are in the ratio 5 : 7. If each saves ₹ 15,000 per month, find their monthly incomes using matrix method. This problem reflects which value ?[4]**

**Sol.** Let the monthly incomes of Aryan and Babban be $3x$ and $4x$ respectively.

Suppose their monthly expenditures are $5y$ and $7y$ respectively.

Since each saves ₹ 15,000 per month.

Monthly saving of Aryan : $3x - 5y = 15{,}000$

Monthly saving of Babban : $4x - 7y = 15{,}000$

The above system of equations can be written in the matrix form as follows :

$$\begin{bmatrix} 3 & -5 \\ 4 & -7 \end{bmatrix}\begin{bmatrix} x \\ y \end{bmatrix} = \begin{bmatrix} 15000 \\ 15000 \end{bmatrix}$$

AX = B, where $A = \begin{bmatrix} 3 & -5 \\ 4 & -7 \end{bmatrix}, X = \begin{bmatrix} x \\ y \end{bmatrix}$ and

$$B = \begin{bmatrix} 15000 \\ 15000 \end{bmatrix}$$

Now, $|A| = \begin{vmatrix} 3 & -5 \\ 4 & -7 \end{vmatrix} = -21-(-20) = -1 \neq 0$

$\Rightarrow$ A is invertible

$$\text{adj } A = \begin{bmatrix} -7 & -4 \\ 5 & 3 \end{bmatrix}^T = \begin{bmatrix} -7 & 5 \\ -4 & 3 \end{bmatrix}$$

So, $A^{-1} = \frac{1}{|A|}\text{adj } A = -1\begin{bmatrix} -7 & 5 \\ -4 & 3 \end{bmatrix} = \begin{bmatrix} 7 & -5 \\ 4 & -3 \end{bmatrix}$

$\therefore \quad X = A^{-1}B$

$$\Rightarrow \begin{bmatrix} x \\ y \end{bmatrix} = \begin{bmatrix} 7 & -5 \\ 4 & -3 \end{bmatrix}\begin{bmatrix} 15000 \\ 15000 \end{bmatrix}$$

$$\Rightarrow \begin{bmatrix} x \\ y \end{bmatrix} = \begin{bmatrix} 105000-75000 \\ 60000-45000 \end{bmatrix}$$

$$\Rightarrow \begin{bmatrix} x \\ y \end{bmatrix} = \begin{bmatrix} 30000 \\ 15000 \end{bmatrix}$$

$\Rightarrow x = 30{,}000$ and $y = 15{,}000$

Therefore,

Monthly income of Aryan = 3 × 30,000 = ₹ 90,000

Monthly income of Babban = 4 × 30,000 = ₹ 1,20,000

**Value :** Saving in good time helps us to survive in bad times. **Ans.**

**9. If $x = a \sin 2t(1 + \cos 2t)$ and $y = b \cos 2t(1 - \cos 2t)$, find the values of $\frac{dy}{dx}$ at $t = \frac{\pi}{4}$ and $t = \frac{\pi}{3}$** **[4]**

**Sol.** We have, $x = a \sin 2t\ (1 + \cos 2t)$

$$\Rightarrow \frac{dx}{dt} = 2a \cos 2t\ (1 + \cos 2t) - 2a \sin^2 2t$$

and, $\quad y = b \cos 2t\ (1 - \cos 2t)$

$$\Rightarrow \frac{dy}{dt} = 2b \cos 2t \sin 2t - 2b \sin 2t\ (1 - \cos 2t)$$

$$\therefore \frac{dy}{dx} = \frac{dy}{dt} \times \frac{dt}{dx}$$

$$= \frac{2b\cos 2t \sin 2t - 2b \sin 2t(1-\cos 2t)}{2a\cos 2t(1+\cos 2t) - 2a\sin^2 2t}$$

$$= \frac{b\,[\cos 2t \sin 2t - \sin 2t(1-\cos 2t)]}{a\,[\cos 2t(1+\cos 2t) - \sin^2 2t]}$$

Since we know that at $t = \frac{\pi}{4}$, $\sin 2t = 1$ and $\cos 2t = 0$,

$$\text{So, } \left(\frac{dy}{dx}\right)_{t=\frac{\pi}{4}} = \frac{b[0.1-1(1-0)]}{a[0(1+0)-1^2]} = \frac{b}{a}$$

Also, we know that at $t = \frac{\pi}{3}$, $\sin 2t = \frac{\sqrt{3}}{2}$ and $\cos 2t = -\frac{1}{2}$

$$\text{So, } \left(\frac{dy}{dx}\right)_{t=\frac{\pi}{3}} = \frac{b\left[\left(-\frac{1}{2}\right)\left(\frac{\sqrt{3}}{2}\right) - \left(\frac{\sqrt{3}}{2}\right)\left(1+\frac{1}{2}\right)\right]}{a\left[\left(-\frac{1}{2}\right)\left(1-\frac{1}{2}\right) - \left(\frac{\sqrt{3}}{2}\right)^2\right]}$$

$$= \frac{b\left[-\frac{\sqrt{3}}{4} - \frac{3\sqrt{3}}{4}\right]}{a\left[-\frac{1}{4} - \frac{3}{4}\right]}$$

$$= \frac{b[-\sqrt{3}]}{a[-1]} = \sqrt{3}\left(\frac{b}{a}\right)$$ **Ans.**

**OR**

**If $y = x^x$ prove that $\frac{d^2y}{dx^2} - \frac{1}{y}\left(\frac{dy}{dx}\right)^2 - \frac{y}{x} = 0$.**

**Sol.** Given, $y = x^x$

Taking log on both sides, we get

$$\log y = \log (x^x)$$

$$\Rightarrow \log y = x \log x$$

On differentiating w.r.t. $x$, we get

$$\frac{1}{y}\cdot\frac{dy}{dx} = \log x + x \times \frac{1}{x} = 1 + \log x$$

$$\Rightarrow \frac{dy}{dx} = x^x\ (1 + \log x)$$

$$\Rightarrow \frac{dy}{dx} = y\ (1 + \log x)$$

Again differentiating w.r.t. $x$, we get

$$\Rightarrow \frac{d^2y}{dx^2} = (1 + \log x)\ \frac{dy}{dx} + y\frac{d}{dx}(1+\log x)$$

$$\Rightarrow \frac{d^2y}{dx^2} = (1 + \log x).x^x(1 + \log x) + x^x \times \frac{1}{x}$$

$$\Rightarrow \frac{d^2y}{dx^2} = x^x(1+\log x)^2 + x^{x-1}$$

$$\Rightarrow \frac{d^2y}{dx^2} = \frac{1}{y}\left(\frac{dy}{dx}\right)^2 + \frac{y}{x}$$

$$\Rightarrow \frac{d^2y}{dx^2} - \frac{1}{y}\left(\frac{dy}{dx}\right)^2 - \frac{y}{x} = 0$$ **Hence Proved.**

**10. Find the values of $p$ and $q$ for which**

$$f(x) = \begin{cases} \dfrac{1-\sin^3 x}{3\cos^2 x} & , \text{ if } x < \dfrac{\pi}{2} \\ p & , \text{ if } x = \dfrac{\pi}{2} \\ \dfrac{q(1-\sin x)}{(\pi-2x)^2} & , \text{ if } x > \dfrac{\pi}{2} \end{cases}$$

**is continuous at $x = \pi/2$.** **[4]**

**Sol.** Given, $f(x) = \begin{cases} \dfrac{1-\sin^3 x}{3\cos^2 x} & , \text{ if } x < \dfrac{\pi}{2} \\ p & , \text{ if } x = \dfrac{\pi}{2} \\ \dfrac{q(1-\sin x)}{(\pi-2x)^2} & , \text{ if } x > \dfrac{\pi}{2} \end{cases}$

$f(x)$ is continuous at $x = \dfrac{\pi}{2}$, then LHL= RHL= $f(a)$

i.e. $\lim\limits_{x\to\frac{\pi}{2}^-} f(x) = \lim\limits_{x\to\frac{\pi}{2}^+} f(x) = f\left(\dfrac{\pi}{2}\right)$

$$\lim_{x\to\frac{\pi}{2}^-} f(x) = \lim_{x\to\frac{\pi}{2}} \left(\frac{1-\sin^3 x}{3\cos^2 x}\right)$$

$$= \lim_{x\to\frac{\pi}{2}} \frac{(1-\sin x)(1+\sin^2 x+\sin x)}{3(1-\sin^2 x)}$$

$$= \lim_{x\to\frac{\pi}{2}} \frac{(1-\sin x)(1+\sin^2 x+\sin x)}{3\,(1+\sin x)\,(1-\sin x)}$$

$$\lim_{x\to\frac{\pi}{2}^-} f(x) = \lim_{x\to\frac{\pi}{2}} \frac{1+\sin^2 x+\sin x}{3(1+\sin x)} = \frac{1+1+1}{3(2)} = \frac{1}{2}$$

Let $x = \dfrac{\pi}{2} + \theta$ as $x \to \dfrac{\pi}{2}, \theta \to 0$

$$\lim_{x\to\frac{\pi}{2}^+} f(x) = \lim_{\theta\to 0} q\frac{\left[1-\sin\left(\frac{\pi}{2}-\theta\right)\right]}{(2\theta)^2} = \frac{q}{4}\lim_{\theta\to 0}\frac{1-\cos\theta}{\theta^2}$$

$$= \frac{q}{4}\lim_{\theta\to 0}\frac{2\sin^2\frac{\theta}{2}}{\theta^2} = \frac{q}{2}\lim_{\theta\to 0}\frac{\sin^2\frac{\theta}{2}}{4\times\left(\frac{\theta}{2}\right)^2} = \frac{q}{8}$$

Now, $\lim\limits_{x\to\frac{\pi}{2}^-} f(x) = \lim\limits_{x\to\frac{\pi}{2}^+} f(x) = f\left(\dfrac{\pi}{2}\right)$

$\Rightarrow \quad \dfrac{1}{2} = p = \dfrac{q}{8}$

$\Rightarrow \quad p = \dfrac{1}{2}$ and $q = 4$ **Ans.**

**11. Show that the equation of normal at any point on the curve $x = 3\cos t - \cos^3 t$ and $y = 3\sin t - \sin^3 t$ is $4(y\cos^3 t - x\sin^3 t) = 3\sin 4t$.** **[4]**

**Sol.** Given, $x = 3\cos t - \cos^3 t$

$\Rightarrow \quad \dfrac{dy}{dt} = -3\sin t + 3\cos^2 t\sin t$

and $\quad y = 3\sin t - \sin^3 t$

$\Rightarrow \quad \dfrac{dy}{dt} = 3\cos t - 3\sin^2 t\cos t$

Slope of the tangent,

$$\frac{dy}{dx} = \frac{\frac{dy}{dt}}{\frac{dx}{dt}} = \frac{3\cos t - 3\sin^2 t\cos t}{-3\sin t + 3\cos^2 t\sin t}$$

$$= \frac{3\cos t[\cos^2 t]}{-3\sin t[\sin^2 t]}$$

$$\Rightarrow \quad \frac{dy}{dx} = \frac{-\cos^3 t}{\sin^3 t}$$

$\therefore$ Slope of the normal $= \dfrac{-dx}{dy} = \dfrac{\sin^3 t}{\cos^3 t}$

The equation of the normal is given by

$$\frac{y-(3\sin t-\sin^3 t)}{x-(3\cos t-\cos^3 t)} = \frac{\sin^3 t}{\cos^3 t}$$

$\Rightarrow \quad y\cos^3 t - 3\sin t\cos^3 t + \sin^3 t\cos^3 t$
$= x\sin^3 t - 3\cos t\sin^3 t + \sin^3 t\cos^3 t$

$\Rightarrow \quad y\cos^3 t - x\sin^3 t = 3(\sin t\cos^3 t - \cos t\sin^3 t)$

$\Rightarrow \quad y\cos^3 t - x\sin^3 t = 3\sin t\cos t(\cos^2 t - \sin^2 t)$

$\Rightarrow \quad y\cos^3 t - x\sin^3 t = \dfrac{3\times 2}{2}\sin t\cos t\cos 2t$

$\Rightarrow \quad y\cos^3 t - x\sin^3 t = \dfrac{3}{2}\sin 2t\cos 2t$

$\Rightarrow \quad y\cos^3 t - x\sin^3 t = \dfrac{3\times 2}{2\times 2}\sin 2t\cos 2t$

$\Rightarrow \quad y\cos^3 t - x\sin^3 t = \dfrac{3}{4}\sin 4t$

$\Rightarrow 4(y\cos^3 t - x\sin^3 t) = 3\sin 4t$ **Hence Proved.**

**12. Find $\displaystyle\int \frac{(3\sin\theta-2)\cos\theta}{5-\cos^2\theta-4\sin\theta}d\theta$** **[4]**

**Sol.** Let $\quad I = \displaystyle\int \frac{(3\sin\theta-2)\cos\theta}{5-\cos^2\theta-4\sin\theta}.d\theta$

$\Rightarrow \quad I = \displaystyle\int \frac{(3\sin\theta-2)\cos\theta}{4+1-\cos^2\theta-4\sin\theta}d\theta$

$\Rightarrow \quad I = \displaystyle\int \frac{(3\sin\theta-2)\cos\theta.d\theta}{4+\sin^2\theta-4\sin\theta}$

$$[\because 1-\cos^2\theta = \sin^2\theta]$$

$$\Rightarrow \quad I = \int \frac{(3\sin\theta - 2)\cos\theta . d\theta}{(\sin\theta - 2)^2}$$

Put $\sin\theta = t$

$$\Rightarrow \quad \cos\theta .\, d\theta = dt$$

$$\therefore \quad I = \int \frac{(3t-2)}{(t-2)^2}\cdot dt$$

Consider,

$$\frac{3t-2}{(t-2)^2} = \frac{A}{(t-2)} + \frac{B}{(t-2)^2}$$

$$3t - 2 = A(t-2) + B$$

On comparing, we get

$$A = 3 \text{ and } B = 4$$

$$I = \int \left( \frac{3}{t-2} + \frac{4}{(t-2)^2} \right) dt$$

$$\Rightarrow \quad I = 3\log|t-2| - \frac{4}{(t-2)} + c$$

$$\Rightarrow \quad I = 3\log|\sin\theta - 2| - \frac{4}{(\sin\theta - 2)} + c$$

**Ans.**

**OR**

**Evaluate $\int_0^{\pi} e^{2x}.\sin\left(\frac{\pi}{4} + x\right) dx$**

**Sol.** Let $I = \int_0^{\pi} e^{2x}.\sin\left(\frac{\pi}{4} + x\right) dx$

$$I = \left[ \sin\left(\frac{\pi}{4} + x\right).\frac{e^{2x}}{2} \right]_0^{\pi} - \int_0^{\pi} \cos\left(\frac{\pi}{4} + x\right)\cdot\frac{e^{2x}}{2}\cdot dx$$

$$\Rightarrow I = \sin\frac{5\pi}{4}\cdot\frac{e^{2\pi}}{2} - \sin\frac{\pi}{4}\times\frac{1}{2} - \frac{1}{2}\left[ \cos\left(\frac{\pi}{4} + x\right).\frac{e^{2x}}{2} \right]_0^{\pi}$$

$$+ \frac{1}{2}\int_0^{\pi} -\sin\left(\frac{\pi}{4} + x\right).\frac{e^{2x}}{2}\cdot dx$$

$$\Rightarrow I = \frac{1}{2}\sin\frac{5\pi}{4}.e^{2\pi} - \frac{1}{2}\sin\frac{\pi}{4}$$

$$-\frac{1}{4}\left[ \cos\frac{5\pi}{4}.e^{2\pi} - \cos\frac{\pi}{4} \right] - \frac{1}{2}\times\frac{1}{2}\int_0^{\pi} \sin\left(\frac{\pi}{4} + x\right).e^{2x}.dx$$

$$\Rightarrow I = \frac{1}{2}\left(-\frac{1}{\sqrt{2}}\right)e^{2\pi} - \frac{1}{2}\times\frac{1}{\sqrt{2}} - \frac{1}{4}\left(\frac{-1}{\sqrt{2}}e^{2\pi} - \frac{1}{\sqrt{2}}\right) - \frac{1}{4}I$$

$$\Rightarrow I + \frac{I}{4} = -\frac{1}{2\sqrt{2}}e^{2\pi} - \frac{1}{2\sqrt{2}} + \frac{1}{4\sqrt{2}}e^{2\pi} + \frac{1}{4\sqrt{2}}$$

$$\Rightarrow \frac{5I}{4} = \frac{-2e^{2\pi} - 2 + e^{2\pi} + 1}{4\sqrt{2}} = \frac{-e^{2\pi} - 1}{4\sqrt{2}}$$

$$\Rightarrow 5I = -\left( \frac{e^{2\pi} + 1}{\sqrt{2}} \right)$$

$$\Rightarrow I = -\frac{1}{5}\left( \frac{e^{2\pi} + 1}{\sqrt{2}} \right)$$

**Ans.**

**13. Find $\int \frac{\sqrt{x}}{\sqrt{a^3 - x^3}} dx.$** **[4]**

**Sol.** Let $I = \int \frac{\sqrt{x}}{\sqrt{a^3 - x^3}}\cdot dx$

Put $x^{3/2} = t$

$$\Rightarrow \quad \frac{3}{2}\sqrt{x}.dx = dt$$

$$\Rightarrow \quad \sqrt{x}.dx = \frac{2}{3}\cdot dt$$

Putting the values in I, we get

$$I = \frac{2}{3}\int \frac{1}{\sqrt{(a^{3/2})^2 - t^2}}\cdot dt$$

$$\Rightarrow \quad I = \frac{2}{3}\sin^{-1}\left(\frac{t}{a^{3/2}}\right) + c$$

$$\Rightarrow \quad I = \frac{2}{3}\sin^{-1}\left(\frac{x}{a}\right)^{3/2} + c$$

**Ans.**

**14. Evaluate : $\int_{-1}^{2} |x^3 - x|\, dx.$** **[4]**

**Sol.** Let $I = \int_{-1}^{2} |x^3 - x|\, dx.$

$$f(x) = x^3 - x = x(x-1)(x+1)$$

The signs of $f(x)$ for the different values are shown in the figure given below :

$$\underset{-\infty}{\longleftarrow} \; - \; \underset{-1}{|} \; + \; \underset{0}{|} \; - \; \underset{1}{|} \; + \; \underset{\infty}{\longrightarrow}$$

$$f(x) > 0 \text{ for all } x \in (-1, 0) \cup (1, 2)$$

$$f(x) < 0 \text{ for all } x \in (0, 1)$$

Therefore,

$$|x^3 - x| = \begin{cases} x^3 - x, & x \in (-1,0) \cup (1,2) \\ -(x^3 - x), & x \in (0,1) \end{cases}$$

$$\therefore I = \int_{-1}^{2} |x^3 - x|\, dx$$

$$= \int_{-1}^{0} |x^3 - x|\, dx + \int_0^1 |x^3 - x|\, dx + \int_1^2 |x^3 - x|\, dx$$

$$= \int_{-1}^{0} (x^3 - x)dx - \int_0^1 (x^3 - x)dx + \int_1^2 (x^3 - x)dx$$

$$= \left[\frac{x^4}{4} - \frac{x^2}{2}\right]_{-1}^{0} - \left[\frac{x^4}{4} - \frac{x^2}{2}\right]_0^1 + \left[\frac{x^4}{4} - \frac{x^2}{2}\right]_1^2$$

$$= -\left(\frac{1}{4}-\frac{1}{2}\right)-\left(\frac{1}{4}-\frac{1}{2}\right)+\left(\frac{16}{4}-\frac{4}{2}\right)-\left(\frac{1}{4}-\frac{1}{2}\right)$$

$$= \frac{3}{4}+(4-2) = \frac{11}{4}$$ **Ans.**

**15. Find the particular solution of the differential equation $(1 - y^2)(1 + \log x)\, dx + 2xy\, dy = 0$, given that $y = 0$ when $x = 1$.** **[4]**

**Sol.** The given differential equation is,

$$(1-y^2)(1+\log x).dx + 2xy\,dy = 0$$

$$\Rightarrow \quad \frac{(1+\log x)}{x}\cdot dx = \frac{-2y}{(1-y^2)}\cdot dy$$

On integrating both side, we have

$$\Rightarrow \quad \int \frac{1+\log x}{x}\cdot dx = \int \frac{-2y}{(1-y^2)}\cdot dy$$

In first integral,

put $\quad 1 + \log x = t$

$$\Rightarrow \quad \frac{1}{x}\cdot dx = dt$$

Also, in second integral,

put $\quad 1 - y^2 = u$

$$\Rightarrow \quad -2y.dy = du$$

$$\therefore \quad \int t.dt = \int \frac{1}{u}.du$$

$$\Rightarrow \quad \frac{t^2}{2} - \log|u| = c$$

$$\Rightarrow \frac{1}{2}(1+\log x)^2 - \log|1-y^2| = c$$

Which is given that $y = 0$ when $x = 1$

So, $\frac{1}{2}(1+\log 1)^2 - \log|1-0^2| = c$

$$\Rightarrow \quad c = \frac{1}{2}$$

$$\therefore \quad \frac{(1+\log x)^2}{2} - \log|1-y^2| = \frac{1}{2}$$

or $(1 + \log x)^2 - 2\log|1 - y^2| = 1$

Which is the required particular solution. **Ans.**

**16. Find the general solution of the following differential equation :**

$$(1+y^2)+(x-e^{\tan^{-1}y})\frac{dy}{dx} = 0$$ **[4]**

**Sol.** The given differential equation is,

$$(1+y^2)+(x-e^{\tan^{-1}y})\frac{dy}{dx} = 0$$

$$\Rightarrow \quad \frac{dx}{dy}+\left(\frac{1}{1+y^2}\right)x = \frac{e^{\tan^{-1}y}}{1+y^2}$$

This is a linear differential equation with

$$P = \frac{1}{1+y^2}$$

and $\quad Q = \dfrac{e^{\tan^{-1}y}}{1+y^2}$

$$\therefore \quad \text{I.F.} = e^{\int \frac{1}{1+y^2}\cdot dx} = e^{\tan^{-1}y}$$

So, the required solution is :

$$xe^{\tan^{-1}y} = \int e^{\tan^{-1}y}\cdot\frac{e^{\tan^{-1}y}}{1+y^2}\cdot dy$$

Put $\quad \tan^{-1}y = t$

$$\Rightarrow \quad \frac{1}{1+y^2}\cdot dy = dt$$

$$\therefore \quad xe^{\tan^{-1}y} = \int e^{2t}.dt$$

$$\Rightarrow \quad xe^{\tan^{-1}y} = \frac{1}{2}e^{2t}+C$$

$$\Rightarrow \quad xe^{\tan^{-1}y} = \frac{1}{2}e^{2\tan^{-1}y}+C$$

$$\Rightarrow \quad x = \frac{1}{2}e^{\tan^{-1}y}+Ce^{-\tan^{-1}y}$$

**Ans.**

**17. Show that the vectors $\vec{a}, \vec{b}$ and $\vec{c}$ are coplanar if $\vec{a}+\vec{b}, \vec{b}+\vec{c}$ and $\vec{c}+\vec{a}$ are coplanar.** **[4]**

**Sol.** As $\vec{a}+\vec{b}, \vec{b}+\vec{c}, \vec{c}+\vec{a}$ are coplanar.

So, $\quad [\vec{a}+\vec{b} \;\; \vec{b}+\vec{c} \;\; \vec{c}+\vec{a}] = 0 \qquad$ ...(i)

Now, if $\vec{a}, \vec{b}$ and $\vec{c}$ are coplanar vectors

then, $[\vec{a} \;\; \vec{b} \;\; \vec{c}] = 0$

Consider $[\vec{a}+\vec{b} \;\; \vec{b}+\vec{c} \;\; \vec{c}+\vec{a}] = 0$

$$\Rightarrow [(\vec{a}+\vec{b})\times(\vec{b}+\vec{c})].(\vec{c}+\vec{a}) = 0$$

$$\Rightarrow [\vec{a}\times\vec{b}+\vec{a}\times\vec{c}+\vec{b}\times\vec{b}+\vec{b}\times\vec{c}].(\vec{c}+\vec{a}) = 0$$

$$\Rightarrow (\vec{a}\times\vec{b}).\vec{c}+(\vec{a}\times\vec{b}).\vec{a}+(\vec{a}\times\vec{c}).\vec{c}+(\vec{a}\times\vec{c}).\vec{a}$$
$$+(\vec{b}\times\vec{c}).\vec{c}+(\vec{b}\times\vec{c}).\vec{a} = 0$$

$$\Rightarrow (\vec{a}\times\vec{b}).\vec{c}+0+0+0+0+(\vec{b}\times\vec{c}).\vec{a} = 0$$

$\Rightarrow [\vec{a}\ \vec{b}\ \vec{c}]+[\vec{b}\ \vec{c}\ \vec{a}]=0$

$\Rightarrow [\vec{a}\ \vec{b}\ \vec{c}]+[\vec{a}\ \vec{b}\ \vec{c}]=0$

$\Rightarrow \quad 2[\vec{a}\ \vec{b}\ \vec{c}]=0$

$\Rightarrow \quad [\vec{a}\ \vec{b}\ \vec{c}]=0$

$\therefore\ \vec{a}, \vec{b}$ and $\vec{c}$ are coplanar if $\vec{a}+\vec{b}, \vec{b}+\vec{c}$ and $\vec{c}+\vec{a}$ are coplanar. **Hence Proved**

**18. Find the vector and cartesian equations of the line through the point (1, 2, – 4) and perpendicular to the two lines.**

$$\vec{r}=(8\hat{i}-19\hat{j}+10\hat{k})+\lambda(3\hat{i}-16\hat{j}+7\hat{k})$$

**and** $\vec{r}=(15\hat{i}+29\hat{j}+5\hat{k})+\mu(3\hat{i}+8\hat{j}-5\hat{k})$ **[4]**

**Sol.** The equations of the given lines are

$\vec{r}=(8\hat{i}-19\hat{j}+10\hat{k})+\lambda(3\hat{i}-16\hat{j}+7\hat{k})$ ......(i)

$\vec{r}=(15\hat{i}+29\hat{j}+5\hat{k})+\mu(3\hat{i}+8\hat{j}-5\hat{k})$ .....(ii)

Normal parallel to (i) is

$\vec{n_1}=3\hat{i}-16\hat{j}+7\hat{k}$

Normal parallel to (ii) is

$\vec{n_2}=3\hat{i}+8\hat{j}-5\hat{k}$

The required line is perpendicular to the given lines. So the normal $\vec{n}$ parallel to the required line perpendicular to $\vec{n_1}$ and $\vec{n_2}$.

$$\therefore\ \vec{n}=\vec{n_1}\times\vec{n_2}=\begin{vmatrix}\hat{i} & \hat{j} & \hat{k}\\ 3 & -16 & 7\\ 3 & 8 & -5\end{vmatrix}=24\hat{i}+36\hat{j}+72\hat{k}$$

Thus, the vector equation of the required line is

$\vec{r}=(\hat{i}+2\hat{j}-4\hat{k})+\gamma(24\hat{i}+36\hat{j}+72\hat{k})$

$\Rightarrow \vec{r}=(\hat{i}+2\hat{j}-4\hat{k})+k(2\hat{i}+3\hat{j}+6\hat{k})$

(where $k = 12\gamma$)

Also, the cartesian equation of the required line is

$$\frac{x-1}{2}=\frac{y-2}{3}=\frac{z+4}{6}$$ **Ans.**

**19. Three persons A, B and C apply for a job of Manager in a Private Company. Chances of their selection (A, B and C) are in the ratio 1 : 2 : 4. The probabilities that A, B and C can introduce changes to improve profits of the company are 0.8, 0.5 and 0.3 respectively. If the change does not take place, find the probability that it is due to the appointment of C.** **[4]**

**Sol.** Let $E_1$, $E_2$ and $E_3$ be the events denoting the selection of A, B and C as managers respectively.

$P(E_1)$ = Probability of selection of A = $\frac{1}{7}$

$P(E_2)$ = Probability of selection of B = $\frac{2}{7}$

$P(E_3)$ = Probability of selection of C = $\frac{4}{7}$

Let A be the event denoting the change not taking place.

$P(A/E_1)$ = Probability that A does not introduce change = 0.2

$P(A/E_2)$ = Probability that B does not introduce change = 0.5

$P(A/E_3)$ = Probability that C does not introduce change = 0.7

$\therefore$ Required probability = $P(E_3/A)$

By Bayes' theorem, we have

$$P(E_3/A)=\frac{P(E_3)P(A/E_3)}{P(E_1)P(A/E_1)+P(E_2)P(A/E_2)+P(E_3)P(A/E_3)}$$

$$=\frac{\frac{4}{7}\times 0.7}{\frac{1}{7}\times 0.2+\frac{2}{7}\times 0.5+\frac{4}{7}\times 0.7}$$

$$=\frac{2.8}{0.2+1+2.8}$$

$$=\frac{2.8}{4}=0.7$$ **Ans.**

**OR**

**A and B throw a pair of dice alternately. A wins the game if he gets a total of 7 and B wins the game if he gets a total of 10. If A starts the game, then find the probability that B wins.**

**Sol.** Total of 7 on the dice can be obtained in the following ways :

(1, 6), (6, 1), (2, 5), (5, 2), (3, 4), (4, 3)

Probability of getting a total of 7 = $\frac{6}{36}=\frac{1}{6}$

Probability of not getting a total of 7 = $1-\frac{1}{6}=\frac{5}{6}$

Total of 10 on the dice can be obtained in the following ways :

(4, 6), (6, 4), (5, 5)

Probability of getting a total of 10

$$=\frac{3}{36}=\frac{1}{12}$$

Probability of not getting a total of 10

$$= 1-\frac{1}{12}=\frac{11}{12}$$

Let E and F be the two events, defined as follows :

E = Getting a total of 7 in a single throw of a dice

F = Getting a total of 10 in a single throw of a dice

$$P(E) = \frac{1}{6}, P(\bar{E})=\frac{5}{6}$$

$$P(F) = \frac{1}{12}, P(\bar{F})=\frac{11}{12}$$

A wins if he gets a total of 7 in 1st, 3rd or 5th.... throws.

Probability of A getting a total of 7 in the1st throw $= \frac{1}{6}$

A will get the 3rd throw if he fails in the 1st throw and B fails in the 2nd throw.

Probability of A getting a total of 7 in the 3rd throw

$$=P(\bar{E})\,P(\bar{F})\,P(E) = \frac{5}{6}\times\frac{11}{12}\times\frac{1}{6}$$

Similarly, probability of getting a total of 7 in the 5th throw = $P(\bar{E})\,P(\bar{F})\,P(\bar{E})\,P(\bar{E})\,P(E)$

$$= \frac{5}{6}\times\frac{11}{12}\times\frac{5}{6}\times\frac{11}{12}\times\frac{1}{6} \text{ and so on}$$

Probability of winning of A

$$=\frac{1}{6}+\left(\frac{5}{6}\times\frac{11}{12}\times\frac{1}{6}\right)+\left(\frac{5}{6}\times\frac{11}{12}\times\frac{5}{6}\times\frac{11}{12}\times\frac{1}{6}\right)+\ldots$$

$$=\frac{\frac{1}{6}}{1-\frac{5}{6}\times\frac{11}{12}}=\frac{12}{17}$$

$\therefore$ Probability of winning of B = 1 – Probability of winning of A

$$= 1-\frac{12}{17}=\frac{5}{17}$$ **Ans.**

## SECTION–C

**20. Let $f$ : N $\to$ N be a function defined as $f(x) = 9x^2 + 6x - 5$. Show that $f$: N $\to$ S, where S is the range of $f$, is invertible. Find the inverse of $f$ and hence find $f^{-1}$ (43) and $f^{-1}$ (163).** **[6]**

**Sol.** Given,

$$f(x) = 9x^2 + 6x - 5$$

Let $x_1, x_2 \in N$ and $f(x_1) = f(x_2)$

$$\Rightarrow \quad 9x_1^2 + 6x_1 - 5 = 9x_2^2 + 6x_2 - 5$$

$$\Rightarrow \quad 9(x_1^2 - x_2^2) + 6(x_1 - x_2) = 0$$

$$\Rightarrow \quad (x_1 - x_2)\,[9(x_1 + x_2) + 6] = 0$$

$$\Rightarrow \quad x_1 - x_2 = 0$$

as $\quad 9x_1 + 9x_2 + 6 \neq 0 \quad [x_1, x_2 \in N]$

$$\Rightarrow \quad x_1 = x_2$$

$\Rightarrow f$ is one-one function.

Let $\quad y = 9x^2 + 6x - 5$

$$\Rightarrow \quad y = (3x + 1)^2 - 1 - 5$$

$$= (3x + 1)^2 - 6$$

$$\Rightarrow \quad (3x + 1)^2 = y + 6$$

$$\Rightarrow \quad 3x + 1 = \sqrt{y+6}$$

$$\Rightarrow \quad x = \frac{\sqrt{y+6}-1}{3} \text{ as } x \in N$$

$$\Rightarrow \quad \sqrt{y+6}-1 > 0$$

$$\Rightarrow \quad y + 6 > 1$$

$$\Rightarrow \quad y > -5 \text{ and } y \in N$$

So, the function is invertible if the range of the function $f(x)$ is {1, 2, 3, .....}.

$f$ : N $\to$ S is onto as codomain = Range

Hence $f$ is invertible.

Therefore, the inverse of the function $f(x)$ is $f^{-1}(y)$, *i.e.*, $x$.

Now, $\quad f^{-1}(y) = \frac{\sqrt{y+6}-1}{3}$

$$\Rightarrow \quad f^{-1}(43) = \frac{\sqrt{43+6}-1}{3} = 2$$

and $\quad f^{-1}(163) = \frac{\sqrt{163+6}-1}{3} = 4$ **Ans.**

**21. Prove that $\begin{vmatrix} yz-x^2 & zx-y^2 & xy-z^2 \\ zx-y^2 & xy-z^2 & yz-x^2 \\ xy-z^2 & yz-x^2 & zx-y^2 \end{vmatrix}$ is divisible by $(x+y+z)$ and hence find the quotient.** **[6]**

**Sol.** Let $\Delta = \begin{vmatrix} yz-x^2 & zx-y^2 & xy-z^2 \\ zx-y^2 & xy-z^2 & yz-x^2 \\ xy-z^2 & yz-x^2 & zx-y^2 \end{vmatrix}$

Applying $C_1 \to C_1 - C_2$, $C_2 \to C_2 - C_3$

$$\Delta = \begin{vmatrix} yz-x^2-zx+y^2 & zx-xy-y^2+z^2 & xy-z^2 \\ zx-xy-y^2+z^2 & xy-yz-z^2+x^2 & yz-x^2 \\ xy-yz-z^2+x^2 & yz-zx-x^2+y^2 & zx-y^2 \end{vmatrix}$$

$$\Delta = \begin{vmatrix} (y-x)(x+y+z) & (z-y)(x+y+z) & xy-z^2 \\ (z-y)(x+y+z) & (x-z)(x+y+z) & yz-x^2 \\ (x-z)(x+y+z) & (y-x)(x+y+z) & zx-y^2 \end{vmatrix}$$

Taking $(x + y + z)$ common from $C_1$ and $C_2$ both

$$\Delta = (x+y+z)^2 \begin{vmatrix} (y-x) & (z-y) & xy-z^2 \\ (z-y) & (x-z) & yz-x^2 \\ (x-z) & (y-x) & zx-y^2 \end{vmatrix}$$

Applying $R_1 \to R_1 + R_2 + R_3$

$$\Delta = (x+y+z)^2 \begin{vmatrix} 0 & 0 & xy+yz+zx-x^2-y^2-z^2 \\ (z-y) & (x-z) & yz-x^2 \\ (x-z) & (y-x) & zx-y^2 \end{vmatrix}$$

Expanding along $R_1$, we get

$$\Delta = (x+y+z)^2 \{ (xy+yz+zx-x^2-y^2-z^2) [(z-y)(y-x)-(x-z)^2]\}$$

$$\Rightarrow \Delta = (x+y+z)^2 \{(xy+yz+zx-x^2-y^2-z^2)(xy+yz+zx-x^2-y^2-z^2)\}$$

$$\Delta = (x+y+z)^2 (xy+zy+zx-x^2-y^2-z^2)^2$$

Hence $\Delta$ is divisible by $(x + y + z)$ and the quotient is $(x + y + z)(xy + yz + zx - x^2 - y^2 - z^2)^2$

**Hence Proved.**

**OR**

**Using elementary transformations, find the inverse of the matrix** $A = \begin{bmatrix} 8 & 4 & 3 \\ 2 & 1 & 1 \\ 1 & 2 & 2 \end{bmatrix}$ **and use it to solve the following system of linear equations :**

$$8x + 4y + 3z = 19$$
$$2x + y + z = 5$$
$$x + 2y + 2z = 7$$

**Sol.** Here, $A = \begin{bmatrix} 8 & 4 & 3 \\ 2 & 1 & 1 \\ 1 & 2 & 2 \end{bmatrix}$

Using A = IA, we have

$$\begin{bmatrix} 8 & 4 & 3 \\ 2 & 1 & 1 \\ 1 & 2 & 2 \end{bmatrix} = \begin{bmatrix} 1 & 0 & 0 \\ 0 & 1 & 0 \\ 0 & 0 & 1 \end{bmatrix} A$$

Applying $R_1 \leftrightarrow R_3$, We get

$$\begin{bmatrix} 1 & 2 & 2 \\ 2 & 1 & 1 \\ 8 & 4 & 3 \end{bmatrix} = \begin{bmatrix} 0 & 0 & 1 \\ 0 & 1 & 0 \\ 1 & 0 & 0 \end{bmatrix} A$$

Applying $R_2 \to R_2 - 2R_1$, $R_3 \to R_3 - 8R_1$, We get

$$\begin{bmatrix} 1 & 2 & 2 \\ 0 & -3 & -3 \\ 0 & -12 & -13 \end{bmatrix} = \begin{bmatrix} 0 & 0 & 1 \\ 0 & 1 & -2 \\ 1 & 0 & -8 \end{bmatrix} A$$

Applying $R_2 \to \frac{R_2}{-3}$, we get

$$\begin{bmatrix} 1 & 2 & 2 \\ 0 & 1 & 1 \\ 0 & -12 & -13 \end{bmatrix} = \begin{bmatrix} 0 & 0 & 1 \\ 0 & -\frac{1}{3} & \frac{2}{3} \\ 1 & 0 & -8 \end{bmatrix} A$$

Applying $R_1 \to R_1 - 2R_2$, we get

$$\begin{bmatrix} 1 & 0 & 0 \\ 0 & 1 & 1 \\ 0 & -12 & -13 \end{bmatrix} = \begin{bmatrix} 0 & \frac{2}{3} & -\frac{1}{3} \\ 0 & -\frac{1}{3} & \frac{2}{3} \\ 1 & 0 & -8 \end{bmatrix} A$$

Applying $R_3 \to R_3 + 12R_2$, we get

$$\begin{bmatrix} 1 & 0 & 0 \\ 0 & 1 & 1 \\ 0 & 0 & -1 \end{bmatrix} = \begin{bmatrix} 0 & \frac{2}{3} & -\frac{1}{3} \\ 0 & -\frac{1}{3} & \frac{2}{3} \\ 1 & -4 & 0 \end{bmatrix} A$$

Applying $R_3 \to -R_3$, we get

$$\begin{bmatrix} 1 & 0 & 0 \\ 0 & 1 & 1 \\ 0 & 0 & 1 \end{bmatrix} = \begin{bmatrix} 0 & \frac{2}{3} & -\frac{1}{3} \\ 0 & -\frac{1}{3} & \frac{2}{3} \\ -1 & 4 & 0 \end{bmatrix} A$$

Applying $R_2 \to R_2 - R_3$, we get

$$\begin{bmatrix} 1 & 0 & 0 \\ 0 & 1 & 0 \\ 0 & 0 & 1 \end{bmatrix} = \begin{bmatrix} 0 & \frac{2}{3} & -\frac{1}{3} \\ 1 & -\frac{13}{3} & \frac{2}{3} \\ -1 & 4 & 0 \end{bmatrix} A$$

Thus, we have

$$A^{-1} = \begin{bmatrix} 0 & \frac{2}{3} & -\frac{1}{3} \\ 1 & -\frac{13}{3} & \frac{2}{3} \\ -1 & 4 & 0 \end{bmatrix}$$

The given system of equations is

$$8x + 4y + 3z = 19$$
$$2x + y + z = 5$$
$$x + 2y + 2z = 7$$

This system of equations can be written as AX = B,

where $A = \begin{bmatrix} 8 & 4 & 3 \\ 2 & 1 & 1 \\ 1 & 2 & 2 \end{bmatrix}$, $X = \begin{bmatrix} x \\ y \\ z \end{bmatrix}$ and $B = \begin{bmatrix} 19 \\ 5 \\ 7 \end{bmatrix}$

$\therefore \quad X = A^{-1}B$

$$\Rightarrow \begin{bmatrix} x \\ y \\ z \end{bmatrix} = \begin{bmatrix} 0 & \frac{2}{3} & -\frac{1}{3} \\ 1 & -\frac{13}{3} & \frac{2}{3} \\ -1 & 4 & 0 \end{bmatrix} \begin{bmatrix} 19 \\ 5 \\ 7 \end{bmatrix}$$

$$\Rightarrow \begin{bmatrix} x \\ y \\ z \end{bmatrix} = \begin{bmatrix} 0+\frac{10}{3}-\frac{7}{3} \\ 19-\frac{65}{3}+\frac{14}{3} \\ -19+20+0 \end{bmatrix} = \begin{bmatrix} 1 \\ 2 \\ 1 \end{bmatrix}$$

$\therefore$ $x = 1, y = 2$ and $z = 1$. **Ans.**

**22. Show that the altitude of the right circular cone of maximum volume that can be inscribed in a sphere of radius $r$ is $\frac{4r}{3}$. Also find maximum volume in terms of volume of the sphere. [6]**

**Sol.** A sphere of fixed radius ($r$) is given.

Let R and $h$ be the radius and the height of the cone respectively.

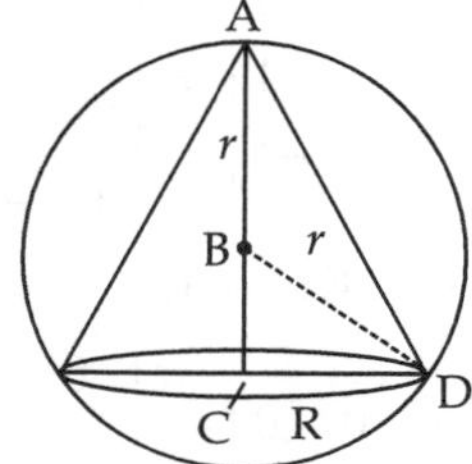


The volume (V) of the cone is given by,

$$V = \frac{1}{3}\pi R^2 h$$

Now, from the right triangle BCD, we have

$$BC = \sqrt{r^2 - R^2}$$

$$\therefore \quad h = r + \sqrt{r^2 - R^2}$$

$$V = \frac{1}{3}\pi R^2 r + \frac{1}{3}\pi R^2\sqrt{r^2 - R^2}$$

$$\frac{dV}{dR} = \frac{2}{3}\pi R r + \frac{2}{3}\pi R\sqrt{r^2 - R^2} + \frac{\pi R^2}{3}\cdot\frac{(-2R)}{2\sqrt{r^2 - R^2}}$$

$$= \frac{2}{3}\pi R r + \frac{2}{3}\pi R\sqrt{r^2 - R^2} - \frac{\pi R^3}{3\sqrt{r^2 - R^2}}$$

$$= \frac{2}{3}\pi R r + \frac{2\pi R(r^2 - R^2) - \pi R^3}{3\sqrt{r^2 - R^2}}$$

$$= \frac{2}{3}\pi R r + \frac{2\pi R r^2 - 3\pi R^3}{3\sqrt{r^2 - R^2}}$$

Now for maximum volume, $\frac{dV}{dR} = 0$

$$\Rightarrow \frac{2\pi r R}{3} = \frac{3\pi R^3 - 2\pi R r^2}{3\sqrt{r^2 - R^2}}$$

$$\Rightarrow 2r\sqrt{r^2 - R^2} = 3R^2 - 2r^2$$

$$\Rightarrow 4r^2(r^2 - R^2) = (3R^2 - 2r^2)^2$$

$$\Rightarrow 4r^4 - 4r^2R^2 = 9R^4 + 4r^4 - 12R^2r^2$$

$$\Rightarrow 9R^4 - 8r^2R^2 = 0$$

$$\Rightarrow 9R^2 = 8r^2$$

$$\Rightarrow R^2 = \frac{8r^2}{9}$$

Now,

$$\frac{d^2V}{dR^2} = \frac{2\pi r}{3} + \frac{3\sqrt{r^2 - R^2}(2\pi r^2 - 9\pi R^2) - (2\pi R r^2 - 3\pi R^3)(-6R)\frac{1}{2\sqrt{r^2 - R^2}}}{9(r^2 - R^2)}$$

$$= \frac{2\pi r}{3} + \frac{3\sqrt{r^2 - R^2}(2\pi r^2 - 9\pi R^2) + (2\pi R r^2 - 3\pi R^3)(6R)\frac{1}{2\sqrt{r^2 - R^2}}}{9(r^2 - R^2)}$$

Now, when $R^2 = \frac{8r^2}{9}$, Clearly $\frac{d^2V}{dR^2} < 0$.

$\therefore$ The volume is maximum when

$$R^2 = \frac{8r^2}{9}.$$

Height of the cone

$$= r + \sqrt{r^2 - \frac{8r^2}{9}} = r + \sqrt{\frac{r^2}{9}} = r + \frac{r}{3} = \frac{4r}{3}.$$

Hence, it can be seen that the altitude of a right circular cone of maximum volume that can be inscribed in a sphere of radius $r$ is $\frac{4r}{3}$.

**Hence Proved.**

Let volume of the sphere be $V_s = \frac{4}{3}\pi r^3$.

$$r = \sqrt[3]{\frac{3V_s}{4\pi}}$$

$\therefore$ Volume of cone, $V = \frac{1}{3}\pi R^2 h$

$$\Rightarrow V = \frac{1}{3}\pi\frac{8r^2}{9}\times\frac{4r}{3} \qquad \left(\because R^2 = \frac{8r^2}{9}\right)$$

$$\Rightarrow V = \frac{32\pi r^3}{81} = \frac{8}{27}\left(\frac{4}{3}\pi r^3\right)$$

$\therefore$ Maximum volume of cone in terms of sphere

$$= \frac{8(\text{Volume of sphere})}{27}$$ **Ans.**

**OR**

**Find the intervals in which $f(x) = \sin 3x - \cos 3x$, $0<x<\pi$, is strictly increasing or strictly decreasing.**

**Sol.** Consider the function

$$f(x) = \sin 3x - \cos 3x$$

$$\Rightarrow \quad f'(x) = 3\cos 3x + 3\sin 3x$$

$$= 3(\sin 3x + \cos 3x)$$

$$= 3\sqrt{2}\left\{\sin 3x \cos\left(\frac{\pi}{4}\right) + \cos 3x \sin\left(\frac{\pi}{4}\right)\right\}$$

$$= 3\sqrt{2}\left\{\sin\left(3x + \frac{\pi}{4}\right)\right\}$$

For the increasing interval $f'(x) > 0$.

$$3\sqrt{2}\left\{\sin\left(3x + \frac{\pi}{4}\right)\right\} > 0$$

$$\sin\left(3x + \frac{\pi}{4}\right) > 0$$

$$\Rightarrow \quad 0 < 3x + \frac{\pi}{4} < \pi$$

$$\Rightarrow \quad \frac{-\pi}{4} < 3x < \frac{3\pi}{4}$$

$$\Rightarrow \quad \frac{-\pi}{12} < x < \frac{\pi}{4}$$

as $0 < x < \pi$ is given

$\Rightarrow 0 < x < \pi/4$

Also, $\sin\left(3x + \frac{\pi}{4}\right) > 0$

when, $2\pi < 3x + \frac{\pi}{4} < 3\pi$

Therefore, intervals in which function is strictly increasing is $0 < x < \frac{\pi}{4}$ and $\frac{7\pi}{12} < x < \frac{11\pi}{12}$.

Similarly, for the decreasing interval $f'(x) < 0$

$$3\sqrt{2}\left\{\sin\left(3x + \frac{\pi}{4}\right)\right\} < 0$$

$$\sin\left(3x + \frac{\pi}{4}\right) < 0$$

$$\Rightarrow \quad \pi < 3x + \frac{\pi}{4} < 2\pi$$

$$\Rightarrow \quad \frac{3\pi}{4} < 3x < \frac{7\pi}{4}$$

$$\Rightarrow \quad \frac{\pi}{4} < x < \frac{7\pi}{12}$$

Also, $\quad \sin\left(3x + \frac{\pi}{4}\right) < 0$

When $\quad 3\pi < 3x + \frac{\pi}{4} < 4\pi,$

$$\Rightarrow \quad \frac{11\pi}{4} < 3x < \frac{15\pi}{4}$$

$$\Rightarrow \quad \frac{11\pi}{12} < x < \frac{15\pi}{12}$$

But $0 < x < \pi$ so, $\frac{11\pi}{12} < x < \pi$

The function is strictly decreasing in $\frac{\pi}{4} < x < \frac{7\pi}{12}$ and $\frac{11\pi}{12} < x < \pi$. **Ans.**

**23. Using integration find the area of the region $\{(x, y) : x^2 + y^2 \le 2ax, y^2 \ge ax, x, y \ge 0\}$ [6]**

**Sol.** We have, $\{(x, y) : x^2 + y^2 \le 2ax, y^2 \ge ax, x, y \ge 0\}$

Consider $\quad x^2 + y^2 = 2ax \quad$ ...(i)

$y^2 = ax \quad$ ...(ii)

$x = 0, y = 0$

Solving equation (i) and (ii), we get

$$x^2 + ax = 2ax$$

$$\Rightarrow \quad x^2 - ax = 0$$

$$\Rightarrow \quad x(x - a) = 0$$

$$\therefore \quad x = 0, a$$

So, points of intersections of (i) and (ii) are (0, 0) and $(a, \pm a)$. Also, equation (i) can be written as, $(x - a)^2 + (y - 0)^2 = a^2$ whose centre is at $(a, 0)$ and radius is of '$a$' units.

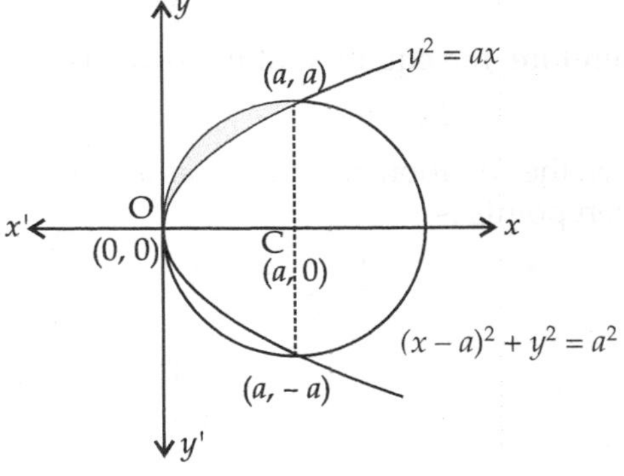


$\therefore$ Require area

$$= \int_0^a y_1 dx - \int_0^a y_2 dx$$

$$= \int_0^a \sqrt{a^2 - (x-a)^2}.dx - \sqrt{a}\int_0^a \sqrt{x}.dx$$

$$= \left[\frac{x-a}{2}\sqrt{a^2 - (x-a)^2} + \frac{a^2}{2}\sin^{-1}\left(\frac{x-a}{a}\right)\right]_0^a$$

$$-\frac{2}{3}\sqrt{a}[x^{3/2}]_0^a$$

$$= [0+0]-\left[0+\frac{a^2}{2}\left(-\frac{\pi}{2}\right)\right]-\frac{2}{3}\sqrt{a}[a^{3/2}-0]$$

$$= \left(\frac{\pi}{4}-\frac{2}{3}\right)a^2 \text{ sq. units}$$ **Ans.**

**24. Find the coordinate of the point P where the line through A(3, – 4, – 5) and B (2, – 3, 1) crosses the plane passing through three points L (2, 2, 1), M (3, 0, 1) and N (4, – 1, 0). Also, find the ratio in which P divides the line segment AB. [6]**

**Sol.** The equation of the plane passing through three given points can be given by

$$\begin{vmatrix} x-2 & y-2 & z-1 \\ x-3 & y-0 & z-1 \\ x-4 & y+1 & z-0 \end{vmatrix} = 0$$

Performing elementary row operations $R_2 \to R_1 - R_2$ and $R_3 \to R_1 - R_3$, we get

$$\begin{vmatrix} x-2 & y-2 & z-1 \\ 3-2 & 0-2 & 0 \\ 4-2 & -1-2 & -1 \end{vmatrix} = 0$$

$$\Rightarrow \begin{vmatrix} x-2 & y-2 & z-1 \\ 1 & -2 & 0 \\ 2 & -3 & -1 \end{vmatrix} = 0$$

Solving the above determinant, we get

$$(x-2)(2-0)-(y-2)(-1-0)+(z-1)(-3+4)=0$$

$$\Rightarrow (2x-4)+(y-2)+(z-1)=0$$

$$\Rightarrow 2x+y+z-7=0$$

Therefore, the equation of the plane is

$$2x+y+z-7=0$$

Now, the equation of the line passing through two given points is

$$\frac{x-3}{2-3} = \frac{y+4}{-3+4} = \frac{z+5}{1+5} = \lambda$$

$$\Rightarrow \frac{x-3}{-1} = \frac{y+4}{1} = \frac{z+5}{6} = \lambda$$

$$\Rightarrow x = (-\lambda+3),\ y = (\lambda-4),\ z = (6\lambda-5)$$

At the point of intersection, these points satisfy the equation of the plane $2x + y + z - 7 = 0$

Putting the values of $x$, $y$ and $z$ in the equation of the plane, we get the value of $\lambda$

$$2(-\lambda+3)+(\lambda-4)+(6\lambda-5)-7 = 0$$

$$\Rightarrow -2\lambda+6+\lambda-4+6\lambda-5-7 = 0$$

$$\Rightarrow 5\lambda = 10$$

$$\Rightarrow \lambda = 2$$

Thus, the point of intersection is P (1, – 2, 7).

Now, let P divide the line AB in the ratio $m : n$.

By the section formula, we have

$$1 = \frac{2m+3n}{m+n}$$

$$\Rightarrow m+2n = 0$$

$$\Rightarrow m = -2n$$

$$\Rightarrow \frac{m}{n} = \frac{-2}{1}$$

Hence, P divides externally the line segment AB in the ratio 2 : 1. **Ans.**

**25. An urn contains 3 white and 6 red balls. Four balls are drawn one by one with replacement from the urn. Find the probability distribution of the number of red balls drawn. Also find mean and variance of the distribution. [6]**

**Sol.** Let X denote the total number of red balls when four balls are drawn one by one with replacement.

$$\text{P (getting a red ball in one draw)} = \frac{6}{9} = \frac{2}{3}$$

$$\text{P (getting a white ball in one draw)} = \frac{1}{3}$$

| X | 0 | 1 | 2 | 3 | 4 |
|---|---|---|---|---|---|
| P(X) | $\left(\frac{1}{3}\right)^4$ | $\frac{2}{3}\left(\frac{1}{3}\right)^3 \cdot {}^4C_1$ | $\left(\frac{2}{3}\right)^2\left(\frac{1}{3}\right)^2 \cdot {}^4C_2$ | $\left(\frac{2}{3}\right)^3\left(\frac{1}{3}\right) \cdot {}^4C_3$ | $\left(\frac{2}{3}\right)^4$ |
| | $\frac{1}{81}$ | $\frac{8}{81}$ | $\frac{24}{81}$ | $\frac{32}{81}$ | $\frac{16}{81}$ |

Using the formula for mean, we have

$$\overline{X} = \Sigma P_i X_i$$

Mean

$$(\overline{X}) = \left(0\times\frac{1}{81}\right)+1\left(\frac{8}{81}\right)+2\left(\frac{24}{81}\right)+3\left(\frac{32}{81}\right)+4\left(\frac{16}{81}\right)$$

$$= \frac{1}{81}(8+48+96+64)$$

$$= \frac{216}{81} = \frac{8}{3}$$

Using the formula for variance, we have

$$\text{Var (X)} = \Sigma P_i X_i^2 - (\Sigma P_i X_i)^2$$

$$\Rightarrow \text{Var (X)} = \left\{\left(0\times\frac{1}{81}\right)+1\left(\frac{8}{81}\right)+4\left(\frac{24}{81}\right)+9\left(\frac{32}{81}\right)\right.$$

$$+16\left(\frac{16}{81}\right)\Big\} - \left(\frac{8}{3}\right)^2$$

$$= \frac{648}{81} - \frac{64}{9}$$

$$= \frac{8}{9}$$

Hence, the mean of the disribution is $\frac{8}{3}$ and the variance of the disribution is $\frac{8}{9}$ **Ans.**

**26. A manufacturer produces two products A and B. Both the products are processed on two different machines. The available capacity of first machine is 12 hours and that of second machine is 9 hours per day. Each unit of product A requires 3 hours on both machines and each unit of product B requires 2 hours on first machine and 1 hour on second machine. Each unit of product A is sold at ₹ 7 profit and B at a profit of ₹ 4. Find the production level per day for maximum profit graphically. [6]**

**Sol.** Let the numbers of units of products A and B to be produced be $x$ and $y$, respectively.

| Product | Machine | |
|---|---|---|
| | 1 ($h$) | II ($h$) |
| A | 3 | 3 |
| B | 2 | 1 |

Total profit : $Z = 7x + 4y$

We have to maximize $Z = 7x + 4y$.

Subject to constraints :

$$3x + 2y \leq 12$$

$$3x + y \leq 9$$

$\Rightarrow$ $x \geq 0$ and $y \geq 0$

The given information can be graphically expressed as follows :

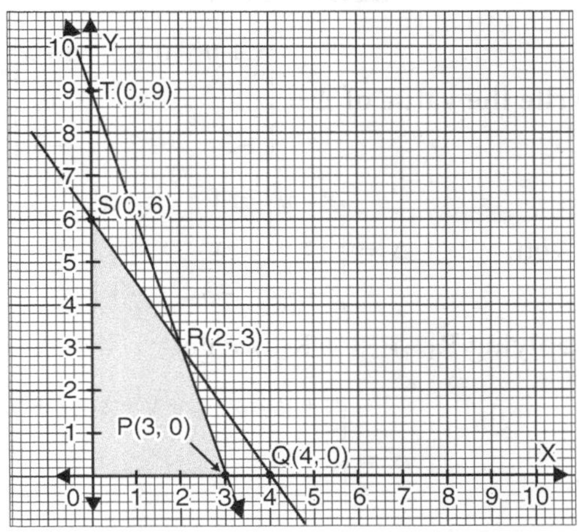


Values of $Z = 7x + 4y$ at the corner points are S follows :

| Corner Points | $Z = 7x + 4y$ |
|---|---|
| S (0,6) | 24 |
| R (2,3) | **26** ← Maximum |
| P (3,0) | 21 |

Therefore, the manufacturer has to produce 2 units of product A and 3 units of product B for the maximum profit of ₹ 26. **Ans.**

All questions are same in Outside Delhi Set II and Set III

# MATHEMATICS
## Outside Delhi [Set-I] 2015

**Time allowed : 3 hours** **Maximum marks : 100**

### SECTION–A

**1. Write the value of** $\Delta = \begin{vmatrix} x+y & y+z & z+x \\ z & x & y \\ -3 & -3 & -3 \end{vmatrix}$ **[1]**

**Sol.** Given, $\Delta = \begin{vmatrix} x+y & y+z & z+x \\ z & x & y \\ -3 & -3 & -3 \end{vmatrix}$

Applying $R_1 \to R_1 + R_2$, we get

$$\Delta = \begin{vmatrix} x+y+z & x+y+z & x+y+z \\ z & x & y \\ -3 & -3 & -3 \end{vmatrix}$$

$$\Delta = -3(x+y+z)\begin{vmatrix} 1 & 1 & 1 \\ z & x & y \\ 1 & 1 & 1 \end{vmatrix}$$

Taking out $(x + y + z)$ and $(-3)$ common from $R_1$ and $R_3$ respectively

$$\Delta = -3\,(x + y + z).\,0$$

[$\because$ $R_1$ and $R_3$ are identical]

$\Rightarrow$ $\Delta = 0$ **Ans.**

**2. Write the sum of the order and degree of the following differential equation :**

$$\frac{d}{dx}\left\{\left(\frac{dy}{dx}\right)^3\right\} = 0$$ **[1]**

**Sol.** Given,

$$\frac{d}{dx}\left\{\left(\frac{dy}{dx}\right)^3\right\} = 0.$$

On differentiating w.r.t. $x$ both sides, we get

$$3\left(\frac{dy}{dx}\right)^2 \times \frac{d^2y}{dx^2} = 0.$$

Since the order and degree of the differential equation is 2 and 1 respectively.

So, the sum of the order and degree is 3. **Ans.**

**3. Write the integrating factor of the following differential equation :**

$$(1+y^2)+(2xy-\cot y)\frac{dy}{dx} = 0.$$ **[1]**

**Sol.** Given,

$$(1 + y^2) + (2xy - \cot y)\frac{dy}{dx} = 0$$

$\Rightarrow$ $(1 + y^2)\,dx + 2xy\,dy = \cot y\,dy$

$\Rightarrow$ $\frac{dx}{dy} + \left(\frac{2y}{1+y^2}\right)x = \frac{\cot y}{1+y^2}$

This is a linear differential equation of the form

$$\frac{dx}{dy} + Rx = S,$$

where $R = \frac{2y}{1+y^2}$

and $S = \frac{\cot y}{1+y^2}$

$\therefore$ Integrating factor $= e^{\int R\,dy} = e^{\int \frac{2y}{1+y^2}dy}$

$= e^{\log(1+y^2)}$

$= 1 + y^2.$ **Ans.**

**4. If $\hat{a}$, $\hat{b}$ and $\hat{c}$ are mutually perpendicular unit vectors, then find the value of $|2\hat{a} + \hat{b} + \hat{c}|$.** **[1]**

**Sol.** Let $x = |2\hat{a} + \hat{b} + \hat{c}|$

$\therefore$ $x^2 = |2\hat{a} + \hat{b} + \hat{c}|^2$

$= 4|\hat{a}|^2 + |\hat{b}|^2 + |\hat{c}|^2 + 2\,(2\,\hat{a}.\hat{b} + \hat{b}.\hat{c} + 2\,\hat{c}.\hat{a})$

Since, $\hat{a}$, $\hat{b}$ and $\hat{c}$ are mutually perpendicular unit vectors.

$\therefore$ $\hat{a}.\hat{b} = \hat{b}.\hat{c} = \hat{c}.\hat{a} = 0$

So, $x^2 = 4 \times 1 + 1 + 1 + 2\,(2 \times 0 + 0 + 2 \times 0)$

$\Rightarrow$ $x^2 = 6$

$\therefore$ $x = \sqrt{6}$

Hence, $|2\hat{a} + \hat{b} + \hat{c}| = \sqrt{6}$. **Ans.**

**5. Write a unit vector perpendicular to both the vectors $\vec{a} = \hat{i} + \hat{j} + \hat{k}$ and $\vec{b} = \hat{i} + \hat{j}$.** **[1]**

**Sol.** A vector perpendicular to both $\vec{a}$ and $\vec{b}$ is $\vec{a} \times \vec{b}$

$\therefore$ Unit vector $\perp$ to $\vec{a}$ and $\vec{b}$ $= \dfrac{\vec{a} \times \vec{b}}{|\vec{a} \times \vec{b}|}$

$$\vec{a} \times \vec{b} = \begin{vmatrix} \hat{i} & \hat{j} & \hat{k} \\ 1 & 1 & 1 \\ 1 & 1 & 0 \end{vmatrix}$$

$$= \hat{i}(0-1) - \hat{j}(0-1) + \hat{k}(1-1)$$

$$= -\hat{i} + \hat{j}$$

$$\Rightarrow \quad |\vec{a} \times \vec{b}| = \sqrt{(-1)^2 + 1^2} = \sqrt{2}$$

$\therefore$ Required unit vector $= \dfrac{-\hat{i} + \hat{j}}{\sqrt{2}}$ **Ans.**

**6. The equations of a line is $5x - 3 = 15y + 7 = 3 - 10z$. Write the direction cosines of the line. [1]**

**Sol.** Given line is $5x - 3 = 15y + 7 = 3 - 10z$

Rewritting the eq. in standard form :

$$5\left(x - \frac{3}{5}\right) = 15\left(y + \frac{7}{15}\right)$$

$$= -10\left(z - \frac{3}{10}\right)$$

*i.e.,*
$$\frac{x - \frac{3}{5}}{\frac{1}{5}} = \frac{y + \frac{7}{15}}{\frac{1}{15}} = \frac{z - \frac{3}{10}}{\frac{-1}{10}}$$

Thus, the direction ratios of the line are

$\frac{1}{5}, \frac{1}{15}, \frac{-1}{10}$ *i.e.,* $6, 2, -3.$

Hence, its direction cosines are

$$\pm\frac{6}{\sqrt{6^2 + 2^2 + (-3)^2}}, \pm\frac{2}{\sqrt{6^2 + 2^2 + (-3)^2}},$$

$$\pm\frac{-3}{\sqrt{6^2 + 2^2 + (-3)^2}} \text{ i.e., } \pm\frac{6}{7}, \pm\frac{2}{7}, \mp\frac{3}{7}$$

*i.e.,* $\frac{6}{7}, \frac{2}{7}, \frac{-3}{7}$ or $\frac{-6}{7}, \frac{-2}{7}, \frac{3}{7}$. **Ans.**

## SECTION–B

**7. To promote the making of toilets for women, an organisation tried to generate awareness through (i) house calls (ii) letters, and (iii) announcements. The cost for each mode per attempt is given below:**

**(i) ₹ 50 (ii) ₹ 20(iii) ₹ 40**

**The number of attempts made in three villages X, Y and Z are given below :**

| | (i) | (ii) | (iii) |
|---|---|---|---|
| X | 400 | 300 | 100 |
| Y | 300 | 250 | 75 |
| Z | 500 | 400 | 150 |

**Find the total cost incurred by the organisation for the three villages separately, using matrices.**

**Write one value generated by the organisation in the society. [4]**

**Sol.** The number of attempts made in three villages X, Y and Z can be represented by the $3 \times 3$ matrix.

$$X = \begin{bmatrix} 400 & 300 & 100 \\ 300 & 250 & 75 \\ 500 & 400 & 150 \end{bmatrix}$$

and the cost for each mode per attempt can be represented by the $3 \times 1$ matrix.

$$Y = \begin{bmatrix} 50 \\ 20 \\ 40 \end{bmatrix}$$

$\therefore$ By matrix multiplication the cost incurred by the organisation for the three villages.

$$XY = \begin{bmatrix} 400 & 300 & 100 \\ 300 & 250 & 75 \\ 500 & 400 & 150 \end{bmatrix}\begin{bmatrix} 50 \\ 20 \\ 40 \end{bmatrix}$$

$$\Rightarrow \quad XY = \begin{bmatrix} 30{,}000 \\ 23{,}000 \\ 39{,}000 \end{bmatrix}$$

Hence the total cost incurred by the organisation for the three villages separately are ₹ 30,000, ₹ 23,000 and ₹ 39,000.

The organisation in the society generated the value of cleanliness for the women welfare. **Ans.**

**8. Solve for $x$ :**

$$\tan^{-1}(x + 1) + \tan^{-1}(x - 1) = \tan^{-1}\frac{8}{31} \quad \textbf{[4]}$$

**Sol.** Given, $\tan^{-1}(x + 1) + \tan^{-1}(x - 1)$

$$= \tan^{-1}\frac{8}{31}$$

$$\Rightarrow \quad \tan^{-1}\frac{x + 1 + x - 1}{1 - (x + 1)(x - 1)} = \tan^{-1}\frac{8}{31}$$

$$\left[\because \tan^{-1}A + \tan^{-1}B = \tan^{-1}\frac{A + B}{1 - AB}\right]$$

$$\Rightarrow \quad \tan^{-1}\frac{2x}{2 - x^2} = \tan^{-1}\frac{8}{31}$$

$$\Rightarrow \quad \frac{2x}{2-x^2} = \frac{8}{31}$$

$$\Rightarrow \quad 62x = 16 - 8x^2$$

$$\Rightarrow \quad 4x^2 + 31x - 8 = 0$$

Solving the quadratic equation, we get

$$(4x-1)(x+8) = 0$$

$$\Rightarrow \quad x = \frac{1}{4} \text{ or } -8.$$

Since, $x = -8$ doesn't satisfy the given equation, so neglecting it.

$$\therefore \quad x = \frac{1}{4}.$$ **Ans.**

**OR**

**Prove the following :**

$$\cot^{-1}\left(\frac{xy+1}{x-y}\right) + \cot^{-1}\left(\frac{yz+1}{y-z}\right) + \cot^{-1}\left(\frac{zx+1}{z-x}\right)$$

$$= 0 \ (0 < xy, yx, zx < 1).$$

**Sol.** L. H. S.

$$= \cot^{-1}\left(\frac{xy+1}{x-y}\right) + \cot^{-1}\left(\frac{yz+1}{y-z}\right) + \cot^{-1}\left(\frac{zx+1}{z-x}\right)$$

$$= \tan^{-1}\left(\frac{x-y}{xy+1}\right) + \tan^{-1}\left(\frac{y-z}{yz+1}\right) + \tan^{-1}\left(\frac{z-x}{zx+1}\right)$$

$$\left[\because \tan^{-1} x = \cot^{-1}\frac{1}{x}\right]$$

$$= \tan^{-1}x - \tan^{-1}y + \tan^{-1}y - \tan^{-1}z + \tan^{-1}z$$

$$- \tan^{-1}x = 0 = \text{R. H. S.}$$

**Hence Proved.**

**9. Using properties of determinants, prove the following:**

$$\begin{vmatrix} a^2 & bc & ac+c^2 \\ a^2+ab & b^2 & ac \\ ab & b^2+bc & c^2 \end{vmatrix} = 4a^2b^2c^2.$$ **[4]**

**Sol.** Let

$$\Delta = \begin{vmatrix} a^2 & bc & ac+c^2 \\ a^2+ab & b^2 & ac \\ ab & b^2+bc & c^2 \end{vmatrix}$$

Taking $a$, $b$ and $c$ common from $C_1$, $C_2$ and $C_3$ respectively.

$$\Delta = abc\begin{vmatrix} a & c & a+c \\ a+b & b & a \\ b & b+c & c \end{vmatrix}$$

Applying $R_1 \rightarrow R_1 + R_2 + R_3$

$$\Delta = abc\begin{vmatrix} 2(a+b) & 2(b+c) & 2(a+c) \\ a+b & b & a \\ b & b+c & c \end{vmatrix}$$

Taking 2 common from $R_1$,

$$\Delta = 2abc\begin{vmatrix} a+b & b+c & a+c \\ a+b & b & a \\ b & b+c & c \end{vmatrix}$$

Applying $R_1 \rightarrow R_1 - R_2$ and $R_3 \rightarrow R_3 - R_1$

$$\Delta = 2abc\begin{vmatrix} 0 & c & c \\ a+b & b & a \\ -a & 0 & -a \end{vmatrix}$$

Now, taking $c$ and $a$ common from $R_1$ and $R_3$

$$\Delta = 2a^2bc^2\begin{vmatrix} 0 & 1 & 1 \\ a+b & b & a \\ -1 & 0 & -1 \end{vmatrix}$$

Expanding along $R_1$,

$$\Delta = 2a^2bc^2\,[0(-b-0) - 1\{-(a+b)+a\} + 1\,(0+b)]$$

$$= 2a^2bc^2\,[0 - (-a-b+a) + b]$$

$$= 2a^2bc^2\,[0 + b + b]$$

$$= 2a^2bc^2\,[2b]$$

$$= 4a^2b^2c^2 = \text{R. H. S.}$$ **Hence Proved.**

**10. Find the adjoint of the matrix :**

$$\mathbf{A} = \begin{pmatrix} -1 & -2 & -2 \\ 2 & 1 & -2 \\ 2 & -2 & 1 \end{pmatrix}$$

**and hence show that A. (adj A) = | A | $I_3$.** **[4]**

**Sol.** We have,

$$A = \begin{pmatrix} -1 & -2 & -2 \\ 2 & 1 & -2 \\ 2 & -2 & 1 \end{pmatrix}$$

$$A_{11} = \begin{vmatrix} 1 & -2 \\ -2 & 1 \end{vmatrix} = 1 - 4 = -3$$

$$A_{12} = -\begin{vmatrix} 2 & -2 \\ 2 & 1 \end{vmatrix} = -(2+4) = -6$$

$$A_{13} = \begin{vmatrix} 2 & 1 \\ 2 & -2 \end{vmatrix} = -4 - 2 = -6$$

$$A_{21} = -\begin{vmatrix} -2 & -2 \\ -2 & 1 \end{vmatrix} = -(-2-4) = 6$$

$$A_{22} = \begin{vmatrix} -1 & -2 \\ 2 & 1 \end{vmatrix} = -1 + 4 = 3$$

$$A_{23} = -\begin{vmatrix} -1 & -2 \\ 2 & -2 \end{vmatrix} = -(2+4) = -6$$

$$A_{31} = \begin{vmatrix} -2 & -2 \\ 1 & -2 \end{vmatrix} = +4+2 = +6$$

$$A_{32} = -\begin{vmatrix} -1 & -2 \\ 2 & -2 \end{vmatrix} = -(2+4) = -6$$

$$A_{33} = \begin{vmatrix} -1 & -2 \\ 2 & 1 \end{vmatrix} = -1+4 = +3$$

$$\text{adj } A = \begin{bmatrix} A_{11} & A_{21} & A_{31} \\ A_{12} & A_{22} & A_{32} \\ A_{13} & A_{23} & A_{33} \end{bmatrix}$$

$$\text{adj } A = \begin{bmatrix} -3 & 6 & 6 \\ -6 & 3 & -6 \\ -6 & -6 & 3 \end{bmatrix}$$

$$|A| = \begin{vmatrix} -1 & -2 & -2 \\ 2 & 1 & -2 \\ 2 & -2 & 1 \end{vmatrix}$$

$$= -1(1-4)+2(2+4)-2(-4-2)$$
$$= -1(-3)+2(6)-2(-6)$$
$$= 3+12+12 = 27$$

Now L. H. S. = A (adj A)

$$= \begin{pmatrix} -1 & -2 & -2 \\ 2 & 1 & -2 \\ 2 & -2 & 1 \end{pmatrix}\begin{pmatrix} -3 & 6 & 6 \\ -6 & 3 & -6 \\ -6 & -6 & 3 \end{pmatrix}$$

$$= \begin{pmatrix} 27 & 0 & 0 \\ 0 & 27 & 0 \\ 0 & 0 & 27 \end{pmatrix}$$

$$= 27\begin{pmatrix} 1 & 0 & 0 \\ 0 & 1 & 0 \\ 0 & 0 & 1 \end{pmatrix}$$

$= |A| . I_3 =$ R. H. S. **Hence Proved.**

**11. Show that the function $f(x) = |x-1| + |x+1|$, for all $x \in$ R, is not differentiable at the points $x = -1$ and $x = 1$.** **[4]**

**Sol.** Given,

$$f(x) = |x-1| + |x+1|$$

$$= \begin{cases} -(x-1)-(x+1) = -2x, & x < -1 \\ -(x-1)+(x+1) = 2, & -1 \le x < 1 \\ (x-1)+(x+1) = 2x, & x \ge 1 \end{cases}$$

Now differentiability at $x = -1$,

$$\text{LHD} = \lim_{h \to 0} \frac{f(-1-h) - f(-1)}{-h}$$
$$= \lim_{h \to 0} \frac{-2(-1-h)-2}{-h}$$
$$= \lim_{h \to 0} \frac{2+2h-2}{-h}$$
$$= \lim_{h \to 0} \frac{2h}{-h}$$
$$= \lim_{h \to 0} -2$$
$$= -2$$

$$\text{RHD} = \lim_{h \to 0} \frac{f(-1+h) - f(-1)}{h}$$
$$= \lim_{h \to 0} \frac{2-2}{h} = \lim_{h \to 0} 0$$
$$= 0$$

Since (LHD) $\neq$ (RHD)

$\therefore f(x)$ is not differentiable at $x = -1$.

Now differentiability at $x = 1$,

$$(\text{LHD at } x = 1) = \lim_{h \to 0} \frac{f(1-h) - f(1)}{-h}$$
$$= \lim_{h \to 0} \frac{2-2}{-h}$$
$$= \lim_{h \to 0} 0$$
$$= 0$$

$$(\text{RHD at } x = 1) = \lim_{h \to 0} \frac{f(1+h) - f(1)}{h}$$
$$= \lim_{h \to 0} \frac{2(1+h)-2}{h}$$
$$= \lim_{h \to 0} \frac{2+2h-2}{h}$$
$$= \lim_{h \to 0} \frac{2h}{h} = \lim_{h \to 0} 2$$
$$= 2$$

Since LHD $\neq$ RHD

$\therefore f(x)$ is not differentiable at $x = 1$.

**Hence Proved.**

**12. If $y = e^{m \sin^{-1} x}$, then show that**

$$\left(1-x^2\right)\frac{d^2y}{dx^2} - x\frac{dy}{dx} - m^2y = 0. \quad \textbf{[4]}$$

**Sol.** Given,

$$y = e^{m \sin^{-1} x}, \quad \text{...(i)}$$

Differentiating both sides w.r.t. $x$, we get

$$\frac{dy}{dx} = e^{m\sin^{-1}x} \times \frac{m}{\sqrt{1-x^2}}$$

$$\Rightarrow \quad \sqrt{1-x^2}\frac{dy}{dx} = m\,e^{m\sin^{-1}x}$$

Again differentiating, w.r.t. $x$, we get

$$\sqrt{1-x^2}\frac{d^2y}{dx^2} + \frac{dy}{dx} \times \left(\frac{-x}{\sqrt{1-x^2}}\right) = \frac{m^2\,e^{m\sin^{-1}x}}{\sqrt{1-x^2}}$$

Multiplying both sides by $\sqrt{1-x^2}$, we get

$$\left(1-x^2\right)\frac{d^2y}{dx^2} - x\frac{dy}{dx} = m^2\,e^{m\sin^{-1}x}$$

$$\Rightarrow \quad \left(1-x^2\right)\frac{d^2y}{dx^2} - x\frac{dy}{dx} - m^2 y = 0 \quad \text{[using (i)]}$$

**Hence Proved.**

**13. If $f(x) = \sqrt{x^2+1}$; $g(x) = \dfrac{x+1}{x^2+1}$ and $h(x) = 2x - 3$, then find $f'\,[h'\,\{g'(x)\}]$.** **[4]**

**Sol.** Given,

$$f(x) = \sqrt{x^2+1},\ g(x) = \frac{x+1}{x^2+1}$$

and $h(x) = 2x - 3$

Now, $$f'(x) = \frac{2x}{2\sqrt{x^2+1}} = \frac{x}{\sqrt{x^2+1}}$$

$$g'(x) = \frac{\left(x^2+1\right)-(x+1)2x}{\left(x^2+1\right)^2} = \frac{1-2x-x^2}{\left(x^2+1\right)^2}$$

and $h'(x) = 2.$

$$\therefore \quad f'[h'\{g'(x)\}] = f'\left[h'\left\{\frac{1-2x-x^2}{\left(x^2+1\right)^2}\right\}\right]$$

$$= f'[2] \qquad [\because h'(x) = 2\,]$$

$$= \frac{2}{\sqrt{2^2+1}} = \frac{2}{\sqrt{5}}.$$

**Ans.**

**14. Evaluate : $\int(3-2x).\sqrt{2+x-x^2}\,dx$.** **[4]**

**Sol.** Let, $$I = \int(3-2x).\sqrt{2+x-x^2}\,dx.$$

Let $$3-2x = \lambda\frac{d}{dx}\left(2+x-x^2\right)+\mu$$

$$\Rightarrow \quad 3-2x = \lambda\,(1-2x)+\mu \qquad \text{...(i)}$$

$$\Rightarrow \quad 3-2x = (-2\lambda)x + \lambda + \mu$$

Equating the coefficients of like terms, we get

$$-2\lambda = -2$$

$$\Rightarrow \quad \lambda = 1$$

and $$\lambda + \mu = 3$$

$$\Rightarrow \quad \mu = 3 - 1 = 2$$

$\therefore$ Equation. (i) becomes $3 - 2x = (1 - 2x) + 2$

$$I = \int\{(1-2x)+2\}\sqrt{2+x-x^2}\,dx$$

$$\Rightarrow \quad I = \int(1-2x)\sqrt{2+x-x^2}\,dx + 2\int\sqrt{2+x-x^2}\,dx$$

$$\Rightarrow \quad I = I_1 + 2I_2 \qquad \text{...(ii)}$$

$$\Rightarrow \quad I_1 = \int(1-2x)\sqrt{2+x-x^2}\,dx$$

Let $$2+x-x^2 = t$$

$$\Rightarrow \quad (1-2x)\,dx = dt$$

$$\therefore \quad I_1 = \int\sqrt{t}\,dt$$

$$\Rightarrow \quad I_1 = \frac{t^{1/2+1}}{\frac{1}{2}+1} + C_1$$

$$\Rightarrow \quad I_1 = \frac{2}{3}t^{3/2} + C_1$$

$$\Rightarrow \quad I_1 = \frac{2}{3}\left(2+x-x^2\right)^{3/2} + C_1 \quad \text{...(iii)}$$

and $$I_2 = \int\sqrt{2+x-x^2}\,dx$$

$$\Rightarrow \quad I_2 = \int\sqrt{2+\frac{1}{4}-\left(x^2-x+\frac{1}{4}\right)}\,dx$$

$$\Rightarrow \quad I_2 = \int\sqrt{\left(\frac{3}{2}\right)^2-\left(x-\frac{1}{2}\right)^2}\,dx$$

$$\Rightarrow \quad I_2 = \frac{1}{2}\left(x-\frac{1}{2}\right)\sqrt{\left(\frac{3}{2}\right)^2-\left(x-\frac{1}{2}\right)^2} + \frac{\frac{9}{4}}{2}\sin^{-1}\left(\frac{x-\frac{1}{2}}{\frac{3}{2}}\right) + C_2$$

$$\Rightarrow \quad I_2 = \frac{1}{2}\left(x-\frac{1}{2}\right)\sqrt{2+x-x^2} + \frac{9}{8}\sin^{-1}\left(\frac{2x-1}{3}\right) + C_2 \quad \text{...(iv)}$$

From (ii), (iii) and (iv), we get

$$I = \frac{2}{3}\left(2+x-x^2\right)^{3/2} + \left(x-\frac{1}{2}\right)\sqrt{2+x-x^2} + \frac{9}{4}\sin^{-1}\left(\frac{2x-1}{3}\right) + C,$$

where

$C = C_1 + C_2.$

**Ans.**

OR

**Evaluate :** $\int \frac{x^2+x+1}{(x^2+1)(x+2)} dx.$

**Sol.** Let $I = \int \frac{x^2+x+1}{(x^2+1)(x+2)} dx$

$$I = \int \left( \frac{x^2+1}{(x^2+1)(x+2)} + \frac{x}{(x^2+1)(x+2)} \right) dx$$

$$\Rightarrow \quad I = \int \frac{1}{x+2} dx + \int \frac{x}{(x^2+1)(x+2)} dx$$

$$\Rightarrow \quad I = \log|x+2| + \int \frac{x}{(x^2+1)(x+2)} dx \quad ...(i)$$

Let $\frac{x}{(x^2+1)(x+2)} = \frac{A}{x+2} + \frac{Bx+C}{x^2+1}$

$$x = A(x^2+1) + (Bx+C)(x+2) \quad ...(ii)$$

Putting $x+2 = 0$

$\Rightarrow$ $x = -2$ in (ii), we get $-2 = 5A$

$$\Rightarrow \quad A = \frac{-2}{5}$$

Putting $x = 0$ and $x = 1$ in (ii), we get

$$0 = A + 2C$$

$$\Rightarrow \quad C = \frac{1}{5}$$

$$1 = 2A + 3B + 3C$$

$$\Rightarrow \quad 3B = \frac{6}{5}$$

$$\Rightarrow \quad B = \frac{2}{5}$$

$$\therefore \quad I = \log|x+2| - \frac{2}{5}\int \frac{dx}{x+2} + \int \frac{\frac{2}{5}x + \frac{1}{5}}{x^2+1} dx$$

$$\Rightarrow \quad I = \log|x+2| - \frac{2}{5}\int \frac{dx}{x+2} + \frac{2}{5}\int \frac{xdx}{x^2+1} + \frac{1}{5}\int \frac{1}{x^2+1} dx$$

$$\Rightarrow \quad I = \log|x+2| - \frac{2}{5}\log|x+2| + \frac{1}{5}\log|x^2+1| + \frac{1}{5}\tan^{-1} x + C$$

$$\therefore \quad I = \frac{3}{5}\log|x+2| + \frac{1}{5}\log|x^2+1| + \frac{1}{5}\tan^{-1} x + C.$$ **Ans.**

**15. Find** $\left( \int_0^{\pi/4} \frac{dx}{\cos^3 x\sqrt{2\sin 2x}}. \right)$. **[4]**

**Sol.** Let $I = \int_0^{\pi/4} \frac{dx}{\cos^3 x\sqrt{2\sin 2x}}.$

Multiplying and dividing denominator by $\cos x$

$$\int_0^{\pi/4} \frac{dx}{\cos^4 x \frac{\sqrt{2\sin 2x}}{\cos x}} = \int_0^{\pi/4} \frac{\sec^4 x\, dx}{\sqrt{\frac{2\sin 2x}{\cos^2 x}}}$$

$$I = \int_0^{\pi/4} \frac{(1+\tan^2 x)\sec^2 x}{2\sqrt{\frac{\sin x\cos x}{\cos^2 x}}} dx$$

[$\because \sin 2x = 2\sin x\cos x$]

$$\Rightarrow \quad I = \int_0^{\pi/4} \frac{(1+\tan^2 x)\sec^2 x}{2\sqrt{\tan x}} dx$$

Putting $\tan x = t$

$\Rightarrow \sec^2 x\, dx = dt$, we get

$$\therefore \quad I = \int_0^1 \frac{(1+t^2)dt}{2\sqrt{t}} \quad \left( \begin{array}{l} \because \text{when } x=0, t=0 \\ \text{and } \quad x=\frac{\pi}{4}, t=1 \end{array} \right)$$

$$\Rightarrow \quad I = \frac{1}{2}\int_0^1 \left( \frac{1}{\sqrt{t}} + t^{\frac{3}{2}} \right) dt$$

$$\Rightarrow \quad I = \frac{1}{2}\left[ 2\sqrt{t} + \frac{2}{5}t^{\frac{5}{2}} \right]_0^1$$

$$\Rightarrow \quad I = \left[ \sqrt{t} + \frac{1}{5}t^{\frac{5}{2}} \right]_0^1$$

$$\Rightarrow \quad I = \left[ \sqrt{1} + \frac{1}{5}1^{\frac{5}{2}} \right] - \left[ \sqrt{0} + \frac{1}{5}0^{\frac{5}{2}} \right]$$

$$\therefore \quad I = \frac{6}{5}.$$ **Ans.**

**16. Find** $\left( \int \frac{\log x}{(x+1)^2} dx. \right)$. **[4]**

**Sol.** Let $I = \int \frac{\log x}{(x+1)^2} dx$

$$= \int \underset{\text{I}}{\log x} . \underset{\text{II}}{\frac{1}{(x+1)^2}} dx$$

Integrating by parts,

$$I = \log x \int \frac{dx}{(x+1)^2} - \int \left( \frac{d}{dx}\log x \int \frac{dx}{(x+1)^2} \right) dx$$

$$\Rightarrow \quad I = -\frac{\log x}{(x+1)} + \int \frac{1}{x(x+1)} dx$$

$$= -\frac{\log x}{(x+1)} + \int \frac{(x+1)-x}{x(x+1)} dx$$

[Add and subtract $x$]

$$\Rightarrow \quad I = -\frac{\log x}{(x+1)} + \int \left(\frac{1}{x} - \frac{1}{x+1}\right) dx$$

$$\Rightarrow \quad I = -\frac{\log x}{(x+1)} + \log|x| - \log|x+1| + C$$

$$\therefore \quad I = \log\left|\frac{x}{x+1}\right| - \frac{\log x}{(x+1)} + C. \qquad \textbf{Ans.}$$

**17. If $\vec{a} = \hat{i} + 2\hat{j} + \hat{k}, \vec{b} = 2\hat{i} + \hat{j}$ and $\vec{c} = 3\hat{i} - 4\hat{j} - 5\hat{k}$ then find a unit vector perpendicular to both of the vectors $(\vec{a} - \vec{b})$ and $(\vec{c} - \vec{b})$.** **[4]**

**Sol.** Here,

$$(\vec{a} - \vec{b}) = \hat{i} + 2\hat{j} + \hat{k} - 2\hat{i} - \hat{j}$$

$$= -\hat{i} + \hat{j} + \hat{k}$$

and

$$(\vec{c} - \vec{b}) = 3\hat{i} - 4\hat{j} - 5\hat{k} - 2\hat{i} - \hat{j}$$

$$= \hat{i} - 5\hat{j} - 5\hat{k}$$

So, required unit vector $\bar{r} = \dfrac{(\vec{a}-\vec{b})\times(\vec{c}-\vec{b})}{\left|(\vec{a}-\vec{b})\times(\vec{c}-\vec{b})\right|}$,

where

$$\vec{r} = (\vec{a} - \vec{b}) \times (\vec{c} - \vec{b})$$

$$= \begin{vmatrix} \hat{i} & \hat{j} & \hat{k} \\ -1 & 1 & 1 \\ 1 & -5 & -5 \end{vmatrix}$$

$$= \hat{i}(-5+5) - \hat{j}(5-1) + \hat{k}\ (5-1)$$

$$= -4\hat{j} + 4\hat{k}$$

Hence, $\hat{r} = \dfrac{4\hat{k} - 4\hat{j}}{\sqrt{4^2 + 4^2}} = \left(\dfrac{\hat{k} - \hat{j}}{\sqrt{2}}\right)$. **Ans.**

**18. Find the equation of a line passing through the point (1, 2, – 4) and perpendicular to two lines**

$$\vec{r} = (8\hat{i} - 19\hat{j} + 10\hat{k}) + \lambda(3\hat{i} - 16\hat{j} + 7\hat{k})$$

**and** $\vec{r} = (15\hat{i} + 29\hat{j} + 5\hat{k}) + \mu(3\hat{i} + 8\hat{j} - 5\hat{k})$. **[4]**

**Sol.** Let the direction ratios of required line be $a, b, c$, Since, the line is perpendicular to

$$\vec{r} = (8\hat{i} - 19\hat{j} + 10\hat{k}) + \lambda(3\hat{i} - 16\hat{j} + 7\hat{k})$$

and $\vec{r} = (15\hat{i} + 29\hat{j} + 5\hat{k}) + \mu(3\hat{i} + 8\hat{j} - 5\hat{k})$

$\therefore \quad 3a - 16b + 7c = 0$

and $\quad 3a + 8b - 5c = 0.$

Solving by cross multiplication, we get

$$\frac{a}{80-56} = \frac{b}{21+15} = \frac{c}{24+48}$$

$$\Rightarrow \quad \frac{a}{24} = \frac{b}{36} = \frac{c}{72}$$

$\therefore$ the direction ratios of line : 2, 3, 6.

Hence, required line through the point (1, 2, – 4) is

$$\vec{r} = \vec{a} + \lambda\vec{b}$$

$$\vec{r} = (\hat{i} + 2\hat{j} - 4\hat{k}) + \lambda(2\hat{i} + 3\hat{j} + 6\hat{k})$$

**Ans.**

**OR**

**Find the equation of the plane passing through the points (–1, 2, 0), (2, 2, –1) and parallel to the line $\dfrac{x-1}{1} = \dfrac{2y+1}{2} = \dfrac{z+1}{-1}$.**

**Sol.** The equation of a plane passing through (–1, 2, 0) is

$$a(x+1) + b(y-2) + c(z-0) = 0 \quad \text{...(i)}$$

It passes through (2, 2, –1)

$$\therefore\ a(2+1) + b(2-2) + c(-1-0) = 0$$

$$3a + 0b - c = 0 \quad \text{...(ii)}$$

The given line is

$$\frac{x-1}{1} = \frac{2y+1}{2} = \frac{z+1}{-1}$$

*i. e.*

$$\frac{x-1}{1} = \frac{y+\frac{1}{2}}{1} = \frac{z+1}{-1}$$

$\therefore$ *d. r'. s* of line are 1, 1, –1

The plane (i) is parallel to the given line

$$a + b - c = 0 \quad \text{...(iii)}$$

Solving (ii) and (iii) by cross multiplication, we get

$$\frac{a}{1} = \frac{b}{2} = \frac{c}{3}$$

*i.e.*, direction ratios of normal to the plane are 1, 2, 3.

$$\therefore \quad 1(x+1) + 2(y-2) + 3(z-0) = 0$$

*i.e.*, $\quad x + 2y + 3z = 3.$ **Ans.**

**19. Three cards are drawn successively with replacement from a well shuffled pack of 52 cards. Find the probability distribution of the**

**number of spades. Hence find the mean of the distribution.** **[4]**

**Sol.** Let X denote the number of spades when three cards are drawn, then, X is a random variable that can take values 0, 1, 2, 3.

Let E be the event when spade card is drawn,

$$p = P(E) = \frac{13}{52} = \frac{1}{4}$$

$$q = 1 - p = 1 - \frac{1}{4} = \frac{3}{4}$$

P(X = 0) = Probability of getting no spade

$$= {}^3C_0\left(\frac{1}{4}\right)^0\left(\frac{3}{4}\right)^3 = \frac{27}{64}$$

P(X = 1) = Probability of getting one spade

$$= {}^3C_1\left(\frac{1}{4}\right)^1\left(\frac{3}{4}\right)^{3-1} = \frac{27}{64}$$

P(X = 2) = Probability of getting two spades

$$= {}^3C_2\left(\frac{1}{4}\right)^2\left(\frac{3}{4}\right)^{3-2} = \frac{9}{64}$$

P(X = 3) = Probability of getting three spades

$$= {}^3C_3\left(\frac{1}{4}\right)^3\left(\frac{3}{4}\right)^{3-3} = \frac{1}{64}$$

Thus, the probability distribution of random variable X is given by

| X | 0 | 1 | 2 | 3 |
|---|---|---|---|---|
| P(X) | $\frac{27}{64}$ | $\frac{27}{64}$ | $\frac{9}{64}$ | $\frac{1}{64}$ |

$\therefore$ Mean $(\overline{X}) = \Sigma XP(X)$

$$= 0\times\frac{27}{64} + 1\times\frac{27}{64} + 2\times\frac{9}{64} + 3\times\frac{1}{64}$$

$$= \frac{48}{64} = \frac{3}{4}.$$ **Ans.**

**OR**

**For 6 trials of an experiment, let X be a binomial variate which satisfies the relation 9P(X = 4) = P(X = 2). Find the probability of success.**

**Solution :** Let $p$ denote the probability of getting success and $q$ be the probability of failure.

Since, $P(x = r) = {}^nC_r\, p^r\, q^{n-r}$

$\therefore$ $P(x = 4) = {}^6C_4\, p^4\, q^{6-4}$

and $P(x = 2) = {}^6C_2\, p^2\, q^{6-2}$

We have $9P(X = 4) = P(X = 2)$

$\Rightarrow$ $9\,{}^6C_4\, p^4\, q^2 = {}^6C_2\, p^2\, q^4$

$\Rightarrow$ $9p^2 = q^2$ $[\because {}^6C_4 = {}^6C_2]$

$\Rightarrow$ $9p^2 = (1-p)^2$ $[\because p + q = 1]$

$\Rightarrow$ $9p^2 = 1^2 + p^2 - 2p$

$\Rightarrow$ $9p^2 - p^2 + 2p - 1 = 0$

$\Rightarrow$ $8p^2 + 2p - 1 = 0$

$\Rightarrow$ $8p^2 + 4p - 2p - 1 = 0$

$\Rightarrow$ $(4p - 1)(2p + 1) = 0$

$\therefore$ $p = \frac{1}{4}, \frac{-1}{2}$

Since probability can not be – ve,

$\therefore$ $p = \frac{1}{4}$

Hence, the probability of success

$= \frac{1}{4}.$ **Ans.**

## SECTION–C

**20. Consider $f : R_+ \to [-9, \infty)$ given by $f(x) = 5x^2 + 6x - 9$. Prove that $f$ is invertible with**

$$f^{-1}(y) = \left(\frac{\sqrt{54+5y} - 3}{5}\right).$$ **[6]**

**Sol.** To prove $f$ is invertible we have to prove that $f$ is one-one and onto.

**For one-one**

Let $x_1, x_2 \in R_+$, then

$f(x_1) = f(x_2)$

$\Rightarrow$ $5x_1^2 + 6x_1 - 9 = 5x_2^2 + 6x_2 - 9$

$\Rightarrow$ $5(x_1^2 - x_2^2) + 6(x_1 - x_2) = 0$

$\Rightarrow$ $(x_1 - x_2)(5x_1 + 5x_2 + 6) = 0$

$\Rightarrow$ $x_1 - x_2 = 0$ as $5x_1 + 5x_2 + 6 \neq 0$

$\Rightarrow$ $x_1 = x_2$

*i.e.,* $f$ is one-one function.

**For onto**

Let $f(x) = y$

$\because$ $y = 5x^2 + 6x - 9$

$\therefore 5x^2 + 6x - (9 + y) = 0$

$$x = \frac{-6 \pm \sqrt{36 + 4\times5(9+y)}}{10}$$

$$= \frac{-6 \pm \sqrt{216 + 20y}}{10}$$

$$= \frac{\pm\sqrt{54+5y} - 3}{5}$$

$$\Rightarrow \quad x = \frac{\sqrt{54+5y}-3}{5} \quad (\because x \in R_+)$$

Clearly $\forall\ y \in [-9, \infty]$, the value of $x \in R_+$

$\Rightarrow f$ is onto function.

Hence $f$ is one-one onto function.

$\Rightarrow f$ is invertible function with

$$f^{-1}(y) = \frac{\sqrt{54+5y}-3}{5}$$ **Hence Proved.**

**OR**

**A binary operation * is defined on the set $x = R - \{-1\}$ by $x * y = x + y + xy, \forall\ x, y \in X$. Check whether * is commutative and associative. Find its identity element and also find the inverse of each element of X.****

**21. Find the value of $p$ for which the curves $x^2 = 9p(9-y)$ and $x^2 = p(y+1)$ cut each other at right angles. [6]**

**Sol.** Given,

$$x^2 = 9p(9-y) \quad ...(i)$$

and $$x^2 = p(y+1) \quad ...(ii)$$

From (i) and (ii), we get

$$9p(9-y) = p(y+1)$$

$$\Rightarrow \quad 81p - 9py = py + p$$

$$\Rightarrow \quad 10py = 80p$$

$$\Rightarrow \quad y = 8$$

$$\therefore \quad x^2 = 9p$$

Now, differentiating (i) and (ii) w.r.t. $x$, we get

$$2x = -9p\frac{dy}{dx}$$

$$\Rightarrow \quad \frac{dy}{dx} = -\frac{2x}{9p}$$

and $$2x = p\frac{dy}{dx}$$

$$\Rightarrow \quad \frac{dy}{dx} = \frac{2x}{p}.$$

Since curves cut each other at right angles

$$\therefore \quad \left(\frac{2x}{p}\right)\left(-\frac{2x}{9p}\right) = -1 \quad [\because m_1m_2 = -1]$$

$$\Rightarrow \quad \frac{-4x^2}{9p^2} = -1$$

$$\Rightarrow \quad \frac{4}{9p^2}(9p) = 1 \quad (\because x^2 = 9p)$$

$$\therefore \quad p = 4.$$ **Ans.**

** Answer is not given due to change in the present syllabus

**22. Using integration, prove that the curves $y^2 = 4x$ and $x^2 = 4y$ divide the area of the square bounded by $x = 0, x = 4, y = 0$ and $y = 4$ into three equal parts. [6]**

**Sol.** Given,

$$y^2 = 4x \quad ...(i)$$

and $$x^2 = 4y \quad ...(ii)$$

Solving (i) and (ii), we get the point of intersection (0, 0) and (4, 4).

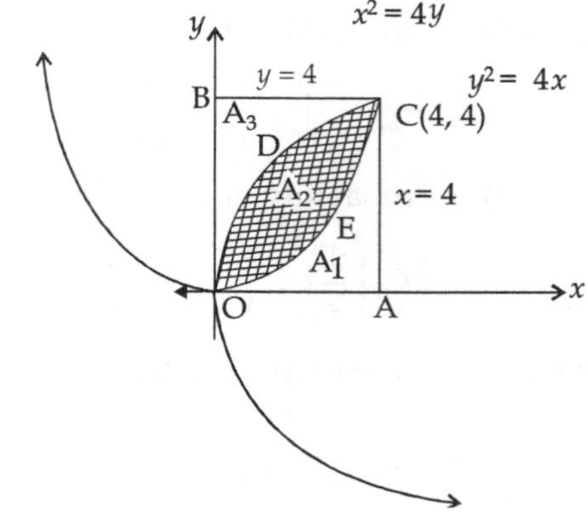


The area of the region OECDO bounded by the given curve $A_2$ = Area under $(y^2 = 4x)$ – Area under $(x^2 = 4y)$

$$= \int_0^4 \left(2\sqrt{x} - \frac{x^2}{4}\right)dx$$

$$= \left[\frac{4}{3}x^{3/2} - \frac{x^3}{12}\right]_0^4$$

$$= \left(\frac{32}{3} - \frac{16}{3}\right) = \frac{16}{3} \text{ sq. units} \quad ...(iii)$$

and the area of the region OACEO

$A_1$ = Area under $x^2 = 4y$

$$= \int_0^4 \left(\frac{x^2}{4}\right)dx = \left[\frac{x^3}{12}\right]_0^4$$

$$= \frac{16}{3} \text{ sq. units} \quad ...(iv)$$

Similarly, the area of the region ODCBO

$$A_3 = \text{Area of square} - (A_1 + A_2)$$

$$= 16 - \left(\frac{16}{3} + \frac{16}{3}\right)$$

$$= 16 - \frac{32}{3}$$

$$= \frac{16}{3} \text{ sq. units} \quad ...(v)$$

From (iii), (iv) and (v), it can be concluded that the given curves divide the area of the square bounded by $x = 0$, $x = 4$, $y = 0$ into three equal parts. **Hence Proved.**

**23. Show that the differential equation $\frac{dy}{dx} = \frac{y^2}{xy - x^2}$ is homogeneous and also solve it. [6]**

**Solution :** We have

$$\frac{dy}{dx} = \frac{y^2}{xy - x^2} \quad ...(i)$$

Let $$f(x, y) = \frac{y^2}{xy - x^2}$$

$$\therefore \quad f(\lambda x, \lambda y) = \frac{\lambda^2 y^2}{\lambda^2 xy - \lambda^2 x^2} = \frac{y^2}{xy - x^2}$$

$$= \lambda^0 f(x, y).$$

Hence, the differential equation is homogeneous.

**Hence Proved.**

Putting $y = vx$ so that $\frac{dy}{dx} = v + x\frac{dv}{dx}$ in (i), we get

$$v + \frac{xdv}{dx} = \frac{v^2 x^2}{vx^2 - x^2}$$

$$\Rightarrow \quad x\frac{dv}{dx} = \frac{v}{v-1}$$

$$\Rightarrow \quad \frac{v-1}{v} dv = \frac{dx}{x}$$

Integrating both sides, we get

$$\Rightarrow \quad \int \frac{v-1}{v} dv = \int \frac{dx}{x}$$

$$\Rightarrow \quad \int \left(1 - \frac{1}{v}\right) dv = \int \frac{dx}{x}$$

$$v - \log|v| = \log|x| + C$$

$$\Rightarrow \quad \frac{y}{x} - \log|y| + \log|x| = \log|x| + C$$

Hence, $x = \frac{y}{\log|y| + C}$ is the required solution of the differential equation. **Ans.**

**OR**

**Find the particular solution of the differential equation $(\tan^{-1}y - x)\, dy = (1 + y^2)\, dx$, given that $x = 1$, when $y = 0$.**

**Sol.** Given,

$$(\tan^{-1} y - x)dy = (1 + y^2)dx$$

$$\Rightarrow \quad \frac{\tan^{-1} y - x}{1 + y^2} dy = dx \quad ...(i)$$

Putting $\tan^{-1} y = t$ and $\frac{1}{1+y^2} = \frac{dt}{dy}$ in (i), we get

$$(t - x)\, dt = dx$$

$$\Rightarrow \quad \frac{dx}{dt} = t - x$$

$$\Rightarrow \quad \frac{dx}{dt} + x = t \quad ...(ii)$$

Here, $$\text{I.F.} = e^{\int 1.dt} = e^t$$

Hence, the solution of the differential equation (ii) is given by

$$x(\text{I.F.}) = \int (\text{I.F.})\, t\, dt$$

$$xe^t = \int e^t t\, dt + C,$$

where C is arbitrary constant

$$\Rightarrow \quad xe^t = te^t - \int e^t dt + C$$

$$\Rightarrow \quad xe^t = te^t - e^t + C$$

$$\Rightarrow xe^{\tan^{-1} y} = e^{\tan^{-1} y}(\tan^{-1} y - 1) + C \quad ...(iii)$$

$[\because t = \tan^{-1} y]$

It is given that $x = 1$ when $y = 0$

So, $$e^0 = e^0(0 - 1) + C$$

$$\Rightarrow \quad C = 2$$

Putting C = 2 in (iii) we get

$$xe^{\tan^{-1} y} = e^{\tan^{-1} y}(\tan^{-1} y - 1) + 2$$

$$\therefore \quad x = \tan^{-1} y + 2e^{\tan y} - 1$$

is the particular solution the given differential equation. **Ans.**

**24. Find the distance of the point P(3, 4, 4) from the point, where the line joining the points A(3, – 4, – 5) and B(2, –3, 1) intersects the plane $2x + y + z = 7$. [6]**

**Sol.** Equation of the line joining the points A(3, – 4, – 5) and B(2, – 3, 1) is

$$\Rightarrow \quad \frac{x-3}{2-3} = \frac{y+4}{-3+4} = \frac{z+5}{1+5}$$

$$\Rightarrow \quad \frac{x-3}{-1} = \frac{y+4}{1} = \frac{z+5}{6} = k.$$

Coordinates of any point on the line are

$$Q\,(3 - k, k - 4, 6k - 5)$$

The given line intersects the plane $2x + y + z = 7$.

So, $2(3 - k) + k - 4 + 6k - 5 = 7$

$\Rightarrow 6 - 2k + k - 4 + 6k - 5 = 7$

$\Rightarrow \quad 5k = 10$

$\Rightarrow \quad k = 2$

$\therefore$ Coordinates of a point where line intersect plane are $(3-k, k-4, 6k-5) = (3-2, 2-4, 12-5) = (1, -2, 7)$

Now, the distance between the point P(3, 4, 4) and A(1, – 2, 7) is given by

$$PQ = \sqrt{(3-1)^2+(4+2)^2+(4-7)^2}$$

$$\Rightarrow \quad PQ = \sqrt{(2)^2+(6)^2+(-3)^2}$$

$$\Rightarrow \quad PQ = \sqrt{4+36+9} = \sqrt{49}$$

$\therefore \quad PQ = 7$ units. **Ans.**

**25. A company manufactures three kinds of calculators : A, B and C in its two factories I and II. The company has got an order for manufacturing at least 6400 calculators of kind A, 4000 of kind B and 4800 of kind C. The daily output of factory I is of 50 calculators of kind A, 50 calculators of kind B, and 30 calculators of kind C. The daily output of factory II is of 40 calculators of kind A, 20 of kind B and 40 of kind C. The cost per day to run factory I is ₹ 12,000 and of factory II is ₹ 15,000. How many days do the two factories have to be in operation to produce the order with the minimum cost? Formulate this problem as an LPP and solve it graphically. [6]**

**Sol.** Let the factories I and II work for $x$ and $y$ number of days respectively.

Thus, the given linear programming problem is Minimize $Z = ₹\ (12000x + 15000y)$

Subject to the constraints

$$50x + 40y \geq 6400$$
$$50x + 20y \geq 4000$$
$$30x + 40y \geq 4800$$
$$x \geq 0$$

and $\quad y \geq 0$

*i.e.* $\quad 5x + 4y \geq 640$

$$5x + 2y \geq 400$$
$$3x + 4y \geq 480$$
$$x \geq 0, y \geq 0.$$

To solve this L. P. P,

Let us consider the equations

$L_1 : 5x + 4y = 640 \quad$ ...(i)

$L_2 : 5x + 2y = 400 \quad$ ...(ii)

$L_3 : 3x + 4y = 480 \quad$ ...(iii)

The point of intersection of $L_1$ and $L_2$ is D(32, 120) and that of $L_1$ and $L_3$ is C (80,60)

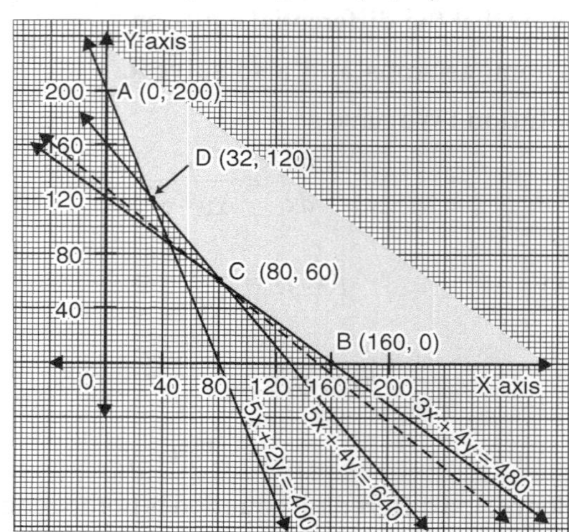


The shaded region is the solution region of the given L. P. P.

| Corner Points | Values of the objective function $Z = 12000x + 15000y$ |
|---|---|
| A (0, 200) | $12000 \times 0 + 15000 \times 200 = 30,00,000$ |
| B (160,0) | $12000 \times 160 + 15000 \times 0 = 19,20,000$ |
| C (80, 60) | $12000 \times 80 + 15000 \times 60 = 18,60,000$ |
| D (32,120) | $12000 \times 32 + 15000 \times 120 = 21,84,000$ |

Out of these values of Z, the minimum value of Z is 18,60,000 at $x = 80$ and $y = 60$.

Since the feasible region is unbounded so we draw the graph of inequality

$$12000\,x + 15000\,y < 1860000$$

*i.e.*, $4x + 5y < 620$

$L : 4x + 5y = 620$

| $x$ | 0 | 155 |
|---|---|---|
| $y$ | 124 | 0 |

We observe that open half one represented by L have no point common with feasible region.

$$Z = 12000 \times 80 + 15000 \times 60$$
$$= ₹\ 18,60,000.$$

Hence, the factories I and II work for 80 and 60 number of days respectively. **Ans.**

**26. In a factory which manufactures bolts, machines A, B and C manufacture respectively 30%, 50% and 20% of the bolts. Of their outputs 3, 4 and 1 percent respectively are defective bolts. A bolt is drawn at random from the product and is found to be defective. Find the probability that this is not manufactured by machine B. [6]**

**Sol.** Let $E_1$, $E_2$, $E_3$ and A be the events defined as below

$E_1$ = the bolt is manufactured by machine A.

$E_2$ = the bolt is manufactured by machine B.

$E_3$ = the bolt is manufactured by machine C.

A = the bolt is defective.

then, $P(E_1)$ = Probability that the bolt drawn is manufactured by machine A = $\frac{30}{100}$

$P(E_2)$ = Probability that the bolt drawn is manufactured by machine B = $\frac{50}{100}$

$P(E_3)$ = Probability that the bolt drawn is manufactured by machine C = $\frac{20}{100}$

$P(A/E_1)$ = Probability that the bolt drawn is defective given that it is manufactured by machine A.

$$P(A/E_1) = \frac{3}{100}$$

Similarly, we have, $P(A/E_2) = \frac{4}{100}$

and $P(A/E_3) = \frac{1}{100}$

Now, using Bayes' theorem

$P(E_2/A)$ = Probability that the bolt is manufactured by machine B given that the bolt drawn is defective.

$$= \frac{P(E_2)\,P(A/E_2)}{P(E_1)P(A/E_1)+P(E_2)P(A/E_2)+P(E_3)P(A/E_3)}$$

$$= \frac{\frac{50}{100}\times\frac{4}{100}}{\frac{30}{100}\times\frac{3}{100}+\frac{50}{100}\times\frac{4}{100}+\frac{20}{100}\times\frac{1}{100}}$$

$$= \frac{200}{90+200+20} = \frac{200}{310} = \frac{20}{31}.$$

Hence, the probability that this is not manufactured by Machine B = $1 - P(E_2/A)$

$$= 1-\frac{20}{31} = \frac{11}{31}$$ **Ans.**

All questions are same in Outside Delhi Set II and Set III

# MATHEMATICS

## Delhi [Set-I] 2015

**Time allowed : 3 hours** **Maximum marks : 100**

### SECTION–A

**1. If $\vec{a} = 7\hat{i} + \hat{j} - 4\hat{k}$ and $\vec{b} = 2\hat{i} + 6\hat{j} + 3\hat{k}$, then find the projection of $\vec{a}$ on $\vec{b}$.** **[1]**

**Sol.** Given,

$$\vec{a} = 7\hat{i}+\hat{j}-4\hat{k}$$

$$\vec{b} = 2\hat{i}+6\hat{j}+3\hat{k}$$

$\therefore$ The projection of $\vec{a}$ on $\vec{b}$ $= \dfrac{\vec{a}.\vec{b}}{|\vec{b}|}$

$$= \frac{(7\times 2)+(1\times 6)+(-4\times 3)}{\sqrt{(2)^2+(6)^2+(3)^2}} = \frac{8}{7}.$$

**Ans.**

**2. Find $\lambda$, if the vectors $\vec{a} = \hat{i} + 3\hat{j} + \hat{k}$, $\vec{b} = 2\hat{i} - \hat{j} - \hat{k}$ and $\vec{c} = \lambda\hat{j} + 3\hat{k}$ are coplanar.** **[1]**

**Sol.** Since, the given vectors are coplanar.

$$[\vec{a}\ \vec{b}\ \vec{c}] = 0$$

Here, $[\vec{a}\ \vec{b}\ \vec{c}] = \begin{vmatrix} 1 & 3 & 1 \\ 2 & -1 & -1 \\ 0 & \lambda & 3 \end{vmatrix} = 0$

Expanding along $R_3$, we get

$0.(-3 + 1) - \lambda(-1 - 2) + 3(-1 - 6) = 0$

$\Rightarrow \quad 3\lambda = 21$

$\therefore \quad \lambda = 7$ **Ans.**

**3. If a line makes angles 90°, 60° and $\theta$ with $x$, $y$ and $z$–axis respectively, where $\theta$ is acute, then find $\theta$.** **[1]**

**Sol.** Given,

$$\alpha = 90°$$
$$\beta = 60°$$
$$\lambda = \theta$$

Let $l$, $m$, $n$ be the direction cosines of the given vector.

Then, $l = \cos\alpha$,

$m = \cos\beta$

and $n = \cos\lambda$

Now, $l^2 + m^2 + n^2 = 1$

$\cos^2\alpha + \cos^2\beta + \cos^2\lambda = 1$

$\Rightarrow \cos^2(90°) + \cos^2(60°) + \cos^2\theta = 1$

$\Rightarrow \quad 0^2 + \left(\dfrac{1}{2}\right)^2 + \cos^2\theta = 1$

$\Rightarrow \quad \cos^2\theta = 1 - \dfrac{1}{4} = \dfrac{3}{4}$

$\Rightarrow \quad \cos\theta = \dfrac{\sqrt{3}}{2}$ ($\because$ $\theta$ is acute)

$\therefore \quad \theta = 30°$ **Ans.**

**4. Write the element $a_{23}$ of $a$ 3 × 3 matrix A = $(a_{ij})$ whose elements $a_{ij}$ are given be $a_{ij} = \dfrac{|i-j|}{2}$.** **[1]**

**Sol.** Given,

$$a_{ij} = \frac{|i-j|}{2}$$

$\therefore \quad a_{23} = \dfrac{|2-3|}{2} = \dfrac{|-1|}{2} = \dfrac{1}{2}$ **Ans.**

**5. Find the differential equation representing the family of curves $v = \dfrac{A}{r} + B$, where A and B are arbitrary constants.** **[1]**

**Sol.** We have,

$$v = \frac{A}{r} + B \quad \text{...(i)}$$

Since, the given equation contains two arbitrary constants, we shall differentiate it two times. Now, differentiating (i) w.r.t. $r$, we get

$$\frac{dv}{dr} = -\frac{A}{r^2} + 0$$

$\Rightarrow \quad r^2\dfrac{dv}{dr} = -A$ ...(ii)

Again, differentiating (ii) w.r.t. $r$, we get

$$r^2 \times \frac{d^2v}{dr^2} + 2r \times \frac{dv}{dr} = 0$$

$\Rightarrow \quad r\dfrac{d^2v}{dr^2} + 2\dfrac{dv}{dr} = 0$

This is the required differential equation representing the family of the given curve. **Ans.**

**6. Find the integrating factor of the differential equation $\left(\dfrac{e^{-2\sqrt{x}}}{\sqrt{x}} - \dfrac{y}{\sqrt{x}}\right)\dfrac{dx}{dy} = 1.$** **[1]**

**Sol.** We have,

$$\left(\frac{e^{-2\sqrt{x}}}{\sqrt{x}} - \frac{y}{\sqrt{x}}\right)\frac{dx}{dy} = 1$$

$$\Rightarrow \quad \frac{dy}{dx} = \frac{e^{-2\sqrt{x}}}{\sqrt{x}} - \frac{y}{\sqrt{x}}$$

$$\Rightarrow \quad \frac{dy}{dx} + \frac{y}{\sqrt{x}} = \frac{e^{-2\sqrt{x}}}{\sqrt{x}},$$

which is a linear differential equation of the form

$$\frac{dy}{dx} + Py = Q,$$

where $P = \frac{1}{\sqrt{x}}$

and $Q = \frac{e^{-2\sqrt{x}}}{\sqrt{x}}$

$$\therefore \quad \text{I. F.} = e^{\int P dx}$$

$$= e^{\int \frac{1}{\sqrt{x}} dx} = e^{2\sqrt{x}}$$ **Ans.**

## SECTION–B

**7. If $A = \begin{bmatrix} 2 & 0 & 1 \\ 2 & 1 & 3 \\ 1 & -1 & 0 \end{bmatrix}$, find $A^2 - 5A + 4I$ and hence find a matrix X such that $A^2 - 5A + 4I + X = 0$.[4]**

**Sol.** We have,

$$A = \begin{bmatrix} 2 & 0 & 1 \\ 2 & 1 & 3 \\ 1 & -1 & 0 \end{bmatrix}$$

$$\Rightarrow \quad A^2 = A.A = \begin{bmatrix} 2 & 0 & 1 \\ 2 & 1 & 3 \\ 1 & -1 & 0 \end{bmatrix}\begin{bmatrix} 2 & 0 & 1 \\ 2 & 1 & 3 \\ 1 & -1 & 0 \end{bmatrix}$$

$$= \begin{bmatrix} 5 & -1 & 2 \\ 9 & -2 & 5 \\ 0 & -1 & -2 \end{bmatrix}$$

$$\Rightarrow \quad -5A = \begin{bmatrix} (-5).2 & (-5).0 & (-5).1 \\ (-5).2 & (-5).1 & (-5).3 \\ (-5).1 & (-5)(-1) & (-5).0 \end{bmatrix}$$

$$= \begin{bmatrix} -10 & 0 & -5 \\ -10 & -5 & -15 \\ -5 & 5 & 0 \end{bmatrix}$$

$$\Rightarrow \quad 4I_3 = 4\begin{bmatrix} 1 & 0 & 0 \\ 0 & 1 & 0 \\ 0 & 0 & 1 \end{bmatrix} = \begin{bmatrix} 4 & 0 & 0 \\ 0 & 4 & 0 \\ 0 & 0 & 4 \end{bmatrix}$$

$$\therefore \quad A^2 - 5A + 4I$$

$$= \begin{bmatrix} 5 & -1 & 2 \\ 9 & -2 & 5 \\ 0 & -1 & -2 \end{bmatrix} + \begin{bmatrix} -10 & 0 & -5 \\ -10 & -5 & -15 \\ -5 & 5 & 0 \end{bmatrix}$$

$$+ \begin{bmatrix} 4 & 0 & 0 \\ 0 & 4 & 0 \\ 0 & 0 & 4 \end{bmatrix} = \begin{bmatrix} -1 & -1 & -3 \\ -1 & -3 & -10 \\ -5 & 4 & 2 \end{bmatrix}$$

Now, $A^2 - 5A + 4I + X = 0$

$$\Rightarrow \quad X = -(A^2 - 5A + 4I)$$

$$\Rightarrow \quad X = (-1)\begin{bmatrix} -1 & -1 & -3 \\ -1 & -3 & -10 \\ -5 & 4 & 2 \end{bmatrix} = \begin{bmatrix} 1 & 1 & 3 \\ 1 & 3 & 10 \\ 5 & -4 & -2 \end{bmatrix}$$

**Ans.**

**OR**

**If $A = \begin{bmatrix} 1 & -2 & 3 \\ 0 & -1 & 4 \\ -2 & 2 & 1 \end{bmatrix}$, find $(A')^{-1}$.**

**Sol.** Given,

$$A = \begin{bmatrix} 1 & -2 & 3 \\ 0 & -1 & 4 \\ -2 & 2 & 1 \end{bmatrix}$$

Now, $$A' = \begin{bmatrix} 1 & 0 & -2 \\ -2 & -1 & 2 \\ 3 & 4 & 1 \end{bmatrix}$$

$$|A'| = \begin{vmatrix} 1 & 0 & -2 \\ -2 & -1 & 2 \\ 3 & 4 & 1 \end{vmatrix}$$

$$= 1(-1-8) - 0(-2-6) - 2(-8+3)$$

$$= -9 + 10 = 1 \neq 0$$

So, A′ is invertible.

$$A'_{11} = -9,\ A'_{12} = 8,\ A'_{13} = -5$$

$$A'_{21} = -8,\ A'_{22} = 7,\ A'_{23} = -4$$

$$A'_{31} = -2,\ A'_{32} = 2,\ A'_{33} = -1$$

$$\text{adj } A' = \begin{bmatrix} A'_{11} & A'_{21} & A'_{31} \\ A'_{12} & A'_{22} & A'_{32} \\ A'_{13} & A'_{23} & A'_{33} \end{bmatrix}$$

$$= \begin{bmatrix} -9 & -8 & -2 \\ 8 & 7 & 2 \\ -5 & -4 & -1 \end{bmatrix}$$

$$\therefore \quad (A')^{-1} = \frac{1}{|A'|}.\text{adj } A'$$

$= \text{adj } A'$ $\quad [\because |A'| = 1]$

$$\Rightarrow \quad (A')^{-1} = \begin{bmatrix} -9 & -8 & -2 \\ 8 & 7 & 2 \\ -5 & -4 & -1 \end{bmatrix}$$ **Ans.**

**8. If $f(x) = \begin{vmatrix} a & -1 & 0 \\ ax & a & -1 \\ ax^2 & ax & a \end{vmatrix}$, using properties of determinants find the value of $f(2x) - f(x)$.** **[4]**

**Sol.** Given,
$$f(x) = \begin{vmatrix} a & -1 & 0 \\ ax & a & -1 \\ ax^2 & ax & a \end{vmatrix}$$

Taking $a$ common from $C_1$

$$\Rightarrow \quad f(x) = a\begin{vmatrix} 1 & -1 & 0 \\ x & a & -1 \\ x^2 & ax & a \end{vmatrix}$$

Applying $C_2 \rightarrow C_2 + C_1$, we get

$$f(x) = a\begin{vmatrix} 1 & 0 & 0 \\ x & x+a & -1 \\ x^2 & x^2+ax & a \end{vmatrix}$$

On expanding along $R_1$,

$$f(x) = a(a^2 + ax + ax + x^2)$$
$$\Rightarrow \quad f(x) = a(a^2 + 2ax + x^2)$$

Also,
$$f(2x) = \begin{vmatrix} a & -1 & 0 \\ 2ax & a & -1 \\ 4ax^2 & 2ax & a \end{vmatrix}$$

$$\Rightarrow \quad f(2x) = a\begin{vmatrix} 1 & -1 & 0 \\ 2x & a & -1 \\ 4x^2 & 2ax & a \end{vmatrix}$$

Applying $C_2 \rightarrow C_2 + C_1$, we get

$$f(2x) = a\begin{vmatrix} 1 & 0 & 0 \\ 2x & 2x+a & -1 \\ 4x^2 & 4x^2+2ax & a \end{vmatrix}$$

On Expanding along $R_1$

$$\Rightarrow \quad f(2x) = a\{a(2x + a) + 4x^2 + 2ax\}$$
$$\Rightarrow \quad f(2x) = a\{4x^2 + a^2 + 4ax\}$$
$$\therefore \quad f(2x) - f(x) = a(4x^2 + a^2 + 4ax - a^2 - 2ax - x^2)$$
$$\Rightarrow \quad = a(3x^2 + 2ax)$$
$$\Rightarrow \quad = ax\,(3x + 2a).$$ **Ans.**

**9. Find : $\int \frac{dx}{\sin x + \sin 2x}$.** **[4]**

**Sol.** Let
$$I = \int \frac{1}{\sin x + \sin 2x}\,dx$$
$$\Rightarrow \quad I = \int \frac{1}{\sin x + 2\sin x \cos x}\,dx$$
$$\Rightarrow \quad I = \int \frac{1}{\sin x(1 + 2\cos x)}\,dx$$
$$\Rightarrow \quad I = \int \frac{\sin x}{\sin^2 x(1 + 2\cos x)}\,dx$$
$$= \int \frac{\sin x\,dx}{(1 - \cos^2 x)\,(1 + 2\cos x)}$$

Putting $\cos x = t, -\sin x\,dx = dt$

$$\Rightarrow \quad \sin x\,dx = -dt$$
$$I = \int \frac{-dt}{(1 - t^2)(1 + 2t)}$$
$$= \int \frac{-1}{(1 - t)(1 + t)(1 + 2t)}\,dt$$

Let
$$\frac{-1}{(1+t)(1-t)(1+2t)} = \frac{A}{(1+t)} + \frac{B}{(1-t)} + \frac{C}{(1+2t)} \quad ...(i)$$

$$\therefore \quad -1 = A\,(1 - t)\,(1 + 2t) + B(1 + t)\,(1 + 2t) + C\,(1 + t)\,(1 - t)$$

Putting $t = 1$ in (i), we get

$$-1 = 6B$$
$$\Rightarrow \quad B = \frac{-1}{6}$$

Putting $t = -1$ in (i), we get

$$-1 = -2A$$
$$\Rightarrow \quad A = \frac{1}{2}$$

Putting $t = -\frac{1}{2}$ in (i), we get

$$-1 = \frac{3}{4}C$$
$$\Rightarrow \quad C = -\frac{4}{3}$$

$$\frac{-1}{(1+t)(1-t)(1+2t)} = \frac{1}{2(1+t)} - \frac{1}{6(1-t)} - \frac{4}{3(1+2t)}$$

$$\Rightarrow \quad I = \int \frac{-dt}{(1+t)(1-t)(1+2t)}$$
$$\Rightarrow \quad I = \frac{1}{2}\int \frac{1}{1+t}\,dt - \frac{1}{6}\int \frac{1}{1-t}\,dt - \frac{4}{3}\int \frac{1}{1+2t}\,dt$$
$$\Rightarrow \quad I = \frac{1}{2}\log|1+t| + \frac{1}{6}\log|1-t|$$

$$-\frac{4}{3\times2}\log|1+2t|+C$$

$$\Rightarrow \quad I = \frac{1}{2}\log|1+\cos x| + \frac{1}{6}\log|1-\cos x|$$

$$-\frac{2}{3}\log|1+2\cos x|+C$$

**Ans.**

**OR**

**Integrate the following w.r.t. $x$ : $\frac{x^2-3x+1}{\sqrt{1-x^2}}$.**

**Sol.** Given,

$$\frac{x^2-3x+1-1+1}{\sqrt{1-x^2}} = -\left[\frac{-x^2+3x-1+1-1}{\sqrt{1-x^2}}\right]$$

$$= -\left[\frac{1-x^2+3x-2}{\sqrt{1-x^2}}\right]$$

$$= -\frac{1-x^2}{\sqrt{1-x^2}} - \frac{3x-2}{\sqrt{1-x^2}}$$

$$= -\sqrt{1-x^2} - \frac{3x-2}{\sqrt{1-x^2}}$$

$$\Rightarrow \int\frac{x^2-3x+1}{\sqrt{1-x^2}}dx = \int\left(-\sqrt{1-x^2}-\frac{3x-2}{\sqrt{1-x^2}}\right)dx$$

$$= -\int\sqrt{1-x^2}\,dx - \int\frac{3x-2}{\sqrt{1-x^2}}dx$$

$$= -\int\sqrt{1-x^2}\,dx - 3\int\frac{x}{\sqrt{1-x^2}}dx + 2\int\frac{1}{\sqrt{1-x^2}}dx$$

$$= -\int\sqrt{1-x^2}\,dx + \frac{3}{2}\int\frac{1}{\sqrt{t}}dt + 2\int\frac{1}{\sqrt{1-x^2}}dx$$

(Putting $1-x^2 = t \Rightarrow -2x\,dx = dt$)

$$= \frac{-x}{2}\sqrt{1-x^2} - \frac{1}{2}\sin^{-1}x + 3\sqrt{1-x^2} + 2\sin^{-1}x + C$$

$$= \frac{-x}{2}\sqrt{1-x^2} + \frac{3}{2}\sin^{-1}x + 3\sqrt{1-x^2} + C$$

**Ans.**

**10. Evaluate : $\int_{-\pi}^{\pi}(\cos ax - \sin bx)^2\,dx$.** **[4]**

**Sol.** Let

$$I = \int_{-\pi}^{\pi}(\cos ax - \sin bx)^2\,dx$$

$$I = \int_{-\pi}^{\pi}(\cos^2 ax + \sin^2 bx - 2\cos ax \sin bx)dx$$

$$I = \int_{-\pi}^{\pi}\cos^2 ax\,dx + \int_{-\pi}^{\pi}\sin^2 bx\,dx - 2\int_{-\pi}^{\pi}\cos ax \sin bx\,dx$$

$$\Rightarrow \quad I = 2\int_0^{\pi}\cos^2 ax\,dx + 2\int_0^{\pi}\sin^2 bx\,dx - 0$$

[Since $\cos^2 ax$ and $\sin^2 bx$ are even functions and $\cos ax \sin bx$ is an odd function]

$$\Rightarrow \quad I = 2\int_0^{\pi}\left(\frac{1+\cos 2ax}{2}\right)dx + 2\int_0^{\pi}\left(\frac{1-\cos 2bx}{2}\right)dx$$

$$\Rightarrow \quad I = \int_0^{\pi}(1+\cos 2ax)dx + \int_0^{\pi}(1-\cos 2bx)\,dx$$

$$\Rightarrow \quad I = \int_0^{\pi}(1+\cos 2ax + 1 - \cos 2bx)dx$$

$$\Rightarrow \quad I = \int_0^{\pi}(2+\cos 2ax - \cos 2bx)dx$$

$$\Rightarrow \quad I = 2[x]_0^{\pi} + \left[\frac{\sin 2ax}{2a}\right]_0^{\pi} - \left[\frac{\sin 2bx}{2b}\right]_0^{\pi}$$

$$\Rightarrow \quad I = 2\pi + \frac{\sin 2a\pi}{2a} - \frac{\sin 2b\pi}{2b}$$

$$\Rightarrow \quad I = 2\pi \text{ if } a, b \in Z$$

**Ans.**

**11. A bag 'A' contains 4 black and 6 red balls and bag 'B' contains 7 black and 3 red balls. A die is thrown. If 1 or 2 appears on it, then bag A is chosen, otherwise bag B. If two balls are drawn at random (without replacement) from the selected bag, find the probability of one of them being red and another black.** **[4]**

**Sol.** Consider the following events :

$E_1$ = Getting 1 or 2 on die.

$E_2$ = Getting 3, 4, 5 or 6 on die.

E = One of the ball drawn is red and another is black.

$$P(E_1) = \frac{2}{6} = \frac{1}{3}$$

and $$P(E_2) = \frac{4}{6} = \frac{2}{3}$$

$P(E/E_1)$ = Probability of drawing a red and a black ball when bag A has been chosen.

$$= P(RB) + P(BR)$$

$$P(E/E_1) = \frac{6}{10}\times\frac{4}{9} + \frac{4}{10}\times\frac{6}{9} = \frac{48}{90} = \frac{8}{15}$$

$P(E/E_2)$ = Probability of drawing a red and a black ball when bag B has been chosen.

$$= \frac{3}{10}\times\frac{7}{9} + \frac{7}{10}\times\frac{3}{9} = \frac{42}{90} = \frac{7}{15}$$

Using the law of total probability, we have

$$P(E) = P(E_1)\,P(E/E_1) + P(E_2)\,P(E/E_2)$$

$$= \frac{1}{3}\times\frac{8}{15} + \frac{2}{3}\times\frac{7}{15} = \frac{22}{45}$$

**Ans.**

OR

**An unbiased coin is tossed 4 times. Find the mean and variance of the number of heads obtained.**

**Sol.** Let X denote the number of heads in the four tosses of the coin, then X is a random variable that can have values 0, 1, 2, 3, 4.

P(X = 0) = Probability of getting no head (TTTT)

$= \frac{1}{16}$

P(X = 1) = Probability of getting one head (HTTT, THTT, TTHT, TTTH)

$= 4\times\frac{1}{16} = \frac{1}{4}$

P(X = 2) = Probability of getting two head (HHTT, HTHT, HTTH, THHT, THTH, TTHH)

$= 6\times\frac{1}{16} = \frac{3}{8}$

P(X = 3) = Probability of getting three head (HHHT, HHTH, HTHH, THHH)

$= 4\times\frac{1}{16} = \frac{1}{4}$

P(X = 4) = Probability of getting four head (HHHH)

$= \frac{1}{16}$

Thus, the probability distribution of random variable X is given by

| X | 0 | 1 | 2 | 3 | 4 |
|---|---|---|---|---|---|
| P(X) | $\frac{1}{16}$ | $\frac{1}{4}$ | $\frac{3}{8}$ | $\frac{1}{4}$ | $\frac{1}{16}$ |

| $x_i$ | $P_i = P(X = x_i)$ | $P_i x_i$ | $P_i x_i^2$ |
|---|---|---|---|
| 0 | $\frac{1}{16}$ | 0 | 0 |
| 1 | $\frac{1}{4}$ | $\frac{1}{4}$ | $\frac{1}{4}$ |
| 2 | $\frac{3}{8}$ | $\frac{3}{4}$ | $\frac{3}{2}$ |
| 3 | $\frac{1}{4}$ | $\frac{3}{4}$ | $\frac{9}{4}$ |
| 4 | $\frac{1}{16}$ | $\frac{1}{4}$ | 1 |
| | | $\Sigma P_i x_i = 2$ | $\Sigma P_i x_i^2 = 5$ |

$\therefore$ Mean $= \overline{X} = \Sigma P_i x_i = 2$

and Var $(x) = \Sigma P_i x_i^2 - (\Sigma P_i x_i)^2 = 5 - 4 = 1$

Hence, Mean = 2 and Variance = 1. **Ans.**

**12. If $\vec{r} = x\hat{i} + y\hat{j} + z\hat{k}$, find $(\vec{r}\times\hat{i}).(\vec{r}\times\hat{j}) + xy$.** **[4]**

**Sol.** Given,

$$\vec{r} = x\hat{i} + y\hat{j} + z\hat{k},$$

Now, $(\vec{r}\times\hat{i}).(\vec{r}\times\hat{j}) + xy$

$= [(x\hat{i} + y\hat{j} + z\hat{k})\times\hat{i}].[(x\hat{i} + y\hat{j} + z\hat{k})\times\hat{j}] + xy$ $[\because \vec{r} = x\hat{i} + y\hat{j} + z\hat{k}]$

$= [(x\hat{i}\times\hat{i} + y\hat{j}\times\hat{i} + z\hat{k}\times\hat{i})].[(x\hat{i}\times\hat{j} + y\hat{j}\times\hat{j} + z\hat{k}\times\hat{j})] + xy$

$\left[\begin{array}{l} \because \hat{i}\times\hat{j} = \hat{k},\ \hat{j}\times\hat{k} = \hat{i}, \\ \hat{k}\times\hat{i} = \hat{j},\ \hat{i}\times\hat{i} = 0 \\ \hat{j}\times\hat{j} = 0,\ \hat{k}\times\hat{k} = 0 \end{array}\right]$

$= (0 - y\hat{k} + z\hat{j}).(x\hat{k} + 0 - z\hat{i}) + xy$

$= (0 - xy + 0) + xy$

$= -xy + xy = 0.$ **Ans.**

**13. Find the distance between the point (– 1, – 5, – 10) and the point of intersection of the line $\frac{x-2}{3} = \frac{y+1}{4} = \frac{z-2}{12}$ and the plane $x - y + z = 5$.** **[4]**

**Sol.** Let $\frac{x-2}{3} = \frac{y+1}{4} = \frac{z-2}{12} = k$

$\Rightarrow$ $x = 3k + 2,$

$y = 4k - 1,$

$z = 12k + 2$

Coordinates of any point on the line are $(3k + 2, 4k - 1, 12k + 2)$.

The point of intersection of the line $\frac{x-2}{3} = \frac{y+1}{4} = \frac{z-2}{12}$ and the plane $x - y + z = 5$ will also be in the form $(3k + 2, 4k - 1, 12k + 2)$ and it will satisfy the equation of plane.

Now, putting $x = 3k + 2$, $y = 4k - 1$

and $z = 12k + 2$ in $x - y + z = 5$,

we get

$3k + 2 - (4k - 1) + 12k + 2 = 5$

$$\Rightarrow \quad 11k + 5 = 5$$
$$\Rightarrow \quad 11k = 0$$
$$\Rightarrow \quad k = 0$$
$$\therefore \quad x = 2,$$
$$y = -1,$$
$$z = 2$$

Hence, the point of intersection of $\frac{x-2}{3} = \frac{y+1}{4} = \frac{z-2}{12}$ and the plane $x - y + z = 5$ is $(2, -1, 2)$.

$\therefore$ Distance between the point $(-1, -5, -10)$ and $(2, -1, 2)$

Using distance formula

$$= \sqrt{(2+1)^2 + (-1+5)^2 + (2+10)^2}$$
$$= \sqrt{3^2 + 4^2 + 12^2}$$
$$= \sqrt{169} = 13 \text{ units}$$

Hence, the distance between the point $(-1, -5, -10)$ and the point of intersection of the line $\frac{x-2}{3} = \frac{y+1}{4} = \frac{z-2}{12}$ and the plane $x - y + z = 5$ is 13 units. **Ans.**

**14. If $\sin[\cot^{-1}(x+1)] = \cos(\tan^{-1} x)$, then find $x$.** **[4]**

**Sol.** Given, $\sin[\cot^{-1}(x+1)] = \cos(\tan^{-1} x)$

$$\sin\left\{\sin^{-1}\frac{1}{\sqrt{1+(x+1)^2}}\right\} = \cos\left\{\cos^{-1}\frac{1}{\sqrt{1+x^2}}\right\}$$

$$\left[\because \cot^{-1} x = \sin^{-1}\frac{1}{\sqrt{1+x^2}} \text{ and } \tan^{-1} x = \cos^{-1}\frac{1}{\sqrt{1+x^2}}\right]$$

$$\Rightarrow \quad \frac{1}{\sqrt{1+(x+1)^2}} = \frac{1}{\sqrt{1+x^2}}$$

$$\Rightarrow \quad \frac{1}{\sqrt{1+x^2+1+2x}} = \frac{1}{\sqrt{1+x^2}}$$

$$\Rightarrow \quad \frac{1}{\sqrt{x^2+2x+2}} = \frac{1}{\sqrt{1+x^2}}$$

$$\Rightarrow \quad \sqrt{1+x^2} = \sqrt{x^2+2x+2}$$

On squaring both sides, we get

$$1 + x^2 = x^2 + 2x + 2$$
$$\Rightarrow \quad 2x + 2 = 1$$
$$\Rightarrow \quad x = \frac{-1}{2}$$
**Ans.**

**OR**

**If $(\tan^{-1} x)^2 + (\cot^{-1} x)^2 = \frac{5\pi^2}{8}$, then find $x$.**

**Sol.** Given, $(\tan^{-1} x)^2 + (\cot^{-1} x)^2 = \frac{5\pi^2}{8}$

$$\Rightarrow \quad (\tan^{-1} x + \cot^{-1} x)^2 - 2\tan^{-1} x \cot^{-1} x = \frac{5\pi^2}{8}$$

$$\Rightarrow \quad \left(\frac{\pi}{2}\right)^2 - 2\tan^{-1} x\left(\frac{\pi}{2} - \tan^{-1} x\right) = \frac{5\pi^2}{8}$$

$$\Rightarrow \quad \frac{\pi^2}{4} - \pi\tan^{-1} x + 2(\tan^{-1} x)^2 = \frac{5\pi^2}{8}$$

$$\Rightarrow \quad 2(\tan^{-1} x)^2 - \pi\tan^{-1} x + \frac{\pi^2}{4} - \frac{5\pi^2}{8} = 0$$

$$\Rightarrow \quad 2(\tan^{-1} x)^2 - \pi\tan^{-1} x - \frac{3\pi^2}{8} = 0$$

Solving the quadratic equation, we get

$$\tan^{-1} x = \frac{\pi \pm \sqrt{\pi^2 + 4\times 2\times\frac{3\pi^2}{8}}}{2\times 2}$$

$$\Rightarrow \quad \tan^{-1} x = \frac{\pi \pm 2\pi}{4}$$

$$\Rightarrow \quad \tan^{-1} x = \frac{3\pi}{4} \text{ or } \tan^{-1} x = \frac{-\pi}{4}$$

$$\Rightarrow \quad x = \tan\frac{3\pi}{4} \text{ or } x = \tan\left(\frac{-\pi}{4}\right)$$

$$\Rightarrow \quad x = -1.$$
**Ans.**

**15. If $y = \tan^{-1}\left(\frac{\sqrt{1+x^2}+\sqrt{1-x^2}}{\sqrt{1+x^2}-\sqrt{1-x^2}}\right)$, $x^2 \leq 1$, then find $\frac{dy}{dx}$.** **[4]**

**Sol.** Given,

$$y = \tan^{-1}\left(\frac{\sqrt{1+x^2}+\sqrt{1-x^2}}{\sqrt{1+x^2}-\sqrt{1-x^2}}\right)$$

Putting $x^2 = \cos 2\theta$ we have

$$y = \tan^{-1}\left(\frac{\sqrt{1+\cos 2\theta}+\sqrt{1-\cos 2\theta}}{\sqrt{1+\cos 2\theta}-\sqrt{1-\cos 2\theta}}\right)$$

$$\Rightarrow \quad y = \tan^{-1}\left(\frac{\sqrt{2\cos^2\theta}+\sqrt{2\sin^2\theta}}{\sqrt{2\cos^2\theta}-\sqrt{2\sin^2\theta}}\right)$$

$$\Rightarrow \quad y = \tan^{-1}\left(\frac{\cos\theta+\sin\theta}{\cos\theta-\sin\theta}\right)$$

Dividing the numerator and denominator by $\cos\theta$, we get

$$y = \tan^{-1}\left(\frac{1+\tan\theta}{1-\tan\theta}\right)$$

$$\Rightarrow \quad y = \tan^{-1}\left(\frac{\tan\frac{\pi}{4}+\tan\theta}{1-\tan\frac{\pi}{4}.\tan\theta}\right)$$

$$\Rightarrow \quad y = \tan^{-1}\left[\tan\left(\frac{\pi}{4}+\theta\right)\right]$$

$$\Rightarrow \quad y = \frac{\pi}{4}+\theta$$

$$\therefore \quad y = \frac{\pi}{4}+\frac{1}{2}\cos^{-1}x^2 \quad (\because x^2 = \cos 2\theta) \quad \text{...(i)}$$

Now, differentiating (i), w.r.t. $x$, we get

$$\frac{dy}{dx} = 0+\frac{1}{2}\times\left(\frac{-1}{\sqrt{1-(x^2)^2}}\right)\times 2x$$

$$\Rightarrow \quad \frac{dy}{dx} = \frac{-x}{\sqrt{1-x^4}}$$

**Ans.**

**16. If $x = a\cos\theta + b\sin\theta$, $y = a\sin\theta - b\cos\theta$, show that $\frac{y^2 d^2y}{dx^2} - \frac{xdy}{dx} + y = 0$.** **[4]**

**Sol.** We have,

$$x = a\cos\theta + b\sin\theta \quad \text{...(i)}$$

$$y = a\sin\theta - b\cos\theta \quad \text{...(ii)}$$

On squaring and adding (i) and (ii), we get

$$x^2+y^2 = (a\cos\theta+b\sin\theta)^2 + (a\sin\theta-b\cos\theta)^2$$

$$= a^2\cos^2\theta + b^2\sin^2\theta + 2ab\cos\theta\sin\theta + a^2\sin^2\theta + b^2\cos^2\theta - 2ab\cos\theta\sin\theta$$

$$= a^2(\cos^2\theta+\sin^2\theta) + b^2(\sin^2\theta+\cos^2\theta)$$

$$\Rightarrow \quad x^2+y^2 = a^2+b^2 \quad \text{...(iii)}$$

Now, differentiating both sides of equation (iii) w.r.t. $x$, we get

$$2x+2y\frac{dy}{dx} = 0$$

$$\Rightarrow \quad 2y\frac{dy}{dx} = -2x$$

$$\Rightarrow \quad \frac{dy}{dx} = -\frac{x}{y} \quad \text{...(iv)}$$

Again, differentiating both sides of (iv) w.r.t. $x$, we get

$$\frac{d^2y}{dx^2} = -\left(\frac{y\times 1 - x\times\frac{dy}{dx}}{y^2}\right)$$

$$\Rightarrow \quad \frac{d^2y}{dx^2} = -\left[\frac{y-x\left(-\frac{x}{y}\right)}{y^2}\right] \quad \text{[from (iv)]}$$

$$\Rightarrow \quad \frac{d^2y}{dx^2} = -\left[\frac{y^2+x^2}{y^3}\right] \quad \text{...(v)}$$

Now, we have

$$\text{L. H. S.} = y^2\frac{d^2y}{dx^2} - x\frac{dy}{dx} + y$$

$$= y^2\left(\frac{-y^2-x^2}{y^3}\right) - x\left(-\frac{x}{y}\right) + y$$

$$= \frac{-y^2-x^2}{y} + \frac{x^2}{y} + y$$

$$= \frac{-y^2-x^2+x^2+y^2}{y}$$

= 0 = R. H. S. **Hence Proved.**

**17. The side of an equilateral triangle is increasing at the rate of 2 cm/s. At what rate is its area increasing when the side of the triangle is 20 cm ?** **[4]**

**Sol.** We know that, Area of an equilateral triangle,

$$A = \frac{\sqrt{3}}{4}a^2,$$

where $a$ = side of an equilateral triangle.

Given $\quad \frac{da}{dt} = 2 \text{ cm/s}$

Now, $\quad \frac{dA}{dt} = \frac{d}{dt}\left(\frac{\sqrt{3}}{4}a^2\right)$

$$= \frac{\sqrt{3}}{4}\times 2\times a\times\frac{da}{dt}$$

$$= \left(\frac{\sqrt{3}}{2}a.\frac{da}{dt}\right)$$

$$= \frac{\sqrt{3}a}{2}.2 = \sqrt{3}a \text{ cm}^2/\text{s}$$

when the side of the triangle is 20 cm.

$$\therefore \quad \left[\frac{dA}{dt}\right]_{a=20} = 20\sqrt{3}\text{ cm}^2/\text{s}$$

Hence, the area is increasing at the rate of $20\sqrt{3}\text{ cm}^2/\text{s}$ when the side of the triangle is 20 cm. **Ans.**

**18. Find $\int (x+3)\sqrt{3-4x-x^2}\,dx$.** **[4]**

**Sol.** Let $I = \int (x+3)\sqrt{3-4x-x^2}\,dx$

Let $x+3 = \lambda \frac{d}{dx}(3-4x-x^2)+\mu$

$\Rightarrow x+3 = \lambda(-4-2x)+\mu$

$\Rightarrow x+3 = (-2\lambda)x - 4\lambda + \mu$

Equating the coefficients of like terms, we get

$-2\lambda = 1$

$\Rightarrow \lambda = \frac{-1}{2}$

and $-4\lambda + \mu = 3$

$\Rightarrow -4\left(\frac{-1}{2}\right)+\mu = 3$

$\Rightarrow \mu = 3-2 = 1$

$\therefore I = \int \left[\frac{1}{2}(4+2x)+1\right]\sqrt{3-4x-x^2}\,dx$

$= \int \frac{1}{2}(4+2x)\sqrt{3-4x-x^2}\,dx + \int \sqrt{3-4x-x^2}\,dx$

$= \frac{1}{2}I_1 + I_2$ ...(i)

Now, $I_1 = \int (4+2x)\sqrt{3-4x-x^2}\,dx$

Let $3-4x-x^2 = t$

$\Rightarrow -(4+2x)dx = dt$

$\therefore I_1 = -\int \sqrt{t}\,dt$

$= -\int t^{\frac{1}{2}}dt$

$= -\frac{2}{3}t^{3/2}+C_1$

$= -\frac{2}{3}(3-4x-x^2)^{3/2}+C_1$ ...(ii)

$(\because t = 3-4x-x^2)$

and $I_2 = \int \sqrt{3-4x-x^2}\,dx$

$= \int \sqrt{3+4-(x^2+4x+4)}\,dx$

$= \int \sqrt{7-(x+2)^2}\,dx = \int \sqrt{(\sqrt{7})^2-(x+2)^2}\,dx$

$= \frac{1}{2}(x+2)\sqrt{(\sqrt{7})^2-(x+2)^2} + \frac{(\sqrt{7})^2}{2}\sin^{-1}\left(\frac{x+2}{\sqrt{7}}\right)+C_2$

$= \frac{1}{2}(x+2)\sqrt{3-4x-x^2}+\frac{7}{2}\sin^{-1}\left(\frac{x+2}{\sqrt{7}}\right)+C_2$ ...(iii)

From (i), (ii) and (iii), we get

$I = \frac{-1}{3}(3-4x-x^2)^{\frac{3}{2}}+\frac{1}{2}(x+2)\sqrt{3-4x-x^2} + \frac{7}{2}\sin^{-1}\left(\frac{x+2}{\sqrt{7}}\right)+C,$

where $C = C_1 + C_2$ **Ans.**

**19. Three schools A, B and C organised a mela for collecting funds for helping the rehabilitation of flood victims. They sold handmade fans, mats and plates from recycled material at a cost of ₹ 25, ₹ 100 and ₹ 50 each. The number of articles sold are given below :**

| Article \ School | A | B | C |
|---|---|---|---|
| Hand-fans | 40 | 25 | 35 |
| Mats | 50 | 40 | 50 |
| Plates | 20 | 30 | 40 |

**Find the fund collected by each school separately by selling the above articles. Also find the total funds collected for the purpose.**

**Write one value generated by the above situation.** **[4]**

**Sol.** The number of articles sold by each school can be represented by the 3 × 3 matrix

$$X = \begin{bmatrix} 40 & 25 & 35 \\ 50 & 40 & 50 \\ 20 & 30 & 40 \end{bmatrix}$$

and the cost of each article can be represented by the 1 × 3 matrix

$$Y = [25\ 100\ 50]$$

$\therefore$ Funds collected by each school separately is given by the matrix multiplication.

$$YX = [25\ 100\ 50]\begin{bmatrix} 40 & 25 & 35 \\ 50 & 40 & 50 \\ 20 & 30 & 40 \end{bmatrix}$$

$$YX = [7000\ 6125\ 7875]$$

Hence, the funds collected by schools A, B and C are ₹ 7,000, ₹ 6,125 and ₹ 7,875 respectively.

The total funds collected for flood victims

= ₹ (7,000 + 6,125 + 7,875)

= ₹ 21,000

The above situation exhibits the helping nature of students. **Ans.**

## SECTION–C

**20. Let N denote the set of all natural numbers and R be the relation on N × N defined by $(a, b)$ R $(c, d)$ if $ad(b + c) = bc(a + d)$. Show that R is an equivalence relation. [6]**

**Sol.** We know that relation R will be an equivalence relation, if we prove it as a reflexive, symmetric and transitive relation.

**(i) Reflexivity :**

Let $(a, b)$, be an arbitrary element of N × N

then, $(a, b) \in N \times N$

$a, b \in N$

$\Rightarrow \quad ab(b + a) = ba(a + b)$

$\Rightarrow \quad (a, b)\ R\ (a, b)$

$\therefore \quad (a, b)\ R\ (a, b)\ \forall\ (a, b) \in N \times N.$

**(ii) Symmetry :**

Let $(a, b)$, $(c, d)$ be an arbitrary element of N × N such that $(a, b)\ R\ (c, d)$

$\Rightarrow \quad ad(b + c) = bc(a + d)$

$\Rightarrow \quad cb(d + a) = da(c + b)$

$\Rightarrow \quad (c, d)\ R\ (a, b)$

$\therefore (a, b)\ R\ (c, d) \Rightarrow (c, d)\ R\ (a, b)\ \forall\ (a, b), (c, d) \in N \times N$

So, R is symmetric on N × N.

**(iii) Transitivity :**

Let $(a, b)$, $(c, d)$, $(e, f)$ be an arbitrary element of N × N such that $(a, b)$ R $(c, d)$ and $(c, d)$ R $(e, f)$, then $(a, b)$ R $(c, d)$

$\Rightarrow \quad ad(b + c) = bc(a + d)$

$\Rightarrow \quad \frac{b+c}{bc} = \frac{a+d}{ad}$

$\Rightarrow \quad \frac{1}{b}+\frac{1}{c} = \frac{1}{a}+\frac{1}{d} \quad$ ...(i)

Also, $(c, d)$ R $(e, f)$

$\Rightarrow \quad cf(d + e) = de(c + f)$

$\Rightarrow \quad \frac{d+e}{de} = \frac{c+f}{cf}$

$\Rightarrow \quad \frac{1}{d}+\frac{1}{e} = \frac{1}{c}+\frac{1}{f} \quad$ ...(ii)

Adding (i) and (ii), we get

$$\left(\frac{1}{b}+\frac{1}{c}\right)+\left(\frac{1}{d}+\frac{1}{e}\right) = \left(\frac{1}{a}+\frac{1}{d}\right)+\left(\frac{1}{c}+\frac{1}{f}\right)$$

$\Rightarrow \quad \frac{1}{b}+\frac{1}{e} = \frac{1}{a}+\frac{1}{f}$

$\Rightarrow \quad \frac{b+e}{be} = \frac{a+f}{af}$

$\Rightarrow \quad af(b + e) = be(a + f)$

$\Rightarrow \quad (a, b)\ R\ (e, f)$

Thus, $(a, b)$ R $(c, d)$ and $(c, d)$ R $(e, f)$

$\Rightarrow (a, b)\ R\ (e, f).\ \forall\ (a, b)\ (c, d)\ (e, f) \in N \times N$

So, R is transitive on N × N.

Hence, R being reflexive, symmetric and transitive, is an equivalence relation on N × N.

**Hence Proved.**

**21. Using integration find the area of the triangle formed by positive $x$-axis and tangent and normal to the circle $x^2 + y^2 = 4$ at $(1, \sqrt{3})$. [6]**

**Solution :** The equation of the given circle is $x^2 + y^2 = 4$. The equation of the normal to the circle at $(1, \sqrt{3})$ is same as the line joining the points $(1, \sqrt{3})$ and $(0, 0)$ which is given by

$$y - 0 = \frac{\sqrt{3}-0}{1-0}(x - 0)$$

$\Rightarrow \quad y = \sqrt{3}x \quad$ ...(i)

So, the slope of the normal is $\sqrt{3}$.

We know that,

slope of normal × slope of tangent = – 1.

$\therefore$ the slope of tangent $= \frac{-1}{\sqrt{3}}$

Now, the equation of the tangent to the circle at $\left(1, \sqrt{3}\right)$ is given by :

$$y - \sqrt{3} = \frac{-1}{\sqrt{3}}(x - 1)$$

$\Rightarrow \quad \sqrt{3}y - 3 = -x + 1$

$\Rightarrow \quad y = \frac{-x+4}{\sqrt{3}} \quad$ ...(ii)

Putting $y = 0$ in (ii), we get $x = 4$.

Thus, AOB is triangle formed by the tangent, normal and the positive $x$-axis.

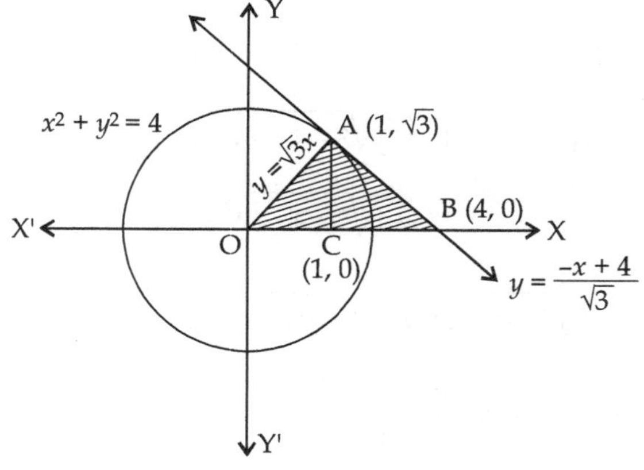


Now, Area of Δ AOB = Area of Δ AOC + Area of Δ ACB.

$= \int_0^1 y\,dx + \int_1^4 y\,dx$

$= \int_0^1 \sqrt{3}\,x\,dx + \int_1^4 \left(\frac{-x+4}{\sqrt{3}}\right)dx$

$= \sqrt{3}\int_0^1 x\,dx - \frac{1}{\sqrt{3}}\int_1^4 x\,dx + \frac{4}{\sqrt{3}}\int_1^4 1.dx$

$= \sqrt{3}\left[\frac{x^2}{2}\right]_0^1 - \frac{1}{\sqrt{3}}\left[\frac{x^2}{2}\right]_1^4 + \frac{4}{\sqrt{3}}[x]_1^4$

$= \sqrt{3}\left(\frac{1}{2}-0\right) - \frac{1}{\sqrt{3}}\left(\frac{16}{2}-\frac{1}{2}\right) + \frac{4}{\sqrt{3}}(4-1)$

$= \frac{\sqrt{3}}{2} - \frac{15}{2\sqrt{3}} + \frac{12}{\sqrt{3}}$

$= \frac{\sqrt{3}}{2} + \frac{9}{2\sqrt{3}} = \frac{\sqrt{3}}{2} + \frac{3\sqrt{3}}{2}$

$= 2\sqrt{3}$ sq. units.

Hence, the area of the tringle so formed is $2\sqrt{3}$ square units. **Ans.**

**OR**

**Evaluate $\int_1^3 (e^{2-3x} + x^2 + 1)\,dx$ as a limit of a sum.**

**Sol.** We have,

$\int_a^b f(x)dx = \lim_{h\to 0} h\ [f(a) + f(a+h) + f(a+2h) + ... + f(a+(n-1)h)]$

where $h = \frac{b-a}{n}$

Here, $a = 1, b = 3$

and $f(x) = e^{2-3x} + x^2 + 1$

$\therefore$ $h = \frac{2}{n}$

$\Rightarrow$ $hn = 2$

Now, $f(a) = f(1)$

$= e^{2-3\times1} + 1^2 + 1$

$f(a+h) = f(1+h)$

$= e^{2-3(1+h)} + (1+h)^2 + 1$

$f(a+2h) = f(1+2h)$

$= e^{2-3(1+2h)} + (1+2h)^2 + 1$

$f(a+(n-1)h) = f(1+(n-1)h)$

$= e^{2-3[1+(n-1)h]} + [1+(n-1)h]^2 + 1$

Adding these equations, we get

$f(a) + f(a+h) + f(a+2h) + ... + f(a+(n-1)h)$

$= [e^{2-3\times1} + 1^2 + 1] + [e^{2-3\times(1+h)} + (1+h)^2 + 1] + [e^{2-3\times(1+2h)}] + (1+2h)^2 + 1 + ... + \{e^{2-3[1+(n-1)h]} + \{1+(n-1)h\}^2 + 1]$

$\therefore \int_1^3 (e^{2-3x} + x^2 + 1)dx$

$= \lim_{h\to 0} h\ [e^2.(e^{-3} + e^{-3(1+h)} + e^{-3(1+2h)} + ... + e]$
$+ \lim_{h\to 0} h\ [1^2 + (1+h)^2 + (1+2h)^2 + ... + \{1+(n-1)h\}^2]$
$+ \lim_{h\to 0} h(1 + 1 + ......\ n \text{ times})$

$= \lim_{h\to 0} h\left\{e^2 \times e^{-3}\frac{(1-e^{-3nh})}{1-e^{-3h}}\right\}$
$+ \lim_{h\to 0} h\,(1 + 1 + ......\ n \text{ times}) + 2h\,(1+2+3+...+(n-1))$
$+ h^2(1^2 + 2^2 + 3^2 + ... + (n-1)^2)$

$+ \lim_{h\to 0} h(n)$ $\left[\because \text{sum to } n \text{ terms of G. P. } S_n = \frac{a(1-r^n)}{1-r}\right]$

$= \lim_{h\to 0} h\left(\frac{e^{-1}(1-e^{-6})}{1-e^{-3h}}\right) + \lim_{h\to 0} h\left[n + 2h \times \frac{(n-1)n}{2} + h^2 \times \frac{(n-1)n(2n-1)}{6}\right] + \lim_{h\to 0} h\,(n)$

$= \lim_{h\to 0} h\left(\frac{e^{-1}(1-e^{-6})}{1-e^{-3h}}\right)$
$+ \lim_{h\to 0}\left[2nh + (nh-h)\,nh + \frac{(nh-h)nh(2nh-h)}{6}\right]$

$= \frac{1}{e}\left(1-\frac{1}{e^6}\right) \times \frac{\lim_{h\to 0} e^{3h}}{3 \times \lim_{h\to 0}\left(\frac{e^{3h}-1}{3h}\right)}$
$+ \lim_{h\to 0}\left[2\times2 + (2-h)\times2 + \frac{(2-h)\times2(2\times2-h)}{6}\right]$

$= \frac{1}{e}\left(1-\frac{1}{e^6}\right) \times \frac{1}{3\times1} + \left(4+4+\frac{8}{3}\right)$

$= \frac{1}{3e}\left(1-\frac{1}{e^6}\right) + \frac{32}{3}$ **Ans.**

**22. Solve the differential equation :**
**$(\tan^{-1} y - x)\,dy = (1 + y^2)\,dx.$** **[6]**

**Sol.** Same as solution Q. 23 (OR)
Set 1 (Outside Delhi) upto equation

$x = \tan^{-1} y - 1 + C\,e^{\tan^{-1}y}$

**OR**

**Find the particular solution of the differential equation $\frac{dy}{dx} = \frac{xy}{x^2+y^2}$ given that $y = 1$, when $x = 0$.**

**Sol.** Given, $$\frac{dy}{dx} = \frac{xy}{x^2+y^2} \quad ...(i)$$

which is a homogeneous differential equation.

Putting $y = vx$

and $$\frac{dy}{dx} = v + x\frac{dv}{dx}$$

$$v + x\frac{dv}{dx} = \frac{vx^2}{x^2+v^2x^2}$$

$$x\frac{dv}{dx} = \frac{v}{1+v^2} - v$$

$$\Rightarrow \quad x\frac{dv}{dx} = \frac{-v^3}{1+v^2}$$

$$\Rightarrow \quad \frac{1+v^2}{v^3}dv = \frac{-1}{x}dx$$

$$\Rightarrow \quad \frac{1}{v^3}dv + \frac{1}{v}dv = \frac{-1}{x}dx$$

Now, integrating both sides, we get

$$\frac{v^{-3+1}}{-3+1} + \log|v| = -\log|x| + C$$

$$\Rightarrow \quad \frac{-1}{2v^2} + \log|v| = -\log|x| + C$$

$$\Rightarrow \quad \frac{-1}{2v^2} + \log|vx| = C$$

$$\Rightarrow \quad \frac{-x^2}{2y^2} + \log|y| = C \quad (\because y = vx)...(ii)$$

Putting $x = 0$, $y = 1$ in (ii)

we get, $C = 0$

Putting $C = 0$ in (ii), we get

$$\log|y| = +\frac{x^2}{2y^2}$$

$$\Rightarrow \quad x^2 = +2y^2 \log|y|$$

Hence, $x^2 = 2y^2 \log|y|$ is the solution of the given equation.

**Ans.**

**23. If lines $\frac{x-1}{2} = \frac{y+1}{3} = \frac{z-1}{4}$ and $\frac{x-3}{1} = \frac{y-k}{2} = \frac{z}{1}$ intersect, then find the value of $k$ and hence find the equation of the plane containing these lines. [6]**

**S** The coordinates of any point on first line are

$$\frac{x-1}{2} = \frac{y+1}{3} = \frac{z-1}{4} = \lambda$$

$$x = 2\lambda + 1,$$

$$y = 3\lambda - 1,$$

$$z = 4\lambda + 1$$

*i.e.*, $(2\lambda + 1, 3\lambda - 1, 4\lambda + 1)$

and the coordinates of any point on second line are

$$\frac{x-3}{1} = \frac{y-k}{2} = \frac{z}{1} = \mu$$

*i.e.*, $$x = \mu + 3,$$

$$y = 2\mu + k,$$

$$z = \mu$$

*i.e.*, $(\mu + 3, 2\mu + k, \mu)$

if these two lines intersect each other, then

$$2\lambda + 1 = \mu + 3, 3\lambda - 1 = 2\mu + k, 4\lambda + 1 = \mu$$

*i.e.* $$2\lambda - \mu = 2, 3\lambda - 2\mu = k + 1, 4\lambda - \mu = -1$$

solving, $2\lambda - \mu = 2$ and $4\lambda - \mu = -1$, we get

$$\lambda = \frac{-3}{2} \text{ and } \mu = -5$$

and substituting the values of $\lambda$ and $\mu$ in $3\lambda - 2\mu = k + 1$, we get

$$k = \frac{9}{2}$$

Now, we have $\vec{b_1} = 2\hat{i} + 3\hat{j} + 4\hat{k}$ and $\vec{b_2} = \hat{i} + 2\hat{j} + \hat{k}$. So, the required plane contains both lines and it passes through a point $\vec{a}$ (1, – 1, 1) and perpendicular vector $\vec{n}$, given by, $\vec{n} = \vec{b_1} \times \vec{b_2}$

$$\vec{n} = \begin{vmatrix} \hat{i} & \hat{j} & \hat{k} \\ 2 & 3 & 4 \\ 1 & 2 & 1 \end{vmatrix}$$

$$= \hat{i}(3-8) - \hat{j}(2-4) + \hat{k}(4-3)$$

$$= -5\hat{i} + 2\hat{j} + \hat{k}$$

$\therefore$ The equation of plane passing through $\vec{a}$ and perpendicular to $\vec{n}$ is given by

$$(\vec{r} - \vec{a}).\vec{n} = 0$$

$$[\vec{r} - (\hat{i} - \hat{j} + \hat{k})].(-5\hat{i} + 2\hat{j} + \hat{k}) = 0$$

$$\Rightarrow \quad \vec{r}(-5\hat{i} + 2\hat{j} + \hat{k}) = (\hat{i} - \hat{j} + \hat{k}).(-5\hat{i} + 2\hat{j} + \hat{k})$$

$$\Rightarrow \quad \vec{r}(-5\hat{i} + 2\hat{j} + \hat{k}) = -6$$

Writing $\vec{r} = x\vec{i} + y\vec{j} + z\vec{k}$

$\Rightarrow \quad 5x - 2y - z - 6 = 0.$ **Ans.**

**24. If A and B are two independent events such that $P(\bar{A} \cap B) = \frac{2}{15}$ and $P(A \cap \bar{B}) = \frac{1}{6}$, then find P(A) and P(B).** **[6]**

**Sol.** Let $P(A) = x$ and $P(B) = y$

$\therefore \quad P(\bar{A}) = 1 - P(A) = 1 - x, P(\bar{B}) = 1 - y$

We have, $P(\bar{A} \cap B) = \frac{2}{15}$ and $(P \cap \bar{B}) = \frac{1}{6}$

Now, $P(\bar{A} \cap B) = \frac{2}{15}$

$\Rightarrow \quad P(\bar{A})\,P(B) = \frac{2}{15}$

$\Rightarrow \quad (1 - x)y = \frac{2}{15}$ ...(i)

Since, A and B are independent events so are $\bar{A}$ and B as well as A and $\bar{B}$.

Given, $P(A \cap \bar{B}) = \frac{1}{6}$

$\Rightarrow \quad P(A)P(\bar{B}) = \frac{1}{6}$

$\Rightarrow \quad x(1 - y) = \frac{1}{6}$ ...(ii)

$\Rightarrow \quad x = \frac{1}{6 - 6y}$

Putting the value of $x$ in eq. (i), we get

$$\left(1 - \frac{1}{6 - 6y}\right)y = \frac{2}{15}$$

$\Rightarrow \quad -90y^2 + 87y = 12$

$\Rightarrow \quad 30y^2 - 29y + 4 = 0$

Solving the quadratic equation, we get

$$y = \frac{4}{5}, y = \frac{1}{6}$$

For $y = \frac{4}{5}$; using (ii), we get

$$x = \frac{1}{6 - 6 \times \frac{4}{5}} = \frac{5}{30 - 24} = \frac{5}{6}$$

For $y = \frac{1}{6}$ using (ii), we get

$$x = \frac{1}{6 - 6 \times \frac{1}{6}} = \frac{1}{5}$$

$\therefore \quad P(A) = \frac{5}{6}, P(B) = \frac{4}{5}$

or $\quad P(A) = \frac{1}{5},$

$P(B) = \frac{1}{6}.$ **Ans.**

**25. Find the local maxima and local minima of the function $f(x) = \sin x - \cos x$, $0 < x < 2\pi$. Also find the local maximum and local minimum values.** **[6]**

**Solution :** We have, $f(x) = \sin x - \cos x$, $0 < x < 2\pi$.

$\Rightarrow \quad f'(x) = \cos x + \sin x$

For local maximum or minimum, we have

$\Rightarrow \quad f'(x) = 0$

$\Rightarrow \quad \cos x + \sin x = 0$

$\Rightarrow \quad \cos x = -\sin x$

$\Rightarrow \quad \tan x = -1$

$\Rightarrow \quad x = \frac{3\pi}{4}$

or $\quad x = \frac{7\pi}{4}$

Thus, $x = \frac{3\pi}{4}$ and $x = \frac{7\pi}{4}$ are possible points of local maximum or minimum.

We have, $f''(x) = \frac{d}{dx}(\cos x + \sin x)$

$= -\sin x + \cos x$

At $x = \frac{3\pi}{4}$, we have

$$f''\left(\frac{3\pi}{4}\right) = -\sin\frac{3\pi}{4} + \cos\frac{3\pi}{4}$$

$$= -\frac{1}{\sqrt{2}} - \frac{1}{\sqrt{2}}$$

$$= -\sqrt{2} < 0$$

$\Rightarrow \quad f''\left(\frac{3\pi}{4}\right) < 0$

So, $x = \frac{3\pi}{4}$ is the point of local maximum.

Local maximum value $= f\left(\frac{3\pi}{4}\right)$

$$= \sin\frac{3\pi}{4} - \cos\frac{3\pi}{4}$$

$$= \frac{1}{\sqrt{2}} + \frac{1}{\sqrt{2}} = \sqrt{2}$$

At $x = \frac{7\pi}{4}$, we have

$$f''\left(\frac{7\pi}{4}\right) = -\sin\frac{7\pi}{4} + \cos\frac{7\pi}{4}$$

$$= \frac{1}{\sqrt{2}} + \frac{1}{\sqrt{2}}$$

$$= \sqrt{2} > 0$$

$$\Rightarrow \quad f''\left(\frac{7\pi}{4}\right) > 0$$

So, $x = \frac{7\pi}{4}$, is the point of local minimum.

Local minimum value $= f\left(\frac{7\pi}{4}\right)$

$$= \sin\frac{7\pi}{4} - \cos\frac{7\pi}{4}$$

$$= -\frac{1}{\sqrt{2}} - \frac{1}{\sqrt{2}} = -\sqrt{2}.$$

**Ans.**

**26. Find graphically, the maximum value of $z = 2x + 5y$, subject to constraints given below : [6]**

$$2x + 4y \leq 8$$

$$3x + y \leq 6$$

$$x + y \leq 4$$

$$x \geq 0,\ y \leq 0$$

**Sol.** We first convert the inequalities into equations to obtain lines

$$2x + 4y = 8 \quad \text{...(i)}$$

$$3x + y = 6 \quad \text{...(ii)}$$

$$x + y = 4 \quad \text{...(iii)}$$

$$x = 0$$

and $y = 0.$

We need to maximize the objective function

$$z = 2x + 5y$$

These lines are drawn and the feasible region of the L.P.P. is the shaded region :

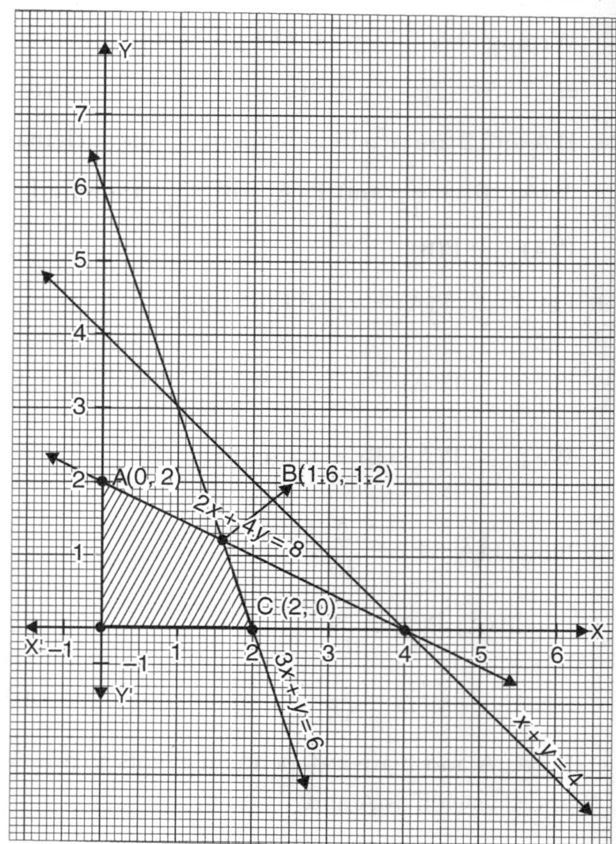


The point of intersection of (i) and (ii) is B (1.6, 1.2)

The coordinates of the corner points of the feasible region are O(0, 0), A(0, 2), B (1.6, 1.2) and C(2, 0).

The value of the objective function at these points are given in the following table :

| Corner Points | Value of the objective function $z = 2x + 5y$ |
|---|---|
| O(0, 0) | $2 \times 0 + 5 \times 0 = 0$ |
| A(0, 2) | $2 \times 0 + 5 \times 2 = 10$ maximum |
| B(1.6, 1.2) | $2 \times 1.6 + 5 \times 1.2 = 9.2$ |
| C(2, 0) | $2 \times 2 + 5 \times 0 = 4$ |

Out of these values of $z$, the maximum value of $z$ is 10 which is attained at the point (0, 2). Thus the maximum value of $z$ is 10. **Ans.**

●●

All questions are same in Delhi Set II and Set III